CALCULUS
THIRD EDITION

CALCULUS
THIRD EDITION

STANLEY I. GROSSMAN
University of Montana

Academic Press, Inc.

(Harcourt Brace Jovanovich, Publishers)
Orlando San Diego San Francisco New York London
Toronto Montreal Sydney Tokyo São Paulo

TO KERSTIN, ERIK, AND AARON

Academic Press, Inc.
Orlando, Florida 32887

United Kingdom Edition Published by Academic Press, Inc. (London) Ltd.
24/28 Oval Road, London NW1 7DX

ISBN: 0-12-304371-9
Library of Congress Catalog Card Number: 83-71160
Printed in the United States of America

Contents

Preface

The study of calculus has been of central importance to scientists throughout most of recorded history. Our present understanding of the subject owes much to Archimedes of Syracuse (287–212 B.C.), who developed what were, for his time, incredibly ingenious techniques for calculating the areas enclosed by a great variety of curves. Archimedes' work led directly to the modern concept of the integral. Much later, toward the end of the seventeenth century, Sir Isaac Newton and Gottfried Leibniz independently showed how to calculate instantaneous rates of change and slopes to curves. This development inspired an enormous variety of mathematical techniques and theorems that could be used to solve problems in a number of diverse and often unrelated fields.

Many of these techniques and theorems are considered to be part of "the calculus." No textbook could discuss all the myriad results in calculus discovered over the past several hundred years. Fortunately, a weeding out process has occurred, and there is now fairly widespread agreement as to what topics properly should be included in a three-semester (or four- to five-quarter) introduction to the subject. This book includes all the standard topics. In the process of writing it, I have worked to achieve certain goals that make the text unique.

EXAMPLES

As a student, I learned calculus from seeing examples and doing exercises. *Calculus, Third Edition*, contains 960 examples—many more than commonly found in standard calculus texts. Each example includes all the algebraic steps needed to complete the solution. As a student, I was infuriated by statements like "it now easily follows that . . ." when it was not at all easy for me. Students have a right to see the "whole hand," so to speak, so that they always know how to get from "a" to "b." In many instances, explanations are highlighted in color to make a step easier to follow.

EXERCISES

The text includes approximately 7000 exercises—including both drill and applied-type problems. More difficult problems are marked with an asterisk (*) and a few especially difficult ones are marked with a double asterisk (**). The exercises provide the most important learning tool in any undergraduate mathematics textbook. I stress to my students that no matter how well they think they understand my lectures or the textbook, they do not really know the material until they have worked problems. A vast difference exists between understanding someone else's solution and solving a new problem by yourself. Learning mathematics without doing problems is about as easy as learning to ski without going to the slopes.

CHAPTER REVIEW EXERCISES

At the end of each chapter, I have provided a collection of review exercises. Any student who can do these exercises can feel confident that he or she understands the material in the chapter.

APPLICATIONS

Calculus *is* applied mathematics. Consequently, this book includes many applied examples and exercises. Many calculus books draw examples exclusively from the physical sciences, even though calculus today is also used in the biological sciences, the social sciences, economics, and business. Thus, the examples and problems in this book, while including a great number from the physical sciences, also cover a wide range of other fields. Moreover, I have included "real-world" data wherever possible to make the examples more meaningful. For example, students are asked to find the escape velocity from Mars, the effective interest rate of a large installment purchase, and the optimal branching angle between two blood vessels. Finally, as most of the world uses the metric system and even the United States is reluctantly following suit, the majority of the applied examples and problems in the book make use of metric units.

OPTIONAL, LONGER APPLICATIONS

By necessity, many applications in a calculus text are short and, as a result, sometimes seem contrived. To solve this problem, I have included a variety of optional sections that discuss important applications in greater detail. There are extensive physical applications in Section 5.9 (Work, Power, and Energy), 7.3 (Periodic Motion), 9.6 (Moments of Inertia and Kinetic Energy), 9.7 (Fluid Pressure), and 21.6 (Vibratory Motion). Two sections, Sections 4.6 and 6.7, contain applications in economics. Finally, new to this edition, is a section (6.8) that describes models of epidemics. In this section, students can learn about a topic that is much discussed in current research in mathematical biology.

REVIEW CHAPTER

Chapter 1 includes general discussions of several topics that are commonly taught in an intermediate algebra–college algebra course, including absolute value and inequalities, circles and lines, and an introduction to functions. This chapter also includes some detailed discussions of graphing techniques, including a unique section on shifting the graphs of functions—an extremely useful procedure.

INTUITION VERSUS RIGOR

Intuition rather than rigor is stressed in the early parts of the book. For example, in the introduction to the limit in Chapter 2, the student's intuition is appealed to in the initial discussion of limits, and the concept is introduced by means of examples, together with detailed tables. I believe that the "ϵ-δ" or "neighborhood" approach to limits can be appreciated only after some feeling for the limit has been developed. However, I also feel strongly that standard definitions must be included, since mathematics depends on rigorous, unambiguous proofs. To resolve this apparent contradiction, I have put much of the one-variable theory in separate optional sections (Sections 2.8, 5.10, and Appendix 5). These sections can be included at any stage, as the instructor sees fit, or omitted without loss of continuity if time is a problem. By the time students reach infinite series, they should have developed enough mathematical sophistication to understand and appreciate some of the subtleties of mathematical proof. Thus, beginning with the discussion of convergence of sequences and series in Chapter 14, I have included ϵ's and δ's whenever necessary. More difficult proofs like the proof of Stokes's theorem, for example, are labeled "optional."

EARLY TRIGONOMETRY

The derivatives of all six trigonometric functions are computed in Section 3.5. Equally important, all six are then used in applications throughout the rest of the book—not just $\sin x$ and $\cos x$. See, for example, the application to blood flow in Example 4.6.10 (on page 246) that makes use of the derivatives of $\csc x$ and $\cot x$.

TRIGONOMETRY REVIEW

High school trigonometry is, of course, a prerequisite for calculus. Many students, however, come to calculus without having seen trigonometry for several years. Much has been forgotten. To fill this gap and to free the instructor from the necessity of taking class time to review trigonometry, I have included a four-section appendix (Appendix 1), which contains all the precalculus trigonometry a student needs.

EXPONENTIAL AND LOGARITHMIC FUNCTIONS

I have introduced an element of choice in the way in which exponential and logarithm functions are first presented. In Section 6.2 a detailed introduction (without calculus) to exponential and logarithmic functions is provided. These functions are also introduced, in Section 6.4, via the definition of the natural logarithm function as an integral. This gives the instructor the option of choosing the introduction more suitable for his or her students, or both approaches may be covered to illustrate interesting relationships among seemingly disparate concepts in mathematics.

USE OF THE HAND-HELD CALCULATOR

Today, most students who study calculus own or have access to a hand-held calculator. As well as being useful in solving computational problems, a calculator can be employed as a learning device. For example, a student will develop a feeling for limits more quickly if he or she can literally *see* the limit being approached. Chapter 2 contains numerous tables that make use of a calculator to illustrate particular limits. At several places in the text a calculator has been used for illustrative purposes. Section 4.8 includes a discussion of Newton's method for finding roots of equations. This method is quite easy to employ with the aid of a calculator. In Section 6.7 Newton's method is used to estimate effective interest rates.

 Examples, problems, and sections employing a hand calculator are marked with the symbol ▥ . Note, however, that the availability of the calculator is *not* a prerequisite for use of this text. The vast majority of problems do not require a calculator and the illustrative tables can be appreciated without independent verification. I recommend that the student who does wish to use a calculator find a calculator with function keys for the three basic trigonometric functions (sin, cos, tan), common (log) and natural (ln) logarithm keys, and an exponential key (usually denoted y^x), to allow easy calculation of powers and roots; in addition, a memory unit (so that numbers can be stored for easy retrieval) is a useful feature.

BIOGRAPHICAL SKETCHES

Mathematics becomes more interesting if one knows something about the historical development of the subject. I try to convince my students that, contrary to what they may believe, many great mathematicians lived interesting and often controversial lives. Thus, to make the subject more interesting and, perhaps, more fun, I have included a number of full-page biographical sketches of mathematicians who helped develop the calculus. In these sketches students will learn about the wonderful inventiveness of Archimedes, the dispute between Newton and Leibniz, the unproven theorem of Fermat, the reactionary behavior of Cauchy, the gift for languages of Hamilton, and the love life of Lagrange. It is my hope that these notes will bring the subject to life.

ANSWERS AND OTHER AIDS

The answers to most odd-numbered exercises appear at the back of the book. In addition, student's manuals containing detailed solutions to all odd-numbered problems are available. Richard Lane at the University of Montana prepared the manual for Chapters 1–14, while Leon Gerber at St. John's University in New York City prepared the student's manual for Chapters 15–21. In addition, Professor Gerber has written an instructor's manual containing detailed solutions to all even-numbered problems.

COMPUTER SUPPLEMENT

As many instructors will want to have their students use a computer in conjunction with their calculus courses, a supplement entitled *Computing for Calculus* has been prepared. This supplement, written by Mark Christensen at Georgia Institute of Technology, includes an introduction to BASIC and programs for implementing the numerical techniques (Newton's method, numerical integration) discussed in the text. It also contains a section on computer graphics.

VECTOR ANALYSIS

Chapter 20 provides a detailed introduction to vector analysis. This chapter contains Green's Theorem, Stokes's Theorem, and the divergence theorem, as well as a discussion of changing variables in multiple integration and the Jacobian. As elsewhere in the text, I have tried to make this more difficult material accessible to the student by providing a large number of simpler examples before tackling general cases. I have included most of the more difficult proofs but have inserted them in a way that allows for their omission without loss of continuity.

COMPLEX NUMBERS

As an added option, I have included a short section (Appendix 6) on complex numbers. While this material is not used in the text, several instructors have indicated that a brief review of this topic is often desirable for freshman students in the natural sciences and engineering.

NUMBERING AND NOTATION IN THE TEXT

Numbering in the book is fairly standard. Within each section, examples, problems, theorems, and equations are numbered consecutively, starting with 1. Reference to an example, problem, theorem, or equation outside the section in which it appears is by chapter, section, and number. Thus, Example 4 in Section 2.3 is called, simply, Example 4 in that section but outside the section is referred to as Example 2.3.4. As already mentioned, the more difficult problems are marked (*) or occasionally (**), and problems where the use of a calculator is advisable are marked ▦ . Sections that are more difficult and can be omitted without loss of continuity and sections that contain specialized applications are labeled "optional." Finally, the ends of examples and proofs of theorems are marked with a ■.

CHANGES IN THE THIRD EDITION

There have been many large and small changes in this edition. A number of reviewers suggested many ways to improve the clarity of the text. As a result of their suggestions, I made literally hundreds of changes in the examples, exercises, theorems, and explanations. The major changes include the following:

- The differentiation and subsequent use of all six trigonometric functions in Chapter 3. Some of the material previously found in Chapter 7 ("The Trigonometric Functions") now appears in Chapters 3, 4, 5, and 6.
- A rearrangement of Chapter 5 ("The Integral") so that antiderivatives are now discussed before definite integrals.
- A rearrangement of Chapter 6 so that inverse functions in general are discussed before exponential and logarithmic functions.
- A discussion of epidemic models in Section 6.8.
- The statement and proof of a uniqueness theorem for Taylor polynomials in Section 13.2. This section contains results that justify procedures for simplifying the computation of certain Taylor polynomials. This important material is absent from most elementary calculus texts.
- A new discussion of the gradient in Section 18.5, which makes this important topic more accessible to students.
- A discussion of simple electric circuits in Sections 21.3 and 21.5.
- A paring of extraneous text and examples. While examples are an extremely important part of any mathematics text, redundant examples make the text harder to teach from. With the help of reviewers, I have cut approximately 10 percent of the examples. I also deleted explanatory material that was considered redundant. The result is a text that, I trust, remains a book for the student but covers the material in a more streamlined fashion.
- Detailed biographical sketches of mathematicians who were important in the development of the calculus.

ACCURACY

The success of a calculus book depends, to a large extent, on its accuracy. A few badly placed typographical errors can turn a good teaching tool into a source of confusion.

Every book has errors. This one has some too (although I'd love to know, as I write this, exactly where they are). Academic Press, however, has gone to considerable lengths to ensure that the book is as error free as possible.

Approximately twenty mathematics professors read portions of my original manuscript. They found a number of errors, which were corrected before the book was typeset. The checking of galley proofs, however, is the most important step in the process of finding and correcting errors. The galleys were checked in the following ways:

1. I read each set of galleys twice.
2. A proofreader compared the galleys to the original manuscript, looking for any discrepancies between the two.

3. A team of two faculty members at a community college read through the galleys, checking both mathematical accuracy and adherence to the manuscript.
4. A team consisting of a faculty member at a university in Florida and a graduate student read the galleys of the textual material.
5. A second team of a faculty member and a graduate student read only the problem sets in the galleys.
6. A faculty member in New York, who prepared the instructor's manual, checked the problem sets for accuracy.

Some of these actions were repeated when the galleys were corrected and turned into page proofs. The result is a book that, while probably not error free, is as clean as could reasonably be expected in its first printing.

ACKNOWLEDGMENTS

Since the first edition of the book appeared in 1977, I have received a large number of helpful comments and suggestions from readers. The following people contributed constructive comments on the first edition: Boyd Benson, Rio Hondo College; Mike Burke, College of San Mateo; Ernest Fandreyer, Fitchburg State College; Bo Green, Abilene Christian University; Floyd F. Helton, College of the Pacific; David Hughes, Abilene Christian University; B. William Irlbeck, Xavier University of Louisiana; Raymond Knodel, Bemidji State University; Jane E. Lewenthal, College of the Pacific; David C. Mayne, Anne Arundel Community College; Charles Miers, University of Victoria; William F. Moss, Georgia Institute of Technology; Rosalie M. O'Mahony, College of San Mateo; Vincent G. Ryan, St. Joseph's College; Ruben D. Schwieger, Sterling College; and John F. Weiler, University of Wisconsin-Stevens Point.

In addition, second edition reviewers included Don Gallagher, Central Oregon State University; David Hansen, Monterey Peninsula College; David Horowitz, Golden West College; John Jewett, Oklahoma State University; Beverly Marshman, University of Waterloo; Alexander Nagel, University of Wisconsin-Madison; Raymond C. Roan, Washington State University; William Rundell, Texas A&M University; Wayne Roberts, Macalester University; Thomas Schwartzbauer, Ohio State University; Houston H. Stokes, University of Illinois-Chicago Circle; and Professor Fredric I. Davis of the United States Naval Academy, who provided the computer sketches in Figures 11.3.4 and 11.3.5.

The following reviewers read all or part of the third edition manuscript and made many very useful suggestions: Duane W. Bailey, Amherst College; Paul F. Baum, Brown University; Don Bellairs, Grossmont College; Neil Berger, University of Illinois-Chicago Circle; Carl Cowen, Purdue University; Daniel Drucker, Wayne State University; Nat Grossman, University of California, Los Angeles; Howard Hamilton, California State University at Sacramento; Robert Lohman, Kent State University; James Maxwell, Oklahoma State University; Frank D. Pederson, Southern Illinois University at Carbondale; Thomas Schaffter, University of Nevada; Lindsey A. Skinner, University of Wisconsin-Milwaukee; Martin Sternstein, Ithaca College; William Ray Wilson, Central Piedmont Community College. The mathematical accuracy was checked by Howard Sherwood, Gwen Sherwood, Michael Taylor, and Kathy Burnett of the University of Central Florida, and by George Coyne and James D. Lange of Valencia Community College.

I am grateful to Charles Bryan at the University of Montana, who suggested the need for the uniqueness theorem for Taylor polynomials that now appears in Section 13.2.

I wish to thank the Addison-Wesley Publishing Company, Inc., for permission to use the material from my book (with William R. Derrick) *Elementary Differential Equations with Applications,* Second Edition, in Chapter 21. I am especially grateful to Leon Gerber at St. John's University in New York City. Professor Gerber prepared the *Student Solutions Manual for Chapters 15–21* and the *Instructor's Manual.* In addition, he made an astonishing number of insightful suggestions that, I am confident, have made this a better book both for students and their instructors.

Finally, I want to thank the editorial staff of Academic Press for providing much needed help and encouragement during the preparation of this third edition.

Stanley I. Grossman

To the Instructor

As an aid to the instructor, the following table indicates the interrelationships of the various parts of the text.

Chapter	Chapter Dependence	Comments
1	—	Section 1.8 provides extremely useful techniques for elementary curve sketching.
2	Chapter 1	Derivatives are first introduced as a rate of change in Section 2.1 with the analytic definition postponed until Section 2.5. Mathematically rigorous (i.e., $\epsilon - \delta$ arguments) definitions of limits are given in Section 2.8. This section is optional.
3	Chapters 1 and 2 (except Section 2.8)	After this chapter students should be able to differentiate any algebraic function as well as the six trigonometric functions.
4	Chapters 1, 2, and 3 (except Section 2.8)	This chapter provides applications of the derivative. Theorems relating to curve sketching depend on the mean value theorem proved in Section 4.2. The first and second derivative tests are treated separately in Sections 4.3 and 4.4, but the two sections together provide a rather complete procedure for sketching a curve. Section 4.8 on Newton's method makes extensive use of the hand calculator.
5	Chapters 1–4	The introduction to the integral. Antiderivatives are discussed in Section 5.2. The Σ notation is discussed in Section 5.3. The two fundamental theorems of calculus are given in Section 5.6. A mean value theorem for integrals and a rigorous proof of the second fundamental theorem of calculus are given in the optional Section 5.10. Section 5.9 is also optional.

Chapter	Chapter Dependence	Comments
6	Section 6.2 is algebra and depends only on Sections 1.7 and 1.8. The rest of the chapter depends on Chapters 1–3, 5, and parts of Chapter 4	Exponential and logarithmic functions are introduced algebraically in Section 6.2 and their derivatives and integrals are computed in Section 6.3. In Section 6.4 ln x is defined as $\int_1^x (1/t)dt$, making use of the second fundamental theorem of calculus. Then, from this definition, the basic properties of logarithmic and exponential functions are derived. If Sections 6.2 and 6.3 are covered, then Section 6.4 can be omitted, and conversely. Or, all three sections can be covered to show students how two seemingly unrelated ideas are really two sides of the same coin. Section 6.6 introduces the simplest differential equations and should be covered in courses with science majors. Section 6.7 is optional but should be covered in courses with economics and business majors. Section 6.8 contains an optional discussion of epidemic models.
7	This chapter presupposes the material in Appendix 1, the review of elementary trigonometry. Sections 7.1–7.3 require the material in Chapters 1–3, 5, and parts of Chapter 4, while Sections 7.4 and 7.5 also depend on Chapter 6	Section 7.3 is optional but should be covered in courses with science majors. Section 7.5 on the inverse hyperbolic functions should be covered if time permits, but is not used in other parts of the text.
8	Chapters 1–3, 5, 6, and 7	Section 8.6 describes the integration of rational functions with linear and quadratic denominators. Section 8.7 extends these results and can be omitted if time is a problem. Section 8.9 on using the integral tables should be covered. This is undoubtedly the most used technique for integration "in the field."
9	Chapters 1–3, 5–8, and parts of Chapter 4	Each section presents a particular geometric or physical concept and is optional.
10	Chapter 1 and Appendix 1	This chapter provides extensive descriptions of the conic sections and the techniques of translation and rotation of axes. No calculus is needed in the chapter.
11	Chapters 1–3, 5, 6, 7, and parts of Chapter 8	Section 11.5 is optional.
12	Chapters 1–3, 5, 6, 7, and parts of Chapter 8	Section 12.2 can be omitted without loss of continuity.
13	Chapters 1–3, 5, 6, 7, and parts of Chapters 8 and 12	Section 13.1 introduces Taylor's theorem as an approximation technique before a discussion of infinite series. Section 13.2 contains an important uniqueness theorem. Section 13.3 makes use of the calculator.

Chapter	Chapter Dependence	Comments
14	Chapters 1–3, 5, 6, 7, and parts of Chapters 8, 12, and 13	All sections in this chapter should be covered.
15	Chapters 1 and 2 only	An introduction to vectors in the plane—includes no calculus.
16	Chapters 1–3 and 5–8, Section 9.2, and parts of Chapter 11, and Chapter 15	An introduction to vector functions in the plane and parametric equations. Sections 16.6–16.8 are optional.
17	Chapters 1–3, 5–8, 15 and 16, Section 9.2, and parts of Chapter 11 and Appendix 3	An extension of the material in Chapters 15 and 16 to three dimensions. Sections 17.6 and 17.8 are optional. Computer-drawn graphs of certain quadric surfaces are given in Section 17.6.
18	Chapters 1–8, 12, and 15–17	An introduction to the calculus of several variables. The gradient is introduced in Section 18.5 as the natural extension of the ordinary derivative, and this is done in such a way as to allow a discussion of functions of n variables (although this discussion is limited to $\mathbb{R}^2$ and $\mathbb{R}^3$). Section 18.9 is optional but should be covered by students of physics and engineering. Section 18.13 on Lagrange multipliers is motivated by the geometry of the gradient. Computer-drawn sketches of some not easily visualized surfaces in $\mathbb{R}^3$ are given in Section 18.1.
19	Chapters 1–9 and parts of Chapters 11, 12, and 15–18	Introduction to multiple integration with an emphasis on applications. Sections 19.3, 19.5, and 19.7 are optional.
20	Chapters 1–8, 15–19, and Appendix 3	A detailed introduction to vector analysis. Sections 20.1–20.3 provide a description of vector analysis in $\mathbb{R}^2$, including a discussion of Green's theorem in the plane from both an analytic and a geometric point of view. Section 20.5 provides an extensive introduction to surface integrals. Section 20.9 provides examples and a proof of the theorem allowing for a change of variables in multiple integration. The material in this chapter is considerably more difficult than the material in other parts of the text. While the proofs of almost all theorems (including Stokes's theorem and the divergence theorem) are given, they can be omitted without loss of continuity.
21	Chapters 1–8	This chapter can be used as the major part of a one-quarter introduction to ordinary differential equations—especially as the fifth quarter or fourth semester of a calculus sequence.

CALCULUS

THIRD EDITION

1 Preliminaries

1.1 SETS OF REAL NUMBERS

In this chapter we present basic information that you will need for your study of calculus. We begin by discussing the **real number system.** Real numbers are discussed in high school mathematics, and we assume that you are familiar with the basic techniques of addition, subtraction, multiplication, division, determining roots, and using exponents.

The real numbers fall into several categories. The **positive integers** (sometimes called the **natural numbers**) are the numbers of counting: 1, 2, 3, 4, 5, . . . (the three dots indicate that the string of numbers goes on indefinitely). The **integers** consist of the positive integers, the negative integers, and the number 0. It is often convenient to represent the integers on a **number line,** as indicated in Figure 1.

FIGURE 1

A **rational number** is a real number that can be written as the quotient of two integers, where the integer in the denominator is not zero:

$$r = \frac{m}{n} \qquad \text{where} \qquad n \neq 0 \tag{1}$$

Every integer n is also a rational number since $n = n/1$.

EXAMPLE 1   The following are rational numbers:

(a) $\frac{1}{2}$ (b) $-\frac{3}{4}$ (c) -5 (d) $0 = \frac{0}{1}$ (e) $-\frac{127}{105}$ (f) $\frac{4521}{7132}$ ■

Any terminating decimal is a rational number, as the following example suggests.

EXAMPLE 2 $r = 0.721$ is a rational number since $r = \frac{721}{1000}$. ■

All rational numbers can be represented in an infinite number of ways. For example,

$$\tfrac{1}{2} = \tfrac{2}{4} = \tfrac{4}{8} = \tfrac{3}{6} = \tfrac{125}{250} = \cdots.$$

Usually, however, a rational number is written as m/n, where m and n have no common factors. That is, we will write the rational number in *lowest terms*.

Any real number that is not rational is called **irrational.** Examples of irrational numbers are $\pi = 3.14159265\ldots$ and $\sqrt{2} = 1.41421356\ldots.$ (A proof that $\sqrt{2}$ is irrational is suggested in Problem 7.)

Every rational number can be written as a *repeating decimal*, whereas no irrational number can be written in this way. For example, $\frac{1}{3} = 0.33333\ldots,$ $\frac{3}{11} = 0.272727$ $\ldots,$ and $\frac{2}{7} = 0.285714285714\ldots$ are examples of rational numbers written as repeating decimals. We will prove that a rational number can be written in this way in Chapter 14 in our discussion of infinite series.

The following theorem will be very useful in later chapters:

Between any two real numbers there is a rational number and an irrational number. (2)

Rational and irrational numbers can be represented on the number line used to depict integers (see Figure 2).

FIGURE 2

The number line can be used to give us a sense of order. We put the number a to the right of the number b if a is greater than b. We then write this inequality as

$$a > b. \tag{3}$$

Similarly, if $b > a$, then a is to the left of b, and we write the inequality as

$$a < b. \tag{4}$$

We use the notation

$$a \leq b \tag{5}$$

to indicate that a is less than or equal to b; that is, $a < b$ or $a = b$. Finally, we write

$$a \geq b \tag{6}$$

to indicate that a is greater than or equal to b.

EXAMPLE 3   The following inequalities are illustrated in the number line in Figure 2.

(a) $2 < 3$ (b) $\frac{1}{3} < \frac{1}{2}$ (c) $\frac{4}{3} < \sqrt{2} < \frac{3}{2}$

(d) $-3 < 1$ (e) $-\sqrt{3} < -1 < 0$ (f) $-4 < -\pi < -3$

The notation $a < c < b$ indicates that c lies between a and b on the number line, as in parts (c), (e), and (f). ■

We will often be interested in **sets** of numbers. A set of numbers is any *well-defined*† collection of numbers. For example, the integers and rational numbers are each sets of numbers. The set of all real numbers is denoted by $\mathbb{R}$. The collection of "large numbers" does not constitute a set since it is not well defined. There is no universal agreement about whether a given number is or is not in this collection. The numbers in a set are called **members,** or **elements,** of that set. If x is an element of the set A, we write $x \in A$ and read this notation as "x is an element of A" or "x belongs to A."

The elements of a set can be written in a bracket notation. For example, the set A of integers between $\frac{1}{2}$ and $\frac{11}{2}$ can be written as

$$A = \{x \in \mathbb{R}: x \text{ is an integer and } \tfrac{1}{2} < x < \tfrac{11}{2}\}.$$

This notation is read as "A is the set of real numbers x such that x is an integer and x is between $\frac{1}{2}$ and $\frac{11}{2}$." Alternatively, we can write

$$A = \{1, 2, 3, 4, 5\}.$$

The **empty set,** denoted by $\varnothing$, is the set containing no elements.

We say that A is a **subset** of B if every element of A is an element of B, and we write

$$A \subseteq B. \tag{7}$$

Two sets are **equal** if they contain the same elements. The **union** of two sets of real numbers A and B, denoted $A \cup B$, is the set of numbers that are in A *or* in B (or in both); in symbols,

$$A \cup B = \{x: x \in A \text{ or } x \in B\}. \tag{8}$$

---

†A set of numbers is **well defined** if we can determine with certainty whether or not a number belongs to the set.

The **intersection** of two sets A and B, denoted $A \cap B$, is the set of numbers in A *and* in B; we have

$$A \cap B = \{x : x \in A \text{ and } x \in B\}. \qquad (9)$$

The sets A and B are said to be **disjoint** if $A \cap B = \varnothing$. That is, A and B are disjoint if they have no elements in common. If A is a set of real numbers, the **complement** of the set A, denoted $\overline{A}$, is the set of real numbers not in A; we have

$$\overline{A} = \{x : x \notin A\}. \qquad (10)$$

The symbol $\notin$ is read, "is not an element of."

EXAMPLE 4 Let $A = \{x : x$ is a positive integer$\}$, $B = \{x : x$ is a positive even integer$\}$, and $C = \{x : x$ is a negative integer or zero$\}$. Then we have the following:

(a) $B \subseteq A$.
(b) $A \cup C = \{x : x$ is an integer$\}$.
(c) $\overline{A} = \{x : x \in \mathbb{R} \text{ and } x$ is not a positive integer$\}$.
(d) $A \cap C = \varnothing$. ∎

We will often be interested in the set of numbers between two numbers a and b. The following definitions are quite important. We assume that $a < b$.

Definition 1 INTERVALS

(i) The **open interval** (a, b) is the set of real numbers between a and b, *not including* the numbers a and b. We have

$$(a, b) = \{x : a < x < b\}. \qquad (11)$$

Note that $a \notin (a, b)$ and $b \notin (a, b)$. The numbers a and b are called **endpoints** of the interval. This is depicted in Figure 3.

a, *b* excluded

a   *b*

FIGURE 3

(ii) The **closed interval** $[a, b]$ is the set of numbers between a and b, *including* the numbers a and b. We have

$$[a, b] = \{x : a \leq x \leq b\}. \qquad (12)$$

Note that $a \in [a, b]$ and $b \in [a, b]$.

As before, the numbers a and b are called **endpoints** of the interval. This situation is depicted in Figure 4.

FIGURE 4

EXAMPLE 5 **(a)** $(-1, 5) = \{x: -1 < x < 5\}$ **(b)** $[0, 8] = \{x: 0 \leq x \leq 8\}$ ■

Sometimes we will need to include one endpoint but not the other.

(iii) The **half-open interval** $[a, b)$ is given by

$$[a, b) = \{x: a \leq x < b\}. \tag{13}$$

We include the endpoint $x = a$ but not the endpoint $x = b$. That is, $a \in [a, b)$ but $b \notin [a, b)$.

(iv) The half-open interval $(a, b]$ is given by

$$(a, b] = \{x: a < x \leq b\}. \tag{14}$$

We have $a \notin (a, b]$ but $b \in (a, b]$. Here b is included but a is not. See Figure 5.

(a)   (b)

FIGURE 5

Intervals may be infinite. We have

$$[a, \infty) = \{x: x \geq a\} \tag{15}$$
$$(a, \infty) = \{x: x > a\} \tag{16}$$
$$(-\infty, a] = \{x: x \leq a\} \tag{17}$$
$$(-\infty, a) = \{x: x < a\} \tag{18}$$
$$(-\infty, \infty) = \mathbb{R} \tag{19}$$

The symbols ∞ and $-\infty$, denoting infinity and minus infinity, respectively, are *not* real numbers and do not obey the usual laws of algebra, but they can be used for notational convenience. This idea is expressed in symbols as

$$[a, \infty) = \{x : a \leq x < \infty\} \quad \text{and} \quad (-\infty, a) = \{x : -\infty < x < a\}.$$

Definition 2 BOUNDEDNESS

(i) The set A is **bounded above** if there is a number M, called an **upper bound** for A, such that $x \leq M$ for every $x \in A$.

(ii) The set A is **bounded below** if there is a number m, called a **lower bound** for A, such that $m \leq x$ for every $x \in A$.

(iii) The set A is **bounded** if it is bounded above and below.

EXAMPLE 6

(a) The interval $[-6, 2]$ is bounded. An upper bound is 2 and a lower bound is -6.

(b) The infinite interval $(0, \infty)$ is bounded below but not above. A lower bound is 0.

(c) The infinite interval $(-\infty, 2]$ is bounded above but not below. An upper bound is 2.

(d) The set of positive integers is bounded below but not above. A lower bound is 1.

(e) The set of rational numbers is neither bounded above nor bounded below.

Note that when a set is bounded above and/or below, the upper or lower bound is not unique. For example, $[-6, 2]$ is bounded above by 2.5, 7, 100, and so on. ■

REMARK. The smallest upper bound that can be found for a set of numbers is called a **least upper bound;** the largest lower bound is called a **greatest lower bound.** An important property of the set of real numbers is that every nonempty set of real numbers that is bounded above has a least upper bound and every nonempty set that is bounded below has a greatest lower bound. This property is called the **completeness axiom** of the real numbers.

PROBLEMS 1.1

In Problems 1–6, insert either $<$ or $>$ in place of the comma so that the resulting inequality is true.

1. $-3, 2$

2. $4, -6$

3. $\pi/3, 1$

4. $\sqrt{2}, \frac{1}{4}$

5. $\sqrt{8}, 3.1$

6. $-10, -3\pi$

*7. Prove that $\sqrt{2}$ is irrational. [*Hint:* Assume that $\sqrt{2}$ is rational. Then $\sqrt{2} = m/n$, where m and n are integers having no common factors. Show that this result implies that $2n^2 = m^2$, and explain why this result indicates that both m and n are even. Finally, explain why this result proves that $\sqrt{2}$ must be irrational.]

8. Let $A = \{\text{positive integers}\}$, $B = \{\text{negative integers}\}$, and $C = \{\text{multiples of } 3\} = \{3n : n \text{ is an integer}\}$. Describe the following:

(a) $A \cup B$

(b) $A \cup B \cup C$ [*Hint:* First define the union of three sets.]

(c) $B \cap C$

(d) $A \cap B$

(e) $\overline{C}$ [*Hint:* Treat C as a subset of $\mathbb{R}$.]

9. Let A = {multiples of 2}, B = {multiples of 3}, and C = {multiples of 5}. Describe the following:
(a) $A \cup C$ (b) $B \cap C$ (c) $A \cup B \cup C$

10. Let $A = (-\infty, 5)$ and $B = [-1, 10]$. Find $A \cup B$ and $A \cap B$.

11. Show that $\overline{A \cap B} = \overline{A} \cup \overline{B}$. [Hint: $C = D$ if and only if $C \subseteq D$ and $D \subseteq C$.]

12. Show that $\overline{A \cup B} = \overline{A} \cap \overline{B}$.

13. Determine whether each of the following sets is bounded, bounded above, bounded below, or neither bounded above nor bounded below.
(a) $(1, 1000)$ (b) $[-10^{37}, 10^{37}]$ (c) $[4, \infty]$
(d) $(-\infty, 6)$ (e) {positive odd integers}
(f) {negative rationals larger than -20}

14. The set $A - B$ is defined by

$$A - B = \{x: x \in A \text{ and } x \notin B\}.$$

That is, $A - B$ is the set of numbers that are in A but not in B. For $A = [1, 3]$ and $B = [2, 4]$, find the following:
(a) $A - B$ (b) $B - A$

15. For the sets in Problem 8, describe the following:
(a) $A - B$ (b) $B - A$ (c) $A - C$ (d) $C - B$

16. For the sets in Problem 9, find the following:
(a) $A - C$ (b) $C - B$ (c) $B - C$

17. For the intervals of Problem 10, find the following:
(a) $A - B$ (b) $B - A$

18. Suppose that $A = (-\infty, -5) \cup (5, \infty)$ and $B = [-10, 3)$. Describe the following sets:
(a) $A - (A - B)$ (b) $A - (B - A)$

19. The set $A \Delta B$ is defined by $A \Delta B = (A - B) \cup (B - A)$. For $A = [1, 3)$ and $B = (2, 5]$, find $A \Delta B$. $A \Delta B$ is called the **symmetric difference** of A and B.

20. Let $A = [0, 5]$, $B = (3, 6]$, and $C = (2, 8)$. Find the following:
(a) $(A \Delta B) \Delta C$ (b) $A \Delta (B \Delta C)$

*21. Show that the following are true.
(a) $A \Delta B = (A \cup B) - (A \cap B)$ (b) $(A \Delta B) \Delta C = A \Delta (B \Delta C)$

22. Under what conditions will the set $A \cup B$ be equal to the set $A \cap B$?

23. Answer true or false.
(a) $\varnothing = 0$ (b) $\varnothing = \{\varnothing\}$ (c) $\varnothing \in \{\varnothing\}$
(d) $\varnothing \subseteq \{\varnothing\}$ (e) $\varnothing \in \varnothing$ (f) $\varnothing \subseteq \varnothing$

1.2 ABSOLUTE VALUE AND INEQUALITIES

Inequalities play a central part in calculus. The following arithmetic facts will prove to be very useful. Let a, b, and c be real numbers.

If $a < b$, then $a + c < b + c$.   (1)

If $a < b$ and $c > 0$, then $ac < bc$.   (2)

If $a < b$ and $c < 0$, then $ac > bc$.   (3)

EXAMPLE 1 To illustrate (3), we note that since $2 < 5$, if we multiply by -1, we obtain $-2 > -5$. This statement is true since -2 is to the right of -5 on the number line. ∎

The rules (1), (2), and (3) can be restated as follows:

1′. Adding a positive or negative number to both sides of an inequality does not change the inequality.
2′. Multiplying both sides of an inequality by a *positive* number does not change the inequality.
3′. Multiplying both sides of an inequality by a *negative* number reverses the inequality.

EXAMPLE 2   Solve the inequalities $-3 < (7 - 2x)/3 \leq 4$.

Solution. In this problem we are asked to solve *two* inequalities:

$$\frac{7 - 2x}{3} \leq 4 \quad \text{and} \quad \frac{7 - 2x}{3} > -3.$$

It is necessary to find the set of numbers for which the inequalities are satisfied. This set is called the **solution set** of the inequalities. We start with

$$-3 < \frac{7 - 2x}{3} \leq 4$$

$$-9 < 7 - 2x \leq 12 \qquad \text{We multiplied by 3.}$$

$$-16 < -2x \leq 5 \qquad \text{We subtracted 7 (added } -7\text{).}$$

$$8 > x \geq -\tfrac{5}{2} \quad \text{or} \quad x \in [-\tfrac{5}{2}, 8) \qquad \text{We multiplied by } -\tfrac{1}{2} \text{ (which reversed the inequalities).}$$

Thus the solution set is the half-open interval $[-\tfrac{5}{2}, 8)$. Note that each of the steps in the computation served to simplify the term containing x. ∎

EXAMPLE 3   Solve the inequality $x^2 - 4x - 12 > 0$.

Solution. Factoring yields $(x - 6)(x + 2) > 0$. In order that the product be positive, the factors must both be positive or both negative. That is, we must have either

$$(x - 6) > 0 \quad \text{and} \quad (x + 2) > 0$$

or

$$(x - 6) < 0 \quad \text{and} \quad (x + 2) < 0.$$

In the first case we have $x > 6$. In the second we must have $x < -2$. Therefore

$$x^2 - 4x - 12 > 0 \quad \text{if} \quad x > 6 \quad \text{or} \quad x < -2,$$

and the solution set is given by $(-\infty, -2) \cup (6, \infty)$. ∎

Definition 1   ABSOLUTE VALUE   The **absolute value** of a number a is the distance from that number to zero and is written $|a|$. See Figure 1.

FIGURE 1

Thus 2 is 2 units from zero, so that $|2| = 2$. The number -3 is 3 units from zero, so that $|-3| = 3$.

We may define:

$$|a| = a \quad \text{if} \quad a \geq 0; \tag{4}$$

$$|a| = -a \quad \text{if} \quad a < 0. \tag{5}$$

The absolute value of a number is a nonnegative number. Note that, for example, $|5| = 5$ and $|-5| = -(-5) = 5$, so that numbers that are negatives of one another have the same absolute value. Another way to calculate absolute value is to observe that

$$(-x)^2 = x^2 = |x|^2 \quad \text{or} \quad |x| = \sqrt{x^2} \tag{6}$$

where, we emphasize, the positive square root is taken. Thus, for example, $|-3| = \sqrt{(-3)^2} = \sqrt{9} = 3$, $|3| = \sqrt{(3)^2} = \sqrt{9} = 3$, and so on.

For all real numbers a and b the following facts can be proven:

$$|-a| = |a| \tag{7}$$

$$|ab| = |a||b| \tag{8}$$

$$|a + b| \leq |a| + |b| \qquad \text{Triangle inequality} \tag{9}$$

Property (7) is obviously true. A proof of property (8) is suggested in Problem 18, and a proof of the triangle inequality is suggested in Problem 22.

In the solution of inequalities by using absolute values, the following property will be very useful:

$$\text{If } a > 0, |x| < a \text{ is equivalent to } -a < x < a. \tag{10}$$

See Figure 2. This indicates that for any x in the open interval $(-a, a)$, the distance from x to 0 is less than a.

FIGURE 2

EXAMPLE 4   Solve the inequality $|x| \leq 3$.

  Solution. $|x| \leq 3$ implies that the distance from x to 0 is less than or equal to 3. Thus $-3 \leq x \leq 3$, and the solution set is the closed interval $[-3, 3]$. See Figure 3. ■

FIGURE 3

EXAMPLE 5   Solve the inequality $|x - 4| < 5$.

  Solution. The distance from $x - 4$ to 0 is less than 5, so $-5 < x - 4 < 5$, and adding 4 to each term, we see that $-1 < x < 9$. Thus the solution set is the open interval $(-1, 9)$. See Figure 4. Another way to think of this set is as the set of all x such that x is within 5 units of 4. ■

FIGURE 4

Another useful property of absolute value is given next.

If $a > 0$, $|x| > a$ is equivalent to $x > a$ or $x < -a$.   **(11)**

Fact (11) is illustrated in Figure 5.

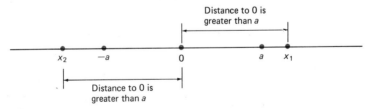

FIGURE 5

EXAMPLE 6   Solve the inequality $|x + 2| \geq 8$.

  Solution. The distance from $x + 2$ to 0 is greater than or equal to 8, so either $x + 2 \geq 8$ or $x + 2 \leq -8$. Hence either $x \geq 6$ or $x \leq -10$. The solution set is $(-\infty, -10] \cup [6, \infty)$. See Figure 6. ■

FIGURE 6

EXAMPLE 7   Solve the inequality $|3x + 4| < 2$.

Solution. We have $-2 < 3x + 4 < 2$. We subtract 4: $-6 < 3x < -2$. Then we divide by 3: $-2 < x < -\frac{2}{3}$, or $x \in (-2, -\frac{2}{3})$. ∎

EXAMPLE 8   Solve the inequality $|5 - 3x| \geq 1$.

Solution. We have either $5 - 3x \geq 1$ or $5 - 3x \leq -1$. In the first case $-3x \geq -4$, which implies that $x \leq \frac{4}{3}$. In the second case $-3x \leq -6$, so $x \geq 2$. The solution set is, therefore, $(-\infty, \frac{4}{3}] \cup [2, \infty)$. ∎

PROBLEMS 1.2

In Problems 1–15, solve the given inequalities and display the solution sets on the number line.

1. $x - 2 < 5$ **2.** $x + 3 > -6$ **3.** $1 \leq 2x + 2 \leq 4$

4. $3x + 6 \geq -3$ **5.** $3 < 2 - 5x < 13$ **6.** $-4 < \dfrac{2x - 4}{3} \leq 7$

7. $1 - 2x < -2$ **8.** $0 < \dfrac{5 - 4x}{2} < 1$ **9.** $1 \leq 3x - 5 < 3$

10. $2 \geq \dfrac{4 - 2x}{5} > -4$ **11.** $\dfrac{1}{x} > 3$ **12.** $\dfrac{4}{3x - 2} \leq -2$

13. $\dfrac{2}{5x} < 4$ **14.** $x^2 - 2x - 8 \leq 0$ **15.** $x^2 - 5x \geq 14$

16. Solve the inequality $a \leq (bx + c)/d < e$, $b, d > 0$.

17. Solve for x:
   (a) $x = |4 - 5|$ **(b)** $x = |-6 - (-2)|$ **(c)** $x = |2| - |-3|$

18. Show that $|xy| = |x||y|$. [*Hint:* Deal with each of four cases separately: (1) $x \geq 0$, $y \geq 0$, (2) $x \geq 0$, $y < 0$, etc.]

19. Show that if $x \geq 0$ and $y \geq 0$, then $|x + y| = |x| + |y|$.

20. If $x > 0$ and $y < 0$, show that $|x + y| < |x| + |y|$.

21. If $x < 0$ and $y < 0$, show that $|x + y| = |x| + |y|$.

22. Using Problems 19–21, prove the triangle inequality (9).

*23. Show that $\big||x| - |y|\big| \leq |x - y|$. [*Hint:* Write $x = (x - y) + y$ and apply the triangle inequality.]

In Problems 24–43, solve the given inequalities and graph the solution sets.

24. $|x| \leq 4$ **25.** $|x| > 2$ **26.** $|x - 2| < 1$
27. $|x + 3| \leq 4$ **28.** $|x + 6| > 3$ **29.** $|2x + 4| < 3$

*30. $|6 - 4x| \geq |x - 2|$ **31.** $\left|\dfrac{8 - 3x}{2}\right| \leq 3$ **32.** $\left|\dfrac{3x + 17}{4}\right| > 9$

33. $x^2 - 4x + 4 \geq 0$ **34.** $x^2 + 2x - 15 < 0$ **35.** $x^2 + 8x + 15 > 0$
36. $3 - \sqrt{x} < 8$

37. $\dfrac{1}{x - 2} > \dfrac{2}{x + 3}$ [*Hint:* Keep track of whether you are multiplying by a positive or negative number.]

38. $\dfrac{2x + 3}{4x - 5} \le -4$

39. $|ax + b| < c, \quad a > 0$

40. $|ax + b| \ge c, \quad a < 0, c > 0$

*41. $x \le |x|$

42. $x > \dfrac{1}{x}$

*43. $|2x| > |5 - 2x|$ [*Hint:* Use equation (6).]

44. Solve the inequalities and graph the solution sets.
   (a) $|2 - x| + |2 + x| \le 10$ **(b)** $|2 - x| + |2 + x| > 6$
   (c) $|2 - x| + |2 + x| \le 4$ **(d)** $|2 - x| + |2 + x| \le 3.99$

45. Use the absolute value function to translate each of the following statements into a single inequality.
   (a) $x \in (-4, 10)$ **(b)** $x \notin (-3, 3)$ **(c)** $x \notin [5, 11]$
   (d) $x \in (-\infty, 2] \cup [9, \infty)$ **(e)** $x \in (-93, 4) \cap (-10, 50)$

46. Write single inequalities that are satisfied by the following:
   (a) All numbers x that are closer to 5 than to 0.
   (b) All numbers y that are closer to -2 than to 2.

47. Show that $\dfrac{s + t + |s - t|}{2}$ equals the maximum of $\{s, t\}$.

48. Show that $\dfrac{s + t - |s - t|}{2}$ equals the minimum of $\{s, t\}$.

49. For what choices of s is $3.72s > 4.06s$?

*50. (a) Suppose that a and b are positive; prove that

$$\sqrt{ab} \le \frac{a + b}{2}.$$

   [*Hint:* Use the fact that $(x - y)^2 \ge 0$ for all real numbers x and y.]
   (b) Use the inequality of part (a) to prove that among all rectangles with an area of 225 cm^2, the one with the shortest perimeter is a square.
   (c) Use the inequality of part (a) to prove that among all rectangles with a perimeter of 300 cm, the one with the largest area is a square.

*51. Show that $AC + BD \le \sqrt{A^2 + B^2}\sqrt{C^2 + D^2}$ for any real numbers A, B, C, and D.

*52. Discuss the following *false proof* that $-1 = 2$:
   Suppose A is a solution to $A = 1 + A^2$. Clearly, A cannot be 0; hence $1 = (1/A) + A$. Use this expression to substitute for 1 in the first conditional equation: $A = [(1/A) + A] + A^2$. Therefore

$$0 = \frac{1}{A} + A^2 \quad \text{and} \quad -\frac{1}{A} = A^2;$$

   hence $A^3 = -1$. We now infer that $A = \sqrt[3]{-1} = -1$. Substituting this value into our original equation, we have

$$-1 = 1 + (-1)^2 = 1 + 1 = 2.$$

   [*Hint:* $(x^2 - x + 1)(x + 1) = x^3 + 1$ is true for all numbers x.]

*53. An estimate for a particular number w is 1.3. Suppose this estimate is accurate to one decimal place; that is, $|w - 1.3| \le 0.05$. Observe that $(1.3)^2 = 1.69$, which rounds off to 1.7. Is the estimate 1.7 for w^2 also accurate to one decimal place? Describe the shortest interval that contains w^2.

*54. The sides of a rectangular piece of paper (a page of this book, for example) are measured. Suppose that we measure to one-decimal-place accuracy and find that the rectangle is 16.1 cm by 23.4 cm; 16.1 times 23.4 equals 376.74, which rounds off to 376.7. Does 376.7 estimate the true area of the rectangle with one-decimal-place accuracy? Explain.

1.3 THE CARTESIAN PLANE

To this point we have been concerned with single numbers. We will now discuss properties of pairs of numbers. An **ordered pair** of numbers consists of two numbers; one is called the **first component** or **element** and the other is called the **second component.** Ordered pairs are written in the form

$$(a, b) \tag{1}$$

where a and b are real numbers. Two ordered pairs are equal if their first components are equal and their second components are equal. Note that $(1, 0)$ and $(1, 1)$ are *different* ordered pairs since their second components are different. Note too that $(1, 2)$ and $(2, 1)$ are different ordered pairs.

Definition 1 CARTESIAN PLANE   The set of all ordered pairs of real numbers is called the **Cartesian plane**† and is denoted $\mathbb{R}^2$. Hence

$$\mathbb{R}^2 = \{(x, y): x \in \mathbb{R} \text{ and } y \in \mathbb{R}\}.$$

Any element (x, y) of the Cartesian plane is called a **point** in the plane.

Any number in $\mathbb{R}$ can be represented graphically on a number line. Analogously, any point (x, y) can be represented graphically as a point in the Cartesian plane. We call the first number in the ordered pair the x**-coordinate** and the second number the y**-coordinate.** Following the technique invented by Descartes, we draw a horizontal line called the x**- axis** and a vertical line perpendicular to it called the y**-axis.** The point at which the two lines meet is called the **origin** and is labeled $(0, 0)$ or 0. With this orientation for the point denoted (a, b), a is the x-coordinate and b is the y-coordinate. Along the x-axis, positive distances are measured to the right of the origin, and along the y-axis, positive distances are measured in the upward direction. This idea is illustrated in Figure 1. The arrows indicate the direction of increasing x and y. A typical point (a, b) is drawn in Figure 1, where $a > 0$ and $b > 0$. We simply measure a units in the positive direction along the x-axis and b units in the positive direction along the y-axis to arrive at the point (a, b). In Figure 2 several different points are depicted. Note that the distinct ordered pairs $(1, 2)$ and $(2, 1)$ represent different points in the plane.

WARNING.   Do not become confused by the fact that (a, b) can represent either a point in the plane or an open interval. The meaning will always be clear from the context.

NOTATION.   With the above construction we will usually refer to the Cartesian plane as the **xy-plane.**

A glance at Figure 3 indicates that the x- and y-axes divide the xy-plane into four regions. These regions are called **quadrants** and are denoted as in the figure. In Figure 2 we see, for example, that $(1, 3)$ is in the first quadrant, $(-1, 3)$ is in the second quadrant, $(-1, -3)$ is in the third quadrant, and $(3, -1)$ is in the fourth quadrant.

---

†See the accompanying biographical sketch.

René Descartes

René Descartes
The Granger Collection

The Cartesian plane is named after the great French mathematician and philosopher René Descartes. Born near the city of Tours in 1596, Descartes received his education first at the Jesuit school at La Flèche and later at Poitier, where he studied law. He had delicate health and, while still in school, developed the habit of spending the greater part of each morning in bed. Later, he considered these morning hours the most productive period of the day.

At the age of 16, Descartes left school and moved to Paris, where he began his study of mathematics. One year later, in 1617, he joined the army of Maurice, Prince of Nassau. He also served with Duke Maximillian I of Bavaria and with the French army at the siege of La Rochelle.

Descartes was not a professional soldier, however, and his periods of military service were broken by periods of travel and study in various European cities. After leaving the army for good, he resettled in Paris to continue his mathematical studies and then moved to Holland where he lived for 20 years.

Much stimulated by the scientists and philosophers he met in France, Holland, and elsewhere, Descartes later became known as the "father of modern philosophy." His statement "Cogito ergo sum" ("I think, therefore I am") played a central role in his philosophical writings.

Descartes's program for philosophical research was enunciated in his famous *Discours de la méthode pour bien conduire sa raison et chercher la vérité dans les sciences* (A Discourse on the Method of Rightly Conducting the Reason and Seeking Truth in the Sciences) published in 1637. This work was accompanied by three appendixes: *La dioptrique* (in which the law of refraction—discovered by Snell—was first published), *Les météores* (which contained the first accurate explanation of the rainbow), and *La géométrie*. *La géométrie*, the third and most famous appendix, took up about a hundred pages of the *Discours*. One of the major achievements of *La géométrie* was that it connected figures of geometry with the equations of algebra. The work established Descartes as the founder of analytic geometry.

In 1649 Descartes was invited to Sweden by Queen Christina. He agreed, reluctantly, but was unable to survive the harsh, Scandinavian winter. He died in Stockholm in early 1650.

We will often need to calculate the distance between two points in the xy-plane. Consider the two points (x_1, y_1) and (x_2, y_2), as in Figure 4. The distance d between (x_1, y_1) and (x_2, y_2) is the length of the line segment PQ. Since PQR is a right triangle, we have, by the Pythagorean theorem,

$$\overline{PQ}^2 = \overline{PR}^2 + \overline{QR}^2.†$$

But $\overline{PR} = |y_2 - y_1|$ and $\overline{QR} = |x_2 - x_1|$, so that $d^2 = \overline{PQ}^2 = \overline{PR}^2 + \overline{QR}^2$, or

$$d^2 = (x_2 - x_1)^2 + (y_2 - y_1)^2.$$

---

†The symbol $\overline{PQ}$ denotes the distance between the points P and Q.

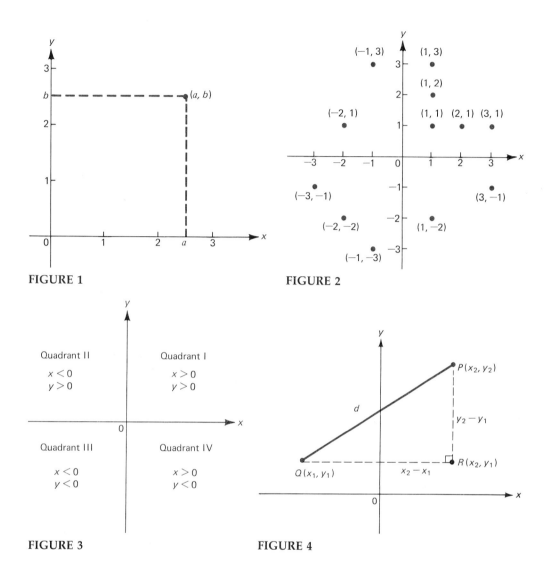

FIGURE 1

FIGURE 2

FIGURE 3

FIGURE 4

Therefore we have the distance formula:

$$d = \sqrt{(x_2 - x_1)^2 + (y_2 - y_1)^2}. \tag{2}$$

EXAMPLE 1 Find the distance between the points $(2, 5)$ and $(-3, 7)$.

Solution. Let $(x_1, y_1) = (2, 5)$ and $(x_2, y_2) = (-3, 7)$, so that from (2) we have

$$d = \sqrt{((-3) - 2)^2 + (7 - 5)^2} = \sqrt{(-5)^2 + 2^2} = \sqrt{29}. \ \blacksquare$$

EXAMPLE 2 Prove that the diagonals of a rectangle have equal lengths.

Solution. We place the rectangle as in Figure 5. The lengths of the two diagonals are $\overline{AC}$ and $\overline{BD}$, But $\overline{AC} = \sqrt{x^2 + y^2}$ and $\overline{BD} = \sqrt{(-x)^2 + y^2}$, so that $\overline{AC} = \overline{BD}. \ \blacksquare$

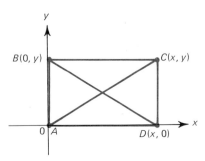

FIGURE 5

Once we have defined a coordinate system with two components, we may discuss **equations in two variables.**

EXAMPLE 3   The following are equations in the two variables x and y:

(a) $y = 3x + 2$
(b) $x^2 + y^2 = 4$

(c) $y = \dfrac{1}{x}$

(d) $x^3y + \sqrt{x + y} = y$ ■

Definition 2   SOLUTION OF AN EQUATION IN TWO VARIABLES   A **solution** to an equation in two variables is a point (x, y) whose coordinates satisfy the equation.

EXAMPLE 4   **(a)** $(1, 5)$ is a solution to the equation $y = 3x + 2$ because

$$\overset{\overset{\displaystyle x}{\downarrow}}{3} \cdot 1 + 2 = \overset{\overset{\displaystyle y}{\downarrow}}{5}.$$

(b) $(\sqrt{3}, 1)$ is a solution to the equation $x^2 + y^2 = 4$ because

$$(\sqrt{3})^2 + 1^2 = 3 + 1 = 4 \quad ■$$
$$\underset{x}{\uparrow} \quad \underset{y}{\uparrow}$$

Definition 3   GRAPH   The **graph** of an equation in two variables is the set of points in the xy-plane whose coordinates satisfy the equation.

In Sections 1.4 and 1.5 we will discuss equations whose graphs are straight lines. In Sections 1.6 we will discuss more general equations and graphs. Now we consider the equation of a circle.

Definition 4   CIRCLE   A **circle** is defined as the set of all points at a given distance from a given point. The given point is called the **center** of the circle, and the common distance from the center is called the **radius.**

EXAMPLE 5 Find the equation of the circle centered at the origin with radius 1.

Solution. If (x, y) is any point on the circle, then the distance from (x, y) to $(0, 0)$ is 1. From (2) we have

$$\sqrt{(x - 0)^2 + (y - 0)^2} = 1,$$

or squaring both sides, we obtain

$$x^2 + y^2 = 1. \tag{3}$$

This circle is sketched in Figure 6. It is called the **unit circle** and was of central importance in your study of the trigonometric functions. (See Appendix 1.2.) ◼

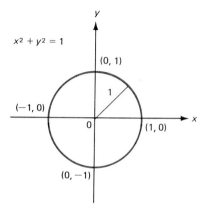

FIGURE 6

We now discuss circles with other centers and other radii.

EXAMPLE 6 Find the equation of the circle centered at $(3, -5)$ with radius 4.

Solution. If (x, y) is on the circle, then the distance from (x, y) to $(3, -5) = 4$, so that

$$\sqrt{(x - 3)^3 + (y + 5)^2} = 4 \qquad \text{or} \qquad (x - 3)^2 + (y + 5)^2 = 16.$$

This circle is sketched in Figure 7a. ◼

In general, the equation of the circle with center at (a, b) and radius r is given by

$$(x - a)^2 + (y - b)^2 = r^2. \tag{4}$$

This circle is sketched in Figure 7b.

EXAMPLE 7 Show that the equation $x^2 - 6x + y^2 + 2y - 17 = 0$ represents a circle, and find its center and radius.

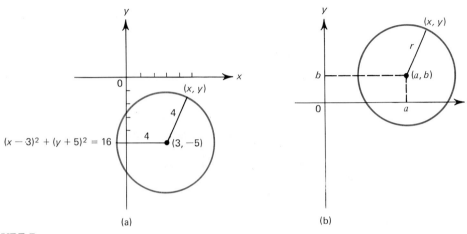

(a)   (b)

FIGURE 7

Solution. We use the technique of completing the squares.† We have

$$x^2 - 6x + y^2 + 2y - 17 = (x^2 - 6x + 9) + (y^2 + 2y + 1) - 9 - 1 - 17$$
$$= (x - 3)^2 + (y + 1)^2 - 27 = 0.$$

This result implies that

$$(x - 3)^2 + (y + 1)^2 = 27,$$

which is the equation of a circle with center at $(3, -1)$ and radius $\sqrt{27}$. ■

We close this section by noting that the graphs of quadratic equations in two variables—that is, equations having the form

$$ax^2 + bxy + cy^2 + dx + ey + f = 0 \qquad (5)$$

—are called **conic sections.** The equation of a circle is a special case of equation (5). Another special case is discussed in Example 1.6.3 (see page 37). We will discuss conic sections in great generality in Chapter 10.

PROBLEMS 1.3

1. Determine the quadrant in which each point lies and draw the point in the xy-plane.

(a) $(3, -2)$ (b) $(-\sqrt{2}, -\sqrt{2})$ (c) $(-\frac{1}{3}, -7)$ (d) $(\pi, 97)$

In Problems 2–8, find the distance between the given points.

2. $(1, 3), (4, 7)$ **3.** $(-7, 2), (4, 3)$ **4.** $(8, -1), (-2, 0)$
5. $(\frac{1}{2}, \frac{1}{3}), (\frac{1}{3}, \frac{1}{2})$ **6.** $(-3, -7), (-1, -2)$ **7.** $(a, b), (b, a)$
8. $(a, b), (0, 0)$

---

†Recall that this technique is sometimes used to solve the general quadratic equation $ax^2 + bx + c = 0$.

9. Confirm that the distance formula gives the answer $|y_2 - y_1|$ as the distance between (a, y_1) and (a, y_2).

10. Find the midpoint of the line segment joining the points:
    (a) (2, 5) and (5, 12)   (b) $(-3, 7)$ and $(4, -2)$

**11. Show that the area A of a triangle with vertices $P_1 = (x_1, y_1)$, $P_2 = (x_2, y_2)$, and $P_3 = (x_3, y_3)$ is given by

$$A = \tfrac{1}{2}|x_1y_2 + x_2y_3 + x_3y_1 - x_1y_3 - x_2y_1 - x_3y_2|.$$

12. Using the result of Problem 11, calculate the area of the triangle with the given vertices.
    (a) (2, 1), (0, 4), (3, −6)   (b) (4, 2), (−1, −5), (7, 3)

13. Let L denote the line segment joining the points $P_1 = (x_1, y_1)$ and $P_2 = (x_2, y_2)$. Show that the midpoint of L is the point

$$P_3 = \left(\frac{x_1 + x_2}{2}, \frac{y_1 + y_2}{2}\right).$$

   [*Hint:* Show that $\overline{P_1P_3} = \overline{P_3P_2}$ and that $\overline{P_1P_2} = \overline{P_1P_3} + \overline{P_3P_2}$.]

*14. Use similar triangles to show that the line segment between (x_1, y_1) and (x_2, y_2) is divided in thirds by

$$Q = \left(\frac{2x_1 + x_2}{3}, \frac{2y_1 + y_2}{3}\right) \quad \text{and} \quad R = \left(\frac{x_1 + 2x_2}{3}, \frac{y_1 + 2y_2}{3}\right).$$

15. Consider the straight-line segment between

   $$P_0 = (x_0, y_0) \quad \text{and} \quad P_1 = (x_1, y_1).$$

   Show that the point $([1 - \lambda]x_0 + \lambda x_1, [1 - \lambda]y_0 + \lambda y_1)$, where λ is a real number between 0 and 1, lies on that line segment.

*16. Show that the area of the parallelogram with vertices at (0, 0), $P_1 = (x_1, y_1)$, $P_2 = (x_2, y_2)$, and $Q = (x_1 + x_2, y_1 + y_2)$ equals $|x_1y_2 - x_2y_1|$.

In Problems 17–25, find the equation of the circle with the given center and radius and sketch its graph.

17. (1, 0), $r = 1$ 18. (0, 1), $r = 1$ 19. (1, 1), $r = \sqrt{2}$
20. (1, −1), $r = 2$ 21. (−1, 4), $r = 5$ 22. $(\tfrac{1}{2}, \tfrac{1}{3})$, $r = \tfrac{1}{2}$
23. $(\pi, 2\pi)$, $r = \sqrt{\pi}$ 24. (4, −5), $r = 7$ 25. (3, −2); $r = 4$

26. Show that the equation $x^2 - 6x + y^2 + 4y - 12 = 0$ is the equation of a circle, and find the circle's center and radius.

*27. Show that the equation $x^2 + ax + y^2 + by + c = 0$ is the equation of a circle if and only if $a^2 + b^2 - 4c > 0$.

*28. Find an equation for the unique circle that contains the points $(0, -2)$, $(6, -12)$, and $(-2, -4)$.

1.4 LINES

Lines will be very important to us in the study of calculus. Two points (x_1, y_1) and (x_2, y_2) determine a line. The *slope* of a line is a measure of the relative rate of change of the x- and y-coordinates of points on the line as we move along the line.

Definition 1 SLOPE Let L denote a line that is not parallel to the y-axis and let $P(x_1, y_1)$ and $Q(x_2, y_2)$ be two points on the line. Then the **slope** of the line, denoted m, is given by

$$m = \text{slope of } L = \frac{y_2 - y_1}{x_2 - x_1} = \frac{\Delta y}{\Delta x}. \tag{1}$$

Here Δy and Δx denote the changes in y and x, respectively. (See Figure 1.) If L is parallel to the y-axis, then the slope is **undefined.†**

In Figure 2 we see that the triangles PQR and STU, whose corresponding sides are parallel, are similar. That means that the ratios of corresponding sides are equal. Thus

$$\frac{y_2 - y_1}{x_2 - x_1} = \frac{y_4 - y_3}{x_4 - x_3}.$$

That is, the same value for the slope is obtained no matter which two points on the line are chosen.

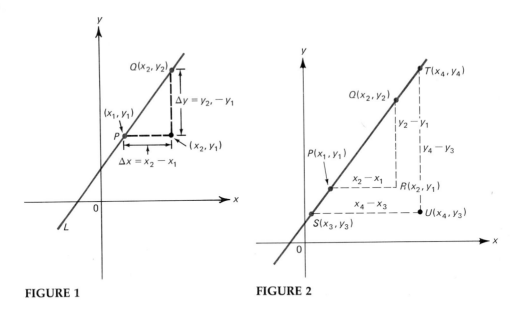

FIGURE 1   **FIGURE 2**

Since the ratio in (1) is unaffected by the choice of points on the line, we may choose P and Q so that $x_2 > x_1$. That is, in going from P to Q, we move in the direction of increasing x. If $y_2 > y_1$ also, then the y values increase and m is positive. If $y_2 < y_1$, then the y values decrease and m is negative. Thus we have the following results:

---

†In some books a line parallel to the y-axis is said to have an **infinite slope.**

> **(i)** If $m > 0$, the graph of the line will point upward as we move from left to right.
> **(ii)** If $m < 0$, the graph of the line will point downward as we move from left to right.

These facts are illustrated in Figure 3.

There are two cases that must be treated separately. In Figure 4a we have drawn the line $y = a$, which is parallel to the x-axis. Here as x changes, y doesn't change at all (since y is equal to the constant a). Therefore $\Delta y / \Delta x = 0/\Delta x = 0$, and we conclude that *lines parallel to the x-axis have a slope of zero*. In Figure 4b we have drawn the line $x = a$, which is parallel to the y-axis. Here when y changes, x doesn't change at all. In this case the slope is undefined, and we see that *lines parallel to the y-axis have an undefined slope*.

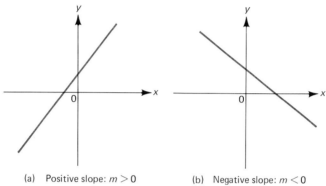

(a)   Positive slope: $m > 0$ (b)   Negative slope: $m < 0$

FIGURE 3

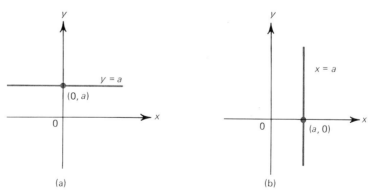

(a)   (b)

FIGURE 4

EXAMPLE 1   Find the slopes of the lines containing the given pairs of points. Then sketch these lines.

(a) $(2, 3)$, $(-1, 4)$ **(b)** $(1, -3)$, $(4, 0)$
(c) $(2, 6)$, $(-1, 6)$ **(d)** $(3, 1)$, $(3, 5)$

Solution.

(a) $m = \dfrac{\Delta y}{\Delta x} = \dfrac{4 - 3}{-1 - 2} = \dfrac{1}{-3} = -\dfrac{1}{3}$

(b) $m = \dfrac{\Delta y}{\Delta x} = \dfrac{0 - (-3)}{4 - 1} = \dfrac{3}{3} = 1$

(c) $m = \dfrac{6 - 6}{-1 - 2} = \dfrac{0}{-3} = 0$

That is, as the x-coordinate changes, the y-coordinate doesn't vary. This line is parallel to the x-axis.

(d) Here the slope is undefined since the line is parallel to the y-axis (the x-coordinates of both points have the same value, 3).

The lines are sketched in Figure 5. ■

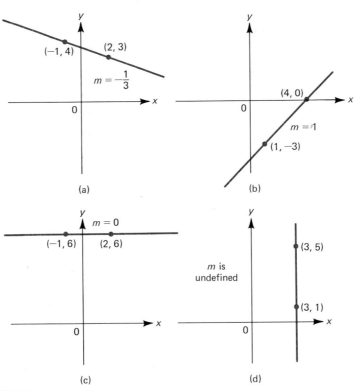

FIGURE 5

Eight different lines, together with their slopes, are sketched in Figure 6.

Definition 2   ANGLE OF INCLINATION   If the line L is not parallel to the x-axis, then its **angle of inclination** θ is the angle between $0°$ and $180°$ that it makes with the positive x-axis. If L is horizontal (parallel to the x-axis), then its angle of inclination is $0°$.

NOTE.   We emphasize that for every line L

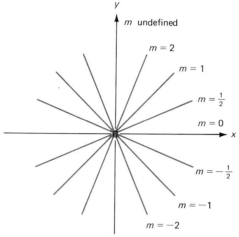

FIGURE 6

$$0° \leq \theta < 180°, \tag{2a}$$

or, in radians,†

$$0 \leq \theta < \pi. \tag{2b}$$

Some angles of inclination are illustrated in Figure 7.

In many places in this text we will use notions from trigonometry. A review of trigonometry appears in Appendix 1. This review contains all the trigonometric facts you will need in your calculus course. In the rest of this text we will assume you are familiar with this material.

We now tie together the notions of slope and angle of inclination.

Theorem 1 If m is the slope of a line L and θ its angle of inclination, then

$$m = \tan \theta \tag{3}$$

Proof

  (i) If L is parallel to the x-axis, then $m = 0$, $\theta = 0$, and $\tan \theta = \tan 0 = 0$ = m.

  (ii) If L is parallel to the y-axis, then m is undefined. But then $\theta = 90° = \pi/2$, and $\tan \theta$ is also undefined.

  (iii) If $m > 0$, then $0° < \theta < 90°$ and from Figure 8a we immediately see that

$$\tan \theta = \frac{\Delta y}{\Delta x} = m.$$

---

†A complete discussion of radian measure is given in Appendix 1.1.

FIGURE 7

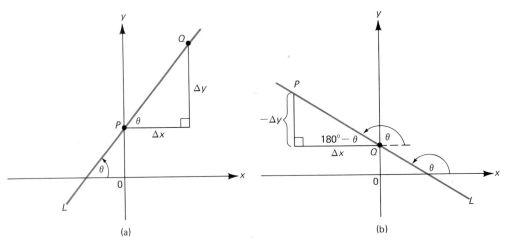

FIGURE 8

(iv) If $m < 0$, then $90° < \theta < 180°$, and as we move from P to Q in Figure 8b, y decreases and $\Delta y < 0$. Thus the length of the line segment joining P to R is $-\Delta y$, and we have

$$\tan(180° - \theta) = \overset{180° \,=\, \pi \text{ radians}}{\tan(\pi - \theta)} = \frac{-\Delta y}{\Delta x} = -m.$$

But

See identities (vi) and (vii)
Appendix A1.2, Table 2

$$\tan(\pi - \theta) = \frac{\sin(\pi - \theta)}{\cos(\pi - \theta)} = \frac{\sin \theta}{-\cos \theta} = -\tan \theta.$$

Thus $-m = -\tan \theta$ or $m = \tan \theta$, and the proof is complete. ∎

REMARK. Lines with large positive or negative slopes have angles of inclination near $90°$ ($= \pi/2$ radians ≈ 1.5708 radians).† These are lines that appear very "steep." Lines with small slopes have angles of inclination near $0°$ or $180°$ (near 0 or π radians).

It follows from elementary geometry that parallel lines have equal angles of inclination (see Figure 9).

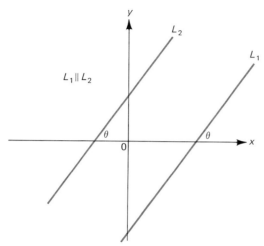

FIGURE 9

Theorem 2 If the lines L_1 and L_2 have slopes m_1 and m_2, respectively, then

$m_1 = m_2$ if and only if‡ L_1 is parallel to L_2. (4)

Proof. If L_1 is parallel to L_2, then $m_1 = \tan \theta = m_2$, where θ is the common angle of inclination. If $m_1 = m_2$, then $\tan \theta_1 = \tan \theta_2$, so that $\theta_1 = \theta_2$ and L_1 is parallel to L_2. ∎

EXAMPLE 2 The line joining the points $(1, -1)$ and $(2, 1)$ is parallel to the line joining the points $(0, 4)$ and $(-2, 0)$ because the slope of each line is 2. ∎

---

†The symbol $\approx$ stands for "is approximately equal to."

‡The words "if and only if" mean that each of the two statements implies the other. For example, the statement "An integer is even if and only if it is a multiple of two" means that (i) if an integer is even, then it is a multiple of two, and (ii) if an integer is a multiple of two, then it is even.

Definition 3 PERPENDICULAR LINES   Two lines L_1 and L_2 are **perpendicular,** or **orthogonal,** if the smallest angle between them is 90°.

Theorem 3   Let L_1 and L_2 be two lines with slopes m_1 and m_2, neither of which is parallel to the y-axis. Then

$$L_1 \text{ is perpendicular to } L_2 \text{ if and only if } m_1 = -\frac{1}{m_2}. \qquad (5)$$

Proof.   A proof is suggested in Problem 57.

EXAMPLE 3   Let the line L_1 contain the two points $(2, -6)$ and $(1, 4)$. Find the slope of a line L_2 that is perpendicular to L_1.

Solution. The slope of $L_1 = m_1 = [4 - (-6)]/(1 - 2) = -10$. Thus $m_2 = -1/(-10) = 1/10$. ∎

PROBLEMS 1.4

In Problems 1–11, find the slope of the line passing through the two given points. Sketch the line.

1. $(1,6), (2,4)$
2. $(-3, 4), (7, 9)$
3. $(-1, -2), (-3, -4)$
4. $(4, 0), (0, 4)$
5. $(-6, 5), (7, -2)$
6. $(1, 7), (-4, 7)$
7. $(2, -3), (5, -3)$
8. $(-2, 4), (-2, 6)$
9. $(0, a), (a, 0), a \neq 0$
10. $(a, b), (b, a), ab \neq 0$
11. $(a, b), (c, d), a \neq c$

In Problems 12–17, determine whether the three points are collinear.

12. $(1, 1), (0, 2), (5, -3)$
13. $(0, -5), (1, -1), (2, 3)$
14. $(2, 1), (3, 2), (2, 3)$
15. $(2, 7), (-3, 4), (7, 10)$
16. $(1, 1), (1, 3), (1, 5)$
17. $(5, -2), (0, -2), (-2, -2)$

In Problems 18–32, find the slope of a line having the given angle of inclination. Use a calculator or table when necessary.

18. 45°
19. 30°
20. 60°
21. 135°
22. 120°
23. 150°
24. 10°
25. 0.5°
26. 179.5°
27. 89.5°
28. 90.5°
29. 37°
30. 161°
31. 110°
32. 73°

In Problems 33–44, find the angle of inclination of a line having the given slope.

33. 1
34. $\sqrt{3}$
35. $-1/\sqrt{3}$
36. $1/\sqrt{3}$
37. 25
38. -37
39. 3.4
40. -0.8
41. 1.5
42. 3.8
43. -2.3
44. -0.005

In Problems 45–53 two pairs of points are given. Determine whether the two lines containing these pairs of points are parallel, perpendicular, or neither.

45. $(1, 8), (2, 9); (1, 2), (0, 1)$
46. $(3, -1), (2, 4); (2, 0), (5, 7)$
47. $(0, 2), (-2, 0); (0, 3), (3, 0)$
48. $(0, 5), (2, -1); (0, 0), (-1, 3)$
49. $(5, 2), (1, 7); (2, 5), (7, 1)$
50. $(1, -2), (2, 4); (4, 1), (-8, 2)$

51. (3, 2), (5, −2); (0, 6), (−5, 6)   **52.** (4, 3), (4, 1); (−2, 4), (−2, 0)
53. (3, 1), (3, 7); (2, 4), (−1, 4)

54. Find three squares that have (0, 7) and (12, 12) as two of the vertices. [*Hint:* It is sufficient to identify a square by listing its vertices.]

55. Suppose $a > 0$ and $a + h > 0$. Show that the straight line through $(a, \sqrt{a})$ and $(a + h, \sqrt{a + h})$ has slope $1/(\sqrt{a + h} + \sqrt{a})$. [*Hint:* $(\sqrt{B} - \sqrt{A})(\sqrt{B} + \sqrt{A}) = B - A$ if $A, B > 0$.]

56. Show that $(x - 3)^2 + (y - 10)^2 = (x + 1)^2 + (y - 5)^2$ is an equation satisfied precisely by the points on the straight line that is the perpendicular bisector of the line segment joining (3, 10) and (−1, 5).

57. Prove Theorem 3 by carrying out the following steps:

 (a) Explain why L_1 and L_2 are perpendicular if and only if, in Figure 10, $\overline{AC}^2 + \overline{CB}^2 = \overline{AB}^2$. [*Hint:* Look at the law of cosines (Problem A1.4.11)]

 (b) Show, using the notation of Figure 10, that $\overline{AC}^2 + \overline{CB}^2 = \overline{AB}^2$ if and only if $c^2 + d^2 = ac + bd$.

 (c) Show that $m_1 = -\dfrac{1}{m_2}$ if and only if $c^2 + d^2 = ac + bc$.

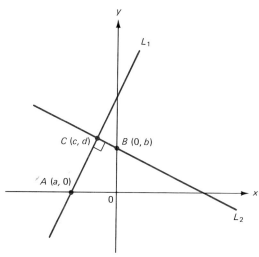

FIGURE 10

1.5 EQUATIONS OF A STRAIGHT LINE

An **equation** of a line is an equation in the variables x and y satisfied by the coordinates of every point on the line and only the points on the line. The *graph* of an equation in x and y is the set of all points drawn in the xy-plane whose coordinates satisfy the equation. If we know two points on a line, then we can calculate an equation of the line.

Definition 1 POINT-SLOPE EQUATION OF A LINE If a line is vertical, then it has the equation $x = a$. If it is not vertical, then, if (x_1, y_1) and (x_2, y_2) are points on the line, it has a slope given by

$$m = \frac{y_2 - y_1}{x_2 - x_1}. \tag{1}$$

Also

$$y_2 - y_1 = m(x_2 - x_1). \qquad \text{We multiplied both sides by } x_2 - x_1. \tag{2}$$

Let (x_1, y_1) be a point on a line with slope m. Then if (x, y) is any other point on the line, its coordinates must satisfy, from (2),

$$y - y_1 = m(x - x_1). \tag{3}$$

Equation (3) is called a **point-slope equation** of the line.

EXAMPLE 1   Find a point-slope equation of the line passing through the points $(-1, -2)$ and $(2, 5)$.

Solution. We first compute

$$m = \frac{5 - (-2)}{2 - (-1)} = \frac{7}{3}.$$

Thus if we choose $(x_1, y_1) = (2, 5)$, a point-slope equation of the line is

$$y - 5 = \tfrac{7}{3}(x - 2).$$

Choosing $(x_1, y_1) = (-1, -2)$, we obtain another point-slope equation of the line:

$$y - (-2) = \tfrac{7}{3}[x - (-1)],$$

or

$$y + 2 = \tfrac{7}{3}(x + 1).$$

This line is sketched in Figure 1. ■

As Example 1 shows, there are many point-slope equations of a line. In fact, there are an infinite number of them—one for each point on the line. A more commonly used equation of a line is given below. First, we define the **y-intercept** of a line to be the y-coordinate of the point at which the line intersects the y-axis.

EXAMPLE 2   Find the y-intercept of the line

$$y + 2 = \tfrac{7}{3}(x + 1). \tag{4}$$

Solution. Any point on the y-axis has x-coordinate 0. Thus setting $x = 0$ in (4), we obtain

$$y + 2 = \tfrac{7}{3}(0 + 1) = \tfrac{7}{3},$$

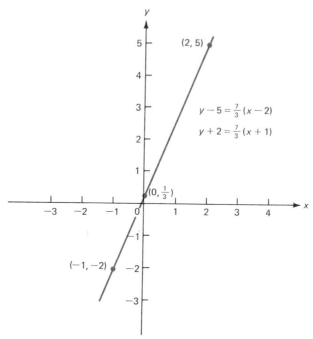

FIGURE 1

or

$$y = \tfrac{7}{3} - 2 = \tfrac{1}{3}.$$

Thus the y-intercept is $\tfrac{1}{3}$. It is indicated in Figure 1. ■

Definition 2 SLOPE-INTERCEPT EQUATION OF A LINE Let m be the slope and b be the y-intercept of a line. Then the **slope-intercept equation** of the line is the equation

$$y = mx + b. \tag{5}$$

EXAMPLE 3 Find the slope-intercept equation of the line passing through $(-1, -2)$ and $(2, 5)$.

Solution. In Example 1 we found that $m = \tfrac{7}{3}$. In Example 2 we found that $b = \tfrac{1}{3}$. Thus inserting these values into equation (5), we find the slope-intercept equation

$$y = \tfrac{7}{3}x + \tfrac{1}{3}.$$

Alternatively, we can start with (see Example 1)

$$y - 5 = \tfrac{7}{3}(x - 2)$$

Then

$$y - 5 = \tfrac{7}{3}x - \tfrac{14}{3}$$
$$y = \tfrac{7}{3}x - \tfrac{14}{3} + 5$$
$$= \tfrac{7}{3}x + \tfrac{1}{3}. \quad \blacksquare$$

There are other ways to represent a straight line. In Example 3 we may write $3y = 7x + 1$, or

$$-7x + 3y = 1. \tag{6}$$

Equation (6) is called a **standard equation** of the line. Another standard equation is $7x - 3y = -1$.

Note that

$$-7x = \frac{x}{-\tfrac{1}{7}} \qquad \text{and} \qquad 3y = \frac{y}{\tfrac{1}{3}}.$$

So

$$\frac{x}{-\tfrac{1}{7}} + \frac{y}{\tfrac{1}{3}} = 1. \tag{7}$$

Equation (7) is called the **intercept form** of the line because both the x- and y-intercepts ($-\tfrac{1}{7}$ and $\tfrac{1}{3}$) are displayed.

EXAMPLE 4 Find the slope-intercept equation of the line passing through the point $(2, 3)$ and parallel to the line whose equation is $y = -3x + 5$.

Solution. Parallel lines have the same slope. The slope of the line $y = -3x + 5$ is -3 since the line is given in its slope-intercept form. Then a point-slope equation of the line is, by (3),

$$y - 3 = -3(x - 2)$$
$$y - 3 = -3x + 6$$
$$y = -3x + 9.$$

The lines are sketched in Figure 2. ◼

EXAMPLE 5 Find the slope-intercept equation of the line passing through $(-1, 3)$ that is perpendicular to the line $2x + 3y = 4$.

Solution. From $2x + 3y = 4$ we obtain

$$3y = -2x + 4,$$

or

$$y = -\tfrac{2}{3}x + \tfrac{4}{3}.$$

Thus the slope of the line $2x + 3y = 4$ is $-\tfrac{2}{3}$, so the line we seek has the slope

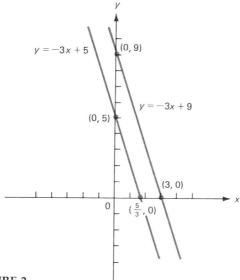

FIGURE 2

$$m = \frac{-1}{-\frac{2}{3}} = \frac{3}{2}.$$

Hence, a point-slope equation of the line passing through $(-1, 3)$ with slope $\frac{3}{2}$ is

$$y - 3 = \tfrac{3}{2}(x + 1)$$
$$y - 3 = \tfrac{3}{2}x + \tfrac{3}{2}$$
$$y = \tfrac{3}{2}x + \tfrac{3}{2} + 3$$
$$= \tfrac{3}{2}x + \tfrac{9}{2}.$$

The two lines are sketched in Figure 3. ■

In Table 1 we summarize some properties of straight lines.

TABLE 1

Equation	Description of Line
$x = a$	Vertical line; x-intercept is a; no y-intercept; no slope
$y = a$	Horizontal line; no x-intercept; y-intercept a; slope $= 0$
$y - y_1 = m(x - x_1)$	*Point-slope form* of line, with slope m and passing through the point (x_1, y_1)
$y = mx + b$	*Slope-intercept form* of line, with slope m and y-intercept b; x-intercept $= -b/m$ if $m \neq 0$
$ax + by = c$	*Standard form;* slope is $-a/b$ if $b \neq 0$; x-intercept is c/a if $a \neq 0$ and y-intercept is c/b if $b \neq 0$
$\dfrac{x}{a} + \dfrac{y}{b} = 1$	*Intercept form:* here it is assumed that $ab \neq 0$; slope is $-b/a$; x-intercept is a and y-intercept is b

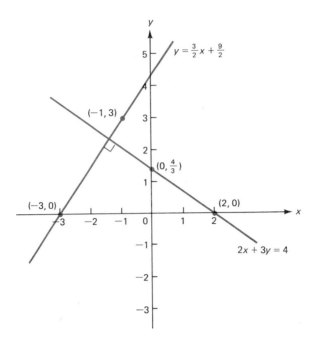

FIGURE 3

There is another type of problem we will encounter.

EXAMPLE 6   Find the point of intersection of the lines $2x + 3y = 7$ and $-x + y = 4$, if one exists.

Solution. The lines have different slopes ($-\frac{2}{3}$ and 1) and are therefore not parallel. So they do have a point of intersection. This point, which we label (a, b), must satisfy both equations. For the first equation we have

$y = -\frac{2}{3}x + \frac{7}{3}$, and for the second $y = x + 4$.

At (a, b),

$$b = -\tfrac{2}{3}a + \tfrac{7}{3} = a + 4 \quad \text{or} \quad \tfrac{5}{3}a = \tfrac{7}{3} - 4 = -\tfrac{5}{3}, \quad \text{and} \quad a = -1.$$

Then $b = a + 4 = 3$, and the point of intersection is $(-1, 3)$. The two lines are sketched in Figure 4. ■

PROBLEMS 1.5

In Problems 1–12, find the slope-intercept form and one point-slope form of the equation of the straight line when either two points on the line or a point and the slope of the line are given. Sketch the graph of the line in the xy-plane.

1. $(1, 2), (3, 6)$ **2.** $(-2, 3), (4, -1)$ **3.** $(3, 7), m = \frac{1}{2}$

4. $(4, -7), m = 0$ **5.** $(-3, -7), m$ undefined   **6.** $(3, -\frac{1}{2}), (\frac{1}{3}, 0)$

7. $(-2, -4), (3, 7)$ **8.** $(5, -1), (8, 2)$ **9.** $(7, -3), m = -\frac{4}{3}$

10. $(-5, 1), m = \frac{3}{7}$ **11.** $(a, b), (c, d), a \neq c$ **12.** $(a, b), m = c$

y

4

$-x + y = 4$

(0, 4)

(−1, 3)   3

$(0, \frac{7}{3})$

2

1

(−4, 0)

$(\frac{7}{2}, 0)$

−4   −3   −2   −1   1   2   3   4

x

−1 $2x + 3y = 7$

−2

−3

FIGURE 4

13. Find the slope-intercept equation of the line parallel to the line $2x + 5y = 6$ that passes through the point $(-1, 1)$.

14. Find the slope-intercept equation of the line parallel to the line $5x - 7y = 3$ that passes through the point $(2, 5)$.

15. Find the standard equation of the line perpendicular to the line $x + 3y = 7$ that passes through the point $(0, 1)$.

16. Find the slope-intercept equation of the line perpendicular to the line $2x - \frac{3}{2}y = 7$ that passes through the point $(-1, 4)$.

17. Find the standard equation of the line perpendicular to the line $ax + by = c$ that passes through the point (α, β). Assume that $a \neq 0$ and $b \neq 0$.

In Problems 18–23, find the point of intersection (if there is one) of the two lines.

18. $x - y = 7$; $2x + 3y = 1$

19. $y - 2x = 4$; $4x - 2y = 6$

20. $4x - 6y = 7$; $6x - 9y = 12$

21. $4x - 6y = 10$; $6x - 9y = 15$

22. $3x + y = 4$; $y - 5x = 2$

23. $3x + 4y = 5$; $6x - 7y = 8$

It is a fact from geometry that the tangent line T to a circle at a given point is perpendicular to the radial line R at that point (see Figure 5). In Problems 24–27, find the slope-intercept equation of the tangent line to the given circle at the given point.

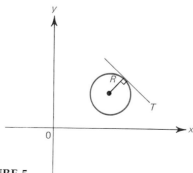

y

R

T

0 x

FIGURE 5

24. $x^2 + y^2 = 1$; $(1/\sqrt{2}, 1/\sqrt{2})$ **–25.** $(x - 1)^2 + (y + 1)^2 = 4$; $(1, 1)$
26. $(x + 2)^2 + (y - 3)^2 = 9$; $(0, 3 + \sqrt{5})$ **27.** $(x - 3)^2 + (y + 2)^2 = 5$; $(4, 0)$

28. (a) Find all points on the line $y = 0$ that are twice as far from $(0, 0)$ as from $(12, 0)$.
   (b) Show that the set of all points in the plane that are twice as far from $(0, 0)$ as from $(12, 0)$ is a circle.

*29. Suppose that $x^2 + y^2 + Ax + By + C = 0$ and $x^2 + y^2 + ax + by + c = 0$ are different circles that meet at two distinct points. Show that the line through those two points of intersection has the equation

$$(A - a)x + (B - b)y + (C - c) = 0.$$

*30. Suppose the point (a, b) is on the circle $x^2 + y^2 = 25$. Show that the line determined by (a, b) and $(5, 0)$ is perpendicular to the line determined by (a, b) and $(-5, 0)$. Assume that $b \neq 0$.

**31. Find the length of the common chord of these circles:
   (a) $(x - a)^2 + (y - b)^2 = R^2$ **(b)** $(x - b)^2 + (y - a)^2 = R^2$

*32. Write an equation for a line that goes through the point $(A, 0)$ and is tangent to the circle $x^2 + y^2 = 1$. Assume $A > 1$.

*33. Find a simple equation satisfied by all points $P = (x, y)$ having the property that the line through P and $(-5, 2)$ is perpendicular to the line through P and $(8, 20)$.

*34. Show that the line with equation $Ax + By = r^2$ is tangent to the circle $x^2 + y^2 = r^2$ at (A, B), assuming that (A, B) lies on the circle.

Let L be a line and let $L_\perp$ denote the line perpendicular to L that passes through a given point P. Then the **distance** from L to P is defined to be the distance between P and the point of intersection of L and $L_\perp$. In Problems 35–39, find the distance between the given line and point.

35. $2x + 3y = -1$; $(0, 0)$ **36.** $3x + y = 7$; $(1, 2)$
37. $5x - 6y = 3$; $(2, \frac{16}{5})$ **38.** $2y - 5x = -2$; $(5, -3)$
39. $6y + 3x = 3$; $(8, -1)$

40. Find the distance between the line $2x - y = 6$ and the point of intersection of the lines $2x - 3y = 1$ and $3x + 6y = 12$.

*41. Show that the distance from the point (x_0, y_0) to the line with equation $Ax + By + C = 0$ is

$$\frac{|Ax_0 + By_0 + C|}{\sqrt{A^2 + B^2}}.$$

*42. The lines $x + 3y = -2$ and $x + 3y = 8$ are parallel. Find the distance between them.

   COMMENTS. Does this problem make sense? Is there a unique reasonable answer? First, state a reasonable definition for the distance between two parallel lines. Second, devise a way to compute that distance in this case. [*Note:* Although computing $8 - (-2)$ involves easy arithmetic, 10 is not a reasonable answer to this problem.]

**43. Suppose we measure the distance between two points in a different way. If $P = (x_1, y_1)$ and $Q = (x_2, y_2)$, then define $d(P, Q) = |x_2 - x_1| + |y_2 - y_1|$. We say that R is "between" P and Q if $d(P, Q) = d(P, R) + d(R, Q)$. Finally, we say that R is "on the the same line as" P and Q if one of the three points is "between" the other two. With these definitions, describe "straight lines." [*Hint:* They're rather fat.]

   NOTE. The distance formula given here is sometimes called the "taxicab" distance function.

1.6 FUNCTIONS

Let us return to the equation of a straight line. For example, if $y = 3x + 5$, then the line can be thought of as the set of all ordered pairs (or points) (x, y) such that $y = 3x + 5$. The important fact here is that for every real number x there is a *unique* real number y such that $y = 3x + 5$ and the ordered pair (x, y) is on the line. We generalize this idea in the following definition.

Definition 1 FUNCTION Let X and Y be sets of real numbers. A **function** f is a rule that assigns to each number x in X a single number $f(x)$ in Y. The number $f(x)$ is called the **image** of the number x. X is called the **domain** of f, written dom f. The set of images of elements of X is called the **range** of f, written *range f* or $f(X)$.

Simply put, a function is a rule that assigns to every x in the domain of f a unique number y in the range of f. We will usually write this as†

$$y = f(x), \tag{1}$$

which is read, "y equals f of x."

To determine the domain of a function (if it is not given explicitly), we need to ask the question "For what values of x does the rule given in equation (1) make sense?" For example, if f is the function written as $f(x) = 1/x$, then since the expression $1/x$ is not defined for $x = 0$, the number 0 is not in the domain of f. However, $1/x$ is defined for any $x \neq 0$, so the domain of f is the set of all real numbers except zero. This result can be written as $\mathbb{R} - \{0\}$.

To determine the range of a function we must ask, "What values do we obtain for y as x takes on all values in dom f?" For example, if f is defined by $f(x) = x^2$, then the domain of f is $\mathbb{R}$, since any real number can be squared. The range of f is the set of nonnegative real numbers, denoted $\mathbb{R}^+$, because the square of any real number is nonnegative, and any nonnegative real number is the square of a real number.

■ WARNING. In writing $y = f(x)$, we must distinguish between the symbol f, which stands for the function rule, and the symbol $f(x)$, which is the *value* the function takes on for a given number x in the domain of f. Here $f(x)$ is a *number* in the range of f.

EXAMPLE 1 Let $f(x) = x^2 - 3x + 1$. Find (a) $f(2)$ and (b) $f(-5)$.

Solution. (a) Since $f(x) = x^2 - 3x + 1$, substituting 2 for x gives us

$$f(2) = 2^2 - (3)(2) + 1 = 4 - 6 + 1 = -1.$$

(b) $f(-5) = (-5)^2 - 3(-5) + 1 = 25 + 15 + 1 = 41.$ ■

Definition 2 GRAPH OF A FUNCTION The **graph** of the function f is the set of ordered pairs $\{(x, f(x)): x \in \text{dom } f\}$.

---

†See the biographical sketch on page 36.

Leonhard Euler

Leonhard Euler
The Granger Collection

The functional notation $y = f(x)$ was first used by the great Swiss mathematician Leonhard Euler (pronounced "oiler") in the *Commentarii Academia Petropolitanae* (*Petersburg Commentaries*), published in 1734–1735.

Born in Basel, Switzerland, Euler's father was a clergyman who hoped that his son would follow him into the ministry. The father was adept at mathematics, however, and together with Johann Bernoulli, instructed young Leonhard in that subject. Euler also studied theology, astronomy, physics, medicine, and several Eastern languages.

In 1727 Euler applied to and was accepted for a chair of medicine and physiology at the St. Petersburg Academy. The day Euler arrived in Russia, however, Catherine I—founder of the Academy—died, and the Academy was plunged into turmoil. By 1730 Euler was pursuing his mathematical career from the chair of natural philosophy. Accepting an invitation from Frederick the Great, Euler went to Berlin in 1741 to head the Prussian Academy. Twenty-five years later, he returned to St. Petersburg, where he died in 1783 at the age of 76.

The most prolific writer in the history of mathematics, Euler found new results in virtually every branch of pure and applied mathematics. Although German was his native language, he wrote mostly in Latin and occasionally in French. His amazing productivity did not decline even when he became totally blind in 1766. During his lifetime, Euler published 530 books and papers. When he died, he left so many unpublished manuscripts that the St. Petersburg Academy was still publishing his work in its *Proceedings* almost half a century later. Euler's work enriched such diverse areas as hydraulics, celestial mechanics, lunar theory, and the theory of music, as well as mathematics.

Euler had a phenomenal memory. As a young man he memorized the entire *Aeneid* by Virgil (in Latin) and many years later could still recite the entire work. He was able to solve astonishingly complex mathematical problems in his head and is said to have solved, again in his head, problems in astronomy that stymied Newton. The French academician François Arago once commented that Euler could calculate without effort "just as men breathe, as eagles sustain themselves in the air."

Euler wrote in a mathematical language that is largely in use today. Among many symbols first used by him are:

$f(x)$ for functional notation
e for the base of the natural logarithm (see Section 6.2)
Σ for the summation sign (see Section 5.3)
i to denote $\sqrt{-1}$ (see Appendix 6)

Euler's textbooks were models of clarity. His texts included the *Introductio in analysin infinitorum* (1748), his *Institutiones calculi differentialis* (1755), and the three-volume *Institutiones calculi integralis* (1768–74). This and others of his works served as models for many of today's mathematics textbooks.

It is said that Euler did for mathematical analysis what Euclid did for geometry. It is no wonder that so many later mathematicians expressed their debt to him.

Thus a point lies on the graph of f if and only if its coordinates satisfy the equation $y = f(x)$.

EXAMPLE 2   The equation $y = f(x) = 3x + 5$ can be thought of as a function, since for every real number x there is a unique real number y that is equal to $3x + 5$. Here the domain of f is $\mathbb{R}$. To calculate the range of f, we note that if y is any real number, then there is a unique number x such that $y = 3x + 5$. To see this, we simply solve for x: $x = (y - 5)/3$. Thus every real number y is in the range of f. In this example the function f is called a **straight-line function.** In general, straight-line functions have the form $y = f(x) = ax + b$, where a and b are real numbers. ■

EXAMPLE 3   Let $y = f(x) = x^2$. As we have seen, this rule constitutes a function whose domain is $\mathbb{R}$ since each real number has a unique square. We also see, as mentioned above, that the range of f equals $\{x: x \geq 0\} = \mathbb{R}^+$, the set of nonnegative real numbers, since the square of any real number is nonnegative. The graph of this function is obtained by plotting all points of the form $(x, y) = (x, x^2)$.† First, we note that because $f(x) = x^2$, $f(x) = f(-x)$ since $(-x)^2 = x^2$. Thus it is only necessary to calculate $f(x)$ for $x \geq 0$. For every $x > 0$ there is a value of $x < 0$ (the number $-x$) that gives the same value of y. In this situation we say that the function is *symmetric* about the y-axis. Some values for $f(x)$ are shown in Table 1. This function, which is graphed in Figure 1, is called a **parabola.** It is a second example of a conic section [see equation (1.3.5)].

TABLE 1

x	0	$\frac{1}{2}$	1	$\frac{3}{2}$	2	$\frac{5}{2}$	3	4	5
$f(x) = x^2$	0	$\frac{1}{4}$	1	$\frac{9}{4}$	4	$\frac{25}{4}$	9	16	25

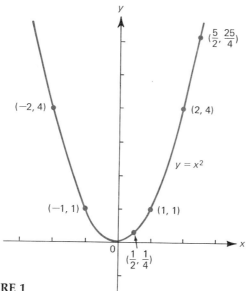

FIGURE 1

†Of course, we can't plot *all* points (there are an infinite number of them). Rather, we plot some sample points and assume they can be connected to obtain the sketch of the graph.

REMARK. We repeat that care must be taken to distinguish between the symbols f and $f(x)$. In this example f is the *rule* that assigns to each real number the square of that number. The symbol $f(x)$ is the *number x^2*. ■

In Example 3 we defined a concept that is worth repeating.

Definition 3   SYMMETRIC FUNCTION   The function given by $y = f(x)$ is **symmetric about the y-axis** if $f(x) = f(-x)$.

A function that is symmetric about the y-axis is also called an **even function.**
    At this point there are essentially three reasons for restricting the domain of a function:

(i) You cannot have zero in a denominator.
(ii) You cannot take the square root (or fourth root, or sixth root, etc.) of a negative number.
(iii) The domain is restricted by the nature of the applied problem under consideration (see Example 7).

We will see a fourth kind of restriction when we discuss logarthmic functions in Section 6.2.

EXAMPLE 4   Let $f(x) = \sqrt{2x - 6}$. Find the domain of f.

Solution. $f(x)$ is defined as long as $2x - 6 \geq 0$ or $2x \geq 6$ or $x \geq 3$.

Thus

$$\operatorname{dom} f = \{x : x \geq 3\} = [3, \infty] ■$$

EXAMPLE 5   Consider the rule $v = g(u) = 1/(u^2 - 1)$. It is clear that $u^2 - 1 = 0$ only when $u = \pm 1$. For a fixed number $u \neq \pm 1$ there is a unique number v that is equal to $1/(u^2 - 1)$. Thus this rule is a function, and

$$\operatorname{dom} g = \{u \in \mathbb{R}: u \neq \pm 1\} = \mathbb{R} - \{1, -1\}.$$

Here we have used the letters u, v, and g instead of x, y, and f to illustrate the fact that there is nothing special about any particular set of letters. Often, we will use the letter t to denote time and the letter s to denote distance. The definition of a function is independent of the letters used to denote the function and variables.
    We now calculate the range of the function. We note that as u become large, $1/(u^2 - 1)$ becomes very small. If $u > 1$ but is close to 1, then $1/(u^2 - 1)$ is very large in the positive direction. For example, $1[(1.0001)^2 - 1] \approx 5000$. If $u < 1$ and u is close to 1, then $u^2 - 1 < 0$ and $1/(u^2 - 1)$ is very large in the negative direction. For example, $1/[(0.9999)^2 - 1] \approx -5000$. Since $u^2 - 1$ can take on any positive real value, so can $1/(u^2 - 1)$. Also, $1/(u^2 - 1)$ is never equal to 0. On the other hand, if $-1 < u < 1$, then $u^2 - 1 < 0$. But $u^2 - 1 \geq -1$, so $1/(u^2 - 1) \leq -1$. Thus range $g = \mathbb{R} - (-1, 0]$. We will not sketch the graph of this function. Methods for doing so will be given in Chapter 4 (Sections 4.3 and 4.4). ■

EXAMPLE 6   Consider the graph of the unit circle $x^2 + y^2 = 1$ given in Figure 2. It is clear that for every real number x in the open interval $(-1, 1)$ there are two values of y given by $y = \pm\sqrt{1 - x^2}$. Hence we do not have a function. We can obtain two separate functions by defining

$$y_1 = f_1(x) = \sqrt{1 - x^2} \qquad \text{and} \qquad y_2 = f_2(x) = -\sqrt{1 - x^2}.$$

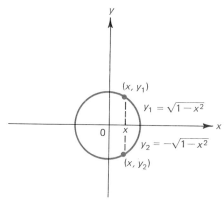

FIGURE 2

Then dom f_1 = dom f_2 = $[-1, 1]$, range f_1 = $[0, 1]$, and range f_2 = $[-1, 0]$. ∎

EXAMPLE 7   The Thunder Power Company has the following rate schedule for electricity consumers:

$6 for the first 30 kilowatthours (kWh) or less each month.

7¢ for each kilowatthour over 30.

(a)   Find a function that gives monthly cost as a function of the number of kilowatthours of electricity used.
(b)   Graph the function.
(c)   What is the cost of 75 kWh in one month?

Solution.   **(a)** Let q denote the number (quantity) of kilowatthours used in a given month, and let $C(q)$ denote the total monthly cost. If $0 \le q \le 30$, then the cost is simply $6, so $C(q) = 6$. If $q > 30$, the number of kilowatthours over 30 is $q - 30$. So from the information given in the statement of the problem, we have

$$C(q) = 6 + 0.07(q - 30) = 6 + 0.07q - (0.07)30 = 6 - 2.10 + 0.07q.$$

cost per
kilowatthours
over 30

number of
kilowatthours
over 30

Thus

$$C(q) = \begin{cases} 6, & 0 \le q \le 30 \\ 3.90 + 0.07q, & q > 30. \end{cases}$$

Note that since a negative number of kilowatt hours cannot be used in a month, we have

$$\text{dom } C \geq \{q : q \geq 0\}.$$

(b) The graph of this cost function is given in Figure 3.

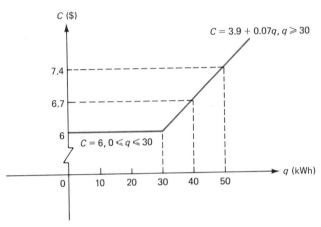

FIGURE 3

(c) $C(75) = 3.90 + 0.07(75) = 3.90 + 5.25 = \$9.15.$ ■

REMARK.   Example 7 gives an example of a function defined in "pieces." It is perfectly legitimate to do so as long as for each x in the domain of f there is a unique y in the range.

EXAMPLE 8   Let

$$f(x) = \begin{cases} 1, & x \geq 0 \\ 2, & x < 0. \end{cases}$$

A graph of this function is given in Figure 4. We have $\text{dom } f = \mathbb{R}$ and range $f = \{1, 2\}.$ ■

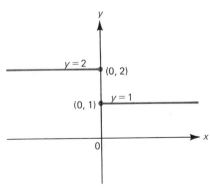

FIGURE 4

EXAMPLE 9 Some people paying federal income tax in the United States do not use standard tax tables. In this case they use Tax Rate Schedules to figure the tax on their net incomes. Tax rates in 1982 for single taxpayers who did not use standard tax tables are given in Table 2.

TABLE 2. FROM 1982 U.S. FEDERAL TAX TABLES

If the amount on Form 1040, line 37 is:		Enter on Form 1040, line 38	
Over —	*But not Over —*		*of the amount over —*
$0	$2,300	—0—	
2,300	3,400	· · · · · · · ·12%	$2,300
3,400	4,400	$132 + 14%	3,400
4,400	6,500	272 + 16%	4,400
6,500	8,500	608 + 17%	6,500
8,500	10,800	948 + 19%	8,500
10,800	12,900	1,385 + 22%	10,800
12,900	15,000	1,847 + 23%	12,900
15,000	18,200	2,330 + 27%	15,000
18,200	23,500	3,194 + 31%	18,200
23,500	28,800	4,837 + 35%	23,500
28,800	34,100	6,692 + 40%	28,800
34,100	41,500	8,812 + 44%	34,100
41,500	· · · · · ·	12,068 + 50%	41,500

Let x denote the net income (in dollars) and $T(x)$ the tax due. Then the *tax function* is defined in pieces. The first five pieces and the last two are given below.

$$T(x) = \begin{cases} 0, & 0 \le x \le 2,300 \\ 0.12(x - 2,300), & 2,300 \le x \le 3,400 \\ 132 + 0.14(x - 3,400), & 3,400 \le x \le 4,400 \\ 272 + 0.16(x - 4,400), & 4,400 \le x \le 6,500 \\ 608 + 0.17(x - 6,500), & 6,500 \le x \le 8,500 \\ \quad \vdots & \quad \vdots \\ 8,812 + 0.44(x - 34,100), & 34,100 \le x \le 41,500 \\ 12,068 + 0.50(x - 41,500), & x \ge 41,500 \quad \blacksquare \end{cases}$$

PROBLEMS 1.6

1. For $f(x) = x^3 - x + 5$, find $f(0)$, $f(1)$, $f(3)$, and $f(-2)$.
2. For $f(x) = 1/(x + 1)$, find $f(0)$, $f(1)$, $f(-2)$, and $f(-5)$.
3. For $f(x) = 1 + \sqrt{x}$, find $f(0)$, $f(1)$, $f(16)$, and $f(25)$.
4. For $f(x) = x/(x - 2)$, find $f(0)$, $f(1)$, $f(-1)$, $f(3)$ and $f(10)$.

In Problems 5–16 an equation involving x and y is given. Determine whether or not y is a function of x.

5. $2x + 3y = 6$

6. $\dfrac{x}{y} = 2$

7. $x^2 - 3y = 4$

8. $x - 3y^2 = 4$

9. $x^2 + y^2 = 4$

10. $x^2 - y^2 = 1$

11. $\sqrt{x + y} = 1$

12. $y^2 + xy + 1 = 0$ [*Hint:* Use the quadratic formula.]

13. $y^3 - x = 0$

14. $y^4 - x = 0$

15. $y = |x|$

16. $y^2 = \dfrac{x}{x + 1}$

17. Explain why the equation $y^n - x = 0$ allows us to write y as a function of x if n is an odd integer but does not allow us to do so if n is an even integer. [*Hint:* First solve Problems 13 and 14.]

In Problems 18–45, find the domain and range of the given function.

18. $y = f(x) = 2x - 3$

19. $s = g(t) = 4t - 5$

20. $y = f(x) = 3x^2 - 1$

21. $v = h(u) = \dfrac{1}{u^2}$

22. $y = f(x) = x^3$

23. $y = f(x) = \dfrac{1}{x + 1}$

24. $s = g(t) = t^2 + 4t + 4$

25. $y = f(x) = \sqrt{x^3 - 1}$

26. $v = h(u) = |u - 2|$

27. $y = f(x) = \dfrac{1}{|x|}$

28. $y = f(x) = \dfrac{1}{|x + 2|}$

29. $y = \begin{cases} x, & x \geq 0 \\ -x, & x < 0 \end{cases}$

30. $y = \begin{cases} x, & x \geq 1 \\ 1, & x < 1 \end{cases}$

31. $y = \begin{cases} x^3, & x > 0 \\ x^2, & x \leq 0 \end{cases}$

32. $y = \dfrac{x^2 + 3x}{x + 3}$

33. $y = \dfrac{x^5 + 5x^2}{x^2}$

34. $y = \dfrac{1}{\sqrt{x - 1}} + 2$

35. $y = \dfrac{1}{x^2 + 1} + 3$

*36. $y = \dfrac{1}{x^2 - 1} + 3$

*37. $y = \dfrac{1}{1 + x^2 + x^4 + x^6}$

*38. $y = \dfrac{x}{x + 1}$

*39. $y = \dfrac{x^2}{x^2 + 3}$

*40. $y = \dfrac{2x + 3}{3x + 4}$

*41. $y = \dfrac{x^n}{x^n + 1}$, n an integer ≥ 1

*42. $y = \sqrt{x^2 - 3x + 2}$

*43. $y = \sqrt[3]{x^2 - 3x + 2}$

44. $y = 3x^2 - 5$, $x \geq 2$

45. $y = \dfrac{1}{x}$, $x < 3$

*46. Let $f(x) = 1/(x - 1)$. Find $f(t^2)$ and $f(3t + 2)$.

47. Let $f(x) = x^2$. Find $f(x + \Delta x)$ and $[f(x + \Delta x) - f(x)]/\Delta x$, where Δx denotes an arbitrary nonzero real number. Simplify your answer.

48. Let $f(x) = \sqrt[3]{x^2 - 1}$. Find $f(1)$, $f(2x)$, $f(3x^2 - 6)$, $f(g(t))$, and $f(x + \Delta x)$.

49. For $f(x) = x/(x^5 + 3)$, calculate $f(-5x)$, $f(1/x)$, $f(x^{10} - 25)$, $f(g(t))$, and $f(x + \Delta x)$.

50. Let $f(x) = \sqrt{x}$. Show, assuming that $\Delta x \neq 0$, that

$$\frac{f(x + \Delta x) - f(x)}{\Delta x} = \frac{1}{\sqrt{x + \Delta x} + \sqrt{x}}.$$

[*Hint:* Multiply and divide by $\sqrt{x + \Delta x} + \sqrt{x}$.]

51. Let $f(x) = |x|/x$. Show that

$$f(x) = \begin{cases} 1, & x > 0 \\ -1, & x < 0 \end{cases}$$

Find the domain and range of f.

52. Describe a computational rule for Fahrenheit temperature as a function of Celsius temperature. [*Hint:* Pure water at sea level boils at 100°C and 212°F, and it freezes at 0°C and 32°F; the graph of this function is a straight line.]

53. Consider the set of all rectangles whose perimeter is 50 cm. For any one rectangle, once its width W is measured, it is possible to do some computations that yield the area of the rectangle. Verify this assertion by producing an explicit expression for area A as a function of width W. Find the domain and the range of your function.

54. (a) Find a straight-line function that gives federal income tax due in 1982 as a function of net income for single taxpayers using the tax tables and earning between $12,900 and $15,000 a year.
(b) Graph this function.
(c) Determine the tax due on an income of $14,000.

55. (a) Find a straight-line function that gives tax due in 1982 as a function of net income for single taxpayers using the tax tables and earning between $28,800 and $34,100 a year.
(b) Graph this function.
(c) Determine the tax due on an income of $32,750.

56. A spotlight shines on a screen; between them is an obstruction that casts a shadow (see Figure 5). Suppose the screen is vertical, 20 m wide by 15 m high, and 50 m from the spotlight. Also suppose the obstruction is a square, 1 m on a side, and is parallel to the screen. Express the area of the shadow as a function of the distance from the light to the obstruction.

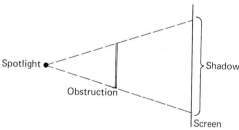

FIGURE 5

*57. A baseball diamond is a square 90 ft long on each side. Casey runs a constant 30 ft/sec whether he hits a ground ball or a home run. Today in his first time at bat, he hit a home run. Write an expression for the function that measures his line-of-sight distance from second base as a function of the time t, in seconds, after he left home plate.

58. Let $f(x)$ be the fifth decimal place of the decimal expansion of x. For example, $f(\frac{1}{64}) = f(0.015625) = 2, f(98.786543210) = 4, f(-78.90123456) = 3$, and so on. Find the domain and range of f.

59. Alec, on vacation in Canada, found that he got a 12% premium on his U.S. money. When he returned, he discovered there was a 12% discount on converting his Canadian money back into U.S. currency. Describe each conversion function, and then show that one is not the inverse of the other. That is, show that after converting both ways, Alex lost money.

1.7 OPERATIONS WITH FUNCTIONS

We begin this section by showing how functions can be added, subtracted, multiplied, and divided.

Definition 1 SUM, DIFFERENCE, PRODUCT, and QUOTIENT FUNCTIONS Let f and g be two functions.

(i) The **sum** $f + g$ is defined by

$$(f + g)(x) = f(x) + g(x), \tag{1}$$

(ii) The **difference** $f - g$ is defined by

$$(f - g)(x) = f(x) - g(x), \tag{2}$$

(iii) The **product** $f \cdot g$ is defined by

$$(f \cdot g)(x) = f(x)g(x), \tag{3}$$

(iv) The **quotient** f/g is defined by

$$\left(\frac{f}{g}\right)(x) = \frac{f(x)}{g(x)}. \tag{4}$$

Furthermore, $f + g$, $f - g$, and $f \cdot g$ are defined for all x for which both f and g are defined. That is,

$$\mathrm{dom}(f + g) = \mathrm{dom}(f - g) = \mathrm{dom}(f \cdot g) = \mathrm{dom}\, f \cap \mathrm{dom}\, g.$$

Finally, f/g is defined whenever both f and g are defined and $g(x) \neq 0$ (so that we do not divide by zero). That is,

$$\mathrm{dom}\left(\frac{f}{g}\right) = \mathrm{dom}\, f \cap \mathrm{dom}\, g - \{x: g(x) = 0\}.$$

EXAMPLE 1 Let $f(x) = \sqrt{x + 1}$ and $g(x) = \sqrt{4 - x^2}$. Since $\mathrm{dom}\, f = [-1, \infty)$ and $\mathrm{dom}\, g = [-2, 2]$, we have $\mathrm{dom}(f + g) = \mathrm{dom}(f - g) = \mathrm{dom}(f \cdot g) = [-1, \infty) \cap [-2, 2] = [-1, 2]$, and $\mathrm{dom}(f/g) = [-1, 2] - \{x: \sqrt{4 - x^2} = 0\} = [-1, 2] - \{-2, 2\} = [-1, 2)$. The functions are

$$(f + g)(x) = \sqrt{x + 1} + \sqrt{4 - x^2}$$
$$(f - g)(x) = \sqrt{x + 1} - \sqrt{4 - x^2}$$
$$(f \cdot g)(x) = \sqrt{x + 1} \cdot \sqrt{4 - x^2} = \sqrt{(x + 1)(4 - x^2)}$$
$$\left(\frac{f}{g}\right)(x) = \frac{\sqrt{x + 1}}{\sqrt{4 - x^2}} = \sqrt{\frac{x + 1}{4 - x^2}}. \quad \blacksquare$$

Definition 1 can lead to some confusion if we are not careful.

EXAMPLE 2   Let $f(x) = \sqrt{x}$ and $g(x) = 2\sqrt{x}$. Then

$$\text{dom } f = [0, \infty), \qquad \text{dom } g = [0, \infty),$$

and

$$\text{dom}(f \cdot g) = \text{dom } f \cap \text{dom } g = [0, \infty).$$

But $(f \cdot g)(x) = \sqrt{x} \cdot 2\sqrt{x} = 2x$, which is defined for every real number x—or is it? The apparent problem here lies in the definition of a function. The function fg, with domain $[0, \infty)$, is defined as the rule $(fg)(x) = 2x$ for $x \geq 0$. This function is *not* the same function as the function defined by $h(x) = 2x$ *without* the restriction that $x \geq 0$. These two functions are not the same since they have different domains. They also have different graphs. These graphs are sketched in Figure 1. We will, for most of the remainder of this book, not worry about such distinctions—but you should be aware that some care has to be taken when dealing with functions. $\blacksquare$

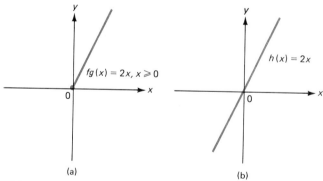

(a)   (b)

FIGURE 1

We will often need to deal with functions of functions.

Definition 2   COMPOSITE FUNCTION   If f and g are functions, then the **composite function** $f \circ g$ is defined by

$$(f \circ g)(x) = f(g(x)) \tag{5}$$

and

$$\mathrm{dom}(f \circ g) = \{x : x \in \mathrm{dom}\ g \text{ and } g(x) \in \mathrm{dom}\ f\}.$$

That is, $(f \circ g)(x)$ is defined for every x such that $g(x)$ and $f(g(x))$ are defined.

EXAMPLE 3   Let $f(x) = \sqrt{x}$ and $g(x) = x^2 + 1$. Then

$$(f \circ g)(x) = f(g(x)) = f(x^2 + 1) = \sqrt{x^2 + 1}$$

and

$$(g \circ f)(x) = g(f(x)) = g(\sqrt{x}) = (\sqrt{x})^2 + 1 = x + 1.$$

Note, for example, that $(f \circ g)(3) = \sqrt{10}$, while $(g \circ f)(3) = 4$. Now dom $f = \mathbb{R}^+$, dom $g = \mathbb{R}$, and we have†

$$\mathrm{dom}(f \circ g) = \{x : g(x) = x^2 + 1 \in \mathrm{dom}\ f\}.$$

But since $x^2 + 1 > 0$, $x^2 + 1 \in \mathrm{dom}\ f$ for every real x, so $\mathrm{dom}(f \circ g) = \mathbb{R}$. On the other hand, $\mathrm{dom}(g \circ f) = \mathbb{R}^+$ since f is only defined for $x \geq 0$. ∎

▨   **WARNING**   As Example 3 suggests, it is *not* true in general that $(f \circ g)(x) = (g \circ f)(x)$.

EXAMPLE 4   Let $f(x) = 3x - 4$ and $g(x) = x^3$. Then

$$(f \circ g)(x) = f(g(x)) = f(x^3) = 3x^3 - 4$$

and

$$(g \circ f)(x) = g(f(x)) = g(3x - 4) = (3x - 4)^3.$$

Here $\mathrm{dom}(f \circ g) = \mathrm{dom}(g \circ f) = \mathbb{R}$. Note that the functions $f \circ g$ and $g \circ f$ are, as in Example 3, quite different. ∎

EXAMPLE 5   Let $f(x) = 4x + 8$ and $g(x) = \frac{1}{4}x - 2$. Then

$$(f \circ g)(x) = f(g(x)) = f(\tfrac{1}{4}x - 2) = 4(\tfrac{1}{4}x - 2) + 8 = x - 8 + 8 = x.$$

Similarly,

$$(g \circ f)(x) = g(f(x)) = g(4x + 8) = \tfrac{1}{4}(4x + 8) - 2 = x + 2 - 2 = x.$$

Thus,

$$(f \circ g)(x) = (g \circ f)(x) = x.$$

When these last equations hold, we say that f and g are *inverse functions*. Inverse functions are discussed in detail in Section 6.1. ∎

---

†$\mathbb{R}^+ = \{x \in \mathbb{R} : x \geq 0\}$

PROBLEMS 1.7

In Problems 1–8 two functions f and g are given. Calculate $f + g$, $f - g$, $f \cdot g$, and f/g, and determine their respective domains.

1. $f(x) = 2x - 5$, $g(x) = -4x$

2. $f(x) = x^2$, $g(x) = x + 1$

3. $f(x) = \sqrt{x + 2}$, $g(x) = \sqrt{2 - x}$

4. $f(x) = x^3 + x$, $g(x) = \dfrac{1}{\sqrt{x + 1}}$

5. $f(x) = 1 + x^5$, $g(x) = 1 - |x|$

6. $f(x) = \sqrt{1 + x}$, $g(x) = \dfrac{1}{x^5}$

7. $f(x) = \sqrt[5]{x + 2}$, $g(x) = \sqrt[4]{x - 3}$

8. $f(x) = \dfrac{x}{x + 1}$, $g(x) = \dfrac{x - 1}{x}$

In Problems 9–20, find $f \circ g$ and $g \circ f$, and determine the domain of each.

9. $f(x) = x + 1$, $g(x) = 2x$

10. $f(x) = x^2$, $g(x) = 2x + 3$

11. $f(x) = 3x + 5$, $g(x) = 5x + 2$

12. $f(x) = \sqrt{x + 1}$, $g(x) = x^4$

13. $f(x) = \dfrac{x}{x + 2}$, $g(x) = \dfrac{x - 1}{x}$

14. $f(x) = |x|$, $g(x) = -x$

15. $f(x) = \sqrt{1 - x}$, $g(x) = \sqrt{x - 1}$

16. $f(x) = \dfrac{|x|}{x}$, $g(x) = x^2$

17. $f(x) = \dfrac{1}{\sqrt{x + 1}}$, $g(x) = x^5$

18. $f(x) = \sqrt[3]{x + 1}$, $g(x) = \sqrt[4]{x - 1}$

19. $f(x) = \begin{cases} x, & x \geq 0 \\ 2x, & x < 0 \end{cases}$, $g(x) = \begin{cases} -3x, & x \geq 0 \\ 5x, & x < 0 \end{cases}$

*20. $f(x) = \begin{cases} x^2, & x > 0 \\ x^3, & x \leq 0 \end{cases}$, $g(x) = \begin{cases} \sqrt{x}, & x > 0 \\ \sqrt{-x}, & x \leq 0 \end{cases}$

21. Let $f(x) = 2x + 4$ and $g(x) = \frac{1}{2}x - 2$. Show that $(f \circ g)(x) = (g \circ f)(x) = x$.

22. For $f(x) = 3x + 2$, find a function g such that $(f \circ g)(x) = (g \circ f)(x) = x$.

23. For $f(x) = x^2$, find two functions g such that $(f \circ g)(x) = x^2 - 10x + 25$.

*24. For $f(x) = ax + b$, find a function g such that $(f \circ g)(x) = (g \circ f)(x) = x$. Assume that $a \neq 0$.

*25. For $f(x) = \sqrt[3]{x - 1}$, find a function g such that $(f \circ g)(x) = (g \circ f)(x) = x$.

26. Let $h(x) = 1/\sqrt{x^2 + 1}$. Determine two functions f and g such that $f \circ g = h$.

27. Let $h(x) = \sqrt[3]{x^2 - 5}$. Determine two functions f and g such that $f \circ g = h$.

28. Let $h(x) = (3x^7 - 5)/8$. Determine two functions f and g such that $f \circ g = h$.

29. Let $k(x) = (1 + \sqrt{x})^{5/7}$. Find dom k. Determine three functions f, g, and h such that $f \circ g \circ h = k$.

30. Let $h(x) = x^2 + x$ and let $f_1(x) = x^2 - x$, $g_1(x) = x + 1$, $f_2(x) = x^2 + 3x + 2$, and $g_2(x) = x - 1$. Show that $f_1 \circ g_1 = f_2 \circ g_2 = h$. This result illustrates that there is often more than one way to write a given function as the composition of two other functions.

*31. Suppose f is a function with domain $[0, \infty)$ and range $[5, \infty)$ such that if $0 \leq x < y$, then $f(x) < f(y)$. Show that there is a function g with domain $[5, \infty)$ and range $[0, \infty)$ such that the following hold:
(a) $f(g(s)) = s$ for all $s \in [5, \infty)$.
(b) $g(f(t)) = t$ for all $t \in [0, \infty)$.

32. Let f and g be the following straight-line functions:

$$f(x) = ax + b, \qquad g(x) = cx + d.$$

Find conditions on a and b in order that $f \circ g = g \circ f$.

*33. For any two subsets A and B of dom f, show the following:
(a) $f(A \cup B) = f(A) \cup f(B)$ (b) $f(A \cap B) \subseteq f(A) \cap f(B)$

34. Each of the following functions satisfies an equation of the form $(f \circ f)(x) = x$ or $(g \circ g \circ g)(x) = x$ or $(h \circ h \circ h \circ h)(x) = x$, and so on. For each function, discover what type of equation is appropriate.

(a) $A(x) = \sqrt[3]{1 - x^3}$

(b) $B(x) = \sqrt[7]{23 - x^7}$

(c) $C(x) = 1 - \dfrac{1}{x}$, dom $C = \mathbb{R} - \{0, 1\}$

(d) $D(x) = \dfrac{1}{1 - x}$, dom $D = \mathbb{R} - \{0, 1\}$

(e) $E(x) = \dfrac{x + 1}{x - 1}$, dom $E = \mathbb{R} - \{1\}$

(f) $F(x) = \dfrac{x - 1}{x + 1}$, dom $F = \mathbb{R} - \{-1, 0, 1\}$

(g) $G(x) = \dfrac{4x - 1}{4x + 2}$, dom $G = \mathbb{R} - \left\{-\dfrac{1}{2}, 0, \dfrac{1}{4}, \dfrac{1}{2}, 1\right\}$

*35. Let $R(x) = 1/x$ and $S(x) = 1 - x$; also let J be the **identity function;** that is, $J(x) = x$. For this problem, suppose each function has domain $\mathbb{R} - \{0, -1\}$. Note that $R \circ R = J = S \circ S$.

(a) Let $T = R \circ S$. Show that $T \circ T = S \circ R$ and $T \circ T \circ T = J$.

(b) Let $U = R \circ S \circ R$. Show that $U = S \circ R \circ S$ and $U \circ U = J$.

(c) Verify that $T \circ R = R \circ T \circ T = T \circ T \circ S = S \circ T$.

(d) Show that the set of six functions $\{J, R, S, T, T \circ T, U\}$ is closed with respect to the operation of composition; that is, if F and G belong to the set, then so does $F \circ G$.

(e) Show that for each function F in the set $\{J, R, S, T, T \circ T, U\}$ there is a unique function G in the set such that $F \circ G = J = G \circ F$.

REMARK.   This set of functions forms what is called a **group.**

*36. Suppose that F and G are functions with the same domain. Describe the graph of the following:

(a) $(y - F(x)) \cdot (y - G(x)) = 0$

(b) $|y - F(x)| + |y - G(x)| = 0$

(c) $(y - F(x))^2 + (y - G(x))^2 = 0$

1.8 SHIFTING THE GRAPHS OF FUNCTIONS (OPTIONAL)

While more advanced methods are needed to obtain the graphs of most functions (without plotting a large number of points), there are some techniques that make it a relatively simple matter to sketch certain functions. As an illustration of what we have in mind, consider the following six functions:

(a) $f(x) = x^2$ (b) $g(x) = x^2 + 1$ (c) $h(x) = x^2 - 1$
(d) $j(x) = (x - 1)^2$ (e) $k(x) = (x + 1)^2$ (f) $l(x) = -x^2$

These functions are all graphed in Figure 1.

   In (a) we use the graph of $y = x^2$ obtained in Figure 1.6.1. For (b) in order to graph $y = x^2 + 1$, we simply add 1 unit to every y value obtained in (a); that is, we shift the graph of $y = x^2$ *up* one unit. Analogously, for (c) we simply shift the graph of $y = x^2$ *down* one unit to obtain the graph of $y = x^2 - 1$.

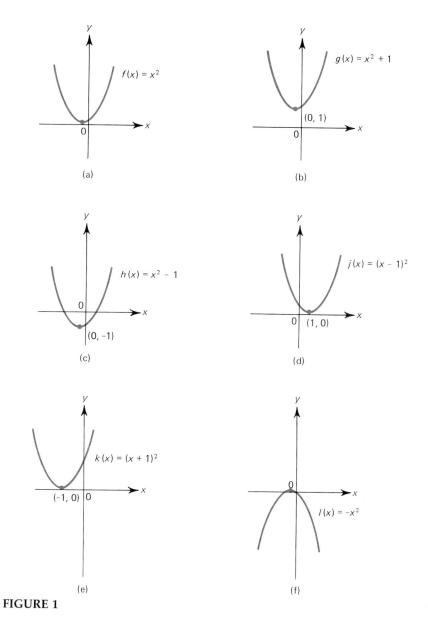

FIGURE 1

The analysis of the graph in (d) is a little trickier. Since, for example, $y = 0$ when $x = 0$ for the function $y = x^2$, then $y = 0$ when $x = 1$ for the function $y = (x - 1)^2$. Similarly, $y = 4$ when $x = -2$ if $y = x^2$, and $y = 4$ when $x = -1$ if $y = (x - 1)^2$. By continuing in this manner, we can see that y values in the graph of $y = x^2$ are the same as y values in the graph of $y = (x - 1)^2$ except that they are achieved one unit later. Some representative values are given in Table 1. Thus we find that the graph of $y = (x - 1)^2$ is the graph of $y = x^2$ *shifted one unit to the right.*

Similarly, in (e) we find that the graph of $y = (x + 1)^2$ is the graph of $y = x^2$ shifted *one unit to the left.* Some values are given in Table 2. Finally, in (f) to obtain the graph of $y = -x^2$, we note that each y value is replaced by its negative, so the graph of $y = -x^2$ is the graph of $y = x^2$ *reflected through the x-axis* (i.e., turned upside down).

TABLE 1

x	x^2	$(x - 1)^2$
−5	25	36
−4	16	25
−3	9	16
−2	4	9
−1	1	4
0	0	1
1	1	0
2	4	1
3	9	4
4	16	9
5	25	16

TABLE 2

x	x^2	$(x + 1)^2$
−5	25	16
−4	16	9
−3	9	4
−2	4	1
−1	1	0
0	0	1
1	1	4
2	4	9
3	9	16
4	16	25
5	25	36

In general, we have the following rules. Let $y = f(x)$.

(i) To obtain the graph of $y = f(x) + c$, shift the graph of $y = f(x)$ *up* c units if $c \geq 0$ and *down* $|c|$ units if $c < 0$.

(ii) To obtain the graph of $y = f(x - c)$, shift the graph of $y = f(x)$ *to the right* c units if $c > 0$ and *to the left* $|c|$ units if $c < 0$.

(iii) To obtain the graph of $y = -f(x)$, *reflect* the graph of $y = f(x)$ *through the* x-*axis*.

(iv) To obtain the graph of $y = f(-x)$, *reflect* the graph of $y = f(x)$ *through the* y-*axis*.

EXAMPLE 1   The graph of $y = \sqrt{x}$ is given in Figure 2a. Then using the above rules, the graphs of $\sqrt{x} + 3$, $\sqrt{x} - 2$, $\sqrt{x - 2}$, $\sqrt{x + 3} = \sqrt{x - (-3)}$, $-\sqrt{x}$, and $\sqrt{-x}$ are given in the other parts of Figure 2. ■

(a)

(b)

(c)

FIGURE 2   (d)   (e)

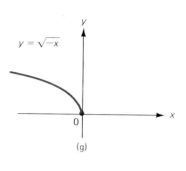

(f)

(g)

FIGURE 2 (Cont.)

EXAMPLE 2   The graph of a certain function $f(x)$ is given in Figure 3a. To obtain the graph of $-f(3 - x)$, we proceed as follows:

   (i) Reflect through the y-axis to obtain the graph of $f(-x)$ (Figure 3b).
   (ii) Shift to the right three units to obtain the graph of $f(-(x - 3)) = f(3 - x)$ (Figure 3c).
   (iii) Reflect through the x-axis to obtain the graph of $y = -f(3 - x)$ (Figure 3d). ■

(a)

(b)

(c)

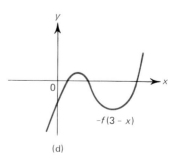

(d)

FIGURE 3

EXAMPLE 3   Graph the parabola $y = x^2 - 10x + 22$.

   Solution. Completing the square, we see that

$$x^2 - 10x + 22 = x^2 - 10x + 25 - 3 = (x - 5)^2 - 3.$$

Thus the graph of $f(x)$ is obtained by shifting the graph of $y = x^2$ to the right five units and then down three units, as in Figure 4. ■

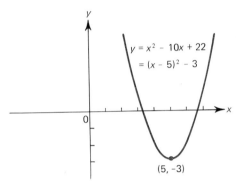

FIGURE 4

PROBLEMS 1.8

1. The graph of $f(x) = x^3$ is given in Figure 5. Sketch the graphs of the following:
 (a) $f(x) = (x - 2)^3$ **(b)** $f(x) = -x^3$ **(c)** $f(x) = (4 - x)^3 + 5$

2. The graph of $f(x) = 1/x$ is given in Figure 6. Sketch the graphs of the following:

 (a) $f(x) = \dfrac{1}{x + 3}$ **(b)** $f(x) = 2 - \dfrac{1}{x}$

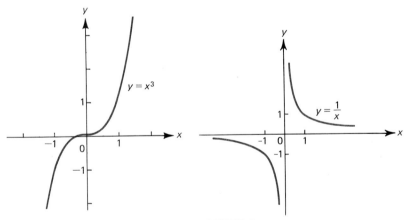

        FIGURE 5   **FIGURE 6**

3. After completing the squares, sketch the graphs of these parabolas:
 (a) $y = x^2 - 4x + 7$ **(b)** $y = x^2 + 8x + 2$
 (c) $y = x^2 + 3x + 4$
 (d) $y = -x^2 + 2x - 3$ [*Hint:* Write $-x^2 + 2x - 3 = -(x^2 - 2x + 3)$.]
 (e) $y = -x^2 - 5x + 8$
In Problems 4–12 the graph of a function is sketched. Obtain the graphs of (a) $f(x - 2)$, (b) $f(x + 3)$, (c) $-f(x)$, (d) $f(-x)$, (e) $f(2 - x) + 3$.

4.

5.

6.

7.

8.

9.

10.

11.

12.

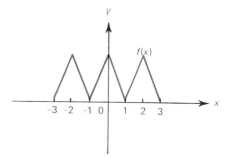

REVIEW QUESTIONS FOR CHAPTER 1

1. Let $A = (-\infty, 2]$, $B = (-4, \infty)$, and $C = [-1, 1]$. Calculate $A \cup B$, $A \cup C$, $B \cup C$, $A \cap B$, $A \cap C$, $B \cap C$, $\overline{A}$, $\overline{B}$, $\overline{C}$, $A - B$, $B - A$, $A - C$, $C - A$, $B - C$, and $C - B$.

In Exercises 2–9, find and graph the solution sets of the given inequalities.

2. $|x| < 2$

3. $|x| \geq 4$

4. $|x - 1| \leq 5$

5. $|x + 3| > 1$

6. $\left| \dfrac{2x - 3}{4} \right| \leq 4$

7. $x^2 + 6x + 9 \geq 0$

8. $x^2 + x - 2 < 0$

9. $\dfrac{4}{x + 3} > \dfrac{1}{x - 2}$

10. Find the distance between these points:
    (a) $(1, 5), (-3, 2)$ (b) $(-6, 1), (-11, -4)$
11. Find the midpoint of the line segment joining the points $(-3, 2)$ and $(5, -8)$.
12. Find the area of the triangle with vertices at $(1, 4)$, $(1, 8)$ and $(7, 8)$.

In Exercises 13–18, find the slope-intercept equation and angle of inclination of a line when either two points on it or its slope and one point are given.

13. $(2, 5), (-1, 3)$ 14. $(-2, 4), m = 3$ 15. $(3, -1), (1, -3)$
16. $(-1, 4), m = 2$ 17. $(1, 4),(1, 7)$ 18. $(3, -8), (-8, -8)$

19. Find the equation of the line parallel to the line $2x - 5y = 6$ that passes through the point $(4, -2)$.
20. Find the equation of the line perpendicular to the line $3x + 2y = 7$ that passes through the point $(-2, 3)$.
21. Find the distance between the line $-x + 3y = 4$ and the point $(2, 3)$.
22. Find the equation of the circle with radius 3 centered at the point $(-1, 2)$.
23. Find the equation of the line tangent to the circle $(x - 1)^2 + y^2 = 10$ at the point $(2, 3)$.

In Exercises 24–32, determine whether the given equation defines a function and, if so, find its domain and range.

24. $4x - 2y = 5$

25. $\dfrac{x^2 - y}{2} = 4$

26. $\dfrac{y}{x} = 1$

27. $(x - 1)^2 + (y - 3)^2 = 4$

28. $y = \sqrt{x + 2}$

29. $3 = \dfrac{1 + x^2 + x^4}{2y}$

30. $y = \dfrac{x}{x^2 + 1}$

31. $y = \dfrac{x}{x^2 - 1}$

32. $y = \sqrt{x^2 - 6}$

33. For $y = f(x) = \sqrt{x^2 - 4}$, calculate $f(2)$, $f(-\sqrt{5})$, $f(x + 4)$, $f(x^3 - 2)$, and $f(-1/x)$.
34. If $y = f(x) = 1/x$, show that for $\Delta x \neq 0$,

$$\frac{f(x + \Delta x) - f(x)}{\Delta x} = \frac{-1}{x(x + \Delta x)}.$$

35. Let $f(x) = \sqrt{x + 1}$ and $g(x) = x^3$. Find $f + g, f - g, f \cdot g, g/f, f \circ g$, and $g \circ f$, and determine their respective domains.
36. Do the same as in Exercise 35 for $f(x) = 1/x$ and $g(x) = x^2 - 4x + 3$.
37. For $f(x) = 4x - 6$, find a function $g(x)$ such that $(f \circ g)(x) = (g \circ f)(x) = x$.
38. The graph of the function $y = f(x)$ is given in Figure 1. Sketch the graphs of $f(x - 3)$, $f(x) - 5, f(-x), -f(x)$, and $4 - f(1 - x)$.
39. Do the same as in Exercise 38 for the function graphed in Figure 2.

FIGURE 1

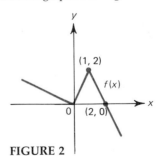

FIGURE 2

2 Limits and Derivatives

2.1 INTRODUCTION TO THE DERIVATIVE

From before the time of the great Greek scientist Archimedes† (287–212 B.C.), mathematicians were concerned with two important problems:

PROBLEM 1   Find the unique tangent line (if one exists) to a given curve at a given point on the curve (see Figure 1).

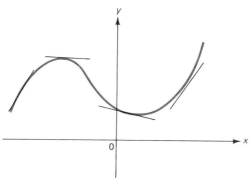

FIGURE 1

---

†We will refer to the work of Archimedes a great deal in Chapter 5.

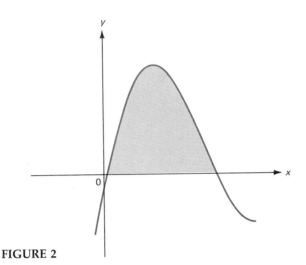

FIGURE 2

PROBLEM 2   Find the area bounded by a given curve and the x-axis (see Figure 2).

The solution to Problem 1 gave rise to what is now termed *differential calculus*, while the solution to Problem 2 gave rise to *integral calculus*.

In this chapter we begin our study of differential calculus. In Chapter 5 we take up Problem 2. We will see that the solutions to both problems are similar; that is, differential and integral calculus are really two closely related parts of the same subject.

The Greeks knew how to find the line tangent to a circle at a given point, using the fact that for a circle the tangent line is perpendicular to the radius at the given point (see Figure 3). The Greeks also discovered how to construct tangent lines to other particular curves, and Archimedes himself devoted a major part of one of his books (*On Spirals*) to the tangent problem for a special curve that is called the *spiral of Archimedes.* †

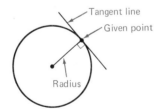

FIGURE 3

However, as more and more curves were studied, it became increasingly difficult to treat the large number of special cases, and a general method was sought to solve all such problems. Unfortunately, these early attempts met with failure, and it wasn't until the independent discoveries of Isaac Newton (1642–1727) and Gottfried Leibniz‡ (1646–1716) that the problem was resolved.

---

†We will discuss the spiral of Archimedes in Chapter 11 (see Example 11.2.9).

‡See the accompanying biographical sketches.

1642–1727

Sir Isaac Newton

Isaac Newton
The Granger Collection

Isaac Newton was born in the small English town of Woolsthorpe on Christmas Day 1642, the year of Galileo's death. His father, a farmer, had died before Isaac was born. His mother remarried when he was three and, thereafter, Isaac was raised by his grandmother. As a boy, Newton showed great cleverness and inventiveness—designing a water clock and a toy gristmill, among other things. One of his uncles, a Cambridge graduate, took an interest in the boy's education, and as a result, Newton entered Trinity College, Cambridge in 1661. His primary interest at that time was chemistry.

Newton's interest in mathematics began with his discovery of two of the great mathematics books of his day: Euclid's *Elements* and Descartes's *La géométrie*. He also became aware of the work of the great scientists who preceded him, including Galileo and Fermat.

By the end of 1664, Newton seems to have mastered all the mathematical knowledge of the time and had begun adding substantially to it. In 1665 he began his study of the rates of change, or *fluxions*, of quantities, such as distances or temperatures that varied continuously. The result of this study was what today we call *differential calculus*.

Newton disliked controversy so much that he delayed the publication of many of his findings for years. An unfortunate result of one of these delays was a conflict with Leibniz over who first discovered calculus. Leibniz made similar discoveries at about the same time as Newton, and to this day there is no universal agreement as to who discovered what first. The conflict stirred up so much ill will that English mathematicians (supporters of Newton) and continental mathematicians (supporters of Leibniz) had virtually no communication for more than a hundred years. English mathematics suffered greatly as a result.

Newton made many of the discoveries that governed physics until the discoveries of Einstein early in this century. In 1679 he used a new measurement of the radius of the earth, together with an analysis of the earth's motion, to formulate his universal law of gravitational attraction. Although he made many other discoveries at that time, he communicated them to no one for five years. In 1684 Edmund Halley (after whom Halley's comet is named) visited Cambridge to discuss his theories of planetary motion with Newton. The conversations with Halley stimulated Newton's interest in celestial mechanics and led him to work out many of the laws that govern the motion of bodies subject to the forces of gravitation. The result of this work was the 1687 publication of Newton's masterpiece, *Philosophiae naturalis principia mathematica* (known as the *Principia*). It was received with great acclaim throughout Europe.

Newton is considered by many the greatest mathematician the world has ever produced. He was the greatest "applied" mathematician, determined by his ability to discover a physical property and analyze it in mathematical terms. Leibniz once said, "Taking mathematics from the beginning of the world to the time when Newton lived, what he did was much the better half." The great English poet Alexander Pope wrote,

> Nature and Nature's laws lay hid in night;
> God said, 'Let Newton be,' and all was light.

Newton, by contrast, was modest about his accomplishments. Late in life he wrote, "If I have seen farther than Descartes, it is because I have stood on the shoulders of giants." All who study mathematics today are standing on Isaac Newton's shoulders.

Gottfried Wilhelm Leibniz

Gottfried Wilhelm Leibniz
The Granger Collection

Born in Leipzig, Germany, Gottfried Wilhelm Leibniz entered the university there at the age of fifteen and received the bachelor's degree before his eighteenth birthday. Truly a Renaissance man, Leibniz taught himself Latin and Greek when he was still a child. At the university he studied law, philosophy, theology, and mathematics. His studies were widespread, and he is considered by many to be the last scholar to have amassed universal knowledge. Leibniz was prepared for the degree of doctor of laws before he was twenty, but the university refused to grant the degree because of his relative youth.

When his law degree was refused, Leibniz moved to Nuremberg. There he wrote a highly regarded essay on teaching law by the historical method and almost immediately thereafter was offered a professorship of law. Leibniz refused this appointment, however, and instead was commissioned by the Elector of Mainz to work on rewriting some statutes. From this time onward, Leibniz worked in the diplomatic service. One of the diplomats Leibniz served was an Elector of Hanover who was a great-grandson of James I of England and who became King George I in 1714.

In the diplomatic service Leibniz traveled widely. In Paris in 1672, he met the great Dutch physicist and mathematician Christiaan Huygens (1629–1695) and persuaded Huygens to tutor him in mathematics. The next year Leibniz had already begun his work on the calculus. He had discovered the fundamental theorem of calculus, knew how to differentiate a wide variety of functions, and had developed much of his notation. On October 29, 1675, he first used the modern integral sign, the elongated S that stood for the Latin word *summa* (sum). Shortly thereafter he was writing derivatives and integrals in much the same way they are written today.

Leibniz's work for the Elector of Hanover gave him much leisure time to pursue subjects that intrigued him. His writing in philosophy made him one of the leading philosophers of his time, and his gift for languages won him fame as a Sanscrit scholar. He devoted a great deal of energy to a scheme for reuniting the Catholic and Protestant churches— although this was less successful than many of his other endeavors. In 1700 he founded the Berlin Academy of Science.

Many of Leibniz's mathematical papers appeared in the journal *Acta eruditorum*, which he cofounded in 1682. This journal contained his work on calculus and led to the bitter controversy with Newton over who first discovered calculus. It seems that Newton made his discoveries first. In fact, Leibniz may have seen some of Newton's work in 1673, although he was not yet sufficiently knowledgeable about analysis and geometry to fully understand what Newton had written. Nevertheless, this possibility remains one of the sources of the great controversy. In any event, Leibniz was the first to publish the important results in the calculus and was the first to use the notation that has now become standard.

In addition to everything else, Leibniz was a scientist, and he and Huygens developed the idea of kinetic energy. Unfortunately, Leibniz's work as a physicist was greatly overshadowed by the physics of Newton, so that his contributions in that area are now largely forgotten.

In 1714 Leibniz was in Hanover while his employer, the Elector of Mainz became King George I of England. Thereafter, Leibniz was largely ignored, and when he died in 1716, his funeral was attended only by his secretary.

Why should we be concerned with finding tangent lines to curves? We will see many, many answers to that question in this book. However, before beginning a discussion of the general theory of tangent lines, let us give one example of the reasoning that led to the development of calculus. The connection of this example to tangent lines will be made soon.

EXAMPLE A boy is standing on a bridge over a highway. The boy drops a rock from a point exactly 100 feet above the roadway. How fast is the rock traveling when it hits?

Solution. Let $s(t)$ denote the height of the rock above the road t seconds after the rock is released. Then it can be shown that,† neglecting air resistance,

$$s(t) = 100 - 16t^2. \tag{1}$$

The rock hits the road when $s(t) = 0$. So from (1) we have

$$0 = 100 - 16t^2$$
$$16t^2 = 100$$
$$t^2 = \tfrac{100}{16}$$
$$t = \sqrt{\tfrac{100}{16}} = \tfrac{10}{4} = \tfrac{5}{2} = 2\tfrac{1}{2} \text{ sec.}$$

We might now reason as follows: The *average velocity* of a moving object is given by

$$\text{average velocity} = \frac{\text{distance traveled}}{\text{elapsed time}}.$$

So in our case

$$\text{average velocity of the rock} = \frac{-100 \text{ ft}}{2.5 \text{ sec}} = -40 \text{ ft/sec.}‡$$

But this result doesn't answer our question. The expression -40 ft/sec represents the average velocity of the rock during the $2\tfrac{1}{2}$ seconds of its flight. When the rock is released by the boy, it isn't moving at all. As it falls, it gains speed until the moment of impact. Certainly, the velocity on impact, after exactly $2\tfrac{1}{2}$ seconds, is greater than the average velocity. But how do we calculate this *instantaneous velocity*, as it is called, after exactly $2\tfrac{1}{2}$ seconds?

Let us begin by sketching, as in Figure 4, the graph of the function $s(t) = 100 - 16t^2$. We do so by plotting the points given in Table 1 and connecting them.

---

†We will derive this equation in Section 5.2 (see Example 5.2.9).
‡The velocity is negative because the rock is falling.

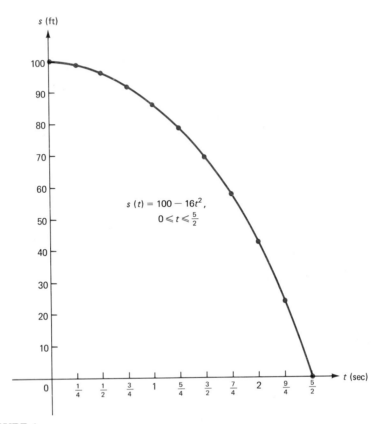

FIGURE 4

TABLE 1

time t	$16t^2$	height $100 - 16t^2$
0	0	100
0.25	1	99
0.5	4	96
0.75	9	91
1	16	84
1.25	25	75
1.5	36	64
1.75	49	51
2	64	36
2.25	81	19
2.5	100	0

Note that $s(t)$ is defined only for $0 \le t \le \frac{5}{2}$, since neither negative time nor negative distance makes any sense in this problem.

Now let us attempt to compute the instantaneous velocity for any value of t in the interval [0, 2.5]. We enlarge the graph in Figure 4 to examine what happens near a point on the curve. The coordinates of any such point are $(t, 100 - 16t^2)$ (see Figure 5).

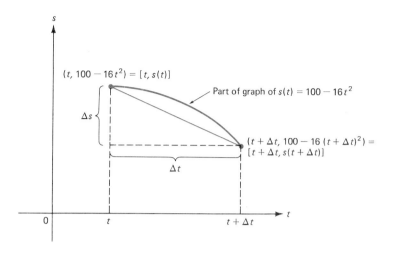

FIGURE 5

The Greek letter Δ is traditionally used to denote changes (differences). Then Δt represents a period of time. Let us compute the average velocity of the rock in the time period t to $t + \Delta t$. We have, as before,

$$\text{average velocity} = \frac{\text{distance traveled}}{\text{elapsed time}} \qquad (2)$$

But in the time interval $[t, t + \Delta t]$ the rock has fallen from a height of $100 - 16t^2$ feet to a height of $100 - 16(t + \Delta t)^2$. Thus from (2) we have

$$\text{average velocity} = \frac{[100 - 16(t + \Delta t)^2] - (100 - 16t^2)}{\Delta t}$$

$$= \frac{[100 - 16(t^2 + 2t\,\Delta t + \Delta t^2)] - 100 + 16t^2}{\Delta t}$$

$$= \frac{100 - 16t^2 - 32t\,\Delta t - 16\,\Delta t^2 - 100 + 16t^2}{\Delta t}$$

$$= \frac{-32t\,\Delta t - 16\,\Delta t^2}{\Delta t} = -32t - 16\,\Delta t. \qquad \text{We divided numerator and denominator by } \Delta t.$$

But we can see from Figure 5 that if Δs denotes the change in the height of the rock, then

$$\text{average velocity between } t \text{ and } t + \Delta t = -32t - 16\,\Delta t$$

$$= \frac{\Delta s}{\Delta t}$$

$$= \text{slope of line joining points } (t, s(t)) \text{ and } (t + \Delta t, s(t + \Delta t)).$$

This line is called a *secant line*, and we see that

$$\text{slope of secant line} = -32t - 16\Delta t. \tag{3}$$

If Δt is very small, then over the time period t to $t + \Delta t$, the velocity changes, but *it does not change very much*. Thus

$$-32t - 16\,\Delta t = \text{average velocity between } t \text{ and } t + \Delta t$$

$$\approx \text{instantaneous velocity at time } t. \tag{4}$$

But as Δt gets smaller and smaller, the approximation in (4) gets better and better. We have

$$\text{instantaneous velocity} = \text{limiting value of average velocity}$$
$$\text{as } \Delta t \text{ approaches } 0. \tag{5}$$

We will spend a good part of this chapter making the statement in (5) more precise. However, we can at this point see that

$$\text{as } \Delta t \text{ approaches } 0, \ -32t - 16\,\Delta t \text{ approaches } -32t. \tag{6}$$

Thus

$$\text{instantaneous velocity at time } t = -32t, \tag{7}$$

and at impact (when $t = 2.5$),

$$\text{instantaneous velocity} = -32(2.5) = -80 \text{ ft/sec } (\approx 54.5 \text{ mi/hr}).$$

The minus sign indicates that the height is decreasing (the rock is going *down*). This is considerably greater than the average velocity of 40 ft/sec.

Now we observe that as Δt becomes smaller, the secant lines approach the line tangent to the curve at the point $(t, 100 - 16t^2)$ (see Figure 6). We therefore see that *the slopes of secant lines approach the slope of the tangent line* as Δt becomes smaller, and we have, from (3),

$$\text{slope of tangent line at point } (t, 100 - 16t^2) = -32t. \tag{8}$$

Or combining (7) and (8), we obtain

$$\text{instantaneous velocity of falling rock} = \text{slope of line tangent to curve}$$
$$s = 100 - 16t^2 \text{ at point}$$
$$(t, 100 - 16t^2).$$

Thus we see that finding the tangent line to a curve has something to do with computing velocity. We will see many other applications of this idea in the chapters that follow.

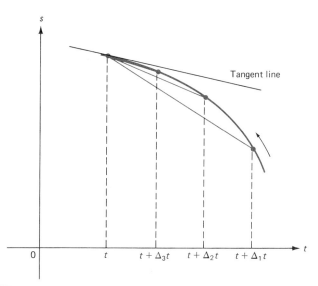

FIGURE 6

In this example we have introduced two important concepts. First, we talked about "limiting values." The notion of a *limit* is central to any study of calculus. We will describe limits in the next three sections.

Second, the slope of the unique tangent line (if one exists) to a point on the graph of a function is called the *derivative* of the function at that point. The calculation and application of derivatives will be the focus of this chapter and Chapters 3 and 4.

2.2 THE CALCULATION OF LIMITS

The notion of a limit, which we will discuss extensively in this chapter, plays a central role in calculus and in much of modern mathematics. However, although mathematics dates back over three thousand years, limits were not really understood until the monumental work of the great French mathematician Augustin-Louis Cauchy† (1789–1857) in the nineteenth century.

In this section we define a limit and show how some limits can be calculated.

EXAMPLE 1   We begin by looking at the function

$$y = f(x) = x^2 + 3. \tag{1}$$

This function is graphed in Figure 1 (see Section 1.8). What happens to $f(x)$ as x gets close to the value $x = 2$? To get an idea, look at Table 1, keeping in mind that x can get close to 2 from the right of 2 and from the left of 2 along the x-axis. It appears from the table that as x gets close to $x = 2$, $f(x) = x^2 + 3$ gets close to 7. This is not surprising since if we now calculate $f(x)$ at $x = 2$, we obtain $f(2) = 2^2 + 3 = 4 + 3 = 7$. In mathematical symbols we write

$$\lim_{x \to 2}(x^2 + 3) = 7.$$

---

†See the accompanying biographical sketch.

Augustin-Louis Cauchy

Augustin-Louis Cauchy
The Granger Collection

Augustin-Louis Cauchy is considered the most out-
standing mathematical analyst of the first half of the
nineteenth century. He was born in Paris in 1789 and
received his early education from his father. In second-
ary school he excelled at classical studies. Entering the
Ecole Polytechnique in 1805, Cauchy greatly impressed
two of the greatest French mathematicians of the time:
Joseph Lagrange (1736–1813) and Simon Laplace
(1749–1827). Although Cauchy studied to be a civil en-
gineer, in 1816 he was persuaded by Lagrange and Laplace to accept a professorship of mathematics at the Ecole
Polytechnique.

Cauchy made many contributions to calculus. In
his 1829 textbook *Leçons sur le calcul différential*, he gave
the first reasonably clear definition of a limit:

*When the successive values attributed to a variable
approach indefinitely a fixed value so as to end by differing
from it by as little as one wishes, this last is called the limit
of all the others.*

Cauchy was the first to define the derivative as the limit of the difference quotient:

$$\frac{\Delta y}{\Delta x} = \frac{f(x + \Delta x) - f(x)}{\Delta x}.$$

He was also responsible for the modern definition of the definite integral as the limit of a sum
(we shall discuss this in Chapter 5).

Cauchy wrote extensively in both pure and applied mathematics. Only Euler wrote
more. He contributed to many areas including real and complex function theory, determinants, probability theory, geometry, wave propogation theory, and infinite series. In the
study of calculus his name appears in the *Cauchy Schwarz inequality*, the *Cauchy mean value
theorem* (Section 12.2), and the *Cauchy root test for convergence of an infinite series* (Section 14.6).

Cauchy is credited with setting a new standard of rigor in mathematical publication.
After Cauchy, it was much more difficult to publish a paper based on intuition; a strict
adherence to formal proof was demanded.

The sheer volume of Cauchy's publication was overwhelming. When the French Academy of Sciences began publishing its journal *Comptes Rendus* in 1835, Cauchy sent his work
there to be published. Soon the printing bill for Cauchy's work alone became so large that
the Academy placed a limit of four pages on each published paper. This rule is still in force
today.

There are some unpleasant stories told of Cauchy. One of the most tragic had to do
with the Norwegian mathematician Niels Henrik Abel (1802–1829). In 1826 Abel, who had
already published some brilliant results, came to Paris in search of an academic position. He
approached Cauchy and gave him an important paper he had just completed. Cauchy
misplaced it. Abel wrote a friend, "Every beginner has a great deal of difficulty in getting
noticed here. I have just finished an extensive treatise on a certain class of transcendental
functions . . . but Mr. Cauchy scarcely deigned to glance at it." While waiting for a suitable
position, Abel lived in an unheated apartment in Paris. In 1829, he died of tuberculosis.

Ironically, a letter offering Abel a professorship of mathematics at the University of Berlin arrived two days after his death.

Cauchy was a political reactionary and a strong supporter of the Bourbons, the French kings who came to power in the years after the French revolution. When King Charles X, the last of the Bourbons, went into exile in 1830, Cauchy was forced to resign his position at the Ecole Polytechnique. He was not allowed to return until 1848—and even then he refused to swear allegiance to the new government. He was also a religious bigot and spent much of his time attempting to convert others to his beliefs.

Cauchy was, however, a courageous defender of academic freedom. In 1843 he published a sharply worded letter in defence of freedom of conscience. This letter was partially responsible for the abolition of the oath of allegiance that Cauchy had so stubbornly refused to sign.

Cauchy died in 1857 at the age of 68 of a bronchial ailment. His last words were spoken to the Archbishop of Paris: "Men pass away, but their deeds abide."

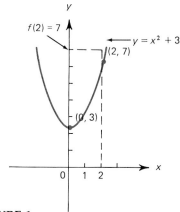

FIGURE 1

TABLE 1

x	$f(x) = x^2 + 3$	x	$f(x) = x^2 + 3$
3	12	1	4
2.5	9.25	1.5	5.25
2.3	8.29	1.7	5.89
2.1	7.41	1.9	6.61
2.05	7.2025	1.95	6.8025
2.01	7.0401	1.99	6.9601
2.001	7.004001	1.999	6.996001
2.0001	7.00040001	1.9999	6.99960001

This notation is read, "The limit as x approaches 2 (or tends to 2) of $x^2 + 3$ is equal to 7."

NOTE.   In order to calculate this limit, we did *not* have to evaluate $x^2 + 3$ at $x = 2$. ■

EXAMPLE 2   What happens to the function $y = f(x) = \sqrt{2x - 6}$ as x gets close to $x = 5$?

Solution. Since when $x = 5$, $\sqrt{2x - 6} = \sqrt{2 \cdot 5 - 6} = \sqrt{10 - 6} = \sqrt{4} = 2$, we might guess that as x gets close to 5, $\sqrt{2x - 6}$ gets close to 2. That this is indeed true is suggested by the computations in Table 2. We write

$$\lim_{x \to 5} \sqrt{2x - 6} = 2. \ \blacksquare$$

TABLE 2

x	$\sqrt{2x - 6}$	x	$\sqrt{2x - 6}$
6.0	2.449489743	4.0	1.414213562
5.5	2.236067977	4.5	1.732050808
5.1	2.049390153	4.9	1.949358869
5.01	2.004993766	4.99	1.994993734
5.001	2.000499938	4.999	1.999499937
5.0001	2.000049999	4.9999	1.999949999

EXAMPLE 3   Consider the function

$$f(x) = \frac{x^2 + 3x + 2}{13x + 8}. \tag{2}$$

To see what happens to this function as x approaches 4, look at Table 3. It seems that $f(x)$ approaches $0.5 = \frac{1}{2}$ as x approaches 4. This result is reasonable since

$$f(4) = \frac{4^2 + 3 \cdot 4 + 2}{13 \cdot 4 + 8} = \frac{16 + 12 + 2}{52 + 8} = \frac{30}{60} = \frac{1}{2}.$$

TABLE 3

x	$f(x) = \dfrac{x^2 + 3x + 2}{13x + 8}$	x	$f(x) = \dfrac{x^2 + 3x + 2}{13x + 8}$
5	0.5753424658	3	0.4255319149
4.5	0.5375939850	3.5	0.4626168224
4.1	0.5075040783	3.9	0.4925042589
4.01	0.5007500416	3.99	0.4992500418
4.001	0.5000750004	3.999	0.4999250004
4.0001	0.5000075000	3.9999	0.4999925000

We write the limit we have discovered as

$$\lim_{x \to 4} \frac{x^2 + 3x + 2}{13x + 8} = \frac{1}{2}. \ \blacksquare$$

EXAMPLE 4   Consider the function

$$f(x) = \frac{x^2 + x}{x}.$$ (3)

Since we cannot divide by zero, this function is defined for every real number except $x = 0$. But $x^2 + x = x(x + 1)$ and $x/x = 1$, so

$$f(x) = \frac{x^2 + x}{x} = \frac{x(x + 1)}{x} = x + 1, \qquad x \neq 0.$$

Let

$$g(x) = x + 1.$$

Then $f(x) = g(x)$ for all $x \neq 0$. However, we emphasize that $f(x)$ and $g(x)$ are *not* the same function since $g(x)$ *is* defined at $x = 0$ while $f(x)$ is not. Nevertheless, for $x \neq 0$, $f(x) = g(x)$. These functions are graphed in Figure 2. It is clear that as x gets close to 0 (without being equal to 0), $f(x)$ gets close to 1 [since $f(x) = x + 1 \approx 0 + 1 = 1$). In mathematical notation we write

$$\lim_{x \to 0} \frac{x^2 + x}{x} = 1.$$ (4)

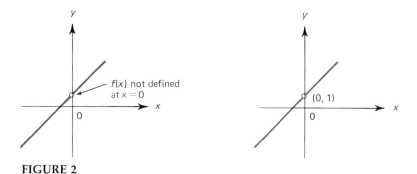

FIGURE 2

It is important to note that for $f(x) = (x^2 + x)/x$, it is still not permissible to set $x = 0$, because this would imply division by zero. However, we now know what happens to this function as x approaches 0. We can see why it is important that we are not required to evaluate $f(x)$ at $x = 0$ when we calculate the limit as x approaches 0. ■

Before giving further examples, we will give a more formal definition of a limit. The definition given below is meant to appeal to your intuition. It is *not* a precise mathematical definition. In this section and the ones that follow, we hope that you will begin to get comfortable with the notion of limits and will acquire some facility in calculating them. Later, at the end of this chapter (in Section 2.8), we will give a formal, mathematically precise definition.

Definition 1   LIMIT   Let L be a real number, and suppose that $f(x)$ is defined on an open interval containing x_0 but not necessarily at x_0 itself. We say that the **limit** as x approaches x_0 of $f(x)$ is L, written

$$\lim_{x \to x_0} f(x) = L,$$
(5)

if, whenever x gets close to x_0 from either side with $x \neq x_0$, $f(x)$ gets close to L.

Here we insist that f be defined on an open interval (see page 4) containing the number x_0 except possibly at x_0 itself. This ensures us that f is defined on both sides of x_0 (see Figure 3). It is important that $f(x)$ get close to L when x gets close to x_0 from either side. In Table 1, for example, $x^2 + 3$ gets close to 7 when x gets close to 2 from the right (the table on the left) and the left (the table on the right). Similarly, when x gets close to 5 from either side, we see that $f(x) = \sqrt{2x - 6}$ gets close to 2.

It should be emphasized that while we do not actually need to know what $f(x_0)$

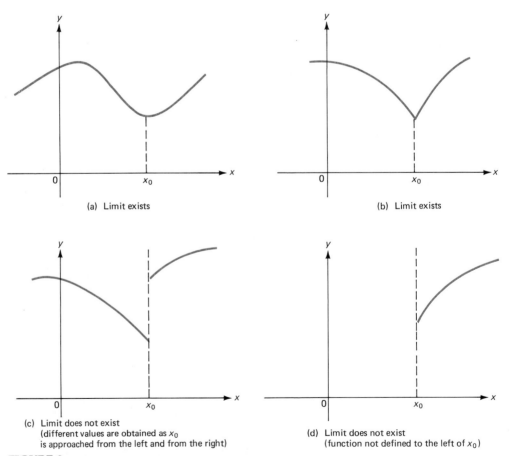

(a) Limit exists

(b) Limit exists

(c) Limit does not exist
(different values are obtained as x_0
is approached from the left and from the right)

(d) Limit does not exist
(function not defined to the left of x_0)

FIGURE 3

is [in fact, $f(x_0)$ need not even exist], it is nevertheless often very helpful to know $f(x_0)$ in the actual computation of $\lim_{x \to x_0} f(x_0)$, since it frequently happens that $\lim_{x \to x_0} f(x)$ indeed equals $f(x_0)$. However, we again emphasize that this result is *not always* the case. In Example 4 we showed that $\lim_{x \to x_0} f(x_0) = 1$ even though $f(0)$ did not exist.

EXAMPLE 5   In the introduction we showed that

$$\lim_{\Delta t \to 0} \frac{[100 - 16(t + \Delta t)^2] - (100 - 16t^2)}{\Delta t} = -32t.$$

The value taken by the function $-32t$ is the slope of the line tangent to the curve $s(t) = 100 - 16t^2$ at the point $(t, 100 - 16t^2)$. ∎

EXAMPLE 6   Calculate

$$\lim_{x \to 0} f(x) = \lim_{x \to 0} \frac{\sqrt{4 + x} - 2}{x}. \tag{6}$$

Solution. Note that if we substitute $x = 0$ in (6), then

$$f(0) = \frac{\sqrt{4 + 0} - 2}{0} = \frac{2 - 2}{0} = \frac{0}{0}$$

which is an undefined expression. In Table 4 we tabulate values of $f(x)$ for x near 0. From the table we conclude that

$$\lim_{x \to 0} \frac{\sqrt{4 + x} - 2}{x} = 0.25.$$

TABLE 4

x	$f(x) = \dfrac{\sqrt{4 + x} - 2}{x}$	x	$f(x) = \dfrac{\sqrt{4 + x} - 2}{x}$
1	0.2360679775	−1	0.2679491924
0.5	0.2426406871	−0.5	0.2583426132
0.1	0.2484567313	−0.1	0.2515823419
0.01	0.2498439448	−0.01	0.2501564457
0.001	0.2499843740	−0.001	0.2500156290
0.0001	0.2499984200	−0.0001	0.2500015900

We can derive this result algebraically:

Multiply and divide by $\sqrt{4 + x} + 2$ 　　　　　　　$(a - b)(a + b) = a^2 - b^2$

$$\lim_{x \to 0} \frac{\sqrt{4 + x} - 2}{x} = \lim_{x \to 0} \frac{(\sqrt{4 + x} - 2)(\sqrt{4 + x} + 2)}{x(\sqrt{4 + x} + 2)} = \lim_{x \to 0} \frac{(\sqrt{4 + x})^2 - 2^2}{x(\sqrt{4 + x} + 2)}$$

$$= \lim_{x \to 0} \frac{(4 + x) - 4}{x(\sqrt{4 + x} + 2)} = \lim_{x \to 0} \frac{x}{x(\sqrt{4 + x} + 2)}.$$

As before, if $x \neq 0$, this last limit is equal to

$$\lim_{x \to 0} \frac{1}{\sqrt{4 + x} + 2} = \frac{1}{\sqrt{4 + 0} + 2} = \frac{1}{4} = 0.25.$$

The technique of multiplying and dividing by a nonzero number to simplify an algebraic expression is one we will use often.

We will see in Section 2.5 (Example 3) that the number $\frac{1}{4}$ is the slope of the tangent to the curve $y = \sqrt{x}$ at the point (4,2). This curve is sketched in Figure 4. ◼

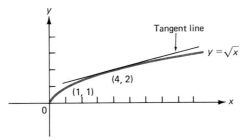

FIGURE 4

EXAMPLE 7   Calculate $\lim_{x \to 0} |x|$.

Solution. From Section 1.2 [see page 9], we have

$$|x| = \begin{cases} x, & x \geq 0 \\ -x, & x \leq 0. \end{cases}$$

If $x > 0$, then $|x| = x$, which tends to 0 as $x \to 0$ from the right of 0. If $x < 0$, then $|x| = -x$, which again tends to 0 as $x \to 0$ from the left of 0. Then since we get the same answer when we approach 0 from the left and from the right, we have

$$\lim_{x \to 0} |x| = 0.$$

This result is pictured in Figure 5. ◼

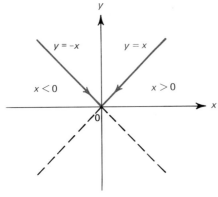

FIGURE 5

EXAMPLE 8 Calculate $\lim\limits_{x \to 0} \dfrac{|x|}{x}$.

Solution. If $x > 0$, then $|x| = x$, so that $|x|/x = x/x = 1$. On the other hand, if $x < 0$, then $|x| = -x$, so that $|x|/x = -x/x = -1$. Note that $|x|/x$ is not defined at $x = 0$. The graph of $|x|/x$ is sketched in Figure 6. In sum, we have

$$\frac{|x|}{x} = \begin{cases} 1, & x > 0 \\ -1, & x < 0. \end{cases}$$

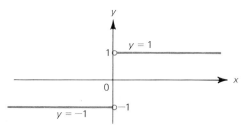

FIGURE 6

From Figure 6 we conclude that $f(x) = |x|/x$ has *no* limit as $x \to 0$; for if $x > 0$, then $f(x)$ remains constant at the value 1, while for $x < 0$, $f(x)$ remains constant at the value -1. Since the value of the limit has to be the same no matter from which direction we approach the value 0, we are left to conclude that there is no limit at 0. Of course, for any other value of x there is a limit. For example, $\lim_{x \to 2} |x|/x = 1$, since near $x = 2$, $|x| = x$ and $|x|/x = 1$. Similarly, $\lim_{x \to -2} |x|/x = -1$. ■

We now turn to a different kind of example.

EXAMPLE 9 Consider the function

$$f(x) = \sqrt{x}.$$

It seems as if $\lim_{x \to 0} \sqrt{x} = 0$, but that is *not* the case since $f(x) = \sqrt{x}$ is *not even defined* for $x < 0$ (we cannot take the square root of a negative number). Therefore as x approaches 0 from the left, $\sqrt{x}$ is not defined, so $\lim_{x \to 0} \sqrt{x}$ *does not exist*. Note also that there is *no* open interval containing 0 in which $f(x)$ is defined. The graph of $f(x) = \sqrt{x}$ is sketched in Figure 7. ■

The result of Example 9 leads to a new definition. Since $\sqrt{x}$ approaches 0 as x approaches 0 from the right, we write

$$\lim_{x \to 0^+} \sqrt{x} = 0.$$

Definition 2 ONE-SIDED LIMITS Let L be a real number.

(i) Suppose that $f(x)$ is defined near x_0 for $x > x_0$ and that as x gets close to x_0 (with $x > x_0$), $f(x)$ gets close to L. Then we say that L is the **right-hand limit** of $f(x)$ as x approaches x_0 and we write

$$\lim_{x \to x_0^+} f(x) = L. \tag{7}$$

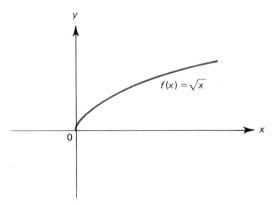

FIGURE 7

(ii) Suppose that $f(x)$ is defined near x_0 for $x < x_0$ and that as x gets close to x_0 (with $x < x_0$), $f(x)$ gets close to L. Then we say that L is the **left-hand limit** of $f(x)$ as x approaches x_0, and we write

$$\lim_{x \to x_0^-} f(x) = L. \tag{8}$$

As before, we stress that these definitions are informal and are only intended to appeal to your intuition. We will give more precise definitions in Section 2.8.

EXAMPLE 10 In Example 8 we discussed the function

$$f(x) = \frac{|x|}{x} = \begin{cases} 1, & x > 0 \\ -1, & x < 0. \end{cases}$$

It follows that

$$\lim_{x \to 0^+} \frac{|x|}{x} = 1 \quad \text{and} \quad \lim_{x \to 0^-} \frac{|x|}{x} = -1. \ \blacksquare$$

EXAMPLE 11 The **greatest integer function** is defined by

$$f(x) = [x]$$

where $[x]$ is the greatest integer smaller than or equal to x. Thus $[3] = 3$, $[\frac{1}{2}] = 0$, $[2.16] = 2$, $[-5.6] = -6$, and so on. A graph of this function is given in Figure 8. From the graph it is evident that

$$\lim_{x \to 2^-} [x] = 1 \quad \text{and} \quad \lim_{x \to 2^+} [x] = 2.$$

Moreover, for any integer n,

$$\lim_{x \to n^-} [x] = n - 1 \quad \text{and} \quad \lim_{x \to n^+} [x] = n. \ \blacksquare$$

In Example 11 we saw that if n is an integer, then $\lim_{x \to n} [x]$ does not exist, since we get different values when $x \to n$ from the left and from the right. For the same

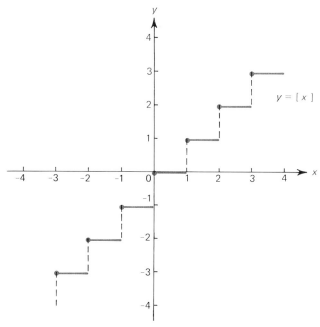

FIGURE 8

reason $\lim_{x \to 0} |x|/x$ does not exist. In general, we have the following theorem, whose proof is given in Section 2.8.

Theorem 1 $\lim_{x \to x_0} f(x) = L$ exists if and only if the following hold:

(i) $\lim_{x \to x_0^+} f(x)$ exists.
(ii) $\lim_{x \to x_0^-} f(x)$ exists.
(iii) $\lim_{x \to x_0^+} f(x) = \lim_{x \to x_0^-} f(x) = L$.

That is, the limit exists if and only if the right- and left-hand limits exist and are equal.

EXAMPLE 12   Let

$$f(x) = \begin{cases} x^2, & x < 1 \\ x^3, & x \geq 1. \end{cases}$$

This function is graphed in Figure 9. Then since for $x > 1$, $f(x) = x^3$, we have

$$\lim_{x \to 1^+} f(x) = \lim_{x \to 1^+} x^3 = 1.$$

Similarly,

$$\lim_{x \to 1^-} f(x) = \lim_{x \to 1^-} x^2 = 1.$$

Since these two limits are equal, we have

$$\lim_{x \to 1} f(x) = 1. \quad \blacksquare$$

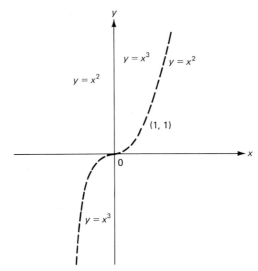

FIGURE 9

EXAMPLE 13   Let

$$f(x) = \begin{cases} x + 1, & x > 0 \\ x - 1, & x < 0. \end{cases}$$

This function is graphed in Figure 10. We immediately see that

$$\lim_{x \to 0^+} f(x) = \lim_{x \to 0^+} (x + 1) = 1$$

and

$$\lim_{x \to 0^-} f(x) = \lim_{x \to 0^-} (x - 1) = -1.$$

Since these limits are different, $\lim_{x \to 0} f(x)$ does not exist.   ∎

PROBLEMS 2.2

1. (a) Draw the graph of the function $f(x) = x + 7$.
   (b) Calculate $f(x)$ for $x = 3, 1, 2.5, 1.5, 2.1, 1.9, 2.01,$ and 1.99.
   (c) Calculate $\lim_{x \to 2}(x + 7)$.
2. (a) Draw the graph of the function $f(x) = x^2 - 4$.
   (b) Calculate $f(x)$ for $x = 2, 0, 1.5, 0.5, 1.1, 0.9, 1.01,$ and 0.99.
   (c) Calculate $\lim_{x \to 1}(x^2 - 4)$.
3. (a) Draw the graph of the function $f(x) = x^2 - 3x + 4$.
   (b) Calculate $f(x)$ for $x = -0.5, -1.5, -0.9, -1.1, -0.99,$ and -1.01.
   (c) Calculate $\lim_{x \to -1}(x^2 - 3x + 4)$.
4. Let $f(x) = (x - 1)(x - 2)/(x - 1)$.
   (a) Explain why $f(x)$ is not defined for $x = 1$.
   (b) Find another function that is equal to f for all $x \neq 1$ and *is* defined for $x = 1$.
   (c) Calculate $\lim_{x \to 1}(x - 1)(x - 2)/(x - 1)$.

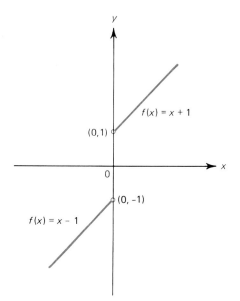

FIGURE 10

5. Let $f(x) = (x^3 - 8)/(x - 2)$.
   (a) Explain why $f(x)$ is not defined at $x = 2$.
   (b) Find another function that is equal to f for all $x \neq 2$ and is defined for $x = 2$. [*Hint:* Calculate $(x^3 - 8)/(x - 2)$ by long division.]
   (c) Calculate $\lim_{x \to 2}(x^3 - 8)/(x - 2)$.
6. Explain why $\lim_{x \to -1} \sqrt{x + 1}$ does not exist.
7. Does $\lim_{x \to -1} \sqrt{x^2 + 1}$ exist? If so, calculate it.

In Problems 8–25, calculate each limit, if it exists, and explain why there is no limit if it does not exist.

8. $\lim_{x \to 5}(x^2 - 6)$

9. $\lim_{x \to 0}(x^3 + 17x + 45)$

10. $\lim_{x \to 0}(-x^5 + 17x^3 + 2x)$

11. $\lim_{x \to 0} \dfrac{1}{x^5 + 6x + 2}$

12. $\lim_{x \to 2}(x^4 - 9)$

13. $\lim_{x \to -1} \dfrac{(x + 1)^2}{x + 1}$

14. $\lim_{x \to -4} \sqrt{x + 4}$

15. $\lim_{x \to 0} \dfrac{x^3}{x^2}$

16. $\lim_{x \to 5} \sqrt{x^2 - 25}$

17. $\lim_{x \to 4} \sqrt{25 - x^2}$

18. $\lim_{x \to 4} \sqrt{x^2 - 25}$

19. $\lim_{x \to 3} \dfrac{x^2 - 4x + 3}{x - 3}$ [*Hint:* Divide.]

20. $\lim_{x \to 2} \sqrt[3]{x^3 - 8}$

21. $\lim_{x \to 2} \sqrt[4]{x^4 - 16}$

22. $\lim_{x \to -2} \dfrac{x^2 + 6x + 8}{x + 2}$

23. $\lim_{x \to 1} \dfrac{x^4 - x}{x^3 - 1}$

24. $\lim_{x \to 1} \dfrac{\sqrt{x} - 1}{x - 1}$ [*Hint:* $a^2 - b^2 = (a + b)(a - b)$.]

25. $\lim\limits_{x \to 2} \dfrac{1 - \sqrt{x/2}}{1 - (x/2)}$

26. Let $f(x) = (x^3 - 6x + 2)/(x^2 + x + 9)$.
  (a) Calculate $f(x)$ for $x = $ 3, 1, 2.5, 1.5, 2.1, 1.9, 2.01, 1.99, 2.001, and 1.999.
  (b) Estimate $\lim_{x \to 2}(x^3 - 6x + 2)/(x^2 + x + 9)$.
  (c) Calculate $f(2)$ and compare it with your estimate.

27. Let $f(x) = \sqrt{x^3 + 13}/(x + 8)$.
  (a) Calculate $f(x)$ for $x = $ -1, -3, -1.5, -2.5, -1.9, -2.1, -1.99, -2.01, -1.999, and -2.001.
  (b) Estimate $\lim_{x \to -2} \sqrt{x^3 + 13}/(x + 8)$.
  (c) Calculate $f(-2)$ and compare it with your estimate.

28. (a) Graph the curve $y = x^2 + 3$.
  (b) Draw (on your graph) the straight line joining the points (1, 4) and (2,7).
  (c) Draw the straight line joining the points (1, 4) and (1.5, 5.25).
  (d) For any real number $h \neq 0$, what is represented by the quotient

$$\frac{[(1 + h)^2 + 3] - 4}{h}?$$

  (e) Calculate $\lim_{h \to 0}\{[(1 + h)^2 + 3] - 4\}/h$.
  (f) What is the slope of the line tangent to the curve $y = x^2 + 3$ at the point (1, 4)?

29. (a) Graph the curve $y = 5 - x^2$.
  (b) Draw (on your graph) the straight line joining the points $(-3, -4)$ and $(-4, -11)$.
  (c) Draw the straight line joining the points $(-3, -4)$ and $(-3.5, -7.25)$.
  (d) For any real number $h \neq 0$, what is represented by the quotient

$$\frac{[5 - (-3 - h)^2] + 4}{-h}?$$

  (e) Calculate $\lim_{h \to 0}\{[5 - (-3 - h)^2] + 4\}/(-h)$.
  (f) What is the slope of the line tangent to the curve $y = 5 - x^2$ at the point $(-3, -4)$?

30. (a) Graph the function $f(x) = |x - 3|$. **(b)** Calculate $\lim_{x \to 3}|x - 3|$.

31. (a) Graph the function $f(x) = |3x - 4|$. **(b)** Calculate $\lim_{x \to 4/3}|3x - 4|$.

32. (a) Graph the function $f(x) = |x + 3|/(x + 3)$.
  (b) Explain why $\lim_{x \to -3}|x + 3|/(x + 3)$ does not exist.
  (c) Calculate $\lim_{x \to 5}|x + 3|/(x + 3)$ and $\lim_{x \to -5}|x + 3|/(x + 3)$.

33. (a) Graph the function $f(x) = |2x - 4|/(2x - 4)$.
  (b) Explain why $\lim_{x \to 2}|2x - 4|/(2x - 4)$ does not exist.
  (c) Calculate $\lim_{x \to 3}|2x - 4|/(2x - 4)$ and $\lim_{x \to 0}|2x - 4|/(2x - 4)$.

*34. Answer true or false. Suppose f is a function such that (1) $\lim_{x \to 7} f(x)$ exists and (2) $f(x) > 4$ for all $x \in \mathbb{R}$.
  (a) $\lim_{x \to 7} f(x) > 4$ **(b)** $\lim_{x \to 7} f(x) > 3.9$
  (c) $\lim_{x \to 7} f(x) > 3.995$ **(d)** $\lim_{x \to 7} f(x) \geq 4$

*35. (a) Find a function A such that

$$\lim_{x \to 2} A(x) = 7 \quad \text{but} \quad A(x) > 7 \quad \text{for all } x.$$

  (b) Find functions B and b such that

$$\lim_{x \to 3} B(x) = \lim_{y \to 3} b(y) \quad \text{but} \quad B(z) = 2 \cdot b(z) \quad \text{for all } z \neq 3.$$

  (c) Find functions C and c such that

$$\lim_{x \to -5} C(x) = \lim_{x \to 7} c(x) \quad \text{but} \quad C(-5) \neq c(7).$$

36. Let $f(x, y) = (2x - 5y)/(3x + 7y)$.
 (a) Find $\lim_{x \to 0}(\lim_{y \to 0} f(x, y))$. **(b)** Find $\lim_{y \to 0}(\lim_{x \to 0} f(x, y))$.
****37.** For each point (x, y) on the curve $y = x^2$, let $A(x)$ be the area of the triangle with vertices $(0, 0)$, $(1, 0)$, and (x, y), and let $B(x)$ be the area of the triangle with vertices $(0, 0)$, $(0, 1)$, and (x, y). Find $\lim_{x \to 0} A(x)/B(x)$ if it exists.
****38.** Compute $\lim_{x \to 0}[-x^2]$ (if it exists) where $[\ \ ]$ denotes the greatest integer function.
****39.** Compute $\lim_{x \to 0} x \cdot [1/x]$ (if it exists).
 40. Merlin strode into calculus class without fanfare and handed the participants the function f, where $f(x) = 7x - 3$ and dom $f = [0, 5)$. Merlin said he would close his eyes and cover his ears while, in turn, each person in the class chose a number s from the domain $[0, 5)$ and then redefined the value $f(s)$ of the function there. When the class had done so, Merlin was tapped on the shoulder. He opened his eyes and ears and said, "Your modification of f is a new function, let's call it g. I don't know what you've done to f, so I can't draw a correct graph of the function g, but I do know that $\lim_{x \to 2} g(x) = 11$." Was Merlin right? Explain.

In Problems 41–55, find the indicated limit, if it exists.

41. $\lim\limits_{x \to 2+} \sqrt{x - 2}$

42. $\lim\limits_{x \to 0+} \sqrt[3]{x}$

43. $\lim\limits_{x \to -2-} \sqrt{x + 2}$

44. $\lim\limits_{x \to -2+} \sqrt{x + 2}$

45. $\lim\limits_{x \to -1-} \sqrt{-1 - x}$

46. $\lim\limits_{x \to -1+} \sqrt{-1 - x}$

47. $\lim\limits_{x \to 1-} \dfrac{|x - 1|}{x - 1}$

48. $\lim\limits_{x \to 1+} \dfrac{|x - 1|}{x - 1}$

49. $\lim\limits_{x \to -2+} \dfrac{3x|x + 2|}{x + 2}$

50. $\lim\limits_{x \to 1+} \sqrt{(x - 1)(x - 2)}$

51. $\lim\limits_{x \to 1-} \sqrt{(x - 1)(x - 2)}$

52. $\lim\limits_{x \to 1} \sqrt{(x - 1)(x - 2)}$

53. $\lim\limits_{x \to (3/2)+} [x]$

54. $\lim\limits_{x \to 2-} [x]$

55. $\lim\limits_{x \to -7+} [x]$

56. Let
$$f(x) = \begin{cases} x^4, & x < 1 \\ x^5, & x \geq 1. \end{cases}$$

Show that $\lim_{x \to 1} f(x) = 1$ by showing that the right- and left-hand limits exist and are equal.

57. Let
$$f(x) = \begin{cases} x - 2, & x > 2 \\ 0, & x \leq 2. \end{cases}$$

Find $\lim_{x \to 2+} f(x)$ and $\lim_{x \to 2-} f(x)$.

58. Let
$$f(x) = \begin{cases} 0, & x < 0 \\ x^2, & 0 \leq x < 2 \\ 4, & x \geq 2. \end{cases}$$

Calculate $\lim_{x \to 0-} f(x)$, $\lim_{x \to 0+} f(x)$, $\lim_{x \to 2-} f(x)$, and $\lim_{x \to 2+} f(x)$.

59. Let
$$f(x) = \begin{cases} |x|, & \text{for } x \leq 2 \\ [x], & \text{for } x > 2. \end{cases}$$

Show that $\lim_{x \to 2} f(x)$ exists.

2.3 THE LIMIT THEOREMS

In this section we state several theorems that will make our calculations of a number of limits a great deal easier. The proofs of these theorems depend on the material in Section 2.8 and are a bit technical. For that reason they are put in an appendix (see Appendix 5).

In Example 2.2.1 we concluded that

$$\lim_{x \to 2}(x^2 + 3) = 2^2 + 3 = 7.$$

That is, the limit of $f(x) = x^2 + 3$ as x tends to 2 is equal to $f(x)$ evaluated at $x = 2$ [i.e., $f(2)$]. Similarly, in Example 2.2.2 we saw that $\lim_{x \to 5} \sqrt{2x - 6} = 2 = f(5)$. However, as we remarked earlier, this process of evaluation (i.e., substituting the value x_0 into $f(x)$ to find a limit as $x \to x_0$) will not always work since $f(x)$ may not even be defined at x_0. Nevertheless, it is true that if f is a polynomial, then it is always possible to calculate the limit by evaluation.

Theorem 1 Let $p(x) = c_0 + c_1x + c_2x^2 + c_3x^3 + \cdots + c_nx^n$ be a *polynomial*, where c_0, $c_1, c_2, c_3, \ldots, c_n$ are real numbers and n is a fixed nonnegative integer. Then

$$\lim_{x \to x_0} p(x) = p(x_0) = c_0 + c_1x_0 + c_2x_0^2 + c_3x_0^3 + \cdots + c_nx_0^n. \tag{1}$$

EXAMPLE 1   Calculate $\lim_{x \to 3}(x^3 - 2x + 6)$.

Solution. $x^3 - 2x + 6$ is a polynomial. Hence

$$\lim_{x \to 3}(x^3 - 2x + 6) = 3^3 - 2 \cdot 3 + 6 = 27 - 6 + 6 = 27. \quad ■$$

EXAMPLE 2   Calculate $\lim_{x \to x_0} 4$ for any real number x_0.

Solution. $f(x) = 4$ is a polynomial (of degree 0). Hence $\lim_{x \to x_0} 4 = P(x_0) = 4$. This result simply states that

the limit of a constant function is that constant.

See Figure 1. In the figure we can see that as x gets close to x_0, $f(x)$ gets close to (is equal to) 4. ■

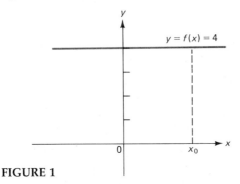

FIGURE 1

Theorem 2 Let c be any real number and suppose that $\lim_{x \to x_0} f(x)$ exists. Then $\lim_{x \to 0} cf(x)$ exists and

$$\lim_{x \to x_0} cf(x) = c \lim_{x \to x_0} f(x). \tag{2}$$

Theorem 2 states that *the limit of a constant times a function is equal to the product of that constant and the limit of the function.*

EXAMPLE 3 Calculate $\lim_{x \to 3} 5(x^3 - 2x + 6)$.

Solution. We can find this limit two ways. We can multiply to find that $5(x^3 - 2x + 6) = 5x^3 - 10x + 30$ and then use Theorem 1. However, in Example 1 we calculated

$$\lim_{x \to 3}(x^3 - 2x + 6) = 27.$$

Therefore, using Theorem 2, we have

$$\lim_{x \to 3} 5(x^3 - 2x + 6) = 5 \lim_{x \to 3}(x^3 - 2x + 6) = 5(27) = 135.$$

We can check this answer.

$$\lim_{x \to 3} 5(x^3 - 2x + 6) = \lim_{x \to 3}(5x^3 - 10x + 30)$$
$$= 5(3^3) - 10(3) + 30 = 5 \cdot 27 - 30 + 30 = 135. \blacksquare$$

Theorem 3 If $\lim_{x \to x_0} f(x)$ and $\lim_{x \to x_0} g(x)$ both exist, then $\lim_{x \to x_0}[f(x) + g(x)]$ exists, and

$$\lim_{x \to x_0}[f(x) + g(x)] = \lim_{x \to x_0} f(x) + \lim_{x \to x_0} g(x). \tag{3}$$

Theorem 3 states that *the limit of the sum of two functions is equal to the sum of their limits.*

EXAMPLE 4 Calculate $\lim_{x \to 0}\{[(x^2 + x)/x] + 4x^3 + 3\}$.

Solution. $\lim_{x \to 0}[(x^2 + x)/x] = 1$ (Example 2.2.4) and $\lim_{x \to 0}[4x^3 + 3] = 4 \cdot 0^3 + 3 = 3$. Hence

$$\lim_{x \to 0}\left(\frac{x^2 + x}{x} + 4x^3 + 3\right) = \lim_{x \to 0}\frac{x^2 + x}{x} + \lim_{x \to 0}(4x^3 + 3) = 1 + 3 = 4. \blacksquare$$

Theorem 4 If $\lim_{x \to x_0} f(x)$ and $\lim_{x \to x_0} g(x)$ both exist, then $\lim_{x \to x_0}[f(x) \cdot g(x)]$ exists, and

$$\lim_{x \to x_0}[f(x) \cdot g(x)] = \left[\lim_{x \to x_0} f(x)\right] \cdot \left[\lim_{x \to x_0} g(x)\right]. \tag{4}$$

Theorem 4 says that *the limit of the product of two functions is the product of their limits.*

EXAMPLE 5 Calculate $\lim_{x\to 13}(\sqrt{x+3})(x-4)$.

Solution. By the methods of the previous section we find that $\lim_{x\to 13}\sqrt{x+3} = \sqrt{13+3} = \sqrt{16} = 4$ and $\lim_{x\to 13}(x-4) = 13-4 = 9$. Therefore

$$\lim_{x\to 13}(\sqrt{x+3})(x-4) = \left[\lim_{x\to 13}\sqrt{x+3}\right]\left[\lim_{x\to 13}(x-4)\right] = 4\cdot 9 = 36.$$

NOTE. In Section 2.7 (Theorem 2.7.2) we will show how limits like $\lim_{x\to 13}\sqrt{x+3}$ can easily by calculated. ∎

EXAMPLE 6 Calculate $\lim_{x\to -1}(x^2-3)^{10}$.

Solution. $\lim_{x\to -1}(x^2-3) = (-1)^2 - 3 = -2$. We now may apply Theorem 4 several times (nine times to be exact):

$$\lim_{x\to -1}(x^2-3)^{10} = \underbrace{\left[\lim_{x\to -1}(x^2-3)\right]\left[\lim_{x\to -1}(x^2-3)\right]\cdots\left[\lim_{x\to -1}(x^2-3)\right]}_{10 \text{ terms}}$$

$$= \underbrace{(-2)(-2)\cdots(-2)}_{10 \text{ terms}} = (-2)^{10} = 1024. \quad\blacksquare$$

The last example can be generalized.

Corollary to Theorem 4 If $\lim_{x\to x_0} f(x)$ exists, then $\lim_{x\to x_0}[f(x)]^n$ exists and

$$\lim_{x\to x_0}[f(x)]^n = \left[\lim_{x\to x_0} f(x)\right]^n, \tag{5}$$

where n is any positive integer.

The limit theorem for quotients is just what you would expect, except that we have to be careful not to divide by zero.

Theorem 5 If $\lim_{x\to x_0} f(x)$ and $\lim_{x\to x_0} g(x)$ both exist, and $\lim_{x\to x_0} g(x) \neq 0$, then $\lim_{x\to x_0} f(x)/g(x)$ exists, and

$$\lim_{x\to x_0}\frac{f(x)}{g(x)} = \frac{\lim_{x\to x_0} f(x)}{\lim_{x\to x_0} g(x)}. \tag{6}$$

Theorem 5 says that *the limit of the quotient of two functions is the quotient of their limits, provided that the limit in the denominator function is not zero.*

EXAMPLE 7 Calculate $\lim_{x\to -2}(x^3-3x+6)/(-x^2+15)$.

Solution.

$$\lim_{x\to -2}(x^3-3x+6) = (-2)^3 - 3(-2) + 6 = -8+6+6 = 4$$

and

$$\lim_{x \to -2}(-x^2 + 15) = -(-2)^2 + 15 = -4 + 15 = 11.$$

Thus

$$\lim_{x \to -2} \frac{x^3 - 3x + 6}{-x^2 + 15} = \frac{\lim_{x \to -2}(x^3 - 3x + 6)}{\lim_{x \to -2}(-x^2 + 15)} = \frac{4}{11}. \blacksquare$$

Definition 1   RATIONAL FUNCTION   A **rational function** $r(x)$ is a function that can be written as the quotient of two polynomials; that is

$$r(x) = \frac{p(x)}{q(x)}, \tag{7}$$

where $p(x)$ and $q(x)$ are both polynomials.

For example, the function

$$r(x) = \frac{x^3 - 3x + 6}{-x^2 + 15},$$

given in Example 7, is a rational function.

Corollary to Theorems 1 and 5   Let $r(x) = p(x)/q(x)$ be a rational function with $q(x_0) \neq 0$. Then

$$\lim_{x \to x_0} r(x) = \lim_{x \to x_0} \frac{p(x)}{q(x)} = \frac{p(x_0)}{q(x_0)} = r(x_0). \tag{8}$$

EXAMPLE 8   Calculate $\lim_{x \to 4}(x^3 - x^2 - 3)/(x^2 - 3x + 5)$.

Solution. Here $q(x) = x^2 - 3x + 5$ and $q(4) = 16 - 12 + 5 = 9 \neq 0$. Therefore

$$\lim_{x \to 4} \frac{x^3 - x^2 - 3}{x^2 - 3x + 5} = \frac{4^3 - 4^2 - 3}{4^2 - 3 \cdot 4 + 5} = \frac{64 - 16 - 3}{16 - 12 + 5} = \frac{45}{9} = 5. \blacksquare$$

EXAMPLE 9   Calculate $\lim_{x \to 1}(x^2 - 1)/(x - 1)$.

Solution. We cannot use the corollary since $q(x) = x - 1$ and $q(1) = 0$. However, for $x \neq 1$

$$\frac{x^2 - 1}{x - 1} = \frac{(x - 1)(x + 1)}{x - 1} = x + 1.$$

So

$$\lim_{x \to 1} \frac{x^2 - 1}{x - 1} = \lim_{x \to 1}(x + 1) \overset{\text{Theorem 1}}{=} 2. \blacksquare$$

PROBLEMS 2.3

In Problems 1–29, use the limit theorems to help calculate the given limits.

1. $\lim\limits_{x\to 3}(x^2 - 2x - 1)$

2. $\lim\limits_{x\to -2}(-x^3 - x^2 - x - 1)$

3. $\lim\limits_{x\to 1} 3$

4. $\lim\limits_{x\to -1}(x^{49} + 1)$

5. $\lim\limits_{x\to 5} 3\sqrt{x - 1}$

6. $\lim\limits_{x\to 3} 5\sqrt{x^2 + 7}$

7. $\lim\limits_{x\to -2} -4\sqrt{x + 3}$

8. $\lim\limits_{x\to 1} 8(x^{100} + 2)$

9. $\lim\limits_{x\to 5}(\sqrt{x - 1} + \sqrt{x^2 - 9})$

10. $\lim\limits_{x\to -2}(1 + x + x^2 + x^3 + \sqrt{x^2 - 3})$

11. $\lim\limits_{x\to 0}(3\sqrt{x + 1} - 5\sqrt{x^2 - 3x + 4})$

12. $\lim\limits_{x\to 27}(\sqrt{x + 9} + \sqrt[3]{x})$

13. $\lim\limits_{x\to 0}(\sqrt{x + 1})(\sqrt{x^2 - 3x + 4})$

14. $\lim\limits_{x\to 5}(\sqrt{x - 1})(\sqrt{x^2 - 9})$

15. $\lim\limits_{x\to 2}(x^3 - 4x^2 + 5x - 3)(x^2 + 17x - 4)$

16. $\lim\limits_{x\to 2}(x^2 - 1)^5$

17. $\lim\limits_{x\to -1}(x^9 + 2)^{53}$

18. $\lim\limits_{x\to 4}(x^2 - x - 10)^7$

19. $\lim\limits_{x\to 0} \dfrac{\sqrt{x + 1}}{\sqrt{x^2 - 3x + 4}}$

20. $\lim\limits_{x\to -2} \dfrac{\sqrt{x^2 - 3}}{1 + x + x^2 + x^3}$

21. $\lim\limits_{x\to 5} \sqrt{\dfrac{x - 1}{x^2 - 9}}$

22. $\lim\limits_{x\to 3} \dfrac{1}{x^3 - 8}$

23. $\lim\limits_{x\to 0} \dfrac{3}{x^5 + 3x^2 + 3}$

24. $\lim\limits_{x\to -4} \dfrac{x^3 - x^2 - x + 1}{x^2 + 3}$

25. $\lim\limits_{x\to 1} \dfrac{x^8 + x^6 + x^4 + x^2 + 1}{x^7 + x^5 + x^3 + x}$

26. $\lim\limits_{x\to 0} \dfrac{x^{28} - 17x^{14} + x^2 - 3}{x^{51} + x^{31} - 23x^2 + 2}$

27. $\lim\limits_{x\to 0} \dfrac{2x^2 + 5x + 1}{3x^5 - 9x + 2}$

28. $\lim\limits_{x\to 0} \dfrac{x^{81} - x^{41} + 3}{23x^4 - 8x^7 + 5}$

29. $\lim\limits_{x\to 2} \dfrac{x^2 - x - 12}{x^2 - 5x + 4}$

*30. Find functions f and g such that

$$\lim\limits_{x\to 7} f(x) = 0 \quad \text{but} \quad \lim\limits_{x\to 7}[f(x) \cdot g(x)] \neq 0.$$

*31. Suppose that $(x + 3)^2 + 41 \leq f(x)$ and $f(x) \leq |x + 3| + 41$ for $-4 \leq x \leq -2$. Find $\lim\limits_{x\to -3} f(x)$ and justify your result. [*Hint*: Sketch the graphs of $y = (x + 3)^2 + 41$ and $y = |x + 3| + 41$ in a short open interval containing -3.]

2.4 INFINITE LIMITS AND LIMITS AT INFINITY

Consider the problem of calculating

$$\lim\limits_{x\to 0} \frac{1}{x^2}.$$

Values of $1/x^2$ for x "near" 0 are given in Table 1. We see that as x gets closer and closer to 0, $f(x) = 1/x^2$ gets larger and larger. In fact, $1/x^2$ grows *without bound* as x approaches 0 from either side. The graph of the function $f(x) = 1/x^2$ is given in Figure 1. In this situation we say that $f(x)$ *tends to infinity as x approaches zero*, and we write

$$\lim\limits_{x\to 0} \frac{1}{x^2} = \infty.$$

TABLE 1

x	x^2	$\dfrac{1}{x^2}$	x	x^2	$\dfrac{1}{x^2}$
1	1	1	-1	1	1
0.5	0.25	4	-0.5	0.25	4
0.1	0.01	100	-0.1	0.01	100
0.01	0.0001	10,000	-0.01	0.0001	10,000
0.001	0.000001	1,000,000	-0.001	0.000001	1,000,000
0.0001	0.00000001	100,000,000	-0.0001	0.00000001	100,000,000

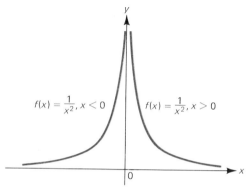

$$f(x) = \frac{1}{x^2}, x < 0 \qquad f(x) = \frac{1}{x^2}, x > 0$$

FIGURE 1

We emphasize that ∞ is the symbol for this behavior. It is *not* a new number.
In general, we have the following definition.

Definition 1   INFINITE LIMIT

(i) If $f(x)$ grows without bound in the positive direction as x gets close to the number x_0 from either side, then we say that $f(x)$ **tends to infinity as x approaches x_0,** and we write

$$\lim_{x \to x_0} f(x) = \infty.$$

(ii) If $f(x)$ grows without bound in the negative direction as x gets close to the number x_0 from either side, then we say that $f(x)$ **tends to minus infinity as x approaches x_0,** and we write

$$\lim_{x \to x_0} f(x) = -\infty.$$

EXAMPLE 1   Since $1/x^2$ grows without bound as $x \to 0$, $-1/x^2$ grows without bound in the negative direction as $x \to 0$. Thus

$$\lim_{x \to 0} -\frac{1}{x^2} = -\infty.$$

The function $-1/x^2$ is sketched in Figure 2. ■

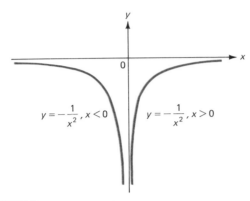

$$y = -\frac{1}{x^2}, x < 0 \qquad y = -\frac{1}{x^2}, x > 0$$

FIGURE 2

EXAMPLE 2   Calculate $\lim_{x \to -1} 1/(x + 1)^2$.

   Solution. The graph of the function $1/(x + 1)^2$ is given in Figure 3. As x gets closer and closer to -1 from either side, $1/(x + 1)^2$ grows without bound. Hence

$$\lim_{x \to -1} \frac{1}{(x + 1)^2} = \infty. \quad ■$$

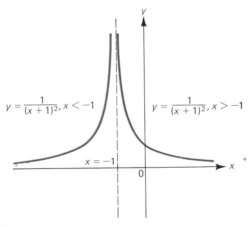

$$y = \frac{1}{(x + 1)^2}, x < -1 \qquad y = \frac{1}{(x + 1)^2}, x > -1$$

$$x = -1$$

FIGURE 3

EXAMPLE 3   Calculate $\lim_{x \to 0} 1/(x^2 + x^3)$.

   Solution. With the aid of a calculator we find values for $1/(x^2 + x^3)$ for x near 0 (see Table 2). We can see that as x gets close to 0, $1/(x^2 + x^3)$ grows without bound. Notice that for x negative, x^3 is also negative. This result poses no problem since for $|x| < 1$, $|x^3| < |x^2|$, so that $x^2 + x^3$ is a positive number. Thus, we have

$$\lim_{x \to 0} \frac{1}{x^2 + x^3} = \infty.$$

TABLE 2

x	$\dfrac{1}{x^2 + x^3}$
0.5	2.66667
0.1	90.90909
0.01	9,900.99
0.001	999,000.999
-0.5	8
-0.1	111.1111
-0.01	10,101.01
-0.001	1,001,001

To calculate this limit without making a table, we first divide the numerator and denominator by x^2. This division can be done since in the calculation of the limit $x \neq 0$, so $x^2 \neq 0$. We then have

$$\lim_{x \to 0} \frac{1}{x^2 + x^3} = \lim_{x \to 0} \frac{1/x^2}{(x^2 + x^3)/x^2} = \lim_{x \to 0} \frac{1/x^2}{1 + x}.$$

Now $\lim_{x \to 0}(1/x^2) = \infty$ and $\lim_{x \to 0}(1 + x) = 1$. Thus the numerator grows without bound while the denominator approaches 1, implying that $1/(x^2 + x^3)$ does tend to infinity. This example illustrates how a difficult calculation can be greatly simplified by a few algebraic manipulations. ∎

EXAMPLE 4   Following the reasoning of Example 3, we find that

$$\lim_{x \to 0} -\frac{1}{x^2 + x^3} = -\infty. \quad \blacksquare$$

EXAMPLE 5   Calculate $\lim_{x \to 0} 1/x$.

Solution. If $x > 0$, then as x gets close to 0, $1/x$ grows without bound in the positive direction. But if $x < 0$, then as x gets close to 0, $1/x$ grows without bound in the negative direction (see Figure 4). Since the behavior of $1/x$ depends on the way in which x approaches 0 (i.e., from the right or from the left), we must conclude that $1/x$ *has no limit* as $x \to 0$, or equivalently, that $\lim_{x \to 0} 1/x$ *does not exist.*

However, if we extend the definition of one-sided limits (Definition 2.2.2) to infinite limits, we easily see from Figure 4 that

$$\lim_{x \to 0^+} \frac{1}{x} = \infty \quad \text{and} \quad \lim_{x \to 0^-} \frac{1}{x} = -\infty. \quad \blacksquare$$

EXAMPLE 6   Calculate $\lim_{x \to 0} -1/(x + x^2)$.

Solution. Before drawing any conclusions, we tabulate, using a calculator, some values of $-1/(x + x^2)$ for x near 0. (Table 3). We see that, as in the previous example, the behavior of $f(x) = -1/(x + x^2)$ as x approaches 0 depends on the way in which

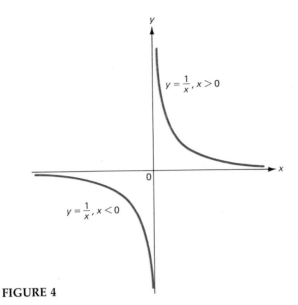

FIGURE 4

x approaches 0. Here if x is positive, then $-1/(x + x^2)$ grows without bound in the negative direction; whereas for x negative, $f(x)$ grows without bound in the positive direction. This result can also be seen by dividing both numerator and denominator by x:

$$\lim_{x \to 0} - \frac{1}{x + x^2} = \lim_{x \to 0} - \frac{1/x}{(x + x^2)/x} = \lim_{x \to 0} - \frac{1/x}{1 + x}.$$

Since the denominator tends to 1, we see that the function $-1/(x + x^2)$ behaves likes the function $-1/x$ near $x = 0$; that is, $\lim_{x \to 0} -1/(x + x^2)$ does not exist.

However, we have

$$\lim_{x \to 0^+} - \frac{1}{x + x^2} = -\infty \quad \text{and} \quad \lim_{x \to 0^-} - \frac{1}{x + x^2} = \infty. \quad \blacksquare$$

TABLE 3

x	$\dfrac{-1}{x + x^2}$
0.5	-1.33333
0.1	-9.09091
0.01	-99.00990
0.001	-999.00100
0.0001	$-9,999.00010$
-0.5	4.00000
-0.1	11.11111
-0.01	101.01010
-0.001	$1,001.00100$
-0.0001	$10,001.00010$

To this point we have considered limits as $x \to x_0$, where x_0 is a real number. But in many important applications it is necessary to determine what happens to $f(x)$ as x becomes very large. For example, we may ask what happens to the function $f(x) = 1/x$ as x becomes large. To illustrate, we again make use of a table (Table 4). It is evident that as x gets large, $1/x$ approaches zero. We then write

$$\lim_{x \to \infty} \frac{1}{x} = 0.$$

Similarly, if x grows large in the negative direction, then $1/x$ approaches 0 (see Table 5); that is,

$$\lim_{x \to -\infty} \frac{1}{x} = 0.$$

TABLE 4

x	$f(x) = \dfrac{1}{x}$
1	1
10	0.1
100	0.01
1,000	0.001
10,000	0.0001

TABLE 5

x	$\dfrac{1}{x}$
-1	-1
-10	-0.1
-100	-0.01
$-1,000$	-0.001
$-10,000$	-0.0001

Look again at Figure 4, which illustrates the behavior of $1/x$ as $x \to \pm\infty$.
These examples suggest the following definition.

Definition 2 LIMITS AT INFINITY

(i) The **limit as x approaches infinity** of $f(x)$ is L, written

$$\lim_{x \to \infty} f(x) = L,$$

if $f(x)$ is defined for all large values of x and if $f(x)$ gets close to L as x increases without bound.

(ii) The **limit as x approaches minus infinity** of $f(x)$ is L, written

$$\lim_{x \to -\infty} f(x) = L,$$

if $f(x)$ is defined for all values of x that are large in the negative direction and $f(x)$ gets close to L as x increases without bound in the negative direction.

We emphasize that the definitions in this section, like the ones in Section 2.2, are *not* precise mathematical statements but are only intended to appeal to your intuition. Precise mathematical definitions will appear in Section 2.8.

The limit theorems of Section 2.3 apply in the same way when $x \to \infty$ or $x \to -\infty$ as they do when $x \to x_0$, where x_0 is a finite number.

EXAMPLE 7 Calculate $\lim_{x\to\infty} 1/x^2$.

> **Solution.** We can calculate this limit in one of three ways. First, we could construct a table of values, as in Table 6. It seems from this table that

$$\lim_{x\to\infty} \frac{1}{x^2} = 0.$$

TABLE 6

x	x^2	$\dfrac{1}{x^2}$
1	1	1
10	100	0.01
100	10,000	0.0001
1,000	1,000,000	0.000001

Second, we could use the corollary to Theorem 2.3.4. Since $\lim_{x\to\infty} 1/x = 0$, we have

$$\lim_{x\to\infty} \frac{1}{x^2} = \lim_{x\to\infty}\left(\frac{1}{x}\right)^2 = \left(\lim_{x\to\infty}\frac{1}{x}\right)^2 = 0^2 = 0.$$

Third, we note that for $x > 1$, $x^2 > x$, so that $0 < 1/x^2 < 1/x$. Then since $1/x \to 0$ as $x \to \infty$, $1/x^2$, which is between 0 and $1/x$, must also approach 0 as $x \to \infty$. This situation is illustrated in Figure 5. ∎

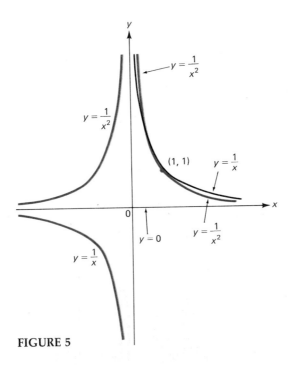

FIGURE 5

The third method used above is an example of the "squeezing" theorem, whose proof is given in Appendix 5.

Theorem 1 **Squeezing Theorem** Let x_0 be a real number, ∞, or $-\infty$.

(i) Suppose the following:
   (a) $f(x) \leq g(x) \leq h(x)$ for all x near x_0.
   (b) $\lim_{x \to x_0} f(x) = \lim_{x \to x_0} h(x) = L$.
   Then

$$\lim_{x \to x_0} g(x) = L.$$

(ii) Suppose the following:
   (a) $f(x) \leq g(x)$ for all x near x_0.
   (b) $\lim_{x \to x_0} f(x) = \infty$.
   Then

$$\lim_{x \to x_0} g(x) = \infty.$$

NOTE: If $x_0 = \infty$, then the phrase "for all x near x_0" means "for all very large x."

This theorem states that if $g(x)$ is "squeezed" between $f(x)$ and $h(x)$ near x_0, and if $f(x)$ and $h(x)$ have the same limit L, then $g(x)$ must also have the limit L. In Example 7

$$f(x) = 0, \qquad g(x) = \frac{1}{x^2}, \qquad \text{and} \qquad h(x) = \frac{1}{x}.$$

If $x > 1$, then $x^2 > x$ and $0 < 1/x^2 < 1/x$, so

$$f(x) \leq g(x) \leq h(x).$$

But $\lim_{x \to \infty} f(x) = \lim_{x \to \infty} 0 = 0$, and $\lim_{x \to \infty} h(x) = \lim_{x \to \infty} 1/x = 0$. Thus $L = 0$, and from the squeezing theorem

$$\lim_{x \to \infty} \frac{1}{x^2} = 0.$$

EXAMPLE 8 Calculate $\lim_{x \to \infty} 1/(x^3 + 8)$.

Solution. For $x > 0$, $x^3 + 8 > x$, and therefore $0 < 1/(x^3 + 8) < 1/x$. Then by the squeezing theorem, since $\lim_{x \to \infty} 1/x = 0$ and $\lim_{x \to \infty} 0 = 0$, we have

$$\lim_{x \to \infty} \frac{1}{x^3 + 8} = 0. \quad \blacksquare$$

EXAMPLE 9 Calculate $\lim_{x \to \infty} x^2$.

Solution. Here as x grows, x^2, which is bigger than x for $x > 1$, grows even faster (see Figure 6). Thus by part (ii) of the squeezing theorem,

$$\lim_{x \to \infty} x^2 = \infty. \quad \blacksquare$$

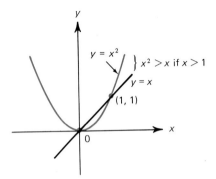

FIGURE 6

EXAMPLE 10 Calculate

$$\lim_{x \to \infty} \frac{2x^2 - 2x + 3}{x^2 + 4x + 4}. \tag{1}$$

Solution. There is an easy way to find this limit, but first we will use a calculator to see if we can find a pattern. From Table 7 we might be led to the conclusion that the limit is 2. To calculate this limit more easily, we could try to use the limit theorem for quotients (Theorem 2.3.5). But

$$\lim_{x \to \infty} 2x^2 - 2x + 3 = \infty, \qquad \lim_{x \to \infty} x^2 + 4x + 4 = \infty,$$

TABLE 7

x	$2x^2 - 2x + 3$	$x^2 + 4x + 4$	$\dfrac{2x^2 - 2x + 3}{x^2 + 4x + 4}$
1	3	9	0.33333
10	183	144	1.27083
100	19,803	10,404	1.90340
1,000	1,998,003	1,004,004	1.99003
10,000	199,980,003	100,040,004	1.99900

and ∞/∞ is an indeterminate quantity.

Fortunately, we may simplify this problem by dividing the numerator and denominator of (1) by x^2. We have

$$\lim_{x \to \infty} \frac{2x^2 - 2x + 3}{x^2 + 4x + 4} = \lim_{x \to \infty} \frac{(2x^2 - 2x + 3)/x^2}{(x^2 + 4x + 4)/x^2}$$

$$= \lim_{x \to \infty} \frac{\dfrac{2x^2}{x^2} - \dfrac{2x}{x^2} + \dfrac{3}{x^2}}{\dfrac{x^2}{x^2} + \dfrac{4x}{x^2} + \dfrac{4}{x^2}} = \lim_{x \to \infty} \frac{2 - \dfrac{2}{x} + \dfrac{3}{x^2}}{1 + \dfrac{4}{x} + \dfrac{4}{x^2}}.$$

But $2/x$, $3/x^2$, $4/x$, and $4/x^2$ all approach 0 as $x \to \infty$. Therefore

$$\lim_{x \to \infty} \frac{2 - \dfrac{2}{x} + \dfrac{3}{x^2}}{1 + \dfrac{4}{x} + \dfrac{4}{x^2}} = \frac{2 - 0 + 0}{1 + 0 + 0} = 2.$$

NOTE. We made use of several limit theorems here. Can you name which ones? ∎

In general, the limits as $x \to \pm\infty$ of rational expressions, like (1), can be found by first dividing numerator and denominator by the highest power of x that appears in the denominator, and then calculating the limit as $x \to \infty$ (or $-\infty$) of both the numerator and denominator.

EXAMPLE 11 Calculate $\lim_{x \to \infty} (x^3 + 3x + 6)/(x^5 + 2x^2 + 9)$.

Solution. Divide by x^5:

$$\lim_{x \to \infty} \frac{x^3 + 3x + 6}{x^5 + 2x^2 + 9} = \frac{\lim\limits_{x \to \infty}\left(\dfrac{1}{x^2} + \dfrac{3}{x^4} + \dfrac{6}{x^5}\right)}{\lim\limits_{x \to \infty}\left(1 + \dfrac{2}{x^3} + \dfrac{9}{x^5}\right)} = \frac{0 + 0 + 0}{1 + 0 + 0} = 0. \quad ∎$$

EXAMPLE 12 Calculate $\lim_{x \to \infty} (x^3 + 2x + 3)/(5x^2 + 1)$.

Solution. Divide by x^2:

$$\lim_{x \to \infty} \frac{x^3 + 2x + 3}{5x^2 + 1} = \lim_{x \to \infty} \frac{x + \dfrac{2}{x} + \dfrac{3}{x^2}}{5 + \dfrac{1}{x^2}}.$$

Here the numerator approaches ∞ and the denominator approaches 5. Thus

$$\lim_{x \to \infty} \frac{x^3 + 2x + 3}{5x^2 + 1} = \infty. \quad ∎$$

In general, if $r(x) = p(x)/q(x)$ and the degree of the polynomial $p(x)$ is greater than the degree of $q(x)$, then $\lim_{x \to \infty} p(x)/q(x) = +\infty$ or $-\infty$.

If the degree of $q(x)$ is greater than the degree of $p(x)$, then $\lim_{x \to \infty} p(x)/q(x) = 0$.

Finally, if the degree of $p(x)$ is equal to the degree of $q(x)$, then $\lim_{x \to \infty} p(x)/q(x) = c_m/d_n$, where c_m is the coefficient of the highest power of x in $p(x)$ and d_n is the coefficient of the highest power of x in $q(x)$. (See Problem 39.)

PROBLEMS 2.4

In Problems 1–35, find each limit (if it exists).

1. $\lim\limits_{x\to 0} \dfrac{1}{x^4}$

2. $\lim\limits_{x\to 0} \dfrac{1}{x^5}$

3. $\lim\limits_{x\to 0} \dfrac{1}{x^6}$

4. $\lim\limits_{x\to 5} \dfrac{1}{(x-5)^2}$

5. $\lim\limits_{x\to 5} \dfrac{1}{(x-5)^3}$

6. $\lim\limits_{x\to -3} \dfrac{x-4}{(x+3)^2}$

7. $\lim\limits_{x\to \pi} \dfrac{\pi}{(x-\pi)^6}$

8. $\lim\limits_{x\to 0} \dfrac{1}{x^4+x^8+x^{12}}$

9. $\lim\limits_{x\to 0} \dfrac{x+x^2}{x^2+x^3}$

10. $\lim\limits_{x\to \infty} \dfrac{1}{x+x^3}$

11. $\lim\limits_{x\to -\infty} \dfrac{x}{1+x}$

12. $\lim\limits_{x\to \infty} \dfrac{1}{1-\sqrt{x}}$

13. $\lim\limits_{x\to \infty} \dfrac{2x}{3x^3+4}$

14. $\lim\limits_{x\to -\infty} \dfrac{2x+3}{3x+2}$

15. $\lim\limits_{x\to \infty} \dfrac{5x-x^2}{3x+x^2}$

16. $\lim\limits_{x\to \infty} \dfrac{1+\sqrt{x}}{1-\sqrt{x}}$

17. $\lim\limits_{x\to \infty} \dfrac{2x^2+3x+5}{3x^2-x+2}$

18. $\lim\limits_{x\to \infty} \dfrac{4x^4+1}{1+5x^4}$

19. $\lim\limits_{x\to \infty} \dfrac{x^5-3x+4}{7x^6+8x^4+2}$

20. $\lim\limits_{x\to \infty} \dfrac{x^8-2x^5+3}{5x^4+3x+1}$

21. $\lim\limits_{x\to \infty} \dfrac{x^8-1}{x^9+1}$

22. $\lim\limits_{x\to -\infty} \dfrac{x^{16}+5x^8+2}{23x^{15}-8x^{13}+x^{11}}$

23. $\lim\limits_{x\to \infty} \dfrac{\sqrt{x}+2}{x+3}$

24. $\lim\limits_{x\to \infty} \dfrac{\sqrt[3]{x}-1}{\sqrt{x}+3}$

25. $\lim\limits_{x\to \infty} \dfrac{3x^{5/3}+2\sqrt{x}-3}{7x^{5/3}-3x+6}$

26. $\lim\limits_{x\to \infty} \dfrac{7x^{1/7}-1}{4x^{1/7}-x^{1/9}}$

*27. $\lim\limits_{x\to \infty} \dfrac{\sqrt{x}}{\sqrt[3]{x}}$

*28. $\lim\limits_{x\to \infty} \dfrac{\sqrt[3]{x}}{\sqrt{x}}$

*29. $\lim\limits_{x\to \infty} \dfrac{\sqrt{x}+1}{\sqrt{x}+3}$

30. $\lim\limits_{x\to 0^+} \dfrac{1}{x^3}$

31. $\lim\limits_{x\to 0^-} \dfrac{1}{x^3}$

32. $\lim\limits_{x\to 1^+} \dfrac{1}{x^2-1}$

33. $\lim\limits_{x\to 1^-} \dfrac{1}{x^2-1}$

34. $\lim\limits_{x\to 1} \dfrac{1}{x^2-1}$

35. $\lim\limits_{x\to 0^+} \dfrac{1}{\sqrt{x}}$

36. We have seen that $\lim_{x\to 0} 1/x^2 = \infty$. How small in absolute value must x be chosen in order that $1/x^2 > 1{,}000{,}000$? 10,000,000? 100,000,000?

37. We have seen that $\lim_{x\to \infty} 1/\sqrt{x} = 0$. How large must x be in order that $1/\sqrt{x} < 0.01$? 0.001? 0.0001?

38. Show that

$$\lim_{x\to -\infty} \frac{2x^5-3x^3+17x+2}{x^5+4x^2+16} = 2.$$

For $x < -2$, how small must x be (i.e., large in the negative direction) in order that this quotient be greater than 1.999?

*39. Let $r(x) = p(x)/q(x)$, where

$$p(x) = c_0 + c_1 x + \cdots + c_m x^m \quad \text{and} \quad q(x) = d_0 + d_1 x + \cdots + d_n x^n.$$

(a) For $m < n$, show that $\lim_{x\to \infty} p(x)/q(x) = 0$. [Hint: Divide top and bottom by x^n.]

(b) For $m > n$, show that $\lim_{x\to \infty} p(x)/q(x) = +\infty$ or $-\infty$.

(c) For $m = n$, show that $\lim_{x\to \infty} p(x)/q(x) = c_m/d_n$.

*40. Explain why we had to stipulate that $x < -2$ in Problem 38.

*41. Compute the following:

(a) $\lim_{x\to\infty}(x - \sqrt{x^2 + 2x})$

(b) $\lim_{x\to\infty}(\sqrt{x + 1} - \sqrt{x})$

*42. Choose a real number c. Show that $\lim_{x\to\infty}(\sqrt{x(x + c)} - x) = c/2$.

2.5 TANGENT LINES AND DERIVATIVES

In Section 2.1 we briefly discussed the notions of the line tangent to a curve at a point and the slope of that tangent line, called the *derivative* of a function at a point. We also showed that the derivative of a function giving the height of a moving object as a function of time is equal to the velocity of that object. In this and the next section we explore these ideas more fully.

Let us begin by describing some of the geometric properties of tangent lines. We recall from our discussion of slope in Section 1.4 that if the straight line $y = mx + b$ has a positive slope ($m > 0$), then y is an *increasing* function of x, and as a point moves along the line from left to right, it also moves *upward*. If the line has a negative slope ($m < 0$), then y is a *decreasing* function of x, and as a point moves along the line from left to right, it also moves *downward*. These ideas are illustrated in Figure 1.

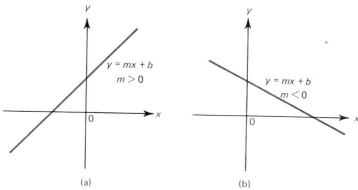

(a)　　　　(b)

FIGURE 1

The central concept in the study of calculus is the concept of a *derivative*. Intuitively, the derivative of a function f at the value x_0 is the slope of the line tangent to the graph of f at the point $(x_0, f(x_0))$. We use the notation $f'(x_0)$ to denote the derivative of f at x_0.

We can make this idea clearer with a picture. Consider the graph of the function f drawn in Figure 2. For each point on this graph there is a unique tangent line. Each of these tangent lines has a slope. At the point $(x_2, f(x_2))$ in Figure 2, for example, the slope of the tangent line is negative, so the derivative of f at x_2, denoted $f'(x_2)$, is negative. Similarly, the slope of the tangent line at $(x_4, f(x_4))$ is positive, so $f'(x_4) > 0$. Since f has a unique tangent line at each point, f has a derivative at each point.

Remember the definition of a function. It is a rule that assigns a unique real number to every number in its domain. For the function graphed in Figure 2 we have assumed that there is a unique tangent line at every point. Thus we have a new

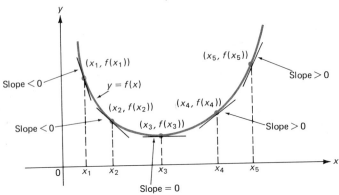

FIGURE 2

function f', called the *derivative* of f, that assigns to each number x_0 a new number $f'(x_0)$. We emphasize that

$$f'(x_0) = \text{slope of tangent line to the graph of } f \text{ at the point } (x_0, f(x_0)).$$

The remainder of this chapter and most of the next one will be concerned with the calculation of derivatives.

Having defined the derivative of a function f, we now begin the task of calculating it. The method we give below is essentially the method of Newton† and Leibniz‡ which resolved the tangent problem posed so long ago by the Greek mathematicians.§

Let us consider the function $y = f(x)$, a part of whose graph is given in Figure 3. To calculate the derivative function, we must calculate the slope of the line tangent

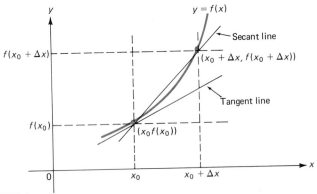

FIGURE 3

---

†*Mathematical Principles of Natural Philosophy (Principia)*, published in 1687.
‡*A New Method for Maxima and Minima, and Also for Tangents, Which is not Obstructed by Irrational Quantities*, published in 1684.
§Actually, this method was first used by the great French mathematician Pierre de Fermat (see the accompanying biographical sketch) (1601–1655) in his *Method of Finding Maxima and Minima* published in 1629. However, Fermat did not explain his procedure satisfactorily, and it was left to Newton and Leibniz to explain the method and apply it to the calculation of tangent lines.

Pierre de Fermat

Pierre de Fermat
The Granger Collection

Pierre de Fermat was born in the French city of Toulouse in approximately 1601. (There were few accurate birth records in the early seventeenth century.) His father was a leather merchant, and what early education Fermat received, he received at home. When Fermat was 30 years old he began working for the Toulouse parliament, first as a lawyer and then as a councillor. Although these jobs were not easy, he still found time to study mathematics on his own. Fermat did have a great deal of correspondence with many of the leading mathematicians of seventeenth century Europe, and perhaps this compensated for his lack of formal mathematical training.

Called the "prince of amateurs" in mathematics, Fermat made contributions to many branches of mathematics, including analytic geometry, infinitesimal analysis, and number theory. One of his discoveries was the formula for finding tangents to the curve, $y = x^n$. As we shall see in Section 3.4, the slope of such a tangent line is nx^{n-1}. For this and other results, Fermat is considered by many to be the discoverer of differential calculus, although his work did not approach the generality of Newton or Leibniz.

Fermat's most important mathematical work was in number theory. One of the major works of Greek mathematics was the *Arithmetica* of Diophantas of Alexandria. In 1621 this work was reintroduced to European mathematicians through the new edition of Claude Gaspard de Bachet (1591–1639). The *Arithmetica* discussed many topics in the theory of numbers, including divisibility, magic squares, and prime numbers. Fermat soon became fascinated with prime numbers and conjectured and sometimes proved facts about them. He conjectured, for example, that all numbers of the form $2^{2^n} + 1$, now called *Fermat numbers*, are always prime. Euler, about a hundred years later, showed this conjecture to be false since $2^{2^5} + 1$ is *not* prime ($2^{2^5} + 1 = 2^{32} + 1 = 4{,}294{,}967{,}297 = 641 \times 6{,}700{,}417$). Many of Fermat's other conjectures, however, are now known to be true.

Fermat is best known for his conjecture:

There are no positive integers a, b, c and n such that $a^n + b^n = c^n$, when $n > 2$.

This conjecture is known as "Fermat's last theorem." Fermat wrote in the margin of his copy of Bachet's *Arithmetica*, "To divide a cube into two cubes, a fourth power, or in general any power whatever into two powers of the same denomination above the second is impossible, and I have assuredly found an admirable proof of this, but the margin is too narrow to contain it." Whether or not Fermat really had a proof of his conjecture is debatable, but as stated above, many of his other assertions have now been proved. In any case, Fermat's marginal remark has frustrated mathematicians for more than three centuries, as no one has been able either to prove the conjecture or to find a counterexample.

Before World War I, the German mathematician Paul Wolfskehl offered a prize of 100,000 marks for the first proof of Fermat's conjecture. Many professional and amateur mathematicians sought the prize in vain. More recently, the conjecture has been verified for large values of n on a computer. But still no general proof has been found.

To this day, mathematics departments around the country (and the world) receive unsolicited "proofs." Some disappointed writers of incorrect "proofs" have accused the mathematical community of a conspiracy to suppress the truth (although the reasons for this alleged conspiracy are not clear). In any event, there have been more incorrect "proofs" published for Fermat's last theorem than for any other mathematical conjecture.

Despite his many contributions to mathematics, Fermat published very little. Instead, he communicated many of his results to his friend Marin Mersenne (1588–1648). As a result, he did not receive credit for many of his discoveries.

to the curve at each point of the curve at which there is a unique tangent line. Let $(x_0, f(x_0))$ be such a point. From now on we will assume that f is defined "near" x_0.

If Δx is a small number (positive or negative), then $x_0 + \Delta x$ will be close to x_0. In moving from x_0 to $x_0 + \Delta x$, the values of f will move from $f(x_0)$ to $f(x_0 + \Delta x)$. Now look at the straight line in Figure 3, called a **secant line,** joining the points $(x_0, f(x_0))$ and $(x_0 + \Delta x, f(x_0 + \Delta x))$. What is its slope? If we define $\Delta y = f(x_0 + \Delta x) - f(x_0)$ and if we use m_s to denote the slope of such a secant line, we have, from Section 1.4,

$$m_s = \frac{\text{change in } y}{\text{change in } x} = \frac{f(x_0 + \Delta x) - f(x_0)}{(x_0 + \Delta x) - x_0} = \frac{f(x_0 + \Delta x) - f(x_0)}{\Delta x} = \frac{\Delta y}{\Delta x}. \tag{1}$$

What does this equation have to do with the slope of the tangent line? The answer is suggested in Figure 4. From this illustration we see that as Δx gets smaller, the secant line gets closer and closer to the tangent line. Put another way, as Δx approaches zero, the slope of the secant line approaches the slope of the tangent line. But the slope of the tangent line at the point $(x_0, f(x_0))$ is the derivative, $f'(x_0)$. We therefore have

$$f'(x_0) = \lim_{\Delta x \to 0} m_s = \lim_{\Delta x \to 0} \frac{\Delta y}{\Delta x} = \lim_{\Delta x \to 0} \frac{f(x_0 + \Delta x) - f(x_0)}{\Delta x} \tag{2}$$

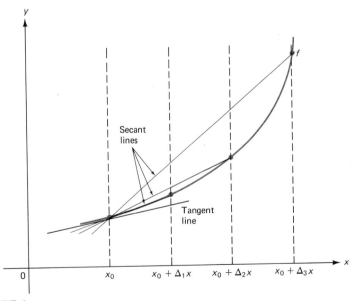

FIGURE 4

Definition 1 DERIVATIVE AT A POINT Let f be defined on an open interval containing the point x_0, and suppose that

$$\lim_{\Delta x \to 0} \frac{f(x_0 + \Delta x) - f(x_0)}{\Delta x}$$

exists and is finite. Then f is said to be **differentiable** at x_0 and the **derivative** of f at x_0, denoted $f'(x_0)$, is given by

$$f'(x_0) = \lim_{\Delta x \to 0} \frac{\Delta y}{\Delta x} = \lim_{\Delta x \to 0} \frac{f(x_0 + \Delta x) - f(x_0)}{\Delta x}. \tag{3}$$

Definition 1 tells us what we mean by the derivative of a function at a point. We next define the derivative function.

Definition 2 THE DERIVATIVE FUNCTION The **derivative** f' of the function f is the **function** defined as follows:

(i) $\mathrm{dom}\, f' = \left\{ x \colon \lim\limits_{\Delta x \to 0} \dfrac{f(x + \Delta x) - f(x)}{\Delta x} \text{ exists and is finite} \right\}$.

(ii) For every x in $\mathrm{dom}\, f'$

$$f'(x) = \lim_{\Delta x \to 0} \frac{f(x + \Delta x) - f(x)}{\Delta x}. \tag{4}$$

That is, f' is the function, defined at every x for which the limit in (4) exists and is finite, that assigns to every x in its domain the derivative $f'(x)$.

REMARK. It follows from Definition 1 that $f'(x)$ can exist only if $f(x)$ is defined. Thus

$$\mathrm{dom}\, f' \subset \mathrm{dom}\, f. \tag{5}$$

We emphasize that the derivative of a function is another function. The value of the derivative at a given number x_0 is the limit obtained in (3).

Note that Definitions 1 and 2 do not say anything about tangent lines. Simply put, f is differentiable at x if the limit in (3) exists. However, we can now formally define what we mean by a tangent line.

Definition 3 TANGENT LINE If the limit in (3) exists and is finite, we say that the graph of the function f has a **tangent line** at the point $(x_0, f(x_0))$. The tangent line is the line passing through the point $(x_0, f(x_0))$ with slope $f'(x_0)$. One equation of this line is

$$y - f(x_0) = f'(x_0)(x - x_0) \tag{6}$$

REMARK. According to this definition, a graph *cannot* have more than one tangent line at the point $(x_0, f(x_0))$ since this would imply that $f'(x_0)$ had more than one value, thereby contradicting the fact that if the limit in (4) exists, then it must be unique.

EXAMPLE 1 Let $y = f(x) = 3x + 5$. Calculate $f'(x)$.

Solution. To solve this problem and the ones that follow, we simply use formula (4). For $f(x) = 3x + 5$,

$$f(x + \Delta x) = 3(x + \Delta x) + 5.$$

Then

$$f'(x) = \lim_{\Delta x \to 0} \frac{f(x + \Delta x) - f(x)}{\Delta x} = \lim_{\Delta x \to 0} \frac{[3(x + \Delta x) + 5] - (3x + 5)}{\Delta x}$$

$$= \lim_{\Delta x \to 0} \frac{3x + 3\Delta x + 5 - 3x - 5}{\Delta x} = \lim_{\Delta x \to 0} \frac{3\Delta x}{\Delta x} = \lim_{\Delta x \to 0} 3 = 3.$$

This answer is not surprising. It simply says that the slope of the line $y = 3x + 5$ is equal to the constant function 3. ∎

Before giving further examples, we introduce the additional symbols dy/dx or df/dx to denote the derivative for $y = f(x)$.

$$f'(x) = \frac{df}{dx} = \frac{dy}{dx} = \lim_{\Delta x \to 0} \frac{\Delta y}{\Delta x} = \lim_{\Delta x \to 0} \frac{f(x + \Delta x) - f(x)}{\Delta x}.$$

The symbol dy/dx is read, "**the derivative of y with respect to x.**" *We emphasize that dy/dx is not a fraction.* At this point the symbols dy and dx have no meaning of their own. (We will define these symbols in Section 3.8.) There are other notations for the derivative. We will often use the symbol y' or $y'(x)$ in place of f' or $f'(x)$. Thus for $y = f(x)$, we may denote the derivative in four different ways:†

$$f'(x) = y'(x) = \frac{df}{dx} = \frac{dy}{dx}.$$

EXAMPLE 2 Calculate the derivative of the function $y = x^2$. What is the equation of the line tangent to the graph of $y = x^2$ at the point $(3, 9)$?

Solution. For $y = f(x) = x^2$, $f(x + \Delta x) = (x + \Delta x)^2$. Then

$$\frac{dy}{dx} = \lim_{\Delta x \to 0} \frac{f(x + \Delta x) - f(x)}{\Delta x} = \lim_{\Delta x \to 0} \frac{(x + \Delta x)^2 - x^2}{\Delta x}$$

$$= \lim_{\Delta x \to 0} \frac{x^2 + 2x\,\Delta x + \Delta x^2 - x^2}{\Delta x} = \lim_{\Delta x \to 0} \frac{2x\,\Delta x + \Delta x^2}{\Delta x}$$

---

†Newton (in England) and Leibniz (in Germany) independently discovered (in the 1670s) the equation for the slope of the tangent line given in this section. Newton used the symbol $\dot{y}$ (read "y dot") and Leibniz used the symbol dy/dx to indicate the derivative.

$$= \lim_{\Delta x \to 0} \frac{\Delta x (2x + \Delta x)}{\Delta x} = \lim_{\Delta x \to 0} (2x + \Delta x) = 2x.$$

At every point of the form $(x, f(x)) = (x, x^2)$, the slope of the line tangent to the curve is $2x$. For $x = 3$, $2x = 6$. Therefore the slope of the tangent line at the point $(3, 9)$ is 6; that is, $f'(3) = 6$.

We can now find the equation of the tangent line since it passes through the point $(3, 9)$ and has the slope 6. We have from (6)

$$y - 9 = 6(x - 3) \quad \text{or} \quad y = 6x - 9.$$

Knowing the derivative of $y = x^2$ helps us to graph the curve. First, we note that x^2 is always positive. If $x < 0$, then $y' = dy/dx = 2x < 0$, so the tangent lines have negative slopes. For $x = 0$, $dy/dx = 2 \cdot 0 = 0$, so the tangent line is horizontal. For $x > 0$, $dy/dx = 2x > 0$, and the tangent lines have positive slopes. Using this information, we obtain the graph of Figure 5. ■

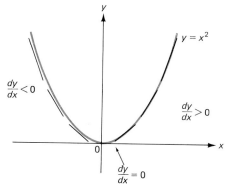

FIGURE 5

EXAMPLE 3   Find the derivative of $y = \sqrt{x}$, and calculate the slope of the tangent line at the point $(4, 2)$.

Solution. We have

$$f'(x) = \lim_{\Delta x \to 0} \frac{f(x + \Delta x) - f(x)}{\Delta x} = \lim_{\Delta x \to 0} \frac{\sqrt{x + \Delta x} - \sqrt{x}}{\Delta x}$$

$$= \lim_{\Delta x \to 0} \frac{(\sqrt{x + \Delta x} - \sqrt{x})(\sqrt{x + \Delta x} + \sqrt{x})}{\Delta x (\sqrt{x + \Delta x} + \sqrt{x})} \qquad \text{We multiplied numerator and denominator by } \sqrt{x + \Delta x} + \sqrt{x}.$$

$$= \lim_{\Delta x \to 0} \frac{(\sqrt{x + \Delta x})^2 - (\sqrt{x})^2}{\Delta x (\sqrt{x + \Delta x} + \sqrt{x})} \qquad \text{Since } (a - b)(a + b) = a^2 - b^2. \\ \text{Here } a = \sqrt{x + \Delta x} \text{ and } b = \sqrt{x}.$$

$$= \lim_{\Delta x \to 0} \frac{(x + \Delta x) - x}{\Delta x (\sqrt{x + \Delta x} + \sqrt{x})} = \lim_{\Delta x \to 0} \frac{\Delta x}{\Delta x (\sqrt{x + \Delta x} + \sqrt{x})}$$

$$= \lim_{\Delta x \to 0} \frac{1}{\sqrt{x + \Delta x} + \sqrt{x}} = \frac{1}{\sqrt{x} + \sqrt{x}} = \frac{1}{2\sqrt{x}}. \tag{7}$$

Thus

$$f'(x) = \frac{d}{dx}\sqrt{x} = \frac{1}{2\sqrt{x}}.$$

At $x = 4$

$$f'(4) = \frac{1}{2\sqrt{4}} = \frac{1}{2 \cdot 2} = \frac{1}{4} = \text{slope of tangent line at } (4, 2).$$

From (6) the equation of the tangent line is given by

$$y - 2 = \tfrac{1}{4}(x - 4) \qquad \text{or} \qquad y = \tfrac{1}{4}x + 1.$$

Recall that in Example 2.2.6 we showed, numerically, that

$$\lim_{\Delta x \to 0} \frac{\sqrt{4 + \Delta x} - \sqrt{4}}{\Delta x} = \frac{1}{4}.$$

Note that although the function $f(x) = \sqrt{x}$ is defined for $x \geq 0$, its derivative $1/(2\sqrt{x})$ is only defined for $x > 0$ since for $x = 0$ the limit taken in (7) does not exist (*explain why*).

We can again make use of the derivative to graph the curve. First note that $\sqrt{x} > 0$ if $x > 0$ and $\sqrt{x}$ is not defined if $x < 0$. Look at the derivative $f'(x) = 1/(2\sqrt{x})$. Since $\sqrt{x} > 0$, $1/(2\sqrt{x}) > 0$, so all tangents to the curve have a positive slope and the function is increasing. But we also have

$$\lim_{x \to \infty} f'(x) = \lim_{x \to \infty} \frac{1}{2\sqrt{x}} = 0.$$

This equation tells us that as x increases, the derivative approaches 0, or equivalently, the tangent lines approach the horizontal (a slope of zero). Thus we conclude that for large values of x the curve $y = \sqrt{x}$ is nearly flat. On the other hand, since $\lim_{x \to 0^+} [1/(2\sqrt{x})] = \infty$, the tangent lines to the curve near $x = 0$ are nearly vertical.† We combine all this information in Figure 6. ∎

EXAMPLE 4 Consider the function $y = |x|$. Since

$$|x| = \begin{cases} x, & x \geq 0 \\ -x, & x \leq 0, \end{cases}$$

we obtain the graph in Figure 7. To see if the graph of f has a tangent line at the point $(0, 0)$, we calculate

---

†Lines with very large slopes are nearly vertical.

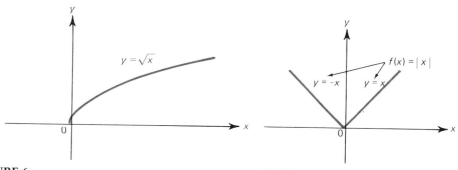

FIGURE 6   **FIGURE 7**

$$f'(0) = \lim_{\Delta x \to 0} \frac{f(0 + \Delta x) - f(0)}{\Delta x} = \lim_{\Delta x \to 0} \frac{|0 + \Delta x| - |0|}{\Delta x} = \lim_{\Delta x \to 0} \frac{|\Delta x|}{\Delta x}.$$

But as we saw in Example 2.2.8, this limit does not exist, so $|x|$ does not have a tangent line at $(0, 0)$. On the other hand, if $x \neq 0$, then the derivative does exist (see Problem 32). ■

EXAMPLE 5   Let

$$f(x) = \begin{cases} x, & x \leq 0 \\ 1 + x, & x > 0. \end{cases}$$

Compute $f'(0)$, if it exists.

Solution The graph of f is given in Figure 8. We have

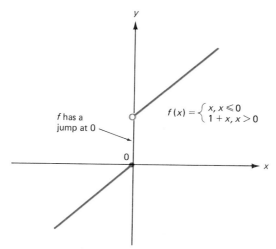

f has a
jump at 0

$f(x) = \begin{cases} x, x \leqslant 0 \\ 1 + x, x > 0 \end{cases}$

FIGURE 8

Since $f(0) = 0$

$$f'(0) = \lim_{\Delta x \to 0} \frac{f(0 + \Delta x) - f(0)}{\Delta x} = \lim_{\Delta x \to 0} \frac{f(\Delta x)}{\Delta x}.$$

Now

$$f(\Delta x) = \begin{cases} \Delta x, & \Delta x \leq 0 \\ 1 + \Delta x, & \Delta x > 0. \end{cases}$$

So

$$\frac{f(\Delta x)}{\Delta x} = \begin{cases} 1, & \text{if } \Delta x \leq 0 \\ \dfrac{1 + \Delta x}{\Delta x}, & \text{if } \Delta x > 0 \end{cases}$$

But

$$\frac{1 + \Delta x}{\Delta x} = 1 + \frac{1}{\Delta x} \qquad \text{and} \qquad \lim_{\Delta x \to 0^+} \frac{1}{\Delta x} = \infty.$$

Thus

$$\lim_{\Delta x \to 0^+} \frac{f(\Delta x)}{\Delta x} = \infty$$

and $f'(x)$ does not exist. [Remember, f is differentiable at x_0 if the limit in (3) exists and is *finite*.] ∎

Almost every function we encounter in this book will be differentiable at every point in its domain. However, there are three commonly encountered situations in which a function will fail to be differentiable at a point.

A function f is not differentiable at a point $x_0 \in \text{dom } f$ if one of the following situations holds:

(i) f has a vertical tangent at x_0 (see Figure 6; $\sqrt{x}$ has a vertical tangent at 0).
(ii) The graph of f comes to a point at x_0 (see Figure 7).
(iii) The graph of f jumps at x_0 (see Figure 8).

Before leaving this section, we give one more definition that we will use often in the rest of the book.

Definition 4 DIFFERENTIABILITY ON AN OPEN INTERVAL   The function f is **differentiable on the open interval** (a, b) if $f'(x)$ exists for every x in (a, b).

EXAMPLE 6   From Example 3 we see that $f(x) = \sqrt{x}$ is differentiable on any interval of the form $(0, b)$, where $b > 0$. ∎

EXAMPLE 7   From Example 2 we see that $f(x) = x^2$ is differentiable on $(-\infty, \infty)$. ∎

PROBLEMS 2.5

1. Determine whether the slopes of the lines in Figure 9 are positive, negative, or zero.

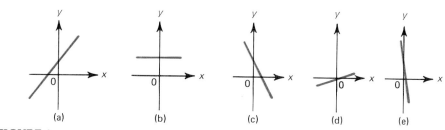

(a)   (b)   (c)   (d)   (e)

FIGURE 9

In Problems 2–12 the graph of a function is given. Several points are listed. At each point, determine whether the function is differentiable, and if so, state whether the derivative appears to be positive, negative, or zero.

7.

8.

9.

10.

11.

12.

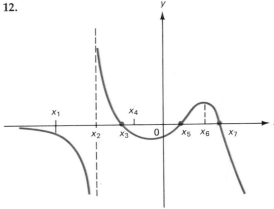

13. Consider the function $f(x) = 3x^2$.
   (a) For $x = 2$, calculate $f(x + \Delta x) = f(2 + \Delta x)$ for $\Delta x = 0.5$, $\Delta x = 0.1$, $\Delta x = 0.01$, $\Delta x = 0.001$, $\Delta x = -0.01$, and $\Delta x = -0.001$.
   (b) Calculate $[f(2 + \Delta x) - f(2)]/\Delta x$ for the values of Δx in part (a) and guess the value of $f'(2)$.
   (c) From the definition, calculate $f'(x)$, use this to compute $f'(2)$, and compare this result with the answer you obtained in part (b).
   (d) What is the equation of the line tangent to the curve at the point $(2, 12)$?
   (e) Using the derivative, sketch the curve.
14. Carry out the steps in Problem 13 for the function $f(x) = 1/x$ at the point $(1, 1)$.
15. Carry out the steps in Problem 13 for the function $f(x) = 5\sqrt{x}$ at the point $(1, 5)$.

In Problem 16–31, find the derivative of the given function and the equation of the tangent line to the curve at the given point.

16. $f(x) = 15x - 2$, $(1, 13)$

17. $f(x) = -4x + 6$, $(3, -6)$

18. $f(x) = (x + 1)^2$, $(1, 4)$

19. $f(x) = 2(x - 3)^2$, $(4, 2)$ $4x - 12$

20. $f(x) = x^2 - 1$, $(-2, 3)$

21. $f(x) = -x^2 + 3x + 5$, $(0, 5)$

22. $f(x) = x^3$, $(-1, -1)$

23. $f(x) = x^3 + 6$, $(-1, 5)$

24. $y = x^3 + x^2 + x + 1$, $(1, 4)$

25. $y = -\sqrt{x} + 3$, $(6, -3)$

26. $y = \sqrt{x} - 2$, $(6, 2)$

27. $y = \sqrt{2x}$, $(8, 4)$

*28. $y = x^{3/2}$, $(1, 1)$ [Hint: $(a^{3/2} - b^{3/2})(a^{3/2} + b^{3/2}) = a^3 - b^3$.]

*29. $y = \dfrac{x + x^2}{\sqrt{x}}$, $(1, 2)$

*30. $y = \dfrac{1}{\sqrt{x}}$, $(4, \frac{1}{2})$

*31. $y = x^4$, $(2, 16)$

32. Let $y = |x|$. Calculate dy/dx for $x \neq 0$.

33. Show from the definition that if $y = mx + b$, then $dy/dx = m$.

34. Show that for any constants a, b, and c

$$\frac{d}{dx}(ax^2 + bx + c) = 2ax + b.$$

35. Show that for any constant a

$$\frac{d}{dx}ax^3 = 3ax^2.$$

36. Calculate the derivative of $ax^3 + bx^2 + cx + d$ for any constants a, b, c, and d.

37. Let $f(x) = x$ and $g(x) = x^2$. For what values of x are the tangents to these curves parallel?

38. Let $f(x) = x^2$ and $g(x) = x^3$. For what values of x are the tangents to these curves parallel?

39. Let $f(x) = ax^2$ and $g(x) = 6x - 5$. For what value of a is the tangent to the graph of f parallel to the tangent to the graph of g when $x = 2$?

*40. Let $f(x) = 1/\sqrt{x}$. For what value of x is the tangent to this curve parallel to the line $x + 8y = 10$?

*41. Let $f(x) = x|x|$. Show that $f'(x) = 2|x|$ for all x.

*42. Let $f(x) = 1/(1 + |x|)$.
  (a) Find $f'(x)$, where f is differentiable.
  (b) Sketch the graph of $y = f(x)$.

*43. Let $f(x) = x[-x^2]$, where $[\ ]$ denotes the greatest integer function. Compute $f'(0)$. (Note that you will obtain a wrong answer if you simplify the difference quotient $[f(0 + \Delta x) - f(0)]/\Delta x$ and then take the shortcut of setting Δx equal to 0.)

**44. Consider

$$\lim_{\Delta x \to 0^+} \frac{f(x + \Delta x) - f(x - \Delta x)}{2\Delta x}$$

as an alternate definition of $f'(x)$. Is this definition equivalent to Definition 1?

*45. A function f is **even** if $f(-x) = f(x)$ and is **odd** if $f(-x) = -f(x)$.
  (a) Suppose that f is an even function that is differentiable everywhere. Show that f' is an odd function.
  (b) Suppose that g is an odd function that is differentiable everywhere. Show that g' is an even function.

*46. Suppose that f is a periodic function that is differentiable everywhere. That is, there is a number ω, called a **period** of f, such that $f(x + \omega) = f(x)$ for all real x. Show that f' has period ω.

*47. (a) Suppose that $|f(x)| \le x^2$ for all x. Show that f is differentiable at 0.
  (b) Suppose that $g(0) = 0$ and $|g(x)| > \sqrt{|x|}$ for all x. Show that g cannot be differentiable at 0.

2.6 THE DERIVATIVE AS A RATE OF CHANGE

In Section 2.1 we saw the relationship between the derivative and the velocity of a falling rock. In this section we see how the derivative represents the rate of change in a variety of interesting situations.

To begin the discussion, we again consider the line given by the equation

$$y = mx + b, \tag{1}$$

where m is the slope and b is the y-intercept (see Figure 1). Let us examine the slope more closely. It is defined as the change in y divided by the change in x:

$$m = \frac{\Delta y}{\Delta x}. \tag{2}$$

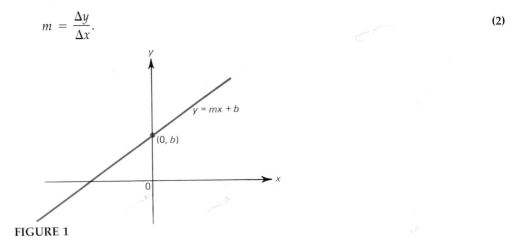

FIGURE 1

Implicit in this definition is the understanding that no matter which two points are chosen on the line, we obtain the same value for this ratio. Thus the slope of a straight line could instead be referred to as *the rate of change of y with respect to x*. It tells us how many units y changes for every one unit that x changes. In fact, instead of following Euclid in defining a straight line as the shortest distance between two points, we could define a straight line as a curve whose rate of change is constant.

EXAMPLE 1 Let $y = 3x - 5$. Then in moving from the point $(1, -2)$ to the point $(2, 1)$ along the line, we see that as x has changed (increased) one unit, y has increased three units, corresponding to the slope $m = 3$. See Figure 2. ∎

We would like to be able to calculate rates of change for functions that are not straight-line functions. We did that in Section 2.1. More generally, suppose that an object is dropped from rest from a given height. The distance s the object has dropped after t seconds (ignoring air resistance) is given by the formula

$$s = \tfrac{1}{2}gt^2 \tag{3}$$

where $g \approx 9.8$ m/sec$^2 \approx 32$ ft/sec^2 is the acceleration due to gravity. We now ask: What is the velocity of the object after 2 sec? To answer this question, we first note that velocity is a rate of change. Whether measured in meters per second, feet per second, or miles per hour, velocity is the ratio of change in distance (meters, feet, miles) to the

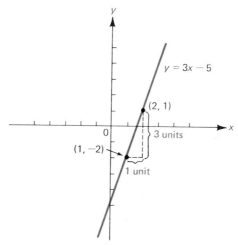

FIGURE 2

change in time (seconds, hours). The discussion in Section 2.1 suggests the following definition.

Definition 1   INSTANTANEOUS VELOCITY   Let $s(t)$ denote the distance traveled by a moving object in t seconds. Then the **instantaneous velocity** after t seconds, denoted ds/dt or $s'(t)$, is given by

$$s'(t) = \frac{ds}{dt} = \lim_{\Delta t \to 0} \frac{\Delta s}{\Delta t} = \lim_{\Delta t \to 0} \frac{s(t + \Delta t) - s(t)}{\Delta t}. \tag{4}$$

NOTE.   The quantity $[s(t + \Delta t) - s(t)]/\Delta t$ is the average velocity of the object between times t and $t + \Delta t$.

As you may have noticed, the velocity ds/dt given by (4) is the derivative of s with respect to t. Although our previous definitions of derivative involved the variables x and y, there is no change in this concept when we insert t in place of x and s in place of y. Thus we can think of a derivative as a velocity, or, more generally, as a rate of change. After Newton discovered the derivative, he used the word "fluxion" instead of velocity in his discussion of a moving object. (In technical terminology, a moving particle is a particle "in flux.")

EXAMPLE 2   If an object has dropped a distance of $\frac{1}{2}gt^2 = 4.9t^2$ meters after t seconds, what is its velocity, ignoring air resistance, after exactly 2 sec?

Solution. $s(t) = 4.9t^2$ and $s(t + \Delta t) = 4.9(t + \Delta t)^2$. Thus

$$\frac{ds}{dt} = \lim_{\Delta t \to 0} \frac{4.9(t + \Delta t)^2 - 4.9t^2}{\Delta t} = 4.9 \lim_{\Delta t \to 0} \frac{(t + \Delta t)^2 - t^2}{\Delta t}$$

$$= 4.9 \lim_{\Delta t \to 0} \frac{t^2 + 2t\,\Delta t + \Delta t^2 - t^2}{\Delta t} = 4.9 \lim_{\Delta t \to 0} \frac{\Delta t(2t + \Delta t)}{\Delta t}$$

$$= 4.9 \lim_{\Delta t \to 0}(2t + \Delta t) = (4.9)(2t) = 9.8t.$$

After 2 sec have elapsed the object is dropping at a velocity of 9.8(2) = 19.6 m/sec. ∎

EXAMPLE 3   Let $P(t)$ denote the population of a colony of bacteria after t hours. If $P(t) = 100 + t^4$, how fast is the population growing after 3 hr?

Solution. This problem is again a "rate of change" problem, even though it does not make sense to speak about "velocity." We are asking for the "instantaneous rate of growth" of the population when $t = 3$ hr. This rate is given by

$$P'(t) = \frac{dP}{dt} = \lim_{\Delta t \to 0} \frac{P(t + \Delta t) - P(t)}{\Delta t}$$

$$= \lim_{\Delta t \to 0} \frac{(100 + (t + \Delta t)^4) - (100 + t^4)}{\Delta t} = \lim_{\Delta t \to 0} \frac{(t + \Delta t)^4 - t^4}{\Delta t} \qquad (5)$$

Note that $[P(t + \Delta t) - P(t)]/\Delta t$ is the average rate of growth between times t and $t + \Delta t$. From the binomial theorem (see Appendix 4),

$$(t + \Delta t)^4 = t^4 + 4t^3\,\Delta t + 6t^2\,\Delta t^2 + 4t\,\Delta t^3 + \Delta t^4. \qquad (6)$$

Inserting (6) into (5), we have (after canceling the t^4 terms)

$$P'(t) = \frac{dP}{dt} = \lim_{\Delta t \to 0} \frac{4t^3\,\Delta t + 6t^2\,\Delta t^2 + 4t\,\Delta t^3 + \Delta t^4}{\Delta t}$$

Dividing numerator and denominator by Δt, we have

$$\frac{dP}{dt} = \lim_{\Delta t \to 0}(4t^3 + 6t^2\,\Delta t + 4t\,\Delta t^2 + \Delta t^3) = 4t^3.$$

When $t = 3$ hr, $dP/dt = 4 \cdot 3^3 = 4 \cdot 27 = 108$, and the population is growing at a rate of 108 individuals per hour. (We know that it is growing because $dP/dt > 0$. It would be declining if dP/dt were negative.) ∎

EXAMPLE 4   The mass of a 4-m-long, nonuniform metal beam varies with the distance along the beam measured from the left end (see Figure 3). The mass μ is given by the formula

$$\mu(x) = x^{3/2} \text{ kilograms.}$$

This equation tells us how much of the mass of the beam is contained in that part of

FIGURE 3

the beam from the left end to the point x units from the left end. For example, at the end $x = 4$, $\mu = 4^{3/2} = 8$ kg, so the entire beam has a mass of 8 kg. On the other hand, when $x = 2$, $\mu = 2^{3/2} = 2\sqrt{2} \approx 2.83$ kg, so the left-hand half of the beam carries only approximately 35% of the mass of the beam ($2\sqrt{2}/8 \approx 0.35$). What is the density ρ of the beam at $x = 1$, $x = 2$, $x = 3$, and $x = 4$ m?

Solution. Density = mass per unit of length, which is expressed in kilograms per meter (kg/m). The density is changing as we move from left to right along the beam. Since the right side carries more of the mass, the density *increases* as we move from left to right. Between $x = 0$ and $x = 2$, the "average" density is ($2\sqrt{2}$ kg)/(2 m) = $\sqrt{2}$ kg/m, while along the entire length of the beam (i.e., between $x = 0$ and $x = 4$) the average density is (8 kg)/(4 m) = 2 kg/m.

The problem asks for the "instantaneous" density at $x = 1$, 2, 3, and 4, or, equivalently, for the instantaneous rate of change of mass per unit of length. We therefore have

$$\rho(x) = \frac{d\mu}{dx} = \lim_{\Delta x \to 0} \frac{\mu(x + \Delta x) - \mu(x)}{\Delta x} = \lim_{\Delta x \to 0} \frac{(x + \Delta x)^{3/2} - x^{3/2}}{\Delta x}. \qquad (7)$$

Since for any numbers a and b, $(a - b)(a + b) = a^2 - b^2$, we have

$$[(x + \Delta x)^{3/2} - x^{3/2}][(x + \Delta x)^{3/2} + x^{3/2}]$$
$$= [(x + \Delta x)^{3/2}]^2 - (x^{3/2})^2$$
$$= (x + \Delta x)^3 - x^3 = x^3 + 3x^2\,\Delta x + 3x\,\Delta x^2 + \Delta x^3 - x^3$$
$$= 3x^2\,\Delta x + 3x\,\Delta x^2 + \Delta x^3.$$

Thus we may multiply numerator and denominator of the last expression in (7) by $(x + \Delta x)^{3/2} + x^{3/2}$ to obtain

$$\frac{d\mu}{dx} = \lim_{\Delta x \to 0} \frac{[(x + \Delta x)^{3/2} - x^{3/2}][(x + \Delta x)^{3/2} + x^{3/2}]}{\Delta x[(x + \Delta x)^{3/2} + x^{3/2}]}$$
$$= \lim_{\Delta x \to 0} \frac{3x^2\,\Delta x + 3x\,\Delta x^2 + \Delta x^3}{\Delta x[(x + \Delta x)^{3/2} + x^{3/2}]} = \lim_{\Delta x \to 0} \frac{3x^2 + 3x\,\Delta x + \Delta x^2}{(x + \Delta x)^{3/2} + x^{3/2}}$$
$$= \frac{3x^2}{x^{3/2} + x^{3/2}} = \frac{3x^2}{2x^{3/2}} = \frac{3}{2}\,x^{1/2} = \frac{3}{2}\,\sqrt{x}.$$

Thus the density ρ is given by the formula

$$\rho(x) = \tfrac{3}{2}\sqrt{x} \text{ kg/m}$$

Therefore

$$\rho(1) = \frac{3}{2} = 1.5, \qquad \rho(2) = \frac{3}{2}\sqrt{2} = \frac{3}{\sqrt{2}} \approx 2.12, \qquad \rho(3) = \frac{3\sqrt{3}}{2} \approx 2.6,$$

and

$$\rho(4) = \frac{3\sqrt{4}}{2} = 3$$

(all in kilograms per meter). We can clearly see how the density increases as we move from left to right. ■

The result of Example 4 is worth mentioning again:

> The density function is the derivative of the mass function.

EXAMPLE 5   **(Marginal Cost).**   A manufacturer buys large quantities of a certain machine replacement part. She finds that her cost depends on the number of cases bought at the same time, and the cost per unit decreases as the number of cases bought increases. She determines that a reasonable model for this situation is given by the formula

$$C(q) = 100 + 5q - 0.01q^2, \qquad 0 \le q \le 250 \tag{8}$$

where q is the number of cases bought (up to 250 cases) and $C(q)$, measured in dollars, is the *total* cost of purchasing q cases. The 100 in (8) is a *fixed* cost, which does not depend on the number of cases bought. What are the incremental and marginal costs for various levels of purchase?

Solution. **Incremental cost** is the cost per additional unit at a given level of purchase. **Marginal cost** is the *rate of change* of the cost with respect to the number of units purchased. These two concepts are different. This difference is like the difference between the average velocity of a falling object over a fixed period of time and the instantaneous velocity of the object at a fixed moment in time. For example, if the manufacturer buys 25 cases, the incremental cost is the cost for one *more* case, that is, the 26th case. This cost is not a constant. We can see this result by calculating that 1 case costs $100 + 5 \cdot 1 - 0.01 = \$104.99$ and two cases cost $100 + 5 \cdot 2 - (0.01)4 = \$109.96$. Thus the incremental cost of buying the second case is $C(2) - C(1) = \$4.97$. On the other hand, for 100 cases it costs

$$100 + 5 \cdot 100 - (0.01)(100)^2 = 600 - 100 = \$500,$$

and it costs

$$100 + 5 \cdot 101 - (0.01)(101)^2 = 100 + 505 - (0.01)(10{,}201)$$
$$= 605 - 102.01 = \$502.99$$

to buy 101 cases. The incremental cost is now $C(101) - C(100) = \$2.99$. The 101st is cheaper than the 2nd.

Next, we compute the marginal cost:

$$\text{marginal cost} = \frac{dC}{dq} = \lim_{\Delta q \to 0} \frac{C(q + \Delta q) - C(q)}{\Delta q}$$

$$= \lim_{\Delta q \to 0} \frac{[100 + 5(q + \Delta q) - 0.01(q + \Delta q)^2] - [100 + 5q - 0.01q^2]}{\Delta q}$$

$$= \lim_{\Delta q \to 0} \frac{5\,\Delta q - 0.01(2q\,\Delta q + \Delta q^2)}{\Delta q} = 5 - 0.02q.$$

Thus at $q = 10$ cases the marginal cost is $5 - 0.2 = \$4.80$; while at $q = 100$ cases the marginal cost is $5 - 2 = \$3$. This result confirms the manufacturer's statement that "the more she buys, the cheaper it gets." ■

PROBLEMS 2.6

In Problems 1–6 distance is given as a function of time. Find the instantaneous velocity at the indicated time.

1. $s = 1 + t + t^2$, $t = 4$

2. $s = t^3 - t^2 + 3$, $t = 5$

3. $s = 1 + \sqrt{2t}$, $t = 8$

4. $s = (1 + t)^2$, $t = 2.5$ [*Hint:* Expand $(1 + t)^2$.]

5. $s = 100t - 5t^2$, $t = 6$

6. $s = t^4 - t^3 + t^2 - t + 5$, $t = 3$

7. Fuel in a rocket burns for $3\frac{1}{2}$ min. In the first t seconds the rocket reaches a height of $70t^2$ feet above the earth (for any t from 0 to 210 sec). What is the velocity of the rocket (in feet per second) after 3 sec? After 10 sec?

8. For the bacteria population of Example 3, how fast is the population growing after 5 hr? After one day? Is this model realistic for large values of t?

9. A colony of bacteria is dying out. It starts with an initial population of 10,000 organisms, and after t days the population $P(t)$ is $10,000 - 2.5t^2$.
   (a) How fast is the population declining after one week? [*Hint:* The minus sign in your answer indicates a decline in the population.]
   (b) After how many days is the colony extinct?

10. The volume of a growing spherical cell is proportional to the cube of the radius of the cell. Let r denote radius and V denote volume; then

$$V = \tfrac{4}{3}\pi r^3. \qquad \frac{dV}{dr} \quad \text{or} \quad V'(r) \quad \text{or} \quad V'$$

What is the rate of growth of the volume with respect to the radius when the radius is $10\ \mu\text{m}$ ($1\ \mu\text{m} = 1$ micrometer $= 1/1{,}000{,}000$ m)?

11. The surface area of a spherical cell is proportional to the square of the radius of the cell: $S = 4\pi r^2$. What is the rate of growth of the surface area as a function of the radius when $r = 10\mu\text{m}$? $20\ \mu\text{m}$?

12. (a) Combining Problems 10 and 11, find an expression for the volume of the cell as a function of the surface area.
   ***(b)** How fast is the volume increasing when the surface area is $100\ \mu\text{m}^2$?

13. The manufacturer in Example 5 finds that her cost function for another machine part is given by

$$C(q) = 100q + 55.$$

Compute her marginal cost.

14. Assume that the cost function of Example 5 is given by

$$C(q) = 200 + 6q - 0.01q^2 + 0.01q^3.$$

(a) Find the marginal cost.

(b) Is the manufacturer better off buying in large quantities?

*15. In Example 5 the difference between the cost of buying 101 units and buying 100 units is given by

$$C(101) - C(100) = \$2.99.$$

However, at $q = 100$, $dC/dq = \$3$. Explain this apparent discrepancy of one cent.

2.7 CONTINUITY

The concept of continuity is one of the central notions in mathematics. Intuitively, a function is continuous at a point if it is defined at that point and if its graph moves unbroken through that point. Figure 1 shows the graphs of six functions, three of which are continuous at x_0 and three of which are not.

There are several equivalent definitions of continuity. The one we give here depends explicitly on the theory of limits developed earlier in this chapter.

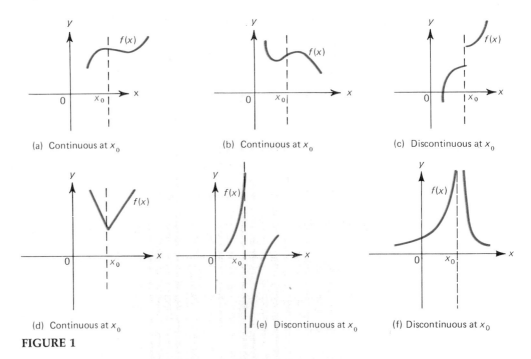

(a) Continuous at x_0

(b) Continuous at x_0

(c) Discontinuous at x_0

(d) Continuous at x_0

(e) Discontinuous at x_0

(f) Discontinuous at x_0

FIGURE 1

Definition 1 CONTINUITY Let $f(x)$ be defined for every x in an interval containing the number x_0.

Then f is **continuous** at x_0 if all of the following three conditions hold:

(i) $f(x_0)$ exists (that is, x_0 is in the domain of f).

(ii) $\lim_{x \to x_0} f(x)$ exists.

(iii) $\lim_{x \to x_0} f(x) = f(x_0)$. (1)

Condition (iii) tells us that if a function f is continuous at x_0, then we can calculate $\lim_{x \to x_0} f(x)$ by evaluation. This is only one of the reasons continuous functions are so important. In the next few chapters we will see that a large majority of the functions we encounter in applications are indeed continuous.

We give an alternative definition of continuity in the problem set of Section 2.8.

EXAMPLE 1   Let $f(x) = x^2$. Then for any real number x_0,

$$\lim_{x \to x_0} f(x) = \lim_{x \to x_0} x^2 = x_0^2 = f(x_0),$$

so f is continuous at every real number (see Figure 2). ■

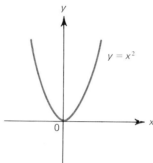

FIGURE 2

EXAMPLE 2   Let $p(x) = c_0 + c_1 x + c_2 x^2 + c_3 x^3 + \cdots + c_n x^n$ be a polynomial. By Theorem 2.3.1,

$$\lim_{x \to x_0} p(x) = p(x_0) \tag{2}$$

for every real number x_0. Therefore

every polynomial is continuous at every real number.

Note that this statement also shows that any constant function is continuous. ■

EXAMPLE 3   Let $r(x) = p(x)/q(x)$ be a rational function [$p(x)$ and $q(x)$ are polynomials]. Then from Theorem 2.3.5 we have, if $q(x_0) \neq 0$, $\lim_{x \to x_0} r(x) = p(x_0)/q(x_0) = r(x_0)$, so

any rational function is continuous at all points x_0 at which the denominator, $q(x_0)$, is nonzero. ■

EXAMPLE 4   Let

$$f(x) = \frac{x^5 + 3x^3 - 4x^2 + 5x - 2}{x^2 - 5x + 6}.$$

Here f is a rational function and therefore is continuous at any x for which the denominator, $x^2 - 5x + 6$, is not zero. Since $x^2 - 5x + 6 = (x - 3)(x - 2) = 0$ only when $x = 2$ or 3, f is continuous at all real numbers except at these two. ∎

Definition 2 DISCONTINUOUS FUNCTION If the function f is not continuous at x_0, then f is said to be **discontinuous** at x_0. Note that f is discontinuous at x_0 if one (or more) of the three conditions given in Definition 1 fails to hold.

Then f is discontinuous at x_0 under any of the following conditions:

(i) $f(x_0)$ does not exist ($x_0 \notin \text{dom } f$).
(ii) $\lim_{x \to x_0} f(x)$ does not exist.
(iii) $\lim_{x \to x_0} f(x)$ exists but is not equal to $f(x_0)$.

REMARK. There are really two somewhat conflicting definitions of the phrase "f is discontinuous at x_0." The first definition is our Definition 2. The second definition requires f to be defined, but not continuous, at x_0. For example, according to this second definition, the function $1/x$ is neither continuous nor discontinuous at 0, since it is not defined for $x = 0$. Analogously, the function $\sqrt{x}$ is not discontinuous at -10 (or any other negative number) because it is not defined there. In this book we will use the first definition exclusively. Thus for us the function $1/x$ is discontinuous at 0 and the function $\sqrt{x}$ is discontinuous at -10†.

As the examples above suggest, most commonly encountered functions are continuous. In this book all the functions we meet will be continuous at every point except in one of the three cases discussed below.

Discontinuity Case 1 *We are dividing by zero.* This case is exemplified by Example 4. Also, $1/x$ is discontinuous at 0.

EXAMPLE 5 Let $f(x) = (x^2 - 1)/(x - 1)$. Although

$$\lim_{x \to 1} \frac{x^2 - 1}{x - 1} = 2,$$

f is discontinuous at 1 since $f(x)$ is not defined there, so condition (i) is violated. This equation provides us with an example of a *removable discontinuity*. The discontinuity can be "removed" in the following way.

Consider the continuous function g, defined by

$$g(x) = \begin{cases} \dfrac{x^2 - 1}{x - 1}, & \text{if } x \neq 1 \\ 2, & \text{if } x = 1. \end{cases}$$

†A more complete discussion of the meaning of the term "discontinuous" can be found in R. C. Buck's *Advanced Calculus* (McGraw-Hill, New York, 1965), pages 84 and 85.

This function is continuous at every real value of x, and $g(x) = f(x)$ for all $x \neq 1$. Notice that any other choice for $g(1)$ would make g discontinuous at 1. For example, if we define

$$g(x) = \begin{cases} \dfrac{x^2 - 1}{x - 1}, & \text{if } x \neq 1 \\ 1, & \text{if } x = 1, \end{cases}$$

then condition (iii) is violated since $\lim_{x \to 1} g(x) = 2$, which is not equal to $g(1)$. ■

NOTE. We will often encounter functions that are defined by two or more "pieces," as above.

Definition 3 REMOVABLE DISCONTINUITY The function f has a **removable discontinuity** at x_0 if $\lim_{x \to x_0} f(x) = c$ exists and is finite but either f is not defined at x_0 or $\lim_{x \to x_0} f(x) \neq f(x_0)$.

NOTE. If f has a removable discontinuity at x_0 and $\lim_{x \to x_0} f(x) = c$, then the function

$$g(x) = \begin{cases} f(x), & x \neq x_0 \\ c, & x = x_0 \end{cases}$$

is equal to f for $x \neq x_0$ and is continuous at x_0.

Discontinuity Case 2 *The function is not defined over a range of values.*

EXAMPLE 6 $f(x) = \sqrt{x}$ is not continuous for $x < 0$ because the square root function is not defined for negative values. ■

Discontinuity Case 3 *The function may be discontinuous if it is defined in pieces.*

EXAMPLE 7 A tomato wholesaler finds that the price of newly harvested tomatoes is 16¢ per pound if he purchases fewer than 100 pounds each day. However, if he purchases at least 100 pounds daily, the price drops to 14¢ per pound. Find the total cost function.

Solution. Let q denote the daily number of pounds bought and C denote the cost; then

$$C(q) = \begin{cases} 0.16q, & \text{if } 0 \leq q < 100 \\ 0.14q, & \text{if } q \geq 100. \end{cases}$$

This function is sketched in Figure 3. It is discontinuous at $q = 100$. Note also that $C(q)$ is not continuous for $q < 0$ because it is not defined for $q < 0$. The discontinuity at $q = 100$ is called a **jump discontinuity**. This term is used when the function "jumps" from one *finite* value to another at a point. ■

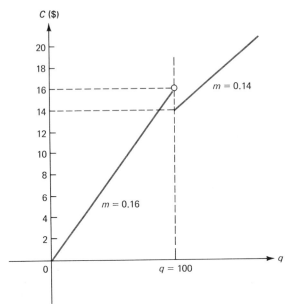

FIGURE 3

EXAMPLE 8   Let

$$f(x) = \begin{cases} x^2, & x \le 1 \\ x^3, & x > 1. \end{cases}$$

In Example 2.2.12 we showed that $\lim_{x \to 1} f(x) = 1$. Thus f is continuous at 1. In fact, since $f(x)$ is equal to either x^2 or x^3 for $x \ne 1$, f is continuous at every real number. This function provides an example of a function defined in pieces that is everywhere continuous. ■

Definition 4   CONTINUITY OVER AN OPEN INTERVAL   A function f is **continuous over** (or **in**) **the open interval** (a, b) if f is continuous at every point in that interval (a may be $-\infty$ and/or b may be $+\infty$).

Definition 5   CONTINUITY OVER A CLOSED INTERVAL   The function f is **continuous** in the **closed interval** $[a, b]$ if the following conditions hold:

   (i) f is continuous at every x in the open interval (a, b).
   (ii) $f(a)$ and $f(b)$ both exist.
   (iii) $\lim_{x \to a+} f(x) = f(a)$ and $\lim_{x \to b-} f(x) = f(b)$.

REMARK.   It may be that f is only defined in the interval $[a, b]$. Then "f is continuous at every point in $[a, b]$" means the usual thing for x in (a, b). However, by "continuity at a" we mean that $\lim_{x \to a+} f(x) = f(a)$, and by "continuity at b" we mean that $\lim_{x \to b-} f(x) = f(b)$. Sometimes this type of continuity at the points a and b is referred to, respectively, as **continuity from the left** and **continuity from the right**.

EXAMPLE 9   Let $f(x) = |x|$. Since for $x > 0$, $f(x) = x$, and for $x < 0$, $f(x) = -x$, f is continuous at every $x \neq 0$. If $x = 0$, then $\lim_{x \to 0} f(x) = \lim_{x \to 0}|x| = 0 = f(0)$, so f is continuous at 0 also (see Figure 4). Thus f is continuous over $(-\infty, \infty)$. ■

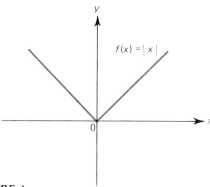

$f(x) = |x|$

FIGURE 4

EXAMPLE 10   Let $f(x) = \sqrt{x}$. Then since $\lim_{x \to 0^+} f(x) = 0 = f(0)$, f is continuous in $[0, \infty)$. Note that $\lim_{x \to 0} f(x)$ does not exist, since $\sqrt{x}$ is not defined for negative values of x. This example illustrates the importance of defining continuity at the endpoints of a closed interval in terms of one-sided limits. ■

EXAMPLE 11   Let $f(x) = [x]$. That is, f is the greatest integer function (see Example 2.2.11). Let us try to calculate $\lim_{x \to 2}[x]$. If $1 < x < 2$, then $[x] = 1$. But if $2 < x < 3$, $[x] = 2$. Hence if we approach $x = 2$ from the right, the functional values approach 2, but if we approach from the left, the functional values approach 1. Therefore $\lim_{x \to 2}[x]$ does not exist, and so $f(x) = [x]$ is discontinuous at $x = 2$. The same argument shows that f is discontinuous at any integer. On the other hand, if x_0 is not an integer, then $[x]$ is continuous at x_0 (why?). Thus f is continuous in any interval (a, b) or $[a, b]$ that does not contain an integer. Note also that $\lim_{x \to 1^+} f(x) = 1 = f(1)$, but $\lim_{x \to 2^-} f(x) = 1 \neq f(2)$, so f is not continuous in $[1, 2]$. ■

The kind of discontinuity exhibited in Example 11, is another example of a jump discontinuity because the function "jumps" from one *finite* value to another, as depicted in Figure 5.

We close this section by citing several theorems that will be very useful in the remainder of the book. The first theorem follows directly from the definition of continuity and the limit theorems of Section 2.3.

Theorem 1   Let the functions f and g be continuous at x_0, and let c be a constant. Then the following functions are continuous at x_0:

  (i) cf.
  (ii) $f + g$.
  (iii) $f \cdot g$.

If, in addition, $g(x_0) \neq 0$, then we have the following:

  (iv) f/g is continuous at x_0.

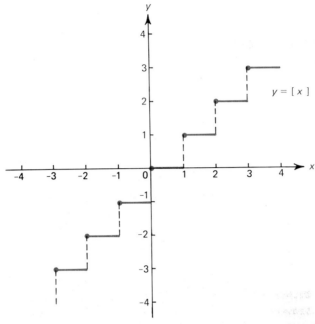

FIGURE 5

To prove part (iv), we have, if $g(x_0) \neq 0$,

$$\lim_{x \to x_0} \frac{f(x)}{g(x)} \overset{\overset{\text{From limit Theorem 5}}{\downarrow}}{=} \frac{\lim_{x \to x_0} f(x)}{\lim_{x \to x_0} g(x)} = \frac{f(x_0)}{g(x_0)} = \left(\frac{f}{g}\right)(x_0)$$

so f/g is continuous at x_0. ∎

The following theorem, whose proof is given in Appendix 5, is very useful for the calculation of limits.

Theorem 2   If f is continuous at a and if $\lim_{x \to x_0} g(x) = a$, then $\lim_{x \to x_0} f(g(x)) = f(a)$. In other words,

$$\lim_{x \to x_0} (f \circ g)(x) = \lim_{x \to x_0} f(g(x)) = f(\lim_{x \to x_0} g(x)). \tag{3}$$

EXAMPLE 12   Calculate $\lim_{x \to 3}[(x^2 - 9)/(x - 3)]^3$.

Solution. Let $f(x) = x^3$ and $g(x) = (x^2 - 9)/(x - 3)$. Then $[(x^2 - 9)/(x - 3)]^3 = f(g(x))$. Since $\lim_{x \to 3} g(x) = 6$, $\lim_{x \to 3} f(g(x)) = 6^3 = 216$. ∎

Next, we show that every differentiable function is continuous. That is, a function f is continuous at any point x_0 at which f is differentiable.

Theorem 3   Let f be differentiable at x_0. Then f is continuous there.

This result appeals to our intuition. For if a function is differentiable at x_0, then its graph has a tangent line at the point $(x_0, f(x_0))$, and the curve seems to move "smoothly" through that point.

Proof. We need to show that $\lim_{x \to x_0} f(x) = f(x_0)$, which is the same as showing that

$$\lim_{x \to x_0}[f(x) - f(x_0)] = 0. \text{ (Why?)}$$

Define $\Delta x = x - x_0$. Then $f(x) = f(x_0 + \Delta x)$, and when $x \to x_0$, we have $\Delta x \to 0$. Then

$$\lim_{x \to x_0}[f(x) - f(x_0)] = \lim_{\Delta x \to 0}[f(x_0 + \Delta x) - f(x_0)]$$

$$= \lim_{\Delta x \to 0} \frac{[f(x_0 + \Delta x) - f(x_0)]\Delta x}{\Delta x} \qquad \text{We multiplied numerator and denominator by } \Delta x.$$

$$= \lim_{\Delta x \to 0} \frac{f(x_0 + \Delta x) - f(x_0)}{\Delta x} \lim_{\Delta x \to 0} \Delta x$$

$$= f'(x_0) \cdot 0 = 0. \quad \blacksquare \qquad \text{Since } f'(x_0) \text{ exists, by assumption.}$$

REMARK. The converse of this theorem is not true. That is, a function that is continuous at a point is *not* necessarily differentiable at that point. For example, we saw in Example 9 that $|x|$ is continuous at 0. However, according to Example 2.5.4, $|x|$ is *not* differentiable at 0.

Two very useful properties of continuous functions that we will need in Chapter 4 are given in the theorems that follow. The proofs of these theorems are beyond the scope of this text. These proofs and the proofs of related results are given in all standard advanced calculus texts.†

Theorem 4 *Upper and Lower Bound Theorem* If f is continuous on the closed, bounded interval $[a, b]$, then f is bounded above and below in that interval. That is, there exist numbers m and M such that

$$m \le f(x) \le M \qquad \text{for every } x \text{ in } [a, b].$$

Moreover, if m is the greatest lower bound (see page 6) for f on $[a, b]$ and M is the least upper bound for f on $[a, b]$, then there exist numbers x_1 and x_2 in $[a, b]$ such that $f(x_1) = m$ and $f(x_2) = M$.

REMARK 1. Theorem 4 is not true, in general, on an open interval (a, b). For example, $f(x) = 1/x$ is continuous on $(0, 1)$, but as $x \to 0^+$, $f(x) \to \infty$, so f is not bounded above on that interval.

REMARK 2. Theorem 4 is also not true, in general, if f is not continuous in $[a, b]$. For example, $f(x) = 1/x$ is not bounded on the interval $[-1, 1]$ (it is not continuous at 0).

---

†See, for example, R. C. Buck, *Advanced Calculus* (McGraw-Hill, New York, 1965), Chapter 2.

Theorem 4 is illustrated in Figure 6 for three different functions.

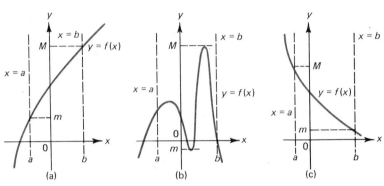

FIGURE 6

Theorem 5 *Intermediate Value Theorem* Let f be continuous on $[a, b]$. Then if c is any number between $f(a)$ and $f(b)$, there is a number $\bar{x}$ in (a, b) such that $f(\bar{x}) = c$.

This theorem has a geometrical interpretation given by Figure 7. The intermediate value theorem could be restated: If the continuous function f takes on the values $f(a)$ and $f(b)$, then it takes on every value in between. The intermediate value theorem can be used to prove many interesting results. It is useful, for example, in finding roots of a polynomial.

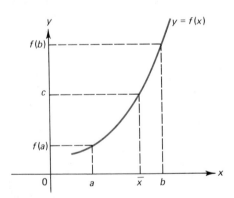

FIGURE 7

EXAMPLE 13 Show that there is a root of $P(x) = x^3 + x^2 + x - 1$ in the interval $[0, 1]$.

Solution. $P(x)$ is continuous (why?). Since $P(0) = -1$ and $P(1) = 2$, there must be a number $\bar{x}$ in $(0, 1)$ such that $P(\bar{x}) = 0$ (since 0 is between -1 and 2). We can do even better. We have

$$P(\tfrac{1}{2}) = \tfrac{1}{8} + \tfrac{1}{4} + \tfrac{1}{2} - 1 = -\tfrac{1}{8} \quad \text{and} \quad P(\tfrac{3}{4}) = \tfrac{27}{64} + \tfrac{9}{16} + \tfrac{3}{4} - 1 = \tfrac{47}{64}.$$

Therefore there is a root between $\tfrac{1}{2}$ and $\tfrac{3}{4}$. Using a calculator, we can narrow the root down further. Since $P(0.543) = -0.002047993$ and $P(0.544) = 0.000925184$, we have located a root between 0.543 and 0.544. This method of "estimating" a root of a polynomial is very useful when used in conjunction with **Newton's method** (Section 4.8). We first use the intermediate value theorem to get "close" to the root (within one decimal place, say) and then use Newton's method to find the root quickly to as many decimal places of accuracy as needed. Doing so, we find one root of $x^3 + x^2 + x - 1$ is $\bar{x} = 0.5436890127$, correct to ten decimal places. ■

PROBLEMS 2.7

In Problems 1–8 find the open interval or intervals in which the given function is continuous.

1. $f(x) = x^{17} - 3x^{15} + 2$

2. $f(x) = x^{1/3}$

3. $f(x) = x^{1/4}$

4. $f(x) = \dfrac{1}{x + 2}$

5. $f(x) = \dfrac{-17x}{x^2 - 1}$

6. $f(x) = \dfrac{1}{(x - 10)^{15}}$

7. $f(x) = \dfrac{2x}{x^3 - 8}$

8. $f(x) = \dfrac{|x + 2|}{x + 2}$

9. For what values of α does the function $f(x) = (x^2 - 4)/(x - \alpha)$ have a removable discontinuity at $x = \alpha$?

10. For what values of α does the function

$$f(x) = \frac{x^3 - 6x^2 + 11x - 6}{x - \alpha}$$

have a removable discontinuity at $x = \alpha$?

*11. Graph the function $f(x) = x[x]$. Find the open intervals in which f is continuous.

12. Show that the function

$$f(x) = \begin{cases} \dfrac{x^3 - 1}{x - 1}, & x \neq 1 \\ 3, & x = 1 \end{cases}$$

is continuous in $(-\infty, \infty)$.

13. For what value of α is the function

$$f(x) = \begin{cases} \dfrac{x^4 - 1}{x - 1}, & x \neq 1 \\ \alpha, & x = 1 \end{cases}$$

continuous at $x = 1$?

14. Let

$$f(x) = \begin{cases} x, & x \neq \text{an integer} \\ x^2, & x = \text{an integer}. \end{cases}$$

Graph the function for $-3 \leq x \leq 3$. For what integer values of x is f continuous?

15. Let

$$f(x) = \begin{cases} 0 & x < 0 \\ x^2, & 0 \leq x < 2 \\ 4, & x \geq 2. \end{cases}$$

Graph the function. Show that f is continuous in $(-\infty, \infty)$.

16. Let

$$f(x) = \begin{cases} x, & x < 0 \\ x^2, & 0 \le x < 1 \\ x^3, & x \ge 1. \end{cases}$$

Graph the function. Show that f is continuous in $(-\infty, \infty)$.

17. Use Theorem 2 to calculate $\lim_{x \to 2}(\sqrt{x} - 1)/(x + 1)$.

18. Use Theorem 2 to calculate $\lim_{x \to -1}(3x^3 + 8x^2 - 9x + 2)^{3/4}$.

19. Let

$$f(x) = \begin{cases} 1, & x \text{ rational} \\ 0, & x \text{ irrational.} \end{cases}$$

Explain why f is not continuous at any point in any interval $[\alpha, \beta]$.

20. A car accelerates from 0 to 80 km/hr in 30 sec. Explain why there is a time between 0 and 30 sec when the car is traveling at exactly 50 km/hr.

*21. Prove that there is a number whose square is 3. [*Hint:* Let $f(x) = x^2$ and apply the intermediate value theorem.]

The function f is **piecewise continuous** in $[a, b]$ if f is continuous at every point in $[a, b]$ except for a finite number of points at which f has a jump or a removable discontinuity. A jump discontinuity is defined in Examples 7 and 11.

22. Show that the function

$$f(x) = \begin{cases} x, & x < 0 \\ 2x, & 0 \le x \le 1 \\ 3x^3, & x > 1 \end{cases}$$

is piecewise continuous over any interval.

23. Is the function $f(x) = (x + 1)/(x - 2)$ piecewise continuous over the interval $[-3, 3]$? Over the interval $[-1, 1]$?

24. Explain why every continuous function is piecewise continuous.

In Problems 25–32 show that the given function is piecewise continuous over the indicated interval. Identify the functions that are continuous.

25. $f(x) = [x] + 1; [-2, 2]$

26. $f(x) = \dfrac{x^2 - 1}{x + 1}; [-2, 2]$

27. $f(x) = \dfrac{[x]}{x}; [1, 3]$

28. $f(x) = \begin{cases} x + x^2, & x \le -1 \\ x^3, & x > -1 \end{cases}; [-2, 0]$

29. $f(x) = \begin{cases} x, & x \text{ an integer} \\ \frac{1}{2}, & x \text{ not an integer} \end{cases}; [-1, 3]$

30. $f(x) = \begin{cases} 2x^2 + 4, & x \le 0 \\ (x - 2)^2, & x > 0 \end{cases}; [-5, 5]$

31. $f(x) = [x]; [-2, 2]$

32. $f(x) = \|[x]\|; [-3, 3]$

33. Let

$$f(x) = \begin{cases} x + 3, & x \le 1 \\ 2x + 5, & x > 1 \end{cases} \quad \text{and} \quad g(x) = \begin{cases} x^2 + 6, & x \le 1 \\ 5x^3 - 1, & x > 1. \end{cases}$$

(a) What is the function $f(x)g(x)$?

(b) Show that $\lim_{x \to 1} f(x)$ does not exist.

(c) Show that $\lim_{x \to 1} g(x)$ does not exist.

(d) Show that $\lim_{x \to 1} f(x)g(x)$ does exist.

34. Find two functions $f(x)$ and $g(x)$ such that $\lim_{x \to x_0} f(x)$ and $\lim_{x \to x_0} g(x)$ do not exist but $\lim_{x \to x_0} f(x)/g(x)$ does exist.

35. Let

$$f(x) = \begin{cases} p(x), & x \le x_0 \\ q(x), & x > x_0, \end{cases}$$

where $p(x)$ and $q(x)$ are polynomials. Under what circumstances will $\lim_{x \to x_0} f(x)$ exist?

*36. Prove that if $f(x_0) > 0$ and f is continuous at x_0, there is a number $\delta > 0$ such that if $x_0 - \delta < x < x_0 + \delta$, then $f(x) > 0$. [*Hint:* Draw a sketch.]

*37. Prove that if $f(x_0) < 0$ and f is continuous at x_0, then there is a number $\delta > 0$ such that if $x_0 - \delta < x < x_0 + \delta$, then $f(x) < 0$.

38. Find the maximum M and the minimum m for the function $f(x) = |x + 3|$ on the interval $[-5, 7]$.

*39. Find M and m for the function

$$f(x) = \frac{x^3 + 3x^2 + 2x + 1}{x + 1}$$

on the interval $[0, 1]$.

40. Give an example of a function defined in $[0, 1]$, bounded in $[0, 1]$, and continuous for $0 < x \le 1$, but discontinuous at $x = 0$.

*41. Suppose that f is continuous in $[a, b]$ with $f'(a) < 0$ and $f'(b) > 0$. Prove that there is a point c in (a, b) such that either $f'(c) = 0$ or $f'(c)$ does not exist.

42. Prove that $f(x) = \sqrt[5]{x}$ is continuous in the interval $[0, 1]$.

*43. Prove that $f(x) = (1 - x^2)^{3/4}$ is continuous in $[-1, 1]$.

*44. Find a function f that is not continuous but such that $|f|$ is continuous.

*45. Discuss continuity, or the lack of it, for the following:

(a) $F(x) = [x] + |x - [x]|$ (b) $G(x) = [x] + (x - [x])^2$

(c) $H(x) = [x] + \sqrt{|x - [x]|}$

*46. Let $P(x) = x^3 + cx + d$. Prove that if $c > 0$, there is exactly one real zero of P.

*47. (a) Sketch the graph of $y = [x] + [-x] + 1$.
   (b) For what values of x is this function continuous?

*48. Suppose that f is continuous on the closed interval I; in addition, suppose that J is a closed interval such that $J \subset I$. Which of the following assertions must be true?

(a) $\min_J f \le \min_I f$ (b) $\min_I f \ge \min_J f$

(c) $\min_J f \le \max_I f$ (d) $\min_I f \ge \max_J f$

(e) $\max_J f \le \min_I f$ (f) $\max_I f \ge \min_J f$

(g) $\max_J f \le \max_I f$ (h) $\max_I f \ge \max_J f$

2.8 THE THEORY OF LIMITS (OPTIONAL)

The value of mathematics lies in its precision. The precision applies not only to its calculations but to the definitions of its terms and the proofs of its results. In mathematics you are not asked to believe something because it seems to be true or because some expert tells you it is true. Rather, you are asked to believe it only after seeing it proved true. On page 68 we provided an intuitive definition of a limit, and in Section 2.3 we stated, but did not prove, a number of very useful limit theorems. At this point you should be comfortable with the notion of a limit. The intuitive definition given in

Section 2.2 is correct. The only problem is that the words "close to" are not precise. What do we mean by "close"? We now answer that question.

The definition of a limit we give next is a precise restatement of what you already know, namely, that as x gets close to x_0, $f(x)$ gets close to L. Before giving this definition, we recall (from Section 1.1) that an *open interval* (a, b) is the set of all points x such that $a < x < b$. A **neighborhood** of the point x_0 is defined as an open interval that contains that point.† Finally, the letters ϵ (epsilon) and δ (delta) used in this section are Greek letters that almost always represent small quantities.

Definition 1 LIMIT Suppose that $f(x)$ is defined in a neighborhood of the point x_0 (a finite number) except possibly at the point x_0 itself. Then

$$\lim_{x \to x_0} f(x) = L$$

if for every $\epsilon > 0$ there is a $\delta > 0$ such that

$$\text{if } 0 < |x - x_0| < \delta, \quad \text{then} \quad |f(x) - L| < \epsilon. \ddagger \tag{1}$$

We must exclude $|x - x_0| = 0$ (i.e., we don't want $x = x_0$) since $f(x)$ may not be defined at x_0.

This formal definition is difficult to understand at first reading. It is worthwhile at this point to draw a picture of what is happening. We note that $0 < |x - x_0| < \delta$ means (from Section 1.2) that $x_0 - \delta < x < x_0 + \delta$, $x \neq x_0$. Similarly, $|f(x) - L| < \epsilon$ means that $L - \epsilon < f(x) < L + \epsilon$. Figure 1 illustrates that no matter how small ϵ is chosen, δ can be made small enough so that $f(x)$ is within ϵ of L.

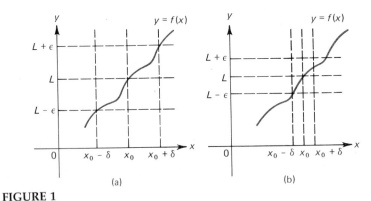

(a)   (b)

FIGURE 1

The best way to illustrate this definition further is to show by example that our intuitive notion of a limit coincides with the formal definition of a limit.

EXAMPLE 1 Show that $\lim_{x \to 2} 3x = 6$.

---

†We can also define a *neighborhood of* ∞ to be an open interval of the form (N, ∞). We will refer to neighborhoods of ∞ in Chapter 12.
‡Here ϵ measures how close $f(x)$ is to L, and δ measures how close x is to x_0.

Solution. Here $x_0 = 2$, $L = 6$, and $f(x) = 3x$. Let $\epsilon > 0$ be given. Then we must show that there is a $\delta > 0$ such that if $0 < |x - 2| < \delta$, then $|3x - 6| < \epsilon$.† There are several ways to show this result. The most direct way is to start out with what we want and work backward. We want $|3x - 6| < \epsilon$. Thus

$$-\epsilon < 3x - 6 < \epsilon$$

$$-\frac{\epsilon}{3} < x - 2 < \frac{\epsilon}{3}, \qquad \text{We divided by 3.}$$

which means that

$$|x - 2| < \frac{\epsilon}{3}.$$

Hence if $|x - 2| < \epsilon/3$, then $|3x - 6| < \epsilon$.

Now we are done. All we need to do is to choose $\delta = \epsilon/3$. Then, as we have seen, if $|x - 2| < \delta$, then $|3x - 6| < \epsilon$. Thus $\lim_{x \to 2} 3x = 6$. For example, if $\epsilon = 0.1 = \frac{1}{10}$, then the choice $\delta = \frac{1}{30}$ will work. This choice is illustrated in Figure 2. It is important to note that this value of δ is *not* unique. Any value of $\delta < \frac{1}{30}$ will work. This fact is not surprising since it merely states that if we start even closer to 2 than $\frac{1}{30}$ of a unit away, $3x$ will still remain within $\frac{1}{10}$ of a unit from 6. ∎

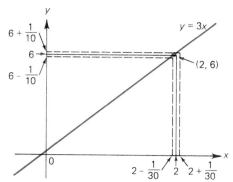

FIGURE 2

Example 1 illustrates the technique for finding a δ that works. We start with the assumption $L - \epsilon < f(x) < L + \epsilon$ and work backward, using elementary algebraic steps, until we obtain an expression of the form $c < x < d$. Then if δ is chosen so that $x_0 - \delta > c$ and $x_0 + \delta < d$, then $|x - x_0| < \delta$ will imply that $c < x < d$. We can usually show that $|f(x) - L| < \epsilon$ by reversing the steps that led to $c < x < d$.

EXAMPLE 2 Show that $\lim_{x \to -1}(4x + 9) = 5$.

---

†Note that since the function $f(x) = 3x$ is defined for all real numbers, it is certainly defined in a neighborhood of $x = 2$.

Solution. Let $\epsilon > 0$ be given. Then we must find a positive number δ such that

$$5 - \epsilon < 4x + 9 < 5 + \epsilon \tag{2}$$

whenever

$$-\delta < x - (-1) < \delta,$$

or

$$-\delta < x + 1 < \delta. \tag{3}$$

Now $4x + 9 = 4(x + 1) + 5$. Thus to find an appropriate choice for δ, we subtract 5 from the inequalities (2) to obtain

$$-\epsilon < 4x + 4 < \epsilon,$$

or after dividing by 4,

$$-\frac{\epsilon}{4} < x + 1 < \frac{\epsilon}{4}.$$

Therefore we see that we may take $\delta = \epsilon/4$ (or anything smaller). This choice of δ will guarantee that $|(4x + 9) - 5| < \epsilon$ whenever $0 < |x + 1| < \delta$. This choice is illustrated in Figure 3. It is now evident that if $\delta = \epsilon/4$, inequality (2) holds if inequality (3) holds. ■

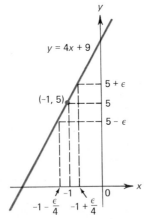

$y = 4x + 9$

$(-1, 5)$

$5 + \epsilon$

5

$5 - \epsilon$

$-1 - \dfrac{\epsilon}{4}$ $-1 + \dfrac{\epsilon}{4}$

FIGURE 3

EXAMPLE 3   Show that $\lim_{x \to 2} x^2 = 4$.

Solution. This example is a bit more involved than the two that preceded it. Let $\epsilon > 0$ be given. We must find a $\delta > 0$ such that

$$-\epsilon < x^2 - 4 < \epsilon$$

whenever

$$-\delta < x - 2 < \delta.$$

We want to show that x^2 is near 4 whenever x is near 2. That is, $|x^2 - 4|$ is small whenever $|x - 2|$ is small. Since x is near 2, we may assume that $1 < x < 3$, so that

$$3 < x + 2 < 5 \qquad \text{and} \qquad |x + 2| < 5.$$

Then

$$|x^2 - 4| = |x + 2||x - 2| < 5|x - 2|. \tag{4}$$

Now choose $\delta = \epsilon/5$. Then if $|x - 2| < \delta$,

$$|x^2 - 4| \underset{\nearrow}{<} 5|x - 2| < 5\delta = 5\,\frac{\epsilon}{5} = \epsilon.$$

From (4)

This result is what we wanted to show (see Figure 4). ■

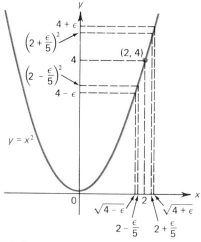

FIGURE 4

The calculation of limits from the definition can be difficult. Fortunately, as we have seen, there are several theorems that make the calculation of limits (at least, in some cases) a relatively simple process. We restate and prove the basic limit theorems in Appendix 5.

We now discuss infinite limits and limits at infinity.

Definition 2   INFINITE LIMITS

(i) $\lim_{x \to x_0} f(x) = \infty$ if for every $N > 0$ there is a $\delta > 0$ such that $f(x) > N$ for all x such that $0 < |x - x_0| < \delta$.

(ii) $\lim_{x \to x_0} f(x) = -\infty$ if for every $N > 0$ there is a $\delta > 0$ such that $f(x) < -N$ for all x such that $0 < |x - x_0| < \delta$.

EXAMPLE 4 Show that $\lim_{x \to 0}(1/x^2) = \infty$.

Solution. Let $N > 0$ be chosen and choose $\delta = 1/\sqrt{N}$. If $0 < |x - 0| < \delta = 1/\sqrt{N}$, then $-1/\sqrt{N} < x < 1/\sqrt{N}$. But if $x < 1/\sqrt{N}$, then $x^2 < (1/\sqrt{N})^2 = 1/N$, so that $1/x^2 > N$. See Figure 5.

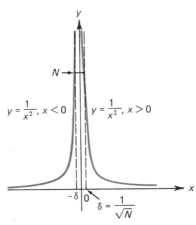

FIGURE 5

NOTE. How did we know that choosing $\delta = 1/\sqrt{N}$ would work? The secret is to start with the inequality you want to end up with and work backward, as in Examples 1, 2, and 3. The inequality $1/x^2 > N$ leads naturally to the inequalities $Nx^2 < 1$, $x^2 < 1/N$, and $x < 1/\sqrt{N}$. Now the choice $\delta = 1/\sqrt{N}$ is obvious. ∎

Definition 3   LIMIT AT INFINITY

(i) $\lim_{x \to \infty} f(x) = L$ if for every $\epsilon > 0$ there is an $N > 0$ such that if $x > N$, then $|f(x) - L| < \epsilon$.

(ii) $\lim_{x \to -\infty} f(x) = L$ if for every $\epsilon > 0$ there is an $N > 0$ such that if $x < -N$, then $|f(x) - L| < \epsilon$.

EXAMPLE 5 Show that $\lim_{x \to -\infty} 1/x = 0$.

Solution. Let $\epsilon > 0$ be given. Then choose $N = 1/\epsilon$. If $x < -N$, then $1/x > -1/N$ and $-\epsilon = -1/N < 1/x < 0$, so $|1/x - 0| < \epsilon$, which is what we had to show. See Figure 6. ∎

EXAMPLE 6 Show that $\lim_{x \to \infty} 1/(\sqrt{x} + 3) = 0$.

Solution. Let $\epsilon > 0$ be given. We want

$$\frac{1}{\sqrt{x} + 3} < \frac{1}{\sqrt{x}} < \epsilon, \qquad \text{or} \qquad 1 < \epsilon \sqrt{x},$$

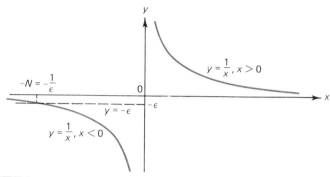

FIGURE 6

or (squaring)

$$1 < \epsilon^2 x.$$

But $\epsilon^2 x > 1$ if $x > 1/\epsilon^2$. Thus let $N = 1/\epsilon^2$. Then for $x > N$, $0 < 1/x < \epsilon^2$, so that

$$\frac{1}{\sqrt{x+3}} < \frac{1}{\sqrt{x}} = \sqrt{\frac{1}{x}} < \sqrt{\epsilon^2} = \epsilon.$$

See Figure 7. ■

FIGURE 7

We next give an example of a function that remains bounded but has no limit as $x \to 0$.

EXAMPLE 7   Let $f(x)$ be defined by

$$f(x) = \begin{cases} 1, & \text{if } x = \dfrac{1}{2^n} \text{ for } n = 1, 2, 3, \ldots \\ x, & \text{otherwise.} \end{cases}$$

A sketch of this function is given in Figure 8.

Since $f(x) = x$ for all values of $x \neq 1/2^n$, it may seem that $f(x) \to 0$ as $x \to 0$. We now show that $\lim_{x \to 0} f(x)$ does not exist. First we show that $\lim_{x \to 0} f(x) \neq 0$. To this end we pick $\epsilon > 0$ such that $\epsilon < 1$. For every $\delta > 0$ there is an n such that $x = 1/2^n < \delta$ (since $1/2^n \to 0$ as n gets very large). Hence no matter how small we choose δ, there will be an x with $0 < x < \delta$ such that $|f(x) - 0| = 1 > \epsilon$. Thus $f(x)$ does not

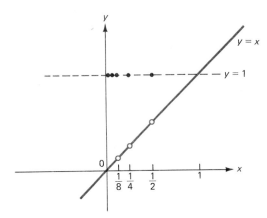

FIGURE 8

tend to 0 as $x \to 0$. But clearly $f(x)$ cannot have any other limit as $x \to 0$ since $|f(x)| < \epsilon$ if $|x| < \epsilon$ except at the points $1/2^n$. Therefore $f(x)$ has no limit as $x \to 0$. However, if $x_0 > 0$, then it is not difficult to show that $\lim_{x \to x_0} f(x) = x_0$ (see Problem 16).

We will, in a moment, turn to the theory of one-sided limits. However, before doing so, we stress that with the material already discussed in this section, it is possible to prove the limit theorems stated in Section 2.3. These proofs are given in Appendix 5.

Definition 4   ONE-SIDED LIMITS

(i) Let f be defined on the open interval (x_0, c) for some number $c > x_0$. Suppose that for every $\epsilon > 0$ there is a $\delta > 0$ such that $x_0 < x < x_0 + \delta$ implies that $|f(x) - L| < \epsilon$. Then L is called the **right-hand limit** of $f(x)$ as $x \to x_0$ and is denoted by

$$\lim_{x \to x_0^+} f(x) = L.$$

(ii) Let f be defined on the open interval (d, x_0) for some number $d < x_0$. If for every $\epsilon > 0$ there is a $\delta > 0$ such that $x_0 - \delta < x < x_0$ implies that $|f(x) - L| < \epsilon$, then L is called the **left-hand limit** of $f(x)$ as $x \to x_0$ and is denoted by

$$\lim_{x \to x_0^-} f(x) = L.$$

These two definitions are illustrated in Figure 9. Note that the functions sketched in the figure do not have a limit as $x \to x_0$.

EXAMPLE 8   Prove that $\lim_{x \to 0^+} \sqrt{x} = 0$.

Solution. Let $\epsilon > 0$ be given. Suppose that $|f(x) - L| = |\sqrt{x} - 0| = \sqrt{x} < \epsilon$. Then $x < \epsilon^2$. This result leads to the obvious choice $\delta = \epsilon^2$. Then if $0 < x < \delta$, $0 < \sqrt{x} < \sqrt{\delta} = \epsilon$, and so $\lim_{x \to 0^+} \sqrt{x} = 0$. ∎

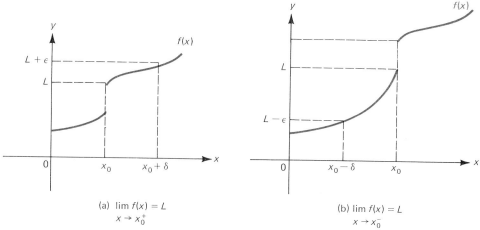

(a) $\lim\limits_{x \to x_0^+} f(x) = L$

(b) $\lim\limits_{x \to x_0^-} f(x) = L$

FIGURE 9

For infinite limits the definitions are similar.

Definition 5   INFINITE ONE-SIDED LIMITS

(i) $\lim_{x \to x_0^+} f(x) = \infty$ means that for every $N > 0$ there is a $\delta > 0$ such that if $x_0 < x < x_0 + \delta$, then $f(x) > N$.

(ii) $\lim_{x \to x_0^-} f(x) = \infty$ if for every $N > 0$ there is a $\delta > 0$ such that if $x_0 - \delta < x < x_0$, then $f(x) > N$.

(iii) $\lim_{x \to x_0^+} f(x) = -\infty$ if for every $N > 0$ there is a $\delta > 0$ such that if $x_0 < x < x_0 + \delta$, then $f(x) < -N$.

(iv) $\lim_{x \to x_0^-} f(x) = -\infty$ if for every $N > 0$ there is a $\delta > 0$ such that if $x_0 - \delta < x < x_0$, then $f(x) < -N$. ■

EXAMPLE 9   Prove that $\lim_{x \to 0^+} 1/x = \infty$ and $\lim_{x \to 0^-} 1/x = -\infty$.

Solution. Let $N > 0$ be given. Then if $\delta = 1/N$, $0 < x < \delta$ implies that $x < 1/N$, or $1/x > N$, which shows that $\lim_{x \to 0^+} 1/x = \infty$. Similarly, if $-\delta < x < 0$, then with $\delta = 1/N$, $x > -\delta = -1/N$ implies that $1/x < -N$ and $\lim_{x \to 0^-} 1/x = -\infty$. ■

We now show that $\lim_{x \to x_0} f(x)$ exists if and only if the right- and left-hand limits exist and are equal. We prove this statement in the case in which the limits are finite. The infinite case is left as a problem (see Problem 19).

Theorem 1 (Theorem 2.2.1) $\lim_{x \to x_0} f(x) = L$ if and only if $\lim_{x \to x_0^+} f(x) = L$ and $\lim_{x \to x_0^-} f(x) = L$.

Proof.   First we show that if $\lim_{x \to x_0} f(x) = L$, then $\lim_{x \to x_0^+} f(x) = L$ and $\lim_{x \to x_0^-} f(x) = L$. To show the right-hand limit, we let $\epsilon > 0$ be given. Then there is a $\delta > 0$ such that if $0 < |x - x_0| < \delta$, then $|f(x) - L| < \epsilon$. In particular, if $x_0 < x < x_0 + \delta$, then $0 < x - x_0 < \delta$, $0 < |x - x_0| < \delta$, and $|f(x) - L| < \epsilon$. Therefore $\lim_{x \to x_0^+} f(x) = L$. The proof for the left-hand limit is similar.

Conversely, suppose that both the right- and left-hand limits exist and are equal to L. For a given $\epsilon > 0$ we choose δ_1 such that if $0 < x - x_0 < \delta_1$, then $|f(x)$

$- L| < \epsilon$. We choose δ_2 such that if $-\delta_2 < x - x_0 < 0$, then $|f(x) - L| < \epsilon$. Let $\delta = \min\{\delta_1, \delta_2\}$. Then, if $|x - x_0| < \delta$, we have $0 < x - x_0 < \delta \le \delta_1$ and $-\delta_2 \le -\delta < x - x_0 < 0$, so that $|f(x) - L| < \epsilon$ whether $x - x_0$ is positive or negative. Hence $\lim_{x \to x_0} f(x) = L$, and the theorem is proved. ∎

EXAMPLE 10 Let

$$f(x) = \begin{cases} x, & x \le 1 \\ x^2, & x > 1. \end{cases}$$

Show that $\lim_{x \to 1} f(x) = 1$.

  Solution. Let $g(x) = x$. For $x \le 1$, $g(x) = f(x)$. Let $h(x) = x^2$. For $x > 1$, $h(x) = f(x)$. Then

$$\lim_{x \to 1^-} f(x) = \lim_{x \to 1^-} g(x) = \lim_{x \to 1^-} x = 1$$

and

$$\lim_{x \to 1^+} f(x) = \lim_{x \to 1^+} h(x) = \lim_{x \to 1^+} x^2 = 1.$$

Since $\lim_{x \to 1^-} f(x) = \lim_{x \to 1^+} f(x) = 1$, Theorem 1 tells us that $\lim_{x \to 1} f(x) = 1$. ∎

PROBLEMS 2.8

In Problems 1–9, verify the given limits directly from Definition 1. In particular, (a) for $\epsilon = \frac{1}{10}$ and (b) $\epsilon = \frac{1}{100}$, what values of δ will ensure that $|f(x) - L| < \epsilon$ if $0 < |x - x_0| < \delta$? (c) Find δ as a function of any ϵ.

1. $\lim_{x \to 1} 7x = 7$
2. $\lim_{x \to 4} (4x - 6) = 10$
3. $\lim_{x \to -2} (5x + 1) = -9$
4. $\lim_{x \to -2} x^2 = 4$
5. $\lim_{x \to 3} (x^2 - 6) = 3$
6. $\lim_{x \to 1} (5 - 2x^2) = 3$
7. $\lim_{x \to -1} (1 + x + x^2) = 1$
8. $\lim_{x \to 0} (4 + 2x - 3x^2) = 4$
   *9. $\lim_{x \to 2} x^3 = 8$

10. (a) Prove from Definition 3 that $\lim_{x \to \infty} 1/\sqrt{x/10} = 0$.
    (b) If $\epsilon = 0.01$, how large must N be so that if $x > N$, then $1/\sqrt{x/10} < \epsilon$?
11. (a) Prove from Definition 2 that $\lim_{x \to 0} 1/x^4 = \infty$.
    (b) How small must δ be chosen so that if $|x| < \delta$, then $1/x^4 > 100{,}000{,}000$?
12. (a) Prove from Definition 3 that $\lim_{x \to -\infty} |1/(x + 3)| = 0$.
    (b) If $\epsilon = 0.01$, how large must N be chosen so that if $x < -N$ then $|1/(x + 3)| < \epsilon$?
13. (a) Prove that $\lim_{x \to \infty} [3 + (4/x^3)] = 3$.
    (b) For $\epsilon = 0.001$, how large must N be chosen so that if $x > N$, then $|[3 + (4/x^3)] - 3| < \epsilon$?
14. (a) Show that if α is any positive integer, then $\lim_{x \to \infty} a/x^\alpha = 0$ for any real number a.
    (b) Prove part (a) when α is a positive rational number.
    (c) Show that $\lim_{x \to \infty} a/(x^\alpha + b) = 0$ for any positive rational number α and any real numbers a and b.
*15. Let r be a real number with $|r| < 1$. Define the **sum of a geometric progression** S_n by

$$S_n = 1 + r + r^2 + \cdots + r^n$$

for each positive integer n.
    (a) Show that $S_n = (1 - r^{n+1})/(1 - r)$. [*Hint:* Compute rS_n and subtract.]
    (b) Show that $\lim_{n \to \infty} S_n = 1/(1 - r)$.

16. Let

$$f(x) = \begin{cases} 1, & \text{if } x = \dfrac{1}{2^n} \text{ for some integer } n \geq 1 \\ x, & \text{otherwise.} \end{cases}$$

Show that if $x_0 \neq 0$, $\lim_{x \to x_0} f(x) = x_0$.

*17. Let

$$f(x) = \begin{cases} 1, & x \text{ rational} \\ 0, & x \text{ irrational.} \end{cases}$$

Show that $\lim_{x \to x_0} f(x)$ does not exist for any real number x_0. [*Hint:* Pick any $\epsilon < 1$. Then show that for any $\delta > 0$ there is an x in $(x_0 - \delta, x_0 + \delta)$ such that $|f(x) - f(x_0)| > \epsilon$.]

18. Let

$$f(x) = \begin{cases} 0, & x < 0 \\ x, & x \geq 0. \end{cases}$$

Prove that $\lim_{x \to 0} f(x) = 0$.

19. Prove that $\lim_{x \to x_0} f(x) = \infty$ if and only if $\lim_{x \to x_0^+} f(x) = \lim_{x \to x_0^-} f(x) = \infty$.

20. Prove, using Definition 4, that $\lim_{x \to 4^+} f(x) = 2$ and $\lim_{x \to 4^-} f(x) = 16$, where

$$f(x) = \begin{cases} x^2, & x < 4 \\ \sqrt{x}, & x \geq 4. \end{cases}$$

21. Show, using Theorem 1, that if

$$f(x) = \begin{cases} 3x + 2, & x \leq 2 \\ 2x^2, & x > 2, \end{cases}$$

then $\lim_{x \to 2} f(x) = 8$.

22. Find a number c such that if

$$f(x) = \begin{cases} 3x, & x \leq -2 \\ cx^2, & x > -2, \end{cases}$$

then $\lim_{x \to -2} f(x)$ exists.

23. Let $f(x) = [x]$. Prove that $\lim_{x \to x_0} f(x)$ exists if and only if x_0 is not an integer.

Consider the alternative definition of continuity:

Definition 6 CONTINUITY ($\epsilon - \delta$ DEFINITION) The function f is **continuous** at x_0 if f is defined in a neighborhood of x_0 and if for every $\epsilon > 0$ there is a $\delta > 0$ such that if $|x - x_0| < \delta$, then $|f(x) - f(x_0)| < \epsilon$.

*24. Using the definition of a limit in this section, show that if f is continuous according to the definition on page 68, then it is continuous according to the definition above.

*25. Show that if f is continuous according to the definition above, then f is continuous according to the definition on page 112.

In Problems 26–33 a function f and numbers x_0 and ϵ are given. Find a number δ such that $|f(x) - f(x_0)| < \epsilon$ whenever $|x - x_0| < \delta$.

26. $f(x) = 2x + 3$; $x_0 = 1$, $\epsilon = 0.1$

27. $f(x) = 3x - 5$; $x_0 = 2$, $\epsilon = 0.01$

28. $f(x) = x^2$; $x_0 = 0$, $\epsilon = 0.1$

29. $f(x) = x^2$; $x_0 = 3$, $\epsilon = 0.1$

30. $f(x) = \dfrac{1}{x}$; $x_0 = 2$, $\epsilon = 0.1$

31. $f(x) = \dfrac{1}{x}$; $x_0 = 1$, $\epsilon = 0.1$

32. $f(x) = \dfrac{1}{x}$; $x_0 = 0.1$, $\epsilon = 0.1$ **33.** $f(x) = \dfrac{1}{x}$; $x_0 = 0.0001$, $\epsilon = 0.1$

REVIEW EXERCISES FOR CHAPTER 2

1. Tabulate values of $f(x) = x^2 - 3x + 6$ for $x = 3, 1, 2.5, 1.5, 2.1, 1.9, 2.01$, and 1.99. What does your table tell you about $\lim_{x\to 2}(x^2 - 3x + 6)$?

2. Tabulate values of $f(x) = x^2 + 10x + 8$ for $x = -4, -2, -3.5, -2.5, -3.1, -2.9, -3.01$, and -2.99. What does your table tell you about $\lim_{x\to -3}(x^2 + 10x + 8)$?

3. Calculate the following:

   (a) $\lim_{x\to 1} (x^3 - 3x + 2)$

   (b) $\lim_{x\to 5} (-x^3 + 17)$

   (c) $\lim_{x\to 3} \dfrac{x^4 - 2x + 1}{x^3 + 3x - 5}$

   (d) $\lim_{x\to -1} \dfrac{x^3 + x^2 + x + 1}{x^4 + x^3 + x^2 + x + 1}$

4. Calculate the following:

   (a) $\lim_{x\to 0} |x + 2|$

   (b) $\lim_{x\to 1} |x - 3|$

   (c) $\lim_{x\to -3} |x + 4|$

   (d) $\lim_{x\to 1} \dfrac{|x|}{x}$

5. Calculate the following:

   (a) $\lim_{x\to 3} \dfrac{(x - 3)(x - 4)}{x - 3}$

   (b) $\lim_{x\to 5} \dfrac{x^2 - 6x + 5}{x - 5}$

6. Do the following limits exist? If not, explain why. If so, calculate them.

   (a) $\lim_{x\to 1} \sqrt{x - 1}$

   (b) $\lim_{x\to 2} \sqrt{x - 1}$

   (c) $\lim_{x\to -3} \dfrac{|x + 2|}{x + 2}$

   (d) $\lim_{x\to -2} \dfrac{|x + 2|}{x + 2}$

   (e) $\lim_{x\to 1} \sqrt[3]{x - 1}$

   (f) $\lim_{x\to 1} \sqrt[4]{x - 1}$

7. Calculate the following:

   (a) $\lim_{x\to 1} 23\sqrt{x - 17}$

   (b) $\lim_{x\to -1} (1 - x + x^2 - x^3 + x^4)$

   (c) $\lim_{x\to 4} \dfrac{x^2 + 9}{x^2 - 9}$

   (d) $\lim_{x\to -1} 5x^{250}$

   (e) $\lim_{x\to -1} 6x^{251}$

   (f) $\lim_{x\to 3} (x^2 + x - 8)^5$

   (g) $\lim_{x\to 0} \dfrac{x^8 - 7x^5 + x^3 - x^2 + 3}{x^{23} - 2x + 9}$

   (h) $\lim_{x\to 0} \dfrac{ax^2 + bx + c}{dx^2 + ex + f}$ (a, b, c, d, e, f are all nonzero real numbers)

8. Calculate the following:

   (a) $\lim_{x\to 1+} \sqrt{x - 1}$

   (b) $\lim_{x\to 1-} \sqrt{x - 1}$

   (c) $\lim_{x\to 1-} \dfrac{1}{x - 1}$

   (d) $\lim_{x\to 1+} \dfrac{1}{x - 1}$

   (e) $\lim_{x\to -2+} \dfrac{x - 2}{x + 2}$

   (f) $\lim_{x\to -2-} \dfrac{x - 2}{x + 2}$

9. Explain why $\lim_{x\to 0}(1/x^3)$ does not exist but $\lim_{x\to 0}(1/x^4)$ does exist.

10. Calculate the following:

(a) $\displaystyle\lim_{x\to2}\frac{x-3}{(x-2)^2}$

(b) $\displaystyle\lim_{x\to-1}\frac{x+10}{(x+1)^{10}}$

11. Calculate the following:

(a) $\displaystyle\lim_{x\to\infty}\frac{1}{x^3}$

(b) $\displaystyle\lim_{x\to\infty}\frac{1}{\sqrt{x+2}}$

(c) $\displaystyle\lim_{x\to\infty}\frac{\sqrt{x}}{x^2+3}$

(d) $\displaystyle\lim_{x\to\infty}\frac{x^3+6x^2+4x+2}{3x^3-9x^2+11}$

(e) $\displaystyle\lim_{x\to\infty}\frac{3x^5-6x^2+3}{x^7-2}$

(f) $\displaystyle\lim_{x\to\infty}\frac{x^7-9}{30x^5+x^4+161}$

12. How large must x be in order that $1/\sqrt[3]{x}<0.01$?

13. Determine whether the derivative appears to be positive, negative, or zero at each point in Figure 1.

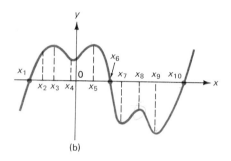

(a) (b)

FIGURE 1

14. Find the equation of the tangent line to the given curve at the given point.

(a) $y=x^2-3$; $(2,1)$

(b) $y=7x^3-8$; $(1,-1)$

(c) $y=\sqrt{x+1}$; $(3,2)$

(d) $y=\dfrac{1}{x+1}$; $(0,1)$

15. Explain why the curve $y=|x+1|$ does not have a derivative at the point $x=-1$.

16. The distance a particle travels is given by $s(t)=t^3+t^2+6$ kilometers after t hours. How fast is it traveling (in kilometers per hour) after 2 hr?

17. The mass of the left x meters of a nonuniform 8-m metal bar is given by $\mu=\sqrt{x+1}$ kilograms (see Figure 2). What is the density of the bar at a point 3 m from the left end?

0   x 8

FIGURE 2

18. The volume of a sphere is $\frac{4}{3}\pi r^3$. How is the volume changing (with respect to changes in r) when the radius r is equal to 2 ft?

19. A slaughterhouse purchases cattle from a ranch at a cost of $C(q)=200+8q-0.02q^2$, where q is the number of head of cattle bought at one time up to a maximum of 150. What is the house's marginal cost as a function of q? Does it pay to buy in large quantities?

20. At what values of x does the function graphed in Figure 3 appear not to have a derivative?

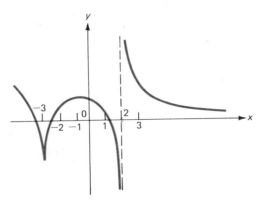

FIGURE 3

21. Show that if

$$f(x) = \begin{cases} 2x + 3, & x \leq -2 \\ x^2 - 5, & x > -2, \end{cases}$$

then $\lim_{x \to -2} f(x)$ exists. Calculate that limit.

22. Let

$$f(x) = \begin{cases} x^2 + 3, & x \leq 3 \\ x^3 + \alpha, & x > 3. \end{cases}$$

For what value of α does $\lim_{x \to 3} f(x)$ exist? Calculate this limit.

In Exercises 23–32, find open intervals in which each of the functions is continuous. Indicate any removable discontinuities.

23. $f(x) = 2\sqrt{x}$

24. $f(x) = 3\sqrt[3]{x}$

25. $f(x) = \dfrac{1}{x - 6}$

26. $f(x) = \dfrac{1}{x^2 - 6}$

27. $f(x) = \dfrac{x}{x^2 - 4}$

28. $f(x) = |x + 2|$

29. $f(x) = \dfrac{|x + 3|}{x + 3}$

30. $f(x) = \dfrac{x^2 - 9}{x + 3}$

31. $f(x) = \dfrac{x + 3}{x^2 - 9}$

32. $f(x) = \dfrac{x^3 + 1}{x + 1}$

The last six exercises rely on the material in Section 2.8. In Exercises 33–37, use the ϵ–δ definition of the limit to verify the given limits.

33. $\lim\limits_{x \to -1} 4x = -4$

34. $\lim\limits_{x \to 3} (x^2 + 2) = 11$

35. $\lim\limits_{x \to \infty} \left(\dfrac{1}{x} + \dfrac{1}{x^2} \right) = 0$

36. $\lim\limits_{x \to \infty} \dfrac{1}{\sqrt[3]{x + 3}} = 0$

37. $\lim\limits_{x \to -3} |x + 4| = 1$

38. Let

$$f(x) = \begin{cases} 1, & x \text{ an integer} \\ 2, & x \text{ not an integer.} \end{cases}$$

Show that $\lim_{x \to \infty} f(x)$ does not exist.

3 More about Derivatives

In the previous chapter we introduced the concepts of the limit and the derivative but found that the calculation of derivatives could be extremely difficult. An even more annoying problem was the seeming necessity to come up with a special technique (like multiplying and dividing by some quantity) each time we took the limit in the process of computing a derivative.

In this chapter we continue the discussion of properties of the derivatives of functions. We will be principally concerned with simplifying the process of a differentiation so that it will no longer be necessary to deal with complicated limits.

In many of the sections of this chapter we will be deriving formulas for calculating derivatives. Formulas are usually not very exciting, and we ask you to bear with us here. By the time you have completed the chapter, you will find that differentiation is not nearly as complicated as it now seems. The work involved in memorizing the appropriate formulas will pay dividends in the chapters to come.

3.1 SOME DIFFERENTIATION FORMULAS

We begin by deriving a formula for calculating the derivative of $y = f(x) = x^n$, where n is a positive integer. First, let us look for a pattern. We have already calculated the following derivatives:

(i) $\frac{d}{dx}x = 1$ (since $y = x = 1 \cdot x + 0$ is the equation of a straight line with slope 1).

(ii) $\frac{d}{dx}x^2 = 2x$ (Example 2.5.2).

(iii) $\frac{d}{dx}x^4 = 4x^3$ (see Example 2.6.3).

Do you see a pattern? The answer is given in Theorem 1.

Theorem 1 If n is a positive integer, then

$$\frac{d}{dx}x^n = nx^{n-1}. \tag{1}$$

Proof. We first note that for any real numbers a and b (such that $a \neq b$),

$$a^2 - b^2 = (a - b)(a + b)$$
$$a^3 - b^3 = (a - b)(a^2 + ab + b^2)$$
$$a^4 - b^4 = (a - b)(a^3 + a^2b + ab^2 + b^3)$$
$$\vdots$$

and for any positive integer n,

$$\frac{a^n - b^n}{a - b} = \frac{(a - b)(a^{n-1} + a^{n-2}b + a^{n-3}b^2 + \cdots + ab^{n-2} + b^{n-1})}{a - b}$$
$$= a^{n-1} + a^{n-2}b + a^{n-3}b^2 + \cdots + ab^{n-2} + b^{n-1}. \tag{2}$$

For each n the equality can be verified by multiplying $(a - b)$ by $a^{n-1} + a^{n-2}b + a^{n-3}b^2 + \cdots + ab^{n-2} + b^{n-1}$ and noting that all terms except a^n and $-b^n$ cancel. Now

$$\frac{d}{dx}x^n = \lim_{\Delta x \to 0}\frac{(x + \Delta x)^n - x^n}{\Delta x} = \lim_{\Delta x \to 0}\frac{(x + \Delta x)^n - x^n}{(x + \Delta x) - x}. \tag{3}$$

Then,

Set $a = x + \Delta x$ and $b = x$ in (2).

$$\lim_{\Delta x \to 0}\frac{(x + \Delta x)^n - x^n}{(x + \Delta x) - x} = \lim_{\Delta x \to 0}[(x + \Delta x)^{n-1} + (x + \Delta x)^{n-2}x + (x + \Delta x)^{n-3}x^2$$
$$+ \cdots + (x + \Delta x)x^{n-2} + x^{n-1}]$$
$$= x^{n-1} + x^{n-2}x + x^{n-3}x^2 + \cdots + x \cdot x^{n-2} + x^{n-1}$$
$$= \underbrace{x^{n-1} + x^{n-1} + \cdots + x^{n-1}}_{n \text{ times}} = nx^{n-1}. \blacksquare$$

EXAMPLE 1 Let $f(x) = x^3$. Then $f'(x) = 3x^2$.

EXAMPLE 2 Let $f(x) = x^{17}$. Then $f'(x) = 17x^{16}$. ▮

We can compute derivatives of more complicated functions by deriving some more general formulas. We first give a theorem that states the obvious fact that a constant function does not change.

Theorem 2 Let $f(x) = c$, a constant. Then $f'(x) = 0$.

NOTE. This result is evident from looking at the graph of a constant function—a horizontal line with a slope of zero (see Figure 1).

FIGURE 1

Proof.

$$f'(x) = \lim_{\Delta x \to 0} \frac{f(x + \Delta x) - f(x)}{\Delta x} = \lim_{\Delta x \to 0} \frac{c - c}{\Delta x}$$

$$= \lim_{\Delta x \to 0} \frac{0}{\Delta x} = \lim_{\Delta x \to 0} 0 = 0. \ ▮$$

The converse of this theorem is also true, namely, that if $f' = 0$ on an interval, then f is a constant function on that interval. The proof of this theorem is more difficult and will be deferred until Section 4.2.

Theorem 3 Let c be a constant. If f is differentiable, then cf is also differentiable, and

$$\frac{d}{dx} cf = c \frac{df}{dx}.$$

(4)

Proof.

$$\frac{d}{dx} cf = \lim_{\Delta x \to 0} \frac{(cf)(x + \Delta x) - (cf)(x)}{\Delta x} = \lim_{\Delta x \to 0} c \left[\frac{f(x + \Delta x) - f(x)}{\Delta x} \right]$$

$$= c \lim_{\Delta x \to 0} \frac{f(x + \Delta x) - f(x)}{\Delta x} = c \frac{df}{dx}.$$

We have used the limit theorem, Theorem 2.3.2, here $(\lim_{x \to x_0} cf = c \lim_{x \to x_0} f)$. ▮

EXAMPLE 3   Compute $\dfrac{d}{dx}(41x^3)$.

Solution.

$$\frac{d}{dx}(41x^3) = 41\frac{d}{dx}x^3 = 41(3x^2) = 123x^2 \ \blacksquare$$

Theorem 4   Let f and g be differentiable. Then $f + g$ is also differentiable, and

$$(f + g)' = \frac{d}{dx}(f + g) = \frac{df}{dx} + \frac{dg}{dx} = f' + g'. \tag{5}$$

That is, *the derivative of the sum of two differentiable functions is the sum of their derivatives.*

Proof.

$$\frac{d}{dx}(f + g) = \lim_{\Delta x \to 0} \frac{(f + g)(x + \Delta x) - (f + g)(x)}{\Delta x}$$

$$= \lim_{\Delta x \to 0} \frac{f(x + \Delta x) + g(x + \Delta x) - f(x) - g(x)}{\Delta x}$$

$$= \lim_{\Delta x \to 0} \left\{ \left[\frac{f(x + \Delta x) - f(x)}{\Delta x}\right] + \left[\frac{g(x + \Delta x) - g(x)}{\Delta x}\right]\right\}$$

By Theorem 3, on page 79
$$\downarrow$$
$$= \lim_{\Delta x \to 0} \frac{f(x + \Delta x) - f(x)}{\Delta x} + \lim_{\Delta x \to 0} \frac{g(x + \Delta x) - g(x)}{\Delta x}$$

$$= \frac{df}{dx} + \frac{dg}{dx} \ \blacksquare$$

EXAMPLE 4   Let $f(x) = 4x^3 + 3\sqrt{x}$. Find $f'(x)$.

Solution.   We saw in Section 2.5, that

$$\frac{d}{dx}\sqrt{x} = \frac{1}{2\sqrt{x}}. \quad \text{(Example 2.5.3)}$$

Thus

By Theorems 3 and 4
$$\downarrow$$
$$\frac{d}{dx}(4x^3 + 3\sqrt{x}) = 4\frac{d}{dx}x^3 + 3\frac{d}{dx}\sqrt{x} = 4 \cdot 3x^2 + \frac{3}{2\sqrt{x}} = 12x^2 + \frac{3}{2\sqrt{x}}. \ \blacksquare$$

We have shown that the derivative of the sum of two functions is the sum of the derivatives of the two functions. It is not too difficult to show that this fact applies to more than two functions. The theorem that states this result is given next. The proof is left as an exercise (see Problem 19).

Theorem 5 Let $f_1, f_2, \ldots, f_n$ be n differentiable functions. Then $f_1 + f_2 + \cdots + f_n$ is differentiable, and

$$(f_1 + f_2 + \cdots + f_n)' = f_1' + f_2' + \cdots + f_n'. \tag{6}$$

EXAMPLE 5 Let $f(x) = 1 + x + x^2 + x^3$. Calculate df/dx.

Solution.

$$\frac{df}{dx} = \frac{d}{dx}1 + \frac{d}{dx}(x) + \frac{d}{dx}(x^2) + \frac{d}{dx}(x^3) = 0 + 1 + 2x + 3x^2$$

$$= 1 + 2x + 3x^2 \blacksquare$$

EXAMPLE 6 Let $s(t) = 17 - 4t^2 + 8t^3 - 9\sqrt{t}$. Calculate ds/dt.

Solution. We find the derivative of each term separately and then add.

(i) $\frac{d}{dt}17 = 0$ since 17 is a constant.

(ii) $\frac{d}{dt}(-4t^2) = -4\frac{d}{dt}t^2 = -4 \cdot 2t = -8t$.

(iii) $\frac{d}{dt}8t^3 = 8\frac{d}{dt}t^3 = 8 \cdot 3t^2 = 24t^2$.

(iv) $\frac{d}{dt}(-9\sqrt{t}) = -9\frac{d}{dt}\sqrt{t} = -9\left(\frac{1}{2\sqrt{t}}\right) = \frac{-9}{2\sqrt{t}}$.

Therefore

$$\frac{ds}{dt} = \frac{d}{dt}(17 - 4t^2 + 8t^3 - 9\sqrt{t}) = \frac{d}{dt}17 - 4\frac{d}{dt}t^2 + 8\frac{d}{dt}t^3 - 9\frac{d}{dt}\sqrt{t}$$

$$= -8t + 24t^2 - \frac{9}{2\sqrt{t}}. \blacksquare$$

Consider the curve depicted in Figure 2. The tangent line T is drawn. The line N, which is perpendicular to T at the point $(x, f(x))$, is called a *normal line.*†

Definition 1 NORMAL LINE A **normal line** to a curve at a point is a line that is perpendicular to the tangent line at that point.

---

†From the Latin *norma*, meaning "square," the carpenter's square. Until the 1830s the English word "normal" meant standing at right angles to the ground.

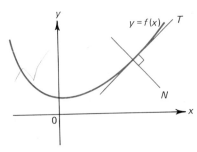

FIGURE 2

If the slope of the tangent line is m and if $m \neq 0$, then from Theorem 1.4.3, the slope of the normal line is $-1/m$. Normal lines are very important in applications, as we will see in some of the later chapters of this book.

EXAMPLE 7 Find the equation of the normal line to the curve $y = x^2$ at the point (2, 4).

Solution. We have $dy/dx = 2x$, and when $x = 2$, $dy/dx = 2 \cdot 2 = 4$. This is the slope of the tangent line at the point (2, 4). Hence the slope of the normal line is $-\frac{1}{4}$, and the normal line has the equation

$$\frac{y - 4}{x - 2} = -\frac{1}{4} \quad \text{or} \quad x + 4y = 18. \blacksquare$$

PROBLEMS 3.1

In Problems 1–10, calculate the derivative of the given function.

1. $f(x) = x^5$
2. $f(x) = 27$
3. $f(x) = 3x^2 + 19x + 2$
4. $g(t) = t^{10} - t^3$
5. $g(t) = t^5 + \sqrt{t}$
6. $g(t) = 1 - t + t^4 - t^7$
7. $h(z) = z^{100} + 100z^{10} + 10$
8. $h(z) = 27z^6 + 3z^5 + 4z$
9. $v(r) = 3r^8 - 8r^6 - 7r^4 + 2r^2 + 3$
10. $v(r) = -3r^{12} + 12r^3$

In Problems 11–17, find the line that is tangent to the given curve at the given point.

11. $y = x^4$; (1, 1)
12. $y = 3x^5 - 3x^3 + 1$; (−1, 1)
13. $y = 2x^7 - x^6 - x^3$; (1, 0)
14. $y = 5x^6 - x^4 + 2x^3$; (1,6)
15. $y = 1 + x + x^2 + x^3 + x^4 + x^5$; (0, 1)
16. $y = 1 - x + 2x^2 - 3x^3 + 4x^4$; (1, 3)
17. $y = x^6 - 6\sqrt{x}$; (1, −5)

18. Show that if $y = x^n$, where $n \geq 2$ is a positive integer, then the tangent line to the curve at the point (0, 0) is the x-axis.
19. **(a)** Prove Theorem 5 when $n = 3$ and $n = 4$.
    *(b) Use mathematical induction to prove Theorem 5 for general n (see Appendix 2).

In Problems 20–24, find the normal line to the given curve at the given point.

20. $y = x^3$; (1, 1)
21. $y = x^7 - 6x^5$; (1, −5)
22. $y = 2x^4 - 3x^3 + 2x + 1$; (2, 13)
23. $y = 4x^{10} + 3x^2 + 6$; (0, 6)
24. $y = x^6 - 6\sqrt{x}$; (1, −5)

25. Find the points on the graph of $y = 2x^3 + 3x^2 - 6x + 1$ where the curve has a horizontal tangent.
26. Let $f(x) = x^3 + 9x^2 + 2x + 2$. Find the values of x for which $f(x) = f'(x)$.
*27. Find the equation of the two lines tangent to the curve $y = x^2 + 3$ that pass through the point $(1, 0)$. [*Hint:* Any point on the curve has the coordinates $(a, a^2 + 3)$. If L is a tangent line at the point $(a, a^2 + 3)$, then the slope of L is $2a$.]
28. Suppose that when an airplane takes off (starting from rest), the distance (in feet) it travels during the first few seconds is given by the formula $s = 1 + 4t + 6t^2$. How fast (in feet per second) is the plane traveling after 10 sec? After 20 sec?
29. A petri dish contains two colonies of bacteria. The population of the first colony is given by $P_1(t) = 1000 + 50t - 20\sqrt{t}$, and the population of the second is given by $P_2(t) = 2000 + 30t^2 - 80t$, where t is measured in hours.
    (a) Find a function that represents the *total* population of the two species.
    (b) What is the instantaneous growth rate of the total population?
    (c) How fast is the total population growing after 4 hr? After 16 hr?
*30. For the model of Problem 29, how fast is the first population growing when the second population is growing at a rate of 160 individuals per hour?
31. Show that the rate of change of the area of a circle with respect to its radius is equal to its circumference.
32. *Stefan's law* for the amount of radiant energy emitted from the surface of a body is given by $R = \sigma T^4$, where R is the rate of emission per unit area, T is the temperature, measured in degrees kelvin (K) and σ is a universal constant called the *Stefan-Boltzmann constant*. † At a temperature of 400 K, what is the instantaneous rate of change of R with respect to the temperature?
*33. A growing grapefruit with a diameter of $2k$ inches has a skin that is $k/12$ inches thick (the skin is included in the diameter of the grapefruit). What is the rate of growth of the volume of the skin (per unit growth in the radius) when the radius of the grapefruit is 3 in.?
34. Where, if ever, is the graph of $y = \sqrt{x}$ parallel to the line $\frac{1}{8}x - 8y = 1$?
*35. Let $f(x) = x^2$ and $g(x) = \frac{1}{3}x^3$. Each vertical line, $x =$ constant, meets the graph of f and the graph of g.
    (a) On what vertical lines do the graphs of f and g have parallel tangents?
    (b) On what horizontal lines do they have parallel tangents?
*36. Consider the graph of $y = ax^2 + bx + c$, $a \neq 0$.
    (a) If a and c have the same sign ($a \cdot c > 0$), then there are exactly two tangents to the graph that pass through the origin. Prove this statement.
    (b) If a and c have opposite signs ($a \cdot c < 0$), then no tangent to the graph passes through the origin. Prove this statement.
    (c) Discuss the remaining case, $c = 0$.
*37. Let $f(x) = (x - 1)^2 + 3 = x^2 - 2x + 4$ and $g(x) = -f(-x) = -x^2 - 2x - 4$. Find each line that is tangent to both of the graphs $y = f(x)$ and $y = g(x)$. [*Note:* A rough sketch indicates that there is at least one such line.]
*38. Let $f(x) = x^2$ and $g(x) = -2x^2 + 28x - 82$. Find each line that is tangent to both of the graphs $y = f(x)$ and $y = g(x)$.
*39. Let $f(x) = x^2$ and $g(x) = -x^2 + 10x - 17$. Find each line that is tangent to both of the graphs $y = f(x)$ and $y = g(x)$.
*40. (a) Show that the normal line to $y = x^2$ at $(3, 9)$ passes through the point $(-3, 10)$.
    (b) Are there any other points on $y = x^2$ such that the normal line there also will pass through $(-3, 10)$? Justify your answer.

---

†$\sigma = 5.67 \times 10^{-8}$ watts/m^2(K)^4.

3.2 THE PRODUCT AND QUOTIENT RULES

In this section we develop some additional rules to simplify the calculation of derivatives. To see why additional rules are needed, consider the problem of calculating the derivatives of

$$f(x) = \sqrt{x}(x^4 + 3) \quad \text{or} \quad g(x) = \frac{x^4 + 3}{\sqrt{x}}.$$

To carry out the calculations from the definition would be very tedious. However, we will shortly see that these calculations can be made rather simple.

Let f and g be two differentiable functions of x. What is the derivative of the product fg? It is easy to be led astray here. Theorem 2.3.4 states that

$$\lim_{x \to x_0} f(x)g(x) = \lim_{x \to x_0} f(x) \lim_{x \to x_0} g(x).$$

However, the derivative of the product is *not* equal to the product of the derivatives. That is,

$$\frac{d}{dx} fg \neq \frac{df}{dx} \cdot \frac{dg}{dx}.$$

Originally, Leibniz, the codiscoverer of the derivative, thought that they were equal. But an easy example shows that this is false. Let $f(x) = x$ and $g(x) = x^2$. Then

$$(fg)(x) = f(x)g(x) = x^3 \quad \text{and} \quad \frac{d}{dx} fg = 3x^2.$$

But

$$\frac{df}{dx} = 1 \quad \text{and} \quad \frac{dg}{dx} = 2x,$$

so that

$$\frac{df}{dx} \cdot \frac{dg}{dx} = 1 \cdot 2x = 2x \neq 3x^2 = \frac{d}{dx} fg.$$

The correct formula, discovered after many false steps by both Leibniz and Newton, is given below.

Theorem 1 **Product Rule** Let f and g be differentiable. Then fg is differentiable, and

$$(fg)' = \frac{d}{dx} fg = f\frac{dg}{dx} + g\frac{df}{dx} = fg' + gf'. \tag{1}$$

Verbally, the product rule says that *the derivative of the product of two functions is equal to the first times the derivative of the second plus the second times the derivative of the first.*

Proof.

$$\frac{d}{dx} fg = \lim_{\Delta x \to 0} \frac{f(x + \Delta x)g(x + \Delta x) - f(x)g(x)}{\Delta x}.$$

This equation doesn't look very much like the derivative of anything. To continue, we will use the trick of adding and subtracting the term $f(x)g(x + \Delta x)$ in the numerator. As we will see, this technique makes everything come out nicely. We have

These are the additional terms.

$$\frac{d}{dx} fg = \lim_{\Delta x \to 0} \frac{f(x + \Delta x)g(x + \Delta x) - \overbrace{f(x)g(x + \Delta x) + f(x)g(x + \Delta x)} - f(x)g(x)}{\Delta x}$$

$$= \lim_{\Delta x \to 0} \frac{g(x + \Delta x)[f(x + \Delta x) - f(x)]}{\Delta x} + \lim_{\Delta x \to 0} \frac{f(x)[g(x + \Delta x) - g(x)]}{\Delta x}$$

$$= \lim_{\Delta x \to 0} g(x + \Delta x) \lim_{\Delta x \to 0} \frac{f(x + \Delta x) - f(x)}{\Delta x} + \lim_{\Delta x \to 0} f(x) \lim_{\Delta x \to 0} \frac{g(x + \Delta x) - g(x)}{\Delta x}$$

$$= g(x) \frac{df}{dx} + f(x) \frac{dg}{dx}.$$

Here we have used the fact that $\lim_{\Delta x \to 0} g(x + \Delta x) = g(x)$. This result follows from the fact that g is differentiable, so that g is continuous by Theorem 2.7.3. As $\Delta x \to 0$, $x + \Delta x \to x$, so by the definition of continuity, $\lim_{x + \Delta x \to x} g(x + \Delta x) = g(x)$. ∎

EXAMPLE 1   Let $h(x) = \sqrt{x}(x^4 + 3)$. Calculate dh/dx.

Solution. $h(x) = f(x)g(x)$, where $f(x) = \sqrt{x}$ and $g(x) = x^4 + 3$. Then

$$\frac{dh}{dx} = f \frac{dg}{dx} + g \frac{df}{dx} = \sqrt{x}(4x^3) + (x^4 + 3)\frac{1}{2\sqrt{x}}.$$

This is the correct answer, but we will use some algebra to simplify the result.

$$\frac{dh}{dx} = \frac{2\sqrt{x}\sqrt{x}(4x^3) + x^4 + 3}{2\sqrt{x}} = \frac{2x(4x^3) + x^4 + 3}{2\sqrt{x}} = \frac{9x^4 + 3}{2\sqrt{x}} \quad ∎$$

EXAMPLE 2   Let $h(t) = (t^2 + 2)(t^3 - 5)$. Compute $h'(t)$.

Solution. We could first multiply through and use the rules given in the previous section. However, it is simpler to compute the derivative directly from the product rule. If $f(t) = t^2 + 2$ and $g(t) = t^3 - 5$, then $h(t) = f(t)g(t)$, and

$$h'(t) = f(t)g'(t) + f'(t)g(t) = (t^2 + 2)(3t^2) + 2t(t^3 - 5).$$

This answer is correct as it stands. We can, if desired, multiply through and combine terms to obtain

$$h'(t) = 3t^4 + 6t^2 + 2t^4 - 10t = 5t^4 + 6t^2 - 10t. \ \blacksquare$$

Having now discussed the product of two functions, we turn to their quotient.

Theorem 2 **Quotient Rule** Let f and g be differentiable at x. Then, if $g(x) \neq 0$, we have

$$\left(\frac{f}{g}\right)' = \frac{d}{dx}\left(\frac{f}{g}\right) = \frac{g(x)(df/dx) - f(x)(dg/dx)}{g^2(x)} = \frac{gf' - fg'}{g^2}. \tag{2}$$

The quotient rule states that *the derivative of the quotient is equal to the denominator times the derivative of the numerator minus the numerator times the derivative of the denominator all over the denominator squared.*

Proof.

$$\frac{d}{dx}\left(\frac{f}{g}\right) = \lim_{\Delta x \to 0} \frac{\frac{f(x+\Delta x)}{g(x+\Delta x)} - \frac{f(x)}{g(x)}}{\Delta x} = \lim_{\Delta x \to 0} \frac{f(x+\Delta x)g(x) - f(x)g(x+\Delta x)}{\Delta x \, g(x+\Delta x)g(x)}$$

These terms sum to zero.

$$= \lim_{\Delta x \to 0} \frac{f(x+\Delta x)g(x) \overbrace{- f(x)g(x) + f(x)g(x)} - f(x)g(x+\Delta x)}{\Delta x \, g(x+\Delta x)g(x)}$$

$$= \lim_{\Delta x \to 0} \frac{\left\{ g(x)\left(\frac{f(x+\Delta x) - f(x)}{\Delta x}\right) - f(x)\left(\frac{g(x+\Delta x) - g(x)}{\Delta x}\right)\right\}}{g(x+\Delta x)g(x)}$$

$$= \frac{\left\{ g(x)\lim_{\Delta x \to 0}\left(\frac{f(x+\Delta x) - f(x)}{\Delta x}\right) - f(x)\lim_{\Delta x \to 0}\left(\frac{g(x+\Delta x) - g(x)}{\Delta x}\right)\right\}}{g(x)\lim_{\Delta x \to 0} g(x+\Delta x)}$$

$$= \frac{g(x)f'(x) - f(x)g'(x)}{[g(x)]^2}$$

Again we used the fact that g is continuous (being differentiable), so that $\lim_{\Delta x \to 0} g(x+\Delta x) = g(x)$. $\blacksquare$

EXAMPLE 3 Let $h(x) = x/(x-1)$. Compute $h'(x)$.

Solution. With $f(x) = x$ and $g(x) = x - 1$, $h(x) = f(x)/g(x)$, and

$$h'(x) = \frac{g(x)f'(x) - f(x)g'(x)}{[g(x)]^2} = \frac{(x-1)(1) - x(1)}{(x-1)^2} = \frac{-1}{(x-1)^2}. \ \blacksquare$$

EXAMPLE 4 Let $h(x) = (x^4 + 3)/\sqrt{x}$. Calculate dh/dx.

Solution. $h(x) = f(x)/g(x)$, where $f(x) = x^4 + 3$ and $g(x) = \sqrt{x}$. Thus

$$\frac{dh}{dx} = \frac{g(x)(df/dx) - f(x)(dg/dx)}{g^2} = \frac{\sqrt{x}(4x^3) - (x^4 + 3)(1/2\sqrt{x})}{(\sqrt{x})^2}$$

$$= \frac{2\sqrt{x}\sqrt{x}(4x^3) - x^4 - 3}{2\sqrt{x}} \cdot \frac{1}{x} = \frac{8x^4 - x^4 - 3}{2x^{3/2}} = \frac{7x^4 - 3}{2x^{3/2}}. \blacksquare$$

In the last section we proved that

$$\frac{d}{dx}x^n = nx^{n-1}$$

if n is a positive integer. With the aid of the quotient rule we can extend this result to the case in which n is a negative integer. [*Note:* When $n = 0$, $x^0 = 1$, which we already know has a derivative of 0.]

Theorem 3   Let $y = x^{-n}$, where n is a positive integer (so that $-n$ is a negative integer). Then

$$\frac{dy}{dx} = \frac{d}{dx}(x^{-n}) = -nx^{-n-1}. \tag{3}$$

Proof. $x^{-n} = 1/x^n$. Then we use the quotient rule (2) to obtain

$$\frac{dy}{dx} = \frac{d}{dx}\left(\frac{1}{x^n}\right) = \frac{x^n \dfrac{d}{dx}(1) - (1)\dfrac{d}{dx}x^n}{(x^n)^2} = \frac{x^n \cdot 0 - nx^{n-1}}{x^{2n}}$$

$$= \frac{-nx^{n-1}}{x^{2n}} = -nx^{(n-1)-2n} = -nx^{-n-1}. \blacksquare$$

NOTE.   We now know that $d(x^n)/dx = nx^{n-1}$ holds for any integer n.

EXAMPLE 5   Let $y = x^{-7}$. Calculate dy/dx.

Solution.   Using (3), we obtain

$$\frac{dy}{dx} = -7x^{-8}. \blacksquare$$

EXAMPLE 6   Let $y = 1/x$. Calculate dy/dx.

Solution. $1/x = x^{-1}$, so that

$$\frac{d}{dx}\left(\frac{1}{x}\right) = \frac{d}{dx}x^{-1} = -1 \cdot x^{-2} = \frac{-1}{x^2}. \blacksquare$$

EXAMPLE 7   Newton's law of universal gravitation states that *the gravitational force between any two particles having masses m_1 and m_2 separated by a distance r is an attraction acting along the line* *joining the particles and has a magnitude*

$$F = G\frac{m_1 m_2}{r^2}, \qquad\qquad (4)$$

where G is a universal constant having the same value for all pairs of particles† (see Figure 1).

FIGURE 1

Two asteroids are approaching each other. The first has a mass of 1000 kg and the second a mass of 3000 kg.

(a) What is the gravitational force between the two asteroids when they are 10 km apart?

(b) How is this force changing at that distance?

Solution.

(a) $r = 10$ km $= 10,000$ m. Then

$$F = G\frac{m_1 m_2}{r^2} = G\frac{(1000)(3000)}{10,000^2} = 0.03G \text{ newton}$$

(b) The rate of change of F when $r = 10,000$ m is $F'(10,000)$. But

$$\frac{dF}{dr} = \frac{d}{dr}Gm_1 m_2 r^{-2} = Gm_1 m_2\frac{d}{dr}r^{-2} = Gm_1 m_2(-2r^{-3}) = \frac{-2Gm_1 m_2}{r^3}$$

$$= \frac{(-2)(1000)(3000)G}{(10,000)^3} = -0.000006G = -6 \times 10^{-6}G \text{ newton per meter}$$

when $r = 10,000$. Here the minus sign indicates that as r increases, the force F decreases, and vice versa. But since the asteroids are approaching one another, r is decreasing, and therefore F is increasing at a rate of $6 \times 10^{-6}G$ newton for every 1 m decrease in r. ∎

PROBLEMS 3.2

In Problems 1–24, find the derivative of the given function. When necessary, use the fact that $(d/dx)(\sqrt{x}) = 1/(2\sqrt{x})$.

1. $f(x) = 2x(x^2 + 1)$

2. $g(t) = t^3(1 + \sqrt{t})$

---

†In metric units $G = 6.673 \times 10^{-11}$ N · m²/kg² (N = newton), where 1 newton is the force that will acceler-ate a 1-kg mass at the rate of 1 m/sec² (1 N = 1 kg/sec²).

3. $s(t) = \dfrac{t^3}{1 + \sqrt{t}}$

4. $f(z) = \dfrac{1 + \sqrt{z}}{z^3}$

5. $f(x) = (1 + x + x^5)(2 - x + x^6)$

6. $f(x) = \dfrac{1 + x + x^5}{2 - x + x^6}$

7. $f(x) = \dfrac{2 - x + x^6}{1 + x + x^5}$

8. $g(t) = (1 + \sqrt{t})(1 - \sqrt{t})$

9. $g(t) = \dfrac{1 + \sqrt{t}}{1 - \sqrt{t}}$

10. $g(t) = \dfrac{1 - \sqrt{t}}{1 + \sqrt{t}}$

11. $p(v) = (v^3 - \sqrt{v})(v^2 + 2\sqrt{v})$

12. $p(v) = \dfrac{v^3 - \sqrt{v}}{v^2 + 2\sqrt{v}}$

13. $p(v) = \dfrac{v^2 + 2\sqrt{v}}{v^3 - \sqrt{v}}$

14. $p(v) = v^{3/2} = v \cdot \sqrt{v}$

15. $p(v) = v^{5/2} = v^2 \cdot v^{1/2}$

16. $p(v) = v^{7/2} = v^3 \cdot v^{1/2}$

17. $m(r) = \dfrac{1}{\sqrt{r}}$

18. $f(x) = \dfrac{1}{x^5 + 3x}$

19. $g(t) = \dfrac{1}{\sqrt{t}}\left(\dfrac{1}{t^4 + 2}\right)$

20. $g(t) = \dfrac{\sqrt{t} - (2/\sqrt{t})}{3t^3 + 4}$

21. $f(x) = \dfrac{1}{x^6}$

22. $g(t) = t^{-100}$

23. $f(x) = \dfrac{5}{7x^3}$

24. $p(v) = \dfrac{v}{v^5}$

In Problems 25–28, find the equation of the tangent line to the curve passing through the given point.

25. $f(x) = 4x(x^5 + 1); (1, 8)$

26. $g(t) = \dfrac{t^2}{1 + \sqrt{t}}; \left(4, \dfrac{16}{3}\right)$

27. $h(u) = \dfrac{1 + \sqrt{u}}{u^2}; (1, 2)$

28. $p(v) = (1 + \sqrt{v})(1 - \sqrt{v}); (1, 0)$

In Problems 29–32, find the equation of the normal line passing through the given point.

29. $f(x) = \dfrac{1}{\sqrt{x}}; (1, 1)$

30. $g(t) = \dfrac{1}{t^7}; (1, 1)$

31. $f(x) = \dfrac{1 - \sqrt{x}}{1 + \sqrt{x}}; (1, 0)$

32. $g(t) = \dfrac{t^3 + 2}{t^2 + 2}; (0, 1)$

33. Let $v(x) = f(x)g(x)h(x)$.
   (a) Using the product rule, show that
   $$\frac{dv}{dx} = f\frac{d(gh)}{dx} + gh\frac{df}{dx}.$$
   (b) Use part (a) to show that
   $$\frac{dv}{dx} = fg\frac{dh}{dx} + fh\frac{dg}{dx} + gh\frac{df}{dx} = f'gh + fg'h + fgh'.$$

In Problems 34–36, use the result of Problem 33 to calculate the derivative of the given function.

34. $v(x) = x(1 + \sqrt{x})(1 - \sqrt{x})$ **35.** $v(x) = (x^2 + 1)(x^3 + 2)(x^4 + 3)$
36. $v(x) = x^{-2}(2 - 3\sqrt{x})(1 + x^3)$

37. Find the derivative of $fg/(f + g)$, where f and g are differentiable functions.
*38. The mass of the earth is 5.983×10^{24} kg. A meteorite with a mass of 10,000 kg is moving toward a collision with the earth.
   (a) When the meteorite is 100 km from the earth, what is the force (in newtons) of gravitational attraction between the meteorite and the earth?
   (b) At that distance, how fast is this force increasing (in newtons per meter) as the meteorite continues on its collision course?
39. According to *Poiseuille's*† *law*, the resistance R of a blood vessel of length l and radius r is given by $R = \alpha l/r^4$, where α is a constant of proportionality determined by the viscosity of blood. Assuming that the length of the vessel is kept constant while the radius increases, how fast is the resistance decreasing (as a function of the radius) when $r = 0.2$ mm?
*40. Find $(d/dv)(v^{n/2})$ where n is a positive integer. [*Hint:* See Problems 14, 15, and 16.]
41. Verify that x^7, $x \cdot x^6$, $x^2 \cdot x^5$, and $x^3 \cdot x^4$ have the same derivative. Do *not* multiply terms together before differentiating.
42. Verify that x^5, x^6/x, x^7/x^2, and x^8/x^3 have the same derivative. Do *not* divide before differentiating.
43. Show that $[F(x)/x^n]' = [xF'(x) - nF(x)]/x^{n+1}$ if F is differentiable.
44. Suppose that $P(x) = (x - a) \cdot (x - b) \cdot (x - c)$. Prove that

$$\frac{P'(x)}{P(x)} = \frac{1}{x - a} + \frac{1}{x - b} + \frac{1}{x - c}.$$

45. Use the product rule to prove that if f is differentiable, then

$$\frac{d}{dx}[f(x)]^2 = 2f(x) \cdot f'(x).$$

46. A beginner's mistake that is not unusual is to compute the derivative of a product as if it equaled the product of the derivatives. If you tackle this problem, you'll discover that this is rarely the case. From the following functions, pick all pairs that satisfy the equation $(F \cdot G)' = F' \cdot G'$.

$$a(x) = 13 \qquad b(x) = x \qquad c(x) = \frac{1}{x}$$

$$d(x) = x + 1 \qquad e(x) = \frac{1}{x - 1}$$

3.3 THE DERIVATIVE OF COMPOSITE FUNCTIONS: THE CHAIN RULE

In this section we derive a result that greatly increases the number of functions whose derivatives can be easily calculated. The idea behind the result is illustrated below. Suppose that $y = f(u)$ is a function of u and $u = g(x)$ is a function of x.‡ Then

---
†Jean Louis Poiseuille (1799–1869) was a French physiologist.
‡So that $y(x) = (f \circ g)(x)$, where $f \circ g$ denotes the composition of f and g. See Section 1.7.

$du/dx = g'(x)$ is the rate of change of u with respect to x, while $dy/du = f'(u)$ is the rate of change of y with respect to u. We now may ask: What is the rate of change of y with respect to x? That is, what is dy/dx?

As an illustration of this idea, suppose that a particle is moving in the xy-plane in such a way that its x-coordinate is given by $x = 3t$, where t stands for time. For example, if t is measured in seconds and x is measured in feet, then in the x direction we suppose that the particle is moving with a velocity of 3 ft/sec. That is, $dx/dt = 3$. In addition, suppose that we know that for every one unit change in the x direction, the particle moves 4 units in the y direction; that is, $dy/dx = 4$ (see Figure 1). Now we ask, What is the velocity of the particle, in feet per second, in the y direction; that is, what is dy/dt? It is clear that for every one unit change in t, x will change 3 ft, and so y will change $4 \cdot 3 = 12$ ft. That is, $dy/dt = 12$. We may write this result as

$$\frac{dy}{dt} = \frac{dy}{dx}\frac{dx}{dt} = 4 \cdot 3 = 12 \text{ ft/sec.} \tag{1}$$

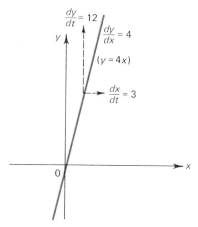

FIGURE 1

This result states simply that if x is changing 3 times as fast as t, and if y is changing 4 times as fast as x, then y is changing $4 \cdot 3 = 12$ times as fast as t.

We now generalize the idea behind this example. Let $y(x) = f(g(x))$. That is, y is the composite function $f \circ g$. It will be very useful to be able to express y' in terms of f' and g'. As will be seen shortly, it can be shown that

$$(f \circ g)'(x) = y'(x) = f'(g(x)) \cdot g'(x). \tag{2}$$

In most of the functions encountered in applications, $g'(x)$ and $f'(g(x))$ exist for almost all real numbers x.

If we write $u = g(x)$, then $f'(g(x)) = f'(u) = dy/du$, $g'(x) = du/dx$, and we can write equation (2) in the form

$$\frac{dy}{dx} = \frac{dy}{du}\frac{du}{dx}. \tag{3}$$

The result given by equation (2) or (3) is called the *chain rule*. Before proving the chain rule, we illustrate its use with some examples.

EXAMPLE 1   Let $u = 3x - 6$ and let $y = 7u + 10$. Then $du/dx = 3$ and $dy/du = 7$, so that

$$\frac{dy}{dx} = \frac{dy}{du}\frac{du}{dx} = 7 \cdot 3 = 21.$$

To calculate dy/dx directly, we compute $y = 7u + 10 = 7(3x - 6) + 10 = 21x - 32$. Therefore

$$\frac{dy}{dx} = 21 = \frac{dy}{du}\frac{du}{dx}. \blacksquare$$

EXAMPLE 2   Let $u = g(x) = \sqrt{x}$ and let $y = f(u) = 2u^2$. Then $y = 2u^2 = 2(\sqrt{x})^2 = 2x$, so that $dy/dx = 2$ by direct calculation. Alternatively, using the chain rule (3), we calculate

$$\frac{du}{dx} = \frac{1}{2\sqrt{x}} \quad \text{and} \quad \frac{dy}{du} = 4u,$$

so that

$$\frac{dy}{dx} = \frac{dy}{du}\frac{du}{dx} = 4u \cdot \frac{1}{2\sqrt{x}} = 4\sqrt{x} \cdot \frac{1}{2\sqrt{x}} = 2. \blacksquare$$

In these two examples we were able to calculate dy/dx directly. It is usually very difficult to do so, and thus the chain rule is very useful. For example, let us calculate dy/dx, where $y = \sqrt{x + x^2}$. If we define $u = x + x^2$, then $y = \sqrt{x + x^2} = \sqrt{u}$. Hence, by the chain rule (3)

$$\frac{dy}{dx} = \frac{dy}{du}\frac{du}{dx} = \frac{1}{2\sqrt{u}}(1 + 2x) = \frac{1}{2\sqrt{x + x^2}}(1 + 2x) = \frac{1 + 2x}{2\sqrt{x + x^2}}.$$

Note that the only other way to calculate this derivative would be to use the original definition (on page 97), which in this example is very difficult.

Now let $u = g(x)$ and $y = f(g(x)) = f(u)$. We assume that for every point x_0 such that $g(x_0)$ is defined, $f(g(x_0))$ is also defined, so it makes sense to talk about the function $f(g(x))$ at x_0.

Theorem 1   *Chain Rule*   Let g and f be differentiable functions such that the above assumptions hold. Then with $u = g(x)$, the composite function $y = (f \circ g)(x) = f(g(x)) = f(u)$ is a differentiable function of x, and

$$\frac{dy}{dx} = \frac{d}{dx}(f \circ g)(x) = f'(g(x))g'(x) = \frac{df}{du}\frac{du}{dx}. \tag{4}$$

Partial Proof. To simplify the proof, we assume that for Δx small, $g(x + \Delta x) - g(x) \neq 0$ (so that we can multiply and divide by it). The case where this assumption is not valid can be treated separately but makes the proof much more complicated. We will not worry about this point here. A complete proof is given in Appendix 5.

$$\frac{d}{dx} f(g(x)) = \lim_{\Delta x \to 0} \frac{f(g(x + \Delta x)) - f(g(x))}{\Delta x}$$

$$= \lim_{\Delta x \to 0} \left[\frac{f(g(x + \Delta x)) - f(g(x))}{\Delta x} \cdot \frac{g(x + \Delta x) - g(x)}{g(x + \Delta x) - g(x)} \right]$$

$$= \lim_{\Delta x \to 0} \frac{f(g(x + \Delta x)) - f(g(x))}{g(x + \Delta x) - g(x)} \cdot \lim_{\Delta x \to 0} \frac{g(x + \Delta x) - g(x)}{\Delta x} \tag{5}$$

The second limit in (5) is $g'(x) = du/dx$. To evaluate the first limit, we note that as $\Delta x \to 0$, $g(x + \Delta x) \to g(x)$, because g, being differentiable, is continuous at x. Then defining the difference $\Delta g = g(x + \Delta x) - g(x)$, we may write

$$g(x + \Delta x) = g(x) + \Delta g. \tag{6}$$

Also, $\Delta g \to 0$ as $\Delta x \to 0$ since

$$\lim_{\Delta x \to 0} \Delta g = \lim_{\Delta x \to 0} [g(x + \Delta x) - g(x)] = \left[\lim_{\Delta x \to 0} g(x + \Delta x) \right] - g(x)$$
$$= g(x) - g(x) = 0.$$

Thus, using (6), we find the first limit of (5) to be

$$\lim_{\Delta g \to 0} \frac{f(g(x) + \Delta g) - f(g(x))}{g(x) + \Delta g - g(x)} = \lim_{\Delta g \to 0} \frac{f(g(x) + \Delta g) - f(g(x))}{\Delta g}$$

$$= f'(g(x)) = f'(u) = \frac{df}{du},$$

and the theorem is proved. ■

REMARK. The procedure in using the chain rule to calculate the derivative dy/dx is to find a function $u(x)$ that has the property that $y(x) = f(u(x))$, where both df/du and du/dx can be calculated without too much difficulty.

EXAMPLE 3   Let $y = f(x) = 1/(x^3 + 1)^5$. Find dy/dx.

Solution. We define $u = g(x) = x^3 + 1$. (There really is no other possibility.) Then $1/(x^3 + 1)^5 = 1/u^5$, and we know how to differentiate $1/u^5$ with respect to u. We have

$$\frac{dy}{du} = \frac{d}{du} \left(\frac{1}{u^5} \right) = \frac{-5}{u^6} \quad \text{and} \quad g'(x) = \frac{d}{dx}(x^3 + 1) = 3x^2.$$

We now use the chain rule to obtain

$$\frac{dy}{dx} = \frac{dy}{du}\frac{du}{dx} = \left(-\frac{5}{u^6}\right)(3x^2) = -\frac{15x^2}{(x^3+1)^6}.$$

After some practice, you can eliminate f and g to obtain, directly,

$$\frac{d}{dx}(x^3+1)^{-5} = -5(x^3+1)^{-6}\frac{d}{dx}(x^3+1) = -5(x^3+1)^{-6}(3x^2) = \frac{-15x^2}{(x^3+1)^6}$$

EXAMPLE 4   Let $y = f(x) = (\sqrt{x}+3)^{15}$. Calculate $dy/dx = f'(x)$.

Solution.

$$\frac{d}{du}u^{15} = 15u^{14}$$

$$\frac{d}{dx}(\sqrt{x}+3)^{15} = 15(\sqrt{x}+3)^{14}\frac{d}{dx}(\sqrt{x}+3)$$

$$= 15(\sqrt{x}+3)^{14}\left(\frac{1}{2\sqrt{x}}\right) = \frac{15(\sqrt{x}+3)^{14}}{2\sqrt{x}}$$

The use of the chain rule may involve other variables and more complicated expressions requiring several of the rules of differentiation. However, it will usually be clear how to proceed.

EXAMPLE 5   Let $s(t) = [(t^3+1)/(t^2-5)]^4$. Find ds/dt.

Solution. $s(t)$ is the fourth power of a function of t. Since we can easily differentiate u^4, we set u equal to this function. That is,

$$u = \frac{t^3+1}{t^2-5}.$$

Then by the quotient rule,

$$\frac{du}{dt} = \frac{(t^2-5)3t^2 - (t^3+1)2t}{(t^2-5)^2} = \frac{3t^4-15t^2-2t^4-2t}{(t^2-5)^2}$$

$$= \frac{t^4-15t^2-2t}{(t^2-5)^2}.$$

Now $s = u^4$ and $ds/du = 4u^3$, so that

$$\frac{ds}{dt} = \frac{ds}{du}\frac{du}{dt} = 4u^3\frac{(t^4-15t^2-2t)}{(t^2-5)^2} = 4\left(\frac{t^3+1}{t^2-5}\right)^3\left[\frac{t^4-15t^2-2t}{(t^2-5)^2}\right]$$

$$= \frac{4t(t^3+1)^3(t^3-15t-2)}{(t^2-5)^5}.$$

There is a formula that can be derived from the chain rule, and it is very useful for calculations. This formula is a generalization of the last two examples.

Theorem 2 **Power Rule** Let $u(x)$ be a differentiable function of x. Then if n is an integer,

$$\frac{d}{dx} u^n = nu^{n-1} \frac{du}{dx}. \tag{7}$$

Alternatively, we may write

$$\frac{d}{dx}[g(x)]^n = n[g(x)]^{n-1}g'(x). \tag{8}$$

Proof. If $f(u) = u^n$, then from the chain rule,

$$\frac{d}{dx} u^n = \frac{df}{du}\frac{du}{dx} = nu^{n-1} \frac{du}{dx}. \quad \blacksquare$$

PROBLEMS 3.3

In Problems 1–25, use the chain rule to find the derivative of the given function.

1. $f(x) = (x + 1)^3$

2. $f(x) = (x^2 - 1)^2$

3. $f(x) = (\sqrt{x} + 2)^4$

4. $f(x) = (x^2 - x^3)^4$

5. $f(x) = (1 + x^6)^6$

6. $f(x) = (1 - x^2 + x^5)^3$

7. $f(x) = (x^2 - 4x + 1)^5$

8. $f(x) = \dfrac{1}{(\sqrt{x} - 3)^4}$

9. $s(t) = \left(\dfrac{t + 1}{t - 1}\right)^3$

10. $s(t) = (\sqrt{t} - t)^7$

11. $g(u) = (u^5 + u^4 + u^3 + u^2 + u + 1)^2$

12. $g(u) = \dfrac{5}{u^3 + u + 1}$

13. $h(y) = (y^2 + 3)^{-4}$

14. $h(y) = (y^3 - \sqrt{y} + 1)^{-17}$

15. $f(x) = (x^2 + 2)^5(x^4 + 3)^3$

16. $f(x) = (x^4 + 1)^{1/2}(x^3 + 3)^4$

17. $s(t) = \dfrac{\sqrt{t^2 + 1}}{(t + 2)^4}$

18. $s(t) = \left(\dfrac{t^4 + 1}{t^4 - 1}\right)^{1/2}$

19. $g(u) = \dfrac{(u^2 + 1)^3(u^2 - 1)^2}{\sqrt{u - 2}}$

20. $g(x) = \dfrac{(x^2 + 1)^2(x^3 + 2)^3}{(x^4 + 3)^{1/2}}$

21. $f(x) = \sqrt{x + \sqrt{1 + \sqrt{x}}}.$ [*Hint:* Use the chain rule twice.]

22. $f(x) = \sqrt{x^2 + \sqrt{1 + x^2}}$

23. $h(y) = (y^{-2} + y^{-3} + y^{-7})^{-5}$

24. $p(s) = \left[\left(1 - \dfrac{1}{s}\right)^{-1} + 4\right]^{-1}$

25. $w(q) = (q^{-7} - q^{-3})^{-1/2}$

26. A missile travels along the path $y = 6(x - 3)^3 + 3x$. When $x = 1$, the missile flies off this path tangentially (i.e., along the tangent line). Where is the missile when $x = 4$? [*Hint:* Find the equation of the tangent line at $x = 1$.]

27. The mass μ of the left x meters of a 4-m metal rod is given by $\mu = 4(1 + \sqrt{x})^3$ kilograms, where x is the distance along the rod measured from the left. What is the density of the rod (in kilograms per meter) when $x = 1$ m?

*28. Let f and g be differentiable functions with $f(g(x)) = x$. Show that

$$f'(g(x)) = \frac{1}{g'(x)}. \tag{9}$$

(This formula is called the *differentiation rule for inverse functions*. This topic is discussed in detail in Section 6.1.)

29. Verify the result of Problem 28 in the following cases.
    (a) $g(x) = 5x$, $f(x) = \frac{1}{5}x$
    (b) $g(x) = 17x - 8$, $f(x) = \frac{1}{17}x + \frac{8}{17}$
    (c) $g(x) = \sqrt{x}$, $f(x) = x^2$
    (d) $g(x) = x^2$ on $(-\infty, 0]$, $f(x) = -\sqrt{x}$

30. Verify that x^{10}, $(x^5)^2$, and $(x^2)^5$ have the same derivatives. [*Hint:* Use the power rule first and then the chain rule. Compare results.]

31. Suppose that $p(x)$ is a polynomial function that is divisible by $(x - 17)^2$; that is, there is a polynomial q such that $p(x) = (x - 17)^2 \cdot q(x)$. Prove that $p'(x)$ is divisible by $x - 17$.

*32. Show that if the polynomial $p(x)$ is divisible by $(ax + b)^n$, where $n > 1$, then $p'(x)$ is divisible by $(ax + b)^{n-1}$.

33. It is possible to obtain the product rule by using just the chain rule, the sum rule, and the fact that $(d/dx)(x^2) = 2x$. Write out such a proof. [*Hint:* Use the fact that $(f + g)^2 = f^2 + 2fg + g^2$.]

34. Differentiate $f(x) \cdot [g(x)]^{-1}$ by using the product rule and the chain rule. Verify that your result equals the derivative of the quotient $f(x)/g(x)$.

**35. Suppose that f is differentiable on $(0, \infty)$ and $f(A \cdot B) = f(A) + f(B)$ for any numbers $A, B > 0$. Prove that $f'(x) = f'(1)/x$ for all $x > 0$.

3.4 THE DERIVATIVE OF A POWER FUNCTION

In this section we derive a general formula for finding the derivative of $y = x^r$, where r is a rational number (remember that a rational number is a number of the form $r = m/n$, where m and n are integers and $n \neq 0$). Such a function is called a **power function.**

If r is a positive or negative integer, then we have shown that

$$\frac{d}{dx} x^r = rx^{r-1}. \tag{1}$$

We will show that formula (1) holds for *all* rational numbers r, and we will suggest an extension of that result to all real numbers. This result will be obtained in several steps.

Theorem 1 Let n be a positive integer. Then the function $f(x) = x^{1/n}$ is differentiable, and

$$\frac{d}{dx} x^{1/n} = \frac{1}{n} x^{(1/n)-1}. \tag{2}$$

Equation (2) is valid whenever $x^{1/n-1}$ is defined.

Proof.   Recall equation (3.1.2):

$$\frac{a^n - b^n}{a - b} = a^{n-1} + a^{n-2}b + a^{n-3}b^2 + \cdots + ab^{n-2} + b^{n-1}. \tag{3}$$

Then

$$\frac{a - b}{a^n - b^n} = \frac{1}{a^{n-1} + a^{n-2}b + a^{n-3}b^2 + \cdots + ab^{n-2} + b^{n-1}}. \tag{4}$$

Let $a = (x + \Delta x)^{1/n}$ and $b = x^{1/n}$. Then

$$a^n - b^n = [(x + \Delta x)^{1/n}]^n - (x^{1/n})^n = (x + \Delta x) - x = \Delta x, \tag{5}$$

and

$$\frac{a - b}{a^n - b^n} = \frac{(x + \Delta x)^{1/n} - x^{1/n}}{[(x + \Delta x)^{1/n}]^n - (x^{1/n})^n} = \frac{(x + \Delta x)^{1/n} - x^{1/n}}{\Delta x} \tag{6}$$

$$= \frac{1}{[(x + \Delta x)^{1/n}]^{n-1} + [(x + \Delta x)^{1/n}]^{n-2}x^{1/n} + \cdots + (x + \Delta x)^{1/n}(x^{1/n})^{n-2} + (x^{1/n})^{n-1}}.$$

Returning to our original problem, we have

$$\frac{d}{dx} x^{1/n} = \lim_{\Delta x \to 0} \frac{(x + \Delta x)^{1/n} - x^{1/n}}{\Delta x}.$$

Then using equation (6), we obtain

$$\lim_{\Delta x \to 0} \frac{(x + \Delta x)^{1/n} - x^{1/n}}{\Delta x} \tag{7}$$

$$= \frac{1}{x^{(n-1)/n} + x^{(n-2)/n}x^{1/n} + x^{(n-3)/n}x^{2/n} + \cdots + x^{1/n}x^{(n-2)/n} + x^{(n-1)/n}}.$$

The denominator contains n terms, each of which is equal to $x^{(n-1)/n}$ (since $x^{(n-2)/n}x^{1/n} = x^{(n-2)/n+(1/n)} = x^{(n-1)/n}$, and so on). Therefore (7) is equal to

$$\frac{1}{nx^{(n-1)/n}} = \frac{1}{nx^{1-(1/n)}} = \frac{1}{n} \cdot \frac{1}{x^{1-(1/n)}} = \frac{1}{n} x^{(1/n)-1},$$

and the theorem is proved. ∎

REMARK.   A much simpler derivation of equation (2) is possible if we assume that $x^{1/n}$ is differentiable. This simpler derivation is suggested in Problem 29. Also an easy proof can be given once we have discussed the derivatives of inverse functions in Section 6.1.

EXAMPLE 1 Let $y = x^{1/5}$. Calculate dy/dx.

Solution.

$$\frac{dy}{dx} = \frac{1}{5}x^{(1/5)-1} = \frac{1}{5}x^{-4/5}. \blacksquare$$

EXAMPLE 2 Let $y = f(x) = (x^3 + 3)^{1/4}$. Find dy/dx.

Solution. We define $u = x^3 + 3$. Then $y = u^{1/4}$, so that by the chain rule,

$$\frac{dy}{dx} = \frac{dy}{du}\frac{du}{dx} = \frac{1}{4}u^{-3/4}(3x^2) = \frac{1}{4}(x^3 + 3)^{-3/4}(3x^2). \blacksquare$$

We now extend Theorem 1 to rational numbers.

Theorem 2 Let $y = x^r$, where $r = m/n$ is a rational number (m and n are integers and $n \neq 0$). Then x^r is differentiable, and

$$\frac{dy}{dx} = \frac{d}{dx}x^r = rx^{r-1}. \tag{8}$$

Equation (8) is valid whenever x^{r-1} is defined.

Proof.

$$\frac{dy}{dx} = \frac{d}{dx}x^r = \frac{d}{dx}x^{m/n} = \frac{d}{dx}(x^{1/n})^m.$$

Let $u = x^{1/n}$. Then $y = u^m$, and

$$\frac{dy}{dx} = \frac{dy}{du}\frac{du}{dx} = mu^{m-1}\cdot\frac{1}{n}x^{(1/n)-1} = m(x^{1/n})^{m-1}\cdot\frac{1}{n}x^{(1-n)/n}$$

$$= \frac{m}{n}(x^{1/n})^{m-1}(x^{1/n})^{1-n} = \frac{m}{n}(x^{1/n})^{m-1+1-n} = \frac{m}{n}x^{(m-n)/n}$$

$$= \frac{m}{n}x^{(m/n)-1} = rx^{r-1},$$

and the theorem is proved. $\blacksquare$

EXAMPLE 3 Let $y = x^{2/3}$. Calculate dy/dx.

Solution.

$$\frac{dy}{dx} = \frac{2}{3}x^{(2/3)-1} = \frac{2}{3}x^{-1/3} = \frac{2}{3x^{1/3}}. \blacksquare$$

EXAMPLE 4 Let $s = (t^3 + 2t + 3)^{11/9}$. Find ds/dt.

Solution. By now you should recognize that the chain rule is called for in this problem. Let $u = t^3 + 2t + 3$. Then $s = u^{11/9}$, so that

$$\frac{ds}{dt} = \frac{ds}{du}\frac{du}{dt} = \frac{11}{9}u^{2/9}(3t^2 + 2) = \frac{11}{9}(t^3 + 2t + 3)^{2/9}(3t^2 + 2). \blacksquare$$

It is true that the formula

$$\frac{d}{dx}x^\alpha = \alpha x^{\alpha-1} \tag{9}$$

holds when α is any real number. We cannot prove this formula here because we don't even know what x^α means if α is not a rational number. Once we have defined the exponential function (in Sections 6.2 and 6.4), a definition of x^α and a proof of (9) will be easy (see Problem 6.3.39). However, we can get an idea of what (9) means if α is irrational by noting that if α is rational, (9) has already been proven; and if α is irrational, then it can be approximated as closely as desired by a rational number (e.g., the irrational number π can be approximated by 3.14, 3.141, 3.1415, 3.14159, etc.).

EXAMPLE 5   Let $y = x^\pi$. Calculate dy/dx.

Solution. Using formula (9), we have $dy/dx = \pi x^{\pi-1}$. ∎

At this point we see that it is possible to differentiate a wide variety of functions. To aid you, we give, in Table 1, a summary of the differentiation rules we have so far discussed. In the notation of the table c stands for an arbitrary constant, and $u(x)$ and $v(x)$ denote differentiable functions.

TABLE 1

Function $y = f(x)$	Its derivative $\dfrac{dy}{dx}$
I. c	$\dfrac{dx}{dx} = 0$
II. $cu(x)$	$\dfrac{d}{dx}cu(x) = c\dfrac{du}{dx}$
III. $u(x) + v(x)$	$\dfrac{d}{dx}(u+v) = \dfrac{du}{dx} + \dfrac{dv}{dx}$
IV. x^r, r a real number	$\dfrac{d}{dx}x^r = rx^{r-1}$
V. $u(x) \cdot v(x)$	$\dfrac{d}{dx}uv = u(x)\dfrac{dv}{dx} + v(x)\dfrac{du}{dx}$
VI. $\dfrac{u(x)}{v(x)}$, $v(x) \neq 0$	$\dfrac{d}{dx}\left[\dfrac{u(x)}{v(x)}\right] = \dfrac{v(x)\dfrac{du}{dx} - u(x)\dfrac{dv}{dx}}{[v(x)]^2}$
VII. $u^r(x)$	$\dfrac{d}{dx}u^r(x) = ru^{r-1}(x)\dfrac{du}{dx}$ (from the chain rule)
VIII. $f(g(x))$	$\dfrac{d}{dx}f(g(x)) = f'(g(x))g'(x)$

PROBLEMS 3.4

In the Problems 1–20, find the derivative of the given function.

1. $f(x) = x^{2/5} + 2x^{1/3}$
3. $f(x) = (x^2 + 1)^{5/3}$

2. $f(x) = 5x^{-2/9}$
4. $f(x) = (x^2 - 1)^{-2/3}$

5. $f(x) = (x^3 + 3)^{1/10}$

6. $f(x) = (x^5 + 2x + 1)^{3/2}$

7. $s(t) = (t^{10} - 2)^{4/7}$

8. $s(t) = (t^3 + t^2 + 1)^{5/8}$

9. $g(u) = \sqrt[3]{u^3 + 3u + 1}$

10. $g(u) = \left(\dfrac{u + 2}{u - 1}\right)^{7/2}$

11. $h(z) = \left(\dfrac{z^2 - 1}{z^2 + 1}\right)^{-7/17}$

12. $h(z) = (z^5 - 1)^{3/4}(z^3 + 2)^{8/9}$

13. $v(r) = (r^2 + 1)^{5/6}(r - 1)^{1/2}$

14. $v(r) = (r + 1)^{1/3}(r - 1)^{1/4}(r + 2)^{1/5}$

15. $f(x) = 3x^{\sqrt{2}}$

16. $f(x) = \sqrt{3}x^{\sqrt{3}}$

17. $s(t) = (t^2 + 1)^{\sqrt{3}}$

18. $s(t) = (t^4 - \pi)^{\pi^2}$

19. $g(u) = (u^{\sqrt{2}} - u^{\sqrt{3}})^4$

20. $g(u) = \dfrac{u^{\sqrt{5}} - 3}{u^{\sqrt{5}} + 1}$

In Problems 21–24, find the equations of the tangent and normal lines to each of the curves at the indicated point.

21. $y = x^{1/2} + x^{3/2}$; $(1, 2)$

22. $y = \dfrac{(x + 2)^3}{(x - 1)^{1/3}}$; $(0, -8)$

23. $y = (x - 1)^{\sqrt{2}}$; $(1, 0)$

24. $y = (x + 11x^2 + 3x^3 + x^4)^{5/4}$; $(1, 32)$

*25. Prove that the tangent line to a circle is perpendicular to a radial line by using the following steps:

(a) Assume that the circle is centered at the origin and show that its equation, for $y \geq 0$, can be written as $y = \sqrt{r^2 - x^2}$, where r is its radius.

(b) Find the equation of the line tangent to the circle at a point (x_0, y_0) on the circle.

(c) Find the equation of the line segment joining $(0, 0)$ to (x_0, y_0).

26. In the study of soil mechanics it is found that the rate of runoff of rainfall from an area to an inlet reaches its maximum after t minutes, where t is given by

$$t = C\left(\frac{L}{mi^2}\right)^{1/3}.$$

Here C is a constant depending on the nature of the soil, L is the distance from the most remote area, m is the slope of the land, and i is the rain intensity (in inches per hour).

(a) What is the instantaneous rate of change of total runoff time as a function of rain intensity if all the other variables are kept constant?

(b) Calculate this rate of change when $L = 1500$ ft, $m = 0.1$, $i = 1.1$ in./hr, and $C = 2$.

27. In the kinetic theory of gases the *root mean square* (rms) speed of gas molecules is related to the average or typical speed of the molecules comprising the gas and is given by

$$v_{rms} = \sqrt{\frac{3P}{\rho}}$$

where P is the pressure (measured in atmospheres, atm) of the gas and ρ is its density (measured in kilograms per cubic meter).

(a) Calculate v_{rms} of hydrogen molecules at 0°C and 1 atm of pressure. At that temperature the density of hydrogen is 8.99×10^{-2} kg/m³.

(b) With P fixed, calculate the instantaneous rate of change of v_{rms} with respect to ρ. (It is known that ρ is inversely proportional to the temperature T with P fixed, so ρ will increase as the gas cools and ρ will decrease as the gas is heated.)

(c) Calculate the instantaneous rate of change of v_{rms} as a function of ρ (with P fixed) for the values of P and ρ given in part (a).

28. In astronomy the *luminosity* of a star is the star's total energy output. Loosely speaking, a star's luminosity is a measure of how bright the star would appear at the surface of the star. The *mass-luminosity relation* gives the approximate luminosity of a star as a function of its mass. It has been found experimentally† that, approximately,

$$\frac{L}{L_0} = \left(\frac{M}{M_0}\right)^r,$$

where L and M are the luminosity and mass of the star and L_0 and M_0 denote the luminosity and mass of our sun. The exponent r depends on the mass of the star, as shown in Table 2.

TABLE 2

Mass range, $\frac{M}{M_0}$	r
1.0–1.4	4.75
1.4–1.7	4.28
1.7–2.5	4.15
2.5–5	3.95
5–10	3.38
10–20	2.80
20–50	2.30
50–100	1.90

(a) In this model how is the luminosity changing as a function of mass when the mass is 2 solar masses?
(b) How is it changing at a mass of 8 solar masses?
(c) How is it changing at a mass of 30 solar masses?
*(d) Write L as a function of M. For what values of M does dL/dM not exist? How would you suggest altering the model to avoid discontinuities in this derivative?
29. Assume that $y(x) = x^{1/n}$ is differentiable.
(a) Show that $x = [y(x)]^n$.
(b) Using the chain rule, differentiate both sides of the equation in part (a) with respect to x.
(c) Using the result of part (b), show that $dy/dx = (1/n)[y(x)]^{1-n}$.
(d) Using part (c), show that $(d/dx)(x^{1/n}) = (1/n)x^{1/n-1}$.

3.5 THE DERIVATIVES OF THE TRIGONOMETRIC FUNCTIONS

In this section we will compute the derivatives of $\sin x$, $\cos x$, and related functions. The material in this section depends on a knowledge of basic trignometry. The background material for this section can be found in the first two sections of Appendix 1. Also, we emphasize that in this section, and for the remainder of this text, we assume, unless otherwise stated, that x (or θ) is measured in radians and that $\text{dom}(\sin x) = \text{dom}(\cos x) = \mathbb{R}$.

---

†These data are based on stellar models computed by D. Ezer and A. Cameron in their paper "Early and main sequence evolution of stars in the range 0.5 to 100 solar masses," *Canadian Journal of Physics*, **45**, 3429–3460 (1967).

To compute the derivative of sin x, we need to know

$$\lim_{\theta \to 0} \frac{\sin \theta}{\theta}.$$

Before doing any formal computations, we illustrate what can happen by means of a table. With θ measured in radians, we obtain Table 1. It seems that the quotient $(\sin \theta)/\theta$ approaches 1 as $\theta \to 0$. We now prove this important result.

TABLE 1

θ (radians)	$\sin \theta$	$\dfrac{\sin \theta}{\theta}$
1	0.8414709848	0.8414709848
0.5	0.4794255386	0.9588510772
0.1	0.0998334166	0.9983341665
0.01	0.0099998333	0.9999833334
0.001	0.0009999998	0.9999998333
0.0001	0.0000999999	0.9999999833

Theorem 1

$$\lim_{\theta \to 0} \frac{\sin \theta}{\theta} = 1 \tag{1}$$

Note that the limit in (1) cannot be calculated by evaluation since $(\sin \theta)/\theta$ is not defined at 0.

REMARK.   Equation (1) is *false* if θ is measured in degrees rather than radians (see Problem 48).

Proof.   We prove the theorem in the case $0 < \theta < \pi/2$. The case $-\pi/2 < \theta < 0$ is similar and is left as an exercise (see Problem 47). Consider Figure 1. From the figure we see that

$$\text{area of } \Delta 0AC < \text{area of sector } 0AC < \text{area of } \Delta 0BC. \tag{2}$$

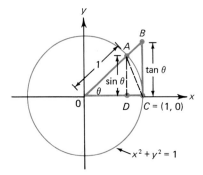

FIGURE 1

To calculate the area of a sector, we use the fact that if $\theta = 2\pi$ radians, then the sector is the area of the entire unit circle, which is π. If an angle of 2π radians corresponds to a sector area of π, then θ radians corresponds to a sector area of $\theta/2$. Thus

$$\text{area of sector } 0AC = \frac{\theta}{2}.$$

Also,

$$\sin \theta = \overline{AD}†$$ Since A is a point on the unit circle (see Appendix A1, Section 2).

$$\tan \theta = \frac{\overline{BC}}{\overline{0C}} = \overline{BC}.$$ Since $\overline{0C} = 1$ (the unit circle has radius 1).

Thus

$$\text{area of } \Delta 0AC = \tfrac{1}{2}(\text{base} \times \text{height}) = \tfrac{1}{2}\overline{0C} \times \overline{AD} = \tfrac{1}{2}\sin \theta$$

and

$$\text{area of } \Delta 0BC = \tfrac{1}{2}\overline{0C} \times \overline{BC} = \tfrac{1}{2}\tan \theta.$$

Thus we obtain, from (2),

$$\tfrac{1}{2}\sin \theta < \tfrac{1}{2}\theta < \tfrac{1}{2}\tan \theta. \tag{3}$$

For $\theta \neq 0$ we multiply the inequalities in (3) by 2 and divide by $\sin \theta$ to obtain

$$1 < \frac{\theta}{\sin \theta} < \frac{1}{\cos \theta}$$

or

$$\cos \theta < \frac{\sin \theta}{\theta} < 1, \qquad 0 < \theta < \frac{\pi}{2}.$$ Since if $0 < a < b$, then $\frac{1}{a} > \frac{1}{b}$. $\tag{4}$

Now since $\sin^2 \theta + \cos^2 \theta = 1$, we see that $\cos \theta = \pm\sqrt{1 - \sin^2 \theta}$. In the interval $0 < \theta < \pi/2$, $\cos \theta > 0$, so $\cos \theta = \sqrt{1 - \sin^2 \theta}$. But from (3),

$$\sin \theta < \theta$$

$$\sin^2 \theta < \theta^2$$

$$\cos \theta = \sqrt{1 - \sin^2 \theta} > \sqrt{1 - \theta^2},$$ We are subtracting a smaller quantity from 1.

---

†Recall that $\overline{AD}$ denotes the length of the line segment joining A to D.

and (4) becomes

$$\sqrt{1 - \theta^2} < \frac{\sin \theta}{\theta} < 1.$$

Now

$$\lim_{\theta \to 0^+} 1 = 1 \quad \text{and} \quad \lim_{\theta \to 0^+} \frac{1}{\sqrt{1 - \theta^2}} = 1.$$

So the squeezing theorem tells us that

$$\lim_{\theta \to 0^+} \frac{\sin \theta}{\theta} = 1.$$

In a similar manner we can show that $\lim_{\theta \to 0^-}[(\sin \theta)/\theta] = 1$, and the proof is complete. Actually, the proof is intuitively reasonable since, in Figure 1, as $\theta \to 0^+$, the length of the line AD and the length of the arc AC (denoted $\widehat{AC}$) get closer and closer together. Then the theorem follows, since $\overline{AD} = \sin \theta$ and $\widehat{AC} = \theta$ ■

We can use Theorem 1 to compute other limits.

EXAMPLE 1 Calculate $\lim_{\theta \to 0}[(\sin 2\theta)/\theta]$.

Solution.

$$\lim_{\theta \to 0} \frac{\sin 2\theta}{\theta} = \lim_{\theta \to 0} \frac{2 \sin 2\theta}{2\theta} = 2 \lim_{\theta \to 0} \frac{\sin 2\theta}{2\theta}.$$

But as $\theta \to 0$, $2\theta \to 0$, so that

$$2 \lim_{\theta \to 0} \frac{\sin 2\theta}{2\theta} = 2 \lim_{2\theta \to 0} \frac{\sin 2\theta}{2\theta} = 2 \cdot 1 = 2. ■$$

EXAMPLE 2 Show that

$$\lim_{\theta \to 0} \frac{\cos \theta - 1}{\theta} = 0. \tag{5}$$

Solution.

$$\lim_{\theta \to 0} \frac{\cos \theta - 1}{\theta} = \lim_{\theta \to 0} \frac{(\cos \theta - 1)(\cos \theta + 1)}{\theta(\cos \theta + 1)} = \lim_{\theta \to 0} \frac{\cos^2 \theta - 1}{\theta(\cos \theta + 1)}$$

$$= \lim_{\theta \to 0} \frac{-\sin^2 \theta}{\theta(\cos \theta + 1)} \quad \text{Since } \sin^2 \theta + \cos^2 \theta = 1.$$

$$= -\lim_{\theta \to 0} \frac{\sin \theta}{\theta} \lim_{\theta \to 0} \frac{\sin \theta}{(\cos \theta + 1)}$$

$$= -1 \cdot \frac{\lim_{\theta \to 0} \sin \theta}{\lim_{\theta \to 0}(\cos \theta + 1)} = -1 \cdot \frac{0}{2} = 0 ■$$

Theorem 2   *Derivative of sin x*

$$\frac{d}{dx} \sin x = \cos x \tag{6}$$

Proof.

$$\frac{d}{dx} \sin x = \lim_{\Delta x \to 0} \frac{\sin(x + \Delta x) - \sin x}{\Delta x}$$

Since $\sin(x + y) = \sin x \cos y + \cos x \sin y$
[see (xiii), Appendix A1, Section 2].

$$= \lim_{\Delta x \to 0} \frac{\sin x \cos \Delta x + \cos x \sin \Delta x - \sin x}{\Delta x}$$

Theorem 2.3.3,

$$= \lim_{\Delta x \to 0} \sin x \left(\frac{\cos \Delta x - 1}{\Delta x} \right) + \lim_{\Delta x \to 0} \cos x \cdot \frac{\sin \Delta x}{\Delta x}$$

Theorem 2.3.2,

$$= \sin x \lim_{\Delta x \to 0} \frac{\cos \Delta x - 1}{\Delta x} + \cos x \lim_{\Delta x \to 0} \frac{\sin \Delta x}{\Delta x}$$

from (5)   from (1)

$$= \sin x \cdot 0 + \cos x \cdot 1 = \cos x \quad \blacksquare$$

Corollary   The function $\sin x$ is continuous at every real number x.

Proof.   Since $\sin x$ is differentiable, it is continuous, by Theorem 2.7.3.   ■

EXAMPLE 3   Compute

$$\frac{d}{dx} \sin x^2.$$

Solution.   Let $u = x^2$ and $f(u) = \sin u$. Then using the chain rule and equation (6), we have

$$\frac{d}{dx} \sin x^2 = \frac{d}{dx} f(u) = \frac{df}{du} \frac{du}{dx} = (\cos u)(2x) = (\cos x^2)2x. \quad \blacksquare$$

The result of Example 3 can be easily generalized.

Theorem 3   If $u(x)$ is a differentiable function of x, then

$$\frac{d}{dx} \sin u = \cos u \frac{du}{dx}. \tag{7}$$

Proof.   Let $f(u) = \sin u$. Then from the chain rule,

$$\frac{d}{dx} f(u) = \frac{df}{du} \frac{du}{dx} = \cos u \frac{du}{dx}. \quad \blacksquare$$

EXAMPLE 4   Calculate

$$\frac{d}{dx} \sin\left(\frac{x + 1}{x^2 - 3}\right).$$

Solution.   Let $u = (x + 1)/(x^2 - 3)$. Then using the quotient rule, we have

$$\frac{du}{dx} = \frac{(x^2 - 3)(1) - (x + 1)(2x)}{(x^2 - 3)^2} = \frac{x^2 - 3 - 2x^2 - 2x}{(x^2 - 3)^2} = \frac{-x^2 - 2x - 3}{(x^2 - 3)^2}.$$

Finally,

$$\frac{d}{dx} \sin\left(\frac{x + 1}{x^2 - 3}\right) = \frac{d}{dx} \sin u = \cos u \frac{du}{dx} = \left[\cos\left(\frac{x + 1}{x^2 - 3}\right)\right]\left[\frac{-(x^2 + 2x + 3)}{(x^2 - 3)^2}\right]. \quad ∎$$

We now turn to the derivative of $\cos x$.

Theorem 4   *Derivative of cos x*

$$\frac{d}{dx} \cos x = -\sin x \tag{8}$$

Proof.   From identity (xi) in Appendix A1, Section 2, $\cos x = \sin[(\pi/2) - x]$, so that

$$\frac{d}{dx} \cos x = \frac{d}{dx} \sin\left(\frac{\pi}{2} - x\right).$$

If $u = (\pi/2) - x$, then $du/dx = -1$, and

Identity (x), Appendix A1.

$$\frac{d}{dx} \sin\left(\frac{\pi}{2} - x\right) = -\cos\left(\frac{\pi}{2} - x\right) = -\sin x. \quad ∎$$

REMARK.   As with $\sin x$, $\cos x$ is evidently continuous at every real number x.

The following result can be proven by using the chain rule, as in Theorem 3.

Theorem 5   If u is a differentiable function of x, then

$$\frac{d}{dx} \cos u = -\sin u \frac{du}{dx}. \tag{9}$$

EXAMPLE 5   Compute $(d/dx)(\cos \sqrt{x})$.

Solution. Let $u = \sqrt{x}$. Then $du/dx = 1/2\sqrt{x}$, and, using (9), we have

$$\frac{d}{dx}\cos\sqrt{x} = \frac{d}{dx}\cos u = -\sin u\frac{du}{dx} = (-\sin\sqrt{x})\left(\frac{1}{2\sqrt{x}}\right) = \frac{-\sin\sqrt{x}}{2\sqrt{x}}. \blacksquare$$

EXAMPLE 6 Compute $(d/dx)(x\sin 2x)$.

Solution. Using the product rule, we obtain

$$\frac{d}{dx}x\sin 2x = x\frac{d}{dx}\sin 2x + 1\sin 2x$$

$$= x\cos 2x\frac{d}{dx}(2x) + \sin 2x = 2x\cos 2x + \sin 2x. \blacksquare$$

We now turn to the derivatives of the other four trigonometric functions. Recall that (these functions are discussed in some detail in the review in Appendix 1.3)

$$\tan x = \frac{\sin x}{\cos x} \qquad \cot x = \frac{\cos x}{\sin x} \qquad \sec x = \frac{1}{\cos x} \qquad \csc x = \frac{1}{\sin x}$$

Theorem 4

$$\frac{d}{dx}\tan x = \sec^2 x \tag{10}$$

$$\frac{d}{dx}\cot x = -\csc^2 x \tag{11}$$

$$\frac{d}{dx}\sec x = \sec x\tan x \tag{12}$$

$$\frac{d}{dx}\csc x = -\csc x\cot x \tag{13}$$

Proof. We will prove formulas (10) and (12) and leave formulas (11) and (13) as exercises (see Problems 49 and 50).
Formula (10):

$$\frac{d}{dx}\tan x = \frac{d}{dx}\frac{\sin x}{\cos x} \overset{\text{Quotient rule}}{=} \frac{\cos x(d/dx)(\sin x) - \sin x(d/dx)(\cos x)}{\cos^2 x}$$

$$= \frac{\cos x(\cos x) - \sin x(-\sin x)}{\cos^2 x} = \frac{\cos^2 x + \sin^2 x}{\cos^2 x} = \frac{1}{\cos^2 x}$$

$$= \left(\frac{1}{\cos x}\right)^2 = \sec^2 x$$

Formula (12):

Power rule
$$\frac{d}{dx}\sec x = \frac{d}{dx}\left(\frac{1}{\cos x}\right) = \frac{d}{dx}(\cos x)^{-1} = (-1)(\cos x)^{-2}\overset{\downarrow}{\frac{d}{dx}}\cos x$$

$$= -\frac{1}{\cos^2 x}(-\sin x) = \frac{\sin x}{\cos^2 x} = \frac{1}{\cos x}\cdot\frac{\sin x}{\cos x} = \sec x\tan x \quad \blacksquare$$

EXAMPLE 7   Compute

$$\frac{d}{dx}\sqrt{\tan x}.$$

Solution.

From (10)
$$\frac{d}{dx}\sqrt{\tan x} = \frac{d}{dx}(\tan x)^{1/2} = \frac{1}{2}(\tan x)^{-1/2}\overset{\downarrow}{\frac{d}{dx}}\tan x = \frac{1}{2\sqrt{\tan x}}\overset{\downarrow}{(\sec^2 x)}$$

$$= \frac{\sec^2 x}{2\sqrt{\tan x}} \quad \blacksquare$$

PROBLEMS 3.5

In Problems 1–30, compute the derivative of the given function.

1. $y = \sin 3x$
2. $y = \cos(x - 3)$
3. $y = x\cos x$

4. $y = \dfrac{\sin x}{x}$
5. $y = \sin(x^3 - 2x + 6)$

6. $y = \dfrac{\sin(x^2 - 3)}{\cos(x^4 + 5)}$
7. $y = \sqrt{\sin x}$
8. $y = (\cos x)^{3/5}$

9. $y = \sin^2 x$
10. $y = \sin^2 x^3$
11. $y = \sin^3 x^2$

*12. $y = \sin\sqrt{\cos x}$
13. $y = \sin^2 x + \cos^2 x$
14. $y = \sin^3 x \cos^4 x$

15. $y = \dfrac{\sin^3 x}{\cos^4 x}$
16. $y = x\cos x - x^2\sin^2 x$
17. $y = x\tan x$

18. $y = \sec x\cos x$
19. $y = \tan x\cot x$
20. $y = \dfrac{\sin x}{\tan x}$

21. $y = \tan^2 x$
22. $y = \sqrt{\csc x}$
23. $y = \sqrt{x}\cot x$

24. $y = \dfrac{1 + \sec x}{1 - \sec x}$
25. $y = \tan^{3/2} x$
26. $y = \sec^3 x$

27. $y = \sec x\tan x$
28. $y = \dfrac{3}{\cot^2 x}$
29. $y = \sqrt{\sin x + \tan x}$

30. $y = \dfrac{\sin x + \cos x}{\tan x}$

In Problems 31–46, calculate the indicated limits.

31. $\lim\limits_{x\to 0} \dfrac{\sin \frac{1}{2}x}{x}$

32. $\lim\limits_{x\to 0} \dfrac{\sin^2 x}{x^2}$

33. $\lim\limits_{x\to 0} \dfrac{\sin^2 4x}{x^2}$ [*Hint:* First find $\lim_{x\to 0}(\sin 4x/x)$.]

34. $\lim\limits_{x\to 0} \dfrac{3x}{\sin 2x}$

35. $\lim\limits_{x\to 0} \dfrac{\sin^7 2x}{4x^7}$ [*Hint:* First find $\lim_{x\to 0}(\sin 2x/x)$.]

36. $\lim\limits_{x\to 0} \dfrac{\sin^2 x}{x}$

37. $\lim\limits_{x\to 0} \dfrac{\sin^2 3x}{4x}$

38. $\lim\limits_{x\to 0} \dfrac{\sin^3 4x}{3x^3}$

39. $\lim\limits_{x\to 0} \dfrac{\sin^3 2x}{x^2}$

40. $\lim\limits_{x\to 0} \dfrac{x^3}{3 \sin^2 2x}$

41. $\lim\limits_{x\to 0} \dfrac{\sin 3x}{\sin 4x}$

42. $\lim\limits_{x\to 0} \dfrac{\sin \frac{1}{2}x}{\sin \frac{1}{5}x}$

43. $\lim\limits_{x\to 0} \dfrac{\sin ax}{\sin bx}$, $ab \neq 0$

44. $\lim\limits_{x\to 0} \dfrac{\cos 2x - 1}{x}$

45. $\lim\limits_{x\to 0} \dfrac{\cos x^2 - 1}{x^4}$

46. $\lim\limits_{x\to 0} \dfrac{\sin x^2}{\sin 5x^2}$

47. Prove Theorem 1 in the case $-\pi/2 < \theta < 0$. [*Hint:* Draw a sketch similar to the sketch in Figure 1 but with θ negative.]

*48. If θ is measured in degrees, show the following:

(a) $\lim\limits_{\theta\to 0^{\circ}} \dfrac{\sin \theta}{\theta} = \dfrac{\pi}{180}$

(b) $\lim\limits_{\theta\to 0^{\circ}} \dfrac{\cos \theta - 1}{\theta} = 0$

In this problem $\sin \theta$ is taken to mean the sine of an angle measured in degrees.

49. Show that

$$\frac{d}{dx}\cot x = \frac{d}{dx}\left(\frac{\cos x}{\sin x}\right) = -\csc^2 x.$$

50. Show that

$$\frac{d}{dx}\csc x = \frac{d}{dx}\left(\frac{1}{\sin x}\right) = -\csc x \cot x$$

*51. Find the derivative of $\tan x$ directly from the limit definition.
(a) Use the fact that $\lim_{x\to 0}(\sin x/x) = 1$ to find $\lim_{x\to 0}(\tan x/x)$.
(b) Use the fact that $\tan(A + B) = (\tan A + \tan B)/(1 - \tan A \cdot \tan B)$ to show that

$$\frac{d}{dx}\tan x = \lim_{\Delta x\to 0} \frac{\tan(x + \Delta x) - \tan x}{\Delta x} = 1 + \tan^2 x = \frac{1}{\cos^2 x}.$$

52. Show the following:
(a) $\sin x/x$ is continuous at every real number except 0.
(b) $\sin x/x$ has a removable discontinuity at 0.

53. Can you explain why the derivatives of $\sec^2 x$ and $\tan^2 x$ are the same? What about $\cot^2 x$ and $\csc^2 x$?

3.6 IMPLICIT DIFFERENTIATION

In most of the problems we have encountered, the variable y was given *explicitly* as a function of the variable x. For example, for each of the functions

$$y = 3x + 6, \quad y = x^2, \quad y = \sqrt{x + 3}, \quad y = 1 + 2x + 4x^3, \quad y = (1 + 8x)^{3/2}$$

the variable y appears alone on the left-hand side. Thus we may say, "You give me an x and I'll tell you the value of $y = f(x)$." One exception to this rule was given in formula (1.3.5) where the variables x and y were given *implicitly* in the equation of the circle centered at $(0, 0)$ with radius r:

$$x^2 + y^2 = r^2. \tag{1}$$

Here x and y are not given separately. In general, we say that x and y are given *implicitly* if neither one is expressed as an explicit function of the other.

NOTE.   This is *not* to say that one variable can*not* be solved explicitly in terms of the other.

EXAMPLE 1   The following are examples in which the variables x and y are given implicitly.

(a) $x^3 + y^3 = 6xy^4$

(b) $(2x^{3/2} + y^{5/3})^{17} - 6y^5 = 2$

(c) $2xy(x + y)^{4/3} = 6x^{17/9}$

(d) $\dfrac{x + y}{\sqrt{x^2 - y^2}} = 16y^5$

(e) $xy = 1$ (Here it is easy to solve for one variable in terms of the other.) ∎

For the example of the circle, it was possible to solve equation (1) explicitly for y in order to calculate dy/dx. However, it is very difficult or impossible to do the same thing for the functions (a)–(d) given in Example 1 (try it!). Nevertheless, the derivative dy/dx *may* exist. Can we calculate it?

To illustrate the answer to this question, let us again return to equation (1):

$$x^2 + y^2 = r^2.$$

If $y > 0$, then

$$y = \sqrt{r^2 - x^2},$$

and

$$\frac{dy}{dx} = \frac{d}{dx}(r^2 - x^2)^{1/2} = \overset{\text{Power rule}}{\underset{\downarrow}{\frac{1}{2}}}(r^2 - x^2)^{-1/2}\frac{d}{dx}(r^2 - x^2) = \frac{-2x}{2\sqrt{r^2 - x^2}}$$

or

$$\frac{dy}{dx} = -\frac{x}{\sqrt{r^2 - x^2}} = -\frac{x}{y}.$$

We now calculate this derivative another way. Assuming that y can be written as a function of x, we can write $y^2 = [f(x)]^2$ for some function $f(x)$ (which we assume to be unknown†). Then by the chain rule,

$$\frac{d}{dx}(y^2) = \frac{d}{dx}[f(x)]^2 = 2f(x) \cdot f'(x) = 2y\frac{dy}{dx}. \tag{2}$$

Now taking the derivatives of both sides of (1) with respect to x and using (2), we obtain

$$2x + 2y\frac{dy}{dx} = \frac{d}{dx}r^2 = 0.\ddagger \tag{3}$$

We can now solve for dy/dx in (3):

$$\frac{dy}{dx} = -\frac{x}{y}. \tag{4}$$

If we do not know y as a function of x, then this is as far as we can go. However, since in this case we may choose $y = \sqrt{r^2 - x^2}$, we may write equation (4) as

$$\frac{dy}{dx} = \frac{-x}{\sqrt{r^2 - x^2}}.$$

NOTE. We should keep in mind that what makes this technique work is that we are *assuming* that y is a differentiable function of x. Thus we may calculate

$$\frac{d}{dx}(y^2) = 2y\frac{dy}{dx}$$

as in the last example.

The method we have used in the above calculation is called **implicit differentiation,** and it is the only way to calculate derivatives when it is impossible to solve for one variable in terms of the other. We begin by assuming that y is a differentiable function of x, without actually having a formula for the function. We proceed to find dy/dx by differentiating and then solving for it algebraically. We illustrate this method with a number of examples.

---

†In this case we know that $f(x) = \sqrt{r^2 - x^2}$ or $-\sqrt{r^2 - x^2}$, but ordinarily (as for the functions in Example 1) $f(x)$ will *really* be unknown.
‡We used the fact that both sides of an equation can always be differentiated provided that the derivatives of the functions involved exist.

EXAMPLE 2 Suppose that

$$x^2 + x^3 = y + y^4. \tag{5}$$

Find dy/dx.

Solution. By the chain rule,

$$\frac{d}{dx} y^4 = 4y^3 \frac{d}{dx} y = 4y^3 \frac{dy}{dx}.$$

Thus we may differentiate both sides of (5) with respect to x to obtain

$$\frac{d}{dx} x^2 + \frac{d}{dx} x^3 = \frac{d}{dx} y + \frac{d}{dx} y^4,$$

or

$$2x + 3x^2 = \frac{dy}{dx} + 4y^3 \frac{dy}{dx} = (1 + 4y^3) \frac{dy}{dx}.$$

Then

$$\frac{dy}{dx} = \frac{2x + 3x^2}{1 + 4y^3}.$$

At the point $(-1, 0)$, for example,†

$$\frac{dy}{dx} = \frac{2(-1) + 3 \cdot 1}{1 + 0} = 1,$$

and the equation of the tangent line at that point is

$$y = x + 1.$$

In this example there was no special reason to calculate dy/dx. We may have just as well been asked to calculate dx/dy.‡ We again use the chain rule to find that

$$\frac{d}{dy} x^2 = 2x \frac{dx}{dy} \qquad \text{and} \qquad \frac{d}{dy} x^3 = 3x^2 \frac{dx}{dy}.$$

Then differentiating both sides of (5) with respect to y yields

$$\frac{d}{dy} x^2 + \frac{d}{dy} x^3 = \frac{d}{dy} y + \frac{d}{dy} y^4,$$

———————————

†You should verify that $(-1, 0)$ is a point on the curve.
‡Assuming, of course, that x is a differentiable function of y.

or

$$2x \frac{dx}{dy} + 3x^2 \frac{dx}{dy} = 1 + 4y^3 \qquad \text{and} \qquad \frac{dx}{dy} = \frac{1 + 4y^3}{2x + 3x^2}.$$

Note that $dx/dy = 1/(dy/dx)$. Although we will not prove this result here it is true that under certain hypotheses $dy/dx = 1/(dx/dy)$ (see Problem 3.3.28, and Section 6.1). ■

EXAMPLE 3   Consider the equation

$$x^2 + y^2 + 4 = 0. \tag{6}$$

If we differentiate implicitly with respect to x, we obtain

$$2x + 2y \frac{dy}{dx} = 0, \qquad \text{or} \qquad \frac{dy}{dx} = -\frac{x}{y}.$$

But this answer is meaningless because there are no real numbers x and y that satisfy Equation (6), so there is no real-valued function implicitly given by (6). What is the meaning of the derivative of a nonexistent function? This example illustrates that implicit differentiation only makes sense when y is a differentiable function of x. ■

EXAMPLE 4   Find the equation of the tangent line to the curve

$$x^4 + y^4 = 17 \tag{7}$$

at the point (2, 1).

Solution. Since

$$\frac{d}{dx} y^4 = 4y^3 \frac{dy}{dx},$$

we may differentiate both sides of (7) with respect to x to obtain

$$\frac{d}{dx} x^4 + \frac{d}{dx} y^4 = \frac{d}{dx} 17, \qquad \text{or} \qquad 4x^3 + 4y^3 \frac{dy}{dx} = 0$$

(the derivative of a constant is zero). Then

$$4y^3 \frac{dy}{dx} = -4x^3 \qquad \text{and} \qquad \frac{dy}{dx} = -\frac{x^3}{y^3}.$$

At the point (2, 1), $dy/dx = -8$, so the equation of the tangent line is

$$\frac{y - 1}{x - 2} = -8, \qquad \text{or} \qquad 8x + y = 17. \; ■$$

EXAMPLE 5   Compute the slope of the line tangent to the curve $\sin xy = x$ at the point (1/2, $\pi/3$).

Solution. We have

$$\frac{d}{dx} \sin xy = \cos xy \left(\frac{d}{dx} xy \right) = \cos xy \left(x \frac{dy}{dx} + y \right)$$

and $(d/dx)x = 1$. Thus

$$\cos xy \left(x \frac{dy}{dx} + y \right) = 1,$$

so that

$$x \frac{dy}{dx} + y = \frac{1}{\cos xy}, \qquad \text{or} \qquad x \frac{dy}{dx} = \frac{1}{\cos xy} - y,$$

and

$$\frac{dy}{dx} = \frac{1}{x} \left(\frac{1}{\cos xy} - y \right).$$

At $(1/2, \pi/3)$, $xy = \pi/6$, $\cos xy = \sqrt{3}/2$, and $1/(\cos xy) = 2/\sqrt{3}$, so that

$$\left. \frac{dy}{dx} \right|_{(1/2, \pi/3)} = 2 \left(\frac{2}{\sqrt{3}} - \frac{\pi}{3} \right) = \frac{4}{\sqrt{3}} - \frac{2\pi}{3}. \quad \blacksquare$$

NOTATION. The notation

$$\left. \frac{dy}{dx} \right|_{(x,y)}$$

stands for the derivative dy/dx evaluated at the point (x, y).

We return briefly to the circle $x^2 + y^2 = r^2$. For $y > 0$ we calculated $dy/dx = -x/y$, which is zero when $x = 0$ (and $y = r$). Thus for $x = 0$ the tangent line has slope zero and is horizontal (see Figure 1). Now let us consider x as a function of y (for $x > 0$) and differentiate implicitly with respect to y to obtain

$$2x \frac{dx}{dy} + 2y \frac{dy}{dy} = \frac{d}{dy} r^2 = 0 \qquad \text{and} \qquad \frac{dx}{dy} = -\frac{y}{x},$$

which is zero at the point $(r, 0)$. If $dx/dy = 0$ at a point, then the tangent line to the curve at that point is parallel to the y-axis. In this case we say that the graph of the function has a *vertical tangent* at the point. This result follows from the same reasoning that shows that the tangent line is parallel to the x-axis at any point at which $dy/dx = 0$. In our example we see that the tangent line to the curve $x^2 + y^2 = r^2$ at point $(0, r)$ is the line $x = r$. This line is vertical, as depicted in Figure 1.

We generalize this example with the following rule:

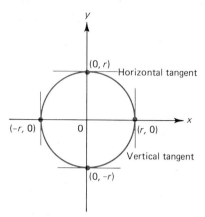

FIGURE 1

(i) If $dy/dx = 0$ at the point (x_0, y_0), then the graph of $y = f(x)$ has a **horizontal tangent** at that point, given by the straight line $y = y_0$.
(ii) If $dx/dy = 0$ at the point (x_0, y_0), then the graph of $y = f(x)$ has a **vertical tangent** at that point given by the straight line $x = x_0$.

NOTE.   In both of these cases $dy/dx \neq 1/(dx/dy)$ since we cannot divide by zero.

PROBLEMS 3.6

In Problems 1–29, find dy/dx and dx/dy by implicit differentiation.

1. $x^3 + y^3 = 3$

2. $x^3 + y^3 = xy$

3. $\sqrt{x} + \sqrt{y} = 2$

4. $xy + x^2y^2 = x^5$

5. $\dfrac{1}{x} + \dfrac{1}{y} = 1$

6. $\dfrac{1}{x^2} - \dfrac{1}{y^2} = x + y$

7. $(x + y)^{1/2} = (x^2 + y)^{1/3}$

8. $xy + x^2y^2 + x^3y^3 = 2$

9. $\dfrac{1}{\sqrt{x^2 + y^2}} = 4$

10. $\sqrt{x^3 + xy + y^3} = 6$

11. $(3xy + 1)^5 = x^2$

12. $\dfrac{x + y}{x - y} = 2$

13. $(x + y)(x - y) = 7$

14. $\sqrt{xy^2 + yx^2} = 0$

15. $\dfrac{x^2 + y^2}{x^2 - y^2} = 4$

16. $x^{3/4} + y^{3/4} = 2$

17. $\dfrac{2xy + 1}{3xy - 1} = 2$

18. $x^{-1/2} + y^{-1/2} = 4$

19. $xy + x^2y^2 = 2$

20. $\dfrac{x}{y} + \dfrac{x^2}{y^2} = 3$

21. $x^{-7/8} + y^{-7/8} = \frac{7}{8}$

22. $x^2 - \sqrt{xy} + y^2 = 6$

23. $(4x^2y^2)^{1/5} = 1$

24. $(\sqrt{x} + \sqrt{y})(\sqrt[3]{x} - \sqrt[3]{y}) = -3$

25. $x^2y^3 + x^3y^2 = xy$

26. $\sin xy = 2$

27. $\cos(x^2 + y^2) = 2x$

28. $\sin x \cos y = y$

29. $\dfrac{\sin x}{\cos y} = \sin(x - y)$

30. Find the equation of the line tangent to the curve $x^5 + y^5 = 2$ at the point $(1, 1)$.

In Problems 31–38, find the points where the given curve has a vertical tangent. Also, find the points where it has a horizontal tangent.

31. $\sqrt{x} + \sqrt{y} = 1$ **32.** $\dfrac{1}{x} + \dfrac{1}{y} = 1$ **33.** $xy = 1$

34. $(x - 3)^2 + (y - 4)^2 = 9$ **35.** $\dfrac{x^2}{a^2} - \dfrac{y^2}{b^2} = 1$ **36.** $\dfrac{x^2}{a^2} + \dfrac{y^2}{b^2} = 1$

37. $y = \dfrac{1}{\sin x}$ **38.** $y \cos x = 3$

39. Use implicit differentiation and the power rule for integer exponents to establish the power rule for rational exponents. Prove that if $f(x) = x^{p/q}$, where p and q are integers, then $f'(x) = x^{p/q-1}$. [*Hint:* If $y = x^{14/3}$, then $y^3 = x^{14}$.]

40. Consider y to be a function of x implicitly specified by

$$\sqrt{\frac{y}{x}} + 4\sqrt{\frac{x}{y}} = \sqrt{18}.$$

Show that $dy/dx = y/x$.

*41. Suppose you're handed a function L such that $L'(x) = 1/x$ for all nonzero x. Also, suppose that you have a function E with the property that $L \circ E = $ identity function. That is, $L(E(y)) = y$ for all real y. Prove that $E' = E$.

**42. Suppose that $[y(x)]^2 = 137/\sqrt{\pi} - x^2 + 2Ax$. Using implicit differentiation, show that each normal to the curve passes through the point $(A, 0)$.

43. Prove that the curves implicitly given by

(i) $3x - 2y + x^3 - x^2y = 0$ and   (ii) $x^2 - 2x + y^2 - 3y = 0$

are perpendicular at the origin (that is, their tangents are perpendicular).

*44. Pick a number m.

(a) Show there are at most two points on $10x^2 + (y/5)^2 = 1$ where the tangent line has slope equal to m.

(b) Show there are at most two points on $x^2 - y^2 = 37$ where the tangent line has slope equal to m.

3.7 HIGHER-ORDER DERIVATIVES

Let $s(t)$ represent the distance a falling object has dropped after t units of time have elapsed. Then as we have seen, the derivative ds/dt evaluated at a time t may be interpreted as the instantaneous velocity of the object at that time. The velocity is the rate of change of position with respect to time. By definition, acceleration is the rate of change of velocity with respect to time. Thus acceleration can be thought of as the derivative of the derivative, or, more simply, as the *second derivative* of the function representing position. If, for example, $s = \frac{1}{2}gt^2$ represents the position of a falling object (as in Section 2.6), then

$$\frac{ds}{dt} = gt \quad \text{and} \quad \frac{d}{dt}\left(\frac{ds}{dt}\right) = g,$$

which is the acceleration due to gravity.

We now generalize these ideas. Let $y = f(x)$ be a differentiable function. Then the derivative $y' = dy/dx = f'$ is also a funciton of x. This new function of x, f', may or may not be a differentiable function. If it is, we call the derivative of f' the **second derivative** of f (i.e., the derivative of the derivative) and denote it by

$$f''. \tag{1}$$

There are other commonly used notations as well. We will write

$$f'' = (f')' = \left(\frac{dy}{dx}\right)' = \frac{d}{dx}\left(\frac{dy}{dx}\right) = \frac{d^2y}{dx^2}. \tag{2}$$

The notations

$$f'' = \frac{d^2y}{dx^2} = y'' \tag{3}$$

will be used interchangeably in this book to denote the second derivative.

Similarly, if f'' exists, it might or might not be differentiable. If it is, then the derivative of f'' is called the **third derivative** of f and is denoted

$$f'''. \tag{4}$$

Alternative notations are

$$f''' = \frac{d^3y}{dx^3} = y'''. \tag{5}$$

We can continue this definition indefinitely as long as each successive derivative is differentiable. After the third derivative we avoid a clumsy succession of primes by using numerals to denote higher derivatives:

$$f^{(4)}, f^{(5)}, f^{(6)}, \ldots . \tag{6}$$

Alternative notations are, for the successive derivatives,

$$f^{(4)} = \frac{d^4y}{dx^4} = y^{(4)} \tag{7}$$

$$f^{(5)} = \frac{d^5y}{dx^5} = y^{(5)} \tag{8}$$

$$\vdots$$

$$f^{(n)} = \frac{d^ny}{dx^n} = y^{(n)}. \tag{9}$$

We emphasize that *each higher order derivative of $y = f(x)$ is a new function of x* (if it exists).

EXAMPLE 1   Let $y = f(x) = x^3$. Then

$$\frac{dy}{dx} = f'(x) = 3x^2.$$

The second derivative is simply the derivative of the first derivative, so

$$\frac{d^2y}{dx^2} = f''(x) = \frac{d}{dx} 3x^2 = 6x.$$

Similarly, the third derivative is the derivative of the second derivative, and we have

$$\frac{d^3y}{dx^3} = f'''(x) = \frac{d}{dx} 6x = 6.$$

Finally,

$$\frac{d^4y}{dx^4} = f^{(4)}(x) = \frac{d}{dx} 6 = 0.$$

Note that for $k \geq 4$, $f^{(k)}(x) = 0$ since the derivative of the zero function is zero. ■

EXAMPLE 2   Let $y = f(x) = 1/x$. Then

$$\frac{dy}{dx} = f'(x) = -\frac{1}{x^2}, \quad \frac{d^2y}{dx^2} = f''(x) = \frac{2}{x^3}, \quad \frac{d^3y}{dx^3} = f'''(x) = -\frac{6}{x^4},$$

and so on. In general, $d^n y/dx^n = f^{(n)}(x) = (-1)^n n!/x^{n+1}$ (see Problem 24). The symbol $n!$, which is read "n factorial," is defined (for the nonnegative integers) by

$$n! = 1 \cdot 2 \cdot 3 \cdot 4 \cdot \cdots \cdot n \quad \text{for} \quad n = 1, 2, 3, \ldots, \tag{10}$$

and

$$0! = 1.$$

For example, $3! = 1 \cdot 2 \cdot 3 = 6$, and $7! = 1 \cdot 2 \cdot 3 \cdot 4 \cdot 5 \cdot 6 \cdot 7 = 5040$. ■

EXAMPLE 3   Let $y = f(x) = \sin x$. Then $dy/dx = f'(x) = \cos x$, $d^2y/dx^2 = f''(x) = -\sin x$, $d^3y/dx^3 = f'''(x) = -\cos x$, and $d^4y/dx^4 = f^{(4)}(x) = \sin x$. That is, $\sin x$ is equal to its fourth derivative and the negative of its second derivative. In Section 7.3 we will use this last fact in our discussion of periodic motion. ■

Although we have now defined derivatives of all orders, almost all of our major applications will involve first and/or second derivatives. Second derivatives are important for several reasons. In Section 4.4, for example, we will show that the sign of the second derivative of a function tells us something about the shape of the graph of

Note that dx does not have to be a small number in accordance with this definition. In fact, dx can take on any value between $-\infty$ and ∞.

The first thing to notice about these definitions is that (since dx is assumed to be nonzero)

$$\frac{dy}{dx} = \frac{f'(x)\,dx}{dx} = f'(x), \tag{7}$$

which is certainly not surprising since dy was chosen so that (7) would be satisfied. We should also note that the definitions here are artificial in the sense that they have been created so that we can manipulate the symbols dx and dy. It must be emphasized that formulas like the chain rule (5) are true not because differentials can automatically be canceled but because they were proven true before we even had such things as differentials around.

EXAMPLE 10 Let $y = x^2$. Calculate the differential dy.

Solution. Since $f'(x) = 2x$, we have $dy = 2x\,dx$. ■

EXAMPLE 11 Let $y = \sqrt{x + 1}$. Find dy.

Solution. Here

$$f'(x) = \frac{1}{2\sqrt{x + 1}} = \frac{dy}{dx},$$

so that

$$dy = \frac{dx}{2\sqrt{x + 1}}. \ ■$$

EXAMPLE 12 Let $y = \sin x^3$. Find dy.

Solution. Here

$$f'(x) = (\cos x^3)3x^2 = 3x^2 \cos x^3 = \frac{dy}{dx}.$$

Thus

$$dy = 3x^2 \cos x^3\,dx. \ ■$$

The differentiation formulas given in Table 3.4.1 can be extended to differential formulas, as shown in Table 4.

We close this section by illustrating the relationship between the increment Δy and the differential dy. Consider the function $y = f(x)$, let the number x_0 be fixed, and suppose that $f'(x_0)$ exists. For any value of Δx we have $dx = \Delta x$. The equation

$$dy = f'(x_0)\,dx \tag{8}$$

TABLE 4

$y = f(x)$	dy
I. c	$dc = 0$
II. $cu(x)$	$d(cu) = c\,du$
III. $u(x) + v(x)$	$d(u + v) = du + dv$
IV. x^r, r real	$d(x^r) = rx^{r-1}\,dx$
V. $u(x) \cdot v(x)$	$d(uv) = u\,dv + v\,du$
VI. $\dfrac{u(x)}{v(x)}$, $v(x) \neq 0$	$d\left(\dfrac{u}{v}\right) = \dfrac{v\,du - u\,dv}{v^2}$
VII. $u^r(x)$	$d[u^r(x)] = ru^{r-1}(x)u'(x)\,dx$
VIII. $f(g(x))$	$d[f(g(x))] = f'(g(x))g'(x)\,dx$
IX. $\sin u(x)$	$d[\sin u(x)] = [\cos u(x)]u'(x)\,dx$
X. $\cos u(x)$	$d[\cos u(x)] = [-\sin u(x)]u'(x)\,dx$

is the equation of a *straight line* with slope $f'(x_0)$ that coincides with the tangent line to the curve $f(x)$. [This line is a straight line in the coordinates (dx, dy) instead of (x, y). The origin in these coordinates is the point $(x_0, f(x_0))$ since $dx = dy = 0$ there.] Now look at Figure 3. We see that $\Delta y = f(x_0 + \Delta x) - f(x_0)$ changes along the curve, while dy changes along the tangent line. In this illustration dy is only close to Δy for small values of $dx = \Delta x$. Tables 1 and 2 given earlier in this section illustrate this phenomenon. In those tables the fourth column is Δy and the fifth column is dy.

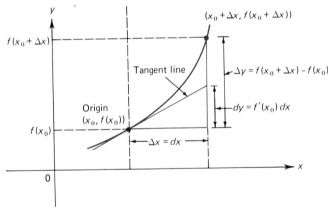

FIGURE 3

PROBLEMS 3.8

1. Let $y = x^3$. For $x = 3$, tabulate values of Δy and dy for $\Delta x = 1, 0.5, 0.1, 0.05, 0.01$, and 0.001. Show how the approximating error decreases as Δx decreases.

2. Let $y = x^2$. For $x = 1$, tabulate values of Δy and dy for $\Delta x = -1, -0.5, -0.1, -0.05, -0.01$, and -0.001. Show how the approximating error decreases as Δx gets closer to zero.

3. Let $y = x^{3/2}$. For $x = 4$, tabulate values of Δy and dy for $\Delta x = 1, -1, 0.5, -0.5, 0.1, -0.1, 0.01$, and -0.01. Show how the approximating error decreases as Δx approaches zero.

In Problems 4–23, calculate an approximate value for the given expression by using differentials.

4. $\sqrt{25.03}$

5. $(0.99)^4$

6. $1 + (4.02)^2$

7. $\dfrac{1}{5.1}$

8. $\dfrac{1}{\sqrt{3.98}}$

9. $\dfrac{1}{35^{1/5}}$

10. $(2.02)^3 - 4(2.02)^2$

11. $(1.03)^{20}$

12. $(0.97)^{5/8}$

13. $(15.95)^{3/4}$

14. $(1.01)^2 + (1.01)^4 + (1.01)^6$

15. $(0.98)^7 - (0.98)^3$

16. $\sqrt[4]{80}$

17. $(5 + \sqrt{3.9})^3$

18. $\cos 58.5°$

19. $\sin\left(\dfrac{\sqrt{\pi}}{2} + 0.1\right)^2$

20. $\tan 44°$

21. $\tan\left(\dfrac{11\pi}{4} + 0.05\right)$

22. $\sec 151°$

23. $\csc 224°$

In Problems 24–42, find the differential of the given function.

24. $y = 3x + 6$

25. $y = 2$

26. $y = x^4$

27. $y = x^{1/3}$

28. $y = (1 + x^2)^4$

29. $y = (1 + x^3)^{1/4}$

30. $y = \dfrac{x + 1}{x - 1}$

31. $y = \dfrac{1}{x}$

32. $y = \sqrt{1 + \sqrt{x}}$

33. $y = \dfrac{1 - \sqrt{x}}{1 + \sqrt{x}}$

34. $y = \dfrac{x + x^2}{1 - x^3}$

35. $y = \sqrt{\dfrac{1 + x}{1 - x}}$

36. $y = (1 + x^3)^{17/2}$

37. $y = \dfrac{1}{x^2 + 2}$

38. $y = (x + 3)^{4/3}(x^2 + 2)^{3/4}$

39. $y = \cos \sqrt{x}$

40. $y = \tan x$

41. $y = \csc x$

42. $y = \sec x$

43. A tank filled with water is in the shape of a right circular cone (Figure 4). The volume of the cone is given by $V = \frac{1}{3}\pi r^2 h$. The radius of the cone is measured to be 2 m, with a maximum error of 0.01 m. The height is exactly 3 m.

FIGURE 4

(a) What is the approximate mass of the water in the tank?

(b) By how much could this calculation be off?

(c) How large a relative error does this represent?

*44. According to **Poiseuille's law,** the resistance R of a blood vessel of length l and radius r is given by $R = \alpha l / r^4$, where α is a constant of proportionality. If l remains fixed, how does a small change in r affect the resistance R?

45. One side of a square field was measured to be 75 ft. What is the relative error in the calculation of the area of the field if the true length of the side is 74.8 ft?

46. What is the largest relative error allowed in the measurement of the radius of a right circular cone if the relative error in the calculation of its volume is to be less than 0.8%?

47. Stefan's law for the emission of radiant energy from the surface of a body is given by $R = \sigma T^4$, where R is the rate of emission per unit area, T is the temperature, measured in degrees kelvin, and σ is a universal constant. A relative error of at most 0.02 in T will result in what maximum relative error in R?

48. A grapefruit with a diameter of 6 in. has a skin $\frac{1}{5}$ of an inch thick (the skin is included in the diameter).

(a) What is the approximate volume of the skin?

(b) What percentage of the total volume is skin?

49. Using the information in Example 8, show that increasing the yield of a nuclear bomb will lead to an ever-decreasing effect on the relative efficiency of the bomb (measured in terms of the distance affected by overpressure).

*50. Justify each of the following estimates.

(a) $1/(1 + x) \approx 1 - x$ if $x \approx 0$.

(b) $\sqrt{1 + x} \approx 1 + \frac{1}{2}x$ if $x \approx 0$.

(c) $(1 + x)^n \approx 1 + nx$ if $x \approx 0$ and $n > 1$ is an integer.

(d) $\sin x \approx x$ if $x \approx 0$.

REVIEW EXERCISES FOR CHAPTER THREE

In Exercises 1–24, calculate dy/dx.

1. $y = 3x + 4$

2. $y = 3x^2 - 6x + 2$

3. $x + y + xy = 3$

4. $y = \dfrac{x + 1}{x - 2}$

5. $y = (3x^2 + 2)\sqrt{x + 1}$

6. $y = x^5 - \sqrt{x}$

7. $x^{3/4} + y^{3/4} = 4$

8. $y = (x^2 - 1)^{1/2}(x^4 - 2)^{1/3}$

9. $y = \sqrt[3]{\dfrac{x + 2}{x - 3}}$

10. $xy^2 + yx^2 = xy$

11. $y = \dfrac{x^2 - 2x + 3}{x^3 + 5x - 6}$

12. $y = \dfrac{3}{x - \sqrt{x^3 - 2}}$

13. $y = (x^3 - 3)^{5/6}(x^3 + 4)^{6/7}$

14. $y = \dfrac{1 + x + x^5}{3 - x^2 + x^7}$

15. $x + \sqrt{x + y} = y^{1/3}$

16. $y = \dfrac{1 - \sqrt[3]{x}}{1 + \sqrt[3]{x}}$

17. $y = (1 + x^3)^{\sqrt{2}}$

18. $(1 + x^2 + y^2)^{3/2} = 6$

19. $y = \dfrac{(x^3 - 1)^{2/3}}{(x^5 + 3)^{3/4}}$

20. $y = \sqrt{1 - \sqrt{1 - \sqrt{x}}}$

21. $y = \sin x^3$

22. $y = \cos \dfrac{1}{x}$

23. $y = \tan x \sec x$

24. $y = \sqrt{1 + \cos x}$

In Exercises 25–30, find the equations of the tangent and normal lines to the given curve at the given point.

25. $y = x^3 - x + 4$; $(1, 4)$

26. $y = \dfrac{\sqrt{x^2 + 9}}{x^2 - 6}$; $\left(4, \dfrac{1}{2}\right)$

27. $y = (x^2 - 4)^2(\sqrt{x} + 3)^{1/2}$; $(1, 18)$

28. $xy^2 - yx^2 = 0$; $(1, 1)$

29. $y = \cos\sqrt{x}$; $\left(\dfrac{\pi^2}{9}, \dfrac{1}{2}\right)$

30. $y = \tan x$; $\left(\dfrac{\pi}{4}, 1\right)$

In Exercises 31–38, calculate the second and third derivatives of the given function.

31. $y = x^7 - 7x^6 + x^3 + 3$ **32.** $y = \sqrt{1 + x}$

33. $y = \dfrac{1}{1 + x}$

34. $y = \dfrac{x + 1}{x - 1}$

35. $y = \dfrac{x^2 - 3}{x^2 + 5}$

36. $y = \sqrt{1 + \sqrt{x}}$

37. $y = \dfrac{\cos x}{x}$

38. $y = \sqrt{\sin x}$

In Exercises 39–46, find the differential dy.

39. $y = 17x^3 - 6$

40. $y = \sqrt{x^2 - 3}$

41. $y = \dfrac{x + 4}{x - 7}$

42. $y = \dfrac{-3}{1 + x}$

43. $y = (\sqrt{x^2 - 3})\sqrt[3]{x^4 + 5}$ **44.** $y = \sqrt{1 + \sqrt{x}}$

45. $y = \cos \dfrac{1}{x}$

46. $y = \sin\sqrt{x + 1}$

47. Using differentials, find an approximate value for (a) $(1.99)^5$, (b) $1/\sqrt[3]{8.02}$.

48. The radius of a sphere is 1 m, with a possible error in measurement of 0.01 m = 1 cm. What is the maximum error you would expect in the calculation of the volume?

4 Applications of the Derivative

4.1 RELATED RATES OF CHANGE

We have seen, beginning in Section 2.6, that the derivative can be interpreted as a rate of change. There are many problems involving two or more variables in which it is necessary to calculate the rate of change of one or more of these variables with respect to time. After giving an example, we will suggest a procedure for handling problems of this type.

EXAMPLE 1   A rope is attached to a pulley mounted on a 15-ft tower. The end of the rope is attached to a heavily loaded cart (see Figure 1). A worker can pull in rope at a rate of 2 ft/sec. How fast is the cart approaching the tower when it is 8 ft from the tower?

Solution. We let s denote the horizontal distance of the cart from the tower and l the length of rope from the top of the tower to the cart, as in Figure 1. Since the speed of the cart is ds/dt, the question asks us to determine ds/dt when $s = 8$ ft. We are told that $dl/dt = -2$.† To calculate ds/dt, we must first find a relationship between s and l. From the Pythagorean theorem we immediately obtain

$$15^2 + s^2 = l^2. \tag{1}$$

---

† dl/dt is negative because the length represented by l is decreasing as the worker pulls in the rope.

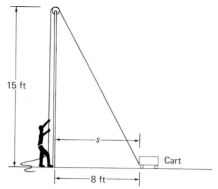

FIGURE 1

We now differentiate (1) implicitly with respect to t to find that

$$\frac{d}{dt}(15^2) + \frac{d}{dt}(s^2) = \frac{d}{dt}(l^2),$$

or

$$0 + 2s\frac{ds}{dt} = 2l\frac{dl}{dt}. \tag{2}$$

When $s = 8$, $l^2 = 15^2 + 8^2 = 225 + 64 = 289$, and $l = 17$. Then inserting $s = 8$, $l = 17$, and $dl/dt = -2$ into equation (2) gives us

$$\frac{ds}{dt} = \frac{17}{8}(-2) = -\frac{17}{4} = -4\frac{1}{4} \text{ ft/sec.}$$

Thus the cart is approaching the tower at a rate of $4\frac{1}{4}$ ft/sec. ■

The solution given above suggests that the following steps be taken to solve a problem involving the rates of change of related variables:

(i) If feasible, draw a picture of what is going on.
(ii) Determine the important variables in the problem and find an equation relating them.
(iii) Differentiate the equation obtained in (ii) with respect to t.
(iv) Solve for the derivative sought.
(v) Evaluate that derivative by substituting given and calculated values of the variables in the problem.
(vi) Interpret your answer in terms of the question posed in the problem.

EXAMPLE 2   An oil storage tank is built in the form of an inverted right circular cone with a height of 6 m and a base radius of 2 m (see Figure 2). Oil is being pumped into the tank at a rate of 2 liters (L)/min $= 0.002$ m³/min (since 1 m³ $= 1000$ L). How fast is the level of the oil rising when the tank is filled to a height of 3 m?

FIGURE 2

Solution. In mathematical terms we are asked to calculate dh/dt when $h = 3$ m, where h denotes the height of the oil at a given time and r the radius of the cone of oil (see Figure 2). The volume of a right circular cone is $V = \frac{1}{3}\pi r^2 h$. From the data given in the problem we see from Figure 2 (using similar triangles) that $h/r = 6/2$, or $r = h/3$. Then

$$V = \frac{1}{3}\pi\left(\frac{h}{3}\right)^2 h = \frac{1}{27}\pi h^3 = \text{volume of oil at height } h.$$

Differentiating with respect to t and using the fact (given to us) that $dV/dt = 0.002$, we obtain

$$\frac{dV}{dt} = \frac{1}{9}\pi h^2 \frac{dh}{dt} = 0.002, \quad \text{or} \quad \frac{dh}{dt} = \frac{9(0.002)}{\pi h^2} = \frac{0.018}{\pi h^2} \text{ m/min.}$$

Then for $h = 3$,

$$\frac{dh}{dt} = \frac{0.018}{\pi \cdot 9} = \frac{0.002}{\pi} \approx 6.37 \times 10^{-4} \text{ m/min,}$$

which is the rate at which the height of the oil is increasing. ∎

EXAMPLE 3   In chemistry an **adiabatic** process is one in which there is no gain or loss of heat. During an adiabatic process the pressure P and volume V of certain gases such as hydrogen or oxygen in a container are related by the formula

$$PV^{1.4} = \text{constant.} \tag{3}$$

At a certain time the volume of hydrogen in a closed container is 4 m³ and the pressure is 0.75 kg/m². Suppose that the volume is increasing at a rate of $\frac{1}{2}$ m³/sec. How fast is the pressure decreasing?

Solution. We use the product rule to differentiate (3) with respect to t.

$$P\frac{d}{dt}(V^{1.4}) + \frac{dP}{dt} \cdot V^{1.4} = 0, \quad P(1.4)V^{0.4}\frac{dV}{dt} + \frac{dP}{dt} \cdot V^{1.4} = 0$$

or

$$\frac{dP}{dt} = -\frac{1.4PV^{0.4}}{V^{1.4}}\frac{dV}{dt} = -1.4\frac{P}{V} \cdot \frac{dV}{dt}.$$

Using the data given in the problem, we find that

$$\frac{dP}{dt} = -\frac{(1.4)(0.75) \text{ kg/m}^2}{4 \text{ m}^3} \cdot (0.5) \text{ m}^3/\text{sec} = -0.13125 \text{ kg/m}^2/\text{sec}.$$

Thus the pressure is decreasing at a rate of 0.13125 kg/m²/sec. ◼

EXAMPLE 4   A ladder 8 m long leans against a wall 4 m high. The lower end of the ladder is pulled away from the wall at a rate of 2 m/sec. How fast is the angle between the top of the ladder and the wall changing when the angle is 60° = $\pi/3$ radians?

Solution. We refer to Figure 3. We are asked to calculate $d\theta/dt$ when $\theta = \pi/3$. From the figure we see that $\sin \theta = x/8$, where x is the distance from the foot of the ladder to the wall. We are told that $dx/dt = 2$. Then using the chain rule, we have

$$\cos \theta \frac{d\theta}{dt} = \frac{d}{dt}\sin \theta = \frac{d}{dt}\left(\frac{x}{8}\right) = \frac{1}{8}\frac{dx}{dt} = \frac{2}{8} = \frac{1}{4}.$$

FIGURE 3

Thus $d\theta/dt = 1/(4 \cos \theta)$, and when $\theta = \pi/3$, $\cos \theta = \frac{1}{2}$, so that $d\theta/dt = 1/4(\frac{1}{2}) = \frac{1}{2}$ radian /sec, which is the rate at which the angle is increasing. ◼

PROBLEMS 4.1

1. Let $xy = 6$. For $dx/dt = 5$, find dy/dt when $x = 3$.
2. Let $x/y = 2$. For $dx/dt = 4$, find dy/dt when $x = 2$.
3. A 10-ft ladder is leaning against the side of a house. As the foot of the ladder is pulled away from the house, the top of the ladder slides down along the side (see Figure 4). The foot of the ladder is pulled away at a rate of 2 ft/sec. How fast is the ladder sliding down when the foot is 8 ft from the house?

FIGURE 4

4. A cylindrical water tank 6 m high with a radius of 2 m is being filled at a rate of 10 L/min. How fast is the water rising when the water level is at a height of 0.5 m? [*Hint:* 1 L = 1000 cm^3, or 1 m^3 = 1000 L.]

5. An airplane at a height of 1000 m is flying horizontally at a velocity of 500 km/hr and passes directly over a civil defense observer. How fast is the plane receding from the observer when it is 1500 m away from the observer?

6. Sand is being dropped in a conical pile at a rate of 15 m^3/min. The height of the pile is always equal to its diameter. How fast is the height increasing when the pile is 7 m high?

7. When helium expands adiabatically, its pressure is related to its volume by the formula $PV^{1.67}$ = constant. At a certain time the volume of the helium in a balloon is 18 m^3 and the pressure is 0.3 kg/m^2. If the pressure is increasing at a rate of 0.01 kg/m^2/sec, how fast is the volume changing? Is the volume increasing or decreasing?

8. A man standing on a pier 15 ft above the water is pulling in his boat by means of a rope attached to the boat's bow. He can pull in the rope at a rate of 5 ft/min. How fast is the boat approaching the foot of the pier when the boat is 20 ft away?

9. A baseball player can run at a top speed of 25 ft/sec. A catcher can throw a ball at a speed of 120 ft/sec. The player attempts to steal third base. The catcher (who is 90 ft from third base) throws the ball toward the third baseman when the player is 30 ft from third base. What is the rate of change of the distance between the ball and the runner at the instant the ball is thrown? (See Figure 5.)

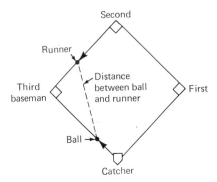

FIGURE 5

10. At 2 P.M. on a certain day ship A is 100 km due north of ship B. At that moment ship A begins to sail due east at a rate of 15 km/hr, while ship B sails due north at a rate of 20 km/hr. How fast is the distance between the two ships changing at 5 P.M.? Is it increasing or decreasing?

11. A storage tank is 20 ft long and its ends are isosceles triangles having bases and altitudes of 3 ft. Water is poured into the tank at a rate of 4 ft^3/min. How fast is the water level rising when the water in the tank is 6 in. deep?

12. Two roads intersect at right angles. A car traveling 80 km/hr reaches the intersection half an hour before a bus that is traveling on the other road at 60 km/hr. How fast is the distance between the car and the bus increasing 1 hr after the bus reaches the intersection? (See Figure 6.)

13. A rock is thrown into a pool of water. A circular wave leaves the point of impact and travels so that its radius increases at a rate of 25 cm/sec. How fast is the circumference of the wave increasing when the radius of the wave is 1 m?

14. The body of a snowman is in the shape of a sphere and is melting at a rate of 2 ft^3/hr. How fast is the radius changing when the body is 3 ft in diameter (assuming that the body stays spherical)?

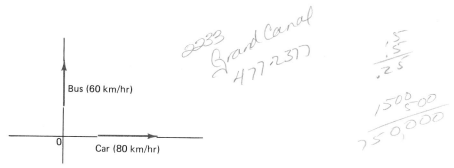

FIGURE 6

15. In Problem 14 how fast is the surface area of the body changing when $d = 3$ ft?

*16. Water is leaking out of the bottom of a hemispherical tank with a radius of 6 m at a rate of 3 m³/hr. If the tank was full at noon, how fast is the height of the water level in the tank changing at 3 P.M.? [*Hint:* The volume of a segment of a sphere of radius r is $\pi h^2[r - (h/3)]$, where h is the height of the segment (see Figure 7).]

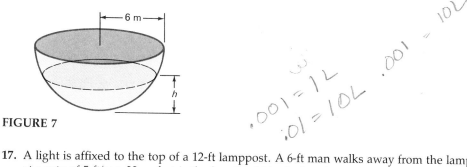

FIGURE 7

17. A light is affixed to the top of a 12-ft lamppost. A 6-ft man walks away from the lamppost at a rate of 5 ft/sec. How fast is the length of his shadow increasing when he is 5 ft away?

18. Bacteria grow in circular colonies. The radius of one colony is increasing at a rate of 4 mm/day. On Wednesday the radius of the colony is 1 mm. How fast is the area of the colony changing one week (i.e., seven days) later?

19. A pill is in the shape of a right circular cylinder with a hemisphere on each end. The height (excluding its hemispherical ends) of the cylinder is half its radius. What is the rate of change of the volume of the pill with respect to the radius of the cylinder when the radius changes?

20. A spherical mothball is dissolving at a rate of 8π cm³/hr. How fast is the radius of the mothball decreasing when the radius is 3 cm?

*21. A particle moves along the parabola $y = x^2$ such that $dx/dt = 3$. Find the rate of change of the distance between the particle and the point (2, 5) when the particle is at $(-1, 1)$; when it is at (3, 9).

*22. In soil mechanics the vertical stress σ_Z in a soil at a point Z feet deep and located r feet (horizontally) from a concentrated surface load of P pounds is given by

$$\sigma_Z = \frac{3P}{2\pi Z^2}\left[1 + \left(\frac{r}{Z}\right)^2\right]^{2.5}.$$

(a) Calculate the vertical stress in a soil at a point 2 ft deep and located 4 ft horizontally from a concentrated surface load of 10,000 lb.

(b) If the load is moved horizontally toward the point at a rate of 1 ft/hr, how is the vertical stress changing when the load is 3 ft from the point?

23. A circular wheel centered at the origin with a radius of 13 cm is rotating in the counterclockwise direction at a constant rate of 10 revolutions per minute (rpm). How fast are x and

y changing when $x = 12$ and $y = 5$? [*Hint:* Express everything in terms of θ, the angle between the radius of the circle and the positive x-axis.]

24. A prison guard tower has a light that is 300 m from a straight wall. Its light revolves at a rate of 5 rpm. How fast is the light beam moving along the wall at a point 300 m along the wall [measured from the point on the wall that is closest (300 m) to the tower]?

25. In Problem 24 how fast is the light moving at a point $300\sqrt{3}$ m along the wall?

26. Two ships I and II meet and then sail off in different directions, keeping an angle of 60° between them (see Figure 8). Ship I travels at 30 km/hr, and ship II travels at 40 km/hr. How fast is the distance between them changing when ship I is 9 km from the meeting point and ship II is 12 km from the meeting point? [*Hint:* Use the law of cosines; see Problem 11, Appendix 1.4.]

FIGURE 8

27. Answer the question of Problem 26 if the angle between the ships' paths is 135°.

28. A bright projection lamp shines from ground level on a vertical screen 25 m away. A man 2 m tall jogs in front of the lamp toward the screen at a constant rate of 3 m/sec. When the man is 15 m from the screen, how fast is the height of his shadow on the screen changing?

*29. A slide projector shines a bright light on a vertical screen 25 m away. A sphere 2 m in diameter is steadily moved in the center of the beam of light toward the screen at a constant rate of 3 m/sec. When the center of the sphere is 15 m from the screen, how fast is the height of its shadow on the screen changing?

30. The amount of fluid flowing across a weir (dam) with a V-shaped notch has been determined to be

$$Q = 2.505\left(\tan \frac{\theta}{2}\right)h^{2.47},$$

where Q is the volume of flow measured in cubic meters per second, h is the height, in meters, of the liquid, measured from the bottom edge of the notch, and θ is the angle of the notch. Assuming that h is kept constant at 2 m while the angle of the weir is increasing, how fast is Q increasing when $\theta = 40°$?

4.2 THE MEAN VALUE THEOREM

This chapter discusses applications. In this section we state and prove one of the most important and applicable theorems of mathematics, the *mean value theorem*. We can use the theorem to prove some important facts about differentiation. We can also use the theorem to obtain information that is useful in graphing a wide variety of functions (as in Section 4.3.) The mean value theorem has many other applications as well. For example, in Chapter 5, our introductory chapter on integral calculus, the mean value theorem will be used to prove another very important theorem—the fundamental theorem of calculus.

Before stating and proving the mean value theorem, we need one preliminary result.

Theorem 1   *Rolle's Theorem†*   Let f be continuous on the closed interval $[a, b]$ and differentiable on the open interval (a, b). If $f(a) = f(b) = 0$, then there exists at least one number c in (a, b) at which

$$f'(c) = 0. \tag{1}$$

REMARK.   The situation described in Rolle's theorem can be depicted as in Figure 1. The proof of the theorem involves describing the situation in the figure.

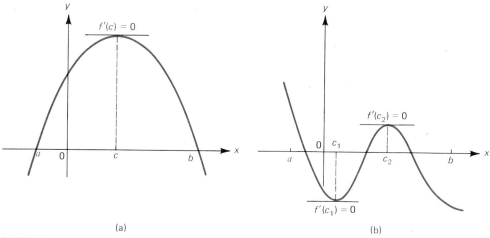

(a)   (b)

FIGURE 1

Proof.   If $f(x) = 0$ for all x in $[a, b]$, then $f'(x) = 0$ in (a, b) and any c in (a, b) will work. If $f(x)$ is not the zero function, then f must become positive or negative in (a, b). Suppose f takes on some positive values in (a, b), as in Figure 1. (Negative values can be handled in the same way.) By Theorem 2.7.4, there is a number x_1 in $[a, b]$ such that $f(x_1) = M$ is the maximum value of $f(x)$ on $[a, b]$. $M > 0$, so x_1 is not equal to a or b since $f(a) = f(b) = 0$ by hypothesis. Therefore since x_1 is in (a, b), $f'(x_1)$ exists [because f is differentiable on (a, b)]. Let us define

$$\epsilon(\Delta x) = \frac{f(x_1 + \Delta x) - f(x_1)}{\Delta x} - f'(x_1). \tag{2}$$

---

†Named after the French mathematician Michel Rolle (1652–1719).

Since

$$f'(x_1) = \lim_{\Delta x \to 0} \frac{f(x_1 + \Delta x) - f(x_1)}{\Delta x}$$

exists, we see that $\epsilon(\Delta x) \to 0$ as $\Delta x \to 0$, and we can rewrite (2) as

$$\frac{f(x_1 + \Delta x) - f(x_1)}{\Delta x} = f'(x_1) + \epsilon(\Delta x). \tag{3}$$

Since f takes its maximum at x_1, $f(x_1 + \Delta x) \leq f(x_1)$, so that

$$f(x_1 + \Delta x) - f(x_1) \leq 0. \tag{4}$$

If $\Delta x > 0$, then from (3) and (4),

$$f'(x_1) + \epsilon(\Delta x) \leq 0, \tag{5}$$

and since $\epsilon(\Delta x) \to 0$ as $\Delta x \to 0$, (5) implies that

$$f'(x_1) \leq 0. \tag{6}$$

Similarly, if $\Delta x < 0$, then from (3),

$$f'(x_1) + \epsilon(\Delta x) \geq 0, \tag{7}$$

which implies that

$$f'(x_1) \geq 0. \tag{8}$$

From (6) and (8) we see that $f'(x_1) = 0$. Setting $c = x_1$ completes the proof. ∎

EXAMPLE 1   Let $f(x) = (x - 1)(x - 2)$. Then since $f(1) = f(2) = 0$, there is a point c in $(1, 2)$ such that $f'(c) = 0$. To find this point, we must differentiate: $f'(x) = (x - 1) + (x - 2) = 2x - 3 = 0$ when $x = \frac{3}{2}$. ∎

EXAMPLE 2   An interesting physical illustration of Rolle's theorem is given by the following situation: Suppose an object (like a ball or rock) is thrown from ground level into the air. Let $s(t)$ denote the height the object reaches at time t. Clearly, $s(0) = 0$ and $s(T) = 0$ at some future time T. (The object hits the ground at time T.) We can then conclude, from Rolle's theorem, that there is a time t_1 such that $s'(t_1) = 0$. That is, there is a time at which the velocity of the object is zero—a physical fact confirmed by experience. ∎

We are now ready to state the mean value theorem.

Theorem 2   *Mean Value Theorem*   Let f be continuous on the closed interval $[a, b]$ and differentiable on the open interval (a, b). Then there exists at least one number c in (a, b) such that

$$f'(c) = \frac{f(b) - f(a)}{b - a}. \tag{9}$$

Before proving the mean value theorem, we describe what it says geometrically. Look at Figure 2. We can see that the expression $[f(b) - f(a)]/[b - a]$ represents the slope of the secant line joining the points $(a, f(a))$ and $(b, f(b))$. Then since $f'(x)$ is the slope of the tangent line to the curve, the mean value theorem states that there is always a number c between a and b such that the slope of the tangent line at the point $(c, f(c))$ is the same as the slope of the secant line; that is, the secant and tangent lines are parallel.

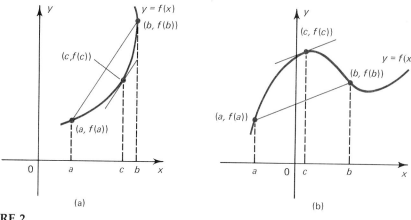

(a)   (b)

FIGURE 2

Proof.   We define the function

$$g(x) = f(x) - \left[\frac{f(b) - f(a)}{b - a}(x - a) + f(a) \right]. \tag{10}$$

The expression

$$y = \frac{f(b) - f(a)}{b - a}(x - a) + f(a)$$

is the equation of a straight line (the secant line discussed above) that passes through the points $(a, f(a))$ and $(b, f(b))$ since when $x = a$, $y = f(a)$, and when $x = b$, $y = [f(b) - f(a)] + f(a) = f(b)$. See Figure 3. The function $g(x)$ represents the vertical difference between the curve $f(x)$ and the secant line. We first observe that since a linear function is everywhere differentiable (being a first-degree polynomial) and since $f(x)$ is continuous on $[a, b]$ and differentiable on (a, b), $g(x)$ is also continuous on $[a, b]$ and differentiable on (a, b). Moreover,

$$g(a) = f(a) - \left[\frac{f(b) - f(a)}{b - a}(a - a) + f(a) \right] = f(a) - f(a) = 0,$$

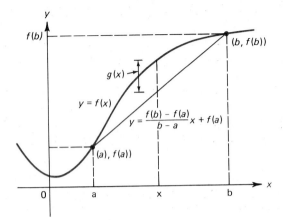

FIGURE 3

and

$$g(b) = f(b) - \left[\frac{f(b) - f(a)}{b - a}(b - a) + f(a)\right]$$

$$= f(b) - [f(b) - f(a) + f(a)] = 0.$$

Thus we may use Rolle's theorem and conclude that there is a number c in (a, b) such that

$$g'(c) = 0.$$

But $g'(x) = f'(x) - \{[f(b) - f(a)]/(b - a)\}$, so that since $g'(c) = 0$,

$$g'(c) = f'(c) - \left[\frac{f(b) - f(a)}{b - a}\right] = 0.$$

That is,

$$f'(c) = \frac{f(b) - f(a)}{b - a},$$

and the proof of the mean value theorem is complete. ∎

EXAMPLE 3   Let $f(x) = x^2$, $a = 2$, and $b = 5$. Then

$$\frac{f(b) - f(a)}{b - a} = \frac{25 - 4}{3} = \frac{21}{3} = 7.$$

By the mean value theorem there is a number c in $(2, 5)$ such that $f'(c) = 7$. Since $f'(x) = 2x$, we have $c = \frac{7}{2}$. (See Figure 4.) ∎

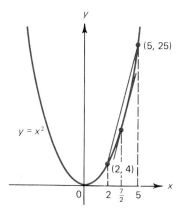

FIGURE 4

EXAMPLE 4   Let $f(x) = \sin x$, $a = 0$, and $b = \pi/2$. Then

$$\frac{\sin b - \sin a}{b - a} = \frac{1}{\pi/2} = \frac{2}{\pi}.$$

There is a number c in $(0, \pi/2)$ such that $f'(c) = \cos c = 2/\pi \approx 0.6366$, and $c \approx 0.8807$. See Figure 5. ∎

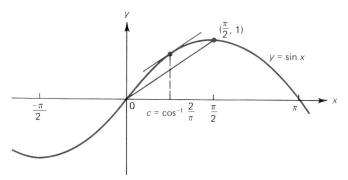

FIGURE 5

EXAMPLE 5   **(The Lorenz Curve)**   We show how the mean value theorem can be used in a practial problem. Consider Table 1, which shows how total income was distributed among Americans in 1955.† In the table the ordering is from smaller incomes to greater incomes. In Figure 6 we have drawn a smooth curve through the points given in Table 1. The curve given in Figure 6 is called a **Lorenz curve.**

Consider the part of the curve between the points $(0.2, 0.05)$ and $(0.4, 0.16)$. We see that the fraction of people between 0.2 and 0.4 (20%) receives $0.11 = 11\%$ of the in-

---

†This table was obtained from P. A. Samuelson, *Economics, An Introductory Analysis*, 4th ed. (McGraw-Hill, New York, 1958), page 69.

TABLE 1

Fraction of people	Fraction of income
0.0	0.0
0.2	0.05
0.4	0.16
0.6	0.33
0.8	0.55
1.0	1.00

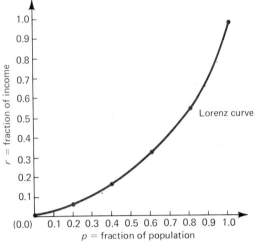

FIGURE 6

come. Thus this part of the population receives less than a fair share (20% of the people should, ideally, receive 20% of the income). The ratio $0.11/0.20 = 0.55$ is a measure of fairness and is easily seen to be equal to the slope of the secant line joining the points $(0.2, 0.05)$ and $(0.4, 0.16)$. Notice that things are better for that 20% of the population between 0.6 and 0.8 because they get $0.22 = 22\%$ $(0.55 - 0.33)$ of the income, with a fairness ratio of $0.22/0.20 = 1.1$. If the Lorenz curve is given by $r = f(p)$, then the slope dr/dp is the "instantaneous" fairness ratio—obtained by letting the percentage of the population (Δp) tend to zero.

   If there is a point on the Lorenz curve with a slope equal to 1, then the p value of that point is called the **equal-share coefficient** (ESC) and is denoted by ϵ. If $\epsilon = 0.55$, for example, it means that 55% of the population have less than an equal share of the income and 45% have more than an equal share.

   It is reasonable to assume that the Lorenz curve is continuous and differentiable. It is also reasonable to assume that the slope of the Lorenz curve is increasing. That is, as we move to the right, equal proportions of the population receive greater proportions of the income (this assumption coincides with the information given in the table). If $r = f(p) = p$, then the slope of the curve is 1, and everyone has an equal share. If $f(p) \neq p$, then by the assumption just stated, $f'(p)$ is an increasing function (as p increases, the fractions of income increase). Thus if there is one value ϵ for which $f'(\epsilon) = 1$, then there is no other value. But $f(0) = 0$, $f(1) = 1$, and by the mean value theorem,

$$1 = \frac{f(1) - f(0)}{1 - 0} = f'(c), \quad \text{where} \quad 0 < c < 1.$$

Thus if $\epsilon = c$, ϵ is the ESC. That is, *every Lorenz curve has an equal-share coefficient.* † ∎

---

†For a much more complete discussion of Lorenz curves and many additional examples, consult Harry M. Schey's paper, "The distribution of resources," UMAP project, Educational Development Center, Newton, Mass. (1978).

EXAMPLE 6 Let $r = p^{3/2}$. Then $dr/dp = \frac{3}{2}p^{1/2}$, which is equal to 1 when $p^{1/2} = \frac{2}{3}$, or $p = \frac{4}{9} \approx 0.44$. T
$\epsilon = 0.44$, which means that 44% of the population receive less than an equal share

We now prove, using the mean value theorem, a theorem stated in Section

Theorem 3 If f is differentiable on (α, β) and if $f'(x) = 0$ for every x, then f
constant function on (α, β).

Proof. Pick any two numbers a and b in (α, β), with $a \neq b$. Then there
number c such that

$$\frac{f(b) - f(a)}{b - a} = f'(c) = 0.$$

Since $b \neq a$, this result can only occur if $f(b) - f(a) = 0$, which implies
$f(b) = f(a)$. Since this result is true for any numbers a and b, f is a const
function. ∎

PROBLEMS 4.2

In Problems 1–11, find a number c that satisfies the conclusion of the mean value theorem
the given function and numbers.

1. $f(x) = x^3$; $a = 1$, $b = 2$

2. $f(x) = \frac{1}{x}$; $a = 1$, $b = 4$

3. $f(x) = \cos x$; $a = 0$, $b = \dfrac{\pi}{2}$

4. $f(x) = \sin 2x$; $a = 0$, $b = \dfrac{\pi}{12}$

5. $f(x) = \tan x$; $a = 0$, $b = \dfrac{\pi}{4}$

6. $f(x) = \sqrt{x}$; $a = 1$, $b = 4$

7. $f(x) = 1 + 2x^2$; $a = -1$, $b = 1$

8. $f(x) = \sqrt[3]{x}$; $a = -8$, $b = -1$

9. $f(x) = x^3 - x$; $a = -1$, $b = 1$

10. $f(x) = (x + 3)(x - 1)(x - 5) + 8$; $a = -3$, $b = 1$

11. $f(x) = \cos(2x) - 2\cos^2 x$; $a = -\dfrac{\pi}{2}$, $b = \dfrac{\pi}{4}$

12. Let $f(x) = |x - 1| - 2$.
  (a) Show that $f(3) = f(-1) = 0$.
  (b) Show that there is no real number c in $(-1, 3)$ such that $f'(c) = 0$.
  (c) Explain why this result does *not* contradict Rolle's theorem.
13. Show that there is no number c that satisfies the mean value theorem in the inter
  $[-1, 1]$ when $f(x) = x^{2/3}$.
14. Let $f(x) = 1/x$. Show that there is no c in the interval $(-1, 2)$ such that

$$f'(c) = \frac{f(2) - f(-1)}{2 - (-1)}.$$

Explain why this result does *not* contradict the mean value theorem.
15. Prove that if the conditions of the mean value theorem hold and if $f(a) = f(b)$, there
  point c in (a, b) such that $f'(c) = 0$.

*16. A function is **one to one** if $f(x) = f(y)$ implies that $x = y$. Use the result of Problem 15 to show that if f is differentiable and $f'(x) \neq 0$ for every real x, then f is one to one.

*17. Let $f(x) = x^3 - 3x^2 + 1$.
   (a) Show that $f(0) = f(3) = 1$.
   (b) Find a number c in $(0, 3)$ such that $f'(c) = 0$.

*18. Let f be differentiable with $f'(x) \leq 2$ for every real number x. Show that there is at most one real number $u \geq 1$ such that $f(u) = u^2$.

*19. Let f be differentiable with $f'(x) \leq 3$ for every real number x. Show that there is at most one real number $u \geq 1$ such that $f(u) = u^3$.

20. A function is called a **contraction** on $[a, b]$ if for every x and y in $[a, b]$,

$$|f(x) - f(y)| \leq \lambda |x - y|,$$

where $\lambda < 1$. Show that if f is differentiable and $|f'(x)| \leq \lambda < 1$ for every x in $[a, b]$, then f is a contraction on $[a, b]$.

21. Show that $f(x) = 1/x$ is a contraction on any interval of the form (a, ∞), where $a > 1$.

22. Show that $f(x) = \sin x$ is not a contraction on $[0, \pi/4]$.

23. For what values of b is the function $f(x) = x^2 + 1$ a contraction on $[0, b]$?

24. Let $P(x) = c_0 + c_1 x + c_2 x^2 + \cdots + c_n x^n$. Prove that if $P'(x_1) = P'(x_2) = 0$ and $P'(x) \neq 0$ for $x_1 < x < x_2$, then there is at most one number x_3 in (x_1, x_2) such that $P(x_3) = 0$.

25. Use the result of Problem 24 to show that the only root of $x^5 - 5x = 0$ in the interval $(-1, 1)$ is $x = 0$.

26. If $f''(x)$ exists for every x, show that between any two points at which $f'(x) = 0$, there is a point at which $f''(x) = 0$.

27. Let

$$f(x) = \begin{cases} x^2, & x \leq 1 \\ ax^3, & x > 1. \end{cases}$$

Is there a choice of α such that the mean value theorem applies to f on the interval $[0, 2]$? Explain.

*28. Consider the parabola $y = f(x) = ax^2 + bx + c$. By the mean value theorem, for any two numbers x_1 and x_2 there is a number x_3 in (x_1, x_2) such that the tangent to the parabola at the point $(x_3, f(x_3))$ is parallel to the secant line joining the points $(x_1, f(x_1))$ and $(x_2, f(x_2))$. Show that $x_3 = (x_1 + x_2)/2$ (see Figure 7).

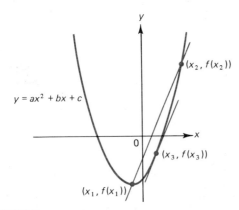

FIGURE 7

*29. Use Rolle's theorem to show that a quadratic polynomial has at most two real roots. [*Hint:* Show that if there are three real roots, then there are two real roots to the equation $P'(x) = 0$ and one root to the equation $P''(x) = 0$, which is impossible.]

*30. Show that a cubic polynomial has at most three real roots. [*Hint:* Follow the steps outlined in Problem 29.]

*31. Use mathematical induction (Appendix 2) to show that an nth degree polynomial has at most n real roots.

32. Use the mean value theorem to show that if f is differentiable and $f'(x) = c$, a nonzero constant, then f is a linear function.

33. Use the mean value theorem to show that if f is differentiable and $f'(x)$ is a linear function, then f is a quadratic polynomial.

34. Use the mean value theorem to show that if f is differentiable and $f'(x)$ is a polynomial of degree n, then $f(x)$ is a polynomial of degree $n + 1$.

35. A Lorenz curve is given by $r = p^n$. Find an expression for the ESC, ϵ_n, in terms of n.

36. Let $r = \frac{1}{2}p^{3/2}(1 + p)$. Compute the ESC to four decimal places.

*37. Let f be differentiable on $[a, b]$. If $f'(a)$ and $f'(b)$ have opposite signs, show that there is at least one number c in (a, b) at which $f'(c) = 0$. Do *not* assume that f' is continuous.

38. Use the result of Problem 37 to show that if f is differentiable on $[a, b]$ and d is any number between $f'(a)$ and $f'(b)$, then there is a number $\bar{x}$ in (a, b) such that $f'(\bar{x}) = d$. Do *not* assume that f' is continuous.

39. Prove **Bernoulli's inequality:** If $\alpha > 1$ and $x > 0$, then

$$(1 + x)^\alpha > 1 + \alpha x.$$

[*Hint:* Define $f(x) = (1 + x)^\alpha - (1 + \alpha x)$. First show that $f(0) = 0$ and then apply the mean value theorem.]

40. Suppose that C and S are two functions such that $C' = -S$ and $S' = C$.
  (a) Prove that $C^2 + S^2$ is constant.
  (b) Prove that $[C(x) - \cos x]^2 + [S(x) - \sin x]^2$ is also constant.
  (c) Suppose $C(0) = 1$ and $S(0) = 0$. Prove that $C(x) = \cos x$ and $S(x) = \sin x$ for all x.

41. Suppose that f has n continuous derivatives on $[a, b]$ and f is equal to zero at $n + 1$ or more points in $[a, b]$. Show that there is at least one number c in (a, b) such that $f^{(n)}(c) = 0$.

*42. Let $P(x) = x^n + ax + b$, $n \geq 0$.
  (a) Show that if n is even, then P cannot have more than two real zeros.
  (b) Show that if n is odd, then P cannot have more than three real zeros.

*43. Pick a real number C. Characterize the set of all polynomial functions P that have the properties (a) $P(1) = C$ and (b) $P(w + x) = P(w) + P(x)$ for all real numbers w and x.

4.3 ELEMENTARY CURVE SKETCHING I: INCREASING AND DECREASING FUNCTIONS AND THE FIRST DERIVATIVE TEST

In Section 2.5 we saw how we could get information about the derivative of a function by looking at the graph of that function and observing where the function is increasing (the tangent line has a positive slope) and decreasing (the tangent line has a negative slope). In Examples 2.5.2 and 2.5.3 we "reversed" the process and showed how information about the derivative of some simple functions could help us to graph those functions. Now that we know how to calculate quite a few derivatives, we will begin the important task of learning how to draw the graphs of a much wider class of functions. In this section we will make use of a theorem about increasing and decreasing functions. This theorem will enable us to get a fairly good picture about the appearance of the graph of the function in question. Later, in Section 4.4 we will use additional facts about derivatives to get even more information about these graphs.

Consider the function whose graph is depicted in Figure 1. In the interval (x_0, x_1) the function increases. In (x_1, x_2) it decreases. From x_2 to x_3 the function again in-

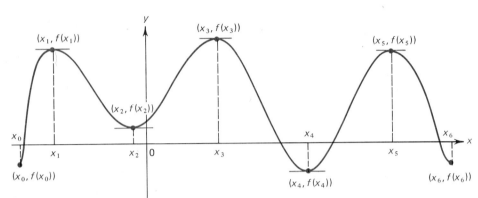

FIGURE 1

creases, and so on. Moreover, when $x = x_1, x_2, x_3, x_4,$ and $x_5,$ the tangent line is horizontal so that the derivative is zero at those points. To be more precise, we have the following definitions.

Definition 1 INCREASING FUNCTION The function f defined on an interval $[a, b]$ is said to be **increasing** on that interval if whenever $a \leq x_1 < x_2 \leq b,$ we have

$$f(x_1) < f(x_2). \tag{1}$$

EXAMPLE 1 In Figure 1 the function is increasing on the intervals $[x_0, x_1], [x_2, x_3],$ and $[x_4, x_5].$ ■

Definition 2 DECREASING FUNCTION The function f defined on an interval $[a, b]$ is said to be **decreasing** on that interval if whenever $a \leq x_1 < x_2 \leq b,$ we have

$$f(x_1) > f(x_2). \tag{2}$$

EXAMPLE 2 In Figure 1 the function is decreasing on the intervals $[x_1, x_2], [x_3, x_4],$ and $[x_5, x_6].$ ■

Definition 3 CRITICAL POINT Let f be defined and continuous on the interval $[a, b]$† and let x_0 be in $(a, b).$ Then the number x_0 is called a **critical point** of f if $f'(x_0) = 0$ *or* if $f'(x_0)$ does not exist.

REMARK. Note that by this definition, a critical point x_0 is *not* an endpoint of the interval $(a, b).$

EXAMPLE 3 In Figure 1 the critical points are $x_1, x_2, x_3, x_4,$ and x_5 because f' evaluated at these points is zero. ■

EXAMPLE 4 Let $f(x) = |x|.$ Then 0 is a critical point since $f'(0)$ does not exist. (See Example 2.5.4.) ■

The critical point x_0 is sometimes called a **critical number.**
In Section 2.5 we inferred that a function was increasing when its derivative was positive and decreasing when its derivative was negative. We now prove it.

---

†See page 116 for the definition of continuity on the closed interval $[a, b].$

Theorem 1 Let f be differentiable for $a < x < b$ and continuous for $a \le x \le b$.

> **(i)** If $f'(x) > 0$ for every x in (a, b), f is increasing on $[a, b]$.
> **(ii)** If $f'(x) < 0$ for every x in (a, b), f is decreasing on $[a, b]$.

Proof. **(i)** Suppose that $a \le x_1 < x_2 \le b$. We must show that $f(x_2) > f(x_1)$. But since f satisfies the hypotheses of the mean value theorem (page 200), we have

$$\frac{f(x_2) - f(x_1)}{x_2 - x_1} = f'(c), \tag{3}$$

where $x_1 < c < x_2$. Thus

$$f(x_2) - f(x_1) = f'(c)(x_2 - x_1). \tag{4}$$

But $f'(c) > 0$ by hypothesis and $x_2 > x_1$, so that from (4) we see that

$$f(x_2) - f(x_1) > 0, \quad \text{or} \quad f(x_2) > f(x_1).$$

(ii) The proof of this part is almost identical to the proof of part (i) except that now $f'(c) < 0$, so that $f(x_2) - f(x_1) < 0$ and $f(x_2) < f(x_1)$. ■

EXAMPLE 5 For what values of x is the function $f(x) = x^2 - 2x + 4$ increasing and decreasing? Sketch the curve.

Solution. $f'(x) = 2x - 2 = 2(x - 1)$. If $x > 1$, $x - 1 > 0$ and $2(x - 1) > 0$. If $x < 1$, $x - 1 < 0$ and $2(x - 1) < 0$. Thus for $x < 1$, f is decreasing, and for $x > 1$, f is increasing. When $x = 1$, $f'(x) = 0$, so that 1 is a critical point. These results are summarized in Table 1. We can use this information to obtain a rough sketch of the function. When $x = 1$, $f(x) = 3$. Table 1 tells us that $f(x)$ decreases until $x = 1$ and increases thereafter. Also, the y-intercept is found by setting $x = 0$ and is at the point $(0, 4)$. We put this information together in Figure 2. We notice that $f(x)$ takes its *minimum value* when $x = 1$. We will shortly see how maximum and minimum values of a function can be found before drawing a sketch of the curve by analyzing critical points more closely. ■

TABLE 1

	$f'(x)$	$f(x)$ is
$-\infty < x < 1$	$-$	decreasing
$x = 1$	0	(critical point)
$1 < x < \infty$	$+$	increasing

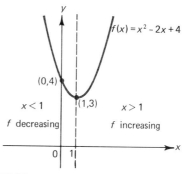

FIGURE 2

EXAMPLE 6   Let $y = x^3 + 3x^2 - 9x - 10$. For what values of x is this function increasing and decreasing? Sketch the curve.

Solution. $dy/dx = 3x^2 + 6x - 9 = 3(x^2 + 2x - 3) = 3(x + 3)(x - 1)$. Now look at Table 2. The critical points are the numbers -3 and 1. At $x = -3$, $y = 17$; at $x = 1$, $y = -15$. Setting $x = 0$, we obtain the y-intercept $y = -10$. Hence the curve passes through the point $(0, -10)$. Using this information, we obtain the graph shown in Figure 3. We notice from the curve that the point $(-3, 17)$ is a maximum point in the sense that "near" $x = -3$, y takes its largest value at $x = -3$. However, there is no *global* (or *absolute*) maximum value for the function since as x increases beyond the value 1, y increases without bound. For example, if $x = 10$, $y = 10^3 + 3 \cdot 10^2 - 9 \cdot 10 - 10 = 1200$, which is much bigger than 17. In this setting the point $(-3, 17)$ is called a *local maximum* (or *relative maximum*) in the sense that the function achieves its maximum value there for points *near* $(-3, 17)$. Similarly, we call the point $(1, -15)$ a *local minimum* (or *relative minimum*). ∎

TABLE 2

	$x + 3$	$x - 1$	$\dfrac{dy}{dx} = 3(x + 3)(x - 1)$	f is
$x < -3$	$-$	$-$	$+$	increasing
$x = -3$	0	-4	0	(critical point)
$-3 < x < 1$	$+$	$-$	$-$	decreasing
$x = 1$	4	0	0	(critical point)
$1 < x$	$+$	$+$	$+$	increasing

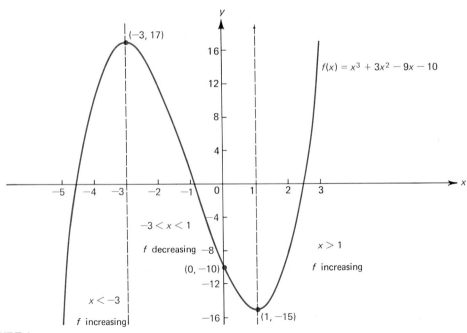

FIGURE 3

In general, we have the following definition.

Definition 4 MAXIMA AND MINIMA

(i) The function f has a **local maximum** at x_0 if there is an open interval (c, d) containing x_0 such that $f(x_0) \geq f(x)$ for every x in (c, d).

(ii) The function f has a **local minimum** at x_0 if there is an open interval (c, d) containing x_0 such that $f(x_0) \leq f(x)$ for every x in (c, d).

(iii) The function f has a **global maximum** at x_0 if $f(x_0) \geq f(x)$ for every x in the domain of f.

(iv) The function f has a **global minimum** at x_0 if $f(x_0) \leq f(x)$ for every x in the domain of f.

NOTE 1.

(i) In Example 6 f has a local, but not a global, maximum at $x = -3$. It also has a local, but not global, minimum at $x = 1$. These results are evident from Figure 3.

(ii) In Example 5 f has a local *and* global minimum at $x = 1$. There is no local or global maximum.

NOTE 2. We should point out that the interval (c, d) in Definition 4 could be very small.

In Figure 4, $f(x)$ has a relative minimum at x_0 and a relative maximum at x_1 even though these are minimum and maximum values of f over very small intervals. This result indicates why calculus methods are superior to other methods for curve sketching. Small "wiggles" such as the two in Figure 4 could easily be missed if we were to try to sketch the curve, say, by plotting points.

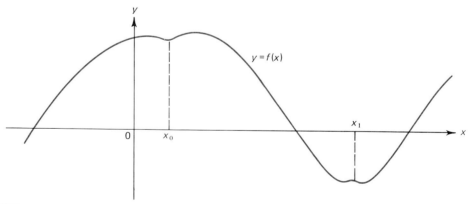

FIGURE 4

As we saw in Examples 5 and 6, whenever we had a local maximum or minimum, we also had a critical point. This is not a coincidence.

Theorem 2 Let f have a local maximum or minimum at x_0. Then x_0 is a critical point of f.

Proof.   Let f have a local maximum or local minimum at x_0. If $f'(x_0)$ does not exist, x_0 is a critical point. Otherwise, we may assume that f has a local maximum or local minimum at x_0 and that $f'(x_0)$ exists. Then it is necessary to show that $f'(x_0) = 0$. But this result follows exactly as in the proof of Rolle's theorem. ■

REMARK.   The converse of this theorem is not true, as the next example shows.

EXAMPLE 7   Let $y = f(x) = x^3$. Then $f'(x) = 3x^2$, which is always positive except at the critical point $x = 0$. If $x < 0$, then $f(x) < 0$; if $x > 0$, then $f(x) > 0$. So the function increases from negative values to zero to positive values at $x = 0$ (see Figure 5) and has neither a local maximum nor a local minimum there. Thus, as this example shows, a critical point may be neither a local maximum nor a local minimum. ■

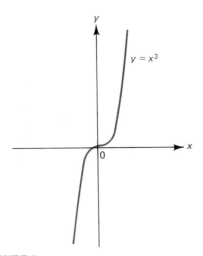

FIGURE 5

Examples 5, 6, and 7 taken together illustrate the fact that at a critical point a function may have a local maximum, a local minimum, or neither. There are two ways to determine when a critical point is a local maximum or minimum. The first of these, called the **first derivative test,** is given in the next theorem. The second is given in the next section.

Theorem 3   *The First Derivative Test*   Let x_0 be a critical point of a function f with x_0 in an open interval (a, b). Suppose that f is continuous for $a \le x \le b$ and differentiable for $a < x < b$, except possibly at x_0 itself.

> **(i)** If $f'(x) > 0$ for $a < x < x_0$ and $f'(x) < 0$ for $x_0 < x < b$, f has a local maximum at x_0.
>
> **(ii)** If $f'(x) < 0$ for $a < x < x_0$ and $f'(x) > 0$ for $x_0 < x < b$, f has a local minimum at x_0.
>
> **(iii)** If $f'(x) < 0$ for $a < x < b$ or $f'(x) > 0$ for $a < x < b$ (except possibly at x_0 itself), f has neither a local maximum nor a local minimum at x_0.

Property (i) says that if f increases to the left of x_0 and decreases to the right of x_0, then f has a local maximum at x_0. Property (ii) says that if f decreases to the left of x_0 and increases to the right of x_0, then f has a local minimum at x_0.

The first derivative test is illustrated in Figure 6.

(i) Local maximum at x_0 (ii) Local minimum at x_0

(iii) Neither local maxima nor local minimum at x_0

FIGURE 6

Proof.

(i) The conditions given imply, by Theorem 1, that f is increasing on $[a, x_0]$ and decreasing on $[x_0, b]$. Thus if $a < x < x_0$, $f(x) < f(x_0)$; if $x_0 < x < b$, then $f(x_0) > f(x)$. Thus by Definition 4, part (i), f has a local maximum at x_0.

(ii) The proof of (ii) is virtually identical to the proof of (i) and is left to the reader.

(iii) Assume that $f'(x) > 0$ in (a, b) (the proof for $f'(x) < 0$ is nearly identical). Then f is increasing in (a, b). Let x be in (a, b). If $x < x_0$, then $f(x) < f(x_0)$; if $x > x_0$, then $f(x) > f(x_0)$. Thus f cannot have either a local maximum or a local minimum at x_0. ∎

This theorem is illustrated again in Figure 7, in which we have drawn three typical tangent lines in each of four cases. We see that f has a local maximum at the critical point x_0 if the curve lies below the tangent lines near that point. Similarly, f has a local minimum if f lies above the tangent lines near that point. Finally, if neither of these "pictures" is valid near x_0, then x_0 has neither a local maximum nor a local minimum.

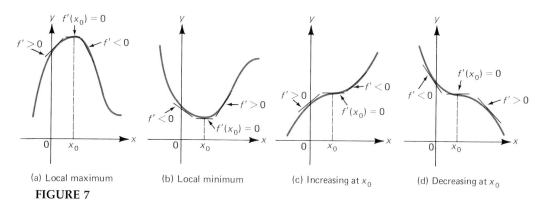

(a) Local maximum   (b) Local minimum   (c) Increasing at x_0 (d) Decreasing at x_0

FIGURE 7

EXAMPLE 5   **(Continued)**   Here $f(x) = x^2 - 2x + 4$ and $f'(x) = 2x - 2 = 2(x - 1)$. Since $f'(x) < 0$ for $x < 1$, $f'(x) > 0$ for $x > 1$, and $f'(1) = 0$, f has a local minimum at $x = 1$ (which is also a global minimum). ∎

$$3(x^2 - 2x - 3) \quad 3(x + 3)(x - 1)$$

EXAMPLE 6   **(Continued)**   Here $f'(x) = 3x^2 - 6x - 9 = 3(x + 3)(x - 1)$. We can use Table 2 to describe the nature of the critical points $x = -3$ and $x = 1$. In $(-4, -2)$, for example, $f'(x) > 0$ for $x < -3$ and $f'(x) < 0$ for $x > -3$. Thus f has a local maximum at $x = -3$. Analogously, in $(0, 2)$, $f'(x) < 0$ for $x < 1$ and $f'(x) > 0$ for $x > 1$. Thus f has a local minimum at $x = 1$. ■

REMARK.   In this example there was nothing special about the intervals $(-4, -2)$ and $(0, 2)$. Any intervals for which the signs of the derivatives went from positive to negative or negative to positive would work.

EXAMPLE 7   **(Continued)**   Here $f'(x) = 3x^2$, which is positive except at $x = 0$. Thus f has neither a local maximum nor a local minimum at 0. ■

EXAMPLE 8   Sketch the curve $y = x^4 - 8x^2$.

Solution. $dy/dx = 4x^3 - 16x = 4x(x^2 - 4) = 4x(x + 2)(x - 2)$. The critical points are $x = 0$ $(y = 0)$, $x = -2$ $(y = -16)$, and $x = 2$ $(y = -16)$. The behavior of the function is given in Table 3. When $x = 0$, $y = 0$, so the curve passes through the origin. Moreover, we can find the x-intercepts—the points on the curve and on the x-axis—by setting $y = 0$ to obtain $0 = x^4 - 8x^2 = x^2(x^2 - 8)$. This expression is equal to zero when $x = 0$ or when $x = \pm\sqrt{8}$. Using all this information, we obtain the curve drawn in Figure 8. ■

TABLE 3

x	$x + 2$	$x - 2$	$f'(x) =$ $4x(x + 2)(x - 2)$	f is
$x < -2$	$-$	$-$	$-$	$-$	decreasing
$x = -2$	$-$	0	$-$	0	(critical point—local minimum)
$-2 < x < 0$	$-$	$+$	$-$	$+$	increasing
$x = 0$	0	$+$	$-$	0	(critical point—local maximum)
$0 < x < 2$	$+$	$+$	$-$	$-$	decreasing
$x = 2$	$+$	$+$	0	0	(critical point—local minimum)
$2 < x$	$+$	$+$	$+$	$+$	increasing

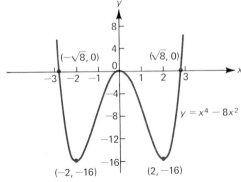

FIGURE 8

EXAMPLE 9   Let $f(x) = x + (1/x)$. When is this function increasing and decreasing? Sketch the curve.

Solution.

$$f'(x) = \frac{d}{dx}x + \frac{d}{dx}x^{-1} = 1 - x^{-2} = 1 - \frac{1}{x^2} = \frac{x^2}{x^2} - \frac{1}{x^2} = \frac{x^2 - 1}{x^2}.$$

First, we notice that the denominator is always positive (for $x \neq 0$). The derivative is not defined for $x = 0$. Thus $f'(x) < 0$ if $x^2 < 1$, and $f'(x) > 0$ if $x^2 > 1$ (see Table 4). Note that 0 is *not* a critical point of f because f is not defined at 0.

TABLE 4

	x^2	$x^2 - 1$	$f'(x) = \dfrac{x^2 - 1}{x^2}$	f is
$x < -1$	>1	$+$	$+$	increasing
$x = -1$	1	0	0	(critical point—local maximum)
$-1 < x < 0$	<1	$-$	$-$	decreasing
$0 < x < 1$	<1	$-$	$-$	decreasing
$x = 1$	1	0	0	(critical point—local minimum)
$1 < x$	>1	$+$	$+$	increasing

To draw the graph, we need more information. We have

$$\lim_{x \to -\infty} \left(x + \frac{1}{x} \right) = \lim_{x \to -\infty} x + \lim_{x \to -\infty} \frac{1}{x} = -\infty,$$

since the second limit is zero. Also,

$$\lim_{x \to \infty} \left(x + \frac{1}{x} \right) = \lim_{x \to \infty} x + \lim_{x \to \infty} \frac{1}{x} = \infty.$$

These equations tell us what happens as x gets large in the positive or negative direction. But what happens near zero? In Example 2.4.5, we saw that $\lim_{x \to 0^+}(1/x) = \infty$ and $\lim_{x \to 0^-}(1/x) = -\infty$. Also, $\lim_{x \to 0}[(x^2 - 1)/x^2] = -\infty$, so the tangent line to the curve becomes vertical as $x \to 0$, since its slope approaches ∞. Finally, as $x \to \infty$, $1/x \to 0$, so that $x + (1/x)$ gets very close to the line $y = x$. This line is called an *asymptote* for the function $y = x + (1/x)$. We now have enough information to sketch the curve (see Figure 9). Here the critical points are the points $(-1, -2)$ and $(1, 2)$. The first is a local maximum and the second is a local minimum. ▪

Definition 5   ASYMPTOTE   An **asymptote** to a curve is a line that the curve approaches as x approaches some fixed number, $+\infty$, or $-\infty$.

In the last example the line $y = x$ is an asymptote since $f(x) - x \to 0$ as $x \to \infty$. Also, the y-axis (the line $x = 0$) is an asymptote since as $x \to 0$ (from the right or left),

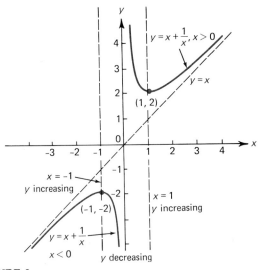

FIGURE 9

the slope of the tangent to the graph of the function approaches $\pm\infty$ and so approaches the vertical line $x = 0$. We will discuss asymptotes in greater detail in the next section.

EXAMPLE 10 Sketch the curve $y = \sin x$.

  Solution. Since $\sin x$ is periodic of period 2π, if we can sketch $\sin x$ for $0 \le x \le 2\pi$, we can obtain the graph of $\sin x$ by extending our sketch. Now $dy/dx = \cos x$. Moreover, $\cos x$ is positive in the first and fourth quadrants, negative in the second and third quadrants, and equal to zero at $\pi/2$ and $3\pi/2$. This information is sum-

TABLE 5

	$f'(x) = \cos x$	$f(x) = \sin x$ is
$0 < x < \pi/2$	$+$	increasing
$x = \pi/2$	0	(critical point—local maximum)
$\pi/2 < x < 3\pi/2$	$-$	decreasing
$x = 3\pi/2$	0	(critical point—local minimum)
$3\pi/2 < x < 2\pi$	$+$	increasing

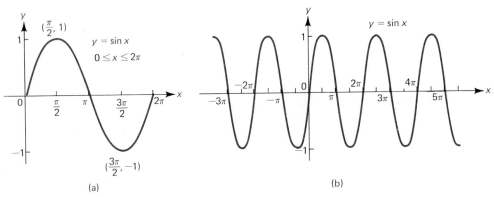

FIGURE 10

marized in Table 5. In addition, $\sin x = 0$ when $x = 0$, π, and 2π. This is all the information we need to obtain the graph of the curve. The sketch in the interval $[0, 2\pi]$ is given in Figure 10a and the extended sketch is given in Figure 10b. ■

PROBLEMS 4.3

In Problems 1–24, (a) find the intervals over which the given function is increasing or decreasing; (b) find all critical points; (c) find the y-intercept (and the x-intercepts if convenient); (d) locate all local maxima and minima by using the first derivative test; (e) sketch the curve.

1. $y = x^2 + x - 30$
2. $y = -x^2 + 2x - 4$
3. $y = 4x^2 - 8$
4. $y = -\frac{3}{2}x^2 + 4x - 7$
5. $y = 2x^3 - 3$
6. $y = x^3 + x$
7. $y = x^3 + x^2$
8. $y = x^3 - x$
9. $y = x^3 - x^2$
10. $y = x^3 - 12x + 10$
11. $y = x^3 - 3x^2 - 45x + 25$
12. $y = 4x^3 - 3x + 2$
13. $y = x^4$
14. $y = x^4 - 18x^2$
15. $y = x^4 - 4x^3 + 4x^2 + 1$
16. $y = x^5$
17. $y = x^3 + 3$
18. $y = x^4 + 1$
19. $y = (x + 1)^{1/3}$
20. $y = (x - 2)^{2/3}$
21. $y = \cos x$
22. $y = \sin 4x$
23. $y = \cos(x/2)$
24. $y = \sin(2x - 5)$

25. Let $f(x) = \sqrt{x + 1}$.
  (a) Show that f is increasing wherever it is defined.
  (b) Show that there are no points on the curve for which $f'(x) = 0$.
  (c) Find the x- and y-intercepts.
  (d) Sketch the curve. What happens as x gets close to -1?
26. Answer the questions of Problem 25 for the function $y = \sqrt[4]{x + 1}$.
27. Answer the questions of Problem 25 for the decreasing function $y = 1/\sqrt{x + 1}$.
28. Sketch the curve $y = x^2 - (1/x^2)$. Be careful near $x = 0$. What function does this curve resemble for large values of x?
29. Sketch the curve $y = (x - 1)/(x - 2)$. Be careful near $x = 2$.
30. Consider the function $y = x/(1 + x)$.
  (a) Show that this function is always increasing except at $x = -1$, where it is not defined.
  (b) What happens as $x \to \pm\infty$?   **(c)** What happens as $x \to -1$?
  (d) Show that $y < 1$ if $x > -1$.   **(e)** Show that $y > 1$ if $x < -1$.
  (f) Sketch the graph.
*31. Suppose that f and g are increasing functions on the interval $[a, b]$.
  (a) Is $f + g$ increasing on $[a, b]$? Explain.
  (b) Is $f \cdot g$ increasing on $[a, b]$? Explain.
  (c) Suppose h is an increasing function on $[0, 1]$ such that $a \le h(0)$ and $h(1) \le b$. Is $f \circ h$ increasing on $[a, b]$? Explain.
32. Let $D(x) = \sqrt{(x - 4)^2 + (3x - 1)^2}$ and $S(x) = (x - 4)^2 + (3x - 1)^2$. Prove the equivalence of the following statements.
  (a) The global minimum of D occurs at $x = c$.
  (b) The global minimum of S occurs at $x = c$.
  Note the practical use of this theoretical result: S is easier to differentiate.
*33. Suppose that $ax^2 + 2bx + c \ge 0$ for each real number x. Show that $b^2 \le ac$.
34. Show that $f(x) = [1/(x + 1) + 1/(x - 1)]$ is decreasing on the interval $(-1, 1)$.
*35. Suppose that $f(x) = (x + 1)^p \cdot (x - 1)^q$, where p and q are integers greater than or equal to 2.
  (a) Show that f' is zero if $x = -1$, $(p - q)/(p + q)$, or 1.
  (b) Describe the local extrema of f. [Note: There are four different cases to analyze, depending on whether p is odd or even and whether q is odd or even.]
*36. Let $f(x) = ax^3 + bx^2 + cx + d$. Prove that f has a local extremum someplace in $(-\infty, \infty)$ if and only if $b^2 > 3ac$.

4.4 ELEMENTARY CURVE SKETCHING II: CONCAVITY AND THE SECOND DERIVATIVE TEST

In this section we continue the discussion of curve sketching begun in Section 4.3. Consider the graph of $f(x) = x^2 - 2x + 4$ given in Figure 1a (see Figure 4.3.2). For $x < 1$, $f'(x) < .0$; at $x = 1$, $f'(x) = 0$; and for $x > 1$, $f'(x) > 0$. Thus the derivative function f' *increases* continuously from negative to positive values. When this situation occurs, the function is said to be *concave up*. Functions that are concave up have the shape of the curve in Figure 1a or 1b.

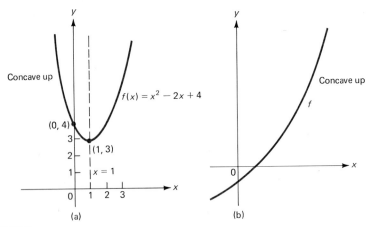

FIGURE 1

Now consider the function $y = -x^2$ whose graph is given in Figure 2a. For $x < 0$, $f'(x) > 0$; at $x = 0$, $f'(x) = 0$; and for $x > 0$, $f'(x) < 0$. Thus the derivative decreases continuously from positive to negative values, and in this case the function is called *concave down*. Functions that are concave down have the shape of the curve in Figure 2a or 2b.

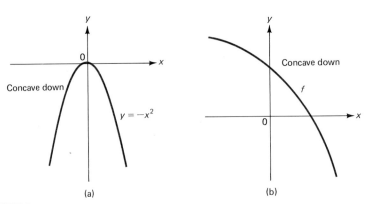

FIGURE 2

The more precise definition of this concept follows.

Definition 1 CONCAVITY Let f be differentiable in the interval (a, b).

(i) f is **concave up** in (a, b) if over that interval f' is an increasing function.
(ii) f is **concave down** in (a, b) if over that interval f' is a decreasing function.

REMARK.   In this definition it is essential that f' exist in (a, b). For example, the function $f(x) = |x|$ has a derivative that goes from negative values (-1) to positive values $(+1)$; yet it has "no concavity" in any interval containing 0.

Many functions will alternate between concave up and concave down.

Definition 2   POINT OF INFLECTION   The point $(x_0, f(x_0))$ is a **point of inflection** of f if f changes its direction of concavity at x_0. That is, it is a point of inflection if there is an interval (c, d) containing x_0 such that one of the following hold:

(i) f is concave down in (c, x_0) and concave up in (x_0, d).
(ii) f is concave up in (c, x_0) and concave down in (x_0, d).

REMARK.   As we will see, f may have a point of inflection at $(x_0, f(x))$ in one of the following two cases:

(i) $f''(x_0) = 0$.
(ii) $f''(x_0)$ does not exist.

EXAMPLE 1   Consider the function graphed in Figure 3. The function is concave down in the intervals (x_0, x_1), (x_2, x_3), and (x_4, x_5) and concave up in the intervals (x_1, x_2) and (x_3, x_4). There are points of inflection at $x_1, x_2, x_3,$ and x_4. ■

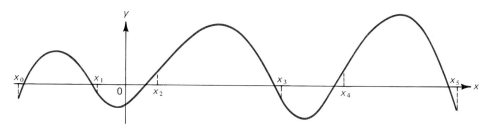

FIGURE 3

Another way to think of concavity is suggested by Figure 4. Here we have drawn in some tangent lines. In Figure 4a, f is concave up and all points on the curve lie *above* the tangent lines. In 4b, f is concave down and all points on the curve lie *below* the

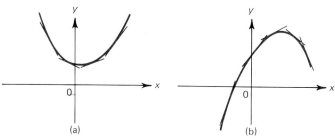

(a)   (b)

FIGURE 4

tangent lines. This information is very useful in curve plotting. If f is concave down in (a, b), then the type of behavior exhibited in Figure 5 is impossible. At the point c the curve lies above the tangent line, but this result is impossible since f is concave down. In general, if a curve is concave up or down in a certain interval, then it cannot "wiggle around" in that interval (i.e. it cannot behave like the curve in Figure 5).

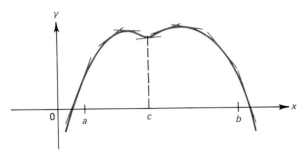

FIGURE 5

Before looking at other examples, let us make some observations. The derivative of f' is f''. If $f'' > 0$, then f' has a positive derivative, and so we can conclude that f' is increasing and f is concave up. If $f'' < 0$, then f' is decreasing and f is concave down. Thus if f'' exists, *we can determine the direction of concavity by simply looking at the sign of the second derivative.* We have therefore proved the following theorem.

Theorem 1.   Suppose that $f''(x)$ exists on the interval (a, b).

(i) If $f''(x) > 0$ in (a, b), f is concave up in (a, b).
(ii) If $f''(x) < 0$ in (a, b), f is concave down in (a, b).

EXAMPLE 2   Let $f(x) = x^3 - 2x$. Then $f'(x) = 3x^2 - 2$ and $f''(x) = 6x$. We see that $f''(x) < 0$ if $x < 0$ and $f''(x) > 0$ if $x > 0$. Thus $(0, 0)$ is a point of inflection. Note that $f''(0) = 0$. The function in sketched in Figure 6. ■

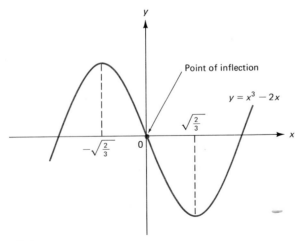

FIGURE 6

EXAMPLE 3   Let $f(x) = x^{3/5}$. Then $f'(x) = \frac{3}{5}x^{-2/5}$ and $f''(x) = -\frac{6}{25}x^{-7/5}$. Here $f''(x) < 0$ for $x > 0$ and $f''(x) > 0$ for $x < 0$. Thus $(0, 0)$ is a point of inflection. Note that both $f'(x)$ and $f''(x)$ become infinite as $x \to 0$; that is, neither $f'(0)$ nor $f''(0)$ exists. The function is sketched in Figure 7. ∎

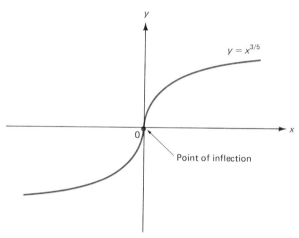

FIGURE 7

Now suppose that f has a point of inflection at $(x_0, f(x_0))$. That means that at x_0, f' goes from increasing to decreasing (Figure 8a) or from decreasing to increasing (Figure 8b). In the first case, f' has a local maximum at x_0, and in the second case f' has a local minimum at x_0. In either case, by Theorem 4.3.2 $(f')' = f'' = 0$ at x_0. This implies the following result.

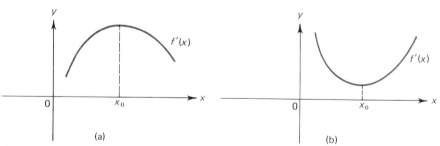

FIGURE 8

Theorem 2.   At a point of inflection $(x_0, f(x_0))$, if $f''(x_0)$ exists, then $f''(x_0) = 0$.

NOTE.   The converse of this theorem is not true in general. That is, if $f''(x_0) = 0$, then x_0 is not necessarily a point of inflection. For example, if $f(x) = x^4$, then $f''(x) = 12x^2 > 0$ for $x \neq 0$. Thus f does not change concavity at 0, so 0 is *not* a point of inflection even though $f''(0) = 0$. (See Problem 64)

Finally, suppose that $f'(x_0) = 0$. If $f''(x_0) > 0$, then the curve is concave up at x_0. A glance at Figure 9a suggests that f has a local minimum at x_0. Similarly, if $f'(x_0) = 0$ and $f''(x_0) < 0$, then f has a local maximum at x_0. It is not difficult to prove these statements (see Problems 61 and 62). In sum, we have the following theorem.

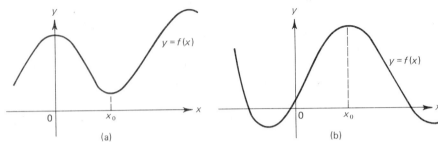

FIGURE 9

Theorem 3 *Second Derivative Test for a Local Maximum or Minimum* Let f be differentiable on an open interval containing x_0 and suppose that $f''(x_0)$ exists.

(i) If $f'(x_0) = 0$ and $f''(x_0) > 0$, f has a local minimum at x_0.
(ii) If $f'(x_0) = 0$ and $f''(x_0) < 0$, f has a local maximum at x_0.

REMARK. If $f'(x_0) = 0$ and $f''(x_0) = 0$, then f may have a local maximum, a local minimum, or neither at x_0. For example, $y = x^3$ has neither at $x = 0$, while $y = x^4$ has a minimum at 0, and $y = -x^4$ has a maximum at 0.

We now give some examples of the technique of curve sketching, making use of information derived from the second derivative.

EXAMPLE 4 If $f(x) = x^2 - 2x + 4$, then $f'(x) = 2x - 2$ and $f''(x) = 2$, which is greater than 0. Thus the curve is concave up for $-\infty < x < \infty$. This result justifies the accuracy of the graph in Figure 1a. The function f has a local (and global) minimum at the point $(1, 3)$. ∎

EXAMPLE 5 If $f(x) = -x^2$, then $f''(x) = -2$, which is less than 0. This curve is concave down for $-\infty < x < \infty$. This result justifies Figure 2a. The function f has a local (and global) maximum at the point $(0, 0)$. ∎

EXAMPLE 6 Let $f(x) = 2x^3 - 3x^2 - 12x + 5$. Sketch the curve.

Solution. $f'(x) = 6x^2 - 6x - 12 = 6(x^2 - x - 2) = 6(x - 2)(x + 1)$, and

$$f''(x) = 12x - 6 = 6(2x - 1).$$

The curve is

concave down if $\quad x < \tfrac{1}{2} \quad$ and $\quad$ concave up if $\quad x > \tfrac{1}{2}.$

The point $x = \tfrac{1}{2}$ $(y = -\tfrac{3}{2})$ is a point of inflection. The critical points are $x = 2$ $(y = -15)$ and $x = -1$ $(y = 12)$. When $x = 2$, $f'' = 6(4 - 1) = 18 > 0$, so that f has a local minimum at $(2, -15)$. When $x = -1$, $f'' = -18 < 0$, so that f has a local maximum at $(-1, 12)$. From an examination of f' we see that f increases up to $x = -1$, decreases for $-1 < x < 2$, and increases thereafter. When $x = 0$, $y = 5$, so the y-intercept is the point $(0, 5)$. We present some of this information in Table 1.

Putting this information all together, we obtain the curve in Figure 10. From this sketch we can see that the graph crosses the x-axis at three places (the three x-intercepts). This result tells us that the cubic equation:

TABLE 1

Critical point x_0	$f''(x_0) = 6(2x - 1)$	Sign of $f''(x_0)$	At x_0 f has a
2	18	+	local minimum
-1	-18	$-$	local maximum

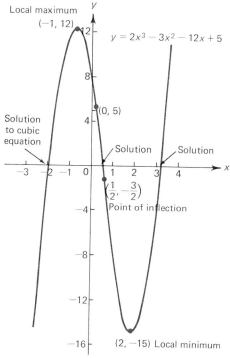

FIGURE 10

$$2x^3 - 3x^2 - 12x + 5 = 0$$

has three real solutions. Moreover, we see that two of these solutions are positive and one is negative. This information, which comes without additional work, can often be very useful. ∎

EXAMPLE 7   Graph the curve $y = x^4 - 4x^3 + 4x^2 + 1$.

Solution. We have

$$\frac{dy}{dx} = 4x^3 - 12x^2 + 8x = 4x(x^2 - 3x + 2) = 4x(x - 1)(x - 2),$$

and

$$\frac{d^2y}{dx^2} = 12x^2 - 24x + 8 = 4(3x^2 - 6x + 2).$$

The second derivative is zero when $3x^2 - 6x + 2 = 0$, or

Quadratic formula

$$x = \frac{6 \pm \sqrt{36 - 24}}{6} = \frac{6 \pm \sqrt{12}}{6} = \frac{3 \pm \sqrt{3}}{3} = 1 \pm \frac{\sqrt{3}}{3} = 1 \pm \frac{1}{\sqrt{3}}.$$

If $x < 1 - (1/\sqrt{3})$, then $d^2y/dx^2 > 0$. If $x > 1 + (1/\sqrt{3})$, then $d^2y/dx^2 > 0$ also. If $1 - (1/\sqrt{3}) < x < 1 + (1/\sqrt{3})$, then $d^2y/dx^2 < 0$. To see this, simply plug in some sample points (try the points $x = 0$, $x = 2$, and $x = 1$). Hence the curve is

concave up for $\quad x < 1 - \dfrac{1}{\sqrt{3}},$

concave down for $\quad 1 - \dfrac{1}{\sqrt{3}} < x < 1 + \dfrac{1}{\sqrt{3}},$

concave up for $\quad x > 1 + \dfrac{1}{\sqrt{3}}.$

The points of inflection are the points $x = 1 - (1/\sqrt{3})$ ($y = \frac{13}{9}$) and $x = 1 + (1/\sqrt{3})$ ($y = \frac{13}{9}$). The critical points are $x = 0$ ($y = 1$), $x = 1$ ($y = 2$), and $x = 2$ ($y = 1$). When $x = 0$, $d^2y/dx^2 = 8$, so there is a local minimum at $(0, 1)$. Similarly, there is a local maximum at $(1, 2)$ and a local minimum at $(2, 1)$. This information is summarized in Table 2. Putting all this information together, we obtain the graph in Figure 11. Since the graph *never* crosses the x-axis, we see immediately that the quartic equation

$$x^4 - 4x^3 + 4x^2 + 1 = 0$$

has *no* real solutions.

TABLE 2

Critical point x_0	$f''(x_0) = 4(3x^2 - 6x + 2)$	Sign of $f''(x_0)$	At x_0 f has a
0	8	+	local minimum
1	-4	$-$	local maximum
2	8	+	local minimum

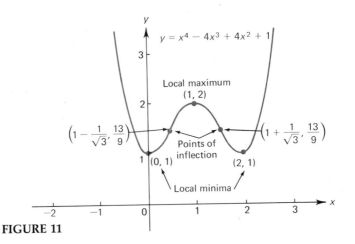

FIGURE 11

The next example shows that in some cases the first derivative test is more useful than the second derivative test.

EXAMPLE 8   Sketch the curve $y = f(x) = x^{2/3}$.

Solution. Since $f'(x) = \frac{2}{3}x^{-1/3} = 2/(3x^{1/3})$, the only critical point is at $x = 0$, since $f'(0)$ is not defined although $f(0)$ is defined. Also, $f''(0)$ is not defined [since $f'(0)$ isn't defined], so the second derivative test won't work. However, since $f'(x) < 0$ for $x < 0$ and $f'(x) > 0$ for $x > 0$, we see that 0 is a local minimum. (It is, in fact, a global minimum.) Also, when $x \neq 0$, $f''(x) < 0$, so the curve is concave down over any interval not containing 0. It is sketched in Figure 12.

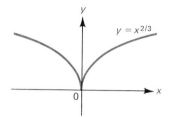

FIGURE 12

NOTE.   The graph of $x^{2/3}$ is said to have a **cusp** at $x = 0$. ∎

We now turn to a different kind of example.

EXAMPLE 9   Let $f(x) = x/(1 + x)$. Graph the function.

Solution. Before doing any calculations, we first note that we will have problems at $x = -1$ since the function is not defined there. We can expect that the function will "blow up" as x approaches -1 from the right or left. Now

$$f'(x) = \frac{d}{dx}\left(\frac{x}{1 + x}\right) = \frac{(1 + x)1 - x}{(1 + x)^2} = \frac{1}{(1 + x)^2}.$$

Therefore f is always increasing (except at $x = -1$, where it is not defined). Also, f has *no* critical points (-1 is *not* a critical point because f is not defined at -1). Now

$$f''(x) = \frac{d}{dx}(1 + x)^{-2} = -\frac{2}{(1 + x)^3}.$$

If $x < -1$, then $1 + x < 0$, $(1 + x)^3 < 0$, and $-2/(1 + x)^3 > 0$. If $x > -1$, then $1 + x > 0$, $(1 + x)^3 > 0$, and $-2/(1 + x)^3 < 0$. Hence f is concave up if $x < -1$ and concave down if $x > -1$.
Observe that

$$\lim_{x \to \infty} \frac{x}{1 + x} = \lim_{x \to \infty} \frac{1}{(1/x) + 1} = 1$$

and

$$\lim_{x \to -\infty} \frac{x}{1 + x} = \lim_{x \to -\infty} \frac{1}{(1/x) + 1} = 1.$$

Therefore as $x \to \pm\infty$, $f(x) \to 1$. The line $y = 1$ is called a **horizontal asymptote** for the function $x/(1 + x)$.

We also observe the following:

If $x < -1$, then $f(x) > 1$ (a negative over a negative, with the numerator more negative than the denominator).

If $-1 < x < 0$, then $f(x) < 0$.

If $x = 0$, then $f(x) = 0$.

If $x > 0$, then $0 < f(x) < 1$ (denominator larger than the numerator).

Also, $\lim_{x \to -1+}[x/(1 + x)] = -\infty$ (since the numerator is negative and the denominator is positive), and $\lim_{x \to -1-}[x/(1 + x)] = \infty$. We next compute

$$\lim_{x \to \pm\infty} f'(x) = \lim_{x \to \pm\infty} \frac{1}{(1 + x)^2} = 0,$$

so the tangent lines become flat (horizontal) as $x \to \infty$. Finally,

$$\lim_{x \to -1} f'(x) = \lim_{x \to -1} \frac{1}{(1 + x)^2} = \infty,$$

so the tangent lines become vertical as $x \to -1$ (from either side). The line $x = -1$ is called a **vertical asymptote**. We put this information all together in Figure 13. ■

FIGURE 13

EXAMPLE 10 Sketch the curve $y = \sec x = 1/\cos x$.

Solution.

Equation (3.5.12)

$$\frac{dy}{dx} = \sec x \tan x = \frac{1}{\cos x}\frac{\sin x}{\cos x} = \frac{\sin x}{\cos^2 x}.$$

Since $\sin 0 = \sin \pi = \sin 2\pi = 0$, we see that $x = 0$, $x = \pi$, and $x = 2\pi$ are critical points in the interval $[0, 2\pi]$. This is the only interval we need to consider because $1/\cos x$, like $\cos x$, is periodic of period 2π. Moreover, since $\cos(\pi/2) = \cos(3\pi/2) = 0$, the lines $x = \pi/2$ and $x = 3\pi/2$ are vertical asymptotes.

Next, we compute

$$\frac{d^2y}{dx^2} = \frac{\cos^2 x(\cos x) - \sin x(2\cos x)(-\sin x)}{\cos^4 x}$$

$$= \frac{\cos^3 x + 2\cos x \sin^2 x}{\cos^4 x} = \frac{\cos^2 x + 2\sin^2 x}{\cos^3 x}.$$

Since the numerator is always positive, the sign of $\cos x$ gives us the sign of the second derivative. We summarize this information in Table 3.

Observing that $\cos x > 0$ for $0 < x < \pi/2$ and $3\pi/2 < x < 5\pi/2$, and $\cos x < 0$ for $\pi/2 < x < 3\pi/2$, we obtain the sketch in Figure 14. Note that there are no points of inflection at $x = \pi/2$ and $x = 3\pi/2$ because f is not defined at these numbers. ■

We now review the kinds of information that are available to help us sketch curves. The following steps should be carried out in order to obtain an accurate picture.

TABLE 3

$f(x) = \sec x = \dfrac{1}{\cos x}$			$f'(x) = \dfrac{\sin x}{\cos^2 x}$	$f''(x) = \dfrac{\cos^2 x + 2\sin^2 x}{\cos^3 x}$	
Interval or point	f'	f	Interval or point	f''	f
0	0	local minimum	$0 \le x < \dfrac{\pi}{2}$	$+$	concave up
$\begin{aligned}0 < x < \pi/2 \\ \pi/2 < x < \pi\end{aligned}$	$+$	increasing	$x = \dfrac{\pi}{2}$	doesn't exist	vertical asymptote
$x = \pi$	0	local maximum	$\dfrac{\pi}{2} < x < \dfrac{3\pi}{2}$	$-$	concave down
$\begin{aligned}\pi < x < 3\pi/2 \\ 3\pi/2 < x < 2\pi\end{aligned}$	$-$	decreasing	$x = \dfrac{3\pi}{2}$	doesn't exist	vertical asymptote
$x = 2\pi$	0	local minimum	$\dfrac{3\pi}{2} < x < \dfrac{5\pi}{2}$	$+$	concave up

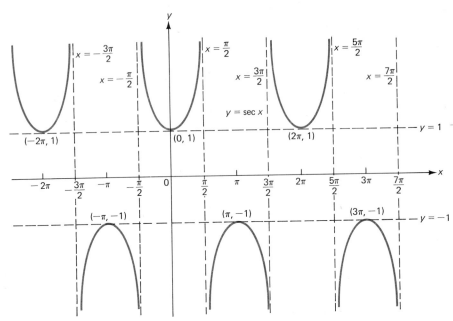

FIGURE 14

To sketch a curve $y = f(x)$, follow these steps:

(i) Calculate the derivative $dy/dx = f'$ and determine where the curve is increasing and decreasing. Find all points at which $f'(x) = 0$ or $f'(x)$ does not exist.

(ii) Calculate the second derivative $f'' = d^2y/dx^2$ and determine where the curve is concave up and concave down.

(iii) Determine local maxima and minima by using either the first or the second derivative test.

(iv) Find all points of inflection.

(v) Determine the y-intercept and (if possible) the x-intercept(s).

(vi) Determine $\lim_{x \to \infty} f(x)$ and $\lim_{x \to -\infty} f(x)$. If either of these is finite, then obtain horizontal asymptotes (as in Example 9).

(vii) If f is not defined at x_0, determine $\lim_{x \to x_0^+} f(x)$ and $\lim_{x \to x_0^-} f(x)$.

(viii) Look for vertical asymptotes. These are the lines of the form $x = x_0$, where $\lim_{x \to x_0^+} f(x) = \infty$ (or $-\infty$) or $\lim_{x \to x_0^-} f(x) = \infty$ (or $-\infty$) and $f(x_0)$ is undefined (see Example 9).

PROBLEMS 4.4

In Problems 1–42, follow the steps outlined in this section to graph the given function.

1. $y = x^2 + x - 30$

2. $y = x^3 - 12x + 10$

3. $y = x^3 - 3x^2 - 45x + 25$

4. $y = 4x^3 - 3x + 2$

5. $y = x^3 + 3$

6. $y = x^4 + 1$

7. $y = 2x^3 - 9x^2 + 12x - 3$

8. $y = \dfrac{x^3}{3} + \dfrac{x^2}{2} - 2x - \dfrac{2}{3}$

9. $y = 2x(x + 4)^3$

10. $y = x^3 - x^2$

11. $y = x^4 - x^3$

12. $y = 3x^5 - 5x^3$

13. $y = \sqrt{1 + x}$

14. $y = \sqrt[3]{1 + x}$

15. $y = x\sqrt{1 + x}$

16. $y = x\sqrt[3]{1 + x}$

17. $y = (x - 2)^2(x + 2)^2$

18. $y = \dfrac{1}{x - 3}$

19. $y = 1 + \dfrac{1}{x^2}$

20. $y = \dfrac{x^2 - 1}{x^2}$

21. $y = \dfrac{1}{(x - 1)(x - 2)}$

22. $y = \dfrac{x - 1}{x - 2}$

23. $y = \dfrac{1}{x^2 - 1}$

24. $y = \dfrac{x}{x^2 - 1}$

25. $y = x - \sqrt[3]{x}$

26. $y = \dfrac{x + 1}{\sqrt{x - 1}}$

27. $y = \sqrt[3]{x^2 - 1}$

28. $y = \sqrt[3]{x^2 + 1}$

29. $y = \dfrac{x^2 + x + 7}{\sqrt{2x + 1}}$

30. $y = \dfrac{4x - 4}{3x + 6}$

31. $y = x^{2/3}(x - 3)$

32. $y = (x - 1)^{1/5}$

33. $y = \dfrac{x^2 - 1}{x^2 + 1}$

*34. $y = \dfrac{x^2 + 1}{x^2 - 1}$

*35. $y = \dfrac{x^2 - 3}{x^2 - 3x + 2}$

36. $y = \cos 3x$

37. $y = \sin(2x + 5)$

*38. $y = \cos^2 x$

*39. $y = \tan x$

40. $y = \csc x$

*41. $y = (x + 3)^4(x - 4)^3$

*42. $y = x^{3/2}(x - 1)^{8/3}$

43. Show that if $f(x) = x^5$, then f has a critical point at $x = 0$ but has neither a local maximum nor local minimum there. Is $x = 0$ a point of inflection?

44. Show that $f(x) = x^2 - 6x + 13$ is always positive. [*Hint:* Graph the curve.]

45. Show that the cubic equation $x^3 + x^2 + 5x - 15 = 0$ has exactly one real root. Is it positive or negative?

46. How many real roots do each of the following cubic equations have?
    (a) $x^3 + 1 = 0$
    (b) $x^3 + x^2 + x + 1 = 0$
    (c) $4x^3 - 3x^2 + 2x - 1 = 0$

47. Show that the equation $x^{11} + x^3 + x + 3 = 0$ has exactly one real root. [*Hint:* Consider its graph.]

48. Show that the quartic equation $x^4 + x^3 - 5x^2 - 6 = 0$ has exactly two real roots. Find open intervals that contain them.

49. Show that the quartic equation $x^4 + 10x^2 + 21 = 0$ has no real roots.

50. Explain why the cubic equation $ax^3 + bx^2 + cx + d = 0$ (a, b, c, d real numbers with $a \neq 0$) always has at least one real root. [*Hint:* Consider its graph.]

51. Explain why the quintic equation $ax^5 + bx^4 + cx^3 + dx^2 + ex + f = 0$, with all coefficients real and $a \neq 0$, has at least one real root.

52. Show graphically that there are an infinite number of solutions to the equation $x = -\tan x$.

53. Show that the solutions of $x = -\tan x$ are critical points of the curve $y = x \sin x$. Use this information to sketch $y = x \sin x$.

A function is **piecewise linear** if it is continuous and its graph consists of a number of straight-line segments. For example, $y = |x|$ is piecewise linear (see Figure 2.2.5). In general, a piecewise linear function looks something like what is shown in Figure 15.

FIGURE 15

54. Show that the function $f(x) = x + |x|$ is piecewise linear and graph it.

55. Graph $f(x) = |x + 3| + x + 3$.

56. Graph $f(x) = |x + 3| + |x - 3| + 2$. How many "pieces" does it take?

57. Graph $f(x) = |x| + |x - 1| + |x - 2| + |x - 3|$. How many "pieces" does it take?

58. Graph $f(x) = |2x - 4| + |3x - 3| + 2x - 7$.

*59. Graph $f(x) = |x^2 - 3|$.

*60. Graph $f(x) = |x^2 - 3| + |x^2 - 1|$.

61. Prove that if $f'(x) = 0$ and $f''(x_0) < 0$, then f has a local maximum at x_0 in some interval $[c, d]$. [*Hint:* Show that near x_0, $f'(x) > 0$ for $x < x_0$ and $f'(x) < 0$ for $x > x_0$.]

62. Explain why if $f'(x_0) = 0$ and $f''(x_0) > 0$, then f has a local minimum at x_0.

*63. A function $f(x)$ satisfies the **differential equation**

$$\frac{df}{dx} = g(x), \qquad f(0) = f_0.$$

The graph of $y = g(x)$ is sketched in Figure 16, and three values for f_0 are shown. Draw a rough sketch of the function $f(x)$ for each of the three possible values for f_0. In your sketch indicate where the function is increasing and decreasing, and indicate maxima, minima, concavity, points of inflection, and asymptotes.

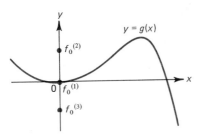

FIGURE 16

64. Show that if $f''(x_0) = 0$ and $f'''(x_0) \neq 0$, then x_0 is a point of inflection.

**65. Let n be a positive integer, $n \geq 2$. Show that the curve

$$\frac{1 + y}{1 - y} = \left(\frac{1 + x}{1 - x}\right)^n$$

has precisely three points of inflection.

*66. Suppose that f, f', and f'' are continuous on $[a, b]$ and $f'(x) = g(x)/h(x)$. Show that if there is a number $c \in [a, b]$ such that $g(c) = 0 \neq h(c)$, then f has a maximum at c if $g'(c)$ and $h(c)$ have opposite signs and f has a minimum at c if $g'(c)$ and $h(c)$ have the same sign.

**67. Suppose that p and q satisfy $p > 1$, $q > 1$, $1/p + 1/q = 1$. Prove that if u, $v \geq 0$, then $uv \leq u^p/p + v^q/q$.

*68. Show that $f(x) = x|x|$ has a point of inflection at $x = 0$. Also show that $f'(0) = 0$. This result provides an example of a point of inflection at x_0 where $f''(x_0)$ does not exist but, also, $|f''(x)|$ does not become infinite as $x \to x_0$.

4.5 THE THEORY OF MAXIMA AND MINIMA

In this section we consider the problem of finding the maximum (largest) and minimum (smallest) values of a function $y = f(x)$ over an interval. In the next section we show how this theory can be used in applications.

Recall the upper and lower bound theorem, which states that a continuous function always has a maximum and minimum value on any closed, bounded interval $[a, b]$. In Section 4.3 we defined what we meant by a function having a local maximum or minimum at a point x_0. Roughly, f has a local maximum at x_0 if among all the values of $f(x)$ in an open interval containing x_0, f takes its largest value at x_0. A similar statement can be made for a local minimum at x_0. We showed (in Theorem 4.3.2) that if f has a local maximum or minimum at x_0, then x_0 is a critical point of f. That is, either $f'(x_0) = 0$ or $f(x_0)$ exists but $f'(x_0)$ does not exist. We further demonstrated (Theorem 4.4.3) the following:

If $f'(x_0) = 0$ and $f''(x_0) < 0$, then f has a local maximum at x_0.

If $f'(x_0) = 0$ and $f''(x_0) > 0$, then f has a local minimum at x_0.†

But there could be more than one local maximum or minimum in an interval. Moreover, there are other ways a maximum or minimum could be reached. Look at Figure 1. The function depicted there has local maxima at x_0, x_2, and x_4 and local minima at x_1, x_3, and x_5. However, in the interval $[a, b]$ the maximum and minimum values of f are taken at none of these critical points. The maximum value of f is taken at $x = b$, and the minimum is taken at $x = a$. Thus it is necessary to check both endpoints as well as all the critical points to find the maximum and minimum.

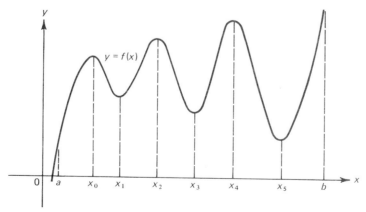

FIGURE 1

---

†Remember also that we may have a local maximum or minimum at x_0 if $f''(x_0)$ is 0 or if it fails to exist.

There is another problem that could arise. Consider the function

$$f(x) = x^{2/3} \quad \text{for} \quad -1 \le x \le 1.$$

This function is graphed in Figure 2. We see that in the interval $[-1, 1]$ f takes on its minimum value at $x = 0$, which is not an endpoint of the interval and at which $f'(0)$ is not zero. In fact, since $f'(x) = 2/(3x^{1/3})$, $f'(0)$ does not even exist. Thus we have the following rule.

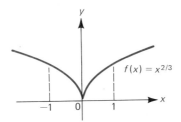

FIGURE 2

> For $f(x)$ continuous on the closed, finite interval $[a, b]$, to find the maximum and minimum values of f over that interval, proceed as follows:
>
> (i) Evaluate f at all critical points in (a, b).
> (ii) Evaluate $f(a)$ and $f(b)$.
>   (a) $f_{\max}[a, b] = $ the largest of the values calculated in (i) and (ii).
>   (b) $f_{\min}[a, b] = $ the smallest of the values calculated in (i) and (ii).

REMARK 1. $f_{\max}[a, b]$ is read as the largest value of f over the interval $[a, b]$; a similar reading is used for $f_{\min}[a, b]$.

REMARK 2. In an infinite interval the function may have neither a maximum nor a minimum. For example, $f(x) = x^2$ has a minimum (zero) but no maximum on $[0, \infty)$. The function $f(x) = x^3$ has neither a maximum nor a minimum in $(-\infty, \infty)$.

EXAMPLE 1 Find the maximum and minimum values of $f(x) = 2x^3 - 3x^2 - 12x + 5$ (a) in the interval $[0, 4]$, and (b) for $x \ge 0$.

Solution. **(a)** $f'(x) = 6x^2 - 6x - 12 = 6(x^2 - x - 2) = 6(x - 2)(x + 1)$. The only place in the interval $[0, 4]$ where $f'(x) = 0$ is at $x = 2$. At $x = 2$, $f(x) = -15$ and this is the only critical point in $[0, 4]$. We also calculate $f(0) = 5$ and $f(4) = 37$. Therefore $f_{\max}[0, 4] = 37$ and $f_{\min}[0, 4] = -15$. This curve is sketched in Figure 3.

(b) We first note that now we are considering the infinite interval $[0, \infty)$, so our upper and lower bound theorem does not apply, and we do not know , in advance, that f has a maximum or minimum on $[0, \infty)$.

From Figure 3 we see that for $x \ge 0$, f takes the minimum value -15 at $x = 2$. However,

$$\lim_{x \to \infty} (2x^3 - 3x^2 - 12x + 5) = \infty,$$

so f has no maximum value on $[0, \infty)$. ∎

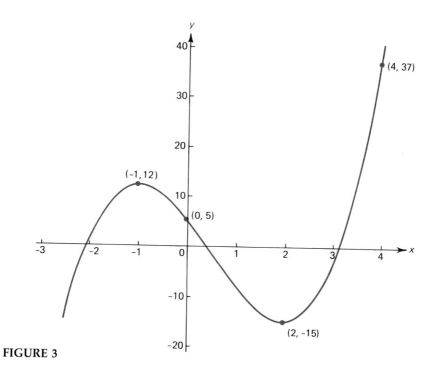

FIGURE 3

EXAMPLE 2   Find the maximum and minimum values of $f(x) = x/(1 + x)$ in the interval $[1, 5]$.

     Solution. $f'(x) = 1/(1 + x)^2$, which is never zero, so there are no critical points.† Since f' is defined everywhere in $[1, 5]$, the maximum and minimum values are taken at the endpoints of the interval. We have $f(1) = \frac{1}{2}$ and $f(5) = \frac{5}{6}$, so $f_{\max}[1, 5] = \frac{5}{6}$ and $f_{\min}[1, 5] = \frac{1}{2}$. This curve is sketched in Figure 4. ∎

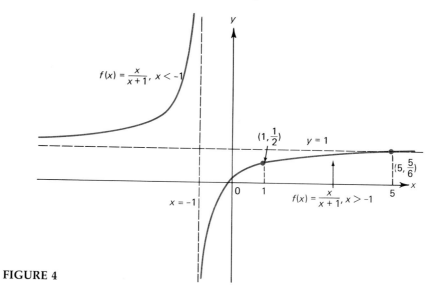

FIGURE 4

---

†$f'(-1)$ does not exist, but -1 is not a critical point because f is not defined at $x = -1$. In any event, we don't have to worry about -1 because it is not in the interval $[1, 5]$.

EXAMPLE 3   Find the maximum and minimum values of $f(x) = -2x\sqrt{x + 3}$ in the interval $[-\frac{5}{2}, 1]$.

Solution.

$$f'(x) = -\frac{2x}{2\sqrt{x + 3}} - 2\sqrt{x + 3} = -\frac{3x + 6}{\sqrt{x + 3}}$$

We see that $f'(x) = 0$ at $x = -2$. For $x = -2$, $f(x) = 4$. Also, -3 is a critical point since $f(-3)$ exists $[f(-3) = 0]$ but $f'(-3)$ does not exist. However, this critical point does not concern us since -3 is not in the interval $[-\frac{5}{2}, 1]$.

At the endpoints,

$$f\left(-\frac{5}{2}\right) = 5\sqrt{\frac{1}{2}} = \frac{5}{\sqrt{2}} \quad \text{and} \quad f(1) = -4.$$

Since

$$\frac{5}{\sqrt{2}} \approx 3.5 < 4, \quad f_{\text{max}}\left[-\frac{5}{2}, 1\right] = 4 \quad \text{and} \quad f_{\text{min}}\left[-\frac{5}{2}, 1\right] = -4.$$

This curve is sketched in Figure 5. ■

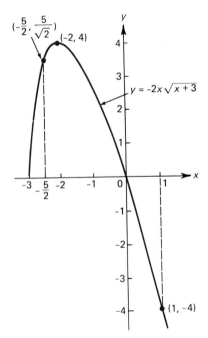

FIGURE 5

EXAMPLE 4   Find the maximum and minimum values of $f(x) = \sin[(x + 2)/3]$ in the interval $[0, \pi]$.

Solution. $f'(x) = \frac{1}{3}\cos[(x + 2)/3] = 0$ when $(x + 2)/3 = \pm\pi/2, \pm3\pi/2, \ldots$ If $(x + 2)/3 = \pi/2$, then $2x + 4 = 3\pi$ and $x = (3\pi - 4)/2 \approx 2.71$. If $(x + 2)/3 = -\pi/2$, then x is negative, so that $x \notin [0, \pi]$. Also, if $(x + 2)/3 = 3\pi/2$, then $2x + 4 = 9\pi$ and $x = (9\pi - 4)/2 \approx 12.14 > \pi$, so $x \notin [0, \pi]$. Thus the only critical point in $[0, \pi]$ is at $(3\pi - 4)/2$. Now $f((3\pi - 4)/2) = \sin(\pi/2) = 1$. At the endpoints, $\sin[(0 + 2)/3] = \sin(2/3) \approx 0.6184$ and $\sin[(\pi + 2)/3] = 0.9898$. Thus $f_{max}[0, \pi] = 1$ and $f_{min}[0, \pi] = 0.6184$. This curve is sketched in Figure 6. ◼

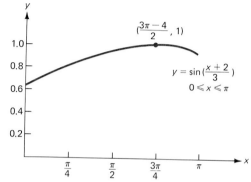

FIGURE 6

EXAMPLE 5   Find the maximum and minimum values of $f(x) = \sqrt[3]{x}$ in $[-1, 1]$.

Solution. $f'(x) = \frac{1}{3}x^{-2/3}$, which is not defined at $x = 0$. There are no other critical points. We therefore check the endpoints and the point $x = 0$. We have $f(1) = 1$, $f(-1) = -1$, and $f(0) = 0$. Therefore $f_{max}[-1, 1] = 1$ and $f_{min}[-1, 1] = -1$. This curve is sketched in Figure 7. ◼

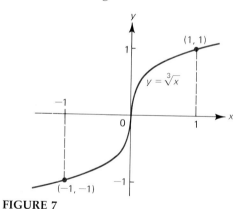

FIGURE 7

PROBLEMS 4.5

In Problems 1–40, find the maximum and minimum values for the given function over the indicated interval.

1. $f(x) = x^2 + x - 30;\ [0, 2]$

2. $f(x) = x^2 + x - 30;\ [-2, 0]$

3. $f(x) = x^3 - 12x + 10;\ [-10, 10]$

4. $f(x) = 4x^3 - 3x + 2;\ [-5, 5]$

5. $f(x) = x^5;\ (-\infty, 1]$

6. $f(x) = x^7;\ [-1, \infty)$

7. $f(x) = x^3 - 3x^2 - 45x + 25;\ [-5, 5]$

8. $f(x) = \dfrac{x^3}{3} + \dfrac{x^2}{2} - 2x - \dfrac{2}{3};\ [-3, 3]$

9. $f(x) = x^{20}$; $[0, 1]$

10. $f(x) = x^4 - 18x^2$; $[-2, 2]$

11. $f(x) = (x + 1)^{1/3}$; $[-2, 7]$

12. $f(x) = (x - 2)^{2/3}$; $[-14, 3]$

13. $f(x) = 2x(x + 4)^3$; $[-1, 1]$

14. $f(x) = x\sqrt[3]{1 + x}$; $[-2, 26]$

15. $f(x) = \dfrac{1}{(x - 1)(x - 2)}$; $[3, 5]$

16. $f(x) = \dfrac{1}{(x - 1)(x - 2)}$; $[-3, 0]$

17. $f(x) = \dfrac{x - 1}{x - 2}$; $[-3, 1]$

18. $f(x) = \dfrac{1}{x^2 - 1}$; $[2, 5]$

19. $f(x) = x^{2/3}(x - 3)$; $[-1, 1]$

20. $f(x) = (x - 1)^{1/5}$; $[-31, 33]$

21. $f(x) = 1 + x + \dfrac{1}{2x^2}$; $\left[\dfrac{1}{2}, \dfrac{3}{2}\right]$

22. $f(x) = \sqrt{x} + \dfrac{1}{\sqrt{x}}$; $\left[\dfrac{1}{2}, \dfrac{3}{2}\right]$

23. $f(x) = (1 + \sqrt{x} + \sqrt[3]{x})^9$; $[0, \infty)$

24. $f(x) = \dfrac{1}{x - 2}$; $[3, \infty)$

25. $f(x) = \begin{cases} x^2, & x \neq 3 \\ 20, & x = 3 \end{cases}$; $[0, 4]$

*26. $f(x) = \begin{cases} x^2, & x \leq 1 \\ x^3, & x \geq 1 \end{cases}$; $[0, 5]$

*27. $f(x) = \begin{cases} x - 3, & x < 1 \\ 2x + 4, & 1 \leq x \leq 3 \\ 3x - 7, & x > 3 \end{cases}$; $[-5, 5]$

28. $f(x) = [x]$; $[-5, 5]$

29. $f(x) = |x^3 - 12x|$; $[-3, 3]$

30. $f(x) = |x^4 - 16x|$; $[-2, 2]$

31. $f(x) = \cos 3x$; $\left[0, \dfrac{\pi}{2}\right]$

32. $f(x) = \sin x \cos x$; $\left[0, \dfrac{\pi}{3}\right]$

33. $f(x) = \sin x^2$; $[0, \pi]$

34. $f(x) = \cos\left(\dfrac{1 - 2x}{3}\right)$; $[-2, 2]$

*35. $f(x) = \sin\left(\dfrac{x}{1 + x}\right)$; $(-1, 1]$

36. $f(x) = \sin x + \cos x$; $[0, 2\pi]$

37. $f(x) = (x - 1)^2 + (2x + 1)^2$; $(-\infty, 0]$

38. $f(x) = (x + 1)^2 + (4x - 5)^2$; $(-\infty, \infty)$

39. $f(x) = \dfrac{1}{\sqrt{1 + x}}$; $[0, \infty)$

40. $f(x) = \dfrac{x}{1 + x}$; $[0, \infty)$

*41. Suppose that $p(x)$ is a polynomial function of degree n such that $p(x) \geq 0$ for all x. Prove that $p(x) + p'(x) + p''(x) + \ldots + p^{(n)}(x) \geq 0$ for all x.

*42. Pick a real number α in the interval $(1, \infty)$. On the interval $[-1, \infty)$ consider the function F, where $F(x) = (1 + x)^{\alpha} - (1 + \alpha x)$.
   (a) Show that F has a unique critical point.
   (b) Show that $F(x) > 0$ unless $x = 0$.
   (c) Prove that $(1 + x)^{\alpha} \geq 1 + \alpha x$ if $x \geq -1$ and $\alpha \geq 1$.

4.6 MAXIMA AND MINIMA: APPLICATIONS

In this section and the next we show how the theory of maxima and minima can be applied to a great variety of problems. Before citing general rules for dealing with these problems, we begin with a simple example that is historically one of the first applications of the theory to a practical problem.

EXAMPLE 1 Suppose that a farmer has 1000 yd of fence that he wishes to use to fence off a rectangular plot along the bank of a river (see Figure 1). What are the dimensions of the maximum area he can enclose?

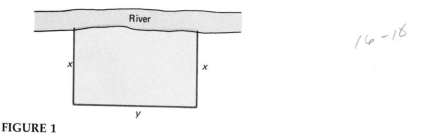

FIGURE 1

Solution. Since one side of the rectangular plot is taken up by the river, the farmer can use the fence for the other three sides. From the figure, the area of the plot is

$$A = xy,$$ (1)

where y is the length of the side parallel to the river. Without other information it would be impossible to solve this problem since the area A is a function of the *two* variables x and y. However, we are also told, since the farmer will obviously use all the available fence, that

$$2x + y = 1000,$$ (2)

or

$$y = 1000 - 2x, \quad \text{for } x \text{ in } [0, 500].$$ (3)

Then

$$A = x(1000 - 2x) = 1000x - 2x^2.$$ (4)

To find the maximum value for A, we set dA/dx equal to zero and use the results of the preceding section. We have

$$\frac{dA}{dx} = 1000 - 4x = 0$$

when $x = 250$. Also,

$$\frac{d^2A}{dx^2} = -4 < 0,$$

so that A is a maximum when $x = 250$. This is depicted in Figure 2.

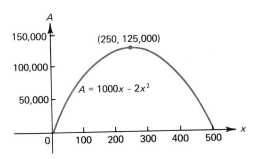

FIGURE 2

When $x = 250$ yd, then $y = 500$ yd and $A = 125,000$ yd^2, which is the maximum area that the farmer can enclose. We note that, as is evident from Figure 2, A is positive when $0 \leq x \leq 500$ and at the endpoints $x = 0$ and $x = 500$ the area is zero, so the local maximum we have obtained is a maximum over the entire interval $[0, 500]$. ■

Example 1 suggests that the following steps be taken to solve a maximum or minimum problem.

(i) Draw a picture if it makes sense to do so.
(ii) Write all the information in the problem in mathematical terms, giving a letter name to each variable.
(iii) Determine the variable that is to be maximized or minimized and write this variable as a function of the other variables in the problem.
(iv) Using information given in the problem, eliminate all variables except one, so that the function to be maximized or minimized is written in terms of *one* of the variables of the problem.
(v) Determine the interval over which this one variable can be defined.
(vi) Follow the steps of the previous section to maximize or minimize the function over this interval (which may be infinite).

EXAMPLE 2   Suppose that the farmer in Example 1 wishes to build his rectangular plot away from the river (so that he must use his fence for all four sides of the rectangle). How large an area can he enclose in this case?

Solution. The situation is now as in Figure 3. The area is again given by $A = xy$, but now

$$2x + 2y = 1000, \tag{5}$$

or

$$y = 500 - x. \tag{6}$$

Hence the problem is to maximize

$$A = x(500 - x) = 500x - x^2, \qquad \text{for } x \text{ in } [0, 500]. \tag{7}$$

FIGURE 3

Now

$$\frac{dA}{dx} = 500 - 2x = 0$$

when $x = 250$. Also, $d^2A/dx^2 = -2$, so that a local maximum is achieved when $x = y = 250$ yd and the answer is $A = 62,500$ yd^2. (Note that the plot is, in this case, a square.) At 0 and 500, $A = 0$, so that $x = 250$ is indeed the maximum. ∎

REMARK 1. The reasoning in Example 2 can be used to prove the following: *For a given perimeter the rectangle containing the greatest area is a square.*

REMARK 2. If we do not require that the plot be rectangular, then we can enclose an even greater area. Although the proof of this fact is beyond the scope of this book, it can be shown that *for a given perimeter the geometric shape with the largest area is a circle.* For example, if the 1000 yd of fence in Example 2 are formed in the shape of a circle, then the circle has a circumference of $2\pi r = 1000$, so that the radius of the circle is $r = 1000/2\pi$ yd. Then $A = \pi r^2 = \pi(1,000,000/4\pi^2) = 1,000,000/4\pi \approx 79,577$ yd^2.

EXAMPLE 3 Find the point on the straight line $x + 2y = 5$ that is closest to the origin.

Solution. The distance D from a point (x, y) to the origin $(0, 0)$ is given by (see Figure 4)

$$D = (x^2 + y^2)^{1/2} \tag{8}$$

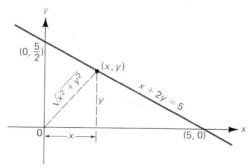

FIGURE 4

But $x = 5 - 2y$, so that

$$D = [(5 - 2y)^2 + y^2]^{1/2} = [25 - 20y + 4y^2 + y^2]^{1/2} = [25 - 20y + 5y^2]^{1/2}. \tag{9}$$

We can simplify our computations by minimizing D^2 instead of D (the minimum values of D and D^2 are taken at the same value of y; see Problem 4.3.32). Let $E = D^2$. Then $E = 25 - 20y + 5y^2$, and

$$\frac{dE}{dy} = -20 + 10y. \tag{10}$$

We immediately see that $dE/dy = 0$ when $y = 2$. When $y < 2$, $dE/dy < 0$, and when $y > 2$, $dE/dy > 0$. Thus by the first derivative test, the minimum occurs at $y = 2$. Then $x = 5 - 2 \cdot 2 = 1$, and at the point $(1, 2)$, $D = \sqrt{5}$.

NOTE. The interval under consideration here is $(-\infty, \infty)$. Explain why it is not necessary to check the endpoints.

EXAMPLE 4   A cylindrical barrel is to be constructed to hold $32\pi \text{m}^3$ of liquid. The cost per square meter of constructing the side of the barrel is half the cost per square meter of constructing the top and bottom. What are the dimensions of the barrel that costs the least to construct?

·*Solution.* Consider Figure 5. Let h be the height of the barrel and let r be the radius of the top and bottom. Then the volume of the barrel† is given by $V = \pi r^2 h = 32\pi \text{m}^3$.

FIGURE 5

Assuming that the cost of constructing the material on the side is k cents per square meter, then the cost of the top and bottom is $2k$ cents per square meter. The area of the top or the bottom is πr^2 square meters, while that of the side is $2\pi rh$ square meters (since the circumference of the side is $2\pi r$). Thus the total cost is given by

$$C = 2\pi r^2(2k) + 2\pi rhk = k(4\pi r^2 + 2\pi rh). \tag{11}$$

To write C as a function of one variable only (which we must do in order to solve the problem), we use $32\pi = \pi r^2 h$ to obtain $h = 32\pi/\pi r^2 = 32/r^2$, so that

$$C = k\left(4\pi r^2 + 2\pi r \cdot \frac{32}{r^2}\right) = k\left(4\pi r^2 + \frac{64\pi}{r}\right) \qquad \text{for} \qquad r > 0. \tag{12}$$

Then since k is a constant,

$$\frac{dC}{dr} = k\left(8\pi r - \frac{64\pi}{r^2}\right).$$

---
†Recall that the volume of a right circular cylinder with radius r and height h is $\pi r^2 h$.

Setting this expression equal to zero, we obtain

$$8\pi r = \frac{64\pi}{r^2}, \qquad 8r^3 = 64, \qquad r^3 = 8, \qquad \text{and} \qquad r = 2 \text{ m}.$$

In addition, $d^2C/dr^2 = k[8\pi + (128\pi/r^3)]$, which is greater than 0 when $r = 2$. Hence there is a local minimum when $r = 2$ m. When $r = 2$, $h = 32/4 = 8$ m. Note that this local minimum is a true minimum since the only endpoint of the interval occurs at $r = 0$, which makes no practical sense since in that case the barrel can hold nothing. Also, from (12) it is apparent that as $r \to \infty$ or $r \to 0^+$, $C = C(r) \to \infty$. ∎

EXAMPLE 5   A cardboard box with a square base and an open top, is to be constructed from a square piece of cardboard 10 cm on a side by cutting out four squares at the corners and folding up the sides. What should be the dimensions of the box in order to make the volume enclosed as large as possible?

Solution. We refer to Figures 6a and 6b. If we cut out squares with sides of x centimeters, the base of the box will have each side equal to $10 - 2x$ centimeters, and the height of the box will be x. Then the volume of the box is given by

$$V = x(10 - 2x)^2, \qquad \text{for } x \text{ in } [0, 5].$$

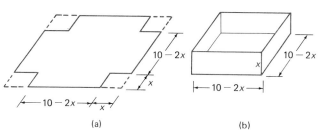

(a)   (b)

FIGURE 6

Now

$$\frac{dV}{dx} = (10 - 2x)^2 + 2x(10 - 2x)(-2) = (10 - 2x)^2 - 4x(10 - 2x),$$

which is equal to zero when

$$4x(10 - 2x) = (10 - 2x)^2.$$

If $(10 - 2x) \neq 0$, then $4x = 10 - 2x$, or $x = \frac{5}{3}$. If $10 - 2x = 0$, then $x = 5$. The second derivative of V is

$$\frac{d^2V}{dx^2} = -8(10 - 2x) + 8x,$$

so that

$$\frac{d^2V}{dx^2} = 40 \qquad \text{when} \qquad x = 5, \qquad \text{and} \qquad \frac{d^2V}{dx^2} = -40 \qquad \text{when} \qquad x = \frac{5}{3}.$$

Thus the maximum occurs when $x = \frac{5}{3}$ cm, and the maximum volume is $V = \frac{2000}{27}$ cm^3. (The minimum volume is zero, which is obtained by setting x equal to 5 cm or 0 cm; i.e., leave a base or a height of zero.) ∎

EXAMPLE 6   In chemistry a **catalyst** is defined as a substance that alters the rate of a chemical reaction without itself undergoing a change; the phenomenon is called **catalysis.** If in a chemical reaction the product of the reaction serves as a catalyst for the reaction, then the process is called **autocatalysis.** Suppose that in the autocatalytic process we start with an amount A of a given substance. Let x be the amount of the product (i.e., the result of the process). It is reasonable to assume that the rate of reaction depends on both the amount of the product x and the amount of remaining substance $(A - x)$. If this rate is given by

$$R = \alpha x(A - x), \tag{13}$$

where α is a known positive constant, for what concentration x is the rate of reaction greatest?

Solution. $R = \alpha(Ax - x^2)$, so that

$$\frac{dR}{dx} = \alpha(A - 2x) = 0 \qquad \text{when} \qquad x = \frac{A}{2}.$$

In addition, $d^2R/dx^2 = -2\alpha < 0$, so a local maximum is reached when $x = A/2$. The endpoints are $x = 0$ and $x = A$, both of which give a reaction rate of zero. Thus we may conclude that the rate of reaction is greatest when exactly half the original substance has been catalyzed. ∎

EXAMPLE 7   A woman is on a lake in a canoe 1 km from the closest point P of a straight shore line (see Figure 7). She wishes to get to a point Q, 10 km along the shore from P. To do so, she paddles to a point R between P and Q and then walks the remaining distance to Q. She can paddle 3 km/hr and walk 5 km/hr. How should she pick the point R to get to Q as quickly as possible?

FIGURE 7

Solution. To simplify matters, we draw Figure 8, where the starting point is placed at the origin. Then as drawn, P has the coordinates $(0, 1)$, R the coordinates $(x, 1)$, and Q the coordinates $(10, 1)$. The problem is to find the value of x (the point where she reaches the shore) that minimizes the total travel time.

The distance from the origin to R is given by

$$d_1 = \sqrt{x^2 + 1}.$$

FIGURE 8

Since (velocity) $\times$ (time) = distance, and the velocity v_1 in this case is 3 km/hr, we have

$$t_1 = \text{time from } 0 \text{ to } R = \frac{d_1}{v_1} = \frac{1}{3}\sqrt{x^2 + 1}. \tag{14}$$

The distance d_2 from R to Q is $10 - x$, and the walking velocity is $v_2 = 5$ km/hr, so that

$$t_2 = \text{time from } R \text{ to } Q = \frac{1}{5}(10 - x).$$

Thus the total time in getting from 0 to Q is given by

$$T = \frac{\sqrt{x^2 + 1}}{3} + \frac{10 - x}{5}. \tag{15}$$

To find the minimum, we calculate dT/dx and set it equal to zero:

$$\frac{dT}{dx} = \frac{x}{3\sqrt{x^2 + 1}} - \frac{1}{5} = 0, \quad \text{or} \quad \frac{x}{3\sqrt{x^2 + 1}} = \frac{1}{5}, \quad \text{or} \quad 5x = 3\sqrt{x^2 + 1}.$$

We now square both sides to obtain

$$25x^2 = 9x^2 + 9, \quad \text{or} \quad x = \pm\frac{3}{4}.$$

We use the value $x = \frac{3}{4}$ since the value $x = -\frac{3}{4}$ does not satisfy $5x = 3\sqrt{x^2 + 1}$, and negative values of x make no sense in this problem. When $x = \frac{3}{4}$, we obtain, from (15),

$$T = \frac{\sqrt{(\frac{9}{16}) + 1}}{3} + \frac{10 - (\frac{3}{4})}{5} = \frac{5}{12} + 2 - \frac{3}{20} = \frac{136}{60} \text{ hr} = 136 \text{ min}.$$

We still have to check the endpoints $x = 0$ (the woman rows straight to the shore) and $x = 10$ (the woman rows directly to her destination). At $x = 0$,

$$T = \tfrac{1}{3} + 2 = 2\tfrac{1}{3} \text{ hr} = 140 \text{ min}.$$

At $x = 10$,

$$T = \frac{\sqrt{101}}{3} \text{ hr} \approx 201 \text{ min.}$$

Thus the minimum time needed is 136 min. ▪

Example 7 can be redone with a different perspective to obtain a very important physical fact.

EXAMPLE 8   According to **Fermat's principle,**† *when light travels through various media, it takes the path that requires the least time.* To illustrate this principle, we suppose that light is traveling from a point beneath the surface of water to a point in the air, above the surface. In each medium (water and air), the light ray will, according to Fermat's principle, travel in a straight line. We assume that the light emanates from a point P one unit of distance below the water surface to a point Q one unit above the surface and two units away in horizontal distance; see Figure 9. In Figure 9 the path from P to Q is not a straight line because the velocity of light in air is not equal to the velocity of light in water. At what point S on the surface of the water between points R and U (in the figure) will the light ray pass in order that it reach Q in minimum time?

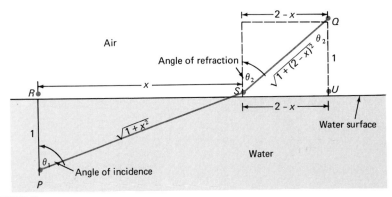

FIGURE 9

Solution. If the distance from R to S is denoted by x, then the distance from S to U is $2 - x$, so that

$$\overline{PS} = \sqrt{1 + x^2} \quad \text{and} \quad \overline{SQ} = \sqrt{1 + (2 - x)^2}. \tag{16}$$

If v_1 and v_2 represent the velocities of the light ray in water and air, respectively, then the total time for the ray to go from P to Q is given by

$$T = \frac{\sqrt{1 + x^2}}{v_1} + \frac{\sqrt{1 + (2 - x)^2}}{v_2}. \tag{17}$$

---

†Pierre de Fermat (1601–1665) was a celebrated French mathematician who helped develop a wide variety of topics of modern mathematics. See his biography on page 95.

Then

$$\frac{dT}{dx} = \frac{x}{v_1\sqrt{1 + x^2}} - \frac{(2 - x)}{v_2\sqrt{1 + (2 - x)^2}}, \tag{18}$$

and

$$\frac{d^2T}{dx^2} = \frac{1}{v_1(1 + x^2)^{3/2}} + \frac{1}{v_2(1 + (2 - x)^2)^{3/2}}.$$

Since $d^2T/dx^2 > 0$, the critical points will all be minima. Setting $dT/dx = 0$ in (18), we obtain

$$\frac{x}{v_1\sqrt{1 + x^2}} = \frac{2 - x}{v_2\sqrt{1 + (2 - x)^2}},$$

or

$$\frac{v_1}{v_2} = \frac{x/\sqrt{1 + x^2}}{(2 - x)/\sqrt{1 + (2 - x)^2}}. \tag{19}$$

But again from Figure 9, if θ_1 and θ_2 denote the angles at which the ray leaves the source and approaches the object, respectively, then

$$\frac{x}{\sqrt{1 + x^2}} = \sin\theta_1 \quad \text{and} \quad \frac{2 - x}{\sqrt{1 + (2 - x)^2}} = \sin\theta_2. \tag{20}$$

Then from (19) the minimum time is achieved when

$$\frac{v_1}{v_2} = \frac{\sin\theta_1}{\sin\theta_2}. \tag{21}$$

The angles θ_1 and θ_2 are called the angles of **incidence** and **refraction,** respectively. Equation (21) is known as **Snell's law,** and it states that, according to Fermat's principle, the ratio of the sines of the angles of incidence and refraction is equal to the ratio of the velocities of light in the two media. Note that if $v_1 = v_2$, then $\theta_1 = \theta_2$, so that the light ray will move in a straight line. ■

In the examples we have given so far, the maximum or minimum did not occur at an endpoint. The next example shows that they can.

EXAMPLE 9 A wire 20 cm long is cut into two pieces. One piece is bent in the shape of an equilateral triangle, and the other is bent in the shape of a circle. How should the wire be cut to maximize the total area enclosed by the shapes? How should it be cut to minimize the total area?

Solution. Let x denote the length of wire used for the equilateral triangle. Then the length of each side is $x/3$, and the circumference of the circle is $20 - x$ (see Figure

10). Using the Pythagorean theorem, we find that the height of the triangle is $\sqrt{3}\,x/6$, so that

$$\text{area of the triangle} = \frac{1}{2}(\text{base}) \times (\text{height}) = \frac{1}{2}\left(\frac{x}{3}\right)\left(\sqrt{3}\frac{x}{6}\right) = \sqrt{3}\frac{x^2}{36}.$$

FIGURE 10

Moreover, since the circumference of the circle is $2\pi r = 20 - x$, we find that the radius of the circle is $(20 - x)/2\pi$, and its area is

$$\pi r^2 = \pi\left(\frac{20 - x}{2\pi}\right)^2 = \frac{(20 - x)^2}{4\pi}.$$

Thus the total area (as a function of x) is

$$A(x) = \frac{\sqrt{3}}{36}x^2 + \frac{(20 - x)^2}{4\pi}, \qquad 0 \le x \le 20.$$

Differentiating, we obtain

$$\frac{dA}{dx} = \frac{\sqrt{3}}{18}x - \frac{(20 - x)}{2\pi} = \left(\frac{\sqrt{3}}{18} + \frac{1}{2\pi}\right)x - \frac{10}{\pi}.$$

This expression is equal to 0 when

$$x = \frac{10/\pi}{(\sqrt{3}/18) + (1/2\pi)} \approx 12.46 \text{ cm.}$$

Also, $d^2A/dx^2 = (\sqrt{3}/18) + (1/2\pi) > 0$, so there is a local minimum at $x \approx 12.46$. For that value of x, $A \approx 11.99$ cm^2. Checking the endpoints of the interval $[0, 20]$, we find that $A(0) = 400/4\pi \approx 31.83$ cm^2, and $A(20) = (\sqrt{3}/36)(400) \approx 19.25$ cm^2. Thus the maximum area is enclosed when we do not cut the wire at all and use it to form a circle. If the wire must be cut, then the function in this problem has no maximum. The local minimum at $x = 12.46$ is indeed the minimum over the entire interval. ■

EXAMPLE 10   Blood is transported in the body by means of arteries, veins, arterioles, and capillaries. This procedure is ideally carried out in such a way as to minimize the energy required† to transport the blood from the heart to the organs and back again. Consider the two blood vessels depicted in Figure 11, where $r_1 > r_2$. There are many ways to try to compute the minimum energy required to pump blood from the point P_1 in the main

---

†Actually, the item to be minimized is the work (energy) required to maintain the system as well as pump the blood. For a more complete discussion of this topic, consult the book by R. Rosen, *Optimality Principles in Biology* (Butterworth, London, 1967).

vessel to the point P_3 in the smaller vessel. We will try to find the minimum total resistance of the blood along the path $P_1 P_2 P_3$.

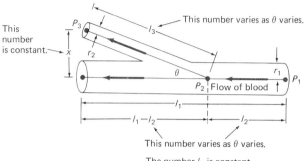

This number is constant.

This number varies as θ varies.

This number varies as θ varies.

The number l_1 is constant.

FIGURE 11

Solution. According to Poiseuille's law (see Problem 3.2.39) the resistance is given by

$$R = \frac{\alpha l}{r^4},$$

where l is the length of the vessel, r is its radius, and α is a constant of proportionality. The problem, then, is to find the "optimal" branching angle θ at which R is a minimum. According to Figure 11, $\sin \theta = x/l_3$, so $l_3 = x/\sin \theta = x \csc \theta$. Also, $x/(l_1 - l_2) = \tan \theta$, so $l_1 - l_2 = x/(\tan \theta) = x \cot \theta$ and $l_2 = l_1 - x \cot \theta$. We now calculate the total resistance along the path $P_1 P_2 P_3$.

$$R_{1,2} = \frac{\alpha l_2}{r_1^{\,4}} \qquad \text{Resistance from } P_1 \text{ to } P_2$$

$$R_{2,3} = \frac{\alpha l_3}{r_2^{\,4}} \qquad \text{Resistance from } P_2 \text{ to } P_3$$

Then $R = R_{1,2} + R_{2,3} = \alpha[(l_2/r_1^{\,4}) + (l_3/r_2^{\,4})]$. But $l_3 = x \csc \theta$ and $l_2 = l_1 - x \cot \theta$, so that

$$R = \alpha \left(\frac{l_1 - x \cot \theta}{r_1^{\,4}} + \frac{x \csc \theta}{r_2^{\,4}} \right).$$

We now simplify matters by using the fact that x is fixed, since x depends on P_1, P_3, and l_1, which are given in the problem, but not on P_2, l_2, l_3, or θ. (P_2, l_2, and l_3 all vary as θ varies, but P_1, P_3, and l_1 do not.) Since R is a function of θ, we can find the minimum value of R by calculating $dR/d\theta$ and setting it to zero. But

$$\frac{dR}{d\theta} = \alpha \left(\frac{x \csc^2 \theta}{r_1^{\,4}} - \frac{x \csc \theta \cot \theta}{r_2^{\,4}} \right)$$

$$= \alpha \left(\frac{x}{\sin^2 \theta \, r_1^{\,4}} - \frac{x \cos \theta}{\sin^2 \theta \, r_2^{\,4}} \right) = \frac{\alpha x}{\sin^2 \theta} \left(\frac{1}{r_1^{\,4}} - \frac{\cos \theta}{r_2^{\,4}} \right).$$

Now $\alpha x/(\sin^2 \theta) \neq 0$, so $dR/d\theta = 0$ when $1/r_1^4 = (\cos \theta)/r_2^4$, or when

$$\cos \theta = \left(\frac{r_2}{r_1}\right)^4.$$

That this value is indeed a minimum can be verified by employing the first derivative test. Thus we can calculate the optimum branching angle by merely considering the ratio of the radii of the blood vessels. In the cited reference by Rosen evidence is given that indicates that branching angles of blood vessels do, in many cases, obey the optimization rule we have just derived. ◼

PROBLEMS 4.6

1. A rectangle has a perimeter of 300 m. What length and width will maximize its area?
2. A farmer wishes to set aside 1 acre of his land for corn and wheat. To keep out the cows, he encloses the field with a fence (costing 50¢ per running foot). In addition, a fence running down the middle of the field is needed, with a cost per foot of $1. Given that 1 acre = 43,560 ft^2, what dimensions should the field have so as to minimize his total cost? The field is rectangular.
3. An isosceles triangle has a perimeter of 24 cm. What are the lengths of the sides of such a triangle that maximize the area of the triangle?
4. Show that among all isosceles triangles with a given perimeter, the area enclosed is greatest when the triangle is equilateral.
5. Find the point on the straight line $2x + 3y = 6$ that is closest to the origin.
6. Find the point on the straight line $3x - 4y = 12$ that is closest to the point (1, 2).
7. Find the point on the circle $x^2 + y^2 = 1$ that is closest to the point (2, 5).
8. Find the points on the hyperbola $x^2 - y^2 = 1$ that are closest to the origin.
9. A wire 35 cm long is cut into two pieces. One piece is bent in the shape of a square, and the other is bent in the shape of a circle. How should the wire be cut to minimize and maximize the total area enclosed by the pieces?
10. Answer Problem 9 if the two pieces are to be formed in the shape of a circle and an equilateral triangle.
11. What is the area of the largest rectangle that can be inscribed in a circle of radius 10?
12. Find the positive number that exceeds its square by the largest amount.
13. Find the two positive numbers whose sum is 20 having the maximum product.
14. A cylindrical tin can is to hold 50 cm^3 of tomato juice. How should the can be constructed in order to minimize the amount of material needed in its construction?
15. A Transylvanian submarine is traveling due east and heading straight for a point P. A battleship from Luxembourg is traveling due south and heading for the same point P. Both ships are traveling at a velocity of 30 km/hr. Initially, their distances from P are 210 km for the submarine and 150 km for the battleship. The range of the submarine's torpedoes is 3 km. How close will the two vessels come? Does the submarine have a chance to torpedo the battleship?
16. Find the dimensions of the right circular cylinder of largest volume that can be inscribed in a sphere of radius 10; of radius r.
17. A Norman window is constructed from a rectangular sheet of glass surmounted by a semicircular sheet of glass. (Stained glass windows are often of this type.) The light that enters through a window is proportional to the area of the window. What are the dimensions of the Norman window having a perimeter of 30 ft that admits the most light?
18. Answer the question of Problem 17 if the glass in the rectangular part of the window is

colored while that of the circular part is clear, given that colored glass admits only half as much light as clear glass.

19. Two saplings, 6 and 8 ft high, are planted 10 ft apart. So that the saplings do not bend, poles the height of the trees are pounded in, then attached to each tree, and a rope tied to the top of each pole is then fixed to the ground (between the two trees) after being pulled taut. How close to the taller tree will the rope be fixed if the total length of the rope is to be minimized?

20. The strength of a rectangular beam is proportional to the product of the breadth and the square of its depth. Find the dimensions of the strongest such beam that can be cut from a cylindrical log of radius $\frac{1}{2}$ m.

21. The stiffness of a rectangular beam is proportional to the product of its breadth and the cube of its depth. Find the dimensions of the stiffest beam that can be cut from a cylindrical log of radius $\frac{1}{2}$ m.

22. A clear rectangle of glass is inserted in a colored semicircular glass window (see Figure 12). The radius of the window is 3 ft and the clear glass passes twice as much light as the colored glass. Find the dimensions of the rectangular insert that passes the maximum light (through the entire window).

FIGURE 12

23. Suppose that the rate of population growth of pigeons in a large city is given by

$$R = 400P^2 - \tfrac{1}{5}P^3,$$

where P is the population of pigeons. For what population level is this rate maximized?

24. The position of a moving object is given by

$$s(t) = 4t - 6t^2 + 6, \qquad t \geq 0.$$

For what value of t is the *velocity* of the object a maximum?

25. In Problem 24, for what value of t is the acceleration a maximum?

26. A ball is thrown in the air from ground level with an initial vertical velocity of 64 ft/sec. How high will the ball go (ignoring air resistance)? After how many seconds will the ball reach its maximum height?

27. **(a)** Show that if two blood vessels have equal radii, then blood resistance is minimized when the branching angle between them is zero (see Example 10).
    (b) Using Table A.4 or a calculator, find the branching angle that minimizes resistance if one blood vessel has twice the diameter of a second.

28. A farmer wishes to divide 20 acres of land along a river into 6 smaller plots by using one fence parallel to the river and 7 fences perpendicular to it. Show that the total amount of fencing is minimized if the sum of the lengths of the 7 cross fences equals the length of the one fence parallel to the river.

29. In water the product of the concentration of hydrogen ions and hydroxl ions is approximately equal to 10^{-14} mol. Find the ratio of hydrogen ions to hydroxl ions that minimizes the *sum* of the concentrations.

30. An oil pipeline is built with two different kinds of tubing. From a point P on one side of a river the line must cross the river and then proceed to a point Q along the bank on the other side (see Figure 13). The tubing used in crossing the river costs 50% more than the tubing that can be used on dry land. The river is $\frac{1}{2}$ km wide, and the point Q is 5 km downriver from P. To what point R across the river must the pipe be directed to minimize the total cost?

FIGURE 13

31. The most important function of the human cough is to increase the velocity of air going out of the windpipe (trachea). Let R_0 denote the "rest radius" of the trachea (i.e., the radius when you are relaxed and not coughing). R_0 is measured in centimeters. Let R be the contracted radius of the trachea during a cough ($R < R_0$); and let V be the average velocity of the air in the trachea when it is contracted to R centimeters. Under some fairly reasonable assumptions regarding the flow of air near the tracheal wall (we assume it is very slow) and the "perfect" elasticity of the wall, we can model the velocity of flow during a cough by the equation†

$$V = \alpha(R_0 - R)R^2 \text{ centimeters per second,}$$

where α is a constant depending on the length of the trachea wall. It is to be hoped that if you are coughing efficiently, your tracheal wall will contract in such a way as to maximize the velocity of air going out of the trachea. Show that V is maximized when the trachea is contracted by one-third of its original radius (so that $R = \frac{2}{3}R_0$). This result has been confirmed, approximately, by X-ray photographs taken during actual coughs.

32. A storage container is constructed in the shape of a V from a rectangular sheet of stainless steel, 6 m long and 2 m wide, by bending the sheet through the middle. What angle should be left between the sides of the container in order to maximize the volume? Assume that triangular pieces, provided free of charge, are welded onto each end.

33. A light is hung over the center of a table 4 m square. The intensity of the light hitting a point on the table is directly proportional to the sine of the angle the path of the light makes with the table and inversely proportional to the distance between the point and the light. How high above the table should the light be placed in order to maximize the light intensity at the corners of the table?

*34. Suppose that a_1, a_2, a_3, and a_4 are four separate estimates of the physical quantity A (e.g., the altitude of a volcanic cone just after an eruption).
   (a) Find an x that minimizes $(a_1 - x)^2 + (a_2 - x)^2$.
   (b) Find a y that minimizes $\sqrt{(a_1 - y)^2 + (a_2 - y)^2 + (a_3 - y)^2}$.
   (c) Find a z that minimizes $\sqrt{(a_1 - z)^2 + (a_2 - z)^2 + (a_3 - z)^2 + (a_4 - z)^2}$.

*35. A farmer has 200 m of fencing. He wishes to construct a given number of rectangular pens by running two fences from east to west and the necessary number of fences from north to south. If the total area of the pens is to be a maximum, what is the total length of the E–W fences and the total length of the N–S fences?

4.7 SOME APPLICATIONS IN ECONOMICS (Optional)

We saw in Chapter 2 (see Example 2.6.5) that the derivative can be used to represent the marginal cost and marginal revenue in producing or selling a given product. The idea of margin can also be applied to other important notions in economics, such as demand, consumption, profit, and savings. The following is a summary of some of the important terms we will need in this section. Other terms will be introduced later.

---

†This equation and a detailed description of this problem appear in Philip Tuchinsky's paper, "The human cough," UMAP Project, Education Development Center, Newton, Mass. (1978).

(i) The **total cost function** gives the cost C of producing q units of a given product. A typical cost function is

$$C = aq^2 + bq + c, \tag{1}$$

where the number c represents the *fixed cost* that will have to be incurred even if nothing is produced (for rent, depreciation, utilities, etc.). Fixed cost is often referred to as *overhead*.

(ii) The **total revenue function** gives the amount R received for selling q units of the product. A typical revenue function is

$$R = aq + bq^2. \tag{2}$$

Revenue can often be calculated by multiplying the price times the number of items sold.

(iii) The **profit function** gives the profit P received when q units of the product are sold. In simple models we will have

$$P = R - C. \tag{3}$$

(iv) The **demand function** expresses the relationship between the unit price that a product can sell for and the number of units that can be sold at that price. Typically, the more units sold, the lower the price. So if p represents the price per unit sold, dp/dq will be negative. A typical demand function is

$$p = a - bq. \tag{4}$$

We now give some examples of how these four functions can be used.

EXAMPLE 1 A toy manufacturer finds that the cost in dollars of producing q copies of a certain doll is given by

$$C = 250 + 3q + 0.01q^2.$$

The dolls can be sold for \$14 each. How many should the manufacturer produce to maximize his profit, assuming that he can sell all he produces?

Solution. Since the price does not vary, the revenue is $14q$ dollars. Thus

$$P = R - C = 14q - (250 + 3q + 0.01q^2)$$
$$= 11q - 0.01q^2 - 250.$$

Then the marginal profit dP/dq is $11 - 0.02q = 0$ when $q = 11/0.02 = 550$ units. Since $d^2P/dq^2 = -0.02$, there is a local maximum at $q = 550$. When $q = 0$, $P = -250$. Also, as q gets very large, the profit becomes negative (since the term $0.01q^2$ becomes larger than the q term). Therefore the maximum profit is $P = 11 \cdot 550 - (0.01)(550)^2 - 250$ $= 6050 - 3025 - 250 = \$2775$, and the answer to the problem is 550 dolls. ■

EXAMPLE 2 A manufacturer of men's shirts figures that her exclusive "Parisian" model will cost $500 for overhead plus $9 for each shirt made. The price she can get for the shirts depends on how exclusive they are. From experience her accountant has estimated the following demand function:

$$p = 30 - 0.2\sqrt{q},$$

where q is the number of shirts sold. How many shirts should the manufacturer produce in order to maximize profit? (Assume that all the shirts produced will be sold.)

Solution. From the information given we have

$$C = 500 + 9q,$$
$$R = (\text{price}) \times (\text{number sold}) = (30 - 0.2\sqrt{q})q = 30q - 0.2q^{3/2},$$

and

$$P = R - C = 21q - 0.2q^{3/2} - 500. \tag{5}$$

Then the marginal profit dP/dq is $21 - 0.3\sqrt{q} = 0$ when $\sqrt{q} = 21/0.3 = 70$ or when $q = 70^2 = 4900$ shirts. Since $d^2P/dq^2 = -0.3/2\sqrt{q}$, which is less than 0, there is a local maximum at $q = 4900$. This result is easily seen to be a true maximum as well and is therefore the answer to the problem. At a production level of 4900,

$$P = 21 \cdot 4900 - (0.2)(4900)^{3/2} - 500$$
$$= 102{,}900 - (0.2)(343{,}000) - 500 = \$33{,}800.$$

Note that if 4900 shirts are sold, then they will be sold at a price of $p = 30 - 0.2(70) = 30 - 14 = \$16$ each. ∎

EXAMPLE 3 A certain manufacturer has a steady demand for 50,000 refrigerators each year. The machines are not made continuously but, rather, in equally sized batches. Production costs are $10,000 to set up the machinery plus $100 for each refrigerator made. In addition, there is a storage (inventory) charge of $2.50 per year for each refrigerator stored.† If the demand is steady throughout the year, how should the manufacturer schedule production runs so as to minimize total costs?

Solution. We have

$$C = \text{total cost} = \text{production cost} + \text{inventory cost} = PC + IC.$$

If batches are in lots of q refrigerators at a time, then there must be $50{,}000/q$ runs each year. The production cost will then be

---

† In economics the total cost is often broken down into **fixed costs** and **variable costs**. Thus total cost = fixed cost + variable cost. In this problem the fixed cost is $10,000 and the variable cost is $100 per refrigerator made + $2.50 per refrigerator stored.

$$PC = \left(\frac{50,000}{q}\right) \cdot 10,000 + (50,000)(100) = \frac{500,000,000}{q} + 5,000,000.$$

Now we calculate inventory costs. It is reasonable to assume that production is scheduled so that one new batch is complete just as the previous batch has run out. Therefore inventory starts at q units and decreases steadily (because of the steady demand) to 0 units. The average number of units in storage at any one time will then be $q/2$ units. Therefore the inventory costs are

$$IC = 2.50\frac{q}{2} = 1.25q.$$

Then we find that

$$C = \frac{500,000,000}{q} + 5,000,000 + 1.25q,$$

and the marginal cost dC/dq is $-(500,000,000/q^2) + 1.25$. This expression is zero when

$$1.25q^2 = 500,000,000,$$

or when

$$q^2 = \frac{500,000,000}{1.25} = 400,000,000 \quad \text{and} \quad q = \pm 20,000.$$

Of course, the value $q = -20,000$ is meaningless. We can then verify that a minimum is indeed reached when $q = 20,000$. If $q = 20,000$, then $50,000/q = 2\frac{1}{2}$, which means that costs will be minimized when there are $2\frac{1}{2}$ runs a year, which works out (practically) to 5 runs every two years. The minimum annual cost is \$5,050,000 (check this result). ∎

Often a person in business will be faced with the choice between increasing or not increasing prices. Generally, the demand function will show that an increase in price will cause a drop in sales. But how much of a drop? If there were a very small drop, then revenue would increase. On the other hand, if a large drop in sales were to result, then revenue would fall. The **average price elasticity of demand** η_{av} (Greek letter eta) is defined as the *relative change* in quantity (q) demanded divided by the relative change in price (p):

$$\text{average price elasticity of demand} = \eta_{av} = -\frac{\Delta q/q}{\Delta p/p}. \tag{6}$$

The relative change in q is the change Δq divided by q, and likewise for the relative change in p. Note that $\Delta q/q$ can also be thought of as the percent change in q. The minus sign in (6) is put there so that η_{av} will turn out to be positive. η_{av} will be positive since if $\Delta p > 0$, Δq will usually be less than 0 (why?).

   In general, if $\eta_{av} < 1$, then the percent decrease in demand is less than the percent increase in price, so that an increase in price will lead to an increase in revenue. If $\eta_{av} > 1$, then the opposite is true. If $\eta_{av} = 1$, then the price increase will not make any difference. The loss in demand will just offset the revenue gained by the increase in price.†

EXAMPLE 4   In Example 2 suppose that the demand function is given by

$$p(q) = 30 - 0.2\sqrt{q}. \tag{7}$$

Calculate the elasticity of demand if the price per shirt is increased from $20 to $22.

   Solution. At a price of $20, we calculate from (7) that $20 = 30 - 0.2\sqrt{q}$, $\sqrt{q} = 10/0.2 = 50$, and $q = 2500$. At $p = 22$, $q = 1600$. Therefore we have $p = 20$, $\Delta p = 2$, $q = 2500$, $\Delta q = -900$, and

$$\eta_{av} = -\frac{(-900)/2500}{2/20} = \frac{90}{25} = 3.6.$$

Since this number is greater than 1, the percent loss in demand (the numerator of η_{av}) is greater than the percent increase in price. Therefore the increase of $2 would result in a net *decrease* in revenue. It also would result in a net decrease in profits. To verify this, we use the profit formula, (5).

$$P(2500) = 21 \cdot 2500 - 0.2(2500)^{3/2} - 500 = \$27{,}000,$$

and

$$P(1600) = 21 \cdot 1600 - 0.2(1600)^{3/2} - 500 = \$20{,}300.$$

There is a decrease in profits of $6700 due to the price increase. ■

   We return to the question whether the person in business should increase or decrease prices. Put another way, will a very small increase in price lead to more or less revenue? To answer this question, let $p = p(q)$ represent the demand function. If $p(q)$ is continuous (as it is assumed to be), a small change Δp in price will be caused by a small change Δq in demand. Then we define the **price elasticity of demand** $\eta(q)$ when the demand is q and the price is $p(q)$ as the limit of η_{av} as $\Delta q \to 0$. That is,

$$\eta(q) = \lim_{\Delta q \to 0} \eta_{av} = \lim_{\Delta q \to 0} -\frac{\Delta q/q}{\Delta p/p} = \lim_{\Delta q \to 0} -\frac{p(q)}{q\,\Delta p/\Delta q} = -\frac{p}{qp'(q)}. \tag{8}$$

†In the terminology of economics a demand curve is **elastic** if $|\eta_{av}| > 1$, is of **unit elasticity** if $|\eta_{av}| = 1$, or is **inelastic** if $|\eta_{av}| < 1$.

EXAMPLE 5 The demand function for a certain electric toaster is $p(q) = 35 - 0.05\sqrt{q}$. Will a rise in price increase or decrease revenue if 10,000 toasters are in demand?

Solution. We calculate η. If $\eta > 1$, then, as before, an increase in price will cause a decrease in revenue, if $\eta < 1$, then an increase in price will cause an increase in revenue. Here

$$p'(q) = -\frac{0.05}{2\sqrt{q}} = -\frac{0.05}{2\sqrt{10{,}000}} = -\frac{0.05}{200} = -0.00025.$$

When $q = 10{,}000$, $p = 35 - 0.05\sqrt{10{,}000} = 30$, and

$$\eta = -\frac{p}{qp'(q)} = \frac{-30}{(10{,}000)(-0.00025)} = \frac{30}{2.5} = 12.$$

Since $\eta > 1$, there will be a *decrease* in revenue if the price is raised. ∎

EXAMPLE 6 Answer the question in Example 5 if 250,000 toasters are in demand.

Solution. When $q = 250{,}000$, $p = 35 - 0.05\sqrt{250{,}000} = 35 - (0.05)(500) = 10$, $p'(250{,}000) = -0.05/2(500) = -0.00005$, and

$$\eta = \frac{-10}{(250{,}000)(-0.00005)} = \frac{-10}{-12.5} = 0.8 < 1.$$

So there would be an increase in revenue if prices were increased. ∎

The notion of elasticity of demand has an interesting interpretation in global economics. Suppose that certain economists in the United States are concerned about an imbalance in the balance of payments. That is, there are more dollars going out than are coming in. To offset this problem, they suggest a devaluation of the dollar. The devaluation would make dollars cheaper abroad, thereby making American goods cheaper to foreign consumers. Therefore there would be an increase in the foreign purchase of U.S. goods, thereby leading to an increase in the number of export dollars flowing back to the United States. Or will something else happen? The economists must be careful. For if the elasticity of demand for American exports is less than 1, more American products would indeed be purchased abroad but there would be a net *decrease* in the value of the dollars paid for these goods. Things are never simple in the area of international trade.

PROBLEMS 4.7

1. The cost of producing q color television sets is given by $C = 5000 + 250q - 0.01q^2$. The revenue received is $R = 400q - 0.02q^2$. Assuming that all sets produced will be sold, how many should be produced so as to maximize the profit?
2. What is the demand function in Problem 1?
3. In Problem 1, at a production level of 10,000 sets, will an increase in price generate an increase or decrease in revenue?

4. Bottles of whiskey cost a distiller in Scotland $2 a bottle to produce. In addition, he has fixed costs of $500. The demand function worldwide for the whiskey is given by $p = 12 - 0.001q$. How many bottles should he produce to maximize his profit?

5. The distiller of Problem 4 now sells exclusively to the United States. He must pay $0.50 per bottle in duty. The demand function does not change. How does he maximize his profit in this case?

*6. The distiller of Problem 4 shifts his sales to France, where an import duty of 20% of the sales price is charged. How does he now maximize his profit?

7. In Problems 4, 5, and 6, determine whether a price increase will result in an increase or decrease in revenue at a production level of 10,000 bottles.

8. Show for any problem of the type we have considered that whenever profit is maximized, the marginal cost and the marginal revenue are equal.

9. A manufacturer of kitchen sinks finds that if she produces q sinks per week, she has fixed costs of $1000, labor and materials costs of $5/sink, and advertising costs of $100\sqrt{q}$. How many sinks should she manufacture weekly to minimize costs?

10. The demand function for the sinks of Problem 9 is $p(q) = 25 - 0.01q$. How many sinks should be manufactured weekly to maximize profits?

11. In Problem 10, at a level of production of 2,000 sinks, will an increase in price lead to an increase or decrease in the revenue?

12. The demand function for a certain manufactured good is $p(q) = 75 - 0.1\sqrt{q} - 0.002q^2$. At what level of production will it not make any difference in the revenue whether the price is increased or decreased?

13. A Detroit manufacturer has a steady annual demand for 50,000 pickup trucks. The trucks are made in batches. The costs of production include a $20,000 setup cost per batch and a cost of $2500 per truck. Inventory charges are $50 per truck per year. How is production to be scheduled to minimize total costs?

14. Suppose that the manufacturer in Problem 13 produces trucks continuously (so that there is only one setup cost), but the demand is not constant and is given by $q = 5000 + (300,000/\sqrt{p})$, where p is the price in dollars and q is the demand. It is assumed that if fewer than 10,000 trucks are sold, the manufacturer will go bankrupt. What should be the price charged for each truck to maximize the total profits? [*Hint:* Find the demand function by writing p as a function of q.]

15. In Problem 14 determine the elasticity function.

16. In Problem 14, at a level of production of 25,000 trucks, will it be profitable to increase prices?

17. A woman has $10,000 to invest in two companies. The return from investing q dollars in company 1 is $4\sqrt{q}$ dollars, and the return from investing q dollars in company 2 is $2\sqrt{q}$ dollars. How should she invest her money so as to maximize her return?

18. The Grosshop Company does public opinion polls. They have observed that the cost of conducting a national survey of n people is

$$C(n) = 25,000 + 0.02(n - 1500)^2.$$

Of course, the more people surveyed, the better are the results (up to a point). They have estimated that the value (in dollars, since better accuracy ensures greater profits) is given by

$$V(n) = 500,000 - 0.1(n - 8,000)^2.$$

If "profit" is defined by $P(n) = V(n) - C(n)$, what is the optimal number of people to be polled (to the nearest person)?

*19. Show that if the demand law is given as $q = q(p)$ (i.e., demand as a function of price rather than vice versa), then

$$\eta(p) \;=\; -\,\frac{pq'(p)}{q(p)}.$$

(This equation expresses the elasticity in terms of price rather than in terms of demand.)

20. Show that if $q(p) = 5/p^4$, then $\eta(p) = 4$.

21. Show, in general, that if $q(p) = a/p^\alpha$, with $\alpha > 0$, then $\eta(p) = \alpha$.

*22. Show that if $p = p(q)$ is the demand function, then at a level of production that maximizes total revenue, the elasticity is 1.

*23. Let the cost function be $C = aq^2 + bq + c$ and the demand function be $p = \alpha - \beta q$, where a, b, c, α, and β are positive.

    (a) At what level of production is profit maximized?

    (b) What is the elasticity?

    (c) At what level of production does it make no difference whether the price is increased or decreased?

4.8 NEWTON'S METHOD FOR SOLVING EQUATIONS

In this section we look at a very different kind of application of the derivative. Consider the equation

$$f(x) = 0, \tag{1}$$

where f is assumed to be differentiable in some interval $[a, b]$. It is often important to calculate the *roots* of equation (1), that is, the values of x that satisfy the equation. For example, if $f(x)$ is a polynomial of degree 5, say, then the roots of $f(x)$ could be as in Figure 1. (We have already used graphs to determine the number of roots of certain polynomials; see pages 222 and 224.)

In the seventeenth century Newton discovered a method for estimating a solution, or root, by defining a sequence of numbers that become successively closer and closer to the root sought. His method is best illustrated graphically. Let $y = f(x)$ as in Figure 2. A number x_0 is chosen arbitrarily. We then locate the point $(x_0, f(x_0))$ on the graph and draw the tangent line to the curve at that point. Next, we follow the tangent

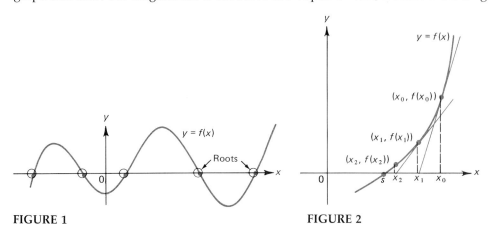

FIGURE 1   **FIGURE 2**

line down until it hits the x-axis. The point of intersection of the tangent line and the x-axis is called x_1. We then repeat the process to arrive at the next point, x_2. On the graph we have labeled the solution to $f(x) = 0$ as s. That is, $f(s) = 0$. For our graph at least, it seems as if the points $x_0, x_1, x_2, \ldots$ are approaching the point $x = s$. In fact, this happens for quite a few functions, and the rate of approach to the solution is quite rapid.

Having briefly looked at a graphical representation of Newton's method, let us next develop a formula for giving us x_1 from x_0, x_2 from x_1, and so on. The slope of the tangent line at the point $(x_0\ f(x_0))$ is $f'(x_0)$. Two points on this line are $(x_1, 0)$ and $(x_0, f(x_0))$. Therefore

$$\frac{0 - f(x_0)}{x_1 - x_0} = f'(x_0). \tag{2}$$

Solving (2) for x_1 gives us

$$x_1 = x_0 - \frac{f(x_0)}{f'(x_0)}. \tag{3}$$

Similarly,

$$x_2 = x_1 - \frac{f(x_1)}{f'(x_1)}. \tag{4}$$

In general, we obtain

$$x_{n+1} = x_n - \frac{f(x_n)}{f'(x_n)}. \tag{5}$$

This last step tells us how to obtain the $(n + 1)$st point if the nth point is given, as long as $f'(x_n) \neq 0$ so that (5) is defined. Thus if we start with a given value x_0, we can obtain $x_1, x_2, x_3, x_4, \ldots$. The formula (5) is called **Newton's formula.** The set of numbers $x_0, x_1, x_2, x_3, \ldots$ is called a **sequence.**† If the numbers in the sequence get closer and closer to a certain number s as n gets larger and larger, then we say that the sequence **converges** to s, and we write

$$\lim_{n \to \infty} x_n = s. \tag{6}$$

Before stating the theorem that guarantees that the sequence defined by (5) converges to a solution of $f(x) = 0$, we give two simple examples.

EXAMPLE 1 Let $r > 1$. Formulate a rule for calculating the square root of r.

---

†We will discuss sequences extensively in Chapter 14.

Solution. We must find an x such that $x = \sqrt{r}$, or $x^2 = r$, or $x^2 - r = 0$. Let $f(x) = x^2 - r$. Then if $f(s) = 0$, s will be a square root of r. [$f(s) = 0$ means that $s^2 - r = 0$, or $s^2 = r$.] By Newton's formula, since $f'(x) = 2x$, we obtain the sequence $x_0, x_1, x_2, \ldots$, where x_0 is arbitrary and

$$x_{n+1} = x_n - \frac{f(x_n)}{f'(x_n)} = x_n - \frac{(x_n{}^2 - r)}{2x_n} = \frac{2x_n{}^2 - x_n{}^2 + r}{2x_n} = \frac{1}{2}\left(x_n + \frac{r}{x_n}\right).^{\dagger} \quad ∎ \quad (7)$$

EXAMPLE 2   Calculate $\sqrt{2}$ by Newton's method.

Solution. In formula (7), $r = 2$, so that

$$x_{n+1} = \frac{1}{2}\left(x_n + \frac{2}{x_n}\right). \tag{8}$$

Using a calculator, we obtain the sequence in Table 1, starting with $x_0 = 1$. We can see here the remarkable accuracy of Newton's method. An answer correct to nine decimal places was obtained after only four steps! We were limited in accuracy only by the fact that our calculator could display only ten digits.   ∎

TABLE 1

n	x_n	$\dfrac{2}{x_n}$	$x_n + \dfrac{2}{x_n}$	x_{n+1} $=\dfrac{1}{2}\left(x_n + \dfrac{2}{x_n}\right)$
0	1.0	2.0	3.0	1.5
1	1.5	1.333333333	2.833333333	1.416666667
2	1.416666667	1.411764706	2.828431373	1.414215686
3	1.414215686	1.414211438	2.828427125	1.414213562
4	1.414213562	1.414213562	2.828427125	1.414213562

We now state a theorem that gives conditions that guarantee that Newton's method will work. The important fact is that we must usually choose a "starting value" x_0 reasonably close to the solution s. The proof of this theorem cannot be given here.‡

Theorem 1.   Suppose that on some interval $[a, b]$, $f(x)$ is defined, f'' exists and is continuous, and the following conditions hold:

    (i) $f(a)$ and $f(b)$ have different signs.
    (ii) $f'(x) \neq 0$ for every x in $[a, b]$.
    (iii) $f''(x)$ does not change sign in $[a, b]$.
    (iv) If $|f'(a)| \leq |f'(b)|$, then $|f(a)|/|f'(a)| \leq b - a$. Or if $|f'(b)| \leq |f'(a)|$, then $|f(b)|/|f'(b)| \leq b - a$.

---

†Note that this formula requires only the basic arithmetic operations of addition, subtraction, multiplication, and division. These are the operations most easily performed in a calculator, which is why many calculators use this formula, or a simple modification of it, to find square roots.

‡For a proof, see Peter Henrici, *Elements of Numerical Analysis* (Wiley, New York, 1969), p. 79.

Then the sequence $x_0, x_1, x_2, x_3, \ldots$ defined by (5) converges to the *unique* solution s of $f(x) = 0$ for every initial choice x_0 in $[a, b]$.

EXAMPLE 3   In Example 2, $f(x) = x^2 - 2$. Let us verify that the conditions of Theorem 1 hold. If $[a, b] = [1, 10]$, for example, then $f(1) = -1$ and $f(10) = 98$, so $f(a)$ and $f(b)$ have different signs. Also, $f'(x) = 2x$, which is nonzero in the interval. Clearly (iii) holds since $f''(x) = 2$, a constant. Finally, $|f'(1)| < |f'(10)|$, and

$$\left|\frac{f(1)}{f'(1)}\right| = \left|-\frac{1}{2}\right| = \frac{1}{2} < b - a = 9.$$

Therefore all conditions are satisfied. ∎

EXAMPLE 4   Formulate a rule for calculating the kth root of a given number r.

Solution. We must find an x such that $x = r^{1/k}$, or $x^k = r$, or $f(x) = x^k - r = 0$. Then $f'(x) = kx^{k-1}$, and

$$x_{n+1} = x_n - \frac{x_n^{\,k} - r}{kx_n^{\,k-1}} = x_n - \frac{1}{k}\frac{x_n^{\,k}}{x_n^{\,k-1}} + \frac{r}{kx_n^{\,k-1}} = \left(1 - \frac{1}{k}\right)x_n + \frac{r}{kx_n^{\,k-1}}. \tag{9}$$
∎

EXAMPLE 5   Calculate $\sqrt[3]{17}$.

Solution. By (9), with $k = 3$ and $r = 17$, we have $x_{n+1} = \frac{2}{3}x_n + (17/3x_n^{\,2})$. Since $\sqrt[3]{17}$ is between 2 and 3, we choose $x_0 = 2$ (3 would do just as well). If $[a, b] = [2, 3]$, then since

$$f(x) = x^3 - 17, \qquad f(2) = -9, \qquad f(3) = 10,$$

and (i) is satisfied. Also, $f'(x) = 3x^2$ and $f''(x) = 6x$, so that (ii) and (iii) are satisfied on $[2, 3]$. Finally, $f'(2) = 12$, $f'(3) = 27$, and

$$\left|\frac{f(2)}{f'(2)}\right| = \left|-\frac{9}{12}\right| = \frac{3}{4} < b - a = 1.$$

Thus we know that our sequence will converge to the unique cube root of 17 in $[2, 3]$. Values of x_n are tabulated in Table 2. The last number is correct to nine decimal places. Again, the rapid convergence of Newton's method is illustrated. ∎

TABLE 2

n	x_n	$\dfrac{2}{3}x_n$	$x_n^{\,2}$	$\dfrac{17}{3x_n^{\,2}}$	$x_{n+1} = \dfrac{2}{3}x_n + \dfrac{17}{3x_n^{\,2}}$
0	2.0	1.333333333	4.0	1.416666667	2.75
1	2.75	1.833333333	7.5625	0.7493112948	2.582644628
2	2.582644628	1.721763085	6.670053275	0.8495684267	2.571331512
3	2.571331512	1.714221008	6.611745745	0.8570605836	2.571281592
4	2.571281592	1.714187728	6.611489025	0.8570938627	2.571281591

EXAMPLE 6   Find the real roots of $p(x) = x^3 + x^2 + 7x - 3$.

Solution. We differentiate to find that $p'(x) = 3x^2 + 2x + 7$. This polynomial has no real roots (explain why), so there are no local maxima or minima. Also, $p''(x) = 6x + 2 = 0$ when $x = -\frac{1}{3}$, so that $(-\frac{1}{3}, -\frac{142}{27})$ is a point of inflection. The graph of $p(x)$ is given in Figure 3. From the graph we see that there is exactly one real root. Also, since $p(0) = -3$ and $p(1) = 6$, the root must lie between 0 and 1. In the interval $[0, 1]$ the conditions of Theorem 1 can be easily verified. Now

$$x_{n+1} = x_n - \frac{p(x_n)}{p'(x_n)} = x_n - \frac{x_n^3 + x_n^2 + 7x_n - 3}{3x_n^2 + 2x_n + 7}$$

$$= \frac{x_n(3x_n^2 + 2x_n + 7) - (x_n^3 + x_n^2 + 7x_n - 3)}{3x_n^2 + 2x_n + 7} = \frac{2x_n^3 + x_n^2 + 3}{3x_n^2 + 2x_n + 7}.$$

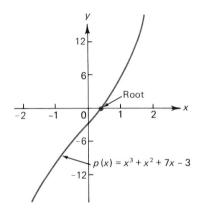

FIGURE 3

If we choose $x_0 = 0$, we obtain the results in Table 3. The root is $s = 0.3970992165$, correct to ten decimal places.

TABLE 3

n	x_n	$2x_n^3 + x_n^2 + 3$	$3x_n^2 + 2x_n + 7$	$x_{n+1} = \dfrac{2x_n^3 + x_n^2 + 3}{3x_n^2 + 2x_n + 7}$
0	0	3.0	7.0	0.4285714286
1	0.4285714286	3.341107872	8.408163265	0.3973647712
2	0.3973647712	3.283385572	8.268425827	0.3970992352
3	0.3970992352	3.282923214	8.267261878	0.3970992164
4	0.3970992164	3.282923182	8.267261796	0.3970992165
5	0.3970992165	3.282923182	8.267261796	0.3970992165

EXAMPLE 7   Find all roots (if any) of the equation $\sin x = x/2$ in the interval $(0, 2\pi]$.

Solution. The graphs of $y = \sin x$ and $y = x/2$ are given in Figure 4. From these it is evident that there is exactly one real root of the equation $\sin x = x/2$ in $(0, 2\pi]$, and it lies in the interval $[\pi/2, \pi]$.

Setting $f(x) = \sin x - x/2$, we have $f'(x) = \cos x - 1/2$, and Newton's method provides the rule

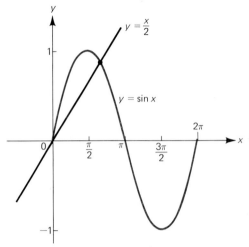

FIGURE 4

$$x_{n+1} = x_n - \frac{f(x_n)}{f'(x_n)} = x_n - \frac{\sin x_n - x_n/2}{\cos x_n - 1/2} = x_n - \frac{2\sin x_n - x_n}{2\cos x_n - 1}.$$

Since we know that the root is in $[\pi/2, \pi]$, we can choose this interval to be our interval. We now verify the four conditions in Theorem 1.

(i) $f(\pi/2) = 1 - \pi/4 > 0$, and $f(\pi) = 0 - \pi/2 < 0$.
(ii) $f'(x) = \cos x - 1/2$, which is negative in $[\pi/2, \pi]$ since $\cos x$ is negative in the second quadrant.
(iii) $f''(x) = -\sin x$, which is negative in $[\pi/2, \pi)$.
(iv) $|f'(\pi)| = 3/2 > |f'(\pi/2)| = 1/2$, and

$$\left|\frac{f(\pi/2)}{f'(\pi/2)}\right| = \left|\frac{1 - (\pi/4)}{1/2}\right| = \left|2 - \frac{\pi}{2}\right| \approx 0.43 < b - a.$$

Thus the Newton iterates will converge for any $x_0 \in [\pi/2, \pi]$.

We start with $x_0 = 2$ and carry out the iteration given in Table 4. We find that the unique solution (correct to ten significant figures) is $x_n = 1.895494267$ radians ($\approx 108.6°$).

TABLE 4

n	x_n	$\sin x_n$	$2\sin x_n - x_n$	$\cos x_n$
0	2	0.9092974268	−0.1814051463	−0.4161468365
1	1.900995594	0.9459777535	−0.0090400871	−0.3242315374
2	1.895511645	0.9477415893	−0.0000284668	−0.3190389944
3	1.895494267	0.9477471335	−0.0000000003	−0.3190225243

n	$2\cos x_n - 1$	$\dfrac{2\sin x_n - x_n}{2\cos x_n - 1}$	$x_{n+1} = x_n - \dfrac{2\sin x_n - x_n}{2\cos x_n - 1}$
0	−1.832293673	0.0990044055	1.900995594
1	−1.648463075	0.0054839488	1.895511645
2	−1.638077989	0.0000173783	1.895494267
3	−1.638045049	0.0000000002	1.895494267

Let us make some final remarks about Newton's method. First, if x_n is a solution to $f(x) = 0$, then

$$x_{n+1} = x_n - \frac{f(x_n)}{f'(x_n)} = x_n - 0 = x_n.$$

This result explains why, in Tables 1, 3, and 4, the value of x_n did not change after the desired accuracy was reached. Second, it is not necessary to check the conditions of Theorem 1 every time Newton's method is used. However, failure to do so could result in one or both of the following problems:

(i) The sequence generated by (5) could grow without bound, never even getting close to a solution.
(ii) The sequence could converge to a solution, but there could be *other* solutions in the interval $[a, b]$ that would be missed. This problem can be avoided if a graph is drawn that gives an idea about where (approximately) the roots are located.

An interesting example of the need to find the root of a polynomial will be given in Example 6.7.14.

PROBLEMS 4.8

In the problems below calculate all answers to as many decimal places of accuracy as are displayed on your calculator.

1. Find an interval over which Theorem 1 applies to the calculation of $\sqrt{90}$. Perform the calculation.
2. Calculate $\sqrt[4]{25}$ by using Newton's method.
3. Calculate $\sqrt[5]{10}$ by using Newton's method.
4. Calculate $\sqrt[6]{100}$ by using Newton's method.
5. Use Newton's method to calculate the roots of $x^2 - 7x + 5 = 0$. It will be necessary to do two separate calculations, using two distinct intervals. Compare this result with the answers obtained by the quadratic formula.
6. Show that there is *no* interval $[a, b]$ for which Theorem 1 applies when $f(x) = x^2 + 5x + 7$. Explain why this must be the case. Try Newton's method with $x_0 = 0$. What happens?
7. Find all solutions of the equation $x^3 - 6x^2 - 15x + 4 = 0$. [*Hint*: Draw a sketch and estimate the roots to the nearest integer. Then use this estimate as your initial choice of x_0 for each of the three roots.]
8. Find all solutions of the equation $x^3 + 14x^2 + 60x + 105 = 0$.
9. Find all solutions of $x^3 - 8x^2 + 2x - 15 = 0$.
*10. Find all solutions of $x^3 + 3x^2 - 24x - 40 = 0$.
11. Using Newton's method, find a formula for finding reciprocals without dividing (the reciprocal of x is $1/x$).
12. Using the formula found in Problem 11, calculate $\frac{1}{7}$ and $\frac{1}{81}$.
13. (a) Show graphically that there is a unique solution to $\cos x = x$ in the interval $[0, \pi/2]$.
    (b) Use Newton's method to find it.
*14. Use Newton's method to find the unique solution to $x = \tan x$ in the interval $(0, 3\pi/2)$.
15. Try to use Newton's method to find a solution to $\sin x = x$ in $[\pi/4, \pi]$. What happens? Which of the hypotheses of Theorem 1 are violated? Why is the method doomed to fail?

REVIEW EXERCISES FOR CHAPTER FOUR

In Exercises 1–4, find the c that satisfies the conclusion of the mean value theorem for the given function and numbers.

1. $f(x) = x^5; a = 0, b = 1$

2. $f(x) = \dfrac{x}{x + 1}; a = 0, b = 2$

3. $f(x) = \sqrt[5]{x}; a = -32, b = -1$

4. $f(x) = \sin x; a = 0, b = \pi$

In Exercises 5–20, graph the given curves by following the eight steps outlined in Section 4.4.

5. $y = x^2 - 3x - 4$

6. $y = x^3 - 3x^2 - 9x + 25$

7. $y = x^5 + 2$

8. $y = \sqrt[3]{x}$

9. $y = \sqrt[4]{x}$

10. $y = \dfrac{1}{x + 2}$

11. $y = x(x - 1)(x - 2)(x - 3)$

12. $y = \dfrac{x + 1}{x + 2}$

13. $y - \dfrac{1}{x^2 - 1}$

14. $y - -x(x - 1)^3$

15. $y = |x - 4|$

16. $y = |x^2 - 4|$

17. $y = \sin\dfrac{x + 1}{2}$

18. $y = \sin^2\left(\dfrac{x}{3}\right)$

19. $y = \cot x$

20. $y = \csc x$

21. A rope is attached to a pulley mounted on top of a 5-m tower. One end of the rope is attached to a heavy mass. If the rope is pulled in at the rate of $1\frac{1}{2}$ m/sec, how fast is the mass approaching the tower when it is 3 m from the base of the tower?

22. A storage tank is in the shape of an inverted right circular cone with radius 6 ft and height 14 ft. If water is pumped into the tank at a rate of 20 gal/min, how fast is the water level rising when it has reached a height of 5 ft? [*Hint:* 1 gal ≈ 0.1337 ft^3.]

23. Find the maximum and minimum values for the following functions over the indicated intervals.
(a) $y = 2x^3 + 9x^2 - 24x + 3; [-2, 5]$ (b) $y = (x - 2)^{1/3}; [0, 4]$
(c) $y = (x + 1)/(x^2 - 4); [-1, 1]$

24. What is the maximum rectangular area that can be enclosed with 800 m of wire fencing?

25. Find the point on the line $2x - 3y = 6$ that is nearest the origin.

26. A cylindrical barrel is to be constructed to hold 128π ft^3 of liquid. The cost per square foot of constructing the side of the barrel is three times that of constructing the top and bottom. What are the dimensions of the barrel that costs the least to construct?

27. The population of a certain species is given by $P(t) = 1000 + 800t + 96t^2 + 12t^3 - t^4$, where t is measured in weeks. After how many weeks is the instantaneous rate of population increase a maximum?

28. A producer of dog food finds that the cost in dollars of producing q cans of dog food is $C = 200 + 0.2q + 0.001q^2$. If the cans sell for $0.35 each, how many cans should be produced to maximize the profit?

29. The cost function for a certain product is $C(q) = 100 + (1000/q)$, and the demand function is $p(q) = 50 - 0.02q$. At what level of production is profit maximized?

30. Calculate the price elasticity of demand for the product of Exercise 29. At what level of production will it make no difference whether prices are increased or decreased?

31. Calculate $\sqrt[6]{135}$ to four decimal places using Newton's method.

32. Use Newton's method to find all roots of $x^3 - 2x^2 + 5x - 8 = 0$.

5 The Integral

5.1 INTRODUCTION

Modern calculus has its origins in two mathematical problems of antiquity. The first of these, the problem of finding the line tangent to a given curve, was, as we have noted, not solved until the seventeenth century. Its solution (by Newton and Leibniz) gave rise to what is known as *differential calculus*. The second of these problems was to find the area enclosed by a given curve. The solution of this problem led to what is now termed *integral calculus*.

It is not known how long scientists have been concerned with finding the area bounded by a curve. In 1858 Henry Rhind, an Egyptologist from Scotland, discovered fragments of a papyrus manuscript written in approximately 1650 B.C., which came to be known as the *Rhind papyrus*. † The Rhind papyrus (see Figure 1) contains 85 problems and was written by the Egyptian scribe Ahmes, who wrote that he copied the problems from an *earlier* manuscript. In Problem 50 Ahmes assumed that the area of a circular field with a diameter of nine units was the same as the area of a square with a side of eight units. If we compare this statement with the correct formula for the area of a circle, we find that

---

†For an interesting discussion of the Rhind papyrus and other similar finds, see C. B. Boyer, *A History of Mathematics* (Wiley, New York, 1968).

FIGURE 1 The Rhind Papyrus (fragment). *Courtesy of the British Museum.*

$$A = \pi r^2 = \pi\left(\frac{9}{2}\right)^2 = \text{(according to Ahmes)} \ 8^2,$$

or

$$\pi \approx \frac{64}{(4.5)^2} = \frac{64}{20.25} \approx 3.16.$$

Thus we see that *before 1650 B.C.* the Egyptians could calculate the area of a circle from the formula

$$A = 3.16r^2.$$

Since $\pi \approx 3.1416$, we see that this result is remarkably close, considering that the Egyptian formula dates back nearly four thousand years!

In ancient Greece there was much interest in obtaining methods for calculating the areas bounded by curves other than circles and rectangles. The problem was solved for a wide variety of curves by Archimedes of Syracuse (287–212 B.C.), who is considered by many to be the greatest mathematician who ever lived.† Archimedes used what he called the *method of exhaustion* to calculate the shaded area A bounded by a parabola. We discuss this method in Section 5.4.

Before discussing the method for computing areas, we will define a new function, called the *antiderivative*. The antiderivative of a function f is a new function F having the property that $dF/dx = f$. We will discuss antiderivatives in the next section.

It is natural to ask, "What do antiderivatives have to do with the computation of area?" The answer is given by the fundamental theorem of calculus (stated on page 309). It is this beautiful result that ties together the great work of Archimedes almost

---

†See the accompanying biographical sketch.

Archimedes of Syracuse

Archimedes of Syracuse
The Granger Collection

Throughout most of the Greek era, Alexandria was the center of mathematical activity. The greatest mathematician of antiquity, however, and perhaps the greatest the world has ever known, was born and died in the Roman town of Syracuse. The son of an astronomer, Archimedes of Syracuse may have been related to King Hieron of Syracuse. When he died in 212 B.C., records indicate that Archimedes was 75 years old, so the likely year of his birth was 287 B.C.

How do we measure greatness? In the case of Archimedes it is easy. The breadth and depth of his discoveries is astonishing. While he did pioneering theoretical work in a great number of scientific areas, he also invented and improved upon many useful mechanical devices.

Archimedes is best known by two stories that are told about him. One of them is related in Plutarch's life of the Roman general Marcellus. During the second Punic War, Syracuse was besieged by Romans led by Marcellus. Archimedes helped to lead the defense of the city by designing catapults with adjustable ranges, cranes that could literally lift enemy ships out of the water, and movable poles that could drop boulders on ships that came within range of the city walls. These devices all made use of the principle of the lever, and Archimedes is said to have claimed that if he were given a lever long enough and a fulcrum on which to rest it, he could move the earth. The story of the siege of Syracuse had an unhappy ending. After Marcellus failed to take Syracuse by a frontal siege, the city fell after a circuitous attack. A Roman soldier was sent by Marcellus to bring Archimedes to him. According to one version of the story, Archimedes had drawn a diagram in the sand and asked the soldier to stand away from it. This angered the soldier, who killed Archimedes with his spear.

The second story about Archimedes concerns the gold crown of King Hieron. Hieron suspected a dishonest goldsmith of filling part of his crown with cheaper silver. The king asked Archimedes for help. Archimedes pondered the problem in his bathtub and hit upon the principle of buoyancy. Absentmindedly, Archimedes leaped out of his bath and ran naked down the street shouting, *"Eureka"* ("I have found it").

The principle of buoyancy was described in Archimedes' treatise *On Floating Bodies:*

> Any solid lighter than a fluid will, if placed in a fluid, be so far immersed that the weight of the solid will be equal to the weight of the fluid displaced.

Archimedes wrote another treatise that could properly be called the first book in mathematical physics: *On the Equilibrium of Planes.* It is concerned with centers of gravity of the triangle and the trapezoid (see Sections 9.4 and 9.5).

At least eight other of Archimedes' treatises have survived. Each work is a model of mathematical rigor, precision, and originality. In his work *On Spirals*, Archimedes computed the area swept out by a spiral (we see how to do this in Section 11.5). In his work *On Conoids and Spheroids*, he computed the area of an ellipse. In his treatise *On the Sphere and the Cylinder,*

Archimedes proved that if a sphere is inscribed in a cylinder the height of which is equal to its diameter, then the ratio of the volumes of the sphere and cylinder is equal to the ratio of their surface areas (3 to 2). He was so pleased with this result that he had a sketch of a sphere inscribed in a cylinder carved on his tombstone.

For the student of calculus, perhaps the most interesting of Archimedes' works is *Quadrature of the Parabola*. In this work Archimedes described a method for estimating the area under a parabola by computing the area under a sum of rectangles. This method, called the *method of exhaustion*, forms the basis for the study of integral calculus. We will describe this method in Section 5.4.

One of the most exciting discoveries of modern times was made in 1906. The Danish scholar J. C. Heiberg had read that at Constantinople (now Istanbul) there was a mathematical palimpsest (a palimpsest is a parchment on which some of the writing has been washed off with new writing appearing in its place). The palimpsest that Heiberg found had been, fortunately, washed off poorly, and he was able to recognize it as a treatise by Archimedes titled, simply, *Method*. By using photographs, Heiberg was able to read most of the 185 leaves of parchment that made up the treatise. The parchment contained a tenth-century copy of the text by Archimedes. *Method* is written in the form of a letter to Archimedes' friend Eratosthenes. In it, Archimedes described the process by which he had discovered many of his theorems. It is a remarkable document because in all his other works he only presented finished, polished results with no indication of how he obtained them. *Method* is a fitting tribute to the genius of this most remarkable of mathematicians.

twenty-three hundred years ago and the much more recent work of Newton and Leibniz. With this theorem we will see that the two ancient problems are really very closely related.

5.2 ANTIDERIVATIVES

In Chapters 2 and 3 we discussed the problem of finding the derivative of a given function. We now consider the problem in reverse.

Definition 1   ANTIDERIVATIVE   Let f be defined on $[a, b]$. Then if there exists a function $y = F(x)$ such that F is continuous on $[a, b]$, differentiable on (a, b), and the derivative of F is f for every x in (a, b), that is, if

$$F'(x) = \frac{dF}{dx} = f(x), \qquad x \text{ in } (a, b) \tag{1}$$

then F is called an **antiderivative,** or **indefinite integral,** of f on the interval $[a, b]$. We write

$$F = \int f(x)\, dx = \int f. \tag{2}$$

This notation is read "$F(x)$ is the integral of $f(x)$ with respect to x," or, simply, "F is an antiderivative of f." If such a function F exists, then f is said to be **integrable,** and the process of calculating an integral is called **integration.** The variable x is called the **variable of integration,** and the function f is called the **integrand.**

NOTATION. The symbol that we use to indicate an integration is a large German s, $\int$, first used by Leibniz in the seventeenth century.†

EXAMPLE 1 Find $\int 3x^2 \, dx$.

Solution. Since $d(x^3)/dx = 3x^2$, we have

$$\int 3x^2 \, dx = x^3.$$

But the derivative of any constant is zero, so that $x^3 + 3$ is also an antiderivative of $3x^2$, as is $x^3 - 5$, $x^3 + 2\pi$, ... In fact, for any constant C, $x^3 + C$ is an antiderivative of $3x^2$. To see this, note that

$$\frac{d}{dx}(x^3 + C) = \frac{d}{dx}(x^3) + \frac{d}{dx}(C) = 3x^2 + 0 = 3x^2. \quad \blacksquare$$

In Example 1 we see that $3x^2$ has an infinite number of antiderivatives. That is why we refer to an *indefinite* integral. If F is an indefinite integral of f, then so is $F + C$ for every constant C, since

$$\frac{d}{dx}(F + C) = \frac{dF}{dx} + \frac{dC}{dx} = f + 0 = f.$$

Of course, you may ask, how do we know we have found all the antiderivatives or indefinite integrals of f? For example, are there any other functions F such that $F'(x) = 3x^2$, other than functions of the form $F(x) = x^3 + C$? The answer is no because of the following theorem.

Theorem 1. If F and G are two differentiable functions that have the same derivative, then they differ by a constant. That is:

If $F'(x) = G'(x)$, then $F(x) - G(x) = C$ for some constant C.

Proof. Let $H(x) = F(x) - G(x)$. Then $H'(x) = F'(x) - G'(x) = 0$. Therefore, by Theorem 4.2.3, $H(x) = F(x) - G(x) = C$, and the theorem is proved. $\blacksquare$

This theorem allows us to define the **most general antiderivative** of f as $F(x) + C$, where F is some antiderivative of f and C is an arbitrary constant. For the remainder of this section we will speak about "the integral," leaving out the word "indefinite," and will look for the most general indefinite integral in our calculations.

---

†The $\int$ was used for s in English books of the seventeenth and eighteenth centuries, too. At the end of a word s was used, and within the word $\int$ was used, as in *hou∫es* (houses).

We now show how some integrals can be calculated. In Section 3.4 we indicated that

$$\frac{d}{dx}x^r = rx^{r-1},\tag{3}$$

where r is a nonzero real number. We therefore have the following theorem.

Theorem 2. If $r \neq -1$, then

$$\int x^r \, dx = \frac{x^{r+1}}{r+1} + C.\tag{4}$$

Proof.

$$\frac{d}{dx}\left[\frac{x^{r+1}}{r+1} + C\right] = \frac{(r+1)x^r}{(r+1)} + 0 = x^r \quad\blacksquare$$

NOTE.   The result does *not* hold for $r = -1$ because the denominator is $-1 + 1 = 0$, and we cannot divide by zero.

EXAMPLE 2   Calculate $\int x^9 \, dx$.

Solution. $r = 9$, so

$$\int x^9 \, dx = \frac{x^{9+1}}{9+1} + C = \frac{x^{10}}{10} + C. \quad\blacksquare$$

EXAMPLE 3   Calculate $\int x^{1/3} \, dx$.

Solution. $r = \frac{1}{3}$, so

$$\int x^{1/3} \, dx = \frac{x^{(1/3)+1}}{(\frac{1}{3})+1} + C = \frac{x^{4/3}}{\frac{4}{3}} + C = \frac{3}{4}x^{4/3} + C. \quad\blacksquare$$

EXAMPLE 4   Calculate $\int (1/\sqrt{t}) \, dt$.

Solution. The only difference between using x and t as the variable of integration is that t instead of x is the variable used in the antiderivative function. Then since $1/\sqrt{t} = t^{-1/2}$,

$$\int \frac{1}{\sqrt{t}} dt = \int t^{-1/2} \, dt = \frac{t^{(-1/2)+1}}{(-\frac{1}{2})+1} + C = \frac{t^{1/2}}{\frac{1}{2}} + C = 2\sqrt{t} + C. \quad\blacksquare$$

In the previous three examples it is easy to *check the answer by differentiating*. This check should always be done since it is always easier to differentiate than to integrate.

Thus differentiation provides a method for verifying the results of calculating an antiderivative.
    As we have already mentioned, Theorem 2 does not hold if $r = -1$. The case $r = -1$ is in some sense more interesting than the cases we have so far discussed. We will analyze this case in great detail in Chapter 6.

    The following theorem follows easily from the analogous results for derivatives.

Theorem 3.   If f and g are integrable† and if k is any constant, then kf and $f + g$ are integrable. Moreover,

$$\int kf(x)\, dx = k \int f(x)\, dx \tag{5}$$

and

$$\int [f(x) + g(x)]\, dx = \int f(x)\, dx + \int g(x)\, dx. \tag{6}$$

Proof.   Let $F = \int f$ and $G = \int g$. Then

$$\frac{d}{dx} kF = k\frac{d}{dx} F = kf,$$

which proves (5). Similarly,

$$\frac{d}{dx}(F + G) = \frac{d}{dx}F + \frac{d}{dx}G = f + g,$$

which proves (6). ∎

REMARK.   This theorem can be extended to more than two functions. For example, it is easy to show that $\int (f + g + h) = \int f + \int g + \int h$.

EXAMPLE 5   Calculate $\int [(3/x^2) + 6x^2]\, dx$.

Solution.

$$\int \left(\frac{3}{x^2} + 6x^2\right) dx \overset{\text{From (6)}}{=} \int \frac{3}{x^2} dx + \int 6x^2 dx \overset{\text{From (5)}}{=} 3\int x^{-2} dx + 6\int x^2 dx$$

$$\overset{\text{From (4)}}{=} \frac{3x^{-2+1}}{-2+1} + \frac{6x^{2+1}}{2+1} + C = -3x^{-1} + 2x^3 + C$$

$$= -\frac{3}{x} + 2x^3 + C$$

† We remind you that f is integrable if f has an antiderivative.

Check.

$$\frac{d}{dx}\left(-\frac{3}{x} + 2x^3 + C\right) = -3(-x^{-2}) + 2 \cdot 3x^2 = \frac{3}{x^2} + 6x^2 \quad \blacksquare$$

EXAMPLE 6   Calculate $\int (\cos x - \sin x)\, dx$.

Solution. Since $(d/dx)(\sin x) = \cos x$ and $(d/dx)(\cos x) = -\sin x$, we have

$$\int (\cos x - \sin x)\, dx = \int \cos x\, dx + \int (-\sin x)\, dx = \sin x + \cos x + C. \quad \blacksquare$$

Let us look more closely at the general integral of a function. Let $f(x) = 2x$. Then $\int f = x^2 + C$. For every value of C we get a different integral. But these integrals are very similar geometrically. For example, the curve $y = x^2 + 1$ is obtained by "shifting" the curve $y = x^2$ up one unit. The curve $y = x^2 + C$ is obtained by shifting the curve $y = x^2$ up C or down $|C|$ units (up if $C > 0$ and down if $C < 0$; see Section 1.8). Some of these curves are plotted in Figure 1. These curves never intersect. To prove this, suppose that (x_0, y_0) is a point on both curves $y = x^2 + A$ and $y = x^2 + B$. Then $y_0 = x_0^2 + A = x_0^2 + B$, which implies that $A = B$. That is, if the two curves have a point in common, then the two curves are the same. Thus if we specify one point through which the integral passes, then we know *the* integral.

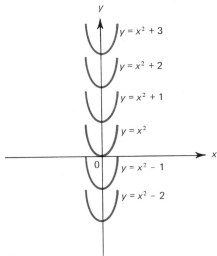

FIGURE 1

EXAMPLE 7   Find $\int 2x\, dx$ that passes through the point $(2, 7)$.

Solution. $y = \int 2x\, dx = x^2 + C$. But when $x = 2$, $y = 7$, so that $7 = 2^2 + C = 4 + C$, or $C = 3$. The solution is the function $x^2 + 3$. $\quad \blacksquare$

We now ask: What functions have antiderivatives? The answer is that all continuous functions have antiderivatives. That is, every continuous function is integrable. We prove the following result in Section 5.10; a graphical indication of the proof is given in Section 5.6.

Theorem 4. If f is continuous on $[a, b]$, then f is integrable on $[a, b]$. That is, there exists a function F such that $F'(x) = f(x)$ at every point x in $[a, b]$. [Here a and b may be $-\infty$ and ∞, respectively, in which case we refer to the interval $(-\infty, \infty)$.]

REMARK. This theorem does *not* say that functions that are not continuous are not integrable.

However, a serious problem remains. Consider the function $y = (1 + x^6)^{1/3}$. This function is certainly continuous (and even differentiable) everywhere. But it is impossible to find a function $F(x)$, written in terms of functions we recognize, that has the property that $F'(x) = (1 + x^6)^{1/3}$. [Try $(1 + x^6)^{4/3}/(\frac{4}{3})$ and see what happens.] This result does not mean that the integral does not exist. It means that there is no way to express this integral in terms of simple functions. In fact, it is perhaps surprising that most continuous functions have antiderivatives that cannot be represented in terms of recognizable functions. We will devote most of Chapter 8 to a discussion of how to find integrals of a wide variety of functions. For the rest we will explore (also in Chapter 8) numerical methods for estimating definite integrals as closely as we wish.

In Section 2.6 we gave several examples of the applicability of the derivative. We close this section by looking at some of these examples from the "reversed" point of view, that of the applicability of the integral.

EXAMPLE 8 A ball is dropped from rest from a certain height. Its velocity after t seconds due to the earth's gravitational field is given by

$$v = 32t,$$

where v is measured in feet per second. How far has the ball fallen after t seconds?

Solution. If $s = s(t)$ represents distance, then

$$\frac{ds}{dt} = v, \quad \text{or} \quad s(t) = \int v(t)\, dt = \int 32t\, dt = 16t^2 + C.$$

But $0 = s(0) = 16 \cdot 0^2 + C$, which implies that $C = 0$. Thus $s(t) = 16t^2$. For example, after 3 sec the ball has fallen $16 \cdot 3^2 = 16(9) = 144$ ft. ∎

We can remember the following rule:

The velocity function is the derivative of the distance function, and the distance function is an antiderivative of the velocity function.

Similarly, since acceleration equals a equals dv/dt, we have:

> The acceleration function is the derivative of the velocity function, and the velocity function is an antiderivative of the acceleration function.

EXAMPLE 9   The acceleration under the pull of the earth's gravity is, approximately, 9.81 m/sec^2. Find a formula for velocity and distance traveled by a particle under the influence of the force of gravitational attraction.

Solution. $v(t) = \int a(t)\, dt = \int 9.81\, dt = 9.81t + C$. Let $v(0)$, the initial velocity of the particle being considered, be denoted by v_0. Then $v(0) = v_0 = (9.81)(0) + C$, or $C = v_0$. Hence

$$v(t) = 9.81t + v_0 \tag{7}$$

and

$$s(t) = \int v(t)\, dt \doteq \int (9.81t + v_0)\, dt = \frac{9.81}{2}t^2 + v_0 t + C.$$

If we let s_0 denote $s(0)$, the initial position of the particle, we obtain

$$s(0) = s_0 = \frac{9.81}{2}(0)^2 + v_0(0) + C, \quad \text{or} \quad C = s_0,$$

so

$$s(t) = \frac{9.81}{2}t^2 + v_0 t + s_0. \tag{8a}$$

This equation is called the **equation of motion** of the particle. Here $s(t)$ is measured in meters. ■

NOTE.   If g is given as $g \approx 32$ ft/sec^2, then (8a) becomes

$$s(t) = 16t^2 + v_0 t + s_0, \tag{8b}$$

since $\frac{1}{2}g \approx 16$. Here $s(t)$ is measured in feet.

EXAMPLE 10   A population is growing with an instantaneous growth rate of $100 + 12\sqrt{t} - 5t$ organisms per hour after t hours. If the initial population is 10,000, what is the population after t hours?

Solution. Since the rate of growth is the derivative of the population size $P(t)$ (see Example 2.6.3), $P(t)$ is an antiderivative of population growth, so

$$P(t) = \int (100 + 12t^{1/2} - 5t)\, dt = 100t + 12\frac{t^{3/2}}{\frac{3}{2}} - \frac{5t^2}{2} + C$$

$$= 100t + 8t^{3/2} - \tfrac{5}{2}t^2 + C.$$

But $P(0) = $ the initial population $= 10,000$, so

$$P(0) = 100(0) + 8(0)^{3/2} - \tfrac{5}{2}(0)^2 + C = 10,000, \quad \text{or} \quad C = 10,000,$$

and

$$P(t) = 100t + 8t^{3/2} - \tfrac{5}{2}t^2 + 10,000$$

organisms after t hours. ∎

EXAMPLE 11 The density ρ of a 4-m nonuniform metal beam, measured in kilograms per meter, is given by

$$\rho(x) = 2\sqrt{x}, \tag{9}$$

where x is the distance along the beam measured from the left end (see Figure 2).

FIGURE 2

(a) What is the mass of the beam up to the point x units from the left?
(b) What is the total mass of the beam?

Solution. (a) Since density is the rate of change of the mass μ (see Example 2.6.4), we have

$$\rho = \frac{d\mu}{dx}, \quad \text{or} \quad \mu(x) = \int \rho(x)\, dx. \tag{10}$$

Then

$$\mu(x) = \int 2\sqrt{x}\, dx = 2 \int x^{1/2}\, dx = 2\frac{x^{3/2}}{\tfrac{3}{2}} + C = \frac{4}{3} x^{3/2} + C.$$

But $\mu(0) = 0$ since there is no mass zero units from the left side of the beam. Hence $C = 0$, and

$$\mu(x) = \tfrac{4}{3} x^{3/2} \text{ kilograms.}$$

(b) When $x = 4$ m, $\mu(x) = \tfrac{4}{3}(4)^{3/2} = \tfrac{4}{3}(8) = \tfrac{32}{3}$ kg, which is the total mass of the beam. ∎

EXAMPLE 12 The marginal revenue that a manufacturer receives for his goods is given by

$$MR = 100 - 0.03q, \quad 0 \le q \le 1000. \tag{11}$$

Find the total revenue function.

Solution. Marginal revenue is similar to marginal cost. (See Example 2.6.5.) Equation (11) tells us that the manufacturer receives less and less per unit the more units he sells. Marginal revenue is the rate of change of money received per unit change in the number of units sold. That is, if R denotes total revenue,

$$\frac{dR}{dq} = MR = 100 - 0.03q. \tag{12}$$

Then

$$R(q) = \int MR = \int (100 - 0.03q)\, dq = 100q - 0.03\frac{q^2}{2} + C.$$

But the manufacturer certainly receives nothing if he sells nothing. Therefore $R(0) = 0 = C$, and

$$R(q) = 100q - 0.03\frac{q^2}{2} \quad \text{for} \quad 0 \le q \le 1000. \quad ∎$$

PROBLEMS 5.2

In Problems 1–17, find the most general antiderivative.

1. $\int dx$

2. $\int x\, dx$

3. $\int a\, dx$, where a is a constant

4. $\int (ax + b)\, dx$, a, b constants

5. $\int (1 + x + x^2 + x^3 + x^4)\, dx$

6. $\int \frac{1}{x^5}\, dx$

7. $\int \frac{4}{x^3}\, dx$

8. $\int \frac{7}{\sqrt{x}}\, dx$

9. $\int \left(\sqrt{x} + \frac{1}{\sqrt{x}}\right) dx$

10. $\int \left(3\sqrt[3]{x} + \frac{3}{\sqrt[3]{x}}\right) dx$

11. $\int (x^{1/2} + x^{3/2} + x^{3/4})\, dx$

12. $\int \left(\frac{4}{3x^4} - \frac{5}{7x^5} + \frac{6}{11x^6}\right) dx$

13. $\int \frac{x^3 + x^2 - x}{x^{3/2}}\, dx$ [*Hint:* Divide through.]

14. $\int \frac{4x^{2/3} - x^{5/8} + x^{17/5}}{5x^3}\, dx$

15. $\int \left(-\frac{17}{x^{13/17}} + \frac{3}{x^{4/9}}\right) dx$

16. $\int (2 \sin x - 3 \cos x)\, dx$

17. $\int \frac{\cos x - 5 \sin x}{4}\, dx$

In Problems 18–24, the derivative of a function and one point on its graph are given. Find the function.

18. $\dfrac{dy}{dx} = x^3 + x^2 - 3$; $(1, 5)$ **19.** $\dfrac{dy}{dx} = 2x(x + 1)$; $(2, 0)$

20. $\dfrac{dy}{dx} = \sqrt[3]{x} + x - \dfrac{1}{3\sqrt[3]{x}}$; $(-1, -8)$ **21.** $y' = 13x^{15/18} - 3$; $(1, 14)$

22. $y' = 1 + x + x^2 + x^3 + x^4 + x^5$; $(0, 83)$

23. $y' = \cos x$; $\left(\dfrac{\pi}{6}, 4\right)$ **24.** $y' = \sin x - \cos x$; $\left(\dfrac{\pi}{4}, \dfrac{3}{\sqrt{2}}\right)$

25. A particle moves with the constant acceleration of 5.8 m/sec². It starts with an initial velocity of 0.2 m/sec and initial position of 25 m. Find the equation of motion of the particle.

26. A bullet is shot from ground level straight up into the air at an initial velocity of 2000 ft/sec. Write a function that tells us the height of the bullet after t seconds, up until the time the bullet hits the earth (assume that the only force acting on the bullet is the force of gravitational attraction, which imparts the constant acceleration $g = 32$ ft/sec²).

27. The acceleration of a moving particle starting at a position 100 m along the x-axis and moving with an initial velocity of $v_0 = 25$ m/min is given by

$$a(t) = 13\sqrt{t} \text{ meters per minute squared.}$$

Find the equation of motion of the particle.

28. The instantaneous rate of change of a population is $50t^2 - 100t^{3/2}$, measured in individuals per year, and the initial population is 25,000.
(a) What is the population after t years?
(b) What is the population after 25 years?

29. The density of a 10-m beam is given by

$$\rho(x) = 3x + 2x^2 - x^{3/2} \text{ kilograms per meter.}$$

What is the total mass (in kilograms) of the beam?

30. The density of a 64-ft beam is given by $\rho(x) = 3x^{5/6}$, measured in pounds per foot. What is the total weight (in pounds) of the beam?

31. The marginal cost function for a certain manufacturer is $MC = 25q - 0.02q^2 + 20$, and it costs $2000 to produce 10 units of a certain product. How much does it cost to produce 100 units? 500 units?

32. The marginal revenue function for a certain product is given by $MR = 3q - 2\sqrt{q} + 10$. How much money will the manufacturer receive if she sells 50 units? 100 units?

33. In Example 12, at what level of production will revenue begin to decrease as additional items are produced? (This level is often called the **point of diminishing returns.**)

*34. Consider the function $F(x) = \int (1/x)\, dx$. We have not yet learned how to integrate $1/x = x^{-1}$, but we can still obtain some information about $F(x)$. Assuming that $F(1) = 0$, use information about the derivatives of F to graph the curve for $x > 0$. [*Hint:* Show that F is always increasing, is concave down, and has a vertical tangent as $x \to 0^+$.]

*35. Find all antiderivatives of $f(x) = |x|$.

5.3 THE Σ NOTATION

In the next section we will see that one way to approximate the area under a curve involves writing a number of sums. There is a simple notation that will be very convenient for us. Consider the sum

$$a_1 + a_2 + a_3 + \cdots + a_n. \tag{1}$$

This sum is written†

$$\sum_{k=1}^{n} a_k, \tag{2}$$

which is read, "the sum of the terms a_k as k goes from 1 to n." In this context Σ is called the **summation sign,** and k is called the **index of summation.**

EXAMPLE 1 Calculate $\sum_{k=1}^{4} k$.

Solution. Here $a_k = k$, so that

$$\sum_{k=1}^{4} k = 1 + 2 + 3 + 4 = 10. \blacksquare$$

EXAMPLE 2 Calculate $\sum_{k=1}^{5} k^2$.

Solution. Here $a_k = k^2$, and

$$\sum_{k=1}^{5} k^2 = 1^2 + 2^2 + 3^2 + 4^2 + 5^2 = 55. \blacksquare$$

EXAMPLE 3 Write the sum $S_8 = 1 - 2 + 3 - 4 + 5 - 6 + 7 - 8$ by using the summation sign.

Solution. Since $1 = (-1)^2$, $-2 = (-1)^3 \cdot 2$, $3 = (-1)^4 \cdot 3$, $\dots$, we have

$$S_8 = \sum_{k=1}^{8} (-1)^{k+1} k. \blacksquare$$

EXAMPLE 4 Write the following sum by using the summation sign:

$$S = \left(\tfrac{1}{8}\right)^2 \tfrac{1}{8} + \left(\tfrac{2}{8}\right)^2 \tfrac{1}{8} + \cdots + \left(\tfrac{7}{8}\right)^2 \tfrac{1}{8} + \left(\tfrac{8}{8}\right)^2 \tfrac{1}{8}.$$

Solution. We have

$$S = \sum_{k=1}^{8} \left(\frac{k}{8}\right)^2 \frac{1}{8} = \sum_{k=1}^{} \left(\frac{1}{8^3}\right) k^2 = \frac{1}{8^3} \sum_{k=1}^{} k^2. \blacksquare$$

In Example 4 we used the following fact:

$$\sum_{k=1}^{n} ca_k = c \sum_{k=1}^{n} a_k$$

---

†The Greek letter Σ (sigma) was first used to denote a sum by the Swiss mathematician Leonhard Euler (1707–1783). A short biography of Euler is given on page 36.

where c is a constant. To see why this is true, observe that

$$\sum_{k=1}^{n} ca_k = ca_1 + ca_2 + ca_3 + \cdots + ca_n = c(a_1 + a_2 + a_3 + \cdots + a_n) = c \sum_{k=1}^{n} a_k.$$

EXAMPLE 5 In the next section we will encounter the sum

$$S = f(x_1)\,\Delta x_1 + f(x_2)\,\Delta x_2 + \cdots + f(x_n)\,\Delta x_n,$$ (3)

which can be written as

$$S = \sum_{k=1}^{n} f(x_k)\,\Delta x_k. \blacksquare$$ (4)

NOTE. We can change the index of summation without changing the sum. For example,

$$\sum_{k=1}^{5} k^2 = \sum_{j=1}^{5} j^2 = \sum_{m=1}^{5} m^2 = 1^2 + 2^2 + 3^2 + 4^2 + 5^2 = 55.$$

PROBLEMS 5.3

In Problems 1–6, evaluate the given sums.

1. $\displaystyle\sum_{k=1}^{4} 2k$ **2.** $\displaystyle\sum_{i=1}^{3} i^3$ **3.** $\displaystyle\sum_{k=0}^{6} 1$

4. $\displaystyle\sum_{k=1}^{8} 3^k$ **5.** $\displaystyle\sum_{i=2}^{5} \frac{i}{i+1}$ **6.** $\displaystyle\sum_{j=5}^{7} \frac{2j+3}{j-2}$

In Problems 7–21, write each sum by using the Σ notation.

7. $1 + 2 + 4 + 8 + 16$

8. $1 - 3 + 9 - 27 + 81 - 243$

9. $\dfrac{2}{3} + \dfrac{3}{4} + \dfrac{4}{5} + \dfrac{5}{6} + \dfrac{6}{7} + \dfrac{7}{8} + \cdots + \dfrac{n}{n+1}$

10. $1 - \dfrac{1}{2!} + \dfrac{1}{3!} - \dfrac{1}{4!} + \dfrac{1}{5!} - \dfrac{1}{6!} + \dfrac{1}{7!}$

11. $1 + 2^{1/2} + 3^{1/3} + 4^{1/4} + 5^{1/5} + \cdots + n^{1/n}$

12. $1 + x^3 + x^6 + x^9 + x^{12} + x^{15} + x^{18} + x^{21}$

13. $x^5 - x^{10} + x^{15} - x^{20} + x^{25} - x^{30} + x^{35} - x^{40}$

14. $-1 + \dfrac{1}{a} - \dfrac{1}{a^2} + \dfrac{1}{a^3} - \dfrac{1}{a^4} + \dfrac{1}{a^5} - \dfrac{1}{a^6} + \dfrac{1}{a^7} - \dfrac{1}{a^8} + \dfrac{1}{a^9}$

*15. $1 \cdot 3 + 3 \cdot 5 + 5 \cdot 7 + 7 \cdot 9 + 9 \cdot 11 + 11 \cdot 13 + 13 \cdot 15 + 15 \cdot 17$

*16. $2^2 \cdot 4 + 3^2 \cdot 6 + 4^2 \cdot 8 + 5^2 \cdot 10 + 6^2 \cdot 12 + 7^2 \cdot 14$

17. $\frac{1}{32}\left(\frac{1}{32}\right)^2 + \frac{1}{32}\left(\frac{2}{32}\right)^2 + \frac{1}{32}\left(\frac{3}{32}\right)^2 + \cdots + \frac{1}{32}\left(\frac{32}{32}\right)^2$

18. $\dfrac{1}{2^n}\left(\dfrac{1}{2^n}\right)^2 + \dfrac{1}{2^n}\left(\dfrac{2}{2^n}\right)^2 + \cdots + \dfrac{1}{2^n}\left(\dfrac{2^n-1}{2^n}\right)^2 + \dfrac{1}{2^n}\left(\dfrac{2^n}{2^n}\right)^2$

19. $\frac{1}{64}\left(\frac{1}{64}\right)^3 + \frac{1}{64}\left(\frac{2}{64}\right)^3 + \frac{1}{64}\left(\frac{3}{64}\right)^3 + \cdots + \frac{1}{64}\left(\frac{64}{64}\right)^3$

20. $0.1 \sin 0.1 + 0.1 \sin 0.2 + 0.1 \sin 0.3 + \cdots + 0.1 \sin 1$

21. $\frac{1}{50}\sqrt{\frac{1}{50}} + \frac{1}{50}\sqrt{\frac{2}{50}} + \frac{1}{50}\sqrt{\frac{3}{50}} + \cdots + \frac{1}{50}\sqrt{\frac{49}{50}} + \frac{1}{50}\sqrt{\frac{50}{50}}$

Observe that there is some flexibility in the Σ notation. For instance, $\Sigma_{i=3}^{8} i$, $\Sigma_{j=3}^{8} j$, $\Sigma_{k=5}^{10} (k-2)$, and $\Sigma_{L=0}^{5} (L+3)$ each equals $3 + 4 + 5 + 6 + 7 + 8$. In Problems 22–24 you are given three expressions; two give the same sum and one is different. Identify the one that does not equal the other two.

22. $\displaystyle\sum_{k=0}^{7} (2k+1), \ \sum_{j=1}^{15} j, \ \sum_{i=2}^{9} (2i-3)$

23. $\displaystyle\sum_{k=1}^{7} k^2, \ \sum_{j=0}^{6} (7-j)^2, \ \sum_{i=1}^{7} (7-i)^2$

24. $\displaystyle\left(\sum_{k=7}^{11} k\right)^4, \ \sum_{m=-11}^{-7} m^4, \ \sum_{n=7}^{11} n^4$

*25. Suppose that g is some function whose domain includes the integers.
   (a) *(Telescoping sum)* Prove that $\Sigma_{k=1}^{n} [g(k) - g(k-1)] = g(n) - g(0)$.
   (b) If we let $g(k) = \frac{1}{2}k \cdot (k+1)$, show that $g(k) - g(k-1) = k$, and part (a) yields $\Sigma_{k=1}^{n} k = n(n+1)/2$. Similarly, if we use $G(k) = \frac{1}{3}k \cdot (k+1) \cdot (k+2)$, show that $G(k) - G(k-1) = k(k+1)$.

*26. Show that $\Sigma_{k=1}^{n} k^2 = n(n+1)(2n+1)/6$. [*Hint:* Find $\Sigma_{k=1}^{n} k(k+1)$ by using the result of Problem 25(b).]

*27. Find a short formula with which to compute $\Sigma_{k=1}^{n} k^3$. [*Hint:* First look at a telescoping sum with $g(k) = \frac{1}{4}k \cdot (k+1) \cdot (k+2) \cdot (k+3)$ and then use formulas for $\Sigma_{k=1}^{n} k$ and $\Sigma_{k=1}^{n} k^2$.]

*28. Find a short formula for computing the sum of the first n odd integers. Show that your formula is correct for all n. [*Hint:* Try a telescoping sum with $g(k) = k^2$.]

*29. Find a short formula by which to compute

$$\tfrac{1}{3} + \tfrac{1}{8} + \tfrac{1}{15} + \tfrac{1}{24} + \tfrac{1}{35} + \cdots + \frac{1}{n^2 - 1} = \frac{1}{1\cdot 3} + \frac{1}{2\cdot 4} + \frac{1}{3\cdot 5} + \frac{1}{4\cdot 6} + \cdots + \frac{1}{n^2 - 1}.$$

*30. Find a short formula for computing

$$\tfrac{1}{2} + \tfrac{1}{6} + \tfrac{1}{12} + \tfrac{1}{20} + \cdots + \frac{1}{n^2 - n} = \frac{1}{1\cdot 2} + \frac{1}{2\cdot 3} + \frac{1}{3\cdot 4} + \frac{1}{4\cdot 5} + \cdots + \frac{1}{(n-1)\cdot n}.$$

$$\left[Hint: \ \frac{1}{k\cdot(k+1)} = \frac{1}{k} - \frac{1}{k+1}.\right]$$

5.4 APPROXIMATIONS TO AREA

In this section we describe a method, first used by Archimedes, for computing the area under a curve. In this method we estimate the area as the sum of areas of rectangles and then show how the estimate can be improved. Finally, we take a limit of these estimates, and it is this limit that is equal to the area sought.

Following Archimedes, we calculate the shaded area A under a parabola (see Figure 1). This region is bounded by the parabola $y = x^2$, the x-axis, and the line

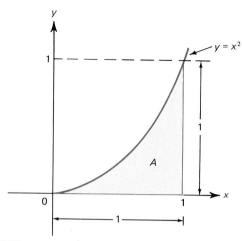

FIGURE 1

$x = 1$. The area is approximated by rectangles under the curve and over the curve (see Figure 2). Since the equation of the parabola is $y = x^2$, the height of each rectangle (which is the y value on the curve) is the square of the x value. If we let s denote the sum of the areas of the rectangles in Figure 2a and S the sum of the areas of the rectangles in Figure 2b, then

$$s < A < S.$$

(1)

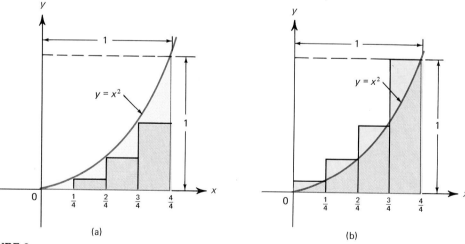

FIGURE 2

Since Archimedes knew how to calculate the area of a rectangle (by multiplying its base by its height), he was able to calculate s and S exactly, thereby obtaining an estimate for the area A. We have

Height of first rectangle in Figure 2a ↘   Length of subinterval ↙   Height of second rectangle in Figure 2a ↙   Height of third rectangle in Figure 2a ↙

$$s = \left(\frac{1}{4}\right)^2\frac{1}{4} \quad + \quad \left(\frac{2}{4}\right)^2\frac{1}{4} \quad + \quad \left(\frac{3}{4}\right)^2\frac{1}{4} \quad = \quad \frac{1}{4}\left[\left(\frac{1}{4}\right)^2 + \left(\frac{2}{4}\right)^2 + \left(\frac{3}{4}\right)^2\right]$$

$$= \frac{1}{4\cdot4^2}(1^2 + 2^2 + 3^2) = \frac{1^2 + 2^2 + 3^2}{4^3} = \frac{14}{64} = \frac{7}{32}.$$

Using the summation notation of the last section, we may write

$$s = \sum_{k=1}^{3}\left(\frac{k}{4}\right)^2\frac{1}{4} = \frac{7}{32}.$$

Similarly,

Height of first rectangle in Figure 2b ↘   Height of second rectangle in Figure 2b ↘   Height of third rectangle in Figure 2b ↘   Height of fourth rectangle in Figure 2b ↘

$$S = \left(\frac{1}{4}\right)^2\frac{1}{4} \quad + \quad \left(\frac{2}{4}\right)^2\frac{1}{4} \quad + \quad \left(\frac{3}{4}\right)^2\frac{1}{4} \quad + \quad \left(\frac{4}{4}\right)^2\frac{1}{4} = \frac{1^2 + 2^2 + 3^2 + 4^2}{4^3}$$

$$= \frac{30}{64} = \frac{15}{32},$$

and we may write

$$S = \sum_{k=1}^{4}\left(\frac{k}{4}\right)^2\frac{1}{4} = \frac{15}{32}.$$

Thus from (1), with $s = \frac{7}{32}$ and $S = \frac{15}{32}$, we obtain

$$0.22 \approx \frac{7}{32} < A < \frac{15}{32} \approx 0.47. \tag{2}$$

It was clear to Archimedes that this estimate could be improved by increasing the number of rectangles so that the error in the estimate becomes smaller. By doubling the number of rectangles, we obtain the approximation depicted in Figure 3. Here

$$s = \left(\frac{1}{8}\right)^2\frac{1}{8} + \left(\frac{2}{8}\right)^2\frac{1}{8} + \cdots + \left(\frac{7}{8}\right)^2\frac{1}{8} = \frac{1^2 + 2^2 + 3^2 + 4^2 + 5^2 + 6^2 + 7^2}{8^3}$$

$$= \frac{140}{512} = \frac{35}{128},$$

and

$$s = \sum_{k=1}^{7}\left(\frac{k}{8}\right)^2\frac{1}{8} = \frac{35}{128}.$$

(a)

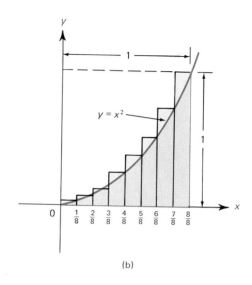
(b)

FIGURE 3

Also,

$$S = \left(\frac{1}{8}\right)^2 \frac{1}{8} + \left(\frac{2}{8}\right)^2 \frac{1}{8} + \cdots + \left(\frac{8}{8}\right)^2 \frac{1}{8}$$

$$= \frac{1^2 + 2^2 + 3^2 + 4^2 + 5^2 + 6^2 + 7^2 + 8^2}{8^3} = \frac{204}{512} = \frac{51}{128},$$

and

$$S = \sum_{k=1}^{8} \left(\frac{k}{8}\right)^2 \frac{1}{8} = \frac{51}{128}.$$

(See Example 5.3.4.) Hence using $s = \frac{35}{128}$ and $S = \frac{51}{128}$ in (1), we have

$$0.27 \approx \frac{35}{128} < A < \frac{51}{128} \approx 0.40. \tag{3}$$

We can continue the process of increasing the number of subintervals to get more and more accurate approximations for A.

Using his method of exhaustion, Archimedes was able to show that with the outer rectangles the assumption $A > \frac{1}{3}$ led to a contradiction. Similarly, using the inner rectangles, he showed that $A < \frac{1}{3}$ led to a contradiction. From this result he concluded that $A = \frac{1}{3}$.

Instead of computing more estimates for A, we will do things in more generality and then prove that A is, indeed, equal to $\frac{1}{3}$. Let $y = f(x)$ be a function that is positive and continuous on the interval $[a, b]$ (see Figure 4). We will now show how the area bounded by this curve, the x-axis, and the lines $x = a$ and $x = b$ can be approximated. The method we will give here is very similar to Archimedes' method of exhaustion.

As before, we begin by dividing the interval $[a, b]$ into a number of smaller

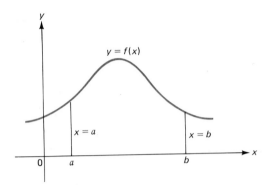

FIGURE 4

subintervals *of equal length*. Such a division is called a **regular partition.** We label the **partition points** $x_0, x_1, x_2, \ldots, x_n$, where $x_0 = a$ and $x_n = b$ and it is assumed that

$$a = x_0 < x_1 < x_2 < \cdots < x_n = b. \tag{4}$$

This partition is illustrated in Figure 5. Since there are n subintervals, each of the same length, and the length of the entire interval is $b - a$, we see that each subinterval has length $(b - a)/n$. We denote this quantity by Δx. We have

$$\Delta x = \frac{b - a}{n}. \tag{5}$$

FIGURE 5

Since the length of the ith subinterval is $x_i - x_{i-1}$, we also have

$$\Delta x = x_i - x_{i-1} = \frac{b - a}{n} \quad \text{for} \quad i = 1, 2, \ldots, n. \tag{6}$$

REMARK.   From (5) or (6) it follows that

$$\lim_{n \to \infty} \Delta x = 0. \tag{7}$$

EXAMPLE 1   In Figure 6 we have sketched four partitions of the interval $[0, 1]$. In (i) $\Delta x = \frac{1}{2} = (1 - 0)/2$; in (ii) $\Delta x = \frac{1}{4}$; in (iii) $\Delta x = \frac{1}{6}$; and in (iv) $\Delta x = \frac{1}{10}$. ∎

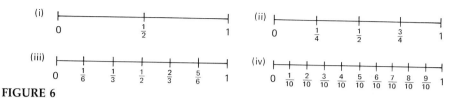

FIGURE 6

NOTATION.   We denote the area in Figure 4 by $A_a{}^b$. This is the area bounded by the curve $y = f(x)$, the x-axis, and the lines $x = a$ and $x = b$.

Suppose that the interval $[a, b]$ is partitioned into n subintervals of equal length. We will approximate $A_a{}^b$ by drawing rectangles whose total area is "close" to the actual area (see Figure 7). We first choose a point *arbitrarily* in each interval $[x_{i-1}, x_i]$ and label it $x_i{}^*$. This technique gives us n points $x_1{}^*, x_2{}^*, \ldots, x_n{}^*$. Next, we locate the points $(x_i{}^*, f(x_i{}^*))$ on the curve for $i = 1, 2, \ldots, n$. The numbers $f(x_i{}^*)$ give us the heights of our n rectangles. The base of each rectangle has length $x_i - x_{i-1} = \Delta x$. This is illustrated in Figure 7. The area A_i of the ith rectangle is the height of the rectangle times its length:

$$A_i = f(x_i{}^*) \, \Delta x. \tag{8}$$

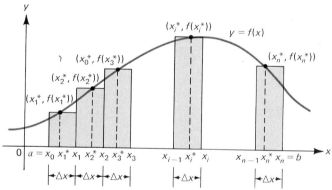

FIGURE 7

Let us take a closer look at the region enclosed by these rectangles. In Figure 8 the shaded regions depict the differences between the region whose area we wish to calculate and the region enclosed by the rectangles. We see that as each rectangle becomes "thinner and thinner," the area of the regions enclosed by the rectangles seems to get closer and closer to the area of the regions under the curve. But the length of the base of each rectangle is Δx, and so if Δx is reasonably small, we have the approximation

$$A_a{}^b \approx A_1 + A_2 + \cdots + A_i + \cdots + A_n = \sum_{i=1}^{n} A_i,$$

or using (8), we have

$$A_a{}^b \approx f(x_1{}^*) \, \Delta x + f(x_2{}^*) \, \Delta x + \cdots + f(x_n{}^*) \, \Delta x$$

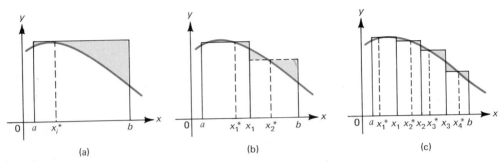

FIGURE 8

and

$$A_a{}^b \approx \sum_{i=1}^{n} f(x_i{}^*) \, \Delta x . \tag{9}$$

Earlier in this section we saw an example of this process at work. There we chose $x_i{}^*$ to be either the right or the left endpoint of the subinterval. Our approximation technique will work for *any* choice of $x_i{}^*$ in $[x_{i-1}, x_i]$. However, it is often easier to deal with endpoints.

EXAMPLE 2 Approximate the area bounded by the curve $y = x^2$ and the x-axis from $x = 0$ to $x = 1$.

Solution. We will partition the interval $[0, 1]$ into n subintervals of equal length and will choose $x_i{}^* = x_i$, the right-hand endpoint of each subinterval. Since $(b - a)/n = (1 - 0)/n = 1/n$, the length of each subinterval is $1/n$ and the partition points are (see Figure 9)

$$0 = \frac{0}{n} < \frac{1}{n} < \frac{2}{n} < \frac{3}{n} < \cdots < \frac{n-1}{n} < \frac{n}{n} = 1.$$

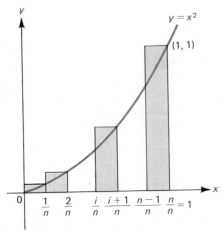

FIGURE 9

Then for $i = 1, 2, \ldots, n$, $f(x_i^*) = f(x_i) = f(i/n) = (i/n)^2$, $\Delta x = 1/n$, and formula (9) becomes

$$A_0^1 \approx \sum_{i=1}^{n} \overbrace{\left(\frac{i}{n}\right)^2}^{f(x_i^*)} \overbrace{\frac{1}{n}}^{\Delta x} = \sum_{i=1}^{n} \frac{i^2}{n^3} = \frac{1}{n^3} \sum_{i=1}^{n} i^2$$

$$= \frac{1}{n^3}(1^2 + 2^2 + \cdots + n^2). \tag{10}$$

In Appendix 2 we prove, using mathematical induction that the sum of the first n squares is given by the formula

$$1^2 + 2^2 + \cdots + n^2 = \sum_{i=1}^{n} i^2 = \frac{n(n + 1)(2n + 1)}{6}. \tag{11}$$

(Or see Problem 5.3.26). You should check this formula by verifying it for a few values of n. Substituting (11) in (10), we have

$$A_0^1 \approx \frac{1}{n^3}\left[\frac{(n)(n + 1)(2n + 1)}{6}\right] = \frac{(n + 1)(2n + 1)}{6n^2}. \tag{12}$$

In Table 1 we tabulate values of this quantity for several different values of n. It seems as if the approximation is tending to the limit $0.3333 \ldots = \frac{1}{3}$ as the number of subintervals grows. Actually, this is easy to prove, since

Divide numerator and denominator by n^2

$$\lim_{n \to \infty} \frac{(n + 1)(2n + 1)}{6n^2} = \lim_{n \to \infty} \frac{2n^2 + 3n + 1}{6n^2} \overset{\downarrow}{=} \lim_{n \to 9} \frac{2 + (3/n) + (1/n^2)}{6} = \frac{2}{6} = \frac{1}{3}. \blacksquare$$

TABLE 1

n	$n + 1$	$2n + 1$	$(n + 1)(2n + 1)$	$6n^2$	$\dfrac{(n + 1)(2n + 1)}{6n^2}$
1	2	3	6	6	1
2	3	5	15	24	0.625
5	6	11	66	150	0.44
10	11	21	231	600	0.385
25	26	51	1,326	3,750	0.3536
50	51	101	5,151	15,000	0.3434
100	101	201	20,301	60,000	0.33835
1,000	1,001	2,001	2,003,001	6,000,000	0.3338335
10,000	10,001	20,001	200,030,001	600,000,000	0.333383335

EXAMPLE 3   Approximate the area bounded by the curve $y = 3x$ and the x-axis from $x = 0$ to $x = 2$.

Solution. The area sought is depicted in Figure 10. Since the area of a triangle is $\frac{1}{2}bh$, where b denotes its base and h its height, we can immediately calculate that $A_0^2 = \frac{1}{2}(2)(6) = 6$. We will show how our approximations lead to the same answer. To

FIGURE 10

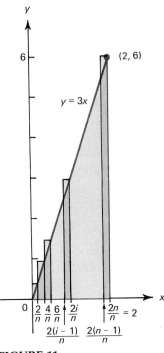

FIGURE 11

do so, we partition $[0, 2]$ into n equal subintervals, each having length $(b - a)/n = 2/n$. The partition points are then

$$0 = \frac{0}{n} < \frac{2}{n} < \frac{4}{n} < \cdots < \frac{2(i - 1)}{n} < \frac{2i}{n} < \cdots < \frac{2(n - 1)}{n} < \frac{2n}{n} = 2.$$

If we choose $x_i^* = 2i/n$, the right-hand endpoint of each subinterval, we obtain the situation depicted in Figure 11. Here $f(x_i^*) = f(2i/n) = 3 \cdot 2i/n = 6i/n$. Then (9) becomes

$$A_0^2 \approx \sum_{i=1}^{n} f(x_i^*) \, \Delta x = \sum_{i=1}^{n} \left(\frac{6i}{n}\right)\frac{2}{n} = \frac{12}{n^2} \sum_{i=1}^{n} i. \tag{13}$$

Now the formula for the sum of the first n integers (proven in Appendix 2 and in Problem 5.3.25) is

$$\sum_{i=1}^{n} i = 1 + 2 + 3 + \cdots + (n - 1) + n = \frac{n(n + 1)}{2}. \tag{14}$$

This formula should also be checked by verifying it for some sample values of n. Using (14) in (13), we obtain

$$A_0^2 \approx \frac{12}{n^2}\left[\frac{n(n + 1)}{2}\right] = \frac{6(n + 1)}{n} = 6\left(1 + \frac{1}{n}\right). \tag{15}$$

It is apparent that these approximations tend to the limit 6. Indeed, we have

Theorem 2 on page 79

$$\lim_{n\to\infty} 6\left(1 + \frac{1}{n}\right) = 6 \lim_{n\to\infty}\left(1 + \frac{1}{n}\right) = 6 \cdot 1 = 6. \blacksquare$$

EXAMPLE 4   Approximate the area bounded by the curve $y = x^3$ and the x-axis from $x = 0$ to $x = 3$.

Solution. We divide the interval into n equal subintervals, each having length $(b - a)/n = 3/n$. The partition points are

$$0 = \frac{0}{n} < \frac{3}{n} < \frac{6}{n} < \frac{9}{n} < \cdots < \frac{3(n-1)}{n} < \frac{3n}{n} = 3.$$

For convenience we choose $x_i{}^* = x_i$ (the right-hand endpoint of the subinterval), so that

$$f(x_i{}^*) = f(x_i) = x_i{}^3 = \left(\frac{3i}{n}\right)^3 = \frac{27i^3}{n^3} \qquad \text{for} \qquad i = 1, 2, \ldots, n$$

(see Figure 12). Then

$$A_0{}^3 \approx \sum_{i=1}^n f(x_i)\,\Delta x = \sum_{i=1}^n \frac{27i^3}{n^3}\cdot\frac{3}{n} = \frac{81}{n^4}\sum_{i=1}^n i^3. \tag{16}$$

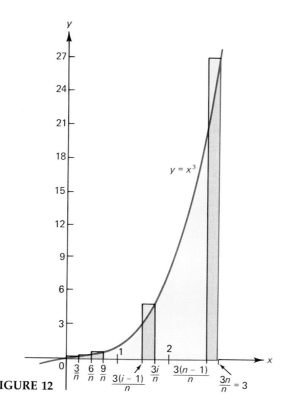

FIGURE 12

Yes, there is a formula for the sum of the first n cubes. In Appendix 2 (see Problem 1) we prove that

$$\sum_{i=1}^{n} i^3 = 1^3 + 2^3 + 3^3 + \cdots + (n-1)^3 + n^3 = \left[\frac{n(n+1)}{2} \right]^2. \tag{17}$$

(See also Problem 5.3.27.) Again you should satisfy yourself that this formula is correct. Substituting (17) in (16), we obtain

$$A_0^3 \approx \frac{81}{n^4} \cdot \frac{n^2(n+1)^2}{4} = \frac{81}{4} \cdot \frac{(n+1)^2}{n^2}. \tag{18}$$

We find that

$$\underset{n \to \infty}{\lim} \overset{\text{Theorem 2, on page 79}}{\frac{81}{4} \cdot \frac{(n+1)^2}{n^2}} \overset{\downarrow}{=} \frac{81}{4} \lim_{n \to \infty} \frac{n^2 + 2n + 1}{n^2} \overset{\text{Divide numerator and denominator by } n^2}{\overset{\downarrow}{=}} \frac{81}{4} \lim_{n \to \infty} \left(1 + \frac{2}{n} + \frac{1}{n^2}\right) = \frac{81}{4}. \blacksquare$$

We will put these calculations on a more formal basis in the next section and show how the techniques we have introduced here can also be used to solve problems that do not involve areas.

PROBLEMS 5.4

Problems 1–4 refer to the function $f(x) = x^2$.
1. Calculate s and S, where the interval $[0, 1]$ is divided into 16 smaller intervals (each having length $\frac{1}{16}$).
2. Calculate s and S, where the interval $[0, 1]$ is divided into 32 equal subintervals.
3. Show that if the interval $[0, 1]$ is divided into n pieces each having length $1/n$, then

$$s = \frac{1}{n^3}[0^2 + 1^2 + \cdots + (n-1)^2] \quad \text{and} \quad S = \frac{1}{n^3}(1^2 + 2^2 + \cdots + n^2).$$

* 4. Without using the explicit formulas in Problem 3, show why the underestimate s and the overestimate S for the area of the region $\{(x, y)|0 \leq x \leq 1, 0 \leq y \leq x^2\}$ differ by $1/n$.

In Problems 5–8, estimate the area bounded by the straight line, the x-axis, and the lines $x = a$ and $x = b$, where a and b are the endpoints of the given interval, by dividing the interval into two, four, and eight subintervals of equal length. Then calculate the actual area by using the fact that the area of a triangle is $\frac{1}{2}bh$ and the area of a rectangle is $b \cdot h$.

5. $y = 7x$; $[0, 4]$
6. $y = 4x$; $[1, 2]$. [*Hint:* To calculate this area exactly, divide the trapezoid you obtain into a triangle and a rectangle, as shown in Figure 13.]
7. $y = 3x + 2$; $[0, 3]$
8. $y = 7x - 3$; $[1, 4]$

In Problems 9–22, (a) estimate the area bounded by the given curve, the x-axis, and the lines $x = a$ and $x = b$ by dividing the interval into a number (given in the braces) of subintervals of equal length and choosing a convenient point in each subinterval; (b) repeat the process by doubling the number of subintervals; (c) obtain a formula for the area enclosed by n estimating rectangles, where n is any positive integer; (d) find the exact area by taking the limit as $n \to \infty$ of the formula obtained in part (c).

FIGURE 13

9. $y = \frac{1}{2}x^2$; [0, 2], {4}

10. $y = 2x^2$; [−1, 1], {3}

11. $y = x^3$; [0, 5], {4}

12. $y = 17x^2$; [0, 8], {5}

13. $y = 3x^3$; [−1, 0], {8}

14. $y = (1 - x)^2$; [0, 1], {6}

15. $y = 1 - x^2$; [0, 1], {4}

16. $y = 1 - x^3$; [0, 1], {5}

17. $y = x^2$; [1, 2], {8}

18. $y = x^3$; [1, 2], {4}

19. $y = x + x^2$; [0, 1], {6}

20. $y = x^3 + 3x$; [0, 1], {5}

21. $y = 1 + 2x + 3x^2$; [0, 1]. {4}

22. $y = 1 + x + x^2 + x^3$; [0, 1], {4}

23. Consider the problem of calculating the area bounded by the curve $y = x^2$, the x-axis, and the lines $x = a$ and $x = b$.

    (a) Divide the interval $[a, b]$ into n subintervals of equal length. That is, $\Delta x = (b - a)/n$. What are the endpoints of each subinterval?

    (b) Letting $x_i{}^* = $ the right-hand endpoint of each subinterval, calculate $f(x_i{}^*) = (x_i{}^*)^2$.

    (c) Use (11) to estimate $A_a{}^b$ for 4, 8, 16, and 32 subintervals.

    *(d) Obtain a formula for the total area under the curve.

*24. Following the steps of Problem 23 and using (17), obtain a formula for the area bounded by the curve $y = x^3$, the x-axis, and the lines $x = a$ and $x = b$, with a and b positive.

*25. Let

$$S_n = \frac{1}{n} \sum_{k=1}^{n} \sqrt{1 - \left(\frac{k}{n}\right)^2}.$$

    (a) Describe a region whose area is estimated by S_n.

    (b) By using general information about the area of the region described in part (a), find $\lim_{n \to \infty} S_n$.

5.5 THE DEFINITE INTEGRAL

In Section 5.4 we computed the area of the region bounded by the graph of a function, the x-axis, and the lines $x = a$ and $x = b$. We did so in three cases: $f(x) = x^2$ (Example 2), $f(x) = 3x$ (Example 3), and $f(x) = x^3$ (Example 4). Our procedure was to write the sum

$$\sum_{i=1}^{n} f(x_i{}^*) \, \Delta x \tag{1}$$

and then see what happened as $n \to \infty$ ($\Delta x \to 0$). In all three examples four conditions held:

(i) f is continuous on $[a, b]$.
(ii) $f(x) \geq 0$ for $x \in [a, b]$.
(iii) All intervals in the partition have equal length. Specifically, $\Delta x = (b - a)/n$ is the length of each subinterval.
(iv) The point x_i^* is taken to be either the left- or right-hand endpoint of the subinterval $[x_{i-1}, x_i]$.

In this section we take the limit of sums like the sum in (1) to define the definite integral. In doing so, we will relax all four of the assumptions listed above. That is, we will use the following assumptions:

(i) f may be discontinuous at some points in $[a, b]$.
(ii) $f(x)$ may be negative for some (or even all) points in $[a, b]$.
(iii) The subintervals $[x_{i-1}, x_i]$ might have different lengths.
(iv) x_i^* can be any number in the interval $[x_{i-1}, x_i]$.

Definition 1   PARTITION   A **partition** P of $[a, b]$ is a division of $[a, b]$ into a number of smaller subintervals. The **partition points** are labeled $x_0, x_1, x_2, \ldots, x_n$, where $x_0 = a$ and $x_n = b$. It is assumed that

$$x_0 < x_1 < x_2 < \cdots < x_n.$$

We define

$$\Delta x_i = x_i - x_{i-1}$$

so that

$$\Delta x_1 = x_1 - x_0, \quad \Delta x_2 = x_2 - x_1, \quad \ldots, \quad \Delta x_n = x_n - x_{n-1}$$

and

$$\sum_{i=1}^{n} \Delta x_i = b - a = \text{total length of the interval.}$$

This is illustrated in Figure 1.

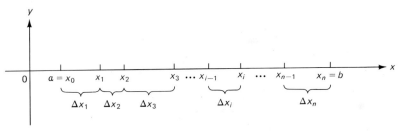

FIGURE 1

EXAMPLE 1   Five partitions of the interval $[0, 1]$ are sketched in Figure 2. ▪

(v) $0 \quad \frac{1}{10} \quad \frac{2}{10} \quad \frac{3}{10} \quad \frac{4}{10} \quad \frac{5}{10} \quad \frac{6}{10} \quad \frac{7}{10} \quad \frac{8}{10} \quad \frac{9}{10} \quad 1$

$$\Delta x_i = \frac{1}{10} \text{ for } i = 1, 2, \dots, 10$$

FIGURE 2

As in Section 5.4, a partition is called **regular** if the partition points are equally spaced. That is:

> The partition is regular if $\Delta x_i = \dfrac{b - a}{n}$ for $i = 1, 2, \dots, n$. $\qquad$ **(2)**

In Figure 2 partitions (i), (ii), and (v) are regular.

In Section 5.4 we examined what happened as $n \to \infty$. Since $\Delta x = (b - a)/n$, $n \to \infty$ implies that $\Delta x \to 0$. If the partition is not regular, we need a new way to describe the fact that the lengths of the subintervals are getting small. We do so by writing

$$\max \Delta x_i \to 0, \qquad\qquad\qquad\qquad\qquad\qquad \textbf{(3)}$$

where $\max \Delta x_i$ is the length of the largest subinterval in the partition.

EXAMPLE 2   In Figure 2 we have the following:

    (i) $\max \Delta x_i = \frac{1}{2}$ **(ii)** $\max \Delta x_i = \frac{1}{4}$ **(iii)** $\max \Delta x_i = \frac{1}{3}$
    (iv) $\max \Delta x_i = \frac{2}{5}$ **(v)** $\max \Delta x_i = \frac{1}{10}$ ▪

NOTE: If P is regular, then

$$\max \Delta x_i = \Delta x = \frac{b - a}{n}. \qquad\qquad\qquad\qquad \textbf{(4)}$$

We can now form sums similar to (1) for any function f defined (but not necessarily continuous or nonnegative) on $[a, b]$.

Step 1. Partition the interval $[a, b]$ by

$$a = x_0 < x_1 < x_2 < \cdots < x_n = b.$$

Step 2. Choose a number x_i^* in each subinterval $[x_{i-1}, x_i]$.

Step 3. Form the sum

$$f(x_1^*) \, \Delta x_1 + f(x_2^*) \, \Delta x_2 + \cdots + f(x_n^*) \, \Delta x_n = \sum_{i=1}^{n} f(x_i^*) \, \Delta x_i. \tag{5}$$

The sum (5) is called a **Riemann sum.†**

For every partition of the interval $[a, b]$ and for every choice of the numbers x_i^*, the Riemann sum (5) is a *real number*. In certain circumstances that number represents an approximation to the area under a curve. So it is natural to ask what happens to the Riemann sum when the lengths of the subintervals get small; that is, what happens to $\sum_{i=1}^{n} f(x_i^*) \, \Delta x_i$ when $\max \Delta x_i \to 0$? We are taking a limit, but it is a special kind of limit.

Let I be a real number. Then we can define the limit of Riemann sums in the following intuitive way.

Definition 2 INTUITIVE DEFINITION OF LIMIT OF RIEMANN SUMS

$$\lim_{\max \Delta x_i \to 0} \sum_{i=1}^{n} f(x_i^*) \, \Delta x_i = L$$

if $\sum_{i=1}^{n} f(x_i^*) \, \Delta x_i$ is as close as we wish to L when $\max \Delta x_i$ is small, no matter how the numbers $x_1^*, x_2^*, \ldots, x_n^*$ are chosen.

We will make this definition more mathematically precise at the end of this section.
We now define one of the most important concepts in our study of calculus.

Definition 3 THE DEFINITE INTEGRAL Let f be defined on $[a, b]$ with $a < b$. Then the **definite integral** of f over the interval $[a, b]$, written $\int_a^b f(x) \, dx$, is given by

$$\int_a^b f(x) \, dx = \lim_{\max \Delta x_i \to 0} \sum_{i=1}^{n} f(x_i^*) \, \Delta x_i, \tag{6}$$

whenever the limit in (6) exists.

The process of calculating an integral is called **integration,** and the numbers a and b are called the **lower** and **upper limits of integration,** respectively. If the limit in (6)

---

†See the accompanying biographical sketch.

Georg Friedrich Riemann

Georg Friedrich Riemann
The Granger Collection

Georg Friedrich Riemann was born in 1826 in the German village of Hanover. His father was a Lutheran pastor. Throughout his life Riemann was exceedingly shy and in frail health. Althought his family was by no means wealthy, Riemann was, nevertheless, able to get a good education, first at the University of Berlin and then at the University of Göttingen. His work at Göttingen culminated with a brilliant thesis in the area of functions of a complex variable (a complex variable is a variable of the form $x + iy$, where x and y are real numbers and $i = \sqrt{-1}$). At Göttingen he worked under the greatest mathematician of the nineteenth century, Karl Friedrich Gauss (1777–1855).

In 1854 Riemann was appointed *Privatdozent* (official but unpaid lecturer) at the University of Göttingen. According to the custom of the day, he was asked to give a probationary lecture. The result was the greatest paper of comparable size ever presented in the history of mathematics. The title of the lecture was "Über die Hypothesen welche der Geometrie zu Grunde liegen" ("On the Hypotheses which Lie at the Foundation of Geometry"). Rather than discussing a specific example, it urged a global view of geometry that revolutionized the study of that subject. After this lecture, and perhaps for the only time in a career that spanned approximately sixty years, Gauss paid compliments to the work of someone else.

Riemann's great paper was presented in 1854 but was not published until 1867. One of the central ideas in the paper was that geometry could be discussed in general curved spaces rather than only in the sphere (which is one specific curved space). Albert Einstein made use of this idea in his general theory of relativity.

Riemann made important contributions to many other areas of mathematics and theoretical physics. Most important for us, he clarified the concept of the definite integral. It is this definition, now known as the *Riemann integral*, that is the basis for the material in this chapter and much of the rest of the book.

In 1859 Riemann became a full professor at the University of Göttingen, having become an assistant professor only two years earlier. The chair that he occupied was previously held by another great German mathematician, Peter Gustav Lejeune Dirichlet (1805–1859). Dirichlet, in turn, had succeeded Gauss. Riemann's career was cut short in 1866, in Italy, where he had gone to seek a cure for tuberculosis. He died at the age of forty.

exists, then f is said to be **integrable on the interval** $[a, b]$. The function $f(x)$ in (6) is called the **integrand.**

The variable x in (6) is called a **dummy variable** since it could be replaced by *any other* variable without changing the value of the integral. This is true because the definite integral is a number. We could, instead, subdivide $[a, b]$ by

$$a = t_0 < t_1 < t_2 < \cdots < t_{n-1} < t_n = b.$$

That is, we are simply renaming the same numbers chosen before. Then we could choose $t_i{}^*$ in $[t_{i-1}, t_i]$ and (6) would become

$$\int_a^b f(t)\, dt = \lim_{\max \Delta t_i \to 0} \sum_{i=1}^n f(t_i{}^*)\, \Delta t_i. \tag{7}$$

This expression is, of course, the same definition as in (6). We can also substitute any other variable for t. Thus

$$\int_a^b f(x)\, dx = \int_a^b f(t)\, dt = \int_a^b f(z)\, dz = \int_a^b f(\text{dummy})\, d(\text{dummy}) \tag{8}$$

REMARK. There is a great difference between the definite integral and indefinite integral, or antiderivative. *The definite integral is a number, while the indefinite integral is a function.* Nevertheless, we use the words "integral," "integrable," and "integration" in both cases. There should not be any confusion between the two uses of the words since the two integrals stand for such different things.

We can tell the difference between a definite and indefinite integral because definite integrals always have lower and upper limits $[\int_a^b f(x)\, dx]$ while indefinite integrals do not. We will show the connection between the two integrals in Section 5.6.

Following this definition, a basic question remains: What functions are integrable? After all, the limit in (6) is usually difficult to obtain (as you could imagine after working through the problems of the previous section), and so it would be nice to know *before* doing any calculations that a given integral does exist. The following theorem, whose proof is beyond the scope of this book,† tells us that a great number of the functions we have already discussed are integrable.

Theorem 1   *Existence of Definite Integrals* Let f be continuous on the interval $[a, b]$. Then f is integrable on $[a, b]$. That is, $\int_a^b f(x)\, dx$ exists.

EXAMPLE 3   The following integrals all exist since the functions being integrated are all continuous on the interval of integration.

(a) $\displaystyle\int_1^7 x^7\, dx$ (b) $\displaystyle\int_1^3 \sqrt{x}\, dx$ (c) $\displaystyle\int_{-10}^{10} |x|\, dx$

(d) $\displaystyle\int_0^3 \frac{1}{\sqrt{(x+1)(x+2)}}\, dx$ (e) $\displaystyle\int_3^5 \frac{1}{x}\, dx$

(f) $\displaystyle\int_2^4 t^{3/5}\, dt$ (g) $\displaystyle\int_0^{1,000,000} s^{100}\, ds$ (h) $\displaystyle\int_{-3}^0 \frac{1}{z-1}\, dz$ ■

There are an infinite number of ways to partition an interval and there are infinitely many numbers in the interval $[x_{i-1}, x_i]$. However, if f is continuous on $[a, b]$, then we know that f is integrable on $[a, b]$, which means that the limit in (6) exists. But if the limit exists, $\sum_{i=1}^n f(x_i{}^*)\, \Delta x_i$ is close to L for max Δx_i small, no matter how $[a, b]$

---

†For a proof, consult any book in advanced calculus, such as R. C. Buck, *Advanced Calculus* (McGraw-Hill, New York, 1965), Chapter 3.

is partitioned or how the $x_i{}^*$'s are chosen. So if f is continuous, we can make things simpler by using a regular partition of $[a, b]$, and, if we wish, we can compute Riemann sums by using the right- or left-hand endpoints of each subinterval, as in Section 5.4. According to equation (4), for a regular partition max $\Delta x_i \to 0$ means $\Delta x \to 0$, or $n \to \infty$, where n is the number of subintervals. Thus using regular partitions, we have the following alternative definition of the definite integral.

Definition 4 ALTERNATIVE DEFINITION OF THE DEFINITE INTEGRAL Let f be continuous on $[a, b]$ with $a < b$. Then

$$\int_a^b f(x)\, dx = \lim_{\Delta x \to 0} \sum_{i=1}^n f(x_i{}^*)\, \Delta x = \lim_{n \to \infty} \sum_{i=1}^n f(x_i{}^*)\, \frac{b-a}{n}. \tag{9}$$

EXAMPLE 4 Calculate (a) $\int_0^1 x^2\, dx$ and (b) $\int_0^3 x^3\, dx$.

Solution. **(a)** From Example 5.4.2 we see that

$$\int_0^1 x^2\, dx = \frac{1}{3}.$$

$\dfrac{x^3}{3} \qquad \dfrac{1-0}{3} = \dfrac{1}{3}$

(b) From Example 5.4.4 we see that

$$\int_0^3 x^3\, dx = \frac{81}{4}. \ \blacksquare$$

$\dfrac{x^4}{4} \qquad \dfrac{3-0}{4} = \dfrac{(3)^4}{4} =$

The following theorem shows how to compute a particularly simple definite interval.

Theorem 2. Let c be a constant. Then

$$\int_a^b c\, dx = c(b-a). \tag{10}$$

In particular, for $c = 1$ we have

$$\int_a^b 1\, dx = \int_a^b dx = b - a. \tag{11}$$

This theorem makes sense geometrically if c is a positive constant. Look at Figure 3. Then $c(b-a)$ is equal to the area of the rectangle under the curve $y = c$ from $x = a$ to $x = b$. The proof of the theorem is left as an exercise (see Problem 25).

To this point we have required that a be less than b in the definition of the integral. The cases $a = b$ and $a > b$ are defined next.

Definition 5 For any real number a,

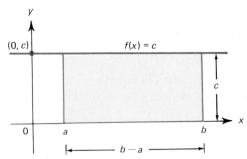

FIGURE 3

$$\int_a^a f(x)\, dx = 0. \tag{12}$$

Note that if $\int_a^a f(x)\, dx$ represents area, then this definition states that the area of a region with a width of zero is zero. (The interval $[a, a]$ has a width of zero.)

Definition 6.   If $a < b$, and $\int_a^b f(x)\, dx$ exists, then

$$\int_b^a f(x)\, dx = -\int_a^b f(x)\, dx. \tag{13}$$

This definition says that reversing the order of integration changes the sign of the integral, thus enabling us to apply Definition 4 when $a > b$.

EXAMPLE 5   Calculate $\int_1^0 x^2\, dx$.

Solution.

$$\int_1^0 x^2\, dx = -\overset{\text{Example 4(a)}}{\underset{\swarrow}{\int_0^1 x^2\, dx}} = -\frac{1}{3} \quad \blacksquare$$

The definite integral does not always represent the area under a curve. The following example illustrates this fact.

EXAMPLE 6   Calculate $\int_{-4}^7 (-2)\, dx$.

Solution. $b - a = 7 - (-4) = 11$, so that

$$\int_{-4}^7 (-2)\, dx = \overset{\text{From Theorem 2}}{\underset{\swarrow}{-2[7 - (-4)]}} = (-2)11 = -22.$$

Here the integral is negative. To see why, look at Figure 4. The area enclosed in the

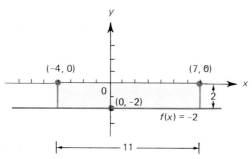

FIGURE 4

figure is equal to 22 square units. However, we see that *the integral treats areas below the x-axis as negative.* Therefore we cannot calculate an area under the curve $y = f(x)$ by simple integration unless $f(x) \geq 0$ on $[a, b]$. To handle the more general case, we define area in terms of the definite integral. ∎

Definition 7   AREA   The **area**† bounded by the function $y = f(x)$, the x-axis, and the lines $x = a$ and $x = b$ (for $a < b$) is denoted by $A_a^{\ b}$ and defined by the formula

$$A_a^{\ b} = \int_a^b |f(x)| \, dx. \tag{14}$$

NOTE.   If $a < b$ and if $f(x)$ is nonnegative for $a \leq x \leq b$, then $|f(x)| = f(x)$, and in this case $A_a^{\ b} = \int_a^b f(x) \, dx$.

EXAMPLE 7   Calculate the area of the region bounded by the curve $y = x^3$, the x-axis, and the lines $x = -1$ and $x = 1$ (see Figure 5).

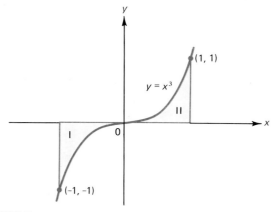

FIGURE 5

---

†If $f(x)$ is negative for some values of x in $[a, b]$, then $A_a^{\ b}$ is sometimes called the *net area of f* over $[a, b]$.

Solution. In $[-1, 0]$, x^3 is negative, so $|x^3| = -x^3$. In $[0, 1]$, x^3 is positive. Thus

$$A_{-1}{}^1 = \int_{-1}^{1} f(x)\,dx, \qquad \text{where} \qquad f(x) = \begin{cases} x^3, & x \ge 0 \\ -x^3, & x < 0. \end{cases}$$

Geometrically, it appears that the area of region I in Figure 5 is equal to the area of region II. To verify this, we have

$$A_{-1}{}^0 = \int_{-1}^{0} (-x^3)\,dx = \lim_{\Delta x \to 0} \sum_{i=1}^{n} [-(x_i{}^*)^3]\,\Delta x.$$

We now choose the partition points

$$-1 = -\frac{n}{n} < -\frac{(n-1)}{n} < \cdots < -\frac{i}{n} < \cdots < -\frac{1}{n} < \frac{0}{n} = 0 \quad \text{and} \quad x_i{}^* = -\frac{i}{n},$$

so that

$$\Delta x = \frac{1}{n}, \qquad (x_i{}^*)^3 = -\frac{i^3}{n^3}, \qquad \text{and} \qquad (x_i{}^*)^3\,\Delta x = -\frac{i^3}{n^4}.$$

Then we have

$$\int_{-1}^{0} -x^3\,dx = \lim_{\Delta x \to 0} \sum_{i=1}^{n} (-x_i{}^*)^3\,\Delta x = \lim_{\Delta x \to 0} -\sum_{i=1}^{n} (x_i{}^*)^3 = -\lim_{n \to \infty} \sum_{i=1}^{n} -\frac{i^3}{n^4}$$

$$\overset{\text{Equation (5.4.17)}}{\underset{\nearrow}{}}$$

$$= \lim_{n \to \infty} \frac{1}{n^4} \sum_{i=1}^{n} i^3 = \lim_{n \to \infty} \frac{1}{n^4} \cdot \frac{n^2(n-1)^2}{4} = \frac{1}{4} \lim_{n \to \infty} \frac{(n-1)^2}{n^2} = \frac{1}{4}.$$

We can, in an analogous manner, calculate $A_0{}^1 = \frac{1}{4}$, so that

$$A_{-1}{}^1 = \tfrac{1}{4} + \tfrac{1}{4} = \tfrac{1}{2}.$$

Note that this result is *not* the same as $\int_{-1}^{1} x^3\,dx$, which is equal to zero! (The "negative" area below the x-axis "cancels" with the positive area above it.) You are urged to verify this result by partitioning the interval $[-1, 1]$ in such a way that areas of rectangles below the x-axis "cancel" with areas of rectangles above it. ∎

WARNING. A fairly common error made by students when they first face the problem of calculating areas by integration is to employ the following *incorrect* reasoning: "If I get a negative answer when calculating an area, then all I need to do is to take the absolute value of my answer to make it right." To see why this reasoning is faulty, consider the problem of computing the area bounded by the line $y = x$ and the x-axis for x between -2 and 1. This area is drawn in Figure 6. It is easy to see that the area of triangle I is $2[A = \frac{1}{2}bh = \frac{1}{2}(2)(2)]$, and the area of triangle II is $\frac{1}{2}$. Thus the total area is $\frac{5}{2}$. However, it is not difficult to show that

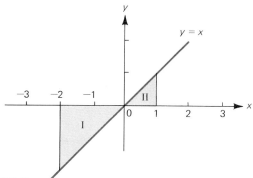

FIGURE 6

$$\int_{-2}^{1} x \, dx = -\frac{3}{2}.$$

This is the answer we would get if we forgot to take the absolute value in $\int_{-1}^{1} |x| \, dx$. Changing the $-\frac{3}{2}$ to $\frac{3}{2}$ will not give us the correct answer.

We now state three theorems that can be very useful for calculating integrals. The proofs of the theorems are not difficult. However, since these will be especially easy to prove after we have discussed the fundamental theorem of calculus in Section 5.6, we will delay the proofs until that section (or see Problems 26, 27, and 28).

Theorem 3 If f is integrable on $[a, b]$ and if $a < c < b$, then f is integrable on $[a, c]$ and on $[c, b]$, and

$$\int_{a}^{b} f = \int_{a}^{c} f + \int_{c}^{b} f. \tag{15}$$

Theorem 3 is obvious if $\int_{a}^{b} f$ represents the area under a curve. In Figure 7 the area under the curve is given by $\int_{a}^{b} f$, which is equal to $A_1 + A_2 = \int_{a}^{c} f + \int_{c}^{b} f$.

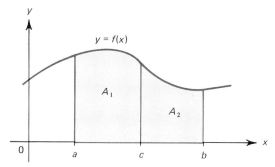

FIGURE 7

EXAMPLE 8 Calculate $\int_{1}^{3} x^3 \, dx$.

Solution. We have already calculated (in Examples 7 and 4)

$$\int_0^1 x^3 \, dx = \frac{1}{4} \quad \text{and} \quad \int_0^3 x^3 \, dx = \frac{81}{4}.$$

Then using (15), we have $\int_0^3 x^3 \, dx = \int_0^1 x^3 \, dx + \int_1^3 x^3 \, dx$, or

$$\int_1^3 x^3 \, dx = \int_0^3 x^3 \, dx - \int_0^1 x^3 \, dx = \frac{81}{4} - \frac{1}{4} = \frac{80}{4} = 20. \ \blacksquare$$

Theorem 4   *Multiplication by a Constant* If f is integrable on $[a, b]$ and if k is any constant, then kf is integrable on $[a, b]$, and

$$\int_a^b kf = k \int_a^b f. \tag{16}$$

EXAMPLE 9   Calculate $\int_1^3 17x^3 \, dx$.

Solution. In Example 8 we saw that $\int_1^3 x^3 \, dx = 20$. Then

$$\int_1^3 17x^3 \, dx = 17 \int_1^3 x^3 \, dx = 17(20) = 340. \ \blacksquare$$

Theorem 5   *The Sum of Two Functions* If the functions f and g are both integrable on $[a, b]$, then $f + g$ is integrable on $[a, b]$, and

$$\int_a^b (f + g) = \int_a^b f + \int_a^b g. \tag{17}$$

That is, *the integral of the sum is the sum of the integrals.*

REMARK. Theorem 5 can be easily extended to the integral of a finite sum of integrable functions.

EXAMPLE 10   Calculate $\int_0^1 (x^2 + x^3) \, dx$.

Solution.

$$\int_0^1 (x^2 + x^3) \, dx = \int_0^1 x^2 \, dx + \int_0^1 x^3 \, dx = \frac{1}{3} + \frac{1}{4} = \frac{7}{12} \ \blacksquare$$

EXAMPLE 11   Calculate $\int_0^1 (x^2 - x^3) \, dx$.

Solution. $\int_0^1 (x^2 - x^3) \, dx = \int_0^1 x^2 \, dx + \int_0^1 (-x^3) \, dx$. But from Theorem 4, $\int_0^1 (-x^3) \, dx = -\int_0^1 x^3 \, dx = -\frac{1}{4}$. Thus

$$\int_0^1 (x^2 - x^3) \, dx = \frac{1}{3} - \frac{1}{4} = \frac{1}{12}. \ \blacksquare$$

The next two theorems give us a way to compare integrals without tedious calculations.

Theorem 6.   If f is integrable on $[a, b]$ and $f \geq 0$ there, then

$$\int_a^b f(x) \, dx \geq 0.$$

Proof.   This result follows since every term in the limit (6) defining the integral is nonnegative.   ■

Theorem 7   *Comparison Theorem*   Let f and g be integrable on $[a, b]$ and suppose that for every x in $[a, b]$,

$$f(x) \leq g(x). \tag{18}$$

Then

$$\int_a^b f(x) \, dx \leq \int_a^b g(x) \, dx. \tag{19}$$

Proof.   Define $h(x) = g(x) - f(x)$. Then by (18), $h(x) \geq 0$ on $[a, b]$. By Theorem 6, $\int_a^b h(x) \, dx \geq 0$. But

$$0 \leq \int_a^b h(x) \, dx = \int_a^b [g(x) - f(x)] \, dx \overset{\text{Theorem 5}}{=} \int_a^b g(x) \, dx + \int_a^b [-f(x)] \, dx$$

$$\overset{\text{Theorem 4}}{=} \int_a^b g(x) \, dx - \int_a^b f(x) \, dx.$$

Therefore $\int_a^b f(x) \, dx \leq \int_a^b g(x) \, dx$.   ■

EXAMPLE 12   On the interval $[0, 1]$, $x^3 \leq x^2$. This inequality tells us (from Theorem 7) that

$$\int_0^1 x^3 \, dx \leq \int_0^1 x^2 \, dx.$$

This is indeed the case since $\int_0^1 x^3 \, dx = \frac{1}{4}$ and $\int_0^1 x^2 \, dx = \frac{1}{3}$.   ■

EXAMPLE 13   Since $x^5 \leq x^7$ on $[2, 5]$, we have

$$\int_2^5 x^5 \, dx \leq \int_2^5 x^7 \, dx.$$ ■

The following theorem is sometimes extremely useful for estimating integrals when calculations are difficult.

Theorem 8 *Upper and Lower Bound Theorem* Suppose that on $[a, b]$

$$m \leq f(x) \leq M \tag{20}$$

for all x in $[a, b]$. That is, m and M are lower and upper bounds, respectively, for the function f on the interval $[a, b]$ (see page 119). Then if f is integrable on $[a, b]$,

$$m(b - a) \leq \int_a^b f(x) \, dx \leq M(b - a). \tag{21}$$

Proof. If we define g by $g(x) = M$, then $f(x) \leq g(x)$, and by the comparison theorem

$$\int_a^b f(x) \, dx \leq \overset{\text{Theorem 2}}{\int_a^b M \, dx} = M(b - a)$$

Similarly,

$$m(b - a) = \int_a^b m \, dx \leq \int_a^b f(x) \, dx,$$

and the theorem is proved. ∎

This result is illustrated in Figure 8.

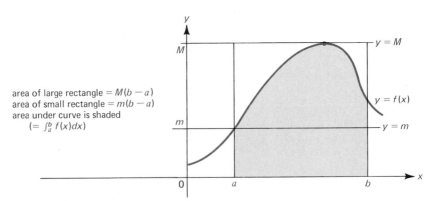

area of large rectangle $= M(b - a)$
area of small rectangle $= m(b - a)$
area under curve is shaded
$\left(= \int_a^b f(x) dx \right)$

FIGURE 8

REMARK. According to Theorem 2.7.4, if f is continuous on $[a, b]$, then f takes minimum and maximum values in that interval, so the numbers m and M do exist. The function $1/\sqrt[3]{x}$, for example, is not continuous and does not have such bounds in the interval $[-1, 1]$.

EXAMPLE 14 Estimate $\displaystyle\int_0^{\pi/6} \sin^5 x \, dx$.

Solution. The function sin x is increasing on $[0, \pi/6]$, so

$$0 = \sin 0 \le \sin x \le \sin \frac{\pi}{6} = \frac{1}{2}.$$

and

$$0 \le \sin^5 x \le \left(\frac{1}{2}\right)^5 = \frac{1}{32} \quad \text{for} \quad 0 \le x \le \frac{\pi}{6}.$$

Thus $m = 0$, $M = \frac{1}{32}$, and

$$0 \le \int_0^{\pi/6} \sin^5 x \, dx \le \frac{1}{32}\left(\frac{\pi}{6} - 0\right) = \frac{\pi}{192} \approx 0.01636. \quad \blacksquare$$

We now show how definite integrals can arise in an application having nothing at all to do with area.

Suppose that a particle is moving with velocity $v = v(t) \ge 0$. If $v(t) = c$, a constant, then we can calculate the distance traveled between the initial time $t = t_0$ and the final time $t = t_f$ by

$$\text{distance} = \text{velocity} \times \text{elapsed time} = v \cdot (t_f - t_0). \tag{22}$$

However, if the velocity is changing (i.e., if the particle is accelerating or decelerating) then formula (22) simply will not work. Still, we can calculate the distance traveled by a method identical to the one we used to calculate areas. The basic idea is simple: Even though $v(t)$ is changing, over a very small period of time it will be almost constant.

We begin by dividing the time interval $[t_0, t_f]$ into n equal subintervals of length $(t_f - t_0)/n$ (see Figure 9). Let t_i^* denote a time in the interval $[t_{i-1}, t_i]$. If n is large enough (i.e., if the lengths of the intervals are small), then the velocity of the particle in the time period $[t_{i-1}, t_i]$ is almost constant, so that it can be approximated by $v(t_i^*)$ in $[t_{i-1}, t_i]$. Let $\Delta t = t_i - t_{i-1}$. If s_i denotes the distance traveled by the particle between the time $t = t_{i-1}$ and the time $t = t_i$, then from (22)

$$s_i \approx v(t_i^*) \, \Delta t. \tag{23}$$

FIGURE 9

The total distance traveled by the particle over the entire time interval $[t_0, t_f]$ is then given (approximately) by

$$s = s_1 + s_2 + \cdots + s_n \approx v(t_1^*) \, \Delta t + v(t_2^*) \, \Delta t + \cdots + v(t_n^*) \, \Delta t$$

$$= \sum_{i=1}^{n} v(t_i^*) \, \Delta t. \tag{24}$$

This should look familiar. Formula (24) is only an approximation because $v(t)$ is not really constant over $[t_{i-1}, t_i]$. However, as the length of the interval becomes smaller and smaller, the approximation gets better and better. We therefore can conclude that

$$s = \lim_{\Delta t \to 0} \sum_{i=1}^{n} v(t_i^*)\, \Delta t = \int_{t_0}^{t_f} v(t)\, dt. \tag{25}$$

Thus we obtain the distance traveled by "adding up" a great number of small distances, just as we obtained the area under a curve by "adding up" a great number of areas of "thin" rectangles. ■

EXAMPLE 15 A race car starting at rest accelerates so that its velocity after t seconds is given by $v(t) = 4t^3$ feet per second (for the first 4 sec).

(a) What is the car's velocity after 3 sec?
(b) How far does the car travel in the first 3 sec?

Solution. (a) $v(3) = 4 \cdot 3^3 = 4(27) = 108$ ft/sec (≈ 73.6 mi/hr).
(b) Here $t_0 = 0$, $t_f = 3$, and $v(t) = 4t^3$. Then

$$s = \int_{t_0}^{t_f} v(t)\, dt = \int_0^3 4t^3\, dt \overset{\text{Theorem 4}}{=} 4\int_0^3 t^3\, dt \overset{\text{Example 4(b)}}{=} 4\left(\frac{81}{4}\right) = 81 \text{ ft.} \ ■$$

There is an interesting interpretation of Theorem 3 in the use of the definite integral in the previous example. If $v(t)$ represents the velocity of a moving object and $v(t) \geq 0$ for $t \in [a, b]$, then

$$\int_a^b v(t)\, dt = \text{distance traveled from time } a \text{ to time } b,$$

$$\int_a^c v(t)\, dt = \text{distance traveled from time } a \text{ to time } c,$$

and

$$\int_c^b v(t)\, dt = \text{distance traveled from time } c \text{ to time } b.$$

It is easy to see why it should be true that if $a < c < b$,

$$\int_a^b v(t)\, dt = \int_a^c v(t)\, dt + \int_c^b v(t)\, dt.$$

In Theorem 1 we stated that all continuous functions are integrable. Many other kinds of functions are integrable as well. On page 122 we defined a *piecewise continuous* function as a function that is continuous at every point in $[a, b]$ except for a finite number of points at which f has a jump discontinuity (a jump discontinuity is a *finite jump*). Thus, for example, the greatest integer function $[x]$ is piecewise continuous on every closed, bounded interval. The following result is true:

If f is piecewise continuous on $[a, b]$, then $\int_a^b f(x)\,dx$ exists.   (26)

We close this section with a formal definition of the limit of Riemann sums. This definition is somewhat similar to the ϵ–δ definition of a limit given in Section 2.8 (Definition 1).

Definition 8   FORMAL DEFINITION OF THE LIMIT OF RIEMANN SUMS   Let L be a real number. Then

$$\lim_{\max \Delta x_i \to 0} \sum_{i=1}^{n} f(x_i{}^*)\,\Delta x_i = L$$

if for every $\epsilon > 0$ there is a number $\delta > 0$ such that for every partition P with $\max \Delta x_i < \delta$,

$$\left| \sum_{i=1}^{n} f(x_i{}^*)\,\Delta x_i - L \right| < \epsilon.$$

That is, if we take any partition of $[a, b]$, with each subinterval of length less than δ, then any Riemann sum $\sum_{i=1}^{n} f(x_i{}^*)\,\Delta x_i$, with $x_i{}^* \in [x_{i-1}, x_i]$, is within ϵ units of L.

PROBLEMS 5.5

In Problems 1–24, calculate the given definite integral by using the definitions, theorems, and examples of this section.

1. $\displaystyle\int_0^4 7x\,dx$

2. $\displaystyle\int_1^2 4x\,dx$

3. $\displaystyle\int_2^5 (3x + 2)\,dx$

4. $\displaystyle\int_1^4 (7t - 3)\,dt$

5. $\displaystyle\int_0^2 \tfrac{1}{2}t^2\,dt$

6. $\displaystyle\int_{-1}^0 t^2\,dt$

7. $\displaystyle\int_{-1}^1 2s^2\,ds$

8. $\displaystyle\int_0^5 s^3\,ds$

9. $\displaystyle\int_0^8 17s^2\,ds$

10. $\displaystyle\int_{-1}^1 z^3\,dz$

11. $\displaystyle\int_1^0 z^3\,dz$

12. $\displaystyle\int_3^0 (2z + 4)\,dz$

13. $\displaystyle\int_0^1 (1 - y^2)\,dy$

14. $\displaystyle\int_0^1 (1 - y^3)\,dy$

15. $\displaystyle\int_5^2 y^3\,dy$

16. $\displaystyle\int_0^1 (1 + x + x^2 + x^3)\,dx$ **17.** $\displaystyle\int_{-2}^1 (x^3 - x^2)\,dx$

18. $\displaystyle\int_1^{-1} (x^3 + 3x)\,dx$

*19. $\displaystyle\int_{-1}^1 |x|\,dx$ [Hint: Draw a picture.]

*20. $\displaystyle\int_{-3}^2 |x + 1|\,dx$

*21. $\displaystyle\int_0^5 [x]\,dx$ [Hint: This is the greatest integer function. To calculate the area, add the areas of the rectangles.]

*22. $\displaystyle\int_{-4}^4 [x]\,dx$

*23. $\displaystyle\int_a^b x^2\,dx$ [Hint: See Problem 5.4.23.]

*24. $\displaystyle\int_a^b x^3\,dx$ [*Hint:* See Problem 5.4.24.]

*25. Prove Theorem 2. [*Hint:* $f(x_i{}^*) = c$ for any number $x_i{}^* \in [a, b]$.]

**26. Give a partial proof of Theorem 3 by dividing the interval $[a, b]$ into $n + m$ subintervals, where the nth partition point is c. [Here it is necessary to assume that both $c - a$ and $b - c$ are rational numbers.]

*27. Prove Theorem 4. [*Hint:* Use the fact that $\lim_{x \to x_0} kf(x) = k \lim_{x \to x_0} f(x)$.]

*28. Prove Theorem 5. [*Hint:* Use the fact that $\lim_{x \to x_0} (f + g) = \lim_{x \to x_0} f + \lim_{x \to x_0} g$.]

29. Explain why $\int_{23}^{47} s^{17}\,ds < \int_{23}^{47} s^{55/3}\,ds$.   30. Explain why $\int_0^1 \sqrt{x}\,dx < \int_0^1 \sqrt[3]{x}\,dx$.

31. Which is greater: $\int_1^2 x\,dx$ or $\int_1^2 \sqrt{x}\,dx$?   32. Which is greater: $\int_{-1}^0 x^{1/3}\,dx$ or $\int_{-1}^0 x^{1/5}\,dx$?

*33. Show that $\int_0^1 \sqrt{1 + x^3}\,dx$ lies between 1 and $\frac{5}{4}$. [*Hint:* Write the integral as the sum of two integrals.]

In Problems 34–43, find upper and lower bounds for the given integrals. Do *not* try to calculate them.

34. $\displaystyle\int_{-1}^1 x^{10}\,dx$ 35. $\displaystyle\int_1^4 4\sqrt{x}\,dx$ 36. $\displaystyle\int_1^8 7x^{1/3}\,dx$

37. $\displaystyle\int_1^9 \frac{1}{\sqrt{x}}\,dx$ 38. $\displaystyle\int_2^3 (x^2 + x^3)\,dx$ 39. $\displaystyle\int_1^{100} \frac{1}{x}\,dx$

40. $\displaystyle\int_{-2}^2 3x^4\,dx$ 41. $\displaystyle\int_0^1 \frac{1}{1 + x^2}\,dx$ 42. $\displaystyle\int_0^{\pi/2} \sin x\,dx$

43. $\displaystyle\int_0^{\pi/3} \cos x\,dx$

44. A ball is dropped from a height of 400 ft. Its velocity after t seconds is given by the formula $v = 32\,t$ feet per second.
    (a) How fast is the ball dropping after 4 sec?
    (b) How far has the ball dropped after 4 sec?
    *(c) After how many seconds will the ball hit the ground?

45. Gravity on a certain planet X is only half that of earth. Then the force of acceleration due to gravity is 16 ft/sec² = 4.91 m/sec² instead of the usual 32 ft/sec² = 9.81 m/sec². Answer the questions in Problem 44 assuming that the ball is dropped on planet X. [*Hint:* Find a new formula for v as a function of t.]

46. A bullet is shot straight into the air with an initial velocity of 500 m/sec. Its velocity after t seconds have elapsed is $v(t) = (500 - 9.81t)$ meters per second.
    (a) After how many seconds will the bullet begin to fall?
    (b) How high will the bullet go?

47. Show that if a, b, and c are any real numbers such that f is integrable on $[a, b]$, $[a, c]$, and $[c, b]$, then

$$\int_a^b f = \int_a^c f + \int_c^b f,$$

where it is *not* required that $a < c < b$. [*Hint:* There are six cases to consider: $a < c < b$, $a < b < c$, $b < a < c$, etc. Use Theorem 3 and Definition 6 to test each case separately.]

*48. Let

$$f(x) = \begin{cases} 1, & \text{if } x \text{ is rational} \\ 0, & \text{if } x \text{ is irrational.} \end{cases}$$

[This function was discussed in Problem 2.7.19 and you should look at that problem again now.]

(a) Explain why, for any partition of the interval $[0, 1]$, it is possible to choose points x_i^* in $[x_{i-1}, x_i]$ such that $f(x_i^*) = 1$.

(b) Explain why it is possible to choose points y_i^* in $[x_{i-1}, x_i]$ such that $f(y_i^*) = 0$.

(c) Show that $\lim_{\Delta x \to 0} [f(x_1^*) \Delta x + f(x_2^*) \Delta x + \cdots + f(x_n^*) \Delta x]$ depends on the choice of the points x_i^*.

(d) Explain why $\int_0^1 f(x)\, dx$ does not exist.

(e) Is f piecewise continuous?

*49. If $f(x)$ is continuous on $[a, b]$, show that

$$\left| \int_a^b f(x)\, dx \right| \leq \int_a^b |f(x)|\, dx.$$

[*Hint:* Show from the definition that $\int_a^b f(x)\, dx \leq \int_a^b |f(x)|\, dx$ and $-\int_a^b f(x)\, dx \leq \int_a^b |f(x)|\, dx$.]

50. Without doing any calculations, explain why the following are true.

(a) $\displaystyle\int_0^{2\pi} \sin x\, dx = \int_0^{2\pi} \cos x\, dx = 0$ (b) $\displaystyle\int_{-\pi/2}^{\pi/2} \sin x\, dx = \int_0^{\pi} \cos x\, dx = 0$

*51. Show that

$$\int_0^n [x]\, dx = \frac{(n-1)\cdot n}{2}$$

if n is a positive integer.

*52. Suppose that $0 \leq a < b$. Evaluate $\int_a^b [x]\, dx$.

*53. Suppose that f is continuous on $[a, b]$. We define the new function g by $g(x) = f(a + b - x)$. Note that g is continuous on $[a, b]$. Prove that $\int_a^b f = \int_a^b g$.

**54. Show that if f is integrable on $[a, b]$, then f is bounded. That is, there is a number $M > 0$ such that $|f(x)| \leq M$ for every $x \in [a, b]$.

5.6 THE FUNDAMENTAL THEOREM OF CALCULUS

In Chapter 2 we saw how the tangent line problem, formulated long ago by Greek mathematicians and not solved until the seventeenth century, gave rise to the modern theory of derivatives. In Sections 5.4 and 5.5 we saw how the area problem whose origins are lost in antiquity gave rise to the idea behind the definite integral. Finally, in Section 5.2 we introduced the indefinite integral, which was seemingly unrelated to the definite integral and was defined as the "inverse" operation to differentiation.

In this section we show how these three operations on functions (the derivative and the two integrals) are intimately related. The remarkable theorem linking these together is called the **fundamental theorem of calculus.**

**Theorem 1   *Fundamental Theorem of Calculus*   Let f be continuous on $[a, b]$. If F is any antiderivative of f on $[a, b]$, then

$$\int_a^b f(t)\, dt = F(b) - F(a). \tag{1}$$

This theorem simply asserts that we may calculate a definite integral by evaluating any antiderivative at the endpoints of the interval of integration and then subtracting.

Proof.   Since f is continuous on $[a, b]$, we know from Theorem 5.5.1 that $\int_a^b f(x)\, dx$ exists. Let F be an antiderivative for f. By the definition of the antiderivative, F is continuous on $[a, b]$ and differentiable on (a, b). Let

$$a = x_0 < x_1 < x_2 < \cdots < x_n = b$$

be a regular partition of $[a, b]$. Consider the subinterval $[x_{i-1}, x_i]$. By the mean value theorem

$$F(x_i) - F(x_{i-1}) = F'(x_i^*)(x_i - x_{i-1}) = F'(x_i^*)\,\Delta x, \tag{2}$$

where $x_{i-1} < x_i^* < x_i$. Thus

$$\sum_{i=1}^{n} [F(x_i) - F(x_{i-1})] = \sum_{i=1}^{n} F'(x_i^*)\,\Delta x. \tag{3}$$

But

$$\begin{aligned}
\sum_{i=1}^{n} [F(x_i) - F(x_{i-1})] &= [F(x_1) - F(x_0)] + [F(x_2) - F(x_1)] \\
&\quad + [F(x_3) - F(x_2)] + \cdots + [F(x_{n-1}) - F(x_{n-2})] \\
&\quad + [F(x_n) - F(x_{n-1})] \\
&= F(x_n) - F(x_0) = F(b) - F(a),
\end{aligned}$$

since all terms except the first and the last "cancel." Thus from (3),

$$F(b) - F(a) = \sum_{i=1}^{n} F'(x_i^*)\,\Delta x,$$

and taking the limit as $\Delta x \to 0$ of both sides, we obtain

$$\lim_{\Delta x \to 0} [F(b) - F(a)] = \lim_{\Delta x \to 0} \sum_{i=1}^{n} F'(x_i^*)\,\Delta x. \tag{4}$$

Since F is an antiderivative of f, we have

$$F'(x_i^*) = f(x_i^*), \tag{5}$$

and inserting (5) into (4) gives us

$$F(b) - F(a) = \lim_{\Delta x \to 0} \sum_{i=1}^{n} f(x_i^*)\,\Delta x. \tag{6}$$

But now we are done, since $\int_a^b f(x)\,dx$ exists and is equal to the limit in the right-hand side of (6), and this limit is independent of the way x_i^*'s are chosen.† Hence from (6),

$$F(b) - F(a) = \int_a^b f(x)\,dx. \quad\blacksquare$$

---

†Remember that this last statement is part of the definition of the definite integral.

EXAMPLE 1   Calculate $\int_0^1 x^2 \, dx$.

Solution. We have seen that $x^3/3$ is an antiderivative for x^2. Thus

$$\int_0^1 x^2 \, dx = \left(\frac{x^3}{3} \text{ evaluated at } x = 1\right) - \left(\frac{x^3}{3} \text{ evaluated at } x = 0\right) = \frac{1}{3} - 0 = \frac{1}{3}. \blacksquare$$

There is a simple notation we will use to avoid writing the words "evaluated at" each time.

NOTATION.

$$F(x)\Big|_a^b = F(b) - F(a).$$

In Example 1, we could have written

$$\int_0^1 x^2 \, dx = \frac{x^3}{3}\Big|_0^1 = \frac{1}{3} - 0 = \frac{1}{3}.$$

REMARK.   It doesn't make any difference which antiderivative we choose to evaluate the definite integral. For example, if C is any constant, then

$$\int_0^1 x^2 \, dx = \left(\frac{x^3}{3} + C\right)\Big|_0^1 = \left(\frac{1}{3} + C\right) - (0 + C) = \frac{1}{3} + C - C = \frac{1}{3}.$$

The constants will always "disappear" in this manner. Thus we will use the "easiest" antiderivative in our evaluation of $\int_a^b f$, which will almost always be the one in which $C = 0$.

EXAMPLE 2   Calculate $\int_0^3 x^3 \, dx$.

Solution. $\int x^3 \, dx = x^4/4 + C$. Then

$$\int_0^3 x^3 \, dx = \frac{x^4}{4}\Big|_0^3 = \frac{3^4}{4} - 0 = \frac{81}{4}. \blacksquare$$

EXAMPLE 3   Calculate $\int_1^2 (3x^4 - x^5) \, dx$.

Solution. $\int (3x^4 - x^5) \, dx = 3x^5/5 - x^6/6 + C$. Thus

$$\int_1^2 (3x^4 - x^5) \, dx = \left(\frac{3x^5}{5} - \frac{x^6}{6}\right)\Big|_1^2 = \left[\frac{3(2)^5}{5} - \frac{2^6}{6}\right] - \left[\frac{3(1)^5}{5} - \frac{1^6}{6}\right]$$

$$= \left(\frac{96}{5} - \frac{64}{6}\right) - \left(\frac{3}{5} - \frac{1}{6}\right) = \left(\frac{576}{30} - \frac{320}{30}\right) - \left(\frac{18}{30} - \frac{5}{30}\right)$$

$$= \frac{243}{30} = \frac{81}{10}. \blacksquare$$

WARNING: It is easy to lose track of minus signs when performing these calculations. Often the minus signs will cancel each other. Be careful!

EXAMPLE 4   Calculate $\int_{-2}^{2} x^2 \, dx$.

Solution.

$$\int_{-2}^{2} x^2 \, dx = \frac{x^3}{3}\Big|_{-2}^{2} = \frac{2^3}{3} - \frac{(-2)^3}{3} = \frac{8}{3} - \left(-\frac{8}{3}\right) = \frac{16}{3} \quad \blacksquare$$

EXAMPLE 5   Calculate

$$\int_{1}^{4} \left(\frac{3}{\sqrt{s}} - 5\sqrt{s}\right) ds.$$

Solution.

$$\int_{1}^{4} \left(\frac{3}{\sqrt{s}} - 5\sqrt{s}\right) ds = \int_{1}^{4} (3s^{-1/2} - 5s^{1/2}) \, ds = \left(6\sqrt{s} - \frac{10}{3}s^{3/2}\right)\Big|_{1}^{4}$$

$$= [6\sqrt{4} - \tfrac{10}{3}(4)^{3/2}] - [6\sqrt{1} - \tfrac{10}{3}(1)^{3/2}]$$

$$= 6(2) - \tfrac{10}{3}(8) - 6 + \tfrac{10}{3} = 6 - \tfrac{70}{3} = -\tfrac{52}{3} \quad \blacksquare$$

EXAMPLE 6   Calculate $\int_{a}^{b} x^r \, dx$, where $r \neq -1$ is a real number.

Solution.

$$\int x^r \, dx = \frac{x^{r+1}}{r+1} + C,$$

so that

$$\int_{a}^{b} x^r \, dx = \frac{x^{r+1}}{r+1}\Big|_{a}^{b} = \frac{1}{r+1}(b^{r+1} - a^{r+1}). \quad \blacksquare \qquad (7)$$

EXAMPLE 7   Calculate $\int_{0}^{\pi/2} \sin x \, dx$.

Solution.

$$\int_{0}^{\pi/2} \sin x \, dx = -\cos x\Big|_{0}^{\pi/2} = -\cos\frac{\pi}{2} - (-\cos 0) = -0 - (-1) = 1 \quad \blacksquare$$

In Section 5.5 we stated three theorems whose proofs were promised here. They were as follows:

(i) $\displaystyle\int_{a}^{b} f = \int_{a}^{c} f + \int_{c}^{b} f$

(ii) $\displaystyle\int_{a}^{b} kf = k \int_{a}^{b} f$

(iii) $\displaystyle\int_{a}^{b} (f + g) = \int_{a}^{b} f + \int_{a}^{b} g$

Let F and G be antiderivatives of f and g, respectively. Then kF is an antiderivative of kf and $F + G$ is an antiderivative of $f + g$ (check this by differentiating). We then have:

(i) $\displaystyle \int_a^c f + \int_c^b f = [F(c) - F(a)] + [F(b) - F(c)] = F(b) - F(a) = \int_a^b f$

(ii) $\displaystyle \int_a^b kf = kF(b) - kF(a) = k[F(b) - F(a)] = k \int_a^b f$

(iii) $\displaystyle \int_a^b (f + g) = (F + G)\Big|_a^b = (F + G)(b) - (F + G)(a)$

$$= [F(b) + G(b)] - [F(a) + G(a)]$$

$$= [F(b) - F(a)] + [G(b) - G(a)] = \int_a^b f + \int_a^b g \quad \blacksquare$$

EXAMPLE 8 Calculate $\int_{-5}^3 |x|\, dx$.

Solution. We know that

$$|x| = \begin{cases} x, & x \geq 0 \\ -x, & x < 0, \end{cases}$$

so that

$$\int |x|\, dx = \begin{cases} \frac{1}{2}x^2, & x \geq 0 \\ -\frac{1}{2}x^2, & x < 0 \end{cases} = \begin{cases} \frac{1}{2}x(x), & x \geq 0 \\ \frac{1}{2}x(-x), & x < 0 \end{cases} = \frac{1}{2}x|x| + C.$$

That is,

$$\int |x|\, dx = \frac{1}{2}x|x| + C.$$

Thus

$$\int_{-5}^3 |x|\, dx = \frac{1}{2}x|x| \Big|_{-5}^3 = \frac{1}{2}(3)(3) - \frac{1}{2}(-5)(5) = \frac{1}{2}(9 + 25) = 17. \quad \blacksquare$$

EXAMPLE 9 Calculate the area bounded by the curve $y = x^3 - 6x^2 + 11x - 6$ and the x-axis.

Solution. We have $y = x^3 - 6x^2 + 11x - 6 = (x - 1)(x - 2)(x - 3)$. The curve is graphed in Figure 1. The desired area is the shaded part of the graph. We know that

$$A = \int_1^3 |x^3 - 6x^2 + 11x - 6|\, dx$$

$$= \int_1^2 (x^3 - 6x^2 + 11x - 6)\, dx + \int_2^3 -(x^3 - 6x^2 + 11x - 6)\, dx$$

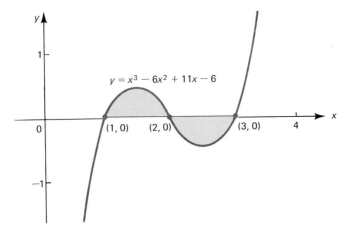

y = x³ − 6x² + 11x − 6

(1, 0)   (2, 0)   (3, 0)

FIGURE 1

$$= \left(\frac{x^4}{4} - 2x^3 + \frac{11x^2}{2} - 6x\right)\Big|_1^2 - \left(\frac{x^4}{4} - 2x^3 + \frac{11x^2}{2} - 6x\right)\Big|_2^3$$

$$= (4 - 16 + 22 - 12 - \tfrac{1}{4} + 2 - \tfrac{11}{2} + 6)$$

$$\quad - (\tfrac{81}{4} - 54 + \tfrac{99}{2} - 18 - 4 + 16 - 22 + 12)$$

$$= \tfrac{1}{2}.$$

Note that

$$\int_1^3 (x^3 - 6x^2 + 11x - 6)\, dx = \left(\frac{x^4}{4} - 2x^3 + \frac{11x^2}{2} - 6x\right)\Big|_1^3 = 0.$$

This example illustrates why, in the process of calculating area, care must be taken so that "positive" areas and "negative" areas don't cancel each other. ■

EXAMPLE 10   Calculate the area bounded by the curve $y = \cos x$, the x-axis, and the lines $x = 0$ and $x = 2\pi$.

Solution. The curve is sketched in Figure 2. We see that $\cos x > 0$ for $0 \le x < \pi/2$, $\cos x < 0$ for $\pi/2 < x < 3\pi/2$, and $\cos x > 0$ for $3\pi/2 < x \le 2\pi$. Thus

$$A = \int_0^{2\pi} |\cos x|\, dx = \int_0^{\pi/2} \cos x\, dx + \int_{\pi/2}^{3\pi/2} -\cos x\, dx + \int_{3\pi/2}^{2\pi} \cos x\, dx$$

$$= \sin x\Big|_0^{\pi/2} - \sin x\Big|_{\pi/2}^{3\pi/2} + \sin x\Big|_{3\pi/2}^{2\pi}$$

$$= \left(\sin\frac{\pi}{2} - \sin 0\right) - \left(\sin\frac{3\pi}{2} - \sin\frac{\pi}{2}\right) + \left(\sin 2\pi - \sin\frac{3\pi}{2}\right)$$

$$= (1 - 0) - (-1 - 1) + (0 - (-1)) = 1 + 2 + 1 + 4.$$

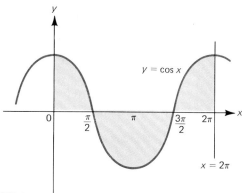

FIGURE 2

Note that

$$\int_0^{2\pi} \cos x \, dx = \sin x \Big|_0^{2\pi} = \sin 2\pi - \sin 0 = 0 - 0 = 0,$$

so that again "positive" and "negative" areas cancel if we are careless. ■

In the statement of the fundamental theorem of calculus we left one important question unanswered. We know that if f has an antiderivative F, then $\int_a^b f(x) \, dx = F(b) - F(a)$. But how do we know that every continuous function has an antiderivative? Theorem 2 below answers that question.

We assume that the function f is continuous on $[a, b]$ (piecewise continuity would do just as well here). Then by Theorem 5.5.1, $\int_a^b f(t) \, dt$ exists. But we have even more than that. If x is any number in $[a, b]$, then f is certainly continuous on the smaller interval $[a, x]$, and so again by Theorem 5.5.1, $\int_a^x f(t) \, dt$ exists. For every value of x in $[a, b]$, this integral is a real number. Now we define a new function G by

$$G(x) = \int_a^x f(t) \, dt. \tag{8}$$

We emphasize that $G(x)$ is the *function* that assigns to every number x in $[a, b]$ the value of the definite integral of f over the interval $[a, x]$. The theorem below tells us that $G(x)$ is really an antiderivative for the function f over the interval $[a, b]$. Its proof is difficult and will be delayed until Section 5.10. However, we will indicate by a graph why the theorem is plausible. The following important result is sometimes called the **second fundamental theorem of calculus.**

Theorem 2   *Second Fundamental Theorem of Calculus*   If f is continuous on $[a, b]$, then the function $G(x) = \int_a^x f(t) \, dt$ is continuous on $[a, b]$, differentiable on (a, b), and for every x in (a, b),

$$G'(x) = f(x). \tag{9}$$

That is, G is an antiderivative of f on the interval $[a, b]$.

Graphical Indication of Proof.   Consider Figure 3, paying particular attention to the various areas under the curve $f(x)$. By definition of the derivative,

$$G'(x) = \lim_{\Delta x \to 0} \frac{G(x + \Delta x) - G(x)}{\Delta x}.$$

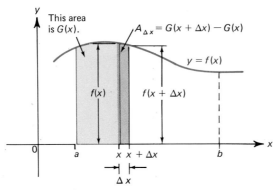

FIGURE 3

Since $G(x) = \int_a^x f(t)\, dt$ = area under the curve $f(x)$ between a and x, and assuming that $f > 0$ on (a, b), we have

$$G(x + \Delta x) = \text{area between } a \text{ and } x + \Delta x$$
$$G(x) = \text{area between } a \text{ and } x$$
$$G(x + \Delta x) - G(x) = \text{area between } x \text{ and } x + \Delta x.$$

This last area is denoted $A_{\Delta x}$ in Figure 3. Now if Δx is small, then x and $x + \Delta x$ are close, and since f is continuous, $f(x + \Delta x)$ is close to $f(x)$. That is, the shaded area $A_{\Delta x}$ is approximately equal to the area of the rectangle with height $f(x)$ and base Δx. We therefore see that

$$G(x + \Delta x) - G(x) = A_{\Delta x} \approx f(x)\, \Delta x.$$

Then for Δx small,

$$\frac{G(x + \Delta x) - G(x)}{\Delta x} \approx \frac{f(x)\, \Delta x}{\Delta x} = f(x).$$

Since as $\Delta x \to 0$, this approximation gets better and better, we may assert that

$$G'(x) = \lim_{\Delta x \to 0} \frac{G(x + \Delta x) - G(x)}{\Delta x} = f(x),$$

which indicates that the derivative of G is f, as we wanted to show. ∎

PROBLEMS 5.6

In Problems 1–25, calculate the given integral.

1. $\int_{-1}^{2} x^4 \, dx$

2. $\int_{2}^{5} 3s^3 \, ds$

3. $\int_{1}^{9} \frac{\sqrt{t}}{2} \, dt$

4. $\int_{1}^{3} (x^2 + 3x + 5) \, dx$

5. $\int_{a}^{b} (c_1 y^2 + c_2 y + c_3) \, dy$

6. $\int_{2}^{4} \left(\frac{1}{z^3} - \frac{1}{z^2} \right) dz$

7. $\int_{1}^{8} \left(\frac{1}{\sqrt[3]{x}} + 7\sqrt[3]{x} \right) dx$

8. $\int_{0}^{1} (1 + \cdot s^8 + s^{16} + s^{32}) \, ds$

9. $\int_{-1}^{1} (p^9 + p^{17}) \, dp$

10. $\int_{-a}^{a} x^{2n+1} \, dx$, where n is a positive integer and a is a real number

11. $\int_{9}^{16} \frac{v+1}{\sqrt{v}} \, dv$ [Hint: Divide.]

12. $\int_{2}^{3} (y-1)(y+2) \, dy$ [Hint: Multiply.]

13. $\int_{1}^{4} \frac{z^2 + 2z + 5}{z^{3/2}} \, dz$

14. $\int_{-2}^{2} (v^2 - 4)(v^5 + 6) \, dv$

15. $\int_{0}^{1} (t^{3/2} - t^{2/3})(t^{4/3} - t^{3/4}) \, dt$

16. $\int_{0}^{1} (\sqrt{x} - \sqrt[3]{x})^2 \, dx$

17. $\int_{1}^{0} s^{100} \, ds$

18. $\int_{-1}^{-2} \frac{1}{s^{100}} \, ds$

19. $\int_{2}^{4} \frac{x^2 + 7x + 6}{x+1} \, dx$ [Hint: Divide.]

20. $\int_{0}^{1} \frac{(x+1)(x-2)^3}{x^2 - x - 2} \, dx$

21. $\int_{-3}^{3} |x - 1| \, dx$

22. $\int_{-5}^{15} |2x + 6| \, dx$

23. $\int_{0}^{\pi/4} 2 \sin x \, dx$

24. $\int_{0}^{\pi/6} (\cos x - \sin x) \, dx$

*25. $\int_{-2}^{5} f(x) \, dx$, where $f(x) = \begin{cases} 1 + x^2, & x \geq 2 \\ 13 - x^3, & x \leq 2 \end{cases}$

In Problems 26–37, find the area bounded by the given curve and the given lines. [Hint: Sketch the curves.]

26. $y = x^2 - 4$; x-axis
27. $y = 4 - x^2$; x-axis
28. $y = x^2 + 2x - 3$; $x = 1$, $x = 3$, x-axis
29. $y = x^2 - 6x + 5$; x-axis
30. $y = x^2 - 15x - 34$; x-axis
31. $y = x^4$; $x = -2$, $x = 4$, x-axis
32. $y = x^3 + 2x^2 - x - 2$; x-axis
33. $y = (x^2 - 1)(x^2 - 4)$; x-axis
34. $y = x^3 + 2x^2 - 13x + 10$; $x = 0$, $x = 3$, x-axis
35. $y = (x - a)(x - b)$, $a < b$; x-axis

36. $y = 2 \sin x$; $x = -\pi/3$, $x = \pi/3$, x-axis

37. $y = \cos x - \sin x$; $x = 0$, $x = \pi$, x-axis

38. An **even function** is a function f that has the property that $f(-x) = f(x)$ for every real number x. [For example, 1, x^2, x^4, $|x|$, and $1/(1 + x^2)$ are all even functions.] Show that if f is even, then

$$\int_{-a}^{a} f = 2 \int_{0}^{a} f$$

for every real number a. Can you explain this fact geometrically?

39. An **odd function** f has the property that $f(-x) = -f(x)$ for every real number x. [For example, x, x^3, and $1/(x^5 + x^7)$ are all odd functions.] Show that if f is odd, then

$$\int_{-a}^{a} f = 0$$

for any real number a. Can you explain this fact geometrically?

40. Calculate the following integrals.

(a) $\displaystyle\int_{-50}^{50} (x + x^3 + x^{17})\, dx$

(b) $\displaystyle\int_{-100}^{100} (1 + x^{1/3} + x^{5/3} + x^{11/3})\, dx$

In Problems 41–44, use Theorem 2 to calculate the derivative $F'(x)$. Then evaluate $F'(x_0)$ for the given x_0.

41. $F(x) = \displaystyle\int_{3}^{x} \frac{dt}{1 + t^3}$; $x_0 = 2$

42. $F(x) = \displaystyle\int_{-1}^{x} \frac{s^2}{s^2 + 5}\, ds$; $x_0 = 3$

43. $F(x) = \displaystyle\int_{0}^{x} \sqrt{\frac{u - 1}{u + 1}}\, du$; $x_0 = 1$

44. $F(x) = \displaystyle\int_{2}^{x} \left(\frac{1 + 2t - 3t^2}{t^3 + t^{7/9}}\right) dt$; $x_0 = 1$

45. The velocity of a moving particle after t seconds is given by $v(t) = t^{3/2} + 16t + 1$, where v is measured in meters per second. How far does the particle travel between $t = 4$ and $t = 9$ sec?

46. A ball is thrown down from a tower with an initial velocity of 25 ft/sec. The tower is 500 ft high.
   (a) How fast is the ball traveling after 3 sec?
   (b) How far does the ball travel in the first 3 sec?
   *(c) How long does it take for the ball to hit the ground? [*Hint:* Use $a = 32$ ft/sec^2.]

*47. The density of a 3-m metal bar is given by $\rho(x) = 1/\sqrt{x + 1}$ kilograms per meter, where x is measured in units from the left.
   (a) What is the total mass of the bar? [*Hint:* It is not difficult to find a function whose derivative is $1/\sqrt{x + 1}$.]
   (b) For what value of x is the mass of the bar from 0 to x units equal to half the total mass of the bar?

48. The marginal revenue a manufacturer receives is given by MR $= 2 - 0.02q + 0.003q^2$ dollars per additional unit sold. How much additional money does the manufacturer receive if he increases sales from 50 to 100 units?

49. The marginal cost to produce a widget is given by MC $= 2 - 0.003q + 0.00005q^2$ dollars. How much does it cost to increase production from 100 to 200 widgets?

50. The population of a species of insects is increasing at a rate of $3000/\sqrt{t}$ individuals per week. How many individuals are added to the population between the 9th and the 25th week?

51. A charged particle enters a linear accelerator. Its velocity increases with a constant acceleration from an initial velocity of 500 m/sec to a velocity of 10,500 m/sec in $\frac{1}{100}$ sec.
   (a) What is the acceleration?
   (b) How far does the particle travel in $\frac{1}{100}$ sec?

If $f(x)$ takes on the n values $x_1, x_2, \ldots x_n$, then the *average* of these values is $(x_1 + x_2 + \cdots + x_n)/n$. That is, we add up the values the function takes and divide the sum by n, the number of points. If f is continuous over the interval $[a, b]$ then the *average value of f over* $[a, b]$ is defined by

$$\text{average value} = \frac{1}{b - a} \int_a^b f(x)\, dx.$$

In Problems 52–63, find the average value of the given function over the given integral.

52. $f(x) = 3x + 5; [1, 2]$

53. $f(x) = C,$ C a constant; $[a, b]$

54. $f(x) = x^2; [0, 2]$

55. $f(x) = x^3; [0, 1]$

56. $f(x) = x^r; [0, 2]$ $(r \neq -1)$

57. $f(x) = x^2 - 2x + 5; [-2, 2]$

58. $f(x) = \sqrt{x} + 5x^3; [0, 4]$

59. $f(x) = 1 + x + x^2 + x^3 + x^4; [0, 1]$

60. $f(x) = \dfrac{1}{x^2}; [1, 3]$

61. $f(x) = x^{2/3} + \dfrac{1}{\sqrt[3]{x}}; [1, 8]$

62. $f(x) = x^{17/11}; [0, 1]$

63. $f(x) = x^{10} - \dfrac{1}{x^{10}}; [1, 2]$

64. The density of a 4-m metal bar is given by $\rho(x) = 1 + x - \sqrt{x}$ kilograms per meter. What is the average density over the length of the entire bar?

65. If the marginal cost of a given product is $50 - 0.05q$ dollars, what is the average cost per unit of production if 200 units are produced?

66. One day the temperature of the air t hours after noon was found to be $60 + 4t - t^2/3$ degrees (Fahrenheit). What was the average temperature between noon and 5 P.M.?

*67. A ball was dropped from rest from a height of 400 ft. What was its average velocity on the way to the ground?

*68. Compute $\int_0^7 \min\{6, 1 + 6x - x^2\}\, dx$.

69. Compute the following integrals.

(a) $\displaystyle \int_{-1}^1 |x|\, dx$

(b) $\displaystyle \int_0^2 |x - 1|\, dx$

(c) $\displaystyle \int_{-5}^{-3} |x + 4|\, dx$

*70. Find all functions f such that **(a)** f' is continuous on $[a, b]$ and **(b)**

$$\frac{d}{dx} \int_a^x f(t)\, dt = \int_a^x \frac{d}{dt} f(t)\, dt \qquad \text{on } [a, b].$$

71. Let $f(x) = -1/x^2$ and $F(x) = 1/x$. Are the following statements true or false?

(a) $F'(x) = f(x)$

(b) $\displaystyle \int f = F$

(c) $\displaystyle \int f = F + \text{constant}$

(d) $\displaystyle \int_{-1}^1 f = F(1) - F(-1)$

(e) $\displaystyle \int_{-1}^1 f$ does not exist

72. Compute

$$\frac{d}{dx} \int_3^x \left(\frac{d}{dt} \cos t \right) dt.$$

*73. Find two nonconstant functions f and g such that

$$\int_0^1 f \cdot g = \left(\int_0^1 f \right) \cdot \left(\int_0^1 g \right).$$

(The real purpose of this exercise is to show you that this relation is unlikely to be true for a randomly chosen pair of functions.)

*74. Suppose that $p > 0$. Prove that

$$\frac{p}{p+1} < \int_0^1 \frac{1}{1+x^p}\, dx < 1.$$

[*Hint:* First show that $1 - x^p < 1/(1 + x^p) < 1$.]

75. Watch carefully. Recall that

$$\frac{d}{dx}\left(-\frac{1}{x}\right) = (-1)\cdot\frac{d}{dx}(x^{-1}) = (-1)(-1)\cdot x^{-2} = \frac{1}{x^2}.$$

Therefore

$$\int_{-3}^3 \frac{1}{x^2}\, dx = -\frac{1}{x}\Big|_{-3}^3 = \left(\frac{-1}{3}\right) - \left(\frac{-1}{-3}\right) = -\frac{2}{3}???$$

But x^2 can't be negative and neither can $1/x^2$. What went wrong to produce a negative value for the integral? [*Hint:* Sketch $y = 1/x^2$ and reread the statement of the fundamental theorem of calculus.]

76. Suppose that a particle is constrained to move only along a straight line with velocity $v(t)$ at time t. Does $\int_a^b v(t)\, dt$ equal the total distance traveled in the time interval $[a, b]$, or does it equal the net distance traveled, that is, the distance traveled in the positive direction minus the distance traveled in the negative direction? Write an integral that computes the other distance.

5.7 INTEGRATION BY SUBSTITUTION

As we saw in the previous section, the problem of finding a definite or indefinite integral is solved if we can find one antiderivative for the function f. However, as we have already mentioned, this task can often be very difficult. So far, the only functions we know how to integrate are functions of the form x^r (where $r \neq -1$), $\sin x$, and $\cos x$.

There are many other functions for which indefinite integrals can be found, but learning to recognize them takes a lot of experience. We shall devote an entire chapter (Chapter 8) to the problem of recognizing the types of functions for which integrals can be expressed in terms of certain elementary functions. In this section we enlarge the class of functions whose integrals can be immediately determined.

We start with the formula

$$\int u^r\, du = \frac{u^{r+1}}{r+1} + C, \quad \text{if} \quad r \neq -1. \tag{1}$$

This formula is the result of Theorem 5.2.2. We can use this formula to prove the following basic result.

Theorem 1 If $u = g(x)$ is a differentiable function of x and $du = g'(x)\, dx$ is its differential, then if $r \neq -1$,

$$\int [g(x)]^r g'(x)\, dx = \int u^r\, du. \qquad (2)$$

Proof. Applying the chain rule to the function $u^{r+1}/(r+1)$, we have

$$\frac{d}{dx}\!\left(\frac{u^{r+1}}{r+1}\right) = \frac{d}{dx}\!\left\{\frac{[g(x)]^{r+1}}{r+1}\right\} = [g(x)]^r g'(x).$$

Thus

$$\int [g(x)]^r g'(x)\, dx = \frac{u^{r+1}}{r+1} + C = \int u^r\, du. \quad \blacksquare$$

Consider the integral

$$\int (1 + x^2)^3\, 2x\, dx. \qquad (3)$$

If we define the function $u = 1 + x^2$, then $du/dx = 2x$, and taking differentials, we have $du = 2x\, dx$. Then $(1 + x^2)^3 = u^3$, and equation (3) becomes

$$\int (1 + x^2)^3\, 2x\, dx = \int u^3\, du = \frac{u^4}{4} + C = \frac{(1 + x^2)^4}{4} + C.$$

In general, to calculate $\int f(x)\, dx$, perform the following steps:

   (i) Make a substitution $u = g(x)$ so that the integral can be expressed in the form $u^r\, du$ (if possible).
   (ii) Calculate $du = g'(x)\, dx$.
   (iii) Write $\int f(x)\, dx$ as $\int u^r\, du$ for some number $r \neq -1$.
   (iv) Integrate using formula (1).
   (v) Substitute $g(x)$ for u to obtain the answer in terms of x.

EXAMPLE 1   Calculate $\int \sqrt{3 + x}\, dx$.

   Solution. Let $u = g(x) = 3 + x$. Then $du = dx$, and

$$\int \overbrace{\sqrt{3 + x}}^{\sqrt{u}}\, \overbrace{dx}^{du} = \int \sqrt{u}\, du = \int u^{1/2}\, du = \frac{2}{3} u^{3/2} + C = \frac{2}{3}(3 + x)^{3/2} + C. \quad \blacksquare$$

EXAMPLE 2   Calculate $\int_0^1 (x^3 - 1)^{11/5} 3x^2\, dx$.

   Solution. Let $u = g(x) = x^3 - 1$. Then $du = 3x^2\, dx$, and

$$\int \overbrace{(x^3 - 1)^{11/5}}^{u^{11/5}}\, \overbrace{3x^2\, dx}^{du} = \int u^{11/5}\, du = \frac{5}{16} u^{16/5} + C = \frac{5}{16}(x^3 - 1)^{16/5} + C.$$

Then

$$\int_0^1 (x^3 - 1)^{11/5}\, 3x^2\, dx = \frac{5}{16}\, (x^3 - 1)^{16/5}\Big|_0^1 = \frac{5}{16}\, [0 - (-1)^{16/5}] = -\frac{5}{16}. \ \blacksquare$$

In the two examples above the integrand was already in the form $\int u^r\, du$ for the appropriate value of r. Sometimes it is not, but we can salvage the problem by multiplying and dividing by an appropriate constant.

EXAMPLE 3   Calculate $\int x\sqrt[3]{1 + x^2}\, dx$.

Solution. Let $u = 1 + x^2$. Then $du = 2x\, dx$. All we have is $x\, dx$. We therefore multiply and divide by 2 to obtain

$$\int x\sqrt[3]{1 + x^2}\, dx = \frac{1}{2}\int \overbrace{\sqrt[3]{1 + x^2}}^{\sqrt[3]{u}}\ \overbrace{2x\, dx}^{du} = \frac{1}{2}\int u^{1/3}\, du$$

$$= \tfrac{1}{2}\cdot\tfrac{3}{4}\, u^{4/3} + C = \tfrac{3}{8}\, (1 + x^2)^{4/3} + C.$$

We were allowed to multiply and divide by the constant 2 because of the theorem that states that $\int kf = k\int f$ for any constant k. $\blacksquare$

$\blacksquare$ WARNING.   *We cannot multiply and divide in the same way by functions that are not constants.* If we try, we get incorrect answers such as

$$\int \sqrt{1 + x^2}\, dx = \frac{1}{2x}\int \sqrt{1 + x^2}\, 2x\, dx \qquad \text{We multiplied and divided by } 2x.$$

$$= \frac{1}{2x}\int u^{1/2}\, du \qquad\qquad \text{Where } u = 1 + x^2.$$

$$= \frac{1}{2x}\cdot\frac{2}{3}\, u^{3/2} + C = \frac{1}{3x}\, (1 + x^2)^{3/2} + C.$$

But the derivative of $(1 + x^2)^{3/2}/3x$ is not even close to $\sqrt{1 + x^2}$. (Check!) It is possible to calculate $\int \sqrt{1 + x^2}\, dx$, but we will have to wait until Chapter 8 to see how to do it.

EXAMPLE 4   Calculate $\int_1^2 s^2/(1 + s^3)^{3/4}\, ds$.

Solution. Let $u = 1 + s^3$. Then $du = 3s^2\, ds$, and we multiply and divide by 3 to obtain

$$\int \frac{s^2}{(1 + s^3)^{3/4}}\, ds = \frac{1}{3}\int \frac{\overbrace{3s^2\, ds}^{du}}{\underbrace{(1 + s^3)^{3/4}}_{u^{-3/4}}} = \frac{1}{3}\int u^{-3/4}\, du = \frac{4}{3}\, u^{1/4} + C$$

$$= \frac{4}{3}(1 + s^3)^{1/4} + C.$$

Therefore

$$\int_1^2 \frac{s^2}{(1 + s^3)^{3/4}} \, ds = \frac{4}{3} (1 + s^3)^{1/4} \Big|_1^2 = \frac{4}{3} (9^{1/4} - 2^{1/4}).$$ ∎

EXAMPLE 5   Calculate $\int [1 + (1/t)]^5/t^2 \, dt$.

Solution. Let $u = 1 + (1/t)$. Then $du = -(1/t^2) \, dt$. We then multiply and divide by -1 to obtain

$$\int \frac{[1 + (1/t)]^5}{t^2} \, dt = -\int \overbrace{\left(1 + \frac{1}{t}\right)^5}^{u^5} \overbrace{\left(-\frac{1}{t^2}\right) \, dt}^{du} = -\int u^5 \, du$$

$$= -\frac{u^6}{6} + C = -\frac{[1 + (1/t)]^6}{6} + C.$$ ∎

EXAMPLE 6   Calculate the area bounded by the curve $y = (1 + x^3)/\sqrt[4]{4x + x^4}$, the x-axis, and the lines $x = 1$ and $x = 3$.

Solution. It is not necessary to draw this curve to see that $(1 + x^3)/\sqrt[4]{4x + x^4} > 0$ for $1 \le x \le 3$. Thus the area is represented by the definite integral

$$\int_1^3 \frac{1 + x^3}{\sqrt[4]{4x + x^4}} \, dx.$$

Let $u = 4x + x^4$. Then $du = (4 + 4x^3) \, dx$, and we multiply and divide by 4 to obtain

$$\int \frac{1 + x^3}{\sqrt[4]{4x + x^4}} \, dx = \frac{1}{4} \int \frac{4(1 + x^3)}{\sqrt[4]{4x + x^4}} \, dx = \frac{1}{4} \int u^{-1/4} \, du$$

$$= \tfrac{1}{3} u^{3/4} + C = \tfrac{1}{3}(4x + x^4)^{3/4} + C.$$

Thus

$$\int_1^3 \frac{1 + x^3}{\sqrt[4]{4x + x^4}} \, dx = \frac{1}{3}(4x + x^4)^{3/4} \Big|_1^3 = \frac{1}{3}[(12 + 81)^{3/4} - (5)^{3/4}]$$

$$= \tfrac{1}{3}(93^{3/4} - 5^{3/4}).$$ ∎

We know that

$$\int \sin u \, du = -\cos u + C \tag{4}$$

and

$$\int \cos u \, du = \sin u + C. \tag{5}$$

These formulas follow immediately from the facts that

$$\frac{d}{du}(-\cos u) = \sin u \quad \text{and} \quad \frac{d}{du}\sin u = \cos u.$$

Thus we can prove the next theorem.

Theorem 2   If $u = g(x)$ is a differentiable function of x and $du = g'(x)\,dx$ is its differential, then

$$\int \sin[g(x)]g'(x)\,dx = \int \sin u\,du = -\cos u + C = -\cos[g(x)] + C \qquad (6)$$

and

$$\int \cos[g(x)]g'(x)\,dx = \int \cos u\,du = \sin u + C = \sin[g(x)] + C. \qquad (7)$$

EXAMPLE 7   Compute $\int \sin x^2(2x)\,dx$.

Solution. If $u = x^2$, then $du = 2x\,dx$, and from (6),

$$\int \overbrace{\sin x^2}^{\sin u}\,\overbrace{(2x)\,dx}^{du} = \int \sin u\,du = -\cos u + C = -\cos x^2 + C. \ \blacksquare$$

EXAMPLE 8   Compute $\int \cos 3x\,dx$.

Solution. If $u = 3x$, then $du = 3dx$, so multiplying and dividing by 3, we obtain

$$\int \cos 3x\,dx = \frac{1}{3}\int \overbrace{(\cos 3x)}^{\cos u}\overbrace{3\,dx}^{du} = \frac{1}{3}\int \cos u\,du$$

$$= \tfrac{1}{3}\sin u + C = \tfrac{1}{3}\sin 3x + C. \ \blacksquare$$

PROBLEMS 5.7

In Problems 1–33, carry out the indicated integration by making an appropriate substitution $u = g(x)$.

1. $\int (1 + x^2)^5\, 2x\,dx$

2. $\int_0^1 (1 + x)^4\,dx$

3. $\int_0^2 (3 - x)^3\,dx$

4. $\int_3^4 \sqrt{5 - t}\,dt$

5. $\int_0^1 (2 - x^4)4x^3\,dx$

6. $\int \frac{2x}{\sqrt{5 + x^2}}\,dx$

7. $\int_0^7 \sqrt{9 + x}\,dx$

8. $\int \sqrt{10 + 3x}\,dx$

9. $\int_0^1 \sqrt{10 - 9x}\,dx$

10. $\int_0^2 x^2\sqrt{1 + x^3}\, dx$ **11.** $\int x^3\sqrt[5]{1 + 3x^4}\, dx$ **12.** $\int_0^2 \frac{t}{(1 + 2t^2)^{5/2}}\, dt$

13. $\int (s^4 + 1)\sqrt{s^5 + 5s}\, ds$ **14.** $\int_{-2}^2 \sqrt{1 + |s|}\, ds$ **15.** $\int_1^2 \frac{t + 3t^2}{\sqrt{t^2 + 2t^3}}\, dt$

16. $\int_0^1 (3w - 2)^{99}\, dw$ **17.** $\int \frac{dx}{\sqrt{x}(1 + \sqrt{x})^5}$ **18.** $\int_1^2 \frac{w + 1}{\sqrt{w^2 + 2w - 1}}\, dw$

19. $\int \frac{[1 + (1/v^2)]^{5/3}}{v^3}\, dv$ **20.** $\int_0^3 \left(\frac{x}{3} - 1\right)^{77}\, dx$

21. $\int (ax + b)\sqrt{ax^2 + 2bx + c}\, dx$

22. $\int (ax^2 + bx + c)\sqrt{2ax^3 + 3bx^2 + 6cx + d}\, dx$

23. $\int \frac{ax + b}{(ax^2 + 2bx + c)^{3/7}}\, dx$ **24.** $\int_0^\alpha t\sqrt{t^2 + \alpha^2}\, dt,\ \alpha > 0$

25. $\int_0^\alpha t^n\sqrt{\alpha^2 + t^{n+1}}\, dt,\ \alpha > 0,\ n \geq 0$ **26.** $\int \frac{s^{n-1}}{\sqrt{a + bs^n}}\, ds$

27. $\int_{-\alpha}^\alpha p^2\sqrt{\alpha^3 - p^3}\, dp,\ \alpha > 0$ **28.** $\int_{-\alpha}^\alpha p^5\sqrt{\alpha^6 - p^6}\, dp,\ \alpha > 0$

29. $\int_{-\alpha}^\alpha p^{7/3}(\alpha^{10/3} - p^{10/3})^{19/7}\, dp$ **30.** $\int_0^{\pi/4} \sin 2x\, dx$

31. $\int x^2 \cos x^3\, dx$ **32.** $\int \frac{\sin\sqrt{x}}{\sqrt{x}}\, dx$

33. $\int_0^{\sqrt{\pi}} x \cos(\pi + x^2)\, dx$

In Problems 34–39, find the areas bounded by the given curves and lines.

34. $y = \sqrt{x + 2}$; x-axis, y-axis, $x = 7$ **35.** $y = x\sqrt{x^2 + 7}$; x-axis, y-axis, $x = 3$

36. $y = x^2\sqrt{\frac{x^3}{3} + 1}$; x-axis, y-axis, $x = 3$

37. $y = \frac{x + 1}{x^3} = \frac{1 + (1/x)}{x^2}$; x-axis, $x = \frac{1}{3}$, $x = \frac{1}{2}$

38. $y = \sin\frac{x}{3}$; x-axis, $x = 0$, $x = \pi$ **39.** $y = x \cos x^2$; x-axis, $x = 0$, $x = \frac{1}{2}\sqrt{\pi}$

40. A particle moves with velocity $v(t) = 1/(3\sqrt{2 + t})$ meters per minute after t minutes. How far does the particle travel between the times $t = 2$ and $t = 7$?

41. The density of a 19-m metal beam is given by $\rho(x) = 1/(3\sqrt[3]{8 + x})$, where x is measured in meters from the left end of the beam and ρ is measured in kilograms per meter.
   (a) What is the mass of the beam?
   ***(b)** For what value x does the interval $[0, x]$ contain exactly half the mass of the beam?

42. The marginal cost incurred in the manufacture of flidgets is given by MC $= 4/\sqrt{q + 4}$. What is the cost incurred by raising production from 60 to 77 units?

43. What is the average velocity of the particle in Problem 40? [*Hint:* See the comment preceding Problem 5.6.52.]

44. What is the average density of the beam in Problem 41?

45. What is the average cost per flidget to manufacture 96 flidgets, where the marginal cost is as given in Problem 42?

5.8 THE AREA BETWEEN TWO CURVES

We introduced the definite integral by calculating the area under a curve. We now show that by using similar methods, we can calculate the area between two curves.

Consider the two functions f and g as graphed in Figure 1. We will calculate the area between the curves $y = f(x)$ and $y = g(x)$ and the lines $x = a$ and $x = b$ by adding up the areas of a large number of rectangles. Note that $f(x) \geq g(x)$ for $a \leq x \leq b$. Let P be a regular partition of the interval $[a, b]$. Then a typical rectangle (see Figure 1) has the area

$$[f(x_i^*) - g(x_i^*)] \, \Delta x,$$

where x_i^* is a point in the interval $[x_{i-1}, x_i]$ and $\Delta x = x_i - x_{i-1}$. Then proceeding exactly as we have proceeded before, we find that

$$\text{area} = \lim_{\Delta x \to 0} \sum_{i=1}^{n} [f(x_i^*) - g(x_i^*)] \, \Delta x = \int_a^b [f(x) - g(x)] \, dx. \qquad (1)$$

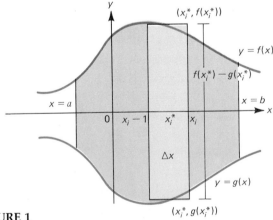

FIGURE 1

This formula is valid as long as $f(x) \geq g(x)$ in $[a, b]$. We will deal with more general cases later in this section.

REMARK. There is no difficulty if one (or both) of the functions takes on negative values in $[a, b]$. If you are uncomfortable with these negative values, simply *shift upward*. If we define $\bar{f}(x) = f(x) + C$ and $\bar{g}(x) = g(x) + C$, where C is a positive constant, then

$$\text{area between } \bar{f} \text{ and } \bar{g} = \int_a^b (\bar{f} - \bar{g}) = \int_a^b [(f + C) - (g + C)] = \int_a^b (f - g)$$
$$= \text{area between } f \text{ and } g.$$

This is illustrated in Figure 2. The shaded areas retain the same area.

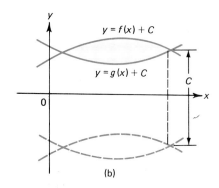

FIGURE 2

EXAMPLE 1   Find the area of the region bounded by $y = x^2$ and $y = 4x$, for x between 0 and 1.

Solution. In any problem of this type it is helpful to draw a graph. In Figure 3 the required area is shaded. We have

$$A = \int_0^1 (4x - x^2)\, dx = \left(2x^2 - \frac{x^3}{3}\right)\Bigg|_0^1 = 2 - \frac{1}{3} = \frac{5}{3}. \quad \blacksquare$$

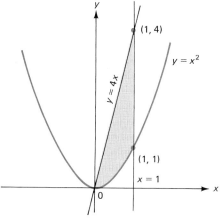

FIGURE 3

EXAMPLE 2   Find the area bounded by the curves $y = x^3$ and $y = \sqrt{x}$.

Solution. This problem only makes sense if the curves intersect at at least two points. (Otherwise we would have to be given other bounding lines.) See Figure 4. We need to find the points of intersection of these two curves. To find them, we set the

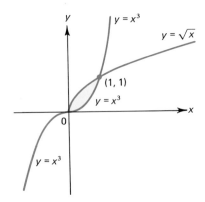

FIGURE 4

two functions equal. If $x^3 = \sqrt{x}$, then $x^6 = x$, or $x^6 - x = x(x^5 - 1) = 0$. This occurs when $x = 0$ and $x = 1$. Then

$$A = \int_0^1 (\sqrt{x} - x^3)\, dx = \left(\frac{2x^{3/2}}{3} - \frac{x^4}{4}\right)\Big|_0^1 = \frac{2}{3} - \frac{1}{4} = \frac{5}{12}. \quad \blacksquare$$

In Examples 1 and 2 note that we have had no trouble deciding which function came first in the expression $f(x) - g(x)$. We always put the larger function first. Thus in $[0, 1]$, $\sqrt{x} \geq x^3$, so that $\sqrt{x}$ comes first.

EXAMPLE 3   Find the area bounded by the two curves $y = x^2 + 3x + 5$ and $y = -x^2 + 5x + 9$.

Solution. We first sketch the two curves (see Figure 5). To find the points of intersection, we set $x^2 + 3x + 5 = -x^2 + 5x + 9$. This leads to the equation $2x^2 - 2x - 4 = 0$, which has roots $x = -1$ and $x = 2$. Thus

$$A = \int_{-1}^2 [(-x^2 + 5x + 9) - (x^2 + 3x + 5)]\, dx$$

$$= \int_{-1}^2 (-2x^2 + 2x + 4)\, dx = \left(-\frac{2x^3}{3} + x^2 + 4x\right)\Big|_{-1}^2 = 9. \quad \blacksquare$$

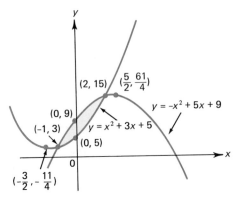

FIGURE 5

EXAMPLE 4   Find the area bounded by the two curves $y = x^2 + 3x + 5$ and $y = -x^2 + 5x + 9$ for x between -1 and 4.

     Solution. This example is more complicated than Example 3. See Figure 6. There are now two areas to be added together. In the first (calculated in Example 3)

$$-x^2 + 5x + 9 \geq x^2 + 3x + 5.$$

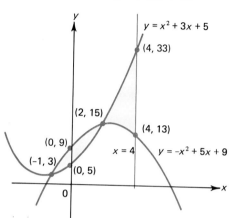

FIGURE 6

In the second

$$x^2 + 3x + 5 \geq -x^2 + 5x + 9.$$

We therefore break the calculation into two parts:

$$A = \int_{-1}^{2} [(-x^2 + 5x + 9) - (x^2 + 3x + 5)]\, dx$$

$$+ \int_{2}^{4} [(x^2 + 3x + 5) - (-x^2 + 5x + 9)]\, dx$$

$$= 9 + \int_{2}^{4} (2x^2 - 2x - 4)\, dx = 9 + \left(\frac{2x^3}{3} - x^2 - 4x\right)\Big|_{2}^{4} = 9 + \frac{52}{3} = \frac{79}{3}.$$

    We note that $(-x^2 + 5x + 9) - (x^2 + 3x + 5)$ in $[-1, 2]$ and $(x^2 + 3x + 5) - (-x^2 + 5x + 9)$ in $[2, 4]$ can be written as $|(x^2 + 3x + 5) - (-x^2 + 5x + 9)|$ in $[-1, 4]$. (Why?) ∎

    We can generalize the result of the last example to obtain the following rule:

The area between the curves $y = f(x)$ and $y = g(x)$ between $x = a$ and $x = b$ ($a < b$) is given by

$$A = \int_{a}^{b} |f(x) - g(x)|\, dx. \tag{2}$$

This rule forces the integrand to be positive, so we cannot run into the problem of adding negative areas (however, it does not make the calculation any easier).

NOTE.   Reread the warning on page 300.

EXAMPLE 5   Compute the area between the curves $y = \sin x$ and $y = \cos x$ for x between 0 and π.

Solution. The two curves are sketched in Figure 7. We find that $\sin x = \cos x = \sqrt{2}/2$ when $x = \pi/4$ and $\sin x = \cos x = -\sqrt{2}/2$ when $x = 5\pi/4$. Also, it is evident from the graph that

$$\cos x > \sin x \qquad \text{for} \qquad 0 \le x < \frac{\pi}{4}$$

and

$$\sin x > \cos x \qquad \text{for } \frac{\pi}{4} < x < \pi.$$

Thus

$$A = \int_0^{\pi/4} (\cos x - \sin x)\, dx + \int_{\pi/4}^{\pi} (\sin x - \cos x)\, dx$$

$$= (\sin x + \cos x)\Big|_0^{\pi/4} + (-\cos x - \sin x)\Big|_{\pi/4}^{\pi}$$

$$= \left(\sin \frac{\pi}{4} + \cos \frac{\pi}{4} - \sin 0 - \cos 0\right) + \left(-\cos \pi - \sin \pi + \cos \frac{\pi}{4} + \sin \frac{\pi}{4}\right)$$

$$= \left(\frac{\sqrt{2}}{2} + \frac{\sqrt{2}}{2} - 0 - 1\right) + \left[-(-1) - (0) + \frac{\sqrt{2}}{2} + \frac{\sqrt{2}}{2}\right]$$

$$= (\sqrt{2} - 1) + (\sqrt{2} + 1) = 2\sqrt{2}. \quad\blacksquare$$

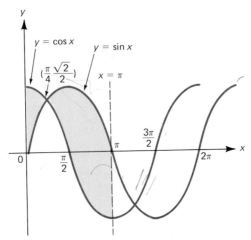

FIGURE 7

EXAMPLE 6   Find the area bounded by the curves $y^2 = 4 + x$ and $x + 2y = 4$.

     Solution. This problem is more complicated than it might at first appear. Look at Figure 8. We can easily calculate that the two curves intersect at $(0, 2)$ and $(12, -4)$ by setting $x = y^2 - 4$ equal to $x = 4 - 2y$ to find that $y = 2$ and $y = -4$. Calculating the area is more complicated. We see that x ranges from -4 to 12. But we can't simply integrate from -4 to 12 with respect to x because the integrand changes. In $[-4, 0]$ a typical rectangle has a height of $\sqrt{4 + x_i^*} - (-\sqrt{4 + x_i^*}) = 2\sqrt{4 + x_i^*}$. But between 0 and 12, a typical rectangle has the height

$$\left(-\frac{x_i^*}{2} + 2\right) - (-\sqrt{4 + x_i^*}) = \sqrt{4 + x_i^*} - \frac{x_i^*}{2} + 2.$$

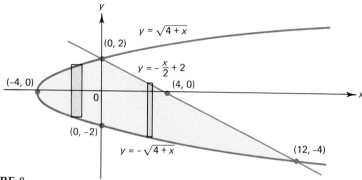

$y = \sqrt{4 + x}$

$(0, 2)$

$y = -\dfrac{x}{2} + 2$

$(-4, 0)$

$(4, 0)$

0

$(0, -2)$

$y = -\sqrt{4 + x}$

$(12, -4)$

FIGURE 8

Thus

$$A = \int_{-4}^{0} 2\sqrt{4 + x}\, dx + \int_{0}^{12} \left(\sqrt{4 + x} - \frac{x}{2} + 2\right) dx$$

$$= \frac{4}{3}(4 + x)^{3/2}\bigg|_{-4}^{0} + \left[\frac{2}{3}(4 + x)^{3/2} - \frac{x^2}{4} + 2x\right]\bigg|_{0}^{12} = \frac{32}{3} + \frac{76}{3} = 36.$$

     There is a much easier way to solve this problem. We are in the habit of writing every functional relationship between x and y in terms of y as a function of x. However, in this problem it is clearly more convenient to treat x as a function of y. It is, for example, an easy matter to obtain the graph of $x = y^2 - 4$. This graph is just like the graph of $y = x^2 - 4$ with the x- and y-axes interchanged. We again draw the graph, with the only change being that everything is labeled as a function of y. This graph is drawn in Figure 9. We now construct our rectangles horizontally intead of vertically by partitioning along the y-axis. The height of a typical rectangle is Δy, and the width is $(4 - 2y_i^*) - (y_i^{*2} - 4)$. We may therefore write

$$A = \int_{-4}^{2} [(4 - 2y) - (y^2 - 4)]\, dy$$

$$= \int_{-4}^{2} (8 - 2y - y^2)\, dy = \left(8y - y^2 - \frac{y^3}{3}\right)\bigg|_{-4}^{2} = 36. \quad \blacksquare$$

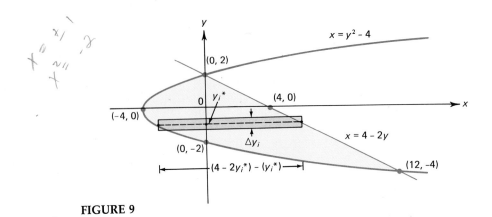

FIGURE 9

In general, if it is easier to treat y as the independent variable, then for $a < b$,

$$A = \int_{y=a}^{y=b} |f(y) - g(y)| \, dy. \qquad\qquad (3)$$

Whether to integrate with respect to x or y is a matter of judgment. In most cases a glance at a sketch of the area being considered will indicate which is preferable.

PROBLEMS 5.8

In Problems 1–22, calculate the area bounded by the given curves and lines. If possible, find the areas by integration with respect to both x and y.

1. $y = x^2$, $y = x$

2. $y = x^2$, $y = x^3$

3. $y = x^2$, $y = x^3$, $x = 3$

4. $y = 2x^2 + 3x + 5$, $y = x^2 + 3x + 6$

5. $y = 3x^2 + 6x + 8$, $y = 2x^2 + 9x + 18$

6. $y = x^2 - 7x + 3$, $y = -x^2 - 4x + 5$

7. $y = 2x$, $y = \sqrt{4x - 24}$, $y = 0$, $y = 10$

8. $y = x^3$, $y = x^3 + x^2 + 6x + 5$

9. $y = x^4 + x - 81$, $y = x$

10. $xy^2 = 1$, $x = 5$, $y = 5$

11. $\sqrt{x} + \sqrt{y} = 4$, $x = 0$, $y = 0$

12. $x + y^2 = 8$, $x + y = 2$

13. $x = y^2$, $x^2 = 6 - 5y$

*14. $xy^2 = 1$, $y = 3 - 2\sqrt{x}$

15. $y = x^{1/3}$, $y = x$

16. $y = \dfrac{2}{x^2} - 1$, $y = x^2$, x-axis

17. $x = y^3 - y^2$, $x = 5y^2$

18. $y = x^3 + x + 1$, $y = x^2 + x + 1$

19. $y = x + 6$, $y = 2x - 2$, $y = 3x + 4$

20. $y = ax + b$, $y = ax^2$, y-axis; $a, b > 0$

21. $y = \sin \dfrac{\pi}{2} x$, $y = x$, $x = 0$, $x = 2$

22. $y = \sin \dfrac{x}{2}$, $y = \cos \dfrac{x}{2}$, $x = 0$, $x = \pi$

23. Find the area of the triangle with vertices at $(1, 6)$, $(2, 4)$, and $(-3, 7)$. [*Hint:* Find the equations of the straight lines forming the sides, and draw a sketch.]

24. Find the area of the triangle with vertices at $(2, 0)$, $(3, 2)$, and $(6, 7)$.

25. Find the area of the triangle with sides given by $x + y = 3$, $2x - y = 4$, $6x + 3y = 3$.

26. Find the area of the triangle with sides given by $y = x - 3$, $y = 2x + 4$, $y = -3x + 8$.

flirtatious

5.9 WORK, POWER, AND ENERGY (OPTIONAL)

$8 - 6 + 5$
$8 + 6 + 5$

Consider a particle acted on by a force. In the simplest case the force is constant, and the particle moves in a straight line in the direction of the force (see Figure 1). In this situation we define the *work W done by the force on the particle* as the product of the magnitude of the force F and the distance s through which the particle travels:

$$W = Fs. \tag{1}$$

FIGURE 1

One unit of work is the work done by a unit force in moving a body a unit distance in the direction of the force. In the metric system the unit of work is 1 newton† meter (N · m), called 1 *joule* (J). In the British system the unit of work is the foot pound‡ (ft · lb).

EXAMPLE 1   A block of mass 10.0 kg is raised 5 m off the ground. How much work is done?

Solution. Force = (mass) × (acceleration). The acceleration here is opposing acceleration due to gravity and is therefore equal to 9.81 m/sec². We have $F = ma = (10 \text{ kg})(9.81 \text{ m/sec}^2) = 98.1$ N. Therefore, $W = Fs = (98.1) \times 5$ m $= 490.5$ J. ■

EXAMPLE 2   How much work is done in lifting a 25-lb weight 5 ft off the ground?

Solution. $W = Fs = 25$ lb $\times 5$ ft $= 125$ ft · lb. ■

Since we will be dealing with mass and force in many problems in this text, we take a moment here to explain units of weight, mass, and force in both the English and metric systems.

In the English system we start with the unit of force, which is the *pound*. (It is also the standard unit of weight.) Since $F = mg$ by Newton's second law of motion, we obtain $m = F/g$. In the English system the unit of mass is the *slug*, and a 1-lb weight (force) has a mass of $(1/32.2)$ lb/(ft/sec²) = 1 slug. Or we can equivalently define a slug as the mass of an object whose acceleration is 1 ft/sec² when it is subjected to a force of 1 lb. This explains why, in Example 2, we did not multiply the weight of 25 lb

---

†1 newton (N) is the force that will accelerate a 1-kg mass at the rate of 1 m/sec²; 1 N = 0.2248 lb.
‡1 Joule (J) = 0.7376 ft · lb, or 1 ft · lb = 1.356 J.

by the acceleration to obtain the force. The weight (in pounds) was *already* given as a force.

In the metric system we start with the standard unit of mass, which is the *kilogram*. In this case to obtain the force, we must multiply the mass by the acceleration:

$$F \text{ (in newtons)} = [\text{mass (in kilograms)}] \times g(= 9.81 \text{ m/sec}^2).$$

In sum, we have the following:

The pound is a force, but not a mass.
The kilogram is a mass but not a force.

We now return to the calculation of work. The case in which the motion of the particle is not in a straight line in the direction of the force is more complicated than the case we discussed in Examples 1 and 2. This case is analyzed in a discussion of vector calculus. However, we can use the integral to handle the case in which the force is variable.

We wish to calculate how much work is done in moving from the pont $x = a$ to the point $x = b$. As usual, we partition the interval into n subintervals, a typical interval being $[x_{i-1}, x_i]$. Now if the force is given by $F(x)$ (which is assumed to be continuous), then over a very small interval the force will be almost constant. Therefore, if x_i^* is a point in the interval $[x_{i-1}, x_i]$, then the work done in moving from x_{i-1} to x_i is, approximately,

$$\Delta W \approx F(x_i^*)(x_i - x_{i-1}) = F(x_i^*) \, \Delta x. \tag{2}$$

Adding the work done over these small intervals and taking the limit as $\Delta x \to 0$ leads to the formula

$$W = \lim_{\Delta x \to 0} \sum_{i=1}^{n} F(x_i^*) \, \Delta x = \int_a^b F(x) \, dx. \tag{3}$$

EXAMPLE 3 A spring is stretched 10 cm by a force of 2 N. How much work is needed to stretch it 50 cm $= \frac{1}{2}$ m (see Figure 2), assuming that the spring satisfies Hooke's law?

$x = 0$

$x = 10$ cm

$F = 2N$

FIGURE 2

Solution. **Hooke's law**† states that the force needed to stretch a spring is proportional to the amount of the spring displaced; that is,

$$F(x) = kx,$$

(4)

where k is a constant of proportionality, called the **spring constant,** and x is the amount displaced. From the information given, we have $2 \text{ N} = k(0.1 \text{ m})$, or $k = 2/0.1 = 20$. Therefore the force is given by $F(x) = 20x$. Then

$$W = \int_0^{1/2} F(x)\, dx = \int_0^{1/2} 20x\, dx = 10x^2 \Big|_0^{1/2} = \frac{10}{4} = \frac{5}{2} \text{ J.}$$

We should point out that Hooke's law is only valid if the spring is not stretched too far. After a certain point the spring will become stretched out of shape, and Hooke's law will fail to hold. ■

EXAMPLE 4   A cylindrical tank 12 ft in diameter and 20 ft high is filled with water. How much work will be done in pumping out the water?

Solution. Look at Figure 3. We define the variable x as the distance to the top of the tank. We then partition the interval $[0, 20]$ in the usual way. Consider the shaded amount of water. The work needed to get it to the top of the tank is its weight (in pounds) times the distance to the top. The density of water is 62.4 lb/ft³. The water layer of width Δx has a volume given by

$$\text{volume} = (\text{area of cross section}) \times (\Delta x)$$
$$= \pi r^2 \Delta x = \pi \left(\tfrac{12}{2}\right)^2 \Delta x = \pi 36 \Delta x \text{ cubic feet.}$$

FIGURE 3

Therefore the weight of the water in the shaded region is equal to

$$\underbrace{(62.4 \text{ lb/ft}^3)}_{\substack{\text{Weight per} \\ \text{cubic foot}}} \underbrace{(36\pi \Delta x) \text{ ft}^3}_{\substack{\text{Volume of} \\ \text{region} = \\ \text{number of} \\ \text{cubic feet}}} = \underbrace{2246.4\pi \Delta x \text{ lb.}}_{\substack{\text{Total weight of} \\ \text{shaded region}}}$$

---

†Named for Robert Hooke (1635–1703), who was an English philosopher, physicist, chemist, and inventor.

If x_i^* is in the interval $[x_{i-1}, x_i]$ then x_i^* is the approximate distance from the water in the layer to the top of the tank. Hence the work needed to get the water in the ith layer to the top is given by $W_i = 2246.4\pi x_i^* \, \Delta x$, and the total work is

$$W = \int_0^{20} 2246.4\pi x \, dx = 1123.2\pi x^2 \Big|_0^{20} = (1123.2)(400)\pi = 449{,}280\pi \text{ ft} \cdot \text{lb.} \quad \blacksquare$$

EXAMPLE 5   A storage tank in the shape of an inverted right circular cone has a radius of 4 m and a height of 8 m. It is filled to a height of 6 m with olive oil (density $= 920$ kg/m^3). To bottle the olive oil, the bottler must first pump it to the top of the tank. How much work is done in accomplishing this task?

Solution. This problem involves a lot of things going on at once. First, we draw a sketch (see Figure 4). It helps if we put in some identifying coordinates.† We place the origin 8 m up and in the center of the cone. Then if the x-axis is the vertical axis and the y-axis is the horizontal axis, we can label various points on the cone as in the figure. Now the "x-axis" in the cone includes the interval $[0, 8]$, and the olive oil fills the interval $[2, 8]$. We partition that last interval in the usual way and look at a layer of oil in the ith subinterval $[x_{i-1}, x_i]$.

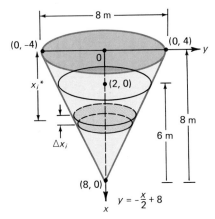

FIGURE 4

The work needed to get this layer of oil to the top is the force needed to overcome gravity times the distance the layer must travel. When Δx is small, this distance is approximately equal to x_i^* for any x_i^* in $[x_{i-1}, x_i]$. What is the force that must be overcome to lift the layer of oil? We have $F = ma$, where $a = 9.81$ m/sec^2. The mass m of the oil in the layer is the volume of the oil times its density $(= 920$ kg/m$^3)$. The volume of the layer is the cross-sectional area times the thickness Δx. Thus the only thing left to calculate is the cross-sectional area. This area is πy_i^2, where y_i is the radius of the layer. We can think of this radius as varying along the straight line from $(8, 0)$ to $(0, 4)$ (see Figure 4 again). The equation of this straight line is

---

†The coordinate system given in Figure 4 may seem a bit strange to you at first. If you are confused, turn the sketch sideways so that the x- and y-axes point in their usual directions.

$$\frac{y-4}{x-0} = -\frac{4}{8}, \quad \text{or} \quad y = -\frac{x}{2} + 4.$$

Then $y_i \approx -(x_i^*/2) + 4$.

Now we put everything together.

$$\text{cross-sectional area of layer} = \pi\left(4 - \frac{x_i^*}{2}\right)^2 \text{ square meters}$$
$$(A = \pi r^2)$$

$$\text{volume of layer} = \pi\overbrace{\left(4 - \frac{x_i^*}{2}\right)^2}^{\text{Area}} \overbrace{\Delta x}^{\text{Thickness}} \text{ cubic meters}$$
$$(V = A \times \text{thickness})$$

$$\text{mass of layer} = \overbrace{920}^{\text{Density}}\pi\overbrace{\left(4 - \frac{x_i^*}{2}\right)^2 \Delta x}^{\text{Volume}} \text{ kilograms}$$
$$(m = \mu V)$$

$$\begin{array}{l}\text{force to overcome force} \\ \text{of gravity to raise layer}\end{array} = \overbrace{(9.81)}^{a}\overbrace{(920)\pi\left(4 - \frac{x_i^*}{2}\right)^2 \Delta x}^{m} \text{ newtons}$$
$$(F = ma)$$

$$= 9025.2\pi\left(4 - \frac{x_i^*}{2}\right)^2 \Delta x \text{ newtons}$$

$$\text{Work to raise layer } x_i^* \text{ m} = x_i^* \overset{s}{\underbrace{9025.2\pi\left(4 - \frac{x_i^*}{2}\right)^2 \Delta x}_{F}} \text{ joules.}$$
$$(W = Fs)$$

Therefore, the total work is given by

$$W = \int_2^8 9025.2\pi\left(4 - \frac{x}{2}\right)^2 x \, dx = 9025.2\pi \int_2^8 \left(16x - 4x^2 + \frac{x^3}{4}\right) dx$$

$$= (9025.2\pi)\left(8x^2 - \frac{4x^3}{3} + \frac{x^4}{16}\right)\Bigg|_2^8 = (9025.2\pi)(63) = 568{,}587.6\pi \text{ J.}$$

You should go through the steps in this long calculation again. The key to solving the problem is to look at a typical layer and determine, using the definition of work, what it takes to move that layer to the top. ■

So far, we have not considered the time spent in doing work. The work it takes to pump oil out of a tank is the same whether it is done in one minute or in one month. Sometimes, however, the rate at which work is done is more important than the actual amount of work done. The rate at which work is done is called **power**. If the work is done at a constant rate, then the power produced in t units of time is $P = W/t$. The **instantaneous power** is the rate of change of work with respect to time:

$$P = \frac{dW}{dt}. \tag{5}$$

In the metric system the unit of power is the *watt* (W),† and

$$1\ W = 1\ J/sec.$$

In the British system the unit of power is 1 ft · lb/sec. Since this unit is very small, another unit called the **horsepower** is often used. We have

$$1\ hp = 550\ ft \cdot lb/sec = 746\ W \approx \tfrac{3}{4}\ kilowatt\ (kW).$$

Sometimes work is expressed in terms of power × time. This convention explains the use of the term kilowatt-hour (kWh), which is the work done by something (such as twenty 50-w light bulbs) working 1 hr at a constant rate of 1kW.

EXAMPLE 6 A man works at a constant rate and performs 129,600 J of work in 4 hr. Then his power output is

$$P = \frac{\text{work done}}{\text{time (in seconds)}} = \frac{129,600\ J}{4(3600)\ sec} = \frac{129,600}{14,400} = 9W \ \blacksquare$$

EXAMPLE 7 The instantaneous power of a machine is $50 - 3\sqrt{t+4}$ watts. How much work does the machine do in 1 min?

Solution. Since $P = dW/dt$, we have

$$W = \int_0^{60} P(t)\ dt = \int_0^{60} (50 - 3\sqrt{t+4})\ dt$$

$$= [50t - 2(t+4)^{3/2}]\Big|_0^{60} = 1992\ J$$

(Here the 60 is for the 60 sec in 1 min.) ■

Let $x(t)$ represent the position of a moving particle at time t, and assume that the particle starts at the origin. Then $x'(t) = v(t)$ (velocity) and $x''(t) = a(t)$ (acceleration). If a is constant, then the velocity at time t is given by $v = \int a\ dt = at + c$. But $v(0) = v_0$, the initial velocity, so that $c = v_0$ and $v = at + v_0$, or

$$a = \frac{v - v_0}{t}. \tag{6}$$

Since velocity changes at a constant rate ($a = $ constant), the average velocity of the particle is simply the average of the initial velocity v_0 and the final velocity v. That is,

†Named after James Watt (1736–1819), the inventor of the steam engine.

$$v_{av} = \frac{v + v_0}{2}.$$

Therefore the distance the particle travels is $x = v_{av} \cdot t$, or

$$x = \frac{v + v_0}{2} t. \tag{7}$$

The work done on the particle in moving from 0 to x is [using (6) and (7) and Newton's second law $F = ma$]

$$W = Fx = max = m\left(\frac{v - v_0}{t}\right)\left(\frac{v + v_0}{2}\right)t = \frac{1}{2}mv^2 - \frac{1}{2}mv_0^2. \tag{8}$$

The product $\frac{1}{2}mv^2$ is called the **kinetic energy** of the particle. It is denoted by the symbol K. We have

$$K = \tfrac{1}{2}mv^2. \tag{9}$$

Equation (8) states the following:

The work done by the force acting on a particle is equal to the change in kinetic energy of the particle.

This result is known as the **work-energy law.**
  We have only shown why the work-energy law is true when a is constant. If a is not constant, we can still illustrate it as follows: From (3) the work done in moving from x_0 to x_1 is

$$W = \int_{x_0}^{x_1} F \, dx.$$

Now

$$a = \frac{dv}{dt} = \frac{dv}{dx}\frac{dx}{dt} \text{ (by the chain rule)} = \frac{dv}{dx}v = v\frac{dv}{dx}. \tag{10}$$

Then

$$W = \int_{x_0}^{x_1} F \, dx = \int_{x_0}^{x_1} ma \, dx = \int_{x_0}^{x_1} mv\frac{dv}{dx} \, dx.$$

We know how to integrate $mv \, dv$ ($\int mv \, dv = mv^2/2$ since m is constant). Using a technique we will develop in Section 8.4, we change the variable of integration. When $x = x_0$, $v = v(x_0) = v_0$; and when $x = x_1$, $v = v(x_1) = v_1$. Furthermore,

$$\frac{dv}{dx} dx = dv.$$

Therefore we have

$$W = \int_{x_0}^{x_1} F \, dx = \int_{x_0}^{x_1} mv \frac{dv}{dx} dx = \int_{v_0}^{v_1} mv \, dv = \frac{1}{2} mv^2 \Big|_{v_0}^{v_1} = \frac{1}{2} mv_1{}^2 - \frac{1}{2} mv_0{}^2$$

$$= \text{change in kinetic energy.}$$

EXAMPLE 8   A block weighing 5 kg slides on a horizontal table with a speed of 0.3 m/sec. It is stopped by a spring with a spring constant $k = \frac{1}{4}$ placed in its path. How much is the spring compressed?

     Solution. We assume (for simplicity) that until the block hits the spring, its velocity is constant. Then its kinetic energy is the constant $\frac{1}{2} mv^2 = \frac{1}{2}(5)(0.3)^2 = 0.45/2$ J. The work done in compressing the spring x units is

$$W = \int_0^x kx \, dx = \frac{kx^2}{2} = \frac{x^2}{8},$$

since $k = \frac{1}{4}$. Therefore equating work to the change in kinetic energy (from 0.45/2 to 0), we have

$$\frac{x^2}{8} = \frac{0.45}{2}, \quad \text{or} \quad x^2 = 1.8, \quad \text{and} \quad x \approx 1.34 \text{ m.} \blacksquare$$

EXAMPLE 9   Neglecting air resistance, find the velocity v_0 at which a missile at the surface of the earth would have to be fired in order to escape the earth's gravitational field.

     Solution. The only force the missile must overcome is the force of gravity. Let δ denote the distance from the center of the earth to its surface ($\delta \approx 6378$ km ≈ 3963 mi). The force of gravity is, by Newton's law of gravitational attraction, inversely proportional to the square of the distance u between the missile and the center of the earth. That is, $F = \alpha/u^2$, where α is a constant of proportionality. To find α, we observe that when $u = \delta$ (so that the missile is at the surface of the earth), $F = mg$ (where m denotes the mass of the missile). Then $F(\delta) = mg = \alpha/\delta^2$, or $\alpha = mg\delta^2$, and

$$F = mg \frac{\delta^2}{u^2}.$$

The work needed to lift the missile from a distance of δ to a height of x is

$$W = \int_\delta^x F(u) \, du = \int_\delta^x mg \frac{\delta^2}{u^2} du = -\frac{mg\delta^2}{u} \Big|_\delta^x = mg\delta^2 \left(\frac{1}{\delta} - \frac{1}{x} \right). \tag{11}$$

     To escape the earth's gravitational pull, the missile must have an initial velocity sufficient to thrust it to an "infinite" height (once a certain height is reached, the missile keeps "rising" indefinitely). We therefore let $x \to \infty$ in (11) to find that the work needed to allow the missile to escape is given by

$$W = mg\delta^2 \frac{1}{\delta} = mg\delta.$$ (12)

Now the missile starts at rest and then moves with a velocity of v_0. The change in kinetic energy is, therefore,

$$\Delta K = \tfrac{1}{2}mv_0^2.$$ (13)

Using the work-energy law, we equate (12) and (13) to obtain

$$v_0 = \sqrt{2g\delta}.$$ (14)

This value of v_0 is called the **escape velocity.** Note that the escape velocity is independent of the mass of the missile. In metric units,

$$v_0 = \sqrt{(2)9.81 \text{ m/sec}^2 \times 6378 \times 10^3 \text{ m}} \approx 11{,}186 \text{ m/sec} \approx 11.2 \text{ km/sec.}$$

In English units,

$$v_0 = \sqrt{(2)32.2 \text{ ft/sec}^2 \times 3963 \times 5280 \text{ ft}} \approx 36{,}709 \text{ ft/sec} \approx 6.95 \text{ mi/sec.} \blacksquare$$

PROBLEMS 5.9

1. How much work is done in vertically lifting a 5-kg rock 2 m?
2. How much work is done in lifting an 8-lb weight 6 ft?
3. A force of 5 lb stretches a spring 6 in. How much work is done in stretching the spring 2 ft?
4. A force of 3 N stretches a spring 50 cm. How much work is done in stretching the spring 3 m?
5. A force of 10 N stretches a spring 25 cm. How much work is done in stretching the spring $1\tfrac{1}{2}$ m?
6. How much work is done to stretch the spring in Problem 5 from 1 to 2 m?
7. A force of 25 lb stretches the spring in Problem 3 how many feet?
8. A force of 8 N stretches the spring in Problem 4 how many meters?
9. A cylindrical tank 10 m high and 10 m in diameter is filled with water. How much work does it take to pump the water out? [*Hint:* The density of water at 0° C is 1000 kg/m³.]
10. A tank in the shape of an inverted cone has a radius of 3 m and a height of 9 m and is filled with water. How much work is needed to pump the water out?
11. How much work is needed in Problem 9 if only half the water is to be pumped out?
*12. How much work is needed in Problem 10 if only half the water is to be pumped out?
13. The cylindrical tank of Problem 9 is now placed 10 m above the ground. How much work is needed to fill the tank from a hose that runs from ground level to the top of the tank?
14. The conical tank of Problem 10 is placed 8 m above the ground. How much work is needed to fill the tank from a hose that runs from ground level to a hole at the vertex (bottom) of the tank?
15. Answer Problem 9 if the tank is filled with olive oil instead of water.

16. Answer Problem 10 if the tank is filled with olive oil instead of water.
17. A conical tank (inverted) has a height of 7 m and a diameter of 6 m (see Figure 5). The tank is filled with ethyl alcohol to a level of 5 m. How much work is needed to pump the alcohol out of the tank? [*Hint:* Density of ethyl alcohol = 810 kg/m³.]

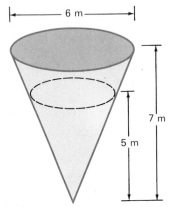

FIGURE 5

18. A 10-ft metal chain weighs $1\frac{1}{2}$ lb/ft. The chain is lifted vertically from one end until the other end is 5 ft off the ground. How much work is done?
19. A 40-ft chain that weighs $1\frac{1}{2}$ lb/ft is hanging from the top of a building. How much work is needed to pull 10 ft of it to the top? [*Hint:* Integrate from 30 to 40 since it is the *last* 10 ft that is being pulled to the top.]
20. A 25-m rope weighing $\frac{3}{4}$ kg/m is hanging from the top of a cliff. How much work is needed to pull 7 m of the rope to the top of the cliff?
21. A swimming pool is 20 ft on each side and 6 ft deep. If it is filled with water, how much work is done by the water as it empties out of a hole in the bottom of the pool?
22. How much work is needed in Problem 21 to fill the pool up to a level of 2 ft?
23. How long will it take a 2-hp motor to do the work required in Problem 22?
24. A 100-m-long chain is attached to a drum. The chain weighs 2 kg/m. If the chain initially does not touch the floor, how much work is done in winding the chain up?
*25. The moon is approximately 380,000 km from the earth. By Newton's law of gravitational attraction, the attraction between two masses is Gm_1m_2/r^2, where G is a universal constant (see Example 3.2.7) and r is the distance between them. If M denotes the mass of the earth, then the mass of the moon is $0.01228M$. In the moon's revolving about the earth, a force is created, called *centrifugal force*, which counteracts the force of gravity and keeps the moon in orbit around the earth.

   A popular science fiction series on television portrays the moon, blown out of orbit by nuclear explosions, wandering aimlessly through space. Using the information above, calculate how much work would be required to accomplish this feat. (You can express your answer in terms of G and M, assuming that the loss of mass is negligible and the final position of the moon is essentially an infinite distance from the earth.)
26. A 40-W light bulb burns for $2\frac{1}{2}$ h. How much work is performed during that time?
27. The instantaneous power of a machine is $P = 30 + 4t - t^2/2$ watts. How much work is performed in 15 sec?
28. The instantaneous power of a machine is $P = t + (30/\sqrt{1 + t})$ watts. How much work is performed in the second minute of its operation?
*29. An arrow weighing 0.0322 lb is shot from a 35-lb draw bow at full draw $d = 16$ in. Assume that the force of the bow (in pounds) is proportional to the draw (in inches). At what

velocity does the arrow leave the bow? [*Hint:* mass in slugs = weight in pounds ÷ acceleration of gravity.]

30. A block weighing 12 kg slides on a horizontal table with a speed of 4 m/sec. It is stopped by a spring, with a spring constant of 2, placed in its path. How much is the spring compressed?

31. The diameter of Mars is 6860 km. The acceleration due to gravity at its surface is 3.92 m/sec^2. What is the escape velocity from Mars?

32. An object is dropped from a distance x_0 meters above the surface of the earth. Assuming that the force of gravity is constant for small distances above the earth, what is the kinetic energy of the object when it hits the earth? What is its final velocity?

33. From what height would you have to drop an automobile in order that it gain the same kinetic energy it has from going 50 mi/hr?

5.10 ADDITIONAL INTEGRATION THEORY (OPTIONAL)

In this chapter we introduced the definite and indefinite integrals. In Chapter 8 we will show how a large variety of integrals can be calculated. The basic theorem underlying these calculations is that if f is continuous in $[a, b]$, then $\int_a^b f(x)\,dx$ exists. The proof of this theorem requires tools from advanced calculus that cannot be developed in an elementary calculus text. Nevertheless, there are interesting results that can be derived by using facts about continuous and differentiable functions that we already know. In this section we will obtain two of these results, assuming only the basic integrability theorem cited above and the intermediate value thorem, restated below.

Theorem 1 *Intermediate Value Theorem (Theorem 2.7.5)* Let f be continuous on $[a, b]$. Then if d is any number between $f(a)$ and $f(b)$, there is a number c in (a, b) such that $f(c) = d$.

We can now state our first result.

Theorem 2 *Mean Value Theorem for Integrals* Suppose that f and g are continuous on $[a, b]$ and that $g(x)$ is never zero on (a, b). Then there exists a number c in (a, b) such that

$$\int_a^b f(x)g(x)\,dx = f(c)\int_a^b g(x)\,dx. \tag{1}$$

Proof. Since f is continuous on $[a, b]$, there exist numbers m and M such that

$$m \le f(x) \le M, \tag{2}$$

and there are numbers x_1 and x_2 in $[a, b]$ such that $f(x_1) = m$ and $f(x_2) = M$ (from Theorem 2.7.4).† Since $g(x) \ne 0$ and g is continuous, we see that either $g(x) >$

---

†Recall from Theorem 2.7.4 that this result applies in the case in which m and M are the greatest lower bound and least upper bound, respectively, for f on $[a, b]$.

0 or $g(x) < 0$ on (a, b). (Otherwise the intermediate value theorem would be contradicted.) Suppose that $g > 0$. The case $g < 0$ is left as a problem (see Problem 1). Then we multiply (2) by $g(x)$ to obtain

$$mg(x) \le f(x)g(x) \le Mg(x),$$

which implies, by the comparison theorem (Theorem 5.5.7), that

$$m \int_a^b g(x)\,dx \le \int_a^b f(x)g(x)\,dx \le M \int_a^b g(x)\,dx. \tag{3}$$

Since $g(x) > 0$, $\int_a^b g(x)\,dx > 0$, and dividing (3) by this integral yields

$$m \le \frac{\displaystyle\int_a^b f(x)g(x)\,dx}{\displaystyle\int_a^b g(x)\,dx} \le M.$$

Since $m = f(x_1)$ and $M = f(x_2)$ are values for $f(x)$, $[\int_a^b f(x)g(x)\,dx]/\int_a^b g(x)\,dx$ is a number between $f(x_1)$ and $f(x_2)$, so that, by the intermediate value theorem, there is a number c such that

$$f(c) = \frac{\displaystyle\int_a^b f(x)g(x)\,dx}{\displaystyle\int_a^b g(x)\,dx},$$

and the theorem is proved. ■

The mean value theorem for integrals is very useful. Some applications are suggested in the problem set. We use the theorem now to prove the second fundamental theorem of calculus.

Theorem 3   *Second Fundamental Theorem of Calculus (Theorem 5.6.2)*   If f is continuous on $[a, b]$, then the function $G(x) = \int_a^x f(t)\,dt$ is continuous on $[a, b]$, differentiable on (a, b), and for every x in (a, b),

$$G'(x) = f(x). \tag{4}$$

That is, G is an antiderivative of f.

Proof.   Let x be in (a, b). Then

Theorem 5.5.3

$$G(x + \Delta x) = \int_a^{x+\Delta x} f(t)\,dt \overset{\downarrow}{=} \int_a^x f(t)\,dt + \int_x^{x+\Delta x} f(t)\,dt. \tag{5}$$

Thus

$$G(x + \Delta x) - G(x) = \int_a^x f(t)\, dt + \int_x^{x+\Delta x} f(t)\, dt - \int_a^x f(t)\, dt$$

$$= \int_x^{x+\Delta x} f(t)\, dt. \tag{6}$$

From the mean value theorem for integrals [using $g(x) \equiv 1$], there is a number c in $(x, x + \Delta x)$ such that

$$\int_x^{x+\Delta x} f(t)\, 1\, dt = f(c) \int_x^{x+\Delta x} 1\, dt = f(c)\, \Delta x. \tag{7}$$

Then using (7) in (6), we see that

$$G(x + \Delta x) - G(x) = f(c)\, \Delta x,$$

or for $\Delta x \ne 0$,

$$\frac{G(x + \Delta x) - G(x)}{\Delta x} = f(c). \tag{8}$$

We now take the limit on both sides of (8) to obtain

$$\lim_{\Delta x \to 0} \frac{G(x + \Delta x) - G(x)}{\Delta x} = \lim_{\Delta x \to 0} f(c), \tag{9}$$

or

$$G'(x) = \lim_{\Delta x \to 0} f(c). \tag{10}$$

But $c \in (x, x + \Delta x)$, so that as $\Delta x \to 0$, $c \to x$. Moreover, since $x \in (a, b)$, f is continuous at x, so

$$\lim_{c \to x} f(c) = f(x). \tag{11}$$

This shows that $G'(x) = f(x)$ for any x in (a, b).

Since a differentiable function is continuous, we see that G is continuous in (a, b). The only thing left to prove is the continuity of G at $x = a$ and $x = b$. That is, we must show (see Definition 2.7.4) that

$$\lim_{x \to a^+} G(x) = G(a) = \int_a^a f(t)\, dt = 0 \tag{12}$$

and

$$\lim_{x \to b^-} G(x) = G(b) = \int_a^b f(t)\, dt. \tag{13}$$

The limit in (12) is equivalent to

$$\lim_{\Delta x \to 0+} \int_a^{a + \Delta x} f(t)\, dt = 0. \tag{14}$$

But $\int_a^{a+\Delta x} f(t)\, dt = f(c)\, \Delta x$, where $a < c < a + \Delta x$. And since f is continuous at a, $\lim_{c \to a+} f(x) = f(a)$ and, as before, $\lim_{\Delta x \to 0+} f(c) = f(a)$. Thus

$$\lim_{\Delta x \to 0+} \int_a^{a + \Delta x} f(t)\, dt = \lim_{\Delta x \to 0+} f(c)\, \Delta x = \lim_{\Delta x \to 0+} f(c) \lim_{\Delta x \to 0+} \Delta x = f(a) \cdot 0 = 0.$$

Thus (12) holds.

To prove (13), we observe that

$$\int_a^b f(t)\, dt - \int_a^{b - \Delta x} f(t)\, dt = \int_{b - \Delta x}^b f(t)\, dt.$$

The same argument used to prove (14) shows that

$$\lim_{\Delta x \to 0+} \int_{b - \Delta x}^b f(t)\, dt = 0.$$

Therefore

$$\lim_{x \to b^-} G(x) = \lim_{\Delta x \to 0+} \int_a^{b - \Delta x} f(t)\, dt = \int_a^b f(t)\, dt = G(b), \tag{15}$$

and (13) is proved. ■

PROBLEMS 5.10

1. Show that the mean value theorem for integrals holds in the case $g < 0$ on (a, b).
2. Prove that for any continuous function $f > 0$,

$$\int_0^\pi f(t) \sin t\, dt \le \int_0^\pi f(t)\, dt.$$

3. Prove that for any continuous function $f > 0$,

$$\int_a^b f(t) \cos^6(t^2 + 1)\, dt \le \int_a^b f(t)\, dt$$

for any real numbers a and b, with $a < b$.
4. Let $G(u) = \int_a^u f(t)\, dt$, where f is continuous. Use the chain rule and Theorem 3 to prove that

$$\frac{dG(u)}{dx} = f(u)\frac{du}{dx}.$$

5. Use the result of Problem 4 to prove that if $b(x)$ is differentiable and f is continuous, then

$$\frac{d}{dx} \int_a^{b(x)} f(t)\, dt = f(b(x))b'(x).$$

6. Use the result of Problem 5 to show that if $a(x)$ is differentiable, then

$$\frac{d}{dx} \int_{a(x)}^{b} f(t)\, dt = -f(a(x))a'(x).$$

7. Using Problems 5 and 6, show that

$$\frac{d}{dx} \int_{a(x)}^{b(x)} f(t)\, dt = f(b(x))b'(x) - f(a(x))a'(x).$$

[*Hint:* Write $\int_{a(x)}^{b(x)} f = \int_{a(x)}^{c} f + \int_{c}^{b(x)} f$.]

In Problems 8–16, use Problem 7 to differentiate the given function

8. $G(x) = \displaystyle\int_x^{x^2} \sin t^2 \, dt$ 9. $G(x) = \displaystyle\int_{\sin x}^{\cos x} t(5 - t)\, dt$ 10. $G(x) = \displaystyle\int_{\sqrt{x}}^{\sqrt[3]{x}} t^2 \, dt$

11. $G(x) = \displaystyle\int_0^{5x} \cos t \, dt$ 12. $G(x) = \displaystyle\int_{\pi/3}^{x^2} \cos t \, dt$ 13. $\displaystyle\int_{2x}^{x^3} \cos t \, dt$

14. $G(x) = \displaystyle\int_3^x \frac{\sin t}{t^2 - 1} \, dt$ 15. $G(x) = \displaystyle\int_5^{2x} \frac{\sin v}{v^2 - 1} \, dv$ *16. $G(x) = \displaystyle\int_{17}^y \frac{\sin x}{x^2 - 1} \, dx$

17. Prove another form of the mean value theorem for integrals: Let f be continuous on $[a, b]$. Then there exists a number c in (a, b) such that $\int_a^b f(t)\, dt = f(c) \cdot (b - a)$.
18. Using the result of Problem 17, prove that $\int_0^{2\pi} \sin t^3 \, dt \leq 2\pi$.
19. Prove that $\int_0^{\pi/2} \sin^8 t \sqrt[3]{\cos t} \, dt \leq \pi/2$.
20. **(a)** Suppose that f is continuous, $f(x) \geq 0$, and $f(x) \neq 0$. Prove that $\int_a^b f(x)\, dx > 0$.
    (b) Use the result of part (a) to show that in the proof of Theorem 2 the inequalities in (3) are strict, so that c is strictly between a and b [unless $f(x)$ is constant, in which case c can be any number between a and b].

REVIEW EXERCISES FOR CHAPTER FIVE

In Exercises 1–12, calculate the given definite or indefinite integral.

1. $\displaystyle\int x^5 \, dx$

2. $\displaystyle\int_0^2 (t^3 + 3t + 5)\, dt$

3. $\displaystyle\int_1^8 \frac{ds}{\sqrt[3]{s}}$

4. $\displaystyle\int \frac{du}{(u + 3)^3}$

5. $\displaystyle\int_0^1 x\sqrt{x^2 + 1} \, dx$

6. $\displaystyle\int \frac{t^2 \, dt}{(t^3 + 3)^{3/4}}$

7. $\displaystyle\int_{-1}^1 (x^2 - 3)(x^5 + 2)\, dx$ 8. $\displaystyle\int_{-3}^4 |s + 2| \, ds$

9. $\displaystyle\int_1^{\sqrt{2}} \frac{[1 - (1/t^2)]^4}{t^3} \, dt$

10. $\displaystyle\int (3ax^2 + 2bx + c)(ax^3 + bx^2 + cx + d)^{-1/9}\, dx$ 11. $\displaystyle\int_0^{\pi/3} (\sin 2x + \cos 3x)\, dx$

12. $\displaystyle\int_0^{6\pi} \left| \sin \frac{x}{3} \right| \, dx$

In Exercises 13–24, find the areas bounded by the given curves and lines.

13. $y = 3x - 7$; x-axis, $x = -2$, $x = 5$ 14. $y = \sqrt{x + 1}$, x-axis, y-axis, $x = 15$

15. $y = -x^2 - x + 2$; x-axis

16. $y = x^3 - 7x^2 + 7x + 15$; x-axis

17. $y = \dfrac{1}{(x + 2)^2}$; x-axis, y-axis, $x = 3$

18. $y = \sin\left(\dfrac{x}{3}\right)$; x-axis, $x = 0$, $x = 6\pi$

19. $y = 3x^3$; $y = x$

20. $x^{1/3} + y^{1/3} = 1$; $x = 0$, $y = 0$

21. $y^2 + x = 0$; $y = 2x + 1$

22. $y = x^3$; $y = x^4$

23. $y = \sin x$; $y = \cos x$; $x = 0$, $x = \pi$

24. $y = \sin 2x$; $y = \cos 3x$; $y = 0$, $y = \dfrac{\pi}{6}$

In Exercises 25–27, calculate the derivative of the given function and evaluate it at the given point.

25. $F(x) = \displaystyle\int_0^x \dfrac{t^3}{\sqrt{t^2 + 17}}\, dt$; $x_0 = 8$ [Recall that $x_0 = x(0)$.]

26. $F(t) = \displaystyle\int_3^t (s^5 - 17s + 8)^{11/3}\, ds$; $t_0 = 1$ **27.** $F(s) = \displaystyle\int_{-1}^s (1 + u^2)^{2001}\, du$; $s_0 = 0$

28. A particle is moving with the velocity $v(t) = t + 1/\sqrt{1 + t}$ meters per second.
   (a) How far does the particle move in the first 15 sec?
   (b) What is the average velocity of the particle?
29. The density of a tree is given by $\rho(h) = 50/\sqrt{h + 1}$ (measured in kilograms per meter), where h denotes the distance (in meters) above the ground. The height of the tree is 24 m.
   (a) What is the total mass of the tree?
   (b) For what value of h does the first h meters of the tree contain half the total mass of the tree?
30. A force of 8 N stretches a spring 30 cm. How much work is done in stretching the spring 60 cm?
31. A tank in the shape of an inverted cone has a radius of 6 ft and a height of 12 ft. It is filled with water. How much work is needed to pump it out?
32. The instantaneous power of a certain machine is given by $P = (t/10) + (t^2/100) + 45\sqrt{t}$ watts, where t is measured in seconds. How much work is done in the first 25 sec of operation?
33. An object is dropped from a height of 500 m. What is the kinetic energy of the object when it hits the earth? What is its final velocity?
34. Compute $(d/dx) \displaystyle\int_{\sqrt[3]{x}}^{\sqrt{x}} (1 + t^2)^{4/5}\, dt$.

6 Exponentials and Logarithms

In this chapter we discuss two of the most important classes of functions of mathematics: the exponential and logarithmic functions. Perhaps you are already familiar with these functions through some previous course. Perhaps you will be seeing these functions for the first time. In any event, we will define these functions and show why they are important by giving a large variety of applications in physics, biology, and economics.

There are two very different ways to define the exponential and logarithmic functions. The first of these starts with the basic rules of exponents with which you are all familiar. The second way starts off with a special integral. In both cases, of course, we are led to the same functions. Both ways of defining these functions can give us insight into how to work with them, and both are given in this text. The first derivation is given in Sections 6.2 and 6.3 and the second in Section 6.4. It is not necessary to cover Section 6.4 if you cover Sections 6.2 and 6.3, and vice versa. However, it is useful to see how two very different approaches lead to the same central idea.

In a sense to be made precise shortly, exponential and logarithmic functions are inverses of one another. For that reason we begin the chapter with a discussion of inverse functions.

6.1 INVERSE FUNCTIONS

Let $y = 2x + 3$. Then we can write x as a function of y: $x = (y - 3)/2$. The functions $2x + 3$ and $(x - 3)/2$ are called *inverse functions*. We have seen other examples of inverse functions. The functions x^3 and $\sqrt[3]{x}$ are inverse functions since $y = x^3$ implies that $x = \sqrt[3]{y}$. In general, we have the following definition.

Definition 1   INVERSE FUNCTIONS   The functions f and g are **inverse functions** if the following conditions hold:

(i) For every x in the domain of g, $g(x)$ is in the domain of f and
$$f(g(x)) = x. \tag{1}$$
(ii) For every x in the domain of f, $f(x)$ is in the domain of g and
$$g(f(x)) = x. \tag{2}$$

In this case we write $f(x) = g^{-1}(x)$ or $g(x) = f^{-1}(x)$, and we say

f is the inverse of g and g is the inverse of f.

In the notation of Section 1.7, we may write

$$(f \circ g)(x) = x \quad\text{and}\quad (g \circ f)(x) = x. \tag{3}$$

EXAMPLE 1   If $f(x) = 2x + 3$ and $g(x) = (x - 3)/2$, then

$$f(g(x)) = 2\left(\frac{x - 3}{2}\right) + 3 = x - 3 + 3 = x$$

and

$$g(f(x)) = \frac{(2x + 3) - 3}{2} = x.$$

So $2x + 3$ and $(x - 3)/2$ are inverse functions. ∎

In general, we find inverse functions by following these steps:

(i) Write y as a function of x.
(ii) Write x as a function of y, if possible, and interchange the letters x and y.

The two functions thus obtained are inverses.

We used the phrase "if possible" in step (ii) above since it is not always possible to solve explicitly for x as a *function* of y.

EXAMPLE 2   Let $y = x^2$. Then $x = \pm\sqrt{y}$. This means that each value of y comes from *two* different

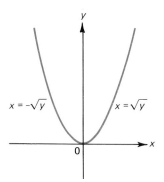

FIGURE 1

values of x. See Figure 1. We recall from Section 1.6 that a function $y = f(x)$ can be thought of as a rule that assigns to each x in its domain a *unique* value of y. Suppose we try to define the inverse of x^2 by taking the positive square root. Then we would like to be able to show that $f = x^2$ and $g = \sqrt{x}$ are inverses. Let $x = -2$. Then $g(f(x)) = \sqrt{(-2)^2} = \sqrt{4} = 2 \neq -2 = x$, so $g(f(x)) \neq x$. ■

To avoid problems like the one just encountered, we make the following definition.

Definition 2   ONE-TO-ONE FUNCTION   The function $y = f(x)$ is **one-to-one** on the interval $[a, b]$, written 1–1, if $x_1, x_2 \in [a, b]$ and $x_1 \neq x_2$ implies that $f(x_1) \neq f(x_2)$. That is, each value of $f(x)$ comes from only one value of x.

The definition of a 1–1 function is illustrated in Figure 2.

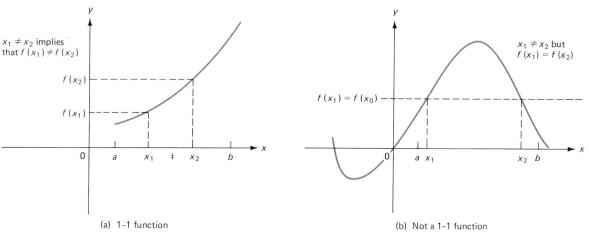

(a) 1–1 function   (b) Not a 1–1 function

FIGURE 2

Theorem 1.   If f is either an increasing or a decreasing function on $[a, b]$, then f is 1–1 on $[a, b]$.

This theorem is illustrated in Figure 3.

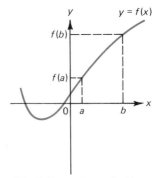

(a) $f(x)$ decreasing on $[a, b]$ (b) $f(x)$ increasing on $[a, b]$

FIGURE 3

Proof. Suppose f is increasing and suppose that $x_1 \neq x_2$. If $x_1 > x_2$, then $f(x_1) > f(x_2)$. If $x_2 > x_1$, then $f(x_2) > f(x_1)$. In either case $f(x_1) \neq f(x_2)$, so f is 1–1. The proof in the case when f is decreasing is similar. ∎

Theorem 2. If f is continuous on $[a, b]$ and has a derivative on (a, b) satisfying $f'(x) \neq 0$ for every x in (a, b), then f is 1–1 on $[a, b]$.

Proof. Suppose there are numbers x_1 and x_2 in $[a, b]$ such that $x_1 \neq x_2$, but $f(x_1) = f(x_2)$. Then, by the mean value theorem, there is a number c in (x_1, x_2) such that

$$0 = f(x_2) - f(x_1) = f'(c)(x_2 - x_1).$$

Since $(x_2 - x_1) \neq 0$, this implies that $f'(c) = 0$, which contradicts the hypothesis that $f'(x) \neq 0$ on (a, b). Thus if $f(x_1) = f(x_2)$, we must have $x_1 = x_2$, and f is 1–1. ∎

REMARK. There are 1–1 functions for which the condition $f'(x) \neq 0$ does not hold. For example, the function x^3 is 1–1 in $[-1, 1]$ but $f'(0) = 0$.

Why are we interested in 1–1 functions? Because if $y = f(x)$ is 1–1, then every value of y comes from a unique value of x, so that we can write x as a function of y, and this new function is the inverse of y. That is, *every 1–1 function has an inverse.*

EXAMPLE 3   Let $f(x) = \sqrt{x}$. Then $f'(x) = 1/(2\sqrt{x}) \neq 0$ for $x > 0$. Thus for $x \geq 0$, f has an inverse, which we find by setting $y = \sqrt{x}$ so that $x = y^2$ and therefore $f^{-1}(x) = x^2$. Note that since dom $f^{-1} = \{x : x \geq 0\}$, f^{-1} is *not* the function discussed in Example 2 (x^2 with domain $\mathbb{R}$). ∎

EXAMPLE 4   Let $f(x) = \sin x$ for $-\pi/2 \leq x \leq \pi/2$. Then $f'(x) = \cos x \neq 0$ in $(-\pi/2, \pi/2)$ so f is 1–1 in $[-\pi/2, \pi/2]$ and therefore has an inverse on $[-\pi/2, \pi/2]$. This inverse function is denoted $\sin^{-1} x$, or arcsin x. For example, $\sin^{-1} 1 = \pi/2$ since $\sin \pi/2 = 1$, and $\sin^{-1} \sqrt{3}/2 = \pi/3$ since $\sin \pi/3 = \sqrt{3}/2$. Note that if we do not restrict the domain of f, then f will not be 1–1 and therefore will not have an inverse. For example, if $\sin x = 1$, then x could be any one of the values $\pi/2, 5\pi/2, 9\pi/2, \ldots$. This is sketched

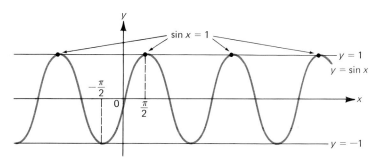

FIGURE 4

in Figure 4. We will discuss the inverse trigonometric functions in great detail in Section 7.2. ■

EXAMPLE 5   Let $f(x) = x^5 + x^3 + x$. Then $f'(x) = 5x^4 + 3x^2 + 1$, which is always positive, so f^{-1} exists. However, there is no way to solve explicitly the equation $y = x^5 + x^3 + x$ for x. This result does not contradict the existence of f^{-1}. It simply illustrates the limitations of our algebraic techniques. ■

It turns out that inverses of continuous functions are continuous and that inverses of differentiable functions with nonzero derivatives are differentiable. The following theorem is proved in Appendix 5.

Theorem 3   *Continuity of Inverse Functions*   Let f be continuous on $[a, b]$ and let $c = f(a)$ and $d = f(b)$. If f has an inverse $g = f^{-1}$, then g is continuous on $[c, d]$ (or $[d, c]$ if $c > d$).

We now prove a theorem that shows a relationship between the derivatives of functions and the derivatives of their inverses.

Theorem 4   *Differentiability of Inverse Functions*   Let $y = f(x)$ be differentiable in some neighborhood† of a point x_0 and suppose that f is 1–1 in a neighborhood of x_0 and that $f'(x_0) \neq 0$. If $x = g(y) = f^{-1}(y)$ is an inverse function of $f(x)$ that is continuous in a neighborhood of $y_0 = f(x_0)$, then $f^{-1}(y) = g(y)$ is differentiable at y_0, and

$$g'(y_0) = \frac{1}{f'(x_0)}, \quad \text{or, equivalently,} \quad \frac{dx}{dy} = \frac{1}{dy/dx}. \tag{4}$$

Proof.   Since

$$\lim_{\Delta x \to 0} \frac{f(x_0 + \Delta x) - f(x_0)}{\Delta x} = f'(x_0) \neq 0,$$

we have

---

†Recall that a neighborhood of x_0 is an open interval containing x_0.

$$\lim_{\Delta x \to 0} \frac{\Delta x}{f(x_0 + \Delta x) - f(x)} = \lim_{\Delta x \to 0} \frac{1}{\dfrac{f(x_0 + \Delta x) - f(x_0)}{\Delta x}}$$

$$\text{Theorem 2.3.5} \to \quad = \frac{\displaystyle\lim_{\Delta x \to 0} 1}{\displaystyle\lim_{\Delta x \to 0} \frac{f(x_0 + \Delta x) - f(x_0)}{\Delta x}} = \frac{1}{f'(x_0)}. \tag{5}$$

Now $x_0 = g(y_0)$ and since $\Delta y = f(x_0 + \Delta x) - f(x_0) = f(x_0 + \Delta x) - y_0$, we have $y_0 + \Delta y = f(x_0 + \Delta x)$, or $x_0 + \Delta x = g(y_0 + \Delta y)$, and $\Delta x = g(y_0 + \Delta y) - g(y_0)$. Thus

$$\overset{\text{Since } g \text{ is continuous at } x_0}{}$$

$$\lim_{\Delta y \to 0} \Delta x = \lim_{\Delta y \to 0} [g(y_0 + \Delta y) - g(y_0)] = 0. \tag{6}$$

Finally, since $y_0 = f(x_0)$, we find that

$$g'(y_0) = \lim_{\Delta y \to 0} \frac{g(y_0 + \Delta y) - g(y_0)}{\Delta y}$$

$$= \lim_{\Delta y \to 0} \frac{(x_0 + \Delta x) - x_0}{f(x_0 + \Delta x) - f(x_0)} = \lim_{\Delta y \to 0} \frac{\Delta x}{f(x_0 + \Delta x) - f(x_0)}$$

$$\overset{\text{From (6)}}{} \qquad \overset{\text{From (5)}}{}$$

$$= \lim_{\Delta x \to 0} \frac{\Delta x}{f(x_0 + \Delta x) - f(x_0)} = \frac{1}{f'(x_0)}.$$

The proof of the theorem is complete. ■

NOTE.   It can be shown (see Problem 25) that if $f'(x_0) \neq 0$ and f' is continuous, then $f(x)$ is 1–1 in a neighborhood of x_0; thus the assumption that $f(x)$ is 1–1 is not really necessary in the statement of the theorem.

REMARK.   It is tempting to prove Theorem 4 by using the chain rule. Since $g(f(x)) = x$, we have

$$\frac{d}{dx} g(f(x)) = \frac{d}{dx} x$$

$$g'(f(x)) f'(x) = 1, \qquad \text{By the chain rule}$$

or

$$g'(y) = g'(f(x)) = \frac{1}{f'(x)}.$$

This derivation is certainly easier, but it is incomplete. Do you see why? In using the chain rule, we assume that $g'(f(x)) = dg/dy$ exists. But this assumption is part of what we need to prove! So we must resort to the longer (but correct) proof.

EXAMPLE 6   Let $y = \sqrt{x}$. Then $dy/dx = 1/(2\sqrt{x})$. The inverse function is $x = g(y) = y^2$ for $y > 0$. Then

$$\frac{dx}{dy} = 2y = 2\sqrt{x} = \frac{1}{1/2\sqrt{x}} = \frac{1}{dy/dx}. \;\blacksquare$$

EXAMPLE 7   Let $y = f(x) = x^2 + 4x + 3$. To find the inverse function, we note that

$$x^2 + 4x + (3 - y) = 0,$$

or

$$x = \overset{\text{Quadratic formula}}{\overset{\downarrow}{\frac{-4 \pm \sqrt{16 - 4(3 - y)}}{2}}} = -2 \pm \frac{\sqrt{4 + 4y}}{2} = -2 \pm \sqrt{1 + y}.$$

The graph of $y = f(x)$ is given in Figure 5. It is apparent that for every value of $y > -1$, there are *two* different values of x. Thus f is not 1–1 and f^{-1} does not exist.

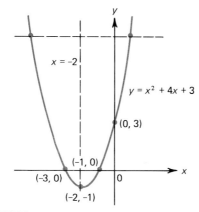

FIGURE 5

However, we may define

$$f_1(x) = x^2 + 4x + 3, \quad x \geq -2, \quad \text{and} \quad f_2(x) = x^2 + 4x + 3, \quad x \leq -2$$

(i.e., we have restricted the domain of f in two different ways to obtain two different functions). Then f_1 and f_2 are both 1–1, and we have the inverse functions

$$g_1(y) = f_1^{-1}(y) = -2 + \sqrt{1 + y} \quad \text{and} \quad g_2(y) = f_2^{-1}(y) = -2 - \sqrt{1 + y}.$$

Note that both $f_1^{-1}(y)$ and $f_2^{-1}(y)$ are not defined for $y < -1$. To verify equation (4), we have, for $x \geq -2$,

$$\frac{dx}{dy} = \frac{dg_1}{dy} = \frac{1}{2\sqrt{1 + y}}.$$

But since $x = -2 + \sqrt{1+y}$, $\sqrt{1+y} = x + 2$. Therefore

$$\frac{1}{2\sqrt{1+y}} = \frac{1}{2x+4} = \frac{1}{dy/dx}.$$

We obtain a similar result for $x \leq -2 \,[x = g_2(y)]$. ∎

EXAMPLE 8 If $y = x^5 + x^3 + x$, then as we saw in Example 5, f^{-1} exists but cannot be explicitly calculated. Nevertheless, we can compute

$$\frac{df^{-1}}{dy} = \frac{dx}{dy} = \frac{1}{dy/dx} = \frac{1}{5x^4 + 3x^2 + 1}.$$

This answer is not exactly what we want since we would like to calculate df^{-1}/dy as a function of y, not of x. Nevertheless, it *is* quite useful since, for example, if we knew a point (x_0, y_0) on the curve $x = f^{-1}(y)$, we could calculate the derivative of f^{-1} at that point. ∎

EXAMPLE 9 Let $y = \sin x$, $-\pi/2 < x < \pi/2$. Then $x = \sin^{-1} y$ exists, and $dx/dy = 1/(dy/dx) = 1/\cos x$. But

$$\sin^2 x + \cos^2 x = 1,$$

so

$$\cos x = \sqrt{1 - \sin^2 x} = \sqrt{1 - y^2}.$$

(Here $\cos x > 0$ because $-\pi/2 < x < \pi/2$.) Thus

$$\frac{d}{dy} \sin^{-1} y = \frac{1}{\sqrt{1-y^2}}.$$ ∎

We close this section by showing the relationship between the graph of a function f and the graph of its inverse function f^{-1}. The graphs of three functions and their inverses are sketched in Figure 6. It appears that the graphs of f and f^{-1} are symmetric about the line $y = x$. Let us see why this observation is true.

Suppose that $y = f(x)$. Then if f^{-1} exists, $x = f^{-1}(y)$. That is, (x, y) is in the graph of f if and only if (y, x) is in the graph of f^{-1}. In Figure 6b, for example, $(\frac{1}{4}, \frac{1}{2})$, $(1, 1)$, $(4, 2)$, and $(9, 3)$ are in the graph of $f(x) = \sqrt{x}$. The points $(\frac{1}{2}, \frac{1}{4})$, $(1, 1)$, $(2, 4)$, and $(3, 9)$ are in the graph of $f^{-1}(x) = x^2$, $x \geq 0$. In Figure 6b if we fold the page along the line $y = x$, we find that the graphs of f and f^{-1} coincide. We say that:

The graphs of f and f^{-1} are **reflections** of one another about the line $y = x$. (7)

This **reflection property,** as it is called, of the graphs of inverse functions enables us immediately to obtain the graph of f^{-1} once the graph of f is known.

(a)

(b)

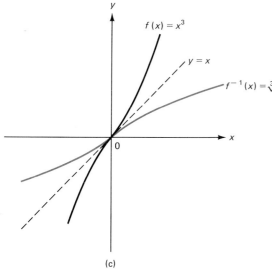

(c)

FIGURE 6

PROBLEMS 6.1

In Problems 1–24, determine intervals over which the given function is 1–1, and determine the inverse function over each interval, if possible. Then use theorem 4 to calculate the derivative of each inverse function.

1. $f(x) = 3x + 5$

2. $f(x) = 17 - 2x$

3. $f(x) = \dfrac{1}{x}$

4. $f(x) = \dfrac{1}{x - 2}$

5. $f(x) = \sqrt{4x + 3}$

6. $f(x) = \sqrt[3]{x + 1}$

7. $f(x) = 1 - x^3$

8. $f(x) = (x + 1)^2$

9. $y = \sin x$ (for $x \in \mathbb{R}$)

10. $y = \cos x$ **11.** $y = \cos\left(\dfrac{x}{4}\right)$ **12.** $y = \dfrac{x}{x+1}$

13. $y = \tan x$ **14.** $y = \sec x$ **15.** $y = x^2 - 5x + 6$
16. $y = x^2 + 2x + 2$ **17.** $y = x^3 + x$
*18. $y = x^4 + 4x^2 + 4$ [*Hint:* Use the quadratic formula.]
19. $y = x^{101}$ **20.** $y = x^9 + x^7 + x^3 + x + 1$

*21. Prove that if $f'(x)$ exists and is continuous in a neighborhood of the point x_0 and if $f'(x_0) \neq 0$, then $f(x)$ is 1–1 in some neighborhood of x_0. [*Hint:* Suppose that $f'(x_0) > 0$. Then in some neighborhood of x_0, $f'(x) > 0$. If x_1 and x_2 are two points in this neighborhood and if $x_1 > x_2$, then $f(x_1) > f(x_2)$. A similar result holds if $f'(x_0) < 0$.]

22. Prove that

$$\frac{d}{dx} x^{1/n} = \frac{1}{n} x^{(1/n)-1}$$

if n is a positive integer. [*Hint:* Let $f(x) = x^n$. Then find f^{-1} and use Theorem 4.]

**23. Suppose that f is strictly increasing and continuous on $[0, a]$. Let $g = f^{-1}$ and suppose that $f(0) = 0$. If $0 \le w \le a$ and $0 \le h \le f(a)$, show that $\int_0^w f + \int_0^h g \ge w \cdot h$.

24. Suppose that f is continuous and strictly increasing with domain $[a, b]$. Prove the following:
 (a) Range of $f = [f(a), f(b)]$ **(b)** f^{-1} exists with domain $[f(a), f(b)]$
 (c) f^{-1} is an increasing function.

6.2 THE EXPONENTIAL AND LOGARITHMIC FUNCTIONS I

Let a be a positive number. Then an **exponential function** is a function of the form

$$f(x) = a^x. \tag{1}$$

where x can be any real number. Before discussing properties of this function, we must explain what we mean by a^x. (We will assume that $a \neq 1$, for if $a = 1$, then $a^x = 1^x = 1$ for every x.) There are several cases to consider.

Case 1. $x = n$, a positive integer. Then

$$a^x = a^n = \underbrace{a \cdot a \cdot a \cdot \cdots \cdot a}_{n \text{ times}}. \tag{2}$$

This expression is, of course, the familiar rule for raising a number to the nth power.

Case 2. $x = -n$, a negative integer. Then

$$a^x = a^{-n} = \frac{1}{a^n}. \tag{3}$$

Case 3. $x = 1/n$, where n is a positive integer. Then

$$a^x = a^{1/n} = \text{the } n\text{th root of } a. \tag{4}$$

That is, $a^{1/n} = b$ is equivalent to $a = b^n$. For example, $64^{1/6} = 2$ since $2^6 = 64$.

In the last two cases we see why it is necessary to require that $a > 0$. If $a = 0$, then $a^{-n} = 1/a^n = 1/0$ is not defined. If $a < 0$, then $a^{1/2}$, $a^{1/4}$, $a^{1/6}$, and so on are not defined [e.g., look at $(-2)^{1/2}$].

Case 4. $x = -1/n$, n is a positive integer. Then

$$a^x = a^{-1/n} = \frac{1}{a^{1/n}}.$$ (5)

Case 5. x is a rational number. Then $x = m/n$, where m and n are integers, and

$$a^x = a^{m/n} = (a^{1/n})^m.$$ (6)

These rules take care of many cases of interest. But how do we interpret expressions like $a^{\sqrt{2}}$ and a^π? To do so, we make use of a fundamental fact of mathematical analysis that any irrational number can be approximated as closely as desired by a rational number.

Case 6. x is an irrational number. Then

$$a^x = \lim_{r \to x} a^r, \qquad r \text{ a rational number.}$$ (7)

EXAMPLE 1   Find an approximate value for 3^π.

Solution. To six decimal places, $\pi = 3.141593$. Using a calculator, we arrive at Table 1. We can continue this calculation to the number of digits displayed on our calculator. To six decimal places, $3^\pi = 31.544281$. ■

TABLE 1

r	3^r
3.0	27.0
3.1	30.1353257
3.14	31.48913565
3.141	31.52374901
3.1415	31.54106996
3.14159	31.54418874
3.141592	31.54425805

(a) power function: x^n (Variable, Constant)
(b) exponential function: a^x (Constant, Variable)

FIGURE 1

Before continuing, it is useful to emphasize the difference between an exponential function and a power function (discussed in Section 3.4). This difference is best illustrated by drawing a picture (see Figure 1). For a power function the exponent is constant, while the number being raised to that power varies. For an exponential function it is the *exponent* that varies, while the number being raised to that exponent is constant. Thus, for example, x^2 is a power function, whereas 2^x is an exponential function. The basic properties of exponential functions that we will need are contained in Theorem 1.

Theorem 1. Let $a > 0$ and let x and y be any real numbers. Then:

(i) $a^x > 0$.

(ii) $a^{-x} = \dfrac{1}{a^x}$.

(iii) $a^{x+y} = a^x \cdot a^y$

(iv) $a^{x-y} = a^x \cdot a^{-y} = \dfrac{a^x}{a^y}$.

(v) $a^0 = a^{x-x} = \dfrac{a^x}{a^x} = 1$.

(vi) $(a^x)^y = a^{xy}$.

(vii) If $a > 1$, then a^x is an increasing function in any interval.

(viii) If $0 < a < 1$, then a^x is a decreasing function in any interval.

(ix) a^x is continuous for every x.

We will not prove this theorem here. The first six properties are familiar if x and y are integers. If x and y are rational numbers, then it is not difficult to show that properties (i)–(vi) hold by using the fact that they hold for integers. Then since (i)–(vi) are true for rational numbers, we can prove them for irrational numbers by taking limits. Properties (vii), (viii), and (ix) will follow easily once we have calculated the derivative of a^x. However, it is easy to see that these facts are true by examining the cases $a > 1$ and $a < 1$ separately and observing what happens as x grows. (Try $a = 2$ and $a = \frac{1}{2}$.) The graphs of $y = 2^x$ and $y = (\frac{1}{2})^x$ are given in Figure 2.

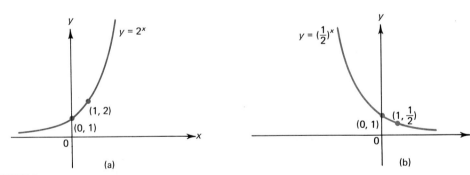

FIGURE 2

If $y = a^x$, we may ask the following question: Which value of x corresponds to a given y?

EXAMPLE 2 Solve for x:
(a) $2^x = 8$ **(b)** $3^x = \frac{1}{9}$ **(c)** $(\frac{1}{2})^x = 8$

Solution. (a) It is clear that $2^x = 8$ if and only if $x = 3$. Similarly, in (b) $3^x = \frac{1}{9}$ if and only if $x = -2$, and in (c), $(\frac{1}{2})^x = 8$ if and only if $x = -3$. $[(\frac{1}{2})^3 = \frac{1}{8}.]$ ∎

From Theorem 1, parts (viii) and (ix), we know that a^x is either an increasing function (if $a > 1$) or a decreasing functions (if $0 < a < 1$). Thus from Theorem 6.1.1 we know that

a^x is 1–1 if $a > 0$ and $a \neq 1$.

We conclude that

a^x has an inverse if $a > 0$ and $a \neq 1$.

The inverse of a^x is called the *logarithm to the base a*.

Definition 1 LOGARITHM TO THE BASE a If $x = a^y$, then the **logarithm to the base a** of x is y, written as

$$y = \log_a x. \tag{8}$$

The relationship between the exponential and logarithmic functions is illustrated in Figure 3. We will discuss the graphs of logarithmic functions later, but the graph of $\log_a x$ can be immediately obtained by turning the graph of $y = a^x$ on its side and flipping it over (i.e., reflecting it through the line $y = x$) so as to interchange the positions of the x- and y-axes (see Figure 3). Figure 3b shows a typical graph of the logarithmic function for $a > 1$. This graph is obtained by reflecting the graph of a^x about the line $y = x$ [see Statement (7), Section 6.1].

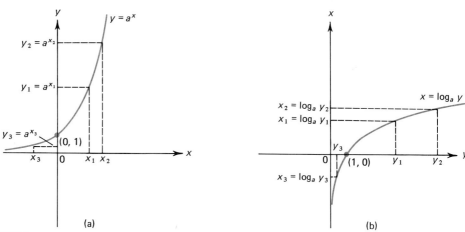

FIGURE 3

The logarithmic function $f(x) = \log_a x$ often causes a lot of confusion when it is first encountered. We emphasize that

$y = \log_a x$ means that $x = a^y$,

and

$y = a^x$ means that $x = \log_a y$.

Think of $y = \log_a x$ as an answer to the question: To what power must a be raised to obtain the number x? This answer immediately implies that

$a^{\log_a x} = x$ for every positive real number x,

or

$\log_a x$ is the power to which we must raise a in order to get x,

and

$\log_a a^x = x$ for every real number x.

Remember that $\log_a x$ is only defined for $x > 0$ since it is impossible to raise a positive number to a power and obtain a number less than or equal to 0.

EXAMPLE 3   In Example 2 we found the following:
(a) $\log_2 8 = 3$ **(b)** $\log_3 \frac{1}{9} = -2$ **(c)** $\log_{1/2} 8 = -3$ ∎

We will draw graphs of several logarithmic functions in Section 6.3 by using properties of their derivatives. Graphs of $y = \log_2 x$ and $y = \log_{1/2} x$ are given in Figure 4.

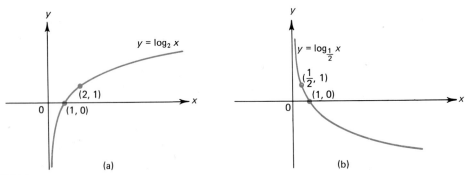

FIGURE 4

Properties of the function $\log_a x$ are summarized in the next theorem.

Theorem 2.   Let $a > 0$ ($a \neq 1$) and let x and y be positive.

(i) $\log_a xy = \log_a x + \log_a y$.

(ii) $\log_a(x/y) = \log_a x - \log_a y$.

(iii) $\log_a 1 = 0$.

(iv) $\log_a(1/x) = -\log_a x$.

(v) $\log_a a = 1$.

(vi) $\log_a x^y = y \log_a x$. Here y can be any real number.

(vii) $\log_a b = 1/(\log_b a)$, $b > 0$, $b \neq 1$.

(viii) If $a > 1$, $\log_a x$ is an increasing function of x for $x > 0$.

(ix) If $0 < a < 1$, $\log_a x$ is a decreasing function of x for $x > 0$.

(x) $f(x) = \log_a x$ is continuous for every $x > 0$.

Proof.

(i) Let $u = \log_a x$ and $v = \log_a y$. Then $a^u = x$ and $a^v = y$, so $xy = a^u \cdot a^v = a^{u+v}$. By definition of the logarithm, we have $\log_a xy = u + v$, and this proves (i).

(ii) With u and v as above, $x/y = a^u/a^v = a^{u-v}$, so $\log_a(x/y) = u - v$, which proves (ii).

(iii) Since $a^0 = 1$, $0 = \log_a 1$.

(iv) Follows from (ii) and (iii),

(v) Since $a^1 = a$, $1 = \log_a a$.

(vi) Let $u = \log_a x$. Then $x = a^u$ and $x^y = (a^u)^y$. Thus $\log_a x^y = uy$, which proves (vi).

(vii) Let $u = \log_a b$ for $b > 0$. Then $a^u = b$. Now taking logarithms with respect to b, we have $\log_b a^u = \log_b b = 1$. But $\log_b a^u = u \log_b a = (\log_a b)(\log_b a) = 1$, which proves (vii).

Properties (viii) and (ix) will follow easily once we have calculated the derivative of $\log_a x$.

(x) You are asked to prove the continuity of $\log_a x$ in Problem 48. ∎

EXAMPLE 4   If $f(x) = a^x$ and $g(x) = \log_a x$, then $f(g(x)) = a^{\log_a x} = x$ and $g(f(x)) = \log_a a^x = x$, so that a^x and $\log_a x$ are indeed inverses. ∎

NOTE.   Example 4 illustrates that $f^{-1}(f(x))$ and $f(f^{-1}(x))$ are *not* the same functions. If $f = a^x$ and $f^{-1} = \log_a x$, then $f(f^{-1}(x)) = a^{\log_a x}$ and $f^{-1}(f(x)) = \log_a a^x$. The first function is defined only when $x > 0$, and the second is defined for *all* x. The two functions have different domains and are therefore different—even though each takes the value x wherever it is defined.

While we have discussed logarithms to any positive base, there are only two bases that are commonly used. Logarithms to base 10 are called **common logarithms** and are very useful for computations since our number system is based on powers of 10. For simplicity we will write $\log_{10} x$ as $\log x$, the base 10 being understood.†

For the remainder of this book we will be concerned with another system of logarithms, called **natural logarithms.** These are logarithms defined by using the base number e.

---

†On scientific calculators there is a key labeled $\boxed{\log}$ or $\boxed{\log x}$. This key refers to $\log_{10} x$.

Definition 2   THE NUMBER e The **number** e is defined by†

$$e = \lim_{n \to \infty} \left(1 + \frac{1}{n} \right)^n. \tag{9}$$

Before discussing this number, we indicate that the limit in (9) exists. We will also prove that it exists in Section 6.4 (Theorem 6.4.7). (A proof is suggested in Problems 49–54.) This limit is different from the ones we encountered in Chapter 2. We can see that the exponent is becoming infinite but that $1 + (1/n)$ is approaching 1. Since 1 to any power is 1 our first guess might be that this limit is equal to 1. Table 2 illustrates what happens as n grows. The last number in the table is e, correct to nine decimal places. For convenience we will use the value 2.72 or 2.718 as an approximation to e. The number e is the second **transcendental** number we have encountered. The first was π. A transcendental number is an irrational number that cannot be written as the root of a polynomial with integer coefficients. (The irrational number $\sqrt{2}$ is not transcendental because it is a root of the polynomial equation $x^2 - 2 = 0$.)

TABLE 2

n	$\dfrac{1}{n}$	$1 + \dfrac{1}{n}$	$\left(1 + \dfrac{1}{n} \right)^n$
1	1	2	2
2	0.5	1.5	2.25
5	0.2	1.2	2.48832
10	0.1	1.1	2.59374246
100	0.01	1.01	2.704813829
1000	0.001	1.001	2.716923932
10,000	0.0001	1.0001	2.718145926
100,000	0.00001	1.00001	2.718268237
1,000,000	0.000001	1.000001	2.718280469
1,000,000,000	10^{-9}	$1 + 10^{-9}$	2.718281828

There are equivalent definitions for e. For example,

$$e = \lim_{n \to -\infty} \left(1 - \frac{1}{n} \right)^{-n}. \tag{10}$$

(See Problem 37.) If $n \to \pm\infty$, then $u = 1/n \to 0$. This leads to another definition for e:

$$e = \lim_{u \to 0} (1 + u)^{1/u}. \tag{11}$$

---

†The number e was discovered by and is named after the Swiss mathematician and physicist Leonhard Euler (1707–1783); see his biographical sketch on page 36.

Two other definitions for e will be given in Sections 6.4 and 14.9.

We will see in the next section why the number e is so useful and important. As we have already stated, logarithms to the base e are called *natural logarithms* and have a special notation:

$$\ln x = \log_e x.\dagger \tag{12}$$

The symbol $\ln x$ should be read, "the natural logarithm of x."

Tables of the exponential functions e^x and e^{-x} are given in Table A.1, and a table of natural logarithms is given in Table A.2 (both at the end of the book).

PROBLEMS 6.2

In Problems 1–32, solve for the unknown variable. (Do not use any table or calculator.)

1. $y = \log_2 4$

2. $y = \log_4 16$

3. $y = \log_{1/3} 27$

4. $y = \frac{1}{3}\log_7(\frac{1}{7})$

5. $y = \pi \log_\pi\left(\frac{1}{\pi^4}\right)$

6. $y = \log_{81} 3$

7. $y = \ln e^5$

8. $y = \ln \frac{1}{e^{3.7}}$

9. $y = \log_{1/4} 2$

10. $y = \log_a a \cdot \log_b b^2 \cdot \log_c c^3$

11. $y = \log_6 36 \log_{25} \frac{1}{5}$

12. $64 = x \log_{1/4} 64$

13. $2^{x^2} = 64$

14. $y = 1.3^{\log_{1.3} 48}$

15. $y = e^{\ln \sqrt{2}}$

16. $y = e^{\ln 14.6}$

17. $y = e^{\ln e^\pi}$

18. $\log_2 x^4 = 4$

19. $\log_x 64 = 3$

20. $\log_x 125 = -3$

21. $\log_x 32 = -5$

22. $y = \log 0.01$

23. $\log x = 10^{-23}$

24. $\log x = 3 \log 2 + 2 \log 3$

25. $\log x = 4 \log \frac{1}{2} - 3 \log \frac{1}{3}$

26. $\log x = a \log b + b \log a$

27. $\log x = \log 1 + \log 2 + \log 3 + \log 4$

28. $\log x = \log 2 - \log 3 + \log 5 - \log 7$

29. $\log x^3 = 2 \log 5 - 3 \log 2$

30. $\log \sqrt{x} = 4 \log 2 - 3 \log \frac{1}{2}$

31. $\ln x^2 - \ln x = \ln 18 - \ln 6$

32. $\log_2 x^{3/2} - \log_2 \sqrt{x} = \log_2 9 - \log_2 3$

In Problems 33–36, simplify the given expressions by using Theorem 2.

33. $\log x^3\sqrt{x}$

34. $\log[\sqrt[3]{x^2 + 3} \, (x^5 - 9)^{1/8}]$

35. $\ln \dfrac{x^5}{(1 + x)^{18}}$

36. $\ln \sqrt[5]{\dfrac{(x + 1)x^2}{(x - 12)}}.$

37. Using a calculator, calculate $[1 - (1/n)]^{-n}$ for $n = 2, 5, 10, 1000, 10,000,$ and $1,000,000$ to illustrate that $\lim_{n \to -\infty}[1 - (1/n)]^{-n} = e$.

*38. Show that if $a > 1$, then $\lim_{x \to 0^+} \log_a x = -\infty$ and $\lim_{x \to \infty} \log_a x = \infty$.

*39. If $0 < a < 1$, show that $\lim_{x \to 0^+} \log_a x = \infty$ and $\lim_{x \to \infty} \log_a x = -\infty$.

40. The acidity of a substance is measured by the concentration of the hydrogen ion $[H^+]$ in the substance. This concentration is usually measured in terms of moles per liter (mol/L). A standard way to describe this acidity is to define the pH of a substance by

$$pH = -\log[H^+].$$

---

†On scientific calculators there is a key labeled ☐ ln ☐ or ☐ lnx ☐. This key refers to $\ln x = \log_e x$.

Distilled water has an approximate H^+ concentration of 10^{-7} mol/L so its pH is $-\log 10^{-7}$ = 7. A substance with a pH under 7 is termed an **acid,** while one with a pH above 7 is called a **base.** Determine the pH of the following substances with the indicated hydrogen ion concentrations.

(a) $[H^+] = 4.2 \times 10^{-6}$ (b) $[H^+] = 8 \times 10^{-6}$ (c) $[H^+] = 0.6 \times 10^{-7}$

* 41. Using a calculator determine which is greater: 999^{1000} or 1000^{999}.

42. A general psychophysical relation was established in 1834 by the physiologist Weber† and given a more precise phrasing later by Fechner.‡ By the **Weber-Fechner law,** $S = c \log(R + d)$, where S is the intensity of a sensation, R is the strength of the stimulus producing it, and c and d are constants. The Greek astronomer Ptolemy catalogued stars according to their visual brightness in six categories, or **magnitudes.** A star of the first magnitude was about $2\frac{1}{2}$ times as bright as a star of the second magnitude, which in turn was about $2\frac{1}{2}$ times as bright as a star of the third magnitude, and so on. Let b_n and b_m denote the apparent brightness of two stars having magnitudes n and m, respectively. Then modern astronomers have established the Weber-Fechner law relating the relative brightness to the difference in magnitudes as

$$(m - n) = 2.5 \log\left(\frac{b_n}{b_m}\right).$$

(a) Using this formula, calculate the ratio of brightness for two stars of the second and fifth magnitudes, respectively.

(b) If star A is five times as bright to the naked eye as star B, what is the difference in their magnitudes?

(c) How much brighter is Sirius (magnitude 1.4) than a star of magnitude 21.5?

(d) The Nova Aquilae in a 2–3-day period in June 1918 increased in brightness about 45,000 times. How many magnitudes did it rise?

*(e) The bright star Castor appears to the naked eye as a single star but can be seen with the aid of a telescope to be really two stars whose magnitudes have been calculated to be 1.97 and 2.95. What is the magnitude of the two combined? [*Hint:* Brightnesses, but not magnitudes, can be added.]

43. The subjective impression of loudness can be described by a Weber-Fechner law. Let I denote the intensity of a sound. The least intense sound that can be heard is $I_0 = 10^{-12}$ watt/m² at a frequency of 1000 cycles/sec (this value is called the **threshold of audibility**). If L denotes the loudness of a sound, measured in decibels,§ then $L = 10 \log(I/I_0)$.

(a) If one sound has twice the intensity of another, what is the ratio of the perceived loudness of the two sounds?

(b) If one sound appears to be twice as loud as another, what is the ratio of their intensities?

(c) Ordinary conversation sounds six times as loud as a low whisper. What is the actual ratio of intensity of their sounds?

44. Natural logarithms can be calculated on a hand calculator even if the calculator does not have an $\boxed{\ln}$ key. If $\frac{1}{2} \leq x \leq \frac{3}{2}$ and if $A \equiv (x - 1)/(x + 1)$, then a good approximation to $\ln x$ is given by

$$\ln x \approx \left[\left(\frac{3A^2}{5} + 1\right) \cdot \frac{A^2}{3} + 1\right]2A, \quad \frac{1}{2} \leq x \leq \frac{3}{2}.$$

(a) Use this formula to calculate $\ln 0.8$ and $\ln 1.2$.

(b) Using facts about logarithms, use the formula to calculate (approximately) $\ln 2 = \ln(\frac{3}{2} \cdot \frac{4}{3})$.

---

†Ernest Weber (1796–1878) was a German physiologist.

‡Gustav Fechner (1801–1887) was a German physicist.

§1 decibel (dB) = $\frac{1}{10}$ bel, named after Alexander Graham Bell (1847–1922), inventor of the telephone.

(c) Using the result of (b), calculate ln 3 and ln 8.

45. The exponential e^x can be estimated for x in $\left[-\frac{1}{2}, \frac{1}{2}\right]$ by the formula.

$$e^x \approx \left(\left\{\left[\left(\frac{x}{5} + 1\right)\frac{x}{4} + 1\right]\frac{x}{3} + 1\right\}\frac{x}{2} + 1\right)x + 1.$$

(a) Calculate an approximate value for $e^{0.13}$.
(b) Calculate an approximate value for $e^{-0.37}$.
(c) Calculate an approximate value for $e^{4.13}$.
(d) Calculate an approximate value for $e^{-2.63}$.   [*Hint:* Use part (b).]
(e) Calculate approximate values for $e^{4.82}$ and $e^{-1.44}$.

46. Show, using logarithms, that $a^b = b^a$, where $a = [1 + (1/n)]^n$ and $b = [1 + (1/n)]^{n+1}$. What does this result prove in the case $n = 1$?

47. The quantity $n! = n(n - 1)(n - 2) \cdots 3 \cdot 2 \cdot 1$ grows very rapidly as n increases. According to **Stirling's formula,** when n is large,

$$n! \approx \sqrt{2\pi n}\left(\frac{n}{e}\right)^n.$$

Use Stirling's formula to estimate 100! and 200!.   [*Hint:* Use common logarithms.]

*48. Prove that if $a > 0$, then the function $y = \log_a x$ is continuous at every $x > 0$.

*49. Suppose that $b > 1$. Show that for $x > -1$, $f(x) > 0$, where $f(x) = (1 + x)^b - (1 + bx)$. [*Hint:* Use the mean value theorem to show that there is a number c in $[0, x]$ such that $f(x) = xb[(1 + c)^{b-1} - 1]$.]

50. Use the result of Problem 49 to show that if $b > 1$ and $x > -1$, then $(1 + x)^b > 1 + bx$.

51. Let $S_n = [1 + (1/n)]^n$ and $T_n = [1 + (1/n)]^{n+1}$. Show that $T_n > S_n$ and $T_n/S_n \to 1$ as $n \to \infty$.

*52. Use the result of Problem 50 to show that $S_{n+1}^{1/n} > S_n^{1/n}$, and conclude that S_n increases as n increases.

*53. Use the result of Problem 50 to show that $(1/T_n)^{1/n} > (1/T_{n-1})^{1/n}$, and conclude that T_n decreases as $n \to \infty$.

54. Explain why the results of Problems 51, 52, and 53 prove that $\lim_{n\to\infty}[1 + (1/n)]^n$ exists and is finite.

*55. Watch carefully. Suppose that $0 < A < B$. Because the logarithm is an increasing function, we have

(a) $\log A < \log B$;   then   **(b)** $10A \cdot \log A < 10B \cdot \log B$,

(c) $\log A^{10A} < \log B^{10B}$,   **(d)** $A^{10A} < B^{10B}$.

On the other hand, we run into trouble with particular choices of A and B. For instance, choose $A = \frac{1}{10}$ and $B = \frac{1}{2}$. Clearly, $0 < A < B$, but $A^{10A} = \left(\frac{1}{10}\right)^1 = \frac{1}{10}$ is greater than $B^{10B} = \left(\frac{1}{2}\right)^5 = \frac{1}{32}$. Where was the first false step made?

**56. Prove that ln x cannot be expressed in the form $p(x)/q(x)$, where p and q are polynomial functions.

6.3 THE DERIVATIVES AND INTEGRALS OF $\log_a x$ and a^x

In this section we calculate the derivatives of $\log_a x$ and a^x.

Theorem 1

$$\frac{d}{dx}\log_a x = \frac{1}{x}\log_a e.$$

(1)

Proof. If $y = \log_a x$, then

$$\frac{dy}{dx} = \lim_{\Delta x \to 0} \frac{\log_a(x + \Delta x) - \log_a x}{\Delta x} \overset{\text{Theorem 6.2.2 (ii).}}{=} \lim_{\Delta x \to 0}\left[\frac{1}{\Delta x} \log_a\left(\frac{x + \Delta x}{x}\right) \right]$$

Multiply and divide by x ⟶   Theorem 6.2.2 (vi),

$$= \lim_{\Delta x \to 0}\left[\frac{x}{x\,\Delta x} \log_a\left(1 + \frac{\Delta x}{x}\right) \right] = \frac{1}{x} \lim_{\Delta x \to 0} \log_a\left(1 + \frac{\Delta x}{x}\right)^{x/\Delta x}.$$

We give a new name to the variable $x/\Delta x$. If $u = \Delta x/x$, then $x/\Delta x = 1/u$, and for each fixed x, $u \to 0$ as $\Delta x \to 0$, so that

$$\lim_{\Delta x \to 0}\left[\frac{1}{x} \log_a\left(1 + \frac{\Delta x}{x}\right)^{x/\Delta x} \right] = \lim_{u \to 0}\left[\frac{1}{x} \log_a(1 + u)^{1/u} \right]$$

$$= \frac{1}{x} \lim_{u \to 0}[\log_a(1 + u)^{1/u}].$$

It follows from the continuity of $\log_a x$, Theorem 2.7.2, and equation (6.2.11), that

$$\lim_{u \to 0} \log_a(1 + u)^{1/u} = \log_a \lim_{u \to 0}(1 + u)^{1/u} = \log_a e.$$

This step completes the proof. ■

Note that since

Theorem 6.2.2 (vii) ⟶

$$\log_a e = \frac{1}{\log_e a} = \frac{1}{\ln a},$$

(1) can be written as

$$\frac{d}{dx} \log_a x = \frac{1}{x \ln a}. \tag{2}$$

If $a = e$, then (1) becomes

$$\frac{d}{dx} \log_e x = \frac{d}{dx} \ln x = \frac{1}{x \ln e} = \frac{1}{x}. \tag{3}$$

The differentiation formula (2) immediately proves parts (viii) and (ix) of Theorem 6.2.2. For if $a > 1$, then $\ln a > 0$, so that $1/(x \ln a) > 0$ (remember that x must always be > 0). Thus $\log_a x$ is an increasing function. On the other hand, if $0 < a <$

1, then $\ln a < 0$, so that $1/(x \ln a) < 0$, and $\log_a x$ is a decreasing function. Also,

$$\frac{d^2}{dx^2} \log_a x = -\frac{1}{x^2 \ln a},$$

so that if $a > 1$, $\log_a x$ is concave down, while if $a < 1$, $\log_a x$ is concave up. We note that $\lim_{x \to \infty} 1/(x \ln a) = 0$, so the graph of $\log_a x$ appears more and more horizontal as $x \to \infty$. Finally, if $a > 1$, $\lim_{x \to 0^+} \log_a x = -\infty$, and if $a < 1$, $\lim_{x \to 0^+} \log_a x = +\infty$, so $\log_a x$ has a vertical asymptote at 0. We draw some typical graphs in Figure 1.

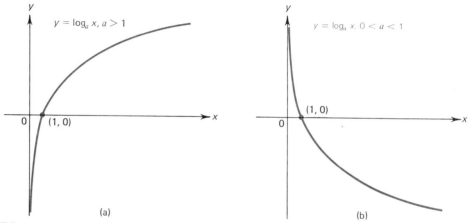

FIGURE 1

Recall that in Section 5.2 (Theorem 5.2.2) we showed that $\int x^r \, dr = [x^{r+1}/(r + 1)] + C$ if $r \neq -1$. We can now handle the case in which $r = -1$. Suppose that $x > 0$; then since $d(\ln x)/dx = 1/x$, we have

$$\int \frac{1}{x} \, dx = \ln x + C \qquad \text{if} \qquad x > 0. \tag{4}$$

Now suppose that $x < 0$. Then $-x > 0$, $\ln(-x)$ is defined, and by the chain rule,

$$\frac{d}{dx} \ln(-x) = \frac{1}{-x} \frac{d}{dx} (-x) = -\frac{1}{x} (-1) = \frac{1}{x},$$

so that

$$\int \frac{1}{x} \, dx = \ln(-x) + C \qquad \text{if} \qquad x < 0. \tag{5}$$

Since

$$|x| = \begin{cases} x, & x > 0 \\ -x, & x < 0, \end{cases}$$

we can put (4) and (5) together to obtain an important integration formula:

$$\int \frac{1}{x} \, dx = \ln|x| + C. \tag{6}$$

We now turn to the differentiation of the exponential function.

Theorem 2.   If $a > 0$

$$\frac{d}{dx} a^x = a^x \ln a. \tag{7}$$

Proof.   We use the fact that a^x is the inverse of $\log_a x$. Let $y = a^x$. Then $y > 0$, $x = \log_a y$, and

$$\frac{dx}{dy} = \frac{1}{y \ln a} \neq 0.$$

Thus by Theorem 6.1.4, a^x is differentiable, and

$$\frac{d}{dx} a^x = \frac{dy}{dx} = \frac{1}{dx/dy} = \frac{1}{1/y \ln a} = y \ln a = a^x \ln a. \quad \blacksquare$$

(Since $y = a^x$)

Using (7), we immediately obtain the integration formula

$$\int a^x \, dx = \frac{1}{\ln a} a^x + C. \tag{8}$$

When $a = e$, (7) becomes

$$\frac{d}{dx} e^x = e^x. \tag{9}$$

This is a remarkable fact. We have shown that *the function e^x is its own derivative!* This is the major reason that the number e is so important. In fact, the constant multiples of e^x are the only continuous functions that have this property. We will exploit the property a great deal in Section 6.6.

From (9) we immediately obtain

$$\int e^x \, dx = e^x + C. \tag{10}$$

We can now draw the graphs of $y = a^x$. If $a > 1$, then

$$\frac{d}{dx} a^x = a^x \ln a > 0,$$

and if $0 < a < 1$,

$$\frac{d}{dx} a^x < 0.$$

Also,

$$\frac{d^2}{dx^2} a^x = \frac{d}{dx} a^x \ln a = (a^x \ln a) \ln a = a^x \ln^2 a > 0.$$

Therefore a^x is always concave up. If $0 < a < 1$, then $\lim_{x \to \infty} a^x = 0$ and $\lim_{x \to -\infty} a^x = \infty$. If $a > 1$, then $\lim_{x \to \infty} a^x = \infty$ and $\lim_{x \to -\infty} a^x = 0$. Since a^x is always positive, we obtain the typical graphs drawn in Figure 2.

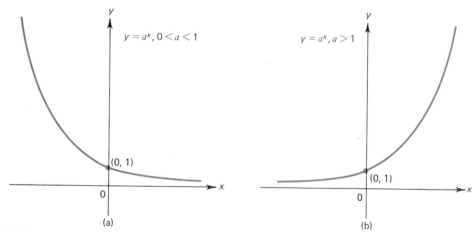

FIGURE 2

We summarize in Table 1 the differentiation and integration formulas obtained in this section.

TABLE 1   SUMMARY OF DIFFERENTIATION AND INTEGRATION FORMULAS

I. $\dfrac{d}{dx} \log_a x = \dfrac{1}{x \ln a} = \dfrac{\log_a e}{x}$

II. $\dfrac{d}{dx} \ln x = \dfrac{1}{x}$

III. $\displaystyle\int \log_a x \, dx = x \log_a x - \dfrac{x}{\ln a} + C.$

IV. $\displaystyle\int \ln x \, dx = x \ln x - x + C$

V. $\displaystyle\int \dfrac{1}{x} dx = \ln|x| + C$

VI. $\dfrac{d}{dx} a^x = a^x \ln a$

VII. $\dfrac{d}{dx} e^x = e^x$

VIII. $\displaystyle\int a^x \, dx = \dfrac{a^x}{\ln a} + C$

IX. $\displaystyle\int e^x \, dx = e^x + C$

NOTE: Formulas III and IV were not proven in this section but were inserted for completeness. They can be easily verified by differentiation. When we discuss integration by parts (in Chapter 8), we will show how these integrals are calculated. You are asked to verify them in Problems 40 and 41.

AN APPLICATION TO THERMODYNAMICS (OPTIONAL)

There is an interesting application of the natural logarithm function in thermodynamics. **Thermodynamics** is the study of the relationship between heat and other forms of energy. **Heat** is the energy that flows from one body to another because of a temperature difference between them. In the metric system the unit of heat energy is the *calorie* (cal). One calorie is the quantity of heat needed to raise the temperature of 1 gram of water from 14.5° to 15.5° C. In terms of mechanical energy, it has been calculated that

$$1 \text{ cal} = 4.186 \text{ joules.}$$

Work, like heat, involves a transfer of energy. We can distinguish between work and heat energy by defining work as energy that passes from one system to another such that a difference of temperature is not involved. To illustrate this difference, rub your hands together vigorously. The heat generated is simply a conversion from mechanical energy to heat energy.

Consider a gas in a cylindrical container with a movable piston (see Figure 3).

FIGURE 3

Heat can flow into or out of the cylinder through the wall, and work can be done on the gas by having the piston compress it, or work can be done by the gas by having the gas expand against the piston. Let P and V denote the pressure and volume of the gas. Suppose the gas expands slightly so that the piston moves a distance Δs. Since pressure is given in terms of force per unit area, the force against the piston is $F = P \times A$, where A is the area of the face of the piston (A is also the cross-sectional area of the interior of the cylinder). Then the work done is $F \times \Delta s$ (see Section 5.9), or

$$\Delta W = PA\,\Delta s = P\Delta V. \tag{11}$$

If the gas expands from an initial volume V_0 to a final value V_f, then (11) yields

$$W = \int_{V_0}^{V_f} P\,dV. \tag{12}$$

The **ideal gas law** relates pressure, volume, and temperature of an ideal gas (i.e., a gas that follows the gas law exactly for all conditions of pressure and temperature; no gas actually does so, but some come close). It is given by

$$PV = nRT, \tag{13}$$

where T is the temperature, n is the number of moles of the gas, and R is a constant.†
Returning to our gas piston system, suppose that the gas expands with no change in the temperature (this phenomenon is called an **isothermal expansion**). The **first law of thermodynamics** states that the change in energy ΔE of any system is given by

$$\Delta E = Q + W, \tag{14}$$

where Q is the heat absorbed by the system and W is the work done on the system (by convention, W is written as negative). In our system no energy is being added (since the temperature does not change), so $\Delta E = 0$ and therefore

$$Q = -W.$$

To calculate W, we use equations (12) and (13) and the fact that the work is negative.

$$W = -\int_{V_0}^{V_f} P \, dV = -\int_{V_0}^{V_f} \frac{nRT}{V} \, dV = -nRT \int_{V_0}^{V_f} \frac{dV}{V} \quad \text{(since T is constant)}$$

$$= -nRT \ln V \Big|_{V_0}^{V_f} = -nRT(\ln V_f - \ln V_0) = -nRT \ln \frac{V_f}{V_0}. \tag{15}$$

EXAMPLE 1   How much heat is absorbed by 1 mol of an ideal gas that doubles in volume at a constant temperature of $0°$ C ($= 273°$ K)?

Solution. Here $n = 1$ and $T = 273°$ K. If V_0 denotes the initial volume, then $V_f = 2V_0$, and

$$W = -273R \ln \frac{2 V_0}{V_0}$$

$$= -273(1.987)(\ln 2) \approx -376 \text{ cal}$$

$$= (-273)(8.314)(\ln 2) \approx -1573 \text{ J}$$

Then the heat energy Q absorbed by the system is $Q = -W = 376$ cal $= 1573$ J. ■

There are many other applications of the logarithmic function in thermodynamics, as you will find by consulting any standard chemistry textbook.

PROBLEMS 6.3

In Problems 1–8, compute the derivative of the given function.

1. $\log_5 x$

2. 5^x

3. $\log_{10} x$

4. 10^x

5. $\log_{1/3} x$

6. $\left(\dfrac{1}{3}\right)^x$

---

†$R = 8.314$ J/(mol)(K) $= 1.987$ cal/(mol)(K).

7. $\log_\pi x$ **8.** π^x

In Problems 9–16 calculate the given definite integral.

9. $\displaystyle\int_1^4 2^x \, dx$

10. $\displaystyle\int_{-1}^3 \left(\frac{1}{2}\right)^x dx$

11. $\displaystyle\int_1^{e^3} \frac{1}{x} \, dx$

12. $\displaystyle\int_1^e \ln x \, dx$

13. $\displaystyle\int_{\log_3 (5 \ln 3)}^{\log_3 (7 \ln 3)} 3^x \, dx$

14. $\displaystyle\int_1^{\ln 4} e^x \, dx$

15. $\displaystyle\int_{\ln 8}^{\ln 23} e^x \, dx$

16. $\displaystyle\int_{\log_{1/4}[-8 \ln(1/2)]}^{\log_{1/4}[-3 \ln(1/2)]} \left(\frac{1}{4}\right)^x dx$

In Problems 17–32, draw a graph of the given function.

17. $y = \ln|x|$

18. $y = \ln(x + 1)$

19. $y = 3 \ln(x - 4)$

20. $y = e^{-x}$

21. $y = \ln(-x)$

22. $y = -2e^{(2 + x)}$

23. $y = e^{3x}$

24. $y = \ln x^4$

25. $y = 3^{2x}$

26. $y = 3 \log_3(x + 1)$

27. $y = \left(\frac{1}{2}\right)^{-x}$

28. $y = \log_{1/2} \dfrac{x}{2}$

29. $y = 10^{x + 3}$

30. $y = \log x^{17/4}$

31. $y = 1 - \ln \sqrt{x}$

32. $y = 7(\log_7 x) - 7$

33. For $x > 0$, express $\int_{-1}^{-x} (1/t) \, dt$ in terms of the natural logarithm.   [*Hint:* Draw a picture.]

34. Do the same as in Problem 33 for $\int_2^x [1/(t - 1)] \, dt$ for $x > 1$.   [Same hint.]

*35. Calculate

$$\lim_{n \to \infty} \left(\frac{1}{n + 1} + \frac{1}{n + 2} + \frac{1}{n + 3} + \cdots + \frac{1}{2n}\right).$$

[*Hint:* Partition the interval [1, 2] into n subintervals. Then this limit represents the integral of some function over the interval [1, 2].]

36. Calculate the heat absorbed by 2 mol of an ideal gas expanding isothermally at a temperature of 373° K when the volume is tripled.

37. Suppose that 500 cal are added to the system depicted in Figure 3 (containing 1 mol) by an external heating device and that the volume of the gas doubles without any change in its temperature of 300° K. How much work is done on the piston?

*38. If 500 cal are externally added to the system of Figure 3 (containing 1 mol of gas) and if the temperature of the gas remains at 273° K, what must be the ratio of V_f to V_0 in order that no heat energy be absorbed by the system?

39. (a) Show that if α is a real number and $x > 0$, then $x^\alpha = e^{\alpha \ln x}$.
   (b) Show, using formulas (3) and (9), that $(d/dx)e^{\alpha \ln x} = (\alpha/x)e^{\alpha \ln x}$.   [*Hint:* Set $u = \alpha \ln x$ and apply the chain rule.]
   (c) Using the results of (a) and (b), show that for any real number α, $(d/dx)x^\alpha = \alpha x^{\alpha - 1}$.

40. Show that $(d/dx)[x \log_a x - x/(\ln a) + C] = \log_a x$ for any constant C, thereby verifying formula III in Table 1.

41. Show that $(d/dx)(x \ln x - x + C) = \ln x$ for any constant C, thereby verifying formula IV in Table 1.

*42. Show that $1 - 1/x \le \ln x \le x - 1$ for all $x > 0$, with equality if and only if $x = 1$.

*43. Suppose that $a > 0$. Define the function A by $A(x) = a^x$. Prove directly from the definition of the derivative that $A'(x) = A'(0) \cdot A(x)$.

*44. Suppose that $0 < q < p$, where p and q are integers. Prove that

$$\lim_{n \to \infty} \sum_{k=1+nq}^{np} \frac{1}{k} = \ln\left(\frac{p}{q}\right).$$

**45. Prove that $2^\alpha < 1 + \alpha$ for all α in $(0, 1)$.   [*Hint:* Let $f(x) = 2^x$ and show that f is concave up.]
 *46. Prove that $x^p \cdot y^q \le px + qy$ if $x, y, p, q > 0$ and $p + q = 1$.
**47. For what values of A is the graph of A^x tangent to the graph of $\log_A x$?

6.4 THE EXPONENTIAL AND LOGARITHMIC FUNCTIONS II

In this section we give alternative definitions of the functions $\ln x$, e^x, a^x, and $\log_a x$. These definitions rely heavily on the second fundamental theorem of calculus which is repeated here.

Theorem 1   **(Theorem 5.6.2).**   If f is continuous on $[a, b]$, then the function $G(x) = \int_a^x f(t)\, dt$ is continuous on $[a, b]$, differentiable on (a, b), and, for every x in (a, b), $G'(x) = f(x)$.

Now let $f(t) = 1/t$. This function is continuous on any interval $[a, b]$ with $a > 0$ and $b > a$.

Definition 1.   THE NATURAL LOGARITHMIC FUNCTION   The **natural logarithm,** denoted $\ln x$, is the function with domain $(0, \infty)$ defined by

$$\ln x = \int_1^x \frac{1}{t}\, dt \qquad \text{for all } x > 0. \tag{1}$$

By Theorem 1, $\ln x$ is continuous and differentiable for $x > 0$, and

$$\frac{d}{dx} \ln x = \frac{1}{x}. \tag{2}$$

NOTE.   We emphasize that dom $\ln x = \{x : x > 0\}$.

Suppose that $x > 1$. Then since $1/t > 0$ for $t > 0$, the function $\ln x$ gives the area under the curve $f(t) = 1/t$ between $t = 1$ and $t = x$. This is illustrated in Figure 1.

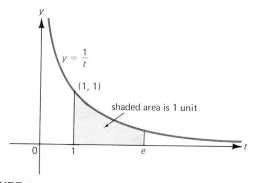

FIGURE 1

Suppose that $x < 0$. Then $-x > 0$, and from the chain rule,

$$\frac{d}{dx} \ln(-x) = \frac{1}{-x} \frac{d}{dx} (-x) = \frac{-1}{-x} = \frac{1}{x}.$$

Since

$$|x| = \begin{cases} x, & x > 0 \\ -x, & x < 0, \end{cases}$$

we obtain the important formulas

$$\frac{d}{dx} \ln|x| = \frac{1}{x} \quad \text{for} \quad x \neq 0 \tag{3}$$

and

$$\int \frac{1}{x} dx = \ln|x| + C. \tag{4}$$

This definition of a logarithmic function is very different from definitions we have seen before. But as we will see, it leads to familiar results. The next theorem gives us the properties we expect of a logarithmic function.

Theorem 2. Let x and y be positive numbers and let c be a real number. Then:

    (i) $\ln 1 = 0$
    (ii) $\ln xy = \ln x + \ln y$
    (iii) $\ln x/y = \ln x - \ln y$
    (iv) $\ln 1/x = -\ln x$
    (v) $\ln x^c = c \ln x$

Proof.
    (i)   ↙ Equation (5.5.12)

$$\ln 1 = \int_1^1 \left(\frac{1}{t}\right) dt = 0$$

    (ii) We have   ↙ Theorem 5.5.3

$$\ln xy = \int_1^{xy} \frac{1}{t} dt = \int_1^x \frac{1}{t} dt + \int_x^{xy} \frac{1}{t} dt. \tag{5}$$

We will evaluate $\int_x^{xy} (1/t)\, dt$. We consider the indefinite integral $\int (1/t)\, dt$ and make the substitution $t = xu$ (see Section 5.7). Then $1/t = 1/xu$, $dt = x\, du$ and $u = t/x$, so that

$$\int \frac{1}{t}\,dt = \int \frac{1}{xu}\,x\,du = \int \frac{du}{u} = \ln|u| + C = \ln\left|\frac{t}{x}\right| + C.$$

Thus

$$\int_x^{xy} \frac{1}{t}\,dt = \ln\left|\frac{t}{x}\right|\Big|_{t=x}^{t=xy} = \ln\left(\frac{xy}{x}\right) - \ln\left(\frac{x}{x}\right) = \ln y - \ln 1 = \ln y. \tag{6}$$

$$\ln 1 = 0 \text{ by (i)}$$

Thus, from (5) and (6),

$$\ln xy = \int_1^x \frac{1}{t}\,dt + \int_x^{xy} \frac{1}{t}\,dt = \ln x + \ln y.$$

(iii)

By (ii)

$$\ln x = \ln\left(\frac{x}{y}\cdot y\right) = \ln\frac{x}{y} + \ln y,$$

so

$$\ln \frac{x}{y} = \ln x - \ln y.$$

(iv)

By (iii)   By (i)

$$\ln \frac{1}{x} = \ln 1 - \ln x = 0 - \ln x = -\ln x$$

(v) Let $u = x^c$. Then $(d/dx)\,u(x) = cx^{c-1}$, and

Chain rule

$$\frac{d}{dx}\ln x^c = \frac{d}{dx}\ln[u(x)] = \frac{1}{u(x)}\frac{d}{dx}u(x)$$

$$= \frac{1}{x^c}\cdot cx^{c-1} = \frac{cx^{c-1}}{x^c} = \frac{c}{x}.$$

Also,

$$\frac{d}{dx}c\ln x = c\frac{d}{dx}\ln x = \frac{c}{x}.$$

We see that the functions $c\ln x$ and $\ln x^c$ have the same derivative. Thus by Theorem 5.2.1, $c\ln x$ and $\ln x^c$ differ by a constant. That is, there is a number k such that

$$\ln x^c = c\ln x + k \quad \text{for every} \quad x > 0. \tag{7}$$

Since (7) holds for every $x > 0$, it holds for $x = 1$. Thus

$$\ln 1^c = c \ln 1 + k,$$

or

By (i)

$$0 = \ln 1 = c \ln 1 + k = c \cdot 0 + k = k. \tag{8}$$

Inserting $k = 0$ into (7), we have

$$\ln x^c = c \ln x.$$

This completes the proof. ■

We now analyze the function $y = \ln x$ in order to obtain its graph. Since $(d/dx)\ln x = 1/x > 0$ if $x > 0$, $\ln x$ is an increasing function. Moreover, $(d^2/dx^2)\ln x = (d/dx)(1/x) = -1/x^2 < 0$, so that $\ln x$ is concave down. Suppose now that $x > 1$. Then from Figure 1 we see that $\ln x > 0$. If $x < 1$, then $1/x > 1$, so that $\ln(1/x) > 0$. But

$$\ln \frac{1}{x} = -\ln x, \qquad \text{so that} \qquad \ln x < 0.$$

Thus

$$\ln x > 0 \quad \text{if} \quad x > 1$$

and

$$\ln x < 0 \quad \text{if} \quad x < 1.$$

Let us estimate $\ln 2$ (very roughly). We have

$$\ln 2 = \int_1^2 \frac{1}{t}\, dt.$$

In the interval $[1, 2]$ the smallest value of $1/t$ is $1/2$ (at $t = 2$), and the largest value is 1 (at $t = 1$). Thus in $[1, 2]$

$$\frac{1}{2} \le \frac{1}{t} \le 1.$$

Then using the comparison theorem (Theorem 5.5.7), we have

$$\int_1^2 \frac{1}{2}\, dt \le \int_1^2 \frac{1}{t}\, dt \le \int_1^2 1\, dt,$$

or

$$\frac{1}{2}t\bigg|_1^2 \le \ln 2 \le t\bigg|_1^2,$$

or

$$\frac{1}{2} \le \ln 2 \le 1. \tag{9}$$

We could evaluate ln 2 more closely (actually $\ln 2 \approx 0.6931\dagger$) but the estimate (9) is all we need at this point.

Using the estimate (9), we can compute $\lim_{x\to\infty} \ln x$ and $\lim_{x\to 0^+} \ln x$. Let $z = 2^n$, where n is a positive integer. Then

$$\ln z = \ln 2^n = n \ln 2 \ge \frac{n}{2}. \tag{10}$$

The last inequality follows from the first inequality in (9). Now let N be given. Let $x = 2^{2(N+1)}$. Then from (10),

$$\ln x \ge \frac{2(N + 1)}{2} = N + 1.$$

Thus for every number N, no matter how large, we can find a number x such that $\ln x > N$. But $\ln x$ is an increasing function, so that if $y > x$, then $\ln y > \ln x > N$. Thus once $\ln x$ gets bigger than N, it stays bigger than N. Since N was arbitrary, we have

$$\lim_{x\to\infty} \ln x = \infty. \tag{11}$$

Finally, if $x \to 0^+$, then $1/x \to \infty$, so $\ln(1/x) \to \infty$. But $\ln(1/x) = -\ln x$, from which we conclude that

$$\lim_{x\to 0^+} \ln x = -\infty. \tag{12}$$

Putting this information all together, we obtain the graph in Figure 2.

The function $\ln x$ is continuous. Since $\ln 1 = 0$ and $\ln 4 = 2 \ln 2 \ge 2 \cdot \frac{1}{2} = 1$, there is, by the intermediate value theorem (Theorem 2.7.5), a number x between 1 and 4 such that $\ln x = 1$. Because $\ln x$ is increasing, the number is unique. This number is one of the most important numbers in mathematics.

Definition 2   THE NUMBER e   The **number e** is defined to be the unique number having the property that $\ln x = 1$. That is,

---

†Values of ln x must be computed numerically. A large number of values of ln x are given in Table A.2.

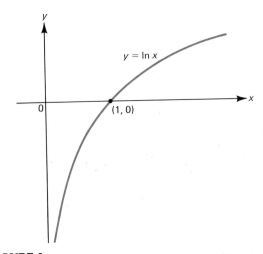

FIGURE 2

$$\ln e = 1. \tag{13}$$

We emphasize that e is now *defined* so that the area under the curve $1/t$ between 1 and e is equal to 1. This is sketched in Figure 3.

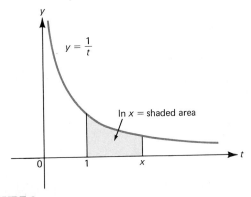

FIGURE 3

Let us now review some of the things we know about the function $\ln x$:

    (i) $\ln x$ is continuous for $x > 0$.
    (ii) $\ln x$ is increasing.
    (iii) As $x \to 0^+$, $\ln x \to -\infty$.
    (iv) As $x \to \infty$, $\ln x \to \infty$.

Putting these facts together, we see that

$$\text{range } \ln x = (-\infty, \infty). \tag{14}$$

Moreover, since $\ln x$ is increasing, $\ln x$ is 1–1 (by Theorem 6.1.1) and therefore it has an inverse. The inverse of $\ln x$ is called the *exponential function*.

Definition 3   THE EXPONENTIAL FUNCTION   The **exponential function** $f(x) = $ **exp(x)** is defined by

$$y = \exp(x) \quad \text{if and only if} \quad x = \ln y. \tag{15}$$

NOTE.   From (14) it follows that dom $\exp(x) = \mathbb{R}$.

The function $\exp(x)$ is called the *inverse* of the logarithmic function $\ln x$ because by (15)

$$\exp(\ln x) = x \quad \text{for } x > 0 \quad \text{and} \quad \ln[\exp(x)] = x \quad \text{for all } x. \tag{16}$$

To see this, we observe that if $y = \ln x$, then $\exp(\ln x) = \exp(y) = x$ by reversing the roles of x and y in (15). Similarly, if $y = \exp(x)$, then $\ln[\exp(x)] = \ln y = x$.

We now prove some facts about the exponential function.

Theorem 3.   Let x, y, and c be real numbers.

(i) $\exp(x) > 0$ for every real number x
(ii) $\exp(0) = 1$
(iii) $\exp(1) = e$
(iv) $\exp(x + y) = \exp(x)\exp(y)$
(v) $\exp(x - y) = \exp(x)/\exp(y)$
(vi) $\exp(-x) = 1/\exp(x)$
(vii) $\exp(cx) = [\exp(x)]^c$
(viii) $\lim\limits_{x \to \infty} \exp(x) = \infty$
(ix) $\lim\limits_{x \to -\infty} \exp(x) = 0$

Proof.

(i) Let $y = \exp(x)$. Then $x = \ln y$, which is defined only if $y > 0$. Thus $y = \exp(x) > 0$.
(ii) Since $\ln 1 = 0$, $\exp(0) = 1$ by (15).
(iii) Since $\ln e = 1$, $\exp(1) = e$ by (15).
(iv) Let $u = \exp(x)$ and $v = \exp(y)$. Then by (15), $x = \ln u$ and $y = \ln v$. Thus

$$x + y = \ln u + \ln v = \overset{\text{Theorem 2(ii)}}{\ln uv},$$

and by (15),

$$uv = \exp(x + y), \quad \text{or} \quad \exp(x)\exp(y) = \exp(x + y).$$

(v) Defining u and v as in (iv), we have

Theorem 2(iii)

$$x - y = \ln u - \ln v = \ln\left(\frac{u}{v}\right),$$

so

$$\frac{u}{v} = \exp(x - y), \quad \text{and} \quad \frac{\exp(x)}{\exp(y)} = \exp(x - y).$$

(vi) $\exp(-x) = \exp(0 - x) = \exp(0)/\exp(x) = 1/\exp(x).$
(vii) Let $u = \exp(x)$. Then $x = \ln u$, and

Theorem 2(v)

$$cx = c \ln u = \ln u^c,$$

so

$$u^c = \exp(cx), \quad \text{and} \quad [\exp(x)]^c = \exp(cx).$$

(viii) In the interval $[1, \frac{3}{2}]$, $1/t \le 1$. Thus

Comparison theorem

$$\int_1^{3/2} \frac{1}{t}\, dt \le \int_1^{3/2} 1\, dt = \frac{1}{2}.$$

Since e is defined by $\int_1^e (1/t)\, dt = 1$, we see that $e \ge \frac{3}{2}$.† Thus

$$\exp(n) = \exp(1 \cdot n) = [\exp(1)]^n = e^n \ge \left(\frac{3}{2}\right)^n.$$

We will see (in Theorem 5) that $(d/dx)\exp(x) > 0$, so that $\exp(x)$ is an increasing function. Thus if $x > n$, then $\exp(x) \ge (\frac{3}{2})^n$, which tends to infinity as $n \to \infty$.
   (ix) By (viii), $\lim_{x\to -\infty} \exp(x) = \lim_{x\to\infty} \exp(-x) = \lim_{x\to\infty} (1/\exp x) = 0$. This completes the proof. ∎

We now show the important connection between the number e and the function $\exp(x)$. First, we need another definition.

Definition 4   THE FUNCTION a^x Let $a > 0$ be a real number and let x be any real number. Then we define

$$a^x = \exp(x \ln a). \tag{17}$$

From the definitions of $\exp(x)$ and a^x, it immediately follows that

†Actually, $e \approx 2.71828$. A large number of values of $\exp(x)$ are given in Table A.1.

$$\ln a^x = x \ln a. \tag{18}$$

We now can prove the following important result.

Theorem 4. For any real number x,

$$\exp(x) = e^x. \tag{19}$$

Proof. By (18),

$$\ln e^x = x \ln e = x.$$

Also, from (16)

$$\ln[\exp(x)] = x.$$

Thus

$$\ln e^x = \ln[\exp(x)]$$

for every real number x. But since $\ln x$ is 1–1, we see that $e^x = \exp(x)$. ∎

From now on we will denote the exponential function by e^x. We can rewrite the facts about e^x given in Theorem 3.

$$
\begin{aligned}
&\textbf{(i)}\ e^x > 0 \text{ for every real number } x &&(20)\\
&\textbf{(ii)}\ e^0 = 1 &&(21)\\
&\textbf{(iii)}\ e^1 = e &&(22)\\
&\textbf{(iv)}\ e^{x+y} = e^x e^y &&(23)\\
&\textbf{(v)}\ e^{x-y} = e^x/e^y &&(24)\\
&\textbf{(vi)}\ e^{-x} = 1/e^x &&(25)\\
&\textbf{(vii)}\ e^{cx} = (e^x)^c &&(26)\\
&\textbf{(viii)}\ \lim_{x\to\infty} e^x = \infty &&(27)\\
&\textbf{(ix)}\ \lim_{x\to-\infty} e^x = 0 &&(28)\\
&\textbf{(x)}\ \ln e^x = x &&(29)\\
&\textbf{(xi)}\ e^{\ln x} = x &&(30)
\end{aligned}
$$

We now compute the derivative of the function a^x.

Theorem 5. Let a be a positive real number. Then

$$\text{(i)}\ \frac{d}{dx} a^x = a^x \ln a.$$

In particular,

$$(ii) \quad \frac{d}{dx} e^x = e^x.$$

That is, *the function e^x is its own derivative!*

Proof.

   (ii) We use the fact that e^x is the inverse of $\ln x$. Let $y = e^x$. Then $y > 0$, $x = \ln y$, and $dx/dy = 1/y \neq 0$. Thus by Theorem 6.1.4, e^x is differentiable, and

$$\frac{d}{dx} e^x = \frac{dy}{dx} = \frac{1}{dx/dy} = \frac{1}{1/y} = y = e^x.$$

(i)

$$\frac{d}{dx} a^x = \overset{\text{From (17)}}{\frac{d}{dx} e^{x \ln a}} = \overset{\text{From (ii) and the chain rule}}{e^{x \ln a} \frac{d}{dx} x \ln a} = e^{x \ln a} \ln a = \overset{\text{From (17)}}{a^x \ln a}. \quad \blacksquare$$

Using Theorem 5, we can prove two additional facts about the function e^x. Since

$$\frac{d}{dx} e^x = e^x > 0 \quad \text{and} \quad \frac{d^2}{dx^2} e^x = \frac{d}{dx} e^x = e^x > 0,$$

we see that

(x) e^x is an increasing function, $\qquad\qquad\qquad\qquad\qquad\qquad$ (31)

and

(xi) the graph of e^x is concave up. $\qquad\qquad\qquad\qquad\qquad\qquad$ (32)

It is now easy to sketch e^x. This is done in Figure 4.

   Using Theorem 5, we can obtain the following integration formulas.

$$\int e^x \, dx = e^x + C \qquad\qquad\qquad\qquad\qquad\qquad (33)$$

and

$$\int a^x \, dx = \frac{a^x}{\ln a} + C. \qquad\qquad\qquad\qquad\qquad\qquad (34)$$

EXAMPLE 1 $\int 2^x \, dx = 2^x/(\ln 2) + C.$ $\blacksquare$

   Finally, we can define other logarithmic functions. Since $(d/dx)a^x = a^x \ln a$, and

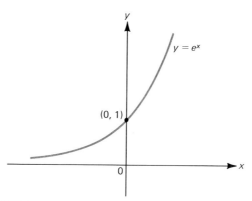

FIGURE 4

$a^x = e^{x \ln a} > 0$, we see that a^x is an increasing function. Thus a^x is 1–1 and it has an inverse. The inverse of a^x is called the *logarithm to the base a*.

Definition 5   LOGARITHM TO THE BASE a Let a be a positive number with $a \neq 1$. Then the **logarithmic function with base a** is defined by

$$y = \log_a x \qquad \text{if and only if} \qquad a^y = x. \tag{35}$$

Theorem 6.   Let $a > 0$ with $a \neq 1$.

> **(i)** $\log_a x = \dfrac{\ln x}{\ln a}$ $\hspace{4cm}$ (36)
>
> **(ii)** $\log_a e = \dfrac{1}{\ln a}$ $\hspace{4cm}$ (37)
>
> **(iii)** $\dfrac{d}{dx} \log_a x = \dfrac{1}{x \ln a} = \dfrac{\log_a e}{x}$ $\hspace{2.5cm}$ (38)

Proof.

(i) Let $y = \log_a x$. Then $x = a^y$ and $\ln x = \ln a^y = \ln e^{y \ln a} = y \ln a$. Thus $y = \ln x / \ln a$.

(ii) From (i)

$$\log_a e = \frac{\ln e}{\ln a} = \frac{1}{\ln a}.$$

(iii)

$$\frac{d}{dx} \log_a x = \frac{d}{dx}\left(\frac{\ln x}{\ln a}\right) = \frac{1}{\ln a}\frac{d}{dx} \ln x = \frac{1}{x \ln a}$$

$$= \frac{1}{x}\frac{1}{\ln a} = \frac{1}{x} \log_a e. \; \blacksquare$$

REMARK 1.   From (i) we see that $\log_e x = \ln x$ since $\ln e = 1$.

REMARK 2.   Logarithms to the base 10, $\log_{10} x$, are called **common logarithms** and are usually denoted $\log x$, with the subscript 10 omitted.

In Section 6.2 we defined the number e by

$$e = \lim_{n \to \infty} \left(1 + \frac{1}{n}\right)^n.$$

We now show that this limit can be obtained by our alternative definition of e.

Theorem 7.

$$e = \lim_{n \to \infty} \left(1 + \frac{1}{n}\right)^n. \tag{39}$$

Proof.   We start with the fact that if $t \in [1, 1 + 1/n]$, then

$$\frac{1}{1 + (1/n)} \le \frac{1}{t} \le 1,$$

so that by the comparison theorem,

$$\int_1^{1+(1/n)} \frac{1}{1+(1/n)}\, dt \le \int_1^{1+(1/n)} \frac{1}{t}\, dt \le \int_1^{1+(1/n)} 1\, dt,$$

or

$$\frac{1}{1+(1/n)}\left[\left(1 + \frac{1}{n}\right) - 1\right] \le \ln\left(1 + \frac{1}{n}\right) \le \left(1 + \frac{1}{n}\right) - 1. \tag{40}$$

Since $(1 + 1/n) - 1 = 1/n$ and $[1/(1 + 1/n)] \cdot 1/n = 1/(n + 1)$, we obtain from (40)

$$\frac{1}{n+1} \le \ln\left(1 + \frac{1}{n}\right) \le \frac{1}{n}. \tag{41}$$

Then, from (41), $(n + 1)\ln(1 + 1/n) \ge 1$ and $n \ln(1 + 1/n) \le 1$. But $(n + 1)\ln(1 + 1/n) = \ln(1 + 1/n)^{n+1} \ge 1$ implies that

$$\left(1 + \frac{1}{n}\right)^{n+1} \ge e. \tag{42}$$

Similarly, $n \ln(1 + 1/n) = \ln(1 + 1/n)^n \le 1$ implies that

$$\left(1 + \frac{1}{n}\right)^n \le e.$$

Thus

$$\left(1 + \frac{1}{n}\right)^n \leq e \leq \left(1 + \frac{1}{n}\right)^{n+1} \qquad \text{for} \qquad n = 1, 2, 3, \ldots. \tag{43}$$

Finally, dividing the right-hand inequality in (43) by $[1 + (1/n)]$ yields

$$\frac{e}{1 + (1/n)} \leq \left(1 + \frac{1}{n}\right)^n \leq e, \tag{44}$$

and since $1 + (1/n) \to 1$ as $n \to \infty$, we see that $e/[1 + (1/n)] \to e$ as $n \to \infty$, and the result now follows from the squeezing theorem (page 89). ∎

REMARK.   The inequalities (43) give us estimates on the size of e. For example, for $n = 10$ we obtain

$$\left(1 + \frac{1}{10}\right)^{10} \leq e \leq \left(1 + \frac{1}{10}\right)^{11},$$

or

$$2.59374246 \leq e \leq 2.853116706.$$

For $n = 1000$ we obtain

$$\left(1 + \frac{1}{1000}\right)^{1000} \leq e \leq \left(1 + \frac{1}{1000}\right)^{1001},$$

or

$$2.716923932 \leq e \leq 2.719640856.$$

In this section we have reproduced all the results about the basic exponential and logarithmic functions obtained in Sections 6.2 and 6.3, but from a completely different point of view. We will use these facts in the next section to differentiate more complicated exponential and logarithmic functions. In Sections 6.6, 6.7, and 6.8 we will show how important those functions are in applications.

PROBLEMS 6.4

In Problems 1–32, solve for the unknown variable. (Do not use any table or calculator.) By convention, the unsubscripted $\log x$ stands for $\log_{10} x$.

1. $y = \log_2 4$

2. $y = \log_4 16$

3. $y = \log_{1/3} 27$

4. $y = \frac{1}{3} \log_7(\frac{1}{7})$

5. $y = \pi \log_\pi \left(\frac{1}{\pi^4}\right)$

6. $y = \log_{81} 3$

7. $y = \ln e^5$

8. $y = \ln \frac{1}{e^{3.7}}$

9. $y = \log_{1/4} 2$

10. $y = \log_a a \cdot \log_b b^2 \cdot \log_c c^3$

11. $y = \log_6 36 \log_{25} \frac{1}{5}$

12. $64 = x \log_{1/4} 64$

13. $2^{x^2} = 64$

14. $y = 1.3^{\log_{1.3} 48}$

15. $y = e^{\ln \sqrt{2}}$

16. $y = e^{\ln 14.6}$

17. $y = e^{\ln e^{\pi}}$

18. $\log_2 x^4 = 4$

19. $\log_x 64 = 3$

20. $\log_x 125 = -3$

21. $\log_x 32 = -5$

22. $y = \log 0.01$

23. $\log x = 10^{-23}$

24. $\log x = 3 \log 2 + 2 \log 3$

25. $\log x = 4 \log \frac{1}{2} - 3 \log \frac{1}{3}$

26. $\log x = a \log b + b \log a$

27. $\log x = \log 1 + \log 2 + \log 3 + \log 4$

28. $\log x = \log 2 - \log 3 + \log 5 - \log 7$

29. $\log x^3 = 2 \log 5 - 3 \log 2$

30. $\log \sqrt{x} = 4 \log 2 - 3 \log \frac{1}{2}$

31. $\ln x^2 - \ln x = \ln 18 - \ln 6$

32. $\log_2 x^{3/2} - \log_2 \sqrt{x} = \log_2 9 - \log_2 3$

In Problems 33–36, simplify the given expressions by using Theorem 2.

33. $\log x^3 \sqrt{x}$

34. $\log[\sqrt[3]{x^2 + 3} \, (x^5 - 9)^{1/8}]$

35. $\ln \dfrac{x^5}{(1 + x)^{18}}$

36. $\ln \sqrt[5]{\dfrac{(x + 1)x^2}{(x - 12)}}.$

37. Using a calculator, calculate $[1 - (1/n)]^{-n}$ for $n = 2, 5, 10, 1000, 10{,}000,$ and $1{,}000{,}000$ to illustrate that $\lim_{n \to -\infty} [1 - (1/n)]^{-n} = e$.

*38. Show that if $a > 1$, then $\lim_{x \to 0^+} \log_a x = -\infty$ and $\lim_{x \to \infty} \log_a x = \infty$.

*39. If $0 < a < 1$, show that $\lim_{x \to 0^+} \log_a x = \infty$ and $\lim_{x \to \infty} \log_a x = -\infty$.

40. The acidity of a substance is measured by the concentration of the hydrogen ion $[H^+]$ in the substance. This concentration is usually measured in terms of moles per liter (mol/L). A standard way to describe this acidity is to define the pH of a substance by

$$pH = -\log[H^+].$$

Distilled water has an approximate H^+ concentration of 10^{-7} mol/L so its pH is $-\log 10^{-7} = 7$. A substance with a pH under 7 is termed an **acid,** while one with a pH above 7 is called a **base.** Determine the pH of the following substances with the indicated hydrogen ion concentrations.

(a) $[H^+] = 4.2 \times 10^{-6}$ **(b)** $[H^+] = 8 \times 10^{-6}$ **(c)** $[H^+] = 0.6 \times 10^{-7}$

* 41. Using a calculator, determine which is greater: 999^{1000} or 1000^{999}.

42. A general psychophysical relation was established in 1834 by the physiologist Weber and given a more precise phrasing later by Fechner. By the **Weber-Fechner law,** $S = c \log(R + d)$, where S is the intensity of a sensation, R is the strength of the stimulus producing it, and c and d are constants. The Greek astronomer Ptolemy catalogued stars according to their visual brightness in six categories, or **magnitudes.** A star of the first magnitude was about $2\frac{1}{2}$ times as bright as a star of the second magnitude, which in turn was about $2\frac{1}{2}$ times as bright as a star of the third magnitude, and so on. Let b_n and b_m denote the apparent brightness of two stars having magnitudes n and m, respectively. Then modern astronomers have established the Weber-Fechner law relating the relative brightness to the difference in magnitudes as

$$(m - n) = 2.5 \log\left(\frac{b_n}{b_m}\right).$$

(a) Using this formula, calculate the ratio of brightness for two stars of the second and fifth magnitudes, respectively.

(b) If star A is five times as bright to the naked eye as star B, what is the difference in their magnitudes?

(c) How much brighter is Sirius (magnitude 1.4) than a star of magnitude 21.5?

(d) The Nova Aquilae in a 2–3-day period in June 1918 increased in brightness about 45,000 times. How many magnitudes did it rise?

*(e) The bright star Castor appears to the naked eye as a single star but can be seen with the aid of a telescope to be really two stars whose magnitudes have been calculated to be 1.97 and 2.95. What is the magnitude of the two combined? [*Hint:* Brightnesses, but not magnitudes, can be added.]

43. The subjective impression of loudness can be described by a Weber-Fechner law. Let I denote the intensity of a sound. The least intense sound that can be heard is $I_0 = 10^{-12} \text{W/m}^2$ at a frequency of 1000 cycles/sec (this value is called the **threshold of audibility**). If L denotes the loudness of a sound, measured in decibels (dB), then $L = 10 \log(I/I_0)$.
(a) If one sound has twice the intensity of another, what is the ratio of the perceived loudness of the two sounds?
(b) If one sound appears to be twice as loud as another, what is the ratio of their intensities?
(c) Ordinary conversation sounds six times as loud as a low whisper. What is the actual ratio of intensity of their sounds?

44. Natural logarithms can be calculated on a hand calculator even if the calculator does not have an $\boxed{\ln}$ key. If $\frac{1}{2} \le x \le \frac{3}{2}$ and if $A \equiv (x-1)/(x+1)$, then a good approximation to $\ln x$ is given by

$$\ln x \approx \left[\left(\frac{3A^2}{5} + 1 \right) \cdot \frac{A^2}{3} + 1 \right] 2A, \qquad \frac{1}{2} \le x \le \frac{3}{2}.$$

(a) Use this formula to calculate $\ln 0.8$ and $\ln 1.2$.
(b) Using facts about logarithms, use the formula to calculate (approximately) $\ln 2 = \ln(\frac{3}{2} \cdot \frac{4}{3})$.
(c) Using the result of (b), calculate $\ln 3$ and $\ln 8$.

45. The exponential e^x can be estimated for x in $[-\frac{1}{2}, \frac{1}{2}]$ by the formula.

$$e^x \approx \left(\left\{ \left[\left(\frac{x}{5} + 1 \right) \frac{x}{4} + 1 \right] \frac{x}{3} + 1 \right\} \frac{x}{2} + 1 \right) x + 1.$$

(a) Calculate an approximate value for $e^{0.13}$.
(b) Calculate an approximate value for $e^{-0.37}$.
(c) Calculate an approximate value for $e^{4.13}$.
(d) Calculate an approximate value for $e^{-2.63}$. [*Hint:* Use part (b).]
(e) Calculate approximate values for $e^{4.82}$ and $e^{-1.44}$.

46. Show, using logarithms, that $a^b = b^a$, where $a = [1 + (1/n)]^n$ and $b = [1 + (1/n)]^{n+1}$. What does this prove in the case $n = 1$?

47. The quantity $n! = n(n-1)(n-2) \cdots 3 \cdot 2 \cdot 1$ grows very rapidly as n increases. According to **Stirling's formula,** when n is large

$$n! \approx \sqrt{2\pi n} \left(\frac{n}{e} \right)^n.$$

Use Stirling's formula to estimate 100! and 200!. [*Hint:* Use common logarithms.]

**48. Prove that $\ln x$ cannot be expressed in the form $p(x)/q(x)$, where p and q are polynomial ·functions.

In Problems 49–56, compute the derivative of the given function.

49. $\log_5 x$ **50.** 5^x **51.** $\log_{10} x$

52. 10^x **53.** $\log_{1/3} x$ **54.** $\left(\frac{1}{3} \right)^x$

55. $\log_\pi x$ **56.** π^x

In Problems 57–64, calculate the given definite integral.

57. $\int_1^4 2^x \, dx$ **58.** $\int_{-1}^3 \left(\frac{1}{2} \right)^x dx$ **59.** $\int_1^{e^3} \frac{1}{x} \, dx$

60. $\displaystyle\int_1^e \ln x \, dx$

61. $\displaystyle\int_{\log_3 (5\ln 3)}^{\log_3 (7\ln 3)} 3^x \, dx$

62. $\displaystyle\int_1^{\ln 4} e^x \, dx$

63. $\displaystyle\int_{\ln 8}^{\ln 23} e^x \, dx$

64. $\displaystyle\int_{\log_{1/4}[-8\ln(1/2)]}^{\log_{1/4}[-3\ln(1/2)]} \left(\frac{1}{4}\right)^x \, dx$

In Problems 65–79, draw a graph of the given function.

65. $y = \ln|x|$

66. $y = \ln(x + 1)$

67. $y = 3\ln(x - 4)$

68. $y = e^{-x}$

69. $y = \ln(-x)$

70. $y = -2e^{(2 + x)}$

71. $y = e^{3x}$

72. $y = \ln x^4$

73. $y = 3^{2x}$

74. $y = 3\log_3(x + 1)$

75. $y = \left(\dfrac{1}{2}\right)^{-x}$

76. $y = \log_{1/2}\dfrac{x}{2}$

77. $y = 10^{x + 3}$

78. $y = \log x^{17/4}$

79. $y = 1 - \ln\sqrt{x}$

**80. For what values of A is the graph of A^x tangent to the graph of $\log_A x$?

6.5 DIFFERENTIATION AND INTEGRATION OF MORE GENERAL EXPONENTIAL AND LOGARITHMIC FUNCTIONS

Once we know how to differentiate the basic functions $\log_a x$ and a^x, we can use the chain rule and implicit differentiation to expand greatly the number of new functions we can differentiate and integrate. We will illustrate this in this section with a number of examples. Since for the rest of the book we will only be concerned with the base e, we will use that base in all our examples.

EXAMPLE 1 Differentiate $y = \ln(x^3 + 3x + 1)$.

Solution. Using the chain rule, we set $u = x^3 + 3x + 1$. Then

$$y = \ln u, \quad \text{and} \quad \frac{dy}{dx} = \frac{dy}{du}\frac{du}{dx} = \frac{1}{u}(3x^2 + 3) = \frac{3x^2 + 3}{x^3 + 3x + 1}.$$

NOTE. $\ln(x^3 + 3x + 1)$ is only defined when $x^3 + 3x + 1 > 0$. ■

EXAMPLE 2 Differentiate $y = \ln(1 + e^x)$.

Solution. Letting $u = 1 + e^x$, we have

$$\frac{dy}{dx} = \frac{dy}{du}\frac{du}{dx} = \frac{1}{u}e^x = \frac{e^x}{1 + e^x}. \quad ■$$

EXAMPLE 3 Find $(d/dx)\ln(\sec x + \tan x)$.

Solution.

$$\frac{d}{dx}\ln(\sec x + \tan x) = \frac{1}{\sec x + \tan x}\cdot\frac{d}{dx}(\sec x + \tan x)$$

$$= \frac{\sec x \tan x + \sec^2 x}{\sec x + \tan x} = \frac{\sec x(\sec x + \tan x)}{\sec x + \tan x} = \sec x. \quad ■$$

Generalizing these three examples, we obtain the rule

$$\frac{d}{dx} \ln u = \frac{1}{u} \frac{du}{dx}.$$ (1)

We must always remember that $\ln u$ is only defined if $u > 0$.
Now consider

$$\frac{d}{dx} \ln|u| \qquad \text{for} \qquad u \neq 0.$$

If $u \neq 0$, then $|u| > 0$ and $\ln|u|$ is defined. There are two cases to consider.

Case 1. $u > 0$. Then

$$|u| = u, \qquad \text{and} \qquad \frac{d}{dx} \ln|u| = \frac{d}{dx} \ln u = \frac{1}{u} \frac{du}{dx}.$$

Case 2. $u < 0$. Then

$$|u| = -u, \quad \text{and} \quad \frac{d}{dx} \ln|u| = \frac{d}{dx} \ln(-u) = \frac{1}{-u} \frac{d(-u)}{dx} = -\frac{1}{u}\left(-\frac{du}{dx}\right) = \frac{1}{u} \frac{du}{dx}.$$

Thus in either case we have

$$\frac{d}{dx} \ln|u| = \frac{1}{u} \frac{du}{dx}, \qquad u \neq 0.$$ (2)

This fact is useful, for now we only have to worry about zero values of u. Negative values of u pose no problem.

Using (2), we immediately obtain [see also equation (6.3.6)]

$$\int \frac{du}{u} = \ln|u| + C.$$ (3)

In other words, *if the numerator is the derivative of the denominator, then the integral of the fraction is the natural logarithm of the absolute value of the denominator.*

EXAMPLE 4 Calculate $\int [2x/(x^2 + 1)] \, dx$.

Solution. Since $2x$ is the derivative of $x^2 + 1$,

$$\int \frac{2x}{x^2 + 1} \, dx = \ln|x^2 + 1| + C = \ln(x^2 + 1) + C$$

(since $x^2 + 1$ is always positive). ■

EXAMPLE 5 Calculate

$$\int \frac{dx}{x \ln x}.$$

Solution. We write the integral as

$$\int \frac{1/x}{\ln x}\, dx .$$

Then the numerator is the derivative of the denominator, and

$$\int \frac{1/x}{\ln x}\, dx = \ln|\ln x| + C. \ \blacksquare$$

EXAMPLE 6 Calculate

$$\int \frac{x^3 + 1}{x^4 + 4x}\, dx .$$

Solution. Here the numerator is not quite the derivative of the denominator. The derivative of the denominator is $4x^3 + 4 = 4(x^3 + 1)$. Thus we need to multiply and divide by 4 to obtain

$$\int \frac{x^3 + 1}{x^4 + 4x}\, dx = \frac{1}{4}\int \frac{4x^3 + 4}{x^4 + 4x}\, dx = \frac{1}{4}\ln|x^4 + 4x| + C. \ \blacksquare$$

EXAMPLE 7 Differentiate $y = e^{2+3x}$.

Solution. If we let $u = 2 + 3x$, then $y = e^u$, and using the chain rule, we obtain

$$\frac{dy}{dx} = \frac{dy}{du}\frac{du}{dx} = e^u \cdot 3 = 3e^{2+3x}. \ \blacksquare$$

EXAMPLE 8 Differentiate $y = e^{\cos^2 x}$.

Solution. If $u = \cos^2 x$, then

$$\frac{du}{dx} = 2 \cos x \frac{d}{dx}(\cos x) = -2 \cos x \sin x .$$

and

$$\frac{dy}{dx} = \frac{dy}{du}\frac{du}{dx} = e^u(-2 \cos x \sin x) = -2 \cos x \sin x(e^{\cos^2 x}). \ \blacksquare$$

We now generalize Examples 7 and 8 to see that

$$\frac{d}{dx} e^u = e^u \frac{du}{dx}. \tag{4}$$

If we integrate (4), we get the rule

$$\int e^u \, du = e^u + C. \tag{5}$$

That is, the integral of e raised to a power times the derivative of the power is simply e to that power.

EXAMPLE 9   Calculate $\int e^{x^3} \cdot 3x^2 \, dx$.

Solution. If $u = x^3$, then $du = 3x^2 \, dx$, so that

$$\int e^{x^3} \cdot 3x^2 \, dx = \int e^u \, du = e^u + C = e^{x^3} + C. \ \blacksquare$$

EXAMPLE 10   Calculate $\int e^{-2x} \, dx$.

Solution. If $u = -2x$, then $du = -2 \, dx$, so we multiply and divide by -2 to obtain

$$\int e^{-2x} \, dx = -\frac{1}{2} \int e^{-2x}(-2) \, dx = -\frac{1}{2} \int e^u \, du = -\frac{1}{2} e^u + C = -\frac{1}{2} e^{-2x} + C.$$

$\blacksquare$

EXAMPLE 11   Calculate $\int (e^{\sqrt{x}}/\sqrt{x}) \, dx$.

Solution. If $u = \sqrt{x}$, then $du = (\frac{1}{2}\sqrt{x}) \, dx$, and we multiply and divide by $\frac{1}{2}$ to obtain

$$\int \frac{e^{\sqrt{x}}}{\sqrt{x}} \, dx = 2 \int \frac{e^{\sqrt{x}}}{2\sqrt{x}} \, dx = 2 \int e^u \, du = 2e^u + C = 2e^{\sqrt{x}} + C. \ \blacksquare$$

We can make use of our ability to integrate and differentiate certain logarithmic and exponential functions to do all the things we have done in earlier chapters—namely, sketch curves, find the area between curves, find maxima and minima, and so on.

EXAMPLE 12   Sketch the curve $y = (\ln^2 x)/x$. [*Note:* By $\ln^2 x$, we mean $(\ln x)^2$.]

Solution. We differentiate to obtain

$$\frac{dy}{dx} = \frac{x \cdot (d/dx)(\ln^2 x) - \ln^2 x}{x^2} = \frac{x \cdot 2(\ln x)(1/x) - \ln^2 x}{x^2} = \frac{2 \ln x - \ln^2 x}{x^2}.$$

The function and its derivative are only defined for $x > 0$. Now $dy/dx = 0$ when

$2 \ln x - \ln^2 x = (\ln x)(2 - \ln x) = 0$. This occurs when $\ln x = 0$ and when $\ln x = 2$. When $\ln x = 0$, $x = 1$. When $\ln x = 2$, $x = e^2 \approx 7.4$. If $0 < x < 1$, then $dy/dx < 0$ and y is decreasing. For $1 < x < e^2$, $dy/dx > 0$ and y is increasing. For $x > e^2$, y is again decreasing. Therefore by the first derivative test, f has a local minimum at $x = 1$ [which gives the point $(1, 0)$] and a local maximum at $x = e^2$ [the point $(e^2, 4/e^2) \approx (7.4, 0.54)$]. To find concavity, we compute

$$\frac{d^2y}{dx^2} = \frac{x^2(d/dx)(2 \ln x - \ln^2 x) - (2 \ln x - \ln^2 x)(2x)}{x^4}$$

$$= \frac{x^2\left(\frac{2}{x} - \frac{2 \ln x}{x}\right) - 4x \ln x + 2x \ln^2 x}{x^4}$$

$$= \frac{2x - 6x \ln x + 2x \ln^2 x}{x^4} = \frac{2 - 6 \ln x + 2 \ln^2 x}{x^3}.$$

To find the points of inflection, we must solve the equation $2 - 6 \ln x + 2 \ln^2 x = 0$. This is a quadratic equation in $\ln x$. We have

$$\ln^2 x - 3 \ln x + 1 = 0, \quad \text{or} \quad \ln x = \frac{3 \pm \sqrt{9 - 4}}{2} = \frac{3 \pm \sqrt{5}}{2}.$$

If $\ln x = (3 + \sqrt{5})/2 \approx 2.62$, then $x \approx e^{2.62} \approx 13.74$. If $\ln x = (3 - \sqrt{5})/2 \approx 0.38$, then $x \approx e^{0.38} \approx 1.46$. These give the two points of inflection. If $x < 1.46$, then the curve is concave up. For $1.46 < x < 13.74$ the curve is concave down, and for $x > 13.74$ the curve is again concave up. Finally, since $\ln x$ seems to increase much more slowly than x itself, it is reasonable to believe that

$$\lim_{x \to 0^+} \frac{\ln^2 x}{x} = \infty \quad \text{and} \quad \lim_{x \to \infty} \frac{\ln^2 x}{x} = 0.$$

We will prove these facts in Chapter 12. Putting all this information together, we obtain the curve in Figure 1. ■

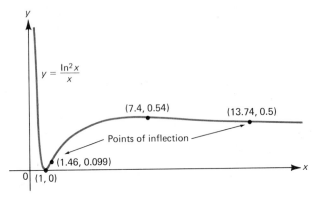

FIGURE 1

EXAMPLE 13 Find the area between the curve $y = e^{x+1}$, the y-axis, the curve $y = x^2$, and the line $x = 1$.

> *Solution.* After graphing, we can depict the shaded area as in Figure 2. Then
>
> $$A = \int_0^1 (e^{x+1} - x^2)\, dx = \left(e^{x+1} - \frac{x^3}{3} \right)\Bigg|_0^1 = e^2 - \frac{1}{3} - e \approx 4.34. \quad \blacksquare$$

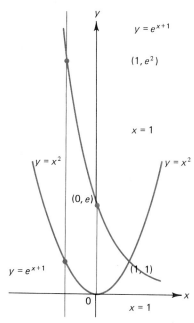

FIGURE 2

EXAMPLE 14 Graph the curve $y = e^{-x} \sin x$.

> *Solution.* First notice that since $-1 \le \sin x \le 1$ and since $e^{-x} \to 0$ as $x \to \infty$, $\lim_{x \to \infty} e^{-x} \sin x = 0$. (Be careful, however. The limit $\lim_{x \to \infty} \sin x$ *does not exist* since $\sin x$ oscillates between -1 and 1 but does not approach any single number.) Also, since $e^{-x} > 0$, $-1 \le \sin x \le 1$ implies that $-e^{-x} \le e^{-x} \sin x \le e^{-x}$, so the graph of the function $e^{-x} \sin x$ stays between the graphs of the functions $-e^{-x}$ and e^{-x}. We then calculate
>
> $$\frac{dy}{dx} = e^{-x} \cos x - e^{-x} \sin x.$$

This expression is zero when $e^{-x} \cos x = e^{-x} \sin x$, or when $\sin x = \cos x$ and $\sin x / \cos x = \tan x = 1$. Since $\tan x = 1$ when $x = \pi/4$, and since $\tan x$ is periodic of period π, we see that the critical points are the points $\pi/4$, $-3\pi/4$, $5\pi/4$, $-7\pi/4$, $9\pi/4$, ... Now $d^2y/dx^2 = -e^{-x} \sin x - e^{-x} \cos x - e^{-x} \cos x + e^{-x} \sin x = -2e^{-x} \cos x$, and there are points of inflection at $x = \pi/2, 3\pi/2, 5\pi/2,$ . . . . When $x = \pi/4$, $d^2y/dx^2 < 0$; and when $x = 5\pi/4$, $\cos x < 0$ (since $5\pi/4$ is in the third quadrant), so that $d^2y/dx^2 > 0$. Hence when $x = \pi/4$, we have a local maximum, and at $x = 5\pi/4$ there is a local minimum. We now draw the curve by using the fact that $e^{-x} \sin x$ always remains between $-e^{-x}$ and e^{-x}. See Figure 3. ■

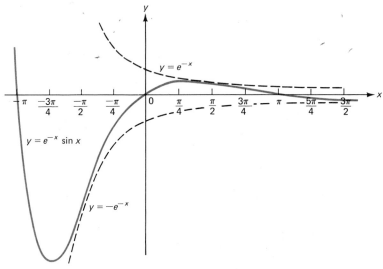

FIGURE 3

Before leaving this section, we show how the use of logarithms can sometimes simplify the process of differentiation. We need only recall the rules given in Theorem 6.2.2 and Theorem 6.4.2.

EXAMPLE 15   Differentiate $y = \sqrt[4]{(x^3 + 1)}/x^{7/9}$.

Solution. We first take the natural logarithm of both sides of the equation:

$$\ln y = \ln\left(\frac{x^3 + 1}{x^{7/9}}\right)^{1/4} = \frac{1}{4}\ln\left(\frac{x^3 + 1}{x^{7/9}}\right) = \frac{1}{4}\ln(x^3 + 1) - \frac{7}{36}\ln x.$$

Now we differentiate implicitly:

$$\frac{1}{y}\frac{dy}{dx} = \frac{3x^2}{4(x^3 + 1)} - \frac{7}{36x},$$

or

$$\frac{dy}{dx} = y\left(\frac{3x^2}{4(x^3 + 1)} - \frac{7}{36x}\right) = \left(\sqrt[4]{\frac{x^3 + 1}{x^{7/9}}}\right)\left(\frac{3x^2}{4(x^3 + 1)} - \frac{7}{36x}\right). ■$$

In Example 15 it was not necessary to use the quotient rule at all. The process we used to simplify the differentiation is called **logarithmic differentiation.**

EXAMPLE 16   Differentiate $y = x^x$.

Solution. If $y = x^x$, then $\ln y = x \ln x$, and

$$\frac{1}{y}\frac{dy}{dx} = x \cdot \frac{1}{x} + \ln x = 1 + \ln x,$$

so that

$$\frac{dy}{dx} = y(1 + \ln x) = x^x(1 + \ln x). \quad \blacksquare$$

PROBLEMS 6.5

In Problems 1–46, perform the indicated integration or differentiation.

1. $\dfrac{d}{dx}\ln(1 + 2x)$

2. $\dfrac{d}{dx}e^{3x+2}$

3. $\dfrac{d}{dx}\ln(1 + x^5)$

4. $\dfrac{d}{dx}\ln\!\left(\dfrac{x}{1 + x}\right)$

5. $\dfrac{d}{dx}e^{1+5x^4}$

6. $\dfrac{d}{dx}e^{x^2+x^3}$

7. $\displaystyle\int \dfrac{1}{2x}\,dx$

8. $\displaystyle\int e^{3x}\,dx$

9. $\displaystyle\int x^2 e^{x^3}\,dx$

10. $\displaystyle\int \dfrac{x}{1 + x^2}\,dx$

11. $\displaystyle\int \dfrac{x^2}{1 + x^3}\,dx$

12. $\displaystyle\int \dfrac{x^{n-1}}{1 + x^n}\,dx$, $n > 0$ an integer

13. $\dfrac{d}{dx}(1 - \ln 3x)$

14. $\dfrac{d}{dx}(1 + \ln x + \ln^2 x + \ln^3 x)$

15. $\displaystyle\int \dfrac{x^3}{x^4 + 25}\,dx$

16. $\displaystyle\int \dfrac{e^{2x}}{1 + e^{2x}}\,dx$

17. $\displaystyle\int \dfrac{e^{\sqrt[3]{x}}}{x^{2/3}}\,dx$

18. $\dfrac{d}{dx}(e^{\sqrt[3]{x}} + 4)$

19. $\dfrac{d}{dx}e^{\ln(x^5+6)}$ [*Hint:* This problem is easier than it looks.]

20. $\dfrac{d}{dx}e^{\ln(1+e^{\ln x})}$

21. $\dfrac{d}{dx}x^{\sqrt{x}}$

22. $\dfrac{d}{dx}x^{2(1-x^3)}$

23. $\dfrac{d}{dx}\!\left(\dfrac{x^2}{3\ln x}\right)$

24. $\displaystyle\int (1 + x + x^2)e^{2x+x^2+(2x^3/3)}\,dx$

25. $\dfrac{d}{dx}e^{\sqrt{x}-1/x}$

26. $\displaystyle\int \dfrac{e^{1/x}}{x^2}\,dx$

27. $\displaystyle\int \dfrac{\ln(1/x)}{x}\,dx$

28. $\displaystyle\int -\dfrac{\ln x}{x}\,dx$

29. $\displaystyle\int_0^{\ln 2} e^{3x}\,dx$

30. $\displaystyle\int_0^3 \dfrac{x^2}{x^3 + 1}\,dx$

31. $\displaystyle\int_{\ln 3}^{\ln 5} \dfrac{e^x}{e^x + 4}\,dx$

32. $\displaystyle\int_1^2 \dfrac{e^{\sqrt{x}}}{8\sqrt{x}}\,dx$

33. $\dfrac{d}{dx}\ln|x^3 + 3|$

34. $\dfrac{d}{dx}\ln|e^x - 8|$

35. $\dfrac{d}{dx}\ln|1 - \ln x|$

36. $\dfrac{d}{dx}\ln\left|\dfrac{x - 2}{x + 3}\right|$

37. $\dfrac{d}{dx}\ln \ln x$

38. $\dfrac{d}{dx} \ln \ln \ln x$

39. $\displaystyle\int_1^3 \dfrac{1}{\ln e^x}\, dx$

40. $\displaystyle\int 2^x e^x\, dx$

41. $\displaystyle\int \dfrac{1}{x(\ln x)^5}\, dx$

42. $\displaystyle\int \dfrac{(\ln x)^{5/9}}{x}\, dx$

43. $\displaystyle\int \dfrac{1}{x \ln x[\ln(\ln x)]}\, dx$

44. $\dfrac{d}{dx} 2e^{2e^{2x}}$

45. $\dfrac{d}{dx} e^{x + e^x}$

*46. $\displaystyle\int \dfrac{1}{e^x + e^{-x} + 2}\, dx$ [*Hint:* Multiply numerator and denominator by e^x and then simplify.]

In Problems 47–56, sketch the given curve, indicating maxima, minima, and points of inflection.

47. $y = e^{x^2}$

48. $y = e^{2x+3}$

49. $y = \ln(1 + e^x)$

50. $y = \ln(1 + x^2)$

51. $y = e^{-x} \cos x$

*52. $y = \dfrac{\ln x}{x}$

53. $y = x \ln x$

54. $y = \ln[\ln(x + 2)]$

*55. $y = e^x \ln x$

56. $y = \dfrac{x}{\ln^2 x}$

In Problems 57–62, find the areas bounded by the given curves and lines.

57. $y = e^{-x}$; $x = 0$, $x = 1$, $y = 2$

58. $y = \ln x$; $x = 1$, $x = 2$, $y = 0$

59. $y = \ln x$; $y = e^x$; $x = 0$, $y = 0$, $x = 3$

60. $y = (\ln x)/x$; $x = 2$, $x = 4$, $y = 0$

61. $y = e^x$; $y = xe^{x^2}$; $x = 0$

*62. $y = \ln(x + 1)$; $y = 3$, $x = 0$

In Problems 63–68, use logarithmic differentiation to evaluate the derivative.

63. $y = \sqrt[5]{\dfrac{x^3 - 3}{1 + x^2}}$

64. $y = \dfrac{x^5(x^4 - 3)}{x^8\sqrt{x^5 + 1}}$

65. $y = (\sqrt{x})(\sqrt[3]{x + 2})(\sqrt[5]{x - 1})$

66. $y = \left[\dfrac{x^3(x + 3)(x^{11/9} + 2)^{1/2}}{(x^5 - 3)\sqrt{x^7 + 2}}\right]^{-9/14}$

67. Show that

$$\dfrac{d}{dx} a^u = a^u \ln a \, \dfrac{du}{dx}.$$

68. Show that

$$\dfrac{d}{dx} \log_a u = \dfrac{1}{u} \dfrac{du}{dx} \log_a e.$$

In Problems 69–82, use the results of Problems 67 and 68 to evaluate the given derivatives and integrals.

69. $\dfrac{d}{dx} 3^{x^2 + 5}$

70. $\dfrac{d}{dx} \log_2(x^3 + 2x - 3)$

71. $\displaystyle\int \dfrac{\log_5 2x}{x}\, dx$

72. $\displaystyle\int_2^3 \dfrac{\sqrt{\log_2 x}}{x}\, dx$

73. $\displaystyle\int \dfrac{3^{\sqrt{x}}}{\sqrt{x}}\, dx$

74. $\dfrac{d}{dx} \log_3(1 + 2^{x^2})$

75. $\displaystyle\int \dfrac{\log_{1/2} x^3}{x}\, dx$

76. $\displaystyle\int_2^4 \left(\dfrac{1}{5}\right)^{x/2}\, dx$

77. $\displaystyle\int \dfrac{3^{\ln x}}{x}\, dx$

78. $\dfrac{d}{dx}\log_3\{\log_{10}[\log_5(x+1)]\}$

79. $\displaystyle\int \dfrac{1}{x\,\log_{14}{}^3 x}\,dx$

80. $\displaystyle\int 13^{\log_{13}e^{2x}}\,dx$

81. $\dfrac{d}{dx}\log_8\dfrac{x^3}{x^4+3}$

82. $\displaystyle\int e^x(7^{-3e^x})\,dx$

83. The revenue a manufacturer receives when q units of a given product are sold is given by

$$R = 0.50q\,(e^{-0.001q}).$$

    (a) What is the marginal revenue when $q = 100$ units?
    (b) At what level of sales is revenue maximized?

84. What is the acceleration after 10 min of a particle whose equation of motion is

$$s(t) = 30 + 3t + 0.01t^2 + \ln t + e^{-2t^2}?$$

    (Time t is measured in minutes and distance is measured in meters.

85. The density of a 5-m bar is given by $\rho(x) = xe^{-x^2/3}$ kilograms per meter, where x is measured in meters from the left.
    (a) At what point is the bar most dense?
    (b) What is the total mass of the bar?

86. Prove that if f is continuous and positive and if $a > 0$, then

$$\int_0^a e^{-t^2} f(t)\,dt \le \int_0^a f(t)\,dt.$$

**87. Suppose that $x > 0$, $y > 0$, and $x + y = 1$. Prove that $x^x + y^y \ge \sqrt{2}$, and discuss conditions for equality.

**88. Suppose that L is a differentiable function such that

$$L(u \cdot v) = L(u) + L(v) \qquad \text{for} \qquad u, v > 0.$$

    How must this function be related to the natural logarithm function $\ln$?

**89. Suppose that E is a differentiable function such that $E(u + v) = E(u) \cdot E(v)$ for all real numbers u and v. How must this function be related to the exponential function e^x?

In Problems 90–92, determine intervals over which the given function is 1–1, and determine the inverse function over each such interval, if possible. Then use Theorem 6.1.4 to calculate the derivative of each inverse function.

90. $y = \dfrac{\ln x}{x}$

91. $y = xe^x$

92. $y = e^{x^2}$

6.6 DIFFERENTIAL EQUATIONS OF EXPONENTIAL GROWTH AND DECAY

In this section we begin to illustrate the great importance of the exponential function e^x in applications. Before citing examples, we will discuss a very basic type of mathematical model.

    Let $y = f(x)$ represent some physical quantity such as the volume of a substance, the population of a certain species, the mass of a decaying radioactive substance, the number of dollars invested in bonds, and so on. Then the growth of $f(x)$ is given by its derivative dy/dx. Thus if $f(x)$ is growing at a constant rate, $dy/dx = k$ and $y = kx + C$; that is, $y = f(x)$ is a straight-line function.

It is sometimes more interesting and more appropriate to consider the **relative rate of growth,** defined by

$$\text{relative rate of growth} = \frac{\text{actual rate of growth}}{\text{size of } f(x)} = \frac{f'(x)}{f(x)} = \frac{dy/dx}{y}. \tag{1}$$

The relative rate of growth indicates the percentage increase or decrease in f. For example, an increase of 100 individuals for a species with a population size of 500 would probably have a significant impact, being an increase of 20%. On the other hand, if the population were 1,000,000, then the addition of 100 would hardly be noticed, being an increase of only 0.01%.

In many applications we are told that the relative rate of growth of the given physical quantity is constant. That is,

$$\frac{dy/dx}{y} = \alpha, \tag{2}$$

or

$$\frac{dy}{dx} = \alpha y, \tag{3}$$

where α is the constant percentage increase or decrease in the quantity.

Another way to view equation (3) is that it tells us that the function is changing at a rate proportional to itself. If the constant of proportionality α is greater than 0, the quantity is increasing; while if $\alpha < 0$, it is decreasing. Equation (3) is called a **differential equation** because it is an equation involving a derivative. Differential equations arise in a great variety of settings in the physical, biological, engineering, and social sciences. It is a vast subject, and many of you will take courses in differential equations after completing your basic calculus sequence. We will encounter differential equations periodically throughout this book.

We return to equation (3). To solve it,† we carry out the following steps:

(i) Multiply both sides of equation (3) by $e^{-\alpha x}$ to obtain

$$\frac{dy}{dx} e^{-\alpha x} - \alpha y e^{-\alpha x} = 0. \tag{4}$$

(ii) Observe that

$$\frac{d}{dx}(e^{-\alpha x} y) = \overset{\text{Product rule}}{\overset{\swarrow}{e^{-\alpha x} \frac{dy}{dx}}} + y(-\alpha e^{-\alpha x}). \tag{5}$$

(iii) Using (5), write (4) as

---

†By a solution to this differential equation, we mean a function f that is differentiable and for which $f'(x) = \alpha f(x)$.

$$\frac{d}{dx}(e^{-\alpha x}y) = 0. \tag{6}$$

(iv) Since the derivative of $ye^{-\alpha x} = 0$, we see that

$$ye^{-\alpha x} = c \tag{7}$$

for some constant c.

(v) Multiply both sides of (7) by $e^{\alpha x}$ to obtain

$$y = ce^{\alpha x}.$$

Thus any solution to (3) can be written as

$$y = ce^{\alpha x}, \tag{8}$$

where c can be any real number. To check this, we simply differentiate:

$$\frac{dy}{dx} = \frac{d}{dx}ce^{\alpha x} = c\frac{d}{dx}e^{\alpha x} = c\alpha e^{\alpha x} = \alpha(ce^{\alpha x}) = \alpha y,$$

so that $y = ce^{\alpha x}$ does satisfy (3). We therefore have proven the next theorem.

Theorem 1   If α is any real number, then there are an infinite number of solutions to the differential equation $y' = \alpha y$; they take the form $y = ce^{\alpha x}$ for any real number α.

If $\alpha > 0$, we say that the quantity described by $f(x)$ is **growing exponentially.** If $\alpha < 0$, it is **decaying exponentially** (see Figure 1). Of course, if $\alpha = 0$, then there is no growth and $f(x)$ remains constant.

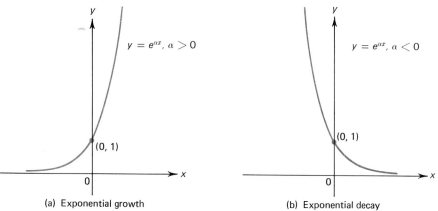

(a) Exponential growth   (b) Exponential decay

FIGURE 1

For a physical problem it would not make sense to have an infinite number of solutions. We can usually get around this difficulty by specifying the value of y for one particular value of x, say $y(x_0) = y_0$. This value is called an **initial condition,** and it will give us a unique solution to the problem. We will see this illustrated in the examples that follow.

EXAMPLE 1 (a) Find all solutions to $dy/dx = 3y$.
(b) Find the solution that satisfies the initial condition $y(0) = 2$.

Solution.

(a) Since $\alpha = 3$, all solutions are of the form

$$y = ce^{3x}.$$

(b) $2 = y(0) = ce^{3\cdot 0} = c \cdot 1 = c$, so $c = 2$ and the unique solution is $y = 2e^{3x}$.

This is illustrated in Figure 2. We can see that while there are indeed an infinite number of solutions, there is only one that passes through the point (0, 2). ■

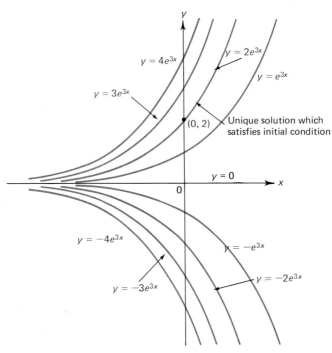

FIGURE 2

EXAMPLE 2 **(Population Growth)** A bacterial population is growing continuously at a rate equal to 10% of its population each day. Its initial size is 10,000 organisms. How many bacteria are present after 10 days? After 30 days?

Solution. Since the percentage growth of the population is 10% = 0.1, we have

$$\frac{dP/dt}{P} = 0.1, \quad \text{or} \quad \frac{dP}{dt} = 0.1P.$$

Here $\alpha = 0.1$, and all solutions have the form

$$P(t) = ce^{0.1t},$$

where t is measured in days. Since $P(0) = 10{,}000$, we have

$$ce^{(0.1)(0)} = c = 10{,}000, \quad \text{and} \quad P(t) = 10{,}000e^{0.1t}.$$

After 10 days $P(10) = 10{,}000e^{(0.1)(10)} = 10{,}000e \approx 27{,}183$, and after 30 days $P(30) = 10{,}000e^{0.1(30)} = 10{,}000e^3 \approx 200{,}855$ bacteria. ■

EXAMPLE 3 **(Newton's Law of Cooling)** **Newton's law of cooling** states that the rate of change of the temperature difference between an object and its surrounding medium is proportional to the temperature difference. If $D(t)$ denotes this temperature difference at time t and if α denotes the constant of proportionality, then we obtain

$$\frac{dD}{dt} = -\alpha D. \tag{9}$$

The minus sign indicates that this difference decreases. (If the object is cooler than the surrounding medium—usually air—it will warm up; if it is hotter, it will cool.) The solution to this differential equation is

$$D(t) = ce^{-\alpha t}.$$

If we denote the initial ($t = 0$) temperature difference by D_0, then

$$D(t) = D_0 e^{-\alpha t} \tag{10}$$

is the formula for the temperature difference for any $t > 0$. Notice that for t large $e^{-\alpha t}$ is very small, so that, as we have all observed, temperature differences tend to die out rather quickly.

We now may ask: In terms of the constant α, how long does it take for the temperature difference to decrease to half its original value?

Solution. The original value is D_0. We are therefore looking for a value of t for which $D(t) = \frac{1}{2}D_0$. That is, $\frac{1}{2}D_0 = D_0 e^{-\alpha t}$, or $e^{-\alpha t} = \frac{1}{2}$. Taking natural logarithms, we obtain

From Table A.2 or a calculator

$$-\alpha t = \ln \frac{1}{2} = -\ln 2 = -0.6931, \quad \text{and} \quad t \approx \frac{0.6931}{\alpha}.$$

Notice that this value of t does *not* depend on the initial temperature difference D_0. ■

EXAMPLE 4 With the air temperature equal to 30°C, an object with an initial temperature of 10°C warmed to 14°C in 1 hr.

(a) What was its temperature after 2 hr?
(b) After how many hours was its temperature 25°C?

Solution. Let $T(t)$ denote the temperature of the object. Then $D(t) = 30 - T(t) = D_0e^{-\alpha t}$ [from (10)]. But $D_0 = 30 - T(0) = 30 - 10 = 20$, so that

$$D(t) = 20e^{-\alpha t}.$$

We are given that $T(1) = 14$, so $D(1) = 30 - T(1) = 16$ and

$$16 = D(1) = 20e^{-\alpha \cdot 1} = 20e^{-\alpha}, \quad \text{or} \quad e^{-\alpha} = 0.8.$$

Thus

$$D(t) = 20e^{-\alpha t} = 20(e^{-\alpha})^t = 20(0.8)^t,$$

and

$$T(t) = 30 - D(t) = 30 - 20(0.8)^t.$$

We can now answer the two questions.

(a) $T(2) = 30 - 20(0.8)^2 = 30 - 20(0.64) = 17.2°C.$
(b) We need to find t such that $T(t) = 25$. That is,

$$25 = 30 - 20(0.8)^t, \quad \text{or} \quad (0.8)^t = \tfrac{1}{4} \quad \text{or} \quad t\ln(0.8) = -\ln 4,$$

and

$$t = \frac{-\ln 4}{\ln(0.8)} \approx \frac{1.3863}{0.2231} \approx 6.2 \text{ hr} = 6 \text{ hr } 12 \text{ min.} \quad\blacksquare$$

EXAMPLE 5 **(Carbon Dating)** Carbon dating is a technique used by archaeologists, geologists, and others who want to estimate the ages of certain artifacts and fossils they uncover. The technique is based on certain properties of the carbon atom. In its natural state the nucleus of the carbon atom ^{12}C has 6 protons and 6 neutrons. An *isotope* of carbon ^{12}C is ^{14}C, which has 2 additional neutrons in its nucleus. ^{14}C is *radioactive*. That is, it emits neutrons until it reaches the stable state ^{12}C. We make the assumption that the ratio of ^{14}C to ^{12}C in the atmosphere is constant. This assumption has been shown experimentally to be approximately valid, for although ^{14}C is being constantly lost through **radioactive decay** (as this process is often termed), new ^{14}C is constantly being produced by the cosmic bombardment of nitrogen in the upper atmosphere. Living plants and animals do not distinguish between ^{12}C and ^{14}C, so at the time of death the ratio of ^{12}C to ^{14}C in an organism is the same as the ratio in the atmosphere. However, this ratio changes after death since ^{14}C is converted to ^{12}C but no further ^{14}C is taken in.

It has been observed that ^{14}C decays at a rate proportional to its mass and that its **half-life** is approximately 5580 years.† That is, if a substance starts with 1 g of ^{14}C,

---

†This number was first determined in 1941 by the American chemist W. S. Libby, who based his calculations on the wood from sequoia trees, whose ages were determined by rings marking years of growth. Libby's method has come to be regarded as the archaeologist's absolute measuring scale. But in truth, this scale is

then 5580 years later it would have $\frac{1}{2}$ g of ^{14}C, the other $\frac{1}{2}$ g having been converted to ^{12}C.

We may now pose a question typically asked by an archaeologist. A fossil is unearthed and it is determined that the amount of ^{14}C present is 40% of what it would be for a similarly sized living organism. What is the approximate age of the fossil?

Solution. Let $M(t)$ denote the mass of ^{14}C present in the fossil. Then since ^{14}C decays at a rate proportional to its mass, we have

$$\frac{dM}{dt} = -\alpha M,$$

where α is the constant of proportionality. Then $M(t) = ce^{-\alpha t}$, where $c = M_0$, the initial amount of ^{14}C present. When $t = 0$, $M(0) = M_0$; when $t = 5580$ years, $M(5580) = \frac{1}{2}M_0$, since half the original amount of ^{14}C has been converted to ^{12}C. We can use this fact to solve for α since we have

$$\tfrac{1}{2}M_0 = M_0 e^{-\alpha \cdot 5580}, \qquad \text{or} \qquad e^{-5580\alpha} = \tfrac{1}{2}.$$

Thus

$$(e^{-\alpha})^{5580} = \tfrac{1}{2}, \qquad \text{or} \qquad e^{-\alpha} = (\tfrac{1}{2})^{1/5580}, \qquad \text{and} \qquad e^{-\alpha t} = (\tfrac{1}{2})^{t/5580},$$

so

$$M(t) = M_0(\tfrac{1}{2})^{t/5580}.$$

Now we are told that after t years (from the death of the fossilized organism to the present) $M(t) = 0.4 M_0$, and we are asked to determine t. Then

$$0.4 M_0 = M_0(\tfrac{1}{2})^{t/5580},$$

and taking natural logarithms (after dividing by M_0), we obtain

$$\ln 0.4 = \frac{t}{5580}\ln\left(\frac{1}{2}\right), \qquad \text{or} \qquad t = \frac{5580 \ln(0.4)}{\ln(\frac{1}{2})} \approx 7376 \text{ years.}$$

The carbon-dating method has been used successfully on numerous occasions. It was this technique that established that the Dead Sea scrolls were prepared and buried about two thousand years ago. ∎

---

flawed. Libby used the assumption that the atmosphere had at all times a constant amount of ^{14}C. Recently, however, the American chemist C. W. Ferguson of the University of Arizona deduced from his study of tree rings in 4000-year-old American giant trees that before 1500 B.C. the radiocarbon content of the atmosphere was considerably higher that it was later. This result implied that objects from the pre-1500 B.C. era were much older than previously believed, because Libby's "clock" allowed for a smaller amount of ^{14}C than actually was present. For example, a find dated at 1800 B.C. was in fact from 2500 B.C. This fact has had a considerable impact on the study of prehistoric times. For a fascinating discussion of this subject, see Gerhard Herm, *The Celts* (St. Martin's Press, New York, 1975), pages 90–92.

PROBLEMS 6.6†

1. The growth rate of a bacteria population is proportional to its size. Initially the population is 10,000, while after 10 days its size is 25,000. What is the population size after 20 days? After 30 days?

2. In Problem 1 suppose instead that the population after 10 days is 6000. What is the population after 20 days? After 30 days?

3. The population of a certain city grows 6% a year. If the population in 1970 was 250,000, what would be the population in 1980? In 2000?

4. When the air temperature is 70°F, an object cools from 170°F to 140°F in $\frac{1}{2}$ hr.
    (a) What will be the temperature after 1 hr?
    (b) When will the temperature be 90°F?   [*Hint:* Use Newton's law of cooling.]

5. A hot coal (temperature 150°C) is immersed in ice water (temperature $-10°C$). After 30 sec the temperature of the coal is 60°C. Assume that the ice water is kept at $-10°C$.
    (a) What is the temperature of the coal after 2 min?
    (b) When will the temperature of the coal be 0°C?

** 6. The president and vice-president sit down for coffee. They are both served a cup of hot black coffee (at the same temperature). The president takes a container of cream and immediately adds it to his coffee, stirs it, and waits. The vice-president waits ten minutes and then adds the same amount of cream (which has been kept cool) to her coffee and stirs it in. Then they both drink. Assuming that the temperature of the cream is lower than that of the air, who drinks the hotter coffee?   [*Hint:* Use Newton's law of cooling. It is necessary to treat each case separately and to keep track of the volumes of coffee, cream, and the coffee-cream mixture.]‡

7. A fossilized leaf contains 70% of a "normal" amount of ^{14}C. How old is the fossil?

8. Forty percent of a radioactive substance disappears in 100 years.
    (a) What is its half-life?
    (b) After how many years will 90% be gone?

9. Salt decomposes in water into sodium $[Na^+]$ and chloride $[Cl^-]$ ions at a rate proportional to its mass. Suppose there were initially 25 kg of salt and 15 kg after 10 hr.
    (a) How much salt would be left after one day?
    (b) After how many hours would there be less than $\frac{1}{2}$ kg of salt left?

10. X rays are absorbed into a uniform, partially opaque body as a function not of time but of penetration distance. The rate of change of the intensity $I(x)$ of the X ray is proportional to the intensity. Here x measures the distance of penetration. The more the X ray penetrates, the lower the intensity is. The constant of proportionality is the density D of the medium being penetrated.
    (a) Formulate a differential equation describing this phenomenon.
    (b) Solve for $I(x)$ in terms of x, D, and the initial (surface) intensity $I(0)$.

11. Radioactive beryllium is sometimes used to date fossils found in deep-sea sediment. The decay of beryllium satisfies the equation

$$\frac{dA}{dt} = -\alpha A, \qquad \text{where} \qquad \alpha = 1.5 \times 10^{-7}.$$

What is the half-life of beryllium?

---

†To complete most of these problems, you will need either to consult Tables A.1 and A.2 or to use a hand calculator with ln and e^x function keys.

‡This is a famous old problem that keeps on popping up (with an ever-changing pair of characters) in books on games and puzzles in mathematics. The problem is hard and has stymied many a mathematician. Do not get frustrated if you cannot solve it. The trick is to write everything down and to keep track of all the variables. The fact that the air is warmer than the cream is critical. It should also be noted that guessing the correct answer is fairly easy. Proving that your guess is correct is what makes the problem difficult.

12. In a certain medical treatment a tracer dye is injected into the pancreas to measure its function rate. A normally active pancreas will secrete 4% of the dye each minute. A physician injects 0.3 g of the dye and 30 min later 0.1 g remains. How much dye would remain if the pancreas were functioning normally?

13. Atmospheric pressure is a function of altitude above sea level and is given by $dP/da = \beta P$, where β is a constant. The pressure is measured in millibars (mbar). At sea level ($a = 0$), $P(0)$ is 1013.25 mbar which means that the atmosphere at sea level will support a column of mercury 1013.25 mm high at a standard temperature of 15°C. At an altitude of $a = 1500$ m, the pressure is 845.6 mbar.
    (a) What is the pressure at $a = 4000$ m?
    (b) What is the pressure at 10 km?
    (c) In California the highest and lowest points are Mount Whitney (4418 m) and Death Valley (86 m below sea level). What is the difference in their atmospheric pressures?
    (d) What is the atmospheric pressure at Mount Everest (elevation 8848 m)?
    (e) At what elevation is the atmospheric pressure equal to 1 mbar?

14. A bacteria population is known to grow exponentially. The following data were collected:

Number of Days	Number of Bacteria
5	936
10	2190
20	11,986

    (a) What was the initial population?
    (b) If the present growth rate were to continue, what would be the population after 60 days?

15. A bacteria population is declining exponentially. The following data were collected:

Number of Hours	Number of Bacteria
12	5969
24	3563
48	1269

    (a) What was the initial population?
    (b) How many bacteria are left after one week?
    (c) When will there be no bacteria left? (i.e., when is $P(t) < 1$?)

16. The models of population growth that we have discussed so far are not always realistic over long periods of time since an exponentially growing population would overrun the earth if it continued to grow unchecked (although exponential growth is often a reasonable approximation to reality over short periods of time). To model the growth situation more accurately, we assume that the average birth rate β is a positive constant [independent of the population $P(t)$] and the average death rate is proportional to the size of the population and is given by $\delta P(t)$, where δ is another positive constant. Then the average population growth is the difference between the birth and death rates and is given by

$$\frac{dP/dt}{P} = \beta - \delta P, \quad or \quad \frac{dP}{dt} = P(\beta - \delta P).$$

This equation is called the **logistic equation,** and the growth exhibited by the population is called **logistic growth.**

(a) Show by differentiation that if $P(0)$ is the initial population, then a solution to the logistic equation is

$$P(t) = \frac{\beta}{\delta + \{[\beta/P0)] - \delta\}e^{-\beta t}}.$$

(b) Show that when $P = \beta/\delta$, $dP/dt = 0$. What does this mean?
(c) Show that $\lim_{t \to \infty} P(t) = \beta/\delta$.
(d) Graph the population curve for $\delta = 0.001$, $\beta = 0.4$, and $P(0) = 300$.
(e) Graph the population curve for $\delta = 0.001$, $\beta = 0.4$, and $P(0) = 500$.
17. Consider the simple electric circuit in Figure 3.

FIGURE 3

(i) An electromotive force (emf) E (volts, V), usually a battery or a generator, drives an electric charge Q (coulombs, C) and produces a current I (amperes, A). A current is a rate of flow of charge: $I = dQ/dt$.

(ii) A resistor of resistance R (ohms, Ω) opposes the current, converting some of the electrical energy to heat. It produces a drop in the voltage given by **Ohm's law:**

$E_R = RI$.

(iii) An inductor of inductance L (henrys, H) opposes any change in the current by causing a drop in the voltage of

$$E_L = L\frac{dI}{dt}.$$

A basic principle of electric circuits is given by **Kirchoff's voltage law,** which states that *the algebraic sum of all voltage drops around a closed circuit is zero.* In the circuit of Figure 3 the resistor and the inductor cause voltage drops of E_R and E_L, respectively. The emf *provides* a voltage of E (which is a voltage *drop* of $-E$). Thus $E_R + E_L - E = 0$, or

$$E = E_R + E_L = RI + L\frac{dI}{dt}.$$

(a) Verify that the solution to the above differential equation is

$$I(t) = \frac{E}{R} + \left[I(0) - \frac{E}{R}\right]e^{-Rt/L},$$

where $I(0)$ is the initial current.
(b) Show that $\lim_{t \to \infty} I(t) = E/R$. Here E/R is called the **steady-state** part of the current and $[I(0) - (E/R)]e^{-Rt/L}$ is called the **transient part** of the current.
18. In Problem 17 let $E = 50$ V, $R = 8\,\Omega$, and $L = 2$ H. Suppose initially that the current is zero.
(a) Calculate the current after 0.05 sec.
(b) Calculate the current after 0.2 sec.
(c) What is the steady-state current?
(d) How long does it take for the current to reach 99.9% of the steady-state current?

6.7 APPLICATIONS IN ECONOMICS (OPTIONAL)

In this section we show in a variety of ways how the exponential function can arise in economics. We begin with the problem of calculating compound interest.

Suppose A_0 dollars are invested in an enterprise (which may be a bank, bonds, a common stock, etc.) with an annual interest rate of i. **Simple interest** is the amount earned on the $\$A_0$ over a period of time. If the $\$A_0$ is invested for t years, then the simple interest I is given by

$$I = A_0 t i. \tag{1}$$

EXAMPLE 1 If $\$1000$ is invested for 5 years with an interest rate of 6%, then $i = 0.06$, and the simple interest earned is

$$I = (\$1000)(5)(0.06) = \$300. \ \blacksquare$$

Compound interest is interest paid on the interest previously earned as well as on the original investment. Suppose that interest is paid annually. Then if $\$A_0$ is invested, the interest after one year is $\$iA_0$, and the original investment is now worth $A_0 + iA_0 = A_0(1 + i)$ dollars. After two years the compound interest paid would be $i[A_0(1 + i)]$ dollars, and the investment would then be worth

$$A_0(1 + i) + iA_0(1 + i) = A_0(1 + i)(1 + i) = A_0(1 + i)^2$$

dollars. Continuing in this fashion, we see that after t years the investment would be worth

$$A(t) = A_0(1 + i)^t \text{ dollars.} \tag{2}$$

We have used the notation $A(t)$ to denote the value of the investment after t years.

EXAMPLE 2 If the interest in Example 1 is compounded annually, then after 5 years the investment is worth

$$A(5) = 1000(1 + 0.06)^5 = 1000(1.33823) = \$1338.23.$$

The actual interest paid is $\$338.23. \ \blacksquare$

In practice, interest is compounded more frequently than annually. If it is paid m times a year, then in each interest period the rate of interest is i/m, and in t years there are tm pay periods. Then, according to formula (2), we obtain the following rule:

The value of an investment of $\$A_0$ compounded m times a year with an annual interest rate of i after t years is

$$A(t) = A_0\left(1 + \frac{i}{m}\right)^{mt}. \tag{3}$$

EXAMPLE 3   If the interest in Example 1 is compounded quarterly (4 times a year), then after 5 years the investment is worth

$$A(5) = 1000\left(1 + \frac{0.06}{4}\right)^{(4)(5)} = 1000(1.015)^{20} = 1000(1.34686) = \$1346.86.$$

The interest paid is now $346.86. ■

It is clear from the above examples that the interest actually paid increases as the number of interest periods increases. We are naturally led to ask: What is the value of an investment if interest is **compounded continuously,** that is, when the number of interest payments approaches infinity? To answer this question, we calculate, from (3),

$$A(t) = \lim_{m\to\infty} A_0\left(1 + \frac{i}{m}\right)^{mt} = A_0 \lim_{m\to\infty}\left[\left(1 + \frac{i}{m}\right)^{m/i}\right]^{it} = A_0\left[\lim_{m\to\infty}\left(1 + \frac{i}{m}\right)^{m/i}\right]^{it}$$

(from the corollary to Theorem 2.3.4). Let $n = m/i$. Then as $m \to \infty$, $n \to \infty$, so that

$$\lim_{m\to\infty}\left(1 + \frac{i}{m}\right)^{m/i} = \lim_{n\to\infty}\left(1 + \frac{1}{n}\right)^{n} = e.$$

Thus

$$A(t) = A_0 e^{it} \tag{4}$$

is the value of an original investment after t years with an interest rate of i compounded continuously

There is another way to arrive at formula (4). In the time period t to $t + \Delta t$, the interest earned is $A(t + \Delta t) - A(t)$. If Δt is small, then the interest paid on $A(t)$ dollars would be [from formula (1)] approximately equal to $A(t)\,\Delta t\,i$. We say "approximately" because $A(t)\,\Delta t\,i$ represents simple interest between t and $t + \Delta t$. However, the difference between this approximation and the actual interest paid is small if Δt is small. Thus

$$A(t + \Delta t) - A(t) \approx A(t)\,\Delta t\,i,$$

or dividing by Δt, we have

$$\frac{A(t + \Delta t) - A(t)}{\Delta t} \approx iA(t).$$

Then taking the limit as $\Delta t \to 0$, we obtain

$$\frac{dA}{dt} = iA(t),$$

and from the last section

$$A(t) = ce^{it}.$$

But $A(0) = A_0$, which tells us that $c = A_0$, and we have again obtained formula (4).

EXAMPLE 4 If the interest in Example 1 is compounded continuously, then

$$A(5) = 1000e^{(0.06)(5)} = 1000e^{0.3} = \$1349.86,$$

and the interest paid is \$349.86. Compare this result to the results in Examples 2 and 3. ▪

EXAMPLE 5 \$5000 is invested in a bond yielding $8\frac{1}{2}\%$ annually. What will the bond be worth in 10 years if interest is compounded continuously?

Solution. $A(t) = A_0 e^{it} = 5000e^{(0.085)(10)} = 5000e^{0.85} = \$11,698.23.$ ▪

EXAMPLE 6 How long does it take for an investment to double if the annual interest rate is 6% compounded continuously?

Solution. We need to determine a value of t such that $A(t) = 2A_0$. That is,

$$A_0 e^{0.06t} = 2A_0, \quad \text{or} \quad e^{0.06t} = 2.$$

Taking natural logarithms, we obtain $0.06t = \ln 2 \approx 0.6931$, and $t \approx 0.6931/0.06 = 11.55$ years. ▪

EXAMPLE 7 What must be the interest rate in order than an investment double in 7 years when interest is compounded continuously?

Solution. We need to determine i such that $A_0 e^{7i} = 2A_0$. Then $e^{7i} = 2$, $7i = \ln 2$, and $i \approx 0.6931/7 = 0.099 = 9.9\%$. ▪

EXAMPLE 8 If money is invested at 8% compounded continuously, what is the effective rate of interest?

Solution. The **effective interest rate** is the actual rate of simple interest received over a 1 year period. If we start with A_0 dollars, there will be $A_0 e^{0.08} \approx A_0(1.0833)$ after 1 year. Thus the effective interest rate is about $8.33 \approx 8\frac{1}{3}\%$. ▪

An extremely important concept in economics is that of the **present value** of money. Since \$1000 invested today will be worth more in the future, say 2 years hence, it follows that \$1000 that will be collected in 2 years is worth *less* today. Put another way, the present value of \$1000 due in 2 years is the amount of money we'd have to invest today (at a given rate of interest) in order to receive \$1000 in 2 years. We can use the compound interest formulas (3) and (4) to calculate present value.

If we invest A_0 dollars today, then after t years our investment is worth $A(t) = A_0[1 + (i/m)]^{mt}$ if compounded periodically and $A(t) = A_0 e^{it}$ if compounded continuously. $A(t)$ dollars after t years has the present value of A_0. Thus present value may be calculated by solving for A_0.

If compounded m times a year,

$$\text{present value} = A_0 = \frac{A(t)}{[1 + (i/m)]^{mt}} = A(t)\left(1 + \frac{i}{m}\right)^{-mt} \tag{5}$$

If compounded continuously,

$$\text{present value} = A_0 = \frac{A(t)}{e^{it}} = A(t)e^{-it}. \tag{6}$$

EXAMPLE 9   What is the present value of $1000 after 5 years at 6% compounded semiannually?

Solution. We have $t = 5$, $A(t) = \$1000$, $i = 0.06$, and $m = 2$, so that

$$A_0 = 1000(1.03)^{-10} = \$744.09.$$

Put another way, we would have to invest $744.09 now at 6% interest to have $1000 in 5 years. ∎

EXAMPLE 10   A manufacturer receives a note promising payment of $25,000 in 3 years. He needs capital now, so he sells the note to a bank who reimburses him based on an annual interest rate of 7% compounded continuously. How much money does the manufacturer receive? (*Note:* The act of purchasing something for its present value is sometimes called **discounting,** and the interest rate i is called the **discount rate**).

Solution. He will receive the present value of the $25,000 based on an interest rate of 7%.

$$A_0 = A(3)e^{-(0.07)3} = 25,000e^{-0.21} = \$20,264.61. ∎$$

EXAMPLE 11   If the manufacturer in Example 10 receives $18,000 from the bank, what is the actual rate of interest he is effectively paying for the loan?

Solution. Here $A_0 = 18,000$, so that

$$18,000 = 25,000e^{-3i}, \quad \text{or} \quad e^{-3i} = \tfrac{18}{25} = 0.72.$$

Then

$$-3i = \ln 0.72 \approx -0.3285, \quad \text{and} \quad i \approx 0.1095 = 10.95\%. ∎$$

In all the above examples we have calculated values of investments assuming that capital was invested at one fixed time. But in many important problems money is invested periodically (e.g., monthly deposits in a bank, annual life insurance premiums, installment loan payments, etc). This type of payment is called an **annuity.** If we deposit a fixed amount B dollars each year, then after 1 year we deposit B dollars; after 2 years the B dollars have now become $B(1 + i)$ (compounded annually), and we deposit another B dollars to obtain $A(2) = B + B(1 + i)$. After 3 years we have $A(3) = B + B(1 + i) + B(1 + i)^2$, and after t years

$$A(t) = B + B(1 + i) + B(1 + i)^2 + \cdots + B(1 + i)^{t-1}$$
$$= B[1 + (1 + i) + (1 + i)^2 + \cdots + (1 + i)^{t-1}].$$

The expression in brackets is the sum of a geometric progression (see Problem 2.8.15 or Section 14.3). Since $1 + a + a^2 + \cdots + a^{n-1} = (1 - a^n)/(1 - a)$, we have

$$A(t) = B\left[\frac{1 - (1 + i)^t}{1 - (1 + i)}\right] = \frac{B[(1 + i)^t - 1]}{i}. \tag{7}$$

If the money is compounded m times during every interval of deposit, then B dollars are worth $B[1 + (i/m)]^m$ dollars after one such interval (from our compound interest formula). Then after t intervals of deposit, we have

$$A(t) = B + B\left(1 + \frac{i}{m}\right)^m + B\left(1 + \frac{i}{m}\right)^{2m} + \cdots + B\left(1 + \frac{i}{m}\right)^{(t-1)m}$$

$$= B\left\{1 + \left(1 + \frac{i}{m}\right)^m + \left[\left(1 + \frac{i}{m}\right)^m\right]^2 + \cdots + \left[\left(1 + \frac{i}{m}\right)^m\right]^{t-1}\right\},$$

or

$$A(t) = B\left\{\frac{[1 + (i/m)]^{mt} - 1}{[1 + (i/m)]^m - 1}\right\}. \tag{8}$$

EXAMPLE 12   If a woman deposits $500 every 6 months and this investment is compounded quarterly at 6%, how much does she have after 10 years?

Solution. Here $B = 500$ and $i = 0.03$ since the interval of deposit is $\frac{1}{2}$ year. Then $m = 2$ (2 payments every $\frac{1}{2}$ year), $t = 20$ (there are 20 semiannual deposits), and

$$A(20) = 500\left\{\frac{[1 + (0.03/2)]^{2(20)} - 1}{[1 + (0.03/2)]^2 - 1}\right\}$$

$$= 500\left(\frac{1.015^{40} - 1}{1.015^2 - 1}\right) \approx \$13,465.98. \quad \blacksquare$$

Suppose money is received, instead of being paid out, at periodic intervals. (For example, you may own an apartment building where monthly rents are received.) What is the present value of such payments? We suppose that an interest rate of i is paid over a fixed time period and assume that $\$B_1$ is expected after the first time period, $\$B_2$ after the second, and so on. Then the present values are $B_1^0 = B_1/(1 + i)$, $B_2^0 = B_2/(1 + i)^2$, and so on, so that the total present value over t time periods is

$$A_0 = \frac{B_1}{1 + i} + \frac{B_2}{(1 + i)^2} + \cdots + \frac{B_t}{(1 + i)^t}. \tag{9}$$

If all payments are equal, then $B_1 = B_2 = \cdots = B_t = B$, and

$$A_0 = B\left[\frac{1}{(1 + i)} + \frac{1}{(1 + i)^2} + \cdots + \frac{1}{(1 + i)^t}\right]$$

$$= \frac{B}{1 + i}\left[1 + \frac{1}{(1 + i)} + \frac{1}{(1 + i)^2} + \cdots + \frac{1}{(1 + i)^{t-1}}\right],$$

or

$$A_0 = \frac{B}{1 + i}\left\{\frac{1 - [1/(1 + i)]^t}{1 - [1/(1 + i)]}\right\} = \frac{B}{i}\left[1 - \left(\frac{1}{1 + i}\right)^t\right]. \tag{10}$$

EXAMPLE 13   What is the present value of an annuity that would pay $3000 a year for 20 years at an interest rate of 6% compounded annually?

Solution. $A(20) = (3000/0.06)(1 - 1/1.06^{20}) = \$34{,}409.76$. ∎

EXAMPLE 14   A new car costs $8000. If you purchase it in 36 monthly installments of $300/month, what is the effective annual interest you are paying?

Solution. The dealer is receiving an annuity of $300/month in exchange for a present value of $8000. Thus in formula (10)

$$A_0 = 8000 = \frac{300}{i}\left[1 - \frac{1}{(1 + i)^{36}}\right].$$

Rewriting, we obtain

$$8000i = 300 - \frac{300}{(1 + i)^{36}}, \qquad 8000i(1 + i)^{36} - 300(1 + i)^{36} + 300 = 0,$$

or dividing by 100,

$$80i(1 + i)^{36} - 3(1 + i)^{36} + 3 = 0.$$

There is, of course, no formula for finding the roots of this 37th-degree polynomial. However, we can estimate the root i by Newton's method (Section 4.8). (Note that $i = 0$ is a root. We seek the first positive root.) Defining

$$P(i) = 80i(1 + i)^{36} - 3(1 + i)^{36} + 3,$$

we have

$$P'(i) = 80(1 + i)^{36} + (80 \cdot 36)i(1 + i)^{35} - 108(1 + i)^{35},$$

and Newton's method yields the sequence

$$i_{n+1} = i_n - \frac{P(i_n)}{P'(i_n)}$$

$$= i_n - \left[\frac{80i_n(1 + i_n)^{36} - 3(1 + i_n)^{36} + 3}{80(1 + i_n)^{36} + 2880i_n(1 + i_n)^{35} - 108(1 + i_n)^{35}} \right],$$

or dividing numerator and denominator by $(1 + i_n)^{35}$ to simplify, we obtain

$$i_{n+1} = i_n - \left\{ \frac{80i_n(1 + i_n) - 3(1 + i_n) + [3/(1 + i_n)^{35}]}{80(1 + i_n) + 2880i_n - 108} \right\}$$

$$= i_n - \left\{ \frac{80i_n^2 + 77i_n - 3 + [3/(1 + i_n)^{35}]}{2960i_n - 28} \right\}.$$

Starting with $i_0 = 0.02$ (representing a monthly rate of 2%), we obtain the values in Table 1. The required interest rate is $i \approx 0.0172 = 1.72\%$ a month (correct to four decimal places). The annual interest is $1.72 \times 12 = 20.64\%$. You would be better off borrowing the money from a bank!

TABLE 1

i_n	$\dfrac{3}{(1 + i_n)^{35}}$	$A,$ $80i_n^2 + 77i_n$ $-3 + \dfrac{3}{(1 + i_n)^{35}}$	$B,$ $2960i_n - 28$	$\dfrac{A}{B}$	$i_{n+1} = i_n - \dfrac{A}{B}$
0.02	1.50008284	0.0720828401	31.2	0.0023103474	0.0176896526
0.0176896526	1.62399129	0.0111284445	24.3613717	0.0004568069	0.0172328456
0.0172328456	1.649712021	0.0003988095	23.00922298	0.0000173325	0.017215513
0.017215513	1.650696151	0.0000005635	22.9579185	0.0000000245	0.0172154884

To conclude this section, we consider the case where income is received continuously, rather than at fixed intervals. Let $a(t)$ denote a **stream of income.** Then the present value of the money received at time t is $a(t)e^{-it}$. Between time t and $t + \Delta t$, the present value of the money received is approximately $a(t)e^{-it} \Delta t$, and under a now familiar argument, the present value of the money received up to the time $t = T$ years is given by

$$\text{present value} = \int_0^T a(t)e^{-it}\, dt. \tag{11}$$

EXAMPLE 15 An investment broker has a choice between investing $100,000 in a growing business or putting it into bonds yielding an 8% annual return, compounded continuously for 20 years. The annual income (received continuously throughout the year) from the business is projected to be $4000(1 + e^{t/10})$ dollars after t years. Where should she put her money?

Solution. Here $a(t) = 4000(1 + e^{0.1t})$. To the broker, the money can earn 8%. Thus the present value of the investment in the business is

$$\int_0^{20} 4000(1 + e^{0.1t})e^{-0.08t} \, dt = 4000 \int_0^{20} (e^{-0.08t} + e^{0.02t}) \, dt$$

$$= 4000 \left\{ \frac{e^{-0.08t}}{-0.08} \bigg|_0^{20} + \frac{e^{0.02t}}{0.02} \bigg|_0^{20} \right\} = 4000 \left(\frac{1 - e^{-1.6}}{0.08} + \frac{e^{0.4} - 1}{0.02} \right)$$

$$\approx 4000(9.9763 + 24.5912) = (4000)(34.5675) = \$138{,}270.$$

Thus if she invests in the business, she can increase the present value of her money by about 38%. Of course, this is under the assumption that there is no greater risk in the business investment. ▪

The notions of discounting and present value have been applied to the management of our dwindling supply of natural resources. For example, what is the value to us now of oil that will be worth more per barrel in the future? Harder to evaluate is the present value of an animal nearing extinction. In order to manage wisely, we must "discount the future" to have a real understanding of what we have now. The more we discount into the future, the more apparent it becomes that what we have is less than we thought we had. There is a very strong interdependence between economic and biological forces in problems of resource management.†

PROBLEMS 6.7

1. A sum of $5000 is invested for 8 years at a return of 7% per year. What simple interest is paid over that period of time?
2. If the money in Problem 1 is compounded annually, what is the investment worth after 8 years?
3. If the money in Problem 1 is compounded monthly, what is the investment worth after 8 years?
4. If the money in Problem 1 is compounded continuously, what is the investment worth after 8 years?
5. Calculate the percentage increase in return on investment if A_0 is invested for 10 years at 6% compounded annually and quarterly.
6. As a gimmick to lure depositors, some banks offer 5% interest compounded continuously in comparison to their competitors who offer $5\frac{1}{8}\%$ compounded annually. Which bank would you choose?
7. Suppose a competitor in Problem 6 now compounds $5\frac{1}{8}\%$ semiannually. Which bank would you choose?
8. Suppose $10,000 is invested in bonds yielding 9% compounded continuously. What will the bonds be worth in 8 years?
9. A certain government bond sells for $750 and can be redeemed for $1000 in 8 years. Assuming continuous compounding, what is the rate of interest paid?
10. How long would it take an investment to increase by half if it is invested at 4% compounded continuously?

---

†The interested reader is urged to look at one or more of these references: (1) S. V. Ciriacy-Wantrup, *Resource Conservation: Economics and Policies* (Univ. of California Press, Berkeley, 1963); (2) C. W. Clark, "Profit maximization and the extinction of animal species," *Journal of Political Economy*, **81**, 950–961 (1973); (3) C. W. Clark, "Mathematical bioeconomics" in *Mathematical Problems in Biology—Victoria Conference* (Springer-Verlag, New York, 1974), pages 29–45.

11. What must be the interest rate in order that an investment triple in 15 years if interest is continuously compounded?

12. If money is invested at 10% compounded continuously, what is the effective interest rate?

13. What is the most a banker should pay for a $10,000 note due in 5 years if he can invest a like amount of money at 9% compounded annually?

14. What is the present value of $5000 in 3 years at 5% compounded quarterly?

15. What is the present value of $8000 in 10 years compounded continuously at 4%?

16. A sales firm receives a cash payment of $8000 for a $15,000 note due in 5 years. Assuming continuous compounding, what interest is it paying?

*17. A real estate dealer owns recreation land that she calculates will appreciate in value according to the formula

$$V(t) = 100,000(1 + e^{\sqrt{0.2}\,t}),$$

where t is measured in years. At any time she sells, she can reinvest the money and be guaranteed a 12% return compounded continuously. When should she sell?

18. If a $2000 bond is bought every year and the money earns 7% compounded annually, how much is available after 20 years?

*19. A man wishes to prepare for his child's education. He calculates that he will need $25,000 in 15 years. He wishes to make monthly deposits into an account that pays 5% compounded quarterly. How large must his payments be if he is to meet his goal?

20. A decision must be made for a capital equipment purchase. For an interest rate of 7% compounded continuously, what is the maximum reasonable purchase price for a machine whose earnings over the next 5 years are projected to be $1000, $1500, $1800, $1300, and $700? Assume a value of zero after 5 years.

21. A novelist writes a novel loaded with adventure and sex, which centers around the story of a five-year-old beagle named Bennette who can turn lead into gold. A publisher is eager to publish this book since it is expected to be a best-seller. The author demands a cash advance based on first year royalties, which are anticipated to be $250,000. Because of the time needed to revise the novel and publish the book, the first royalty payments are not due for three years. The publisher must pay 11% compounded continuously on any money he pays out now in anticipation of future earnings. If the author's demand is to be met, what is a reasonable cash advance?

*22. A broker invests $10,000 in second mortgages and receives $100/month for 20 years. At the end of that time his investment is worthless. What is his effective rate of return on capital?

23. An investment counselor has the opportunity to purchase an apartment building that yields, after expenses, $6,000 a month. The purchase price is $750,000. She calculates that she will hold the property for 20 years, after which it will begin to deteriorate and thereafter be essentially worthless. Assume that she could earn 6% compounded monthly in another venture.

(a) Calculate the present value of the investment in the building.

(b) Should she make the purchase?

[*Hint:* Here $i = \frac{1}{2}\%$ and t is the number of months.]

24. Answer the questions in Problem 23 assuming that the counselor can realize a return of 12% compounded monthly.

*25. In Problems 23 and 24, for what annual interest rate (compounded monthly) will it make no difference whether or not the counselor invests in the building?

*26. A new refrigerator costs $400. At 24 monthly installments of $20, what is the effective annual interest being paid?

27. In order to achieve an effective annual interest of 6% in Problem 26, what monthly payment should be paid?

28. A $30,000 mortgage is obtained for a house with monthly payments of $230 spread over a 30-year period. What is the interest rate?

29. What would be the monthly payments on a $25,000, 25-year mortgage at $7\frac{1}{2}\%$?

30. What is the present value of an annuity if the stream of income is $a(t) = 10,000(1 + e^{0.13t})$ after 10 years, assuming continuous compounding at 8%?

6.8 A MODEL FOR EPIDEMICS (OPTIONAL)

In recent years many mathematicians and biologists have attempted to find reasonable mathematical models to describe the growth of an epidemic in a population. One such model has been used by the Center for Disease Control in Atlanta, Georgia, to help form a public-testing policy to limit the spread of gonorrhea.† In this section we provide a simple model to describe what may happen in an epidemic.

An epidemic is the spread of an infectious disease through a community, affecting a significant proportion of the population. The epidemic may begin when a certain number of infected individuals invade the community. The epidemic could result, for example, from new people moving to the community or old residents returning from a trip. We make the following assumptions:

(i) Everyone in the community is, initially, susceptible to the disease. That is, no one is immune.

(ii) The disease is spread only by direct contact between a susceptible person (hereafter called a **susceptible**) and an infected person (called an **infective**).

(iii) Everyone who has had the disease and has *recovered* is immune, although such a person may still be an infective. The people who die from the disease are also considered to be in the **recovered class** in this model.

(iv) After the disease has been introduced into the community, the total population of the community remains fixed and is denoted N.

(v) The infectives are introduced into the community (i.e., the epidemic "starts") at time $t = 0$.

To model the spread of the disease, we define three variables:

$x(t)$ is the number of susceptibles at time t.

$y(t)$ is the number of infectives at time t.

$z(t)$ is the number of recovered persons at time t.

Then by assumption (iv), we have

$$x(t) + y(t) + z(t) = N. \tag{1}$$

Also, we have

$$y(0) = \text{number of initial infectives.} \tag{2}$$

---

†J. A. Yorke, H. W. Hethcote, and A. Nold, "Dynamics and control of the transmission of gonorrhea," *Sexually Transmitted Diseases*, **5** (No. 2), 51–56 (1978).

$$x(0) = N - y(0), \tag{3}$$

and

$$z(0) = 0. \tag{4}$$

Equation (4) states the obvious fact that no one has yet recovered or died from the disease at the time the epidemic begins.

To get a better idea of what is going on, look at Figure 1. There we see that a susceptible can become infective and an infected person can recover. These are the only possibilities. For reasons that are obvious, the model we are describing is, in the literature, usually referred to as an **SIR model.**

FIGURE 1

How does the disease spread? Evidently, the disease will spread more rapidly, (i.e., susceptibles will become infectives) if we increase the number of infectives or the number of susceptibles. Thus it is reasonable to assume that the rate of change of the number of susceptibles is proportional both to the number of susceptibles and to the number of infectives. In mathematical terms we have

$$x'(t) = -\alpha x(t)y(t), \tag{5}$$

where α is a constant of proportionality and the minus sign indicates that the number of susceptibles is decreasing. Equation (5) is called the **law of mass action.**

On the other hand, it is reasonable to assume that the rate at which people recover or die from the disease is proportional to the number of infectives. This assumption gives us the equation

$$z'(t) = \beta y(t). \tag{6}$$

In Equations (5) and (6) α is often called the **infection rate** and β is called the **removal rate.**

A major question about any epidemic is, can it be controlled? It is clear that the quantity $y'(t)$ is a measure of how bad the epidemic is. The bigger $y'(t)$ is, the greater is the number of people becoming infected. We can say that the epidemic has been **controlled** if at some point $y'(t) \leq 0$. Certainly, at some point $y'(t)$ must become negative. This follows from the fact that the population size N is fixed. In the worst possible case everyone becomes infected. When this outcome occurs, $y'(t)$ will be negative as infectives die or recover, and there cannot possibly be any new infectives (there's no one left to become infected). Thus the epidemic is really controlled if $y'(t) \leq 0$ for every $t > 0$. That is, no more people are, at the onset, entering the class of infectives than are leaving it.

We begin our analysis by dividing equation (5) by equation (6):

$$\frac{x'(t)}{z'(t)} = \frac{-\alpha x(t)y(t)}{\beta y(t)} = \frac{-\alpha}{\beta} x(t). \tag{7}$$

Rearranging the terms in (7) yields

$$\frac{x'(t)}{x(t)} = \frac{-\alpha}{\beta} z'(t),$$

and after integrating both sides, we have

$$\int \frac{x'(t)}{x(t)} dt = \frac{-\alpha}{\beta} \int z'(t) \, dt + C,$$

or

$$\ln x(t) = \frac{-\alpha}{\beta} z(t) + C. \tag{8}$$

Then setting $t = 0$ in (8) and noting that $z(0) = 0$, we have

$$\ln x(0) = 0 + C, \qquad \text{or} \qquad C = \ln x(0).$$

Thus

$$\ln x(t) = \frac{-\alpha}{\beta} z(t) + \ln x(0),$$

or

$$\ln x(t) - \ln x(0) = \frac{-\alpha}{\beta} z(t),$$

or

$$\ln \frac{x(t)}{x(0)} = \frac{-\alpha}{\beta} z(t), \qquad \text{or} \qquad \frac{x(t)}{x(0)} = e^{(-\alpha/\beta)z(t)}.$$

Finally,

$$x(t) = x(0)e^{(-\alpha/\beta)z(t)}. \tag{9}$$

Then we find that

$$x'(t) = x(0)\left(\frac{-\alpha}{\beta}\right)e^{(-\alpha/\beta)z(t)}[z'(t)].$$

Since $-\alpha/\beta < 0$, $x(0) > 0$, and $z'(t) \geq 0$ [from 6], we find that, for every $t \geq 0$,

$$x'(t) \leq 0.† \tag{10}$$

Suppose that

$$\beta > \alpha x(0). \tag{11}$$

Then $x(0) \leq \beta/\alpha$, and since $x(t)$ is decreasing, by (10),

$$x(t) \leq \frac{\beta}{\alpha} \qquad \text{for every} \qquad t \geq 0. \tag{12}$$

From (1),

$$y(t) = N - x(t) - z(t),$$

so that

$$y'(t) = -x'(t) - z'(t) = \alpha x(t)y(t) - \beta y(t),$$

or

$$y'(t) = [\alpha x(t) - \beta]y(t). \tag{13}$$

Now we can answer our question. For if $\beta \geq \alpha x(0)$, then by (12), $\alpha x(t) - \beta < 0$ for every $t \geq 0$, and using this fact in (13) tells us that

$$y'(t) \leq 0 \qquad \text{for every} \qquad t \geq 0.$$

That is:

> The epidemic will be controlled at the start if the condition $x(0) < \beta/\alpha$ holds.

Often the term β/α is called the **relative removal rate.** Then our result can be stated as follows:

> An epidemic will not ensue if the initial number of susceptibles (N minus the number of initial infectives) does not exceed the relative removal rate.

EXAMPLE 1 Let $N = 1000$, $y(0) = 50$, $\alpha = 0.0001$, and $\beta = 0.01$. Then $x(0) = 950$ and $\beta/\alpha = 100$, so $x(0) > \beta/\alpha$, and an epidemic will ensue. ∎

---

†This fact is obvious from the model, since the number of susceptibles decreases.

EXAMPLE 2   Let $N = 1000$, $y(0) = 50$, $\alpha = 0.0001$, and $\beta = 0.1$. Then $x(0) = 950$, $\beta/\alpha = 1000$, and an epidemic will not ensue. ∎

We now ask another question. Suppose that $x(0) > \beta/\alpha$ and an epidemic does ensue. How many people will eventually become infected? This question is more difficult to answer. Suppose that the last person becomes infected at time T. Then the total number of people who have become infected up until and including time T is given by

$$W = y(T) + z(T). \tag{14}$$

The number W is called the **extent** of the epidemic. Equation (14) can be explained by noting that every person who was ever infected is either still infected or has died or recovered. If the epidemic has ended, then

$$x'(t) = 0 \qquad \text{for} \qquad t \geq T. \tag{15}$$

But by (5)

$$x'(T) = -\alpha x(T)y(T) = 0. \tag{16}$$

If $x(T) = 0$, then $y(T) + z(T) = N$, and everyone has become infected. If $x(T) > 0$, then (16) implies that $y(T) = 0$. Thus in this more interesting case

$$W = z(T).$$

Now in the interval $[0, T]$,

$$z'(t) = \beta y(t) = \beta[N - x(t) - z(t)]$$
$$= \beta[N - x(0)e^{(-\alpha/\beta)z(t)} - z(t)].$$

But $z'(T) = \beta y(T) = 0$, so that

$$0 = \beta[N - x(0)e^{(-\alpha/\beta)z(T)} - z(T)]. \tag{17}$$

Thus, using $z(T) = W$, we can divide both sides of (17) by β to obtain

$$N - x(0)e^{(-\alpha/\beta)W} - W = 0. \tag{18}$$

If α and β are > 0, it is impossible to find an explicit solution for W in terms of α, β, and N. The best we can do is solve it numerically. This practice is a reasonable thing to do using the powerful Newton's method (see Section 4.8). ∎

EXAMPLE 3   Let $N = 1000$, $y(0) = 50$, $\alpha = 0.0001$, and $\beta = 0.09$. Compute the extent of the epidemic.

Solution. We first note that $x(0) = 950$ and $\beta/\alpha = 900$, so that $x(0) > \beta/\alpha$, and an epidemic will ensue. Since $\alpha/\beta = 1/900$, we must find a root of the equation

$$f(W) = 1000 - 950e^{-W/900} - W = 0. \tag{19}$$

Then

$$f'(W) = \frac{950}{900} e^{-W/900} - 1,$$

and by Newton's method (see Section 4.8) we obtain the iterates

$$W_{n+1} = W_n - \frac{f(W_n)}{f'(W_n)},$$

or

$$W_{n+1} = W_n - \frac{1000 - 950e^{-W_n/900} - W_n}{(950/900)e^{-W_n/900} - 1}. \tag{20}$$

We have no idea, initially, what W is [although $W > y(0) = 50$], so we start with a guess: $W_0 = 100$. We then obtain the iterates given in Table 1. After 5 iterations we obtain the value $W_5 = 370.7732035$, which is correct to ten significant figures, as can be verified by plugging it into equation (19). Thus $W \approx 371$, which means that by the time the epidemic has run its course, 371 individuals have been infected. Of course, we cannot say how many of these remain infected, recover, or die. ∎

TABLE 1

n	W_n	$e^{-W_n/900}$	(a) $1000 - 950e^{-W_n/900} - W_n$	(b) $\frac{950}{900}e^{-W_n/900} - 1$	$\frac{(a)}{(b)}$	$W_{n+1} = W_n - \frac{(a)}{(b)}$
0	100	0.8948393168	49.90264904	−0.0554473878	−900	1000
1	1000	0.3291929878	−312.7333384	−0.6525185129	479.2712119	520.7287882
2	520.7287882	0.560689758	−53.38405829	−0.4081608110	130.7917293	389.9370589
3	389.9370589	0.6483896843	−5.907258934	−0.3155886666	18.71822267	371.2188362
4	371.2188362	0.6620161196	−0.1341497754	−0.3012052071	0.4453766808	370.7734595
5	370.7734595	0.6623438079	−0.0000770206	−0.3008593139	0.000256002	370.7732035
6	370.7732035	0.6623439963	0.0000000002	−0.3008591150	−0.0000000006	370.7732035

Before leaving this section we mention some of the limitations of this model. The constants α and β will often vary with time. Also, the recovered individuals may, after a time, lose their immunity (or never acquire it) and reenter the susceptible state. This situation could give rise to periodic epidemics in which individuals get the disease many times. This lack of immunity holds for many diseases, notably gonorrhea and certain types of flu. Finally, there may be many factors other than relative population sizes that control the sizes of the three classes. Such factors might include weather, the available food supply, living conditions, and the presence of other diseases in the community. However, even a simple model like this one can give us the kind of insight needed to study more complicated situations. In a few cases this has been done with great success.

PROBLEMS 6.8

In Problems 1–5, values for N, $y(0)$, α, and β are given. Determine whether an epidemic will occur, and if so, find the extent of the epidemic.

1. $N = 1000$, $y(0) = 100$, $\alpha = 0.001$, $\beta = 0.4$
2. $N = 1000$, $y(0) = 100$, $\alpha = 0.0001$, $\beta = 0.1$
3. $N = 10{,}000$, $y(0) = 1500$, $\alpha = 10^{-5}$, $\beta = 0.2$
4. $N = 10{,}000$, $y(0) = 1500$, $\alpha = 10^{-5}$, $\beta = 0.02$
5. $N = 25{,}000$, $y(0) = 5000$, $\alpha = 10^{-5}$, $\beta = 0.1$

6. Let $f(W)$ be given by (19). Show that there is exactly one *positive* value of W for which $f(W) = 0$. [*Hint:* Show that $f(0) > 0$ and that $f''(W) < 0$ for all W and that $f(W) < 0$ if W is sufficiently large.]

REVIEW EXERCISES FOR CHAPTER SIX

In Exercises 1–6, solve for the given variable.

1. $y = \log_3 9$ 2. $y = \log_{1/3} 9$ 3. $4 = \log_x \frac{1}{16}$
4. $y = e^{\ln 17.2}$ 5. $\log x = 10^{-9}$ 6. $\log_x 32 = -5$

7. Simplify

$$\ln \left[\frac{\sqrt{x^3 - 1}(x^5 + 1)^{4/3}}{\sqrt{(x - 3)(x - 2)}} \right]^{-5/6}.$$

8. If $y = 3 \ln x$, what happens to y if x doubles?
9. Sketch the curve $y = \ln |x + 1|$.   10. Sketch the curve $y = e^{x+2}$.

In Exercises 11–25, calculate the given derivative or integral.

11. $\dfrac{d}{dx} \ln(1 + x^2)$ 12. $\displaystyle\int \dfrac{3x}{1 + x^2}\, dx$ 13. $\dfrac{d}{dx} e^{x^2 + 2x + 1}$

14. $\dfrac{d}{dx} e^{\sqrt{x^3 - 3}}$ 15. $\displaystyle\int \dfrac{e^{\sqrt{x}}}{5\sqrt{x}}\, dx$ 16. $\displaystyle\int \dfrac{1}{3x \ln x}\, dx$

17. $\dfrac{d}{dx} \ln |e^x - 5|$ 18. $\dfrac{d}{dx} \ln[x + \ln(x + 3)]$ 19. $\displaystyle\int \dfrac{e^{-1/x^2}}{x^3}\, dx$

20. $\dfrac{d}{dx} x^{x+1}$ 21. $\dfrac{d}{dx} \left[\dfrac{\sqrt[3]{(x + 5)}(x - 5)}{\sqrt{(x + 1)}(x + 2)} \right]$ 22. $\displaystyle\int \dfrac{4^{\ln x}}{x}\, dx$

23. $\displaystyle\int \dfrac{1}{x} \sqrt{\log_5 x}\, dx$ 24. $\dfrac{d}{dx} \log_3 2^{x+5}$ 25. $\dfrac{d}{dx} e^{\ln(x^3 + 1)}$

In Exercises 26–29, sketch the given curve, indicating local maxima and minima and points of inflection.

26. $y = xe^{x^2+1}$

27. $y = \dfrac{x}{\ln x}$

28. $y = xe^{-x}$

29. $y = \ln |\ln x|$

30. The relative annual rate of growth of a population is 15%. If the initial population is 10,000, what is the population after 5 years? After 10 years?

31. In Exercise 30, how long will it take for the population to double?

32. When a cake is taken out of the oven, its temperature is 125°C. Room temperature is 23°C. The temperature of the cake is 80°C after 10 min.
 (a) What will be its temperature after 20 min?
 (b) How long will the cake take to cool to 25°C?

33. A fossil contains 35% of the normal amount of ^{14}C. What is its approximate age?

34. What is the half-life of an exponentially decaying substance that loses 20% of its mass in one week?

35. How long will it take the substance in Exercise 34 to lose 75% of its mass? 95% of its mass?

36. A sum of $10,000 is invested at an interest rate of 6% compounded quarterly. What is the investment worth in 8 years?

37. What is the investment in Exercise 36 worth if interest is compounded continuously?

38. What is the present value of a $5000 note payable in 4 years based on 7% compounded semiannually?

39. What is the present value of the note in Exercise 38 compounded continuously?

40. What is the present value of an annuity yielding $4000(1 + e^{0.15t})$ dollars over the next 10 years, assuming continuous compounding at 9%?

In Exercises 41–48, determine intervals over which the given function is 1–1, and determine the inverse function over each such interval, if possible. Then calculate the derivative of each inverse function.

41. $f(x) = \sqrt{2x - 5}$

42. $f(x) = \sqrt[3]{x - 2}$

43. $f(x) = e^{x+4}$

44. $f(x) = \ln(4x - 3)$

45. $f(x) = \sin 3x$

46. $f(x) = \sqrt[3]{e^x + 2}$

47. $y = x^2 + 5x + 6$

48. $y = x^4 + 3x^2 + 2$

7 More on Trigonometric Functions and the Hyperbolic Functions

7.1 INTEGRATION OF TRIGONOMETRIC FUNCTIONS

In Section 3.5 we computed the derivatives of the six trigonometric functions. The following theorem follows from the formulas obtained in Section 3.5 and the chain rule. Keep in mind that, as in Section 3.5, if x, θ, or u represents an angle, then x, θ, or u is measured in *radians*.

Theorem 1. Suppose $u(x)$ is a differentiable function of x and $u(x)$ is measured in radians.

$$\textbf{(i) } \frac{d}{dx} \sin u = \cos u \, \frac{du}{dx}$$

$$\textbf{(ii) } \frac{d}{dx} \cos u = -\sin u \, \frac{du}{dx}$$

$$\textbf{(iii) } \frac{d}{dx} \tan u = \sec^2 u \, \frac{du}{dx}$$

$$\textbf{(iv) } \frac{d}{dx} \cot u = -\csc^2 u \, \frac{du}{dx}$$

$$\textbf{(v) } \frac{d}{dx} \sec u = \sec u \tan u \, \frac{du}{dx}$$

$$\textbf{(vi) } \frac{d}{dx} \csc u = -\csc u \cot u \, \frac{du}{dx}$$

We can reverse the process of differentiation to calculate the integrals of certain trigonometric functions. The formulas in Table 1 follow from Theorem 1.

TABLE 1

(i) $\int \cos u \, du = \sin u + C$

(iv) $\int \csc^2 u \, du = -\cot u + C$

(ii) $\int \sin u \, du = -\cos u + C$

(v) $\int \sec u \tan u \, du = \sec u + C$

(iii) $\int \sec^2 u \, du = \tan u + C$

(vi) $\int \csc u \cot u \, du = -\csc u + C$

In this section we will integrate functions that can be written in one of the above forms. In Chapter 8 we will see how other trigonometric functions can be integrated.

EXAMPLE 1   Calculate $\int 2 \cos 2x \, dx$.

Solution. If $u = 2x$, then $du = 2 \, dx$, so that

$$\int 2 \cos 2x \, dx = \int \cos u \, du = \sin u + C = \sin 2x + C. \ \blacksquare$$

EXAMPLE 2   Calculate $\int x^2 \sec x^3 \tan x^3 \, dx$.

Solution. If $u = x^3$, then $du = 3x^2 \, dx$, so we multiply and divide by 3 to obtain

$$\frac{1}{3} \int \sec x^3 \tan x^3 \cdot 3x^2 \, dx = \frac{1}{3} \int \sec u \tan u \, du = \frac{1}{3} \sec u + C = \frac{1}{3} \sec x^3 + C. \ \blacksquare$$

EXAMPLE 3   Calculate $\int_0^{\pi/3} \tan^5 x \sec^2 x \, dx$.

Solution. If we set $u = \tan x$, then $du = \sec^2 x \, dx$, so that

$$\int \tan^5 x \sec^2 x \, dx \int u^5 \, du = \frac{u^6}{6} + C = \frac{\tan^6 x}{6} + C,$$

and

$$\int_0^{\pi/3} \tan^5 x \sec^2 x \, dx = \left. \frac{\tan^6 x}{6} \right|_0^{\pi/3} = \frac{(\sqrt{3})^6}{6} = \frac{27}{6} = \frac{9}{2}. \ \blacksquare$$

$$\swarrow \tan \frac{\pi}{3} = \sqrt{3}$$

EXAMPLE 4   Calculate $\int \tan x \, dx$.

Solution.

$$\int \tan x \, dx = \int \frac{\sin x}{\cos x} \, dx.$$

If $u = \cos x$, then $du = -\sin x \, dx$, so that

$$\int \frac{\sin x \, dx}{\cos x} = -\int \frac{-\sin x \, dx}{\cos x} = -\int \frac{du}{u} = -\ln|u| + C = -\ln|\cos x| + C.$$

Since $-\ln x = \ln(1/x)$, we also have

$$\int \frac{\sin x}{\cos x} \, dx = \ln|\sec x| + C. \quad \blacksquare$$

EXAMPLE 5 Calculate $\int_0^{\pi/2} \sin^2 x \, dx$.

Solution. By identity (xxi) in Appendix 1.2,

$$\int_0^{\pi/2} \sin^2 x \, dx = \int_0^{\pi/2} \frac{1 - \cos 2x}{2} \, dx = \int_0^{\pi/2} \frac{1 \, dx}{2} - \int_0^{\pi/2} \frac{\cos 2x}{2} \, dx$$

$$= \frac{x}{2}\Big|_0^{\pi/2} - \frac{\sin 2x}{4}\Big|_0^{\pi/2} = \frac{\pi}{4}. \quad \blacksquare$$

EXAMPLE 6 Calculate the area of the region in the first quadrant bounded by $y = \sin x$, $y = \cos x$, and the y-axis.

Solution. The area requested is drawn in Figure 1. The curves intersect when $\sin x = \cos x$ or $\tan x = 1$, and $x = \pi/4$. Then

$$A = \int_0^{\pi/4} (\cos x - \sin x) \, dx = (\sin x + \cos x)\Big|_0^{\pi/4} = \sqrt{2} - 1. \quad \blacksquare$$

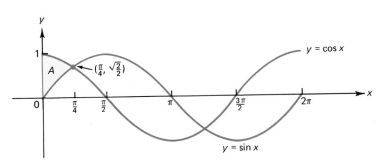

FIGURE 1

PROBLEMS 7.1

In Problems 1–36, calculate the integrals.

1. $\int \sin 3x \, dx$

2. $\int \cos \frac{x}{5} \, dx$

3. $\int_0^{\pi/12} 3 \cos 2x \, dx$

4. $\int_{-\pi/8}^{\pi/8} \frac{\sin 4x}{2} \, dx$

5. $\int \sin 2x \cos 2x \, dx$

6. $\int \cos^3 x \sin x \, dx$

7. $\displaystyle\int_0^{\pi/6} \sqrt{\sin x}\, \cos x\, dx$

8. $\displaystyle\int \sin^{3/2} \frac{x}{2} \cos \frac{x}{2}\, dx$

9. $\displaystyle\int \sec 4x \tan 4x\, dx$

10. $\displaystyle\int_0^{3\pi/4} \tan \frac{x}{3}\, dx$

11. $\displaystyle\int_0^{\pi/9} (\sin^2 x + \cos^2 x)\, dx$

12. $\displaystyle\int_0^{\pi/2} \sec^2(2 + x) \tan(2 + x)\, dx$

13. $\displaystyle\int \sqrt{\cos x}\, \sin x\, dx$

14. $\displaystyle\int_0^{\pi/10} \sin 5x \cos 5x\, dx$

15. $\displaystyle\int x \csc^2 x^2\, dx$

16. $\displaystyle\int \frac{\csc \sqrt{x} \cot \sqrt{x}}{\sqrt{x}}\, dx$

17. $\displaystyle\int \cos^{1/3} 2x \sin 2x\, dx$

18. $\displaystyle\int 3 \sin^2 \frac{x}{2} \cos \frac{x}{2}\, dx$

19. $\displaystyle\int (\sec x)^{6/5} \tan x\, dx$

20. $\displaystyle\int_0^{\pi/4} (\sec^2 x + \tan^2 x)\, dx$ [Hint: Use the identity $\tan^2 x = \sec^2 x - 1$.]

21. $\displaystyle\int x \cot x^2 \csc x^2\, dx$

22. $\displaystyle\int_0^1 \sec x \cos x\, dx$

23. $\displaystyle\int_0^{\pi/4} \sec^5 x \tan x\, dx$

24. $\displaystyle\int \csc^{3/2} x \cot x\, dx$

25. $\displaystyle\int \sec^2 x \cot x\, dx$ $\left[\text{Hint: } \cot x = \frac{1}{\tan x}.\right]$

26. $\displaystyle\int_{\pi/4}^{\pi/2} \frac{1 + \cot^2 x}{\csc^2 x}\, dx$

*27. $\displaystyle\int \frac{1}{\sqrt{\csc x \tan x}}\, dx$

28. $\displaystyle\int \sin^2 5x \csc 5x\, dx$

*29. $\displaystyle\int \frac{\cos^2 \sqrt{x} \sec \sqrt{x}}{\sqrt{x}}\, dx$

30. $\displaystyle\int \cot \frac{x}{8}\, dx$

31. $\displaystyle\int \frac{\cos x}{1 + \sin x}\, dx$

32. $\displaystyle\int x \tan x^2\, dx$

33. $\displaystyle\int \cot(2 + x)\, dx$

34. $\displaystyle\int e^{\cos x} \sin x\, dx$

35. $\displaystyle\int e^{-2 \tan x} \sec^2 x\, dx$

36. $\displaystyle\int \sin(\cos x) \sin x\, dx$

*37. Calculate $\int \sec x\, dx$. [Hint: Multiply the numerator and denominator by $\sec x + \tan x$.]

*38. Calculate $\int \csc x\, dx$.

39. To calculate $\int \sin x \cos x\, dx$, we first set $u = \sin x$. Then

$$\int \sin x \cos x\, dx = \int u\, du = \frac{u^2}{2} + C = \frac{\sin^2 x}{2} + C.$$

Next, we set $u = \cos x$, so that $du = -\sin x\, dx$ and

$$\int \sin x \cos x\, dx = -\int \cos x(-\sin x)\, dx = -\int u\, du = -\frac{u^2}{2} + C = -\frac{\cos^2 x}{2} + C.$$

Can you explain this apparent discrepancy?

40. What is wrong with the following calculation?

$$\int_0^\pi \sec^2 x \, dx = \tan x \Big|_0^\pi = \tan \pi - \tan 0 = 0$$

41. Calculate the area bounded by the curve $y = \sin x$, the x-axis, the line $x = 0$, and the line $x = \pi$.

42. Calculate the area bounded by the curve $y = \cot x$, the x-axis, the line $x = \pi/3$, and the line $x = \pi/2$.

43. Calculate the area bounded by one arch of the curve $y = \sin^2 x$ and the x-axis.

44. Calculate the area bounded by the curve $y = \sec x$, the x- and y-axes, and the line $x = \pi/4$.

45. Calculate the area bounded by the x-axis and one arch of the curve $y = \cos^2(x/3)$.

46. Calculate the area bounded by $y = \sin x$, $y = \cos x$, and the line $x = \pi$ for $x \le \pi$.

47. Calculate the area bounded by the curve $y = \tan x$, the y-axis, and the line $y = 1$.

48. Calculate the area bounded by the curve $y = \csc^2 x$, the lines $x = \pi/4$ and $x = \pi/2$, and the x-axis.

49. Calculate the area bounded by the curves $y = \sin^2 x$ and $y = \cos^2 x$ and the y-axis, $0 \le x \le \pi/4$.

50. Find one of the areas bounded by the curve $y = \csc^2 x$ and the line $y = 2$.

51. A manufacturer of machine parts finds that her marginal cost (in dollars) per unit fluctuates and is given by MC $= 7.5 + 2 \cos(\pi q/500)$, where q is the number of units manufactured. What is the total cost of producing 250 units if her fixed cost (overhead) is $50?

52. The density of a nonuniform 4-m metal bar is given by

$$\rho = 0.8 + \tan\left(1 + \frac{x}{8}\right)$$

(measured in kilograms per meter). What is the total mass of the bar?

53. (a) Compute $\int [\cos\theta/(1 + \sin\theta)] \, d\theta$.   **(b)** Compute $\int (\sin\theta/\cos\theta) \, d\theta$.
   (c) Show that $\cos\theta/(1 + \sin\theta) = (1 - \sin\theta)/\cos\theta$.
   (d) Use the results of (a), (b), and (c) to compute $\int \sec\theta \, d\theta = \int (1/\cos\theta) \, d\theta$.

54. (a) Show that $\sin\theta/(1 - \cos\theta) = (1 + \cos\theta)/\sin\theta$.
   (b) Use the result of (a) to compute $\int \csc\theta \, d\theta = \int (1/\sin\theta) \, d\theta$.

55. (a) Compute $(d/dx)(e^x \sin x)$.   **(b)** Compute $(d/dx)(e^x \cos x)$.
   (c) Compute $(d/dx)[e^x(\sin x + \cos x)]$.   **(d)** Compute $(d/dx)[e^x(\sin x - \cos x)]$.
   (e) Use the result of part (d) to find $\int e^x \sin x \, dx$.
   (f) Use the result of part (c) to find $\int e^x \cos x \, dx$.

7.2 THE INVERSE TRIGONOMETRIC FUNCTIONS

Let $y = \sin x$. Can we solve for x as a function of y? Using what we now know, the answer is no. Suppose that $\sin x = \frac{1}{2}$. Then x could be $\pi/6$, or it could be $5\pi/6$, or it could be $13\pi/6$, and so on. In fact, if y is any number in the interval $[-1, 1]$, then there are an *infinite* number of values of x for which $\sin x = y$. This is illustrated in Figure 1. At the circled points, $\sin x = y_0$. We can eliminate this problem by restricting x to lie in a certain interval, say $[-\pi/2, \pi/2]$. In the interval $[-\pi/2, \pi/2]$, $\sin x$ is 1–1 (see page 351). Thus, as in Figure 1, for each value of y in $[-1, 1]$, there is a unique value x in $[-\pi/2, \pi/2]$ such that $\sin x = y$. Then from the discussion in Section 6.1, $\sin x$ has an inverse.

Definition 1   THE INVERSE SINE FUNCTION   The **inverse sine function** is the

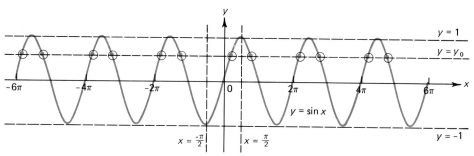

FIGURE 1

function that assigns to each number x in $[-1, 1]$ the unique number y in $[-\pi/2, \pi/2]$ such that $x = \sin y$. We write

$$y = \sin^{-1} x. \tag{1}$$

NOTE. The -1 appearing in (1) does *not* mean $1/(\sin x)$, which is equal to $(\sin x)^{-1} = \csc x$.

Another commonly used notation for the inverse sine function is

$$y = \arcsin x. \tag{2}$$

EXAMPLE 1   Calculate $\sin^{-1} x$ for the following:

(a) $x = 0$ (b) $x = 1$ (c) $x = \frac{1}{2}$ (d) $x = -1$

Solution.

(a) $\sin^{-1} 0 = 0$ since $\sin 0 = 0$.
(b) $\sin^{-1} 1 = \pi/2$ since $\sin \pi/2 = 1$.
(c) $\sin^{-1} \frac{1}{2} = \pi/6$ since $\sin \pi/6 = \frac{1}{2}$.
(d) $\sin^{-1}(-1) = -\pi/2$ since $\sin(-\pi/2) = -1$. ■

We emphasize that the function $y = \sin^{-1} x = \arcsin x$ is only defined if we restrict y to lie in $[-\pi/2, \pi/2]$. Note, for the moment, that if x is in $[-1, 1]$ and if y is in $[-\pi/2, \pi/2]$, then $\sin(\sin^{-1} x) = x$ and $\sin^{-1}(\sin y) = y$. (This result follows from Definition 1 and is consistent with the definition of an inverse function). However, it is not true in general. For example, let $y = 2\pi$. Then $x = \sin y = 0$ and $y = \sin^{-1} x = \sin^{-1}(\sin y) = \sin^{-1}(0) = 0$, which is not equal to the original value of y. We now differentiate $y = \sin^{-1} x$ for y in $[-\pi/2, \pi/2]$.

Theorem 1.   In the open interval $(-1, 1)$, $y = \sin^{-1} x = \arcsin x$ is differentiable, and

$$\frac{d}{dx} \sin^{-1} x = \frac{d}{dx} \arcsin x = \frac{1}{\sqrt{1 - x^2}}, \quad -1 < x < 1. \tag{3}$$

Proof. Let $y = \sin^{-1} x$ with $x \in (-1, 1)$. Then $x = \sin y$, and

$$\frac{dx}{dy} = \cos y.$$

Since $x \neq \pm 1$, $y \neq \pm \pi/2$, so $\cos y \neq 0$ and $dx/dy \neq 0$ for $-1 < x < 1$. Thus from Theorem 6.1.4,

$$\frac{dy}{dx} = \frac{1}{dx/dy} = \frac{1}{\cos y}. \tag{4}$$

In the interval $(-\pi/2, \pi/2)$, $\cos y > 0$. Thus

$$\cos y = \overset{\text{Since } \sin^2 y + \cos^2 y = 1}{\sqrt{1 - \sin^2 y}}$$

and

$$\frac{dy}{dx} = \frac{1}{\cos y} = \frac{1}{\sqrt{1 - \sin^2 y}} = \overset{\text{Since } x = \sin y}{\frac{1}{\sqrt{1 - x^2}}}. \quad\blacksquare$$

To graph $y = \sin^{-1} x$ for x in $(-1, 1)$, we first observe that $dy/dx = 1/\sqrt{1 - x^2} > 0$, so that the function is increasing. There are no critical points. In addition,

$$\frac{d^2y}{dx^2} = \frac{d}{dx}(1 - x^2)^{-1/2} = \frac{x}{(1 - x^2)^{3/2}},$$

which is negative for $x < 0$ and positive for $x > 0$. The origin is a point of inflection. The graph is given in Figure 2b; next to it (in Figure 2a) we place, as a frame of reference, the graph of $y = \sin x$.

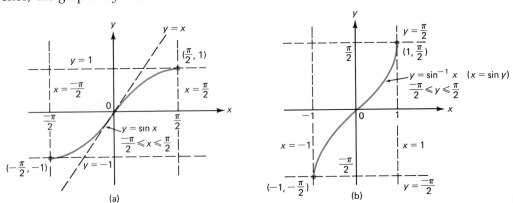

FIGURE 2

Alternatively, note [see statement (6.1.7)] that the graph of $\sin^{-1} x$ is the reflection about the line $y = x$ of the graph of $\sin x$.

For the remainder of this chapter we will use the notation $y = \sin^{-1} x$ because it expresses more clearly the fact that the function is the inverse of the sine function.

Next, consider the graph of $\cos x$ in Figure 3(a). If $y = \cos x$ and x is restricted to lie in the interval $[0, \pi]$, then for every y in $[-1, 1]$, there is a unique x such that $\cos x = y$. That is, $\cos x$ is 1–1 in the interval $[0, \pi]$.

Definition 2 INVERSE COSINE FUNCTION The **inverse cosine function** is the function that assigns to each number x in $[-1, 1]$, the unique number y in $[0, \pi]$ such that $x = \cos y$. We write $y = \cos^{-1} x$.

NOTE. $\cos^{-1} x$ is also written as $\arccos x$.

EXAMPLE 2 Calculate $y = \cos^{-1} x$ for the following:

(a) $x = 0$ (b) $x = 1$ (c) $x = \frac{1}{2}$ (d) $x = -1$

Solution.

(a) $\cos^{-1} 0 = \pi/2$ since $\cos \pi/2 = 0$.
(b) $\cos^{-1} 1 = 0$ since $\cos 0 = 1$.
(c) $\cos^{-1} \frac{1}{2} = \pi/3$ since $\cos \pi/3 = \frac{1}{2}$.
(d) $\cos^{-1}(-1) = \pi$ since $\cos \pi = -1$. ■

Theorem 2.

$$\frac{d}{dx} \cos^{-1} x = -\frac{1}{\sqrt{1 - x^2}}. \qquad -1 < x < 1. \tag{5}$$

Proof. If $y = \cos^{-1} x$, then $x = \cos y$ and $1 = (-\sin y)\, dy/dx$, or

$$\frac{dy}{dx} = -\frac{1}{\sin y} = -\frac{1}{\sqrt{1 - x^2}}$$

($\sin y = +\sqrt{1 - \cos^2 y}$ because $0 \le y \le \pi$). Note that dy/dx exists according to Theorem 6.1.4. ■

The graphs of $y = \cos x$ and $y = \cos^{-1} x$ are given in Figure 3.

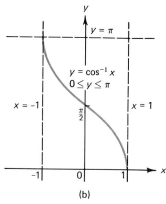

(a)   (b)

FIGURE 3

We next consider the graph of $y = \tan x$ given in Figure 4(a). The function $\tan x$ can take on values in the interval $(-\infty, \infty)$. To get a unique x for a given y, we restrict x to the interval $(-\pi/2, \pi 2)$.

Definition 3   INVERSE TANGENT FUNCTION   The **inverse tangent function** is the function that assigns to each real number x the unique number y in $(-\pi/2, \pi 2)$ such that $x = \tan y$. We write $y = \tan^{-1} x = \arctan x$.

Theorem 3.

$$\frac{d}{dx} \tan^{-1} x = \frac{1}{1 + x^2} \qquad -\infty < x < \infty. \tag{6}$$

Proof.   If $y = \tan^{-1} x$, then $x = \tan y$ and $1 = (\sec^2 y) \, dy/dx$, or

See Formula (1) in Appendix 1.3

$$\frac{dy}{dx} = \frac{1}{\sec^2 y} = \frac{1}{1 + \tan^2 y} = \frac{1}{1 + x^2}.$$

Note that dy/dx exists according to Theorem 6.1.4. ■

The graphs of $\tan x$ and $\tan^{-1} x$ are given in Figure 4.

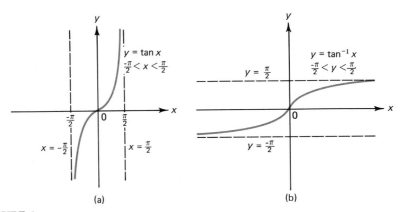

(a)   (b)

FIGURE 4

There are, of course, three other inverse trigonometric functions. These arise less frequently in applications, and so we will delay their introduction until the problems.

We summarize in Table 1 the formulas that can be obtained from the results of this section. The last three rows will be derived in the problem set.

REMARK.   In Table 1 it may seem surprising that $\int du/\sqrt{1 - u^2} = \sin^{-1} u + C$ but $-\int du/\sqrt{1 - u^2} = \cos^{-1} u + C$. This result is explained by noting that $\sin^{-1} u + \cos^{-1} u$ is a constant function (see Problem 59).

EXAMPLE 3   Calculate dy/dx for $y = \sin^{-1} x^2$.

TABLE 1

$$\frac{d}{dx} \sin^{-1} u = \frac{1}{\sqrt{1 - u^2}} \frac{du}{dx} \qquad \int \frac{du}{\sqrt{1 - u^2}} = \sin^{-1} u + C$$

$$\frac{d}{dx} \cos^{-1} u = -\frac{1}{\sqrt{1 - u^2}} \frac{du}{dx} \qquad -\int \frac{du}{\sqrt{1 - u^2}} = \cos^{-1} u + C$$

$$\frac{d}{dx} \tan^{-1} u = \frac{1}{1 + u^2} \frac{du}{dx} \qquad \int \frac{du}{1 + u^2} = \tan^{-1} u + C$$

$$\frac{d}{dx} \cot^{-1} u = -\frac{1}{1 + u^2} \frac{du}{dx} \qquad -\int \frac{du}{1 + u^2} = \cot^{-1} u + C$$

$$\frac{d}{dx} \sec^{-1} u = \frac{1}{|u|\sqrt{u^2 - 1}} \frac{du}{dx} \qquad \int \frac{du}{u\sqrt{u^2 - 1}} = \sec^{-1} |u| + C$$

$$\frac{d}{dx} \csc^{-1} u = -\frac{1}{|u|\sqrt{u^2 - 1}} \frac{du}{dx} \qquad -\int \frac{du}{u\sqrt{u^2 - 1}} = \csc^{-1} |u| + C$$

Solution. If $u = x^2$, then $du/dx = 2x$, and using the chain rule we have

$$\frac{dy}{dx} = \frac{dy}{du} \frac{du}{dx} = \frac{1}{\sqrt{1 - u^2}} \cdot 2x = \frac{2x}{\sqrt{1 - x^4}}. \quad ■$$

EXAMPLE 4   Calculate $(d/dx)[\cos^{-1}(\ln x)]$.

Solution. First note that this function is only defined for $1/e \le x \le e$ (so that $-1 \le \ln x \le 1$). Then for $u = \ln x$ and $1/e < x < e$,

$$\frac{du}{dx} = \frac{1}{x} \qquad \text{and} \qquad \frac{d}{dx} \cos^{-1} \ln x = -\frac{1}{\sqrt{1 - \ln^2 x}} \cdot \frac{1}{x}. \quad ■$$

EXAMPLE 5   Calculate $\int dx/(a^2 + x^2)$.

Solution. We want to write the integral in the form

$$\int \frac{du}{1 + u^2}$$

since we can integrate this. To do so, we factor the a^2 term in the denominator to obtain

$$\int \frac{dx}{a^2 + x^2} = \frac{1}{a^2} \int \frac{dx}{1 + (x^2/a^2)}.$$

Let $u = x/a$; then $du = (1/a)\, dx$, so that

$$\frac{1}{a^2} \int \frac{dx}{1 + (x^2/a^2)} = \frac{1}{a} \int \frac{(1/a)\, dx}{1 + (x/a)^2} = \frac{1}{a} \int \frac{du}{1 + u^2} = \frac{1}{a} \tan^{-1} u + C$$

$$= \frac{1}{a} \tan^{-1} \frac{x}{a} + C. \quad ■$$

EXAMPLE 6   Calculate $\int (x^2/\sqrt{1-x^6})\,dx$.

Solution. If $u = x^3$, then $du = 3x^2\,dx$, so we multiply and divide by 3 to obtain

$$\int \frac{x^2}{\sqrt{1-x^6}}\,dx = \frac{1}{3}\int \frac{3x^2}{\sqrt{1-(x^3)^2}}\,dx = \frac{1}{3}\int \frac{du}{\sqrt{1-u^2}} = \frac{1}{3}\sin^{-1} u + C$$

$$= \frac{1}{3}\sin^{-1} x^3 + C. \quad\blacksquare$$

In some applications it is necessary to evaluate expressions like $\sin(\tan^{-1} 3)$. To do so, we draw a triangle with an angle whose tangent is 3. See Figure 5. If $\tan \theta = 3$, then with the sides as in the figure, we use the Pythagorean theorem to find that the hypotenuse must be $\sqrt{10}$. Then $\sin \theta = 3/\sqrt{10}$, $\cos \theta = 1/\sqrt{10}$, and so on.

EXAMPLE 7   Calculate $\tan(\cos^{-1}\frac{3}{7})$.

Solution. By using the triangle in Figure 6, we see that $\tan(\cos^{-1}\frac{3}{7}) = \tan \theta = \sqrt{40}/3 = 2\sqrt{10}/3$. $\quad\blacksquare$

EXAMPLE 8   Calculate $\cos(\sin^{-1} x)$.

Solution. Look at Figure 7. We see that $\cos(\sin^{-1} x) = \cos \theta = \sqrt{1-x^2}$. $\quad\blacksquare$

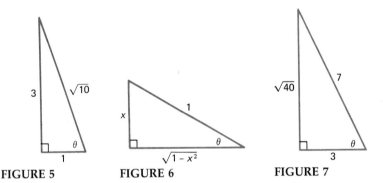

FIGURE 5   FIGURE 6   FIGURE 7

EXAMPLE 9   A department store wishes to place a large advertising display near its main entrance. The display is a rectangular sign 7 ft tall that is mounted on a wall and whose bottom edge is 8 ft from the floor (see Figure 8). The advertising manager wishes to place the entrance walkway at a distance x feet from the wall to maximize the "viewing angle" θ. It is assumed that the eye level of the average customer is 5 ft from the floor. How far from the wall should the walkway be placed?

Solution. We write θ as a function of x and then find the value of x that maximizes the function. From Figure 8, we have

$$\tan \theta_1 = \frac{3}{x} \tag{7}$$

and

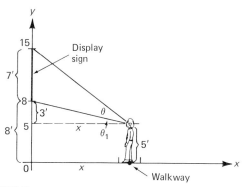

FIGURE 8

$$\tan(\theta_1 + \theta) = \frac{10}{x}.$$
(8)

We have

From (7)

$$\theta_1 = \tan^{-1}\frac{3}{x},$$

and

From (8)

$$\theta_1 + \theta = \tan^{-1}\frac{10}{x}.$$
(9)

Then

$$\theta = (\theta_1 + \theta) - \theta_1 = \tan^{-1}\frac{10}{x} - \tan^{-1}\frac{3}{x},$$

and

$$\frac{d\theta}{dx} = \frac{1}{1 + (10/x)^2}\frac{d}{dx}\left(\frac{10}{x}\right) - \frac{1}{1 + (3/x)^2}\frac{d}{dx}\left(\frac{3}{x}\right)$$

$$= \frac{-10/x^2}{1 + (100/x^2)} - \frac{-3/x^2}{1 + (9/x^2)}$$

Multiply numerator and denominator by x^2.

$$= \frac{-10}{x^2 + 100} + \frac{3}{x^2 + 9}.$$
(10)

Setting $d\theta/dx = 0$ and adding the fractions, we have

$$\frac{d\theta}{dx} = \frac{-10(x^2 + 9) + 3(x^2 + 100)}{(x^2 + 100)(x^2 + 9)} = 0,$$

or

$$-10(x^2 + 9) + 3(x^2 + 100) = 0,$$

or

$$-10x^2 - 90 + 3x^2 + 300 = 0,$$

or

$$-7x^2 + 210 = 0, \quad \text{and} \quad x = \pm\sqrt{30}.$$

Since x must be positive, we see that the only critical point occurs at $x = \sqrt{30}$. If $x < \sqrt{30}$, $-7x^2 + 210 > 0$; and when $x > \sqrt{30}$, $-7x^2 + 210 < 0$. Thus $d\theta/dx > 0$ when $x < \sqrt{30}$, and $d\theta/dx < 0$ when $x > \sqrt{30}$, so that by the first derivative test, $\theta(x)$ has a local maximum at $x = \sqrt{30}$. Thus the walkway should be placed $\sqrt{30} \approx 5.477$ ft from the wall, and the maximum viewing angle is

$$\theta(\sqrt{30}) = \tan^{-1}\left(\frac{10}{\sqrt{30}}\right) - \tan^{-1}\left(\frac{3}{\sqrt{30}}\right)$$

$$\approx \tan^{-1} 1.82574 - \tan^{-1} 0.54772$$

$$\approx 61.29° - 28.71° = 32.58°.$$

Note that as $x \to 0$ or $x \to \infty$, $\theta \to 0$, so that this angle is a true maximum. ∎

PROBLEMS 7.2

In Problems 1–18, calculate the indicated values. Do not use a calculator.

1. $\sin^{-1}\dfrac{\sqrt{3}}{2}$

2. $\cos^{-1}\left(-\dfrac{\sqrt{3}}{2}\right)$

3. $\sin^{-1}\left(-\dfrac{1}{2}\right)$

4. $\tan^{-1}(-1)$

5. $\tan^{-1}\dfrac{1}{\sqrt{3}}$

6. $\tan^{-1}\left(-\dfrac{1}{\sqrt{3}}\right)$

7. $\sin^{-1}\left(-\dfrac{\sqrt{2}}{2}\right)$

8. $\tan^{-1}\cos(-5\pi)$

9. $\tan^{-1}(\sin 30\pi)$

10. $\sin(\cos^{-1}\frac{3}{5})$
11. $\cos[\sin^{-1}(-\frac{3}{5})]$
12. $\tan(\sin^{-1}\frac{3}{5})$
13. $\sin(\tan^{-1}\frac{3}{5})$
14. $\sin[\tan^{-1}(-5)]$
*15. $\tan(\sin^{-1}5)$ [*Hint:* Watch out.]
16. $\sin(\tan^{-1}x)$
17. $\sin(\cos^{-1}x)$
18. $\tan(\sin^{-1}x)$

19. Show that if $-1 \le x \le 1$, then $\sin(\cos^{-1}x) = \cos(\sin^{-1}x)$.
20. Show that $\lim_{x\to\infty} \tan^{-1}x = \pi/2$ and $\lim_{x\to-\infty} \tan^{-1}x = -\pi/2$.

In Problems 21–54, calculate the given derivative or integral.

21. $\dfrac{d}{dx}\sin^{-1}3x$

*22. $\dfrac{d}{dx}\sin^{-1}(1 + x^2)$

23. $\dfrac{d}{dx}\sin^{-1}\sqrt{x}$

24. $\dfrac{d}{dx} \cos^{-1}(1 - 2x)$

25. $\dfrac{d}{dx} \cos^{-1}(x^3 + x)$

26. $\dfrac{d}{dx} \cos^{-1}\dfrac{x^3 + 1}{x^5}$

27. $\dfrac{d}{dx} \tan^{-1}\dfrac{x}{2}$

28. $\dfrac{d}{dx} \tan^{-1}\sqrt[3]{x}$

29. $\dfrac{d}{dx} \tan^{-1}(x - 4)^2$

30. $\displaystyle\int \dfrac{dx}{\sqrt{1 - 9x^2}}$

31. $\displaystyle\int \dfrac{dx}{\sqrt{9 - 25x^2}}$

32. $\displaystyle\int \dfrac{dx}{4x^2 + 1}$

33. $\displaystyle\int \dfrac{x}{x^4 + 1}\,dx$

34. $\displaystyle\int \dfrac{x^3}{x^8 + 4}\,dx$

35. $\displaystyle\int \dfrac{\sqrt{x}}{x^3 + 1}\,dx$

36. $\displaystyle\int \dfrac{\sqrt{x}}{\sqrt{1 - x^3}}\,dx$

37. $\dfrac{d}{dx} \tan^{-1} \ln x$

38. $\dfrac{d}{dx} \ln(\tan^{-1} x)$

39. $\dfrac{d}{dx} \sqrt{\sin^{-1} x + \cos^{-1} x}$

40. $\displaystyle\int_0^{\sqrt{3}/2} \dfrac{dx}{\sqrt{1 - x^2}}$

41. $\displaystyle\int \dfrac{e^{-x}}{4 + e^{-2x}}\,dx$

42. $\displaystyle\int \dfrac{\sin x}{\sqrt{4 - \cos^2 x}}\,dx$

43. $\dfrac{d}{dx} \ln \sin^{-1}(e^{-x})$

44. $\dfrac{d}{dx} \tan^{-1}\dfrac{1}{x}$

45. $\dfrac{d}{dx} x^2 \cos^{-1}(1 - x)$

46. $\displaystyle\int \dfrac{dx}{x^2 + 2x + 2}$ [*Hint:* Use the fact that $(x + 1)^2 = x^2 + 2x + 1$.]

47. $\displaystyle\int \dfrac{dx}{x^2 - 2x + 2}$

48. $\displaystyle\int_0^2 \dfrac{dx}{1 + x^2}$

49. $\displaystyle\int_0^1 \dfrac{dx}{\sqrt{1 - x^2}}$

50. $\displaystyle\int \dfrac{\sin^{-1} x}{\sqrt{1 - x^2}}\,dx$

51. $\displaystyle\int \dfrac{\cos^{-1} x}{\sqrt{1 - x^2}}\,dx$

52. $\displaystyle\int_0^1 \dfrac{\sqrt{\tan^{-1} x}}{1 + x^2}\,dx$

53. $\displaystyle\int \dfrac{\sec x \tan x}{1 + \sec^2 x}$

54. $\displaystyle\int \dfrac{\sec^2 x}{\sqrt{1 - \tan^2 x}}\,dx$

55. Find the area bounded by the curve $y = 1/(1 + x^2)$, the x- and y-axes, and the line $x = 1$.

56. Find the area bounded by the curve $y = 1/\sqrt{1 - x^2}$, the x-axis, the y-axis, and the line $x = \frac{1}{2}$.

57. Find the area bounded by the curve $y = -1/\sqrt{1 - x^2}$ and the line $y = -2$. [*Hint:* Sketch the curve.]

58. Show that

$$\int \dfrac{dx}{\sqrt{a^2 - x^2}} = \sin^{-1}\dfrac{x}{|a|} + C.$$

*59. Show that $\sin^{-1} x + \cos^{-1} x = \pi/2$. [*Hint:* First show by differentiation that $\sin^{-1} x + \cos^{-1} x = C$. Then find C by evaluating at one value of x.]

60. Show that

$$\dfrac{d}{dx}(x \sin^{-1} x + \sqrt{1 - x^2}) = \sin^{-1} x.$$

This gives us the integral of $\sin^{-1} x$.

61. Show that

$$\dfrac{d}{dx}(x \cos^{-1} x - \sqrt{1 - x^2}) = \cos^{-1} x.$$

62. Show that

$$\frac{d}{dx}\left[x \tan^{-1} x - \frac{1}{2} \ln(1 + x^2)\right] = \tan^{-1} x.$$

63. **Definition of cot^{-1} x.** Consider the graph of $y = \cot x$ in Figure 1(b) in Appendix 1.3. Show that if $y = \cot x$, then for every y in $(-\infty, \infty)$ there is a unique x in $(0, \pi)$ such that $y = \cot x$. Then define the function $y = \cot^{-1} x$ by $y = \cot^{-1} x$ if $x = \cot y$, $0 < y < \pi$.

64. Calculate the following:

    (a) $\cot^{-1} 0$ **(b)** $\cot^{-1}(-1)$

    (c) $\cot^{-1}(-\sqrt{3})$ **(d)** $\cot^{-1}\left(\dfrac{1}{\sqrt{3}}\right)$

65. Show that

$$\frac{d}{dx} \cot^{-1} x = -\frac{1}{1 + x^2}.$$

66. Graph the function $y = \cot^{-1} x$, $0 < y < \pi$.

67. **Definition of sec^{-1} x.** Consider the graph of $y = \sec x$ in Figure 1(c) in Appendix 1.3. Show that if $y = \sec x$, then for every y in $(-\infty, -1]$ or $[1, \infty)$ there is a unique x in $(\pi/2, \pi]$ or $[0, \pi/2)$ such that $y = \sec x$. Explain why the value $x = \pi/2$ must be excluded. Then define the function $y = \sec^{-1} x$ by $y = \sec^{-1} x$ if $x = \sec y$, $0 \le y < \pi/2$ or $\pi/2 < y \le \pi$.†

68. Show that $\sec^{-1} x = \cos^{-1}(1/x)$, for $|x| \ge 1$.

69. Calculate the following:

    (a) $\sec^{-1} 2$ **(b)** $\sec^{-1}(-2)$

    (c) $\sec^{-1} \sqrt{2}$ **(d)** $\sec^{-1}\left(\dfrac{2}{\sqrt{3}}\right)$

    (e) $\sec^{-1} 1$ **(f)** $\sec^{-1}(-1)$

70. **(a)** Show that if $x \ge 1$, then

$$\frac{d}{dx} \sec^{-1} x = \frac{1}{x\sqrt{x^2 - 1}}.$$

    (b) Show that if $x \le -1$, then

$$\frac{d}{dx} \sec^{-1} x = -\frac{1}{x\sqrt{x^2 - 1}}.$$

    (c) Conclude that

$$\frac{d}{dx} \sec^{-1} x = \frac{1}{|x|\sqrt{x^2 - 1}}.‡$$

    (d) What is the domain of the function $y' = (d/dx)(\sec^{-1} x)$?

71. Graph the function $y = \sec^{-1} x$, $0 \le y < \pi/2$ and $\pi/2 < y \le \pi$.

---

†This problem suggests one way to define $\sec^{-1} x$. An alternative way is to restrict the range of $\sec^{-1} x$ to the intervals $[-\pi, -\pi/2)$ and $[0, \pi/2)$. Then for every y in $(-\infty, -1]$ or $[1, \infty)$ there is a unique x in $[-\pi, -\pi/2)$ or $[0, \pi/2)$ such that $y = \sec x$.

‡With the alternative definition of $\sec^{-1} x$ given in the previous footnote, it is not difficult to show that

$$\frac{d}{dx} \sec^{-1} x = \frac{1}{x\sqrt{x^2 - 1}}.$$

Some mathematicians prefer this alternative definition because it is then not necessary to keep track of the absolute value in the derivative of $\sec^{-1} x$.

72. **Definition of $\csc^{-1} x$.** Consider the graph of the function $y = \csc x$ in Figure 1(d) in Appendix 1.3. Show that if $y = \csc x$, then for every y in $(-\infty, -1]$ or $[1, \infty)$, there is a unique x in $[-\pi/2, 0)$ or $(0, \pi/2]$ such that $y = \csc x$. Explain why the value $x = 0$ must be excluded. Then define the function $y = \csc^{-1} x$ by $y = \csc^{-1} x$ if $x = \csc y$, $-\pi/2 \leq y < 0$ or $0 < y \leq \pi/2$.

73. Show that $\csc^{-1} x = \sin^{-1}(1/x)$, for $|x| \geq 1$.

74. Calculate the following:
    (a) $\csc^{-1} 2$
    (b) $\csc^{-1}(-2)$
    (c) $\csc^{-1} \sqrt{2}$
    (d) $\csc^{-1}\left(\dfrac{2}{\sqrt{3}}\right)$
    (e) $\csc^{-1} 1$
    (f) $\csc^{-1}(-1)$

75. (a) Show that if $x \geq 1$, then
$$\frac{d}{dx} \csc^{-1} x = -\frac{1}{x\sqrt{x^2 - 1}}.$$

    (b) Show that if $x \leq -1$, then
$$\frac{d}{dx} \csc^{-1} x = \frac{1}{x\sqrt{x^2 - 1}}.$$

    (c) Conclude that
$$\frac{d}{dx} \csc^{-1} x = -\frac{1}{|x|\sqrt{x^2 - 1}}.$$

76. Calculate the following:
    (a) $\sin(\csc^{-1} 4)$
    (b) $\sec(\cot^{-1} 8)$
    (c) $\cos[\sec^{-1}(-10)]$
    (d) $\cot[\sec^{-1}(-3)]$
    (e) $\tan(\csc^{-1} 4)$
    (f) $\sin(\cot^{-1} \frac{1}{5})$

77. Verify the integral formulas in Table 1.

78. Show by differentiation that $\tan^{-1} x + \cot^{-1} x = \pi/2$.

In Problems 79–94, calculate the given derivative or integral.

79. $\dfrac{d}{dx} \sec^{-1}(4x + 2)$

80. $\dfrac{d}{dx} \cot^{-1}(x^2 + 5)$

81. $\dfrac{d}{dx} \csc^{-1} \sqrt{x}$

82. $\dfrac{d}{dx} x^2 \sec^{-1} x^2$

83. $\displaystyle\int_{-6}^{-3\sqrt{2}} \frac{dx}{x\sqrt{x^2 - 9}}$

84. $\displaystyle\int_{2/\sqrt{3}}^{2} \frac{dx}{x\sqrt{x^2 - 1}}$

85. $\dfrac{d}{dx} \cot^{-1}(e^x)$

86. $\dfrac{d}{dx} \csc^{-1} \ln x$

*87. $\displaystyle\int \frac{dx}{x\sqrt{x^2 - a^2}}$

88. $\displaystyle\int \frac{dx}{1 + (3 - x)^2}$

89. $\displaystyle\int_{0}^{\ln 2} \frac{e^x \, dx}{e^x\sqrt{e^{2x} - 1}}$

*90. $\displaystyle\int \frac{dx}{x\sqrt{x - 1}}$ [*Hint:* Let $u = \sqrt{x}$.]

91. $\dfrac{d}{dx} \sec^{-1}(\cot x)$

92. $\dfrac{d}{dx} \cot^{-1}(\sec x)$

*93. $\dfrac{d}{dx} \sin^{-1}(\cot^{-1} \ln x)$

94. $\displaystyle\int_{0}^{\pi/4} \frac{\sin x}{1 + \cos^2 x} \, dx$

95. Show that
$$\int \frac{dx}{x\sqrt{x^2 - a^2}} = \frac{1}{a} \sec^{-1}\left|\frac{x}{a}\right| + C = \frac{1}{a} \cos^{-1}\left|\frac{a}{x}\right| + C.$$

*96. Show that $\sec^{-1} x + \csc^{-1} x$ is constant.

*97. Prove that $\cot^{-1}[(x + 1)/(x - 1)] + \cot^{-1} x$ is a constant if $x > 1$ and is a different constant for $x < 1$. Then find the constants. [*Hint:* Differentiate and see what happens.]

98. Answer the question posed in Example 9 if the display sign is 5 ft high and its bottom edge is 6 ft from the floor.

99. A man stands on top of a vertical cliff, 100 m above a lake. He watches a man in a rowboat move directly away from the foot of the cliff at the rate of 30 m/min. How fast is the angle of depression of his line of sight changing when the boat is 60 m from the foot of the cliff? [*Hint:* See Figure 9.]

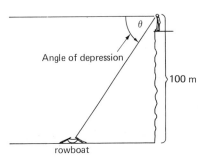

θ

Angle of depression

100 m

rowboat

FIGURE 9

100. A lighthouse containing a revolving beacon light is on a small island 3 km from the nearest point Q on a straight shoreline. The light from the spotlight moves along the shore as the beacon revolves. At a point 1 km from Q the light is sweeping along the shoreline at a rate of 40 km/min. How fast (in revolutions per minute) is the beacon revolving?

101. A visitor to the Kennedy Space Center in Florida watches a rocket blast off. The visitor is standing 3 km from the blast site. At a given moment the rocket makes an angle of 45° with her line of sight. The visitor estimates that this angle is changing at a rate of 20° per second. If the rocket is traveling vertically, what is its velocity at the given moment? [*Hint:* Convert everything to radians.]

102. Find the minimum of the function $\tan^{-1}(1/x) + \cot^{-1}(1/x) = f(x)$ on the interval $[1, \sqrt{3}]$.

*103. Let $A(x) = \cot^{-1} x$ and $B(x) = \tan^{-1}(1/x)$.
  (a) Verify that $A(1) = B(1)$.
  (b) Compare $A'(x)$ with $B'(x)$.
  (c) Discuss and graph the function $A(x) - B(x)$.

104. Compute $(d/dx)[\sin^{-1} x] \cdot (\sin x)]$.

*105. Let $f(x) = \sin^{-1}[(x^2 - 8)/8]$ and $g(x) = 2 \sin^{-1}(x/4) - \pi/2$.
  (a) Verify that $f(0) = g(0)$.
  (b) Find the maximal domain for each of the functions.
  (c) Find $f'(x)$ and $g'(x)$; compare them.
  (d) Find all x such that $f(x) = g(x)$.

*106. Discuss the following "proof" that $\tan^{-1}[(1 + x)/(1 - x)] = \pi/4 + \tan^{-1} x$.

  (a) $\dfrac{d}{dx} \tan^{-1}\left(\dfrac{1 + x}{1 - x}\right) = \dfrac{1}{1 + \left(\dfrac{1 + x}{1 - x}\right)^2} \cdot \dfrac{2}{(1 - x)^2} = \dfrac{2}{(1 - x)^2 + (1 + x)^2}$

$= \dfrac{1}{1 + x^2} = \dfrac{d}{dx} \tan^{-1} x.$

  (b) $\tan^{-1}[(1 + x)/(1 - x)] - \tan^{-1} x$ is constant; evaluating this expression for $x = 0$, we get $\tan^{-1} 1 - \tan^{-1} 0 = \pi/4 - 0 = \pi/4$. [*Hint:* Notice that $\tan^{-1} \sqrt{3} = \pi/3$, whereas $\tan^{-1}[(1 + \sqrt{3})/(1 - \sqrt{3})] = -5\pi/12$; include this result in your discussion.]

7.3 PERIODIC MOTION (OPTIONAL)

Periodic motion is a very common occurrence in physical or biological† settings. A simple example is given by the back and forth motion of a pendulum. Another is given by the oscillating size of the population of an animal species. Yet another is given by the rise and fall of average daily temperature in a fixed location over a time span of several years.

It often happens that the equations describing periodic, or **harmonic**,‡ motion (as it is often called) involve the sine and cosine functions. To see how such an equation could arise, we consider a model for the horizontal motion of a spring.

Consider a mass m attached to a spring resting on a horizontal frictionless surface. If the mass is pulled out beyond its equilibrium (resting) position, then the spring will generally exert a restoring force F, which is proportional to distance pulled and is in the opposite direction. This phenomenon can be expressed by

$$F = -kx, \tag{1}$$

where k is a positive number (called the **spring constant;** see page 335). The minus sign indicates that the force acts in such a way as to bring the mass back to its equilibrium position. This is illustrated in Figure 1. The restoring force of the spring is the only force acting on the mass, if we ignore air resistance, since we have assumed that there is no friction. By Newton's second law of motion the sum of the forces acting on the mass is given by $F = ma$, where a is the acceleration of the particle. But $a = d^2x/dt^2$. This leads to the equation $m\, d^2x/dt^2 = -kx$, or

$$m\frac{d^2x}{dt^2} + kx = 0. \tag{2}$$

FIGURE 1

Since k and m are positive numbers, we can write $k/m = \omega^2$, where ω is a positive number. Then if we divide (2) through by m, we obtain

---

†There is a great variety of periodic or rhythmic phenomena in biology. For example, diurnal (daily) or *circadian* rhythms have been studied for more than one hundred years. The interested reader should consult the collection of papers in *Circadian Clocks*, Jürgen Aschoff, ed. (North-Holland Publ. Co., Amsterdam, 1965) or *Biological Rhythms in Human and Animal Physiology* by G. G. Luce (Dover, New York, 1971).

‡The term "harmonic" is applied to expressions containing certain combinations of the sine and cosine functions.

$$\frac{d^2x}{dt^2} + \omega^2 x(t) = 0. \tag{3}$$

Equation (3) is called the **equation of the harmonic oscillator.** It is another kind of differential equation (we discussed an easier differential equation in Section 6.6). The equation of the harmonic oscillator arises in a great number of physical and biological applications. It can be shown† that any solution to equation (3) can be written in the form

$$x(t) = A \cos(\omega t + \delta), \tag{4}$$

where A and δ are constants. We can check that a function of the form (4) is a solution of (3) by differentiating:

$$x'(t) = -Aw \sin(\omega t + \delta) \quad \text{and} \quad x''(t) = -A\omega^2 \cos(\omega t + \delta) = -\omega^2 x(t),$$

so that

$$x'' + \omega^2 x = -\omega^2 x + \omega^2 x = 0.$$

Whatever the values A and δ, equation (4) tells us that the mass vibrates back and forth with a period T of $2\pi/\omega$ since

$$A \cos\left[\omega\left(t + \frac{2\pi}{\omega}\right) + \delta\right] = A \cos(\omega t + 2\pi + \delta) = A \cos(\omega t + \delta).$$

If t is measured in seconds, then the mass oscillates once every $2\pi/\omega$ seconds and therefore oscillates $\omega/2\pi$ times in 1 sec. In this case we say that the **frequency** f is $\omega/2\pi$ cycles per second.‡ In terms of m and k, since $\omega^2 = k/m$, the period, measured in seconds, is

$$T = \frac{2\pi}{\omega} = \frac{2\pi}{\sqrt{k/m}} = 2\pi\sqrt{\frac{m}{k}}, \tag{5}$$

and the frequency is

$$f = \frac{1}{T} = \frac{\omega}{2\pi} = \frac{1}{2\pi}\sqrt{\frac{k}{m}}. \tag{6}$$

The constant A has a simple physical meaning. It is the **amplitude** of the oscillation [since $-1 \le \cos x \le 1$, $-A \le A \cos(\omega t + \delta) \le A$]. The amplitude is the

---

†For a derivation of Equation (4) see W. R. Derrick and S. I. Grossman, *Elementary Differential Equations with Applications,* 2nd ed., Addison-Wesley, Reading, Mass., 1981.

‡Cycles per second are also called *hertz* (Hz), named after the German physicist Heinrich Rudolph Hertz (1857–1894).

maximum value of the displacement from equilibrium. The number δ, which can always be taken to be between $-\pi/2$ and $\pi/2$, is called the **phase constant.** Since $\cos(\omega t + \delta) = \cos\{\omega[t + (\delta/\omega)]\}$, we see from Section 1.8 that this constant has the simple effect of shifting the cosine curve δ/ω units to the left if δ is positive and $|\delta/\omega|$ units to the right if δ is negative (see Figure 2).If $\delta = 0$, we obtain the cosine function itself. If $\delta = -\pi/2$, then we obtain the sine function since

$$\cos\left(x - \frac{\pi}{2}\right) = \cos\left(\frac{\pi}{2} - x\right) = \sin x.$$

Identity (x), in Appendix 1.2

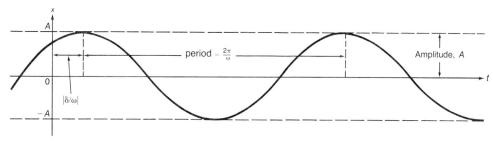

FIGURE 2

To obtain the constants A and δ, we must have further information, as the next example illustrates.

EXAMPLE 1   A 2-kg mass is attached to a spring with spring constant 8 and is pulled out a distance of 25 cm and then released. Find the equation of motion of the mass on the spring. Where is the mass after $\frac{1}{2}$ sec? After 5 sec? What are the period and frequency of oscillation?

Solution. Here $m = 2$ and $k = 8$, so $\omega = \sqrt{k/m} = 2$. The equation of motion is $x(t) = A\cos(2t + \delta)$. When $t = 0$, $x(0) = 0.25$ m. At the instant the mass is released, it is not moving, so that its velocity is zero. But $v(t) = x'(t) = -2A\sin(2t + \delta)$. Thus we obtain

$$x(0) = A\cos\delta = 0.25 \quad \text{and} \quad x'(0) = -2A\sin\delta = 0.$$

Since $-2A \neq 0$, we must have $\sin\delta = 0$ or $\delta = \sin^{-1}0 = 0$. Then $x(0) = A\cos 0 = A = 0.25 = \frac{1}{4}$. Hence the equation of motion is

$$x(t) = \frac{\cos 2t}{4}.$$

If $t = \frac{1}{2}$, then

$$x(t) = \frac{\cos 1}{4} \approx 0.135 \text{ m}.$$

If $t = 5$, then $x(t) = (\cos 10)/4 \approx -0.21$ m. Thus after 5 sec the spring is compressed 21 cm. Finally, $T = 2\pi/\omega = \pi$ and $f = (1/\pi)$ cycle/sec $= 1/\pi$ Hz. ■

EXAMPLE 2   What is the maximum speed of the mass in Example 1?

Solution. In general, if $x(t) = A \cos(\omega t + \delta)$, then $v = dx/dt = -A\omega \sin(\omega t + \delta)$. Since $-1 \leq \sin(\omega t + \delta) \leq 1$, we see that $v_{max} = A\omega$. In this problem $A = \frac{1}{4}$ and $\omega = 2$, so that $v_{max} = \frac{1}{2}$ m/sec. ∎

EXAMPLE 3   In a **predator-prey relationship** the population growth of the predator species is proportional to the population of the prey species, while the rate of decline of the prey species is proportional to the population of the predator species. Find the populations of each species as a function of time.

Solution. Let $x(t)$ represent the population of the predator species and $y(t)$ the population of the prey species. Then

$$\frac{dx}{dt} = k_1 y \tag{7}$$

and

$$\frac{dy}{dt} = -k_2 x, \tag{8}$$

where k_1 and k_2 are positive constants.

To solve these equations, we differentiate (7) with respect to t:

$$\overset{\text{From (8)}}{\frac{d^2x}{dt^2} = k_1 \frac{dy}{dt}} = k_1(-k_2 x), \qquad \text{or} \qquad \frac{d^2x}{dt^2} + k_1 k_2 x = 0.$$

If we set $k_1 k_2 = \omega^2$, then we obtain the equation of the harmonic oscillator. Thus

$$x(t) = A \cos(\omega t + \delta). \tag{9}$$

Then since

$$\overset{\text{From (7)}}{y(t) = (1/k_1)\, dx/dt},$$

We have

$$y(t) = -\frac{A\omega}{k_1} \sin(\omega t + \delta). \; ∎ \tag{10}$$

EXAMPLE 4   In Example 3 suppose that $k_1 = k_2 = 1$ and initially the population of the predator species is 500 while that of the prey species is 2000. Find the populations of both species as a function of time if t is measured in weeks.

Solution. Here $\omega = \sqrt{k_1 k_2} = 1$. Then, from equations (9) and (10),

$$x(0) = A \cos \delta = 500 \qquad \text{and} \qquad y(0) = -A \sin \delta = 2000$$

(since $k_1 = 1$). Then $A = 500/\cos \delta$, so that $y(0) = (-500/\cos \delta) \sin \delta = 2000$, or $\tan \delta = -4$. And since

$$\tan(-x) = \frac{\sin(-x)}{\cos(-x)} = -\frac{\sin x}{\cos x} = -\tan x,$$

$$\delta = \tan^{-1}(-4) \overset{\text{Calculator}}{\approx} -76° \approx -1.326 \text{ radians}$$

$$A = \frac{500}{\cos \delta} = \frac{500}{\cos[\tan^{-1}(-4)]} \overset{\text{Figure 3}}{=} \frac{500}{1/\sqrt{17}} \approx 2061.6$$

FIGURE 3

Thus the population equations are

$$x(t) = 2061.6 \cos(t - 1.326)$$

and

$$y(t) = -2061.6 \sin(t - 1.326).$$

Note that the population of the prey species is zero when $t = 1.326$ weeks, which means that the prey becomes extinct after that period of time.

REMARK.   This model is, of course, highly simplistic. It is more reasonable to assume that the rate of growth of the prey species also depends on, for example, the size of the prey species. Other factors, such as weather, time of year, and availability of food, also affect population growth. More complicated models are too complex to be discussed here. Nevertheless, even models as simple as this one can sometimes be used to help analyze complicated biological phenomena. ■

PROBLEMS 7.3

1. Solve the equation of the harmonic oscillator when $\omega = 1$, $x(0) = 0$, and $x'(0) = 1$.
2. Find the equation of motion of a 3-kg mass attached to a spring that slides on a horizontal frictionless surface given that the amplitude of oscillation is 6, the spring constant is 2, and $x(1) = 3$.
3. A 5-kg mass attached to a spring oscillates with an amplitude of 8 m and a phase constant of $\frac{1}{4}$ and is displaced $4\sqrt{2}$ m when $t = 2$ sec. What are the period and frequency of the oscillations? What is the spring constant?
4. A mass attached to a spring oscillates, and it is known that initially $x(0) = 0$ and $v(0) = 10$ m/sec. If the period of oscillation is 5 sec, find the equation of motion of the mass.

*5. Find the equation of motion of a mass attached to a spring on a horizontal frictionless surface with initial position x_0, initial velocity v_0, and period T. [*Hint:* Write everything in terms of x_0, v_0, and T and assume that the mass will not move if $x_0 = 0$; that is, assume that equilibrium is at 0.]

6. For $x(t) = A \cos(\omega t + \delta)$, find the following:
   (a) The time at which the velocity is a maximum.
   (b) The time at which the velocity is zero.
   (c) The time at which the acceleration is a maximum.
   (d) The time at which the acceleration is zero.

7. Show that the solution of the equation of the harmonic oscillator can also be written

$$x(t) = c_1 \cos \omega t + c_2 \sin \omega t$$

   for appropriate constants c_1 and c_2.

8. Suppose $k_1 = 8$ and $k_2 = 2$ in Example 3. Find the population equations of the predator and prey species if the initial populations of the two species are 1000 and 10,000, respectively. When is the prey species extinct? Assume that time is measured in weeks.

9. Answer the questions of Problem 8 for $k_1 = 2$ and $k_2 = 8$ (with the same initial populations).

7.4 THE HYPERBOLIC FUNCTIONS

In many physical applications (especially those involving differential equations) functions arise that are combinations of e^x and e^{-x}. In fact, this happens so often that certain of the combinations that occur most frequently have been given special names.

Definition 1 THE HYPERBOLIC SINE AND COSINE FUNCTIONS   The **hyperbolic cosine function** is defined by

$$\cosh x = \frac{e^x + e^{-x}}{2}. \tag{1}$$

The domain of $\cosh x$ is $(-\infty, \infty)$.

   The **hyperbolic sine function** is defined by

$$\sinh x = \frac{e^x - e^{-x}}{2}. \tag{2}$$

The domain of $\sinh x$ is $(-\infty, \infty)$.†

   Since $e^{-x} > 0$ for all x, we see that for every x

$$\cosh x > \sinh x. \tag{3}$$

_____

†$\cosh x$ is pronounced "kosh" x and $\sinh x$ is pronounced "cinch" x.

Theorem 1

$$\frac{d}{dx} \cosh x = \sinh x \qquad\qquad (4)$$

and

$$\frac{d}{dx} \sinh x = \cosh x. \qquad\qquad (5)$$

Proof.

$$\frac{d}{dx} \cosh x = \frac{d}{dx}\left(\frac{e^x + e^{-x}}{2}\right) = \frac{e^x - e^{-x}}{2} = \sinh x.$$

The proof for the derivative of sinh is just as easy. ■

It is not very difficult to draw the graphs of these functions. We immediately see that $\sinh x < 0$ if $x < 0$ and $\sinh x > 0$ if $x > 0$. If $x = 0$, $\sinh x = 0$ and $\cosh x = 1$. Therefore $\cosh x$ is decreasing if $x < 0$ and $\cosh x$ is increasing if $x > 0$, and the point $x = 0$ is a critical point for $\cosh x$. In addition,

$$\frac{d^2}{dx^2}(\cosh x) = \cosh x > 0,$$

so that $\cosh x$ is concave up and the point $(0, 1)$ is a minimum. This information is all depicted in Figure 1b.

Since

$$\frac{d}{dx} \sinh x = \cosh x > 0,$$

$\sinh x$ is always increasing (and has no critical points). We have

$$\frac{d^2}{dx^2} \sinh x = \sinh x$$

so that $\sinh x$ is concave down for $x < 0$ and concave up for $x > 0$. At $(0, 0)$ there is a point of inflection. This figure is drawn in Figure 1a.

Consider the expression $\cosh^2 x - \sinh^2 x$. From Definition 1, it is easy to show that

$$\cosh^2 x - \sinh^2 x = 1. \qquad\qquad (6)$$

(See Problem 1.)

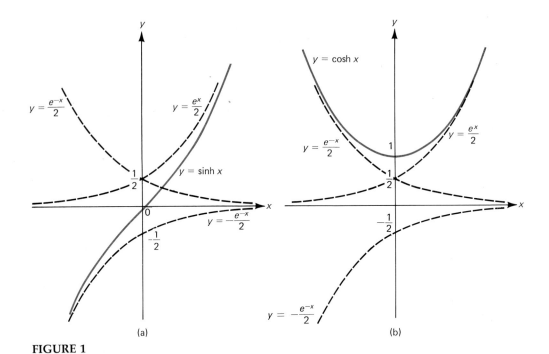

FIGURE 1

The trigonometric, or circular, functions $\sin \theta$ and $\cos \theta$ are defined with reference to the unit circle whose equation is $x^2 + y^2 = 1$ (see Figure 1 in Appendix 1.2). That is, if (x, y) is a point on the unit circle and θ is the angle between the x-axis and the radial line joining $(0, 0)$ to (x, y), then by definition $x = \cos \theta$ and $y = \sin \theta$.

Now the equation $x^2 - y^2 = 1$ is the equation of a hyperbola, sometimes called the *unit* hyperbola (see Section 10.3), whose graph is given in Figure 2. If $x = \cosh \theta$ and $y = \sinh \theta$ for some number θ, then from (6), (x, y) is a point on this hyperbola. This is the reason $\cosh x$ and $\sinh x$ are called *hyperbolic* functions.

From Theorem 1 we obtain the differentiation and integration formulas shown in Table 1.

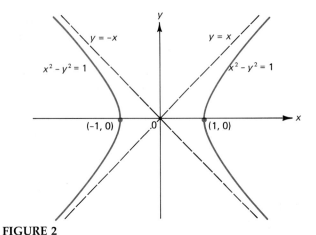

FIGURE 2

TABLE 1

$\dfrac{d}{dx}\cosh u = \sinh u\,\dfrac{du}{dx}$	$\displaystyle\int \sinh u\,du = \cosh u + C$
$\dfrac{d}{dx}\sinh u = \cosh u\,\dfrac{du}{dx}$	$\displaystyle\int \cosh u\,du = \sinh u + C$

EXAMPLE 1 Calculate $(d/dx)(\cosh \sqrt{x^3 - 1})$.

Solution. If $u = \sqrt{x^3 - 1}$, then

$$\frac{du}{dx} = \frac{3x^2}{2\sqrt{x^3 - 1}},$$

and

$$\frac{d}{dx}\cosh \sqrt{x^3 - 1} = \sinh u\,\frac{du}{dx} = (\sinh \sqrt{x^3 - 1})\,\frac{3x^2}{2\sqrt{x^3 - 1}}. \ \blacksquare$$

EXAMPLE 2 Calculate $\int x \sinh 3x^2\,dx$.

Solution. If $u = 3x^2$, then $du = 6x\,dx$, so that

$$\int x \sinh 3x^2\,dx = \frac{1}{6}\int \sinh 3x^2 (6x)\,dx = \frac{1}{6}\int \sinh u\,du$$

$$= \frac{1}{6}\cosh u + C = \frac{1}{6}\cosh 3x^2 + C. \ \blacksquare$$

EXAMPLE 3 A **catenary** is the curve formed by a uniform flexible cable hanging from two points under the influence of its own weight. If the lowest point of the catenary is at the point $(0, a)$, then its equation (which we will not derive here) is given by

$$y = a \cosh \frac{x}{a}, \qquad a > 0.$$

This curve is graphed in Figure 3. ■

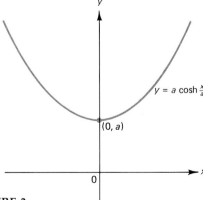

$y = a \cosh \frac{x}{a}$

$(0, a)$

FIGURE 3

EXAMPLE 4   **Symbiosis** is defined as the relationship of two or more different organisms in a close association that may be of benefit to each. If two species are living in a symbiotic relationship, then we may suppose that the rate of growth of the first population is proportional to the size of the second, and vice versa. If $x(t)$ and $y(t)$ denote the populations of the two species, respectively, then we have

$$\frac{dx}{dt} = k_1 y(t) \tag{7}$$

and

$$\frac{dy}{dt} = k_2 x(t), \tag{8}$$

where k_1 and k_2 are positive constants (compare this model with the model in Example 7.3.3). We then differentiate (7) to obtain

$$\frac{d^2x}{dt^2} = k_1 \frac{dy}{dt} \overset{\text{From (8)}}{=} k_1 k_2 x.$$

If we define $\omega^2 = k_1 k_2 > 0$, then we have

$$\frac{d^2x}{dt^2} - \omega^2 x(t) = 0. \tag{9}$$

The only difference between equation (9) and the equation of harmonic motion [equation (7.3.3)] is the minus sign before the $\omega^2 x$ term. It is easy to see that any function of the form

$$x(t) = c_1 \sinh \omega t + c_2 \cosh \omega t, \tag{10}$$

where c_1 and c_2 are constants, is a solution to (9) (see Problem 2). Although we will not prove this result here, these are the only solutions to (9). Also, from (7) we see that $y(t) = (1/k_1)\, dx/dt$, so that

$$y(t) = \frac{1}{k_1}(\omega c_1 \cosh \omega t + \omega c_2 \sinh \omega t),$$

or

$$y(t) = \frac{\omega}{k_1}(c_1 \cosh \omega t + c_2 \sinh \omega t). \tag{11}$$

This is the best we can do unless further information is provided. ∎

EXAMPLE 5   Find the populations as functions of t of the species in Example 4 if $k_1 = 4$, $k_2 = 1$, and initially, $x(0) = 500$, $y(0) = 1000$, and t is measured in months. What are the populations after $\frac{1}{2}$ month?

Solution. We have $\omega^2 = k_1 k_2 = 4$, so that $\omega = 2$. Then

$$x(t) = c_1 \sinh 2t + c_2 \cosh 2t \quad \text{and} \quad y(t) = \tfrac{1}{2}(c_1 \cosh 2t + c_2 \sinh 2t).$$

When $t = 0$, $\sinh 0 = 0$ and $\cosh 0 = 1$, so that

$$x(0) = c_2 = 500, \quad y(0) = \tfrac{1}{2}c_1 = 1000, \quad \text{and } c_1 = 2000.$$

Thus

$$x(t) = 2000 \sinh 2t + 500 \cosh 2t \quad \text{and} \quad y(t) = 1000 \cosh 2t + 250 \sinh 2t.$$

We then calculate $\sinh 1 \approx 1.175$ and $\cosh 1 \approx 1.543$, so that $x(\tfrac{1}{2}) \approx 3122$ and $y(\tfrac{1}{2}) \approx 1837$. ■

After looking at the several examples in this section and in Section 7.3, we see that the trigonometric and hyperbolic functions describe different types of behavior: *Trigonometric functions describe oscillatory behavior, while hyperbolic functions describe situations in which there is exponential growth or decay (or an exponential approach to a steady state).*

As you might expect, there are four other hyperbolic functions. They do not arise nearly as often in applications as the two we have already discussed.

Definition 2 OTHER HYPERBOLIC FUNCTIONS

(i) $\tanh x = \dfrac{\sinh x}{\cosh x} = \dfrac{e^x - e^{-x}}{e^x + e^{-x}}.$

The domain of $\tanh x$ is $(-\infty, \infty)$.

(ii) $\coth x = \dfrac{1}{\tanh x} = \dfrac{\cosh x}{\sinh x} = \dfrac{e^x + e^{-x}}{e^x - e^{-x}}.$

The domain of $\coth x$ is all x except $x = 0$.

(iii) $\operatorname{sech} x = \dfrac{1}{\cosh x} = \dfrac{2}{e^x + e^{-x}}.$

The domain of $\operatorname{sech} x$ is $(-\infty, \infty)$.

(iv) $\operatorname{csch} x = \dfrac{1}{\sinh x} = \dfrac{2}{e^x - e^{-x}}.$

The domain of $\operatorname{csch} x$ is all x except $x = 0$.

The graphs of these four functions are given in Figure 4.

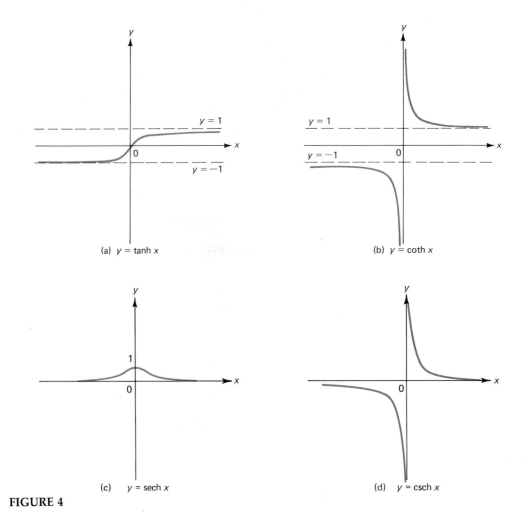

(a) $y = \tanh x$

(b) $y = \coth x$

(c) $y = \operatorname{sech} x$

(d) $y = \operatorname{csch} x$

FIGURE 4

Integrals and derivatives of these four hyperbolic functions are given in Table 2. The proofs of these facts are left as problems.

TABLE 2

$$\frac{d}{dx} \tanh u = \operatorname{sech}^2 u \, \frac{du}{dx} \qquad \int \operatorname{sech}^2 u \, du = \tanh u + C$$

$$\frac{d}{dx} \coth u = -\operatorname{csch}^2 u \, \frac{du}{dx} \qquad \int \operatorname{csch}^2 u \, du = -\coth u + C$$

$$\frac{d}{dx} \operatorname{sech} u = -\operatorname{sech} u \tanh u \, \frac{du}{dx} \qquad \int \operatorname{sech} u \tanh u = -\operatorname{sech} u + C$$

$$\frac{d}{dx} \operatorname{csch} u = -\operatorname{csch} u \coth u \, \frac{du}{dx} \qquad \int \operatorname{csch} u \coth u = -\operatorname{csch} u + C$$

PROBLEMS 7.4

1. Show from the definitions that $\cosh^2 x - \sinh^2 x = 1$.
2. Verify that (10) is a solution to equation (9) for all real numbers c_1 and c_2.
3. Show that $(d/dx)(\tanh x) = \operatorname{sech}^2 x$.   [*Hint:* Use $\tanh x = (\sinh x)/(\cosh x)$.]
4. Show that $(d/dx)(\coth x) = -\operatorname{csch}^2 x$.
5. Show that $(d/dx)(\operatorname{sech} x) = -\operatorname{sech} x \tanh x$.
6. Show that $(d/dx)(\operatorname{csch} x) = -\operatorname{csch} x \coth x$.

In Problems 7–30, calculate the given derivatives and integrals.

7. $\dfrac{d}{dx} \sinh(4x + 2)$

8. $\dfrac{d}{dx} \cosh \dfrac{1}{x}$

9. $\dfrac{d}{dx} \operatorname{csch}(\ln x)$

10. $\displaystyle\int \sinh 2x \, dx$

11. $\displaystyle\int \dfrac{\operatorname{sech}^2 \sqrt{x}}{\sqrt{x}} \, dx$

12. $\displaystyle\int \dfrac{\operatorname{csch} \ln x \, \coth \ln x}{x} \, dx$

13. $\dfrac{d}{dx} \sin(\sinh x)$

14. $\dfrac{d}{dx} \sinh(\sin x)$

15. $\dfrac{d}{dx} e^{\tanh^2 x}$

16. $\displaystyle\int e^{\sinh x} \cosh x \, dx$

*17. $\displaystyle\int \cosh^3 x \, dx$ [*Hint:* Use Problem 1.]

18. $\displaystyle\int \tanh x \, dx$

19. $\displaystyle\int \coth x \, dx$

20. $\displaystyle\int \dfrac{\operatorname{csch}^2 x \, dx}{\sqrt[3]{\coth x}}$

21. $\displaystyle\int \dfrac{\sinh(1/x)}{x^2} \, dx$

22. $\dfrac{d}{dx} \tanh(\operatorname{csch} x)$

*23. $\dfrac{d}{dx} \sin^{-1}(\cosh x)$ [*Hint:* Be careful.]

24. $\dfrac{d}{dx} \tanh(\tan^{-1} x)$

25. $\displaystyle\int \dfrac{\cosh 2x}{1 + \sinh 2x} \, dx$

*26. $\displaystyle\int \dfrac{[\cos^{-1}(\tanh x)]\operatorname{sech}^2 x}{\sqrt{1 - \tanh^2 x}} \, dx$

27. $\displaystyle\int \operatorname{sech}^{11/3} x \tanh x \, dx$

*28. $\dfrac{d}{dx} \cosh x^x$

*29. $\dfrac{d}{dx} x^{\cosh x}$

*30. $\dfrac{d}{dx} \sinh x^{\cosh x}$

31. Find the area bounded by the catenary $y = a \cosh(x/a)$, the x-axis, and the lines $x = -a$ and $x = a$.
32. Show that $e^x = \cosh x + \sinh x$.
33. Show that $e^{-x} = \cosh x - \sinh x$.
34. Use the result of Problem 32 to prove that

$$(\sinh x + \cosh x)^n = \sinh nx + \cosh nx.$$

35. Use the result of Problem 33 to prove that

$$(\cosh x - \sinh x)^n = \cosh nx - \sinh nx.$$

36. Show that $\sinh(x + y) = \sinh x \cosh y + \cosh x \sinh y$. [*Hint:* $\sinh(x + y) = (e^{x+y} - e^{-(x+y)})/2 = (e^x e^y - e^{-x} e^{-y})/2$. Now write $\sinh x \cosh y + \cosh x \sinh y$ in terms of exponentials.]
37. Show that $\cosh(x + y) = \cosh x \cosh y + \sinh x \sinh y$.
38. Show that $\sinh(-x) = -\sinh x$.
39. Show that $\cosh(-x) = \cosh x$.
40. Using Problems 36–39, derive formulas for $\sinh(x - y)$ and $\cosh(x - y)$.
41. Show that $\sinh 2x = 2 \sinh x \cosh x$.
42. Show that $\cosh 2x = \cosh^2 x + \sinh^2 x = 2 \sinh^2 x + 1$.
43. Using Equation (6) show the following:
    (a) $\operatorname{sech}^2 x + \tanh^2 x = 1$ **(b)** $\coth^2 x - \operatorname{csch}^2 x = 1$
44. Using Equation (6), find $\sinh x$ if $\cosh x = 8$.
45. Find $\cosh x$ if $\sinh x = -\frac{2}{5}$.
46. Using Problem 43, calculate $\operatorname{sech} x$ if $\tanh x = \frac{1}{2}$.
47. Calculate $\tanh x$ if $\operatorname{sech} x = \frac{3}{8}$.
48. Calculate $\coth x$ if $\operatorname{csch} x = 3$.
49. Calculate $\operatorname{csch} x$ if $\coth x = \frac{3}{2}$.
*50. If $\sinh x = \frac{2}{3}$, calculate the other five hyperbolic functions.
*51. If $\operatorname{csch} x = 5$, calculate the other five hyperbolic functions.

7.5 THE INVERSE HYPERBOLIC FUNCTIONS (OPTIONAL)

From Figure 7.4.1, we see that for every real number y there is a unique x such that $y = \sinh x$. Thus $\sinh x$ is 1–1 and therefore has an inverse function.

Definition 1   INVERSE HYPERBOLIC SINE   The **inverse hyperbolic sine function** is defined by $y = \sinh^{-1} x$ if and only if $x = \sinh y$. The domain of $y = \sinh^{-1} x$ is $(-\infty, \infty)$.

NOTE.   In contrast to the definitions of the inverse trigonometric functions, with the inverse hyperbolic sine function there is no need to restrict y to a fixed interval since the function $\sinh x$ is 1–1.

Also from Figure 7.4.1 we see that for each $y \geq 1$ there are two values of x for which $y = \cosh x$: one positive, one negative. Thus we have the following definition.

Definition 2   INVERSE HYPERBOLIC COSINE   The **inverse hyperbolic cosine function** is defined by $y = \cosh^{-1} x$ if and only if $\cosh y = x$, $y \geq 0$. The domain of $\cosh^{-1} x$ is the interval $[1, \infty)$.

REMARK.   Here we need to restrict y to be nonnegative so that for each x there will be a *unique* y such that $x = \cosh y$ (otherwise, there would be two values for $\cosh^{-1} x$, and $\cosh^{-1} x$ would then *not* be a function).

Theorem 1.

$$\textbf{(i)} \quad \frac{d}{dx} \sinh^{-1} x = \frac{1}{\sqrt{x^2 + 1}} \tag{1}$$

(ii) $\dfrac{d}{dx} \cosh^{-1} x = \dfrac{1}{\sqrt{x^2 - 1}}$, $x > 1$ **(2)**

Proof.

(i) If $y = \sinh^{-1} x$, then $x = \sinh y$, so that

$$1 = \cosh y \,\frac{dy}{dx} \quad \text{and} \quad \frac{dy}{dx} = \frac{1}{\cosh y}.$$

But $\cosh^2 y - \sinh^2 y = 1$, so that $\cosh y = \sqrt{1 + \sinh^2 y} = \sqrt{1 + x^2}$, which proves (i). Note that dy/dx exists according to Theorem 6.4.1.

(ii) If $y = \cosh^{-1} x$, then $x = \cosh y$ and

$$1 = \sinh y \,\frac{dy}{dx}, \quad \text{or} \quad \frac{dy}{dx} = \frac{1}{\sinh y}.$$

But $\sinh y = \pm\sqrt{\cosh^2 y - 1}$. Since $y \geq 0$, $\sinh y \geq 0$, so that

$$\sinh y = \sqrt{\cosh^2 y - 1} = \sqrt{x^2 - 1},$$

which proves (ii). ■

The graphs of $y = \sinh^{-1} x$ and $y = \cosh^{-1} x$ are given in Figure 1.

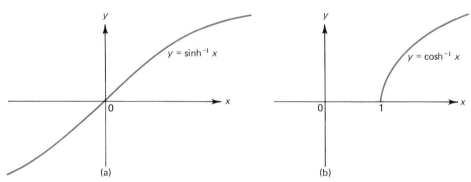

FIGURE 1

From the graph of $y = \tanh x$ in Figure 7.4.4(a), we see that for every y in $(-1, 1)$ there is an x such that $y = \tanh x$.

Definition 3 INVERSE HYPERBOLIC TANGENT   The **inverse hyperbolic tangent function** is defined by $y = \tanh^{-1} x$ if and only if $x = \tanh y$. The domain of $\tanh^{-1} x$ is $(-1, 1)$.

Theorem 2.

$$\frac{d}{dx} \tanh^{-1} x = \frac{1}{1 - x^2}, \qquad -1 < x < 1$$

Proof.   If $y = \tanh^{-1} x$, then $x = \tanh y$, and

$$1 = \operatorname{sech}^2 y \, \frac{dy}{dx}, \qquad \text{or} \qquad \frac{dy}{dx} = \frac{1}{\operatorname{sech}^2 y} = \frac{1}{1 - \tanh^2 y} = \frac{1}{1 - x^2}$$

[see Problem 7.4.43(a)]. Again, dy/dx exists according to Theorem 6.4.1. ■

A sketch of $y = \tanh^{-1} x$ appears in Figure 2.

We summarize in Table 1 the rules we have discovered in this section. The derivation of the last formula is suggested in Problem 35.

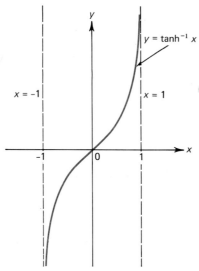

$y = \tanh^{-1} x$

$x = -1$ $x = 1$

FIGURE 2

TABLE 1

$\dfrac{d}{dx} \sinh^{-1} u = \dfrac{1}{\sqrt{u^2 + 1}} \dfrac{du}{dx}$	$\displaystyle\int \dfrac{du}{\sqrt{u^2 + 1}} = \sinh^{-1} u + C$
$\dfrac{d}{dx} \cosh^{-1} u = \dfrac{1}{\sqrt{u^2 - 1}} \dfrac{du}{dx}, \; u > 1$	$\displaystyle\int \dfrac{du}{\sqrt{u^2 - 1}} = \cosh^{-1} u + C, \; u > 1$
$\dfrac{d}{dx} \tanh^{-1} u = \dfrac{1}{1 - u^2} \dfrac{du}{dx}, \; \lvert u \rvert < 1$	$\displaystyle\int \dfrac{du}{1 - u^2} = \begin{cases} \tanh^{-1} u + C, \lvert u \rvert < 1 \\ \tanh^{-1} (1/u) + C, \lvert u \rvert > 1 \end{cases}$

EXAMPLE 1   Calculate $(d/dx)[\cosh^{-1}(x^2 + e^x)]$.

Solution. If $u = x^2 + e^x$, then $du/dx = 2x + e^x$, and

$$\frac{d}{dx} \cosh^{-1}(x^2 + e^x) = \frac{2x + e^x}{\sqrt{(x^2 + e^x)^2 - 1}}.$$ ■

EXAMPLE 2   Calculate $(d/dx)(\tanh^{-1} \tan x)$.

Solution. If $u = \tan x$, $du/dx = \sec^2 x$, so that

$$\frac{d}{dx} \tanh^{-1} \tan x = \frac{\sec^2 x}{1 - \tan^2 x}.$$ ■

There are three other inverse hyperbolic functions but these are rarely used, so we will not discuss them here. In fact, the inverse hyperbolic functions occur rarely in applications and are usually used only as devices to help us to integrate, as we will see in Chapter 8.

PROBLEMS 7.5

In Problems 1–12, calculate the derivative of the given function.

1. $y = \sinh^{-1}(3x + 2)$
2. $y = \cosh^{-1} \sqrt{x}$
3. $y = \tanh^{-1} \ln x$
4. $y = \sinh^{-1} \sec x$
5. $y = \sec \sinh^{-1} x$
6. $y = \sinh^{-1}(\cosh x)$

7. $y = \cosh(\sinh^{-1} x)$
8. $y = \tanh^{-1} \dfrac{1}{x^2}$

9. $y = \sqrt{\sinh^{-1} x}$
10. $y = [\sinh^{-1}(x + 1)]\cosh^{-1}(x + 2)$

11. $y = \dfrac{\sinh^{-1} x}{\cosh^{-1} x}$
12. $y = \cos(\sinh^{-1} x)$

13. Show that

$$\frac{d}{dx} \sinh^{-1} \frac{x}{a} = \frac{1}{\sqrt{x^2 + a^2}} \quad \text{for every } x.$$

14. Show that

$$\frac{d}{dx} \cosh^{-1} \frac{x}{a} = \frac{1}{\sqrt{x^2 - a^2}} \quad \text{for} \quad \frac{x}{a} > 1.$$

15. Show that

$$\frac{d}{dx} \tanh^{-1} \frac{x}{a} = \frac{a}{a^2 - x^2} \quad \text{for} \quad x^2 < a^2,$$

so that

$$\int \frac{du}{a^2 - u^2} = \frac{1}{a} \tanh^{-1} \frac{u}{a} + C.$$

*16. Show that $\sinh^{-1} x = \ln(x + \sqrt{1 + x^2})$. [*Hint:* differentiate both functions and show that they have the same derivative. This means that they differ by a constant. Then show that this constant is zero by evaluating both functions at $x = 0$.]

*17. Show that $\cosh^{-1} x = \ln(x + \sqrt{x^2 - 1})$.

*18. Show that $\tanh^{-1} x = \frac{1}{2} \ln[(1 + x)/(1 - x)]$.

19. Use Problems 16, 17, and 18 to evaluate the following:

    (a) $\sinh^{-1} 4$ **(b)** $\cosh^{-1} 3$ **(c)** $\tanh^{-1} \frac{1}{2}$ **(d)** $\sinh^{-1}(-1)$

20. Use the results of Problems 13 and 16 to show that

$$\int \frac{du}{\sqrt{u^2 + a^2}} = \ln(u + \sqrt{a^2 + u^2}) + C.$$

21. Use the results of Problems 14 and 17 to show that for $u > a$

$$\int \frac{du}{\sqrt{u^2 - a^2}} = \ln(u + \sqrt{u^2 - a^2}) + C.$$

22. Use the results of Problems 15 and 18 to show that for $u^2 < a^2$

$$\int \frac{a\,du}{a^2 - u^2} = \frac{1}{2} \ln\left(\frac{a + u}{a - u}\right) + C.$$

In Problems 23–34, calculate the integrals.

23. $\displaystyle\int \frac{x}{\sqrt{x^4 - 4}}\,dx$ **24.** $\displaystyle\int \frac{x}{\sqrt{x^4 + 4}}\,dx$ **25.** $\displaystyle\int \frac{x}{4 - x^4}\,dx$

26. $\displaystyle\int \frac{e^x}{\sqrt{1 + e^{2x}}}\,dx$ **27.** $\displaystyle\int \frac{e^x}{\sqrt{e^{2x} - 9}}\,dx$ **28.** $\displaystyle\int \frac{e^x}{16 - e^{2x}}\,dx$

29. $\displaystyle\int \frac{\cos x}{\sqrt{1 + \sin^2 x}}\,dx$ **30.** $\displaystyle\int \frac{\sin x}{\sqrt{1 + \cos^2 x}}\,dx$ **31.** $\displaystyle\int \frac{\sec x \tan x}{\sqrt{4 + \sec^2 x}}\,dx$

32. $\displaystyle\int \frac{dx}{x\sqrt{\ln^2 x - 25}}$ **33.** $\displaystyle\int \frac{dx}{\sqrt{x^2 + x}}$ **34.** $\displaystyle\int \frac{7}{4 - 3x^2}\,dx$

                                          [*Hint:* Complete the square.]

35. For $|x| > 1$, show that

$$\int \frac{dx}{1 - x^2} = \tanh^{-1} \frac{1}{x} + C. \quad [\textit{Hint: } \text{Find the derivative of } \tanh^{-1} 1/x \text{ for } |x| > 1.]$$

REVIEW EXERCISES FOR CHAPTER SEVEN

1. Graph the curve $y = 4 \tan(5x + 1)$. What is its period?

2. Graph the following curves:

    (a) $y = 4 \sin(2\pi x + 4)$ **(b)** $y = -3 \cos\left[\left(\frac{x}{3}\right) + 2\right]$

In Exercises 3–26, calculate the derivative or integral.

3. $\dfrac{d}{dx} \sin(8x^2 + 2)$ **4.** $\dfrac{d}{dx} \csc \sqrt{x}$ **5.** $\dfrac{d}{dx} \sin^{-1}(e^x + 1)$

6. $\dfrac{d}{dx} \tan^{-1} \sqrt{\cos x}$ **7.** $\dfrac{d}{dx}(\sin x + \cos^{-1} x)^{1/3}$ **8.** $\dfrac{d}{dx} \ln |\tan x|$

9. $\displaystyle\int_0^{5\pi/6} \tan \frac{x}{5}\,dx$ **10.** $\displaystyle\int \sin 3x \cos 3x\,dx$ **11.** $\displaystyle\int \frac{\sec \sqrt{x} \tan \sqrt{x}}{\sqrt{x}}\,dx$

12. $\displaystyle\int \frac{dx}{\sqrt{1-x^2}}$

13. $\displaystyle\int \frac{3x\,dx}{x^4+1}$

14. $\displaystyle\int \frac{x^2}{\sqrt{1-x^6}}\,dx$

15. $\displaystyle\frac{d}{dx}\tan^{-1}\frac{1}{x^2}$

16. $\displaystyle\frac{d}{dx}\cot\!\left(\frac{1+x}{1-x}\right)$

17. $\displaystyle\frac{d}{dx}\,xe^x\sin^{-1}x$

18. $\displaystyle\int \frac{\sec^2(1/x)}{x^2}\,dx$

19. $\displaystyle\int e^x(\csc^2 e^x)\,dx$

20. $\displaystyle\int \frac{e^x}{\sqrt{1-e^{2x}}}\,dx$

21. $\displaystyle\frac{d}{dx}\tan(\cos x)$

22. $\displaystyle\frac{d}{dx}\ln|\sec x + \tan x|$

23. $\displaystyle\frac{d}{dx}\frac{\sin x}{\sin^{-1}x}$

24. $\displaystyle\int \frac{e^{5x}}{1+e^{10x}}\,dx$

25. $\displaystyle\int \sin^2(3x+2)\,dx$

26. $\displaystyle\int \cot\frac{x}{5}\,dx$

27. Calculate the following:
   (a) $\sin(\cos^{-1}\frac{7}{10})$ **(b)** $\tan[\sec^{-1}(-4)]$ **(c)** $\csc(\cot^{-1}\frac{2}{3})$

28. A 2-kg mass is attached to a spring with a spring constant $k = 2$. The spring slides on a horizontal frictionless surface. The mass is initially displaced 50 cm from the equilibrium position of the spring and is then released. Find the equation of motion of the mass for all future time.

In Problems 29–46, calculate the derivative or integral.

29. $\displaystyle\frac{d}{dx}\sinh(x^2+3x)$

30. $\displaystyle\frac{d}{dx}\operatorname{csch}\sqrt[3]{x}$

31. $\displaystyle\frac{d}{dx}e^{\operatorname{sech}^3 x}$

32. $\displaystyle\int \coth x\,dx$

33. $\displaystyle\int x^3\sinh x^4\,dx$

34. $\displaystyle\int \frac{\operatorname{sech}^2\ln x}{x}\,dx$

35. $\displaystyle\frac{d}{dx}\sin(\cosh x)$

36. $\displaystyle\frac{d}{dx}\cos^{-1}(\operatorname{sech}^9 x)$

37. $\displaystyle\frac{d}{dx}\coth(\cot x)$

38. $\displaystyle\int \frac{\operatorname{csch}^2 x}{40+3\coth x}\,dx$

39. $\displaystyle\int \operatorname{csch}^{4/3}x\coth x\,dx$

40. $\displaystyle\int \frac{\cosh(1/x^5)}{x^6}\,dx$

41. $\displaystyle\frac{d}{dx}\sinh^{-1}(\cos x)$

42. $\displaystyle\frac{d}{dx}\cosh^{-1}\frac{1}{x}$

43. $\displaystyle\frac{d}{dx}\tanh^{-1}\sqrt{1+x}$

44. $\displaystyle\int \frac{x}{9-x^4}\,dx$

45. $\displaystyle\int \frac{\sec^2 x}{\sqrt{\tan^2 x - 4}}\,dx$

46. $\displaystyle\int \frac{e^{4x}}{\sqrt{16+e^{8x}}}\,dx$

47. Find $\sinh x$ if $\cosh x = 2$.
48. Find $\operatorname{sech} x$ if $\tanh x = \frac{3}{7}$.

8 Techniques of Integration

In Chapters 3, 6, and 7 we developed methods for differentiating almost any conceivable function we could write down. However, as we have seen, the process of integration is much more difficult. There are many harmless looking functions that are continuous (and so have indefinite integrals) but for which it is impossible to express indefinite integrals in terms of the functions we have so far discussed. The functions e^{x^2} and $\sin(1/\sqrt{x})$ fall into this category. We simply do not have enough functions and must do the best we can with algebraic functions (i.e., functions like $[\sqrt{x^2 - 1}/(x^4 - 2\sqrt{x})]^{3/5}$), exponential and logarithmic functions, trigonometric functions and their inverses, and the hyperbolic functions (which are really exponential functions). It is true that most continuous functions do not have integrals expressible in terms of the functions listed above.

The aim of this chapter is to show how to recognize classes of functions that can be integrated in the sense of actually finding a recognizable antiderivative. There are many techniques which have evolved over a period of decades, and we will present the more general ones. This chapter consists primarily of a set of techniques. Each of these was discovered in response to the need to calculate a particular integral. There is no short, easy way to learn these techniques, and it will be necessary to do great numbers of problems.†

---

†Actually, there are theorems that give us ways to determine whether a function can be integrated in terms of the functions listed above. For an interesting discussion of this topic, see the paper by D. G. Mead, "Integration," *American Mathematical Monthly*, **68**, 152–156 (1961).

After you have completed the first eight sections of this chapter, it will be apparent that there are many integrals for which none of our techniques will yield an answer. For such cases we can still evaluate the definite integral, since the definite integral of any continuous function can be calculated to any number of decimal places of accuracy by a variety of numerical techniques. Some of these techniques will be discussed in Section 8.10.

8.1 REVIEW OF THE BASIC FORMULAS OF INTEGRATION

We give below the integration formulas that were developed in Chapters 5, 6, and 7.

$$\int u^n \, du = \frac{u^{n+1}}{n+1} + C \qquad (n \neq -1) \tag{1}$$

$$\int \frac{1}{u} \, du = \ln |u| + C \tag{2}$$

$$\int a^u \, du = \frac{1}{\ln a} a^u + C \tag{3}$$

$$\int e^u \, du = e^u + C \tag{4}$$

$$\int \cos u \, du = \sin u = C \tag{5}$$

$$\int \sin u \, du = -\cos u + C \tag{6}$$

$$\int \sec^2 u \, du = \tan u + C \tag{7}$$

$$\int \csc^2 u \, du = -\cot u + C \tag{8}$$

$$\int \sec u \tan u \, du = \sec u + C \tag{9}$$

$$\int \csc u \cot u \, du = -\csc u + C \tag{10}$$

$$\int \tan u \, du = -\ln |\cos u| + C = \ln |\sec u| + C \qquad \text{(see Example 7.1.4)} \tag{11}$$

$$\int \cot u \, du = \ln |\sin u| + C \tag{12}$$

$$\int \sec u \, du = \ln |\sec u + \tan u| + C \qquad \text{(see Example 6.5.3)†} \tag{13}$$

---

†This integral seems to have been pulled out of a hat. For a fascinating discussion of the original derivation of $\int \sec \theta \, d\theta$, see V. Frederick Rickey and Philip M. Tuchinsky, "An application of geography to mathematics: History of the integral of the secant," *Mathematics Magazine*, **53** (No. 3), 162–166 (May 1980).

$$\int \csc u \, du = -\ln |\csc u + \cot u| + C \quad \text{(see Problems 7.1.37 and 7.1.38)} \quad \textbf{(14)}$$

$$\int \frac{du}{\sqrt{1 - u^2}} = \sin^{-1} u + C = -\cos^{-1} u + C$$
$$\text{(see Theorems 7.2.1 and 7.2.2)} \quad \textbf{(15)}$$

$$\int \frac{du}{1 + u^2} = \tan^{-1} u + C = -\cot^{-1} u + C$$
$$\text{(see Theorem 7.2.3 and Problem 7.2.65)} \quad \textbf{(16)}$$

$$\int \frac{du}{a^2 + u^2} = \frac{1}{a} \tan^{-1} \frac{u}{a} + C \quad \text{(see Example 7.2.5)} \quad \textbf{(17)}$$

$$\int \frac{du}{u\sqrt{u^2 - 1}} = \sec^{-1} |u| + C \quad \text{(see Problem 7.2.70)} \quad \textbf{(18)}$$

$$\int \sinh u \, du = \cosh u + C \quad \textbf{(19)}$$

$$\int \cosh u \, du = \sinh u + C \quad \textbf{(20)}$$

$$\int \text{sech}^2 u \, du = \tanh u + C \quad \textbf{(21)}$$

$$\int \text{csch}^2 u \, du = -\coth u + C \quad \textbf{(22)}$$

$$\int \text{sech} \, u \tanh u \, du = -\text{sech} \, u + C \quad \textbf{(23)}$$

$$\int \text{csch} \, u \coth u \, du = -\text{csch} \, u + C \quad \textbf{(24)}$$

$$\int \frac{du}{\sqrt{u^2 + 1}} = \sinh^{-1} u + C = \ln |u + \sqrt{1 + u^2}| + C \text{ (see Theorem 7.5.1)} \quad \textbf{(25)}$$

$$\int \frac{du}{\sqrt{u^2 - 1}} = \cosh^{-1} u + C = \ln |u + \sqrt{u^2 - 1}| + C \quad (u > 1) \quad \textbf{(26)}$$

$$\int \frac{du}{1 - u^2} = \tanh^{-1} u + C = \frac{1}{2} \ln \left| \frac{1 + u}{1 - u} \right| + C \quad (u^2 < 1)$$
$$\text{(see Theorem 7.5.2)} \quad \textbf{(27)}$$

A much more complete table of integrals appears in Table A.4 at the end of the book.

8.2 INTEGRATION BY PARTS

Of all the methods to be discussed in this chapter, the most powerful is called **integration by parts;** it is derived from the product rule of differentiation (written in terms of differentials):

$$d(uv) = u\, dv + v\, du. \tag{1}$$

Integrating both sides of (1), we obtain

$$uv = \int u\, dv + \int v\, du,$$

or, rearranging terms,

$$\int u\, dv = uv - \int v\, du. \tag{2}$$

As we will see, the idea in using integration by parts is to rewrite an expression that is difficult to integrate in terms of an expression for which an integral is readily obtainable.

EXAMPLE 1   Calculate $\int xe^x\, dx$.

Solution. We cannot integrate xe^x directly because the x term gets in the way. However, if we set $u = x$ and $dv = e^x\, dx$, then $du = dx$, $v = \int e^x\, dx = e^x$, and

$$\int \overset{u}{x}\overset{dv}{e^x\, dx} = \int u\, dv = uv - \int v\, du = \overset{u\,v}{xe^x} - \int \overset{v\,du}{e^x\, dx} = xe^x - e^x + C.$$

This result can be checked by differentiation. Note how the x term "disappeared." This example is typical of the type of problem that can be solved by integration by parts. ■

EXAMPLE 2   Calculate $\int_0^{\pi/2} x \cos x\, dx$.

Solution. Again we could integrate directly were it not for the x term. We set $u = x$ (so that $du = dx$, causing the x term to "vanish") and $dv = \cos x\, dx$. Then $v = \sin x$ and

$$\int_0^{\pi/2} \overset{u}{x}\, \overset{dv}{\cos x\, dx} = \int_0^{\pi/2} u\, dv = uv \Big|_0^{\pi/2} - \int_0^{\pi/2} v\, du$$

$$= \overset{u}{x}\, \overset{v}{\sin x} \Big|_0^{\pi/2} - \int_0^{\pi/2} \overset{v}{\sin x}\, \overset{du}{dx} = \frac{\pi}{2} + \cos x \Big|_0^{\pi/2} = \frac{\pi}{2} - 1. \quad ■$$

EXAMPLE 3   Calculate $\int \ln x\, dx$.

Solution. There are two terms here, $\ln x$ and dx. If we are to integrate by parts, the only choices we have for u and dv are $u = \ln x$ and $dv = dx$. Then $du = (1/x)\, dx$, $v = x$, and

$$\int \overbrace{\ln x}^{u} \overbrace{dx}^{dv} = \int u\,dv = uv - \int v\,du = \overset{v}{x} \overset{u}{\ln x} - \int \overset{v}{x} \cdot \overbrace{\frac{1}{x}\,dx}^{du} = x\ln x - x + C.$$

NOTE. In many of the integrations involving ln x, we may take $u = \ln x$ so that $du = (1/x)\,dx$ and the ln term "vanishes."

EXAMPLE 4   Calculate $\int_0^1 x^3 e^{x^2}\,dx$.

Solution. Here are several choices for u and dv. The possibilities are shown in Table 1. From the table we see that while there are several choices for u and dv, only one works. The others fail either because (a) dv cannot be readily integrated to find v or (b) $\int v\,du$ is not any easier to integrate than the integral we started with. Hence to complete the problem, we set $u = x^2$ and $dv = xe^{x^2}\,dx$ to obtain

$$\int_0^1 x^3 e^{x^2}\,dx = \int_0^1 \overset{u}{x^2} \cdot \overbrace{xe^{x^2}\,dx}^{dv} = \overset{u}{x^2}\,\overset{v}{\frac{e^{x^2}}{2}}\Big|_0^1 - \int_0^1 \overset{v}{\frac{1}{2}e^{x^2}} \cdot \overbrace{2x\,dx}^{du}$$

$$= \frac{e}{2} - \int_0^1 xe^{x^2}\,dx = \frac{e}{2} - \frac{1}{2}e^{x^2}\Big|_0^1 = \frac{e}{2} - \frac{e}{2} + \frac{1}{2} = \frac{1}{2}.\ \blacksquare$$

TABLE 1

u	dv	du	v	$uv - \int v\,du$	Comments
x^3	$e^{x^2}\,dx$	$3x^2\,dx$	$\int dv = \int e^{x^2}\,dx$ (We're stuck.)	—	Try something else.
e^{x^2}	$x^3\,dx$	$2xe^{x^2}\,dx$	$\dfrac{x^4}{4}$	$\dfrac{x^4 e^{x^2}}{4} - \dfrac{1}{2}\int x^5 e^{x^2}\,dx$	We're worse off than when we started.
xe^{x^2}	$x^2\,dx$	$e^{x^2}(1 + 2x^2)\,dx$	$\dfrac{x^3}{3}$	$\dfrac{x^4 e^{x^2}}{3} - \dfrac{1}{3}\int x^3 e^{x^2}(1 + 2x^2)\,dx$	Ditto.
x^2	$xe^{x^2}\,dx$	$2x\,dx$	$\dfrac{1}{2}e^{x^2}$	$\dfrac{x^2 e^{x^2}}{2} - \int xe^{x^2}\,dx$	We can integrate this.
$x^2 e^{x^2}$	$x\,dx$	$e^{x^2}(2x^3 + 2x)\,dx$	$\dfrac{x^2}{2}$	$\dfrac{x^4}{2}e^{x^2} - \int e^{x^2}(x^5 + x^3)\,dx$	What a mess!
x	$x^2 e^{x^2}\,dx$	dx	$\int x^2 e^{x^2}\,dx$	—	Try something else.
$x^3 e^{x^2}$	dx	$e^{x^2}(2x^4 + 3x^2)\,dx$	x	$x^4 e^{x^2} - \int e^{x^2}(2x^5 + 3x^3)\,dx$	This is the worst of all.

Example 4 points out two things to look for in choosing u and dv:

> **(i)** It must be possible to evaluate $\int dv$.
> **(ii)** $\int v\, du$ should be easier to evaluate than $\int u\, dv$.

EXAMPLE 5 Calculate $\int x^2 e^{-x}\, dx$.

Solution. We need to get rid of the x^2 term. Let $u = x^2$ and $dv = e^{-x}\, dx$. Then $du = 2x\, dx$ and $v = -e^{-x}$, so that

$$\int x^2 e^{-x}\, dx = -x^2 e^{-x} + 2 \overset{-2\int -e^{-x}\,dx}{\int xe^{-x}\, dx}.$$

We see that $\int xe^{-x}\, dx$ is simpler than $\int x^2 e^{-x}\, dx$, but it is still necessary to integrate by parts once more. Setting $u = x$ and $dv = e^{-x}\, dx$, we have $du = dx$ and $v = -e^{-x}$, so that

$$2\int xe^{-x}\, dx = -2xe^{-x} + 2\int e^{-x}\, dx = -2xe^{-x} - 2e^{-x} + C.$$

Therefore

$$\int x^2 e^{-x}\, dx = -x^2 e^{-x} - 2xe^{-x} - 2e^{-x} + C. \ \blacksquare$$

EXAMPLE 6 Calculate $\int \sin^{-1} x\, dx$.

Solution. There is no other choice but to set $u = \sin^{-1} x$ and $dv = dx$. Then $du = dx/\sqrt{1 - x^2}$, $v = x$, and

$$\int \sin^{-1} x\, dx = x\sin^{-1} x - \int \frac{x}{\sqrt{1 - x^2}}\, dx = x\sin^{-1} x + \sqrt{1 - x^2} + C. \ \blacksquare$$

EXAMPLE 7 Evaluate $\int e^x \sin x\, dx$.

Solution. We set $I = \int e^x \sin x\, dx$. (The reason for this will become apparent as we work the problem.) Let $u = e^x$ and $dv = \sin x\, dx$. Then $du = e^x\, dx$, $v = -\cos x$, and

$$I = -e^x \cos x + \int e^x \cos x\, dx.$$

It seems as if we have not accomplished anything, but this is not the case. We now set $u = e^x$ and $dv = \cos x\, dx$, so that $du = e^x\, dx$, $v = \sin x$, and

$$I = -e^x \cos x + e^x \sin x - \int e^x \sin x\, dx = e^x(\sin x - \cos x) - I,$$

or

$$2I = e^x(\sin x - \cos x),$$

and

$$I = \int e^x \sin x \, dx = \frac{e^x(\sin x - \cos x)}{2} + C. \ \blacksquare$$

EXAMPLE 8   Evaluate $\int \sec^3 x \, dx$.

Solution. Let $I = \int \sec^3 x \, dx$. We know that $(d/dx)\tan x = \sec^2 x$. We set $u = \sec x$ and $dv = \sec^2 x \, dx$, so that

$$du = \sec x \tan x \, dx, \qquad v = \tan x,$$

and

$$I = \sec x \tan x - \int \sec x \tan^2 x \, dx. \tag{3}$$

Now

$$\tan^2 x = \sec^2 x - 1, \tag{4}$$

and by inserting (4) into (3), we obtain

$$I = \sec x \tan x - \int \sec x (\sec^2 x - 1) dx = \sec x \tan x - \int \sec^3 x \, dx + \int \sec x \, dx$$

$$= \sec x \tan x - I + \ln |\sec x + \tan x|,$$

or

$$2I = \sec x \tan x + \ln |\sec x + \tan x|,$$

and

$$I = \int \sec^3 x \, dx = \frac{1}{2}(\sec x \tan x + \ln |\sec x + \tan x|) + C.$$

This integral comes up frequently in applications, as we will see in Chapter 9. It's worth memorizing. $\blacksquare$

We can use the technique of integration by parts to obtain **reduction formulas.** A reduction formula allows us to evaluate a whole class of integrals by reducing each integral in the class to a simpler one. This idea is best illustrated by an example.

EXAMPLE 9 Prove that

$$\int \sin^n x \, dx = -\frac{1}{n} \sin^{n-1} x \cos x + \frac{n-1}{n} \int \sin^{n-2} x \, dx, \tag{5}$$

where $n > 1$ is a positive integer.

Solution. Let $u = \sin^{n-1} x$ and $dv = \sin x \, dx$. Then

$$du = (n-1)\sin^{n-2} x \cos x \, dx, \quad v = -\cos x,$$

and

$$I = \int \sin^n x \, dx = -\sin^{n-1} x \cos x + (n-1) \int \sin^{n-2} x \cos^2 x \, dx.$$

Using $\cos^2 x = 1 - \sin^2 x$, we obtain

$$I = -\sin^{n-1} x \cos x + (n-1) \int (\sin^{n-2} x)(1 - \sin^2 x) \, dx$$

$$= -\sin^{n-1} x \cos x + (n-1) \int \sin^{n-2} x \, dx - (n-1) \int \sin^n x \, dx$$

$$= -\sin^{n-1} x \cos x + (n-1) \int \sin^{n-2} x \, dx - (n-1)I,$$

or

$$I + (n-1)I = nI = -\sin^{n-1} x \cos x + (n-1) \int \sin^{n-2} x \, dx,$$

and

$$I = -\frac{1}{n} \sin^{n-1} x \cos x + \frac{n-1}{n} \int \sin^{n-2} x \, dx. \quad \blacksquare$$

REMARK. As Example 9 indicates, we can integrate $\sin^n x$ if we know how to integrate $\sin^{n-2} x$. By using the reduction formula (5) repeatedly, if necessary, we can integrate any positive integral power of $\sin x$ since we can integrate 1 and $\sin x$ (because even powers of $\sin x$ will "reduce" to 1 and odd powers will "reduce" to $\sin x$).

EXAMPLE 10 Evaluate $\int \sin^3 x \, dx$.

Solution. Using (5) with $n = 3$, we have

$$\int \sin^3 x \, dx = -\frac{1}{3} \sin^2 x \cos x + \frac{2}{3} \int \sin x \, dx = -\frac{1}{3} \sin^2 x \cos x - \frac{2}{3} \cos x + C. \quad \blacksquare$$

There are many other reduction formulas. Several are given in the table of integrals (Table A4). You are asked to derive four of them in Problems 34, 36, and 38.

PROBLEMS 8.2

In Problems 1–27, evaluate the given integral.

1. $\displaystyle\int xe^{3x}\,dx$

2. $\displaystyle\int_0^1 xe^{-7x}\,dx$

3. $\displaystyle\int_0^4 x^2 e^{x/4}\,dx$

4. $\displaystyle\int_1^2 x\ln x\,dx$

5. $\displaystyle\int x^7\ln x\,dx$

6. $\displaystyle\int_0^{\pi/2} x\sin x\,dx$

7. $\displaystyle\int x^2\cos 2x\,dx$

8. $\displaystyle\int_0^5 x\sqrt{3x+1}\,dx$

9. $\displaystyle\int x\sqrt{1-\frac{x}{2}}\,dx$

10. $\displaystyle\int \cos^{-1}x\,dx$

11. $\displaystyle\int_0^{1/2} \tan^{-1}2x\,dx$

12. $\displaystyle\int x\sinh x\,dx$

13. $\displaystyle\int_0^1 x^2\cosh 2x\,dx$

14. $\displaystyle\int x5^x\,dx$

15. $\displaystyle\int_0^1 \sin^{-1}x\,dx$

*16. $\displaystyle\int x\tan^{-1}x\,dx$

17. $\displaystyle\int \cos(\ln x)\,dx$

18. $\displaystyle\int e^x\cos x\,dx$

19. $\displaystyle\int e^{3x}\sin 2x\,dx$

*20. $\displaystyle\int e^{ax}\sin bx\,dx,\, a,b\neq 0$

21. $\displaystyle\int_0^1 x^5 e^{x^3}\,dx$

22. $\displaystyle\int_1^2 \ln^2 x\,dx$

23. $\displaystyle\int \sin^4 x\,dx$

*24. $\displaystyle\int e^{ax}\cos bx\,dx,\, a,b\neq 0$

*25. $\displaystyle\int x^2\tan^{-1}x\,dx$

*26. $\displaystyle\int \sinh 2x\cosh 3x\,dx$

*27. $\displaystyle\int \sinh 2x\sinh 5x\,dx$

*28. Show that if $a\neq 0$,

$$\int x^3\sin ax\,dx = \left(\frac{3x^2}{a^2}-\frac{6}{a^4}\right)\sin ax + \left(\frac{6x}{a^3}-\frac{x^3}{a}\right)\cos ax + C.$$

*29. Using mathematical induction (Appendix 2) and integration by parts, show that if n is a positive integer and $a\neq 0$,

$$\int x^n e^{ax}\,dx = \frac{e^{ax}}{a}\left[x^n - \frac{nx^{n-1}}{a} + \frac{n(n-1)x^{n-2}}{a^2} - \cdots + \frac{(-1)^n n!}{a^n}\right] + C,$$

where $n! = n(n-1)(n-2)\cdots\cdots 3\cdot 2\cdot 1$.

*30. Show that if $a\neq 0$,

$$\int x^2\sinh ax\,dx = \left(\frac{x^2}{a}+\frac{2}{a^3}\right)\cosh ax - \frac{2x}{a^2}\sinh ax + C.$$

31. The density of a 5-m bar is given by $\rho = xe^{-x}$ kilograms per meter, where x is measured in meters from the left. What is the total mass of the bar?

32. The marginal cost of a certain product fluctuates according to the rule

$$MC = 20 + e^{-0.01q}\sin q,$$

where q is the number of items produced. What is the total cost of producing 100 units?

33. (a) Graph the curve $y = xe^{-x}$. [*Hint:* Use the fact (which we will prove in Chapter 12) that $\lim_{x\to\infty} xe^{-x} = 0$.]
   (b) Calculate the area bounded by the curve $y = xe^{-x}$, the x-axis, and the lines $x = -1$ and $x = 1$.
34. Prove the reduction formula

$$\int \cos^n x \, dx = \frac{1}{n} \cos^{n-1} x \sin x + \frac{n-1}{n} \int \cos^{n-2} x \, dx,$$

where $n > 1$ is an integer.
35. Use the result of Problem 34 to evaluate $\int \cos^3 x \, dx$.
*36. Show that

$$\int \sec^n x \, dx = \frac{1}{n-1} \tan x \sec^{n-2} x + \frac{n-2}{n-1} \int \sec^{n-2} x \, dx,$$

where $n > 1$ is an integer.

37. Use the result of Problem 36 to evaluate $\int \sec^5 x \, dx$.
38. (a) Show that $\int x^n \cos x \, dx = x^n \sin x - n \int x^{n-1} \sin x \, dx$, where $n > 1$.
   (b) Show that $\int x^k \sin x \, dx = -x^k \cos x + k \int x^{k-1} \cos x \, dx$.
39. Use the results of Problem 38 to evaluate $\int x^3 \cos x \, dx$.
**40. Suppose that f and g are continuous on $[a, b]$ and f has a continuous derivative that is never zero. Show that there is at least one choice of c such that $a \le c \le b$ and

$$\int_a^b fg = f(a) \int_a^c g + f(b) \int_c^b g.$$

*41. For $-\pi/2 < x < \pi/2$, let $T_n(x) = \int_0^x \tan^n \theta \, d\theta$.
   (a) Find explicit formulas for $T_0(x)$ and $T_1(x)$.
   (b) Find a recursion formula connecting $T_n(x)$ with $T_{n-2}(x)$. [*Hint:* $\tan^2 \theta = \sec^2 \theta - 1$.]
   (c) Use your answer to part (b) to find $T_2(x)$ and $T_3(x)$.
*42. For $0 < x < \pi$, let $C_n(x) = \int_{\pi/2}^x \cot^n \theta \, d\theta$.
   (a) Find explicit formulas for $C_0(x)$ and $C_1(x)$.
   (b) Find a recursion formula connecting $C_n(x)$ with $C_{n-2}(x)$. [*Hint:* $\cot^2 \theta = \csc^2 \theta - 1$.]
   (c) Use your answer to part (b) to find $C_2(x)$ and $C_3(x)$.
*43. Let

$$I_n(x) = \int_0^x \frac{t^n}{\sqrt{t^2 + 1}} \, dt.$$

   (a) Prove that $I_0(x) = \ln(x + \sqrt{x^2 + 1})$.
   (b) Find $I_1(x)$.
   (c) Find a recursion relation for $I_n(x)$ of the form $I_n(x) = [P_n(x)]\sqrt{x^2 + 1} + C_n I_{n-2}(x)$, where $P_n(x)$ is a polynomial function and C_n is a constant.
*44. Let $G_n(x) = \int_0^x e^{-t} t^{n-1} \, dt$.
   (a) Find a recursion relation connecting G_{n+1} with G_n.
   (b) Find explicit expressions for $G_1(x)$, $G_2(x)$, and $G_3(x)$.
*45. Let $J_n(x) = \int (x^n/\sqrt{x^k + 1}) \, dx$. Find a recursion relation connecting J_n with J_{n-k}.
*46. What is wrong with the following reasoning?

$$\int \cot x \, dx = \int \csc x \frac{d}{dx}(\sin x) \, dx = \csc x \sin x - \int \sin x \frac{d}{dx}(\csc x) \, dx = 1 + \int \cot x \, dx.$$

Therefore $0 = 1$.

8.3 INTEGRALS OF CERTAIN TRIGONOMETRIC FUNCTIONS

In this section we show how to integrate some classes of trigonometric functions. They are (with m and n positive integers):

(i) $\int \sin^m x \cos^n x \, dx$, either m or n odd

(ii) $\int \sin^m x \cos^n x \, dx$, both m and n even

(iii) $\int \sin mx \cos nx \, dx$

(iv) $\int \sin mx \sin nx \, dx$

(v) $\int \cos mx \cos nx \, dx$

(vi) $\int \tan^n x \, dx$

(vii) $\int \sec^n x \, dx$

(viii) $\int \tan^m x \sec^n x \, dx$, n even

(ix) $\int \tan^m x \sec^n x \, dx$, n and m odd

(x) $\int \tan^m x \sec^n x \, dx$, n odd, m even

We will see that each case can be treated fairly easily by making use of some trigonometric identity.

Case (i). $\int \sin^m x \cos^n x \, dx$, where either m or n is an odd integer.†

EXAMPLE 1   Calculate $\int \sin^3 x \cos^{4/3} x \, dx$.

Solution. Write $\sin^3 x = (\sin^2 x) \sin x = (1 - \cos^2 x) \sin x$. Then

$$\int \sin^3 x \cos^{4/3} x \, dx = \int (1 - \cos^2 x) \sin x \cos^{4/3} x \, dx$$

$$= \int \cos^{4/3} x \sin x \, dx - \int \cos^{10/3} x \sin x \, dx$$

$$= -\tfrac{3}{7} \cos^{7/3} x + \tfrac{3}{13} \cos^{13/3} x + C. \ \blacksquare$$

---

†Here if m is an odd integer, then n need not be an integer, and vice versa.

In Case (i), by breaking off $\sin^2 x$ terms (if m is odd) or $\cos^2 x$ terms (if n is odd), we can always obtain integrals of the form $\int \cos^p x \sin x \, dx$ or $\int \sin^q x \cos x \, dx$, and these can be integrated directly since

$$\int \cos^p x \sin x \, dx = -\frac{\cos^{p+1} x}{p+1} + C \qquad \text{and} \qquad \int \sin^q x \cos x \, dx = \frac{\sin^{q+1} x}{q+1} + C.$$

Case (ii). $\int \sin^m x \cos^n x \, dx$, both m and n are even. We use the identities

$$\cos^2 x = \frac{1 + \cos 2x}{2}, \qquad \sin^2 x = \frac{1 - \cos 2x}{2} \tag{1}$$

[See (xx) and (xxi) in Table 2 in Appendix 1.2]

EXAMPLE 2 Calculate $\int_0^{\pi/2} \sin^2 x \, dx$.

Solution.

$$\int_0^{\pi/2} \sin^2 x \, dx = \frac{1}{2} \int_0^{\pi/2} (1 - \cos 2x) \, dx = \left(\frac{x}{2} - \frac{\sin 2x}{4} \right) \Big|_0^{\pi/2} = \frac{\pi}{4} \quad \blacksquare$$

EXAMPLE 3 Calculate $\int_0^{\pi/2} \sin^2 x \cos^4 x \, dx$.

Solution.

$$\int_0^{\pi/2} \sin^2 x \cos^4 x \, dx = \int_0^{\pi/2} \overbrace{\frac{1 - \cos 2x}{2}}^{\sin^2 x} (\cos^2 x)^2 \, dx$$

$$= \frac{1}{2} \int_0^{\pi/2} (1 - \cos 2x) \overbrace{\left(\frac{1 + \cos 2x}{2} \right)^2}^{\cos^2 x} \, dx$$

$$= \frac{1}{2} \cdot \frac{1}{4} \int_0^{\pi/2} (1 - \cos 2x)(1 + 2\cos 2x + \cos^2 2x) \, dx$$

$$= \frac{1}{8} \int_0^{\pi/2} (1 + \cos 2x - \cos^2 2x - \cos^3 2x) \, dx$$

$$= \frac{1}{8} \int_0^{\pi/2} (1 + \cos 2x) \, dx - \frac{1}{8} \int_0^{\pi/2} \cos^2 2x \, dx - \frac{1}{8} \int_0^{\pi/2} \cos^3 2x \, dx$$

$$= \frac{1}{8} \left(x + \frac{\sin 2x}{2} \right) \Big|_0^{\pi/2} - \frac{1}{8} \int_0^{\pi/2} \overbrace{\frac{1 + \cos 4x}{2}}^{\cos^2 2x} \, dx$$

$$-\frac{1}{8}\int_0^{\pi/2} \overbrace{(1 - \sin^2 2x)}^{\cos^2 2x;\ see\ Case\ (i)} \cos 2x\ dx$$

$$= \frac{\pi}{16} - \frac{1}{16}\left(x + \frac{\sin 4x}{4}\right)\Big|_0^{\pi/2} - \frac{1}{8}\int_0^{\pi/2}(\cos 2x - \sin^2 2x \cos 2x)\ dx$$

$$= \frac{\pi}{16} - \frac{\pi}{32} - \frac{1}{8}\left(\frac{\sin 2x}{2} - \frac{\sin^3 2x}{6}\right)\Big|_0^{\pi/2} = \frac{\pi}{32} - 0 = \frac{\pi}{32}\ \blacksquare$$

Cases (iii), (iv), and (v).   We use the three identities

$$\sin mx \cos nx = \tfrac{1}{2}\{\sin[(m + n)x] + \sin[(m - n)x]\}, \tag{2}$$

$$\sin mx \sin nx = \tfrac{1}{2}\{\cos[(m - n)x] - \cos[(m + n)x]\}, \tag{3}$$

$$\cos mx \cos nx = \tfrac{1}{2}\{\cos[(m - n)x] + \cos[(m + n)x]\}. \tag{4}$$

These identities are easily obtained from the addition formulas (see Problems 28, 29, 30).

EXAMPLE 4   Calculate $\int \sin 3x \cos 7x\ dx$.

Solution.

$$\int \sin 3x \cos 7x\ dx \overset{\underset{\text{From (2)}}{\downarrow}}{=} \frac{1}{2}\int [\sin(3 + 7)x + \sin(3 - 7)x]\ dx$$

$$= \frac{1}{2}\int [\sin 10x + \overset{\underset{\sin(-x) = -\sin x}{\downarrow}}{\sin(-4x)}]\ dx = \frac{1}{2}\int (\sin 10x - \sin 4x)\ dx$$

$$= \frac{1}{2}\left(\frac{\cos 4x}{4} - \frac{\cos 10x}{10}\right) + C\ \blacksquare$$

Other integrals in Cases (iii), (iv), and (v) are handled just as easily.

Case (vi). $\int \tan^n x\ dx$. We use the identity $1 + \tan^2 x = \sec^2 x$.

EXAMPLE 5   Calculate $\int \tan^2 x\ dx$.

Solution. $\int \tan^2 x\ dx = \int (\sec^2 x - 1)\ dx = \tan x - x + C$. $\blacksquare$

EXAMPLE 6   Calculate $\int \tan^5 x\ dx$.

Solution.

$$\int \tan^5 x\ dx = \int \tan^3 x \tan^2 x\ dx = \int \tan^3 x \overbrace{(\sec^2 x - 1)}^{\tan^2 x}\ dx$$

$$= \int \tan^3 x \overbrace{\sec^2 x \, dx}^{d(\tan x)} - \int \tan^3 x \, dx$$

$$= \frac{\tan^4 x}{4} - \int \tan x \overbrace{(\sec^2 x - 1)}^{\tan^2 x} \, dx$$

$$= \frac{\tan^4 x}{4} - \frac{\tan^2 x}{2} + \int \tan x \, dx$$

$$= \frac{\tan^4 x}{4} - \frac{\tan^2 x}{2} - \ln |\cos x| + C \quad \blacksquare$$

Case (vii). $\int \sec^n x \, dx$. We use the same identity as in Case (vi).

EXAMPLE 7   Calculate $\int_0^{\pi/4} \sec^4 x \, dx$.

Solution.

$$\int_0^{\pi/4} \sec^4 x \, dx = \int_0^{\pi/4} \sec^2 x (1 + \tan^2 x) \, dx$$

$$= \int_0^{\pi/4} \sec^2 x \, dx + \int_0^{\pi/4} \tan^2 x \sec^2 x \, dx$$

$$= \tan x \Big|_0^{\pi/4} + \frac{\tan^3 x}{3} \Big|_0^{\pi/4} = \frac{4}{3} \quad \blacksquare$$

NOTE.   There are two other ways to integrate $\sec^n x \, dx$. We can use integration by parts, as in Example 8.2.8, or we can prove and use the recursion formula given in Problem 8.2.36. There is also a recursion formula for computing $\int \tan^n x \, dx$. This formula is given in Problem 54.

Case (viii). $\int \tan^m x \sec^n x \, dx$, n even. Use $1 + \tan^2 x = \sec^2 x$ and break off one $\sec^2 x$ term, since $d(\tan x) = \sec^2 x \, dx$.

EXAMPLE 8   Calculate $\int_0^{\pi/3} \tan^7 x \sec^4 x \, dx$.

Solution.

$$\int_0^{\pi/3} \tan^7 x \sec^2 x \sec^2 x \, dx = \int_0^{\pi/3} (\tan^7 x)(1 + \tan^2 x) \sec^2 x \, dx$$

$$= \int_0^{\pi/3} \tan^7 x \sec^2 x \, dx + \int_0^{\pi/3} \tan^9 x \sec^2 x \, dx$$

$$= \left(\frac{\tan^8 x}{8} + \frac{\tan^{10} x}{10} \right) \Big|_0^{\pi/3} = \frac{\sqrt{3}^8}{8} + \frac{\sqrt{3}^{10}}{10} = \frac{3^4}{8} + \frac{3^5}{10}$$

$$= \frac{81}{8} + \frac{243}{10} = \frac{1377}{40} \quad \blacksquare$$

Case (ix). $\int \tan^m x \sec^n x \, dx$, m, n odd. In this case, break off a term of the form $\sec x \tan x$, since $d(\sec x) = \sec x \tan x \, dx$.

EXAMPLE 9 $\int \tan^3 x \sec^5 x \, dx = \int \tan^2 x \sec^4 x \cdot \sec x \tan x \, dx$. Since $\sec x \tan x$ is the derivative of $\sec x$, we write everything in terms of $\sec x$:

$$\int \tan^2 x \sec^4 x \cdot \sec x \tan x \, dx = \int (\sec^2 x - 1)\sec^4 x \cdot \sec x \tan x \, dx$$

$$= \int (\sec^6 x - \sec^4 x)\sec x \tan x \, dx$$

$$= \frac{\sec^7 x}{7} - \frac{\sec^5 x}{5} + C. \quad \blacksquare$$

Case (x). $\int \tan^m x \sec^n x \, dx$, m even, n odd. In this case we can write everything in terms of $\sec x$ and proceed as in Case (vii).

EXAMPLE 10   Calculate $\int \tan^2 x \sec x \, dx$.

Solution.

$$\int \tan^2 x \sec x \, dx = \int (\sec^2 x - 1)\sec x \, dx = \int \sec^3 x \, dx - \int \sec x \, dx$$
$$\text{(From Example 8.2.8)}$$
$$= \tfrac{1}{2}(\sec x \tan x + \ln |\sec x + \tan x|) - \ln |\sec x + \tan x| + C$$
$$= \tfrac{1}{2}[\sec x \tan x - \ln |\sec x + \tan x|] + C \quad \blacksquare$$

PROBLEMS 8.3

In Problems 1–27, calculate the integral.

1. $\int_0^\pi \sin^3 x \cos^2 x \, dx$

2. $\int \cos^3 x \sqrt{\sin x} \, dx$

3. $\int_0^{\pi/4} \cos^2 x \, dx$

4. $\int \sin^4 x \, dx$

5. $\int \cos^4 2x \, dx$

6. $\int_0^{\pi/2} \sin^2 3x \cos^4 3x \, dx$

7. $\int \sin 2x \cos 3x \, dx$

8. $\int_0^{\pi/2} \sin \frac{x}{2} \cos \frac{x}{3} \, dx$

9. $\int \sin \sqrt{2}x \cos \frac{x}{\sqrt{2}} \, dx$

10. $\int_0^\pi \sin 2x \sin \frac{x}{2} \, dx$

11. $\int \cos 10x \cos 100x \, dx$

12. $\int_0^{\pi/8} \tan^2 2x \, dx$

13. $\int \sec^3 5x \, dx$

14. $\int_0^{\pi/6} \sec^3 2x \tan 2x \, dx$

15. $\int_0^\pi \sin^6 x \cos^4 x \, dx$

16. $\int \cos^{10} x \sin^7 x \, dx$

17. $\int_0^{\pi/4} \sec^3 x \tan^3 x \, dx$

18. $\int \sec^6 \frac{x}{2} \, dx$

19. $\int \sin^5 x \cos^2 x \, dx$

20. $\int \cos^5 x \sin^2 x \, dx$

21. $\int_0^{\pi/2} \tan^5 \frac{x}{2} \, dx$

22. $\displaystyle\int_0^{\pi/3} \sec^6 x \, dx$

23. $\displaystyle\int \tan^3 x \sec^4 x \, dx$

24. $\displaystyle\int \frac{\sin^5 x}{\cos^2 x} \, dx$

25. $\displaystyle\int_0^{\pi/4} \sin^2 x \tan x \, dx$

26. $\displaystyle\int \sec x \sin^3 x \, dx$

27. $\displaystyle\int \cos^3 x \sqrt{\csc x} \, dx$

28. Show that $\sin mx \cos nx = \frac{1}{2}\{\sin[(m+n)x] + \sin[(m-n)x]\}$.   [*Hint:* Add the identities $\sin(a+b) = \sin a \cos b + \cos a \sin b$ and $\sin(a-b) = \sin a \cos b - \cos a \sin b$.]

29. Prove identity (3).   **30.** Prove identity (4).

31. Show that

$$\int \sin mx \sin nx \, dx = \frac{1}{2}\left\{\frac{\sin[(m-n)x]}{m-n} - \frac{\sin[(m+n)x]}{m+n}\right\} + C$$

if $m \neq \pm n$.

32. Show that

$$\int \cos mx \cos nx \, dx = \frac{1}{2}\left\{\frac{\sin[(m-n)x]}{m-n} + \frac{\sin[(m+n)x]}{m+n}\right\} + C.$$

33. Show that $\int_0^{2\pi} \sin mx \cos nx \, dx = 0$ for any integers m and n.

34. Show that $\int_0^{2\pi} \cos mx \cos nx \, dx = 0$ if $m \neq \pm n$.

35. Show that $\int_0^{2\pi} \sin mx \sin nx \, dx = 0$ if $m \neq \pm n$.

Using the identity $1 + \cot^2 x = \csc^2 x$, we can evaluate integrals of the form $\int \cot^n x \, dx$, $\int \csc^n x \, dx$, and $\int \cot^m x \csc^n x \, dx$. We need only use the methods outlined for integrals of type (vi)–(x). In Problems 36–44, evaluate the integral.

36. $\displaystyle\int \cot^2 x \, dx$

37. $\displaystyle\int \cot^5 x \, dx$

38. $\displaystyle\int_{\pi/4}^{\pi/2} \csc^4 x \, dx$

39. $\displaystyle\int \csc^3 x \, dx$

40. $\displaystyle\int_{\pi/6}^{\pi/2} \cot^7 x \csc^4 x \, dx$

41. $\displaystyle\int \cot^3 x \csc^5 x \, dx$

42. $\displaystyle\int \cot^2 x \csc^6 x \, dx$

43. $\displaystyle\int \csc^4 2x \, dx$

44. $\displaystyle\int \cot^2 x \csc x \, dx$

45. Show that

$$\int_0^{\pi/2} \sin^2 x \, dx = \frac{\pi}{4} = \frac{1}{2} \cdot \frac{\pi}{2}.$$

46. Show that

$$\int_0^{\pi/2} \sin^4 x \, dx = \frac{3}{4} \cdot \frac{1}{2} \cdot \frac{\pi}{2}.$$

47. Show that

$$\int_0^{\pi/2} \sin^6 x \, dx = \frac{5}{6} \cdot \frac{3}{4} \cdot \frac{1}{2} \cdot \frac{\pi}{2}.$$

48. By using the reduction formula of Example 8.2.9, show that

$$\int_0^{\pi/2} \sin^{2n} x \, dx = \frac{2n-1}{2n} \int_0^{\pi/2} \sin^{2n-2} x \, dx.$$

*49. Using Problems 45–48, show that

$$\int_0^{\pi/2} \sin^{2n} x \, dx = \frac{2n-1}{2n} \cdot \frac{2n-3}{2n-2} \cdot \frac{2n-5}{2n-4} \cdots \cdots \frac{5}{6} \cdot \frac{3}{4} \cdot \frac{1}{2} \cdot \frac{\pi}{2}.$$

This is called a **Wallis product.**

*50. Show that

$$\int_0^{\pi/2} \sin^{2n+1} x\, dx = \frac{2n}{2n+1}\cdot\frac{2n-2}{2n-1}\cdot\frac{2n-4}{2n-3}\cdot\ldots\cdot\frac{4}{5}\cdot\frac{2}{3}.$$

This is another Wallis product.

51. Show that $\left|\int_0^{\pi/2}\sin^{2n} x\, dx\right| \le \pi/2$.

*52. Show, using the reduction formula of Example 8.2.9, that

$$\lim_{n\to\infty}\frac{\displaystyle\int_0^{\pi/2}\sin^{2n} x\, dx}{\displaystyle\int_0^{\pi/2}\sin^{2n+1} x\, dx} = 1.$$

*53. Divide the result of Problem 49 by the result of Problem 50 and use Problem 52 to derive the *Wallis formula*

$$\frac{\pi}{2} = \lim_{n\to\infty}\frac{2}{1}\cdot\frac{2}{3}\cdot\frac{4}{3}\cdot\frac{4}{5}\cdot\frac{6}{5}\cdot\frac{6}{7}\cdot\ldots\cdot\frac{2n}{2n-1}\cdot\frac{2n}{2n+1}.$$

54. Derive the formula

$$\int \tan^n ax\, dx = \frac{\tan^{n-1} ax}{(n-1)a} - \int \tan^{n-2} ax\, dx.$$

*55. (a) Find $\int (\cos^2 x - \sin^2 x)\, dx$ without using double-angle formulas such as $\cos(2x) = \cos^2 x - \sin^2 x$.
   (b) Find $\int \cos^2 x\, dx$ and $\int \sin^2 x\, dx$. [*Hint:* $\int \cos^2 x\, dx - \int \sin^2 x\, dx$ was found in part (a) and $\int \cos^2 x\, dx + \int \sin^2 x\, dx$ is easy.]

**56. Compute

$$I_k = \int_0^{\pi/2}\frac{\sin^2(k\theta)}{\sin\theta}\, d\theta,$$

where k is a positive integer. [*Hint:* Compute I_1 and I_2; then compute $I_{k+1} - I_k$; guess a generalized result and use mathematical induction to prove that your guess is correct.]

8.4 THE IDEA BEHIND INTEGRATION BY SUBSTITUTION

In Section 5.7 we saw how a substitution could be made to change $\int f(x)\, dx$ into an integral of the form $\int u'\, du$. There are many different kinds of substitutions that will allow us to integrate more easily. Some of these will be illustrated in Sections 8.5 and 8.8. In this section we will describe the steps involved in making a substitution in an integral. We begin with two examples.

EXAMPLE 1 Compute $\int_0^1 \sqrt{1-x^2}\, dx$. We make the substitution $x = \sin u$ (the reason for making this substitution will be clear shortly). In converting from an integral in x into an integral in which u is the variable of integration, we must change three things:

(i) Change *the integrand*: $\sqrt{1-x^2} = \sqrt{1-\sin^2 u} = \sqrt{\cos^2 u} = \cos u$.†
(ii) Change dx: Since $dx/du = (d/du)\sin u = \cos u$, we have

†When we substitute $x = \sin u$, we mean that $u = \sin^{-1} x$. By the definition of the inverse sine function on page 430, $-\pi/2 \le u \le \pi/2$. On the interval $[-\pi/2, \pi/2]$, $\cos u \ge 0$, which is why we take the positive square root of $\cos^2 u$ here.

$$dx = \cos u \, du.$$

(iii) Change *the limits of integration*: x ranges between 0 and 1. When $x = 0$, $\sin u = 0$ and $u = \sin^{-1} 0 = 0$; when $x = 1$, $\sin u = 1$ and $u = \sin^{-1} 1 = \pi/2$.

Thus

$$\int_{x=0}^{1} \sqrt{1 - x^2} \, dx = \int_{u=0}^{\pi/2} \cos u \cdot \cos u \, du = \int_{0}^{\pi/2} \cos^2 u \, du = \frac{1}{2}\int_{0}^{\pi/2} (1 + \cos 2u) \, du$$

$$= \frac{1}{2}\left(u + \frac{\sin 2u}{2}\right)\Big|_{0}^{\pi/2} = \frac{1}{2}\left(\frac{\pi}{2} + 0\right) = \frac{\pi}{4}.$$

NOTE: This result should not be surprising since $\int_0^1 \sqrt{1 - x^2} \, dx$ is the area enclosed by the part of the unit circle lying in the first quadrant. ∎

EXAMPLE 2 Compute $\int_0^2 x^2 \sqrt{1 + x^3} \, dx$.

Solution. We make the substitution $1 + x^3 = u$, or $x = \sqrt[3]{u - 1}$. If we do so, we must change everything in the integral from an integral with respect to x to an integral with respect to u. Again we must change three things:

(i) Change *the integrand*: If $x = (u - 1)^{1/3}$,

$$x^2 \sqrt{1 + x^3} = (u - 1)^{2/3} \sqrt{u} = (u - 1)^{2/3} u^{1/2}.$$

(ii) Change dx:

$$\frac{dx}{du} = \frac{1}{3}(u - 1)^{-2/3}, \quad \text{so} \quad dx = \tfrac{1}{3}(u - 1)^{-2/3} \, du$$

(iii) Change *the limits of integration*: When $x = 0$, $(u - 1)^{1/3} = 0$, or $u - 1 = 0$, or $u = 1$. When $x = 2$, $(u - 1)^{1/3} = 2$, or $u - 1 = 2^3 = 8$, and $u = 9$.

Then

$$\int_0^2 x^2 \sqrt{1 + x^3} \, dx = \int_1^9 (u - 1)^{2/3} u^{1/2} \cdot \frac{1}{3}(u - 1)^{-2/3} \, du = \frac{1}{3}\int_1^9 u^{1/2} \, du$$

$$= \frac{1}{3}\cdot\frac{2}{3} u^{3/2}\Big|_1^9 = \frac{2}{9}(9^{3/2} - 1) = \frac{2}{9}(27 - 1) = \frac{52}{9}. \blacksquare$$

We now generalize the technique used in Examples 1 and 2. We emphasize that here *we are discussing what to do once a substitution has been chosen*. Finding the substitution that works is usually a much more difficult problem. We will discuss how to choose reasonable substitutions for a wide variety of integrals in Sections 8.5 and 8.8.
Consider the definite integral

$$\int_a^b f(x)\, dx \tag{1}$$

Let g be a differentiable, 1–1 function. Suppose we wish to make the substitution $x = g(u)$ in order to change the integral with respect to x to an integral with respect to u. We want to do so because with the proper choice of the function g, it will be easier to carry out the integration.

Since g is 1–1, g has an inverse function. Let

$$c = g^{-1}(a) \quad \text{and} \quad d = g^{-1}(b), \tag{2}$$

so that

$$a = g(c) \quad \text{and} \quad b = g(d). \tag{3}$$

Let F be an antiderivative for f. Then, by the chain rule,

$$\frac{d}{du} F(g(u)) = F'(g(u))g'(u) = f(g(u))g'(u). \tag{4}$$

Then, using the fundamental theorem of calculus twice, we obtain

$$\int_c^d f(g(u))g'(u)\, du = \int_c^d \left[\frac{d}{du} F(g(u))\right] du$$

$$= F(g(u))\Big|_{u=c}^{u=d} = F(g(d)) - F(g(c))$$

$$= F(b) - F(a) = \int_a^b f(x)\, dx.$$

Summarizing, we have

$$\int_a^b f(x)\, dx = \int_c^d f(g(u))g'(u)\, du, \tag{5}$$

where $c = g^{-1}(a)$ and $d = g^{-1}(b)$.

Thus to change the integral in (1) from an integral with respect to the variable x to an integral with respect to the variable u, we must do three things:

(i) Replace $f(x)$ by $f(g(u))$.
(ii) Replace dx by $g'(u)\, du$.
(iii) Replace a and b by $c = g^{-1}(a)$ and $d = g^{-1}(b)$.

EXAMPLE 3   Consider

$$\int_1^{64} \frac{x+1}{\sqrt{x}(1 + \sqrt[3]{x})}\, dx.$$

If we make the substitution $x = u^6$, then $\sqrt{x} = u^3$, $\sqrt[3]{x} = u^2$, and $dx = 6u^5\,du$. When $x = 1$, $u = x^{1/6} = 1$; when $x = 64$, $u = 2$. Thus

$$\int_1^{64} \frac{x+1}{\sqrt{x}(1+\sqrt[3]{x})}\,dx = \int_1^2 \underbrace{\frac{u^6+1}{u^3(1+u^2)}}_{\substack{x+1\\ \overline{\sqrt{x}(1+\sqrt[3]{x})}}} \cdot \overbrace{6u^5\,du}^{dx} = 6\int_1^2 \frac{u^8+u^2}{1+u^2}\,du.$$

This integral can be computed by noting that $(u^8 + u^2)/(1 + u^2) = u^6 - u^4 + u^2$. ■

It is important to remember that in substituting a new variable in an integral, *it is necessary to change everything—including the limits of integration.*

PROBLEMS 8.4

In Problems 1–14, transform the given integral to an integral with respect to a new variable by making the indicated substitution. Do not try to integrate.

1. $\displaystyle\int_0^3 e^{x^4}\,dx;\; x^4 = u$

2. $\displaystyle\int_1^e x^2 \ln x\,dx;\; \ln x = u$

3. $\displaystyle\int_0^1 \sqrt{1 + x^5}\,dx;\; 1 + x^5 = z$

4. $\displaystyle\int_0^1 \sqrt{1 + x^2}\,dx;\; x = \tan\theta$

5. $\displaystyle\int_0^1 \frac{dx}{3 + \sqrt{x}};\; \sqrt{x} = v$

6. $\displaystyle\int_1^4 \frac{x^{1/3}}{2 + x^{2/3}}\,dx;\; x^{1/3} = y$

7. $\displaystyle\int_0^1 (1 - x^2)^{3/2}\,dx;\; x = \sin\theta$

8. $\displaystyle\int_{-2}^1 x^2\sqrt{x+3}\,dx;\; \sqrt{x+3} = s$

9. $\displaystyle\int_0^4 x^3\sqrt{16 - x^2}\,dx;\; \sqrt{16 - x^2} = y$

10. $\displaystyle\int_1^2 \frac{dx}{x^3(9 - x^3)^{1/3}};\; v = \left(\frac{9 - x^3}{x^3}\right)^{1/3}$

11. $\displaystyle\int_0^3 x^3\sqrt{9 - x^2}\,dx;\; x = 3\cos\theta$

12. $\displaystyle\int_1^{\sqrt{2}} (x^2 - 1)^{5/2}\,dx;\; x = \sec\theta$

*13. $\displaystyle\int_0^1 \sqrt{1 + x^2}\,dx;\; x = \sinh t$

14. $\displaystyle\int_0^1 \frac{x^7}{1 + x^2}\,dx;\; x = \tan\theta$

*15. Let

$$I = \int_{-1}^1 \frac{1}{1 + x^2}\,dx.$$

Because $1/(1 + x^2) > \frac{1}{2}$ for all $x \in (-1, 1)$, we see that $I > 1$. However, if we try the change of variable $u = 1/x$, then $du = (-1/x^2)\,dx$, $dx = (-1/u^2)\,du$, and

$$I = \int_{-1}^1 \frac{1}{1 + x^2}\,dx = \int_{1/-1}^{1/1} \frac{1}{1 + (1/u^2)}\left(-\frac{1}{u^2}\right)du = -\int_{-1}^1 \frac{1}{1 + u^2}\,du = -I;$$

but the only number satisfying $I = -I$ is $I = 0$. What went wrong?

16. Prove that if f is continuous on $[a, b]$, then $\int_a^b f(x)\,dx = \int_a^b f(a + b - x)\,dx$.

*17. Discuss the following procedure used to compute $\int_{-1}^7 (3 + 5x^2)x^2\,dx$. Let $x^2 = t$; then $2x\,dx = dt$. Since $t = 1$ when $x = -1$ and $t = 49$ when $x = 7$, we evaluate the integral

$$\int_1^{49} (3 + 5t)\sqrt{t}\,\frac{1}{2}\,dt = \int_1^{49}\left(\frac{3}{2}t^{1/2} + \frac{5}{2}t^{3/2}\right)dt = (t^{3/2} + t^{5/2})\Big|_{t=1}^{49}$$

$$= (343 + 16{,}807) - (1 + 1) = 17{,}148.$$

[*Note:* $\int_{-1}^{7} (3x^2 + 5x^4)\,dx = 17{,}152.$]

**18. Suppose that f is continuous on $[a, b]$ and $f[(a + b)/2 + x] = f[(a + b)/2 - x]$. Prove that $\int_a^b xf(x)\,dx = [(a + b)/2] \int_a^b f(x)\,dx$.

8.5 INTEGRALS INVOLVING $\sqrt{a^2 - x^2}$, $\sqrt{a^2 + x^2}$, AND $\sqrt{x^2 - a^2}$: TRIGONOMETRIC SUBSTITUTIONS

We may simplify a large number of complicated-looking integrals by making an appropriate trigonometric substitution. Our technique makes use of the following two identities, each of which we have used before:

$$\sin^2\theta + \cos^2\theta = 1 \tag{1}$$

$$1 + \tan^2\theta = \sec^2\theta. \tag{2}$$

We use these identities in Table 1.

TABLE 1 USEFUL TRIGONOMETRIC SUBSTITUTIONS

Term	Substitution	Term Becomes	
$a^2 - x^2$	$x = a\sin\theta$	$a^2 - a^2\sin^2\theta = a^2(1 - \sin^2\theta) = a^2\cos^2\theta$	(3)
$a^2 + x^2$	$x = a\tan\theta$	$a^2 + a^2\tan^2\theta = a^2(1 + \tan^2\theta) = a^2\sec^2\theta$	(4)
$x^2 - a^2$	$x = a\sec\theta$	$a^2\sec^2\theta - a^2 = a^2(\sec^2\theta - 1) = a^2\tan^2\theta$	(5)

EXAMPLE 1 Calculate $\int_0^1 \sqrt{4 - x^2}\,dx$.

Solution. Since $4 - x^2 = 2^2 - x^2$, we make the substitution $x = 2\sin\theta$. Then $dx = 2\cos\theta\,d\theta$ and

$$\sqrt{4 - x^2} = \sqrt{4 - 4\sin^2\theta} = \sqrt{4(1 - \sin^2\theta)} = \sqrt{4\cos^2\theta} = 2\cos\theta. \tag{6}$$

For the same reason as used in Example 8.4.1 (see the footnote), we took the positive square root of $4\cos^2\theta$. When $x = 2\sin\theta = 0$, $\theta = \sin^{-1}0 = 0$. When $x = 2\sin\theta = 1$, $\sin\theta = \frac{1}{2}$ and $\theta = \sin^{-1}\frac{1}{2} = \pi/6$. Then

$$\int_0^1 \sqrt{4 - x^2}\,dx = \int_0^{\pi/6} (2\cos\theta)(2\cos\theta)\,d\theta = 4\int_0^{\pi/6} \cos^2\theta\,d\theta$$

$$= 2\int_0^{\pi/6} (1 + \cos 2\theta)\,d\theta = 2\left(\theta + \frac{\sin 2\theta}{2}\right)\Big|_0^{\pi/6}$$

$$= 2\left(\frac{\pi}{6} + \frac{\sqrt{3}}{4}\right) = \frac{\pi}{3} + \frac{\sqrt{3}}{2}. \blacksquare$$

EXAMPLE 2   Calculate $\int \sqrt{4 - x^2}\, dx$.

Solution. This problem is more complicated than the one before. Setting $x = 2 \sin \theta$, we obtain

$$\int \sqrt{4 - x^2}\, dx = \int 4 \cos^2 \theta\, d\theta = 2\left(\theta + \frac{\sin 2\theta}{2}\right) = 2\theta + \sin 2\theta + C.$$

But here we need an answer that is a function of x, not θ. If $x = 2 \sin \theta$, then $\theta = \sin^{-1} x/2$. Also,

$$\sin 2\theta = 2 \sin \theta \cos \theta.$$

Now, from the triangle in Figure 1 with $\sin \theta = x/2$, we see that $\cos \theta = \sqrt{4 - x^2}/2$.

FIGURE 1

Thus

$$\int \sqrt{4 - x^2}\, dx = 2\theta + \sin 2\theta = 2\theta + 2 \sin \theta \cos \theta = 2 \sin^{-1} \frac{x}{2} + 2\left(\frac{x}{2}\right) \frac{\sqrt{4 - x^2}}{2}$$

$$= 2 \sin^{-1} \frac{x}{2} + \frac{x}{2} \sqrt{4 - x^2} + C.$$

This result can be checked by differentiation. Note that

$$\left. \left(2 \sin^{-1} \frac{x}{2} + \frac{x}{2} \sqrt{4 - x^2}\right) \right|_0^1 = 2\left(\frac{\pi}{6}\right) + \frac{1}{2} \sqrt{3} = \frac{\pi}{3} + \frac{\sqrt{3}}{2},$$

as in Example 1. ■

Examples 1 and 2 illustrate that *when we are making substitutions, it is easier to evaluate definite integrals, since in those cases it is not necessary to write the answer in terms of the original variable.* In calculating indefinite integrals of this sort, we often find it helpful to draw a triangle to yield a function of x for the answer.

EXAMPLE 3   Calculate $\int_0^4 x^3 \sqrt{16 - x^2}\, dx$.

Solution. Let $x = 4 \sin \theta$. Then $\sqrt{16 - x^2} = \sqrt{16 - 16 \sin^2 \theta} = \sqrt{16 \cos^2 \theta}$† $= 4 \cos \theta$, $dx = 4 \cos \theta\, d\theta$ and since $\theta = \sin^{-1}(x/4)$, the limits of integration become $\sin^{-1} 0 = 0$ and $\sin^{-1}(4/4) = \pi/2$. Then

---

†We take the positive square root of $16 \cos^2 \theta$ for the reason given in the footnote on page 478.

$$\int_0^4 x^3 \sqrt{16 - x^2}\, dx = 4^5 \int_0^{\pi/2} \sin^3 \theta \cos^2 \theta\, d\theta = 1024 \int_0^{\pi/2} (1 - \cos^2 \theta) \cos^2 \theta \sin \theta\, d\theta$$

$$= 1024 \int_0^{\pi/2} (\cos^2 \theta - \cos^4 \theta) \sin \theta\, d\theta$$

$$= 1024 \left(\frac{\cos^5 \theta}{5} - \frac{\cos^3 \theta}{3} \right) \Big|_0^{\pi/2}$$

$$= 1024(\tfrac{1}{3} - \tfrac{1}{5}) = 1024 \cdot \tfrac{2}{15} = \tfrac{2048}{15}. \ \blacksquare$$

EXAMPLE 4   Calculate $\int \sqrt{x^2 - a^2}\, dx$, assuming that $|x| > a > 0$.

Solution. Let $x = a \sec \theta$. Then $dx = a \sec \theta \tan \theta\, d\theta$ and

$$\sqrt{x^2 - a^2} = \sqrt{a^2 \sec^2 \theta - a^2} = \sqrt{a^2 \tan^2 \theta} = \pm a \tan \theta.$$

The plus sign is taken if $\tan \theta > 0$ and the minus sign is taken if $\tan \theta < 0$. We assume for the moment that $x > a$. By choosing $x = a \sec \theta$, we mean that $\theta = \sec^{-1}(x/a)$. By the definition of the inverse secant function on page 440, $0 \le \theta < \pi/2$, or $\pi/2 < \theta \le \pi$. With $x > a$, $x/a > 0$ and $\sec^{-1}(x/a)$ is in $[0, \pi/2)$. Thus $\tan \theta > 0$.

Since $x = a \sec \theta$, $a^2 \tan^2 \theta = a^2(\sec^2 \theta - 1) = a^2 \sec^2 \theta - a^2 = x^2 - a^2$, so that

$$\tan \theta = \frac{\sqrt{x^2 - a^2}}{a} \qquad \text{and} \qquad \sec \theta = \frac{x}{a}.$$

Thus

$$\int \sqrt{x^2 - a^2}\, dx = a^2 \int \tan^2 \theta \sec \theta\, d\theta = a^2 \int (\sec^2 \theta - 1) \sec \theta\, d\theta$$

$$= a^2 \int (\sec^3 \theta - \sec \theta)\, d\theta = \frac{a^2}{2} (\sec \theta \tan \theta + \ln |\sec \theta + \tan \theta|)$$

$$- a^2 \ln |\sec \theta + \tan \theta| \qquad \text{(see Example 8.2.8)}$$

$$= \frac{a^2}{2} (\sec \theta \tan \theta - \ln |\sec \theta + \tan \theta|)$$

$$= \frac{a^2}{2} \left(\frac{x}{a} \frac{\sqrt{x^2 - a^2}}{a} - \ln \left| \frac{x}{a} + \frac{\sqrt{x^2 - a^2}}{a} \right| \right) + C$$

for $x > a$.

It is not difficult to verify (see Problems 61 and 62) that the same answer is obtained in the case in which $x < -a < 0$. Thus

$$\int \sqrt{x^2 - a^2}\, dx = \frac{a^2}{2} \left(\frac{x}{a} \frac{\sqrt{x^2 - a^2}}{a} - \ln \left| \frac{x + \sqrt{x^2 - a^2}}{a} \right| \right) + C,$$

or

$$\int \sqrt{x^2 - a^2}\, dx = \frac{x\sqrt{x^2 - a^2}}{2} - \frac{a^2}{2} \ln \left| x + \sqrt{x^2 - a^2} \right| + C.$$

Note that $(a^2/2) \ln \left| [(x + \sqrt{x^2 - a^2})/a] \right| = (a^2/2) \ln |x + \sqrt{x^2 - a^2}| - (a^2/2) \ln |a|$, and we included the constant term $(a^2/2) \ln |a|$ in the general constant C. ∎

EXAMPLE 5 Calculate $\int dx/(a^2 + x^2)^{3/2}$, $a > 0$.

Solution. Let $x = a \tan \theta$. Then $dx = a \sec^2 \theta\, d\theta$ and

$$(a^2 + x^2)^{3/2} = (\sqrt{a^2 + x^2})^3 = (\sqrt{a^2 + a^2 \tan^2 \theta})^3 = (\sqrt{a^2 \sec^2 \theta})^3$$
$$= (\pm a \sec \theta)^3 = \pm a^3 \sec^3 \theta.$$

By choosing $x = a \tan \theta$, we mean that $\tan \theta = x/a$ and $\theta = \tan^{-1}(x/a)$. By the definition of the inverse tangent function (on page 434), $-\pi/2 < \theta < \pi/2$. Thus since $\sec \theta = 1/\cos \theta > 0$ in $(-\pi/2, \pi/2)$, we take the positive value of $a \sec \theta$. Hence

$$\int \frac{dx}{(a^2 + x^2)^{3/2}} = \int \frac{a \sec^2 \theta\, d\theta}{a^3 \sec^3 \theta} = \frac{1}{a^2} \int \frac{d\theta}{\sec \theta}$$
$$= \frac{1}{a^2} \int \cos \theta\, d\theta = \frac{\sin \theta}{a^2} + C.$$

From the triangle in Figure 2, if $x/a = \tan \theta$, then $\sin \theta = x/\sqrt{a^2 + x^2}$, so that

$$\int \frac{dx}{(a^2 + x^2)^{3/2}} = \frac{x}{a^2\sqrt{a^2 + x^2}} + C. \quad ∎$$

FIGURE 2

EXAMPLE 6 Compute $\int dx/(x^2\sqrt{x^2 + 5})$.

Solution. This example is more complicated than others we have done, but our procedure will still work. The term $x^2 + 5$ is of the form $x^2 + a^2$ with $a = \sqrt{5}$. We thus choose $x = \sqrt{5} \tan \theta$, so that $dx = \sqrt{5} \sec^2 \theta\, d\theta$ and $x^2\sqrt{x^2 + 5} = 5 \tan^2 \theta \times \sqrt{5 \tan^2 \theta + 5} = 5 \tan^2 \theta \sqrt{5} \sec^2 \theta = 5\sqrt{5} \tan^2 \theta \sec \theta$. We choose the positive square root of $5 \sec^2 \theta$ for the same reason as given in Example 5. Thus

$$\int \frac{dx}{x^2\sqrt{x^2 + 5}} = \int \frac{\sqrt{5} \sec^2 \theta\, d\theta}{5\sqrt{5} \tan^2 \theta \sec \theta} = \frac{1}{5} \int \frac{\sec \theta}{\tan^2 \theta}\, d\theta.$$

Now we can simplify this expression by writing everything in terms of sines and cosines:

$$\frac{1}{5}\int \frac{\sec\theta}{\tan^2\theta}\,d\theta = \frac{1}{5}\int \sec\theta\cot^2\theta\,d\theta = \frac{1}{5}\int \frac{1}{\cos\theta}\cdot\frac{\cos^2\theta}{\sin^2\theta}\,d\theta = \frac{1}{5}\int \frac{\cos\theta}{\sin^2\theta}\,d\theta$$

$$= \frac{1}{5}\left(\frac{-1}{\sin\theta}\right) = -\frac{1}{5}\csc\theta + C.$$

If $x = \sqrt{5}\tan\theta$, then $\tan\theta = x/\sqrt{5}$, and from the triangle in Figure 3, $\csc\theta = \sqrt{x^2 + 5}/x$. Thus

$$\int \frac{dx}{x^2\sqrt{x^2 + 5}} = -\frac{\sqrt{x^2 + 5}}{5x} + C.$$

In problems this complicated the answer should be checked by differentiation. ■

FIGURE 3

EXAMPLE 7   Calculate $\int_0^{1/3} \sqrt{4 - 9x^2}\,dx$.

Solution. Let $u = 3x$. Then $du = 3\,dx$ so that $dx = \frac{1}{3}\,du$; $u = 0$ when $x = 0$; $u = 1$ when $x = \frac{1}{3}$; and

$$\int_0^{1/3} \sqrt{4 - 9x^2}\,dx = \frac{1}{3}\int_0^1 \sqrt{4 - u^2}\,du.$$

Now set $u = 2\sin\theta$. Then

$$\frac{1}{3}\int_0^1 \sqrt{4 - u^2}\,du = \frac{1}{3}\int_0^{\pi/6} 4\cos^2\theta\,d\theta = \frac{2}{3}\int_0^{\pi/6} (1 + \cos 2\theta)\,d\theta$$

$$= \frac{2}{3}\left(\theta + \frac{\sin 2\theta}{2}\right)\Big|_0^{\pi/6} = \frac{2}{3}\left(\frac{\pi}{6} + \frac{\sqrt{3}}{4}\right) = \frac{\pi}{9} + \frac{\sqrt{3}}{6}.\quad ■$$

We close this section by noting that while the substitutions described in this section will usually lead to a solution to the problem, they are not always the easiest ones to use. There will sometimes be a "better" substitution—better in the sense that it will give us something we can integrate much more quickly. We will see an example of this in Section 8.8 (see Example 8.8.6).

PROBLEMS 8.5

In Problems 1–54, evaluate the given integral by making an appropriate trigonometric substitution.

1. $\displaystyle\int_0^1 \frac{x}{\sqrt{4 - x^2}}\,dx$

2. $\displaystyle\int \frac{dx}{\sqrt{4 - x^2}}$

3. $\displaystyle\int \frac{x^2}{\sqrt{4 - x^2}}\,dx$

4. $\int_0^1 (4 - x^2)^{3/2}\, dx$

5. $\int_0^1 x^3 \sqrt{4 - x^2}\, dx$

6. $\int \dfrac{x}{\sqrt{x^2 - 4}}\, dx$

7. $\int \dfrac{dx}{\sqrt{x^2 - 4}}$

8. $\int \dfrac{x^2}{\sqrt{x^2 - 4}}\, dx$

9. $\int_{4/\sqrt{3}}^4 \sqrt{x^2 - 4}\, dx$

*10. $\int x^3 \sqrt{x^2 - 4}\, dx$

11. $\int_0^2 \dfrac{x}{\sqrt{4 + x^2}}\, dx$

12. $\int \dfrac{dx}{\sqrt{4 + x^2}}$

13. $\int \dfrac{dx}{4 + x^2}$

14. $\int_0^2 \dfrac{x^2}{\sqrt{4 + x^2}}\, dx$

*15. $\int (4 + x^2)^{3/2}\, dx$

16. $\int_0^2 x^3 \sqrt{4 + x^2}\, dx$

17. $\int_0^{\sqrt{3}} \dfrac{x^5}{1 + x^2}\, dx$

18. $\int_{\sqrt{2}}^2 (x^2 - 2)^{1/2}\, dx$

19. $\int_0^{\sqrt{2}} \dfrac{dx}{2 + x^2}$

20. $\int_0^{1/\sqrt{2}} \dfrac{x^3}{2 + x^2}\, dx$

21. $\int_2^4 \sqrt{4x^2 - 9}\, dx$

*22. $\int_1^3 \dfrac{dx}{x^4 \sqrt{x^2 + 3}}$

23. $\int_4^8 \dfrac{x^2}{(x^2 - 4)^{3/2}}\, dx$

24. $\int_1^4 \dfrac{dx}{x^4 \sqrt{x^2 + 16}}$

25. $\int_4^5 \dfrac{dx}{(x^2 - 9)^{3/2}}$

26. $\int \sqrt{a^2 - x^2}\, dx$

27. $\int x^2 \sqrt{a^2 - x^2}\, dx$

28. $\int \sqrt{a^2 + x^2}\, dx$

29. $\int \sqrt{1 + a^2 x^2}\, dx$

30. $\int \sqrt{1 - a^2 x^2}\, dx$

31. $\int \dfrac{\sqrt{5x^2 - 9}}{x}\, dx$

32. $\int \dfrac{\sqrt{5x^2 + 9}}{x}\, dx$

33. $\int \dfrac{\sqrt{9 - 5x^2}}{x}\, dx$

34. $\int (a^2 - x^2)^{3/2}\, dx$

35. $\int \dfrac{dx}{x\sqrt{x^2 - a^2}}$

36. $\int \dfrac{dx}{x^3 \sqrt{x^2 - a^2}}$

37. $\int x^3(x^2 - a^2)^{3/2}\, dx$

*38. $\int \dfrac{(x^2 - a^2)^{3/2}}{x^3}\, dx$

39. $\int \dfrac{x^3}{\sqrt{a^2 - x^2}}\, dx$

40. $\int \dfrac{dx}{x\sqrt{a^2 - x^2}}$

41. $\int \dfrac{dx}{x^3 \sqrt{a^2 - x^2}}$

42. $\int \dfrac{dx}{(a^2 - x^2)^{3/2}}$

43. $\int \dfrac{dx}{(x^2 - a^2)^{3/2}}$

44. $\int x^2(a^2 - x^2)^{3/2}\, dx$

45. $\int \dfrac{dx}{x\sqrt{x^2 + a^2}}$

46. $\int \dfrac{\sqrt{x^2 + a^2}}{x}\, dx$

47. $\int \dfrac{x^2}{(x^2 + a^2)^{3/2}}\, dx$

*48. $\int \dfrac{dx}{x^3(x^2 + a^2)^{3/2}}$

49. $\int \dfrac{x^2}{x^2 + a^2}\, dx$

50. $\int \dfrac{dx}{x^3(x^2 + a^2)}$

51. $\int \dfrac{dx}{x^2(x^2 + a^2)^2}$

52. $\int \dfrac{dx}{\sqrt{a^2 - b^2 x^2}}$

54. $\int \dfrac{dx}{\sqrt{a^2 + b^2 x^2}}$

54. $\int \dfrac{dx}{\sqrt{b^2 x^2 - a^2}}$

55. Prove by integration that $\cosh^{-1} x = \ln |x + \sqrt{x^2 - 1}|$ if $x \geq 1$.

56. Prove that
$$\tanh^{-1} x = \frac{1}{2}\ln \frac{1 + x}{1 - x}$$
if $-1 < x < 1$.

57. The density of a 5-m bar is given by $\rho(x) = \sqrt{50 - 2x^2}$. Find the total mass of the bar.
58. The marginal cost of a certain product is given by $MC(q) = 20 + (30/\sqrt{5 + 4q^2})$ cents per unit. What is the total cost of producing 100 units?
59. (a) Graph the curve $y = \sqrt{1 - x^2}$ for $0 \le x \le 1$.
    (b) Find the area bounded by the curve and the x- and y-axes.
60. (a) Graph the curve $y = \sqrt{9 + x^2}$ for $x \ge 0$.
    (b) Find the area bounded by the curve, the x- and y-axes, and the line $x = 4$.
61. In Example 4, assume that $x < -a < 0$.
    (a) Show that if $x = a \sec\theta$, then $a\tan\theta < 0$.
    (b) Show that

$$\int \sqrt{x^2 - a^2}\, dx = -\frac{a^2}{2}\left[\frac{x}{a}\left(\frac{-\sqrt{x^2 - a^2}}{a}\right) - \ln\left|\frac{x}{a} - \frac{\sqrt{x^2 - a^2}}{a}\right|\right] + C$$

$$= \frac{a^2}{2}\left[\frac{x}{a}\frac{\sqrt{x^2 - a^2}}{a} + \ln\left|\frac{x - \sqrt{x^2 - a^2}}{a}\right|\right] + C.$$

62. (a) Show, using properties of logarithms, that

$$-\ln\left|\frac{x + \sqrt{x^2 - a^2}}{a}\right| = \ln\left|\frac{x - \sqrt{x^2 - a^2}}{a}\right| + C.$$

[Hint: Multiply and divide by $x - \sqrt{x^2 - a^2}$.]
    (b) Using the results of Problem 61 and part (a), show that

$$\int \sqrt{x^2 - a^2}\, dx = \frac{a^2}{2}\left(\frac{x}{a}\frac{\sqrt{x^2 - a^2}}{a} - \ln\left|\frac{x + \sqrt{x^2 - a^2}}{a}\right|\right) + C$$

for $x < -a < 0$.
*63. Compute the following integral without making any substitutions:

$$\int_0^1 (\sqrt{2 - x^2} - \sqrt{2x - x^2})\, dx.$$

[Hint: Find a region whose area is computed by this integral.]
64. (a) Show that

$$\int \frac{dx}{\sqrt{x^2 + 1}} = \ln(x + \sqrt{1 + x^2}) + C_1.$$

Recall that

$$\int \frac{dx}{\sqrt{x^2 + 1}} = \sinh^{-1} x + C_2 \quad \text{(see page 456)}.$$

    (b) Show that $\sinh^{-1} x = \ln(x + \sqrt{1 + x^2})$.

8.6 THE INTEGRATION OF RATIONAL FUNCTIONS I: LINEAR AND QUADRATIC DENOMINATORS

In this section and the next one, we will show how, at least in theory, every rational function can be integrated. Recall that a rational function is a quotient of the form $p(x)/q(x)$, where p and q are both polynomials. In this section we will deal with the case in which $q(x)$ is either a linear or a quadratic polynomial.

Case 1. $q(x)$ is a linear function. That is

$$q(x) = ax + b.$$ (1)

We illustrate what can happen.

EXAMPLE 1   Compute $\int dx/(ax + b)$.

Solution. Let $u = ax + b$. Then $du = a\,dx$ and

$$\int \frac{dx}{ax + b} = \frac{1}{a} \int \frac{a\,dx}{ax + b} = \frac{1}{a} \int \frac{du}{u} = \frac{1}{a} \ln |u| + C,$$

or

$$\int \frac{dx}{ax + b} = \frac{1}{a} \ln |ax + b| + C. \quad\blacksquare$$ (2)

Thus we can integrate $1/q(x)$ if $q(x) = ax + b$. To integrate $p(x)/(ax + b)$, we simply divide. If the degree of $p(x) = n$, then

$$\frac{p(x)}{ax + b} = \frac{c}{ax + b} + r(x),$$ (3)

where $r(x)$ is a polynomial of degree $n - 1$. Instead of proving formula (3), we give an example.

EXAMPLE 2   Compute $\int [(x^3 - x^2 + 4x - 5)/(x - 2)]\,dx$.

Solution. We divide:

$$
\begin{array}{r}
x^2 + x + 6 \\
x - 2\overline{)x^3 - x^2 + 4x - 5} \\
\underline{x^3 - 2x^2 } \\
x^2 + 4x \\
\underline{x^2 - 2x } \\
6x - 5 \\
\underline{6x - 12} \\
7
\end{array}
$$

So

$$\frac{x^3 - x^2 + 4x - 5}{x - 2} = x^2 + x + 6 + \frac{7}{x - 2},$$

and

$$\int \frac{x^3 - x^2 + 4x - 5}{x - 2}\, dx = \int \left(x^2 + x + 6 + \frac{7}{x - 2} \right) dx$$

$$= \frac{x^3}{3} + \frac{x^2}{2} + 6x + 7 \ln |x - 2| + C. \quad \blacksquare$$

As Examples 1 and 2 illustrate, we can integrate any rational function with a linear denominator.

Case 2. $q(x)$ is a quadratic function.   That is,

$$q(x) = px^2 + qx + r.$$

This case is a bit more complicated. We need two basic integrals. In Example 7.2.5 we showed that

$$\int \frac{dx}{a^2 + x^2} = \frac{1}{a} \tan^{-1}\frac{x}{a} + C. \tag{4}$$

The other integral is computed below.

EXAMPLE 3   Compute $\int dx/(a^2 - x^2)$, $a \neq 0$.

Solution. This integral can be computed in several ways. One way is to make the substitution $x = a \sin \theta$. However, we will solve it in another way, and by doing so, we will illustrate an important technique used in integrating rational functions. We start by writing

$$\frac{1}{a^2 - x^2} = \frac{1}{(a - x)(a + x)}.$$

We will find constants A and B such that

$$\frac{1}{(a - x)(a + x)} = \frac{A}{a - x} + \frac{B}{a + x} \tag{5}$$

If (5) holds, then by adding fractions, we obtain

$$\frac{1}{(a - x)(a + x)} = \frac{A(a + x) + B(a - x)}{(a - x)(a + x)} = \frac{(A - B)x + (A + B)a}{(a - x)(a + x)}. \tag{6}$$

Equation (6) can hold only if the numerators are the same for *every* x. That is, for every x,

$$(A - B)x + (A + B)a = 1. \tag{7}$$

We will encounter equations like (7) quite frequently. To find the numbers A and B, we observe that the polynomial $(A - B)x + (A + B)a$ is equal to the polynomial 1. Two polynomials are equal if and only if the coefficients of like powers of x (including $x^0 = 1$) are equal. Since

$$1 = 0 \cdot x + 1 \cdot 1 = (A - B) \cdot x + [(A + B)a] \cdot 1,$$

we see that

$$A - B = 0 \quad \text{and} \quad (A + B)a = 1.$$

The first of these equations tells us that $A = B$. From the second we see that

$$(A + B)a = (A + A)a = 2Aa = 1,$$

or

$$A = \frac{1}{2a} = B.$$

Thus from (7) we obtain

$$\int \frac{dx}{a^2 - x^2} = \int \frac{dx}{(a - x)(a + x)} = \int \left[\frac{(1/2a)}{a - x} + \frac{(1/2a)}{a + x} \right] dx$$

$$= -\frac{1}{2a} \ln |a - x| + \frac{1}{2a} \ln |a + x| + C$$

$$= \frac{1}{2a} (\ln |a + x| - \ln |a - x|) + C.$$

or

$$\int \frac{dx}{a^2 - x^2} = \frac{1}{2a} \ln \left| \frac{a + x}{a - x} \right| + C. \quad \blacksquare \tag{8}$$

From (8), we obtain

$$\int \frac{dx}{x^2 - a^2} = -\int \frac{dx}{a^2 - x^2} = -\frac{1}{2a} \ln \left| \frac{a + x}{a - x} \right| \underset{|-u| = |u|}{=} -\frac{1}{2a} \ln \left| \frac{x + a}{x - a} \right| \underset{-\ln u = \ln \frac{1}{u}}{=} \frac{1}{2a} \ln \left| \frac{x - a}{x + a} \right|$$

or

$$\int \frac{dx}{x^2 - a^2} = \frac{1}{2a} \ln \left| \frac{x - a}{x + a} \right| + C. \tag{9}$$

The method we have just used is called the **method of partial fractions.**

We now know how to integrate $1/(a^2 + x^2)$ and $1/(a^2 - x^2)$. To deal with $1/(px^2 + qx + r)$, we either factor $px^2 + qx + r$, if possible, or we complete the square. We have the following rule:

To compute $\int dx/(px^2 + qx + r)$, we proceed as follows:

(i) Factor the quadratic term in the denominator.
(ii) Use the method of partial fractions to complete the integration.

Or if the denominator cannot easily be factored:

(i) Factor the number p from the denominator.
(ii) Complete the square.
(iii) Make an appropriate substitution to write the denominator as $a^2 - u^2$ or $a^2 + u^2$.
(iv) Use formula (4), (8), or (9) to complete the integration.

EXAMPLE 4 Compute $\int dx/(x^2 + 4x - 5)$.

Solution.

$$\frac{1}{x^2 + 4x - 5} = \frac{1}{(x + 5)(x - 1)} = \frac{A}{x + 5} + \frac{B}{x - 1}. \tag{10}$$

We can solve for A and B as in Example 3. But there is a quicker way to obtain these values. Multiplying both sides of the right-hand equation in (10) by $(x + 5)(x - 1)$, we obtain

$$1 = A(x - 1) + B(x + 5) \tag{11}$$

Setting $x = -5$ in (11) yields

$$1 = -6A, \quad \text{or} \quad A = -\tfrac{1}{6}.$$

Setting $x = 1$ in (11) leads to

$$1 = 6B, \quad \text{or} \quad B = \tfrac{1}{6}.$$

Thus

$$\int \frac{dx}{x^2 + 4x - 5} = \frac{1}{6} \int \left(\frac{-1}{x + 5} + \frac{1}{x - 1} \right) dx = \frac{1}{6}(-\ln |x + 5| + \ln |x - 1|)$$

$$= \frac{1}{6} \ln \left| \frac{x - 1}{x + 5} \right| + C.$$

We can also solve this problem by first completing the square and then using equation (9):

$$\int \frac{dx}{x^2 + 4x - 5} = \int \frac{dx}{(x + 2)^2 - 4 - 5} = \int \frac{dx}{(x + 2)^2 - 9}.$$

Let $u = x + 2$; then $du = dx$ and

$$\int \frac{dx}{x^2 + 4x - 5} = \int \frac{du}{u^2 - 9} = \int \frac{du}{u^2 - 3^2}.$$

Then from equation (9) this expression is equal to

$$\frac{1}{2 \cdot 3} \ln \left| \frac{u - 3}{u + 3} \right| = \frac{1}{6} \ln \left| \frac{(x + 2) - 3}{(x + 2) + 3} \right| = \frac{1}{6} \ln \left| \frac{x - 1}{x + 5} \right| + C. \ ■$$

REMARK. As the last example indicates, there are sometimes two (or more) ways to integrate a rational function with a quadratic denominator. Which method should you use? The answer is always to choose the quickest method. If the denominator is easily factored, then partial fractions is usually the fastest way to proceed. The advantage of the other method is that it always works, whether or not the denominator can be factored.

EXAMPLE 5   Compute $\int dx/(2x^2 - 4x + 12)$.

Solution.

$$\int \frac{dx}{2x^2 - 4x + 12} = \frac{1}{2} \int \frac{dx}{x^2 - 2x + 6} = \frac{1}{2} \int \frac{dx}{(x - 1)^2 - 1 + 6} = \frac{1}{2} \int \frac{dx}{(x - 1)^2 + 5}.$$

Let $u = x - 1$. Then $du = dx$ and the last term is equal to

$$\frac{1}{2} \int \frac{du}{u^2 + 5} = \frac{1}{2} \int \frac{du}{u^2 + (\sqrt{5})^2}.$$

From equation (4), this is equal to

$$\frac{1}{2} \left(\frac{1}{\sqrt{5}} \right) \tan^{-1} \frac{u}{\sqrt{5}} = \frac{1}{2\sqrt{5}} \tan^{-1} \frac{x - 1}{\sqrt{5}} + C. \ ■$$

We can integrate $1/(px^2 + qx + r)$ by completing the square in the denominator or by using partial fractions. If $p(x)$ is a linear polynomial, then we can integrate $p(x)/q(x)$ by partial fractions or by using the procedure given in the following example.

EXAMPLE 6   Compute $\int (x + 8)/(x^2 + 6x - 7) \, dx$.

Solution. We first note that

$$\frac{d}{dx}(x^2 + 6x - 7) = 2x + 6.$$

Thus if we can put $2x + 6$ in the numerator and let $u = x^2 + 6x - 7$, we will have an expression of the form du/u, which is easily integrated. Let's see how this can be done:

$$\int \frac{x + 8}{x^2 + 6x - 7}\, dx \overset{\substack{\text{Multiply and divide by 2} \\ \downarrow}}{=} \frac{1}{2}\int \frac{2x + 16}{x^2 + 6x - 7}\, dx \overset{\substack{\text{We want } 2x + 6 \text{ in the numerator} \\ \swarrow}}{=} \frac{1}{2}\int \frac{2x + 6 + 10}{x^2 + 6x - 7}\, dx$$

$$= \frac{1}{2}\int \frac{2x + 6}{x^2 + 6x - 7}\, dx + \frac{1}{2}\int \frac{10}{x^2 + 6x - 7}\, dx$$

$$= \frac{1}{2}\ln|x^2 + 6x - 7| + 5\int \frac{dx}{x^2 + 6x - 7}$$

$$= \frac{1}{2}\ln|x^2 + 6x - 7| + 5\int \frac{dx}{(x + 3)^2 - 9 - 7}$$

$$= \frac{1}{2}\ln|x^2 + 6x - 7| + 5\int \frac{dx}{(x + 3)^2 - 16}.$$

$$\int \frac{dx}{(x + 3)^2 - 16} \overset{\substack{u = x + 3 \\ du = dx \\ \downarrow}}{=} \int \frac{du}{u^2 - 16} \overset{\substack{\text{By (9)} \\ \swarrow}}{=} \frac{1}{8}\ln\left|\frac{u - 4}{u + 4}\right|$$

$$= \frac{1}{8}\ln\left|\frac{(x + 3) - 4}{(x + 3) + 4}\right| = \frac{1}{8}\ln\left|\frac{x - 1}{x + 7}\right|.$$

So

$$\int \frac{x + 8}{x^2 + 6x - 7}\, dx = \frac{1}{2}\ln|x^2 + 6x - 7| + \frac{5}{8}\ln\left|\frac{x - 1}{x + 7}\right| + C.$$

REMARK 1.   This problem can also be solved by partial fractions. Note that

$$\frac{x + 8}{x^2 + 6x - 7} = \frac{x + 8}{(x + 7)(x - 1)} = -\frac{1}{8}\left(\frac{1}{x + 7}\right) + \frac{9}{8}\left(\frac{1}{x - 1}\right)$$

(see Problem 43). We solved the problem the way we did to illustrate a method that always works. ∎

REMARK 2.   Let $n = $ degree of $p(x)$ with $n \geq 2$. If $q(x)$ is a quadratic polynomial, then we can divide to obtain

$$\frac{p(x)}{q(x)} = \frac{ax + b}{q(x)} + r(x), \tag{12}$$

Where a and b are constants and $r(x)$ is a polynomial of degree $n - 2$. For example,

$$\int \frac{x^5 + 3x^4 - x^3 + 9x^2 - 4x + 12}{x^2 + 2x + 5} \, dx$$

$$= \int \left(x^3 + x^2 - 8x + 20 + \frac{-4x - 88}{x^2 + 2x + 5} \right) dx$$

$$= \int \left[x^3 + x^2 - 8x + 20 - 2\left(\frac{2x + 2}{x^2 + 2x + 5} \right) + \frac{-84}{(x + 1)^2 + 4} \right] dx$$

From (4)
$$\downarrow$$

$$= \frac{x^4}{4} + \frac{x^3}{3} - 4x^2 + 20x - 2 \ln(x^2 + 2x + 5) - 42 \tan^{-1}\left(\frac{x + 1}{2} \right) + C.$$

As the preceding examples indicate, we can integrate any rational function with a linear or quadratic denominator. In the next section we will show how other rational functions can be integrated.

PROBLEMS 8.6

In Problems 1–36, calculate the given integral.

1. $\int \dfrac{dx}{2x - 5}$

2. $\int \dfrac{dx}{5x + 10}$

3. $\int \dfrac{dx}{3x + 11}$

4. $\int \dfrac{dx}{3x - 11}$

5. $\int \dfrac{dx}{x^2 + 9}$

6. $\int \dfrac{dx}{2x^2 + 8}$

7. $\int \dfrac{dx}{x^2 - 16}$

8. $\int \dfrac{dx}{x^2 - 12}$

9. $\int \dfrac{dx}{x^2 + 12}$

10. $\int \dfrac{x + 1}{x - 1} dx$ $\left[Hint: \dfrac{x + 1}{x - 1} = 1 + \dfrac{2}{x - 1} \right]$

11. $\int \dfrac{x - 1}{x + 1} dx$

12. $\int \dfrac{2x + 3}{x - 4} dx$

13. $\int \dfrac{x^2 + 4x + 5}{x - 2} dx$

14. $\int \dfrac{x^3 - x}{2x + 6} dx$

15. $\int \dfrac{x^4 - x^3 + 3x^2 - 2x + 7}{x + 2} dx$

16. $\int \dfrac{x + 1}{x^2 + 1} dx$

17. $\int \dfrac{2x - 3}{x^2 - 4} dx$

18. $\int \dfrac{3x + 5}{x^2 + 9} dx$

19. $\int \dfrac{dx}{(x - 1)(x - 4)}$

20. $\int \dfrac{dx}{(x + 3)(x - 7)}$

21. $\int \dfrac{dx}{(x - a)(x - b)}, a \neq b$

22. $\int \dfrac{2x - 3}{(x - 1)(x - 4)} dx$

23. $\int \dfrac{3x - 5}{(x + 3)(x - 7)} dx$

24. $\int \dfrac{cx + d}{(x - a)(x - b)} dx, a \neq b, \\ c \neq 0$

25. $\int \dfrac{x^2 - 1}{x^2 + 1} dx$

26. $\int \dfrac{x - 1}{x^2 - 6x + 16} dx$

27. $\int \dfrac{x^2 - 1}{x^2 - 6x + 25} dx$

28. $\int \dfrac{x+2}{x^2-6x-7}\,dx$ **29.** $\int \dfrac{x^3-3}{x^2-6x-7}\,dx$ **30.** $\int \dfrac{x^3-x+2}{x^2-6x-7}\,dx$

31. $\int \dfrac{x^5-1}{x^2-1}\,dx$ **32.** $\int \dfrac{x^3-1}{x^2+x+1}\,dx$ **33.** $\int \dfrac{x^3-x^2+2x+5}{x^2+x+1}\,dx$

34. $\int \dfrac{x^4-2x^3+x-3}{2x^2-8x+10}\,dx$ **35.** $\int \dfrac{x^2-x+2}{3x^2+6x-15}\,dx$ **36.** $\int \dfrac{x^2+3x+4}{x^2+x}\,dx$

37. Calculate $\int \sin x/(\cos^2 x + 4\cos x + 4)\,dx$. [*Hint:* Make the substitution $u = \cos x$.]

38. Calculate $\int \cos x/(\sin^2 x - 5\sin x + 6)\,dx$.

39. Calculate the area bounded by the curve $y = (1+x^2)/(1+x)^2$, the x- and y-axes, and the line $x = 1$.

40. Calculate the area bounded by the curve $y = 1/(x^2+x)$, the y-axis, and the lines $x = 1$ and $x = 2$.

41. Show that if $r > (q^2/4)$, then

$$\int \frac{dx}{x^2+qx+r} = \frac{2}{\sqrt{4r-q^2}} \tan^{-1} \frac{2x+q}{\sqrt{4r-q}} + C.$$

42. (a) Show that if $r < (q^2/4)$, then

$$\int \frac{dx}{x^2+qx+r} = \frac{1}{\sqrt{q^2-4r}} \ln\left|\frac{\sqrt{q^2-4r}+2x+q}{\sqrt{q^2-4r}-2x-q}\right| + C.$$

(b) What happens if $r = q^2/4$?

43. (a) Show that

$$\frac{x+8}{x^2+6x-7} = \frac{x+8}{(x+7)(x-1)} = -\frac{1}{8}\left(\frac{1}{x+7}\right) + \frac{9}{8}\left(\frac{1}{x-1}\right).$$

(b) Using part (a), show that

$$\int \frac{x+8}{x^2+6x-7}\,dx = \frac{9}{8}\ln|x-1| - \frac{1}{8}\ln|x+7| + C.$$

(c) Show, using properties of the logarithm, that

$$\frac{1}{2}\ln|x^2+6x-7| + \frac{5}{8}\ln\left|\frac{x-1}{x+7}\right| = \frac{9}{8}\ln|x-1| - \frac{1}{8}\ln|x+7|.$$

[*Hint:* $\ln|x^2+6x-7| = \ln|(x+7)(x-1)|$.]

8.7 THE INTEGRATION OF RATIONAL FUNCTIONS II: THE METHOD OF PARTIAL FRACTIONS

In the last section we saw how to integrate $p(x)/q(x)$, where $q(x)$ is a linear or a quadratic polynomial. In this section we use the technique of partial fractions to compute the integral of other rational functions. We used this technique in Example 8.6.3 in our computation of $\int dx/(a^2-x^2)$. This technique is really a technique in algebra. The idea is to use algebra to write $p(x)/q(x)$ as a sum of rational functions, each having a denominator that is a power of either a linear or a quadratic polynomial. Then we use the techniques in Section 8.6 to complete the integration.

Before doing examples, we cite some results from algebra. A quadratic polynomial is called **irreducible** if it cannot be factored into linear factors. That means that the polynomial $px^2 + qx + r$ is irreducible† if the quadratic equation $px^2 + qx + r = 0$ has no real roots; that is, if

$$q^2 - 4pr < 0. \tag{1}$$

EXAMPLE 1   The polynomial $2x^2 + 4x + 3$ is irreducible because $4^2 - 4(2)(3) = 16 - 24 = -8 < 0.$ ■

EXAMPLE 2   The polynomial $x^2 - 5x + 6$ is not irreducible because $(-5)^2 - 4(1)(6) = 25 - 24 = 1 > 0$. Note that $x^2 - 5x + 6 = (x - 3)(x - 2)$, which is a product of linear factors. ■

The following theorem from algebra will be very helpful to us. We will not prove it, but we will indicate instead how it can be used.

Theorem 1   Let $r(x) = p(x)/q(x)$ be a rational function with degree of $p(x) <$ degree of $q(x)$. Then

$$\frac{p(x)}{q(x)} = R_1 + R_2 + \cdots + R_n, \tag{2}$$

where each R_i is a rational function whose denominator is a positive integral power of either a linear polynomial of the form $ax + b$ or an irreducible quadratic polynomial of the form $px^2 + qx + r$. The formula (2) is called the **partial fraction decomposition** of the function $p(x)/q(x)$.

We emphasize that according to Theorem 1, we can integrate (at least theoretically) any rational function by writing it as a sum of terms we already know how to integrate. To use Theorem 1, we provide a four-step procedure for carrying out the partial fraction decomposition of any rational function. To simplify our discussion, we will assume that if the degree of $q(x)$ is n, then the coefficient of the x^n term, called the **leading coefficient,** is 1. If it is not, we can factor out the coefficient to make the leading coefficient equal to 1. For example,

$$\frac{x^3 - x^2 + 2}{4x^4 - 7x^3 + 6x^2 - 2x + 3} = \frac{1}{4}\left[\frac{x^3 - x^2 + 2}{x^4 - \frac{7}{4}x^3 + \frac{3}{2}x^2 - \frac{1}{2}x + \frac{3}{4}}\right],$$

and we can integrate the function in brackets and then multiply by $\frac{1}{4}$ at the end. Here then is our procedure.

---

†Keep in mind that there is a difference between linear factors of multiplicity two [such as $(x + 1)^2$] and irreducible quadratic factors (such as $x^2 + 1$). To emphasize this point, we note that

$$\int \frac{dx}{(x + 1)^2} = -\frac{1}{x + 1} + C \quad \text{and} \quad \int \frac{dx}{x^2 + 1} = \tan^{-1} x + C.$$

PROCEDURE FOR INTEGRATING $p(x)/q(x)$

Step 1. If degree of $p(x) \geq$ degree of $q(x)$, first divide to obtain

$$\frac{p(x)}{q(x)} = s(x) + \frac{t(x)}{q(x)}, \tag{3}$$

where s, t, and q are polynomials and degree of $t(x) <$ degree of $q(x)$.

Step 2. Factor $q(x)$ into linear and/or irreducible quadratic factors.† A linear factor will have the form $x - a_i$ [where a_i is a root of $q(x)$], and a quadratic factor will have the form $x^2 + bx + c$. This follows from the fact that the leading coefficient of $q(x)$ is 1. **(4)**

Step 3. For each factor of the form $(x - a_i)^k$, the partial fraction decomposition (2) contains a sum of k partial fractions

$$\frac{A_1}{x - a_i} + \frac{A_2}{(x - a_i)^2} + \cdots + \frac{A_k}{(x - a_i)^k}. \tag{5}$$

In particular, if $k = 1$, then the decomposition has one term of the form

$$\frac{A}{x - a_i}. \tag{6}$$

Step 4. For each factor of the form $(x^2 + bx + c)^m$, the partial fraction decomposition (2) contains a sum of k partial fractions

$$\frac{A_1 x + B_1}{x^2 + bx + c} + \frac{A_2 x + B_2}{(x^2 + bx + c)^2} + \cdots + \frac{A_m x + B_m}{(x^2 + bx + c)^m}. \tag{7}$$

In particular, if $m = 1$, then the decomposition has one term of the form

$$\frac{Ax + B}{x^2 + ax + b}. \tag{8}$$

We illustrate this technique with a number of examples.

EXAMPLE 3 Compute $\int dx/[(x - 1)(x - 2)(x - 3)]$.

Solution. By (6),

---

†This step is the hardest step since there are no rules for factoring high-degree polynomials. In every example we give, such a factoring will be possible. But it would be impossible to factor exactly a polynomial of degree 5, say, if its coefficients were randomly chosen. Thus we say that "theoretically" an antiderivative can be found for any rational function. In practice, this step may be impossible.

$$\frac{1}{(x-1)(x-2)(x-3)} = \frac{A}{x-1} + \frac{B}{x-2} + \frac{C}{x-3}. \tag{9}$$

There are several ways to find A, B, and C. The simplest one is as follows: Multiply both sides of (9) by $(x-1)(x-2)(x-3)$. Then

$$1 = A(x-2)(x-3) + B(x-1)(x-3) + C(x-1)(x-2). \tag{10}$$

Equation (10) holds for every value of x. Setting $x = 1$ makes the last two terms in (10) equal to 0, which gives us the equation

$$1 = A(-1)(-2), \quad \text{or} \quad A = \tfrac{1}{2}.$$

Similarly, setting $x = 2$ gives us

$$1 = B(1)(-1), \quad \text{or} \quad B = -1.$$

Setting $x = 3$ leads to

$$1 = C(2)(1), \quad \text{or} \quad C = \tfrac{1}{2}.$$

Thus

$$\frac{1}{(x-1)(x-2)(x-3)} = \frac{\tfrac{1}{2}}{x-1} + \frac{-1}{x-2} + \frac{\tfrac{1}{2}}{x-3}, \tag{11}$$

so that

$$\int \frac{dx}{(x-1)(x-2)(x-3)} = \frac{1}{2}\ln|x-1| - \ln|x-2| + \frac{1}{2}\ln|x-3| + C.$$

We can also combine these terms to obtain

$$\ln\left|\frac{\sqrt{(x-1)(x-3)}}{x-2}\right| + C. \ \blacksquare$$

EXAMPLE 4   Calculate $\int dx/[x(x-1)^2]$.

Solution. From (5) we have

$$\frac{1}{x(x-1)^2} = \frac{A}{x} + \frac{B}{x-1} + \frac{C}{(x-1)^2}. \tag{12}$$

Multiplying both sides of (12) by $x(x-1)^2$, we obtain

$$1 = A(x-1)^2 + Bx(x-1) + Cx. \tag{13}$$

Setting $x = 0$ in (13) gives us $1 = A$. Setting $x = 1$ in (13) leads to $1 = C$. Finally, since

$A = C = 1$, (13) can be rewritten as

$$1 = (x - 1)^2 + Bx(x - 1) + x,$$

and setting $x = 2$ (any number other than 0 or 1 will work), we obtain

$$1 = 1 + 2B + 2, \quad \text{or} \quad 2B = -2 \quad \text{and} \quad B = -1.$$

Thus

$$\int \frac{dx}{x(x - 1)^2} = \int \left[\frac{1}{x} - \frac{1}{x - 1} + \frac{1}{(x - 1)^2} \right] dx$$

$$= \ln |x| - \ln |x - 1| - \frac{1}{x - 1} + C$$

$$= \ln \left| \frac{x}{x - 1} \right| - \frac{1}{x - 1} + C. \ \blacksquare$$

EXAMPLE 5 Compute $\int dx/(x^3 + x)$.

Solution. From (8),

$$\frac{1}{x^3 + x} = \frac{1}{x(x^2 + 1)} = \frac{A_1}{x} + \frac{A_2 x + B_2}{x^2 + 1} \tag{14}$$

since $x^2 + 1$ is irreducible. Multiplying both sides of (14) by $x(x^2 + 1)$ and setting $x = 0$ yields

$$1 = A_1(x^2 + 1) + (A_2 x + B_2)x \quad \text{and} \quad 1 = A_1.$$

Thus

$$1 - (x^2 + 1) = A_2 x^2 + B_2 x, \quad \text{or} \quad -x^2 = A_2 x^2 + B_2 x.$$

Dividing by x, we have

$$-x = A_2 x + B_2,$$

and again setting $x = 0$, we obtain

$$B_2 = 0.$$

Thus $-x = A_2 x$, or $A_2 = -1$, and

$$\int \frac{dx}{x(x^2 + 1)} = \int \left[\frac{1}{x} - \frac{x}{x^2 + 1} \right] dx = \ln |x| - \frac{1}{2} \ln |x^2 + 1|$$

$$= \ln \left| \frac{x}{\sqrt{x^2 + 1}} \right| + C. \quad \blacksquare$$

EXAMPLE 6 Compute $\int [(2x^5 - 3x^4 + 3x^3 + 4x^2 - 2x + 5)/(x^4 + 5x^2 + 6)] \, dx$.

Solution. Since the degree of the numerator is greater than the degree of the denominator, we divide to obtain

$$\frac{2x^5 - 3x^4 + 3x^3 + 4x^2 - 2x + 5}{x^4 + 5x^2 + 6} = 2x - 3 + \frac{-7x^3 + 19x^2 - 14x + 23}{x^4 + 5x^2 + 6}.$$

Now $x^4 + 5x^2 + 6 = (x^2 + 2)(x^2 + 3)$ and both terms are irreducible, so

$$\frac{-7x^3 + 19x^2 - 14x + 23}{(x^2 + 2)(x^2 + 3)} = \frac{A_1 x + B_1}{x^2 + 2} + \frac{A_2 x + B_2}{x^2 + 3}. \qquad (15)$$

Multiplying both sides of (15) by $(x^2 + 2)(x^2 + 3)$ yields

$$-7x^3 + 19x^2 - 14x + 23 = (A_1 x + B_1)(x^2 + 3) + (A_2 x + B_2)(x^2 + 2)$$
$$= (A_1 + A_2)x^3 + (B_1 + B_2)x^2$$
$$+ (3A_1 + 2A_2)x + (3B_1 + 2B_2).$$

Then equating the coefficients of like powers of x, we have

$$A_1 + A_2 = -7, \qquad B_1 + B_2 = 19,$$
$$3A_1 + 2A_2 = -14, \qquad 3B_1 + 2B_2 = 23.$$

The solutions to these systems of equations are

$$A_1 = 0, \qquad A_2 = -7, \qquad B_1 = -15, \qquad B_2 = 34.$$

Thus

$$\int \frac{2x^5 - 3x^4 + 3x^3 + 4x^2 - 2x + 5}{x^2 + 5x^2 + 6} \, dx = \int \left(2x - 3 + \frac{-15}{x^2 + 2} + \frac{-7x + 34}{x^2 + 3} \right) dx.$$

Now

$$\int \frac{-7x + 34}{x^2 + 3} \, dx = -\frac{7}{2} \int \frac{2x - (68/7)}{x^2 + 3} \, dx$$

$$= -\frac{7}{2} \int \frac{2x}{x^2 + 3} \, dx + 34 \int \frac{dx}{x^2 + 3}$$

$$= -\frac{7}{2} \ln(x^2 + 3) + \frac{34}{\sqrt{3}} \tan^{-1} \frac{x}{\sqrt{3}}.$$

So finally, we obtain

$$\int \frac{2x^5 - 3x^4 + 3x^3 + 4x^2 - 2x + 5}{x^4 + 5x^2 + 6} \, dx$$

$$= x^2 - 3x - \frac{15}{\sqrt{2}} \tan^{-1} \frac{x}{\sqrt{2}} - \frac{7}{2} \ln(x^2 + 3) + \frac{34}{\sqrt{3}} \tan^{-1} \frac{x}{\sqrt{3}} + C. \quad \blacksquare$$

There is one case that we have not considered, that of repeated irreducible quadratic factors. We illustrate this case with an example.

EXAMPLE 7   Compute $\int dx/(x^2 + 2x + 10)^2$.

Solution. The polynomial $x^2 + 2x + 10 = (x + 1)^2 + 9$ is irreducible. We make the substitution $(x + 1) = 3 \tan \theta$. Then

$$dx = 3 \sec^2 \theta \, d\theta$$

and

$$(x^2 + 2x + 10)^2 = [(x + 1)^2 + 9]^2 = (9 \tan^2 \theta + 9)^2 = (9 \sec^2 \theta)^2 = 81 \sec^4 \theta.$$

Thus

$$\int \frac{dx}{(x^2 + 2x + 10)^2} = \int \frac{3 \sec^2 \theta}{81 \sec^4 \theta} \, d\theta = \frac{1}{27} \int \frac{1}{\sec^2 \theta} \, d\theta = \frac{1}{27} \int \cos^2 \theta \, d\theta$$

$$= \frac{1}{54} \int (1 + \cos 2\theta) \, d\theta$$

$$= \frac{1}{54}\left(\theta + \frac{\sin 2\theta}{2}\right) = \frac{1}{54}(\theta + \sin \theta \cos \theta)$$

$$= \frac{1}{54}\left[\tan^{-1}\left(\frac{x + 1}{3}\right) + \overbrace{\frac{x + 1}{\sqrt{(x + 1)^2 + 9}}}^{\sin \theta} \cdot \overbrace{\frac{3}{\sqrt{(x + 1)^2 + 9}}}^{\cos \theta}\right]$$

Figure 1

$$= \frac{1}{54}\left[\tan^{-1}\left(\frac{x + 1}{3}\right) + \frac{3(x + 1)}{x^2 + 2x + 10}\right] + C. \quad \blacksquare$$

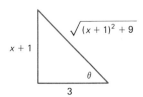

FIGURE 1

We close this section with one additional example and an interesting observation.

EXAMPLE 8   Compute $\int [(x^5 + 1)/(x^6 + 2x)]\, dx$.

       Solution. This problem could be solved by partial fractions *if* we could factor $x^6 + 2x$. But even then we would have to do a lot of algebra. Instead, we multiply numerator and denominator by x^4 to obtain

$$\int \frac{x^5 + 1}{x^6 + 2x}\, dx = \int \frac{x^9 + x^4}{x^{10} + 2x^5}\, dx = \frac{1}{10} \int \frac{10x^9 + 10x^4}{x^{10} + 2x^5}\, dx = \frac{1}{10} \ln |x^{10} + 2x^5| + C. \quad \blacksquare$$

       Why did this technique work? In a recent paper† J. E. Nymann at the University of Texas in El Paso showed the following: To evaluate

$$\int \frac{ax^n + b}{x(cx^n + d)}\, dx, \qquad \text{where} \qquad ad - bc \neq 0, \tag{16}$$

we can always find a real number r such that

$$\int \frac{ax^n + b}{x(cx^n + d)} \cdot \frac{x^r}{x^r}\, dx = \frac{1}{k} \int \frac{du}{u} \tag{17}$$

for some constant k. You are asked to verify this result in Problem 38.

       NOTE. This technique works *only* for integrals that can be written in the form of equation (16).

PROBLEMS 8.7

In Problems 1–34, calculate the integral.

1. $\displaystyle \int_{-1}^{0} \frac{dx}{(x - 1)(x - 2)(x - 3)}$

2. $\displaystyle \int \frac{x + 2}{(x - 1)(x - 2)(x - 3)}\, dx$

3. $\displaystyle \int \frac{x^2 + 3x + 4}{(x - 1)(x - 2)(x - 3)}\, dx$

4. $\displaystyle \int \frac{x^3 + 3x^2 + 2x + 1}{(x - 1)(x - 2)(x - 3)}\, dx$

5. $\displaystyle \int \frac{x^5}{x^3 - x}\, dx$

6. $\displaystyle \int \frac{x}{(x - 4)^3}\, dx$

7. $\displaystyle \int \frac{x^4}{(x - 4)^3}\, dx$

8. $\displaystyle \int \frac{x^{10}}{x^2(x + 3)}\, dx$

9. $\displaystyle \int \frac{dx}{x^2(x + 2)^2}$

10. $\displaystyle \int \frac{dx}{x^2(x - 1)}$

11. $\displaystyle \int \frac{x^3 + x^2 - 2}{x^4}\, dx$

12. $\displaystyle \int \frac{x^3}{(x - 1)^3}\, dx$

13. $\displaystyle \int \frac{dx}{x^4 - 5x^2 + 4}$

14. $\displaystyle \int \frac{x}{x^4 - 18x^2 + 81}\, dx$

---

†J. E. Nymann, "An alternative for partial fractions (part of the time)," *Two-Year College Mathematics Journal,* **14** (No. 1), 60–61 (January 1983).

15. $\displaystyle\int \frac{x^4 - x^3 + x^2 - 7x + 2}{x^3 + x^2 - 14x - 24}\, dx$

16. $\displaystyle\int \frac{x^2 + 1}{(x^2 - 1)^2}\, dx$

17. $\displaystyle\int_1^2 \frac{x^4}{16x - x^3}\, dx$

18. $\displaystyle\int_0^1 \frac{dx}{(x - 2)^2(x - 3)^2}$

19. $\displaystyle\int \frac{x^2 + 2}{x(x - 1)^2(x + 1)}\, dx$

20. $\displaystyle\int \frac{x^3}{x^4 - x^2}\, dx$

21. $\displaystyle\int \frac{dx}{(x + 2)(x^2 + 1)}$

22. $\displaystyle\int \frac{4x + 3}{(x^2 + 1)(x^2 + 2)}\, dx$

23. $\displaystyle\int \frac{x^3 + x^2 + 3}{(x^2 + 2)^2}\, dx$

24. $\displaystyle\int \frac{x + 4}{(x^2 + 4x + 5)^2}\, dx$

25. $\displaystyle\int \frac{x^3}{x^4 - 8x^2 + 16}\, dx$

26. $\displaystyle\int \frac{x}{(x + 1)^2(x^2 + 1)}\, dx$

27. $\displaystyle\int \frac{x}{x^4 - 1}\, dx$

28. $\displaystyle\int \frac{x^2 - 2}{(x^2 + 1)^2}\, dx$

29. $\displaystyle\int \frac{x^3 + 5}{(x^2 - x + \frac{1}{2})(x^2 - 6x + 10)}\, dx$

30. $\displaystyle\int \frac{dx}{x^4 + x^2}$

31. $\displaystyle\int \frac{x^4 + 1}{x^5 + x^3}\, dx$

32. $\displaystyle\int \frac{x^2 + 1}{(x^2 + 16)^3}\, dx$

33. $\displaystyle\int \frac{x^3 + 1}{(x^2 + 4x + 13)^2}\, dx$

**34. $\displaystyle\int \frac{(2x^3 - 4x^2 + 70x - 35)}{(x^2 - 2x + 17)(x^2 - 10x + 26)}\, dx$

35. Calculate $\int_0^{\pi/2} [(\cos x)/(\sin^2 x + 2 \sin x + 2)]\, dx$.   [*Hint:* Let $u = \sin x$.]

36. Calculate $\int_0^{\pi} [(\sin x)/(\cos^2 x - 4 \cos x + 13)^2]\, dx$.

37. Calculate the area bounded by the curve $y = x^3/(x^2 + 1)^3$, the x- and y-axes, and the line $x = 1$.

*38. Suppose that $ad - bc \neq 0$. Show that if

$$r = \frac{ad - bc(n + 1)}{bc - ad} \qquad \text{and} \qquad k = \frac{cdn}{ad - bc},$$

then

$$\int \frac{ax^n + b}{x(cx^n + d)}\, dx = \int \frac{ax^{n+r} + bx^r}{cx^{n+r+1} + dx^{r+1}}\, dx = \frac{1}{k}\int \frac{du}{u},$$

where $u = cx^{n+r+1} + dx^{r+1}$.

In Problems 39–42, use the result of Problem 38 to evaluate the given integral.

39. $\displaystyle\int \frac{x^2 - 1}{x^3 - 2x}\, dx$

40. $\displaystyle\int \frac{x^6 + 1}{x^7 - x}\, dx$

41. $\displaystyle\int \frac{3x^4 + 4}{2x^5 + 5x}\, dx$

42. $\displaystyle\int \frac{5x^{10} + 8}{-x^{11} + 2x}\, dx$

8.8 OTHER SUBSTITUTIONS

In addition to the integrals discussed in Section 8.5, there are many other types of expressions that can be integrated by making an appropriate substitution. Unfortunately, many of these substitutions work only for a very small class of integrals. Even when a function can be integrated by substitution, there is usually no obvious way to find the substitution that works, and it is often necessary to resort to trial and error. In this section we provide a few examples of substitutions that do work.

When an integrand contains a term of the form $[f(x)]^{m/n}$, then the substitution

$$u = f(x)^{1/n} \tag{1}$$

will often be useful.

EXAMPLE 1 Calculate $\int dx/(2 + \sqrt{x})$.

Solution. Let $u = x^{1/2}$. Then $x = u^2$ and $dx = 2u\,du$, so

$$\int \frac{dx}{2 + \sqrt{x}} = \int \frac{2u}{2 + u}\,du \overset{\text{Divide } 2u \text{ by } 2 + u}{=} \int \left(2 - \frac{4}{u + 2}\right)du$$

$$= 2u - 4\ln|u + 2| = 2\sqrt{x} - 4\ln(\sqrt{x} + 2) + C. \quad \blacksquare$$

If more than one rational power of x appears, make a substitution of the form $u = x^{1/n}$, where n is chosen to be the least common multiple of the denominators of the fractional powers that appear, so that you end up with a rational function of u that can be integrated by the method of partial fractions.

EXAMPLE 2 Calculate

$$\int \frac{x - 1}{6\sqrt{x}(4 + \sqrt[3]{x})}\,dx.$$

Solution. The integrand contains terms of the form $x^{1/2}$ and $x^{1/3}$. If we choose $u = x^{1/6}$, then $x^{1/2} = u^3$, $x^{1/3} = u^2$, and since $x = u^6$, $dx = 6u^5\,du$. Thus everything will be expressed in powers of u. We have

$$\int \frac{x - 1}{6\sqrt{x}(4 + \sqrt[3]{x})}\,dx = \int \frac{(u^6 - 1)6u^5}{6u^3(4 + u^2)}\,du = \int \frac{u^{11} - u^5}{u^5 + 4u^3}\,du = \int \frac{u^8 - u^2}{u^2 + 4}\,du.$$

Since the degree of the numerator is greater than the degree of the denominator, we divide:

$$\int \frac{u^8 - u^2}{u^2 + 4}\,du = \int \left(u^6 - 4u^4 + 16u^2 - 65 + \frac{260}{u^2 + 4}\right)du$$

$$= \frac{u^7}{7} - \frac{4u^5}{5} + \frac{16u^3}{3} - 65u + 130\tan^{-1}\frac{u}{2}$$

$$= \frac{x^{7/6}}{7} - \frac{4x^{5/6}}{5} + \frac{16x^{1/2}}{3} - 65x^{1/6} + 130\tan^{-1}\left(\frac{x^{1/6}}{2}\right) + C. \quad \blacksquare$$

EXAMPLE 3   Calculate $\int_{-1}^{0} x\sqrt[3]{x+1}\, dx$.

Solution. Let $u = (x+1)^{1/3}$. Then $u^3 = x+1$, or $x = u^3 - 1$, and $dx = 3u^2\, du$. When $x = -1$, $u = 0$, and when $x = 0$, $u = 1$. Thus

$$\int_{-1}^{0} x\sqrt[3]{x+1}\, dx = \int_{0}^{1} (u^3 - 1)u \cdot 3u^2\, du = \int_{0}^{1} (3u^6 - 3u^3)\, du$$

$$= \frac{3}{7}u^7 - \frac{3}{4}u^4 \Big|_{0}^{1} = -\frac{9}{28}.$$

This problem could also be solved by the method of integration by parts or by making the substitution $u = x + 1$. ▪

Most rational functions of $\sin x$ and $\cos x$ can be integrated with the use of a special trigonometric substitution.

EXAMPLE 4   Calculate $\int dx/(2 + \cos x)$.

Solution. Let $u = \tan(x/2)$.† Then $x = 2\tan^{-1} u$, and

$$dx = \frac{2}{1+u^2}\, du. \tag{2}$$

From Figure 1

$$\sin\frac{x}{2} = \frac{u}{\sqrt{1+u^2}} \qquad \text{and} \qquad \cos\frac{x}{2} = \frac{1}{\sqrt{1+u^2}}.$$

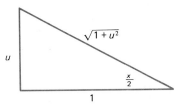

FIGURE 1

Using the double angle formulas, $\sin 2x = 2\sin x\cos x$ and $\cos 2x = \cos^2 x - \sin^2 x$, we obtain

$$\sin x = 2\sin\frac{x}{2}\cos\frac{x}{2} = \frac{2u}{1+u^2} \tag{3}$$

---

†The reason for making this seemingly arbitrary substitution will soon become apparent. It is certainly not obvious that it will work before we try it.

and

$$\cos x = \cos^2\frac{x}{2} - \sin^2\frac{x}{2} = \frac{1 - u^2}{1 + u^2}.$$ (4)

Thus using (2) and (4), we obtain

$$\int \frac{dx}{2 + \cos x} = \int \frac{2/(1 + u^2)}{2 + [(1 - u^2)/(1 + u^2)]}\,du = \int \frac{2}{3 + u^2}\,du$$

$$= \frac{2}{\sqrt{3}}\tan^{-1}\frac{u}{\sqrt{3}} = \frac{2}{\sqrt{3}}\tan^{-1}\left[\frac{\tan(x/2)}{\sqrt{3}}\right] + C. \ \blacksquare$$

EXAMPLE 5   Calculate $\int_0^{\pi/2} dx/(\sin x + \cos x)$.

Solution. Let $u = \tan x/2$. Then using (2), (3), and (4), we obtain

$$\int_0^{\pi/2} \frac{dx}{\sin x + \cos x} = \int_0^1 \frac{2/(1 + u^2)}{[2u/(1 + u^2)] + [(1 - u^2)/(1 + u^2)]}\,du$$

$$= \int_0^1 \frac{2\,du}{1 + 2u - u^2} = \int_0^1 \frac{2\,du}{1 - (u^2 - 2u)}$$

$$= \int_0^1 \frac{2\,du}{1 + 1 - (u^2 - 2u + 1)} = \int_0^1 \frac{2\,du}{2 - (u - 1)^2}.$$

Now let $u - 1 = \sqrt{2}\sin\theta$ so that $du = \sqrt{2}\cos\theta\,d\theta$, $\theta = 0$ when $u = 1$, and $\theta = \sin^{-1}(-1/\sqrt{2}) = -\pi/4$ when $u = 0$. Then

$$\int_0^1 \frac{2\,du}{2 - (u - 1)^2} = \int_{-\pi/4}^0 \frac{2\sqrt{2}\cos\theta}{2\cos^2\theta}\,d\theta = \sqrt{2}\int_{-\pi/4}^0 \sec\theta\,d\theta$$

$$= \sqrt{2}\ln|\sec\theta + \tan\theta|\,\Big|_{-\pi/4}^0$$

$$= \sqrt{2}[\ln 1 - \ln(\sqrt{2} - 1)] \approx 1.25. \ \blacksquare$$

We saw in Section 8.5 that a variety of functions could be integrated by making an appropriate trigonometric substitution. In many cases, however, a different substitution will work much better.

EXAMPLE 6   Compute $\int_0^1 [x^3/(3 + x^2)^{5/2}]\,dx$.

Solution. An obvious thing to try is $x = \sqrt{3}\tan\theta$. However, let's try the substitution $u = 3 + x^2$ first. Then $du = 2x\,dx$ and $x^2 = u - 3$, so that

$$x^3\,dx = x^2 \cdot x\,dx = (u - 3)\tfrac{1}{2}\,du$$

and

$$\int_0^1 \frac{x^3}{(3 + x^2)^{5/2}} \, dx = \frac{1}{2} \int_3^4 \frac{u - 3}{u^{5/2}} \, du = \frac{1}{2} \int_3^4 (u^{-3/2} - 3u^{-5/2}) \, du$$

$$= \frac{1}{2}(-2u^{-1/2} + 2u^{-3/2})\Big|_3^4 = (-u^{-1/2} + u^{-3/2})\Big|_3^4$$

$$= -\frac{1}{2} + \frac{1}{8} + \frac{1}{\sqrt{3}} - \frac{1}{3\sqrt{3}} = \frac{2}{3\sqrt{3}} - \frac{3}{8}.$$

This substitution turns out to be much easier than the substitution $x = \sqrt{3} \tan \theta$. Try it. ■

PROBLEMS 8.8

In Problems 1–34, calculate the given integrals by making an appropriate substitution.

1. $\displaystyle\int \frac{\sqrt{x + 2} - 1}{\sqrt{x + 2} + 1} \, dx$

2. $\displaystyle\int_{-1}^8 \frac{\sqrt[3]{x} - 1}{\sqrt[3]{x} + 1} \, dx$

3. $\displaystyle\int \frac{dx}{4 + 5\sqrt{x}}$

4. $\displaystyle\int_1^{32} \frac{dx}{x(1 + x^{1/5})}$

5. $\displaystyle\int x\sqrt{1 + x} \, dx$

6. $\displaystyle\int \frac{1 + x}{\sqrt{1 - x}} \, dx$

7. $\displaystyle\int_0^{13} x(1 + 2x)^{2/3} \, dx$

8. $\displaystyle\int_0^1 x^8(1 - x^3)^{1/3} \, dx$

9. $\displaystyle\int \frac{dx}{x^3(16 - x^3)^{1/3}}$

10. $\displaystyle\int_0^1 x^8(1 - x^3)^{6/5} \, dx$

11. $\displaystyle\int_0^2 \frac{3x^5}{(1 + x^3)^{3/2}} \, dx$

12. $\displaystyle\int \frac{x - 4}{\sqrt{x}(1 + \sqrt[3]{x})} \, dx$

13. $\displaystyle\int \frac{3x^2 - x}{\sqrt{1 + x}} \, dx$

14. $\displaystyle\int_0^1 \frac{dx}{1 + x^{1/5}}$

15. $\displaystyle\int_0^1 \frac{dx}{x^{1/3} + x^{1/4}}$

16. $\displaystyle\int \frac{x^{1/2}}{x^{1/3} + x^{1/4}} \, dx$

*17. $\displaystyle\int \frac{\sqrt{x}}{1 + x^{2/3}} \, dx$ [*Hint:* Factor $u^4 + 1$ into a product of irreducible quadratic factors.]

18. $\displaystyle\int \frac{x^3 - x}{\sqrt{x^2 - 2}} \, dx$

*19. $\displaystyle\int \frac{\sqrt[3]{1 - x}}{1 + x} \, dx$

20. $\displaystyle\int \frac{x}{(x - 1)^4} \, dx$

21. $\displaystyle\int \frac{x^{2/3} + 1}{x^{2/3} - 1} \, dx$

*22. $\displaystyle\int \frac{x^{3/4} - 16}{x^{3/4} + 16} \, dx$

23. $\displaystyle\int_0^1 x^2\sqrt{2 + 3x} \, dx$

24. $\displaystyle\int \frac{x^2}{\sqrt{2 + 3x}} \, dx$

25. $\displaystyle\int \frac{1 + x^{1/2}}{x^{5/6} + x^{7/6}} \, dx$

26. $\displaystyle\int \frac{dx}{\sqrt{e^{2x} - 1}}$

27. $\displaystyle\int \frac{dx}{1 + \sin x}$

28. $\displaystyle\int_0^{\pi/2} \frac{dx}{1 + \sin x + \cos x}$

29. $\displaystyle\int_0^{\pi/3} \frac{dx}{3 + 2\cos x}$

30. $\displaystyle\int \frac{dx}{\sin x + \tan x}$

31. $\displaystyle\int \frac{dx}{\sin 2x + \cos 2x}$

32. $\displaystyle\int \frac{2 - \sin x}{2 + \sin x} \, dx$

33. $\displaystyle\int \frac{\csc x \, dx}{3\csc x + 2\cot x + 2}$

34. $\displaystyle\int \frac{dx}{10\sec x - 6}$

35. Calculate $\int dx/x\sqrt{x^2 + 4x - 5}$.   [*Hint:* Try the substitution $u = 1/x$.]

36. Calculate $\int dx/x^3\sqrt{1 + x + x^2}$.

Problems 37–50 were given in Section 8.5. Redo them now by making a substitution that is *not* trigonometric.

37. $\int_0^1 \dfrac{x}{\sqrt{4 - x^2}}\, dx$ **38.** $\int \dfrac{x}{\sqrt{x^2 - 4}}\, dx$ **39.** $\int_0^1 x^3 \sqrt{4 - x^2}\, dx$

40. $\int_0^{\sqrt{3}} \dfrac{x^5}{1 + x^2}\, dx$ **41.** $\int x^3 \sqrt{x^2 - 4}\, dx$ **42.** $\int_0^2 x^3 \sqrt{4 + x^2}\, dx$

43. $\int \dfrac{\sqrt{5x^2 - 9}}{x}\, dx$ **44.** $\int_0^{1/\sqrt{2}} \dfrac{x^3}{2 + x^2}\, dx$ **45.** $\int \dfrac{\sqrt{9 - 5x^2}}{x}\, dx$

46. $\int \dfrac{\sqrt{5x^2 + 9}}{x}\, dx$ **47.** $\int x^3(x^2 - a^2)^{3/2}\, dx$ **48.** $\int \dfrac{dx}{x\sqrt{a^2 - x^2}}$

49. $\int \dfrac{x^3}{\sqrt{a^2 - x^2}}\, dx$ **50.** $\int \dfrac{\sqrt{x^2 + a^2}}{x}\, dx$

51. Compute $\int 1/(e^x + 1)\, dx$ by making an appropriate substitution and then integrating by partial fractions.

52. Consider $\int x^2 \sqrt{x^2 - 1}\, dx$. We can integrate this expression by making the substitution $x = \sec \theta$. Show that other "reasonable" nontrigonometric substitutions will *not* work.   [*Hint:* Let $u = \sqrt{x^2 - 1}$ and see what happens.]

*53. Use *Euler's substitution*, $t = x + \sqrt{x^2 + a}$, to find $\int dx/\sqrt{x^2 + 5}$.

*54. Compute $\int_0^\pi \ln \sin \theta \, d\theta$.

*55. Compute $\int \sqrt{\sec^2 x + A}\, dx$, $A \geq 0$.   [*Hint:* Start with the substitution $u = \sec x$. Two other substitutions are needed.]

*56. Suppose $a < b$. Evaluate $\int_a^b \sqrt{(x - a)(b - x)}\, dx$.   [*Hint:* Consider the graph of $y = \sqrt{(x - a)(b - x)}$.]

*57. Apply the substitution $x = \tan^{-1}[(e^{eu} - e^{-eu})/2]$ to find $\int \sec x \, dx$.

*58. Apply the substitution $x = \cot^{-1}(\sinh \theta)$ to find $\int \csc x \, dx$.

*59. Apply the substitution $x = \tan^{-1}(\sinh \theta)$ to find $\int \sec^3 x \, dx$.

8.9 USING THE INTEGRAL TABLES

There are many types of integrals that are encountered in applications. For that reason extensive tables of integrals have been made available. One of the most complete can be found in *Table of Integrals, Series and Products, Corrected and Enlarged Edition*, by I. S. Gradshteyn and I. M. Ryzhik (New York: Academic Press, 1979). It gives 150 pages of indefinite integrals and 400 pages of definite integrals. A more modest table is Table A.4 at the back of this book.

In the first eight sections of this chapter we saw how to integrate a wide variety of functions. All the integrals given in the tables can be obtained by one or more of our methods. However, a great deal of time can be saved by using the tables when one of the more common integrals is encountered. We illustrate this technique with several examples.

EXAMPLE 1   Use the integral table to find $\int dx/\sqrt{5x^2 + 8}$.

Solution.

$\overset{\text{Entry 22 in Table A.4}}{\searrow}$

$$\int \frac{du}{\sqrt{u^2 + a^2}} = \ln \left| u + \sqrt{u^2 + a^2} \right| + C \tag{1}$$

If $u = \sqrt{5x^2} = \sqrt{5}x$ and $a = \sqrt{8}$, then $\sqrt{u^2 + a^2} = \sqrt{5x^2 + 8}$. Also, $du = \sqrt{5}\, dx$. Thus multiplying and dividing the integrand by $\sqrt{5}$ and using (1), we obtain

$$\int \frac{dx}{\sqrt{5x^2 + 8}} = \frac{1}{\sqrt{5}} \int \frac{\sqrt{5}\,dx}{\sqrt{5x^2 + 8}} = \frac{1}{\sqrt{5}} \int \frac{du}{\sqrt{u^2 + 8}}$$

$$= \frac{1}{\sqrt{5}} \ln|u + \sqrt{u^2 + 8}| = \frac{1}{\sqrt{5}} \ln|\sqrt{5}x + \sqrt{5x^2 + 8}| + C. \blacksquare$$

EXAMPLE 2 Use the integral table to find $\int dx/(4 - 3\sin 5x)$.

Solution.

Entry 107 in Table A.4

$$\int \frac{du}{p + q\sin u} = \frac{2}{\sqrt{p^2 - q^2}} \tan^{-1}\left[\frac{p\tan(u/2) + q}{\sqrt{p^2 - q^2}}\right] + C, \qquad |p| > |q| \tag{2}$$

Here $p = 4$, $q = -3$, $|4| > |-3|$ and $u = 5x$. Then $du = 5\,dx$, so that

$$\int \frac{dx}{4 - 3\sin 5x} = \frac{1}{5} \int \frac{5\,dx}{4 - 3\sin 5x} = \frac{1}{5} \int \frac{du}{4 - 3\sin u}$$

$$= \frac{1}{5} \frac{2}{\sqrt{16 - 9}} \tan^{-1}\left[\frac{4\tan(u/2) - 3}{\sqrt{16 - 9}}\right]$$

$$= \frac{2}{5\sqrt{7}} \tan^{-1}\left[\frac{4\tan(5x/2) - 3}{\sqrt{7}}\right] + C. \blacksquare$$

EXAMPLE 3 Use the integral table to find $\int xe^{-5x^2}\sin 4x^2\,dx$.

Solution.

Entry 168

$$\int e^{au}\sin bu\,du = \frac{e^{au}}{a^2 + b^2}(a\sin bu - b\cos bu) + C \tag{3}$$

If $u = x^2$, then $du = 2x\,dx$, so that with $a = -5$ and $b = 4$,

$$\int xe^{-5x^2}\sin 4x^2\,dx = \frac{1}{2}\int e^{-5x^2}\sin 4x^2(2x\,dx)$$

$$= \frac{1}{2}\int e^{-5u}\sin 4u\,du = \frac{1}{2}\left(\frac{e^{-5u}}{25 + 16}\right)(-5\sin 4u - 4\cos 4u)$$

$$= \frac{-e^{-5x^2}}{82}(5\sin 4x^2 + 4\cos 4x^2) + C. \blacksquare$$

EXAMPLE 4 Use the table of integrals to find $\int \sin^4 3x\,dx$.

Solution.

Entry 109 and Example 8.2.9,

$$\int \sin^n au\,du = -\frac{1}{an}\sin^{n-1}au\cos au + \frac{n-1}{n}\int \sin^{n-2}au\,du. \tag{4}$$

So, using (4) twice, we have

$$\int \sin^4 3x \, dx = -\frac{1}{12} \sin^3 3x \cos 3x + \frac{3}{4} \int \sin^2 3x \, dx$$

$$= -\frac{1}{12} \sin^3 3x \cos 3x + \frac{3}{4}\left[-\frac{1}{6} \sin 3x \cos 3x + \frac{1}{2} \overbrace{\int dx}^{\sin^{2-2} x \, = \, 1} \right]$$

$$= -\tfrac{1}{12} \sin^3 3x \cos 3x - \tfrac{1}{8} \sin 3x \cos 3x + \tfrac{3}{8}x + C. \ \blacksquare$$

EXAMPLE 5   Use the table of integrals to find $\int x^3 e^{-2x} \, dx$.

Solution.

$$\int u^n e^{au} \, dx \overset{\text{Entry 166}}{=} \frac{u^n e^{au}}{a} - \frac{n}{a} \int u^{n-1} e^{au} \, du \tag{5}$$

Here $a = -2$ and $u = x$, and using (5) three times, we obtain

$$\int x^3 e^{-2x} \, dx = \frac{x^3 e^{-2x}}{-2} + \frac{3}{2} \int x^2 e^{-2x} \, dx$$

$$= -\frac{x^3 e^{-2x}}{2} + \frac{3}{2}\left(\frac{x^2 e^{-2x}}{-2} + \frac{2}{2} \int x e^{-2x} \, dx\right)$$

$$= -\frac{x^3 e^{-2x}}{2} - \frac{3}{4}x^2 e^{-2x} + \frac{3}{2}\left(\frac{x e^{-2x}}{-2} + \frac{1}{2} \int e^{-2x} \, dx\right)$$

$$= -\frac{x^3 e^{-2x}}{2} - \frac{3}{4}x^2 e^{-2x} - \frac{3}{4}x e^{-2x} - \frac{3}{8}e^{-2x} + C$$

$$= -e^{-2x}\left(\frac{x^3}{2} + \frac{3x^2}{4} + \frac{3}{4}x + \frac{3}{8}\right) + C. \ \blacksquare$$

PROBLEMS 8.9

In the following problems use the table of integrals (Table A.4) to compute the integral.

1. $\int \dfrac{dx}{9 - 4x^2}$

2. $\int \dfrac{dx}{4x^2 - 9}$

3. $\int \dfrac{dx}{9 + 4x^2}$

4. $\int \dfrac{dx}{\sqrt{16 + 3x^2}}$

5. $\int \sqrt{16 - 3x^2} \, dx$

6. $\int \dfrac{x}{\sqrt{16 - 3x^2}} \, dx$

7. $\int \dfrac{x}{\sqrt{16 + 3x^2}} \, dx$

8. $\int \dfrac{\sqrt{10 - 2x^2}}{x} \, dx$

9. $\int \dfrac{\sqrt{10 + 2x^2}}{x} \, dx$

10. $\int \dfrac{\sqrt{25x^2 - 4}}{3x} \, dx$

11. $\int \dfrac{dx}{x\sqrt{2x^2 - 9}}$

12. $\int \dfrac{4x^2}{\sqrt{9 + 4x^2}} \, dx$

13. $\int \dfrac{dx}{(4x^2 - 16)^{3/2}}$

14. $\int \dfrac{x^2}{\sqrt{2x^2 - 5}}\,dx$

*15. $\int \dfrac{dx}{x^2\sqrt{9x^2 - 1}}$

16. $\int \dfrac{x}{3 + 2x}\,dx$

17. $\int \dfrac{x^2}{3 + 2x}\,dx$

18. $\int \dfrac{dx}{x(3 + 2x)}$

19. $\int \dfrac{dx}{x^2(3 + 2x)}$

20. $\int x\sqrt{3 + 2x}\,dx$

21. $\int \tan^4 \dfrac{x}{3}\,dx$

22. $\int \dfrac{\sqrt{3 + 2x}}{x}\,dx$

23. $\int \sin^3 x \cos^4 x\,dx$

24. $\int \cos^4 2x\,dx$

25. $\int x^2 \cos 4x^3 \cos 5x^3\,dx$

26. $\int x \sec^4 x^2\,dx$

27. $\int x \cos^{-1} 2x\,dx$

28. $\int \sin 3x \cos 4x\,dx$

*29. $\int x^3 \sin^{-1} \dfrac{x}{4}\,dx$

30. $\int x^{14} \sin x^3\,dx$

31. $\int xe^{-x}\,dx$

32. $\int x \tan^{-1} x^2\,dx$

33. $\int x^2 e^{4x^3}\,dx$

34. $\int x^5 \tan^{-1} x^3\,dx$

*35. $\int x\sqrt{3x - 4x^2}\,dx$

36. $\int x^3 e^{-x^2}\,dx$

37. $\int x^2 e^{4x^3} \cos 7x^3\,dx$

38. $\int \dfrac{x^3}{\sqrt{7x^2 - x^4}}\,dx$

8.10 NUMERICAL INTEGRATION

Consider the problem of evaluating

$$\int_0^1 \sqrt{1 + x^3}\,dx \quad \text{or} \quad \int_0^1 e^{x^2}\,dx.$$

Since both $\sqrt{1 + x^3}$ and e^{x^2} are continuous in $[0, 1]$, we know that both the definite integrals given here exist. They represent the areas under the curves $y = \sqrt{1 + x^3}$ and $y = e^{x^2}$ for x between 0 and 1. The problem is that none of the methods we have studied (or any other method for that matter) will enable us to find the antiderivative of $\sqrt{1 + x^3}$ or e^{x^2} because neither antiderivative can be expressed in terms of the functions we know.

In fact, there are a great number of continuous functions for which an antiderivative cannot be expressed in terms of functions we know. In those cases we cannot use the fundamental theorem of calculus to evaluate a definite integral. Nevertheless, it may be very important to approximate the value of such an integral. For that reason many methods have been devised to approximate the value of a definite integral to as many decimal places as are deemed necessary. All these techniques come under the heading of **numerical integration.** We will not discuss this vast subject in great generality here. Rather, we will introduce two reasonably effective methods for estimating a definite integral: the **trapezoidal rule** and **Simpson's rule.**†

---

†One reasonably elementary book that gives a more complete discussion of numerical integration is by Conte and deBoor, *Elementary Numerical Analysis: An Algorithmic Approach*, 3rd edition (McGraw-Hill, New York, 1980).

Consider the problem of calculating

$$\int_a^b f(x)\, dx.\tag{1}$$

We know that

$$\int_a^b f(x)\, dx = \lim_{\Delta x \to 0}[f(x_1{}^*)\,\Delta x + f(x_2{}^*)\,\Delta x + \cdots + f(x_n{}^*)\,\Delta x]$$

$$= \lim_{\Delta x \to 0}\sum_{i=1}^{n} f(x_i{}^*)\,\Delta x.\tag{2}$$

In other words, when the lengths of the subintervals in a partition of $[a, b]$ are small, the sum in the right-hand side of (2) gives us a crude approximation to the integral. Here the area is approximated by a sum of areas of rectangles. We saw some examples of this type of approximation in Sections 5.4 and 5.5. We now develop a more efficient way to approximate the integral.

TRAPEZOIDAL RULE

Let f be as in Figure 1 and let us partition the interval $[a, b]$ by the equally spaced points

$$a = x_0 < x_1 < x_2 < \cdots < x_{i-1} < x_i < \cdots < x_n = b,\tag{3}$$

where $x_i - x_{i-1} = \Delta x = (b - a)/n$. In Figure 1 we have indicated that the area under the curve can be approximated by the sum of the areas of n trapezoids. One typical

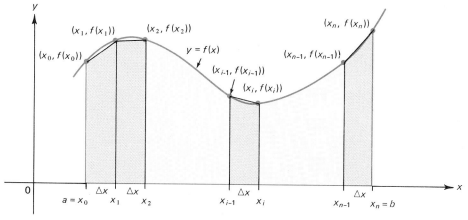

FIGURE 1

trapezoid is sketched in Figure 2. The area of the trapezoid is the area of the rectangle plus the area of the triangle. But the area of the rectangle is $f(x_i)\,\Delta x$, and the area of the triangle is $\frac{1}{2}[f(x_{i-1}) - f(x_i)]\,\Delta x$, so that

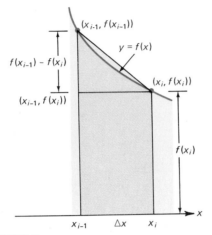

FIGURE 2

$$\text{area of trapezoid} = \overbrace{f(x_i)\,\Delta x}^{\text{Area of rectangle}} + \overbrace{\tfrac{1}{2}[f(x_{i-1}) - f(x_i)]\,\Delta x}^{\text{Area of triangle}}$$

$$= \tfrac{1}{2}[f(x_{i-1}) + f(x_i)]\,\Delta x. \dagger \qquad (4)$$

Then

$$\int_a^b f(x)\,dx \approx \text{sum of the areas of the trapezoids}$$

$$= \tfrac{1}{2}[f(x_0) + f(x_1)]\,\Delta x + \tfrac{1}{2}[f(x_1) + f(x_2)]\,\Delta x + \cdots$$

$$+ \tfrac{1}{2}[f(x_{n-2}) + f(x_{n-1})]\,\Delta x + \tfrac{1}{2}[f(x_{n-1}) + f(x_n)]\,\Delta x,$$

or

$$\int_a^b f(x)\,dx \approx \tfrac{1}{2}\Delta x[f(x_0) + 2f(x_1) + 2f(x_2) + \cdots + 2f(x_{n-1}) + f(x_n)]. \qquad (5)$$

The approximation formula (5) is called the **trapezoidal rule** for numerical integration. Note that since $\Delta x = (b - a)/n$, we can write (5) as

$$\int_a^b f(x)\,dx \approx \frac{b - a}{2n}[f(x_0) + 2f(x_1) + 2f(x_2) + \cdots + 2f(x_{n-1}) + f(x_n)]. \qquad (6)$$

---

†Note that this expression is the same as the average of the area of the rectangle R_{i-1} whose height is $f(x_{i-1})$ (the left-hand endpoint) and the area of the rectangle R_i whose height is $f(x_i)$ (the right-hand endpoint). That is,

$$\tfrac{1}{2}[(\text{area of } R_{i-1}) + (\text{area of } R_i)] = \tfrac{1}{2}[f(x_{i-1})\,\Delta x + f(x_i)\,\Delta x] = \tfrac{1}{2}[f(x_{i-1}) + f(x_i)]\,\Delta x.$$

EXAMPLE 1   Estimate $\int_1^2 (1/x)\, dx$ by using the trapezoidal rule first with $n = 5$ and then with $n = 10$.

Solution.

(i) Here $n = 5$ and

$$\Delta x = \frac{b - a}{n} = \frac{2 - 1}{5} = \frac{1}{5} = 0.2.$$

Then $x_0 = 1$, $x_1 = 1.2$, $x_2 = 1.4$, $x_3 = 1.6$, $x_4 = 1.8$, and $x_5 = 2$. From (5)

$$\int_1^2 \frac{1}{x}\, dx \approx \frac{1}{2}\, \Delta x [f(x_0) + 2f(x_1) + 2f(x_2) + 2f(x_3) + 2f(x_4) + f(x_5)]$$

$$= \frac{0.2}{2}\left(\frac{1}{1} + \frac{2}{1.2} + \frac{2}{1.4} + \frac{2}{1.6} + \frac{2}{1.8} + \frac{1}{2}\right)$$

$$\approx 0.1(1 + 1.6667 + 1.4286 + 1.25 + 1.111 + 0.5)$$

$$= 0.1(6.9564) = 0.6956.$$

(ii) Now $n = 10$ and $\Delta x = 1/10 = 0.1$, so that $x_0 = 1$, $x_1 = 1.1, \ldots, x_9 = 1.9$, and $x_{10} = 2$. Thus

$$\int_1^2 \frac{1}{x}\, dx \approx \frac{1}{2}\,(0.1)\left[1 + \frac{2}{1.1} + \frac{2}{1.2} + \frac{2}{1.3} + \frac{2}{1.4} + \frac{2}{1.5} + \frac{2}{1.6} + \frac{2}{1.7} + \frac{2}{1.8} + \frac{2}{1.9} + \frac{1}{2}\right]$$

$$\approx 0.05[1 + 1.8182 + 1.6667 + 1.5385 + 1.4286 + 1.3333 + 1.25$$

$$+ 1.1765 + 1.1111 + 1.0526 + 0.5]$$

$$= 0.05[13.8755] = 0.6938.$$

We can check our calculations by integrating:

$$\int_1^2 \frac{1}{x}\, dx = \ln x \Big|_1^2 = \ln 2 - \ln 1 = \ln 2 \approx 0.6931. \quad \blacksquare$$

We can see that by increasing the number of intervals, we increase the accuracy of our answer. This, of course, is not surprising. However, we are naturally led to ask what kind of accuracy we can expect by using the trapezoidal rule. In general, there are two kinds of errors encountered when we use a numerical method to integrate. The first kind is the kind we have already encountered—the error obtained by approximating the curve between the points $(x_{i-1}, f(x_{i-1}))$ and $(x_i, f(x_i))$ by the straight line joining those points. Since we now consider the function at a finite or *discrete* number of points, the error incurred by this approximation is called **discretization error.** However, there is another kind of error we will always encounter. As we saw in Example 1, we rounded our calculations to four decimal places. Each such "rounding" led to an error in our calculation. The accumulated effect of this rounding is called **round-off error.** Note that as we increase the number of intervals in our calculation, we improve the accuracy of our approximation to the area under the curve. This, evidently, has the effect of reducing the discretization error. On the other hand, an increase in the number of subintervals leads to an increase in the number of computations, which, in turn, leads to an increase in the accumulated round-off error. In fact, there is a delicate balance between these two types of errors and often there is an "optimal" number of intervals to be chosen so as to minimize the total error. Round-off error depends on the type of device used for the computations (pencil and paper, slide rule, hand calculator, computer, etc.) and will not be discussed further here. However, we can give a formula for estimating the discretization error incurred in using the trapezoidal rule.

Let the sum in (5) be denoted by T and let ϵ_n^T denote the discretization error:

$$\epsilon_n^T = \int_a^b f(x)\, dx - T$$

when n subintervals are used.

Theorem 1   Let f be continuous on $[a, b]$ and suppose that f' and f'' exist on $[a, b]$. Then there is a number c in (a, b) such that

$$\epsilon_n^T = -\frac{(b-a)}{12}(\Delta x)^2 f''(c). \tag{7}$$

The proof of this theorem is beyond the scope of this book but it can be found in most standard numerical analysis texts, including the one cited earlier in this section.

Corollary 1   *Error Bound for Trapezoidal Rule*   If $|f''(x)| \le M$ for all x in $[a, b]$, then

$$|\epsilon_n^T| \le M\frac{(b-a)^3}{12n^2}. \tag{8}$$

Proof. From (7)

$$|\epsilon_n^T| = \frac{b-a}{12}\Delta x^2\, |f''(c)| \le \frac{b-a}{12}\left(\frac{b-a}{n}\right)^2 M = \frac{M(b-a)^3}{12n^2}. \quad\blacksquare$$

REMARK. The expression in (8) gives the maximum possible value of the error. Often, as in the next example, the actual error will be quite a bit less. You may think of the error bound as a *guarantee* that the error will not be any greater. Accurate estimates of the error are more difficult to obtain.

EXAMPLE 2   Find a bound on the discretization error incurred when estimating $\int_1^2 (1/x)\, dx$ by using the trapezoidal rule with n subintervals.

Solution. $f(x) = 1/x$, $f'(x) = -1/x^2$, and $f''(x) = 2/x^3$. Hence $f''(x)$ is bounded above by 2 for x in $[1, 2]$. Then from (8)

$$|\epsilon_n^T| \leq \frac{2(2-1)^3}{12n^2} = \frac{1}{6n^2}.$$

For example, for $n = 5$ we calculated $\int_1^2 (1/x)\, dx \approx 0.6956$. Then the actual error is

$$\epsilon_n^T \approx 0.6931 - 0.6956 = -0.0025.$$

This result compares with a maximum possible error of $1/6n^2 = 1/(6 \cdot 25) = 1/150 \approx 0.0067$. For $n = 10$ the actual error is

$$\epsilon_{10}^T \approx 0.6931 - 0.6938 = -0.0007.$$

This result compares with a maximum possible error of $1/6n^2 = 1/600 \approx 0.0017$. Hence we see, in this example at least, that the error bound in (8) is a crude estimate of the actual error. Nevertheless, even this crude bound allows us to estimate the accuracy of our calculation in the cases where we can*not* check our answers by computing an antiderivative. Of course, these are the only cases of interest since we would not use a numerical technique if we could calculate the answer exactly. ■

EXAMPLE 3   Use the trapezoidal rule to estimate $\int_0^2 e^{x^2}\, dx$ with a maximum error of 1.

Solution. We must choose n large enough so that $|\epsilon_n^T| \leq 1$. For $f(x) = e^{x^2}$, we have $f'(x) = 2xe^{x^2}$ and $f''(x) = (2 + 4x^2)e^{x^2}$. Since this function is an increasing function, its maximum over the interval $[0, 2]$ occurs at 2. Then $M = f''(2) = 18e^4 \approx 983$. Hence from (8)

$$|\epsilon_n^T| \leq \frac{M(b-a)^3}{12n^2} \leq \frac{(983)2^3}{12n^2} \approx \frac{655}{n^2}.$$

We need $655/n^2 \leq 1$, or $n^2 \geq 655$, or $n \geq \sqrt{655}$. The smallest n that meets this requirement is $n = 26$. Hence we use the trapezoidal rule with $n = 26$ and $\Delta x = (b-a)/n = 2/26 = 1/13$. We have $x_0 = 0$, $x_1 = 1/13$, $x_2 = 2/13, \ldots, x_{25} = 25/13$, and $x_{26} = 26/13 = 2$. Then

$$\int_0^2 e^{x^2}\, dx \approx \frac{1}{2} \cdot \frac{1}{13}[e^0 + 2e^{(1/13)^2} + 2e^{(2/13)^2} + \cdots + 2e^{(25/13)^2} + e^{(26/13)^2}]$$

$$\approx \frac{1}{26}(1 + 2.012 + 2.048 + 2.109 + 2.199 + 2.319 + 2.475 + 2.673$$

$$+ 2.921 + 3.230 + 3.614 + 4.092 + 4.689 + 5.437 + 6.378 + 7.572$$

$$+ 9.097 + 11.059 + 13.603 + 16.933 + 21.328 + 27.184 + 35.060$$

$$+ 45.756 + 60.427 + 80.751 + 54.598)$$

$$= \frac{1}{26}(430.564) \approx 16.560.\dagger$$

---

†Values of the function $\int_0^x e^{t^2}\, dt$ have been tabulated. To six decimal places, the correct value of $\int_0^2 e^{t^2}\, dt$ is 16.452627. Thus our answer is actually correct to within 0.11.

This answer is correct to within 1 unit. The next method we discuss will enable us to calculate this integral with greater accuracy and less work. ■

SIMPSON'S RULE

We now derive a second method for estimating a definite integral. Look at the three sketches in Figure 3. In Figure 3a the area under the curve $y = f(x)$ over the interval $[x_i, x_{i+2}]$ is approximated by rectangles, where the height of each rectangle is the value of the function at an endpoint of an interval. In Figure 3b we have depicted the trapezoidal approximation to this area. The "top" of the first trapezoid is the straight line joining the *two* points $(x_i, f(x_i))$ and $(x_{i+1}, f(x_{i+1}))$ and the "top" of the second trapezoid is given in an analogous manner. In Figure 3c we are approximating the required area by drawing a figure whose "top" is the parabola passing through the *three* points $(x_i, f(x_i))$, $(x_{i+1}, f(x_{i+1}))$, and $(x_{i+2}, f(x_{i+2}))$. As we will see, this method will give us a better approximation to the area under the curve. First, we need to calculate the area depicted in Figure 3c.

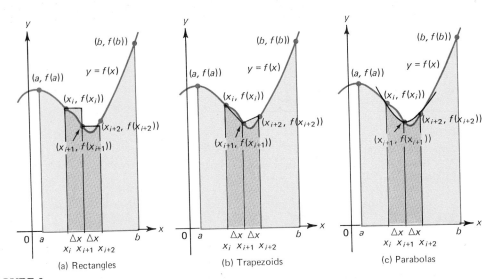

FIGURE 3

Theorem 2 The area A_{i+2} bounded by the parabola† passing through the points $(x_i, f(x_i))$, $(x_{i+1}, f(x_{i+1}))$, and $(x_{i+2}, f(x_{i+2}))$, the lines $x = x_i$ and $x = x_{i+2}$, and the x-axis (where $x_{i+1} - x_i = x_{i+2} - x_{i+1} = \Delta x$) is given by

$$A_{i+2} = \tfrac{1}{3} \Delta x [f(x_i) + 4f(x_{i+1}) + f(x_{i+2})]. \tag{9}$$

Proof (Optional). First, we observe that the parabola can be written in the form

$$y = ax^2 + bx + c \tag{10}$$

---

†In Figure 3c we assume that f is positive over the interval $[x_i, x_{i+2}]$. The method we are about to develop does *not* require that f be positive. However, the method is easier to motivate if we make this assumption.

where a, b, and c are real numbers. Now the area under this parabola is given by

$$A_{i+2} = \int_{x_i}^{x_i+2} (ax^2 + bx + c)\, dx = \int_{x_i}^{x_i+2\Delta x} (ax^2 + bx + c)\, dx$$

$$= \left(\frac{ax^3}{3} + \frac{bx^2}{2} + cx\right)\Big|_{x_i}^{x_i+2\Delta x}$$

$$= \tfrac{1}{3}a(x_i + 2\,\Delta x)^3 + \tfrac{1}{2}b(x_i + 2\,\Delta x)^2 + c(x_i + 2\,\Delta x) - \tfrac{1}{3}ax_i^3 - \tfrac{1}{2}bx_i^2 - cx_i$$

$$= \tfrac{1}{3}a(x_i^3 + 6x_i^2\,\Delta x + 12x_i\,\Delta x^2 + 8\,\Delta x^3) + \tfrac{1}{2}b(x_i^2 + 4x_i\,\Delta x + 4\,\Delta x^2)$$

$$\quad + c(x_i + 2\,\Delta x) - \tfrac{1}{3}ax_i^3 - \tfrac{1}{2}bx_i^2 - cx_i$$

$$= \tfrac{1}{3}a\,\Delta x(6x_i^2 + 12x_i\,\Delta x + 8\,\Delta x^2) + \tfrac{1}{2}b\,\Delta x(4x_i + 4\,\Delta x) + 2c\,\Delta x$$

$$= \tfrac{1}{3}a\,\Delta x(6x_i^2 + 12x_i\,\Delta x + 8\,\Delta x^2) + \tfrac{1}{3}b\,\Delta x(6x_i + 6\,\Delta x) + \tfrac{1}{3}(6c\,\Delta x)$$

$$\underbrace{\qquad\qquad}_{\text{We multiplied and divided by 3/2}} \underbrace{\qquad}_{\text{We multiplied and divided by 3}}$$

$$= \tfrac{1}{3}\,\Delta x[a(6x_i^2 + 12x_i\,\Delta x + 8\,\Delta x^2) + b(6x_i + 6\,\Delta x) + 6c]. \tag{11}$$

Now using the fact that the parabola passes through the points $(x_i, f(x_i))$, $(x_{i+1}, f(x_{i+1}))$, and $(x_{i+2}, f(x_{i+2}))$, we have

$$f(x_i) = ax_i^2 + bx_i + c$$

$$f(x_{i+1}) = a(x_i + \Delta x)^2 + b(x_i + \Delta x) + c$$

$$= a(x_i^2 + 2x_i\,\Delta x + \Delta x^2) + b(x_i + \Delta x) + c$$

$$f(x_{i+2}) = a(x_i + 2\,\Delta x)^2 + b(x_i + 2\,\Delta x) + c$$

$$= a(x_i^2 + 4x_i\,\Delta x + 4\,\Delta x^2) + b(x_i + 2\,\Delta x) + c.$$

Finally,

$$f(x_i) + 4f(x_{i+1}) + f(x_{i+2})$$

$$= a(6x_i^2 + 12x_i\,\Delta x + 8\,\Delta x^2) + b(6x_i + 6\,\Delta x) + 6c. \tag{12}$$

Comparing (12) with (11) gives us (9), and the proof is complete. ∎

Now suppose that the interval $[a, b]$ is divided into $2n$ subintervals of equal lengths $\Delta x = (b - a)/2n$. Then from (9) we have

$$\int_a^b f(x)\, dx \approx A_2 + A_4 + A_6 + \cdots + A_{2n}$$

$$= \tfrac{1}{3}\,\Delta x[f(x_0) + 4f(x_1) + f(x_2)] + \tfrac{1}{3}\,\Delta x[f(x_2) + 4f(x_3) + f(x_4)]$$

$$+ \cdots + \tfrac{1}{3}\,\Delta x[f(x_{2n-2}) + 4f(x_{2n-1}) + f(x_{2n})],$$

or

$$\int_a^b f(x)\, dx \approx \tfrac{1}{3} \Delta x[f(x_0) + 4f(x_1) + 2f(x_2) + 4f(x_3) + 2f(x_4)$$

$$+ \cdots + 2f(x_{2n-2}) + 4f(x_{2n-1}) + f(x_{2n})]. \tag{13}$$

The approximation in (13) is called **Simpson's rule†** (or the **parabolic rule**) for approximating a definite integral. From (13) we see that there is a bit more work needed to estimate an integral by using Simpson's rule than by using the trapezoidal rule with the same number of subintervals. However, the discretization error in Simpson's rule is usually a good deal less, as is suggested in the following theorem, whose proof can be found in the text cited earlier in this section.

REMARK. It is important to note that the formula for the trapezoidal rule has n subintervals and the formula for Simpson's rule has $2n$ subintervals. With $2n$ subintervals we have n parabolas, the areas under which we must compute. That is why we use the number $2n$ instead of n in using Simpson's rule.

Theorem 3 Let f be continuous on $[a, b]$ and suppose that f', f'', f''', and $f^{(4)}$ all exist on $[a, b]$. Then the discretization error ϵ_{2n}^S of Simpson's rule (13), using $2n$ equally spaced subintervals of length $\Delta x = (b - a)/2n$, is given by

$$\epsilon_{2n}^S = -\frac{(b-a)}{180}(\Delta x)^4 f^{(4)}(c), \tag{14}$$

where c is some number in the open interval (a, b).

Corollary 2 *Error Bound for Simpson's Rule* If $|f^{(4)}(x)| \le M$ for all x in $[a, b]$, then

$$|\epsilon_{2n}^S| \le \frac{M(b-a)^5}{2880n^4}. \tag{15}$$

Proof.

$$|\epsilon_{2n}^S| = \frac{b-a}{180}(\Delta x)^4 |f^{(4)}(c)| \le \left(\frac{b-a}{180}\right)\left(\frac{b-a}{2n}\right)^4 M$$

$$\le \frac{M(b-a)^5}{(180)(16n^4)} = \frac{M(b-a)^5}{2880n^4}. \quad \blacksquare$$

EXAMPLE 4 Use Simpson's rule to estimate $\int_1^2 (1/x)\, dx$ by using 10 subintervals. What is the maximum error in your estimate? Compare this error with the exact answer of $\ln 2$.

Solution. Here we have $\Delta x = \tfrac{1}{10}$ and $n = 5$ ($2n = 10$), so that from (13)

---

†Named after the British mathematician Thomas Simpson (1710–1761), who published the result in his *Mathematical Dissertations on Physical and Analytical Subjects* in 1743.

$$\int_1^2 \frac{1}{x}\,dx \approx \frac{1}{30}\left(\frac{1}{1} + \frac{4}{1.1} + \frac{2}{1.2} + \frac{4}{1.3} + \frac{2}{1.4} + \frac{4}{1.5} + \frac{2}{1.6} + \frac{4}{1.7} + \frac{2}{1.8} + \frac{4}{1.9} + \frac{1}{2}\right)$$

$$= \frac{1}{30}(1 + 3.636364 + 1.666667 + 3.076923 + 1.428571 + 2.666667$$

$$+ 1.25 + 2.352941 + 1.111111 + 2.105263 + 0.5)$$

$$= \frac{1}{30}(20.794507) \approx 0.693150.$$

To six decimal places, ln 2 = 0.693147, so our answer is very accurate indeed. To calculate the maximum possible error, we first need to calculate $f^{(4)}$. But $f(x) = 1/x$, $f'(x) = -1/x^2$, $f''(x) = 2/x^3$, $f'''(x) = -6/x^4$, and $f^{(4)}(x) = 24/x^5$. Over the interval [1, 2], $|24/x^5| \le 24$, so that $M = 24$. Then we use formula (15) with $M = 24$ and $n = 5$ (so that $2n = 10$) to obtain

$$|\epsilon_{2n}^S| \le \frac{24}{(2880)5^4} = \frac{24}{(2880)(625)} = \frac{24}{1,800,000} \approx 0.0000133.$$

Our actual error is $0.693150 - 0.693147 = 0.000003$, which is about one-fourth the maximum possible discretization error. Notice that in this example Simpson's rule gives a far more accurate answer than the trapezoidal rule using the same number of subintervals (ten). ■

EXAMPLE 5   Use Simpson's rule to estimate $\int_0^2 e^{x^2}\,dx$ with a maximum error of 0.1.

Solution. If $f(x) = e^{x^2}$, we have already calculated (in Example 3) that $f''(x) = (2 + 4x^2)e^{x^2}$. Then $f'''(x) = (12x + 8x^3)e^{x^2}$ and $f^{(4)}(x) = (12 + 48x^2 + 16x^4)e^{x^2}$. This is an increasing function for x in [0, 2], so that $M = f^{(4)}(2) = 460e^4 \approx 25{,}115$. Since $(b - a)^5 = 2^5 = 32$, we must choose n such that

$$|\epsilon_{2n}^S| \le \frac{M(b - a)^5}{2880n^4} \approx \frac{(25{,}115)(32)}{2880n^4} \approx \frac{279}{n^4} < 0.1 = \frac{1}{10}.$$

We then need $n^4 > 2790$, or $n > 2790^{1/4}$. The smallest integer n that satisfies this inequality is $n = 8$. Thus to obtain the required accuracy, we use Simpson's rule with $2n = 16$ subintervals. Then $\Delta x = (b - a)/2n = \frac{2}{16} = \frac{1}{8}$, and we have

$$\int_0^2 e^{x^2}\,dx \approx \tfrac{1}{3} \cdot \tfrac{1}{8}(e^0 + 4e^{(1/8)^2} + 2e^{(2/8)^2} + 4e^{(3/8)^2} + \cdots + 2e^{(14/8)^2}$$

$$+ 4e^{(15/8)^2} + e^{(16/8)^2})$$

$$\approx \tfrac{1}{24}(1 + 4.0630 + 2.1290 + 4.6040 + 2.5681 + 5.9116 + 3.5101$$

$$+ 8.6014 + 5.4366 + 14.1812 + 9.5415 + 26.4940 + 18.9755$$

$$+ 56.0879 + 42.7619 + 134.5478 + 54.5982)$$

$$\approx \tfrac{1}{24}(395.0118) \approx 16.4588.$$

This answer is correct to within one-tenth of a unit.† Notice how in the calculation of $\int_0^2 e^{x^2}\, dx$, Simpson's rule give us more accuracy with fewer calculations than the trapezoidal rule does. ■

There are many other methods that can be used to approximate definite integrals. For example, there are methods in which the points $x_0, x_1, \ldots, x_n$ are *not* equally spaced. One such method is called **Gaussian quadrature.** We will not discuss this very useful method here except to note that it can be found in any introductory book on numerical analysis. Finally, as we will discuss in Chapters 13 and 14, techniques using infinite series can be used to approximate certain definite integrals.

We conclude by noting that every definite integral that is known to exist can be evaluated to any number of decimal places of accuracy if one is supplied with the appropriate calculating tool. The problems at the end of this section can all be done reasonably quickly by using a scientific hand calculator. For more accuracy it may be necessary to evaluate a function at hundreds, or even thousands, of points. This problem is a manageable one only if you have access to a high-speed computer or a programmable calculator. If, in fact, you do have such access, you should write a computer program to estimate an integral using Simpson's rule and then use it to calculate each of the integrals in the problem set to at least six decimal places of accuracy.

PROBLEMS 8.10

In Problems 1–15, estimate the given definite integral by using (i) the trapezoidal rule and (ii) Simpson's rule over the given number of intervals. Then (iii) use error formula (8) to obtain a bound for the error of the trapezoidal approximation and (iv) use error formula (15) to obtain a bound for the error of the approximation using Simpson's rule. Then (v) calculate the integral exactly. Finally, compare the actual errors in your computations with the maximum possible errors found in (iii) and (iv).

1. $\int_0^1 x\, dx$; 4 intervals

2. $\int_{-2}^2 x\, dx$; 6 intervals

3. $\int_0^1 x^2\, dx$; 4 intervals

4. $\int_0^1 e^x\, dx$; 4 intervals

5. $\int_0^2 e^x\, dx$; 6 intervals

6. $\int_1^2 \frac{1}{x^2}\, dx$; 6 intervals

7. $\int_1^2 \sqrt{x}\, dx$; 8 intervals

8. $\int_0^3 \frac{1}{\sqrt{1+x}}\, dx$; 6 intervals

9. $\int_0^{\pi/2} \sin x\, dx$; 4 intervals

10. $\int_0^{\pi/2} \cos x\, dx$; 4 intervals

11. $\int_0^{\pi/3} \tan x\, dx$; 8 intervals

12. $\int_2^5 \frac{x}{\sqrt{x^2+1}}\, dx$; 6 intervals

13. $\int_1^e \ln x\, dx$; 6 intervals

14. $\int_0^1 \frac{1}{1+x^2}\, dx$; 10 intervals

15. $\int_0^1 e^x \sin x\, dx$; 8 intervals

---

†Since the correct value is 16.452627 (correct to six decimal places), our answer is really correct to within 0.007 units.

In Problems 16–30, approximate the given integral by using (i) the trapezoidal rule and (ii) Simpson's rule, with the indicated number of subintervals.

16. $\int_0^1 \sqrt{x + x^2}\, dx$; 4 intervals

17. $\int_1^2 \frac{\sin x}{x}\, dx$; 6 intervals

18. $\int_1^3 \frac{x}{\sin x}\, dx$; 6 intervals

19. $\int_0^1 e^{\sqrt{x}}\, dx$; 6 intervals

20. $\int_0^1 e^{x^3}\, dx$; 8 intervals

21. $\int_0^\pi \sin x^2\, dx$; 8 intervals

22. $\int_1^2 \sqrt{\ln x}\, dx$; 10 intervals

23. $\int_{-1}^1 e^{-x^2}\, dx$; 10 intervals

24. $\int_0^1 \sqrt{1 + x^3}\, dx$; 10 intervals

25. $\int_0^1 \frac{dx}{\sqrt{1 + x^3}}$; 10 intervals

26. $\int_0^1 xe^{x^3}\, dx$; 10 intervals

27. $\int_0^1 \ln(1 + e^x)\, dx$; 8 intervals

28. $\int_1^3 \frac{x^2}{\sqrt[3]{1 + x}}\, dx$; 10 intervals

29. $\int_0^1 \sinh x^2\, dx$; 10 intervals

30. $\int_0^{\pi/4} \sin(\tan x)\, dx$; 10 intervals

In Problems 31–36, find a bound for the discretization error by using the trapezoidal rule and Simpson's rule.

31. the integral of Problem 20

32. the integral of Problem 21

33. the integral of Problem 23

34. the integral of Problem 24

35. the integral of Problem 26

36. the integral of Problem 27

37. The integral $(1/\sqrt{2\pi}) \int_{-a}^{a} e^{-x^2/2}\, dx$ is very important in probability theory. Using Simpson's rule, estimate $(1/\sqrt{2\pi}) \int_{-1}^{1} e^{-x^2/2}\, dx$ with an error of less than 0.01. [*Hint:* Show that $\int_{-a}^{a} e^{-x^2/2}\, dx = 2 \int_0^a e^{-x^2/2}\, dx$.]

38. Estimate $(1/\sqrt{2\pi}) \int_{-5}^{5} e^{-x^2/2}\, dx$ with an error of less than 0.01.

*39. **(a)** Estimate $(1/\sqrt{2\pi}) \int_{-50}^{50} e^{-x^2/2}\, dx$ with an error of less than 0.1.
   (b) Can you guess what happens to $(1/\sqrt{2\pi}) \int_{-N}^{N} e^{-x^2/2}\, dx$ as N grows without bound?

40. We know that

$$\int_0^1 \frac{dx}{1 + x^2} = \tan^{-1} x \bigg|_0^1 = \frac{\pi}{4}.$$

Thus

$$\pi = 4 \int_0^1 \frac{dx}{1 + x^2}.$$

(a) With Simpson's rule how many subintervals does it take to estimate π with an error of less than 0.0001?
(b) Using this number of subintervals, give your estimate of π.

41. How many subintervals would it take to estimate ln 2 by using the trapezoidal rule applied to the integral $\int_1^2 (1/x)\, dx$ with an error of less than 10^{-10}?

42. How many subintervals would it take to perform the estimate in Problem 41 by using Simpson's rule?

*43. The function $J_{1/2}(x) = \sqrt{2/\pi x}\,\sin x$ is called the **Bessel function of order 1/2** and occurs frequently in applications in physics and engineering. Estimate $\int_{1/2}^{1} J_{1/2}(x)\,dx$ with an error of less than 0.01.

44. Show that Simpson's rule provides the exact answer for $\int_a^b P_3(x)\,dx$ if P_3 is a polynomial of degree three or less.

REVIEW EXERCISES FOR CHAPTER EIGHT

In Exercises 1–86, use one or more of the techniques of this chapter to calculate the given integral. Do not use tables.

1. $\displaystyle\int_0^1 \frac{x^2}{\sqrt{4 + x^2}}\,dx$

2. $\displaystyle\int x\sqrt{x + 2}\,dx$

3. $\displaystyle\int \frac{x^2 + 3}{(x^2 + 1)^2}\,dx$

4. $\displaystyle\int \frac{dx}{(x - 1)(x + 3)}$

5. $\displaystyle\int_0^{\pi/2} \cos^3 x \sin^2 x\,dx$

6. $\displaystyle\int_0^1 \frac{\sqrt{x}}{1 + \sqrt{x}}\,dx$

7. $\displaystyle\int xe^{-2x}\,dx$

8. $\displaystyle\int \sin 4x \cos 5x\,dx$

9. $\displaystyle\int \tan^5 x\,dx$

10. $\displaystyle\int_0^{\pi/4} \sec^2 x \tan^3 x\,dx$

11. $\displaystyle\int_0^1 \frac{dx}{1 + \sqrt[3]{x}}$

12. $\displaystyle\int_0^1 (1 + x^2)^{3/2}\,dx$

13. $\displaystyle\int_{\sqrt{2}}^2 \frac{x^2}{(x^2 - 1)^{3/2}}\,dx$

14. $\displaystyle\int \frac{dx}{\cos x - \sin x}$

15. $\displaystyle\int e^{2x} \sin 2x\,dx$

16. $\displaystyle\int_1^2 x^3 \ln x\,dx$

17. $\displaystyle\int \sin^3 x \cos^3 x\,dx$

18. $\displaystyle\int \sec^4 x \tan^2 x\,dx$

19. $\displaystyle\int \csc^3 x \cot^3 x\,dx$

20. $\displaystyle\int \frac{dx}{(x - 1)(x^2 + x + 1)}$

21. $\displaystyle\int \frac{x^5 - 1}{x^3 - x}\,dx$

22. $\displaystyle\int \frac{x^2}{(x + 1)^2(x - 1)^2}\,dx$

23. $\displaystyle\int x \sinh x\,dx$

24. $\displaystyle\int \frac{dx}{x^2 - 4x + 5}$

25. $\displaystyle\int \frac{dx}{x^2 - 4x + 4}$

26. $\displaystyle\int \sin 5x \sin 6x\,dx$

27. $\displaystyle\int \cos 6x \cos 7x\,dx$

28. $\displaystyle\int \frac{1 + x + x^2}{(x^2 + 2x + 3)^2}\,dx$

29. $\displaystyle\int \frac{x^2 + 2}{(x - 3)(x + 4)(x - 5)}\,dx$

30. $\displaystyle\int_0^3 \frac{x^2}{\sqrt{9 + x^2}}\,dx$

31. $\displaystyle\int x^3\sqrt{x^2 - 4}\,dx$

32. $\displaystyle\int x \cos^{-1} x\,dx$

33. $\displaystyle\int \sec^4 x\,dx$

34. $\displaystyle\int \frac{x^3}{(x + 1)^4}\,dx$

35. $\displaystyle\int \frac{(x + 1)^3}{x^3}\,dx$

36. $\displaystyle\int_0^{\pi/4} \sin x \tan^2 x\,dx$

37. $\displaystyle\int_0^{\pi/2} \sin^6 x\,dx$

38. $\displaystyle\int \sec^4 x \tan^3 x\,dx$

39. $\displaystyle\int_0^1 (1 + x^2)^{-1/2}\,dx$

40. $\displaystyle\int \frac{dx}{2 - \cos x}$

41. $\displaystyle\int \frac{dx}{1 + 5 \sin x}$

42. $\displaystyle\int_0^1 x^3(1 - x)^{2/3}\,dx$

43. $\displaystyle\int \frac{2x^3}{(1 + x^4)^{4/3}}\,dx$

44. $\displaystyle\int \frac{1 - x}{1 + \sqrt{x}}\,dx$

45. $\displaystyle\int \frac{x - 2x^3}{\sqrt{2 + 3x}}\, dx$

46. $\displaystyle\int \sin(\ln x)\, dx$

47. $\displaystyle\int e^{-x} \cos \frac{x}{3}\, dx$

48. $\displaystyle\int \sinh x \cosh 2x\, dx$

49. $\displaystyle\int \cosh x \cosh 3x\, dx$

50. $\displaystyle\int \frac{dx}{2 \sec x - 1}$

51. $\displaystyle\int x^2 e^{-x}\, dx$

52. $\displaystyle\int x^{3/2} \ln x\, dx$

53. $\displaystyle\int \frac{\sin \sqrt{x} \cos \sqrt{x}}{\sqrt{x}}\, dx$

54. $\displaystyle\int \frac{dx}{\sqrt{4x^2 - 9}}$

55. $\displaystyle\int \frac{dx}{x\sqrt{9 - 4x^2}}$

56. $\displaystyle\int \frac{x}{(2x + 3)^2}\, dx$

57. $\displaystyle\int \frac{\sqrt{x^2 - 16}}{x^2}\, dx$

58. $\displaystyle\int \frac{dx}{x\sqrt{4x - x^2}}$

59. $\displaystyle\int x \sin^{3/4} 2x^2 \cos 2x^2\, dx$

60. $\displaystyle\int_0^{\pi/2} \sin^6 x\, dx$

61. $\displaystyle\int_0^{\pi/2} \sin^7 x\, dx$

62. $\displaystyle\int_0^{\pi/2} \cos^5 x\, dx$

63. $\displaystyle\int \sinh^2 3x\, dx$

64. $\displaystyle\int x^5 \ln 4x\, dx$

65. $\displaystyle\int x^2 e^{-3x^3} \sin 2x^3\, dx$

66. $\displaystyle\int \csc^4 2x \cot 2x\, dx$

67. $\displaystyle\int \sec^4 3x\, dx$

68. $\displaystyle\int \operatorname{sech} 4x\, dx$

69. $\displaystyle\int e^{2x} \sinh 3x\, dx$

70. $\displaystyle\int e^{-x} \cosh 5x\, dx$

71. $\displaystyle\int \tanh^2 3x\, dx$

72. $\displaystyle\int \frac{dx}{1 + \cos 2x}$

73. $\displaystyle\int \frac{dx}{4 - 3 \cos 2x}$

74. $\displaystyle\int \frac{dx}{1 - \sin 2x}$

75. $\displaystyle\int \frac{dx}{1 + \sin 2x}$

76. $\displaystyle\int x^3 \sin 4x\, dx$

77. $\displaystyle\int x^3 \cos 4x\, dx$

78. $\displaystyle\int x(x^2 - 1)^{7/3}\, dx$

79. $\displaystyle\int \frac{x^5}{9 + x^{12}}\, dx$

80. $\displaystyle\int \frac{x^5}{9 - x^{12}}\, dx$

81. $\displaystyle\int x^5 \sqrt{9 + x^{12}}\, dx$

82. $\displaystyle\int \frac{dx}{x^2 + 4x + 13}$

83. $\displaystyle\int \frac{x^2 - 3}{x^3 - 1}\, dx$

84. $\displaystyle\int \frac{x^4}{(x - 1)(x^2 + 2)}\, dx$

85. $\displaystyle\int \frac{2x - 3}{(x - 1)^2}\, dx$

86. $\displaystyle\int \frac{x + 2}{(x^2 + 1)(x^2 + 4)}\, dx$

In Exercises 87–96, evaluate the integral by using an integral table.

87. $\displaystyle\int \frac{dx}{16 - x^2}$

88. $\displaystyle\int \frac{dx}{\sqrt{25 - 4x^2}}$

89. $\displaystyle\int \frac{x}{\sqrt{25 - 9x^2}}\, dx$

90. $\displaystyle\int \frac{dx}{x\sqrt{4x^2 - 25}}$

91. $\displaystyle\int \frac{9x^2}{\sqrt{16 + 9x^2}}\, dx$

92. $\displaystyle\int \sin^4 3x\, dx$

93. $\displaystyle\int x^3 \cos^{-1} 3x\, dx$

94. $\displaystyle\int x^4 e^{-2x}\, dx$

95. $\displaystyle\int x^2 \tan^{-1} x^3\, dx$

96. $\displaystyle\int x \sin 2x^2 \sin 3x^2\, dx$

In Exercises 97–102, use the trapezoidal rule (T) or Simpson's rule (S) to estimate the given integral with the given number of subintervals.

97. $\int_0^1 e^{x^3} dx$; T, $n = 4$

98. $\int_0^1 e^{x^3} dx$; S, $n = 4$

99. $\int_0^1 \dfrac{dx}{\sqrt{1 + x^4}}$; T, $n = 6$

100. $\int_0^1 \dfrac{dx}{\sqrt{1 + x^4}}$; S, $n = 6$

101. $\int_0^{\pi/2} \cos \sqrt{x}\, dx$; S, $n = 6$

102. $\int_{\pi/6}^{\pi/2} \ln(\sin x)\, dx$; S, $n = 8$

103. How many subintervals are needed in Exercise 97 to obtain a discretization error less than 0.01?

104. How many subintervals are needed in Exercise 98 to obtain a discretization error less than 0.00001?

105. Use Simpson's rule to estimate $\int_1^2 (1/x^2)\, dx$ with an error of less than 0.0001. Compare your answer with the actual answer, which is easily obtained by integration.

106. Answer the questions in Exercise 105 for the integral $\int_1^2 \ln x\, dx$.

9 Further Applications of the Definite Integral

In this chapter we continue the discussion of applications of the integral begun in Chapter 5. The first applications are geometric, and the last are applications in physics. After completing this chapter, you should have a sense of the great usefulness of the integral as a tool in many different branches of science.

9.1 VOLUMES

In this section we show how the definite integral can be used to calculate the volumes of certain solid figures. We begin by deriving the familiar formula for the volume of a right circular cone. Such a cone is depicted in Figure 1. To calculate its volume, we place the cone with its vertex at the origin and its central axis along the y-axis (see Figure 2). The line joining the points $(0, 0)$ and (r, h) has the equation $y = (h/r)x$, and we may think of the cone as the solid obtained by rotating the region bounded by this line, the y-axis, and the line $y = h$ around the y-axis. In this context the cone is called a **solid of revolution.**

We will calculate the volume in several steps, using two different methods.

METHOD 1: DISK METHOD

Step 1: Partition an appropriate interval. We form an equally spaced partition of the interval along the y-axis between $y = 0$ and $y = h$ by using the partition points

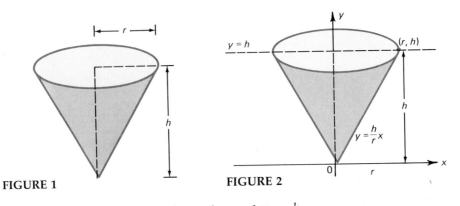

FIGURE 1   **FIGURE 2**

$$0 = y_0 < y_1 < y_2 < \cdots < y_{n-1} < y_n = h,$$

where the width of each subinterval given by $\Delta y = y_i - y_{i-1}$ is assumed to be small. This has the effect of dividing the cone into "slices," or **disks,** parallel to the x-axis (see Figure 3).

FIGURE 3

Step 2: Approximate the volume of each disk.   As drawn in Figure 4, each disk is, for Δy small, very close to a right circular cylinder with radius $x_i = (r/h)y_i$ and height Δy. It is not quite a cylinder because at the top the radius is $(r/h)y_i$, while at the bottom the radius is $(r/h)y_{i-1}$. However, if Δy is small, this difference will be small.

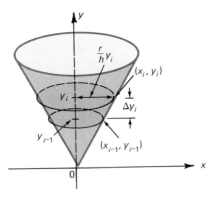

FIGURE 4

The volume of a cylinder with radius R and height H is $\pi R^2 H$. Thus if ΔV_i denotes the volume of the ith disk, we have

$$\Delta V_i \approx \pi x_i^2 \, \Delta y = \pi \left(\frac{r}{h} y_i\right)^2 \Delta y = \frac{\pi r^2}{h^2} y_i^2 \, \Delta y.$$

Step 3: *Add up the volumes of the disks to obtain an approximation for the volume of the cone.* Then

$$V = \Delta V_1 + \Delta V_2 + \cdots + \Delta V_n \approx \frac{\pi r^2}{h^2}(y_1{}^2\,\Delta y + y_2{}^2\,\Delta y + \cdots + y_n{}^2\,\Delta y)$$

$$= \frac{\pi r^2}{h^2}\sum_{i=1}^{n} y_i{}^2\,\Delta y. \tag{1}$$

We emphasize that the expression in (1) is an approximation to the volume. It provides the exact volume of the object drawn in Figure 5.

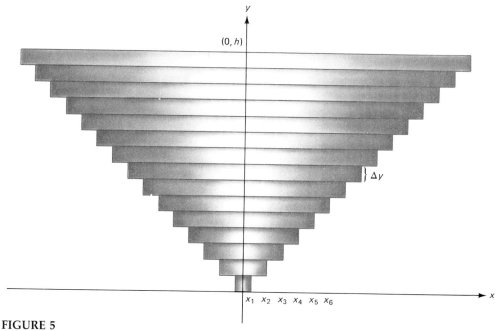

FIGURE 5

Step 4: *Take the limit as* $\Delta y \to 0$. The approximations in (1) get better and better as $\Delta y \to 0$. Thus we have

$$V = \lim_{\Delta y \to 0} \frac{\pi r^2}{h^2}\sum_{i=1}^{n} y_i{}^2\,\Delta y = \frac{\pi r^2}{h^2}\lim_{\Delta y \to 0}\sum_{i=1}^{n} y_i{}^2\,\Delta y$$

$$\underset{\text{Equation (5.5.9)}}{} = \frac{\pi r^2}{h^2}\int_0^h y^2\,dy = \frac{\pi r^2}{h^2}\left.\frac{y^3}{3}\right|_0^h = \frac{1}{3}\pi r^2 h,$$

which is the formula you might have seen earlier for the volume of a cone.

METHOD 2: SHELL METHOD

We now calculate the same volume in a different way.

Step 1: *We form an equally spaced partition of the x-axis between $x = 0$ and $x = r$ by using the partition points*

$$0 = x_0 < x_1 < x_2 < \cdots < x_{n-1} < x_n = r,$$

where the length of each subinterval, given by $\Delta x = x_i - x_{i-1}$, is assumed to be small. The situation is depicted in Figure 6. This has the effect of dividing the cone into what are called **cylindrical shells** (a cylindrical shell has the shape of an empty tin can with the ends removed). A typical cylindrical shell is drawn in Figure 7.

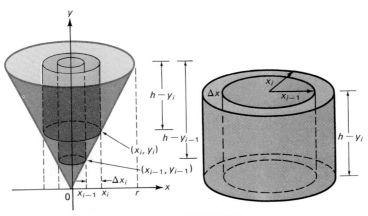

FIGURE 6   **FIGURE 7**

Step 2: *Approximate the volume of each cylindrical shell.* First, we note that a cylindrical shell is an "outer" cylinder with an "inner" cylinder removed. In Figure 6 the "shell" between x_{i-1} and x_i is not quite the same as the shell in Figure 7 because the bottom is not flat. To visualize this, note that the inner cylinder has height $h - y_{i-1}$, and the outer cylinder has height $h - y_i$ (see Figure 8). If Δx is small, then the difference in the volume between the cylindrical shell in Figure 7 and the shell in Figure 8b will be small. Thus we use the volume of the cylindrical shell as an approximation to the volume of the actual shell obtained in the partition.

We have

volume of cylindrical shell =

volume of outer cylinder minus volume of inner cylinder.

The volume of a cylinder is $\pi r^2 h$, where r is its radius and h is its height. From Figure 7 we see that

volume of outer cylinder $= \pi x_i^2 (h - y_i)$

and

volume of inner cylinder $= \pi x_{i-1}^2 (h - y_i)$.

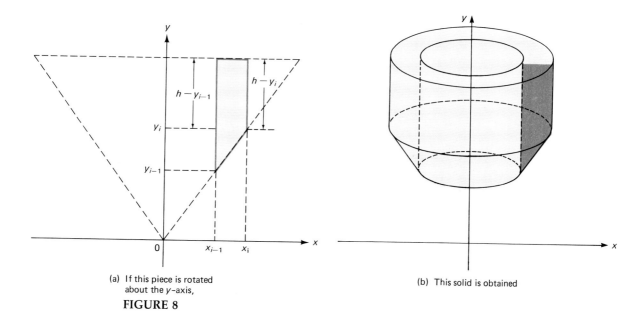

(a) If this piece is rotated about the y-axis,

(b) This solid is obtained

FIGURE 8

Thus if ΔV_i denotes the volume of the ith shell (sketched in Figure 8b), then

$$\Delta V_i \approx \text{volume of } i\text{th cylindrical shell} = \pi[x_i^2(h - y_i) - x_{i-1}^2(h - y_i)]$$
$$= \pi(x_i^2 - x_{i-1}^2)(h - y_i)$$
$$= \pi(x_i + x_{i-1})(x_i - x_{i-1})(h - y_i)$$
$$= \pi(x_i + x_{i-1})(h - y_i)\,\Delta x = 2\pi\left(\frac{x_i + x_{i-1}}{2}\right)(h - y_i)\,\Delta x$$

or

$$\Delta V_i \approx 2\pi\left(\frac{x_i + x_{i-1}}{2}\right)(h - y_i)\,\Delta x. \tag{2}$$

The right side of equation (2) gives the volume of the ith cylindrical shell exactly. Since the average of the numbers x_i and x_{i-1} is $(x_i + x_{i-1})/2$, we have the following result:

volume of a cylindrical shell $= 2\pi$(average radius)(height)(thickness).

Now if Δx is small, then $x_{i-1} = x_i - \Delta x \approx x_i$, so

$$\text{average radius} = \frac{x_i + x_{i-1}}{2} \approx \frac{x_i + x_i}{2} = x_i,$$

and (2) becomes

$$\Delta V_i \approx 2\pi x_i (h - y_i)\, \Delta x. \tag{3}$$

Since $y = (h/r)x$, $y_i = (h/r)x_i$, and we obtain, from (3),

$$\Delta V_i \approx 2\pi x_i \left[h - \frac{h}{r}x_i \right]\Delta x. \tag{4}$$

Step 3: *Add up the volumes of the cylindrical shells to obtain an approximation to the volume of the cone.*

$$V \approx \Delta V_1 + \Delta V_2 + \cdots + \Delta V_n = \sum_{i=1}^{n} \Delta V_i$$

$$\approx \sum_{i=1}^{n} 2\pi x_i \left(h - \frac{h}{r}x_i \right)\Delta x.$$

Step 4: *Take the limit as $\Delta x \to 0$.*

$$V = \lim_{\Delta x \to 0} \sum_{i=1}^{n} 2\pi x_i \left(h - \frac{h}{r}x_i \right)\Delta x$$

$$= \int_0^r 2\pi x\left(h - \frac{h}{r}x \right)dx = 2\pi h \int_0^r \left(x - \frac{x^2}{r} \right) dx$$

$$= 2\pi h \left(\frac{x^2}{2} - \frac{x^3}{3r} \right)\Big|_0^r = 2\pi h \left(\frac{r^2}{2} - \frac{r^2}{3} \right) = 2\pi h \frac{r^2}{6} = \frac{1}{3}\pi r^2 h,$$

as before.

The two methods used above can be applied in a wide variety of problems. When the curve $y = f(x)$ is revolved about an axis, we obtain a **solid of revolution** whose volume can be obtained by the **disk method** or the **cylindrical shell method**. Before giving additional examples, however, we summarize the procedure that we used here. We will use this procedure in every application in this chapter. We call our procedure the **integration process.**

THE INTEGRATION PROCESS FOR APPLICATIONS

To obtain a quantity Q, follow these steps:

(i) Break the quantity Q, defined in the interval $[a, b]$, into a number of small pieces, denoted ΔQ_i, so that

$$Q = \sum_{i=1}^{n} \Delta Q_i.$$

Do so by partitioning either the x-axis or the y-axis.

(ii) Select an arbitrary piece and work the problem on that piece, obtaining an approximate value: $\Delta Q_i \approx q(x_i{}^*)\, \Delta x.$

(iii) Add the results for all the pieces:

$$Q \approx \sum_{i=1}^{n} q(x_i{}^*) \, \Delta x.$$

(iv) Take the limit of this sum to obtain a definite integral:

$$Q = \lim_{\Delta x \to 0} \sum_{i=1}^{n} q(x_i{}^*) \, \Delta x = \lim_{n \to \infty} \sum_{i=1}^{n} q(x_i{}^*) \, \Delta x.$$

(v) Evaluate the definite integral to obtain the answer: $Q = \int_a^b q(x) \, dx$.

As we solve more and more problems by using the integration process for applications, we will obtain a number of formulas. But you should keep in mind that *it is the process that is important, not the formulas.* Once you have mastered the process, you will be able to solve an astonishingly wide variety of problems that once seemed very difficult or even impossible.

EXAMPLE 1   The region bounded by the curve $y = x^2$, the x-axis, and the line $x = 2$ is revolved about the x-axis. Calculate the volume of the solid generated.

Solution. The solid is depicted in Figure 9. By partitioning the x-axis, we obtain disks. The volume of a typical disk is given by

$$\Delta V_i \approx \pi r^2 h = \pi y_i{}^2 \, \Delta x = \pi x_i{}^4 \, \Delta x,$$

so that

$$V = \int_0^2 \pi x^4 \, dx = \frac{\pi x^5}{5} \Big|_0^2 = \frac{32\pi}{5}.$$

Using cylindrical shells, we have, using Figure 10 and equation (3),

$$\Delta V_i \approx 2\pi y_i (2 - x_i) \, \Delta y = 2\pi y_i (2 - \sqrt{y_i}) \, \Delta y,$$

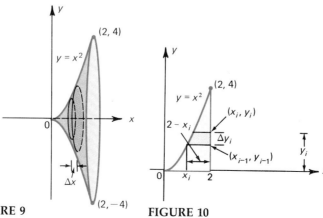

FIGURE 9   **FIGURE 10**

so that

$$V = \int_0^4 2\pi(2 - \sqrt{y})y \, dy = 2\pi \int_0^4 (2y - y^{3/2}) \, dy = 2\pi\left(y^2 - \frac{2}{5}y^{5/2}\right)\Big|_0^4$$

$$= 2\pi\left(16 - \frac{64}{5}\right) = \frac{32\pi}{5}. \quad \blacksquare$$

It is, of course, not necessary to use both methods to calculate one volume. Typically, one method will be easier to use than the other in a given problem, and deciding which method to use will be a matter of judgment. The calculation of a given volume follows fairly easily once one of the axes has been partitioned and the volume of a typical disk or shell calculated, provided that the function involved is easy to integrate. We give two definitions that are used in the calculation of volumes.

Definition 1 VOLUME BY DISK METHOD   Let f be continuous and nonnegative on $[a, b]$. (See Figure 11a.) Then the **volume** of the solid generated by revolving the area below the graph of f about the x-axis between $x = a$ and $x = b$ (see Figure 11b) is defined by

$$V = \int_a^b \pi[f(x)]^2 \, dx. \tag{5}$$

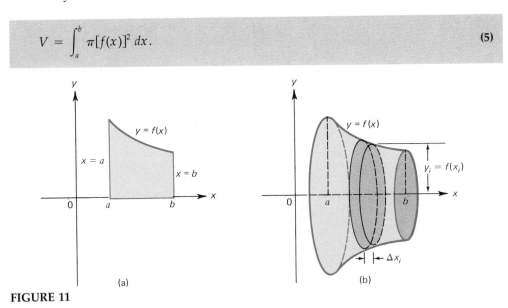

(a)

(b)

FIGURE 11

Definition 2 VOLUME BY CYLINDRICAL SHELL METHOD   If the area in Figure 11(a) is revolved about the y-axis (see Figure 12), then the **volume** of the solid generated is defined by

$$V = \int_a^b 2\pi x f(x) \, dx. \tag{6}$$

WARNING: Formulas (5) and (6) are given as guidelines. They are useless if you forget how they were obtained. The important thing is the integration process, not a formula that comes as a result of using that process.

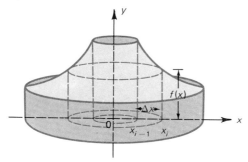

FIGURE 12

EXAMPLE 2 Calculate the volume of a sphere of radius r.

Solution. It is not difficult to see that we can generate the sphere of radius r by rotating the semicircle and the points interior to it pictured in Figure 13 around the x-axis. Since $x^2 + y^2 = r^2$ is the equation of the circle of radius r centered at the origin, the equation of the semicircle in the first and second quadrants is given by $y = \sqrt{r^2 - x^2}$. Then by formula (5)

$$V = \pi \int_{-r}^{r} (\sqrt{r^2 - x^2})^2 \, dx = \pi \int_{-r}^{r} (r^2 - x^2) \, dx = \pi\left(r^2 x - \frac{x^3}{3}\right)\Big|_{-r}^{r} = \frac{4}{3}\pi r^3.$$

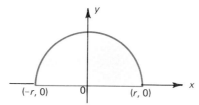

FIGURE 13

The same answer is obtained by rotating the semicircle in the first and fourth quadrants around the y-axis and using formula (6). ■

EXAMPLE 3 The circular disk $(x - 2)^2 + y^2 \leq 1$ is rotated around the y-axis. Find the volume of the "doughnut" shaped region generated. (This solid is called a *torus.*)

Solution. The circular disk $(x - 2)^2 + y^2 \leq 1$ consists of the circle centered at $(2, 0)$ with radius 1 and all points interior to it (see Figure 14a). In the upper part of the circle (i.e., in the first quadrant) $y = \sqrt{1 - (x - 2)^2}$, while in the lower part of the circle y is negative and is equal to $-\sqrt{1 - (x - 2)^2}$. Thus the height of a typical cylindrical shell is equal to

$$\sqrt{1 - (x_i - 2)^2} - [-\sqrt{1 - (x_i - 2)^2}] = 2\sqrt{1 - (x_i - 2)^2}.$$

Thus

$$\Delta V_i \approx 2\pi x_i [2\sqrt{1 - (x_i - 2)^2}] \, \Delta x$$

and

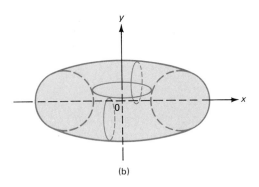

(a)

(b)

FIGURE 14

$$V \approx \sum_{i=1}^{n} 2\pi x_i [2\sqrt{1 - (x_i - 2)^2}]\, \Delta x.$$

So taking the limit as $\Delta x \to 0$, we have

$$V = \lim_{\Delta x \to 0} \sum_{i=1}^{n} 2\pi x_i [2\sqrt{1 - (x_i - 2)^2}]\, \Delta x$$

and

$$V = \int_{1}^{3} 2\pi x [2\sqrt{1 - (x - 2)^2}]\, dx = 4\pi \int_{1}^{3} x\sqrt{1 - (x - 2)^2}\, dx.$$

To integrate this, we make the substitution $x - 2 = \sin \theta$. Then

$$dx = \cos \theta \, d\theta, \qquad x = 2 + \sin \theta,$$

and

$$\sqrt{1 - (x - 2)^2} = \sqrt{1 - \sin^2 \theta} = \cos \theta.$$

When $x = 1$, $\sin \theta = 1 - 2 = -1$ and $\theta = -\pi/2$. When $x = 3$, $\sin \theta = 1$ and $\theta = \pi/2$. Thus

$$V = 4\pi \int_{-\pi/2}^{\pi/2} (2 + \sin \theta) \cos^2 \theta \, d\theta$$

$$= 8\pi \int_{-\pi/2}^{\pi/2} \cos^2 \theta \, d\theta + 4\pi \int_{-\pi/2}^{\pi/2} \cos^2 \theta \sin \theta \, d\theta$$

$$= 4\pi \int_{-\pi/2}^{\pi/2} (1 + \cos 2\theta) \, d\theta - \frac{4\pi}{3} \cos^3 \theta \Big|_{-\pi/2}^{\pi/2}$$

$$= 4\pi \left(\theta + \frac{\sin 2\theta}{2} \right) \Big|_{-\pi/2}^{\pi/2} - 0 = 4\pi^2. \ \blacksquare$$

EXAMPLE 4   Find the volume generated when the area under one loop of the curve $y = \sin x$ between $x = 0$ and $x = \pi$ is rotated about (a) the x-axis and (b) the y-axis.

Solution. **(a)**   The area to be rotated is drawn in Figure 15a. From formula (5) the volume of the solid drawn in Figure 15b is given by

$$V = \pi \int_0^\pi \sin^2 x \, dx = \frac{\pi}{2} \int_0^\pi (1 - \cos 2x) \, dx = \frac{\pi}{2}\left(x - \frac{\sin 2x}{2}\right)\Big|_0^\pi = \frac{\pi^2}{2}.$$

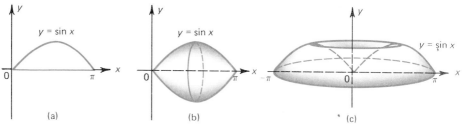

(a)   (b)   (c)

FIGURE 15

(b)   From formula (6) the volume of the solid drawn in Figure 15c is given by

$$V = 2\pi \int_0^\pi x \sin x \, dx.$$

We can integrate this expression by parts. Let $u = x$ and $dv = \sin x \, dx$. Then $du = dx$, $v = -\cos x$, and

$$V = 2\pi\left(-x \cos x\Big|_0^\pi + \int_0^\pi \cos x \, dx\right) = 2\pi\left(\pi + \sin x\Big|_0^\pi\right) = 2\pi^2. \ \blacksquare$$

EXAMPLE 5   The area bounded by the curve $y = x^2 + 2$ and the line $y = x + 8$ is rotated around the x-axis. Find the volume of the solid generated.

Solution. The area is depicted in Figure 16a, and the solid generated is depicted in Figure 16c. The points of intersection of the two curves are easily found to be $(-2, 6)$ and $(3, 11)$. When the strip depicted in Figure 16a is rotated around the x-axis, it generates a circular ring, as in Figure 16b. The volume of this ring is the volume of the large disk (radius $x_i + 8$) minus the volume of the smaller disk (radius $x_i^2 + 2$). Thus

$$\Delta V_i = \pi(x_i + 8)^2 \, \Delta x - \pi(x_i^2 + 2)^2 \, \Delta x,$$

and the total volume is given by

$$V = \pi \int_{-2}^3 [(x + 8)^2 - (x^2 + 2)^2] \, dx$$

$$= \pi \int_{-2}^3 (x^2 + 16x + 64 - x^4 - 4x^2 - 4) \, dx$$

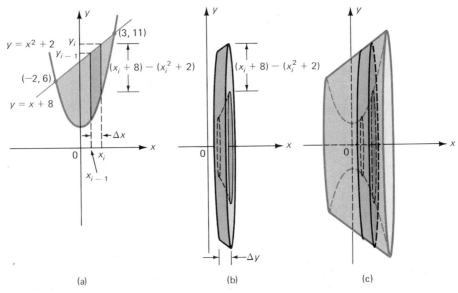

FIGURE 16

$$= \pi \int_{-2}^{3} (-x^4 - 3x^2 + 16x + 60)\, dx$$

$$= \pi \left(-\frac{x^5}{5} - x^3 + 8x^2 + 60x \right) \Big|_{-2}^{3} = 250\pi. \quad \blacksquare$$

The above example can be generalized to obtain the following definition (see Figure 17).

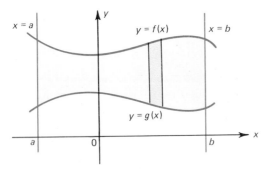

FIGURE 17

Definition 3   Let f and g be continuous and nonnegative on $[a, b]$ with $f(x) \geq g(x)$ ≥ 0 for every x in $[a, b]$. Then the **volume** generated by rotating the region bounded by the curves and lines $y = f(x)$, $y = g(x)$, $x = a$, and $x = b$ around the x-axis is defined by

$$V = \pi \int_{a}^{b} \{ [f(x)]^2 - [g(x)]^2 \}\, dx . \tag{7}$$

REMARK.   In (7) note that we have the difference of squares, $f^2 - g^2$, *not* the square of the difference, $(f - g)^2$.

The techniques of this section can also be used to calculate the volumes of certain solids in which parallel cross sections all have the same basic shape. By a **cross section** we mean the figure (in two dimensions) obtained by slicing the solid with a plane. By slicing the figure with parallel planes, we obtain a number of disks. (Think of cutting an egg into many thin, parallel slices. Each cross section has the shape of an oval.) Each "slice" has a volume approximately equal to the area of the cross section times its thickness. To calculate the volume of the solid, we simply "add up" the volumes of the separate slices by the now familiar process of integration.

EXAMPLE 6   The base of a certain solid is the circular disk $x^2 + y^2 \le 4$ in the xy-plane. Each plane perpendicular to the x-axis cuts the solid in an equilateral triangle. Find the volume of the solid.

Solution.   The solid is drawn in Figure 18a. If the solid is sliced along the x-axis, the volume of a typical slice is $A_i \, \Delta x$, where A_i is the area of an equilateral triangle.

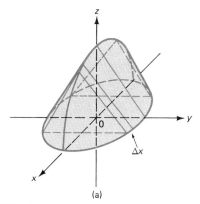

(a)

FIGURE 18

One side of the triangle has length

$$\sqrt{4 - x_i^2} - (-\sqrt{4 - x_i^2}) = 2\sqrt{4 - x_i^2}$$

(according to Figure 18b). Using Figure 19, we easily calculate that the area of an equilateral triangle with side s is given by $A = (\sqrt{3}/4)s^2$. Thus the area A_i above is

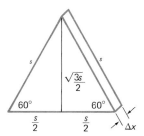

FIGURE 19

given by

$$A_i = \frac{\sqrt{3}}{4}(4)(4 - x_i^2) = \sqrt{3}(4 - x_i^2), \quad \text{and} \quad \Delta V_i \approx \sqrt{3}(4 - x_i^2)\, \Delta x,$$

so that

$$V = \sqrt{3} \int_{-2}^{2} (4 - x^2)\, dx = \sqrt{3}\left(4x - \frac{x^3}{3}\right)\Big|_{-2}^{2} = \frac{32\sqrt{3}}{3}. \quad \blacksquare$$

PROBLEMS 9.1

In Problems 1–20, find the volume generated when the region bounded by the given curves and lines is rotated about the x-axis.

1. $y = x$, $x = 2$, $y = 0$
2. $y = 2x + 3$, $x = 1$, $x = 4$, $y = 0$
3. $y = 2x^2$, $x = 1$, $y = 0$
4. $y = \sqrt{x + 1}$, $x = 1$, $x = 5$, $y = 0$
5. $y = 4 - x$, $x = 0$, $y = 0$
6. $y = x^3$, $x = 2$, $y = 1$
7. $y = \cos x$, $x = \pi/2$, $y = 0$, $x = 0$
8. $y = \tan x$, $x = \pi/4$, $y = 0$
9. $y = e^x$, $x = -1$, $x = 1$, $y = 0$
10. $y = \sqrt{1 - x^3}$, $x = 0$, $y = 0$
11. $y = (1 - x^2)^{1/4}$, $x = 0$, $y = 0$
*12. $x = y - y^2$, $x = 0$
13. $y = \ln x$, $x = 1$, $x = e$, $y = 0$
14. $y = x^2$, $y = x^3$
15. $y = \sin x$, $y = \cos x$, $0 \le x \le \pi/4$
16. $y = \sec x$, $x = 0$, $x = \pi/3$, $y = 0$
17. $y = xe^x$, $x = 1$, $y = 0$
18. $y = \sin^{-1} x$, $x = 1$, $y = 0$

19. $y = \dfrac{1}{x}$, $x = 1$, $x = 2$, $y = 0$
20. $y = \sinh^{1/2} x$, $x = 1$, $y = 0$

In Problems 21–32, find the volumes generated when the region bounded by the given curves and lines is rotated about the y-axis.

21. $x + y = 3$, $x = 0$, $y = 0$
22. $y = \frac{1}{2}x^2$, $x = 1$, $y = 0$
23. $y = x$, $y = 1 - x$, $x = 0$
24. $y = x^{1/3}$, $x = 1$, $y = 0$
25. $y = \ln x$, $x = 1$, $x = 3$, $y = 0$
26. $y = \cos x$, $x = \pi/2$, $y = 0$
27. $y = \sqrt{x}$, $y = x^2$
28. $y = x^2$, $y = x^3$
29. $y = \sin x$, $y = \cos x$, $0 \le x \le \pi/4$
30. $x = y^2 + 1$, $x = 2$
31. $y = e^{x^2}$, $x = 0$, $x = 1$, $y = 0$
32. $y = \cos(x^2)$, $x = 0$, $x = \sqrt{\pi/6}$, $y = 0$

33. Calculate the volume of the sphere of radius r by rotating the interior of the semicircle $x = \sqrt{r^2 - y^2}$ about the y-axis.

34. The circle centered at $(a, 0)$ with radius r $(r < a)$ is rotated about the y-axis. Show that the volume of the torus so generated is equal to $2\pi^2 ar^2$.

35. The equation of an ellipse centered at the origin is given by

$$\frac{x^2}{a^2} + \frac{y^2}{b^2} = 1.$$

Calculate the volume of the "football" shaped solid generated by rotating this ellipse about the x-axis.

36. Find the volume generated by rotating the triangle with vertices at $(1, 1)$, $(2, 3)$, and $(3, 2)$ about the x-axis.

37. Find the volume generated if the triangle of Problem 36 is rotated about the y-axis.
38. What is the volume generated when the area bounded by the line $y = x$ and the curve $y = x^2$ is rotated about the line $x = 2$? About the line $y = 2$?
**39. Let (a_1, b_1), (a_2, b_2), and (a_3, b_3) be three points in the first quadrant. Calculate the volume generated when the triangle with vertices at these points is rotated about the x-axis.
40. Calculate the volume generated by rotating the area bounded by $y = \sqrt{x}$, $x = 1$, and $y = 0$ about the line $x = 4$.
41. Find the volume generated by rotating the area bounded by $y = \sin x$, $x = 0$, $x = \pi/2$, and $y = 0$ about the line $y = 2$.
*42. A water tank in the shape of a sphere has a radius of 10 m. What is the mass of water in the tank if the tank is filled to a depth of 2 m? [*Hint:* The density of water is approximately 1000 kg/m^3.]
43. The base of a certain solid is a circle with radius 2, while each cross section perpendicular to the base of the solid is a square. Find the volume of the solid.
44. Find the volume of the solid in Problem 43 if perpendicular cross sections are isosceles triangles with height 1.
45. Find the volume of the solid in Problem 43 if perpendicular cross sections are semicircles.
*46. A right circular cylinder of radius 2 m is cut by two planes. The first is parallel to the base of the cylinder, while the second makes an angle of 45° with the base. The two planes intersect at a diameter of the circular cross section of the cylinder, thereby cutting out a wedge. What is the volume of this wedge? [*Hint:* Look at Figure 20. The volume can be calculated two ways: by taking (a) cross sections parallel to the base or (b) cross sections perpendicular to it.]

FIGURE 20

47. Calculate the volume of the wedge in Problem 46 if the second plane intersects the first at an angle of 60°.
48. Calculate the volume of the wedge in Problem 46 if the second plane intersects the first at an angle of α degrees.
49. The base of a certain solid is the region between the line $y = x$ and the curve $y = x^2$. Find the volume of the solid if cross sections perpendicular to the x-axis are:
    (a) Squares.
    (b) Isosceles right triangles with hypotenuses in the xy-plane.
    (c) Semicircles.
*50. A water tank is in the shape of the solid generated by rotating the region bounded by $y = x^2$, $x = 0$, and $y = 4$ around the y-axis. The tank is filled with water. Find the work needed to pump all the water to the top of the tank. (Assume that distance is measured in meters.)
*51. A storage tank filled with olive oil is in the shape of the solid generated by rotating the region bounded by $y = \ln x$, $y = 0$, $y = 3$, and $x = 0$ about the y-axis. Find the work needed to pump the oil to the top of the tank. (The density of olive oil is 920 kg/m^3.)

9.2 ARC LENGTH

In this section we show how to calculate the length of a curve in the xy-plane (called a **plane curve**). A typical curve is sketched in Figure 1. The problem is to calculate the length of the curve from $x = a$ to $x = b$. Such a problem arises in many interesting physical examples. For example, if an object is thrown or shot into the air, then its path as a function of time is a parabola opening downward. To calculate the length of the path traveled by the object, one must find the length of the parabola from some initial time t_0 to a final time t_f.

To find the length of the curve in Figure 1, we go through our integration process. We begin by dividing the curve into small pieces. We do so by partitioning the interval $[a, b]$ into n subintervals of equal length. We then find the approximate length Δs_i of the arc in each subinterval, and add up these lengths. (This technique is, by now, very familiar.) The appearance of the curve near a typical subinterval is sketched in Figure 2. If Δx is small, and if $f(x)$ is continuous in the interval $[x_{i-1}, x_i]$, then the length of the curve between x_{i-1} and x_i, which we will denote by Δs_i, can be approximated by the straight-line segment joining the points $(x_{i-1}, f(x_{i-1}))$ and $(x_i, f(x_i))$. The length of this line segment is, by the Pythagorean theorem, given by

$$\sqrt{\Delta x^2 + \Delta y_i{}^2}$$

Thus

$$\Delta s_i \approx \sqrt{\Delta x^2 + \Delta y_i{}^2}. \tag{1}$$

FIGURE 1   **FIGURE 2**

Now $\Delta y_i = f(x_i) - f(x_{i-1})$. If f is continuous on $[a, b]$ and differentiable on (a, b), then f is continuous in the smaller interval $[x_{i-1}, x_i]$ and differentiable on (x_{i-1}, x_i). By the mean value theorem (Theorem 4.2.2), there is a number x_i^* in (x_{i-1}, x_i) such that

$$\Delta y_i = f(x_i) - f(x_{i-1}) = f'(x_i^*)(x_i - x_{i-1}) = f'(x_i^*)\, \Delta x. \tag{2}$$

Then inserting (2) into (1), we have

$$\Delta s_i \approx \sqrt{\Delta x^2 + [f'(x_i^*)]^2\, \Delta x^2}$$
$$= \sqrt{\Delta x^2 (1 + [f'(x_i^*)]^2)} = \sqrt{1 + [f'(x_i^*)]^2}\, \Delta x,$$

so that

$$s = \sum_{i=1}^{n} \Delta s_i \approx \sum_{i=1}^{n} \sqrt{1 + [f'(x_i{}^*)]^2}\, \Delta x.$$

If f' is also continuous in $[a, b]$, then $(f')^2$ is continuous in $[a, b]$, and $\sqrt{1 + [f'(x)]^2}$ is continuous in $[a, b]$. Thus $\int_a^b \sqrt{1 + [f'(x)]^2}\, dx$ exists. But

$$\int_a^b \sqrt{1 + [f'(x)]^2}\, dx = \lim_{\Delta x \to 0} \sum_{i=1}^{n} \sqrt{1 + [f'(x_i{}^*)]^2}\, \Delta x.$$

This derivation is used as motivation for the following definition.

Definition 1 LENGTH OF A CURVE (ARC LENGTH) Suppose that f and f' are continuous in $[a, b]$. Then the **length of the curve (arc length)** $f(x)$ between $x = a$ and $x = b$ is defined by

$$s = \int_a^b \sqrt{1 + [f'(x)]^2}\, dx. \tag{3}$$

This integral is often written as $\int_0^s ds$, where $ds = \sqrt{1 + [f'(x)]^2}\, dx$ denotes the **differential of arc length.**

EXAMPLE 1 Calculate the length of the straight line $y = x$ between $x = 0$ and $x = 1$.

Solution. We easily see that this distance is simply the straight-line distance between $(0, 0)$ and $(1, 1)$ and is equal to $\sqrt{2}$. However, we will also calculate it by using formula (3), mainly to illustrate for ourselves the formula's validity. Since $f'(x) = 1$, we have

$$s = \int_0^1 \sqrt{1 + 1^2}\, dx = \int_0^1 \sqrt{2}\, dx = \sqrt{2}. \ \blacksquare$$

EXAMPLE 2 Calculate the length of the arc of the curve $y = x^2$ between $x = 0$ and $x = 2$.

Solution. Since $f'(x) = 2x$, $[f'(x)]^2 = 4x^2$, and

$$s = \int_0^2 \sqrt{1 + 4x^2}\, dx = 2 \int_0^2 \sqrt{x^2 + \tfrac{1}{4}}\, dx.$$

Let $x = \tfrac{1}{2} \tan \theta$. Then

$$s = 2 \int_0^{\tan^{-1}(4)} \sqrt{\tfrac{1}{4} \tan^2 \theta + \tfrac{1}{4}} \cdot \tfrac{1}{2} \sec^2 \theta\, d\theta = \tfrac{1}{2} \int_0^{\tan^{-1}(4)} \sec^3 \theta\, d\theta$$

$$= \tfrac{1}{4}(\ln |\sec \theta + \tan \theta| + \sec \theta \tan \theta)\Big|_0^{\tan^{-1}(4)} \qquad \text{(from Example 8.2.8).} \tag{4}$$

When $\theta = \tan^{-1} 4$, $\tan \theta = 4$ and $\sec \theta = \sqrt{17}$ (since $1 + \tan^2 \theta = \sec^2 \theta$, we have $1 + 4^2 = \sec^2 \theta$). Thus from (4) we find that $s = \tfrac{1}{4}[\ln(\sqrt{17} + 4) + 4\sqrt{17}]$. $\blacksquare$

EXAMPLE 3 Calculate the length of the arc of the curve $y = 3 + x^{2/3}$ between $x = 1$ and $x = 8$.

Solution. $f'(x) = (2/3)x^{-1/3}$, so that

$$s = \int_1^8 \sqrt{1 + \frac{4}{9x^{2/3}}}\, dx = \int_1^8 \sqrt{\frac{4 + 9x^{2/3}}{9x^{2/3}}}\, dx = \int_1^8 \frac{\sqrt{4 + 9x^{2/3}}}{3x^{1/3}}\, dx.$$

Let $u = 4 + 9x^{2/3}$. Then $du = (9)(\frac{2}{3})x^{-1/3}\, dx = (6/x^{1/3})\, dx$. When $x = 1$, $u = 13$, and when $x = 8$, $u = 40$. Then, since $\frac{1}{3} = (\frac{1}{18})(6)$, we have

$$s = \frac{1}{18}\int_1^8 \sqrt{4 + 9x^{2/3}}\,\frac{6}{x^{1/3}}\, dx = \frac{1}{18}\int_{13}^{40} \sqrt{u}\, du = \left(\frac{1}{18}\right)\left(\frac{2}{3}\right) u^{3/2}\Big|_{13}^{40}$$

$$= \frac{1}{27}(40^{3/2} - 13^{3/2}).\ \blacksquare$$

EXAMPLE 4 Calculate the length of the loop of the sine function between $x = 0$ and $x = \pi/2$.

Solution. The arc whose length is requested is sketched in Figure 3. Since $f'(x) = \cos x$, we have

$$s = \int_0^{\pi/2} \sqrt{1 + \cos^2 x}\, dx.$$

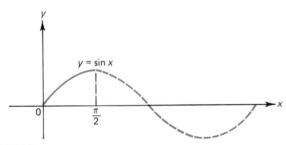

FIGURE 3

Unfortunately, there is no way to find an antiderivative for $\sqrt{1 + \cos^2 x}$. We can estimate this integral by using either the trapezoidal rule or Simpson's rule. To two decimal places, the value of the integral is 1.93. $\blacksquare$

We close this section by noting that the calculation of arc length is usually more difficult than the calculation of volume because it is often difficult or impossible to find an antiderivative for $\sqrt{1 + [f'(x)]^2}$. This fact illustrates one of the many reasons why numerical techniques are so important.

PROBLEMS 9.2

In Problems 1–10, calculate the length of the arc of the curve $y = f(x)$ over the given interval.

1. $y = 3x + 4;\ [1, 5]$

2. $y = 2 + \frac{1}{2}x^2;\ [0, 1]$

3. $y = x^{3/2};\ [0, 1]$

4. $y = \frac{1}{3}(x^2 + 2)^{3/2};\ [0, 3]$

5. $y = \frac{1}{6}\left(x^3 + \frac{3}{x}\right)$; [1, 3] $\qquad$ **6.** $y = \frac{1}{3}(x^{3/2} - 3\sqrt{x})$; [1, 4]

7. $y = \frac{x^4}{4} + \frac{1}{8x^2}$; [1, 2] $\qquad$ ***8.** $y = 2\ln(1 + x)$; [0, 1]

9. $y = e^x$; [1, 2] [*Hint:* Use the substitution $1 + e^{2x} = u^2$.]
10. $y = \cosh x$; [−1, 1] [*Hint:* Write $\cosh x$ in exponential form.]

*11. (a) Using the technique of this section, show that the circumference of the unit circle $x^2 + y^2 = 1$ is 2π. [*Hint:* Show that the length of the arc of the circle in the first quadrant between $x = 0$ and $x = 1/\sqrt{2}$ is $\pi/4$.]
  (b) Explain why this calculation cannot be used to find the length of the circumference because it *makes use* of the fact that the length of the circumference of a circle of radius r is $2\pi r$.
*12. (a) Show that the circumference of the circle $(x - a)^2 + (y - b)^2 = r^2$ is $2\pi r$.
  (b) Answer part (b) of Problem 11.
*13. Find the length of the curve $x^{2/3} + y^{2/3} = 1$. [*Hint:* Write y as a function of x and determine the values of x for which the function is defined.]
14. Find the length of the arc of $y = e^{-x}$ between $x = 0$ and $x = 1$.
15. Find the length of the arc of the curve $(y + 1)^2 = 4x^3$ between $x = 0$ and $x = 1$.

In Problems 16–24, find a definite integral that is equal to the length of the arc of the given curve over the given interval. Do not try to evaluate the integral.

16. $y = \cos x$; [0, $\pi/2$] $\qquad$ **17.** $y^2 = (1 + x^2)^3$, $y \geq 0$; [0, 1]
18. $y = \tan x$; [0, $\pi/4$] $\qquad$ **19.** $y = x \sinh x$; [0, 1]
20. $y = \sin^{-1} x$; [0, 1] $\qquad$ **21.** $y = e^{x^2}$; [0, 4]
22. $y = e^x \sin x$; [$\pi/6$, π] $\qquad$ **23.** $y = \cosh(1 + x^2)$; [−1, 1]
24. $y = \tan^{-1} e^x$; [3, 6]

9.3 SURFACE AREA

In Section 9.1 we discussed the notion of a solid of revolution obtained by rotating a curve around a given line. In this section we show how to calculate the lateral surface area of such a solid. The technique we use is very similar to the one we used in our calculation of arc length in Section 9.2.

Suppose that the curve in Figure 1a is rotated around the x-axis. We again partition the interval [a, b]. When we rotate the "subarc" of the curve $f(x)$ in the interval [x_{i-1}, x_i] around the x-axis, we obtain a thin surface like the one depicted in Figure 1b. The idea is, as before, to estimate the surface area of this thin surface, add up corresponding areas, and then take the limit as $\Delta x \to 0$. When we do so we will have the *lateral surface area* of the surface. This surface area excludes the circular areas at the ends. You can think of the lateral surface area as the area of the sides of a barrel. The areas of the top and bottom of the barrel are excluded.

Suppose f is a constant function. That is, $f(x) \equiv c$. Then the curve $y = f(x) = c$ is a horizontal straight line. The area of the surface of the strip in Figure 1b would be $2\pi c \, \Delta x$ since the surface is a right circular cylinder with radius c and width Δx.

If $f(x)$ is not constant, then the surface is no longer a cylinder. However, if Δx is small, it has, approximately, the shape of a cylinder. The "width" of this surface is

(a) (b)

FIGURE 1

now Δs_i, the length of the curve $f(x)$ between x_{i-1} and x_i. Its radius varies.

We assume, as in Section 9.2, that f is continuous and has a continuous derivative on $[a, b]$. From Theorem 2.7.4 there are numbers m and M and numbers $\bar{x}_1$ and $\bar{x}_2$ in $[a, b]$ such that

$$f(\bar{x}_1) = m \leq f(x) \leq M = f(\bar{x}_2)$$

on $[x_{i-1}, x_i]$. This means that the radius of the cylindrical region in Figure 1b varies between the values m and M. Thus if ΔS_i denotes the area of the part of the surface between x_{i-1} and x_i, we have

$$2\pi m \, \Delta s_i \leq \Delta S_i \leq 2\pi M \, \Delta s_i,$$

or

$$m \leq \frac{\Delta S_i}{2\pi \, \Delta s_i} \leq M.$$

We see that $\Delta S_i/(2\pi \, \Delta s_i)$ is a number between m and M. Since f is continuous on $[a, b]$, it takes on every value between m and M according to the intermediate value theorem (Theorem 2.7.5). Thus there is a number x_i^{**} in (x_{i-1}, x_i) such that

$$\frac{\Delta S_i}{2\pi \, \Delta s_i} = f(x_i^{**}),$$

or

$$\Delta S_i = 2\pi f(x_i^{**}) \, \Delta s_i. \tag{1}$$

Also, from Section 9.2 there is a number x_i^* in (x_{i-1}, x_i) such that

$$\Delta s_i \approx \sqrt{1 + [f'(x_i^*)]^2} \, \Delta x. \tag{2}$$

Suppose that Δx is very small. Then since both x_i^{**} and x_i^* are in (x_{i-1}, x_i),

$x_i^{**} \approx x_i^{*}$, and by the continuity of f,

$$f(x_i^{**}) \approx f(x_i^{*}).\qquad(3)$$

Then, combining (1), (2), and (3), we obtain

$$\Delta S_i \approx 2\pi f(x_i^{*})\sqrt{1 + [f'(x_i^{*})]^2}\,\Delta x,$$

and

$$S = \sum_{i=1}^{n} \Delta S_i \approx \sum_{i=1}^{n} 2\pi f(x_i^{*})\sqrt{1 + [f'(x_i^{*})]^2}\,\Delta x.$$

Thus

$$S = \lim_{\Delta x \to 0} \sum_{i=1}^{n} 2\pi f(x_i^{*})\sqrt{1 + [f'(x_i^{*})]^2}\,\Delta x$$

This result suggests the following definition.

Definition 1 LATERAL SURFACE AREA (ROTATION AROUND x-AXIS) Suppose that $f \geq 0$ on $[a, b]$ and f' is continuous on $[a, b]$. Then

$$\textbf{lateral surface area} = S = \int_a^b 2\pi f(x)\sqrt{1 + [f'(x)]^2}\,dx.\qquad(4)$$

EXAMPLE 1 Calculate the lateral surface area of a right circular cone with radius r and height h.

Solution. We place the cone so that its vertex is at the origin and its central axis is along the x-axis (see Figure 2). We see that the cone is generated by revolving the region under the line $y = (r/h)x$ about the x-axis. Then from formula (4)

$$S = \int_0^h 2\pi\left(\frac{r}{h}x\right)\sqrt{1 + \left(\frac{r}{h}\right)^2}\,dx = 2\pi\frac{r}{h}\sqrt{1 + \left(\frac{r}{h}\right)^2}\int_0^h x\,dx$$

$$= 2\pi\frac{r}{h}\sqrt{\frac{h^2 + r^2}{h^2}} \cdot \frac{h^2}{2} = \pi r\sqrt{r^2 + h^2}. \qquad\blacksquare$$

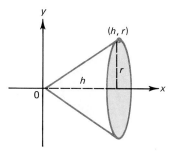

FIGURE 2

EXAMPLE 2   The region under the curve $y = \frac{1}{2}x^2$ between $x = 0$ and $x = 1$ is rotated about the x-axis. Find the lateral surface area of the solid generated.

Solution. The solid is pictured in Figure 3. Since $f(x) = \frac{1}{2}x^2$, $f'(x) = x$, and

$$S = \int_0^1 (2\pi)(\tfrac{1}{2}x^2)\sqrt{1 + x^2}\, dx.$$

Let $x = \tan\theta$. Then

$$S = \pi \int_0^{\pi/4} \tan^2\theta \sec^3\theta\, d\theta.$$

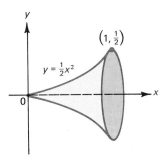

FIGURE 3

We can complete the integration by using the methods of Section 8.3. When we do so, we obtain

$$S = \pi\left(-\frac{\ln|\sec\theta + \tan\theta|}{8} - \frac{\sec\theta\tan\theta}{8} + \frac{\sec^3\theta\tan\theta}{4}\right)\Bigg|_0^{\pi/4}$$

$$= \pi\left[-\frac{\ln(\sqrt{2} + 1)}{8} + \frac{3\sqrt{2}}{8}\right]. \quad\blacksquare$$

Suppose that the area between curve $x = g(y)$ and the y-axis is rotated about the y-axis between $y = c$ and $y = d$. Then we have the following definition.

Definition 2   LATERAL SURFACE AREA (ROTATION AROUND y-AXIS)   Suppose that $g \ge 0$ on $[c, d]$ and g' is continuous on $[c, d]$. Then

$$\textbf{lateral surface area} = S = \int_c^d 2\pi g(y)\sqrt{1 + [g'(y)]^2}\, dy. \tag{5}$$

EXAMPLE 3   Calculate the lateral surface area of the solid obtained by revolving the curve $y = x^{1/3}$ about the y-axis between $x = 0$ and $x = 8$.

Solution. The situation is depicted in Figure 4. Here $y = x^{1/3}$, $x = g(y) = y^3$, and $g'(y) = 3y^2$. If x is in $[0, 8]$, then y is in $[0, 2]$, and we find that

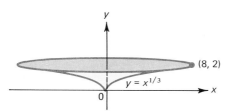

FIGURE 4

From (5)

$$S \stackrel{\swarrow}{=} 2\pi \int_0^2 y^3 \sqrt{1 + 9y^4} \, dy = \frac{2\pi}{36} \int_0^2 (1 + 9y^4)^{1/2} 36y^3 \, dy$$

$$= \frac{\pi}{27}(1 + 9y^4)^{3/2} \Big|_0^2 = \frac{\pi}{27}(145^{3/2} - 1). \blacksquare$$

PROBLEMS 9.3

In Problems 1–12, find the lateral surface area of the solid generated when the area below the given curve over the given interval is rotated about the x-axis.

1. $y = \sqrt{x}$; $[1, 4]$ **2.** $y = 2\sqrt{x}$; $[9, 25]$ **3.** $y = 2x$; $[0, 1]$

4. $y = x + 1$; $[1, 4]$ **5.** $y = \alpha x + \beta$; $[a, b]$ **6.** $y = e^{-x}$; $[0, 1]$

7. $y = \sqrt{2 - x}$; $[0, 1]$ **8.** $y = \cos x$; $\left[0, \dfrac{\pi}{2}\right]$ **9.** $y = \dfrac{x^3}{6} + \dfrac{1}{2x}$; $[1, 2]$

10. $y = \cosh x$; $[0, 1]$ ***11.** $y = \dfrac{1}{x}$; $[1, 2]$ **12.** $y = \dfrac{3}{2}x^{2/3}$; $[1, 8]$

13. Show that the surface area of a sphere of radius r is $4\pi r^2$. [*Hint:* Revolve the area in the first-quadrant part of the disk $x^2 + y^2 \le r^2$ around the x-axis to obtain the lateral surface area of the hemisphere of radius r, and then multiply by 2.]

14. Calculate the total surface area of the solid obtained by revolving about the y-axis the region bounded by the curve $y = \ln x$, the x-axis, and the line $x = e$.

*15. The region enclosed by the ellipse $(x^2/a^2) + (y^2/b^2) = 1$ is revolved about the x-axis. Find the surface area of the resulting football-shaped solid. [Note that there are really two cases: If $a > b$, we indeed get a "football," but if $a < b$, we get something that looks more like a Frisbee.]

16. The area below the line $y = 2x + 3$ and above the x-axis between $x = -1$ and $x = 1$ is revolved about the line $y = -1$. Find the resulting surface area. [*Hint:* Be careful.]

17. The arc in the fourth quadrant of the circle $x^2 + (y - 1)^2 = 16$ is revolved about the y-axis. Find the resulting lateral surface area.

*18. When a circle is revolved about a line that does not intersect it, the resulting solid is called a **torus**. Calculate the surface area of the torus obtained by rotating the circle $x^2 + y^2 = r^2$ about the line $x = q$ (where $0 < r < q$).

19. On the earth a **parallel** is a line around the planet that runs parallel to the equator. A **zone** is any region between two parallels. Show that if we assume that the earth is spherical,†

---

†Of course, the earth is not a perfect sphere but is rather flattened at the poles and takes the shape of what is termed an *oblate spheroid.* However, it is very close to a sphere; the difference between its equatorial diameter and its polar diameter is only about one-third of 1% of the equatorial diameter.

then the area of any zone is given by $S = \pi h d$, where d is the diameter of the earth ($\sim 13{,}300$ km) and h is the distance between the parallels. Note that this area is independent of the position of the zone of the sphere.

9.4 CENTER OF MASS AND THE FIRST MOMENT

In the physical applications given in earlier chapters we treated moving objects as though they were particles, having mass but not size. However, in the real world objects are made up of many particles. If a moving object is rotating or vibrating, for example, then different particles in the object may exhibit different kinds of behavior. Fortunately, we can deal with this kind of situation by representing the motion of the object as the motion of a special point, called the **center of mass** of that object. In general, we can describe the motion of an entire system if we can describe the motion of its center of mass and the motion of the system around that center of mass.

We can derive a simple equation for the center of mass by considering the motion of a seesaw. Consider a system containing two point masses m_1 and m_2, located at distances x_1 and x_2 from a reference point, which we label 0. In Figure 1 we have indicated two masses, one of which is located to the left of 0 so that its "distance" is treated as negative. We place the two masses on opposite ends of a seesaw as in Figure 2 and label the pivot point P. From experience we know that the farther away a mass is from the pivot point, the more likely the mass will "control" the motion. To put this idea more precisely, we consider the quantity $m_1(x_1 - P)$, measured in kilogram meters. By the principle of the lever, the seesaw will exactly balance if

$$m_1(x_1 - P) + m_2(x_2 - P) = 0. \tag{1}$$

FIGURE 1

FIGURE 2

The quantity on the left-hand side of (1) is called the **first moment of the system around the line $x = P$**. If $P = 0$, then the first moment is equal to $m_1 x_1 + m_2 x_2$ and is called the first moment about the y-axis (the line $x = 0$). When the first moment is zero, the seesaw will balance, and we say that the system is in **equilibrium.** Solving equation (3) for P, we obtain

$$P = \frac{m_1 x_1 + m_2 x_2}{m_1 + m_2}. \tag{2}$$

The point P is called the **center of mass** of the two masses and is denoted $\bar{x}$. That is, from (2)

$$\bar{x} = \frac{m_1 x_1 + m_2 x_2}{m_1 + m_2}. \tag{3}$$

We have seen that the seesaw will balance if the pivot is at its center of mass. (Of course, in this derivation we ignored the mass of the board of the seesaw.)

If the system contains n particles, where $n > 2$, then we can, in an analogous manner, define the center of mass by

$$\bar{x} = \frac{m_1 x_1 + m_2 x_2 + \cdots + m_n x_n}{m_1 + m_2 + \cdots + m_n} = \frac{\text{first moment of the system about the } y\text{-axis}}{\text{total mass of the system}} \tag{4}$$

EXAMPLE 1 Five particles of masses 2, 1, 4, 3, and 5 kg, respectively, are located on the x-axis with x values 3, -2, 4, 7, and -3 m. Calculate the center of mass of the system.

Solution.

$$\bar{x} = \frac{m_1 x_1 + m_2 x_2 + m_3 x_3 + m_4 x_4 + m_5 x_5}{m_1 + m_2 + m_3 + m_4 + m_5}$$

$$= \frac{2(3) + 1(-2) + 4(4) + 3(7) + 5(-3)}{2 + 1 + 4 + 3 + 5} = \frac{26}{15} \text{ m.} \quad \blacksquare$$

Now consider a rigid, horizontal rod, with variable density $\rho(x)$, that extends along the x-axis from $x = a$ to $x = b$. We can think of the rod as a system of very closely packed particles. In fact, the number of particles (atoms) is so large, and the spacing between them so small, that we can think of the body as having a continuous distribution of mass (i.e., having a continuous density function). We first calculate the first moment of the rod and begin by partitioning the interval $[a, b]$:

$$a = x_0 < x_1 < x_2 < \cdots < x_n = b.$$

If $\Delta x = x_i - x_{i-1}$ is small, the density $\rho(x)$ (which is assumed to be continuous) is almost constant in $[x_{i-1}, x_1]$. Then the mass Δm_i of the "subrod" over the subinterval $[x_{i-1}, x_i]$ is given by

Mass = density $\times$ length
$$\Delta m_i \approx \rho(x_i) \, \Delta x. \tag{5}$$

What we have done here is to approximate the continuous rod by a system of n point masses situated at the n partition points $x_1, x_2, \ldots, x_n$. (See Figure 3.) Com-

FIGURE 3

bining equations (4) and (5), we see that the center of mass of this approximating system is given by

$$\bar{x}_{approx} = \frac{x_1\rho(x_1)\,\Delta x + x_2\rho(x_2)\,\Delta x + \cdots + x_n\rho(x_n)\,\Delta x}{\rho(x_1)\,\Delta x + \rho(x_2)\,\Delta x + \cdots + \rho(x_n)\,\Delta x} = \frac{\sum_{i=1}^{n} x_i\rho(x_i)\,\Delta x}{\sum_{i=1}^{n} \rho(x_i)\,\Delta x}. \tag{6}$$

Now as the lengths of the individual sections of the partition approach zero, $\bar{x}_{approx}$ approaches the center of mass of the continuous rod. That is,

$$\bar{x} = \lim_{\Delta x \to 0} \bar{x}_{approx} = \frac{\displaystyle\int_a^b x\rho(x)\,dx}{\displaystyle\int_a^b \rho(x)\,dx}. \tag{7}$$

As we saw in Chapter 5 (see p. 275), $\int_a^b \rho(x)\,dx$ = the mass of the rod. We define $\int_a^b x\rho(x)\,dx$ to be the **first moment of the rod around the y-axis** (the line $x = 0$). We can then rewrite (7) as

$$\bar{x} = x-\text{coordinate of the center of mass}$$
$$= \frac{\text{first moment around the }y\text{-axis}}{\text{total mass}}. \tag{8}$$

EXAMPLE 2   The density ρ of a 4-m nonuniform metal beam is given by $\rho(x) = 2\sqrt{x}$ kilograms per meter, where x is the distance along the beam. Find the center of mass of the beam.

Solution. We place the beam along the x-axis from $x = 0$ to $x = 4$. (See Figure 4.) Then

$$\bar{x} = \frac{2\displaystyle\int_0^4 x\sqrt{x}\,dx}{2\displaystyle\int_0^4 \sqrt{x}\,dx} = \frac{\frac{4}{5}x^{5/2}\Big|_0^4}{\frac{4}{3}x^{3/2}\Big|_0^4} = \frac{\frac{4}{5}\cdot 32}{\frac{4}{3}\cdot 8} = \frac{12}{5}\text{ m.}$$

Here the first moment is $\frac{128}{5}$ kg · m and the total mass of the beam is $\frac{32}{3}$ kg. ■

FIGURE 4

PROBLEMS 9.4

In Problems 1–4, find the first moment around the origin and the center of mass of the system of masses m_i located at the points P_i on the x-axis, where m_i is measured in kilograms and the x units are meters.

1. $m_1 = 4$, $m_2 = 6$; $P_1 = (3, 0)$, $P_2 = (-5, 0)$
2. $m_1 = 3$, $m_2 = 4$, $m_3 = 7$; $P_1 = (2, 0)$, $P_2 = (-3, 0)$, $P_3 = (1, 0)$
3. $m_1 = 2$, $m_2 = 8$, $m_3 = 5$, $m_4 = 3$; $P_1 = (-6, 0)$, $P_2 = (2, 0)$, $P_3 = (20, 0)$, $P_4 = (-1, 0)$
4. $m_1 = 4$, $m_2 = 1$, $m_3 = 6$, $m_4 = 3$, $m_5 = 10$; $P_1 = (4, 0)$, $P_2 = (10, 0)$, $P_3 = (-10, 0)$, $P_4 = (5, 0)$, $P_5 = (-7, 0)$

In Problems 5–16 a rod of variable density $\rho(x)$ lies along the x-axis in the given interval $[a, b]$. Find the first moment of the rod about zero, the total mass of the rod, and its center of mass.

5. $\rho(x) = x$; $[0, 1]$

6. $\rho(x) = x^{1/3}$; $[1, 8]$

7. $\rho(x) = x^3$; $[1, 2]$

8. $\rho(x) = \dfrac{1}{x}$; $[1, 3]$

9. $\rho(x) = 2 + \sin x$; $[0, 2\pi]$
10. $\rho(x) = e^x$; $[0, 1]$
11. $\rho(x) = \ln x$; $[1, 4]$
12. $\rho(x) = xe^x$; $[0, 1]$
13. $\rho(x) = x \ln x$; $[1, e]$
14. $\rho(x) = e^x \sin x$; $[0, \pi/2]$

15. $\rho(x) = \dfrac{1}{(x + 1)(x + 2)}$; $[0, 3]$

16. $\rho(x) = \dfrac{1}{\sqrt{1 - x^2}}$; $\left[0, \dfrac{1}{2}\right]$

*17. A rod with total mass m of variable density $\rho(x)$ lies along the x-axis between $x = a$ and $x = b$. The rod is split into two pieces at $x = c$; let $\bar{x}_1$ denote the center of mass of the first piece over the interval $[a, c]$, let $\bar{x}_2$ denote the center of mass of the second piece over the interval $[c, b]$, and let $\bar{x}$ denote the center of mass of the entire (unsplit) rod. Show that

$$\bar{x} = \frac{m_1 \bar{x}_1 + m_2 \bar{x}_2}{m_1 + m_2}.$$

18. Suppose the density of a rod is $\rho(x) = x^2$ between $x = 0$ and $x = 3$ and the rod is split at $x = 2$. Verify the result of Problem 17.

9.5 THE CENTROID OF A PLANE REGION

We began the last section by considering the center of mass of a finite number of point masses located along the x-axis. We found that the center of mass $\bar{x}$ was given by

$$\bar{x} = \frac{\text{first moment of the system around the } y\text{-axis}}{\text{total mass}} \tag{1}$$

We denote the first moment of the system around the y-axis by M_y, and we use the Greek letter μ (mu) to denote the total mass ($m_1 + m_2 + \cdots + m_n$) of the system. Then we have

$$\bar{x} = \frac{M_y}{\mu}. \tag{2}$$

Now suppose, instead, that the n masses m_i are located at the points (x_i, y_i) in the xy-plane. Then we calculate the center of mass by treating separately the x- and y-coordinates of the points at which the masses are located. For the x-coordinates we simply use formula (2). We then define the first moment of the system around the x-axis (the line $y = 0$) by

$$M_x = m_1 y_1 + m_2 y_2 + \cdots + m_n y_n = \sum_{i=1}^{n} m_i y_i, \tag{3}$$

and we have

$$\bar{y} = \frac{M_x}{\mu}. \tag{4}$$

The center of mass of the n particles is then located at the point $(\bar{x}, \bar{y})$. We can think of *the center of mass as a balancing point*. To picture this, imagine that the n masses are placed on a thin, weightless plate. Then the plate will balance on a pivot if the pivot is located at the point $(\bar{x}, \bar{y})$. This is illustrated in Figure 1.

FIGURE 1

EXAMPLE 1 Calculate the center of mass of the system of three particles with masses $m_1 = 10$ g, $m_2 = 5$ g, and $m_3 = 8$ g located, respectively, at the points $(2, -1)$, $(3, 2)$, and $(-6, 1)$ (measured in centimeters).

Solution. We calculate $M_y = 10(2) + 5(3) + 8(-6) = -13$ g · cm, $M_x = 10(-1) + 5(2) + 8(1) = 8$ g · cm, and $\mu = 10 + 5 + 8 = 23$ g. Then $\bar{x} = M_y/\mu = -\frac{13}{23}$ cm, $\bar{y} = M_x/\mu = \frac{8}{23}$ cm, and the center of mass is located at the point $(-\frac{13}{23}, \frac{8}{23})$. ∎

If the masses in the preceding discussion all have the same mass m, then

$$M_y = mx_1 + mx_2 + \cdots + mx_n = m(x_1 + x_2 + \cdots + x_n) = m \sum_{i=1}^{n} x_i,$$

$$M_x = my_1 + my_2 + \cdots + my_n = m(y_1 + y_2 + \cdots + y_n) = m \sum_{i=1}^{n} y_i,$$

and $\mu = m + m + \cdots + m = nm$, so that

$$\bar{x} = \frac{m(x_1 + x_2 + \cdots + x_n)}{nm} = \frac{x_1 + x_2 + \cdots + x_n}{n} = \frac{\sum_{i=1}^{n} x_i}{n} \tag{5}$$

and

$$\bar{y} = \frac{m(y_1 + y_2 + \cdots + y_n)}{nm} = \frac{y_1 + y_2 + \cdots + y_n}{n} = \frac{\sum_{i=1}^{n} y_i}{n} \tag{6}$$

We see that $\bar{x}$ is simply the arithmetic average of the x values, $\bar{y}$ is the arithmetic average of the y values, and the center of mass *does not depend* on the common mass value m. Since in this setting the actual value of the mass is irrelevant, we give the center of mass another name: the **centroid**. *The centroid is defined as the center of mass in the case of equal masses.*

Now let R denote a region in the plane, which we can think of as a thin sheet of material with *uniform* (i.e., constant) area density ρ. With this interpretation, the region is called a **plane lamina** and ρ is measured in kg/m^2, g/cm^2, lb/ft^2, or lb/in^2. (See Figure 2.) If $\rho = 1$, then the total mass of the lamina is equal to the area of the region R. We now calculate the centroid of such a lamina. We can think of the center of mass $(\bar{x}, \bar{y})$ as the pivot point at which the lamina would exactly balance.

Suppose, first, that R is the region bounded by the graph of a continuous, nonnegative function f, the x-axis, and the lines $x = a$ and $x = b$ (see Figure 3). We partition the interval $[a, b]$ in the usual manner:

$$a = x_0 < x_1 < x_2 < \cdots < x_n = b.$$

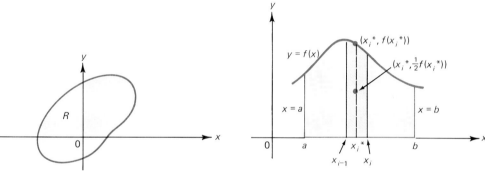

FIGURE 2   **FIGURE 3**

The shaded region in Figure 3 is roughly rectangular in shape. Since density is constant, the center of mass of the rectangular lamina lies in the center of the rectangle, which is at the point $(x_i^*, \frac{1}{2}f(x_i^*))$, where $x_i^* = (x_{i-1} + x_i)/2$. The area of the rectangular lamina is, approximately, $f(x_i^*)(x_i - x_{i-1}) = f(x_i^*)\,\Delta x$, and the mass of the lamina is

$$\underset{\text{Mass}}{\Delta\mu_i} \approx \underset{\text{density} \times \text{area}}{\rho f(x_i^*)\,\Delta x}.$$

The first moment of this ith sublamina around the y-axis (mass times average distance to the y-axis) is then given by

$$\Delta(M_y)_i \approx \Delta\mu_i x_i^* \approx \rho x_i^* f(x_i^*)\,\Delta x,$$

and the first moment of the entire lamina around the y-axis is defined to be

$$M_y = \lim_{\Delta x \to 0} \sum_{i=1}^{n} \rho x_i^* f(x_i^*) \, \Delta x = \rho \int_a^b x f(x) \, dx. \tag{7}$$

Similarly, the first moment of the ith sublamina around the x-axis (mass times average distance to the x-axis) is given by

$$\Delta(M_x)_i \approx \overbrace{[\rho f(x_i^*) \, \Delta x]}^{\substack{\text{Mass} \\ \overline{\text{Density} \times \text{Area}}}} \cdot \overbrace{\left[\frac{1}{2} f(x_i^*)\right]}^{\substack{\text{Average distance} \\ \text{to } x\text{-axis}}} = \frac{\rho}{2} [f(x_i^*)]^2 \, \Delta x,$$

and the first moment of the entire lamina around the x-axis is defined to be

$$M_x = \lim_{\Delta x \to 0} \sum_{i=1}^{n} \frac{\rho}{2} [f(x_i^*)]^2 \, \Delta x = \frac{\rho}{2} \int_a^b [f(x)]^2 \, dx. \tag{8}$$

Since ρ is constant, the mass μ of the lamina is equal to ρ times the area of the region, or

$$\mu = \rho \int_a^b f(x) \, dx.$$

Definition 1   CENTER OF MASS   We define the **center of mass** $(\bar{x}, \bar{y})$ of the lamina by

$$\bar{x} = \frac{M_y}{\mu} = \frac{\rho \displaystyle\int_a^b x f(x) \, dx}{\rho \displaystyle\int_a^b f(x) \, dx} = \frac{\displaystyle\int_a^b x f(x) \, dx}{\displaystyle\int_a^b f(x) \, dx} \tag{9}$$

and

$$\bar{y} = \frac{M_x}{\mu} = \frac{\dfrac{\rho}{2} \displaystyle\int_a^b [f(x)]^2 \, dx}{\rho \displaystyle\int_a^b f(x) \, dx} = \frac{\dfrac{1}{2} \displaystyle\int_a^b [f(x)]^2 \, dx}{\displaystyle\int_a^b f(x) \, dx}. \tag{10}$$

Equations (9) and (10) tell us that if the lamina is uniform (i.e., if it has a constant density), then its center of mass depends only on the region and not on the density. As before, we then call the center of mass the **centroid** of the plane region R, and we can define the **first moments** of the region R around the x- and y-axes by, respectively,

$$M_y = \rho \int_a^b x f(x) \, dx \quad \text{and} \quad M_x = \frac{\rho}{2} \int_a^b [f(x)]^2 \, dx, \tag{11}$$

$$\bar{x} = \frac{M_y}{\rho A} \quad \text{and} \quad \bar{y} = \frac{M_x}{\rho A}, \tag{12}$$

where A denotes the area of the region.

REMARK.   Since $\bar{x}$ and $\bar{y}$ do not depend on the density, we can compute $\bar{x}$ and $\bar{y}$ most easily by setting $\rho = 1$ in (11) and (12).

EXAMPLE 2   Calculate the centroid of the region in the first quadrant bounded by the curve $y = x^2$, the x-axis, and the line $x = 1$.

Solution. We have

$$\overset{\text{Setting } \rho = 1}{\swarrow}$$

$$M_y = \int_0^1 x^3 \, dx = \frac{1}{4}, \qquad M_x = \frac{1}{2} \int_0^1 x^4 \, dx = \frac{1}{10},$$

and $A = \int_0^1 x^2 \, dx = \frac{1}{3}$. Then

$$\bar{x} = \frac{M_y}{A} = \frac{\frac{1}{4}}{\frac{1}{3}} = \frac{3}{4} \quad \text{and} \quad \bar{y} = \frac{\frac{1}{10}}{\frac{1}{3}} = \frac{3}{10}.$$

The centroid is the point $(\frac{3}{4}, \frac{3}{10})$. This is depicted in Figure 4. ∎

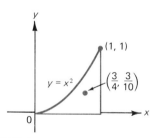

FIGURE 4

EXAMPLE 3   Find the centroid of the region bounded by the curve $y = 3x^2 - 6x + 4$, the x- and y-axes, and the line $x = 2$.

Solution. We have

$$\overset{\text{Setting } \rho = 1}{\swarrow}$$

$$M_y = \int_0^2 (3x^3 - 6x^2 + 4x) \, dx = 4,$$

$$M_x = \frac{1}{2} \int_0^2 (3x^2 - 6x + 4)^2 \, dx$$

$$= \frac{1}{2} \int_0^2 (9x^4 - 36x^3 + 60x^2 - 48x + 16) \, dx = \frac{24}{5},$$

and

$$A = \int_0^2 (3x^2 - 6x + 4)\, dx = 4,$$

so that $(\bar{x}, \bar{y}) = (1, \frac{6}{5})$. This is sketched in Figure 5. Note that in this case the centroid lies *outside* the region under consideration. ∎

FIGURE 5

We now generalize these results. Let the region R be bounded by the two curves $f(x)$ and $g(x)$, where both f and g are continuous and $f \geq g$. Then the situation is as depicted in Figure 6. If x ranges over the interval $[a, b]$, then we again partition the interval and find that the centroid of a typical rectangular subregion is, approximately, $(\bar{x}_i, \bar{y}_i) = (x_i^*, \frac{1}{2}(f(x_i^*) + g(x_i^*)))$, where $x_i^* = \frac{1}{2}[x_{i-1} + x_i]$. Then replacing the function f in equation (7) by $f - g$, we find that

$$\bar{x} = \frac{M_y}{\rho A} \quad \text{and} \quad \bar{y} = \frac{M_x}{\rho A}, \tag{13}$$

where

$$M_y = \rho \int_a^b x[f(x) - g(x)]\, dx, \tag{14}$$

$$M_x = \frac{\rho}{2} \int_a^b [f(x) + g(x)][f(x) - g(x)]\, dx$$

$$= \frac{\rho}{2} \int_a^b \{[f(x)]^2 - [g(x)]^2\}\, dx, \tag{15}$$

and

$$A = \int_a^b [f(x) - g(x)]\, dx. \tag{16}$$

The expression for M_x follows from the fact that the area of the ith subregion is approximately equal to $\Delta x[f(x_i^*) - g(x_i^*)]$.

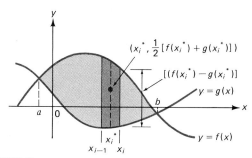

FIGURE 6

REMARK. Again we may set $\rho = 1$ in (13), (14), and (15) to simplify computations.

EXAMPLE 4 Find the centroid of the region bounded by the curves $y = x^3$ and $y = \sqrt{x}$.

Solution. The two curves intersect at $(0, 0)$ and $(1, 1)$. Since $\sqrt{x} \geq x^3$ on $[0, 1]$, we have

$$M_y = \int_0^1 x(\sqrt{x} - x^3)\, dx = \int_0^1 (x^{3/2} - x^4)\, dx = \frac{1}{5},$$

where *Setting $\rho = 1$* (annotation)

$$M_x = \frac{1}{2}\int_0^1 (x - x^6)\, dx = \frac{5}{28},$$

and

$$A = \int_0^1 (\sqrt{x} - x^3)\, dx = \frac{5}{12},$$

so that $\bar{x} = \frac{12}{25}$ and $\bar{y} = \frac{3}{7}$. This is sketched in Figure 7. ■

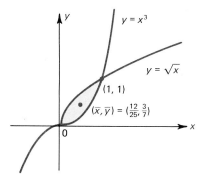

FIGURE 7

EXAMPLE 5 Calculate the centroid of the region bounded by $y = \sin x$, $y = \cos x$, and the lines $x = \pi/2$ and $x = \pi$.

Solution. The region is sketched in Figure 8. We have

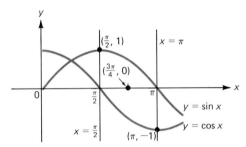

FIGURE 8

Setting $\rho = 1$

$$M_y = \int_{\pi/2}^{\pi} x(\sin x - \cos x)\, dx$$

Integrating by parts

$$= -x(\cos x + \sin x)\Big|_{\pi/2}^{\pi} + \int_{\pi/2}^{\pi} (\cos x + \sin x)\, dx$$

$$= \frac{3\pi}{2} + (\sin x - \cos x)\Big|_{\pi/2}^{\pi} = \frac{3\pi}{2},$$

Identity (xvi) in Appendix A1.2

$$M_x = \frac{1}{2}\int_{\pi/2}^{\pi} (\sin^2 x - \cos^2 x)\, dx = -\frac{1}{2}\int_{\pi/2}^{\pi} \cos 2x\, dx = 0,$$

and

$$A = \int_{\pi/2}^{\pi} (\sin x - \cos x)\, dx = 2,$$

so that $(\bar{x}, \bar{y}) = (3\pi/4, 0)$. ∎

The following theorem whose proof is suggested in Problems 39 and 40, can be very useful in the calculation of centroids.

Theorem 1.

(i) If the plane region R is symmetric about the line $x = c$, then $\bar{x} = c$.
(ii) If the plane region R is symmetric about the line $y = d$, then $\bar{y} = d$.

EXAMPLE 6   Calculate the centroid of the semicircle $y = \sqrt{16 - x^2}$.

Solution. The semicircle (half the circle $x^2 + y^2 = 16$) is sketched in Figure 9. Since the region is symmetric about the y-axis (the line $x = 0$), we have $\bar{x} = 0$. We then calculate

$$M_x = \frac{1}{2}\int_{-4}^{4} (16 - x^2)\, dx = \frac{128}{3},$$

and since $A = \frac{1}{2}\pi r^2 = 8\pi$, we find that $\bar{y} = 16/3\pi$. ∎

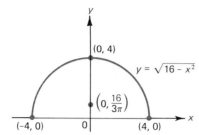

FIGURE 9

PROBLEMS 9.5

In Problems 1–4, find the center of mass of the system of masses m_i located at the points P_i, where each m_i is measured in grams and x and y units are centimeters.

1. $m_1 = 4$, $m_2 = 6$; $P_1 = (3, 4)$, $P_2 = (-5, 3)$
2. $m_1 = 3$, $m_2 = 4$, $m_3 = 7$; $P_1 = (2, 3)$, $P_2 = (-3, -4)$, $P_3 = (1, 2)$
3. $m_1 = 2$, $m_2 = 8$, $m_3 = 5$, $m_4 = 3$; $P_1 = (-6, 2)$, $P_2 = (2, 3)$, $P_3 = (20, -5)$, $P_4 = (-1, 8)$
4. $m_1 = 4$, $m_2 = 1$, $m_3 = 6$, $m_4 = 3$, $m_5 = 10$; $P_1 = (4, 7)$, $P_2 = (10, -3)$, $P_3 = (-10, 0)$, $P_4 = (0, 5)$, $P_5 = (6, -6)$

In Problems 5–20, calculate the centroid of the plane region bounded by the given curve and the x-axis over the indicated interval.

5. $y = 2x + 3$; $[0, 1]$ **6.** $y = x^3$; $[1, 2]$ **7.** $y = x^{1/3}$; $[0, 1]$
8. $y = x^4$; $[-1, 2]$ **9.** $y = \sqrt{10 + x}$; $[-1, 6]$ ***10.** $y = x\sqrt{x^2 + 9}$; $[0, 3]$

11. $y = \dfrac{1}{x + 1}$; $[0, 1]$ **12.** $y = e^x$; $[0, 1]$

13. $y = \dfrac{1}{(x + 1)(x + 2)}$; $[0, 1]$ **14.** $y = \sin 2x$; $\left[0, \dfrac{\pi}{4}\right]$

15. $y = \cos 3x$; $[0, \pi/9]$ **16.** $y = \sin^2 x$; $[0, \pi/2]$ **17.** $y = \ln x$; $[1, 2]$

18. $y = x^{3/5}$; $[0, 32]$ **19.** $y = \dfrac{x}{x - 1}$; $[-1, 0]$ **20.** $y = \dfrac{1}{x^2 + 1}$; $[0, 1]$

21. Find the centroid of the region in the first quadrant bounded by the curve $x = y^2 + 1$ and the line $y = 1$.
*22. Find the centroid of the region in the first quadrant bounded by the curve $x^{2/3} + y^{2/3} = 1$.

In Problems 23–34, find the centroids of the plane regions bounded by the given curves and lines.

23. $y = x^2$, $y = 4x$, $x = 0$, $x = 1$ **24.** $y = x^2 + 3x + 5$, $y = 2x^2 - x + 8$
25. $y = 2x^5 + 3$, $y = 32x + 3$, $x \geq 0$ **26.** $y^2 = 4 + x$, $x + 2y = 4$
27. $y = x^2$, $y = x^3$, $x = 3$, $x \geq 1$ **28.** $xy = 1$, $x = 5$, $y = 5$
29. $x + y^2 = 8$, $x + y = 2$ **30.** $y = x^{1/3}$, $y = x$ (in the first quadrant)
31. $y = \sin x$, $y = \cos x$, $x = 0$, $x = \pi/4$ **32.** $y = e^{2x}$, $y = e^{-x}$, $x = 2$
33. $y = \ln x$, $y = e^x$; $[1, 2]$ **34.** $y = \ln x$, $y = e^{-x}$; $[2, 3]$

35. Show that the centroid of the ellipse $(x^2/a^2) + (y^2/b^2) = 1$ is the origin.
36. Use Theorem 1 to calculate the centroid of the region bounded by the curve $y = x^8$ and the line $y = 1$.

37. Use Theorem 1 to calculate the centroid of the region bounded by $y = \cos x$ and the x-axis in the interval $[\pi/2, 3\pi/2]$.

38. Find the center of mass of the plane lamina in the first quadrant bounded by the curve $y = x^2$, the y-axis, and the line $y = 1$, if the area density at any point (x, y) in the lamina is given by $\rho(x, y) = \sqrt{1 - x}$ g/cm^2. (Assume that units on the x-axis are centimeters.)

*39. Prove Theorem 1(i). [*Hint:* Divide the interval $[a, b]$ into $2n$ subintervals, half of which lie to the left of c and half of which lie to the right. Then treat the line $x = c$ as if it were the y-axis and show that $\bar{x} = 0$.]

*40. Prove Theorem 1(ii).

41. Prove the following special case of the **first theorem of Pappus: If the plane region R is revolved about the y-axis and if the y-axis does not intersect R, then the volume generated is equal to the product of the area of R and the length of the circumference of the circle traced by the centroid of R. [*Hint:* The length of the circumference of the circle traced by the centroid is $2\pi\bar{x}$, so it is necessary to show that $V = 2\pi\bar{x}A$. To prove this, write down the formula for $\bar{x}$ and A and compare $2\pi\bar{x}A$ with the formula for calculating V by the method of cylindrical shells.]

**42. Prove that the statement of the first theorem of Pappus in Problem 41 holds if R is revolved about any line that does not intersect it.

43. Use the first theorem of Pappus to calculate the volume of the torus generated by rotating the circle $(x - a)^2 + y^2 = r^2$ $(r < a)$ about the y-axis. (Compare this with Problem 9.1.34.)

44. Use the first theorem of Pappus to calculate the volume of the solid generated by rotating the triangle with vertices $(-1, 2)$, $(1, 2)$, and $(0, 4)$ about the x-axis.

*45. Use the first theorem of Pappus to calculate the volume of the torus generated by rotating the unit circle about the line $y = 4 - x$.

*46. Use the first theorem of Pappus to calculate the volume of the "elliptical torus" generated by rotating the ellipse $(x^2/a^2) + (y^2/b^2) = 1$ about the line $y = 3a$. Assume that $3a > b > 0$.

9.6 MOMENTS OF INERTIA AND KINETIC ENERGY (OPTIONAL)

In Section 9.4 we defined the first moment around the y-axis (the line $x = 0$) of n masses by

$$M_y = m_1 x_1 + m_2 x_2 + \cdots + m_n x_n = \sum_{i=1}^{n} m_i x_i.$$

There we used the fact that the force exerted by an object on a pivot point is proportional to the mass of the object times its distance from the pivot.

For certain types of problems other kinds of moments are more useful. We define the **second moment** of the system of n masses around the y-axis by

$$I_y = m_1 x_1^2 + m_2 x_2^2 + \cdots + m_n x_n^2 = \sum_{i=1}^{n} m_i x_i^2. \tag{1}$$

EXAMPLE 1   Calculate the second moment I_y of the system of three masses $m_1 = 5$ kg, $m_2 = 3$ kg, and $m_3 = 4$ kg, situated along the x-axis at the points $(3, 0)$, $(-2, 0)$, and $(4, 0)$, respectively, where the units along the x-axis are meters.

Solution. $I_y = 5(3)^2 + 3(-2)^2 + 4(4)^2 = 121 \text{ kg} \cdot \text{m}^2$. ◼

Now consider the motion of a rotating particle of mass m at a distance r from an axis of rotation. (See Figure 1.) The **angular velocity** ω of the particle is measured in radians per unit time. Thus, for example, if $\omega = 4\pi$ and the unit of time is the second, then the particle makes two complete rotations in one second ($4\pi = 2 \cdot 2\pi$) since there are 2π radians in an angle of one revolution. The velocity of the particle is then given by

$$v = \omega r, \tag{2}$$

since ωr is the distance traveled in one unit of time. Furthermore, the kinetic energy (see p. 339) is given by

$$K = \tfrac{1}{2}mv^2 = \tfrac{1}{2}mr^2\omega^2. \tag{3}$$

FIGURE 1

Suppose that a rotating rigid body contains n particles of masses m_i at distances r_i, respectively, from the axis of rotation. Then the total kinetic energy of the body is given by

$$K = \frac{1}{2}(m_1r_1^2 + m_2r_2^2 + \cdots + m_nr_n^2)\omega^2 = \frac{1}{2}\left(\sum_{i=1}^{n} m_ir_i^2\right)\omega^2. \tag{4}$$

The quantity

$$I = m_1r_1^2 + m_2r_2^2 + \cdots + m_nr_n^2 = \sum_{i=1}^{n} m_ir_i^2 \tag{5}$$

is called the **rotational inertia,** or **moment of inertia,** of the body about the axis of rotation. Note that this quantity is equal to the second moment defined by (1) and depends on the choice of axis of rotation as well as the shape of the body and the distribution of masses along it.

EXAMPLE 2 A dumbell consists of two 10-kg weights separated by a 1-m rod. Neglecting the weight of the rod and treating the weights as point masses, calculate the moment of inertia of the dumbbell, first, about an axis perpendicular to it and passing through the center of the rod and, second, about an axis passing through one of the weights.

Solution. The situation is depicted in Figure 2. We can think of the axis in both cases as a line perpendicular to the plane of the page. In the first case $m_1 = m_2 = 10$ and $r_1 = r_2 = \frac{1}{2}$, so that $I = 10(\frac{1}{2})^2 + 10(\frac{1}{2})^2 = 5 \text{ kg} \cdot \text{m}^2$. In the second case we assume

FIGURE 2

the axis passes through the mass on the left (the other case leads to the same result). Then $m_1 = m_2 = 10$, $r_1 = 0$, and $r_2 = 1$, so that $I = 10(1)^2 = 10$ kg·m². We see that the rotational inertia is twice as great about an axis through one end as it is about an axis through the center of the dumbbell. ∎

The ideas above can easily be generalized to calculate the moment of inertia of a "continuous" body rotating about a given axis.†

EXAMPLE 3   A rigid 4-m rod spins about its left end 50 times a minute. Its density is given by $\rho(x) = (1 + \sqrt{x})$ kg/m. Find its kinetic energy.

Solution. The rod is sketched in Figure 3. From equation (4) the kinetic energy is given by $K = \frac{1}{2}I\omega^2$. In this problem $\omega = 50(2\pi) = 100\pi$ radians/min. If we partition the interval $[0, 4]$, we may treat each of the n subintervals as point masses. The mass of the ith subinterval is approximately equal to $\rho(x_i)(x_{i-1} - x_i) = (1 + \sqrt{x_i})\,\Delta x_i$. Then the moment of inertia of the ith subinterval is given by

$$\Delta I_i \approx m_i r_i^2 = [(1 + \sqrt{x_i})\,\Delta x]x_i^2,$$

so that

$$I \approx \sum_{i=1}^n (1 + \sqrt{x_i})x_i^2\,\Delta x,$$

and letting $\Delta x \to 0$, we obtain

$$I = \lim_{\Delta x \to 0} \sum_{i=1}^n (1 + \sqrt{x_i})x_i^2\,\Delta x$$

$$= \int_0^4 (1 + \sqrt{x})x^2\,dx = \int_0^4 (x^2 + x^{5/2})\,dx = \frac{1216}{21}\text{ kg·m}^2.$$

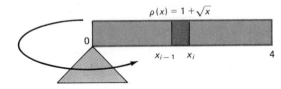

FIGURE 3

---

†For example, the moment of inertia of the earth about its polar axis is, approximately, $I = 8.08 \times 10^{37}$ kg·m².

Then

$$K = \frac{1}{2}I\omega^2 = \frac{1}{2} \cdot \frac{1216}{21}(100\pi)^2 = \frac{6,080,000}{21}\pi^2 \text{ kg} \cdot \text{m}^2/\text{min}^2$$

$$\approx 794 \text{ N} \cdot \text{m} = 794 \text{ J.†} \quad \blacksquare$$

The technique of the above example can be used to calculate moments of inertia for two-dimensional regions.

EXAMPLE 4 The region bounded by the curve $y = \sqrt{x}$, the x-axis, and the line $x = 4$ is rotated about the y-axis. Find the moment of inertia if the region has the constant area density of 3 kg/m^2.

Solution. The region is sketched in Figure 4. We partition the interval $[0, 4]$. We can then think of the rotating body as a system of n rotating "rectangularly shaped" masses. The mass m_i of the ith subregion is equal to its area times the density, or

$$\Delta m_i = 3\sqrt{x_i}\,\Delta x.$$

Also, we see from Figure 4 that $r_i \approx x_i$, so that

$$\Delta I_i = \Delta m_i r_i^2 = 3x_i^{5/2}\,\Delta x.$$

Thus from (5)

$$I \approx \sum_{i=1}^{n} 3x_i^{5/2}\,\Delta x,$$

and

$$I = \lim_{\Delta x \to 0} \sum_{i=1}^{n} 3x_i^{5/2}\,\Delta x = 3\int_0^4 x^{5/2}\,dx = \frac{768}{7} \text{ kg} \cdot \text{m}^2. \quad \blacksquare$$

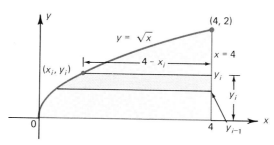

FIGURE 4

---

†1 N = 1 kg $\cdot$ m/sec^2 = 3600 kg $\cdot$ m/min^2; 1 joule (J) = 1 N $\cdot$ m.

EXAMPLE 5   Calculate the moment of inertia when the region in Example 4 is rotated about the x-axis.

Solution. The region is the same as before, but now we must use horizontal strips to obtain a rotating strip of almost constant radius. This is illustrated in Figure 5. We partition the interval $[0, 2]$ along the y-axis to obtain n horizontal rectangular subregions. The mass of each subregion is given by

$$\Delta m_i = \text{area} \times \text{density} = 3(4 - x_i)\,\Delta y = 3(4 - y_i^2)\,\Delta y.$$

Since $r_i = y_i$, we have

$$\Delta I_i = \Delta m_i r_i^2 = 3(4 - y_i^2)y_i^2\,\Delta y,$$

and

$$I = 3\int_0^2 (4 - y^2)y^2\,dy = \frac{64}{5}\ \text{kg}\cdot\text{m}^2. \ \blacksquare$$

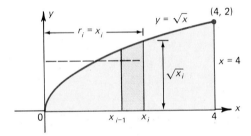

FIGURE 5

The technique shown in Example 5 can be used in some problems where density is not constant.

EXAMPLE 6   A plane lamina takes the shape of the region bounded by $y = \sin x$, the x-axis, and the line $x = \pi/2$. At any point (x, y) on the lamina, the density is given by $\rho(x, y) = x^2$ g/cm^2. The lamina is rotated about the y-axis. Calculate the moment of inertia if the x- and y-units are centimeters.

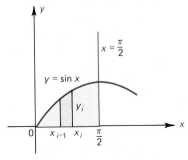

FIGURE 6

Solution. The lamina is sketched in Figure 6. Here we have

$$\Delta m_i = \text{area} \times \text{density} \approx y_i \, \Delta x \cdot x_i^2 = x_i^2 \sin x_i \, \Delta x,$$

and $r_i = x_i$, so that $\Delta m_i r_i^2 = x_i^4 \sin x_i \, \Delta x$ and

$$I = \int_0^{\pi/2} x^4 \sin x \, dx$$

After integrating by parts four times

$$= (-x^4 \cos x + 4x^3 \sin x + 12x^2 \cos x - 24x \sin x - 24 \cos x)\Big|_0^{\pi/2}$$

$$= \left(\frac{\pi^3}{2} - 12\pi + 24\right) \text{g} \cdot \text{cm}^2. \quad \blacksquare$$

We conclude this section by defining another interesting physical quantity. If one particle rotates about an axis, then we have seen that $I = mr^2$, or

$$r = \sqrt{\frac{I}{m}}. \tag{6}$$

This quantity is called the **radius of gyration.** If there are more particles, we may define the radius of gyration by using Equation (6). If a plane lamina is rotated about an axis and if r denotes its radius of gyration, then in its rotation the lamina exhibits the same behavior as if it were a point mass located r units from the axis.

EXAMPLE 7 Calculate the radius of gyration of the region in Example 4.

Solution. We have $r = \sqrt{I/m}$, where $I = (768/7) \text{ kg} \cdot \text{m}^2$. Also, the mass of the region is the product of its density and its area, or

$$m = 3 \int_0^4 \sqrt{x} \, dx = 3 \cdot \frac{16}{3} = 16 \text{ kg}.$$

Then

$$r = \sqrt{\frac{768/7 \text{ kg} \cdot \text{m}^2}{16} \frac{}{\text{kg}}} = \sqrt{\frac{48}{7}} \text{ m}. \quad \blacksquare$$

PROBLEMS 9.6

In Problems 1–3 masses are located on the x-axis at the points P_i. Find the moment of inertia about the y-axis.

1. $m_1 = 3$ kg, $m_2 = 2$ kg; $P_1 = (4, 0)$, $P_2 = (-6, 0)$ (measured in meters)
2. $m_1 = 4$ g, $m_2 = 1$ g, $m_3 = 5$ g; $P_1 = (-3, 0)$, $P_2 = (-4, 0)$, $P_3 = (7, 0)$ (measured in centimeters)

3. $m_1 = 2$ kg, $m_2 = 3$ kg, $m_3 = 42$ kg, $m_4 = 3$ kg; $P_1 = (4, 0)$, $P_2 = (3, 0)$, $P_3 = (-4, 0)$, $P_4 = (-3, 0)$ (measured in meters)

4. A dumbbell is composed of two 8-kg masses separated by a bar 1 m long. Ignoring the weight of the bar and treating the masses as point masses, calculate the moment of inertia of the dumbbell about an axis perpendicular to it and passing through:
   (a) The center of the rod.
   (b) A point one-third of the way along the rod.
5. Find the kinetic energy in each case of Problem 4 if the dumbbell rotates 25 times per minute.
6. Find the radius of gyration in each case of Problem 4.
7. A dumbbell is unevenly weighted with masses of 5 and 7 kg, respectively, separated by a 1-m bar. Ignoring the weight of the bar and treating the masses as point masses, calculate the moment of inertia of the dumbbell about an axis perpendicular to it and passing through:
   (a) The center of the rod.   (b) The 5-kg mass.   (c) The 7-kg mass.
   (d) A point one-quarter of the way along the rod, closer to the 5-kg mass.
8. Find the kinetic energy in each case of Problem 7 if the dumbbell rotates 10 times per second.
9. Find the radius of gyration in each case of Problem 7.
10. A 3-m rod with density $\sqrt{16 - x^2}$ kilograms per meter (x being measured from the "left" end) is rotated about its left end. Find the moment of inertia and radius of gyration.
11. Find the kinetic energy if the rod in Problem 10 rotates 20 times per second.

In Problems 12–31, find the moment of inertia and the radius of gyration when the region bounded by the given curves and lines with given constant area density ρ is rotated about the given axis.

12. The rectangle bounded by $x = 0$, $x = 2$, $y = 0$, and $y = 3$; $\rho = 2$; x-axis.
13. The region of Problem 12 about the y-axis.
14. $y = x^2$, $y = 0$, $x = 1$; $\rho = 3$; x-axis.
15. The region of Problem 14 about the y-axis.
16. The region of Problem 14 about the line $x = 2$.
17. The region of Problem 14 about the line $y = -3$.
18. $y = \cos x$, $x = 0$, $y = 0$, $x = \pi/2$; $\rho = 2$; x-axis.
19. $y = \cos x$, $x = 0$, $y = 0$, $x = \pi/2$; $\rho = 2$; y-axis.
20. $y = 1/x$, $y = 0$, $x = 1$, $x = 2$; $\rho = 4$; x-axis.
21. $y = 1/x$, $y = 0$, $x = 1$, $x = 2$; $\rho = 4$; y-axis.
22. $y = e^x$, $y = 0$, $x = 0$, $x = 1$; $\rho = 5$; x-axis.
23. $y = e^x$, $y = 0$, $x = 0$, $x = 1$; $\rho = 5$; y-axis.
24. $\sqrt{x} + \sqrt{y} = 1$, $y = 0$, $x = 0$; $\rho = 1$; x-axis.
25. The circle $(x - 2)^2 + y^2 = 1$; $\rho = 2$; y-axis.
26. The upper half of the ellipse $(x^2/a^2) + (y^2/b^2) = 1$; $\rho = 4$; y-axis.
27. The circle $(x - 2)^2 + y^2 = 1$; $\rho = 2$; the line $x = -4$.
28. The circle $(x - 2)^2 + y^2 = 1$; $\rho = 3$; the line $y = 5$.
29. $x = y^2$, $x = 0$, $y = 2$; $\rho = 7$; x-axis.
30. $x = y^2$, $x = 0$, $y = 2$; $\rho = 7$; y-axis.
31. The quarter circle $x^2 + y^2 = 1$ in the first quadrant; $\rho = 3$; the line $y = -x$.

32. An aluminum triangle with constant density ρ and vertices at $(0, 0)$, $(0, 4)$, and $(-3, 0)$ is spun with angular velocity ω about the x-axis. Calculate its kinetic energy (assume that the units are kilograms and meters).
33. Calculate the kinetic energy of the triangle in Problem 32 if it is spun about the y-axis.
34. Calculate the kinetic energy in Problems 12, 14, 18, 20, 22, 24, 25, 29, and 31 if the region rotates 8 times per second (assume that the units are kilograms and meters).

9.7 FLUID PRESSURE (OPTIONAL)

In this section we show how the integral can be used to calculate fluid pressures. By a **fluid** we mean a substance that can flow. This definition includes both liquids and gases. It is reasonable only to talk about forces applied to the *surface* of a fluid. If the fluid is at rest, then such forces can be applied only perpendicular to the surface (Figure 1a), since if a force were applied tangentially (i.e., from the side, as in Figure 1b), then layers of fluid would simply slide over one another, and the fluid would no longer be at rest, as we have assumed.

(a)   (b)

FIGURE 1

When a force is applied perpendicular to the surface of a fluid it is called a **normal force.** In this case, we define the **pressure** P to be the magnitude of the normal force per unit of area. Commonly used units of pressure are lb/ft^2, lb/in.^{2}, N/m^2, and N/cm^2. For example, the air pressure at sea level is 14.7 lb/in.^{2} = 1.013×10^5 N/m^2.

Now suppose that a thin horizontal lamina is submerged at a certain depth in a fluid and we wish to calculate the total force exerted on the lamina by the column of the fluid on one side of the lamina. The force is given by

$$F = PA,$$

(1)

where P is the pressure and A is area. Then

$$P = \frac{F}{A} = \frac{mg}{A},$$

(2)

where g is the acceleration due to gravity. Over a unit of area, the mass of a fluid is given by the product of its density ρ (in lb/in.^{3}, lb/ft^3, kg/m^3 or g/cm^3) times the depth d of the column of fluid.

If the mass is measured in slugs (1 slug = 1 lb/g), then

$$P = \rho d.$$

(3)

If mass if measured in kilograms, then

$$P = \rho d g,\dagger$$

(4)

---

†Recall that the forces are measured in pounds (in the English system) and newtons (in the metric system). To obtain force from kilogram measurements, we need to multiply by g. Thus if water density is given in terms of pounds per cubic foot, no multiplication by g is necessary to obtain force since pounds already represent force. However, if the density is given in terms of kilograms per cubic meter, multiplication by g is necessary to obtain force.

where $g = 9.81$ m/sec^2. Note that 1 kg $\cdot$ m/sec$^2 = 1$ N.

REMARK. It is a physical fact that when an object is submerged in a fluid, the pressure at any depth is the same in every direction (up, down, sideways).

EXAMPLE 1   A swimming pool with length 8 m, width 5 m, and depth 3 m is filled with water. What is the force exerted on the bottom of the pool by the weight of the water?

Solution. The density of water (at 0°C) is, approximately,

$$\rho = 1000 \text{ kg/m}^3.$$

From (4)

$$P = \rho dg = (1000 \text{ kg/m}^3)(3 \text{ m})(9.81 \text{ m/sec}^2) = 29{,}430 \, \frac{\text{kg} \cdot \text{m/sec}^2}{\text{m}^2}$$

$$= 29{,}430 \text{ N/m}^2.$$

Then using (1), we have

$$F = PA = (29{,}430 \text{ N/m}^2)(5 \times 8 \text{ m}^2) = 1{,}177{,}200 \text{ N.} \quad \blacksquare$$

EXAMPLE 2   A rectangular sheet of metal 10 ft × 14 ft is submerged horizontally in water to a depth of 12 ft. What is the force exerted on one side of the sheet of metal by the weight of the water?

Solution. The density of water at 0°C is, approximately,

$$\rho = 62.4 \text{ lb/ft}^3.$$

Using (3), we have

$$P = \rho d = (62.4 \text{ lb/ft}^3)(12 \text{ ft}) = 748.8 \text{ lb/ft}^2.$$

Then

$$F = PA = (748.8 \text{ lb/ft}^2)(10 \times 14 \text{ ft}^2) = 104{,}832 \text{ lb.} \quad \blacksquare$$

EXAMPLE 3   **Archimedes' principle**† states that a body immersed in a fluid is buoyed up by a force equal to the weight of the fluid displaced. In the case of a floating object Archimedes' principle states that a floating body must displace its own weight in water. To demonstrate Archimedes' principle, we let a block of width w and area A (see Figure 2) be submerged a distance h in a liquid of density ρ. By equation (4) the force $F_\uparrow$ on the bottom of the block points up and is given by

$$F_\uparrow = PA = \rho g(w + h)A.$$

---

†It was this discovery that is reputed to have led Archimedes to jump from his bath and run home naked, shouting "Eureka" ("I have found it."). See Archimedes' biography on page 267.

Water surface

FIGURE 2

Similarly, the force $F_\downarrow$ on the top of the block points down and is given by

$$F_\downarrow = PA = \rho g h A.$$

The net (or resultant) force is given by

$$F_\uparrow - F_\downarrow = \rho g (w + h)A - \rho g h A = \rho g w A = m_l g,$$

where m_l is the total mass of the liquid displaced by the block (since the mass of the block is its volume wA times its density ρ). Thus the block undergoes an upward force equal to the weight of the water it displaces. ■

In the problems discussed above, we calculated the force of the liquid on a plane surface. That is, water pressure was applied uniformly over a flat surface. However, if a body is submerged in a fluid, then the lateral pressure (pressure on the wall or side of the body) varies with the depth of the fluid so that formulas (3) or (4) cannot be used directly.

EXAMPLE 4 A trough 3 m in length and filled with water has a cross section in the shape of a trapezoid with lower base 2 m, upper base 4 m, and depth $1\frac{1}{2}$ m. Find the total force due to water pressure on one end of the trough.

Solution. One end of the trough is sketched in Figure 3. To facilitate our calculations, we put in x- and y-coordinates as shown. Then we partition the interval $[0, \frac{3}{2}]$ along the y-axis. The force against the ith strip (shaded) is equal to the pressure along the strip times the area of the strip, and (since pressure is fairly constant on the strip if Δy is small) we have

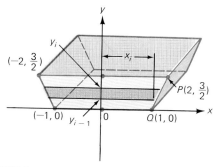

FIGURE 3

From (4)

$$\Delta F_i = P_i A_i \approx \rho d_i g A_i.$$

But $\rho = 1000$ kg/m^3, $d_i \approx \frac{3}{2} - y_i$, $g = 9.81$ m/sec^2, and $A_i \approx 2x_i \, \Delta y$. Now we calculate x as a function of y on the edge of the trapezoid. The line PQ on the edge passes through the points $(2, \frac{3}{2})$ and $(1, 0)$, leading to the equation

$$\frac{y}{x-1} = \frac{-\frac{3}{2}}{-1} = \frac{3}{2}, \quad \text{or} \quad x = \tfrac{2}{3}y + 1.$$

Thus

$$\Delta F_i = \rho g \left(\tfrac{3}{2} - y_i\right)(2x_i) \, \Delta y = 2\rho g \left(\tfrac{3}{2} - y_i\right)\left(\tfrac{2}{3}y_i + 1\right) \Delta y$$

$$= \frac{\rho g}{3}(3 - 2y_i)(2y_i + 3) \, \Delta y = \frac{\rho g}{3}(9 - 4y_i^2) \, \Delta y.$$

Finally, we let Δy approach zero to obtain

$$F = \frac{\rho g}{3} \int_0^{3/2} (9 - 4y^2) \, dy = \frac{\rho g}{3}(9) = 3\rho g = 3(1000)(9.81) = 29{,}430 \text{ N.} \quad \blacksquare$$

EXAMPLE 5   A cylindrical barrel whose end has a diameter of 4 ft is submerged in seawater (density approximately 64.3 lb/ft^3 at 0°C). Find the total force due to water pressure at one end if the center of the barrel is at a depth of 12 ft.

Solution.   One end of the barrel is sketched in Figure 4. We introduce a coordinate system that places the origin at the center of the barrel. If we now partition the "y-axis" as shown, then the force on a typical horizontal strip is given by

$$\Delta F_i = P_i A_i \approx \rho d_i A_i \qquad \text{[from (3)]}.$$

FIGURE 4

But $\rho = 64.3$ lb/ft^3, $d_i \approx 12 - y_i$ and $A_i \approx$ (length of strip) $\times$ (Δy). Since the equation of the circular end is $x^2 + y^2 = 4$, we have $x = \sqrt{4 - y^2}$ to the right of the y-axis and $x = -\sqrt{4 - y^2}$ to the left. Thus the length of the strip is approximately equal to $\sqrt{4 - y_i^2} - (-\sqrt{4 - y_i^2}) = 2\sqrt{4 - y_i^2}$, and we find that $A_i \approx 2\sqrt{4 - y_i^2}\ \Delta y$. Then

$$\Delta F_i \approx 2\rho(12 - y_i)\sqrt{4 - y_i^2}\ \Delta y$$

and

$$F = 2\rho \int_{-2}^{2} (12 - y)\sqrt{4 - y^2}\ dy = 24\rho \int_{-2}^{2} \sqrt{4 - y^2}\ dy - 2\rho \int_{-2}^{2} y\sqrt{4 - y^2}\ dy.$$

The first integral is the area of the semicircle $x = \sqrt{4 - y^2}$ and is equal to $\frac{1}{2}\pi r^2 = \frac{1}{2}\pi(2)^2 = 2\pi$ (or the integral can be calculated by making the substitution $y = \sin\theta$). Thus

$$F = (24\rho)(2\pi) + \frac{2\rho}{3}(4 - y^2)^{3/2}\Big|_{-2}^{2} = (48\pi)\rho + 0$$

$$= (48\pi)(64.3) = 3086.4\pi\ \text{lb} \approx 9696\ \text{lb.} \quad\blacksquare$$

The two examples above illustrate how to go about solving problems involving fluid pressure. We will not cite a general formula since the type of problems encountered can vary so widely. Once a convenient coordinate system is introduced, it is usually not too difficult to obtain the answer.

PROBLEMS 9.7

1. A circular swimming pool has a radius of 6 m and is 2 m deep. Find the force due to water pressure on the bottom of the pool when the pool is filled with water.
2. Find the force in Problem 1 if the pool is filled with seawater ($\rho \approx 1030$ kg/m^3 at 0°C).
3. A tube of mercury ($\rho = 13.6$ gm/cm^3) is inverted over a beaker of mercury as shown in Figure 5. The pressures P_C and P_D at the points C and D are the same because the points are at the same height. Let P_A denote atmospheric pressure.

FIGURE 5

(a) Find a formula giving P_A as a function of h, the height of the column of mercury.
(b) For $h = 76$ cm, show that $P_A = 1.01 \times 10^5$ N/m^2. [*Hint:* Use $g = 981$ cm/sec^2.] This value of P_A is called **one standard atmosphere.**

4. A rectangular swimming pool 5 m wide and 7 m long has a bottom that slopes at a constant rate, being $\frac{1}{2}$ m deep at the shallow end and 4 m deep at the deep end. When the pool is filled with water, find the force due to liquid pressure on the bottom of the pool.

5. A trough filled with water has a cross section in the shape of an equilateral triangle with 5-ft sides (the wide end is up). What is the force due to water pressure on one end of the trough?

6. In Problem 5, find the force when the trough is filled to a depth of **(a)** 1 ft, **(b)** 2 ft, **(c)** 3 ft.

7. A trough filled with ethyl alcohol ($\rho = 810$ kg/m³) has a cross section in the shape of a semicircle with a radius of 1 m (the open end is up). What is the force due to liquid pressure on one end of the trough?

*8. In Problem 7, find the force when the trough is filled to a depth of **(a)** 25 cm, **(b)** 50 cm, **(c)** 75 cm.

9. The dam across a certain river has the shape of the parabola $y = x^2/25$. The river is 25 m wide at the top. Find the force due to water pressure on a face of the dam that is vertical.

*10. Suppose the dam across the river in Problem 9 has a face that slopes upstream from the water surface at an angle of 45°. Find the force due to liquid pressure on it. [*Hint:* First use Problem 9 to find the depth of the water.]

*11. Find the force in Problem 10 if the dam slopes at an angle of 30°.

12. An elliptical plate is submerged in seawater so that the center of the ellipse is at a depth of 6 m. The equation of the ellipse is $(x^2/4) + (y^2/9) = 1$. Find the force due to water pressure on the plate.

13. A cask with sides in the shape of the trapezoid of Example 4 is submerged in seawater so that the top of the trapezoid is level with the surface of the water. Find the force due to liquid pressure on one end of the cask.

14. The gate of a dam is in the shape of an isosceles triangle with a 6-ft base and 4-ft sides. The upper vertex of the triangle is at a depth of 22 ft. Find the force on the gate when the gate is closed.

*15. Find the force on one side of the trough in Example 4.

16. Find the force on the sides of the pool (at the deeper and shallower ends) in Problem 4.

17. Find the force on one side of the trough in Problem 5 if the trough is 10 ft long.

18. The face of a dam is in the shape of one arc of the curve $y = 50 \sin(\pi x/100)$ for $x \in [-100, 0]$. The surface of the water is 1 m above the top of the dam. Find the force due to water pressure on the face of the dam.

19. A plate is in the shape of the region bounded by $y = e^{-x}$, $x = 0$, $x = 5$, and $y = 0$. The plate is submerged in seawater so that its bottom edge is at a depth of 10 m. Find the force due to water pressure on one side of the plate.

*20. A dam has a circular gate of radius 1 m. The gate can withstand a force of 20,000 N. To what height can the dam be filled?

*21. Prove that the total force against one side of a vertical submerged region R is equal to the product of the pressure at the centroid of R times the area of R.

In Problems 22–27 a plate in the shape of the given region is submerged in water with its centroid at the given depth. Use Problem 21 to calculate the force due to liquid pressure.

22. The region bounded by $y = 1/(x + 1)$, $y = 0$, $x = 0$, $x = 1$; $d = 10$ m.

23. $y = x^{1/3}$, $y = 0$, $x = 1$; $d = 15$ m. 24. $y = e^x$, $y = 0$, $x = 0$, $x = 2$; $d = 25$ ft.

25. $y = x^4$, $y = 0$, $x = 2$; $d = 30$ ft.

26. $y = 1/(x^2 + 1)$, $y = 0$, $x = 0$, $x = 1$; $d = 8$ m.

27. $y = \ln x$, $y = 0$, $x = 1$, $x = 2$; $d = 20$ ft.

REVIEW EXERCISES FOR CHAPTER NINE

In Exercises 1–8, find the volume of the solid generated when the region bounded by the given curves and lines is rotated about the given axis.

1. $y = 3x + 2$; $x = 0$; $y = 0$; $x = 2$; x-axis
2. $y = \sqrt{2x + 1}$; $x = 0$; $y = 0$; $x = 4$; x-axis
3. $y = x^5$; $x = 1$; $y = 0$; x-axis
4. $y = x^{1/4}$; $x = 1$; $y = 0$; y-axis
5. $y = \sin 2x$; $x = 0$; $x = \pi/4$; x-axis
6. $y = \cos x$; $x = 0$; $x = \pi/2$; y-axis
7. $(x - 2)^2 + y^2 \leq 1$; y-axis
8. $x^2 + (y + 3)^2 \leq 1$; x-axis

9. The base of a certain solid is a circle of radius 3, while each perpendicular cross section is a square. What is the volume of the solid?

10. Find the volume of the solid in Exercise 9 if each perpendicular cross section is an equilateral triangle.

In Exercises 11–14, calculate the length of each given arc.

11. $y = 2x + 3$; x in $[2, 7]$
12. $y = x^2$; x in $[0, 2]$
13. $y = \frac{2}{3}x^{3/2}$; x in $[0, 1]$
14. $y = \ln(x^2 - 1)$; x in $[2, 3]$

15. Find a definite integral that is equal to the length of the curve $y = x \sin x$ between $x = 0$ and $x = \pi/2$.

In Exercises 16–19, find the lateral surface area of the solid generated when the area below the given curve over the given interval is rotated about the x-axis.

16. $y = \sqrt{x}$; $[0, 2]$
17. $y = e^x$; $[0, 1]$
18. $y = \cos x$; $[\pi/2, \pi]$
19. $y = -1/x$; $[1, 2]$

20. The region enclosed by the ellipse $x^2 + (y^2/4) = 1$ is revolved about the y-axis. Find the surface area of the resulting football-shaped solid.

21. Find the surface area if the region of Exercise 20 is revolved about the x-axis.

22. Find the first moment around the origin and the center of mass of the system of masses $m_1 = 2$ kg, $m_2 = 5$ kg, and $m_3 = 8$ kg located along the x-axis at the points $P_1 = (4, 0)$, $P_2 = (-9, 0)$, $P_3 = (2, 0)$.

23. Find the center of mass of a rod lying along the x-axis in the interval $[0, 3]$ with density $\rho(x) = x^2$.

24. Find the center of mass of a rod lying along the x-axis in the interval $[1, 4]$ with density $\rho(x) = 1/x^3$.

25. Find the center of mass of the system of masses $m_1 = 3$ g, $m_2 = 7$ g, $m_3 = 4$ g, and $m_4 = 8$ g located at the points $P_1 = (-2, 3)$, $P_2 = (4, 6)$, $P_3 = (3, -7)$, $P_4 = (0, -1)$, where x and y values are measured in centimeters.

26. Calculate the centroid of the plane region bounded by $y = x^2$, $y = 0$, $x = 1$, $x = 3$.

27. Calculate the centroid of the region bounded by $y = \sqrt{1 + x}$, $y = 0$, $x = 0$, $x = 3$.

28. Calculate the centroid of the region bounded by $y = \cos x$, $y = 0$, $x = 0$, $x = \pi/2$.

29. Calculate the centroid of the region bounded by $y = x^2 + 5x + 6$ and $y = x + 3$.

30. Calculate the centroid of the region bounded by $y = e^x$, $y = e^{-x}$, and $x = 1$.

31. Calculate the moment of inertia about the y-axis of the system of masses $m_1 = 3$ kg, $m_2 = 8$ kg, and $m_3 = 5$ kg located at the points $P_1 = (4, 0)$, $P_2 = (-6, 0)$, $P_3 = (-2, 0)$.

32. A dumbbell is composed of a 3-kg mass and a 5-kg mass separated by a 1-m bar. Ignoring the weight of the bar and treating the masses as point masses, calculate the moment of inertia of the dumbbell about an axis perpendicular to it and passing through:
    (a) The center of the bar.
    (b) The 5-kg mass.
    (c) A point one-third of the way along the bar, closer to the 3-kg mass.

33. Calculate the radius of gyration in each case of Exercise 32.

34. Suppose the dumbbell in Problem 32 rotates 15 times per second. Calculate the kinetic energy in each of the three cases.

35. Find the moment of inertia and the radius of gyration when the region bounded by $y = x^3$, $y = 0$, and $x = 1$ is rotated about the x-axis, assuming a constant area density of 2 kg/m^2.

36. Find the moment of inertia and the radius of gyration if the region in Exercise 35 is rotated about the y-axis.

37. In Exercise 35 assume that the region is rotated 25 times per second. Calculate the kinetic energy.

38. The region bounded by $y = 1 - x$ and the x- and y-axes is rotated about the x-axis. Assuming a constant area density of 5 g/cm², calculate the moment of inertia and the radius of gyration.

39. Calculate the kinetic energy in Exercise 38 if the region is rotated 15 times per minute.

39. The unit disk $x^2 + y^2 \leq 1$ is rotated about the line $x = 2$. Calculate the moment of inertia and the radius of gyration. Assume that $\rho = 1$.

41. Solve Exercise 40 if the axis of rotation is the line $y = -5$.

42. Show that the second moment of a circle of radius r about an axis b units from its center is $\pi r^2(\frac{1}{4}r^2 + b^2)$.

43. Find the force due to water pressure on the bottom of a rectangular swimming pool 5 m long, 4 m wide, and 2 m deep when it is filled with water.

44. Find the force due to water pressure on the bottom of the pool in Exercise 43 if the bottom slopes downward from a depth of 1 m at one end to a depth of 3 m at the other.

45. A trough filled with water is 5 m long and has a cross section in the shape of a trapezoid 3 m wide at the bottom, $1\frac{1}{2}$ m wide at the top, and 2 m deep. What is the force due to water pressure on one end of the trough?

46. Answer the questions in Exercises 44 and 45 if the trough is filled with ethyl alcohol (density $\rho = 810$ kg/m³).

47. In Exercise 45, find the force if the trough is filled with water to a depth of **(a)** $\frac{1}{2}$ m, **(b)** 1 m, **(c)** $1\frac{1}{2}$ m.

48. A cylindrical barrel whose end has a diameter of 2 m is submerged in seawater (density $\rho = 1030$ kg/m³). Find the total force due to water pressure at one end if the center of the barrel is at a depth of 15 m.

10 Topics in Analytic Geometry

In this chapter we will discuss the general second-degree equation

$$Ax^2 + Bxy + Cy^2 + Dx + Ey + F = 0. \tag{1}$$

In particular, we will analyze three special cases of equation (1): the ellipse, the parabola, and the hyperbola. Each of these three curves (plus the circle, which is a special kind of ellipse) can be obtained as the intersection of a plane with a right circular cone (see Figure A). That is why these curves are called **conic sections.** We will also show that the graph of most second-degree equations given by (1) is either a circle, an ellipse, a parabola, or a hyperbola. The exceptional cases, called **degenerate conic sections,** consists of equations whose graphs are empty, a single point, a line, or a pair of lines.

10.1 THE ELLIPSE AND TRANSLATION OF AXES

Definition 1   THE ELLIPSE   An **ellipse** is the set of points (x, y) such that the sum of the distances from (x, y) to two given points is fixed. Each of the two points is called a **focus** of the ellipse.

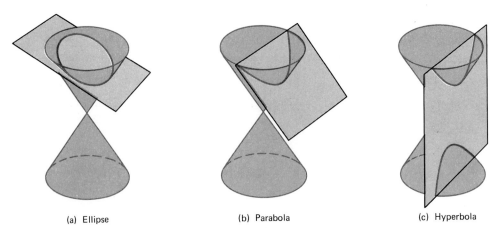

(a) Ellipse   (b) Parabola   (c) Hyperbola

FIGURE A

We now calculate the equation of an ellipse. To simplify our calculations, we assume that the foci F_1 and F_2 are the points $(-c, 0)$ and $(c, 0)$. (See Figure 1.) If (x, y) is on the ellipse, then by the definition of the ellipse,

$$\overline{F_1P} + \overline{F_2P} = 2a,\tag{1}$$

where $2a > 2c$ ($2c$ is the distance between the foci and $2a$ is the given sum of the distances). Let (x, y) be a point on the ellipse. Then

$$\sqrt{(x + c)^2 + y^2} + \sqrt{(x - c)^2 + y^2} = 2a, \qquad \text{From (1)}$$

or

$$\sqrt{(x + c)^2 + y^2} = 2a - \sqrt{(x - c)^2 + y^2}$$

$$(x + c)^2 + y^2 = 4a^2 - 4a\sqrt{(x - c)^2 + y^2} + (x - c)^2 + y^2 \qquad \text{We squared both sides.}$$

$$4xc = 4a^2 - 4a\sqrt{(x - c)^2 + y^2}. \qquad \text{We simplified.}$$

$$\sqrt{(x - c)^2 + y^2} = a - \frac{c}{a}x \qquad \begin{array}{l}\text{We divided by}\\ 4a \text{ and rearranged}\\ \text{terms.}\end{array}$$

$$(x - c)^2 + y^2 = a^2 - 2cx + \frac{c^2}{a^2}x^2. \qquad \begin{array}{l}\text{We squared}\\ \text{again.}\end{array}$$

Then

$$x^2 - 2xc + c^2 + y^2 = a^2 - 2cx + \frac{c^2}{a^2}x^2,$$

or

$$x^2\left(1 - \frac{c^2}{a^2}\right) + y^2 = a^2 - c^2,$$

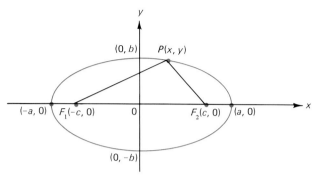

FIGURE 1

or

$$x^2\left(\frac{a^2 - c^2}{a^2}\right) + y^2 = a^2 - c^2,$$

and after dividing both sides by $a^2 - c^2$, which is positive by assumption, we have

$$\frac{x^2}{a^2} + \frac{y^2}{a^2 - c^2} = 1.$$

Finally, we define the positive number b by $b^2 = a^2 - c^2$ to obtain the standard equation of the ellipse.

Definition 2 STANDARD EQUATION OF THE ELLIPSE The **standard equation of the ellipse** is given by

$$\frac{x^2}{a^2} + \frac{y^2}{b^2} = 1. \tag{2}$$

Here

$$a^2 = b^2 + c^2,$$

where $(c, 0)$ and $(-c, 0)$ are the foci. Note that this ellipse is symmetric about both the x- and y-axes. Here, since $a > b$, the line segment from $(-a, 0)$ to $(a, 0)$ is called the **major axis** and the line segment from $(0, -b)$ to $(0, b)$ is called the **minor axis.** The point $(0, 0)$, which is at the intersection of the axes, is called the **center** of the ellipse. The points $(a, 0)$ and $(-a, 0)$ are called the **vertices** of the ellipse. In general, the vertices of an ellipse are the endpoints of the major axis.

EXAMPLE 1 Find the equation of the ellipse with foci at $(-3, 0)$ and $(3, 0)$ and with $a = 5$.

Solution. $c = 3$ and $a = 5$, so that $b^2 = a^2 - c^2 = 16$, and we obtain

$$\frac{x^2}{25} + \frac{y^2}{16} = 1.$$

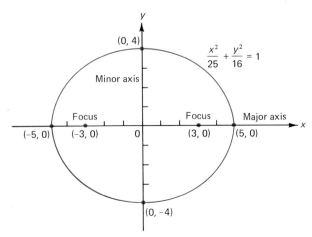

FIGURE 2

The ellipse is sketched in Figure 2. ◪

We can reverse the roles of x and y in the preceding discussion. Suppose that the foci are at $(0, c)$ and $(0, -c)$ on the y-axis. Then if the fixed sum of the distances is given as $2b$, we obtain, using similar reasoning,

$$\frac{x^2}{a^2} + \frac{y^2}{b^2} = 1,$$

where $a^2 = b^2 - c^2$. Now the major axis is on the y-axis, the minor axis is on the x-axis, and the vertices are $(0, b)$ and $(0, -b)$. In general, if the ellipse is given by (2), then we have the equations given in Table 1.

EXAMPLE 2 Discuss the curve $9x^2 + 4y^2 = 36$.

Solution. Dividing both sides by 36, we obtain

$$\frac{x^2}{4} + \frac{y^2}{9} = 1.$$

Here $a = 2$, $b = 3$, and the major axis is on the y-axis. Since $c^2 = 9 - 4 = 5$, the foci are at $(0, \sqrt{5})$ and $(0, -\sqrt{5})$. This curve is sketched in Figure 3. ◪

EXAMPLE 3 Find the equation of the ellipse with foci at $(-5, 0)$ and $(5, 0)$ and whose minor axis is the line segment extending from $(0, -3)$ to $(0, 3)$.

Solution. We have $c = 5$ and $b = 3$, so that,

$$a^2 = b^2 + c^2 = 34,$$

and the equation of the ellipse is

TABLE 1   STANDARD ELLIPSES

Equation	Description	Picture
$\dfrac{x^2}{a^2} + \dfrac{y^2}{b^2} = 1,\ a > b$	Ellipse with major axis on x-axis; $a^2 = b^2 + c^2$	
$\dfrac{x^2}{a^2} + \dfrac{y^2}{b^2} = 1,\ b > a$	Ellipse with major axis on y-axis; $b^2 = a^2 + c^2$	
$\dfrac{x^2}{a^2} + \dfrac{y^2}{b^2} = 1,\ b = a$	Circle with radius $a\ (= b)$	
$\dfrac{x^2}{a^2} + \dfrac{y^2}{b^2} = 0$	Degenerate ellipse; single point $(0, 0)$	
$\dfrac{x^2}{a^2} + \dfrac{y^2}{b^2} = -1$	Degenerate ellipse; graph is empty	

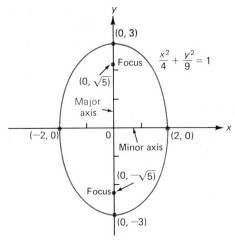

FIGURE 3

$$\frac{x^2}{34} + \frac{y^2}{9} = 1.$$

It is sketched in Figure 4. ■

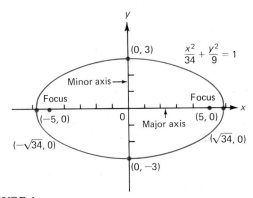

FIGURE 4

Definition 3   ECCENTRICITY   The **eccentricity** e of an ellipse is defined by

$$e = \frac{c}{a} \quad \text{if} \quad a \geq b \tag{3}$$

and

$$e = \frac{c}{b} \quad \text{if} \quad b \geq a. \tag{4}$$

NOTE.   If $a \geq b$, then the length of the major axis is $2a$; if $b \geq a$, the length of the major axis is $2b$. In both cases the distance between the foci is $2c$. Since $2c/2a = c/a$ and $2c/2b = c/b$, we have

$$e = \frac{\text{distance between foci}}{\text{length of major axis}} = \frac{\text{distance between foci}}{\text{distance between vertices}}. \tag{5}$$

The eccentricity of an ellipse is a measure of the shape of the ellipse and is always a number in the interval $[0, 1]$. If $e = 0$, then the ellipse is a circle, since in that case $c = 0$, so that $a^2 = b^2$ and the foci now coincide and are the center of the circle. As e approaches 1, the ellipse becomes progressively flatter and approaches the major axis: the straight-line segment from $(-a, 0)$ to $(a, 0)$ if $a > b$, and from $(0, -b)$ to $(0, b)$ if $b > a$. In general,

the larger the eccentricity, the flatter the ellipse.

In Example 1, $e = 3/5 = 0.6$. In Example 2, $c^2 = b^2 - a^2 = 5$, so that $e = \sqrt{5}/3 \approx 0.74536$. In Example 3, $e = 5/\sqrt{34} \approx 0.85749$.

We now turn to a different question. What happens if the center of the ellipse $(x^2/a^2) + (y^2/b^2) = 1$ is shifted from $(0, 0)$ to a new point (x_0, y_0)? Consider the equation

$$\frac{(x - x_0)^2}{a^2} + \frac{(y - y_0)^2}{b^2} = 1. \tag{6}$$

If we define two new variables by

$$x' = x - x_0 \quad \text{and} \quad y' = y - y_0, \tag{7}$$

then (6) becomes

$$\frac{(x')^2}{a^2} + \frac{(y')^2}{b^2} = 1. \tag{8}$$

This expression is the equation of an ellipse centered at the origin in the new coordinate system (x', y'). But $(x', y') = (0, 0)$ implies $(x - x_0, y - y_0) = (0, 0)$, or $x = x_0$ and $y = y_0$. That is, equation (6) is the equation of the "shifted" ellipse. See Figure 5. We have performed what is called a **translation of axes.** That is, we moved (or **translated**) the x- and y-axes to new positions so that they intersect at the point (x_0, y_0).

EXAMPLE 4   Find the equation of the ellipse centered at $(4, -2)$ with foci at $(1, -2)$ and $(7, -2)$ and minor axis joining the points $(4, 0)$ and $(4, -4)$.

Solution. The points are sketched in Figure 6a. Here $c = (7 - 1)/2 = 3$, $b = [0 - (-4)]/2 = 2$, and $a^2 = b^2 + c^2 = 13$. Thus from (6) we have

$$\frac{(x - 4)^2}{13} + \frac{(y + 2)^2}{4} = 1.$$

This ellipse is sketched in Figure 6b. Its major axis is on the line $y = -2$ and its minor axis is on the line $x = 4$. Its eccentricity is $3/\sqrt{13} \approx 0.83205$. ∎

FIGURE 5

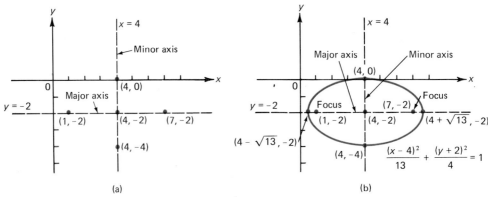

FIGURE 6

EXAMPLE 5   Discuss the curve given by

$$9x^2 + 36x + 4y^2 - 8y + 4 = 0.$$

Solution. We write this expression as

$$9(x^2 + 4x) + 4(y^2 - 2y) = -4.$$

Then after completing the squares, we obtain

$$9(x^2 + 4x + \overset{\text{Added}}{4}) - \overset{\text{Subtracted } 9 \cdot 4}{36} + 4(y^2 - 2y + \overset{\text{Added}}{1}) - \overset{\text{Subtracted } 4 \cdot 1}{4} = -4,$$

or

$$9(x + 2)^2 + 4(y - 1)^2 = 36,$$

and, dividing both sides by 36,

$$\frac{(x + 2)^2}{4} + \frac{(y - 1)^2}{9} = 1.$$

This is the equation of an ellipse centered at $(-2, 1)$. Here $a = 2$, $b = 3$, and $c = \sqrt{b^2 - a^2} = \sqrt{5}$. The foci are therefore at $(-2, 1 - \sqrt{5})$ and $(-2, 1 + \sqrt{5})$. The major axis is on the line $x = -2$ and the minor axis is on the line $y = 1$. The eccentricity is $\sqrt{5}/3 \approx 0.74536$. The ellipse is sketched in Figure 7. ∎

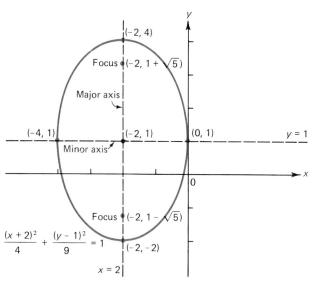

FIGURE 7

PROBLEMS 10.1

In Problems 1–18 the equation of an ellipse (or circle) is given. Find its center, foci, vertices, major and minor axes, and eccentricity. Then sketch it.

1. $\dfrac{x^2}{16} + \dfrac{y^2}{25} = 1$ 　　　　**2.** $\dfrac{x^2}{25} + \dfrac{y^2}{16} = 1$ 　　　　**3.** $x^2 + \dfrac{y^2}{9} = 1$

4. $\dfrac{x^2}{9} + y^2 = 1$ 　　　　**5.** $x^2 + 4y^2 = 16$ 　　　　**6.** $4x^2 + y^2 = 16$

7. $\dfrac{(x-1)^2}{16} + \dfrac{(y+3)^2}{25} = 1$ 　　　　**8.** $\dfrac{(x+3)^2}{25} + \dfrac{(y-1)^2}{16} = 1$

9. $2x^2 + 2y^2 = 2$ 　　　　　　　　　**10.** $4x^2 + y^2 = 9$
11. $x^2 + 4y^2 = 9$ 　　　　　　　　　**12.** $4(x-3)^2 + (y-7)^2 = 9$
13. $4x^2 + 8x + y^2 + 6y = 3$ 　　　　**14.** $x^2 + 6x + 4y^2 + 8y = 3$
15. $4x^2 + 8x + y^2 - 6y = 3$ 　　　　**16.** $x^2 + 2x + y^2 + 2y = 7$
17. $3x^2 + 12x + 8y^2 - 4y = 20$ 　　**18.** $2x^2 - 3x + 4y^2 + 5y = 37$

19. Find the equation of an ellipse with foci at $(0, 4)$ and $(0, -4)$ and vertices at $(0, 5)$ and $(0, -5)$.
20. Find the equation of the ellipse with vertices at $(2, 0)$ and $(-2, 0)$ and eccentricity 0.8.
21. Find the equation of two ellipses with center at $(-1, 4)$ that have the same shape as the ellipse of Problem 19 and that contain the same area.
***22.** Find the equation of the "ellipse" centered at the origin that is symmetric with respect to both the x- and y-axes and that passes through the points $(1, 2)$ and $(-1, -4)$.
23. Find the equation of the line tangent to the ellipse $2x^2 + 3y^2 = 14$ at the point $(1, 2)$.
***24.** Find two values of the number c such that the line $x + 3y = c$ is tangent to the ellipse $6x^2 + y^2 = 24$. Find the two points of tangency.

25. Show that the graph of the equation $x^2 + 2x + 2y^2 + 12y = c$ is:
    (a) An ellipse if $c > -19$.
    (b) A single point if $c = -19$.
    (c) Empty if $c < -19$.
*26. Find conditions on the numbers a, b, and c in order that the graph of the equation $x^2 + ax + 2y^2 + by = c$ be (a) an ellipse, (b) a single point, (c) empty.
27. Compute the area enclosed by the ellipse

$$\frac{x^2}{a^2} + \frac{y^2}{b^2} = 1.$$

10.2 THE PARABOLA

Definition 1   THE PARABOLA   A **parabola** is the set of points (x, y) equidistant from a fixed point and a fixed line that does not contain the fixed point. The fixed point is called the **focus,** and the fixed line is called the **directrix.**

To calculate the equation of a parabola, we begin by placing the axes so that the focus is the point $(0, c)$ and the directrix is the line $y = -c$. (See Figure 1.) If $P = (x, y)$ is a point on the parabola, then we have

$$\sqrt{x^2 + (y - c)^2} = |y + c|,$$

or squaring,

$$x^2 + (y - c)^2 = (y + c)^2,$$

which reduces to

$$x^2 = 4cy. \tag{1}$$

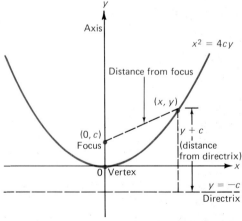

FIGURE 1

We see that the parabola given by (1) is symmetric about the y-axis. This line is called the **axis** of the parabola. Note that the axis contains the focus and is perpendicular to the directrix. The point at which the axis and the parabola intersect is called the **vertex.** The vertex is equidistant from the focus and the directrix.

EXAMPLE 1   Describe the parabola given by $x^2 = 12y$.

Solution. Here $4c = 12$, so that $c = 3$, the focus is the point $(0, 3)$, and the directrix is the line $y = -3$. The axis of the parabola is the y-axis and the vertex is the origin. The curve is sketched in Figure 2. ∎

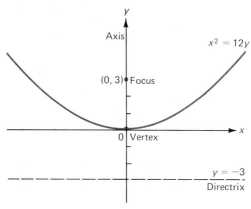

FIGURE 2

EXAMPLE 2   Describe the parabola given by $x^2 = -8y$.

Solution. Here $4c = -8$, so that $c = -2$, and the focus is $(0, -2)$, the directrix is the line $y = 2$, and the curve opens downward, as shown in Figure 3. ∎

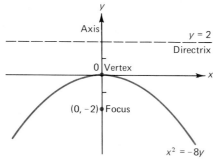

FIGURE 3

In general,

the parabola described by $x^2 = 4cy$ opens upward if $c > 0$ and opens downward if $c < 0$.

To verify this statement, note that

$$y = \frac{x^2}{4c}, \qquad y' = \frac{x}{2c}, \qquad \text{and} \qquad y'' = \frac{1}{2c}.$$

Thus the graph of $x^2 = 4cy$ is concave up if $c > 0$ and is concave down if $c < 0$.

As with the ellipse, we can exchange the role of x and y. We then obtain the following:

> The equation of the parabola with focus at $(c, 0)$ and directrix the line $x = -c$ is
>
> $$y^2 = 4cx. \tag{2}$$
>
> If $c > 0$, the parabola opens to the right; and if $c < 0$, the parabola opens to the left.

EXAMPLE 3   Describe the parabola $y^2 = 16x$.

Solution. Here $4c = 16$, so that $c = 4$, and the focus is $(4, 0)$, the directrix is the line $x = -4$, the axis is the x-axis (since the parabola is symmetric about the x-axis), and the vertex is the origin. The curve is sketched in Figure 4. ∎

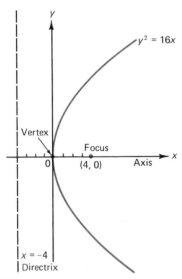

FIGURE 4

In Table 1 we list the **standard parabolas.**

All the parabolas we have so far drawn have had their vertices at the origin. Other parabolas can be obtained by a simple translation of axes. The parabolas

$$(x - x_0)^2 = 4c(y - y_0) \tag{3}$$

and

$$(y - y_0)^2 = 4c(x - x_0) \tag{4}$$

TABLE 1   STANDARD PARABOLAS

Equation	Description	Picture
$x^2 = 4cy$	Focus: $(0, c)$ Directrix: $y = -c$ Axis: y-axis Vertex: $(0, 0)$ Curve opens upward if $c > 0$ and downward if $c < 0$	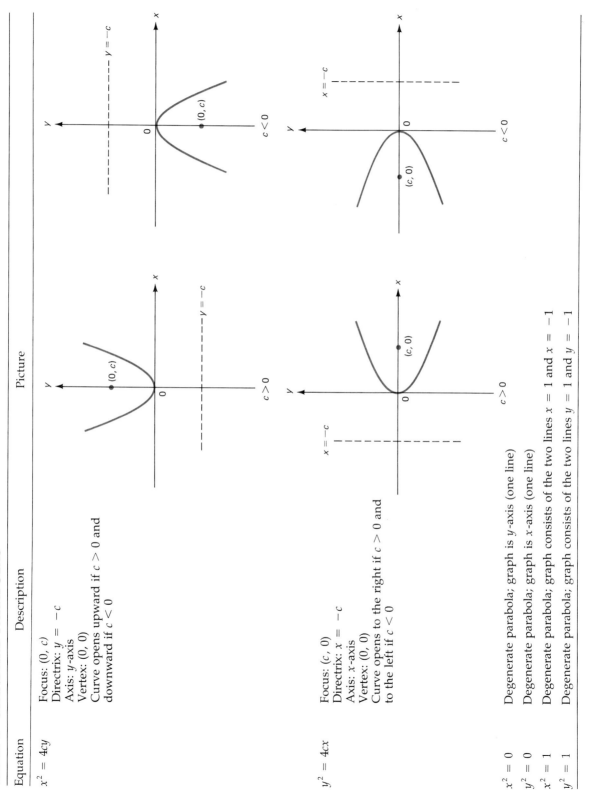
$y^2 = 4cx$	Focus: $(c, 0)$ Directrix: $x = -c$ Axis: x-axis Vertex: $(0, 0)$ Curve opens to the right if $c > 0$ and to the left if $c < 0$	
$x^2 = 0$	Degenerate parabola; graph is y-axis (one line)	
$y^2 = 0$	Degenerate parabola; graph is x-axis (one line)	
$x^2 = 1$	Degenerate parabola; graph consists of the two lines $x = 1$ and $x = -1$	
$y^2 = 1$	Degenerate parabola; graph consists of the two lines $y = 1$ and $y = -1$	

have vertices at the point (x_0, y_0).

EXAMPLE 4   Describe the parabola $(y - 2)^2 = -8(x + 3)$.

   Solution. This parabola has its vertex at $(-3, 2)$. It is obtained by shifting the parabola $y^2 = -8x$ three units to the left and two units up (see Section 1.8). The focus and directrix of $y^2 = -8x$ are $(-2, 0)$ and $x = 2$. Hence after translation, the focus and directrix of $(y - 2)^2 = -8(x + 3)$ are $(-5, 2)$ and $x = -1$. The axis of this curve is $y = 2$. The two parabolas are sketched in Figure 5. ■

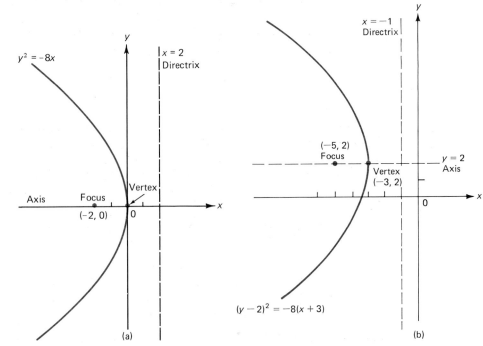

(a)   (b)

FIGURE 5

EXAMPLE 5   Describe the curve $x^2 - 4x + 2y + 10 = 0$.

   Solution. We first complete the square:

$$x^2 - 4x + 2y + 10 = (x - 2)^2 - 4 + 2y + 10 = 0,$$

or

$$(x - 2)^2 = -2(y + 3).$$

This expression is the equation of a parabola with vertex at $(2, -3)$. Since $4c = -2$, $c = -\frac{1}{2}$, and the focus is $(2, -3 - \frac{1}{2}) = (2, -\frac{7}{2})$. The directrix is the line $y = -3 - (-\frac{1}{2}) = -\frac{5}{2}$, and the axis is the line $x = 2$. The curve is sketched in Figure 6. ■

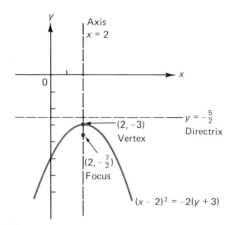

FIGURE 6

PROBLEMS 10.2

In Problems 1–18 the equation of a parabola is given. Find its focus, directrix, axis, and vertex. Then sketch it.

1. $x^2 = 16y$

2. $y^2 = 16x$

3. $x^2 = -16y$

4. $y^2 = -16x$

5. $2x^2 = 3y$

6. $2y^2 = 3x$

7. $4x^2 = -9y$

8. $7y^2 = -20x$

9. $(x - 1)^2 = -16(y + 3)$

10. $(y - 1)^2 = -16(x + 3)$

11. $x^2 + 4y = 9$

12. $(x + 1)^2 + 25y = 50$

13. $x^2 + 2x + y + 1 = 0$

14. $x + y - y^2 = 4$

15. $x^2 + 4x + y = 0$

16. $y^2 + 4y + x = 0$

17. $x^2 + 4x - y = 0$

18. $y^2 + 4y - x = 0$

19. Find the equation of the parabola with focus $(0, 4)$ and directrix the line $y = -4$.

20. Find the equation of the parabola with focus $(-3, 0)$ and directrix the line $x = 3$.

21. Find the equation of the parabola obtained when the parabola of Problem 20 is shifted so that its vertex is at the point $(-2, 5)$.

*22. Find the equation of the parabola with vertex at $(1, 2)$ and directrix the line $x = y$. [*Hint:* See Problem 1.5.41.]

23. The parabola in Problem 20 is translated so that its vertex is now at $(3, -1)$. Find its new focus and directrix.

24. Find the equation of the line tangent to the parabola $x^2 = 9y$ at the point $(3, 1)$.

*25. Show that among all points lying on a parabola, the point closest to the focus is the vertex.

*26. Let (x_0, y_0) be a point on a parabola. Let L_1 denote the line passing through (x_0, y_0) and the focus, and let L_2 denote the line through (x_0, y_0) and parallel to the axis of the parabola. Show that the tangent line to the parabola passing through (x_0, y_0) makes equal angles with L_1 and L_2.

**27. An asteroid orbits the earth making a parabolic orbit around it; that is, the earth is at the focus of the parabola. The asteroid is 150,000 km from the earth at a point where the line from the asteroid to the earth makes an angle of 30° with the axis of the parabola. How close will the asteroid come to the earth? [*Hint:* See Problem 25.]

**28. Answer the question in Problem 27 if the angle is 57°.

29. The **latus rectum** of a parabola is the chord passing through the focus that is perpendicular to the axis (see Figure 7). Compute the length of the latus rectum of the parabola $y^2 = 6x$.

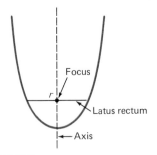

FIGURE 7

*30. Show that if a parabola has the equation $x^2 = 4cy$ or $y^2 = 4cx$, then the length of its latus rectum is $|4c|$.

10.3 THE HYPERBOLA

Definition 1   THE HYPERBOLA   A **hyperbola** is a set of points (x, y) with the property that the positive difference between the distances from (x, y) and each of two given (distinct) points is a constant. Each of the two given points is called a **focus** of the hyperbola.

To calculate the equation of a hyperbola, we place the axes so that the foci are the points $(c, 0)$ and $(-c, 0)$ and the difference of the distances from a point (x, y) on the hyperbola to the foci is equal to $2a > 0$. Note that from the triangle in Figure 1, it follows that $c > a$. (Explain why.) Then if (x, y) is a point on the hyperbola,

$$\overline{PF_2} - \overline{PF_1} = 2a,$$

or

$$\sqrt{(x + c)^2 + y^2} - \sqrt{(x - c)^2 + y^2} = 2a$$

(here we assumed that $\overline{PF_2} > \overline{PF_1}$ so that $\overline{PF_2} - \overline{PF_1}$ gives us a positive distance). Then we obtain, successively,

$$\sqrt{(x + c)^2 + y^2} = \sqrt{(x - c)^2 + y^2} + 2a$$

$$(x + c)^2 + y^2 = (x - c)^2 + y^2 + 4a\sqrt{(x - c)^2 + y^2} + 4a^2 \qquad \text{We squared.}$$

$$4cx - 4a^2 = 4a\sqrt{(x - c)^2 + y^2} \qquad \text{We simplified.}$$

$$\frac{c}{a}x - a = \sqrt{(x - c)^2 + y^2} \qquad \begin{array}{l}\text{We divided}\\\text{by } 4a.\end{array}$$

$$\frac{c^2}{a^2}x^2 - 2cx + a^2 = (x - c)^2 + y^2, \qquad \begin{array}{l}\text{We squared}\\\text{again.}\end{array}$$

or

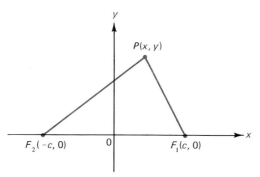

FIGURE 1

$$\left(\frac{c^2}{a^2} - 1\right)x^2 - y^2 = c^2 - a^2. \tag{1}$$

Finally, since $c > a$, $c^2 - a^2 > 0$, and we can define the positive number b by

$$b^2 = c^2 - a^2. \tag{2}$$

Then dividing both sides of (1) by $c^2 - a^2$, we obtain

$$\frac{x^2}{a^2} - \frac{y^2}{b^2} = 1. \tag{3}$$

A similar derivation shows that if we assume that $\overline{PF_1} > \overline{PF_2}$, we also obtain equation (3).

These two cases correspond to the right and left branches of the hyperbola sketched in Figure 2. Note that the hyperbola given by (3) is symmetric about both the x- and y-axes. The midpoint of the line segment joining the foci is called the **center** of the hyperbola. The **transverse axis** of the hyperbola is the line containing the foci. The **vertices** of the hyperbola are the points of intersection of the hyperbola with its

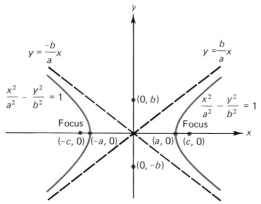

FIGURE 2

transverse axis. The hyperbola given by (3) is sketched in Figure 2. The hyperbola (3) has the asymptotes $y = \pm(b/a)x$. To prove this, we first note that for $|x| > a$,

$$y = \pm\frac{b}{a}\sqrt{x^2 - a^2}.$$

Then we find that

$$\lim_{x\to\infty} \left(\frac{b}{a}\sqrt{x^2 - a^2} - \frac{b}{a}x\right) = \frac{b}{a}\lim_{x\to\infty}(\sqrt{x^2 - a^2} - x)$$

$$= \frac{b}{a}\lim_{x\to\infty}\frac{(\sqrt{x^2 - a^2} - x)(\sqrt{x^2 - a^2} + x)}{\sqrt{x^2 - a^2} + x}$$

$$= \frac{b}{a}\lim_{x\to\infty}\frac{(x^2 - a^2) - x^2}{\sqrt{x^2 - a^2} + x}$$

$$= \frac{b}{a}\lim_{x\to\infty}\frac{-a^2}{\sqrt{x^2 - a^2} + x} = 0.$$

Similarly,

$$\lim_{x\to\infty}\left[-\frac{b}{a}\sqrt{x^2 - a^2} - \left(-\frac{b}{a}x\right)\right] = 0.$$

EXAMPLE 1   Find the equation of the hyperbola with foci at $(4, 0)$ and $(-4, 0)$ and with $a = 3$.

Solution. Here $a^2 = 9$ and $b^2 = c^2 - a^2 = 16 - 9 = 7$, so that the equation is

$$\frac{x^2}{9} - \frac{y^2}{7} = 1.$$

The asymptotes are $y = \pm(b/a)x = \pm(\sqrt{7}/3)x$. The curve is sketched in Figure 3. ■

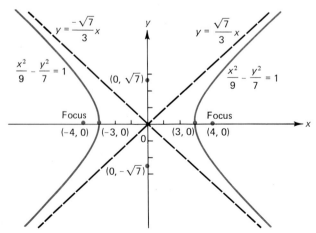

FIGURE 3

EXAMPLE 2 Discuss the curve given by $x^2 - 4y^2 = 9$.

Solution. Dividing by 9, we obtain $(x^2/9) - (4y^2/9) = 1$, or

$$\frac{x^2}{9} - \frac{y^2}{(\frac{3}{2})^2} = 1.$$

This is the equation of a hyperbola with $a = 3$ and $b = \frac{3}{2}$. Then $c^2 = a^2 + b^2 = 9 + 9/4 = 45/4$, so that the foci are $(\sqrt{45}/2, 0)$ and $(-\sqrt{45}/2, 0)$. The vertices are $(3, 0)$ and $(-3, 0)$, and the asymptotes are $y = \pm\frac{1}{2}x$. The curve is sketched in Figure 4. ■

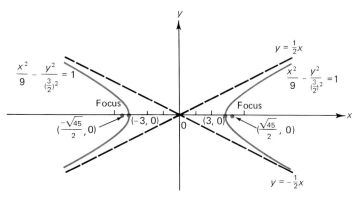

FIGURE 4

As with the ellipse and parabola, the roles of x and y can be reversed. The graph of the equation

$$\frac{y^2}{a^2} - \frac{x^2}{b^2} = 1 \tag{4}$$

is sketched in Figure 5. In (4) the transverse axis is on the y-axis. From (2) we have $c^2 = a^2 + b^2$, and the foci are at $(0, c)$ and $(0, -c)$. The vertices of the hyperbola given by (4) are the points $(0, a)$ and $(0, -a)$. The asymptotes are the lines $x = \pm(b/a)y$.

EXAMPLE 3 Discuss the curve given by $4y^2 - 9x^2 = 36$.

Solution. Dividing by 36, we obtain

$$\frac{y^2}{9} - \frac{x^2}{4} = 1.$$

Hence $a = 3$, $b = 2$, and $c^2 = 9 + 4 = 13$, so that $c = \sqrt{13}$. The foci are $(0, \sqrt{13})$ and $(0, -\sqrt{13})$, and the vertices are $(0, 3)$ and $(0, -3)$. The asymptotes are the lines $x = \frac{2}{3}y$ and $x = -\frac{2}{3}y$. This curve is sketched in Figure 6. ■

The hyperbolas we have sketched to this point are **standard hyperbolas.** Our results are summarized in Table 1.

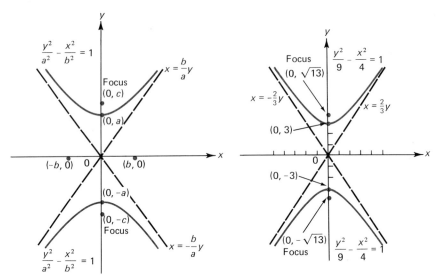

FIGURE 5   **FIGURE 6**

TABLE 1   STANDARD HYPERBOLAS

Equations	Description	Picture
$\dfrac{x^2}{a^2} - \dfrac{y^2}{b^2} = 1$	Hyperbola with foci at $(c, 0)$ and $(-c, 0)$, where $c^2 = a^2 + b^2$; transverse axis is the x-axis; center at origin; asymptotes $y = \pm(b/a)x$; curve opens to right and left	See Figure 2
$\dfrac{y^2}{a^2} - \dfrac{x^2}{b^2} = 1$	Hyperbola with foci at $(0, c)$ and $(0, -c)$, where $c^2 = a^2 + b^2$; transverse axis is the y-axis; center at origin; asymptotes $x = \pm(b/a)y$; curve opens at top and bottom	See Figure 5
$\dfrac{x^2}{a^2} - \dfrac{y^2}{b^2} = 0$	Degenerate hyperbola; graph consists of two lines: $y = \pm(b/a)x$	

The hyperbolas we have so far discussed have had their centers at the origin. Other hyperbolas can be obtained by a translation of the axes. The hyperbolas

$$\frac{(x - x_0)^2}{a^2} - \frac{(y - y_0)^2}{b^2} = 1 \tag{5}$$

and

$$\frac{(y - y_0)^2}{a^2} - \frac{(x - x_0)^2}{b^2} = 1 \tag{6}$$

have centers at the point (x_0, y_0).

EXAMPLE 4   Describe the hyperbola $(y - 2)^2/9 - (x + 3)^2/4 = 1$.

    Solution. This curve is the hyperbola of Example 3 shifted three units to the left and two units up, so that its center is at $(-3, 2)$. It is sketched in Figure 7. ∎

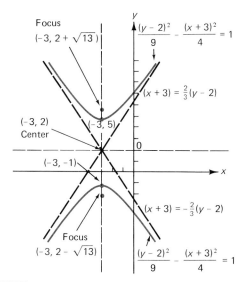

FIGURE 7

EXAMPLE 5   Describe the curve $x^2 - 4y^2 - 4x - 8y - 9 = 0$.

    Solution. We have

$$(x^2 - 4x) - (4y^2 + 8y) - 9 = 0,$$

or completing the squares,

$$(x - 2)^2 - 4 - 4[(y + 1)^2 - 1] - 9 = 0,$$

or,

$$(x - 2)^2 - 4(y + 1)^2 = 9,$$

and

$$\frac{(x - 2)^2}{9} - \frac{(y + 1)^2}{(\frac{3}{2})^2} = 1.$$

This is the equation of a hyperbola with center at $(2, -1)$. Except for the translation, it is the hyperbola of Example 2 and is sketched in Figure 8. ∎

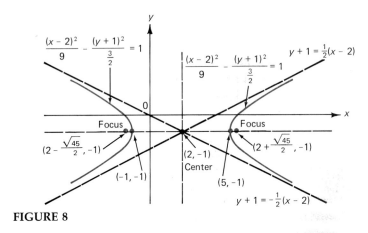

FIGURE 8

PROBLEMS 10.3

In Problems 1–20 the equation of a hyperbola is given. Find its foci, transverse axis, center, vertices, and asymptotes. Then sketch it.

1. $\dfrac{x^2}{16} - \dfrac{y^2}{25} = 1$

2. $\dfrac{y^2}{16} - \dfrac{x^2}{25} = 1$

3. $\dfrac{y^2}{25} - \dfrac{x^2}{16} = 1$

4. $\dfrac{x^2}{25} - \dfrac{y^2}{16} = 1$

5. $y^2 - x^2 = 1$

6. $x^2 - y^2 = 1$

7. $x^2 - 4y^2 = 9$

8. $4y^2 - x^2 = 9$

9. $y^2 - 4x^2 = 9$

10. $4x^2 - y^2 = 9$

11. $2x^2 - 3y^2 = 4$

12. $3x^2 - 2y^2 = 4$

13. $2y^2 - 3x^2 = 4$

14. $3y^2 - 2x^2 = 4$

15. $(x - 1)^2 - 4(y + 2)^2 = 4$

16. $\dfrac{(y + 3)^2}{4} - \dfrac{(x - 2)^2}{9} = 1$

17. $4x^2 + 8x - y^2 - 6y = 21$

18. $-4x^2 - 8x + y^2 - 6y = 20$

19. $2x^2 - 16x - 3y^2 + 12y = 45$

20. $2y^2 - 16y - 3x^2 + 12x = 45$

21. Find the equation of the hyperbola with foci at $(5, 0)$ and $(-5, 0)$ and vertices $(4, 0)$ and $(-4, 0)$.

22. Find the equation of the hyperbola with foci at $(0, 5)$ and $(0, -5)$ and vertices at $(0, 4)$ and $(0, -4)$.

23. Find the equation of the hyperbola with center at $(0, 0)$ and vertices at $(2, 0)$ and $(-2, 0)$ that is asymptotic to the lines $y = \pm 3x$.

24. Find the equation of the hyperbola obtained by shifting the hyperbola of Problem 22 so that its center is at $(4, -3)$.

25. The **eccentricity** e of a hyperbola is defined by

$$e = \frac{\text{distance between foci}}{\text{distance between vertices}}.$$

Show that the eccentricity of any hyperbola is greater than 1.

26. For the hyperbola $x^2/a^2 - y^2/b^2 = 1$, show that $e = \sqrt{a^2 + b^2}/a = \sqrt{1 + (b/a)^2}$.

In Problems 27–36 find the eccentricity of the given hyperbola.

27. the hyperbola of Problem 1   **28.** the hyperbola of Problem 4
29. the hyperbola of Problem 5   **30.** the hyperbola of Problem 8
31. the hyperbola of Problem 9   **32.** the hyperbola of Problem 12
33. the hyperbola of Problem 13   **34.** the hyperbola of Problem 16
35. the hyperbola of Problem 19   **36.** the hyperbola of Problem 20

37. Find the equation of the hyperbola obtained by shifting the hyperbola of Problem 17 two units down and five units to the right.

*38. Find the equation of the hyperbola with center at $(0, 0)$ and axis parallel to the x-axis that passes through the points $(1, 2)$ and $(5, 12)$.

39. Find the equation of the curve having the property that the difference of the distances from a point on the curve to the points $(1, -2)$ and $(4, 3)$ is 5.

*40. A tangent line to the hyperbola $9y^2 - 16x^2 = 25$ passes through the point $(1, 0)$. Find the point or points where this line is tangent to the hyperbola.

41. Show that the graph of the curve $x^2 + 4x - 3y^2 + 6y = c$ is **(a)** a hyperbola if $c \neq -1$, **(b)** a pair of striaght lines if $c = -1$. **(c)** If $c = -1$, find the equations of the lines.

42. Find conditions relating the numbers a, b, and c such that the graph of the equation $2x^2 + ax - 3y^2 + by = c$ is **(a)** a hyperbola, **(b)** a pair of straight lines.

10.4 SECOND-DEGREE EQUATIONS AND ROTATION OF AXES

The curves we have so far considered in this chapter have had their axes parallel to the two coordinate axes. This is not always the case. The curves sketched in Figure 1 are, respectively, an ellipse, a parabola, and a hyperbola. To obtain curves like those pictured, we must rotate the coordinate axes through an appropriate angle. How do we do so? Suppose that the x- and y-axes are rotated through an angle of θ with respect to the origin (see Figure 2). Let $P(x, y)$ represent a typical point in the coordinates x and y. We now seek a representation of that point in the "new" coordinates x' and y'. From Figure 2 we see that

$$x = \overline{0A} = \overline{0P} \cos(\theta + \alpha)$$

and

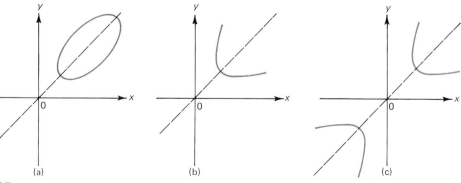

(a)   (b)   (c)

FIGURE 1

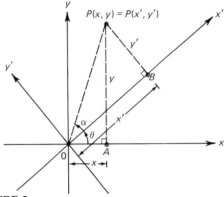

FIGURE 2

$$y = \overline{AP} = \overline{OP} \sin(\theta + \alpha).$$

But

$$\cos(\theta + \alpha) = \cos\theta\cos\alpha - \sin\theta\sin\alpha,$$

and

$$\sin(\theta + \alpha) = \sin\alpha\cos\theta + \cos\alpha\sin\theta.$$

Thus

$$x = \overline{OP}(\cos\theta\cos\alpha - \sin\theta\sin\alpha) \tag{1}$$

and

$$y = \overline{OP}(\sin\alpha\cos\theta + \cos\alpha\sin\theta). \tag{2}$$

Now from the right triangle $0BP$ we find that

$$\sin\alpha = \frac{y'}{\overline{OP}} \qquad \text{or} \qquad \overline{OP}\sin\alpha = y'.$$

Similarly,

$$\cos\alpha = \frac{x'}{\overline{OP}} \qquad \text{or} \qquad \overline{OP}\cos\alpha = x'.$$

Substituting these last expressions into (1) and (2) yields

$$x = x'\cos\theta - y'\sin\theta \tag{3}$$

$$y = x'\sin\theta + y'\cos\theta. \tag{4}$$

Equations (3) and (4) can be solved simultaneously to express the "new" coordinates x' and y' in terms of the "old" coordinates x and y. Switching from (x', y') to (x, y) is the same as rotating through an angle of $-\theta$. Since $\cos(-\theta) = \cos\theta$ and $\sin(-\theta) = -\sin\theta$, we obtain, from (3) and (4),

$$x' = x\cos\theta + y\sin\theta \tag{5}$$

$$y' = -x\sin\theta + y\cos\theta. \tag{6}$$

EXAMPLE 1 Find the equation of the curve obtained from the graph of $x^2 - xy + y^2 = 10$ by rotating the axes through an angle of $\pi/4$.

Solution. Since $\cos\pi/4 = \sin\pi/4 = 1/\sqrt{2}$, we obtain, from equations (3) and (4),

$$x = \frac{x' - y'}{\sqrt{2}} \quad \text{and} \quad y = \frac{x' + y'}{\sqrt{2}}.$$

Substitution of these into the equation $x^2 - xy + y^2 = 10$ yields

$$\left(\frac{x' - y'}{\sqrt{2}}\right)^2 - \left(\frac{x' - y'}{\sqrt{2}}\right)\left(\frac{x' + y'}{\sqrt{2}}\right) + \left(\frac{x' + y'}{\sqrt{2}}\right)^2 = 10,$$

or

$$\frac{(x')^2 - 2x'y' + (y')^2}{2} - \left[\frac{(x')^2 - (y')^2}{2}\right] + \frac{(x')^2 + 2x'y' + (y')^2}{2} = 10,$$

which after simplification yields

$$\frac{(x')^2 + (3y')^2}{2} = 10, \quad \text{or} \quad (x')^2 + 3(y')^2 = 20,$$

and, finally,

$$\frac{(x')^2}{20} + \frac{(y')^2}{\frac{20}{3}} = 1.$$

In the new coordinates x' and y' this is the equation of an ellipse with $a = \sqrt{20}$, $b = \sqrt{\frac{20}{3}}$, and $c = \sqrt{20 - \frac{20}{3}} = \sqrt{\frac{40}{3}}$, and foci at $(\sqrt{\frac{40}{3}}, 0)$ and $(-\sqrt{\frac{40}{3}}, 0)$. [This ellipse can be obtained by rotating the ellipse $(x^2/20) + (y^2/\frac{20}{3}) = 1$ through an angle of $\pi/4$.] It is sketched in Figure 3. We can obtain the foci in the coordinates (x, y) by using equations (3) and (4). Since

$$x = \frac{x' - y'}{\sqrt{2}} \quad \text{and} \quad y = \frac{x' + y'}{\sqrt{2}},$$

we find that $(\sqrt{\frac{40}{3}}, 0)$ in the coordinates (x', y') comes from $(\sqrt{\frac{20}{3}}, \sqrt{\frac{20}{3}})$ in the coordi-

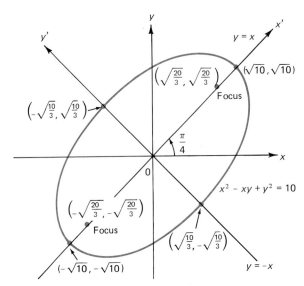

FIGURE 3

nates (x, y) and $(-\sqrt{\frac{40}{3}}, 0)$ comes from $(-\sqrt{\frac{20}{3}}, -\sqrt{\frac{20}{3}})$. The major axis is on the line $y = x$. ∎

EXAMPLE 2 Find the equation of the curve obtained from the graph of $xy = 1$ by rotating the axes through an angle of $\pi/4$.

  Solution. As in Example 1, we have $x = (x' - y')/\sqrt{2}$ and $y = (x' + y')/\sqrt{2}$, so that

$$1 = xy = \left(\frac{x' - y'}{\sqrt{2}}\right)\left(\frac{x' + y'}{\sqrt{2}}\right) = \frac{(x')^2}{2} - \frac{(y')^2}{2}.$$

Thus in the new (rotated) coordinate system we obtain the hyperbola $[(x')^2/2] - [(y')^2/2] = 1$. This curve is sketched in Figure 4. We may say that the equation $xy = 1$ is really the equation of a "standard" hyperbola rotated through an angle of $\pi/4$. ∎

  The last two examples illustrate the fact that the equations of our three basic curves can take forms other than the standard forms given by equations (10.1.2) and (10.1.6), (10.2.1)–(10.2.4), and (10.3.3)–(10.3.6). It turns out that any second-degree equation $Ax^2 + Bxy + Cy^2 + Dx + Ey + F = 0$ can be written in the standard form whose graph is a circle, an ellipse, a parabola, a hyperbola, or a degenerate form such as a line, a pair of lines, a point, or an empty set of points. We will not discuss this fact further except to state the following theorem. (You are asked to prove the theorem in Problem 24.)

Theorem 1. Consider the second-degree equation

$$Ax^2 + Bxy + Cy^2 + Dx + Ey + F = 0. \tag{7}$$

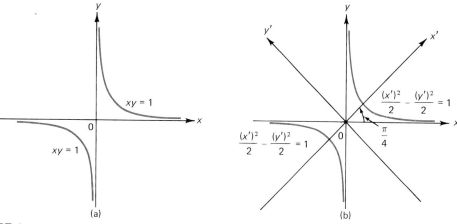

FIGURE 4

(i) If $B^2 - 4AC = 0$, then (7) is the equation of a parabola, a line, or two parallel lines or is imaginary.†
(ii) If $B^2 - 4AC < 0$, then (7) is the equation of a circle, an ellipse, or a single point or is imaginary.
(iii) If $B^2 - 4AC > 0$, then (7) is the equation of a hyperbola or two intersecting lines.

EXAMPLE 3   Determine the type of curve represented by the equation

$$16x^2 - 24xy + 9y^2 + 100x - 200y + 100 = 0. \qquad (8)$$

Then write the equation in a standard form by finding an appropriate translation and rotation of axes.

Solution. Here $A = 16$, $B = -24$, and $C = 9$, so that $B^2 - 4AC = (24)^2 - 4(16)(9) = 576 - 576 = 0$. Thus the equation represents a parabola (or a degenerate form of a parabola). To write it in a standard form, we first rotate the axes through an appropriate angle θ to eliminate the xy term in (8). To determine θ, we substitute

$$x = x' \cos \theta - y' \sin \theta \qquad \text{and} \qquad y = x' \sin \theta + y' \cos \theta$$

in equation (8). We then obtain

$$16(x' \cos \theta - y' \sin \theta)^2 - 24(x' \cos \theta - y' \sin \theta)(x' \sin \theta + y' \cos \theta)$$
$$+ 9(x' \sin \theta + y' \cos \theta)^2 + 100(x' \cos \theta - y' \sin \theta)$$
$$- 200(x' \sin \theta + y' \cos \theta) + 100 = 0.$$

---

†By "imaginary" we mean that the graph contains no real points; for example, $x^2 + 2xy + y^2 + 1 = 0$ satisfies $B^2 - 4AC = 0$; but there are no real values of x and y that satisfy $0 = x^2 + 2xy + y^2 + 1 = (x + y)^2 + 1$. (Explain why.)

The idea now is to choose θ so that the coefficient of the term $x'y'$ is zero. After simplification we find that this coefficient is given by

$$-24(\cos^2 \theta - \sin^2 \theta) - 14 \sin \theta \cos \theta.$$

Setting this expression equal to zero and using the fact that $\cos 2\theta = \cos^2 \theta - \sin^2 \theta$ and $\sin 2\theta = 2 \sin \theta \cos \theta$, we obtain

$$24 \cos 2\theta + 7 \sin 2\theta = 0,$$

or dividing by $\cos 2\theta$ and rearranging terms,

$$\tan 2\theta = -\frac{24}{7}.$$

If $\tan 2\theta = -\frac{24}{7}$, then $\cos 2\theta = -\frac{7}{25}$,† and also

$$\cos \theta = \sqrt{\frac{1 + \cos 2\theta}{2}} = \sqrt{\frac{9}{25}} = \frac{3}{5}.$$

Furthermore, $\sin \theta = \frac{4}{5}$. At this point we do not need to know θ but only the values of $\cos \theta$ and $\sin \theta$. Next,

$$x = x' \cos \theta - y' \sin \theta = \tfrac{1}{5}(3x' - 4y')$$

and

$$y = x' \sin \theta + y' \cos \theta = \tfrac{1}{5}(4x' + 3y').$$

We substitute these expressions into (8) to obtain (after simplification)

$$25(y')^2 - 200y' - 100x' + 100 = 0.$$

To further simplify, we divide by 25 and complete the square:

$$(y')^2 - 8y' - 4x' + 4 = 0,$$

or

$$(y' - 4)^2 - 16 - 4x' + 4 = 0,$$

or

$$(y' - 4)^2 = 4x' + 12 = 4(x' + 3).$$

---

†Alternatively, we could choose $\cos 2\theta = \frac{7}{25}$, yielding $\cos \theta = \frac{4}{5}$ and $\sin \theta = -\frac{3}{5}$. This, however, is merely a 90° rotation of the coordinate axes obtained with the choice $\cos 2\theta = -\frac{7}{25}$.

This is the equation of a parabola with vertex at $(-3, 4)$ in the new (rotated) coordinate system. To sketch the parabola, we need to calculate the angle of rotation given by $\theta = \cos^{-1}\frac{3}{5} \approx 0.927 \approx 53°$. The curve is sketched in Figure 5. ▪

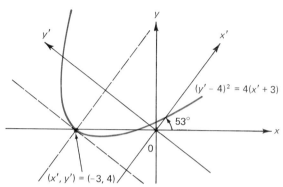

FIGURE 5

We note here that the xy term in (7) can always be eliminated by a rotation of axes through an angle θ, where θ is given by the equation

$$\cot 2\theta = \frac{A - C}{B}. \tag{9}$$

The proof of this fact follows exactly as in the derivation of the angle of rotation in Example 3 and is left as an exercise. (See Problem 21.)

We conclude by pointing out that instead of rotating axes, we can rotate curves, keeping the axes fixed. Note the following important fact:

Rotating a curve through an angle θ has the effect of rotating the axes through an angle $-\theta$.

PROBLEMS 10.4

1. Describe the curve obtained from the graph of $4x^2 - 2xy + 4y^2 = 45$ by rotating the axes through an angle of $\pi/4$.
2. Describe the curve obtained from the graph of $x^2 + 2\sqrt{3}xy - y^2 = 4$ by rotating the axes through an angle of $\pi/6$.
3. What is the equation of the line obtained from the line $2x - 3y = 6$ by rotating the axes through an angle of $\pi/6$?
4. Find the equation of the line obtained from the line $ax + by + c = 0$ by rotating the axes through an angle θ.
5. Find the equation of a parabola obtained from the parabola $y^2 = -12x$ if the axes are rotated until the axis of the parabola coincides with the line $y = \sqrt{3}x$.
*6. Find the equation of the ellipse whose major axis is on the line $y = -x$, which is obtained by rotating the ellipse $(x^2/25) + (y^2/16) = 1$. [*Hint:* Find the angle through which the axes must be rotated to accomplish this.]

In Problems 7–20, find a rotation of coordinate axes in which the given equation written in the new coordinates has no xy term. Describe each curve and then sketch it.

7. $4x^2 + 4xy + y^2 = 9$ **8.** $4x^2 + 4xy - y^2 = 9$

9. $3x^2 - 2xy - 5 = 0$ **10.** $xy = 2$

11. $xy = a, \, a > 0$ **12.** $xy = a, \, a < 0$

13. $4x^2 + 4xy + y^2 + 20x - 10y = 0$ **14.** $x^2 + 4xy + 4y^2 - 6 = 0$

15. $2x^2 + xy + y^2 = 4$ **16.** $9x^2 + 6xy + y^2 + 10x - 30y = 0$

17. $3x^2 - 6xy + 5y^2 = 36$ **18.** $x^2 - 3xy + 4y^2 = 1$

19. $3y^2 - 4xy + 30y - 20x + 40 = 0$ **20.** $6x^2 + 5xy - 6y^2 + 7 = 0$

21. Show that if $A \neq C$, the xy term in the second-degree equation (7) will be eliminated by rotation through an angle θ if θ is given by

$$\cot 2\theta = \frac{A - C}{B}.$$

22. Show that if $A = C$ in Problem 21, then the xy term will be eliminated by a rotation through an angle of either $\pi/4$ or $-\pi/4$.

*23. Suppose that a rotation converts $Ax^2 + Bxy + Cy^2$ into $A'(x')^2 + B'(x'y') + C'(y')^2$.
   (a) Show that $A + C = A' + C'$ **(b)** Show that $B^2 - 4AC = (B')^2 - 4A'C'$.

*24. Use the result of Problem 23 to prove Theorem 1. [*Hint:* Rotate the axes so that $B' = 0$.]

25. Show that, for every constant $k \neq 0$, $xy = k$ is the equation of a hyperbola.

REVIEW EXERCISES FOR CHAPTER TEN

In Problems 1–14, identify the type of conic. If it is an ellipse (or circle), give its foci, center, vertices, major and minor axes, and eccentricity. If it is a parabola, give its focus, directrix, axis, and vertex. If it is a hyperbola, give its foci, transverse axis, center, vertices, asymptotes, and eccentricity. Finally, sketch the curve (if it is not degenerate).

1. $\dfrac{x^2}{9} + \dfrac{y^2}{16} = 1$ **2.** $\dfrac{x^2}{9} - \dfrac{y^2}{16} = 1$ **3.** $\dfrac{x^2}{9} - \dfrac{y}{16} = 0$

4. $\dfrac{y^2}{16} - \dfrac{x}{9} = 0$ **5.** $\dfrac{y^2}{9} - \dfrac{x^2}{16} = 1$ **6.** $\dfrac{x^2}{16} + \dfrac{y^2}{9} = 1$

7. $\dfrac{(x-1)^2}{4} + \dfrac{(y+1)^2}{9} = 1$ **8.** $\dfrac{(x+2)^2}{25} - \dfrac{(y+3)^2}{4} = 1$

9. $\dfrac{(x+2)^2}{25} + \dfrac{(y-5)^2}{25} = 0$ **10.** $x^2 + 2x + y^2 + 2y = 0$

11. $x^2 + 2x - y^2 + 2y = 0$ **12.** $x^2 + 2x - 2y = 0$

13. $4x^2 + 4x + 3y^2 + 24y = 5$ **14.** $-3x^2 + 6x + 2y^2 + 4y = 6$

15. Find the equation of an ellipse with foci at $(3, 0)$ and $(-3, 0)$ and eccentricity 0.6.

16. Find the equation of the parabola with focus $(3, 0)$ and directrix the line $x = -4$.

17. Find the equation of the hyperbola with foci $(0, 3)$ and $(0, -3)$ and vertices $(0, 2)$ and $(0, -2)$.

18. Find the equation of the hyperbola centered at $(0, 0)$ with vertices at $(0, 3)$ and $(0, -3)$ that is asymptotic to the lines $y = \pm 5x$.

19. Describe the curve obtained from the graph of $2x^2 + xy + 2y^2 = 10$. [*Hint:* Rotate through an angle of $\pi/4$.]

20. Find the equation of the line obtained by rotating the line $2x - 3y = 7$ through an angle of $\pi/3$.

In Problems 21–26, find a rotation of coordinate axes in which the given equation written in the new coordinates has no xy term. Describe each curve and then sketch it.

21. $x^2 + 4xy + 4y^2 = 9$

22. $x^2 + 4xy - 4y^2 = 9$

23. $xy = -3$

24. $4x^2 + 3xy + y^2 = 1$

25. $-2x^2 + 3xy + 4y^2 = 5$

26. $x^2 + 2xy + y^2 = 6$

11 Polar Coordinates

11.1 THE POLAR COORDINATE SYSTEM

In Section 1.3 we introduced the Cartesian plane (the xy-plane), and up to this moment we have represented points in the plane by their x- and y-coordinates. There are many other ways to represent points in the plane, the most important of which is called the **polar coordinate system.**

We begin by choosing a fixed point, which we label O, and a ray (half line) that extends in one direction from O, which we label OA. The fixed point is called the **pole** (or **origin**), and the ray OA is called the **polar axis.** In Figure 1 the polar axis is drawn as a horizontal ray that extends indefinitely to the right (just like the positive x-axis).

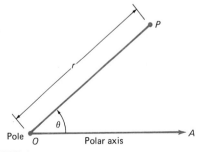

FIGURE 1

This representation is a matter of convention, although it would be correct to have the polar axis extend in any direction.

If $P \neq 0$ is any other point in the plane, let r denote the distance between O and P, and let θ represent the angle (in radians) between OA and OP, measured counterclockwise from OA to OP. With this representation we initially use the convention that $0 \leq \theta < 2\pi$ and $r > 0$ (except at the pole). Then every point in the plane, except the pole, can be represented by a pair of numbers (r, θ), where $r > 0$ and $0 \leq \theta < 2\pi$. The pole can be represented as $(0, \theta)$ for any number θ. The representation $P = (r, \theta)$ is called **the polar representation** of the point, and r and θ are called the **polar coordinates** of P. Some typical points are sketched in Figure 2.

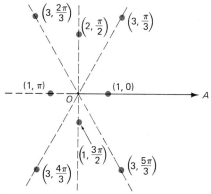

FIGURE 2

If we are given two numbers r and θ, with $r \geq 0$ and $0 \leq \theta < 2\pi$, then we can draw the point $P = (r, \theta)$ in the plane. But what do we do if either $r < 0$ or $\theta > 2\pi$? To avoid this difficulty, we simply extend our definition. Let (r, θ) be any pair of real numbers. To locate the point $P = (r, \theta)$, we first rotate the polar axis through an angle of θ in the counterclockwise direction if θ is positive and in the clockwise direction if θ is negative. Let us call this new ray OB. Then if $r > 0$, the point P is placed on the ray OB, r units from the pole. If $r < 0$, the ray OB is extended backward through the pole and the point P is then located $|r|$ units from the pole along this extended ray.

EXAMPLE 1   Plot the following points:

(a) $\left(2, \dfrac{8\pi}{3}\right)$ (b) $\left(-1, \dfrac{\pi}{4}\right)$ (c) $\left(3, -22\dfrac{1}{2}\pi\right)$ (d) $\left(-2, -\dfrac{\pi}{2}\right)$

Solution. The four points are plotted in Figure 3.

In (a), since $8\pi/3 = 2\pi + (2\pi/3)$, the line OB makes an angle of $2\pi/3$ with OA and the point P lies 2 units along this line.

In (b) we rotate $\pi/4$ units and then extend OB backward since $r < 0$. The point P is located 1 unit along this extended line.

In (c) we rotate in a clockwise direction and find that the line OB makes an angle of $\pi/2$ with the polar axis since $22\frac{1}{2}\pi = 11(2\pi) + (\pi/2)$. The point P is located 3 units along OB.

In (d) we rotate OA in a clockwise direction $\pi/2$ radians and then extend OB

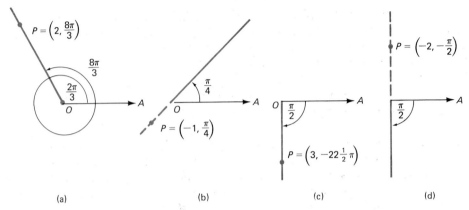

(a)   (b)   (c)   (d)

FIGURE 3

backward (and therefore upward) since $r < 0$. The point P is then located 2 units along this line. ∎

We immediately notice that if we do not restrict r and θ, then there are many (in fact, an infinite number of) representations for each point in the plane. For example, in Figure 3d we see that $(-2, -\pi/2) = (2, \pi/2) = (2, (\pi/2) + 2n\pi)$ for any integer n. In general,

$$P = (r, \theta) = (-r, \theta + \pi) = (r, \theta + 2n\pi) \quad \text{for} \quad n = \pm 1, \pm 2, \pm 3, \ldots$$

(see Problem 36). Actually, these ordered pairs are not equal. The equal sign indicates that the points they represent are the same. Therefore, with this understanding we will write $(r_1, \theta_1) = (r_2, \theta_2)$ in the rest of this chapter when the points they represent are the same, even though r_1 may not be equal to r_2, and θ_1 may not be equal to θ_2. However, to avoid difficulties, if a point $P \neq 0$ is given, then we can write $P = (r, \theta)$ in polar coordinates in a unique way if we specify that $r > 0$ and $0 \leq \theta < 2\pi$. If P is the pole O, then $P = (0, \theta)$ for any real number θ, and there is no unique representation unless we specify a value for θ.

What is the relationship between polar and rectangular (Cartesian) coordinates? To find it, we place the pole at the origin in the xy-plane with the polar axis along the x-axis as in Figure 4. Let $P = (x, y) = (r, \theta)$ be a point in the plane. Then as is evident from Figure 4, we have

From polar to rectangular coordinates

$$\cos \theta = \frac{x}{r} \quad \text{and} \quad \sin \theta = \frac{y}{r}.$$

or

$$x = r \cos \theta \quad \text{and} \quad y = r \sin \theta. \tag{1}$$

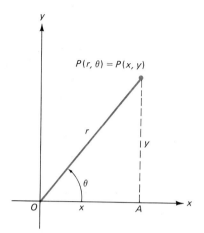

FIGURE 4

From rectangular to polar coordinates

Similarly, from the Pythagorean theorem we see that $x^2 + y^2 = r^2$, so that if we specify that $r > 0$, we have

$$r = \sqrt{x^2 + y^2}. \tag{2}$$

Finally, we can calculate θ from x and y by the relation

$$\tan \theta = \frac{y}{x} \quad \text{if} \quad x \neq 0. \tag{3}$$

Before citing examples, we note that these conversion formulas have only been illustrated for $r > 0$ and $0 \leq \theta < \pi/2$. The fact that $0 \leq \theta < \pi/2$ in Figure 4 is irrelevant since, using the circular definition of $\sin x$ and $\cos x$ in Appendix 1 (Section A1.2), we see that $x/r = \cos \theta$ and $y/r = \sin \theta$ for any real number θ.

WARNING.   Formula (3) does *not* determine θ uniquely. The signs of x and y must be taken into account. For example, for $(1, 1)$ and $(-1, -1)$, $\tan \theta = 1$. But in the first case $\theta = \pi/4$ and in the second case $\theta = 5\pi/4$ (see Figure 5).

To deal with the case $r < 0$, we note the identity

$$(-r, \theta) = (r, \theta + \pi). \tag{4}$$

The proof is left as an exercise (see Problem 36).

With these formulas we can always convert from polar to rectangular coordinates in a unique way. To convert from rectangular to polar coordinates in a unique way, we must specify that $r > 0$ and $0 \leq \theta < 2\pi$.

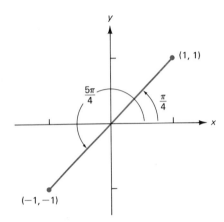

FIGURE 5

EXAMPLE 2   Convert from polar to rectangular coordinates:

(a) $\left(3, \dfrac{\pi}{6}\right)$ (b) $\left(4, \dfrac{2\pi}{3}\right)$ (c) $\left(-6, \dfrac{\pi}{4}\right)$ (d) $(2, 0)$ (e) $(1, -\pi)$

Solution. (a) $x = 3 \cos \pi/6$ and $y = 3 \sin \pi/6$, so that $(x, y) = (3\sqrt{3}/2, 3/2)$.
(b) $x = 4 \cos 2\pi/3 = -4 \cos \pi/3 = -4/2 = -2$ and $y = 4 \sin 2\pi/3 = 4 \sin \pi/3 = 4\sqrt{3}/2 = 2\sqrt{3}$, so that $(x, y) = (-2, 2\sqrt{3})$.
(c) $(-6, \pi/4) = (6, (\pi/4) + \pi) = (6, 5\pi/4)$ [from (4)], and we have $x = 6 \cos 5\pi/4 = 6(-\sqrt{2}/2) = -3\sqrt{2}$, $y = 6 \sin 5\pi/4 = -3\sqrt{2}$, and $(x, y) = (-3\sqrt{2}, -3\sqrt{2})$.
(d) $x = 2 \cos 0 = 2$ and $y = 2 \sin 0 = 0$, so that $(x, y) = (2, 0)$.
(e) Since $(1, -\pi) = (1, -\pi + 2\pi) = (1, \pi)$, we have $x = \cos \pi = -1$, $y = \sin \pi = 0$, and $(x, y) = (-1, 0)$.

These five points are sketched in Figure 6. ■

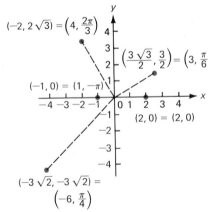

FIGURE 6

EXAMPLE 3   Convert from rectangular to polar coordinates (with $r > 0$ and $0 \le \theta < 2\pi$):

(a) $(1, \sqrt{3})$ (b) $(2, -2)$ (c) $(-4\sqrt{3}, -4)$

Solution. **(a)** $r = \sqrt{x^2 + y^2} = \sqrt{1^2 + \sqrt{3}^2} = \sqrt{1 + 3} = 2$ and $\tan\theta = \sqrt{3}/1 = \sqrt{3}$. Since $(1, \sqrt{3})$ is in the first quadrant, we have $\theta = \pi/3$, so that $(r, \theta) = (2, \pi/3)$.
   (b) $r = \sqrt{(2)^2 + (-2)^2} = \sqrt{8} = 2\sqrt{2}$ and $\tan\theta = -2/2 = -1$. Since $(2, -2)$ is in the fourth quadrant, $\theta = 7\pi/4$ and $(r, \theta) = (2\sqrt{2}, 7\pi/4)$.
   (c) $r = \sqrt{(-4\sqrt{3})^2 + (-4)^2} = \sqrt{48 + 16} = 8$ and $\tan\theta = -4/(-4\sqrt{3}) = 1/\sqrt{3}$. Since $(-4\sqrt{3}, -4)$ is in the third quadrant, we have $\theta = 7\pi/6$ and $(r, \theta) = (8, 7\pi/6)$.

These three points are sketched in Figure 7. ■

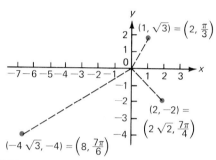

FIGURE 7

PROBLEMS 11.1

In Problems 1–18 a point is given in polar coordinates. Write the point in rectangular coordinates and plot it in the xy-plane, showing both representations.

1. $(3, 0)$ **2.** $(4, \pi/4)$ **3.** $(-5, 0)$
4. $(-4, \pi/4)$ **5.** $(6, 7\pi/6)$ **6.** $(-3, -\pi/6)$
7. $(5, \pi)$ **8.** $(5, -\pi)$ **9.** $(-5, -\pi)$
10. $(2, 3\pi/4)$ **11.** $(-2, 3\pi/4)$ **12.** $(-2, 5\pi/4)$
13. $(2, 7\pi/4)$ **14.** $(-2, -\pi/4)$ **15.** $(1, 3\pi/2)$
16. $(1, -\pi/2)$ **17.** $(-1, \pi/2)$ **18.** $(-1, -3\pi/2)$

In Problems 19–35 a point is given in rectangular coordinates. Write the point in polar coordinates with $r > 0$ and $0 \le \theta < 2\pi$. Then plot the point with both representations in the xy-plane.

19. $(3, 0)$ **20.** $(7, 0)$ **21.** $(-3, 0)$
22. $(1, 1)$ **23.** $(-1, -1)$ **24.** $(1, -1)$
25. $(-1, 1)$ **26.** $(0, 1)$ [*Hint:* Draw a sketch first.]
27. $(0, -1)$ **28.** $(2, 2\sqrt{3})$ **29.** $(-2, 2\sqrt{3})$
30. $(2, -2\sqrt{3})$ **31.** $(-2, -2\sqrt{3})$ **32.** $(2\sqrt{3}, 2)$
33. $(2\sqrt{3}, -2)$ **34.** $(-2\sqrt{3}, 2)$ **35.** $(-2\sqrt{3}, -2)$

36. Show that $(-r, \theta) = (r, \theta + \pi)$. [*Hint:* Draw a sketch.]

11.2 GRAPHING IN POLAR COORDINATES

In rectangular coordinates we define the graph of the equation $y = f(x)$, or, more generally, $F(x, y) = 0$, as the set of points (x, y) whose coordinates satisfy the equation. In polar coordinates, however, we must be careful, since each point in the plane has an infinite number of representations.

Definition 1 GRAPH IN POLAR COORDINATES The **graph** of an equation written in polar coordinates r and θ consists of those points P having at least one representation $P = (r, \theta)$ whose coordinates satisfy the equation.

EXAMPLE 1 Find the graph of the polar equation $r = 1$.

Solution. The set of points for which $r = 1$ is simply the set of points one unit from the pole, which is the definition of the unit circle. Thus the graph of $r = 1$ is the unit circle. To see this in another way, note that $r = \sqrt{x^2 + y^2} = 1$ implies that $x^2 + y^2 = 1$. This curve is sketched in Figure 1, together with the curves $r = 2$ (the circle of radius 2), $r = 3$, and $r = \frac{1}{2}$. ■

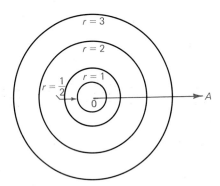

FIGURE 1

REMARK. Every point on the circle $r = 1$ has a representation for which $r \neq 1$ since the point $(-1, \theta)$ is on the circle for any real number θ. This presents no problem since $(-1, \theta) = (1, \theta + \pi)$.

EXAMPLE 2 Sketch the curve $\theta = \pi/4$.

Solution. The curve $\theta = \pi/4$ is the straight line passing through the pole making an angle of $\pi/4$ with the polar axis (see Figure 2). It extends in both directions because it contains points of the form $(-r, \pi/4) = (r, 5\pi/4)$. To see in another way that this graph is a straight line, note that $y/x = \tan \theta = \tan \pi/4 = 1$, so that $y = x$ in the rectangular representation of the line. ■

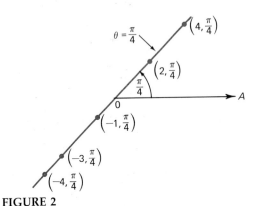

FIGURE 2

Examples 1 and 2 give us some very useful information. Previously, we sketched graphs by representing a typical point $P = (x_0, y_0)$ as the intersection of the line $x = x_0$ with the line $y = y_0$ (Figure 3a). In polar coordinates we can represent the point $P = (r_0, \theta_0)$ as the intersection of the circle $r = r_0$ with the ray $\theta = \theta_0$ $(r \geq 0)$ (Figure 3b).

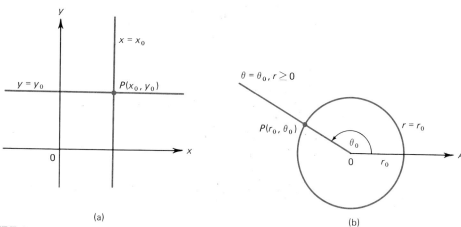

(a)

(b)

FIGURE 3

We now consider the graphs of some more general curves in polar coordinates. To aid us in obtaining sketches of these curves, we cite three rules that are often useful.

RULES OF SYMMETRY

(i) If in a polar equation θ can be replaced by $-\theta$ without changing the equation, then the polar graph is symmetric about the polar axis (see Figure 4a).

(ii) If θ can be replaced by $\pi - \theta$ without changing the equation, then the polar graph is symmetric about the line $\theta = \pi/2$ (see Figure 4b).

(iii) If r can be replaced by $-r$, or, equivalently, if θ can be replaced by $\theta + \pi$, without changing the equation, then the polar graph is symmetric about the pole (see Figure 4c).

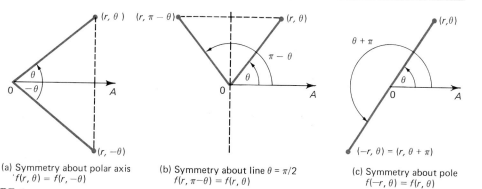

(a) Symmetry about polar axis
$f(r, \theta) = f(r, -\theta)$

(b) Symmetry about line $\theta = \pi/2$
$f(r, \pi-\theta) = f(r, \theta)$

(c) Symmetry about pole
$f(-r, \theta) = f(r, \theta)$

FIGURE 4

EXAMPLE 3   Sketch the curve $r = \cos\theta$.

   Solution. Since $\cos(-\theta) = \cos\theta$, rule (i) implies that the graph is symmetric about the polar axis, so we need only consider values of θ between 0 and π. In Table 1 we tabulate r for some typical values of θ in $[0, \pi]$. From this table we can plot the curve for $0 \le \theta < \pi$. This is done in Figure 5. Here we have used the fact that

$$\left(-\frac{1}{2}, \frac{2\pi}{3}\right) = \left(\frac{1}{2}, \frac{5\pi}{3}\right), \qquad \left(-\frac{\sqrt{2}}{2}, \frac{3\pi}{4}\right) = \left(\frac{\sqrt{2}}{2}, \frac{7\pi}{4}\right),$$

and

$$\left(-\frac{\sqrt{3}}{2}, \frac{5\pi}{6}\right) = \left(\frac{\sqrt{3}}{2}, \frac{11\pi}{6}\right).$$

   From the sketch in Figure 5 it appears that the graph is a circle. To see this analytically, we convert to rectangular coordinates. Since $r = \cos\theta$, we have, for $r \ne 0$,

$$r^2 = r\cos\theta$$

($r = 0$ is just the origin, which is already known to be on the graph), or

$$x^2 + y^2 = x.$$

TABLE 1

θ	0	$\dfrac{\pi}{6}$	$\dfrac{\pi}{4}$	$\dfrac{\pi}{3}$	$\dfrac{\pi}{2}$	$\dfrac{2\pi}{3}$	$\dfrac{3\pi}{4}$	$\dfrac{5\pi}{6}$	π
$r = \cos\theta$	1	$\dfrac{\sqrt{3}}{2}$	$\dfrac{\sqrt{2}}{2}$	$\dfrac{1}{2}$	0	$-\dfrac{1}{2}$	$-\dfrac{\sqrt{2}}{2}$	$-\dfrac{\sqrt{3}}{2}$	-1

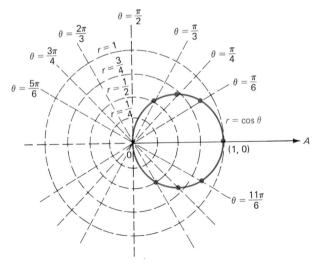

FIGURE 5

This equation can be written as

$$x^2 + y^2 - x = 0.$$

Then completing the square, we obtain $(x - \frac{1}{2})^2 + y^2 = \frac{1}{4}$, which is the equation of a circle centered at $(\frac{1}{2}, 0)$ with radius $\frac{1}{2}$. ■

In general, the graph of the equation $r = a \cos \theta + b \sin \theta$ is a circle for any real numbers a and b. (See Problems 66, 67, and 68.)

EXAMPLE 4   Sketch the curve $r = 1 + \sin \theta$.

Solution. Since $\sin(\pi - \theta) = \sin \theta$, the curve is symmetric about the line $\theta = \pi/2$. We therefore need only consider values of θ in $[0, \pi/2]$ and $[3\pi/2, 2\pi]$. Typical values of the function are given in Table 2. We use these values in Figure 6 and then use symmetry to reflect the curve about the line $\theta = \pi/2$. The heart-shaped curve we have sketched is called a **cardioid,** from the Greek *kardia,* meaning ''heart.'' ■

TABLE 2

θ	0	$\dfrac{\pi}{6}$	$\dfrac{\pi}{4}$	$\dfrac{\pi}{3}$	$\dfrac{\pi}{2}$	$\dfrac{3\pi}{2}$	$\dfrac{5\pi}{3}$	$\dfrac{7\pi}{4}$	$\dfrac{11\pi}{6}$
$r = 1 + \sin \theta$	1	$\dfrac{3}{2}$	$1 + \dfrac{\sqrt{2}}{2}$	$1 + \dfrac{\sqrt{3}}{2}$	2	0	$1 - \dfrac{\sqrt{3}}{2}$	$1 - \dfrac{\sqrt{2}}{2}$	$\dfrac{1}{2}$

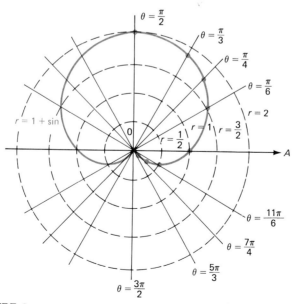

FIGURE 6

EXAMPLE 5   Sketch the curve $r = 3 - 2 \cos \theta$.

Solution. The curve is symmetric about the polar axis, and we therefore calculate values of r for θ in $[0, \pi]$ (Table 3). We then obtain the graph sketched in Figure 7. The curve sketched in Figure 7 is called a **limaçon.**† ▪

TABLE 3

θ	0	$\dfrac{\pi}{6}$	$\dfrac{\pi}{4}$	$\dfrac{\pi}{3}$	$\dfrac{\pi}{2}$	$\dfrac{2\pi}{3}$	$\dfrac{3\pi}{4}$	$\dfrac{5\pi}{6}$	π
$r = 3 - 2\cos\theta$	1	$3 - \sqrt{3}$	$3 - \sqrt{2}$	2	3	4	$3 + \sqrt{2}$	$3 + \sqrt{3}$	5

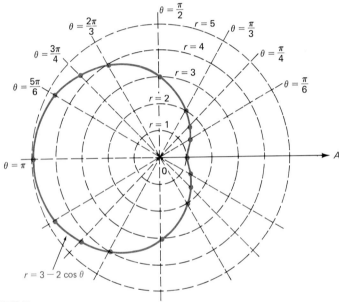

FIGURE 7

EXAMPLE 6   Sketch the curve $r = 1 + 2\cos\theta$.

Solution. This curve is also symmetric about the polar axis, and we therefore tabulate the function for θ in $[0, \pi]$ (Table 4).

Using the facts that

$$\left(1 - \sqrt{2}, \frac{3\pi}{4}\right) = \left(\sqrt{2} - 1, \frac{7\pi}{4}\right), \qquad \left(1 - \sqrt{3}, \frac{5\pi}{6}\right) = \left(\sqrt{3} - 1, \frac{11\pi}{6}\right),$$

and

$$(-1, \pi) = (1, 2\pi) = (1, 0),$$

together with the symmetry around the polar axis, we obtain the graph in Figure 8. This curve is also called a limaçon. ▪

†From the Latin word *limax,* meaning "snail." The curve is also referred to as *Pascal's limaçon* since it was discovered by Etienne Pascal (1588–1640), father of the famous French mathematician Blaise Pascal (1623–1662).

TABLE 4

θ	0	$\dfrac{\pi}{6}$	$\dfrac{\pi}{4}$	$\dfrac{\pi}{3}$	$\dfrac{\pi}{2}$	$\dfrac{2\pi}{3}$	$\dfrac{3\pi}{4}$	$\dfrac{5\pi}{6}$	π
$r = 1 + 2\cos\theta$	3	$1+\sqrt{3}$	$1+\sqrt{2}$	2	1	0	$1-\sqrt{2}$	$1-\sqrt{3}$	-1

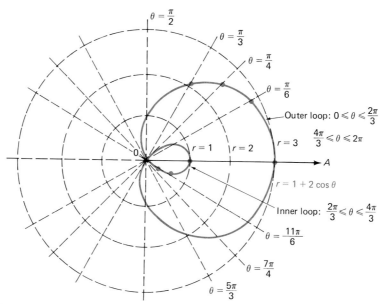

FIGURE 8

We can generalize the results of the last three examples.

> The graph of the equation
>
> $$r = a \pm b\cos\theta \quad \text{or} \quad r = a \pm b\sin\theta, \quad a > 0, b > 0,$$
>
> is a limaçon, which can take one of three possible shapes:
>
>   **(i)** If $a = b$, we obtain a cardioid as in Figure 6.
>   **(ii)** If $a > b$, we obtain a limaçon having the appearance of the curve in Figure 7.
>   **(iii)** If $b > a$, we obtain a limaçon with a loop as in Figure 8.

EXAMPLE 7   Sketch the curve $r = 3\cos 2\theta$.

Solution. We first test for symmetry. Replacing θ by $-\theta$, we have $3\cos 2(-\theta) = 3\cos 2\theta$, implying that the curve is symmetric about the polar axis. Similarly, replacing θ by $\pi - \theta$ yields the equations

$$3\cos 2(\pi - \theta) = 3\cos(2\pi - 2\theta) = 3\cos(-2\theta) = 3\cos 2\theta,$$

so that the curve is symmetric about the line $\theta = \pi/2$. Finally, replacing θ by $\pi + \theta$, we obtain

$$3 \cos 2(\pi + \theta) = 3 \cos(2\pi + 2\theta) = 3 \cos 2\theta,$$

so that the curve is also symmetric about the pole. Thus it is only necessary to sketch the curve for $0 \leq \theta \leq \pi/2$ and then reflect about the polar axis, the line $x = \pi/2$, and the pole. Values for r and θ are given in Table 5. Now

$$\left(-\frac{3}{2}, \frac{\pi}{3}\right) = \left(\frac{3}{2}, \frac{4\pi}{3}\right), \qquad \left(\frac{-3\sqrt{3}}{2}, \frac{5\pi}{12}\right) = \left(\frac{3\sqrt{3}}{2}, \frac{17\pi}{12}\right),$$

and

$$\left(-3, \frac{\pi}{2}\right) = \left(3, \frac{3\pi}{2}\right).$$

Then using symmetry, we obtain the curve sketched in Figure 9. This curve is called a **four-leafed rose.** ■

TABLE 5

θ	0	$\dfrac{\pi}{12}$	$\dfrac{\pi}{8}$	$\dfrac{\pi}{6}$	$\dfrac{\pi}{4}$	$\dfrac{\pi}{3}$	$\dfrac{5\pi}{12}$	$\dfrac{\pi}{2}$
$r = 3 \cos 2\theta$	3	$\dfrac{3\sqrt{3}}{2}$	$\dfrac{3\sqrt{2}}{2}$	$\dfrac{3}{2}$	0	$-\dfrac{3}{2}$	$-\dfrac{3\sqrt{3}}{2}$	-3

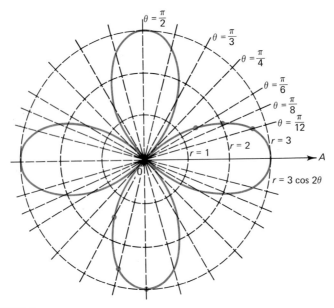

FIGURE 9

In general, any curve of the form

$$r = a \cos 2\theta \qquad \text{or} \qquad r = a \sin 2\theta$$

is a *four-leafed rose*.

EXAMPLE 8 Sketch the curve $r^2 = 4 \sin 2\theta$.

Solution. Two things should be immediately evident from the equation. First, since $(-r)^2 = r^2$, the graph is symmetric about the pole. Second, since $r^2 \geq 0$, the function is only defined for values of θ such that $\sin 2\theta \geq 0$. If θ is restricted to the interval $[0, 2\pi]$, then $\sin 2\theta \geq 0$ if and only if θ is in $[0, \pi/2]$ or $[\pi, 3\pi/2]$. Then using Table 6 and the symmetry about the pole, we obtain the graph sketched in Figure 10. This curve is called a **lemniscate.†**

TABLE 6

θ	0	$\dfrac{\pi}{12}$	$\dfrac{\pi}{8}$	$\dfrac{\pi}{6}$	$\dfrac{\pi}{4}$	$\dfrac{\pi}{3}$	$\dfrac{5\pi}{12}$	$\dfrac{\pi}{2}$
$r = \sqrt{4 \sin 2\theta}$	0	$\sqrt{2}$	$2^{3/4}$	$\sqrt{2} \cdot 3^{1/4}$	2	$\sqrt{2} \cdot 3^{1/4}$	$\sqrt{2}$	0

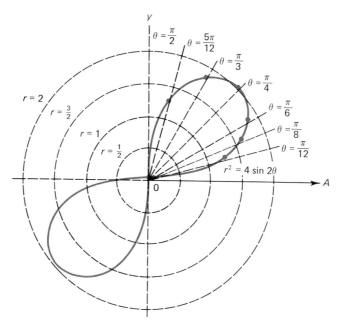

FIGURE 10

---

†From the Greek *lemniskos* and the Latin *lemniscus,* meaning "knotted ribbon." Actually, this curve was originally called the *lemniscate of Bernoulli,* named after the Swiss mathematician Jacques Bernoulli (1654–1705), who described the curve in the *Acta Eruditorum* published in 1694.

In general, any curve with the equation

$$r^2 = a \sin b\theta \qquad \text{or} \qquad r^2 = a \cos b\theta$$

is a *lemniscate*.

EXAMPLE 9   Sketch the curve $r = \theta$, $\theta \geq 0$.

Solution. This curve has no symmetry. To sketch its graph, we note that as we move around the pole, θ increases as r increases. The graph is sketched in Figure 11, and the curve depicted is called the **spiral of Archimedes.**† It intersects the polar axis when $\theta = 2n\pi$, n an integer, and it intersects the line $\theta = \pi/2$ when $\theta = [(2n + 1)/2]\pi$. ∎

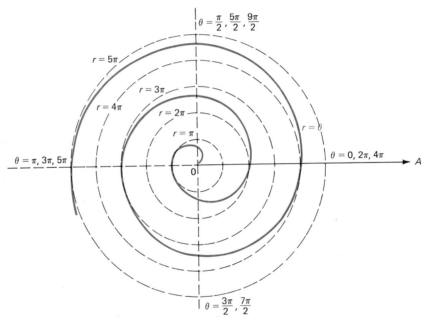

FIGURE 11

PROBLEMS 11.2

In Problems 1–65, sketch the graph of the given equation, indicating any symmetry about the polar axis, the line $\theta = \pi/2$, and/or the pole.

1. $r = 5$ **2.** $r = 7$ **3.** $r = -4$
4. $\theta = 3\pi/8$ **5.** $\theta = -\pi/6$ **6.** $\theta = 13\pi/5$
7. $r = 5 \sin \theta$ **8.** $r = 5 \cos \theta$ **9.** $r = -5 \sin \theta$

---

†Archimedes devoted a great deal of study to the spiral that can be described by the polar equation $r = a\theta$. He used it initially in his attempt to solve the ancient problem of trisecting the angle. Later he calculated part of its area by his method of exhaustion. He described his work on this subject in one of his most important works, entitled *On Spirals*.

10. $r = -5 \cos \theta$ 　　　11. $r = 5 \cos \theta + 5 \sin \theta$ 　　12. $r = -5 \cos \theta + 5 \sin \theta$
13. $r = 5 \cos \theta - 5 \sin \theta$ 　14. $r = -5 \cos \theta - 5 \sin \theta$ 　15. $r = 2 + 2 \sin \theta$
16. $r = 2 + 2 \cos \theta$ 　　　17. $r = 2 - 2 \sin \theta$ 　　　18. $r = 2 - 2 \cos \theta$
19. $r = -2 + 2 \sin \theta$ 　　20. $r = -2 + 2 \cos \theta$ 　　21. $r = -2 - 2 \sin \theta$
22. $r = -2 - 2 \cos \theta$ 　　23. $r = 1 + 3 \sin \theta$ 　　　24. $r = 3 + \sin \theta$
25. $r = -2 + 4 \cos \theta$ 　　26. $r = -4 + 2 \cos \theta$ 　　27. $r = -3 - 4 \cos \theta$
28. $r = -4 - 3 \cos \theta$ 　　29. $r = -3 - 4 \sin \theta$ 　　30. $r = -4 - 3 \sin \theta$
31. $r = 4 - 3 \cos \theta$ 　　　32. $r = 3 - 4 \cos \theta$ 　　　33. $r = 4 + 3 \sin \theta$
34. $r = 3 + 4 \sin \theta$ 　　　35. $r = 3 \sin 2\theta$ 　　　　36. $r = -3 \cos 2\theta$
37. $r = -3 \sin 2\theta$ 　　　　　　　　　　　　　　　　　38. $r = 3 \cos 2\theta$
39. $r = 5 \sin 3\theta$ (This curve is called a **three-leafed rose**.)
40. $r = 5 \cos 3\theta$ 　　　　41. $r = -5 \sin 3\theta$ 　　　42. $r = -5 \cos 3\theta$
43. $r = 2 \cos 4\theta$ (This curve is called an **eight-leafed rose**.)
44. $r = 2 \sin 4\theta$ 　　　　　　　　　　　　　　　　　45. $r = -2 \cos 4\theta$
46. $r = 3\theta, \theta \geq 0$ (46–48 are all spirals of Archimedes.)
47. $r = -5\theta, \theta \geq 0$ 　　　　　　　　　　　　48. $r = \theta/2, \theta \geq 0$
49. $r = e^{\theta}$ (This curve is called a **logarithmic spiral** since $\ln r = \theta$.)
50. $r = e^{\theta/2}$ 　　　　　51. $r = e^{3\theta}$ 　　　　52. $r^2 = \cos 2\theta$
53. $r^2 = \sin 2\theta$ 　　　　54. $r^2 = -\cos 2\theta$ 　　55. $r^2 = -\sin 2\theta$
56. $r^2 = 4 \sin 2\theta$ 　　　57. $r^2 = -25 \cos 2\theta$ 　58. $r^2 = -25 \sin 2\theta$
59. $r = \sin \theta \tan \theta$ (This curve is called a **cissoid**.)
60. $r = 2 - 3 \sec \theta$ (This curve is called a **conchoid**.) 　61. $r = 4 + 3 \csc \theta$
62. $r^2 = \theta$ (This curve is called a **parabolic spiral**.)
63. $(r + 1)^2 = 3\theta$ 　　　64. $r = |\sin \theta|$ 　　　　65. $r = |\cos \theta|$

66. By translating into rectangular coordinates and completing the square, show that for any real number a the graph of the equation $r = a \cos \theta$ is a circle in the xy-plane with center at $(a/2, 0)$ and radius $|a/2|$.
67. Show that the graph of the equation $r = b \sin \theta$ in the xy-plane is a circle with radius $|b/2|$ centered at $(0, b/2)$.
68. Show that the graph of the equation $r = a \cos \theta + b \sin \theta$ in the xy-plane is a circle centered at $(a/2, b/2)$ with radius $\sqrt{a^2 + b^2}/2$.
69. Find the polar equation of the circle centered at $(-2, \frac{3}{2})$ with radius $\frac{5}{2}$.
70. Show that the polar equation $r \cos \theta = a$ is the equation of a vertical line for any real number a.
71. Show that the polar equation $r \sin \theta = a$ is the equation of a horizontal line for any real number a.
72. Show that the polar equation $r(a \cos \theta + b \sin \theta) = c$ with a, b, c real numbers is the equation of a straight line, if $a^2 + b^2 \neq 0$.
73. Sketch the graphs:
   (a) $r \sin \theta = 3$ 　　　**(b)** $r \cos \theta = -2$ 　　　**(c)** $3r \cos \theta = 8$
   (d) $r(2 \sin \theta - 3 \cos \theta) = 4$ 　　　**(e)** $r(-5 \sin \theta + 10 \cos \theta) = 20$

11.3 POINTS OF INTERSECTION OF GRAPHS OF POLAR EQUATIONS

Suppose that we have two functions given in rectangular coordinates: $y = f(x)$ and $y = g(x)$. To find the points of intersection of the graphs of these equations, we simply find the roots of the equation $f(x) = g(x)$. This method works because every point in the xy-plane has a unique representation. However, as we have already seen, the same is not true in polar coordinates. Therefore the "analytic" method described above will usually not locate all the points of intersection of the graphs of $r = f(\theta)$ and $r = g(\theta)$, since the graphs can intersect for different representations of the points of intersection.

EXAMPLE 1   Find all points of intersection of the curves $r = \sin\theta$ and $r = \cos\theta$.

Solution. We first try our "old" method of setting $\sin\theta = \cos\theta$. Then $\tan\theta = 1$, which means that $\theta = \pi/4 + n\pi$ for $n = 0, \pm 1, \pm 2, \pm 3, \ldots$. Thus

$$r(\theta) = \left(\frac{1}{\sqrt{2}}, \frac{\pi}{4}\right)$$

is one point of intersection.

Next, we draw the graphs of these two curves. From the discussion of the last section we know that $r = \sin\theta$ and $r = \cos\theta$ are the equations of circles whose graphs are sketched in Figure 1. Evidently, the pole is also a point of intersection. How did we miss it? The answer is simply that the pole can be represented as $(0, \theta)$ for any real number θ. On the curve $r = \sin\theta$, $r = 0$ when $\theta = n\pi$; and on $r = \cos\theta$, $r = 0$ when $\theta = (n + \frac{1}{2})\pi$. Since these two representations of the pole are not the same, our old method will not locate this point of intersection. ▪

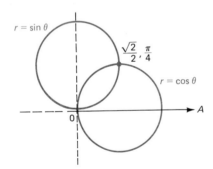

FIGURE 1

In general, to find all points of intersection of two polar equations, it is useful to sketch their graphs.

EXAMPLE 2   Find all points of intersection of the graphs of the equations $r = 3$ and $r = 3\cos 2\theta$.

Solution. Setting $3 = 3\cos 2\theta$, we have $\cos 2\theta = 1$, or $2\theta = 2n\pi$, and $\theta = n\pi$. If $0 \le \theta < 2\pi$, we have the points of intersection $(3, 0)$ and $(3, \pi)$. But, again, some points are missed. The equation $r = 3\cos 2\theta$ is the equation of a four-leafed rose (see Example 11.2.7). The two graphs are sketched in Figure 2. For $0 \le \theta < 2\pi$ we see that there are really four points of intersection. The ones we missed were $(-3, \pi/2) = (3, 3\pi/2)$ and $(-3, 3\pi/2) = (3, \pi/2)$. Thus the points $(3, \pi/2)$ and $(3, 3\pi/2)$ are also points of intersection of the two curves. ▪

EXAMPLE 3   Find the points of intersection of the graphs of the equations $r = 1 + \sin\theta$ and $r = 1 + 2\cos\theta$.

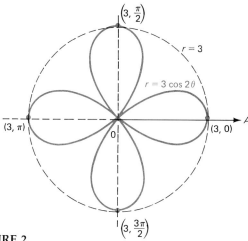

FIGURE 2

Solution. We first set $1 + \sin \theta = 1 + 2 \cos \theta$ to obtain $\tan \theta = 2$, or $\theta = \tan^{-1} 2$. When $\theta = \tan^{-1} 2$, $\sin \theta = 2/\sqrt{5}$ and $\cos \theta = 1/\sqrt{5}$ (draw a triangle to verify this). To check for other points of intersection, we sketch the two curves (see Examples 11.2.4 and 11.2.6) in Figure 3. Again we find that the pole is a point of intersection. In addition, the curves intersect at the point $(1, 0)$ since when $\theta = \pi$, $r = 1 + 2 \cos \theta = 1 - 2 = -1$ and $(-1, \pi) = (1, 2\pi) = (1, 0)$. Thus three points of intersection are at $(1 + (2/\sqrt{5}), \tan^{-1} 2)$, $(1, 0)$, and O. But, surprisingly, these are not all. In Figure 4 the two curves are plotted by computer on the same axes. In Figure 5 a "blown up" picture of the curves near the origin is given. In this blowup we can see that two points were missed.

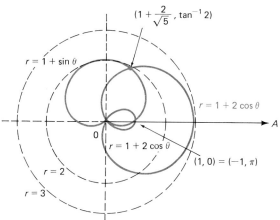

FIGURE 3

These two additional points of intersection fall within the width of the colored lines in Figure 3 (and so are not visible on our sketch). To find them, we first note that $\theta = \pi + \tan^{-1} 2$ (in the third quadrant) satisfies $\tan \theta = 2$ since the tangent function is periodic of period π. Then

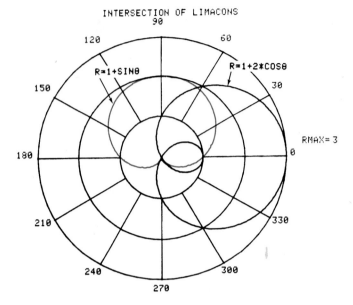

R=1+SIN0 AND R=1+2*COS0 PLOTTED ON SAME AXES
FIGURE 4

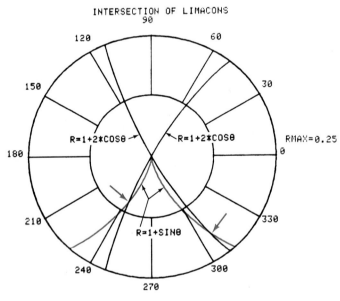

R=1+SIN0 AND R=1+2*COS0 PLOTTED ON SAME AXES
FIGURE 5

$$\sin(\pi + \tan^{-1} 2) = -\sin(\tan^{-1} 2) = \frac{-2}{\sqrt{5}}$$

and

$$\cos(\pi + \tan^{-1} 2) = -\cos(\tan^{-1} 2) = \frac{-1}{\sqrt{5}}.$$

Thus $r = 1 + \sin\theta = 1 + 2\cos\theta = 1 - 2/\sqrt{5}$, and we see that

$$(1 - (2/\sqrt{5}), \pi + \tan^{-1} 2)$$

is a fourth point of intersection.

It is more difficult to find the fifth point of intersection. We begin with the identity

$$(r, \theta) = (-r, \theta + \pi).$$

This suggests that we consider the possibility that the two curves intersect at a point (r, θ) for which $r = 1 + \sin\theta$ and $-r = 1 + 2\cos(\theta + \pi)$. (Explain why!) But then $r = -1 - 2\cos(\theta + \pi) = -1 + 2\cos\theta$, and we are led to the equation

$$1 + \sin\theta = -1 + 2\cos\theta, \tag{1}$$

or

$$2 + \sin\theta = 2\cos\theta,$$

and squaring, we have

$$4 + 4\sin\theta + \sin^2\theta = 4\cos^2\theta = 4 - 4\sin^2\theta,$$

or

$$0 = 5\sin^2\theta + 4\sin\theta = \sin\theta(5\sin\theta + 4),$$

with solutions $\sin\theta = 0$ and $\sin\theta = -\frac{4}{5}$. But if $\sin\theta = 0$, we again obtain the point $(1, 0)$. If $\sin\theta = -\frac{4}{5}$, then θ is in the third or fourth quadrant. To find out which, we again start with Equation (1) to obtain

$$\sin^2\theta = (-2 + 2\cos\theta)^2 = 4 - 8\cos\theta + 4\cos^2\theta,$$

or using $\sin^2\theta = 1 - \cos^2\theta$,

$$0 = 5\cos^2\theta - 8\cos\theta + 3 = (5\cos\theta - 3)(\cos\theta - 1)$$

with solutions $\cos\theta = 1$ and $\cos\theta = \frac{3}{5}$. Hence $\cos\theta = \frac{3}{5}$, θ is in the fourth quadrant, and $\theta = -\sin^{-1}\frac{4}{5} = -\cos^{-1}\frac{3}{5}$, giving us the last point of intersection, $(\frac{1}{5}, -\sin^{-1}\frac{4}{5})$. [Note that this point is the same point as $(-\frac{1}{5}, \pi - \cos^{-1}\frac{3}{5})$.] ∎

From Example 3 it is evident that finding all points of intersection can be very tricky. A method for determining all points of intersection without sketching the graphs is outlined in Problems 22 and 23.

PROBLEMS 11.3

In Problems 1–20, find all points of intersection of the graphs of the two polar equations.

1. $r = 1, r = \sin\theta$
2. $r = \frac{1}{2}, r = \cos\theta$
3. $r = 2(1 - \sin\theta), r = 2(1 - \cos\theta)$
4. $r = 1 - \sin\theta, r = 1 + \cos\theta$
5. $r = 3, \theta = \pi/4$ (Be careful!)
6. $r = -2, \theta = \pi/3$
7. $r = -2 + 2\cos\theta, r = -4 + 2\sin\theta$
8. $r = \sqrt{2}\sin\theta, r = \cos\theta$
9. $r = \sqrt{3}\sin\theta, r = \cos\theta$
10. $r^2 = \sin 2\theta, r^2 = \cos 2\theta$
11. $r = \frac{3}{2}, r = 3\cos 2\theta$
12. $r = 2, r = 4\sin 2\theta$
13. $r^2 = \cos\theta, r = \cos\theta$
14. $r^2 = \sin\theta, r = \sin\theta$
15. $r = \theta, r^2 = \theta, r, \theta \geq 0, 0 \leq r \leq 2$
16. $r = 5\sin 3\theta, r = 5\cos 3\theta$
17. $r = \sin 2\theta, r = \cos 2\theta$
18. $r = \pi/3, r = 2\theta$
19. $r = \sin\theta, r = \sec\theta$
20. $r = |\sin\theta|, r = |\cos\theta|$

*21. Show that all points of intersection of the two curves $r = a\csc\theta$ and $r^2 = 40 - 39\cos 7\theta$ lie on a straight line parallel to the polar axis.

*22. If the graphs of the equations $r = f(\theta)$ and $r = g(\theta)$ intersect at a point $(r, \theta) \neq 0$, explain why one of the following three equations must hold for r and θ:
   (i) $r = f(\theta), r = g(\theta)$.
   (ii) $r = f(\theta), r = g(\theta + 2n\pi)$, for some integer n.
   (iii) $r = f(\theta), -r = g(\theta + (2n + 1)\pi)$ for some integer n.

23. Show that if $r = f(\theta)$ and $r = g(\theta)$ intersect at the pole, then there exist numbers θ_1 and θ_2 such that $f(\theta_1) = g(\theta_2) = 0$.

Using the results of Problems 22 and 23, we can find all points of intersection of the graphs of $r = f(\theta)$ and $r = g(\theta)$ by checking the pole and solving equations (i), (ii), and (iii). In Problems 24–34, use this method to find all points of intersection.

24. $r = 1 - \sin\theta, r = 1 + \sqrt{3}\cos\theta$
25. $r = \tan 2\theta, r = 1$
26. $r = 3\tan\theta\sin\theta, r = 3\cos\theta$
27. $r = 8(1 + \sin\theta), r = \dfrac{6}{1 - \sin\theta}$
28. $r = 4(1 + \cos\theta), r = \dfrac{3}{1 - \cos\theta}$
29. $r = \dfrac{1}{2}\tan\theta, r = \dfrac{1}{\sqrt{3}}\sin\theta$
30. $r^2 = \sin\theta, r = \csc\theta$
31. $r^2 = -\cos\theta, r = \sec\theta$
32. $r = \csc\theta, r = \cot\theta$
33. $r = \theta, r = 2\theta, \theta \geq 0$
34. $r = 2\theta, r = 3\theta, \theta \geq 0$

11.4 DERIVATIVES AND TANGENT LINES

As we have seen repeatedly in this text, the derivative of a function tells us a great deal about the behavior of that function. If $r = f(\theta)$, then $dr/d\theta$, which we can easily calculate, tells us how r changes when θ changes. However, other applications of the derivative cannot be immediately extended to polar coordinates.

Suppose we wish to find the slope of the tangent line to a curve given in polar coordinates. We can immediately see that this slope is *not* given by $dr/d\theta$. For example, let $r = c$, a constant. Then $dr/d\theta = 0$, but zero is not the slope of the tangent to the curve $r = c$ since $r = c$ is the equation of a circle. To calculate the slope of the tangent line, we must translate into rectangular coordinates and then calculate dy/dx.

From Section 11.1 we have

$$x = r \cos \theta \qquad \text{and} \qquad y = r \sin \theta.$$

Since we are assuming that r is a function of θ, we may also consider x and y to be functions of θ. Then using the product rule and differentiating with respect to θ, we obtain

$$\frac{dx}{d\theta} = \frac{dr}{d\theta} \cos \theta - r \sin \theta \tag{1}$$

and

$$\frac{dy}{d\theta} = \frac{dr}{d\theta} \sin \theta + r \cos \theta. \tag{2}$$

By the chain rule, $dy/dx = (dy/d\theta)/(dx/d\theta)$, so that

$$\frac{dy}{dx} = \frac{(dr/d\theta)\sin \theta + r \cos \theta}{(dr/d\theta)\cos \theta - r \sin \theta}. \tag{3}$$

EXAMPLE 1 Calculate the slope of the line tangent to the limaçon $r = 1 + 2 \cos \theta$ at the point $(2, \pi/3)$ (see Example 11.2.6).

Solution. Here $dr/d\theta = -2 \sin \theta = -\sqrt{3}$ when $\theta = \pi/3$. Then substituting $\theta = \pi/3$ and $r = 2$ into (3), we obtain

$$\frac{dy}{dx} = \frac{(-\sqrt{3})(\sqrt{3}/2) + (2)(1/2)}{(-\sqrt{3})(1/2) - (2)(\sqrt{3}/2)} = \frac{1}{3\sqrt{3}}. \quad \blacksquare$$

EXAMPLE 2 Find the equation of the line tangent to the four-leafed rose $r = 3 \cos 2\theta$ at the point $(3/2, \pi/6)$ (see Example 11.2.7).

Solution. When $\theta = \pi/6$, $dr/d\theta = -6 \sin 2\theta = -6 \sin(\pi/3) = -6(\sqrt{3}/2) = -3\sqrt{3}$, $r = 3 \cos 2\theta = 3 \cos(\pi/3) = 3 \cdot \frac{1}{2} = \frac{3}{2}$, $\sin \theta = \frac{1}{2}$, and $\cos \theta = \sqrt{3}/2$. Thus substitution of $r = \frac{3}{2}$ and $\theta = \pi/6$ into equation (3) yields

$$\frac{dy}{dx} = \frac{(-3\sqrt{3})(1/2) + (3/2)(\sqrt{3}/2)}{(-3\sqrt{3})(\sqrt{3}/2) - (3/2)(1/2)} = \frac{-3\sqrt{3}/4}{-21/4} = \frac{\sqrt{3}}{7}.$$

At $(3/2, \pi/6)$, $x = r \cos \theta = 3\sqrt{3}/4$ and $y = r \sin \theta = \frac{3}{4}$. Thus

$$\frac{y - (3/4)}{x - (3\sqrt{3}/4)} = \frac{\sqrt{3}}{7},$$

which reduces to $7y - \sqrt{3}x = 3$. $\blacksquare$

As these two examples illustrate, the calculation of dy/dx can be tedious. Using some elementary geometry, we can simplify our work. From Theorem 1.4.1, we see that dy/dx can be thought of as the tangent of the angle α that the tangent line makes with the positive x-axis, or, in polar coordinates, the tangent of the angle it makes with the polar axis (see Figure 1). Then equation (3) can be written

$$\tan \alpha = \frac{(dr/d\theta)\sin \theta + r \cos \theta}{(dr/d\theta)\cos \theta - r \sin \theta}. \tag{4}$$

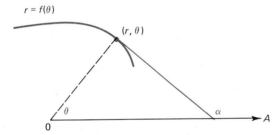

FIGURE 1

If $\cos \theta \neq 0$, then we can divide numerator and denominator of (4) by $\cos \theta$ to obtain

$$\tan \alpha = \frac{(dr/d\theta)\tan \theta + r}{(dr/d\theta) - r \tan \theta}. \tag{5}$$

Now let β denote the angle, measured counterclockwise, from the line $\theta =$ constant to the tangent line (see Figure 2). From elementary geometry we have $\pi = \theta + \beta + (\pi - \alpha)$, or

$$\beta = \alpha - \theta. \tag{6}$$

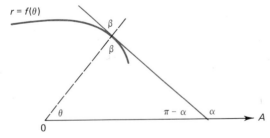

FIGURE 2

Recalling the formula (see Problem A.1.3.22)

$$\tan(A - B) = \frac{\tan A - \tan B}{1 + \tan A \tan B}, \tag{7}$$

we have

$$\tan \beta = \tan(\alpha - \theta) = \frac{\tan \alpha - \tan \theta}{1 + \tan \alpha \tan \theta}. \tag{8}$$

Substituting (5) into (8), we obtain

$$\tan \beta = \frac{\dfrac{(dr/d\theta)\tan \theta + r}{(dr/d\theta) - r \tan \theta} - \tan \theta}{1 + \dfrac{(dr/d\theta)\tan \theta + r}{(dr/d\theta) - r \tan \theta} \tan \theta}$$

Then multiplying numerator and denominator by $(dr/d\theta) - r \tan \theta$, we have

$$\tan \beta = \frac{(dr/d\theta)\tan \theta + r - (dr/d\theta)\tan \theta + r \tan^2 \theta}{(dr/d\theta) - r \tan \theta + (dr/d\theta)\tan^2 \theta + r \tan \theta}$$

$$= \frac{r(1 + \tan^2 \theta)}{(dr/d\theta)(1 + \tan^2 \theta)} = \frac{r}{dr/d\theta},$$

and finally,

$$\tan \beta = \tan(\alpha - \theta) = \frac{r}{dr/d\theta}. \tag{9}$$

Often it will be more convenient to calculate $\tan \beta$ in a given problem. The next example illustrates how the slope of the tangent line may be calculated by concentrating on the angle β.

EXAMPLE 3   Calculate the slope of the line tangent to the limaçon $r = 1 + 2 \sin \theta$ at the point $(2, \pi/6)$.

Solution. We have $dr/d\theta = 2 \cos \theta = \sqrt{3}$ when $\theta = \pi/6$. Then from (9),

$$\tan \beta = \frac{r}{dr/d\theta} = \frac{2}{\sqrt{3}}.$$

But we can rearrange equation (8) and solve for $\tan \alpha$ to obtain

$$\tan \alpha = \frac{\tan \theta + \tan \beta}{1 - \tan \theta \tan \beta}. \tag{10}$$

Then substituting $\tan \theta = \tan \pi/6 = 1/\sqrt{3}$ and $\tan \beta = 2/\sqrt{3}$ into (10), we have

$$\frac{dy}{dx} = \tan \alpha = \frac{(1/\sqrt{3}) + (2/\sqrt{3})}{1 - (1/\sqrt{3}) \cdot (2/\sqrt{3})} = 3\sqrt{3}. \ \blacksquare$$

EXAMPLE 4   Calculate the slope of the line tangent to the curve $r = \theta$ at the point $(\pi/4, \pi/4)$.

Solution. Since $dr/d\theta = 1$, $\tan \beta = r/(dr/d\theta) = \pi/4$. Then from (10),

$$\frac{dy}{dx} = \tan \alpha = \frac{\tan(\pi/4) + \pi/4}{1 - \tan(\pi/4) \cdot \pi/4} = \frac{1 + (\pi/4)}{1 - (\pi/4)} = \frac{4 + \pi}{4 - \pi}. \quad \blacksquare$$

As a further application of the usefulness of the angle β, suppose that $r = f(\theta)$ and that θ^* denotes a number for which $f(\theta^*) = 0$. Then the curve passes through the pole when $\theta = \theta^*$. Thus the angle between the line $\theta = \theta^*$ and the tangent to the curve at $(0, \theta^*)$ is 0 or, put another way, the line $\theta = \theta^*$ *is* a tangent line to the curve $r = f(\theta)$ at the pole (see Figure 3). This all follows from (9) since $\tan \beta = 0/(dr/d\theta) = 0$.

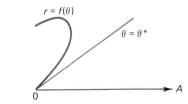

FIGURE 3

In general, the following rule is useful for sketching the graphs of equations in polar coordinates.

> If $\theta_1, \theta_2, \ldots, \theta_n$ are roots of $f(\theta) = 0$, then there is a branch of the graph of $r = f(\theta)$ that is tangent to the line $\theta = \theta_i$ at the pole for each i, $i = 1, 2, \ldots, n$.

EXAMPLE 5 In Example 11.2.4, we sketched the graph of the cardioid $r = 1 + \sin \theta$. We can improve that sketch by noting that since $r = 0$ when $\theta = 3\pi/2$ (so that $\sin \theta = -1$), the cardioid is tangent to the line $\theta = 3\pi/2$ at the pole. This curve is sketched in Figure 4. ◼

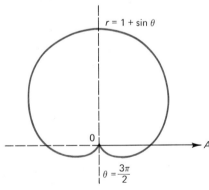

FIGURE 4

EXAMPLE 6 In Example 11.2.7 we sketched the graph of the four-leafed rose $r = 3 \cos 2\theta$. We have $r = 0$ when $\theta = \pi/4$ and $3\pi/4$ (so that $\cos 2\theta = 0$). Therefore the four-leafed rose has branches that are tangent to the lines $\theta = \pi/4$ and $\theta = 3\pi/4$ at the pole. This curve is sketched in Figure 5. ■

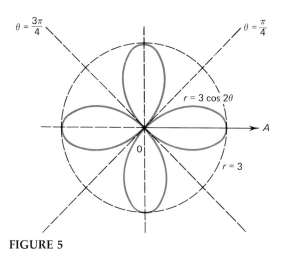

FIGURE 5

PROBLEMS 11.4

In Problems 1–10, calculate the slope of the line tangent to the given curve at the given point by using equation (3) or (4).

1. $r = 1$; $(1, \pi/2)$

2. $r = 3$; $(3, 0)$

3. $r = \sin \theta$; $(1/\sqrt{2}, \pi/4)$

4. $r = 2 \cos \theta$; $(1, \pi/3)$

5. $r = 2 - 3 \sin \theta$; $(1/2, \pi/6)$

6. $r = 2 + 2 \cos \theta$; $(3, \pi/3)$

7. $r = 2\theta$; $(\pi, \pi/2)$

8. $r = 3 \cos 2\theta$; $(0, \pi/4)$

9. $r = -4 \sin 2\theta$; $(2\sqrt{3}, 5\pi/3)$

10. $r = e^{\theta}$; $(e^{\pi/2}, \pi/2)$

In Problems 11–24, calculate the tangent of the angle β between the radial line $\theta = $ constant and the tangent line at the given point and then use that calculation to find the slope of the tangent line.

11. $r = 3 \sin \theta$; $(3, \pi/2)$

12. $r = -2 \cos \theta$; $(\sqrt{3}, 5\pi/6)$

13. $r = 4 - 3 \cos \theta$; $(5/2, \pi/3)$

14. $r = 5 \sin 3\theta$; $(5/\sqrt{2}, \pi/12)$

15. $r = e^{\theta/2}$; $(e^{\pi/4}, \pi/2)$

16. $r = 3\theta$; $(3\pi/2, \pi/2)$

17. $r = \sin \theta \tan \theta$; $(1/\sqrt{2}, \pi/4)$

18. $r = 2 \cos 4\theta$; $(1, \pi/12)$

19. $r^2 = \cos 2\theta$; $(1/\sqrt{2}, \pi/6)$

20. $r = 2 + 3 \sec \theta$; $(8, \pi/3)$

21. $r = 9 - 3 \csc \theta$; $(3, \pi/6)$

22. $r^2 = \theta$; $(\sqrt{\pi}/2, \pi/4)$

23. $r = 7 \cos 7\theta$; $(7/\sqrt{2}, \pi/28)$

24. $r = 2 \tan \theta$; $(2, \pi/4)$

25. The curve $r = ae^{b\theta}$, where a and b are nonzero real numbers, is called a **logarithmic spiral**. Show that the angle β between the tangent line and the radial line joining the pole and the point of tangency is a constant. Use this information to sketch the curve.

*26. Using the half angle formula, show that for any real number $a \neq 0$, $\tan \beta = \tan \theta/2$ at any point on the cardioid $r = a(1 - \cos \theta)$.

*27. Find all relative maxima and minima of y for the function $r = 1 + 2 \cos \theta$.

28. Find all relative maxima and minima of y for the function $r = 3 \cos 2\theta$.

*29. Find all relative maxima and minima of y for the function $r = 1 - \sin \theta$.

*30. Let $r = f(\theta)$ and $r = g(\theta)$ be the equations of two curves that intersect at the point $P = (r_0, \theta_0)$. Let β_1 and β_2 denote, respectively, the angles from the line $\theta = \theta_0$ to the tangent lines $(T_1$ and $T_2)$ to the two curves at P. We define the angle γ between the two curves at P as the angle measured counterclockwise from T_1 to T_2. Show that

$$\gamma = \beta_2 - \beta_1 \quad \text{or} \quad \gamma = \beta_2 - \beta_1 + \pi$$

31. Show that in Problem 30

$$\tan \gamma = \left| \frac{\tan \beta_1 - \tan \beta_2}{1 + \tan \beta_1 \tan \beta_2} \right|.$$

In Problems 32–39, use the results of Problems 30 and 31 to calculate (to the nearest degree) the angle between the given curves at all nonpolar points of intersection. (Use a calculator.)

32. $r = \sin \theta; r = \cos \theta$
33. $r = 5 \sin 2\theta; r = 5 \cos 2\theta$
34. $r = 1 + \sin \theta; r = 1 - \cos \theta$
35. $r = 2; r = 4 \cos \theta$
36. $r = 1 + \sin \theta; r = 1 + 2 \cos \theta$
37. $r = \sin \theta; r = \sec \theta$
38. $r = 1 - \sin \theta; r = 1 + \cos \theta$
39. $r^2 = \cos \theta; r = \cos \theta$

40. Show that the curves $r = a \cos \theta$ and $r = b \sin \theta$ intersect at right angles for any nonzero real numbers a and b.
41. Show that the curves $r^2 = a^2 \sec 2\theta$ and $r^2 = b^2 \csc 2\theta$ intersect at right angles.
42. Show that, except at the pole, the curves $r = a(1 - \cos \theta)$ and $r = a(1 + \cos \theta)$ intersect at right angles.

In Problems 43–56, find all values θ_i for which the given curve has a branch that is tangent to the line $\theta = \theta_i$ at the pole. Then sketch the curve.

43. $r = \sin \theta$
44. $r = \cos \theta$
45. $r = \sin 2\theta$
46. $r = \cos 2\theta$
47. $r = a \sin 2\theta, a > 0$
48. $r = a \cos 2\theta, a > 0$
49. $r^2 = \sin 2\theta$
50. $r^2 = \cos 2\theta$
51. $r^2 = a \sin 2\theta$
52. $r^2 = a \cos 2\theta; a > 0$
53. $r = 4 \sin 3\theta$
54. $r = a(1 - \sin \theta)$
55. $r = 1 + 2 \sin \theta$
56. $r = 7(\cos 2\theta - \sqrt{3} \sin 2\theta)$

11.5 AREAS IN POLAR COORDINATES

In this section we derive a method for calculating the area of a region whose boundary is given in polar coordinates. We begin by recalling the formula for the area of a sector. Consider the sector bounded by the lines $\theta = \theta_1$ and $\theta = \theta_2$ and the circle $r = r_0$ (the shaded region in Figure 1). Let φ denote the angle (given in radians) between these two lines ($\varphi = \theta_2 - \theta_1$) and let A denote the area of the sector. Then since the area of the circle is πr_0^2, we have the proportion

$$\frac{\text{number of radians in sector}}{\text{number of radians in circle}} = \frac{\text{area of sector}}{\text{area of circle}},$$

or

$$\frac{\varphi}{2\pi} = \frac{A}{\pi r_0^2}, \tag{1}$$

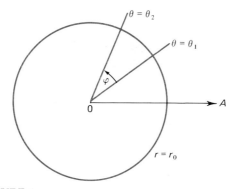

FIGURE 1

or

$$A = \tfrac{1}{2}r_0^2\varphi. \tag{2}$$

Now consider the plane region bounded by the continuous polar curve $r = f(\theta)$ and the lines $\theta = \alpha$ and $\theta = \beta$ (see Figure 2). To calculate the area of this region, we divide the region into n equal subregions by the rays $\alpha = \theta_0 < \theta_1 < \theta_2 < \cdots < \theta_{n-1} < \theta_n = \beta$. We use the notation $\Delta\theta$ to denote the angle between any two successive lines:

$$\Delta\theta = \theta_i - \theta_{i-1}.$$

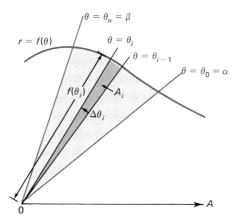

FIGURE 2

Let r_i and R_i be the smallest and largest values of $f(\theta)$ for $\theta_{i-1} \le \theta \le \theta_i$, and let ΔA_i denote the area of the ith subregion. Then

$$\tfrac{1}{2}r_i^2\,\Delta\theta \le \Delta A_i \le \tfrac{1}{2}R_i^2\,\Delta\theta. \tag{3}$$

Since $r = f(\theta)$ is continuous, so is $r^2 = f^2(\theta)$ and so is the function $g(\theta) = \tfrac{1}{2}f^2(\theta)\,\Delta\theta$ (for $\Delta\theta$ fixed). The function $g(\theta)$ takes on the values $\tfrac{1}{2}r_i^2\,\Delta\theta$ and $\tfrac{1}{2}R_i^2\,\Delta\theta$. Since ΔA_i is between these values, there is, by the intermediate value theorem, a number θ_i^* in (θ_{i-1}, θ_i) such that $g(\theta_i^*) = \Delta A_i$. But

$$\Delta A_i = g(\theta_i^*) = \tfrac{1}{2}[f(\theta_i^*)]^2 \, \Delta\theta.$$

Then using a familiar argument, we have

$$A = \lim_{\Delta\theta \to 0} \sum_{i=1}^{n} \Delta A_i = \lim_{\Delta\theta \to 0} \sum_{i=1}^{n} \frac{1}{2}[f(\theta_i^*)]^2 \, \Delta\theta,$$

or

$$A = \frac{1}{2} \int_{\alpha}^{\beta} [f(\theta)]^2 \, d\theta.$$

EXAMPLE 1   Calculate the area of the region enclosed by the graph of the cardioid $r = 1 + \sin\theta$.

Solution. The curve is symmetric about the line $\theta = \pi/2$ and is sketched in Figure 3. (See Example 11.2.4.) Thus we need only calculate the area for θ in $[-\pi/2, \pi/2]$ and then multiply it by 2. We have

$$\text{half the area} = \frac{1}{2} \int_{-\pi/2}^{\pi/2} (1 + \sin\theta)^2 \, d\theta = \frac{1}{2} \int_{-\pi/2}^{\pi/2} (1 + 2\sin\theta + \sin^2\theta) \, d\theta$$

$$= \frac{1}{2}\,\theta\,\Big|_{-\pi/2}^{\pi/2} + \int_{-\pi/2}^{\pi/2} \sin\theta \, d\theta + \frac{1}{4} \int_{-\pi/2}^{\pi/2} (1 - \cos 2\theta) \, d\theta$$

$$= \frac{\pi}{2} + 0 + \frac{\pi}{4} = \frac{3\pi}{4},$$

and the total area $A = 2 \cdot 3\pi/4 = 3\pi/2$. ∎

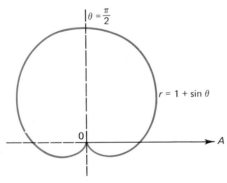

FIGURE 3

EXAMPLE 2   Find the area enclosed by the smaller loop of the limaçon $r = 1 + 2\cos\theta$.

Solution. The curve is sketched in Figure 4 (see Example 11.2.6). The curve is

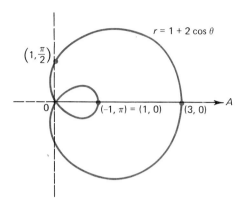

FIGURE 4

symmetric about the polar axis and the inner loop is determined by the interval $(2\pi/3, 4\pi/3)$ (see Table 11.2.4). Then

$$A = \frac{1}{2} \int_{2\pi/3}^{4\pi/3} (1 + 2\cos\theta)^2 \, d\theta = \frac{1}{2} \int_{2\pi/3}^{4\pi/3} (1 + 4\cos\theta + 4\cos^2\theta) \, d\theta$$

$$= \frac{1}{2} \left[\theta + 4\sin\theta + 2\left(\theta + \frac{\sin 2\theta}{2}\right) \right] \Big|_{2\pi/3}^{4\pi/3} = \frac{2\pi - 3\sqrt{3}}{2}. \quad \blacksquare$$

EXAMPLE 3   Find the area inside the circle $r = 5\sin\theta$ and outside the limaçon $r = 2 + \sin\theta$.

Solution. The region is sketched in Figure 5. The two curves intersect whenever $2 + \sin\theta = 5\sin\theta$, or $\sin\theta = \frac{1}{2}$, so that $\theta = \pi/6$ and $\theta = 5\pi/6$. The area of the region can be calculated by calculating the area of the limaçon between $\pi/6$ and $5\pi/6$ and subtracting it from the area of the circle for θ in that interval. By symmetry, we need only calculate the area between $\theta = \pi/6$ and $\theta = \pi/2$ and then multiply it by 2. We therefore have

$$A = \int_{\pi/6}^{\pi/2} (5\sin\theta)^2 \, d\theta - \int_{\pi/6}^{\pi/2} (2 + \sin\theta)^2 \, d\theta$$

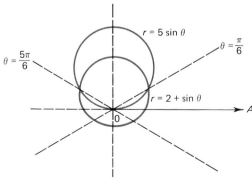

FIGURE 5

(we have already multiplied by 2)

$$= \int_{\pi/6}^{\pi/2} (25 \sin^2 \theta - 4 - 4 \sin \theta - \sin^2 \theta) \, d\theta$$

$$= \int_{\pi/6}^{\pi/2} [12(1 - \cos 2\theta) - 4 - 4 \sin \theta] \, d\theta$$

$$= (8\theta - 6 \sin 2\theta + 4 \cos \theta) \Big|_{\pi/6}^{\pi/2} = \frac{8\pi}{3} + \sqrt{3}. \ \blacksquare$$

PROBLEMS 11.5

In Problems 1–10, calculate the area of the smallest region enclosed by the curve and the two lines.

1. $r = \theta; \theta = 0; \theta = \pi/2; 0 \le r \le \pi/2$ **2.** $r = 3\theta; \theta = 0; \theta = \pi$
3. $r = 2 \cos \theta; \theta = -\pi/2; \theta = \pi/2$ **4.** $r = a \sin \theta; \theta = 0; \theta = \pi$
5. $r = a \cos 5\theta; \theta = -\pi/10; \theta = \pi/10$ **6.** $r = a \sin 3\theta; \theta = 0; \theta = \pi/6$
7. $r = 2\theta^2; \theta = 0; \theta = \pi/2$ **8.** $r = e^\theta; \theta = 0; \theta = 2\pi/3$
9. $r = 5\theta^4; \theta = 0; \theta = \pi$ **10.** $r = 1/\theta; \theta = \pi/6; \theta = \pi/2$

In Problems 11–26, sketch the given curve and find the entire area enclosed by it. Where appropriate, assume that $a > 0$.

11. $r = a \sin \theta$ **12.** $r = a \cos \theta$ **13.** $r = a(\cos \theta + \sin \theta)$
14. $r = a(\cos \theta - \sin \theta)$ **15.** $r = a(\sin \theta - \cos \theta)$ **16.** $r = 3 + 2 \sin \theta$
17. $r = 3 + 2 \cos \theta$ **18.** $r = 3 - 2 \sin \theta$ **19.** $r = 3 - 2 \cos \theta$
20. $r = a \cos 2\theta$ **21.** $r = a \sin 2\theta$ **22.** $r = a \sin 3\theta$
23. $r = a \cos 5\theta$ **24.** $r^2 = a^2 \sin 2\theta$ **25.** $r^2 = a \cos 2\theta$
26. $r = \sqrt{\cos \theta}$

In Problems 27–40, sketch the given region and find its area.

27. The smaller loop of the limaçon $r = 2 - 3 \sin \theta$.
28. The smaller loop of the limaçon $r = 2 + 3 \cos \theta$.
29. The smaller loop of the limaçon $r = a + b \sin \theta, b > a > 0$.
30. The area outside the smaller loop and inside the larger loop of the limaçon in Problem 27.
31. The area outside the smaller loop and inside the larger loop of the limaçon of Problem 28.
32. The area outside the smaller loop and inside the larger loop of the limaçon of Problem 29.
33. The region bounded by the graphs of $r = 2 \sin \theta, r = 2 + 2 \sin \theta, \theta = 0,$ and $\theta = \pi$.
34. The region inside the cardioid $r = a(1 + \sin \theta)$ and outside the circle $r = a, a > 0$.
*35. The region outside the circle $r = a \cos \theta$ and inside the lemniscate $r^2 = a^2 \sin 2\theta$.
36. The region inside the two circles $r = 3 \sin \theta$ and $r = 3 \cos \theta$.
37. The region enclosed by the graphs of $r = 4 \sin 2\theta$ and $r = 4 \cos 2\theta$.
*38. The region inside the circle $r = \cos \theta$ and inside the four-leafed rose $r = \cos 2\theta$.
39. The region enclosed by one "petal" of the rose $r = a \cos n\theta, a > 0, n \ge 1$ a fixed integer.
*40. The region enclosed by both the circles $r = a \sin \theta$ and $r = b \cos \theta, a > 0$ and $b > 0$.

41. A roller coaster encloses a region bounded by the graph of the equation $r^2 = 9 \sin 2\theta$. If r is measured in meters, what is the area of this region?

REVIEW EXERCISES FOR CHAPTER ELEVEN

In Exercises 1–6, convert from polar to rectangular coordinates.

1. $(2, 0)$ **2.** $(3, \pi/6)$ **3.** $(-7, \pi/2)$
4. $(3, 23\pi/3)$ **5.** $(-1, -\pi/2)$ **6.** $(2, 11\pi/12)$

In Exercises 7–12, convert from rectangular to polar coordinates with $r > 0$ and $0 \le \theta < 2\pi$.

7. $(2, 0)$ **8.** $(1, \sqrt{3})$ **9.** $(\sqrt{3}, -1)$
10. $(6, 6)$ **11.** $(-6, -6)$ **12.** $(-6, 6)$

In Exercises 13–26, sketch the given equation, indicating any symmetry about the polar axis, the line $\theta = \pi/2$, and/or the pole.

13. $r = 8$ **14.** $\theta = \pi/3$ **15.** $r = 2\cos\theta$
16. $r = 3 - 3\sin\theta$ **17.** $r = 3 - 2\sin\theta$ **18.** $r = 2 - 3\sin\theta$
19. $r = 5\cos 2\theta$ **20.** $r^2 = 4\sin 2\theta$ **21.** $r = 3\sin 4\theta$
22. $r = 3\theta,\ \theta \ge 0$ **23.** $r = e^{2\theta}$ **24.** $r^2 = 4\theta,\ r \ge 0$
25. $r\sin\theta = 4$ **26.** $r\cos\theta = -2$

27. Find, in rectangular coordinates, the equation of the circle

$$r = 3\sin\theta + 4\cos\theta.$$

28. Find the polar equation of the circle centered at $(\frac{5}{2}, -6)$ with radius $\frac{13}{2}$.

In Exercises 29–36, find all points of intersection of the graphs of the two polar equations.

29. $r = 1;\ r = \cos\theta$ **30.** $r = 2 + \cos\theta;\ r = 2 - \sin\theta$
31. $r^2 = \sin\theta;\ r = \sin\theta$ **32.** $r^2 = 4\cos 2\theta;\ r^2 = 4\sin 2\theta$
33. $r = \theta;\ r^2 = 4\theta;\ \theta,\ r \ge 0$ **34.** $r = \sin\theta;\ r = \sqrt{3}\cos\theta$
35. $r = \sqrt{3}\tan 2\theta;\ r = 1$ **36.** $r = 2\cot\theta;\ r = 8\cos\theta$

In Exercises 37–46, calculate the slope of the line tangent to the given curve at the given point by two methods; directly from formula (11.4.3) and indirectly by first calculating $\tan\beta$.

37. $r = 1;\ (1, 3\pi/2)$ **38.** $r = \sin\theta;\ (1/2, \pi/6)$
39. $r = 4\theta;\ (\pi, \pi/4)$ **40.** $r = 3\sin 2\theta;\ (3/\sqrt{2}, \pi/8)$
41. $r = 3 + 3\sin\theta;\ (9/2, \pi/6)$ **42.** $r = e^{\theta/3};\ (e^{\pi/6}, \pi/2)$
43. $r^2 = \sin 2\theta;\ (1, \pi/4)$ **44.** $r = 4\sin 3\theta;\ (2\sqrt{3}, \pi/9)$
45. $r = 10\sin 10\theta;\ (10, \pi/20)$ **46.** $r = 3\tan\theta;\ (3, \pi/4)$

47. Calculate, to the nearest degree, the angle formed by the curves $r = 2\sin\theta$ and $r = 2\cos\theta$ at each of their nonpolar points of intersection.

48. Repeat the calculation of Exercise 47 for the curves $r = 3 - 3\cos\theta$ and $r = 3 + 3\sin\theta$.

49. Find all numbers θ_i for which the curve $r = 2\sin 2\theta$ has a branch that is tangent to the line $\theta = \theta_i$ at the pole. Then sketch the curve.

50. Repeat Exercise 49 for the curve $r = -4(1 - 2\cos\theta)$.

In Exercises 51–61, find the area of the given region.

51. The region bounded by $r = \theta$, $\theta = \pi/6$, and $\theta = \pi/2$, $\pi/6 \le r \le \pi/2$.
52. The region bounded by $r = a\cos 4\theta$, $\theta = 0$, and $\theta = \pi/12$.
53. The region enclosed by the circle $r = a(\sin\theta + \cos\theta)$, $a > 0$.
54. The region enclosed by the limaçon $r = 4 + 3\cos\theta$.

55. The region enclosed by the cardioid $r = -2 - 2 \sin \theta$.
56. The region enclosed by the smaller loop of the limaçon $r = 3 - 4 \cos \theta$.
57. The region outside the smaller loop and inside the larger loop of the limaçon in Exercise 56.
58. The region inside the cardioid $r = 4(1 + \cos \theta)$ and outside the circle $r = 4$.
59. The region inside the two circles $r = 2 \sin \theta$ and $r = 2 \cos \theta$.
60. The region enclosed by the lemniscate $r^2 = 36 \cos 2\theta$.
61. The region enclosed by one petal of the rose $r = 4 \sin 10\theta$.

12 Indeterminate Forms and Improper Integrals

12.1 THE INDETERMINATE FORM 0/0 AND L'HÔPITAL'S RULE

In earlier chapters we encountered quotients of the form $f(x)/g(x)$. In limit Theorem 2.3.5, we showed that if the limits $\lim_{x \to x_0} f(x)$ and $\lim_{x \to x_0} g(x)$ both exist, then

$$\lim_{x \to x_0} \frac{f(x)}{g(x)} = \frac{\lim_{x \to x_0} f(x)}{\lim_{x \to x_0} g(x)}, \tag{1}$$

provided that $\lim_{x \to x_0} g(x) \neq 0$.

However, this last condition is frequently an obstacle in important applications. For example, in Section 3.5 we showed that $\lim_{x \to 0}[(\sin x)/x] = 1$. We did so despite the fact that $\lim_{x \to 0} x = 0$ so that rule (1) did not apply. As another example, we saw (in Example 2.2.4) that

$$\lim_{x \to 0} \frac{x(x + 1)}{x} = 1,$$

again despite the fact that $\lim_{x \to 0} x = 0$.

In general, we have the following definition.

Definition 1 INDETERMINATE FORM 0/0 Let f and g be two functions having the property that $\lim_{x \to x_0} f(x) = 0$ and $\lim_{x \to x_0} g(x) = 0$. Then the function f/g has **the indeterminate form 0/0** at x_0.

We now give a rule for finding $\lim_{x \to x_0}[f(x)/g(x)]$ in certain cases when f/g has the indeterminate form 0/0 at x_0. The proof of this result will be given in Section 12.2 and other indeterminate forms will be considered in Section 12.3.

Theorem 1. *L'Hôpital's† Rule for the Indeterminate Form 0/0.*

Let x_0 be a real number, $+\infty$, or $-\infty$, and let f and g be two functions that satisfy the following:

    (i) f and g are differentiable at every point in a neighborhood‡ of x_0, except possibly at x_0 itself.
    (ii) $\lim_{x \to x_0} f(x) = 0$ and $\lim_{x \to x_0} g(x) = 0$.
    (iii) $\lim_{x \to x_0}[f'(x)/g'(x)] = L$, where L is a real number, $+\infty$, or $-\infty$.

Then

$$\lim_{x \to x_0} \frac{f(x)}{g(x)} = L.$$

That is, under the conditions of the theorem, the limit of the quotient of the two functions is equal to the limit of the quotient of their derivatives.§ Furthermore, Theorem 1 is also true for right- and left-hand limits.

EXAMPLE 1 Calculate $\lim_{x \to 1}(x^3 - 5x^2 + 6x - 2)/(x^5 - 4x^4 + 7x^2 - 9x + 5)$.

Solution. We first note that

$$\lim_{x \to 1}(x^3 - 5x^2 + 6x - 2) = 0 \quad \text{and} \quad \lim_{x \to 1}(x^5 - 4x^4 + 7x^2 - 9x + 5) = 0,$$

so that the indicated limit has the form 0/0. Then applying L'Hôpital's rule, we have

$$\lim_{x \to 1} \frac{x^3 - 5x^2 + 6x - 2}{x^5 - 4x^4 + 7x^2 - 9x + 5} = \lim_{x \to 1} \frac{3x^2 - 10x + 6}{5x^4 - 16x^3 + 14x - 9}$$

$$= \frac{-1}{-6} = \frac{1}{6}. \quad \blacksquare$$

---

†This theorem is named after the French mathematician Marquis de L'Hôpital (1661–1704). L'Hôpital included this theorem in a book, considered to be the first calculus textbook ever written, published in 1696. Actually, the theorem was first proven by the great Swiss mathematician Jean Bernoulli (1667–1748), who was one of L'Hôpital's tutors and who sent L'Hôpital the proof in a letter in 1694.

‡On page 124 we defined a neighborhood of a point x_0 as an open interval containing x_0, and we defined a neighborhood of ∞ as an open interval of the form (a, ∞) for some real number a.

§We emphasize that we are taking the quotient of the two derivatives (f'/g'), *not* the derivative of the quotient $((f/g)')$.

EXAMPLE 2 Calculate $\lim_{x\to 0^+}[\sqrt{x}/(\sin 3\sqrt{x})]$.

Solution. Since $\sqrt{x}$ is only defined for $x \geq 0$ and since we are considering only a right-hand limit, a neighborhood of 0 is an interval of the form $(0, b)$. In this context the hypotheses of Theorem 1 hold, and we have

$$\lim_{x\to 0^+} \frac{\sqrt{x}}{\sin 3\sqrt{x}} = \lim_{x\to 0^+} \frac{1/2\sqrt{x}}{(\cos 3\sqrt{x})(3/2\sqrt{x})} = \lim_{x\to 0^+} \frac{1}{3\cos 3\sqrt{x}} = \frac{1}{3}. \ \blacksquare$$

EXAMPLE 3 Calculate $\lim_{x\to 0}[(\sin x)/x^3]$.

Solution.

$$\lim_{x\to 0} \frac{\sin x}{x^3} = \lim_{x\to 0} \frac{\cos x}{3x^2} = \infty$$

NOTE. We cannot apply L'Hôpital's rule to the last expression because $\lim_{x\to 0} \cos x = 1 \neq 0$. $\blacksquare$

EXAMPLE 4 Calculate $\lim_{x\to 1}(\ln x)/(x - 1)$.

Solution. We note that $\lim_{x\to 1} \ln x = 0 = \lim_{x\to 1}(x - 1)$, so that L'Hôpital's rule applies. Then

$$\lim_{x\to 1} \frac{\ln x}{x - 1} = \lim_{x\to 1} \frac{1/x}{1} = \lim_{x\to 1} \frac{1}{x} = 1. \ \blacksquare$$

EXAMPLE 5 Calculate $\lim_{x\to\infty}[1 + (1/x)]^x$.

Solution. First, note that the expression $\lim_{x\to\infty}[1 + (1/x)]^x$ was used in Section 6.2 to define the number e. We cannot apply L'Hôpital's rule directly since the indicated limit does not have the form of a quotient. Let us define $y = [1 + (1/x)]^x$. Then

$$\ln y = x \ln\left(1 + \frac{1}{x}\right) = \frac{\ln[1 + (1/x)]}{1/x}.$$

Since $\lim_{x\to\infty} \ln[1 + (1/x)] = 0$ and $\lim_{x\to\infty} 1/x = 0$, we can now apply L'Hôpital's rule to obtain

$$\lim_{x\to\infty} \ln y = \lim_{x\to\infty} \frac{\ln[1 + (1/x)]}{1/x} = \lim_{x\to\infty} \frac{\frac{1}{1 + (1/x)} \cdot \frac{-1}{x^2}}{-1/x^2}$$

$$= \lim_{x\to\infty} \frac{1}{1 + (1/x)} = 1.$$

Hence $\ln y \to 1$ as $x \to \infty$. Thus

$$y = e^{\ln y} \to \overset{\text{Since } e^x \text{ is continuous}}{e^1} = e$$

(which is what we expected). $\blacksquare$

EXAMPLE 6   Calculate

$$\lim_{x \to -2} \frac{3x^3 + 16x^2 + 28x + 16}{x^5 + 4x^4 + 4x^3 + 3x^2 + 12x + 12}.$$

Solution. First, we note that

$$\lim_{x \to -2} (3x^3 + 16x^2 + 28x + 16) = 0$$

and

$$\lim_{x \to -2} (x^5 + 4x^4 + 4x^3 + 3x^2 + 12x + 12) = 0,$$

so that

Differentiate top and bottom

$$\lim_{x \to -2} \frac{3x^3 + 16x^2 + 28x + 16}{x^5 + 4x^4 + 4x^3 + 3x^2 + 12x + 12} = \lim_{x \to -2} \frac{9x^2 + 32x + 28}{5x^4 + 16x^3 + 12x^2 + 6x + 12}.$$

But

$$\lim_{x \to -2} (9x^2 + 32x + 28) = 0$$

and

$$\lim_{x \to -2} (5x^4 + 16x^3 + 12x^2 + 6x + 12) = 0,$$

so we simply apply L'Hôpital's rule again:

$$= \lim_{x \to -2} \frac{18x + 32}{20x^3 + 48x^2 + 24x + 6} = \frac{-4}{-10} = \frac{2}{5}. \blacksquare$$

WARNING: Do not try to apply L'Hôpital's rule when either the numerator or denominator of f/g has a finite, nonzero limit at x_0. For example, we easily see that

$$\lim_{x \to 0} \frac{x}{1 + \sin x} = \frac{\lim_{x \to 0} x}{\lim_{x \to 0}(1 + \sin x)} = \frac{0}{1} = 0.$$

But if we try to apply L'Hôpital's rule, we obtain

$$\lim_{x \to 0} \frac{x}{1 + \sin x} = \lim_{x \to 0} \frac{1}{\cos x} = 1,$$

which is an incorrect result.

PROBLEMS 12.1

In Problems 1–26, calculate the given limit.

1. $\lim_{x \to 0} \dfrac{1 - \cos x}{x}$

2. $\lim_{x \to \pi/2} \dfrac{x - (\pi/2)}{\cos x}$

3. $\lim_{x \to 0^+} \dfrac{\sin x}{\sqrt{x}}$

4. $\lim\limits_{x\to 0} \dfrac{\sin x}{x^5}$

5. $\lim\limits_{x\to 1} \dfrac{x^4 - x^3 + x^2 - 1}{x^3 - x^2 + x - 1}$

6. $\lim\limits_{x\to\infty} \dfrac{1/x}{\ln[1 + (1/x)]}$

7. $\lim\limits_{x\to 0} \dfrac{e^x - 1}{x(3 + x)}$

8. $\lim\limits_{x\to 0} \dfrac{x + \sin 5x}{x - \sin 5x}$

9. $\lim\limits_{x\to 0} \dfrac{x + \sin ax}{x - \sin ax}, \ a \neq 1$

10. $\lim\limits_{x\to 0} \dfrac{x - \sin x}{x^3}$

11. $\lim\limits_{x\to 1^-} \dfrac{x^2 - 1}{\sqrt{1 - x}}$

12. $\lim\limits_{x\to 0} \dfrac{3 + x - 3e^x}{x(2 + 5e^x)}$

*13. $\lim\limits_{x\to 1^-} \dfrac{\sqrt{1 - x^3}}{\sqrt{1 - x^4}}$

*14. $\lim\limits_{x\to 0} \left(\dfrac{2}{\sin x} - \dfrac{2}{x}\right)$

15. $\lim\limits_{x\to 0^+} \dfrac{4^x - 3^x}{\sqrt{x}}$

16. $\lim\limits_{x\to \pi/2} \dfrac{\cos x}{\sin 2x}$

17. $\lim\limits_{x\to 0} (1 + x)^{2/x}$

18. $\lim\limits_{x\to 0} \dfrac{\tan^{-1} x}{x}$

19. $\lim\limits_{x\to 0} \dfrac{x^2}{\sin^{-1} x}$

20. $\lim\limits_{x\to 0} \dfrac{\sin^{-1} x - x}{5x^2}$

*21. $\lim\limits_{x\to 0} \dfrac{\sinh x - \sin x}{\sin^3 x}$

22. $\lim\limits_{x\to 0} (x + e^x)^{1/x}$

23. $\lim\limits_{x\to 4\pi} \dfrac{(x - 4\pi)^2}{\ln \cos x}$

24. $\lim\limits_{x\to -\infty} \dfrac{e^x}{2^x}$

25. $\lim\limits_{x\to x_0} \dfrac{\sqrt{x} - \sqrt{x_0}}{x - x_0}$

26. $\lim\limits_{x\to 0} \dfrac{\tanh ax}{\tanh bx}$

27. Show that if a and b are real numbers, $b \neq 0$, and if $x_0 > 0$, then

$$\lim_{x\to x_0} \frac{x^a - x_0^a}{x^b - x_0^b} = \frac{a}{b} x_0^{a-b}.$$

*28. Calculate $\lim_{x\to 0} (\int_0^x \cos t^2 \, dt)/(\int_0^x e^{t^2} \, dt)$. [*Hint:* Use the second fundamental theorem of calculus (Theorem 5.6.2).]

29. Calculate $\lim_{x\to 0^+} (\int_0^x \sin t^3 \, dt)/x^4$.

30. **(a)** Show that the sum of the **geometric progression,** S_n, is given by

$$S_n = 1 + a + a^2 + \cdots + a^n = \frac{1 - a^{n+1}}{1 - a}.$$

(b) Calculate $\lim_{a\to 1} S_n$ and compare this with the sum of the first $n + 1$ terms of the geometric progression with $a = 1$.

31. Show that $\lim_{x\to 0} (1 + x)^{a/x} = e^a$ for any real number $a \neq 0$.

32. Calculate $\lim_{x\to\infty} x \sin(1/x)$.

33. Show that $\lim_{x\to\infty} x^a \sin(1/x) = 0$ if $0 < a < 1$.

34. Show that $\lim_{x\to\infty} x^a \sin(1/x) = +\infty$ if $a > 1$.

35. In Problem 6.6.17, we discussed the simple electric circuit shown in Figure 1 and found that the current at any time t is given by $I(t) = (E/R) + [I(0) - (E/R)]e^{-Rt/L}$. Assuming that $I(0) = 0$, find, for fixed t, E, and L, $\lim_{R\to 0^+} I(t)$.

FIGURE 1

36. Suppose that f is a function continuous on the interval $[a, b]$. Construct a new function A defined on $(a, b]$ by

$$A(x) = \frac{1}{x - a} \int_a^x f(t)\, dt.$$

That is, $A(x)$ is the average value of f on the interval $[a, x]$. Find $\lim_{x \to a^+} A(x)$.

*37. Compute

$$\lim_{\epsilon \to 0} \int \frac{\dfrac{1}{x - a - \epsilon} - \dfrac{1}{x - a}}{\epsilon}\, dx.$$

*38. Fix two positive real numbers a and b and let $f(x) = \int_a^b t^x\, dt$. Show that $\lim_{x \to -1} f(x)$ exists and obtain a simple expression for it.

12.2 PROOF OF L'HÔPITAL'S RULE (OPTIONAL)

In this section we will prove L'Hôpital's rule for the indeterminate form 0/0. Before doing so, however, we need to prove another kind of mean value theorem that will be useful here and in Chapter 13.

Theorem 1 *Cauchy† Mean Value Theorem*   Suppose that the two functions f and g are continuous in the closed interval $[a, b]$ and differentiable in the open interval (a, b). Suppose further that $g'(x) \neq 0$ for x in (a, b). Then there exists at least one number c in (a, b) such that

$$\frac{f(b) - f(a)}{g(b) - g(a)} = \frac{f'(c)}{g'(c)}. \tag{1}$$

Proof.   Let

$$h(x) = [g(b) - g(a)][f(x) - f(a)] - [g(x) - g(a)][f(b) - f(a)].$$

Then clearly $h(a) = h(b) = 0$. Since h is continuous in $[a, b]$ and differentiable in (a, b), we can apply Rolle's theorem (Theorem 4.2.1), which tells us that there exists at least one number c in (a, b) such that $h'(c) = 0$. But

$$h'(c) = [g(b) - g(a)][f'(c)] - [g'(c)][f(b) - f(a)] = 0. \tag{2}$$

Then $f'(c)[g(b) - g(a)] = g'(c)[f(b) - f(a)]$. Since $g'(c) \neq 0$ by assumption, we can divide by it and by $[g(b) - g(a)]$ to obtain equation (1). ■

NOTE.   It is *not* necessary to assume that $g(b) - g(a) \neq 0$ since this fact is implied by the hypotheses (see Problem 1).

We can now prove L'Hôpital's rule for the indeterminate form 0/0.

Theorem 2.   Let x_0 be a real number, $+\infty$, or $-\infty$ and let f and g be two functions that satisfy the following:

---

†See the biography of Cauchy on page 64.

(i) f and g are differentiable at every point in a neighborhood of x_0, except possibly at x_0 itself.

(ii) $\lim_{x \to x_0} f(x) = 0$ and $\lim_{x \to x_0} g(x) = 0$.

(iii) $\lim_{x \to x_0} [f'(x)/g'(x)] = L$, where L is a real number, $+\infty$, or $-\infty$.

Then $\lim_{x \to x_0} [f(x)/g(x)] = L$.

Furthermore, the theorem is true for right- and left-hand limits as well.

REMARK. The idea of the proof is simple. The proof, however, is not so simple because of a number of small details. Here's the idea: Since $\lim_{x \to x_0} f(x) = 0 = \lim_{x \to x_0} g(x)$, we assume that $f(x_0) = g(x_0) = 0$. Then if $x > x_0$, we find from the Cauchy mean value theorem that there is a number c in (x_0, x) such that

$$\frac{f(x)}{g(x)} = \frac{f(x) - 0}{g(x) - 0} = \frac{f(x) - f(x_0)}{g(x) - g(x_0)} = \frac{f'(c)}{g'(c)} \to L \quad \text{as} \quad x \to x_0,$$

since as $x \to x_0$, $c \to x_0$ (because c is squeezed between x_0 and x). Now to the details.

Proof. We prove the theorem first in the case in which $x \to x_0^+$, where x_0 is a real number.

Step 1. Choose a positive number $\Delta_1 x$ such that f and g are differentiable in $(x_0, x_0 + \Delta_1 x)$ and continuous in $(x_0, x_0 + \Delta_1 x]$.   We can do this by (i).

Step 2. Choose $\Delta_2 x$ such that $f'(x)/g'(x)$ exists in $(x_0, x_0 + \Delta_2 x)$.   We can do this because $\lim_{x \to x_0^+} [f'(x)/g'(x)]$ exists.

Step 3. Note that $g'(x) \neq 0$ for x in $(x_0, x_0 + \Delta_2 x)$.   Since $f'(x)/g'(x)$ exists (by Step 2) in $(x_0, x_0 + \Delta_2 x)$.

Step 4. Define $\Delta x = \min(\Delta_1 x, \Delta_2 x)$.

Step 5. Since $\lim_{x \to x_0} f(x) = 0$, $f(x_0) = 0$, f is not defined at x_0 or $f(x_0) = k \neq 0$. In the latter cases we simply redefine f to be zero at x_0. Similarly, either $g(x_0) = 0$ or g can be defined to be zero at x_0. Then by hypothesis (ii), both functions are continuous in the closed interval $[x_0, x_0 + \Delta x]$.

Step 6. Apply the Cauchy mean value theorem to f and g in the interval $(x_0, x_0 + \Delta x)$. There is a number $c_{\Delta x}$ in $(x_0, x_0 + \Delta x)$† such that

$$\frac{f'(c_{\Delta x})}{g'(c_{\Delta x})} = \frac{f(x_0 + \Delta x) - f(x_0)}{g(x_0 + \Delta x) - g(x_0)} = \frac{f(x_0 + \Delta x)}{g(x_0 + \Delta x)}.$$

Step 7. Observe that $c_{\Delta x} \to x_0^+$ as $\Delta x \to 0^+$, so

$$\lim_{x \to x_0^+} \frac{f(x)}{g(x)} = \lim_{\Delta x \to 0^+} \frac{f(x_0 + \Delta x)}{g(x_0 + \Delta x)} = \lim_{\Delta x \to 0^+} \frac{f'(c_{\Delta x})}{g'(c_{\Delta x})}$$

$$= \lim_{c_{\Delta x} \to x_0^+} \frac{f'(c_{\Delta x})}{g'(c_{\Delta x})} = \lim_{x \to x_0^+} \frac{f'(x)}{g'(x)} = L.$$

---

†The notation $c_{\Delta x}$ implies that the number depends on the choice of Δx.

This step completes the proof in the case $x \to x_0{}^+$.

If $x \to x_0{}^-$, then we obtain the same proof by considering the interval $(x_0 - \Delta x, x_0)$. If $x \to x_0$, then $x \to x_0{}^-$ and $x \to x_0{}^+$. Finally, for the case in which $x \to \infty$, let $y = 1/x$. Then $x = 1/y$, $dx/dy = -1/y^2$, and by the chain rule,

$$f'(x) = \left[f\left(\frac{1}{y}\right) \right]' = f'\left(\frac{1}{y}\right) \frac{d}{dy}\left(\frac{1}{y}\right) = f'\left(\frac{1}{y}\right)\left(-\frac{1}{y^2}\right).$$

Thus

$$\lim_{x \to \infty} \frac{f'(x)}{g'(x)} = \lim_{y \to 0^+} \frac{f'(1/y)(-1/y^2)}{g'(1/y)(-1/y^2)} = \lim_{y \to 0^+} \frac{f'(1/y)}{g'(1/y)}$$

By the case proven above

$$= \lim_{y \to 0^+} \frac{f(1/y)}{g(1/y)} = \lim_{x \to \infty} \frac{f(x)}{g(x)},$$

and the case where $x \to \infty$ is proved. The case in which $x \to -\infty$ is handled in a similar manner. ∎

PROBLEM 12.2

1. Use the mean value theorem (Theorem 4.2.2) to show that under the hypotheses of Theorem 1, $g(b) \neq g(a)$.

12.3 OTHER INDETERMINATE FORMS

There are other situations in which $\lim_{x \to x_0}[f(x)/g(x)]$ cannot be evaluated directly. One important case is defined below.

Definition 1 INDETERMINATE FORM ∞/∞. Let f and g be two functions having the property that $\lim_{x \to x_0} f(x) = \pm\infty$ and $\lim_{x \to x_0} g(x) = \pm\infty$, where x_0 is real, $+\infty$, or $-\infty$. Then the function f/g has the **indeterminate form** ∞/∞ at x_0.

Theorem 1 *L'Hôpital's Rule for the Indeterminate Form ∞/∞.* let x_0 be a real number, $+\infty$, or $-\infty$ and let f and g be two functions that satisfy the following:

(i) f and g are differentiable at every point in a neighborhood of x_0, except possibly at x_0 itself.
(ii) $\lim_{x \to x_0} f(x) = \pm\infty$ and $\lim_{x \to x_0} g(x) = \pm\infty$.
(iii) $\lim_{x \to x_0}[f'(x)/g'(x)] = L$, where L is a real number, $+\infty$, or $-\infty$.

Then

$$\lim_{x \to x_0} \frac{f(x)}{g(x)} = L. \tag{1}$$

The result is also true for right- and left-hand limits.

The proof of this theorem is more complicated than the proof of the earlier version of L'Hôpital's rule and is omitted.†

EXAMPLE 1 Calculate $\lim_{x\to\infty}(e^x/x)$.

Solution. Since $e^x \to \infty$ as $x \to \infty$, L'Hôpital's rule applies, and we have

$$\lim_{x\to\infty}\frac{e^x}{x} = \lim_{x\to\infty}\frac{e^x}{1} = \infty. \ \blacksquare$$

EXAMPLE 2 Calculate $\lim_{x\to0^+}(x \ln x)$.

Solution. We write $x \ln x = (\ln x)/(1/x)$. Then since $\lim_{x\to0^+} \ln x = -\infty$ and $\lim_{x\to0^+}(1/x) = \infty$, we have

$$\lim_{x\to0^+}(x \ln x) = \lim_{x\to0^+}\frac{\ln x}{1/x} = \lim_{x\to0^+}\frac{1/x}{-1/x^2} = \lim_{x\to0^+}(-x) = 0. \ \blacksquare$$

EXAMPLE 3 Calculate $\lim_{x\to0^+}x^{ax}$, $a > 0$.

Solution. If $y = x^{ax}$, then $\ln y = ax \ln x$, and

$$\lim_{x\to0^+} \ln y = \lim_{x\to0^+}(ax \ln x) = \lim_{x\to0^+}\frac{\ln x}{1/ax} = \lim_{x\to0^+}\frac{1/x}{-1/ax^2} = \lim_{x\to0^+}(-ax) = 0.$$

Since $\ln y \to 0$, we have $y = x^{ax} \to 1$ as $x \to 0^+$. $\blacksquare$

EXAMPLE 4 Compute $\lim_{x\to\infty}3x/(5x + 2 \sin x)$.

Solution. By dividing numerator and denominator by x, we obtain

$$\lim_{x\to\infty}\frac{3x}{5x + 2 \sin x} = \lim_{x\to\infty}\frac{3}{5 + (\sin x)/x} = \frac{3}{5}.$$

Also, $\lim_{x\to\infty} 3x = \lim_{x\to\infty}(5x + 2 \sin x) = \infty$. However, if we try to apply L'Hôpital's rule, we obtain

$$\lim_{x\to\infty}\frac{3x}{5x + 2 \sin x} \stackrel{?}{=} \lim_{x\to\infty}\frac{3}{5 + 2 \cos x}$$

which *does not exist.* This is one example of a situation in which $\lim_{x\to x_0} f(x)/g(x)$ exists, but $\lim_{x\to x_0} f'(x)/g'(x)$ does not exist. It illustrates the fact that in order to apply L'Hôpital's rule we must know that $\lim_{x\to x_0} f'(x)/g'(x)$ exists. Otherwise, the rule cannot be applied. $\blacksquare$

EXAMPLE 5 Calculate $\lim_{x\to\infty}x^a e^{-bx}$ for any real number a and any positive real number b.

†For a proof, consult R. C. Buck and F. Buck, *Advanced Calculus*, 3rd ed. (McGraw-Hill, New York, 1978), pages 121–122.

Solution.

Case 1. $a = 0$. Then

$$\lim_{x \to \infty} x^a e^{-bx} = \lim_{x \to \infty} x^0 e^{-bx} = \lim_{x \to \infty} \frac{1}{e^{bx}} = 0.$$

Case 2. $a \neq 0$. Let $y = x^a e^{-bx}$. Then

$$\ln y = \ln x^a + \ln e^{-bx} = a \ln x - bx = x\left(a \frac{\ln x}{x} - b\right).$$

Now

$$\lim_{x \to \infty} \frac{\ln x}{x} = \lim_{x \to \infty} \frac{1/x}{1} = \lim_{x \to \infty} \frac{1}{x} = 0.$$

This means that

$$\lim_{x \to \infty} \left(\frac{a \ln x}{x} - b\right) = -b < 0 \qquad \text{and} \qquad \lim_{x \to \infty} x\left(\frac{a \ln x}{x} - b\right) = -\infty.$$

Thus $\ln y \to -\infty$, which means that $y \to 0$ as $x \to \infty$. Therefore

$$\lim_{x \to \infty} x^a e^{-bx} = 0 \qquad \text{if} \qquad b > 0. \tag{2}$$

This result is very interesting and useful. It tells us that *the exponential function e^x grows much faster than any power function.* ■

Now let $P_n(x)$ be a polynomial of degree n. That is,

$$P_n(x) = a_n x^n + a_{n-1}x^{n-1} + a_{n-2}x^{n-2} + \cdots + a_2 x^2 + a_1 x + a_0.$$

From (2) we see that, for $k = 0, 1, 2, \ldots, n$,

$$\lim_{x \to \infty} a_k x^k e^{-bx} = a_k \lim_{x \to \infty} x^k e^{-bx} = a_k \cdot 0 = 0.$$

Thus we have the following theorem.

Theorem 2. If $b > 0$, then

$$\lim_{x \to \infty} [P_n(x)e^{-bx}] = 0 \qquad \text{for all values of } n. \tag{3}$$

We will make use of this theorem in later chapters.

There are other indeterminate forms that can be dealt with by applying L'Hôpital's rule. For example, in Example 2 we calculated $\lim_{x \to 0^+}(x \ln x)$. Since

$\lim_{x \to 0^+} x = 0$ and $\lim_{x \to 0^+} \ln x = -\infty$, this indeterminate expression is really of the form $0 \cdot \infty$. An indeterminate form of this type can often be treated by putting it into one of the forms $0/0$ or ∞/∞.

In Table 1 we give a list of indeterminate forms. We stress, however, that only the indeterminate forms $0/0$ and ∞/∞ can be evaluated directly. In all other cases it is necessary to bring the expression into the form $0/0$ or ∞/∞. This is done either by an algebraic manipulation (as in Example 2) or by taking logarithms (as in Example 3).

TABLE 1 INDETERMINATE FORMS

Indeterminate Form	Example
$\dfrac{0}{0}$	$\lim\limits_{x \to 0} \dfrac{\sin x}{x}$ (Section 3.5)
$\dfrac{\infty}{\infty}$	$\lim\limits_{x \to \infty} \dfrac{2x^2 - 2x + 3}{x^2 + 4x + 4}$ (Example 2.4.10)
$0 \cdot \infty$	$\lim\limits_{x \to 0^+} x \ln x$ (Example 2)
$\infty - \infty$	$\lim\limits_{x \to 0^+} (\csc x - \cot x)$ (Example 6)
0^0	$\lim\limits_{x \to 0^+} x^{ax}$ (Example 3)
∞^0	$\lim\limits_{x \to \infty} x^{1/x}$ (Example 7)
1^∞	$\lim\limits_{x \to \infty} \left(1 + \dfrac{1}{x}\right)^x$ (Example 12.1.5)

EXAMPLE 6   Calculate $\lim_{x \to 0^+} (\csc x - \cot x)$.

Solution. We first note that

$$\lim_{x \to 0^+} \csc x = \lim_{x \to 0^+} \frac{1}{\sin x} = \infty$$

and

$$\lim_{x \to 0^+} \cot x = \lim_{x \to 0^+} \frac{\cos x}{\sin x} = \infty.$$

We now write

$$\lim_{x \to 0^+} (\csc x - \cot x) = \lim_{x \to 0^+} \frac{1}{\sin x} - \frac{\cos x}{\sin x}$$

$$= \lim_{x \to 0^+} \frac{1 - \cos x}{\sin x}$$

L'Hôpital's rule for $\dfrac{0}{0}$ ↘

$$= \lim_{x \to 0^+} \frac{\sin x}{\cos x} = 0. \ \blacksquare$$

EXAMPLE 7 Calculate $\lim_{x\to\infty} x^{1/x}$.

Solution. This expression is of the form ∞^0. We set $y = x^{1/x}$. Then $\ln y = (1/x) \ln x$, and

$$\lim_{x\to\infty} \ln y = \lim_{x\to\infty} \frac{\ln x}{x} = \lim_{x\to\infty} \frac{1/x}{1} = 0,$$

so that $\ln y \to 0$ and $y = x^{1/x} \to 1$. ■

The technique used in Example 7 can often be used to evaluate limits of expressions having the form

$$f(x)^{g(x)}.$$

REMARK. In using this technique, we make use of the fact that e^x is continuous. Without that we couldn't conclude, for example, that $y \to 1$ just because $\ln y \to 0$. But because of the continuity of e^x, we may conclude that

$$\lim_{x\to\infty} y = \lim_{x\to\infty} e^{\ln y} = e^{\lim_{x\to\infty} \ln y} = e^0 = 1.$$

PROBLEMS 12.3

In Problems 1–32, evaluate the given limit.

1. $\lim_{x\to\infty} \dfrac{x^3 + 3x + 4}{2x^3 - 4x + 2}$

2. $\lim_{x\to\infty} \dfrac{4x^{5/2} + 3\sqrt{x} - 10}{3x^{5/2} - 8x^{3/2} + 45x^2}$

3. $\lim_{x\to\infty} \dfrac{\ln x}{\sqrt{x}}$

4. $\lim_{x\to\pi/2^-} \dfrac{\sec x}{\tan x}$

5. $\lim_{x\to\infty} \dfrac{\ln x}{x^a}, a > 0$

6. $\lim_{x\to0^+} \dfrac{\ln(\sin x)}{\ln(\tan x)}$

*7. $\lim_{x\to0^+} \dfrac{\ln x}{\cot x}$

8. $\lim_{x\to\pi/2^-} \dfrac{\tan 2x}{\tan x}$

9. $\lim_{x\to\infty} \left(x \tan \dfrac{1}{x}\right)$

10. $\lim_{x\to0^+} x^{-1/x}$

*11. $\lim_{x\to\infty} \dfrac{x}{e^{\sqrt{x}}}$

*12. $\lim_{x\to\infty} (xe^{-x^a}), a > 0$

13. $\lim_{x\to0^+} x^{\sin x}$

14. $\lim_{x\to\pi/2^+} \left(x - \dfrac{\pi}{2}\right)^{\cos x}$

15. $\lim_{x\to0^+} (x \ln \sin x)$

16. $\lim_{x\to0^+} (\sin x)^x$

17. $\lim_{x\to\infty} \left(1 + \dfrac{1}{2x}\right)^{x^2}$

*18. $\lim_{x\to\infty} \left(1 + \dfrac{1}{ax}\right)^{x^a}, a > 0$ [*Hint:* Consider separately the cases $a < 1, a = 1, a > 1$.]

19. $\lim_{x\to0^+} (1 + \sinh x)^{1/x}$

20. $\lim_{x\to0^+} (1 + \sinh x)^{a/x}, a > 0$

21. $\lim\limits_{x \to 0+} (\csc 2x \, \sin 3x)$

22. $\lim\limits_{x \to (\pi/2)} (\sec x \, \cos 3x)$

*23. $\lim\limits_{x \to 0+} (\sin x)^{(\sin x) - x}$

24. $\lim\limits_{x \to \pi/2 -} (\cos x)^{\sec x}$

25. $\lim\limits_{x \to 2} x^{1/(2-x)}$

26. $\lim\limits_{x \to \infty} (1 + 5x)^{e^{-x}}$

27. $\lim\limits_{x \to \infty} \left(\cos \dfrac{1}{x}\right)^x$

28. $\lim\limits_{x \to 3} \left[\dfrac{1}{x - 3} - \dfrac{1}{(x - 3)^2}\right]$

29. $\lim\limits_{x \to 1-} \left(\dfrac{1}{1 - x} - \dfrac{1}{\ln x}\right)$

30. $\lim\limits_{x \to \infty} (\sqrt{x^2 + x} - x)$

31. $\lim\limits_{x \to 2+} \left(\dfrac{x}{x - 2} - \dfrac{1}{\ln(x - 1)}\right)$

32. $\lim\limits_{x \to \infty} (x^3 - \sqrt{x^6 - 5x^3 + 3})$

*33. Since $\lim_{x \to \infty} \sin x$ does not exist (but oscillates continuously between -1 and $+1$), it is evident that $\lim_{x \to \infty} e^{-\sin x}$ does not exist. But

$$\lim_{x \to \infty} e^{-\sin x} = \lim_{x \to \infty} \frac{2x + \sin 2x}{(2x + \sin 2x)e^{\sin x}}.$$

Now show the following successively:

(a) $2 + 2 \cos 2x = 4 \cos^2 x$. [*Hint:* $\cos 2x = 2 \cos^2 x - 1$.]

(b) $\lim\limits_{x \to \infty} \dfrac{4 \cos x}{(2x + 4 \cos x + \sin 2x)e^{\sin x}} = 0.$

Then apply L'Hôpital's rule [using (a) and (b)] to show that:

(c) $\lim\limits_{x \to \infty} \dfrac{2x + \sin 2x}{(2x + \sin 2x)e^{\sin x}} = 0.$

Thus $\lim_{x \to \infty} e^{-\sin x} = 0$. Where have we been led astray?

34. Show that $\lim_{x \to 0}(e^{-1/x^2}/x) = 0$. [*Hint:* Write $\lim_{x \to 0}(e^{-1/x^2}/x) = \lim_{x \to 0}[(1/x)/e^{1/x^2}]$.]

35. Show that for every integer n, $\lim_{x \to 0}(e^{-1/x^2}/x^n) = 0$.

*36. Use the result of Problem 35 to show that the function

$$f(x) = \begin{cases} e^{-1/x^2}, & x \neq 0 \\ 0, & x = 0 \end{cases}$$

has continuous derivatives of all orders at every real number x. [*Hint:* Show that $f^{(n)}(x)$ is a sum of terms of the form $ax^{-m}e^{-1/x^2}$, where $n + 2 \leq m \leq 3n$.]

*37. Compute:

(a) $\lim\limits_{x \to 0+} x^{-k/\ln x}, \ k > 0$

(b) $\lim\limits_{x \to 0+} x^{-k/\sqrt{-\ln x}}, \ k > 0$

*38. Let $g(x) = x \ln x/(x^2 - 1)$. Compute the following:

(a) $\lim\limits_{x \to 0+} g(x)$

(b) $\lim\limits_{x \to 1} g(x)$

39. (a) Show that $\lim_{x \to \infty}(x + \cos x)/(x + \sin x) = 1$.

(b) Explain why $\lim_{x \to \infty}(1 - \sin x)/(1 + \cos x)$ does not exist.

Do the results of parts (a) and (b) show that L'Hôpital's rule is not always true?

12.4 IMPROPER INTEGRALS

In our introduction to the definite integral in Chapter 5, we cited Theorem 5.5.1, which states that $\int_a^b f(x) \, dx$ exists if $f(x)$ is continuous in the closed interval $[a, b]$. However, in many interesting applications one of two situations occurs: either (1) a or b is infinite or (2) f becomes infinite at one or more values in the interval $[a, b]$. If one of these situations occurs, we say that the integral in question is an **improper integral.** In this section we will learn how to evaluate these two different types of improper integrals.

Improper Integral—Type 1 $b = +\infty$, $a = -\infty$, or both.

Before dealing with the general case, we give an example.

EXAMPLE 1   Calculate the area in the first quadrant under the curve $y = e^{-x}$.

Solution. The region in question is sketched in Figure 1. For the first time we are asked to calculate the area of a region that stretches an infinite distance. To deal with this situation, we calculate the area from 0 to N, where N is some "large" number, and then we see what happens as $N \to \infty$. We have

$$\text{area from 0 to } N = A_0^N = \int_0^N e^{-x}\, dx = -e^{-x}\Big|_0^N = 1 - e^{-N}.$$

Then

$$\text{total area} = \lim_{N\to\infty} A_0^N = \lim_{N\to\infty}(1 - e^{-N}) = 1.$$

Thus the total area in the first quadrant under the curve $y = e^{-x}$ is 1. ■

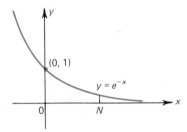

FIGURE 1

Example 1 leads to the following general definitions.

Definition 1   IMPROPER INTEGRAL—TYPE 1: INFINITE LIMIT OF INTEGRATION

(i) Let a be a real number and let f be a function having the property that $\int_a^N f(x)\, dx$ exists for every real number $N \ge a$. Then we define the **improper integral**

$$\int_a^\infty f(x)\, dx = \lim_{N\to\infty} \int_a^N f(x)\, dx \tag{1}$$

provided that this limit exists.

If $\int_a^\infty f(x)\, dx$ exists and is finite, we say that the improper integral is **convergent.** If the limit in (1) does not exist, or if it exists and is infinite, then we say that the improper integral is **divergent.**

(ii) If $\int_{-N}^b f(x)\, dx$ exists for every real number N such that $-N \le b$, we define

$$\int_{-\infty}^{b} f(x)\,dx = \lim_{N\to\infty} \int_{-N}^{b} f(x)\,dx \tag{2}$$

whenever the limit exists. We define the terms "convergent" and "divergent" as in (i).
(iii) If $\int_{-M}^{0} f(x)\,dx$ and $\int_{0}^{N} f(x)\,dx$ exist for every real N and M, then we define

$$\int_{-\infty}^{\infty} f(x)\,dx = \lim_{N\to\infty} \int_{0}^{N} f(x)\,dx + \lim_{M\to\infty} \int_{-M}^{0} f(x)\,dx \tag{3}$$

whenever both of these limits exist.

EXAMPLE 2 Evaluate $\int_{1}^{\infty} (1/x)\,dx$.

Solution. $\int_{1}^{N} (1/x)\,dx = \ln x \big|_{1}^{N} = \ln N - \ln 1 = \ln N$. But $\lim_{N\to\infty} \ln N = \infty$, so that the improper integral is divergent. ■

EXAMPLE 3 Evaluate $\int_{0}^{\infty} e^x\,dx$.

Solution. $\int_{0}^{N} e^x\,dx = e^N - 1$, which approaches ∞ as $N\to\infty$, so this improper integral is also divergent. ■

EXAMPLE 4 Evaluate $\int_{-\infty}^{0} e^x\,dx$.

Solution. $\int_{-N}^{0} e^x\,dx = 1 - e^{-N}$, which approaches 1 as $N\to\infty$, so that $\int_{-\infty}^{0} e^x\,dx$ is convergent and is equal to 1. ■

EXAMPLE 5 Evaluate $\int_{0}^{\infty} \cos x\,dx$.

Solution. $\int_{0}^{N} \cos x\,dx = \sin x \big|_{0}^{N} = \sin N$, which has no limit as $N\to\infty$, so that the improper integral $\int_{0}^{\infty} \cos x\,dx$ diverges. This makes sense graphically. In Figure 2 the integral in question is the area of the shaded region, where the areas below the x-axis are treated as negative. Thus the total area has no finite value because we must keep on adding and subtracting areas of equal size. ■

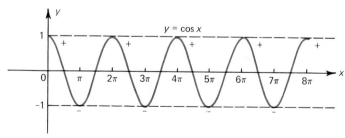

FIGURE 2

EXAMPLE 6 Evaluate $\int_{-\infty}^{\infty} xe^{-x^2}\,dx$.

Solution. The function xe^{-x^2} is sketched in Figure 3. $\int_{0}^{N} xe^{-x^2}\,dx = -\tfrac{1}{2}e^{-x^2}\big|_{0}^{N} = \tfrac{1}{2}(1 - e^{-N^2}) \to \tfrac{1}{2}$ as $N\to\infty$, and $\int_{-M}^{0} xe^{-x^2}\,dx = -\tfrac{1}{2}e^{-x^2}\big|_{-M}^{0} = \tfrac{1}{2}(e^{-M^2} - 1) \to -\tfrac{1}{2}$ as $M\to\infty$. Thus since both limits exist,

FIGURE 3

$$\int_{-\infty}^{\infty} xe^{-x^2}\, dx = \int_{0}^{\infty} xe^{-x^2}\, dx + \int_{-\infty}^{0} xe^{-x^2}\, dx = \frac{1}{2} - \frac{1}{2} = 0. \blacksquare$$

EXAMPLE 7   Calculate $\int_{-\infty}^{\infty} x^3\, dx$.

Solution. The area to be calculated is sketched in Figure 4. Since $\lim_{N\to\infty} \int_{0}^{N} x^3\, dx = \lim_{N\to\infty}(N^4/4) = \infty$, the integral diverges. Note, however, that

$$\lim_{N\to\infty} \int_{-N}^{N} x^3\, dx = \lim_{N\to\infty} \frac{x^4}{4}\bigg|_{-N}^{N} = \lim_{N\to\infty}\left(\frac{N^4}{4} - \frac{N^4}{4}\right) = \lim_{N\to\infty} 0 = 0,$$

which explains why we *do not* define $\int_{-\infty}^{\infty} f(x)\, dx$ as $\lim_{N\to\infty} \int_{-N}^{N} f(x)\, dx$. The problem here is that $\int_{0}^{\infty} x^3\, dx = \infty$ and $\int_{-\infty}^{0} x^3\, dx = -\infty$. We simply cannot "cancel off" infinite terms. The expression $\infty - \infty$ is not defined. $\blacksquare$

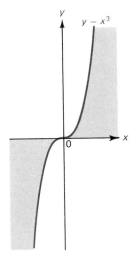

FIGURE 4

EXAMPLE 8   Evaluate

$$\int_{1}^{\infty} \frac{1}{x^k}\, dx, \qquad k > 0,\ k \neq 1$$

Solution.

$$\int_{1}^{N} \frac{1}{x^k}\, dx = \int_{1}^{N} x^{-k}\, dx = \frac{x^{-k+1}}{-k+1}\bigg|_{1}^{N} = \frac{1}{k-1}(1 - N^{1-k})$$

If $0 < k < 1$, then $N^{1-k} \to \infty$ as $N \to \infty$ and the integral diverges. If $k > 1$, then $N^{1-k} = 1/N^{k-1} \to 0$ as $N \to \infty$ and the integral converges. Combining this result with that of Example 2, we have

$$\int_1^\infty \frac{1}{x^k}\,dx \quad \begin{cases} \text{diverges if } 0 < k \le 1, \\ \text{converges to } 1/(k-1) \text{ if } k > 1. \end{cases} \quad \blacksquare$$

EXAMPLE 9 Evaluate

$$\int_{-\infty}^\infty \frac{dx}{x^2 - 4x + 9}.$$

Solution.

$$\int_0^N \frac{dx}{x^2 - 4x + 9} = \int_0^N \frac{dx}{(x-2)^2 + 5} = \frac{1}{\sqrt{5}} \tan^{-1} \frac{x-2}{\sqrt{5}} \Big|_0^N$$

$$= \frac{1}{\sqrt{5}} \left[\tan^{-1} \frac{N-2}{\sqrt{5}} - \tan^{-1}\left(-\frac{2}{\sqrt{5}}\right) \right]$$

and $\lim_{N \to \infty} \tan^{-1}[(N-2)/\sqrt{5}] = \pi/2$. Thus

$$\int_0^\infty \frac{dx}{x^2 - 4x + 5} = \frac{1}{\sqrt{5}} \left[\frac{\pi}{2} - \tan^{-1}\left(-\frac{2}{\sqrt{5}}\right) \right].$$

Similarly,

$$\int_{-M}^0 \frac{dx}{x^2 - 4x + 9} = \frac{1}{\sqrt{5}} \tan^{-1} \frac{x-2}{\sqrt{5}} \Big|_{-M}^0$$

$$= \frac{1}{\sqrt{5}} \left[\tan^{-1}\left(-\frac{2}{\sqrt{5}}\right) - \tan^{-1} \frac{-M-2}{\sqrt{5}} \right].$$

and

$$\int_{-\infty}^0 \frac{dx}{x^2 - 4x + 5} = \frac{1}{\sqrt{5}} \left[\tan^{-1}\left(-\frac{2}{\sqrt{5}}\right) - \left(-\frac{\pi}{2}\right) \right],$$

Thus

$$\int_{-\infty}^\infty \frac{dx}{x^2 - 4x + 5} = \int_0^\infty \frac{dx}{x^2 - 4x + 5} + \int_{-\infty}^0 \frac{dx}{x^2 - 4x + 5}$$

$$= \frac{1}{\sqrt{5}}\left(\frac{\pi}{2} + \frac{\pi}{2} \right) = \frac{\pi}{\sqrt{5}}. \quad \blacksquare$$

Improper Integral—Type 2: Integral becomes infinite in $[a, b]$

We now turn to the second type of improper integral in which the integrand becomes infinite in the closed interval $[a, b]$. As before, we begin with an example.

EXAMPLE 10 Calculate the area in the first quadrant under the curve $y = 1/\sqrt{x}$ between $x = 0$ and $x = 1$.

   Solution. The region in question is sketched in Figure 5. As in Example 1, the region is infinite. However, if ϵ is a small, positive real number, then we can calculate $\int_\epsilon^1 (1/\sqrt{x}) \, dx$ and see what happens as $\epsilon \to 0^+$. We have

$$A_\epsilon^1 = \int_\epsilon^1 \frac{1}{\sqrt{x}} \, dx = 2\sqrt{x}\,\Big|_\epsilon^1 = 2(1 - \sqrt{\epsilon}).$$

Then

$$\text{total area} = \lim_{\epsilon \to 0^+} A_\epsilon^1 = \lim_{\epsilon \to 0^+} 2(1 - \sqrt{\epsilon}) = 2. \ \blacksquare$$

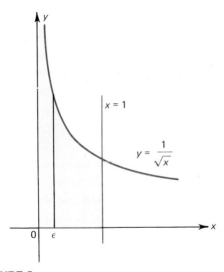

FIGURE 5

   Before citing the general definition, we remark that if f is continuous on $(a, b]$ but $f(x) \to +\infty$ or $-\infty$ as $x \to a^+$, then the graph of f has a **vertical asymptote** at $x = a$. That is, as $x \to a^+$, the graph of $y = f(x)$ approaches the vertical line $x = a$. In Example 10 we see this illustrated for $a = 0$.

Definition 2 IMPROPER INTEGRAL—TYPE 2: INTEGRAND BECOMES INFINITE IN $[a, b]$ Let a and b be finite numbers.

   (i) If $\int_{a+\epsilon}^b f(x) \, dx$ exists for every ϵ in $(0, b - a]$ and if f has a vertical asymptote at $x = a$, then

$$\int_a^b f(x) \, dx = \lim_{\epsilon \to 0^+} \int_{a+\epsilon}^b f(x) \, dx \qquad \qquad (4)$$

provided that the limit exists.

(ii) If $\int_a^{b-\epsilon} f(x)\,dx$ exists for every ϵ in $(0,\, b - a]$ and if f has a vertical asymptote at $x = b$, then

$$\int_a^b f(x)\,dx = \lim_{\epsilon \to 0^+} \int_a^{b-\epsilon} f(x)\,dx \tag{5}$$

provided that the limit exists.

(iii) If for c in $(a,\, b)$ f has a vertical asymptote at $x = c$ and if the integrals $\int_a^{c-\epsilon_1} f(x)\,dx$ and $\int_{c+\epsilon_2}^b f(x)\,dx$ exist for ϵ_1 in $(0,\, c - a]$ and ϵ_2 in $(0,\, b - c]$, then

$$\int_a^b f(x)\,dx = \lim_{\epsilon_1 \to 0^+} \int_a^{c-\epsilon_1} f(x)\,dx + \lim_{\epsilon_2 \to 0^+} \int_{c+\epsilon_2}^b f(x)\,dx \tag{6}$$

provided that both of these integrals exist.

For each of the cases (i), (ii), and (iii), the improper integral is **convergent** if the appropriate limit or limits exist and are finite. Otherwise, it is **divergent**.

The three cases of Type 2 are illustrated in Figure 6. Of course, we could have just as well drawn the curves approaching $-\infty$ instead of $+\infty$, but the basic idea is the same.

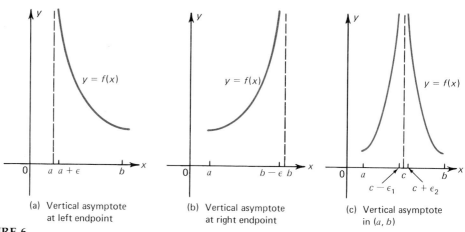

(a) Vertical asymptote at left endpoint

(b) Vertical asymptote at right endpoint

(c) Vertical asymptote in (a, b)

FIGURE 6

In Example 10 we found that $\int_0^1 (1/\sqrt{x})\,dx$ is a converging improper integral.

EXAMPLE 11   Evaluate $\int_0^1 (1/x)\,dx$.

Solution. $\int_\epsilon^1 (1/x)\,dx = -\ln \epsilon$, which approaches ∞ as $\epsilon \to 0^+$, so that $\int_0^1 (1/x)\,dx$ diverges. ∎

EXAMPLE 12   Evaluate $\int_0^a (1/x^k)\,dx$, where $k > 0$, $a > 0$.

Solution.

$$\int_\epsilon^a \left(\frac{1}{x^k}\right) dx = \int_\epsilon^a x^{-k} dx = \frac{x^{-k+1}}{-k+1}\Big|_\epsilon^a = \frac{1}{1-k}(a^{1-k} - \epsilon^{1-k})$$

If $k < 1$, then $1 - k > 0$, so $\lim_{\epsilon \to 0^+} \epsilon^{1-k} = 0$. If $k > 1$, then $k - 1 > 0$ and $\lim_{\epsilon \to 0^+} \epsilon^{1-k}$
$= \lim_{\epsilon \to 0^+}(1/\epsilon^{k-1}) = \infty$. Thus adding in the case $k = 1$ from Example 11, we have

$$\int_0^a \frac{dx}{x^k} \begin{cases} \text{diverges} & \text{if } k \geq 1, \\ = \dfrac{a^{1-k}}{1-k} & \text{if } 0 < k < 1. \ \blacksquare \end{cases} \tag{7}$$

EXAMPLE 13   Calculate

$$\int_0^2 \frac{dx}{(x-2)^{4/5}}.$$

Solution. The integrand is not defined at $x = 2$. We have

$$\int_0^2 \frac{dx}{(x-2)^{4/5}} = \lim_{\epsilon \to 0^+} \int_0^{2-\epsilon} \frac{dx}{(x-2)^{4/5}} = \lim_{\epsilon \to 0^+} 5(x-2)^{1/5}\Big|_0^{2-\epsilon}$$

$$= \lim_{\epsilon \to 0^+} 5[(\ \epsilon)^{1/5} - (-2)^{1/5}] = 5 \cdot 2^{1/5}. \ \blacksquare$$

EXAMPLE 14   Calculate

$$\int_0^3 \frac{dx}{(x-1)^3}$$

Solution. See Figure 7. In this problem the integrand has a vertical asymptote at $x = 1$, so we must consider two improper integrals:

$$\int_0^3 \frac{dx}{(x-1)^3} = \int_0^1 \frac{dx}{(x-1)^3} + \int_1^3 \frac{dx}{(x-1)^3}.$$

Let $y = x - 1$. Then $dy = dx$, and

$$\int_0^3 \frac{dx}{(x-1)^3} = \int_{-1}^2 \frac{dy}{y^3} = \int_{-1}^0 \frac{dy}{y^3} + \int_0^2 \frac{dy}{y^3}.$$

Using (7), we see that the last integral diverges, since $3 > 1$. Thus $\int_0^3 (dx/(x-1)^3)$ diverges.

NOTE.   If we had not noticed the discontinuity at $x = 1$, we might have blindly integrated to obtain

$$\int_0^3 \frac{dx}{(x-1)^3} = -\frac{1}{2(x-1)^2}\Big|_0^3 = \frac{1}{2}\left(1 - \frac{1}{4}\right) = \frac{3}{8},$$

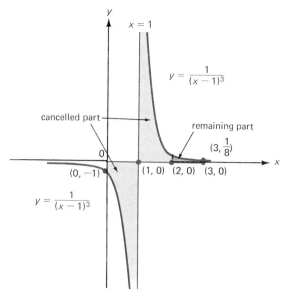

FIGURE 7

which is, of course, wrong. The reason that we obtain this finite answer is that two infinite areas have been canceled (as illustrated in Figure 7). ■

EXAMPLE 15 Calculate

$$\int_{-1}^{1} \frac{dx}{\sqrt{1 - x^2}}.$$

Solution. Here we have vertical asymptotes at both endpoints. Before integrating, we make the substitution $x = \sin \theta$. Then

$$\int_{-1}^{1} \frac{dx}{\sqrt{1 - x^2}} = \int_{-\pi/2}^{\pi/2} \frac{\cos \theta \, d\theta}{\cos \theta} = \pi.$$

Here we have transformed an improper integral into an ordinary one. This conversion is legitimate as long as the vertical asymptotes occur at the endpoints. Note that if we integrate directly, we get $\sin^{-1}(1 - \epsilon_1)$ and $\sin^{-1}(-1 + \epsilon_2)$, which tend, respectively, to $\pi/2$ and $-\pi/2$ as ϵ_1 and $\epsilon_2 \to 0^+$. ■

EXAMPLE 16 Calculate $\int_{-1}^{8} (1/\sqrt[3]{x}) \, dx$.

Solution. Since we have a vertical asymptote at $x = 0$, we write this expression as the sum of two integrals:

$$\int_{-1}^{8} \frac{1}{\sqrt[3]{x}} dx = \int_{-1}^{0} \frac{1}{\sqrt[3]{x}} dx + \int_{0}^{8} \frac{1}{\sqrt[3]{x}} dx$$

$$= \lim_{\epsilon_1 \to 0^+} \int_{-1}^{-\epsilon_1} x^{-1/3} dx + \lim_{\epsilon_2 \to 0^+} \int_{\epsilon_2}^{8} x^{-1/3} dx$$

$$= \lim_{\epsilon_1 \to 0^+} \frac{3}{2} x^{2/3} \bigg|_{-1}^{-\epsilon_1} + \lim_{\epsilon_2 \to 0^+} \frac{3}{2} x^{2/3} \bigg|_{\epsilon_2}^{8}$$

$$= \lim_{\epsilon_1 \to 0^+} \frac{3}{2}[(-\epsilon_1)^{2/3} - 1] + \lim_{\epsilon_2 \to 0^+} \frac{3}{2}[4 - \epsilon_2^{3/2}]$$

$$= -\frac{3}{2} + 6 = \frac{9}{2}. \quad \blacksquare$$

PROBLEMS 12.4

In Problems 1–45, determine whether the given improper integral converges or diverges. If it converges, calculate its value.

1. $\int_0^\infty e^{-2x}\, dx$

2. $\int_{-\infty}^4 e^{3x}\, dx$

3. $\int_{-\infty}^\infty e^{-0.01x}\, dx$

4. $\int_0^\infty xe^{-2x}\, dx$

5. $\int_{-\infty}^\infty x^3 e^{-x^4}\, dx$

6. $\int_{-\infty}^\infty x^2 e^{-x^3}\, dx$

7. $\int_{-\infty}^1 \frac{dx}{\sqrt{4-x}}$

8. $\int_{-\infty}^1 \frac{dx}{(4-x)^{3/2}}$

9. $\int_{10}^\infty \frac{dx}{x^2 - 1}$

10. $\int_0^\infty \frac{dx}{1 + x^2}$

11. $\int_a^\infty \frac{dx}{b^2 + x^2},\ a > 0$

12. $\int_0^\infty \frac{2x}{x^2 + 1}\, dx$

13. $\int_0^\infty \frac{1}{x^3}\, dx$

14. $\int_0^\infty \frac{1}{\sqrt{x}}\, dx$

15. $\int_0^1 \frac{dx}{x^{3/2}}$

16. $\int_0^{\pi/2} \tan x\, dx$

17. $\int_2^3 \frac{dx}{\sqrt{x-2}}$

18. $\int_2^3 \frac{dx}{(x-2)^{1/3}}$

19. $\int_2^3 \frac{dx}{(x-2)^{4/3}}$

20. $\int_0^{\pi/4} \csc x\, dx$

21. $\int_0^{\pi/2} \sec x\, dx$

22. $\int_0^2 \frac{dx}{(x-1)^{1/3}}$

23. $\int_0^2 \frac{x\, dx}{(x^2-1)^{1/3}}$

24. $\int_{-2}^1 \frac{dx}{x^2 - 1}$

25. $\int_1^\infty \frac{dx}{x \ln x}$

26. $\int_0^\infty e^{-x} \sin x\, dx$

27. $\int_{-\infty}^0 e^x \cos x\, dx$

28. $\int_0^\infty e^{-ax} \sin bx\, dx,\ a > 0$

29. $\int_0^5 \frac{dx}{\sqrt{25 - x^2}}$

30. $\int_0^\infty \frac{x^5}{e^{x^6}}\, dx$

31. $\int_0^2 \frac{dx}{x - 1}$

32. $\int_{-2}^0 \frac{dx}{(x + 2)^5}$

33. $\int_{-2}^0 \frac{dx}{(x + 2)^{1/5}}$

34. $\int_0^\pi \tan x\, dx$

35. $\int_2^4 \frac{dx}{(x - 3)^7}$

36. $\int_2^4 \frac{dx}{(x - 3)^{1/7}}$

*37. $\int_0^{\pi/2} \frac{dx}{1 - \sin x}$

*38. $\int_0^\pi \frac{dx}{1 - \cos x}$

39. $\int_{-\infty}^\infty \frac{x^2}{x^2 + 1}\, dx$

40. $\int_5^\infty \frac{dx}{x^2 - 6x + 8}$

41. $\int_1^\infty \frac{dx}{x^2 - 6x + 8}$

42. $\int_{-\infty}^0 \frac{x}{(x^2 + 1)^{5/2}}\, dx$

43. $\int_5^\infty \frac{dx}{x (\ln x)^2}$

44. $\int_0^1 \ln x\, dx$

45. $\int_0^\infty \frac{e^{-2\sqrt{x}}}{\sqrt{x}}\, dx$

46. In Example 2 we showed that the area under the curve $y = 1/x$ for $x \geq 1$ is infinite.
    (a) Show that if this curve is revolved about the x-axis, then the volume of the resulting solid of revolution is finite, and calculate this volume.
    (b) Calculate the lateral surface area.
47. Find the volume of the solid generated when the curve $y = e^{-x}$ for $x \geq 0$ is rotated about the x-axis.
48. Find the volume generated when the curve in Problem 47 is rotated about the y-axis.
49. Find the lateral surface area of the solid in Problem 47.
50. Find the centroid of the region bounded by $y = e^{-x}$ and the x- and y-axes.
*51. Show by mathematical induction (Appendix 2) and integration by parts that for any integer $n > 0$, $\int_0^\infty x^n e^{-x}\, dx = n!$
*52. Calculate $\int_0^\infty x^n e^{-ax}\, dx$ for $a > 0$.
*53. Let f and g be two continuous functions such that $f(x) \geq g(x) \geq 0$ for $x \geq a$. Show the following:
    (a) If $\int_a^\infty f(x)\, dx$ converges, then $\int_a^\infty g(x)\, dx$ converges.
    (b) If $\int_a^\infty g(x)\, dx$ diverges, then $\int_a^\infty f(x)$ diverges.
    These two results are called **comparison theorems**. [Hint: For (a) show that for N large, $\int_a^N g(x)\, dx \leq \int_a^N f(x)\, dx \leq L$, where $L = \int_a^\infty f(x)\, dx$. For (b) show that $\int_a^N f(x)\, dx \geq \int_a^N g(x)\, dx$, which approaches ∞ as $N \to \infty$.]
54. Use the result of Problem 53 to show that $\int_4^\infty (1/\sqrt{x^3 + 1 + \sin x})\, dx$ converges.
55. Show that $\int_1^\infty (\sqrt{1 + x^{1/8}}/x^{3/4})\, dx$ diverges.
56. Show that $\int_1^\infty [1/\ln(1 + x)]\, dx$ diverges.
57. Show that $\int_{-\infty}^\infty e^{-x^2}\, dx$ converges.
58. Show that $\int_1^\infty [(\sin^2 x)/x^2]\, dx$ converges.
*59. Show that $\int_0^1 (1/\sqrt{\sin x})\, dx$ converges. [Hint: Show that $\sin x \geq kx$ for some constant k so that $1/\sqrt{\sin x} \leq 1/\sqrt{kx}$, which has a convergent integral on $[0, 1]$.]

A **probability density function** is a function f whose domain is the set of all real numbers such that

(i) $f(x) \geq 0$ for all x in $\mathbb{R}$,
(ii) $\int_{-\infty}^\infty f(x)\, dx = 1$.

Such functions are very useful in probability theory.

60. Show that

$$f(x) = \begin{cases} 0, & x < a \\ \dfrac{1}{b - a}, & a \leq x \leq b \\ 0, & x \geq b \end{cases}$$

is a probability density function. This function is called a **uniform** density function.
61. Show that

$$f(x) = \begin{cases} \dfrac{1}{a} e^{-x/a}, & x \geq 0 \\ 0, & x < 0, \end{cases}$$

for $a > 0$, is a probability density function. This function is called the **exponential** density function.
62. Show that $\int_{-\infty}^\infty (1/\sqrt{2\pi})e^{-t^2/2}\, dt$ converges. The integrand here is called the **unit normal** density function. Tables of $\int_{-\infty}^\infty (1/\sqrt{2\pi})e^{-t^2/2}\, dt$ are available in most books on probability theory.

The **expected value** or **mean** μ associated with the probability density function f is defined by

$$\mu = \int_{-\infty}^{\infty} xf(x)\, dx.$$

That is, the expected value is the first moment of the probability density function. We can think of the expected value as a weighted average of various probabilities.

63. Calculate the expected value associated with the probability density function given in Problem 60.
64. Calculate the expected value associated with the probability density function given in Problem 61.
65. Calculate the expected value associated with the probability density function given in Problem 62.

The **variance** σ^2 associated with the probability density function f is defined by

$$\sigma^2 = \int_{-\infty}^{\infty} (x - \mu)^2 f(x)\, dx.$$

We can think of the variance as the "average" square of the deviation from the mean. We define the **standard deviation** σ associated with the function f to be the square root of the variance σ^2.

66. Calculate the variance and standard deviation associated with the probability density function given in Problem 60.
67. Calculate the variance and standard deviation associated with the probability density function given in Problem 61.
68. Show that the variance and standard deviation associated with the probability density function given in Problem 62 are finite.
*69. Compute $\int_0^1 [1 - (1 - x)^N]/x\, dx$, where N is a positive integer.
*70. Let N be a positive integer.
   (a) Show that $I_N = \int_0^1 (\ln t)^N\, dt$ converges.
   (b) Evaluate I_N.
*71. Prove:
   (a) $\int_0^1 \ln t\, dt < (1/N) \sum_{k=1}^N \ln(k/N) < \int_{1/N}^1 \ln t\, dt$ for all positive integers N.
   (b) $\lim_{N\to\infty}(1/N) \sum_{k=1}^N \ln(k/N) = \int_0^1 \ln t\, dt$.
   (c) $\lim_{N\to\infty}(\sqrt[N]{N!}/N) = (1/e)$.
*72. Suppose that α is a positive real number; define $\Gamma(\alpha) = \int_0^\infty x^{\alpha-1}e^{-x}\, dx$. Prove that $\Gamma(\alpha + 1) = \alpha\Gamma(\alpha)$. [See Problem 51 for the case when α is an integer.]
*73. Let $H(r) = \int_0^\infty [(1 - e^{-t})/t^r]\, dt$, where $1 < r < 2$. Express H in terms of the function Γ defined in Problem 72.
*74. Using only the fact that $\int_0^\infty [1/(1 + t^2)]\, dt$ exists (i.e., is finite), find the solution, A, to the equation

$$\int_0^A \frac{1}{1 + t^2}\, dt = \frac{1}{2}\int_0^\infty \frac{1}{1 + t^2}\, dt = \int_A^\infty \frac{1}{1 + t^2}\, dt.$$

*75. The **Laplace transform** of a function $f(t)$, denoted $\hat{f}(s)$, is defined by

$$\hat{f}(s) = \int_0^\infty e^{-st} f(t)\, dt.$$

Suppose that $\int_0^\infty |f(t)|\, dt$ converges. Show that $\hat{f}(s)$ exists and is finite for every nonnegative real number s.

REVIEW EXERCISES FOR CHAPTER TWELVE

In Exercises 1–20, evaluate the given limit.

1. $\lim\limits_{x \to 0} \dfrac{\sin 3x}{2x}$

2. $\lim\limits_{x \to \infty} \dfrac{\ln x}{x^2}$

3. $\lim\limits_{x \to 0} \dfrac{2x^3 + 3x + 4}{x - x^3}$

4. $\lim\limits_{x \to \pi/4} \dfrac{\cos 2x}{(\pi/4) - x}$

5. $\lim\limits_{x \to \infty} 3xe^{-x}$

6. $\lim\limits_{x \to 0} \dfrac{1 - e^x}{x^2 + x}$

7. $\lim\limits_{x \to 0+} \dfrac{\sqrt{x}}{\sin x}$

8. $\lim\limits_{x \to 0+} (\sin x)^{2x}$

9. $\lim\limits_{x \to \pi/4-} [(\pi/4) - x]^{\cos 2x}$

10. $\lim\limits_{x \to 0+} x^{3x}$

11. $\lim\limits_{x \to 0} \dfrac{\sin x}{x^3}$

12. $\lim\limits_{x \to 0} \dfrac{x - \sin x}{2x^3}$

13. $\lim\limits_{x \to \pi/2-} \dfrac{\tan x}{\sec x}$

14. $\lim\limits_{x \to 0+} \dfrac{\ln(\tan x)}{\ln(\sin x)}$

15. $\lim\limits_{x \to \infty} (x - \sqrt{x^2 + x})$

16. $\lim\limits_{x \to \infty} (x^4 - \sqrt{x^8 + x^4 + 1})$

17. $\lim\limits_{x \to \infty} \dfrac{e^x}{2^x}$

18. $\lim\limits_{x \to \infty} \dfrac{e^x}{4^x}$

19. $\lim\limits_{x \to 0} x^{10}e^{-1/x^2}$

20. $\lim\limits_{x \to 0+} \dfrac{\int_0^x e^{t^3}\,dt}{\int_0^x e^{t^2}\,dt}$

In Exercises 21–40, determine whether the given improper integral converges or diverges. If it converges, calculate its value.

21. $\displaystyle\int_0^\infty e^{-3x}\,dx$

22. $\displaystyle\int_{-\infty}^2 e^{50x}\,dx$

23. $\displaystyle\int_0^\infty x^3 e^{-7x}\,dx$

24. $\displaystyle\int_{-\infty}^\infty x^2 e^{-x^3}\,dx$

25. $\displaystyle\int_{50}^\infty \dfrac{dx}{x^2 - 4}$

26. $\displaystyle\int_0^1 \dfrac{dx}{\sqrt[4]{x}}$

27. $\displaystyle\int_0^1 \dfrac{dx}{x^4}$

28. $\displaystyle\int_{-\infty}^\infty \dfrac{dx}{x^3}$

29. $\displaystyle\int_{-\infty}^0 \dfrac{dx}{(1 - x)^{5/2}}$

30. $\displaystyle\int_3^4 \dfrac{dx}{(x - 3)^{6/5}}$

31. $\displaystyle\int_0^{\pi/2} \csc x\,dx$

32. $\displaystyle\int_0^4 \dfrac{dx}{x - 2}$

33. $\displaystyle\int_0^4 \dfrac{dx}{\sqrt{x - 2}}$

34. $\displaystyle\int_3^5 \dfrac{dx}{(x - 4)^{10}}$

35. $\displaystyle\int_0^\infty \dfrac{x^3}{x^3 + 1}\,dx$

36. $\displaystyle\int_0^e \ln x\,dx$

37. $\displaystyle\int_2^\infty \dfrac{dx}{(x - 1)^{1/100}}$

38. $\displaystyle\int_0^\infty e^{-x} \sin 2x\,dx$

39. $\displaystyle\int_0^\infty \dfrac{dx}{x^2 - 3x + 2}$

40. $\displaystyle\int_{-\infty}^0 e^{2x} \cos 3x\,dx$

41. Find the volume generated when the region in the first quadrant bounded by $y = 1/(x + 1)^2$ is rotated about the x-axis.

42. Find the volume generated if the region in Exercise 41 is rotated about the y-axis.

43. Find the centroid of the region in the first quadrant bounded by $y = 1/(1 + x)^3$.

44. Show that $\int_0^{\pi/2} (1/\sqrt{\cos x})\,dx$ converges.

45. Show that $\int_{10}^\infty (1/\ln\sqrt{2 + x})\,dx$ diverges.

46. Show that the function

$$f(x) = \begin{cases} \frac{1}{5}e^{-x/5}, & x \geq 0 \\ 0, & x < 0 \end{cases}$$

is a probability density function.

47. Calculate the expected value associated with the density function given in Exercise 46.

48. Calculate the variance and standard deviation associated with the density function given in Exercise 46.

13 Taylor Polynomials and Approximation

Many functions arising in applications are difficult to deal with. A continuous function, for example, may take a complicated form, or it may take a simple form that, nevertheless, cannot be integrated.

For this reason mathematicians and physicists have developed methods for approximating certain functions by other functions that are much easier to handle. Some of the easiest functions to deal with are the polynomials, since in addition to having other useful properties, they can be differentiated and integrated any number of times and still remain polynomials. In this chapter we will show how certain continuous functions can be approximated by polynomials.

13.1 TAYLOR'S THEOREM AND TAYLOR POLYNOMIALS

In this section we show how a function can be approximated as closely as desired by a polynomial, provided that the function possesses a sufficient number of derivatives.

We begin by reminding you of the factorial notation defined for all positive integers n:

$$n! = n(n-1)(n-2) \cdots 3 \cdot 2 \cdot 1.$$

That is, $n!$ is the product of the first n positive integers. For example, $3! = 3 \cdot 2 \cdot 1 = 6$ and $5! = 5 \cdot 4 \cdot 3 \cdot 2 \cdot 1 = 120$. By convention, we define $0!$ to be equal to 1.

Definition 1   TAYLOR† POLYNOMIAL   Let the function f and its first n derivatives exist on the closed interval $[a, b]$. Then, for $x \in [a, b]$, the nth degree **Taylor polynomial** of f at a is the nth degree polynomial $P_n(x)$, given by

$$P_n(x) = f(a) + \frac{f'(a)}{1!}(x - a) + \frac{f''(a)}{2!}(x - a)^2 + \frac{f'''(a)}{3!}(x - a)^3 + \cdots$$

$$+ \frac{f^{(n)}(a)}{n!}(x - a)^n = \sum_{k=0}^{n} \frac{f^{(k)}(a)(x - a)^k}{k!}. \ddagger \quad (1)$$

EXAMPLE 1   Calculate the fifth-degree Taylor polynomial of $f(x) = \sin x$ at 0.

Solution. We have $f(x) = \sin x$, $f'(x) = \cos x$, $f''(x) = -\sin x$, $f'''(x) = -\cos x$, $f^{(4)}(x) = \sin x$, and $f^{(5)}(x) = \cos x$. Then $f(0) = 0$, $f'(0) = 1$, $f''(0) = 0$, $f'''(0) = -1$, $f^{(4)}(0) = 0$, $f^{(5)}(0) = 1$, and we obtain

$$P_5(x) = f(0) + \frac{f'(0)}{1!}x + \frac{f''(0)}{2!}x^2 + \frac{f'''(0)}{3!}x^3 + \frac{f^{(4)}(0)}{4!}x^4 + \frac{f^{(5)}(0)}{5!}x^5$$

$$= x - \frac{x^3}{3!} + \frac{x^5}{5!} = x - \frac{x^3}{6} + \frac{x^5}{120}. \ \blacksquare$$

Definition 2   REMAINDER TERM   Let $P_n(x)$ be the nth degree Taylor polynomial of the function f. Then the **remainder term**, denoted $R_n(x)$, is given by

$$R_n(x) = f(x) - P_n(x). \quad (2)$$

Why do we study Taylor polynomials? Because of the following remarkable result that tells us that a Taylor polynomial provides a good approximation to a function f.

Theorem 1.   *Taylor's Theorem (Taylor's Formula with Remainder)*   Suppose that $f^{n+1}(x)$ exists on the closed interval $[a, b]$. Let x be any number in $[a, b]$. Then there is a number $c§$ in (a, x) such that

$$R_n(x) = \frac{f^{(n+1)}(c)}{(n + 1)!}(x - a)^{n+1}. \quad (3)$$

---

†Named after the English mathematician Brook Taylor (1685–1731), who published what we now call *Taylor's formula* in *Methodus Incrementorum* in 1715. There was a considerable controversy over whether Taylor's discovery was, in fact, a plagiarism of an earlier result of the Swiss mathematician Jean Bernoulli (1667–1748).

‡In this notation we have $f^{(0)}(a) = f(a)$.

§c depends on x.

The expression in (3) is called **Lagrange's form of the remainder.†** Using (3), we can write Taylor's formula as

$$f(x) = f(a) + \frac{f'(a)}{1!}(x - a) + \frac{f''(a)}{2!}(x - a)^2 + \cdots$$

$$+ \frac{f^{(n)}(a)}{n!}(x - a)^n + \frac{f^{(n+1)}(c)}{(n + 1)!}(x - a)^{n+1} \quad \text{(4)}$$

REMARK. In Section 13.2 we will show that the Taylor polynomial $P_n(x)$ is *unique* in a sense to be made precise later.

In Section 13.2 we will prove Taylor's formula. Also, we will show that under certain reasonable assumptions, the remainder term $R_n(x)$ given by (3) actually approaches 0 as $n \to \infty$. Thus as n increases, $P_n(x)$ is an increasingly useful approximation to the function f over the interval $[a, b]$.

In the remainder of this section we will calculate some Taylor polynomials.

EXAMPLE 2　Calculate the fifth-degree Taylor polynomial of $f(x) = \sin x$ at $\pi/6$.

Solution. Using the derivatives found in Example 1, we have $f(\pi/6) = 1/2$, $f'(\pi/6) = \sqrt{3}/2$, $f''(\pi/6) = -1/2$, $f'''(\pi/6) = -\sqrt{3}/2$, $f^{(4)}(\pi/6) = 1/2$, and $f^{(5)}(\pi/6) = \sqrt{3}/2$, so that in this case

$$P_5(x) = \frac{1}{2} + \frac{\sqrt{3}}{2}\left(x - \frac{\pi}{6}\right) - \frac{1}{2}\frac{[x - (\pi/6)]^2}{2!} - \frac{\sqrt{3}}{2}\frac{[x - (\pi/6)]^3}{3!}$$

$$+ \frac{1}{2}\frac{[x - (\pi/6)]^4}{4!} + \frac{\sqrt{3}}{2}\frac{[x - (\pi/6)]^5}{5!}$$

$$= \frac{1}{2} + \frac{\sqrt{3}}{2}\left(x - \frac{\pi}{6}\right) - \frac{1}{4}\left(x - \frac{\pi}{6}\right)^2 - \frac{\sqrt{3}}{12}\left(x - \frac{\pi}{6}\right)^3$$

$$+ \frac{1}{48}\left(x - \frac{\pi}{6}\right)^4 + \frac{\sqrt{3}}{240}\left(x - \frac{\pi}{6}\right)^5. \quad \blacksquare$$

Examples 1 and 2 illustrate that in many cases it is easiest to calculate the Taylor polynomial at 0. In this situation we have

$$P_n(x) = f(0) + f'(0)x + \frac{f''(0)}{2!}x^2 + \cdots + \frac{f^{(n)}(0)}{n!}x^n. \quad \text{(5)}$$

EXAMPLE 3　Find the eighth-degree Taylor polynomial of $f(x) = e^x$ at 0.

Solution. Here $f(x) = f'(x) = f''(x) = \cdots = f^{(8)}(x) = e^x$, and $e^0 = 1$, so that

---

†See the accompanying biographical sketch of Lagrange.

Joseph Louis Lagrange

Joseph Louis Lagrange
The Granger Collection

Joseph Louis Lagrange was one of the two greatest mathematicians of the eighteenth century—the other being Leonhard Euler (see page 36). Born in 1736 in Turin, Italy, Lagrange was the youngest of eleven children of French and Italian parents and the only one to survive to adulthood. Educated in Turin, Joseph Louis became a professor of mathematics in the military academy there when he was still quite young.

Lagrange's early publications established his reputation. When Euler left his post at the court of Frederick the Great in Berlin in 1766, he recommended that Lagrange be appointed his successor. Accepting Euler's advice, Frederick wrote to Lagrange that "the greatest king in Europe" wished to invite to his court "the greatest mathematician in Europe." Lagrange accepted and remained in Berlin for twenty years. Afterwards, he accepted a post at the École Polytechnique in France.

Lagrange had a deep influence on nineteenth and twentieth century mathematics. He is perhaps best known as the first great mathematician to attempt to make calculus mathematically rigorous. His major work in this area was his 1797 paper "Théorie des fonctions analytiques contenant les principes du calcul différentiel." In this work, Lagrange tried to make calculus more logical—rather than more useful. His key idea was to represent a function $f(x)$ by a Taylor series. For example, we can write $1/(1 - x) = 1 + x + x^2 + \cdots + x^n + \cdots$ (a result that can be obtained by long division). Lagrange multiplied the coefficient of x^n by $n!$ and called the result the nth *derived function* of $1/(1 - x)$ at $x = 0$. This is the origin of the word *derivative*. The notation $f'(x)$, $f''(x)$, ... was first used by Lagrange as was the form of the remainder term (3).

Lagrange is known for much else as well. Beginning in the 1750s, he invented the calculus of variations. He made significant contributions to ordinary differential equations, partial differential equations, numerical analysis, number theory, and algebra. In 1788 he published his *Mécanique Analytique*, which contained the equations of motion of a dynamical system. Today these equations are known as *Lagrange's equations*.

Lagrange lived in France during the French revolution. In 1790 he was placed on a committee to reform weights and measures and later became the head of a related committee that, in 1799, recommended the adoption of the system that we know today as the *metric system*. Despite his work for the revolution, however, Lagrange was disgusted by its cruelties. After the great French chemist Lavoisier was guillotined, Lagrange exclaimed, "It took the mob only a moment to remove his head; a century will not suffice to reproduce it."

In his later years, Lagrange was often lonely and depressed. When he was 56, the 17-year-old daughter of his friend the astronomer P. C. Lemonier was so moved by his unhappiness that she proposed to him. The resulting marriage apparently turned out to be ideal for both.

Perhaps the greatest tribute to Lagrange was given by Napoleon Bonaparte, who said, "Lagrange is the lofty pyramid of the mathematical sciences."

$$P_8(x) = 1 + x + \frac{x^2}{2!} + \frac{x^3}{3!} + \frac{x^4}{4!} + \frac{x^5}{5!} + \frac{x^6}{6!} + \frac{x^7}{7!} + \frac{x^8}{8!} = \sum_{k=0}^{8} \frac{x^k}{k!}.$$

EXAMPLE 4   We can extend Example 1 to see that for $f(x) = \sin x$ at 0, the Taylor polynomials for different values of n are given by

$$P_0(x) = 0$$

$$P_1(x) = x = P_2(x)$$

$$P_3(x) = x - \frac{x^3}{3!} = P_4(x)$$

$$P_5(x) = x - \frac{x^3}{3!} + \frac{x^5}{5!} = P_6(x)$$

$$P_7(x) = x - \frac{x^3}{3!} + \frac{x^5}{5!} - \frac{x^7}{7!} = P_8(x)$$

$$\vdots$$

$$P_{2n+1}(x) = x - \frac{x^3}{3!} + \frac{x^5}{5!} - \cdots + \frac{(-1)^n x^{2n+1}}{(2n+1)!} = \sum_{k=0}^{n} \frac{(-1)^k x^{2k+1}}{(2k+1)!} = P_{2n+2}(x).$$

In Figure 1 we reproduce a computer-drawn graph showing how, with smaller values of x in the interval $[0, 2\pi]$, the Taylor polynomials get closer and closer to the function $\sin x$ as n increases. We will say more about this phenomenon in Section 13.3. ■

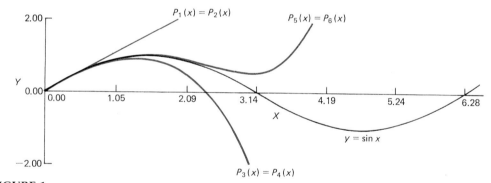

FIGURE 1

EXAMPLE 5   Find the fifth-degree Taylor polynomial of $f(x) = 1/(1 - x)$ at 0.

Solution. Here

$$f(x) = \frac{1}{1-x}, \qquad f'(x) = \frac{1}{(1-x)^2}, \qquad f''(x) = \frac{2}{(1-x)^3},$$

$$f'''(x) = \frac{6}{(1-x)^4}, \qquad f^{(4)}(x) = \frac{24}{(1-x)^5}, \qquad f^{(5)}(x) = \frac{120}{(1-x)^6}.$$

Thus $f(0) = 1, f'(0) = 1, f''(0) = 2, f'''(0) = 6, f^{(4)}(0) = 24, f^{(5)}(x) = 120$, and

$$P_5(x) = 1 + x + \frac{2x^2}{2!} + \frac{6x^3}{3!} + \frac{24x^4}{4!} + \frac{120x^5}{5!}$$

$$= 1 + x + x^2 + x^3 + x^4 + x^5 = \sum_{k=0}^{5} x^k.$$

In Figure 2 we reproduce a computer-drawn sketch of $1/(1-x)$ and its first four Taylor polynomials. ■

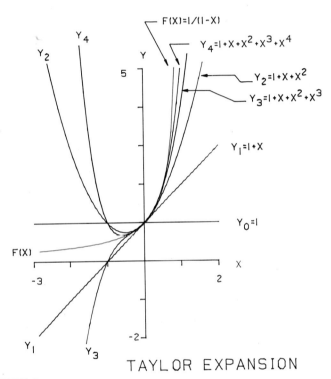

TAYLOR EXPANSION

FIGURE 2

Note that in Examples 1, 2, and 3 the given function had continuous derivatives of all orders defined for all real numbers. In Example 5, $f(x)$ is defined over intervals of the form $[-b, b]$, where $b < 1$. Thus Taylor's theorem does *not* apply in any interval containing 1. *It is always necessary to check whether the hypotheses of Taylor's theorem hold over a given interval.*

Before leaving this section, we observe that *the nth-degree Taylor polynomial at a of a function is the polynomial that agrees with the function and each of its first n derivatives at a.* This follows immediately from (1):

$$P_n(a) = f(a)$$

$$P_n'(a) = \left[f'(a) + f''(a)(x - a) + f'''(a)\frac{(x - a)^2}{2!} + \cdots + \frac{f^{(n)}(a)}{(n - 1)!}(x - a)^{n-1} \right]\Bigg|_{x=a}$$

$$= f'(a),$$

$$P_n''(a) = \left[f''(a) + f'''(a)(x - a) + \cdots + \frac{f^{(n)}(a)}{(n-2)!}(x-a)^{n-2} \right]\Bigg|_{x=a} = f''(a),$$

and so on. In particular, since the $(n+1)$st derivative of an nth-degree polynomial is zero, we find that *if $Q(x)$ is a polynomial of degree n, then $P_n(x) = Q(x)$.* This follows immediately from the fact that the remainder term given by (3) will be zero since $Q^{(n+1)}(c) = 0$.

EXAMPLE 6 Let $Q(x) = 3x^4 + 2x^3 - 4x^2 + 5x - 8$. Compute $P_4(x)$ at 0.

Solution. We have

$$Q(0) = -8$$
$$Q'(0) = (12x^3 + 6x^2 - 8x + 5)|_{x=0} = 5,$$
$$Q''(0) = (36x^2 + 12x - 8)|_{x=0} = -8,$$
$$Q'''(0) = (72x + 12)|_{x=0} = 12,$$
$$Q^{(4)}(0) = 72.$$

Therefore,

$$P_4(x) = -8 + 5x - \frac{8x^2}{2!} + \frac{12x^3}{3!} + \frac{72x^4}{4!} = -8 + 5x - 4x^2 + 2x^3 + 3x^4 = Q(x),$$

as expected. ■

PROBLEMS 13.1

In Problems 1–26, find the Taylor polynomial of the given degree n for the function f at the number a.

1. $f(x) = \cos x$; $a = \pi/4$; $n = 6$

2. $f(x) = \sqrt{x}$; $a = 1$; $n = 4$

3. $f(x) = \ln x$; $a = e$; $n = 5$

4. $f(x) = \ln(1 + x)$; $a = 0$; $n = 5$

5. $f(x) = 1/x$; $a = 1$; $n = 4$

6. $f(x) = \dfrac{1}{(1+x)}$; $a = 0$; $n = 5$

7. $f(x) = \tan x$; $a = 0$; $n = 4$

8. $f(x) = \tan^{-1} x$; $a = 0$; $n = 6$

9. $f(x) = \tan x$; $a = \pi$; $n = 4$

10. $f(x) = \tan^{-1} x$; $a = 1$; $n = 6$

11. $f(x) = \dfrac{1}{(1+x^2)}$; $a = 0$; $n = 4$

12. $f(x) = \dfrac{1}{\sqrt{x}}$; $a = 4$; $n = 3$

13. $f(x) = \sinh x$; $a = 0$; $n = 4$

14. $f(x) = \cosh x$; $a = 0$; $n = 4$

15. $f(x) = \ln \sin x$; $a = \pi/2$; $n = 3$

16. $f(x) = \ln \cos x$; $a = 0$; $n = 3$

17. $f(x) = \dfrac{1}{\sqrt{4-x}}$; $a = 0$; $n = 4$

18. $f(x) = \dfrac{1}{\sqrt{4-x}}$; $a = 3$; $n = 4$

19. $f(x) = e^{\alpha x}$; $a = 0$; $n = 6$; α real

20. $f(x) = \sin \alpha x$; $a = 0$; $n = 6$; α real

21. $f(x) = \sin^{-1} x$; $a = 0$; $n = 3$

22. $f(x) = 1 + x + x^2$; $a = 0$; $n = 10$

23. $f(x) = a_0 + a_1x + a_2x^2 + a_3x^3$, $a = 1$, $n = 10$
24. $f(x) = e^{x^2}$; $a = 0$; $n = 4$ **25.** $f(x) = \sin x^2$; $a = 0$; $n = 4$
26. $f(x) = \cos x^2$; $a = 0$; $n = 4$

27. Show that the nth-degree Taylor polynomial of $f(x) = 1/(1-x)$ at 0 is given by

$$P_n(x) = 1 + x + x^2 + \cdots + x^n = \sum_{k=0}^{n} x^k.$$

13.2 A PROOF OF TAYLOR'S THEOREM, ESTIMATES ON THE REMAINDER TERM, AND A UNIQUENESS THEOREM (OPTIONAL)

In this section we begin by proving Taylor's formula with remainder, as stated on page 668.

Proof of Taylor's Theorem

We show that if $f^{(n+1)}(x)$ exists in $[a, b]$, then for any number $x \in [a, b]$, there is a number c in $[a, x]$ such that

$$f(x) = f(a) + \frac{f'(a)}{1!}(x-a) + \frac{f''(a)}{2!}(x-a)^2 + \cdots + \frac{f^{(n)}(a)}{n!}(x-a)^n + R_n(x), \tag{1}$$

where

$$R_n(x) = \frac{f^{(n+1)}(c)}{(n+1)!}(x-a)^{n+1}. \tag{2}$$

Let $x \in [a, b]$ be fixed. We define the new function $h(t)$ by

$$h(t) = f(x) - f(t) - f'(t)(x-t) - \frac{f''(t)}{2!}(x-t)^2 - \cdots$$
$$- \frac{f^{(n)}(t)}{n!}(x-t)^n - \frac{R_n(x)(x-t)^{n+1}}{(x-a)^{n+1}} \tag{3}$$

Then

$$h(x) = f(x) - f(x) - f'(x)(x-x) - \frac{f''(x)}{2!}(x-x)^2 - \cdots$$
$$- \frac{f^{(n)}(x)}{n!}(x-x)^n - \frac{R_n(x)(x-x)^{n+1}}{(x-a)^{n+1}} = 0$$

and

$$h(a) = f(x) - f(a) - f'(a)(x - a) - \frac{f''(a)}{2!}(x - a)^2 - \cdots$$

$$- \frac{f^{(n)}(a)}{n!}(x - a)^n - \frac{R_n(x)(x - a)^{n+1}}{(x - a)^{n+1}}$$

$$= f(x) - P_n(x) - R_n(x) = R_n(x) - R_n(x) = 0.$$

Since $f^{(n+1)}$ exists, $f^{(n)}$ is differentiable so that h, being a sum of products of differentiable functions, is also differentiable for t in (a, x). Remember that x is fixed so h is a function of t only.

Recall Rolle's theorem (see page 199), which states that if h is continuous on $[a, b]$, differentiable on (a, b), and $h(a) = h(b) = 0$, then there is at least one number c in (a, b) such that $h'(c) = 0$. We see that the conditions of Rolle's theorem hold in the interval $[a, x]$ so that there is a number c in (a, x) with $h'(c) = 0$. In Problems 1–3 you are asked to show that

$$h'(t) = \frac{-f^{(n+1)}(t)(x - t)^n}{n!} + \frac{(n + 1)R_n(x)(x - t)^n}{(x - a)^{n+1}}$$

Then, setting $t = c$, we obtain

$$0 = h'(c) = \frac{-f^{(n+1)}(c)(x - c)^n}{n!} + \frac{(n + 1)R_n(x)(x - c)^n}{(x - a)^{n+1}}$$

Finally, dividing the equations above through by $(x - c)^n$ and solving for $R_n(x)$, we obtain

$$R_n(x) = \frac{f^{(n+1)}(c)(x - a)^n}{(n + 1)n!} = \frac{f^{(n+1)}(c)(x - a)^{n+1}}{(n + 1)!}.$$

This is what we wanted to prove. ∎

REMARK. If you go over the proof, you may observe that we didn't need to assume that $x > a$; if we replace the interval $[a, x]$ with the interval $[x, a]$, then all results are still valid.

In a remark on page 669 we said that the Taylor polynomial was unique. We now make that statement more precise.

First, we note that if $f^{(n+1)}$ is continuous in $[a, b]$ and $a < c < x$ [where c is the c that appears in the formula for $R_n(x)$], we have

$$\lim_{x \to a} f^{(n+1)}(c) = f^{(n+1)}(a). \tag{4}$$

Then

From (2)

$$\lim_{x \to a} \frac{R_n(x)}{(x-a)^n} = \lim_{x \to a} \frac{f^{(n+1)}(c)}{(n+1)!} \frac{(x-a)^{n+1}}{(x-a)^n} = \frac{1}{(n+1)!} \lim_{x \to a} f^{(n+1)}(c) \lim_{x \to a} (x-a)$$

From (4)

$$= \frac{1}{(n+1)!} f^{(n+1)}(a) \cdot 0 = 0.$$

That is,

$$\lim_{x \to a} \frac{R_n(x)}{(x-a)^n} = 0. \tag{5}$$

We can now state our uniqueness result.

Theorem 1. *Uniqueness of the Taylor Polynomial*   If f has $n + 1$ continuous derivatives, then the Taylor polynomial $P_n(x)$ is the *only* nth-degree polynomial whose remainder term satisfies (5).

Proof.   Suppose that there is another nth-degree polynomial $Q_n(x)$ such that

$$f(x) = P_n(x) + R_n(x) = Q_n(x) + S_n(x),$$

where

$$\lim_{x \to a} \frac{R_n(x)}{(x-a)^n} = 0, \quad \text{and} \quad \lim_{x \to a} \frac{S_n(x)}{(x-a)^n} = 0.$$

Let

$$D(x) = P_n(x) - Q_n(x) = S_n(x) - R_n(x).$$

We will show that $D(x) = 0$, which will imply that $P_n(x) = Q_n(x)$. This will show that $P_n(x)$ is unique. Since $D(x)$ is the difference of nth-degree polynomials, $D(x)$ is an nth-degree polynomial, and it can be written in the form

$$D(x) = b_0 + b_1(x-a) + b_2(x-a)^2 + \cdots + b_n(x-a)^n.$$

Now

$$\lim_{x \to a} \frac{D(x)}{(x-a)^n} = \lim_{x \to a} \frac{S_n(x)}{(x-a)^n} - \lim_{x \to a} \frac{R_n(x)}{(x-a)^n} = 0 - 0 = 0.$$

Similarly, if $0 \le m < n$,

Multiply and divide by $(x-a)^{n-m}$

$$\lim_{x \to a} \frac{D(x)}{(x-a)^m} = \lim_{x \to a} \frac{(x-a)^{n-m} D(x)}{(x-a)^n}$$

$$= \lim_{x \to a} (x - a)^{n-m} \lim_{x \to a} \frac{D(x)}{(x - a)^n} = 0 \cdot 0 = 0,$$

and we have

$$\lim_{x \to a} \frac{D(x)}{(x - a)^m} = 0 \qquad \text{for} \qquad m = 0, 1, 2, \ldots, n. \tag{6}$$

To complete the proof, we note that

By (6) with $m = 0$

$$b_0 = D(a) = \lim_{x \to a} D(x) = 0.$$

Then

$$D(x) = b_1(x - a) + b_2(x - a)^2 + \cdots + b_n(x - a)^n,$$

so that

$$\frac{D(x)}{x - a} = b_1 + b_2(x - a) + \cdots + b_n(x - a)^{n-1}$$

and

By (6) with $m = 1$

$$b_1 = \lim_{x \to a} \frac{D(x)}{x - a} = 0.$$

Suppose we have shown that $b_0 = b_1 = \cdots = b_k = 0$, where $k < n$. Then

$$D(x) = b_{k+1}(x - a)^{k+1} + b_{k+2}(x - a)^{k+2} + \cdots + b_n(x - a)^n,$$

so that

$$\frac{D(x)}{(x - a)^{k+1}} = b_{k+1} + b_{k+2}(x - a) + \cdots + b_n(x - a)^{n-k-1}$$

and

By (6) with $m = k + 1 \le n$

$$b_{k+1} = \lim_{x \to a} \frac{D(x)}{(x - a)^{k+1}} = 0.$$

This shows that $b_0 = b_1 = b_2 = \cdots = b_n = 0$, which means that $D(x) = 0$ for every x in $[a, b]$, so that $P_n(x) = Q_n(x)$ for every x in $[a, b]$. Thus the Taylor polynomial is unique. ∎

We now prove three results that are very useful for applications.

Theorem 2. If f has $n + 1$ continuous derivatives, there exists a positive number M_n such that

$$|R_n(x)| \le M_n \frac{|x-a|^{n+1}}{(n+1)!}$$

for all x in $[a, b]$. Here M_n is an upper bound for the $(n+1)$st derivative of f in the interval $[a, b]$.

Proof.   Since $f^{(n+1)}$ is continuous on $[a, b]$, it is bounded above and below on that interval (from Theorem 2.7.4). That is, there is a number M_n such that $|f^{(n+1)}(x)| \le M_n$ for every x in $[a, b]$. Since

$$R_n(x) = f^{(n+1)}(c)\frac{(x-a)^{n+1}}{(n+1)!},$$

with c in (a, x), we see that

$$|R_n(x)| = |f^{(n+1)}(c)|\frac{(x-a)^{n+1}}{(n+1)!} \le M_n\frac{(x-a)^{n+1}}{(n+1)!},$$

and the theorem is proved. ■

In the next section we will show how this bound on the magnitude of the remainder term can be used to find some very interesting approximations. Since many of these examples use the Taylor polynomial at 0, we restate the results of this section for that special case: If $f, f', \ldots, f^{(n+1)}$ are continuous in $[0, b]$, then for any x in $[0, b]$,

$$f(x) = f(0) + f'(0) + f''(0)\frac{x^2}{2!} + \cdots + f^{(n)}(0)\frac{x^n}{n!} + R_n(x) = \sum_{k=0}^{n} \frac{f^{(k)}(0)x^k}{k!} + R_n(x),$$

and there is a number M_n such that

$$|R_n(x)| \le M_n \frac{|x|^{n+1}}{(n+1)!}.$$

We close this section by showing that if there is a number K such that $|f^{(n)}(x)| \le K$ for every x in $[a, b]$ and for every positive integer n, then the remainder terms actually approach zero as n increases. First, we need the following result.

Theorem 3.   Let x be any real number. Then $x^n/n! \to 0$ as $n \to \infty$.

Proof.   Let N be an integer such that $N > 2|x|$. Then $|x| < N/2$, and for $n > N$ and $k > 0$,

$$\frac{|x|}{N+k} < \frac{N}{2(N+k)} = \frac{1}{2}\left(\frac{N}{N+k}\right) < \frac{1}{2}.$$

Then

$$\frac{x^n}{n!} \le \frac{|x|^n}{n!} = \overbrace{\frac{|x| \cdot |x| \cdot \cdots \cdot |x|}{1 \cdot 2 \cdot 3 \cdot \cdots \cdot N}}^{N \text{ times}} \cdot \underbrace{\frac{|x|}{N+1} \cdot \frac{|x|}{N+2} \cdot \cdots \cdot \frac{|x|}{n!}}_{\substack{n - N \text{ terms, each of} \\ \text{which is } < \frac{1}{2}}} < \frac{|x|^N}{N!} \left(\frac{1}{2}\right)^{n-N}$$

$$= \frac{|x|^N}{N!} \cdot 2^N \cdot \left(\frac{1}{2}\right)^n.$$

Since x and N are fixed, we define $M = (|x|^N/N!) \cdot 2^N$, so that for each $n > N$, $0 < |x|^n/n! < M(\frac{1}{2})^n$, which approaches zero as $n \to \infty$.† Thus by the squeezing theorem (Theorem 2.4.1), we see that $x^n/n! \to 0$. ∎

Theorem 4. Suppose that $f^{(n)}(x)$ is defined for all $n \ge 1$ and $|f^{(n)}(x)| \le K$ for all $x \in [a, b]$. Then

$$|R_n(x)| \to 0 \qquad \text{as} \qquad n \to \infty.$$

Proof. We have

$$|R_n(x)| = \left| f^{(n+1)}(c) \frac{(x-a)^{n+1}}{(n+1)!} \right| = \left| f^{(n+1)}(c) \right| \frac{|x-a|^{n+1}}{(n+1)!}$$

$$\le K \frac{|x-a|^{n+1}}{(n+1)!}.$$

But by Theorem 3 this last term approaches zero as $n \to \infty$. Thus the theorem is proved. ∎

PROBLEMS 13.2

1. Show that $\dfrac{d}{dt}[f'(t)(x-t)] = -f'(t) + f''(t)(x-t)$.

2. Show that for $1 \le k \le n$,

$$\frac{d}{dt}\left[\frac{f^{(k)}(t)}{k!}(x-t)^k\right] = \frac{-f^{(k)}(t)}{(k-1)!}(x-t)^{k-1} + \frac{f^{(k+1)}(t)}{k!}(x-t)^k$$

3. Use the results of Problems 1 and 2 to show that

$$h'(t) = \frac{-f^{(n+1)}(t)}{n!}(x-t)^n + \frac{(n+1)R_n(x)(x-t)^n}{(x-a)^{n+1}}$$

where $h(t)$ is given by equation (3).

*4. Prove the following: Let f and its first $n-1$ derivatives exist in an open interval containing a, and suppose that $f^{(n)}(a)$ exists. Then $f(x) = P_n(x) + R(x)$, where $P_n(x)$ is the nth-degree Taylor polynomial centered at a and

$$\lim_{x \to a} \frac{R(x)}{(x-a)^n} = 0.$$

---

†This fact is proved in Section 14.1 (Theorem 14.1.1).

13.3 APPROXIMATION USING TAYLOR POLYNOMIALS

In this section we show how Taylor's formula can be used as a tool for making approximations. In many of the examples that follow, results have been obtained by making use of a hand calculator. If a hand calculator is available, we suggest that you use it to check the computations.

For convenience, we summarize the results of the preceding two sections.

Let $f, f', f'', \ldots, f^{(n+1)}$ be continuous on $[a, b]$. Then for any x in $[a, b]$,

$$f(x) = f(a) + f'(a)(x - a) + f''(a)\frac{(x - a)^2}{2!} + \cdots + f^{(n)}(a)\frac{(x - a)^n}{n!} + R_n(x)$$

$$= \sum_{k=0}^{n} \frac{f^{(k)}(a)(x - a)^k}{k!} + R_n(x), \tag{1}$$

where

$$R_n(x) = f^{(n+1)}(c)\frac{(x - a)^{n+1}}{(n + 1)!}, \tag{2}$$

for some number c (which depends on x) in the interval (a, x). Moreover, there exists a positive number M_n such that for every x in $[a, x]$,

$$|R_n(x)| \leq M_n\frac{(x - a)^{n+1}}{(n + 1)!}, \tag{3}$$

where M_n is the maximum value of $|f^{(n+1)}(x)|$ on the interval $[a, b]$.

We stress that (3) provides an upper bound for the error. In many cases the actual error [the difference $|f(x) - P_n(x)|$] will be considerably less than $M_n(x - a)^{n+1}/(n + 1)!$.

EXAMPLE 1 In Example 13.1.1, we found that the fifth-degree Taylor polynomial of $f(x) = \sin x$ at 0 is $P_5(x) = x - (x^3/3!) + (x^5/5!)$. We then have

$$\sin x = x - \frac{x^3}{3!} + \frac{x^5}{5!} + R_5(x), \tag{4}$$

where

$$R_5(x) = \frac{f^{(6)}(c)(x - 0)^6}{6!}.$$

But $f^{(6)}(c) = \sin^{(6)}(c) = -\sin c$ and $|-\sin c| \leq 1$. Thus for x in $[0, 1]$,

$$|R_5(x)| \leq \frac{1(x - 0)^6}{6!} = \frac{x^6}{720}.$$

For example, suppose we wish to calculate $\sin(\pi/10)$. From (4)

$$\sin\frac{\pi}{10} = \frac{\pi}{10} - \frac{\pi^3}{3!10^3} + \frac{\pi^5}{5!10^5} + R_n(x)$$

with

$$\left|R_5\left(\frac{\pi}{10}\right)\right| \le \frac{(\pi/10)^6}{720} \approx 0.00000134.$$

We find that

$$\sin\frac{\pi}{10} \approx \frac{\pi}{10} - \frac{1}{3!}\frac{\pi^3}{10^3} + \frac{1}{5!}\frac{\pi^5}{10^5} \approx 0.3141593 - 0.0051677 + 0.0000255$$

$$= 0.3090171.$$

The actual value of $\sin \pi/10 = \sin 18° = 0.3090170$, correct to seven decimal places, so our actual error is 0.0000001, which is quite a bit less than 0.0000013. This illustrates the fact that the actual error (the value of the remainder term) is often quite a bit smaller than the theoretical upper bound on the error given by formula (3). ∎

REMARK. In Example 1 the fifth-degree Taylor polynomial is also the sixth-degree Taylor polynomial [since $\sin^{(6)}(0) = -\sin 0 = 0$]. Thus we can use the error estimate for P_6. Since $|\sin^{(7)}(c)| = |-\cos c| \le 1$, we have

$$R_6(x) \le \frac{x^7}{7!} = \frac{x^7}{5040}.$$

If $x = \pi/10$, we obtain

$$\left|R_6\left(\frac{\pi}{10}\right)\right| \le \frac{(\pi/10)^7}{5040} \approx 0.0000000599.$$

Note that, to ten decimal places,

$$\frac{\pi}{10} - \frac{(\pi/10)^3}{3!} + \frac{(\pi/10)^5}{5!} = 0.3090170542,$$

and to ten decimal places,

$$\sin\frac{\pi}{10} = 0.3090169944,$$

with an actual error of 0.0000000598. Now we see that our estimate on the remainder term is really quite accurate.

EXAMPLE 2 Use a Taylor polynomial to estimate $e^{0.3}$ with an error of less than 0.0001.

Solution. For convenience, choose the interval $[0, 1]$. Then on $[0, 1]$, e^x and all its derivatives have a maximum value of $e^1 = e$. Then for any n, if we use a Taylor polynomial at 0 of degree n, we have

$$|R_n(0.3)| \le e \cdot \frac{(0.3)^{n+1}}{(n + 1)!}.$$

Since $e \approx 2.71828\ldots$, we use the bound $e < 2.72$. We must choose n so that $(e)(0.3)^{n+1}/(n + 1)! < (2.72)(0.3)^{n+1}/(n + 1)! < 0.0001$. For $n = 3$, $(2.72)(0.3)^4/4! \approx 0.00092$, while for $n = 4$, $(2.72)(0.3)^5/5! \approx 0.0000551 < 0.00006$. Thus we choose a fourth-degree Taylor polynomial for our approximation, and we know in advance that $|e^{0.3} - P_4(0.3)| < 0.00006$. We have

$$P_4(x) = 1 + x + \frac{x^2}{2!} + \frac{x^3}{3!} + \frac{x^4}{4!}$$

(see Example 13.1.3). Then

$$P_4(0.3) = 1 + 0.3 + \frac{(0.3)^2}{2!} + \frac{(0.3)^3}{3!} + \frac{(0.3)^4}{4!}$$

$$\approx 1 + 0.3 + 0.045 + 0.0045 + 0.00034 = 1.34984.$$

The actual value of $e^{0.3}$ is 1.34986, correct to five decimal places. Thus the error in our calculation is, approximately, 0.00002, one-third the calculated upper bound on the error. ∎

EXAMPLE 3 Compute $\cos[(\pi/4) + 0.1]$ with an error of less than 0.00001.

Solution. In this problem it is clearly convenient to use the Taylor expansion at $\pi/4$. Since all derivatives of $\cos x$ are bounded by 1, we have, for any n,

$$|R_n(x)| \le \frac{[x - (\pi/4)]^{n+1}}{(n + 1)!},$$

so that $R_n[(\pi/4) + 0.1] \le (0.1)^{n+1}/(n + 1)!$. If $n = 2$, then $(0.1)^3/3! = 0.00017$, and for $n = 3$, $(0.1)^4/4! = 0.000004 < 0.00001$. Thus we need to calculate $P_3(x)$ at $\pi/4$. But

$$P_3(x) = \cos\frac{\pi}{4} - \left(\sin\frac{\pi}{4}\right)\left(x - \frac{\pi}{4}\right) - \left(\cos\frac{\pi}{4}\right)\frac{[x - (\pi/4)]^2}{2} + \left(\sin\frac{\pi}{4}\right)\frac{[x - (\pi/4)]^3}{6}$$

$$= \frac{1}{\sqrt{2}}\left\{1 - \left(x - \frac{\pi}{4}\right) - \frac{[x - (\pi/4)]^2}{2} + \frac{[x - (\pi/4)]^3}{6}\right\},$$

and for $x = (\pi/4) + 0.1$,

$$\cos\left(\frac{\pi}{4} + 0.1\right) \approx P_3\left(\frac{\pi}{4} + 0.1\right) = \frac{1}{\sqrt{2}}\left[1 - 0.1 - \frac{(0.1)^2}{2} + \frac{(0.1)^3}{6}\right] \approx 0.63298.$$

This is correct to five decimal places. ∎

We now consider a more general example.

THE LOGARITHM [ln(1 + x)]

In Section 14.3 we will show that for any number $u \neq 1$, the formula for the sum of a **geometric progression** is given by†

$$1 + u + u^2 + \cdots + u^n = \frac{1 - u^{n+1}}{1 - u} = \frac{1}{1 - u} - \frac{u^{n+1}}{1 - u}. \tag{5}$$

This leads to the expression

$$\frac{1}{1 - u} = 1 + u + u^2 + \cdots + u^n + \frac{u^{n+1}}{1 - u}. \tag{6}$$

Setting $u = -t$ in (6) gives us

$$\frac{1}{1 + t} = 1 - t + t^2 - t^3 + \cdots + (-1)^n t^n + (-1)^{n+1} \frac{t^{n+1}}{1 + t}. \tag{7}$$

Integration of both sides of (7) from 0 to x yields

$$\ln(1 + x) = x - \frac{x^2}{2} + \frac{x^3}{3} - \cdots + (-1)^n \frac{x^{n+1}}{n + 1} + (-1)^{n+1} \int_0^x \frac{t^{n+1}}{t + 1} \, dt, \tag{8}$$

which is valid if $x > -1$ [since $\ln(1 + x)$ is not defined for $x \leq -1$].
   Now let

$$R_{n+1}(x) = (-1)^{n+1} \int_0^x \frac{t^{n+1}}{t + 1} \, dt.$$

We will show two things: First, we will prove that, for $-1 < x \leq 1$,

$$\lim_{n \to \infty} R_{n+1}(x) = 0.$$

This will ensure that the polynomial given in (8) provides an increasingly good approximation to $\ln(1 + x)$ as n increases. Second, we will show that for every $n \geq 1$,

$$\lim_{x \to 0} \frac{R_{n+1}(x)}{x^{n+1}} = 0.$$

---

†You were asked to prove this result in Problem 12.1.30.

Then, according to the uniqueness theorem (Theorem 13.2.1), we can conclude that the polynomial in (8) is *the* $(n + 1)$st degree Taylor polynomial for $\ln(1 + x)$ in the interval $-1 < x \leq 1$.

There are two cases to consider.

Case 1. $0 \leq x \leq 1$. Then for t in $[0, 1]$, $1/(1 + t) \leq 1$, so that

$$|R_{n+1}(x)| = \int_0^x \frac{t^{n+1}}{t + 1} \, dt \leq \int_0^x t^{n+1} \, dt = \frac{x^{n+2}}{n + 2} \leq \frac{1}{n + 2}, \tag{9}$$

which approaches 0 as $n \to \infty$.

Moreover, from (9),

$$\lim_{x \to 0} \frac{|R_{n+1}(x)|}{x^{n+1}} \leq \lim_{x \to 0+} \frac{x^{n+2}}{x^{n+1}(n + 2)} = \lim_{x \to 0+} \frac{x}{n + 2} = 0.$$

(with annotations: "$x > 0$" pointing to the first expression, "Remember, n is fixed" pointing to the second)

Case 2. $-1 < x < 0$. First, we note that

$$(-1)^{n+1} \int_0^x \frac{t^{n+1}}{t + 1} \, dt = (-1)^n \int_x^0 \frac{t^{n+1}}{t + 1} \, dt.$$

In this latter integral t is in the interval $[x, 0]$, where $-1 < x < 0$ (see Figure 1). Then we have $1 \geq 1 + t \geq 1 + x > 0$, so that

$$1 \leq \frac{1}{1 + t} \leq \frac{1}{1 + x}$$

FIGURE 1

and

$$|R_{n+1}(x)| \leq \int_x^0 \frac{|t|^{n+1}}{|t + 1|} \, dt \leq \int_x^0 \frac{(-t)^{n+1}}{(x + 1)} \, dt \quad \text{Since } |t| = -t \text{ in } [x, 0]$$

$$= \frac{1}{x + 1} \int_x^0 (-t)^{n+1} \, dt = -\frac{1}{x + 1} \frac{(-t)^{n+2}}{(n + 2)} \bigg|_x^0 = \frac{(-x)^{n+2}}{(1 + x)(n + 2)}.$$

Thus

$$|R_{n+1}(x)| \leq \frac{(-x)^{n+2}}{(1 + x)(n + 2)} \leq \frac{1}{(1 + x)(n + 2)}, \tag{10}$$

since $-1 < x < 0$ implies that $0 < -x < 1$, and we have $|R_n(x)| \to 0$ as $n \to \infty$.

Moreover, from (10),

$$\lim_{x \to 0} \frac{|R_{n+1}(x)|}{x^{n+1}} \leq \lim_{x \to 0+} \frac{(-x)^{n+2}}{(-x)^{n+1}(n + 2)} = \lim_{x \to 0+} \frac{-x}{n + 2} = 0.$$

(with annotations: "$-x > 0$" pointing to the first expression, "n is fixed" pointing to the second)

We conclude that the Taylor polynomial

$$x - \frac{x^2}{2} + \frac{x^3}{3} - \cdots + (-1)^n \frac{x^{n+1}}{n+1}$$

is a good approximation to $\ln(1 + x)$ for sufficiently large n, provided that $-1 < x \leq 1$.

EXAMPLE 4 Calculate $\ln 1.4$ with an error of less than 0.001.

Solution. Here $x = 0.4$, and from (9) we need to find an n such that $(0.4)^{n+2}/(n+2) < 0.001$. We have $(0.4)^5/5 \approx 0.00205$ and $(0.4)^6/6 \approx 0.00068$, so choosing $n = 4$ ($n + 2 = 6$), we obtain the Taylor polynomial [from (8)]

$$P_5(x) = x - \frac{x^2}{2} + \frac{x^3}{3} - \frac{x^4}{4} + \frac{x^5}{5}$$

[remember, from (8) the last term in $P_{n+1}(x)$ is $(-1)^n x^{n+1}/(n+1)$], and

$$\ln 1.4 \approx P_5(0.4) = 0.4 - \frac{(0.4)^2}{2} + \frac{(0.4)^3}{3} - \frac{(0.4)^4}{4} + \frac{(0.4)^5}{5}$$

$$\approx 0.4 - 0.08 + 0.02133 - 0.0064 + 0.00205 = 0.33698.$$

The actual value of $\ln 1.4$ is 0.33647, correct to five decimal places, so that the error is $0.33698 - 0.33647 = 0.00051$. This error is slightly less than the maximum possible error of 0.00068, and so in this case our error bound is fairly sharp. Note that the error bound (9) used in this problem is *not* the same as the error bound given by (3). To obtain the bound (3), we would have to compute the sixth derivative of $\ln(1 + x)$. ◼

EXAMPLE 5 Calculate $\ln 0.9$ with an error of less than 0.001.

Solution. Here we need to find $\ln(1 + x)$ for $x = -0.1$. Using (10), we have

$$|R_{n+1}(x)| \leq \frac{(-x)^{n+2}}{(1+x)(n+2)} = \frac{(0.1)^{n+2}}{0.9(n+2)}.$$

For $n = 1$, $(0.1)^3/(0.9)(3) = 0.00037$, so that we need only evaluate $P_{n+1}(-0.1) = P_2(-0.1) = (-0.1) - [(-0.1)^2/2] = -0.10500$. The value of $\ln 0.9$, correct to five decimal places, is -0.10536, so our actual error of 0.00036 and our maximum possible error of 0.00037 almost coincide. ◼

THE ARC TANGENT ($\tan^{-1} x$)

If in Equation (6) we substitute $u = -t^2$, then for any number t we obtain

$$\frac{1}{1+t^2} = 1 - t^2 + t^4 - t^6 + \cdots + (-1)^n t^{2n} + (-1)^{n+1} \frac{t^{2(n+1)}}{1+t^2}. \tag{11}$$

Integration of both sides of (11) from 0 to x yields

$$\tan^{-1} x = x - \frac{x^3}{3} + \frac{x^5}{5} - \frac{x^7}{7} + \cdots + \frac{(-1)^n x^{2n+1}}{2n+1} + (-1)^{n+1} \int_0^x \frac{t^{2(n+1)}}{1+t^2} \, dt. \tag{12}$$

This equation is valid for any real number $x \geq 0$.

The polynomial in (12) is a polynomial of degree $2n + 1$. Since $1/(1 + t^2) \leq 1$, we have

$$|R_{2n+1}(x)| = \int_0^x \frac{t^{2(n+1)}}{1+t^2} \, dt \leq \int_0^x t^{2(n+1)} \, dt = \frac{x^{2n+3}}{2n+3}. \tag{13}$$

Thus

$$\lim_{x \to 0} \frac{|R_{2n+1}(x)|}{x^{2n+1}} = \lim_{x \to 0} \frac{x^2}{2n+3} = 0,$$

so that again by the uniqueness theorem, the polynomial in (12) is the Taylor polynomial for $\tan^{-1} x$ at 0. We also see, from (13), that

$$R_{2n+1}(x) \to 0 \quad \text{as} \quad n \to \infty \quad \text{if} \quad |x| \leq 1.$$

We conclude that the Taylor polynomial

$$x - \frac{x^3}{3} + \frac{x^5}{5} + \cdots + \frac{(-1)^n x^{2n+1}}{2n+1}$$

provides a good approximation to $\tan^{-1} x$ for sufficiently large n, provided that $|x| \leq 1$.

Equation (12) give us a formula for calculating π.† Setting $x = 1$, we have

$$\frac{\pi}{4} = \tan^{-1} 1 \approx 1 - \frac{1}{3} + \frac{1}{5} - \frac{1}{7} + \frac{1}{9} - \frac{1}{11} + \cdots + \frac{(-1)^n}{2n+1}.$$

The error here [given by (13)], is bounded by $1/(2n + 3)$, which approaches zero very slowly. To get an error less than 0.001, we would need $1/(2n + 3) < 1/1000$, or $2n + 3 > 1000$, and $n \geq 499$. For example, $1 - \frac{1}{3} + \frac{1}{5} - \frac{1}{7} + \frac{1}{9} - \frac{1}{11} + \frac{1}{13} - \frac{1}{15} = 0.75427$, while $\pi/4$ is 0.78540. A better way to approximate π is suggested in Problems 29 and 30.

---

†This formula was discovered by the Scottish mathematician James Gregory (1638–1675) and was first published in 1712.

▦ PROBLEMS 13.3

In Problems 1–10, find a bound for $|R_n(x)|$ for x in the given interval, where $P_n(x)$ is a Taylor polynomial of degree n having terms of the form $(x - a)^k$.

1. $f(x) = \sin x; a = \pi/4; n = 6; x \in [0, \pi/2]$
2. $f(x) = \sqrt{x}; a = 1; n = 4; x \in [\frac{1}{4}, 4]$

3. $f(x) = \dfrac{1}{x}; a = 1; n = 4; x \in \left[\dfrac{1}{2}, 2\right]$

4. $f(x) = \tan x; a = 0; n = 4; x \in [0, \pi/4]$

5. $f(x) = \dfrac{1}{\sqrt{x}}; a = 5; n = 5; x \in \left[\dfrac{19}{4}, \dfrac{21}{4}\right]$

6. $f(x) = \sinh x; a = 0; n = 4; x \in [0, 1]$
7. $f(x) = \ln \cos x; a = 0; n = 3; x \in [0, \pi/6]$
8. $f(x) = e^{ax}; a = 0; n = 4; x \in [0, 1]$
9. $f(x) = e^{x^2}; a = 0; n = 4; x \in [0, \frac{1}{3}]$
10. $f(x) = \sin x^2; a = 0; n = 4; x \in [0, \pi/4]$

In Problems 11–26, use a Taylor polynomial to estimate the given number with the given degree of accuracy.

11. $\sin\left(\dfrac{\pi}{6} + 0.2\right)$; error < 0.001

12. $\sin 33°$; error < 0.001 [*Hint:* Convert to radians.]

13. $\tan\left(\dfrac{\pi}{4} + 0.1\right)$; error < 0.01

14. e; error < 0.0001 [*Hint:* You may assume that $2 < e < 3$.]
15. e^{-1}; error < 0.001 **16.** $\ln 2$; error < 0.1
17. $\ln 1.5$; error < 0.001 **18.** $\ln 0.5$; error < 0.0001
19. e^3; error < 0.01 [*Hint:* See Problem 14.]
20. $\tan^{-1} 0.5$; error < 0.001
*21. $\sinh \frac{1}{2}$; error < 0.01 ***22.** $\cosh \frac{1}{2}$; error < 0.01
23. $\sin 100°$; error < 0.001 **24.** $\cos 195°$; error < 0.001

25. $\dfrac{1}{\sqrt{1.1}}$; error < 0.001 ***26.** $\ln \cos 0.3$; error < 0.01

27. Use the result of Problem 9 to estimate $\int_0^{1/3} e^{x^2}\, dx$. What is the maximum error of your estimate?
28. Use the result of Problem 7 to estimate $\int_0^{\pi/6} \ln \cos x\, dx$. What is the maximum error of your estimate?
*29. Use the formula $\tan(A \pm B) = (\tan A \pm \tan B)/(1 \mp \tan A \tan B)$ to prove that $\tan(4 \tan^{-1} 1/5 - \tan^{-1} 1/239) = 1$.   [*Hint:* First calculate $\tan 2(\tan^{-1} 1/5)$ and $\tan (4 \tan^{-1} 1/5)$.] This implies that $4 \tan^{-1} 1/5 - \tan^{-1} 1/239 = \pi/4$.
*30. Use the result of Problem 29 to calculate π to five decimal places.
31. Let $f(x) = (1 + x)^n$, where n is a positive integer.
    (a) Show that $f^{(n+1)}(x) = 0$.
    (b) Show that

$$f(x) = 1 + nx + \frac{n(n-1)}{2!}x^2 + \frac{n(n-1)(n-2)}{3!}x^3 + \cdots + x^n.$$

This result is called the **binomial theorem.** (Also see Appendix 4.)

32. Use the binomial theorem to calculate **(a)** $(1.2)^4$, **(b)** $(0.8)^5$, **(c)** $(1.03)^4$.

33. Let $f(x) = (1 + x)^\alpha$, where α is any real number.

    (a) Show that

$$f(x) = 1 + \alpha x + \frac{\alpha(\alpha - 1)}{2!}x^2 + \frac{\alpha(\alpha - 1)(\alpha - 2)}{3!}x^3 + \cdots$$

$$+ \frac{\alpha(\alpha - 1)(\alpha - 2)\cdots(\alpha - n + 1)}{n!}x^n + R_n(x).$$

    (b) Show that if $|x| < 1$, then $R_n(x) \to 0$ as $n \to \infty$. This result is called the **general binomial expansion. If $x = -1$, the result is true for $\alpha > -1$. If $x = 1$, it is true for $\alpha > 0$.

34. Use the result of Problem 33 to calculate the following numbers to four decimal places of accuracy.

    (a) $\sqrt{1.2}$ **(b)** $(0.9)^{3/4}$ **(c)** $(1.8)^{1/4}$

    (d) $\dfrac{1}{\sqrt[3]{1.01}}$ **(e)** $2^{5/3}$ **(f)** $(0.4)^{1.6}$

REVIEW EXERCISES FOR CHAPTER THIRTEEN

In Exercises 1–8, find the Taylor polynomial of the given degree at the number a.

1. $f(x) = e^x$; $a = 0$; $n = 3$ **2.** $f(x) = \ln x$; $a = 1$, $n = 4$

3. $f(x) = \sin x$; $a = \pi/6$; $n = 3$ **4.** $f(x) = \cos x$; $a = \pi/2$; $n = 5$

5. $f(x) = \cot x$; $a = \pi/2$; $n = 4$ **6.** $f(x) = \sinh x$; $a = 0$; $n = 3$

7. $f(x) = x^3 - x^2 + 2x + 3$; $a = 0$; $n = 8$ **8.** $f(x) = e^{-x^2}$; $a = 0$; $n = 5$

In Exercises 9–13, find a bound for $|R_n(x)|$ for x in the given interval.

9. $f(x) = \cos x$; $a = \pi/6$; $n = 5$; $x \in [0, \pi/2]$

10. $f(x) = \sqrt[3]{x}$; $a = 1$; $n = 4$; $x \in [\frac{7}{8}, \frac{9}{8}]$

11. $f(x) = e^x$; $a = 0$; $n = 6$; $x \in [-\ln e, \ln e]$

12. $f(x) = \cot x$; $a = \pi/2$; $n = 2$; $x \in [\pi/4, 3\pi/4]$

13. $f(x) = e^{-x^2}$; $a = 0$; $n = 4$; $x \in [-1, 1]$

In Exercises 14–20, use a Taylor polynomial to estimate the given number with the given degree of accuracy.

14. $\cos\left(\dfrac{\pi}{3} + 0.1\right)$; error < 0.001

15. $\cos 43°$; error < 0.001

16. $\cot\left(\dfrac{\pi}{4} + 0.1\right)$; error < 0.001

*17. $\ln 2$; error < 0.0001 [*Hint:* Look at $\ln[(1 + x)/(1 - x)]$.]

18. e^2; error < 0.0001

19. $\tan^{-1} 0.3$; error < 0.0001

20. $\ln \sin\left(\dfrac{\pi}{2} + 0.2\right)$; error < 0.01

21. Use a Taylor polynomial of degree 4 to estimate $\int_0^{1/2} \cos x^2 \, dx$. What is the maximum error of your estimate?

22. Use a Taylor polynomial of degree 4 to estimate $\int_0^1 e^{-x^3} \, dx$. What is the maximum error of your estimate?

14 Sequences and Series

In Chapter 13 we saw how a wide variety of functions could be approximated by polynomials. A polynomial in x is a *finite* sum of terms of the form $a_k x^k$. In this chapter we will discuss what we mean by an *infinite* sum. The infinite sums we describe are called *infinite series*. We will discuss the theory of infinite series and will show how it can give us a great deal of information about a wide variety of functions.

14.1 SEQUENCES OF REAL NUMBERS

According to a popular dictionary,† a *sequence* is "the following of one thing after another." In mathematics we could define a sequence intuitively as a succession of numbers that never terminates. The numbers in the sequence are called the *terms* of the sequence. In a sequence there is one term for each positive integer.

EXAMPLE 1 Consider the sequence

$$\underset{\substack{\text{1st}\\\text{term}}}{\frac{1}{2}}, \quad \underset{\substack{\text{2nd}\\\text{term}}}{\frac{1}{4}}, \quad \underset{\substack{\text{3rd}\\\text{term}}}{\frac{1}{8}}, \quad \underset{\substack{\text{4th}\\\text{term}}}{\frac{1}{16}}, \quad \underset{\substack{\text{5th}\\\text{term}}}{\frac{1}{32}}, \cdots, \underset{\substack{n\text{th}\\\text{term}}}{\frac{1}{2^n}}, \cdots$$

---

†*The Random House Dictionary* (Ballantine Books, New York, 1978).

We see that there is one term for each positive integer. The terms in this sequence form an infinite set of real numbers, which we write as

$$A = \left\{ \frac{1}{2}, \frac{1}{4}, \frac{1}{8}, \ldots, \frac{1}{2^n}, \ldots \right\}. \tag{1}$$

That is, the set A consists of all numbers of the form $1/2^n$, where n is a positive integer. There is another way to describe this set. We define the function f by the rule $f(n) = 1/2^n$, where the domain of f is the set of positive integers. Then the set A is precisely the set of values taken by the function f. ∎

In general, we have the following formal definition.

Definition 1   SEQUENCE   A **sequence** of real numbers is a function whose domain is the set of positive integers. The values taken by the function are called **terms** of the sequence.

NOTATION.   We will often denote the terms of a sequence by a_n. Thus if the function given in Definition 1 is f, then $a_n = f(n)$. With this notation, *we can denote the set of values taken by the sequence by* $\{a_n\}$. Also, we will use n, m, and so on as integer variables and x, y, and so on as real variables.

EXAMPLE 2   The following are sequences of real numbers:

(a) $\{a_n\} = \left\{ \dfrac{1}{n} \right\}$ (b) $\{a_n\} = \{\sqrt{n}\}$ (c) $\{a_n\} = \left\{ \dfrac{1}{n!} \right\}$

(d) $\{a_n\} = \{\sin n\}$ (e) $\{a_n\} = \left\{ \dfrac{e^n}{n!} \right\}$ (f) $\{a_n\} = \left\{ \dfrac{n-1}{n} \right\}$ ∎

We sometimes denote a sequence by writing out the values $\{a_1, a_2, a_3, \ldots\}$.

EXAMPLE 3   We write out the values of the sequences in Example 2:

(a) $\left\{ 1, \dfrac{1}{2}, \dfrac{1}{3}, \dfrac{1}{4}, \ldots, \dfrac{1}{n}, \ldots \right\}$

(b) $\{1, \sqrt{2}, \sqrt{3}, \sqrt{4}, \ldots, \sqrt{n}, \ldots\}$

(c) $\left\{ 1, \dfrac{1}{2}, \dfrac{1}{6}, \dfrac{1}{24}, \ldots, \dfrac{1}{n!}, \ldots \right\}$

(d) $\{\sin 1, \sin 2, \sin 3, \sin 4, \ldots, \sin n, \ldots\}$

(e) $\left\{ e, \dfrac{e^2}{2}, \dfrac{e^3}{6}, \dfrac{e^4}{24}, \ldots, \dfrac{e^n}{n!}, \ldots \right\}$

(f) $\left\{ 0, \dfrac{1}{2}, \dfrac{2}{3}, \dfrac{3}{4}, \ldots, \dfrac{n-1}{n}, \ldots \right\}$ ∎

Because a sequence is a function, it has a graph. In Figure 1 we draw part of the graphs of four of the sequences in Example 3.

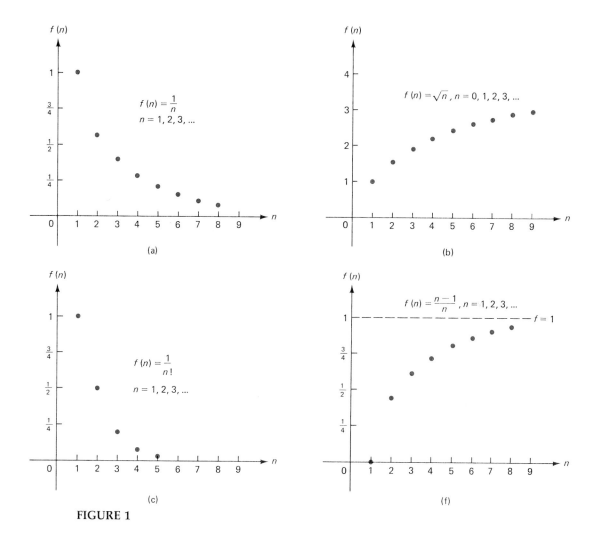

FIGURE 1

EXAMPLE 4   Find the general term a_n of the sequence

$$\{-1,\ 1\ -1,\ 1,\ -1,\ 1,\ -1,\ \dots\ \}.$$

Solution. We see that $a_1 = -1$, $a_2 = 1$, $a_3 = -1$, $a_4 = 1$, ... Hence

$$a_n = \begin{cases} -1, & \text{if } n \text{ is odd} \\ 1, & \text{if } n \text{ is even}. \end{cases}$$

A more concise way to write this term is

$$a_n = (-1)^n.$$

We draw the graph of this sequence in Figure 2. ▨

It is evident that as n gets large, the numbers $1/n$ get small. We can write

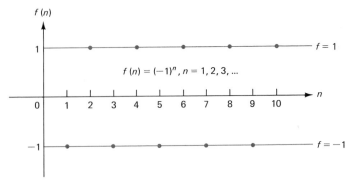

FIGURE 2

$$\lim_{n \to \infty} \frac{1}{n} = 0.$$

This is also suggested by the graph in Figure 1a. Similarly, it is not hard to show that as n gets large, $(n - 1)/n$ gets close to 1. We write

$$\lim_{n \to \infty} \frac{n - 1}{n} = 1.$$

This is illustrated in Figure 1f.

On the other hand, it is clear that $a_n = (-1)^n$ does not get close to any one number as n increases. It simply oscillates back and forth between the numbers $+1$ and -1. This is illustrated in Figure 2.

For the remainder of this section we will be concerned with calculating the limit of a sequence as $n \to \infty$. Since a sequence is a special type of function, our formal definition of the limit of a sequence is going to be very similar to the definition of $\lim_{x \to \infty} f(x)$.

Definition 2   FINITE LIMIT OF A SEQUENCE   A sequence $\{a_n\}$† has the limit L if for every $\epsilon > 0$ there exists an integer $N > 0$ such that if $n \geq N$, then $|a_n - L| < \epsilon$. We write

$$\lim_{n \to \infty} a_n = L. \qquad (2)$$

Intuitively, this definition states that $a_n \to L$ if as n increases without bound, a_n gets arbitrarily close to L (see pages 87 and 128). We illustrate this definition in Figure 3.

Definition 3   INFINITE LIMIT OF A SEQUENCE   The sequence $\{a_n\}$ has the limit ∞ if for every positive number M there is an integer $N > 0$ such that if $n > N$, then $a_n > M$. In this case we write

---

†To be precise, $\{a_n\}$ denotes the set of values taken by the sequence. There is a difference between the sequence, which is a function f, and the values $a_n = f(n)$ taken by this function. However, because it is more convenient to write down the values the sequence takes, we will, from now on, use the notation $\{a_n\}$ to denote a sequence.

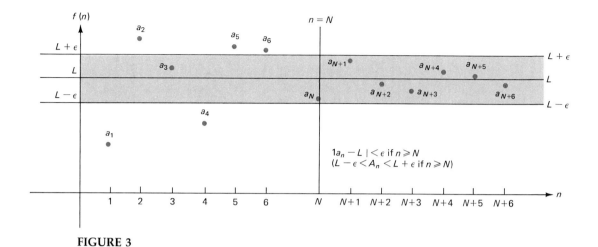

FIGURE 3

$$\lim_{n \to \infty} a_n = \infty.$$

Intuitively, $\lim_{n \to \infty} a_n = \infty$ means that as n increases without bound, a_n also increases without bound.

The theorem below gives us a very useful result.

Theorem 1   Let r be a real number. Then

$$\lim_{n \to \infty} r^n = 0 \qquad \text{if} \qquad |r| < 1$$

and

$$\lim_{n \to \infty} |r^n| = \infty \qquad \text{if} \qquad |r| > 1.$$

Proof.

Case 1 $r = 0$. Then $r^n = 0$, and the sequence has the limit 0.

Case 2 $0 < |r| < 1$.   For a given $\epsilon > 0$, choose N such that

$$N > \frac{\ln \epsilon}{\ln |r|}.$$

Note that since $|r| < 1$, $\ln|r| < 0$. Now if $n > N$,

$$n > \frac{\ln \epsilon}{\ln |r|} \qquad \text{and} \qquad n \ln|r| < \ln \epsilon.$$

The second inequality follows from the fact that $\ln|r|$ is negative, and multiplying both sides of an inequality by a negative number reverses the inequality. Thus

$$\ln|r^n - 0| = \ln|r^n| = \ln|r|^n = n \ln|r| < \ln \epsilon.$$

Since $\ln|r^n - 0| < \ln \epsilon$ and $\ln x$ is an increasing function, we conclude that $|r^n - 0| < \epsilon$. Thus according to the definition of a finite limit of a sequence,

$$\lim_{n \to \infty} r^n = 0.$$

Case 3 $|r| > 1$. Let $M > 0$ be given. Choose $N > \ln M / \ln|r|$. Then if $n > N$,

$$\ln|r^n| = n \ln|r| \overset{n > N}{>} N \ln|r| > \left(\frac{\ln M}{\ln|r|} \right)(\ln|r|) = \ln M,$$

so that

$$|r^n| > M \quad \text{if} \quad n > N.$$

From the definition of an infinite limit of a sequence, we see that

$$\lim_{n \to \infty} |r^n| = \infty. \quad \blacksquare$$

In Sections 2.3 and 2.8 we stated a number of limit theorems. In Section 2.4 and in Chapter 12 we saw how to evaluate limits as $x \to \infty$. All these results can be applied when n, rather than x, approaches infinity. The only difference is that as n grows, it takes on integer values. For convenience, we state without proof the major limit theorems we need for sequences. The proofs are similar to the proofs given in Appendix 5.

Theorem 2 Suppose that $\lim_{n \to \infty} a_n$ and $\lim_{n \to \infty} b_n$ both exist and are finite.

(i) $\lim_{n \to \infty} \alpha a_n = \alpha \lim_{n \to \infty} a_n$ for any number α.   **(3)**

(ii) $\lim_{n \to \infty}(a_n + b_n) = \lim_{n \to \infty} a_n + \lim_{n \to \infty} b_n$.   **(4)**

(iii) $\lim_{n \to \infty} a_n b_n = (\lim_{n \to \infty} a_n)(\lim_{n \to \infty} b_n)$.   **(5)**

(iv) If $\lim_{n \to \infty} b_n \neq 0$, then

$$\lim_{n \to \infty} \frac{a_n}{b_n} = \frac{\lim_{n \to \infty} a_n}{\lim_{n \to \infty} b_n}. \qquad (6)$$

Theorem 3 *Continuity Theorem* Suppose that L is finite and $\lim_{n \to \infty} a_n = L$. If f is continuous in an open interval containing L, then

$$\lim_{n \to \infty} f(a_n) = f(\lim_{n \to \infty} a_n) = f(L). \qquad (7)$$

Theorem 4 *Squeezing Theorem* Suppose that $\lim_{n \to \infty} a_n = \lim_{n \to \infty} b_n = L$ and that $\{c_n\}$ is a sequence having the property that for $n > N$ (a positive integer), $a_n \leq c_n \leq b_n$. Then

$$\lim_{n \to \infty} c_n = L. \qquad (8)$$

We now give a central definition in the theory of sequences.

Definition 4 CONVERGENCE AND DIVERGENCE OF A SEQUENCE If the limit in (2) exists and if L is finite, we say that the sequence **converges** or is **convergent.** Otherwise, we say that the sequence **diverges** or is **divergent.**

EXAMPLE 5 The sequence $\{1/2^n\}$ is convergent since, by Theorem 1, $\lim_{n \to \infty} 1/2^n = \lim_{n \to \infty} (1/2)^n = 0$. ◼

EXAMPLE 6 The sequence $\{r^n\}$ is divergent for $r > 1$ since $\lim_{n \to \infty} r^n = \infty$ if $r > 1$. ◼

EXAMPLE 7 The sequence $\{(-1)^n\}$ is divergent since the values a_n alternate between -1 and $+1$ but do not stay close to any fixed number as n becomes large. ◼

Since we have a large body of theory and experience behind us in the calculation of ordinary limits, we would like to make use of that experience to calculate limits of sequences. The following theorem, whose proof is left as a problem (see Problem 34), is extremely useful.

Theorem 5 Suppose that $\lim_{x \to \infty} f(x) = L$, a finite number, ∞, or $-\infty$. If f is defined for every positive integer, then the limit of the sequence $\{a_n\} = \{f(n)\}$ is also equal to L. That is, $\lim_{x \to \infty} f(x) = \lim_{n \to \infty} a_n = L$.

EXAMPLE 8 Calculate $\lim_{n \to \infty} 1/n^2$.

Solution. Since $\lim_{x \to \infty} 1/x^2 = 0$, we have $\lim_{n \to \infty} 1/n^2 = 0$ (by Theorem 5). ◼

EXAMPLE 9 Does the sequence $\{e^n/n\}$ converge or diverge?

Solution. Since $\lim_{x \to \infty} e^x/x = \lim_{x \to \infty} e^x/1$ (by L'Hôpital's rule) $= \infty$, we find that the sequence diverges. ◼

REMARK. It should be emphasized that Theorem 5 does *not* say that if $\lim_{x \to \infty} f(x)$ does not exist, then $\{a_n\} = \{f(n)\}$ diverges. For example, let

$$f(x) = \sin \pi x.$$

Then $\lim_{x \to \infty} f(x)$ does not exist, but $\lim_{n \to \infty} f(n) = \lim_{n \to \infty} \sin \pi n = 0$ since $\sin \pi n = 0$ for every integer n.

EXAMPLE 10 Let $\{a_n\} = \{[1 + (1/n)]^n\}$. Does this sequence converge or diverge?

Solution. Since $\lim_{x \to \infty} [1 + (1/x)]^x = e$, we see that a_n converges to the limit e. ◼

EXAMPLE 11 Determine the convergence or divergence of the sequence $\{(\ln n)/n\}$.

Solution. $\lim_{x \to \infty} [(\ln x)/x] = \lim_{x \to \infty} [(1/x)/1] = 0$ by L'Hôpital's rule, so that the sequence converges to 0. ◼

EXAMPLE 12 Let $p(x) = c_0 + c_1 x + \cdots + c_m x^m$ and $q(x) = d_0 + d_1 x + \cdots + d_r x^r$. In Problem 2.4.39 we showed that if the rational function $r(x) = p(x)/q(x)$, then, if $c_m d_r \neq 0$,

$$\lim_{x \to \infty} \frac{p(x)}{q(x)} = \begin{cases} 0, & \text{if } m < r \\ \pm\infty, & \text{if } m > r \\ \dfrac{c_m}{d_r}, & \text{if } m = r. \end{cases}$$

Thus the sequence $\{p(n)/q(n)\}$ converges to 0 if $m < r$, converges to c_m/d_r if $m = r$, and diverges if $m > r$. ■

EXAMPLE 13 Does the sequence $\{(5n^3 + 2n^2 + 1)/(2n^3 + 3n + 4)$ converge or diverge?

Solution. Here $m = r = 3$, so that by the result of Example 12 the sequence converges to $c_3/d_3 = \frac{5}{2}$. ■

EXAMPLE 14 Does the sequence $\{n^{1/n}\}$ converge or diverge?

Solution. Since $\lim_{x \to \infty} x^{1/x} = 1$ (see Example 12.3.7), the sequence converges to 1. ■

EXAMPLE 15 Determine the convergence or divergence of the sequence $\{\sin \alpha n/n^\beta\}$, where α is a real number and $\beta > 0$.

Solution. Since $-1 \le \sin \alpha x \le 1$, we see that

$$-\frac{1}{x^\beta} \le \frac{\sin \alpha x}{x^\beta} \le \frac{1}{x^\beta} \qquad \text{for any } x > 0.$$

But $\pm \lim_{x \to \infty} 1/x^\beta = 0$, and so by the squeezing theorem (Theorem 2.4.1), $\lim_{x \to \infty}[(\sin \alpha x)/x^\beta] = 0$. Therefore the sequence $\{(\sin \alpha n)/n^\beta\}$ converges to 0. ■

As in Example 15, the squeezing theorem can often be used to calculate the limit of a sequence.

PROBLEMS 14.1

In Problems 1–9, find the first five terms of the given sequence.

1. $\left\{\dfrac{1}{3^n}\right\}$
2. $\left\{\dfrac{n + 1}{n}\right\}$
3. $\left\{1 - \dfrac{1}{4^n}\right\}$

4. $\{\sqrt[3]{n}\}$
5. $\{e^{1/n}\}$
6. $\{n \cos n\}$

7. $\{\sin n\pi\}$
8. $\{\cos n\pi\}$
9. $\left\{\sin \dfrac{n\pi}{2}\right\}$

In Problems 10–27, determine whether the given sequence is convergent or divergent. If it is convergent, find its limit.

10. $\left\{\dfrac{3}{n}\right\}$
11. $\left\{\dfrac{1}{\sqrt{n}}\right\}$
12. $\left\{\dfrac{n + 1}{n^{5/2}}\right\}$

13. $\{\sin n\}$ **14.** $\{\sin n\pi\}$ **15.** $\left\{\cos\left(n + \dfrac{\pi}{2}\right)\right\}$

16. $\left\{\dfrac{n^5 + 3n^2 + 1}{n^6 + 4n}\right\}$ **17.** $\left\{\dfrac{4n^5 - 3}{7n^5 + n^2 + 2}\right\}$ **18.** $\left\{\left(1 + \dfrac{4}{n}\right)^n\right\}$

19. $\left\{\left(1 + \dfrac{1}{4n}\right)^n\right\}$ **20.** $\left\{\dfrac{\sqrt{n}}{\ln n}\right\}$

21. $\{\sqrt{n + 3} - \sqrt{n}\}$ [*Hint:* Multiply and divide by $\sqrt{n + 3} + \sqrt{n}$.]

22. $\left\{\dfrac{2^n}{n!}\right\}$ **23.** $\left\{\dfrac{\alpha^n}{n!}\right\}$ (α real) **24.** $\left\{\dfrac{4}{\sqrt{n^2 + 3} - n}\right\}$

25. $\left\{\dfrac{(-1)^n n^3}{n^3 + 1}\right\}$ **26.** $\{(-1)^n \cos n\pi\}$ **27.** $\left\{\dfrac{(-1)^n}{\sqrt{n}}\right\}$

In Problems 28–33, find the general term a_n of the given sequence.

28. $\{1, -2, 3, -4, 5, -6, \ldots\}$ **29.** $\{1, 2 \cdot 5, 3 \cdot 5^2, 4 \cdot 5^3, 5 \cdot 5^4, \ldots\}$

30. $\{\frac{1}{2}, \frac{2}{3}, \frac{3}{4}, \frac{4}{5}, \frac{5}{6}, \ldots\}$ **31.** $\{\frac{1}{2}, \frac{3}{4}, \frac{7}{8}, \frac{15}{16}, \frac{31}{32}, \ldots\}$

*32. $\{\frac{1}{3}, \frac{2}{5}, \frac{3}{7}, \frac{4}{9}, \frac{5}{11}, \ldots\}$ **33.** $\{1, -\frac{1}{3}, \frac{1}{9}, -\frac{1}{27}, \ldots\}$

*34. Prove Theorem 5. [*Hint:* Use Definition 2.8.3.]

35. Show that if $\{a_n\}$ and $\{b_n\}$ are two sequences such that $|a_n| \le |b_n|$ for each n, then if $|b_n|$ converges to 0, $|a_n|$ also converges to 0. [*Hint:* Use the squeezing theorem.]

36. Use the result of Problem 35 to show that the sequence $\{(a \sin bn + c \cos dn)/n^{p^2}\}$ converges to 0 for any real numbers a, b, c, d, and $p \ne 0$.

37. Prove that if $|r| < 1$, then the sequence $\{nr^n\}$ converges to 0.

*38. Show that if $\{a_n\}$ converges, then the $\lim_{n\to\infty} a_n$ is unique. [*Hint:* Assume that $\lim_{n\to\infty} a_n = L$, $\lim_{n\to\infty} a_n = M$, and $L \ne M$. Then choose $\epsilon = \frac{1}{2}|L - M|$ to show that Definition 2 is violated.]

*39. Suppose that $\{a_n\}$ is a sequence such that $a_{n+1} = a_n - a_n^2$, where a_0 is given. Find all values of a_0 for which $\lim_{n\to\infty} a_n = 0$.

**40. Suppose that $a_{n+1} = 1/(2 + a_n)$. For what choices of a_0 does the sequence $\{a_k\}$ diverge?

*41. Prove or disprove: If $\lim_{n\to\infty} n^p a_n = 0$, then there exists an $\epsilon > 0$ such that $\lim_{n\to\infty} n^{p+\epsilon} a_n$ also exists (i.e., is finite).

42. Show that $a_n = [1 + (\alpha/n)]^n$ converges to e^α as $n \to \infty$.

14.2 BOUNDED AND MONOTONIC SEQUENCES

There are certain kinds of sequences that have special properties worthy of mention.

Definition 1   BOUNDEDNESS

(i) The sequence $\{a_n\}$ is **bounded above** if there is a number M_1 such that

$$a_n \le M_1 \tag{1}$$

for every positive integer n.

(ii) It is **bounded below** if there is a number M_2 such that

$$M_2 \le a_n \tag{2}$$

for every positive integer n.

(iii) It is **bounded** if there is a number $M > 0$ such that

$$|a_n| \leq M$$

for every positive integer n.

The numbers M_1, M_2, and M are called, respectively, an **upper bound,** a **lower bound,** and a **bound** for $\{a_n\}$.

(iv) If the sequence is not bounded, it is called **unbounded.**

REMARK.   If $\{a_n\}$ is bounded above and below, then it is bounded. Simply set $M = \max\{|M_1|, |M_2|\}$.

EXAMPLE 1   The sequence $\{\sin n\}$ has the upper bound of 1, the lower bound of -1, and the bound of 1 since $-1 \leq \sin n \leq 1$ for every n. Of course, any number greater than 1 is also a bound. ■

EXAMPLE 2   The sequence $\{(-1)^n\}$ has the upper bound 1, the lower bound -1, and the bound 1. ■

EXAMPLE 3   The sequence $\{2^n\}$ is bounded below by 2 but has no upper bound and so is unbounded. ■

EXAMPLE 4   The sequence $\{(-1)^n 2^n\}$ is bounded neither below nor above. ■

It turns out that the following statement is true:

Every convergent sequence is bounded.

Theorem 1.   If the sequence $\{a_n\}$ is convergent, then it is bounded.

Proof.   Before giving the technical details, we remark that the idea behind the proof is easy. For if $\lim_{n \to \infty} a_n = L$, then a_n is close to the finite number L if n is large. Thus, for example, $|a_n| \leq |L| + 1$ if n is large enough. Since a_n is a real number for every n, the first few terms of the sequence are bounded, and these two facts give us a bound for the entire sequence.

Now to the details: Let $\epsilon = 1$. Then there is an $N > 0$ such that (according to Definition 14.1.2)

$$|a_n - L| < 1 \quad \text{if} \quad n \geq N. \tag{3}$$

Let

$$K = \max\{|a_1|, |a_2|, \ldots, |a_N|\}. \tag{4}$$

Since each a_n is finite, K, being the maximum of a finite number of terms, is also finite. Now let

$$M = \max\{|L| + 1, K\}. \tag{5}$$

It follows from (4) that if $n \leq N$, then $|a_n| \leq K$. If $n \geq N$, then from (3), $|a_n| < |L| + 1$; so in either case $|a_n| \leq M$, and the theorem is proved. ■

Sometimes it is difficult to find a bound for a convergent sequence.

EXAMPLE 5   Find an M such that $5^n/n! \leq M$.

Solution. We know from Theorem 13.2.3 that $\lim_{n \to \infty} x^n/n! = 0$ for every real number x. In particular, $\{5^n/n!\}$ is convergent and therefore must be bounded. Perhaps the easiest way to find the bound is to tabulate a few values, as in Table 1. It is clear from the table that the maximum value of a_n occurs at $n = 4$ or $n = 5$ and is equal to 26.04. Of course, any number larger than 26.04 is also a bound for the sequence. ■

TABLE 1

n	$\dfrac{5^n}{n!}$	n	$\dfrac{5^n}{n!}$
1	5	7	15.5
2	12.5	8	9.69
3	20.83	9	5.38
4	26.04	10	2.69
5	26.04	20	0.000039
6	21.7		

Theorem 1 is useful in another way. Since every convergent sequence is bounded, it follows that:

> **Every unbounded sequence is divergent.**

EXAMPLE 6   The following sequences are divergent since they are clearly unbounded:

(a) $\{\ln \ln n\}$ (starting at $n = 2$)   **(b)** $\{n \sin n\}$ **(c)** $\{(-\sqrt{2})^n\}$ ■

The converse of Theorem 1 is *not* true. That is, it is not true that every bounded sequence is convergent. For example, the sequences $\{(-1)^n\}$ and $\{\sin n\}$ are both bounded *and* divergent. Since boundedness alone does not ensure convergence, we need some other property. We investigate this idea now.

Definition 2   MONOTONICITY

(i) The sequence $\{a_n\}$ is **monotone increasing** if $a_n \leq a_{n+1}$ for every $n \geq 1$.
(ii) The sequence $\{a_n\}$ is **monotone decreasing** if $a_n \geq a_{n+1}$ for every $n \geq 1$.
(iii) The sequence $\{a_n\}$ is **monotonic** if it is either monotone increasing or monotone decreasing.

Definition 3   STRICT MONOTONICITY

(i) The sequence $\{a_n\}$ is **strictly increasing** if $a_n < a_{n+1}$ for every $n \geq 1$.

(ii) The sequence $\{a_n\}$ is **strictly decreasing** if $a_n > a_{n+1}$ for every $n \geq 1$.

(iii) The sequence $\{a_n\}$ is **strictly monotonic** if it is either strictly increasing or strictly decreasing.

EXAMPLE 7   The sequence $\{1/2^n\}$ is strictly decreasing since $1/2^n > 1/2^{n+1}$ for every n. ∎

EXAMPLE 8   Determine whether the sequence $\{2n/(3n + 2)\}$ is increasing, decreasing, or not monotonic.

Solution. If we write out the first few terms of the sequence,. we find that $\{2n/(3n + 2)\} = \{\frac{2}{5}, \frac{4}{8}, \frac{6}{11}, \frac{8}{14}, \frac{10}{17}, \frac{12}{20}, \ldots \}$. Since these terms are strictly increasing, we suspect that $\{2n/(3n + 2)\}$ is an increasing sequence. To check this, we try to verify that $a_n < a_{n+1}$. We have

$$a_{n+1} = \frac{2(n + 1)}{3(n + 1) + 2} = \frac{2n + 2}{3n + 5}.$$

Then $a_n < a_{n+1}$ implies that

$$\frac{2n}{3n + 2} < \frac{2n + 2}{3n + 5}.$$

Multiplying both sides of this inequality by $(3n + 2)(3n + 5)$, we obtain

$$(2n)(3n + 5) < (2n + 2)(3n + 2), \qquad \text{or} \qquad 6n^2 + 10n < 6n^2 + 10n + 4.$$

Since this last inequality is obviously true for all $n \geq 1$, we can reverse our steps to conclude that $a_n < a_{n+1}$, and the sequence is strictly increasing. ∎

EXAMPLE 9   Determine whether the sequence $\{(\ln n)/n\}$, $n > 1$, is increasing, decreasing, or not monotonic.

Solution. Let $f(x) = (\ln x)/x$. Then $f'(x) = [x(1/x) - (\ln x)1]/x^2 = (1 - \ln x)/x^2$. If $x > e$, then $\ln x > 1$ and $f'(x) < 0$. Thus the sequence $\{(\ln n)/n\}$ is decreasing for $n \geq 3$. However, $(\ln 1)/1 = 0 < (\ln 2)/2 \approx 0.35$, so initially, the sequence is increasing. Thus the sequence is not monotone. It is decreasing if we start with $n = 3$. ∎

EXAMPLE 10   The sequence $\{[n/4]\}$ is increasing but not strictly increasing. Here $[x]$ is the "greatest integer" function (see Example 2.2.11). The first twelve terms are 0, 0, 0, 1, 1, 1, 1, 2, 2, 2, 2, 3. For example, $a_9 = [\frac{9}{4}] = 2$. ∎

EXAMPLE 11   The sequence $\{(-1)^n\}$ is not monotonic since successive terms oscillate between $+1$ and -1. ∎

In all the examples we have given, a divergent sequence diverges for one of two reasons: It goes to infinity (it is unbounded) or it oscillates [like $(-1)^n$, which oscillates between -1 and 1]. But if a sequence is bounded, it does not go to infinity. And if it is monotone, then it does not oscillate. Thus the following theorem should not be surprising.

Theorem 2

A bounded monotonic sequence is convergent.

Proof. We will prove this theorem for the case in which the sequence $\{a_n\}$ is increasing. The proof of the other case is similar. Since $\{a_n\}$ is bounded, there is a number M such that $a_n \leq M$ for every n. Let L be the smallest such upper bound. Now let $\epsilon > 0$ be given. Then there is a number $N > 0$ such that $a_N > L - \epsilon$. If this were not true, then we would have $a_n \leq L - \epsilon$ for all $n \geq 1$. Then $L - \epsilon$ would be an upper bound for $\{a_n\}$, and since $L - \epsilon < L$, this would contradict the choice of L as the smallest such upper bound. Since $\{a_n\}$ is increasing, we have, for $n \geq N$,

$$L - \epsilon < a_N \leq a_n \leq L < L + \epsilon. \tag{6}$$

But the inequalities in (6) imply that $|a_n - L| \leq \epsilon$ for $n \geq N$, which proves, according to the definition of convergence, that $\lim_{n\to\infty} a_n = L$. ∎

The number L is called the **least upper bound** for the sequence $\{a_n\}$. It is an axiom of the real number system that every set of real numbers that is bounded above has a least upper bound and that every set of real numbers that is bounded below has a **greatest lower bound.** This axiom is called the **completeness axiom** and is of paramount importance in theoretical mathematical analysis. We discussed the completeness axiom earlier on page 6.

We have actually proved a stronger result. Namely, that *if the sequence $\{a_n\}$ is bounded above and increasing, then it converges to its least upper bound. Similarly, if $\{a_n\}$ is bounded below and decreasing, then it converges to its greatest lower bound.*

EXAMPLE 12 In Example 8 we saw that the sequence $\{2n/(3n+2)\}$ is strictly increasing. Also, since $2n/(3n+2) < 3n/(3n+2) < 3n/3n = 1$, we see that $\{a_n\}$ is also bounded, so that by Theorem 2, $\{a_n\}$ is convergent. We easily find that $\lim_{n\to\infty} 2n/(3n+2) = \frac{2}{3}$. ∎

PROBLEMS 14.2

In Problems 1–12, determine whether the given sequence is bounded or unbounded. If it is bounded, find the smallest bound for $|a_n|$.

1. $\left\{\dfrac{1}{n+1}\right\}$
2. $\{\sin n\pi\}$
3. $\{\cos n\pi\}$
4. $\{\sqrt{n}\sin n\}$
5. $\left\{\dfrac{2^n}{1+2^n}\right\}$
6. $\left\{\dfrac{2^n+1}{2^n}\right\}$
7. $\left\{\dfrac{1}{n!}\right\}$
8. $\left\{\dfrac{3^n}{n!}\right\}$
9. $\left\{\dfrac{n^2}{n!}\right\}$
10. $\left\{\dfrac{2n}{2^n}\right\}$
11. $\left\{\dfrac{\ln n}{n}\right\}$
*12. $\{ne^{-n}\}$

13. Show that for $n > 2^{10}$, $n^{10}/n! > (n+1)^{10}/(n+1)!$, and use this result to conclude that $\{n^{10}/n!\}$ is bounded.

In Problems 14–28, determine whether the given sequence is monotone increasing, strictly increasing, monotone decreasing, strictly decreasing, or not monotonic.

14. $\{\sin n\pi\}$

15. $\left\{\dfrac{3^n}{2 + 3^n}\right\}$

16. $\left\{\left(\dfrac{n}{25}\right)^{1/3}\right\}$

17. $\{n + (-1)^n\sqrt{n}\}$

18. $\left\{\dfrac{\sqrt{n + 1}}{n}\right\}$

19. $\left\{\dfrac{n!}{n^n}\right\}$

20. $\left\{\dfrac{n^n}{n!}\right\}$

21. $\left\{\dfrac{2n!}{1 \cdot 3 \cdot 5 \cdot 7 \cdots (2n - 1)}\right\}$

*22. $\{n + \cos n\}$

23. $\left\{\dfrac{2^{2n}}{n!}\right\}$

24. $\left\{\dfrac{\sqrt{n} - 1}{n}\right\}$

25. $\left\{\dfrac{n - 1}{n + 1}\right\}$

26. $\left\{\ln\left(\dfrac{3n}{n + 1}\right)\right\}$

27. $\{\ln n - \ln(n + 2)\}$

28. $\left\{\left(1 + \dfrac{3}{n}\right)^{1/n}\right\}$

*29. Show that the sequence $\{(2^n + 3^n)^{1/n}\}$ is convergent.

*30. Show that $\{(a^n + b^n)^{1/n}\}$ is convergent for any positive real numbers a and b. [*Hint:* First do Problem 29. Then treat the cases $a = b$ and $a \neq b$ separately.]

*31. Show that the sequence $\{n!/n^n\}$ is bounded. [*Hint:* Show that $n!/n^n > (n + 1)!/(n + 1)^n$ for sufficiently large n.]

32. Prove that the sequence $\{n!/n^n\}$ converges. [*Hint:* Use the result of Problem 31.]

33. Use Theorem 2 to show that $\{\ln n - \ln(n + 4)\}$ converges.

34. Show that the sequence of Problem 21 is convergent.

35. Show that the sequence $\{2 \cdot 5 \cdot 8 \cdot 11 \cdots (3n - 1)/3^n n!\}$ is convergent.

14.3 GEOMETRIC SERIES

Consider the sum

$$S_7 = 1 + 2 + 4 + 8 + 16 + 32 + 64 + 128.$$

This can be written as

$$S_7 = 1 + 2 + 2^2 + 2^3 + 2^4 + 2^5 + 2^6 + 2^7 = \sum_{k=0}^{7} 2^k.$$

GEOMETRIC PROGRESSION
In general, the sum of a **geometric progression** is a sum of the form

$$S_n = 1 + r + r^2 + r^3 + \cdots + r^{n-1} + r^n = \sum_{k=0}^{n} r^k, \tag{1}$$

where r is a real number and n is a fixed positive integer.

We now obtain a formula for the sum in (1).

Theorem 1 If $r \neq 1$, the sum of a geometric progression (1) is given by

$$S_n = \frac{1 - r^{n+1}}{1 - r}.$$

(2)

Proof. We write

$$S_n = 1 + r + r^2 + r^3 + \cdots + r^{n-1} + r^n$$

(3)

and then multiply both sides of (3) by r:

$$rS_n = r + r^2 + r^3 + r^4 + \cdots + r^n + r^{n+1}.$$

(4)

We now subtract (4) from (3) and note that all terms except the first and the last cancel:

$$S_n - rS_n = 1 - r^{n+1},$$

or

$$(1 - r)S_n = 1 - r^{n+1}.$$

(5)

Finally, we divide both sides of (5) by $1 - r$ (which is nonzero) to obtain equation (2). ∎

NOTE. If $r = 1$, we obtain

$$S_n = \overbrace{1 + 1 + \cdots + 1}^{n + 1 \text{ terms}} = n + 1.$$

EXAMPLE 1 Calculate $S_7 = 1 + 2 + 4 + 8 + 16 + 32 + 64 + 128$, using formula (2).

Solution. Here $r = 2$ and $n = 7$, so that

$$S_7 = \frac{1 - 2^8}{1 - 2} = 2^8 - 1 = 256 - 1 = 255. \quad \blacksquare$$

EXAMPLE 2 Calculate $\sum_{k=0}^{10} \left(\frac{1}{2}\right)^k$.

Solution. Here $r = \frac{1}{2}$ and $n = 10$, so that

$$S_{10} = \frac{1 - \left(\frac{1}{2}\right)^{11}}{1 - \frac{1}{2}} = \frac{1 - \frac{1}{2048}}{\frac{1}{2}} = 2\left(\frac{2047}{2048}\right) = \frac{2047}{1024}. \quad \blacksquare$$

EXAMPLE 3 Calculate

$$S_6 = 1 - \frac{2}{3} + \left(\frac{2}{3}\right)^2 - \left(\frac{2}{3}\right)^3 + \left(\frac{2}{3}\right)^4 - \left(\frac{2}{3}\right)^5 + \left(\frac{2}{3}\right)^6 = \sum_{k=0}^{6} \left(-\frac{2}{3}\right)^k.$$

Solution. Here $r = -\frac{2}{3}$ and $n = 6$, so that

$$S_6 = \frac{1 - (-\frac{2}{3})^7}{1 - (-\frac{2}{3})} = \frac{1 + \frac{128}{2187}}{\frac{5}{3}} = \frac{3}{5}\left(\frac{2315}{2187}\right) = \frac{463}{729}. \blacksquare$$

EXAMPLE 4   Calculate the sum $1 + b^2 + b^4 + b^6 + \cdots + b^{20} = \Sigma_{k=0}^{10}\, b^{2k}$ for $b \neq \pm 1$.

Solution. Note that the sum can be written $1 + b^2 + (b^2)^2 + (b^2)^3 + \cdots + (b^2)^{10}$. Here $r = b^2 \neq 1$ and $n = 10$, so that

$$S_{10} = \frac{1 - (b^2)^{11}}{1 - b^2} = \frac{b^{22} - 1}{b^2 - 1}. \blacksquare$$

The sum of a geometric progression is the sum of a finite number of terms. We now see what happens if the number of terms is infinite. Consider the sum

$$S = 1 + \frac{1}{2} + \frac{1}{4} + \frac{1}{8} + \frac{1}{16} + \cdots = \sum_{k=0}^{\infty} \left(\frac{1}{2}\right)^k. \tag{6}$$

What can such a sum mean? We will give a formal definition in a moment. For now, let us show why it is reasonable to say that $S = 2$. Let $S_n = \Sigma_{k=0}^{n} \left(\frac{1}{2}\right)^k = 1 + \frac{1}{2} + \frac{1}{4} + \cdots + \left(\frac{1}{2}\right)^n$. Then

$$S_n = \frac{1 - (\frac{1}{2})^{n+1}}{1 - \frac{1}{2}} = 2\left[1 - \left(\frac{1}{2}\right)^{n+1}\right].$$

Thus for any n (no matter how large), $1 \leq S_n < 2$. Hence the numbers S_n are bounded. Also, since $S_{n+1} = S_n + (\frac{1}{2})^{n+1} > S_n$, the numbers S_n are monotone increasing. Thus the sequence $\{S_n\}$ converges. But

$$S = \lim_{n\to\infty} S_n.$$

Thus S has a finite sum. To compute it, we note that

$$S = \lim_{n\to\infty} S_n = \lim_{n\to\infty} 2[1 - (\tfrac{1}{2})^{n+1}] = 2 \lim_{n\to\infty} [1 - (\tfrac{1}{2})^{n+1}] = 2$$

since $\lim_{n\to\infty} (\tfrac{1}{2})^{n+1} = 0$.

GEOMETRIC SERIES
The infinite sum $\Sigma_{k=0}^{\infty} (\tfrac{1}{2})^k$ is called a *geometric series*. In general, a **geometric series** is an infinite sum of the form

$$S = \sum_{k=0}^{\infty} r^k = 1 + r + r^2 + r^3 + \cdots. \tag{7}$$

CONVERGENCE AND DIVERGENCE OF A GEOMETRIC SERIES
Let $S_n = \Sigma_{k=0}^{n} r^k$. Then we say that the geometric series **converges** if $\lim_{n\to\infty} S_n$ exists and is finite. Otherwise, the series is said to **diverge**.

EXAMPLE 5   Let $r = 1$. Then

$$S_n = \sum_{k=0}^{n} 1^k = \sum_{k=0}^{n} 1 = \underbrace{1 + 1 + \cdots + 1}_{n+1 \text{ times}} = n + 1.$$

Since $\lim_{n \to \infty}(n + 1) = \infty$, the series $\sum_{k=0}^{\infty} 1^k$ diverges. ■

EXAMPLE 6   Let $r = -2$. Then

$$S_n = \sum_{k=0}^{n} (-2)^k = \frac{1 - (-2)^{n+1}}{1 - (-2)} = \frac{1}{3}[1 - (-2)^{n+1}].$$

But $(-2)^{n+1} = (-1)^{n+1}(2^{n+1}) = \pm 2^{n+1}$. As $n \to \infty$, $2^{n+1} \to \infty$. Thus the series $\sum_{k=0}^{\infty} (-2)^{k+1}$ diverges. ■

Theorem 2

> Let $S = \sum_{k=0}^{\infty} r^k$ be a geometric series.
>
> **(i)** The series converges to
>
> $$\frac{1}{1 - r} \qquad \text{if} \qquad |r| < 1.$$
>
> **(ii)** The series diverges if $|r| \geq 1$.

Proof.   **(i)** If $|r| < 1$, then $\lim_{n \to \infty} r^{n+1} = 0$. Thus

$$S = \lim_{n \to \infty} S_n = \lim_{n \to \infty} \frac{1 - r^{n+1}}{1 - r} = \frac{1}{1 - r} \lim_{n \to \infty} (1 - r^{n+1})$$

$$= \frac{1}{1 - r} (1 - 0) = \frac{1}{1 - r}.$$

(ii) If $|r| > 1$, then $\lim_{n \to \infty} |r|^{n+1} = \infty$. Thus $1 - r^{n+1}$ does not have a finite limit and the series diverges. Finally, if $r = 1$, then the series diverges, by Example 5, and if $r = -1$, then S_n alternates between the numbers 0 and 1, so that the series diverges. ■

EXAMPLE 7 $1 - \frac{2}{3} + (\frac{2}{3})^2 - \cdots = \sum_{k=0}^{\infty} (-\frac{2}{3})^k = 1/[1 - (-\frac{2}{3})] = 1/(\frac{5}{3}) = \frac{3}{5}.$ ■

EXAMPLE 8

$$1 + \frac{\pi}{4} + \left(\frac{\pi}{4}\right)^2 + \left(\frac{\pi}{4}\right)^3 + \cdots = \sum_{k=0}^{\infty} \left(\frac{\pi}{4}\right)^k = \frac{1}{1 - (\pi/4)}$$

$$= \frac{4}{4 - \pi} \approx 4.66. ■$$

In the rest of this chapter we will discuss a great number of infinite sums. We conclude this section by showing how the geometric series can be used to resolve an ancient paradox.

Some of the early work on limits was motivated by unresolved questions that had been posed by Greek mathematicians. For example, the fifth century B.C. philosopher and mathematician Zeno (ca. 495–435 B.C.) posed four problems that came to be known as **Zeno's paradoxes.** In the second of these Zeno argued that the legendary Greek hero Achilles could never overtake a tortoise. Suppose that the tortoise starts 100 yd ahead and that Achilles can run ten times as fast as the tortoise. Then when Achilles has run 100 yd, the tortoise has run 10 yd, and when Achilles has covered this distance, the tortoise is still a yard ahead; and so on. It seems that the tortoise will stay ahead!

We can view this seeming paradox in another way, which is equally contradictory of common sense. Suppose that a man is standing a certain distance, say 10 ft, from a door (see Figure 1). Using Zeno's reasoning, we may claim that it is impossible for the man to walk to the door. In order to reach the door, the man must walk half the distance (5 ft) to the door. He then reaches point ① on Figure 1. From point ①, 5 ft from the door, he must again walk halfway ($2\frac{1}{2}$ ft) to the door, to point ②. Continuing in this manner, no matter how close he comes to the door, he must walk halfway to the door and halfway from there and halfway from there, . . . , and so on. Thus no matter how close the man gets to the door, he still has half of some remaining distance to cover. It seems that the man will never actually reach the door. Of course, this contradicts our common sense. But where is the flaw in Zeno's reasoning?

FIGURE 1

It took more than two thousand years for mathematicians to provide a satisfactory answer to this question, and in order to do so, they had to use the notion of a limit. Intuitively, we sense that Zeno's man is indeed covering an infinite number of intervals in his walk toward the door, but each interval is over a shorter and shorter distance and, therefore, takes less and less time. Indeed, the time necessary to walk over each succeeding interval "approaches" the limit zero, thus allowing the man to reach the door. Let us prove that the man can indeed reach the door in finite time.

Suppose the man in Figure 1 starts walking toward the door at the fixed velocity of 5 ft/sec. Let us calculate the time it takes him to walk to the door, using Zeno's argument. Since (velocity) × (time) = distance, we have $t = $ distance/velocity, where t stands for time. Thus it takes the man $t = (5 \text{ ft})/(5 \text{ ft/sec}) = 1$ sec to walk to the point 5 ft from the door (recall that he starts 10 ft from the door). To walk to the next point, $2\frac{1}{2}$ ft from the door, takes $(2\frac{1}{2} \text{ ft})/(5 \text{ ft/sec}) = \frac{1}{2}$ sec. The next point takes $(1\frac{1}{4} \text{ ft})/(5 \text{ ft/sec}) = \frac{1}{4}$ sec to reach. It is clear that to reach succeeding points, each half the distance to the door from the preceding point, the man will take $\frac{1}{8}$ sec, $\frac{1}{16}$ sec, . . . , $(\frac{1}{2})^n$

sec, ... Thus the total time he takes to walk to the door is

$$t = 1 + \tfrac{1}{2} + \tfrac{1}{4} + \tfrac{1}{8} + \tfrac{1}{16} + \cdots = 2 \text{ sec,}$$

since this is nothing but the sum of a geometric series with $r = \tfrac{1}{2}$. Hence the man will reach the door in 2 sec, as is certainly not surprising. Therefore we see that with the concept of an infinite sum, Zeno's paradox is really no paradox at all.

In Problem 27 we ask you to "explain" the seeming paradox in the original version of Zeno's paradox: the race between Achilles and the tortoise.

PROBLEMS 14.3

In Problems 1–11, calculate the sum of the given geometric progression.

1. $1 + 3 + 9 + 27 + 81 + 243$

2. $1 + \dfrac{1}{4} + \dfrac{1}{16} + \cdots + \dfrac{1}{4^8}$

3. $1 - 5 + 25 - 125 + 625 - 3125$

4. $0.2 + 0.2^2 + 0.2^3 + \cdots + 0.2^9$

5. $0.3^2 - 0.3^3 + 0.3^4 - 0.3^5 + 0.3^6 - 0.3^7 + 0.3^8$

6. $1 + b^3 + b^6 + b^9 + b^{12} + b^{15} + b^{18} + b^{21}$

7. $1 - \dfrac{1}{b^2} + \dfrac{1}{b^4} - \dfrac{1}{b^6} + \dfrac{1}{b^8} - \dfrac{1}{b^{10}} + \dfrac{1}{b^{12}} - \dfrac{1}{b^{14}}$

8. $\pi - \pi^3 + \pi^5 - \pi^7 + \pi^9 - \pi^{11} + \pi^{13}$

9. $1 + \sqrt{2} + 2 + 2^{3/2} + 4 + 2^{5/2} + 8 + 2^{7/2} + 16$

10. $1 - \dfrac{1}{\sqrt{3}} + \dfrac{1}{3} - \dfrac{1}{3\sqrt{3}} + \dfrac{1}{9} - \dfrac{1}{9\sqrt{3}} + \dfrac{1}{27} - \dfrac{1}{27\sqrt{3}} + \dfrac{1}{81}$

11. $-16 + 64 - 256 + 1024 - 4096$

12. A bacteria population initially contains 1000 organisms and each bacterium produces two live bacteria every 2 hr. How many organisms will be alive after 12 hr if none of the bacteria die during the growth period?

In Problems 13–22, calculate the sum of the given geometric series.

13. $1 + \dfrac{1}{4} + \dfrac{1}{4^2} + \dfrac{1}{4^3} + \cdots$

14. $1 - \tfrac{1}{2} + \tfrac{1}{4} - \tfrac{1}{8} + \tfrac{1}{16} - \cdots$

15. $1 + \tfrac{1}{10} + \tfrac{1}{100} + \tfrac{1}{1000} + \cdots$

16. $1 - \tfrac{1}{10} + \tfrac{1}{100} - \tfrac{1}{1000} + \cdots$

17. $1 + \dfrac{1}{\pi} + \dfrac{1}{\pi^2} + \dfrac{1}{\pi^3} + \cdots$

18. $1 + 0.7 + 0.7^2 + 0.7^3 + \cdots$

19. $1 - 0.62 + 0.62^2 - 0.62^3 + 0.62^4 - \cdots$

20. $\tfrac{1}{4} + \tfrac{1}{16} + \tfrac{1}{64} + \cdots$ [*Hint:* Factor out the term $\tfrac{1}{4}$.]

21. $\tfrac{3}{5} - \tfrac{3}{25} + \tfrac{3}{125} - \cdots$

22. $\tfrac{1}{9} + \tfrac{1}{27} + \tfrac{1}{81} + \cdots$

23. How large must n be in order that $\left(\tfrac{1}{2}\right)^n < 0.01$?

24. How large must n be in order that $(0.8)^n < 0.01$?

25. How large must n be in order that $(0.99)^n < 0.01$?

26. If $x > 1$, show that

$$1 + \frac{1}{x} + \frac{1}{x^2} + \frac{1}{x^3} + \cdots = \frac{x}{x - 1}.$$

*****27.** Suppose that in the original version of Zeno's paradox, the tortoise is moving at a rate of 1 km/hr while Achilles is running at a rate of 201 km/hr. Give the tortoise a 40-km head start.
(a) Show, using the arguments of this section, that Achilles will really overtake the tortoise.
(b) How long will it take?

14.4 INFINITE SERIES

In Section 14.3 we defined the geometric series $\sum_{k=0}^{\infty} r^k$ and showed that if $|r| < 1$, the series converges to $1/(1 - r)$. Let us again look at what we did. If S_n denotes the sum of the first $n + 1$ terms of the geometric series, then

$$S_n = 1 + r + r^2 + \cdots + r^n = \frac{1 - r^{n+1}}{1 - r}, \qquad r \neq 1. \tag{1}$$

For each n we obtain the number S_n, and therefore we can define a new sequence $\{S_n\}$ to be the sequence of **partial sums** of the geometric series. If $|r| < 1$, then

$$\lim_{n \to \infty} S_n = \lim_{n \to \infty} \frac{1 - r^{n+1}}{1 - r} = \frac{1}{1 - r}.$$

That is, the convergence of the geometric series is implied by the convergence of the sequence of partial sums $\{S_n\}$.

We now give a more general definition of these concepts.

Definition 1   INFINITE SERIES

Let $\{a_n\}$ be a sequence. Then the infinite sum

$$\sum_{k=1}^{\infty} a_k = a_1 + a_2 + a_3 + \cdots + a_n + \cdots \tag{2}$$

is called an **infinite series** (or, simply, **series**). Each a_k in (2) is called a **term** of the series. The **partial sums** of the series are given by

$$S_n = \sum_{k=1}^{n} a_k.$$

The term S_n is called the **nth partial sum** of the series. If the sequence of partial sums $\{S_n\}$ converges to L, then we say that the infinite series $\sum_{k=1}^{\infty} a_k$ **converges** to L, and we write

$$\sum_{k=1}^{\infty} a_k = L. \tag{3}$$

Otherwise, we say that the series $\sum_{k=1}^{\infty} a_k$ **diverges**.

REMARK. Occasionally a series will be written with the first term other than a_1. For example, $\sum_{k=0}^{\infty} (\frac{1}{2})^k$ and $\sum_{k=2}^{\infty} 1/(\ln k)$ are both examples of infinite series. In the second case we must start with $k = 2$ since $1/(\ln 1)$ is not defined.

EXAMPLE 1   We can write the number $\frac{1}{3}$ as

$$\frac{1}{3} = 0.33333\ldots = \frac{3}{10} + \frac{3}{100} + \frac{3}{1000} + \cdots + \frac{3}{10^n} + \cdots. \tag{4}$$

This expression is an infinite series. Here $a_n = 3/10^n$ and

$$S_n = \frac{3}{10} + \frac{3}{100} + \cdots + \frac{3}{10^n} = \overbrace{0.333\ldots3}^{n \text{ places}}.$$

We can formally prove that this sum converges by noting that

$$S = \frac{3}{10}\left(1 + \frac{1}{10} + \frac{1}{100} + \cdots\right) = \frac{3}{10}\sum_{k=0}^{\infty}\left(\frac{1}{10}\right)^k$$

$$\underset{\text{By Theorem 14.3.2}}{\swarrow}$$

$$= \frac{3}{10}\left[\frac{1}{1 - (\frac{1}{10})}\right] = \frac{3}{10}\left(\frac{1}{\frac{9}{10}}\right) = \frac{3}{10}\cdot\frac{10}{9} = \frac{3}{9} = \frac{1}{3}. \quad\blacksquare$$

As a matter of fact, any decimal number x can be thought of as a convergent infinite series, for if $x = 0.\,a_1a_2a_3\ldots a_n\ldots$, then

$$x = \frac{a_1}{10} + \frac{a_2}{100} + \frac{a_3}{1000} + \cdots + \frac{a_n}{10^n} + \cdots = \sum_{k=1}^{\infty}\frac{a_k}{10^k}.\dagger$$

EXAMPLE 2   Express the **repeating decimal** $0.123123123\ldots$ as a rational number (the quotient of two integers).

Solution.

$$0.123123123\ldots = 0.123 + 0.000123 + 0.000000123 + \cdots$$

$$= \frac{123}{10^3} + \frac{123}{10^6} + \frac{123}{10^9} + \cdots = \frac{123}{10^3}\left[1 + \frac{1}{10^3} + \frac{1}{(10^3)^2} + \cdots\right]$$

$$= \frac{123}{1000}\sum_{k=0}^{\infty}\left(\frac{1}{1000}\right)^k = \frac{123}{1000}\left[\frac{1}{1 - (1/1000)}\right] = \frac{123}{1000}\cdot\frac{1}{999/1000}$$

$\dagger$Since $0 \le a_k < 10$,

$$\sum_{k=1}^{\infty}\frac{a_k}{10^k} < \sum_{k=1}^{\infty}\frac{10}{10^k} = \sum_{k=1}^{\infty}\frac{1}{10^{k-1}} = 1 + \frac{1}{10} + \left(\frac{1}{10}\right)^2 + \cdots = \frac{1}{1 - \frac{1}{10}} = \frac{10}{9}.$$

In Section 14.5 we will prove the comparison test. Once we have this test, the inequality given above implies that $\sum_{k=1}^{\infty}(a_k/10^k)$ converges.

$$= \frac{123}{1000} \cdot \frac{1000}{999} = \frac{123}{999} = \frac{41}{333}. \quad \blacksquare$$

In general, we can use the geometric series to write any repeating decimal in the form of a fraction by using the technique of Example 1 or 2. In fact, *the rational numbers are exactly those real numbers that can be written as repeating decimals*. Repeating decimals include numbers like $3 = 3.00000\ldots$ and $\frac{1}{4} = 0.25 = 0.25000000\ldots$.

EXAMPLE 3   **Telescoping Series**   Consider the infinite series $\sum_{k=1}^{\infty} 1/k(k+1)$. We write the first three partial sums:

$$S_1 = \sum_{k=1}^{1} \frac{1}{k(k+1)} = \frac{1}{1 \cdot 2} = \frac{1}{2} = 1 - \frac{1}{2},$$

$$S_2 = \sum_{k=1}^{2} \frac{1}{k(k+1)} = \frac{1}{1 \cdot 2} + \frac{1}{2 \cdot 3} = \frac{1}{2} + \frac{1}{6} = \frac{2}{3} = 1 - \frac{1}{3},$$

$$S_3 = \sum_{k=1}^{3} \frac{1}{k(k+1)} = \frac{1}{1 \cdot 2} + \frac{1}{2 \cdot 3} + \frac{1}{3 \cdot 4} = \frac{1}{2} + \frac{1}{6} + \frac{1}{12} = \frac{3}{4} = 1 - \frac{1}{4}.$$

We can use partial fractions to rewrite the general term as

$$a_k = \frac{1}{k(k+1)} = \frac{1}{k} - \frac{1}{k+1},$$

from which we can get a better view of the nth partial sum:

$$S_n = \left(\frac{1}{1} - \frac{1}{2}\right) + \left(\frac{1}{2} - \frac{1}{3}\right) + \left(\frac{1}{3} - \frac{1}{4}\right) + \cdots + \left(\frac{1}{n-1} - \frac{1}{n}\right) + \left(\frac{1}{n} - \frac{1}{n+1}\right)$$

$$= 1 - \frac{1}{n+1},$$

because all other terms cancel. Since $\lim_{n\to\infty} S_n = \lim_{n\to\infty}\{1 - [1/(n-1)]\} = 1$, we see that

$$\sum_{k=1}^{\infty} \frac{1}{k(k+1)} = 1.$$

When, as here, alternate terms cancel, we say that the series is a **telescoping series**. $\blacksquare$

REMARK.   *Often, it is not possible to calculate the exact sum of an infinite series, even if it can be shown that the series converges.*

EXAMPLE 4   Consider the series

$$\sum_{k=1}^{\infty} \frac{1}{k} = 1 + \frac{1}{2} + \frac{1}{3} + \frac{1}{4} + \cdots + \frac{1}{n} + \cdots. \tag{5}$$

This series is called the **harmonic series.** Although $a_n = 1/n \to 0$ as $n \to \infty$, it is not difficult to show that the harmonic series diverges. To see this, we write

2 terms 4 terms 8 terms

$$\sum_{k=1}^{\infty} \frac{1}{k} = 1 + \frac{1}{2} + \underbrace{\left(\frac{1}{3} + \frac{1}{4}\right)}_{> \frac{1}{2}} + \underbrace{\left(\frac{1}{5} + \frac{1}{6} + \frac{1}{7} + \frac{1}{8}\right)}_{> \frac{1}{2}} + \underbrace{\left(\frac{1}{9} + \cdots + \frac{1}{16}\right)}_{> \frac{1}{2}} + \cdots.$$

Here we have written the terms in groups containing 2^n numbers. Note that $\frac{1}{3} + \frac{1}{4} > \frac{2}{4} = \frac{1}{2}, \frac{1}{5} + \frac{1}{6} + \frac{1}{7} + \frac{1}{8} > \frac{1}{8} + \frac{1}{8} + \frac{1}{8} + \frac{1}{8} = \frac{1}{2}$, and so on. Thus $\sum_{k=1}^{\infty} 1/k > 1 + \frac{1}{2} + \frac{1}{2} + \cdots$, and the series diverges. ▨

WARNING.   Example 4 clearly shows that even though the sequence $\{a_n\}$ converges to 0, the series Σa_n may, in fact, diverge. That is, if $a_n \to 0$, then $\sum_{k=1}^{\infty} a_k$ may or may not converge. Some additional test is needed to determine convergence or divergence.

It is often difficult to determine whether a series converges or diverges. For that reason a number of techniques have been developed to make it easier to do so. We will present some easy facts here, and then we will develop additional techniques in the three sections that follow.

Theorem 1   Let c be a constant. Suppose that $\sum_{k=1}^{\infty} a_k$ and $\sum_{k=1}^{\infty} b_k$ both converge. Then $\sum_{k=1}^{\infty} (a_k + b_k)$ and $\sum_{k=1}^{\infty} ca_k$ converge, and

(i) $\displaystyle\sum_{k=1}^{\infty} (a_k + b_k) = \sum_{k=1}^{\infty} a_k + \sum_{k=1}^{\infty} b_k$,   **(6)**

(ii) $\displaystyle\sum_{k=1}^{\infty} ca_k = c \sum_{k=1}^{\infty} a_k$.   **(7)**

This theorem should not be surprising. Since the sum in a series is the limit of a sequence (the sequence of partial sums), the first part, for example, simply restates the fact that the limit of the sum is the sum of the limits. This is Theorem 14.1.2(ii).

Proof.

(i) Let $S = \sum_{k=1}^{\infty} a_k$ and $T = \sum_{k=1}^{\infty} b_k$. The partial sums are given by $S_n = \sum_{k=1}^{n} a_k$ and $T_n = \sum_{k=1}^{n} b_k$. Then

$$\sum_{k=1}^{\infty} (a_k + b_k) = \lim_{n \to \infty} \sum_{k=1}^{n} (a_k + b_k) = \lim_{n \to \infty} \left(\sum_{k=1}^{n} a_k + \sum_{k=1}^{n} b_k\right) = \lim_{n \to \infty} (S_n + T_n)$$

$$= \lim_{n \to \infty} S_n + \lim_{n \to \infty} T_n = S + T = \sum_{k=1}^{\infty} a_k + \sum_{k=1}^{\infty} b_k.$$

(ii) $\displaystyle\sum_{k=1}^{\infty} ca_k = \lim_{n \to \infty} \sum_{k=1}^{n} ca_k = \lim_{n \to \infty} c \sum_{k=1}^{n} a_k = \lim_{n \to \infty} cS_n$

$$= c \lim_{n \to \infty} S_n = cS = c \sum_{k=1}^{\infty} a_k. \quad ▨$$

EXAMPLE 5   Show that $\sum_{k=1}^{\infty} \{[1/k(k+1)] + (\frac{5}{6})^k\}$ converges.

Solution. This follows since $\sum_{k=1}^{\infty} 1/k(k+1)$ converges (Example 3) and $\sum_{k=1}^{\infty} (\frac{5}{6})^k$ converges because $\sum_{k=1}^{\infty} (\frac{5}{6})^k = \sum_{k=0}^{\infty} (\frac{5}{6})^k - (\frac{5}{6})^0$ [we added and subtracted the term $(\frac{5}{6})^0 = 1$] $= 1/(1 - \frac{5}{6}) - 1 = 5.$ ■

EXAMPLE 6   Does $\sum_{k=1}^{\infty} 1/50k$ converge or diverge?

Solution. We show that the series diverges by assuming that it converges to obtain a contradiction. If $\sum_{k=1}^{\infty} 1/50k$ did converge, then $50 \sum_{k=1}^{\infty} 1/50k$ would also converge by Theorem 1. But then $50 \sum_{k=1}^{\infty} 1/50k = \sum_{k=1}^{\infty} 50 \cdot 1/50k = \sum_{k=1}^{\infty} 1/k$, and this series is the harmonic series, which we know diverges. Hence $\sum_{k=1}^{\infty} 1/50k$ diverges. ■

Another useful test is given by the following theorem and corollary.

Theorem 2   If $\sum_{k=1}^{\infty} a_k$ converges, then $\lim_{n \to \infty} a_n = 0$.

Proof. Let $S = \sum_{k=1}^{\infty} a_k$. Then the partial sums S_n and S_{n-1} are given by

$$S_n = \sum_{k=1}^{n} a_k = a_1 + a_2 + \cdots + a_{n-1} + a_n$$

and

$$S_{n-1} = \sum_{k=1}^{n-1} a_k = a_1 + a_2 + \cdots + a_{n-1},$$

so that

$$S_n - S_{n-1} = a_n.$$

Then

$$\lim_{n \to \infty} a_n = \lim_{n \to \infty} (S_n - S_{n-1}) = \lim_{n \to \infty} S_n - \lim_{n \to \infty} S_{n-1} = S - S = 0. \;■$$

We have already seen that the converse of this theorem is false. The convergence of $\{a_n\}$ to 0 does *not* imply that $\sum_{k=1}^{\infty} a_k$ converges. For example, the harmonic series does not converge, but the sequence $\{1/n\}$ does converge to zero.

Corollary

If $\{a_n\}$ does not converge to 0, then $\sum_{k=1}^{\infty} a_k$ diverges.

EXAMPLE 7 $\sum_{k=1}^{\infty} (-1)^k$ diverges since the sequence $\{(-1)^k\}$ does not converge to zero. ■

EXAMPLE 8 $\sum_{k=1}^{\infty} k/(k+100)$ diverges since $\lim_{n \to \infty} a_n = \lim_{n \to \infty} n/(n+100) = 1 \neq 0$. ■

PROBLEMS 14.4

In Problems 1–15, a convergent infinite series is given. Find its sum.

1. $\displaystyle\sum_{k=0}^{\infty} \frac{1}{4^k}$

2. $\displaystyle\sum_{k=0}^{\infty} \left(-\frac{2}{3}\right)^k$

3. $\displaystyle\sum_{k=2}^{\infty} \frac{1}{2^k}$

4. $\displaystyle\sum_{k=1}^{\infty} \frac{1}{2^{k-1}}$

5. $\displaystyle\sum_{k=-3}^{\infty} \frac{1}{2^{k+3}}$

6. $\displaystyle\sum_{k=3}^{\infty} \left(\frac{2}{3}\right)^k$

7. $\displaystyle\sum_{k=0}^{\infty} \frac{100}{5^k}$

8. $\displaystyle\sum_{k=0}^{\infty} \frac{5}{100^k}$

9. $\displaystyle\sum_{k=2}^{\infty} \frac{1}{k(k+1)}$

10. $\displaystyle\sum_{k=3}^{\infty} \frac{1}{k(k-1)}$

11. $\displaystyle\sum_{k=0}^{\infty} \frac{1}{(k+1)(k+2)}$

12. $\displaystyle\sum_{k=-1}^{\infty} \frac{1}{(k+3)(k+4)}$

13. $\displaystyle\sum_{k=2}^{\infty} \frac{2^{k+3}}{3^k}$

14. $\displaystyle\sum_{k=2}^{\infty} \frac{2^{k+4}}{3^{k-1}}$

15. $\displaystyle\sum_{k=4}^{\infty} \frac{5^{k-2}}{6^{k+1}}$

In Problems 16–24, write the repeating decimals as rational numbers.

16. $0.666\ldots$

17. $0.353535\ldots$

18. $0.282828\ldots$

19. $0.717171\ldots$

20. $0.214214214\ldots$

21. $0.501501501\ldots$

22. $0.124242424\ldots$

23. $0.11362362362\ldots$

24. $0.513651365136\ldots$

25. Give a new proof, using the corollary to Theorem 2, that the geometric series diverges if $|r| \geq 1$.

In Problems 26–30, use Theorem 1 to calculate the sum of the convergent series.

26. $\displaystyle\sum_{k=0}^{\infty} \left[\frac{1}{2^k} + \frac{1}{5^k}\right]$

27. $\displaystyle\sum_{k=1}^{\infty} \left[\frac{1}{k(k+1)} + \frac{1}{(k+1)(k+2)}\right]$

28. $\displaystyle\sum_{k=0}^{\infty} \left[\frac{3}{5^k} - \frac{7}{4^k}\right]$

29. $\displaystyle\sum_{k=1}^{\infty} \left[\frac{8}{5^k} - \frac{7}{(k+3)(k+4)}\right]$

30. $\displaystyle\sum_{k=3}^{\infty} \left[\frac{12 \cdot 2^{k+1}}{3^{k-2}} - \frac{15 \cdot 3^{k+1}}{4^{k+2}}\right]$

31. Show that for any nonzero real numbers a and b, $\sum_{k=1}^{\infty} a/bk$ diverges.

32. Show that if the sequences $\{a_k\}$ and $\{b_k\}$ differ only for a finite number of terms, then $\sum_{k=1}^{\infty} a_k$ and $\sum_{k=1}^{\infty} b_k$ either both converge or both diverge.

33. Use the result of Problem 32 to show that $\sum_{k=1}^{\infty} 1/(k+6)$ diverges.

*34. Show that if $\sum_{k=1}^{\infty} a_k$ converges and $\sum_{k=1}^{\infty} b_k$ diverges, then $\sum_{k=1}^{\infty} (a_k + b_k)$ diverges. [*Hint:* Assume that $\sum_{k=1}^{\infty} (a_k + b_k)$ converges and then show that this leads to a contradiction of Theorem 1.]

35. Use the result of Problem 34 to show that $\sum_{k=1}^{\infty} (3/2^k + 2 \cdot 5^k)$ diverges.

36. Give an example in which $\sum_{k=1}^{\infty} a_k$ and $\sum_{k=1}^{\infty} b_k$ both diverge but $\sum_{k=1}^{\infty} (a_k + b_k)$ converges.

37. Use the geometric series to show that

$$\frac{1}{1+x} = \sum_{k=0}^{\infty} (-1)^k x^k$$

for any real number x with $|x| < 1$.

38. Show that $1/(1 + x^2) = \sum_{k=0}^{\infty} (-1)^k x^{2k}$ if $|x| < 1$.

*39. At what time between 1 P.M. and 2 P.M. is the minute hand of a clock exactly over the hour hand? [*Hint:* The minute hand moves 12 times as fast as the hour hand. Start at 1:00 P.M. When the minute hand has reached 1, the hour hand points to $1 + \frac{1}{12}$; when the minute hand has reached $1 + \frac{1}{12}$, the hour hand has reached $1 + \frac{1}{12} + \frac{1}{12} \cdot \frac{1}{12}$; etc. Now add up the geometric series.]

*40. At what time between 7 A.M. and 8 A.M. is the minute hand exactly over the hour hand?

41. A ball is dropped from a height of 8 m. Each time it hits the ground, it rebounds to a height of two-thirds the height from which it fell. Find the total distance traveled by the ball until it comes to rest (i.e., until it stops bouncing).

*42. Pick a_0 and a_1. For $n \geq 2$, compute a_n recursively so that $n(n - 1)a_n = (n - 1)(n - 2)a_{n-1} - (n - 3)a_{n-2}$. Evaluate $\Sigma_{n=0}^{\infty} a_n$.

14.5 SERIES WITH NONNEGATIVE TERMS I: TWO COMPARISON TESTS AND THE INTEGRAL TEST

In this section and the next one we consider series of the form $\Sigma_{k=1}^{\infty} a_k$, where each a_k is nonnegative. Such series are often easier to handle than others. One fact is easy to prove. The sequence $\{S_n\}$ of partial sums is a monotone increasing sequence since $S_{n+1} = S_n + a_{n+1}$ and $a_{n+1} \geq 0$ for every n. Then if $\{S_n\}$ is bounded, it is convergent by Theorem 14.2.2, and we have the following theorem:

Theorem 1   An infinite series of nonnegative terms is convergent if and only if its sequence of partial sums is bounded.

EXAMPLE 1   Show that $\Sigma_{k=1}^{\infty} 1/k^2$ is convergent.

Solution. We group the terms as follows:

$$\sum_{k=1}^{\infty} \frac{1}{k^2} = \frac{1}{1^2} + \overbrace{\frac{1}{2^2} + \frac{1}{3^2}}^{2 \text{ terms}} + \overbrace{\frac{1}{4^2} + \frac{1}{5^2} + \frac{1}{6^2} + \frac{1}{7^2}}^{4 \text{ terms}} + \overbrace{\frac{1}{8^2} + \cdots + \frac{1}{15^2}}^{8 \text{ terms}} + \cdots$$

$$\leq 1 + \overbrace{\frac{1}{2^2} + \frac{1}{2^2}}^{2 \text{ terms}} + \overbrace{\frac{1}{4^2} + \frac{1}{4^2} + \frac{1}{4^2} + \frac{1}{4^2}}^{4 \text{ terms}} + \overbrace{\frac{1}{8^2} + \cdots + \frac{1}{8^2}}^{8 \text{ terms}} + \cdots$$

$$= 1 + \frac{2}{2^2} + \frac{4}{4^2} + \frac{8}{8^2} + \cdots = 1 + \frac{1}{2} + \frac{1}{4} + \frac{1}{8} + \cdots$$

$$= \sum_{k=0}^{\infty} \frac{1}{2^k} = 2.$$

Thus the sequence of partial sums is bounded by 2 and is therefore convergent. ∎

With Theorem 1 the convergence or divergence of a series of nonnegative terms depends on whether or not its partial sums are bounded. There are several tests that can be used to determine whether or not the sequence of partial sums of a series is bounded. We will deal with these one at a time.

Theorem 2 *Comparison Test* Let $\Sigma_{k=1}^{\infty} a_k$ be a series with $a_k \geq 0$ for every k.

> **(i)** If there exists a convergent series $\Sigma_{k=1}^{\infty} b_k$ and a number N such that $a_k \leq b_k$ for every $k \geq N$, then $\Sigma_{k=1}^{\infty} a_k$ converges.
>
> **(ii)** If there exists a divergent series $\Sigma_{k=1}^{\infty} c_k$ and a number N such that $a_k \geq c_k \geq 0$ for every $k \geq N$, then $\Sigma_{k=1}^{\infty} a_k$ diverges.

Proof. In either case the sum of the first N terms is finite, so we need only consider the series $\Sigma_{k=N+1}^{\infty} a_k$ since if this is convergent or divergent, then the addition of a finite number of terms does not affect the convergence or divergence.

(i) $\Sigma_{k=N+1}^{\infty} b_k$ is a nonnegative series (since $b_k \geq a_k \geq 0$ for $k > N$) and is convergent. Thus the partial sums $T_n = \Sigma_{k=N+1}^{n} b_k$ are bounded. If $S_n = \Sigma_{k=N+1}^{n} a_k$, then $S_n \leq T_n$, and so the partial sums of $\Sigma_{k=N+1}^{\infty} a_k$ are also bounded, implying that $\Sigma_{k=N+1}^{\infty} a_k$ is convergent.

(ii) Let $U_n = \Sigma_{k=N+1}^{n} c_k$. By Theorem 1 these partial sums are unbounded since $\Sigma_{k=N+1}^{\infty} c_k$ diverges. Since in this case $S_n \geq U_n$, the partial sums of $\Sigma_{k=N+1}^{\infty} a_k$ are also unbounded, and the series $\Sigma_{k=N+1}^{\infty} a_k$ diverges. ∎

REMARK. One fact mentioned in the proof of (i) is important enough to state again: *If for some positive integer N, $\Sigma_{k=N+1}^{\infty} a_k$ converges, then $\Sigma_{k=1}^{\infty} a_k$ also converges. If $\Sigma_{k=N+1}^{\infty} a_k$ diverges, then $\Sigma_{k=1}^{\infty} a_k$ diverges. That is, the addition of a finite number of terms does not affect convergence or divergence.*

EXAMPLE 2 Determine whether $\Sigma_{k=1}^{\infty} 1/\sqrt{k}$ converges or diverges.

Solution. Since $1/\sqrt{k} \geq 1/k$ for $k \geq 1$, and since $\Sigma_{k=1}^{\infty} 1/k$ diverges, we see that by the comparison test, $\Sigma_{k=1}^{\infty} 1/\sqrt{k}$ diverges. ∎

EXAMPLE 3 Determine whether $\Sigma_{k=1}^{\infty} 1/k!$ converges or diverges.

Solution. If $k \geq 4$, $k! \geq 2^k$. To see this, note that $4! = 24$ and $2^4 = 16$. Then $5! = 5 \cdot 24$ and $2^5 = 2 \cdot 16$ and since $5 > 2$, $5! > 2^5$, and so on. Then since $\Sigma_{k=1}^{\infty} 1/2^k$ converges, we see that $\Sigma_{k=1}^{\infty} 1/k!$ converges. In fact, as we will show in Section 14.9, it converges to $e - 1$. That is,

$$e = 1 + 1 + \frac{1}{2!} + \frac{1}{3!} + \frac{1}{4!} + \cdots. \quad ∎ \tag{1}$$

Theorem 3 *The Integral Test* Let f be a function that is continuous, positive, and decreasing for all $x \geq 1$. Then the series

$$\sum_{k=1}^{\infty} f(k) = f(1) + f(2) + f(3) + \cdots + f(n) + \cdots \tag{2}$$

converges if $\int_1^{\infty} f(x)\, dx$ converges, and diverges if $\int_1^{n} f(x)\, dx \to \infty$ as $n \to \infty$.

Proof.   The idea behind this proof is fairly easy. Take a look at Figure 1. Comparing areas, we immediately see that

$$f(2) + f(3) + \cdots + f(n) \le \int_1^n f(x)\,dx \le f(1) + f(2) + \cdots + f(n-1).$$

(a)

(b)

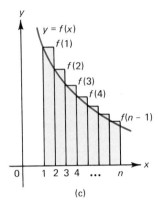

(c)

FIGURE 1

If $\lim_{n\to\infty} \int_1^n f(x)\,dx$ is finite, then the partial sums $[f(2) + f(3) + \cdots + f(n)]$ are bounded and the series converges. On the other hand, if $\lim_{n\to\infty} \int_1^n f(x)\,dx = \infty$, then the partial sums $[f(1) + f(2) + \cdots + f(n-1)]$ are unbounded and the series diverges. ∎

EXAMPLE 4   Consider the series $\sum_{k=1}^{\infty} 1/k^{\alpha}$ with $\alpha > 0$. We have already seen that this series diverges for $\alpha = 1$ (the harmonic series) and converges for $\alpha = 2$ (Example 1). Now let $f(x) = 1/x^{\alpha}$. Then for $\alpha \ne 1$

$$\int_1^n f(x)\,dx = \int_1^n \frac{1}{x^{\alpha}}\,dx = \frac{x^{1-\alpha}}{1-\alpha}\bigg|_1^n = \frac{1}{1-\alpha}(n^{1-\alpha} - 1).$$

This last expression converges to $1/(\alpha - 1)$ if $\alpha > 1$ and diverges if $\alpha < 1$. For $\alpha = 1$

$$\int_1^n f(x)\,dx = \int_1^n \frac{1}{x}\,dx = \ln x\bigg|_1^n = \ln n,$$

which diverges. (This is another proof that the harmonic series diverges.) Hence

$$\sum_{k=1}^{\infty} \frac{1}{k^{\alpha}} \quad \begin{cases} \text{diverges if } \alpha \le 1, \\ \text{converges if } \alpha > 1. \end{cases} \; \blacksquare$$

EXAMPLE 5   Determine whether $\sum_{k=1}^{\infty} (\ln k)/k^2$ converges or diverges.

Solution.   We easily see, using L'Hôpital's rule, that

$$\lim_{x\to\infty} \frac{\ln x}{\sqrt{x}} = \lim_{x\to\infty} \frac{1/x}{1/2\sqrt{x}} = 2\lim_{x\to\infty} \frac{\sqrt{x}}{x} = 2\lim_{x\to\infty} \frac{1}{\sqrt{x}} = 0,$$

so that for k sufficiently large, $\ln k \leq \sqrt{k}$. Thus

$$\frac{\ln k}{k^2} \leq \frac{\sqrt{k}}{k^2} = \frac{1}{k^{3/2}}.$$

But $\sum_{k=1}^{\infty} 1/k^{3/2}$ converges by the result of Example 4, and therefore by the comparison test, $\sum_{k=1}^{\infty} (\ln k)/k^2$ also converges.

NOTE. The integral test can also be used directly here since $\int_1^{\infty} \ln x/x^2 \, dx$ can be integrated by parts with $u = \ln x$. ■

EXAMPLE 6 Determine whether $\sum_{k=1}^{\infty} 1/[k \ln(k + 5)]$ converges or diverges.

Solution. First, we note that $1/[k \ln(k + 5)] > 1/[(k + 5) \ln(k + 5)]$. Also,

$$\int_1^n \frac{dx}{(x + 5) \ln(x + 5)} = \ln \ln(x + 5) \Big|_1^n = \ln \ln(n + 5) - \ln \ln 6,$$

which diverges, so that $\sum_{k=1}^{\infty} 1/[k \ln(k + 5)]$ also diverges. ■

We now give another test that is an extension of the comparison test.

Theorem 4 *The Limit Comparison Test* Let $\sum_{k=1}^{\infty} a_k$ and $\sum_{k=1}^{\infty} b_k$ be series with positive terms.

If there is a number $c > 0$ such that

$$\lim_{k \to \infty} \frac{a_k}{b_k} = c, \tag{3}$$

then either both series converge or both series diverge.

Proof. We have $\lim_{k \to \infty}(a_k/b_k) = c > 0$. In the definition of a limit on page 692, let $\epsilon = c/2$. Then there is a number $N > 0$ such that

$$\left|\frac{a_k}{b_k} - c\right| < \frac{c}{2} \quad \text{if} \quad k \geq N. \tag{4}$$

Equation (4) is equivalent to

$$-\frac{c}{2} < \frac{a_k}{b_k} - c < \frac{c}{2},$$

or

$$\frac{c}{2} < \frac{a_k}{b_k} < \frac{3c}{2}. \tag{5}$$

From the right inequality in (5), we obtain

$$a_k < \frac{3c}{2} b_k. \tag{6}$$

If $\Sigma \, b_k$ is convergent, then so is $(3c/2) \, \Sigma \, b_k = \Sigma \, (3c/2)b_k$. Thus from (6) and the comparison test, $\Sigma \, a_k$ is convergent. From the left inequality in (5), we have

$$a_k > \frac{c}{2} b_k. \tag{7}$$

If $\Sigma \, b_k$ is divergent, then so is $(c/2) \, \Sigma \, b_k = \Sigma \, (c/2) \, b_k$. Then using (7) and the comparison test, we find that $\Sigma \, a_k$ is also divergent. Thus if $\Sigma \, b_k$ is convergent, then $\Sigma \, a_k$ is convergent; and if $\Sigma \, b_k$ is divergent, then $\Sigma \, a_k$ is divergent. This proves the theorem. ∎

EXAMPLE 7   Show that $\Sigma_{k=1}^{\infty} 1/(ak^2 + bk + c)$ is convergent, where a, b, and c are positive real numbers.

Solution. We know from Example 1 that $\Sigma_{k=1}^{\infty} 1/k^2$ is convergent. If

$$a_k = \frac{1}{k^2} \quad \text{and} \quad b_k = \frac{1}{ak^2 + bk + c},$$

then

$$\frac{a_k}{b_k} = \frac{1/k^2}{1/(ak^2 + bk + c)} = \frac{ak^2 + bk + c}{k^2} = a + \frac{b}{k} + \frac{c}{k^2},$$

so that $\lim_{k \to \infty} (a_k/b_k) = a > 0$. Thus by the limit comparison test,

$$\sum_{k=1}^{\infty} \frac{1}{ak^2 + bk + c}$$

is convergent. ∎

EXAMPLE 8   Determine whether $\Sigma_{k=1}^{\infty} (k + 1)/(k + 3)(k + 5)$ converges or diverges.

Solution. For k large, $k + 1 \approx k$ (since $(k + 1)/k \to 1$ as $k \to \infty$), $k + 3 \approx k$, and $k + 5 \approx k$. Thus

$$\frac{k + 1}{(k + 3)(k + 5)} \approx \frac{k}{k^2} = \frac{1}{k},$$

and $\Sigma_{k=1}^{\infty} 1/k$ diverges. Therefore we suspect that $\Sigma_{k=1}^{\infty} (k + 1)/(k + 3)(k + 5)$ also diverges. This, of course, is not a proof, but it helps us to know what answer we should get. Let $a_k = 1/k$ (this is a term of a natural series with which to compare). If $b_k = (k + 1)/(k + 3)(k + 5)$, then

$$\frac{a_k}{b_k} = \frac{1/k}{(k-1)/(k+3)(k+5)} = \frac{(k+3)(k+5)}{(k)(k+1)}$$

$$= \frac{k^2 + 8k + 15}{k^2 + k} = \frac{1 + (8/k) + (15/k^2)}{1 + (1/k)} \to 1 \quad \text{as} \quad k \to \infty.$$

Thus by the limit comparison test, $\sum_{k=1}^{\infty} (k+1)/(k+3)(k+5)$ diverges because $\sum_{k=1}^{\infty} (1/k)$ diverges. ■

PROBLEMS 14.5

In Problems 1–33, determine the convergence or divergence of the given series.

1. $\displaystyle\sum_{k=1}^{\infty} \frac{1}{k^2 + 1}$

2. $\displaystyle\sum_{k=10}^{\infty} \frac{1}{k(k-3)}$

3. $\displaystyle\sum_{k=4}^{\infty} \frac{1}{5k + 50}$

4. $\displaystyle\sum_{k=1}^{\infty} \frac{1}{\sqrt{k^2 + 2k}}$

5. $\displaystyle\sum_{k=1}^{\infty} \frac{1}{\sqrt{k^3 + 1}}$

6. $\displaystyle\sum_{k=1}^{\infty} \frac{\ln k}{k^3}$

7. $\displaystyle\sum_{k=2}^{\infty} \frac{1}{k^2 + 1}$

8. $\displaystyle\sum_{k=2}^{\infty} \frac{4}{k \ln k}$

9. $\displaystyle\sum_{k=0}^{\infty} ke^{-k}$

10. $\displaystyle\sum_{k=3}^{\infty} k^3 e^{-k^4}$

11. $\displaystyle\sum_{k=5}^{\infty} \frac{1}{k(\ln k)^3}$

12. $\displaystyle\sum_{k=4}^{\infty} \frac{1}{k^2 \sqrt{\ln k}}$

13. $\displaystyle\sum_{k=1}^{\infty} \frac{1}{(3k-1)^{3/2}}$

14. $\displaystyle\sum_{k=1}^{\infty} \frac{1}{\sqrt{k^2 + 3}}$

15. $\displaystyle\sum_{k=2}^{\infty} \frac{1}{k \sqrt{\ln k}}$

16. $\displaystyle\sum_{k=1}^{\infty} \frac{1}{50 + \sqrt{k}}$

17. $\displaystyle\sum_{k=3}^{\infty} \left(\frac{k}{k+1}\right)^k$

18. $\displaystyle\sum_{k=1}^{\infty} \left(\frac{k}{k+1}\right)^{1/k}$

19. $\displaystyle\sum_{k=4}^{\infty} \frac{1}{k \ln \ln k}$

20. $\displaystyle\sum_{k=1}^{\infty} \sin \frac{1}{k}$

21. $\displaystyle\sum_{k=1}^{\infty} \frac{1}{(k+2)\sqrt{\ln(k+1)}}$

22. $\displaystyle\sum_{k=1}^{\infty} \frac{e^{1/k}}{k^2}$

23. $\displaystyle\sum_{k=1}^{\infty} \frac{\tan^{-1} k}{1 + k^2}$

24. $\displaystyle\sum_{k=1}^{\infty} \operatorname{sech} k$

25. $\displaystyle\sum_{k=10}^{\infty} \frac{1}{k(\ln k)(\ln \ln k)}$

26. $\displaystyle\sum_{k=1}^{\infty} \frac{k^2}{50k}$

27. $\displaystyle\sum_{k=1}^{\infty} \frac{1}{\cosh^2 k}$

28. $\displaystyle\sum_{k=1}^{\infty} \tan^{-1} k$

*29. $\displaystyle\sum_{k=2}^{\infty} \frac{1}{\sqrt{k} \ln^{10} k}$

30. $\displaystyle\sum_{k=1}^{\infty} \frac{\sqrt{k}}{3k^2 + 2k + 20}$

31. $\displaystyle\sum_{k=1}^{\infty} \frac{(k+1)^{7/8}}{k^3 + k^2 + 3}$

32. $\displaystyle\sum_{k=1}^{\infty} \frac{1}{(k+1)\ln(k+1)}$

33. $\displaystyle\sum_{k=1}^{\infty} \frac{k^5 + 2k^4 + 3k + 7}{k^6 + 3k^4 + 2k^2 + 1}$

*34. Let $p(x)$ be a polynomial of degree n with positive coefficients and let $q(x)$ be a polynomial of degree $\leq n + 1$ with positive coefficients. Show that $\sum_{k=1}^{\infty} p(k)/q(k)$ diverges.

*35. With $p(x)$ as in Problem 34 and $r(x)$ a polynomial of degree $\geq n + 2$ with positive coefficients, show that $\sum_{k=1}^{\infty} p(k)/r(k)$ converges.

36. Determine whether $\sum_{n=1}^{\infty} (1/n)^{1 + (1/n)}$ converges or diverges.

37. Let $S_n = \ln n! = \ln 2 + \ln 3 + \cdots + \ln n$. By calculating $\int_1^n \ln x \, dx$ and comparing areas, as in the proof of the integral test, show that for $n \geq 2$

$$\ln(n-1)! < n \ln n - n + 1 < \ln n!$$

and that

$$(n - 1)! < n^n e^{-(n-1)} < n!.$$

*38. Let $S_n = \Sigma_{k=1}^{n} 1/k$. Show that

$$\ln(n + 1) < S_n < \ln n + 1.$$

[*Hint:* Use the inequality $1/(k + 1) \leq 1/x \leq 1/k$ if $0 < k \leq x \leq k + 1$, integrate, and add, as in the proof of the integral test.]

39. Let $S_n = \Sigma_{k=1}^{n} 1/k$.

(a) Show that the sequence $\{S_n - \ln(n + 1)\}$ is increasing.

*(b) Show that this sequence is bounded by 1. [*Hint:* $S_n - \ln(n + 1) =$ the sum of the areas of "triangular" shaped regions in the region. Move those "triangles" over so that they all fit into the first rectangle.]

(c) Show that $\lim_{n\to\infty} [S_n - \ln(n + 1)]$ exists. This limit is denoted by γ, which is called the **Euler constant:**

$$\gamma = \lim_{n\to\infty} \left[1 + \frac{1}{2} + \frac{1}{3} + \cdots + \frac{1}{n} - \ln(n + 1) \right]$$

$$- \lim_{n\to\infty} \left(1 + \frac{1}{2} + \frac{1}{3} + \cdots + \frac{1}{n} - \ln n \right).$$

This number arises in physical applications. To seven decimal places, $\gamma = 0.5772157$.

14.6 SERIES WITH NONNEGATIVE TERMS II: THE RATIO AND ROOT TESTS

In this section we discuss two more tests that can be used to determine whether an infinite series converges or diverges. The first of these, the ratio test, is useful in a wide variety of applications.

Theorem 1 *The Ratio Test* Let $\Sigma_{k=1}^{\infty} a_k$ be a series with $a_k > 0$ for every k, and suppose that

$$\lim_{n\to\infty} \frac{a_{n+1}}{a_n} = L. \tag{1}$$

(i) If $L < 1$, $\Sigma_{k=1}^{\infty} a_k$ converges.

(ii) If $L > 1$, $\Sigma_{k=1}^{\infty} a_k$ diverges.

(iii) If $L = 1$, $\Sigma_{k=1}^{\infty} a_k$ may converge or diverge and the ratio test is inconclusive; some other test must be used.

Proof.

(i) Pick $\epsilon > 0$ such that $L + \epsilon < 1$. By the definition of the limit in (1), there is a number $N > 0$ such that if $n \geq N$, we have

$$\frac{a_{n+1}}{a_n} < L + \epsilon.$$

Then

$$a_{n+1} < a_n(L + \epsilon), \qquad a_{n+2} < a_{n+1}(L + \epsilon) < a_n(L + \epsilon)^2,$$

and

$$a_{n+k} < a_n(L + \epsilon)^k \tag{2}$$

for each $k \geq 1$ and each $n \geq N$. In particular, for $k \geq N$ we use (2) to obtain

$$a_k = a_{(k-N)+N} \leq a_N(L + \epsilon)^{k-N}.$$

Then

$$S_n = \sum_{k=N}^{n} a_k \leq \sum_{k=N}^{n} a_N(L + \epsilon)^{k-N} = \frac{a_N}{(L + \epsilon)^N} \sum_{k=N}^{n} (L + \epsilon)^k.$$

But since $L + \epsilon < 1$, $\sum_{k=0}^{\infty}(L + \epsilon)^k = 1/[1 - (L + \epsilon)]$ (since this last sum is the sum of a geometric series). Thus

$$S_n \leq \frac{a_N}{(L + \epsilon)^N} \cdot \frac{1}{1 - (L + \epsilon)}$$

and so the partial sums of $\sum_{k=N}^{\infty} a_k$ are bounded, implying that $\sum_{k=N}^{\infty} a_k$ converges. Thus $\sum_{k=1}^{\infty} a_k = \sum_{k=1}^{N-1} a_k + \sum_{k=N}^{\infty} a_k$ also converges.

(ii) If $1 < L < \infty$, pick ϵ such that $L - \epsilon > 1$. Then for $n \geq N$, the same proof as before (with the inequalities reversed) shows that

$$a_k \geq a_N(L - \epsilon)^{k-N}$$

and that

$$S_n = \sum_{k=N}^{n} a_k > \frac{a_N}{(L - \epsilon)^N} \sum_{k=N}^{n} (L - \epsilon)^k.$$

But since $L - \epsilon > 1$, $\sum_{k=N}^{\infty}(L - \epsilon)^k$ diverges, so that the partial sums S_n are unbounded and $\sum_{k=N}^{\infty} a_k$ diverges. The proof in the case $L = \infty$ is suggested in Problem 33.

(iii) To illustrate (iii), we show that $L = 1$ can occur for a converging or diverging series.

(a) The harmonic series $\sum_{k=1}^{\infty} 1/k$ diverges. But

$$\lim_{n \to \infty} \frac{a_{n+1}}{a_n} = \lim_{n \to \infty} \frac{1/(n + 1)}{1/n} = \lim_{n \to \infty} \frac{n}{n + 1} = 1.$$

(b) The series $\sum_{k=1}^{\infty} 1/k^2$ converges. Here

$$\lim_{n\to\infty}\frac{a_{n+1}}{a_n} = \lim_{n\to\infty}\frac{1/(n+1)^2}{1/n^2} = \lim_{n\to\infty}\left(\frac{n}{n+1}\right)^2 = 1. \blacksquare$$

REMARK. The ratio test is very useful. But in those cases where $L = 1$, we must try another test to determine whether the series converges or diverges.

EXAMPLE 1 We have used the comparison test to show that $\Sigma_{k=1}^{\infty} 1/k!$ converges. Using the ratio test, we find that

$$\lim_{n\to\infty}\frac{a_{n+1}}{a_n} = \lim_{n\to\infty}\frac{1/(n+1)!}{1/n!} = \lim_{n\to\infty}\frac{n!}{(n+1)!} = \lim_{n\to\infty}\frac{1}{n+1} = 0 < 1,$$

so that the series converges. $\blacksquare$

EXAMPLE 2 Determine whether the series $\Sigma_{k=0}^{\infty}(100)^k/k!$ converges or diverges.

Solution. Here

$$\lim_{n\to\infty}\frac{a_{n+1}}{a_n} = \lim_{n\to\infty}\frac{(100)^{n+1}/(n+1)!}{(100)^n/n!} = \lim_{n\to\infty}\frac{100}{n+1} = 0,$$

so that the series converges. $\blacksquare$

EXAMPLE 3 Determine whether the series $\Sigma_{k=1}^{\infty} k^k/k!$ converges or diverges.

Solution.

$$\lim_{n\to\infty}\frac{a_{n+1}}{a_n} = \lim_{n\to\infty}\frac{[(n+1)^{n+1}/(n+1)!]}{n^n/n!} = \lim_{n\to\infty}\frac{(n+1)^{n+1}}{(n+1)n^n}$$

$$= \lim_{n\to\infty}\left(\frac{n+1}{n}\right)^n = \lim_{n\to\infty}\left(1+\frac{1}{n}\right)^n = e > 1,$$

so that the series diverges. $\blacksquare$

EXAMPLE 4 Determine whether the series $\Sigma_{k=1}^{\infty}(k+1)/[k(k+2)]$ converges or diverges.

Solution. Here

$$\lim_{n\to\infty}\frac{a_{n+1}}{a_n} = \lim_{n\to\infty}\frac{(n+2)/(n+1)(n+3)}{(n+1)/n(n+2)} = \lim_{n\to\infty}\frac{n(n+2)^2}{(n+1)^2(n+3)}$$

$$= \lim_{n\to\infty}\frac{n^3+4n^2+4n}{n^3+5n^2+7n+3} = 1. \qquad \text{By Example 14.1.12}$$

Thus the ratio test fails. However, $\lim_{k\to\infty}[(k+1)/k(k+2)]/(1/k) = 1$, so that $\Sigma_{k=1}^{\infty}(k+1)/[k(k+2)]$ diverges by the limit comparison test. $\blacksquare$

Theorem 2 *The Root Test* Let $\Sigma_{k=1}^{\infty} a_k$ be a series with $a_k > 0$ and suppose that $\lim_{n\to\infty}(a_n)^{1/n} = R.$

> **(i)** If $R < 1$, $\sum_{k=1}^{\infty} a_k$ converges.
> **(ii)** If $R > 1$, $\sum_{k=1}^{\infty} a_k$ diverges.
> **(iii)** If $R = 1$, the series either converges or diverges, and no conclusions can be drawn from this test.

The proof of this theorem is similar to the proof of the ratio test and is left as an exercise (see Problems 27–29).

EXAMPLE 5   Determine whether $\sum_{k=2}^{\infty} 1/(\ln k)^k$ converges or diverges.

Solution. Note first that we start at $k = 2$ since $1/(\ln 1)^1$ is not defined.

$$\lim_{n \to \infty}\left[\frac{1}{(\ln n)^n}\right]^{1/n} = \lim_{n \to \infty}\frac{1}{\ln n} = 0,$$

so that the series converges.   ■

EXAMPLE 6   Determine whether the series $\sum_{k=1}^{\infty}(k^k/3^{4k+5})$ converges or diverges.

Solution. $\lim_{n \to \infty}(n^n/3^{4n+5})^{1/n} = \lim_{n \to \infty}(n/3^{4+5/n}) = \infty$, since $\lim_{n \to \infty} 3^{4+5/n} = 3^4 = 81$. Thus the series diverges.   ■

EXAMPLE 7   Determine whether the series $\sum_{k=1}^{\infty} (1/2 + 1/k)^k$ converges or diverges.

Solution.

$$\lim_{n \to \infty}\left[\left(\frac{1}{2} + \frac{1}{n}\right)^n\right]^{1/n} = \lim_{n \to \infty}\left(\frac{1}{2} + \frac{1}{n}\right) = \frac{1}{2} < 1,$$

so the series converges.   ■

PROBLEMS 14.6

In Problems 1–25, determine whether the given series converges or diverges.

1. $\sum_{k=1}^{\infty} \dfrac{2^k}{k^2}$ **2.** $\sum_{k=1}^{\infty} \dfrac{5^k}{k^5}$ **3.** $\sum_{k=1}^{\infty} \dfrac{r^k}{k^r}, \ 0 < r < 1$

4. $\sum_{k=1}^{\infty} \dfrac{r^k}{k^r}, \ r > 1$ **5.** $\sum_{k=2}^{\infty} \dfrac{k!}{k^k}$ **6.** $\sum_{k=1}^{\infty} \dfrac{k^k}{(2k)!}$

7. $\sum_{k=1}^{\infty} \dfrac{e^k}{k^5}$ **8.** $\sum_{k=1}^{\infty} \dfrac{e^k}{k!}$ **9.** $\sum_{k=1}^{\infty} \dfrac{k^{2/3}}{10^k}$

10. $\sum_{k=1}^{\infty} \dfrac{3^k + k}{k! + 2}$ **11.** $\sum_{k=2}^{\infty} \dfrac{k}{(\ln k)^k}$ **12.** $\sum_{k=1}^{\infty} \dfrac{4^k}{k^3}$

13. $\sum_{k=2}^{\infty} \left(1 + \dfrac{1}{k}\right)^k$ **14.** $\sum_{k=1}^{\infty} \dfrac{\sqrt{k} \ln k}{k^3 + 1}$ **15.** $\sum_{k=1}^{\infty} \dfrac{3^{4k+5}}{k^k}$

16. $\sum_{k=1}^{\infty} \dfrac{a^{mk+b}}{k^k}, \ a > 1, \ b$ real   **17.** $\sum_{k=1}^{\infty} \dfrac{k^k}{a^{mk+b}}, \ a > 1, \ b$ real   **18.** $\sum_{k=1}^{\infty} \dfrac{k^6 5^k}{(k + 1)!}$

19. $\displaystyle\sum_{k=1}^{\infty} \frac{k^2 k!}{(2k)!}$

20. $\displaystyle\sum_{k=1}^{\infty} \frac{(2k)!}{k^2 k!}$

*21. $\displaystyle\sum_{k=1}^{\infty} \left(\frac{k!}{k^k}\right)^k$

22. $\displaystyle\sum_{k=1}^{\infty} \left(\frac{k^k}{k!}\right)^k$

23. $\displaystyle\sum_{k=2}^{\infty} \frac{e^k}{(\ln k)^k}$

24. $\displaystyle\sum_{k=1}^{\infty} \frac{(\ln k)^k}{k^2}$

25. $\displaystyle\sum_{k=1}^{\infty} \left(\frac{k}{3k+2}\right)^k$

26. Show that $\Sigma_{k=0}^{\infty} x^k/k!$ converges for every real number x.

*27. Prove part (i) of the root test (Theorem 2).   [*Hint:* If $R < 1$, choose $\epsilon > 0$ so that $R + \epsilon < 1$. Show that there is an N such that if $n \geq N$, then $a_n < (R + \epsilon)^n$. Then complete the proof by comparing $\Sigma\, a_k$ with the sum of a geometric series.]

28. Prove part (ii) of the root test.   [*Hint:* Follow the steps of Problem 27.]

29. Show that if $a_n^{1/n} \to 1$, then $\Sigma_{k=1}^{\infty} a_k$ may converge or diverge.   [*Hint:* Consider $\Sigma\, 1/k$ and $\Sigma\, 1/k^2$.]

30. Prove that $k!/k^k \to 0$ as $k \to \infty$.

31. Let $a_k = 3/k^2$ if k is even and $a_k = 1/k^2$ if k is odd. Show that $\lim_{n\to\infty} (a_{n+1}/a_n)$ does not exist, but $\Sigma_{k=1}^{\infty} a_k$ converges.

32. Construct a series of positive terms for which $\lim_{n\to\infty} (a_{n+1}/a_n)$ does not exist but for which $\Sigma_{k=1}^{\infty} a_k$ diverges.

33. Prove that if $a_n > 0$ and $\lim_{n\to\infty} (a_{n+1}/a_n) = \infty$, then $\Sigma_{k=1}^{\infty} a_k$ diverges.   [*Hint:* Show that for N sufficiently large, $a_k \geq 2^{k-N} a_n$.]

14.7 ABSOLUTE AND CONDITIONAL CONVERGENCE: ALTERNATING SERIES

In Sections 14.5 and 14.6 all the series we dealt with had positive terms. In this section we consider special types of series that have positive and negative terms.

Definition 1   ABSOLUTE CONVERGENCE   The series $\Sigma_{k=1}^{\infty} a_k$ is said to **converge absolutely** if the series $\Sigma_{k=1}^{\infty} |a_k|$ converges.

EXAMPLE 1   The series

$$\sum_{k=1}^{\infty} \frac{(-1)^{k+1}}{k^2} = \frac{1}{1^2} - \frac{1}{2^2} + \frac{1}{3^2} - \frac{1}{4^2} + \cdots$$

converges absolutely because $\Sigma_{k=1}^{\infty} |(-1)^{k+1}/k^2| = \Sigma_{k=1}^{\infty} 1/k^2$ converges.   ∎

EXAMPLE 2   The series

$$\sum_{k=1}^{\infty} \frac{(-1)^{k+1}}{k} = \frac{1}{1} - \frac{1}{2} + \frac{1}{3} - \frac{1}{4} + \frac{1}{5} + \cdots$$

does not converge absolutely because $\Sigma_{k=1}^{\infty} 1/k$ diverges.   ∎

The importance of absolute convergence is given in the theorem below.

Theorem 1.   If $\Sigma_{k=1}^{\infty} |a_k|$ converges, then $\Sigma_{k=1}^{\infty} a_k$ also converges. That is:

> Absolute convergence implies convergence.

REMARK.   The converse of this theorem is false. That is, there are series that are convergent but not absolutely convergent. We will see examples of this phenomenon shortly.

Proof.   Since $a_k \leq |a_k|$, we have

$$0 \leq a_k + |a_k| \leq 2|a_k|.$$

Since $\Sigma_{k=1}^{\infty} |a_k|$ converges, we see that $\Sigma_{k=1}^{\infty} (a_k + |a_k|)$ converges by the comparison test. Then since $a_k = (a_k + |a_k|) - |a_k|$, $\Sigma_{k=1}^{\infty} a_k$ converges because it is the sum of two convergent series. ∎

EXAMPLE 3   The series $\Sigma_{k=1}^{\infty} (-1)^{k+1}/k^2$ considered in Example 1 converges since it converges absolutely. ∎

Definition 2   ALTERNATING SERIES   A series in which successive terms have opposite signs is called an **alternating series**.

EXAMPLE 4   The series

$$\sum_{k=1}^{\infty} \frac{(-1)^{k+1}}{k} = 1 - \frac{1}{2} + \frac{1}{3} - \frac{1}{4} + \frac{1}{5} - \frac{1}{6} + \cdots$$

is an alternating series. ∎

EXAMPLE 5   The series $1 + \frac{1}{2} - \frac{1}{3} - \frac{1}{4} + \frac{1}{5} + \frac{1}{6} - \cdots$ is not an alternating series because two successive terms have the same sign. ∎

Let us consider the series of Example 4:

$$S = 1 - \tfrac{1}{2} + \tfrac{1}{3} - \tfrac{1}{4} + \tfrac{1}{5} - \tfrac{1}{6} + \cdots.$$

Calculating successive partial sums, we find that

$$S_1 = 1, \qquad S_2 = \tfrac{1}{2}, \qquad S_3 = \tfrac{5}{6}, \qquad S_4 = \tfrac{7}{12}, \qquad S_5 = \tfrac{47}{60}, \qquad \cdots.$$

It is clear that this series is not diverging to infinity (indeed, $\frac{1}{2} \leq S_n \leq 1$) and that the partial sums are getting "narrowed down." At this point it is reasonable to suspect that the series converges. But it does *not* converge absolutely (since the series of absolute values is the harmonic series), and we cannot use any of the tests of the previous section since the terms are not nonnegative. The result we need is given in the theorem below.

Theorem 2   *Alternating Series Test*   Let $\{a_k\}$ be a decreasing sequence of positive numbers such that $\lim_{k \to \infty} a_k = 0$. Then the alternating series $\Sigma_{k=1}^{\infty} (-1)^{k+1} a_k = a_1 - a_2 + a_3 - a_4 + \cdots$ converges.

Proof. Looking at the odd-numbered partial sums of this series, we find that

$$S_{2n+1} = (a_1 - a_2) + (a_3 - a_4) + (a_5 - a_6) + \cdots + (a_{2n-1} - a_{2n}) + a_{2n+1}.$$

Since $\{a_k\}$ is decreasing, all the terms in parentheses are nonnegative, so that $S_{2n+1} \geq 0$ for every n. Moreover,

$$S_{2n+3} = S_{2n+1} - a_{2n+2} + a_{2n+3} = S_{2n+1} - (a_{2n+2} - a_{2n+3}),$$

and since $a_{2n+2} - a_{2n+3} \geq 0$, we have

$$S_{2n+3} \leq S_{2n+1}.$$

Hence the sequence of odd-numbered partial sums is bounded below by 0 and is decreasing and is therefore convergent by Theorem 14.2.2 Thus S_{2n+1} converges to some limit L. Now let us consider the sequence of even-numbered partial sums. We find that $S_{2n+2} = S_{2n+1} - a_{2n+2}$ and since $a_{2n+2} \to 0$,

$$\lim_{n\to\infty} S_{2n+2} = \lim_{n\to\infty} S_{2n+1} - \lim_{n\to\infty} a_{2n+2} = L - 0 - L,$$

so that the even partial sums also converge to L. Since both the odd and even sums converge to L, we see that the partial sums converge to L, and the proof is complete. ∎

EXAMPLE 6 The following alternating series are convergent by the alternating series test:

(a) $1 - \dfrac{1}{2} + \dfrac{1}{3} - \dfrac{1}{4} + \dfrac{1}{5} - \dfrac{1}{6} + \cdots$

(b) $1 - \dfrac{1}{\sqrt{2}} + \dfrac{1}{\sqrt{3}} - \dfrac{1}{\sqrt{4}} + \dfrac{1}{\sqrt{5}} - \dfrac{1}{\sqrt{6}} + \dfrac{1}{\sqrt{7}} - \cdots$

(c) $\dfrac{1}{\ln 2} - \dfrac{1}{\ln 3} + \dfrac{1}{\ln 4} - \dfrac{1}{\ln 5} + \dfrac{1}{\ln 6} - \cdots$

(d) $1 - \dfrac{1}{2} + \dfrac{1}{2^2} - \dfrac{1}{2^3} + \dfrac{1}{2^4} - \dfrac{1}{2^5} + \dfrac{1}{2^6} - \dfrac{1}{2^7} + \cdots$ ∎

Definition 3 CONDITIONAL CONVERGENCE An alternating series is said to be **conditionally convergent** if it is convergent but not absolutely convergent.

In Example 6 all the series are conditionally convergent except the last one, which is absolutely convergent.

It is not difficult to estimate the sum of a convergent alternating series. We again consider the series

$$S = 1 - \tfrac{1}{2} + \tfrac{1}{3} - \tfrac{1}{4} + \tfrac{1}{5} - \cdots.$$

Suppose we wish to approximate S by its nth partial sum S_n. Then

$$S - S_n = \pm\left(\frac{1}{n+1} - \frac{1}{n+2} + \frac{1}{n+3} - \frac{1}{n+4} + \cdots\right) = R_n.$$

But we can estimate the remainder term R_n:

$$|R_n| = \left|\left[\frac{1}{n+1} - \left(\frac{1}{n+2} - \frac{1}{n+3}\right) - \left(\frac{1}{n+4} - \frac{1}{n+5}\right) - \cdots\right]\right| \le \frac{1}{n+1}.$$

That is, the error is less than the first term that we left out! For example, $|S - S_{20}| \le \frac{1}{21} \approx 0.0476$.

In general, we have the following result, whose proof is left as an exercise (see Problem 31).

Theorem 3 If $S = \sum_{k=1}^{\infty}(-1)^{k+1}a_k$ is a convergent alternating series with monotone decreasing terms, then for any n,

$$|S - S_n| \le |a_{n+1}|. \tag{1}$$

EXAMPLE 7 The series

$$\sum_{k=1}^{\infty}\frac{(-1)^{k+1}}{\ln(k+1)} = \frac{1}{\ln 2} - \frac{1}{\ln 3} + \frac{1}{\ln 4} - \frac{1}{\ln 5} + \cdots$$

can be approximated by S_n with an error of less than $1/\ln(n+2)$. For example, with $n = 10$, $1/\ln(n+2) = 1/\ln 12 \approx 0.4$. Hence the sum

$$\sum_{k=1}^{\infty}\frac{(-1)^{k+1}}{\ln(k+1)} = \frac{1}{\ln 2} - \frac{1}{\ln 3} + \cdots$$

can be approximated by

$$S_{10} = \frac{1}{\ln 2} - \frac{1}{\ln 3} + \frac{1}{\ln 4} - \frac{1}{\ln 5} + \frac{1}{\ln 6} - \frac{1}{\ln 7} + \frac{1}{\ln 8} - \frac{1}{\ln 9} + \frac{1}{\ln 10} - \frac{1}{\ln 11}$$

$$\approx 0.7197,$$

with an error of less than 0.4. ∎

By modifying Theorem 3 we can significantly improve on the last result.

Theorem 4 Suppose that the hypotheses of Theorem 3 hold and that, in addition, the sequence $\{|a_n - a_{n+1}|\}$ is monotone decreasing. Let $T_n = S_{n-1} - (-1)^n \frac{1}{2}a_n$. Then

$$|S - T_n| \le \frac{1}{2}|a_n - a_{n+1}|. \tag{2}$$

This result follows from Theorem 3 and is also left as an exercise (see Problem 45).

EXAMPLE 8   We can improve the estimate in Example 7. We may approximate $\sum_{k=1}^{\infty}(-1)^{k+1}/\ln(k+1)$ by

$$\frac{1}{\ln 2} - \frac{1}{\ln 3} + \frac{1}{\ln 4} - \frac{1}{\ln 5} + \frac{1}{\ln 6} - \frac{1}{\ln 7} + \frac{1}{\ln 8} - \frac{1}{\ln 9} + \frac{1}{\ln 10} - \frac{1}{2\ln 11} \approx 0.9282.$$

With $n = 10$ (so that $n + 1 = 11$),

$$T_{10} = S_9 - \frac{1}{2}\left(\frac{1}{\ln 11}\right),$$

which is precisely the sum given above. Thus

$$|S - T_{10}| < \frac{1}{2}|a_{10} - a_{11}| = \frac{1}{2}\left(\frac{1}{\ln 11} - \frac{1}{\ln 12}\right) \approx 0.0073.$$

This result is a considerable improvement.

Note that in order to justify this result, we must verify that $|a_n - a_{n+1}|$ is monotone decreasing. But here

$$|a_n - a_{n+1}| = \frac{1}{\ln(n+1)} - \frac{1}{\ln(n+2)}.$$

Let

$$f(x) = \frac{1}{\ln(x+1)} - \frac{1}{\ln(x+2)}.$$

Then

$$f'(x) = -\frac{1}{(x+1)\ln^2(x+1)} + \frac{1}{(x+2)\ln^2(x+2)} < 0.$$

Thus f is a decreasing function, which shows that $f(n+1) < f(n)$. ∎

There is one fascinating fact about an alternating series that is conditionally but not absolutely convergent:

> By reordering the terms of a conditionally convergent alternating series, the new series of rearranged terms can be made to converge to any real number.

Let us illustrate this fact with the series

$$S = 1 - \tfrac{1}{2} + \tfrac{1}{3} - \tfrac{1}{4} + \tfrac{1}{5} - \tfrac{1}{6} + \cdots.$$

The odd-numbered terms sum to a divergent series:

$$1 + \tfrac{1}{3} + \tfrac{1}{5} + \tfrac{1}{7} + \cdots. \tag{3}$$

The even-numbered terms are likewise a divergent series:

$$-\tfrac{1}{2} - \tfrac{1}{4} - \tfrac{1}{6} - \cdots. \tag{4}$$

If either of these series converged, then the other one would too (by Theorem 14.4.1(i), and then the entire series would be absolutely convergent (which we know to be false). Now choose any real number, say 1.5. Then:

(i) Choose enough terms from the series (3) so that the sum exceeds 1.5. We can do so since the series diverges.

$$1 + \tfrac{1}{3} + \tfrac{1}{5} = 1.53333 \ldots.$$

(ii) Add enough negative terms from (4) so that the sum is now just under 1.5.

$$1 + \tfrac{1}{3} + \tfrac{1}{5} - \tfrac{1}{2} = 1.0333 \ldots.$$

(iii) Add more terms from (3) until 1.5 is exceeded.

$$1 + \tfrac{1}{3} + \tfrac{1}{5} - \tfrac{1}{2} + \tfrac{1}{7} + \tfrac{1}{9} + \tfrac{1}{11} + \tfrac{1}{13} + \tfrac{1}{15} = 1.5218.$$

(iv) Again subtract terms from (4) until the sum is under 1.5.

$$1 + \tfrac{1}{3} + \tfrac{1}{5} - \tfrac{1}{2} + \tfrac{1}{7} + \tfrac{1}{9} + \tfrac{1}{11} + \tfrac{1}{13} + \tfrac{1}{15} - \tfrac{1}{4} = 1.2718.$$

We continue the process to "converge" to 1.5. Since the terms in each series are decreasing to 0, the amount above or below 1.5 will approach 0 and the partial sums converge.

We will indicate in Section 14.10 that without rearranging, we have

$$\sum_{k=1}^{\infty} \frac{(-1)^{k+1}}{k} = 1 - \frac{1}{2} + \frac{1}{3} - \frac{1}{4} + \frac{1}{5} - \frac{1}{6} + \cdots = \ln 2 \approx 0.693147. \tag{5}$$

We again illustrate what can happen when the terms in the series (5) are rearranged.† Consider the series

$$S^* = 1 - \tfrac{1}{2} - \tfrac{1}{4} + \tfrac{1}{3} - \tfrac{1}{6} - \tfrac{1}{8} + \tfrac{1}{5} - \tfrac{1}{10} - \tfrac{1}{12} + \cdots. \tag{6}$$

This series is the rearrangement of the series (5) in which each odd term is followed by two even terms. Inserting parentheses, we have

$$S^* = (1 - \tfrac{1}{2}) - \tfrac{1}{4} + (\tfrac{1}{3} - \tfrac{1}{6}) - \tfrac{1}{8} + (\tfrac{1}{5} - \tfrac{1}{10}) - \tfrac{1}{12} + \cdots$$
$$= \tfrac{1}{2} - \tfrac{1}{4} + \tfrac{1}{6} - \tfrac{1}{8} + \tfrac{1}{10} - \tfrac{1}{12} + \cdots$$
$$= \tfrac{1}{2}(1 - \tfrac{1}{2} + \tfrac{1}{3} - \tfrac{1}{4} + \tfrac{1}{5} - \tfrac{1}{6} + \cdots) = \tfrac{1}{2}\ln 2.$$

---

†This example is cited in the paper "Rearranging the alternating harmonic series" by C. C. Cowen, K. R. Davidson, and R. P. Kaufman, *American Mathematics Monthly*, **87**, 817–819, (December 1980).

TABLE 1 TESTS OF CONVERGENCE

Test	First discussed on page	Description	Examples and Comments
Convergence test for a geometric series	705	$\sum_{k=0}^{\infty} r^k$ converges to $1/(1-r)$ if $|r| < 1$ and diverges if $|r| > 1$	$\sum_{k=0}^{\infty} (\tfrac{1}{2})^k$ converges to 2; $\sum_{k=0}^{\infty} 2^k$ diverges
Look at the terms of the series— the limit test	712	If $|a_k|$ does not converge to 0, then $\sum a_n$ diverges	If $a_k \to 0$, then $\sum_0^{\infty} a_k$ may converge ($\sum_{k=0}^{\infty} 1/k^2$) or it may not (the harmonic series $\sum_{k=0}^{\infty} 1/k$)
Comparison test	715	If $0 \le a_k \le b_k$ and $\sum b_k$ converges, then $\sum a_k$ converges. If $a_k \ge b_k \ge 0$ and $\sum b_k$ diverges, then $\sum a_k$ diverges	It is not necessary that $a_k \le b_k$ or $a_k \ge b_k$ for *all* k, only for $k \ge N$ for some integer N; convergence or divergence of a series is not affected by the values of the first few terms
Integral test	715	If $a_k = f(k) \ge 0$, then $\sum_{k=1}^{\infty} a_k$ converges if $\int_1^{\infty} f(x)\,dx$ converges and $\sum_{k=1}^{\infty} a_k$ diverges if $\int_1^{\infty} f(x)\,dx$ diverges	Use this test whenever $f(x)$ can easily be integrated
$\sum_{k=1}^{\infty} 1/k^{\alpha}$	716	$\sum_{k=1}^{\infty} 1/k^{\alpha}$ diverges if $0 \le \alpha \le 1$ and converges if $\alpha > 1$	
Limit comparison test	717	If $a_k \ge 0$, $b_k \ge 0$ and there is a number c such that $\lim_{k \to \infty} a_k/b_k = c$, then either both series converge or both series diverge	Use the limit comparison test when a series $\sum b_k$ can be found such that (a) it is known whether $\sum b_k$ converges or diverges and (b) it appears that a_k/b_k has an easily computed limit; (b) will be true, for instance, when $a_k = 1/p(k)$ and $b_k = 1/q(k)$ where $p(k)$ and $q(k)$ are polynomials

Using results in the paper cited on page 729, we can show that

$$1 + \tfrac{1}{3} - \tfrac{1}{2} + \tfrac{1}{5} + \tfrac{1}{7} - \tfrac{1}{4} + \tfrac{1}{9} + \tfrac{1}{10} - \tfrac{1}{6} + \cdots$$

converges to $\tfrac{3}{2} \ln 2$ and that

$$1 - \tfrac{1}{2} - \tfrac{1}{4} - \tfrac{1}{6} + \tfrac{1}{3} - \tfrac{1}{8} - \tfrac{1}{10} - \tfrac{1}{12} + \tfrac{1}{5} - \cdots$$

converges to $\ln 2 - \tfrac{1}{2} \ln 3$.

TABLE 1   TESTS OF CONVERGENCE (Continued)

Test	First discussed on page	Description	Examples and Comments
Ratio test	720	If $a_k > 0$ and $\lim_{n\to\infty} a_{n+1}/a_n = L$, then $\sum_{k=1}^{\infty} a_k$ converges if $L < 1$ and diverges when $L > 1$	This is often the easiest test to apply; note that if $L = 1$, then the series may either converge $(\sum 1/k^2)$ or diverge $(\sum 1/k)$
Root test	722	If $a_k > 0$ and $\lim_{n\to\infty}(a_n)^{1/n} = R$, then $\sum_{k=1}^{\infty} a_k$ converges if $R < 1$ and diverges if $R > 1$	If $R = 1$, the series may either converge $(\sum 1/k^2)$ or diverge $(\sum 1/k)$; the root test is the hardest test to apply; it is most useful when a_k is something raised to the kth power $[\sum 1/(\ln k)^k$, for example]
Alternating series test	725	$\sum (-1)^k a_k$ with $a_k \geq 0$ converges if (a) $a_k \to 0$ as $k \to \infty$ and (b) $\{a_k\}$ is a decreasing sequence; also, $\sum (-1)^k a_k$ diverges if $\lim_{k\to\infty} a_k \neq 0$	This test can only be applied when the terms are alternately positive and negative; if there are two or more positive (or negative) terms in a row, then try another test
Absolute convergence test for a series with both positive and negative terms	724	$\sum a_k$ converges absolutely if $\sum |a_k|$ converges	To determine whether $\sum |a_k|$ converges, try any of the tests that apply to series with nonnegative terms

REMARK.   It should be noted that *any* rearrangement of the terms of an *absolutely converging* series converges to the same number.

We close this section by providing in Table 1 a summary of the convergence tests we have discussed.

PROBLEMS 14.7

In Problems 1–30, determine whether the given series is absolutely convergent, conditionally convergent, or divergent.

1. $\sum_{k=1}^{\infty} (-1)^k$

2. $\sum_{k=1}^{\infty} \frac{(-1)^{k+1}}{2k}$

3. $\sum_{k=2}^{\infty} \frac{(-1)^k}{k \ln k}$

4. $\sum_{k=1}^{\infty} \frac{(-1)^k}{k^{3/2}}$

5. $\sum_{k=2}^{\infty} \frac{(-1)^k k}{\ln k}$

6. $\sum_{k=1}^{\infty} \frac{(-1)^k \ln k}{k}$

7. $\sum_{k=1}^{\infty} \frac{(-1)^{k+1}}{5k - 4}$

8. $\sum_{k=1}^{\infty} \sin \frac{k\pi}{2}$

9. $\sum_{k=0}^{\infty} \cos \frac{k\pi}{2}$

10. $\displaystyle\sum_{k=1}^{\infty} \frac{(-3)^k}{k!}$

11. $\displaystyle\sum_{k=1}^{\infty} \frac{k!}{(-3)^k}$

12. $\displaystyle\sum_{k=1}^{\infty} \frac{(-2)^k}{k^2}$

13. $\displaystyle\sum_{k=1}^{\infty} \frac{k^2}{(-2)^k}$

14. $\displaystyle\sum_{k=2}^{\infty} \frac{(-1)^{k+1}}{\sqrt{k(k-1)}}$

15. $\displaystyle\sum_{k=2}^{\infty} \frac{(-1)^k k^2}{k^3+1}$

16. $\displaystyle\sum_{k=1}^{\infty} \frac{\cos(k\pi/6)}{k^2}$

17. $\displaystyle\sum_{k=3}^{\infty} \frac{\sin(k\pi/7)}{k^3}$

18. $\displaystyle\sum_{k=1}^{\infty} \frac{(-1)^k(k+2)}{k(k+1)}$

19. $\displaystyle\sum_{k=2}^{\infty} \frac{(-1)^k k(k+1)}{(k+2)^3}$

20. $\displaystyle\sum_{k=2}^{\infty} \frac{(-1)^k k(k+1)}{(k+2)^4}$

21. $\displaystyle\sum_{k=1}^{\infty} \frac{(-1)^k 2^k}{k}$

22. $\displaystyle\sum_{k=1}^{\infty} \frac{(-1)^{k+1}}{k!}$

23. $\displaystyle\sum_{k=1}^{\infty} \frac{(-1)^k k^k}{k!}$

24. $\displaystyle\sum_{k=1}^{\infty} \frac{(-1)^k \sqrt{k}}{k+3}$

25. $\displaystyle\sum_{k=2}^{\infty} \frac{(-1)^k(k^2+3)}{k^3+4}$

26. $\displaystyle\sum_{k=2}^{\infty} \frac{(-1)^k}{\sqrt[3]{\ln k}}$

27. $\displaystyle\sum_{k=1}^{\infty} \frac{(-1)^k k^2}{4+k^2}$

28. $\displaystyle\sum_{k=1}^{\infty} (-1)^k \left(1+\frac{1}{k}\right)^k$

29. $\displaystyle\sum_{k=2}^{\infty} \frac{(-1)^k}{k\sqrt{\ln k}}$

30. $\displaystyle\sum_{k=2}^{\infty} \frac{(-1)^k k^3}{k^3+2k^2+k-1}$

*31. Prove Theorem 3. [*Hint:* Assume that the odd-numbered terms are positive. Show that the sequence $\{S_{2n}\}$ is increasing and that $S_{2n} < S_{2n+2} < S$ for all $n \geq 1$. Then show that the sequence of odd-numbered partial sums is decreasing and that $S < S_{2n+1} < S_{2n-1}$ for all $n \geq 1$. Conclude that (a) $0 < S - S_{2n} < a_{2n+1}$ for all $n \geq 1$ and that (b) $0 < S_{2n-1} - S < -a_{2n}$. Use inequalities (a) and (b) to prove the theorem.]

In Problems 32–37, use the result of Theorem 3 or Theorem 4 to estimate the given sum to within the indicated accuracy.

32. $\displaystyle\sum_{k=1}^{\infty} \frac{(-1)^{k+1}}{k!}$; error < 0.001

33. $\displaystyle\sum_{k=1}^{\infty} \frac{(-1)^{k+1}}{k^2}$; error < 0.01

34. $\displaystyle\sum_{k=1}^{\infty} \frac{(-1)^{k+1}}{k^4}$; error < 0.0001

35. $\displaystyle\sum_{k=2}^{\infty} \frac{(-1)^{k+1}}{k \ln k}$; error < 0.05

36. $\displaystyle\sum_{k=1}^{\infty} \frac{(-1)^{k+1}}{k^k}$; error < 0.0001

37. $\displaystyle\sum_{k=1}^{\infty} \frac{(-1)^{k+1}}{\sqrt{k}}$; error < 0.1

38. Find the first ten terms of a rearrangement of the series $\sum_{k=1}^{\infty}(-1)^{k+1}/k$ that converges to 0.

39. Find the first ten terms of a rearrangement of the series $\sum_{k=1}^{\infty}(-1)^{k+1}/k$ that converges to 0.3.

40. Explain why there is no rearrangement of the series $\sum_{k=1}^{\infty}(-1)^k/k^2$ that converges to -1.

41. Prove that if $\sum_{k=1}^{\infty} a_k$ is a convergent series of nonzero terms, then $\sum_{k=1}^{\infty} 1/|a_k|$ diverges.

42. Show that if $\sum_{k=1}^{\infty} a_k$ is absolutely convergent, then $\sum_{k=1}^{\infty} a_k^{\,p}$ is convergent for any integer $p \geq 1$.

43. Give an example of a sequence $\{a_k\}$ such that $\sum_{k=1}^{\infty} a_k^2$ converges but $\sum_{k=1}^{\infty} a_k$ diverges.

*44. Give an example of a sequence $\{a_k\}$ such that $\sum_{k=1}^{\infty} a_k$ converges but $\sum_{k=1}^{\infty} a_k^3$ diverges.

*45. Prove Theorem 4. [*Hint:* Write the series as $S = \frac{1}{2}a_1 + \frac{1}{2}(a_1 - a_2) - \frac{1}{2}(a_2 - a_3) + \frac{1}{2}(a_3 - a_4) - \cdots = \frac{1}{2}a_1 + \sum_{k=1}^{\infty}(-1)^{k+1}(a_k - a_{k+1})/2$. Then apply Theorem 3.]

14.8 POWER SERIES

In previous sections we discussed infinite series of real numbers. Here we discuss series of functions.

Definition 1   POWER SERIES

(i) A **power series** in x is a series of the form

$$\sum_{k=0}^{\infty} a_k x^k = a_0 + a_1 x + a_2 x^2 + \cdots + a_n x^n + \cdots.$$ (1)

(ii) A power series in $(x - x_0)$ is a series of the form

$$\sum_{k=0}^{\infty} a_k (x - x_0)^k = a_0 + a_1 (x - x_0) + a_2 (x - x_0)^2$$

$$+ \cdots + a_n (x - x_0)^n + \cdots,$$ (2)

where x_0 is a real number.

A power series in $(x - x_0)$ can be converted to a power series in u by the change of variables $u = x - x_0$. Then $\Sigma_{k=0}^{\infty} a_k (x - x_0)^k = \Sigma_{k=0}^{\infty} a_k u^k$. For example, consider

$$\sum_{k=0}^{\infty} \frac{(x - 3)^k}{k!}$$ (3)

If $u = x - 3$, then the power series in $(x - 3)$ given by (3) can be written as

$$\sum_{k=0}^{\infty} \frac{u^k}{k!},$$

which is a power series in u.

Definition 2   CONVERGENCE AND DIVERGENCE OF A POWER SERIES

(i) A power series is said to **converge** at x if the series of real numbers $\Sigma_{k=0}^{\infty} a_k x^k$ converges. Otherwise, it is said to **diverge** at x.

(ii) A power series is said to converge in a set D of a real numbers if it converges for every real number x in D.

EXAMPLE 1   For what real numbers does the power series

$$\sum_{k=0}^{\infty} \frac{x^k}{3^k} = 1 + \frac{x}{3} + \frac{x^2}{3^2} + \frac{x^3}{3^3} + \cdots$$

converge?

Solution. The nth term in this series is $x^n/3^n$. Using the ratio test, we find that

$$\lim_{n \to \infty} \frac{|a_{n+1}|}{|a_n|} = \lim_{n \to \infty} \frac{|x^{n+1}|/3^{n+1}}{|x^n|/3^n} = \lim_{n \to \infty} \left| \frac{x}{3} \right| = \left| \frac{x}{3} \right|.$$

We put in the absolute value bars since the ratio test only applies to a series of *positive* terms. However, as this example shows, we can use the ratio test to test for the absolute convergence of any series of nonzero terms by inserting absolute value bars, thereby making all the terms positive.

Thus the power series converges absolutely if $|x/3| < 1$ or $|x| < 3$ and diverges if $|x| > 3$. The case $|x| = 3$ has to be treated separately. For $x = 3$

$$\sum_{k=0}^{\infty} \frac{x^k}{3^k} = \sum_{k=0}^{\infty} \frac{3^k}{3^k} = \sum_{k=0}^{\infty} 1^k,$$

which diverges. For $x = -3$

$$\sum_{k=0}^{\infty} \frac{x^k}{3^k} = \sum_{k=0}^{\infty} (-1)^k,$$

which also diverges. Thus the series converges in the open interval $(-3, 3)$. We will show in Theorem 1 that since the series diverges for $x = 3$, it diverges for $|x| > 3$, so that conditional convergence at any x for which $|x| > 3$ is ruled out. ■

EXAMPLE 2 For what values of x does the series $\sum_{k=0}^{\infty} x^k/(k + 1)$ converge?

Solution. Here $a_n = x^n/(n + 1)$ so that

$$\lim_{n \to \infty} \frac{|a_{n+1}|}{|a_n|} = \lim_{n \to \infty} \left| \frac{x^{n+1}/(n + 2)}{x^n/(n + 1)} \right| = |x| \lim_{n \to \infty} \frac{n + 1}{n + 2} \overset{\underset{\lim_{n \to \infty} \frac{n+1}{n+2} = 1}{\downarrow}}{=} |x|.$$

Thus the series converges absolutely for $|x| < 1$ and diverges for $|x| > 1$. If $x = 1$, then

$$\sum_{k=0}^{\infty} \frac{x^k}{k + 1} = \sum_{k=0}^{\infty} \frac{1}{k + 1},$$

which diverges since this series is the harmonic series. If $x = -1$, then

$$\sum_{k=0}^{\infty} \frac{x^k}{k + 1} = \sum_{k=0}^{\infty} \frac{(-1)^k}{k + 1} = 1 - \frac{1}{2} + \frac{1}{3} - \frac{1}{4} + \cdots,$$

which converges conditionally by the alternating series test (see Example 14.7.6(a)). In sum, the power series $\sum_{k=0}^{\infty} x^k/(k + 1)$ converges in the half-open interval $[-1, 1)$. ■

The following theorem is of great importance in determining the range of values over which a power series converges.

Theorem 1

(i) If $\sum_{k=0}^{\infty} a_k x^k$ converges at x_0, $x_0 \neq 0$, then it converges absolutely at all x such that $|x| < |x_0|$.

(ii) If $\sum_{k=0}^{\infty} a_k x^k$ diverges at x_0, then it diverges at all x such that $|x| > |x_0|$.

Proof.

(i) Since $\Sigma_{k=0}^{\infty} a_k x_0^k$ converges, $a_k x_0^k \to 0$ as $k \to \infty$ by Theorem 14.4.2. This implies that for all k sufficiently large, $|a_k x_0^k| < 1$. Then if $|x| < |x_0|$ and if k is sufficiently large,

$$|a_k x^k| = \left| a_k \frac{x_0^k x^k}{x_0^k} \right| = |a_k x_0^k| \left| \frac{x}{x_0} \right|^k < \left| \frac{x}{x_0} \right|^k.$$

Since $|x| < |x_0|$, $|x/x_0| < 1$, and the geometric series $\Sigma_{k=0}^{\infty} |x/x_0|^k$ converges. Thus $\Sigma_{k=0}^{\infty} |a_k x^k|$ converges by the comparison test.

(ii) Suppose $|x| > |x_0|$ and $\Sigma_{k=0}^{\infty} a_k x_0^k$ diverges. If $\Sigma_{k=0}^{\infty} a_k x^k$ did converge, then by part (i), $\Sigma_{k=0}^{\infty} a_k x_0^k$ would also converge. This contradiction completes the proof of the theorem. ■

Theorem 1 is very useful for it enables us to place all power series in one of three categories:

Definition 3 RADIUS OF CONVERGENCE

Category 1: $\Sigma_{k=0}^{\infty} a_k x^k$ converges only at 0.

Category 2: $\Sigma_{k=0}^{\infty} a_k x^k$ converges for all real numbers.

Category 3: There exists a positive real number R, called the **radius of convergence** of the power series, such that $\Sigma_{k=0}^{\infty} a_k x^k$ converges if $|x| < R$ and diverges if $|x| > R$. At $x = R$ and at $x = -R$, the series may converge or diverge.

We can extend the notion of radius of convergence to Categories 1 and 2:

1. In Category 1 we say that the radius of convergence is 0.
2. In Category 2 we say that the radius of convergence is ∞.

NOTE. The series in Examples 1 and 2 both fall into Category 3. In Example 1, $R = 3$; and in Example 2, $R = 1$.

EXAMPLE 3 For what values of x does the series $\Sigma_{k=0}^{\infty} k! x^k$ converge?

Solution. Here

$$\lim_{n \to \infty} \left| \frac{a_{n+1}}{a_n} \right| = \lim_{n \to \infty} \left| \frac{(n+1)! x^{n+1}}{n! x^n} \right| = |x| \lim_{n \to \infty} (n+1) = \infty,$$

so that if $x \neq 0$, the series diverges. Thus $R = 0$ and the series falls into Category 1. ■

EXAMPLE 4 For what values of x does the series $\Sigma_{k=0}^{\infty} x^k/k!$ converge?

Solution. Here

$$\lim_{n \to \infty} \left| \frac{a_{n+1}}{a_n} \right| = \lim_{n \to \infty} \left| \frac{x^{n+1}/(n+1)!}{x^n/n!} \right| = |x| \lim_{n \to \infty} \frac{1}{n+1} = 0,$$

so that the series converges for every real number x. Here $R = \infty$ and the series falls into Category 2. ■

In going through these examples, we find that there is an easy way to calculate the radius of convergence. The proof of the following theorem is left as an exercise (see Problems 36 and 37).

Theorem 2.   Consider the power series $\sum_{k=0}^{\infty} a_k x^k$ and suppose that $\lim_{n \to \infty} |a_{n+1}/a_n|$ exists and is equal to L or that $\lim_{n \to \infty} |a_n|^{1/n}$ exists and is equal to L.

(i) If $L = \infty$, then $R = 0$ and the series falls into Category 1.
(ii) If $L = 0$, then $R = \infty$ and the series falls into Category 2.
(iii) If $0 < L < \infty$, then $R = 1/L$ and the series falls into Category 3.

Definition 4   INTERVAL OF CONVERGENCE   The **interval of convergence** of a power series is the interval over which the power series converges.

Using Theorem 2, we can calculate the interval of convergence of a power series in one or two steps:

(i) Calculate R. If $R = 0$, the series converges only at 0; and if $R = \infty$, the interval of convergence is $(-\infty, \infty)$.
(ii) If $0 < R < \infty$, check the values $x = -R$ and $x = R$. Then the interval of convergence is $(-R, R)$, $[-R, R)$, $(-R, R]$, or $[-R, R]$, depending on the convergence or divergence of the series at $x = R$ and $x = -R$.

NOTE.   In Example 1, the interval of convergence is $(-3, 3)$ and in Example 2 the interval of convergence is $[-1, 1)$.

EXAMPLE 5   Find the radius of convergence and interval of convergence of the power series $\sum_{k=0}^{\infty} 2^k x^k / \ln(k + 2)$.

Solution. Here $a_n = 2^n / \ln(n + 2)$ and

$$L = \lim_{n \to \infty} \left| \frac{a_{n+1}}{a_n} \right| = \lim_{n \to \infty} \left| \frac{2^{n+1}/\ln(n + 3)}{2^n/\ln(n + 2)} \right| = 2 \lim_{n \to \infty} \frac{\ln(n + 2)}{\ln(n + 3)} = 2.$$

Thus $R = 1/L = \frac{1}{2}$. For $x = \frac{1}{2}$,

$$\sum_{k=0}^{\infty} \frac{2^k x^k}{\ln(k + 2)} = \sum_{k=0}^{\infty} \frac{1}{\ln(k + 2)},$$

which diverges by comparison with the harmonic series since $1/\ln(k + 2) > 1/(k + 2)$ if $k \geq 1$. If $x = -\frac{1}{2}$, then

$$\sum_{k=0}^{\infty} \frac{2^k x^k}{\ln(k + 2)} = \sum_{k=0}^{\infty} \frac{(-1)^k}{\ln(k + 2)},$$

which converges by the alternating series test. Thus the interval of convergence is $\left[-\frac{1}{2}, \frac{1}{2}\right)$. ∎

EXAMPLE 6 Find the radius of convergence and interval of convergence of the power series $\sum_{k=0}^{\infty} (-1)^k (x - 3)^k/(k + 1)^2$.

Solution. We make the substitution $u = x - 3$. The series then becomes $\sum_{k=0}^{\infty} (-1)^k u^k/(k + 1)^2$, and

$$L = \lim_{n \to \infty} \left| \frac{a_{n+1}}{a_n} \right| = \lim_{n \to \infty} \frac{(n + 1)^2}{(n + 2)^2} = 1,$$

so that $R = 1$. If $u = -1$,

$$\sum_{k=0}^{\infty} \frac{(-1)^k u^k}{(k + 1)^2} = \sum_{k=0}^{\infty} \frac{1}{(k + 1)^2},$$

which converges. If $u = 1$,

$$\sum_{k=0}^{\infty} \frac{(-1)^k u^k}{(k + 1)^2} = \sum_{k=0}^{\infty} \frac{(-1)^k}{(k + 1)^2},$$

which also converges. Thus the interval of convergence of the series $\sum_{k=0}^{\infty}(-1)^k u^k/(k + 1)^2$ is $[-1, 1]$. Since $u = x - 3$, the original series converges for $-1 \le x - 3 \le 1$, or $2 \le x \le 4$. Hence the interval of convergence is $[2, 4]$. ∎

EXAMPLE 7 Find the radius of convergence and interval of convergence of the power series $\sum_{k=0}^{\infty} x^{2k} = 1 + x^2 + x^4 + \cdots$.

Solution.

$$\sum_{k=0}^{\infty} x^{2k} = 1 + 0 \cdot x + 1 \cdot x^2 + 0 \cdot x^3 + 1 \cdot x^4 + 0 \cdot x^5 + 1 \cdot x^6 + \cdots.$$

This example illustrates the pitfalls of blindly applying formulas. We have $a_0 = 1$, $a_1 = 0$, $a_2 = 1$, $a_3 = 0$, ... Thus the ratio a_{n+1}/a_n is 0 if n is even and is undefined if n is odd. The simplest thing to do here is to apply the ratio test directly. The ratio of consecutive terms is $x^{2k+2}/x^{2k} = x^2$. Thus the series converges if $|x| < 1$, diverges if $|x| > 1$, and the radius of convergence is 1. If $x = \pm 1$, then $x^2 = 1$ and the series diverges. Finally, the interval of convergence is $(-1, 1)$. ∎

PROBLEMS 14.8

In Problems 1–33, find the radius of convergence and interval of convergence of the given power series.

1. $\sum_{k=0}^{\infty} \frac{x^k}{6^k}$
2. $\sum_{k=0}^{\infty} \frac{(-1)^k x^k}{8^k}$
3. $\sum_{k=0}^{\infty} \frac{(x + 1)^k}{3^k}$

4. $\displaystyle\sum_{k=0}^{\infty} \frac{(-1)^k(x-3)^k}{4^k}$

5. $\displaystyle\sum_{k=0}^{\infty} (3x)^k$

6. $\displaystyle\sum_{k=0}^{\infty} \frac{x^k}{k^2+1}$

7. $\displaystyle\sum_{k=0}^{\infty} \frac{(x-1)^k}{k^3+3}$

8. $\displaystyle\sum_{k=2}^{\infty} \frac{x^k}{(\ln k)^2}$

9. $\displaystyle\sum_{k=0}^{\infty} \frac{(x+17)^k}{k!}$

10. $\displaystyle\sum_{k=2}^{\infty} \frac{x^k}{k \ln k}$

11. $\displaystyle\sum_{k=0}^{\infty} x^{2k}$

12. $\displaystyle\sum_{k=1}^{\infty} \frac{x^{2k}}{k}$

13. $\displaystyle\sum_{k=1}^{\infty} \frac{(-1)^k x^{2k}}{k^k}$

14. $\displaystyle\sum_{k=1}^{\infty} \frac{kx^k}{\ln(k+1)}$

15. $\displaystyle\sum_{k=0}^{\infty} \frac{(-1)^k kx^k}{\sqrt{k+1}}$

16. $\displaystyle\sum_{k=1}^{\infty} \frac{x^k}{k^k}$

*17. $\displaystyle\sum_{k=2}^{\infty} \frac{x^k}{(\ln k)^k}$ [*Hint:* Use the root test.]

*18. $\displaystyle\sum_{k=1}^{\infty} \frac{3^k x^k}{k^5}$

19. $\displaystyle\sum_{k=1}^{\infty} \frac{(-2x)^k}{k^4}$

20. $\displaystyle\sum_{k=0}^{\infty} \frac{(2x+3)^k}{k!}$

21. $\displaystyle\sum_{k=0}^{\infty} \frac{(2x+3)^k}{5^k}$

22. $\displaystyle\sum_{k=0}^{\infty} \frac{(3x-5)^k}{3^{2k}}$

23. $\displaystyle\sum_{k=0}^{\infty} \left(\frac{k}{15}\right)^k x^k$

24. $\displaystyle\sum_{k=0}^{\infty} (-1)^k x^{2k}$

25. $\displaystyle\sum_{k=0}^{\infty} (-1)^k x^{2k+1}$

26. $\displaystyle\sum_{k=1}^{\infty} \frac{(\ln k)(x+3)^k}{k+1}$

*27. $\displaystyle\sum_{k=1}^{\infty} k^k(x+1)^k$

*28. $\displaystyle\sum_{k=1}^{\infty} \frac{k^k}{k!}x^k$ [*Hint:* See Example 14.6.3, and use Stirling's formula given in Problem 6.2.47.]

29. $\displaystyle\sum_{k=0}^{\infty} \frac{(x+10)^k}{(k+1)3^k}$

*30. $\displaystyle\sum_{k=1}^{\infty} \frac{k!}{k^k}x^k$

31. $\displaystyle\sum_{k=0}^{\infty} [1+(-1)^k]x^k$

32. $\displaystyle\sum_{k=0}^{\infty} \frac{[1+(-1)^k]}{k!}x^k$

33. $\displaystyle\sum_{k=1}^{\infty} \frac{[1+(-1)^k]}{k}x^k$

34. Show that the interval of convergence of the power series $\sum_{k=0}^{\infty} (ax+b)^k/c^k$ with $a>0$ and $c>0$ is $((-c-b)/a, (c-b)/a)$.

35. Prove that if the interval of convergence of a power series is $(a, b]$, then the power series is conditionally convergent at b.

36. Prove the ratio limit part of Theorem 2.   [*Hint:* Show that if $|x|<1/L$, then the series converges absolutely by applying the ratio test. Then show that if $|x|>1/L$, the series diverges.]

37. Show that if $\lim_{n\to\infty} |a_n|^{1/n} = L$, then the radius of convergence of $\sum_{k=0}^{\infty} a_k x^k$ is $1/L$.

38. Show that if the radius of convergence of $\sum_{k=0}^{\infty} a_k x^k$ is R, and if $m>0$ is an integer, then the radius of convergence of the power series $\sum_{k=0}^{\infty} a_k x^{mk}$ is $R^{1/m}$.

14.9 DIFFERENTIATION AND INTEGRATION OF POWER SERIES

Consider the power series

$$\sum_{k=0}^{\infty} a_k(x-x_0)^k = a_0 + a_1(x-x_0) + a_2(x-x_0)^2 + \cdots \tag{1}$$

with a given interval of convergence I. For each x in I we may define a new function f by

$$f(x) = \sum_{k=0}^{\infty} a_k(x - x_0)^k. \tag{2}$$

As we will see in this section and in Section 14.10, many familiar functions can be written as power series. In this section we will discuss some properties of a function given in the form of equation (2).

EXAMPLE 1   We know that

$$\sum_{k=0}^{\infty} x^k = 1 + x + x^2 + \cdots = \frac{1}{1 - x} \quad \text{if} \quad |x| < 1. \tag{3}$$

Thus the function $1/(1 - x)$ for $|x| < 1$ can be defined by

$$f(x) = \frac{1}{1 - x} = \sum_{k=0}^{\infty} x^k \quad \text{if} \quad |x| < 1. \; \blacksquare$$

EXAMPLE 2   Substituting x^4 for x in (3) leads to the equality

$$f(x) = \frac{1}{1 - x^4} = \sum_{k=0}^{\infty} x^{4k} = 1 + x^4 + x^8 + x^{12} + \cdots \quad \text{if} \quad |x| < 1. \; \blacksquare$$

EXAMPLE 3   Substituting $-x$ for x in (3) leads to the equality

$$f(x) = \frac{1}{1 - (-x)} = \frac{1}{1 + x} = \sum_{k=0}^{\infty} (-1)^k x^k$$

$$= 1 - x + x^2 - x^3 + x^4 - \cdots \quad \text{if} \quad |x| < 1. \; \blacksquare$$

Once we see that certain functions can be written as power series, the next question that naturally arises is whether such functions can be differentiated and integrated. The remarkable theorem given next ensures that every function represented as a power series can be differentiated and integrated at any x such that $|x| < R$, the radius of convergence. Moreover, we see how the derivative and integral can be calculated. The proof of this theorem is long (but not conceptually difficult) and so is omitted.†

Theorem 1.   Let the power series $\sum_{k=0}^{\infty} a_k x^k$ have the radius of convergence $R > 0$. Let

$$f(x) = \sum_{k=0}^{\infty} a_k x^k = a_0 + a_1 x + a_2 x^2 + \cdots \quad \text{for} \quad |x| < R.$$

Then for $|x| < R$ we have the following:

(i) $f(x)$ is continuous.
(ii) The derivative $f'(x)$ exists, and

---

†See R. C. Buck *Advanced Calculus*, McGraw-Hill, New York, 1965, p. 198.

$$f'(x) = \frac{d}{dx}a_0 + \frac{d}{dx}a_1x + \frac{d}{dx}a_2x^2 + \cdots$$

$$= a_1 + 2a_2x + 3a_3x^2 + \cdots = \sum_{k=1}^{\infty} ka_k x^{k-1}.$$

(iii) The antiderivative $\int f(x)\, dx$ exists and

$$\int f(x)\, dx = \int a_0\, dx + \int a_1 x\, dx + \int a_2 x^2\, dx + \cdots$$

$$= a_0 x + a_1 \frac{x^2}{2} + a_2 \frac{x^3}{3} + \cdots + C = \sum_{k=0}^{\infty} a_k \frac{x^{k+1}}{k+1} + C.$$

Moreover, the two series $\sum_{k=1}^{\infty} ka_k x^{k-1}$ and $\sum_{k=0}^{\infty} a_k x^{k+1}/(k+1)$ both have the radius of convergence R.

Simply put, this theorem says that the derivative of a converging power series is the series of derivatives of its terms and that the integral of a converging power series is the series of integrals of its terms.

A more concise statement of Theorem 1 is: *A power series may be differentiated and integrated term by term within its radius of convergence.*

EXAMPLE 4 From Example 3 we have

$$\frac{1}{1+x} = 1 - x + x^2 - x^3 + \cdots = \sum_{k=0}^{\infty} (-1)^k x^k \tag{4}$$

for $|x| < 1$. Substituting $u = x + 1$, we have $x = u - 1$ and $1/(1 + x) = 1/u$. If $-1 < x < 1$, then $0 < u < 2$, and we obtain

$$\frac{1}{u} = 1 - (u - 1) + (u - 1)^2 - (u - 1)^3 + \cdots = \sum_{k=0}^{\infty} (-1)^k (u - 1)^k$$

for $0 < u < 2$. Integration then yields

$$\ln u = \int \frac{du}{u} = u - \frac{(u-1)^2}{2} + \frac{(u-1)^3}{3} - \cdots + C.$$

Since $\ln 1 = 0$, we immediately find that $C = -1$, so that

$$\ln u = \sum_{k=0}^{\infty} (-1)^k \frac{(u-1)^{k+1}}{k+1} \tag{5}$$

for $0 < u < 2$. Here we have expressed the logarithmic function defined on the interval $(0, 2)$ as a power series. ∎

EXAMPLE 5   The series

$$f(x) = 1 + x + \frac{x^2}{2!} + \frac{x^3}{3!} + \cdots = \sum_{k=0}^{\infty} \frac{x^k}{k!} \tag{6}$$

converges for every real number x (i.e., $R = \infty$; see Example 14.8.4). But

$$f'(x) = \frac{d}{dx} 1 + \frac{d}{dx} x + \frac{d}{dx} \frac{x^2}{2!} + \cdots = 1 + x + \frac{x^2}{2!} + \cdots = f(x).$$

Thus f satisfies the differential equation

$$f' = f,$$

and so from the discussion in Section 6.6, we find that

$$f(x) = ce^x \tag{7}$$

for some constant c. Then substituting $x = 0$ into equations (6) and (7) yields

$$f(0) = 1 = ce^0 = c,$$

so that $f(x) = e^x$. We have obtained the important expansion that is valid for any real number x:

$$e^x = 1 + x + \frac{x^2}{2!} + \frac{x^3}{3!} + \cdots = \sum_{k=0}^{\infty} \frac{x^k}{k!}. \tag{8}$$

For example, if we substitute the value $x = 1$ into (8), we obtain partial sum approximations for $e = 1 + 1 + 1/2! + 1/3! + \cdots$ (see Table 1). The last value $(\sum_{k=0}^{8} 1/k!)$ is correct to five decimal places.

TABLE 1

n	0	1	2	3	4	5	6	7	8
$S_n = \sum_{k=0}^{n} \frac{1}{k!}$	1	2	2.5	2.66667	2.70833	2.71667	2.71806	2.71825	2.71828

EXAMPLE 6   Substituting $x = -x$ in (8), we obtain

$$e^{-x} = 1 - x + \frac{x^2}{2!} - \frac{x^3}{3!} + \cdots = \sum_{k=0}^{\infty} (-1)^k \frac{x^k}{k!}. \tag{9}$$

Since this is an alternating series if $x > 0$, Theorem 14.7.3 tells us that the error

$|S - S_n|$ in approximating e^{-x} for $x > 0$ is bounded by $|a_{n+1}| = x^{n+1}/(n + 1)!$.† For example, to calculate e^{-1} with an error of less than 0.0001, we must have $|S - S_n| \leq 1/(n + 1)! < 0.0001 = 1/10{,}000$, or $(n + 1)! > 10{,}000$. If $n = 7$, $(n + 1)! = 8! = 40{,}320$, so that $\Sigma_{k=0}^{7} (-1)^k/k!$ will approximate e^{-1} correct to four decimal places. We obtain

$$e^{-1} \approx 1 - 1 + \frac{1}{2!} - \frac{1}{3!} + \frac{1}{4!} - \frac{1}{5!} + \frac{1}{6!} - \frac{1}{7!}$$

$$= \frac{1}{2} - \frac{1}{6} + \frac{1}{24} - \frac{1}{120} + \frac{1}{720} - \frac{1}{5040}$$

$$= 0.5 - 0.16667 + 0.04167 - 0.00833 + 0.00139 - 0.0002$$

$$\approx 0.36786.$$

Note that $e^{-1} \approx 0.36788$ correct to five decimal places. ∎

EXAMPLE 7   Consider the series

$$f(x) = 1 - \frac{x^2}{2!} + \frac{x^4}{4!} - \frac{x^6}{6!} + \cdots = \sum_{k=0}^{\infty} (-1)^k \frac{x^{2k}}{(2k)!}. \qquad (10)$$

It is easy to see that $R = \infty$ since the series is absolutely convergent for every x by comparison with the series (8) for e^x. [The series (8) is larger than the series (10) since it contains the terms $x^n/n!$ for n both even and odd, not just for n even, as in (10).] Differentiating, we obtain

$$f'(x) = -x + \frac{x^3}{3!} - \frac{x^5}{5!} + \frac{x^7}{7!} - \cdots = \sum_{k=0}^{\infty} (-1)^{k+1} \frac{x^{2k+1}}{(2k + 1)!}. \qquad (11)$$

Since this series has a radius of convergences $R = \infty$, we can differentiate once more to obtain

$$f''(x) = -1 + \frac{x^2}{2!} - \frac{x^4}{4!} + \frac{x^6}{6!} - \cdots = \sum_{k=0}^{\infty} \frac{(-1)^{k+1}x^{2k}}{(2k)!} = -f(x).$$

Thus we see that f satisfies the differential equation

$$f'' + f = 0. \qquad (12)$$

Moreover, from equations (10) and (11) we see that

$$f(0) = 1 \quad \text{and} \quad f'(0) = 0. \qquad (13)$$

We discussed this differential equation in Section 7.3. It is easily seen that the function $f(x) = \cos x$ satisfies equation (12) together with the conditions (13). In fact, although

---

†We can apply Theorem 14.7.3 here because the terms in the sequence $\{x^n/n!\}$ are monotone decreasing as long as $x < n + 1$.

we do not prove it here, it is the only function that does so. Thus we have

$$\cos x = 1 - \frac{x^2}{2!} + \frac{x^4}{4!} - \frac{x^6}{6!} + \cdots = \sum_{k=0}^{\infty} (-1)^k \frac{x^{2k}}{(2k)!}. \tag{14}$$

Since

$$\frac{d}{dx} \cos x = -\sin x,$$

we obtain, from (14), the series (after multiplying both sides by -1)

$$\sin x = x - \frac{x^3}{3!} + \frac{x^5}{5!} - \frac{x^7}{7!} + \cdots = \sum_{k=0}^{\infty} (-1)^k \frac{x^{2k+1}}{(2k+1)!}. \blacksquare \tag{15}$$

Power series expansions can be very useful for approximate integration.

EXAMPLE 8   Calculate $\int_0^1 e^{-t^2}\, dt$ with an error < 0.0001.

Solution. Substituting t^2 for x in (9), we find that

$$e^{-t^2} = 1 - t^2 + \frac{t^4}{2!} - \frac{t^6}{3!} + \cdots = \sum_{k=0}^{\infty} (-1)^k \frac{t^{2k}}{k!}. \tag{16}$$

Then we integrate to find that

$$\int_0^x e^{-t^2}\, dt = x - \frac{x^3}{3} + \frac{x^5}{5 \cdot 2!} - \frac{x^7}{7 \cdot 3!} + \cdots = \sum_{k=0}^{\infty} \frac{(-1)^k x^{2k+1}}{(2k+1)k!}. \tag{17}$$

The error $|S - S_n|$ is bounded by

$$|a_{n+1}| = \left| \frac{x^{2(n+1)+1}}{[2(n+1)+1](n+1)!} \right|.\dagger$$

In our example $x = 1$, so we need to choose n so that

$$\frac{1}{(2n+3)(n+1)!} < 0.0001, \qquad \text{or} \qquad (2n+3)(n+1)! > 10{,}000.$$

If $n = 6$, then $(2n+3)(n+1)! = (15)(7!) = 75{,}600$. With this choice of n ($n = 5$ is too small), we obtain

$$\int_0^1 e^{-t^2}\, dt \approx 1 - \frac{1}{3} + \frac{1}{5 \cdot 2!} - \frac{1}{7 \cdot 3!} + \frac{1}{9 \cdot 4!} - \frac{1}{11 \cdot 5!} + \frac{1}{13 \cdot 6!}$$

$$\approx 1 - 0.33333 + 0.1 - 0.02381 + 0.00463 - 0.00076 + 0.00011$$

$$= 0.74684,$$

---

†As in Example 6, we can apply Theorem 14.7.3 as long as the terms are decreasing. Here $a_{n-1} < a_n$ if $x^2 < (2n+3)(n+1)/(2n+1)$. This holds for every n when $x = 1$.

and to four decimal places,

$$\int_0^1 e^{-t^2}\, dt = 0.7468. \quad \blacksquare$$

In Section 8.10 we discussed the trapezoidal rule and Simpson's rule as examples of some more general ways to calculate definite integrals that do not require the existence of a series expansion of the function being integrated. However, as shown in Example 8 on page 743, a power series provides an easy method of numerical integration when the power series representation of a function is readily obtainable.

PROBLEMS 14.9

1. By substituting x^2 for x in (4), find a series expansion for $1/(1 + x^2)$ that is valid for $|x| < 1$.
2. Integrate the series obtained in Problem 1 to obtain a series expansion for $\tan^{-1} x$.
3. Use the result of Problem 2 to obtain an estimate of π that is correct to two decimal places.
4. Use the series expansion for $\ln x$ to calculate the following to two decimal places of accuracy:
   (a) $\ln 0.5$ (b) $\ln 1.6$

In Problems 5–13, estimate the given integral to within the given accuracy.

5. $\displaystyle\int_0^1 e^{-t^2}\, dt$; error < 0.01 6. $\displaystyle\int_0^1 e^{-t^3}\, dt$; error < 0.001

7. $\displaystyle\int_0^{1/2} \cos t^2\, dt$; error < 0.001 8. $\displaystyle\int_0^{1/2} \sin t^2\, dt$; error < 0.0001

9. $\displaystyle\int_0^1 t^2 e^{-t^2}\, dt$; error < 0.01 [*Hint:* The series expansion of $t^2 e^{-t^2}$ is obtained by multiplying each term of the series expansion of e^{-t^2} by t^2.]

10. $\displaystyle\int_0^{1/4} t^5 e^{-t^5}\, dt$; error < 0.0001 11. $\displaystyle\int_0^1 \cos \sqrt{t}\, dt$; error < 0.01

12. $\displaystyle\int_0^1 t \sin \sqrt{t}\, dt$; error < 0.001 13. $\displaystyle\int_0^{1/2} \frac{dt}{1 + t^8}$; error < 0.0001

14. Find a series expansion for xe^x that is valid for all real values of x.
15. Use the result of Problem 14 to find a series expansion for $\int_0^x te^t\, dt$.
16. Use the result of Problem 15 to show that $\sum_{k=0}^{\infty} 1/(k + 2)k! = 1$.
17. Find a power series expansion for $\int_0^x [\ln(1 + t)/t]\, dt$.
18. Expand $1/x$ as a power series of the form $\sum_{k=0}^{\infty} a_k(x - 1)^k$. What is the interval of convergence of this series?
*19. Define the function $J(x)$ by $J(x) = \sum_{k=0}^{\infty} [(-1)^k/(k!)^2](x/2)^{2k}$.
   (a) What is the interval of convergence of this series?
   (b) Show that $J(x)$ satisfies the differential equation

$$x^2 J''(x) + xJ'(x) + x^2 J(x) = 0.$$

The function $J(x)$ is called a *Bessel function of order zero*.†

---

†Named after the German physicist and mathematician Wilhelm Bessel (1784–1846), who used the function in his study of planetary motion. The Bessel functions of various orders arise in many applications in modern physics and engineering.

14.10 TAYLOR AND MACLAURIN SERIES

In the last two sections we used the fact that within its interval of convergence, the function

$$f(x) = \sum_{k=0}^{\infty} a_k(x - x_0)^k$$

is differentiable and integrable. In this section we look more closely at the coefficients a_k and show that they can be represented in terms of derivatives of the function f.

We begin with the case $x_0 = 0$ and assume that $R > 0$, so that the theorem on power series differentiation applies. We have

$$f(x) = \sum_{k=0}^{\infty} a_k x^k = a_0 + a_1 x + a_2 x^2 + \cdots + a_n x^n + \cdots, \tag{1}$$

and clearly,

$$f(0) = a_0 + 0 + 0 + \cdots + 0 + \cdots = a_0. \tag{2}$$

If we differentiate (1), we obtain

$$f'(x) = \sum_{k=1}^{\infty} k a_k x^{k-1} = a_1 + 2a_2 x + 3a_3 x^2 + \cdots + n a_n x^{n-1} + \cdots \tag{3}$$

and

$$f'(0) = a_1. \tag{4}$$

Continuing to differentiate, we obtain

$$f''(x) = \sum_{k=2}^{\infty} k(k - 1)a_k x^{k-2}$$

$$= 2a_2 + 3 \cdot 2a_3 x + 4 \cdot 3a_4 x^2 + \cdots + n(n - 1)a_n x^{n-2} + \cdots$$

and

$$f''(0) = 2a_2,$$

or

$$a_2 = \frac{f''(0)}{2} = \frac{f''(0)}{2!}. \tag{5}$$

Similarly,

$$f'''(x) = \sum_{k=3}^{\infty} k(k - 1)(k - 2)a_k x^{k-3}$$

$$= 3 \cdot 2a_3 + 4 \cdot 3 \cdot 2a_4 x + 5 \cdot 4 \cdot 3a_5 x^2 + \cdots + n(n-1)(n-2)a_n x^{n-3} + \cdots$$

and

$$f'''(0) = 3 \cdot 2a_3,$$

or

$$a_3 = \frac{f'''(0)}{3 \cdot 2} = \frac{f'''(0)}{3!}. \tag{6}$$

It is not difficult to see that this pattern continues and that for every positive integer n

$$a_n = \frac{f^{(n)}(0)}{n!}. \tag{7}$$

For $n = 0$ we use the convention $0! = 1$ and $f^{(0)}(x) = f(x)$. Then formula (7) holds for every n, and we have the following:

If

$$f(x) = \sum_{k=0}^{\infty} a_k x^k,$$

then

$$f(x) = \sum_{k=0}^{\infty} \frac{f^{(k)}(0)}{k!} x^k$$

$$= f(0) + f'(0)x + f''(0)\frac{x^2}{2!} + \cdots + f^{(n)}(0)\frac{x^n}{n!} + \cdots \tag{8}$$

for every x in the interval of convergence.

In the general case, if

$$f(x) = \sum_{k=0}^{\infty} a_k (x - x_0)^k$$

$$= a_0 + a_1(x - x_0) + a_2(x - x_0)^2 + \cdots + a_n(x - x_0)^n + \cdots, \tag{9}$$

then

$$f(x_0) = a_0,$$

and differentiating as before, we find that

$$a_n = \frac{f^{(n)}(x_0)}{n!}. \tag{10}$$

Thus we have the following: If

$$f(x) = \sum_{k=0}^{\infty} a_k (x - x_0)^k,$$

then

$$\begin{aligned} f(x) &= \sum_{k=0}^{\infty} \frac{f^{(k)}(x_0)}{k!}(x - x_0)^k \\ &= f(x_0) + f'(x_0)(x - x_0) + f''(x_0)\frac{(x - x_0)^2}{2!} + \cdots \\ &\quad + f^{(n)}(x_0)\frac{(x - x_0)^n}{n!} + \cdots \end{aligned} \tag{11}$$

for every x in the interval of convergence.

Definition 1 TAYLOR AND MACLAURIN SERIES The series in (11) is called the **Taylor series**† of the function f at x_0. The special case $x_0 = 0$ in (8) is called a **Maclaurin series**.‡ We see that the first n terms of the Taylor series of a function are simply the Taylor polynomial described in Section 13.1.

■ WARNING: We have shown here that *if $f(x) = \sum_{k=0}^{\infty} a_k(x - x_0)^k$, then f is infinitely differentiable* (i.e., f has derivatives of all orders) and that the series for f is the Taylor series (or Maclaurin series if $x_0 = 0$) of f. What we have *not* shown is that if f is infinitely differentiable at x_0, then f has a Taylor series expansion at x_0. In general, this last statement is false, as we will see in Example 3 on page 749.

EXAMPLE 1 Find the Maclaurin series for e^x.

Solution. If $f(x) = e^x$, then $f(0) = f'(0) = \cdots = f^{(k)}(0) = 1$, and

$$e^x = \sum_{k=0}^{\infty} \frac{x^k}{k!} = 1 + x + \frac{x^2}{2!} + \frac{x^3}{3!} + \cdots + \frac{x^n}{n!} + \cdots. \tag{12}$$

This series is the series we obtained in Example 14.9.5. It is important to note here that

---

†The history of the Taylor series is somewhat muddled. It has been claimed that the basis for its development was found in India before 1550! (Taylor published the result in 1715.) For an interesting discussion of this controversy, see the paper by C. T. Rajagopal and T. V. Vedamurthi, "On the Hindu proof of Gregory's series," *Scripta Mathematica* **17**, 65–74 (1951).

‡See the accompanying biographical sketch.

Colin Maclaurin

Considered the finest British mathematician of the generation after Newton, Colin Maclaurin was certainly one of the best mathematicians of the eighteenth century.

Born in Scotland, Maclaurin was a mathematical prodigy and entered Glasgow University at the age of eleven. By the age of nineteen he was a professor of mathematics in Aberdeen and later obtained a post at the University of Edinburgh.

Maclaurin is best known for the term *Maclaurin series*, which is the Taylor series in the case $x_0 = 0$. He used this series in his 1742 work, *Treatise of Fluxions*. (Maclaurin acknowledged that the series had first been used by Taylor in 1715.) The *Treatise of Fluxions* was most significant in that it presented the first logical description of Newton's method of fluxions. This work was written to defend Newton from the attacks of the powerful Bishop George Berkeley (1685–1753). Berkeley was troubled (as are many of today's calculus students) by the idea of a quotient that takes the form 0/0. This, of course, is what we obtain when we take a derivative. Berkeley wrote:

And what are these fluxions? The velocities of evanescent increments. And what are these same evanescent increments? They are neither finite quantities nor quantities infinitely small nor yet nothing. May we not call them ghosts of departed quantities?

Maclaurin answered Berkeley using geometric arguments. Later, Newton's calculus was put on an even firmer footing by the work of Lagrange in 1797 (see page 670).

Maclaurin made many other contributions to mathematics—especially in the areas of geometry and algebra. He published his *Geometria organica* when only 21 years old. His posthumous work *Treatise of Algebra,* published in 1748, contained many important results, including the well-known *Cramer's rule* for solving a system of equations (Cramer published the result in 1750).

In 1745, when "Bonnie Prince Charlie" marched against Edinburgh, Maclaurin helped defend the city. When the city fell, Maclaurin escaped, fleeing to York, where he died in 1746 at the age of 48.

what this example shows is that *if* e^x has a Maclaurin series expansion, then the series must be the series (12). It does not show that e^x actually does have such a series expansion. To prove that the series in (12) is really equal to e^x, we differentiate, as in Example 14.9.5, and use the fact that the only continuous function that satisfies

$$f'(x) = f(x), \qquad f(0) = 1,$$

is the function e^x. ∎

EXAMPLE 2   Assuming that the function $f(x) = \cos x$ can be written as a Maclaurin series, find that series.

Solution. If $f(x) = \cos x$, then $f(0) = 1$, $f'(0) = 0$, $f''(0) = -1$, $f'''(0) = 0$, $f^{(4)}(0) = 1$, and so on, so that if

$$\cos x = \sum_{k=0}^{\infty} a_k x^k,$$

then

$$\cos x = f(0) + f'(0) + \frac{f''(0)x^2}{2!} + \frac{f'''(0)x^3}{3!} + \frac{f^{(4)}(0)x^4}{4!} + \cdots,$$

or

$$\cos x = 1 - \frac{x^2}{2!} + \frac{x^4}{4!} - \frac{x^6}{6!} + \cdots = \sum_{k=0}^{\infty} \frac{(-1)^k x^{2k}}{(2k)!}. \tag{13}$$

This series is the series found in Example 14.9.7.

NOTE. Again, this does not prove that the equality in (13) is correct. It only shows that *if* $\cos x$ has a Maclaurin expansion, then the expansion must be given by (13). We will show that $\cos x$ has a Maclaurin series in Example 5. ∎

EXAMPLE 3   Let

$$f(x) = \begin{cases} e^{-1/x^2}, & \text{if } x \neq 0 \\ 0, & \text{if } x = 0. \end{cases}$$

Find a Maclaurin expansion for f if one exists.

Solution. First, we note that since $\lim_{x \to 0} e^{-1/x^2} = 0$, f is continuous. Now recall from Example 12.3.5, that $\lim_{x \to \infty} x^a e^{-bx} = 0$ if $b > 0$. Let $y = 1/x^2$. Then as $x \to 0$, $y \to \infty$. Also, $1/x^n = (1/x^2)^{n/2}$, so that $\lim_{x \to 0} (e^{-1/x^2}/x^n) = \lim_{y \to \infty} y^{n/2} e^{-y} = 0$ by Example 12.3.5.

Now for $x \neq 0$, $f'(x) = (2/x^3)e^{-1/x^2} \to 0$ as $x \to 0$, so that f' is continuous at 0. Similarly, $f''(x) = [(4/x^6) - (6/x^4)]e^{-1/x^2}$, which also approaches 0 as $x \to 0$ by the limit result above. In fact, *every* derivative of f is continuous and $f^{(n)}(0) = 0$ for every n. (You were asked to prove this result in Problem 12.3.36.) Thus f is infinitely differentiable, and *if* it had a Maclaurin series that represented the function, then we would have

$$f(x) = f(0) + f'(0)x + f''(0)\frac{x^2}{2!} + \cdots.$$

But $f(0) = f'(0) = f''(0) = \cdots = 0$, so that the Maclaurin series would be the zero series. But since f is obviously not the zero function, we can only conclude that there is *no* Maclaurin series that represents f at any point other than 0. ∎

Example 3 illustrates that infinite differentiability is not sufficient to guarantee that a given function can be represented by its Taylor series. Something more is needed.

Definition 2   ANALYTIC FUNCTION   We say that a function f is **analytic** at x_0 if f can be represented by a Taylor series in some neighborhood of x_0.

We see that the functions e^x and $\cos x$ are analytic at 0, while the function

$$f(x) = \begin{cases} e^{-1/x^2}, & x \neq 0 \\ 0, & x = 0 \end{cases}$$

is not. A condition that guarantees analyticity of an infinitely differentiable function is given below.

Theorem 1. Suppose that the function f has continuous derivatives of all orders in a neighborhood $N(x_0)$ of the number x_0.

Then f is analytic at x_0 if and only if

$$\lim_{n \to \infty} R_n(x) = \lim_{n \to \infty} \frac{f^{(n+1)}(c_n)}{(n+1)!}(x - x_0)^{n+1} = 0 \qquad (14)$$

for every x in $N(x_0)$ where c_n is between x_0 and x.

REMARK. The expression between the equal signs in (14) is simply the remainder term given by Taylor's theorem (see page 668).

Proof. The hypotheses of Taylor's theorem apply, so that we can write, for any n,

$$f(x) = P_n(x) + R_n(x), \qquad (15)$$

where $P_n(x)$ is the nth degree Taylor polynomial for f. To show that f is analytic, we must show that

$$\lim_{n \to \infty} P_n(x) = f(x) \qquad (16)$$

for every x in $N(x_0)$. But if x is in $N(x_0)$, we obtain, from (14) and (15),

$$\lim_{n \to \infty} P_n(x) = \lim_{n \to \infty} [f(x) - R_n(x)] = f(x) - \lim_{n \to \infty} R_n(x) = f(x) - 0 = f(x).$$

Conversely, if f is analytic, then $f(x) = \lim_{n \to \infty} P_n(x)$ so $R_n(x) \to 0$ as $n \to \infty$. ∎

EXAMPLE 4 If $f(x) = e^x$, then $f^{(n)}(x) = e^x$, and

$$\lim_{n \to \infty} \left| \frac{f^{(n+1)}(c_n)}{(n+1)!}(x - 0)^{n+1} \right| = \lim_{n \to \infty} \frac{e^{c_n}|x|^{n+1}}{(n+1)!} \overset{0 < c_n < |x|}{\leq} e^{|x|} \lim_{n \to \infty} \frac{|x|^{n+1}}{(n+1)!} \to 0,$$

since $|x|^{n+1}/(n+1)!$ is the $(n+2)$nd term in the converging power series $\sum_{k=0}^{\infty} |x|^k/k!$ and the terms in a converging power series $\to 0$ by the Theorem 14.4.2. Since this result is true for any $x \in \mathbb{R}$, we may take $N = (-\infty, \infty)$ to conclude that the series (12) is valid for every real number x. ∎

EXAMPLE 5   Let $f(x) = \cos x$. Since all derivatives of $\cos x$ are equal to $\pm \sin x$ or $\pm \cos x$, we see that $|f^{(n+1)}(c_n)| \leq 1$. Then for $x_0 = 0$, $|R_n(x)| \leq |x|^{n+1}/(n+1)!$, which $\to 0$ as $n \to \infty$, so that the series (13) is also valid for every real number x. ■

EXAMPLE 6   It is evident that for the function in Example 3, $R_n(x) \nrightarrow 0$ if $x \neq 0$. This follows from the fact that $R_n(x) = f(x) - P_n(x) = e^{-1/x^2} - 0 = e^{-1/x^2} \neq 0$ if $x \neq 0$. ■

EXAMPLE 7   Find the Taylor expansion for $f(x) = \ln x$ at $x = 1$.

Solution. Since $f'(x) = 1/x$, $f''(x) = -1/x^2$, $f'''(x) = 2/x^3$, $f^{(4)}(x) = -6/x^4, \ldots,$ $f^{(n)}(x) = (-1)^{n+1}(n-1)!/x^n$, we find that $f(1) = 0$, $f'(1) = 1$, $f''(1) = -1$, $f'''(1) = 2$, $f^{(4)}(1) = -6, \ldots, f^{(n)}(1) = (-1)^{n+1}(n-1)!$. Then wherever valid,

$$\ln x = \sum_{k=0}^{\infty} f^{(k)}(1)\frac{(x-1)^k}{k!}$$

$$= 0 + (x-1) - \frac{(x-1)^2}{2} + \frac{2(x-1)^3}{3!} - \frac{3!(x-1)^4}{4!} + \frac{4!(x-1)^5}{5!} + \cdots,$$

or

$$\ln x = (x-1) - \frac{(x-1)^2}{2} + \frac{(x-1)^3}{3} - \frac{(x-1)^4}{4} + \cdots$$

$$= \sum_{k=1}^{\infty} \frac{(-1)^{k+1}(x-1)^k}{k}. \quad ■ \tag{17}$$

In Section 13.3 [see equation (13.3.8)] we showed that

$$\ln(1+x) = \sum_{k=1}^{n+1} \frac{(-1)^{k+1}x^k}{k} + R_{n+1}(x), \tag{18}$$

where $R_{n+1}(x) \to 0$ as $n \to \infty$ whenever $-1 < x \leq 1$ [see equations (13.3.9) and (13.3.10)]. From (18) we see that

$$\ln u = \ln[1 + (u-1)] = \sum_{k=1}^{n+1} \frac{(-1)^{k+1}(u-1)^k}{k} + R_{n+1}(u-1), \tag{19}$$

where $R_{n+1}(u-1) \to 0$ as $n \to \infty$ whenever $-1 < u - 1 \leq 1$, or $0 < u \leq 2$. But this implies that the series (17) converges to $\ln(x)$ for $0 < x \leq 2$. When $x = 2$, we obtain, from (17),

$$\ln 2 = 1 - \frac{1}{2} + \frac{1}{3} - \frac{1}{4} + \frac{1}{5} - \cdots = \sum_{k=1}^{\infty} \frac{(-1)^{k+1}}{k}. \tag{20}$$

EXAMPLE 8   Find a Taylor series for $f(x) = \sin x$ at $x = \pi/3$.

Solution. Here $f(\pi/3) = \sqrt{3}/2$, $f'(\pi/3) = 1/2$, $f''(\pi/3) = -\sqrt{3}/2$, $f'''(\pi/3) = -1/2$, and so on, so that

$$\sin x = \frac{\sqrt{3}}{2} + \frac{1}{2}\left(x - \frac{\pi}{3}\right) - \frac{\sqrt{3}}{2}\frac{[x - (\pi/3)]^2}{2!} - \frac{1}{2}\frac{[x - (\pi/3)]^3}{3!}$$

$$+ \frac{\sqrt{3}}{2}\frac{[x - (\pi/3)]^4}{4!} + \cdots .$$

The proof that this series is valid for every real number x is similar to the proof in Example 5 and is therefore omitted. ∎

We provide here a list of useful Maclaurin series:

$$e^x = \sum_{k=0}^{\infty} \frac{x^k}{k!} = 1 + x + \frac{x^2}{2!} + \frac{x^3}{3!} + \cdots \tag{21}$$

$$\cos x = \sum_{k=0}^{\infty} \frac{(-1)^k x^{2k}}{(2k)!} = 1 - \frac{x^2}{2!} + \frac{x^4}{4!} - \frac{x^6}{6!} + \cdots \tag{22}$$

$$\sin x = \sum_{k=0}^{\infty} \frac{(-1)^k x^{2k+1}}{(2k+1)!} = x - \frac{x^3}{3!} + \frac{x^5}{5!} - \frac{x^7}{7!} + \cdots \tag{23}$$

$$\cosh x = \sum_{k=0}^{\infty} \frac{x^{2k}}{(2k)!} = 1 + \frac{x^2}{2!} + \frac{x^4}{4!} + \frac{x^6}{6!} + \cdots \tag{24}$$

$$\sinh x = \sum_{k=0}^{\infty} \frac{x^{2k+1}}{(2k+1)!} = x + \frac{x^3}{3!} + \frac{x^5}{5!} + \frac{x^7}{7!} + \cdots \tag{25}$$

You are asked to prove, in Problems 1 and 2, that the series (23), (24), and (25) are valid for every real number x.

BINOMIAL SERIES

We close this section by deriving another series that is quite useful. Let $f(x) = (1 + x)^r$, where r is a real number not equal to an integer. We have

$$f'(x) = r(1 + x)^{r-1},$$
$$f''(x) = r(r - 1)(1 + x)^{r-2},$$
$$f'''(x) = r(r - 1)(r - 2)(1 + x)^{r-3},$$
$$\vdots$$
$$f^{(n)}(x) = r(r - 1)(r - 2) \cdots (r - n + 1)(1 + x)^{r-n}.$$

Note that since r is not an integer, $r - n$ is never equal to 0, and all derivatives exist and are nonzero as long as $x \neq -1$. Then

$$f(0) = 1,$$
$$f'(0) = r,$$

$$f''(0) = r(r - 1),$$
$$\vdots$$
$$f^{(n)}(0) = r(r - 1) \cdots (r - n + 1),$$

and we can write

$$(1 + x)^r = 1 + rx + \frac{r(r - 1)}{2!}x^2 + \frac{r(r - 1)(r - 2)}{3!}x^3 + \cdots$$

$$+ \frac{r(r - 1) \cdots (r - n + 1)}{n!}x^n + \cdots \tag{26}$$

$$= 1 + \sum_{k=1}^{\infty} \frac{r(r - 1) \cdots (r - k + 1)}{k!}x^k$$

The series (26) is called the **binomial series.**

Some applications of the binomial series are given in Problems 25–28.

PROBLEMS 14.10

1. Prove that the series (23) represents $\sin x$ for all real x.
2. **(a)** Prove that the series (24) represents $\cosh x$ for all real x.
   (b) Use the fact that $\sinh x = (d/dx) \cosh x$ to derive the series in (25).
3. Find the Taylor series for e^x at 1.
4. Find the Maclaurin series for e^{-x}.
5. Find the Taylor series for $\cos x$ at $\pi/4$.
6. Find the Taylor series for $\sinh x$ at $\ln 2$.
7. Find the Maclaurin series for $e^{\alpha x}$, α real.
8. Find the Maclaurin series for xe^x.
9. Find the Maclaurin series for $x^2 e^{-x^2}$.
10. Find the Maclaurin series for $(\sin x)/x$.
11. Find the Taylor series for e^x at $x = -1$.
12. Find the Maclaurin series for $\sin^2 x$. [*Hint:* $\sin^2 x = (1 - \cos 2x)/2$.]
13. Find the Taylor series for $(x - 1)\ln x$ at 1. Over what interval is this representation valid?
14. Find the first three nonzero terms of the Maclaurin series for $\tan x$. What is its interval of convergence?
15. Find the first four terms of the Taylor series for $\csc x$ at $\pi/2$. What is its interval of convergence?
16. Find the first three nonzero terms of the Maclaurin series for $\ln |\cos x|$. What is its interval of convergence? [*Hint:* $\int \tan x \, dx = -\ln |\cos x|$.]
17. Find the Taylor series of $\sqrt{x}$ at $x = 4$. What is its radius of convergence?
18. Find the Maclaurin series of $\sin^{-1} x$. What is its radius of convergence? [*Hint:* Expand $1/\sqrt{1 - x^2}$ and integrate.]
19. Use the Maclaurin series for $\sin x$ to obtain the Maclaurin series for $\sin x^2$.
20. Find the Maclaurin series for $\cos x^2$.
21. Differentiate the Maclaurin series for $\sin x$ and show that it is equal to the Maclaurin series for $\cos x$.
22. Differentiate the Maclaurin series for $\sinh x$ and show that it is equal to the Maclaurin series for $\cosh x$.

23. Using the fact that if f has a Taylor series at x_0, then the Taylor series is given by (11), show that $1/(1-x) = 1 + x + x^2 + \cdots$ is the Taylor series for $1/(1-x)$ when $|x| < 1$.

24. Find the Maclaurin series for $\tan^{-1} x$. What is its interval of convergence? [*Hint:* Integrate the series for $1/(1+x^2)$.]

25. Show that for any real number r

$$1 + \frac{r}{2} + \frac{r(r-1)}{2^2 2!} + \cdots + \frac{r(r-1)\cdots(r-n+1)}{2^n n!} + \cdots = \left(\frac{3}{2}\right)^r.$$

26. Use Equation (26) to find a power series representation for $\sqrt[4]{1+x}$.

27. Use the result of Problem 26 to find a power series representation for $\sqrt[4]{1+x^3}$.

28. Use the result of Problem 27 to estimate $\int_0^{0.5} \sqrt[4]{1+x^3}\, dx$ to four significant figures.

29. Using the technique suggested in Problems 26–28, estimate $\int_0^{1/4} (1 + \sqrt{x})^{3/5}\, dx$ to four significant figures.

30. The **error function** (which arises in mathematical statistics) is defined by

$$\text{erf}(x) = \frac{2}{\sqrt{\pi}} \int_0^x e^{-t^2}\, dt.$$

(a) Find a Maclaurin series for $\text{erf}(x)$ by integrating the Maclaurin series for e^{-x^2}.

(b) Use the series obtained in (a) to estimate, with an error < 0.0001, $\text{erf}(1)$ and $\text{erf}(\frac{1}{2})$.

31. The **complementary error function** is defined by

$$\text{erfc}(x) = 1 - \text{erf}(x) = 1 - \frac{2}{\sqrt{\pi}} \int_0^x e^{-t^2}\, dt = \frac{2}{\sqrt{\pi}} \int_x^\infty e^{-t^2}\, dt.$$

Find a Maclaurin series for $\text{erfc}(x)$ and use it to estimate $\text{erfc}(1)$ and $\text{erfc}(\frac{1}{2})$ with a maximum error of 0.0001. Note that for large values of x, $\text{erfc}(x)$ can be estimated by integrating the last integral by parts.

*32. The **sine integral** is defined by

$$\text{Si}(x) = \int_0^x \frac{\sin t}{t}\, dt.$$

(a) Show that $\text{Si}(x)$ is defined and continuous for all real x.

(b) Find a Maclaurin series expansion for $\text{Si}(x)$.

(c) Estimate $\text{Si}(1)$ and $\text{Si}(\frac{1}{2})$ with a maximum error of 0.0001.

REVIEW EXERCISES FOR CHAPTER FOURTEEN

1. Find the first five terms of the sequence $\{(n-2)/n\}$.

2. Find the first seven terms of the sequence $\{n^2 \sin n\}$.

In Exercises 3–8, determine whether the given sequence is convergent or divergent. If it is convergent, find its limit.

3. $\left\{\dfrac{-7}{n}\right\}$

4. $\{\cos \pi n\}$

5. $\left\{\dfrac{\ln n}{\sqrt{n}}\right\}$

6. $\left\{\dfrac{7^n}{n!}\right\}$

7. $\left\{\left(1 - \dfrac{2}{n}\right)^n\right\}$

8. $\left\{\dfrac{3}{\sqrt{n^2+8}-n}\right\}$

9. Find the general term of the sequence $\frac{1}{8}, \frac{3}{16}, \frac{5}{32}, \frac{7}{64}, \ldots$.

10. Find the general term of the sequence $1, -\frac{1}{5}, \frac{1}{25}, -\frac{1}{125}, \frac{1}{625}, \ldots$.

In Exercises 11–19, determine whether the given sequence is bounded or unbounded and increasing, decreasing, or not monotonic.

11. $\{\sqrt{n} \cos n\}$

12. $\left\{\dfrac{3}{n+2}\right\}$

13. $\left\{\dfrac{2^n}{1+2^n}\right\}$

14. $\left\{\dfrac{n^n}{n!}\right\}$

15. $\left\{\dfrac{n!}{n^n}\right\}$

16. $\left\{\dfrac{\sqrt{n}+1}{n}\right\}$

17. $\left\{\left(1-\dfrac{1}{n}\right)^{1/n}\right\}$

18. $\left\{\dfrac{n-7}{n+4}\right\}$

19. $\{(3^n+5^n)^{1/n}\}$

20. Write by using the Σ notation:
$$1 - 2^{1/2} + 3^{1/3} - 4^{1/4} + 5^{1/5} - \cdots + (-1)^{n+1}\, n^{1/n}.$$

21. Evaluate $\Sigma_{k=2}^{10} 4^k$.

22. Evaluate $\Sigma_{k=1}^{\infty} 1/3^k$.

23. Evaluate $\Sigma_{k=3}^{\infty}[(\tfrac{3}{4})^k - (\tfrac{2}{5})^k]$.

24. Evaluate $\Sigma_{k=2}^{\infty} 1/k(k-1)$.

25. Write $0.79797979\ldots$ as a rational number.

26. Write $0.142314231423\ldots$ as a rational number.

27. At what time between 9 P.M. and 10 P.M. is the minute hand of a clock exactly over the hour hand?

In Exercises 28–40, determine the convergence or divergence of the given series.

28. $\displaystyle\sum_{k=1}^{\infty} \frac{1}{k^3-5}$

29. $\displaystyle\sum_{k=5}^{\infty} \frac{1}{k(k+6)}$

30. $\displaystyle\sum_{k=1}^{\infty} \frac{1}{\sqrt{k^3+4}}$

31. $\displaystyle\sum_{k=2}^{\infty} \frac{3}{\ln k}$

32. $\displaystyle\sum_{k=4}^{\infty} \frac{1}{\sqrt[3]{k^3+50}}$

33. $\displaystyle\sum_{k=1}^{\infty} \frac{r^k}{k^r},\ 0<r<1$

34. $\displaystyle\sum_{k=1}^{\infty} \frac{k!}{k^k}$

35. $\displaystyle\sum_{k=2}^{\infty} \frac{10^k}{k^5}$

36. $\displaystyle\sum_{k=1}^{\infty} \frac{k^{6/5}}{8^k}$

37. $\displaystyle\sum_{k=1}^{\infty} \frac{\sqrt{k}\ln(k+3)}{k^2+2}$

38. $\displaystyle\sum_{k=1}^{\infty} \operatorname{csch} k$

39. $\displaystyle\sum_{k=2}^{\infty} \frac{e^{1/k}}{k^{3/2}}$

40. $\displaystyle\sum_{k=1}^{\infty} \frac{k(k+6)}{(k+1)(k+3)(k+5)}$

In Exercises 41–52, determine whether the given alternating series is absolutely convergent, conditionally convergent, or divergent.

41. $\displaystyle\sum_{k=1}^{\infty} \frac{(-1)^{k+1}}{50k}$

42. $\displaystyle\sum_{k=2}^{\infty} \frac{(-1)^k\sqrt{k}}{\ln k}$

43. $\displaystyle\sum_{k=2}^{\infty} \frac{(-1)^{k+1}}{\sqrt{k(k-1)}}$

44. $\displaystyle\sum_{k=2}^{\infty} \frac{(-1)^k k^2}{k^3+1}$

45. $\displaystyle\sum_{k=2}^{\infty} \frac{(-1)^k k^2}{k^4+1}$

46. $\displaystyle\sum_{k=2}^{\infty} \frac{(-1)^k k^3}{k^3+1}$

47. $\displaystyle\sum_{k=3}^{\infty} \frac{(-1)^k(k+2)(k+3)}{(k+1)^3}$

48. $\displaystyle\sum_{k=2}^{\infty} \frac{(-1)^k 3^k}{3^k}$

49. $\displaystyle\sum_{k=1}^{\infty} \frac{(-1)^k k^k}{k!}$

50. $\displaystyle\sum_{k=1}^{\infty} \frac{(-1)^k k^4}{k^4+20k^3+17k+2}$

51. $\displaystyle\sum_{k=1}^{\infty} (-1)^k\left(1+\frac{1}{k}\right)^k$

52. $\displaystyle\sum_{k=1}^{\infty} \frac{(-1)^k k!}{k^k}$

53. Calculate $\Sigma_{k=1}^{\infty}(-1)^{k+1}/k^3$ with an error of less than 0.001.

54. Calculate $\Sigma_{k=0}^{\infty}(-1)^k/k!$ with an error of less than 0.0001.

55. Find the first ten terms of a rearrangement of the series $\Sigma_{k=1}^{\infty}(-1)^{k+1}/k$ that converges to $\tfrac{1}{2}$.

In Exercises 56–67, find the radius of convergence and interval of convergence of the given power series.

56. $\displaystyle\sum_{k=0}^{\infty} \frac{x^k}{3^k}$

57. $\displaystyle\sum_{k=0}^{\infty} \frac{(-1)^k x^k}{3^k}$

58. $\displaystyle\sum_{k=0}^{\infty} \frac{x^k}{k^2 + 2}$

59. $\displaystyle\sum_{k=1}^{\infty} \frac{x^k}{k^k}$

60. $\displaystyle\sum_{k=2}^{\infty} \frac{x^k}{(2 \ln k)^k}$

61. $\displaystyle\sum_{k=0}^{\infty} \frac{(3x + 5)^k}{k!}$

62. $\displaystyle\sum_{k=0}^{\infty} \frac{(3x - 5)^k}{3^k}$

63. $\displaystyle\sum_{k=0}^{\infty} \left(\frac{k}{6}\right)^k x^k$

64. $\displaystyle\sum_{k=0}^{\infty} (-1)^k x^{3k}$

65. $\displaystyle\sum_{k=1}^{\infty} \frac{\ln k (x - 1)^k}{k + 2}$

66. $\displaystyle\sum_{k=1}^{\infty} \frac{[1 - (-1)^{k+1}]}{k} x^k$

67. $\displaystyle\sum_{k=0}^{\infty} \frac{(x + 8)^k}{(k + 1)3^k}$

68. Estimate $\int_0^{1/2} e^{-t^4} \, dt$ with an error of less than 0.00001.

69. Estimate $\int_0^{1/2} \sin t^2 \, dt$ with an error of less than 0.001.

70. Estimate $\int_0^{1/2} t^3 e^{-t^3} \, dt$ with an error of less than 0.001.

71. Estimate $\int_0^{1/2} [1/(1 + t^4)] \, dt$ with an error of less than 0.00001.

72. Find the Maclaurin series for $x^2 e^x$.

73. Find the Taylor series for e^x at $\ln 2$.

74. Find the Maclaurin series for $\cos^2 x$. [*Hint:* $\cos^2 x = (1 + \cos 2x)/2$.]

75. Find the Maclaurin series for $\sin \alpha x$, α real.

15 Vectors in the Plane

With this chapter we begin a new subject in the study of calculus—the study of vectors and vector functions. This will lead us, in Chapter 18, to the study of functions of two and more variables, and it is a sharp departure from the first 14 chapters of this book.

The modern study of vectors began essentially with the work of the great Irish mathematician Sir William Rowan Hamilton (1805–1865) who worked with what he called quaternions.† After Hamilton's death, his work on quaternions was supplanted by the more adaptable work on vector analysis by the American mathematician and physicist Josiah Willard Gibbs (1839–1903) and the general treatment of ordered n-tuples by the German mathematician Hermann Grassman (1809–1877).

Throughout Hamilton's life and for the remainder of the nineteenth century, there was considerable debate over the usefulness of quaternions and vectors. At the end of the century, the great British physicist Lord Kelvin wrote that quaternions "... although beautifully ingenious, have been an unmixed evil to those who have touched them in any way ... vectors ... have never been of the slightest use to any creature."

But Kelvin was wrong. Today nearly all branches of classical and modern physics are represented using the language of vectors. Vectors are also used with increasing frequency in the social and biological sciences. Quaternions, too, have

---

†See the accompanying biographical sketch.

Sir William Rowan Hamilton

Sir William Rowan Hamilton
The Granger Collection

Born in Dublin in 1805, where he spent most of his life, William Rowan Hamilton was without question Ireland's greatest mathematician. Hamilton's father (an attorney) and mother died when he was a small boy. His uncle, a linguist, took over the boy's education. By his fifth birthday, Hamilton could read English, Hebrew, Latin, and Greek. By his 13th birthday he had mastered not only the languages of continental Europe, but also Sanscrit, Chinese, Persian, Arabic, Malay, Hindi, Bengali, and several others as well. Hamilton liked to write poetry, both as a child and as an adult, and his friends included the great English poets Samuel Taylor Coleridge and William Wordsworth. Hamilton's poetry was considered so bad, however, that it is fortunate that he developed other interests—especially in mathematics.

Although he enjoyed mathematics as a young boy, Hamilton's interest was greatly enhanced by a chance meeting at the age of 15 with Zerah Colburn, the American lightning calculator. Shortly afterwards, Hamilton began to read important mathematical books of the time. In 1823, at the age of 18, he discovered an error in Simon Laplace's *Mécanique céleste* and wrote an impressive paper on the subject. A year later he entered Trinity College in Dublin.

Hamilton's university career was astonishing. At the age of 21, while still an undergraduate, he had so impressed the faculty that he was appointed Royal Astronomer of Ireland and Professor of Astronomy at the University. Shortly thereafter, he wrote what is now considered a classical work on optics. Using only mathematical theory, he predicted conical refraction in certain types of crystals. Later this theory was confirmed by physicists. Largely because of this work, Hamilton was knighted in 1835.

Hamilton's first great purely mathematical paper appeared in 1833. In this work he described an algebraic way to manipulate pairs of real numbers. This work gives rules that are used today to add, subtract, multiply, and divide complex numbers. At first, however, Hamilton was unable to devise a multiplication for triples or n-tuples of numbers for $n > 2$. For 10 years he pondered this problem, and it is said that he solved it in an inspiration while walking on the Brougham Bridge in Dublin in 1843. The key was to discard the familiar commutative property of multiplication. The new objects he created were called *quaternions*, which were the precursors of what we now call vectors.

For the rest of his life, Hamilton spent most of his time developing the algebra of quaternions. He felt that they would have revolutionary significance in mathematical physics. His monumental work on this subject, *Treatise on Quaternions*, was published in 1853. Thereafter, he worked on an enlarged work, *Elements of Quaternions*. Although Hamilton died in 1865 before his *Elements* was completed, the work was published by his son in 1866.

Students of mathematics and physics know Hamilton in a variety of other contexts. In mathematical physics, for example, one encounters the Hamiltonian function, which often represents the total energy in a system and the Hamilton-Jacobi differential equations of dynamics. In matrix theory, the Cayley-Hamilton theorem states that every matrix satisfies its own characteristic equation.

Despite the great work he was doing, Hamilton's final years were a torment to him. His wife was a semi-invalid and he was plagued by alcoholism. It is therefore gratifying to point out that during these last years, the newly formed American National Academy of Sciences elected Sir William Rowan Hamilton to be its first foreign associate.

recently been used in physics—in particle theory and other areas.

In this chapter we will explore properties of vectors in the plane. When going through this material, the reader should keep in mind that, like most important discoveries, vectors have been a source of great controversy—a controversy that was not resolved until well into the twentieth century.†

15.1 VECTORS AND VECTOR OPERATIONS

In many applications of mathematics to the physical and biological sciences and engineering, scientists are concerned with entities that have both magnitude (length) and direction. Examples include the notions of force, velocity, acceleration, and momentum. It is frequently useful to express these quantities geometrically.

Let P and Q be two different points in the plane. Then the **directed line segment** from P to Q, denoted by $\overrightarrow{PQ}$, is the straight-line segment that extends from P to Q (see Figure 1a). Note that the directed line segments $\overrightarrow{PQ}$ and $\overrightarrow{QP}$ are different since they point in opposite directions. (Figure 1b).

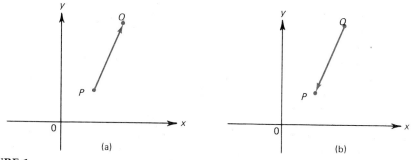

(a)   (b)

FIGURE 1

The point P in the directed line segment $\overrightarrow{PQ}$ is called the **initial point** of the segment and the point Q is called the **terminal point.** The two important properties of a directed line segment are its magnitude (length) and its direction. If two directed line segments $\overrightarrow{PQ}$ and $\overrightarrow{RS}$ have the same magnitude and direction, we say that they are **equivalent** no matter where they are located with respect to the origin. The directed line segments in Figure 2 are all equivalent.

Definition 1   GEOMETRIC DEFINITION OF A VECTOR   The set of all directed line segments equivalent to a given directed line segment is called a **vector.** Any directed line segment in that set is called a **representation** of the vector.

REMARK.   The directed line segments in Figure 2 are all representations of the same vector.

---

†For interesting discussions of the development of modern vector analysis, consult the book by M. J. Crowe, *A History of Vector Analysis* (Univ. of Notre Dame Press, Notre Dame, 1967), or Morris Kline's excellent book *Mathematical Thought from Ancient to Modern Times* (Oxford Univ. Press, New York, 1972), Chapter 32.

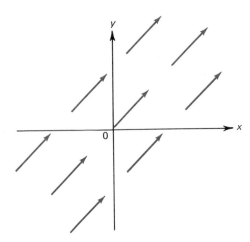

FIGURE 2

NOTATION.   We will denote vectors by lowercase boldface letters such as **v**, **w**, **a**, **b**.

From Definition 1 we see that a given vector **v** can be represented in many different ways. In fact, let $\overrightarrow{PQ}$ be a representation of **v**. Then without changing magnitude or direction, we can move $\overrightarrow{PQ}$ in a parallel way so that its initial point is shifted to the origin. We then obtain the directed line segment $\overrightarrow{OR}$, which is another representation of the vector **v** (see Figure 3). Now suppose that R has the Cartesian coordinates (a, b). Then we can describe the directed line segment $\overrightarrow{OR}$ by the coordinates (a, b). That is, $\overrightarrow{OR}$ is the directed line segment with initial point $(0, 0)$ and terminal point (a, b). Since one representation of a vector is as good as another, we can write the vector **v** as (a, b). In sum, we see that a vector can be thought of as a point in the xy-plane.

Definition 2   ALGEBRAIC DEFINITION OF A VECTOR   A **vector v** in the xy-plane is an ordered pair of real numbers (a, b). The numbers a and b are called the **components** of the vector **v**. The **zero vector** is the vector $(0, 0)$ and is denoted **0**.

Definition 3   SCALAR   Since we will often have to distinguish between real numbers and vectors (which are pairs of real numbers), we will use the term **scalar†** to denote a real number.

Definition 4   MAGNITUDE OF A VECTOR   Since a vector is really a set of equivalent directed line segments, we define the **magnitude** or **length** of a vector as the length of any one of its representations.

---

†The term "scalar" originated with Hamilton. His definition of the quaternion included what he called a *real part* and an *imaginary part*. In his paper, "On Quaternions, or on a New System of Imaginaries in Algebra," in *Philosophical Magazine*, 3rd Ser., **25**, 26–27 (1844), he wrote

> The algebraically *real* part may receive . . . all values contained on the one *scale* of progression of numbers from negative to positive infinity; we shall call it therefore the *scalar part*, or simply the *scalar* of the quaternion. . . .

Moreover, in the same paper Hamilton went on to define the imaginary part of his quaternion as the *vector* part. Although this was not the first usage of the word *vector*, it was the first time it was used in the context of Definitions 1 and 2. In fact, it is fair to say that the paper from which the above quotation was taken marks the beginning of modern vector analysis.

FIGURE 3

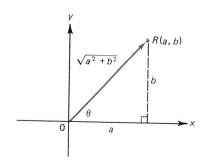

FIGURE 4

Using the representation $\overrightarrow{OR}$, and writing the vector $\mathbf{v} = (a, b)$, we find that

$$|\mathbf{v}| = \text{magnitude of } \mathbf{v} = \sqrt{a^2 + b^2}. \tag{1}$$

This follows from the Pythagorean theorem (see Figure 4). We have used the notation $|\mathbf{v}|$ to denote the magnitude of $\mathbf{v}$. Note that $|\mathbf{v}|$ *is a scalar.*

EXAMPLE 1   Calculate the magnitudes of the vectors (a) $(2, 2)$; (b) $(2, 2\sqrt{3})$; (c) $(-2\sqrt{3}, 2)$; (d) $(-3, -3)$; (e) $(6, -6)$.

Solution.

(a) $|\mathbf{v}| = \sqrt{2^2 + 2^2} = \sqrt{8} = 2\sqrt{2}$
(b) $|\mathbf{v}| = \sqrt{2^2 + (2\sqrt{3})^2} = 4$
(c) $|\mathbf{v}| = \sqrt{(-2\sqrt{3})^2 + 2^2} = 4$
(d) $|\mathbf{v}| = \sqrt{(-3)^2 + (-3)^2} = \sqrt{18} = 3\sqrt{2}$
(e) $|\mathbf{v}| = \sqrt{6^2 + (-6)^2} = \sqrt{72} = 6\sqrt{2}$ ■

Definition 5 DIRECTION OF A VECTOR We now define the **direction** of the vector $\mathbf{v} = (a, b)$ to be the angle θ, measured in radians, that the vector makes with the positive x-axis. By convention, we choose θ such that $0 \le \theta < 2\pi$.

It follows from Figure 4 that if $a \ne 0$, then

$$\tan \theta = \frac{b}{a}. \tag{2}$$

REMARK 1. The zero vector has a magnitude of 0. Since the initial and terminal points coincide, we say that *the zero vector has no direction.*

REMARK 2. If the initial point of $\mathbf{v}$ is not at the origin, then the direction of $\mathbf{v}$ is defined as the direction of the vector equivalent to $\mathbf{v}$ whose initial point *is* at the origin.

REMARK 3. It follows from Remark 2 that *parallel vectors have the same direction.*

EXAMPLE 2   Calculate the directions of the vectors in Example 1.

Solution. We depict these five vectors in Figure 5.

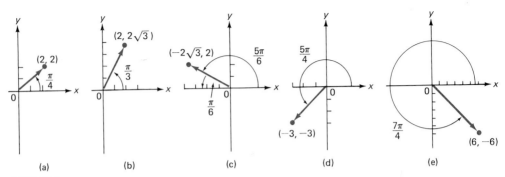

FIGURE 5

(a) Here **v** is in the first quadrant, and since $\tan \theta = 2/2 = 1$, $\theta = \pi/4$.

(b) Here $\theta = \tan^{-1} 2\sqrt{3}/2 = \tan^{-1} \sqrt{3} = \pi/3$.

(c) We see that **v** is in the second quadrant, and since $\tan^{-1} 2/(2\sqrt{3}) = \tan^{-1} 1/\sqrt{3} = \pi/6$, we see from the figure that $\theta = \pi - (\pi/6) = 5\pi/6$.

(d) Here **v** is in the third quadrant, and since $\tan^{-1} 1 = \pi/4$, $\theta = \pi + (\pi/4) = 5\pi/4$.

(e) Since **v** is in the fourth quadrant, and since $\tan^{-1}(-1) = -\pi/4$, $\theta = 2\pi - (\pi/4) = 7\pi/4$. ◼

Let $\mathbf{v} = (a, b)$. Then as we have seen, **v** can be represented in many different ways. For example, the representation of **v** with the initial point (c, d) has the terminal point $(c + a, d + b)$. This is depicted in Figure 6. It is easy to show that the directed line segment $\overrightarrow{PQ}$ in Figure 6 has the same magnitude and direction as the vector **v** (see Problems 48 and 49).

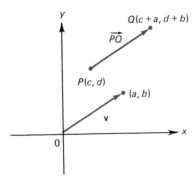

FIGURE 6

EXAMPLE 3   Find a representation of the vector $(2, -1)$ whose initial point is the point $P = (5, -4)$.

Solution. Let $Q = (5 + 2, -4 - 1) = (7, -5)$. Then $\overrightarrow{PQ}$ is a representation of the vector $(2, -1)$. This is illustrated in Figure 7. ◼

We now turn to the question of adding vectors and multiplying them by scalars.

Definition 6 ADDITION AND SCALAR MULTIPLICATION OF A VECTOR Let $\mathbf{u} = (a_1, b_1)$ and $\mathbf{v} = (a_2, b_2)$ be two vectors in the plane and let α be a scalar. Then we define

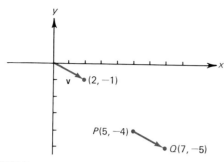

FIGURE 7

(i) $\mathbf{u} + \mathbf{v} = (a_1 + a_2, b_1 + b_2)$
(ii) $\alpha\mathbf{u} = (\alpha a_1, \alpha b_1)$
(iii) $-\mathbf{v} = (-1)\mathbf{v} = (-a_2, -b_2)$
(iv) $\mathbf{u} - \mathbf{v} = \mathbf{u} + (-\mathbf{v}) = (a_1 - a_2, b_1 - b_2)$

We have

> To add two vectors, we add their corresponding components, and to multiply a vector by a scalar, we multiply each of its components by that scalar.

EXAMPLE 4 Let $\mathbf{u} = (1, 3)$ and $\mathbf{v} = (-2, 4)$. Calculate (a) $\mathbf{u} + \mathbf{v}$; (b) $3\mathbf{u}$; (c) $-\mathbf{v}$; (d) $\mathbf{u} - \mathbf{v}$; and (e) $-3\mathbf{u} + 5\mathbf{v}$.

Solution.

(a) $\mathbf{u} + \mathbf{v} = (1 + (-2), 3 + 4) = (-1, 7)$
(b) $3\mathbf{u} = 3(1, 3) = (3, 9)$
(c) $-\mathbf{v} = (-1)(-2, 4) = (2, -4)$
(d) $\mathbf{u} - \mathbf{v} = \mathbf{u} + (-\mathbf{v}) = (1 + 2, 3 - 4) = (3, -1)$
(e) $-3\mathbf{u} + 5\mathbf{v} = (-3, -9) + (-10, 20) = (-13, 11)$ ■

There are interesting geometric interpretations of vector addition and scalar multiplication. First, let $\mathbf{v} = (a, b)$ and let α be any scalar. Then

$$|\alpha\mathbf{v}| = |(\alpha a, \alpha b)| = \sqrt{\alpha^2 a^2 + \alpha^2 b^2} = |\alpha|\sqrt{a^2 + b^2} = |\alpha||\mathbf{v}|.$$

> That is, multiplying a vector by a scalar has the effect of multiplying the length of the vector by the absolute value of that scalar.

Moreover, if $\alpha > 0$, then $\alpha\mathbf{v}$ is in the same quadrant as $\mathbf{v}$ and, since $\tan^{-1}(\alpha b/\alpha a) = \tan^{-1}(b/a)$, the direction of $\alpha\mathbf{v}$ is the same as the direction of $\mathbf{v}$. If $\alpha < 0$, then the direction of $\alpha\mathbf{v}$ is equal to the direction of $\mathbf{v}$ plus π (which is the direction of $-\mathbf{v}$).

EXAMPLE 5 Let $\mathbf{v} = (1, 1)$. Then $|\mathbf{v}| = \sqrt{1 + 1} = \sqrt{2}$ and $|2\mathbf{v}| = |(2, 2)| = \sqrt{2^2 + 2^2} = \sqrt{8} = 2\sqrt{2} = 2|\mathbf{v}|$. Also, $|-2\mathbf{v}| = \sqrt{(-2)^2 + (-2)^2} = 2\sqrt{2} = 2|\mathbf{v}|$. Moreover, the direction of $2\mathbf{v}$ is $\pi/4$, while the direction of $-2\mathbf{v}$ is $5\pi/4$. This is illustrated in Figure 8. ■

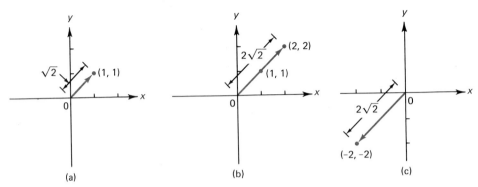

FIGURE 8

Now suppose we add the vectors $\mathbf{u} = (a_1, b_1)$ and $\mathbf{v} = (a_2, b_2)$, as in Figure 9. From the figure we see that the vector $\mathbf{u} + \mathbf{v} = (a_1 + a_2, b_1 + b_2)$ can be obtained by shifting the representation of the vector $\mathbf{v}$ so that its initial point coincides with the terminal point (a_1, b_1) of the vector $\mathbf{v}$. We can therefore obtain the vector $\mathbf{u} + \mathbf{v}$ by drawing a parallelogram with one vertex at the origin and sides $\mathbf{u}$ and $\mathbf{v}$. Then $\mathbf{u} + \mathbf{v}$ is the vector that points from the origin along the diagonal of the parallelogram.

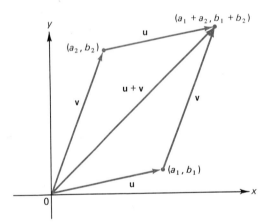

FIGURE 9

NOTE.   Since a straight line is the shortest distance between two points, it immediately follows from Figure 9 that

$$|\mathbf{u} + \mathbf{v}| \le |\mathbf{u}| + |\mathbf{v}|.$$

For obvious reasons this inequality is called the **triangle inequality.**

We can also obtain a geometric representation of the vector $\mathbf{u} - \mathbf{v}$. Since $\mathbf{u} = \mathbf{u} - \mathbf{v} + \mathbf{v}$, the vector $\mathbf{u} - \mathbf{v}$ is the vector that must be added to $\mathbf{v}$ to obtain $\mathbf{u}$. This is illustrated in Figure 10a.

The following theorem lists several properties that hold for any vectors, $\mathbf{u}$, $\mathbf{v}$, and $\mathbf{w}$ and any scalars α and β. Since the proof is easy, we leave it as an exercise (see Problem 58). Some parts of this theorem have already been proven.

FIGURE 10

Theorem 1. Let **u**, **v**, and **w** be any three vectors in the plane, let α and β be scalars, and let **0** denote the zero vector.

(i) $\mathbf{u} + \mathbf{v} = \mathbf{v} + \mathbf{u}$ **(ii)** $\mathbf{u} + (\mathbf{v} + \mathbf{w}) = (\mathbf{u} + \mathbf{v}) + \mathbf{w}$
(iii) $\mathbf{v} + \mathbf{0} = \mathbf{v}$
(iv) $0\mathbf{v} = \mathbf{0}$ (here the 0 on the left is the scalar zero)
(v) $\alpha\mathbf{0} = \mathbf{0}$ **(vi)** $(\alpha\beta)\mathbf{v} = \alpha(\beta\mathbf{v})$
(vii) $\mathbf{v} + (-\mathbf{v}) = \mathbf{0}$ **(viii)** $1\mathbf{v} = \mathbf{v}$
(ix) $(\alpha + \beta)\mathbf{v} = \alpha\mathbf{v} + \beta\mathbf{v}$ **(x)** $\alpha(\mathbf{u} + \mathbf{v}) = \alpha\mathbf{u} + \alpha\mathbf{v}$
(xi) $|\alpha\mathbf{v}| = |\alpha||\mathbf{v}|$ **(xii)** $|\mathbf{u} + \mathbf{v}| \le |\mathbf{u}| + |\mathbf{v}|$

Many of the properties above can be illustrated geometrically. For example, rule (i), which is called the **commutative law for vector addition,** is illustrated in Figure 11. Similarly, rule (ii), which is called the **associative law for vector addition,** is illustrated in Figure 12.

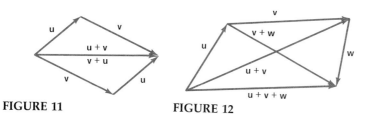

FIGURE 11   **FIGURE 12**

When a set of vectors together with a set of scalars and the operations of addition and scalar multiplication have the properties given in Theorem 1(i)–(x), we say that the vectors form a **vector space.** The set of vectors of the form (a, b), where a and b are real numbers, is denoted $\mathbb{R}^2$. We will not discuss properties of abstract vector spaces here, except to say that all abstract vector spaces have properties very similar to the properties of the vector space $\mathbb{R}^2$.

There are two special vectors in $\mathbb{R}^2$ that allow us to represent other vectors in $\mathbb{R}^2$ in a convenient way. We will denote the vector $(1, 0)$ by the vector symbol **i** and the vector $(0, 1)$ by the vector symbol **j** (see Figure 13).† If (a, b) denotes any other vector in $\mathbb{R}^2$, then since $(a, b) = a(1, 0) + b(0, 1)$, we may write

$$\mathbf{v} = (a, b) = a\mathbf{i} + b\mathbf{j}. \tag{3}$$

---

†The symbols **i** and **j** were first used by Hamilton. He defined his quarternion as a quantity of the form $a + b\mathbf{i} + c\mathbf{j} + d\mathbf{k}$, where a was the "scalar part" and $b\mathbf{i} + c\mathbf{j} + d\mathbf{k}$ the "vector part." In Section 17.2 we will write vectors in space in the form $b\mathbf{i} + c\mathbf{j} + d\mathbf{k}$.

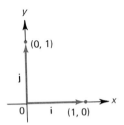

FIGURE 13

Moreover, any vector in $\mathbb{R}^2$ can be represented in a unique way in the form $a\mathbf{i} + b\mathbf{j}$ since the representation of (a, b) as a point in the plane is unique. (Put another way, a point in the xy-plane has one and only one x-coordinate and one and only one y-coordinate.) Thus Theorem 1 holds with this new representation as well.

When the vector $\mathbf{v}$ is written in the form $\mathbf{v} = a\mathbf{i} + b\mathbf{j}$, we way that $\mathbf{v}$ *is resolved into its horizontal and vertical components*, since a is the horizontal component of $\mathbf{v}$ while b is its vertical component. The vectors $\mathbf{i}$ and $\mathbf{j}$ are called **basis vectors** for the vector space $\mathbb{R}^2$.

Now suppose that a vector $\mathbf{v}$ can be represented by the directed line segment $\overrightarrow{PQ}$, where $P = (a_1, b_1)$ and $Q = (a_2, b_2)$. (See Figure 14.) If we label the point (a_2, b_1) as R, then we immediately see that

$$\mathbf{v} = \overrightarrow{PQ} = \overrightarrow{PR} + \overrightarrow{RQ}. \tag{4}$$

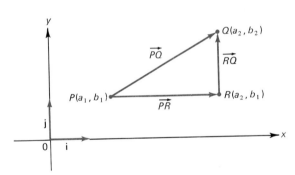

FIGURE 14

If $a_2 \geq a_1$, then the length of $\overrightarrow{PR}$ is $a_2 - a_1$, and since $\overrightarrow{PR}$ has the same direction as $\mathbf{i}$ (since they are parallel), we can write

$$\overrightarrow{PR} = (a_2 - a_1)\mathbf{i}. \tag{5}$$

If $a_2 < a_1$, then the length of $\overrightarrow{PR}$ is $a_1 - a_2$, but then $\overrightarrow{PR}$ has the same direction as $-\mathbf{i}$ so $\overrightarrow{PR} = (a_1 - a_2)(-\mathbf{i}) = (a_2 - a_1)\mathbf{i}$ again. Similarly,

$$\overrightarrow{RQ} = (b_2 - b_1)\mathbf{j}, \tag{6}$$

and we may write [using (4), (5), and (6)]

$$\mathbf{v} = (a_2 - a_1)\mathbf{i} + (b_2 - b_1)\mathbf{j}. \tag{7}$$

EXAMPLE 6   Resolve the vector represented by the directed line segment from $(-2, 3)$ to $(1, 5)$ into its vertical and horizontal components.

Solution. Using (7), we have

$$\mathbf{v} = (a_2 - a_1)\mathbf{i} + (b_2 - b_1)\mathbf{j} = [1 - (-2)]\mathbf{i} + (5 - 3)\mathbf{j} = 3\mathbf{i} + 2\mathbf{j}. \ \blacksquare$$

We conclude this section by defining a kind of vector that is very useful in certain types of applications.

Definition 7   UNIT VECTOR   A **unit vector u** is a vector that has length 1.

EXAMPLE 7   The vector $\mathbf{u} = (1/2)\mathbf{i} + (\sqrt{3}/2)\mathbf{j}$ is a unit vector since

$$|\mathbf{u}| = \sqrt{\left(\frac{1}{2}\right)^2 + \left(\frac{\sqrt{3}}{2}\right)^2} = \sqrt{\frac{1}{4} + \frac{3}{4}} = 1. \ \blacksquare$$

Let $\mathbf{u} = a\mathbf{i} + b\mathbf{j}$ be a unit vector. Then $|\mathbf{u}| = \sqrt{a^2 + b^2} = 1$, so $a^2 + b^2 = 1$ and $\mathbf{u}$ is a point on the unit circle (see Figure 15). If θ is the direction of $\mathbf{u}$, then we immediately see that $a = \cos\theta$ and $b = \sin\theta$. Thus any unit vector $\mathbf{u}$ can be written in the form

$$\mathbf{u} = (\cos\theta)\mathbf{i} + (\sin\theta)\mathbf{j} \tag{8}$$

where θ is the direction of $\mathbf{u}$.

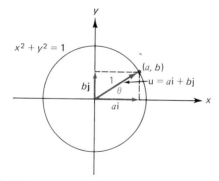

FIGURE 15

EXAMPLE 8   The unit vector $\mathbf{u} = (1/2)\mathbf{i} + (\sqrt{3}/2)\mathbf{j}$ of Example 7 can be written in the form (8) with $\theta = \cos^{-1}(1/2) = \pi/3$. Note that since $\cos\theta = 1/2$ and $\sin\theta = \sqrt{3}/2$, θ is in the first quadrant. We need this fact to conclude that $\theta = \pi/3$. It is also true that $\cos 5\pi/3 = 1/2$, but $5\pi/3$ is in the fourth quadrant. $\blacksquare$

Finally:

Let $\mathbf{v}$ be any nonzero vector. Then $\mathbf{u} = \mathbf{v}/|\mathbf{v}|$ is the unit vector having the same direction as $\mathbf{v}$.

(See Problem 31.)

EXAMPLE 9   Find the unit vector having the same direction as $\mathbf{v} = 2\mathbf{i} - 3\mathbf{j}$.

Solution. Here $|\mathbf{v}| = \sqrt{4 + 9} = \sqrt{13}$, so $\mathbf{u} = \mathbf{v}/|\mathbf{v}| = (2/\sqrt{13})\mathbf{i} - (3/\sqrt{13})\mathbf{j}$ is the required unit vector. ■

EXAMPLE 10   Find the vector $\mathbf{v}$ whose direction is $5\pi/4$ and whose magnitude is 7.

Solution. A unit vector $\mathbf{u}$ with direction $5\pi/4$ is given by

$$\mathbf{u} = \left(\cos\frac{5\pi}{4}\right)\mathbf{i} + \left(\sin\frac{5\pi}{4}\right)\mathbf{j} = -\frac{1}{\sqrt{2}}\mathbf{i} - \frac{1}{\sqrt{2}}\mathbf{j}.$$

Then $\mathbf{v} = 7\mathbf{u} = -(7/\sqrt{2})\mathbf{i} - (7/\sqrt{2})\mathbf{j}$. This vector is sketched in Figure 16a. In Figure 16b we have translated $\mathbf{v}$ so that it points toward the origin. This representation of $\mathbf{v}$ will be useful in Section 15.3. ■

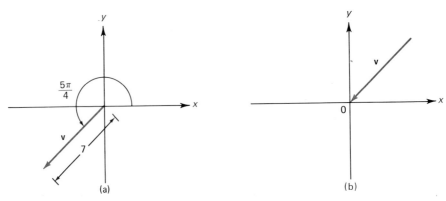

FIGURE 16

We conclude this section with a summary of properties of vectors, given in Table 1.

TABLE 1

Object	Intuitive definition	Expression in terms of components if $\mathbf{u} = u_1\mathbf{i} + u_2\mathbf{j} = (u_1, u_2)$ and $\mathbf{v} = v_1\mathbf{i} + v_2\mathbf{j} = (v_1, v_2)$
Vector $\mathbf{v}$	Magnitude and direction ↗ $\mathbf{v}$	$v_1\mathbf{i} + v_2\mathbf{j}$ or (v_1, v_2)
$|\mathbf{v}|$	Magnitude or length of $\mathbf{v}$	$\sqrt{v_1^2 + v_2^2}$
$\alpha\mathbf{v}$	↗$\mathbf{v}$ ↗$\alpha\mathbf{v}$ (Here $\alpha = 2$)	$\alpha v_1\mathbf{i} + \alpha v_2\mathbf{j}$ or $(\alpha v_1, \alpha v_2)$
$-\mathbf{v}$	↗$\mathbf{v}$ ↙$-\mathbf{v}$	$-v_1\mathbf{i} - v_2\mathbf{j}$ or $(-v_1, -v_2)$ or $-(v_1, v_2)$
$\mathbf{u} + \mathbf{v}$	$\mathbf{u} + \mathbf{v}$	$(u_1 + v_1)\mathbf{i} + (u_2 + v_2)\mathbf{j}$ or $(u_1 + v_1, u_2 + v_2)$
$\mathbf{u} - \mathbf{v}$	$\mathbf{u} - \mathbf{v}$	$(u_1 - v_1)\mathbf{i} + (u_2 - v_2)\mathbf{j}$ or $(u_1 - v_1, u_2 - v_2)$

PROBLEMS 15.1

In Problems 1–6 a vector **v** and a point P are given. Find a point Q such that the directed line segment $\overrightarrow{PQ}$ is a representation of **v**. Sketch **v** and $\overrightarrow{PQ}$.

1. $\mathbf{v} = (2, 5); P = (1, -2)$

2. $\mathbf{v} = (5, 8); P = (3, 8)$

3. $\mathbf{v} = (-3, 7); P = (7, -3)$

4. $\mathbf{v} = -\mathbf{i} - 7\mathbf{j}; P = (0, 1)$

5. $\mathbf{v} = 5\mathbf{i} - 3\mathbf{j}; P = (-7, -2)$

6. $\mathbf{v} = e\mathbf{i} + \pi\mathbf{j}; P = (\pi, \sqrt{2})$

In Problems 7–18, find the magnitude and direction of the given vector.

7. $\mathbf{v} = (4, 4)$

8. $\mathbf{v} = (-4, 4)$

9. $\mathbf{v} = (4, -4)$

10. $\mathbf{v} = (-4, -4)$

11. $\mathbf{v} = (\sqrt{3}, 1)$

12. $\mathbf{v} = (1, \sqrt{3})$

13. $\mathbf{v} = (-1, \sqrt{3})$

14. $\mathbf{v} = (1, -\sqrt{3})$

15. $\mathbf{v} = (-1, -\sqrt{3})$

16. $\mathbf{v} = (1, 2)$

17. $\mathbf{v} = (-5, 8)$

18. $\mathbf{v} = (11, -14)$

In Problems 19–26, write in the form $a\mathbf{i} + b\mathbf{j}$ the vector **v** that is represented by $\overrightarrow{PQ}$. Sketch $\overrightarrow{PQ}$ and **v**.

19. $P = (1, 2); Q = (1, 3)$

20. $P = (2, 4); Q = (-7, 4)$

21. $P = (5, 2); Q = (-1, 3)$

22. $P = (8, -2); Q = (-3, -3)$

23. $P = (7, -1); Q = (-2, 4)$

24. $P = (3, -6); Q = (8, 0)$

25. $P = (-3, -8); Q = (-8, -3)$

26. $P = (2, 4); Q = (-4, -2)$

27. Let $\mathbf{u} = (2, 3)$ and $\mathbf{v} = (-5, 4)$. Find the following:

    (a) $3\mathbf{u}$ **(b)** $\mathbf{u} + \mathbf{v}$ **(c)** $\mathbf{v} - \mathbf{u}$ **(d)** $2\mathbf{u} - 7\mathbf{v}$

    Sketch these vectors.

28. Let $\mathbf{u} = 2\mathbf{i} - 3\mathbf{j}$ and $\mathbf{v} = -4\mathbf{i} + 6\mathbf{j}$. Find the following:

    (a) $\mathbf{u} + \mathbf{v}$ **(b)** $\mathbf{u} - \mathbf{v}$ **(c)** $3\mathbf{u}$

    (d) $-7\mathbf{v}$ **(e)** $8\mathbf{u} - 3\mathbf{v}$ **(f)** $4\mathbf{v} - 6\mathbf{u}$

    Sketch these vectors.

29. Show that the vectors **i** and **j** are unit vectors.

30. Show that the vector $(1/\sqrt{2})\mathbf{i} + (1/\sqrt{2})\mathbf{j}$ is a unit vector.

31. Show that if $\mathbf{v} = a\mathbf{i} + b\mathbf{j} \neq 0$, then $\mathbf{u} = (a/\sqrt{a^2 + b^2})\mathbf{i} + (b/\sqrt{a^2 + b^2})\mathbf{j}$ is a unit vector having the same direction as **v**.

In Problems 32–37, find a unit vector having the same direction as the given vector.

32. $\mathbf{v} = 2\mathbf{i} + 3\mathbf{j}$

33. $\mathbf{v} = \mathbf{i} - \mathbf{j}$

34. $\mathbf{v} = (3, 4)$

35. $\mathbf{v} = (3, -4)$

36. $\mathbf{v} = -3\mathbf{i} + 4\mathbf{j}$

37. $\mathbf{v} = (a, a), a \neq 0$

38. If $\mathbf{v} = a\mathbf{i} + b\mathbf{j} \neq 0$, show that $a/\sqrt{a^2 + b^2} = \cos\theta$ and $b/\sqrt{a^2 + b^2} = \sin\theta$, where θ is the direction of **v**.

39. For $\mathbf{v} = 2\mathbf{i} - 3\mathbf{j}$, find $\sin\theta$ and $\cos\theta$.

40. For $\mathbf{v} = -3\mathbf{i} + 8\mathbf{j}$, find $\sin\theta$ and $\cos\theta$.

A vector **v** has a direction opposite to that of a vector **u** if $|\text{direction } \mathbf{v} - \text{direction } \mathbf{u}| = \pi$. In Problems 41–46, find a unit vector **u** that has a direction opposite the direction of the given vector **v**.

41. $\mathbf{v} = \mathbf{i} + \mathbf{j}$

42. $\mathbf{v} = 2\mathbf{i} - 3\mathbf{j}$

43. $\mathbf{v} = (-3, 4)$

44. $\mathbf{v} = (-2, 3)$

45. $\mathbf{v} = -3\mathbf{i} - 4\mathbf{j}$

46. $\mathbf{v} = (8, -3)$

47. Let $\mathbf{u} = 2\mathbf{i} - 3\mathbf{j}$ and $\mathbf{v} = -\mathbf{i} + 2\mathbf{j}$. Find a unit vector having the same direction as the following:

    (a) $\mathbf{u} + \mathbf{v}$ **(b)** $2\mathbf{u} - 3\mathbf{v}$ **(c)** $3\mathbf{u} + 8\mathbf{v}$

48. Let $P = (c, d)$ and $Q = (c + a, d + b)$. Show that the magnitude of $\overrightarrow{PQ}$ is $\sqrt{a^2 + b^2}$.

49. Show that the direction of $\overrightarrow{PQ}$ in Problem 48 is the same as the direction of the vector (a, b). [*Hint:* If $R = (a, b)$, show that the line passing through the points P and Q is parallel to the line passing through the points 0 and R.]

In Problems 50–57, find a vector **v** having the given magnitude and direction.   [*Hint:* See Example 10.]

50. $|\mathbf{v}| = 3$; $\theta = \pi/6$ **51.** $|\mathbf{v}| = 8$; $\theta = \pi/3$ **52.** $|\mathbf{v}| = 7$; $\theta = \pi$
53. $|\mathbf{v}| = 4$; $\theta = \pi/2$ **54.** $|\mathbf{v}| = 1$; $\theta = \pi/4$ **55.** $|\mathbf{v}| = 6$; $\theta = 2\pi/3$
56. $|\mathbf{v}| = 8$; $\theta = 3\pi/2$ **57.** $|\mathbf{v}| = 6$; $\theta = 11\pi/6$

58. Prove Theorem 1. [*Hint:* Use the definitions of addition and scalar multiplication of vectors.]

59. Show algebraically (i.e., strictly from the definitions of vector addition and magnitude) that for any two vectors **u** and **v**, $|\mathbf{u} + \mathbf{v}| \leq |\mathbf{u}| + |\mathbf{v}|$.

60. Show that if neither **u** nor **v** is the zero vector, then $|\mathbf{u} + \mathbf{v}| = |\mathbf{u}| + |\mathbf{v}|$ if and only if **u** is a positive scalar multiple of **v**.

15.2 THE DOT PRODUCT

In Section 15.1 we showed how a vector could be multiplied by a scalar but not how two vectors could be multiplied. Actually, there are several ways to define the product of two vectors, and in this section we will discuss one of them. We will discuss a second product operation in Section 17.4.

Definition 1 DOT PRODUCT   Let $\mathbf{u} = (a_1, b_1) = a_1\mathbf{i} + b_1\mathbf{j}$ and $\mathbf{v} = (a_2, b_2) = a_2\mathbf{i} + b_2\mathbf{j}$. Then the **dot product** of **u** and **v**, denoted $\mathbf{u} \cdot \mathbf{v}$, is defined by

$$\mathbf{u} \cdot \mathbf{v} = a_1 a_2 + b_1 b_2. \qquad (1)$$

REMARK.   The dot product of two vectors is a *scalar*. For this reason the dot product is often called the **scalar product.** It is also called the **inner product.**

EXAMPLE 1   If $\mathbf{u} = (1, 3)$ and $\mathbf{v} = (4, -7)$, then

$$\mathbf{u} \cdot \mathbf{v} = 1(4) + 3(-7) = 4 - 21 = -17. \quad \blacksquare$$

Theorem 1.   For any vectors **u**, **v**, **w**, and scalar α,

(i) $\mathbf{u} \cdot \mathbf{v} = \mathbf{v} \cdot \mathbf{u}$ (ii) $(\mathbf{u} + \mathbf{v}) \cdot \mathbf{w} = \mathbf{u} \cdot \mathbf{w} + \mathbf{v} \cdot \mathbf{w}$
(iii) $(\alpha\mathbf{u}) \cdot \mathbf{v} = \alpha(\mathbf{u} \cdot \mathbf{v})$ (iv) $\mathbf{u} \cdot \mathbf{u} \geq 0$ and $\mathbf{u} \cdot \mathbf{u} = 0$ if and only if $\mathbf{u} = 0$
(v) $|\mathbf{u}| = \sqrt{\mathbf{u} \cdot \mathbf{u}}$

Proof.   Let $\mathbf{u} = (u_1, u_2)$, $\mathbf{v} = (v_1, v_2)$, and $\mathbf{w} = (w_1, w_2)$.

(i) $\mathbf{u} \cdot \mathbf{v} = u_1 v_1 + u_2 v_2 = v_1 u_1 + v_2 u_2 = \mathbf{v} \cdot \mathbf{u}$
(ii) $(\mathbf{u} + \mathbf{v}) \cdot \mathbf{w} = (u_1 + v_1, u_2 + v_2) \cdot (w_1, w_2) = (u_1 + v_1)w_1 + (u_2 + v_2)w_2$
$\qquad\qquad = u_1 w_1 + u_2 w_2 + v_1 w_1 + v_2 w_2 = \mathbf{u} \cdot \mathbf{w} + \mathbf{v} \cdot \mathbf{w}$
(iii) $(\alpha\mathbf{u}) \cdot \mathbf{v} = (\alpha u_1, \alpha u_2) \cdot (v_1, v_2) = \alpha u_1 v_1 + \alpha u_2 v_2 = \alpha(u_1 v_1 + u_2 v_2) = \alpha(\mathbf{u} \cdot \mathbf{v})$
(iv) $\mathbf{u} \cdot \mathbf{u} = u_1^2 + u_2^2 \geq 0$ and $\mathbf{u} \cdot \mathbf{u} = 0$ if and only if $u_1 = u_2 = 0$.
(v) $\sqrt{\mathbf{u} \cdot \mathbf{u}} = \sqrt{(u_1, u_2) \cdot (u_1, u_2)} = \sqrt{u_1^2 + u_2^2} = |\mathbf{u}|$ $\blacksquare$

The dot product is useful in a wide variety of applications. An interesting one follows.

Definition 2   ANGLE BETWEEN TWO VECTORS   Let **u** and **v** be two nonzero vectors. Then the **angle** φ between **u** and **v** is defined to be the smallest angle† between the representations of **u** and **v** that have the origin as their initial points. If **u** $= \alpha$**v** for some scalar α, then we define $\varphi = 0$ if $\alpha > 0$ and $\varphi = \pi$ if $\alpha < 0$.

Theorem 2.   Let **u** and **v** be two nonzero vectors. Then if φ is the angle between them,

$$\cos \varphi = \frac{\mathbf{u} \cdot \mathbf{v}}{|\mathbf{u}|\,|\mathbf{v}|}. \tag{2}$$

Proof.   The law of cosines (see Problem 11, Appendix 1.4) states that in the triangle of Figure 1,

$$c^2 = a^2 + b^2 - 2ab \cos C. \tag{3}$$

We now place the representations of **u** and **v** with initial points at the origin so that **u** $= (a_1, b_1)$ and **v** $= (a_2, b_2)$ (see Figure 2). Then from the law of cosines,

$$|\mathbf{v} - \mathbf{u}|^2 = |\mathbf{v}|^2 + |\mathbf{u}|^2 - 2|\mathbf{u}||\mathbf{v}| \cos \varphi.$$

But using Theorem 1 several times, we have

$$|\mathbf{v} - \mathbf{u}|^2 = (\mathbf{v} - \mathbf{u}) \cdot (\mathbf{v} - \mathbf{u}) = \mathbf{v} \cdot \mathbf{v} - 2\mathbf{u} \cdot \mathbf{v} + \mathbf{u} \cdot \mathbf{u} = |\mathbf{v}|^2 - 2\mathbf{u} \cdot \mathbf{v} + |\mathbf{u}|^2.$$

Thus after simplification, we obtain

$$-2\mathbf{u} \cdot \mathbf{v} = -2|\mathbf{u}||\mathbf{v}| \cos \varphi,$$

from which the theorem follows.   ■

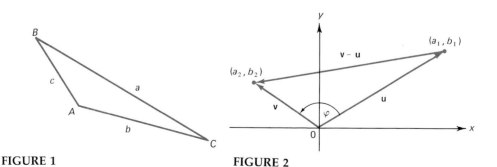

FIGURE 1   **FIGURE 2**

---

†The smallest angle will be in the interval $[0, \pi]$.

REMARK.   Using Theorem 2, we could define the dot product $\mathbf{u} \cdot \mathbf{v}$ by

$$\mathbf{u} \cdot \mathbf{v} = |\mathbf{u}||\mathbf{v}| \cos \varphi. \tag{4}$$

EXAMPLE 2   Find the cosine of the angle between the vectors $\mathbf{u} = 2\mathbf{i} + 3\mathbf{j}$ and $\mathbf{v} = -7\mathbf{i} + \mathbf{j}$.

Solution. $\mathbf{u} \cdot \mathbf{v} = -14 + 3 = -11$, $|\mathbf{u}| = \sqrt{2^2 + 3^2} = \sqrt{13}$, and $|\mathbf{v}| = \sqrt{(-7)^2 + 1^2} = \sqrt{50}$, so

$$\cos \varphi = \frac{\mathbf{u} \cdot \mathbf{v}}{|\mathbf{u}||\mathbf{v}|} = \frac{-11}{\sqrt{13}\sqrt{50}} = \frac{-11}{\sqrt{650}} \approx -0.4315. \quad \blacksquare$$

Definition 3   PARALLEL VECTORS   Two nonzero vectors $\mathbf{u}$ and $\mathbf{v}$ are **parallel** if the angle between them is 0 or π.

EXAMPLE 3   Show that the vectors $\mathbf{u} = (2, -3)$ and $\mathbf{v} = (-4, 6)$ are parallel.

Solution.

$$\cos \varphi = \frac{\mathbf{u} \cdot \mathbf{v}}{|\mathbf{u}||\mathbf{v}|} = \frac{-8 - 18}{\sqrt{13}\sqrt{52}} = \frac{-26}{\sqrt{13}(2\sqrt{13})} = \frac{-26}{2(13)} = -1,$$

so $\varphi = \pi$. $\blacksquare$

Theorem 3.   If $\mathbf{u} \neq 0$, then $\mathbf{v} = \alpha\mathbf{u}$ for some nonzero constant α if and only if $\mathbf{u}$ and $\mathbf{v}$ are parallel.

Proof.   This follows from the last part of Definition 2 (see also Problem 43). $\blacksquare$

Definition 4   ORTHOGONAL VECTORS   The nonzero vectors $\mathbf{u}$ and $\mathbf{v}$ are called **orthogonal** (or **perpendicular**) if the angle between them is $\pi/2$.

EXAMPLE 4   Show that the vectors $\mathbf{u} = 3\mathbf{i} - 4\mathbf{j}$ and $\mathbf{v} = 4\mathbf{i} + 3\mathbf{j}$ are orthogonal.

Solution. $\mathbf{u} \cdot \mathbf{v} = 3 \cdot 4 - 4 \cdot 3 = 0$. This implies that $\cos \varphi = (\mathbf{u} \cdot \mathbf{v})/(|\mathbf{u}||\mathbf{v}|) = 0$. Since φ is in the interval $[0, \pi]$, $\varphi = \pi/2$.

Theorem 4.   The nonzero vectors $\mathbf{u}$ and $\mathbf{v}$ are orthogonal if and only if $\mathbf{u} \cdot \mathbf{v} = 0$.

Proof.   This proof is also easy and is left as an exercise (see Problem 44). $\blacksquare$

REMARK.   The condition $\mathbf{u} \cdot \mathbf{v} = 0$ is often given as the *definition* of orthogonal.

A number of interesting problems involve the notion of the *projection* of one vector onto another. Before defining this term, we prove the following theorem.

Theorem 5.   Let $\mathbf{v}$ be a nonzero vector. Then for any other vector $\mathbf{u}$, the vector

$$\mathbf{w} = \mathbf{u} - [(\mathbf{u} \cdot \mathbf{v})/|\mathbf{v}|^2]\mathbf{v} \text{ is orthogonal to } \mathbf{v}.$$

Proof.

$$\mathbf{w} \cdot \mathbf{v} = \left(\mathbf{u} - \frac{(\mathbf{u} \cdot \mathbf{v})\mathbf{v}}{|\mathbf{v}|^2} \right) \cdot \mathbf{v} = \mathbf{u} \cdot \mathbf{v} - \frac{(\mathbf{u} \cdot \mathbf{v})(\mathbf{v} \cdot \mathbf{v})}{|\mathbf{v}|^2}$$

$$= \mathbf{u} \cdot \mathbf{v} - \frac{(\mathbf{u} \cdot \mathbf{v})|\mathbf{v}|^2}{|\mathbf{v}|^2} = \mathbf{u} \cdot \mathbf{v} - \mathbf{u} \cdot \mathbf{v} = 0$$

The vectors **u**, **v**, and **w** are illustrated in Figure 3. ■

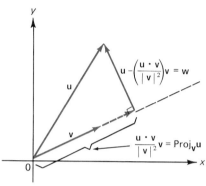

FIGURE 3

Definition 5   PROJECTION   Let **u** and **v** be nonzero vectors. Then the **projection of u onto v** is a vector, denoted Proj$_\mathbf{v}$ **u**, which is defined by

$$\text{Proj}_\mathbf{v}\, \mathbf{u} = \frac{\mathbf{u} \cdot \mathbf{v}}{\mathbf{v} \cdot \mathbf{v}}\mathbf{v} = \frac{\mathbf{u} \cdot \mathbf{v}}{|\mathbf{v}|^2}\mathbf{v} = \left(\frac{\mathbf{u} \cdot \mathbf{v}}{|\mathbf{v}|} \right)\frac{\mathbf{v}}{|\mathbf{v}|}. \tag{5}$$

The **component** *of* **u** *in the direction* **v** *is* $\mathbf{u} \cdot \mathbf{v}/|\mathbf{v}|$. $\tag{6}$

Note that $\mathbf{v}/|\mathbf{v}|$ is a unit vector in the direction of **v**.

REMARK 1.   From Figure 3 and the fact that cos $\varphi = (\mathbf{u} \cdot \mathbf{v})/(|\mathbf{u}||\mathbf{v}|)$, we find that

v and Proj$_\mathbf{v}$ **u** have

       (i) the same direction if $\mathbf{u} \cdot \mathbf{v} > 0$ and
       (ii) opposite directions if $\mathbf{u} \cdot \mathbf{v} < 0$.

REMARK 2.   Proj$_\mathbf{v}$ **u** can be thought of as the "**v**-component" of the vector **u**. We will see an illustration of this in our discussion of force in the next section.

REMARK 3.   If **u** and **v** are orthogonal, then $\mathbf{u} \cdot \mathbf{v} = 0$, so that Proj$_\mathbf{v}$ **u** = **0**.

REMARK 4.   An alternative definition of projection is: If **u** and **v** are nonzero vectors, then Proj$_\mathbf{v}$ **u** is the unique vector having the properties

> **(i)** Proj$_\mathbf{v}$ $\mathbf{u}$ is parallel to $\mathbf{v}$ and   **(ii)** $\mathbf{u}$ − Proj$_\mathbf{v}$ $\mathbf{u}$ is orthogonal to $\mathbf{v}$.

EXAMPLE 5   Let $\mathbf{u} = 2\mathbf{i} + 3\mathbf{j}$ and $\mathbf{v} = \mathbf{i} + \mathbf{j}$. Calculate Proj$_\mathbf{v}$ $\mathbf{u}$.

Solution. Proj$_\mathbf{v}$ $\mathbf{u} = (\mathbf{u} \cdot \mathbf{v})\mathbf{v}/|\mathbf{v}|^2 = [5/(\sqrt{2})^2]\mathbf{v} = (5/2)\mathbf{i} + (5/2)\mathbf{j}$. This is illustrated in Figure 4. ■

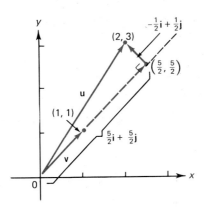

FIGURE 4

EXAMPLE 6   Let $\mathbf{u} = 2\mathbf{i} - 3\mathbf{j}$ and $\mathbf{v} = \mathbf{i} + \mathbf{j}$. Find Proj$_\mathbf{v}$ $\mathbf{u}$.

Solution. Here $(\mathbf{u} \cdot \mathbf{v})/|\mathbf{v}|^2 = -\frac{1}{2}$, so that Proj$_\mathbf{v}$ $\mathbf{u} = -\frac{1}{2}\mathbf{i} - \frac{1}{2}\mathbf{j}$. This is illustrated in Figure 5. ■

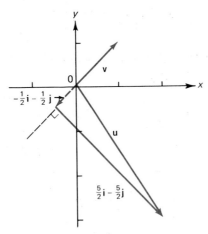

FIGURE 5

PROBLEMS 15.2

In Problems 1–10, calculate both the dot product of the two vectors and the cosine of the angle between them.

1. $\mathbf{u} = \mathbf{i} + \mathbf{j}$; $\mathbf{v} = \mathbf{i} - \mathbf{j}$

2. $\mathbf{u} = 3\mathbf{i}$; $\mathbf{v} = -7\mathbf{j}$

3. $\mathbf{u} = -5\mathbf{i}$; $\mathbf{v} = 18\mathbf{j}$

4. $\mathbf{u} = \alpha\mathbf{i}$; $\mathbf{v} = \beta\mathbf{j}$; α, β real and nonzero

5. $\mathbf{u} = 2\mathbf{i} + 5\mathbf{j}$; $\mathbf{v} = 5\mathbf{i} + 2\mathbf{j}$

6. $\mathbf{u} = 2\mathbf{i} + 5\mathbf{j}$; $\mathbf{v} = 5\mathbf{i} - 3\mathbf{j}$

7. $\mathbf{u} = -3\mathbf{i} + 4\mathbf{j}$; $\mathbf{v} = -2\mathbf{i} - 7\mathbf{j}$

8. $\mathbf{u} = 4\mathbf{i} + 5\mathbf{j}$; $\mathbf{v} = 7\mathbf{i} - 4\mathbf{j}$

9. $\mathbf{u} = 11\mathbf{i} - 8\mathbf{j}$; $\mathbf{v} = 4\mathbf{i} - 7\mathbf{j}$

10. $\mathbf{u} = -13\mathbf{i} + 8\mathbf{j}$; $\mathbf{v} = 2\mathbf{i} + 11\mathbf{j}$

11. Show that for any nonzero real numbers α and β, the vectors $\mathbf{u} = \alpha\mathbf{i} + \beta\mathbf{j}$ and $\mathbf{v} = \beta\mathbf{i} - \alpha\mathbf{j}$ are orthogonal.

12. Let $\mathbf{u}$, $\mathbf{v}$, and $\mathbf{w}$ denote three arbitrary vectors. Explain why the product $\mathbf{u} \cdot \mathbf{v} \cdot \mathbf{w}$ is *not defined*.

In Problems 13–20, determine whether the given vectors are orthogonal, parallel, or neither. Then sketch each pair.

13. $\mathbf{u} = 3\mathbf{i} + 5\mathbf{j}$; $\mathbf{v} = -6\mathbf{i} - 10\mathbf{j}$

14. $\mathbf{u} = 2\mathbf{i} + 3\mathbf{j}$; $\mathbf{v} = 6\mathbf{i} - 4\mathbf{j}$

15. $\mathbf{u} = 2\mathbf{i} + 3\mathbf{j}$; $\mathbf{v} = 6\mathbf{i} + 4\mathbf{j}$

16. $\mathbf{u} = 2\mathbf{i} + 3\mathbf{j}$; $\mathbf{v} = -6\mathbf{i} + 4\mathbf{j}$

17. $\mathbf{u} = 7\mathbf{i}$; $\mathbf{v} = -23\mathbf{j}$

18. $\mathbf{u} = 2\mathbf{i} - 6\mathbf{j}$; $\mathbf{v} = -\mathbf{i} + 3\mathbf{j}$

19. $\mathbf{u} = \mathbf{i} + \mathbf{j}$; $\mathbf{v} = \alpha\mathbf{i} + \alpha\mathbf{j}$; α real

20. $\mathbf{u} = -2\mathbf{i} + 3\mathbf{j}$; $\mathbf{v} = -\mathbf{i} + 2\mathbf{j}$

21. Let $\mathbf{u} = 3\mathbf{i} + 4\mathbf{j}$ and $\mathbf{v} = \mathbf{i} + \alpha\mathbf{j}$.
   (a) Determine α such that $\mathbf{u}$ and $\mathbf{v}$ are orthogonal.
   (b) Determine α such that $\mathbf{u}$ and $\mathbf{v}$ are parallel.
   (c) Determine α such that the angle between $\mathbf{u}$ and $\mathbf{v}$ is $\pi/4$.
   (d) Determine α such that the angle between $\mathbf{u}$ and $\mathbf{v}$ is $\pi/3$.

22. Let $\mathbf{u} = -2\mathbf{i} + 5\mathbf{j}$ and $\mathbf{v} = \alpha\mathbf{i} - 2\mathbf{j}$.
   (a) Determine α such that $\mathbf{u}$ and $\mathbf{v}$ are orthogonal.
   (b) Determine α such that $\mathbf{u}$ and $\mathbf{v}$ are parallel.
   (c) Determine α such that the angle between $\mathbf{u}$ and $\mathbf{v}$ is $2\pi/3$.
   (d) Determine α such that the angle between $\mathbf{u}$ and $\mathbf{v}$ is $\pi/3$.

23. In Problem 21 show that there is no value of α for which $\mathbf{u}$ and $\mathbf{v}$ have opposite directions.

24. In Problem 22 show that there is no value of α for which $\mathbf{u}$ and $\mathbf{v}$ have the same direction.

In Problems 25–38, calculate $\text{Proj}_{\mathbf{v}}\,\mathbf{u}$.

25. $\mathbf{u} = 3\mathbf{i}$; $\mathbf{v} = \mathbf{i} + \mathbf{j}$

26. $\mathbf{u} = -5\mathbf{j}$; $\mathbf{v} = \mathbf{i} + \mathbf{j}$

27. $\mathbf{u} = 2\mathbf{i} + \mathbf{j}$; $\mathbf{v} = \mathbf{i} - 2\mathbf{j}$

28. $\mathbf{u} = 2\mathbf{i} + 3\mathbf{j}$; $\mathbf{v} = 4\mathbf{i} + \mathbf{j}$

29. $\mathbf{u} = \mathbf{i} + \mathbf{j}$; $\mathbf{v} = 2\mathbf{i} - 3\mathbf{j}$

30. $\mathbf{u} = \mathbf{i} + \mathbf{j}$; $\mathbf{v} = 2\mathbf{i} + 3\mathbf{j}$

31. $\mathbf{u} = 4\mathbf{i} + 5\mathbf{j}$; $\mathbf{v} = 2\mathbf{i} + 4\mathbf{j}$

32. $\mathbf{u} = 4\mathbf{i} + 5\mathbf{j}$; $\mathbf{v} = 2\mathbf{i} - 4\mathbf{j}$

33. $\mathbf{u} = -4\mathbf{i} + 5\mathbf{j}$; $\mathbf{v} = 2\mathbf{i} - 4\mathbf{j}$

34. $\mathbf{u} = -4\mathbf{i} - 5\mathbf{j}$; $\mathbf{v} = -2\mathbf{i} - 4\mathbf{j}$

35. $\mathbf{u} = \alpha\mathbf{i} + \beta\mathbf{j}$; $\mathbf{v} = \mathbf{i} + \mathbf{j}$

36. $\mathbf{u} = \mathbf{i} + \mathbf{j}$; $\mathbf{v} = \alpha\mathbf{i} + \beta\mathbf{j}$

37. $\mathbf{u} = \alpha\mathbf{i} - \beta\mathbf{j}$; $\mathbf{v} = \mathbf{i} + \mathbf{j}$

38. $\mathbf{u} = \alpha\mathbf{i} - \beta\mathbf{j}$; $\mathbf{v} = -\mathbf{i} + \mathbf{j}$

39. Let $\mathbf{u} = a_1\mathbf{i} + b_1\mathbf{j}$ and $\mathbf{v} = a_2\mathbf{i} + b_2\mathbf{j}$. Give a condition on a_1, b_1, a_2, and b_2 that will ensure that $\mathbf{v}$ and $\text{Proj}_{\mathbf{v}}\,\mathbf{u}$ have the same direction.

40. In Problem 39, give a condition that will ensure that $\mathbf{v}$ and $\text{Proj}_{\mathbf{v}}\,\mathbf{u}$ have opposite directions.

41. Let $P = (2, 3)$, $Q = (5, 7)$, $R = (2, -3)$, and $S = (1, 2)$. Calculate $\text{Proj}_{\overrightarrow{PQ}}\,\overrightarrow{RS}$ and $\text{Proj}_{\overrightarrow{RS}}\,\overrightarrow{PQ}$.

42. Let $P = (-1, 3)$, $Q = (2, 4)$, $R = (-6, -2)$, and $S = (3, 0)$. Calculate $\text{Proj}_{\overrightarrow{PQ}}\,\overrightarrow{RS}$ and $\text{Proj}_{\overrightarrow{RS}}\,\overrightarrow{PQ}$.

43. Prove that the nonzero vectors $\mathbf{u}$ and $\mathbf{v}$ are parallel if and only if $\mathbf{v} = \alpha\mathbf{u}$ for some nonzero constant α. [*Hint:* Show that $\cos \varphi = \pm 1$ if and only if $\mathbf{v} = \alpha\mathbf{u}$.]

44. Prove that the nonzero vectors $\mathbf{u}$ and $\mathbf{v}$ are orthogonal if and only if $\mathbf{u} \cdot \mathbf{v} = 0$.

45. Show that the vector $\mathbf{v} = a\mathbf{i} + b\mathbf{j}$ is orthogonal to the line $ax + by + c = 0$.

46. Show that the vector $\mathbf{u} = b\mathbf{i} - a\mathbf{j}$ is parallel to the line $ax + by + c = 0$.

47. A triangle has vertices $(1, 3)$, $(4, -2)$, and $(-3, 6)$. Find the cosine of each of its angles.

48. A triangle has vertices (a_1, b_1), (a_2, b_2), and (a_3, b_3). Find a formula for the cosines of each of its angles.

*49. The **Cauchy-Schwarz inequality** states that for any real numbers a_1, a_2, b_1, and b_2,

$$\left| \sum_{k=1}^{2} a_k b_k \right| \leq \left(\sum_{k=1}^{2} a_k^2 \right)^{1/2} \left(\sum_{k=1}^{2} b_k^2 \right)^{1/2}.$$

Use the dot product to prove this formula. Under what circumstances can the inequality be replaced by an equality?   [*Hint:* Show that $|\mathbf{a} \cdot \mathbf{b}| \le |\mathbf{a}| \, |\mathbf{b}|$.]

50. Prove that the shortest distance between a point and a line is measured along a line through the point and perpendicular to the line.

51. Find the distance between $P = (2, 3)$ and the line through the points $Q = (-1, 7)$ and $R = (3, 5)$.   [*Hint:* Draw a picture and use the Pythagorean theorem.]

52. Find the distance between $(3, 7)$ and the line along the vector $\mathbf{v} = 2\mathbf{i} - 3\mathbf{j}$ that passes through the origin.

53. Use the dot product and the Cauchy-Schwarz inequality (Problem 49) to prove the triangle inequality: $|\mathbf{u} + \mathbf{v}| \le |\mathbf{u}| + |\mathbf{v}|$.

15.3 SOME APPLICATIONS OF VECTORS (OPTIONAL)

In this section we discuss some elementary applications of vectors. Instead of attempting to give a set of rules for solving problems involving vectors, we will give a variety of examples. Further examples of the use of vectors will be given in the next chapter, after we discuss the differentiation of vector functions.

EXAMPLE 1   Show that the midpoints of the sides of a quadrilateral are the vertices of a parallelogram.

Solution. The situation is sketched in Figure 1. Using the fact that B, D, F, and H are midpoints, we have

$$\overrightarrow{BD} = \overrightarrow{BC} + \overrightarrow{CD} = \tfrac{1}{2}\overrightarrow{AC} + \tfrac{1}{2}\overrightarrow{CE} = \tfrac{1}{2}(\overrightarrow{AC} + \overrightarrow{CE}) = \tfrac{1}{2}\overrightarrow{AE}$$

and

$$\overrightarrow{HF} = \overrightarrow{HG} + \overrightarrow{GF} = \tfrac{1}{2}\overrightarrow{AG} + \tfrac{1}{2}\overrightarrow{GE} = \tfrac{1}{2}(\overrightarrow{AG} + \overrightarrow{GE}) = \tfrac{1}{2}\overrightarrow{AE}.$$

Thus $\overrightarrow{BD} = \overrightarrow{HF}$. Similarly, $\overrightarrow{HB} = \overrightarrow{FD}$, so $BDFH$ is a parallelogram.   ∎

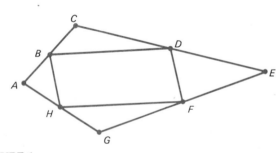

FIGURE 1

EXAMPLE 2   Show that the line segment joining the midpoints of two sides of a triangle has the length of half the third side and is parallel to it.

Solution. We refer to Figure 2. From the figure we see that

$$\overrightarrow{ST} = \overrightarrow{PT} - \overrightarrow{PS}.$$

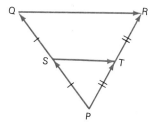

FIGURE 2

But $\overrightarrow{PT} = \frac{1}{2}\overrightarrow{PR}$ and $\overrightarrow{PS} = \frac{1}{2}\overrightarrow{PQ}$. Thus

$$\overrightarrow{ST} = \frac{1}{2}\overrightarrow{PR} - \frac{1}{2}\overrightarrow{PQ} = \frac{1}{2}(\overrightarrow{PR} - \overrightarrow{PQ}) = \frac{1}{2}\overrightarrow{QR}.$$

Hence from Theorem 15.2.3, $\overrightarrow{ST}$ is parallel to $\overrightarrow{QR}$. Moreover $|\overrightarrow{ST}| = \frac{1}{2}|\overrightarrow{QR}|$. This is what we wanted to show. ∎

In earlier chapters we solved problems involving the notion of force. We started such problems with a statement like "a force of x newtons is applied to. . . ." Implicit in such a problem is the idea that a force with a certain magnitude (measured in pounds or newtons) is exerted in a certain direction. In this context force can (and should) be thought of as a vector.

If more than one force is applied to an object, then we define the **resultant** of the forces applied to the object to be the *vector sum* of these forces. We can think of the resultant as the *net* applied force.

EXAMPLE 3   A force of 3 N† is applied to the left side of an object, one of 4 N is applied from the bottom, and a force of 7 N is applied from an angle of $\pi/4$ to the horizontal. What is the resultant of forces applied to the object?

Solution. The forces are indicated in Figure 3. We write each force as a magnitude times a unit vector in the indicated direction. For convenience we can think of the center of the object as being at the origin. Then $\mathbf{F}_1 = 3\mathbf{i}$; $\mathbf{F}_2 = 4\mathbf{j}$; and $\mathbf{F}_3 = -(7/\sqrt{2})(\mathbf{i} + \mathbf{j})$. This last vector follows from the fact that the vector $-(1/\sqrt{2})(\mathbf{i} + \mathbf{j})$ is a unit vector pointing toward the origin making an angle of $\pi/4$ with the x-axis (see Example 15.1.10). Then the resultant is given by

$$\mathbf{F} = \mathbf{F}_1 + \mathbf{F}_2 + \mathbf{F}_3 = \left(3 - \frac{7}{\sqrt{2}}\right)\mathbf{i} + \left(4 - \frac{7}{\sqrt{2}}\right)\mathbf{j}.$$

The magnitude of $\mathbf{F}$ is

$$|\mathbf{F}| = \sqrt{\left(3 - \frac{7}{\sqrt{2}}\right)^2 + \left(4 - \frac{7}{\sqrt{2}}\right)^2} = \sqrt{74 - \frac{98}{\sqrt{2}}} \approx 2.17 \text{ N}.$$

The direction θ can be calculated by first finding the unit vector in the direction of $\mathbf{F}$:

---

†Units of force are discussed on page 333.

FIGURE 3

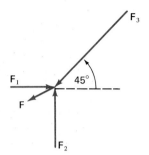

FIGURE 4

$$\frac{\mathbf{F}}{|\mathbf{F}|} = \frac{3 - (7/\sqrt{2})}{\sqrt{74 - (98/\sqrt{2})}}\mathbf{i} + \frac{4 - (7/\sqrt{2})}{\sqrt{74 - (98/\sqrt{2})}}\mathbf{j}$$

$$= (\cos\theta)\mathbf{i} + (\sin\theta)\mathbf{j}.$$

Then

$$\cos\theta = \frac{3 - (7/\sqrt{2})}{\sqrt{74 - (98/\sqrt{2})}} \approx -0.8990 \quad \text{and} \quad \sin\theta = \frac{4 - (7/\sqrt{2})}{\sqrt{74 - (98/\sqrt{2})}} \approx -0.4379.$$

This means that θ is in the third quadrant, and $\theta \approx 3.5949 \approx 206°$ (or $-154°$). This is illustrated in Figure 4. ■

Recall from Section 5.9 that the work done by a force F in moving an object a distance d is given by

$$W = Fd \tag{1}$$

where units of work are newton-meters or foot-pounds. In formula (1) it is assumed that the force is applied in the same direction as the direction of motion. However, this is not always the case. In general, we may define

$$W = (\text{component of } \mathbf{F} \text{ in the direction of motion}) \times (\text{distance moved}). \tag{2}$$

If the object moves from P to Q, then the distance moved is $|\overrightarrow{PQ}|$. The vector $\mathbf{d}$, one of whose representations is $\overrightarrow{PQ}$, is called a **displacement vector.** Then from Equation (6) on page 773,

$$\text{component of } \mathbf{F} \text{ in direction of motion} = \frac{\mathbf{F} \cdot \mathbf{d}}{|\mathbf{d}|}. \tag{3}$$

Finally, combining (2) and (3), we obtain

$$W = \frac{\mathbf{F} \cdot \mathbf{d}}{|\mathbf{d}|}|\mathbf{d}| = \mathbf{F} \cdot \mathbf{d}. \tag{4}$$

That is, *the work done is the dot product of the force* **F** *and the displacement vector* **d**. Note that if **F** acts in the direction **d** and if φ denotes the angle (which is zero) between **F** and **d**, then $\mathbf{F} \cdot \mathbf{d} = |\mathbf{F}|\,|\mathbf{d}|\,\cos\varphi = |\mathbf{F}|\,|\mathbf{d}|\,\cos 0 = |\mathbf{F}|\,|\mathbf{d}|$, which is formula (1).

EXAMPLE 4   A force of 4 N has the direction $\pi/3$. What is the work done in moving an object from the point (1, 2) to the point (5, 4), where distances are measured in meters?

   Solution.   A unit vector with direction $\pi/3$ is given by $\mathbf{u} = (\cos \pi/3)\mathbf{i} + (\sin \pi/3)\mathbf{j}$ $= (1/2)\mathbf{i} + (\sqrt{3}/2)\mathbf{j}$. Thus $\mathbf{F} = 4\mathbf{u} = 2\mathbf{i} + 2\sqrt{3}\mathbf{j}$. The displacement vector **d** is given by $(5 - 1)\mathbf{i} + (4 - 2)\mathbf{j} = 4\mathbf{i} + 2\mathbf{j}$. Thus

$$W = \mathbf{F} \cdot \mathbf{d} = (2\mathbf{i} + 2\sqrt{3}\mathbf{j}) \cdot (4\mathbf{i} + 2\mathbf{j}) = (8 + 4\sqrt{3}) \approx 14.93 \text{ N} \cdot \text{m}.$$

The component of **F** in the direction of motion is sketched in Figure 5.   ∎

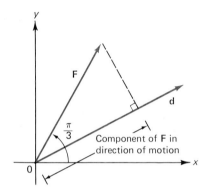

FIGURE 5

PROBLEMS 15.3

1. Show that the diagonals of a parallelogram bisect each other.
2. Show that the diagonals of a rhombus are orthogonal.
3. Show that in a trapezoid the line segment that joins the midpoints of the two sides that are not parallel is parallel to the parallel sides and has a length equal to the average of the lengths of the parallel sides.
4. Use vector methods to show that the angles opposite the equal sides in an isosceles triangle are equal.
*5. Prove that the medians of a triangle intersect at a fixed point that is two-thirds the distance from each vertex to the opposite side.
6. Use the result of Problem 5 to find the point of intersection of the medians of the triangle with vertices at (3, 4), (2, −1), and (−3, 2).

In Problems 7–15, find the resultant of the forces acting on an object. Then find the force vector that must be applied so that the object will remain at rest.

7. 2 N (from right), 5 N (from above)
8. 2 N (from left), 5 N (from below)
9. 3 N (from left), 5 N (from right), 3 N (from above)
10. 10 lb (from right), 8 lb (from below)
11. 5 lb (from above), 4 lb (from direction $\pi/6$)
12. 6 N (from left), 4 N (from direction $\pi/4$), 2 N (from direction $\pi/3$)

13. 2 N (from above), 3 N (from direction $3\pi/4$)
14. 5 N (from direction $\pi/3$), 5 N (from direction $2\pi/3$)
15. 7 N (from direction $\pi/6$), 7 N (from direction $\pi/3$), 14 N (from direction $5\pi/4$)

In Problems 16–24, find the work done when the force with given magnitude and direction moves an object from P to Q. All distances are measured in meters. (Note that work can be negative.)

16. $|\mathbf{F}| = 3$ N; $\theta = 0$; $P = (2, 3)$; $Q = (1, 7)$
17. $|\mathbf{F}| = 2$ N; $\theta = \pi/2$; $P = (5, 7)$; $Q = (1, 1)$
18. $|\mathbf{F}| = 6$ N; $\theta = \pi/4$; $P = (2, 3)$; $Q = (-1, 4)$
19. $|\mathbf{F}| = 4$ N; $\theta = \pi/6$; $P = (-1, 2)$; $Q = (3, 4)$
20. $|\mathbf{F}| = 7$ N; $\theta = 2\pi/3$; $P = (4, -3)$; $Q = (1, 0)$
21. $|\mathbf{F}| = 3$ N; $\theta = 3\pi/4$; $P = (2, 1)$; $Q = (1, 2)$
22. $|\mathbf{F}| = 6$ N; $\theta = \pi$; $P = (3, -8)$; $Q = (5, 10)$
23. $|\mathbf{F}| = 4$ N; θ is the direction of $2\mathbf{i} + 3\mathbf{j}$; $P = (2, 0)$; $Q = (-1, 3)$
24. $|\mathbf{F}| = 5$ N; θ is the direction of $-3\mathbf{i} + 2\mathbf{j}$; $P = (1, 3)$; $Q = (4, -6)$
25. Two tugboats are towing a barge (see Figure 6). Tugboat 1 pulls with a force of 500 N at an angle of 20° with the horizontal. Tugboat 2 pulls with a force of x newtons at an angle of 30°. The barge moves horizontally (i.e., $\theta = 0$). Find x.

FIGURE 6

26. Answer the question of Problem 25 if the angles are 50° and 75°, respectively, and all other data remain the same.
27. In Problem 25 how much work is done by each tugboat in moving the barge a distance of 750 m?
28. In Problem 26 how much work is done by each tugboat in moving the barge a distance of 2 km?

REVIEW EXERCISES FOR CHAPTER FIFTEEN

In Exercises 1–6, find the magnitude and direction of the given vector.

1. $\mathbf{v} = (3, 3)$
2. $\mathbf{v} = -3\mathbf{i} + 3\mathbf{j}$
3. $\mathbf{v} = (2, -2\sqrt{3})$
4. $\mathbf{v} = (\sqrt{3}, 1)$
5. $\mathbf{v} = -12\mathbf{i} - 12\mathbf{j}$
6. $\mathbf{v} = \mathbf{i} + 4\mathbf{j}$

In Exercises 7–10, write the vector $\mathbf{v}$ that is represented by $\overrightarrow{PQ}$ in the form $a\mathbf{i} + b\mathbf{j}$.

7. $P = (2, 3)$; $Q = (4, 5)$
8. $P = (1, -2)$; $Q = (7, 12)$
9. $P = (-1, -6)$; $Q = (3, -4)$
10. $P = (-1, 3)$; $Q = (3, -1)$

11. Let $\mathbf{u} = (2, 1)$ and $\mathbf{v} = (-3, 4)$. Find the following:
  (a) $5\mathbf{u}$
  (b) $\mathbf{u} - \mathbf{v}$
  (c) $-8\mathbf{u} + 5\mathbf{v}$

12. Let $\mathbf{u} = -4\mathbf{i} + \mathbf{j}$ and $\mathbf{v} = -3\mathbf{i} - 4\mathbf{j}$. Find the following:
  (a) $-3\mathbf{v}$ **(b)** $\mathbf{u} + \mathbf{v}$ **(c)** $3\mathbf{u} - 6\mathbf{v}$

In Exercises 13–19, find a unit vector having the same direction as the given vector.

13. $\mathbf{v} = \mathbf{i} + \mathbf{j}$ **14.** $\mathbf{v} = -\mathbf{i} + \mathbf{j}$ **15.** $\mathbf{v} = 2\mathbf{i} + 5\mathbf{j}$
16. $\mathbf{v} = -7\mathbf{i} + 3\mathbf{j}$ **17.** $\mathbf{v} = 3\mathbf{i} + 4\mathbf{j}$ **18.** $\mathbf{v} = -2\mathbf{i} - 2\mathbf{j}$
19. $\mathbf{v} = a\mathbf{i} - a\mathbf{j}$

20. For $\mathbf{v} = 4\mathbf{i} - 7\mathbf{j}$, find $\sin\theta$ and $\cos\theta$, where θ is the direction of $\mathbf{v}$,
21. Find a unit vector with direction opposite to that of $\mathbf{v} = 5\mathbf{i} + 2\mathbf{j}$.
22. Find two unit vectors orthogonal to $\mathbf{v} = \mathbf{i} - \mathbf{j}$.
23. Find a unit vector with direction opposite to that of $\mathbf{v} = 10\mathbf{i} - 7\mathbf{j}$.

In Exercises 24–27, find a vector $\mathbf{v}$ having the given magnitude and direction.

24. $|\mathbf{v}| = 2; \theta = \pi/3$ **25.** $|\mathbf{v}| = 1; \theta = \pi/2$
26. $|\mathbf{v}| = 4; \theta = \pi$ **27.** $|\mathbf{v}| = 7; \theta = 5\pi/6$

In Exercises 28–31, calculate the dot product of the two vectors and the cosine of the angle between them.

28. $\mathbf{u} = \mathbf{i} - \mathbf{j}; \mathbf{v} = \mathbf{i} + 2\mathbf{j}$ **29.** $\mathbf{u} = -4\mathbf{i}; \mathbf{v} = 11\mathbf{j}$
30. $\mathbf{u} = 4\mathbf{i} - 7\mathbf{j}; \mathbf{v} = 5\mathbf{i} + 6\mathbf{j}$ **31.** $\mathbf{u} = -\mathbf{i} - 2\mathbf{j}; \mathbf{v} = 4\mathbf{i} + 5\mathbf{j}$

In Exercises 32–37, determine whether the given vectors are orthogonal, parallel, or neither. Then sketch each pair.

32. $\mathbf{u} = 2\mathbf{i} - 6\mathbf{j}; \mathbf{v} = -\mathbf{i} + 3\mathbf{j}$ **33.** $\mathbf{u} = 4\mathbf{i} - 5\mathbf{j}; \mathbf{v} = 5\mathbf{i} - 4\mathbf{j}$
34. $\mathbf{u} = 4\mathbf{i} - 5\mathbf{j}; \mathbf{v} = -5\mathbf{i} + 4\mathbf{j}$ **35.** $\mathbf{u} = -7\mathbf{i} - 7\mathbf{j}; \mathbf{v} = \mathbf{i} + \mathbf{j}$
36. $\mathbf{u} = -7\mathbf{i} - 7\mathbf{j}; \mathbf{v} = -\mathbf{i} + \mathbf{j}$ **37.** $\mathbf{u} = -7\mathbf{i} - 7\mathbf{j}; \mathbf{v} = -\mathbf{i} - \mathbf{j}$

38. Let $\mathbf{u} = 2\mathbf{i} + 3\mathbf{j}$ and $\mathbf{v} = 4\mathbf{i} + \alpha\mathbf{j}$.
  (a) Determine α such that $\mathbf{u}$ and $\mathbf{v}$ are orthogonal.
  (b) Determine α such that $\mathbf{u}$ and $\mathbf{v}$ are parallel.
  (c) Determine α such that the angle between $\mathbf{u}$ and $\mathbf{v}$ is $\pi/4$.
  (d) Determine α such that the angle between $\mathbf{u}$ and $\mathbf{v}$ is $\pi/6$.

In Exercises 39–44, calculate $\text{Proj}_{\mathbf{v}}\,\mathbf{u}$.

39. $\mathbf{u} = 14\mathbf{i}; \mathbf{v} = \mathbf{i} + \mathbf{j}$ **40.** $\mathbf{u} = 14\mathbf{i}; \mathbf{v} = \mathbf{i} - \mathbf{j}$
41. $\mathbf{u} = 3\mathbf{i} - 2\mathbf{j}; \mathbf{v} = 3\mathbf{i} + 2\mathbf{j}$ **42.** $\mathbf{u} = 3\mathbf{i} + 2\mathbf{j}; \mathbf{v} = \mathbf{i} - 3\mathbf{j}$
43. $\mathbf{u} = 2\mathbf{i} - 5\mathbf{j}; \mathbf{v} = -3\mathbf{i} - 7\mathbf{j}$ **44.** $\mathbf{u} = 4\mathbf{i} - 5\mathbf{j}; \mathbf{v} = -3\mathbf{i} - \mathbf{j}$

45. Let $P = (3, -2)$, $Q = (4, 7)$, $R = (-1, 3)$, and $S = (2, -1)$. Calculate $\text{Proj}_{\overrightarrow{PQ}}\,\overrightarrow{RS}$ and $\text{Proj}_{\overrightarrow{RS}}\,\overrightarrow{PQ}$.

In Exercises 46–48, calculate the resultant of the forces acting on an object.

46. 3 N (from left), 2 N (from below)
47. 5 N (from right), 2 N (from left), 3 N (from above)
48. 2 N (from direction $\pi/4$), 5 N (from left), 4 N (from direction $2\pi/3$), 6 N (from direction $3\pi/4$)

In Exercises 49–52, find the work done when the force with given magnitude and direction moves an object from P to Q. All distances are measured in meters.

49. $|\mathbf{F}| = 2$ N; $\theta = \pi/4$; $P = (1, 6)$; $Q = (2, 4)$
50. $|\mathbf{F}| = 3$ N; $\theta = \pi/2$; $P = (3, -5)$; $Q = (2, 7)$
51. $|\mathbf{F}| = 11$ N; $\theta = \pi/6$; $P = (-1, -2)$; $Q = (-7, -4)$
52. $|\mathbf{F}| = 8$ N; $\theta = 2\pi/3$; $P = (-1, 4)$; $Q = (5, -6)$

16 Vector Functions, Vector Differentiation, and Parametric Equations

16.1 VECTOR FUNCTIONS AND PARAMETRIC EQUATIONS

In Chapter 15 we considered vectors (in the plane) that could be written as

$$\mathbf{v} = (a, b) = a\mathbf{i} + b\mathbf{j}. \tag{1}$$

In this chapter we see what happens when the numbers a and b in (1) are replaced by functions $f_1(t)$ and $f_2(t)$.

Definition 1 VECTOR FUNCTION   Let f_1 and f_2 be functions of the real variable t. Then for all values of t for which $f_1(t)$ and $f_2(t)$ are defined, we define the **vector-valued function f** by

$$\mathbf{f}(t) = (f_1(t), f_2(t)) = f_1(t)\mathbf{i} + f_2(t)\mathbf{j}. \tag{2}$$

The **domain** of $\mathbf{f}$ is the intersection of the domains of f_1 and f_2.

REMARK.   For simplicity we will refer to vector-valued functions as **vector functions**.

EXAMPLE 1   Let $\mathbf{f}(t) = f_1(t)\mathbf{i} + f_2(t)\mathbf{j} = (1/t)\mathbf{i} + \sqrt{t + 1}\mathbf{j}$. Find the domain of $\mathbf{f}$.

Solution. The domain of **f** is the set of all t for which f_1 and f_2 are defined. Since $f_1(t)$ is defined for $t \neq 0$ and $f_2(t)$ is defined for $t \geq -1$, we see that the domain of **f** is the set $\{t: t \geq -1 \text{ and } t \neq 0\}$. ∎

Let **f** be a vector function. Then for each t in the domain of **f**, the endpoint of the vector $f_1(t)\mathbf{i} + f_2(t)\mathbf{j}$ is a point (x, y) in the xy-plane, where

$$x = f_1(t) \qquad \text{and} \qquad y = f_2(t). \tag{3}$$

Definition 2 PLANE CURVES AND PARAMETRIC EQUATIONS Suppose that the interval $[a, b]$ is in the domain of the function **f** and that both f_1 and f_2 are continuous in $[a, b]$. Then the set of points $(f_1(t), f_2(t))$ for $a \leq t \leq b$ is called a **plane curve** C. Equation (2) is called the **vector equation** of C, while equations (3) are called the **parametric equations** or **parametric representation** of C. In this context the variable t is called a **parameter**.

REMARK. We will usually refer to a plane curve simply as a **curve**.

EXAMPLE 2 Describe the curve given by the vector equation

$$\mathbf{f}(t) = (\cos t)\mathbf{i} + (\sin t)\mathbf{j}, \qquad 0 \leq t \leq 2\pi. \tag{4}$$

Solution. We first see that for every t, $|\mathbf{f}(t)| = 1$ since $|\mathbf{f}(t)| = \sqrt{\cos^2 t + \sin^2 t} = 1$. Moreover, if we write the curve in its parametric representation, we find that

$$x = \cos t, \qquad y = \sin t, \tag{5}$$

and since $\cos^2 t + \sin^2 t = 1$, we have

$$x^2 + y^2 = 1,$$

which is, of course, the equation of the unit circle. This curve is sketched in Figure 1. Note that in the sketch the parameter t represents both the length of the arc from $(1, 0)$ to the endpoint of the vector and the angle (measured in radians) the vector makes with the positive x-axis. The representation $x^2 + y^2 = 1$ is called the *Cartesian equation* of the curve given by (5).

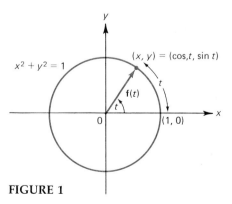

FIGURE 1

REMARK: As t increases from 0 to 2π, we move around the unit circle in the counterclockwise direction. If t were instead restricted to the range $0 \le t \le \pi$, then we would not get the entire circle. Rather, we would stop at the point $(\cos \pi, \sin \pi) = (-1, 0)$, which would give us the upper semicircle only. ▮

Definition 3 CARTESIAN EQUATION OF A PLANE CURVE   A **Cartesian equation** of the curve $f(t) = x(t)\mathbf{i} + y(t)\mathbf{j}$ is an equation relating the variables x and y only.

EXAMPLE 3 Describe and sketch the curve given parametrically by $x = t + 3$, $y = t^2 - t + 2$.

Solution. With problems of this type the easiest thing to do is to write t as a function of x or y, if possible. Since $x = t + 3$, we immediately see that $t = x - 3$ and $y = t^2 - t + 2 = (x - 3)^2 - (x - 3) + 2 = x^2 - 7x + 14$. This is the Cartesian equation of the curve and is the equation of a parabola. It is sketched in Figure 2. Note that in the Cartesian equation of the parabola, the parameter t does not appear. ▮

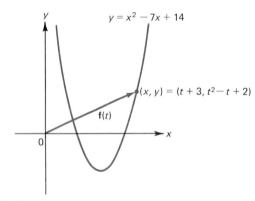

$y = x^2 - 7x + 14$

$(x, y) = (t + 3, t^2 - t + 2)$

$f(t)$

FIGURE 2

EXAMPLE 4 Describe and sketch the curve given by the vector equation

$$\mathbf{f}(r) = (1 - r^4)\mathbf{i} + r^2\mathbf{j}. \tag{6}$$

Solution. First, we note that the parameter in this problem is r instead of t. This makes absolutely no difference since, as in the case of the variable of integration, the parameter is a "dummy" variable. Now to get a feeling for the shape of this curve, we display in Table 1, values of x and y for various values of r. Plotting some of these points leads to the sketch in Figure 3. To write the Cartesian equation for this curve, we square both sides of the equation $y = r^2$ to obtain $y^2 = r^4$ and $x = 1 - r^4 = 1 - y^2$, which is the equation of the parabola sketched in Figure 4. Note that this curve is *not* the same as the curve sketched in Figure 3 since the parametric representation $y = r^2$ requires that y be nonnegative. Thus the curve described by (6) is only the *upper half* of the parabola described by the equation $x = 1 - y^2$.

TABLE 1

r	0	$\pm\frac{1}{2}$	± 1	$\pm\frac{3}{2}$	± 2
$x = 1 - r^4$	1	$\frac{15}{16}$	0	$-\frac{65}{16}$	-15
$y = r^2$	0	$\frac{1}{4}$	1	$\frac{9}{4}$	4

▮

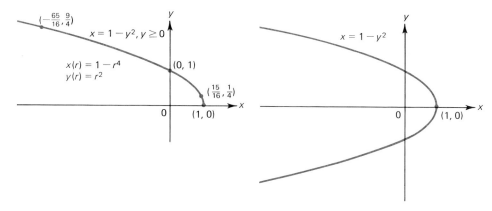

FIGURE 3   **FIGURE 4**

We now obtain a parametric representation of a line L. Let $P = (x_1, y_1)$ and $Q = (x_2, y_2)$ be two points on the line and let R be a third point on the line (see Figure 5). Since the vectors $\overrightarrow{PQ}$ and $\overrightarrow{PR}$ have the same or opposite directions, they are parallel and by Theorem 15.2.3, we have

$$\overrightarrow{PR} = t\,\overrightarrow{PQ} \tag{7}$$

for some real number t (t is positive if $\overrightarrow{PQ}$ and $\overrightarrow{PR}$ have the same direction and t is negative if $\overrightarrow{PQ}$ and $\overrightarrow{PR}$ have opposite directions). Then from Figure 5 we see that

$$\overrightarrow{OR} = \overrightarrow{OP} + \overrightarrow{PR} = \overrightarrow{OP} + t\,\overrightarrow{PQ}. \tag{8}$$

But $\overrightarrow{PQ} = (x_2 - x_1)\mathbf{i} + (y_2 - y_1)\mathbf{j}$ and $\overrightarrow{OP} = x_1\mathbf{i} + y_1\mathbf{j}$, so from (8), if $R = (x, y)$, we obtain

$$x\mathbf{i} + y\mathbf{j} = x_1\mathbf{i} + y_1\mathbf{j} + t(x_2 - x_1)\mathbf{i} + t(y_2 - y_1)\mathbf{j}$$

or

$$x\mathbf{i} + y\mathbf{j} = [x_1 + t(x_2 - x_1)]\mathbf{i} + [y_1 + t(y_2 - y_1)]\mathbf{j}.$$

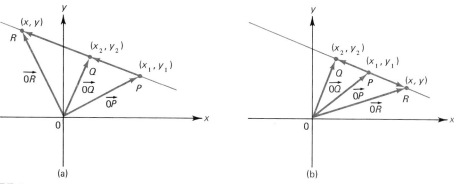

(a)   (b)

FIGURE 5

Thus a parametric equation of the line passing through the points (x_1, y_1) and (x_2, y_2) is

$$x = x_1 + t(x_2 - x_1) \quad \text{and} \quad y = y_1 + t(y_2 - y_1). \tag{9}$$

EXAMPLE 5   Find a parametric representation of the line passing through the points $P = (2, 5)$ and $Q = (-3, 9)$.

Solution. Using (9), we immediately find that

$$x = 2 - 5t \quad \text{and} \quad y = 5 + 4t. \tag{10}$$

Note that the problem asked for *a* parametric representation of the line, since there are several such representations. For example, reversing the roles of P and Q, we obtain the representation [from the new $P = (-3, 9)$ and $Q = (2, 5)$]

$$x = -3 + 5s \quad \text{and} \quad y = 9 - 4s. \tag{11}$$

Although these representations look different, they do represent the same line, as is seen by inserting a few sample values for t and s. For example, if $t = 1$, from (10) we obtain the point $(-3, 9)$. This point is obtained from (11) by setting $s = 0$.

To obtain another parametric representation, we first observe that the slope-intercept form of the line passing through the points $(2, 5)$ and $(-3, 9)$ is given by

$$\frac{y - 5}{x - 2} = \frac{9 - 5}{-3 - 2} = -\frac{4}{5}, \quad \text{or} \quad y = -\frac{4}{5}x + \frac{33}{5}.$$

Then setting $t = x$, we obtain the "slope-intercept" parametrization

$$x = t, \quad y = -\tfrac{4}{5}t + \tfrac{33}{5}. \ \blacksquare$$

The following example shows how vector analysis can be useful in a simple physical problem.

EXAMPLE 6   A cannonball shot from a cannon has an initial velocity of 600 m/sec. The muzzle of the cannon is inclined at an angle of $30°$. Ignoring air resistance, determine the path of the cannonball.

Solution. We place the x- and y-axes so that the mouth of the cannon is at the origin (see Figure 6). The velocity vector $\mathbf{v}$ can be resolved into its vertical and horizontal components:

$$\mathbf{v} = v_x \mathbf{i} + v_y \mathbf{j}.$$

A unit vector in the direction of $\mathbf{v}$ is

$$\mathbf{u} = (\cos 30°)\mathbf{i} + (\sin 30°)\mathbf{j} = \frac{\sqrt{3}}{2}\mathbf{i} + \frac{1}{2}\mathbf{j}$$

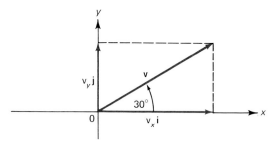

FIGURE 6

so initially,

$$\mathbf{v} = |\mathbf{v}|\mathbf{u} = 600\mathbf{u} = 300\sqrt{3}\mathbf{i} + 300\mathbf{j}.$$

The scalar $|\mathbf{v}|$ is called the **speed** of the cannonball. Thus initially, $v_x = 300\sqrt{3}$ m/sec (the initial speed in the horizontal direction) and $v_y = 300$ m/sec (the initial speed in the vertical direction). Now the vertical acceleration (due to gravity) is

$$a_y = -9.81 \text{ m/sec}^2 \quad \text{and} \quad v_y = \int a_y \, dt = -9.81t + C.$$

Since, initially, $v_y(0) = 300$, we have $C = 300$ and

$$v_y = -9.81t + 300.$$

Then

$$y(t) = \int v_y \, dt = -\frac{9.81t^2}{2} + 300t + C_1$$

and since $y(0) = 0$ (we start at the origin), we find that

$$y(t) = -\frac{9.81t^2}{2} + 300t.$$

To calculate the x-component of the position vector, we note that, ignoring air resistance, the velocity in the horizontal direction is constant;† that is,

$$v_x = 300\sqrt{3} \text{ m/sec.}$$

Then $x(t) = \int v_x \, dt = 300\sqrt{3}t + C_2$, and since $x(0) = 0$, we obtain

$$x = 300\sqrt{3}t.$$

Thus the position vector describing the location of the cannonball is

---

†There are no forces acting on the ball in the horizontal direction since the force of gravity acts only in the vertical direction.

$$\mathbf{s}(t) = x(t)\mathbf{i} + y(t)\mathbf{j} = 300\sqrt{3}t\mathbf{i} + \left(300t - \frac{9.81t^2}{2}\right)\mathbf{j}.$$

To obtain the Cartesian equation of this curve, we start with

$$x = 300\sqrt{3}t$$

so that

$$t = \frac{x}{300\sqrt{3}}$$

and

$$y = 300\left(\frac{x}{300\sqrt{3}}\right) - \frac{9.81}{2}\frac{x^2}{(300\sqrt{3})^2} = \frac{x}{\sqrt{3}} - \frac{9.81}{540,000}x^2.$$

This parabola is sketched in Figure 7. ▪

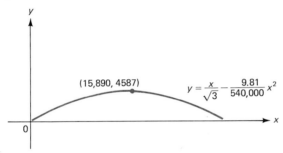

$$y = \frac{x}{\sqrt{3}} - \frac{9.81}{540,000}x^2$$

(15,890, 4587)

FIGURE 7

We close this section by describing, parametrically, a curve that historically has been of great interest. Suppose that a wheel of radius r is rolling in a straight line without slipping. Let P be a fixed point on the wheel a distance s from the center. As the wheel moves, the curve traced by the point P is called a **trochoid**† and if $s = r$ (i.e., if P is on the circumference of the wheel), then the curve traced is called a **cycloid**.‡

To simplify our computations, we place the wheel so that it rolls in a clockwise fashion on the x-axis (see Figure 8a). Suppose that P moves through an angle of α radians to reach its position in Figure 8b. Then the new position vector $\overrightarrow{OP}$ is given by

$$\overrightarrow{OP} = \overrightarrow{OR} + \overrightarrow{RC} + \overrightarrow{CP}. \tag{12}$$

---

†From the Greek word "trochos," meaning "wheel."

‡From the Greek word "kyklos," meaning "circle." The cycloid was a source of great controversy in the seventeenth century. Many of its properties were discovered by the French mathematician Gilles Personne de Roberval (1602–1675), although the curve was first discussed by Galileo. Unfortunately, Roberval, for unknown reasons, did not publish his discoveries concerning the cycloid, which meant that he lost credit for most of them. The ensuing arguments over who discovered what were so bitter that the cycloid became known as the "Helen of geometers" (after Helen of Troy—the source of the intense jealousy that led to the Trojan war).

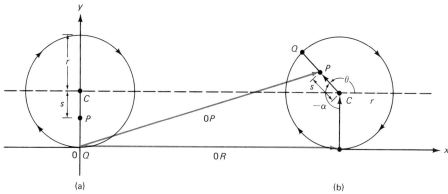

(a)   (b)

FIGURE 8

The length of $\overrightarrow{0R}$ is the length of the arc of the circle that is generated by moving through an angle of α radians and is equal to α times the radius of the circle $= \alpha r$. Thus

$$\overrightarrow{0R} = \alpha r \mathbf{i}.$$

Clearly, $\overrightarrow{RC} = r\mathbf{j}$, and if θ denotes the angle the vector $\overrightarrow{CP}$ makes with the positive x-axis, then

$$\overrightarrow{CP} = s[(\cos \theta)\mathbf{i} + (\sin \theta)\mathbf{j}].$$

Hence using (12), we have

$$\begin{aligned} \overrightarrow{0P} &= \alpha r \mathbf{i} + r\mathbf{j} + s(\cos \theta)\mathbf{i} + s(\sin \theta)\mathbf{j} \\ &= [\alpha r + s(\cos \theta)]\mathbf{i} + [r + s(\sin \theta)]\mathbf{j}. \end{aligned} \tag{13}$$

But since $\alpha + \theta = 3\pi/2$, we have $\theta = 3\pi/2 - \alpha$, so

$$\cos \theta = \cos\left(\frac{3\pi}{2} - \alpha\right) = -\sin \alpha, \qquad \sin \theta = \sin\left(\frac{3\pi}{2} - \alpha\right) = -\cos \alpha,$$

and the equation of the trochoid becomes

$$x\mathbf{i} + y\mathbf{j} = \overrightarrow{0P} = (r\alpha - s \sin \alpha)\mathbf{i} + (r - s \cos \alpha)\mathbf{j} \tag{14}$$

or

$$x = r\alpha - s \sin \alpha, \qquad y = r - s \cos \alpha. \tag{15}$$

For example, if $r = 3$ and $s = 2$, (15) becomes

$$x = 3\alpha - 2 \sin \alpha, \qquad y = 3 - 2 \cos \alpha.$$

A computer-drawn sketch of this curve is given in Figure 9.

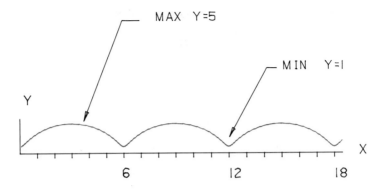

NOTE: ONE UNIT = 3.14159

FIGURE 9

When $r = s$, P is on the circumference of the wheel, and we obtain the cycloid given parametrically by

$$x = r(\alpha - \sin \alpha), \qquad y = r(1 - \cos \alpha). \tag{16}$$

Note that $y \geq 0$ in (16) and $y = 0$ when $\cos \alpha = 1$, which occurs when α is a multiple of 2π. For these values of α, $x = r\alpha = 2\pi nr$. Moreover, y takes on its maximum value $2r$ when $\alpha = \pi, 3\pi, 5\pi, \ldots$. With this information, we can sketch the cycloid, as in Figure 10.

REMARK. Here is one instance in which it is much easier to represent a curve parametrically. If you doubt it, try to write the Cartesian equation of the trochoid [equation (14)] or the simpler cycloid (see Problem 49). You will quickly learn to appreciate the advantages of parametric representation. In fact, while it is possible to find the Cartesian equation of the trochoid, the parametric representation is much easier to use. ■

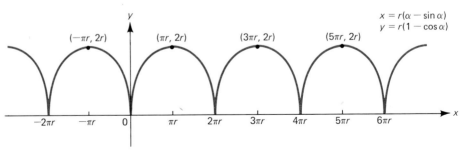

FIGURE 10

PROBLEMS 16.1

In Problems 1–8, find the domain of each vector-valued function.

1. $\mathbf{f}(t) = \dfrac{1}{t}\mathbf{i} + \dfrac{1}{t-1}\mathbf{j}$

2. $\mathbf{f}(t) = \sqrt{t}\,\mathbf{i} + \dfrac{1}{t}\mathbf{j}$

3. $\mathbf{f}(s) = \dfrac{1}{s^2-1}\mathbf{i} + (s^2-1)\mathbf{j}$

4. $\mathbf{f}(u) = e^{1/u}\mathbf{i} + e^{-1/(u+1)}\mathbf{j}$

5. $\mathbf{f}(s) = (\ln s)\mathbf{i} + \ln(1-s)\mathbf{j}$

6. $\mathbf{f}(r) = (\sin r)\mathbf{i} + r\mathbf{j}$

7. $\mathbf{f}(t) = (\sec t)\mathbf{i} + (\csc t)\mathbf{j}$

8. $\mathbf{f}(w) = (\tan w)\mathbf{i} + (\cot w)\mathbf{j}$

In Problems 9–24, find the Cartesian equation of each curve and then sketch the curve in the xy-plane.

9. $\mathbf{f}(t) = t^2\mathbf{i} + 2t\mathbf{j}$

10. $\mathbf{f}(t) = (2t-3)\mathbf{i} + t^2\mathbf{j}$

11. $\mathbf{f}(t) = t^2\mathbf{i} + t^3\mathbf{j}$

12. $\mathbf{f}(t) = 3(\sin t)\mathbf{i} + 3(\cos t)\mathbf{j}$

13. $\mathbf{f}(t) = (2t-1)\mathbf{i} + (4t+3)\mathbf{j}$

14. $\mathbf{f}(t) = 2(\cosh t)\mathbf{i} + 2(\sinh t)\mathbf{j}$

15. $\mathbf{f}(t) = (t^4 + t^2 + 1)\mathbf{i} + t^2\mathbf{j}$

16. $\mathbf{f}(t) = t^2\mathbf{i} + t^8\mathbf{j}$

17. $\mathbf{f}(t) = t^3\mathbf{i} + (t^9 - 1)\mathbf{j}$

18. $\mathbf{f}(t) = t\mathbf{i} + e^t\mathbf{j}$

19. $\mathbf{f}(t) = e^t\mathbf{i} + t^2\mathbf{j}$

20. $\mathbf{f}(t) = (t^2 + t - 3)\mathbf{i} + \sqrt{t}\,\mathbf{j}$

*21. $\mathbf{f}(t) = e^t(\sin t)\mathbf{i} + e^t(\cos t)\mathbf{j}$ [*Hint:* Show that $|\mathbf{f}(t)| = e^t$.]

22. $\mathbf{f}(t) = \mathbf{i} + (\tan t)\mathbf{j}$

23. $\mathbf{f}(t) = e^t\mathbf{i} + e^{2t}\mathbf{j}$

*24. $\mathbf{f}(t) = \dfrac{6t}{1+t^3}\mathbf{i} + \dfrac{6t^2}{1+t^3}\mathbf{j}$

In Problems 25–30, find two parametric representations of the straight line that passes through the given points.

25. $(2, 4)$; $(1, 6)$

26. $(-3, 2)$; $(0, 4)$

27. $(3, 5)$; $(-1, -7)$

28. $(4, 6)$; $(7, 9)$

29. $(-2, 3)$; $(4, 7)$

30. $(-4, 0)$; $(3, -2)$

31. The equation $(x^2/a^2) + (y^2/b^2) = 1$ is the equation of an ellipse (Section 10.1). Show that the curve given by the vector equation $\mathbf{f}(t) = a(\cos t)\mathbf{i} + b(\sin t)\mathbf{j}$ is an ellipse.

32. A cannonball is shot upward from ground level at an angle of 45° with an initial speed of 1300 ft/sec. Find a parametric representation of the path of the cannonball. Then find the Cartesian equation of this path.

33. How many feet (horizontally) does the cannonball in Problem 32 travel before it hits the ground?

34. How many meters (horizontally) does the cannonball in Example 6 travel before it hits the ground?

35. An object is thrown down from the top of a 150-m building at an angle of 30° (below the horizontal) with an initial velocity of 100 m/sec. Determine a parametric representation of the path of the object. [*Hint:* Draw a picture.]

36. When the object in Problem 35 hits the ground, how far is it from the base of the building?

37. A point is located 25 cm from the center of a wheel 1 m in diameter. Find the parametric representation of the curve traced by that point as the wheel rolls.

38. Answer the question in Problem 37 if the point is located on the circumference of the wheel.

*39. A **hypocycloid**† is a curve generated by the motion of a point P on the circumference of a circle that rolls internally, without slipping, on a larger circle (see Figure 11). Assume that

---

†The "hypo" comes from the Greek "hupo," meaning "under."

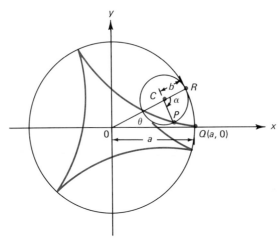

FIGURE 11

the radius of the large circle is a, while that of the smaller circle is b. If θ is the angle in Figure 11, show that a parametric representation of the hypocycloid is

$$x = (a - b)\cos\theta + b\cos\left(\frac{a - b}{b}\right)\theta \quad \text{and} \quad y = (a - b)\sin\theta - b\sin\left(\frac{a - b}{b}\right)\theta.$$

[*Hint:* First show that $a\theta = b\alpha$.]

40. If $a = 4b$ in Problem 39, then the curve generated is called a **hypocycloid of four cusps.** Sketch this curve.

*41. Show that the Cartesian equation of the hypocycloid of four cusps is

$$x^{2/3} + y^{2/3} = a^{2/3}.$$

[*Hint:* Use the identities $\cos 3\theta = 4\cos^3\theta - 3\cos\theta$ and $\sin 3\theta = 3\sin\theta - 4\sin^3\theta$ and the fact that $a - b = 3b$ and $(a - b)/b = 3$.]

42. If $a = nb$ in Problem 39, then the curve generated is called a **hypocycloid of n cusps.** Sketch the hypocycloid of seven cusps.

43. Show that the hypocycloid of two cusps is a straight-line segment.

44. Calculate the area bounded by the hypocycloid of four cusps. [*Hint:* Use the result of Problem 41.]

*45. An **epicycloid†** is a curve generated by the motion of a point on the circumference of a circle that rolls externally, without slipping, on a fixed circle. Show that the parametric represent-ation of an epicycloid is given by

$$x = (a + b)\cos\theta - b\cos\left(\frac{a + b}{b}\right)\theta \quad \text{and} \quad y = (a + b)\sin\theta - b\sin\left(\frac{a + b}{b}\right)\theta,$$

where a and b are as in Problem 39.

46. If $a = 4b$, show that the epicycloid generated has the sketch given in Figure 12.

47. Sketch the epicycloid in the case $a = 5b$.

48. Sketch the epicycloid in the case $a = 8b$.

*49. Show that the Cartesian equation of the cycloid is

$$x = r\cos^{-1}\left(\frac{r - y}{y}\right) - \sqrt{2ry - y^2}.$$

―――――――――

†From the Greek "epi," meaning "upon" or "over."

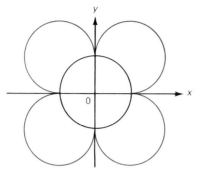

FIGURE 12

50. Show that a parametric representation of the hyperbola $(x^2/a^2) - (y^2/b^2) = 1$ is given by

$x = a \sec \theta, \qquad y = b \tan \theta.$

51. Use the substitution $\tan \theta = \sin t$ to show that the lemniscate $r^2 = \cos 2\theta$ has the cartesian parametric representation $x = \cos t/(1 + \sin^2 t)$, $y = \sin t \cos t/(1 + \sin^2 t)$.

16.2 THE EQUATION OF THE TANGENT LINE TO A PARAMETRIC CURVE

Suppose that $x = f_1(t)$ and $y = f_2(t)$ is the parametric representation of a curve C. We would like to be able to calculate the equation of the line tangent to the curve without having to determine the Cartesian equation of the curve. However, there are complications that might occur since the curve could intersect itself. This happens if there are two numbers $t_1 \neq t_2$ such that $f_1(t_1) = f_1(t_2)$ and $f_2(t_1) = f_2(t_2)$. Thus there are three possibilities. At a given point the curve could have the following:

(i) a unique tangent
(ii) no tangent
(iii) two or more tangents

This is illustrated in Figure 1. A condition that ensures that there is at least one tangent

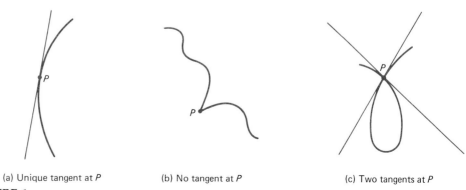

(a) Unique tangent at P (b) No tangent at P (c) Two tangents at P

FIGURE 1

line at each point is that

$$f_1' \text{ and } f_2' \text{ exist and } [f_1'(t)]^2 + [f_2'(t)]^2 \neq 0. \tag{1}$$

REMARK.   Condition (1) simply states that the derivatives of f_1 and f_2 are not zero at the same value of t.

Theorem 1   Let (x_0, y_0) be on the curve C given by $x = f_1(t)$ and $y = f_2(t)$. If the curve passes through (x_0, y_0) when $t = t_0$,† then the slope m of the line tangent to C at (x_0, y_0) is given by

$$m = \lim_{t \to t_0} \frac{f_2'(t)}{f_1'(t)}, \tag{2}$$

provided that this limit exists.

   Proof.   We refer to Figure 2. From the figure we see that

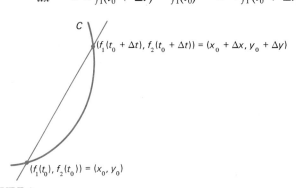

$$\frac{dy}{dx} = \lim_{\Delta t \to 0} \frac{f_2(t_0 + \Delta t) - f_2(t_0)}{f_1(t_0 + \Delta t) - f_1(t_0)} \overset{\text{L'Hôpital's rule}}{=} \lim_{\Delta t \to 0} \frac{f_2'(t_0 + \Delta t)}{f_1'(t_0 + \Delta t)} \overset{\text{let } t = t_0 + \Delta t}{=} \lim_{t \to t_0} \frac{f_2'(t)}{f_1'(t)} \blacksquare$$

FIGURE 2

Corollary.   If $f_2'(t_0)$ exists and $f_1'(t_0) \neq 0$, then

$$m = \frac{f_2'(t_0)}{f_1'(t_0)} \tag{3}$$

   Proof.

$$\frac{dy}{dx} = \lim_{\Delta t \to 0} \frac{f_2(t_0 + \Delta t) - f_2(t_0)}{f_1(t_0 + \Delta t) - f_1(t_0)} = \lim_{\Delta t \to 0} \frac{[f_2(t_0 + \Delta t) - f_2(t_0)]/\Delta t}{[f_1(t_0 + \Delta t) - f_1(t_0)]/\Delta t} \overset{\text{limit theorem for quotients}}{=} \frac{f_2'(t_0)}{f_1'(t_0)}. \blacksquare$$

---

†The curve may pass through the point (x_0, y_0) for other values of t as well, and therefore it may have other tangent lines at that point, as in Figure 1c.

Using the corollary, we find that the tangent line to C at t_0 is given by

$$\frac{y - y_0}{x - x_0} = \frac{f_2'(t_0)}{f_1'(t_0)},$$

or after simplification,

$$f_2'(t_0)(x - x_0) - f_1'(t_0)(y - y_0) = 0. \tag{4}$$

Note that if $f_1'(t_0) = 0$, then equation (4) is still valid, and we obtain

$$f_2'(t_0)(x - x_0) = 0. \tag{5}$$

Since by assumption $[f_1'(t_0)]^2 + [f_2'(t_0)]^2 \neq 0$, the assumption $f_1'(t_0) = 0$ implies that $f_2'(t_0) \neq 0$. Thus we can divide by $f_2'(t_0)$ in (5) to obtain the equation

$$x - x_0 = 0, \quad \text{or} \quad x = x_0.$$

In this case the tangent line is vertical. If $f_2'(t_0) = 0$, then C has a horizontal tangent at t_0.

In sum, if condition (1) holds, then

(i) C has a vertical tangent line at t_0 if

$$\left.\frac{dx}{dt}\right|_{t=t_0} = f_1'(t_0) = 0 \quad \text{and} \quad f_2'(t_0) \neq 0.$$

(ii) C has a horizontal tangent line at t_0 if

$$\left.\frac{dy}{dt}\right|_{t=t_0} = f_2'(t_0) = 0 \quad \text{and} \quad f_1'(t_0) \neq 0.$$

(6)

EXAMPLE 1   Find the equation of a line (or lines) tangent to the curve $x = e^t$, $y = e^{-t}$ at the point $(1, 1)$.

Solution. The point $(1, 1)$ is reached only when $t = 0$. Then

$$\frac{dy}{dx} = \frac{d(e^{-t})/dt}{d(e^t)/dt} = \left.-\frac{e^{-t}}{e^t}\right|_{t=0} = -1,$$

and the equation of the tangent line is

$$\frac{y - 1}{x - 1} = -1, \quad \text{or} \quad y = -x + 2. \quad \blacksquare$$

EXAMPLE 2   Find the equation of the line tangent to the curve

$$x = 2t^3 - 15t^2 + 24t + 7, \qquad y = t^2 + t + 1 \qquad \text{at} \quad t = 2.$$

Solution. Here $f_1'(2) = (6t^2 - 30t + 24)|_{t=2} = -12$ and $f_2'(2) = (2t + 1)|_{t=2} = 5$. When $t = 2$, $x = 11$ and $y = 7$. Then using (4), we obtain

$$5(x - 11) + 12(y - 7) = 0, \qquad \text{or} \qquad 5x + 12y = 139. \quad \blacksquare$$

EXAMPLE 3   Find all horizontal and vertical tangents to the curve of Example 2.

Solution. There are vertical tangents when

$$f_1'(t) = 6t^2 - 30t + 24 = 0 = 6(t^2 - 5t + 4) = 6(t - 4)(t - 1),$$

or when $t = 1$ and $t = 4$, since $f_2'(t)$ is nonzero at these points. When $t = 1$, $x = 18$; and when $t = 4$, $x = -9$. Thus the vertical tangents are the lines $x = 18$ and $x = -9$. There is a horizontal tangent when $2t + 1 = 0$, or $t = -\frac{1}{2}$. When $t = -\frac{1}{2}$, $y = \frac{3}{4}$; and the line $y = \frac{3}{4}$ is a horizontal tangent. $\quad \blacksquare$

EXAMPLE 4   Find all points at which the curve $x = 2 + 7 \cos \theta$, $y = 8 + 3 \sin \theta$ has a vertical or horizontal tangent.

Solution. We find that $f_1'(\theta) = x'(\theta) = -7 \sin \theta$, which is 0 when $\theta = n\pi$ for some integer n. If $\theta = n\pi$, then

$$x = \begin{cases} 9, & n \text{ even} \\ -5, & n \text{ odd} \end{cases} \qquad \text{and} \qquad y = 8,$$

so there are vertical tangents at $(9, 8)$ and $(-5, 8)$. Similarly, $f_2'(\theta) = y'(\theta) = 3 \cos \theta$, which is 0 when $\theta = (n + \frac{1}{2})\pi$ for some integer n. If $\theta = (n + \frac{1}{2})\pi$, then

$$x = 2 \qquad \text{and} \qquad y = \begin{cases} 11, & n \text{ even} \\ 5, & n \text{ odd}, \end{cases}$$

so the curve has horizontal tangents at $(2, 11)$ and $(2, 5)$. You should verify that when $f_1'(\theta) = 0$, $f_2'(\theta) \neq 0$, and vice versa.

REMARK.   It is not difficult to verify that x and y satisfy the equation

$$\frac{(x - 2)^2}{49} + \frac{(y - 8)^2}{9} = 1.$$

This is the equation of an ellipse centered at $(2, 8)$. $\quad \blacksquare$

PROBLEMS 16.2

In Problems 1–16, find the slope of the line tangent to the given curve for the given value of the parameter.

1. $x = t^2 - 2$; $y = 4t$; $t = 3$
2. $x = t^3$; $y = t^4 - 5$; $t = -1$
3. $x = t^2$; $y = \sqrt{1 - t}$; $t = \frac{1}{2}$
4. $x = t + 4$; $y = t^3 - t + 4$; $t = 2$
5. $x = e^{2t}$; $y = e^{-2t}$; $t = 1$
6. $x = \cos \theta$; $y = \sin \theta$; $\theta = \pi/4$

7. $x = \cos 2\theta$; $y = \sin 2\theta$; $\theta = \pi/4$
8. $x = \tan \theta$; $y = \sec \theta$; $\theta = \pi/4$
9. $x = \sec \theta$; $y = \tan \theta$; $\theta = \pi/4$
10. $x = \cosh t$; $y = \sinh t$; $t = 0$
11. $x = \sqrt{1 - \sin \theta}$; $y = \sqrt{1 + \cos \theta}$; $\theta = 0$
12. $x = 8 \cos \theta$; $y = -3 \sin \theta$; $\theta = 2\pi/3$
13. $x = \cos^3 \theta$; $y = \sin^3 \theta$; $\theta = \pi/6$
14. $x = \cos^3 \theta$; $y = \sin^3 \theta$; $\theta = \pi/2$.

15. $x = \theta$; $y = \dfrac{1}{\theta}$; $\theta = \dfrac{\pi}{4}$
16. $x = t \cosh t$; $y = \dfrac{\tanh t}{1 + t}$; $t = 0$

In Problems 17–22, find the equation of the tangent line.

17. the curve of Problem 1
18. the curve of Problem 4
19. the curve of Problem 5
20. the curve of Problem 7
21. the curve of Problem 11
22. the curve of Problem 13

In Problems 23–32, find the points [in the form (x, y)] at which the given curves have vertical and horizontal tangents.

23. $x = t^2 - 1$; $y = t^2 - 4$
24. $x = 2 \cos \theta$; $y = 3 \sin \theta$
25. $x = \sin 3\theta$; $y = \cos 5\theta$
26. $x = \sin \theta + \cos \theta$; $y = \sin \theta - \cos \theta$
27. $x = e^{3t}$; $y = e^{-5t}$
28. $x = \sinh t$; $y = \cosh t$
29. $x = \cosh t$; $y = \sinh t$
*30. $x = \theta \sin \theta$; $y = \theta \cos \theta$

31. $x = \dfrac{1}{\sqrt{1 - t^2}}$; $y = \sqrt{1 - t^2}$
32. $x = \ln(1 + t^2)$; $y = \ln(1 + t^3)$

33. Suppose a curve is given in polar coordinates by the equation $r = f(\theta)$. Show that the curve can then be given parametrically by

$$x = f(\theta)\cos \theta \qquad \text{and} \qquad y = f(\theta)\sin \theta.$$

In Problems 34–41, use the result of Problem 33 to determine the slope of the tangent line to the given curve for the given value of θ.

34. $r = 5 \sin \theta$; $\theta = \pi/6$
35. $r = 5 \cos \theta + 5 \sin \theta$; $\theta = \pi/4$
36. $r = -4 + 2 \cos \theta$; $\theta = \pi/3$
37. $r = 4 + 3 \sin \theta$; $\theta = 2\pi/3$
38. $r = 3 \sin 2\theta$; $\theta = \pi/6$
39. $r = 5 \sin 3\theta$; $\theta = \pi/4$
40. $r = e^{\theta/2}$; $\theta = 0$
41. $r^2 = \cos 2\theta$; $\theta = \pi/6$

42. Find the equation of the two tangents to the curve

$$x = t^3 - 2t^2 - 3t + 11, \qquad y = t^2 - 2t - 5$$

at the point $(11, -2)$.

*43. Calculate the equations of the three lines that are tangent to the curve $x = t^3 - 2t^2 - t + 3$, $y = 2t - t^2 - (2/t)$ at the point $(1, -1)$.

44. Calculate the slope of the tangent to the cycloid given in equations (16.1.16) and show that the tangent is vertical when α is a multiple of 2π.

45. For what values of α does the trochoid given by equations (16.1.15) have a vertical tangent?

16.3 THE DIFFERENTIATION AND INTEGRATION OF A VECTOR FUNCTION

In Section 16.2 we showed how the derivative dy/dx could be calculated when x and y were given parametrically in terms of t. In this section we will show how to calculate the derivative of a vector function.

Since a derivative is a special kind of limit, we first need to define the limit of a vector function. This definition is what you might expect.

Definition 1   LIMIT OF A VECTOR FUNCTION   Let $\mathbf{f}(t) = f_1(t)\mathbf{i} + f_2(t)\mathbf{j}$. Let t_0 be any real number, $+\infty$, or $-\infty$. If $\lim_{t \to t_0} f_1(t)$ and $\lim_{t \to t_0} f_2(t)$ both exist, then we define

$$\lim_{t \to t_0} \mathbf{f}(t) = \left[\lim_{t \to t_0} f_1(t)\right]\mathbf{i} + \left[\lim_{t \to t_0} f_2(t)\right]\mathbf{j}. \tag{1}$$

That is, *the limit of a vector function is determined by the limits of its component functions.* Thus in order to calculate the limit of a vector function, it is only necessary to calculate two ordinary limits.

EXAMPLE 1   Let $\mathbf{f}(t) = [(\sin t)/t]\mathbf{i} + [\ln(3 + t)]\mathbf{j}$. Calculate $\lim_{t \to 0} \mathbf{f}(t)$.

Solution.

$$\lim_{t \to 0} \mathbf{f}(t) = \left[\lim_{t \to 0} \frac{\sin t}{t}\right]\mathbf{i} + \left[\lim_{t \to 0} \ln(3 + t)\right]\mathbf{j} = \mathbf{i} + (\ln 3)\mathbf{j} \ \blacksquare$$

Definition 2   CONTINUITY OF A VECTOR FUNCTION $\mathbf{f}$ is **continuous** at t_0 if the component functions f_1 and f_2 are continuous at t_0. Thus, $\mathbf{f}$ is continuous at t_0 if

(i) $\mathbf{f}$ is defined at t_0.
(ii) $\lim_{t \to t_0} \mathbf{f}(t)$ exists.
(iii) $\lim_{t \to t_0} \mathbf{f}(t) = \mathbf{f}(t_0)$.

Definition 3   DERIVATIVE OF A VECTOR FUNCTION   Let f be defined at t. Then $\mathbf{f}$ is **differentiable** at t if

$$\lim_{\Delta t \to 0} \frac{\mathbf{f}(t + \Delta t) - \mathbf{f}(t)}{\Delta t} \tag{2}$$

exists and is finite. The vector function $\mathbf{f}'$ defined by

$$\mathbf{f}'(t) = \frac{d\mathbf{f}}{dt} = \lim_{\Delta t \to 0} \frac{\mathbf{f}(t + \Delta t) - \mathbf{f}(t)}{\Delta t} \tag{3}$$

is called the **derivative** of $\mathbf{f}$, and the domain of $\mathbf{f}'$ is the set of all t such that the limit in (2) exists.

Definition 4   DIFFERENTIABILITY IN AN OPEN INTERVAL   The vector function $\mathbf{f}$ is **differentiable** on the open interval I if $\mathbf{f}'(t)$ exists for every t in I.

Before giving examples of the calculation of derivatives, we prove a theorem that makes this calculation no more difficult than the calculation of "ordinary" derivatives.

Theorem 1 If $\mathbf{f}(t) = f_1(t)\mathbf{i} + f_2(t)\mathbf{j}$, then at any value t for which $f_1'(t)$ and $f_2'(t)$ exist,

$$\mathbf{f}'(t) = f_1'(t)\mathbf{i} + f_2'(t)\mathbf{j}. \tag{4}$$

That is, the derivative of a vector function is determined by the derivatives of its component functions.

Proof.

$$
\begin{aligned}
\mathbf{f}'(t) &= \lim_{\Delta t \to 0} \frac{\mathbf{f}(t + \Delta t) - \mathbf{f}(t)}{\Delta t} \\[2mm]
&= \lim_{\Delta t \to 0} \frac{[f_1(t + \Delta t)\mathbf{i} + f_2(t + \Delta t)\mathbf{j}] - [f_1(t)\mathbf{i} + f_2(t)\mathbf{j}]}{\Delta t} \\[2mm]
&= \lim_{\Delta t \to 0} \frac{[f_1(t + \Delta t) - f_1(t)]\mathbf{i} + [f_2(t + \Delta t) - f_2(t)]\mathbf{j}}{\Delta t} \\[2mm]
&= \lim_{\Delta t \to 0} \left[\frac{f_1(t + \Delta t) - f_1(t)}{\Delta t} \right]\mathbf{i} + \lim_{\Delta t \to 0} \left[\frac{f_2(t + \Delta t) - f_2(t)}{\Delta t} \right]\mathbf{j} \\[2mm]
&= f_1'(t)\mathbf{i} + f_2'(t)\mathbf{j} \quad ■
\end{aligned}
$$

EXAMPLE 2 Let $\mathbf{f}(t) = (\cos t)\mathbf{i} + e^{2t}\mathbf{j}$. Calculate $\mathbf{f}'(t)$.

Solution.

$$\mathbf{f}'(t) = \frac{d}{dt}(\cos t)\mathbf{i} + \frac{d}{dt}e^{2t}\mathbf{j} = -(\sin t)\mathbf{i} + 2e^{2t}\mathbf{j} \quad ■$$

Once we know how to calculate the first derivative of $\mathbf{f}$, we can calculate higher derivatives as well.

Definition 5 SECOND DERIVATIVE If the function $\mathbf{f}'$ is differentiable at t, we define the **second derivative** of $\mathbf{f}$ to be the derivative of $\mathbf{f}'$. That is,

$$\mathbf{f}'' = (\mathbf{f}')'. \tag{5}$$

EXAMPLE 3 Let $\mathbf{f}(t) = (\ln t)\mathbf{i} + (1/t)\mathbf{j}$. Calculate $\mathbf{f}''(t)$.

Solution. $\mathbf{f}'(t) = (1/t)\mathbf{i} - (1/t^2)\mathbf{j}$, so

$$\mathbf{f}''(t) = -\frac{1}{t^2}\mathbf{i} + \frac{2}{t^3}\mathbf{j}.$$

Note that $\mathbf{f}'$ and $\mathbf{f}''$ are defined for all $t > 0$. ($\mathbf{f}'$ and $\mathbf{f}''$ are not defined for $t \leq 0$ because $\ln t$ is not defined for $t \leq 0$.) ■

GEOMETRIC INTERPRETATION OF $\mathbf{f}'$ We now seek a geometric interpretation for $\mathbf{f}'$. As we have seen, the set of vectors $\mathbf{f}(t) = f_1(t)\mathbf{i} + f_2(t)\mathbf{j}$ for t in the domain of

f form a curve C in the plane (see Figure 1a). For fixed t, $\mathbf{f}(t)$ is the vector drawn in this figure. If φ denotes the direction of $\mathbf{f}'$, then [see equation (15.1.2)]

$$\tan \varphi = \frac{f_2'(t)}{f_1'(t)} = \frac{dy/dt}{dx/dt} = \frac{dy}{dx}. \tag{6}$$

The last step is justified by the chain rule. This result implies that $\mathbf{f}'(t)$ is tangent to the curve $\mathbf{f}$ at the point $(f_1(t), f_2(t))$ (see Figure 1b). This is a natural extension of the usual notion of a tangent line to the case of a vector function.

To see this in another way, we see in Figure 2 that the vector $\mathbf{f}(t + \Delta t) - \mathbf{f}(t)$ is a secant vector whose direction approaches that of the tangent vector as $\Delta t \to 0$.

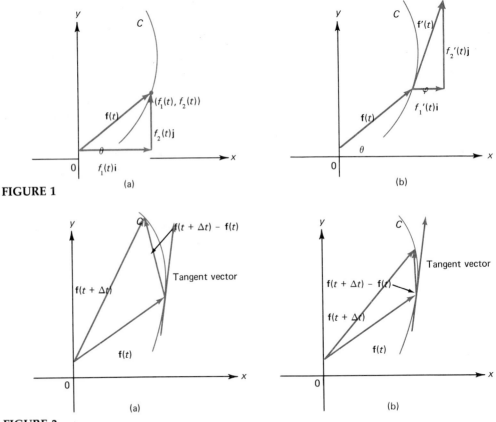

FIGURE 1

FIGURE 2

Definition 6   UNIT TANGENT VECTOR   It is sometimes useful to calculate a **unit tangent vector** to a curve C. This vector is a tangent vector with a magnitude of 1. The unit tangent vector is usually denoted $\mathbf{T}$ and can be calculated by the formula

$$\mathbf{T}(t) = \frac{\mathbf{f}'(t)}{|\mathbf{f}'(t)|} \tag{7}$$

for any number t so long as $\mathbf{f}'(t) \neq 0$. This follows since $\mathbf{f}'(t)$ is a tangent vector and $\mathbf{f}'/|\mathbf{f}'|$ is a unit vector.

EXAMPLE 4 Find the unit tangent vector to the curve $\mathbf{f} = (\ln t)\mathbf{i} + (1/t)\mathbf{j}$ at $t = 1$.

Solution. $\mathbf{f}'(t) = (1/t)\mathbf{i} - (1/t^2)\mathbf{j} = \mathbf{i} - \mathbf{j}$ when $t = 1$. Since $|\mathbf{f}'| = |\mathbf{i} - \mathbf{j}| = \sqrt{2}$, we find that

$$\mathbf{T} = \frac{1}{\sqrt{2}}\mathbf{i} - \frac{1}{\sqrt{2}}\mathbf{j}. \quad \blacksquare$$

We now turn to the integration of vector functions.

Definition 7   INTEGRAL OF A VECTOR FUNCTION

(i) Let $\mathbf{f}(t) = f_1(t)\mathbf{i} + f_2(t)\mathbf{j}$ and suppose that the component functions f_1 and f_2 have antiderivatives. Then we define the **antiderivative,** or **indefinite integral,** of $\mathbf{f}$ by

$$\int \mathbf{f}(t)\, dt = \left(\int f_1(t)\, dt \right)\mathbf{i} + \left(\int f_2(t)\, dt \right)\mathbf{j} + \mathbf{C}, \tag{8}$$

where $\mathbf{C}$ is a constant vector of integration.

(ii) If f_1 and f_2 are integrable over the interval $[a, b]$, then we define the **definite integral** of $\mathbf{f}$ by

$$\int_a^b \mathbf{f}(t)\, dt = \left(\int_a^b f_1(t)\, dt \right)\mathbf{i} + \left(\int_a^b f_2(t)\, dt \right)\mathbf{j}. \tag{9}$$

REMARK 1.   The antiderivative (or indefinite integral) of $\mathbf{f}$ is a new vector function and is *not unique* since $\int f_1(t)\, dt$ and $\int f_2(t)\, dt$ are not unique.

REMARK 2.   The definite integral of $\mathbf{f}$ is a constant vector since $\int_a^b f_1(t)\, dt$ and $\int_a^b f_2(t)\, dt$ are constants.

EXAMPLE 5 Let $\mathbf{f}(t) = (\cos t)\mathbf{i} + (\sin t)\mathbf{j}$. Calculate the following:

(a) $\displaystyle\int \mathbf{f}(t)\, dt$ **(b)** $\displaystyle\int_0^{\pi/2} \mathbf{f}(t)\, dt$

Solution.

(a) $\displaystyle\int \mathbf{f}(t)\, dt = \left(\int \cos t\, dt \right)\mathbf{i} + \left(\int \sin t\, dt \right)\mathbf{j} = (\sin t + C_1)\mathbf{i} + (-\cos t + C_2)\mathbf{j}$

$$= (\sin t)\mathbf{i} - (\cos t)\mathbf{j} + \mathbf{C},$$

where $\mathbf{C} = C_1\mathbf{i} + C_2\mathbf{j}$ is a constant vector.

(b) $\displaystyle\int_0^{\pi/2} \mathbf{f}(t)\, dt = \left(\sin t \Big|_0^{\pi/2} \right)\mathbf{i} + \left(-\cos t \Big|_0^{\pi/2} \right)\mathbf{j} = \mathbf{i} + \mathbf{j}$ ■

EXAMPLE 6 For $\mathbf{f}'(t) = t\mathbf{i} + t^3\mathbf{j}$ and $\mathbf{f}(0) = \mathbf{i} - 2\mathbf{j}$, calculate $\mathbf{f}(t)$.

Solution. We first find that

$$\mathbf{f}(t) = \int \mathbf{f}'(t)\, dt = \left(\frac{t^2}{2} + C_1\right)\mathbf{i} + \left(\frac{t^4}{4} + C_2\right)\mathbf{j} = \frac{t^2}{2}\mathbf{i} + \frac{t^4}{4}\mathbf{j} + \mathbf{C}. \tag{10}$$

To evaluate $\mathbf{C}$, we substitute the value $t = 0$ into (10) to find that $\mathbf{f}(0) = \mathbf{C} = \mathbf{i} - 2\mathbf{j}$, so that

$$\mathbf{f}(t) = \frac{t^2}{2}\mathbf{i} + \frac{t^4}{4}\mathbf{j} + \mathbf{i} - 2\mathbf{j} = \left(\frac{t^2}{2} + 1\right)\mathbf{i} + \left(\frac{t^4}{4} - 2\right)\mathbf{j}. \ \blacksquare$$

PROBLEMS 16.3

In Problems 1–10, calculate the first and second derivatives of the given vector function.

1. $\mathbf{f}(t) = t\mathbf{i} - t^5\mathbf{j}$

2. $\mathbf{f}(t) = (1 + t^2)\mathbf{i} + \frac{2}{t}\mathbf{j}$

3. $\mathbf{f}(t) = (\sin 2t)\mathbf{i} + (\cos 3t)\mathbf{j}$

4. $\mathbf{f}(t) = \frac{t}{1+t}\mathbf{i} - \frac{1}{\sqrt{t}}\mathbf{j}$

5. $\mathbf{f}(t) = (\ln t)\mathbf{i} + e^{3t}\mathbf{j}$

6. $\mathbf{f}(t) = e^t(\sin t)\mathbf{i} + e^t(\cos t)\mathbf{j}$

7. $\mathbf{f}(t) = (\tan t)\mathbf{i} + (\sec t)\mathbf{j}$

8. $\mathbf{f}(t) = (\tan^{-1} t)\mathbf{i} + (\sin^{-1} t)\mathbf{j}$

9. $\mathbf{f}(t) = (\ln \cos t)\mathbf{i} + (\ln \sin t)\mathbf{j}$

10. $\mathbf{f}(t) = (\cosh t)\mathbf{i} + (\sinh t)\mathbf{j}$

In Problems 11–20, find the unit tangent vector to the given curve for the given value of t.

11. $\mathbf{f}(t) = t^2\mathbf{i} + t^3\mathbf{j}; t = 1$

12. $\mathbf{f}(t) = t\mathbf{i} + \frac{1}{t}\mathbf{j}; t = 1$

13. $\mathbf{f}(t) = (\cos t)\mathbf{i} + (\sin t)\mathbf{j}; t = 0$

14. $\mathbf{f}(t) = (\cos t)\mathbf{i} + (\sin t)\mathbf{j}; t = \pi/2$

15. $\mathbf{f}(t) = (\cos t)\mathbf{i} + (\sin t)\mathbf{j}; t = \pi/4$

16. $\mathbf{f}(t) = (\cos t)\mathbf{i} + (\sin t)\mathbf{j}; t = 3\pi/4$

17. $\mathbf{f}(t) = (\tan t)\mathbf{i} + (\sec t)\mathbf{j}; t = 0$

18. $\mathbf{f}(t) = (\ln t)\mathbf{i} + e^{2t}\mathbf{j}; t = 1$

19. $\mathbf{f}(t) = \frac{t}{t+1}\mathbf{i} + \frac{t+1}{t}\mathbf{j}; t = 2$

20. $\mathbf{f}(t) = \frac{t+1}{t}\mathbf{i} + \frac{t}{t+1}\mathbf{j}; t = 2$

In Problems 21–28, calculate the indicated integral.

21. $\int_0^2 (t^2\mathbf{i} + t^4\mathbf{j})\, dt$

22. $\int [(\sin 2t)\mathbf{i} + e^t\mathbf{j}]\, dt$

23. $\int (t^{-1/2}\mathbf{i} + t^{1/2}\mathbf{j})\, dt$

24. $\int_0^{\pi/4} [(\cos 2t)\mathbf{i} - (\sin 2t)\mathbf{j}]\, dt$

25. $\int_0^1 [(\sinh t)\mathbf{i} - (\cosh t)\mathbf{j}]\, dt$

26. $\int_1^e \left[\left(\frac{1}{t}\right)\mathbf{i} - \left(\frac{3}{t}\right)\mathbf{j}\right] dt$

27. $\int [(\ln t)\mathbf{i} + te^t\mathbf{j}]\, dt$

28. $\int [(\tan t)\mathbf{i} + (\sec t)\mathbf{j}]\, dt$

29. Let $\mathbf{f}'(t) = t^3\mathbf{i} - t^5\mathbf{j}$ and let $\mathbf{f}(0) = 2\mathbf{i} + 5\mathbf{j}$. Find $\mathbf{f}(t)$.

30. Let $\mathbf{f}'(t) = (1/\sqrt{t})\mathbf{i} + \sqrt{t}\mathbf{j}$ and let $\mathbf{f}(1) = -2\mathbf{i} + \mathbf{j}$. Find $\mathbf{f}(t)$.

31. Let $\mathbf{f}'(t) = (\cos t)\mathbf{i} + (\sin t)\mathbf{j}$ and let $\mathbf{f}(\pi/2) = \mathbf{i}$. Find $\mathbf{f}(t)$.

32. Find $\mathbf{f}(t)$ in Problem 31 for $\mathbf{f}(\pi/2) = \mathbf{j}$.
33. The ellipse $(x^2/a^2) + (y^2/b^2) = 1$ can be written parametrically as $x = a \cos \theta$, $y = b \sin \theta$. Find a unit tangent vector to the ellipse at $\theta = \pi/4$.
34. Find a unit tangent vector to the cycloid $\mathbf{f}(\alpha) = r(\alpha - \sin \alpha)\mathbf{i} + r(1 - \cos \alpha)\mathbf{j}$ for $\alpha = 0$.
35. Find a unit tangent vector to the cycloid of Problem 34 for $\alpha = \pi/2$.
36. Find a unit tangent vector to the cycloid of Problem 34 for $\alpha = \pi/3$.
37. Find, for $\theta = \pi/6$, a unit tangent vector to the hypocycloid of Problem 16.1.39, assuming that $a = 5$ and $b = 2$.

16.4 SOME DIFFERENTIATION FORMULAS

In Section 3.1 we gave a number of rules for differentiation. Many of these rules carry over with little change to the differentiation of a vector function.

Theorem 1   Let $\mathbf{f}$ and $\mathbf{g}$ be vector functions that are differentiable in an interval I. Let the scalar function h be differentiable in I. Finally, let α be a scalar and let $\mathbf{v}$ be a constant vector. Then, on I, we have

(i) $\mathbf{f} + \mathbf{g}$ is differentiable and

$$\frac{d}{dt}(\mathbf{f} + \mathbf{g}) = \frac{d\mathbf{f}}{dt} + \frac{d\mathbf{g}}{dt} = \mathbf{f}' + \mathbf{g}'. \tag{1}$$

(ii) $\alpha\mathbf{f}$ is differentiable and

$$\frac{d}{dt}\alpha\mathbf{f} = \alpha\frac{d\mathbf{f}}{dt} = \alpha\mathbf{f}'. \tag{2}$$

(iii) $\mathbf{v} \cdot \mathbf{f}$ is differentiable and

$$\frac{d}{dt}\mathbf{v} \cdot \mathbf{f} = \mathbf{v} \cdot \frac{d\mathbf{f}}{dt} = \mathbf{v} \cdot \mathbf{f}' \tag{3}$$

(iv) $h\mathbf{f}$ is differentiable and

$$\frac{d}{dt}h\mathbf{f} = h\frac{d\mathbf{f}}{dt} + \frac{dh}{dt}\mathbf{f} = h\mathbf{f}' + h'\mathbf{f}. \tag{4}$$

(v) $\mathbf{f} \cdot \mathbf{g}$ is differentiable and

$$\frac{d}{dt}\mathbf{f} \cdot \mathbf{g} = \mathbf{f} \cdot \frac{d\mathbf{g}}{dt} + \frac{d\mathbf{f}}{dt} \cdot \mathbf{g} = \mathbf{f} \cdot \mathbf{g}' + \mathbf{f}' \cdot \mathbf{g}. \tag{5}$$

Proof.

(i) $\dfrac{d}{dt}(\mathbf{f} + \mathbf{g}) = \dfrac{d}{dt}(f_1\mathbf{i} + f_2\mathbf{j} + g_1\mathbf{i} + g_2\mathbf{j})$

$\qquad\qquad = \dfrac{d}{dt}[(f_1 + g_1)\mathbf{i} + (f_2 + g_2)\mathbf{j}] = (f_1 + g_1)'\mathbf{i} + (f_2 + g_2)'\mathbf{j}$

$\qquad\qquad = (f_1' + g_1')\mathbf{i} + (f_2' + g_2')\mathbf{j}$

$\qquad\qquad = (f_1'\mathbf{i} + f_2'\mathbf{j}) + (g_1'\mathbf{i} + g_2'\mathbf{j}) = \mathbf{f}' + \mathbf{g}'$

(ii) $\dfrac{d}{dt}(\alpha\mathbf{f}) = \dfrac{d}{dt}(\alpha f_1\mathbf{i} + \alpha f_2\mathbf{j}) = \alpha f_1'\mathbf{i} + \alpha f_2'\mathbf{j} = \alpha(f_1'\mathbf{i} + f_2'\mathbf{j}) = \alpha\mathbf{f}'$

(iii) Let $\mathbf{v} = v_1\mathbf{i} + v_2\mathbf{j}$, where v_1 and v_2 are constants. Then

$\qquad \mathbf{v} \cdot \mathbf{f} = v_1 f_1 + v_2 f_2 \qquad$ and $\qquad (\mathbf{v} \cdot \mathbf{f})' = v_1 f_1' + v_2 f_2' = \mathbf{v} \cdot \mathbf{f}'.$

(iv) $\dfrac{d}{dt}h\mathbf{f} = \dfrac{d}{dt}(hf_1\mathbf{i} + hf_2\mathbf{j}) = (hf_1)'\mathbf{i} + (hf_2)'\mathbf{j} = (hf_1' + h'f_1)\mathbf{i} + (hf_2' + h'f_2)\mathbf{j}$

$\qquad\qquad = h(f_1'\mathbf{i} + f_2'\mathbf{j}) + h'(f_1\mathbf{i} + f_2\mathbf{j}) = h\mathbf{f}' + h'\mathbf{f}$

(v) $\mathbf{f} \cdot \mathbf{g} = (f_1\mathbf{i} + f_2\mathbf{j}) \cdot (g_1\mathbf{i} + g_2\mathbf{j}) = f_1 g_1 + f_2 g_2$, so that

$\qquad \dfrac{d}{dt}(\mathbf{f} \cdot \mathbf{g}) = f_1 g_1' + f_1' g_1 + f_2 g_2' + f_2' g_2 = f_1 g_1' + f_2 g_2' + f_1' g_1 + f_2' g_2$

$\qquad\qquad = (f_1\mathbf{i} + f_2\mathbf{j}) \cdot (g_1'\mathbf{i} + g_2'\mathbf{j}) + (f_1'\mathbf{i} + f_2'\mathbf{j}) \cdot (g_1\mathbf{i} + g_2\mathbf{j})$

$\qquad\qquad = \mathbf{f} \cdot \mathbf{g}' + \mathbf{f}' \cdot \mathbf{g}.$ ∎

EXAMPLE 1   Let $\mathbf{f}(t) = t\mathbf{i} + t^3\mathbf{j}$, $\mathbf{g}(t) = (\cos t)\mathbf{i} + (\sin t)\mathbf{j}$ and $\mathbf{v} = 2\mathbf{i} - 3\mathbf{j}$. Calculate (a) $(\mathbf{f} + \mathbf{g})'$, (b) $(\mathbf{v} \cdot \mathbf{f})'$, and (c) $(\mathbf{f} \cdot \mathbf{g})'$.

Solution.

(a) $(\mathbf{f} + \mathbf{g})' = \mathbf{f}' + \mathbf{g}' = (\mathbf{i} + 3t^2\mathbf{j}) + [-(\sin t)\mathbf{i} + (\cos t)\mathbf{j}]$

$\qquad\qquad = (1 - \sin t)\mathbf{i} + (3t^2 + \cos t)\mathbf{j}$

(b) $(\mathbf{v} \cdot \mathbf{f})' = \mathbf{v} \cdot \mathbf{f}' = (2\mathbf{i} - 3\mathbf{j}) \cdot (\mathbf{i} + 3t^2\mathbf{j}) = 2 - 9t^2$

(c) $(\mathbf{f} \cdot \mathbf{g})' = \mathbf{f} \cdot \mathbf{g}' + \mathbf{f}' \cdot \mathbf{g}$

$\qquad\qquad = (t\mathbf{i} + t^3\mathbf{j}) \cdot [-(\sin t)\mathbf{i} + (\cos t)\mathbf{j}] + (\mathbf{i} + 3t^2\mathbf{j}) \cdot [(\cos t)\mathbf{i} + (\sin t)\mathbf{j}]$

$\qquad\qquad = -t \sin t + t^3 \cos t + \cos t + 3t^2 \sin t$

$\qquad\qquad = (\cos t)(t^3 + 1) + (\sin t)(3t^2 - t)$ ∎

EXAMPLE 2   Let $\mathbf{f}(t) = (\cos t)\mathbf{i} + (\sin t)\mathbf{j}$. Calculate $\mathbf{f} \cdot \mathbf{f}'$.

Solution.

$$\mathbf{f} \cdot \mathbf{f}' = [(\cos t)\mathbf{i} + (\sin t)\mathbf{j}] \cdot [-(\sin t)\mathbf{i} + (\cos t)\mathbf{j}]$$
$$= -\cos t \sin t + \sin t \cos t = 0 \quad \blacksquare$$

Example 2 can be generalized to the following interesting result.

Theorem 2 Let $\mathbf{f}(t) = f_1\mathbf{i} + f_2\mathbf{j}$ be a differentiable vector function such that $|\mathbf{f}(t)| = \sqrt{f_1^2(t) + f_2^2(t)}$ is constant. Then

$$\mathbf{f} \cdot \mathbf{f}' = 0. \tag{6}$$

Proof. Suppose $|\mathbf{f}(t)| = C$, a constant. Then

$$\mathbf{f} \cdot \mathbf{f} = f_1^2 + f_2^2 = |\mathbf{f}|^2 = C^2,$$

so

$$\frac{d}{dt}(\mathbf{f} \cdot \mathbf{f}) = \frac{d}{dt}C^2 = 0.$$

But

$$\frac{d}{dt}(\mathbf{f} \cdot \mathbf{f}) = \mathbf{f} \cdot \mathbf{f}' + \mathbf{f}' \cdot \mathbf{f} = 2\mathbf{f} \cdot \mathbf{f}' = 0,$$

so

$$\mathbf{f} \cdot \mathbf{f}' = 0. \quad \blacksquare$$

NOTE. In Example 2, $|\mathbf{f}(t)| = \sqrt{\cos^2 t + \sin^2 t} = 1$ for every t.

UNIT NORMAL VECTOR There is an interesting geometric application of Theorem 2. Let $\mathbf{f}(t)$ be a differentiable vector function. For all t for which $\mathbf{f}'(t) \neq 0$, we let $\mathbf{T}(t)$ denote the unit tangent vector to the curve $\mathbf{f}(t)$. Then since $|\mathbf{T}(t)| = 1$ by the definition of a *unit* tangent vector, we have, from Theorem 2 [assuming that $\mathbf{T}'(t)$ exists],

$$\mathbf{T}(t) \cdot \mathbf{T}'(t) = 0. \tag{7}$$

That is, $\mathbf{T}'(t)$ is *orthogonal* to $\mathbf{T}(t)$. Recalling from Section 3.1 (p. 141) that a line perpendicular to a tangent line is called a **normal line,** we call the vector $\mathbf{T}'$ a **normal vector** to the curve $\mathbf{f}$. Finally, whenever $\mathbf{T}'(t) \neq 0$, we can define the **unit normal vector** to the curve $\mathbf{f}$ at t as

$$\mathbf{n}(t) = \frac{\mathbf{T}'(t)}{|\mathbf{T}'(t)|}. \tag{8}$$

EXAMPLE 3   Calculate a unit normal vector to the curve $\mathbf{f}(t) = (\cos t)\mathbf{i} + (\sin t)\mathbf{j}$ at $t = \pi/4$.

Solution. First, we calculate $\mathbf{f}'(t) = -(\sin t)\mathbf{i} + (\cos t)\mathbf{j}$, and since $|\mathbf{f}'(t)| = 1$, we find that

$$\mathbf{T}(t) = \frac{\mathbf{f}'(t)}{|\mathbf{f}'(t)|} = -(\sin t)\mathbf{i} + (\cos t)\mathbf{j}.$$

Then

$$\mathbf{T}'(t) = -(\cos t)\mathbf{i} - (\sin t)\mathbf{j} = \mathbf{n}(t)$$

since $|\mathbf{T}'(t)| = 1$, so that at $t = \pi/4$,

$$\mathbf{n} = -\frac{1}{\sqrt{2}}\mathbf{i} - \frac{1}{\sqrt{2}}\mathbf{j}.$$

This is sketched in Figure 1 [remember that $\mathbf{f}(t) = (\cos t)\mathbf{i} + (\sin t)\mathbf{j}$ is the parametric equation of the unit circle]. The reason the vector $\mathbf{n}$ points inward is that it is the negative of the position vector at $t = \pi/4$. ∎

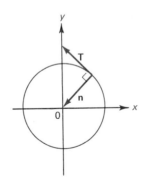

FIGURE 1

EXAMPLE 4   Calculate a unit normal vector to the curve $\mathbf{f}(t) = [(t^3/3) - t]\mathbf{i} + t^2\mathbf{j}$ at $t = 3$.

Solution. Here $\mathbf{f}'(t) = (t^2 - 1)\mathbf{i} + 2t\mathbf{j}$ and

$$|\mathbf{f}'(t)| = \sqrt{(t^2 - 1)^2 + 4t^2} = \sqrt{t^4 - 2t^2 + 1 + 4t^2} = \sqrt{t^4 + 2t^2 + 1} = t^2 + 1.$$

Thus

$$\mathbf{T}(t) = \frac{\mathbf{f}'(t)}{|\mathbf{f}'(t)|} = \frac{t^2 - 1}{t^2 + 1}\mathbf{i} + \frac{2t}{t^2 + 1}\mathbf{j}.$$

Then

$$\mathbf{T}'(t) = \frac{d}{dt}\left(\frac{t^2 - 1}{t^2 + 1}\right)\mathbf{i} + \frac{d}{dt}\left(\frac{2t}{t^2 + 1}\right)\mathbf{j} = \frac{4t}{(t^2 + 1)^2}\mathbf{i} + \frac{2 - 2t^2}{(t^2 + 1)^2}\mathbf{j}.$$

Finally,

$$|\mathbf{T}'(t)| = \left\{ \left[\frac{4t}{(t^2 + 1)^2} \right]^2 + \left[\frac{2 - 2t^2}{(t^2 + 1)^2} \right]^2 \right\}^{1/2} = \frac{1}{(t^2 + 1)^2} (16t^2 + 4 - 8t^2 + 4t^4)^{1/2}$$

$$= \frac{1}{(t^2 + 1)^2} \sqrt{4t^4 + 8t^2 + 4} = \frac{2}{(t^2 + 1)^2} \sqrt{t^4 + 2t^2 + 1}$$

$$= \frac{2(t^2 + 1)}{(t^2 + 1)^2} = \frac{2}{t^2 + 1},$$

so that

$$\mathbf{n}(t) = \frac{\mathbf{T}'(t)}{|\mathbf{T}'(t)|} = \frac{t^2 + 1}{2} \left[\frac{4t}{(t^2 + 1)^2} \mathbf{i} + \frac{2 - 2t^2}{(t^2 + 1)^2} \mathbf{j} \right] = \frac{2t}{t^2 + 1} \mathbf{i} + \frac{1 - t^2}{t^2 + 1} \mathbf{j}.$$

At $t = 3$, $\mathbf{T}(t) = \frac{4}{5}\mathbf{i} + \frac{3}{5}\mathbf{j}$ and $\mathbf{n} = \frac{3}{5}\mathbf{i} - \frac{4}{5}\mathbf{j}$. Note that $\mathbf{T}(3) \cdot \mathbf{n}(3) = 0$. This is sketched in Figure 2. ∎

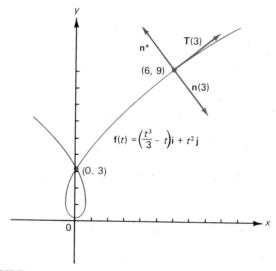

FIGURE 2

In Figure 2 we see that there are two vectors perpendicular to $\mathbf{T}$ at the point P: one pointing "inward" ($\mathbf{n}$) and one pointing "outward" ($\mathbf{n}^*$). In general, the situation is as shown in Figure 3. How do we know which of the two vectors is $\mathbf{n}$? If φ is the direction of $\mathbf{T}$, then since $\mathbf{T}$ is a unit vector,

$$\mathbf{T} = (\cos \varphi)\mathbf{i} + (\sin \varphi)\mathbf{j},$$

so that

$$\frac{d\mathbf{T}}{dt} = \frac{d\mathbf{T}}{d\varphi} \frac{d\varphi}{dt}$$

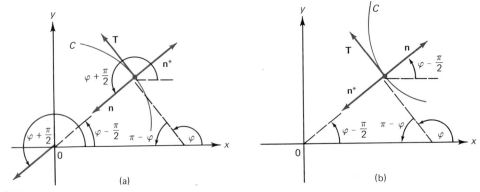

FIGURE 3

and

$$\left|\frac{d\mathbf{T}}{dt}\right| = \left|\frac{d\mathbf{T}}{d\varphi}\right|\left|\frac{d\varphi}{dt}\right| = \left|\frac{d\varphi}{dt}\right| \quad \left(\text{since }\left|\frac{d\mathbf{T}}{d\varphi}\right| = 1\right).$$

Then

$$\mathbf{n}(t) = \frac{d\mathbf{T}/dt}{|d\mathbf{T}/dt|} = \frac{d\mathbf{T}}{d\varphi}\frac{d\varphi/dt}{|d\varphi/dt|} = \frac{d\varphi/dt}{|d\varphi/dt|}[-(\sin\varphi)\mathbf{i} + (\cos\varphi)\mathbf{j}].$$

Since $d\varphi/dt$ is a scalar, $(d\varphi/dt)/|d\varphi/dt| = \pm 1$. It is $+1$ in Figure 3a since φ increases as t increases so that $d\varphi/dt > 0$, while it is -1 in Figure 3b since φ decreases as t increases. In Figure 3a, then,

$$\mathbf{n}(t) = -(\sin\varphi)\mathbf{i} + (\cos\varphi)\mathbf{j} = \cos[\varphi + (\pi/2)]\mathbf{i} + \sin[\varphi + (\pi/2)]\mathbf{j}.$$

That is, the direction of $\mathbf{n}$ is $\varphi + (\pi/2)$ and $\mathbf{n}$ must point inward (on the concave side of C) as in Figure 3a. In the other case, $(d\varphi/dt)/|d\varphi/dt| = -1$, so

$$\mathbf{n}(t) = (\sin\varphi)\mathbf{i} - (\cos\varphi)\mathbf{j} = \cos[\varphi - (\pi/2)]\mathbf{i} + \sin[\varphi - (\pi/2)]\mathbf{j}.$$

That is, the direction of $\mathbf{n}$ is $\varphi - (\pi/2)$ and so $\mathbf{n}$ points as in Figure 3b. In either case:

> *The unit normal vector **n** always points in the direction of the concave side of the curve.*

PROBLEMS 16.4

In Problems 1–15, calculate the indicated derivative.

1. $\dfrac{d}{dt}\{[2t\mathbf{i} + (\cos t)\mathbf{j}] + [(\tan t)\mathbf{i} - (\sec t)\mathbf{j}]\}$

2. $\dfrac{d}{dt}\{e^t[(\cos t)\mathbf{i} + (\sin t)\mathbf{j}]\}$

3. $\dfrac{d}{dt}\left[(3\mathbf{i} - 4\mathbf{j}) \cdot \left(\sqrt{t}\,\mathbf{i} - \dfrac{1}{\sqrt{t}}\mathbf{j}\right)\right]$

4. $\dfrac{d}{dt}\left[(t^3\mathbf{i} - t^5\mathbf{j}) \cdot \left(-\dfrac{1}{t^3}\mathbf{i} + \dfrac{1}{t^5}\mathbf{j}\right)\right]$

5. $\dfrac{d}{dt}[(t^3\mathbf{i} - t^5\mathbf{j}) + (-t^3\mathbf{i} + t^5\mathbf{j})]$

6. $\dfrac{d}{dt}\{[(\sinh t)\mathbf{i} + (\cosh t)\mathbf{j}] \cdot (-2\mathbf{i} + 7\mathbf{j})\}$

7. $\dfrac{d}{dt}\{3 \sin t[(\sinh t)\mathbf{i} + (\cosh t)\mathbf{j}]\}$

8. $\dfrac{d}{dt}\{[(\sinh t)\mathbf{i} + (\cosh t)\mathbf{j}] \cdot [\sin t)\mathbf{i} + (\cos t)\mathbf{j}]\}$

9. $\dfrac{d}{dt}\left[\left(\dfrac{t+1}{t}\mathbf{i} + \dfrac{t}{t+1}\mathbf{j}\right) \cdot \left(t^{10}\mathbf{i} - \dfrac{1}{t^{10}}\mathbf{j}\right)\right]$

10. $\dfrac{d}{dt}\{[e^{2t}(\cos t)\mathbf{i} + e^{2t}(\sin t)\mathbf{j}] \cdot (t\mathbf{i} + t\mathbf{j})$

11. $\dfrac{d}{dt}\{[(\ln t)\mathbf{i} + (\ln t^3)\mathbf{j}] \cdot [(\tanh t)\mathbf{i} + (\operatorname{sech} t)\mathbf{j}]\}$

12. $\dfrac{d}{dt}\{[(\sin^{-1} t)\mathbf{i} + (\cos^{-1} t)\mathbf{j}] \cdot (2\mathbf{i} - 10\mathbf{j})\}$

13. $\dfrac{d}{dt}\{[\sin^{-1} t)\mathbf{i} + (\cos^{-1} t)\mathbf{j}] + [(\cos t)\mathbf{i} + (\sin t)\mathbf{j}]\}$

14. $\dfrac{d}{dt}\{(\cos t)[(\sin^{-1} t)\mathbf{i} + (\cos^{-1} t)\mathbf{j}]\}$

15. $\dfrac{d}{dt}\{[(\sin^{-1} t)\mathbf{i} + (\cos^{-1} t)\mathbf{j}] \cdot [(\tan^{-1} t)\mathbf{i} + (\cos t)\mathbf{j}]\}$

In Problems 16–29, find $\mathbf{T}(t)$, $\mathbf{n}(t)$, and the particular vectors $\mathbf{T}$ and $\mathbf{n}$ when $t = t_0$. Then sketch the curve near $t = t_0$ and include the vectors $\mathbf{T}$ and $\mathbf{n}$ in your sketch.

16. $\mathbf{f} = (\cos 3t)\mathbf{i} + (\sin 3t)\mathbf{j}; t = 0$

17. $\mathbf{f} = (\cos 5t)\mathbf{i} + (\sin 5t)\mathbf{j}; t = \pi/2$

18. $\mathbf{f} = 2(\cos 4t)\mathbf{i} + 2(\sin 4t)\mathbf{j}; t = \pi/4$

19. $\mathbf{f} = -3(\cos 10t)\mathbf{i} - 3(\sin 10t)\mathbf{j}; t = \pi$

20. $\mathbf{f} = 8(\cos t)\mathbf{i} + 8(\sin t)\mathbf{j}; t = \pi/4$

21. $\mathbf{f} = 4t\mathbf{i} + 2t^2\mathbf{j}; t = 1$

22. $\mathbf{f} = (2 + 3t)\mathbf{i} + (8 - 5t)\mathbf{j}; t = 3$

23. $\mathbf{f} = (4 - 7t)\mathbf{i} + (-3 + 5t)\mathbf{j}; t = -5$

24. $\mathbf{f} = (a + bt)\mathbf{i} + (c + dt)\mathbf{j}, a, b, c, d$ real; $t = t_0$

25. $\mathbf{f} = 2t\mathbf{i} + (e^{-t} + e^{t})\mathbf{j}; t = 0$

*26. $\mathbf{f} = (t - \cos t)\mathbf{i} + (1 - \sin t)\mathbf{j}; t = \pi/2$

*27. $\mathbf{f} = (t - \cos t)\mathbf{i} + (1 - \sin t)\mathbf{j}; t = \pi$

*28. $\mathbf{f} = (t - \cos t)\mathbf{i} + (1 - \sin t)\mathbf{j}; t = \pi/4$

29. $\mathbf{f} = (\ln \sin t)\mathbf{i} + (\ln \cos t)\mathbf{j}; t = \pi/6$

16.5 ARC LENGTH REVISITED

In Section 9.2 we derived a formula for the length of an arc of the curve $y = f(x)$ between $x = a$ and $x = b$ [equation (9.2.3)). We now derive a formula for arc length in a more general setting. Let the curve C be given parametrically by

$$x = f_1(t), \qquad y = f_2(t). \tag{1}$$

We will assume in this section that f_1' and f_2' exist. Let t_0 be a fixed number, which fixes a point $P_0 = (x_0, y_0) = (f_1(t_0), f_2(t_0))$ on the curve (see Figure 1). The arrows in Figure 1 indicate the direction in which a point moves along the curve as t increases. We define the function $s(t)$ by

$s(t) = $ length along the curve C from $(f_1(t_0), f_2(t_0))$ to $(f_1(t), f_2(t))$.

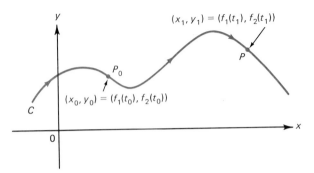

FIGURE 1

The following theorem allows us to calculate the length of a curve given parametrically.

Theorem 1 Suppose that f_1' and f_2' are continuous in the interval $[t_0, t_1]$. Then $s(t)$ is a differentiable function of t for $t \in [t_0, t_1]$ and

$$\frac{ds}{dt} = \sqrt{[f_1'(t)]^2 + [f_2'(t)]^2} = \sqrt{\left(\frac{dx}{dt}\right)^2 + \left(\frac{dy}{dt}\right)^2}. \tag{2}$$

Sketch of Proof. The proof of this theorem is quite difficult. However, it is possible to give an intuitive idea of what is going on. Consider an arc of the curve between t and $t + \Delta t$ (see Figure 2). First, we note that as $\Delta t \to 0$, the ratio of the length of the secant line L to the length of the arc Δs between the points $(x(t_0), y(t_0))$ and $(x(t_0 + \Delta t), y(t_0 + \Delta t))$ approaches 1. That is,

$$\lim_{\Delta t \to 0} \frac{\Delta s}{L} = 1. \tag{3}$$

Then since $L = \sqrt{\Delta x^2 + \Delta y^2}$, we have

$$\frac{\Delta s}{\Delta t} = \left(\frac{\Delta s}{L}\right)\left(\frac{L}{\Delta t}\right) = \left(\frac{\Delta s}{L}\right)\left(\frac{\sqrt{\Delta x^2 + \Delta y^2}}{\Delta t}\right) = \left(\frac{\Delta s}{L}\right)\sqrt{\left(\frac{\Delta x}{\Delta t}\right)^2 + \left(\frac{\Delta y}{\Delta t}\right)^2}$$

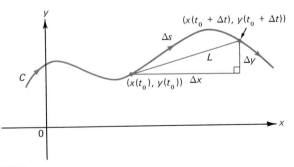

FIGURE 2

and

$$\frac{ds}{dt} = \lim_{\Delta t \to 0} \frac{\Delta s}{\Delta t} = \lim_{\Delta t \to 0} \frac{\Delta s}{L} \cdot \lim_{\Delta t \to 0} \sqrt{\left(\frac{\Delta x}{\Delta t}\right)^2 + \left(\frac{\Delta y}{\Delta t}\right)^2} = \sqrt{\left(\frac{dx}{dt}\right)^2 + \left(\frac{dy}{dt}\right)^2}.$$

The last step can be justified under the assumption that dx/dt and dy/dt are continuous, but its proof is beyond the scope of this book.† ∎

We use equation (2) to define the length of the arc from t_0 to t_1.

Definition 1 ARC LENGTH. Suppose that f_1' and f_2' are continuous in the interval $[t_0, t_1]$. Then the **arc length** of the curve $(f_1(t), f_2(t))$ is given by

$$s(t_1) = \text{length of arc from } t_0 \text{ to } t_1$$

$$= \int_{t_0}^{t_1} \left(\frac{ds}{dt}\right) dt = \int_{t_0}^{t_1} \sqrt{\left(\frac{dx}{dt}\right)^2 + \left(\frac{dy}{dt}\right)^2}\, dt. \tag{4}$$

REMARK. If we can write y as a differentiable function of x, then (4) follows from the definition of arc length in Chapter 9 [see equation (9.2.3)]. To see this, we start with

$$s = \int_a^b \sqrt{1 + \left(\frac{dy}{dx}\right)^2}\, dx,$$

where $x(t_0) = a$ and $x(t_1) = b$. Then by the chain rule,

$$\frac{dy}{dx} = \frac{dy/dt}{dx/dt} \qquad \text{and} \qquad dx = \frac{dx}{dt}\, dt.$$

Thus

$$s = \int_a^b \sqrt{1 + \left(\frac{dy}{dx}\right)^2}\, dx = \int_{t_0}^{t_1} \sqrt{1 + \left(\frac{dy/dt}{dx/dt}\right)^2}\, \frac{dx}{dt}\, dt$$

$$= \int_{t_0}^{t_1} \sqrt{\frac{(dx/dt)^2 + (dy/dt)^2}{(dx/dt)^2}}\, \frac{dx}{dt}\, dt = \int_{t_0}^{t_1} \left(\frac{\sqrt{(dx/dt)^2 + (dy/dt)^2}}{dx/dt}\right) \frac{dx}{dt}\, dt \quad \left(\text{if } \frac{dx}{dt} > 0\right)$$

$$= \int_{t_0}^{t_1} \sqrt{\left(\frac{dx}{dt}\right)^2 + \left(\frac{dy}{dt}\right)^2}\, dt.$$

We emphasize that although this last result may be interesting, it does not prove the theorem because we cannot always write y as a function of x. For example, if $x = \cos t$ and $y = \sin t$, then $x^2 + y^2 = 1$ and $y = \pm\sqrt{1 - x^2}$, which is not a function of x since for every x in $(-1, 1)$ there are two values of y.

---

†See, for example, R. C. Buck, *Advanced Calculus* (McGraw-Hill, New York, 1965), page 321.

Equation (4) can be rewritten in a slightly different form. Using the chain rule (applied to differentials), we have

$$\frac{ds}{dt} \, dt = ds, \tag{5}$$

and so (4) becomes

$$s(t_1) = \int_{t_0}^{t_1} ds, \tag{6}$$

where

$$ds = \sqrt{\left(\frac{dx}{dt}\right)^2 + \left(\frac{dy}{dt}\right)^2} \, dt. \tag{7}$$

In this context the variable s is called the **parameter of arc length.** We will discuss this further in the next section. Note, however, that while x measures distance along the horizontal axis and y measures distance along the vertical axis, s measures distance *along the curve* given parametrically by equations (1).

EXAMPLE 1   Calculate the length of the curve $x = \cos t$, $y = \sin t$ in the interval $[0, 2\pi]$.

Solution.

$$ds = \sqrt{\left(\frac{dx}{dt}\right)^2 + \left(\frac{dy}{dt}\right)^2} \, dt = \sqrt{(-\sin t)^2 + (\cos t)^2} \, dt = dt,$$

so that

$$s = \int_0^{2\pi} ds = \int_0^{2\pi} dt = 2\pi.$$

NOTE.   Since $x = \cos t$, $y = \sin t$ for t in $[0, 2\pi]$ is a parametric representation of the unit circle, we have verified the accuracy of formula (4) in this instance, since the circumference of the unit circle is 2π. ∎

EXAMPLE 2   Let the curve C be given by $x = t^2$ and $y = t^3$. Calculate the length of the arc from $t = 0$ to $t = 3$.

Solution. Here

$$\sqrt{\left(\frac{dx}{dt}\right)^2 + \left(\frac{dy}{dt}\right)^2} = \sqrt{(2t)^2 + (3t^2)^2} = \sqrt{4t^2 + 9t^4} = t\sqrt{4 + 9t^2},$$

so that

$$s = \int_0^3 t\sqrt{4 + 9t^2}\, dt = \frac{1}{27}(4 + 9t^2)^{3/2}\Big|_0^3 = \frac{85^{3/2} - 8}{27}. \quad \blacksquare$$

Suppose now that a curve is given in polar coordinates by $r = f(\theta)$. Since $x = r\cos\theta$ and $y = r\sin\theta$ (equation (11.1.1)), we have

$$x = f(\theta)\cos\theta \qquad \text{and} \qquad y = f(\theta)\sin\theta,$$

so that

$$\frac{dx}{d\theta} = f'(\theta)\cos\theta - f(\theta)\sin\theta \qquad \text{and} \qquad \frac{dy}{d\theta} = f'(\theta)\sin\theta + f(\theta)\cos\theta.$$

Therefore

$$\left(\frac{dx}{d\theta}\right)^2 + \left(\frac{dy}{d\theta}\right)^2 = [f'(\theta)]^2\cos^2\theta - 2f'(\theta)f(\theta)\sin\theta\cos\theta + [f(\theta)]^2\sin^2\theta$$

$$+ [f'(\theta)]^2\sin^2\theta + 2f'(\theta)f(\theta)\sin\theta\cos\theta + [f(\theta)]^2\cos^2\theta$$

$$= [f(\theta)]^2 + [f'(\theta)]^2 = r^2 + \left(\frac{dr}{d\theta}\right)^2.$$

Thus the formula for arc length in polar coordinates becomes

$$s = \int_{\theta = \theta_0}^{\theta = \theta_1} \sqrt{r^2 + \left(\frac{dr}{d\theta}\right)^2}\, d\theta. \tag{8}$$

EXAMPLE 3   Calculate the arc length of the cardioid $r = a(1 + \cos\theta)$, $a > 0$.

Solution. The curve is sketched in Figure 3. Since the curve is symmetric about the polar axis, we calculate the arc length between $\theta = 0$ and $\theta = \pi$ and then multiply the result by 2. We have

$$\frac{s}{2} = \int_0^\pi \sqrt{r^2 + \left(\frac{dr}{d\theta}\right)^2}\, d\theta = \int_0^\pi a\sqrt{(1 + \cos\theta)^2 + \sin^2\theta}\, d\theta$$

$$= a\int_0^\pi \sqrt{1 + 2\cos\theta + \cos^2\theta + \sin^2\theta}\, d\theta$$

$$= a\int_0^\pi \sqrt{(2 + 2\cos\theta)}\, d\theta = a\int_0^\pi \sqrt{\frac{4 + 4\cos\theta}{2}}\, d\theta$$

$$= 2a\int_0^\pi \sqrt{\frac{1 + \cos\theta}{2}}\, d\theta = 2a\int_0^\pi \cos\frac{\theta}{2}\, d\theta = 4a\sin\frac{\theta}{2}\Big|_0^\pi = 4a.$$

The total arc length is therefore $8a$.

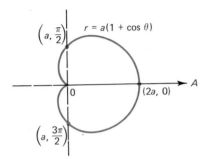

FIGURE 3

REMARK.   If we ignored symmetry and tried to integrate from 0 to 2π, we would obtain

$$s = 2a \int_0^{2\pi} \sqrt{\frac{1 + \cos\theta}{2}}\, d\theta = 2a \int_0^{2\pi} \cos\frac{\theta}{2}\, d\theta = 4a \sin\frac{\theta}{2}\Big|_0^{2\pi} = 0.$$

The problem is that $\cos\theta/2 \ge 0$ for $0 \le \theta \le \pi$ and $\cos\theta/2 \le 0$ for $\pi \le \theta \le 2\pi$. We see that $\sqrt{(1 + \cos\theta/2}} = \cos(\theta/2)$ for $0 \le \theta \le \pi$ and $\sqrt{(1 + \cos\theta)/2} = -\cos(\theta/2)$ for $\pi \le \theta \le 2\pi$. Thus it would be necessary to write

$$s = 2a \int_0^{2\pi} \left|\cos\frac{\theta}{2}\right| d\theta = 2a\left(\int_0^{\pi} \cos\frac{\theta}{2}\, d\theta - \int_{\pi}^{2\pi} \cos\frac{\theta}{2}\, d\theta\right)$$

$$= 4a\left(\sin\frac{\theta}{2}\Big|_0^{\pi} - \sin\frac{\theta}{2}\Big|_{\pi}^{2\pi}\right) = 4a[1 - (-1)] = 8a,$$

as before. In other problems the reader is cautioned to pay attention to signs when taking square roots. ■

There is a more concise way to write our formula for arc length by using vector notation. Let the curve C be given by

$$\mathbf{f}(t) = f_1(t)\mathbf{i} + f_2(t)\mathbf{j}. \tag{9}$$

Then

$$\mathbf{f}'(t) = f_1'(t)\mathbf{i} + f_2'(t)\mathbf{j}$$

and

$$|\mathbf{f}'(t)| = \sqrt{[f_1'(t)]^2 + [f_2'(t)]^2} = \frac{ds}{dt}, \tag{10}$$

so that the length of the arc between t_0 and t_1 is given by

$$s = \int_{t_0}^{t_1} |\mathbf{f}'(t)|\, dt. \tag{11}$$

EXAMPLE 4 Calculate the length of the arc of the curve $\mathbf{f}(t) = (2t - t^2)\mathbf{i} + \frac{8}{3}t^{3/2}\mathbf{j}$ between $t = 1$ and $t = 3$.

Solution. $\mathbf{f}'(t) = (2 - 2t)\mathbf{i} + 4\sqrt{t}\,\mathbf{j}$ and

$$|\mathbf{f}'(t)| = \sqrt{4 - 8t + 4t^2 + 16t} = 2\sqrt{t^2 + 2t + 1} = 2(t + 1),$$

so

$$s = \int_1^3 2(t + 1)\,dt = (t^2 + 2t)\Big|_1^3 = 12. \quad \blacksquare$$

PROBLEMS 16.5

In Problems 1–20, find the length of the arc over the given interval or the length of the closed curve.

1. $x = t^3; y = t^2; 1 \le t \le 4$

2. $x = \cos 2\theta; y = \sin 2\theta; 0 \le \theta \le \pi/2$

3. $x = t^3 + 1; y = 3t^2 + 2; 0 \le t \le 2$

4. $x = 1 + t; y = (1 + t)^{3/2}; 0 \le t \le 1$

5. $x = \dfrac{1}{\sqrt{t + 1}}; y = \dfrac{t}{2(t + 1)}; 0 \le t \le 4$

6. $x = e^t \cos t; y = e^t \sin t; 0 \le t \le \pi/2$

7. $x = \sin^2 t; y = \cos^2 t; 0 \le t \le \pi/2$

8. The hypocycloid of four cusps $x = a \cos^3 \theta$, $y = a \sin^3 \theta$, $a > 0$. [*Hint:* Calculate the length in the first quadrant and multiply by 4.]

*9. The cardioid $r = a(1 + \sin \theta)$. [*Hint:*

$$\int \sqrt{1 + \sin \theta}\,d\theta = \int \sqrt{1 + \sin \theta} \cdot \frac{\sqrt{1 - \sin \theta}}{\sqrt{1 - \sin \theta}}\,d\theta.$$

Pay attention to signs.]

10. One arc of the cycloid $x = a(\theta - \sin \theta)$, $y = a(1 - \cos \theta)$, $a > 0$.

11. $x = t^3; y = t^2; -1 \le t \le 1$ [*Hint:* $\sqrt{t^2} = -t$ for $t < 0$.]

12. $r = a \sin \theta; 0 \le \theta \le \pi/2$, $a > 0$

13. $r = a \cos \theta; 0 \le \theta \le \pi$, $a > 0$

*14. $r = a\theta; 0 \le \theta \le 2\pi$, $a > 0$

15. $r = e^\theta; 0 \le \theta \le 3$

16. $r = \theta^2; 0 \le \theta \le \pi$

17. $r = 6\cos^2(\theta/2); 0 \le \theta \le \pi/2$

18. $r = \sin^3(\theta/3); 0 \le \theta \le \pi/2$ [*Hint:* $\sin^2(\theta/3) = \frac{1}{2}(1 - \cos(2\theta/3))$.]

19. $\mathbf{f}(t) = e^t(\sin t)\mathbf{i} + e^t(\cos t)\mathbf{j}; 0 \le t \le \pi/2$

20. $\mathbf{f}(t) = 3(\cos \theta)\mathbf{i} + 3(\sin \theta)\mathbf{j}; 0 \le \theta \le 2\pi$

21. The parametric representation of the ellipse $(x^2/a^2) + (y^2/b^2) = 1$ is given by $x = a \cos \theta$, $y = b \sin \theta$. Find an integral that represents the length of the circumference of an ellipse but do not try to evaluate it. The integral you obtain is called an *elliptic integral* and it arises in a variety of physical applications. It cannot be integrated (except numerically) unless $a = b$.

*22. A tack is stuck in the front tire of a bicycle wheel with a diameter of 1 m. What is the total distance traveled by the tack if the bicycle moves a total of 30π m?

16.6 ARC LENGTH AS A PARAMETER

In many problems it is convenient to use the arc length s as a parameter. We can think of the vector function $\mathbf{f}$ as the *position vector* of a particle moving in the xy-plane. Then if $P_0 = (x_0, y_0)$ is a fixed point on the curve C described by the vector function $\mathbf{f}$, we may write

$$\mathbf{f}(s) = x(s)\mathbf{i} + y(s)\mathbf{j}, \tag{1}$$

where s is the distance along the curve measured from P_0 in the direction of increasing s. In this way we can determine the x- and y-components of the position vector as we move s units *along the curve*.

EXAMPLE 1 Write the vector $\mathbf{f}(t) = (\cos t)\mathbf{i} + (\sin t)\mathbf{j}$ (which describes the unit circle) with arc length as a parameter. Take $P_0 = (1, 0)$.

Solution. We have $ds/dt = 1$ (from Example 16.5.1), so that since $(1, 0)$ is reached when $t = 0$, we find that

$$s = \int_0^t ds = \int_0^t \frac{ds}{du}\, du = \int_0^t du = t.$$

Thus we may write

$$\mathbf{f}(s) = (\cos s)\mathbf{i} + (\sin s)\mathbf{j}.$$

For example, if we begin at the point $(1, 0)$ and move π units along the unit circle (which is half the unit circle), then we move to the point $(\cos \pi, \sin \pi) = (-1, 0)$. This is what we would expect. See Figure 1. ∎

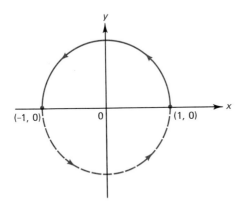

FIGURE 1

EXAMPLE 2 Let $\mathbf{f}(t) = (2t - t^2)\mathbf{i} + \frac{8}{3}t^{3/2}\mathbf{j}$, $t \geq 0$. Write this curve with arc length as a parameter.

Solution. Suppose that the fixed point is $P_0 = (0, 0)$ when $t = 0$. Then from Example 16.5.4 we have

$$\frac{ds}{dt} = 2(t + 1),$$

so that

$$s = \int_0^t \frac{ds}{du} \, du = \int_0^t 2(u + 1) \, du = t^2 + 2t.$$

This leads to the equations

$$t^2 + 2t - s = 0 \quad \text{and} \quad t = \frac{-2 + \sqrt{4 + 4s}}{2} = \sqrt{1 + s} - 1.$$

We took the positive square root here since it is assumed that t starts at 0 and increases. Then

$$x = 2t - t^2 = 4\sqrt{1 + s} - 4 - s$$
$$y = \tfrac{8}{3}t^{3/2} = \tfrac{8}{3}(\sqrt{1 + s} - 1)^{3/2},$$

and we obtain

$$\mathbf{f}(s) = (4\sqrt{1 + s} - 4 - s)\mathbf{i} + \tfrac{8}{3}(\sqrt{1 + s} - 1)^{3/2}\mathbf{j}. \quad \blacksquare$$

As Example 2 illustrates, writing $\mathbf{f}$ explicitly with arc length as a parameter can be tedious (or, more often, impossible).

There is an interesting and important relationship between position vectors, tangent vectors, and normal vectors that becomes apparent when we use s as a parameter.

Theorem 1 If the curve C is parametrized by $\mathbf{f}(s) = x(s)\mathbf{i} + y(s)\mathbf{j}$, where s is arc length and x and y have continuous derivatives, then the unit tangent vector $\mathbf{T}$ is given by

$$\mathbf{T}(s) = \frac{d\mathbf{f}}{ds}. \tag{2}$$

Proof. With *any* parametrization of C, the unit tangent vector is given by [see equation (16.3.7)]

$$\mathbf{T}(t) = \frac{\mathbf{f}'(t)}{|\mathbf{f}'(t)|}.$$

So choosing $t = s$ yields

$$\mathbf{T}(s) = \frac{d\mathbf{f}/ds}{|d\mathbf{f}/ds|}.$$

But from equation (16.5.10), $|\mathbf{f}'(t)| = ds/dt$, so

$$\left|\frac{d\mathbf{f}}{ds}\right| = \left|\frac{d\mathbf{f}/dt}{ds/dt}\right| = \left|\frac{\mathbf{f}'(t)}{ds/dt}\right| = 1,$$

and the proof is complete. ■

Theorem 1 is quite useful in that it provides a check of our calculation of the parametrization in terms of arc length. For if

$$\mathbf{f}(s) = x(s)\mathbf{i} + y(s)\mathbf{j},$$

then

$$\mathbf{T} = \frac{d\mathbf{f}}{ds} = \frac{dx}{ds}\mathbf{i} + \frac{dy}{ds}\mathbf{j}.$$

But $|\mathbf{T}| = 1$, so that $|\mathbf{T}|^2 = 1$, which implies that

$$\left(\frac{dx}{ds}\right)^2 + \left(\frac{dy}{ds}\right)^2 = 1. \tag{3}$$

We can apply this result in Example 2. We have

$$x(s) = 4\sqrt{1+s} - 4 - s \qquad \text{and} \qquad y(s) = \tfrac{8}{3}(\sqrt{1+s} - 1)^{3/2},$$

so

$$\frac{dx}{ds} = \frac{2}{\sqrt{1+s}} - 1, \qquad \frac{dy}{ds} = \frac{2}{\sqrt{1+s}}(\sqrt{1+s} - 1)^{1/2},$$

and

$$\left(\frac{dx}{ds}\right)^2 + \left(\frac{dy}{ds}\right)^2 = \frac{4}{1+s} - \frac{4}{\sqrt{1+s}} + 1 + \frac{4}{1+s}(\sqrt{1+s} - 1) = 1,$$

as expected.

PROBLEMS 16.6

In the following problems, find parametric equations in terms of the arc length s measured from the point reached when $t = 0$. Verify your solution by using formula (3).

1. $\mathbf{f} = 3t^2\mathbf{i} + 2t^3\mathbf{j}$
2. $\mathbf{f} = t^3\mathbf{i} + t^2\mathbf{j}$
3. $\mathbf{f} = (t^3 + 1)\mathbf{i} + (t^2 - 1)\mathbf{j}$
4. $\mathbf{f} = (3t^2 + a)\mathbf{i} + (2t^3 + b)\mathbf{j}$
5. $\mathbf{f} = 3(\cos\theta)\mathbf{i} + 3(\sin\theta)\mathbf{j}$
6. $\mathbf{f} = a(\sin\theta)\mathbf{i} + a(\cos\theta)\mathbf{j}$
7. $\mathbf{f} = a(\cos\theta)\mathbf{i} + a(\sin\theta)\mathbf{j}$
8. $\mathbf{f} = 3(\cos t + t\sin t)\mathbf{i} + 3(\sin t - t\cos t)\mathbf{j}$
9. $\mathbf{f} = (a + b\cos\theta)\mathbf{i} + (c + b\sin\theta)\mathbf{j}$
10. $\mathbf{f} = ae^t(\cos t)\mathbf{i} + ae^t(\sin t)\mathbf{j}$
11. One cusp of the hypocycloid of four cusps

$$\mathbf{f} = a(\cos^3\theta)\mathbf{i} + a(\sin^3\theta)\mathbf{j}, \qquad 0 \le \theta \le \frac{\pi}{2}, \quad a > 0.$$

*12. The cycloid $x = a(\theta - \sin\theta)$, $y = a(1 - \cos\theta)$, $a > 0$.

16.7 VELOCITY, ACCELERATION, FORCE, AND MOMENTUM (OPTIONAL)

Suppose that an object is moving in the plane. Then we can describe its motion parametrically by the vector function

$$\mathbf{f}(t) = f_1(t)\mathbf{i} + f_2(t)\mathbf{j}. \tag{1}$$

In this context **f** is called the **position vector** of the object, and the curve described by **f** is called the **trajectory** of the object. We then have the following definition.

Definition 1   VELOCITY AND ACCELERATION VECTOR If $\mathbf{f}'$ and $\mathbf{f}''$ exist, then

   (i) $\mathbf{v}(t) = \mathbf{f}'(t) = f_1'(t)\mathbf{i} + f_2'(t)\mathbf{j}$ (2)

is called the **velocity vector** of the moving object at time t.

   (ii) $\mathbf{a}(t) = \dfrac{d\mathbf{v}}{dt} = \mathbf{f}''(t) = f_1''(t)\mathbf{i} + f_2''(t)\mathbf{j}$ (3)

is called the **acceleration vector** of the object.

This definition is, of course, not surprising. It simply extends to the vector case our notion of velocity as the derivative of position and acceleration as the derivative of velocity.

Definition 2   SPEED AND ACCELERATION SCALAR
   (i) The **speed** $v(t)$ of a moving object is the magnitude of the velocity vector.
   (ii) The **acceleration scalar** $a(t)$ is the magnitude of the acceleration vector.

REMARK 1.   Since we have already shown that $|\mathbf{f}'(t)| = ds/dt$ [equation (16.5.10)], we have, since $v(t) = |\mathbf{v}(t)| = |\mathbf{f}'(t)|$,

$$v(t) = \frac{ds}{dt}. \tag{4}$$

REMARK 2.   Although $\mathbf{a}(t)$ is the derivative of $\mathbf{v}(t)$, it is *not true* in general that $a(t)$ is the derivative of the speed $v(t)$. For example, consider the motion along the unit circle given by

$$\mathbf{f}(t) = (\cos t)\mathbf{i} + (\sin t)\mathbf{j}.$$

Then

$$\mathbf{v}(t) = -(\sin t)\mathbf{i} + (\cos t)\mathbf{j} \quad \text{and} \quad \mathbf{a}(t) = -(\cos t)\mathbf{i} - (\sin t)\mathbf{j}.$$

But $v(t) = |\mathbf{v}(t)| = 1$, so that $dv/dt = 0$. But $a(t) = |\mathbf{a}(t)| = 1$, which is, evidently, not equal to dv/dt.

REMARK 3.   It follows from Theorem 16.4.2 that *if speed is constant, then the velocity and acceleration vectors are orthogonal.*

EXAMPLE 1   A particle is moving along the circle with the position vector $\mathbf{f} = 3(\cos 2t)\mathbf{i} + 3(\sin 2t)\mathbf{j}$. Calculate $\mathbf{v}(t)$, $\mathbf{a}(t)$, $v(t)$, and $a(t)$, and find the velocity and acceleration vectors when $t = \pi/6$. Assume that distance is measured in meters.

Solution. Here $\mathbf{v}(t) = \mathbf{f}'(t) = -6(\sin 2t)\mathbf{i} + 6(\cos 2t)\mathbf{j}$ and $\mathbf{a}(t) = -12(\cos 2t)\mathbf{i} - 12(\sin 2t)\mathbf{j}$. Then $v(t) = |\mathbf{v}(t)| = \sqrt{36 \sin^2 2t + 36 \cos^2 2t} = 6$ m/sec and $a(t) = |\mathbf{a}(t)| = 12$ m/sec^2. Finally, $\mathbf{v}(\pi/6) = -3\sqrt{3}\mathbf{i} + 3\mathbf{j}$ and $\mathbf{a}(\pi/6) = -6\mathbf{i} - 6\sqrt{3}\mathbf{j}$. We sketch these vectors in Figure 1. Note that $\mathbf{a}(t) \cdot \mathbf{v}(t) = 0$. This follows from Theorem 16.4.2 and the fact that $|\mathbf{v}(t)|$ is constant.   ∎

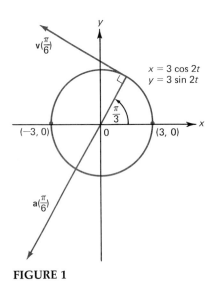

FIGURE 1

We now calculate the vectors that describe the motion of an object in a vertical plane. First, we make the simplifying assumption that the only force acting on the object is the force of gravity. We ignore, for example, the frictional force due to air resistance. The force of gravity is directed vertically downward. There is no force acting in the horizontal direction. Thus the acceleration vector is given by

$$\mathbf{a} = -g\mathbf{j}, \tag{5}$$

where the constant $g = 9.81$ m/sec^2 $= 32.2$ ft/sec^2. In addition,

$$\mathbf{F} = m\mathbf{a} = -mg\mathbf{j},$$

where m is the mass of the object. To obtain the velocity vector $\mathbf{v}(t)$, we simply integrate (5) to obtain

$$\mathbf{v}(t) = -gt\mathbf{j} + \mathbf{C}, \tag{6}$$

where $\mathbf{C}$ is a constant vector. But from (6) we find that

$$\mathbf{v}(0) = \mathbf{C},$$

so that $\mathbf{C}$ is the initial velocity vector $\mathbf{v}_0$ (the velocity at $t = 0$) and (6) becomes

$$\mathbf{v}(t) = -gt\mathbf{j} + \mathbf{v}_0. \tag{7}$$

Integrating (7), we obtain the position vector

$$\mathbf{f}(t) = -\tfrac{1}{2}gt^2\mathbf{j} + \mathbf{v}_0 t + \mathbf{D}, \tag{8}$$

where $\mathbf{D}$ is another constant vector. Evaluating (8) at $t = 0$ yields $\mathbf{D} = \mathbf{f}(0)$ = the initial position vector $\mathbf{f}_0$, so (8) becomes

$$\mathbf{f}(t) = -\tfrac{1}{2}gt^2\mathbf{j} + \mathbf{v}_0 t + \mathbf{f}_0. \tag{9}$$

EXAMPLE 2 A cannon whose muzzle is tilted upward at an angle of 30° shoots a ball at an initial velocity of 600 m/sec.

(a) Find the position vector at all times $t \geq 0$.
(b) How much time does the ball spend in the air?
(c) How far does the cannonball travel?
(d) How high does the ball get?
(e) How far from the cannon does the ball land?
(f) What is the speed of the ball at the time of impact with the earth? Assume that the mouth of the cannon is at ground level and ignore air resistance.

Solution.

(a) From Example 16.1.6 we find, resolving the initial velocity vector into its vertical and horizontal components, that $\mathbf{v}_0 = 300\sqrt{3}\mathbf{i} + 300\mathbf{j}$. Moreover, if we place the origin so that it coincides with the mouth of the cannon, then the initial position vector $\mathbf{f}_0$ is $(0, 0) = 0\mathbf{i} + 0\mathbf{j}$. From (9)

$$\mathbf{f}(t) = -\tfrac{1}{2}gt^2\mathbf{j} + 300\sqrt{3}t\mathbf{i} + 300t\mathbf{j} + \mathbf{0}$$
$$= 300\sqrt{3}t\mathbf{i} + (300t - \tfrac{1}{2}gt^2)\mathbf{j}.$$

(b) The ball hits the ground when the vertical component of the position vector is zero, that is, when $300t - \tfrac{1}{2}gt^2 = 0$, which occurs at $t = 600/g$ seconds.

(c) The total distance traveled is

$$s = \int_0^{600/g} \left(\frac{ds}{dt}\right) dt.$$

But

$$\frac{ds}{dt} = v(t) = |\mathbf{v}(t)| = |\mathbf{f}'(t)| = |300\sqrt{3}\mathbf{i} + (300 - gt)\mathbf{j}|$$
$$= \sqrt{270{,}000 + (300 - gt)^2}$$

Thus using formula (16.5.4), we have

$$s = \int_0^{600/g} \left(\frac{ds}{dt}\right) dt = \int_0^{600/g} \sqrt{270,000 + (300 - gt)^2}\, dt \text{ meters}$$

Let $300 - gt = \sqrt{270,000} \tan \theta = 300\sqrt{3} \tan \theta$. Then

$$-g\, dt = 300\sqrt{3} \sec^2 \theta\, d\theta$$

and

$$dt = \left(\frac{-300\sqrt{3}}{g}\right) \sec^2 \theta\, d\theta.$$

Also,

$$\sqrt{270,000 + (300 - gt)^2} = \sqrt{270,000(1 + \tan^2 \theta)} = 300\sqrt{3} \sec \theta.$$

When $t = 0$, $300 = 300\sqrt{3} \tan \theta$, $\tan \theta = 1/\sqrt{3}$, and $\theta = \pi/6$. When $t = 600/g$, $\tan \theta = -1/\sqrt{3}$ and $\theta = -\pi/6$. Thus

$$s = -\frac{1}{g} \int_{\pi/6}^{-\pi/6} (300\sqrt{3})(300\sqrt{3}) \sec^3 \theta\, d\theta$$

$$= \frac{270,000}{g} \int_{-\pi/6}^{\pi/6} \sec^3 \theta\, d\theta = \frac{540,000}{g} \int_0^{\pi/6} \sec^3 \theta\, d\theta$$

Example 8.2.8

$$= \frac{270,000}{g} \left. (\ln |\sec \theta + \tan \theta| + \sec \theta \tan \theta) \right|_0^{\pi/6}$$

$$= \frac{270,000}{g} \left(\ln \left| \frac{2}{\sqrt{3}} + \frac{1}{\sqrt{3}} \right| + \frac{2}{\sqrt{3}} \frac{1}{\sqrt{3}} \right) = \frac{270,000}{g} \left(\ln \sqrt{3} + \frac{2}{3} \right) \approx 33.467 \text{ km.}$$

(d) The maximum height is achieved when $dy/dt = 0$. That is, when $300 - gt = 0$ or when $t = 300/g$. For that value of t, $y_{max} = 45,000/g \approx 4587.2 \text{ m} = 4.5872$ km.

(e) In $600/g$ seconds, the x-component of $\mathbf{f}$ increases from 0 to $(300\sqrt{3}) \cdot (600/g)$ $\approx 31,780.7 \text{ m} \approx 31.78$ km.

(f) Since speed $= v(t) = \sqrt{270,000 + (300 - gt)^2}$ meters per second, we find that upon impact $t = 600/g$ seconds, so that

$$v\left(\frac{600}{g}\right) = \sqrt{270,000 + \left(300 - g \cdot \frac{600}{g}\right)^2} = \sqrt{270,000 + 90,000}$$

$$= \sqrt{360,000} = 600 \text{ m/sec.}$$

These results are illustrated in Figure 2. ∎

FIGURE 2

There is another convenient application of vectors applied to the motion of an object.

Definition 3 MOMENTUM   The **momentum P** of a particle at any time t is a vector defined as the product of the mass m of the particle and its velocity $\mathbf{v}$. That is,

$$\mathbf{P} = m\mathbf{v}. \tag{10}$$

NEWTON'S SECOND LAW OF MOTION   Newton's second law of motion states that the *rate of change of momentum of a moving object is proportional to the resultant force and is in the direction of that force*. That is,

$$\mathbf{F} = \frac{d\mathbf{P}}{dt}. \tag{11}$$

If the mass of the object is constant, then using (10) and (11), we have the familiar law

$$\mathbf{F} = \frac{d}{dt}m\mathbf{v} = m\frac{d\mathbf{v}}{dt} = m\mathbf{a}. \tag{12}$$

If mass is not constant, then (12) becomes

$$\mathbf{F} = \frac{d}{dt}m\mathbf{v} = m\frac{d\mathbf{v}}{dt} + \mathbf{v}\frac{dm}{dt}. \tag{13}$$

EXAMPLE 3   The cannonball of Example 2 has a mass of 8 kg. Find the force acting on the ball at any time t.

Solution. Here m is constant, so that $\mathbf{F} = m\mathbf{a} = (8\text{ kg})(-g\mathbf{j}) = -8g\mathbf{j}$, which is a force with a magnitude of $8g$ newtons acting vertically downward. There is no force in the horizontal direction. ∎

EXAMPLE 4   An object of mass m moves in the elliptical orbit given by $\mathbf{f}(t) = a(\cos \alpha t)\mathbf{i} + b(\sin \alpha t)\mathbf{j}$. Find the force acting on the object at any time t.

Solution. We easily find that $\mathbf{a}(t) = \mathbf{f}''(t) = -\alpha^2 a(\cos \alpha t)\mathbf{i} - \alpha^2 b(\sin \alpha t)\mathbf{j} = -\alpha^2 \mathbf{f}(t)$. Since m is constant, $\mathbf{F} = m\mathbf{a} = -m\alpha^2 \mathbf{f}(t)$. Thus the force always acts in the direction opposite to the direction of the position vector (thereby pointing *toward* the origin) and has a magnitude proportional to the distance of the object from the origin. Such a force is called a **central force** (see Figure 3). ■

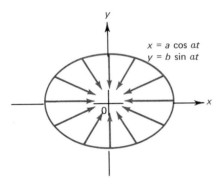

$$x = a \cos at$$
$$y = b \sin at$$

FIGURE 3

If the vector sum of forces acting on a system is zero, then, since $0 = \mathbf{F} = \mathbf{P}'(t)$, we find that the *total momentum of the system is constant*. This fact is called the **principle of the conservation of linear momentum.** We now see how that principle can be applied to the motion of a rocket.

Let the mass of a rocket be denoted by m, and suppose that the rocket is traveling with velocity $\mathbf{v}(t)$. The rocket is propelled by gas emissions. We assume that the exhaust velocity of the gases ejected from the rocket, relative to the rocket, is $\mathbf{u}(t)$. Then the total velocity of the gas (relative to the earth, say) is $\mathbf{u} + \mathbf{v}$. The rate of change of momentum on the rocket is

$$\mathbf{P}'(t) = \frac{d}{dt}(m\mathbf{v}) = m\frac{d\mathbf{v}}{dt} + \mathbf{v}\frac{dm}{dt}.$$

As the rocket loses mass, the exhaust gains it, so the rate of change of momentum of the gases is given by

$$-(\mathbf{u} + \mathbf{v})\frac{dm}{dt}.$$

Here dm/dt is the rate at which exhaust gas accumulates. The total force acting on the rocket-gas system is the sum of the rates of change of momentum for each of the two components of the system; that is,

$$\mathbf{F} = m\frac{d\mathbf{v}}{dt} + \mathbf{v}\frac{dm}{dt} - \mathbf{u}\frac{dm}{dt} - \mathbf{v}\frac{dm}{dt} = m\frac{d\mathbf{v}}{dt} - \mathbf{u}\frac{dm}{dt},$$

or

$$\mathbf{F} + \mathbf{u}\frac{dm}{dt} = m\frac{d\mathbf{v}}{dt}. \tag{14}$$

The term $\mathbf{u}\, dm/dt$ is called the **thrust** of the rocket. Note that $dm/dt < 0$ (since the rocket is losing mass), so that the thrust has the direction *opposite* to that of the velocity $\mathbf{u}$ of the exhaust gases.

Now if the rocket is free of any gravitational field, then there are no external forces acting on the system, so $\mathbf{F} = 0$ and (14) becomes

$$\mathbf{u}\frac{dm}{dt} = m\frac{d\mathbf{v}}{dt}. \tag{15}$$

If $\mathbf{u}$ is constant, then (15) becomes

$$\frac{d\mathbf{v}}{dt} = \frac{\mathbf{u}}{m}\frac{dm}{dt},$$

and integration yields

$$\mathbf{v}(t) - \mathbf{v}_0 = \mathbf{u}\int_{m_0}^{m}\frac{dm}{m} = \mathbf{u}\ln m\Big|_{m_0}^{m},$$

or

$$\mathbf{v} - \mathbf{v}_0 = -\mathbf{u}\ln\left(\frac{m_0}{m}\right). \tag{16}$$

Here $\mathbf{v}_0$ is the initial velocity vector and m_0 is the initial mass of the rocket and the fuel combined. Thus once the rocket has escaped all gravitational fields, we can calculate its velocity at any time if we know the velocity of the escaping gases and the proportion of gases that have been expelled.

EXAMPLE 5 A 1000-kg rocket carrying 2000 kg of fuel is motionless in space. Its engine starts and its exhaust velocity is 0.7 km/sec. What is its speed when all its fuel is consumed?

Solution. Here $\mathbf{v}_0 = \mathbf{0}$. Then

$$|\mathbf{v}| = |\mathbf{u}|\ln\frac{m_0}{m} = 0.7\ln\frac{3000}{1000} = 0.7\ln 3 \approx 0.769 \text{ km/sec}$$

$$\approx 2769 \text{ km/hr} \ (\approx 1720 \text{ mi/hr}).\ \blacksquare$$

EXAMPLE 6 The calculation above was made in the absence of gravitational forces. Let us now assume that the rocket is fired upward from the earth and assume that gas is ejected at the constant rate of 100 kg/sec. Then the force propelling the rocket upward is equal to the thrust and is given by

$$F_{\text{thrust}} = \left|\mathbf{u}\frac{dm}{dt}\right| = |(700 \text{ m/sec})(-100 \text{ kg/sec})| = 70{,}000 \text{ N}.$$

The thrust is opposed by the force of gravity:

$$F_{\text{grav}} = -mg = (-3000 \text{ kg})(9.81 \text{ m/sec}^2) = -29{,}430 \text{ N}.$$

(Note that we have treated $\mathbf{u}$ and $\mathbf{F}$ as scalars since we are assuming motion in only one direction: straight up.) We see that the initial net upward force is given by

$$F_{\text{thrust}} + F_{\text{grav}} = 70{,}000 - 29{,}430 = 40{,}570 \text{ N}.$$

Finally, just before all the fuel is used up, the force due to gravity is

$$F_{\text{grav}} = (-1000 \text{ kg})(9.81 \text{ m/sec}^2) = -9810 \text{ N},$$

so the net force is $70{,}000 - 9810 = 60{,}190$ N. Of course, when the fuel burns out, the rocket will fall back to earth.

REMARK.   In this example we assumed, as usual, that the earth's gravitational attraction could be represented by the constant $-mg$; that is, the rocket did not go high enough to escape or partially escape the effect of the earth's gravitational field. ∎

PROBLEMS 16.7

In Problems 1–14, the position vector of a moving particle is given. For the indicated value of t, calculate the velocity vector, the acceleration vector, the speed, and the acceleration scalar. Then sketch the portion of the trajectory showing the velocity and acceleration vectors.

1. $\mathbf{f} = (\cos 3t)\mathbf{i} + (\sin 3t)\mathbf{j}; \ t = 0$
2. $\mathbf{f} = (\cos 5t)\mathbf{i} + (\sin 5t)\mathbf{j}; \ t = \pi/2$
3. $\mathbf{f} = 2(\cos 4t)\mathbf{i} + 2(\sin 4t)\mathbf{j}; \ t = \pi/6$
4. $\mathbf{f} = -3(\cos 10t)\mathbf{i} - 3(\sin 10t)\mathbf{j}; \ t = \pi$
5. $\mathbf{f} = 4t\mathbf{i} + 2t^2\mathbf{j}; \ t = 1$
6. $\mathbf{f} = (2 + 3t)\mathbf{i} + (8 - 5t)\mathbf{j}; \ t = 3$
7. $\mathbf{f} = (4 - 7t)\mathbf{i} + (-3 + 5t)\mathbf{j}; \ t = -5$
8. $\mathbf{f} = (a + bt)\mathbf{i} + (c + dt)\mathbf{j}; \ t = t_0, \ a, \ b, \ c, \ d, \text{ real}$
9. $\mathbf{f} = 2t\mathbf{i} + (e^{-t} + e^t)\mathbf{j}; \ t = 0$ 10. $\mathbf{f} = \cosh t\mathbf{i} + 2t\mathbf{j}; \ t = 2$
11. $\mathbf{f} = (t - \cos t)\mathbf{i} + (1 - \sin t)\mathbf{j}; \ t = \pi/2$
12. $\mathbf{f} = (t - \cos t)\mathbf{i} + (1 - \sin t)\mathbf{j}; \ t = \pi$
13. $\mathbf{f} = (t - \cos t)\mathbf{i} + (1 - \sin t)\mathbf{j}; \ t = \pi/4$
14. $\mathbf{f} = (\ln \sin t)\mathbf{i} + (\ln \cos t)\mathbf{j}; \ t = \pi/6$
*15. A bullet is shot from a gun with an initial velocity of 1200 m/sec. The gun is inclined at an angle of 45°. Find (a) the total distance traveled by the bullet, (b) the horizontal distance traveled by the bullet, (c) the maximum height it reaches, and (d) the speed of the bullet at impact. Assume that the gun is held at ground level.
*16. Answer the questions of Problem 15 if the angle of inclination of the gun is 60° and the initial velocity is 3000 ft/sec.
17. A man is standing at the top of a 200-m building. He throws a ball horizontally with an initial speed of 20 m/sec.
    (a) How far does the ball travel in its path?
    (b) How far from the base of the building does the ball hit the ground?
    (c) At what angle with the horizontal does the ball hit the ground?
    (d) With what speed does the ball hit the ground?
18. An airplane, flying horizontally at a height of 1500 m and with a speed of 450 km/hr, releases a bomb.
    (a) How long does it take the bomb to hit the ground?
    (b) How far does it travel in the horizontal direction?
    (c) What is its speed of impact?

19. A girl is standing 20 ft from a tall building. She throws a ball at an angle of 60° toward the building with an initial speed of 60 ft/sec. Her hand is 4 ft above the ground when she releases the ball. How high up the building does the ball hit?

*20. A man can throw a football a maximum distance of 60 yd. What is the maximum speed, in feet per second, that he can throw the football?

21. A 2000-kg rocket carrying 3000 kg of fuel is initially motionless in space. Its exhaust velocity is 1 km/sec. What is its velocity when all its fuel is consumed?

22. The rocket in Problem 21 is launched from the earth, and gas is emitted at a constant rate of 75 kg/sec.
    (a) Find the net initial force acting on the rocket.
    (b) Find the net force just before all the fuel is expended.

23. In Problem 22, what is the minimum thrust needed to get the rocket off the ground?

16.8 CURVATURE AND THE ACCELERATION VECTOR (OPTIONAL)

The derivative dy/dx of a curve $y = f(x)$ measures the rate of change of the vertical component of the curve with respect to the horizontal component. As we have seen, the derivative ds/dt represents the change in the length of the arc traced out by the vector $\mathbf{f} = x(t)\mathbf{i} + y(t)\mathbf{j}$ as t increases. Another quantity of interest is the rate of change of the direction of the curve with respect to the length of the curve. That is, how much does the direction change for every one-unit change in the arc length? We are thus led to the following definitions.

Definition 1   CURVATURE
   (i) Let the curve C be given by the differentiable vector function $\mathbf{f}(t) = f_1(t)\mathbf{i} + f_2(t)\mathbf{j}$. Let $\varphi(t)$ denote the direction of $\mathbf{f}'(t)$. Then the **curvature** of C, denoted $\kappa(t)$, is the absolute value of the rate of change of direction with respect to arc length; that is,

$$\kappa(t) = \left| \frac{d\varphi}{ds} \right|. \tag{1}$$

Note that $\kappa(t) \geq 0$.
   (ii) The **radius of curvature** $\rho(t)$ is defined by

$$\rho(t) = \frac{1}{\kappa(t)} \quad \text{if} \quad \kappa(t) > 0. \tag{2}$$

REMARK 1.   The curvature is a measure of *how fast* the curve turns as we move along it.

REMARK 2.   If $\kappa(t) = 0$, we say that the radius of curvature is *infinite*. To understand this idea, note that if $\kappa(t) = 0$, then the "curve" does not bend and so is a straight line (see Example 2). A straight line can be thought of as an arc of a circle with *infinite* radius.

In Chapter 17 we will need a definition of curvature that does not depend on the angle φ.

Theorem 1   If $\mathbf{T}(t)$ denotes the unit tangent vector to $\mathbf{f}$, then

$$\kappa(t) = \left|\frac{d\mathbf{T}}{ds}\right|. \tag{3}$$

Proof.   By the chain rule (which applies just as well to vector-valued functions),

$$\frac{d\mathbf{T}}{ds} = \frac{d\mathbf{T}}{d\varphi}\frac{d\varphi}{ds}.$$

But since φ is the direction of $\mathbf{f}'$, and therefore also the direction of $\mathbf{T}$, we have

$$\mathbf{T} = (\cos\varphi)\mathbf{i} + (\sin\varphi)\mathbf{j},$$

so that

$$\frac{d\mathbf{T}}{d\varphi} = -(\sin\varphi)\mathbf{i} + (\cos\varphi)\mathbf{j} \quad \text{and} \quad \left|\frac{d\mathbf{T}}{d\varphi}\right| = \sqrt{\sin^2\varphi + \cos^2\varphi} = 1.$$

Thus

$$\left|\frac{d\mathbf{T}}{ds}\right| = \left|\frac{d\mathbf{T}}{d\varphi}\right|\left|\frac{d\varphi}{ds}\right| = 1\left|\frac{d\varphi}{ds}\right| = \kappa(t),$$

and the theorem is proved. ■

We now derive an easier way to calculate $\kappa(t)$.

Theorem 2   With the curve C given as in Definition 1, the curvature of C is given by the formula

$$\kappa(t) = \frac{|(dx/dt)(d^2y/dt^2) - (dy/dt)(d^2x/dt^2)|}{[(dx/dt)^2 + (dy/dt)^2]^{3/2}} = \frac{|x'y'' - y'x''|}{[(x')^2 + (y')^2]^{3/2}} \tag{4}$$

where $x(t) = f_1(t)$ and $y(t) = f_2(t)$.

Proof.   By the chain rule,

$$\frac{d\varphi}{ds} = \frac{d\varphi}{dt}\frac{dt}{ds} = \frac{d\varphi/dt}{ds/dt}. \tag{5}$$

From equation (16.5.10),

$$\frac{ds}{dt} = \sqrt{\left(\frac{dx}{dt}\right)^2 + \left(\frac{dy}{dt}\right)^2}, \tag{6}$$

and from equation (16.3.6), we obtain

$$\tan \varphi = \frac{dy/dt}{dx/dt}, \tag{7}$$

or

$$\varphi = \tan^{-1}\frac{dy/dt}{dx/dt}. \tag{8}$$

Differentiating both sides of (8) with respect to t, we obtain

$$\frac{d\varphi}{dt} = \frac{1}{1 + \left(\frac{dy/dt}{dx/dt}\right)^2} \frac{d}{dt}\left(\frac{dy/dt}{dx/dt}\right) = \frac{(dx/dt)^2}{\left(\frac{dx}{dt}\right)^2 + \left(\frac{dy}{dt}\right)^2} \cdot \frac{\frac{dx}{dt}\frac{d^2y}{dt^2} - \frac{dy}{dt}\frac{d^2x}{dt^2}}{(dx/dt)^2} \tag{9}$$

Substitution of (6) and (9) into (5) completes the proof of the theorem. ▨

EXAMPLE 1 We certainly expect that the curvature of a circle is constant and that its radius of curvature is its radius. Show that this is true.

Solution. The circle of radius r centered at the origin is given parametrically by

$$x = r \cos t, \qquad y = r \sin t.$$

Then $dx/dt = -r \sin t$, $d^2x/dt^2 = -r \cos t$, $dy/dt = r \cos t$, and $d^2y/dt^2 = -r \sin t$, so from (4),

$$\kappa(t) = \frac{|r^2 \sin^2 t + r^2 \cos^2 t|}{[r^2 \sin^2 t + r^2 \cos^2 t]^{3/2}} = \frac{r^2}{r^3} = \frac{1}{r}$$

and

$$\rho(t) = \frac{1}{\kappa(t)} = r,$$

as expected. ▨

EXAMPLE 2 Show that for a straight line $\kappa(t) = 0$.

Solution. A line can be represented parametrically [see equation (16.1.9)] by $x = x_1 + t(x_2 - x_1)$ and $y = y_1 + t(y_2 - y_1)$. Then $dx/dt = x_2 - x_1$, $d^2x/dt^2 = 0$, $dy/dt = y_2 - y_1$, and $d^2y/dt^2 = 0$. Substitution of these values into (4) immediately yields $\kappa(t) = 0$. ▨

EXAMPLE 3 Find the curvature and radius of curvature of the curve given parametrically by $x = (t^3/3) - t$ and $y = t^2$ (see Example 16.4.4) at $t = 2\sqrt{2}$.

Solution. We have $dx/dt = t^2 - 1$, $d^2x/dt^2 = 2t$, $dy/dt = 2t$, and $d^2y/dt^2 = 2$, so

$$\kappa(t) = \frac{|2(t^2 - 1) - 2t(2t)|}{[(t^2 - 1)^2 + (2t)^2]^{3/2}} = \frac{2(t^2 + 1)}{(t^4 + 2t^2 + 1)^{3/2}} = \frac{2(t^2 + 1)}{(t^2 + 1)^3} = \frac{2}{(t^2 + 1)^2}.$$

Then $\kappa(2\sqrt{2}) = 2/81$, and the radius of curvature $\rho(2\sqrt{2}) = 1/\kappa(2\sqrt{2}) = 81/2$. A portion of this curve near $t = 2\sqrt{2}$ is sketched in Figure 1. For reference purposes the unit tangent vector **T** at $t = 2\sqrt{2}$ is included. ∎

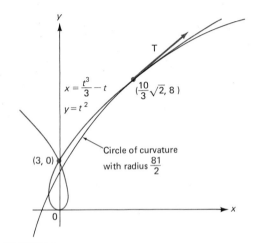

FIGURE 1

If we are given the Cartesian equation of a curve, $y = f(x)$, then the formula for the curvature, now denoted by $\kappa(x)$, is simpler. Proceeding as before, we have

$$\frac{d\varphi}{dx} = \frac{d\varphi}{ds}\frac{ds}{dx}, \qquad \text{or} \qquad \frac{d\varphi}{ds} = \frac{d\varphi/dx}{ds/dx}.$$

From equation (9.2.3) we have

$$\frac{ds}{dx} = \sqrt{1 + \left(\frac{dy}{dx}\right)^2}.$$

Also, since $\tan\varphi = dy/dx$, we have

$$\varphi = \tan^{-1}\frac{dy}{dx} \qquad \text{and} \qquad \frac{d\varphi}{dx} = \frac{1}{1 + (dy/dx)^2} \cdot \frac{d^2y}{dx^2},$$

so that

$$\kappa(x) = \left|\frac{d\varphi}{dx}\right| = \frac{|d^2y/dx^2|}{[1 + (dy/dx)^2]^{3/2}}. \tag{10}$$

EXAMPLE 4   Calculate the curvature and radius of curvature of the curve $y = 3/x$ at the point $(1, 3)$.

Solution. Here $dy/dx = -3/x^2$, $d^2y/dx^2 = 6/x^3$, and

$$\kappa(x) = \frac{|6/x^3|}{[1 + (9/x^4)]^{3/2}} = \frac{|6/x^3|}{[(9 + x^4)/x^4]^{3/2}} = \frac{|6x^3|}{(9 + x^4)^{3/2}}.$$

When $x = 1$, $\kappa(1) = 6/10\sqrt{10} = 3\sqrt{10}/50 \approx 0.19$. Then $\rho(1) = 50/3\sqrt{10} \approx 5.27$. The curve is sketched in Figure 2. ■

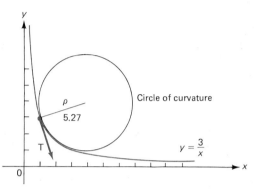

FIGURE 2

NOTE 1.   In Figures 1 and 2 the circle with radius of curvature $\rho(t)$ that lies on the concave side of C is called the **circle of curvature,** or **osculating circle.**

NOTE 2.   The curve $y = 3/x$ is concave up for $x > 0$. Graphically, it seems as if the curve is "flattening out" as x gets larger. This is easy to prove since $\lim_{x\to\infty} \kappa(x) = 0$. In fact, it can be shown that if any curve is concave up (or down) for x sufficiently large, then $\lim_{x\to\infty} \kappa(x) = 0$.

TANGENTIAL AND NORMAL COMPONENTS OF ACCELERATION

There is an interesting relationship between curvature and acceleration vectors. If a particle is moving along the curve C with position vector

$$\mathbf{f}(t) = x(t)\mathbf{i} + y(t)\mathbf{j},$$

then from the last section the acceleration vector is given by

$$\mathbf{a}(t) = \frac{d^2x}{dt^2}\mathbf{i} + \frac{d^2y}{dt^2}\mathbf{j}. \tag{11}$$

The representation (11) resolves **a** into its horizontal and vertical components. However, there is another representation that is often more useful. Imagine yourself driving on the highway. If the car in which you are riding accelerates forward, you are pressed to the back of your seat. If it turns sharply to one side, you are thrown to the other. Both motions are due to acceleration. The second force is related to the rate at which the car turns, which is, of course, related to the curvature of the road. Thus we would like to express the acceleration vector as a component in the direction of motion and a component somehow related to the curvature of the path. How do we do so? A glance at either Figure 1 or Figure 2 reveals the answer. The radial line of the circle of curvature is perpendicular to the unit tangent vector $\mathbf{T}$ since $\mathbf{T}$ is tangent to the circle of curvature, and in a circle tangent and radial lines at a point are orthogonal. (See Example 16.4.3.) But a vector that is perpendicular to $\mathbf{T}$ has the direction of the unit normal vector $\mathbf{n}$. Thus the component of acceleration in the direction $\mathbf{n}$ will be a measure of the acceleration due to turning. We would like to write

$$\mathbf{a} = a_T\mathbf{T} + a_n\mathbf{n}, \tag{12}$$

where a_T and a_n are, respectively, the components of $\mathbf{a}$ in the tangential and normal directions. Before calculating a_T and a_n, we need the following result.

Theorem 3 For a curve C, let $\mathbf{T}$ and $\mathbf{n}$ be the unit tangent and unit normal vectors, respectively, and let φ be as before. Then

(i) $$\frac{d\mathbf{T}}{d\varphi} = \pm\mathbf{n} \tag{13}$$

and

(ii) $$\frac{d\mathbf{T}}{ds} = \kappa\mathbf{n}, \tag{14}$$

where κ is the curvature of C. In (13) the plus sign is taken if φ increases as s increases and the minus sign is taken if φ decreases as s increases.

Proof.
(i) By the definition of $\mathbf{T}$,

$$\mathbf{T} = (\cos\varphi)\mathbf{i} + (\sin\varphi)\mathbf{j}.$$

Then

$$\frac{d\mathbf{T}}{d\varphi} = -(\sin\varphi)\mathbf{i} + (\cos\varphi)\mathbf{j} = \pm\mathbf{n}†$$

since $\mathbf{T}\cdot(d\mathbf{T}/d\varphi) = 0$ and $|-(\sin\varphi)\mathbf{i} + (\cos\varphi)\mathbf{j}| = 1$.
(ii) If $d\varphi/ds > 0$, then

$$\frac{d\mathbf{T}}{ds} = \frac{d\mathbf{T}}{d\varphi}\frac{d\varphi}{ds} = \mathbf{n}\kappa$$

---

†To see why $d\mathbf{T}/d\varphi = -\mathbf{n}$ if $d\varphi/ds < 0$, look again at the discussion on page 808. There we showed that if $d\varphi/ds < 0$ (Figure 16.4.3b), then

$$\mathbf{n} = \sin\varphi\mathbf{i} - \cos\varphi\mathbf{j} = -[-\sin\varphi\mathbf{i} + \cos\varphi\mathbf{j}] = -\frac{d\mathbf{T}}{d\varphi}.$$

from the definition of κ. If $d\varphi/ds < 0$, then

$$\frac{d\mathbf{T}}{ds} = \frac{d\mathbf{T}}{d\varphi}\frac{d\varphi}{ds} = (-\mathbf{n})(-\kappa) = \kappa\mathbf{n}. \quad \blacksquare$$

We can now show how to write $\mathbf{a}(t)$ in its tangential and normal components.

Theorem 4   Let the curve C be given by the vector function $\mathbf{f}(t)$. Then the acceleration vector $\mathbf{a}(t) = \mathbf{f}''(t)$ can be written as

$$\mathbf{a} = \frac{d^2s}{dt^2}\mathbf{T} + \left(\frac{ds}{dt}\right)^2 \kappa\mathbf{n} \tag{15}$$

where $\mathbf{T}$ and $\mathbf{n}$ are the unit tangent and unit normal vectors to the curve, s is the arc length measured from the point reached when $t = 0$, and κ is the curvature.

NOTE.   Since $v = ds/dt$ is the speed, we can write (15) as

$$\mathbf{a} = \frac{dv}{dt}\mathbf{T} + v^2\kappa\mathbf{n} = \frac{dv}{dt}\mathbf{T} + \frac{v^2}{\rho}\mathbf{n}. \tag{16}$$

Thus the tangential component of acceleration is $a_{\mathrm{T}} = dv/dt$ and the normal component of acceleration is $a_{\mathrm{n}} = v^2/\rho$.

Proof of Theorem 4.   We have

$$\mathbf{v} = \frac{d\mathbf{f}}{dt} = \frac{d\mathbf{f}}{ds}\frac{ds}{dt} = \mathbf{T}\frac{ds}{dt}$$

(from Theorem 16.6.1, $\mathbf{T} = d\mathbf{f}/ds$). Then

$$\mathbf{a} = \frac{d\mathbf{v}}{dt} = \frac{d}{dt}\left(\mathbf{T}\frac{ds}{dt}\right) = \frac{d^2s}{dt^2}\mathbf{T} + \frac{ds}{dt}\frac{d\mathbf{T}}{dt}.$$

But

$$\frac{ds}{dt}\frac{d\mathbf{T}}{dt} = \frac{ds}{dt}\left(\frac{d\mathbf{T}}{ds}\frac{ds}{dt}\right) = \left(\frac{ds}{dt}\right)^2\frac{d\mathbf{T}}{ds} = \left(\frac{ds}{dt}\right)^2 \kappa\mathbf{n}.$$

The last step follows from Theorem 3, and the theorem is proved. $\blacksquare$

The result given by (15) or (16) is very important in physics. If an object of constant mass m is traveling along a trajectory, then the force acting to keep the object on that trajectory is given by

$$\mathbf{F} = m\mathbf{a} = m\frac{dv}{dt}\mathbf{T} + mv^2\kappa\mathbf{n}. \tag{17}$$

The term $mv^2 \kappa$ is the magnitude of the force necessary to keep the object from "moving off" the trajectory because of the force exerted on the object caused by turning.

EXAMPLE 5   A 1500-kg race car is driven at a speed of 150 km/hr on a circular race track of radius 50 m. What frictional force must be exerted by the tires on the road surface to keep the car from skidding?

Solution. The frictional force exerted by the tires must be equal to the component of the force (due to acceleration) normal to the circular race track. That is,

$$F = mv^2 \kappa = \frac{mv^2}{\rho} = (1500 \text{ kg}) \frac{(150{,}000 \text{ m})^2}{(3600 \text{ sec})^2} \cdot \frac{1}{50 \text{ m}}$$

$$= 52{,}083\tfrac{1}{3}(\text{kg})(\text{m})/\text{sec}^2 = 52{,}083\tfrac{1}{3} \text{ N.} \quad \blacksquare$$

EXAMPLE 6   Let the car of Example 5 have the **coefficient of friction** μ. That is, the maximum frictional force that can be exerted by the car on the road surface is μmg, where mg is the **normal force** of the car on the road (the force of the car on the road due to gravity). What is the minimum value μ can take in order that the car not slide off the road?

Solution. We must have $\mu mg \geq 52{,}083\tfrac{1}{3}$ N. But $\mu mg = \mu(9.81)(1500)$, so we obtain

$$\mu \geq \frac{52{,}083\tfrac{1}{3}}{(9.81)(1500)} \approx 3.54. \quad \blacksquare$$

PROBLEMS 16.8

In Problems 1–20, find the curvature and radius of curvature for each curve. Sketch the unit tangent vector and the circle of curvature.

1. $f = 2 \cos t i + 2 \sin t j; \; t = \pi/4$
2. $f = 2 \cos t i + 2 \sin t j; \; t = \pi/27$
3. $f = t i + t^2 j; \; t = 1$
4. $f = 3 \sin t i + 4 \cos t j; \; t = 0$
5. $f = 3 \sin t i + 4 \cos t j; \; t = \pi/2$
6. $f = 3 \sin t i + 4 \cos t j; \; t = \pi/4$
7. $f = (\cos t + t \sin t)i + (\sin t - t \cos t)j; \; t = \pi/6$
8. $y = x^2; \; (0, 0)$
9. $y = x^2; \; (1, 1)$
10. $xy = 1; \; (1, 1)$
11. $y = e^x; \; (0, 1)$
12. $y = e^x; \; (1, e)$
13. $y = \ln x; \; (1, 0)$
14. $y = \cos x; \; (\pi/3, \tfrac{1}{2})$
15. $y = ax^2 + bx + c; \; (0, c), a \neq 0$
16. $y = \ln \cos x; \; (\pi/4, \ln (1/\sqrt{2}))$
17. $y = \sqrt{1 - x^2}; \; (0, 1)$
*18. $y = \sin^{-1} x; \; (1, \pi/2)$
19. $x = \cos y; \; (0, \pi/2)$
20. $x = y^3; \; (1, 1)$

21. At what point on the parabola $y = ax^2$ is the curvature a maximum?
22. At what point on the curve $y = \ln x$ is the curvature a maximum?
23. For what value of t in the interval $[0, \pi/2]$ is the curvature of the curve $f(t) = a(\cos^3 t)i + a(\sin^3 t)j$ a minimum? For what value is it a maximum?
*24. Let $r = f(\theta)$ be the equation of a curve in polar coordinates. Show that

$$\kappa(\theta) = \frac{|r^2 + 2(dr/d\theta)^2 - r\,d^2r/d\theta^2|}{[r^2 + (dr/d\theta)^2]^{3/2}}.$$

In Problems 25–30, use the result of Problem 24 to calculate the curvature. In all cases assume that $a > 0$.

25. $r = a \sin 2\theta$; $\theta = \pi/8$

26. $r = a\theta$; $\theta = 1$

27. $r = 2a \cos \theta$; $\theta = \pi/3$

28. $r = a(1 + \sin \theta)$; $\theta = \pi/2$

29. $r = a(1 - \cos \theta)$; $\theta = \pi$

30. $r = e^{a\theta}$; $\theta = 1$

In Problems 31–38, find the tangential and normal components of acceleration for each of the given position vectors.

31. $\mathbf{f} = (\cos 2t)\mathbf{i} + (\sin 2t)\mathbf{j}$

32. $\mathbf{f} = 2(\cos t)\mathbf{i} + 3(\sin t)\mathbf{j}$

33. $\mathbf{f} = t\mathbf{i} + t^2\mathbf{j}$

34. $\mathbf{f} = t\mathbf{i} + (\cos t)\mathbf{j}$

35. $\mathbf{f} = (t^3 - 3t)\mathbf{i} + (t^2 - 1)\mathbf{j}$

36. $\mathbf{f} = e^{-t}\mathbf{i} + e^t\mathbf{j}$

37. $\mathbf{f} = t^2\mathbf{i} + t^3\mathbf{j}$

38. $\mathbf{f} = (\sin t^2)\mathbf{i} + (\cos t^2)\mathbf{j}$

39. Show that if a particle is moving at a constant speed, then the tangential component of acceleration is zero.

40. Suppose that the driver of the car of Examples 5 and 6 reduces his speed by a factor of M. Show that the frictional force needed to keep the car from skidding is reduced by a factor of M^2.

41. A truck traveling at 80 km/hr and weighing 10,000 kg is moving on a curved stretch of track. The equation of the curved section is the parabola $y = x^2 - x$ meters. What is the frictional force exerted by the wheels of the truck on the track at the "point" $(0, 0)$?

42. If the coefficient of friction for the truck in Problem 41 is 2.5, what is the maximum speed it can achieve at the point $(0, 0)$ without going off the road?

43. If the race car of Example 5 is placed on a track with half the radius of the original one, how much slower would it have to be driven so as not to increase the normal component of acceleration?

*44. A woman swings a rope attached to a bucket containing 3 kg of water. The pail rotates in the vertical plane in a circular path with a radius of 1 m. What is the smallest number of revolutions that must be made every minute in order that the water stay in the pail? [*Hint:* Calculate the pressure of the water on the bottom of the pail. This can be determined by first calculating the normal force of the motion. Then the water will stay in the bucket if this normal force exceeds the force due to gravity.]

REVIEW EXERCISES FOR CHAPTER SIXTEEN

In Exercises 1–8, find the Cartesian equation of each curve and then sketch the curve in the xy-plane.

1. $\mathbf{f}(t) = t\mathbf{i} + 2t\mathbf{j}$

2. $\mathbf{f}(t) = (2t - 6)\mathbf{i} + t^2\mathbf{j}$

3. $\mathbf{f}(t) = t^2\mathbf{i} + (2t - 6)\mathbf{j}$

4. $\mathbf{f}(t) = t^2\mathbf{i} + t^4\mathbf{j}$

5. $\mathbf{f}(t) = (\cos 4t)\mathbf{i} + (\sin 4t)\mathbf{j}$

6. $\mathbf{f}(t) = 4(\sin t)\mathbf{i} + 9(\sin t)\mathbf{j}$

7. $\mathbf{f}(t) = t^6\mathbf{i} + t^2\mathbf{j}$

8. $\mathbf{f}(t) = e^t(\cos t)\mathbf{i} + e^t(\sin t)\mathbf{j}$

In Exercises 9–16, find the slope of the line tangent to the given curve for the given value of the parameter, and then find all points at which the curve has vertical and horizontal tangents.

9. $x = t^3$; $y = 6t$; $t = 1$

10. $x = t^7$; $y = t^8 - 5$; $t = 2$

11. $x = \sin 5\theta$; $y = \cos 5\theta$; $\theta = \pi/3$

12. $x = \cos^2 \theta$; $y = -3\theta$; $\theta = \pi/4$

13. $x = \cosh t$; $y = \sinh t$; $t = 0$

14. $x = \dfrac{2}{\theta}$; $y = -3\theta$; $\theta = 10\pi$

15. $x = 3 \cos \theta$; $y = 4 \sin \theta$; $\theta = \pi/3$

16. $x = 3 \cos \theta$; $y = -4 \sin \theta$; $\theta = 2\pi/3$

17. Find the slope of the line tangent to the polar curve $r = -3 \cos \theta$ for $\theta = \pi/3$.
18. Find the slope of the line tangent to the polar curve $r = \sin 2\theta$ for $\theta = \pi/8$.
19. Find the slope of the line tangent to the spiral of Archimedes $r = \theta$ for $\theta = \pi$.
20. Calculate the equation of the line that is tangent to the curve $x = t^2 - t - 2$, $y = t^2 - t + 1$ at the point $(4, 7)$.

In Exercises 21–24, find the first and second derivatives of the given vector functions.

21. $\mathbf{f}(t) = 2t\mathbf{i} - t^3\mathbf{j}$

22. $\mathbf{f}(t) = \left(\dfrac{1}{t^2}\right)\mathbf{i} + \sqrt[3]{t}\,\mathbf{j}$

23. $\mathbf{f}(t) = (\cos 5t)\mathbf{i} + 2(\sin 4t)\mathbf{j}$

24. $\mathbf{f}(t) = (\tan t)\mathbf{i} + (\cot t)\mathbf{j}$

In Exercises 25–30, find the unit tangent and unit normal vectors to the given curve for the given value of t.

25. $\mathbf{f}(t) = t^4\mathbf{i} + t^5\mathbf{j}$; $t = 1$

26. $\mathbf{f}(t) = (\cos 2t)\mathbf{i} + (\sin 2t)\mathbf{j}$; $t = \pi/6$

27. $\mathbf{f}(t) = (\sin 5t)\mathbf{i} + (\cos 5t)\mathbf{j}$; $t = \pi/30$

28. $\mathbf{f}(t) = (\cosh 2t)\mathbf{i} + (\sinh 2t)\mathbf{j}$; $t = 0$

29. $\mathbf{f}(t) = (\ln t)\mathbf{i} + \sqrt{t}\,\mathbf{j}$; $t = 1$

30. $\mathbf{f}(t) = (\tan t)\mathbf{i} + (\cot t)\mathbf{j}$; $t = \pi/3$

In Exercises 31–34, calculate the integral.

31. $\displaystyle\int_0^3 (t^3\mathbf{i} + t^5\mathbf{j})\, dt$

32. $\displaystyle\int [(\cos 3t)\mathbf{i} + (\sin 3t)\mathbf{j}]\, dt$

33. $\displaystyle\int (\sqrt{t}\,\mathbf{i} + \sqrt[3]{t}\,\mathbf{j})\, dt$

34. $\displaystyle\int_0^{\pi/3} [(\cos t)\mathbf{i} + (\tan t)\mathbf{j}]\, dt$

35. For $\mathbf{f}'(t) = t^7\mathbf{i} - t^6\mathbf{j}$ and for $\mathbf{f}(0) = -\mathbf{i} + 3\mathbf{j}$, find $\mathbf{f}(t)$.

In Exercises 36–40, calculate the derivative.

36. $\dfrac{d}{dt}[(2t\mathbf{i} + \sqrt{t}\,\mathbf{j}) \cdot (4\mathbf{i} - 3\mathbf{j})]$

37. $\dfrac{d}{dt}\{[(\cos t)\mathbf{i} + (\sin t)\mathbf{j}] \cdot [(\cosh t)\mathbf{i} - (\sinh t)\mathbf{j}]\}$

38. $\dfrac{d}{dt}[(e^t)(t^2\mathbf{i} - t^{3/2}\mathbf{j})]$

39. $\dfrac{d}{dt}\left[\left(\dfrac{t}{1+t}\mathbf{i} - \dfrac{t+1}{t}\mathbf{j}\right) \cdot (e^t\mathbf{i} + e^{-t}\mathbf{j})\right]$

40. $\dfrac{d}{dt}\{(2\mathbf{i} - 11\mathbf{j}) \cdot [-(\tan t)\mathbf{i} + (\sec t)\mathbf{j}]\}$

In Exercises 41–46, find the length of the arc over the given interval or the length of the closed curve.

41. $x = \cos 4\theta$; $y = \sin 4\theta$; $0 \le \theta \le \pi/12$
42. $x = e^t \sin t$; $y = e^t \cos t$; $0 \le t \le \pi/2$
43. $r = 2(1 + \cos \theta)$
44. $r = 5 \sin \theta$
45. $r = \theta^2$; $1 \le \theta \le 5$
46. $r = 2\theta$; $0 \le \theta \le 20\pi$

In Exercises 47–50, find parametric equations in terms of the arc length s measured from the point reached when $t = 0$. Verify your answer by using equation (16.6.3).

47. $\mathbf{f} = 3t\mathbf{i} + 4t^{3/2}\mathbf{j}$
48. $\mathbf{f} = \frac{2}{9}t^{9/2}\mathbf{i} + \frac{1}{3}t^3\mathbf{j}$
49. $\mathbf{f} = 2(\cos 3t)\mathbf{i} + 2(\sin 3t)\mathbf{j}$
50. $\mathbf{f} = e^t(\sin t)\mathbf{i} + e^t(\cos t)\mathbf{j}$

In Exercises 51–56 the position vector of a moving particle is given. For the indicated value of t, calculate the velocity vector, the acceleration vector, the speed, and the acceleration scalar. Then sketch a portion of the trajectory showing the velocity and acceleration vectors.

51. $\mathbf{f} = (\cos 2t)\mathbf{i} + (\sin 2t)\mathbf{j}; \ t = \pi/6$

52. $\mathbf{f} = 6t\mathbf{i} + 2t^3\mathbf{j}; \ t = 1$

53. $\mathbf{f} = (2^t + e^{-t})\mathbf{i} + 2t\mathbf{j}; \ t = 0$

54. $\mathbf{f} = (1 - \cos t)\mathbf{i} + (t - \sin t)\mathbf{j}; \ t = \pi/2$

55. $\mathbf{f} = 2(\sinh t)\mathbf{i} + 4t\mathbf{j}; \ t = 1$

56. $\mathbf{f} = (3 + 5t)\mathbf{i} + (2 + 8t)\mathbf{j}; \ t = 6$

57. A boy throws a ball into the air with an initial speed of 30 m/sec. The ball is thrown from shoulder level (1.5 m above the ground) at an angle of 30° with the horizontal. Ignoring air resistance, find the following:

(a) The total distance traveled by the ball.

(b) The horizontal distance traveled by the ball.

(c) The maximum height the ball reaches.

(d) The speed of the ball at impact with the ground.

(e) The angle at which the path of the ball meets the ground at the point of impact.

58. A 1500-kg rocket carrying 1000 kg of fuel starts from a position motionless in space. Its exhaust velocity is 1.5 km/sec. What is its velocity when all its fuel is consumed?

59. The rocket in Exercise 58 is launched from the earth, and gas is emitted at a constant rate of 40 kg/sec.

(a) Find the net initial force acting on the rocket.

(b) Find the net force just before all the fuel is expended.

60. In Exercise 59, what is the minimum thrust needed to get the rocket off the ground?

In Exercises 61–70, find the curvature and radius of curvature for each curve. Sketch the unit tangent vector and the circle of curvature.

61. $\mathbf{f} = (\cos 2t)\mathbf{i} + (\sin 2t)\mathbf{j}; \ t = \pi/3$

62. $\mathbf{f} = t^2\mathbf{i} + 2t\mathbf{j}; \ t = 2$

63. $\mathbf{f} = 4(\cos t)\mathbf{i} + 9(\sin t)\mathbf{j}; \ t = \pi/4$

64. $y = 2x^2; \ (0, 0)$

65. $xy = 1; \ (2, \frac{1}{2})$

66. $y = e^{-x}; \ (1, 1/e)$

67. $y = \sqrt{x}; \ (4, 2)$

68. $r = 3 \cos 2\theta; \ \theta = \pi/6$

69. $r = 1 + \sin \theta; \ \theta = \pi/2$

70. $r = 3\theta; \ \theta = \pi$

In Exercises 71–74, find the tangential and normal components of acceleration for each of the given position vectors.

71. $\mathbf{f} = 2(\sin t)\mathbf{i} + 2(\cos t)\mathbf{j}$

72. $\mathbf{f} = 4(\cos t)\mathbf{i} + 9(\sin t)\mathbf{j}$

73. $\mathbf{f} = 3t^2\mathbf{i} + 2t^3\mathbf{j}$

74. $\mathbf{f} = (\cos t^2)\mathbf{i} + (\sin t^2)\mathbf{j}$

75. A 1300-kg race car is driven at a speed of 175 km/hr on a circular race track of radius 65 m. What frictional force must be exerted by the tires on the road surface to keep the car from skidding?

17 Vectors in Space

In Chapters 15 and 16 we developed a theory of vectors in the plane. In this chapter we will extend that development to vectors in space and will find that there are great similarities between the two notions.

17.1 THE RECTANGULAR COORDINATE SYSTEM IN SPACE

In Section 1.3 we showed how any point in a plane can be represented as an ordered pair of real numbers. It is not surprising, then, that any point in space can be represented by an **ordered triple** of real numbers

$$(a, b, c), \tag{1}$$

where a, b, and c are real numbers.

Definition 1 THREE-DIMENSIONAL SPACE $\mathbb{R}^3$ The set of ordered triples of the form (1) is called **real three-dimensional space** and is denoted $\mathbb{R}^3$.

There are many ways to represent a point in $\mathbb{R}^3$. Two other methods will be given in Section 17.8. However, the most common representation (given in Definition 1) is very similar to the representation of a point in the plane by its x- and y-coordinates.

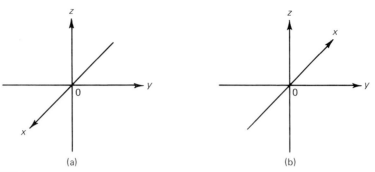

(a)   (b)

FIGURE 1

We begin, as before, by choosing a point in $\mathbb{R}^3$ and calling it the **origin,** denoted by 0. Then we draw three mutually perpendicular axes, called the **coordinate axes,** which we label the x-**axis,** the y-**axis,** and the z-**axis.** These axes can be selected in a variety of ways, but the most common selection has the x- and y-axes drawn horizontally with the z-axis vertical. On each axis we choose a positive direction and measure distance along each axis as the number of units in this positive direction measured from the origin.

The two basic systems of drawing these axes are depicted in Figure 1. If the axes are placed as in Figure 1a, then the system is called a **right-handed system;** if they are placed as in Figure 1b, the system is a **left-handed system.** In the figures the arrows indicate the positive directions on the axes. The reason for this choice of terms is as follows: In a right-handed system, if you place your right hand so that your index finger points in the positive direction of the x-axis while your middle finger points in the positive direction of the y-axis, then your thumb will point in the positive direction of the z-axis. This is illustrated in Figure 2. For a left-handed system the same rule will work for your left hand. For the remainder of this text we will follow common practice and depict the coordinate axes using a right-handed system.

If you have trouble visualizing the placement of these axes, do the following. Face any uncluttered corner (on the floor) of the room in which you are sitting. Call the corner the origin. Then the x-axis lies along the floor, along the wall, and to your left; the y-axis lies along the floor, along the wall, and to your right; and the z-axis lies along the vertical intersection of the two perpendicular walls. This is illustrated in Figure 3.

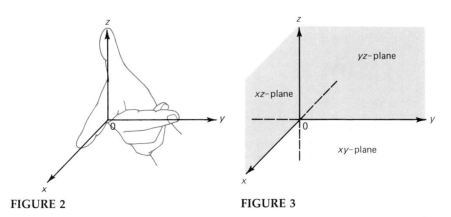

FIGURE 2   **FIGURE 3**

The three axes in our system determine three **coordinate planes** that we will call the *xy*-**plane,** the *xz*-**plane,** and the *yz*-**plane.** The *xy*-plane contains the *x*- and *y*-axes and is simply the plane with which we have been dealing in most of this book. The *xz*- and *yz*-planes can be thought of in a similar way.

Having built our structure of coordinate axes and planes, we can describe any point *P* in space in a unique way:

$$P = (x, y, z), \tag{2}$$

where the first coordinate *x* is the distance from the *yz*-plane to *P* (measured in the positive direction of the *x*-axis), the second coordinate *y* is the distance from the *xz*-plane to *P* (measured in the positive direction of the *y*-axis), and the third coordinate *z* is the distance from the *xy*-plane to *P* (measured in the positive direction of the *z*-axis). Thus, for example, any point in the *xy*-plane has *z*-coordinate 0; any point in the *xz*-plane has *y*-coordinate 0; and any point in the *yz*-plane has *x*-coordinate 0. Some representative points are sketched in Figure 4.

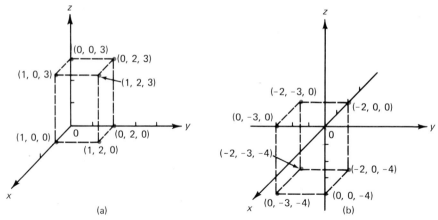

FIGURE 4

In this system the three coordinate planes divide $\mathbb{R}^3$ into eight **octants,** just as in $\mathbb{R}^2$ the two coordinate axes divide the plane into four quadrants. The first octant is always chosen to be the one in which the three coordinates are positive.

The coordinate system we have just established is often referred to as the **rectangular coordinate system,** or the **Cartesian coordinate system.** Once we are comfortable with the notion of depicting a point in this system, then we can generalize many of our ideas from the plane.

Theorem 1.   Let $P = (x_1, y_1, z_1)$ and $Q = (x_2, y_2, z_2)$ be two points in space. Then the distance $\overline{PQ}$ between *P* and *Q* is given by

$$\overline{PQ} = \sqrt{(x_1 - x_2)^2 + (y_1 - y_2)^2 + (z_1 - z_2)^2}. \tag{3}$$

Proof.   The two points are sketched in Figure 5. From the Pythagorean theorem,

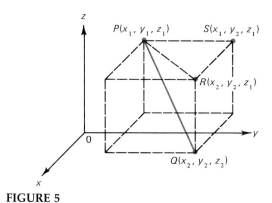

FIGURE 5

since the line segments PR and RQ are perpendicular, the triangle PQR is a right triangle and

$$\overline{PQ}^2 = \overline{PR}^2 + \overline{RQ}^2. \tag{4}$$

But using the Pythagorean theorem again, we have

$$\overline{PR}^2 = \overline{PS}^2 + \overline{SR}^2, \tag{5}$$

so combining (4) and (5) yields

$$\overline{PQ}^2 = \overline{PS}^2 + \overline{SR}^2 + \overline{RQ}^2. \tag{6}$$

Since the x- and z-coordinates of P and S are equal,

$$\overline{PS}^2 = (y_2 - y_1)^2. \tag{7}$$

Similarly,

$$\overline{RS}^2 = (x_2 - x_1)^2 \tag{8}$$

and

$$\overline{RQ}^2 = (z_2 - z_1)^2. \tag{9}$$

Thus using (7), (8), and (9) in (6) yields

$$\overline{PQ}^2 = (x_2 - x_1)^2 + (y_2 - y_1)^2 + (z_2 - z_1)^2,$$

and the proof is complete. ■

EXAMPLE 1 Calculate the distance between the points $(3, -1, 6)$ and $(-2, 3, 5)$.

Solution. $\overline{PQ} = \sqrt{[3 - (-2)]^2 + [-1 - 3]^2 + (6 - 5)^2} = \sqrt{42}.$ ■

Definition 2 GRAPH IN $\mathbb{R}^3$ The **graph** of an equation in $\mathbb{R}^3$ is the set of all points in $\mathbb{R}^3$ whose coordinates satisfy the equation.

One of our first examples of a graph in $\mathbb{R}^2$ was the graph of the unit circle $x^2 + y^2 = 1$. This example can easily be generalized.

Definition 3 SPHERE   A **sphere** is the set of points in space at a given distance from a given point. The given point is called the **center** of the sphere, and the given distance is called the **radius** of the sphere.

EXAMPLE 2   Suppose that the center of a sphere is the origin (0, 0, 0) and the radius of the sphere is 1. Let (x, y, z) be a point on the sphere. Then from (3)

$$1 = \sqrt{(x - 0)^2 + (y - 0)^2 + (z - 0)^2}.$$

Simplifying and squaring, we obtain

$$x^2 + y^2 + z^2 = 1, \tag{10}$$

which is the equation of the **unit sphere.** ■

In general, if the center of a sphere is (a, b, c), the radius is r, and if (x, y, z) is a point on the sphere, we obtain

$$r = \sqrt{(x - a)^2 + (y - b)^2 + (z - c)^2},$$

or

$$(x - a)^2 + (y - b)^2 + (z - c)^2 = r^2. \tag{11}$$

This is sketched in Figure 6.

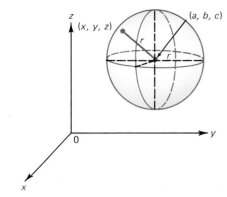

FIGURE 6

EXAMPLE 3   Find the equation of the sphere with center at $(1, -3, 2)$ and radius 5.

Solution. From (11) we obtain

$$(x - 1)^2 + (y + 3)^2 + (z - 2)^2 = 25. \ \blacksquare$$

EXAMPLE 4   Show that

$$x^2 - 6x + y^2 + 2y + z^2 + 10z + 5 = 0 \tag{12}$$

is the equation of a sphere, and find its center and radius.

Solution. Completing the squares, we obtain

$$x^2 - 6x = (x - 3)^2 - 9$$
$$y^2 + 2y = (y + 1)^2 - 1$$
$$z^2 + 10z = (z + 5)^2 - 25,$$

so that

$$0 = x^2 - 6x + y^2 + 2y + z^2 + 10z + 5$$
$$= (x - 3)^2 - 9 + (y + 1)^2 - 1 + (z + 5)^2 - 25 + 5$$

and

$$(x - 3)^2 + (y + 1)^2 + (z + 5)^2 = 30,$$

which is the equation of a sphere with center $(3, -1, -5)$ and radius $\sqrt{30}$. ■

■   WARNING: Not every second-degree equation in a form similar to (12) is the equation of a sphere. For example, if the number 5 in (12) is replaced by 40, we obtain

$$(x - 3)^2 + (y + 1)^2 + (z + 5)^2 = -5.$$

Clearly, the sum of squares cannot be negative, so there are *no* points in $\mathbb{R}^3$ that satisfy this equation. On the other hand, if we replaced the 5 by 35, we would obtain

$$(x - 3)^2 + (y + 1)^2 + (z + 5)^2 = 0.$$

This equation can hold only when $x = 3$, $y = -1$, and $z = -5$. In this case the graph of the equation contains the single point $(3, -1, -5)$.

PROBLEMS 17.1

In Problems 1–15, sketch the given point in $\mathbb{R}^3$.

1. $(1, 4, 2)$ **2.** $(3, -2, 1)$ **3.** $(-1, 5, 7)$
4. $(8, -2, 3)$ **5.** $(-2, 1, -2)$ **6.** $(1, -2, 1)$
7. $(3, 2, -5)$ **8.** $(-2, -3, -8)$ **9.** $(2, 0, 4)$
10. $(-3, -8, 0)$ **11.** $(0, 4, 7)$ **12.** $(1, 3, 0)$
13. $(3, 0, 0)$ **14.** $(0, 8, 0)$ **15.** $(0, 0, -7)$

In Problems 16–25, find the distance between the two points.

16. $(8, 1, 6)$; $(8, 1, 4)$ **17.** $(3, -4, 3)$; $(3, 2, 5)$
18. $(3, -4, 7)$; $(3, -4, 9)$ **19.** $(-2, 1, 3)$; $(4, 1, 3)$
20. $(2, -7, 5)$; $(8, -7, -1)$ **21.** $(1, 3, -2)$; $(4, 7, -2)$
22. $(3, 1, 2)$; $(1, 2, 3)$ **23.** $(5, -6, 4)$; $(3, 11, -2)$
24. $(-1, -7, -2)$; $(-4, 3, -5)$ **25.** $(8, -2, -3)$; $(-7, -5, 1)$

26. Find the equation of the sphere with center at $(2, -1, 4)$ and radius 2.
27. Find the equation of the sphere with center at $(-1, 8, -3)$ and radius $\sqrt{5}$.
28. Show that the equation $x^2 + y^2 + z^2 - 4x - 4y + 8z + 8 = 0$ is the equation of a sphere, and find its center and radius.
29. Do the same for the equation $x^2 + y^2 + z^2 + 3x - y + 2z - 1 = 0$.
30. Find a number α such that the equation $x^2 + y^2 + z^2 - 2x + 8y - 5z + \alpha = 0$ has exactly one solution.
31. In Problem 30 give a condition on α in order that the equation have no solution.
32. The equation $x^2 + y^2 + z^2 + ax + by + cz + d = 0$ is a second-degree equation in three variables. Let $e = d - (a/2)^2 - (b/2)^2 - (c/2)^2$. Show that the equation **(a)** is the equation of a sphere if $e < 0$, **(b)** has exactly one solution if $e = 0$, and **(c)** has no solutions if $e > 0$.
33. Three points P, Q, and R are **collinear** if they lie on the same straight line. Show that, in $\mathbb{R}^2$, P, Q, and R are collinear if $\overline{PR} = \overline{PQ} + \overline{QR}$ or $\overline{PQ} = \overline{PR} + \overline{RQ}$ or $\overline{QR} = \overline{QP} + \overline{PR}$. Use this last fact in $\mathbb{R}^3$ to show that the points $(-1, -1, -1)$, $(5, 8, 2)$, and $(-3, -4, -2)$ are collinear.
34. Show that the points $(3, 0, 1)$, $(0, -4, 0)$, and $(6, 4, 2)$ are collinear.
*35. Let $P = (x_1, y_1, z_1)$ and $Q = (x_2, y_2, z_2)$. Show that the midpoint of PQ is the point $R = ((x_1 + x_2)/2, (y_1 + y_2)/2, (z_1 + z_2)/2)$. [*Hint:* Show that P, Q, and R are collinear and that $\overline{PR} = \overline{RQ}$.]
36. Find the midpoint of the line joining the points $(2, -1, 4)$ and $(5, 7, -3)$.
37. Find the equation of the sphere that has a diameter with endpoints $(3, 1, -2)$ and $(4, 1, 6)$. [*Hint:* Find the center and radius of the sphere by using the result of Problem 35.]
38. One sphere is said to be **inscribed** in a second sphere if it has a smaller radius and the same center as the second sphere. Find the equation of a sphere of radius 1 inscribed in the sphere given by $x^2 + y^2 + z^2 - 2x - 4y + z - 2 = 0$.
39. Find the volumes of the spheres of Problems 37 and 38.
40. Find the surface areas of the spheres of Problems 37 and 38.

17.2 VECTORS IN $\mathbb{R}^3$

In Chapter 15 we developed properties of vectors in the plane $\mathbb{R}^2$. Given the similarity between the coordinate systems in $\mathbb{R}^2$ and $\mathbb{R}^3$, it should come as no surprise to learn that vectors in $\mathbb{R}^2$ and $\mathbb{R}^3$ have very similar structures. In this section we will develop the notion of a vector in space. The development will closely follow the development in Sections 15.1 and 15.2 and, therefore, some of the details will be omitted.

Let P and Q be two distinct points in $\mathbb{R}^3$. Then the **directed line segment** $\overrightarrow{PQ}$ is the straight line segment that extends from P to Q. Two directed line segments are **equivalent** if they have the same magnitude and direction. A **vector** in $\mathbb{R}^3$ is the set of all directed line segments equivalent to a given directed line segment, and any directed line segment $\overrightarrow{PQ}$ in that set is called a **representation** of the vector.

So far, our definitions are identical. For convenience we will choose P to be the origin and label the endpoint of the vector R, so that the vector $\mathbf{v} = \overrightarrow{OR}$ can be described by the coordinates (x, y, z) of the point R. Then the **magnitude** of $\mathbf{v} = |\mathbf{v}| = \sqrt{x^2 + y^2 + z^2}$ (from Theorem 17.1.1).

EXAMPLE 1 Let $\mathbf{v} = (1, 3, -2)$. Find $|\mathbf{v}|$.

Solution. $|\mathbf{v}| = \sqrt{1^2 + 3^2 + (-2)^2} = \sqrt{14}$. ■

Let $\mathbf{u} = (x_1, y_1, z_1)$ and $\mathbf{v} = (x_2, y_2, z_2)$ be two vectors and let α be a real number (scalar). Then we define

$$\mathbf{u} + \mathbf{v} = (x_1 + x_2, y_1 + y_2, z_1 + z_2)$$

and

$$\alpha\mathbf{u} = (\alpha x_1, \alpha y_1, \alpha z_1).$$

This is the same definition of vector addition and scalar multiplication we had before and is illustrated in Figure 1.

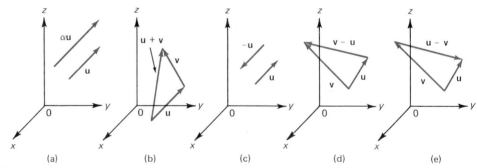

(a)   (b)   (c)   (d)   (e)

FIGURE 1

EXAMPLE 2 Let $\mathbf{u} = (2, 3, -1)$ and $\mathbf{v} = (-6, 2, 4)$. Find (a) $\mathbf{u} + \mathbf{v}$, (b) $3\mathbf{v}$, (c) $-\mathbf{u}$, and (d) $4\mathbf{u} - 3\mathbf{v}$.

Solution.

(a) $\mathbf{u} + \mathbf{v} = (2 - 6, 3 + 2, -1 + 4) = (-4, 5, 3)$
(b) $3\mathbf{v} = (-18, 6, 12)$
(c) $-\mathbf{u} = (-2, -3, 1)$
(d) $4\mathbf{u} - 3\mathbf{v} = (8, 12, -4) - (-18, 6, 12) = (26, 6, -16)$ ■

The following theorem extends to three dimensions the results of Theorem 15.1.1. Its proof is easy and is left as an exercise (Problem 54).

Theorem 1. Let $\mathbf{u}$, $\mathbf{v}$, and $\mathbf{w}$ be any three vectors in space, let α and β be scalars, and let $\mathbf{0}$ denote the zero vector $(0, 0, 0)$.

(i) u + v = v + u	**(ii) u + (v + w) = (u + v) + w**
(iii) v + 0 = v	**(iv) 0v = 0**
(v) α0 = 0	**(vi) $(\alpha\beta)$v = $\alpha(\beta$v$)$**
(vii) v + (−v) = 0	**(viii) (1)v = v**
(ix) $(\alpha + \beta)$v = αv + βv	**(x) α(u + v) = αu + αv**
(xi) $\|\alpha$v$\| = \|\alpha\| \|$v$\|$	**(xii) $\|$u + v$\| \leq \|$u$\| + \|$v$\|$**

A **unit vector u** is a vector with magnitude 1. If **v** is any nonzero vector, then **u** = **v**/|**v**| is a unit vector having the same direction as **v**.

EXAMPLE 3   Find a unit vector having the same direction as **v** = (2, 4, −3).

Solution. Since $|\mathbf{v}| = \sqrt{2^2 + 4^2 + (-3)^2} = \sqrt{29}$, we have

$$\mathbf{u} = \left(\frac{2}{\sqrt{29}}, \frac{4}{\sqrt{29}}, \frac{-3}{\sqrt{29}}\right). \quad \blacksquare$$

We can now formally define the direction of a vector in $\mathbb{R}^3$. We cannot define it to be the angle θ the vector makes with the positive x-axis, since, for example, if $0 < \theta < \pi/2$, then there are an *infinite number* of unit vectors making the angle θ with the positive x-axis, and these together form a cone (see Figure 2).

Definition 1   DIRECTION OF A VECTOR   The **direction** of a nonzero vector **v** in $\mathbb{R}^3$ is defined to be the unit vector **u** = **v**/|**v**|.

REMARK.   We could have defined the direction of a vector **v** in $\mathbb{R}^2$ in this way. For if **u** = **v**/|**v**|, then **u** = $(\cos \theta, \sin \theta)$, where θ is the direction of **v** (according to the $\mathbb{R}^2$ definition).

DIRECTION COSINES   It would still be useful to define the direction of a vector in terms of some angles. Let **v** be the vector $\overrightarrow{OP}$ depicted in Figure 3. We define α to be the angle between **v** and the positive x-axis, β the angle between **v** and the positive

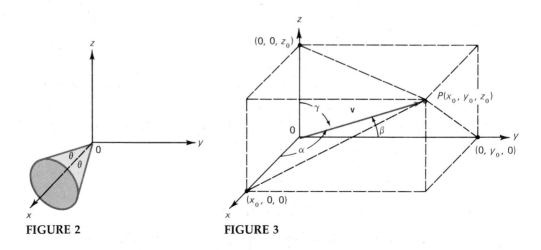

FIGURE 2   **FIGURE 3**

y-axis, and γ the angle between **v** and the positive z-axis. The angles α, β, and γ are called the **direction angles** of the vector **v**. Then from Figure 3

$$\cos \alpha = \frac{x_0}{|\mathbf{v}|}, \qquad \cos \beta = \frac{y_0}{|\mathbf{v}|}, \qquad \cos \gamma = \frac{z_0}{|\mathbf{v}|}. \tag{1}$$

If **v** is a unit vector, then $|\mathbf{v}| = 1$ and

$$\cos \alpha = x_0, \qquad \cos \beta = y_0, \qquad \cos \gamma = z_0. \tag{2}$$

By definition, each of these three angles lies between 0 and π. The cosines of these angles are called the **direction cosines** of the vector **v**.

Note from (1) that

$$\cos^2 \alpha + \cos^2 \beta + \cos^2 \gamma = \frac{x_0^2 + y_0^2 + z_0^2}{|\mathbf{v}|^2} = \frac{x_0^2 + y_0^2 + z_0^2}{x_0^2 + y_0^2 + z_0^2} = 1. \tag{3}$$

If α, β, and γ are any three numbers between 0 and π such that condition (3) is satisfied, then they uniquely determine a unit vector given by $\mathbf{u} = (\cos \alpha, \cos \beta, \cos \gamma)$.

REMARK.   If $\mathbf{v} = (a, b, c)$ and $|\mathbf{v}| \neq 0$, then the numbers a, b, and c are called **direction numbers** of the vector **v**.

EXAMPLE 4   Find the direction cosines of the vector $\mathbf{v} = (4, -1, 6)$.

Solution. The direction of **v** is $\mathbf{v}/|\mathbf{v}| = \mathbf{v}/\sqrt{53} = (4/\sqrt{53}, -1/\sqrt{53}, 6/\sqrt{53})$. Then $\cos \alpha = 4/\sqrt{53} \approx 0.5494$, $\cos \beta = -1/\sqrt{53} \approx -0.1374$, and $\cos \gamma = 6/\sqrt{53} \approx 0.8242$. From these we use a calculator to obtain $\alpha \approx 56.7° \approx 0.989$ radian, $\beta \approx 97.9° \approx 1.71$ radians, and $\gamma = 34.5° \approx 0.602$ radian. The vector, along with the angles α, β, and γ, is sketched in Figure 4. ■

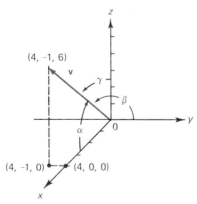

FIGURE 4

EXAMPLE 5   Find a vector **v** of magnitude 7 whose direction cosines are $1/\sqrt{6}$, $1/\sqrt{3}$, and $1/\sqrt{2}$.

NOTE.   We can solve this problem because $(1/\sqrt{6})^2 + (1/\sqrt{3})^2 + (1/\sqrt{2})^2 = 1$.

Solution. Let $\mathbf{u} = (1/\sqrt{6}, 1/\sqrt{3}, 1/\sqrt{2})$. Then $\mathbf{u}$ is a unit vector since $|\mathbf{u}| = 1$. Thus the direction of $\mathbf{v}$ is given by $\mathbf{u}$, and so

$$\mathbf{v} = |\mathbf{v}|\mathbf{u} = 7\mathbf{u} = \left(\frac{7}{\sqrt{6}}, \frac{7}{\sqrt{3}}, \frac{7}{\sqrt{2}}\right). \quad \blacksquare$$

It is interesting to note that if $\mathbf{v}$ in $\mathbb{R}^2$ is written

$$\mathbf{v} = (\cos\theta)\mathbf{i} + (\sin\theta)\mathbf{j},$$

where θ is the direction of $\mathbf{v}$, then $\cos\theta$ and $\sin\theta$ are the direction cosines of $\mathbf{v}$. Here $\alpha = \theta$ and we define β to be the angle that $\mathbf{v}$ makes with the y-axis (see Figure 5). Then $\beta = (\pi/2) - \alpha$, so that

$$\cos\beta = \cos\left(\frac{\pi}{2} - \alpha\right) = \sin\alpha,$$

and $\mathbf{v}$ can be written in the "direction cosine" form

$$\mathbf{v} = (\cos\alpha)\mathbf{i} + (\cos\beta)\mathbf{j}.$$

In Chapter 15 we showed how any vector in the plane can be written in terms of the basis vectors $\mathbf{i}$ and $\mathbf{j}$. To extend this idea to $\mathbb{R}^3$, we define

$$\mathbf{i} = (1, 0, 0), \quad \mathbf{j} = (0, 1, 0), \quad \mathbf{k} = (0, 0, 1) \tag{4}$$

Here $\mathbf{i}$, $\mathbf{j}$, and $\mathbf{k}$ are unit vectors. The vector $\mathbf{i}$ lies along the x-axis, $\mathbf{j}$ along the y-axis, and $\mathbf{k}$ along the z-axis. These vectors are sketched in Figure 6. If $\mathbf{v} = (x, y, z)$ is any vector in $\mathbb{R}^3$, then

$$\mathbf{v} = (x, y, z) = (x, 0, 0) + (0, y, 0) + (0, 0, z) = x\mathbf{i} + y\mathbf{j} + z\mathbf{k}. \tag{5}$$

That is, *any vector* $\mathbf{v}$ *in* $\mathbb{R}^3$ *can be written in a unique way in terms of the vectors* $\mathbf{i}$, $\mathbf{j}$, *and* $\mathbf{k}$.

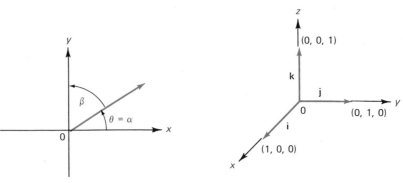

FIGURE 5   **FIGURE 6**

Let $P = (a_1, b_1, c_1)$ and $Q = (a_2, b_2, c_2)$. Then as in Section 15.1, the vector $\mathbf{v} = \overrightarrow{PQ}$ can be written

$$\mathbf{v} = (a_2 - a_1)\mathbf{i} + (b_2 - b_1)\mathbf{j} + (c_2 - c_1)\mathbf{k} \tag{6}$$

(see Problem 55).

EXAMPLE 6   Find a vector in space that can be represented by the directed line segment from $(2, -1, 4)$ to $(5, 1, -3)$.

Solution. $\mathbf{v} = (5 - 2)\mathbf{i} + [1 - (-1)]\mathbf{j} + (-3 - 4)\mathbf{k} = 3\mathbf{i} + 2\mathbf{j} - 7\mathbf{k}.$ ▪

We now turn to the notion of dot product (or scalar product) in $\mathbb{R}^3$.

Definition 2   DOT PRODUCT   If $\mathbf{u} = x_1\mathbf{i} + y_1\mathbf{j} + z_1\mathbf{k}$ and $\mathbf{v} = x_2\mathbf{i} + y_2\mathbf{j} + z_2\mathbf{k}$, then we define the **dot product** (or **scalar product** or **inner product**) by

$$\mathbf{u} \cdot \mathbf{v} = x_1x_2 + y_1y_2 + z_1z_2. \tag{7}$$

As before, the dot product of two vectors is a *scalar*. Note that $\mathbf{i} \cdot \mathbf{i} = 1$, $\mathbf{j} \cdot \mathbf{j} = 1$, $\mathbf{k} \cdot \mathbf{k} = 1$, $\mathbf{i} \cdot \mathbf{j} = 0$, $\mathbf{j} \cdot \mathbf{k} = 0$, and $\mathbf{i} \cdot \mathbf{k} = 0$.

EXAMPLE 7   For $\mathbf{u} = 2\mathbf{i} - 3\mathbf{j} - 4\mathbf{k}$ and $\mathbf{v} = -3\mathbf{i} + \mathbf{j} - 2\mathbf{k}$, calculate $\mathbf{u} \cdot \mathbf{v}$.

Solution. $\mathbf{u} \cdot \mathbf{v} = 2(-3) + (-3)(1) + (-4)(-2) = -1.$ ▪

Theorem 2.   For any vectors $\mathbf{u}$, $\mathbf{v}$, and $\mathbf{w}$ in space, and for any scalar α, we have

> **(i)** $\mathbf{u} \cdot \mathbf{v} = \mathbf{v} \cdot \mathbf{u}$
> **(ii)** $(\mathbf{u} + \mathbf{v}) \cdot \mathbf{w} = \mathbf{u} \cdot \mathbf{w} + \mathbf{v} \cdot \mathbf{w}$
> **(iii)** $(\alpha\mathbf{u}) \cdot \mathbf{v} = \alpha(\mathbf{u} \cdot \mathbf{v})$
> **(iv)** $\mathbf{u} \cdot \mathbf{u} \geq 0$, and $\mathbf{u} \cdot \mathbf{u} = 0$ if and only if $\mathbf{u} = \mathbf{0}$
> **(v)** $|\mathbf{u}| = \sqrt{\mathbf{u} \cdot \mathbf{u}}$

Proof.   The proof is almost identical to the proof of Theorem 15.2.1 and is left as an exercise (see Problem 56). ▪

Theorem 3.   If φ denotes the angle between two nonzero vectors $\mathbf{u}$ and $\mathbf{v}$, we have

$$\cos \varphi = \frac{\mathbf{u} \cdot \mathbf{v}}{|\mathbf{u}|\,|\mathbf{v}|}. \tag{8}$$

Proof.   The proof is almost identical to the proof of Theorem 15.2.2 and is left as an exercise (see Problem 57). For an interesting corollary to this theorem, see Problem 61. ▪

EXAMPLE 8   Calculate the cosine of the angle between $\mathbf{u} = 3\mathbf{i} - \mathbf{j} + 2\mathbf{k}$ and $\mathbf{v} = 4\mathbf{i} + 3\mathbf{j} - \mathbf{k}$.

Solution. $\mathbf{u} \cdot \mathbf{v} = 7$, $|\mathbf{u}| = \sqrt{14}$, and $|\mathbf{v}| = \sqrt{26}$, so that $\cos \varphi = 7/\sqrt{(14)(26)} = 7/\sqrt{364} \approx 0.3669$ and $\varphi = 68.5° \approx 1.2$ radians. ∎

Definition 3   PARALLEL AND ORTHOGONAL VECTORS

    (i) Two nonzero vectors **u** and **v** are **parallel** if the angle between them is 0 or π.

    (ii) Two nonzero vectors **u** and **v** are **orthogonal** (or **perpendicular**) if the angle between them is $\pi/2$.

Theorem 4.

    (i) If $\mathbf{u} \neq 0$, then **u** and **v** are parallel if and only if $\mathbf{v} = \alpha \mathbf{u}$ for some constant α.

    (ii) If **u** and **v** are nonzero, then **u** and **v** are orthogonal if and only if $\mathbf{u} \cdot \mathbf{v} = 0$.

Proof.   Again the proof is easy and is left as an exercise (see Problem 58). ∎

EXAMPLE 9   Show that the vectors $\mathbf{u} = \mathbf{i} + 3\mathbf{j} - 4\mathbf{k}$ and $\mathbf{v} = -2\mathbf{i} - 6\mathbf{j} + 8\mathbf{k}$ are parallel.

Solution.   Here $(\mathbf{u} \cdot \mathbf{v})/(|\mathbf{u}||\mathbf{v}|) = -52/(\sqrt{26}\sqrt{104}) = -52/(\sqrt{26} \cdot 2\sqrt{26}) = -1$, so $\cos \theta = -1$, $\theta = \pi$, and **u** and **v** are parallel (but have opposite directions). An easier way to see this result is to note that $\mathbf{v} = -2\mathbf{u}$ so that by Theorem 4, **u** and **v** are parallel. This is sketched in Figure 7. ∎

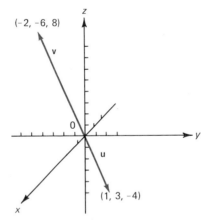

FIGURE 7

    We now turn to the definition of the projection of one vector on another. First, we state the theorem that is the analog of Theorem 15.2.5 (and has an identical proof).

Theorem 5.   Let **v** be a nonzero vector. Then for any other vector **u**,

$$\mathbf{w} = \mathbf{u} - \frac{\mathbf{u} \cdot \mathbf{v}}{|\mathbf{v}|^2}\mathbf{v}$$

is orthogonal to **v**.

Definition 4 PROJECTION   Let **u** and **v** be nonzero vectors. Then the **projection**†
of **u** onto **v**, denoted Proj$_\text{v}$ **u**, is defined by

$$\text{Proj}_\text{v}\ \mathbf{u} = \frac{\mathbf{u} \cdot \mathbf{v}}{\mathbf{v} \cdot \mathbf{v}}\mathbf{v} = \frac{\mathbf{u} \cdot \mathbf{v}}{|\mathbf{v}|^2}\mathbf{v} = \left(\frac{\mathbf{u} \cdot \mathbf{v}}{|\mathbf{v}|}\right)\frac{\mathbf{v}}{|\mathbf{v}|}. \tag{9}$$

The **component** of **u** in the direction **v** is given by $(\mathbf{u} \cdot \mathbf{v})/|\mathbf{v}|$.

EXAMPLE 10   Let $\mathbf{u} = 2\mathbf{i} + 3\mathbf{j} + \mathbf{k}$ and $\mathbf{v} = \mathbf{i} + 2\mathbf{j} - 6\mathbf{k}$. Find Proj$_\text{v}$ **u**.

Solution. Here $(\mathbf{u} \cdot \mathbf{v})/|\mathbf{v}|^2 = \frac{2}{41}$, so

$$\text{Proj}_\text{v}\ \mathbf{u} = \frac{2}{41}\mathbf{i} + \frac{4}{41}\mathbf{j} - \frac{12}{41}\mathbf{k}.$$

The component of **u** in the direction **v** is $(\mathbf{u} \cdot \mathbf{v})/|\mathbf{v}| = 2/\sqrt{41}$.   ■

Note that, as in the planar case, Proj$_\text{v}$ **u** is a vector that has the same direction as
v if $\mathbf{u} \cdot \mathbf{v} > 0$ and the direction opposite to that of **v** if $\mathbf{u} \cdot \mathbf{v} < 0$.

PROBLEMS 17.2

In Problems 1–20, find the magnitude and the direction cosines of the given vector.

1. $\mathbf{v} = 3\mathbf{j}$	**2.** $\mathbf{v} = -3\mathbf{i}$	**3.** $\mathbf{v} = 14\mathbf{k}$
4. $\mathbf{v} = -8\mathbf{j}$	**5.** $\mathbf{v} = 4\mathbf{i} - \mathbf{j}$	**6.** $\mathbf{v} = \mathbf{i} + 2\mathbf{k}$
7. $\mathbf{v} = -2\mathbf{i} + 3\mathbf{j}$	**8.** $\mathbf{v} = \mathbf{i} + \mathbf{j} + \mathbf{k}$	**9.** $\mathbf{v} = \mathbf{i} - \mathbf{j} + \mathbf{k}$
10. $\mathbf{v} = \mathbf{i} + \mathbf{j} - \mathbf{k}$	**11.** $\mathbf{v} = -\mathbf{i} + \mathbf{j} + \mathbf{k}$	**12.** $\mathbf{v} = \mathbf{i} - \mathbf{j} - \mathbf{k}$
13. $\mathbf{v} = -\mathbf{i} + \mathbf{j} - \mathbf{k}$	**14.** $\mathbf{v} = -\mathbf{i} - \mathbf{j} + \mathbf{k}$	**15.** $\mathbf{v} = -\mathbf{i} - \mathbf{j} - \mathbf{k}$
16. $\mathbf{v} = 2\mathbf{i} + 5\mathbf{j} - 7\mathbf{k}$	**17.** $\mathbf{v} = -7\mathbf{i} + 2\mathbf{j} - 13\mathbf{k}$	**18.** $\mathbf{v} = \mathbf{i} + 7\mathbf{j} - 7\mathbf{k}$
19. $\mathbf{v} = -3\mathbf{i} - 3\mathbf{j} + 8\mathbf{k}$	**20.** $\mathbf{v} = -2\mathbf{i} - 3\mathbf{j} - 4\mathbf{k}$	

21. The three direction angles of a certain unit vector are the same and are between 0 and $\pi/2$. What is the vector?

22. Find a vector of magnitude 12 that has the same direction as the vector of Problem 21.

23. Show that there is no unit vector whose direction angles are $\pi/6$, $\pi/3$, and $\pi/4$.

24. Let $P = (2, 1, 4)$ and $Q = (3, -2, 8)$. Find a unit vector in the direction of $\overrightarrow{PQ}$.

25. Let $P = (-3, 1, 7)$ and $Q = (8, 1, 7)$. Find a unit vector whose direction is opposite that of $\overrightarrow{PQ}$.

26. In Problem 25 find all points R that $\overrightarrow{PR} \perp \overrightarrow{PQ}$.

*27. Show that the set of points that satisfy the condition of Problem 26 and the condition $|\overrightarrow{PR}| = 1$ form a circle.

In Problems 28–46 let $\mathbf{u} = 2\mathbf{i} - 3\mathbf{j} + 4\mathbf{k}$, $\mathbf{v} = -2\mathbf{i} - 3\mathbf{j} + 5\mathbf{k}$, $\mathbf{w} = \mathbf{i} - 7\mathbf{j} + 3\mathbf{k}$, and $\mathbf{t} = 3\mathbf{i} + 4\mathbf{j} + 5\mathbf{k}$.

28. Calculate $\mathbf{u} + \mathbf{v}$.	**29.** Calculate $2\mathbf{u} - 3\mathbf{v}$.
30. Calculate $-18\mathbf{u}$.	**31.** Calculate $\mathbf{w} - \mathbf{u} - \mathbf{v}$.

---

†The projection vector in $\mathbb{R}^2$ is sketched in Figure 15.2.3. The derivation and a sketch of a projection vector in $\mathbb{R}^3$ are virtually the same.

32. Calculate $\mathbf{t} + 3\mathbf{w} - \mathbf{v}$.

33. Calculate $2\mathbf{u} - 7\mathbf{w} + 5\mathbf{v}$.

34. Calculate $2\mathbf{v} + 7\mathbf{t} - \mathbf{w}$.

35. Calculate $\mathbf{u} \cdot \mathbf{v}$.

36. Calculate $|\mathbf{w}|$.

37. Calculate $\mathbf{u} \cdot \mathbf{w} - \mathbf{w} \cdot \mathbf{t}$.

38. Calculate the angle between $\mathbf{u}$ and $\mathbf{w}$.

39. Calculate the angle between $\mathbf{t}$ and $\mathbf{w}$.

40. Calculate the angle between $\mathbf{v}$ and $\mathbf{t}$.

41. Calculate $\text{Proj}_{\mathbf{v}}\, \mathbf{u}$.

42. Calculate $\text{Proj}_{\mathbf{u}}\, \mathbf{v}$.

43. Calculate $\text{Proj}_{\mathbf{t}}\, \mathbf{w}$.

44. Calculate $\text{Proj}_{\mathbf{w}}\, \mathbf{t}$.

45. Calculate $\text{Proj}_{\mathbf{w}}\, \mathbf{u}$.

46. Calculate $\text{Proj}_{\mathbf{t}}\, \mathbf{v}$.

47. Find the distance between the point $P = (2, 1, 3)$ and the line passing through the points $Q = (-1, 1, 2)$ and $R = (6, 0, 1)$. [*Hint:* See Problem 15.2.50.]

48. Find the distance from the point $P = (1, 0, 1)$ to the line passing through the points $Q = (2, 3, -1)$ and $R = (6, 1, -3)$.

49. Show that the points $P = (3, 5, 6)$, $Q = (1, 2, 7)$, and $R = (6, 1, 0)$ are the vertices of a right triangle.

50. Show that the points $P = (3, 2, -1)$, $Q = (4, 1, 6)$, $R = (7, -2, 3)$, and $S = (8, -3, 10)$ are the vertices of a parallelogram.

*51. A polyhedron in space with exactly four vertices is called a **tetrahedron** (see Figure 8). Let $\mathbf{P}$ represent the vector $\overrightarrow{OP}$, $\mathbf{Q}$ the vector $\overrightarrow{OQ}$, and so on. A line is drawn from each vertex to the centroid of the opposite side. Show that these four lines meet at the endpoint of the vector

$$\mathbf{v} = \frac{\mathbf{P} + \mathbf{Q} + \mathbf{R} + \mathbf{S}}{4}.$$

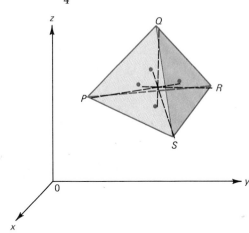

FIGURE 8

52. A force of 3 N acts in the direction of the vector with direction cosines $(1/\sqrt{6}, 1/\sqrt{3}, 1/\sqrt{2})$. Find the work done in moving the object from the point $(1, 2, 3)$ to the point $(2, 8, 11)$, where distance is measured in meters. [*Hint:* See Example 15.3.4.]

53. Find the work done when a force of 3 N acting in the direction of the vector $\mathbf{v} = \mathbf{i} + \mathbf{j} - \mathbf{k}$ moves an object from $(-1, 3, 4)$ to $(3, 7, -2)$. Again distance is measured in meters.

54. Prove Theorem 1.

55. Prove that formula (6) is correct. [*Hint:* Follow the steps leading to formula (15.1.7).]

56. Prove Theorem 2.

57. Prove Theorem 3.

58. Prove Theorem 4.

*59. Let PQR be a triangle in $\mathbb{R}^3$. Show that if a force of N newtons of constant direction moves an object around the triangle, then the total work done by that force is zero.

*60. Find the angle between the diagonal of a cube and the diagonal of one of its faces.

61. **(a)** Use Theorem 3 to prove the **Cauchy-Schwarz inequality:**

$$\left(\sum_{i=1}^{3} u_i v_i \right)^2 \leq \sum_{i=1}^{3} u_i^2 \sum_{i=1}^{3} v_i^2, \tag{10}$$

where the u_i's and v_i's are real numbers.

(b) Show that equality holds in (10) if and only if at least one of the vectors (u_1, u_2, u_3) and (v_1, v_2, v_3) is a multiple of the other.

62. Use the dot product to find two unit vectors perpendicular to the vectors $(1, 2, 3)$ and $(-4, 1, 5)$.

63. Find two unit vectors perpendicular to $(-2, 0, 4)$ and $(3, -2, -1)$.

17.3 LINES IN $\mathbb{R}^3$

In Section 1.5 we derived the equation of a line in the plane. In that section we showed that the equation of the line could be determined if we knew either (i) two points on the line, or (ii) one point on the line and the direction (slope) of the line. Moreover, in Section 16.1 we showed how the equation of a line could be written parametrically. In $\mathbb{R}^3$ our intuition tells us that the basic ideas are the same. Since two points determine a line, we should be able to calculate the equation of a line in space if we know two points on it. Alternatively, if we know one point and the direction of a line, we should also be able to find its equation.

We begin with two points $P = (x_1, y_1, z_1)$ and $Q = (x_2, y_2, z_2)$ on a line L. A vector parallel to L is a vector with representation $\mathbf{v} = \overrightarrow{PQ}$, or [from formula (17.2.6)]

$$\mathbf{v} = (x_2 - x_1)\mathbf{i} + (y_2 - y_1)\mathbf{j} + (z_2 - z_1)\mathbf{k}. \tag{1}$$

Now let $R = (x, y, z)$ be another point on the line. Then $\overrightarrow{PR}$ is parallel to $\overrightarrow{PQ}$, which is parallel to $\mathbf{v}$, so that by Theorem 17.2.4(i),

$$\overrightarrow{PR} = t\mathbf{v} \tag{2}$$

for some real number t. Now look at Figure 1. From the figure we have (in each of the three possible cases)

$$\overrightarrow{OR} = \overrightarrow{OP} + \overrightarrow{PR}, \tag{3}$$

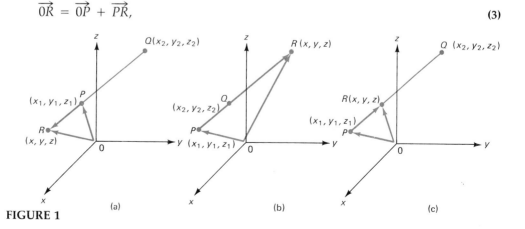

FIGURE 1

and combining (2) and (3)

$$\overrightarrow{PR} = \overrightarrow{OR} - \overrightarrow{OP} = t\mathbf{v}.$$

Thus

$$\overrightarrow{OR} = \overrightarrow{OP} + t\mathbf{v}, \tag{4}$$

or

$$(x, y, z) = (x_1, y_1, z_1) + t(x_2 - x_1, y_2 - y_1, z_2 - z_1). \tag{4a}$$

Equation (4) or (4a) is called the **vector equation** of the line L. For if R is on L, then (4) is satisfied for some real number t. Conversely, if (4) is satisfied, then reversing our steps, we see that $\overrightarrow{PR}$ is parallel to $\mathbf{v}$, which means that R is on L.

NOTE.   Since $\mathbf{v} = \overrightarrow{PQ} = \overrightarrow{OQ} - \overrightarrow{OP}$, (4) can be rewritten as

$$\overrightarrow{OR} = \overrightarrow{OP} + t(\overrightarrow{OQ} - \overrightarrow{OP}),$$

or

$$\overrightarrow{OR} = (1 - t)\overrightarrow{OP} + t\overrightarrow{OQ}. \tag{5}$$

The vector equation (5) is sometimes very useful.

If we write out the components of equation (4), we obtain

$$x\mathbf{i} + y\mathbf{j} + z\mathbf{k} = x_1\mathbf{i} + y_1\mathbf{j} + z_1\mathbf{k} + t(x_2 - x_1)\mathbf{i} + t(y_2 - y_1)\mathbf{j} + t(z_2 - z_1)\mathbf{k},$$

or

$$\begin{aligned} x &= x_1 + t(x_2 - x_1), \\ y &= y_1 + t(y_2 - y_1), \\ z &= z_1 + t(z_2 - z_1). \end{aligned} \tag{6}$$

The equations (6) are called the **parametric equations** of a line.

Finally, solving for t in (6), and defining $x_2 - x_1 = a$, $y_2 - y_1 = b$, and $z_2 - z_1 = c$, we find that

$$\frac{x - x_1}{a} = \frac{y - y_1}{b} = \frac{z - z_1}{c}. \tag{7}$$

The equations (7) are called the **symmetric equations** of the line.

Here a, b, and c are direction numbers of the vector $\mathbf{v}$. Of course, equations (7) are valid only if a, b, and c are nonzero.

EXAMPLE 1 Find a vector equation, parametric equations, and symmetric equations of the line L passing through the points $P = (2, -1, 6)$ and $Q = (3, 1, -2)$.

Solution. First, we calculate

$$\mathbf{v} = (3 - 2)\mathbf{i} + [1 - (-1)]\mathbf{j} + (-2 - 6)\mathbf{k} = \mathbf{i} + 2\mathbf{j} - 8\mathbf{k}.$$

Then from (4), if $R = (x, y, z)$ is on the line,

$$\overrightarrow{OR} = x\mathbf{i} + y\mathbf{j} + z\mathbf{k} = \overrightarrow{OP} + t\mathbf{v} = 2\mathbf{i} - \mathbf{j} + 6\mathbf{k} + t(\mathbf{i} + 2\mathbf{j} - 8\mathbf{k}),$$

or

$$x = 2 + t, \qquad y = -1 + 2t, \qquad z = 6 - 8t.$$

Finally, since $a = 1$, $b = 2$, and $c = -8$, we find the symmetric equations

$$\frac{x - 2}{1} = \frac{y + 1}{2} = \frac{z - 6}{-8}.$$

To check this, we verify that $(2, -1, 6)$ and $(3, 1, -2)$ are indeed on the line. We have

$$\frac{2 - 2}{1} = \frac{-1 + 1}{2} = \frac{6 - 6}{-8} = 0$$

and

$$\frac{3 - 2}{1} = \frac{1 + 1}{2} = \frac{-2 - 6}{-8} = 1.$$

Other points on the line can be found. For example, if $t = 3$, we obtain

$$3 = \frac{x - 2}{1} = \frac{y + 1}{2} = \frac{z - 6}{-8},$$

which yields the point $(5, 5, -18)$. ∎

EXAMPLE 2 Find symmetric equations of the line passing through the point $(1, -2, 4)$ and parallel to the vector $\mathbf{v} = \mathbf{i} + \mathbf{j} - \mathbf{k}$.

Solution. We simply choose $\overrightarrow{OP} = \mathbf{i} - 2\mathbf{j} + 4\mathbf{k}$ and $\mathbf{v}$ as above. Then $a = 1$, $b = 1$, $c = -1$, and we obtain

$$\frac{x - 1}{1} = \frac{y + 2}{1} = \frac{z - 4}{-1}. \quad ∎$$

What happens if one of the direction numbers a, b, or c is zero?

EXAMPLE 3 Find symmetric equations of the line containing the points $P = (3, 4, -1)$ and $Q = (-2, 4, 6)$.

Solution. Here $\mathbf{v} = -5\mathbf{i} + 7\mathbf{k}$, and $a = -5$, $b = 0$, $c = 7$. Then a parametric representation of the line is

$$x = 3 - 5t, \qquad y = 4, \qquad z = -1 + 7t.$$

Solving for t, we find that

$$\frac{x - 3}{-5} = \frac{z + 1}{7} \qquad \text{and} \qquad y = 4.$$

The equation $y = 4$ is the equation of a plane parallel to the xz-plane, so we have obtained an equation of a line in that plane. ■

EXAMPLE 4 Find symmetric equations of the line in the xy-plane that passes through the points $(x_1, y_1, 0)$ and $(x_2, y_2, 0)$.

Solution. Here $\mathbf{v} = (x_2 - x_1)\mathbf{i} + (y_2 - y_1)\mathbf{j}$, and we obtain

$$\frac{x - x_1}{x_2 - x_1} = \frac{y - y_1}{y_2 - y_1} \qquad \text{and} \qquad z = 0.$$

We can rewrite this as

$$y - y_1 = \left(\frac{y_2 - y_1}{x_2 - x_1}\right)(x - x_1).$$

Here $(y_2 - y_1)/(x_2 - x_1) = m$, the slope of the line, and when $x = 0$, $y = y_1 - [(y_2 - y_1)/(x_2 - x_1)]x_1 = b$, the y-intercept of the line. That is, $y = mx + b$, which is the slope-intercept form of a line in the xy-plane. Thus we see that the symmetric equations of a line in space are really a generalization of the equation of a line in the plane. ■

Now, what happens if two of the direction numbers are zero?

EXAMPLE 5 Find the symmetric equations of the line passing through the points $P = (2, 3, -2)$ and $Q = (2, -1, -2)$.

Solution. Here $\mathbf{v} = -4\mathbf{j}$, so that $a = 0$, $b = -4$, and $c = 0$. A parametric representation of the line is, by equations (6), given by

$$x = 2, \qquad y = 3 - 4t, \qquad z = -2.$$

Now $x = 2$ is the equation of a plane parallel to the yz-plane, while $z = -2$ is the equation of a plane parallel to the xy-plane. Their intersection is the line $x = 2$, $z = $

-2, which is parallel to the y-axis. In fact, the equation $y = 3 - 4t$, says, essentially, that y can take on any value (while x and z remain fixed). ◼

The results of Examples 1, 3, and 5 are summarized in Theorem 1.

Theorem 1.   Let L be a line passing through the point (x_1, y_1, z_1) and parallel to the vector $\mathbf{v} = a\mathbf{i} + b\mathbf{j} + c\mathbf{k}$. Then symmetric equations of the line are as follows:

(i) $$\frac{x - x_1}{a} = \frac{y - y_1}{b} = \frac{z - z_1}{c}$$

if a, b, and c are all nonzero.

(ii) $$x = x_1, \qquad \frac{y - y_1}{b} = \frac{z - z_1}{c}$$

if $a = 0$. Then the line is parallel to the yz-plane. If either b or $c = 0$, but $a \neq 0$, similar results hold.

(iii) $$x = x_1, \qquad y = y_1, \qquad z = z_1 + ct$$

if a and b are 0. Then the line is parallel to the z-axis. If a and c or b and c are 0, similar results hold.

◼   WARNING: The parametric or symmetric equations of a line are *not* unique. To see this, simply choose two other points on the line.

EXAMPLE 6   In Example 1, the line contains the point $(5, 5, -18)$. Choose $P = (5, 5, -18)$ and $Q = (3, 1, -2)$. We find that $\mathbf{v} = -2\mathbf{i} - 4\mathbf{j} + 16\mathbf{k}$, so that

$$x = 5 - 2t, \qquad y = 5 - 4t, \qquad z = -18 + 16t.$$

[Note that if $t = \frac{3}{2}$, we obtain $(x, y, z) = (2, -1, 6)$.] The symmetric equations are now

$$\frac{x - 5}{-2} = \frac{y - 5}{-4} = \frac{z + 18}{16}. \quad ◼$$

PROBLEMS 17.3

In Problems 1–22, find a vector equation, parametric equations, and symmetric equations of the indicated line.

1. Containing $(2, 1, 3)$ and $(1, 2, -1)$.
2. Containing $(1, -1, 1)$ and $(-1, 1, -1)$.
3. Containing $(1, 3, 2)$ and $(2, 4, -2)$.
4. Containing $(-2, 4, 5)$ and $(3, 7, 2)$.
5. Containing $(-4, 1, 3)$ and $(-4, 0, 1)$.
6. Containing $(2, 3, -4)$ and $(2, 0, -4)$.
7. Containing $(1, 2, 3)$ and $(3, 2, 1)$.
8. Containing $(7, 1, 3)$ and $(-1, -2, 3)$.
9. Containing $(1, 2, 4)$ and $(1, 2, 7)$.
10. Containing $(-3, -1, -6)$ and $(-3, 1, 6)$.
11. Containing $(2, 2, 1)$ and parallel to $2\mathbf{i} - \mathbf{j} - \mathbf{k}$.
12. Containing $(-1, -6, 2)$ and parallel to $4\mathbf{i} + \mathbf{j} - 3\mathbf{k}$.

13. Containing $(1, 0, 3)$ and parallel to $\mathbf{i} - \mathbf{j}$.
14. Containing $(2, 1, -4)$ and parallel to $\mathbf{i} + 4\mathbf{k}$.
15. Containing $(-1, -2, 5)$ and parallel to $-3\mathbf{j} + 7\mathbf{k}$.
16. Containing $(-2, 3, -2)$ and parallel to $4\mathbf{k}$.
17. Containing $(-1, -3, 1)$ and parallel to $-7\mathbf{j}$.
18. Containing $(2, 1, 5)$ and parallel to $3\mathbf{i}$.
19. Containing (a, b, c) and parallel to $d\mathbf{i} + e\mathbf{j}$, $d, e \neq 0$.
20. Containing (a, b, c) and parallel to $d\mathbf{k}$, $d \neq 0$.
21. Containing $(4, 1, -6)$ and parallel to $(x - 2)/3 = (y + 1)/6 = (z - 5)/2$.
22. Containing $(3, 1, -2)$ and parallel to $(x + 1)/3 = (y + 3)/2 = (z - 2)/(-4)$.

23. Let L_1 be given by

$$\frac{x - x_1}{a_1} = \frac{y - y_1}{b_1} = \frac{z - z_1}{c_1}$$

and L_2 be given by

$$\frac{x - x_2}{a_2} = \frac{y - y_2}{b_2} = \frac{z - z_2}{c_2}.$$

Show that a direction vector of L_1 is orthogonal to a direction vector of L_2 if and only if $a_1 a_2 + b_1 b_2 + c_1 c_2 = 0$.

24. Show that direction vectors of the lines

$$L_1: \frac{x - 3}{2} = \frac{y + 1}{4} = \frac{z - 2}{-1} \quad \text{and} \quad L_2: \frac{x - 3}{5} = \frac{y + 1}{-2} = \frac{z - 3}{2}$$

are orthogonal.

25. Show that the lines

$$L_1: \frac{x - 1}{1} = \frac{y + 3}{2} = \frac{z + 3}{3} \quad \text{and} \quad L_2: \frac{x - 3}{3} = \frac{y - 1}{6} = \frac{z - 3}{9}$$

are equations of the same straight line.

In the plane, two lines that are not parallel have exactly one point of intersection. In $\mathbb{R}^3$, this is not the case. For example, the lines $L_1: x = 2, y = 3$ (parallel to the z-axis) and $L_2: x = 1$, $z = 3$ (parallel to the y-axis) are not parallel and have no points in common. It takes a bit of work to determine whether two lines in $\mathbb{R}^3$ do have a point in common (they usually do not).

26. Determine whether the lines

$$L_1: \quad x = 1 + t, \quad y = -3 + 2t, \quad z = -2 - t$$

and

$$L_2: \quad x = 17 + 3s, \quad y = 4 + s, \quad z = -8 - s$$

have a point of intersection. [*Hint:* If (x, y, z) is a point common to both lines, then $x = 1 + t = 17 + 3s, y = -3 + 2t = 4 + s, z = -2 - t = -8 - s$. Find (if possible) numbers s and t that satisfy all three of these equations.

27. Determine whether the lines

$$x = 2 - t, \quad y = 1 + t, \quad z = -2 - t$$

and

$$x = 1 + s, \quad y = -2s, \quad z = 3 + 2s$$

have a point in common.

In Problems 28–33, determine whether the given pair of lines has a point of intersection. If so, find it.

28. $L_1: x = 2 + t, y = -1 + 2t, z = 3 + 4t; L_2: x = 9 + s, y = -2 - s, z = 1 - 2s$

29. $L_1: x = 3 + 2t, y = 2 - t, z = 1 + t; L_2: x = 4 - s, y = -2 + 3s, z = 2 + 2s$

30. $L_1: \dfrac{x - 4}{-3} = \dfrac{y - 1}{7} = \dfrac{z + 2}{-8}; L_2: \dfrac{x - 5}{1} = \dfrac{y - 3}{-1} = \dfrac{z - 1}{2}$

31. $L_1: \dfrac{x - 2}{-5} = \dfrac{y - 1}{1} = \dfrac{z - 3}{4}; L_2: \dfrac{x + 3}{4} = \dfrac{y - 2}{-1} = \dfrac{z - 7}{6}$

32. $L_1: x = 4 - t, y = 7 + 5t, z = 2 - 3t; L_2: x = 1 + 2s, y = 6 - 2s, z = 10 + 3s$

33. $L_1: x = 1 + t, y = 2 - t, z = 3t; L_2: x = 3s, y = 2 - s, z = 2 + s$

34. Let L be given in its vector form $\overrightarrow{OR} = \overrightarrow{OP} + t\mathbf{v}$. Find a number t such that $\overrightarrow{OR}$ is perpendicular to $\mathbf{v}$.

35. Use the result of Problem 34 to find the distance between the line L (containing P and parallel to $\mathbf{v}$) and the origin.
(a) $P = (2, 1, -4); \mathbf{v} = \mathbf{i} + \mathbf{j} + \mathbf{k}$
(b) $P = (1, 2, -3); \mathbf{v} = 3\mathbf{i} - \mathbf{j} - \mathbf{k}$
(c) $P = (-1, -4, 2); \mathbf{v} = -\mathbf{i} + \mathbf{j} + 2\mathbf{k}$

*36. Show that the lines $L_1: x = x_1 + a_1t, y = y_1 + b_1t, z = z_1 + c_1t$ and $L_2: x = x_2 + a_2s, y = y_2 + b_2s, z = z_2 + c_2s$ have a point in common or are parallel if and only if the determinant

$$\begin{vmatrix} a_1 & a_2 & x_1 - x_2 \\ b_1 & b_2 & y_1 - y_2 \\ c_1 & c_2 & z_1 - z_2 \end{vmatrix} = 0.$$

37. Apply the result of Problem 36 to the lines in Problem 28.

38. Do the same for the lines in Problem 29.

17.4 THE CROSS PRODUCT OF TWO VECTORS

To this point the only product of vectors we have considered has been the dot or scalar product. We now define a new product, called the *cross product*† (or *vector product*), which is defined only in $\mathbb{R}^3$.

Definition 1 CROSS PRODUCT Let $\mathbf{u} = a_1\mathbf{i} + b_1\mathbf{j} + c_1\mathbf{k}$ and $\mathbf{v} = a_2\mathbf{i} + b_2\mathbf{j} + c_2\mathbf{k}$. Then the **cross product (vector product)** of $\mathbf{u}$ and $\mathbf{v}$, denoted $\mathbf{u} \times \mathbf{v}$, is a new vector defined by

$$\mathbf{u} \times \mathbf{v} = (b_1c_2 - c_1b_2)\mathbf{i} + (c_1a_2 - a_1c_2)\mathbf{j} + (a_1b_2 - b_1a_2)\mathbf{k}. \tag{1}$$

Note that *the result of the cross product is a vector, while the result of the dot product is a scalar*.

Here the cross product seems to have been defined somewhat arbitrarily. There

---

†The cross product was defined by Hamilton in one of a series of papers discussing his quaternions, which were published in *Philosophical Magazine* between the years 1844 and 1850.

are obviously many ways to define a vector product. Why was this definition chosen? We will answer that question in this section by demonstrating some of the properties of the cross product and illustrating some of its uses.

EXAMPLE 1   Let $u = i - j + 2k$ and $v = 2i + 3j - 4k$. Calculate $w = u \times v$.

Solution. Using formula (1), we have

$$w = [(-1)(-4) - (2)(3)]i + [(2)(2) - (1)(-4)]j + [(1)(3) - (-1)(2)]k$$

$$= -2i + 8j + 5k. \blacksquare$$

NOTE.   In this example $u \cdot w = v \cdot w = 0$. That is, $u \times v$ is orthogonal to both u and v. As we will shortly see, the cross product of u and v is always orthogonal to both u and v.

Before continuing our discussion of the uses of the cross product, we remark that there is an easy way to remember how to calculate $u \times v$ if you are familiar with the elementary properties of 3×3 determinants. If you are not, we suggest that you turn to Appendix 3 where these properties are discussed.

Theorem 1.

$$u \times v = \begin{vmatrix} i & j & k \\ a_1 & b_1 & c_1 \\ a_2 & b_2 & c_2 \end{vmatrix}^\dagger$$

Proof.

$$\begin{vmatrix} i & j & k \\ a_1 & b_1 & c_1 \\ a_2 & b_2 & c_2 \end{vmatrix} = i \begin{vmatrix} b_1 & c_1 \\ b_2 & c_2 \end{vmatrix} - j \begin{vmatrix} a_1 & c_1 \\ a_2 & c_2 \end{vmatrix} + k \begin{vmatrix} a_1 & b_1 \\ a_2 & b_2 \end{vmatrix} \qquad (2)$$

$$= (b_1 c_2 - c_1 b_2)i + (a_2 c_1 - a_1 c_2)j + (a_1 b_2 - b_1 a_2)k,$$

which is equal to $u \times v$ according to Definition 1. $\blacksquare$

EXAMPLE 2   Calculate $u \times v$, where $u = 2i + 4j - 5k$ and $v = -3i - 2j + k$.

Solution.

$$u \times v = \begin{vmatrix} i & j & k \\ 2 & 4 & -5 \\ -3 & -2 & 1 \end{vmatrix} = (4 - 10)i - (2 - 15)j + (-4 + 12)k$$

$$= -6i + 13j + 8k. \blacksquare$$

The following theorem summarizes some properties of the cross product.

---

†The determinant is defined as a real number, not a vector. This use of the determinant notation is simply a convenient way to denote the cross product.

Theorem 2.   Let **u**, **v**, and **w** be vectors in $\mathbb{R}^3$, and let α be a scalar.

> **(i)** $\mathbf{u} \times \mathbf{0} = \mathbf{0} = \mathbf{0} \times \mathbf{u}$.
> **(ii)** $\mathbf{u} \times \mathbf{v} = -(\mathbf{v} \times \mathbf{u})$.
> **(iii)** $(\alpha\mathbf{u} \times \mathbf{v}) = \alpha(\mathbf{u} \times \mathbf{v})$.
> **(iv)** $\mathbf{u} \times (\mathbf{v} + \mathbf{w}) = (\mathbf{u} \times \mathbf{v}) + (\mathbf{u} \times \mathbf{w})$.
> **(v)** $(\mathbf{u} \times \mathbf{v}) \cdot \mathbf{w} = \mathbf{u} \cdot (\mathbf{v} \times \mathbf{w})$. (This product is called the **scalar triple product** of **u**, **v**, and **w**.)
> **(vi)** $\mathbf{u} \cdot (\mathbf{u} \times \mathbf{v}) = \mathbf{v} \cdot (\mathbf{u} \times \mathbf{v}) = 0$. (That is, $\mathbf{u} \times \mathbf{v}$ is orthogonal to both **u** and **v**.)
> **(vii)** If **u** and **v** are parallel, then $\mathbf{u} \times \mathbf{v} = \mathbf{0}$.

Proof.

(i) Let $\mathbf{u} = a_1\mathbf{i} + b_1\mathbf{j} + c_1\mathbf{k}$. Then

$$\mathbf{u} \times \mathbf{0} = \begin{vmatrix} \mathbf{i} & \mathbf{j} & \mathbf{k} \\ a_1 & b_1 & c_1 \\ 0 & 0 & 0 \end{vmatrix} = 0\mathbf{i} + 0\mathbf{j} + 0\mathbf{k} = \mathbf{0}.$$

Similarly, $\mathbf{0} \times \mathbf{u} = \mathbf{0}$.

(ii) Let $\mathbf{v} = a_2\mathbf{i} + b_2\mathbf{j} + c_2\mathbf{k}$. Then

$$\mathbf{u} \times \mathbf{v} = \begin{vmatrix} \mathbf{i} & \mathbf{j} & \mathbf{k} \\ a_1 & b_1 & c_1 \\ a_2 & b_2 & c_2 \end{vmatrix} = -\begin{vmatrix} \mathbf{i} & \mathbf{j} & \mathbf{k} \\ a_2 & b_2 & c_2 \\ a_1 & b_1 & c_1 \end{vmatrix} = -(\mathbf{v} \times \mathbf{u}),$$

since interchanging the rows of a determinant has the effect of multiplying the determinant by -1 [see Theorem 5(iii), Appendix 3].

(iii)

$$(\alpha\mathbf{u}) \times \mathbf{v} = \begin{vmatrix} \mathbf{i} & \mathbf{j} & \mathbf{k} \\ \alpha a_1 & \alpha b_1 & \alpha c_1 \\ a_2 & b_2 & c_2 \end{vmatrix} = \alpha\begin{vmatrix} \mathbf{i} & \mathbf{j} & \mathbf{k} \\ a_1 & b_1 & c_1 \\ a_2 & b_2 & c_2 \end{vmatrix} = \alpha(\mathbf{u} \times \mathbf{v})$$

The second equality follows from Theorem 5(iv) in Appendix 3.

(iv) Let $\mathbf{w} = a_3\mathbf{i} + b_3\mathbf{j} + c_3\mathbf{k}$. Then

$$\mathbf{u} \times (\mathbf{v} + \mathbf{w}) = \begin{vmatrix} \mathbf{i} & \mathbf{j} & \mathbf{k} \\ a_1 & b_1 & c_1 \\ a_2 + a_3 & b_2 + b_3 & c_2 + c_3 \end{vmatrix} = \begin{vmatrix} \mathbf{i} & \mathbf{j} & \mathbf{k} \\ a_1 & b_1 & c_1 \\ a_2 & b_2 & c_2 \end{vmatrix} + \begin{vmatrix} \mathbf{i} & \mathbf{j} & \mathbf{k} \\ a_1 & b_1 & c_1 \\ a_3 & b_3 & c_3 \end{vmatrix}$$

$$= (\mathbf{u} \times \mathbf{v}) + (\mathbf{u} \times \mathbf{w}).$$

The second equality is easily verified by direct calculation.

(v) $(\mathbf{u} \times \mathbf{v}) \cdot \mathbf{w} = [(b_1c_2 - c_1b_2)\mathbf{i} + (c_1a_2 - a_1c_2)\mathbf{j} + (a_1b_2 - b_1a_2)\mathbf{k}]$
$\cdot (a_3\mathbf{i} + b_3\mathbf{j} + c_3\mathbf{k})$
$= b_1c_2a_3 - c_1b_2a_3 + c_1a_2b_3 - a_1c_2b_3 + a_1b_2c_3 - b_1a_2c_3.$

We can easily show that $\mathbf{u} \cdot (\mathbf{v} \times \mathbf{w})$ is equal to the same expression (see Problem 35).

NOTE.   For an interesting geometric intrepretation of the scalar triple product, see Problem 41.

(vi) We know that $\mathbf{u} \cdot (\mathbf{u} \times \mathbf{v}) = (\mathbf{u} \times \mathbf{v}) \cdot \mathbf{u}$ [since the dot product is commutative—see Theorem 17.2.2(i)]. But from parts (ii) and (v) of this theorem,

$$(\mathbf{u} \times \mathbf{v}) \cdot \mathbf{u} = \mathbf{u} \cdot (\mathbf{v} \times \mathbf{u}) = \mathbf{u} \cdot (-\mathbf{u} \times \mathbf{v}) = -\mathbf{u} \cdot (\mathbf{u} \times \mathbf{v}).$$

Thus $\mathbf{u} \cdot (\mathbf{u} \times \mathbf{v}) = -\mathbf{u} \cdot (\mathbf{u} \times \mathbf{v})$, which can only occur if $\mathbf{u} \cdot (\mathbf{u} \times \mathbf{v}) = 0$. A similar computation shows that $\mathbf{v} \cdot (\mathbf{u} \times \mathbf{v}) = 0$.
(vii) If $\mathbf{u}$ and $\mathbf{v}$ are parallel, then $\mathbf{v} = \alpha\mathbf{u}$ for some scalar α [from Theorem 17.2.4(i)], so that

$$\mathbf{u} \times \mathbf{v} = \begin{vmatrix} \mathbf{i} & \mathbf{j} & \mathbf{k} \\ a_1 & b_1 & c_1 \\ \alpha a_1 & \alpha b_1 & \alpha c_1 \end{vmatrix} = 0$$

since the third row is a multiple of the second row [see Theorem 5(ii) in Appendix 3]. ∎

NOTE.   We could have proved this theorem without using determinants, but the proof would have involved many more computations.

What happens when we take cross products of the basis vectors $\mathbf{i}, \mathbf{j}, \mathbf{k}$? It is easy to verify the following:

$$\mathbf{i} \times \mathbf{i} = \mathbf{j} \times \mathbf{j} = \mathbf{k} \times \mathbf{k} = 0, \tag{3}$$
$$\mathbf{i} \times \mathbf{j} = \mathbf{k}, \quad \mathbf{k} \times \mathbf{i} = \mathbf{j}, \quad \mathbf{j} \times \mathbf{k} = \mathbf{i},$$
$$\mathbf{j} \times \mathbf{i} = -\mathbf{k}, \quad \mathbf{i} \times \mathbf{k} = -\mathbf{j}, \quad \mathbf{k} \times \mathbf{j} = -\mathbf{i}.$$

To remember these results, consider the circle in Figure 1. The cross product of two consecutive vectors in the clockwise direction is positive, while the cross product of two consecutive vectors in the counterclockwise direction is negative. Note that the formulas above show that the cross product is *not* associative, since, for example, $\mathbf{i} \times (\mathbf{i} \times \mathbf{j}) = \mathbf{i} \times \mathbf{k} = -\mathbf{j}$ while $(\mathbf{i} \times \mathbf{i}) \times \mathbf{j} = 0 \times \mathbf{j} = 0$, so that

FIGURE 1

$$\mathbf{i} \times (\mathbf{i} \times \mathbf{j}) \neq (\mathbf{i} \times \mathbf{i}) \times \mathbf{j}.$$

In general,

$$\mathbf{u} \times (\mathbf{v} \times \mathbf{w}) \neq (\mathbf{u} \times \mathbf{v}) \times \mathbf{w}.$$

EXAMPLE 3   Calculate $(3\mathbf{i} + 4\mathbf{k}) \times (2\mathbf{i} - 3\mathbf{j})$.

Solution. This is a good example of the usefulness of Theorem 2 and the formulas (3). We have

$$(3\mathbf{i} + 4\mathbf{k}) \times (2\mathbf{i} - 3\mathbf{j}) = (3 \cdot 2)(\mathbf{i} \times \mathbf{i}) + (4 \cdot 2)(\mathbf{k} \times \mathbf{i}) - (3 \cdot 3)(\mathbf{i} \times \mathbf{j})$$
$$+ 4(-3)(\mathbf{k} \times \mathbf{j})$$
$$= 0 + 8\mathbf{j} - 9\mathbf{k} + 12\mathbf{i} = 12\mathbf{i} + 8\mathbf{j} - 9\mathbf{k}. \quad \blacksquare$$

EXAMPLE 4   Find a line whose direction vector is orthogonal to the direction vectors of the lines $(x - 1)/3 = (y + 6)/4 = (z - 2)/-2$ and $(x + 2)/-3 = (y - 3)/4 = (z + 1)/1$ and that passes through the point $(2, -1, 1)$.

Solution. The directions of these lines are

$$\mathbf{v}_1 = 3\mathbf{i} + 4\mathbf{j} - 2\mathbf{k} \quad \text{and} \quad \mathbf{v}_2 = -3\mathbf{i} + 4\mathbf{j} + \mathbf{k}.$$

A vector orthogonal to these vectors is

$$\mathbf{w} = \mathbf{v}_1 \times \mathbf{v}_2 = \begin{vmatrix} \mathbf{i} & \mathbf{j} & \mathbf{k} \\ 3 & 4 & -2 \\ -3 & 4 & 1 \end{vmatrix} = 12\mathbf{i} + 3\mathbf{j} + 24\mathbf{k}.$$

Then symmetric equations of a line satisfying the requested conditions are given by

$$L_1: \quad \frac{x - 2}{12} = \frac{y + 1}{3} = \frac{z - 1}{24}.$$

NOTE. $\mathbf{w}_1 = \mathbf{v}_2 \times \mathbf{v}_1 = -(\mathbf{v}_1 \times \mathbf{v}_2)$ is also orthogonal to $\mathbf{v}_1$ and $\mathbf{v}_2$, so symmetric equations of another line are given by

$$L_2: \quad \frac{x - 2}{-12} = \frac{y + 1}{-3} = \frac{z - 1}{-24}.$$

However, L_1 and L_2 are really the same line. (Explain why.) $\blacksquare$

The preceding example leads to a basic question. We know that $\mathbf{u} \times \mathbf{v}$ is a vector orthogonal to $\mathbf{u}$ and $\mathbf{v}$. But there are always *two* unit vectors orthogonal to $\mathbf{u}$ and $\mathbf{v}$ (see Figure 2). The vectors $\mathbf{n}$ and $-\mathbf{n}$ ($\mathbf{n}$ stands for normal, of course) are both orthogonal to $\mathbf{u}$ and $\mathbf{v}$. Which one is in the direction of $\mathbf{u} \times \mathbf{v}$? The answer is given by the **right-hand rule.** If the right hand is placed so that the index finger points in the direction of $\mathbf{u}$ while the middle finger points in the direction of $\mathbf{v}$, then the thumb points in the direction of $\mathbf{u} \times \mathbf{v}$ (see Figure 3).

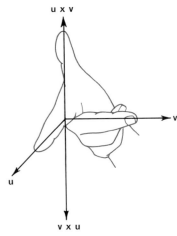

FIGURE 2   **FIGURE 3**

Having discussed the direction of the vector **u** × **v**, we now turn to a discussion of its magnitude.

Theorem 3.   If φ is the angle between **u** and **v**, then

$$|\mathbf{u} \times \mathbf{v}| = |\mathbf{u}||\mathbf{v}| \sin \varphi. \qquad (4)$$

Proof.   It is easy to show (by comparing components) that

$$|\mathbf{u} \times \mathbf{v}|^2 = |\mathbf{u}|^2|\mathbf{v}|^2 - (\mathbf{u} \cdot \mathbf{v})^2 \qquad (5)$$

(see Problem 36). Then since $(\mathbf{u} \cdot \mathbf{v})^2 = |\mathbf{u}|^2|\mathbf{v}|^2 \cos^2 \varphi$ (from Theorem 17.2.3),

$$|\mathbf{u} \times \mathbf{v}|^2 = |\mathbf{u}|^2|\mathbf{v}|^2 - |\mathbf{u}|^2|\mathbf{v}|^2 \cos^2 \varphi = |\mathbf{u}|^2|\mathbf{v}|^2(1 - \cos^2 \varphi)$$
$$= |\mathbf{u}|^2|\mathbf{v}|^2 \sin^2 \varphi,$$

and the theorem follows after taking square roots of both sides. ∎

There is an interesting geometric interpretation of Theorem 3. The vectors **u** and **v** are sketched in Figure 4 and can be thought of as two adjacent sides of a parallelogram. Then from elementary geometry we see that

$$\text{area of the parallelogram} = |\mathbf{u}||\mathbf{v}| \sin \varphi = |\mathbf{u} \times \mathbf{v}|. \qquad (6)$$

EXAMPLE 5   Find the area of a parallelogram with consecutive vertices at $P = (1, 3, -2)$, $Q = (2, 1, 4)$, and $R = (-3, 1, 6)$.

Solution.   One such parallelogram is sketched in Figure 5 (there are two others). We have

FIGURE 4

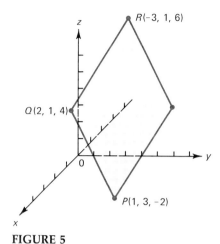

FIGURE 5

$$\text{area} = |\overrightarrow{PQ} \times \overrightarrow{QR}| = |(\mathbf{i} - 2\mathbf{j} + 6\mathbf{k}) \times (-5\mathbf{i} + 2\mathbf{k})|$$

$$= \begin{vmatrix} \mathbf{i} & \mathbf{j} & \mathbf{k} \\ 1 & -2 & 6 \\ -5 & 0 & 2 \end{vmatrix} = |-4\mathbf{i} - 32\mathbf{j} - 10\mathbf{k}| = \sqrt{1140} \text{ square units.} \blacksquare$$

PROBLEMS 17.4

In Problems 1–20, find the cross product $\mathbf{u} \times \mathbf{v}$.

1. $\mathbf{u} = \mathbf{i} - 2\mathbf{j}; \mathbf{v} = 3\mathbf{k}$
2. $\mathbf{u} = 3\mathbf{i} - 7\mathbf{j}; \mathbf{v} = \mathbf{i} + \mathbf{k}$
3. $\mathbf{u} = \mathbf{i} - \mathbf{j}; \mathbf{v} = \mathbf{j} + \mathbf{k}$
4. $\mathbf{u} = -7\mathbf{k}; \mathbf{v} = \mathbf{j} + 2\mathbf{k}$
5. $\mathbf{u} = -2\mathbf{i} + 3\mathbf{j}; \mathbf{v} = 7\mathbf{i} + 4\mathbf{k}$
6. $\mathbf{u} = a\mathbf{i} + b\mathbf{j}; \mathbf{v} = c\mathbf{i} + d\mathbf{j}$
7. $\mathbf{u} = a\mathbf{i} + b\mathbf{k}; \mathbf{v} = c\mathbf{i} + d\mathbf{k}$
8. $\mathbf{u} = a\mathbf{j} + b\mathbf{k}; \mathbf{v} = c\mathbf{i} + d\mathbf{k}$
9. $\mathbf{u} = 2\mathbf{i} - 3\mathbf{j} + \mathbf{k}; \mathbf{v} = \mathbf{i} + 2\mathbf{j} + \mathbf{k}$
10. $\mathbf{u} = 3\mathbf{i} - 4\mathbf{j} + 2\mathbf{k}; \mathbf{v} = 6\mathbf{i} - 3\mathbf{j} + 5\mathbf{k}$
11. $\mathbf{u} = -3\mathbf{i} - 2\mathbf{j} + \mathbf{k}; \mathbf{v} = 6\mathbf{i} + 4\mathbf{j} - 2\mathbf{k}$
12. $\mathbf{u} = \mathbf{i} + 7\mathbf{j} - 3\mathbf{k}; \mathbf{v} = -\mathbf{i} - 7\mathbf{j} + 3\mathbf{k}$
13. $\mathbf{u} = \mathbf{i} - 7\mathbf{j} - 3\mathbf{k}; \mathbf{v} = -\mathbf{i} + 7\mathbf{j} - 3\mathbf{k}$
14. $\mathbf{u} = 2\mathbf{i} - 3\mathbf{j} + 5\mathbf{k}; \mathbf{v} = 3\mathbf{i} - \mathbf{j} - \mathbf{k}$
15. $\mathbf{u} = 10\mathbf{i} + 7\mathbf{j} - 3\mathbf{k}; \mathbf{v} = -3\mathbf{i} + 4\mathbf{j} - 3\mathbf{k}$
16. $\mathbf{u} = 2\mathbf{i} + 4\mathbf{j} - 6\mathbf{k}; \mathbf{v} = -\mathbf{i} - \mathbf{j} + 3\mathbf{k}$
17. $\mathbf{u} = 2\mathbf{i} - \mathbf{j} + \mathbf{k}; \mathbf{v} = 4\mathbf{i} + 2\mathbf{j} + 2\mathbf{k}$
18. $\mathbf{u} = 3\mathbf{i} - \mathbf{j} + 8\mathbf{k}; \mathbf{v} = \mathbf{i} + \mathbf{j} - 4\mathbf{k}$
19. $\mathbf{u} = a\mathbf{i} + a\mathbf{j} + a\mathbf{k}; \mathbf{v} = b\mathbf{i} + b\mathbf{j} + b\mathbf{k}$
20. $\mathbf{u} = a\mathbf{i} + b\mathbf{j} + c\mathbf{k}; \mathbf{v} = a\mathbf{i} + b\mathbf{j} - c\mathbf{k}$

21. Find two unit vectors orthogonal to both $\mathbf{u} = 2\mathbf{i} - 3\mathbf{j}$ and $\mathbf{v} = 4\mathbf{j} + 3\mathbf{k}$.
22. Find two unit vectors orthogonal to both $\mathbf{u} = \mathbf{i} + \mathbf{j} + \mathbf{k}$ and $\mathbf{v} = \mathbf{i} - \mathbf{j} - \mathbf{k}$.
23. Use the cross product to find the sine of the angle φ between the vectors $\mathbf{u} = 2\mathbf{i} + \mathbf{j} - \mathbf{k}$ and $\mathbf{v} = -3\mathbf{i} - 2\mathbf{j} + 4\mathbf{k}$.
24. Use the dot product to calculate the cosine of the angle between the vectors of Problem 23. Then show that for the values you have calculated, $\sin^2 \varphi + \cos^2 \varphi = 1$.

In Problems 25–28, find a line whose direction vector is orthogonal to the direction vectors of the two given lines and that passes through the given point.

25. $\dfrac{x + 2}{-3} = \dfrac{y - 1}{4} = \dfrac{z}{-5}; \dfrac{x - 3}{7} = \dfrac{y + 2}{-2} = \dfrac{z - 8}{3}; (1, -3, 2)$

26. $\dfrac{x-2}{-4} = \dfrac{y+3}{-7} = \dfrac{z+1}{3}; \dfrac{x+2}{3} = \dfrac{y-5}{-4} = \dfrac{z+3}{-2}; (-4, 7, 3)$

27. $x = 3 - 2t, y = 4 + 3t, z = -7 + 5t; x = -2 + 4s, y = 3 - 2s, z = 3 + s; (-2, 3, 4)$
28. $x = 4 + 10t, y = -4 - 8t, z = 3 + 7t; x = -2t, y = 1 + 4t, z = -7 - 3t; (4, 6, 0)$

In Problems 29–34, find the area of a parallelogram with the given adjacent vertices.

29. $(1, -2, 3); (2, 0, 1); (0, 4, 0)$ 30. $(-2, 1, 1); (2, 2, 3); (-1, -2, 4)$
31. $(-2, 1, 0); (1, 4, 2); (-3, 1, 5)$ 32. $(7, -2, -3); (-4, 1, 6); (5, -2, 3)$
33. $(a, 0, 0); (0, b, 0); (0, 0, c)$ 34. $(a, b, 0); (a, 0, b); (0, a, b)$

35. Show that if $\mathbf{u} = (a_1, b_1, c_1)$, $\mathbf{v} = (a_2, b_2, c_2)$, and $\mathbf{w} = (a_3, b_3, c_3)$, then

$$\mathbf{u} \cdot (\mathbf{v} \times \mathbf{w}) = \begin{vmatrix} a_1 & b_1 & c_1 \\ a_2 & b_2 & c_2 \\ a_3 & b_3 & c_3 \end{vmatrix}.$$

36. Show that $|\mathbf{u} \times \mathbf{v}|^2 = |\mathbf{u}|^2|\mathbf{v}|^2 - (\mathbf{u} \cdot \mathbf{v})^2$. [*Hint:* Write out in terms of components.]
37. Show that the area of the triangle PQR is given by $A = \frac{1}{2}|\overrightarrow{PQ} \times \overrightarrow{QR}|$.
38. Use the result of Problem 37 to calculate the area of the triangle with vertices at $(2, 1, -4)$, $(1, 7, 2)$, and $(3, -2, 3)$.
39. Calculate the area of the triangle with vertices at $(3, 1, 7)$, $(2, -3, 4)$, and $(7, -2, 4)$.
40. Calculate the area of the triangle with vertices at $(1, 0, 0)$, $(0, 1, 0)$, and $(0, 0, 1)$. Sketch this triangle.
*41. Let $\mathbf{u}$, $\mathbf{v}$, and $\mathbf{w}$ be three vectors that are not in the same plane. Then they form the sides of a *parallelepiped* in space (see Figure 6). Prove that the volume of the parallelepiped is given by $V = |(\mathbf{u} \times \mathbf{v}) \cdot \mathbf{w}|$. [*Hint:* The area of the base is $|\mathbf{u} \times \mathbf{v}|$.]

FIGURE 6

42. Calculate the volume of the parallelepiped determined by the vectors $\mathbf{u} = 2\mathbf{i} - \mathbf{j} + \mathbf{k}$, $\mathbf{v} = 3\mathbf{i} + 2\mathbf{j} - 2\mathbf{k}$, $\mathbf{w} = 3\mathbf{i} + 2\mathbf{j}$.
43. Calculate the volume of the parallelepiped determined by the vectors $\mathbf{i} - \mathbf{j}$, $3\mathbf{i} + 2\mathbf{k}$, $-7\mathbf{j} + 3\mathbf{k}$.
44. Calculate the volume of the parallelepiped determined by the vectors $\overrightarrow{PQ}$, $\overrightarrow{PR}$, and $\overrightarrow{PS}$, where $P = (2, 1, -1)$, $Q = (-3, 1, 4)$, $R = (-1, 0, 2)$, and $S = (-3, -1, 5)$.
*45. Calculate the distance between the lines

$$L_1: \dfrac{x-2}{3} = \dfrac{y-5}{2} = \dfrac{z-1}{-1} \quad \text{and} \quad L_2: \dfrac{x-4}{-4} = \dfrac{y-5}{4} = \dfrac{z+2}{1}.$$

[*Hint:* The distance is measured along a vector $\mathbf{v}$ that is perpendicular to both L_1 and L_2. Let P be a point on L_1 and Q a point on L_2. Then the length of the projection of $\overrightarrow{PQ}$ on $\mathbf{v}$ is the distance between the lines, measured along a vector that is perpendicular to them both. See Figure 7.]

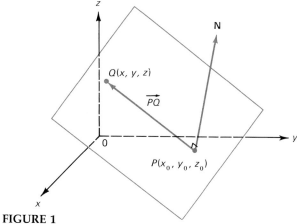

FIGURE 7

*46. Find the distance between the lines

$$L_1: \quad \frac{x+2}{3} = \frac{y-7}{-4} = \frac{z-2}{2} \quad \text{and} \quad L_2: \quad \frac{x-1}{-3} = \frac{y+2}{4} = \frac{z+1}{1}.$$

*47. Find the distance between the lines $x = 2 - 3t$, $y = 1 + 2t$, $z = -2 - t$, and $x = 1 + 4s$, $y = -2 - s$, $z = 3 + s$.

*48. Find the distance between the lines $x = -2 - 5t$, $y = -3 - 2t$, $z = 1 + 4t$, and $x = 2 + 3s$, $y = -1 + s$, $z = 3s$.

17.5 PLANES

In Section 17.3 we derived the equation of a line in space by specifying a point on the line and a vector *parallel* to this line. We can derive the equation of a plane in space by specifying a point in the plane and a vector orthogonal to every vector in the plane. This orthogonal vector is called a **normal vector** and is denoted by **N**. (See Figure 1.)

FIGURE 1

Definition 1 PLANE Let P be a point in space and let $\mathbf{N}$ be a given nonzero vector. Then the set of all points Q for which $\overrightarrow{PQ}$ and $\mathbf{N}$ are orthogonal constitutes a **plane** in $\mathbb{R}^3$.

NOTATION: We will usually denote a plane by the symbol Π.

Let $P = (x_0, y_0, z_0)$ and $\mathbf{N} = a\mathbf{i} + b\mathbf{j} + c\mathbf{k}$. Then if $Q = (x, y, z)$,

$$\overrightarrow{PQ} = (x - x_0)\mathbf{i} + (y - y_0)\mathbf{j} + (z - z_0)\mathbf{k}.$$

If $\overrightarrow{PQ} \perp \mathbf{N}$, then $\overrightarrow{PQ} \cdot \mathbf{N} = 0$. But this implies that

$$a(x - x_0) + b(y - y_0) + c(z - z_0) = 0. \tag{1}$$

A more common way to write the equation of a plane is easily derived from (1):

$$ax + by + cz = d, $$

where

$$d = ax_0 + by_0 + cz_0 = \overrightarrow{OP} \cdot \mathbf{N}. \tag{2}$$

EXAMPLE 1 Find an equation of the plane Π passing through the point $(2, 5, 1)$ and normal to the vector $\mathbf{N} = \mathbf{i} - 2\mathbf{j} + 3\mathbf{k}$.

Solution. From (1) we immediately obtain

$$(x - 2) - 2(y - 5) + 3(z - 1) = 0,$$

or

$$x - 2y + 3z = -5. \tag{3}$$

This plane is sketched in Figure 2. ∎

REMARK. The plane can be sketched by setting $x = y = 0$ in (3) to obtain $(0, 0, -\frac{5}{3})$, $x = z = 0$ to obtain $(0, \frac{5}{2}, 0)$, and $y = z = 0$ to obtain $(-5, 0, 0)$. These three points all lie on the plane.

The three coordinate planes are easily represented as follows:

(i) The xy-plane. This plane passes through the origin $(0, 0, 0)$ and any vector lying along the z-axis is normal to it. The simplest such vector is $\mathbf{k}$. Thus from (1) we obtain

$$0(x - 0) + 0(y - 0) + 1(z - 0) = 0,$$

which yields

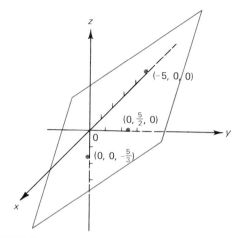

FIGURE 2

$$z = 0 \tag{4}$$

as the equation of the xy-plane. (This result should not be very surprising.)

REMARK.   The equation $z = 0$ is really a shorthand notation for the equation of the xy-plane. The full notation is

the xy-plane $= \{(x, y, z) : z = 0\}$.

The shorthand notation is fine as long as we don't lose sight of the fact that we are in $\mathbb{R}^3$.

(ii) **The xz-plane** has the equation

$$y = 0. \tag{5}$$

(iii) **The yz-plane** has the equation

$$x = 0. \tag{6}$$

Three points that are not collinear determine a plane since they determine two nonparallel vectors that intersect at a point.

EXAMPLE 2   Find an equation of the plane Π passing through the points $P = (1, 2, 1)$, $Q = (-2, 3, -1)$, and $R = (1, 0, 4)$.

Solution. The vectors $\overrightarrow{PQ} = -3\mathbf{i} + \mathbf{j} - 2\mathbf{k}$ and $\overrightarrow{QR} = 3\mathbf{i} - 3\mathbf{j} + 5\mathbf{k}$ lie on the plane and are therefore orthogonal to the normal vector. Thus

$$\mathbf{N} = \overrightarrow{PQ} \times \overrightarrow{QR} = \begin{vmatrix} \mathbf{i} & \mathbf{j} & \mathbf{k} \\ -3 & 1 & -2 \\ 3 & -3 & 5 \end{vmatrix} = -\mathbf{i} + 9\mathbf{j} + 6\mathbf{k},$$

and we obtain

$$\text{II:} \quad -(x - 1) + 9(y - 2) + 6(z - 1) = 0,$$

or

$$-x + 9y + 6z = 23.$$

Note that if we choose another point, say Q, we get the equation

$$-(x + 2) + 9(y - 3) + 6(z + 1) = 0,$$

which reduces to

$$-x + 9y + 6z = 23.$$

The plane is sketched in Figure 3. ■

Definition 2 PARALLEL PLANES Two planes are **parallel** if their normal vectors are parallel.

Two parallel planes are drawn in Figure 4.

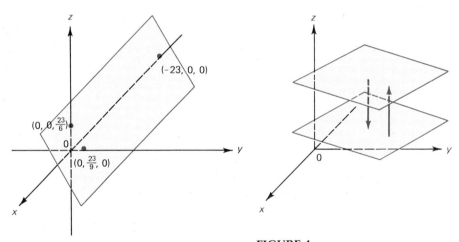

FIGURE 3   FIGURE 4

EXAMPLE 3   The planes II_1: $2x + 3y - z = 3$ and II_2: $-4x - 6y + 2z = 8$ are parallel since $\mathbf{N}_1 = 2\mathbf{i} + 3\mathbf{j} - \mathbf{k}$, $\mathbf{N}_2 = -4\mathbf{i} - 6\mathbf{j} + 2\mathbf{k}$, and $\mathbf{N}_2 = -2\mathbf{N}_1$ (and $\mathbf{N}_1 \times \mathbf{N}_2 = 0$). ■

If two distinct planes are not parallel, then they intersect in a straight line.

EXAMPLE 4   Find all points of intersection of the planes $2x - y - z = 3$ and $x + 2y + 3z = 7$.

Solution. When the planes intersect, we have

$$2x - y - z = 3$$

and

$$x + 2y + 3z = 7.$$

Multiplying the first equation by 2 and adding it to the second, we obtain

$$
\begin{array}{r}
4x - 2y - 2z = 6 \\
\underline{x + 2y + 3z = 7} \\
5x \qquad\quad + z = 13
\end{array}
$$

or $z = -5x + 13$. Then from the first equation

$$y = 2x - z - 3 = 2x - (-5x + 13) - 3 = 7x - 16.$$

Then setting $x = t$, we obtain the parametric representation of the line of intersection:

$$x = t, \qquad y = -16 + 7t, \qquad z = 13 - 5t.$$

This line is sketched in Figure 5. Note that this line is orthogonal to both normal vectors. ◾

We conclude this section by indicating how the distance from a plane to a point can be calculated. Look at Figure 6. If Q is the point, then the required distance is the distance measured along a line orthogonal to Π. That is, the shortest distance is obtained by "dropping a perpendicular" from the point to the plane, which is done by calculating (for any point P on the plane)

$$D = |\text{Proj}_{\mathbf{N}}\, \overrightarrow{PQ}| = \frac{|\overrightarrow{PQ} \cdot \mathbf{N}|}{|\mathbf{N}|}. \tag{7}$$

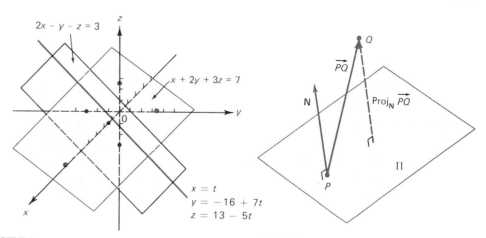

FIGURE 5   **FIGURE 6**

EXAMPLE 5 Find the distance D between the plane $2x - y + 3z = 6$ and the point $Q = (3, 5, -7)$.

Solution. One point on the plane is $P = (3, 0, 0)$ and $\mathbf{N} = 2\mathbf{i} - \mathbf{j} + 3\mathbf{k}$. Then $\overrightarrow{PQ} = 5\mathbf{j} - 7\mathbf{k}$, $|\overrightarrow{PQ} \cdot \mathbf{N}| = 26$ and $|\mathbf{N}| = \sqrt{14}$, so that

$$D = \frac{26}{\sqrt{14}}. \quad \blacksquare$$

EXAMPLE 6 Find the distance from the plane $ax + by + cz = d$ to the origin.

Solution. If $a \neq 0$, one point on the plane is $P = (d/a, 0, 0)$. [If $a = 0$ but $b \neq 0$, a point on the plane is $(0, d/b, 0)$, leading to the same result.] Then $\mathbf{N} = a\mathbf{i} + b\mathbf{j} + c\mathbf{k}$ and $\overrightarrow{PO} = (d/a)\mathbf{i}$, so that $|\overrightarrow{OP} \cdot \mathbf{N}| = |d|$, $\mathbf{N} = \sqrt{a^2 + b^2 + c^2}$, and

$$D = \frac{|d|}{\sqrt{a^2 + b^2 + c^2}}. \quad \blacksquare$$

PROBLEMS 17.5

In Problems 1–16, find the equation of the plane and sketch it.

1. $P = (0, 0, 0)$; $\mathbf{N} = \mathbf{i}$
2. $P = (0, 0, 0)$; $\mathbf{N} = \mathbf{j}$
3. $P = (0, 0, 0)$; $\mathbf{N} = \mathbf{k}$
4. $P = (1, 2, 3)$; $\mathbf{N} = \mathbf{i} + \mathbf{j}$
5. $P = (1, 2, 3)$; $\mathbf{N} = \mathbf{i} + \mathbf{k}$
6. $P = (1, 2, 3)$; $\mathbf{N} = \mathbf{j} + \mathbf{k}$
7. $P = (2, -1, 6)$; $\mathbf{N} = 3\mathbf{i} - \mathbf{j} + 2\mathbf{k}$
8. $P = (-4, -7, 5)$; $\mathbf{N} = -3\mathbf{i} - 4\mathbf{j} + \mathbf{k}$
9. $P = (-3, 11, 2)$; $\mathbf{N} = 4\mathbf{i} + \mathbf{j} - 7\mathbf{k}$
10. $P = (3, -2, 5)$; $\mathbf{N} = 2\mathbf{i} - 7\mathbf{j} - 8\mathbf{k}$
11. $P = (4, -7, -3)$; $\mathbf{N} = -\mathbf{i} - \mathbf{j} - \mathbf{k}$
12. $P = (8, 1, 0)$; $\mathbf{N} = -7\mathbf{i} + \mathbf{j} + 2\mathbf{k}$
13. containing $(1, 2, -4)$, $(2, 3, 7)$, and $(4, -1, 3)$
14. containing $(-7, 1, 0)$, $(2, -1, 3)$, and $(4, 1, 6)$
15. containing $(1, 0, 0)$, $(0, 1, 0)$, and $(0, 0, 1)$
16. containing $(2, 3, -2)$, $(4, -1, -1)$, and $(3, 1, 2)$

Two planes are **orthogonal** if their normal vectors are orthogonal. In Problems 17–23, determine whether the given planes are parallel, orthogonal, coincident (i.e., the same), or none of these.

17. $\Pi_1: x + y + z = 2$; $\Pi_2: 2x + 2y + 2z = 4$
18. $\Pi_1: x - y + z = 3$; $\Pi_2: -3x + 3y - 3z = -9$
19. $\Pi_1: 2x - y + z = 3$; $\Pi_2: x + y - z = 7$
20. $\Pi_1: 2x - y + z = 3$; $\Pi_2: x + y + z = 3$
21. $\Pi_1: 3x - 2y + 7z = 4$; $\Pi_2: -2x + 4y + 2z = 16$
22. $\Pi_1: -4x + 4y - 6z = 7$; $\Pi_2: 2x - 2y + 3z = -3$
23. $\Pi_1: -4x + 4y - 6z = 6$; $\Pi_2: 2x - 2y + 3z = -3$

In Problems 24–26, find the equation of the set of all points of intersection of the two planes.

24. $\Pi_1: x - y + z = 2$; $\Pi_2: 2x - 3y + 4z = 7$
25. $\Pi_1: 3x - y + 4z = 3$; $\Pi_2: -4x - 2y + 7z = 8$
26. $\Pi_1: -2x - y + 17z = 4$; $\Pi_2: 2x - y - z = -7$

In Problems 27–30, find the distance from the given point to the given plane.

27. $(2, -1, 4)$; $3x - y + 7z = 2$
28. $(4, 0, 1)$; $2x - y + 8z = 3$
29. $(-7, -2, -1)$; $-2x + 8z = -5$
30. $(-3, 0, 2)$; $-3x + y + 5z = 0$

31. Prove that the distance between the plane $ax + by + cz = d$ and the point (x_0, y_0, z_0) is given by

$$D = \frac{|ax_0 + by_0 + cz_0 - d|}{\sqrt{a^2 + b^2 + c^2}}.$$

The *angle between two planes* is defined to be the acute angle between their normal vectors. In Problems 32–34, find the angle between the two planes described in the indicated problem.

32. Problem 24   **33.** Problem 25   **34.** Problem 26

*35. Let **u** and **v** be two nonparallel vectors in a plane Π. Show that if **w** is any other vector in Π, then there exist scalars α and β such that

$$\mathbf{w} = \alpha\mathbf{u} + \beta\mathbf{v}.$$

This expression is called the **parametric representation** of the plane Π. [*Hint:* Draw a parallelogram in which $\alpha\mathbf{u}$ and $\beta\mathbf{v}$ form adjacent sides and the diagonal vector is **w**.]

36. Three vectors **u**, **v**, and **w** are called **coplanar** if they all lie in the same plane Π. Show that if **u**, **v**, and **w** all pass through the origin, then they are coplanar if and only if the scalar triple product equals zero:

$$\mathbf{u} \cdot (\mathbf{v} \times \mathbf{w}) = 0.$$

In Problems 37–41, determine whether the three given position vectors (i.e., one endpoint at the origin) are coplanar. If they are coplanar, find the equation of the plane containing them.

37. $\mathbf{u} = 2\mathbf{i} - 3\mathbf{j} + 4\mathbf{k}$; $\mathbf{v} = 7\mathbf{i} - 2\mathbf{j} + 3\mathbf{k}$; $\mathbf{w} = 9\mathbf{i} - 5\mathbf{j} + 7\mathbf{k}$
38. $\mathbf{u} = -3\mathbf{i} + \mathbf{j} + 8\mathbf{k}$; $\mathbf{v} = -2\mathbf{i} - 3\mathbf{j} + 5\mathbf{k}$; $\mathbf{w} = 2\mathbf{i} + 14\mathbf{j} - 4\mathbf{k}$
39. $\mathbf{u} = 2\mathbf{i} + \mathbf{j} - 2\mathbf{k}$; $\mathbf{v} = 2\mathbf{i} - \mathbf{j} - 2\mathbf{k}$; $\mathbf{w} = 2\mathbf{i} - \mathbf{j} + 2\mathbf{k}$
40. $\mathbf{u} = 3\mathbf{i} - 2\mathbf{j} + \mathbf{k}$; $\mathbf{v} = \mathbf{i} + \mathbf{j} - 5\mathbf{k}$; $\mathbf{w} = -\mathbf{i} + 5\mathbf{j} - 16\mathbf{k}$
41. $\mathbf{u} = 2\mathbf{i} - \mathbf{j} - \mathbf{k}$; $\mathbf{v} = 4\mathbf{i} + 3\mathbf{j} + 2\mathbf{k}$; $\mathbf{w} = 6\mathbf{i} + 7\mathbf{j} + 5\mathbf{k}$

*42. Let $P(x_1, y_1, z_1)$, $Q(x_2, y_2, z_2)$, and $R(x_3, y_3, z_3)$ be three points in $\mathbb{R}^3$ that are not collinear. Show that an equation of the plane passing through the three points is

$$\begin{vmatrix} x & y & z & 1 \\ x_1 & y_1 & z_1 & 1 \\ x_2 & y_2 & z_2 & 1 \\ x_3 & y_3 & z_3 & 1 \end{vmatrix} = 0.$$

43. Let $a_1x + b_1y + c_1z = d_1$, $a_2x + b_2y + c_2z = d_2$, and $a_3x + b_3y + c_3z = d_3$ be equations of three planes. Show that the planes have a unique point in common if

$$\begin{vmatrix} a_1 & b_1 & c_1 \\ a_2 & b_2 & c_2 \\ a_3 & b_3 & c_3 \end{vmatrix} \neq 0.$$

17.6 QUADRIC SURFACES

A **surface** in space is defined as the set of points in $\mathbb{R}^3$ satisfying the equation $F(x, y, z) = 0$. For example, the equation

$$F(x, y, z) = x^2 + y^2 + z^2 - 1 = 0 \tag{1}$$

is the equation of the unit sphere, as we saw in Section 17.1. In this section we will take a brief look at some of the most commonly encountered surfaces in $\mathbb{R}^3$. We will take a more detailed look at general surfaces in $\mathbb{R}^3$ in Chapter 18.

Having already discussed the sphere, we turn our attention to the cylinder.

Definition 1   CYLINDER   Let a line L and a plane curve C be given. A **cylinder** is the surface generated when a line parallel to L moves around C, remaining parallel to L.

The line L is called the **generatrix** of the cylinder, and the curve C is called its **directrix**.

EXAMPLE 1   Let L be the z-axis and C the circle $x^2 + y^2 = a^2$ in the xy-plane. Sketch the cylinder.

Solution. As we move a line along the circle $x^2 + y^2 = a^2$ and parallel to the z-axis, we obtain the **right circular cylinder** $x^2 + y^2 = a^2$ sketched in Figure 1a. A computer-drawn sketch of the right circular cylinder $x^2 + y^2 = 4$ is given in Figure 1b. ■

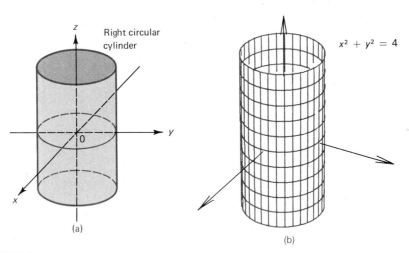

FIGURE 1

REMARK. We can write the equation $x^2 + y^2 = 4$ as

$$x^2 + y^2 + 0 \cdot z^2 = 4.$$

This illustrates that we are talking about a cylinder in $\mathbb{R}^3$ rather than the circle $x^2 + y^2 = 4$ in $\mathbb{R}^2$.

EXAMPLE 2   Suppose L is the x-axis and C is given by $y = z^2$. Sketch the resulting cylinder.

Solution. The curve $y = z^2$ is a parabola in the yz-plane. As we move along it, parallel to the x-axis, we obtain the **parabolic cylinder** sketched in Figure 2. ■

EXAMPLE 3   Suppose L is the x-axis and C is given by $y^2 + (z^2/4) = 1$. Sketch the resulting cylinder.

Solution. $y^2 + (z^2/4) = 1$ is an ellipse in the yz-plane. As we move along it parallel to the x-axis, we obtain the **elliptic cylinder** sketched in Figure 3. ■

As we saw from the discussion in Chapter 10, a second-degree equation in the variables x and y forms a circle, parabola, ellipse, or hyperbola (or a degenerate form of one of these such as a single point or a straight line or a pair of straight lines) in the plane. In $\mathbb{R}^3$ we have the following definition.

Definition 2   QUADRIC SURFACE   A **quadric surface** in $\mathbb{R}^3$ is the graph of a second-degree equation in the variables x, y, and z. Such an equation takes the form

FIGURE 2

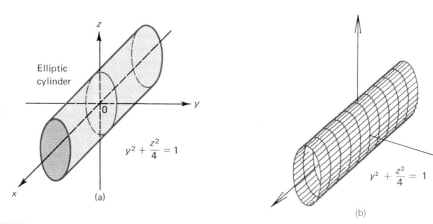

FIGURE 3

$$Ax^2 + By^2 + Cz^2 + Dxy + Exz + Fyz + Gx + Hy + Jz + K = 0. \tag{2}$$

We have already seen sketches of several quadric surfaces. We list below the standard forms† of eleven types of nondegenerate‡ quadric surfaces. Although we will not prove this result here, any quadric surface can be written in one of these forms by a translation or rotation of the coordinate axes. Here is the list:

1. sphere
2. right circular cylinder
3. parabolic cylinder
4. elliptic cylinder
5. hyperbolic cylinder
6. ellipsoid
7. hyperboloid of one sheet

---

†A standard form is one in which the surface has its center or vertex at the origin and its axes parallel to the coordinate axes.

‡By degenerate we again mean a surface whose graph contains a finite number of points (or none) or consists of a pair of planes. We saw examples of degenerate spheres (zero or one point) in Section 17.1.

8. hyperboloid of two sheets
9. elliptic paraboloid
10. hyperbolic paraboloid
11. elliptic cone

We have already seen sketches of surfaces 1, 2, 3, and 4.

5. **The hyperbolic cylinder:** $(y^2/b^2) - (x^2/a^2) = 1$. This is the equation of a hyperbola in the xy-plane. See Figure 4a. A computer-drawn sketch of $(y^2/4) - (x^2/9) = 1$ is given in Figure 4b.

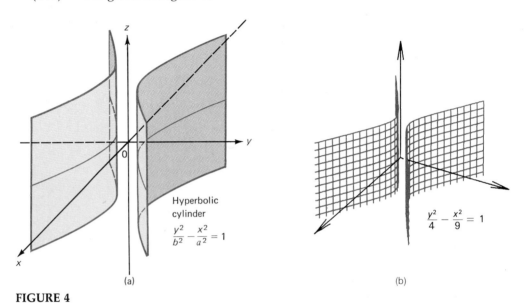

(a)   (b)

FIGURE 4

We can best describe the remaining six surfaces by looking at **cross sections** parallel to a given coordinate plane. For example, in the unit sphere $x^2 + y^2 + z^2 = 1$, cross sections parallel to the xy-plane are circles. To see this, let $z = c$, where $-1 < c < 1$ (this is a plane parallel to the xy-plane). Then

$$x^2 + y^2 = 1 - c^2,$$

which is the equation of a circle in the xy-plane.

6. **The ellipsoid:** $(x^2/a^2) + (y^2/b^2) + (z^2/c^2) = 1$. This surface is sketched in Figure 5a. It is a closed "watermelon-shaped" surface. Cross sections parallel to the xy-plane, the xz-plane, and the yz-plane are all ellipses. A computer-drawn sketch of the ellipsoid $(x^2/4) + (y^2/9) + (z^2/16) = 1$ is given in Figure 5b.

NOTE. If the ellipse $(x^2/a^2) + (y^2/b^2) = 1$ is revolved about the x-axis, the resulting surface is the ellipsoid $(x^2/a^2) + (y^2/b^2) + (z^2/c^2) = 1$. What do you get if the ellipse is rotated around the y-axis? (See Problem 35.)

7. **The hyperboloid of one sheet:** $(x^2/a^2) + (y^2/b^2) - (z^2/c^2) = 1$. This surface is sketched in Figure 6a. Cross sections parallel to the xy-plane are ellipses. Cross

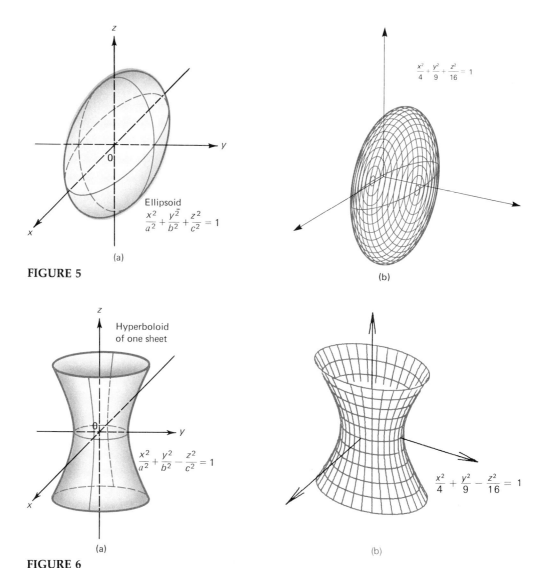

Ellipsoid
$$\frac{x^2}{a^2} + \frac{y^2}{b^2} + \frac{z^2}{c^2} = 1$$

(a)

FIGURE 5

$$\frac{x^2}{4} + \frac{y^2}{9} + \frac{z^2}{16} = 1$$

(b)

Hyperboloid
of one sheet

$$\frac{x^2}{a^2} + \frac{y^2}{b^2} - \frac{z^2}{c^2} = 1$$

(a)

$$\frac{x^2}{4} + \frac{y^2}{9} - \frac{z^2}{16} = 1$$

(b)

FIGURE 6

sections parallel to the xz-plane and the yz-plane are hyperbolas. A computer-drawn sketch of $(x^2/4) + (y^2/9) - (z^2/16) = 1$ is given in Figure 6b.

8. **The hyperboloid of two sheets:** $(z^2/c^2) - (x^2/a^2) - (y^2/b^2) = 1$. This surface is sketched in Figure 7a. Cross sections are the same as those for the hyperboloid of one sheet. Note that the equation implies that $|z| \geq |c|$ (explain why). A computer-drawn sketch of $(z^2/4) - (x^2/4) - (y^2/9) = 1$ is given in Figure 7b.

9. **The elliptic paraboloid:** $z = (x^2/a^2) + (y^2/b^2)$. This surface is sketched in Figure 8a. For each positive fixed z, $(x^2/a^2) + (y^2/b^2) = z$ is the equation of an ellipse. Hence cross sections parallel to the xy-plane are ellipses. If x or y is fixed, then we obtain parabolas. Hence cross sections parallel to the xz- or yz-planes are parabolas. A computer-drawn sketch of $z = (x^2/4) + (y^2/9)$ is given in Figure 8b.

10. **The hyperbolic paraboloid:** $z = (x^2/a^2) - (y^2/b^2)$. This surface is sketched in Figure 9a. For each fixed z we obtain a hyperbola parallel to the xy-plane. Hence cross

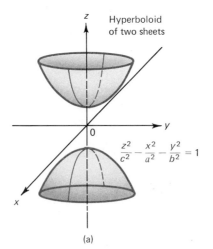

Hyperboloid of two sheets

$$\frac{z^2}{c^2} - \frac{x^2}{a^2} - \frac{y^2}{b^2} = 1$$

(a)

$$\frac{z^2}{4} - \frac{x^2}{4} - \frac{y^2}{9} = 1$$

(b)

FIGURE 7

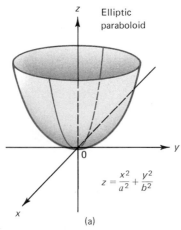

Elliptic paraboloid

$$z = \frac{x^2}{a^2} + \frac{y^2}{b^2}$$

(a)

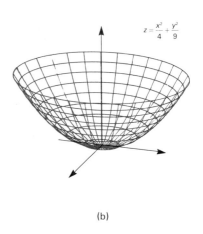

$$z = \frac{x^2}{4} + \frac{y^2}{9}$$

(b)

FIGURE 8

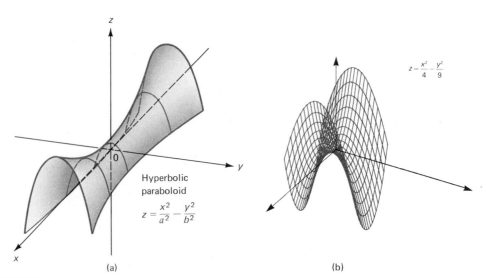

Hyperbolic paraboloid

$$z = \frac{x^2}{a^2} - \frac{y^2}{b^2}$$

(a)

$$z = \frac{x^2}{4} - \frac{y^2}{9}$$

(b)

FIGURE 9

sections parallel to the xy-plane are hyperbolas. If x or y is fixed, we obtain parabolas. Thus cross sections parallel to the xz- and yz-planes are parabolas. The shape of the graph suggests why the hyperbolic paraboloid is often called a **saddle surface.** A computer-drawn sketch of $z = (x^2/4) - (y^2/9)$ is given in Figure 9b.

11. **The elliptic cone:** $(x^2/a^2) + (y^2/b^2) = z^2$. This surface is sketched in Figure 10a. We get one cone for $z > 0$ and another for $z < 0$. Cross sections cut by planes not passing through the origin are either parabolas, circles, ellipses, or hyperbolas. That is, the cross sections are the **conic sections** discussed in Chapter 10. If $a = b$, we obtain the equation of a **circular cone.** A computer drawn sketch of $(x^2/4) + (y^2/9) = z^2$ is given in Figure 10b.

(a)

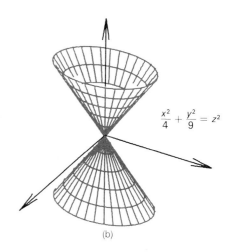

(b)

FIGURE 10

EXAMPLE 4   Describe the surface given by $x^2 - 4y^2 - 9z^2 = 25$.

Solution. Dividing by 25, we obtain

$$\frac{x^2}{25} - \frac{4y^2}{25} - \frac{9z^2}{25} = 1, \quad \text{or} \quad \frac{x^2}{25} - \frac{y^2}{(5/2)^2} - \frac{z^2}{(5/3)^2} = 1.$$

This is the equation of a hyperboloid of two sheets. Note that here cross sections parallel to the yz-plane are ellipses. ∎

EXAMPLE 5   Describe the surface given by $4x - 4y^2 - z^2 = 0$.

Solution. Dividing by 4, we obtain

$$x = y^2 + \frac{z^2}{4}.$$

This is the equation of an elliptic paraboloid. The only difference between this surface and the ones described under surface 9 is that now cross sections parallel to the yz-plane are ellipses while cross sections parallel to the other coordinate planes are parabolas. This surface is sketched in Figure 11. ∎

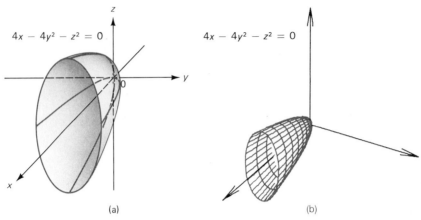

$4x - 4y^2 - z^2 = 0$

$4x - 4y^2 - z^2 = 0$

(a)   (b)

FIGURE 11

EXAMPLE 6   Describe the surface $x^2 + 8x - 2y^2 + 8y + z^2 = 0$.

Solution. Completing the squares, we obtain

$$(x + 4)^2 - 2(y - 2)^2 + z^2 = 8,$$

or dividing by 8,

$$\frac{(x + 4)^2}{8} - \frac{(y - 2)^2}{4} + \frac{z^2}{8} = 1.$$

This surface is a hyperboloid of one sheet, which, however, is centered at $(-4, 2, 0)$ instead of the origin. Moreover, cross sections parallel to the xz-plane are ellipses, while those parallel to the other coordinate planes are hyperbolas. ∎

As a rule, to identify a quadric surface with x, y, and z terms, completing the square will generally resolve the problem. If there are terms of the form cxy, dyz, or exz present, then it is necessary to rotate the axes, a technique that we will not discuss here except to note that it is similar to the $\mathbb{R}^2$ technique discussed in Section 10.4.

■   **WARNING:** Watch out for degenerate surfaces. For example, there are clearly no points that satisfy

$$x^2 + \frac{y^2}{4} + \frac{z^2}{9} + 1 = 0.$$

PROBLEMS 17.6

In Problems 1–7, draw a sketch of the given cylinder. Here the directrix C is given. The generatrix L is the axis of the variable missing in the equation.

1. $y = \sin x$ **2.** $z = \sin y$ **3.** $y = \cos z$
4. $y = \cosh x$ **5.** $z = x^3$ **6.** $z = |y|$
7. $x^2 + y^2 + 2y = 0$ [*Hint:* Complete the square.]

In Problems 8–34, identify the quadric surface and sketch it.

8. $x^2 + y^2 = 4$

9. $y^2 + z^2 = 4$

10. $x^2 + z^2 = 4$

11. $\dfrac{x^2}{4} - \dfrac{y^2}{9} = 1$

12. $x^2 + 4z^2 = 1$

13. $y^2 - 2y + 4z^2 = 1$

14. $x^2 - z^2 = 1$

15. $3x^2 - 4y^2 = 4$

16. $x^2 + y^2 + z^2 = 1$

17. $x^2 + y^2 - z^2 = 1$

18. $y^2 + 2y - z^2 + x^2 = 1$

19. $x^2 + 2y^2 + 3z^2 = 4$

20. $x + 2y^2 + 3z^2 = 4$

21. $x^2 - 2y - 3z^2 = 4$

22. $x^2 + 4x + y^2 + 6y - z^2 - 8z = 2$

23. $4x - x^2 + y^2 + z^2 = 0$

24. $5x^2 + 7y^2 + 8z^2 = 8z$

25. $x^2 - y^2 - z^2 = 1$

26. $x^2 - 2y^2 - 3z^2 = 4$

27. $z^2 - x^2 - y^2 = 2$

28. $y^2 - 3x^2 - 3z^2 = 27$

29. $x^2 + y^2 + z^2 = 2(x + y + z)$

30. $4x^2 + 4y^2 + 16z^2 = 16$

31. $4y^2 - 4x^2 + 8z^2 = 16$

32. $z + x^2 - y^2 = 0$

33. $-x^2 + 2x + y^2 - 6y = z$

34. $x + y + z = x^2 + z^2$

35. Identify and sketch the surface generated when the ellipse $(x^2/a^2) + (y^2/b^2) = 1$ is revolved about the y-axis.

17.7 VECTOR FUNCTIONS AND PARAMETRIC EQUATIONS IN $\mathbb{R}^3$

Having seen how vectors in $\mathbb{R}^2$ generalize to vectors in $\mathbb{R}^3$, we can imagine properties of vector functions in $\mathbb{R}^3$. We will describe these briefly, as almost all their properties have already been proved in Chapter 16.

Definition 1 VECTOR FUNCTIONS IN $\mathbb{R}^3$ Let f_1, f_2, and f_3 be functions of the real variable t. Then for all values of t for which $f_1(t)$, $f_2(t)$, and $f_3(t)$ are defined, we define the **vector-valued function f of a real variable** t by

$$\mathbf{f}(t) = (f_1(t), f_2(t), f_3(t)) = f_1(t)\mathbf{i} + f_2(t)\mathbf{j} + f_3(t)\mathbf{k}. \tag{1}$$

If f_1, f_2, and f_3 are continuous over an interval I, then as t varies over I, the set of points traced out by the end of the vector $\mathbf{f}$ is called a **curve** in space.

EXAMPLE 1   Sketch the curve $\mathbf{f}(t) = (\cos t)\mathbf{i} + 4(\sin t)\mathbf{j} + t\mathbf{k}$.

Solution. Here $x = \cos t$ and $y = 4 \sin t$, so eliminating t, we obtain $x^2 + (y^2/16) = 1$, which is the equation of an ellipse in the xy-plane. Since $z = t$ increases as t increases, the curve is a spiral that climbs up the side of an elliptical cylinder, as sketched in Figure 1. This curve is called an **elliptical helix.** ■

Definition 2 LIMIT   If $\mathbf{f}(t) = f_1(t)\mathbf{i} + f_2(t)\mathbf{j} + f_3(t)\mathbf{k}$, and if $\lim_{t \to t_0} f_1(t)$, $\lim_{t \to t_0} f_2(t)$, and $\lim_{t \to t_0} f_3(t)$ exist, then we define

$$\lim_{t \to t_0} \mathbf{f}(t) = \left[\lim_{t \to t_0} f_1(t)\right]\mathbf{i} + \left[\lim_{t \to t_0} f_2(t)\right]\mathbf{j} + \left[\lim_{t \to t_0} f_3(t)\right]\mathbf{k}. \tag{2}$$

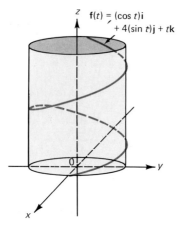

FIGURE 1

Definition 3   DERIVATIVE

$$\frac{d\,\mathbf{f}(t)}{dt} = \mathbf{f}'(t) = \lim_{\Delta t \to 0} \frac{\mathbf{f}(t + \Delta t) - \mathbf{f}(t)}{\Delta t}, \tag{3}$$

provided that this limit exists.

Theorem 1. $\mathbf{f}$ is differentiable if and only if its component functions f_1, f_2, and f_3 are differentiable, and in that case

$$\mathbf{f}'(t) = f_1'(t)\mathbf{i} + f_2'(t)\mathbf{j} + f_3'(t)\mathbf{k}. \tag{4}$$

EXAMPLE 2   Find the derivative of $\mathbf{f}(t) = (\cos t)\mathbf{i} + (\sin t)\mathbf{j} + t\mathbf{k}$.

Solution. $\mathbf{f}'(t) = -(\sin t)\mathbf{i} + (\cos t)\mathbf{j} + \mathbf{k}.$ ■

The following theorem is new.

Theorem 2.   If $\mathbf{f}$ and $\mathbf{g}$ are differentiable, then $\mathbf{f} \times \mathbf{g}$ is differentiable and

$$(\mathbf{f} \times \mathbf{g})' = (\mathbf{f}' \times \mathbf{g}) + (\mathbf{f} \times \mathbf{g}'). \tag{5}$$

Proof.   The proof is an easy application of the product rule of differentiation and is left as an exercise (see Problem 25). ■

NOTE.   Since $\mathbf{g} \times \mathbf{f} = -(\mathbf{f} \times \mathbf{g})$, it follows that $(\mathbf{f} \times \mathbf{g})' = -(\mathbf{g} \times \mathbf{f})'$. Thus the order in which things are written is important in (5).

EXAMPLE 3   Let $\mathbf{f}(t) = (\cos t)\mathbf{i} + (\sin t)\mathbf{j} + t\mathbf{k}$ and $\mathbf{g}(t) = t^2\mathbf{i} + t^3\mathbf{j} + t^4\mathbf{k}$. Calculate

$$\frac{d}{dt}(\mathbf{f} \times \mathbf{g}).$$

Solution.

$$(\mathbf{f} \times \mathbf{g})' = (\mathbf{f}' \times \mathbf{g}) + (\mathbf{f} \times \mathbf{g}')$$

$$= [-(\sin t)\mathbf{i} + (\cos t)\mathbf{j} + \mathbf{k}] \times (t^2\mathbf{i} + t^3\mathbf{j} + t^4\mathbf{k})$$

$$+ [(\cos t)\mathbf{i} + (\sin t)\mathbf{j} + t\mathbf{k}] \times (2t\mathbf{i} + 3t^2\mathbf{j} + 4t^3\mathbf{k})$$

$$= \begin{vmatrix} \mathbf{i} & \mathbf{j} & \mathbf{k} \\ -\sin t & \cos t & 1 \\ t^2 & t^3 & t^4 \end{vmatrix} + \begin{vmatrix} \mathbf{i} & \mathbf{j} & \mathbf{k} \\ \cos t & \sin t & t \\ 2t & 3t^2 & 4t^3 \end{vmatrix}$$

$$= (t^4 \cos t - t^3)\mathbf{i} + (t^2 + t^4 \sin t)\mathbf{j} - (t^3 \sin t + t^2 \cos t)\mathbf{k}$$

$$+ (4t^3 \sin t - 3t^3)\mathbf{i} + (2t^2 - 4t^3 \cos t)\mathbf{j} + (3t^2 \cos t - 2t \sin t)\mathbf{k}$$

$$= (4t^3 \sin t + t^4 \cos t - 4t^3)\mathbf{i} + (t^4 \sin t - 4t^3 \cos t + 3t^2)\mathbf{j}$$

$$+ [2t^2 \cos t - (t^3 + 2t)\sin t]\mathbf{k}. \quad \blacksquare$$

Facts about arc length and tangent vectors hold exactly as in $\mathbb{R}^2$. Proofs of the parts of the following theorem are identical to the proofs given in Chapter 16.

Theorem 3. Let $\mathbf{f}$ have a continuous derivative and let $\mathbf{T}(t)$ and $s(t)$ denote the unit tangent vector and arc length, respectively.

   (i) If $\mathbf{f}'(t) \neq \mathbf{0}$, $\mathbf{T}(t) = \mathbf{f}'(t)/|\mathbf{f}'(t)|$ [see equation (16.3.7)].
   (ii) If $|\mathbf{f}(t)|$ is constant over an interval I, then $\mathbf{f} \cdot \mathbf{f}' = 0$ on I (see Theorem 16.4.2).
   (iii) $ds/dt = |\mathbf{f}'(t)|$ [see Theorem 16.5.1 and equation (16.5.10)].
   (iv) $s(t_1) = \int_{t_0}^{t_1} |\mathbf{f}'(t)| \, dt$ [see equation (16.5.4)]; here $s(t)$ measures arc length from a reference point $P_0 = (x(t_0), y(t_0), z(t_0))$.
   (v) $\mathbf{T} = d\mathbf{f}/ds$ (see Theorem 16.6.1).

REMARK.   Formula (iv) can be derived as in Section 16.5 if we start with an intuitive notion of arc length. However, formula (iv) can be used as the *definition* of arc length. The precise definition is given in Definition 4.

Definition 4   ARC LENGTH   Suppose f has a continuous derivative in the interval $[t_0, b]$ and suppose that for every t_1 in $[t_0, b]$, $\int_{t_0}^{t_1} |\mathbf{f}'(t)| \, dt$ exists. Then $\mathbf{f}$ is said to be **rectifiable** in the interval $[t_0, b]$, and the **arc length** of the curve $\mathbf{f}(t)$ in the interval $[t_0, t_1]$ is given by

$$s(t_1) = \int_{t_0}^{t_1} |\mathbf{f}'(t)| \, dt,$$

where $t_0 \leq t_1 \leq b$.

EXAMPLE 4   Let $\mathbf{f}(t) = (\cos t)\mathbf{i} + (\sin t)\mathbf{j} + t\mathbf{k}$. (This surface is called a **circular helix.**)

   (a) Calculate $\mathbf{T}(t)$ at $t = \pi/3$.
   (b) Find the length of the arc from $t = 0$ to $t = 4$.

Solution. **(a)** $\mathbf{f}'(t) = -(\sin t)\mathbf{i} + (\cos t)\mathbf{j} + \mathbf{k}$ and

$$|\mathbf{f}'(t)| = \sqrt{\sin^2 t + \cos^2 t + 1} = \sqrt{2},$$

so that

$$\mathbf{T}(t) = -\frac{\sin t}{\sqrt{2}}\mathbf{i} + \frac{\cos t}{\sqrt{2}}\mathbf{j} + \frac{1}{\sqrt{2}}\mathbf{k}.$$

At $t = \pi/3$,

$$\mathbf{T}\left(\frac{\pi}{3}\right) = -\frac{\sqrt{3}}{2\sqrt{2}}\mathbf{i} + \frac{1}{2\sqrt{2}}\mathbf{j} + \frac{1}{\sqrt{2}}\mathbf{k}.$$

(b) Since $ds/dt = |\mathbf{f}'(t)| = \sqrt{2}$,

$$s(4) = \int_0^4 \sqrt{2}\, dt = 4\sqrt{2}. \quad \blacksquare$$

The curvature $\kappa(t)$ and the unit normal vector $\mathbf{n}(t)$ are defined differently in $\mathbb{R}^3$. They would have to be. For one thing, there is a whole "plane-full" of unit vectors orthogonal to $\mathbf{T}$. Also, curvature cannot be defined as $|d\varphi/ds|$ since there are now three angles that define the direction of C. Therefore, we use an alternative definition of curvature (see Theorem 16.8.1).

Definition 5   CURVATURE   If $\mathbf{f}$ has a continuous derivative, then the **curvature** of $\mathbf{f}$ is given by

$$\kappa(t) = \left|\frac{d\mathbf{T}}{ds}\right|. \tag{6}$$

REMARK.   By Theorem 16.8.1, equation (6) is equivalent in $\mathbb{R}^2$ to the definition $\kappa(t) = |d\varphi/ds|$, which was our $\mathbb{R}^2$ definition.

Definition 6   PRINCIPAL UNIT NORMAL VECTOR   For any value of t for which $\kappa(t) \neq 0$, the **principal unit normal vector n** is defined by

$$\mathbf{n}(t) = \frac{1}{\kappa(t)}\frac{d\mathbf{T}}{ds}. \tag{7}$$

REMARK.   Having defined a unit normal vector, we must show that it is orthogonal to the unit tangent vector in order to justify its name.

Theorem 4. $\mathbf{n}(t) \perp \mathbf{T}(t)$.

Proof.   Since $1 = \mathbf{T} \cdot \mathbf{T}$, we differentiate both sides with respect to s to obtain

$$0 = \mathbf{T} \cdot \frac{d\mathbf{T}}{ds} + \frac{d\mathbf{T}}{ds} \cdot \mathbf{T} = 2\mathbf{T} \cdot \frac{d\mathbf{T}}{ds} = (2\kappa)\mathbf{T} \cdot \mathbf{n}.\ \blacksquare$$

EXAMPLE 5   Find the curvature and principal unit normal vector, $\kappa(t)$ and $\mathbf{n}(t)$, for the curve of Example 4 at $t = \pi/3$.

Solution. Since $ds/dt = \sqrt{2}$, $s = \sqrt{2}t$ and $t = s/\sqrt{2}$, so that

$$\mathbf{T}(s) = -\frac{\sin(s/\sqrt{2})}{\sqrt{2}}\mathbf{i} + \frac{\cos(s/\sqrt{2})}{\sqrt{2}}\mathbf{j} + \frac{1}{\sqrt{2}}\mathbf{k}$$

and

$$\frac{d\mathbf{T}}{ds} = -\frac{\cos(s/\sqrt{2})}{2}\mathbf{i} - \frac{\sin(s/\sqrt{2})}{2}\mathbf{j}.$$

Hence

$$\kappa(t) = \sqrt{\frac{\cos^2(s/\sqrt{2})}{4} + \frac{\sin^2(s/\sqrt{2})}{4}} = \frac{1}{2}$$

and

$$\mathbf{n}(t) = \frac{1}{\kappa(t)}\frac{d\mathbf{T}}{ds} = -(\cos t)\mathbf{i} - (\sin t)\mathbf{j}.$$

At $t = \pi/3$,

$$\mathbf{n}\left(\frac{\pi}{3}\right) = -\frac{1}{2}\mathbf{i} - \frac{\sqrt{3}}{2}\mathbf{j}.\ \blacksquare$$

REMARK.   When it is convenient to solve for t in terms of s, we can calculate $d\mathbf{T}/ds$ directly by the relation

$$\frac{d\mathbf{T}}{ds} = \frac{d\mathbf{T}/dt}{ds/dt}. \tag{8}$$

There is a third vector that is often useful in applications.

Definition 7   BINORMAL VECTOR   The **binormal vector B** to the curve **f** is defined by

$$\mathbf{B} = \mathbf{T} \times \mathbf{n}. \tag{9}$$

From this definition we see that **B** is orthogonal to both **T** and **n**. Moreover, since

$\mathbf{T} \perp \mathbf{n}$, the angle θ between $\mathbf{T}$ and $\mathbf{n}$ is $\pi/2$, and

$$|\mathbf{B}| = |\mathbf{T} \times \mathbf{n}| = |\mathbf{T}|\,|\mathbf{n}|\sin\theta = 1,$$

so that $\mathbf{B}$ is a unit vector. Also, $\mathbf{T}$, $\mathbf{n}$, and $\mathbf{B}$ form a right-handed system.

EXAMPLE 6   Find the binormal vector to the curve of Example 5 at $t = \pi/3$.

Solution.

$$\mathbf{B} = \mathbf{T} \times \mathbf{n} = \begin{vmatrix} \mathbf{i} & \mathbf{j} & \mathbf{k} \\ -\dfrac{\sin t}{\sqrt{2}} & \dfrac{\cos t}{\sqrt{2}} & \dfrac{1}{\sqrt{2}} \\ -\cos t & -\sin t & 0 \end{vmatrix} = \dfrac{\sin t}{\sqrt{2}}\mathbf{i} - \dfrac{\cos t}{\sqrt{2}}\mathbf{j} + \dfrac{1}{\sqrt{2}}\mathbf{k}$$

When $t = \pi/3$, $\mathbf{B} = (\sqrt{3}/2\sqrt{2})\mathbf{i} - (1/2\sqrt{2})\mathbf{j} + (1/\sqrt{2})\mathbf{k}$. The vectors $\mathbf{T}$, $\mathbf{n}$, and $\mathbf{B}$ are sketched in Figure 2. ■

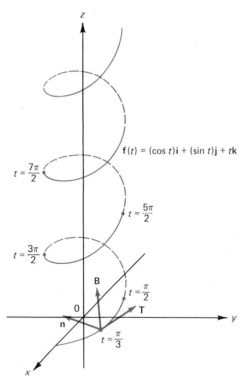

FIGURE 2

Our definitions of velocity and acceleration vectors remain unchanged in $\mathbb{R}^3$. Suppose that an object is moving in space. Then we can describe its motion parametrically by the vector function

$$\mathbf{f}(t) = f_1(t)\mathbf{i} + f_2(t)\mathbf{j} + f_3(t)\mathbf{k}. \qquad (10)$$

In this context $\mathbf{f}$ is again called the **position vector** of the object, and the curve described by **f** is called the **trajectory** of the point.

Definition 8   VELOCITY AND ACCELERATION VECTORS   If **f**′ and **f**″ exist, then

(i) $\mathbf{v}(t) = \mathbf{f}'(t) = f_1'(t)\mathbf{i} + f_2'(t)\mathbf{j} + f_3'(t)\mathbf{k}$ (11)

is called the **velocity vector.**

(ii) $\mathbf{a}(t) = \dfrac{d\mathbf{v}}{dt} = \mathbf{f}''(t) = f_1''(t)\mathbf{i} + f_2''(t)\mathbf{j} + f_3''(t)\mathbf{k}$ (12)

is called the **acceleration vector.**

(iii)   The **speed** of the object $v(t)$ is the magnitude of the velocity vector:

$v(t) = |\mathbf{v}(t)|.$ (13)

(iv)   The **acceleration scalar** $a(t)$ is the magnitude of the acceleration vector:

$a(t) = |\mathbf{a}(t)|.$ (14)

EXAMPLE 7   Find the velocity, acceleration, speed, and acceleration scalar of an object whose position vector is given by $\mathbf{f}(t) = (\cos t)\mathbf{i} + (\sin t)\mathbf{j} + t^3\mathbf{k}$.

Solution.

$\mathbf{v}(t) = -(\sin t)\mathbf{i} + (\cos t)\mathbf{j} + 3t^2\mathbf{k}$

$\mathbf{a}(t) = -(\cos t)\mathbf{i} - (\sin t)\mathbf{j} + 6t\mathbf{k}$

$v(t) = \sqrt{\sin^2 t + \cos^2 t + (3t^2)^2} = \sqrt{1 + 9t^4}$

$a(t) = \sqrt{\cos^2 t + \sin^2 t + (6t)^2} = \sqrt{1 + 36t^2}$

For example, at $t = 1$, $v(1) = \sqrt{10}$ and $a(1) = \sqrt{37}$. ∎

As in the $\mathbb{R}^2$ case, the acceleration vector $\mathbf{a}(t)$ can be resolved into its tangential and normal components.

Theorem 5.

$$\mathbf{a}(t) = \frac{d^2s}{dt^2}\mathbf{T} + \left(\frac{ds}{dt}\right)^2 \kappa\mathbf{n}.$$

Proof. $\mathbf{v}(t) = (d\mathbf{f}/dt) = (d\mathbf{f}/ds)(ds/dt) = \mathbf{T}(ds/dt)$ [from Theorem 3(v)]. Then

$$\mathbf{a}(t) = \frac{d\mathbf{v}}{dt} = \mathbf{T}\frac{d^2s}{dt^2} + \frac{d\mathbf{T}}{dt}\frac{ds}{dt} = \mathbf{T}\frac{d^2s}{dt^2} + \left(\frac{d\mathbf{T}}{ds}\frac{ds}{dt}\right)\left(\frac{ds}{dt}\right)$$

$$= \frac{d^2s}{dt^2}\mathbf{T} + \kappa\mathbf{n}\left(\frac{ds}{dt}\right)^2,$$

and the theorem is proved. ■

Using this result, we can prove the next theorem.

Theorem 6.

$$\kappa = \frac{|\mathbf{f}' \times \mathbf{f}''|}{|\mathbf{f}'|^3}.$$

Proof. $\mathbf{f}' \times \mathbf{f}'' = (\mathbf{T}\, ds/dt) \times [(d^2s/dt^2)\mathbf{T} + \kappa\mathbf{n}(ds/dt)^2]$. Now $\mathbf{T} \times \mathbf{T} = 0$ and $\mathbf{T} \times \mathbf{n} = \mathbf{B}$. Thus using Theorem 17.4.2(iv), we have

$$\mathbf{f}' \times \mathbf{f}'' = \left(\frac{ds}{dt}\right)^3 \kappa\mathbf{B}.$$

Then taking absolute values and using the fact that $|ds/dt| = |\mathbf{f}'|$ yields

$$|\mathbf{f}' \times \mathbf{f}''| = \kappa|\mathbf{f}'|^3,$$

from which the result follows. ■

Theorem 6 is useful for calculating curvature when it is not easy to write $\mathbf{f}$ in terms of the arc length parameter s.

EXAMPLE 8   Calculate the curvature of $\mathbf{f} = t^2\mathbf{i} + t^3\mathbf{j} + t^4\mathbf{k}$ at $t = 1$.

Solution. Here

$$\mathbf{f}' = 2t\mathbf{i} + 3t^2\mathbf{j} + 4t^3\mathbf{k}$$

$$|\mathbf{f}'| = \sqrt{4t^2 + 9t^4 + 16t^6} = t\sqrt{4 + 9t^2 + 16t^4}$$

$$\mathbf{f}'' = 2\mathbf{i} + 6t\mathbf{j} + 12t^2\mathbf{k}$$

$$\mathbf{f}' \times \mathbf{f}'' = \begin{vmatrix} \mathbf{i} & \mathbf{j} & \mathbf{k} \\ 2t & 3t^2 & 4t^3 \\ 2 & 6t & 12t^2 \end{vmatrix} = 12t^4\mathbf{i} - 16t^3\mathbf{j} + 6t^2\mathbf{k}$$

$$|\mathbf{f}' \times \mathbf{f}''| = \sqrt{144t^8 + 256t^6 + 36t^4} = t^2\sqrt{144t^4 + 256t^2 + 36}.$$

Thus

$$\kappa = \frac{\sqrt{144t^4 + 256t^2 + 36}}{t(4 + 9t^2 + 16t^4)^{3/2}}.$$

At $t = 1$, $\kappa = \sqrt{436}/29^{3/2} \approx 0.1337$. ■

PROBLEMS 17.7

In Problems 1–6, find the unit tangent vector **T** for the given value of t.

1. $\mathbf{f}(t) = t\mathbf{i} + t^2\mathbf{j} + t^3\mathbf{k}; \ t = 1$
2. $\mathbf{f}(t) = t^3\mathbf{i} + t^5\mathbf{j} + t^7\mathbf{k}; \ t = 1$
3. $\mathbf{f}(t) = t\mathbf{i} + e^t\mathbf{j} + e^{-t}\mathbf{k}; \ t = 0$
4. $\mathbf{f}(t) = t^2\mathbf{i} + t^2\mathbf{j} + t^{5/2}\mathbf{k}; \ t = 4$
5. $\mathbf{f}(t) = 4(\cos 2t)\mathbf{i} + 9(\sin 2t)\mathbf{j} + t\mathbf{k}; \ t = \pi/4$
6. $\mathbf{f}(t) = (\cosh t)\mathbf{i} + (\sinh t)\mathbf{j} + t^2\mathbf{k}; \ t = 0$

7. Find the arc length of the curve $\mathbf{f} = 2(\cos 3t)\mathbf{i} + 2(\sin 3t)\mathbf{j} + t^2\mathbf{k}$ between $t = 0$ and $t = 10$.

8. Find the arc length of the curve $\mathbf{f} = e^t(\cos 2t)\mathbf{i} + e^t(\sin 2t)\mathbf{j} + e^t\mathbf{k}$ between $t = 1$ and $t = 4$.

9. Find the arc length of the curve $\mathbf{f} = \frac{2}{3}t^3\mathbf{i} + (1 + t^{9/2})\mathbf{j} + (1 - t^{9/2})\mathbf{k}$ between $t = 0$ and $t = 2$.

In Problems 10–15, find the velocity vector, the speed, the acceleration vector, and the acceleration scalar for the given value of t.

10. $\mathbf{f} = t^2\mathbf{i} + t^3\mathbf{j} + t^4\mathbf{k}; \ t = 2$
11. $\mathbf{f} = (\cos t)\mathbf{i} + (\sin t)\mathbf{j} + t^4\mathbf{k}; \ t = 1$

12. $\mathbf{f} = (\ln t)\mathbf{i} + \dfrac{1}{t}\mathbf{j} + \dfrac{1}{t^2}\mathbf{k}; \ t = 1$
13. $\mathbf{f} = (\cosh t)\mathbf{i} + (\sinh t)\mathbf{j} + t\mathbf{k}; \ t = 0$

14. $\mathbf{f} = e^t\mathbf{i} + e^{-t}\mathbf{j} + \sqrt{t}\,\mathbf{k}; \ t = 4$
15. $\mathbf{f} = \sqrt{t}\,\mathbf{i} + \sqrt[3]{t}\,\mathbf{j} + \sqrt[4]{t}\,\mathbf{k}; \ t = 1$

In Problems 16–23, find the unit tangent vector **T**, the unit normal vector **n**, the binormal vector **B**, and the curvature κ at the given value of t. Check that $\mathbf{n} \cdot \mathbf{T} = 0$.

16. $\mathbf{f} = a(\sin t)\mathbf{i} + a(\cos t)\mathbf{j} + t\mathbf{k}; \ t = \pi/4; \ a > 0$
17. $\mathbf{f} = a(\sin t)\mathbf{i} + a(\cos t)\mathbf{j} + t\mathbf{k}; \ t = \pi/6; \ a > 0$
18. $\mathbf{f} = a(\cos t)\mathbf{i} + b(\sin t)\mathbf{j} + t\mathbf{k}; \ t = 0; \ a > 0, b > 0, a \neq b$
19. $\mathbf{f} = a(\cos t)\mathbf{i} + b(\sin t)\mathbf{j} + t\mathbf{k}; \ t = \pi/2; \ a > 0, b > 0, a \neq b$
20. $\mathbf{f} = t\mathbf{i} + t^2\mathbf{j} + t^3\mathbf{k}; \ t = 0$
21. $\mathbf{f} = e^t(\cos 2t)\mathbf{i} + e^t(\sin 2t)\mathbf{j} + e^t\mathbf{k}; \ t = 0$
22. $\mathbf{f} = e^t(\cos 2t)\mathbf{i} + e^t(\sin 2t)\mathbf{j} + e^t\mathbf{k}; \ t = \pi/4$
23. $\mathbf{f} = \mathbf{i} + t\mathbf{j} + t^2\mathbf{k}; \ t = 1$

24. Let $\mathbf{f} = a(\cos t)\mathbf{i} + a(\sin t)\mathbf{j} + t\mathbf{k}$ (a circular helix). Show that the angle between the unit tangent vector **T** and the z-axis is constant.

25. Show, by writing out the component functions, that $(\mathbf{f} \times \mathbf{g})' = (\mathbf{f}' \times \mathbf{g}) + (\mathbf{f} \times \mathbf{g}')$.

26. Show that the curvature of a straight line in space is zero.

*27. Let **B** be the binormal vector to a curve **f**, and let s denote arc length. Show that if $d\mathbf{B}/ds \neq 0$, then $d\mathbf{B}/ds \perp \mathbf{B}$ and $d\mathbf{B}/ds \perp \mathbf{T}$. [*Hint:* Differentiate $\mathbf{B} \cdot \mathbf{T} = 0$ and $\mathbf{B} \cdot \mathbf{B} = 1$.]

*28. Since, from Problem 27, $d\mathbf{B}/ds$ is orthogonal to **B** and **T**, it must be parallel to **n**. Hence

$$\frac{d\mathbf{B}}{ds} = -\tau\mathbf{n}$$

for some number τ. This number is called the **torsion** of the curve C and is a measure of how much the curve twists. Show that

$$\frac{d\mathbf{n}}{ds} = -\kappa\mathbf{T} + \tau\mathbf{B}.$$

**29. Show that

$$\mathbf{f}'''(t) = \left[\frac{d^3s}{dt^3} - \left(\frac{ds}{dt}\right)^3\kappa^2\right]\mathbf{T} + \left[3\left(\frac{d^2s}{dt^2}\right)\frac{ds}{dt}\kappa + \left(\frac{ds}{dt}\right)^2\frac{d\kappa}{dt}\right]\mathbf{n} + \left(\frac{ds}{dt}\right)^3\kappa\tau\mathbf{B}.$$

**30. Use the result of Problem 29 to show that

$$\tau = \frac{\mathbf{f}''' \cdot (\mathbf{f}' \times \mathbf{f}'')}{\kappa^2 (ds/dt)^6} = \frac{\mathbf{f}''' \cdot (\mathbf{f}' \times \mathbf{f}'')}{|\mathbf{f}' \times \mathbf{f}''|^2}.$$

*31. Calculate the torsion of $\mathbf{f} = (\cos t)\mathbf{i} + (\sin t)\mathbf{j} + t\mathbf{k}$ at $t = \pi/6$.

*32. Calculate the torsion of $\mathbf{f} = e^t(\cos t)\mathbf{i} + e^t(\sin t)\mathbf{j} + e^t\mathbf{k}$ at $t = 0$.

17.8 CYLINDRICAL AND SPHERICAL COORDINATES

In this chapter, so far, we have represented points using rectangular (Cartesian) coordinates. However, there are many ways to represent points in space, some of which are quite bizarre. In this section we will briefly introduce two common ways to represent points in space. The first is the generalization of the polar coordinate system in the plane.

CYLINDRICAL COORDINATES

In the **cylindrical coordinate system** a point P is given by

$$P = (r, \theta, z), \tag{1}$$

where $r \geq 0$, $0 \leq \theta < 2\pi$, r and θ are polar coordinates of the projection of P onto the xy-plane, called the **polar plane,** and z is the distance (measured in the positive direction) of this plane from P (see Figure 1). In this figure $\overrightarrow{0Q}$ is the projection of $\overrightarrow{0P}$ on the xy-plane.

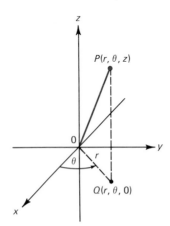

FIGURE 1

EXAMPLE 1   Discuss the graphs of the equations in cylindrical coordinates:
(a) $r = c, c > 0$ (b) $\theta = c$ (c) $z = c$
(d) $r_1 \leq r \leq r_2,\ \theta_1 \leq \theta \leq \theta_2,\ z_1 \leq z \leq z_2$

Solution. (a) If $r = c$, a constant, then θ and z can vary freely, and we obtain

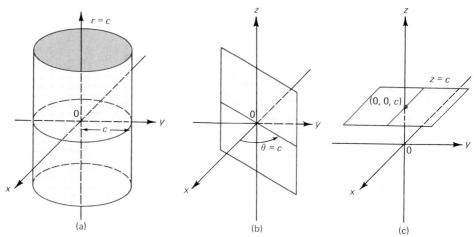

(a)   (b)   (c)

FIGURE 2

a right circular cylinder with radius c. (See Figure 2a.) This is the analog of the circle whose equation in polar coordinates is $r = c$ and is the reason the system is called the cylindrical coordinate system.

(b) If $\theta = c$, we obtain a half plane through the z-axis (see Figure 2b).

(c) If $z = c$, we obtain a plane parallel to the polar plane. This plane is sketched in Figure 2c.

(d) $r_1 \leq r \leq r_2$ gives the region between the cylinders $r = r_1$ and $r = r_2$. $\theta_1 \leq \theta \leq \theta_2$ is the wedge-shaped region between the half planes $\theta = \theta_1$ and $\theta = \theta_2$. Finally, $z_1 \leq z \leq z_2$ is the "slice" of space between the planes $z = z_1$ and $z = z_2$. Putting these results together, we get the rectangularly shaped solid in Figure 3. ∎

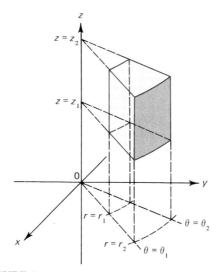

FIGURE 3

To translate from cylindrical to rectangular coordinates and back again, we have the formulas

$$x = r \cos \theta, \qquad y = r \sin \theta, \qquad z = z \tag{2}$$

and

$$r = \sqrt{x^2 + y^2}, \qquad \tan \theta = \frac{y}{x}, \qquad z = z. \tag{3}$$

These need no proof as they follow from the formulas for polar coordinates in Chapter 11.

EXAMPLE 2 Convert $P = (2, \pi/3, -5)$ from cylindrical to rectangular coordinates.

Solution. $x = r \cos \theta = 2 \cos(\pi/3) = 1$, $y = r \sin \theta = 2 \sin(\pi/3) = \sqrt{3}$, and $z = -5$. Thus in rectangular coordinates $P = (1, \sqrt{3}, -5)$. ∎

EXAMPLE 3 Convert $(-3, 3, 7)$ from rectangular to cylindrical coordinates.

Solution. $r = \sqrt{3^2 + 3^2} = 3\sqrt{2}$, $\theta = \tan^{-1}(3/-3) = \tan^{-1}(-1) = 3\pi/4$ [since $(-3, 3)$ is in the second quadrant], and $z = 7$, so $(-3, 3, 7) = (3\sqrt{2}, 3\pi/4, 7)$ in cylindrical coordinates. ∎

SPHERICAL COORDINATES

The second new coordinate system in space is the **spherical coordinate system.** This system is not a generalization of any system in the plane. A typical point P in space is represented as

$$P = (\rho, \theta, \varphi) \tag{4}$$

where $\rho \geq 0$, $0 \leq \theta < 2\pi$, $0 \leq \varphi \leq \pi$, and

ρ is the (positive) distance between the point and the origin.

θ is the same as in cylindrical coordinates.

φ is the angle between $\overrightarrow{OP}$ and the positive z-axis.

This system is illustrated in Figure 4.

EXAMPLE 4 Discuss the graphs of the equations in spherical coordinates:
(a) $\rho = c, c > 0$ **(b)** $\theta = c$ **(c)** $\varphi = c$
(d) $\rho_1 \leq \rho \leq \rho_2$, $\theta_1 \leq \theta \leq \theta_2$, $\varphi_1 \leq \varphi \leq \varphi_2$

Solution. **(a)** The set of points of which $\rho = c$ is the set of points c units from the origin. This set constitutes a sphere centered at the origin with radius c and is what gives the coordinate system its name. This surface is sketched in Figure 5a.

(b) As with cylindrical coordinates, the graph is a half plane containing the z-axis. See Figure 5b.

FIGURE 4

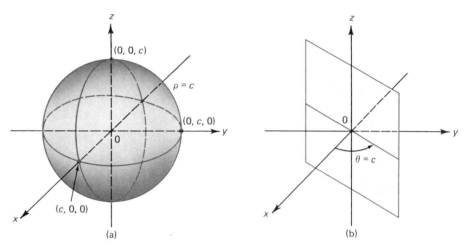

FIGURE 5

(c) If $0 < c < \pi/2$, then the graph of $\varphi = c$ is obtained by rotating the vector $\overrightarrow{OP}$ around the z-axis. This yields the circular cone sketched in Figure 6a. If $\pi/2 < c < \pi$, then we obtain the cone of Figure 6b. Finally, if $c = 0$, we obtain the positive z-axis; if $c = \pi$, we obtain the negative z-axis; and if $c = \pi/2$, we obtain the xy-plane.

(d) $\rho_1 \le \rho \le \rho_2$ gives the region between the sphere $\rho = \rho_1$ and the sphere $\rho = \rho_2$. $\theta_1 \le \theta \le \theta_2$ consists, as before, of the wedge-shaped region between the half planes $\theta = \theta_1$ and $\theta = \theta_2$. Finally, $\varphi_1 \le \varphi \le \varphi_2$ yields the region between the cones $\varphi = \varphi_1$ and $\varphi = \varphi_2$. Putting these results together, we obtain the solid (called a **spherical wedge**) sketched in Figure 7. ▮

To convert from spherical to rectangular coordinates, we have, from Figure 4,

$$x = \overline{0Q} \cos \theta \quad \text{and} \quad y = \overline{0Q} \sin \theta.$$

But from Figure 4

$$\overline{0Q} = \overline{PR} = \rho \sin \varphi, \tag{5}$$

(a)

(b)

FIGURE 6

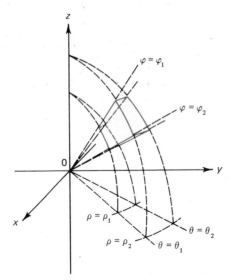

FIGURE 7

so that

$$x = \rho \sin \varphi \cos \theta \qquad (6)$$
$$y = \rho \sin \varphi \sin \theta. \qquad (7)$$

Finally, from triangle $0PR$ we have

$$z = \rho \cos \varphi. \qquad (8)$$

To convert to spherical coordinates from rectangular coordinates, we have

$$\overline{OP} = \sqrt{x^2 + y^2 + z^2} = \rho, \tag{9}$$

so that from (8)

$$\cos \varphi = \frac{z}{\rho} = \frac{z}{\sqrt{x^2 + y^2 + z^2}}.$$

Knowing ρ and φ, we can then calculate θ from equation (6) and (7).

EXAMPLE 5 Convert $(4, \pi/6, \pi/4)$ from spherical to rectangular coordinates.

Solution.

$$x = \rho \sin \varphi \cos \theta = 4 \sin\frac{\pi}{4}\cos\frac{\pi}{6} = 4\frac{\sqrt{2}}{2}\frac{\sqrt{3}}{2} = \sqrt{6},$$

$$y = \rho \sin \varphi \sin \theta = 4 \sin\frac{\pi}{4}\sin\frac{\pi}{6} = 4\frac{\sqrt{2}}{2}\frac{1}{2} = \sqrt{2},$$

and

$$z = \rho \cos \varphi = 4 \cos\frac{\pi}{4} = 2\sqrt{2}.$$

Therefore, in rectangular coordinates $(4, \pi/6, \pi/4)$ is $(\sqrt{6}, \sqrt{2}, 2\sqrt{2})$. ■

EXAMPLE 6 Convert $(1, \sqrt{3}, -2)$ from rectangular to spherical coordinates.

Solution. $\rho = \sqrt{x^2 + y^2 + z^2} = \sqrt{1 + 3 + 4} = \sqrt{8} = 2\sqrt{2}$. Then from (8)

$$\cos \varphi = \frac{z}{\rho} = \frac{-2}{2\sqrt{2}} = -\frac{1}{\sqrt{2}},$$

so that $\varphi = 3\pi/4$. Finally, from (6)

$$\cos \theta = \frac{x}{\rho \sin \varphi} = \frac{1}{(2\sqrt{2})(1/\sqrt{2})} = \frac{1}{2},$$

so that $\theta = \pi/3$, and we find that $(1, \sqrt{3}, -2)$ is $(2\sqrt{2}, \pi/3, 3\pi/4)$ in spherical coordinates. ■

We will not discuss the graphs of different kinds of surfaces given in cylindrical and spherical coordinates as that would take us too far afield. Both coordinate systems are useful in a wide variety of physical applications. For example, points on the earth and its interior are much more easily described by using a spherical coordinate system than a rectangular system.

We will return to cylindrical and spherical coordinates in Chapter 19 in our discussion of multiple integration. For convenience, we reproduce below the formulas for changing from one coordinate system to another that we derived in this section.

CONVERSION FORMULAS

(i) From cylindrical to rectangular coordinates:

$$x = r \cos \theta$$
$$y = r \sin \theta$$
$$z = z.$$

(ii) From rectangular to cylindrical coordinates:

$$r = \sqrt{x^2 + y^2}$$
$$\tan \theta = \frac{y}{x}, \qquad 0 \le \theta < 2\pi$$
$$z = z.$$

(iii) From spherical to rectangular coordinates:

$$x = \rho \sin \varphi \cos \theta$$
$$y = \rho \sin \varphi \sin \theta$$
$$z = \rho \cos \varphi.$$

(iv) From rectangular to spherical coordinates:

$$\rho = \sqrt{x^2 + y^2 + z^2}$$
$$\varphi = \cos^{-1}\frac{z}{\rho}, \qquad 0 \le \varphi \le \pi$$
$$\cos \theta = \frac{x}{\rho \sin \varphi} \quad \text{or} \quad \sin \theta = \frac{y}{\rho \sin \varphi}, \qquad 0 \le \theta < 2\pi.$$

PROBLEMS 17.8

In Problems 1–9, convert from cylindrical to rectangular coordinates.

1. $(2, \pi/3, 5)$ **2.** $(1, 0, -3)$ **3.** $(8, 2\pi/3, 1)$
4. $(4, \pi/2, -7)$ **5.** $(3, 3\pi/4, 2)$ **6.** $(10, 5\pi/3, -3)$
7. $(10, \pi, -3)$ **8.** $(13, 3\pi/2, 4)$ **9.** $(7, 5\pi/4, 2)$

In Problems 10–18, convert from rectangular to cylindrical coordinates.

10. $(1, 0, 0)$ **11.** $(0, 1, 0)$ **12.** $(0, 0, 1)$

13. $(1, 1, 2)$ **14.** $(-1, 1, 4)$ **15.** $(2, 2\sqrt{3}, -5)$
16. $(2\sqrt{3}, -2, 8)$ **17.** $(2, -2\sqrt{3}, 4)$ **18.** $(-2\sqrt{3}, -2, 1)$

In Problems 19–27, convert from spherical to rectangular coordinates.

19. $(2, 0, \pi/3)$ **20.** $(4, \pi/4, \pi/4)$ **21.** $(6, \pi/2, \pi/3)$
22. $(3, \pi/6, 5\pi/6)$ **23.** $(7, 7\pi/4, 3\pi/4)$ **24.** $(3, \pi/2, \pi/2)$
25. $(4, \pi/3, 2\pi/3)$ **26.** $(4, 2\pi/3, \pi/3)$ **27.** $(5, 11\pi/6, 5\pi/6)$

In Problems 28–36, convert from rectangular to spherical coordinates.

28. $(1, 1, 0)$ **29.** $(1, 1, \sqrt{2})$ **30.** $(1, -1, \sqrt{2})$
31. $(-1, -1, \sqrt{2})$ **32.** $(1, -\sqrt{3}, 2)$ **33.** $(-\sqrt{3}, -1, 2)$
34. $(2, \sqrt{3}, 4)$ **35.** $(-2, \sqrt{3}, -4)$ **36.** $(-\sqrt{3}, -2, -4)$

37. Write the equation $x^2 + y^2 + z^2 = 25$ in cylindrical and spherical coordinates.
38. Write the plane $ax + by + cz = d$ in cylindrical and spherical coordinates.
39. Write the equation $r = 9\sin\theta$ in rectangular coordinates.
40. Write the equation $r^2\sin 2\theta = z^3$ in rectangular coordinates.
41. Write the surface $x^2 + y^2 - z^2 = 1$ in cylindrical and spherical coordinates.
42. Write the surface $x^2 + 4y^2 + 4z^2 = 1$ in cylindrical and spherical coordinates.
43. Write the equation $z = r^2$ in rectangular coordinates.
44. Write the equation $\rho\cos\varphi = 1$ in rectangular coordinates.
45. Write the equation $\rho^2\sin\varphi\cos\varphi = 1$ in rectangular coordinates.
46. Write the equation $\rho = \sin\theta\cos\varphi$ in rectangular coordinates.
47. Write the equation $z = r^2\sin 2\theta$ in rectangular coordinates.
48. Write the equation $\rho = 2\cot\theta$ in rectangular coordinates.
49. Write the equation $x^2 + (y - 3)^2 + z^2 = 9$ in spherical coordinates.
50. Using the dot product, show that $\cos\varphi = z/\rho$.

REVIEW EXERCISES FOR CHAPTER SEVENTEEN

In Exercises 1–4, sketch the two given points and then find the distance between them.

1. $(3, 1, 2); (-1, -3, -4)$ **2.** $(4, -1, 7); (-5, 1, 3)$
3. $(-2, 4, -8); (0, 0, 6)$ **4.** $(2, -7, 0); (0, 5, -8)$

5. Find the equation of the sphere centered at $(-1, 4, 2)$ with radius 3.
6. Show that $x^2 + y^2 + z^2 - 4x + 8y + 10z = 12$ is the equation of a sphere, and find its center and radius.
7. Show that the points $(1, 3, 0)$, $(3, -1, -2)$, and $(-1, 7, 2)$ are collinear.

In Exercises 8–11, find the magnitude and direction cosines of the given vector.

8. $\mathbf{v} = 2\mathbf{i} - \mathbf{k}$ **9.** $\mathbf{v} = 3\mathbf{j} + 11\mathbf{k}$
10. $\mathbf{v} = \mathbf{i} - 2\mathbf{j} - 3\mathbf{k}$ **11.** $\mathbf{v} = -4\mathbf{i} + \mathbf{j} + 6\mathbf{k}$

12. Find a unit vector in the direction of $\overrightarrow{PQ}$, where $P = (3, -1, 2)$ and $Q = (-4, 1, 7)$.
13. Find a unit vector whose direction is opposite to that of $\overrightarrow{PQ}$, where $P = (1, -3, 0)$ and $Q = (-7, 1, -4)$.

In Exercises 14–21, let $\mathbf{u} = \mathbf{i} - 2\mathbf{j} + 3\mathbf{k}$, $\mathbf{v} = -3\mathbf{i} + 2\mathbf{j} + 5\mathbf{k}$, and $\mathbf{w} = 2\mathbf{i} - 4\mathbf{j} + \mathbf{k}$. Calculate the following:

14. $\mathbf{u} - \mathbf{v}$ **15.** $3\mathbf{v} + 5\mathbf{w}$ **16.** $\text{Proj}_{\mathbf{v}}\,\mathbf{w}$
17. $\text{Proj}_{\mathbf{w}}\,\mathbf{u}$ **18.** $2\mathbf{u} - 4\mathbf{v} + 7\mathbf{w}$ **19.** $\mathbf{u}\cdot\mathbf{w} - \mathbf{w}\cdot\mathbf{v}$
20. the angle between $\mathbf{u}$ and $\mathbf{v}$ **21.** the angle between $\mathbf{v}$ and $\mathbf{w}$

22. Find the distance from the point $P = (3, -1, 2)$ to the line passing through the points $Q = (-2, -1, 6)$ and $R = (0, 1, -8)$.
23. Find the work done when a force of 4 N acting in the direction of the vector $\mathbf{v} = -\mathbf{i} + \mathbf{j} + \mathbf{k}$ moves an object from $(2, 1, -6)$ to $(3, 5, 8)$ (distance in meters).

In Exercises 24–27, find a vector equation, parametric equations, and symmetric equations of the given line.

24. Containing $(3, -1, 4)$ and $(-1, 6, 2)$.
25. Containing $(-4, 1, 0)$ and $(3, 0, 7)$.
26. Containing $(3, 1, 2)$ and parallel to $3\mathbf{i} - \mathbf{j} - \mathbf{k}$.
27. Containing $(1, -2, -3)$ and parallel to the line $(x + 1)/5 = (y - 2)/-3 = (z - 4)/2$.

28. Show that the lines L_1: $x = 3 - 2t$, $y = 4 + t$, $z = -2 + 7t$ and L_2: $x = -3 + s$, $y = 2 - 4s$, $z = 1 + 6s$ have no points of intersection.
29. Find the distance from the origin to the line passing through the point $(3, 1, 5)$ and having the direction $\mathbf{v} = 2\mathbf{i} - \mathbf{j} + \mathbf{k}$.

In Exercises 30–33, find the cross product $\mathbf{u} \times \mathbf{v}$.

30. $\mathbf{u} = 3\mathbf{i} - \mathbf{j}$; $\mathbf{v} = 2\mathbf{i} + 4\mathbf{k}$ 31. $\mathbf{u} = 7\mathbf{j}$; $\mathbf{v} = \mathbf{i} - \mathbf{k}$
32. $\mathbf{u} = 4\mathbf{i} - \mathbf{j} + 7\mathbf{k}$; $\mathbf{v} = -7\mathbf{i} + \mathbf{j} - 2\mathbf{k}$ 33. $\mathbf{u} = -2\mathbf{i} + 3\mathbf{j} - 4\mathbf{k}$; $\mathbf{v} = -3\mathbf{i} + \mathbf{j} - 10\mathbf{k}$

34. Find two unit vectors orthogonal to both $\mathbf{u} = \mathbf{i} - \mathbf{j} + 3\mathbf{k}$ and $\mathbf{v} = -2\mathbf{i} - 3\mathbf{j} + 4\mathbf{k}$.
35. Find the equation of the line passing through $(-1, 2, 4)$ and orthogonal to L_1: $(x - 1)/4 = (y + 6)/3 = z/-2$ and L_2: $(x + 3)/5 = (y - 1)/1 = (z + 3)/4$.
36. Calculate the area of a parallelogram with the adjacent vertices $(1, 4, -2)$, $(-3, 1, 6)$, and $(1, -2, 3)$.
37. Calculate the area of a triangle with vertices at $(2, 1, 3)$, $(-4, 1, 7)$, and $(-1, -1, 3)$.
38. Calculate the volume of the parallelepiped determined by the vectors $\mathbf{i} + \mathbf{j}$, $2\mathbf{i} - 3\mathbf{k}$, and $2\mathbf{j} + 7\mathbf{k}$.

In Exercises 39–41, find the equation of the plane containing the given point and orthogonal to the given normal vector.

39. $P = (1, 3, -2)$; $\mathbf{N} = \mathbf{i} + \mathbf{k}$ 40. $P = (1, -4, 6)$; $\mathbf{N} = 2\mathbf{j} - 3\mathbf{k}$
41. $P = (-4, 1, 6)$; $\mathbf{N} = 2\mathbf{i} - 3\mathbf{j} + 5\mathbf{k}$

42. Find the equation of the plane containing the points $(-2, 4, 1)$, $(3, -7, 5)$, and $(-1, -2, -1)$.
43. Find all points of intersection of the planes Π_1: $-x + y + z = 3$ and Π_2: $-4x + 2y - 7z = 5$.
44. Find all points of intersection of the planes Π_1: $-4x + 6y + 8z = 12$ and Π_2: $2x - 3y - 4z = 5$.
45. Find all points of intersection of the planes Π_1: $3x - y + 4z = 8$ and Π_2: $-3x - y - 11z = 0$.
46. Find the distance from $(1, -2, 3)$ to the plane $2x - y - z = 6$.
47. Find the angle between the planes of Exercise 43.
48. Show that the position vectors $\mathbf{u} = \mathbf{i} - 2\mathbf{j} + \mathbf{k}$, $\mathbf{v} = 3\mathbf{i} + 2\mathbf{j} - 3\mathbf{k}$, and $\mathbf{w} = 9\mathbf{i} - 2\mathbf{j} - 3\mathbf{k}$ are coplanar, and find the equation of the plane containing them.

In Exercises 49–51, draw a sketch of the given cylinder. The directrix C is given. The generatrix L is the axis of the variable missing in the equation.

49. $x = \cos y$ 50. $y = z^2$ 51. $z = \sqrt[3]{x}$

In Exercises 52–58, identify the quadric surface and sketch it.

52. $x^2 + y^2 = 9$

53. $x^2 - \dfrac{y^2}{4} + \dfrac{z^2}{9} = 1$

54. $x^2 - \dfrac{y^2}{4} - \dfrac{z^2}{9} = 1$

55. $-9x^2 + 16y^2 - 9z^2 = 25$

56. $4x^2 + y^2 + 4z^2 - 8y = 0$

57. $x = \dfrac{y^2}{4} - \dfrac{z^2}{9}$

58. $y^2 + z^2 = x^2$

In Exercises 59–61, find the unit tangent vector **T** for the given value for t.

59. $\mathbf{f}(t) = t^3\mathbf{i} - \sqrt{t}\,\mathbf{j} + 2t\,\mathbf{k};\ t = 1$

60. $\mathbf{f}(t) = e^t\mathbf{i} + e^{-t}\mathbf{j} + t^3\mathbf{k};\ t = 0$

61. $\mathbf{f}(t) = (\cos 3t)\mathbf{i} - (\sin 3t)\mathbf{j} + t\,\mathbf{k};\ t = \pi/9$

62. Find the arc length of the curve $\mathbf{f} = 3(\sin 2t)\mathbf{i} + 3(\cos 2t)\mathbf{j} + t^2\mathbf{k}$ between $t = 0$ and $t = 5$.

63. The position vector of a moving particle is given by $\mathbf{f} = t^3\mathbf{i} + (\cos t)\mathbf{j} - (\sin t)\mathbf{k}$. Find the velocity, the speed, the acceleration vector, and the acceleration scalar when $t = \pi/4$.

64. Answer the questions in Exercise 63 for $\mathbf{f} = e^t\mathbf{i} + (\ln t)\mathbf{j} - t^{3/2}\mathbf{k}$ at $t = 1$.

In Exercises 65–67, find the unit tangent vector **T**, the unit normal vector **n**, the binormal vector **B**, and the curvature κ for the given value of t.

65. $\mathbf{f} = 2(\cos t)\mathbf{i} + 2(\sin t)\mathbf{j} + t\,\mathbf{k};\ t = \pi/6$ **66.** $\mathbf{f} = t^2\mathbf{i} + t^3\mathbf{j} - t\,\mathbf{k};\ t = 0$

67. $\mathbf{f} = e^t\mathbf{i} + e^t(\cos t)\mathbf{j} + e^t(\sin t)\mathbf{k};\ t = 0$

68. Find the torsion of the function in Exercise 66 at $t = 0$.

In Exercises 69–76, convert from one coordinate system to another as indicated.

69. $(3,\ \pi/6,\ -1)$, cylindrical to rectangular   **70.** $(2,\ 2,\ -4)$, rectangular to cylindrical

71. $(2,\ 2\pi/3,\ 4)$, cylindrical to rectangular   **72.** $(-2,\ 2\sqrt{3},\ -4)$ rectangular to cylindrical

73. $(3,\ \pi/3,\ \pi/4)$, spherical to rectangular   **74.** $(2,\ 7\pi/3,\ 2\pi/3)$, spherical to rectangular

75. $(-1,\ 1,\ -\sqrt{2})$, rectangular to spherical   **76.** $(2,\ -\sqrt{3},\ 4)$, rectangular to spherical

77. Write the equation $x^2 + y^2 + z^2 = 25$ in cylindrical and spherical coordinates.

78. Write the equation $r^2 \cos 2\theta = z^3$ in rectangular coordinates.

79. Write the equation $x^2 - y^2 + z^2 = 1$ in cylindrical and spherical coordinates.

80. Write the equation $z = r^3$ in rectangular coordinates.

81. Write the equation $\rho \sin \varphi = 1$ in rectangular coordinates.

18 Differentiation of Functions of Two and Three Variables

For most of the functions we have so far encountered in this book, we have been able to write $y = f(x)$. This means that we could write the variable y explicitly in terms of the single variable x. However, in a great variety of applications it is necessary to write the quantity of interest in terms of two or more variables. We have already encountered this situation. For example, the volume of a right circular cylinder is given by

$$V = \pi r^2 h,$$

where r is the radius of the cylinder and h is its height. That is, V is a function of the *two* variables r and h.

As a second example, we discussed [see equation (6.3.13), the **ideal gas law,** which relates pressure, volume, and temperature for an ideal gas: We have

$$PV = nRT,$$

where P is the pressure of the gas, V is the volume, T is the absolute temperature (i.e., in degrees kelvin), n is the number of moles of the gas, and R is a constant. Solving for P, we find that

$$P = \frac{nRT}{V}.$$

That is, we can write P as a function of the *three* variables n, T, and V.

As a third example, we saw in Problem 3.2.39 that according to Poiseuille's law, the resistance R of a blood vessel of length l and radius r is given by

$$R = \frac{\alpha l}{r^4},$$

where α is a constant of proportionality. If l and r are allowed to vary, then R is written as a function of the *two* variables l and r.

As a final example, let the vector $\mathbf{v}$ be given by

$$\mathbf{v} = x\mathbf{i} + y\mathbf{j} + z\mathbf{k}.$$

Then the magnitude of $\mathbf{v}$, $|\mathbf{v}|$, is given by

$$|\mathbf{v}| = \sqrt{x^2 + y^2 + z^2}.$$

That is, the magnitude of $\mathbf{v}$ is written as a function of the *three* variables, x, y, and z.

There are many examples like the four cited above. It is probably fair to say that very few physical, biological, or economic quantities can be properly expressed in terms of one variable alone. Often we write these quantities in terms of one variable simply because functions of only one variable are the easiest functions to handle.

In this chapter we will see how many of the operations we have studied in our discussion of the "one-variable" calculus can be extended to functions of two and three variables. We will begin by discussing the basic notions of limits and continuity and will go on to differentiation and applications of differentiation. In Chapter 19 we will discuss the integration of functions of two and three variables.

18.1 FUNCTIONS OF TWO AND THREE VARIABLES

In Chapters 16 and 17 we discussed the notion of vector-valued functions. For example, if $\mathbf{f}(t) = f_1(t)\mathbf{i} + f_2(t)\mathbf{j}$, then for every t in the domain of $\mathbf{f}$, we obtain a vector $\mathbf{f}(t)$. In defining functions of two or more variables, we obtain a somewhat reversed situation. For example, in the formula for the volume of a right circular cylinder, we have

$$V = \pi r^2 h. \tag{1}$$

Here the volume V is written as a function of the two variables r and h. Put another way, for every ordered pair of positive real numbers (r, h), there is a unique positive number V such that $V = \pi r^2 h$. To indicate the dependence of V on the variables r and h, we write $V(r, h)$. That is, V can be thought of as a function that assigns a positive real number to every ordered pair of real numbers. Thus we see what we meant when we said that the situation is "reversed." Instead of having a vector function of one variable (a scalar), we have a scalar function of an ordered pair of variables (which is a vector in $\mathbb{R}^2$). We now give a definition of a function of two variables.

Definition 1 FUNCTION OF TWO VARIABLES   Let D be a subset of $\mathbb{R}^2$. Then a **real-valued function of two variables** f is a rule that assigns to each point (x, y) in D

a unique real number that we denote $f(x, y)$. The set D is called the **domain** of f. The set $\{f(x, y): (x, y) \in D\}$, which is the set of values the function f takes on, is called the **range** of f.

EXAMPLE 1   Let $D = \mathbb{R}^2$. For each point $(x, y) \in \mathbb{R}^2$, we assign the number $f(x, y) = x^2 + y^4$. Since $x^2 + y^4 \geq 0$ for every pair of real numbers (x, y), we see that the range of f is $\mathbb{R}^+$, the set of nonnegative real numbers. ■

When the domain D is not given, we will take the domain of f to be the largest subset of $\mathbb{R}^2$ for which the expression $f(x, y)$ makes sense.

EXAMPLE 2   Find the domain and range of the function f given by $f(x, y) = \sqrt{4 - x^2 - y^2}$, and find $f(0, 1)$ and $f(1, -1)$.

Solution. Clearly, f is defined when the expression under the square root sign is nonnegative. Thus $D = \{(x, y): x^2 + y^2 \leq 4\}$. This is the disk† centered at the origin with radius 2. It is sketched in Figure 1. Since x^2 and y^2 are nonnegative, $4 - x^2 - y^2$ is largest when $x = y = 0$. Thus the largest value of $\sqrt{4 - x^2 - y^2} = \sqrt{4} = 2$. Since $x^2 + y^2 \leq 4$, the smallest value of $4 - x^2 - y^2$ is 0, taken when $x^2 + y^2 = 4$ (at all points on the circle $x^2 + y^2 = 4$). Thus the range of f is the closed interval $[0, 2]$. Finally,

$$f(0, 1) = \sqrt{4 - 0^2 - 1^2} = \sqrt{3} \qquad \text{and} \qquad f(-1, 1) = \sqrt{4 - (-1)^2 - 1^2} = \sqrt{2}.$$ ■

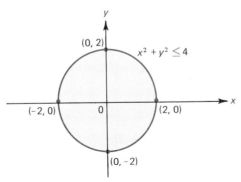

FIGURE 1

REMARK.   We emphasize that the domain of f is a subset of $\mathbb{R}^2$, while the range is a subset of $\mathbb{R}$, the real numbers.

CAUTION.   In Figure 1 we sketched the *domain* of the function. We did *not* sketch its graph. The graph of a function of two or more variables is more complicated and will be discussed shortly.

In Chapter 1 we wrote $y = f(x)$. That is, we used the letter y to denote the value of a function of one variable. Here we can use the letter z (or any other letter for that matter) to denote the value taken by f, which is now a function of two variables. We then write

---

†The *disk* of radius r is the circle of radius r together with all points interior to that circle.

$$z = f(x, y). \tag{2}$$

EXAMPLE 3   Find the domain and range of the function

$$z = f(x, y) = \frac{\sqrt{x^2 + y^2 - 9}}{x - y}.$$

Solution. $\sqrt{x^2 + y^2 - 9}$ is defined only when $x^2 + y^2 \geq 9$. This is the circle $x^2 + y^2 = 9$ with radius 3 and its exterior. However, $f(x, y)$ is also not defined when $x - y = 0$ or when $y = x$. Thus all points on the line $y = x$ are excluded from the domain of f. We have the domain of $f = \{(x, y): x^2 + y^2 \geq 9 \text{ and } x \neq y\}$. This domain is sketched in Figure 2. The points on the dotted line are excluded.

We now calculate the range of f. We first note that $\sqrt{x^2 + y^2 - 9}$ can take any nonnegative value. Since $x - y$ can be positive or negative, we see that $\sqrt{x^2 + y^2 - 9}/(x - y)$ can take on any real value, so the range of f is $\mathbb{R}$. Note that $f(x, y) = 0$ at any point on the circle except at the points $(3/\sqrt{2}, 3/\sqrt{2})$ and $(-3/\sqrt{2}, -3/\sqrt{2})$. The function is not defined at these last two points. ∎

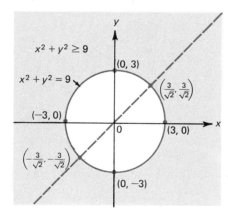

FIGURE 2

EXAMPLE 4   Let $z = f(x, y) = 3/\sqrt{x^2 - y}$. Find the domain and range of f.

Solution. $f(x, y)$ is defined if $x^2 - y > 0$ or $y < x^2$. This is the region in the plane "outside" the parabola $y = x^2$ and is sketched in Figure 3. Note that the parabola $y = x^2$ is *not* in the domain of f. The range of f is the set of positive real numbers since when $x^2 - y > 0$, $3/\sqrt{x^2 - y}$ can take on any positive real number. ∎

EXAMPLE 5   Find the domain and range of $f(x, y) = \tan^{-1}(y/x)$.

Solution. $\tan^{-1}(y/x)$ is defined so long as $x \neq 0$. Hence the domain of $f = \{(x, y): x \neq 0\}$. From the definition of $\tan^{-1} x$ in Section 7.2, we find that the range of f is the open interval $(-\pi/2, \pi/2)$. ∎

We now turn to the definition of a function of three variables.

Definition 2   FUNCTION OF THREE VARIABLES   Let D be a subset of $\mathbb{R}^3$. Then a **real-valued function of three variables** f is a rule that assigns to each point (x, y, z)

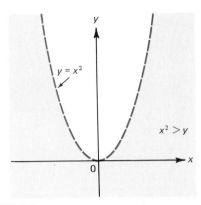

FIGURE 3

in D a unique real number that we denote $f(x, y, z)$. The set D is called the **domain** of f, and the set $\{f(x, y, z): (x, y, z) \in D\}$, which is the set of values the function f takes on, is called the **range** of f.

NOTE. We will often use the letter w to denote the values that a function of three variables takes. We then have

$$w = f(x, y, z). \tag{3}$$

EXAMPLE 6 Let $w = f(x, y, z) = \sqrt{1 - x^2 - (y^2/4) - (z^2/9)}$. Find the domain and range of f. Calculate $f(0, 1, 1)$.

Solution. $f(x, y, z)$ is defined if $1 - x^2 - (y^2/4) - (z^2/9) \geq 0$, which occurs if $x^2 + (y^2/4) + (z^2/9) \leq 1$. From Section 17.6 we see that the equation $x^2 + (y^2/4) + (z^2/9) = 1$ is the equation of an ellipsoid. Thus the domain of f is the set of points (x, y, z) in $\mathbb{R}^3$ that are on and interior to this ellipsoid (sketched in Figure 4). The range of f is

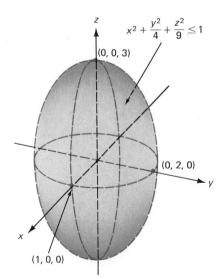

FIGURE 4

the closed interval $[0, 1]$. This follows because $x^2 + (y^2/4) + (z^2/9) \geq 0$, so that $1 - x^2 - (y^2/4) - (z^2/9)$ is maximized when $x = y = z = 0$. Finally,

$$f(0, 1, 1) = \sqrt{1 - 0^2 - \frac{1^2}{4} - \frac{1^2}{9}} = \sqrt{1 - \frac{1}{4} - \frac{1}{9}} = \sqrt{1 - \frac{13}{36}} = \frac{\sqrt{23}}{6}.$$

NOTE.   Figure 4 is a sketch of the domain of f, not a graph of f itself. ■

EXAMPLE 7   Let $w = f(x, y, z) = \ln(x + 2y + z - 3)$. Find the domain and range of f. Calculate $f(1, 2, 3)$.

Solution. $f(x, y, z)$ is defined when $x + 2y + z - 3 > 0$, which occurs when $x + 2y + z > 3$. The equation $x + 2y + z = 3$ is the equation of a plane. Thus the domain of f is the *half space* above (but not including) this plane. The plane is sketched in Figure 5. The range of f is $\mathbb{R}$. Finally, $f(1, 2, 3) = \ln(1 + 4 + 3 - 3) = \ln 5 \approx 1.61$.
■

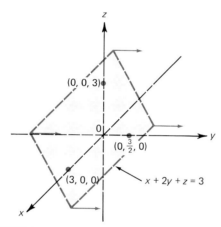

FIGURE 5

We will not here define a function of more than three variables, since such a definition would be very similar to Definitions 1 and 2.

We now turn to a discussion of the graph of a function. Recall from Section 1.6 that the graph of a function of one variable f is the set of all points (x, y) in the plane such that $y = f(x)$. Using this definition as our model, we have the following definition.

Definition 3   GRAPH OF A FUNCTION OF TWO VARIABLES   The **graph** of a function f of two variables x and y is the set of all points (x, y, z) in $\mathbb{R}^3$ such that $z = f(x, y)$. The graph of a function of two variables is called a **surface** in $\mathbb{R}^3$.

EXAMPLE 8   Sketch the graph of the function

$$z = f(x, y) = \sqrt{1 - x^2 - y^2}. \tag{4}$$

Solution.   We first note that $z \geq 0$. Then squaring both sides of (4), we have

$z^2 = 1 - x^2 - y^2$, or $x^2 + y^2 + z^2 = 1$. This equation is, of course, the equation of the unit sphere. However, since $z \geq 0$, the graph of f is the hemisphere sketched in Figure 6. ■

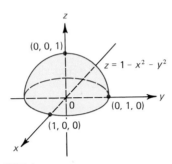

FIGURE 6

It is often very difficult to sketch the graph of a function $z = f(x, y)$ since, except for the quadric surfaces we discussed in Section 17.6, we really do not have a vast "catalog" of surfaces to which to refer. Moreover, the techniques of curve plotting in three dimensions are tedious, to say the least, and plotting points in space will, except in the most trivial of cases, not get us very far. The best we can usually do is to describe **cross sections** of the surface that lie in planes parallel to the coordinate planes. This will give an idea of what the surface looks like. Then if we have access to a computer with graphing capabilities, we can obtain a computer-drawn sketch of the surface. We illustrate this process with two examples.

EXAMPLE 9   Obtain a sketch of the surface

$$z = f(x, y) = x^3 + 3x^2 - y^2 - 9x + 2y - 10. \tag{5}$$

Solution. The xz-plane has the equation $y = 0$. Planes parallel to the xz-plane have the equation $y = c$, where c is a constant. Setting $y = c$ in (5), we have

$$z = x^3 + 3x^2 - 9x - 10 + (-c^2 + 2c)$$
$$= x^3 + 3x^2 - 9x - 10 + k,$$

where $k = -c^2 + 2c$ is a constant. We obtained a graph of the curve $y = x^3 + 3x^2 - 9x - 10$ in the xy-plane in Example 4.3.6 (see Figure 4.3.3).

Using the material in Section 1.8, we can obtain the graph of $z = x^3 + 3x^2 - 9x - 10 + k$ by shifting the graph of $z = x^3 + 3x^2 - 9x - 10$ up or down k units. For several values of k, cross sections lying in planes parallel to the xz-plane are given in Figure 7a.

REMARK.   The maximum value of the function $g(c) = -c^2 + 2c$ is 1, taken when $c = 1$ (you should verify this). Thus 1 is the largest value k can assume in the parallel cross sections, so $-10 + k \leq -9$, and the "highest" cross section is $z = x^3 + 3x^2 - 9x - 9$.

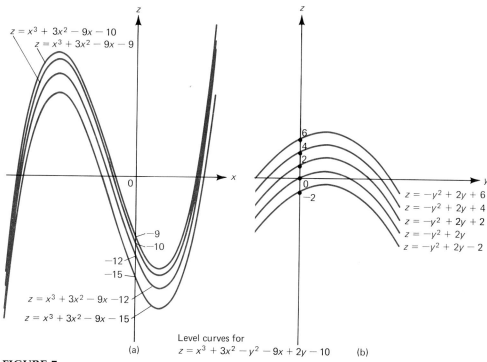

FIGURE 7

The yz-plane has the equation $x = 0$. Planes parallel to the yz-plane have the equation $x = c$, where c is a constant. Setting $x = c$ in (5), we obtain

$$z = -y^2 + 2y + (c^3 + 3c^2 - 9c - 10) = -y^2 + 2y + k.$$

For several values of k, cross sections parallel to the yz-plane are drawn in Figure 7b. Note that this k can take on any real value. In Figure 8 we provide a computer-drawn sketch of the surface. ▪

EXAMPLE 10   Obtain a sketch of the surface $z = \sin x + e^y$.

Solution. Cross sections lying in planes parallel to the xz-plane have the equation $z = \sin x + e^c = \sin x + k$, where $k > 0$. Some of the cross sections are sketched in Figure 9a. Cross sections lying in planes parallel to the yz-plane have the equation $z = \sin c + e^y = e^y + k$, where $-1 \le k \le 1$. Some of these cross sections are sketched in Figure 9b. In Figure 10 we provide a computer-drawn sketch of the surface. ▪

REMARK.   The curves $y = \sin x$ and $y = e^x$ are easy for us to sketch. Nevertheless, if we combine them as in the last example, the surface is difficult to sketch. For less "obvious" cross sections, graphs can be exceedingly difficult to obtain.

As we have stated, except for the standard quadric surfaces described in Section 17.6, it is extremely difficult to sketch three-dimensional graphs by hand. However, virtually every modern computer (and microcomputer) has graphing capabilities. In

$$z = x^3 + 3x^2 - y^2 - 9x + 2y - 10$$

FIGURE 8

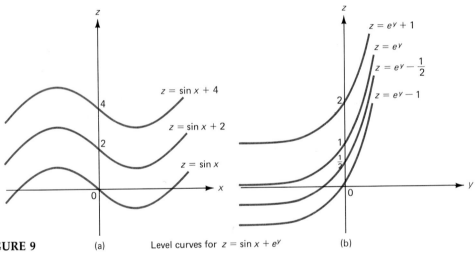

FIGURE 9   (a)   Level curves for $z = \sin x + e^y$ (b)

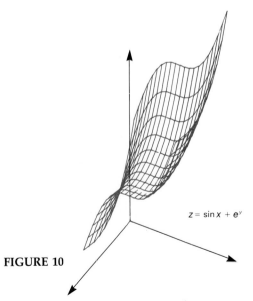

$$z = \sin x + e^y$$

FIGURE 10

Figure 11 we provide microcomputer-drawn sketches of the graphs of four different functions of two variables.

The situation becomes much more complicated when we try to sketch the graph of a function of three variables $w = f(x, y, z)$. We would need *four dimensions* to sketch such a surface. Being human, we are limited to three dimensions, and we see that we have reached the point where our comfortable three-dimensional geometry fails us.

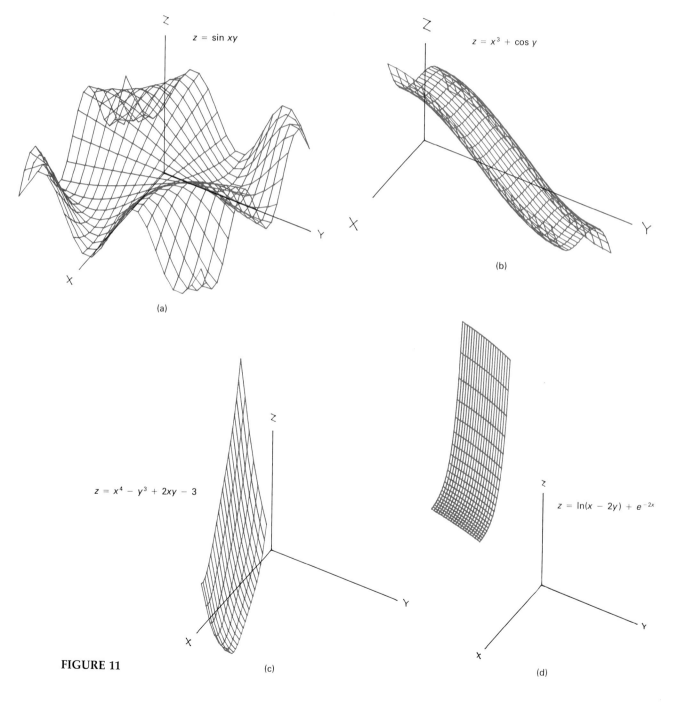

FIGURE 11

We are *not* saying that curves and surfaces in four-dimensional space do not exist; they do—only that we are not able to sketch them.

We will not spend much time in this text discussing spaces of dimension higher than three. To give you a taste of the subject, we can define four-dimensional space $\mathbb{R}^4$ as the set of all "points" or *four-tuples* (x, y, z, w), where x, y, z, and w are real numbers. Then the graph of a function $w = f(x, y, z)$ is the set of points (x, y, z, w) in $\mathbb{R}^4$ with $w = f(x, y, z)$. *It is important to note that the only difference between a graph of a function of two or three variables is our ability to sketch one (sometimes) but not the other.* If we do not insist on being able to sketch things, then there is really no essential difference between $\mathbb{R}^3$ or $\mathbb{R}^4$ (or $\mathbb{R}^5$, $\mathbb{R}^6$, . . . , for that matter). For the remainder of this chapter we will focus on $\mathbb{R}^2$ and $\mathbb{R}^3$, keeping in mind that all our definitions apply to higher dimensions as well.

We have shown that it is usually very difficult to sketch the graph of a function of two variables. Fortunately, there is a way to describe the graph of such a function in two dimensions. The idea for what we are about to do comes from cartographers (mapmakers). Cartographers have the problem of indicating three-dimensional features (such as mountains and valleys) on a two-dimensional surface. They solve the problem by drawing a **contour map (topographic map),** which is a map in which points of constant elevation are joined to form curves, called **contour curves.** The closer together these contour curves are drawn, the steeper is the terrain. A portion of a typical contour map is sketched in Figure 12.

FIGURE 12

LEVEL CURVES

We can use the same idea to depict the function $z = f(x, y)$ graphically. If z is fixed, then the equation $f(x, y) = z$ is the equation of a curve in the xy-plane, called a **level curve.**

We can think of a level curve as the projection of a cross section lying in a plane parallel to the xy-plane. Each value of z gives us such a curve. In other words, a level curve is the projection of the intersection of the surface $z = f(x, y)$ with the plane $z = c$. This idea is best illustrated with some examples.

EXAMPLE 11   Sketch the level curves of $z = x^2 + y^2$.

Solution. If $z > 0$, then $z = a^2$ for some positive number $a > 0$. Hence all level curves are circles of the form $x^2 + y^2 = a^2$. The number a^2 can be thought of as the "elevation" of points on a level curve. Some of these curves are sketched in Figure 13. Each level curve encloses a projection of a "slice" of the actual graph of the function in three dimensions. In this case each circle is the projection onto the xy-plane of a part of the surface in space. Actually, this example is especially simple because we can, without much difficulty, sketch the graph in space. From Section 17.6 the equation $z = x^2 + y^2$ is the equation of an elliptic paraboloid (actually, a circular paraboloid). It is sketched in Figure 14. In this easy case we can see that if we slice this surface parallel to the xy-plane, we obtain the circles whose projections onto the xy-plane are the level curves. In most cases, of course, we will not be able to sketch the graph in space easily, so we will have to rely on our "contour map" sketch in $\mathbb{R}^2$. ■

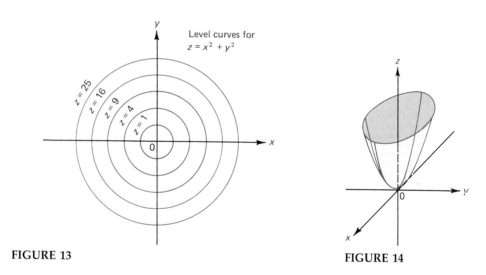

FIGURE 13   FIGURE 14

There are some interesting applications of level curves in the sciences and economics. Three are given below.

(i) Let $T(x, y)$ denote the temperature at a point (x, y) in the xy-plane. The level curves $T(x, y) = c$ are called **isothermal curves.** All points on such a curve have the same temperature.

(ii) Let $V(x, y)$ denote the voltage (or potential) at a point in the xy-plane. The level curves $V(x, y) = c$ are called **equipotential curves.** All points on such a curve have the same voltage.

(iii) A manufacturer makes two products. Let x and y denote the number of units of the first and second products produced during a given year. If $P(x, y)$ denotes the profit the manufacturer receives each year, then the level curves $P(x, y) = c$ are **constant profit curves.** All points on such a curve yield the same profit.

We conclude this section by noting that the idea behind level curves can be used to describe functions of three variables of the form $w = f(x, y, z)$. For each fixed w, the equation $f(x, y, z) = w$ is the equation of a surface in space, called a **level surface.** Since surfaces are so difficult to draw, we will not pursue the notion of level surfaces here.

PROBLEMS 18.1

In Problems 1–35, find the domain and range of the indicated function.

1. $f(x, y) = \sqrt{x^2 + y^2}$

2. $f(x, y) = \sqrt{1 + x + y}$

3. $f(x, y) = \dfrac{x}{y}$

4. $f(x, y) = \sqrt{1 - x^2 - 4y^2}$

5. $f(x, y) = \sqrt{1 - x^2 + 4y^2}$

6. $f(x, y) = \sin(x + y)$

7. $f(x, y) = e^x + e^y$

8. $f(x, y) = \dfrac{1}{(x^2 - y^2)^{3/2}}$

9. $f(x, y) = \tan(x - y)$

10. $f(x, y) = \sqrt{\dfrac{x + y}{x - y}}$

11. $f(x, y) = \sqrt{\dfrac{x - y}{x + y}}$

12. $f(x, y) = \sin^{-1}(x + y)$

13. $f(x, y) = \cos^{-1}(x - y)$

14. $f(x, y) = \dfrac{y}{|x|}$

15. $f(x, y) = \dfrac{x^2 - y^2}{x + y}$

16. $f(x, y) = \ln(1 + x^2 - y^2)$

*17. $f(x, y) = \left(\dfrac{x}{2y}\right) + \left(\dfrac{2y}{x}\right)$

18. $f(x, y, z) = x + y + z$

19. $f(x, y, z) = \sqrt{x + y + z}$

20. $f(x, y, z) = \dfrac{1}{\sqrt{x^2 + y^2 + z^2}}$

21. $f(x, y, z) = \dfrac{1}{\sqrt{x^2 - y^2 + z^2}}$

22. $f(x, y, z) = \dfrac{1}{\sqrt{x^2 - y^2 - z^2}}$

23. $f(x, y, z) = \sqrt{-x^2 - y^2 - z^2}$

24. $f(x, y, z) = \ln(x - 2y - 3z + 4)$

25. $f(x, y, z) = \dfrac{xy}{z}$

26. $f(x, y, z) = \sin(x + y - z)$

27. $f(x, y, z) = \sin^{-1}(x + y - z)$

28. $f(x, y, z) = \ln(x + y - z)$

29. $f(x, y, z) = \tan^{-1}\left(\dfrac{x + z}{y}\right)$

30. $f(x, y, z) = e^{xy + z}$

31. $f(x, y, z) = \dfrac{e^x + e^y}{e^z}$

32. $f(x, y, z) = xyz$

33. $f(x, y, z) = \dfrac{1}{xyz}$

34. $f(x, y, z) = \dfrac{x}{y + z}$

35. $f(x, y, z) = \sin x + \cos y + \sin z$

In Problems 36–42, sketch the graph of the given function.

36. $z = 4x^2 + 4y^2$

37. $y = x^2 + 4z^2$

38. $x = 4z^2 - 4y^2$

39. $z = x^2 - 4y^2$

40. $z = \sqrt{x^2 + 4y^2 + 4}$

41. $x = \sqrt{x^2 - 4z^2 + 4}$

42. $x = \sqrt{4 - z^2 - 4y^2}$

In Problems 43–50, describe the level curves of the given function and sketch these curves for the given values of z.

43. $z = \sqrt{1 + x + y}; z = 0, 1, 5, 10$

44. $z = \dfrac{x}{y}; z = 1, 3, 5, -1, -3$

45. $z = \sqrt{1 - x^2 - 4y^2}; z = 0, \frac{1}{4}, \frac{1}{2}, 1$

46. $z = \sqrt{1 + x^2 - y}; z = 0, 1, 2, 5$

47. $z = \cos^{-1}(x - y); z = 0, \dfrac{\pi}{6}, \dfrac{\pi}{3}, \dfrac{\pi}{2}$

48. $z = \sqrt{\dfrac{x + y}{x - y}}; z = 0, 1, 2, 5$

49. $z = \tan(x + y); z = 0, 1, -1, \sqrt{3}$

50. $z = \tan^{-1}(x - y^2); z = 0, \pi/6, \pi/4$

51. The temperature T at any point on an object in the plane is given by $T(x, y) = 20 + x^2 + 4y^2$. Sketch the isothermal curves for $T = 50$, $T = 60$, and $T = 70$ degrees.

52. The voltage at a point (x, y) on a metal plate placed in the xy-plane is given by $V(x, y) = \sqrt{1 - 4x^2 - 9y^2}$. Sketch the equipotential curves for $V = 1.0$ V, $V = 0.5$ V, and $V = 0.25$ V.

53. A manufacturer earns $P(x, y) = 100 + 2x^2 + 3y^2$ dollars each year for producing x and y units, respectively, of two products. Sketch the constant profit curves for $P = \$100$, $P = \$200$, and $P = \$1000$.

In Problems 54–63 a function of two variables is given. Match the function to one of the ten computer-drawn graphs given below.

(a)

(b)

(c)

(d)

(e)

(f)

(g)

(h)

(i)

(j)

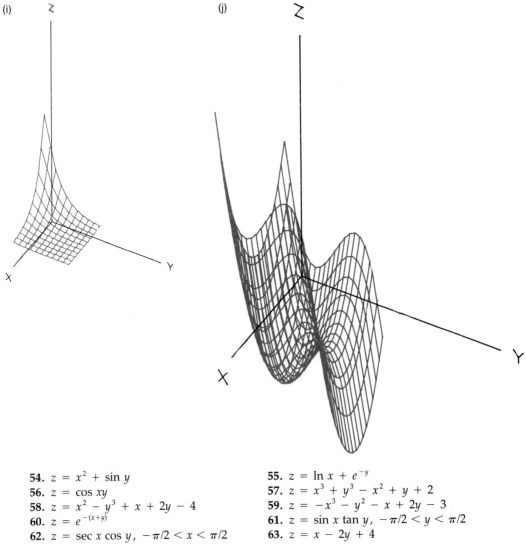

54. $z = x^2 + \sin y$

55. $z = \ln x + e^{-y}$

56. $z = \cos xy$

57. $z = x^3 + y^3 - x^2 + y + 2$

58. $z = x^2 - y^3 + x + 2y - 4$

59. $z = -x^3 - y^2 - x + 2y - 3$

60. $z = e^{-(x+y)}$

61. $z = \sin x \tan y,\ -\pi/2 < y < \pi/2$

62. $z = \sec x \cos y,\ -\pi/2 < x < \pi/2$

63. $z = x - 2y + 4$

18.2 LIMITS AND CONTINUITY

In this section we discuss the fundamental concepts of limits and continuity for functions of two and three variables. Recall that in the definition of a limit in Chapter 2†, the notions of an open and closed interval were fundamental. We could say, for

---

†The formal definition of a limit was given on page 124. We reproduce it here. Suppose that $f(x)$ is defined in a neighborhood of the point x_0 (a finite number) except possibly at the point x_0 itself. Then

$$\lim_{x \to x_0} f(x) = L$$

if for every $\epsilon > 0$, there is a $\delta > 0$, such that if

$$0 < |x - x_0| < \delta, \quad \text{then} \quad |f(x) - L| < \epsilon.$$

example, that x was close to x_0 if $|x - x_0|$ was sufficiently small or, equivalently, if x was contained in a small open interval (neighborhood) centered at x_0. It is interesting to see how these ideas extend to $\mathbb{R}^2$ or $\mathbb{R}^3$.

Let (x_0, y_0) be a point in $\mathbb{R}^2$. What do we mean by the equation

$$|(x, y) - (x_0, y_0)| = r? \tag{1}$$

Since (x, y) and (x_0, y_0) are vectors in $\mathbb{R}^2$,

$$|(x, y) - (x_0, y_0)| = |(x - x_0, y - y_0)| = \overset{\text{Equation (15.1.1)}}{\sqrt{(x - x_0)^2 + (y - y_0)^2}}. \tag{2}$$

Inserting (2) in (1) and squaring both sides yields

$$(x - x_0)^2 + (y - y_0)^2 = r^2, \tag{3}$$

which is the equation of the circle of radius r centered at (x_0, y_0). Then the set of points whose coordinates (x, y) satisfy the inequality

$$|(x, y) - (x_0, y_0)| < r \tag{4}$$

is the set of all points in $\mathbb{R}^2$ interior to the circle given by (3). This set is sketched in Figure 1a. Similarly, the inequality

$$|(x, y) - (x_0, y_0)| \leq r \tag{5}$$

describes the set of all points interior to and on the circle given by (3). This set is sketched in Figure 1b. This leads to the following definition.

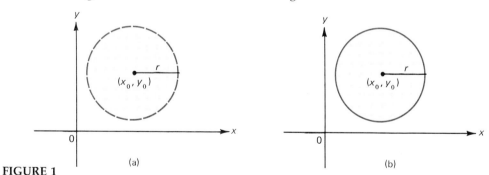

FIGURE 1

Definition 1

(i) The **open disk** D_r centered at (x_0, y_0) with radius r is the subset of $\mathbb{R}^2$ given by

$$\{(x, y): |(x, y) - (x_0, y_0)| < r\}.$$

(ii) The **closed disk** centered at (x_0, y_0) with radius r is the subset of $\mathbb{R}^2$ given by

$$\{(x, y): |(x, y) - (x_0, y_0)| \leq r\}.$$

(iii) The **boundary** of the open or closed disk defined in (i) or (ii) is the circle

$$\{(x, y): |(x, y) - (x_0, y_0)| = r\}.$$

(iv) A **neighborhood** of a point (x_0, y_0) in $\mathbb{R}^2$ is an open disk centered at (x_0, y_0).

REMARK.   In this definition the words "open" and "closed" have meanings very similar to their meanings in the terms "open interval" and "closed interval." An open interval does not contain its endpoints. An open disk does not contain any point on its boundary. Similarly, a closed interval contains all its boundary points, as does a closed disk.

With these definitions it is easy to define a limit of a function of two variables. Intuitively, we say that $f(x, y)$ approaches the limit L as (x, y) approaches (x_0, y_0) if $f(x, y)$ gets arbitrarily "close" to L as (x, y) approaches (x_0, y_0) along any **path**.† We define this notion precisely below.

Definition 2   DEFINITION OF A LIMIT   Let $f(x, y)$ be defined in a neighborhood of (x_0, y_0) but not necessarily at (x_0, y_0) itself. Then the **limit** of $f(x, y)$ as (x, y) approaches (x_0, y_0) *is L*, written

$$\lim_{(x,y)\to(x_0,y_0)} f(x, y) = L, \tag{6}$$

if for every number $\epsilon > 0$, there is a number $\delta > 0$ such that $|f(x, y) - L| < \epsilon$ for every $(x, y) \neq (x_0, y_0)$ in the open disk centered at (x_0, y_0) with radius δ.

REMARK.   This definition is illustrated in Figure 2.

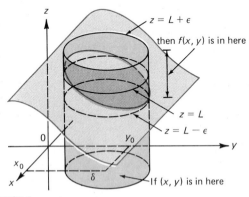

FIGURE 2

EXAMPLE 1   Show, using the definition of a limit, that $\lim_{(x,y)\to(1,2)}(3x + 2y) = 7$.

Solution. Let $\epsilon > 0$ be given. We need to choose a $\delta > 0$ such that $|3x + 2y - 7| < \epsilon$ if $0 < \sqrt{(x - 1)^2 + (y - 2)^2} < \delta$. We start with

---

†A path is another name for a curve joining (x, y) to (x_0, y_0).

Triangle inequality on page 9

$$|3x + 2y - 7| = |3x - 3 + 2y - 4| \leq |3x - 3| + |2y - 4|$$

Explain why

$$= |3(x - 1)| + |2(y - 2)| = 3|x - 1| + 2|y - 2|. \tag{7}$$

Now

$$|x - 1| = \sqrt{(x - 1)^2} \leq \sqrt{(x - 1)^2 + (y - 2)^2}$$

and

$$|y - 2| = \sqrt{(y - 2)^2} \leq \sqrt{(x - 1)^2 + (y - 2)^2}.$$

So from (7)

$$|3x + 2y - 7| \leq 3\sqrt{(x - 1)^2 + (y - 2)^2} + 2\sqrt{(x - 1)^2 + (y - 2)^2}$$

$$= 5\sqrt{(x - 1)^2 + (y - 2)^2}. \tag{8}$$

Now we want $|3x + 2y - 7| < \epsilon$. We choose $\delta = \epsilon/5$. Then from (8)

$$|3x + 2y - 7| \leq 5\sqrt{(x - 1)^2 + (y - 2)^2} < 5\delta = 5\left(\frac{\epsilon}{5}\right) = \epsilon,$$

and the requested limit is shown. ∎

   In the intuitive definition of a limit which preceded Definition 2, we used the phrase "if $f(x, y)$ gets arbitrarily close to L as (x, y) approaches (x_0, y_0) *along any path.*" Recall that $\lim_{x \to x_0} f(x) = L$ only if $\lim_{x \to x_0^+} f(x) = \lim_{x \to x_0^-} f(x) = L$. That is, the limit exists only if we get the same value from either side of x_0. In $\mathbb{R}^2$ the situation is more complicated because (x, y) can approach (x_0, y_0) not just along two but along an *infinite number* of paths. Some of these are illustrated in Figure 3. Thus the only way we can verify a limit is by making use of Definition 2 or some appropriate limit theorem that can be proven directly from Definition 2. We illustrate the kinds of problems we can encounter in the next two examples.

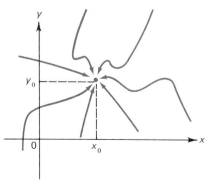

FIGURE 3

EXAMPLE 2   Let $f(x, y) = (y^2 - x^2)/(y^2 + x^2)$ for $(x, y) \neq (0, 0)$. We will show that $\lim_{(x,y)\to(0,0)} f(x, y)$ does not exist. There are an infinite number of approaches to the origin. For example, if we approach along the x-axis, then $y = 0$ and, if the indicated limit exists,

$$\lim_{(x,y)\to(0,0)} \frac{y^2 - x^2}{y^2 + x^2} = \lim_{(x,y)\to(0,0)} \frac{-x^2}{x^2} = \lim_{(x,y)\to(0,0)} -1 = -1.$$

On the other hand, if we approach along the y-axis, then $x = 0$ and

$$\lim_{(x,y)\to(0,0)} \frac{y^2 - x^2}{y^2 + x^2} = \lim_{(x,y)\to(0,0)} \frac{y^2}{y^2} = \lim_{(x,y)\to(0,0)} 1 = 1.$$

Thus we get different answers depending on how we approach the origin. To prove that the limit cannot exist, we note that we have shown that in any open disk centerd at the origin, there are points at which f takes on the values $+1$ and -1. Hence f cannot have a limit as $(x, y) \to (0, 0)$. ∎

Example 2 leads to the following general rule:

RULE FOR NONEXISTENCE OF A LIMIT

If we get two or more different values for $\lim_{(x,y)\to(x_0,y_0)} f(x, y)$ as we approach (x_0, y_0) along different paths, then $\lim_{(x,y)\to(x_0,y_0)} f(x, y)$ does not exist.

EXAMPLE 3   Let $f(x, y) = xy^2/(x^2 + y^4)$. We will show that $\lim_{(x,y)\to(0,0)} f(x, y)$ does not exist. First, let us approach the origin along a straight line passing through the origin that is not the y-axis. Then $y = mx$ and

$$f(x, y) = \frac{x(m^2 x^2)}{x^2 + m^4 x^4} = \frac{m^2 x}{1 + m^2 x^2},$$

which approaches 0 as $x \to 0$. Thus along every straight line, $f(x, y) \to 0$ as $(x, y) \to (0, 0)$. But if we approach $(0, 0)$ along the parabola $x = y^2$, then

$$f(x, y) = \frac{y^2(y^2)}{y^4 + y^4} = \frac{1}{2},$$

so that along this parabola, $f(x, y) \to \frac{1}{2}$ as $(x, y) \to (0, 0)$, and the limit does not exist. ∎

EXAMPLE 4   Prove that $\lim_{(x,y)\to(0,0)} [xy^2/(x^2 + y^2)] = 0$.

Solution. It is easy to show that this limit is zero along any straight line passing through the origin. But as we have seen, this is not enough. We must rely on our definition. Let $\epsilon > 0$ be given; then we must show that there is a $\delta > 0$ such that if $0 < \sqrt{x^2 + y^2} < \delta$, then

$$|f(x, y)| = \left| \frac{xy^2}{x^2 + y^2} \right| < \epsilon.$$

But $|x| \leq \sqrt{x^2 + y^2}$ and $y^2 \leq x^2 + y^2$, so

$$\left| \frac{xy^2}{x^2 + y^2} \right| \leq \frac{(\sqrt{x^2 + y^2})(x^2 + y^2)}{x^2 + y^2} = \sqrt{x^2 + y^2}.$$

Hence if we choose $\delta = \epsilon$, we will have $|f(x, y)| < \epsilon$ if $\sqrt{x^2 + y^2} < \delta$. ■

REMARK.   This example illustrates the fact that while the existence of different limits for different paths implies that no limit exists, only the definition or an appropriate limit theorem can be used to prove that a limit does exist.

As Examples 1 and 4 illustrate, it is often tedious to calculate limits from the definition. Fortunately, just as in the case of functions of a single variable, there are a number of theorems that greatly facilitate the calculation of limits. As you will recall, one of our important definitions stated that if f is continuous, then $\lim_{x \to x_0} f(x) = f(x_0)$. We now define continuity of a function of two variables.

Definition 3   CONTINUITY

(i) Let $f(x, y)$ be defined at every point (x, y) in a neighborhood of (x_0, y_0). Then f is **continuous** at (x_0, y_0) if all of the following conditions hold:

   (a) $f(x_0, y_0)$ exists [i.e., (x_0, y_0) is in the domain of f].
   (b) $\lim_{(x,y) \to (x_0, y_0)} f(x, y)$ exists.
   (c) $\lim_{(x,y) \to (x_0, y_0)} f(x, y) = f(x_0, y_0)$.

(ii) If one or more of these three conditions fail to hold, then f is said to be **discontinuous** at (x_0, y_0).
(iii) f is **continuous in a subset** S of $\mathbb{R}^2$ if f is continuous at every point (x, y) in S.

REMARK.   Condition (c) tells us that if a function f is continuous at (x_0, y_0), then we can calculate $\lim_{(x,y) \to (x_0, y_0)} f(x, y)$ by evaluation of f at (x_0, y_0).

EXAMPLE 5   Let $f(x, y) = xy^2/(x^2 + y^2)$. Then f is discontinuous at $(0, 0)$ because $f(0, 0)$ is not defined. ■

EXAMPLE 6   Let

$$f(x, y) = \begin{cases} \dfrac{xy^2}{x^2 + y^4}, & (x, y) \neq (0, 0) \\ 0, & (x, y) = (0, 0). \end{cases}$$

Here f is defined at $(0, 0)$ but it is still discontinuous there because $\lim_{(x,y) \to (0,0)} f(x, y)$ does not exist (see Example 3). ■

EXAMPLE 7   Let

$$f(x, y) = \begin{cases} \dfrac{xy^2}{x^2 + y^2}, & (x, y) \neq (0, 0) \\ 0, & (x, y) = (0, 0). \end{cases}$$

Here f is continuous at $(0, 0)$, according to Example 4. ■

EXAMPLE 8   Let

$$f(x, y) = \begin{cases} \dfrac{xy^2}{x^2 + y^2}, & (x, y) \neq (0, 0) \\ 1, & (x, y) = (0, 0). \end{cases}$$

Then f is discontinuous at $(0, 0)$ because $\lim_{(x,y)\to(0,0)} f(x, y) = 0 \neq f(0, 0) = 1$, so condition (c) is violated. ■

Naturally, we would like to know what functions are continuous. We can answer this question if we look at the continuous functions of one variable. We start with polynomials.

Definition 4   POLYNOMIAL AND RATIONAL FUNCTION

(i) A **polynomial** $p(x, y)$ in the two variables x and y is a finite sum of terms of the form $Ax^m y^n$, where m and n are positive integers and A is a real number.
(ii) A **rational function** $r(x, y)$ in the two variables x and y is a function that can be written as the quotient of two polynomials: $r(x, y) = p(x, y)/q(x, y)$.

EXAMPLE 9 $p(x, y) = 5x^5y^2 + 12xy^9 - 37x^{82}y^5 + x + 4y - 6$ is a polynomial. ■

EXAMPLE 10

$$r(x, y) = \frac{8x^3y^7 - 7x^2y^4 + xy - 2y}{1 - 3y^3 + 7x^2y^2 + 18yx^7}$$

is a rational function. ■

The limit theorems we stated in Section 2.3 and proved in Appendix 5 can be extended, with minor modifications, to functions of two variables. We will not state them here. However, by using them it is not difficult to prove the following theorem about continuous functions.

Theorem 1

(i) Any polynomial p is continuous at any point in $\mathbb{R}^2$.
(ii) Any rational function $r = p/q$ is continuous at any point (x_0, y_0) for which $q(x_0, y_0) \neq 0$. It is discontinuous when $q(x_0, y_0) = 0$ because it is then not defined at (x_0, y_0).
(iii) If f and g are continuous at (x_0, y_0), then $f + g$, $f - g$, and $f \cdot g$ are continuous at (x_0, y_0).†

---

†$(f + g)(x, y) = f(x, y) + g(x, y)$, just as in the case of functions of one variable. Similarly, $(f \cdot g)(x, y) = [f(x, y)][g(x, y)]$, and $(f/g)(x, y) = f(x, y)/g(x, y)$.

(iv) If f and g are continuous at (x_0, y_0) and if $g(x_0, y_0) \neq 0$, then f/g is continuous at (x_0, y_0).

(v) If f is continous at (x_0, y_0) and if h is a function of one variable that is continuous at $f(x_0, y_0)$, then the composite function $h \circ f$, defined by $(h \circ f)(x, y) = h(f(x, y))$, is continuous at (x_0, y_0).

EXAMPLE 11 Calculate $\lim_{(x,y)\to(4,1)}(x^3y^2 - 4xy)/(x + 6xy^3)$.

Solution. $(x^3y^2 - 4xy)/(x + 6xy^3)$ is continuous at $(4, 1)$ so we can calculate the limit by evaluation. We have

$$\lim_{(x,y)\to(4,1)} \frac{x^3y^2 - 4xy}{x + 6xy^3} = \frac{x^3y^2 - 4xy}{x + 6xy^3}\bigg|_{(4,1)} = \frac{64 \cdot 1 - 4 \cdot 4 \cdot 1}{4 + 6 \cdot 4 \cdot 1} = \frac{48}{28} = \frac{12}{7}. \blacksquare$$

EXAMPLE 12 Calculate $\lim_{(x,y)\to(\pi/6, \pi/3)} \sin(x + y)$.

Solution. $f(x, y) = x + y$ is continuous, as is $h(x) = \sin x$. Then by (v), $(h \circ f)(x, y) = \sin(x + y)$ is continuous, and therefore

$$\lim_{(x,y)\to(\pi/6, \pi/3)} \sin(x + y) = \sin\left(\frac{\pi}{6} + \frac{\pi}{3}\right) = \sin\left(\frac{\pi}{2}\right) = 1. \blacksquare$$

EXAMPLE 13 For what values of (x, y) is the function

$$f(x, y) = \frac{x^3 + 7xy^5 - 8y^6}{x^2 - y^2}$$

continuous?

Solution. Since f is a rational function in x and y, f is continuous as long as the denominator is not zero, that is, when $y \neq \pm x$. Thus f is continuous at every point (x, y) in $\mathbb{R}^2$ not on one of the lines $y = x$ and $y = -x$. $\blacksquare$

All the ideas in this section can be generalized to functions of three variables. A **neighborhood** of radius r of the point (x_0, y_0, z_0) in $\mathbb{R}^3$ consists of all points in $\mathbb{R}^3$ interior to the sphere

$$(x - x_0)^2 + (y - y_0)^2 + (z - z_0)^2 = r^2.$$

That is, the neighborhood (also called an **open ball**) is described by

$$\{(x, y, z): (x - x_0)^2 + (y - y_0)^2 + (z - z_0)^2 < r^2\}.$$

A typical neighborhood is sketched in Figure 4. The **closed ball** is described by

$$\{(x, y, z): (x - x_0)^2 + (y - y_0)^2 + (z - z_0)^2 \leq r^2\}.$$

With this definition, the definition of the limit $\lim_{(x,y,z)\to(x_0,y_0,z_0)} f(x, y, z) = L$ is analogous to that given in Definition 2. All other definitions, remarks, and theorems given in this section apply to functions of three (or more) variables.

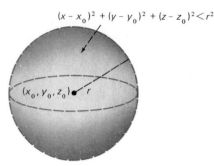

$$(x - x_0)^2 + (y - y_0)^2 + (z - z_0)^2 < r^2$$

FIGURE 4

PROBLEMS 18.2

In Problems 1–5, sketch the indicated region.

1. The open disk centered at $(3, 0)$ with radius 2.
2. The open disk centered at $(3, 2)$ with radius 3.
3. The closed disk centered at $(-1, 1)$ with radius 1.
4. The open ball centered at $(1, 0, 0)$ with radius 1.
5. The closed ball centered at $(0, 1, 1)$ with radius 2.

In Problems 6–11, use Definition 2 to verify the indicated limit.

6. $\lim\limits_{(x,y)\to(1,2)} (3x + y) = 5$

7. $\lim\limits_{(x,y)\to(3,-1)} (x - 7y) = 10$

8. $\lim\limits_{(x,y)\to(5,-2)} (ax + by) = 5a - 2b$

*9. $\lim\limits_{(x,y)\to(1,1)} \dfrac{x}{y} = 1$

10. $\lim\limits_{(x,y)\to(0,0)} \dfrac{2x^2 y}{x^2 + y^2} = 0$

11. $\lim\limits_{(x,y)\to(4,1)} (x^2 + 3y^2) = 19$

In Problems 12–21, show that the given limit does not exist.

12. $\lim\limits_{(x,y)\to(0,0)} \dfrac{x + y}{x - y}$

13. $\lim\limits_{(x,y)\to(0,0)} \dfrac{xy}{x^2 - y^2}$

14. $\lim\limits_{(x,y)\to(0,0)} \dfrac{xy}{x^2 + y^2}$

15. $\lim\limits_{(x,y)\to(0,0)} \dfrac{xy^3}{x^4 + y^4}$

16. $\lim\limits_{(x,y)\to(0,0)} \dfrac{xy}{x^3 + y^3}$

17. $\lim\limits_{(x,y)\to(0,0)} \dfrac{(x^2 + y^2)^2}{x^4 + y^4}$

18. $\lim\limits_{(x,y)\to(0,0)} \dfrac{x^2 - 2y}{y^2 + 2x}$

19. $\lim\limits_{(x,y)\to(0,0)} \dfrac{ax^2 + by}{cy^2 + dx}, a, b, c, d > 0$

20. $\lim\limits_{(x,y,z)\to(0,0,0)} \dfrac{xy + 2xz + 3yz}{x^2 + y^2 + z^2}$

21. $\lim\limits_{(x,y,z)\to(0,0,0)} \dfrac{xyz}{x^3 + y^3 + z^3}$

In Problems 22–25, show that the indicated limit exists and calculate it.

22. $\lim\limits_{(x,y)\to(0,0)} \dfrac{3xy}{\sqrt{x^2 + y^2}}$

23. $\lim\limits_{(x,y)\to(0,0)} \dfrac{5x^2 y^2}{x^4 + y^2}$

24. $\displaystyle\lim_{(x,y)\to(0,0)} \frac{x^3 + y^3}{x^2 + y^2}$

25. $\displaystyle\lim_{(x,y,z)\to(0,0,0)} \frac{yx^2 + z^3}{x^2 + y^2 + z^2}$

In Problems 26–35, calculate the indicated limit.

26. $\displaystyle\lim_{(x,y)\to(-1,2)} (xy + 4y^2x^3)$

27. $\displaystyle\lim_{(x,y)\to(-1,2)} \frac{4x^3y^2 - 2xy^5 + 7y - 1}{3xy - y^4 + 3x^3}$

28. $\displaystyle\lim_{(x,y)\to(-4,3)} \frac{1 + xy}{1 - xy}$

29. $\displaystyle\lim_{(x,y)\to(\pi,\pi/3)} \ln(x + y)$

30. $\displaystyle\lim_{(x,y)\to(1,2)} \ln(1 + e^{x+y})$

31. $\displaystyle\lim_{(x,y)\to(2,5)} \sinh\left(\frac{x + 1}{y - 2}\right)$

*32. $\displaystyle\lim_{(x,y)\to(1,1)} \frac{x - y}{x^2 - y^2}$

33. $\displaystyle\lim_{(x,y)\to(2,2)} \frac{x^3 - 2xy + 3x^2 - 2y}{x^2y + 4y^2 - 6x^2 + 24y}$

34. $\displaystyle\lim_{(x,y,z)\to(1,1,3)} \frac{xy^2 - 4xz^2 + 5yz}{3z^2 - 8z^3y^7x^4 + 7x - y + 2}$

35. $\displaystyle\lim_{(x,y,z)\to(4,1,3)} \ln(x - yz + 4x^3y^5z)$

In Problems 36–49, describe the maximum region over which the given function is continuous.

36. $f(x, y) = \sqrt{x - y}$

37. $f(x, y) = \dfrac{x^3 + 4xy^6 - 7x^4}{x^2 + y^2}$

38. $f(x, y) = \dfrac{x^3 + 4xy^6 - 7x^4}{x^3 - y^3}$

*39. $f(x, y) = \dfrac{xy^3 - 17x^2y^5 + 8x^3y}{xy + 3y - 4x - 12}$

40. $f(x, y) = \ln(3x + 2y + 6)$

41. $f(x, y) = \tan^{-1}(x - y)$

42. $f(x, y) = e^{xy+2}$

43. $f(x, y) = \dfrac{x^3 - 1 + 3y^5x^2}{1 - xy}$

44. $f(x, y) = \dfrac{x}{\sqrt{1 - (x^2/4) - y^2}}$

45. $f(x, y, z) = e^{(xy + yz - \sqrt{x})}$

46. $f(x, y, z) = \dfrac{xyz^2 + yzx^2 - 3x^3yz^5}{x - y + 2z + 4}$

47. $f(x, y, z) = y \ln(xz)$

48. $f(x, y) = \cos^{-1}(x^2 - y)$

49. $f(x, y, z) = \dfrac{1}{\sqrt{1 - x^2 - y^2 - z^2}}$

50. Find a function $g(x)$ such that the function

$$f(x, y) = \begin{cases} \dfrac{x^2 - y^2}{x - y}, & x \neq y \\ g(x), & x = y \end{cases}$$

is continuous at every point in $\mathbb{R}^2$.

51. Find a number c such that the function

$$f(x, y) = \begin{cases} \dfrac{3xy}{\sqrt{x^2 + y^2}}, & (x, y) \neq (0, 0) \\ c, & (x, y) = (0, 0) \end{cases}$$

is continuous at the origin.

52. Find a number c such that

$$f(x, y) = \begin{cases} \dfrac{xy}{|x| + |y|}, & \text{if } (x, y) \neq (0, 0) \\ c, & \text{if } (x, y) = (0, 0) \end{cases}$$

is continuous at the origin.

*53. Discuss the continuity at the origin of

$$f(x, y, z) = \begin{cases} \dfrac{yz - x^2}{x^2 + y^2 + z^2}, & \text{if } (x, y, z) \neq (0, 0, 0) \\ 0, & \text{if } (x, y, z) = (0, 0, 0). \end{cases}$$

18.3 PARTIAL DERIVATIVES

In this section we show one of the ways a function of several variables can be differentiated. The idea is simple. Let $z = f(x, y)$. If we keep one of the variables, say y, fixed, then f can be treated as a function of x only and we can calculate the derivative (if it exists) of f with respect to x. This new function is called the *partial derivative of f with respect to x* and is denoted $\partial f / \partial x$.† Before giving a more formal definition, we give an example.

EXAMPLE 1   Let $z = f(x, y) = x^2 y + \sin xy^2$. Calculate $\partial f / \partial x$.

Solution. Treating y as if it were a constant, we have

$$\frac{\partial f}{\partial x} = \frac{\partial}{\partial x}(x^2 y + \sin xy^2) = \frac{\partial}{\partial x} x^2 y + \frac{\partial}{\partial x} \sin xy^2 = 2xy + y^2 \cos xy^2. \quad \blacksquare$$

Definition 1   PARTIAL DERIVATIVES   Let $z = f(x, y)$.

(i) The **partial derivative of f with respect to** x is the function

$$\frac{\partial z}{\partial x} = \frac{\partial f}{\partial x} = \lim_{\Delta x \to 0} \frac{f(x + \Delta x, y) - f(x, y)}{\Delta x}. \tag{1}$$

$\partial f / \partial x$ is defined at every point (x, y) in the domain of f such that the limit (1) exists.

(ii) The **partial derivative of f with respect to** y is the function

$$\frac{\partial z}{\partial y} = \frac{\partial f}{\partial y} = \lim_{\Delta y \to 0} \frac{f(x, y + \Delta y) - f(x, y)}{\Delta y}. \tag{2}$$

---

†The symbol ∂ of partial derivatives is not a letter from any alphabet but an invented mathematical symbol, which may be read "partial." Historically, the difference between an ordinary and partial derivative was not recognized at first, and the same symbol d was used for both. The symbol ∂ was introduced in the eighteenth century by the mathematicians Alexis Fontaine des Bertins (1705–1771), Leonhard Euler (1707–1783), Alexis-Claude Clairaut (1713–1765), and Jean Le Rond d'Alembert (1717–1783) and was used in their development of the theory of partial differentiation.

$\partial f/\partial y$ is defined at every point (x, y) in the domain of f such that the limit (2) exists.

REMARK 1. This definition allows us to calculate partial derivatives in the same way we calculate ordinary derivatives by allowing only one of the variables to vary. It also allows us to use all the formulas from one-variable calculus.

REMARK 2. The partial derivatives $\partial f/\partial x$ and $\partial f/\partial y$ give us the rate of change of f as each of the variables x and y change with the other one held fixed. They do *not* tell us how f changes when x and y change simultaneously. We will discuss this different topic in Section 18.5.

REMARK 3. It should be emphasized that while the functions $\partial f/\partial x$ and $\partial f/\partial y$ are computed with one of the variables held constant, each is a function of both variables.

EXAMPLE 2 Let $f(x, y) = \sqrt{x + y^2}$. Calculate $\partial f/\partial x$ and $\partial f/\partial y$.

Solution.

$$\frac{\partial f}{\partial x} = \frac{1}{2\sqrt{x + y^2}} \frac{\partial}{\partial x}(x + y^2) = \frac{1}{2\sqrt{x + y^2}}(1 + 0) = \frac{1}{2\sqrt{x + y^2}}$$

since we are treating y as a constant.

$$\frac{\partial f}{\partial y} = \frac{1}{2\sqrt{x + y^2}} \frac{\partial}{\partial y}(x + y^2) = \frac{1}{2\sqrt{x + y^2}}(0 + 2y) = \frac{y}{\sqrt{x + y^2}}$$

since we are treating x as a constant. ■

EXAMPLE 3 Let $f(x, y) = (1 + x^2 + y^5)^{4/3}$. Calculate $\partial f/\partial x$ and $\partial f/\partial y$ at the point $(3, 1)$.

Solution.

$$\frac{\partial f}{\partial x} = \frac{4}{3}(1 + x^2 + y^5)^{1/3} \frac{\partial}{\partial x}(1 + x^2 + y^5)$$

$$= \frac{4}{3}(1 + x^2 + y^5)^{1/3} \cdot 2x = \frac{8x}{3}(1 + x^2 + y^5)^{1/3}.$$

At $(3, 1)$,

$$\frac{\partial f}{\partial x} = \frac{(8)(3)}{3}(1 + 3^2 + 1^5)^{1/3} = 8\sqrt[3]{11}.$$

Also,

$$\frac{\partial f}{\partial y} = \frac{4}{3}(1 + x^2 + y^5)^{1/3} \frac{\partial}{\partial y}(1 + x^2 + y^5)$$

$$= \frac{4}{3}(1 + x^2 + y^5)^{1/3} \cdot 5y^4 = \frac{20y^4}{3}(1 + x^2 + y^5)^{1/3}.$$

At (3, 1),

$$\frac{\partial f}{\partial y} = \frac{20}{3} \sqrt[3]{11}. \quad \blacksquare$$

We now obtain a geometric interpretation of the partial derivative. Let $z = f(x, y)$. As we saw in Section 18.1, this is the equation of a surface in $\mathbb{R}^3$. To obtain $\partial z/\partial x$, we hold y fixed at some constant value y_0. The equation $y = y_0$ is a plane in space parallel to the xz-plane (whose equation is $y = 0$). Thus if y is constant, $\partial z/\partial x$ is the rate of change of f with respect to x as x changes along the curve C, which is at the intersection of the surface $z = f(x, y)$ and the plane $y = y_0$. This is illustrated in Figure 1. To be more precise, if (x_0, y_0, z_0) is a point on the surface $z = f(x, y)$, then $\partial z/\partial x$ evaluated at (x_0, y_0) is the slope of the line tangent to the surface at the point (x_0, y_0, z_0) that lies in the plane $y = y_0$. Analogously, $\partial z/\partial y$ evaluated at (x_0, y_0) is the slope of the line tangent to the surface at the point (x_0, y_0, z_0) that lies in the plane $x = x_0$ (since x is held fixed in order to calculate $\partial z/\partial y$). This is illustrated in Figure 2.

There are other ways to denote partial derivatives. We will often write

$$f_x = \frac{\partial f}{\partial x} \quad \text{and} \quad f_y = \frac{\partial f}{\partial y}. \tag{3}$$

If f is a function of other variables, say s and t, then we may write $\partial f/\partial s = f_s$ and $\partial f/\partial t = f_t$.

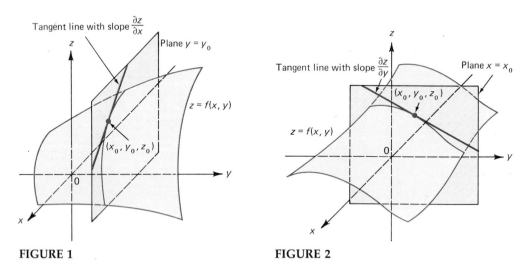

FIGURE 1   FIGURE 2

EXAMPLE 4   As we calculated in Section 9.1, the volume of a cone of radius r and height h is given by

$$V = \tfrac{1}{3}\pi r^2 h.$$

Then the rate of change of V with respect to r (with h fixed) is given by

$$\frac{\partial V}{\partial r} = V_r = \frac{\partial}{\partial r}\left(\frac{1}{3}\pi r^2 h\right) = \frac{2}{3}\pi r h,$$

and the rate of change of V with respect to h (with r fixed) is given by

$$\frac{\partial V}{\partial h} = V_h = \frac{\partial}{\partial h}\left(\frac{1}{3}\pi r^2 h\right) = \frac{1}{3}\pi r^2. \ \blacksquare$$

We now turn to the question of finding partial derivatives of functions of three variables.

Definition 2 PARTIAL DERIVATIVES Let $w = f(x, y, z)$ be defined in a neighborhood of the point (x, y, z).

   (i) The **partial derivative of f with respect to** x is the function

$$\frac{\partial w}{\partial x} = \frac{\partial f}{\partial x} = f_x = \lim_{\Delta x \to 0}\frac{f(x + \Delta x, y, z) - f(x, y, z)}{\Delta x}. \tag{4}$$

f_x is defined at every point (x, y, z) in the domain of f at which the limit in (4) exists.
   (ii) The **partial derivative of f with respect to** y is the function

$$\frac{\partial w}{\partial y} = \frac{\partial f}{\partial y} = f_y = \lim_{\Delta y \to 0}\frac{f(x, y + \Delta y, z) - f(x, y, z)}{\Delta y}. \tag{5}$$

f_y is defined at every point (x, y, z) in the domain of f at which the limit in (5) exists.
   (iii) The **partial derivative of f with respect to** z is the function

$$\frac{\partial w}{\partial z} = \frac{\partial f}{\partial z} = f_z = \lim_{\Delta z \to 0}\frac{f(x, y, z + \Delta z) - f(x, y, z)}{\Delta z}. \tag{6}$$

f_z is defined at any point (x, y, z) in the domain of f at which the limit in (6) exists.

REMARK. As can be seen from Definition 2, to calculate $\partial f/\partial x$, we simply treat y and z as constants and then calculate an ordinary derivative.

EXAMPLE 5 Let $w = f(x, y, z) = xz + e^{y^2 z} + \sqrt{xy^2 z^3}$. Calculate $\partial w/\partial x$, $\partial w/\partial y$, and $\partial w/\partial z$.

   Solution. To calculate $\partial w/\partial x$, we keep y and z fixed. Then

$$\frac{\partial w}{\partial x} = \frac{\partial f}{\partial x} = f_x = \frac{\partial}{\partial x}xz + \frac{\partial}{\partial x}e^{y^2 z} + \frac{1}{2\sqrt{xy^2 z^3}}\frac{\partial}{\partial x}(xy^2 z^3)$$

$$= z + 0 + \frac{y^2 z^3}{2\sqrt{xy^2 z^3}} = z + \frac{y^2 z^3}{2\sqrt{xy^2 z^3}}.$$

To calculate $\partial w/\partial y$, we keep x and z fixed. Then

$$\frac{\partial w}{\partial y} = \frac{\partial f}{\partial y} = f_y = \frac{\partial}{\partial y}xz + e^{y^2z}\frac{\partial}{\partial y}(y^2z) + \frac{1}{2\sqrt{xy^2z^3}}\frac{\partial}{\partial y}(xy^2z^3)$$

$$= 0 + 2yze^{y^2z} + \frac{2xyz^3}{2\sqrt{xy^2z^3}} = 2yze^{y^2z} + \frac{xyz^3}{\sqrt{xy^2z^3}}.$$

To calculate $\partial w/\partial z$, we keep x and y fixed. Then

$$\frac{\partial w}{\partial z} = \frac{\partial f}{\partial z} = f_z = x + e^{y^2z}\frac{\partial}{\partial z}y^2z + \frac{1}{2\sqrt{xy^2z^3}}\frac{\partial}{\partial z}(xy^2z^3)$$

$$= x + y^2e^{y^2z} + \frac{3xy^2z^2}{2\sqrt{xy^2z^3}}. \blacksquare$$

EXAMPLE 6   Let C denote the oxygen consumption (per unit weight) of a fur-bearing animal, let T denote its internal body temperature (in degrees Celsius), let t denote the outside temperature of its fur, and let w denote its weight (in kilograms). It has been experimentally determined that if T is considerably larger than t, then a reasonable model for the oxygen consumption of the animal is given by

$$C = \frac{5(T - t)}{2w^{2/3}}.$$

Calculate (a) C_T, (b) C_t, and (c) C_w.

Solution.

(a) $C_T = \dfrac{\partial C}{\partial T} = \dfrac{\partial}{\partial T}\left(\dfrac{5T}{2w^{2/3}} - \dfrac{5t}{2w^{2/3}}\right) = \dfrac{5}{2w^{2/3}}$

(b) $C_t = -\dfrac{5}{2w^{2/3}}$

(c) $C_w = \dfrac{\partial}{\partial w}\dfrac{5}{2}(T - t)w^{-2/3} = -\dfrac{5}{3}(T - t)w^{-5/3}$

Note that (a) and (b) imply that since $w > 0$, an increase in internal body temperature leads to an increase in oxygen consumption (if t and w do not change) while an increase in fur temperature leads to a decrease in oxygen consumption (with T and w held constant). Does this result make sense intuitively? Furthermore, if T and t are held constant, then since $T > t$, an increase in the animal's weight will lead to a decrease in its oxygen consumption. $\blacksquare$

We can use Definitions 1 and 2 to develop a definition of partial derivatives of functions of four or more variables. Although we will not give a formal definition, the general idea is to hold all the variables but one constant and then perform an "ordinary" differentiation.

In Section 2.7 we showed that a differentiable function is continuous. The situation is more complicated for a function of two variables. In fact, there are functions that are discontinuous at a point but for which all partial derivatives exist at that point.

EXAMPLE 7   Let

$$f(x, y) = \begin{cases} \dfrac{xy}{x^2 + y^2}, & (x\,y) \neq (0, 0) \\ 0, & (x, y) = (0, 0). \end{cases}$$

Show that f_x and f_y exist at $(0, 0)$ but that f is not continuous there.

Solution.   We first show that $\lim_{(x,y)\to(0,0)} f(x, y)$ does not exist, so that f cannot be continuous at $(0, 0)$. To show that, we first let $(x, y) \to (0, 0)$ along the line $y = x$. Then

$$\frac{xy}{x^2 + y^2} = \frac{x^2}{x^2 + x^2} = \frac{1}{2},$$

so if $\lim_{(x,y)\to(0,0)} f(x, y)$ existed, it would have to equal $\frac{1}{2}$. But if we now let $(x, y) \to (0, 0)$ along the line $y = -x$, we have

$$\frac{xy}{x^2 + y^2} = \frac{-x^2}{x^2 + x^2} = -\frac{1}{2},$$

so along this line the limit is $-\frac{1}{2}$. Hence the limit does not exist. On the other hand, we have

$$f_x(0, 0) = \lim_{\Delta x \to 0} \frac{f(0 + \Delta x, 0) - f(0, 0)}{\Delta x} = \lim_{\Delta x \to 0} \frac{\dfrac{(0 + \Delta x) \cdot 0}{\Delta x^2 + 0^2}}{\Delta x}$$

$$= \lim_{\Delta x \to 0} \frac{0}{\Delta x} = \lim_{\Delta x \to 0} 0 = 0.$$

Similarly, $f_y(0, 0) = 0$. Hence both f_x and f_y exist at $(0, 0)$ (and are equal to zero) even though f is not continuous there. ■

In Section 18.5 we will show a relationship between continuity and partial derivatives and show that a certain kind of differentiability does imply continuity.

PROBLEMS 18.3

In Problems 1–12, calculate $\partial z/\partial x$ and $\partial z/\partial y$.

1. $z = x^2 y$
2. $z = xy^2$
3. $z = 3e^{xy^3}$
4. $z = \sin(x^2 + y^3)$
5. $z = 4x/y^5$
6. $z = e^y \tan x$
7. $z = \ln(x^3 y^5 - 2)$
8. $z = \sqrt{xy + 2y^3}$
9. $z = (x + 5y \sin x)^{4/3}$
10. $z = \sinh(2x - y)$
*11. $z = \sin^{-1}\left(\dfrac{x^2 - y^2}{x^2 + y^2}\right)$
12. $z = \sec xy$

In Problems 13–20, calculate the value of the given partial derivative at the given point.

13. $f(x, y) = x^3 - y^4; f_x(1, -1)$

14. $f(x, y) = \ln(x^2 + y^4); f_y(3, 1)$

15. $f(x, y) = \sin(x + y); f_x(\pi/6, \pi/3)$

16. $f(x, y) = e^{\sqrt{x^2 + y}}; f_y(0, 4)$

17. $f(x, y) = \sinh(x - y); f_x(3, 3)$

18. $f(x, y) = \sqrt[3]{x^2 y} - y^2 x^5; f_y(-2, 4)$

19. $f(x, y) = \dfrac{x^2 - y^2}{x^2 + y^2}; f_y(2, -3)$

20. $f(x, y) = \tan^{-1}\dfrac{y}{x}; f_x(4, 4)$

In Problems 21–29, calculate $\partial w/\partial x$, $\partial w/\partial y$, and $\partial w/\partial z$.

21. $w = xyz$

22. $w = \sqrt{x + y + z}$

23. $w = \dfrac{x + y}{z}$

24. $w = \dfrac{x^2 - y^2 + z^2}{x^2 + y^2 + z^2}$

25. $w = \ln(x^3 + y^2 + z)$

26. $w = e^{x + 2y + 3x}$

27. $w = \sin(xyz)$

28. $w = \tan^{-1}\dfrac{xz}{y}$

29. $w = \cosh\sqrt{x + 2y + 5z}$

In Problems 30–39, calculate the value of the given partial derivative at the given point.

30. $f(x, y, z) = xyz; f_x(2, 3, 4)$

31. $f(x, y, z) = \sqrt{x + 2y + 3z}; f_y(2, -1, 3)$

32. $f(x, y, z) = \dfrac{x - y}{z}; f_z(-3, -1, 2)$

33. $f(x, y, z) = \sin(z^2 - y^2 + x); f_y(0, 1, 0)$

34. $f(x, y, z) = \ln(x + 2y + 3z); f_z(2, 2, 5)$

35. $f(x, y, z) = \tan^{-1}\dfrac{xy}{z}; f_x(1, 2, -2)$

36. $f(x, y, z) = \dfrac{y^3 - z^5}{x^2 y + z}; f_y(4, 0, 1)$

37. $f(x, y, z) = \sqrt{\dfrac{x + y - z}{x + y + z}}; f_z(1, 1, 1)$

38. $f(x, y, z) = e^{xy}(\cosh z - \sinh z); f_z(2, 3, 0)$

39. $f(x, y, z) = \sqrt{x^2 + y^2 + z^2}; f_x(a, b, c)$

40. Let $f(x, y) = \begin{cases} \dfrac{x + y}{x - y}, & (x, y) \neq (0, 0) \\ 0, & (x, y) = (0, 0). \end{cases}$

(a) Show that f is not continuous at $(0, 0)$. [*Hint:* Show that $\lim_{(x, y) \to (0, 0)} f(x, y)$ does not exist.]
(b) Do $f_x(0, 0)$ and $f_y(0, 0)$ exist?

41. A **partial differential equation** is an equation involving partial derivatives. Show that the function $z = f(x, y) = e^{(x + \sqrt{3}y)/4} - 4x - 2y - 4 - 2\sqrt{3}$ satisfies the partial differential equation $\partial z/\partial x + \sqrt{3}\,\partial z/\partial y - z = 4x + 2y$.

42. Show that the function $f(x, y) = z = e^{x + (5/4)y} - \frac{7}{2}e^{y/2} + \frac{7}{2}$ is a solution to the partial differential equation $3f_x - 4f_y + 2f = 7$ that also satisfies $f(x, 0) = e^x$.

43. Find the equation of the line tangent to the surface $z = x^3 - 4y^3$ at the point $(1, -1, 5)$ that (a) lies in the plane $x = 1$; (b) lies in the plane $y = -1$.

44. Find the equation of the line tangent to the surface $z = \tan^{-1}(y/x)$ at the point $(\sqrt{3}, 1, \pi/6)$ that (a) lies in the plane $x = \sqrt{3}$; (b) lies in the plane $y = 1$.

45. Find the equation of the line tangent to the surface $x^2 + 4y^2 + 4z^2 = 9$ that lies on the plane $y = 1$ at the point $(1, 1, 1)$.

46. The ideal gas law states that $P = nRT/V$. Assume that the number of moles n of an ideal gas and the temperature T of the gas are held constant at the values 10 and 20°C

($= 293$ °K), respectively. What is the rate of change of the pressure, as a function of the volume, when the volume of the gas is 2 liters?

47. For any ideal gas show that $(\partial V/\partial T)(\partial T/\partial P)(\partial P/\partial V) = -1$.

48. In Section 6.7 we showed that the present value of an annuity for which B dollars are to be paid every year for t years is given by

$$A_0 = \frac{B}{i}\left[1 - \left(\frac{1}{1+i}\right)^t\right]$$

[see equation (6.7.10)].

(**a**) If t is fixed, how does the present value of the annuity change as the rate of interest changes?

(**b**) How is the present value changing with respect to i if $B = \$500$ and $i = 6\%$?

(**c**) If i is fixed, how does the present value of the annuity change as the number of years during which the payments are made is increased?

49. A fur-bearing animal weighing 10 kg has a constant internal body temperature of 23°C. Using the model of Example 6, if the outside temperature is dropping, how is the oxygen consumption of the animal changing when the outside temperature of its fur is 5°C?

50. The cost to a manufacturer of producing x units of product A and y units of product B is given (in dollars) by

$$C(x, y) = \frac{50}{2+x} + \frac{125}{(3+y)^2}.$$

Calculate the marginal cost of each of the two products.

51. The revenue received from the manufacturer of Problem 50 is given by

$$R(x, y) = \ln(1 + 50x + 75y) + \sqrt{1 + 40x + 125y}.$$

Calculate the marginal revenue from each of the two products.

52. If a particle is falling in a fluid, then according to *Stokes' law* the velocity of the particle is given by

$$V = \frac{2g}{9}\,(\rho_P - \rho_f)\frac{r^2}{\eta},$$

where g is the acceleration due to gravity, ρ_P is the density of the particle, ρ_f is the density of the fluid, r is the radius of the particle (in centimeters), and η is the absolute viscosity of the liquid. Calculate $V\rho_P$, $V\rho_f$, V_r, and V_η.

53. Let $f(x, y) = g(x)h(y)$, where g and h are differentiable functions of a single variable. Show that the partial derivatives $\partial f/\partial x$ and $\partial f/\partial y$ exist and that

$$\frac{\partial f}{\partial x} = g'(x)h(y) \qquad \text{and} \qquad \frac{\partial f}{\partial y} = g(x)h'(y).$$

54. Let g, h, and k be differentiable and let $f(x, y, z) = g(x)h(y)k(z)$. Show that $\partial f/\partial x$, $\partial f/\partial y$, and $\partial f/\partial z$ all exist and

$$\frac{\partial f}{\partial x} = g'(x)h(y)k(z), \qquad \frac{\partial f}{\partial y} = g(x)h'(y)k(z), \qquad \frac{\partial f}{\partial z} = g(x)h(y)k'(z).$$

In Problems 55–58, assume that $\partial f/\partial x$ and $\partial g/\partial x$ exist at every point (x, y) in $\mathbb{R}^2$.

55. Prove that $(\partial/\partial x)(f + g)$ exists and

$$\frac{\partial}{\partial x}(f + g) = \frac{\partial f}{\partial x} + \frac{\partial g}{\partial x}.$$

56. Prove that $(\partial/\partial x)(af)$ exists for every constant a and $(\partial/\partial x)(af) = a(\partial f/\partial x)$.

57. Prove that $(\partial/\partial x)(fg)$ exists and is equal to $f(\partial g/\partial x) + g(\partial f/\partial x)$.

58. If $\partial g/\partial x \neq 0$, prove that $(\partial/\partial x)(f/g)$ exists and is equal to $[g(\partial f/\partial x) - f(\partial g/\partial x)]/g^2$.

18.4 HIGHER-ORDER PARTIAL DERIVATIVES

We have seen that if $y = f(x)$, then

$$y' = \frac{df}{dx} \quad \text{and} \quad y'' = \frac{d^2f}{dx^2} = \frac{d}{dx}\left(\frac{df}{dx}\right).$$

That is, the second derivative of f is the derivative of the first derivative of f. Analogously, if $z = f(x, y)$, then we can differentiate each of the two "first" partial derivatives $\partial f/\partial x$ and $\partial f/\partial y$ with respect to both x and y to obtain four **second partial derivatives** as follows:

Definition 1   SECOND PARTIAL DERIVATIVES

(i) Differentiate twice with respect to x:

$$\frac{\partial^2 z}{\partial x^2} = \frac{\partial^2 f}{\partial x^2} = f_{xx} = \frac{\partial}{\partial x}\left(\frac{\partial f}{\partial x}\right). \tag{1}$$

(ii) Differentiate first with respect to x and then with respect to y:

$$\frac{\partial^2 z}{\partial y\, \partial x} = \frac{\partial^2 f}{\partial y\, \partial x} = f_{xy} = \frac{\partial}{\partial y}\left(\frac{\partial f}{\partial x}\right). \tag{2}$$

(iii) Differentiate first with respect to y and then with respect to x:

$$\frac{\partial^2 z}{\partial x\, \partial y} = \frac{\partial^2 f}{\partial x\, \partial y} = f_{yx} = \frac{\partial}{\partial x}\left(\frac{\partial f}{\partial y}\right). \tag{3}$$

(iv) Differentiate twice with respect to y:

$$\frac{\partial^2 z}{\partial y^2} = \frac{\partial^2 f}{\partial y^2} = f_{yy} = \frac{\partial}{\partial y}\left(\frac{\partial f}{\partial y}\right). \tag{4}$$

REMARK 1.   The derivatives $\partial^2 f/\partial x\, \partial y$ and $\partial^2 f/\partial y\, \partial x$ are called the **mixed second partials.**

REMARK 2.   It is much easier to denote the second partials by f_{xx}, f_{xy}, f_{yx}, and f_{yy}. We will therefore use this notation for the remainder of this section. Note that the symbol f_{xy} indicates that we differentiate first with respect to x and then with respect to y.

EXAMPLE 1   Let $z = f(x, y) = x^3y^2 - xy^5$. Calculate the four second partial derivatives.

Solution.   We have $f_x = 3x^2y^2 - y^5$ and $f_y = 2x^3y - 5xy^4$.

(a) $f_{xx} = \dfrac{\partial}{\partial x}(f_x) = 6xy^2$

(b) $f_{xy} = \dfrac{\partial}{\partial y}(f_x) = 6x^2y - 5y^4$

(c) $f_{yx} = \dfrac{\partial}{\partial x}(f_y) = 6x^2y - 5y^4$

(d) $f_{yy} = \dfrac{\partial}{\partial y}(f_y) = 2x^3 - 20xy^3$ ∎

In Example 1 we saw that $f_{xy} = f_{yx}$. This result is no accident, as we see by the following theorem whose proof can be found in any intermediate calculus text.†

Theorem 1   Suppose that f, f_x, f_y, f_{xy}, and f_{yx} are all continuous at (x_0, y_0). Then

$$f_{xy}(x_0, y_0) = f_{yx}(x_0, y_0). \tag{5}$$

This result is often referred to as the **equality of mixed partials.**‡

The definition of second partial derivatives and the theorem on the equality of mixed partials are easily extended to functions of three variables. If $w = f(x, y, z)$, then we have the nine second partial derivatives (assuming that they exist):

$$\frac{\partial^2 f}{\partial x^2} = f_{xx}, \qquad \frac{\partial^2 f}{\partial y\, \partial x} = f_{xy}, \qquad \frac{\partial^2 f}{\partial z\, \partial x} = f_{xz},$$

$$\frac{\partial^2 f}{\partial x\, \partial y} = f_{yx}, \qquad \frac{\partial^2 f}{\partial y^2} = f_{yy}, \qquad \frac{\partial^2 f}{\partial z\, \partial y} = f_{yz},$$

$$\frac{\partial^2 f}{\partial x\, \partial z} = f_{zx}, \qquad \frac{\partial^2 f}{\partial y\, \partial z} = f_{zy}, \qquad \frac{\partial^2 f}{\partial z^2} = f_{zz}.$$

Theorem 2   If f, f_x, f_y, f_z, and all six mixed partials are continuous at a point (x_0, y_0, z_0), then at that point

$$f_{xy} = f_{yx}, \qquad f_{xz} = f_{zx}, \qquad f_{yz} = f_{zy}.$$

†See, for example, S. Grossman, *Calculus, Part 2* (Academic Press, Orlando, 1982), page 461.
‡This theorem was first stated by Euler in a 1734 paper devoted to a problem in hydrodynamics.

EXAMPLE 2   Let $f(x, y, z) = xy^3 - zx^5 + x^2yz$. Calculate all nine second partial derivatives and show that all three pairs of mixed partials are equal.

Solution. We have

$$f_x = y^3 - 5zx^4 + 2xyz,$$
$$f_y = 3xy^2 + x^2z,$$

and

$$f_z = -x^5 + x^2y.$$

Then

$$f_{xx} = -20zx^3 + 2yz, \qquad f_{yy} = 6xy, \qquad f_{zz} = 0,$$

$$f_{xy} = \frac{\partial}{\partial y}(y^3 - 5zx^4 + 2xyz) = 3y^2 + 2xz,$$

$$f_{yx} = \frac{\partial}{\partial x}(3xy^2 + x^2z) = 3y^2 + 2xz,$$

$$f_{xz} = \frac{\partial}{\partial z}(y^3 - 5zx^4 + 2xyz) = -5x^4 + 2xy,$$

$$f_{zx} = \frac{\partial}{\partial x}(-x^5 + x^2y) = -5x^4 + 2xy,$$

$$f_{yz} = \frac{\partial}{\partial z}(3xy^2 + x^2z) = x^2,$$

$$f_{zy} = \frac{\partial}{\partial y}(-x^5 + x^2y) = x^2. \quad \blacksquare$$

We conclude this section by pointing out that we can easily define partial derivatives of orders higher than two. For example,

$$f_{zyx} = \frac{\partial^3 f}{\partial x\, \partial y\, \partial z} = \frac{\partial}{\partial x}\left(\frac{\partial^2 f}{\partial y\, \partial z}\right) = \frac{\partial}{\partial x}(f_{zy}).$$

EXAMPLE 3   Calculate f_{xxx}, f_{xzy}, f_{yxz}, and f_{yxzx} for the function of Example 2.

Solution. We easily obtain the three third partial derivatives:

$$f_{xxx} = \frac{\partial}{\partial x}(f_{xx}) = \frac{\partial}{\partial x}(-20zx^3 + 2yz) = -60zx^2$$

$$f_{xzy} = \frac{\partial}{\partial y}(f_{xz}) = \frac{\partial}{\partial y}(-5x^4 + 2xy) = 2x$$

$$f_{yxz} = \frac{\partial}{\partial z}(f_{yx}) = \frac{\partial}{\partial z}(3y^2 + 2xz) = 2x.$$

Note that $f_{xzy} = f_{yxz}$. This again is no accident and follows from the generalization of Theorem 2 to mixed third partial derivatives. Finally, the fourth partial derivative f_{yxzx} is given by

$$f_{yxzx} = \frac{\partial}{\partial x}(f_{yxz}) = \frac{\partial}{\partial x}(2x) = 2. \quad \blacksquare$$

PROBLEMS 18.4

In Problems 1–12, calculate the four second partial derivatives and show that the mixed partials are equal.

1. $f(x, y) = x^2 y$
2. $f(x, y) = xy^2$
3. $f(x, y) = 3e^{xy^3}$
4. $f(x, y) = \sin(x^2 + y^3)$
5. $f(x, y) = \dfrac{4x}{y^5}$
6. $f(x, y) = e^y \tan x$
7. $f(x, y) = \ln(x^3 y^5 - 2)$
8. $f(x, y) = \sqrt{xy + 2y^3}$
9. $f(x, y) = (x + 5y \sin x)^{4/3}$
10. $f(x, y) = \sinh(2x - y)$
*11. $f(x, y) = \sin^{-1}\left(\dfrac{x^2 - y^2}{x^2 + y^2}\right)$
12. $f(x, y) = \sec xy$

In Problems 13–21, calculate the nine second partial derivatives and show that the three pairs of mixed partials are equal.

13. $f(x, y, z) = xyz$
14. $f(x, y, z) = x^2 y^3 z^4$
15. $f(x, y, z) = \dfrac{x + y}{z}$
16. $f(x, y, z) = \sin(x + 2y + z^2)$
17. $f(x, y, z) = \tan^{-1}\dfrac{xz}{y}$
18. $f(x, y, z) = \cos xyz$
19. $f(x, y, z) = e^{3xy} \cos z$
20. $f(x, y, z) = \ln(xy + z)$
21. $f(x, y, z) = \cosh \sqrt{x + yz}$

22. How many third partial derivatives are there for a function of (a) two variables; (b) three variables?
23. How many fourth partial derivatives are there for a function of (a) two variables; (b) three variables?
*24. How many nth partial derivatives are there for a function of (a) two variables; (b) three variables?

In Problems 25–30, calculate the given partial derivative.

25. $f(x, y) = x^2 y^3 + 2y$; f_{xyx}
26. $f(x, y) = \sin(2xy^4)$; f_{xyy}
27. $f(x, y) = \ln(3x - 2y)$; f_{yxy}
28. $f(x, y, z) = x^2 y + y^2 z - 3\sqrt{xz}$; f_{xyz}
29. $f(x, y, z) = \cos(x + 2y + 3z)$; f_{zzx}
30. $f(x, y, z) = e^{xy} \sin z$; f_{zxyx}

31. Consider a string that is tightly stretched between two fixed points 0 and L on the x-axis.

The string is pulled back and released at a time $t = 0$, causing it to vibrate. One position of the string is sketched in Figure 1. Let $y(x, t)$ denote the height of the string at any time $t \geq 0$ and at any point x in the interval $[0, L]$. It can be shown that $y(x, t)$ satisfies the partial differential equation†

$$\frac{\partial^2 y}{\partial t^2} = c^2 \frac{\partial^2 y}{\partial x^2},$$

where c is a constant. This equation is called the one-dimensional **wave equation.** Show that

$$y(x, t) = \tfrac{1}{2}[(x - ct)^2 + (x + ct)^2]$$

is a solution to the wave equation, where c is any constant.

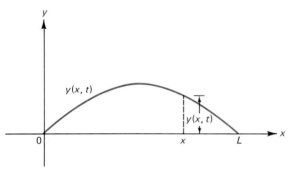

FIGURE 1

32. Consider a cylindrical rod composed of a uniform heat-conducting material of length l and radius r and assume that heat can enter and leave the rod only through its ends. Let $T(x, t)$ denote the absolute temperature at time t at a point x units along the rod (see Figure 2). Then it can be shown‡ that T satisfies the partial differential equation

$$\frac{\partial T}{\partial t} = \delta \frac{\partial^2 T}{\partial x^2},$$

where δ is a positive constant called the **diffusivity** of the rod. This equation is called the **heat equation** (or **diffusion equation**). Show that the function

$$T(x, t) = e^{-\alpha^2 \delta t} \sin \alpha x$$

satisfies the heat equation for any constant α.

FIGURE 2

33. Show that $T(x, y) = (1/\sqrt{t})e^{-x^2/4\delta t}$ satisfies the heat equation.
34. Find constants α and β such that $T(x, t) = e^{\alpha x + \beta t}$ satisfies the heat equation.
35. One of the most important partial differential equations of mathematical physics is **Laplace's equation** *in* $\mathbb{R}^2$, given by

---

†For a derivation of this equation, see W. Derrick and S. Grossman, *Elementary Differential Equations with Applications*, 2nd ed. (Addison-Wesley, Reading, Mass., 1981).

‡See Derrick and Grossman, *Elementary Differential Equations*.

$$\frac{\partial^2 f}{\partial x^2} + \frac{\partial^2 f}{\partial y^2} = 0.$$

Laplace's equation can be used to model steady state heat flow in a closed, bounded region in $\mathbb{R}^2$. Show that the function $f(x, y) = x^2 - y^2$ satisfies Laplace's equation.

36. Show that $f(x, y) = \tan^{-1}(y/x)$ satisfies Laplace's equation.
37. Show that $f(x, y) = \ln(x^2 + y^2)$ satisfies Laplace's equation.
38. Show that $f(x, y) = \sin x \sinh y$ satisfies Laplace's equation.
39. Laplace's equation in $\mathbb{R}^3$ is given by

$$\frac{\partial^2 f}{\partial x^2} + \frac{\partial^2 f}{\partial y^2} + \frac{\partial^2 f}{\partial z^2} = 0.$$

Show that $f(x, y, z) = x^2 + y^2 - 2z^2$ satisfies Laplace's equation in $\mathbb{R}^3$.

18.5 DIFFERENTIABILITY AND THE GRADIENT

In this section we discuss the notion of the differentiability of a function of several variables. There are several ways to introduce this subject and the way we have chosen is designed to illustrate the great similarities between differentiation of functions of one variable and differentiation of functions of several variables.

We begin with a function of one variable,

$$y = f(x).$$

If f is differentiable, then

$$f'(x) = \frac{dy}{dx} = \lim_{\Delta x \to 0} \frac{\Delta y}{\Delta x}. \tag{1}$$

Then if we define the new function $\epsilon(\Delta x)$ by

$$\epsilon(\Delta x) = \frac{\Delta y}{\Delta x} - f'(x), \tag{2}$$

we have

$$\lim_{\Delta x \to 0} \epsilon(\Delta x) = \lim_{\Delta x \to 0} \left(\frac{\Delta y}{\Delta x} - f'(x) \right) = \lim_{\Delta x \to 0} \frac{\Delta y}{\Delta x} - f'(x)$$
$$= f'(x) - f'(x) = 0. \tag{3}$$

Multiplying both sides of (2) by Δx and rearranging terms, we obtain

$$\Delta y = f'(x) \, \Delta x + \epsilon(\Delta x) \, \Delta x.$$

Note that here Δy depends on both Δx and x. Finally, since $\Delta y = f(x + \Delta x) - f(x)$, we obtain

$$f(x + \Delta x) - f(x) = f'(x)\,\Delta x + \epsilon(\Delta x)\,\Delta x. \tag{4}$$

Why did we do all this? We did so in order to be able to state the following alternative definition of differentiability of a function f of one variable.

Definition 1 ALTERNATIVE DEFINITION OF DIFFERENTIABILITY OF A FUNCTION OF ONE VARIABLE   Let f be a function of one variable. Then f is **differentiable** at a number x if there is a function $f'(x)$ and a function $g(\Delta x)$ such that

$$f(x + \Delta x) - f(x) = f'(x)\,\Delta x + g(\Delta x), \tag{5}$$

where $\lim_{\Delta x \to 0}[g(\Delta x)/\Delta x] = 0$.

We will soon show how the definition (5) can be extended to a function of two or more variables. First, we give a definition.

Definition 2 DIFFERENTIABILITY OF A FUNCTION OF TWO VARIABLES   Let f be a real-valued function of two variables that is defined in a neighborhood of a point (x, y) and such that $f_x(x, y)$ and $f_y(x, y)$ exist. Then f is **differentiable** at (x, y) if there exist functions $\epsilon_1(\Delta x, \Delta y)$ and $\epsilon_2(\Delta x, \Delta y)$ such that

$$f(x + \Delta x, y + \Delta y) - f(x, y) = f_x(x, y)\,\Delta x + f_y(x, y)\,\Delta y$$
$$+ \epsilon_1(\Delta x, \Delta y)\,\Delta x + \epsilon_2(\Delta x, \Delta y)\,\Delta y, \tag{6}$$

where

$$\lim_{(\Delta x, \Delta y) \to (0,0)} \epsilon_1(\Delta x, \Delta y) = 0 \quad \text{and} \quad \lim_{(\Delta x, \Delta y) \to (0,0)} \epsilon_2(\Delta x, \Delta y) = 0. \tag{7}$$

EXAMPLE 1   Let $f(x, y) = xy$. Show that f is differentiable at every point (x, y) in $\mathbb{R}^2$.

Solution.

$$f(x + \Delta x, y + \Delta y) - f(x, y) = (x + \Delta x)(y + \Delta y) - xy$$
$$= xy + y\,\Delta x + x\,\Delta y + \Delta x\,\Delta y - xy$$
$$= y\,\Delta x + x\,\Delta y + \Delta x\,\Delta y$$

Now $f_x = y$ and $f_y = x$, so we have

$$f(x + \Delta x, y + \Delta y) - f(x, y) = f_x(x, y)\,\Delta x + f_y(x, y)\,\Delta y + \Delta y\,\Delta x + 0 \cdot \Delta y.$$

Setting $\epsilon_1(\Delta x, \Delta y) = \Delta y$ and $\epsilon_2(\Delta x, \Delta y) = 0$, we see that

$$\lim_{(\Delta x, \Delta y) \to (0,0)} \epsilon_1(\Delta x, \Delta y) = \lim_{(\Delta x, \Delta y) \to (0,0)} \epsilon_2(\Delta x, \Delta y) = 0.$$

This result shows that $f(x, y) = xy$ is differentiable at every point in $\mathbb{R}^2$. ∎

We now rewrite our definition of differentiability in a more compact form. Since a point (x, y) is a vector in $\mathbb{R}^2$, we will write (as we have done before) $\mathbf{x} = (x, y)$. Then if $z = f(x, y)$, we can simply write

$$z = f(\mathbf{x}).$$

Similarly, if $w = f(x, y, z)$, we may write

$$w = f(\mathbf{x}),$$

where $\mathbf{x}$ is the vector (x, y, z). With this notation we may use the symbol $\Delta \mathbf{x}$ to denote the vector $(\Delta x, \Delta y)$ in $\mathbb{R}^2$ or $(\Delta x, \Delta y, \Delta z)$ in $\mathbb{R}^3$.

Next, we write

$$g(\Delta \mathbf{x}) = \epsilon_1(\Delta x, \Delta y)\, \Delta x + \epsilon_2(\Delta x, \Delta y)\, \Delta y. \tag{8}$$

Note that $(\Delta x, \Delta y) \to (0, 0)$ can be written in the compact form $\Delta \mathbf{x} \to \mathbf{0}$. Then if the conditions (7) hold, we see that

$$\lim_{\Delta \mathbf{x} \to 0} \frac{|g(\Delta \mathbf{x})|}{|\Delta \mathbf{x}|} \overset{\nearrow |\Delta \mathbf{x}| = \sqrt{\Delta x^2 + \Delta y^2}}{\leq} \lim_{\Delta \mathbf{x} \to 0} |\epsilon_1(\Delta x, \Delta y)| \frac{|\Delta x|}{\sqrt{\Delta x^2 + \Delta y^2}} + \lim_{\Delta \mathbf{x} \to 0} |\epsilon_2(\Delta x, \Delta y)| \frac{|\Delta y|}{\sqrt{\Delta x^2 + \Delta y^2}}$$

$$\overset{\frac{|\Delta x|}{\sqrt{\Delta x^2 + \Delta y^2}} \leq 1}{\underset{\nearrow}{\leq}} \lim_{\Delta \mathbf{x} \to 0} |\epsilon_1(\Delta x, \Delta y)| + \lim_{\Delta \mathbf{x} \to 0} |\epsilon_2(\Delta x, \Delta y)| = 0 + 0 = 0.$$

Finally, we have the following important definition.

Definition 3   THE GRADIENT   Let f be a function of two variables such that f_x and f_y exist at a point $\mathbf{x} = (x, y)$. Then the **gradient** of f at $\mathbf{x}$, denoted $\nabla f(\mathbf{x})$, is given by

$$\nabla f(\mathbf{x}) = f_x(x, y)\mathbf{i} + f_y(x, y)\mathbf{j}. \tag{9}$$

Note that the gradient of f is a **vector function.** That is, for every point $\mathbf{x}$ in $\mathbb{R}^2$ for which $\nabla f(\mathbf{x})$ is defined, we see that $\nabla f(\mathbf{x})$ is a vector in $\mathbb{R}^2$.

REMARK.   The gradient of f is denoted ∇f, which is read "del" f. This symbol, an inverted Greek delta, was first used in the 1850s, although the name "del" first appeared in print only in 1901. The symbol ∇ is also called *nabla*. This name is used because someone once suggested to the Scottish mathematician Peter Guthrie Tait (1831–1901) that ∇ looks like an Assyrian harp, the Assyrian name of which is nabla.†
Tait, incidentally, was one of the mathematicians who helped carry on Hamilton's development of the theory of quaternions and vectors in the nineteenth century.

---

†Fortunately, most (but certainly not all) of the mathematical terms currently in use have more to do with the objects they describe. We might further point out that ∇ is also called *atled*, which is delta spelled backward (no kidding).

EXAMPLE 2 In Example 1, $f(x, y) = xy$, $f_x = y$, and $f_y = x$, so that

$$\nabla f(\mathbf{x}) = \nabla f(x, y) = y\mathbf{i} + x\mathbf{j}. \ \blacksquare$$

Using this new notation, we observe that

$$\nabla f(\mathbf{x}) \cdot \Delta\mathbf{x} = (f_x\mathbf{i} + f_y\mathbf{j}) \cdot (\Delta x\mathbf{i} + \Delta y\mathbf{j}) = f_x(x, y)\,\Delta x + f_y(x, y)\,\Delta y.$$

Also,

$$f(x + \Delta x, y + \Delta y) = f(\mathbf{x} + \Delta\mathbf{x}).$$

Thus we have the following definition, which is implied by Definition 2.

Definition 4 DIFFERENTIABILITY Let f be a function of two variables that is defined in a neighborhood of a point $\mathbf{x} = (x, y)$. Let $\Delta\mathbf{x} = (\Delta x, \Delta y)$. If $f_x(x, y)$ and $f_y(x, y)$ exist, then f is **differentiable** at $\mathbf{x}$ if there is a function g such that

$$f(\mathbf{x} + \Delta\mathbf{x}) - f(\mathbf{x}) = \nabla f(\mathbf{x}) \cdot \Delta\mathbf{x} + g(\Delta\mathbf{x}), \tag{10}$$

where

$$\lim_{\Delta x \to 0} \frac{g(\Delta\mathbf{x})}{|\Delta\mathbf{x}|} = 0. \tag{11}$$

REMARK. Although formulas (5) and (10) look very similar, there are two fundamental differences. First, f' is a scalar, while ∇f is a vector. Second, $f'(x)\,\Delta x$ is an ordinary product of real numbers, while $\nabla f(\mathbf{x}) \cdot \Delta\mathbf{x}$ is a dot product of vectors.

There are two reasons for giving you this new definition. First, it illustrates that the gradient of a function of two variables is the natural extension of the derivative of a function of one variable. Second, the definition (10), (11) can be used to define the notion of differentiability for a function of three or more variables as well. We will say more about this subject shortly.

One important question remains: What functions are differentiable? A partial answer is given in Theorem 1.

Theorem 1 Let f, f_x, and f_y be defined and continuous in a neighborhood of $\mathbf{x} = (x, y)$. Then f is differentiable at $\mathbf{x}$.

REMARK 1. It is essential that f_x and f_y be continuous in this theorem. In Example 18.3.6 we saw an example of a function for which f_x and f_y exist at $(0, 0)$ but f itself is not continuous there. As we will see in Theorem 2, if f is differentiable at a point, then it is continuous there. Thus the function in Example 18.3.6 is not differentiable at $(0, 0)$ even though $f_x(0, 0)$ and $f_y(0, 0)$ exist.

REMARK 2. If the hypotheses of Theorem 1 are satisfied, then f is said to be **continuously differentiable** at $\mathbf{x}$.

We will postpone the proof until the end of the section.

EXAMPLE 3 Let $z = f(x, y) = \sin xy^2 + e^{x^2y^3}$. Show that f is differentiable and calculate ∇f. Find $\nabla f(1, 1)$.

Solution. $\partial f/\partial x = y^2 \cos xy^2 + 2xy^3 e^{x^2y^3}$ and $\partial f/\partial y = 2xy \cos xy^2 + 3x^2y^2 e^{x^2y^3}$. Since $\partial f/\partial x$ and $\partial f/\partial y$ are continuous, f is differentiable and

$$\nabla f(x, y) = (y^2 \cos xy^2 + 2xy^3 e^{x^2y^3})\mathbf{i} + (2xy \cos xy^2 + 3x^2y^2 e^{x^2y^3})\mathbf{j}.$$

At $(1, 1)$, $\nabla f(1, 1) = (\cos 1 + 2e)\mathbf{i} + (2 \cos 1 + 3e)\mathbf{j}$. ∎

In Section 18.3 we showed that the existence of all of its partial derivatives at a point does *not* ensure that a function is continuous at that point. However, differentiability (according to Definition 4) does ensure continuity.

Theorem 2 If f is differentiable at $\mathbf{x}_0 = (x_0, y_0)$, then f is continuous at $\mathbf{x}_0$.

Proof. We must show that $\lim_{(x,y)\to(x_0,y_0)} f(x, y) = f(x_0, y_0)$, or, equivalently, $\lim_{\mathbf{x}\to\mathbf{x}_0} f(\mathbf{x}) = f(\mathbf{x}_0)$. But if we define $\Delta\mathbf{x}$ by $\Delta\mathbf{x} = \mathbf{x} - \mathbf{x}_0$, this is the same as showing that

$$\lim_{\Delta\mathbf{x}\to 0} f(\mathbf{x}_0 + \Delta\mathbf{x}) = f(\mathbf{x}_0). \tag{12}$$

Since f is differentiable at $\mathbf{x}_0$,

$$f(\mathbf{x}_0 + \Delta\mathbf{x}) - f(\mathbf{x}_0) = \nabla f(\mathbf{x}_0) \cdot \Delta\mathbf{x} + g(\Delta\mathbf{x}). \tag{13}$$

But as $\Delta\mathbf{x} \to 0$, both terms on the right-hand side of (13) approach zero, so

$$\lim_{\Delta\mathbf{x}\to 0} [f(\mathbf{x}_0 + \Delta\mathbf{x}) - f(\mathbf{x}_0)] = 0,$$

which means that (12) holds and the theorem is proved. ∎

The converse to this theorem is false, as it is in one-variable calculus. That is, there are functions that are continuous, but not differentiable, at a given point. For example, the function

$$f(x, y) = \sqrt[3]{x} + \sqrt[3]{y}$$

is continuous at any point (x, y) in $\mathbb{R}^2$. But

$$\nabla f(x, y) = \frac{1}{3x^{2/3}}\mathbf{i} + \frac{1}{3y^{2/3}}\mathbf{j},$$

so f is not differentiable at any point (x, y) for which either x or y is zero. That is, $\nabla f(x, y)$ is not defined on the x- and y-axes. Hence f is not differentiable along these axes.

In Section 3.1 we showed that

$$(f + g)' = f' + g' \qquad \text{and} \qquad (\alpha f)' = \alpha f';$$

that is, the derivative of the sum of two functions is the sum of the derivatives of the two functions and the derivative of a scalar multiple of a function is the scalar times the derivative of the function. These results can be extended to the gradient vector.

Theorem 3   Let f and g be differentiable in a neighborhood of $\mathbf{x} = (x, y)$. Then for every scalar α, αf and $f + g$ are differentiable at $\mathbf{x}$, and

(i) $\nabla(\alpha f) = \alpha \, \nabla f$, and

(ii) $\nabla(f + g) = \nabla f + \nabla g$.

Proof.

   (i) From the definition of differentiability (Definition 4), there is a function $h_1(\Delta \mathbf{x})$ such that

$$f(\mathbf{x} + \Delta \mathbf{x}) - f(\mathbf{x}) = \nabla f(\mathbf{x}) \cdot \Delta \mathbf{x} + h_1(\Delta \mathbf{x}),$$

where $\lim_{\Delta \mathbf{x} \to 0} [h_1(\Delta \mathbf{x})/|\Delta \mathbf{x}|] = 0$. Thus $\alpha f(\mathbf{x} + \Delta \mathbf{x}) - \alpha f(\mathbf{x}) = \alpha \, \nabla f(\mathbf{x}) \cdot \Delta \mathbf{x} + \alpha h_1(\Delta \mathbf{x})$, and

$$\lim_{\Delta \mathbf{x} \to 0} \frac{\alpha h_1(\Delta \mathbf{x})}{|\Delta \mathbf{x}|} = \alpha \lim_{\Delta \mathbf{x} \to 0} \frac{h_1(\Delta \mathbf{x})}{|\Delta \mathbf{x}|} = \alpha 0 = 0.$$

But

$$\overset{\displaystyle \swarrow \, \alpha \frac{\partial f}{\partial x} = \frac{\partial(\alpha f)}{\partial x}}{\alpha \, \nabla f(\mathbf{x}) = \alpha(f_x \mathbf{i} + f_y \mathbf{j}) = (\alpha f)_x \mathbf{i} + (\alpha f)_y \mathbf{j} = \nabla(\alpha f).}$$

Thus

$$\alpha f(\mathbf{x} + \Delta \mathbf{x}) - \alpha f(\mathbf{x}) = \nabla \alpha f(\mathbf{x}) \cdot \Delta \mathbf{x} + \alpha h_1(\Delta \mathbf{x}),$$

which shows that αf is differentiable and $\nabla(\alpha f) = \alpha \nabla f$.
   (ii) As above, there is a function $h_2(\Delta \mathbf{x})$ such that $g(\mathbf{x} + \Delta \mathbf{x}) - g(\mathbf{x}) = \nabla g(\mathbf{x}) \cdot \Delta \mathbf{x} + h_2(\Delta \mathbf{x})$, where $\lim_{\Delta \mathbf{x} \to 0} [h_2(\Delta \mathbf{x})/|\Delta \mathbf{x}|] = 0$. Thus

$$\begin{aligned}
(f + g)(\mathbf{x} + \Delta \mathbf{x}) - (f + g)(\mathbf{x}) &= [f(\mathbf{x} + \Delta \mathbf{x}) + g(\mathbf{x} + \Delta \mathbf{x})] - [f(\mathbf{x}) + g(\mathbf{x})] \\
&= [f(\mathbf{x} + \Delta \mathbf{x}) - f(\mathbf{x})] + [g(\mathbf{x} + \Delta \mathbf{x}) - g(\mathbf{x})] \\
&= \nabla f(\mathbf{x}) \cdot \Delta \mathbf{x} + h_1(\Delta \mathbf{x}) + \nabla g(\mathbf{x}) \cdot \Delta \mathbf{x} + h_2(\Delta \mathbf{x}) \\
&= [\nabla f(\mathbf{x}) + \nabla g(\mathbf{x})] \cdot \Delta \mathbf{x} + [h_1(\Delta \mathbf{x}) + h_2(\Delta \mathbf{x})],
\end{aligned}$$

where

$$\lim_{\Delta x \to 0} \frac{[h_1(\Delta x) + h_2(\Delta x)]}{|\Delta x|} = 0.$$

To complete the proof, we observe that

$$\nabla f(\mathbf{x}) + \nabla g(\mathbf{x}) = (f_x \mathbf{i} + f_y \mathbf{j}) + (g_x \mathbf{i} + g_y \mathbf{j}) = (f_x + g_x)\mathbf{i} + (f_y + g_y)\mathbf{j}$$

$$\frac{\partial}{\partial x}(f + g) = \frac{\partial f}{\partial x} + \frac{\partial g}{\partial x}$$

$$= (f + g)_x \mathbf{i} + (f + g)_y \mathbf{j} = \nabla(f + g).$$

Thus $f + g$ is differentiable and $\nabla(f + g) = \nabla f + \nabla g$. ∎

REMARK. Any function that satisfies conditions (i) and (ii) of Theorem 3 is called a **linear mapping** or a **linear operator.** Linear operators play an extremely important role in advanced mathematics.

All the definitions and theorems in this section hold for functions of three or more variables. We give the equivalent results for functions of three variables below.

Definition 5 THE GRADIENT Let f be a scalar function of three variables such that f_x, f_y, and f_z exist at a point $\mathbf{x} = (x, y, z)$. Then the **gradient** of f at $\mathbf{x}$, denoted $\nabla f(\mathbf{x})$, is given by the vector

$$\nabla f(\mathbf{x}) = f_x(x, y, z)\mathbf{i} + f_y(x, y, z)\mathbf{j} + f_z(x, y, z)\mathbf{k}. \tag{14}$$

Definition 6 DIFFERENTIABILITY Let f be a function of three variables that is defined in a neighborhood of $\mathbf{x} = (x, y, z)$, and let $\Delta x = (\Delta x, \Delta y, \Delta z)$. If $f_x(x, y, z)$, $f_y(x, y, z)$, and $f_z(x, y, z)$ exist, then f is **differentiable** at $\mathbf{x}$ if there is a function g such that

$$f(\mathbf{x} + \Delta \mathbf{x}) - f(\mathbf{x}) = \nabla f \cdot \Delta \mathbf{x} + g(\Delta \mathbf{x}),$$

where

$$\lim_{|\Delta x| \to 0} \frac{g(\Delta x)}{|\Delta x|} = 0.$$

Equivalently, we can write

$$f(x + \Delta x, y + \Delta y, z + \Delta z) - f(x, y, z) = f_x(x, y, z)\,\Delta x + f_y(x, y, z)\,\Delta y$$
$$+ f_z(x, y, z)\,\Delta z + g(\Delta x, \Delta y, \Delta z),$$

where

$$\lim_{(\Delta x, \Delta y, \Delta z) \to (0,0,0)} \frac{g(\Delta x, \Delta y, \Delta z)}{\sqrt{\Delta x^2 + \Delta y^2 + \Delta z^2}} = 0.$$

Theorem 1' If f, f_x, f_y, and f_z exist and are continuous in a neighborhood of $\mathbf{x} = (x, y, z)$, then f is differentiable at $\mathbf{x}$.

Theorem 2' Let f be a function of three variables that is differentiable at $\mathbf{x}_0$. Then f is continuous at $\mathbf{x}_0$.

EXAMPLE 4 Let $f(x, y, z) = xy^2z^3$. Show that f is differentiable at any point $\mathbf{x}_0$, calculate ∇f, and find $\nabla f(3, -1, 2)$.

Solution. $\partial f/\partial x = y^2z^3$, $\partial f/\partial y = 2xyz^3$, and $\partial f/\partial z = 3xy^2z^2$. Since f, $\partial f/\partial x$, $\partial f/\partial y$, and $\partial f/\partial z$ are all continuous, we know that f is differentiable and that

$$\nabla f = y^2z^3\mathbf{i} + 2xyz^3\mathbf{j} + 3xy^2z^2\mathbf{k}$$

and

$$\nabla f(3, -1, 2) = 8\mathbf{i} - 48\mathbf{j} + 36\mathbf{k}. \ \blacksquare$$

Theorem 3' Let f and g be differentiable in a neighborhood of $\mathbf{x} = (x, y, z)$. Then for any scalar α, αf and $f + g$ are differentiable at $\mathbf{x}$, and

(i) $\nabla(\alpha f) = \alpha \nabla f$, and

(ii) $\nabla(f + g) = \nabla f + \nabla g$.

We conclude this section with a proof of Theorem 1. The proof of Theorem 1' is similar.

Proof of Theorem 1. We begin by restating the mean value theorem for a function f of one variable.

Mean Value Theorem Let f be continuous on $[a, b]$ and differentiable on (a, b). Then there is a number c in (a, b) such that

$$f(b) - f(a) = f'(c)(b - a).$$

Now we have assumed that f, f_x, and f_y are all continuous in a neighborhood N of $\mathbf{x} = (x, y)$. Choose $\Delta\mathbf{x}$ so small that $\mathbf{x} + \Delta\mathbf{x}$ is in N. Then

$$\Delta f(\mathbf{x}) = f(x + \Delta x, y + \Delta y) - f(x, y)$$

This term was added and subtracted

$$= [f(x + \Delta x, y + \Delta y) - \overbrace{f(x + \Delta x, y)] + [f(x + \Delta x, y)} - f(x, y)]. \quad \textbf{(15)}$$

If $x + \Delta x$ is fixed, then $f(x + \Delta x, y)$ is a function of y that is continuous and differentiable in the interval $[y, y + \Delta y]$. Hence by the mean value theorem there is a number c_2 between y and $y + \Delta y$ such that

$$f(x + \Delta x, y + \Delta y) - f(x + \Delta x, y) = f_y(x + \Delta x, c_2)[(y + \Delta y) - y]$$
$$= f_y(x + \Delta x, c_2) \Delta y. \tag{16}$$

Similarly, with y fixed, $f(x, y)$ is a function of x only, and we obtain

$$f(x + \Delta x, y) - f(x, y) = f_x(c_1, y) \Delta x, \tag{17}$$

where c_1 is between x and $x + \Delta x$. Thus using (16) and (17) in (15), we have

$$\Delta f(\mathbf{x}) = f_x(c_1, y) \Delta x + f_y(x + \Delta x, c_2) \Delta y. \tag{18}$$

Now both f_x and f_y are continuous at $\mathbf{x} = (x, y)$, so since c_1 is between x and $x + \Delta x$ and c_2 is between y and $y + \Delta y$, we obtain

$$\lim_{\Delta \mathbf{x} \to 0} f_x(c_1, y) = f_x(x, y) = f_x(\mathbf{x}) \tag{19}$$

and

$$\lim_{\Delta \mathbf{x} \to 0} f_y(x + \Delta x, c_2) = f_y(x, y) = f_y(\mathbf{x}). \tag{20}$$

Let

$$\epsilon_1(\Delta \mathbf{x}) = f_x(c_1, y) - f_x(x, y). \tag{21}$$

From (19) it follows that

$$\lim_{|\Delta \mathbf{x}| \to 0} \epsilon_1(\Delta \mathbf{x}) = 0. \tag{22}$$

Similarly, if

$$\epsilon_2(\Delta \mathbf{x}) = f_y(x + \Delta x, c_2) - f_y(x, y), \tag{23}$$

then

$$\lim_{|\Delta \mathbf{x}| \to 0} \epsilon_2(\Delta \mathbf{x}) = 0. \tag{24}$$

Now define

$$g(\Delta \mathbf{x}) = \epsilon_1(\Delta \mathbf{x}) \Delta x + \epsilon_2(\Delta \mathbf{x}) \Delta y. \tag{25}$$

From (22) and (24) it follows that

$$\lim_{|\Delta \mathbf{x}| \to 0} \frac{g(\Delta \mathbf{x})}{|\Delta \mathbf{x}|} = 0. \tag{26}$$

Finally, since

$$f_x(c_1, y) = f_x(x, y) + \epsilon_1(\Delta x) \qquad \text{From (21)} \qquad \qquad \textbf{(27)}$$

and

$$f_y(x + \Delta x, c_2) = f_y(x, y) + \epsilon_2(\Delta x), \qquad \text{From (23)} \qquad \qquad \textbf{(28)}$$

we may substitute (27) and (28) into (18) to obtain

$$\Delta f(\mathbf{x}) = f(\mathbf{x} + \Delta \mathbf{x}) - f(\mathbf{x}) = [f_x(\mathbf{x}) + \epsilon_1(\Delta \mathbf{x})]\,\Delta x + [f_y(\mathbf{x}) + \epsilon_2(\Delta \mathbf{x})]\,\Delta y$$
$$= f_x(\mathbf{x})\,\Delta x + f_y(\mathbf{x})\,\Delta y + g(\Delta \mathbf{x}) = (f_x \mathbf{i} + f_y \mathbf{j}) \cdot (\Delta \mathbf{x}) + g(\Delta \mathbf{x}),$$

where $\lim_{\Delta x \to 0}[g(\Delta \mathbf{x})/|\Delta \mathbf{x}|] \to 0$, and the proof is (at last) complete. ∎

PROBLEMS 18.5

1. Let $f(x, y) = x^2 + y^2$. Show, by using Definition 2, that f is differentiable at any point in $\mathbb{R}^2$.

2. Let $f(x, y) = x^2 y^2$. Show, by using Definition 2, that f is differentiable at any point in $\mathbb{R}^2$.

3. Let $f(x, y) =$ be any polynomial in the variables x and y. Show that f is differentiable.

In Problems 4–24, calculate the gradient of the given function. If a point is also given, evaluate the gradient at that point.

4. $f(x, y) = (x + y)^2$

5. $f(x, y) = e^{\sqrt{xy}};\ (1, 1)$

6. $f(x, y) = \cos(x - y);\ (\pi/2, \pi/4)$

7. $f(x, y) = \ln(2x - y + 1)$

8. $f(x, y) = \sqrt{x^2 + y^3}$

9. $f(x, y) = \tan^{-1}\dfrac{y}{x};\ (3, 3)$

10. $f(x, y) = y\tan(y - x)$

11. $f(x, y) = x^2 \sinh y$

12. $f(x, y) = \sec(x + 3y);\ (0, 1)$

13. $f(x, y) = \dfrac{x - y}{x + y};\ (3, 1)$

14. $f(x, y) = \dfrac{x^2 - y^2}{x^2 + y^2}$

15. $f(x, y) = \dfrac{e^{x^2} - e^{-y^2}}{3y}$

16. $f(x, y, z) = xyz;\ (1, 2, 3)$

17. $f(x, y, z) = \sin x \cos y \tan z;\ (\pi/6, \pi/4, \pi/3)$

18. $f(x, y, z) = \dfrac{x^2 - y^2 + z^2}{3xy};\ (1, 2, 0)$

19. $f(x, y, z) = x \ln y - z \ln x$

20. $f(x, y, z) = xy^2 + y^2 z^3;\ (2, 3, -1)$

21. $f(x, y, z) = (y - z)e^{x + 2y + 3z};\ (-4, -1, 3)$

22. $f(x, y, z) = x \sin y \ln z;\ (1, 0, 1)$

23. $f(x, y, z) = \dfrac{x - z}{\sqrt{1 - y^2 + x^2}};\ (0, 0, 1)$

24. $f(x, y, z) = x \cosh z - y \sin x$

25. Show that if f and g are differentiable functions of three variables, then

$$\nabla(f + g) = \nabla f + \nabla g.$$

26. Show that if f and g are differentiable functions of three variables, then fg is differentiable and

$$\nabla(fg) = f(\nabla g) + g(\nabla f).$$

*27. Show that $\nabla f = \mathbf{0}$ if and only if f is constant.

*28. Show that if $\nabla f = \nabla g$, then there is a constant c for which $f(x, y) = g(x, y) + c$. [*Hint:* Use the result of Problem 27.]

*29. What is the most general function f such that $\nabla f(\mathbf{x}) = \mathbf{x}$ for every $\mathbf{x}$ in $\mathbb{R}^2$?

*30. Let $f(x, y) = \begin{cases} (x^2 + y^2) \sin \dfrac{1}{\sqrt{x^2 + y^2}}, & (x, y) \neq (0, 0) \\ 0, & (x, y) = (0, 0). \end{cases}$

    (a) Calculate $f_x(0, 0)$ and $f_y(0, 0)$.

    (b) Explain why f_x and f_y are *not* continuous at $(0, 0)$.

    (c) Show that f is differentiable at $(0, 0)$.

31. Suppose that f is a differentiable function of one variable and g is a differentiable function of three variables. Show that $f \circ g$ is differentiable and $\nabla f \circ g = f'(g) \nabla g$.

18.6 THE CHAIN RULE

In this section we derive the chain rule for functions of two and three variables. Let us recall the chain rule for the composition of two functions of one variable:

Let $y = f(u)$ and $u = g(x)$ and assume that f and g are differentiable. Then

$$\frac{dy}{dx} = \frac{dy}{du}\frac{du}{dx} = f'(g(x))g'(x). \tag{1}$$

If $z = f(x, y)$ is a function of two variables, then there are two versions of the chain rule.

Theorem 1 CHAIN RULE Let $z = f(x, y)$ be differentiable and suppose that $x = x(t)$ and $y = y(t)$. Assume further that dx/dt and dy/dt exist and are continuous. Then z can be written as a function of the parameter t, and

$$\frac{dz}{dt} = \frac{\partial z}{\partial x}\frac{dx}{dt} + \frac{\partial z}{\partial y}\frac{dy}{dt} = f_x\frac{dx}{dt} + f_y\frac{dy}{dt}. \tag{2}$$

We can also write this result using our gradient notation. If $\mathbf{g}(t) = x(t)\mathbf{i} + y(t)\mathbf{j}$, then $\mathbf{g}'(t) = (dx/dt)\mathbf{i} + (dy/dt)\mathbf{j}$, and (2) can be written as

$$\frac{d}{dt}f(x(t), y(t)) = (f \circ \mathbf{g})'(t) = [f(\mathbf{g}(t))]' = \nabla f \cdot \mathbf{g}'(t). \tag{3}$$

Theorem 2 CHAIN RULE Let $z = f(x, y)$ be differentiable and suppose that x and y are functions of the two variables r and s. That is, $x = x(r, s)$ and $y = y(r, s)$.

Suppose further that $\partial x/\partial r$, $\partial x/\partial s$, $\partial y/\partial r$, and $\partial y/\partial s$ all exist and are continuous. Then z can be written as a function of r and s, and

$$\frac{\partial z}{\partial r} = \frac{\partial z}{\partial x}\frac{\partial x}{\partial r} + \frac{\partial z}{\partial y}\frac{\partial y}{\partial r} \tag{4}$$

$$\frac{\partial z}{\partial s} = \frac{\partial z}{\partial x}\frac{\partial x}{\partial s} + \frac{\partial z}{\partial y}\frac{\partial y}{\partial s}. \tag{5}$$

We will leave the proofs of these theorems until the end of this section.

EXAMPLE 1   Let $z = f(x, y) = xy^2$. Let $x = \cos t$ and $y = \sin t$. Calculate dz/dt.

Solution.

$$\frac{dz}{dt} = \frac{\partial z}{\partial x}\frac{dx}{dt} + \frac{\partial z}{\partial y}\frac{dy}{dt} = y^2(-\sin t) + 2xy(\cos t)$$

$$= (\sin^2 t)(-\sin t) + 2(\cos t)(\sin t)(\cos t)$$

$$= 2\sin t \cos^2 t - \sin^3 t$$

We can calculate this result another way. Since $z = xy^2$, we have $z = (\cos t)(\sin^2 t)$. Then

$$\frac{dz}{dt} = (\cos t)(2 \sin t \cos t) + (\sin^2 t)(-\sin t)$$

$$= 2 \sin t \cos^2 t - \sin^3 t. \ \blacksquare$$

EXAMPLE 2   Let $z = f(x, y) = \sin xy^2$. Suppose that $x = r/s$ and $y = e^{r-s}$. Calculate $\partial z/\partial r$ and $\partial z/\partial s$.

Solution.

$$\frac{\partial z}{\partial r} = \frac{\partial z}{\partial x}\frac{\partial x}{\partial r} + \frac{\partial z}{\partial y}\frac{\partial y}{\partial r} = (y^2 \cos xy^2)\frac{1}{s} + (2xy \cos xy^2)e^{r-s}$$

$$= \frac{e^{2(r-s)}\cos[(r/s)e^{2(r-s)}]}{s} + \frac{2r}{s}\left\{\cos\left[\frac{r}{s}e^{2(r-s)}\right]\right\}e^{2(r-s)}$$

and

$$\frac{\partial z}{\partial s} = \frac{\partial z}{\partial x}\frac{\partial x}{\partial s} + \frac{\partial z}{\partial y}\frac{\partial y}{\partial s} = (y^2 \cos xy^2)\frac{-r}{s^2} + (2xy \cos xy^2)(-e^{r-s})$$

$$= \frac{-re^{2(r-s)}\cos[(r/s)e^{2(r-s)}]}{s^2} - \frac{2r}{s}\left\{\cos\left[\frac{r}{s}e^{2(r-s)}\right]\right\}e^{2(r-s)}. \ \blacksquare$$

The chain rules given in Theorem 1 and Theorem 2 can easily be extended to functions of three or more variables.

Theorem 1' Let $w = f(x, y, z)$ be a differentiable function. If $x = x(t)$, $y = y(t)$, $z = z(t)$, and if dx/dt, dy/dt, and dz/dt exist and are continuous, then

$$\frac{dw}{dt} = \frac{\partial w}{\partial x}\frac{dx}{dt} + \frac{\partial w}{\partial y}\frac{dy}{dt} + \frac{\partial w}{\partial z}\frac{dz}{dt}. \tag{6}$$

Theorem 2' Let $w = f(x, y, z)$ be a differentiable function and let $x = x(r, s)$, $y = y(r, s)$, and $z = z(r, s)$. Then if all indicated partial derivatives exist and are continuous, we have

$$\frac{\partial w}{\partial r} = \frac{\partial w}{\partial x}\frac{\partial x}{\partial r} + \frac{\partial w}{\partial y}\frac{\partial y}{\partial r} + \frac{\partial w}{\partial z}\frac{\partial z}{\partial r} \tag{7}$$

and

$$\frac{\partial w}{\partial s} = \frac{\partial w}{\partial x}\frac{\partial x}{\partial s} + \frac{\partial w}{\partial y}\frac{\partial y}{\partial s} + \frac{\partial w}{\partial z}\frac{\partial z}{\partial s}. \tag{8}$$

Theorem 3 Let $w = f(x, y, z)$ be a differentiable function and let $x = x(r, s, t)$, $y = y(r, s, t)$, and $z = z(r, s, t)$. Then if all indicated partial derivatives exist and are continuous, we have

$$\begin{aligned}
\frac{\partial w}{\partial r} &= \frac{\partial w}{\partial x}\frac{\partial x}{\partial r} + \frac{\partial w}{\partial y}\frac{\partial y}{\partial r} + \frac{\partial w}{\partial z}\frac{\partial z}{\partial r} \\
\frac{\partial w}{\partial s} &= \frac{\partial w}{\partial x}\frac{\partial x}{\partial s} + \frac{\partial w}{\partial y}\frac{\partial y}{\partial s} + \frac{\partial w}{\partial z}\frac{\partial z}{\partial s} \\
\frac{\partial w}{\partial t} &= \frac{\partial w}{\partial x}\frac{\partial x}{\partial t} + \frac{\partial w}{\partial y}\frac{\partial y}{\partial t} + \frac{\partial w}{\partial z}\frac{\partial z}{\partial t}.
\end{aligned} \tag{9}$$

We saw in Section 4.1 that the chain rule is useful to solve "related rates" types of problems. The chain rules for functions of two or more variables are useful in a similar way.

EXAMPLE 3 The radius of a right circular cone is increasing at a rate of 3 cm/sec, while its height is increasing at a rate of 5 cm/sec. How fast is the volume increasing when $r = 15$ cm and $h = 25$ cm?

Solution. We are asked to find dV/dt, where $V = \frac{1}{3}\pi r^2 h$. But

$$\begin{aligned}
\frac{dV}{dt} &= \frac{\partial V}{\partial r}\frac{dr}{dt} + \frac{\partial V}{\partial h}\frac{dh}{dt} = \frac{2}{3}\pi r h(3) + \frac{1}{3}\pi r^2(5) \\
&= 2\pi r h + \tfrac{5}{3}\pi r^2 = \pi[2 \cdot 15 \cdot 25 + \tfrac{5}{3}(15)^2] \\
&= 1125\pi \text{ cm}^3/\text{sec.} \quad \blacksquare
\end{aligned}$$

EXAMPLE 4 According to the ideal gas law, the pressure, volume, and absolute temperature of n moles of an ideal gas are related by

$$P\dot{V} = nRT,$$

where R is a constant. Suppose that the volume of an ideal gas is increasing at a rate of 10 cm³/min and the pressure is decreasing at a rate of 0.3 N/cm²/min (see Section 9.7 for a discussion of pressure). How is the temperature of the gas changing when the volume of 5 mol of a gas is 100 cm³ and the pressure is 2 N/cm²?

Solution. We have $T = PV/nR$, where $n = 5$. Then

$$\frac{dT}{dt} = \frac{\partial T}{\partial P}\frac{dP}{dt} + \frac{\partial T}{\partial V}\frac{dV}{dt} = \frac{V}{nR}(-0.3) + \frac{P}{nR}(10) = \frac{100}{5R}(-0.3) + \frac{2}{5R}(10)$$

$$= \frac{-2}{R} \text{ °K/min.} \blacksquare$$

We conclude this section with a proof of Theorem 2. Theorem 1 follows easily from Theorem 2. (Explain why.)

Proof of Theorem 2. We will show that

$$\frac{\partial z}{\partial r} = \frac{\partial z}{\partial x}\frac{\partial x}{\partial r} + \frac{\partial z}{\partial y}\frac{\partial y}{\partial r}.$$

Equation (5) follows in an identical manner. Since $x = x(r, s)$ and $y = y(r, s)$, a change Δr in r will cause a change Δx in x and a change Δy in y. We may therefore write

$$\Delta x = x(r + \Delta r, s) - x(r, s)$$

and

$$\Delta y = y(r + \Delta r, s) - y(r, s).$$

Since z is differentiable, we may write (Definition 18.5.2)

$$\Delta z = \frac{\partial z}{\partial x}\Delta x + \frac{\partial z}{\partial y}\Delta y + \epsilon_1(\Delta x, \Delta y)\,\Delta x + \epsilon_2(\Delta x, \Delta y)\,\Delta y, \tag{10}$$

where

$$\lim_{(\Delta x, \Delta y)\to(0,0)} \epsilon_1(\Delta x, \Delta y) = \lim_{(\Delta x, \Delta y)\to(0,0)} \epsilon_2(\Delta x, \Delta y) = 0.$$

Then dividing both sides of (10) by Δr and taking limits, we obtain

$$\lim_{\Delta r\to 0}\frac{\Delta z}{\Delta r} = \lim_{\Delta r\to 0}\left[\frac{\partial z}{\partial x}\frac{\Delta x}{\Delta r} + \frac{\partial z}{\partial y}\frac{\Delta y}{\Delta r} + \epsilon_1(\Delta x, \Delta y)\frac{\Delta x}{\Delta r} + \epsilon_2(\Delta x, \Delta y)\frac{\Delta y}{\Delta r}\right]. \tag{11}$$

Since x and y are continuous functions of r, we have

$$\lim_{\Delta r \to 0} \Delta x = 0 \quad \text{and} \quad \lim_{\Delta r \to 0} \Delta y = 0,$$

so that

$$\lim_{\Delta r \to 0} \epsilon_1(\Delta x, \Delta y) = \lim_{(\Delta x, \Delta y) \to (0,0)} \epsilon_1(\Delta x, \Delta y) = 0$$

and

$$\lim_{\Delta r \to 0} \epsilon_2(\Delta x, \Delta y) = \lim_{(\Delta x, \Delta y) \to (0,0)} \epsilon_2(\Delta x, \Delta y) = 0.$$

Thus the limits in (11) become

$$\frac{\partial z}{\partial r} = \frac{\partial z}{\partial x}\frac{\partial x}{\partial r} + \frac{\partial z}{\partial y}\frac{\partial y}{\partial r} + 0 \cdot \frac{\partial x}{\partial r} + 0 \cdot \frac{\partial y}{\partial r},$$

and the theorem is proved. ∎

PROBLEMS 18.6

In Problems 1–11, use the chain rule to calculate dz/dt or dw/dt. Check your answer by first writing z or w as a function of t and then differentiating.

1. $z = xy, x = e^t, y = e^{2t}$

2. $z = x^2 + y^2, x = \cos t, y = \sin t$

3. $z = \dfrac{y}{x}, x = t^2, y = t^3$

4. $z = e^x \sin y, x = \sqrt{t}, y = \sqrt[3]{t}$

5. $z = \tan^{-1}\dfrac{y}{x}, x = \cos 3t, y = \sin 5t$

6. $z = \sinh(x - 2y), x = 2t^4, y = t^2 + 1$

7. $w = x^2 + y^2 + z^2, x = \cos t, y = \sin t, z = t$

8. $w = xy - yz + zx, x = e^t, y = e^{2t}, z = e^{3t}$

9. $w = \dfrac{x + y}{z}, x = t, y = t^2, z = t^3$

10. $w = \sin(x + 2y + 3z), x = \tan t, y = \sec t, z = t^5$

11. $w = \ln(2x - 3y + 4z), x = e^t, y = \ln t, z = \cosh t$

In Problems 12–26, use the chain rule to calculate the indicated partial derivatives.

12. $z = xy; x = r + s; y = r - s; \partial z/\partial r$ and $\partial z/\partial s$

13. $z = x^2 + y^2; x = \cos(r + s); y = \sin(r - s); \partial z/\partial r$ and $\partial z/\partial s$

14. $z = \dfrac{y}{x}; x = e^r; y = e^s; \dfrac{\partial z}{\partial r}$ and $\dfrac{\partial z}{\partial s}$

15. $z = \sin\dfrac{y}{x}; x = \dfrac{r}{s}; y = \dfrac{s}{r}; \dfrac{\partial z}{\partial r}$ and $\dfrac{\partial z}{\partial s}$

16. $z = \dfrac{e^{x+y}}{e^{x-y}}; x = \ln rs; y = \ln\dfrac{r}{s}; \dfrac{\partial z}{\partial r}$ and $\dfrac{\partial z}{\partial s}$

17. $z = x^2y^3$; $x = r - s^2$; $y = 2s + r$; $\partial z/\partial r$ and $\partial z/\partial s$

18. $w = x + y + z$; $x = rs$; $y = r + s$; $z = r - s$; $\partial w/\partial r$ and $\partial w/\partial s$

19. $w = \dfrac{xy}{z}$; $x = r$, $y = s$, $z = t$; $\dfrac{\partial w}{\partial r}$, $\dfrac{\partial w}{\partial s}$, and $\dfrac{\partial w}{\partial t}$

20. $w = \dfrac{xy}{z}$; $x = r + s$, $y = t - r$, $z = s + 2t$; $\dfrac{\partial w}{\partial r}$, $\dfrac{\partial w}{\partial s}$ and $\dfrac{\partial w}{\partial t}$

21. $w = \sin xyz$; $x = s^2r$, $y = r^2s$, $z = r - s$; $\partial w/\partial r$ and $\partial w/\partial s$

22. $w = \sinh(x + 2y + 3z)$; $x = \sqrt{r + s}$, $y = \sqrt[3]{s - t}$, $z = \dfrac{1}{r + t}$, $\dfrac{\partial w}{\partial r}$, $\dfrac{\partial w}{\partial s}$, and $\dfrac{\partial w}{\partial t}$

23. $w = x^2y + yz^2$; $x = rst$, $y = \dfrac{rs}{t}$, $z = \dfrac{1}{rst}$; $\dfrac{\partial w}{\partial r}$, $\dfrac{\partial w}{\partial s}$, and $\dfrac{\partial w}{\partial t}$

24. $w = \ln(x + 2y + 3z)$; $x = rt^3 + s$; $y = t - s^5$, $z = e^{r+s}$; $\partial w/\partial r$, $\partial w/\partial s$, and $\partial w/\partial t$

25. $w = e^{xy/z}$; $x = r^2 + t^2$, $y = s^2 - t^2$, $z = r^2 + s^2$; $\partial w/\partial r$, $\partial w/\partial s$, and $\partial w/\partial t$

*26. $u = xy + w^2 - z^3$; $x = t + r - q$, $y = q^2 + s^2 - t + r$, $z = \dfrac{qr + st}{r^4}$, $w = \dfrac{r - s}{t + q}$; $\dfrac{\partial u}{\partial r}$, $\dfrac{\partial u}{\partial s}$,

$\dfrac{\partial u}{\partial t}$, and $\dfrac{\partial u}{\partial q}$

27. The radius of a right circular cone is increasing at a rate of 7 in./min, while its height is decreasing at a rate of 20 in./min. How fast is the volume changing when $r = 45$ in. and $h = 100$ in.? Is the volume increasing or decreasing?

28. The radius of a right circular cylinder is decreasing at a rate of 12 cm/sec, while its height is increasing at a rate of 25 cm/sec. How is the volume changing when $r = 180$ cm and $h = 500$ cm? Is the volume increasing or decreasing?

29. The volume of 10 mol of an ideal gas is decreasing at a rate of 25 cm³/min and its temperature is increasing at a rate of 1°C/min. How fast is the pressure changing when $V = 1000$ cm³ and $P = 3$ N/cm²? Is the pressure increasing or decreasing? Leave your answer in terms of R.

30. The pressure of 8 mol of an ideal gas is decreasing at a rate of 0.4 N/cm²/min, while the temperature is decreasing at a rate of 0.5 °K/min. How fast is the volume of the gas changing when $V = 1000$ cm³ and $P = 3$ N/cm²? Is the volume increasing or decreasing? [*Hint:* Use the value $R = 8.314$ J/mol °K.]

31. The angle A of a triangle ABC is increasing at a rate of 3°/sec, the side AB is increasing at a rate of 1 cm/sec, and the side AC is decreasing at a rate of 2 cm/sec. How fast is the side BC changing when $A = 30°$, $AB = 10$ cm, and $AC = 24$ cm? Is the length of BC increasing or decreasing? [*Hint:* Use the law of cosines and convert to radians.]

32. The wave equation (see Problem 18.4.31) is the partial differential equation

$$\frac{\partial^2 y}{\partial t^2} = c^2 \frac{\partial^2 y}{\partial x^2}.$$

Show that if f is any differentiable function, then

$$y(x, t) = \tfrac{1}{2}[f(x - ct) + f(x + ct)]$$

is a solution to this equation.

33. Let $z = f(x, y)$ be differentiable. Write the expression $(\partial z/\partial x)^2 + (\partial z/\partial y)^2$ in terms of polar coordinates r and θ.

34. Let $w = f(x, y, z)$ be differentiable and let $x = r\cos\theta$, $y = r\sin\theta$, and $z = t$. Calculate $\partial w/\partial r$, $\partial w/\partial\theta$, and $\partial w/\partial t$. (These are cylindrical coordinates.)

35. Let $w = f(x, y, z)$ be differentiable and let $x = \rho \sin \varphi \cos \theta$, $y = \rho \sin \varphi \sin \theta$, and $z = \rho \cos \varphi$. Calculate $\partial w / \partial \rho$, $\partial w / \partial \varphi$, and $\partial w / \partial \theta$. (These are spherical coordinates.)

*36. Let $z = f(x, y)$, let $x = x(r, s)$, and let $y = y(r, s)$. Show that

$$\frac{\partial^2 z}{\partial r^2} = \frac{\partial}{\partial x}\left(\frac{\partial z}{\partial x}\frac{\partial x}{\partial r} + \frac{\partial z}{\partial y}\frac{\partial y}{\partial r}\right)\left(\frac{\partial x}{\partial r}\right) + \frac{\partial}{\partial y}\left(\frac{\partial z}{\partial x}\frac{\partial x}{\partial r} + \frac{\partial z}{\partial y}\frac{\partial y}{\partial r}\right)\left(\frac{\partial y}{\partial r}\right).$$

**37. Laplace's equation (see Problem 18.4.35) is the partial differential equation

$$\frac{\partial^2 f}{\partial x^2} + \frac{\partial^2 f}{\partial y^2} = 0.$$

If we write (x, y) in polar coordinates ($x = r \cos \theta$, $y = r \sin \theta$), show that Laplace's equation becomes

$$\frac{\partial^2 f}{\partial r^2} + \frac{1}{r^2}\frac{\partial^2 f}{\partial \theta^2} + \frac{1}{r}\frac{\partial f}{\partial r} = 0.$$

[*Hint:* Write $\partial^2 f / \partial x^2$ and $\partial^2 f / \partial y^2$ in terms of r and θ, using the result of Problem 36.]

18.7 TANGENT PLANES, NORMAL LINES, AND GRADIENTS

Let $z = f(x, y)$ be a function of two variables. As we have seen, the graph of f is a surface in $\mathbb{R}^3$. More generally, the graph of the equation $F(x, y, z) = 0$ is a surface in $\mathbb{R}^3$. The surface $F(x, y, z) = 0$ is called **differentiable** at a point (x_0, y_0, z_0) if $\partial F / \partial x$, $\partial F / \partial y$, and $\partial F / \partial z$ all exist and are continuous at (x_0, y_0, z_0). In $\mathbb{R}^2$ a differentiable curve has a unique tangent line at each point. In $\mathbb{R}^3$ a differentiable surface in $\mathbb{R}^3$ has a unique tangent plane at each point at which $\partial F / \partial x$, $\partial F / \partial y$, and $\partial F / \partial z$ are not all zero. We will formally define what we mean by a tangent plane to a surface after a bit, although it should be easy enough to visualize (see Figure 1). We note here that not every surface has a tangent plane at every point. For example, the cone $z = \sqrt{x^2 + y^2}$ has no tangent plane at the origin (see Figure 2).

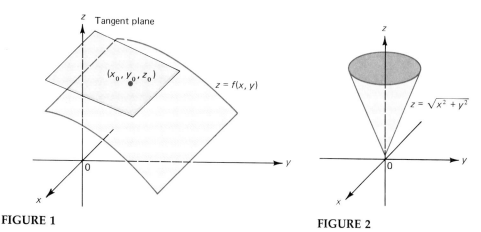

FIGURE 1 **FIGURE 2**

Assume that the surface S given by $F(x, y, z) = 0$ is differentiable. Let C be any curve lying on S. That is, C can be given parametrically by $\mathbf{g}(t) = x(t)\mathbf{i} + y(t)\mathbf{j} +$

$z(t)\mathbf{k}$. (Recall from Definition 17.7.1, the definition of a curve in $\mathbb{R}^3$.) Then for points on the curve, $F(x, y, z)$ can be written as a function of t, and from the vector form of the chain rule [equation (18.6.3)] we have

$$F'(t) = \nabla F \cdot \mathbf{g}'(t). \tag{1}$$

But since $F(x(t), y(t), z(t)) = 0$ for all t [since $(x(t), y(t), z(t))$ is on S], we see that $F'(t) = 0$ for all t. But $\mathbf{g}'(t)$ is tangent to the curve C for every number t. Thus (1) implies the following:

> The gradient of F at a point $\mathbf{x}_0 = (x_0, y_0, z_0)$ on S is orthogonal to the tangent vector at $\mathbf{x}_0$ to any curve C remaining on S and passing through $\mathbf{x}_0$.

This statement is illustrated in Figure 3.

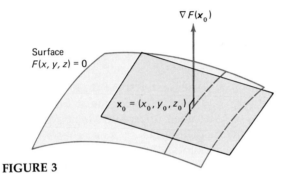

FIGURE 3

Thus if we think of all the vectors tangent to a surface at a point $\mathbf{x}_0$ as constituting a plane, then $\nabla F(\mathbf{x}_0)$ is a *normal* vector to that plane. This motivates the following definition.

Definition 1   TANGENT PLANE AND NORMAL LINE   Let F be differentiable at $\mathbf{x}_0 = (x_0, y_0, z_0)$ and let the surface S be defined by $F(x, y, z) = 0$.

   (i) The **tangent plane** to S at (x_0, y_0, z_0) is the plane passing through the point (x_0, y_0, z_0) with normal vector $\nabla F(\mathbf{x}_0)$.
   (ii) The **normal line** to S at $\mathbf{x}_0$ is the line passing through $\mathbf{x}_0$ having the same direction as $\nabla F(\mathbf{x}_0)$.

EXAMPLE 1   Find the equation of the tangent plane and symmetric equations of the normal line to the ellipsoid $x^2 + (y^2/4) + (z^2/9) = 3$ at the point $(1, 2, 3)$.

   Solution. Since $F(x, y, z) = x^2 + (y^2/4) + (z^2/9) - 3 = 0$, we have

$$\nabla F = \frac{\partial F}{\partial x}\mathbf{i} + \frac{\partial F}{\partial y}\mathbf{j} + \frac{\partial F}{\partial z}\mathbf{k} = 2x\mathbf{i} + \frac{y}{2}\mathbf{j} + \frac{2z}{9}\mathbf{k}.$$

Then $\nabla F(1, 2, 3) = 2\mathbf{i} + \mathbf{j} + \frac{2}{3}\mathbf{k}$, and the equation of the tangent plane is

$$2(x - 1) + (y - 2) + \tfrac{2}{3}(z - 3) = 0,$$

or

$$2x + y + \tfrac{2}{3}z = 6.$$

The normal line is given by

$$\frac{x - 1}{2} = y - 2 = \tfrac{3}{2}(z - 3). \quad \blacksquare$$

The situation is even simpler if we can write the surface in the form $z = f(x, y)$. That is, the surface is the graph of a function of two variables. Then $F(x, y, z) = f(x, y) - z = 0$, so that

$$F_x = f_x, \qquad F_y = f_y, \qquad F_z = -1,$$

and the normal vector $\mathbf{N}$ to the tangent plane is

$$\mathbf{N} = f_x(x_0, y_0)\mathbf{i} + f_y(x_0, y_0)\mathbf{j} - \mathbf{k}. \tag{2}$$

REMARK.   One interesting consequence of this fact is that if $z = f(x, y)$ and if $\nabla f(x_0, y_0) = \mathbf{0}$, then *the tangent plane to the surface at $(x_0, y_0, f(x_0, y_0))$ is parallel to the xy-plane (i.e., it is horizontal)*. This occurs because at $(x_0, y_0, f(x_0, y_0))$, $\mathbf{N} = (\partial f/\partial x)\mathbf{i} + (\partial f/\partial y)\mathbf{j} - \mathbf{k} = \nabla f - \mathbf{k} = -\mathbf{k}$. Thus the z-axis is normal to the tangent plane.

EXAMPLE 2   Find the tangent plane and normal line to the surface $z = x^3 y^5$ at the point $(2, 1, 8)$.

Solution. $\mathbf{N} = (\partial z/\partial x)\mathbf{i} + (\partial z/\partial y)\mathbf{j} - \mathbf{k} = 3x^2 y^5\mathbf{i} + 5x^3 y^4\mathbf{j} - \mathbf{k} = 12\mathbf{i} + 40\mathbf{j} - \mathbf{k}$ at $(2, 1, 8)$. Then the tangent plane is given by

$$12(x - 2) + 40(y - 1) - (z - 8) = 0,$$

or

$$12x + 40y - z = 56.$$

Symmetric equations of the normal line are

$$\frac{x - 2}{12} = \frac{y - 1}{40} = \frac{z - 8}{-1}. \quad \blacksquare$$

We can write the equation of the tangent plane to a surface $z = f(x, y)$ so that it looks like the equation of the tangent line to a curve in $\mathbb{R}^2$. This will further illustrate the connection between the derivative of a function of one variable and the gradient. Recall from Section 17.5 that if P is a point on a plane and $\mathbf{N}$ is a normal vector, then if Q denotes any other point on the plane, the equation of the plane can be written

$$\overrightarrow{PQ} \cdot \mathbf{N} = 0. \tag{3}$$

In this case, since $z = f(x, y)$, a point on the surface takes the form $(x, y, z) = (x, y, f(x, y))$. Then since $\mathbf{N} = f_x \mathbf{i} + f_y \mathbf{j} - \mathbf{k}$, the equation of the tangent plane at $(x_0, y_0, f(x_0, y_0))$ becomes, using (3),

$$\begin{aligned}
0 &= [(x, y, z) - (x_0, y_0, z_0)] \cdot (f_x, f_y, -1) \\
&= (x - x_0, y - y_0, z - z_0) \cdot (f_x, f_y, -1) \\
&= (x - x_0)f_x + (y - y_0)f_y - (z - z_0).
\end{aligned} \tag{4}$$

We can rewrite (4) as

$$z = f(x_0, y_0) + (x - x_0)f_x + (y - y_0)f_y. \tag{5}$$

Denote (x_0, y_0) by $\mathbf{x}_0$ and (x, y) by $\mathbf{x}$. Then (5) can be written as

$$z = f(\mathbf{x}_0) + (\mathbf{x} - \mathbf{x}_0) \cdot \nabla f(\mathbf{x}_0). \tag{6}$$

Recall that if $y = f(x)$ is differentiable at x_0, then the equation of the tangent line to the curve at the point $(x_0, f(x_0))$ is given by

$$\frac{y - f(x_0)}{x - x_0} = f'(x_0),$$

or

$$y = f(x_0) + (x - x_0)f'(x_0). \tag{7}$$

This similarity between (6) and (7) illustrates quite vividly the important of the gradient vector of a function of several variables as the generalization of the derivative of a function of one variable.

PROBLEMS 18.7

In Problems 1–16, find the equation of the tangent plane and symmetric equations of the normal line to the given surface at the given point.

1. $x^2 + y^2 + z^2 = 1$; $(1, 0, 0)$
2. $x^2 + y^2 + z^2 = 1$; $(0, 1, 0)$
3. $x^2 + y^2 + z^2 = 1$; $(0, 0, 1)$
4. $x^2 - y^2 + z^2 = 1$; $(1, 1, 1)$
5. $\dfrac{x^2}{a^2} + \dfrac{y^2}{b^2} + \dfrac{z^2}{c^2} = 3$; (a, b, c)
6. $\dfrac{x^2}{a^2} + \dfrac{y^2}{b^2} + \dfrac{z^2}{c^2} = 3$; $(-a, b, -c)$
7. $x^{1/2} + y^{1/2} + z^{1/2} = 6$; $(4, 1, 9)$
8. $ax + by + cz = d$; $\left(\dfrac{1}{a}, \dfrac{1}{b}, \dfrac{d-2}{c}\right)$

9. $xyz = 4$; $(1, 2, 2)$
10. $xy^2 - yz^2 + zx^2 = 1$; $(1, 1, 1)$
11. $4x^2 - y^2 - 5z^2 = 15$; $(3, 1, -2)$
12. $xe^y - ye^z = 1$; $(1, 0, 0)$
13. $\sin xy - 2 \cos yz = 0$; $(\pi/2, 1, \pi/3)$
14. $x^2 + y^2 + 4x + 2y + 8z = 7$; $(2, -3, -1)$

15. $e^{xyz} = 5$; $(1, 1, \ln 5)$
16. $\sqrt{\dfrac{x + y}{z - 1}} = 1$; $(1, 1, 3)$

In Problems 17–24, write the equation of the tangent plane in the form (6) and find the symmetric equations of the normal line to the given surface.

17. $z = xy^2$; $(1, 1, 1)$
18. $z = \ln(x - 2y)$; $(3, 1, 0)$

19. $z = \sin(2x + 5y)$; $(\pi/8, \pi/20, 1)$
20. $z = \sqrt{\dfrac{x + y}{x - y}}$; $(5, 4, 3)$

21. $z = \tan^{-1}\dfrac{y}{x}$; $\left(-2, 2, -\dfrac{\pi}{4}\right)$
22. $z = \sinh xy^2$; $(0, 3, 0)$

23. $z = \sec(x - y)$; $(\pi/2, \pi/6, 2)$
24. $z = e^x \cos y + e^y \cos x$; $(\pi/2, 0, e^{\pi/2})$

*25. Find the two points of intersection of the surface $z = x^2 + y^2$ and the line
$$\frac{x - 3}{1} = \frac{y + 1}{-1} = \frac{z + 2}{-2}.$$

26. At each of the points of intersection found in Problem 25, find the cosine of the angle between the given line and the normal line to the surface.

*27. Show that every line normal to the surface of a sphere passes through the center of the sphere.

28. Show that every line normal to the cone $z^2 = ax^2 + ay^2$ intersects the z-axis.

29. Let f be a differentiable function of one variable and let $z = yf(y/x)$. Show that all tangent planes to the surface defined by this equation have a point in common.

30. The *angle between two surfaces* at a point of intersection is defined to be the angle between their normal lines. Show that if two surfaces $F(x, y, z) = 0$ and $G(x, y, z) = 0$ intersect at right angles at a point $\mathbf{x}_0$, then
$$\nabla F(\mathbf{x}_0) \cdot \nabla G(\mathbf{x}_0) = 0.$$

31. Show that the sum of the squares of the x-, y-, and z-intercepts of any plane tangent to the surface $x^{2/3} + y^{2/3} + z^{2/3} = a^{2/3}$ is constant.

32. The equation $F(x, y) = 0$ defines a curve in $\mathbb{R}^2$. Show that if F is differentiable, then $\nabla F(x, y)$ is normal to the curve at every point.

33. Use the result of Problem 32 to find the equation of the tangent line to a curve $F(x, y) = 0$ at a point (x_0, y_0).

In Problems 34–39, use the results of Problems 32 and 33 to find a normal vector and the equation of the tangent line to the curve at the given point.

34. $xy = 5$; $(1, 5)$
35. $x^2 + xy + y^2 + 3x - 5y = 16$; $(1, -2)$

36. $\dfrac{x + y}{x - y} = 7$; $(4, 3)$
37. $xe^{xy} = 1$; $(1, 0)$

38. $\dfrac{x^2}{4} + \dfrac{y^2}{16} = 1$; $(\sqrt{2}, 2\sqrt{2})$
39. $\tan(x + y) = 1$; $(\pi/4, 0)$

40. Show that at any point (x, y), $y \neq 0$, the curve $x/y = a$ is orthogonal to the curve $x^2 + y^2 = r^2$ for any constants a and r.

18.8 DIRECTIONAL DERIVATIVES AND THE GRADIENT

Let us take another look at the partial derivatives $\partial f/\partial x$ and $\partial f/\partial y$ of the function $z = f(x, y)$. We have

$$\frac{\partial f}{\partial x}(x_0, y_0) = \lim_{\Delta x \to 0} \frac{f(x_0 + \Delta x, y_0) - f(x_0, y_0)}{\Delta x}. \tag{1}$$

This measures the rate of change of f as we approach the point (x_0, y_0) along a vector parallel to the x-axis [since $(x_0 + \Delta x, y_0) - (x_0, y_0) = (\Delta x, 0) = \Delta x \mathbf{i}$]. Similarly

$$\frac{\partial f}{\partial y}(x_0, y_0) = \lim_{\Delta y \to 0} \frac{f(x_0, y_0 + \Delta y) - f(x_0, y_0)}{\Delta y} \tag{2}$$

measures the rate of change of f as we approach the point (x_0, y_0) along a vector parallel to the y-axis.

It is frequently of interest to compute the rate of change of f as we approach (x_0, y_0) along a vector that is not parallel to one of the coordinate axes. The situation is depicted in Figure 1. Suppose that (x, y) approaches the fixed point (x_0, y_0) along the line segment joining them, and let t denote the distance between the two points. We want to determine the relative rate of change in f with respect to a change in t.

Let $\mathbf{u}$ denote a unit vector with the initial point at (x_0, y_0) and parallel to $\overrightarrow{PQ}$ (see Figure 2). Since $\mathbf{u}$ and $\overrightarrow{PQ}$ are parallel, there is, by Theorem 15.2.3, a value of t such that

$$\overrightarrow{PQ} = t\mathbf{u}. \tag{3}$$

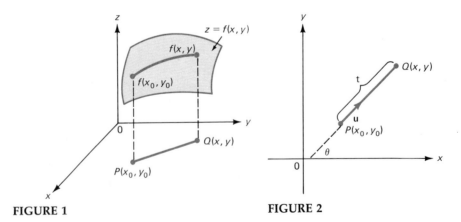

FIGURE 1   FIGURE 2

Note that $t > 0$ if $\mathbf{u}$ and $\overrightarrow{PQ}$ have the same direction and $t < 0$ if $\mathbf{u}$ and $\overrightarrow{PQ}$ have opposite directions. Now

$$\overrightarrow{PQ} = (x - x_0)\mathbf{i} + (y - y_0)\mathbf{j}, \tag{4}$$

and since $\mathbf{u}$ is a unit vector, we have

$$\mathbf{u} = \cos\theta\mathbf{i} + \sin\theta\mathbf{j}, \tag{5}$$

where θ is the direction of $\mathbf{u}$. Thus insterting (4) and (5) into (3), we have

$$(x - x_0)\mathbf{i} + (y - y_0)\mathbf{j} = t\cos\theta\mathbf{i} + t\sin\theta\mathbf{j},$$

or

$$x = x_0 + t\cos\theta$$
$$y = y_0 + t\sin\theta. \tag{6}$$

The equations (6) are the parametric equations of the line passing through P and Q. Using (6), we have

$$z = f(x, y) = f(x_0 + t\cos\theta, y_0 + t\sin\theta). \tag{7}$$

Remember that θ is fixed—it is the direction of approach. Thus $(x, y) \to (x_0, y_0)$ along $\overrightarrow{PQ}$ is equivalent to $t \to 0$ in (7). Hence to compute the instantaneous rate of change of f as $(x, y) \to (x_0, y_0)$ along the vector $\overrightarrow{PQ}$, we need to compute dz/dt. But by the chain rule,

$$\frac{dz}{dt} = \frac{\partial f}{\partial x}(x, y)\frac{dx}{dt} + \frac{\partial f}{\partial y}(x, y)\frac{dy}{dt},$$

or

$$\frac{dz}{dt} = f_x(x, y)\cos\theta + f_y(x, y)\sin\theta, \tag{8}$$

and

$$\frac{dz}{dt} = [f_x(x_0 + t\cos\theta, y_0 + t\sin\theta)]\cos\theta$$
$$+ [f_y(x_0 + t\cos\theta, y_0 + t\sin\theta)]\sin\theta. \tag{9}$$

If we set $t = 0$ in (9), we obtain the instantaneous rate of change of f in the direction $\overrightarrow{PQ}$ at the point (x_0, y_0). That is,

$$\left.\frac{dz}{dt}\right|_{t=0} = f_x(x_0, y_0)\cos\theta + f_y(x_0, y_0)\sin\theta. \tag{10}$$

But (10) can be written [using (5)] as

$$\left.\frac{dz}{dt}\right|_{t=0} = \nabla f(x_0, y_0) \cdot \mathbf{u}. \tag{11}$$

This leads to the following definition.

Definition 1   DIRECTIONAL DERIVATIVE   Let f be differentiable at a point $\mathbf{x}_0 = (x_0, y_0)$ in $\mathbb{R}^2$ and let $\mathbf{u}$ be a unit vector. Then the **directional derivative of f in the direction u**, denoted $f_{\mathbf{u}}'(\mathbf{x}_0)$, is given by

$$f_{\mathbf{u}}'(\mathbf{x}_0) = \nabla f(\mathbf{x}_0) \cdot \mathbf{u}. \tag{12}$$

REMARK 1.   Note that if $\mathbf{u} = \mathbf{i}$, then $\nabla f \cdot \mathbf{u} = \partial f / \partial x$ and (12) reduces to the partial derivative $\partial f / \partial x$. Similarly, if $\mathbf{u} = \mathbf{j}$, then (12) reduces to $\partial f / \partial y$.

REMARK 2.   Definition 1 makes sense if f is a funtion of three variables. Then, of course, $\mathbf{u}$ is a unit vector in $\mathbb{R}^3$.

REMARK 3.   There is another definition of the directional derivative. It is given by

$$f_{\mathbf{u}}'(\mathbf{x}_0) = \lim_{h \to 0} \frac{f(\mathbf{x}_0 + h\mathbf{u}) - f(\mathbf{x}_0)}{h}. \tag{13}$$

It can be shown that if the limit in (13) exists, it is equal to $\nabla f(\mathbf{x}_0) \cdot \mathbf{u}$ if f is differentiable.

EXAMPLE 1   Let $z = f(x, y) = xy^2$. Calculate the directional derivative of f in the direction of the vector $\mathbf{v} = 2\mathbf{i} + 3\mathbf{j}$ at the point $(4, -1)$.

Solution. A unit vector in the direction $\mathbf{v}$ is $\mathbf{u} = (2/\sqrt{13})\mathbf{i} + (3/\sqrt{13})\mathbf{j}$. Also, $\nabla f = y^2\mathbf{i} + 2xy\mathbf{j}$. Thus

$$f_{\mathbf{u}}'(x, y) = \nabla f(\mathbf{x}) \cdot \mathbf{u} = \frac{2y^2}{\sqrt{13}} + \frac{6xy}{\sqrt{13}} = \frac{2y^2 + 6xy}{\sqrt{13}}.$$

At $(4, -1)$, $f_{\mathbf{u}}'(4, -1) = -22/\sqrt{13}$. ∎

EXAMPLE 2   Let $f(x, y, z) = x \ln y - e^{xz^3}$. Calculate the directional derivative of f in the direction of the vector $\mathbf{v} = \mathbf{i} - \mathbf{j} + 3\mathbf{k}$. Evaluate this derivative at the point $(-5, 1, -2)$.

Solution. A unit vector in the direction of $\mathbf{v}$ is $\mathbf{u} = (1/\sqrt{11})\mathbf{i} - (1/\sqrt{11})\mathbf{j} + (3/\sqrt{11})\mathbf{k}$, and

$$\nabla f = (\ln y - z^3 e^{xz^3})\mathbf{i} + \frac{x}{y}\mathbf{j} - 3xz^2 e^{xz^3}\mathbf{k}.$$

Thus

$$f_{\mathbf{u}}'(\mathbf{x}) = \nabla f(\mathbf{x}) \cdot \mathbf{u} = \frac{\ln y - z^3 e^{xz^3} - (x/y) - 9xz^2 e^{xz^3}}{\sqrt{11}},$$

and at $(-5, 1, -2)$

$$f_u'(-5, 1, -2) = \frac{5 + 188e^{40}}{\sqrt{11}}. \quad \blacksquare$$

There is an interesting geometric interpretation of the directional derivative. By Definitions 15.2.5 and 17.2.4 the projection of ∇f on $\mathbf{u}$ is given by

$$\text{Proj}_u \nabla f = \frac{\nabla f \cdot \mathbf{u}}{|\mathbf{u}|^2} \mathbf{u},$$

and since $\mathbf{u}$ is a unit vector, the component of ∇f in the direction $\mathbf{u}$ is given by

$$\frac{\nabla f \cdot \mathbf{u}}{|\mathbf{u}|^2} = \nabla f \cdot \mathbf{u}.$$

Thus the *directional derivative of f in the direction* $\mathbf{u}$ *is the component of the gradient of f in the direction* $\mathbf{u}$. This is illustrated in Figure 3.

Component of $\nabla f(\mathbf{x})$ in direction u

FIGURE 3

We now derive another remarkable property of the gradient. Recall that $\mathbf{u} \cdot \mathbf{v} = |\mathbf{u}| \, |\mathbf{v}| \cos \theta$, where θ is the smallest angle between the vectors $\mathbf{u}$ and $\mathbf{v}$. Thus the directional derivative of f in the direction $\mathbf{u}$ can be written as

$$f_u'(\mathbf{x}) = \nabla f(\mathbf{x}) \cdot \mathbf{u} = |\nabla f(\mathbf{x})| \, |\mathbf{u}| \cos \theta, \tag{14}$$

or since $\mathbf{u}$ is a unit vector,

$$f_u'(\mathbf{x}) = |\nabla f(\mathbf{x})| \cos \theta.$$

Now $\cos \theta = 1$ when $\theta = 0$, which occurs when $\mathbf{u}$ has the direction of ∇f. Similarly, $\cos \theta = -1$ when $\theta = \pi$, which occurs when $\mathbf{u}$ has the direction of $-\nabla f$. Also, $\cos \theta = 0$ when $\theta = \pi/2$. Thus since $-1 \le \cos \theta \le 1$, equation (14) implies the following important result.

Theorem 1   Let f be differentiable. Then f increases most rapidly in the direction of its gradient and decreases most rapidly in the direction opposite to that of its gradient. It changes least in a direction perpendicular to its gradient.

EXAMPLE 3   Consider the sphere $x^2 + y^2 + z^2 = 1$. We can write the upper half of this sphere (i.e., the upper hemisphere) as $z = f(x, y) = \sqrt{1 - x^2 - y^2}$. Then

$$\nabla f = \frac{-x}{\sqrt{1 - x^2 - y^2}}\mathbf{i} + \frac{-y}{\sqrt{1 - x^2 - y^2}}\mathbf{j}.$$

The direction of ∇f is

$$\tan^{-1}\left(\frac{-y/\sqrt{1 - x^2 - y^2}}{-x/\sqrt{1 - x^2 - y^2}}\right) = \tan^{-1}\left(\frac{-y}{-x}\right).$$

But $\tan^{-1}(-y/-x)$ is the direction of the vector $-x\mathbf{i} - y\mathbf{j}$, which points from (x, y) to $(0, 0)$. Thus if we start at a point (x, y, z) on the sphere, the path of steepest ascent (increase) is a great circle passing through the point $(0, 0, 1)$ (called the *north pole* of the sphere). ■

EXAMPLE 4   The distribution of voltage on a metal plate is given by

$$V = 50 - x^2 - 4y^2.$$

(a)   At the point $(1, -2)$, in what direction does the voltage increase most rapidly?
(b)   In what direction does it decrease most rapidly?
(c)   What is the magnitude of this increase or decrease?
(d)   In what direction does it change least?

Solution. $\nabla V = V_x\mathbf{i} + V_y\mathbf{j} = -2x\mathbf{i} - 8y\mathbf{j}$. At $(1, -2)$, $\nabla V = -2\mathbf{i} + 16\mathbf{j}$.

(a)   The voltage increases most rapidly as we move in the direction of $-2\mathbf{i} + 16\mathbf{j}$.
(b)   It decreases most rapidly in the direction of $2\mathbf{i} - 16\mathbf{j}$.
(c)   The magnitude of the increase or decrease is $\sqrt{2^2 + 16^2} = \sqrt{260}$.
(d)   A unit vector perpendicular to ∇V is $(16\mathbf{i} + 2\mathbf{j})/\sqrt{260}$. The voltage changes least in this or the opposite direction. ■

EXAMPLE 5   In Example 4, describe the path of a particle that starts at the point $(1, -2)$ and moves in the direction of greatest voltage increase.

Solution. The path of the particle will be that of the gradient. If the particle follows the path $\mathbf{f}(t) = x(t)\mathbf{i} + y(t)\mathbf{j}$, then since the direction of the path is $\mathbf{f}'(t) = x'(t)\mathbf{i} + y'(t)\mathbf{j}$ and since this direction is also given by $\nabla V = -2x\mathbf{i} - 8y\mathbf{j}$, we must have

$$x'(t) = -2x(t) \qquad \text{and} \qquad y'(t) = -8y(t).$$

From Section 6.6 (see page 401), the solutions to these differential equations are

$$x(t) = c_1 e^{-2t} \qquad \text{and} \qquad y(t) = c_2 e^{-8t}.$$

But $x(0) = 1$ and $y(0) = -2$, so

$$x(t) = e^{-2t} \qquad \text{and} \qquad y(t) = -2e^{-8t}.$$

Then since $e^{-8t} = (e^{-2t})^4$, we see that the particle moves along the path

$$y = -2x^4.$$

REMARK. Technically, a direction is a unit vector, so we should choose the direction $(-2x\mathbf{i} - 8y\mathbf{j})/\sqrt{4x^2 + 64y^2}$ in our computations. But this choice would not change the final answer. A method for obtaining the answer by using unit vectors is suggested in Problem 27. ■

One other fact about directional derivatives and gradients should be mentioned here. Since the gradient vector is normal to the curve $f(x, y) = C$, for any constant C, we say that the directional derivative of f in the direction of the gradient is the **normal derivative** of f and is denoted df/dn. We then have, from equation (14),

$$\text{normal derivative} = \frac{df}{dn} = |\nabla f|. \tag{15}$$

EXAMPLE 6   Let $f(x, y) = xy^2$. Calculate the normal derivative. Evaluate df/dn at the point $(3, -2)$.

Solution. $\nabla f = y^2\mathbf{i} + 2xy\mathbf{j}$. Then $df/dn = |\nabla f| = \sqrt{y^4 + 4x^2y^2}$. At $(3, -2)$, $df/dn = \sqrt{16 + 144} = \sqrt{160}$. ■

PROBLEMS 18.8

In Problems 1–15, calculate the directional derivative of the given function at the given point in the direction of the given vector **v**.

1. $f(x, y) = xy$ at $(2, 3)$; $\mathbf{v} = \mathbf{i} + 3\mathbf{j}$
2. $f(x, y) = 2x^2 - 3y^2$ at $(1, -1)$; $\mathbf{v} = -\mathbf{i} + 2\mathbf{j}$
3. $f(x, y) = \ln(x + 3y)$ at $(2, 4)$; $\mathbf{v} = \mathbf{i} + \mathbf{j}$
4. $f(x, y) = ax^2 + by^2$ at (c, d); $\mathbf{v} = \alpha\mathbf{i} + \beta\mathbf{j}$

5. $f(x, y) = \tan^{-1}\dfrac{y}{x}$ at $(2, 2)$; $\mathbf{v} = 3\mathbf{i} - 2\mathbf{j}$

6. $f(x, y) = \dfrac{x - y}{x + y}$ at $(4, 3)$; $\mathbf{v} = -\mathbf{i} - 2\mathbf{j}$

7. $f(x, y) = xe^y + ye^x$ at $(1, 2)$; $\mathbf{v} = \mathbf{i} + \mathbf{j}$
8. $f(x, y) = \sin(2x + 3y)$ at $(\pi/12, \pi/9)$; $\mathbf{v} = -2\mathbf{j} + 3\mathbf{j}$
9. $f(x, y, z) = xy + yz + xz$ at $(1, 1, 1)$; $\mathbf{v} = \mathbf{i} + \mathbf{j} + \mathbf{k}$
10. $f(x, y, z) = xy^3z^5$ at $(-3, -1, 2)$; $\mathbf{v} = -\mathbf{i} - 2\mathbf{j} + \mathbf{k}$
11. $f(x, y, z) = \ln(x + 2y + 3z)$ at $(1, 2, 0)$; $\mathbf{v} = 2\mathbf{i} + \mathbf{j} - \mathbf{k}$
12. $f(x, y, z) = xe^{yz}$ at $(2, 0, -4)$; $\mathbf{v} = -\mathbf{i} + 2\mathbf{j} + 5\mathbf{k}$
13. $f(x, y, z) = x^2y^3 + z\sqrt{x}$ at $(1, -2, 3)$; $\mathbf{v} = 5\mathbf{j} + \mathbf{k}$
14. $f(x, y, z) = e^{-(x^2+y^2+z^2)}$ at $(1, 1, 1)$; $\mathbf{v} = \mathbf{i} + 3\mathbf{j} - 5\mathbf{k}$

15. $f(x, y, z) = \dfrac{1}{\sqrt{x^2 + y^2 + z^2}}$ at $(-1, 2, 3)$; $\mathbf{v} = \mathbf{i} - \mathbf{j} + \mathbf{k}$

16. The voltage (potential) at any point on a metal structure is given by

$$v(x, y, z) = \frac{1}{0.02 + \sqrt{x^2 + y^2 + z^2}}.$$

At the point $(1, -1, 2)$, in what direction does the voltage increase most rapidly?

17. The temperature at any point in a solid metal ball centered at the origin is given by

$$T(x, y, z) = 100e^{-(x^2+y^2+z^2)}.$$

(a) Where is the ball hottest?

(b) Show that at any point (x, y, z) on the ball, the direction of greatest increase in temperature is a vector pointing toward the origin.

18. The temperature distribution of a ball centered at the origin is given by

$$T(x, y, z) = \frac{100}{x^2 + y^2 + z^2 + 1}.$$

(a) Where is the ball hottest?

(b) Find the direction of greatest decrease of temperature at the point $(3, -1, 2)$.

(c) Find the direction of greatest increase in temperature. Does this vector point toward the origin?

19. The temperature distribution on a plate is given by

$$T(x, y) = 1 - \frac{x^2}{a^2} - \frac{y^2}{b^2}.$$

Find the path of a heat-seeking particle (i.e., a particle that always moves in the direction of greatest increase in temperature) if it starts at the point (a, b).

20. Find the path of the particle in Problem 19 if it starts at the point $(-a, b)$.

21. The height of a mountain is given by $h(x, y) = 3000 - 2x^2 - y^2$, where the x-axis points north, the y-axis points east, and all distances are measured in meters. Suppose that a mountain climber is at the point $(30, -20, 800)$.

(a) If the climber moves in the southwest direction, will he or she ascend or descend?

(b) In what direction should the climber move so as to ascend most rapidly?

*22. Prove that if $w = f(x, y, z)$ is differentiable at $\mathbf{x}$, then all the first partials exist at $\mathbf{x}$, and

$$\nabla f(\mathbf{x}) = f_x(\mathbf{x})\mathbf{i} + f_y(\mathbf{x})\mathbf{j} + f_z(\mathbf{x})\mathbf{k}.$$

(This result is the converse to Theorem 18.5.1.)

In Problems 23–26, calculate the normal derivative at the given point.

23. $f(x, y) = x + 2y$ at $(1, 4)$

24. $f(x, y) = e^{x+3y}$ at $(1, 0)$

25. $f(x, y) = \tan^{-1}\frac{y}{x}$ at $(-1, -1)$

26. $f(x, y) = \sqrt{\frac{x - y}{x + y}}$ at $(3, 1)$

27. We refer to Example 5.

(a) Show that a unit vector in the direction of motion is given by

$$\frac{x'}{\sqrt{x'^2 + y'^2}}\mathbf{i} + \frac{y'}{\sqrt{x'^2 + y'^2}}\mathbf{j}.$$

(b) Show that a unit vector having the direction of the gradient is

$$\frac{-2x}{\sqrt{4x^2 + 64y^2}}\mathbf{i} - \frac{8y}{\sqrt{4x^2 + 64y^2}}\mathbf{j}.$$

(c) By equating coordinates in (a) and (b), show that

$$\frac{y'(t)}{y(t)} = \frac{4x'(t)}{x(t)}.$$

(d) Integrate both sides of the equation in (c) to show that $\ln|y'(t)| = \ln[x(t)]^4 + C$.

(e) Use (d) to show that $y(t) = k[x(t)]^4$, where $k = \pm e^c$.

(f) Show that $k = -2$ in (e) by using the point $(1, -2)$ on the curve.

18.9 CONSERVATIVE VECTOR FIELDS AND THE GRADIENT (OPTIONAL)

In this brief section we show how the gradient can arise in a fundamental way in physics. Suppose that for every $\mathbf{x}$ in $\mathbb{R}^2$ (or $\mathbb{R}^3$), $\mathbf{F}(\mathbf{x})$ is another vector in $\mathbb{R}^2$ (or $\mathbb{R}^3$). Then $\mathbf{F}$ is really a vector function of a vector. We call such a function a **vector field.**† An important question that can arise is: Does there exist a function $f(\mathbf{x})$ such that $\nabla f(\mathbf{x}) = \mathbf{F}(\mathbf{x})$? That is, is $\mathbf{F}$ the gradient of some function f? In Section 18.11 we answer this question. However, in this section we will assume that there exists a function f such that

$$\mathbf{F} = \nabla f \tag{1}$$

If $\mathbf{F} = -\nabla f$ for some function f, then $\mathbf{F}$ is said to be a **conservative vector field** and f is called a **potential function** for $\mathbf{F}$. The reason for this terminology will be made clear shortly. [If $\mathbf{F} = \nabla f$, then $\mathbf{F} = -\nabla(-f)$, so that the introduction of the minus sign does not cause any problem.]

Now let $\mathbf{g}(t) = x(t)\mathbf{i} + y(t)\mathbf{j}$ be a differentiable curve and suppose that a particle of mass m moves along it. Suppose further that the force acting on the particle at any time t is given by $\mathbf{F}(\mathbf{x}(t))$, where $\mathbf{F}$ is assumed to be a conservative vector field. By Newton's second law,

$$\mathbf{F}(\mathbf{x}(t)) = m\mathbf{a}(t) = m\mathbf{x}''(t). \tag{2}$$

But since $\mathbf{F}$ is conservative, (1) implies that $\mathbf{F}(\mathbf{x}) = -\nabla f(\mathbf{x})$ for some differentiable function f. Then we have

$$-\nabla f(\mathbf{x}(t)) = m\mathbf{x}''(t),$$

or

$$m\mathbf{x}'' + \nabla f(\mathbf{x}) = \mathbf{0}. \tag{3}$$

We now take the dot product of both sides of (3) with $\mathbf{x}'$ to obtain

$$m\mathbf{x}' \cdot \mathbf{x}'' + \nabla f(\mathbf{x}) \cdot \mathbf{x}' = 0. \tag{4}$$

†We will discuss vector fields in more detail in Section 20.1.

But by the product rule,

$$\frac{d}{dx}|\mathbf{x}'(t)|^2 = \frac{d}{dt}(\mathbf{x}'(t) \cdot \mathbf{x}'(t)) = \mathbf{x}'(t) \cdot \mathbf{x}''(t) + \mathbf{x}''(t) \cdot \mathbf{x}'(t) = 2\mathbf{x}'(t) \cdot \mathbf{x}''(t), \qquad (5)$$

and by equation (18.6.3),

$$\frac{d}{dt}f(\mathbf{x}(t)) = \nabla f(\mathbf{x}(t)) \cdot \mathbf{x}'(t) \qquad (6)$$

Using (5) and (6) in (4), we obtain

$$\frac{d}{dt}\left[\frac{1}{2}m|\mathbf{x}'|^2 + f(\mathbf{x}(t))\right] = 0,$$

which implies that

$$\tfrac{1}{2}m|\mathbf{x}'|^2 + f(\mathbf{x}(t)) = C, \qquad (7)$$

where C is a constant. This is one of the versions of the **law of conservation of energy**. The term $\frac{1}{2}m|\mathbf{x}'|^2 = \frac{1}{2}m|\mathbf{v}|^2$ is called the **kinetic energy** of the particle, and the term $f(\mathbf{x}(t))$ is called the **potential energy** of the particle. Equation (7) tells us simply that if the force function $\mathbf{F}$ is conservative, then the total energy of the system is constant and, moreover, the potential function f of $\mathbf{F}$ represents the potential energy of the system.

The **principle of the conservation of energy** states that energy may be transformed from one form to another but cannot be created or destroyed; that is, the total energy is constant. Thus it seems reasonable that force fields in classical physics are conservative (although since the work of Einstein, it has been found that energy can be transformed into mass and vice versa, so that there are forces that are not conservative). One example of a conservative force is given by the force of gravitational attraction. Let m_1 represent the mass of a (relatively) fixed object in space and let m_2 denote the mass of an object moving near the fixed object. Then the magnitude of the gravitational force between the objects is given by (see Example 3.2.7)

$$|\mathbf{F}| = G\frac{m_1 m_2}{r^2}, \qquad (8)$$

where r is the distance between the objects and G is a universal constant. If we assume that the first object is at the origin, then we may denote the position of the second object by $\mathbf{x}(t)$, and then since $r = |\mathbf{x}|$, (8) can be written

$$|\mathbf{F}| = \frac{Gm_1 m_2}{|\mathbf{x}|^2}. \qquad (9)$$

Also, the force acts toward the origin, that is, in the direction opposite to that of the positive vector $\mathbf{x}$. Therefore

$$\text{direction of } \mathbf{F} = -\frac{\mathbf{x}}{|\mathbf{x}|}, \tag{10}$$

so from (9) and (10),

$$\mathbf{F}(t) = \frac{\alpha \mathbf{x}(t)}{|\mathbf{x}(t)|^3}, \tag{11}$$

where $\alpha = -Gm_1m_2$. We now prove that $\mathbf{F}$ is conservative. To show this, we must come up with a function f such that $\mathbf{F} = -\nabla f$. Here we will pull f "out of a hat." In Section 18.11 we will show you how to construct such an f (if one exists). Let

$$f(\mathbf{x}) = \frac{\alpha}{|\mathbf{x}|}.$$

Then $f(\mathbf{x}) = \alpha/\sqrt{x^2 + y^2 + z^2}$ and

$$-\nabla f(\mathbf{x}) = -\alpha\left[\frac{-x}{(x^2 + y^2 + z^2)^{3/2}}\mathbf{i} + \frac{-y}{(x^2 + y^2 + z^2)^{3/2}}\mathbf{j} + \frac{-z}{(x^2 + y^2 + z^2)^{3/2}}\mathbf{k}\right]$$

$$= \frac{\alpha}{|\mathbf{x}|^3}(x\mathbf{i} + y\mathbf{j} + z\mathbf{k}) = \frac{\alpha\mathbf{x}}{|\mathbf{x}|^3} = \mathbf{F}.$$

Thus $\mathbf{F}$ is conservative, so that, with respect to gravitational forces, the law of conservation of energy holds.

PROBLEMS 18.9

1. Show that the force given by $\mathbf{F}(\mathbf{x}) = -\alpha\mathbf{x}/|\mathbf{x}|^2$ is conservative by finding a potential function for it.
2. Do the same for the force given by $\mathbf{F}(\mathbf{x}) = -\alpha\mathbf{x}/|\mathbf{x}|^4$.
3. If α is a constant, show that the force $\mathbf{F}(\mathbf{x}) = -\alpha\mathbf{x}/|\mathbf{x}|^k$ is conservative.
4. Show that the force $\mathbf{F}(x, y) = y\mathbf{i} + x\mathbf{j}$ is conservative.
5. Show that if $\mathbf{F}$ and $\mathbf{G}$ are conservative, then $\alpha\mathbf{F} + \beta\mathbf{G}$ is also conservative for any constants α and β.
*6. Show that the force $\mathbf{F}(x, y) = y\mathbf{i} - x\mathbf{j}$ is *not* conservative. [*Hint:* Assume that $\mathbf{F} = -\nabla f$, so that $-y = \partial f/\partial x$ and $x = \partial f/\partial y$. Then integrate to obtain a contradiction.]

18.10 THE TOTAL DIFFERENTIAL AND APPROXIMATION

In Section 3.8 we used the notions of increments and differentials to approximate a function. We used the fact that if Δx was small, then

$$f(x + \Delta x) - f(x) = \Delta y \approx f'(x)\, \Delta x. \tag{1}$$

We also defined the differential dy by

$$dy = f'(x)\, dx = f'(x)\, \Delta x \qquad\qquad (2)$$

(since dx is defined to be equal to Δx). Note that in (2) it is not required that Δx be small.

We now extend these ideas to functions of two or three variables.

Definition 1   INCREMENT AND TOTAL DIFFERENTIAL   Let $f = f(\mathbf{x})$ be a function of two or three variables and let $\Delta \mathbf{x} = (\Delta x, \Delta y)$ or $(\Delta x, \Delta y, \Delta z)$.

(i) The **increment of** f, denoted Δf, is defined by

$$\Delta f = f(\mathbf{x} + \Delta \mathbf{x}) - f(\mathbf{x}). \qquad\qquad (3)$$

(ii) The **total differential of** f, denoted df, is given by

$$df = \nabla f(\mathbf{x}) \cdot \Delta \mathbf{x}. \qquad\qquad (4)$$

Note that equation (4) is very similar in form to equation (2).

REMARK 1.   If f is a function of two variables, then (3) and (4) become

$$\Delta f = f(x + \Delta x, y + \Delta y) - f(x, y), \qquad\qquad (5)$$

and the total differential is

$$df = f_x(x, y)\, \Delta x + f_y(x, y)\, \Delta y. \qquad\qquad (6)$$

REMARK 2.   If f is a function of three variables, then (3) and (4) become

$$\Delta f = f(x + \Delta x, y + \Delta y, z + \Delta z) - f(x, y, z) \qquad\qquad (7)$$

and

$$df = f_x(x, y, z)\, \Delta x + f_y(x, y, z)\, \Delta y + f_z(x, y, z)\, \Delta z. \qquad\qquad (8)$$

REMARK 3.   Note that in the definition of the total differential, it is *not* required that $|\Delta \mathbf{x}|$ be small.

From Theorems 18.5.1 and 18.5.1′ and the definition of differentiability, we see that if $|\Delta \mathbf{x}|$ is small and if f is differentiable, then

$$\Delta f \approx df. \qquad\qquad (9)$$

We can use the relation (9) to approximate functions of several variables in much the same way that we used the relation (1) to approximate the values of functions of one variable.

EXAMPLE 1   Use the total differential to estimate $\sqrt{(2.98)^2 + (4.03)^2}$.

Solution. Let $f(x, y) = \sqrt{x^2 + y^2}$. Then we are asked to calculate $f(2.98, 4.03)$. We know that $f(3, 4) = \sqrt{3^2 + 4^2} = 5$. Thus we need to calculate $f(3 - 0.02, 4 + 0.03)$. Now, at $(3, 4)$,

$$\nabla f(\mathbf{x}) = \frac{x}{\sqrt{x^2 + y^2}}\mathbf{i} + \frac{y}{\sqrt{x^2 + y^2}}\mathbf{j} = \frac{3}{5}\mathbf{i} + \frac{4}{5}\mathbf{j}.$$

Then using (6), we have

$$df = \frac{3}{5}\Delta x + \frac{4}{5}\Delta y = (0.6)(-0.02) + (0.8)(0.03) = 0.012.$$

Hence

$$f(3 - 0.02, 4 + 0.03) - f(3, 4) = \Delta f \approx df = 0.012,$$

so

$$f(2.98, 4.03) \approx f(3, 4) + 0.012 = 5.012.$$

The exact value of $\sqrt{(2.98)^2 + (4.03)^2}$ is $\sqrt{8.8804 + 16.2409} = \sqrt{25.1213} \approx 5.012115$, so that $\Delta f \approx 0.012115$ and our approximation is very good indeed.

EXAMPLE 2 The radius of a cone is measured to be 15 cm and the height of the cone is measured to be 25 cm. There is a maximum error of ± 0.02 cm in the measurement of the radius and ± 0.05 cm in the measurement of the height. (a) What is the approximate volume of the cone? (b) What is the maximum error in the calculation of the volume?

Solution. **(a)** $V = \frac{1}{3}\pi r^2 h \approx \frac{1}{3}\pi(15)^2 25 = 1875\pi \text{ cm}^3 \approx 5890.5 \text{ cm}^3$.
(b) $\nabla V = V_r\mathbf{i} + V_h\mathbf{j} = \frac{2}{3}\pi rh\mathbf{i} + \frac{1}{3}\pi r^2\mathbf{j} = \pi(250\mathbf{i} + 75\mathbf{j})$. Then choosing $\Delta x = 0.02$ and $\Delta y = 0.05$ to find the maximum error, we have

$$\Delta V \approx dV = \nabla V \cdot \Delta \mathbf{x} = \pi[250(0.02) + 75(0.05)] = \pi(5 + 3.75)$$
$$= 8.75\pi \approx 27.5 \text{ cm}^3.$$

Thus the maximum error in the calculation is, approximately, 27.5 cm³, which means that

$$5890.5 - 27.5 < V < 5890.5 + 27.5,$$

or

$$5863 \text{ cm}^3 < V < 5918 \text{ cm}^3.$$

Note that an error of 27.5 cm³ is only a *relative error* of $27.5/5890.5 \approx 0.0047$, which is a very small relative error (see p. 185 for a discussion of relative error). ∎

EXAMPLE 3 A cylindrical tin can has an inside radius of 5 cm and a height of 12 cm. The thickness of the tin is 0.2 cm. Estimate the amount of tin needed to construct the can (including its ends).

    Solution. We need to estimate the difference between the "outer" and "inner" volumes of the can. We have $V = \pi r^2 h$. The inner volume is $\pi(5^2)(12) = 300\pi$ cm³, and the outer volume is $\pi(5.2)^2(12.4)$. The difference is

$$\Delta V = \pi(5.2)^2(12.4) - 300\pi \approx dV.$$

Since $\nabla V = 2\pi r h \mathbf{i} + \pi r^2 \mathbf{j} = \pi(120\mathbf{i} + 25\mathbf{j})$, we have

$$dV = \pi(120(0.2) + 25(0.4)) = 34\pi.$$

Thus the amount of tin needed is, approximately, 34π cm³ ≈ 106.8 cm³. ∎

PROBLEMS 18.10

In Problems 1–12, calculate the total differential df.

1. $f(x, y) = xy^3$

2. $f(x, y) = \tan^{-1}\dfrac{y}{x}$

3. $f(x, y) = \sqrt{\dfrac{x - y}{x + y}}$

4. $f(x, y) = xe^y$

5. $f(x, y) = \ln(2x + 3y)$

6. $f(x, y) = \sin(x - 4y)$

7. $f(x, y, z) = xy^2z^5$

8. $f(x, y, z) = \dfrac{xy}{z}$

9. $f(x, y, z) = \ln(x + 2y + 3z)$

10. $f(x, y, z) = \sec xy - \tan z$

11. $f(x, y, z) = \cosh(xy - z)$

12. $f(x, y, z) = \dfrac{x - z}{y + 3x}$

13. Let $f(x, y) = xy^2$.
    (a) Calculate explicitly the difference $\Delta f - df$.
    (b) Verify your answer by calculating $\Delta f - df$ at the point $(1, 2)$, where $\Delta x = -0.01$ and $\Delta y = 0.03$.

*14. Repeat the steps of Problem 13 for the function $f(x, y) = x^3y^2$.

In Problems 15–23, use the total differential to estimate the given number.

15. $3.01/5.99$

16. $19.8\sqrt{65}$

17. $\sqrt{35.6}\,\sqrt[3]{64.08}$

18. $(2.01)^4(3.04)^7 - (2.01)(3.04)^9$

19. $\sqrt{\dfrac{5.02 - 3.96}{5.02 + 3.96}}$

20. $((4.95)^2 + (7.02))^{1/5}$

21. $\dfrac{(3.02)(1.97)}{\sqrt{8.95}}$

22. $\sin\left(\dfrac{11\pi}{24}\right)\cos\left(\dfrac{13\pi}{36}\right)$

23. $(7.92)\sqrt{5.01 - (0.98)^2}$

24. The radius and height of a cylinder are, approximately, 10 cm and 20 cm, respectively. The maximum errors in approximation are ± 0.03 cm and ± 0.07 cm.
(a) What is the approximate volume of the cylinder?
(b) What is the approximate maximum error in this calculation?

25. Two sides of a triangular piece of land were measured to be 50 m and 110 m; there was an error of at most ± 0.3 m in each measurement. The angle between the sides was measured to be $60°$ with an error of at most $\pm 1°$.
(a) Using the law of cosines, find the approximate length of the third side of the triangle.
(b) What is the approximate maximum error of your measurement? [*Hint:* It will be necessary to convert $1°$ to radians.]

26. How much wood is contained in the sides of a rectangular box with sides of inside measurements 1 m, 1.2 m, and 1.6 m if the thickness of the wood making up the sides is 5 cm ($= 0.05$ m)?

27. When three resistors are connected in parallel, the total resistance R (measured in ohms) is given by

$$\frac{1}{R} = \frac{1}{r_1} + \frac{1}{r_2} + \frac{1}{r_3},$$

where r_1, r_2, and r_3 are the three separate resistances. Let $r_1 = 6 \pm 0.1\ \Omega$, $r_2 = 8 \pm 0.03\ \Omega$, and $r_3 = 12 \pm 0.15\ \Omega$.
(a) Estimate R.
(b) Find an approximate value for the maximum error in your estimate.

28. The volume of 10 mol of an ideal gas was calculated to be 500 cm^3 at a temperature of 40°C ($= 313$ °K). The maximum error in each measurement n, V, and T was $\frac{1}{2}\%$.
(a) Calculate the approximate pressure of the gas (in newtons per square centimeter).
(b) Find the approximate maximum error in your computation. [*Hint:* Recall that according to the ideal gas law, $PV = nRT$.]

18.11 EXACT VECTOR FIELDS OR HOW TO OBTAIN A FUNCTION FROM ITS GRADIENT

If $z = f(x, y)$ is differentiable, then the gradient of f exists and

$$\nabla f = f_x(x, y)\mathbf{i} + f_y(x, y)\mathbf{j}. \tag{1}$$

We can write (1) as

$$\nabla f = P(x, y)\mathbf{i} + Q(x, y)\mathbf{j}. \tag{2}$$

In this section we will try to reverse this process. This corresponds roughly to finding an antiderivative of a function of one variable. We will ask the following question: Given a *vector field* of the form

$$\mathbf{F}(x, y) = P(x, y)\mathbf{i} + Q(x, y)\mathbf{j}, \tag{3}$$

does there exist a differentiable function f such that $\nabla f = \mathbf{F}$?

Before answering this question in general, we will look at two examples that illustrate what can happen.

EXAMPLE 1   Let $\mathbf{F} = 9x^2y^2\mathbf{i} + (6x^3y + 2y)\mathbf{j}$. Show that there is a function f such that $\nabla f = \mathbf{F}$.

Solution. If there is such an f, then

$$\frac{\partial f}{\partial x} = 9x^2y^2 \quad \text{and} \quad \frac{\partial f}{\partial y} = 6x^3y + 2y. \tag{4}$$

Integrating both sides of the first of these equations *with respect to x*, we obtain

$$f(x, y) = 3x^3y^2 + g(y). \tag{5}$$

The term $g(y)$ is a *constant function of integration* since

$$\frac{\partial}{\partial x}g(y) = 0.$$

Next we differentiate (5) *with respect to y* and use the second equation of (4):

$$\frac{\partial f}{\partial y} = 6x^3y + g'(y) = 6x^3y + 2y.$$

Thus $g'(y) = 2y$ and $g(y) = y^2 + C$. Hence for any constant C,

$$f(x, y) = 3x^3y^2 + y^2 + C$$

is a function that satisfies $\nabla f = \mathbf{F}$. (This result should be checked.) ∎

REMARK.   When we integrate a function of x and y with respect to x, the "constant" of integration is always a function of y. Analogously, when we integrate a function of x and y with respect to y, the "constant" of integration is always a function of x.

EXAMPLE 2   Show that there is *no* function f that satisfies

$$\nabla f = y\mathbf{i} - x\mathbf{j}.$$

Solution. Suppose that $\nabla f = y\mathbf{i} - x\mathbf{j}$. Then $\partial f/\partial x = y$ and $\partial f/\partial y = -x$. Integration of the first of these equations with respect to x yields

$$f(x, y) = yx + g(y),$$

so that

$$\frac{\partial f}{\partial y} = x + g'(y) = -x.$$

This implies that $g'(y) = -2x$, which is impossible since g and g' are functions of y only. Hence there is no function f whose gradient is $y\mathbf{i} - x\mathbf{j}$. ∎

We now state a general result that tells us whether **F** is a gradient of some function f (i.e., whether **F** is conservative). Half its proof is difficult and is omitted.† The other half is suggested in Problem 24.

Theorem 1 Let $\mathbf{F}(x, y) = P(x, y)\mathbf{i} + Q(x, y)\mathbf{j}$ and suppose that P, Q, $\partial P/\partial y$, and $\partial Q/\partial x$ are continuous in an open disk D centered at (x, y). Then in D, **F** is the gradient of a function f if and only if

$$\frac{\partial P}{\partial y} = \frac{\partial Q}{\partial x}. \tag{6}$$

Definition 1 EXACT VECTOR FIELD We say that the vector field $P(x, y)\mathbf{i} + Q(x, y)\mathbf{j}$ is **exact** if condition (6) holds, that is, if $\partial P/\partial y = \partial Q/\partial x$.

EXAMPLE 3 Show that the vector field

$$\mathbf{F}(x, y) = \left(4x^3y^3 + \frac{1}{x}\right)\mathbf{i} + \left(3x^4y^2 - \frac{1}{y}\right)\mathbf{j}$$

is exact, and find all functions f for which $\nabla f = \mathbf{F}$.

Solution. $P(x, y) = 4x^3y^3 + (1/x)$ and $Q(x, y) = 3x^4y^2 - (1/y)$, so

$$\frac{\partial P}{\partial y} = 12x^3y^2 = \frac{\partial Q}{\partial x}.$$

If $\nabla f = \mathbf{F}$, then $\partial f/\partial x = P$, so

$$f(x, y) = \int\left(4x^3y^3 + \frac{1}{x}\right) dx = x^4y^3 + \ln|x| + g(y).$$

Differentiating with respect to y, we have

$$Q = \frac{\partial f}{\partial y} = 3x^4y^2 + g'(y) = 3x^4y^2 - \frac{1}{y}.$$

Thus $g'(y) = -1/y$, $g(y) = -\ln|y| + C$, and finally,

$$f(x, y) = x^4y^3 + \ln|x| - \ln|y| + C = x^4y^3 + \ln\left|\frac{x}{y}\right| + C.$$

This answer should be checked. Note that the presence of the constant C indicates that the potential function f is not unique. ∎

---

†For a proof see R. C. Buck, *Advanced Calculus* (McGraw-Hill, New York, 1965), page 425.

We close this section by pointing out that it is now easy to check whether a force field **F** is conservative. It is necessary only to check whether or not **F** is exact.

PROBLEMS 18.11

In Problems 1–14, test for exactness. If the given vector field is exact, find all functions f for which $\nabla f = \mathbf{F}$.

1. $\mathbf{F}(x, y) = 2xy\mathbf{i} + (x^2 + 1)\mathbf{j}$
2. $\mathbf{F}(x, y) = (4x^3 - ye^{xy})\mathbf{i} + (\tan y - xe^{xy})\mathbf{j}$
3. $\mathbf{F}(x, y) = (4x^2 - 4y^2)\mathbf{i} + (8xy - \ln y)\mathbf{j}$
4. $\mathbf{F}(x, y) = [x \cos(x + y) + \sin(x + y)]\mathbf{i} + x \cos(x + y)]\mathbf{j}$
5. $\mathbf{F}(x, y) = 2x(\cos y)\mathbf{i} + x^2(\sin y)\mathbf{j}$

6. $\mathbf{F}(x, y) = \left[\dfrac{\ln(\ln y)}{x} + \dfrac{2}{3}xy^3\right]\mathbf{i} + \left(\dfrac{\ln x}{y \ln y} + x^2y^2\right)\mathbf{j}$

7. $\mathbf{F}(x, y) = (x - y \cos x)\mathbf{i} - (\sin x)\mathbf{j}$
8. $\mathbf{F}(x, y) = e^{x^2y}\mathbf{i} + e^{x^2y}\mathbf{j}$
9. $\mathbf{F}(x, y) = (3x \ln x + x^5 - y)\mathbf{i} - x\mathbf{j}$

10. $\mathbf{F}(x, y) = \left[\left(\dfrac{1}{x^2}\right) + y^2\right]\mathbf{i} + (2xy)\mathbf{j}$

11. $\mathbf{F}(x, y) = \left(\tan^2 x - \dfrac{y}{x^2 + y^2}\right)\mathbf{i} + \left(\dfrac{x}{x^2 + y^2} - e^y\right)\mathbf{j}$

12. $\mathbf{F}(x, y) = [\cos(x + 3y) - x^4]\mathbf{i} + \left[3 \cos(x + 3y) - \left(\dfrac{2}{y}\right)\right]\mathbf{j}$

13. $\mathbf{F}(x, y) = (x^2 + y^2 + 1)\mathbf{i} - (xy + y)\mathbf{j}$

14. $\mathbf{F}(x, y) = \left[-\left(\dfrac{1}{x^3}\right) + 4x^3y\right]\mathbf{i} + (\sin y + \sqrt{y} + x^4)\mathbf{j}$

Consider the vector field

$$\mathbf{F}(x, y, z) = P(x, y, z)\mathbf{i} + Q(x, y, z)\mathbf{j} + R(x, y, z)\mathbf{k}.$$

F is exact (i.e., conservative) if there is a differentiable function f such that

$$\nabla f(x, y, z) = \mathbf{F}(x, y, z).$$

If $P, Q, R, \partial P/\partial y, \partial P/\partial z, \partial Q/\partial x, \partial Q/\partial z, \partial R/\partial x$, and $\partial R/\partial y$ are continuous, then, analogously to Theorem 1, **F** is exact if and only if

$$\frac{\partial P}{\partial y} = \frac{\partial Q}{\partial x}, \qquad \frac{\partial R}{\partial x} = \frac{\partial P}{\partial z}, \qquad \frac{\partial Q}{\partial z} = \frac{\partial R}{\partial y}.$$

In Problems 15–20, use the result above to test for exactness. If **F** is exact, find all functions f for which $\nabla f = \mathbf{F}$.

15. $\mathbf{F} = \mathbf{i} + \mathbf{j} + \mathbf{k}$
16. $\mathbf{F} = yz\mathbf{i} + xz\mathbf{j} + xy\mathbf{k}$

17. $\mathbf{F} = \left[\left(\dfrac{y}{z}\right) + x^2\right]\mathbf{i} + \left[\left(\dfrac{x}{z}\right) - \sin y\right]\mathbf{j} + \left[\cos z - \left(\dfrac{xy}{z^2}\right)\right]\mathbf{k}$

18. $\mathbf{F} = \left(\dfrac{1}{x + 2y + 3z} - 3x\right)\mathbf{i} + \left(\dfrac{2}{x + 2y + 3z} + y^2\right)\mathbf{j} + \left(\dfrac{3}{x + 2y + 3z} - \tan^{-1} z\right)\mathbf{k}$

19. $\mathbf{F} = (e^{yz} + y)\mathbf{i} + (xze^{yz} - x)\mathbf{j} + (xye^{yz} + 2z)\mathbf{k}$
20. $\mathbf{F} = (2xy^3 + x + z)\mathbf{i} + (3x^2y^2 - y)\mathbf{j} + (x + \sin z)\mathbf{k}$

*21. A particle is moving in space with position vector $\mathbf{x}(t)$ and is subjected to a force $\mathbf{F}$ of magnitude $\alpha/|\mathbf{x}(t)|^3$ (α a constant) directed toward the origin. Show that the force is conservative and find all potential functions for $\mathbf{F}$.

*22. Answer the questions in Problem 21 if the magnitude of the force is $\alpha/|\mathbf{x}(t)|^k$ for some positive integer k.

*23. Suppose that a moving particle is subjected to a force of constant magnitude that always points toward the origin. Show that the force is conservative. [*Hint:* If the path is given by $\mathbf{x}(t) = x(t)\mathbf{i} + y(t)\mathbf{j}$, find a unit vector that points toward the origin.]

24. We omitted the proof of Theorem 1 but half of it is easy to prove. Show that if in Theorem 1 $\mathbf{F} = \nabla f$, then

$$\frac{\partial P}{\partial y} = \frac{\partial Q}{\partial x}.$$

[*Hint:* Use Theorem 18.4.1.]

18.12 MAXIMA AND MINIMA FOR A FUNCTION OF TWO VARIABLES

In Section 4.5 we discussed methods for obtaining maximum and minimum values for a function of one variable. The first basic fact we used was that if f is continuous on a closed bounded interval $[a, b]$, then f takes on a maximum and minimum value in $[a, b]$. We then defined a critical point to be a number x at which $f'(x) = 0$ or for which $f(x)$ exists but $f'(x)$ does not. Finally, we defined local maxima and minima (which occur at critical points) and gave conditions on first and second derivatives that ensured that a critical point was a local maximum or minimum.

The theory for functions of two variables is more complicated, but some of the basic ideas are the same. The first thing we need to know is that a function *has* a maximum or minimum. The proof of the following theorem can be found in any advanced calculus book.†

Theorem 1. Let f be a function of two variables that is continuous on the closed disk D. Then there exist points $\mathbf{x}_1$ and $\mathbf{x}_2$ in D such that

$$m = f(\mathbf{x}_1) \le f(\mathbf{x}) \le f(\mathbf{x}_2) = M \tag{1}$$

for any other point $\mathbf{x}$ in D. The numbers m and M are called, respectively, the **global minimum** and the **global maximum** for the function f over the disk D.

We now define what we mean by local minima and maxima.

Definition 1 DEFINITION OF LOCAL MAXIMA AND MINIMA Let f be defined in a neighborhood of a point $\mathbf{x}_0 = (x_0, y_0)$.

---

†See, for example, Buck, *Advanced Calculus*, page 74.

(i) f has a **local maximum** at x_0 if there is a neighborhood N_1 of x_0 such that for every point x in N_1,

$$f(x) \leq f(x_0). \tag{2}$$

(ii) f has a **local minimum** at x_0 if there is a neighborhood N_2 of x_0 such that for every point x in N_2,

$$f(x) \geq f(x_0). \tag{3}$$

(iii) If f has a local maximum or minimum at x_0, then x_0 is called an **extreme point** of f.

NOTE: Compare this definition with the definitions on page 211.

A rough sketch of a function with several local maxima is given in Figure 1. The following theorem is a natural generalization of Theorem 4.3.2.

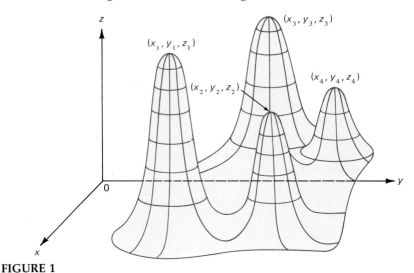

FIGURE 1

Theorem 2. Let f be differentiable at x_0 and suppose that x_0 is an extreme point of f. Then

$$\nabla f(x_0) = 0. \tag{4}$$

Proof. We must show that $f_x(x_0, y_0)$ and $f_y(x_0, y_0) = 0$. To prove that $f_x(x_0, y_0) = 0$, define a function h by

$$h(x) = f(x, y_0).$$

Then $h(x_0) = f(x_0, y_0)$ and $h'(x_0) = f_x(x_0, y_0)$. But since f has a local maximum or (minimum) at (x_0, y_0), h has a local maximum (or minimum) at x_0, so that by

Theorem 4.3.2, $h'(x_0) = 0$. Thus $f_x(x_0, y_0) = 0$. In a very similar way we can show that $f_y(x_0, y_0) = 0$ by defining $h(y) = f(x_0, y)$, and the theorem is proved. ▪

Definition 2   CRITICAL POINT $\mathbf{x}_0$ is a **critical point** of f if f is differentiable at $\mathbf{x}_0$ and $\nabla f(\mathbf{x}_0) = \mathbf{0}$.

REMARK 1.   Definitions 1 and 2 and Theorems 1 and 2 hold for functions of three variables with very little change.

REMARK 2.   By the remark on page 957, we see that *if $\mathbf{x}_0$ is a critical point of f, then the tangent plane to the surface $z = f(x, y)$ is horizontal* (i.e., parallel to the xy-plane).

REMARK 3.   Note that this definition differs from our definition of a critical point of a function of one variable in that, for a function of one variable, x_0 is a critical point if $f(x_0)$ is defined but $f'(x_0)$ does not exist.

As in the case of a function of one variable, the fact that $\mathbf{x}_0$ is a critical point of f does not guarantee that $\mathbf{x}_0$ is an extreme point of f, as Example 3 below illustrates.

EXAMPLE 1   Let $f(x, y) = 1 + x^2 + 3y^2$. Then $\nabla f = 2x\mathbf{i} + 6y\mathbf{j}$, which is zero only when $(x, y) = (0, 0)$. Thus $(0, 0)$ is the only critical point, and it is clearly a local (and global) minimum. This is illustrated in Figure 2. ▪

EXAMPLE 2   Let $f(x, y) = 1 - x^2 - 3y^2$. Then $\nabla f = -2x\mathbf{i} - 6y\mathbf{j}$, which is zero only at the origin. In this case $(0, 0)$ is a local (and global) maximum for f. See Figure 3. ▪

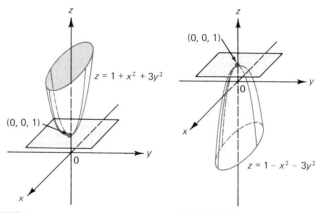

FIGURE 2   **FIGURE 3**

EXAMPLE 3   Let $f(x, y) = x^2 - y^2$. Then $\nabla f = 2x\mathbf{i} - 2y\mathbf{j}$, which again is zero only at $(0, 0)$. But $(0, 0)$ is *neither* a local maximum nor a local minimum for f. To see this, we simply note that f can take positive and negative values in any neighborhood of $(0, 0)$ since $f(x, y) > 0$ if $|x| > |y|$ and $f(x, y) < 0$ if $|x| < |y|$. This is illustrated by Figure 4. The figure sketched in Figure 4 is called a **hyperbolic paraboloid,** or **saddle surface** (see Figure 17.6.9). ▪

We have the following definition.

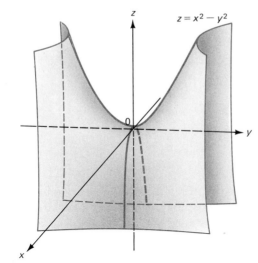

FIGURE 4

Definition 3   SADDLE POINT   If x_0 is a critical point of f but is not an extreme point of f, then x_0 is called a **saddle point** of f.

The three examples above illustrate that a critical point of f may be a local maximum, a local minimum, or a saddle point.

EXAMPLE 4   Let $f(x, y) = \sqrt{x^2 + y^2}$. Then it is obvious that f has a local (and global) minimum at $(0, 0)$. But

$$\nabla f = \frac{x}{\sqrt{x^2 + y^2}}\mathbf{i} + \frac{y}{\sqrt{x^2 + y^2}}\mathbf{j},$$

so that $(0, 0)$ is *not* a critical point since $\nabla f(0, 0)$ *does not exist*. We see this clearly in Figure 5. The graph of f (which is a cone) does not have a tangent plane at $(0, 0)$. ■

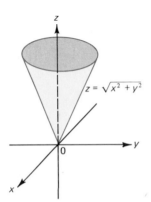

FIGURE 5

Examples 1, 2, and 3 indicate that more is needed to determine whether a critical point is an extreme point (in most cases it will not be at all obvious). You might suspect that the answer has something to do with the signs of the second partial derivatives of f, as in the case of functions of one variable. The following theorem gives the answer. Although we could give its proof by using results we already know, the proof is very tedious, so we will omit it. Its proof can be found (as usual) in many more advanced calculus texts.†

Theorem 3 *Second Derivatives Test* Let f and all its first and second partial derivatives be continuous in a neighborhood of the critical point (x_0, y_0). Let

$$D(x, y) = f_{xx}(x, y)f_{yy}(x, y) - [f_{xy}(x, y)]^2, \tag{5}$$

and let D denote $D(x_0, y_0)$.

> **(i)** If $D > 0$ and $f_{xx}(x_0, y_0) > 0$, then f has a local minimum at (x_0, y_0).
> **(ii)** If $D > 0$ and $f_{xx}(x_0, y_0) < 0$, then f has a local maximum at (x_0, y_0).
> **(iii)** If $D < 0$, then (x_0, y_0) is a saddle point of f.
> **(iv)** If $D = 0$, then any of the preceding alternatives is possible.

REMARK 1.   Using a 2×2 determinant, we can write (5) as

$$D \equiv \begin{vmatrix} f_{xx} & f_{xy} \\ f_{yx} & f_{yy} \end{vmatrix}.$$

REMARK 2.   This theorem does not provide the full answer to the questions of what are the extreme points of f in $\mathbb{R}^2$. It is still necessary to check points at which ∇f does not exist (as in Example 4).

EXAMPLE 5   Let $f(x, y) = 1 + x^2 + 3y^2$. Then as we saw in Example 1, $(0, 0)$ is the only critical point of f. But $f_{xx} = 2$, $f_{yy} = 6$, and $f_{xy} = 0$, so $D(0, 0) = 12$ and $f_{xx} > 0$, which *proves* that f has a local minimum at $(0, 0)$. ■

EXAMPLE 6   Let $f(x, y) = -x^2 - y^2 + 2x + 4y + 5$. Determine the nature of the critical points of f.

Solution. $\nabla f = (-2x + 2)\mathbf{i} + (-2y + 4)\mathbf{j}$. We see that $\nabla f = \mathbf{0}$ when

$$-2x + 2 = 0 \quad \text{and} \quad -2y + 4 = 0,$$

which occurs only at the point $(1, 2)$. Also $f_{xx} = -2$, $f_{yy} = -2$, and $f_{xy} = 0$, so $D = 4$ and $f_{xx} < 0$, which implies that there is a local maximum at $(1, 2)$. At $(1, 2)$, $f = 10$. ■

EXAMPLE 7   Let $f(x, y) = 2x^3 - 24xy + 16y^3$. Determine the nature of the critical points of f.

Solution. We have $\nabla f(x, y) = (6x^2 - 24y)\mathbf{i} + (-24x + 48y^2)\mathbf{j}$. If $\nabla f = \mathbf{0}$, we have

---

†See, for example, S. Grossman, *Calculus Part 2* (Academic Press, Orlando, 1982), page 479.

$$6(x^2 - 4y) = 0 \quad \text{and} \quad -24(x - 2y^2) = 0.$$

Clearly, one critical point is (0, 0). To obtain another, we must solve the simultaneous equations

$$x^2 - 4y = 0$$
$$x - 2y^2 = 0.$$

The second equation tells us that $x = 2y^2$. Substituting this expression into the first equation yields

$$4y^4 - 4y = 0,$$

with solutions $y = 0$ and $y = 1$, giving us the critical points (0, 0) and (2, 1). Now $f_{xx} = 12x$, $f_{yy} = 96y$, and $f_{xy} = -24$, so

$$D(x, y) = (12x)(96y) - 24^2 = 1152xy - 576.$$

Then $D(0, 0) = -576$ and $D(2, 1) = 1728$. We find that (0, 0) is a saddle point, and since $f_{xx}(2, 1) > 0$, (2, 1) is a local minimum. ◼

EXAMPLE 8   Let $f(x, y) = 4x^2 - 4xy + y^2 + 5$. Determine the nature of the critical points of f.

Solution. $\nabla f = (8x - 4y)\mathbf{i} + (-4x + 2y)\mathbf{j}$. We see that $\nabla f = \mathbf{0}$ when

$$8x - 4y = 0 \quad \text{and} \quad -4x + 2y = 0.$$

This occurs if (x, y) is any point on the line $y = 2x$. Thus in this case there are an *infinite number* of critical points. Now $f_{xx} = 8$, $f_{yy} = 2$, and $f_{xy} = -4$, so $D = 8 \cdot 2 - 16 = 0$. It is impossible to determine the nature of the critical points of x by the second derivatives test. However, since we can find that

$$f(x, y) = (2x - y)^2 + 5,$$

we see that $f(x, y) \geq 5$ and $f(x, y) = 5$ when $y = 2x$. Thus *every point* on the line $y = 2x$ is a local (and global) minimum. ◼

EXAMPLE 9   If $f(x, y) = -(4x^2 - 4xy + y^2) + 5$, then following the reasoning of Example 8, it is easy to see that *every point* on the line $y = 2x$ is a local maximum but that $D(x, y) = 0$ at any such point. ◼

EXAMPLE 10   Let $f(x, y) = x^3 - y^3$. Then $\nabla f = 3x^2\mathbf{i} - 3y^2\mathbf{j}$, and the only critical point is (0, 0). But $f_{xx} = 6x$, $f_{yy} = -6y$, and $f_{xy} = 0$, so $D(x, y) = -36xy$ and $D(0, 0) = 0$. In this case it is clear that since f can take on positive and negative values in any neighborhood of (0, 0), (0, 0) is a saddle point of f. ◼

REMARK.   Examples 8, 9, and 10 illustrate that any of the three cases can occur when $D = 0$.

EXAMPLE 11   A rectangular wooden box with an open top is to contain α cubic centimeters, where α is a given positive number. Ignoring the thickness of the wood, how is the box to be constructed so as to use the smallest amount of wood?

Solution. If the dimensions of the box are x, y, and z, then

$$V = xyz = \alpha > 0,$$

so we have $x > 0$, $y > 0$, and $z > 0$. We must minimize the area of the sides of the box, where the area is given by

$$A = xy + 2xz + 2yz$$

(see Figure 6). But $z = \alpha/xy$, so the problem becomes one of minimizing $A(x, y)$, where

$$A(x, y) = xy + 2x\frac{\alpha}{xy} + 2y\frac{\alpha}{xy} = xy + \frac{2\alpha}{y} + \frac{2\alpha}{x}.$$

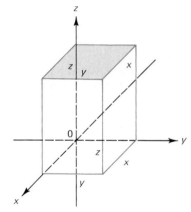

FIGURE 6

Now

$$\nabla A(x, y) = \left(y - \frac{2\alpha}{x^2} \right)\mathbf{i} + \left(x - \frac{2\alpha}{y^2} \right)\mathbf{j}.$$

This is zero when

$$y = \frac{2\alpha}{x^2} \qquad \text{and} \qquad x = \frac{2\alpha}{y^2},$$

or

$$yx^2 = 2\alpha = xy^2.$$

One obvious point is (0, 0), which is ruled out since $x > 0$ and $y > 0$. Also, (0, 0) is not a critical point since ∇A is not defined at (0, 0). Now $xy^2 = x^2 y$ implies that $xy(y - x) = 0$. Since $xy \neq 0$, we have $y = x$, so

$$yx^2 = x^3 = 2\alpha$$

and

$$x = y = \sqrt[3]{2\alpha}.$$

Thus $(\sqrt[3]{2\alpha}, \sqrt[3]{2\alpha})$ is the only critical point.

We can verify that the point $(\sqrt[3]{2\alpha}, \sqrt[3]{2\alpha})$ does indeed give us a local minimum, since $D(\sqrt[3]{2\alpha}, \sqrt[3]{2\alpha}) > 0$ and

$$A_{xx}(\sqrt[3]{2\alpha}, \sqrt[3]{2\alpha}) = \left.\frac{4\alpha}{x^3}\right|_{\sqrt[3]{2\alpha}} = \frac{4\alpha}{2\alpha} = 2,$$

which is greater than 0. Finally, we obtain

$$z = \frac{\alpha}{xy} = \frac{\alpha}{x^2} = \frac{\alpha x}{x^3} = \frac{\alpha x}{2\alpha} = \frac{x}{2},$$

so we can minimize the amount of wood needed by building a box with a square base and a height equal to half its length (or width). Note that since $(\sqrt[3]{2\alpha}, \sqrt[3]{2\alpha})$ is the only critical point, it is easy to check that there is a global minimum there. ∎

PROBLEMS 18.12

In Problems 1–18, determine the nature of the critical points of the given function.

1. $f(x, y) = 7x^2 - 8xy + 3y^2 + 1$
2. $f(x, y) = x^2 + y^3 - 3xy$
3. $f(x, y) = x^2 + 3y^2 + 4x - 6y + 3$
4. $f(x, y) = x^2 + y^2 + 4xy + 6y - 3$
5. $f(x, y) = x^2 + y^2 + 4x - 2y + 3$
6. $f(x, y) = xy^2 + x^2 y - 3xy$
7. $f(x, y) = x^3 + 3xy^2 + 3y^2 - 15x + 2$
8. $f(x, y) = x^3 + y^3 - 3xy$

9. $f(x, y) = \dfrac{1}{y} - \dfrac{1}{x} - 4x + y$
10. $f(x, y) = \dfrac{1}{x} + \dfrac{2}{y} + 2x + y + 1$

11. $f(x, y) = x^2 - xy + y^2 + 2x + 2y$
*12. $f(x, y) = xy + \dfrac{8}{x} + \dfrac{1}{y}$

13. $f(x, y) = (4 - x - y)xy$
*14. $f(x, y) = \sin x + \sin y + \sin(x + y)$

15. $f(x, y) = 2x^2 + y^2 + \dfrac{2}{x^2 y}$
16. $f(x, y) = 4x^2 + 12xy + 9y^2 + 25$

17. $f(x, y) = x^{25} - y^{25}$
18. $f(x, y) = \tan xy$

19. Find three positive numbers whose sum is 50 and such that their product is a maximum.
20. Find three positive numbers x, y, and z whose sum is 50 such that the product $xy^2 z^3$ is a maximum.

21. Find three numbers whose sum is 50 and the sum of whose squares is a minimum.
22. Use the methods of this section to find the minimum distance from the point $(1, -1, 2)$ to the plane $x + 2y - z = 4$. [*Hint:* Express in terms of x and y the distance between $(1, -1, 2)$ and a point (x, y, z) on the plane.]
23. Find the minimum distance between the point $(2, 0, 1)$ and the plane $-x + y + z = 8$.
24. What is the maximum volume of an open-top rectangular box that can be built from α square meters of wood? Assume that any size board can be obtained without waste.
*25. Find the dimensions of the rectangular box of maximum volume that can be inscribed in the ellipsoid

$$\frac{x^2}{a^2} + \frac{y^2}{b^2} + \frac{z^2}{c^2} = 1$$

whose faces are parallel to the coordinate planes.
26. Show that the rectangular box inscribed in a sphere that encloses the greatest volume has the dimensions of a cube.
27. A company uses two types of raw material, I and II, for its product. If it uses x units of I and y units of II, it can produce U units of the finished item, where

$$U(x, y) = 8xy + 32x + 40y - 4x^2 - 6y^2.$$

Each unit of I costs \$10 and each unit of II costs \$4. Each unit of the product can be sold for \$40. How can the company maximize its profits?
*28. Show that the function $e^{(x^2 + y^2)/y} + (11/x) + (5/y) + \sin x^2 y$ has a global minimum in the first quadrant.

18.13 CONSTRAINED MAXIMA AND MINIMA —LAGRANGE MULTIPLIERS

In the previous section we saw how to find the maximum and minimum of a function of two variables by taking gradients and applying a second derivative test. It often happens that there are side conditions (or **constraints**) attached to a problem. For example, we have been asked to find the shortest distance from a point (x_0, y_0) to a line $y = mx + b$. We could write this problem as follows:

Minimize the function: $z = \sqrt{(x - x_0)^2 + (y - y_0)^2}$

subject to the constraint: $y - mx - b = 0$.

As another example, suppose that a region of space containing a sphere is heated and a function $w = T(x, y, z)$ gives the temperature of every point of the region. Then if the sphere is given by $x^2 + y^2 + z^2 = r^2$, and if we wish to find the hottest point on the sphere, we have the following problem:

Maximize: $w = T(x, y, z)$

subject to the constraint: $x^2 + y^2 + z^2 - r^2 = 0$.

We now generalize these two examples. Let f and g be functions of two variables. Then we can formulate a **constrained maximization** (or **minimization**) problem as follows:

> Maximize (or minimize): $z = f(x, y)$ **(1)**
>
> subject to the constraint: $g(x, y) = 0.$ **(2)**

If f and g are functions of three variables, we have the following problem:

> Maximize (or minimize): $w = f(x, y, z)$ **(3)**
>
> subject to the constraint: $g(x, y, z) = 0.$ **(4)**

We now develop a method for dealing with problems of the type (1), (2), or (3), (4). Let C be a curve in $\mathbb{R}^2$ or $\mathbb{R}^3$ given parametrically by the differentiable function $\mathbf{F}(t)$. That is, C is given by

$$\mathbf{F}(t) = x(t)\mathbf{i} + y(t)\mathbf{j} \qquad (\text{in } \mathbb{R}^2)$$

or

$$\mathbf{F}(t) = x(t)\mathbf{i} + y(t)\mathbf{j} + z(t)\mathbf{k} \qquad (\text{in } \mathbb{R}^3).$$

Let $f(\mathbf{x})$ denote the function (of two or three variables) that is to be maximized.

Theorem 1.   Suppose that f is differentiable at a point $\mathbf{x}_0$ and that among all points on a curve C, f takes its maximum (or minimum) value at $\mathbf{x}_0$. Then $\nabla f(\mathbf{x}_0)$ is orthogonal to C at $\mathbf{x}_0$. That is, since $\mathbf{F}'(t)$ is tangent to C, if $\mathbf{x}_0 = \mathbf{F}(t_0)$, then

> $$\nabla f(\mathbf{x}_0) \cdot \mathbf{F}'(t_0) = 0. \qquad \textbf{(5)}$$

Proof.   For $\mathbf{x}$ on C, $\mathbf{x} = \mathbf{F}(t)$, so that the composite function $f(\mathbf{F}(t))$ has a maximum (or minimum) at t_0. Therefore its derivative at t_0 is 0. By the chain rule,

$$\frac{d}{dt} f(\mathbf{F}(t)) = \nabla f(\mathbf{F}(t)) \cdot \mathbf{F}'(t),$$

and at t_0

$$0 = \nabla f(\mathbf{F}(t_0)) \cdot \mathbf{F}'(t_0) = \nabla f(\mathbf{x}_0) \cdot \mathbf{F}'(t_0),$$

and the theorem is proved.   ∎

For f as a function of two variables, the result of Theorem 1 is illustrated in Figure 1.

We can use Theorem 1 to make an interesting geometric observation.† Suppose that, subject to the constraint $g(x, y) = 0$, f takes its maximum (or minimum) at the

---

†This observation was first made by the French mathematician Joseph-Louis Lagrange (1736–1813). (See the biography of Lagrange on page 670).

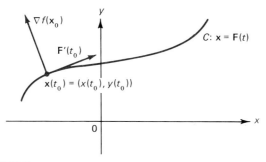

FIGURE 1

point $\mathbf{x}_0 = (x_0, y_0)$. The equation $g(x, y) = 0$ determines a curve C in the xy-plane, and by Theorem 1, $\nabla f(x_0, y_0)$ is orthogonal to C at (x_0, y_0). But from Section 18.7, $\nabla g(x_0, y_0)$ is also orthogonal to C at (x_0, y_0). Thus we see that

$\nabla g(x_0, y_0)$ and $\nabla f(x_0, y_0)$ are parallel.

Hence there is a number λ such that

$$\nabla f(x_0, y_0) = \lambda \nabla g(x_0, y_0).$$

We can extend this observation to the following rule, which applies equally well to functions of three or more variables:

If, subject to the constraint $g(\mathbf{x}) = 0$, f takes its maximum (or minimum) value at a point $\mathbf{x}_0$, then there is a number λ such that

$$\nabla f(\mathbf{x}_0) = \lambda \nabla g(\mathbf{x}_0). \tag{6}$$

The number λ is called a **Lagrange multiplier.** We will illustrate the Lagrange multiplier technique with a number of examples.

EXAMPLE 1 Find the maximum and minimum values of $f(x, y) = xy^2$ subject to the condition $x^2 + y^2 = 1$.

Solution. We have $g(x, y) = x^2 + y^2 - 1 = 0$. Then

$$\nabla f = y^2 \mathbf{i} + 2xy \mathbf{j} \quad \text{and} \quad \nabla g = 2x \mathbf{i} + 2y \mathbf{j}.$$

At a maximizing or minimizing point we have

$$\nabla f = \lambda \nabla g,$$

or

$$y^2 \mathbf{i} + 2xy \mathbf{j} = 2x\lambda \mathbf{i} + 2y\lambda \mathbf{j},$$

which leads to the equations

$$y^2 = 2x\lambda$$
$$2xy = 2y\lambda.$$

Multiplying the first equation by y and the second by x, we obtain

$$y^3 = 2xy\lambda$$
$$2x^2y = 2xy\lambda,$$

or

$$y^3 = 2x^2y.$$

But $x^2 = 1 - y^2$, so

$$y^3 = 2(1 - y^2)y = 2y - 2y^3,$$

or

$$3y^3 = 2y.$$

The solutions to this last equation are $y = 0$ and $y = \pm\sqrt{\frac{2}{3}}$. This leads to the six points

$$(1, 0),\ (-1, 0),\ \left(\frac{1}{\sqrt{3}}, \sqrt{\frac{2}{3}}\right),\ \left(-\frac{1}{\sqrt{3}}, \sqrt{\frac{2}{3}}\right),\ \left(\frac{1}{\sqrt{3}}, -\sqrt{\frac{2}{3}}\right),\ \left(-\frac{1}{\sqrt{3}}, -\sqrt{\frac{2}{3}}\right).$$

Evaluating $f(x, y) = xy^2$ at these points, we have

$$f(1, 0) = f(-1, 0) = 0,\qquad f\left(\frac{1}{\sqrt{3}}, \sqrt{\frac{2}{3}}\right) = f\left(\frac{1}{\sqrt{3}}, -\sqrt{\frac{2}{3}}\right) = \frac{2}{3\sqrt{3}},$$

and

$$f\left(-\frac{1}{\sqrt{3}}, \sqrt{\frac{2}{3}}\right) = f\left(-\frac{1}{\sqrt{3}}, -\sqrt{\frac{2}{3}}\right) = -\frac{2}{3\sqrt{3}}.$$

Thus the maximum value of f is $2/(3\sqrt{3})$ and the minimum value of f is $-2/(3\sqrt{3})$. Note that there is neither a maximum nor a minimum at $(1, 0)$ and $(-1, 0)$ even though $\nabla f(1, 0) = 0\,\nabla g(1, 0)$ and $\nabla f(-1, 0) = 0\,\nabla g(-1, 0) = \mathbf{0}$. ∎

EXAMPLE 2 Find the points on the sphere $x^2 + y^2 + z^2 = 1$ closest to and farthest from the point $(1, 2, 3)$.

Solution. We wish to minimize and maximize $f(x, y, z) = (x - 1)^2 + (y - 2)^2 + (z - 3)^2$ subject to $g(x, y, z) = x^2 + y^2 + z^2 - 1 = 0$. We have $\nabla f(x, y, z) = 2(x - 1)\mathbf{i} + 2(y - 2)\mathbf{j} + 2(z - 3)\mathbf{k}$, and $\nabla g(x, y, z) = 2x\mathbf{i} + 2y\mathbf{j} + 2z\mathbf{k}$. Condition

(6) implies that at a maximizing point, $\nabla f = \lambda \nabla g$, so

$$2(x - 1) = 2x\lambda$$
$$2(y - 2) = 2y\lambda$$
$$2(z - 3) = 2z\lambda.$$

If $\lambda \neq 1$, then we find that

$$x - 1 = x\lambda, \quad \text{or} \quad x - x\lambda = 1, \quad \text{or} \quad x(1 - \lambda) = 1,$$

and

$$x = \frac{1}{1 - \lambda}.$$

Similarly, we obtain

$$y = \frac{2}{1 - \lambda} \quad \text{and} \quad z = \frac{3}{1 - \lambda}.$$

Then

$$1 = x^2 + y^2 + z^2 = \frac{1}{(1 - \lambda)^2}(1^2 + 2^2 + 3^2) = \frac{14}{(1 - \lambda)^2},$$

so that

$$(1 - \lambda)^2 = 14, \quad (1 - \lambda) = \pm\sqrt{14},$$

and

$$\lambda = 1 \pm \sqrt{14}.$$

If $\lambda = 1 + \sqrt{14}$, then $(x, y, z) = (-1/\sqrt{14}, -2/\sqrt{14}, -3/\sqrt{14})$. If $\lambda = 1 - \sqrt{14}$, then $(x, y, z) = (1/\sqrt{14}, 2/\sqrt{14}, 3/\sqrt{14})$. Finally, evaluation shows us that f is maximized at $(-1/\sqrt{14}, -2/\sqrt{14}, -3/\sqrt{14})$ and that f is minimized at $(1/\sqrt{14}, 2/\sqrt{14}, 3/\sqrt{14})$. ∎

We note that we can use Lagrange multipliers in $\mathbb{R}^3$ if there are two or more constraint equations. Suppose, for example, we wish to maximize (or minimize) $w = f(x, y, z)$ subject to the constraints

$$g(x, y, z) = 0 \tag{7}$$

and

$$h(x, y, z) = 0. \tag{8}$$

Each of the equations (7) and (8) represents a surface in $\mathbb{R}^3$, and their intersection forms a curve in $\mathbb{R}^3$. By an argument very similar to the one we used earlier (but applied in $\mathbb{R}^3$ instead of $\mathbb{R}^2$), we find that if f is maximized (or minimized) at (x_0, y_0, z_0), then $\nabla f(x_0, y_0, z_0)$ is in the plane determined by $\nabla g(x_0, y_0, z_0)$ and $\nabla h(x_0, y_0, z_0)$. Thus there are numbers λ and μ such that

$$\nabla f(x_0, y_0, z_0) = \lambda \nabla g(x_0, y_0, z_0) + \mu \nabla h(x_0, y_0, z_0) \tag{9}$$

(see Problem 17.5.35). Formula (9) is the generalization of formula (6) in the case of two constraints.

EXAMPLE 3　　Find the maximum value of $w = xyz$ among all points (x, y, z) lying on the line of intersection of planes $x + y + z = 30$ and $x + y - z = 0$.

　　Solution. Setting $f(x, y, z) = xyz$, $g(x, y, z) = x + y + z - 30$, and $h(x, y, z) = x + y - z$, we obtain

$$\nabla f = yz\mathbf{i} + xz\mathbf{j} + xy\mathbf{k}$$

$$\nabla g = \mathbf{i} + \mathbf{j} + \mathbf{k}$$

$$\nabla h = \mathbf{i} + \mathbf{j} - \mathbf{k},$$

and using equation (9) to obtain the maximum, we obtain the equations

$$yz = \lambda + \mu$$

$$xz = \lambda + \mu$$

$$xy = \lambda - \mu.$$

Multiplying the three equations by x, y, and z, respectively, we find that

$$xyz = (\lambda + \mu)x$$

$$xyz = (\lambda + \mu)y$$

$$xyz = (\lambda - \mu)z.$$

If $\lambda + \mu = 0$, then $yz = 0$ and $xyz = 0$, which is not a maximum value since xyz can be positive (for example, $x = 8$, $y = 7$, $z = 15$ is in the constraint set and $xyz = 840$). Thus we can divide the first two equations by $\lambda + \mu$ to find that

$$x = y.$$

Since $x + y - z = 0$, we have $2x - z = 0$, or $z = 2x$. But then

$$30 = x + y + z = x + x + 2x = 4x,$$

or

$$x = \tfrac{15}{2}.$$

Then

$$y = \tfrac{15}{2}, \quad z = 15,$$

and the maximum value of xyz occurs at $(\tfrac{15}{2}, \tfrac{15}{2}, 15)$ and is equal to $(\tfrac{15}{2})(\tfrac{15}{2})15 = 843\tfrac{3}{4}$. ∎

We conclude this section with two observations. First, while the outlined steps make the method of Lagrange multipliers seem easy, it should be noted that solving three nonlinear equations in three unknowns or four such equations in four unknowns often entails very involved algebraic manipulations. Second, no method is given for determining whether a solution found actually yields a maximum, a minimum, or neither. Fortunately, in many practical applications the existence of a maximum or a minimum can readily be inferred from the nature of the particular problem.

PROBLEMS 18.13

1. Use Lagrange multipliers to find the minimum distance from the point $(1, 2)$ to the line $2x + 3y = 5$.
2. Use Lagrange multipliers to find the minimum distance from the point $(3, -2)$ to the line $y = 2 - x$.
3. Use Lagrange multipliers to find the minimum distance from the point $(1, -1, 2)$ to the plane $x + y - z = 3$.
4. Use Lagrange multipliers to find the minimum distance from the point $(3, 0, 1)$ to the plane $2x - y + 4z = 5$.
5. Use Lagrange multipliers to find the minimum distance from the plane $ax + by + cz = d$ to the origin.
*6. Find the maximum and minimum values of $x^2 + y^2$ subject to the condition $x^3 + y^3 = 6xy$.
7. Find the maximum and minimum values of $2x^2 + xy + y^2 - 2y$ subject to the condition $y = 2x - 1$.
8. Find the maximum and minimum values of $x^2 + y^2 + z^2$ subject to the condition $z^2 = x^2 - 1$.
9. Find the maximum and minimum values of $x^3 + y^3 + z^3$ if (x, y, z) lies on the sphere $x^2 + y^2 + z^2 = 4$.
10. Find the maximum and minimum values of $x + y + z$ if (x, y, z) lies on the sphere $x^2 + y^2 + z^2 = 1$.
11. Find the maximum and minimum values of xyz if (x, y, z) is on the ellipsoid $x^2 + (y^2/4) + (z^2/9) = 1$.
12. Solve Problem 11 if (x, y, z) is on the ellipsoid $(x^2/a^2) + (y^2/b^2) + (z^2/c^2) = 1$.
*13. Minimize the function $x^2 + y^2 + z^2$ for (x, y, z) on the planes $3x - y + z = 6$ and $x + 2y + 2z = 2$.
14. Find the minimum values of $x^3 + y^3 + z^3$ for (x, y, z) on the planes $x + y + z = 2$ and $x + y - z = 3$.
15. Find the maximum and minimum distances from the origin to a point on the ellipse $(x^2/a^2) + (y^2/b^2) = 1$.
16. Find the maximum and minimum distances from the origin to a point on the ellipsoid $(x^2/a^2) + (y^2/b^2) + (z^2/c^2) = 1$.

*17. Find the maximum value of $x_1 + x_2 + x_3 + x_4$ subject to $x_1^2 + x_2^2 + x_3^2 + x_4^2 = 1$. [*Hint:* Use the obvious generalization of Lagrange multipliers to functions of four variables.]

*18. Find the maximum value of $x_1 + x_2 + \cdots + x_n$ subject to $x_1^2 + x_2^2 + \cdots + x_n^2 = 1$.

19. Using Lagrange multipliers, show that among all rectangles with the same perimeter, the square encloses the greatest area.

*20. Show that among all triangles having the same perimeter, the equilateral triangle has the greatest area. [*Hint:* The area is $\sqrt{s(s-a)(s-b)(s-c)}$, where a, b, and c are the sides and $s = (a + b + c)/2$.]

21. Find the maximum and minimum values of xyz subject to $x^2 + z^2 = 1$ and $x = y$.

22. Find the volume of the largest rectangular parallelepiped that can be inscribed in the ellipsoid $x^2 + 4y^2 + 9z^2 = 9$.

23. A silo is in the shape of a cylinder topped with a cone (see Figure 2). The radius of each is 6 m, and the total surface area is 200 m² (excluding the base). What are the heights of the cylinder and cone that maximize the volume enclosed by the silo?

FIGURE 2

24. Prove the *general arithmetic-geometric inequality*: If $p + q + r = 1$, then $x^p y^q z^r \leq px + qy + rz$ with equality if and only if $x = y = z$.

25. The base of an open-top rectangular box costs $3 per square meter to construct, while the sides cost only $1 per square meter. Find the dimensions of the box of greatest volume that can be constructed for $36.

26. A manufacturing company has three plants I, II, and III, which produce x, y, and z units, respectively, of a certain product. The annual revenue from this production is given by

$$R(x, y, z) = 6xyz^2 - 400{,}000x - 400{,}000y - 400{,}000z.$$

If the company is to produce 1000 units annually, how should it allocate production so as to maximize profits?

27. A firm has $250,000 to spend on labor and raw materials. The output of the firm is αxy, where α is a constant and x and y are, respectively, the quantity of labor and raw materials consumed. The unit price of hiring labor is $5000, and the unit price of raw materials is $2500. Find the ratio of x to y that maximizes output.

28. The temperature of a point (x, y, z) on the unit sphere is given by $T(x, y, z) = xy + yz$. What is the hottest point on the sphere?

REVIEW EXERCISES FOR CHAPTER EIGHTEEN

In Exercises 1–6, find the domain and range of the indicated function.

1. $f(x, y) = \sqrt{x^2 - y^2}$

2. $f(x, y) = \dfrac{1}{\sqrt{x^2 + y^2}}$

3. $f(x, y) = \cos(x + 3y)$

4. $f(x, y, z) = \sqrt{1 - x^2 - y^2 - z^2}$

5. $f(x, y, z) = \dfrac{1}{\sqrt{x^2 + y^2 + z^2 - 1}}$

6. $f(x, y) = \ln(x - y + 4z - 3)$

In Exercises 7–10, describe the level curves of the given function and sketch these curves for the given values of z.

7. $z = \sqrt{1 - x - y}$; $z = 0, 1, 3, 8$

8. $z = \sqrt{1 - y^2 + x}$; $z = 0, 2, 4, 7$

9. $z = \ln(x - 3y)$; $z = 0, 1, 2, 3$

10. $z = \dfrac{x^2 + y^2}{x^2 - y^2}$; $z = 1, 3, 6$

11. Sketch the open disk centered at $(-1, 2)$ with radius 4.

12. Sketch the closed ball centered at $(1, 2, 3)$ with radius 2.

13. Show that $\lim\limits_{(x,y)\to(0,0)} \dfrac{xy}{y^2 - x^2}$ does not exist.

14. Show that $\lim\limits_{(x,y)\to(0,0)} \dfrac{y^2 - 2x}{y^2 + 2x}$ does not exist.

15. Show that $\lim\limits_{(x,y)\to(0,0)} \dfrac{4xy^3}{x^2 + y^4} = 0$.

16. Show that $\lim\limits_{(x,y,z)\to(0,0,0)} \dfrac{zy^2 + x^3}{x^2 + y^2 + z^2} = 0$.

17. Calculate $\lim\limits_{(x,y)\to(1,-2)} \dfrac{1 + x^2y}{2 - y}$.

18. Calculate $\lim\limits_{(x,y)\to(1,4)} \ln(1 + x + \sqrt{y})$.

19. Calculate $\lim\limits_{(x,y,z)\to(2,-1,1)} \dfrac{x^2 - yz^3}{1 + xyz - 2y^5}$.

20. Find the maximum region over which the function $f(x, y) = \ln(1 - 2x + 3y)$ is continuous.

21. Find the maximum region over which the function $f(x, y, z) = \ln(x - y - z + 4)$ is continuous.

22. Find the maximum region over which the function

$$f(x, y, z) = \dfrac{1}{\sqrt{1 - x^2 + y^2 - z^2}}$$

is continuous.

23. Find a number c such that the function

$$f(x, y) = \begin{cases} \dfrac{-2xy}{\sqrt{x^2 + y^2}}, & (x, y) \neq (0, 0) \\ c, & (x, y) = (0, 0) \end{cases}$$

is continuous at the origin.

In Exercises 24–35, calculate all first partial derivatives.

24. $f(x, y) = \dfrac{y}{x}$

25. $f(x, y) = \cos(x - 3y)$

26. $f(x, y) = \dfrac{1}{\sqrt{x^2 - y^2}}$

27. $f(x, y) = \tan^{-1} \dfrac{y}{1 + x}$

28. $f(x, y, z) = \ln(x - y + 4z)$

29. $f(x, y, z) = \dfrac{1}{\sqrt{x^2 + y^2 + z^2}}$

30. $f(x, y) = \cosh\dfrac{y}{x^2}$

31. $f(x, y, z) = \sec\dfrac{x - y}{z}$

32. $f(x, y, z) = (x^2y - y^3z^5 + x\sqrt{z})^{2/3}$

33. $f(x, y, z) = \dfrac{x^2 - y^3}{y^3 + z^4}$

34. $f(x, y, z, w) = \dfrac{x - z + w}{y + 2w - x}$

35. $f(x, y, z, w) = e^{(x-w)/(y+z)}$

In Exercises 36–41, calculate all second partial derivatives and show that all pairs of mixed partials are equal.

36. $f(x, y) = xy^3$

37. $f(x, y) = \tan^{-1}\dfrac{y}{x}$

38. $f(x, y) = \sqrt{x^2 - y^2}$

39. $f(x, y) = \dfrac{x + y}{x - y}$

40. $f(x, y, z) = \ln(2 - 3x + 4y - 7z)$

41. $f(x, y, z) = \dfrac{1}{\sqrt{1 - x^2 - y^2 - z^2}}$

42. Let $f(x, y, z) = x^2y^3 - zx^5$. Calculate f_{yzx}. **43.** Let $f(x, y, z) = (x - y)/z$. Calculate f_{zxxyz}.

In Exercises 44–51, calculate the gradient of the given function at the given point.

44. $f(x, y) = x^2 - y^3; (1, 2)$

45. $f(x, y) = \tan^{-1}\dfrac{y}{x}; (-1, -1)$

46. $f(x, y) = \dfrac{x - y}{x + y}; (3, 2)$

47. $f(x, y) = \cos(x - 2y); \left(\dfrac{\pi}{2}, \dfrac{\pi}{6}\right)$

48. $f(x, y, z) = xy + yz^3; (1, 2, -1)$

49. $f(x, y, z) = \dfrac{x - y}{3z}; (2, 1, 4)$

50. $f(x, y, z) = \dfrac{1}{\sqrt{x^2 + y^2 + z^2}}; (a, b, c)$

51. $f(x, y, z) = e^{-(x^2+y^3+z^4)}; (0, -1, 1)$

In Exercises 52–60, use the chain rule to calculate the indicated derivative.

52. $z = 2xy; x = \cos t, y = \sin t; dz/dt$

53. $z \sin^{-1}\dfrac{y}{x}; x = 1 + t; y = t^2; \dfrac{dz}{dt}$

54. $w = \ln(1 - x - 2y + 3z); x = e^t \sin t, y = e^t \cos t, z = t^2; dw/dt$

55. $z = \dfrac{y}{x}; x = r - s; y = r + s; \dfrac{\partial z}{\partial s}$

56. $z = xy^3; x = \dfrac{r}{s}; y = \dfrac{s^2}{r}; \dfrac{\partial z}{\partial r}$

57. $z = \sin(x - y); x = e^{r+s}, y = e^{r-s}; \partial z/\partial s$

58. $w = xyz; x = rs, y = \dfrac{r}{s}, z = s^2r^3; \dfrac{\partial w}{\partial r}$ and $\dfrac{\partial w}{\partial s}$

59. $w = x^3y + y^3z; x = rst; y = \dfrac{rs}{t}, z = \dfrac{rt}{s}; \dfrac{\partial w}{\partial s}$ and $\dfrac{\partial w}{\partial t}$

60. $w = \ln(x + 2y + 5z); x = e^{r+s+t}, y = \sqrt{rst^2}, z = \dfrac{1}{\sqrt{r + s + t}}; \dfrac{\partial w}{\partial r}$ and $\dfrac{\partial w}{\partial t}$

In Exercises 61–66, find the equation of the tangent plane and symmetric equations of the normal line to the given surface at the given point.

61. $x^2 + y^2 + z^2 = 3; (1, 1, 1)$

62. $x^{1/2} + y^{3/2} + z^{1/2} = 3; (1, 0, 4)$

63. $3x - y + 5z = 15; (-1, 2, 4)$

64. $xy^2 - yz^3 = 0, (1, 1, 1)$

65. $xyz = 6; (-2, 1, -3)$

66. $\sqrt{\dfrac{x - y}{y + z}} = \dfrac{1}{2}; (2, 1, 3)$

In Exercises 67–72, calculate the directional derivative of the given function at the given point, in the direction of the given vector **v**.

67. $f(x, y) = \dfrac{y}{x}$ at $(1, 2)$; $\mathbf{v} = \mathbf{i} - \mathbf{j}$

68. $f(x, y) = 3x^2 - 4xy$ at $(3, -1)$; $\mathbf{v} = 2\mathbf{i} + 5\mathbf{j}$

69. $f(x, y) = \tan^{-1}\dfrac{y}{x}$ at $(1, -1)$; $\mathbf{v} = -3\mathbf{i} + 2\mathbf{j}$

70. $f(x, y, z) = xy^2 - zy^3$ at $(1, 2, 3)$; $\mathbf{v} = \mathbf{i} - \mathbf{j} + 2\mathbf{k}$

71. $f(x, y, z) = \dfrac{1}{\sqrt{x^2 + y^2 + z^2}}$ at $(1, -1, 2)$; $\mathbf{v} = -2\mathbf{i} + \mathbf{j} - 3\mathbf{k}$

72. $f(x, y, z) = e^{-(x+y^2-xz)}$ at $(1, 0, -1)$; $\mathbf{v} = 2\mathbf{i} + 5\mathbf{j} + \mathbf{k}$

73. Show that the force $\mathbf{f}(\mathbf{x}) = -3\mathbf{x}/|\mathbf{x}|^5$ is conservative and find a potential function for $\mathbf{f}$.

74. Show that the force $\mathbf{f}(x, y) = y^2\mathbf{i} - x^2\mathbf{j}$ is not conservative.

In Exercises 75–80, calculate the total differential df.

75. $f(x, y) = x^3y^2$

76. $f(x, y) = \cos^{-1}\dfrac{y}{x}$

77. $f(x, y) = \sqrt{\dfrac{x + 1}{y - 1}}$

78. $f(x, y, z) = xy^5z^3$

79. $f(x, y, z) = \ln(x - y + 4z)$

80. $f(x, y) = \left(\dfrac{x - y}{y + z}\right)^{1/3}$

In Exercises 81–83, estimate the given number by using the total differential.

81. $4.03/6.97$

82. $\sqrt{(4.97)^2 + (12.02)^2}$

83. $\sqrt{3.97}(10.05 - 1.03)^{3/2}$

84. How much wood is contained in the sides of a rectangular box with sides of inside measurements 1.5 m, 1.3 m, and 2 m if the thickness of the wood making up the sides is 3 cm ($= 0.03$ m)?

In Exercises 85–90, test for exactness. If the vector field is exact, find all functions f for which $\nabla f = \mathbf{F}$.

85. $\mathbf{F}(x, y) = (y^2 + 1)\mathbf{i} + (2xy)\mathbf{j}$

86. $\mathbf{F}(x, y) = -(\sin y)\mathbf{i} + (y - x\cos y)\mathbf{j}$

87. $\mathbf{F}(x, y) = (2xy)\mathbf{i} + \left(\dfrac{1}{y^2} - x^2\right)\mathbf{j}$

88. $\mathbf{F}(x, y) = (\sin x + \sqrt{x} + y^4)\mathbf{i} + \left(\dfrac{-1}{y^3} + 4y^3x\right)\mathbf{j}$

89. $\mathbf{F}(x, y, z) = (y^3z^5 + 3x)\mathbf{i} + (-4\sqrt{y} + 3xy^2z^5)\mathbf{j} + (\sin z + 5xy^3z^4)\mathbf{k}$

90. $\mathbf{F}(x, y, z) = \left(x + \dfrac{1}{z}\right)\mathbf{i} + \left(y - \dfrac{1}{z}\right)\mathbf{j} + \left(z + \dfrac{y - x}{z^2}\right)\mathbf{k}$

In Exercises 91–96, determine the nature of the critical points of the given function.

91. $f(x, y) = 6x^2 + 14y^2 - 16xy + 2$ **92.** $f(x, y) = x^5 - y^5$

93. $f(x, y) = \dfrac{1}{y} + \dfrac{2}{x} + 2y + x + 4$ **94.** $f(x, y) = 49 - 16x^2 + 24xy - 9y^2$

95. $f(x, y) = x^2 + y^2 + \dfrac{2}{xy^2}$ **96.** $f(x, y) = \cot xy$

97. Find the minimum distance from the point $(2, -1, 4)$ to the plane $x - y + 3z = 7$.

98. What is the smallest amount of wood needed to build an open-top rectangular box enclosing a volume of 25 m³?

99. What is the maximum volume of an open-top rectangular box that can be built from 10 m² of wood?

100. Show that there is no maximum or minimum value of xy^3z^2 if (x, y, z) lies on the plane $x - y + 2z = 2$.

101. What are the dimensions of the rectangular parallelepiped of maximum volume that can be inscribed in the ellipsoid $x^2 + 9y^2 + 4z^2 = 36$?

102. Minimize the function $x^2 + y^2 + z^2$ for (x, y, z) on the planes $2x + y + z = 2$ and $x - y - 3z = 4$.

19 Multiple Integration

In Chapter 5 we introduced the concept of the definite integral of a function of one variable. In this chapter we show how the notion of the definite integral can be extended to functions of several variables. In particular, we will discuss the double integral of a function of two variables and the triple integral of a function of three variables.

Multiple integrals were developed essentially by Euler (whom we have referred to earlier). Euler had a clear conception of the double integral over a bounded region enclosed by arcs, and in a paper published in 1769,† he gave a procedure for evaluating double integrals by repeated integration. We will discuss Euler's technique in Section 19.2. Multiple integrals were first used by Newton in his work (which appeared in the *Principia*) on the gravitational attraction exerted by spheres on particles. It should be mentioned, however, that Newton had only a geometric interpretation of a multiple integral. The more precise analytical definitions had to await the work of Euler and, later, Lagrange.

We begin our discussion of multiple integration with an introduction to the double integral. Our development will closely parallel the development of the definite integral in Chapter 5.

---

†*Novi Comm. Acad. Sci. Petrop.*, **14**, 72–103 (1769).

19.1 VOLUME UNDER A SURFACE AND THE DOUBLE INTEGRAL

In our development of the definite integral in Chapter 5 we began by calculating the area under a curve $y = f(x)$ (and above the x-axis) for x in the interval $[a, b]$. We initially assumed that, on $[a, b]$, $f(x) \geq 0$. We carry out a similar development by obtaining an expression which represents a volume in $\mathbb{R}^3$.

We begin by considering an especially simple case. Let R denote the rectangle in $\mathbb{R}^2$ given by

$$R = \{(x, y): a \leq x \leq b \text{ and } c \leq y \leq d\}. \tag{1}$$

This rectangle is sketched in Figure 1. Let $z = f(x, y)$ be a continuous function that is nonnegative over R. That is, $f(x, y) \geq 0$ for every (x, y) in R. We now ask: What is the volume "under" the surface $z = f(x, y)$ and "over" the rectangle R? The volume requested is sketched in Figure 2.

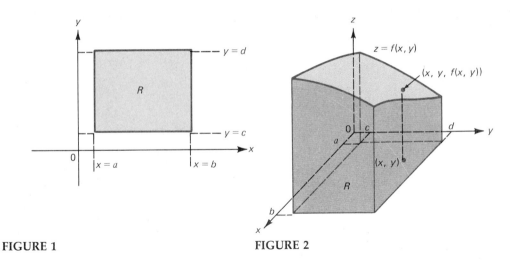

FIGURE 1   **FIGURE 2**

We will calculate this volume in much the same way we calculated the area under a curve in Chapter 5. We begin by "partitioning" the rectangle.

Step 1. Form a **regular partition** (i.e., all subintervals have the same length) of the intervals $[a, b]$ and $[c, d]$:

$$a = x_0 < x_1 < x_2 < \cdots < x_{n-1} < x_n = b, \tag{2}$$

$$c = y_0 < y_1 < y_2 < \cdots < y_{m-1} < y_m = d. \tag{3}$$

We then define

$$\Delta x = x_i - x_{i-1} = \frac{b - a}{n} \tag{4}$$

$$\Delta y = y_j - y_{j-1} = \frac{d - c}{m}, \tag{5}$$

and define the subrectangles R_{ij} by

$$R_{ij} = \{(x, y): x_{i-1} \le x \le x_i \text{ and } y_{j-1} \le y \le y_j\} \tag{6}$$

for $i = 1, 2, \ldots, n$ and $j = 1, 2, \ldots, m$. This is sketched in Figure 3. Note that there are nm subrectangles R_{ij} covering the rectangle R.

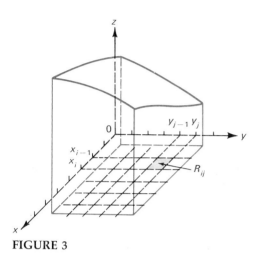

FIGURE 3

REMARK. In defining the definite integral, we allowed for the possibility that the subintervals had different lengths (see p. 292). However, we observed on page 297 that we obtained the same result by using regular partitions whenever $\int_a^b f(x)\,dx$ exists. This is true if f is continuous. Now we are assuming that $z = f(x, y)$ is continuous. Under this assumption it is sufficient to consider only regular partitions. Since these are easier to deal with, we will work exclusively with regular partitions in the remainder of this book.

Step 2. Estimate the volume under the surface and over each subrectangle.
   Let $(x_i{}^*, y_j{}^*)$ be a point in R_{ij}. Then the volume V_{ij} under the surface and over R_{ij} is approximated by

$$V_{ij} \approx f(x_i{}^*, y_j{}^*)\,\Delta x\,\Delta y = f(x_i{}^*, y_j{}^*)\,\Delta A, \tag{7}$$

where $\Delta A = \Delta x\,\Delta y$ is the area of R_{ij}. The expression on the right-hand side of (7) is simply the volume of the parallelepiped (three-dimensional box) with base R_{ij} and height $f(x_i{}^*, y_j{}^*)$. This volume corresponds to the approximate area $A_i \approx f(x_i{}^*)\,\Delta x_i$ that we used on page 285. Unless $f(x, y)$ is a plane parallel to the xy-plane, the expression $f(x_i{}^*, y_j{}^*)\,\Delta A$ will not in general be equal to the volume under the surface S. But if Δx and Δy are small, the approximation will be a good one. The difference between the actual V_{ij} and the approximate volume given in (7) is illustrated in Figure 4.

Step 3. Add up the approximate volumes to obtain an approximation to the total volume sought.
   The total volume is

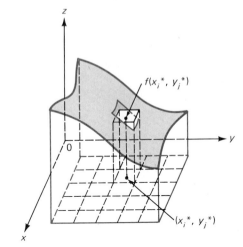

FIGURE 4

$$V = V_{11} + V_{12} + \cdots + V_{1m} + V_{21} + V_{22} + \cdots V_{2m}$$
$$+ \cdots + V_{n1} + V_{n2} + \cdots + V_{nm}. \qquad \textbf{(8)}$$

To simplify notation, we use the summation sign Σ introduced in Section 5.3. Since we are summing over two variables i and j, we need two such signs:

$$V = \sum_{i=1}^{n} \sum_{j=1}^{m} V_{ij}. \qquad \textbf{(9)}$$

The expression in (9) is called a **double sum.** If we "write out"† the expression in (9), we obtain the expression in (8). Then combining (7) and (9), we have

$$V \approx \sum_{i=1}^{n} \sum_{j=1}^{m} f(x_i{}^*, y_j{}^*)\, \Delta A. \qquad \textbf{(10)}$$

Step 4.   Take a limit as both Δx and Δy approach zero.
To indicate that this is happening, we define

$$\Delta s = \sqrt{(\Delta x)^2 + (\Delta y)^2}.$$

Geometrically, Δs is the length of a diagonal of the rectangle R_{ij} whose sides have lengths Δx and Δy (see Figure 5). As $\Delta s \to 0$, the number of subrectangles R_{ij} increases without bound and the area of each R_{ij} approaches zero. This implies that the volume approximation given by (7) is getting closer and closer to the

---

†This writing out is done by summing over j first and then over i. For example,

$$\sum_{i=1}^{3} \sum_{j=1}^{4} a_{ij} = \sum_{i=1}^{3} (a_{i1} + a_{i2} + a_{i3} + a_{i4})$$

$$= a_{11} + a_{12} + a_{13} + a_{14} + a_{21} + a_{22} + a_{23} + a_{24} + a_{31} + a_{32} + a_{33} + a_{34}.$$

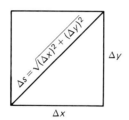

FIGURE 5

"true" volume over R_{ij}. Thus the approximation (10) gets better and better as $\Delta s \to 0$, which enables us to write

$$V = \lim_{\Delta s \to 0} \sum_{i=1}^{n} \sum_{j=1}^{m} f(x_i{}^*, y_j{}^*) \, \Delta A. \tag{11}$$

EXAMPLE 1 Calculate the volume under the plane $z = x + 2y$ and over the rectangle $R = \{(x, y): 1 \le x \le 2 \text{ and } 3 \le y \le 5\}$.

Solution. The solid whose volume we wish to calculate is sketched in Figure 6.

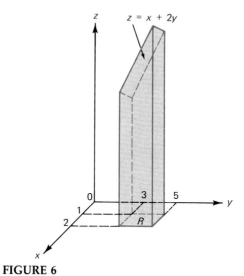

FIGURE 6

Step 1

For simplicity, we partition each of the intervals $[1, 2]$ and $[3, 5]$ into n subintervals of equal length (i.e., $m = n$):

$$1 = x_0 < x_1 < \cdots < x_n = 2$$
$$3 = y_0 < y_1 < \cdots < y_n = 5,$$

where

$$x_i = 1 + \frac{i}{n}, \qquad \Delta x = \frac{1}{n}$$

and

$$y_j = 3 + \frac{2j}{n}, \qquad \Delta y = \frac{2}{n}.$$

Step 2

Then choosing $x_i^* = x_i$ and $y_j^* = y_j$, we obtain

$$V_{ij} \approx f(x_i^*, y_j^*)\, \Delta A = (x_i + 2y_j)\, \Delta x\, \Delta y$$

$$= \left[\left(1 + \frac{i}{n}\right) + 2\left(3 + \frac{2j}{n}\right) \right] \frac{1}{n} \cdot \frac{2}{n}$$

$$= \left(7 + \frac{i}{n} + \frac{4j}{n}\right) \frac{2}{n^2}.$$

Step 3

$$V = \sum_{i=1}^{n} \sum_{j=1}^{n} V_{ij} \approx \sum_{i=1}^{n} \sum_{j=1}^{n} \left(\frac{14}{n^2} + \frac{2i}{n^3} + \frac{8j}{n^3} \right)$$

$$= \underbrace{\sum_{i=1}^{n} \sum_{j=1}^{n} \frac{14}{n^2}}_{①} + \underbrace{\sum_{i=1}^{n} \sum_{j=1}^{n} \frac{2i}{n^3}}_{②} + \underbrace{\sum_{i=1}^{n} \sum_{j=1}^{n} \frac{8j}{n^3}}_{③}.$$

It is not difficult to evaluate each of these double sums. There are n^2 terms in each sum. Since $14/n^2$ does not depend on i or j, we evaluate the sum ① by simply adding up the term $14/n^2$ a total of n^2 times. Thus

$$\sum_{i=1}^{n} \sum_{j=1}^{n} \frac{14}{n^2} = n^2 \left(\frac{14}{n^2} \right) = 14.$$

Next, if we set $i = 1$ in ③, then we have $\sum_{j=1}^{n} 8j/n^3$. Similarly, setting $i = 2, 3, 4, \ldots, n$ in ③ yields $\sum_{j=1}^{n} 8j/n^3$. Thus in ③ we obtain the term $\left(\sum_{j=1}^{n} 8j/n^3\right)$ n times. But

$$\sum_{j=1}^{n} \frac{8j}{n^3} = \frac{8}{n^3} \sum_{j=1}^{n} j = \frac{8}{n^3}(1 + 2 + \cdots + n)$$

Equation 5.4.14

$$\searrow \quad = \frac{8}{n^3} \left[\frac{n(n+1)}{2} \right]$$

$$= \frac{4(n+1)}{n^2}.$$

Thus

$$\sum_{i=1}^{n} \sum_{j=1}^{n} \frac{8j}{n^3} = n \left\{ \sum_{j=1}^{n} \frac{8j}{n^3} \right\} = n \left[\frac{4(n+1)}{n^2} \right] = \frac{4(n+1)}{n}.$$

To calculate ②, we use the same argument as in ③:

$$\sum_{j=1}^{n} \frac{2i}{n^3} = n \left(\frac{2i}{n^3} \right) = \frac{2i}{n^2},$$

so that

$$\sum_{i=1}^{n} \sum_{j=1}^{n} \frac{2i}{n^3} = \sum_{i=1}^{n} \frac{2i}{n^2} = \frac{2}{n^2} \sum_{i=1}^{n} i = \frac{2}{n^2} \left[\frac{n(n+1)}{2} \right] = \frac{n+1}{n}.$$

Finally, we have

$$\sum_{i=1}^{n} \sum_{j=1}^{n} V_{ij} \approx 14 + \frac{4(n+1)}{n} + \frac{n+1}{n}.$$

Step 4

Now as $\Delta s \to 0$, both Δx and Δy approach 0, so $n = (b-a)/\Delta x \to \infty$. Thus

$$V = \lim_{\Delta s \to 0} \sum_{i=1}^{n} \sum_{j=1}^{n} f(x_i{}^*, y_j{}^*) \, \Delta A = \lim_{n \to \infty} \sum_{i=1}^{n} \sum_{j=1}^{n} f(x_i{}^*, y_j{}^*) \, \Delta A$$

$$= \lim_{n \to \infty} \left[14 + 4 \left(\frac{n+1}{n} \right) + \frac{n+1}{n} \right] = 14 + 4 + 1 = 19. \quad \blacksquare$$

The calculation we just made was very tedious. Instead of making other calculations like this one, we will define the double integral and, in Section 19.2, show how double integrals can be easily calculated.

Definition 1 THE DOUBLE INTEGRAL Let $z = f(x, y)$ and let the rectangle R be given by (1). Let $\Delta A = \Delta x \, \Delta y$. Suppose that

$$\lim_{\Delta s \to 0} \sum_{i=1}^{n} \sum_{j=1}^{m} f(x_i{}^*, y_j{}^*) \, \Delta A$$

exists and is independent of the way in which the points $(x_i{}^*, y_j{}^*)$ are chosen. Then the **double integral of f over R**, written $\iint_R f(x, y) \, dA$, is defined by

$$\iint_R f(x, y) \, dA = \lim_{\Delta s \to 0} \sum_{i=1}^{n} \sum_{j=1}^{m} f(x_i{}^*, y_j{}^*) \, \Delta A. \tag{12}$$

If the limit in (12) exists, then the function f is said to be **integrable** over R.

We observe that this definition says nothing about volumes (just as the definition of the definite integral in Section 5.5 says nothing about areas). For example, if $f(x, y)$

takes on negative values in R, then the limit in (12) will not represent the volume under the surface $z = f(x, y)$. However, the limit in (12) may still exist, and in that case f will be integrable over R.

NOTE. $\iint_R f(x, y) \, dA$ is a number, not a function. This is analogous to the fact that the definite integral $\int_a^b f(x) \, dx$ is a number. We will not encounter indefinite double integrals in this book.

As we already stated, we will not calculate any other double integrals in this section but will wait until Section 19.2 to see how these calculations can be made simple. We should note, however, that the result of Example 1 can now be restated as

$$\iint_R (x + 2y) \, dA = 19,$$

where R is the rectangle $\{(x, y): 1 \le x \le 2 \text{ and } 3 \le y \le 5\}$.

What functions are integrable over a rectangle R? The following theorem is the double integral analog of Theorem 5.5.1.

Theorem 1 *Existence of the Double Integral Over a Rectangle* If f is continuous on R, then f is integrable over R.

We will not give proofs of the theorems stated in this section regarding double integrals. The proofs are similar to, but more complicated than, the analogous proofs for theorems on definite integrals stated in Section 5.5. The proofs of all these theorems can be found in any standard advanced calculus text.†

We now turn to the question of defining double integrals over regions in $\mathbb{R}^2$ that are not rectangular. We will denote a region in $\mathbb{R}^2$ by Ω. The two types of regions in which we will be most interested are illustrated in Figure 7. In this figure g_1, g_2, h_1, and h_2 denote continuous functions. A more general region Ω is sketched in Figure 8. We assume that the region is bounded. This means that there is a number M such that for

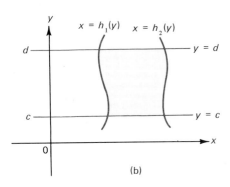

FIGURE 7

---

†See, for example, R. C. Buck, *Advanced Calculus* (McGraw-Hill, New York, 1965).

every (x, y) in Ω, $|(x, y)| = \sqrt{x^2 + y^2} \leq M$. Since Ω is bounded, we can draw a rectangle R around it. Let f be defined over Ω. We then define a new function F by

$$F(x, y) = \begin{cases} f(x, y), & \text{for } (x, y) \text{ in } \Omega \\ 0, & \text{for } (x, y) \text{ in } R \text{ but not in } \Omega. \end{cases} \tag{13}$$

Definition 2 INTEGRABILITY OVER A REGION Let f be defined for (x, y) in Ω and let F be defined by (13). Then we write

$$\iint_\Omega f(x, y)\, dA = \iint_R F(x, y)\, dA \tag{14}$$

if the integral on the right exists. In this case we say that f is **integrable** over Ω.

REMARK.   If we divide R into nm subrectangles, as in Figure 9, then we can see what is happening. For each subrectangle R_{ij} that lies entirely in Ω, $F = f$, so the volume of the "parallelepiped" above R_{ij} is given by

$$V_{ij} \approx f(x_i{}^*, y_j{}^*)\, \Delta x\, \Delta y = F(x_i{}^*, y_j{}^*)\, \Delta x\, \Delta y.$$

However, if R_{ij} is in R but not in Ω, then $F = 0$, so

$$V_{ij} \approx F(x_i{}^*, y_j{}^*)\, \Delta x\, \Delta y = 0.$$

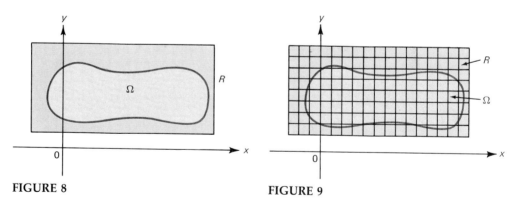

FIGURE 8   FIGURE 9

Finally, if R_{ij} is partly in Ω and partly outside of Ω, then there is no real problem since, as $\Delta s \to 0$, the sum of the volumes above these rectangles (along the boundary of Ω) will approach zero—unless the boundary of Ω is very complicated indeed. Thus we see that the limit of the sum of the volumes of the "parallelepipeds" above R is the same as the limit of the sum of the volumes of the "parallelepipeds" above Ω. This should help explain the "reasonableness" of expression (14).

Theorem 2 *Existence of the Double Integral Over a More General Region*   Let Ω be one of the regions depicted in Figure 7 where the functions g_1 and g_2 or h_1 and h_2 are continuous. Let F be defined by (13). If f is continuous over Ω, then f is integrable over Ω and its integral is given by (14).

REMARK 1.   There are some regions Ω that are so complicated that there are functions continuous but not integrable over Ω. We will not concern ourselves with such regions in this book.

REMARK 2.   *If f is nonnegative and integrable over Ω, then*

$$\iint\limits_{\Omega} f(x, y) \, dA$$

is defined as the volume under the surface $z = f(x, y)$ and over the region Ω.

REMARK 3.   *If the function $f(x, y) = 1$ is integrable over Ω, then*

$$\iint\limits_{\Omega} 1 \, dA = \iint\limits_{\Omega} dA \tag{15}$$

is equal to the area of the region Ω. To see this, note that

$$V_{ij} \approx f(x_i^*, y_j^*) \, \Delta A = \Delta A,$$

so the double integral (15) is the limit of the sum of areas of rectangles in Ω.

We close this section by stating five theorems about double integrals. Each one is analogous to a theorem about definite integrals.

Theorem 3   If f is integrable over Ω, then for any constant c, cf is integrable over Ω, and

$$\iint\limits_{\Omega} cf(x, y) \, dA = c \iint\limits_{\Omega} f(x, y) \, dA \tag{16}$$

(see Theorem 5.5.4).

Theorem 4   If f and g are integrable over Ω, then $f + g$ is integrable over Ω, and

$$\iint\limits_{\Omega} [f(x, y) + g(x, y)] \, dA = \iint\limits_{\Omega} f(x, y) \, dA + \iint\limits_{\Omega} g(x, y) \, dA \tag{17}$$

(see Theorem 5.5.5).

Theorem 5   If f is integrable over Ω_1 and Ω_2, where Ω_1 and Ω_2 have no points in common except perhaps those of their common boundary, then f is integrable over $\Omega = \Omega_1 \cup \Omega_2$, and

$$\iint\limits_{\Omega} f(x, y) \, dA = \iint\limits_{\Omega_1} f(x, y) \, dA + \iint\limits_{\Omega_2} f(x, y) \, dA$$

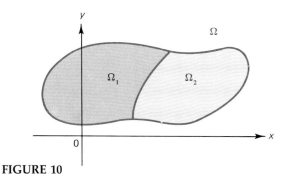

FIGURE 10

(see Theorem 5.5.3). A typical region Ω is depicted in Figure 10.

Theorem 6 If f and g are integrable over Ω and $f(x, y) \le g(x, y)$ for every (x, y) in Ω, then

$$\iint_\Omega f(x, y) \, dA \le \iint_\Omega g(x, y) \, dA \tag{18}$$

(see Theorem 5.5.7).

Theorem 7 Let f be integrable over Ω. Suppose that there exist constants m and M such that

$$m \le f(x, y) \le M \tag{19}$$

for every (x, y) in Ω. If A_Ω denotes the area of Ω, then

$$mA_\Omega \le \iint_\Omega f(x, y) \, dA \le MA_\Omega \tag{20}$$

(see Theorem 5.5.8).

Theorem 7 can be useful for estimating double integrals.

EXAMPLE 2 Let Ω be the rectangle $\{(x, y): a \le x \le b \text{ and } c \le y \le d\}$. Find upper and lower bounds for

$$\iint_\Omega \sin(x - 3y^3) \, dA.$$

Solution. Since $-1 \le \sin(x - 3y^3) \le 1$, and since $A_\Omega = (b - a)(d - c)$, we have, using (20),

$$-(b - a)(d - c) \le \iint_\Omega \sin(x - 3y^3) \, dA \le (b - a)(d - c). \quad ■$$

EXAMPLE 3   Let Ω be the disk $\{(x, y): x^2 + y^2 \leq 1\}$. Find upper and lower bounds for

$$\iint_{\Omega} \frac{1}{1 + x^2 + y^2} \, dA.$$

Solution. Since $0 \leq x^2 + y^2 \leq 1$ in Ω, we easily see that

$$\frac{1}{2} \leq \frac{1}{1 + x^2 + y^2} \leq 1.$$

Since the area of the disk is π, we have

$$\frac{\pi}{2} \leq \iint_{\Omega} \frac{1}{1 + x^2 + y^2} \, dA \leq \pi.$$

In fact, it can be shown (see Problem 19.4.25) that the value of the integral is $\pi \ln 2 \approx 0.693\pi$. ∎

PROBLEMS 19.1

In Problems 1–8, let Ω denote the rectangle $\{(x, y): 0 \leq x \leq 3 \text{ and } 1 \leq y \leq 2\}$. Use the technique employed in Example 1 to calculate the given double integral. Use Theorem 3 and/or 4 where appropriate.

1. $\iint_{\Omega}(2x + 3y) \, dA$ 2. $\iint_{\Omega}(x - y) \, dA$
3. $\iint_{\Omega}(y - x) \, dA$ 4. $\iint_{\Omega}(ax + by + c) \, dA$
5. $\iint_{\Omega}(x^2 + y^2) \, dA$ 6. $\iint_{\Omega}(x^2 - y^2) \, dA$
   [*Hint:* Use formula (5.4.11)]
7. $\iint_{\Omega}(2x^2 + 3y^2) \, dA$ 8. $\iint_{\Omega}(ax^2 + by) \, dA$

In Problems 9–14, let Ω denote the rectangle $\{(x, y): -1 \leq x \leq 0 \text{ and } -2 \leq y \leq 3\}$. Calculate the double integral.

9. $\iint_{\Omega}(x + y) \, dA$ 10. $\iint_{\Omega}(3x - y) \, dA$
11. $\iint_{\Omega}(y - 2x) \, dA$ 12. $\iint_{\Omega}(x^2 + 2y^2) \, dA$
13. $\iint_{\Omega}(y^2 - x^2) \, dA$ 14. $\iint_{\Omega}(3x^2 - 5y^2) \, dA$

In Problems 15–19, use Theorem 7 to obtain upper and lower bounds for the given integral.

15. $\iint_{\Omega}(x^5y^2 + xy) \, dA$, where Ω is the rectangle $\{(x, y): 0 \leq x \leq 1 \text{ and } 1 \leq y \leq 2\}$.
16. $\iint_{\Omega} e^{-(x^2+y^2)} \, dA$, where Ω is the disk $x^2 + y^2 \leq 4$.
*17. $\iint_{\Omega}[(x - y)/(4 - x^2 - y^2)] \, dA$, where Ω is the disk $x^2 + y^2 \leq 1$.
18. $\iint_{\Omega} \cos(\sqrt{|x|} - \sqrt{|y|}) \, dA$, where Ω is the region of Problem 17.
19. $\iint_{\Omega} \ln(1 + x + y) \, dA$, where Ω is the region bounded by the lines $y = x$, $y = 1 - x$, and the x-axis.

*20. Let Ω be one of the regions depicted in Figure 7. Which is greater:

$$\iint_{\Omega} e^{(x^2+y^2)} \, dA \qquad \text{or} \qquad \iint_{\Omega} (x^2 + y^2) \, dA?$$

19.2 THE CALCULATION OF DOUBLE INTEGRALS

In this section we derive an easy method for calculating $\iint_\Omega f(x, y)\, dx\, dy$, where Ω is one of the regions depicted in Figure 19.1.7.

We begin, as in Section 19.1, by considering

$$\iint_R f(x, y)\, dA, \tag{1}$$

where R is the rectangle

$$R = \{(x, y): a \le x \le b \text{ and } c \le y \le d\}. \tag{2}$$

If $z = f(x, y) \ge 0$ for (x, y) in R, then the double integral in (1) is the volume under the surface $z = f(x, y)$ and over the rectangle R in the xy-plane. We now calculate this volume by partitioning the x-axis taking "slices" parallel to the yz-plane. This is illustrated in Figure 1. We can approximate the volume by adding up the volumes of the various "slices." The face of each "slice" lies in the plane $x = x_i$, and the volume of the ith slice is approximately equal to the area of its face times its thickness Δx. What is the area of the face? If x is fixed, then $z = f(x, y)$ can be thought of as a curve lying in the plane $x = x_i$. Thus the area of the ith face is the area bounded by this curve, the y-axis, and the lines $y = c$ and $y = d$. This area is sketched in Figure 2. If $f(x_i, y)$ is a continuous function of y, then the area of the ith face, denoted A_i, is given by

$$A_i = \int_c^d f(x_i, y)\, dy.$$

FIGURE 1

FIGURE 2

By treating x_i as a constant, we can compute A_i as an ordinary definite integral, where the variable is y. Note, too, that $A(x) = \int_c^d f(x, y)\, dy$ is a function of x only and can therefore be integrated as in Chapter 5. Then the volume of the ith slice is approximated by

$$V_i \approx \left\{ \int_c^d f(x_i, y)\, dy \right\} \Delta x$$

so that, adding up these "subvolumes" and taking the limit as Δx approaches zero, we obtain

$$V = \int_a^b \left\{ \int_c^d f(x, y)\, dy \right\} dx = \int_a^b A(x)\, dx. \tag{3}$$

Definition 1   REPEATED INTEGRAL   The expression in (3) is called a **repeated integral** or **iterated integral**. Since we also have

$$V = \iint_R f(x, y)\, dA,$$

we obtain

$$\iint_R f(x, y)\, dA = \int_a^b \left\{ \int_c^d f(x, y)\, dy \right\} dx. \tag{4}$$

REMARK 1.   Usually we will write equation (4) without braces. We then have

$$\iint_R f(x, y)\, dA = \int_a^b \int_c^d f(x, y)\, dy\, dx. \tag{5}$$

REMARK 2.   We should emphasize that the first integration in $\int_a^b \int_c^d f(x, y)\, dy\, dx$ is performed by treating x as a constant.

Similarly, if we instead begin by partitioning the y-axis, we find that the area of the face of a "slice" lying in the plane $y = y_i$ is given by

$$A_i = \int_a^b f(x, y_i)\, dx,$$

where now A_i is an integral in the variable x. Thus as before,

$$V = \int_c^d \left\{ \int_a^b f(x, y)\, dx \right\} dy, \tag{6}$$

and

$$\iint_R f(x, y)\, dA = \int_c^d \int_a^b f(x, y)\, dx\, dy. \tag{7}$$

EXAMPLE 1   Calculate the volume under the plane $z = x + 2y$ and over the rectangle

$R = \{(x, y): 1 \le x \le 2 \text{ and } 3 \le y \le 5\}.$

Solution. We calculated this volume in Example 19.1.1. Using equation (5), we have

$$V = \iint_R (x + 2y) \, dA = \int_1^2 \left[\int_3^5 (x + 2y) \, dy \right] dx\dagger$$

$$= \int_1^2 \left[(xy + y^2) \Big|_{y=3}^{y=5} \right] dx = \int_1^2 [(5x + 25) - (3x + 9)] \, dx$$

$$= \int_1^2 (2x + 16) \, dx = (x^2 + 16x) \Big|_1^2 = 19.$$

Similarly, using equation (7), we have

$$V = \int_3^5 \left\{ \int_1^2 (x + 2y) \, dx \right\} dy = \int_3^5 \left[\left(\frac{x^2}{2} + 2yx \right) \Big|_{x=1}^{x=2} \right] dy$$

$$= \int_3^5 \left\{ (2 + 4y) - \left(\frac{1}{2} + 2y \right) \right\} dy = \int_3^5 \left(2y + \frac{3}{2} \right) dy$$

$$= \left(y^2 + \frac{3}{2} y \right) \Big|_3^5 = 19. \quad \blacksquare$$

EXAMPLE 2 Calculate the volume of the region beneath the surface $z = xy^2 + y^3$ and over the rectangle $R = \{(x, y): 0 \le x \le 2 \text{ and } 1 \le y \le 3\}$.

Solution. A computer-drawn sketch of this region is given in Figure 3. Using equation (5), we have

$$V = \int_0^2 \int_1^3 (xy^2 + y^3) \, dy \, dx = \int_0^2 \left[\left(\frac{xy^3}{3} + \frac{y^4}{4} \right) \Big|_1^3 \right] dx$$

$$= \int_0^2 \left[\left(9x + \frac{81}{4} \right) - \left(\frac{x}{3} + \frac{1}{4} \right) \right] dx = \int_0^2 \left(\frac{26}{3} x + 20 \right) dx$$

$$= \left(\frac{13x^2}{3} + 20x \right) \Big|_0^2 = \frac{52}{3} + 40 = \frac{172}{3}.$$

You should verify that the same answer is obtained by using equation (7). $\blacksquare$

We now extend our results to more general regions. Let

$$\Omega = \{(x, y): a \le x \le b \text{ and } g_1(x) \le y \le g_2(x)\}. \tag{8}$$

---

†Remember, in computing the bracketed integral, we treat x as a constant.

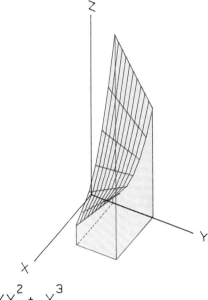

$$Z = XY^2 + Y^3$$

FIGURE 3

This region is sketched in Figure 4. We assume that for every x in $[a, b]$,

$$g_1(x) \le g_2(x). \tag{9}$$

If we partition the x-axis as before, then we obtain slices lying in the planes $x = x_i$, a typical one of which is sketched in Figure 5. Then

$$A_i = \int_{g_1(x_i)}^{g_2(x_i)} f(x_i, y) \, dy, \qquad V_i \approx \left\{ \int_{g_1(x_i)}^{g_2(x_i)} f(x_i, y) \, dy \right\} \Delta x,$$

and

$$V = \iint_\Omega f(x, y) \, dA = \int_a^b \int_{g_1(x)}^{g_2(x)} f(x, y) \, dy \, dx. \tag{10}$$

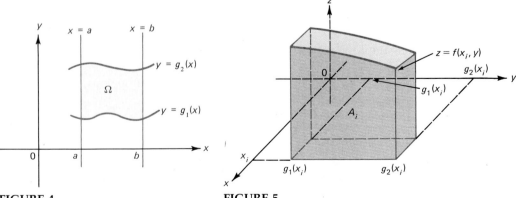

FIGURE 4   **FIGURE 5**

Similarly, let

$$\Omega = \{(x, y): h_1(y) \le x \le h_2(y) \text{ and } c \le y \le d\} \tag{11}$$

(see Figure 6). Then

$$V = \int_c^d \int_{h_1(y)}^{h_2(y)} f(x, y) \, dx \, dy. \tag{12}$$

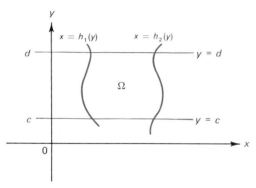

FIGURE 6

We summarize these results in the following theorem.

Theorem 1 Let f be continuous over a region Ω given by equation (8) or (11).

(i) If Ω is of the form (8), where g_1 and g_2 are continuous, then

$$\iint_\Omega f(x, y) \, dA = \int_a^b \int_{g_1(x)}^{g_2(x)} f(x, y) \, dy \, dx.$$

(ii) If Ω is of the form (11), where h_1 and h_2 are continuous, then

$$\iint_\Omega f(x, y) \, dA = \int_c^d \int_{h_1(y)}^{h_2(y)} f(x, y) \, dx \, dy.$$

REMARK 1.   We have not actually proved this theorem here but have merely indicated why it should be so. A rigorous proof can be found in any advanced calculus text.

REMARK 2.   Note that this theorem says nothing about volume. It can be used to calculate any double integral if the hypotheses of the theorem are satisfied and if each function being integrated has an antiderivative that can be written in terms of elementary functions.

REMARK 3.   Many regions are of the form (8) or (11). In addition, almost all regions that arise in practical applications can be broken into a finite number of regions of the form (8) or (11), and the integration can be carried out using Theorem 19.1.5.

EXAMPLE 3   Find the volume of the solid under the surface $z = x^2 + y^2$ and lying above the region

$$\Omega = \{(x, y): 0 \le x \le 1 \text{ and } x^2 \le y \le \sqrt{x}\}.$$

Solution. Ω is sketched in Figure 7. We see that $0 \le x \le 1$ and $x^2 \le y \le \sqrt{x}$. Then using (10), we have

$$V = \int_0^1 \int_{x^2}^{\sqrt{x}} (x^2 + y^2)\, dy\, dx = \int_0^1 \left\{ \left(x^2 y + \frac{y^3}{3} \right) \Big|_{x^2}^{\sqrt{x}} \right\} dx$$

$$= \int_0^1 \left\{ \left(x^2\sqrt{x} + \frac{(\sqrt{x})^3}{3} \right) - \left(x^2 \cdot x^2 + \frac{(x^2)^3}{3} \right) \right\} dx$$

$$= \int_0^1 \left(x^{5/2} + \frac{x^{3/2}}{3} - x^4 - \frac{x^6}{3} \right) dx$$

$$= \left(\frac{2x^{7/2}}{7} + \frac{2x^{5/2}}{15} - \frac{x^5}{5} - \frac{x^7}{21} \right) \Big|_0^1 = \frac{2}{7} + \frac{2}{15} - \frac{1}{5} - \frac{1}{21} = \frac{18}{105}.$$

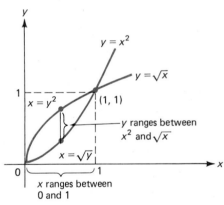

FIGURE 7

We can calculate this integral in another way. We note that x varies between the curves $x = y^2$ and $x = \sqrt{y}$. Then using (12), since $0 \le y \le 1$ and $y^2 \le x \le \sqrt{y}$, we have

$$V = \int_0^1 \int_{y^2}^{\sqrt{y}} (x^2 + y^2)\, dx\, dy,$$

which is easily seen to be equal to 18/105. ∎

EXAMPLE 4   Let $f(x, y) = x^2 y$. Calculate the integral of f over the region bounded by the x-axis and the semicircle $x^2 + y^2 = 4$, $y \ge 0$.

Solution. The region of integration is sketched in Figure 8. Using equation (8), we see that $0 \le y \le \sqrt{4 - x^2}$, $-2 \le x \le 2$, so that, integrating first with respect to y, we obtain

$$\iint_\Omega x^2 y \, dA = \int_{-2}^{2} \int_{0}^{\sqrt{4-x^2}} x^2 y \, dy \, dx$$

$$= \int_{-2}^{2} \left\{ \frac{x^2 y^2}{2} \Big|_{0}^{\sqrt{4-x^2}} \right\} dx = \int_{-2}^{2} \frac{x^2(4 - x^2)}{2} \, dx$$

$$= \int_{-2}^{2} \left(2x^2 - \frac{x^4}{2} \right) dx = \left(\frac{2x^3}{3} - \frac{x^5}{10} \right) \Big|_{-2}^{2} = \frac{64}{15}.$$

We can also use equation (11) and integrate first with respect to x. Then $-\sqrt{4 - y^2} \le x \le \sqrt{4 - y^2}$, $0 \le y \le 2$, and

$$V = \int_{0}^{2} \int_{-\sqrt{4-y^2}}^{\sqrt{4-y^2}} x^2 y \, dx \, dy = \int_{0}^{2} \left(\frac{x^3 y}{3} \Big|_{-\sqrt{4-y^2}}^{\sqrt{4-y^2}} \right) dy$$

$$= \int_{0}^{2} \frac{2}{3} (4 - y^2)^{3/2} \, y \, dy = \frac{-2}{15} (4 - y^2)^{5/2} \Big|_{0}^{2} = \frac{64}{15}. \quad \blacksquare$$

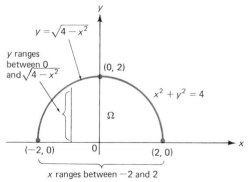

FIGURE 8

REVERSING THE ORDER OF INTEGRATION

EXAMPLE 5   Evaluate $\int_{1}^{2} \int_{1}^{x^2} (x/y) \, dy \, dx$.

Solution.

$$\int_{1}^{2} \int_{1}^{x^2} \frac{x}{y} \, dy \, dx = \int_{1}^{2} \left\{ x \ln y \Big|_{1}^{x^2} \right\} dx = \int_{1}^{2} x \ln x^2 \, dx = \int_{1}^{2} 2x \ln x \, dx$$

It is necessary to use integration by parts to complete the problem. Setting $u = \ln x$ and $dv = 2x \, dx$, we have $du = (1/x) \, dx$, $v = x^2$, and

$$\int_{1}^{2} 2x \ln x \, dx = x^2 \ln x \Big|_{1}^{2} - \int_{1}^{2} x \, dx = 4 \ln 2 - \frac{x^2}{2} \Big|_{1}^{2} = 4 \ln 2 - \frac{3}{2}.$$

There is an easier way to calculate the double integral. We simply **reverse the order of integration.** The region of integration is sketched in Figure 9. If we want to integrate first with respect to x, we note that we can describe the region by

$$\Omega = \{(x, y): \sqrt{y} \le x \le 2 \text{ and } 1 \le y \le 4\}.$$

Then

$$\int_1^2 \int_1^{x^2} \frac{x}{y}\, dy\, dx = \iint_\Omega \frac{x}{y}\, dA = \int_1^4 \int_{\sqrt{y}}^2 \frac{x}{y}\, dx\, dy = \int_1^4 \left\{ \frac{x^2}{2y}\Big|_{\sqrt{y}}^2 \right\} dy$$

$$= \int_1^4 \left(\frac{2}{y} - \frac{1}{2}\right) dy = \left(2 \ln y - \frac{y}{2}\right)\Big|_1^4 = 2 \ln 4 - \frac{3}{2}$$

$$= 4 \ln 2 - \frac{3}{2}.$$

Note that in this case it is easier to integrate first with respect to x. ■

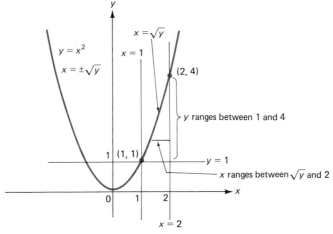

FIGURE 9

The technique used in Example 5 suggests the following:

> When changing the order of integration, first sketch the region of integration in the xy-plane.

REMARK. Why is it legitimate to reverse the order of integration? There is a theorem† that asserts the following: Suppose that the region Ω can be written as

$$\Omega = \{(x, y): a \le x \le b, g_1(x) \le y \le g_2(x)\} = \{(x, y): c \le y \le d, h_1(y) \le x \le h_2(y)\}.$$

---

†This result is a special case of **Fubini's theorem**, which is proved in most advanced calculus books.

Then if f is continuous on Ω,

$$\iint\limits_{\Omega} f(x, y) \, dA = \int_a^b \int_{g_1(x)}^{g_2(x)} f(x, y) \, dy \, dx = \int_c^d \int_{h_1(y)}^{h_2(y)} f(x, y) \, dx \, dy.$$

We will not prove this result here.

EXAMPLE 6   Compute $\int_0^2 \int_y^2 e^{x^2} \, dx \, dy$.

Solution. The region of integration is sketched in Figure 10. We first observe that the double integral cannot be evaluated directly since it is impossible to find an antiderivative for e^{x^2}. Instead, we reverse the order of integration. From Figure 10 we see that Ω can be written as $0 \le y \le x$, $0 \le x \le 2$, so

$$\int_0^2 \int_y^2 e^{x^2} \, dx \, dy = \iint\limits_{\Omega} e^{x^2} \, dA = \int_0^2 \int_0^x e^{x^2} \, dy \, dx = \int_0^2 \left(ye^{x^2} \Big|_{y=0}^{y=x} \right) dx = \int_0^2 xe^{x^2} \, dx$$

$$= \frac{1}{2} e^{x^2} \Big|_0^2 = \frac{1}{2}(e^4 - 1). \quad \blacksquare$$

EXAMPLE 7   Reverse the order of integration in the iterated integral $\int_0^1 \int_{\sqrt{y}}^2 f(x, y) \, dx \, dy$.

Solution. The region of integration is sketched in Figure 11. This region is divided into two subregions Ω_1 and Ω_2. What happens if we integrate first with respect to y? In Ω_1, $0 \le y \le x^2$. In Ω_2, $0 \le y \le 1$. Thus

$$\int_0^1 \int_{\sqrt{y}}^2 f(x, y) \, dx \, dy = \iint\limits_{\Omega} f(x, y) \, dA \overset{\text{Theorem 19.1.5}}{=} \iint\limits_{\Omega_1} f(x, y) \, dA + \iint\limits_{\Omega_2} f(x, y) \, dA$$

$$= \int_0^1 \int_0^{x^2} f(x, y) \, dy \, dx + \int_1^2 \int_0^1 f(x, y) \, dy \, dx. \quad \blacksquare$$

FIGURE 10

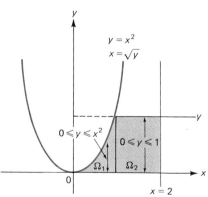

FIGURE 11

EXAMPLE 8   Find the volume in the first octant bounded by the three coordinate planes and the surface $z = 1/(1 + x + 3y)^3$.

Solution. The solid here extends over the infinite region $\{(x, y): 0 \le x \le \infty$ and $0 \le y \le \infty\}$. Thus

$$V = \int_0^\infty \int_0^\infty \frac{1}{(1 + x + 3y)^3} \, dx \, dy = \int_0^\infty \lim_{N \to \infty}\left(-\frac{1}{2(1 + x + 3y)^2}\Big|_0^N \right) dy$$

$$= \int_0^\infty \frac{1}{2(1 + 3y)^2} \, dy = \lim_{N \to \infty}\left(-\frac{1}{6(1 + 3y)}\right)\Big|_0^N = \frac{1}{6}.$$

Note that improper double integrals can be treated in the same way that we treat improper "single" integrals. ∎

EXAMPLE 9   Find the volume of the solid bounded by the coordinate planes and the plane $2x + y + z = 2$.

Solution. We have $z = 2 - 2x - y$ and this expression must be integrated over the region in the xy-plane bounded by the line $2x + y = 2$ (obtained when $z = 0$) and the x- and y-axes. See Figure 12. We therefore have

$$V = \int_0^1 \int_0^{2-2x} (2 - 2x - y) \, dy \, dx = \int_0^1 \left\{ \left(2y - 2xy - \frac{y^2}{2}\right)\Big|_0^{2-2x} \right\} dx$$

$$= \int_0^1 \left\{ 2(2 - 2x) - 2x(2 - 2x) - \frac{(2 - 2x)^2}{2} \right\} dx$$

$$= \int_0^1 (2x^2 - 4x + 2) \, dx = \left(\frac{2x^3}{3} - 2x^2 + 2x \right)\Big|_0^1 = \frac{2}{3}. \; ∎$$

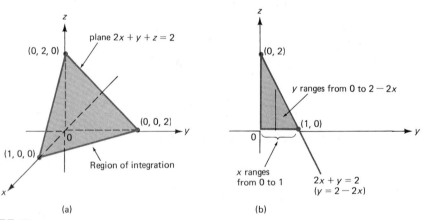

FIGURE 12

EXAMPLE 10   Find the volume of the solid bounded by the circular paraboloid $x = y^2 + z^2$ and the plane $x = 1$.

Solution. The solid is sketched in Figure 13. There are many ways to obtain its volume. Perhaps the best way to do so is to get another perspective on the picture. In Figure 14 we redraw Figure 13 with x as the vertical axis. The volume of the indicated element is $(1 - x) \, \Delta y \, \Delta z$, where $x = y^2 + z^2$. We can take advantage of symmetry to

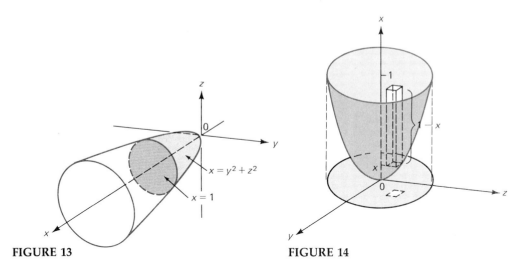

FIGURE 13   **FIGURE 14**

write

$$V = 4\int_0^1 \int_0^{\sqrt{1-z^2}} (1 - x)\, dy\, dz = 4\int_0^1 \int_0^{\sqrt{1-z^2}} (1 - y^2 - z^2)\, dy\, dz$$

$$= 4\int_0^1 \left(y - \frac{y^3}{3} - yz^2 \right)\Bigg|_{y=0}^{y=\sqrt{1-z^2}} dz$$

$$= 4\int_0^1 \left\{ (1 - z^2)^{1/2} - \frac{(1 - z^2)^{3/2}}{3} - z^2(1 - z^2)^{1/2} \right\} dz.$$

We can integrate this by making the substitution $z = \sin\theta$. Then

$$V = 4\int_0^{\pi/2} \left(\cos\theta - \frac{\cos^3\theta}{3} - \sin^2\theta\cos\theta \right) \cos\theta\, d\theta$$

$$= 4\int_0^{\pi/2} \left[\cos\theta\underbrace{(1 - \sin^2\theta)}_{=\,\cos^2\theta} - \frac{\cos^3\theta}{3} \right] \cos\theta\, d\theta$$

$$= \frac{8}{3}\int_0^{\pi/2} \cos^4\theta\, d\theta = \frac{2}{3}\int_0^{\pi/2} (1 + \cos 2\theta)^2\, d\theta$$

$$= \frac{2}{3}\int_0^{\pi/2} (1 + 2\cos 2\theta + \cos^2 2\theta)\, d\theta$$

$$= \frac{2}{3}\int_0^{\pi/2} \left(1 + 2\cos 2\theta + \frac{1}{2} + \frac{1}{2}\cos 4\theta \right) d\theta$$

$$= \frac{2}{3}\cdot\frac{3}{2}\cdot\frac{\pi}{2} = \frac{\pi}{2}. \ \blacksquare$$

REMARK.   We will derive a much easier way to obtain this answer in Section 19.4 (see Example 19.4.7).

PROBLEMS 19.2

In Problems 1–23, evaluate the given double integral.

1. $\displaystyle\int_0^1\int_0^2 xy^2\,dx\,dy$

2. $\displaystyle\int_{-1}^3\int_2^4 (x^2-y^3)\,dy\,dx$

3. $\displaystyle\int_2^5\int_0^4 e^{(x-y)}\,dx\,dy$

4. $\displaystyle\int_0^1\int_{x^2}^x x^3y\,dy\,dx$

5. $\displaystyle\int_2^4\int_{1+y}^{2+3y} (x-y^2)\,dx\,dy$

6. $\displaystyle\int_{\pi/4}^{\pi/3}\int_{\sin x}^{\cos x} (x+2y)\,dy\,dx$

7. $\displaystyle\int_0^3\int_{-\sqrt{9-y^2}}^{\sqrt{9-y^2}} x^2y\,dx\,dy$

8. $\displaystyle\int_1^2\int_{y^5}^{3y^5} \frac{1}{x}\,dx\,dy$

9. $\iint_\Omega(x^2+y^2)\,dA$, where $\Omega=\{(x,y): 1\le x\le 2$ and $-1\le y\le 1\}$.

10. $\iint_\Omega 2xy\,dA$, where $\Omega=\{(x,y): 0\le x\le 4$ and $1\le y\le 3\}$.

11. $\iint_\Omega(x-y)^2\,dA$, where $\Omega=\{(x,y): -2\le x\le 2$ and $0\le y\le 1\}$.

12. $\iint_\Omega\sin(2x+3y)\,dA$, where $\Omega=\{(x,y): 0\le x\le \pi/6$ and $0\le y\le \pi/18\}$.

13. $\iint_\Omega xe^{(x^2+y)}\,dA$, where Ω is the region of Problem 10.

14. $\iint_\Omega(x-y^2)\,dA$, where Ω is the region in the first quadrant bounded by the x-axis, the y-axis, and the unit circle.

15. $\iint_\Omega(x^2+y)\,dA$, where Ω is the region of Problem 14.

16. $\iint_\Omega(x^3-y^3)\,dA$, where Ω is the region of Problem 14.

17. $\iint_\Omega(x+2y)\,dA$, where Ω is the triangular region bounded by the lines $y=x$, $y=1-x$, and the y-axis.

18. $\iint_\Omega e^{x+2y}\,dA$, where Ω is the region of Problem 17.

19. $\iint_\Omega(x^2+y)\,dA$, where Ω is the region in the first quadrant between the parabolas $y=x^2$ and $y=1-x^2$.

20. $\iint_\Omega(1/\sqrt{y})\,dA$, where Ω is the region of Problem 19.

21. $\iint_\Omega(y/\sqrt{x^2+y^2})\,dA$, where $\Omega=\{(x,y): 1\le x\le y$ and $1\le y\le 2\}$.

22. $\iint_\Omega[e^{-y}/(1+x^2)]\,dA$, where Ω is the first quadrant.

23. $\iint_\Omega(x+y)e^{-(x+y)}\,dA$, where Ω is the first quadrant.

In Problems 24–33, (a) sketch the region over which the integral is taken. Then (b) change the order of integration, and (c) evaluate the given integral.

24. $\displaystyle\int_0^2\int_{-1}^3 dx\,dy$

25. $\displaystyle\int_0^4\int_{-5}^8 (x+y)\,dy\,dx$

26. $\displaystyle\int_2^4\int_1^y \frac{y^3}{x^3}\,dx\,dy$

27. $\displaystyle\int_0^1\int_0^x dy\,dx$

28. $\displaystyle\int_0^1\int_x^1 dy\,dx$

29. $\displaystyle\int_0^{\pi/2}\int_0^{\cos y} y\,dx\,dy$

30. $\displaystyle\int_0^2\int_0^{\sqrt{4-y^2}} (4-x^2)^{3/2}\,dx\,dy$

31. $\displaystyle\int_0^1\int_{\sqrt{x}}^{\sqrt[3]{x}} (1+y^6)\,dy\,dx$

32. $\displaystyle\int_0^1\int_{\sqrt{y}}^1 \sqrt{3-x^3}\,dx\,dy$

33. $\displaystyle\int_0^\infty\int_x^\infty \frac{1}{(1+y^2)^{7/5}}\,dy\,dx$

34. Show that if both integrals exist, then

$$\int_0^\infty\int_0^x f(x,y)\,dy\,dx = \int_0^\infty\int_y^\infty f(x,y)\,dx\,dy.$$

[*Hint:* Draw a picture.]

In Problems 35–44, find the volume of the given solid.

35. The solid bounded by the plane $x + y + z = 3$ and the three coordinate planes.
*36. The solid bounded by the planes $x = 0$, $z = 0$, $x + 2y + z = 6$, and $x - 2y + z = 6$.
37. The solid bounded by the cylinders $x^2 + y^2 = 4$ and $y^2 + z^2 = 4$.
38. The solid bounded by the cylinder $x^2 + z^2 = 1$ and the planes $y = 0$ and $y = 2$.
39. The ellipsoid $x^2 + 4y^2 + 9z^2 = 36$.
40. The solid bounded above by the sphere $x^2 + y^2 + z^2 = 9$ and below by the plane $z = \sqrt{5}$.
41. The solid bounded by the planes $y = 0$, $y = x$, and the cylinder $x + z^2 = 2$.
*42. The solid bounded by the parabolic cylinder $x = z^2$ and the planes $y = 1$, $y = 5$, $z = 1$, and $x = 0$.
*43. The solid bounded by the paraboloid $y = x^2 + z^2$ and the plane $x + y = 3$.
44. The solid bounded by the surface $z = e^{-(x+y)}$ and the three coordinate planes.

*45. Use a double integral to find the area of each of the regions bounded by the x-axis and the curves $y = x^3 + 1$ and $y = 3 - x^2$.
46. Use a double integral to find the area in the first quadrant bounded by the curves $y = x^{1/m}$ and $y = x^{1/n}$, where m and n are positive and $n > m$.
47. Let $f(x, y) = g(x)h(y)$, where g and h are continuous. Let Ω be the rectangle $\{(x, y): a \le x \le b \text{ and } c \le y \le d\}$. Show that

$$\iint_{\Omega} f(x, y) \, dA = \left\{\int_a^b g(x) \, dx\right\} \cdot \left\{\int_c^d h(y) \, dy\right\}.$$

48. Sketch the solid whose volume is given by

$$V = \int_1^3 \int_0^2 (x + 3y) \, dx \, dy.$$

*49. Sketch the solid whose volume is given by

$$V = \int_0^1 \int_{x^2}^{\sqrt{x}} \sqrt{x^2 + y^2} \, dy \, dx.$$

19.3 DENSITY, MASS, AND CENTER OF MASS (OPTIONAL)

Let $\rho(x, y)$ denote the density of a plane object (like a thin lamina, for example). Suppose that the object occupies a region Ω in the xy-plane. Then the mass of a small rectangle of sides Δx and Δy centered at the point (x, y) is approximated by

$$\rho(x, y) \, \Delta x \, \Delta y = \rho(x, y) \, \Delta A. \tag{1}$$

and the total mass of the object is

$$\mu = \iint_{\Omega} \rho(x, y) \, dA. \tag{2}$$

Compare this formula with the formula for the mass of an object lying along the x-axis with density $\rho(x)$ [see equation (5.5.10)].

In Sections 9.4 and 9.5 we showed how to calculate the first moment and center of mass of an object around the x- and y-axes. For example, we defined

$$M_y = \int_a^b x\rho(x)\, dx \tag{3}$$

to be the first moment about the y-axis when we had a system of masses distributed along the x-axis. Similarly, we calculated the x-coordinate of the center of mass of the object to be

$$\bar{x} = \frac{\int_b^a x\rho(x)\, dx}{\int_a^b \rho(x)\, dx} = \frac{\text{first moment about } y\text{-axis}}{\text{mass}} = \frac{M_y}{\mu} \tag{4}$$

[see equation (9.4.7)].

When, in Section 9.5, we calculated the centroid of a plane region, we found that it was necessary to assume that the region had a constant area density ρ. However, by using double integrals, we can get away from this restriction. Consider the plane region whose mass is given by (2). Then we define

$$M_y = \text{first moment around } y\text{-axis} = \iint_\Omega x\rho(x, y)\, dA. \tag{5}$$

Look at Figure 1. The first moment about the y-axis of a small rectangle centered at (x, y) is given by

$$x_i\rho(x_i{}^*, y_j{}^*)\, \Delta x\, \Delta y, \tag{6}$$

and if we add up these moments for all such "subrectangles" and take a limit, we arrive at equation (5). Finally, we define the **center of mass** of the plane region to be the point $(\bar{x}, \bar{y})$, where

$$\bar{x} = \frac{M_y}{\mu} = \frac{\iint_\Omega x\rho(x, y)\, dA}{\iint_\Omega \rho(x, y)\, dA} \tag{7}$$

and

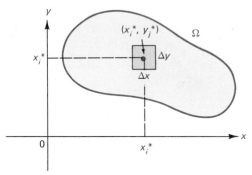

FIGURE 1

$$\bar{y} = \frac{M_x}{\mu} = \frac{\iint_\Omega y\rho(x,y)\,dA}{\iint_\Omega \rho(x,y)\,dA}. \tag{8}$$

These formulas are derived in essentially the same way that we derived formulas (9.5.9) and (9.5.10). As before, if ρ is constant, the center of mass of a region is called the **centroid** of that region.

EXAMPLE 1   A plane lamina has the shape of the triangle bounded by the lines $y = x$, $y = 2 - x$, and the x-axis. Its density function is given by $\rho(x, y) = 1 + 2x + y$. Distance is measured in meters, and mass is measured in kilograms. Find the mass and center of mass of the lamina.

Solution.   The region is sketched in Figure 2. The mass is given by

$$\mu = \iint_\Omega \rho(x,y)\,dA = \int_0^1 \int_y^{2-y} (1 + 2x + y)\,dx\,dy$$

$$= \int_0^1 \left\{ (x + x^2 + xy)\Big|_y^{2-y} \right\} dy = \int_0^1 (6 - 4y - 2y^2)\,dy$$

$$= \left(6y - 2y^2 - \frac{2y^3}{3} \right)\Big|_0^1 = \frac{10}{3}\ \text{kg.}$$

Then

$$M_y = \int_0^1 \int_y^{2-y} x(1 + 2x + y)\,dx\,dy$$

$$= \int_0^1 \int_y^{2-y} (x + 2x^2 + xy)\,dx\,dy = \int_0^1 \left\{ \left(\frac{x^2}{2} + \frac{2x^3}{3} + \frac{x^2 y}{2} \right)\Big|_y^{2-y} \right\} dy$$

$$= \int_0^1 \left(\frac{22}{3} - 8y + 2y^2 - \frac{4y^3}{3} \right) dy$$

$$= \left(\frac{22}{3}y - 4y^2 + \frac{2y^3}{3} - \frac{y^4}{3} \right)\Big|_0^1 = \frac{11}{3}\ \text{kg} \cdot \text{m.}$$

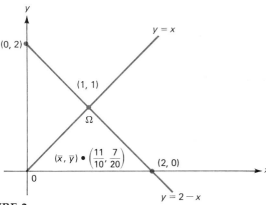

FIGURE 2

$$M_x = \int_0^1 \int_y^{2-y} y(1 + 2x + y)\, dx\, dy = \int_0^1 \left\{ (xy + x^2y + xy^2) \Big|_y^{2-y} \right\} dy$$

$$= \int_0^1 (6y - 4y^2 - 2y^3)\, dy = \left(3y^2 - \frac{4y^3}{3} - \frac{y^4}{2} \right) \Big|_0^1 = \frac{7}{6}\, \text{kg} \cdot \text{m}.$$

Thus

$$\bar{x} = \frac{M_y}{\mu} = \frac{11/3}{10/3} = \frac{11}{10}\, \text{m} \quad \text{and} \quad \bar{y} = \frac{M_x}{\mu} = \frac{7/6}{10/3} = \frac{7}{20}\, \text{m}. \quad \blacksquare$$

In Problems 9.5.41 and 9.5.42, we stated the **first theorem of Pappus.**† We prove this result here.

Theorem 1 *First Theorem of Pappus* Suppose that the plane region Ω is revolved about a line L in the xy-plane that does not intersect it. Then the volume generated is equal to the product of the area of Ω and the length of the circumference of the circle traced by the centroid of Ω.

Proof. We construct a coordinate system, placing the y-axis so that it coincides with the line L and Ω is in the first quadrant. The situation is then as depicted in Figure 3. We form a regular partition of the region Ω and choose the point (x_i^*, y_j^*) to be a point in the subrectangle Ω_{ij}. If Ω_{ij} is revolved about L (the y-axis), it forms a ring, as depicted in Figure 4. The volume of the ring is, approximately,

$$V_{ij} \approx (\text{circumference of circle with radius } x_i^*) \times (\text{thickness of ring})$$

$$\times (\text{height of ring})$$

$$= 2\pi x_i^* \, \Delta x \, \Delta y = 2\pi x_i^* \, \Delta A,$$

and using a familiar limiting argument, we have

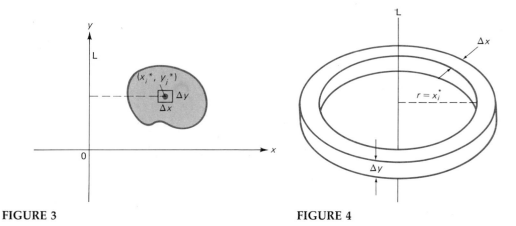

FIGURE 3 **FIGURE 4**

Pappus of Alexandria

Born during the reign of the Roman emperor Diocletian (A.D. 284 – 305), Pappus of Alexandria lived approximately 600 years after the time of Euclid and Archimedes and devoted much of his life attempting to revitalize interest in the traditional study of Greek geometry.

Pappus' greatest work was his *Mathematical Collection*, written between A.D. 320 and 340. This work is significant for three reasons. First, in it are collected works of more than 30 different mathematicians of antiquity. We owe much of our knowledge of Greek geometry to the *Collection*. Second, it provides alternative proofs of the results of the greatest of the Greeks, including Euclid and Archimedes. Third, the *Collection* contains a variety of discoveries not found in any earlier work.

The *Collection* comprises eight books, each one containing a variety of interesting and sometimes amusing results. In Book V, for example, Pappus showed that if two regular polygons have equal perimeters, then the one with the greater number of sides has the larger area. Pappus used this result to suggest the great wisdom of bees, as bees construct their hives using hexagonal (6-sided) cells, rather than square or triangular ones.

Book VII contains the theorem proved in this section. It is also the book that is the most important in the history of mathematics in that it contains the *Treasury of Analysis*. The *Treasury* is a collection of mathematical facts that, together with Euclid's *Elements*, claimed to contain the material the practicing mathematician in the fourth century needed to know. Although mathematicians wrote in Greek for another thousand or so years, no follower wrote a work of equal significance.

$$V = \lim_{\substack{\Delta x \to 0 \\ \Delta y \to 0}} \sum_{i=1}^{n} \sum_{j=1}^{m} V_{ij} = \iint_{\Omega} 2\pi x \, dA = 2\pi \iint_{\Omega} x \, dA. \tag{9}$$

Now we can think of Ω as a thin lamina with constant density $\rho = 1$. Then from (7) (with $\rho = 1$),

$$\bar{x} = \frac{\iint_{\Omega} x \, dA}{\iint_{\Omega} dA} \overset{\text{From (9)}}{=} \frac{V/2\pi}{\iint_{\Omega} dA} \overset{\text{Formula (19.1.15)}}{=} \frac{V/2\pi}{\text{area of } \Omega},$$

or

$$V = (2\pi\bar{x})(\text{area of } \Omega)$$

$$= \left(\begin{array}{c} \text{length of the circumference of the circle} \\ \text{traced by the centroid of } \Omega \end{array} \right) \times (\text{area of } \Omega). \quad \blacksquare$$

EXAMPLE 2 Use the first theorem of Pappus to calculate the volume of the torus generated by rotating the circle $(x - a)^2 + y^2 = r^2$ $(r < a)$ about the y-axis.

Solution. The circle and the torus are sketched in Figure 5. The area of the circle is πr^2. The radius of the circle traced by the centroid $(a, 0)$ is a, and the circumference is $2\pi a$. Thus

$$V = (2\pi a)\pi r^2 = 2\pi^2 a r^2. \quad \blacksquare$$

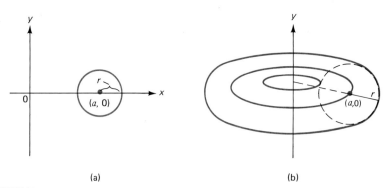

(a)   (b)

FIGURE 5

PROBLEMS 19.3

In Problems 1–12, find the mass and center of mass of an object that lies in the given region with the given area density function.

1. $\Omega = \{(x, y): 1 \le x \le 2, -1 \le y \le 1\}; \rho(x, y) = x^2 + y^2$.
2. $\Omega = \{(x, y): 0 \le x \le 4, 1 \le y \le 3\}; \rho(x, y) = 2xy$.
3. $\Omega = \{(x, y): 0 \le x \le \pi/6, 0 \le y \le \pi/18\}; \rho(x, y) = \sin(2x + 3y)$.
4. $\Omega = \{(x, y): -2 \le x \le 2, 0 \le y \le 1\}; \rho(x, y) = (x - y)^2$.
5. Ω is the region of Problem 2; $\rho(x, y) = xe^{x-y}$.
6. Ω is the quarter of the unit circle lying in the first quadrant; $\rho(x, y) = x + y^2$.
7. Ω is the region of Problem 6; $\rho(x, y) = x^2 + y$.
8. Ω is the region of Problem 6; $\rho(x, y) = x^3 + y^3$.
9. Ω is the triangular region bounded by the lines $y = x$, $y = 1 - x$, and the x-axis; $\rho(x, y) = x + 2y$.
10. Ω is the region of Problem 9; $\rho(x, y) = e^{x+2y}$.
11. Ω is the first quadrant; $\rho(x, y) = e^{-y}/(1 + x)^3$.
12. Ω is the first quadrant; $\rho(x, y) = (x + y)e^{-(x+y)}$

13. Use the first theorem of Pappus to calculate the volume of the torus generated by rotating the unit circle about the line $y = 4 - x$.
14. Use the first theorem of Pappus to calculate the volume of the solid generated by rotating the triangle with vertices $(-1, 2)$, $(1, 2)$, and $(0, 4)$ about the x-axis.
15. Use the first theorem of Pappus to calculate the volume of the "elliptical torus" generated by rotating the ellipse $(x^2/a^2) + (y^2/b^2) = 1$ about the line $y = 3a$. Assume that $3a > b > 0$.

19.4 DOUBLE INTEGRALS IN POLAR COORDINATES

In this section we will see how to evaluate double integrals of functions in the form $z = f(r, \theta)$, where r and θ denote the polar coordinates of a point in the plane.

Let $z = f(r, \theta)$ and let Ω denote the "polar rectangle"

$$\theta_1 \le \theta \le \theta_2, \qquad r_1 \le r \le r_2. \tag{1}$$

This region is sketched in Figure 1. We will calculate the volume of the solid between the surface $z = f(r, \theta)$ and the region Ω. We partition Ω by small "polar rectangles"

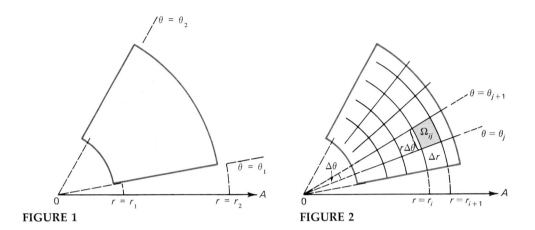

FIGURE 1   **FIGURE 2**

and calculate the volume over such a region (see Figure 2). The volume of the part of the solid over the region Ω_{ij} is given, approximately, by

$$f(r_i,\ \theta_j)A_{ij},\tag{2}$$

where A_{ij} is the area of Ω_{ij}. Recall from Section 11.5 that if $r = f(\theta)$, then the area bounded by the lines $\theta = \alpha$, $\theta = \beta$, and the curve $r = f(\theta)$ is given by

$$A = \int_{\alpha}^{\beta} \frac{1}{2}[f(\theta)]^2\ d\theta.\tag{3}$$

Thus

$$A_{ij} = \frac{1}{2}\int_{\theta_j}^{\theta_{j+1}} (r_{i+1}^2 - r_i^2)\ d\theta$$

$$= \frac{1}{2}(r_{i+1}^2 - r_i^2)\ \theta\Big|_{\theta_j}^{\theta_{j+1}} = \frac{1}{2}(r_{i+1}^2 - r_i^2)(\theta_{j+1} - \theta_j)$$

$$= \frac{1}{2}(r_{i+1} + r_i)(r_{i+1} - r_i)(\theta_{j+1} - \theta_j)$$

$$= \frac{1}{2}(r_{i+1} + r_i)\ \Delta r\ \Delta\theta.$$

But if Δr is small, then $r_{i+1} \approx r_i$, and we have

$$A_{ij} \approx \tfrac{1}{2}(2r_i)\ \Delta r\ \Delta\theta = r_i\ \Delta r\ \Delta\theta.$$

Then

$$V_{ij} \approx f(r_i,\ \theta_j)A_{ij} \approx f(r_i,\ \theta_j)r_i\ \Delta r\ \Delta\theta,$$

so that, adding up the individual volumes and taking a limit, we obtain

$$V = \iint_\Omega f(r, \theta)r \, dr \, d\theta. \tag{4}$$

NOTE. Do not forget the extra r in the above formula.

EXAMPLE 1 Find the volume enclosed by the sphere $x^2 + y^2 + z^2 = a^2$.

 Solution. We will calculate the volume enclosed by the hemisphere $z = \sqrt{a^2 - x^2 - y^2}$ and then multiply by two. To do so, we first note that this volume is the volume of the solid under the hemisphere and above the disk $x^2 + y^2 \le a^2$. We use polar coordinates, since in polar coordinates $x^2 + y^2 = (r \cos \theta)^2 + (r \sin \theta^2) = r^2$. On the disk, $0 \le r \le a$ and $0 \le \theta \le 2\pi$ (see Figure 3). Then $z = \sqrt{a^2 - (x^2 + y^2)} = \sqrt{a^2 - r^2}$, so by (4)

$$V = \int_0^{2\pi} \int_0^a \sqrt{a^2 - r^2}\, r \, dr \, d\theta = \int_0^{2\pi} \left\{ -\frac{1}{3}(a^2 - r^2)^{3/2} \Big|_0^a \right\} d\theta$$

$$= \int_0^{2\pi} \frac{1}{3} a^3 \, d\theta = \frac{2\pi a^3}{3}.$$

Thus the volume of the sphere is $2(2\pi a^3/3) = (4/3)\pi a^3$. ∎

EXAMPLE 2 Find the volume of the solid bounded above by the surface $z = 3 + r$ and below by the region enclosed by the cardioid $r = 1 + \sin \theta$.

 Solution. The cardioid is sketched in Figure 4 and can be described by

$$\Omega = \{(r, \theta): 0 \le \theta \le 2\pi \text{ and } 0 \le r \le 1 + \sin \theta\}.$$

Then from (4)

$$V = \int_0^{2\pi} \int_0^{1 + \sin \theta} (3 + r)r \, dr \, d\theta = \int_0^{2\pi} \left\{ \left(\frac{3r^2}{2} + \frac{r^3}{3} \right) \Big|_0^{1 + \sin \theta} \right\} d\theta$$

$$= \int_0^{2\pi} \left\{ \frac{3}{2}(1 + 2 \sin \theta + \sin^2 \theta) + \frac{1}{3}(1 + 3 \sin \theta + 3 \sin^2 \theta + \sin^3 \theta) \right\} d\theta$$

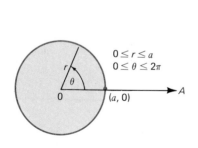

$$0 \le r \le a$$
$$0 \le \theta \le 2\pi$$

FIGURE 3

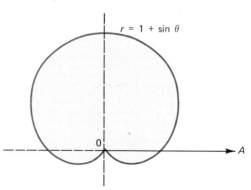

$r = 1 + \sin \theta$

FIGURE 4

$$= \int_0^{2\pi} \left(\frac{11}{6} + 4 \sin \theta + \frac{5}{2} \sin^2 \theta + \frac{\sin^3 \theta}{3} \right) d\theta$$

$$= \int_0^{2\pi} \left\{ \frac{11}{6} + 4 \sin \theta + \frac{5}{4}(1 - \cos 2\theta) + \frac{\sin \theta}{3}(1 - \cos^2 \theta) \right\} d\theta$$

$$= \left(\frac{11}{6}\theta - 4 \cos \theta + \frac{5}{4}\theta - \frac{5}{8}\sin 2\theta - \frac{\cos \theta}{3} + \frac{\cos^3 \theta}{9} \right)\bigg|_0^{2\pi} = \frac{37}{6}\pi. \ \blacksquare$$

REMARK.   We can also use this technique to calculate areas. If $f(r, \theta) = 1$, then

$$\iint_\Omega r \, dr \, d\theta = \text{area of } \Omega. \tag{5}$$

(See Remark 3 following Theorem 19.1.2.)

EXAMPLE 3   Calculate the area enclosed by the cardioid $r = 1 + \sin \theta$.

Solution.

$$A = \int_0^{2\pi} \int_0^{1+\sin \theta} r \, dr \, d\theta = \int_0^{2\pi} \left\{ \frac{r^2}{2}\bigg|_0^{1+\sin \theta} \right\} d\theta$$

$$= \frac{1}{2} \int_0^{2\pi} (1 + 2 \sin \theta + \sin^2 \theta) \, d\theta = \frac{1}{2} \int_0^{2\pi} \left(1 + 2 \sin \theta + \frac{1 - \cos 2\theta}{2} \right) d\theta$$

$$= \frac{1}{2} \left(\frac{3\theta}{2} - 2 \cos \theta - \frac{\sin 2\theta}{4} \right)\bigg|_0^{2\pi} = \frac{3\pi}{2} \ \blacksquare$$

As we will see, it is often very useful to write a double integral in terms of polar coordinates. Let $z = f(x, y)$ be a function defined over a region Ω. Then using polar coordinates, we can write

$$z = f(r \cos \theta, r \sin \theta), \tag{6}$$

and we can also describe Ω in terms of polar coordinates. The volume of the solid under f and over Ω is the same whether we use rectangular or polar coordinates. Thus writing the volume in both rectangular and polar coordinates, we obtain the useful **change-of-variables formula**

$$\iint_\Omega f(x, y) \, dA = \iint_\Omega f(r \cos \theta, r \sin \theta) r \, dr \, d\theta. \tag{7}$$

EXAMPLE 4   The density at any point on a semicircular plane lamina is proportional to the square of the distance from the point to the center of the circle. Find the mass of the lamina.

Solution.   We have $\rho(x, y) = \alpha(x^2 + y^2) = \alpha r^2$ and $\Omega = \{(r, \theta): 0 \le r \le a \text{ and } 0 \le \theta \le \pi\}$, where a is the radius of the circle. Then

$$\mu = \int_0^\pi \int_0^a (\alpha r^2) r \, dr \, d\theta = \int_0^\pi \left\{ \frac{\alpha r^4}{4}\bigg|_0^a \right\} d\theta = \frac{\alpha a^4}{4} \int_0^\pi d\theta = \frac{\alpha \pi a^4}{4}.$$

This double integral can be computed without using polar coordinates, but the computation is much more tedious. Try it! ■

EXAMPLE 5 Find the volume of the solid bounded by the xy-plane, the cylinder $x^2 + y^2 = 4$, and the paraboloid $z = 2(x^2 + y^2)$.

Solution. The volume requested is the volume under the surface $z = 2(x^2 + y^2) = 2r^2$ and above the circle $x^2 + y^2 = 4$. Thus

$$V = \int_0^{2\pi} \int_0^2 2r^2 \cdot r \, dr \, d\theta = \int_0^{2\pi} \left\{ \frac{r^4}{2} \Big|_0^2 \right\} d\theta = 16\pi. \quad ■$$

EXAMPLE 6 In probability theory one of the most important integrals that is encountered is the integral

$$\int_{-\infty}^{\infty} e^{-x^2} \, dx.$$

We now show how a combination of double integrals and polar coordinates can be used to evaluate it. Let

$$I = \int_0^{\infty} e^{-x^2} \, dx.$$

Then by symmetry $\int_{-\infty}^{\infty} e^{-x^2} \, dx = 2I$. Thus we need only to evaluate I. But since any dummy variable can be used in a definite integral, we also have

$$I = \int_0^{\infty} e^{-y^2} \, dy.$$

Thus

$$I^2 = \left(\int_0^{\infty} e^{-x^2} \, dx \right)\left(\int_0^{\infty} e^{-y^2} \, dy \right),$$

and from the result of Problem 19.2.47,†

$$I^2 = \int_0^{\infty} \int_0^{\infty} e^{-x^2} e^{-y^2} \, dx \, dy = \int_0^{\infty} \int_0^{\infty} e^{-(x^2+y^2)} \, dx \, dy = \iint_{\Omega} e^{-(x^2+y^2)} \, dA,$$

where Ω denotes the first quadrant. In polar coordinates the first quadrant can be written as

---

†Problem 19.2.47 really does not apply to improper integrals such as this one. The answer we obtain is correct, but additional theory is needed to justify this next step. The needed result is best left to a course in advanced calculus.

$$\Omega = \left\{ (r, \theta): 0 \le r < \infty \text{ and } 0 \le \theta \le \frac{\pi}{2} \right\}.$$

Thus since $x^2 + y^2 = r^2$, we obtain

$$I^2 = \int_0^{\pi/2} \int_0^\infty e^{-r^2} r \, dr \, d\theta = \int_0^{\pi/2} \left(\lim_{N \to \infty} \int_0^N e^{-r^2} r \, dr \right) d\theta$$

$$= \int_0^{\pi/2} \left(\lim_{N \to \infty} -\frac{1}{2} e^{-r^2} \Big|_0^N \right) d\theta = \frac{1}{2} \int_0^{\pi/2} d\theta = \frac{\pi}{4}.$$

Hence $I^2 = \pi/4$, so $I = \sqrt{\pi}/2$, and

$$\int_{-\infty}^\infty e^{-x^2} \, dx = 2I = \sqrt{\pi}.$$

By making the substitution $u = x/\sqrt{2}$, it is easy to show that

$$\int_{-\infty}^\infty e^{-x^2/2} \, dx = \sqrt{2\pi}.$$

The function $\rho(x) = (1/\sqrt{2\pi})e^{-x^2/2}$ is called the **density function for the unit normal distribution.** We have just shown that $\int_{-\infty}^\infty \rho(x) \, dx = 1$. ∎

EXAMPLE 7 Find the volume of the solid bounded by the circular paraboloid $x = y^2 + z^2$ and the plane $x = 1$.

Solution. This problem is the same as the one in Example 19.2.10. In that example we found that

$$V = \iint_\Omega (1 - y^2 - z^2) \, dy \, dz,$$

where Ω is the circle $y^2 + z^2 = 1$. But this integral is easily evaluated by using polar coordinates (with y and z in place of x and y in the polar coordinate formulas). We have

$$V = \int_0^{2\pi} \int_0^1 (1 - r^2) r \, dr \, d\theta$$

$$= \int_0^{2\pi} \left\{ \left(\frac{r^2}{2} - \frac{r^4}{4} \right) \Big|_0^1 \right\} d\theta = \int_0^{2\pi} \frac{1}{4} d\theta = \frac{\pi}{2}. ∎$$

PROBLEMS 19.4

In Problems 1–5, calculate the volume under the given surface that lies over the given region Ω.

1. $z = r$; Ω is the circle of radius a.

2. $z = r^n$; n is a positive integer; Ω is the circle of radius a.

3. $z = 3 - r$; Ω is the circle $r = 2 \cos \theta$.
4. $z = r^2$; Ω is the cardioid $r = 4(1 - \cos \theta)$.
5. $z = r^3$; Ω is the region enclosed by the spiral of Archimedes $r = a\theta$ and the polar axis for θ between 0 and 2π.

In Problems 6–15, calculate the area of the region enclosed by the given curve or curves.

6. $r = 1 - \cos \theta$ **7.** $r = 4(1 + \cos \theta)$
8. $r = 1 + 2 \cos \theta$ (outer loop)   **9.** $r = 3 - 2 \sin \theta$
10. $r^2 = \cos 2\theta$ **11.** $r^2 = 4 \sin 2\theta$
12. $r = a + b \sin \theta$, $a > b > 0$ **13.** $r = \tan \theta$ and the line $\theta = \pi/4$
14. Outside the circle $r = 6$ and inside the cardioid $r = 4(1 + \sin \theta)$.
15. Inside the cardioid $r = 2(1 + \cos \theta)$ but outside the circle $r = 2$.

16. Find the volume of the solid bounded above by the sphere $x^2 + y^2 + z^2 = 4a^2$, below by the xy-plane, and on the sides by the cylinder $x^2 + y^2 = a^2$.
17. Find the area of the region interior to the curve $(x^2 + y^2)^3 = 9y^2$.
18. Find the volume of the solid bounded by the cone $x^2 + y^2 = z^2$ and the cylinder $x^2 + y^2 = 4y$.
19. Find the volume of the solid bounded by the cone $z^2 = x^2 + y^2$ and the paraboloid $2z = x^2 + y^2$.
20. Find the volume of the solid bounded by the cylinder $x^2 + y^2 = 9$ and the hyperboloid $x^2 + y^2 - z^2 = 1$.
21. Find the volume of the solid centered at the origin that is bounded above by the surface $z = e^{-(x^2+y^2)}$ and below by the unit circle.
22. Find the centroid of the region bounded by $r = \cos \theta + 2 \sin \theta$.
23. Find the centroid of the region bounded by the limaçon $r = 3 + \sin \theta$.
24. Find the centroid of the region bounded by the limaçon $r = a + b \cos \theta$, $a > b > 0$.
25. Show that $\iint_\Omega 1/(1 + x^2 + y^2) \, dA = \pi \ln 2$, where Ω is the unit disk.

19.5 SURFACE AREA

In Section 9.2 we found a formula for the length of a plane curve [given by $y = f(x)$] by showing that the length of a small "piece" of the curve was approximately equal to $\sqrt{1 + [f'(x)]^2} \, \Delta x$. We now define the area of a surface $z = f(x, y)$ that lies over a region Ω in the xy-plane. We define it by analogy with the arc length formula

Definition 1 LATERAL SURFACE AREA Let f be continuous with continuous partial derivatives in the region Ω in the xy-plane. Then the **lateral surface area** σ of the graph of f over Ω is defined by

$$\sigma = \iint_\Omega \sqrt{1 + f_x^2(x, y) + f_y^2(x, y)} \, dA. \tag{1}$$

REMARK 1. The assumption that f is continuously differentiable over Ω ensures that the integral in (1) exists.

REMARK 2. We will show why formula (1) makes intuitive sense at the end of this section.

REMARK 3. We emphasize that formula (1) is a definition. A definition is not something that we have to prove, of course, but it is something that we have to live with. It will be comforting, therefore, to use our definition to evaluate areas where we know what we want the answer to be.

EXAMPLE 1 Calculate the lateral surface area cut from the cylinder $y^2 + z^2 = 9$ by the planes $x = 0$ and $x = 4$.

Solution. The surface is a right circular cylinder with height 4 and radius 3 whose axis lies along the x-axis. Thus

$$S = 2\pi rh = 24\pi.$$

We now solve the problem by using formula (1). The surface area consists of two equal parts, one for $z > 0$ and one for $z < 0$. We calculate the surface area for $z > 0$ and multiply the result by 2. We have, for $z > 0$, $z = f(x, y) = \sqrt{9 - y^2}$, so $f_x = 0$, $f_y = -y/\sqrt{9 - y^2}$, and $f_y{}^2 = y^2/(9 - y^2)$. Thus since $0 \le x \le 4$ and $-3 \le y \le 3$, we have

$$S = 2\int_0^4 \int_{-3}^3 \sqrt{1 + \frac{y^2}{9 - y^2}}\, dy\, dx = 4\int_0^4 \int_0^3 \sqrt{\frac{9}{9 - y^2}}\, dy\, dx$$

$$= 12\int_0^4 \left(\int_0^3 \frac{dy}{\sqrt{9 - y^2}}\right) dx.$$

Setting $y = 3\sin\theta$, we have

$$S = 12\int_0^4 \left(\int_0^{\pi/2} \frac{3\cos\theta}{3\cos\theta}\, d\theta\right) dx = 12\int_0^4 \frac{\pi}{2}\, dx = 12(4)\left(\frac{\pi}{2}\right) = 24\pi. \quad\blacksquare$$

We now compute surface area in problems where we don't know the answer in advance.

EXAMPLE 2 Calculate the lateral surface area cut from the cylinder $y^2 + z^2 = 9$ by the planes $x = 0$, $x = 1$, $y = 0$, and $y = 2$.

Solution. The required area is sketched in Figure 1. It consists of two equal parts: one for $z > 0$ and one for $z < 0$. We will calculate the surface area for $z > 0$ and multiply the result by 2. For $z > 0$, we have $z = \sqrt{9 - y^2}$. Ω is the rectangle $\{(x, y): 0 \le x \le 1 \text{ and } 0 \le y \le 2\}$. Then

$$f_x = 0 \quad\text{and}\quad f_y = -\frac{y}{\sqrt{9 - y^2}},$$

so that

$$\sigma = \int_0^2 \int_0^1 \sqrt{1 + 0^2 + \left(\frac{-y}{\sqrt{9 - y^2}}\right)^2}\, dx\, dy = \int_0^2 \int_0^1 \sqrt{1 + \frac{y^2}{9 - y^2}}\, dx\, dy$$

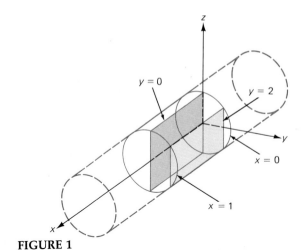

FIGURE 1

$$= 3 \int_0^2 \int_0^1 \frac{1}{\sqrt{9 - y^2}} \, dx \, dy = 3 \int_0^2 \left(\frac{x}{\sqrt{9 - y^2}} \Big|_0^1 \right) dy = 3 \int_0^2 \frac{dy}{\sqrt{9 - y^2}}$$

$$= 3 \sin^{-1} \frac{y}{3} \Big|_0^2 = 3 \sin^{-1} \frac{2}{3}.$$

Thus the total surface area is $6 \sin^{-1} \frac{2}{3} \approx 4.38$. ∎

EXAMPLE 3   Find the lateral surface area of the circular paraboloid $z = x^2 + y^2$ between the xy-plane and the plane $z = 9$.

Solution. The surface area requested is sketched in Figure 2. The region Ω is the disk $x^2 + y^2 \leq 9$. We have

$$f_x = 2x \qquad \text{and} \qquad f_y = 2y,$$

so

$$\sigma = \iint_\Omega \sqrt{1 + 4x^2 + 4y^2} \, dA.$$

Clearly, this problem calls for the use of polar coordinates. We have

$$\sigma = \int_0^{2\pi} \int_0^3 \sqrt{1 + 4r^2} \, r \, dr \, d\theta = \int_0^{2\pi} \left\{ \frac{1}{12} (1 + 4r^2)^{3/2} \Big|_0^3 \right\} d\theta$$

$$= \frac{\pi}{6} (37^{3/2} - 1) \approx 117.3$$ ∎

EXAMPLE 4   Calculate the area of the part of the surface $z = x^3 + y^4$ that lies over the square $\{(x, y): 0 \leq x \leq 1, 0 \leq y \leq 1\}$.

Solution. $f_x = 3x^2$ and $f_y = 4y^3$, so

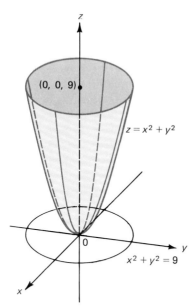

FIGURE 2

$$\sigma = \int_0^1 \int_0^1 \sqrt{1 + 9x^4 + 16y^6} \ dx \ dy.$$

However, this is as far as we can go unless we resort to numerical techniques to approximate this double integral. As with ordinary definite integrals, many double integrals cannot be integrated in terms of functions that we know. However, there are a great number of techniques for approximating a double integral numerically that parallel the techniques discussed in Section 8.10. ■

Derivation of the Formula for Surface Area. We begin by calculating the surface area $\Delta\sigma$ over a rectangle ΔR with sides Δx and Δy. The situation is depicted in Figure 3, in which it is assumed that $f(x, y) > 0$ for (x, y) in Ω. We assume that f has continuous partial derivatives over R. If Δx and Δy are small, then the region $PQSR$ in space has, approximately, the shape of a parallelogram. Thus by equation 17.4.6,

$$\Delta\sigma \approx \text{area of parallelogram} = |\overrightarrow{PQ} \times \overrightarrow{PR}|. \tag{2}$$

Now

$$\begin{aligned}
\overrightarrow{PQ} &= (x + \Delta x, y, f(x + \Delta x, y)) - (x, y, f(x, y)) \\
&= (\Delta x, 0, f(x + \Delta x, y) - f(x, y)).
\end{aligned}$$

But if Δx is small, then

$$\frac{f(x + \Delta x, y) - f(x, y)}{\Delta x} \approx f_x(x, y),$$

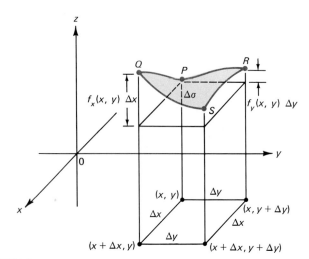

FIGURE 3

so

$$f(x + \Delta x, y) - f(x, y) \approx f_x(x, y) \, \Delta x$$

and

$$\overrightarrow{PQ} \approx (\Delta x, 0, f_x(x, y) \, \Delta x). \tag{3}$$

Similarly,

$$\overrightarrow{PR} = (x, y + \Delta y, f(x, y + \Delta y)) - (x, y, f(x, y))$$
$$= (0, \Delta y, f(x, y + \Delta y) - f(x, y)),$$

and if Δy is small,

$$\overrightarrow{PR} \approx (0, \Delta y, f_y(x, y) \, \Delta y). \tag{4}$$

Thus from (3) and (4),

$$\overrightarrow{PQ} \times \overrightarrow{PQ} \approx \begin{vmatrix} \mathbf{i} & \mathbf{j} & \mathbf{k} \\ \Delta x & 0 & f_x(x, y) \, \Delta x \\ 0 & \Delta y & f_y(x, y) \, \Delta y \end{vmatrix}$$
$$= -f_x(x, y) \, \Delta x \, \Delta y \mathbf{i} - f_y(x, y) \, \Delta x \, \Delta y \mathbf{j} + \Delta x \, \Delta y \mathbf{k}$$
$$= (-f_x(x, y)\mathbf{i} - f_y(x, y)\mathbf{j} + \mathbf{k}) \, \Delta x \, \Delta y,$$

so that from (2)

$$\Delta \sigma \approx \sqrt{f_x^2(x, y) + f_y^2(x, y) + 1} \overbrace{\Delta x \, \Delta y}^{= \Delta A}. \tag{5}$$

Finally, adding up the surface area over rectangles that partition Ω and taking a limit yields

$$\sigma = \iint_\Omega \sqrt{1 + f_x^{\,2}(x, y) + f_y^{\,2}(x, y)}\, dA. \quad\blacksquare \tag{1}$$

REMARK. We caution that equation (1) has *not* been proved. However, if f has continuous partial derivatives, it certainly is plausible.

PROBLEMS 19.5

In Problems 1–9, find the area of the part of the surface that lies over the given region.

1. $z = x + 2y$; $\Omega = \{(x, y): 0 \le x \le y, 0 \le y \le 2\}$
2. $z = 4x + 7y$; $\Omega =$ region between $y = x^2$ and $y = x^5$
3. $z = ax + by$; $\Omega =$ upper half of unit circle
4. $z = y^2$; $\Omega = \{(x, y): 0 \le x \le 2, 0 \le y \le 4\}$
*5. $z = 3 + x^{2/3}$; $\Omega = \{(x, y): -1 \le x \le 1, 1 \le y \le 2\}$
6. $z = (x^4/4) + (1/8x^2)$; $\Omega = \{(x, y): 1 \le x \le 2, 0 \le y \le 5\}$
7. $z = \frac{1}{3}(y^2 + 2)^{3/2}$; $\Omega = \{(x, y): -4 \le x \le 7, 0 \le y \le 3\}$
8. $z = 2\ln(1 + y)$; $\Omega = \{(x, y): 0 \le x \le 2, 0 \le y \le 1\}$
*9. $(z + 1)^2 = 4x^3$; $\Omega = \{(x, y): 0 \le x \le 1, 0 \le y \le 2\}$

*10. Calculate the lateral surface area of the cylinder $y^{2/3} + z^{2/3} = 1$ for x in the interval $[0, 2]$.
*11. Find the surface area of the hemisphere $x^2 + y^2 + z^2 = a^2$, $z \ge 0$.
*12. Find the surface area of the part of the sphere $x^2 + y^2 + z^2 = a^2$ that is also inside the cylinder $x^2 + y^2 = ay$.
*13. Find the area of the surface in the first octant cut from the cylinder $x^2 + y^2 = 16$ by the plane $y = z$.
14. Find the area of the portion of the sphere $x^2 + y^2 + z^2 = 16z$ lying within the circular paraboloid $z = x^2 + y^2$.
15. Find the area of the surface cut from the hyperbolic paraboloid $4z = x^2 - y^2$ by the cylinder $x^2 + y^2 = 16$.
*16. Let $z = f(x, y)$ be the equation of a plane (i.e., $z = ax + by + c$). Show that over the region Ω, the area of the plane is given by

$$\sigma = \iint_\Omega \sec \gamma\, dA,$$

where γ is the angle between the normal vector $\mathbf{N}$ to the plane and the positive z-axis. [*Hint:* Show, using the dot product, that

$$\cos \gamma = \frac{\mathbf{N} \cdot \mathbf{k}}{|\mathbf{N}|} = \frac{1}{\sqrt{1 + a^2 + b^2}}.]$$

In Problems 17–20, find a double integral that represents the area of the given surface over the given region. Do *not* try to evaluate the integral.

17. $z = x^3 + y^3$; Ω is the unit circle.
18. $z = \ln(x + 2y)$; $\Omega = \{(x, y): 0 \le x \le 1, 0 \le y \le 4\}$.
19. $z = \sqrt{1 + x + y}$; Ω is the triangle bounded by $y = x$, $y = 4 - x$, and the y-axis.
20. $z = e^{x-y}$; Ω is the ellipse $4x^2 + 9y^2 = 36$.

*21. Find a double integral that represents the surface area of the ellipsoid $(x^2/a^2) + (y^2/b^2) + (z^2/c^2) = 1$.

19.6 THE TRIPLE INTEGRAL

In this section we discuss the idea behind the triple integral of a function of three variables $f(x, y, z)$ over a region S in $\mathbb{R}^3$. This is really a simple extension of the double integral. For that reason we will omit a number of technical details.

We start with a parallelepiped π in $\mathbb{R}^3$, which can be written as

$$\pi = \{(x, y, z): a_1 \le x \le a_2, b_1 \le y \le b_2, c_1 \le z \le c_2\} \tag{1}$$

and is sketched in Figure 1. We construct regular partitions of the three intervals $[a_1, a_2]$, $[b_1, b_2]$, and $[c_1, c_2]$:

$$a_1 = x_0 < x_1 < \cdots < x_n = a_2$$
$$b_1 = y_0 < y_1 < \cdots < y_m = b_2$$
$$c_1 = z_0 < z_1 < \cdots < z_p = c_2,$$

to obtain nmp "boxes." The volume of a typical box B_{ijk} is given by

$$\Delta V = \Delta x \, \Delta y \, \Delta z, \tag{2}$$

where $\Delta x = x_i - x_{i-1}$, $\Delta y = y_j - y_{j-1}$, and $\Delta z = z_k - z_{k-1}$. We then form the sum

$$\sum_{i=1}^{n} \sum_{j=1}^{m} \sum_{k=1}^{p} f(x_i^*, y_j^*, z_k^*) \, \Delta V, \tag{3}$$

where (x_i^*, y_j^*, z_k^*) is in B_{ijk}.

We now define

$$\Delta u = \sqrt{\Delta x^2 + \Delta y^2 + \Delta z^2}.$$

Geometrically, Δu is the length of a diagonal of a rectangular solid with sides having lengths Δx, Δy, and Δz, respectively. We see that as $\Delta u \to 0$, Δx, Δy, and Δz approach 0, and the volume of each box tends to zero. We then take the limit as $\Delta u \to 0$.

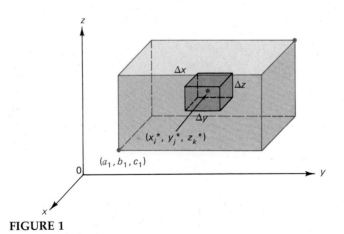

FIGURE 1

Definition 1   THE TRIPLE INTEGRAL   Let $w = f(x, y, z)$ and let the parallelepiped π be given by (1). Suppose that

$$\lim_{\Delta u \to 0} \sum_{i=1}^{n} \sum_{j=1}^{m} \sum_{k=1}^{p} f(x_i^*, y_j^*, z_k^*) \, \Delta x \, \Delta y \, \Delta z$$

exists and is independent of the way in which the points (x_i^*, y_j^*, z_k^*) are chosen. Then the **triple integral** of f over π, written $\iiint_\pi f(x, y, z) \, dV$, is defined by

$$\iiint_\pi f(x, y, z) \, dV = \lim_{\Delta u \to 0} \sum_{i=1}^{n} \sum_{j=1}^{m} \sum_{k=1}^{p} f(x_i^*, y_j^*, z_k^*) \, \Delta V \tag{4}$$

As with double integrals, we can write triple integrals as iterated (or repeated) integrals. If π is defined by (1), we have

$$\iiint_\pi f(x, y, z) \, dV = \int_{a_1}^{a_2} \int_{b_1}^{b_2} \int_{c_1}^{c_2} f(x, y, z) \, dz \, dy \, dx. \tag{5}$$

EXAMPLE 1   Evaluate $\iiint_\pi xy \cos yz \, dV$, where π is the parallelepiped

$$\left\{(x, y, z): 0 \le x \le 1, 0 \le y \le 1, 0 \le z \le \frac{\pi}{2}\right\}.$$

Solution.

$$\iiint_\pi xy \cos yz \, dV = \int_0^1 \int_0^1 \int_0^{\pi/2} xy \cos yz \, dz \, dy \, dx$$

$$= \int_0^1 \int_0^1 \left\{ xy \cdot \frac{1}{y} \sin yz \Big|_0^{\pi/2} \right\} dy \, dx = \int_0^1 \int_0^1 x \sin\left(\frac{\pi}{2} y\right) dy \, dx$$

$$= \int_0^1 \left\{ -\frac{2}{\pi} x \cos \frac{\pi}{2} y \Big|_0^1 \right\} dx = \int_0^1 \frac{2}{\pi} x \, dx = \frac{x^2}{\pi} \Big|_0^1 = \frac{1}{\pi} \quad \blacksquare$$

THE TRIPLE INTEGRAL OVER A MORE GENERAL REGION

We now define the triple integral over a more general region S. We assume that S is bounded. Then we can enclose S in a parallelepiped π and define a new function F by

$$F(x, y, z) = \begin{cases} f(x, y, z), & \text{if } (x, y, z) \text{ is in } S \\ 0, & \text{if } (x, y, z) \text{ is in } \pi \text{ but not in } S. \end{cases}$$

We then define

$$\iiint_S f(x, y, z) \, dV = \iiint_\pi F(x, y, z) \, dV. \tag{6}$$

REMARK 1.   If f is continuous over S and if S is a region of the type we will discuss below, then $\iiint_S f(x, y, z)\, dV$ will exist. The proof of this fact is beyond the scope of this text but can be found in any advanced calculus text.

REMARK 2. If $f \geq 0$ on S, then the triple integral $\iiint_S f(x, y, z)\, dV$ represents the "volume" in four-dimensional space $\mathbb{R}^4$ of the region bounded above by f and below by S. We cannot, of course, draw this volume, but otherwise the theory of volumes carries over to four (and more) dimensions.

Now let S take the form

$$S = \{(x, y, z): a_1 \leq x \leq a_2,\ g_1(x) \leq y \leq g_2(x),\ h_1(x, y) \leq z \leq h_2(x, y)\}. \tag{7}$$

What does such a solid look like? We first note that the equations $z = h_1(x, y)$ and $z = h_2(x, y)$ are the equations of surfaces in $\mathbb{R}^3$. The equations $y = g_1(x)$ and $y = g_2(x)$ are equations of cylinders in $\mathbb{R}^3$, and the equations $x = a_1$ and $x = a_2$ are equations of planes in $\mathbb{R}^3$. The solid S is sketched in Figure 2. We assume that $g_1, g_2, h_1,$ and h_2 are continuous. If f is continuous, then $\iiint_S f(x, y, z)\, dV$ will exist and

$$\iiint_S f(x, y, z)\, dV = \int_{a_1}^{a_2} \int_{g_1(x)}^{g_2(x)} \int_{h_1(x, y)}^{h_2(x, y)} f(x, y, z)\, dz\, dy\, dx. \tag{8}$$

Note the similarity between equation (8) and equation (19.2.10).

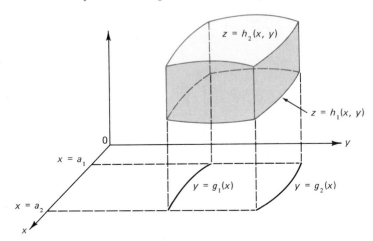

FIGURE 2

EXAMPLE 2 Evaluate $\iiint_S 2x^3 y^2 z\, dV$, where S is the region

$$\{(x, y, z): 0 \leq x \leq 1,\ x^2 \leq y \leq x,\ x - y \leq z \leq x + y\}.$$

Solution.

$$\iiint_S 2x^3 y^2 z\, dV = \int_0^1 \int_{x^2}^x \int_{x-y}^{x+y} 2x^3 y^2 z\, dz\, dy\, dx$$

$$= \int_0^1 \int_{x^2}^x \left\{ x^3 y^2 z^2 \Big|_{x-y}^{x+y} \right\} dy\, dx$$

$$= \int_0^1 \int_{x^2}^x x^3 y^2 [(x + y)^2 - (x - y)^2] \, dy \, dx$$

$$= \int_0^1 \int_{x^2}^x 4x^4 y^3 \, dy \, dx = \int_0^1 \left\{ x^4 y^4 \Big|_{x^2}^x \right\} dx$$

$$= \int_0^1 (x^8 - x^{12}) \, dx = \frac{1}{9} - \frac{1}{13} = \frac{4}{117}. \quad \blacksquare$$

Many of the applications we saw for the double integral can be extended to the triple integral. We present three of them below.

I. VOLUME

Let the region S be defined by (7). Then, since $\Delta V = \Delta x \, \Delta y \, \Delta z$ represents the volume of a "box" in S, when we add up the volumes of these boxes and take a limit, we obtain the total volume of S. That is,

$$\text{volume of } S = \iiint_S dV. \tag{9}$$

EXAMPLE 3 Calculate the volume of the region of Example 2.

Solution.

$$V = \int_0^1 \int_{x^2}^x \int_{x-y}^{x+y} dz \, dy \, dx = \int_0^1 \int_{x^2}^x \left\{ z \Big|_{x-y}^{x+y} \right\} dy \, dx$$

$$= \int_0^1 \int_{x^2}^x 2y \, dy \, dx = \int_0^1 \left\{ y^2 \Big|_{x^2}^x \right\} dx = \int_0^1 (x^2 - x^4) \, dx$$

$$= \frac{1}{3} - \frac{1}{5} = \frac{2}{15} \quad \blacksquare$$

EXAMPLE 4 Find the volume of the tetrahedron formed by the planes $x = 0$, $y = 0$, $z = 0$, and $x + (y/2) + (z/4) = 1$.

Solution. The tetrahedron is sketched in Figure 3. We see that z ranges from 0 to the plane $x + (y/2) + (z/4) = 1$, or $z = 4[1 - x - (y/2)]$. This last plane intersects the xy-plane in a line whose equation (obtained by setting $z = 0$) is given by $0 = 1 - x - (y/2)$ or $y = 2(1 - x)$, so that y ranges from 0 to $2(1 - x)$ (see Figure 4). Finally, this line intersects the x-axis at the point $(1, 0, 0)$, so that x ranges from 0 to 1, and we have

$$V = \int_0^1 \int_0^{2(1-x)} \int_0^{4(1-x-y/2)} dz \, dy \, dx = \int_0^1 \int_0^{2(1-x)} 4\left(1 - x - \frac{y}{2}\right) dy \, dx$$

$$= 4 \int_0^1 \left\{ \left(y(1 - x) - \frac{y^2}{4} \right) \Big|_0^{2(1-x)} \right\} dx = 4 \int_0^1 [2(1 - x)^2 - (1 - x)^2] \, dx$$

$$= -\frac{4}{3}(1 - x)^3 \Big|_0^1 = \frac{4}{3}.$$

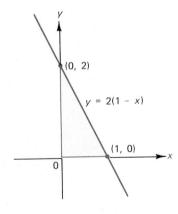

FIGURE 3   **FIGURE 4**

REMARK.   It was not necessary to integrate in the order z, then y, then x. We could have written, for example,

$$0 \le x \le 1 - \frac{y}{2} - \frac{z}{4}.$$

The intersection of this plane with the yz-plane is the line $0 = 1 - (y/2) - (z/4)$, or $z = 4[1 - (y/2)]$. The intersection of this line with the y-axis occurs at the point $(0, 2, 0)$. Thus

$$V = \int_0^2 \int_0^{4(1-y/2)} \int_0^{1-y/2-z/4} dx \, dz \, dy$$

$$= \int_0^2 \int_0^{4(1-y/2)} \left(1 - \frac{y}{2} - \frac{z}{4}\right) dz \, dy = \int_0^2 \left\{ \left[\left(1 - \frac{y}{2}\right)z - \frac{z^2}{8} \right] \Big|_0^{4(1-y/2)} \right\} dy$$

$$= \int_0^2 \left\{ 4\left(1 - \frac{y}{2}\right)^2 - \frac{16[1 - (y/2)]^2}{8} \right\} dy = 2 \int_0^2 \left(1 - \frac{y}{2}\right)^2 dy$$

$$= -\frac{4}{3}\left(1 - \frac{y}{2}\right)^3 \Big|_0^2 = \frac{4}{3}.$$

We could also integrate in any of four other orders (xyz, yzx, yxz, and zxy) to obtain the same result. ■

II. DENSITY AND MASS

Let the function $\rho(x, y, z)$ denote the density (in kilograms per cubic meter, say) of a solid S in $\mathbb{R}^3$. Then for a "box" of sides Δx, Δy, and Δz, the approximate mass of the box will be equal to $\rho(x_i^*, y_j^*, z_k^*) \, \Delta x \, \Delta y \, \Delta z = \rho(x_i^*, y_j^*, z_k^*) \, \Delta V$ if Δx, Δy, and Δz are small. We then obtain

$$\text{total mass of } S = \mu(S) = \iiint_S \rho(x, y, z) \, dV \qquad\qquad \textbf{(10)}$$

EXAMPLE 5   The density of the solid of Example 2 is given by $\rho(x, y, z) = x + 2y + 4z$ kg/m³. Calculate the total mass of the solid.

Solution.

$$\mu(S) = \int_0^1 \int_{x^2}^x \int_{x-y}^{x+y} (x + 2y + 4z)\, dz\, dy\, dx = \int_0^1 \int_{x^2}^x \left\{ [(x + 2y)z + 2z^2] \Big|_{x-y}^{x+y} \right\} dy\, dx$$

$$= \int_0^1 \int_{x^2}^x (10xy + 4y^2)\, dy\, dx = \int_0^1 \left\{ 5xy^2 + \frac{4y^3}{3} \right\} \Big|_{x^2}^x dx$$

$$= \int_0^1 \left(5x^3 - 5x^5 + \frac{4}{3}x^3 - \frac{4}{3}x^6 \right) dx = \int_0^1 \left(\frac{19}{3}x^3 - 5x^5 - \frac{4}{3}x^6 \right) dx$$

$$= \frac{19}{12} - \frac{5}{6} - \frac{4}{21} = \tfrac{47}{84} \text{ kg} \quad \blacksquare$$

III. FIRST MOMENTS AND CENTER OF MASS

In $\mathbb{R}^3$ we use the symbol M_{yz} to denote the first moment with respect to the yz-plane. Similarly, M_{xz} denotes the first moment with respect to the xz-plane, and M_{xy} denotes the first moment with respect to the xy-plane. Since the distance from a point (x, y, z) to the yz-plane is x, and so on, we may use familiar reasoning to obtain

$$M_{yz} = \iiint_S x\rho(x, y, z)\, dV \tag{11}$$

$$M_{xz} = \iiint_S y\rho(x, y, z)\, dV \tag{12}$$

$$M_{xy} = \iiint_S z\rho(x, y, z)\, dV. \tag{13}$$

The **center of mass** of S is then given by

$$(\bar{x}, \bar{y}, \bar{z}) = \left(\frac{M_{yz}}{\mu}, \frac{M_{xz}}{\mu}, \frac{M_{xy}}{\mu} \right), \tag{14}$$

where μ denotes the mass of S. If $\rho(x, y, z)$ is constant, then as in $\mathbb{R}^1$ and $\mathbb{R}^2$, the center of mass is called the **centroid**.

EXAMPLE 6   Find the center of mass of the solid in Example 5.

Solution. We have already found that $\mu = \tfrac{47}{84}$. We next calculate the moments.

$$M_{yz} = \int_0^1 \int_{x^2}^x \int_{x-y}^{x+y} x(x + 2y + 4z)\, dz\, dy\, dx$$

$$= \int_0^1 \int_{x^2}^x \left\{ ((x^2 + 2xy)z + 2xz^2) \Big|_{x-y}^{x+y} \right\} dy \, dx$$

$$= \int_0^1 \int_{x^2}^x (10x^2y + 4xy^2) \, dy \, dx = \int_0^1 \left\{ \left(5x^2y^2 + \frac{4xy^3}{3} \right) \Big|_{x^2}^x \right\} dx$$

$$= \int_0^1 \left(5x^4 - 5x^6 + \frac{4}{3}x^4 - \frac{4}{3}x^7 \right) dx = 1 - \frac{5}{7} + \frac{4}{15} - \frac{1}{6} = \frac{27}{70}$$

$$M_{xz} = \int_0^1 \int_{x^2}^x \int_{x-y}^{x+y} y(x + 2y + 4z) \, dz \, dy \, dx$$

$$= \int_0^1 \int_{x^2}^x \left\{ [(xy + 2y^2)z + 2yz^2] \Big|_{x-y}^{x+y} \right\} dy \, dx$$

$$= \int_0^1 \int_{x^2}^x (10xy^2 + 4y^3) \, dy \, dx = \int_0^1 \left\{ \left(\frac{10}{3}xy^3 + y^4 \right) \Big|_{x^2}^x \right\} dx$$

$$= \int_0^1 \left(\frac{10}{3}x^4 - \frac{10}{3}x^7 + x^4 - x^8 \right) dx = \frac{2}{3} - \frac{5}{12} + \frac{1}{5} - \frac{1}{9} = \frac{61}{180}$$

$$M_{xy} = \int_0^1 \int_{x^2}^x \int_{x-y}^{x+y} z(x + 2y + 4z) \, dz \, dy \, dx$$

$$= \int_0^1 \int_{x^2}^x \left\{ \left[(x + 2y)\frac{z^2}{2} + \frac{4z^3}{3} \right] \Big|_{x-y}^{x+y} \right\} dy \, dz$$

$$= \int_0^1 \int_{x^2}^x \left[(x + 2y)(2xy) + \frac{4}{3}(6x^2y + 2y^3) \right] dy \, dx$$

$$= \int_0^1 \int_{x^2}^x \left(10x^2y + 4xy^2 + \frac{8}{3}y^3 \right) dy \, dx$$

$$= \int_0^1 \left\{ \left(5x^2y^2 + \frac{4}{3}xy^3 + \frac{2}{3}y^4 \right) \Big|_{x^2}^x \right\} dx$$

$$= \int_0^1 \left(5x^4 - 5x^6 + \frac{4}{3}x^4 - \frac{4}{3}x^7 + \frac{2}{3}x^4 - \frac{2}{3}x^8 \right) dx$$

$$= \int_0^1 \left(7x^4 - 5x^6 - \frac{4}{3}x^7 - \frac{2}{3}x^8 \right) dx = \frac{7}{5} - \frac{5}{7} - \frac{1}{6} - \frac{2}{27} = \frac{841}{1890}$$

Thus

$$(\bar{x}, \bar{y}, \bar{z}) = \left(\frac{M_{yz}}{\mu}, \frac{M_{xz}}{\mu}, \frac{M_{xy}}{\mu} \right) = \left(\frac{27/70}{47/84}, \frac{61/180}{47/84}, \frac{841/1890}{47/84} \right)$$

$$= \left(\frac{162}{235}, \frac{427}{705}, \frac{1682}{2115} \right) \approx (0.689, 0.606, 0.795).$$

Here distances are measured in meters. ■

PROBLEMS 19.6

In Problems 1–7, evaluate the repeated triple integral.

1. $\displaystyle\int_0^1 \int_0^y \int_0^x y \; dz \; dx \; dy$

2. $\displaystyle\int_0^2 \int_{-z}^z \int_{y-z}^{y+z} 2xz \; dx \; dy \; dz$

3. $\displaystyle\int_{a_1}^{a_2} \int_{b_1}^{b_2} \int_{c_1}^{c_2} dy \; dx \; dz$

4. $\displaystyle\int_0^{\pi/2} \int_0^{\pi/2} \int_0^z \sin\left(\frac{x}{z}\right) dx \; dz \; dy$

5. $\displaystyle\int_1^2 \int_{1-y}^{1+y} \int_0^{yz} 6xyz \; dx \; dz \; dy$

6. $\displaystyle\int_0^1 \int_0^{\sqrt{1-x^2}} \int_0^x yz \; dz \; dy \; dx$

7. $\displaystyle\int_0^1 \int_{-\sqrt{1-x^2}}^{\sqrt{1-x^2}} \int_{-\sqrt{1-x^2-y^2}}^{\sqrt{1-x^2-y^2}} z^2 \; dz \; dy \; dx$ [*Hint:* Use polar coordinates.]

*8. Change the order of integration in Problem 1 and write the integral in the form (a) $\int_?^? \int_?^? \int_?^? y \; dx \; dy \; dz$; (b) $\int_?^? \int_?^? \int_?^? y \; dy \; dz \; dx$. [*Hint:* Sketch the region in Problem 1 from the given limits and then find the new limits directly from the figure.]

*9. Write the integral of Problem 6 in the form (a) $\int_?^? \int_?^? \int_?^? yz \; dx \; dy \; dz$; (b) $\int_?^? \int_?^? \int_?^? yz \; dy \; dz \; dx$.

*10. Write the integral of Problem 7 in the form $\int_?^? \int_?^? \int_?^? z^2 \; dx \; dz \; dy$.

In Problems 11–18, find the volume of the given solid.

11. The tetrahedron with vertices at the points $(0, 0, 0)$, $(1, 0, 0)$, $(0, 1, 0)$, and $(0, 0, 1)$.

12. The tetrahedron with vertices at the points $(0, 0, 0)$, $(a, 0, 0)$, $(0, b, 0)$, and $(0, 0, c)$.

13. The solid in the first octant bounded by the cylinder $x^2 + z^2 = 9$, the plane $x + y = 4$, and the three coordinate planes.

*14. The solid bounded by the planes $x - 2y + 4z = 4$, $-2x + 3y - z = 6$, $x = 0$, and $y = 0$.

15. The solid bounded above by the sphere $x^2 + y^2 + z^2 = 16$ and below by the plane $z = 2$.

16. The solid bounded by the parabolic cylinder $z = 5 - x^2$ and the planes $z = y$ and $z = 2y$ that lies in the half space $y \geq 0$.

17. The solid bounded by the ellipsoid $(x^2/a^2) + (y^2/b^2) + (z^2/c^2) = 1$.

18. The solid bounded by the elliptic cylinder $9x^2 + y^2 = 9$ and the planes $z = 0$ and $x + y + 9z = 9$.

In Problems 19–23, find the mass and center of mass of the given solid having the given density.

19. the solid of Problem 11; $\rho(x, y, z) = x$

20. the solid of Problem 12; $\rho(x, y, z) = x^2 + y^2 + z^2$

21. the solid of Problem 13; $\rho(x, y, z) = z$

22. the solid of Problem 15; $\rho(x, y, z) = x + 2y + 4z$

23. the solid of Problem 17; $\rho(x, y, z) = \alpha x^2 + \beta y^2 + \gamma z^2$

24. Find the mass of a cube whose side has a length of k units if its density at any point is proportional to the distance from a given face of the cube.

25. Find the centroid of the tetrahedron of Problem 12.

26. Find the centroid of the ellipsoid of Problem 17.

27. Find the centroid of the solid of Problem 16.

28. Find the centroid of the solid of Problem 18.

29. The solid S lies in the first octant and is bounded by the planes $z = 0$, $y = 1$, $x = y$, and the hyperboloid $z = xy$. Its density is given by $\rho(x, y, z) = 1 + 2z$. (a) Find the center of mass of S. (b) Show that the center of mass lies inside S.

19.7 THE TRIPLE INTEGRAL IN CYLINDRICAL AND SPHERICAL COORDINATES

In this section we show how triple integrals can be written by using cylindrical and spherical coordinates.

I. CYLINDRICAL COORDINATES

Recall from Section 17.8 that the cylindrical coordinates of a point in $\mathbb{R}^3$ are (r, θ, z), where r and θ are the polar coordinates of the projection of the point onto the xy-plane and z is the usual z-coordinate. In order to calculate an integral of a region given in cylindrical coordinates, we go through a procedure very similar to the one we used to write double integrals in polar coordinates in Section 19.4.

Consider the "cylindrical parallelepiped" given by

$$\pi_c = \{(r, \theta, z): r_1 \leq r \leq r_2; \; \theta_1 \leq \theta \leq \theta_2; \; z_1 \leq z \leq z_2\}.$$

This solid is sketched in Figure 1. If we partition the z-axis for z in $[z_1, z_2]$, we obtain "slices." For a fixed z the area of the face of a slice of π_c is, according to equation (19.4.5), given by

$$A_i = \int_{\theta_1}^{\theta_2} \int_{r_1}^{r_2} r \, dr \, d\theta. \tag{1}$$

The volume of a slice is, by (1), given by

$$V_i = A_i \, \Delta z = \left\{ \int_{\theta_1}^{\theta_2} \int_{r_1}^{r_2} r \, dr \, d\theta \right\} \Delta z.$$

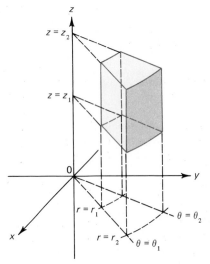

FIGURE 1

Then adding these volumes and taking the limit as before, we obtain

$$\text{volume of } \pi_c = \int_{z_1}^{z_2} \int_{\theta_1}^{\theta_2} \int_{r_1}^{r_2} r \, dr \, d\theta \, dz. \tag{2}$$

In general, let the region S be given in cylindrical coordinates by

$$S = \{(r, \theta, z): \theta_1 \leq \theta \leq \theta_2, 0 \leq g_1(\theta) \leq r \leq g_2(\theta), h_1(r, \theta) \leq z \leq h_2(r, \theta)\}, \tag{3}$$

and let f be a function of r, θ, and z. Then the triple integral of f over S is given by

$$\iiint_S f = \int_{\theta_1}^{\theta_2} \int_{g_1(\theta)}^{g_2(\theta)} \int_{h_1(r, \theta)}^{h_2(r, \theta)} f(r, \theta, z) r \, dz \, dr \, d\theta. \tag{4}$$

EXAMPLE 1   Find the mass of a solid bounded by the cylinder $r = \sin \theta$, the planes $z = 0$, $\theta = 0$, $\theta = \pi/3$, and the cone $z = r$, if the density is given by $\rho(r, \theta, z) = 4r$.

Solution. We first note that the solid may be written as

$$S = \{(r, \theta, z): 0 \leq \theta \leq \frac{\pi}{3}, 0 \leq r \leq \sin \theta, 0 \leq z \leq r\}.$$

Then

$$\mu = \int_0^{\pi/3} \int_0^{\sin \theta} \int_0^r (4r) r \, dz \, dr \, d\theta$$

$$= \int_0^{\pi/3} \int_0^{\sin \theta} \left\{ 4r^2 z \Big|_0^r \right\} dr \, d\theta = \int_0^{\pi/3} \int_0^{\sin \theta} 4r^3 \, dr \, d\theta$$

$$= \int_0^{\pi/3} \left\{ r^4 \Big|_0^{\sin \theta} \right\} d\theta = \int_0^{\pi/3} \sin^4 \theta \, d\theta = \int_0^{\pi/3} \left(\frac{1 - \cos 2\theta}{2} \right)^2 d\theta$$

$$= \frac{1}{4} \int_0^{\pi/3} \left(1 - 2 \cos 2\theta + \frac{1 + \cos 4\theta}{2} \right) d\theta$$

$$= \frac{1}{4} \left(\frac{3\theta}{2} - \sin 2\theta + \frac{\sin 4\theta}{8} \right) \Big|_0^{\pi/3} = \frac{1}{4} \left(\frac{\pi}{2} - \frac{\sqrt{3}}{2} - \frac{\sqrt{3}}{16} \right)$$

$$= \frac{\pi}{8} - \frac{9\sqrt{3}}{64}. \quad \blacksquare$$

The formula for conversion from rectangular to cylindrical coordinates is virtually identical to formula (19.4.7). Let S be a region in $\mathbb{R}^3$. Since $x = r \cos \theta$, $y = r \sin \theta$, and $z = z$, we have

$$\iiint_S f(x, y, z) \, dV = \iiint_S f(r \cos \theta, r \sin \theta, z) r \, dz \, dr \, d\theta. \tag{5}$$

Given in rectangular   Given in cylindrical
coordinates   coordinates

REMARK.   Again, do not forget the extra r when you convert to cylindrical coordinates.

EXAMPLE 2   Find the mass of the solid bounded by the paraboloid $z = x^2 + y^2$ and the plane $z = 4$ if the density at any point is proportional to the distance from the point to the z-axis.

Solution. The solid is sketched in Figure 2. The density is given by $\rho(x, y, z) = \alpha\sqrt{x^2 + y^2}$, where α is a constant of proportionality. Thus since the solid may be written as

$$S = \{(x, y, z): -2 \le x \le 2, -\sqrt{4 - x^2} \le y \le \sqrt{4 - x^2}, x^2 + y^2 \le z \le 4\},$$

we have

$$\mu = \iiint_S \alpha\sqrt{x^2 + y^2}\, dV = \int_{-2}^{2} \int_{-\sqrt{4 - x^2}}^{\sqrt{4 - x^2}} \int_{x^2 + y^2}^{4} \alpha\sqrt{x^2 + y^2}\, dz\, dy\, dx.$$

We write this expression in cylindrical coordinates, using the fact that $\alpha\sqrt{x^2 + y^2} = \alpha r$. We note that the largest value of r is 2, since at the "top" of the solid, $r^2 = x^2 + y^2 = 4$. Then

$$\mu = \int_0^{2\pi} \int_0^{2} \int_{r^2}^{4} (\alpha r) r\, dz\, dr\, d\theta = \int_0^{2\pi} \int_0^{2} \alpha r^2(4 - r^2)\, dr\, d\theta$$

$$= \alpha \int_0^{2\pi} \left\{ \left(\frac{4r^3}{3} - \frac{r^5}{5} \right) \Big|_0^2 \right\} d\theta = \frac{64\alpha}{15} \int_0^{2\pi} d\theta = \frac{128}{15}\pi\alpha. \blacksquare$$

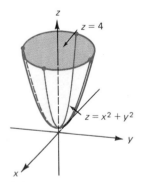

$z = 4$

$z = x^2 + y^2$

FIGURE 2

II. SPHERICAL COORDINATES

Recall from Section 17.8 that a point P in $\mathbb{R}^3$ can be written in the spherical coordinates (ρ, θ, φ), where $\rho \ge 0, 0 \le \theta < 2\pi, 0 \le \varphi \le \pi$. Here ρ is the distance between the point and the origin, θ is the same as in cylindrical coordinates, and φ is the angle between $\overrightarrow{OP}$ and the positive z-axis. Consider the "spherical parallelepiped"

$$\pi_s = \{(\rho, \theta, \varphi): \rho_1 \le \rho \le \rho_2, \theta_1 \le \theta \le \theta_2, \varphi_1 \le \varphi \le \varphi_2\}. \tag{6}$$

This solid is sketched in Figure 3. To approximate the volume of π_s, we partition the intervals $[\rho_1, \rho_2]$, $[\theta_1, \theta_2]$, and $[\varphi_1, \varphi_2]$. This partition gives us a number of "spherical boxes," one of which is sketched in Figure 4. The length of an arc of a circle is given by

$$L = r\theta, \tag{7}$$

where r is the radius of the circle and θ is the angle that "cuts off" the arc [see Appendix 1.1, equation (1.1.8)]. In Figure 4 one side of the spherical box is $\Delta\rho$. Since $\rho = \rho_i$ is the equation of a sphere, we find, from (7), that the length of a second side is $\rho_i \Delta\varphi$. Finally, the length of the third side is $\rho_i \sin \varphi_k \Delta\theta$. This follows from equation (7) and equation (17.8.5). Thus the volume of the box in Figure 4 is given, approximately, by

$$V_i \approx (\Delta\rho)(\rho_i \, \Delta\varphi)(\rho_i \sin \varphi_k \, \Delta\theta),$$

and using a familiar argument, we have

$$\text{volume of } \pi_s = \iiint\limits_{\pi_s} \rho^2 \sin \varphi \, d\rho \, d\varphi \, d\theta = \int_{\theta_1}^{\theta_2} \int_{\varphi_1}^{\varphi_2} \int_{\rho_1}^{\rho_2} \rho^2 \sin \varphi \, d\rho \, d\varphi \, d\theta. \tag{8}$$

If f is a function of the variables ρ, θ, and φ, we have

$$\iiint\limits_{\pi_s} f = \iiint\limits_{\pi_s} f(\rho, \theta, \varphi)\rho^2 \sin \varphi \, d\rho \, d\varphi \, d\theta. \tag{9}$$

More generally, let the region S be defined in spherical coordinates by

$$S = \{(\rho, \theta, \varphi): \theta_1 \le \theta \le \theta_2, g_1(\theta) \le \varphi \le g_2(\theta), h_1(\theta, \varphi) \le \rho \le h_2(\theta, \varphi)\}. \tag{10}$$

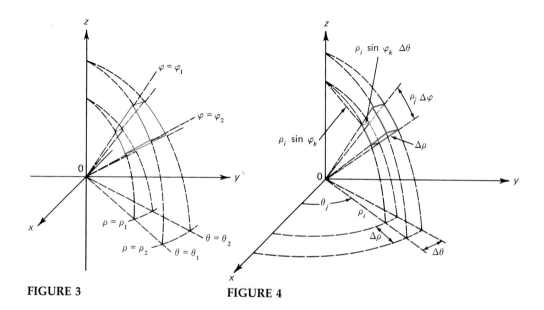

FIGURE 3   FIGURE 4

Then

$$\iiint_S f = \int_{\theta_1}^{\theta_2} \int_{g_1(\theta)}^{g_2(\theta)} \int_{h_1(\theta,\varphi)}^{h_2(\theta,\varphi)} f(\rho,\theta,\varphi)\rho^2 \sin\varphi \, d\rho \, d\varphi \, d\theta. \tag{11}$$

Recall that to convert from rectangular to spherical coordinates, we have the formulas [see equations (17.8.6), (17.8.7), and (17.8.8)]

$$x = \rho \sin\varphi \cos\theta, \qquad y = \rho \sin\varphi \sin\theta, \qquad z = \rho \cos\varphi.$$

Thus to convert from a triple integral in rectangular coordinates to a triple integral in spherical coordinates, we have

$$\iiint_S f(x,y,z)\,dV = \iiint_S f(\rho\sin\varphi\cos\theta,\rho\sin\varphi\sin\theta,\rho\cos\varphi)\rho^2 \sin\varphi \, d\rho \, d\varphi \, d\theta.$$

Given in rectangular
coordinates

Given in spherical coordinates

EXAMPLE 3 Calculate the volume enclosed by the sphere $x^2 + y^2 + z^2 = a^2$.

Solution. In rectangular coordinates, since

$$S = \{(x,y,z): -a \le x \le a, -\sqrt{a^2 - x^2} \le y \le \sqrt{a^2 - x^2},$$
$$-\sqrt{a^2 - x^2 - y^2} \le z \le \sqrt{a^2 - x^2 - y^2}\},$$

we can write the volume as

$$V = \int_{-a}^{a} \int_{-\sqrt{a^2-x^2}}^{\sqrt{a^2-x^2}} \int_{-\sqrt{a^2-x^2-y^2}}^{\sqrt{a^2-x^2-y^2}} dz \, dy \, dx.$$

This expression is very tedious to calculate. So instead, we note that the region enclosed by a sphere can be represented in spherical coordinates as

$$S = \{(\rho,\theta,\varphi): 0 \le \rho \le a, 0 \le \theta \le 2\pi, 0 \le \varphi \le \pi\}.$$

Thus

$$V = \int_0^{2\pi} \int_0^{\pi} \int_0^{a} \rho^2 \sin\varphi \, d\rho \, d\varphi \, d\theta = \int_0^{2\pi} \int_0^{\pi} \frac{a^3}{3} \sin\varphi \, d\varphi \, d\theta$$

$$= \frac{a^3}{3} \int_0^{2\pi} \left\{ -\cos\varphi \Big|_0^{\pi} \right\} d\theta = \frac{2a^3}{3} \int_0^{2\pi} d\theta = \frac{4\pi a^3}{3}. \quad \blacksquare$$

EXAMPLE 4 Find the mass of the sphere in Example 3 if its density at a point is proportional to the distance from the point to the origin.

Solution. We have density $= \alpha\sqrt{x^2 + y^2 + z^2} = \alpha\rho$. Thus

$$\mu = \int_0^{2\pi} \int_0^{\pi} \int_0^a \alpha\rho(\rho^2 \sin \varphi)\, d\rho\, d\varphi\, d\theta = \frac{\alpha a^4}{4} \int_0^{2\pi} \int_0^{\pi} \sin \varphi\, d\varphi\, d\theta$$

$$= \frac{\alpha a^4}{4} (4\pi) = \pi a^4 \alpha. \quad\blacksquare$$

EXAMPLE 5   Find the centroid of the "ice-cream-cone-shaped" region below the sphere $x^2 + y^2 + z^2 = z$ and above the cone $z^2 = x^2 + y^2$.

Solution. The region is sketched in Figure 5. Since $x^2 + y^2 + z^2 = \rho^2$ and $z = \rho \cos \varphi$, the sphere $\{(x, y, z): x^2 + y^2 + z^2 = z\}$ can be written in spherical coordinates as

$$\rho^2 = \rho \cos \varphi, \quad \text{or} \quad \rho = \cos \varphi.$$

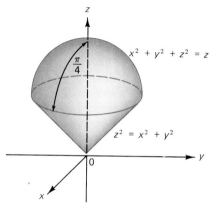

FIGURE 5

Since $x^2 + y^2 = \rho^2 \sin^2 \varphi \cos^2 \theta + \rho^2 \sin^2 \varphi \sin^2 \theta = \rho^2 \sin^2 \varphi$, the equation of the cone is

$$\rho^2 \cos^2 \varphi = \rho^2 \sin^2 \varphi, \quad \text{or} \quad \cos^2 \varphi = \sin^2 \varphi,$$

and since $0 \le \varphi \le \pi$, the equation is

$$\varphi = \frac{\pi}{4}.$$

Thus, assuming that the density of the region is the constant k, we have

$$\mu = \int_0^{2\pi} \int_0^{\pi/4} \int_0^{\cos \varphi} k\rho^2 \sin \varphi\, d\rho\, d\varphi\, d\theta$$

$$= k \int_0^{2\pi} \int_0^{\pi/4} \sin \varphi \left\{ \frac{\rho^3}{3} \Big|_0^{\cos \varphi} \right\} d\varphi\, d\theta = k \int_0^{2\pi} \int_0^{\pi/4} \frac{\cos^3 \varphi}{3} \sin \varphi\, d\varphi\, d\theta$$

$$= k \int_0^{2\pi} \left\{ -\frac{\cos^4 \varphi}{12} \Big|_0^{\pi/4} \right\} d\theta = \frac{1}{12} k \int_0^{2\pi} \frac{3}{4}\, d\theta = \frac{\pi k}{8}.$$

Since $x = \rho \sin \varphi \cos \theta$, we have

$$M_{yz} = k \int_0^{2\pi} \int_0^{\pi/4} \int_0^{\cos \varphi} (\rho \sin \varphi \cos \theta)\rho^2 \sin \varphi \, d\rho \, d\varphi \, d\theta$$

$$= k \int_0^{2\pi} \cos \theta \, d\theta \int_0^{\pi/4} \int_0^{\cos \varphi} \rho^3 \sin^2 \varphi \, d\rho \, d\varphi = 0,$$

since

$$\int_0^{2\pi} \cos \theta \, d\theta = 0$$

(an unsurprising result because of symmetry). Similarly, since $y = \rho \sin \varphi \sin \theta$,

$$M_{xz} = k \int_0^{2\pi} \int_0^{\pi/4} \int_0^{\cos \varphi} (\rho \sin \varphi \sin \theta)\rho^2 \sin \varphi \, d\rho \, d\varphi \, d\theta = 0$$

(again because of symmetry). Finally, since $z = \rho \cos \varphi$,

$$M_{xy} = k \int_0^{2\pi} \int_0^{\pi/4} \int_0^{\cos \varphi} (\rho \cos \varphi)\rho^2 \sin \varphi \, d\rho \, d\varphi \, d\theta$$

$$= \frac{k}{4} \int_0^{2\pi} \int_0^{\pi/4} \cos^5 \varphi \sin \varphi \, d\varphi \, d\theta = \frac{k}{4} \int_0^{2\pi} \left\{ -\frac{\cos^6 \varphi}{6} \Big|_0^{\pi/4} \right\} d\theta$$

$$= \frac{1}{24} \cdot \frac{7}{8} k \int_0^{2\pi} d\theta = \frac{7\pi k}{96}.$$

Thus $\bar{x} = \bar{y} = 0$ and $\bar{z} = M_{xy}/\mu = (7\pi k/96)/(\pi k/8) = 7/12.$ ∎

PROBLEMS 19.7

Solve Problems 1–11 by using cylindrical coordinates.

1. Find the volume of the region inside both the sphere $x^2 + y^2 + z^2 = 4$ and the cylinder $(x - 1)^2 + y^2 = 1$.
2. Find the centroid of the region of Problem 1.
*3. Suppose the density of the region of Problem 1 is proportional to the square of the distance to the xy-plane and is measured in kilograms per cubic meter. Find the center of mass of the region.
4. Find the volume of the solid bounded above by the paraboloid $z = 4 - x^2 - y^2$ and below by the xy-plane.
5. Find the center of mass of the solid of Problem 4 if the density is proportional to the distance to the xy-plane.
6. Find the volume of the solid bounded by the plane $z = y$ and the paraboloid $z = x^2 + y^2$.
7. Find the centroid of the region of Problem 6.
8. Find the volume of the solid bounded by the two cones $z^2 = x^2 + y^2$ and $z^2 = 16x^2 + 16y^2$ between $z = 0$ and $z = 2$.
9. Find the center of mass of the solid in Problem 8 if the density at any point is proportional to the distance to the z-axis.

10. Find the volume of the solid bounded by the hyperboloid of two sheets $z^2 - x^2 - y^2 = 1$ and the cone $z^2 = x^2 + y^2$ for $0 \le z \le a$.

11. Evaluate $\displaystyle\int_0^1 \int_0^{\sqrt{1-x^2}} \int_{\sqrt{x^2+y^2}}^{\sqrt{1-x^2-y^2}} z^3 \, dz \, dy \, dx$.

Solve Problems 12–21 by using spherical coordinates.

12. Find the mass of the unit sphere if the density at any point is proportional to the distance to the boundary of the sphere.

13. Find the volume of the solid inside the sphere $x^2 + y^2 + z^2 = 4$ and outside the cone $z^2 = x^2 + y^2$.

14. Find the center of mass of the solid of Problem 12 if the density at a point is proportional to the distance from the origin.

15. A solid fills the space between two concentric spheres centered at the origin. The radii of the spheres are a and b, where $0 < a < b$. Find the mass of the solid if the density at each point is inversely proportional to its distance from the origin.

16. Find the center of mass of the object of Problem 13 if the density at a point is proportional to the square of the distance from the origin.

17. Find the volume of one of the smaller wedges cut from the unit sphere by two planes that meet at a diameter at an angle of $\pi/6$.

18. Find the mass of a wedge cut from a sphere of radius a by two planes that meet at a diameter at an angle of b radians if the density at any point on the wedge is proportional to the distance to that diameter.

19. Evaluate $\displaystyle\int_{-3}^{3} \int_{-\sqrt{9-x^2}}^{\sqrt{9-x^2}} \int_{-\sqrt{9-x^2-y^2}}^{\sqrt{9-x^2-y^2}} (x^2 + y^2 + z^2)^{3/2} \, dz \, dy \, dx$.

20. Evaluate $\displaystyle\int_0^1 \int_0^{\sqrt{1-x^2}} \int_0^{\sqrt{1-x^2-y^2}} (1/\sqrt{x^2 + y^2 + z^2}) \, dz \, dy \, dx$.

21. Evaluate $\displaystyle\int_0^a \int_0^{\sqrt{a^2-x^2}} \int_0^{\sqrt{a^2-x^2-y^2}} \frac{z^3}{(x^2 + y^2)^{1/2}} \, dz \, dy \, dx$.

*22. A "sphere" in $\mathbb{R}^4$ has the equation $x^2 + y^2 + z^2 + w^2 = a^2$.

(a) Explain why it is reasonable that the volume of this "sphere" be given by

$$V = \int_{-a}^{a} \int_{-\sqrt{a^2-x^2}}^{\sqrt{a^2-x^2}} \int_{-\sqrt{a^2-x^2-y^2}}^{\sqrt{a^2-x^2-y^2}} \int_{-\sqrt{a^2-x^2-y^2-z^2}}^{\sqrt{a^2-x^2-y^2-z^2}} dw \, dz \, dy \, dx$$

$$= 16 \int_0^a \int_0^{\sqrt{a^2-x^2}} \int_0^{\sqrt{a^2-x^2-y^2}} \int_0^{\sqrt{a^2-x^2-y^2-z^2}} dw \, dz \, dy \, dx.$$

(b) Using spherical coordinates, evaluate the integral in (a). [*Hint:* The first integral is easy and reduces the "quadruple" integral to a triple integral.]

REVIEW EXERCISES FOR CHAPTER NINETEEN

In Exercises 1–8, evaluate the integral.

1. $\displaystyle\int_0^1 \int_0^2 x^2 y \, dx \, dy$

2. $\displaystyle\int_0^1 \int_{x^2}^{x} xy^3 \, dy \, dx$

3. $\displaystyle\int_2^4 \int_{1+y}^{2+5y} (x - y^2) \, dx \, dy$

4. $\displaystyle\int_0^4 \int_{-\sqrt{16-x^2}}^{\sqrt{16-x^2}} 4y \, dy \, dx$

5. $\iint_\Omega (y - x^2) \, dx \, dy$, where $\Omega = \{(x, y): -3 \le x \le 3, 0 \le y \le 2\}$.
6. $\iint_\Omega (y - x^2) \, dx \, dy$, where Ω is the region in the first quadrant bounded by the x-axis, the y-axis, and the circle $x^2 + y^2 = 4$.
7. $\iint_\Omega (x + y^2) \, dx \, dy$, where Ω is the region in the first quadrant bounded by the x-axis, the y-axis, and the unit circle.
8. $\iint_\Omega (2x + y)e^{-(x+y)} \, dx \, dy$, where Ω is the first quadrant.

9. Find upper and lower bounds for $\iint_\Omega \sin(x^5 + y^{12}) \, dA$, where Ω is the region in the first quadrant bounded by the x-axis, the y-axis, and the circle $x^2 + y^2 = 4$.
10. Change the order of integration of $\int_2^5 \int_1^x 3x^2 y \, dy \, dx$ and evaluate.
11. Change the order of integration of $\int_0^3 \int_0^{\sqrt{9-x^2}} (9 - y^2)^{3/2} \, dy \, dx$ and evaluate.
12. Change the order of integration of $\int_0^\infty \int_x^\infty f(x, y) \, dy \, dx$.
13. Find the volume of the solid bounded by the plane $x + 2y + 3z = 6$ and the three coordinate planes.
14. Find the volume of the solid bounded by the cylinder $y^2 + z^2 = 4$ and the planes $x = 0$ and $x = 3$.
15. Find the volume enclosed by the ellipsoid $4x^2 + y^2 + 25z^2 = 100$.
16. Find the volume of the solid bounded by the paraboloid $x = y^2 + z^2$ and the plane $y = x - 2$.
17. Find the center of mass of the unit disk if the density at a point is proportional to the distance to the boundary of the disk.
18. Find the center of mass of the region $\Omega = \{(x, y): 0 \le x \le 3, 1 \le y \le 5\}$ if $\rho(x, y) = 3x^2 y$.
19. Find the center of mass of the triangular region bounded by the lines $y = x$, $y = 2 - x$, and the y-axis if $\rho(x, y) = 3xy + y^3$.
20. Calculate the volume under the surface $z = r^5$ and over the circle of radius 6 centered at the origin.
21. Calculate the area of the region enclosed by the curve $r = 2(1 + \sin \theta)$.
22. Calculate the area of the region inside the cardioid $r = 3(1 + \sin \theta)$ and outside the circle $r = 2$.
23. Find the volume of the solid bounded by the cone $z^2 = x^2 + y^2$ and the paraboloid $4z = x^2 + y^2$.
24. Find the centroid of the region bounded by the right-hand loop of the lemniscate $r^2 = 4 \cos 2\theta$.
25. Find the area of the part of the plane $z = 3x + 5y$ over the region bounded by the curves $x = y^2$ and $x = y^3$.
26. Find the area of the part of the surface $z = (y^4/4) + (1/8y^2)$ over the rectangle $\{(x, y): 0 \le x \le 3, 2 \le y \le 4\}$.
27. Find the surface area of the hemisphere $x^2 + y^2 + z^2 = 16$, $z \ge 0$.
28. Evaluate $\int_0^1 \int_0^y \int_0^x x^2 \, dz \, dx \, dy$.   29. Evaluate $\int_1^2 \int_{2-x}^{2+x} \int_0^{xz} 12xyz \, dy \, dz \, dx$.
30. Change the order of integration in Exercise 28 and write the integral in the following forms:
    (a) $\int_?^? \int_?^? \int_?^? x^2 \, dx \, dy \, dz$ (b) $\int_?^? \int_?^? \int_?^? x^2 \, dy \, dz \, dx$
31. Find the volume of the tetrahedron with vertices at $(0, 0, 0)$, $(2, 0, 0)$, $(0, 1, 0)$, and $(0, 0, 3)$.
32. Find the volume of the solid bounded by the parabolic cylinder $z = 4 - y^2$ and the planes $z = x$ and $z = 2x$ that lies in the half plane $z \ge 0$.
33. Find the mass of the solid of Exercise 31 if the density function is $\rho(x, y, z) = 2x + y^2$.
34. Find the center of mass of the solid bounded by the ellipsoid $4x^2 + y^2 + 25z^2 = 100$ if the density function is $\rho(x, y, z) = z^2$.
35. Find the centroid of the solid in Exercise 32.
36. Find the volume of the region inside both the sphere $x^2 + y^2 + z^2 = a^2$ and the cylinder $x^2 + [y - (a/2)]^2 = (a/2)^2$.
37. Find the centroid of the solid of Exercise 36.
38. The density of the region of Exercise 36 is proportional to the square of the distance to the xy-plane. Find the center of mass of the region.

39. Find the volume of the solid bounded by the plane $z = x$ and the paraboloid $z = x^2 + y^2$.

40. Evaluate $\displaystyle\int_{-2}^{2} \int_{-\sqrt{4-x^2}}^{\sqrt{4-x^2}} \int_{\sqrt{x^2+y^2}}^{\sqrt{4-x^2-y^2}} z^2 \, dz \, dy \, dx$.

41. Find the volume of the solid inside the sphere $x^2 + y^2 + z^2 = 9$ and outside the cone $z^2 = x^2 + y^2$.

42. Find the center of mass of the solid in Exercise 41 if the density at a point is proportional to its distance to the origin.

43. Find the volume of a wedge cut from a sphere of radius 2 by two planes that meet at a diameter at an angle of $\pi/3$.

44. Evaluate $\displaystyle\int_{0}^{2} \int_{0}^{\sqrt{4-x^2}} \int_{0}^{\sqrt{4-x^2-y^2}} \frac{z^2}{(x^2 + y^2)^{1/2}} \, dz \, dy \, dx$.

20 Introduction to Vector Analysis

20.1 VECTOR FIELDS

We wrote about vector fields briefly in our discussion of potential functions and the gradient in Section 18.9. The notion of a vector field is central to the subject of **vector analysis.** We provide formal definitions now.

Definition 1   VECTOR FIELD IN $\mathbb{R}^2$ Let Ω be a region in $\mathbb{R}^2$. Then $\mathbf{F}$ is a **vector field** in $\mathbb{R}^2$ if $\mathbf{F}$ assigns to every $\mathbf{x}$ in Ω a unique vector $\mathbf{F}(\mathbf{x})$ in $\mathbb{R}^2$.

REMARK.   Simply put, a vector field in $\mathbb{R}^2$ is a function whose domain is a subset of $\mathbb{R}^2$ and whose range is a subset of $\mathbb{R}^2$. Two vector fields in $\mathbb{R}^2$ are sketched in Figure 1. The meaning of this sketch is that to every point $\mathbf{x}$ in Ω a unique vector $\mathbf{F}(\mathbf{x})$ is assigned. That is, the function value $\mathbf{F}(\mathbf{x})$ is represented by an arrow with $\mathbf{x}$ at the "tail" of the arrow.

EXAMPLE 1   **(Gradient Field in $\mathbb{R}^2$)**   Let $z = f(x, y)$ be a differentiable function. Then $\mathbf{F} = \nabla f = f_x \mathbf{i} + f_y \mathbf{j}$ is a vector field. ◼

Definition 1 can be extended, in an obvious way, to $\mathbb{R}^3$.

Definition 2   VECTOR FIELD IN $\mathbb{R}^3$ Let S be a region in $\mathbb{R}^3$. Then $\mathbf{F}$ is a **vector field** in $\mathbb{R}^3$ if $\mathbf{F}$ assigns to each vector $\mathbf{x}$ in S a unique vector $\mathbf{F}(\mathbf{x})$ in $\mathbb{R}^3$.

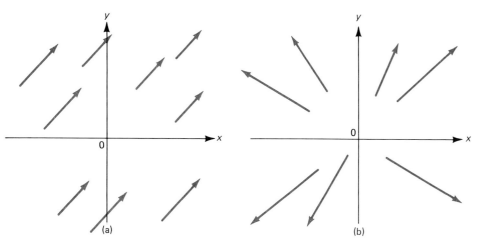

FIGURE 1

REMARK.   A vector field in $\mathbb{R}^3$ is a function whose domain and range are subsets of $\mathbb{R}^3$.

Two vector fields in $\mathbb{R}^3$ are sketched in Figure 2.

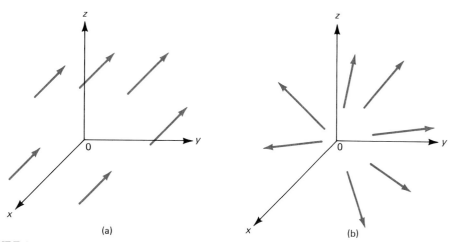

FIGURE 2

EXAMPLE 2   **(Gradient Field in $\mathbb{R}^3$)**   If $u = f(x, y, z)$ is differentiable, then $\mathbf{F} = \nabla f = f_x \mathbf{i} + f_y \mathbf{j} + f_z \mathbf{k}$ is a vector field. It is called a **gradient vector field.** ■

EXAMPLE 3   From Section 18.9, if $\mathbf{F} = -\nabla f$ for some differentiable function $z = f(x, y)$ or $w = f(x, y, z)$, then $\mathbf{F}$ is called a **conservative vector field.** ■

EXAMPLE 4   **(Gravitational Field)**   Let m_1 represent the mass of a (relatively) fixed object in space and let m_2 denote the mass of an object moving near the fixed object. Then the magnitude of the gravitational force between the objects is given by (see Example 3.2.7)

$$|\mathbf{F}| = G\frac{m_1 m_2}{r^2}, \tag{1}$$

where r is the distance between the objects and G is a universal constant. If we assume that the first object is at the origin, then we may denote the position of the second object by $\mathbf{x}(t)$, and then, since $r = |\mathbf{x}|$, (1) can be written

$$|\mathbf{F}| = \frac{Gm_1 m_2}{|\mathbf{x}|^2}. \tag{2}$$

Also, the force acts toward the origin, that is, in the direction opposite to that of the position vector $\mathbf{x}$. Therefore

$$\text{direction of } \mathbf{F} = -\frac{\mathbf{x}}{|\mathbf{x}|}, \tag{3}$$

so from (2) and (3),

$$\mathbf{F}(t) = \frac{\alpha \mathbf{x}(t)}{|\mathbf{x}(t)|^3}, \tag{4}$$

where $\alpha = -Gm_1 m_2$.

The vector field (4) is called a **gravitational field.** As we saw in Section 18.9, such a field is conservative. We can sketch this vector field without much difficulty because for every $\mathbf{x} \neq 0 \in \mathbb{R}^3$, $\mathbf{F}(\mathbf{x})$ points toward the origin. The sketch appears in Figure 3. ▨

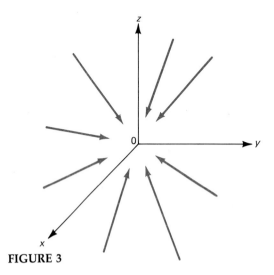

FIGURE 3

Vector fields arise in a great number of physical applications. There are, for example, mechanical force fields, magnetic fields, electric fields, velocity fields, and direction fields.

EXAMPLE 5   Let $\mathbf{F}$ be the vector field in $\mathbb{R}^2$ given by

$$\mathbf{F} = \frac{2(x^2 - y^2 - 1)}{[(x + 1)^2 + y^2][(x - 1)^2 + y^2]}\mathbf{i} + \frac{4xy}{[(x + 1)^2 + y^2][(x - 1)^2 + y^2]}\mathbf{j}. \tag{5}$$

It can be shown that $\mathbf{F}$ is the electric field in the plane caused by two infinite straight wires that are perpendicular to the xy-plane and that pass through the points $(1, 0)$ and $(-1, 0)$. It is assumed that the wires are uniformly charged with electricity and that the charge of one is opposite to the charge of the other. This field is sketched in Figure 4. (Also, see Problem 25.) ■

EXAMPLE 6 **(Velocity Field of a Rotating Body)** Suppose a body is rotating in space around an axis of rotation (see Figure 5). We define a vector $\mathbf{w}$ as follows: $\mathbf{w}$ has the direction of the axis of rotation such that if we look from the initial point of $\mathbf{w}$ to its terminal point, the motion appears clockwise. We now define

$$\textbf{angular speed } \omega = \frac{\text{tangential speed of a point } \mathbf{x}(t)}{\text{distance from } \mathbf{x} \text{ to the axis of rotation}}. \tag{6}$$

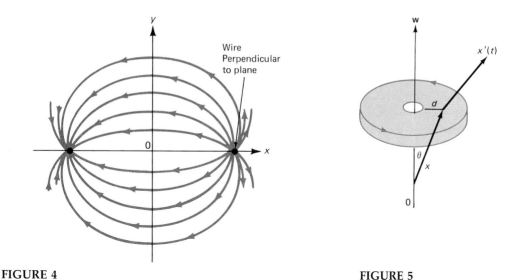

FIGURE 4   **FIGURE 5**

Note that if $\mathbf{x}(t)$ represents the position vector, then the velocity vector is $\mathbf{x}'(t)$ and its speed is $|\mathbf{x}'(t)|$. Thus if d denotes the distance of a point $\mathbf{x}$ from the axis of rotation, we have, from (6),

$$\omega = \frac{|\mathbf{x}'(t)|}{d}. \tag{7}$$

Then we define the magnitude of $\mathbf{w}$ to be ω. That is,

$$|\mathbf{w}| = \omega. \tag{8}$$

Let θ denote the angle between $\mathbf{x}$ and $\mathbf{w}$. Then from Figure 5

$$\sin \theta = \frac{d}{|\mathbf{x}|},$$

or

$$d = |\mathbf{x}| \sin \theta,$$

and

$$\omega d = |\mathbf{x}| \, \omega \sin \theta = |\mathbf{x}| \, |\mathbf{w}| \sin \theta = |\mathbf{x} \times \mathbf{w}|.$$

Thus from (7)

$$|\mathbf{x}'(t)| = |\mathbf{x} \times \mathbf{w}|. \tag{9}$$

Now, again from Figure 5, we see that the velocity vector $\mathbf{x}'$ is perpendicular to both $\mathbf{x}$ and $\mathbf{w}$, so that the direction of $\mathbf{v} = \mathbf{x}'$ is the direction of $\mathbf{w} \times \mathbf{x}$. Combining this result with (9), we obtain

$$\mathbf{v} = \mathbf{w} \times \mathbf{x}.$$

We can write this **velocity field** as

$$\mathbf{v}(x, y, z) = \mathbf{w} \times (x\mathbf{i} + y\mathbf{j} + z\mathbf{k}). \tag{10}$$

This field is sketched in Figure 6. ■

FIGURE 6

EXAMPLE 7 Suppose that the axis of rotation in Example 6 is the y-axis and that $\mathbf{w}$ points in the positive y-direction. Then

$$\mathbf{w} = \omega\mathbf{j} \quad \text{and} \quad \mathbf{w} \times \mathbf{x} = \begin{vmatrix} \mathbf{i} & \mathbf{j} & \mathbf{k} \\ 0 & \omega & 0 \\ x & y & z \end{vmatrix} = \omega z\mathbf{i} - \omega x\mathbf{k} = \omega(z\mathbf{i} - x\mathbf{k}). \quad \blacksquare$$

NOTE. In Examples 6 and 7, ω is not necessarily constant. If ω is constant, the body is said to have *constant angular speed*.

PROBLEMS 20.1

In Problems 1–21, compute the gradient vector field of the given function.

1. $f(x, y) = (x + y)^2$

2. $f(x, y) = e^{\sqrt{xy}}$

3. $f(x, y) = \cos(x - y)$

4. $f(x, y) = \ln(2x - y + 1)$

5. $f(x, y) = \sqrt{x^2 + y^3}$

6. $f(x, y) = \tan^{-1} \dfrac{y}{x}$

7. $f(x, y) = y \tan(y - x)$

8. $f(x, y) = x^2 \sinh y$

9. $f(x, y) = \sec(x + 3y)$

10. $f(x, y) = \dfrac{x - y}{x + y}$

11. $f(x, y) = \dfrac{x^2 - y^2}{x^2 + y^2}$

12. $f(x, y) = \dfrac{e^{x^2} - e^{-y^2}}{3y}$

13. $f(x, y, z) = xyz$

14. $f(x, y, z) = \sin x \cos y \tan z$

15. $f(x, y, z) = \dfrac{x^2 - y^2 + z^2}{3xy}$

16. $f(x, y, z) = x \ln y - z \ln x$

17. $f(x, y, z) = xy^2 + y^2 z^3$

18. $f(x, y, z) = (y - z)e^{x + 2y + 3z}$

19. $f(x, y, z) = x \sin y \ln z$

20. $f(x, y, z) = \dfrac{x - z}{\sqrt{1 - y^2 + x^2}}$

21. $f(x, y, z) = x \cosh z - y \sin x$

22. Define a gravitational field in $\mathbb{R}^2$ and show that it is conservative.

23. Find the velocity field of a rotating body whose axis of rotation is the z-axis with $\mathbf{w}$ pointing in the direction of the positive z-axis. Assume that ω is constant.

24. Answer the question of Problem 23 after replacing the z-axis with the x-axis.

25. Show that the force field given in Example 5 is the gradient of the function

$$F(x, y) = \ln \frac{\sqrt{(x - 1)^2 + y^2}}{\sqrt{(x + 1)^2 + y^2}}.$$

20.2 WORK, LINE INTEGRALS IN THE PLANE, AND INDEPENDENCE OF PATH

In Section 15.3 we showed that if a constant force $\mathbf{F}$ is applied to a particle that moves along a vector $\mathbf{d}$, then the work done by the force on the particle is given by

$$W = \mathbf{F} \cdot \mathbf{d}. \tag{1}$$

We now calculate the work done when a particle moves along a curve C. In doing so, we will define an important concept in applied mathematics—the *line integral*.

Suppose that a curve in the plane is given parametrically by

$$C: \quad \mathbf{x}(t) = f(t)\mathbf{i} + g(t)\mathbf{j}, \tag{2}$$

or

$$C: \quad \mathbf{x}(t) = x(t)\mathbf{i} + y(t)\mathbf{j}. \tag{2'}$$

If a force is applied to a particle moving along C, then such a force will have magnitude and direction, so the force will be a vector function of the vector $\mathbf{x}(t)$. That is, the force will be a vector field. We write

$$\mathbf{F}(\mathbf{x}) = \mathbf{F}(x, y) = P(x, y)\mathbf{i} + Q(x, y)\mathbf{j} \tag{3}$$

where P and Q are scalar-valued functions. The problem is to determine the work done when a particle moves on C from a point $\mathbf{x}(a)$ to a point $\mathbf{x}(b)$ subject to the force $\mathbf{F}$ given by (3). We will assume in our discussion that the curve C is *smooth* or *piecewise smooth*. By that we mean that the functions $f(t)$ and $g(t)$ in (2) are continuously differentiable or that f' and g' exist and are piecewise continuous (see p. 122 for the definition of piecewise continuity).

A typical curve C is sketched in Figure 1. Let

$$W(t) = \text{work done in moving from } \mathbf{x}(a) \text{ to } \mathbf{x}(t). \tag{4}$$

Then the work done in moving from $\mathbf{x}(t)$ to $\mathbf{x}(t + \Delta t)$ is given by

$$W(t + \Delta t) - W(t). \tag{5}$$

Now if Δt is small, then the part of the curve between $\mathbf{x}(t)$ and $\mathbf{x}(t + \Delta t)$ is "close" to a straight line and so can be approximated by the vector

$$\mathbf{x}(t + \Delta t) - \mathbf{x}(t). \tag{6}$$

If Δt is small and if $\mathbf{F}(\mathbf{x})$ is continuous, then the force aplied between $\mathbf{x}(t)$ and $\mathbf{x}(t + \Delta t)$ is approximately equal to $\mathbf{F}(\mathbf{x}(t))$. Thus by (1), if Δt is small, then

$$W(t + \Delta t) - W(t) \approx \mathbf{F}(\mathbf{x}(t)) \bullet [\mathbf{x}(t + \Delta t) - \mathbf{x}(t)]. \tag{7}$$

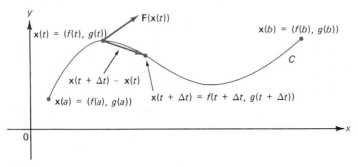

FIGURE 1

We divide both sides of (7) by Δt and take the limit as $\Delta t \to 0$ to obtain

$$\lim_{\Delta t \to 0} \frac{W(t + \Delta t) - W(t)}{\Delta t} = \lim_{\Delta t \to 0} \left\{ \mathbf{F}(\mathbf{x}(t)) \cdot \frac{[\mathbf{x}(t + \Delta t) - \mathbf{x}(t)]}{\Delta t} \right\},$$

or

$$W'(t) = \mathbf{F}(\mathbf{x}(t)) \cdot \mathbf{x}'(t). \tag{8}$$

But

$$W(a) = 0$$

and

$$W(b) = \text{total work done on the particle.}$$

Thus

$$W = \text{total work done} = W(b) - W(a) = \int_a^b W'(t)\, dt,$$

or

$$W = \int_a^b \mathbf{F}(\mathbf{x}(t)) \cdot \mathbf{x}'(t)\, dt. \tag{9}$$

We write equation (9) as

$$W = \int_C \mathbf{F}(\mathbf{x}) \cdot d\mathbf{x}. \tag{10}$$

The symbol $\int_C$ is read "the integral along the curve C." The integral in (10) is called a *line integral of* $\mathbf{F}$ *over* C.

REMARK 1. If C lies along the x-axis, then C is given by $\mathbf{x}(t) = x(t)\mathbf{i} + 0\mathbf{j}$† and $\mathbf{x}'(t) = x'(t)\mathbf{i}$, so that (10) becomes

$$\int_{[a,\, b]} F(x)\, dx = \int_a^b F(x)\, dx,$$

which is our usual definite integral.

REMARK 2. Since $\mathbf{F}$ is given by (3), we can write equation (9) as

---

†We have substituted $x(t)$ for $f(t)$ and 0 for $g(t)$ here.

$$W = \int_a^b [P(x(t), y(t))\, x'(t) + Q(x(t), y(t))\, y'(t)]\, dt. \tag{11}$$

EXAMPLE 1   A particle is moving along the parabola $y = x^2$ subject to a force given by the vector field $2xy\mathbf{i} + (x^2 + y^2)\mathbf{j}$. How much work is done in moving from the point (1, 1) to the point (3, 9) if forces are measured in newtons and distances are measured in meters?

Solution. The curve C is given parametrically by

$$\mathbf{x}(t) = t\mathbf{i} + t^2\mathbf{j} \qquad \text{between} \qquad t = 1 \quad \text{and} \quad t = 3.$$

We therefore have $f(t) = t$ and $g(t) = t^2$. Then $P = 2xy = 2t^3$ and $Q = x^2 + y^2 = t^2 + t^4$. Also, $f'(t) = 1$ and $g'(t) = 2t$, so by (11)

$$W = \int_1^3 [(2t^3)1 + (t^2 + t^4)(2t)]\, dt = \int_1^3 (4t^3 + 2t^5)\, dt = 322\tfrac{2}{3}\ \text{J.} \quad\blacksquare$$

EXAMPLE 2   Two electrical charges of like polarity (i.e., both positive or both negative) will repel each other. If a charge of α coulombs is placed at the origin and a charge of 1 coulomb of the same polarity is at the point (x, y), then the force of repulsion is given by

$$F(x, y) = \frac{\alpha x}{(x^2 + y^2)^{3/2}}\mathbf{i} + \frac{\alpha y}{(x^2 + y^2)^{3/2}}\mathbf{j}.$$

How much work is done by the force on the 1-coulomb charge as the charge moves on the straight line from (1, 0) to (3, −2)?

Solution. The straight line is the line $y = 1 - x$, or parametrically,

$$\mathbf{x}(t) = t\mathbf{i} + (1 - t)\mathbf{j} \qquad \text{for} \qquad 1 \le t \le 3.$$

Then

$$P = \frac{\alpha t}{[(t^2 + (1 - t)^2]^{3/2}} = \frac{\alpha t}{(2t^2 - 2t + 1)^{3/2}}$$

and

$$Q = \frac{\alpha(1 - t)}{[t^2 + (1 - t)^2]^{3/2}} = \frac{\alpha(1 - t)}{(2t^2 - 2t + 1)^{3/2}}.$$

Then since $f'(t) = 1$ and $g'(t) = -1$,

$$W = \alpha \int_1^3 \left\{ \frac{t}{(2t^2 - 2t + 1)^{3/2}}(1) + \frac{(1 - t)}{(2t^2 - 2t + 1)^{3/2}}(-1) \right\} dt$$

$$= \alpha \int_1^3 \frac{2t - 1}{(2t^2 - 2t + 1)^{3/2}}\, dt = \frac{\alpha}{2} \int_1^3 \frac{4t - 2}{(2t^2 - 2t + 1)^{3/2}}\, dt$$

$$= -\alpha (2t^2 - 2t + 1)^{-1/2} \Big|_1^3 = \alpha \left(1 - \frac{1}{\sqrt{13}} \right). \quad \blacksquare$$

EXAMPLE 3   How much work is done by the force on the charge in Example 2 if it moves in the counterclockwise direction along the semicircle $x^2 + y^2 = a^2$, $y \geq 0$?

Solution. Here

$$\mathbf{x}(t) = a(\cos t)\mathbf{i} + a(\sin t)\mathbf{j}, \qquad 0 \leq t \leq \pi.$$

Then

$$P = \frac{\alpha x}{(x^2 + y^2)^{3/2}} = \frac{a\alpha \cos t}{(a^2 \cos^2 t + a^2 \sin^2 t)^{3/2}} = \frac{a\alpha \cos t}{a^3} = \frac{\alpha \cos t}{a^2}$$

and

$$Q = \frac{\alpha y}{(x^2 + y^2)^{3/2}} = \frac{\alpha \sin t}{a^2}.$$

Since $f'(t) = -a \sin t$ and $g'(t) = a \cos t$,

$$W = \frac{\alpha}{a} \int_0^\pi [\cos t(-\sin t) + \sin t(\cos t)] \, dt = 0.$$

This result is no surprise since in this example the force $\mathbf{F}$ and the direction of the curve $\mathbf{x}'$ are orthogonal. Thus the component of the force in the direction of motion is zero, which means that the force does no work. ■

We used the notion of work to motivate the discussion of the line integral. We now give a general definition of the line integral in the plane.

Definition 1   LINE INTEGRAL IN THE PLANE   Let P and Q be continuous on a set S containing the smooth (or piecewise smooth) curve C given by

$$C: \quad \mathbf{x}(t) = f(t)\mathbf{i} + g(t)\mathbf{j}, \qquad t \in [a, b].$$

Let the vector field $\mathbf{F}$ be given by

$$\mathbf{F}(x, y) = P(x, y)\mathbf{i} + Q(x, y)\mathbf{j}.$$

Then the **line integral** of $\mathbf{F}$ over C is given by

$$\int_C \mathbf{F}(\mathbf{x}) \cdot \mathbf{dx} = \int_a^b [P(f(t), g(t))f'(t) + Q(f(t), g(t))g'(t)] \, dt. \qquad (12)$$

REMARK.   If C is piecewise smooth but not smooth, then C is made up of a number of "sections," each of which is smooth. Since C is continuous, these sections are joined. Some typical piecewise smooth curves are sketched in Figure 2. If C is made up of the n smooth curves $C_1, C_2, \ldots, C_n$, then

$$\int_C \mathbf{F}(\mathbf{x}) \cdot d\mathbf{x} = \int_{C_1} \mathbf{F}(\mathbf{x}) \cdot d\mathbf{x} + \int_{C_2} \mathbf{F}(\mathbf{x}) \cdot d\mathbf{x} + \cdots + \int_{C_n} \mathbf{F}(\mathbf{x}) \cdot d\mathbf{x}. \qquad (13)$$

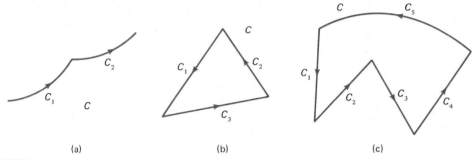

(a)   (b)   (c)

FIGURE 2

EXAMPLE 4   Calculate $\int_C \mathbf{F}(\mathbf{x}) \cdot d\mathbf{x}$, where $\mathbf{F}(x, y) = xy\mathbf{i} + ye^x\mathbf{j}$ and C is the rectangle joining the points $(0, 0)$, $(2, 0)$, $(2, 1)$, and $(0, 1)$ if C is traversed in the counterclockwise direction.

Solution. The rectangle is sketched in Figure 3, and it is made up of four smooth curves (straight lines). We have

$$C_1: \quad \mathbf{x}(t) = t\mathbf{i}, \qquad\qquad 0 \le t \le 2$$
$$C_2: \quad \mathbf{x}(t) = 2\mathbf{i} + t\mathbf{j}, \qquad 0 \le t \le 1$$
$$C_3: \quad \mathbf{x}(t) = (2 - t)\mathbf{i} + \mathbf{j}, \qquad 0 \le t \le 2$$
$$C_4: \quad \mathbf{x}(t) = (1 - t)\mathbf{j}, \qquad 0 \le t \le 1.$$

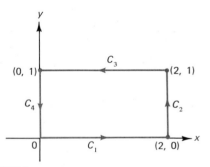

FIGURE 3

Note, for example, that on C_3, $t = 0$ corresponds to the point $2\mathbf{i} + \mathbf{j} = (2, 1)$ and $t = 2$ corresponds to $(2 - 2)\mathbf{i} + \mathbf{j} = (0, 1)$. Thus as t increases, we do move along C_3 in the direction indicated by the arrow in Figure 3. This illustrates why our parametrization of the rectangle is correct.

Now

$$\int_C \mathbf{F}(\mathbf{x}) \cdot d\mathbf{x} = \int_{C_1} \mathbf{F}(\mathbf{x}) \cdot d\mathbf{x} + \int_{C_2} \mathbf{F}(\mathbf{x}) \cdot d\mathbf{x} + \int_{C_3} \mathbf{F}(\mathbf{x}) \cdot d\mathbf{x} + \int_{C_4} \mathbf{F}(\mathbf{x}) \cdot d\mathbf{x}.$$

On C_1, $x = t$ and $y = 0$, so that $xy = 0$, $ye^x = 0$, and

$$\int_{C_1} \mathbf{F}(\mathbf{x}) \cdot d\mathbf{x} = 0.$$

On C_2, $x = 2$, $y = t$, $x' = 0$, $y' = 1$, and $ye^x = te^2$, so that

$$\int_{C_2} \mathbf{F}(\mathbf{x}) \cdot d\mathbf{x} = \int_0^1 te^2 \, dt = \frac{e^2}{2}.$$

On C_3, $x = (2 - t)$, $y = 1$, $x' = -1$, $y' = 0$, and $xy = 2 - t$, so that

$$\int_{C_3} \mathbf{F}(\mathbf{x}) \cdot d\mathbf{x} = \int_0^2 (2 - t)(-1) \, dt = -2.$$

On C_4, $x = 0$, $y = 1 - t$, $x' = 0$, $y' = -1$, $xy = 0$, and $ye^x = 1 - t$, so that

$$\int_{C_4} \mathbf{F}(\mathbf{x}) \cdot d\mathbf{x} = \int_0^1 (1 - t)(-1) \, dt = -\frac{1}{2}.$$

Thus

$$\int_C \mathbf{F}(\mathbf{x}) \cdot d\mathbf{x} = 0 + \frac{e^2}{2} - 2 - \frac{1}{2} = \frac{e^2}{2} - \frac{5}{2} \approx 1.2. \quad ■$$

There are certain conditions under which the calculation of a line integral becomes very easy. We first illustrate what we have in mind with an example.

EXAMPLE 5 Let $\mathbf{F}(x, y) = y\mathbf{i} + x\mathbf{j}$. Calculate $\int_C \mathbf{F}(\mathbf{x}) \cdot d\mathbf{x}$, where C is as follows:

(a) The straight line from $(0, 0)$ to $(1, 1)$.
(b) The parabola $y = x^2$ from $(0, 0)$ to $(1, 1)$.
(c) The curve $\mathbf{x}(t) = t^{3/2}\mathbf{i} + t^5\mathbf{j}$ from $(0, 0)$ to $(1, 1)$.

Solution. (a) C is given by $\mathbf{x}(t) = t\mathbf{i} + t\mathbf{j}$. Then $\mathbf{x}'(t) = \mathbf{i} + \mathbf{j}$ and $\mathbf{F} \cdot \mathbf{x}' = (t\mathbf{i} + t\mathbf{j}) \cdot (\mathbf{i} + \mathbf{j}) = 2t$, so

$$\int_C \mathbf{F}(\mathbf{x}) \cdot d\mathbf{x} = \int_0^1 2t \, dt = 1.$$

(b) C is given by $\mathbf{x}(t) = t\mathbf{i} + t^2\mathbf{j}$. Then $\mathbf{x}'(t) = \mathbf{i} + 2t\mathbf{j}$ and $\mathbf{F} \cdot \mathbf{x}' = (t^2\mathbf{i} + t\mathbf{j}) \cdot (\mathbf{i} + 2t\mathbf{j}) = 3t^2$, so

$$\int_C \mathbf{F}(\mathbf{x}) \cdot d\mathbf{x} = \int_0^1 3t^2 \, dt = 1.$$

(c) Here $\mathbf{x}'(t) = \frac{3}{2}\sqrt{t}\,\mathbf{i} + 5t^4\mathbf{j}$ and $\mathbf{F}(x, y) = t^5\mathbf{i} + t^{3/2}\mathbf{j}$, so $\mathbf{F} \cdot \mathbf{x}' = (t^5\mathbf{i} + t^{3/2}\mathbf{j}) \cdot$
$(\frac{3}{2}\sqrt{t}\,\mathbf{i} + 5t^4\mathbf{j}) = \frac{3}{2}t^{11/2} + 5t^{11/2} = \frac{13}{2}t^{11/2}$. Then

$$\int_C \mathbf{F} \cdot \mathbf{x}'\, dt = \int_0^1 \tfrac{13}{2}t^{11/2}\, dt = 1. \; \blacksquare$$

Before stating our next result, we make several definitions.

Definition 2 CONNECTED SET A set Ω in $\mathbb{R}^2$ is **connected** if any two points in Ω can be joined by a piecewise smooth curve lying entirely in Ω.

Definition 3 OPEN SET A set Ω in $\mathbb{R}^2$ is **open** if for every point $\mathbf{x} \in \Omega$, there is an open disk D with center at $\mathbf{x}$ that is contained in Ω.

Definition 4 REGION A **region** Ω in $\mathbb{R}^2$ is an open, connected set.

These definitions are illustrated in Figure 4.

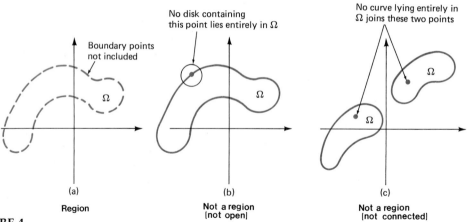

FIGURE 4

(a) Region

(b) Not a region (not open)

(c) Not a region (not connected)

INDEPENDENCE OF PATH

In Example 5 we saw that on three very different curves, we obtained the same answer as we moved between the points $(0, 0)$ and $(1, 1)$. In fact, as we will show in a moment, we will get the same answer if we integrate along any piecewise smooth curve C joining these two points. When this happens, we say that the line integral is **independent of the path.** A condition that ensures that a line integral is independent of the path over which it is integrated is given below.

Theorem 1 Let $\mathbf{F}$ be continuous in a region Ω in $\mathbb{R}^2$. Then $\mathbf{F}$ is the gradient of a differentiable function f if and only if for any piecewise smooth curve C lying in Ω, the line integral $\int_C \mathbf{F}(\mathbf{x}) \cdot d\mathbf{x}$ is independent of the path.

Proof. We first assume that $\mathbf{F}(\mathbf{x}) = \nabla f$ for some differentiable function f. Recall the vector form of the chain rule:

$$\frac{d}{dt} f(\mathbf{x}(t)) = \nabla f(\mathbf{x}(t)) \cdot \mathbf{x}'(t) \tag{14}$$

[see equation (18.6.3)]. Now suppose that C is given by $\mathbf{x}(t)$: $a \le t \le b$, $\mathbf{x}(a) = \mathbf{x}_0$, and $\mathbf{x}(b) = \mathbf{x}_1$. We will assume that C is smooth. Otherwise, we could write the line integral in the form (13) and treat the integral over each smooth curve C_i separately. Using (14), we have

$$\int_C \mathbf{F}(\mathbf{x}) \cdot d\mathbf{x} = \int_a^b \mathbf{F}(\mathbf{x}(t)) \cdot \mathbf{x}'(t)\, dt \overset{\mathbf{F} = \nabla f}{=} \int_a^b \nabla f(\mathbf{x}(t)) \cdot \mathbf{x}'(t)\, dt$$

$$= \int_a^b \frac{d}{dt} f(\mathbf{x}(t))\, dt = f(\mathbf{x}(b)) - f(\mathbf{x}(a)) = f(\mathbf{x}_1) - f(\mathbf{x}_0).$$

This proves that the line integral is independent of the path since $f(\mathbf{x}_1) - f(\mathbf{x}_0)$ does not depend on the particular curve chosen.

We now assume that $\int_C \mathbf{F}(\mathbf{x}) \cdot d\mathbf{x}$ is independent of the path and prove that $\mathbf{F} = \nabla f$ for some differentiable function f. Let $\mathbf{x}_0$ be a fixed point in Ω and let $\mathbf{x}$ be any other point in Ω. Since Ω is connected, there is at least one piecewise smooth path C joining $\mathbf{x}_0$ and $\mathbf{x}$, with C wholly contained in Ω. We define a function f by

$$f(\mathbf{x}) = \int_C \mathbf{F}(\mathbf{x}) \cdot d\mathbf{x}.$$

This function is well defined because, by hypothesis, $\int_C \mathbf{F}(\mathbf{x}) \cdot d\mathbf{x}$ is the same no matter what path is chosen between $\mathbf{x}_0$ and $\mathbf{x}$. Write $\mathbf{x}_0 = (x_0, y_0)$ and $\mathbf{x} = (x, y)$. Since Ω is open, there is an open disk D centered at (x, y) that is contained in Ω. Choose $\Delta x > 0$ such that $(x + \Delta x, y) \in D$, and let C_1 be the horizontal line segment joining (x, y) to $(x + \Delta x, y)$. The situation is depicted in Figure 5. We see that $C \cup C_1$ is a path joining (x_0, y_0) to $(x + \Delta x, y)$, so that

$$f(x + \Delta x, y) = \int_C \mathbf{F}(\mathbf{x}) \cdot d\mathbf{x} + \int_{C_1} \mathbf{F}(\mathbf{x}) \cdot d\mathbf{x} = f(x, y) + \int_{C_1} \mathbf{F}(\mathbf{x}) \cdot d\mathbf{x}. \tag{15}$$

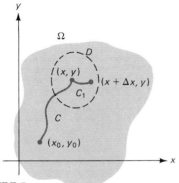

FIGURE 5

A parametrization for C_1 (which is a horizontal line) is

$$\mathbf{x}(t) = (x + t\,\Delta x, y), \qquad 0 \le t \le 1$$

and

$$\mathbf{x}'(t) = (\Delta x, 0).$$

Suppose that $\mathbf{F}(\mathbf{x}) = P(x, y)\mathbf{i} + Q(x, y)\mathbf{j}$. Then

$$\mathbf{F}(\mathbf{x}) \cdot \mathbf{dx} = P(x, y)\,\Delta x. \tag{16}$$

From (15) and (16) we compute

$$f(x + \Delta x, y) - f(x, y) = \int_{C_1} \mathbf{F}(\mathbf{x}) \cdot \mathbf{dx} = \int_0^1 P(x + t\,\Delta x, y)\,\Delta x\,dt,$$

so that

$$\frac{f(x + \Delta x, y) - f(x, y)}{\Delta x} = \int_0^1 P(x + t\,\Delta x, y)\,dt. \tag{17}$$

In (17), $P(x + t\,\Delta x, y)$ is a function of one variable (t), so we may apply the mean value theorem for integrals (Theorem 5.10.2) to see that there is a number $\bar{x}$ with $x < \bar{x} < x + \Delta x$ such that

$$\int_0^1 P(x + t\,\Delta x, y)\,dt = P(\bar{x}, y)\int_0^1 dt = P(\bar{x}, y).$$

Thus taking the limits as $\Delta x \to 0$ on both sides of (17), we have

$$\frac{\partial f(\mathbf{x})}{\partial x} = \lim_{\Delta x \to 0} \frac{f(x + \Delta x, y) - f(x, y)}{\Delta x} = \lim_{\Delta x \to 0} P(\bar{x}, y) \overset{\downarrow}{=} P(x, y).$$

P is continuous and $\bar{x} \to x$ as $\Delta x \to 0$

In a similar manner, we can show that $(\partial f/\partial y)(\mathbf{x}) = Q(x, y)$. This shows that $\mathbf{F} = \nabla f$, and the proof is complete. ∎

In proving this theorem, we also proved the following corollary.

Corollary 1 Suppose that $\mathbf{F}$ is a continuous in a region Ω in $\mathbb{R}^2$ and $\mathbf{F} = \nabla f$ for some differentiable function f. Then for any piecewise smooth curve C in Ω starting at the point $\mathbf{x}_0$ and ending at the point $\mathbf{x}_1$,

$$\int_C \mathbf{F}(\mathbf{x}) \cdot \mathbf{dx} = f(\mathbf{x}_1) - f(\mathbf{x}_0); \tag{18}$$

that is, the value of the integral depends only on the endpoints of the path.

REMARK 1.   This corollary is really the line integral analog of the fundamental the-orem of calculus. It says that we can evaluate the line integral of a gradient field by evaluating at two points the function for which **F** is the gradient.

REMARK 2.   In Theorem 1 it is important that **F** be continuous on Ω, not only on C. (See Problem 29.)

EXAMPLE 5 (Continued).   Since $y\mathbf{i} + x\mathbf{j} = \nabla(xy)$, we immediately find that for any curve C starting at (0, 0) and ending at (1, 1)

$$\int_C \mathbf{F(x)} \cdot \mathbf{dx} = xy\big|_{(1,1)} - xy\big|_{(0,0)} = 1 - 0 = 1. \quad\blacksquare$$

Using Theorem 18.11.1 we have the following obvious corollary to Theorem 1.

Corollary 2   Let $\mathbf{F}(x, y) = P(x, y)\mathbf{i} + Q(x, y)\mathbf{j}$. Then if P, Q, $\partial P/\partial y$, and $\partial Q/\partial x$ are all continuous on a region containing C and if $\partial P/\partial y = \partial Q/\partial x$, then $\int_C \mathbf{F(x)} \cdot \mathbf{dx}$ is independent of the path.

EXAMPLE 6   Let $\mathbf{F}(x, y) = [4x^3y^3 + (1/x)]\mathbf{i} + [3x^4y^2 - (1/y)]\mathbf{j}$. Calculate $\int_C \mathbf{F(x)} \cdot \mathbf{dx}$ for any smooth curve C, not crossing the x- or y-axis, starting at (1, 1) and ending at (2, 3).

Solution.   We showed in Example 18.11.3 that **F** is exact and that $\mathbf{F} = \nabla f$, where $f(x, y) = x^4y^3 + \ln|x/y|$. Thus $\int_C \mathbf{F(x)} \cdot \mathbf{dx}$ is independent of the path, and

$$\int_C \mathbf{F(x)} \cdot \mathbf{dx} = f(2, 3) - f(1, 1) = (16)(27) + \ln\left(\frac{2}{3}\right) - 1 = 431 + \ln\left(\frac{2}{3}\right). \quad\blacksquare$$

EXAMPLE 7   Let $\mathbf{F(x)} = x\mathbf{i} + (x - y)\mathbf{j}$. Let C_1 be the part of the curve $y = x^2$ that connects (0, 0) to (1, 1), and let C_2 be the part of the curve $y = x^3$ that connects these two points. Compute $\int_{C_1} \mathbf{F(x)} \cdot \mathbf{dx}$ and $\int_{C_2} \mathbf{F(x)} \cdot \mathbf{dx}$.

Solution.   Along C_1, $\mathbf{x}(t) = t\mathbf{i} + t^2\mathbf{j}$ and $\mathbf{x}'(t) = \mathbf{i} + 2t\mathbf{j}$ for $0 \le t \le 1$, so that

$$\int_{C_1} \mathbf{F(x)} \cdot \mathbf{dx} = \int_0^1 [t \cdot 1 + (t - t^2)2t]\, dt = \int_0^1 (t + 2t^2 - 2t^3)\, dt$$

$$= \left(\frac{t^2}{2} + \frac{2t^3}{3} - \frac{t^4}{2}\right)\bigg|_0^1 = \frac{2}{3}.$$

Along C_2, $\mathbf{x}(t) = t\mathbf{i} + t^3\mathbf{j}$ and $\mathbf{x}' = \mathbf{i} + 3t^2\mathbf{j}$ for $0 \le t \le 1$, so

$$\int_{C_2} \mathbf{F(x)} \cdot \mathbf{dx} = \int_0^1 [t \cdot 1 + (t - t^3)(3t^2)]\, dt = \int_0^1 (t + 3t^3 - 3t^5)\, dt$$

$$= \left(\frac{t^2}{2} + \frac{3}{4}t^4 - \frac{1}{2}t^6\right)\bigg|_0^1 = \frac{3}{4}.$$

We see that $\int_C \mathbf{F(x)} \cdot \mathbf{dx}$ is *not* independent of the path. Note that here $\partial P/\partial y = 0$ and $\partial Q/\partial x = 1$, so **F** is *not* exact. $\quad\blacksquare$

CH. 20 INTRODUCTION TO VECTOR ANALYSIS

There is another important consequence of Theorem 1. Let C be a closed curve (i.e., $\mathbf{x}_0 = \mathbf{x}_1$).

> If $\mathbf{F}$ is continuous in a region Ω and if $\mathbf{F}$ is the gradient of a differentiable function f, then for any closed curve C lying in Ω,
>
> $$\int_C \mathbf{F}(\mathbf{x}) \cdot d\mathbf{x} = f(\mathbf{x}_1) - f(\mathbf{x}_0) = 0.$$

There is an interesting physical interpretation of this fact:

> *The work done by a conservative force field as it moves a particle completely around a closed path is zero.†*

This also follows from the more general fact that if $\mathbf{F}$ is conservative, then the work done in moving an object between two points depends only on the points, and not on the path taken between the points.

PROBLEMS 20.2

In Problems 1–14, calculate $\int_C \mathbf{F}(\mathbf{x}) \cdot d\mathbf{x}$.

1. $\mathbf{F}(x, y) = x^2\mathbf{i} + y^2\mathbf{j}$; C is the straight line from $(0, 0)$ to $(2, 4)$.
2. $\mathbf{F}(x, y) = x^2\mathbf{i} + y^2\mathbf{j}$; C is the parabola $y = x^2$ from $(0, 0)$ to $(2, 4)$.
3. $\mathbf{F}(x, y) = xy\mathbf{i} + (y - x)\mathbf{j}$; C is the line $y = 2x - 4$ from $(1, -2)$ to $(2, 0)$.
4. $\mathbf{F}(x, y) = xy\mathbf{i} + (y - x)\mathbf{j}$; C is the curve $y = \sqrt{x}$ from $(0, 0)$ to $(1, 1)$.
5. $\mathbf{F}(x, y) = xy\mathbf{i} + (y - x)\mathbf{j}$; C is the unit circle in the counterclockwise direction.
6. $\mathbf{F}(x, y) = xy\mathbf{i} + (y - x)\mathbf{j}$; C is the triangle joining the points $(0, 0)$, $(0, 1)$, and $(1, 0)$ in the counterclockwise direction.
7. $\mathbf{F}(x, y) = xy\mathbf{i} + (y - x)\mathbf{j}$; C is the triangle joining the points $(0, 0)$, $(1, 0)$, and $(1, 1)$ in the counterclockwise direction.
8. $\mathbf{F}(x, y) = e^x\mathbf{i} + e^y\mathbf{j}$; C is the curve of Problem 4.
9. $\mathbf{F}(x, y) = (x^2 + 2y)\mathbf{i} - y^2\mathbf{j}$; C is the part of the ellipse $x^2 + 9y^2 = 9$ joining the points $(0, -1)$ and $(0, 1)$ in the clockwise direction.
10. $\mathbf{F}(x, y) = (\cos x)\mathbf{i} - (\sin y)\mathbf{j}$; C is the curve of Problem 9.
11. $\mathbf{F}(x, y) = e^{x+y}\mathbf{i} + e^{x-y}\mathbf{j}$; C is the curve of Problem 6.
12. $\mathbf{F}(x, y) = e^{x+y}\mathbf{i} + e^{x-y}\mathbf{j}$; C is the curve of Problem 7.
13. $\mathbf{F}(x, y) = (y/x^2)\mathbf{i} + (x/y^2)\mathbf{j}$; C is the straight line segment from $(2, 1)$ to $(4, 6)$.
14. $\mathbf{F}(x, y) = (\ln x)\mathbf{i} + (\ln y)\mathbf{j}$; C is the curve $\mathbf{x}(t) = 2t\mathbf{i} + t^3\mathbf{j}$ for $1 \le t \le 4$.

In Problems 15–19 forces are given in newtons and distances are given in meters.

15. Calculate the work done when a force field $\mathbf{F}(x, y) = x^3\mathbf{i} + xy\mathbf{j}$ moves a particle from the point $(0, 1)$ to the point $(1, e^{\pi/2})$ along the curve $\mathbf{x}(t) = (\sin t)\mathbf{i} + e^t\mathbf{j}$.
16. Calculate the work done when the force field $\mathbf{F}(x, y) = xy\mathbf{i} + (2x^3 - y)\mathbf{j}$ moves a particle around the unit circle in the counterclockwise direction.

---

†This is true in general when the closed path is contained in a region over which the force field is continuous.

17. What is the work done if the particle in Problem 16 is moved in the clockwise direction?

18. Calculate the work done by the force field $\mathbf{F}(x, y) = -y^2 x \mathbf{i} + 2x \mathbf{j}$ when a particle is moved around the ellipse $(x^2/a^2) + (y^2/b^2) = 1$ in the counterclockwise direction.

19. Calculate the work done by the force field $\mathbf{F}(x, y) = 2xy \mathbf{i} + y^2 \mathbf{j}$ when a particle is moved around the triangle of Problem 7.

20. What is the work done on the 1-coulomb particle of Example 2 if it moves along the line $y = 2x - 3$ from $(1, -1)$ to $(2, 1)$?

In Problems 21–28, show that $\mathbf{F}$ is exact and use Theorem 1 to calculate $\int_C \mathbf{F}(\mathbf{x}) \cdot d\mathbf{x}$, where C is any smooth curve starting at $\mathbf{x}_0$ and ending at $\mathbf{x}_1$.

21. $\mathbf{F}(x, y) = 2xy \mathbf{i} + (x^2 + 1)\mathbf{j};\ \mathbf{x}_0 = (0, 1),\ \mathbf{x}_1 = (2, 3)$
22. $\mathbf{F}(x, y) = (4x^2 - 4y^2)\mathbf{i} + (\ln y - 8xy)\mathbf{j};\ \mathbf{x}_0 = (-1, 1),\ \mathbf{x}_1 = (4, e)$
23. $\mathbf{F}(x, y) = [x \cos(x + y) + \sin(x + y)]\mathbf{i} + x \cos(x + y)\mathbf{j};\ \mathbf{x}_0 = (0, 0),\ \mathbf{x}_1 = (\pi/6, \pi/3)$

24. $\mathbf{F}(x, y) = \left[\dfrac{1}{x^2} + y^2\right]\mathbf{i} - 2xy \mathbf{j};\ \mathbf{x}_0 = (1, 4),\ \mathbf{x}_1 = (3, 2)$

25. $\mathbf{F}(x, y) = 2x(\cos y)\mathbf{i} - x^2(\sin y)\mathbf{j};\ \mathbf{x}_0 = (0, \pi/2),\ \mathbf{x}_1 = (\pi/2, 0)$
26. $\mathbf{F}(x, y) = (2xy^3 - 2)\mathbf{i} + (3x^2 y^2 + \cos y)\mathbf{j};\ \mathbf{x}_0 = (1, 0),\ \mathbf{x}_1 = (0, -\pi)$
27. $\mathbf{F}(x, y) = e^y \mathbf{i} + xe^y \mathbf{j};\ \mathbf{x}_0 = (0, 0),\ \mathbf{x}_1 = (5, 7)$
28. $\mathbf{F}(x, y) = (\cosh x)(\cosh y)\mathbf{i} + (\sinh x)(\sinh y)\mathbf{j};\ \mathbf{x}_0 = (0, 0),\ \mathbf{x}_1 = (1, 2)$

*29. Let the force field $\mathbf{F}$ be given by

$$\mathbf{F}(x, y) = \frac{y}{x^2 + y^2}\mathbf{i} - \frac{x}{x^2 + y^2}\mathbf{j}.$$

(a) Show that $\mathbf{F}$ is exact.
(b) Calculate $\int_C \mathbf{F}(\mathbf{x}) \cdot d\mathbf{x}$, where C is the unit circle in the counterclockwise direction.
(c) Explain why $\int_C \mathbf{F}(\mathbf{x}) \cdot d\mathbf{x} \neq 0$. Does this result contradict the last fact stated in this section?

20.3 GREEN'S THEOREM IN THE PLANE

In this section we state a result that gives an important relationship between line integrals and double integrals.

Let Ω be a region in the plane (a typical region is sketched in Figure 1). The curve (indicated by the arrows) that goes around the edge of Ω in the direction that keeps

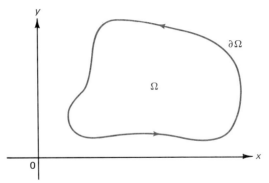

FIGURE 1

Ω on the left (sometimes called the *counterclockwise* direction) is called the **boundary of** Ω and is denoted $\partial\Omega$. Let

$$\mathbf{F}(x, y) = P(x, y)\mathbf{i} + Q(x, y)\mathbf{j}.$$

If the curve $\partial\Omega$ is given by

$$\partial\Omega: \quad \mathbf{x}(t) = x(t)\mathbf{i} + y(t)\mathbf{j},$$

we can write

$$\mathbf{F}(\mathbf{x}) \cdot \mathbf{dx} = P\, dx + Q\, dy. \tag{1}$$

We then denote the line integral of $\mathbf{F}$ around $\partial\Omega$ by

$$\oint_{\partial\Omega} P\, dx + Q\, dy. \tag{2}$$

The symbol $\oint_{\partial\Omega}$ indicates that $\partial\Omega$ is a closed curve around which we integrate in the counterclockwise direction (the direction of the arrow).

Definition 1 SIMPLE CURVE A curve C is called **simple** if it does not cross itself. That is, suppose C is given by

$$C: \quad \mathbf{x}(t) = f(t)\mathbf{i} + g(t)\mathbf{j}, \quad t \in [a, b].$$

Then C is simple if and only if $\mathbf{x}(t_1) \neq \mathbf{x}(t_2)$ whenever $t_1 \neq t_2$.
This notion is illustrated in Figure 2.

Definition 2 SIMPLY CONNECTED REGION A region Ω in the xy-plane is called **simply connected** if it has the following property: If C is a simple closed curve contained in Ω, then every point in the region enclosed by C is also in Ω. Intuitively, a region is simply connected if it has no holes.
We illustrate this definition in Figure 3.

We are now ready to state the principal result of this section.

Theorem 1 *Green's Theorem† in the Plane* Let Ω be a simply connected region in the xy-plane bounded by a piecewise smooth curve $\partial\Omega$. Let P and Q be continuous with continuous first partials in an open disk containing Ω. Then

---

†Named after George Green (1793–1841), a British mathematician and physicist who wrote an essay in 1828 on electricity and magnetism that contained this important theorem. Green was the self-educated son of a baker. His 1828 essay was published for private circulation. It was largely overlooked until it was rediscovered by Lord Kelvin in 1846. The theorem was independently discovered by the Russian mathematician Michel Ostrogradski (1801–1861), and to this day the theorem is known in the Soviet Union as *Ostrogradski's theorem.*

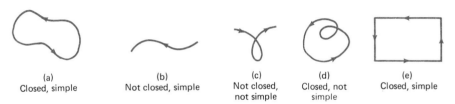

(a)
Closed, simple

(b)
Not closed, simple

(c)
Not closed,
not simple

(d)
Closed, not
simple

(e)
Closed, simple

FIGURE 2

Every point on this
line segment is removed

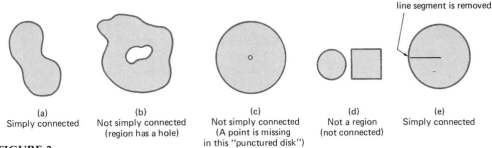

(a)
Simply connected

(b)
Not simply connected
(region has a hole)

(c)
Not simply connected
(A point is missing
in this "punctured disk")

(d)
Not a region
(not connected)

(e)
Simply connected

FIGURE 3

$$\oint_{\partial\Omega} P \, dx + Q \, dy = \iint_{\Omega} \left(\frac{\partial Q}{\partial x} - \frac{\partial P}{\partial y} \right) dx \, dy. \tag{3}$$

REMARK.   This theorem shows how the line integral of a function around the boundary of a region is related to a double integral over that region.

Partial Proof of Green's Theorem.   We prove Green's theorem in the case in which Ω takes the simple form given in Figure 4. The region Ω can be written

$$\{(x, y): a \le x \le b, \, g_1(x) \le y \le g_2(x)\}, \tag{4}$$

and

$$\{(x, y): c \le y \le d, \, h_1(y) \le x \le h_2(y)\}. \tag{5}$$

We first calculate

$$\iint_{\Omega} \frac{\partial P}{\partial y} \, dx \, dy = \int_a^b \left\{ \int_{g_1(x)}^{g_2(x)} \frac{\partial P}{\partial y} \, dy \right\} dx = \int_a^b \left\{ P(x, y) \Big|_{y=g_1(x)}^{y=g_2(x)} \right\} dx$$

$$= \int_a^b [P(x, g_2(x)) - P(x, g_1(x))] \, dx. \tag{6}$$

Now

$$\oint_{\partial\Omega} P \, dx = \int_{C_1} P \, dx + \int_{C_2} P \, dx, \tag{7}$$

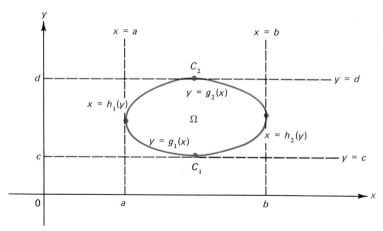

FIGURE 4

where C_1 is the graph of $g_1(x)$ from $x = a$ to $x = b$ and C_2 is the graph of $g_2(x)$ from $x = b$ to $x = a$ (note the order). Since on C_1, $y = g_1(x)$, we have

$$\int_{C_1} P(x, y) \, dx = \int_a^b P(x, g_1(x)) \, dx. \tag{8}$$

Similarly,

$$\int_{C_2} P(x, y) \, dx = \int_b^a P(x, g_2(x)) \, dx = -\int_a^b P(x, g_2(x)) \, dx. \tag{9}$$

Thus

$$\oint_{\partial\Omega} P \, dx = \int_a^b P(x, g_1(x)) \, dx - \int_a^b P(x, g_2(x)) \, dx$$

$$= -\int_a^b [P(x, g_2(x)) - P(x, g_1(x))] \, dx. \tag{10}$$

Comparing (6) and (10), we find that

$$\iint_\Omega -\frac{\partial P}{\partial y} \, dx \, dy = \oint_{\partial\Omega} P \, dx. \tag{11}$$

Similarly, using the representation (5), we have

$$\iint_\Omega \frac{\partial Q}{\partial x} \, dx \, dy = \int_c^d \left\{ \int_{h_1(y)}^{h_2(y)} \frac{\partial Q}{\partial x} \, dx \right\} dy$$

$$= \int_c^d [Q(h_2(y), y) - Q(h_1(y), y)] \, dy,$$

which by analogous reasoning yields the equation

$$\iint_\Omega \frac{\partial Q}{\partial x}\, dx\, dy = \oint_{\partial\Omega} Q\, dy. \tag{12}$$

Adding (11) and (12) completes the proof of the theorem in the special case that Ω can be written in the form (4) and (5).† ∎

REMARK. The relationship between independence of path and Green's theorem is given in the diagram below. Here $\mathbf{F}(x, y) = P(x, y)\mathbf{i} + Q(x, y)\mathbf{j}$.

$\int_c \mathbf{F} \cdot d\mathbf{x}$ is independent of path

$\Longleftrightarrow$ If and only if

there is an f such that $\nabla f = \mathbf{F}$

$\frac{\partial P}{\partial y} = \frac{\partial Q}{\partial x}$

$\oint_c \mathbf{F} \cdot d\mathbf{x} = 0$ around all simple closed curves $\Leftrightarrow$ The hypotheses of Green's theorem apply to the region

EXAMPLE 1 Evaluate $\oint_{\partial\Omega} xy\, dx + (x - y)\, dy$, where Ω is the rectangle $\{(x, y): 0 \le x \le 1, 1 \le y \le 3\}$.

Solution. $P(x, y) = xy$, $Q(x, y) = x - y$, $\partial Q/\partial x = 1$, and $\partial P/\partial y = x$, so

$$\oint_{\partial\Omega} xy\, dx + (x - y)\, dy = \int_0^1 \int_1^3 (1 - x)\, dy\, dx = \int_0^1 \left\{ (1 - x)y \Big|_1^3 \right\} dx$$

$$= \int_0^1 2(1 - x)\, dx = 1. \ \blacksquare$$

EXAMPLE 2 Evaluate $\oint_C (x^3 + y^3)\, dx + (2y^3 - x^3)\, dy$, where C is the unit circle.

Solution. We first note that $C = \partial\Omega$, where Ω is the unit disk. Next, we have $(\partial Q/\partial x) - (\partial P/\partial y) = -3x^2 - 3y^2 = -3(x^2 + y^2)$. Thus

$$\oint_C (x^3 + y^3)\, dx + (2y^3 - x^3)\, dy = -3 \iint_{\text{unit disk}} (x^2 + y^2)\, dx\, dy$$

Converting to polar coordinates

$$= -3 \int_0^{2\pi} \int_0^1 r^2 \cdot r\, dr\, d\theta$$

$$= -3 \int_0^{2\pi} \left\{ \frac{r^4}{4} \Big|_0^1 \right\} d\theta$$

$$= -\frac{3}{4} \int_0^{2\pi} d\theta = -\frac{3\pi}{2}. \ \blacksquare$$

Green's theorem can be useful for calculating area. Recall that

† For a proof of Green's theorem for more general regions, see R. C. Buck, *Advanced Calculus* (McGraw-Hill, New York, 1965), page 408.

area enclosed by $\Omega = \iint\limits_{\Omega} dA.$ (13)

But by Green's theorem,

$$\iint\limits_{\Omega} dA = \oint\limits_{\partial\Omega} x \, dy = \oint\limits_{\partial\Omega} (-y) \, dx = \frac{1}{2} \oint\limits_{\partial\Omega} [(-y) \, dx + x \, dy]$$ (14)

(explain why). Any of the line integrals in (14) can be used to calculate area.

EXAMPLE 3   Use Green's theorem to calculate the area enclosed by the ellipse $(x^2/a^2) + (y^2/b^2) = 1$.

Solution. The ellipse can be written parametrically as

$$\mathbf{x}(t) = a(\cos t)\mathbf{i} + b(\sin t)\mathbf{j}, \qquad 0 \le t \le 2\pi.$$

Then using the first line integral in (14), we obtain

$$A = \oint\limits_{\partial\Omega} x \, dy = \int_0^{2\pi} (a \cos t) \frac{d}{dt} (b \sin t) \, dt$$

$$= \int_0^{2\pi} (a \cos t) b \cos t \, dt = ab \int_0^{2\pi} \cos^2 t \, dt$$

$$= \frac{ab}{2} \int_0^{2\pi} (1 + \cos 2t) \, dt = \pi ab.$$

Note how much easier this calculation is than the direct evaluation of $\iint_A dx \, dy$, where A denotes the area enclosed by the ellipse. ▪

There are two very interesting and important vector interpretations of Green's theorem.

Definition 3   CURL AND DIVERGENCE   Let $\mathbf{F}(x, y) = P(x, y)\mathbf{i} + Q(x, y)\mathbf{j}$ be a vector field in the plane.

(i) The **curl** of $\mathbf{F}$ is given by

$$\text{curl } \mathbf{F} = \frac{\partial Q}{\partial x} - \frac{\partial P}{\partial y}.$$ (15)

(ii) The **divergence** of $\mathbf{F}$, denoted div $\mathbf{F}$, is given by

$$\text{div } \mathbf{F} = \frac{\partial P}{\partial x} + \frac{\partial Q}{\partial y}.$$ (16)

Before explaining these terms more fully, we will write Green's theorem in two equivalent vector forms.

First, recall from equation (16.6.2) that if $\mathbf{T}$ denotes the unit tangent vector to a curve $\mathbf{x}(t) = x(t)\mathbf{i} + y(t)\mathbf{j}$ and if s denotes the parameter of arc length, then

$$\mathbf{T} = \frac{d\mathbf{x}}{ds},$$

or

$$d\mathbf{x} = \mathbf{T}\, ds. \tag{17}$$

Now let $\mathbf{T}$ denote the unit tangent vector to $\partial\Omega$. Then if $\mathbf{F} = P\mathbf{i} + Q\mathbf{j}$, we can write, using (17),

$$\oint_{\partial\Omega} P\, dx + Q\, dy = \oint_{\partial\Omega} \mathbf{F} \cdot d\mathbf{x} = \oint_{\partial\Omega} \mathbf{F} \cdot \mathbf{T}\, ds. \tag{18}$$

Then applying Green's theorem and using (15) and (18), we obtain the following theorem.

Theorem 2 *First Vector Form of Green's Theorem†* Under the hypotheses of Theorem 1,

$$\oint_{\partial\Omega} \mathbf{F} \cdot \mathbf{T}\, ds = \iint_{\Omega} \text{curl } \mathbf{F}\, dx\, dy. \tag{19}$$

Now from the expression

$$\oint_{\partial\Omega} P\, dx + Q\, dy = \iint_{\Omega} \left(\frac{\partial Q}{\partial x} - \frac{\partial P}{\partial y} \right) dx\, dy,$$

we may replace P by $-Q$ and Q by P to obtain

$$\oint_{\partial\Omega} -Q\, dx + P\, dy = \iint_{\Omega} \left(\frac{\partial P}{\partial x} + \frac{\partial Q}{\partial y} \right) dx\, dy. \tag{20}$$

If $dx\, \mathbf{i} + dy\, \mathbf{j}$ represents the vector $\mathbf{T}\, ds$, then $dy\, \mathbf{i} - dx\, \mathbf{j}$ represents the vector $\mathbf{n}\, ds$, where $\mathbf{n}$ is the unit normal vector to the curve. This is easy to see since $(dx\, \mathbf{i} + dy\, \mathbf{j}) \cdot (dy\, \mathbf{i} - dx\, \mathbf{j}) = 0$ and both $\mathbf{T}\, ds$ and $\mathbf{n}\, ds$ have the same magnitude. Thus the left-hand side of (20) can be written

$$\oint_{\partial\Omega} -Q\, dx + P\, dy = \oint_{\partial\Omega} \mathbf{F} \cdot \mathbf{n}\, ds. \tag{21}$$

---

†This form of Green's theorem is sometimes called **Stokes's theorem in the plane.**

Using (16) and (21) in (20) yields the next theorem.

Theorem 3 *Second Vector Form of Green's Theorem†* Under the hypotheses of Theorem 1,

$$\oint_{\partial \Omega} \mathbf{F} \cdot \mathbf{n} \, ds = \iint_{\Omega} \text{div } \mathbf{F} \, dx \, dy. \tag{22}$$

CIRCULATION AND FLUX

There are interesting physical interpretations of the two vector forms of Green's theorem. Let $\mathbf{F}(x, y)$ denote the direction and rate of flow of a fluid at a point (x, y) in the plane. The integral

$$\oint_{\partial \Omega} \mathbf{F} \cdot \mathbf{T} \, ds$$

is the integral of the component of the flow in the direction tangent to the boundary of Ω and is called the **circulation** of $\mathbf{F}$ around the boundary of Ω (see Figure 5). If Ω is small, then curl $\mathbf{F}$ is nearly a constant, and by (19) we have

$$\oint_{\partial \Omega} \mathbf{F} \cdot \mathbf{T} \, ds = \iint_{\Omega} \text{curl } \mathbf{F} \, dx \, dy \approx \text{curl } \mathbf{F}(x, y) \iint_{\Omega} dx \, dy$$

$$= [\text{curl } \mathbf{F}(x, y)](\text{area of } \Omega). \tag{23}$$

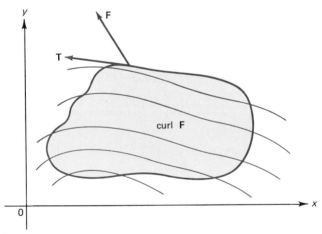

FIGURE 5

From (23) it appears that the curl represents the circulation per unit area at the point (x, y). If curl $\mathbf{F} = 0$ for every (x, y) in Ω, then the fluid flow $\mathbf{F}$ is called **irrotational.**

---

†This form of Green's theorem is sometimes called the **divergence theorem in the plane.**

Now $\oint_{\partial\Omega} \mathbf{F} \cdot \mathbf{n} \, ds$ is the component of flow in the direction of the outward normal to $\partial\Omega$ and is called the **flux** across $\partial\Omega$ (see Figure 6). The flux is the rate at which fluid is flowing across the boundary of Ω from inside Ω. If Ω is small, then using (22), we have

$$\oint_{\partial\Omega} \mathbf{F} \cdot \mathbf{n} \, ds = \iint_{\Omega} \text{div } \mathbf{F} \, dx \, dy \approx [\text{div } \mathbf{F}(x, y)](\text{area of } \Omega).\tag{24}$$

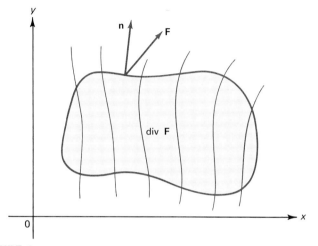

FIGURE 6

Thus the divergence of $\mathbf{F}$ represents the net rate of flow away from (x, y). If div $\mathbf{F} = 0$ for every (x, y) in Ω, then the flow $\mathbf{F}$ is called **incompressible,** because if div $\mathbf{F} = 0$, then the net flow into the region is 0. That is, fluid is not accumulating. The accumulation of fluid in a fixed region would cause it to become compressed.

Before leaving this section we note that there are other ways to write the curl and divergence, and therefore there are other ways to write Green's theorem in the plane. Very loosely speaking, we can think of the gradient symbol ∇ as the vector function $(\partial/\partial x, \partial/\partial y)$. Then if $\mathbf{F} = (P, Q)$,

$$\nabla \times \mathbf{F} = \begin{vmatrix} \mathbf{i} & \mathbf{j} & \mathbf{k} \\ \dfrac{\partial}{\partial x} & \dfrac{\partial}{\partial y} & 0 \\ P & Q & 0 \end{vmatrix} = \left(\dfrac{\partial Q}{\partial x} - \dfrac{\partial P}{\partial y} \right) \mathbf{k} = \text{curl } \mathbf{F} \, \mathbf{k}.$$

Thus curl $\mathbf{F} = (\text{curl } \mathbf{F} \, \mathbf{k}) \cdot \mathbf{k} = (\nabla \times \mathbf{F}) \cdot \mathbf{k}$ (since $\mathbf{k} \cdot \mathbf{k} = 1$). For this reason curl $\mathbf{F}$ is often written (with the $\mathbf{k}$ omitted) as

$$\text{curl } \mathbf{F} = \nabla \times \mathbf{F}.\tag{25}$$

Similarly,

$$\nabla \cdot \mathbf{F} = \left(\frac{\partial}{\partial x}, \frac{\partial}{\partial y}\right) \cdot (P, Q) = \frac{\partial P}{\partial x} + \frac{\partial Q}{\partial y} = \text{div } \mathbf{F},$$

and we write

$$\text{div } \mathbf{F} = \nabla \cdot \mathbf{F}. \tag{26}$$

Then using (25), (26), and Theorems 2 and 3, we have the following forms of Green's theorem:

$$\oint_{\partial\Omega} \mathbf{F} \cdot \mathbf{T}\, ds = \iint_{\Omega} (\nabla \times \mathbf{F}) \cdot \mathbf{k}\, dx\, dy \tag{27}$$

and

$$\oint_{\partial\Omega} \mathbf{F} \cdot \mathbf{n}\, ds = \iint_{\Omega} \nabla \cdot \mathbf{F}\, dx\, dy. \tag{28}$$

PROBLEMS 20.3

In Problems 1–15, find the line integral by using Green's theorem.

1. $\oint_{\partial\Omega} 3y\, dx + 5x\, dy$; $\Omega = \{(x, y): 0 \le x \le 1, 0 \le y \le 1\}$.
2. $\oint_{\partial\Omega} ay\, dx + bx\, dy$; Ω is the region of Problem 1.
3. $\oint_{\partial\Omega} e^x \cos y\, dx + e^x \sin y\, dy$; Ω is the region enclosed by the triangle with vertices at $(0, 0)$, $(1, 0)$, and $(0, 1)$.
4. The integral of Problem 3, where Ω is the region enclosed by the triangle with vertices at $(0, 0)$, $(1, 1)$, and $(1, 0)$.
5. The integral of Problem 3, where Ω is the region enclosed by the rectangle with vertices at $(0, 0)$, $(2, 0)$, $(2, 1)$, and $(0, 1)$.
6. $\oint_{\partial\Omega} 2xy\, dx + x^2\, dy$; Ω is the unit disk.
7. $\oint_{\partial\Omega}(x^2 + y^2)\, dx - 2xy\, dy$; Ω is the unit disk.
8. $\oint_{\partial\Omega}(1/y)\, dx + (1/x)\, dy$; Ω is the region bounded by the lines $y = 1$ and $x = 16$ and the curve $y = \sqrt{x}$.
9. $\oint_{\partial\Omega} \cos y\, dx + \cos x\, dy$; Ω is the region enclosed by the rectangle $\{(x, y): 0 \le x \le \pi/4, 0 \le y \le \pi/3\}$.
10. $\oint_{\partial\Omega} x^2 y\, dx - xy^2\, dy$; Ω is the disk $x^2 + y^2 \le 9$.
11. $\oint_{\partial\Omega} y \ln x\, dy$; $\Omega = \{(x, y): 1 \le y \le 3, e^y \le x \le e^{y^3}\}$.
12. $\oint_{\partial\Omega} \sqrt{1 + y^2}\, dx$; $\Omega = \{(x, y): -1 \le y \le 1, y^2 \le x \le 1\}$.
13. $\oint_{\partial\Omega} ay\, dx + bx\, dy$; Ω is a region of the type (4), (5).
14. $\oint_{\partial\Omega} e^x \sin y\, dx + e^x \cos y\, dy$; Ω is the region enclosed by the ellipse $(x^2/a^2) + (y^2/b^2) = 1$.
15. $\oint_{\partial\Omega}(-4x/\sqrt{1 + y^2})\, dx + (2x^2 y/(1 + y^2)^{3/2})\, dy$; Ω is a region of the type (4), (5).

16. Use one of the line integrals in (14) to calculate the area of the circle $\mathbf{x}(t) = a(\cos t)\mathbf{i} + a(\sin t)\mathbf{j}$.
17. Use Green's theorem to calculate the area enclosed by the triangle with vertices at (a_1, b_1), (a_2, b_2), and (a_3, b_3), assuming that the three points are not collinear.

18. Use Green's theorem to calculate the area of the quadrilateral with vertices at $(0, 0)$, $(2, 1)$, $(-1, 3)$, and $(4, 4)$.

19. Use Green's theorem to calculate the area of the quadrilateral with vertices at (a_1, b_1), (a_2, b_2), (a_3, b_3), and (a_4, b_4), assuming that no three of the points are collinear and that no point is within the triangle whose vertices are the other three points.

In Problems 20–26, calculate (a) curl $\mathbf{F}$, (b) $\oint_{\partial\Omega} \mathbf{F} \cdot \mathbf{T} \, ds$, (c) div $\mathbf{F}$, and (d) $\oint_{\partial\Omega} \mathbf{F} \cdot \mathbf{n} \, ds$.

20. $\mathbf{F}(x, y) = x^2\mathbf{i} + y^2\mathbf{j}$; Ω is the region of Problem 1.
21. $\mathbf{F}(x, y) = y^2\mathbf{i} + x^2\mathbf{j}$; Ω is the region of Problem 1.
22. $\mathbf{F}(x, y) = ay\mathbf{i} + bx\mathbf{j}$; Ω is the region of Problem 1.
23. $\mathbf{F}(x, y) = y^3\mathbf{i} + x^3\mathbf{j}$; Ω is the unit disk.
24. $\mathbf{F}(x, y) = x\mathbf{i} + y\mathbf{j}$; Ω is the unit disk.
25. $\mathbf{F}(x, y) = y\mathbf{i} - x\mathbf{j}$; Ω is the unit disk.
26. $\mathbf{F}(x, y) = xy\mathbf{i} + (y^2 - x^2)\mathbf{j}$; Ω is the region of Problem 3.

27. Let $\partial\Omega$ be the ellipse $(x^2/a^2) + (y^2/b^2) = 1$. Let $\mathbf{F}(x, y)$ be the vector field $-x\mathbf{i} - y\mathbf{j}$ that, at any point (x, y), points toward the origin. Show that $\oint_{\partial\Omega} \mathbf{F} \cdot \mathbf{T} \, ds = 0$.
*28. Let Ω be the disk $x^2 + y^2 \leq a^2$. Show that $\oint_{\partial\Omega} \alpha\sqrt{x^2 + y^2} \, dx + \beta\sqrt{x^2 + y^2} \, dy = 0$.
29. Let Ω be as in Problem 28 and suppose that g is continuously differentiable. Show that $\oint_{\partial\Omega} \alpha g(x^2 + y^2) \, dx + \beta g(x^2 + y^2) \, dy = 0$.
30. Show that the vector flow $\mathbf{F}(x, y) = x\mathbf{i} + y\mathbf{j}$ is irrotational.
31. Show that the vector flow $\mathbf{F}(x, y) = \sin x \, e^x\mathbf{i} + y^{5/2}\mathbf{j}$ is irrotational.
32. Show that for any continuously differentiable functions f and g, the vector flow $\mathbf{F}(x, y) = f(x)\mathbf{i} + g(y)\mathbf{j}$ is irrotational.
33. Show that the vector flow $\mathbf{F}(x, y) = y\sqrt{x^2 + y^2}\mathbf{i} - x\sqrt{x^2 + y^2}\mathbf{j}$ is incompressible.
34. Let g be as in Problem 29. Show that the vector flow $\mathbf{F}(x, y) = -yg(x^2 + y^2)\mathbf{i} + xg(x^2 + y^2)\mathbf{j}$ is incompressible.
35. Let $\mathbf{F}(x, y) = y/(x^2 + y^2)\mathbf{i} - x/(x^2 + y^2)\mathbf{j}$.
   (a) Show that curl $\mathbf{F} = 0$.
   (b) Show that $\oint_C \mathbf{F} \cdot \mathbf{T} \, ds \neq 0$ if C is the unit circle oriented counterclockwise.
   (c) Explain why the results of (a) and (b) do not contradict Theorem 2.

20.4 LINE INTEGRALS IN SPACE

In this section we begin a development that will extend the results of the last two sections to $\mathbb{R}^3$. We start with the notion of a line integral in space, which is the natural generalization of a line integral in the plane.

Suppose that a curve C in space is given parametrically as

$$C: \quad \mathbf{x}(t) = f(t)\mathbf{i} + g(t)\mathbf{j} + h(t)\mathbf{k}, \tag{1}$$

or

$$C: \quad \mathbf{x}(t) = x(t)\mathbf{i} + y(t)\mathbf{j} + z(t)\mathbf{k}. \tag{2}$$

As in Section 20.2, we assume that a force $\mathbf{F}$ is applied to a particle moving along the curve. Now $\mathbf{F} = \mathbf{F}(x, y, z)$ is a vector field in $\mathbb{R}^3$, and we can write

$$\mathbf{F}(\mathbf{x}) = \mathbf{F}(x, y, z) = P(x, y, z)\mathbf{i} + Q(x, y, z)\mathbf{j} + R(x, y, z)\mathbf{k}. \tag{3}$$

As before, we will assume that the curve C is *smooth* or *piecewise smooth*. That is, the functions $f(t)$, $g(t)$, and $h(t)$ are continuously differentiable, or f', g', and h' exist and are piecewise continuous.

Following the development in Section 20.2 [see equation (20.2.7)], we find that the work done in moving along the curve from $\mathbf{x}(t)$ to $\mathbf{x}(t + \Delta t)$ is, if Δt is small, given approximately by

$$W(t + \Delta t) - W(t) \approx \mathbf{F}(\mathbf{x}(t)) \cdot [\mathbf{x}(t + \Delta t) - \mathbf{x}(t)], \tag{4}$$

so that, as before,

$$W'(t) = \mathbf{F}(\mathbf{x}(t)) \cdot \mathbf{x}'(t) \tag{5}$$

and

$$W = \text{total work done} = \int_a^b \mathbf{F}(\mathbf{x}(t)) \cdot \mathbf{x}'(t) \, dt, \tag{6}$$

which we write as

$$W = \int_C \mathbf{F}(\mathbf{x}) \cdot d\mathbf{x}. \tag{7}$$

We see that there is no essential difference between the formula for work in $\mathbb{R}^2$ and the formula for work in $\mathbb{R}^3$.

EXAMPLE 1   A particle moves along the elliptical helix (see Example 17.7.1) $\mathbf{x}(t) = (\cos t)\mathbf{i} + 4(\sin t)\mathbf{j} + t\mathbf{k}$. It is subject to a force given by the vector field $x^2\mathbf{i} + y^2\mathbf{j} + 2xyz\mathbf{k}$. How much work is done in moving from the point $(1, 0, 0)$ to the point $(0, 4, \pi/2)$ if forces are measured in newtons and distances are measured in meters?

Solution. The curve C is given by

$$\mathbf{x}(t) = (\cos t)\mathbf{i} + (4 \sin t)\mathbf{j} + t\mathbf{k}, \qquad 0 \le t \le \frac{\pi}{2}.$$

Then

$$\mathbf{x}'(t) = (-\sin t)\mathbf{i} + 4(\cos t)\mathbf{j} + \mathbf{k}$$
$$\mathbf{F}(\mathbf{x}(t)) = (\cos^2 t)\mathbf{i} + 16(\sin^2 t)\mathbf{j} + 8t(\sin t)(\cos t)\mathbf{k}$$

and

$$\mathbf{F}(\mathbf{x}(t)) \cdot \mathbf{x}'(t) = -\sin t \cos^2 t + 64 \cos t \sin^2 t + 8t \sin t \cos t.$$

Thus from (6)

$$W = \int_0^{\pi/2} (-\sin t \cos^2 t + 64 \cos t \sin^2 t + 8t \sin t \cos t) \, dt$$

$$= \left(\frac{\cos^3 t}{3} + \frac{64 \sin^3 t}{3}\right)\Big|_0^{\pi/2} + \int_0^{\pi/2} 8t \sin t \cos t \, dt = 21 + \int_0^{\pi/2} 8t \sin t \cos t \, dt.$$

Now

$$\int 8t \sin t \cos t \, dt = 4 \int t \sin 2t \, dt.$$

We integrate by parts. Let $u = t$ and $dv = \sin 2t \, dt$; then $du = dt$, $v = (-\cos 2t)/2$, and

$$\int_0^{\pi/2} 8t \sin t \cos t \, dt = -4t\frac{\cos 2t}{2}\Big|_0^{\pi/2} + 4 \int_0^{\pi/2} \frac{\cos 2t}{2} \, dt = \pi.$$

Thus

$$W = 21 + \pi \text{ joules.} \quad \blacksquare$$

A **region** S in space is, as in $\mathbb{R}^2$, an open, connected set. The region is **simply connected** if it has no holes.

We now define the line integral in space.

Definition 1   LINE INTEGRAL IN SPACE   Suppose that P, Q, and R are continuous on a region S in $\mathbb{R}^3$ containing the smooth or piecewise smooth curve C given by

$$C: \quad \mathbf{x}(t) = f(t)\mathbf{i} + g(t)\mathbf{j} + h(t)\mathbf{k}, \qquad t \in [a, b].$$

Let the vector field $\mathbf{F}$ be given by

$$\mathbf{F}(x, y, z) = P(x, y, z)\mathbf{i} + Q(x, y, z)\mathbf{j} + R(x, y, z)\mathbf{k}.$$

Then the **line integral** of $\mathbf{F}$ over C is given by

$$\int_C \mathbf{F}(\mathbf{x}) \cdot d\mathbf{x} = \int_a^b [P(f(t), g(t), h(t))f'(t) + Q(f(t), g(t), h(t))g'(t)$$
$$+ R(f(t), g(t), h(t))h'(t)] \, dt. \quad \textbf{(8)}$$

As in the plane, the computation of a line integral in space is very easy if the integral is **independent of the path.** The following theorem generalizes Theorem 20.2.1. Its proof is essentially identical to the proof of that theorem.

Theorem 1.   Let $\mathbf{F}$ be continuous on a region S in $\mathbb{R}^3$. Then $\mathbf{F}$ is the gradient of a function f if and only if the following conditions hold:

(i) For any piecewise smooth curve C lying in S, the line integral

$$\int_C \mathbf{F}(\mathbf{x}) \cdot d\mathbf{x}$$

is independent of the path.

(ii) For any piecewise smooth curve C in S starting at the point $\mathbf{x}_0$ and ending at the point $\mathbf{x}_1$,

$$\int_C \mathbf{F}(\mathbf{x}) \cdot d\mathbf{x} = f(\mathbf{x}_1) - f(\mathbf{x}_0). \tag{9}$$

EXAMPLE 2   Let $\mathbf{F}(x, y, z) = y^2z^3\mathbf{i} + 2xyz^3\mathbf{j} + 3xy^2z^2\mathbf{k}$. Compute $\int_C \mathbf{F}(\mathbf{x}) \cdot d\mathbf{x}$, where C is any curve starting at $(1, 2, 3)$ and ending at $(-1, 4, 1)$.

Solution. In Example 18.5.4 we saw that $\mathbf{F} = \nabla f$, where $f = xy^2z^3$. Thus from (9)

$$\int_C \mathbf{F}(\mathbf{x}) \cdot d\mathbf{x} = f(-1, 4, 1) - f(1, 2, 3) = -16 - 108 = -124. \quad \blacksquare$$

In Section 18.11 we saw that $\mathbf{F}(x, y) = P(x, y)\mathbf{i} + Q(x, y)\mathbf{j}$ is the gradient of a function f if $\partial P/\partial y = \partial Q/\partial x$. If $\mathbf{F}(x, y, z) = P(x, y, z)\mathbf{i} + Q(x, y, z)\mathbf{j} + R(x, y, z)\mathbf{k}$, then there is a condition that can be used to check whether there is a differentiable function f such that $\mathbf{F} = \nabla f$. We give this result without proof.†

Theorem 2.   Let $\mathbf{F}(x, y, z) = P(x, y, z)\mathbf{i} + Q(x, y, z)\mathbf{j} + R(x, y, z)\mathbf{k}$ and suppose that $P, Q, R, \partial P/\partial y, \partial P/\partial z, \partial Q/\partial x, \partial Q/\partial z, \partial R/\partial x$, and $\partial R/\partial y$ are continuous in a simply connected region. Then $\mathbf{F}$ is the gradient of a differentiable function f if and only if

$$\frac{\partial P}{\partial y} = \frac{\partial Q}{\partial x}, \qquad \frac{\partial R}{\partial x} = \frac{\partial P}{\partial z}, \qquad \frac{\partial Q}{\partial z} = \frac{\partial R}{\partial y}. \tag{10}$$

If these conditions hold, $\mathbf{F}$ is said to be **exact**.

EXAMPLE 3   Let $\mathbf{F}(x, y, z) = yz\mathbf{i} + xz\mathbf{j} + xy\mathbf{k}$. Then $\partial P/\partial y = z = \partial Q/\partial x$, $\partial R/\partial x = y = \partial P/\partial z$, and $\partial Q/\partial z = x = \partial R/\partial y$. Thus $\mathbf{F}$ is exact, and there is a differentiable function f such that $\mathbf{F} = \nabla f$, so that

$$f_x = yz, \qquad f_y = xz, \qquad f_z = xy.$$

Then

$$f(x, y, z) = \int f_x \, dx = \int yz \, dx = xyz + g(y, z)$$

for some differentiable function g of y and z only. (This means that g is a constant function *with respect to x*.) Hence

$$f_y = xz + g_y(y, z) = xz,$$

---

†For a proof, see R. C. Buck, *Advanced Calculus* (McGraw-Hill, New York, 1965), page 424.

so $g_y(y, z) = 0$ and $g(y, z) = h(z)$, where h is a differentiable function of z only. Then

$$f(x, y, z) = xyz + h(z)$$

and

$$f_z(x, y, z) = xy + h'(z) = xy,$$

so $h'(z) = 0$ and $h(z) = C$, a constant. Finally, we have

$$f(x, y, z) = xyz + C,$$

and it is easily verified that $\nabla f = \mathbf{F}$. ■

EXAMPLE 4   Let $\mathbf{F}(x, y, z) = xy\mathbf{i} + yz\mathbf{j} + xz\mathbf{k}$. Then $\partial P/\partial y = x$ and $\partial Q/\partial x = 0$, so $\mathbf{F}$ is not exact. ■

Once we know how to determine whether a vector field is exact, we can use the following fact, which follows immediately from Theorem 1.

> *The work done by a conservative force field as it moves a particle completely around a closed path in $\mathbb{R}^3$ is zero.*

Before leaving this section we note that line integrals in space are often written in a different way. If $\mathbf{F}(x, y, z) = P(x, y, z)\mathbf{i} + Q(x, y, z)\mathbf{j} + R(x, y, z)\mathbf{k}$, and if we write $\mathbf{dx} = dx\mathbf{i} + dy\mathbf{j} + dz\mathbf{k}$, then

$$\int_C \mathbf{F(x)} \cdot \mathbf{dx} = \int_C (P\,dx + Q\,dy + R\,dz). \tag{11}$$

We emphasize that formula (11) does not tell us anything new; it simply provides a different notation. For example, in Example 2

$$\int_C \mathbf{F(x)} \cdot \mathbf{dx} = \int_C (y^2 z^3\,dx + 2xyz^3\,dy + 3xy^2 z^2\,dz),$$

and we still need formula (8) to evaluate the integral.

PROBLEMS 20.4

In Problems 1–4, evaluate $\int_C \mathbf{F(x)} \cdot \mathbf{dx}$.

1. $\mathbf{F}(x, y, z) = x\mathbf{i} + y\mathbf{j} + z\mathbf{k}$; C is the curve $\mathbf{x}(t) = t\mathbf{i} + t^2\mathbf{j} + t^3\mathbf{k}$ from $(0, 0, 0)$ to $(1, 1, 1)$.
2. $\mathbf{F}(x, y, z) = 2xz\mathbf{i} - xy\mathbf{j} + yz^2\mathbf{k}$; C is the curve of Problem 1.
3. $\mathbf{F}(x, y, z) = x^2\mathbf{i} + y^2\mathbf{j} + z^2\mathbf{k}$; C is the helix $\mathbf{x}(t) = (\cos t)\mathbf{i} + (\sin t)\mathbf{j} + t\mathbf{k}$ from $(1, 0, 0)$ to $(0, 1, \pi/2)$.
4. $\mathbf{F}(x, y, z) = yz\mathbf{i} + xz\mathbf{j} + xy\mathbf{k}$; C is the curve of Problem 3.

In Problems 5–7, show that $\mathbf{F}$ is exact and use Theorem 1 to evaluate $\int_C \mathbf{F(x)} \cdot d\mathbf{x}$ when C is any piecewise smooth curve joining the two given points.

5. $\mathbf{F}(x, y, z) = y^2z^4\mathbf{i} + 2xyz^4\mathbf{j} + 4xy^2z^3\mathbf{k}$ from $(0, 0, 0)$ to $(3, 2, 1)$.
6. $\mathbf{F}(x, y, z) = [(y/z) + x^2]\mathbf{i} + [(x/z) - \sin y]\mathbf{j} + [\cos z - (xy/z^2)]\mathbf{k}$ from $(0, \pi/3, \pi/2)$ to $(1, \pi/2, \pi)$.
7. $\mathbf{F}(x, y, z) = [1/(x + 2y + 3z) - 3x]\mathbf{i} + [2/(x + 2y + 3z) + y^2]\mathbf{j} + [3/(x + 2y + 3z)]\mathbf{k}$ from $(1, 0, 0)$ to $(2, 3, 4)$.
8. Compute $\int_{(2,1,2)}^{(-1,0,4)} \mathbf{F(x)} \cdot d\mathbf{x}$, where $\mathbf{F(x)} = (yz + 2)\mathbf{i} + (xz - 3)\mathbf{j} + (xy + 5)\mathbf{k}$.
9. Suppose that $\mathbf{F(x)} = P(x, y, z)\mathbf{i} + Q(x, y, z)\mathbf{j} + R(x, y, z)\mathbf{k}$ is continuously differentiable and the gradient of a differentiable function. Show that

$$\frac{\partial P}{\partial y} = \frac{\partial Q}{\partial x}, \quad \frac{\partial P}{\partial z} = \frac{\partial R}{\partial x}, \quad \frac{\partial Q}{\partial x} = \frac{\partial R}{\partial y}.$$

20.5 SURFACE INTEGRALS

In Section 20.3 we discussed Green's theorem, which gave a relationship between a line integral over a closed curve and a double integral over a region enclosed by that curve. In this section we discuss the notion of a surface integral. This will enable us, in Section 20.7, to extend Green's theorem to integrals over regions in space.

Recall from Section 17.6 that a surface in space is defined as the set of points satisfying the equation

$$G(x, y, z) = 0. \tag{1}$$

In this section we will begin by considering surfaces that can be written in the form

$$z = f(x, y) \tag{2}$$

for some function f. Note that (2) is a special case of (1), as can be seen by defining $G(x, y, z) = z - f(x, y)$ so that $z = f(x, y)$ is equivalent to $G(x, y, z) = 0$.

Definition 1 SMOOTH SURFACE   The surface is called **smooth** at a point (x_0, y_0, z_0) if $\partial f/\partial x$ and $\partial f/\partial y$ are continuous at (x_0, y_0). If the surface is smooth at all points in the domain of f, we speak of it as a **smooth surface.** That is, if f is continuously differentiable, then the surface is smooth.

A surface integral is very much like a double integral. Suppose that the surface $z = f(x, y)$ is smooth for (x, y) in a bounded region Ω in the xy-plane. Since $f(x, y)$ is continuous on Ω, we find from Definitions 19.1.1 and 19.1.2 and Theorem 19.1.1, that

$$\iint_\Omega f(x, y)\, dA = \lim_{\Delta s \to 0} \sum_{i=1}^{n}\sum_{j=1}^{m} f(x_i{}^*, y_j{}^*)\, \Delta x \Delta y \tag{3}$$

exists, where $\Delta s = \sqrt{\Delta x^2 + \Delta y^2}$ and the limit is independent of the way in which the points (x_i^*, y_j^*) are chosen in the rectangle R_{ij}. We stress that in (3) the quantity $\Delta x \, \Delta y$ represents the *area* of the rectangle R_{ij}.

The double integral (3) is an integral over a region in the plane. A *surface integral,* which we will soon define, is an integral over a surface in space. Suppose we wish to integrate the function $F(x, y, z)$ over the surface S given by $z = f(x, y)$ where $(x, y) \in \Omega$ and, as before, Ω is a bounded region in the xy-plane. Such a surface is sketched in Figure 1.

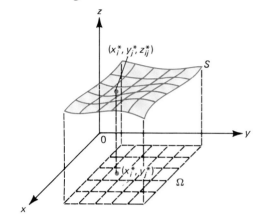

FIGURE 1

We partition Ω into rectangles (and parts of rectangles) as before. This procedure provides a partition of S into mn "subsurfaces" S_{ij}, where

$$S_{ij} = \{(x, y, z): z = f(x, y) \text{ and } (x, y) \in R_{ij}\}. \tag{4}$$

We choose a point in each S_{ij}. Such a point will have the form (x_i^*, y_j^*, z_{ij}^*), where $z_{ij}^* = f(x_i^*, y_j^*)$ and $(x_i^*, y_j^*) \in R_{ij}$. We let $\Delta\sigma_{ij}$ denote the surface area of S_{ij}. This is analogous to the notation $\Delta x \, \Delta y$ as the area of the rectangle R_{ij}. Then we write the double sum

$$\sum_{i=1}^{n}\sum_{j=1}^{m} F(x_i^*, y_j^*, z_{ij}^*) \, \Delta\sigma_{ij}$$

and consider

$$\lim_{\Delta s \to 0} \sum_{i=1}^{n}\sum_{j=1}^{m} F(x_i^*, y_j^*, z_{ij}^*) \, \Delta\sigma_{ij}. \tag{5}$$

Definition 2 INTEGRAL OVER A SURFACE   Suppose the limit in (5) exists and is independent of the way the surface S is partitioned and the way in which the points (x_i^*, y_j^*, z_{ij}^*) are chosen in S_{ij}. Then F is said to be **integrable** over S and the **surface integral** of F over S, denoted $\iint_S F(x, y, z) \, d\sigma$, is given by

$$\iint_S F(x, y, z) \, d\sigma = \lim_{\Delta s \to 0} \sum_{i=1}^{n}\sum_{j=1}^{m} F(x_i^*, y_j^*, z_{ij}^*) \, \Delta\sigma_{ij}. \tag{6}$$

There are three central questions that are raised by this definition.

(i) Under what conditions does the surface integral of F over S exist?
(ii) How can we compute the surface integral if it does exist?
(iii) What is a surface integral good for?

We will answer the first two questions now and will answer the third one by giving some applications of the surface integral later in the section.

The key to evaluating the limit in (6) is to note that, from formula (19.5.5),

$$\Delta\sigma_{ij} \approx \sqrt{f_x{}^2(x_i{}^*, y_j{}^*) + f_y{}^2(x_i{}^*, y_j{}^*) + 1}\; \Delta x\, \Delta y. \tag{7}$$

Then inserting (7) into the limit in (6) and noting that $z_{ij}{}^* = f(x_i{}^*, y_j{}^*)$, we have

$$\lim_{\Delta s \to 0} \sum_{i=1}^{n}\sum_{j=1}^{m} F(x_i{}^*, y_j{}^*, z_{ij}{}^*)\, \Delta\sigma_{ij}$$

$$= \lim_{\Delta s \to 0} \sum_{i=1}^{n}\sum_{j=1}^{m} F(x_i{}^*, y_j{}^*, f(x_i{}^*, y_j{}^*))\sqrt{f_x{}^2(x_i{}^*, y_j{}^*) + f_y{}^2(x_i{}^*, y_j{}^*) + 1}\; \Delta x\, \Delta y. \tag{8}$$

Now if $z = f(x, y)$ is a smooth surface over Ω, then f_x and f_y are continuous over Ω, so that $\sqrt{f_x{}^2 + f_y{}^2 + 1}$ is also continuous over Ω. Furthermore, if $F(x, y, z)$ is continuous for (x, y, z) on S, then by Theorem 19.1.1 and Definitions 19.1.1 and 19.1.2,

$$\iint_\Omega F(x, y, f(x, y))\sqrt{f_x{}^2(x, y) + f_y{}^2(x, y) + 1}\; dA$$

exists and is equal to the right-hand limit in (8). We therefore have the following important result.

Theorem 1. Let $S: z = f(x, y)$ be a smooth surface for (x, y) in the bounded region Ω in the xy-plane. Then if F is continuous on S, F is integrable over S, and

$$\iint_S F(x, y, z)\, d\sigma = \iint_\Omega F(x, y, f(x, y))\sqrt{f_x{}^2(x, y) + f_y{}^2(x, y) + 1}\; dA. \tag{9}$$

REMARK 1. Using (9), we can compute a surface integral over S by transforming it into an ordinary double integral over the region Ω that is the projection of S into the xy-plane.

REMARK 2. If $F(x, y, z) = 1$ in (9), then (9) reduces to

$$\sigma = \iint_S d\sigma = \iint_\Omega \sqrt{f_x{}^2(x, y) + f_y{}^2(x, y) + 1}\; dA. \tag{10}$$

This is the formula for surface area given in Section 19.5.

EXAMPLE 1 Compute $\iint_S (x + 2y + 3z)\, d\sigma$, where S is the part of the plane $2x - y + z = 3$, that lies above the triangular region Ω in the xy-plane bounded by the x- and y-axes and the line $y = 1 - 2x$.

Solution. We have $z = f(x, y) = 3 - 2x + y$, so that $f_x = -2$, $f_y = 1$, and $\sqrt{f_x^2(x, y) + f_y^2(x, y) + 1} = \sqrt{6}$. The region Ω is sketched in Figure 2. Then from (9)

$$I = \iint_S (x + 2y + 3z)\, d\sigma = \sqrt{6} \iint_\Omega [x + 2y + 3(3 - 2x + y)]\, dA$$

$$= \sqrt{6} \int_0^{1/2} \int_0^{1-2x} (9 - 5x + 5y)\, dy\, dx = \sqrt{6} \int_0^{1/2} \left(9y - 5xy + \frac{5y^2}{2}\right)\Big|_0^{1-2x} dx$$

$$= \sqrt{6} \int_0^{1/2} \left[9(1 - 2x) - 5x(1 - 2x) + \frac{5}{2}(1 - 2x)^2\right] dx$$

$$= \sqrt{6} \int_0^{1/2} \left(20x^2 - 33x + \frac{23}{2}\right) dx = \frac{59}{24}\sqrt{6}. \quad \blacksquare$$

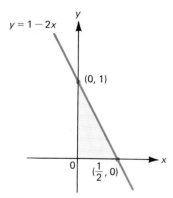

FIGURE 2

EXAMPLE 2 Compute $\iint_S (x^2 + y^2 + 3z^2)\, d\sigma$, where S is the part of the circular paraboloid $z = x^2 + y^2$ with $x^2 + y^2 \le 9$.

Solution. Here $f(x, y) = x^2 + y^2$, so that $f_x = 2x$, $f_y = 2y$, and $d\sigma = \sqrt{1 + 4x^2 + 4y^2}\, dx\, dy$. Thus

$$I = \iint_S (x^2 + y^2 + 3z^2)\, d\sigma = \iint_\Omega [x^2 + y^2 + 3(x^2 + y^2)^2]\sqrt{1 + 4x^2 + 4y^2}\, dA,$$

where Ω is the disk (in the xy-plane) $x^2 + y^2 \le 9$. The problem is greatly simplified by the use of polar coordinates. We have, using $x^2 + y^2 = r^2$,

$$I = \int_0^{2\pi} \int_0^3 [r^2 + 3(r^2)^2]\sqrt{1 + 4r^2}\, r\, dr\, d\theta$$

$$= \int_0^{2\pi} \int_0^3 (r^3 + 3r^5)\sqrt{1 + 4r^2}\, dr\, d\theta$$

$$= 2\pi \int_0^3 (r^3 + 3r^5)\sqrt{1 + 4r^2}\, dr.$$

There are several ways to complete the evaluation of this integral. One way is to integrate by parts twice (start with $u = r^2 + 3r^4$). Another way is to make the substitution $r = \frac{1}{2}\tan\varphi$. The result is

$$I = 2\pi\left\{21(37)^{3/2} + \frac{1}{120}[1 - 55(37)^{5/2}] + \frac{1}{280}(37^{7/2} - 1)\right\} \approx 12{,}629.4. \quad\blacksquare$$

EXAMPLE 3 Evaluate $\iint_S (x + y + z)\, d\sigma$, where S is the part of the surface $z = x^3 + y^4$ lying over the square $\{(x, y): 0 \le x \le 1, 0 \le y \le 1\}$.

Solution. We have $f_x = 3x^2$, $f_y = 4y^3$, and $d\sigma = \sqrt{1 + 9x^4 + 16y^6}\, dx\, dy$, so

$$\iint_S (x + y + z)\, d\sigma = \int_0^1\int_0^1 (x + x^3 + y + y^4)\sqrt{1 + 9x^4 + 16y^6}\, dx\, dy.$$

This is as far as we can go because there is no way to compute this integral directly. The best we can do is to use numerical integration to approximate the answer. $\quad\blacksquare$

REMARK. Example 3 is typical. Like the computations of arc length and surface area, the direct evaluation of a surface integral will often be impossible since it will involve integrals of functions for which antiderivatives cannot be found.

There are many applications for which it is necessary to compute a surface integral. For example, suppose that for (x, y) in a region Ω in the xy-plane, $z = f(x, y)$ is the equation of a thin metallic surface in space. Suppose further that the density of the surface varies and is given by $\rho(x, y, z)$ for (x, y, z) on the surface. Our problem is to compute the total mass of the sheet. We do so by first partitioning Ω in the usual way. This procedure leads to a partition of the metallic surface into sub-surfaces, as before. We again denote this subsurface S_{ij}:

$$S_{ij} = \{(x, y, z): z = f(x, y) \text{ and } (x, y) \in R_{ij}\}.$$

If $\Delta\sigma_{ij}$ denotes the surface area of S_{ij}, if $\Delta\sigma_{ij}$ is small, and if $\rho(x, y, z)$ is continuous, then $\rho(x, y, z)$ is approximately constant on S_{ij}. If (x_i^*, y_j^*) is a point in R_{ij}, then $(x_i^*, y_j^*, z_{ij}^*) = (x_i^*, y_j^*, f(x_i^*, y_j^*))$ is a point on S_{ij}, and

$$\rho(x, y, z) \approx \rho(x_i^*, y_j^*, f(x_i^*, y_j^*)) \tag{11}$$

for (x, y, z) on S_{ij}. Now if area density is constant, then the mass μ of an object with area σ and density ρ is given by

$$\mu = \rho\sigma. \tag{12}$$

Thus if μ_{ij} denotes the mass of S_{ij}, then combining (11) and (12), we have

$$\mu_{ij} \approx \rho(x_i^*, y_j^*, f(x_i^*, y_j^*))\, \Delta\sigma_{ij}, \tag{13}$$

and adding up the masses of the *mn* "subsurfaces" S_{ij}, we obtain

$$\mu \approx \sum_{i=1}^{n} \sum_{j=1}^{m} \rho(x_i^*, y_j^*, f(x_i^*, y_j^*)) \, \Delta\sigma_{ij}. \tag{14}$$

Finally, we let $\Delta s \to 0$, as before. This gives us

$$\mu = \lim_{\Delta s \to 0} \sum_{i=1}^{n} \sum_{j=1}^{m} \rho(x_i^*, y_j^*, f(x_i^*, y_j^*)) \, \Delta\sigma_{ij} = \iint_{S} \rho(x, y, f(x, y)) \, d\sigma. \tag{15}$$

EXAMPLE 4 A metallic dome has the shape of a hemisphere centered at the origin with radius 4 m. Its area density at a point (x, y, z) in space is given by $\rho(x, y, z) = 25 - x^2 - y^2$ kilograms per square meter. Find the total mass of the dome.

Solution. The equation of the hemisphere is $x^2 + y^2 + z^2 = 16$ with $z \geq 0$. Thus $z = f(x, y) = \sqrt{16 - x^2 - y^2}$ (for $x^2 + y^2 \leq 16$),

$$f_x = \frac{-x}{\sqrt{16 - x^2 - y^2}}, \qquad f_y = \frac{-y}{\sqrt{16 - x^2 - y^2}},$$

and

$$\sqrt{f_x^2 + f_y^2 + 1} = \sqrt{\frac{x^2}{16 - x^2 - y^2} + \frac{y^2}{16 - x^2 - y^2} + 1}$$

$$= \sqrt{\frac{x^2 + y^2 + (16 - x^2 - y^2)}{16 - x^2 - y^2}} = \frac{4}{\sqrt{16 - x^2 - y^2}}.$$

Then using polar coordinates, we have

$$\mu = \iint_{S} \rho(x, y, z) \, d\sigma = \int_{0}^{2\pi} \int_{0}^{4} (25 - r^2)\left(\frac{4}{\sqrt{16 - r^2}}\right) r \, dr \, d\theta$$

Ω is the circle centered at $(0, 0)$ with radius 4

$$= 2\pi \int_{0}^{4} (25 - r^2)\left(\frac{4}{\sqrt{16 - r^2}}\right) r \, dr \, d\theta.$$

Now let $16 - r^2 = u^2$, so that $25 - r^2 = 9 + u^2$ and $r \, dr = -u \, du$. Then the integral becomes

$$2\pi \int_{4}^{0} (9 + u^2)\left(\frac{4}{u}\right)(-u \, du) = 8\pi \int_{0}^{4} (9 + u^2) \, du$$

$$= (8\pi)\left(\frac{172}{3}\right) = \frac{1376\pi}{3} \approx 1441 \text{ kg.} \ \blacksquare$$

We now derive another way to represent a surface integral. Let $z = f(x, y)$ be a smooth surface in space for $(x, y) \in \Omega$. As before, the surface can be written

$$G(x, y, z) = 0,$$

where

$$G(x, y, z) = z - f(x, y). \tag{16}$$

Definition 3 OUTWARD UNIT NORMAL VECTOR   We know from Section 18.7 that the gradient vector ∇G is orthogonal or normal to the surface at every point on the surface. Then if $\nabla G \neq 0$,

$$\mathbf{n} = \frac{\nabla G}{|\nabla G|} \tag{17}$$

is a unit normal vector. It is called the **outward unit normal vector.** In general, if $G(x, y, z) = 0$ is any surface in space, then at any point at which ∇G exists and is nonzero, the vector defined by (17) is called the outward unit normal vector.

REMARK 1.   There are two unit vectors normal to a smooth surface at any point. Which one is the outward normal vector? This is a difficult question to answer because, depending on how the surface is defined, we can obtain either one. To see why, note that the surface $G(x, y, z) = 0$ can also be written as $-G(x, y, z) = 0$. In the first case $\mathbf{n} = \nabla G / |\nabla G|$, while in the second $\mathbf{n} = \nabla(-G)/|\nabla(-G)| = -\nabla G/|-\nabla G| = -\nabla G/|\nabla G|$. The second "outward" normal is the negative of the first and so points in the opposite direction. To resolve this difficulty, we must define the *orientation* of a surface. This topic is discussed in advanced calculus books. For our purposes we will write closed surfaces so that $\mathbf{n}$ points outside the surface. We will say more about the direction of $\mathbf{n}$ for other kinds of surfaces in Section 20.7.

REMARK 2.   We can think of the direction of $\mathbf{n}$ in the following way. If $G(x, y, z) = 0$, then the outward normal vector $\mathbf{n}$ points from the surface $G(x, y, z) = 0$ toward the surface $G(x, y, z) = \epsilon$ for every $\epsilon > 0$. This occurs because ∇G points in the direction of increasing G (see Theorem 18.8.1). For example, the unit outward normal vector $\mathbf{n}$ to the unit sphere $x^2 + y^2 + z^2 - 1 = 0$ points toward the sphere of radius 2: $x^2 + y^2 + z^2 - 1 = 3$. Evidently, $\mathbf{n}$ points away from the center of the sphere.

Now from (16), $\nabla G = G_x \mathbf{i} + G_y \mathbf{j} + G_z \mathbf{k} = -f_x \mathbf{i} - f_y \mathbf{j} + \mathbf{k}$, so that

$$\mathbf{n} = \frac{\nabla G}{|\nabla G|} = \frac{-f_x \mathbf{i} - f_y \mathbf{j} + \mathbf{k}}{\sqrt{f_x^2 + f_y^2 + 1}}. \tag{18}$$

We now define the angle γ to be the acute angle between $\mathbf{n}$ and the positive z-axis. This angle is depicted in Figure 3. Since $\mathbf{k}$ is a unit vector having the direction of the positive z-axis, we have, from Theorem 17.2.3,

$$\cos \gamma = \frac{\mathbf{n} \cdot \mathbf{k}}{|\mathbf{n}| |\mathbf{k}|} = \frac{1}{\sqrt{f_x^2 + f_y^2 + 1}} \tag{19}$$

and

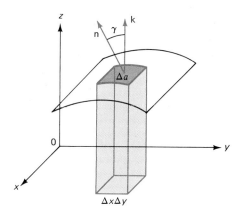

FIGURE 3

$$\sec \gamma = \sqrt{f_x^2 + f_y^2 + 1}. \tag{20}$$

Hence we have

$$\iint_S F(x, y, z)\, d\sigma = \iint_\Omega F(x, y, f(x, y))\, \sec \gamma(x, y)\, dx\, dy. \tag{21}$$

EXAMPLE 5   Compute the surface integral of Example 1 by using formula (21).

Solution. Here S is the plane $2x - y + z = 3$. From Section 17.5 we know that $\mathbf{N} = 2\mathbf{i} - \mathbf{j} + \mathbf{k}$ is normal to this surface, so $\mathbf{n} = \mathbf{N}/|\mathbf{N}| = (1/\sqrt{6})(2\mathbf{i} - \mathbf{j} + \mathbf{k})$ is the outward unit normal vector. Then $\cos \gamma = \mathbf{n} \cdot \mathbf{k} = 1/\sqrt{6}$ and $\sec \gamma = \sqrt{6}$, so

$$\iint_S (x + 2y + 3z)\, d\sigma = \iint_\Omega [(x + 2y) + 3(3 - 2x + y)]\sqrt{6}\, dx\, dy,$$

and we can complete the integration as before. ■

In all the preceding discussion we assumed that the surface was written as $z = f(x, y)$. Suppose instead that the surface is given by

$$S: x = g(y, z), \tag{22}$$

where $(y, z) \in \Omega$, a region in the yz-plane, or

$$S: y = h(x, z), \tag{23}$$

where $(x, z) \in \Omega$, a region in the xz-plane. Let α and β be defined as the acute angles between the unit outward normal to S and the positive x-axis and y-axis, respectively (see Figure 4). Then we have the following:

(i) If the surface is given by (22), then,

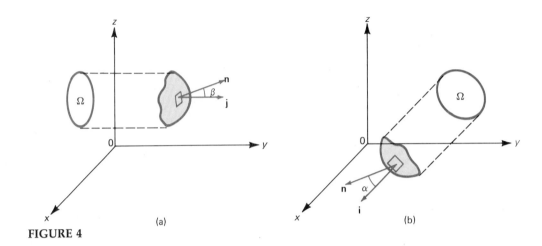

FIGURE 4

$$\iint_S F(x, y, z)\, d\sigma = \iint_\Omega F(g(y, z), y, z)\sqrt{g_y^{\,2} + g_z^{\,2} + 1}\, dy\, dz \qquad (24)$$

$$= \iint_\Omega F(g(y, z), y, x) \sec \alpha\, dy\, dz.$$

(ii) If the surface is given by (23), then

$$\iint_S F(x, y, z)\, d\sigma = \iint_\Omega F(x, h(x, z), z)\sqrt{h_x^{\,2} + h_z^{\,2} + 1}\, dx\, dz \qquad (25)$$

$$= \iint_\Omega F(x, h(x, z), z) \sec \beta\, dx\, dz.$$

EXAMPLE 6 Compute $\iint_S y\, d\sigma$, where S is the hyperbolic paraboloid $x = y^2 - (z^2/2)$ for $0 \le y \le \sqrt{2}$ and $0 \le z \le 1$.

  Solution. Here $g(y, z) = y^2 - (z^2/2)$, $g_y = 2y$, $g_z = -z$, and $\sqrt{g_y^{\,2} + g_z^{\,2} + 1} = \sqrt{4y^2 + z^2 + 1}$, so

$$I = \iint_S y\, d\sigma = \int_0^1 \int_0^{\sqrt{2}} y\sqrt{4y^2 + z^2 + 1}\, dy\, dz = \frac{1}{12}\int_0^1 (4y^2 + z^2 + 1)^{3/2}\Big|_{y=0}^{y=\sqrt{2}} dz$$

$$= \frac{1}{12}\left[\int_0^1 (z^2 + 9)^{3/2}\, dz - \int_0^1 (z^2 + 1)^{3/2}\, dz\right].$$

These integrals can be integrated by making the substitutions $z = 3\tan\theta$ and $z = \tan\theta$, respectively. We then obtain

$$\iint_S y\, d\sigma = \frac{1}{12}\left[\frac{47}{8}\sqrt{10} + \frac{243}{8}\ln\!\left(\frac{1 + \sqrt{10}}{9}\right) - \frac{19}{32}\sqrt{2} - \frac{3}{32}\ln(1 + \sqrt{2})\right]. \quad \blacksquare$$

FLUX

We now give another application of the surface integral and represent the integral in yet another way. Suppose that a smooth surface is immersed in a continuous vector field. For example, this could be a permeable surface immersed in a fluid or a conductor surrounded by an electric field. We denote the vector field by $\mathbf{v}(x, y, z)$ and assume that it is a flow field; that is, we assume that something is moving. For example, the fluid could be in motion. As a particle moves through the fluid, it traces out a **streamline.** One streamline is illustrated in Figure 5. As the particle moves from P_1 to P_2 to P_3, it traces out a path that is determined by the vectors $\mathbf{v}(x, y, z)$ at each point through which it passes. That is, $\mathbf{v}$ is tangent to the streamline at every point. Thus if the particle is at the point (x_0, y_0, z_0), then $\mathbf{v}(x_0, y_0, z_0)$ determines where the particle will be in the next instant of time. Loosely speaking, the *flux* of a vector field is determined by the number of streamlines that cut through the surface in a direction normal to the surface. For an electric field the flux density is proportional to the number of lines of force that cut through the surface.

FIGURE 5

A typical surface immersed in a vector field is sketched in Figure 6a. In Figure 6b we have sketched the flux across a small subsurface. The flux across a point on the surface can be thought of as the component of the vector field in the direction normal to the surface at that point. From Definition 17.2.4, this **normal component,** as it is called, of the vector field is given by

$$\frac{\mathbf{v} \cdot \mathbf{n}}{|\mathbf{n}|} = \mathbf{v} \cdot \mathbf{n}$$

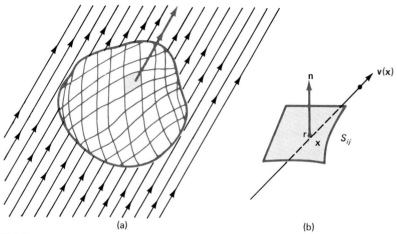

(a)   (b)

FIGURE 6

since $|\mathbf{n}| = 1$. If the subsurface S_{ij} is very small, then because $\mathbf{v}$ is continuous, $\mathbf{v} \cdot \mathbf{n}$ will vary little over S_{ij} and the flux across S_{ij} will be approximately equal to $\mathbf{v}(x_i^*, y_j^*, f(x_i^*, y_j^*)) \cdot \mathbf{n}$ times the surface area of S_{ij}. Thus

$$\text{flux across } S_{ij} \approx \mathbf{v}(x_i^*, y_j^*, f(x_i^*, y_j^*)) \cdot \mathbf{n} \, \Delta\sigma_{ij}. \tag{26}$$

Then

$$\text{total flux} \approx \sum_{i=1}^{n} \sum_{j=1}^{m} \mathbf{v}(x_i^*, y_j^*, f(x_i^*, y_j^*)) \cdot \mathbf{n} \, \Delta\sigma_{ij}. \tag{27}$$

This result leads to the following formal definition.

Definition 4.   FLUX OF A VECTOR FIELD OVER A SURFACE   Let S be a smooth surface with the vector field $\mathbf{v}(x, y, z)$ continuous on S. Then the **flux** of $\mathbf{v}$ over S is defined by

$$\text{flux of } \mathbf{v} \text{ over } S = \iint_S \mathbf{v} \cdot \mathbf{n} \, d\sigma. \tag{28}$$

The following theorem tells us how flux can be computed.

Theorem 2   Let $S: z = f(x, y)$ for $(x, y) \in \Omega$ be a smooth surface and let $\mathbf{n}(x, y, z)$ be its outward unit normal vector. Let $\mathbf{v} = \mathbf{v}(x, y, z)$ be a vector field continuous on S. If

$$\mathbf{v}(x, y, z) = P(x, y, z)\mathbf{i} + Q(x, y, z)\mathbf{j} + R(x, y, z)\mathbf{k}, \tag{29}$$

then

$$\text{flux of } \mathbf{v} \text{ over } S = \iint_S \mathbf{v} \cdot \mathbf{n} \, d\sigma = \iint_\Omega (-Pf_x - Qf_y + R) \, dx \, dy. \tag{30}$$

Proof.   Using (18) and (19), we find that

$$\mathbf{v} \cdot \mathbf{n} = (P\mathbf{i} + Q\mathbf{j} + R\mathbf{k}) \cdot \left(\frac{-f_x\mathbf{i} - f_y\mathbf{j} + \mathbf{k}}{\sqrt{f_x^2 + f_y^2 + 1}} \right)$$

$$= \frac{1}{\sqrt{f_x^2 + f_y^2 + 1}} (-Pf_x - Qf_y + R)$$

$$= \frac{-Pf_x - Qf_y + R}{\sec \gamma}. \tag{31}$$

Now from (21)

$$\iint_S \mathbf{v} \cdot \mathbf{n} \, d\sigma = \iint_\Omega \mathbf{v} \cdot \mathbf{n} \sec \gamma \, dx \, dy. \tag{32}$$

But from (31)

$$\mathbf{v} \cdot \mathbf{n} \sec \gamma = -Pf_x - Qf_y + R. \tag{33}$$

Inserting (33) into the right-hand integral in (32) completes the proof of the theorem. ■

EXAMPLE 7 Find the flux across the conical surface $z = \sqrt{x^2 + y^2}$, $x^2 + y^2 \neq 0$, where $x^2 + y^2 \leq 1$, $x \geq 0$, $y \geq 0$, if the velocity field is given by $\mathbf{v} = x^2\mathbf{i} + y^2\mathbf{j} + z\mathbf{k}$.

Solution. We have $f_x = x/\sqrt{x^2 + y^2}$ and $f_y = y/\sqrt{x^2 + y^2}$, so

$$\text{flux} = \iint_{\Omega} (-Pf_x - Qf_y + R)\, dx\, dy$$

$$= \iint_{\Omega} \left(-\frac{x^3}{\sqrt{x^2 + y^2}} - \frac{y^3}{\sqrt{x^2 + y^2}} + \sqrt{x^2 + y^2} \right) dx\, dy$$

Using polar coordinates

$$= \int_0^{\pi/2} \int_0^1 \left(-\frac{r^3 \cos^3 \theta}{r} - \frac{r^3 \sin^3 \theta}{r} + r \right) r\, dr\, d\theta$$

$$= \int_0^{\pi/2} \int_0^1 [-r^3(\cos^3 \theta + \sin^3 \theta) + r^2]\, dr\, d\theta$$

$$= \int_0^{\pi/2} \left\{ \left[-\frac{r^4}{4}(\cos^3 \theta + \sin^3 \theta) + \frac{r^3}{3} \right]\Big|_0^1 \right\} d\theta$$

$$= \int_0^{\pi/2} \frac{1}{3}\, d\theta - \frac{1}{4}\int_0^{\pi/2} (\cos^3 \theta + \sin^3 \theta)\, d\theta = \frac{\pi}{6} - \frac{1}{3}. \; ■$$

EXAMPLE 8 Compute the flux of the vector field $\mathbf{v}(x, y, z) = x^2 y\mathbf{i} - 2yz\mathbf{j} + x^3 y^2\mathbf{k}$ over the surface of the rectangular solid S:

$$0 \leq x \leq 1, \quad 0 \leq y \leq 2, \quad 0 \leq z \leq 3.$$

Solution. The surface S is sketched in Figure 7. This surface is not smooth, but it is piecewise smooth. That is, it is smooth on each face. So we compute the surface integral over each face and add up the results. The work is summarized in Table 1. For example, consider the face $x = 1$. This face is in a plane parallel to the yz-plane, so that $\Delta\sigma_{ij} = \Delta y\, \Delta z$ and

TABLE 1

Face	Outward unit normal $\mathbf{n}$	$\mathbf{v} \cdot \mathbf{n}$	$\iint_{\text{face}} (\mathbf{v} \cdot \mathbf{n})\, d\sigma = $ flux along face
$x = 0$	$-\mathbf{i}$	$-x^2 y = 0$	0, since $x = 0$
$x = 1$	$\mathbf{i}$	$x^2 y = y$	6
$y = 0$	$-\mathbf{j}$	$2yz = 0$	0, since $y = 0$
$y = 2$	$\mathbf{j}$	$-2yz = -4z$	-18
$z = 0$	$-\mathbf{k}$	$-x^3 y^2$	$-\frac{2}{3}$
$z = 3$	$\mathbf{k}$	$x^3 y^2$	$\frac{2}{3}$

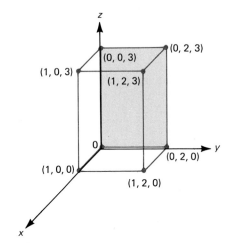

FIGURE 7

$$\iint\limits_{x=1} (\mathbf{v} \bullet \mathbf{n}) \, d\sigma = \int_0^3 \int_0^2 y \, dy \, dz = 6.$$

The surface $y = 2$ is in a plane parallel to the xz-plane, so, analogously, $\Delta\sigma_{ij} = \Delta x \, \Delta z$ and

$$\iint\limits_{y=2} (\mathbf{v} \bullet \mathbf{n}) \, d\sigma = \int_0^3 \int_0^1 -4z \, dx \, dz = -18.$$

Similarly,

$$\iint\limits_{z=0} (\mathbf{v} \bullet \mathbf{n}) \, d\sigma = \int_0^2 \int_0^1 -x^3y^2 \, dx \, dy = -\frac{2}{3}$$

and

$$\iint\limits_{z=3} (\mathbf{v} \bullet \mathbf{n}) \, d\sigma = \int_0^2 \int_0^1 x^3y^2 \, dx \, dy = \frac{2}{3}.$$

Thus the total flux is $6 - 18 - \frac{2}{3} + \frac{2}{3} = -12.$ ∎

Formulas (21), (24), and (25) gives us a way to represent the **differential** $d\sigma$. We have

$$d\sigma = \sec \gamma \, dx \, dy, \tag{34}$$
$$d\sigma = \sec \beta \, dx \, dz, \tag{35}$$

and

$$d\sigma = \sec \alpha \, dy \, dz. \tag{36}$$

We have already seen [formula (19)] that

$$\cos \gamma = \mathbf{n} \cdot \mathbf{k}. \tag{37}$$

Similarly, it is easy to show that

$$\cos \beta = \mathbf{n} \cdot \mathbf{j} \tag{38}$$

and

$$\cos \alpha = \mathbf{n} \cdot \mathbf{i}. \tag{39}$$

Thus

$$\mathbf{n} = (\cos \alpha)\mathbf{i} + (\cos \beta)\mathbf{j} + (\cos \gamma)\mathbf{k}, \tag{40}$$

and if $\mathbf{v} = P\mathbf{i} + Q\mathbf{j} + R\mathbf{k}$, we can write the flux as

$$\iint_S \mathbf{v} \cdot \mathbf{n} \, d\sigma = \iint_S (P \cos \alpha + Q \cos \beta + R \cos \gamma) \, d\sigma. \tag{41}$$

PROBLEMS 20.5

In Problems 1–17, evaluate the surface integral over the given surface.†

1. $\iint_S x \, d\sigma$, where $S: z = x^2$, $0 \le x \le 1$, $0 \le y \le 2$.
2. $\iint_S y \, d\sigma$, where S is as in Problem 1.
3. $\iint_S x^2 \, d\sigma$, where S is as in Problem 1.
4. $\iint_S (x^2 - 2y^2) \, d\sigma$, where S is as in Problem 1.
5. $\iint_S \sqrt{1 + 4z} \, d\sigma$, where S is as in Problem 1.
*6. $\iint_S x \, d\sigma$, where S is the hemisphere $x^2 + y^2 + z^2 = 4$, $z \ge 0$, $x^2 + y^2 \le 4$.
*7. $\iint_S xy \, d\sigma$, where S is as in Problem 6.
8. $\iint_S (x + y) \, d\sigma$, where S is the plane $x + 2y - 3z = 4$; $0 \le x \le 1$, $1 \le y \le 2$.
9. $\iint_S yz \, d\sigma$, where S is as in Problem 8.
10. $\iint_S z^2 \, d\sigma$, where S is as in Problem 8.
*11. $\iint_S (x^2 + y^2 + z^2) \, d\sigma$, where S is the part of the plane $x - y = 4$ that lies inside the cylinder $y^2 + z^2 = 4$.
12. $\iint_S \cos z \, d\sigma$, where S is the plane $2x + 3y + z = 1$ for $0 \le x \le 1$ and $-1 \le y \le 2$.
*13. $\iint_S z \, d\sigma$, where S is the tetrahedron bounded by the coordinate planes and the plane $4x + 8y + 2z = 16$.
14. $\iint_S |x| \, d\sigma$, where S is the hemisphere $x^2 + y^2 + z^2 = 4$; $x \ge 0$, $y^2 + z^2 \le 4$.
15. $\iint_S z \, d\sigma$, where S is the surface of Problem 14.
16. $\iint_S z^2 \, d\sigma$, where S is the hemisphere $x^2 + y^2 + z^2 = 9$, $y \ge 0$, $x^2 + z^2 \le 9$.
17. $\iint_S x^2 \, d\sigma$, where S is the surface of Problem 16.

---

†Note that there may be two answers if it is not clear which of the two normals points outward.

18. Find the mass of a triangular metallic sheet with corners at $(1, 0, 0)$, $(0, 1, 0)$, and $(0, 0, 1)$ if its density is constant.

19. Find the mass of the sheet of Problem 18 if its density is proportional to x^2.

20. Find the mass of a metallic sheet in the shape of the hemisphere $x^2 + y^2 + z^2 = 9$, $z \geq 0$, $x^2 + y^2 \leq 9$, if its density is proportional to its distance from the origin.

21. **(a)** Show that $\iint_S (x^2 + y^2) \, d\sigma$ over the hemisphere $x^2 + y^2 + z^2 = r^2$, $z \geq 0$, $x^2 + y^2 \leq r^2$, is equal to $\frac{4}{3}\pi r^4$.
    (b) Show that $\iint_S x^2 \, d\sigma = \iint_S y^2 \, d\sigma = \iint_S z^2 \, d\sigma = \frac{1}{3}\iint_S (x^2 + y^2 + z^2) \, d\sigma$.
    (c) Explain why the last integral is equal to $\frac{2}{3}\pi r^4$, without performing any integration.
    (d) Use (c) to explain why $\iint_S (x^2 + y^2) \, d\sigma = \frac{4}{3}\pi r^4$, without performing any integration.

In Problems 22–30, compute the flux $\iint_S \mathbf{v} \cdot \mathbf{n} \, d\sigma$ for the given surface lying in the given vector field.

22. $S: z = xy$; $0 \leq x \leq 1$, $0 \leq y \leq 2$; $\mathbf{v} = x^2 y \mathbf{i} - z \mathbf{j}$
23. $S: z = 4 - x - y$; $x \geq 0$, $y \geq 0$, $z \geq 0$; $\mathbf{v} = -3x \mathbf{i} - y \mathbf{j} + 3z \mathbf{k}$
24. $S: x^2 + y^2 + z^2 = 1$; $z \geq 0$; $\mathbf{v} = x \mathbf{i} + y \mathbf{j} + z \mathbf{k}$
25. $S: x^2 + y^2 + z^2 = 1$; $y \geq 0$; $\mathbf{v} = x \mathbf{i} + y \mathbf{j} + z \mathbf{k}$
26. $S: x^2 + y^2 + z^2 = 1$; $x \leq 0$; $\mathbf{v} = x \mathbf{i} + y \mathbf{j} + z \mathbf{k}$
27. $S: z = \sqrt{x^2 + y^2}$; $x^2 + y^2 \leq 1$; $\mathbf{v} = x \mathbf{i} - y \mathbf{j} + xy \mathbf{k}$
28. $S: x = \sqrt{y^2 + z^2}$; $y^2 + z^2 \leq 1$; $\mathbf{v} = y \mathbf{i} - z \mathbf{j} + yz \mathbf{k}$
*29. $S:$ region bounded by $y = 1$ and $y = \sqrt{x^2 + z^2}$; $x^2 + z^2 \leq 1$; $\mathbf{v} = x \mathbf{i} - z \mathbf{j} + xz \mathbf{k}$
30. $S:$ unit sphere; $\mathbf{v} = x \mathbf{i} + y \mathbf{j} + z \mathbf{k}$

31. Show that if $\mathbf{v} = a \mathbf{i} + b \mathbf{j} + c \mathbf{k}$, where a, b, and c are constants, then $\iint_S \mathbf{v} \cdot \mathbf{n} \, d\sigma = 0$, where S is the sphere $x^2 + y^2 + z^2 = r^2$.

32. Find the flux of the vector field $\mathbf{v} = xz \mathbf{i} + y^2 \mathbf{j} - xy^3 z^2 \mathbf{k}$ over the rectangular solid of Example 8.

33. Find the flux of the vector field of Problem 32 over the tetrahedron formed by the coordinate planes and the plane $x + y + z = 1$.

34. If all integrals exist, show that $\iint_S [F(x, y, z) + G(x, y, z)] \, d\sigma = \iint_S F(x, y, z) \, d\sigma + \iint_S G(x, y, z) \, d\sigma$.

20.6 DIVERGENCE AND CURL OF A VECTOR FIELD IN $\mathbb{R}^3$

In Section 20.3 we defined the divergence of $F(x, y) = P(x, y)\mathbf{i} + Q(x, y)\mathbf{j}$ by

$$\text{div } \mathbf{F} = \frac{\partial P}{\partial x} + \frac{\partial Q}{\partial y} \tag{1}$$

and the curl of $\mathbf{F}$ by

$$\text{curl } \mathbf{F} = \frac{\partial Q}{\partial x} - \frac{\partial P}{\partial y}. \tag{2}$$

We also wrote

$$\text{div } \mathbf{F} = \nabla \cdot \mathbf{F} \tag{3}$$

and

$$\text{curl } \mathbf{F} = \nabla \times \mathbf{F}, \tag{4}$$

where ∇, the gradient symbol, is regarded as the vector $(\partial/\partial x, \partial/\partial y)$. Moreover, we used both div $\mathbf{F}$ and curl $\mathbf{F}$ to give alternative versions of Green's theorem in the plane.

In this section we define the divergence and curl of a vector field in $\mathbb{R}^3$, and in Sections 20.7 and 20.8 we will show how these are used in the statement of two very important theorems about surface integrals.

Let the function $F(x, y, z)$ be given. Then the gradient of F is given by

$$\nabla F = \frac{\partial F}{\partial x}\mathbf{i} + \frac{\partial F}{\partial y}\mathbf{j} + \frac{\partial F}{\partial z}\mathbf{k}. \tag{5}$$

We can think of the gradient as a function that takes a differentiable function of (x, y, z) into a vector field in $\mathbb{R}^3$. We write this function, symbolically, as

$$\nabla = \frac{\partial}{\partial x}\mathbf{i} + \frac{\partial}{\partial y}\mathbf{j} + \frac{\partial}{\partial z}\mathbf{k}. \tag{6}$$

Of course, the "thing" written in (6) has no real meaning in itself. However, it is a useful device for writing things down. For example, (5) can be written as

$$\nabla F = \left(\frac{\partial}{\partial x}\mathbf{i} + \frac{\partial}{\partial y}\mathbf{j} + \frac{\partial}{\partial z}\mathbf{k}\right)F = \frac{\partial F}{\partial x}\mathbf{i} + \frac{\partial F}{\partial y}\mathbf{j} + \frac{\partial F}{\partial z}\mathbf{k}.$$

We now define the divergence and curl of a vector field $\mathbf{F}$ in $\mathbb{R}^3$ given by

$$\mathbf{F}(x, y, z) = P(x, y, z)\mathbf{i} + Q(x, y, z)\mathbf{j} + R(x, y, z)\mathbf{k} \tag{7}$$

Definition 1   DIVERGENCE AND CURL   Let the vector field $\mathbf{F}$ be given by (7), where P, Q, and R are differentiable. Then the **divergence** of $\mathbf{F}$ (div $\mathbf{F}$) and **curl** of $\mathbf{F}$ (curl $\mathbf{F}$) are given by

$$\text{div } \mathbf{F} = \frac{\partial P}{\partial x} + \frac{\partial Q}{\partial y} + \frac{\partial R}{\partial z} \tag{8}$$

and

$$\text{curl } \mathbf{F} = \left(\frac{\partial R}{\partial y} - \frac{\partial Q}{\partial z}\right)\mathbf{i} + \left(\frac{\partial P}{\partial z} - \frac{\partial R}{\partial x}\right)\mathbf{j} + \left(\frac{\partial Q}{\partial x} - \frac{\partial P}{\partial y}\right)\mathbf{k}. \tag{9}$$

NOTE.   div $\mathbf{F}$ is a scalar function and curl $\mathbf{F}$ is a vector field.

Before giving examples of divergence and curl, we derive an easy way to remember how to compute them.

Theorem 1 Let ∇ be given by (6) and let the differentiable vector field $\mathbf{F}$ be given by (7). Then

(i) div $\mathbf{F} = \nabla \cdot \mathbf{F}$ (10)

and

(ii) curl $\mathbf{F} = \nabla \times \mathbf{F}$.   (11)

Proof.

(i) $\quad \nabla \cdot \mathbf{F} = \left(\dfrac{\partial}{\partial x}\mathbf{i} + \dfrac{\partial}{\partial y}\mathbf{j} + \dfrac{\partial}{\partial z}\mathbf{k} \right) \cdot (P\mathbf{i} + Q\mathbf{j} + R\mathbf{k})$

$\qquad\qquad = \dfrac{\partial P}{\partial x} + \dfrac{\partial Q}{\partial y} + \dfrac{\partial R}{\partial z} = \text{div } \mathbf{F}$

(ii) $\quad \nabla \times \mathbf{F} = \begin{vmatrix} \mathbf{i} & \mathbf{j} & \mathbf{k} \\[4pt] \dfrac{\partial}{\partial x} & \dfrac{\partial}{\partial y} & \dfrac{\partial}{\partial z} \\[6pt] P & Q & R \end{vmatrix}$

$\qquad\qquad = \left(\dfrac{\partial R}{\partial y} - \dfrac{\partial Q}{\partial z} \right)\mathbf{i} + \left(\dfrac{\partial P}{\partial z} - \dfrac{\partial R}{\partial x} \right)\mathbf{j} + \left(\dfrac{\partial Q}{\partial x} - \dfrac{\partial P}{\partial y} \right)\mathbf{k} = \text{curl } \mathbf{F} \quad\blacksquare$

EXAMPLE 1 Compute the divergence and curl of $\mathbf{F}(x, y, z) = xy\mathbf{i} + (z^2 - 2y)\mathbf{j} + \cos yz\,\mathbf{k}$.

Solution.

$\text{div } \mathbf{F} = \dfrac{\partial}{\partial x}(xy) + \dfrac{\partial}{\partial y}(z^2 - 2y) + \dfrac{\partial}{\partial z}(\cos yz)$

$\qquad\quad = y - 2 - y \sin yz$

and

$\text{curl } \mathbf{F} = \begin{vmatrix} \mathbf{i} & \mathbf{j} & \mathbf{k} \\[4pt] \dfrac{\partial}{\partial x} & \dfrac{\partial}{\partial y} & \dfrac{\partial}{\partial z} \\[6pt] xy & z^2 - 2y & \cos yz \end{vmatrix}$

$\qquad\quad = \left[\dfrac{\partial}{\partial y}\cos yz - \dfrac{\partial}{\partial z}(z^2 - 2y) \right]\mathbf{i} + \left(\dfrac{\partial}{\partial z}xy - \dfrac{\partial}{\partial x}\cos yz \right)\mathbf{j}$

$\qquad\qquad + \left[\dfrac{\partial}{\partial x}(z^2 - 2y) - \dfrac{\partial}{\partial y}xy \right]\mathbf{k}$

$\qquad\quad = (-z \sin yz - 2z)\mathbf{i} - x\mathbf{k}. \quad\blacksquare$

EXAMPLE 2   Compute the divergence and curl of

$$F(x, y, z) = yz\,\mathbf{i} + xz\,\mathbf{j} + xy\,\mathbf{k}.$$

Solution.

$$\text{div } \mathbf{F} = \frac{\partial}{\partial x}yz + \frac{\partial}{\partial y}xz + \frac{\partial}{\partial z}xy = 0$$

$$\text{curl } \mathbf{F} = \begin{vmatrix} \mathbf{i} & \mathbf{j} & \mathbf{k} \\ \dfrac{\partial}{\partial x} & \dfrac{\partial}{\partial y} & \dfrac{\partial}{\partial z} \\ yz & xz & xy \end{vmatrix}$$

$$= \left(\frac{\partial}{\partial y}yx - \frac{\partial}{\partial z}xz\right)\mathbf{i} + \left(\frac{\partial}{\partial z}yz - \frac{\partial}{\partial x}xy\right)\mathbf{j} + \left(\frac{\partial}{\partial x}xz - \frac{\partial}{\partial y}yz\right)\mathbf{k}$$

$$= (x - x)\mathbf{i} + (y - y)\mathbf{j} + (z - z)\mathbf{k} = \mathbf{0} \quad \blacksquare$$

As in $\mathbb{R}^2$, the curl of a vector field **F** represents the circulation per unit area at the point (x, y, z). If curl $\mathbf{F} = \mathbf{0}$ for every (x, y, z) in some region W in $\mathbb{R}^3$, then the fluid flow **F** is called **irrotational**. The divergence of **F** at a point (x, y, z) represents the net rate of flow away from (x, y, z). If div $\mathbf{F} = 0$ for every (x, y, z) in W, then the flow **F** is called **incompressible**. The vector field in Example 2 is both irrotational and incompressible.

Recall from Theorem 20.4.2 that the vector field $\mathbf{F} = P\mathbf{i} + Q\mathbf{j} + R\mathbf{k}$ is the gradient of a function f if and only if

$$\frac{\partial P}{\partial y} = \frac{\partial Q}{\partial x}, \qquad \frac{\partial R}{\partial x} = \frac{\partial P}{\partial z}, \qquad \frac{\partial Q}{\partial z} = \frac{\partial R}{\partial y}. \tag{12}$$

Using (12) and (9), we obtain the following interesting result.

Theorem 2   The differentiable vector field **F** is the gradient of a function f if and only if curl $\mathbf{F} = \mathbf{0}$.

PROBLEMS 20.6

In Problems 1–10, compute the divergence and curl of the given vector field.

1. $\mathbf{F}(x, y, z) = x^2\mathbf{i} + y^2\mathbf{j} + z^2\mathbf{k}$
2. $\mathbf{F}(x, y, z) = (\sin y)\mathbf{i} + (\sin z)\mathbf{j} + (\sin x)\mathbf{k}$
3. $\mathbf{F}(x, y, z) = a\mathbf{i} + b\mathbf{j} + c\mathbf{k}; a, b, c$ constants
4. $\mathbf{F}(x, y, z) = \sqrt{1 + x^2 + y^2}\,\mathbf{i} + \sqrt{1 + x^2 + y^2}\,\mathbf{j} + z^4\mathbf{k}$
5. $\mathbf{F}(x, y, z) = xy\mathbf{i} + yz\mathbf{j} + xz\mathbf{k}$
6. $\mathbf{F}(x, y, z) = (y^2 + z^2)\mathbf{i} + (x^2 + z^2)\mathbf{j} + (x^2 + y^2)\mathbf{k}$
7. $\mathbf{F}(x, y, z) = e^{yz}\mathbf{i} + e^{xz}\mathbf{j} + e^{xy}\mathbf{k}$
8. $\mathbf{F}(x, y, z) = e^{xy}\mathbf{i} + e^{yz}\mathbf{j} + e^{xz}\mathbf{k}$
9. $\mathbf{F}(x, y, z) = \dfrac{x}{y}\mathbf{i} + \dfrac{y}{z}\mathbf{j} + \dfrac{z}{x}\mathbf{k}$

10. $F(x, y, z) = \sqrt{y + z}\,i + \sqrt{x + z}\,j + \sqrt{x + y}\,k$

11. Let f, g, and h be differentiable functions of two variables. Show that the vector field $F(x, y, z) = f(y, z)i + g(x, z)j + h(x, y)k$ is incompressible.

In Problems 12–18, assume that all given functions are differentiable.

12. Show that $\text{div}(F + G) = \text{div } F + \text{div } G$.
13. If $f = f(x, y, z)$ is a scalar function, show that $\text{div } (fF) = f \text{ div } F + \nabla f \cdot F$.
14. Show that $\text{curl}(F + G) = \text{curl } F + \text{curl } G$.
15. If f is as in Problem 13, show that $\text{curl}(fF) = f \text{ curl } F + \nabla f \times F$.
16. If f is as in Problem 13, show that curl grad $f = \text{curl}(\nabla f) = 0$.
17. Show that div curl $F = 0$.
18. Show that $\text{div}(F \times G) = G \cdot \text{curl } F - F \cdot \text{curl } G$.

The **Laplacian** of a twice-differentiable scalar function $f = f(x, y, z)$, denoted $\nabla^2 f$, is defined by

$$\text{Laplacian of } f = \nabla^2 f = \frac{\partial^2 f}{\partial x^2} + \frac{\partial^2 f}{\partial y^2} + \frac{\partial^2 f}{\partial x^2}. \tag{13}$$

In Problems 19–22, compute $\nabla^2 f$.

19. $f(x, y, z) = xyz$

20. $f(x, y, z) = x^2 + y^2 + z^2$

21. $f(x, y, z) = \dfrac{1}{\sqrt{x^2 + y^2 + z^2}}$

22. $f(x, y, z) = 2x^2 + 5y^2 + 3z^2$

23. A function that satisfies the equation $\nabla^2 f = 0$, called **Laplace's equation,** is called **harmonic.** Which of the functions in Problems 19–22 are harmonic?
24. Show that $\nabla^2 f = \text{div (grad } f)$.
25. The **Laplacian** of a vector field F is given by $\nabla^2 F = \nabla^2 Pi + \nabla^2 Qj + \nabla^2 Rk$, where $F = Pi + Qj + Rk$. Show that curl curl $F = \nabla$ div $F - \nabla^2 F$.
26. Let F be the vector field of Problem 1. Verify that div curl $F = 0$.
27. Let f be the function given in Problem 21. Verify that curl grad $f = 0$.
28. It is true, although we will not attempt to prove it, that if div $F = 0$, then $F = \text{curl } G$ for some vector field G. Let

$$F = xi + \frac{y}{2}j - \frac{3}{2}zk.$$

(a) Verify that div $F = 0$.
(b) Find a vector field G such that $F = \text{curl } G$.
29. If curl $F = 0$, then $F = \nabla f$ for some function f. Let $F = 4xyz\,i + 2x^2 z\,j + 2x^2 y\,k$.
(a) Verify that curl $F = 0$.
(b) Find a function f such that $F = \nabla f$.

20.7 STOKES'S THEOREM

In Section 20.3 we discussed Green's theorem in the plane, one vector form of which was given by (see Theorem 20.3.2)

$$\oint_{\partial\Omega} F \cdot T\, dx = \iint_{\Omega} \text{curl } F\, dx\, dy, \tag{1}$$

where $\partial\Omega$ is the piecewise smooth boundary of a region Ω in the xy-plane. We now generalize this result.

If C is a closed curve enclosing the region Ω in the plane, then by *traversing C in the positive sense* we mean moving around C so that Ω is always on the left. This motion corresponds to moving in a counterclockwise direction (see Figure 1).

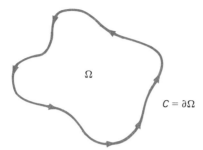

FIGURE 1

REMARK. Like the definition of the outward unit normal vector, the words "counterclockwise" and "left" are ambiguous when applied to closed curves in $\mathbb{R}^3$. To see why, picture a circle in a plane in $\mathbb{R}^3$. You can look at the circle from either side of the plane. The counterclockwise direction from one side would appear as the clockwise direction from the other side. (Take a large piece of translucent paper, place it between yourself and a friend, and ask him or her to draw a circle in the counterclockwise direction. It will appear clockwise to you.) Thus the "positive sense" in which a curve is traversed is arbitrary. However, once we have chosen a positive or counterclockwise direction, the outward unit normal vector can be unambiguously defined: At any point on a smooth surface, we choose a tangent vector $\mathbf{u}$ lying in the unique tangent plane at that point. We then obtain a second tangent vector $\mathbf{v}$ by rotating $\mathbf{u}$ 90° in the counterclockwise direction while remaining in the tangent plane. Finally, we define the direction of $\mathbf{n}$ to be the direction of $\mathbf{u} \times \mathbf{v}$ (so that the vectors $\mathbf{u}$, $\mathbf{v}$, and $\mathbf{n}$ form a right-handed system).

Let $S: z = f(x, y)$ be a smooth surface for $(x, y) \in \Omega$, where Ω is bounded. We assume that the boundary of S, denoted ∂S, is a piecewise smooth, simple closed curve in $\mathbb{R}^3$. The positive direction on ∂S corresponds to the positive direction of $\partial\Omega$, where $\partial\Omega$ is the projection of ∂S into the xy-plane. This orientation is illustrated in Figure 2.

We now state the second major result in vector calculus (the first was Green's theorem).

Theorem 1 *Stokes's Theorem.*† Let $\mathbf{F}(x, y, z) = P(x, y, z)\mathbf{i} + Q(x, y, z)\mathbf{j} + R(x, y, z)\mathbf{k}$ be continuously differentiable on a bounded region W in space that contains the smooth surface S, and let ∂S be a piecewise smooth, simple closed curve traversed in the positive sense. Then

---

†Named after the English mathematician and physicist Sir G. G. Stokes (1819–1903). Stokes is also known as one of the first to discuss the notion of uniform convergence (in 1848).

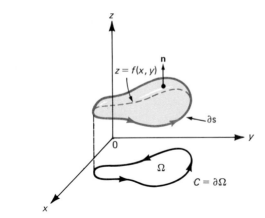

FIGURE 2

$$\oint_{\partial S} \mathbf{F} \cdot \mathbf{T} \, ds = \iint_{S} \text{curl } \mathbf{F} \cdot \mathbf{n} \, d\sigma, \tag{2}$$

where **T** is the unit tangent vector to the curve ∂S at a point (x, y, z) and **n** is the outward unit normal vector to the surface S at the point (x, y, z).

The proof of this theorem is difficult and is left to the end of this section.

REMARK 1.   Note the similarity between Stokes's theorem and the first vector form of Green's theorem.

REMARK 2.   If ∂S is given parametrically by $\mathbf{x}(t) = x(t)\mathbf{i} + y(t)\mathbf{j} + z(t)\mathbf{k}$, then by Theorem 17.7.3(v),

$$\mathbf{T} = \frac{d\mathbf{x}}{ds}, \quad \text{or} \quad d\mathbf{x} = \mathbf{T} \, ds. \tag{3}$$

Thus Stokes's theorem can be written as

$$\oint_{\partial S} \mathbf{F} \cdot d\mathbf{x} = \iint_{S} \text{curl } \mathbf{F} \cdot \mathbf{n} \, d\sigma. \tag{4}$$

REMARK 3.   Using the notation given in equation (20.4.11), we can also write Stokes's theorem as

$$\oint_{\partial S} P \, dx + Q \, dy + R \, dz = \iint_{S} \text{curl } \mathbf{F} \cdot \mathbf{n} \, d\sigma. \tag{5}$$

EXAMPLE 1   Use Stokes's theorem to evaluate $\oint_{C} \mathbf{F} \cdot d\mathbf{x}$, where $\mathbf{F}(x, y, z) = (z - 2y)\mathbf{i} + (3x - 4y)\mathbf{j} + (z + 3y)\mathbf{k}$ and C is the unit circle in the plane $z = 2$.

Solution. C is given parametrically by $\mathbf{x}(t) = (\cos t)\mathbf{i} + (\sin t)\mathbf{j} + 2\mathbf{k}$. Clearly, C bounds the unit disk $x^2 + y^2 \leq 1$, $z = 2$ in the plane $z = 2$. (It bounds many surfaces—for example, a hemisphere with this circle as its base. But the disk is the simplest one to use.) Then

$$\oint_C \mathbf{F} \cdot \mathbf{dx} = \iint_S \text{curl } \mathbf{F} \cdot \mathbf{n} \, d\sigma.$$

We compute

$$\text{curl } \mathbf{F} = \begin{vmatrix} \mathbf{i} & \mathbf{j} & \mathbf{k} \\ \dfrac{\partial}{\partial x} & \dfrac{\partial}{\partial y} & \dfrac{\partial}{\partial z} \\ z - 2y & 3x - 4y & z + 3y \end{vmatrix} = 3\mathbf{i} + \mathbf{j} + 5\mathbf{k},$$

and $\mathbf{n} = \mathbf{k}$ (since $\mathbf{k}$ is normal to any vector lying in a plane parallel to the xy-plane). Thus

$$\text{curl } \mathbf{F} \cdot \mathbf{n} = 5,$$

so that

$$\oint_C \mathbf{F} \cdot \mathbf{dx} = 5 \iint_S d\sigma.$$

But $\iint_S d\sigma$ = the surface area of S = the area of the unit disk = π. Thus

$$\oint_C \mathbf{F} \cdot \mathbf{dx} = 5\pi. \quad \blacksquare$$

EXAMPLE 2 Compute $\oint_C \mathbf{F} \cdot \mathbf{dx}$, where $\mathbf{F}$ is as in Example 1 and C is the boundary of the triangle joining the points $(1, 0, 0)$, $(0, 1, 0)$, and $(0, 0, 1)$.

Solution. We already have found that curl $\mathbf{F} = 3\mathbf{i} + \mathbf{j} + 5\mathbf{k}$. The curve C lies in a plane. Two vectors on the plane are $\mathbf{i} - \mathbf{j}$ and $\mathbf{j} - \mathbf{k}$ (explain why). A normal vector to the plane is therefore given by

$$\mathbf{N} = (\mathbf{i} - \mathbf{j}) \times (\mathbf{j} - \mathbf{k}) = \begin{vmatrix} \mathbf{i} & \mathbf{j} & \mathbf{k} \\ 1 & -1 & 0 \\ 0 & 1 & -1 \end{vmatrix} = \mathbf{i} + \mathbf{j} + \mathbf{k},$$

so that $\mathbf{n} = (1/\sqrt{3})(\mathbf{i} + \mathbf{j} + \mathbf{k})$ is the required outward unit normal vector. Thus curl $\mathbf{F} \cdot \mathbf{n} = 9/\sqrt{3} = 3\sqrt{3}$, and we have

$$\oint_C \mathbf{F} \cdot \mathbf{dx} = 3\sqrt{3} \iint_S d\sigma = 3\sqrt{3} \times \text{(area of the triangle)}.$$

The triangle is an equilateral triangle with the length of one side given by

$$\ell = \text{distance between } (1, 0, 0) \text{ and } (0, 1, 0) = \sqrt{2}.$$

Then the area of the triangle is easily computed to be $\sqrt{3}/2$. Thus

$$\oint_C \mathbf{F} \cdot \mathbf{dx} = (3\sqrt{3})\left(\frac{\sqrt{3}}{2}\right) = \frac{9}{2}. \quad \blacksquare$$

REMARK. In the preceding example tangent vectors to the surface (a plane) are vectors lying on the plane. The vector $\mathbf{j} - \mathbf{k}$ is obtained from the vector $\mathbf{i} - \mathbf{j}$ by rotating $\mathbf{i} - \mathbf{j}$ 120° in the counterclockwise direction when viewed from the side of the plane that does not contain the origin. This procedure leads to $\mathbf{n} = (1/\sqrt{3})(\mathbf{i} + \mathbf{j} + \mathbf{k})$. If we chose the "other" counterclockwise direction, then we would obtain $\mathbf{n} = -(1/\sqrt{3})(\mathbf{i} + \mathbf{j} + \mathbf{k})$ and $\oint_C \mathbf{F} \cdot \mathbf{dx} = -\frac{9}{2}$. Which answer is correct? Both, because the answer depends on our choice of a counterclockwise direction. Without defining the orientation of a surface, this is the best we can do. There will always be two possible answers unless either the counterclockwise direction or the direction of the outward unit normal vector is specified in advance. We will not worry about this problem anymore in this text.

EXAMPLE 3   Verify Stokes's theorem for the surface $z = x^2 + y^2$, $x^2 + y^2 \le 1$, $\mathbf{F}(x, y, z) = y^2\mathbf{i} + x\mathbf{j} + z^2\mathbf{k}$.

Solution. The surface is a circular paraboloid with boundary curve given by the circle $x^2 + y^2 = 1$, $z = 1$ (see Figure 3). This curve can be given parametrically by

$$\mathbf{x}(t) = (\cos t)\mathbf{i} + (\sin t)\mathbf{j} + \mathbf{k}, \qquad 0 \le t \le 2\pi. \tag{6}$$

Thus $\mathbf{x}'(t) = -\sin t\mathbf{i} + \cos t\mathbf{j}$ and $\mathbf{F} \cdot \mathbf{dx} = (-\sin^3 t + \cos^2 t)\,dt$, so that

$$\oint_{\partial S} \mathbf{F} \cdot \mathbf{dx} = \int_0^{2\pi} (-\sin^3 t + \cos^2 t)\,dt = \pi.$$

Now

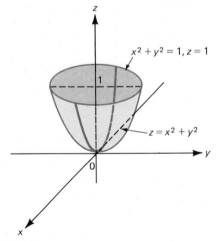

FIGURE 3

$$\text{curl } \mathbf{F} = \begin{vmatrix} \mathbf{i} & \mathbf{j} & \mathbf{k} \\ \dfrac{\partial}{\partial x} & \dfrac{\partial}{\partial y} & \dfrac{\partial}{\partial z} \\ y^2 & x & z^2 \end{vmatrix} = (1 - 2y)\mathbf{k},$$

so that using formula (20.5.30),

$$\iint_S \text{curl } \mathbf{F} \cdot \mathbf{n} \, d\sigma = \iint_\Omega (1 - 2y) \, dx \, dy.$$

Using polar coordinates, we have $y = r \sin \theta$ and

$$\iint_S \text{curl } \mathbf{F} \cdot \mathbf{n} \, d\sigma = \int_0^{2\pi} \int_0^1 (1 - 2r \sin \theta) r \, dr \, d\theta$$

$$= \int_0^{2\pi} \int_0^1 (r - 2r^2 \sin \theta) \, dr \, d\theta = \int_0^{2\pi} \left\{ \left(\frac{r^2}{2} - \frac{2r^3}{3} \sin \theta \right) \Big|_0^1 \right\} d\theta$$

$$= \int_0^{2\pi} \left(\frac{1}{2} - \frac{2}{3} \sin \theta \right) d\theta = \pi. \quad \blacksquare$$

From Theorem 20.6.2, in a simply connected region, $\mathbf{F}$ is the gradient of a function f if and only if curl $\mathbf{F} = \mathbf{0}$. This result provides another proof of the fact that

$$\oint_C \mathbf{F} \cdot \mathbf{dx} = 0$$

if $\mathbf{F}$ is the gradient of a function f, since

$$\oint_C \mathbf{F} \cdot \mathbf{dx} = \iint_S \text{curl } \mathbf{F} \cdot \mathbf{n} \, d\sigma = 0,$$

where S is any smooth surface whose boundary is C.

We can combine several results to obtain the following theorem.

Theorem 2. Let $\mathbf{F} = P(x, y, z)\mathbf{i} + Q(x, y, z)\mathbf{j} + R(x, y, z)\mathbf{k}$ be continuously differentiable on a simply connected region S in space. Then the following conditions are equivalent. That is, if one is true, all are true.

    (i) $\mathbf{F}$ is the gradient of a differentiable function f.

    (ii) $\dfrac{\partial P}{\partial y} = \dfrac{\partial Q}{\partial x}, \dfrac{\partial R}{\partial x} = \dfrac{\partial P}{\partial z}$, and $\dfrac{\partial Q}{\partial z} = \dfrac{\partial R}{\partial y}$.

    (iii) $\int_C \mathbf{F} \cdot \mathbf{dx} = 0$ for every piecewise smooth, simple closed curve C lying in S.

    (iv) $\int_C \mathbf{F} \cdot \mathbf{dx}$ is independent of path.

    (v) Curl $\mathbf{F} = \mathbf{0}$.

There are many interesting physical results that follow from Stokes's theorem.

One of them is given here. Suppose a steady current is flowing in a wire with an electric current density given by the vector field **i**. It is well known in physics that such a current will set up a magnetic field, which is usually denoted **B**. In Figure 4 we see a collection of compass needles near a wire carrying (a) no current and (b) a very strong current. A special case of one of the famous Maxwell equations† states that

$$\text{curl } \mathbf{B} = \mathbf{i}. \tag{7}$$

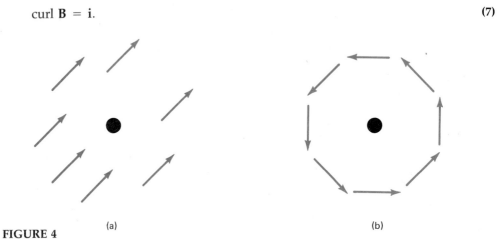

(a)   (b)

FIGURE 4

From this equation we can deduce an important physical law. Let S be a surface with smooth boundary ∂S. Then $\oint_{\partial S} \mathbf{B} \cdot \mathbf{T}\, ds$ is defined as the **circulation** of the magnetic field around ∂S. By Stokes's theorem

$$\oint_{\partial S} \mathbf{B} \cdot \mathbf{T}\, ds = \iint_{S} \text{curl } \mathbf{B} \cdot \mathbf{n}\, d\sigma.$$

But by (7), $\iint_{S} \text{curl } \mathbf{B} \cdot \mathbf{n}\, d\sigma = \iint_{S} \mathbf{i} \cdot \mathbf{n}\, d\sigma$, so

$$\oint_{\partial S} \mathbf{B} \cdot \mathbf{T}\, ds = \iint_{S} \mathbf{i} \cdot \mathbf{n}\, d\sigma. \tag{8}$$

In other words, (8) states that *the total current flowing through an electric field in a surface S is equal to the circulation of the magnetic field induced by* **i** *around the boundary of S*. This important result is known as **Ampère's law.**

We close this section with a proof of Stokes's theorem.

Proof of Stokes's Theorem (Optional).   We have $\mathbf{F} = P\mathbf{i} + Q\mathbf{j} + R\mathbf{k}$ and

$$\text{curl } \mathbf{F} = \left(\frac{\partial R}{\partial y} - \frac{\partial Q}{\partial z}\right)\mathbf{i} + \left(\frac{\partial P}{\partial z} - \frac{\partial R}{\partial x}\right)\mathbf{j} + \left(\frac{\partial Q}{\partial x} - \frac{\partial P}{\partial y}\right)\mathbf{k}. \tag{9}$$

The surface S is given by $S: z = f(x, y)$ for (x, y) in Ω. Then by Theorem 20.5.2,

†James Maxwell (1831–1879) was a British physicist. He formulated four equations, known as **Maxwell's equations,** which were supposed to explain all electromagnetic phenomena.

$$\iint\limits_S \text{curl } \mathbf{F} \cdot \mathbf{n} \, d\sigma = \iint\limits_\Omega \left[\left(\frac{\partial R}{\partial y} - \frac{\partial Q}{\partial z} \right)\left(-\frac{\partial f}{\partial x} \right) + \left(\frac{\partial P}{\partial z} - \frac{\partial R}{\partial x} \right)\left(-\frac{\partial f}{\partial y} \right) \right.$$
$$\left. + \left(\frac{\partial Q}{\partial x} - \frac{\partial P}{\partial y} \right) \right] dx \, dy. \quad \textbf{(10)}$$

Suppose that $\mathbf{x}(t) = x(t)\mathbf{i} + y(t)\mathbf{j}$ for $t_0 \le t \le t_1$ parametrizes $\partial\Omega$. Then ∂S is parametrized by

$$\mathbf{x}(t) = x(t)\mathbf{i} + y(t)\mathbf{j} + f(x(t), y(t))\mathbf{k}, \qquad t_0 \le t \le t_1.$$

Thus

$$\oint\limits_{\partial S} \mathbf{F} \cdot \mathbf{T} \, ds = \oint\limits_{\partial S} \mathbf{F} \cdot d\mathbf{x} = \int_{t_0}^{t_1} \left(P \frac{dx}{dt} + Q \frac{dy}{dt} + R \frac{dz}{dt} \right) dt. \quad \textbf{(11)}$$

By the chain rule,

$$\frac{dz}{dt} = \frac{\partial z}{\partial x}\frac{dx}{dt} + \frac{\partial z}{\partial y}\frac{dy}{dt}. \quad \textbf{(12)}$$

Substituting (12) into (11) yields

$$\oint\limits_{\partial S} \mathbf{F} \cdot \mathbf{T} \, ds = \int_{t_0}^{t_1}\left[\left(P + R\frac{\partial z}{\partial x} \right)\frac{dx}{dt} + \left(Q + R\frac{\partial z}{\partial y} \right)\frac{dy}{dt} \right] dt$$
$$= \oint\limits_{\partial\Omega}\left(P + R\frac{\partial z}{\partial x} \right) dx + \left(Q + R\frac{\partial z}{\partial y} \right) dy. \quad \textbf{(13)}$$

We apply Green's theorem to (13) to obtain

$$\oint\limits_{\partial S}\mathbf{F} \cdot \mathbf{T} \, ds = \iint\limits_\Omega \left\{ \frac{\partial[Q + R(\partial z/\partial y)]}{\partial x} - \frac{\partial[P + R(\partial z/\partial x)]}{\partial y} \right\} dx \, dy. \quad \textbf{(14)}$$

We now apply the chain rule again. This is a bit complicated. For example,

$$\frac{\partial}{\partial x}Q(x, y, z) = \frac{\partial Q}{\partial x} + \frac{\partial Q}{\partial y}\frac{\partial y}{\partial x} + \frac{\partial Q}{\partial z}\frac{\partial z}{\partial x} = \frac{\partial Q}{\partial x} + \frac{\partial Q}{\partial z}\frac{\partial z}{\partial x},$$

since y is not a function of x (so that $\partial y/\partial x = 0$). Also

$$\frac{\partial}{\partial x}\left(R\frac{\partial z}{\partial y} \right) = \left[\frac{\partial}{\partial x}R(x, y, z) \right]\frac{\partial z}{\partial y} + R\frac{\partial}{\partial x}\left\{ \frac{\partial[z(x, y)]}{\partial y} \right\}$$
$$= \frac{\partial R}{\partial x}\frac{\partial z}{\partial y} + \frac{\partial R}{\partial z}\frac{\partial z}{\partial x}\frac{\partial z}{\partial y} + R\frac{\partial^2 z}{\partial x \, \partial y} + R\frac{\partial^2 z}{\partial y^2}\frac{\partial y}{\partial x},$$

and the last term is zero because y is not a function of x. Differentiating inside the right-hand integral in (14), we obtain

$$\oint_{\partial S} \mathbf{F} \cdot \mathbf{T} \, ds = \iint_{\Omega} \left[\left(\frac{\partial Q}{\partial x} + \frac{\partial Q}{\partial z} \frac{\partial z}{\partial x} + \frac{\partial R}{\partial x} \frac{\partial z}{\partial y} + \frac{\partial R}{\partial z} \frac{\partial z}{\partial x} \frac{\partial z}{\partial y} + R \frac{\partial^2 z}{\partial x \, \partial y} \right) \right.$$

$$\left. - \left(\frac{\partial P}{\partial y} + \frac{\partial P}{\partial z} \frac{\partial z}{\partial y} + \frac{\partial R}{\partial y} \frac{\partial z}{\partial x} + \frac{\partial R}{\partial z} \frac{\partial z}{\partial y} \frac{\partial z}{\partial x} + R \frac{\partial^2 z}{\partial y \, \partial x} \right) \right] dz \, dy \text{ (15)}$$

After terms cancel

$$= \iint_{\Omega} \left(\frac{\partial Q}{\partial x} + \frac{\partial Q}{\partial z} \frac{\partial z}{\partial x} + \frac{\partial R}{\partial x} \frac{\partial z}{\partial y} - \frac{\partial P}{\partial y} - \frac{\partial P}{\partial z} \frac{\partial z}{\partial y} - \frac{\partial R}{\partial y} \frac{\partial z}{\partial x} \right) dx \, dy.$$

After rearranging and noting that $\partial f/\partial x = \partial z/\partial x$ and $\partial f/\partial y = \partial z/\partial y$, we see that the integrals in (10) and (15) are identical. Thus

$$\iint_S \text{curl } \mathbf{F} \cdot \mathbf{n} \, d\sigma = \oint_{\partial S} \mathbf{F} \cdot \mathbf{T} \, ds. \quad \blacksquare$$

NOTE.   We have proved this theorem only under the simplifying assumption that the surface can be written in the form $z = f(x, y)$. The proof for more general surfaces is best left to an advanced calculus book.

PROBLEMS 20.7

In Problems 1–8, evaluate the line integral by using Stokes's theorem.

1. $\oint_C \mathbf{F} \cdot \mathbf{dx}$, where $\mathbf{F}(x, y, z) = (x + y)\mathbf{i} + (z - 2x + y)\mathbf{j} + (y - z)\mathbf{k}$ and C is the unit circle in the plane $z = 5$.
2. $\oint_C \mathbf{F} \cdot \mathbf{dx}$, where $\mathbf{F}(x, y, z) = ax\mathbf{i} + by\mathbf{j} + cz\mathbf{k}$ and C is the curve of Problem 1.
3. $\oint_C \mathbf{F} \cdot \mathbf{dx}$, where $\mathbf{F}$ is as in Example 1 and C is the boundary of the triangle joining the points $(2, 0, 0)$, $(0, 2, 0)$, and $(0, 0, 2)$.
4. $\oint_C \mathbf{F} \cdot \mathbf{dx}$, where $\mathbf{F}$ is as in Example 2 and C is the boundary of the triangle joining the points $(d, 0, 0)$, $(0, d, 0)$, and $(0, 0, d)$.
5. $\oint_C y^2 x^3 \, dx + 2xyz^3 \, dy + 3xy^2z^2 \, dz$, where C is given parametrically by $\mathbf{x}(t) = 2 \cos t\,\mathbf{i} + 3\mathbf{j} + 2 \sin t\,\mathbf{k}$, $0 \le t \le 2\pi$.
6. $\oint_C \mathbf{F} \cdot \mathbf{dx}$, where $\mathbf{F} = 2y(x - z)\mathbf{i} + (x^2 + z^2)\mathbf{j} + y^3\mathbf{k}$, C is the square $0 \le x \le 3$, $0 \le y \le 3$, $z = 4$.
7. $\oint_C e^x \, dx + x \sin y \, dy + (y^2 - x^2) \, dz$, where C is the equilateral triangle formed by the intersection of the plane $x + y + z = 3$ with the three coordinate planes.
8. $\oint_C \mathbf{F} \cdot \mathbf{T} \, ds$, where $\mathbf{F} = -3y\mathbf{i} + 3x\mathbf{j} + \mathbf{k}$ and C is the circle $x^2 + y^2 = 1$, $z = 3$.
9. Show that Green's theorem is really a special case of Stokes's theorem.

In Problems 10–13, verify Stokes's theorem for the given S and $\mathbf{F}$.

10. $\mathbf{F} = z^2\mathbf{i} + x^2\mathbf{j} + y^2\mathbf{k}$; S is the part of the plane $x + y + z = 1$ lying in the first octant (i.e., $x \ge 0$, $y \ge 0$, $z \ge 0$).
11. $\mathbf{F} = y^2\mathbf{i} + z^2\mathbf{j} + x^2\mathbf{k}$ and S is the part of the plane $x + 2y + 3z = 6$ lying in the first octant.
12. $\mathbf{F} = 2y\mathbf{i} + x^2\mathbf{j} + 3x\mathbf{k}$; S is the hemisphere $x^2 + y^2 + z^2 = 16$, $z \ge 0$.
13. $\mathbf{F} = y\mathbf{i} - x\mathbf{j} - z\mathbf{k}$; S is the circle $x^2 + y^2 \le 9$, $z = 2$.
14. Let S be a sphere. Use Stokes's theorem to show that $\iint_S \text{curl } \mathbf{F} \cdot \mathbf{n} \, d\sigma = 0$.

*15. Let S be a smooth closed surface bounding a simply connected region. Show that $\iint_S \operatorname{curl} \mathbf{F} \cdot \mathbf{n} \, d\sigma = 0$ for any continuously differentiable vector field $\mathbf{F}$.

*16. Let $\mathbf{E}$ be an electric field and let $\mathbf{B}(t)$ be a magnetic field in space induced by $\mathbf{E}$. One of Maxwell's equations states that $\operatorname{curl} \mathbf{E} = -\partial \mathbf{B}/\partial t$. Let S be a surface in space with smooth boundary C. We define

$$\text{voltage drop around } C = \oint_C \mathbf{E} \cdot \mathbf{dx}.$$

Prove **Faraday's law,** which states that the voltage drop around C is equal to the time rate of decrease of the magnetic flux through S. [*Warning:* At one point it is necessary to interchange the order of differentiation and integration. A proof that this manipulation is "legal" requires techniques from advanced calculus; so just assume that it can be done.]

20.8 THE DIVERGENCE THEOREM

In Section 20.3 we gave the second vector form of Green's theorem (Theorem 20.3.3):

$$\oint_{\partial\Omega} \mathbf{F} \cdot \mathbf{n} \, ds = \iint_{\Omega} \operatorname{div} \mathbf{F} \, dx \, dy, \tag{1}$$

where $\partial\Omega$ is the piecewise smooth boundary of a region Ω in the xy-plane. In this section we extend this theorem to obtain the third major result in vector integral calculus: the **divergence theorem.**

Green's theorem gives us a relationship between a line integral in $\mathbb{R}^2$ around a closed curve and a double integral over the region in $\mathbb{R}^2$ enclosed by that curve. Stokes's theorem shows a relationship between line integral in $\mathbb{R}^3$ around a closed curve and a surface integral over a surface that has the closed curve as a boundary. As we will see, the divergence theorem gives us a relationship between a surface integral over a closed surface and a triple integral over the solid bounded by that surface.

Let S be a surface that forms the complete boundary of a solid W in space. A typical region and its boundary are sketched in Figure 1. We will assume that S is smooth or piecewise smooth. This assumption ensures that $\mathbf{n}(x, y, z)$, the outward unit normal to S, is continuous or piecewise continuous as a function of (x, y, z).

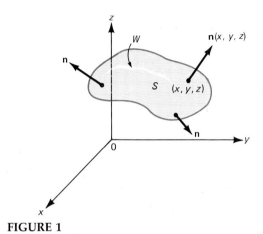

FIGURE 1

Theorem 1   *The Divergence Theorem.* †   Let W be a solid in $\mathbb{R}^3$ totally bounded by the smooth or piecewise smooth surface S. Let $\mathbf{F}$ be a smooth vector field on W, and let $\mathbf{n}$ denote the outward unit normal to S. Then

$$\iint_S \mathbf{F} \cdot \mathbf{n} \, d\sigma = \iiint_W \operatorname{div} \mathbf{F} \, dx \, dy \, dz. \tag{2}$$

REMARK 1.   Just as Green's theorem transforms a line integral to an ordinary double integral, the divergence theorem transforms a surface integral to an ordinary triple integral.

REMARK 2.   If $\mathbf{F} = P\mathbf{i} + Q\mathbf{j} + R\mathbf{k}$, then using equation (20.5.41), we can write the divergence theorem in the form

$$\iint_S (P \cos \alpha + Q \cos \beta + R \cos \gamma) \, d\sigma = \iiint_W \operatorname{div} \mathbf{F} \, dx \, dy \, dz, \tag{3}$$

where α, β, and γ are the acute angles that $\mathbf{n}$ makes with the positive coordinate axes.

Proof of the Divergence Theorem.   In the proof of Green's theorem we assumed (see Figure 20.3.4) that the region Ω could be enclosed by two curves—an upper curve $y = g_2(x)$ and a lower curve $y = g_1(x)$ or a left curve $x = h_1(y)$ and a right curve $x = h_2(y)$. This is equivalent to saying that any line parallel to the x- or y-axis crosses $\partial\Omega$ in at most two points. In order to prove the divergence theorem, we assume much the same thing. That is, we assume that W can be enclosed by an upper surface and a lower surface so that any line parallel to one of the coordinate axes crosses S in at most two points. This assumption is not necessary; smoothness of S is sufficient. But it allows us to give a reasonably simple proof.

From (3) it is sufficient to prove that

$$\iint_S P \cos \alpha \, d\sigma = \iiint_W \frac{\partial P}{\partial x} \, dx \, dy \, dz, \tag{4}$$

$$\iint_S Q \cos \beta \, d\sigma = \iiint_W \frac{\partial Q}{\partial y} \, dx \, dy \, dz, \tag{5}$$

and

$$\iint_S R \cos \gamma \, d\sigma = \iiint_W \frac{\partial R}{\partial z} \, dx \, dy \, dz. \tag{6}$$

The proofs of (4), (5), and (6) are virtually identical, so we will prove (6) only and

---

†This theorem is also known as Gauss's theorem, named after the German mathematician Carl Friedrich Gauss (1777–1855). See the accompanying biographical sketch.

Carl Friedrich Gauss

Carl Friedrich Gauss
The Granger Collection

The greatest mathematician of the nineteenth century, Carl Friedrich Gauss is considered one of the three greatest mathematicians of all time—the others being Archimedes and Newton.

Gauss was born in Brunswick, Germany, in 1777. His father, a hard working laborer who was exceptionally stubborn and did not believe in formal education, did what he could to keep Gauss from appropriate schooling. Fortunately for Carl (and for mathematics), his mother, while uneducated herself, encouraged her son in his studies and took considerable pride in his achievements until her death at the age of 97.

Gauss was a child prodigy. At the age of three, he found an error in his father's bookkeeping. A famous story tells of Carl, age 10, as a student in the local Brunswick school. The teacher there was known to assign tasks to keep his pupils busy. One day he asked his students to add the numbers from 1 to 100. Almost at once, Carl placed his slate face down with the words, "There it is." Afterwards, the teacher found that Gauss was the only one with the correct answer, 5050. Gauss had noticed that the numbers could be arranged in 50 pairs, each with the sum 101 (1 + 100, 2 + 99, and so on) and 50×101 = 5050. Later in life, Gauss joked that he could add before he could speak.

When Gauss was 15, the Duke of Brunswick noticed him and became his patron. The Duke helped him enter Brunswick College in 1795 and, three years later, to enter the university at Göttingen. Undecided between careers in mathematics and philosophy, Gauss chose mathematics after two remarkable discoveries. First, he invented the method of least squares a decade before the result was published by Legendre. Second, a month before his 19th birthday, he solved a problem whose solution had been sought for more than two thousand years. Gauss showed how to construct, using compass and ruler, a regular polygon with the number of sides not a multiple of 2, 3, or 5. On March 30, 1796, the day of this discovery, he began a diary, which contained as its first entry rules for construction of a 17-sided regular polygon. The diary, which contains 146 statements of results in only 19 pages, is one of the most important documents in the history of mathematics.

After a short period at Göttingen, Gauss went to the University of Helmstädt and, in 1798 at the age of 20, wrote his now famous doctoral dissertation. In it he gave the first mathematically rigorous proof of the fundamental theorem of algebra—that every polynomial of degree n, has, counting multiplicities, exactly n roots. Many mathematicians, including Euler, Newton, and Lagrange, had attempted to prove this result.

Gauss made a great number of discoveries in physics as well as in mathematics. For example, in 1801 he used a new procedure to calculate, from very little data, the orbit of the planetoid Ceres. In 1833, he invented the electromagnetic telegraph with his colleague Wilhelm Weber (1804–1891). While he did brilliant work in astronomy and electricity, however, it was Gauss's mathematical output that was astonishing. He made fundamental contributions to algebra and geometry. In 1811, he discovered a result that led to the development of complex variable theory by Cauchy. He is encountered in courses in matrix theory in the Gauss–Jordan method of elimination. Students of numerical analysis study Gaussian quadrature—a technique for numerical integration.

Gauss became a professor of mathematics at Göttingen in 1807 and remained in that post until his death in 1855. Even after his death, his mathematical spirit remained to haunt nineteenth century mathematicians. Often it turned out that an important new result was discovered earlier by Gauss and could be found in his unpublished notes.

In his mathematical writings, Gauss was a perfectionist and is probably the last mathematician who knew everything in his subject. Claiming that a cathedral was not a cathedral until the last piece of scaffolding was removed, he endeavored to make each of his published works complete, concise, and polished. He used a seal that pictured a tree carrying only a few fruit together with the motto *pauca sed matura* (few, but ripe). But Gauss also believed that mathematics must reflect the real world. At his death, Gauss was honored by a commemorative medal on which was inscribed "George V. King of Hanover to the Prince of Mathematicians."

leave (4) and (5) to you. By the assumption stated above, we can think of S as "walnut shaped." That is, S consists of two surfaces: an upper surface S_2: $z = f_2(x, y)$ and a lower surface S_1: $z = f_1(x, y)$. This is depicted in Figure 2. We then have

$$\iint_S R \cos \gamma \, d\sigma = \iint_{S_1} R \cos \gamma \, d\sigma + \iint_{S_2} R \cos \gamma \, d\sigma.$$

Now from equation (20.5.21),

$$\iint_{S_2} R(x, y, z) \cos \gamma \, d\sigma = \iint_\Omega R(x, y, f_2(x, y)) \cos \gamma \sec \gamma \, dx \, dy$$

$$= \iint_\Omega R(x, y, f_2(x, y)) \, dx \, dy, \tag{7}$$

where Ω is the projection of S_2 into the xy-plane. On S_1 the situation is a bit different because the outer normal $\mathbf{n}$ on S_1 points downward and we assumed in deriving equation (20.5.21), that $\mathbf{n}$ points upward (so that the angle γ between $\mathbf{n}$ and the positive z-axis is an acute angle). We solve this problem by using $-\mathbf{n}$ in (20.5.21) so that on S_1 we have

$$\iint_{S_1} R(x, y, z) \cos \gamma \, d\sigma = -\iint_\Omega R(x, y, f_1(x, y)) \cos \gamma \sec \gamma \, dx \, dy$$

$$= -\iint_\Omega R(x, y, f_1(x, y)) \, dx \, dy.$$

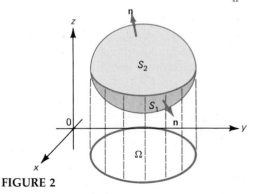

FIGURE 2

Thus

$$\iint_S R \cos \gamma \, ds = \iint_\Omega [R(x, y, f_2(x, y)) - R(x, y, f_1(x, y))] \, dx \, dy$$

$$= \iint_\Omega \left(\int_{f_1(x, y)}^{f_2(x, y)} \frac{\partial R}{\partial z} \, dz \right) dx \, dy = \iiint_W \frac{\partial R}{\partial z} \, dz \, dx \, dy,$$

and under the restrictions already mentioned, this completes the proof. ∎

EXAMPLE 1 Compute $\iint_S \mathbf{F} \cdot \mathbf{n} \, d\sigma$, where $\mathbf{F}(x, y, z) = x^2\mathbf{i} + 2y\mathbf{j} + 4z^2\mathbf{k}$ and S is the surface of the cylinder $x^2 + y^2 \le 4$, $0 \le z \le 2$.

Solution. div $\mathbf{F} = 2x + 2 + 8z$, so

$$\iint_S \mathbf{F} \cdot \mathbf{n} \, ds = \iiint_W (2x + 2 + 8z) \, dz \, dx \, dy$$

$$= \iint_{x^2+y^2\le4} \int_0^2 (2x + 2 + 8z) \, dz \, dx \, dy$$

$$= \iint_{x^2+y^2\le4} \left[(2xz + 2z + 4z^2) \Big|_{z=0}^{z=2} \right] dx \, dy$$

$$= \iint_{x^2+y^2\le4} (4x + 20) \, dx \, dy = \int_0^{2\pi} \int_0^2 (4r \cos \theta + 20)r \, dr \, d\theta$$

$$= 4 \int_0^{2\pi} \int_0^2 (r^2 \cos \theta + 5r) \, dr \, d\theta = 4 \int_0^{2\pi} \left(\frac{8}{3} \cos \theta + 10 \right) d\theta = 80\pi. \ \blacksquare$$

EXAMPLE 2 Compute $\iint_S \mathbf{F} \cdot \mathbf{n} \, d\sigma$, where $\mathbf{F} = 2xy\mathbf{i} + 3y\mathbf{j} + 2z\mathbf{k}$ and S is the boundary of the solid bounded by the three coordinate planes and the plane $x + y + z = 1$.

Solution. div $\mathbf{F} = 2y + 3 + 2 = 2y + 5$, so

$$\iint_S \mathbf{F} \cdot \mathbf{n} \, d\sigma = \iiint_W (2y + 5) \, dx \, dy \, dz,$$

$$\int_0^1 \int_0^{1-y} \int_0^{1-x-y} (2y + 5) \, dz \, dx \, dy = \int_0^1 \int_0^{1-y} (2y + 5)(1 - x - y) \, dx \, dy$$

$$= \int_0^1 \left[-(2y + 5) \frac{(1 - x - y)^2}{2} \Big|_{x=0}^{x=1-y} \right] dy$$

$$= \frac{1}{2} \int_0^1 (2y + 5)(1 - y)^2 \, dy = \frac{11}{12}. \ \blacksquare$$

EXAMPLE 3 Compute $\iint_S \mathbf{F} \cdot \mathbf{n} \, d\sigma$, where $\mathbf{F} = (5x + \sin y \tan z)\mathbf{i} + (y^2 - e^{x-2\cos z^3})\mathbf{j} + (4xy)^{3/5}\mathbf{k}$ and S is the boundary of the solid bounded by the parabolic cylinder $y = 9 - x^2$, the plane $y + z = 1$, and the xy- and xz-planes.

Solution. The solid is sketched in Figure 3. It would be difficult to compute this

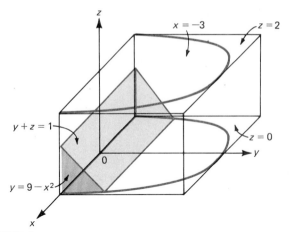

FIGURE 3

surface integral directly, but with the divergence theorem it becomes relatively easy. Since div $\mathbf{F} = 5 + 2y$, we have

$$\iint_S \mathbf{F} \cdot \mathbf{n} \, d\sigma = \iiint_W (5 + 2y) \, dx \, dy \, dz$$

$$= \int_0^1 \int_{-\sqrt{9-y}}^{\sqrt{9-y}} \int_0^{1-y} (5 + 2y) \, dz \, dx \, dy$$

$$= 2 \int_0^1 \int_0^{\sqrt{9-y}} (5 + 2y)(1 \quad y) \, dx \, dy$$

$$= 2 \int_0^1 \int_0^{\sqrt{9-y}} (5 - 3y - 2y^2) \, dx \, dy$$

$$= 2 \int_0^1 (5 - 3y - 2y^2) \sqrt{9 - y} \, dy \approx 16.66. \quad \blacksquare$$

At the end of Section 20.3 we stated that the divergence of $\mathbf{F}$ at a point (x, y) represents the net rate of flow away from (x, y). We can see this more clearly in light of the divergence theorem. Let us divide this region W into a set of small subregions W_i with boundaries S_i. Since div $\mathbf{F}$ is continuous, if W_i is sufficiently small, div $\mathbf{F}$ is approximately constant, and

$$\iiint_{W_i} \text{div } \mathbf{F} \, dx \, dy \, dz \approx \text{div } \mathbf{F} \iiint_{W_i} dx \, dy \, dz$$

$$= (\text{div } \mathbf{F}) \text{ volume } W_i. \tag{8}$$

But

$$\iiint_{W_i} \text{div } \mathbf{F} \, dx \, dy \, dz = \iint_{S_i} \mathbf{F} \cdot \mathbf{T} \, ds, \tag{9}$$

which is the flux of **F** through S_i. Thus by (8) and (9), at a point (x, y, z)

$$\text{div } \mathbf{F}(x, y, z) \approx \frac{\iiint_{W_i} \text{div } \mathbf{F} \, dx \, dy \, dz}{\text{volume of } W_i} = \frac{\iint_{S_i} \mathbf{F} \cdot \mathbf{T} \, ds}{\text{volume of } W}.$$

This last term can be thought of as *flux per unit volume* or net average rate of flow away from (x, y, z).

PROBLEMS 20.8

In Problems 1–15, evaluate the surface integral by using the divergence theorem.

1. $\iint_S \mathbf{F} \cdot \mathbf{n} \, d\sigma$, where $\mathbf{F} = x\mathbf{i} + y\mathbf{j} + z\mathbf{k}$ and S is the unit sphere.
2. $\iint_S \mathbf{F} \cdot \mathbf{n} \, d\sigma$, where $\mathbf{F} = x^2\mathbf{i} + y^2\mathbf{j} + z^2\mathbf{k}$ and S is the unit sphere.
3. $\iint_S \mathbf{F} \cdot \mathbf{n} \, d\sigma$, where $\mathbf{F}$ is as in Problem 1 and S is the cylinder $x^2 + y^2 \leq 4$, $0 \leq z \leq 3$.
4. $\iint_S \mathbf{F} \cdot \mathbf{n} \, d\sigma$, where $\mathbf{F} = y\mathbf{i} + z\mathbf{j} + x\mathbf{k}$ and S is as in Problem 3.
5. $\iint_S \mathbf{F} \cdot \mathbf{n} \, d\sigma$, where $\mathbf{F} = (y^2 + z^2)^{3/2}\mathbf{i} + [\sin(x^2 - z^5)^{4/3}]\mathbf{j} + e^{x^2 - y^3}\mathbf{k}$ and S is the ellipsoid

$$\frac{x^2}{a^2} + \frac{y^2}{b^2} + \frac{z^2}{c^2} = 1, \qquad abc \neq 0.$$

6. $\iint_S \mathbf{F} \cdot \mathbf{n} \, d\sigma$, where $\mathbf{F} = x\mathbf{i} + y\mathbf{j} + z\mathbf{k}$ and S is the surface of the unit cube $0 \leq x \leq 1$, $0 \leq y \leq 1$, $0 \leq z \leq 1$.
7. $\iint_S \mathbf{F} \cdot \mathbf{n} \, d\sigma$, where $\mathbf{F} = x^2\mathbf{i} + y^2\mathbf{j} - xy\mathbf{k}$ and S is as in Problem 6.
8. $\iint_S \mathbf{F} \cdot \mathbf{n} \, d\sigma$, where $\mathbf{F} = xyz\mathbf{i} + yz\mathbf{j} + z\mathbf{k}$ and S is as in Problem 6.
9. $\iint_S \mathbf{F} \cdot \mathbf{n} \, d\sigma$, where $\mathbf{F} = 2x\mathbf{i} + 3y\mathbf{j} + z\mathbf{k}$ and S is the boundary of the hemisphere $x^2 + y^2 + z^2 = 9$, $z \geq 0$.
10. $\iint_S \mathbf{F} \cdot \mathbf{n} \, d\sigma$, where $\mathbf{F} = x^2\mathbf{i} + y^2\mathbf{j} + z^2\mathbf{k}$ and S is as in Problem 9.
11. $\iint_S \mathbf{F} \cdot \mathbf{n} \, d\sigma$, where $\mathbf{F} = xy\mathbf{i} + y^2\mathbf{j} + yz\mathbf{k}$ and S is the boundary of the tetrahedron with vertices at $(0, 0, 0)$, $(1, 0, 0)$, $(0, 1, 0)$, and $(0, 0, 1)$.
12. $\iint_S \mathbf{F} \cdot \mathbf{n} \, d\sigma$, where $\mathbf{F} = y^2\mathbf{i} + x^2\mathbf{j} + z^2\mathbf{k}$ and S is as in Problem 11.
13. $\iint_S \mathbf{F} \cdot \mathbf{n} \, d\sigma$, where $\mathbf{F} = x(1 - \sin y)\mathbf{i} + (y - \cos y)\mathbf{j} + z\mathbf{k}$ and S is as in Problem 11.
14. $\iint_S \mathbf{F} \cdot \mathbf{n} \, d\sigma$, where $\mathbf{F} = x\mathbf{i} + y\mathbf{j} + z\mathbf{k}$ and S is the surface of the region bounded by the parabolic cylinder $z = 1 - y^2$, the plane $x + z = 2$, and the xy- and yz-planes.
15. $\iint_S \mathbf{F} \cdot \mathbf{n} \, d\sigma$, where $\mathbf{F} = (x^2 + e^{y\cos z})\mathbf{i} + (xy - \tan z^{1/3})\mathbf{j} + (x - y^{3/5})^{2/9}\mathbf{k}$ and S is as in Problem 14.

16. If $\mathbf{F}$ is twice continuously differentiable, prove that $\iint_S \text{curl } \mathbf{F} \cdot \mathbf{n} \, d\sigma = 0$ for any smooth closed surface S.
17. Show that $\iint_S \mathbf{F} \cdot \mathbf{n} \, d\sigma = 0$ if $\mathbf{F}$ is a constant vector field and S is a closed smooth surface.
18. If $\mathbf{F} = x\mathbf{i} + y\mathbf{j} + z\mathbf{k}$ and W is a solid with closed smooth boundary S, show that

$$\text{volume of } W = \frac{1}{3} \iint_S \mathbf{F} \cdot \mathbf{n} \, d\sigma.$$

19. Suppose that a vector field $\mathbf{v}$ is tangent to a closed smooth surface S, where S is the boundary of a solid W. Show that $\iiint_W \text{div } \mathbf{v} \, dx \, dy \, dz = 0$.
20. One of Maxwell's equations states that an electric field $\mathbf{E}$ in space satisfies div $\mathbf{E} = (1/\epsilon_0)\rho$, where ϵ_0 is a constant and ρ is the charge density. Show that the flux of the displacement $\mathbf{D} = \epsilon_0\mathbf{E}$ across a closed surface is equal to the charge q inside the surface. This last result is known as *Gauss's law*.

20.9 CHANGING VARIABLES IN MULTIPLE INTEGRALS AND THE JACOBIAN

In many places in this and the preceding chapter we found it useful to evaluate a double integral by first converting to polar coordinates. In Section 19.4 we proved that [see equation (19.4.7)]

$$\iint_\Omega f(x, y)\, dx\, dy = \iint_\Omega f(\cos\theta,\, r\sin\theta) r\, dr\, d\theta. \tag{1}$$

In Section 19.7 we showed how to convert an ordinary triple integral from an integral in Cartesian coordinates to an integral in cyclindrical or spherical coordinates.

In this section we show how, under certain conditions, it is possible to convert from one set of coordinates to another in double and triple integrals. Much of what we do is a generalization of the formula for changing variables in ordinary definite integrals. Let us recall that formula now. If $x = g(u)$ is a differentiable one-to-one function, then [equation (8.4.5)]

$$\int_b^a f(x)\, dx = \int_c^d f(g(u))\, g'(u)\, du, \tag{2}$$

where $c = g^{-1}(a)$ and $d = g^{-1}(b)$.

Now suppose we wish to change the variables of integration in the double integral

$$\iint_\Omega f(x, y)\, dx\, dy. \tag{3}$$

We assume that the new variables are called u and v and that they are related to the old variables x and y by the relations

$$x = g(u, v) \qquad \text{and} \qquad y = h(u, v). \tag{4}$$

The functional relationship described by (4) is called a **mapping** from the uv-plane into the xy-plane. We will assume that there is a region Σ in the uv-plane that gets mapped **onto** Ω by the mapping described by (4) and the mapping is one-to-one. That is:

(i) For every $(x, y) \in \Omega$, there is a $(u, v) \in \Sigma$ such that $x = g(u, v)$ and $y = h(u, v)$.
(ii) If $g(u_1, v_1) = g(u_2, v_2)$ and $h(u_1, v_1) = h(u_2, v_2)$, then $u_1 = u_2$ and $v_1 = v_2$.

Condition (ii) is the natural extension of the definition of one-to-one for functions of two variables. We illustrate what is going on in Figure 1.

EXAMPLE 1   Let $x = u^2 - v^2$ and $y = 3uv$. We will compute the image of the vertical line $u = k$ in the uv-plane under the mapping. Substituting k for u in the equations given above, we have

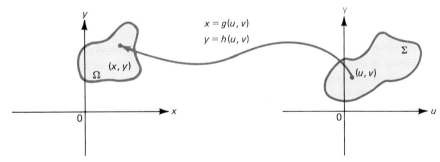

FIGURE 1

$$x = k^2 - v^2 \qquad \text{and} \qquad y = 3kv.$$

Thus $v = y/3k$ and

$$x = k^2 - \frac{y^2}{9k^2}. \tag{5}$$

For every $k \neq 0$, equation (5) is the equation of a parabola in the xy-plane. Moreover, if $v = c$, a constant, then $x = u^2 - c^2$, $y = 3uc$, $u = y/3c$, and

$$x = \frac{y^2}{9c^2} - c^2, \tag{6}$$

which is also a parabola for $c \neq 0$. Thus the functions given above map straight lines into parabolas. Some of these curves are sketched in Figure 2. ∎

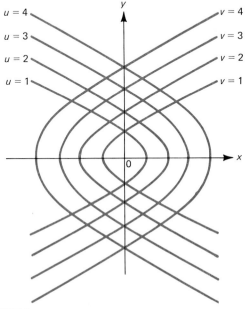

FIGURE 2

Definition 1 JACOBIAN† Let $x = g(u, v)$ and $y = h(u, v)$ be differentiable. Then the **Jacobian** of x and y with respect to u and v, denoted $\partial(x, y)/\partial(u, v)$, is given by

$$\frac{\partial(x, y)}{\partial(u, v)} = \frac{\partial x}{\partial u}\frac{\partial y}{\partial v} - \frac{\partial x}{\partial v}\frac{\partial y}{\partial u} = \begin{vmatrix} \dfrac{\partial x}{\partial u} & \dfrac{\partial x}{\partial v} \\[2mm] \dfrac{\partial y}{\partial u} & \dfrac{\partial y}{\partial v} \end{vmatrix}. \tag{7}$$

EXAMPLE 2 If $x = u^2 - v^2$ and $y = 3uv$ as in Example 1, then $\partial x/\partial u = 2u$, $\partial x/\partial v = -2v$, $\partial y/\partial u = 3v$, and $\partial y/\partial v = 3u$, so

$$\begin{vmatrix} \dfrac{\partial x}{\partial u} & \dfrac{\partial x}{\partial v} \\[2mm] \dfrac{\partial y}{\partial u} & \dfrac{\partial y}{\partial v} \end{vmatrix} = \begin{vmatrix} 2u & -2v \\[2mm] 3v & 3u \end{vmatrix} = 6(u^2 + v^2). \ \blacksquare$$

EXAMPLE 3 Let $x = r \cos \theta$ and $y = r \sin \theta$. Then

$$\frac{\partial x}{\partial r} = \cos \theta, \qquad \frac{\partial x}{\partial \theta} = -r \sin \theta, \qquad \frac{\partial y}{\partial r} = \sin \theta, \qquad \frac{\partial y}{\partial \theta} = r \cos \theta,$$

so

$$\frac{\partial(x, y)}{\partial(u, v)} = \begin{vmatrix} \cos \theta & -r \sin \theta \\ \sin \theta & r \cos \theta \end{vmatrix}$$

$$= r \cos^2 \theta + r \sin^2 \theta = r(\sin^2 \theta + \cos^2 \theta) = r. \ \blacksquare$$

To obtain our main result we need assumptions (i) and (ii) cited earlier. We need also to assume the following:

(iii) $C_1 = \partial\Omega$ and $C_2 = \partial\Sigma$ are simple closed curves, and as (u, v) moves once about C_2 in the positive direction, $(x, y) = (g(u, v), h(u, v))$ moves once around C_1 in the positive or negative direction.

(iv) All second order partial derivatives are continuous.

Theorem 1 *Change of Variables in a Double Integral* If assumptions (i), (ii), (iii), and (iv) hold, then

$$\iint_\Omega f(x, y)\, dx\, dy = \pm \iint_\Sigma f[g(u, v), h(u, v)]\frac{\partial(x, y)}{\partial(u, v)}\, du\, dv. \tag{8}$$

---

†Named after the German mathematician Carl Gustav Jacob Jacobi (1804–1851). See the accompanying biographical sketch.

Carl Gustav Jacob Jacobi
Historical Pictures Service, Chicago

The son of a prosperous banker, Carl Gustav Jacob Jacobi was born in Potsdam, Germany, in 1804. He was educated at the University of Berlin, where he received his doctorate in 1825. In 1827, he was appointed Extraordinary Professor of Mathematics at the University of Königsberg. Jacobi taught at Königsberg until 1842, when he returned to Berlin under a pension from the Prussian government. He remained in Berlin until his death in 1851.

A prolific writer of mathematical treatises, Jacobi was best known in his time for his results in the theory of elliptic functions. Today, however, he is most remembered for his work on determinants. He was one of the two most creative developers of determinant theory, the other being Cauchy. In 1829, Jacobi published a paper on algebra that contained the notation for the Jacobian that we use today. In 1841 he published an extensive treatise titled *De determinantibus functionalibus*, which was devoted to results about the Jacobian. Jacobi showed the relationship between the Jacobian of functions of several variables and the derivative of a function of one variable. He also showed that n functions of n variables are linearly independent if and only if their Jacobian is not identically zero.

In addition to being a fine mathematician, Jacobi was considered the greatest teacher of mathematics of his generation. He inspired and influenced an astonishing number of students. To dissuade his students from mastering great amounts of mathematics before setting off to do their own research, Jacobi often remarked, "Your father would never have married, and you would not be born, if he had insisted on knowing all the girls in the world before marrying one."

Jacobi believed strongly in research in pure mathematics and frequently defended it against the claim that research should always be applicable to something. He once said, "The real end of science is the honor of the human mind."

The plus (minus) sign is taken if as (u, v) moves around C_2 in the positive direction $(x, y) = (g(u, v), h(u, v))$ moves around C_1 in the positive (negative) direction.

Proof. Define $F(x, y) = \int_{x_0}^{x} f(t, y)\, dt$. By the fundamental theorem of calculus, F is continuous and

$$\frac{\partial F}{\partial x} = f(x, y). \tag{9}$$

Then by Green's theorem

$$\iint_\Omega f(x, y)\, dx\, dy = \iint_\Omega \frac{\partial F}{\partial x}\, dx\, dy = \oint_{C_1} F\, dy. \tag{10}$$

We wish to write $\oint_{C_1} F\, dy$ as a line integral in the uv-plane. Let $\mathbf{u} = u(t)\mathbf{i} + v(t)\mathbf{j}$ be a parametric representation of C_2 in the uv-plane for $a \le t \le b$. Then by assumption (iii), $\mathbf{x} = x(t)\mathbf{i} + y(t)\mathbf{j} = g(u(t), v(t))\mathbf{i} + h(u(t), v(t))\mathbf{j}$, $a \le t \le b$, is a parametric representation for C_1. The only difference is that this representation may traverse C_1 in either the positive or the negative direction. Thus

$$\oint_{C_1} F(x, y) \, dy = \int_a^b F(g(u(t), v(t)), h(u(t), v(t))) \frac{dy}{dt} \, dt. \tag{11}$$

Now by the chain rule and the fact that $y = h(u, v)$,

$$\frac{dy}{dt} = \frac{\partial h}{\partial u} u'(t) + \frac{\partial h}{\partial v} v'(t). \tag{12}$$

To simplify notation, let $\overline{F}(u, v) = F(g(u, v), h(u, v))$. Then if we substitute (12) into (11), we obtain

$$\oint_{C_1} F(x, y) \, dy = \int_a^b \overline{F}(u(t), v(t)) \left[\frac{\partial h}{\partial u} u'(t) + \frac{\partial h}{\partial v} v'(t) \right] dt$$

$$= \int_a^b \left[\overline{F} \frac{\partial h}{\partial u} u'(t) + \overline{F} \frac{\partial h}{\partial v} v'(t) \right] dt. \tag{13}$$

Using the definition of the line integral, we can write (13) as

$$\oint_{C_1} F(x, y) \, dy = \pm \oint_{C_1} \overline{F} \frac{\partial h}{\partial u} \, du + \overline{F} \frac{\partial h}{\partial v} \, dv, \tag{14}$$

where the $\pm$ depends on whether (x, y) traverses C_1 in the same direction as, or in the direction opposite to, that in which (u, v) traverses C_2 as t goes from a to b. Let

$$P = \overline{F} \frac{\partial h}{\partial u} \quad \text{and} \quad Q = \overline{F} \frac{\partial h}{\partial v}. \tag{15}$$

Then by Green's theorem

$$\oint_{C_2} P \, du + Q \, dv = \iint_\Sigma \left(\frac{\partial Q}{\partial u} - \frac{\partial P}{\partial v} \right) du \, dv. \tag{16}$$

But from the product rule

$$\frac{\partial Q}{\partial u} = \frac{\partial \overline{F}}{\partial u} \frac{\partial h}{\partial v} + \overline{F} \frac{\partial^2 h}{\partial u \, \partial v} \tag{17}$$

and

$$\frac{\partial P}{\partial v} = \frac{\partial \overline{F}}{\partial v} \frac{\partial h}{\partial u} + \overline{F} \frac{\partial^2 h}{\partial v \, \partial u}, \tag{18}$$

so that from (14)–(18),

Equal because mixed
partials are continuous

$$\oint_{C_1} F\, dy = \pm \iint_{\Sigma} \left(\frac{\partial \overline{F}}{\partial u} \frac{\partial h}{\partial v} + \overline{F} \frac{\partial^2 h}{\partial u\, \partial v} - \frac{\partial \overline{F}}{\partial v} \frac{\partial h}{\partial u} - \overline{F} \frac{\partial^2 h}{\partial v\, \partial u} \right) du\, dv$$

$$= \pm \iint_{\Sigma} \left(\frac{\partial \overline{F}}{\partial u} \frac{\partial h}{\partial v} - \frac{\partial \overline{F}}{\partial v} \frac{\partial h}{\partial u} \right) du\, dv. \qquad (19)$$

But

$$\frac{\partial \overline{F}}{\partial u} = \frac{\partial \overline{F}}{\partial x} \frac{\partial x}{\partial u} + \frac{\partial \overline{F}}{\partial y} \frac{\partial y}{\partial u} = \frac{\partial \overline{F}}{\partial x} \frac{\partial g}{\partial u} + \frac{\partial \overline{F}}{\partial y} \frac{\partial h}{\partial u},$$

and similarly for $\partial \overline{F}/\partial v$. Thus

$$\oint_{C_1} F\, dy = \pm \iint_{\Sigma} \left\{ \left(\frac{\partial \overline{F}}{\partial x} \frac{\partial g}{\partial u} + \frac{\partial \overline{F}}{\partial y} \frac{\partial h}{\partial u} \right) \frac{\partial h}{\partial v} - \left(\frac{\partial \overline{F}}{\partial x} \frac{\partial g}{\partial v} + \frac{\partial \overline{F}}{\partial y} \frac{\partial h}{\partial v} \right) \frac{\partial h}{\partial u} \right\} du\, dv$$

$$= \pm \iint_{\Sigma} \frac{\partial \overline{F}}{\partial x} \left(\frac{\partial g}{\partial u} \frac{\partial h}{\partial v} - \frac{\partial g}{\partial v} \frac{\partial h}{\partial u} \right) du\, dv.$$

But $\partial \overline{F}/\partial x = f(g(u, v), h(u, v))$ and $(\partial g/\partial u)(\partial h/\partial v) - (\partial g/\partial v)(\partial h/\partial u) = \partial(x, y)/\partial(u, v)$. Thus

$$\oint_{C_1} F\, dy = \pm \iint_{\Sigma} f(g(u, v), h(u, v)) \frac{\partial(x, y)}{\partial(u, v)} du\, dv. \qquad (20)$$

Combining (10) with (20) completes the proof. ■

REMARK. Conditions (i) and (ii) given on page 1122 are often difficult to verify. It can be shown that these conditions hold if all partial derivatives are continuous in Ω and the Jacobian $\partial(x, y)/\partial(u, v)$ is not zero on Ω.

EXAMPLE 4   We can use Theorem 1 to obtain the polar coordinate formula (1) very easily. For if $x = r \cos \theta$ and $y = r \sin \theta$, then $\partial(x, y)/\partial(r, \theta) = r$ by Example 3, and (1) follows immediately from (8). ■

EXAMPLE 5   Let Ω be the region in the upper half of the xy-plane bounded by the parabolas $y^2 = 9 - 9x$ and $y^2 = 9 + 9x$ and by the x-axis. Compute $\iint_{\Omega} (x + y)\, dx\, dy$ by making the change of variables $x = u^2 - v^2$, $y = 3uv$.

Solution. We saw in Example 1 that this mapping takes straight lines in the uv-plane into parabolas in the xy-plane. For example, if $y^2 = 9 - 9x$, then $y^2 = 9u^2v^2 = 9 - 9x$, or $u^2v^2 = 1 - x = 1 - u^2 + v^2$, or $u^2 + u^2v^2 = 1 + v^2 = u^2(1 + v^2)$, and $u^2 = 1$, so $u = \pm 1$. Similarly, if $y^2 = 9 + 9x$, then $v = \pm 1$. Since, in Ω, $y \geq 0$, we must have $uv \geq 0$, so that u and v have the same sign. We will choose $u = v = 1$ for reasons to be made clear shortly. Note also that if $u = 0$, then $y = 0$ and $x = -v^2 \leq 0$, so that the positive v-axis in the uv-plane is mapped into the negative x-axis in the xy-plane. Similarly, the positive u-axis in the uv-plane is mapped into the positive

x-axis in the xy-plane. The situation is sketched in Figure 3. The reason we chose $u = v = 1$ is that moving around Σ in the positive direction corresponds to moving around Ω in the positive direction. [Try it: Take the path $v = 0$ (the u-axis) to $u = 1$ to $v = 1$ to $u = 0$.] Thus the integral around the "parabolic" region given in the problem can be reduced to an integral around a square. Also, as we computed in Example 2,

$$\frac{\partial(x, y)}{\partial(u, v)} = 6(u^2 + v^2).$$

Finally, $x + y = u^2 + 3uv - v^2$, so

$$\iint_\Omega (x + y) \, dx \, dy = 6 \int_0^1 \int_0^1 (u^2 + 3uv - v^2)(u^2 + v^2) \, du \, dv$$

$$= 6 \int_0^1 \int_0^1 (u^4 + 3u^3v + 3uv^3 - v^4) \, du \, dv = \frac{9}{2}. \quad \blacksquare$$

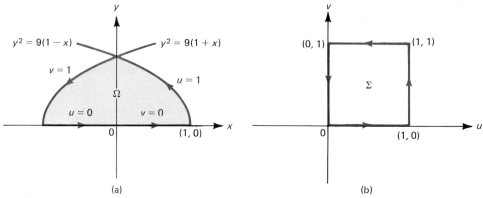

(a)   (b)

FIGURE 3

For triple integrals there is a result analogous to the one we have proven for double integrals. We will state this result without proof.

Let

$$x = g(u, v, w) \qquad y = h(u, v, w), \qquad z = j(u, v, w). \tag{21}$$

We define the **Jacobian** of the transformation from a region U in a uvw-space to a region W in xyz-space as

$$\text{Jacobian} = \frac{\partial(x, y, z)}{\partial(u, v, w)} = \begin{vmatrix} \dfrac{\partial x}{\partial u} & \dfrac{\partial x}{\partial v} & \dfrac{\partial x}{\partial w} \\[2mm] \dfrac{\partial y}{\partial u} & \dfrac{\partial y}{\partial v} & \dfrac{\partial y}{\partial w} \\[2mm] \dfrac{\partial z}{\partial u} & \dfrac{\partial z}{\partial v} & \dfrac{\partial z}{\partial w} \end{vmatrix}. \tag{22}$$

Then under hypotheses similar to the ones made in Theorem 1, we have the following theorem.

Theorem 2 *Change of Variables in a Triple Integral*

$$\iiint_W F(x, y, z)\, dx\, dy\, dz =$$
$$\pm \iiint_U F(g(u, v, w), h(u, v, w), j(u, v, w)) \frac{\partial(x, y, z)}{\partial(u, v, w)}\, du\, dv\, dw$$

EXAMPLE 6   Let $x = r \cos \theta$, $y = r \sin \theta$, and $z = z$. These are cylindrical coordinates. Then

$$\frac{\partial(x, y, z)}{\partial(r, \theta, z)} = \begin{vmatrix} \dfrac{\partial x}{\partial r} & \dfrac{\partial x}{\partial \theta} & \dfrac{\partial x}{\partial z} \\[2mm] \dfrac{\partial y}{\partial r} & \dfrac{\partial y}{\partial \theta} & \dfrac{\partial y}{\partial z} \\[2mm] \dfrac{\partial z}{\partial r} & \dfrac{\partial z}{\partial \theta} & \dfrac{\partial z}{\partial z} \end{vmatrix} = \begin{vmatrix} \cos \theta & -r \sin \theta & 0 \\ \sin \theta & r \cos \theta & 0 \\ 0 & 0 & 1 \end{vmatrix} = r(\cos^2\theta + \sin^2 \theta) = r,$$

so

$$\iiint_W f(x, y, z)\, dx\, dy\, dz = \iiint_U f(r \cos \theta, r \sin \theta, z) r\, dz\, dr\, d\theta.$$

This is equation 19.7.5. ∎

EXAMPLE 7   Let $x = \rho \sin \varphi \cos \theta$, $y = \rho \sin \varphi \sin \theta$, and $z = \rho \cos \varphi$. These are spherical coordinates (see Section 17.8). Then

$$\frac{\partial(x, y, z)}{\partial(\rho, \varphi, \theta)} = \begin{vmatrix} \dfrac{\partial x}{\partial \rho} & \dfrac{\partial x}{\partial \varphi} & \dfrac{\partial x}{\partial \theta} \\[2mm] \dfrac{\partial y}{\partial \rho} & \dfrac{\partial y}{\partial \varphi} & \dfrac{\partial y}{\partial \theta} \\[2mm] \dfrac{\partial z}{\partial \rho} & \dfrac{\partial z}{\partial \varphi} & \dfrac{\partial z}{\partial \theta} \end{vmatrix} = \begin{vmatrix} \sin \varphi \cos \theta & \rho \cos \varphi \cos \theta & -\rho \sin \varphi \sin \theta \\ \sin \varphi \sin \theta & \rho \cos \varphi \sin \theta & \rho \sin \varphi \cos \theta \\ \cos \varphi & -\rho \sin \varphi & 0 \end{vmatrix}$$

Expanding in the last row
↓
$$= \cos \varphi \begin{vmatrix} \rho \cos \varphi \cos \theta & -\rho \sin \varphi \sin \theta \\ \rho \cos \varphi \sin \theta & \rho \sin \varphi \cos \theta \end{vmatrix}$$
$$+ \rho \sin \theta \begin{vmatrix} \sin \varphi \cos \theta & -\rho \sin \varphi \sin \theta \\ \sin \varphi \sin \theta & \rho \sin \varphi \cos \theta \end{vmatrix}$$
$$= \cos \varphi (\rho^2 \cos \varphi \sin \varphi \cos^2 \theta + \rho^2 \sin \varphi \cos \varphi \sin^2 \theta)$$
$$+ \rho \sin \varphi (\rho \sin^2 \varphi \cos^2 \theta + \rho \sin^2 \varphi \sin^2 \theta)$$

$$= \cos\varphi(\rho^2 \sin\varphi \cos\varphi) + \rho \sin\varphi(\rho \sin^2\varphi)$$

$$= \rho^2 \sin\varphi(\cos^2\varphi + \sin^2\varphi) = \rho^2 \sin\varphi.$$

Thus

$$\iiint_W f(x, y, z)\, dx\, dy\, dz =$$

$$\iiint_u f(\rho \sin\varphi \cos\theta, \rho \sin\varphi \sin\theta, \rho \cos\theta)\rho^2 \sin\varphi\, d\rho\, d\varphi\, d\theta. \quad \blacksquare$$

This is formula (19.7.12).

PROBLEMS 20.9

In Problems 1–20, compute the Jacobian of the given transformation.

1. $x = u + v$, $y = u - v$

2. $x = u^2 - v^2$, $y = u^2 + v^2$

3. $x = u^2 - v^2$, $y = 2uv$

4. $x = \sin u$, $y = \cos v$

5. $x = u + 3v - 1$, $y = 2u + 4v + 6$

6. $x = v - 2u$, $y = u + 2v$

7. $x = au + bv$, $y = bu - av$

8. $x = e^v$, $y = e^u$

9. $x = ue^v$, $y = ve^u$

*10. $x = u^v$, $y = v^u$

11. $x = \ln(u + v)$, $y = \ln uv$

12. $x = \tan u$, $y = \sec v$

13. $x = u \sec v$, $y = v \csc u$

14. $x = u \ln v$, $y = v \ln u$

15. $x = u + v + w$, $y = u - v - w$, $z = -u + v + w$

16. $x = au + bv + cw$, $y = au - bv - cw$, $z = -au + bv + cw$

17. $x = u^2 + v^2 + w^2$, $y = u + v + w$, $z = uvw$

18. $x = u \sin v$, $y = v \cos w$, $z = w \sin u$ **19.** $x = e^u$, $y = e^v$, $z = e^w$

20. $x = u \ln(v + w)$, $y = v \ln(u + w)$, $z = w \ln(u + v)$

In Problems 21–25, transform the integral in (x, y) to an integral in (u, v) by using the given transformation. You need not evaluate the integral.

21. $\int_0^1 \int_y^1 xy\, dx\, dy$; $x = u - v$, $y = u + v$.

22. $\iint_\Omega e^{(x+y)/(x-y)}\, dx\, dy$, where Ω is the region in the first quadrant between the lines $x + y = 1$ and $x + y = 2$; $x = u + v$, $y = u - v$.

23. $\iint_\Omega y\, dx\, dy$, where Ω is the region $7x^2 + 6\sqrt{3}x(y - 1) + 13(y - 1)^2 \le 16$; use the transformation $x = \sqrt{3}u + (\frac{1}{2})v$, $y = 1 - u + (\sqrt{3}/2)v$.

24. $\int_0^1 \int_0^x (x^2 + y^2)\, dx\, dy$; $x = v$, $y = u$.

*25. $\iint_\Omega (y - x)\, dy\, dx$, Ω is the region bounded by $y = 2$, $y = x$, and $x = -y^2$; $x = v - u^2$, $y = u + v$.

26. Let Ω be the region in the first quadrant of the xy-plane bounded by the hyperbolas $xy = 1$, $xy = 2$, and the lines $x = y$ and $x = 4y$. Compute $\iint_\Omega x^2 y^2\, dx\, dy$ by setting $x = u$ and $y = u/v$.

27. Let W be the solid enclosed by the ellipsoid $(x^2/a^2) + (y^2/b^2) + (z^2/c^2) = 1$. Then

$$\text{volume of } W = \iiint_W dx\, dy\, dz.$$

Compute this volume by making the transformation $x = au$, $y = bv$, $z = cw$. [*Hint:* In uvw-space you'll obtain a sphere.]

28. Compute $\iiint_W (xy + xz + yz)\, dx\, dy\, dz$, where W is the region of Problem 27.

REVIEW EXERCISES FOR CHAPTER TWENTY

In Exercises 1–6, compute the gradient vector of the given function.

1. $f(x, y) = (x + y)^3$

2. $f(x, y) = \sin(x + 2y)$

3. $f(x, y) = \sqrt{xy}$

4. $f(x, y) = \dfrac{x + y}{x - y}$

5. $f(x, y, z) = x^2 + y^2 + z^2$

6. $f(x, y, z) = xyz$

In Exercises 7 and 8, sketch the given vector field.

7. $\mathbf{F}(x, y) = (x^2 + y^2)\mathbf{i} - 2xy\mathbf{j}$

8. $\mathbf{F}(x, y, z) = x\mathbf{i} - y\mathbf{j} - z\mathbf{k}$

In Exercises 9–11, calculate $\int_C \mathbf{F}(\mathbf{x}) \cdot d\mathbf{x}$.

9. $\mathbf{F}(x, y) = x^2\mathbf{i} + y^2\mathbf{j}$; C is the curve $y = x^{3/2}$ from $(0, 0)$ to $(1, 1)$.

10. $\mathbf{F}(x, y) = x^2y\mathbf{i} - xy^2\mathbf{j}$; C is the unit circle in the counterclockwise direction.

11. $\mathbf{F}(x, y) = 3xy\mathbf{i} - y\mathbf{j}$; C is the triangle joining the points $(0, 0)$, $(1, 1)$, and $(0, 1)$ in the counterclockwise direction.

12. Calculate the work done when the force field $\mathbf{F}(x, y) = x^2y\mathbf{i} + (y^3 + x^3)\mathbf{j}$ moves a particle around the unit circle in the counterclockwise direction.

13. Calculate the work done when the force field $\mathbf{F}(x, y) = 3(x - y)\mathbf{i} + x^5\mathbf{j}$ moves a particle around the triangle of Exercise 11.

14. Show that $\mathbf{F}(x, y) = 3x^2y^2\mathbf{i} + 2x^3y\mathbf{j}$ is exact, and calculate $\int_C \mathbf{F}(\mathbf{x}) \cdot d\mathbf{x}$, where C starts at $(1, 2)$ and ends at $(3, -1)$.

15. Show that $\mathbf{F}(x, y) = e^{xy}(1 + xy)\mathbf{i} + x^2e^{xy}\mathbf{j}$ is exact, and calculate $\int_C \mathbf{F}(\mathbf{x}) \cdot d\mathbf{x}$, where C starts at $(-1, 0)$ and ends at $(0, 1)$.

16. Evaluate $\oint_{\partial\Omega} 2y\, dx + 4x\, dy$, where $\Omega = \{(x, y): 0 \le x \le 2, 0 \le y \le 2\}$.

17. Evaluate $\oint_{\partial\Omega} x^2y\, dx + xy^2\, dy$, where Ω is the region enclosed by the triangle of Exercise 11.

18. Evaluate $\oint_{\partial\Omega}(x^3 + y^3)\, dx + (x^2y^2)\, dy$, where Ω is the unit disk.

19. Evaluate $\oint_{\partial\Omega}\sqrt{1 + x^2}\, dy$; $\Omega = \{(x, y): -1 \le x \le 1, x^2 \le y \le 1\}$.

20. Evaluate

$$\oint_{\partial\Omega} \frac{2x^2 + y^2 - xy}{(x^2 + y^2)^{1/2}}\, dx + \frac{xy - x^2 - 2y^2}{(x^2 + y^2)^{1/2}}\, dy,$$

where Ω is a region of the type (20.3.4) or (20.3.5) that does not contain the origin.

21. Let $\mathbf{F}(x, y) = xy^2\mathbf{i} + x^2y\mathbf{j}$ and let Ω denote the disk of radius 2 centered at $(0, 0)$. Calculate (a) curl $\mathbf{F}$, (b) $\oint_{\partial\Omega} \mathbf{F} \cdot \mathbf{T}\, ds$, (c) div $\mathbf{F}$, and (d) $\oint_{\partial\Omega} \mathbf{F} \cdot \mathbf{n}\, ds$.

22. Do the same for $\mathbf{F}(x, y) = y^2\mathbf{i} - x^2\mathbf{j}$, where Ω is the region enclosed by the triangle of Exercise 11.

23. Show that the vector flow $\mathbf{F}(x, y) = (\cos x^2)\mathbf{i} + e^y\mathbf{j}$ is irrotational.

24. Show that the vector flow $\mathbf{F}(x, y) = -(x^2y + y^3)\mathbf{i} + (x^3 + xy^2)\mathbf{j}$ is incompressible.

In Exercises 25 and 26, evaluate $\int_C \mathbf{F}(\mathbf{x}) \cdot d\mathbf{x}$.

25. $\mathbf{F}(x, y, z) = x\mathbf{i} + y\mathbf{j} + z\mathbf{k}$; C is the curve $\mathbf{x}(t) = t^3\mathbf{i} + t^2\mathbf{j} + t\mathbf{k}$ from $(0, 0, 0)$ to $(1, 1, 1)$.

26. $\mathbf{F}(x, y, z) = x^2\mathbf{i} + y^2\mathbf{j} + z^2\mathbf{k}$; C is the helix $\mathbf{x}(t) = (\sin t)\mathbf{i} + (\cos t)\mathbf{j} + 2t\mathbf{k}$ from $(0, 1, 0)$ to $(1, 0, \pi)$.

27. Show that $\mathbf{F}(x, y, z) = -(y/z)\mathbf{i} - (x/z)\mathbf{j} + (xy/z^2)\mathbf{k}$ is exact, and use that fact to evaluate $\int_C \mathbf{F}(\mathbf{x}) \cdot d\mathbf{x}$, where C is a piecewise smooth curve joining the points $(1, 1, 1)$ and $(2, -1, 3)$ and not crossing the xy-plane.

In Exercises 28–33, evaluate the surface integral over the given surface.

28. $\iint_S y \, d\sigma$, where $S: z = y^2, \ 0 \le x \le 2, \ 0 \le y \le 1$.

29. $\iint_S (x^2 + y^2) \, d\sigma$, where S is as in Exercise 28.

30. $\iint_S y^2 \, d\sigma$, where S is the hemisphere $x^2 + y^2 + z^2 = 9, \ z \ge 0, \ x^2 + y^2 \le 9$.

31. $\iint_S xz \, d\sigma$, where S is as in Exercise 30.

32. $\iint_S y \, d\sigma$, where S is the boundary of the tetrahedron bounded by the coordinate planes and the plane $x + y + z = 1$.

33. $\iint_S (x^2 + y^2 + z^2) \, d\sigma$, where S is the part of the plane $y - x = 3$ that lies inside the cylinder $y^2 + z^2 = 9$.

34. Find the mass of a triangular metallic sheet with corners at $(1, 0, 0)$, $(0, 1, 0)$, and $(0, 0, 1)$ if its density is proportional to y^2.

35. Find the mass of a metallic sheet in the shape of the hemisphere $x^2 + y^2 + z^2 = 1, \ y \ge 0$, $x^2 + z^2 \le 1$, if its density is proportional to the distance from the y-axis.

In Exercises 36–39, compute the flux $\iint_S \mathbf{v} \cdot \mathbf{n} \, d\sigma$ for the given surface lying in the given vector field.

36. $S: z = 2xy; \ 0 \le x \le 1, \ 0 \le y \le 4; \ \mathbf{v} = xy^2 \mathbf{i} - 2z\mathbf{j}$

37. $S: x^2 + y^2 + z^2 = 1; \ z \ge 0; \ \mathbf{v} = x\mathbf{i} + 2y\mathbf{j} + 3z\mathbf{k}$

38. $S: y = \sqrt{x^2 + z^2}; \ x^2 + z^2 \le 4; \ \mathbf{v} = x\mathbf{i} - xz\mathbf{j} + z\mathbf{k}$

39. $S:$ unit sphere; $\mathbf{v} = x^2 \mathbf{i} + y^2 \mathbf{j} + z^2 \mathbf{k}$

In Exercises 39–45, compute the divergence and curl of the given vector field.

40. $\mathbf{F}(x, y, z) = x\mathbf{i} + y\mathbf{j} + z\mathbf{k}$

41. $\mathbf{F}(x, y, z) = (x - y)\mathbf{i} + (y - z)\mathbf{j} + (z - x)\mathbf{k}$

42. $\mathbf{F}(x, y, z) = yz\mathbf{i} + xz\mathbf{j} + xy\mathbf{k}$

43. $\mathbf{F}(x, y, z) = \ln x\mathbf{i} + \ln y\mathbf{j} + \ln z\mathbf{k}$

44. $\mathbf{F}(x, y, z) = e^{yz}\mathbf{i} + e^{xz}\mathbf{j} + e^{xy}\mathbf{k}$

45. $\mathbf{F}(x, y, z) = \cos y\mathbf{i} + \cos x\mathbf{j} + \cos z\mathbf{k}$

In Exercises 46–48, evaluate the line integral by using Stokes's theorem.

46. $\oint_C \mathbf{F} \cdot d\mathbf{x}$, where $\mathbf{F}(x, y, z) = (x + 2y)\mathbf{i} + (y - 3z)\mathbf{j} + (z - x)\mathbf{k}$ and C is the unit circle in the plane $z = 2$.

47. $\oint_C \mathbf{F} \cdot d\mathbf{x}$, where $\mathbf{F} = x\mathbf{i} + y\mathbf{j} + z\mathbf{k}$ and C is the boundary of the triangle joining the points $(1, 0, 0)$, $(0, 1, 0)$, and $(0, 0, 1)$.

48. $\oint_C \mathbf{F} \cdot d\mathbf{x}$, where $\mathbf{F} = y^2 \mathbf{i} + x^2 \mathbf{j} + z^2 \mathbf{k}$ and C is the boundary of the part of the plane $x + y + z = 1$ lying in the first octant.

49. Compute $\iint_S \operatorname{curl} \mathbf{F} \cdot \mathbf{n} \, d\sigma$, where S is the unit sphere and $\mathbf{F} = e^{xy}\mathbf{i} + \tan^{-1} z\mathbf{j} + (x + y + z)^{7/3} z^2 \mathbf{k}$.

In Exercises 50–55, evaluate the surface integral by using the divergence theorem.

50. $\iint_S \mathbf{F} \cdot \mathbf{n} \, d\sigma$, where $\mathbf{F} = x\mathbf{i} + 2y\mathbf{j} + 3z\mathbf{k}$ and S is the unit sphere.

51. $\iint_S \mathbf{F} \cdot \mathbf{n} \, d\sigma$, where $\mathbf{F} = ax\mathbf{i} + by\mathbf{j} + cz\mathbf{k}$ and S is the unit sphere.

52. $\iint_S \mathbf{F} \cdot \mathbf{n} \, d\sigma$, where $\mathbf{F}$ is as in Exercise 50 and S is the cylinder $x^2 + y^2 = 9, \ 0 \le z \le 6$.

53. $\iint_S \mathbf{F} \cdot \mathbf{n} \, d\sigma$, where $\mathbf{F} = (y^3 - z)\mathbf{i} + x^2 e^z \mathbf{j} + \sin xy\mathbf{k}$ and S is the ellipsoid $(x^2/4) + (y^2/16) + (z^2/25) = 1$.

54. $\iint_S \mathbf{F} \cdot \mathbf{n} \, d\sigma$, where $\mathbf{F}$ is as in Exercise 50 and S is the boundary of the unit cube.

55. $\iint_S \mathbf{F} \cdot \mathbf{n} \, d\sigma$, where $\mathbf{F} = xz\mathbf{i} + xy\mathbf{j} + xyz\mathbf{k}$ and S is the unit cube in the first octant.

In Exercises 56–63, compute the Jacobian of the given transformation.

56. $x = u + 2v, \ y = 2u - v$

57. $x = u^3 - v^3, \ y = u^3 + v^3$

58. $x = u \ln v, \ y = v \ln u$

59. $x = ve^u, \ y = ue^v$

60. $x = \dfrac{u}{v}, \ y = \dfrac{v}{u}$

61. $x = v \tan u, \ y = \tan uv$

62. $x = u + v + w, y = u - 2v + 3w, z = -2u + v - 5w$

63. $x = vw, y = uw, z = uv$

64. Transform the integral $\int_0^1 \int_x^1 xy \, dx \, dy$ by making the transformation $x = u + v, y = u - v$.

65. Transform $\iint_\Omega e^{(x-y)/(x+y)} \, dx \, dy$, where Ω is the region in the first quadrant between the lines $x + y = 2$ and $x + y = 3$, by making the transformation $u = x - y$ and $v = x + y$. Then evaluate the integral.

21 Ordinary Differential Equations

21.1 INTRODUCTION

Many of the basic laws of the physical sciences and, more recently, of the biological and social sciences are formulated in terms of mathematical relations involving certain known and unknown quantities and their derivatives. Such relations are called **differential equations.**

We discussed some very simple differential equations in Sections 6.6 and 7.3. In this chapter we will take a brief look at some other types of differential equations. In this section we categorize the types of differential equations that may be encountered.

The most obvious classification is based on the nature of the derivative (or derivatives) in the equation. A differential equation involving only ordinary derivatives (derivatives of functions of one variable) is called an **ordinary differential equation.** The equations in Sections 6.6 and 7.3 are all ordinary differential equations. A differential equation containing partial derivatives is called a **partial differential equation.** Examples of partial differential equations are given in Problems 18.3.41, 18.3.42, 18.4.31, 18.4.32, 18.4.35, and 18.4.39.

In this chapter we will consider only the solutions of certain ordinary differential equations, which we will simply refer to as differential equations, dropping the word "ordinary." Procedures for solving all but the most trivial partial differential equations are beyond the scope of this elementary discussion.

The **order** of a differential equation is the order of the highest-order derivative appearing in the equation.

EXAMPLE 1   The following are examples of differential equations with indicated orders.

(a) $dy/dx = 3x$ (first order).
(b) $x''(t) + 4x'(t) - x(t) = \sin t$ (second order).
(c) $(dy/dx)^5 - 2e^y = 6 \cos x$ (first order).
(d) $(y^{(4)})^2 - 2y''' + y' - 6y = \sqrt{x}$ (fourth order). ∎

In Example 6.6.1, we solved the differential equation

$$\frac{dy}{dx} = 3y, \tag{1}$$

$$y(0) = 2. \tag{2}$$

A solution to (1), (2) is defined to be a function that

(i) is differentiable,
(ii) satisfies the differential equation (1) on some open interval containing 0, and
(iii) satisfies the initial condition (2).

We showed that the only solution to (1) that satisfied (2) was given by

$$y = 2e^{3x}. \tag{3}$$

The system (1), (2) is called an **initial-value problem.** In general, for a certain class of initial-value problems there is a unique solution.† This statement makes sense intuitively. For instance, in Example 6.6.2 we found the following system of differential equations governing the population growth of a certain species of bacteria:

$$\frac{dP}{dt} = 0.1P \tag{4}$$

such that

$$P(0) = 10,000. \tag{5}$$

It is reasonable to expect that with the initial population given by (5) and the assumption that population growth is governed by the equation (4), the population in the future will be completely determined. That is, there is one and only one way to write the population $P(t)$, which is a way of saying that the initial-value problem (4), (5) has a unique solution.

There are other types of problems in which other than initial conditions are given. For example, let a string be held taut between two points 1 m apart, as in Figure 1. Suppose the string is then plucked so that it begins to vibrate. Let $v(x, t)$ denote the height of the string at a distance x units from the left-hand endpoint and at a time t,

---

†Theorems on the existence and uniqueness of solutions to a wide class of ordinary differential equations can be found in *Calculus, Part 2* by S. Grossman (Academic Press, Orlando, 1982), Appendix 5.

FIGURE 1

where $0 \leq x \leq 1$ and $t \geq 0$. Then since the left- and right-hand endpoints are held fixed, we have the conditions

$$v(0, t) = v(1, t) = 0 \qquad \text{for all } t \geq 0. \tag{6}$$

The conditions (6) are called **boundary conditions,** and if we write a differential equation governing the motion of the string (see Problem 18.4.31), then the equation, together with the boundary conditions (6), is called a **boundary value problem.** In general, a boundary value problem is a differential equation together with values given at two or more points, while an initial-value problem is a differential equation with values given at only one point.

EXAMPLE 2   The problem

$$y'' + 2y' + 3y = 0; \qquad y(0) = 1, \quad y(1) = 2$$

is a boundary value problem. The problem

$$y'' + 2y' + 3y = 0; \qquad y(0) = 1, \quad y'(0) = 2$$

is an initial-value problem. ∎

In this chapter we will consider only simple first- and second-order ordinary differential equations and some simple initial-value problems. For a more complete discussion the reader is referred to an introductory book on differential equations.

PROBLEMS 21.1

In Problems 1–6, state the order of the differential equation.

1. $x'(t) + 2x(t) = t$

2. $\dfrac{d^2y}{dx^2} + 4\left(\dfrac{dy}{dx}\right)^3 - y = x^3$

3. $y''' + y = 0$

4. $\left(\dfrac{dx}{dt}\right)^5 = x^4$

5. $x'' - x^2 = 2x^{(4)}$

6. $[x''(t)]^4 - [x'(t)]^{3/2} = 2x(t) + e^t$

7. Verify that $y_1 = 2 \sin x$ and $y_2 = -4 \cos x$ are solutions to $y'' + y = 0$.
8. Verify that $y = (x/2)e^x$ is a solution to $y'' - y = e^x$.
9. Verify that $y = 3x/(4x - 3)$ is the solution to the initial-value problem

$$x^2 \frac{dy}{dx} + y^2 = 0; \qquad y(1) = 3.$$

10. Verify that $y = -\frac{1}{2}x^2e^{-x}$ is a solution to the equation $y''' - 3y' - 2y = 3e^{-x}$.
11. Verify that $y = e^{3x}$ satisfies the initial-value problem

$$y'' - 6y' + 9y = 0, \qquad y(0) = 1, \quad y'(0) = 3.$$

[*Hint:* First check that the equation is satisfied and then verify that the two initial conditions are also satisfied.]

*12. Verify that $y = e^{-x/2}[\cos(\sqrt{3}/2)x + (7/\sqrt{3})\sin(\sqrt{3}/2)x]$ is the solution to the initial-value problem

$$y'' + y' + y = 0, \qquad y(0) = 1, \quad y'(0) = 3.$$

21.2 FIRST-ORDER EQUATIONS—SEPARATION OF VARIABLES

Consider the first-order differential equation

$$\frac{dy}{dx} = f(x, y).　\tag{1}$$

Suppose that $f(x, y)$ can be written as

$$f(x, y) = \frac{g(x)}{h(y)},\tag{2}$$

where g and h are each functions of only one variable. Then (1) can be written

$$h(y)\frac{dy}{dx} = g(x),$$

and integrating both sides with respect to x, we have

$$\int h(y)\, dy = \int h(y)\frac{dy}{dx}\, dx = \int g(x)\, dx + C.\tag{3}$$

The method of solution suggested in (3) is called the method of **separation of variables.** In general, if a differential equation can be written in the form $h(y)\, dy = g(x)\, dx$, then direct integration of both sides (if possible) will produce a family of solutions.

EXAMPLE 1   Solve the differential equation $dy/dx = 4y$.

Solution. We solved equations like this one in Section 6.6. We have $f(x, y) = 4y$, and we can write

$$\frac{dy}{y} = 4\, dx,$$

so

$$\int \frac{dy}{y} = \int 4\, dx,$$

or

$$\ln |y| = 4x + C,$$

or

$$|y| = e^{4x+C} = e^{4x}e^{C},$$

which can be written

$$y = ke^{4x},$$

where k is any real number except 0. In addition, the constant function $y \equiv 0$ is also a solution. ∎

EXAMPLE 2   Solve the initial-value problem $dy/dx = y^2(1 + x^2)$, $y(0) = 1$.

      Solution. We have $f(x, y) = y^2(1 + x^2) = g(x)/h(y)$, where $g(x) = 1 + x^2$ and $h(y) = 1/y^2$. Then, successively,

$$\frac{dy}{y^2} = (1 + x^2)\, dx, \qquad \int \frac{dy}{y^2} = \int (1 + x^2)\, dx, \qquad -\frac{1}{y} = x + \frac{x^3}{3} + C,$$

or

$$y = -\frac{1}{x + (x^3/3) + C}.$$

For every number C this expression is a solution to the differential equation. Moreover, the constant function $y \equiv 0$ is also a solution. When $x = 0$, $y = 1$, so

$$1 = y(0) = -\frac{1}{0 + 0 + C} = -\frac{1}{C},$$

implying that $C = -1$, and we obtain the unique solution to the initial-value problem:

$$y = -\frac{1}{x + (x^3/3) - 1}.$$

REMARK.   This solution (like any solution to a differential equation) can be checked by differentiation. You should *always* carry out this check. To check, we have

$$\frac{dy}{dx} = \frac{1}{[x + (x^3/3) - 1]^2}(1 + x^2) = \left\{\frac{-1}{[x + (x^3/3) - 1]}\right\}^2 (1 + x^2)$$

$$= y^2(1 + x^2).$$

Also, $y(0) = -1/(0 + 0 - 1) = 1$. ∎

EXAMPLE 3 Solve the differential equation $dx/dt = t\sqrt{1 - x^2}$.

Solution. We have

$$\frac{dx}{\sqrt{1 - x^2}} = t\,dt, \qquad \text{or} \qquad \int \frac{dx}{\sqrt{1 - x^2}} = \int t\,dt.$$

Integration yields

$$\sin^{-1} x = \frac{t^2}{2} + C, \qquad \text{or} \qquad x = \sin\!\left(\frac{t^2}{2} + C\right).$$

This answer should be checked by differentiation, using the fact that $\cos[(t^2/2) + C] = \sqrt{1 - \sin^2[(t^2/2) + C]}$. Note that, as in Example 1, there are an infinite number of solutions to the equation. We can obtain a unique solution by specifying an initial condition. For example, if we have $x(0) = \frac{1}{2}$, then $\frac{1}{2} = \sin(0 + C) = \sin C$, and $C = \sin^{-1}\frac{1}{2} = \pi/6$. Then the unique solution would be

$$x(t) = \sin\!\left(\frac{t^2}{2} + \frac{\pi}{6}\right). \quad\blacksquare$$

EXAMPLE 4 **(Logistic Growth)** Let $P(t)$ denote the population of a species at time t. The **growth rate** of the population is defined as the growth in the population divided by the size of the population. Thus, for example, if the birth rate is 3.2 per 100 and the death rate is 1.8 per 100, then the growth rate is $3.2 - 1.8 = 1.4$ per $100 = 1.4/100 = 0.014$. We then write $dP/dt = 0.014P$.

Suppose that in a given population the average birth rate is a positive constant β. It is reasonable to assume that the average death rate is proportional to the number of individuals in the population. Greater populations mean greater crowding and more competition for food and territory. We call this constant of proportionality δ (which is greater than 0). Thus

$$\text{growth rate} = \frac{\text{growth in population}}{\text{population size}} = \frac{dP/dt}{P} \overset{\text{From the discussion above}}{=} \beta - \delta P$$

or

$$\frac{dP}{dt} = P(\beta - \delta P). \tag{4}$$

This differential equation, together with the condition

$$P(0) = P_0 \qquad \text{(the initial population)}, \tag{5}$$

is an initial-value problem. To solve, we have

$$\frac{dP}{P(\beta - \delta P)} = dt,$$

or

$$\int \frac{dP}{P(\beta - \delta P)} = \int dt = t + C. \tag{6}$$

To calculate the integral on the left, we use partial fractions. We have (verify this)

$$\frac{1}{P(\beta - \delta P)} = \frac{1}{\beta P} + \frac{\delta}{\beta(\beta - \delta P)},$$

so

$$\int \frac{dP}{P(\beta - \delta P)} = \int \frac{1}{\beta} \frac{dP}{P} + \frac{\delta}{\beta} \int \frac{dP}{\beta - \delta P}$$

$$= \frac{1}{\beta} \ln P - \frac{1}{\beta} \ln(\beta - \delta P) = t + C,$$

or

$$\ln\left(\frac{P}{\beta - \delta P}\right)^{1/\beta} = t + C.$$

Thus

$$\left(\frac{P}{\beta - \delta P}\right)^{1/\beta} = e^{t+C} = e^{C}e^{t} = C_1 e^{t},$$

so that

$$\frac{P}{\beta - \delta P} = C_2 e^{\beta t} \qquad (\text{where } C_2 = C_1^{\beta}). \tag{7}$$

Using the initial condition (5), we have

$$\frac{P_0}{\beta - \delta P_0} = C_2 e^{0} = C_2,$$

so

$$C_2 = \frac{P_0}{\beta - \delta P_0}. \tag{8}$$

Finally, with the insertion of (8) into (7) it follows (after some algebra) that

$$P(t) = \frac{\beta}{\delta + [(\beta/P_0) - \delta]e^{-\beta t}}. \tag{9}$$

Equation (4) is called the **logistic equation,** and we have shown that the solution to the logistic equation with initial population P_0 is given by (9). Sketches of the growth governed by the logistic equation are given in Figure 1. ■

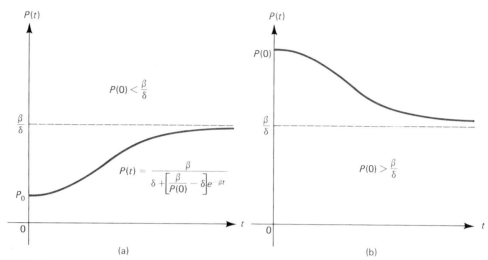

FIGURE 1

We close this section by noting that most differential equations, even of the first order, cannot be solved by elementary methods (although there are other techniques of solution besides the technique of separation of variables). However, even when an equation cannot be solved in a "closed form" (i.e., by writing one variable in terms of another), numerical techniques usually exist for calculating a solution to as many decimal places of accuracy as needed.

In the next section we discuss a wide class of first-order equations for which closed-form solutions can be found.

PROBLEMS 21.2

In Problems 1–12, solve the given differential equation by the method of separation of variables. If an initial condition is given, find the unique solution to the initial-value problem.

1. $\dfrac{dy}{dx} = -7x$

2. $\dfrac{dy}{dx} = e^{x+y}$; $y(0) = 2$

3. $\dfrac{dx}{dt} = \sin x \cos t$; $x\left(\dfrac{\pi}{2}\right) = 3$

4. $\dfrac{dy}{dx} = \dfrac{1}{y^3}$

5. $x^3(y^2 - 1)\dfrac{dy}{dx} = (x + 3)y^5$

6. $\dfrac{dy}{dx} = 2x^2y^2$; $y(1) = 2$

7. $\dfrac{dx}{dt} = e^x \sin t$; $x(0) = 1$

8. $\dfrac{dy}{dx} = 1 + y^2$; $y(0) = 1$

9. $\dfrac{dx}{dt} = x(1 - \cos 2t)$; $x(0) = 1$

*10. $\dfrac{dx}{dt} = x^n t^m$; n, m integers   [*Hint:* Consider the cases $n = 1$, $m = -1$ separately.]

11. $\dfrac{dy}{dx} = x^2(1 + y^2)$

12. $\dfrac{dx}{dt} = \dfrac{e^x t}{e^x + t^2 e^x}$

13. In the logistic equation of Example 4, assume that $\delta = 0.00005$, $\beta = 0.2$, and $P_0 = 20{,}000$. Calculate $P(10)$.

14. The population of a certain species of bacteria is growing according to the logistic equation. Determine the **equilibrium population** P_e, given by

$$P_e = \lim_{t \to \infty} P(t).$$

15. The population of Australia was 12,100,000 in 1968 and 13,268,000 in 1973. Assuming that the growth rate is proportional to the size of the population, estimate the population in 1978, 1983, and 1988.

16. An object of mass m that falls from rest, starting at a point near the earth's surface, is subjected to two forces: a downward force mg and a resisting force proportional to the square of the velocity of the body. Thus

$$F = ma = m\dfrac{dv}{dt} = mg - \alpha v^2,$$

where α is a constant of proportionality.
   (a) Find $v(t)$ as a function of t. [*Hint:* Use the fact that $v(0) = 0$.]
   (b) Show that the velocity does not increase indefinitely but approaches the equilibrium value $\sqrt{mg/\alpha}$. This last value is called **terminal velocity.**

17. The economist Vilfredo Pareto (1848–1923) discovered that the rate of decrease of the number of people y in a stable economy having an income of at least x dollars is directly proportional to the number of such people and inversely proportional to their income. Obtain an expression (**Pareto's law**) for y in terms of x.

*18. On a certain day it began to snow early in the morning and the snow continued to fall at a constant rate. The velocity at which a snowplow is able to clear a road is inversely proportional to the height of the accumulated snow. The snowplow started at 11 A.M. and cleared 4 mi by 2 P.M. By 5 P.M., it cleared another 2 mi. When did it start snowing?

19. A large open cistern filled with water has the shape of a hemisphere with radius 25 ft. The bowl has a circular hole of radius 1 ft in the bottom. By **Torricelli's law,† water will flow out of the hole with the same speed it would attain in falling freely from the level of the water to the hole. How long will it take for all the water to flow from the cistern?

**20. In Problem 19, find the shape of the cistern that would ensure that the water level drops at a constant rate.

21.3 FIRST-ORDER LINEAR EQUATIONS

An nth-order differential equation is called **linear** if it can be written in the form

$$\dfrac{d^n y}{dx^n} + a_{n-1}(x)\dfrac{d^{n-1} y}{dx^{n-1}} + \cdots + a_1(x)\dfrac{dy}{dx} + a_0(x)y = f(x). \tag{1}$$

---

†Evangelista Torricelli (1608–1647) was an Italian physicist.

For example, the most general first-order linear equation takes the form

$$\frac{dy}{dx} + a(x)y = f(x), \tag{2}$$

while a second-order linear equation can be written as

$$\frac{d^2y}{dx^2} + a(x)\frac{dy}{dx} + b(x)y = f(x). \tag{3}$$

In this section we will discuss first-order linear equations; we will discuss second-order linear equations in Sections 21.4 and 21.5.

One nice fact about linear equations is given by the following theorem.†

Theorem 1   Let $x_0, y_0, y_1, \ldots, y_{n-1}$ be real numbers and let $a_0, a_1, \ldots, a_{n-1}$ be continuous. Then there is a unique solution $y = f(x)$ to the nth-order linear differential equation (1) that satisfies

$$f(x_0) = y_0, \qquad f'(x_0) = y_1, \qquad \ldots, \qquad f^{(n-1)}(x_0) = y_{n-1}. \tag{4}$$

That is, if we specify n initial conditions, there exists a unique solution.

We say that the linear equation (1) is **homogeneous** if $f(x) = 0$ for every number x in the domain of f. Otherwise, we say that the equation is **nonhomogeneous.** If the functions $a_0(x), a_1(x), \ldots, a_{n-1}(x)$ are constant, then the equation is said to have **constant coefficients.** Otherwise, it is said to have **variable coefficients.** It turns out that we can solve by integration *all* first-order linear equations and all nth-order linear homogeneous and nonhomogeneous equations with constant coefficients. In this section we show how solutions to linear equations in the form (2) can be explicitly calculated. We do this in three cases.

Case 1   Constant coefficients, homogeneous. Then (2) can be written

$$\frac{dy}{dx} + ay = 0, \tag{5}$$

where a is a constant, or

$$\frac{dy}{dx} = -ay.$$

Solutions (from Section 6.6) are given by

$$y = Ce^{-ax} \tag{6}$$

†For a proof, see Derrick and Grossman, *Elementary Differential Equations with Applications,* Second Edition, Addison-Wesley, Reading, MA., 1981, Chapter 10.

for any constant C. Many examples of this type of equation were given in Section 6.6.

Case 2 Constant coefficients, nonhomogeneous. Then (2) can be written

$$\frac{dy}{dx} + ay = f(x), \tag{7}$$

where a is a constant. It is now impossible to separate the variables. However, equation (7) can be solved by multiplying both sides of (7) by an **integrating factor.** We first note that

$$\frac{d}{dx}(e^{ax}y) = e^{ax}\frac{dy}{dx} + ae^{ax}y = e^{ax}\left(\frac{dy}{dx} + ay\right). \tag{8}$$

Thus if we multiply both sides of (7) by e^{ax}, we obtain

$$e^{ax}\left(\frac{dy}{dx} + ay\right) = e^{ax}f(x),$$

or using (8),

$$\frac{d}{dx}(e^{ax}y) = e^{ax}f(x),$$

and upon integration,

$$e^{ax}y = \int e^{ax}f(x)\,dx + C,$$

where $\int e^{ax}f(x)\,dx$ denotes one particular antiderivative. This leads to the general solution

$$y = e^{-ax}\int e^{ax}f(x)\,dx + Ce^{-ax}. \tag{9}$$

The term e^{ax} is called an integrating factor for (7) because it allows us, after multiplication, to solve the equation by integration.

EXAMPLE 1 Find all solutions to

$$\frac{dy}{dx} + 3y = x. \tag{10}$$

Solution. We multiply both sides of the equation by e^{3x}. Then

$$e^{3x}\left(\frac{dy}{dx} + 3y\right) = xe^{3x}, \quad \text{or} \quad \frac{d}{dx}(e^{3x}y) = xe^{3x},$$

and

$$e^{3x}y = \int xe^{3x}\,dx.$$

But setting $u = x$ and $dv = e^{3x}\,dx$, we find that $du = dx$, $v = \frac{1}{3}e^{3x}$, and

$$\int xe^{3x}\,dx = \frac{x}{3}e^{3x} - \frac{1}{3}\int e^{3x}\,dx = \frac{x}{3}e^{3x} - \frac{1}{9}e^{3x} + C,$$

so

$$e^{3x}y = \frac{x}{3}e^{3x} - \frac{1}{9}e^{3x} + C,$$

and

$$y = \frac{x}{3} - \frac{1}{9} + Ce^{-3x}. \tag{11}$$

This answer should be checked by differentiation. ∎

Let us take a closer look at the answer we just obtained. The answer (11) is given in two parts. The functions Ce^{-3x} (where C is an arbitrary real number) are all solutions to the homogeneous equation

$$\frac{dy}{dx} + 3y = 0,$$

while $(x/3) - \frac{1}{9}$ is one particular solution to the nonhomogeneous problem (10). You should verify this statement. That this result is no accident is suggested by the following theorem.

Theorem 2

(i) Let y_1 and y_2 be two (nonzero) solutions to the homogeneous equation

$$\frac{dy}{dx} + a(x)y = 0. \tag{12}$$

Then for some constant α,

$$y_1(x) = \alpha y_2(x) \qquad \text{for every real number } x. \tag{13}$$

(ii) Let y_p and y_q be two solutions to the nonhomogeneous equation (2). Then

$$y = y_p - y_q$$

is a solution to the homogeneous equation (12).

REMARK 1. Part (i) of the theorem states that if you know one nonzero solution to the homogeneous equation (12), then you know them all.

REMARK 2. Part (ii) of the theorem is important because it states, in effect, that if you know *one* solution to the nonhomogeneous equation (2) and all the solutions to the homogeneous equation (12), then you also know all solutions to the nonhomogeneous equation. This follows since if y_p is the known solution to (2) and if y_q is any other solution to (2), then since $y_q - y_p$ is a solution y_h to (12), we can write

$$y_q - y_p = y_h, \quad \text{or} \quad y_q = y_p + y_h.$$

That is, any other solution to (2) can be written as the sum of the one known solution to (2) and some solution to (12). Thus, for example, in (11), we have $y_p = (x/3) - \frac{1}{9}$, while all solutions to $(dy/dx) + 3y = 0$ are given by $y_h = Ce^{-3x}$. Since, as we will see, nonhomogeneous equations are harder to solve than homogeneous ones, it is nice to know that we need only find *one* solution to a given linear, nonhomogeneous equation.

Proof of Theorem 2.

(i) Assume that $y_2(x) \neq 0$ for every x. Then

$$\left(\frac{y_1}{y_2}\right)' = \frac{y_2 y_1' - y_1 y_2'}{y_2^2} \overset{y_1' = -ay_1}{=} \frac{y_2(-ay_1) - y_1(-ay_2)}{y_2^2} = 0,$$

which implies that y_1/y_2 is constant, so $y_1/y_2 = \alpha$, or $y_1(x) = \alpha y_2(x)$.

Note that we may assume that one of the solutions y_1 or y_2 is never zero. For if both are identically zero, then $y_1(x) = \alpha y_2(x)$ for every number α. On the other hand, if y_2, say, is not identically zero, then it can never be equal to zero. For if $y_2(x_0) = 0$, for some number x_0, then, by the uniqueness theorem (Theorem 1), y_2 is the zero function since the zero function satisfies (12) and the initial condition $y_2(x_0) = 0$.

(ii) let $y = y_p - y_q$. Then

$$\frac{dy}{dx} + a(x)y = \frac{d}{dx}(y_p - y_q) + a(x)(y_p - y_q)$$

$$= \frac{dy_p}{dx} - \frac{dy_q}{dx} + a(x)y_p - a(x)y_q$$

$$= \left[\frac{dy_p}{dx} + a(x)y_p\right] - \left[\frac{dy_q}{dx} + a(x)y_q\right]$$

$$= f(x) - f(x) = 0,$$

since y_p and y_q both solve (2). ∎

EXAMPLE 2 Find the solution to $(dy/dx) + 2y = e^{-5x}$ that satisfies $y(0) = 4$.

Solution. We multiply both sides of the equation by e^{2x} to obtain

$$e^{2x}\left(\frac{dy}{dx} + 2y\right) = e^{-3x}, \quad \text{or} \quad \frac{d}{dx}(e^{2x}y) = e^{-3x},$$

and

$$e^{2x}y = \int e^{-3x}\, dx = -\frac{1}{3}e^{-3x} + C.$$

Then

$$y = -\tfrac{1}{3}e^{-5x} + Ce^{-2x}.$$

Note that $-\tfrac{1}{3}e^{-5x}$ is one solution to $(dy/dx) + 2y = e^{-5x}$, while Ce^{-2x} represents all solutions to $(dy/dx) + 2y = 0$. Finally, we have

$$4 = y(0) = -\tfrac{1}{3} + C, \quad \text{or} \quad C = \tfrac{13}{3},$$

and the solution to the initial-value problem is

$$y = -\tfrac{1}{3}e^{-5x} + \tfrac{13}{3}e^{-2x}. \quad \blacksquare$$

We now turn to the final case.

Case 3   Variable coefficients, nonhomogeneous. We first note the following facts, the first of which follows from the fundamental theorem of calculus (assuming that a is continuous):

(i) $\dfrac{d}{dx}\displaystyle\int a(x)\, dx = a(x).$ (14)

[This holds for *any* antiderivative $\int a(x)\, dx$.]

(ii) $\dfrac{d}{dx}e^{\int a(x)dx} = e^{\int a(x)dx}\dfrac{d}{dx}\displaystyle\int a(x)\, dx = a(x)e^{\int a(x)dx}.$ (15)

Now consider the equation (2),

$$\frac{dy}{dx} + a(x)y = f(x).$$

We multiply both sides by the integrating factor $e^{\int a(x)dx}$. Then we have

$$e^{\int a(x)dx}\frac{dy}{dx} + a(x)e^{\int a(x)dx}y = e^{\int a(x)dx}f(x).$$ (16)

Now we note that, from (15),

$$\frac{d}{dx}\, y e^{\int a(x)dx} = y\frac{d}{dx}\, e^{\int a(x)dx} + \frac{dy}{dx}\, e^{\int a(x)dx}$$

$$= a(x)e^{\int a(x)dx}\, y + \frac{dy}{dx}\, e^{\int a(x)dx}$$

$$= \text{the left-hand side of (16).}$$

Thus from (16)

$$\frac{d}{dx}\,[e^{\int a(x)dx}\, y] = e^{\int a(x)dx}\, f(x),$$

or, integrating,

$$e^{\int a(x)dx}\, y = \int e^{\int a(x)dx}\, f(x)\, dx + C,$$

and

$$y = e^{-\int a(x)dx} \int e^{\int a(x)dx}\, f(x)\, dx + Ce^{-\int a(x)dx}. \qquad (17)$$

It is probably a waste of time to try to memorize the complicated-looking formula (17). Rather, it is important to remember that multiplication by the integrating factor $e^{\int a(x)dx}$ will always enable you to reduce the problem of solving a differential equation to the problem of calculating an integral.

EXAMPLE 3   Find all solutions to the equation

$$\frac{dy}{dx} + \frac{4}{x}\, y = 3x^2.$$

Solution. Here $a(x) = (4/x)$, $\int a(x)\, dx = 4\ln x = \ln x^4$, and $e^{\int a(x)dx} = e^{\ln x^4} = x^4$ (since $e^{\ln u} = u$ for all $u > 0$). Thus we can multiply both sides of the equation by the integrating factor x^4 to obtain

$$x^4\frac{dy}{dx} + 4x^3 y = 3x^6.$$

But

$$\frac{d}{dx}\,(yx^4) = \frac{dy}{dx}\,(x^4) + y(4x^3),$$

and our equation has become

$$\frac{d}{dx}(yx^4) = 3x^6.$$

Then we integrate to find that

$$yx^4 = \frac{3x^7}{7} + C, \quad \text{or} \quad y = \frac{3x^3}{7} + \frac{C}{x^4}.$$

Note that $3x^3/7$ is one solution to the nonhomogeneous equation $(dy/dx) + (4/x)y = 3x^2$, while C/x^4 represents all solutions to the homogeneous equation $(dy/dx) + (4/x)y = 0$. This should be checked by differentiation. ∎

EXAMPLE 4 Find the solution to the initial-value problem $dy/dx = x^2 - 3x^2y$ that satisfies $y(1) = 2$.

Solution. We write the equation in the form

$$\frac{dy}{dx} + 3x^2y = x^2,$$

so $a(x) = 3x^2$, $\int a(x)\, dx = x^3$, and $e^{\int a(x)dx} = e^{x^3}$. Then we have, after multiplication by e^{x^3},

$$\frac{d}{dx}(ye^{x^3}) = x^2 e^{x^3}$$

and

$$ye^{x^3} = \int x^2 e^{x^3}\, dx = \frac{1}{3}e^{x^3} + C,$$

so

$$y = \tfrac{1}{3} + Ce^{-x^3}.$$

Then

$$2 = y(1) = \frac{1}{3} + Ce^{-1} = \frac{1}{3} + \frac{C}{e},$$

or

$$\frac{C}{e} = \frac{5}{3} \quad \text{and} \quad C = \frac{5e}{3}.$$

Thus the solution to the initial-value problem is given by

$$y = \tfrac{1}{3} + (\tfrac{5}{3}e)(e^{-x^3}) = \tfrac{1}{3} + \tfrac{5}{3}e^{1-x^3}. \quad \blacksquare$$

SIMPLE ELECTRIC CIRCUITS

As an application of the material in this section, we will consider simple electric circuits containing a resistor and an inductor or capacitor in series with a source of electromotive force (emf). Such circuits are shown in Figures 1a and 1b, and their action can be understood very easily without any special knowledge of electricity.

(i) An electromotive force (emf) E (volts, V), usually a battery or generator, drives an electric charge Q (coulombs, C) and produces a current I (amperes, A). The current is defined as the rate of flow of the charge, and we can write

$$I = \frac{dQ}{dt}. \tag{18}$$

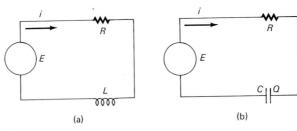

(a)   (b)

FIGURE 1

(ii) A resistor of resistance R (ohms, Ω) is a component of the circuit that opposes the current, dissipating the energy in the form of heat. It produces a drop in voltage given by **Ohm's law:**

$$E_R = RI. \tag{19}$$

(iii) An inductor of inductance L (henrys, H) opposes any change in current by producing a voltage drop of

$$E_L = L\frac{dI}{dt}. \tag{20}$$

(iv) A capacitor of capacitance C (farads, F) stores charge. In so doing, it resists the flow of further charge, causing a drop in the voltage of

$$E_C = \frac{Q}{C}. \tag{21}$$

The quantities R, L, and C are usually constants associated with the particular components in the circuit; E may be a constant or a function of time. The fundamental principle guiding such circuits is given by **Kirchhoff's voltage law.**

> *The algebraic sum of all voltage drops around a closed circuit is zero.*

In the circuit of Figure 1a the resistor and the inductor cause voltage drops of E_R and E_L, respectively. The emf, however, provides a voltage of E (i.e., a voltage drop of $-E$). Thus Kirchhoff's voltage law yields

$$E_R + E_L - E = 0.$$

Transposing E to the other side of the equation and using equations (19) and (20) to replace E_R and E_L, we have

$$L\frac{dI}{dt} + RI = E. \tag{22}$$

The following two examples illustrate the use of equation (22) in analyzing the circuit shown in Figure 1a.

EXAMPLE 5 An inductance of 2 H and a resistance of 10 Ω are connected in series with an emf of 100 V. If the current is zero when $t = 0$, what is the current at the end of 0.1 sec?

Solution. Since $L = 2$, $R = 10$, and $E = 100$, equation (22) and the initial current yield the initial-value problem:

$$2\frac{dI}{dt} + 10I = 100, \qquad I(0) = 0. \tag{23}$$

Dividing both sides of (23) by 2, we note that the resulting linear first-order equation has e^{5t} as an integrating factor; that is,

$$\frac{d}{dt}(e^{5t}I) = e^{5t}\left(\frac{dI}{dt} + 5I\right) = 50e^{5t}. \tag{24}$$

Integrating both sides of equation (24), we get

$$e^{5t}I(t) = 10e^{5t} + C,$$

or

$$I(t) = 10 + Ce^{-5t}. \tag{25}$$

Setting $t = 0$ in (25) and using the initial condition $I(0) = 0$, we have

$$0 = I(0) = 10 + C, \tag{26}$$

which implies that $C = -10$. Substituting this value into (26), we obtain an equation for the current at all times t:

$$I(t) = 10(1 - e^{-5t}).$$

Thus when $t = 0.1$, we have

$$I(0.1) = 10(1 - e^{-0.5}) = 3.93 \text{ A.} \quad \blacksquare$$

EXAMPLE 6 Suppose that the emf $E = 100 \sin 60t$ volts but all other values remain the same as those given in Example 5. Then equation (22) becomes

$$2\frac{dI}{dt} + 10I = 100 \sin 60t, \qquad I(0) = 0.$$

Again, dividing by 2 and multiplying both sides by the integrating factor e^{5t}, we have

$$\frac{d}{dt}(e^{5t}I) = e^{5t}\left(\frac{dI}{dt} + 5I\right) = 50e^{5t} \sin 60t. \qquad (27)$$

Integrating both sides of equation (27) and using formula 168 in Table A.4, we obtain

$$I(t) = e^{-5t}\left[50 \int (\sin 60t)e^{5t}\, dt + C\right]$$

$$= e^{-5t}\left[50e^{5t}\left(\frac{5 \sin 60t - 60 \cos 60t}{3625}\right) + C\right]$$

$$= \frac{2 \sin 60t - 24 \cos 60t}{29} + Ce^{-5t}.$$

Thus setting $t = 0$, we find that $C = \frac{24}{29}$ and

$$I(0.1) = \frac{2 \sin 6 - 24 \cos 6}{29} + \frac{24}{29}e^{-0.5} = -0.31 \text{ A.} \quad \blacksquare$$

For the circuit in Figure 1b we have $E_R + E_C - E = 0$, or

$$RI + \frac{Q}{C} = E.$$

Using the fact that $I = dQ/dt$, we obtain the linear first-order equation

$$R\frac{dQ}{dt} + \frac{Q}{C} = E. \qquad (28)$$

The next example illustrates how to use equation (28).

EXAMPLE 7 A resistance of 2000 Ω and a capacitance of 5×10^{-6} F are connected in series with an emf of 100 V. What is the current at $t = 0.1$ sec if $I(0) = 0.01$ A?

Solution. Setting $R = 2000$, $C = 5 \times 10^{-6}$, and $E = 100$ in (28), we have

$$2000 \frac{dQ}{dt} + \frac{1,000,000}{5} Q = 100,$$

or, dividing both sides by 2000,

$$\frac{dQ}{dt} + 100Q = \frac{1}{20}. \tag{29}$$

Then we can determine $Q(0)$ since

$$\tfrac{1}{20} = Q'(0) + 100Q(0) = I(0) + 100Q(0).$$

Thus

$$Q(0) = \tfrac{1}{100}[\tfrac{1}{20} - I(0)] = \tfrac{1}{100}[\tfrac{1}{20} - \tfrac{1}{100}] = \tfrac{1}{100}(\tfrac{4}{100}) = 4 \times 10^{-4} \text{ C.} \tag{30}$$

Multiplying both sides of (29) by the integrating factor e^{100t}, we get

$$\frac{d}{dt} (e^{100t}Q) = \frac{e^{100t}}{20},$$

and integrating this equation yields

$$e^{100t}Q = \frac{e^{100t}}{2000} + C.$$

Dividing both sides by e^{100t} gives us

$$Q(t) = \tfrac{1}{2000} + Ce^{-100t},$$

and setting $t = 0$, we find that $C = Q(0) - \tfrac{1}{2000} = 4 \times 10^{-4} - 5 \times 10^{-4} = -10^{-4}$. Thus the charge at all times t is

$$Q(t) = \frac{5 - e^{-100t}}{10^4},$$

and the current is

$$I(t) = Q'(t) = \tfrac{1}{100}e^{-100t}.$$

Thus $I(0.1) = 10^{-2}e^{-10} \approx 4.54 \times 10^{-7}$ A. ∎

PROBLEMS 21.3

In Problems 1–14, find all solutions to the given equation. If an initial condition is given, find the particular solution that satisfies that condition.

1. $\dfrac{dy}{dx} = 4x$

2. $\dfrac{dy}{dx} + xy = 0; \, y(0) = 2$

3. $\dfrac{dx}{dt} + x = \sin t; x(0) = 1$

4. $\dfrac{dx}{dt} + x = \dfrac{1}{1 + e^{2t}}$

*5. $\dfrac{dy}{dx} - y \ln x = x^x$

6. $\dfrac{dy}{dx} + (\tan x)y = 2x \sec x; y\left(\dfrac{\pi}{4}\right) = 1$

7. $\dfrac{dx}{dt} - ax = be^{at}; a, b$ constant

8. $\dfrac{dx}{dt} - ax = be^{ct}; x(0) = d; a, b, c, d$ constants

9. $\dfrac{dy}{dx} = x + 2y \tan 2x$

10. $\dfrac{dy}{dx} = 2y + x^2 e^{2x}$

11. $(x^2 + 1)\dfrac{dy}{dx} + 2xy = x; y(0) = 1$

12. $\dfrac{dy}{dx} + \dfrac{y}{x^2} = \dfrac{3}{x^2}; y(1) = 2$

13. $\dfrac{dx}{dt} + \dfrac{4t}{t^2 + 1}x = 3t; x(0) = 4$

14. $\dfrac{ds}{dt} + s \tan t = \cos t; s\left(\dfrac{\pi}{3}\right) = \dfrac{1}{2}$

*15. Solve the equation $y - x(dy/dx) = (dy/dx) y^2 e^y$ by reversing the roles of x and y (i.e., by writing x as a function of y). [*Hint:* Use the fact (from Section 6.1) that under certain conditions $dx/dy = 1/(dy/dx)$.]

16. Use the technique of Problem 15 to solve the equation $dy/dx = -1/(x - e^{-y})$.

17. A chemical substance is injected into the bloodstream of a patient in the treatment of a disease. Let $A(t)$ denote the amount of the chemical in the bloodstream at time t. Suppose that the chemical is injected at a rate of C milligrams (mg) per minute and that it diffuses out of the blood at a rate that is proportional to the amount of the chemical present, with constant of proportionality r. **(a)** Write a first-order differential equation that is satisfied by $A(t)$. Assume that initially the chemical is not present in the bloodstream. **(b)** Solve this differential equation. **(c)** Determine the equilibrium amount of the chemical present in the bloodstream.

In Problems 18–19, assume that the RL circuit shown in Figure 1a has the given resistance, inductance, emf, and initial current. Find an expression for the current at all times t and calculate the current after 1 sec.

18. $R = 8 \, \Omega, L = 1 \, H, E = 6 \, V, I(0) = 1 \, A$

19. $R = 50 \, \Omega, L = 2 \, H, E = 100 \, V, I(0) = 0 \, A$

In Problems 20–22, use the given resistance, capacitance, emf, and initial charge in the RC circuit shown in Figure 1b. Find an expression for the charge at all times t.

20. $R = 1 \, \Omega, C = 1 \, F, E = 12 \, V, Q(0) = 0 \, C$

21. $R = 10 \, \Omega, C = 0.001 \, F, E = 10 \cos 60t$ volts, $Q(0) = 0 \, C$

22. $R = 1 \, \Omega, C = 0.01 \, F, E = \sin 60t$ volts, $Q(0) = 0 \, C$

*23. Solve the problem in Example 7 with an emf of $100 \sin 120\pi t$ volts.

24. An inductance of 1 H and a resistance of 2 Ω are connected in series with a battery of $6e^{-0.0001t}$ volts. No current is flowing initially. When will the current measure 0.5 A?

25. A variable resistance $R = 1/(5 + t)$ ohms and a capacitance of 5×10^{-6} F are connected in series with an emf of 100 V. If $Q(0) = 0$, what is the charge on the capacitor after 1 microsecond (10^{-6} sec)?

26. In the RC circuit (Figure 1b) with constant voltage E, how long will it take the current to decrease to one-half its original value?

27. Suppose that the voltage in an RC circuit is $E(t) = E_0 \cos \omega t$, where $2\pi/\omega$ is the period of the cycle. Assuming that the initial charge is zero, what are the charge and current as functions of R, C, ω, and t?

28. Show that the current in Problem 27 consists of two parts: a steady state term that has a period of $2\pi/\omega$ and a transient term that tends to zero as t increases.

29. In Problem 28, show that if R is small, then the transient term can be quite large for small values of t. (This is why fuses can blow when a switch is flipped.)

21.4 SECOND-ORDER LINEAR, HOMOGENEOUS EQUATIONS WITH CONSTANT COEFFICIENTS

The most general second-order linear differential equation is

$$y''(x) + a(x)y'(x) + b(x)y(x) = f(x). \tag{1}$$

Unlike the analogous first-order equation (21.3.2), it is not possible, in general, to find solutions to (1) by integration. However, it is always possible to solve (1) if the functions $a(x)$ and $b(x)$ are constants. We will therefore consider the constant-coefficient equation

$$y''(x) + ay'(x) + by(x) = f(x) \tag{2}$$

and the related homogeneous equation

$$y''(x) + ay'(x) + by(x) = 0. \tag{3}$$

In this section we will show how solutions to (3) can be found. In Section 21.5 we will deal with the nonhomogeneous equation (2).

The following theorem is central.

Theorem 1 Let $y_1(x)$ and $y_2(x)$ be solutions to the homogeneous equation (3).

(i) Then for any constants c_1 and c_2, $c_1y_1 + c_2y_2$ is also a solution to (3).

(ii) Let $y(x)$ be any other solution to (3). Then if y_2 is not a constant multiple of y_1 and $y_1 \neq 0$, there exist constants k_1 and k_2 such that

$$y(x) = k_1y_1(x) + k_2y_2(x) \tag{4}$$

for every x at which $y_1(x)$ and $y_2(x)$ are defined.

REMARK 1.   The fact in (i) is referred to as the **principle of superposition.** The expression $c_1y_1 + c_2y_2$ is called a **linear combination** of the functions y_1 and y_2. The principle of superposition states that any linear combination of solutions to (3) is again a solution to (3).

REMARK 2.   Part (ii) tells us that if we know two solutions to (3) that are "independent" in the sense that one is not a constant multiple of the other,† then we

---

†y_2 is a constant multiple of y_1 if there exists a constant c such that $y_2(x) = cy_1(x)$ for every x for which $y_1(x)$ is defined.

know them all; for any other solution can be written as a linear combination of these two independent solutions.

The proof of the first part of the theorem is easy and is left as an exercise (see Problem 21). The proof of the second part of the theorem relies on the uniqueness theorem for initial-value problems (Theorem 21.3.1). However, the proof would take us too far afield and so is omitted.†

We now turn to the problem of finding two independent solutions to (3). Recall that for the analogous first-order equation $y' + ay = 0$, the general solution is $y(x) = Ce^{-ax}$. It is then reasonable to "guess" that there may be a solution to (3) of the form $y(x) = e^{\lambda x}$ for some number λ. Setting $y(x) = e^{\lambda x}$ in (3), we obtain

$$y' = \lambda e^{\lambda x}, \qquad y'' = \lambda^2 e^{\lambda x}.$$

and

$$y'' + ay' + by = (\lambda^2 + a\lambda + b)e^{\lambda x}.$$

We see that $y = e^{\lambda x}$ will be a solution to (3) if and only if

$$\lambda^2 + a\lambda + b = 0. \tag{5}$$

Equation (5) is called the **auxiliary equation** of the homogeneous differential equation (3). From the quadratic formula, equation (5) has the roots

$$\lambda_1 = \frac{-a + \sqrt{a^2 - 4b}}{2} \qquad \text{and} \qquad \lambda_2 = \frac{-a - \sqrt{a^2 - 4b}}{2}. \tag{6}$$

There are three possibilities: $a^2 - 4b > 0$, $a^2 - 4b = 0$, and $a^2 - 4b < 0$.

Case 1

TWO REAL ROOTS

If $a^2 - 4b > 0$, then λ_1 and λ_2 are two distinct real numbers and $y_1(x) = e^{\lambda_1 x}$ and $y_2 = e^{\lambda_2 x}$ are independent solutions since, clearly, $e^{\lambda_2 x}$ is not a constant multiple of $e^{\lambda_1 x}$ if $\lambda_1 \neq \lambda_2$. The general solution to (3) is then given by

$$y(x) = c_1 e^{\lambda_1 x} + c_2 e^{\lambda_2 x}, \tag{7}$$

where c_1 and c_2 denote arbitrary constants.

EXAMPLE 1 Find the general solution to

---

†For a proof, see Derrick and Grossman, *Elementary Differential Equations*, 2nd ed., Chapter 3.

$$y''(x) + 2y'(x) - 8y(x) = 0.$$

Solution. The auxiliary equation is

$$\lambda^2 + 2\lambda - 8 = 0, \quad \text{or} \quad (\lambda + 4)(\lambda - 2) = 0,$$

with roots $\lambda = -4$ and $\lambda = 2$. Thus two independent solutions are $y_1(x) = e^{-4x}$ and $y_2(x) = e^{2x}$, so the general solution is

$$y(x) = c_1 e^{-4x} + c_2 e^{2x}. \quad \blacksquare$$

EXAMPLE 2   Find the solution to the equation in Example 1 that satisfies $y(0) = 1$ and $y'(0) = 3$.

Solution. Since $y(x) = c_1 e^{-4x} + c_2 e^{2x}$,

$$y'(x) = -4c_1 e^{-4x} + 2c_2 e^{2x},$$

so

$$y(0) = c_1 + c_2 = 1$$

and

$$y'(0) = -4c_1 + 2c_2 = 3.$$

Then using the techniques of Appendix 3, we easily find the solution

$$c_1 = -\tfrac{1}{6}. \quad c_2 = \tfrac{7}{6},$$

so the solution to the initial-value problem is

$$y(x) = -\tfrac{1}{6}e^{-4x} + \tfrac{7}{6}e^{2x}.$$

This answer should be checked by differentiation. $\quad \blacksquare$

Case 2

REAL REPEATED ROOT

If $a^2 - 4b = 0$, then the roots in (6) are both equal to $-a/2$. One solution is, therefore,

$$y_1(x) = e^{-(a/2)x}. \tag{8}$$

Another, independent solution is given by

$$y_2(x) = xe^{-(a/2)x}. \tag{9}$$

We prove this by differentiation. We have

$$y_2'(x) = \left(1 - \frac{a}{2}x\right) e^{-(a/2)x}$$

and

$$y_2''(x) = \left(\frac{a^2}{4}x - a\right) e^{-(a/2)x},$$

so that

$$y'' + ay' + by = e^{-(a/2)x}\left[\left(\frac{a^2}{4}x - a\right) + a\left(1 - \frac{a}{2}x\right) + bx\right]$$

$$= e^{-(a/2)x}\left(-\frac{a^2}{4}x + bx\right).$$

But $a^2 - 4b = 0$, so $a^2/4 = b$ and

$$e^{-(a/2)x}\left(-\frac{a^2}{4}x + bx\right) = 0,$$

which completes the demonstration. Finally, in Case 2 we see that the general solution is given by

$$y(x) = c_1 e^{-(a/2)x} + c_2 x e^{(a/2)x}. \tag{10}$$

EXAMPLE 3   Find the general solution of

$$y''(x) - 6y'(x) + 9y(x) = 0.$$

Solution. Here $\lambda^2 - 6\lambda + 9 = 0 = (\lambda - 3)^2$, so the only root is $\lambda = 3$. Thus the general solution is given [using (10)] by

$$y(x) = c_1 e^{3x} + c_2 x e^{3x}. \quad \blacksquare$$

EXAMPLE 4   Find the solution of the equation in Example 3 that satisfies the initial conditions $y(0) = 2$ and $y'(0) = -3$.

Solution. We have $y'(x) = 3c_1 e^{3x} + c_2(1 + 3x)e^{3x}$. Thus

$$y(0) = c_1 = 2 \quad \text{and} \quad y'(0) = 3c_1 + c_2 = -3,$$

so that $c_1 = 2$, $c_2 = -9$, and the solution to the initial-value problem is

$$y(x) = 2e^{3x} - 9xe^{3x} = e^{3x}(2 - 9x). \quad \blacksquare$$

Case 3

TWO COMPLEX ROOTS

If $a^2 - 4b < 0$, then the roots of (6) are not real numbers. Rather than digress with a discussion of complex numbers, we will simply tell you the two independent solutions. Let

$$\alpha = -\frac{a}{2} \quad \text{and} \quad \beta = \frac{\sqrt{4b - a^2}}{2}. \tag{11}$$

By assumption, $4b - a^2 > 0$, so that both α and β are real numbers. Then two independent solutions of (3) are given by

$$y_1(x) = e^{\alpha x} \cos \beta x \quad \text{and} \quad y_2(x) = e^{\alpha x} \sin \beta x, \tag{12}$$

and the general solution is

$$y(x) = e^{\alpha x}(c_1 \cos \beta x + c_2 \sin \beta x). \tag{13}$$

We prove that $e^{\alpha x} \cos \beta x$ is a solution of $y'' + ay' + by = 0$. The proof for $e^{\alpha x} \sin \beta x$ is similar. If

$$y(x) = e^{\alpha x} \cos \beta x,$$

then

$$y'(x) = e^{\alpha x}(\alpha \cos \beta x - \beta \sin \beta x)$$

and

$$y''(x) = e^{\alpha x}[(\alpha^2 - \beta^2) \cos \beta x - 2\alpha\beta \sin \beta x].$$

We have

$$y'' + ay' + by = e^{\alpha x}[(\alpha^2 - \beta^2 + a\alpha + b) \cos \beta x - (2\alpha\beta + a\beta) \sin \beta x].$$

To show that $e^{\alpha x} \cos \beta x$ is a solution, we must show that

$$\alpha^2 - \beta^2 + a\alpha + b = 0 \quad \text{and} \quad 2\alpha\beta + a\beta = 0.$$

But from (11), $\alpha^2 = a^2/4$ and $\beta^2 = (4b - a^2)/4$, so

$$\alpha^2 - \beta^2 + a\alpha + b = \frac{a^2}{4} - b + \frac{a^2}{4} - \frac{a^2}{2} + b = 0.$$

Similarly,

$$2\alpha\beta + a\beta = \beta(2\alpha + a) = \beta\left[2\left(-\frac{a}{2}\right) + a\right] = \beta \cdot 0 = 0.$$

EXAMPLE 5   Find all solutions to the equation of the harmonic oscillator

$$y'' + y = 0.$$

Solution. We discussed this equation in Section 7.3. Since $a = 0$ and $b = 1$, $a^2 - 4b = -4 < 0$. Thus $\alpha = -a/2 = 0$, and $\beta = \sqrt{4b - a^2}/2 = \sqrt{4}/2 = 1$, so that the general solution is, according to equation (13), given by

$$y(x) = c_1 \cos x + c_2 \sin x.$$

This is what we found in Section 7.3. ∎

EXAMPLE 6   Find the general solution to

$$y'' + y' + y = 0.$$

Solution. We have $\lambda^2 + \lambda + 1 = 0$ and $a^2 - 4b = 1 - 4 = -3$. Thus if we set

$$\alpha = -\frac{a}{2} = -\frac{1}{2} \quad \text{and} \quad \beta = \frac{\sqrt{4b - a^2}}{2} = \frac{\sqrt{4 - 1}}{2} = \frac{\sqrt{3}}{2},$$

we obtain the general solution

$$y(x) = e^{-x/2}\left(c_1 \cos \frac{\sqrt{3}}{2}x + c_2 \sin \frac{\sqrt{3}}{2}x\right). \quad ∎$$

EXAMPLE 7   Find the solution of the equation of Example 6 that satisfies $y(0) = 1$ and $y'(0) = 3$.

Solution. We have

$$y'(x) = e^{-x/2}\left[\left(-\frac{1}{2}c_1 + \frac{\sqrt{3}}{2}c_2\right)\cos\frac{\sqrt{3}}{2}x + \left(-\frac{\sqrt{3}}{2}c_1 - \frac{1}{2}c_2\right)\sin\frac{\sqrt{3}}{2}x\right].$$

Thus

$$y(0) = c_1 = 1$$

and

$$y'(0) = -\frac{1}{2}c_1 + \frac{\sqrt{3}}{2}c_2 = 3,$$

with solutions $c_1 = 1$ and $c_2 = 7/\sqrt{3}$. Thus the solution to the initial-value problem is

$$y(x) = e^{-x/2}\left(\cos\frac{\sqrt{3}}{2}x + \frac{7}{\sqrt{3}}\sin\frac{\sqrt{3}}{2}x\right). \blacksquare$$

PROBLEMS 21.4

In Problems 1–20, find the general solution of each equation. When initial conditions are specified, determine the particular solution that satisfies them.

1. $y'' - 4y = 0$

2. $y'' + y' - 3y = 0; y(0) = 0, y'(0) = 1$

3. $y'' + 2y' + 2y = 0$

4. $y'' - 3y' + 2y = 0$

5. $8y'' + 4y' + y = 0; y(0) = 0, y'(0) = 1$

6. $y'' + 8y' + 16y = 0$

7. $y'' + 5y' + 6y = 0; y(0) = 1, y'(0) = 2$

8. $y'' + y' + 7y = 0$

9. $4y'' + 20y' + 25y = 0; y(0) = 1, y'(0) = 2$

10. $y'' + y' + 2y = 0$

11. $y'' - y' - 6y = 0; y(0) = -1; y'(0) = 1$

12. $y'' - 5y' = 0$

13. $y'' + 4y = 0; y(\pi/4) = 1, y'(\pi/4) = 3$

14. $y'' - 10y' + 25y = 0; y(0) = 2, y'(0) = -1$

15. $y'' + 17y' = 0; y(0) = 1, y'(0) = 0$

16. $y'' + 2\pi y' + \pi^2 y = 0; y(1) = 1, y'(1) = 1/\pi$

17. $y'' - 13y' + 42y = 0$

18. $y'' + 4y' + 6y = 0$

19. $y'' = y; y(0) = 2, y'(0) = -3$

20. $y'' - y' + y = 0; y(0) = 3, y'(0) = 7$

21. Prove Theorem 1(i). [*Hint:* Let $y = c_1 y_1 + c_2 y_2$. Differentiate twice and show that y satisfies equation (3), using the fact that y_1 and y_2 satisfy equation (3).]

21.5 SECOND-ORDER NONHOMOGENEOUS EQUATIONS WITH CONSTANT COEFFICIENTS: THE METHOD OF UNDETERMINED COEFFICIENTS

In this section we present a method for finding a particular solution to the non-homogeneous equation

$$y''(x) + ay'(x) + by(x) = f(x). \tag{1}$$

We need only find one solution to (1) since if y_p and y_q are solutions to (1), then $y_p - y_q$ is a solution to the homogeneous equation

$$y''(x) + ay'(x) + by(x) = 0. \tag{2}$$

To see this, we note that

$$(y_p - y_q)'' + a(y_p - y_q)' + b(y_p - y_q) = y_p'' - y_q'' + ay_p' - ay_q' + by_p - by_q$$
$$= (y_p'' + ay_p' + by_p) - (y_q'' + ay_q' + by_q)$$
$$= f - f = 0.$$

Thus the general solution to (1) can be written as the sum of one particular solution y_p to (1) plus the general solution to the homogeneous equation (2).

Let $P_n(x)$ be a polynomial of degree n. We will show how to find a solution to (1) if $f(x)$ takes one of the following forms:

(i) $P_n(x)$ (3)

(ii) $P_n(x)e^{ax}$ (4)

(iii) $P_n(x)e^{ax} \sin bx$ or $P_n(x)e^{ax} \cos bx$ (5)

The technique we will use is to "guess" that there is a solution to (1) in the same basic "form" as $f(x)$ and then substitute this "guessed" solution into (1) to determine the unknown coefficients. This technique is called the **method of undetermined coefficients** and is best illustrated with examples. There is a more general method for finding solutions to (1) for an arbitrary f, but that method, called **variation of constants,** is more complicated and is best left to a book on differential equations.

EXAMPLE 1   Find the general solution to

$$y'' + y = x^2.$$ (6)

Solution.   Since x^2 is a polynomial of degree 2, we "guess" that there is a solution to (6) that is a polynomial of degree 2. Let

$$y_p(x) = ax^2 + bx + c.$$

Then

$$y_p' = 2ax + b, \qquad y_p'' = 2a,$$

and

$$y_p'' + y_p = ax^2 + bx + c + 2a.$$

If y_p is a solution to (6), then we must have

$$ax^2 + bx + c + 2a = x^2,$$

which implies that $a = 1$, $b = 0$, and $c + 2a = 0$, so that $a = 1$, $b = 0$, $c = -2$, and

$$y_p(x) = x^2 - 2.$$

You should verify that this is indeed a solution to (6).
Since the general solution to $y'' + y = 0$ is

$$c_1 \cos x + c_2 \sin x,$$

the general solution to (6) is

$$y(x) = c_1 \cos x + c_2 \sin x + x^2 - 2. \quad \blacksquare$$

EXAMPLE 2   Find the solution to

$$y'' - y' - 6y = e^x \cos x \tag{7}$$

that satisfies $y(0) = 1$ and $y'(0) = 0$.

   Solution. We first find a particular solution to (7). We guess that there is a solution of the form

$$y_p(x) = ae^x \sin x + be^x \cos x.$$

Then

$$y_p'(x) = ae^x(\sin x + \cos x) + be^x(\cos x - \sin x)$$

and

$$y_p''(x) = 2ae^x \cos x - 2be^x \sin x.$$

Substitution of these into (7) yields

$$e^x(2a \cos x - 2b \sin x) - e^x[(a - b) \sin x + (a + b) \cos x]$$
$$- 6e^x(a \sin x + b \cos x) = e^x \cos x.$$

Dividing by e^x and equating coefficients of $\sin x$ and $\cos x$ yields

$$2a - (a + b) - 6b = 1,$$
$$-2b - (a - b) - 6a = 0,$$

or

$$a - 7b = 1,$$
$$-7a - b = 0,$$

with solutions $a = \frac{1}{50}$ and $b = -\frac{7}{50}$. Thus

$$y_p(x) = \frac{e^x}{50}(\sin x - 7 \cos x).$$

This result should be checked by differentiation.
   The general solution to the homogeneous equation

$$y'' - y' - 6y = 0$$

is

$$c_1 e^{3x} + c_2 e^{-2x},$$

so the general solution to (7) is

$$y(x) = c_1 e^{3x} + c_2 e^{-2x} + \frac{e^x}{50}(\sin x - 7 \cos x).$$

Then

$$y'(x) = 3c_1 e^{3x} - 2c_2 e^{-2x} + \frac{e^x}{50}(8 \sin x - 6 \cos x),$$

and

$$1 = y(0) = c_1 + c_2 - \tfrac{7}{50},$$
$$0 = y'(0) = 3c_1 - 2c_2 - \tfrac{6}{50},$$

with solutions

$$c_1 = \tfrac{24}{50} \quad \text{and} \quad c_2 = \tfrac{33}{50}.$$

Thus the solution to the initial-value problem is

$$y(x) = \tfrac{1}{50}[24e^{3x} + 33e^{-2x} + e^x(\sin x - 7 \cos x)]. \quad \blacksquare$$

EXAMPLE 3 Find the general solution of

$$y'' - 4y' + 4y = xe^{3x}. \tag{8}$$

Solution. Since x is a polynomial of degree 1, we seek a solution of the form

$$y_p(x) = (ax + b)e^{3x}.$$

Then

$$y_p' = (3ax + 3b + a)e^{3x} \quad \text{and} \quad y_p'' = (9ax + 9b + 6a)e^{3x},$$

so

$$y'' - 4y' + 4y = e^{3x}(ax + b + 2a) = xe^{3x}.$$

Thus

$$a = 1,$$
$$b + 2a = 0,$$

and $a = 1$ and $b = -2$, so that

$$y_p(x) = (x - 2)e^{3x}.$$

Finally, the general solution to the related homogeneous equation $y'' - 4y' + 4y = 0$ is $c_1 e^{2x} + c_2 x e^{2x}$, so the general solution to (8) is

$$y(x) = c_1 e^{2x} + c_2 x e^{2x} + (x - 2)e^{3x}. \blacksquare$$

EXAMPLE 4   Find a particular solution to

$$y'' - y = 3e^{-x}. \tag{9}$$

Solution. We first note that the general solution to the related homogeneous equation

$$y'' - y = 0 \tag{10}$$

is $c_1 e^x + c_2 e^{-x}$. Thus $3e^{-x}$ is a solution to (10). If we try to determine a solution to (9) in the form $y_p(x) = ae^{-x}$, we find that

$$y_p' = -ae^{-x}, \qquad y_p'' = ae^{-x},$$

and

$$y_p'' - y_p = ae^{-x} - ae^{-x} = 0.$$

But this result should have been obvious anyway since ae^{-x} is a solution to (10) and, therefore, could not be a solution to (9). To find a solution to (9), we try the trick that worked for solving Case 2 of the homogeneous equation. We look for a solution to (10) in the form

$$y_p(x) = axe^{-x}.$$

Then

$$y_p'(x) = a(1 - x)e^{-x} \qquad \text{and} \qquad y_p''(x) = a(x - 2)e^{-x},$$

so

$$y_p'' - y_p = a(-2)e^{-x} = 3e^{-x}.$$

This implies that $a = -\frac{3}{2}$, so a particular solution is given by

$$y_p(x) = -\frac{3}{2}xe^{-x}. \blacksquare$$

We now describe in detail the method of undetermined coefficients to obtain a solution to the nonhomogeneous equation (1) when $f(x)$ is of the form (3), (4), or (5).

Case 1   No term in $y_p(x)$ (as given in Table 1) is a solution of the related homogeneous equation (2). Then a particular solution of (1) will have a form according to Table 1.

TABLE 1

$f(x)$	$y_p(x)$
$P_n(x)$	$a_0 + a_1x + a_2x^2 + \cdots + a_nx^n$
$P_n(x)e^{ax}$	$(a_0 + a_1x + a_2x^2 + \cdots + a_nx^n)e^{ax}$
$\left.\begin{array}{l} P_n(x)e^{ax}\sin bx \\ \text{or} \\ P_n(x)e^{ax}\cos bx \end{array}\right\}$	$(a_0 + a_1x + a_2x^2 + \cdots + a_nx^n)e^{ax}\sin bx$ $+ (b_0 + b_1x + b_2x^2 + \cdots + b_nx^n)e^{ax}\cos bx$

Case 2 If any term of $y_p(x)$ is a solution of (2), then multiply the appropriate function $y_p(x)$ of Case 1 by x^k, where k is the smallest integer having the property that no term of $x^k y_p(x)$ is a solution of (2).

MORE ON SIMPLE ELECTRIC CIRCUITS

We will make use of the concepts developed in Section 21.3 and the methods of this chapter to study a simple electric circuit containing a resistor, an inductor, and a capacitor in series with an electromotive force (Figure 1). Suppose that R, L, C, and E are constants. Applying Kirchhoff's law, we obtain

$$L\frac{dI}{dt} + RI + \frac{Q}{C} = E. \tag{11}$$

FIGURE 1

Since $dQ/dt = I$, we differentiate (11) to get the second-order homogeneous differential equation

$$L\frac{d^2I}{dt^2} + R\frac{dI}{dt} + \frac{I}{C} = 0. \tag{12}$$

To solve this equation, we note that the auxiliary equation

$$\lambda^2 + \frac{R}{L}\lambda + \frac{1}{CL} = 0$$

has the following roots:

$$\lambda_1 = \frac{-R + \sqrt{R^2 - 4L/C}}{2L}, \qquad \lambda_2 = \frac{-R - \sqrt{R^2 - 4L/C}}{2L} \tag{13}$$

Equation (12) may now be solved using the methods of this section.

EXAMPLE 5   Let $L = 1$ H, $R = 100\ \Omega$, $C = 10^{-4}$ F, and $E = 1000$ V in the circuit shown in Figure 1. Suppose that no charge is present and no current is flowing at time $t = 0$ when E is applied. Here $R^2 - 4L/C = 10{,}000 - 4 \cdot 10^4 = -30{,}000$ so we have case 3 with $\alpha = -R/2L = -50$ and $\beta = \sqrt{30{,}000}/2 = 50\sqrt{3}$. Thus the general solution to (12) is

$$I(t) = e^{-50t}(c_1 \cos 50\sqrt{3}t + c_2 \sin 50\sqrt{3}t).$$

Applying the initial condition $I(0) = 0$, we have $c_1 = 0$. Hence

$$I(t) = c_2 e^{-50t} \sin 50\sqrt{3}t \quad \text{and} \quad I'(t) = c_2 e^{-50t}(50\sqrt{3} \cos 50\sqrt{3}t - 50 \sin 50\sqrt{3}t).$$

To establish the value of c_2, we must make use of equation (11) and the initial condition $Q(0) = 0$. Now

$$Q(t) = C\left(E - L\frac{dI}{dt} - RI\right)$$

$$= 10^{-4}[1000 - c_2 e^{-50t}(50\sqrt{3} \cos 50\sqrt{3}t - 50 \sin 50\sqrt{3}t + 100 \sin 50\sqrt{3}t)]$$

$$\overset{\displaystyle 50 \cdot 10^{-4} = \tfrac{1}{200}}{\swarrow}$$

$$= \frac{1}{10} - \frac{c_2}{200} e^{-50t}(\sin 50\sqrt{3}t + \sqrt{3} \cos 50\sqrt{3}t).$$

Thus

$$Q(0) = \frac{1}{10} - \frac{c_2}{200}\sqrt{3} = 0 \quad \text{so} \quad c_2 = 20/\sqrt{3}.$$

Finally,

$$Q(t) = \frac{1}{10} - \frac{1}{10\sqrt{3}} e^{-50t}(\sin 50\sqrt{3}t + \sqrt{3} \cos 50\sqrt{3}t)$$

and

$$I(t) = \frac{20}{\sqrt{3}} e^{-50t} \sin 50\sqrt{3}t.$$

From these equations we observe that the current will rapidly damp out and that the charge will rapidly approach its steady state value of $\frac{1}{10}$ C. ∎

EXAMPLE 6   Let the inductance, resistance, and capacitance in Example 5 remain the same, but suppose $E = 962 \sin 60t$. By equation (11) we have

$$\frac{dI}{dt} + 100I + 10^4 Q = 962 \sin 60t, \tag{14}$$

and converting (14) so that all expressions are in terms of $Q(t)$, we obtain

$$\frac{d^2Q}{dt^2} + 100\frac{dQ}{dt} + 10^4Q = 962 \sin 60t. \tag{15}$$

It is evident that equation (15) has a particular solution of the form

$$Q_p(t) = A_1 \sin 60t + A_2 \cos 60t. \tag{16}$$

To determine the values A_1 and A_2, we substitute (16) into (14), obtaining the simultaneous equations

$$6400A_1 - 6000A_2 = 962$$

$$6000A_1 + 6400A_2 = 0.$$

Thus $A_1 = \frac{2}{25}$, $A_2 = -\frac{3}{40}$, and since the general solution of the homogeneous equation is the same as that of equation (12), the general solution of equation (11) is

$$Q(t) = e^{-50t}(c_1 \cos 50\sqrt{3}t + c_2 \sin 50\sqrt{3}t) + \tfrac{2}{25} \sin 60t - \tfrac{3}{40} \cos 60t. \tag{17}$$

Differentiating (17), we obtain

$$I(t) = 50e^{-50t}[(\sqrt{3}c_2 - c_1) \cos 50\sqrt{3}t - (c_2 + \sqrt{3}c_1) \sin 50\sqrt{3}t]$$

$$+ \tfrac{24}{5} \cos 60t + \tfrac{9}{2} \sin 60t.$$

Setting $t = 0$ and using the initial conditions, we obtain the simultaneous equations

$$c_1 = \tfrac{3}{400}$$

$$50(\sqrt{3}c_2 - c_1) = -\tfrac{12}{25}.$$

Therefore, $c_1 = \tfrac{3}{40}$ and $c_2 = -21/1000\sqrt{3}$. ∎

PROBLEMS 21.5

In Problems 1–15, find the general solution of each given differential equation. If initial conditions are given, find the particular solution that satisfies them.

1. $y'' + 6y = 4 \cos x$
2. $y'' + y' - 6y = 10e^{-x}$; $y(0) = 0$, $y'(0) = 2$
3. $y'' + 3y' + 2y = 3e^{4x}$
4. $y'' - 2y' + y = -2e^{3x}$
5. $y'' + y' = 4x^2$; $y(0) = 3$, $y'(0) = -1$
6. $y'' + y = 1 + x + x^2$
7. $y'' - 6y' + 9y = 4xe^{3x}$; $y(0) = 1$, $y'(0) = 0$
8. $y'' + y = \cos x$
9. $y'' - 2y' + y = 2e^x$
10. $y'' - 9y' + 20y = 40x$; $y(0) = 2$, $y'(0) = 3$
11. $y'' + 16y = 8 \cos 4x$; $y(\pi/2) = 0$, $y'(\pi/2) = 1$
12. $y'' + y' = x^2 - x^3 + 2$
*13. $y'' + 4y' + 5y = \sinh 2x \sin x$
*14. $y'' + y' + y = e^{-x/2} \cos (\sqrt{3}/2)x$
15. $y'' - 2y' + 3y = e^x \sin \sqrt{2}x$

16. Show that if y_1 is a solution to $y'' + ay' + by = f_1(x)$ and if y_2 is a solution to $y'' + ay' + by = f_2(x)$, then $y_1 + y_2$ is a solution to $y'' + ay' + by = f_1(x) + f_2(x)$.

In Problems 17–20, use the result of Problem 16 to obtain a particular solution to the given equation.

17. $y'' + y = 3 - 4 \cos x$

18. $y'' - 2y' - 3y = e^{2x} + x - x^2$

19. $y'' + 9y = 2 \sin 3x + x^2$

20. $y'' + 6y' + 9y = xe^x + \cos x$

21. Let $f(x)$ be a polynomial of degree n. Show that for any constants a and $b \neq 0$, there is a solution to $y'' + ay' + by = f(x)$ that is a polynomial of degree n.

22. In Example 5, let $L = 10$ H, $R = 250$ Ω, $C = 10^{-3}$ F, and $E = 900$ V. With the same assumptions, calculate the current and charge for all values of $t \geq 0$.

23. In Problem 22, suppose instead that $E = 50 \cos 30t$. Find $Q(t)$ for $t \geq 0$.

In Problems 24 and 25, find the steady state current in the RLC circuit of Figure 1.

24. $L = 10$ H, $R = 40$ Ω, $C = 0.025$ F, $E = 100 \cos 5t$ volts

25. $L = 5$ H, $R = 10$ Ω, $C = 0.1$ F, $E = 25 \sin t$ volts

26. Find the transient current in the RLC circuit of Figure 1, using the data in Problem 24.

27. Repeat Problem 26 with the data in Problem 25.

21.6 VIBRATORY MOTION (OPTIONAL)

Differential equations were used for the first time to describe the motion of particles. As a simple example, consider the motion of a mass m attached to a coiled spring, the upper end of which is securely fastened (see Figure 1).

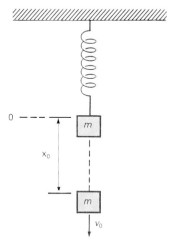

FIGURE 1

We have denoted the equilibrium or rest position of the mass on the spring by the number 0. Suppose that the mass is given an initial displacement x_0 and an initial velocity v_0. We now describe the motion of the mass on the spring. By Hooke's law (see Example 5.9.3) the magnitude of the force exerted on the mass by the spring is proportional to the distance the spring is displaced and is directed toward the equilibrium position. That is,

$$F = -kx, \tag{1}$$

where k is a constant of proportionality. But by Newton's second law,

$$F = ma = m\frac{d^2x}{dt^2}. \tag{2}$$

If we assume that there is no other force acting on the mass, then combining (1) and (2), we obtain

$$m\frac{d^2x}{dt^2} = -kx, \tag{3}$$

or

$$m\frac{d^2x}{dt^2} + kx = 0,$$

which can be written

$$x'' + \frac{k}{m}x = 0. \tag{4}$$

The auxiliary equation for (4) is

$$\lambda^2 + \frac{k}{m} = 0,$$

and since $k/m > 0$, we obtain the general solution

$$x(t) = c_1 \cos\sqrt{\frac{k}{m}}\,t + c_2 \sin\sqrt{\frac{k}{m}}\,t.$$

Then

$$x'(t) = -c_1\sqrt{\frac{k}{m}} \sin\sqrt{\frac{k}{m}}\,t + c_2\sqrt{\frac{k}{m}} \cos\sqrt{\frac{k}{m}}\,t.$$

But $x(0) = x_0$ and $x'(0) = v_0$, so

$$x(0) = x_0 = c_1 \cos 0 + c_2 \sin 0 = c_1$$

and

$$x'(0) = v_0 = -c_1 \sqrt{\frac{k}{m}} \sin 0 + c_2 \sqrt{\frac{k}{m}} \cos 0 = c_2 \sqrt{\frac{k}{m}}.$$

Thus $c_1 = x_0$, $c_2 = v_0/\sqrt{k/m} = v_0\sqrt{m/k}$, and we obtain the equation of periodic motion:

$$x(t) = x_0 \cos\sqrt{\frac{k}{m}}\,t + v_0\sqrt{\frac{m}{k}} \sin\sqrt{\frac{k}{m}}\,t. \tag{5}$$

We see from (5) that the motion of the mass is periodic with period

$$P = 2\pi\sqrt{\frac{m}{k}} \tag{6}$$

and **frequency** f (the frequency is the number of complete oscillations per unit time), given by

$$f = \frac{\sqrt{k/m}}{2\pi}. \tag{7}$$

It is not difficult to show (see Problem 1) that the maximum value taken by x, called the **amplitude** of the motion (and denoted by A), is given by

$$A = \sqrt{x_0^2 + v_0^2\frac{m}{k}}. \tag{8}$$

Note that while the amplitude of the motion depends on the initial displacement and velocity, the period and frequency of the motion do not. These latter depend only on the spring constant k and the mass m.

EXAMPLE 1   A spring, fixed at its upper end, supports a 5-kg mass that stretches the spring 50 cm. Find the equation of motion of the mass if it is drawn to a position 20 cm below its equilibrium position and released.

Solution. The force exerted by the mass on the spring is given by $F = mg = 5g$, where g is the acceleration due to gravity. Then by Hooke's law, we have

$$F = -kx, \quad \text{or} \quad 5g = -k(-50),$$

and

$$k = \frac{g}{10},$$

so

$$\sqrt{\frac{k}{m}} = \sqrt{\frac{g/10}{5}} = \sqrt{\frac{g}{50}}.$$

Since $x_0 = 20$ and $v_0 = 0$, we obtain, from (5),

$$x(t) = 20 \cos \sqrt{\frac{g}{50}}\, t.$$

Also, $P = 2\pi \sqrt{50/g}$, $f = (1/2\pi)\sqrt{g/50}$, and $A = 20$ cm. Note here that we must use the value $g = 981$ cm/sec^2 since our unit of length is the centimeter. ∎

DAMPED HARMONIC MOTION

We now turn to a more practical situation. In the above discussion we made the assumption that there were no external forces acting on the mass. This assumption is, however, not very realistic. We must deal with other forces such as friction in the spring and air resistance, which tend to slow things down. Such forces are called **damping forces.** It is reasonable to assume that the magnitude of the damping force is proportional to the speed of the mass (e.g., the slower the motion, the smaller the air resistance). Thus to equation (3) we add the term $\mu\, dx/dt$, where μ, called the **damping constant,** depends on all the external factors. We then obtain the equation

$$m \frac{d^2 x}{dt^2} = -kx - \mu \frac{dx}{dt},$$

or

$$x'' + \frac{\mu}{m} x' + \frac{k}{m} x = 0; \qquad x(0) = x_0, \quad x'(0) = v_0. \tag{9}$$

The auxiliary equation for (9) is

$$\lambda^2 + \frac{\mu}{m}\lambda + \frac{k}{m} = 0, \tag{10}$$

and

$$\lambda = \frac{-\mu \pm \sqrt{\mu^2 - 4mk}}{2m}. \tag{11}$$

There are three cases to consider.

Case 1 ($\mu^2 > 4mk$)   Then both roots of (10) are negative and we obtain the general solution to (9):

$$x(t) = c_1 e^{\lambda_1 t} + c_2 e^{\lambda_2 t}, \tag{12}$$

where $\lambda_1 = (-\mu + \sqrt{\mu^2 - 4mk})/2m$ and $\lambda_2 = (-\mu - \sqrt{\mu^2 - 4mk})/2m$. Clearly, $x(t) \to 0$ as $t \to \infty$, and the mass quickly returns to its equilibrium position. (See Figure 2.)

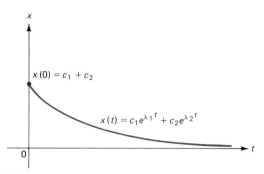

FIGURE 2

Case 2 ($\mu^2 = 4mk$) Then $\lambda_1 = \lambda_2 = -\mu/2m$ and

$$x(t) = c_1 e^{(-\mu/2m)t} + c_2 t e^{(-\mu/2m)t},\qquad(13)$$

and, again, $x(t) \to 0$ as $t \to \infty$.

Case 3 ($\mu^2 < 4mk$) Then the general solution is

$$x(t) = e^{(-\mu/2m)t}\left(c_1 \cos \frac{\sqrt{4mk - \mu^2}}{2m}t + c_2 \sin \frac{\sqrt{4mk - \mu^2}}{2m}t\right).\qquad(14)$$

This motion is more interesting and is called **damped harmonic motion.** The mass oscillates about its equilibrium position but with an ever-decreasing amplitude. The term $e^{(-\mu/2m)t}$ is called a **damping factor.** This motion is illustrated in Figure 3.

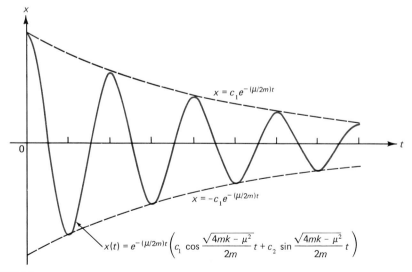

FIGURE 3

EXAMPLE 2 In Example 1, assume that there are damping forces and that $\mu = 2$. Find the new equation of motion.

Solution. We have $\mu^2 - 4mk = 4 - 4(5)(g/10) = 4 - 2g$, which is less than 0 (since $g = 981$). Thus since $\mu/2m = 2/10 = 0.2$, we obtain

$$x(t) = e^{-0.2t}\left(c_1 \cos \frac{\sqrt{2g-4}}{10}t + c_2 \sin \frac{\sqrt{2g-4}}{10}t\right).$$

Then

$$x'(t) = e^{-0.2t}\left[\left(c_2\frac{\sqrt{2g-4}}{10} - 0.2c_1\right)\cos\frac{\sqrt{2g-4}}{10}t\right.$$
$$\left. + \left(-\frac{\sqrt{2g-4}}{10}c_1 - 0.2c_2\right)\sin\frac{\sqrt{2g-4}}{10}t\right].$$

Thus

$$20 = x(0) = c_1$$

and

$$0 = x'(0) = c_2\frac{\sqrt{2g-4}}{10} - 0.2c_1,$$

and we obtain

$$c_1 = 20, \qquad c_2 = \frac{40}{\sqrt{2g-4}},$$

and

$$x(t) = e^{-0.2t}\left(20\cos\frac{\sqrt{2g-4}}{10}t + \frac{40}{\sqrt{2g-4}}\sin\frac{\sqrt{2g-4}}{10}t\right).$$

This motion is sketched in Figure 4. ■

FORCED VIBRATIONS

We now turn to another situation. The motion of the mass considered in the two cases above is determined by the inherent forces of the spring-weight system and the natural forces acting on the system. Accordingly, the vibrations are called **free** or **natural vibrations.** We will now assume that the system is also subject to an external periodic force $\alpha \sin \omega t$, perhaps due to the motion of the object to which the upper end of the spring is attached (see Figure 1). In this case the mass will undergo **forced vibrations.**

Equation (9) may be replaced by the nonhomogeneous second-order differential equation

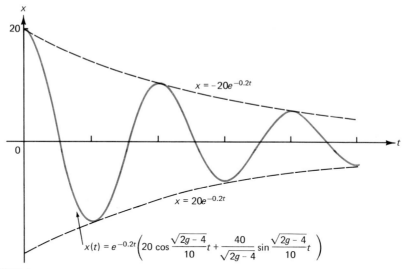

$$x(t) = e^{-0.2t}\left(20 \cos \frac{\sqrt{2g-4}}{10}t + \frac{40}{\sqrt{2g-4}} \sin \frac{\sqrt{2g-4}}{10}t\right)$$

FIGURE 4

$$m\frac{d^2x}{dt^2} = -kx - \mu\frac{dx}{dt} + \alpha \sin \omega t,$$

which we write in the form

$$\frac{d^2x}{dt^2} + \frac{\mu}{m}\frac{dx}{dt} + \frac{k}{m}x = \frac{\alpha}{m} \sin \omega t. \tag{15}$$

By the method of undetermined coefficients, we know that $x(t)$ has a particular solution of the form

$$x_p(t) = A_1 \cos \omega t + A_2 \sin \omega t. \tag{16}$$

Substituting this function into (15) yields the simultaneous equations

$$\left(\frac{k}{m} - \omega^2\right)A_1 + \frac{\mu\omega}{m}A_2 = 0$$

$$-\frac{\mu\omega}{m}A_1 + \left(\frac{k}{m} - \omega^2\right)A_2 = \frac{\alpha}{m}, \tag{17}$$

from which we obtain

$$A_1 = \frac{-\alpha\mu\omega}{(k - m\omega^2)^2 + (\mu\omega)^2} \quad \text{and} \quad A_2 = \frac{\alpha(k - m\omega^2)}{(k - m\omega^2)^2 + (\mu\omega)^2}, \tag{18}$$

To find the general solution of equation (15), we add the general solution of the homogeneous equation to the particular solution (16). The nature of the general solution again depends on the discriminant $\sqrt{\mu^2 - 4mk}$. If $\mu^2 \geq 4mk$, we are super-imposing the periodic function (16) on either equation (12) or (13). Since (12) and (13) damp out as t increases, the motion for large t will be very close to (16). On the other hand, if $\mu^2 < 4mk$, we will suppose initially that $\mu = 0$. Then the general solution for $\sqrt{k/m} \neq \omega$ is

$$x(t) = c_1 \cos\sqrt{\frac{k}{m}}\,t + c_2 \sin\sqrt{\frac{k}{m}}\,t + \frac{\alpha}{m[(k/m) - \omega^2]} \sin \omega t. \tag{19}$$

Note that the effect of the sinusoidal force $\alpha \sin \omega t$ is merely to superimpose itself on the simple harmonic motion of the system. The period of this added motion is the same as that of the external force, with constant amplitude for each fixed ω, but the amplitude approaches infinity as ω approaches $\sqrt{k/m}$. If $\omega = \sqrt{k/m}$, then the particular solution must have the form (from Case 2 on page 1166)

$$x_p(t) = A_1 t \cos \omega t + A_2 t \sin \omega t. \tag{20}$$

Substituting (20) into (15) yields $A_1 = -\alpha/2\omega$ and $A_2 = 0$, so we have the general solution

$$x(t) = c_1 \cos \omega t + c_2 \sin \omega t - \frac{\alpha}{2\omega} t \cos \omega t. \tag{21}$$

We note that as t increases, the vibrations will get larger without bound. Of course, this situation cannot arise in practice, since some resistance to the motion will always be present ($\mu \neq 0$). However, it is evident that if the resistance is small and the external forcing motion agrees with the natural vibration, then the displacement x may become so large that the elastic limit of the spring will be exceeded and a permanent set or distortion will occur. If $\omega = \sqrt{k/m}$, we say the external force is in **resonance** with the vibrating mass. The phenomenon of resonance is of great importance in engineering. It is necessary in some cases, as with our spring problem, to avoid a resonant condition so that no undue stresses will occur; on the other hand, resonance is desirable in many radio circuit problems.

PROBLEMS 21.6

1. Let $x(t)$ be given by (5).
   (a) Show that $x'(t) = 0$ when $\tan\sqrt{k/m}\,t = (v_0/x_0)\sqrt{m/k}$.
   (b) Show that for t as given in (a),

$$\sin\sqrt{\frac{k}{m}}\,t = \frac{v_0\sqrt{m/k}}{\sqrt{x_0^2 + v_0^2(m/k)}} \quad \text{and} \quad \cos\sqrt{\frac{k}{m}}\,t = \frac{x_0}{\sqrt{x_0^2 + v_0^2(m/k)}}.$$

   (c) Show that the maximum value of $x(t)$ is $\sqrt{x_0^2 + v_0^2(m/k)}$.
2. In Example 2, for what minimum damping constant would the mass fail to oscillate about its equilibrium?
3. One end of a rubber band is fixed at a point A. A 2-kg mass, attached to the other end, stretches the rubber band vertically to the point B in such a way that the length AB is 32 cm greater than the natural length of the band. If the mass is further drawn to a position 16 cm

below B and released, what will be its velocity (if we neglect resistance) as it passes the position B?

4. If in Problem 3 the mass is released at a position 16 cm above B, what will be its velocity as it passes 2 cm above B?

*5. A cylindrical block of wood of radius and height 1 ft and weighing 12.48 lb floats with its axis vertical in water (62.4 lb/ft^3). If it is depressed so that the surface of the water is tangent to the block and is then released, what will be its period of vibration and equation of motion? Neglect resistance. [*Hint:* The upward force on the block is equal to the weight of the water displaced by the block.]

*6. A cubical block of wood, 1 ft on a side, is depressed so that its upper face lies along the surface of the water and is then released. The period of vibration is found to be 1 sec. Neglecting resistance, what is the weight of the block of wood (in pounds)?

7. A 10-g mass suspended from a spring vibrates freely, the resistance being numerically equal to half the velocity (in meters per second) at any instant. The period of the motion is 8 sec. What is the spring constant (in grams per second per second)?

8. A weight w (pounds) is suspended from a spring whose constant is 10 lb/ft. The motion of the weight is subject to a resistance (pounds) numerically equal to half the velocity (feet per second). If the motion is to have a 1-sec period, what are the possible values of w?

*9. A 50-g mass is hanging at rest on a spring that is stretched 10 cm by the weight. The upper end of the spring is given the periodic force $10 \sin 2t$ gram meters per second per second and the resistance has a magnitude $100\sqrt{17}$ (g/second) times the velocity (in meters per second). Find the equation of motion of the mass.

*10. Suppose in Problem 9 that the upper end of the spring is given, instead, the force $1 - e^{-\sqrt{g}t}$ gram meters per second per second. Find the displacement of the mass after 0.1 sec.

REVIEW EXERCISES FOR CHAPTER TWENTY-ONE

In Exercises 1–27, find the general solution to the given differential equation. If initial conditions are given, find the unique solution to the initial-value problem.

1. $\dfrac{dy}{dx} = 3x$

2. $\dfrac{dy}{dx} = e^{x-y}; \; y(0) = 4$

3. $\dfrac{dx}{dt} = e^x \cos t; \; x(0) = 3$

4. $\dfrac{dx}{dt} = x^{13}t^{11}$

5. $\dfrac{dy}{dx} + 3y = \cos x; \; y(0) = 1$

6. $\dfrac{dx}{dt} + 3x = \dfrac{1}{1 + e^{3t}}$

7. $\dfrac{dx}{dt} = 3x + t^3 e^{3t}; \; x(1) = 2$

8. $\dfrac{dy}{dx} + y \cot x = \sin x; \; y\left(\dfrac{\pi}{6}\right) = \dfrac{1}{2}$

9. $y'' - 5y' + 4y = 0$

10. $y'' - 9y' + 14y = 0; \; y(0) = 2, \; y'(0) = 1$

11. $y'' - 9y = 0$

12. $y'' + 9y = 0$

13. $y'' + 6y' + 9y = 0$

14. $y'' + 8y' + 16y = 0; \; y(0) = -1, \; y'(0) = 3$

15. $y'' - 2y' + 2y = 0; \; y(0) = 0, \; y'(0) = 1$

16. $y'' + 8y' = 0; \; y(0) = 2; \; y'(0) = -3$

17. $y'' + 4y = 2 \sin x$

18. $y'' + y' - 12y = 4e^{2x}; \; y(0) = 1, \; y'(0) = -1$

19. $y'' + y' + y = e^{-x/2} \sin\left(\dfrac{\sqrt{3}}{2}\right)x$

20. $y'' + 4y = 6x \cos 2x; \; y(0) = 1, \; y'(0) = 0$

21. $y'' + y = x^3 - x$

22. $y'' - 2y' + y = e^{-x}$

23. $y'' - 8y' + 16y = e^{4x}; y(0) = 3, y'(0) = 1$
24. $y'' - 2y' + 3y = e^x \cos \sqrt{2}x$ **25.** $y'' + y = x + e^x + \sin x$
26. $y'' + 16y = \cos 4x + x^2 - 3$ **27.** $y'' + y = 0; y(0) = y'(0) = 0$

28. A spring, fixed at its upper end, supports a 10-kg mass that stretches the spring 60 cm. Find the equation of motion of the mass if it is drawn to a position 10 cm below its equilibrium position and released with an initial velocity of 5 cm/sec upward.

29. What are the period, frequency, and amplitude of the motion of the mass in Exercise 28?

30. Find the equation of motion of the mass of Exercise 28 if it is subjected to damping forces having the damping constant $\mu = 3$.

31. In Exercise 30, for what minimum damping constant would the mass fail to oscillate about its equilibrium?

32. For what value of ω will the external force of 10 sin ωt newtons produce resonance in the spring of Exercise 31?

Appendix 1
Review of Trigonometry

1.1 ANGLES AND RADIAN MEASURE

We begin by drawing the unit circle—the circle with radius 1 centered at the origin (see Figure 1). **Angles** are measured starting at the positive x-axis. An angle is positive if it is measured in the counterclockwise direction, and it is negative if it is measured in the clockwise direction. We measure angles in degrees, using the fact that the circle contains 360°. Then we can describe any angle by comparison with the circle. Some angles are depicted in Figure 2. In Figure 2e we obtained the angle $-90°$ by moving

FIGURE 1

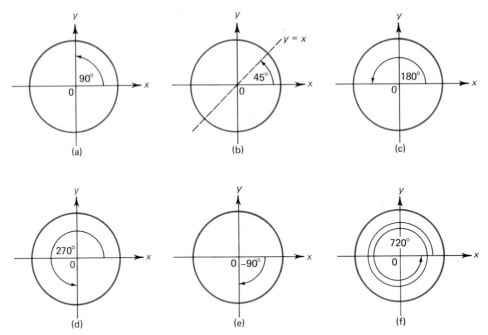

FIGURE 2

in the negative (clockwise) direction. In part (f) we obtained an angle of 720° by moving around the circle twice in the counterclockwise direction.

Figure 2 illustrates the great advantage of using circles rather than triangles to measure angles. Any angle in a triangle must be between 0° and 180°. In a circle there is no such restriction.

There is another way to measure angles, which, in many instances, is more useful than measurement in degrees. Let R denote the radial line which makes an angle of θ with the positive x-axis. See Figure 3. Let (x, y) denote the point at which this radial line intersects the unit circle. Then the **radian measure** of the angle is the length of the arc of the unit circle from the point $(1, 0)$ to the point (x, y).

Since the circumference of a circle is $2\pi r$, where r is its radius, the circumference of the unit circle is 2π. Thus

$$360° = 2\pi \text{ radians.} \tag{1}$$

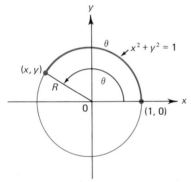

FIGURE 3

Then since $180° = \frac{1}{2}(360°)$, $180° = \frac{1}{2}(2\pi) = \pi$ radians. In general, θ (in degrees) is to $360°$ as θ (in radians) is to 2π radians. Thus $(\theta/360)(\text{degrees}) = (\theta/2\pi)(\text{radians})$, or

$$\theta(\text{degrees}) = \frac{180}{\pi}\theta(\text{radians}) \tag{2}$$

and

$$\theta(\text{radians}) = \frac{\pi}{180}\theta(\text{degrees}). \tag{3}$$

We can calculate that 1 radian $= 180/\pi \approx 57.3°$ and $1° = \pi/180 \approx 0.0175$ radians. Representative values of θ in degrees and radians are given in Table 1.

TABLE 1

θ (degrees)	0	90	180	270	360	45	30	60	-90	135	120	720
θ (radians)	0	$\dfrac{\pi}{2}$	π	$\dfrac{3\pi}{2}$	2π	$\dfrac{\pi}{4}$	$\dfrac{\pi}{6}$	$\dfrac{\pi}{3}$	$\dfrac{-\pi}{2}$	$\dfrac{3\pi}{4}$	$\dfrac{2\pi}{3}$	4π

The radian measure of an angle does not refer to "degrees" but, instead, refers to distance measured along an arc of the unit circle. This is an advantage when discussing trigonometric functions that arise in applications having nothing at all to do with angles (see Section 7.3).

Let C_r denote the circle of radius r centered at the origin (see Figure 4). If $0P$ denotes a radial line as pictured in the figure, then $0P$ cuts an arc from C_r of length L. Let θ be the positive angle between $0P$ and the positive x-axis. If $\theta = 360°$, then $L = 2\pi r$. If $\theta = 180°$, then $L = \pi r$. In fact, it is evident from the figure that

$$\frac{\theta°}{360°} = \frac{L}{2\pi r}, \tag{4}$$

or

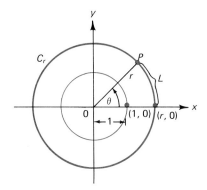

FIGURE 4

$$\theta° = 360° \frac{L}{2\pi r}. \tag{5}$$

If we measure θ in radians, then (4) becomes

$$\frac{\theta}{2\pi} = \frac{L}{2\pi r}, \tag{6}$$

or

$$\theta = \frac{2\pi L}{2\pi r} = \frac{L}{r}. \tag{7}$$

Finally, rewriting (7), we obtain

$$L = r\theta. \tag{8}$$

That is, if θ is measured in radians, then the angle θ "cuts" from the circle of radius r centered at the origin an arc of length $r\theta$. Note that if $r = 1$, then (8) reduces to $L = \theta$, which is the definition of the radian measure of an angle.

EXAMPLE 1 What is the length of an arc cut from the circle of radius 4 centered at the origin by an angle of (a) 45°, (b) 60°, (c) 270°?

Solution. From (8) we find that $L = 4\theta$, where θ is the radian measure of the angle. We therefore have the following:

(a) $L = 4 \cdot \dfrac{\pi}{4} = \pi$ (b) $L = 4 \cdot \dfrac{\pi}{3} = \dfrac{4\pi}{3}$ (c) $L = 4 \cdot \dfrac{3\pi}{2} = 6\pi$ ∎

PROBLEMS 1.1

In Problems 1–6, convert from degrees to radians.

1. $\theta = 150°$ 2. $\theta = -45°$ 3. $\theta = 300°$
4. $\theta = 72°$ 5. $\theta = 144°$ 6. $\theta = 1080°$

In Problems 7–12, convert from radians to degrees.

7. $\pi/12$ 8. $7\pi/12$ 9. $\pi/8$
10. 3π 11. $-(\pi/3)$ 12. $5\pi/4$

13. Let C denote the circle of radius 2 centered at the origin. If a radial line cuts an arc of length π [starting at the point (2, 0)], what is the angle (in degrees) between this line and the positive x-axis?
14. If the radial line in Problem 13 makes an angle of 75° with the positive x-axis, what is the length of the arc it cuts from the circle?

1.2 THE TRIGONOMETRIC FUNCTIONS AND BASIC IDENTITIES

We again begin with the unit circle (see Figure 1). An angle θ uniquely determines a point (x, y) where the radial line intersects the circle. We then define

$$\text{cosine } \theta = x \quad \text{and} \quad \text{sine } \theta = y. \tag{1}$$

These are the two basic trigonometric (or circular) functions, usually written $\cos \theta$ and $\sin \theta$. Since the equation of the circle is $x^2 + y^2 = 1$, we see that $\sin \theta$ and $\cos \theta$ satisfy the equation

$$\sin^2 \theta + \cos^2 \theta = 1. \tag{2}$$

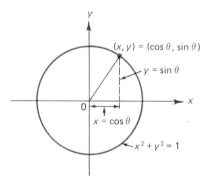

FIGURE 1

We emphasize that $\cos \theta$ is the x-coordinate of the point (x, y) and $\sin \theta$ is the y-coordinate. As θ varies, $\cos \theta$ and $\sin \theta$ oscillate between $+1$ and -1. For example, if $\theta = 0$, then the radial line intersects the circle at the point $(1, 0)$, and we have $\cos 0 = 1$ and $\sin 0 = 0$. If $\theta = 90° = \pi/2$, then the radial line intersects the circle at the point $(0, 1)$, and we have $\cos 90° = \cos \pi/2 = 0$ and $\sin 90° = \sin \pi/2 = 1$. If $\theta = 45° = \pi/4$, then the radial line is the line $y = x$. Since $x^2 + y^2 = 1$ and $y = x$ at the point of intersection, we see that $x^2 + x^2 = 2x^2 = 1$, or $x = y = 1/\sqrt{2} = \sqrt{2}/2$. Thus $\cos 45° = \cos \pi/4 = \sqrt{2}/2$ and $\sin 45° = \sin \pi/4 = \sqrt{2}/2$. It will be shown in Appendix 1.4 that $\cos 30° = \cos \pi/6 = \sqrt{3}/2$, $\sin 30° = \sin \pi/6 = \frac{1}{2}$, $\cos 60° = \cos \pi/3 = \frac{1}{2}$, and $\sin 60° = \sin \pi/3 = \sqrt{3}/2$. The most commonly used values of $\cos \theta$ and $\sin \theta$ are given in Table 1.

Some basic facts about the functions $\sin \theta$ and $\cos \theta$ can be derived by simply looking at the graph of the unit circle. First, we note that if we add $360°$ to the angle θ in Figure 1, then we end up with the same point (x, y) on the circle. Thus

$$\cos(\theta + 360°) = \cos(\theta + 2\pi) = \cos \theta \tag{3}$$

and

TABLE 1

θ	0	$\dfrac{\pi}{6}$	$\dfrac{\pi}{4}$	$\dfrac{\pi}{3}$	$\dfrac{\pi}{2}$	π	$\dfrac{3\pi}{2}$	2π
$\cos\theta$	1	$\dfrac{\sqrt{3}}{2}$	$\dfrac{\sqrt{2}}{2}$	$\dfrac{1}{2}$	0	-1	0	1
$\sin\theta$	0	$\dfrac{1}{2}$	$\dfrac{\sqrt{2}}{2}$	$\dfrac{\sqrt{3}}{2}$	1	0	-1	0

$$\sin(\theta + 360°) = \sin(\theta + 2\pi) = \sin\theta. \tag{4}$$

In general, if α is the *smallest* positive number such that $f(x + \alpha) = f(x)$, we say that f is **periodic** of **period** α. Thus from (3) and (4) we see that the functions $\cos\theta$ and $\sin\theta$ are periodic of period 2π.

A glance at Figure 2 tells us the sign of the two basic functions. With all the information above we can draw a sketch of $y = \cos\theta$ and $y = \sin\theta$. This sketch is shown in Figures 3 and 4.

FIGURE 2

FIGURE 3

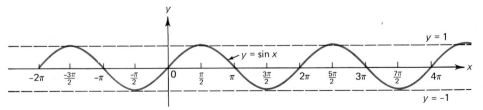

FIGURE 4

Now look at Figure 5a. We see that in a comparison of θ and $-\theta$, the x-coordinates are the same while the y-coordinates have opposite signs. This suggests that

$$\cos(-\theta) = \cos\theta \tag{5}$$

and

$$\sin(-\theta) = -\sin\theta. \tag{6}$$

To obtain another identity, we add $180° = \pi$ to θ (Figure 5b). Then the x- and y-coordinates of $\theta + \pi$ have signs opposite to those of the x- and y-coordinates of θ. Thus

$$\cos(\theta + 180°) = \cos(\theta + \pi) = -\cos\theta \tag{7}$$

and

$$\sin(\theta + 180°) = \sin(\theta + \pi) = -\sin\theta. \tag{8}$$

Several other identities can be obtained by simply glancing at a graph of the unit circle.

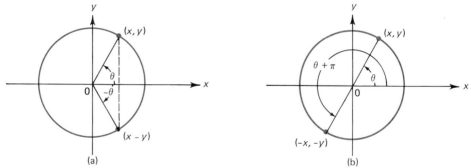

(a)   (b)

FIGURE 5

We now obtain another identity that is very useful in computations.

Theorem 1

$$\cos(\theta + \varphi) = \cos\theta \cos\varphi - \sin\theta \sin\varphi.$$

Proof.   We prove the theorem in the case θ and φ are between 0 and $\pi/2$. We will leave another case as a problem (see Problem 16). From Figure 6 we see that the arc P_1P_3 has the same length as the arc P_2P_4 (they are both equal to the radian measure of the angle $\theta + \varphi$). Then the distance from P_1 to P_3 is the same as the distance from P_2 to P_4. Using the distance formula [see equation (1.3.2)], we obtain

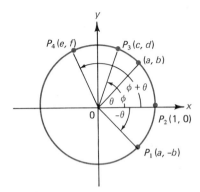

FIGURE 6

$$\overline{P_1P_3}\dagger = \sqrt{(c-a)^2 + (d+b)^2} = \sqrt{(e-1)^2 + f^2} = \overline{P_2P_4}. \tag{9}$$

But

$$\cos\theta = a, \quad \cos\varphi = c, \quad \cos(\theta + \varphi) = e,$$
$$\sin\theta = b, \quad \sin\varphi = d, \quad \sin(\theta + \varphi) = f. \tag{10}$$

Then we square both sides of (9):

$$c^2 - 2ac + a^2 + d^2 + 2bd + b^2 = e^2 - 2e + 1 + f^2.$$

Since $a^2 + b^2 = c^2 + d^2 = e^2 + f^2 = 1$ (why?), we have

$$-2ac + 2bd + 2 = -2e + 2,$$

or

$$e = ac - bd. \tag{11}$$

Substituting (10) into (11) proves the theorem. ∎

There are many other identities that can be proved by using Theorem 1. We indicate in Table 2 some of the identities we will find useful in other parts of this text. The proofs of these identities are suggested in the problems. We use x and y instead of θ and φ in this table. These identities will be very useful to us when we discuss techniques of integration in Chapter 8.

---

†The symbol $\overline{P_1P_3}$ denotes the distance between the points P_1 and P_3.

TABLE 2   BASIC IDENTITIES INVOLVING cos x AND sin x

(i) $\sin^2 x + \cos^2 x = 1$
(ii) $\cos(-x) = \cos x$
(iii) $\sin(-x) = -\sin x$
(iv) $\cos(x + \pi) = -\cos x$
(v) $\sin(x + \pi) = -\sin x$
(vi) $\cos(\pi - x) = -\cos x$
(vii) $\sin(\pi - x) = \sin x$

(viii) $\cos\left(\dfrac{\pi}{2} + x\right) = -\sin x$

(ix) $\sin\left(\dfrac{\pi}{2} + x\right) = \cos x$

(x) $\cos\left(\dfrac{\pi}{2} - x\right) = \sin x$

(xi) $\sin\left(\dfrac{\pi}{2} - x\right) = \cos x$

(xii) $\cos(x + y) = \cos x \cos y - \sin x \sin y$
(xiii) $\sin(x + y) = \sin x \cos y + \cos x \sin y$
(xiv) $\cos(x - y) = \cos x \cos y + \sin x \sin y$
(xv) $\sin(x - y) = \sin x \cos y - \cos x \sin y$
(xvi) $\cos 2x = \cos^2 x - \sin^2 x = 2\cos^2 x - 1 = 1 - 2\sin^2 x$
(xvii) $\sin 2x = 2\sin x \cos x$

(xviii) $\cos\dfrac{x}{2} = \pm\sqrt{\dfrac{1 + \cos x}{2}}$

(xix) $\sin\dfrac{x}{2} = \pm\sqrt{\dfrac{1 - \cos x}{2}}$ } The **half angle** formulas

(xx) $\cos^2 x = \dfrac{1 + \cos 2x}{2}$

(xxi) $\sin^2 x = \dfrac{1 - \cos 2x}{2}$

(xxii) $\cos x - \cos y = 2\sin\dfrac{x + y}{2}\sin\dfrac{y - x}{2}$

(xxiii) $\sin x - \sin y = 2\sin\dfrac{x - y}{2}\cos\dfrac{x + y}{2}$

PROBLEMS 1.2

In Problems 1–15, use the basic identities to calculate $\sin\theta$ and $\cos\theta$.

1. $\theta = 6\pi$

2. $\theta = -30°$

3. $\theta = 7\pi/6$

4. $\theta = 5\pi/6$

5. $\theta = 75°$

6. $\theta = 15°$

7. $\theta = 13\pi/12$

8. $\theta = -150°$

9. $\theta = -\pi/12$

10. $\theta = \pi/8$

11. $\theta = \pi/16$ [*Hint:* Use the result of Problem 10.]

12. $\theta = 3\pi/8$

13. $\theta = 67\frac{1}{2}°$

14. $\theta = 7\pi/24$

15. $\theta = -7\frac{1}{2}°$

16. Prove Theorem 1 in the case $0° < \theta < 90°$ and $90° < \varphi < 180°$.
17. Prove that $\cos[(\pi/2) + x] = -\sin x$.
18. Show that $\sin[(\pi/2) + x] = \cos x$. [*Hint:* Use Problem 17 to show that $\sin[x + (\pi/2)] = -\cos(x + \pi)$ and then use identity (iv).]
19. Use Problems 17 and 18 to show that $\sin(x + y) = \sin x \cos y + \cos x \sin y$. [*Hint:* Start with $\sin(x + y) = -\cos[(\pi/2) + x + y]$ and then apply Theorem 1.]
20. Prove identities (xiv) and (xv). [*Hint:* Use Theorem 1, Problem 19, and identities (ii) and (iii).]
21. Prove identities (vi), (vii), (x), and (xi). [*Hint:* Use Problem 20.]
22. Prove identities (xvi) and (xvii). [*Hint:* Use Theorem 1 and Problem 19.]
23. Prove that $\cos(x/2) = \pm\sqrt{(1 + \cos x)/2}$. [*Hint:* Use identity (xvi) to show that $\cos x = 2\cos^2(x/2) - 1$.]
24. Prove identity (xix). [*Hint:* Use identity (i) and Problem 23.]
25. Prove identities (xx) and (xxi). [*Hint:* Use identities (xviii) and (xix).]
26. Prove that $\cos x - \cos y = 2\sin[(x + y)/2]\sin[(y - x)/2]$. [*Hint:* Expand the right side by using identities (xiii), (xv), and (xix).]
27. Prove identity (xxiii).
28. Graph the function $y = 3\sin x$. The greatest value a periodic function takes is called the **amplitude** of the function. Show that in this case the amplitude is equal to 3.
29. Graph the function $y = -2\cos x$. What is the amplitude?
30. Show that the function $y = \sin 2x$ is periodic of period π. Graph the function.
31. Show that the function $y = 4\cos(x/2)$ is periodic of period 4π. What is its amplitude? Graph the curve.
32. Graph the curve $y = 3\sin(x/3)$.
33. Graph the curve $y = \sin(x - 1)$. [*Hint:* See Section 1.8.] Show that its period is 2π.
34. Graph the curve $y = 2\sin[(x/2) + 3]$. What is its period? What is its amplitude?
35. Graph the curve $y = 3\cos(3x - \frac{1}{3})$. What is its period? What is its amplitude?

1.3 OTHER TRIGONOMETRIC FUNCTIONS

Besides the two functions we have already discussed, there are four other trigonometric functions, which can be defined in terms of $\sin x$ and $\cos x$:

(i) tangent $x = \tan x = \dfrac{\sin x}{\cos x}$ for $\cos x \neq 0$

(ii) cotangent $x = \cot x = \dfrac{\cos x}{\sin x} = \dfrac{1}{\tan x}$ for $\sin x \neq 0$

(iii) secant $x = \sec x = \dfrac{1}{\cos x}$ for $\cos x \neq 0$

(iv) cosecant $x = \csc x = \dfrac{1}{\sin x}$ for $\sin x \neq 0$

Each of these four functions grows without bound as x approaches certain values. When $x \to 0^+$, $\cot x$ and $\csc x$ approach $+\infty$; and when $x \to 0^-$, $\cot x$ and $\csc x$

approach $-\infty$. This is true because $\sin x > 0$ for x near 0 and positive, and $\sin x < 0$ for x near 0 and negative (see Figure 4 in Appendix 1.2). Also, $\cos x$ is near 1 for x near 0. Similarly, we obtain

$$\lim_{x \to \pi/2^-} \tan x = +\infty, \qquad \lim_{x \to \pi/2^+} \tan x = -\infty,$$

$$\lim_{x \to \pi/2^-} \sec x = +\infty, \qquad \lim_{x \to \pi/2^+} \sec x = -\infty,$$

These facts hold because $\cos x$ is positive for $x < \pi/2$ and x near $\pi/2$, and $\cos x$ is negative for $x > \pi/2$ and x near $\pi/2$. (See Figure 3 in Appendix 1.2).
    We also observe that

$$\tan 0 = \frac{\sin 0}{\cos 0} = \frac{0}{1} = 0 \qquad \text{and} \qquad \cot \frac{\pi}{2} = \frac{\cos(\pi/2)}{\sin(\pi/2)} = \frac{0}{1} = 0.$$

We note that since $-1 \le \sin x \le 1$ and $-1 \le \cos x \le 1$, we have $|\sec x| \ge 1$ and $|\csc x| \ge 1$. That is, $\sec x$ and $\csc x$ can never take values in the open interval $(-1, 1)$. In addition,

$$\tan(x + \pi) = \frac{\sin(x + \pi)}{\cos(x + \pi)} = \frac{-\sin x}{-\cos x} = \frac{\sin x}{\cos x} = \tan x,$$

so that $\tan x$ is periodic of period π. Similarly, $\cot x$ is periodic of period π. Also,

$$\sec(x + 2\pi) = \frac{1}{\cos(x + 2\pi)} = \frac{1}{\cos x} = \sec x,$$

so that $\sec x$ (and $\csc x$) are periodic of period 2π.
    Values of $\tan x$, $\cot x$, $\sec x$, and $\csc x$ are given in Table 1. Putting this information all together and using our knowledge of the functions $\sin x$ and $\cos x$, we obtain the graphs given in Figure 1.

TABLE 1

x	0	$\dfrac{\pi}{6}$	$\dfrac{\pi}{4}$	$\dfrac{\pi}{3}$	$\dfrac{\pi}{2}$	π	$\dfrac{3\pi}{2}$	2π
$\tan x$	0	$\dfrac{1}{\sqrt{3}}$	1	$\sqrt{3}$	undefined	0	undefined	0
$\cot x$	undefined	$\sqrt{3}$	1	$\dfrac{1}{\sqrt{3}}$	0	undefined	0	undefined
$\sec x$	1	$\dfrac{2}{\sqrt{3}}$	$\sqrt{2}$	2	undefined	-1	undefined	1
$\csc x$	undefined	2	$\sqrt{2}$	$\dfrac{2}{\sqrt{3}}$	1	undefined	-1	undefined

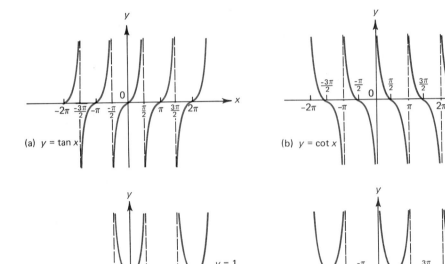

FIGURE 1

There are many identities involving the four functions introduced in this section. However, for our purposes there are two that will prove especially useful:

$$1 + \tan^2 x = \sec^2 x, \tag{1}$$

$$1 + \cot^2 x = \csc^2 x. \tag{2}$$

Both can be obtained by starting with the identity

$$\sin^2 x + \cos^2 x = 1$$

and then dividing both sides by $\cos^2 x$ to obtain (1) and by $\sin^2 x$ to obtain (2).

PROBLEMS 1.3

In Problems 1–15, calculate $\tan x$, $\cot x$, $\sec x$, and $\csc x$.

1. $x = 6\pi$ **2.** $x = -30°$ **3.** $x = 7\pi/6$
4. $x = 5\pi/6$ **5.** $x = 75°$ **6.** $x = 15°$
7. $x = 13\pi/12$ **8.** $x = -150°$ **9.** $x = -\pi/12$
10. $x = \pi/8$ **11.** $x = \pi/16$ **12.** $x = 3\pi/8$
13. $x = 67\frac{1}{2}°$ **14.** $x = 7\pi/24$ **15.** $x = -7\frac{1}{2}°$

In Problems 16–21, find the period of the given function and sketch the graph of the function.

16. $y = \tan 2x$ **17.** $y = 3 \sec x/3$ **18.** $y = 4 \cot(4x + 1)$

19. $y = 2 \csc 6x$ **20.** $y = -2 \tan\left(\dfrac{x}{5} + 1\right)$ **21.** $y = 8 \sec(5x + 5)$

22. Using the corresponding formulas for sine and cosine, show that

$$\tan(x - y) = \frac{\tan x - \tan y}{1 + \tan x \tan y}.$$

23. Show that

$$\tan(x + y) = \frac{\tan x + \tan y}{1 - \tan x \tan y}.$$

24. Show that

$$\tan \frac{x}{2} = \frac{1 - \cos x}{\sin x} = \frac{\sin x}{1 + \cos x}.$$

1.4 TRIANGLES

In many elementary courses in trigonometry the six trigonometric functions are introduced in terms of the ratios of sides of a right triangle. Consider the angle θ in the right triangle in Figure 1. The side opposite θ is labeled "op," the side adjacent to θ (which is not the hypotenuse) is labeled "a," and the hypotenuse (the side opposite the right angle) is labeled "h." Then we define

$$\sin \theta = \frac{\text{opposite}}{\text{hypotenuse}} = \frac{op}{h}, \qquad \cos \theta = \frac{\text{adjacent}}{\text{hypotenuse}} = \frac{a}{h},$$

$$\tan \theta = \frac{\text{opposite}}{\text{adjacent}} = \frac{op}{a}, \qquad \cot \theta = \frac{\text{adjacent}}{\text{opposite}} = \frac{a}{op},$$

$$\sec \theta = \frac{\text{hypotenuse}}{\text{adjacent}} = \frac{h}{a}, \qquad \csc \theta = \frac{\text{hypotenuse}}{\text{opposite}} = \frac{h}{op}.$$

FIGURE 1

Of course, these definitions are limited to angles between 0° and 90° (since the sum of the angles of a triangle is 180° and the right angle is 90°).

We now show that these "triangular" definitions give the same values as the "circular" definitions given earlier. It is only necessary to show this for the functions $\sin \theta$ and $\cos \theta$, since the other four functions are defined in terms of them. To verify this for $\sin \theta$ and $\cos \theta$, we place the triangle as in Figure 2. We then draw the circle with radius h that is centered at the origin and draw the unit circle. The triangles $0AB$

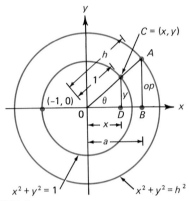

FIGURE 2

and $0CD$ are similar (since they have the same angles). Therefore the ratios of corresponding sides are equal. This fact tells us that

$$\frac{op}{h} = \frac{y}{1} \quad \text{and} \quad \frac{a}{h} = \frac{x}{1}.$$

But op/h is the triangular definition of $\sin \theta$, while $y/1 = y$ is the circular definition of $\sin \theta$. Thus the two definitions lead to the same function. In a similar fashion, we see that the two definitions of $\cos \theta$ lead to the same function.

Let L be any straight line. Its slope is the tangent of the angle θ that the line makes with the positive x-axis. To see this we look at the line parallel to the given line that passes through the origin, and we draw the unit circle around that new line (see Figure 3). Then the slope of the new line (which is parallel to L) that contains the points $(0, 0)$ and $(\cos \theta, \sin \theta)$ is given by

$$m = \frac{\Delta y}{\Delta x} = \frac{\sin \theta - 0}{\cos \theta - 0} = \tan \theta.$$

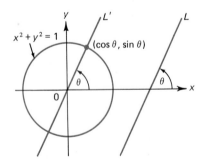

FIGURE 3

Triangles are often useful for computations of values of trigonometric functions. For example, we can use a triangle to prove that $\sin 30° = \frac{1}{2}$. Look at the equilateral triangle in Figure 4. Let the sides of the triangle have lengths of 1 unit and let BD be the angle bisector of angle B, which is also the perpendicular bisector of side BD (this can be proven since the triangles ABD and DBC are congruent). The length of side BD

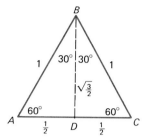

FIGURE 4

is, from the Pythagorean theorem, equal to $\sqrt{3}/2$. Then

$$\sin 30° = \frac{op}{h} = \frac{\frac{1}{2}}{1} = \frac{1}{2}, \quad \cos 30° = \frac{a}{h} = \frac{\sqrt{3}/2}{1} = \frac{\sqrt{3}}{2},$$

and so on. Another use of triangles is given in Example 1.

EXAMPLE 1 If $\sin \theta = \frac{3}{5}$, calculate $\cos \theta$, $\tan \theta$, $\cot \theta$, $\sec \theta$, and $\csc \theta$.

Solution. We draw a triangle (see Figure 5). Since $\sin \theta = op/h = \frac{3}{5}$, we set $op = 3$ and $h = 5$. Then $a = \sqrt{5^2 - 3^2} = 4$, and from the triangle, $\cos \theta = \frac{4}{5}$, $\tan \theta = \frac{3}{4}$, $\cot \theta = \frac{4}{3}$, $\sec \theta = \frac{5}{4}$, and $\csc \theta = \frac{5}{3}$. There is another possible answer. The function $\sin \theta$ is positive in the second quadrant. In that quadrant $\cos \theta$ is negative, $\tan \theta$ is negative, $\cot \theta$ is negative, $\sec \theta$ is negative, and $\csc \theta$ is positive. Thus another possible set of answers is $\cos \theta = -\frac{4}{5}$, $\tan \theta = -\frac{3}{4}$, $\cot \theta = -\frac{4}{3}$, $\sec \theta = -\frac{5}{4}$, and $\csc \theta = \frac{5}{3}$. In order to get a unique answer, we must indicate the quadrant to which θ belongs. ■

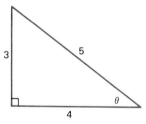

FIGURE 5

The procedure outlined in Example 1 will be useful in Chapter 8. There are many other things that can be discussed by using triangles and are covered in any trigonometry course. Two interesting rules, the *law of cosines* and the *law of sines*, are stated in Problems 11 and 13.

PROBLEMS 1.4

In Problems 1–10, the value of one of the six trigonometric functions is given. Find the values of the other five functions in the indicated quadrant.

1. $\cos \theta = \frac{5}{11}$; first quadrant

2. $\tan \theta = 3$; first quadrant

3. $\sec \theta = 2$; fourth quadrant

4. $\cot \theta = -1$; second quadrant

5. $\csc \theta = 5$; second quadrant
      6. $\sin \theta = -\frac{2}{3}$; third quadrant
7. $\sin \theta = -\frac{2}{3}$; fourth quadrant
      8. $\sec \theta = -7$; second quadrant
9. $\tan \theta = 10$; third quadrant
      10. $\cot \theta = 3$; first quadrant

*11. Let A, B, and C be the vertices of a triangle and let a, b, and c denote the corresponding opposite sides (see Figure 6a). Then the **law of cosines** states that

$$c^2 = a^2 + b^2 - 2ab \cos C.$$

Prove this result. [*Hint:* Place the triangle as in Figure 6b. Use the circular definition of the functions $\sin x$ and $\cos x$ to show that the coordinates of the vertex A are $(b \cos C, b \sin C)$. Then use the distance formula to find $c = \overline{AB}$.]

(a)

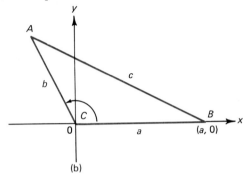
(b)

FIGURE 6

12. The sides of a triangle are 2, 5, and 6. Use the law of cosines to calculate the angles of the triangle.

13. Let ABC be the vertices of a triangle. The **law of sines** states that

$$\frac{a}{\sin A} = \frac{b}{\sin B} = \frac{c}{\sin C}.$$

Prove this result. [*Hint:* Drop a perpendicular line from A to the side a (see Figure 7a or 7b). Show that, in either case,

$$\frac{h}{c} = \sin B \qquad \text{and} \qquad \frac{h}{b} = \sin C$$

(in Figure 7b we need the fact that $\sin(\pi - C) = \sin C$). Use these facts to complete the proof.]

(a)

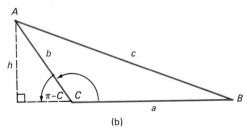
(b)

FIGURE 7

14. Two angles of a triangle are 23° and 85°. The side between them has a length of 5. Use the law of sines to find the lengths of the other two sides.

Appendix 2
Mathematical Induction

Mathematical induction[†] is the name given to an elementary logical principle that can be used to prove a certain type of mathematical statement. Typically, we use mathematical induction to prove that a certain statement or equation holds for every positive integer. For example, we may need to prove that $2^n > n$ for all integers $n \geq 1$.

To do so, we proceed in two steps:

(i) We prove that the statement is true for some integer N (usually $N = 1$).
(ii) We *assume* that the statement is true for an integer k and then *prove* that it is true for the integer $k + 1$.

If we can complete these two steps, then we will have demonstrated the validity of the statement for *all* positive integers greater than or equal to N. To convince you of this fact, we reason as follows: Since the statement is true for N [by step (i)] it is true for the integer $N + 1$ [by step (ii)]. Then it is also true for the integer $(N + 1) + 1 = N + 2$ [again by step (ii)], and so on. We now demonstrate the procedure with some examples.

[†]This technique was first used in a mathematical proof by the great French mathematician Pierre de Fermat (1601–1665). See his biography on page 95.

EXAMPLE 1   Show that $2^n > n$ for all integers $n \geq 1$.

Solution.
(i) If $n = 1$, then $2^n = 2^1 = 2 > 1 = n$, so $2^n > n$ for $n = 1$.
(ii) Assume that $2^k > k$, where $k \geq 1$ is an integer. Then

$$\text{Since } 2^k > k$$
$$2^{k+1} = 2 \cdot 2^k = 2^k + 2^k \overset{\downarrow}{>} k + k \geq k + 1.$$

This completes the proof since we have shown that $2^1 > 1$, which implies, by step (ii), that $2^2 > 2$, so that, again by step (ii), $2^3 > 3$, so that $2^4 > 4$, and so on. ∎

EXAMPLE 2   Use mathematical induction to prove the formula for the sum of the first n positive integers:

$$1 + 2 + 3 + \cdots + n = \frac{n(n + 1)}{2}. \tag{1}$$

Solution
(i) If $n = 1$, then the sum of the first one integer is 1. But $(1)(1 + 1)/2 = 1$, so that equation (1) holds in the case in which $n = 1$.
(ii) Assume that (1) holds for $n = k$; that is,

$$1 + 2 + 3 + \cdots + k = \frac{k(k + 1)}{2}.$$

We must now show that it holds for $n = k + 1$. That is, we must show that

$$1 + 2 + 3 + \cdots + k + (k + 1) = \frac{(k + 1)(k + 2)}{2}.$$

But

$$1 + 2 + 3 + \cdots + k + (k + 1) = (1 + 2 + 3 + \cdots + k) + (k + 1)$$
$$= \frac{k(k + 1)}{2} + (k + 1)$$
$$= \frac{k(k + 1) + 2(k + 1)}{2} = \frac{(k + 1)(k + 2)}{2},$$

and the proof is complete. ∎

You may wish to try a few examples to illustrate that formula (1) really works. For example,

$$1 + 2 + 3 + 4 + 5 + 6 + 7 + 8 + 9 + 10 = \frac{10(11)}{2} = 55.$$

EXAMPLE 3 Use mathematical induction to prove the formula for the sum of the squares of the first n positive integers:

$$1^2 + 2^2 + 3^2 + \cdots + n^2 = \frac{n(n + 1)(2n + 1)}{6}. \tag{2}$$

Solution

(i) Since $1(1 + 1)(2 \cdot 1 + 1)/6 = 1 = 1^2$, equation (2) is valid for $n = 1$.

(ii) Suppose that equation (2) holds for $n = k$; that is,

$$1^2 + 2^2 + 3^2 + \cdots + k^2 = \frac{k(k + 1)(2k + 1)}{6}.$$

Then to prove that (2) is true for $n = k + 1$, we have

$$1^2 + 2^2 + 3^2 + \cdots + k^2 + (k + 1)^2 = \frac{k(k + 1)(2k + 1)}{6} + (k + 1)^2$$

$$= \frac{k(k + 1)(2k + 1) + 6(k + 1)^2}{6}$$

$$= \frac{k + 1}{6}[k(2k + 1) + 6(k + 1)]$$

$$= \frac{k + 1}{6}(2k^2 + 7k + 6) = \frac{k + 1}{6}[(k + 2)(2k + 3)]$$

$$= \frac{(k + 1)(k + 2)[2(k + 1) + 1]}{6},$$

which is equation (2) for $n = k + 1$, and the proof is complete. ◼

Again you may wish to experiment with this formula. For example,

$$1^2 + 2^2 + 3^2 + 4^2 + 5^2 + 6^2 + 7^2 = \frac{7(7 + 1)(2 \cdot 7 + 1)}{6} = \frac{7 \cdot 8 \cdot 15}{6} = 140.$$

EXAMPLE 4 For $a \neq 1$, use mathematical induction to prove the formula for the sum of a geometric progression:

$$1 + a + a^2 + \cdots + a^n = \frac{1 - a^{n+1}}{1 - a}. \tag{3}$$

Solution.

(i) If $n = 0$, then

$$\frac{1 - a^{0+1}}{1 - a} = \frac{1 - a}{1 - a} = 1.$$

Thus equation (3) holds for $n = 0$. (We use $n = 0$ instead of $n = 1$ since $a^0 = 1$ is the first term.)

(ii) Assume that (3) holds for $n = k$; that is,

$$1 + a + a^2 + \cdots + a^k = \frac{1 - a^{k+1}}{1 - a}.$$

Then

$$1 + a + a^2 + \cdots + a^k + a^{k+1} = \frac{1 - a^{k+1}}{1 - a} + a^{k+1}$$

$$= \frac{1 - a^{k+1} + (1 - a)a^{k+1}}{1 - a} = \frac{1 - a^{k+2}}{1 - a},$$

so that equation (3) also holds for $n = k + 1$, and the proof is complete. ∎

EXAMPLE 5 Let $f_1, f_2, \ldots, f_n$ be differentiable functions. Use mathematical induction to prove that

$$\frac{d}{dx}(f_1 + f_2 + \cdots + f_n) = \frac{df_1}{dx} + \frac{df_2}{dx} + \cdots + \frac{df_n}{dx}. \tag{4}$$

Solution.

(i) For $n = 2$, equation (4) was demonstrated in the proof of Theorem 3.1.4.

(ii) Assume that equation (4) is valid for $n = k$; that is,

$$\frac{d}{dx}(f_1 + f_2 + \cdots + f_k) = \frac{df_1}{dx} + \frac{df_2}{dx} + \cdots + \frac{df_k}{dx}.$$

Let $g(x) = f_1(x) + f_2(x) + \cdots + f_k(x)$. Then

$$\frac{d}{dx}(f_1 + f_2 + \cdots + f_k + f_{k+1}) = \frac{d}{dx}(g + f_{k+1})$$

$$= \frac{dg}{dx} + \frac{df_{k+1}}{dx} \quad \text{(by the case } n = 2\text{)}$$

$$= \frac{d}{dx}(f_1 + f_2 + \cdots + f_k) + \frac{df_{k+1}}{dx}$$

$$= \frac{df_1}{dx} + \frac{df_2}{dx} + \cdots + \frac{df_k}{dx} + \frac{df_{k+1}}{dx},$$

which is equation (4) in the case $n = k + 1$, and the theorem is proved. ∎

PROBLEMS

1. Use mathematical induction to prove that the sum of the cubes of the first n positive integers is given by

$$1^3 + 2^3 + 3^3 + \cdots + n^3 = \frac{n^2(n + 1)^2}{4}. \tag{5}$$

2. Let the functions $f_1, f_2, \ldots, f_n$ be integrable on $[0, 1]$. Show that $f_1 + f_2 + \cdots + f_n$ is integrable on $[0, 1]$ and that

$$\int_0^1 [f_1(x) + f_2(x) + \cdots + f_n(x)]\, dx = \int_0^1 f_1(x)\, dx + \int_0^1 f_2(x)\, dx + \cdots + \int_0^1 f_n(x)\, dx.$$

3. Use mathematical induction to prove that the nth derivative of the nth-order polynomial

$$P_n(x) = x^n + a_{n-1}x^{n-1} + a_{n-2}x^{n-2} + \cdots + a_1 x^1 + a_0$$

is equal to $n!$ $[n! = n(n - 1)(n - 2) \cdots 3 \cdot 2 \cdot 1]$.

4. Show that if $a \neq 1$,

$$1 + 2a + 3a^2 + \cdots + na^{n-1} = \frac{1 - (n + 1)a^n + na^{n+1}}{(1 - a)^2}.$$

*5. Prove, using mathematical induction, that there are exactly 2^n subsets of a set containing n elements.

6. Use mathematical induction to prove that

$$\ln(a_1 a_2 a_3 \ldots a_n) = \ln a_1 + \ln a_2 + \cdots + \ln a_n,$$

if $a_k > 0$ for $k = 1, 2, \ldots, n$.

7. Let $\mathbf{u}, \mathbf{v}_1, \mathbf{v}_2, \ldots, \mathbf{v}_n$ be $n + 1$ vectors in $\mathbb{R}^2$. Prove that

$$\mathbf{u} \cdot (\mathbf{v}_1 + \mathbf{v}_2 + \cdots + \mathbf{v}_n) = \mathbf{u} \cdot \mathbf{v}_1 + \mathbf{u} \cdot \mathbf{v}_2 + \cdots + \mathbf{u} \cdot \mathbf{v}_n.$$

Appendix 3
Determinants

In several parts of this book we made use of determinants. In this appendix we will show how determinants arise and will discuss their uses.

We begin by considering the system of two linear equations in two unknowns:

$$a_{11}x_1 + a_{12}x_2 = b_1,$$
$$a_{21}x_1 + a_{22}x_2 = b_2. \tag{1}$$

For simplicity we assume that the constants a_{11}, a_{12}, a_{21}, and a_{22} are all nonzero (otherwise, the system can be solved directly). To solve the system (1), we reduce it to one equation in one unknown. To accomplish this, we multiply the first equation by a_{22} and the second by a_{12} to obtain

$$a_{11}a_{22}x_1 + a_{22}a_{12}x_2 = a_{22}b_1,$$
$$a_{12}a_{21}x_1 + a_{22}a_{12}x_2 = a_{12}b_2. \tag{2}$$

Then subtracting the second equation from the first, we have

$$(a_{11}a_{22} - a_{12}a_{21})\, x_1 = a_{22}b_1 - a_{12}b_2. \tag{3}$$

Now we define the quantity

$$D = a_{11}a_{22} - a_{12}a_{21}. \tag{4}$$

If $D \neq 0$, then (3) yields

$$x_1 = \frac{a_{22}b_1 - a_{12}b_2}{D},$$

(5)

and x_2 may be obtained by substituting this value of x_1 into either of the equations of (1). *Thus if $D \neq 0$, the system (1) has a unique solution.*

On the other hand, suppose that $D = 0$. Then $a_{11}a_{22} = a_{12}a_{21}$, and equation (3) leads to the equation

$$0 = a_{22}b_1 - a_{12}b_2.$$

Either this equation is true or it is false. If it is false, that is, if $a_{22}b_1 - a_{12}b_2 \neq 0$, then the system (1) has *no* solution. If the equation is true, that is, $a_{22}b_1 - a_{12}b_2 = 0$, then the second equation of (2) is a multiple of the first and (1) consists essentially of only one equation. In this case we may choose x_1 arbitrarily and calculate the corresponding value of x_2, which means that there are an *infinite* number of solutions. In sum, we have shown that *if $D = 0$, then the system (1) has either no solution or an infinite number of solutions.*

These facts are easily visualized geometrically by noting that (1) consists of the equations of two straight lines. A solution of the system is a point of intersection of the two lines. If the slopes are different, then $D \neq 0$ and the two lines intersect at a single point, which is the unique solution. It is easy to show (see Problem 21) that $D = 0$ if and only if the slopes of the two lines are the same. If $D = 0$, either we have two parallel lines and no solution, since the lines never intersect, or both equations yield the same line and every point on this line is a solution. These results are illustrated in Figure 1.

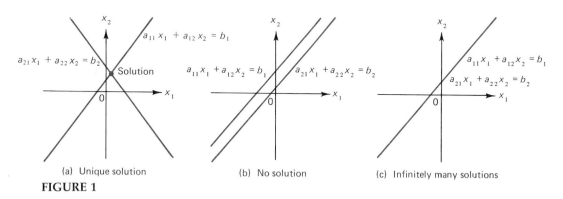

(a) Unique solution   (b) No solution   (c) Infinitely many solutions

FIGURE 1

EXAMPLE 1   Consider the following systems of equations:

      (i) $2x_1 + 3x_2 = 12$ **(ii)** $x_1 + 3x_2 = 3$ **(iii)** $x_1 + 3x_2 = 3$
            $x_1 + x_2 = 5$ $3x_1 + 9x_2 = 8$ $3x_1 + 9x_2 = 9$

In system (i), $D = 2 \cdot 1 - 3 \cdot 1 = -1 \neq 0$, so there is a unique solution, which is easily found to be $x_1 = 3$, $x_2 = 2$. In system (ii), $D = 1 \cdot 9 - 3 \cdot 3 = 0$. Multiplying the first equation by 3 and then subtracting this from the second equation, we obtain the equation $0 = -1$, which is impossible. Thus there is no solution. In (iii), $D = 1 \cdot 9 - 3 \cdot 3 = 0$. But now the second equation is simply three times the first equation. If x_2 is arbitrary, then $x_1 = 3 - 3x_2$, and there are an infinite number of solutions. ■

Returning again to the system (1), we define the **determinant of the system** as

$$D = a_{11}a_{22} - a_{12}a_{21}. \tag{6}$$

For convenience of notation we denote the determinant by writing the coefficients of the system in a square array:

$$D = \begin{vmatrix} a_{11} & a_{12} \\ a_{21} & a_{22} \end{vmatrix} = a_{11}a_{22} - a_{12}a_{21}. \tag{7}$$

Therefore, a 2×2 determinant is the product of the two elements in the upper-left-to-lower-right diagonal minus the product of the other two elements.

We have proved the following theorem.

Theorem 1 For the 2×2 system (1), there is a unique solution if and only if the determinant D is not equal to zero. If $D = 0$, then there is either no solution or an infinite number of solutions.

Let us now consider the general system of n equations in n unknowns:

$$\begin{aligned}
a_{11}x_1 + a_{12}x_2 + \cdots + a_{1n}x_n &= b_1, \\
a_{21}x_1 + a_{22}x_2 + \cdots + a_{2n}x_n &= b_2, \\
&\ \ \vdots \\
a_{n1}x_1 + a_{n2}x_2 + \cdots + a_{nn}x_n &= b_n,
\end{aligned} \tag{8}$$

and define the determinant of such a system in order to obtain a theorem like the one above for $n \times n$ systems. We begin by defining the determinant of a 3×3 system:

$$D = \begin{vmatrix} a_{11} & a_{12} & a_{13} \\ a_{21} & a_{22} & a_{23} \\ a_{31} & a_{32} & a_{33} \end{vmatrix} = a_{11}\begin{vmatrix} a_{22} & a_{23} \\ a_{32} & a_{33} \end{vmatrix} - a_{12}\begin{vmatrix} a_{21} & a_{23} \\ a_{31} & a_{33} \end{vmatrix} + a_{13}\begin{vmatrix} a_{21} & a_{22} \\ a_{31} & a_{32} \end{vmatrix}. \tag{9}$$

We see that to calculate a 3×3 determinant, it is necessary to calculate three 2×2 determinants.

EXAMPLE 2

$$\begin{vmatrix} 3 & 5 & 2 \\ 4 & 2 & 3 \\ -1 & 2 & 4 \end{vmatrix} = 3\begin{vmatrix} 2 & 3 \\ 2 & 4 \end{vmatrix} - 5\begin{vmatrix} 4 & 3 \\ -1 & 4 \end{vmatrix} + 2\begin{vmatrix} 4 & 2 \\ -1 & 2 \end{vmatrix}$$

$$= 3 \cdot 2 - 5 \cdot 19 + 2 \cdot 10 = -69. \ ■$$

The general definition of the determinant of the $n \times n$ system of equations (8) is simply an extension of this procedure:

$$D = \begin{vmatrix} a_{11} & a_{12} & \cdots & a_{1n} \\ a_{21} & a_{22} & \cdots & a_{2n} \\ \vdots & \vdots & & \vdots \\ a_{n1} & a_{n2} & \cdots & a_{nn} \end{vmatrix} = a_{11}A_{11} - a_{12}A_{12} + \cdots + (-1)^{n+1}a_{1n}A_{1n}, \qquad (10)$$

where A_{1j} is the $(n-1) \times (n-1)$ determinant obtained by crossing out the first row and jth column of the original $n \times n$ determinant. Thus an $n \times n$ determinant can be obtained by calculating n $(n-1) \times (n-1)$ determinants. Note that in definition (10) the signs alternate. The signs of the n^2 $(n-1) \times (n-1)$ determinants can easily be illustrated by the following schematic diagram:

$$\begin{vmatrix} + & - & + & - & + & - & \cdots \\ - & + & - & + & - & + & \cdots \\ + & - & + & - & + & - & \cdots \\ - & + & - & + & - & + & \cdots \\ + & - & + & - & + & - & \cdots \\ \vdots & \vdots & \vdots & \vdots & \vdots & \vdots & \ddots \end{vmatrix}$$

EXAMPLE 3
$$\begin{vmatrix} 1 & 3 & 5 & 2 \\ 0 & -1 & 3 & 4 \\ 2 & 1 & 9 & 6 \\ 3 & 2 & 4 & 8 \end{vmatrix} = 1\begin{vmatrix} -1 & 3 & 4 \\ 1 & 9 & 6 \\ 2 & 4 & 8 \end{vmatrix} - 3\begin{vmatrix} 0 & 3 & 4 \\ 2 & 9 & 6 \\ 3 & 4 & 8 \end{vmatrix} + 5\begin{vmatrix} 0 & -1 & 4 \\ 2 & 1 & 6 \\ 3 & 2 & 8 \end{vmatrix} - 2\begin{vmatrix} 0 & -1 & 3 \\ 2 & 1 & 9 \\ 3 & 2 & 4 \end{vmatrix}$$

$$= 1(-92) - 3(-70) + 5(2) - 2(-16) = 160.$$

(The values in parentheses are obtained by calculating the four 3×3 determinants.) ■

The reason for considering determinants of systems of n equations in n unknowns is that Theorem 1 also holds for these systems (although this fact will not be proven here).†

Theorem 2 For the system (8) there is a unique solution if and only if the determinant D, defined by (10), is not zero. If $D = 0$, then there is either no solution or an infinite number of solutions.

---

†For a proof, see S. I. Grossman, *Elementary Linear Algebra*, 2nd ed. (Wadsworth, Belmont, Calif., 1984), Chapter 3.

It is clear that calculating determinants by formula (10) can be extremely tedious, especially if $n \geq 5$. For that reason techniques are available for greatly simplifying these calculations. Some of these techniques are described in the theorems below. The proofs of these theorems can be found in the text cited in the footnote.

We begin with the result that states that the determinant can be obtained by expanding in any row.

Theorem 3 For any i, $i = 1, 2, \ldots n$,

$$D = \begin{vmatrix} a_{11} & a_{12} & \cdots & a_{1n} \\ a_{21} & a_{22} & \cdots & a_{2n} \\ \vdots & \vdots & & \vdots \\ a_{n1} & a_{n2} & \cdots & a_{nn} \end{vmatrix} = (-1)^{i+1}a_{i1}A_{i1} + (-1)^{i+2}a_{i2}A_{i2} + \cdots + (-1)^{i+n}a_{in}A_{in},$$

where A_{ij} is the $(n-1) \times (n-1)$ determinant obtained by crossing off the ith row and jth column of D. Notice that the signs in the expansion of a determinant alternate.

EXAMPLE 4 Calculate $\begin{vmatrix} 3 & 5 & 2 \\ 4 & 2 & 3 \\ -1 & 2 & 4 \end{vmatrix}$ by expanding in the second row (see Example 2).

Solution.

$$\begin{vmatrix} 3 & 5 & 2 \\ 4 & 2 & 3 \\ -1 & 2 & 4 \end{vmatrix} = (-1)^{2+1}(4)\begin{vmatrix} 5 & 2 \\ 2 & 4 \end{vmatrix} + (-1)^{2+2}(2)\begin{vmatrix} 3 & 2 \\ -1 & 4 \end{vmatrix} + (-1)^{2+3}(3)\begin{vmatrix} 3 & 5 \\ -1 & 2 \end{vmatrix}$$

$$= -4(16) + 2(14) - 3(11) = -69.$$

We remark that we can also get the same result by expanding in the third row of D. ∎

Theorem 4 For any j, $j = 1, 2, \ldots, n$,

$$D = \begin{vmatrix} a_{11} & a_{12} & \cdots & a_{1n} \\ a_{21} & a_{22} & \cdots & a_{2n} \\ \vdots & \vdots & & \vdots \\ a_{n1} & a_{n2} & \cdots & a_{nn} \end{vmatrix} = (-1)^{1+j}a_{1j}A_{1j} + (-1)^{2+j}a_{2j}A_{2j} + \cdots + (-1)^{n+j}a_{nj}A_{nj},$$

where A_{ij} is as defined in Theorem 3.

EXAMPLE 5   Calculate $D = \begin{vmatrix} 3 & 5 & 2 \\ 4 & 2 & 3 \\ -1 & 2 & 4 \end{vmatrix}$ by expanding in the third column.

Solution.

$$\begin{vmatrix} 3 & 5 & 2 \\ 4 & 2 & 3 \\ -1 & 2 & 4 \end{vmatrix} = (-1)^{1+3}(2)\begin{vmatrix} 4 & 2 \\ -1 & 2 \end{vmatrix} + (-1)^{2+3}(3)\begin{vmatrix} 3 & 5 \\ -1 & 2 \end{vmatrix} + (-1)^{3+3}(4)\begin{vmatrix} 3 & 5 \\ 4 & 2 \end{vmatrix}$$

$$= 2(10) - 3(11) + 4(-14) = -69. \quad\blacksquare$$

Theorem 5   Let $D = \begin{vmatrix} a_{11} & a_{12} & \cdots & a_{1n} \\ a_{21} & a_{22} & \cdots & a_{2n} \\ \vdots & \vdots & & \vdots \\ a_{n1} & a_{n2} & \cdots & a_{nn} \end{vmatrix}$.

(i) If any row or column of D is zero, then $D = 0$.
(ii) If any row (column) is a multiple of any other row (column), then $D = 0$.
(iii) Interchanging any two rows (columns) of D has the effect of multiplying D by -1.
(iv) Multiplying a row (column) of D by a constant α has the effect of multiplying D by α.
(v) If any row (column) of D is multiplied by a constant and added to a different row (column) of D, then D is unchanged.

EXAMPLE 6   Calculate $D = \begin{vmatrix} 2 & 1 & 4 & 3 \\ 3 & 1 & -2 & -1 \\ 14 & -2 & 0 & 6 \\ 6 & 2 & -4 & -2 \end{vmatrix}$

Solution. $D = 0$ according to (ii) since the fourth row is twice the second row. This can easily be verified. $\quad\blacksquare$

EXAMPLE 7   Calculate $D = \begin{vmatrix} 1 & 14 & 3 \\ 2 & 28 & -2 \\ 0 & -42 & 1 \end{vmatrix}$

Solution. According to (iv), we may divide the second column by 14, which has the effect of dividing D by 14. Then

$$\frac{D}{14} = \begin{vmatrix} 1 & 1 & 3 \\ 2 & 2 & -2 \\ 0 & -3 & 1 \end{vmatrix},$$

or

$$D = \begin{vmatrix} 1 & 1 & 3 \\ 14\cdot 2 & 2 & -2 \\ 0 & -3 & 1 \end{vmatrix} = 14(-24) = -336. \ \blacksquare$$

EXAMPLE 8 Calculate $D = \begin{vmatrix} 0 & -42 & 1 \\ 2 & 28 & -2 \\ 1 & 14 & 3 \end{vmatrix}$.

Solution. D is obtained from the determinant of Example 7 by interchanging the first and third rows. Hence, according to (iii), $D = -(-336) = 336.$ ∎

The results in Theorem 5 can be used to simplify the calculation of determinants.

EXAMPLE 9 Calculate $D = \begin{vmatrix} 1 & 3 & 5 & 2 \\ 0 & -1 & 3 & 4 \\ 2 & 1 & 9 & 6 \\ 3 & 2 & 4 & 8 \end{vmatrix}$.

Solution. This determinant was calculated in Example 3. The idea is to use Theorem 5 to make the evaluation of the determinant almost trivial. We begin by multiplying the first row by -2 and adding it to the third row. By (v), this manipulation will leave the determinant unchanged.

$$D = \begin{vmatrix} 1 & 3 & 5 & 2 \\ 0 & -1 & 3 & 4 \\ 2+(-2)1 & 1+(-2)3 & 9+(-2)5 & 6+(-2)2 \\ 3 & 2 & 4 & 8 \end{vmatrix}$$

$$= \begin{vmatrix} 1 & 3 & 5 & 2 \\ 0 & -1 & 3 & 4 \\ 0 & -5 & -1 & 2 \\ 3 & 2 & 4 & 8 \end{vmatrix}$$

Now we multiply the first row by -3 and add it to the fourth row:

$$D = \begin{vmatrix} 1 & 3 & 5 & 2 \\ 0 & -1 & 3 & 4 \\ 0 & -5 & -1 & 2 \\ 0 & -7 & -11 & 2 \end{vmatrix}.$$

We now expand D by its first column:

$$D = 1 \begin{vmatrix} -1 & 3 & 4 \\ -5 & -1 & 2 \\ -7 & -11 & 2 \end{vmatrix} - 0 \begin{vmatrix} 3 & 5 & 2 \\ -5 & -1 & 2 \\ -7 & -11 & 2 \end{vmatrix} + 0 \begin{vmatrix} 3 & 5 & 2 \\ -1 & 3 & 4 \\ -7 & -11 & 2 \end{vmatrix} + 0 \begin{vmatrix} 3 & 5 & 2 \\ -1 & 3 & 4 \\ -5 & -1 & 2 \end{vmatrix}$$

$$= \begin{vmatrix} -1 & 3 & 4 \\ -5 & -1 & 2 \\ -7 & -11 & 2 \end{vmatrix},$$

which is a 3×3 determinant. We can calculate it by expansion or we can reduce further. By Theorem 5, parts (iv) and (v),

$$\begin{vmatrix} -1 & 3 & 4 \\ -5 & -1 & 2 \\ -7 & -11 & 2 \end{vmatrix} = - \begin{vmatrix} 1 & -3 & -4 \\ -5 & -1 & 2 \\ -7 & -11 & 2 \end{vmatrix} = - \begin{vmatrix} 1 & -3 & -4 \\ 0 & -16 & -18 \\ 0 & -32 & -26 \end{vmatrix} = - \begin{vmatrix} -16 & -18 \\ -32 & -26 \end{vmatrix}$$

$$= -[(-16)(-26) - (-18)(-32)] = 160.$$

In the second step we multiplied the first row by 5 and added it to the second, and multiplied the first row by 7 and added it to the third. Note how we were able to reduce the problem to the calculation of a single 2×2 determinant. ∎

There is one further result about determinants that will be useful in Chapter 17.

Theorem 6 Let $D = \begin{vmatrix} a_{11} & a_{12} & \cdots & a_{1n} \\ a_{21} & a_{22} & \cdots & a_{2n} \\ \vdots & \vdots & & \vdots \\ a_{i1} + b_{i1} & a_{i2} + b_{i2} & \cdots & a_{in} + b_{in} \\ \vdots & \vdots & & \vdots \\ a_{n1} & a_{n2} & \cdots & a_{nn} \end{vmatrix}$. Then

$$D = \begin{vmatrix} a_{11} & a_{12} & \cdots & a_{1n} \\ a_{21} & a_{22} & \cdots & a_{2n} \\ \vdots & \vdots & & \vdots \\ a_{i1} & a_{i2} & \cdots & a_{in} \\ \vdots & \vdots & & \vdots \\ a_{n1} & a_{n2} & \cdots & a_{nn} \end{vmatrix} + \begin{vmatrix} a_{11} & a_{12} & \cdots & a_{1n} \\ a_{21} & 22 & \cdots & a_{2n} \\ \vdots & \vdots & & \vdots \\ b_{i1} & b_{i2} & \cdots & b_{in} \\ \vdots & \vdots & & \vdots \\ a_{n1} & a_{n2} & \cdots & a_{nn} \end{vmatrix}. \tag{11}$$

EXAMPLE 10 To illustrate Theorem 6, we note that

$$
\begin{vmatrix} 2 & 1 & 4 \\ 3+5 & 2-3 & 1+2 \\ 0 & -4 & 2 \end{vmatrix} = \begin{vmatrix} 2 & 1 & 4 \\ 3 & 2 & 1 \\ 0 & -4 & 2 \end{vmatrix} + \begin{vmatrix} 2 & 1 & 4 \\ 5 & -3 & 2 \\ 0 & -4 & 2 \end{vmatrix}
$$

$$ = -38 - 86 = -124. \quad \blacksquare $$

We conclude this appendix by showing how determinants can be used to obtain the unique solution (if one exists) of the system (8) of n equations in n unknowns. We define the determinants

$$
D_1 = \begin{vmatrix} b_1 & a_{12} & \cdots & a_{1n} \\ b_2 & a_{22} & \cdots & a_{2n} \\ \vdots & \vdots & & \vdots \\ b_n & a_{n2} & \cdots & a_{nn} \end{vmatrix},
$$

$$
D_2 = \begin{vmatrix} a_{11} & b_1 & a_{13} & \cdots & a_{1n} \\ a_{21} & b_2 & a_{23} & \cdots & a_{2n} \\ \vdots & \vdots & \vdots & & \vdots \\ a_{n1} & b_n & a_{n3} & \cdots & a_{nn} \end{vmatrix}, \ldots ,
$$

$$
D_k = \begin{vmatrix} a_{11} & a_{12} & \cdots & a_{1,k-1} & b_1 & a_{1,k+1} & \cdots & a_{1n} \\ a_{21} & a_{22} & \cdots & a_{2,k-1} & b_2 & a_{2,k+1} & \cdots & a_{2n} \\ \vdots & \vdots & & \vdots & \vdots & \vdots & & \vdots \\ a_{n1} & a_{n2} & \cdots & a_{n,k-1} & b_n & a_{n,k+1} & \cdots & a_{nn} \end{vmatrix}, \ldots ,
$$

$$
D_n = \begin{vmatrix} a_{11} & a_{12} & \cdots & a_{1,n-1} & b_1 \\ a_{21} & a_{22} & \cdots & a_{2,n-1} & b_2 \\ \vdots & \vdots & & \vdots & \vdots \\ a_{n1} & a_{n2} & \cdots & a_{n,n-1} & b_n \end{vmatrix}, \tag{12}
$$

obtained by replacing the kth column of D by the column

$$
\begin{pmatrix} b_1 \\ b_2 \\ \vdots \\ b_n \end{pmatrix}.
$$

Then we have the following theorem, known as **Cramer's rule.**

Theorem 7 *Cramer's Rule* Let D and D_k, $k = 1, 2, \ldots, n$, be given as in (10) and (12). If $D \neq 0$, then the unique solution to the system (8) is given by the values

$$x_1 = \frac{D_1}{D}, \qquad x_2 = \frac{D_2}{D}, \qquad \cdots, \qquad x_n = \frac{D_n}{D}. \tag{13}$$

EXAMPLE 11   Consider the system

$$2x_1 + 4x_2 - x_3 = -5,$$
$$-4x_1 + 3x_2 + 5x_3 = 14,$$
$$6x_1 - 3x_2 - 2x_3 = 5.$$

We have

$$D = \begin{vmatrix} 2 & 4 & -1 \\ -4 & 3 & 5 \\ 6 & -3 & -2 \end{vmatrix} = 112, \qquad D_1 = \begin{vmatrix} -5 & 4 & -1 \\ 14 & 3 & 5 \\ 5 & -3 & -2 \end{vmatrix} = 224,$$

$$D_2 = \begin{vmatrix} 2 & -5 & -1 \\ -4 & 14 & 5 \\ 6 & 5 & -2 \end{vmatrix} = -112, \qquad D_3 = \begin{vmatrix} 2 & 4 & -5 \\ -4 & 3 & 14 \\ 6 & -3 & 5 \end{vmatrix} = 560.$$

Therefore

$$x_1 = \frac{D_1}{D} = 2, \qquad x_2 = \frac{D_2}{D} = -1, \qquad x_3 = \frac{D_3}{D} = 5. \quad \blacksquare$$

NOTE: As a general rule, avoid Cramer's rule if $n > 3$. There is too much work involved.

PROBLEMS

For each of the 2 × 2 systems in Problems 1–8, calculate the determinant D. If $D \neq 0$, find the unique solution. If $D = 0$, determine whether there is no solution or an infinite number of solutions.

1. $2x_1 + 4x_2 = 6$
   $\quad x_1 + x_2 = 3$

2. $2x_1 + 4x_2 = 6$
   $\quad x_1 + 2x_2 = 5$

3. $2x_1 + 4x_2 = 6$
   $\quad x_1 + 2x_2 = 3$

4. $6x_1 - 3x_2 = 3$
   $\quad -2x_1 + x_2 = -1$

5. $6x_1 - 3x_2 = 3$
   $\quad -2x_1 + x_2 = 1$

6. $6x_1 - 3x_2 = 3$
   $\quad -2x_1 + 2x_2 = -1$

7. $2x_1 + 5x_2 = 0$
   $\quad 3x_1 - 7x_2 = 0$

8. $2x_1 - 3x_2 = 0$
   $\quad -4x_1 + 6x_2 = 0$

In Problems 9 – 20, calculate the determinant.

9. $\begin{vmatrix} 1 & 2 & 3 \\ 6 & -1 & 4 \\ 2 & 0 & 6 \end{vmatrix}$

10. $\begin{vmatrix} 4 & -1 & 0 \\ 2 & 1 & 7 \\ -2 & 3 & 4 \end{vmatrix}$

11. $\begin{vmatrix} 7 & 2 & 3 \\ 0 & 4 & 1 \\ 0 & 0 & 5 \end{vmatrix}$

12. $\begin{vmatrix} 1 & -1 & 4 \\ 3 & -2 & 1 \\ 5 & 1 & 7 \end{vmatrix}$ 　　**13.** $\begin{vmatrix} 4 & 2 & 7 \\ 1 & 5 & 3 \\ -1 & 1 & 4 \end{vmatrix}$ 　　**14.** $\begin{vmatrix} -1 & 0 & 4 \\ 7 & 3 & 2 \\ 4 & 1 & 5 \end{vmatrix}$

15. $\begin{vmatrix} 1 & 4 & 7 & 2 \\ 0 & 5 & 8 & 1 \\ 0 & 0 & -3 & 4 \\ 0 & 0 & 0 & 8 \end{vmatrix}$ 　　**16.** $\begin{vmatrix} a_1 & a_2 & a_3 & a_4 \\ 0 & b_1 & b_2 & b_3 \\ 0 & 0 & c_1 & c_2 \\ 0 & 0 & 0 & d_1 \end{vmatrix}$ 　　**17.** $\begin{vmatrix} 2 & 1 & 3 & 4 \\ 3 & -2 & 5 & 1 \\ 4 & 0 & 4 & 5 \\ 2 & 1 & 7 & -4 \end{vmatrix}$

18. $\begin{vmatrix} 1 & 3 & -1 & 7 \\ -2 & 5 & 2 & 8 \\ -3 & 7 & 3 & 3 \\ 5 & 0 & -5 & 11 \end{vmatrix}$ 　　**19.** $\begin{vmatrix} 2 & 3 & 1 & 4 \\ 2 & 2 & 4 & 6 \\ 3 & -1 & -2 & 4 \\ 4 & 2 & -3 & -5 \end{vmatrix}$ 　　**20.** $\begin{vmatrix} 1 & 0 & 2 & 3 & 1 \\ 0 & 4 & -1 & -2 & 3 \\ 2 & 1 & 0 & -1 & 1 \\ -3 & 2 & 2 & 0 & 5 \\ 0 & 3 & 6 & 1 & -3 \end{vmatrix}$

21. Show that two lines in system (1) have the same slope if and only if the determinant of the system is zero.

In Problems 22–28, solve the system by using Cramer's rule.

22. $\begin{aligned} 3x_1 - x_2 &= 13 \\ -4x_1 + 6x_2 &= -8 \end{aligned}$ 　　**23.** $\begin{aligned} 2x_1 + 6x_2 + 3x_3 &= 9 \\ -3x_1 - 17x_2 - x_3 &= 4 \\ 4x_1 + 3x_2 + x_3 &= -7 \end{aligned}$

24. $\begin{aligned} 2x_1 + x_3 &= 0 \\ 3x_1 - 2x_2 + 2x_3 &= -4 \\ 4x_1 - 5x_2 &= 3 \end{aligned}$ 　　**25.** $\begin{aligned} x_1 - 8x_2 - x_3 &= -1 \\ -x_1 + 4x_2 + x_3 &= 3 \\ 3x_1 - 2x_2 + 6x_3 &= 5 \end{aligned}$

26. $\begin{aligned} x_1 + 2x_2 - x_3 - 4x_4 &= 1 \\ -x_1 + 2x_3 + 6x_4 &= 5 \\ -4x_2 - 2x_3 - 8x_4 &= -8 \\ 3x_1 - 2x_2 + 5x_4 &= 3 \end{aligned}$ 　　**27.** $\begin{aligned} 2x_1 + 5x_2 - 3x_3 + 2x_4 &= 3 \\ -x_1 - 3x_2 + 2x_3 - x_4 &= -1 \\ -3x_1 + 4x_2 + 8x_3 - 2x_4 &= 4 \\ 6x_1 - x_2 - 6x_3 + 4x_4 &= 2 \end{aligned}$

28. $\begin{aligned} x_1 - 2x_2 + x_3 &= 0 \\ 2x_1 - x_2 - 3x_3 &= 0 \\ 5x_1 + 7x_2 - 8x_3 &= 0 \end{aligned}$

Appendix 4
The Binomial Theorem

The binomial theorem provides a useful device for multiplying expressions of the form $(x + y)^n$, where n is a positive integer. You are all familiar with the expression

$$(x + y)^2 = x^2 + 2xy + y^2.$$

In addition, it is not difficult to show that

$$(x + y)^3 = x^3 + 3x^2y + 3xy^2 + y^3.$$

To calculate larger powers of $x + y$, we first define the **binomial coefficient** $\binom{n}{k}$ for n and k positive integers by

$$\binom{n}{k} = \frac{n(n-1)(n-2)\cdots(n-k+1)}{k(k-1)\cdots 3\cdot 2\cdot 1} = \frac{n!}{k!(n-k)!}, \tag{1}$$

where $n! = n(n-1)(n-2)\cdots 3\cdot 2\cdot 1$, and, by convention, $0! = 1$.

EXAMPLE 1 Evaluate (a) $\binom{4}{2}$, (b) $\binom{8}{5}$, (c) $\binom{7}{0}$.

Solution.

(a) $\binom{4}{2} = 4!/2!2! = (4\cdot 3\cdot 2)/(2\cdot 2) = 6$.

(b) $\binom{8}{5} = 8!/5!3! = (8\cdot 7\cdot 6\cdot 5!)/5!3! = (8\cdot 7\cdot 6)/6 = 56$.

(c) $\binom{7}{0} = 7!/7!0! = 1/0! = 1/1 = 1$. ■

Theorem 1 *The Binomial Theorem* Let n be a positive integer. Then

$$(x + y)^n = x^n + \binom{n}{1}x^{n-1}y + \binom{n}{2}x^{n-2}y^2 + \cdots + \binom{n}{n-1}xy^{n-1} + y^n, \qquad (2)$$

or more concisely,

$$(x + y)^n = \sum_{j=0}^{n} \binom{n}{j}x^{n-j}y^j. \qquad (3)$$

Proof.† We prove this theorem by mathematical induction (see Appendix 2).

(i) If $n = 1$, then

$$\sum_{j=0}^{n} \binom{n}{j}x^{n-j}y^j = \sum_{j=0}^{1} \binom{1}{j}x^{1-j}y^j = \binom{1}{0}x^1y^0 + \binom{1}{1}x^0y^1 = x + y,$$

implying the validity of equation (3) for $n = 1$.

(ii) We assume that equation (3) holds for $n = k$ and prove it for $n = k + 1$. By assumption, we have

$$(x + y)^k = \sum_{j=0}^{k} \binom{k}{j}x^{k-j}y^j. \qquad (4)$$

Then

$$(x + y)^{k+1} = (x + y)^k(x + y) = \left[\sum_{j=0}^{k} \binom{k}{j}x^{k-j}y^j\right](x + y),$$

or

$$(x + y)^{k+1} = \sum_{j=0}^{k} \binom{k}{j}x^{k+1-j}y^j + \sum_{j=0}^{k} \binom{k}{j}x^{k-j}y^{j+1}. \qquad (5)$$

We now write out the sums in (5):

$$(x + y)^{k+1} = \binom{k}{0}x^{k+1} + \binom{k}{1}x^ky + \binom{k}{2}x^{k-1}y^2 + \cdots + \binom{k}{j}x^{k+1-j}y^j$$
$$+ \binom{k}{j+1}x^{k-j}y^{j+1} + \cdots + \binom{k}{k}xy^k \qquad (6)$$
$$+ \binom{k}{0}x^ky + \binom{k}{1}x^{k-1}y^2 + \binom{k}{2}x^{k-2}y^3 + \cdots + \binom{k}{j-1}x^{k+1-j}y^j$$
$$+ \binom{k}{j}x^{k-j}y^{j+1} + \cdots + \binom{k}{k}y^{k+1}.$$

---

†The proof is difficult and may be omitted without loss of continuity. Another (simpler) proof, which makes use of Taylor's theorem, is suggested in Problem 13.3.31.

In (6) we see that there are two terms containing the expressions $x^{k+1-j}y^j$ (for $j = 1, 2, \ldots, k$), namely,

$$\binom{k}{j}x^{k+1-j}y^j + \binom{k}{j-1}x^{k+1-j}y^j. \tag{7}$$

But

$$\binom{k}{j} + \binom{k}{j-1} = \frac{k(k-1)\cdots(k-j+2)(k-j+1)}{j!}$$

$$+ \frac{k(k-1)\cdots(k-j+2)}{(j-1)!}. \tag{8}$$

We multiply the second term in (8) by j/j to obtain

$$\binom{k}{j} + \binom{k}{j-1} = \frac{k(k-1)\cdots(k-j+2)}{j!}[(k-j+1)+j] \tag{9}$$

[since $j(j-1)! = j!$]

$$= \frac{(k+1)k(k-1)\cdots(k-j+2)}{j!}$$

$$= \frac{(k+1)k\cdots(k-j+2)(k+1-j)!}{j!(k+1-j)!} = \binom{k+j}{j}.$$

Hence

$$\binom{k}{j}x^{k+1-j}y^j + \binom{k}{j-1}x^{k+1-j}y^j = \binom{k+1}{j}x^{k+1-j}y^j.$$

Finally from (6), (7), and (9), we have

$$(x+y)^{k+1} = \sum_{j=0}^{k+1}\binom{k+1}{j}x^{k+1-j}y^j,$$

which is equation (3) for $n = k + 1$, and the theorem is proved. ∎

EXAMPLE 2 Calculate $(x+y)^5$.

Solution.

$$(x+y)^5 = \sum_{k=0}^{5}\binom{5}{k}x^{5-k}y^k$$

$$= \binom{5}{0}x^5 + \binom{5}{1}x^4y + \binom{5}{2}x^3y^2 + \binom{5}{3}x^2y^3 + \binom{5}{4}xy^4 + \binom{5}{5}y^5$$

$$= x^5 + 5x^4y + 10x^3y^2 + 10x^2y^3 + 5xy^4 + y^5 \quad \blacksquare$$

EXAMPLE 3   Find the coefficient of the term containing x^3y^6 in the expansion of $(x + y)^9$.

   Solution. In (3) we obtain the term x^3y^6 by setting $j = 6$ (so that $9 - j = 3$). The coefficient is

$$\binom{9}{6} = \frac{9!}{6!3!} = \frac{9 \cdot 8 \cdot 7}{3!} = \frac{9 \cdot 8 \cdot 7}{3 \cdot 2} = 84. \quad \blacksquare$$

EXAMPLE 4   Calculate $(2x - 3y)^4$.

   Solution.

$$(2x - 3y)^4 = \binom{4}{0}(2x)^4 + \binom{4}{1}(2x)^3(-3y)^1 + \binom{4}{2}(2x)^2(-3y)^2$$

$$+ \binom{4}{3}(2x)^1(-3y)^3 + \binom{4}{4}(-3y)^4$$

$$= 16x^4 + 4(8x^3)(-3y) + 6(4x^2)(9y^2) + 4(2x)(-27y^3) + 81y^4$$

$$= 16x^4 - 96x^3y + 216x^2y^2 - 216xy^3 + 81y^4 \quad \blacksquare$$

PROBLEMS

In Problems 1–8, calculate the binomial coefficients.

1. $\binom{5}{3}$ **2.** $\binom{7}{4}$ **3.** $\binom{9}{2}$

4. $\binom{10}{5}$ **5.** $\binom{11}{3}$ **6.** $\binom{20}{0}$

7. $\binom{41}{41}$ **8.** $\binom{12}{7}$

9. Prove that $\binom{n}{k} = \binom{n}{n-k}$ for any integers $0 \le k \le n$.

10. Calculate $(x + y)^6$.   **11.** Calculate $(a + b)^7$.

12. Calculate $(u - w)^6$.   **13.** Calculate $(x - 2y)^5$.

14. Calculate $(x^2 - y^3)^4$.   **15.** Calculate $(ax - by)^5$.

16. Calculate $(x^n + y^n)^5$.   **17.** Calculate $[(x/2) + (y/3)]^5$.

18. Find the coefficient of x^5y^7 in the expansion of $(x + y)^{12}$.

19. Find the coefficient of x^8y^3 in the expansion of $(x + y)^{11}$.

20. Show that in the expansion of $(x + y)^n$ the coefficient of x^ky^{n-k} is equal to the coefficient of $x^{n-k}y^k$.

21. Show that for any integer n

$$\binom{n}{0} + \binom{n}{1} + \binom{n}{2} + \cdots + \binom{n}{n} = 2^n.$$

[*Hint:* Expand $(1 + 1)^n$.]

22. Show that for any integer n,

$$\binom{n}{0} - \binom{n}{1} + \binom{n}{2} - \binom{n}{3} + \cdots + (-1)^n\binom{n}{n} = 0.$$

[*Hint:* Expand $(1 - 1)^n$.]

Appendix 5
The Proofs of Some Theorems on Limits, Continuity, and Differentiation

In this appendix we will provide rigorous proofs of several theorems discussed in Chapters 2 and 3. These proofs require a knowledge of the material in Section 2.8. In the proofs of the limit theorems we will assume that x_0 and L are real numbers. The cases of infinite limits and limits at infinity will be discussed in the problem set.

Theorem 1   *(Theorem 2.3.2)*   Let c be any real number and suppose that $\lim_{x \to x_0} f(x)$ exists. Then

$$\lim_{x \to x_0} cf(x) = c \lim_{x \to x_0} f(x).$$

Proof.   *Case (i):* $c = 0$. Then since $\lim_{x \to x_0} 0 = 0$ (see Problem 1), we have

$$\lim_{x \to x_0} cf(x) = \lim_{x \to x_0} 0 = 0 = 0 \cdot \lim_{x \to x_0} f(x).$$

Case (ii): $c \neq 0$. Let $\lim_{x \to x_0} f(x) = L$ and let $\epsilon > 0$ be given. Choose $\delta > 0$ so that if $0 < |x - x_0| < \delta$, then $|f(x) - L| < \epsilon/|c|$ [we can always do this since $\lim_{x \to x_0} f(x) = L$]. Then

$$|cf(x) - cL| = |c(f(x) - L)| = |c| \, |f(x) - L| < |c| \frac{\epsilon}{|c|} = \epsilon$$

if $0 < |x - x_0| < \delta$, and the theorem is proved. ∎

Theorem 2 *(Theorem 2.3.3)* If $\lim_{x\to x_0} f(x)$ and $\lim_{x\to x_0} g(x)$ both exist (and are finite), then

$$\lim_{x\to x_0} [f(x) + g(x)] = \lim_{x\to x_0} f(x) + \lim_{x\to x_0} g(x).$$

Proof. Let $\lim_{x\to x_0} f(x) = L_1$ and $\lim_{x\to x_0} g(x) = L_2$. For a given $\epsilon > 0$, choose δ_1 such that if $0 < |x - x_0| < \delta_1$, then $|f(x) - L_1| < \epsilon/2$; and choose δ_2 such that if $0 < |x - x_0| < \delta_2$, then $|g(x) - L_2| < \epsilon/2$. Let δ be the smaller of δ_1 and δ_2 (denoted $\delta = \min\{\delta_1, \delta_2\}$). Then if $0 < |x - x_0| < \delta$,

$$|(f(x) + g(x)) - (L_1 + L_2)| = |(f(x) - L_1) + (g(x) - L_2)|$$

$$\leq |f(x) - L_1| + |g(x) - L_2| = \frac{\epsilon}{2} + \frac{\epsilon}{2} = \epsilon,$$

and the theorem is proved. (The last step follows from the triangle inequality; see Section 1.2.) ∎

REMARK. It is easy to extend Theorem 2 to a finite sum. We have, if all indicated limits exist,

$$\lim_{x\to x_0} [f_1(x) + f_2(x) + \cdots + f_n(x)] = \lim_{x\to x_0} f_1(x) + \lim_{x\to x_0} f_2(x) + \cdots + \lim_{x\to x_0} f_n(x).$$

Theorem 3 *(Theorem 2.3.4)* If $\lim_{x\to x_0} f(x)$ and $\lim_{x\to x_0} g(x)$ both exist, then

$$\lim_{x\to x_0} f(x) \cdot g(x) = \left[\lim_{x\to x_0} f(x)\right]\left[\lim_{x\to x_0} g(x)\right].$$

Proof. Let $L_1 = \lim_{x\to x_0} f(x)$ and $L_2 = \lim_{x\to x_0} g(x)$. Let $\epsilon > 0$ be given. We choose four δ's as follows:

(i) Choose $\delta_1 > 0$ such that if $0 < |x - x_0| < \delta_1$, then $|f(x) - L_1| < 1$ so that $|f(x)| < |L_1| + 1$.
(ii) Choose δ_2 such that if $0 < |x - x_0| < \delta_2$, then

$$|g(x) - L_2| < \frac{\epsilon}{2}\left(\frac{1}{|L_1| + 1}\right).$$

(iii) Choose $\delta_3 > 0$ such that if $0 < |x - x_0| < \delta_3$, then

$$|f(x) - L_1| < \frac{\epsilon}{2}\left(\frac{1}{|L_2| + 1}\right).$$

(iv) Choose $\delta = \min\{\delta_1, \delta_2, \delta_3\}$.

We now show that if $0 < |x - x_0| < \delta$, then $|f(x)g(x) - L_1L_2| < \epsilon$. This step will complete the proof of the theorem. We have, for $0 < |x - x_0| < \delta$,

$$|f(x)g(x) - L_1 L_2| = |f(x)g(x) - f(x)L_2 + f(x)L_2 - L_1 L_2|$$
$$\leq |f(x)g(x) - f(x)L_2| + |f(x)L_2 - L_1 L_2| \quad \text{(by the triangle}$$
$$= |f(x)|\,|g(x) - L_2| + |L_2|\,|f(x) - L_1| \qquad \text{inequality)}$$
$$< (|L_1| + 1)\left(\frac{\epsilon}{2}\right)\left(\frac{1}{|L_1| + 1}\right) + |L_2|\left(\frac{\epsilon}{2}\right)\left(\frac{1}{|L_2| + 1}\right)$$
$$< \frac{\epsilon}{2} + \frac{\epsilon}{2} = \epsilon. \quad \blacksquare$$

Corollary 1   If $\lim_{x \to x_0} f(x)$ exists and n is a positive integer, then

$$\lim_{x \to x_0} (f(x))^n = \left(\lim_{x \to x_0} f(x)\right)^n.$$

Proof.

$$\lim_{x \to x_0} f^2(x) = \lim_{x \to x_0} f(x) \cdot \lim_{x \to x_0} f(x) = \left[\lim_{x \to x_0} f(x)\right]^2$$

For $n > 2$, simply use this argument as many times as necessary (i.e., use mathematical induction). $\blacksquare$

Corollary 2   *(**Theorem 2.3.1**)*   Let $P(x) = c_0 + c_1 x + c_2 x^2 + \cdots + c_n x^n$ be a polynomial, where $c_0, c_1, c_2, \ldots, c_n$ are real numbers. Then

$$\lim_{x \to x_0} P(x) = P(x_0) = c_0 + c_1 x_0 + c_2 x_0^2 + \cdots + c_n x_0^n.$$

Proof.   We use the fact that $\lim_{x \to x_0} c_0 = c_0$ and $\lim_{x \to x_0} x = x_0$ (see Problems 1 and 3). Then using Theorems 1, 2, and 3 and Corollary 1, we obtain

$$\lim_{x \to x_0} P(x) = \lim_{x \to x_0} (c_0 + c_1 x + c_2 x^2 + \cdots + c_n x^n)$$
$$= \lim_{x \to x_0} c_0 + c_1 \lim_{x \to x_0} x + c_2 \left(\lim_{x \to x_0} x\right)^2 + \cdots + c_n \left(\lim_{x \to x_0} x\right)^n$$
$$= c_0 + c_1 x_0 + c_2 x_0^2 + c_3 x_0^3 + \cdots + c_n x_0^n = P(x_0). \quad \blacksquare$$

Theorem 4   *(**Theorem 2.3.5**)*   If $\lim_{x \to x_0} f(x)$ and $\lim_{x \to x_0} g(x)$ both exist and $\lim_{x \to x_0} g(x) \neq 0$, then

$$\lim_{x \to x_0} \frac{f(x)}{g(x)} = \frac{\lim_{x \to x_0} f(x)}{\lim_{x \to x_0} g(x)}.$$

Proof.   As before, let $L_1 = \lim_{x \to x_0} f(x)$ and $L_2 = \lim_{x \to x_0} g(x)$. Since $L_2 \neq 0$, we show first that $\lim_{x \to x_0} 1/g(x) = 1/L_2$. We must show that for a given $\epsilon > 0$, there is a $\delta > 0$ such that when $|x - x_0| < \delta$,

$$\left| \frac{1}{g(x)} - \frac{1}{L_2} \right| < \epsilon. \tag{1}$$

But the inequality (1) is equivalent to

$$\left| \frac{L_2 - g(x)}{L_2 g(x)} \right| < \epsilon. \tag{2}$$

Select δ_1 such that $0 < |x - x_0| < \delta_1$ implies that $|g(x) - L_2| < |L_2|/2$. Then

$$|L_2| = |(L_2 - g(x)) + g(x)| \le |L_2 - g(x)| + |g(x)| < \tfrac{1}{2}|L_2| + |g(x)|,$$

and so $|g(x)| > \tfrac{1}{2}|L_2|$. Then for $0 < |x - x_0| < \delta_1$,

$$|L_2 g(x)| = |L_2| \, |g(x)| > |L_2| \left| \frac{L_2}{2} \right| = \frac{L_2^2}{2},$$

or $1/|L_2 g(x)| < 2/L_2^2$. Similarly, there is a δ_2 such that if $0 < |x - x_0| < \delta_2$, then

$$|L_2 - g(x)| < \frac{L_2^2}{2} \epsilon.$$

Now choose $\delta = \min\{\delta_1, \delta_2\}$. Then for $0 < |x - x_0| < \delta$,

$$\left| \frac{L_2 - g(x)}{L_2 g(x)} \right| = \frac{1}{|L_2 g(x)|} \cdot |L_2 - g(x)| < \frac{2}{L_2^2} \cdot \frac{L_2^2}{2} \epsilon = \epsilon,$$

which shows that $\lim_{x \to x_0} 1/g(x) = 1/L_2$. Finally, we obtain from this result and Theorem 3

$$\lim_{x \to x_0} \frac{f(x)}{g(x)} = \lim_{x \to x_0} f(x) \cdot \lim_{x \to x_0} \frac{1}{g(x)} = L_1 \cdot \frac{1}{L_2} = \frac{L_1}{L_2},$$

and Theorem 4 is proved. ∎

Corollary 3   Let $r(x) = p(x)/q(x)$ be a rational function (the quotient of two polynomials). Then if $q(x_0) \ne 0$,

$$\lim_{x \to x_0} r(x) = r(x_0) = \frac{p(x_0)}{q(x_0)}.$$

Proof.   From Theorem 4 and Corollary 2 of Theorem 3,

$$\lim_{x \to x_0} r(x) = \lim_{x \to x_0} \frac{p(x)}{q(x)} = \frac{\lim_{x \to x_0} p(x)}{\lim_{x \to x_0} q(x)} = \frac{p(x_0)}{q(x_0)} = r(x_0). \ \blacksquare$$

Theorem 5   *(Theorem 2.4.1: Squeezing Theorem)*   Suppose that $f(x) \le g(x) \le h(x)$ for

x in a neighborhood of x_0 and $\lim_{x \to x_0} f(x) = \lim_{x \to x_0} h(x) = L$, where x_0 may be $+\infty$ or $-\infty$. Then

$$\lim_{x \to x_0} g(x) = L.$$

Proof. We prove this theorem in the case that x_0 is finite. For the case $x_0 = +\infty$ or $-\infty$, see Problem 4. Let $\epsilon > 0$ be given. Choose δ_1 such that $|f(x) - L| < \epsilon$ if $0 < |x - x_0| < \delta_1$, and choose δ_2 so that $|h(x) - L| < \epsilon$ if $0 < |x - x_0| < \delta_2$. Let the inequalities $f(x) \le g(x) \le h(x)$ hold for x in some open interval (a, b) (recall the definition of a neighborhood). Then since x_0 is not one of the numbers a or b, there exists a δ_3 such that $a < x_0 - \delta_3 < x_0 < x_0 + \delta_3 < b$, and therefore, for $0 < |x - x_0| < \delta_3$, $f(x) \le g(x) \le h(x)$. Now let $\delta = \min\{\delta_1, \delta_2, \delta_3\}$. Then if $0 < |x - x_0| < \delta$,

$$g(x) - L \le h(x) - L < \epsilon.$$

Since $|f(x) - L| < \epsilon$, we have $f(x) - L > -\epsilon$. Then

$$g(x) - L \ge f(x) - L > -\epsilon.$$

Putting this information all together, we obtain

$$|g(x) - L| < \epsilon,$$

and the squeezing theorem is proved. ◼

We have continually talked about *the* limit. We have not yet proven that if a limit exists, then it is unique. We do so now.

Theorem 6 If $\lim_{x \to x_0} f(x)$ exists, then it is unique.

Proof. Suppose that $\lim_{x \to x_0} f(x) = L_1$ and $\lim_{x \to x_0} f(x) = L_2$. Let $\epsilon > 0$ be given. There are positive numbers δ_1 and δ_2 such that if $0 < |x - x_0| < \delta = \min\{\delta_1, \delta_2\}$, then $|f(x) - L_1| < \epsilon$ and $|f(x) - L_2| < \epsilon$. Then

$$|L_1 - L_2| = |L_1 - f(x) + f(x) - L_2| \le |L_1 - f(x)| + |f(x) - L_2|$$
$$= |f(x) - L_1| + |L_2 - f(x)| < \epsilon + \epsilon = 2\epsilon.$$

Thus $|L_1 - L_2| < 2\epsilon$ for *every* positive number ϵ. This can only happen if $L_1 = L_2$ and the theorem is proved. ◼

We now turn to an important limit theorem involving continuity. First, we recall our definition of continuity.

Definition 1 (DEFINITION 2.7.1) Let $f(x)$ be defined for every x in an open interval containing the number x_0. Then f is **continuous** at x_0 if all of the following three conditions hold:

    (i) $f(x_0)$ exists (that is, x_0 is in the domain of f).
    (ii) $\lim_{x \to x_0} f(x)$ exists.
    (iii) $\lim_{x \to x_0} f(x) = f(x_0)$.

Using our $\epsilon - \delta$ definition of a limit, we can give a second definition of continuity. This second definition will enable us to prove our important limit theorem.

Definition 2   CONTINUITY   The function f is **continuous** at x_0 if f is defined in a neighborhood of x_0 and, for every $\epsilon > 0$, there is a $\delta > 0$ such that if $|x - x_0| < \delta$, then $|f(x) - f(x_0)| < \epsilon$.

Before using these two definitions, we show that they are equivalent. That is, if f is continuous according to Definition 1, then it is continuous according to Definition 2, and vice versa.

Theorem 7   Definitions 1 and 2 are equivalent.

    Proof.
        (i) We first show that if a function is continuous according to Definition 1, then it is continuous according to Definition 2. Suppose that f is defined in a neighborhood of x_0 and that $\lim_{x \to x_0} f(x) = f(x_0)$. Let $\epsilon > 0$ be given. Then by the definition of limit, there is a δ such that if $0 < |x - x_0| < \delta$, then $|f(x) - f(x_0)| < \epsilon$. Further, $|f(x_0) - f(x_0)| = 0 < \epsilon$. Thus f is defined in a neighborhood of x_0, and for every $\epsilon > 0$ there is a $\delta > 0$ such that if $|x - x_0| < \delta$, then $|f(x) - f(x_0)| < \epsilon$. Hence f is continuous in the sense of Definition 2.
        (ii) Conversely, if we assume that f is continuous at x_0 in the sense of Definition 2, then f is defined in a neighborhood of x_0, and for every $\epsilon > 0$ there is a $\delta > 0$ such that $|x - x_0| < \delta$ implies $|f(x) - f(x_0)| < \epsilon$. Then by the definition of limit, $\lim_{x \to x_0} f(x) = f(x_0)$, and this statement is Definition 1. ∎

We now state and prove our limit theorem.

Theorem 8   *(Theorem 2.7.2)*   If f is continuous at a and if $\lim_{x \to x_0} g(x) = a$, then $\lim_{x \to x_0} f(g(x)) = f(a)$.

    Proof.   Let $\epsilon > 0$ be given. We must show that there exists a $\delta > 0$ such that if $|x - x_0| < \delta$, then $|f(g(x)) - f(a)| < \epsilon$. We do this in two steps.
        (i) Since f is continuous at a, by Definition 2, there is a $\delta_1 > 0$ such that if $|x - a| < \delta_1$, then $|f(x) - f(a)| < \epsilon$.
        (ii) Since $\lim_{x \to x_0} g(x) = a$, there is a $\delta > 0$ such that if $0 < |x - x_0| < \delta$, then $|g(x) - a| < \delta_1$. But if $|g(x) - a| < \delta_1$, then, from (i), $|f(g(x)) - f(a)| < \epsilon$, and the theorem is proved. ∎

In Section 3.3 we gave a partial proof of the chain rule (on p. 153). In this section we give a complete proof of this important theorem. We begin with an important observation.

By the definition of the derivative,

$$f'(x) = \frac{dy}{dx} = \lim_{\Delta x \to 0} \frac{\Delta y}{\Delta x}. \tag{3}$$

This equation implies that if Δx is small, then $\Delta y/\Delta x$ is close to dy/dx. For the number x fixed, we define

$$\epsilon(\Delta x) = \frac{dy}{dx} - \frac{\Delta y}{\Delta x} = f'(x) - \frac{\Delta y}{\Delta x}. \tag{4}$$

Then from (3)

$$\lim_{\Delta x \to 0} \epsilon(\Delta x) = \lim_{\Delta x \to 0} \left(\frac{dy}{dx} - \frac{\Delta y}{\Delta x} \right) = 0.$$

Now multiplying both sides of (4) by Δx and rearranging terms, we obtain

$$\Delta y = f'(x)\,\Delta x - \epsilon(\Delta x) \cdot \Delta x \tag{5}$$

where $\lim_{\Delta x \to 0} \epsilon(\Delta x) = 0$. We have looked at this expression before in a different form. Since Δy is an increment and $f'(x)\,\Delta x$ is the differential dy, the expression $\epsilon(\Delta x) \cdot \Delta x$ is the difference between the increment and the differential (see Section 3.8).

We are now ready to prove the chain rule.

Theorem 9 *Chain Rule* Let g and f be differentiable functions such that for every point x_0 at which $g(x)$ is defined, $f(g(x_0))$ is also defined. Then with $u = g(x)$, the composite function $y = (f \circ g)(x) = f(g(x)) = f(u)$ is a differentiable function of x, and

$$\frac{dy}{dx} = \frac{d}{dx}(f \circ g)(x) = \frac{d}{dx} f(g(x)) = f'(g(x))g'(x) = \frac{df}{du}\frac{du}{dx}.$$

Proof.

$$\frac{d}{dx} f(g(x)) = \lim_{\Delta x \to 0} \frac{f(g(x + \Delta x)) - f(g(x))}{\Delta x}$$

if this limit exists. Since f can be written as a function of u, we can rewrite (5) as

$$\Delta f = f'(u)\,\Delta u - \epsilon(\Delta u)\,\Delta u, \tag{6}$$

where $\epsilon(\Delta u) \to 0$ as $\Delta u \to 0$, $\Delta u = g(x + \Delta x) - g(x)$, and $f'(u) = f'(g(x))$. Then using (6), we obtain

$$f(g(x + \Delta x)) - f(g(x)) = \Delta f = f'(u)\,\Delta u - \epsilon(\Delta u)\,\Delta u, \tag{7}$$

where $\epsilon(\Delta u) \to 0$ as $\Delta u \to 0$. Dividing (7) by Δx, we obtain

$$\frac{f(g(x + \Delta x)) - f(g(x))}{\Delta x} = f'(u)\frac{\Delta u}{\Delta x} - \epsilon(\Delta u)\frac{\Delta u}{\Delta x}.$$

Finally, we take limits as $\Delta x \to 0$. Since $u = g(x)$ and g is differentiable, $\lim_{\Delta x \to 0} \Delta u/\Delta x = du/dx = g'(x)$. Also, $\Delta u = g(x + \Delta x) - g(x)$. Since g is differentiable at x, it is continuous there, and $\lim_{\Delta x \to 0} g(x + \Delta x) = g(x)$. Or alternatively, $\lim_{\Delta x \to 0} \Delta u = \lim_{\Delta x \to 0} [g(x + \Delta x) - g(x)] = 0$. Thus as $\Delta x \to 0$, $\Delta u \to 0$ and $\lim_{\Delta x \to 0} \epsilon(\Delta u) = 0$. Hence

$$\lim_{\Delta x \to 0} \epsilon(\Delta u) \cdot \frac{\Delta u}{\Delta x} = \lim_{\Delta x \to 0} \epsilon(\Delta u) \lim_{\Delta x \to 0} \frac{\Delta u}{\Delta x} = 0 \cdot \frac{du}{dx} = 0.$$

Putting all this information together, we obtain

$$\lim_{\Delta x \to 0} \frac{f(g(x + \Delta x)) - f(g(x))}{\Delta x} = \lim_{\Delta x \to 0} f'(u) \frac{\Delta u}{\Delta x} - \lim_{\Delta x \to 0} \epsilon(\Delta u) \frac{\Delta u}{\Delta x}$$

$$= f'(g(x))g'(x) - 0,$$

and the chain rule is proved. ∎

In Section 6.1 we stated an important theorem about inverse functions. We prove that theorem here.

Theorem 10 *(Theorem 6.1.3)* Let f be continuous on $[a, b]$ and let $c = f(a)$ and $d = f(b)$. If f has an inverse $g = f^{-1}$, then g is continuous on $[c, d]$ (or $[d, c]$ if $c > d$).

Proof.
(i) We first show that f is monotone on $[a, b]$. Since f^{-1} exists, f is 1–1 on $[a, b]$. Suppose there are numbers x_1, x_2, and x_3 in $[a, b]$ such that $x_1 < x_2 < x_3$ and $f(x_1) < f(x_2)$ but $f(x_3) < f(x_2)$ (so that f is neither increasing nor decreasing). The situation is depicted in Figure 1a.

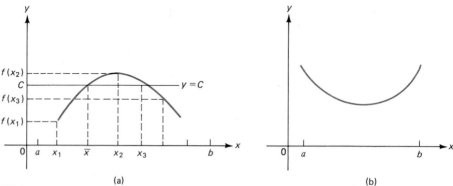

(a)   (b)

FIGURE 1

Let c be a number between $f(x_3)$ and $f(x_2)$. By the intermediate value theorem there is a number $\bar{x}$ in (x_1, x_2) such that $f(\bar{x}) = c$. Similarly, there is a number x^* in (x_2, x_3) such that $f(x^*) = c$. Since $\bar{x} \neq x^*$ but $f(\bar{x}) = f(x^*) = c$, we have obtained a contradiction of the fact that f is 1–1. A similar argument rules out the graph in Figure 1b. Thus we conclude that f is monotone.

(ii) We now show that $g = f^{-1}$ is monotone on $[c, d]$. Suppose that f is increasing (a similar argument works if f is decreasing), and suppose that y_1, $y_2 \in (c, d)$, with $y_1 < y_2$. There are numbers x_1 and x_2 in (a, b) such that $y_1 = f(x_1)$ and $y_2 = f(x_2)$. Suppose that $g(y_1) \geq g(y_2)$. Then since f is increasing,

$$f(\overbrace{g(y_1)}^{x_1}) \geq f(\overbrace{g(y_2)}^{x_2})$$

so that $y_1 = f(x_1) \geq f(x_2) = y_2$. This result contradicts the fact that $y_1 < y_2$. Thus g is also increasing. Thus we have shown the following:

(a) If f is increasing, f^{-1} is increasing.
(b) If f is decreasing, f^{-1} is decreasing.

(iii) We show that $g = f^{-1}$ is continuous in $[c, d]$. That is, we must show that for every $\bar{y} \in [c, d]$,

$$\lim_{y \to \bar{y}} g(y) = g(\bar{y}), \tag{8}$$

where the limit in (8) is a one-sided limit if $\bar{y} = c$ or d. We will assume that $\bar{y} \in (c, d)$ and leave the endpoints as an exercise. Then (8) is equivalent to the following: For every $\epsilon > 0$, there is a number $\delta > 0$ such that

$$\text{if} \quad |y - \bar{y}| < \delta, \quad \text{then} \quad |g(y) - g(\bar{y})| < \epsilon,$$

or equivalently,

$$\text{if} \quad \bar{y} - \delta < y < \bar{y} + \delta \quad \text{then} \quad g(\bar{y}) - \epsilon < g(y) < g(\bar{y}) + \epsilon. \tag{9}$$

Choose $\epsilon > 0$ and assume that $a < g(\bar{y}) - \epsilon < g(\bar{y}) + \epsilon < b$ (this is no restriction). Let $\bar{x} = g(\bar{y})$, so that $f(\bar{x}) = \bar{y}$. Let

$$\delta_1 = \bar{y} - f(\bar{x} - \epsilon), \quad \delta_2 = f(\bar{x} + \epsilon) - \bar{y}, \quad \delta = \min\{\delta_1, \delta_2\}. \tag{10}$$

Since f is increasing, δ_1, δ_2, and therefore δ are all positive. Suppose that $\bar{y} - \delta < y < \bar{y} + \delta$. Then from (10)

$$\bar{y} - \delta_1 \leq \bar{y} - \delta < y < \bar{y} + \delta \leq \bar{y} + \delta_2,$$

and

$$f(\bar{x} - \epsilon) < y < f(\bar{x} + \epsilon),$$

which implies, since g is increasing, that

$$g[f(\bar{x} - \epsilon)] < g(y) < g[f(\bar{x} + \epsilon)].$$

But since $g = f^{-1}$, we have

$$\bar{x} - \epsilon < g(y) < \bar{x} + \epsilon.$$

This completes the proof. ■

PROBLEMS

1. Prove from the definition of a limit that for any finite real number x_0, $\lim_{x \to x_0} c = c$ for any constant c.
2. Prove that $\lim_{x \to \infty} c = c$.
3. Prove that $\lim_{x \to x_0} x = x_0$ for every real number x_0.
4. Prove that if $f(x) \le g(x) \le h(x)$ for all x larger than some number N, and $\lim_{x \to \infty} f(x) = \lim_{x \to \infty} h(x) = L$, then $\lim_{x \to \infty} g(x) = L$. [*Hint:* Treat the cases $L < \infty$ and $L = \infty$ separately.]
5. Prove that $\lim_{x \to \infty} cf(x) = c \lim_{x \to \infty} f(x)$.
6. Prove that if $\lim_{x \to \infty} f(x)$ and $\lim_{x \to \infty} g(x)$ exist and are finite, then

$$\lim_{x \to \infty} [f(x) + g(x)] = \lim_{x \to \infty} f(x) + \lim_{x \to \infty} g(x).$$

*7. Give an example to show that the result of Problem 6 is false in general if the limits are not assumed to be finite. [*Hint:* Find $f(x)$ and $g(x)$ so that $\lim_{x \to \infty} f(x) = \lim_{x \to \infty} g(x) = \infty$ but that $f(x) - g(x)$ is a constant.]
8. Show that if $\lim_{x \to \infty} f(x)$ and $\lim_{x \to \infty} g(x)$ are finite, then

$$\lim_{x \to \infty} f(x)g(x) = \left[\lim_{x \to \infty} f(x) \right]\left[\lim_{x \to \infty} g(x) \right].$$

*9. Give an example to show that the result of Problem 8 is false in general if the limits are not finite. [*Hint:* Find $f(x)$ and $g(x)$ such that $\lim_{x \to \infty} f(x) = \infty$, $\lim_{x \to \infty} g(x) = 0$, and $f(x) \cdot g(x)$ is a constant. Note that the expression $0 \cdot \infty$ is not defined.]
10. Show that if $\lim_{x \to \infty} f(x)$ is finite and $\lim_{x \to \infty} g(x)$ is finite and nonzero, then

$$\lim_{x \to \infty} \frac{f(x)}{g(x)} = \frac{\lim_{x \to \infty} f(x)}{\lim_{x \to \infty} g(x)}.$$

*11. Give an example to show that the result of Problem 10 is false in general if the limits are not assumed to be finite.
12. Prove that if $\lim_{x \to \infty} f(x)$ exists, then it is unique.

Appendix 6
Complex Numbers

In algebra we encounter the problem of finding the roots of the polynomial

$$\lambda^2 + a\lambda + b = 0. \tag{1}$$

To find the roots, we use the quadratic formula to obtain

$$\lambda = \frac{-a \pm \sqrt{a^2 - 4b}}{2}. \tag{2}$$

If $a^2 - 4b > 0$, there are two real roots. If $a^2 - 4b = 0$, we obtain the single root (of multiplicity 2) $\lambda = -a/2$. To deal with the case $a^2 - 4b < 0$, we introduce the **imaginary number**†

$$i = \sqrt{-1}. \tag{3}$$

---

†You should not be troubled by the term "imaginary." It's just a name. The British mathematician Alfred North Whitehead, in the chapter on imaginary numbers in his *Introduction to Mathematics*, wrote:

At this point it may be useful to observe that a certain type of intellect is always worrying itself and others by discussion as to the applicability of technical terms. Are the incommensurable numbers properly called numbers? Are the positive and negative numbers really numbers? Are the imaginary numbers imaginary, and are they numbers?—are types of such futile questions. Now, it cannot be too clearly understood that, in science, technical terms are names arbitrarily assigned, like Christian names to children. There can be no question of the names being right or wrong. They may be judicious or injudicious; for they can sometimes be so arranged as to be easy to remember, or so as to suggest relevant and important ideas. But the essential principle involved was quite clearly enunciated in Wonderland to Alice by Humpty Dumpty, when he told her, apropos of his use of words, 'I pay them extra and make them mean what I like'. So we will not bother as to whether imaginary numbers are imaginary, or as to whether they are numbers, but will take the phrase as the arbitrary name of a certain mathematical idea, which we will now endeavour to make plain.

Then for $a^2 - 4b < 0$

$$\sqrt{a^2 - 4b} = \sqrt{(4b - a^2)(-1)} = \sqrt{4b - a^2}\sqrt{-1} = \sqrt{4b - a^2}\,i,$$

and the two roots of (1) are given by

$$\lambda_1 = -\frac{a}{2} + \frac{\sqrt{4b - a^2}}{2}\,i \quad \text{and} \quad \lambda_2 = -\frac{a}{2} - \frac{\sqrt{4b - a^2}}{2}\,i.$$

EXAMPLE 1   Find the roots of the quadratic equation $\lambda^2 + 2\lambda + 5 = 0$.

Solution. We have $a = 2$, $b = 5$, and $a^2 - 4b = -16$. Thus $\sqrt{a^2 - 4b} = \sqrt{-16} = \sqrt{16}\sqrt{-1} = 4i$, and the roots are

$$\lambda_1 = \frac{-2 + 4i}{2} = -1 + 2i \quad \text{and} \quad \lambda_2 = -1 - 2i. \ \blacksquare$$

Definition 1   COMPLEX NUMBER   A **complex number** is a number of the form

$$z = \alpha + i\beta \tag{4}$$

where α and β are real numbers, α is called the **real part** of z and is denoted Re z. β is called the **imaginary part** of z and is denoted Im z. Representation (4) is sometimes called the **Cartesian form** of the complex number z.

REMARK.   If $\beta = 0$ in equation (4), then $z = \alpha$ is a real number. In this context we can regard the set of real numbers as a subset of the set of complex numbers.

EXAMPLE 2   In Example 1, Re $\lambda_1 = -1$ and Im $\lambda_1 = 2$. ■

We can add and multiply complex numbers by using the standard rules of algebra.

EXAMPLE 3   Let $z = 2 + 3i$ and $w = 5 - 4i$. Calculate (a) $z + w$, (b) $3w - 5z$, and (c) zw.

Solution.
(a) $z + w = (2 + 3i) + (5 - 4i) = (2 + 5) + (3 - 4)i = 7 - i$.
(b) $3w = 3(5 - 4i) = 15 - 12i$, $5z = 10 + 15i$, and $3w - 5z = (15 - 12i) - (10 + 15i) = (15 - 10) + i(-12 - 15) = 5 - 27i$.
(c) $zw = (2 + 3i)(5 - 4i) = (2)(5) + 2(-4i) + (3i)(5) + (3i)(-4i) = 10 - 8i + 15i - 12i^2 = 10 + 7i + 12 = 22 + 7i$. Here we used the fact that $i^2 = -1$. ■

We can plot a complex number z in the xy-plane by plotting Re z along the x-axis and Im z along the y-axis. Thus each complex number can be thought of as a point in the xy-plane. With this representation the xy-plane is called the **complex plane.** Some representative points are plotted in Figure 1.

If $z = \alpha + i\beta$, then we define the **conjugate** of z, denoted $\bar{z}$, by

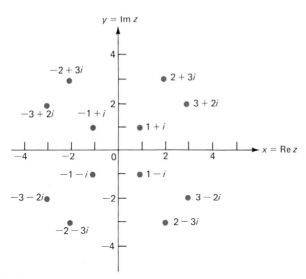

FIGURE 1

$$\bar{z} = \alpha - i\beta. \tag{5}$$

Figure 2 depicts a representative value of z and $\bar{z}$.

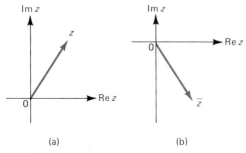

(a)   (b)

FIGURE 2

EXAMPLE 4   Compute the conjugate of (a) $1 + i$, (b) $3 - 4i$, (c) $-7 + 5i$, and (d) -3.

Solution. (a) $\overline{1 + i} = 1 - i$. (b) $\overline{3 - 4i} = 3 + 4i$. (c) $\overline{-7 + 5i} = -7 - 5i$. (d) $\overline{-3} = -3$. ■

It is not difficult to show (see Problem 35) that

$$\bar{z} = z \qquad \text{if and only if } z \text{ is real.} \tag{6}$$

If $z = \beta i$ with β real, then z is said to be **pure imaginary.** We can then show (see Problem 36) that

$$\bar{z} = -z \qquad \text{if and only if } z \text{ is pure imaginary.} \tag{7}$$

Let $p_n(x) = a_0 + a_1x + a_2x^2 + \cdots + a_nx^n$ be a polynomial with real coefficients. Then it can be shown (see Problem 41) that the complex roots of the equation $p_n(x) = 0$ occur in complex conjugate pairs. That is, if z is a root of $p_n(x) = 0$, then so is $\bar{z}$. We saw this fact illustrated in Example 1 in the case in which $n = 2$.

For $z = \alpha + i\beta$ we define the **magnitude** of z, denoted $|z|$, by

$$|z| = \sqrt{\alpha^2 + \beta^2}, \tag{8}$$

and we define the **argument** of z, denoted arg z, as the angle θ between the line $0z$ and the positive x-axis. From Figure 3 we see that $r = |z|$ is the distance from z to the origin, and

$$\theta = \arg z = \tan^{-1}\frac{\beta}{\alpha}. \tag{9}$$

By convention, we always choose a value of $\tan^{-1}\beta/\alpha$ that lies in the interval

$$-\pi < \theta \le \pi. \tag{10}$$

From Figure 4 we see that

$$|\bar{z}| = |z| \tag{11}$$

and

$$\arg \bar{z} = -\arg z. \tag{12}$$

FIGURE 3

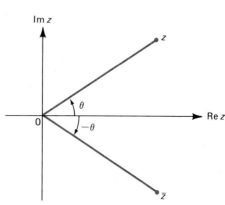

FIGURE 4

We can use $|z|$ and arg z to describe what is often a more convenient way to represent complex numbers.† From Figure 3 it is evident that if $z = \alpha + i\beta$, $r = |z|$, and $\theta = \arg z$, then

$$\alpha = r \cos \theta \quad \text{and} \quad \beta = r \sin \theta. \tag{13}$$

We will see at the end of this appendix that

$$e^{i\theta} = \cos \theta + i \sin \theta. \tag{14}$$

Since $\cos(-\theta) = \cos \theta$ and $\sin(-\theta) = -\sin \theta$, we also have

$$e^{-i\theta} = \cos(-\theta) + i \sin(-\theta) = \cos \theta - i \sin \theta. \tag{14'}$$

Formula (14) is called **Euler's formula**.‡ Using Euler's formula and equation (13), we have

$$z = \alpha + i\beta = r \cos \theta + ir \sin \theta = r(\cos \theta + i \sin \theta),$$

or

$$z = re^{i\theta}. \tag{15}$$

Representation (15) is called the **polar form** of the complex number z.

EXAMPLE 5 Determine the polar forms of the following complex numbers: (a) 1 (b) -1 (c) i (d) $1 + i$ (e) $-1 - \sqrt{3}$ (f) $-2 + 7i$

Solution. The six points are plotted in Figure 5.

(a) From Figure 5a it is clear that arg $1 = 0$. Since Re $1 = 1$, we see that, in polar form,

$$1 = 1e^{i0} = 1e^0 = 1.$$

(b) Since $\arg(-1) = \pi$ (Figure 5b) and $|-1| = 1$, we have

$$-1 = 1e^{\pi i} = e^{i\pi}.$$

(c) From Figure 5c we see that arg $i = \pi/2$. Since $|i| = \sqrt{0^2 + 1^2} = 1$, it follows that

---

†If you have already studied polar coordinates in Chapter 11, you will find this new representation very familiar.
‡Named for the Swiss mathematician Leonhard Euler (1707–1783). See his biographical sketch on page 36.

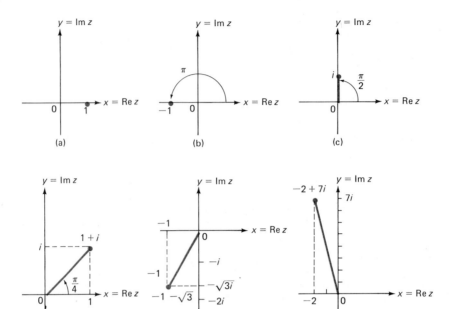

FIGURE 5

$i = e^{i\pi/2}$.

(d) $\arg(1 + i) = \tan^{-1} 1/1 = \pi/4$, and $|1 + i| = \sqrt{1^2 + 1^2} = \sqrt{2}$, so that

$$1 + i = \sqrt{2}\, e^{i\pi/4}.$$

(e) Here $\tan^{-1} \beta/\alpha = \tan^{-1} \sqrt{3} = \pi/3$. However, $\arg z$ is in the third quadrant, so $\theta = (\pi/3) + \pi = 4\pi/3$. Also, $|-1 - \sqrt{3}i| = \sqrt{1^2 + (\sqrt{3})^2} = \sqrt{1 + 3} = 2$, so that $-1 - \sqrt{3} = 2e^{4\pi i/3}$.

(f) To compute this complex number, we need a calculator. A calculator indicates that

$$\arg z = \tan^{-1}(-\tfrac{7}{2}) = \tan^{-1}(-3.5) \approx -1.2925.$$

But $\tan^{-1} x$ is defined as a number in the interval $(-\pi/2, \pi/2)$. Since from Figure 5f θ is in the second quadrant, we see that $\arg z = \tan^{-1}(-3.5) + \pi \approx 1.8491$. Next, we see that

$$|-2 + 7i| = \sqrt{(-2)^2 + 7^2} = \sqrt{53}.$$

Hence

$$-2 + 7i \approx \sqrt{53}\, e^{1.8491i}. \quad\blacksquare$$

EXAMPLE 6   Convert the following complex numbers from polar to Cartesian form: (a) $2e^{i\pi/3}$, (b) $4e^{3\pi i/2}$.

Solution.
(a) $e^{i\pi/3} = \cos\pi/3 + i\sin\pi/3 = \frac{1}{2} + (\sqrt{3}/2)i$. Thus $2e^{i\pi/3} = 1 + \sqrt{3}i$.
(b) $e^{3\pi i/2} = \cos 3\pi/2 + i\sin 3\pi/2 = 0 + i(-1) = -i$. Thus $4e^{3\pi i/2} = -4i$. ▪

If $\theta = \arg z$, then by equation (12), $\arg \bar{z} = -\theta$. Thus since $|\bar{z}| = |z|$, we have the following:

$$\text{If } z = re^{i\theta}, \text{ then } \bar{z} = re^{-i\theta}. \tag{16}$$

Suppose we write a complex number in its polar form $z = re^{i\theta}$. Then

$$z^n = (re^{i\theta})^n = r^n(e^{i\theta})^n = r^n e^{in\theta} = r^n(\cos n\theta + i\sin n\theta). \tag{17}$$

Formula (17) is useful for a variety of computations. In particular, when $r = |z| = 1$, we obtain the **De Moivre formula:**†

$$(\cos\theta + i\sin\theta)^n = \cos n\theta + i\sin n\theta. \tag{18}$$

EXAMPLE 7 Compute $(1 + i)^5$.

Solution. In Example 5(d) we showed that $1 + i = \sqrt{2}\,e^{\pi i/4}$. Then

$$(1 + i)^5 = (\sqrt{2}\,e^{\pi i/4})^5 = (\sqrt{2})^5 e^{5\pi i/4} = 4\sqrt{2}\left(\cos\frac{5\pi}{4} + i\sin\frac{5\pi}{4}\right)$$

$$= 4\sqrt{2}\left(-\frac{1}{\sqrt{2}} - \frac{1}{\sqrt{2}}i\right) = -4 - 4i.$$

This can be checked by direct calculation. If the direct calculation seems no more difficult, then try to compute $(1 + i)^{20}$ directly. Proceeding as above, we obtain

$$(1 + i)^{20} = (\sqrt{2})^{20}e^{20\pi i/4} = 2^{10}(\cos 5\pi + i\sin 5\pi)$$

$$= 2^{10}(-1 + 0) = -1024. ▪$$

Proof of Euler's Formula.‡ We will show that

$$e^{i\theta} = \cos\theta + i + \sin\theta \tag{19}$$

by using power series. If you have not covered Chapter 14, skip the proof. We have

---

†Abraham De Moivre (1667–1754) was a French mathematician well known for his work in probability theory, infinite series, and trigonometry. He was so highly regarded that Newton often told those who came to him with questions on mathematics. "Go to M. De Moivre; he knows these things better than I do."
‡If time permits.

$$e^x = 1 + x + \frac{x^2}{2!} + \frac{x^3}{3!} + \cdots, \dagger \tag{20}$$

$$\sin x = x - \frac{x^3}{3!} + \frac{x^5}{5!} - \cdots, \tag{21}$$

$$\cos x = 1 - \frac{x^2}{2!} + \frac{x^4}{4!} - \cdots. \tag{22}$$

Then

$$e^{i\theta} = 1 + (i\theta) + \frac{(i\theta)^2}{2!} + \frac{(i\theta)^3}{3!} + \frac{(i\theta)^4}{4!} + \frac{(i\theta)^5}{5!} + \cdots. \tag{23}$$

Now $i^2 = -1$, $i^3 = -i$, $i^4 = 1$, $i^5 = i$, and so on. Thus (23) can be written

$$e^{i\theta} = 1 + i\theta - \frac{\theta^2}{2!} - \frac{i\theta^3}{3!} + \frac{\theta^4}{4!} + \frac{i\theta^5}{5!} - \cdots$$

$$= \left(1 - \frac{\theta^2}{2!} + \frac{\theta^4}{4!} - \cdots \right) + i\left(\theta - \frac{\theta^3}{3!} + \frac{\theta^5}{5!} - \cdots \right)$$

$$= \cos\theta + i\sin\theta.$$

This completes the proof. ■

PROBLEMS

In Problems 1–5, perform the indicated operation.

1. $(2 - 3i) + (7 - 4i)$ **2.** $3(4 + i) - 5(-3 + 6i)$
3. $(1 + i)(1 - i)$ **4.** $(2 - 3i)(4 + 7i)$
5. $(-3 + 2i)(7 + 3i)$

In Problems 6–15, convert the complex number to its polar form.

6. $5i$ **7.** $5 + 5i$ **8.** $-2 - 2i$ **9.** $3 - 3i$
10. $2 + 2\sqrt{3}i$ **11.** $3\sqrt{3} + 3i$ **12.** $1 - \sqrt{3}i$ **13.** $4\sqrt{3} - 4i$
14. $-6\sqrt{3} - 6i$ **15.** $-1 - \sqrt{3}i$

In Problems 16–25, convert from polar to Cartesian form.

16. $e^{3\pi i}$ **17.** $2e^{-7\pi i}$ **18.** $\frac{1}{2}e^{3\pi i/4}$ **19.** $\frac{1}{2}e^{-3\pi i/4}$
20. $6e^{\pi i/6}$ **21.** $4e^{5\pi i/6}$ **22.** $4e^{-5\pi i/6}$ **23.** $3e^{-2\pi i/3}$
24. $\sqrt{3}e^{23\pi i/4}$ **25.** e^i

In Problems 26–34, compute the conjugate of the given number.

26. $3 - 4i$ **27.** $4 + 6i$ **28.** $-3 + 8i$ **29.** $-7i$

---

†Although we will not prove it here, these series expansions are also valid when x is a complex number.

30. 16 **31.** $2e^{\pi i/7}$ **32.** $4e^{3\pi i/5}$ **33.** $3e^{-4\pi i/11}$

34. $e^{0.012i}$

35. Show that $z = \alpha + i\beta$ is real if and only if $z = \bar{z}$. [*Hint:* if $z = \bar{z}$, show that $\beta = 0$.]

36. Show that $z = \alpha + i\beta$ is pure imaginary if and only if $z = -\bar{z}$. [*Hint:* if $z = -\bar{z}$, show that $\alpha = 0$.]

37. For any complex number z, show that $z\bar{z} = |z|^2$.

38. Show that the circle of radius 1 centered at the origin (the *unit circle*) is the set of points in the complex plane that satisfy $|z| = 1$.

39. For any complex number z_0 and real number a, describe $\{z: |z - z_0| = a\}$.

40. Describe $\{z: |z - z_0| \le a\}$, where z_0 and a are as in Problem 39.

*41. Let $p(\lambda) = \lambda^n + a_{n-1}\lambda^{n-1} + a_{n-2}\lambda^{n-2} + \cdots + a_1\lambda + a_0$ with $a_0, a_1, \ldots, a_{n-1}$ real numbers. Show that if $p(z) = 0$, then $p(\bar{z}) = 0$. That is, *The roots of polynomials with real coefficients occur in complex conjugate pairs.*

42. Derive expressions for $\cos 4\theta$ and $\sin 4\theta$ by comparing the De Moivre formula and the expansion of $(\cos \theta + i \sin \theta)^4$.

*43. Prove De Moivre's formula by mathematical induction. [*Hint:* Recall the trigonometric identities $\cos(x + y) = \cos x \cos y - \sin x \sin y$ and $\sin(x + y) = \sin x \cos y + \cos x \sin y$.]

TABLE A.1   EXPONENTIAL FUNCTIONS

x	e^x	e^{-x}	x	e^x	e^{-x}
0.00	1.0000	1.0000	3.0	20.086	0.0498
0.05	1.0513	0.9512	3.1	22.198	0.0450
0.10	1.1052	0.9048	3.2	24.533	0.0408
0.15	1.1618	0.8607	3.3	27.113	0.0369
0.20	1.2214	0.8187	3.4	29.964	0.0334
0.25	1.2840	0.7788	3.5	33.115	0.0302
0.30	1.3499	0.7408	3.6	36.598	0.0273
0.35	1.4191	0.7047	3.7	40.447	0.0247
0.40	1.4918	0.6703	3.8	44.701	0.0224
0.45	1.5683	0.6376	3.9	49.402	0.0202
0.50	1.6487	0.6065	4.0	54.598	0.0183
0.55	1.7333	0.5769	4.1	60.340	0.0166
0.60	1.8221	0.5488	4.2	66.686	0.0150
0.65	1.9155	0.5220	4.3	73.700	0.0136
0.70	2.0138	0.4966	4.4	81.451	0.0123
0.75	2.1170	0.4724	4.5	90.017	0.0111
0.80	2.2255	0.4493	4.6	99.484	0.0101
0.85	2.3396	0.4274	4.7	109.95	0.0091
0.90	2.4596	0.4066	4.8	121.51	0.0082
0.95	2.5857	0.3867	4.9	134.29	0.0074
1.0	2.7183	0.3679	5.0	148.41	0.0067
1.1	3.0042	0.3329	5.1	164.02	0.0061
1.2	3.3201	0.3012	5.2	181.27	0.0055
1.3	3.6693	0.2725	5.3	200.34	0.0050
1.4	4.0552	0.2466	5.4	221.41	0.0045
1.5	4.4817	0.2231	5.5	244.69	0.0041
1.6	4.9530	0.2019	5.6	270.43	0.0037
1.7	5.4739	0.1827	5.7	298.87	0.0033
1.8	6.0496	0.1653	5.8	330.30	0.0030
1.9	6.6859	0.1496	5.9	365.04	0.0027
2.0	7.3891	0.1353	6.0	403.43	0.0025
2.1	8.1662	0.1225	6.5	665.14	0.0015
2.2	9.0250	0.1108	7.0	1096.6	0.0009
2.3	9.9742	0.1003	7.5	1808.0	0.0006
2.4	11.023	0.0907	8.0	2981.0	0.0003
2.5	12.182	0.0821	8.5	4914.8	0.0002
2.6	13.464	0.0743	9.0	8103.1	0.0001
2.7	14.880	0.0672	9.5	13,360	0.00007
2.8	16.445	0.0608	10.0	22,026	0.00004
2.9	18.174	0.0550			

TABLE A.2 NATURAL LOGARITHMS

n	$\log_e n$	n	$\log_e n$	n	$\log_e n$
0.0	—	4.5	1.5041	9.0	2.1972
0.1	-2.3026	4.6	1.5261	9.1	2.2083
0.2	-1.6094	4.7	1.5476	9.2	2.2192
0.3	-1.2040	4.8	1.5686	9.3	2.2300
0.4	-0.9163	4.9	1.5892	9.4	2.2407
0.5	-0.6931	5.0	1.6094	9.5	2.2513
0.6	-0.5108	5.1	1.6292	9.6	2.2618
0.7	-0.3567	5.2	1.6487	9.7	2.2721
0.8	-0.2231	5.3	1.6677	9.8	2.2824
0.9	-0.1054	5.4	1.6864	9.9	2.2925
1.0	0.0000	5.5	1.7047	10	2.3026
1.1	0.0953	5.6	1.7228	11	2.3979
1.2	0.1823	5.7	1.7405	12	2.4849
1.3	0.2624	5.8	1.7579	13	2.5649
1.4	0.3365	5.9	1.7750	14	2.6391
1.5	0.4055	6.0	1.7918	15	2.7081
1.6	0.4700	6.1	1.8083	16	2.7726
1.7	0.5306	6.2	1.8245	17	2.8332
1.8	0.5878	6.3	1.8405	18	2.8904
1.9	0.6419	6.4	1.8563	19	2.9444
2.0	0.6931	6.5	1.8718	20	2.9957
2.1	0.7419	6.6	1.8871	25	3.2189
2.2	0.7885	6.7	1.9021	30	3.4012
2.3	0.8329	6.8	1.9169	35	3.5553
2.4	0.8755	6.9	1.9315	40	3.6889
2.5	0.9163	7.0	1.9459	45	3.8067
2.6	0.9555	7.1	1.9601	50	3.9120
2.7	0.9933	7.2	1.9741	55	4.0073
2.8	1.0296	7.3	1.9879	60	4.0943
2.9	1.0647	7.4	2.0015	65	4.1744
3.0	1.0986	7.5	2.0149	70	4.2485
3.1	1.1314	7.6	2.0281	75	4.3175
3.2	1.1632	7.7	2.0142	80	4.3820
3.3	1.1939	7.8	2.0541	85	4.4427
3.4	1.2238	7.9	2.0669	90	4.4998
3.5	1.2528	8.0	2.0794	95	4.5539
3.6	1.2809	8.1	2.0919	100	4.6052
3.7	1.3083	8.2	2.1041	200	5.2983
3.8	1.3350	8.3	2.1163	300	5.7038
3.9	1.3610	8.4	2.1282	400	5.9915
4.0	1.3863	8.5	2.1401	500	6.2146
4.1	1.4110	8.6	2.1518	600	6.3069
4.2	1.4351	8.7	2.1633	700	6.5511
4.3	1.4586	8.8	2.1748	800	6.6846
4.4	1.4816	8.9	2.1861	900	6.8024

TABLE A.3 HYPERBOLIC FUNCTIONS

x	$\sinh x$	$\cosh x$	$\tanh x$	x	$\sinh x$	$\cosh x$	$\tanh x$
0.0	0.00000	1.0000	0.00000	3.0	10.018	10.068	0.99505
0.1	0.10017	1.0050	0.09967	3.1	11.076	11.122	0.99595
0.2	0.20134	1.0201	0.19738	3.2	12.246	12.287	0.99668
0.3	0.30452	1.0453	0.29131	3.3	13.538	13.575	0.99728
0.4	0.41075	1.0811	0.37995	3.4	14.965	14.999	0.99777
0.5	0.52110	1.1276	0.46212	3.5	16.543	16.573	0.99818
0.6	0.63665	1.1855	0.53705	3.6	18.285	18.313	0.99851
0.7	0.75858	1.2552	0.60437	3.7	20.211	20.236	0.99878
0.8	0.88811	1.3374	0.66404	3.8	22.339	22.362	0.99900
0.9	1.0265	1.4331	0.71630	3.9	24.691	24.711	0.99918
1.0	1.1752	1.5431	0.76159	4.0	27.290	27.308	0.99933
1.1	1.3356	1.6685	0.80050	4.1	30.162	30.178	0.99945
1.2	1.5095	1.8107	0.83365	4.2	33.336	33.351	0.99955
1.3	1.6984	1.9709	0.86172	4.3	36.843	36.857	0.99963
1.4	1.9043	2.1509	0.88535	4.4	40.719	40.732	0.99970
1.5	2.1293	2.3524	0.90515	4.5	45.003	45.014	0.99975
1.6	2.3756	2.5775	0.92167	4.6	49.737	49.747	0.99980
1.7	2.6456	2.8283	0.93541	4.7	54.969	54.978	0.99983
1.8	2.9422	3.1075	0.94681	4.8	60.751	60.759	0.99986
1.9	3.2682	3.4177	0.95624	4.9	67.141	67.149	0.99989
2.0	3.6269	3.7622	0.96403	5.0	74.203	74.210	0.99991
2.1	4.0219	4.1443	0.97045	5.1	82.008	82.014	0.99993
2.2	4.4571	4.5679	0.97574	5.2	90.633	90.639	0.99994
2.3	4.9370	5.0372	0.98010	5.3	100.17	100.17	0.99995
2.4	5.4662	5.5569	0.98367	5.4	110.70	110.71	0.99996
2.5	6.0502	6.1323	0.98661	5.5	122.34	122.35	0.99997
2.6	6.6947	6.7690	0.98903	5.6	135.21	135.22	0.99997
2.7	7.4063	7.4735	0.99101	5.7	149.43	149.44	0.99998
2.8	8.1919	8.2527	0.99263	5.8	165.15	165.15	0.99998
2.9	9.0596	9.1146	0.99396	5.9	182.52	182.52	0.99998

Note: When x is beyond the limits of the table (that is, when $x > 5.9$), $\sinh x$ and $\cosh x$ may be obtained to five significant figures by the formula $\cosh x = \sinh x = \frac{1}{2}e^x$; furthermore, for such values of x, $\tanh x = 1.0000$.

TABLE A.4 INTEGRALS[a]

STANDARD FORMS

1. $\int a\,dx = ax + C$

2. $\int af(x)\,dx = a\int f(x)\,dx + C$

3. $\int u\,dv = uv - \int v\,du$ (integration by parts)

4. $\int u^n\,du = \dfrac{u^{n+1}}{n+1} + C, \quad n \neq -1$

5. $\int \dfrac{du}{u} = \ln u$ if $u > 0$ or $\ln(-u)$ if $u < 0$
$= \ln|u| + C$

6. $\int e^u\,du = e^u + C$

7. $\int a^u\,du = \int e^{u\ln a}\,du$
$= \dfrac{e^{u\ln a}}{\ln a} = \dfrac{a^u}{\ln a} + C, \quad a > 0, a \neq 1$

8. $\int \sin u\,du = -\cos u + C$

9. $\int \cos u\,du = \sin u + C$

10. $\int \tan u\,du = \ln|\sec u| = -\ln|\cos u| + C$

11. $\int \cos u\,du = \ln|\sin u| + C$

12. $\int \sec u\,du = \ln|\sec u + \tan u|$
$= \ln\left|\tan\left|\dfrac{u}{2} + \dfrac{\pi}{4}\right|\right| + C$

13. $\int \csc u\,du = \ln|\csc u - \cot u| = \ln\left|\tan\dfrac{u}{2}\right| + C$

14. $\int \sec^2 u\,du = \tan u + C$

15. $\int \csc^2 u\,du = -\cot u + C$

16. $\int \sec u \tan u\,du = \sec u + C$

17. $\int \csc u \cot u\,du = -\csc u + C$

18. $\int \dfrac{du}{u^2 + a^2} = \dfrac{1}{a}\tan^{-1}\dfrac{u}{a} + C$

19. $\int \dfrac{du}{u^2 - a^2} = \dfrac{1}{2a}\ln\left|\dfrac{u-a}{u+a}\right| + C$
$= -\dfrac{1}{a}\coth^{-1}\dfrac{u}{a} + C, \quad u^2 > a^2$

20. $\int \dfrac{du}{a^2 - u^2} = \dfrac{1}{2a}\ln\left|\dfrac{a+u}{a-u}\right| + C$
$= \dfrac{1}{a}\tanh^{-1}\dfrac{u}{a} + C, \quad u^2 < a^2$

21. $\int \dfrac{du}{\sqrt{a^2 - u^2}} = \sin^{-1}\dfrac{u}{|a|} + C$

22. $\int \dfrac{du}{\sqrt{u^2 + a^2}} = \ln(u + \sqrt{u^2 + a^2}) + C$

23. $\int \dfrac{du}{\sqrt{u^2 - a^2}} = \ln|u + \sqrt{u^2 - a^2}| + C$

24. $\int \dfrac{du}{u\sqrt{u^2 - a^2}} = \dfrac{1}{|a|}\sec^{-1}\left|\dfrac{u}{a}\right| + C$

25. $\int \dfrac{du}{u\sqrt{u^2 + a^2}} = -\dfrac{1}{a}\ln\left|\dfrac{a + \sqrt{u^2 + a^2}}{u}\right| + C$

26. $\int \dfrac{du}{u\sqrt{a^2 - u^2}} = -\dfrac{1}{a}\ln\left|\dfrac{a + \sqrt{a^2 - u^2}}{u}\right| + C$

[a]All angles are measured in radians.

INTEGRALS INVOLVING $au + b$

27. $\displaystyle\int \frac{du}{au+b} = \frac{1}{a}\ln|au+b| + C$

28. $\displaystyle\int \frac{u\,du}{au+b} = \frac{u}{a} - \frac{b}{a^2}\ln|au+b| + C$

29. $\displaystyle\int \frac{u^2\,du}{au+b} = \frac{(au+b)^2}{2a^3} - \frac{2b(au+b)}{a^3} + \frac{b^2}{a^3}\ln|au+b| + C$

30. $\displaystyle\int \frac{du}{u(au+b)} = \frac{1}{b}\ln\left|\frac{u}{au+b}\right| + C$

31. $\displaystyle\int \frac{du}{u^2(au+b)} = -\frac{1}{bu} + \frac{a}{b^2}\ln\left|\frac{au+b}{u}\right| + C$

32. $\displaystyle\int \frac{du}{(au+b)^2} = \frac{-1}{a(au+b)} + C$

33. $\displaystyle\int \frac{u\,du}{(au+b)^2} = \frac{b}{a^2(au+b)} + \frac{1}{a^2}\ln|au+b| + C$

34. $\displaystyle\int \frac{du}{u(au+b)^2} = \frac{1}{b(au+b)} + \frac{1}{b^2}\ln\left|\frac{u}{au+b}\right| + C$

35. $\displaystyle\int (au+b)^n\,du = \frac{(au+b)^{n+1}}{(n+1)a} + C,\ n \neq -1$

36. $\displaystyle\int u(au+b)^n\,du = \frac{(au+b)^{n+2}}{(n+2)a^2} - \frac{b(au+b)^{n+1}}{(n+1)a^2} + C,\quad n \neq -1, -2$

37. $\displaystyle\int u^m(au+b)^n\,du = \begin{cases} \dfrac{u^{m+1}(au+b)^n}{m+n+1} + \dfrac{nb}{m+n+1}\displaystyle\int u^m(au+b)^{n-1}\,du \\[3ex] \dfrac{u^m(au+b)^{n+1}}{(m+n+1)a} - \dfrac{mb}{(m+n+1)a}\displaystyle\int u^{m-1}(au+b)^n\,du \\[3ex] \dfrac{-u^{m+1}(au+b)^{n+1}}{(n+1)b} + \dfrac{m+n+2}{(n+1)b}\displaystyle\int u^m(au+b)^{n+1}\,du \end{cases}$

INTEGRALS INVOLVING $\sqrt{au+b}$

38. $\displaystyle\int \frac{du}{\sqrt{au+b}} = \frac{2\sqrt{au+b}}{a} + C$

39. $\displaystyle\int \frac{u\,du}{\sqrt{au+b}} = \frac{2(au-2b)}{3a^2}\sqrt{au+b} + C$

40. $\displaystyle\int \frac{du}{u\sqrt{au+b}} = \begin{cases} \dfrac{1}{\sqrt{b}}\ln\left|\dfrac{\sqrt{au+b}-\sqrt{b}}{\sqrt{au+b}+\sqrt{b}}\right| + C,\quad b > 0 \\[3ex] \dfrac{2}{\sqrt{-b}}\tan^{-1}\sqrt{\dfrac{au+b}{-b}} + C,\quad b < 0 \end{cases}$

41. $\displaystyle\int \sqrt{au+b}\,du = \frac{2\sqrt{(au+b)^3}}{3a} + C$

42. $\displaystyle\int u\sqrt{au+b}\,du = \frac{2(3au-2b)}{15a^2}\sqrt{(au+b)^3} + C$

43. $\displaystyle\int \frac{\sqrt{au+b}}{u}\,du = 2\sqrt{au+b} + b\int \frac{du}{u\sqrt{au+b}}$ (See 40.)

INTEGRALS INVOLVING $u^2 + a^2$

44. $\displaystyle\int \frac{du}{u^2+a^2} = \frac{1}{a}\tan^{-1}\frac{u}{a} + C$

45. $\displaystyle\int \frac{u\,du}{u^2+a^2} = \frac{1}{2}\ln(u^2+a^2) + C$

46. $\displaystyle\int \frac{u^2\,du}{u^2+a^2} = u - a\tan^{-1}\frac{u}{a} + C$

47. $\displaystyle\int \frac{du}{u(u^2+a^2)} = \frac{1}{2a^2}\ln\left(\frac{u^2}{u^2+a^2}\right) + C$

48. $\displaystyle\int \frac{du}{u^2(u^2 + a^2)} = -\frac{1}{a^2 u} - \frac{1}{a^3} \tan^{-1} \frac{u}{a} + C$

49. $\displaystyle\int \frac{du}{(u^2 + a^2)^n} = \frac{u}{2(n-1)a^2(u^2 + a^2)^{n-1}} + \frac{2n-3}{(2n-2)a^2} \int \frac{du}{(u^2 + a^2)^{n-1}}$

50. $\displaystyle\int \frac{u\,du}{(u^2 + a^2)^n} = \frac{-1}{2(n-1)(u^2 + a^2)^{n-1}} + C, \; n \neq 1$

51. $\displaystyle\int \frac{du}{u(u^2 + a^2)^n} = \frac{1}{2(n-1)a^2(u^2 + a^2)^{n-1}} + \frac{1}{a^2} \int \frac{du}{u(u^2 + a^2)^{n-1}}, \; n \neq 1$

INTEGRALS INVOLVING $u^2 - a^2$, $u^2 > a^2$

52. $\displaystyle\int \frac{du}{u^2 - a^2} = \frac{1}{2a} \ln\left|\frac{u-a}{u+a}\right| + C$

53. $\displaystyle\int \frac{u\,du}{u^2 - a^2} = \frac{1}{2} \ln(u^2 - a^2) + C$

54. $\displaystyle\int \frac{u^2\,du}{u^2 - a^2} = u + \frac{a}{2} \ln\left|\frac{u-a}{u+a}\right| + C$

55. $\displaystyle\int \frac{du}{u(u^2 - a^2)} = \frac{1}{2a^2} \ln\left|\frac{u^2 - a^2}{u^2}\right| + C$

56. $\displaystyle\int \frac{du}{u^2(u^2 - a^2)} = \frac{1}{a^2 u} + \frac{1}{2a^3} \ln\left|\frac{u-a}{u+a}\right| + C$

57. $\displaystyle\int \frac{du}{(u^2 - a^2)^2} = \frac{-u}{2a^2(u^2 - a^2)} - \frac{1}{4a^3} \ln\left|\frac{u-a}{u+a}\right| + C$

58. $\displaystyle\int \frac{du}{(u^2 - a^2)^n} = \frac{-u}{2(n-1)a^2(u^2 - a^2)^{n-1}} - \frac{2n-3}{(2n-2)a^2} \int \frac{du}{(u^2 - a^2)^{n-1}}$

59. $\displaystyle\int \frac{u\,du}{(u^2 - a^2)^n} = \frac{-1}{2(n-1)(u^2 - a^2)^{n-1}} + C$

60. $\displaystyle\int \frac{du}{u(u^2 - a^2)^n} = \frac{-1}{2(n-1)a^2(u^2 - a^2)^{n-1}} - \frac{1}{a^2} \int \frac{du}{u(u^2 - a^2)^{n-1}}$

INTEGRALS INVOLVING $a^2 - u^2$, $u^2 < a^2$

61. $\displaystyle\int \frac{du}{a^2 - u^2} = \frac{1}{2a} \ln\left|\frac{a+u}{a-u}\right| + C \quad \text{or} \quad \frac{1}{a} \tanh^{-1} \frac{u}{a} + C$

62. $\displaystyle\int \frac{u\,du}{a^2 - u^2} = -\frac{1}{2} \ln|a^2 - u^2| + C$

63. $\displaystyle\int \frac{u^2\,du}{a^2 - u^2} = -u + \frac{a}{2} \ln\left|\frac{a+u}{a-u}\right| + C$

64. $\displaystyle\int \frac{du}{u(a^2 - u^2)} = \frac{1}{2a^2} \ln\left|\frac{u^2}{a^2 - u^2}\right| + C$

65. $\displaystyle\int \frac{du}{(a^2 - u^2)^2} = \frac{u}{2a^2(a^2 - u^2)} + \frac{1}{4a^3} \ln\left|\frac{a+u}{a-u}\right| + C$

66. $\displaystyle\int \frac{u\,du}{(a^2 - u^2)^2} = \frac{1}{2(a^2 - u^2)} + C$

INTEGRALS INVOLVING $\sqrt{u^2 + a^2}$

67. $\displaystyle\int \frac{du}{\sqrt{u^2 + a^2}} = \ln(u + \sqrt{u^2 + a^2}) + C \quad \text{or} \quad \sinh^{-1} \frac{u}{|a|} + C$

68. $\int \dfrac{u\,du}{\sqrt{u^2 + a^2}} = \sqrt{u^2 + a^2} + C$

69. $\int \dfrac{u^2\,du}{\sqrt{u^2 + a^2}} = \dfrac{u\sqrt{u^2 + a^2}}{2} - \dfrac{a^2}{2}\ln(u + \sqrt{u^2 + a^2}) + C$

70. $\int \dfrac{du}{u\sqrt{u^2 + a^2}} = -\dfrac{1}{a}\ln\left|\dfrac{a + \sqrt{u^2 + a^2}}{u}\right| + C$

71. $\int \sqrt{u^2 + a^2}\,du = \dfrac{u\sqrt{u^2 + a^2}}{2} + \dfrac{a^2}{2}\ln(u + \sqrt{u^2 + a^2}) + C$

72. $\int u\sqrt{u^2 + a^2}\,du = \dfrac{(u^2 + a^2)^{3/2}}{3} + C$

73. $\int u^2\sqrt{u^2 + a^2}\,du = \dfrac{u(u^2 + a^2)^{3/2}}{4} - \dfrac{a^2 u\sqrt{u^2 + a^2}}{8} - \dfrac{a^4}{8}\ln(u + \sqrt{u^2 + a^2}) + C$

74. $\int \dfrac{\sqrt{u^2 + a^2}}{u}\,du = \sqrt{u^2 + a^2} - a\ln\left|\dfrac{a + \sqrt{u^2 + a^2}}{u}\right| + C$

75. $\int \dfrac{\sqrt{u^2 + a^2}}{u^2}\,du = -\dfrac{\sqrt{u^2 + a^2}}{u} + \ln(u + \sqrt{u^2 + a^2}) + C$

INTEGRALS INVOLVING $\sqrt{u^2 - a^2}$

76. $\int \dfrac{du}{\sqrt{u^2 - a^2}} = \ln|u + \sqrt{u^2 - a^2}| + C$ $\qquad$ **77.** $\int \dfrac{u\,du}{\sqrt{u^2 - a^2}} = \sqrt{u^2 - a^2} + C$

78. $\int \dfrac{u^2\,du}{\sqrt{u^2 - a^2}} = \dfrac{u\sqrt{u^2 - a^2}}{2} + \dfrac{a^2}{2}\ln|u + \sqrt{u^2 - a^2}| + C$

79. $\int \dfrac{du}{u\sqrt{u^2 - a^2}} = \dfrac{1}{|a|}\sec^{-1}\left|\dfrac{u}{a}\right| + C$

80. $\int \sqrt{u^2 - a^2}\,du = \dfrac{u\sqrt{u^2 - a^2}}{2} - \dfrac{a^2}{2}\ln|u + \sqrt{u^2 - a^2}| + C$

81. $\int u\sqrt{u^2 - a^2}\,du = \dfrac{(u^2 - a^2)^{3/2}}{3} + C$

82. $\int u^2\sqrt{u^2 - a^2}\,du = \dfrac{u(u^2 - a^2)^{3/2}}{4} + \dfrac{a^2 u\sqrt{u^2 - a^2}}{8} - \dfrac{a^4}{8}\ln|u + \sqrt{u^2 - a^2}| + C$

83. $\int \dfrac{\sqrt{u^2 - a^2}}{u}\,du = \sqrt{u^2 - a^2} - |a|\sec^{-1}\left|\dfrac{u}{a}\right| + C$

84. $\int \dfrac{\sqrt{u^2 - a^2}}{u^2}\,du = -\dfrac{\sqrt{u^2 - a^2}}{u} + \ln|u + \sqrt{u^2 - a^2}| + C$

85. $\int \dfrac{du}{(u^2 - a^2)^{3/2}} = -\dfrac{u}{a^2\sqrt{u^2 - a^2}} + C$

INTEGRALS INVOLVING $\sqrt{a^2 - u^2}$

86. $\int \dfrac{du}{\sqrt{a^2 - u^2}} = \sin^{-1}\dfrac{u}{|a|} + C$ $\qquad$ **87.** $\int \dfrac{u\,du}{\sqrt{a^2 - u^2}} = -\sqrt{a^2 - u^2} + C$

88. $\displaystyle\int \frac{u^2\,du}{\sqrt{a^2-u^2}} = -\frac{u\sqrt{a^2-u^2}}{2} + \frac{a^2}{2}\sin^{-1}\frac{u}{|a|} + C$ **89.** $\displaystyle\int \frac{du}{u\sqrt{a^2-u^2}} = -\frac{1}{a}\ln\left|\frac{a+\sqrt{a^2-u^2}}{u}\right| + C$

90. $\displaystyle\int \frac{du}{u^2\sqrt{a^2-u^2}} = -\frac{\sqrt{a^2-u^2}}{a^2 u} + C$

91. $\displaystyle\int \sqrt{a^2-u^2}\,du = \frac{u\sqrt{a^2-u^2}}{2} + \frac{a^2}{2}\sin^{-1}\frac{u}{|a|} + C$ **92.** $\displaystyle\int u\sqrt{a^2-u^2}\,du = -\frac{(a^2-u^2)^{3/2}}{3} + C$

93. $\displaystyle\int u^2\sqrt{a^2-u^2}\,du = -\frac{u(a^2-u^2)^{3/2}}{4} + \frac{a^2 u\sqrt{a^2-u^2}}{8} + \frac{a^4}{8}\sin^{-1}\frac{u}{|a|} + C$

94. $\displaystyle\int \frac{\sqrt{a^2-u^2}}{u}\,du = \sqrt{a^2-u^2} - a\ln\left|\frac{a+\sqrt{a^2-u^2}}{u}\right| + C$

95. $\displaystyle\int \frac{\sqrt{a^2-u^2}}{u^2}\,du = -\frac{\sqrt{a^2-u^2}}{u} - \sin^{-1}\frac{u}{|a|} + C$

INTEGRALS INVOLVING THE TRIGONOMETRIC FUNCTIONS

96. $\displaystyle\int \sin au\,du = -\frac{\cos au}{a} + C$ **97.** $\displaystyle\int u\sin au\,du = \frac{\sin au}{a^2} - \frac{u\cos au}{a} + C$

98. $\displaystyle\int u^2\sin au\,du = \frac{2u}{a^2}\sin au + \left(\frac{2}{a^3} - \frac{u^2}{a}\right)\cos au + C$

99. $\displaystyle\int \frac{du}{\sin au} = \frac{1}{a}\ln(\csc au - \cot au)$ **100.** $\displaystyle\int \sin^2 au\,du = \frac{u}{2} - \frac{\sin 2au}{4a} + C$

$\displaystyle\qquad\qquad = \frac{1}{a}\ln\left|\tan\frac{au}{2}\right| + C$

101. $\displaystyle\int u\sin^2 au\,du = \frac{u^2}{4} - \frac{u\sin 2au}{4a} - \frac{\cos 2au}{8a^2} + C$ **102.** $\displaystyle\int \frac{du}{\sin^2 au} = -\frac{1}{a}\cot au + C$

103. $\displaystyle\int \sin pu\sin qu\,du = \frac{\sin(p-q)u}{2(p-q)} - \frac{\sin(p+q)u}{2(p+q)} + C, \quad p \neq \pm q$

104. $\displaystyle\int \frac{du}{1-\sin au} = \frac{1}{a}\tan\left(\frac{\pi}{4} + \frac{au}{2}\right) + C$

105. $\displaystyle\int \frac{u\,du}{1-\sin au} = \frac{u}{a}\tan\left(\frac{\pi}{4} + \frac{au}{2}\right) + \frac{2}{a^2}\ln\left|\sin\left(\frac{\pi}{4} - \frac{au}{2}\right)\right| + C$

106. $\displaystyle\int \frac{du}{1+\sin au} = -\frac{1}{a}\tan\left(\frac{\pi}{4} - \frac{au}{2}\right) + C$

107. $\displaystyle\int \frac{du}{p+q\sin au} = \begin{cases} \dfrac{2}{a\sqrt{p^2-q^2}}\tan^{-1}\dfrac{p\tan\frac{1}{2}au + q}{\sqrt{p^2-q^2}} + C, & |p| > |q| \\[2ex] \dfrac{1}{a\sqrt{q^2-p^2}}\ln\left|\dfrac{p\tan\frac{1}{2}au + q - \sqrt{q^2-p^2}}{p\tan\frac{1}{2}au + q + \sqrt{q^2-p^2}}\right| + C, & |p| < |q| \end{cases}$

108. $\displaystyle\int u^m\sin au\,du = -\frac{u^m\cos au}{a} + \frac{mu^{m-1}\sin au}{a^2} - \frac{m(m-1)}{a^2}\int u^{m-2}\sin au\,du$

109. $\displaystyle \int \sin^n au\ du = -\frac{\sin^{n-1} au\ \cos au}{an} + \frac{n-1}{n} \int \sin^{n-2} au\ du$

110. $\displaystyle \int \frac{du}{\sin^n au} = \frac{-\cos au}{a(n-1)\sin^{n-1} au} + \frac{n-2}{n-1} \int \frac{du}{\sin^{n-2} au},\quad n \neq 1$

111. $\displaystyle \int \cos au\ du = \frac{\sin au}{a} + C$ **112.** $\displaystyle \int u \cos au\ du = \frac{\cos au}{a^2} + \frac{u \sin au}{a} + C$

113. $\displaystyle \int u^2 \cos au\ du = \frac{2u}{a^2} \cos au + \left(\frac{u^2}{a} - \frac{2}{a^3}\right)\sin au + C$

114. $\displaystyle \int \frac{du}{\cos au} = \frac{1}{a} \ln(\sec au + \tan au) = \frac{1}{a} \ln\left|\tan\left(\frac{\pi}{4} + \frac{au}{2}\right)\right| + C$

115. $\displaystyle \int \cos^2 au\ du = \frac{u}{2} + \frac{\sin 2au}{4a} + C$ **116.** $\displaystyle \int u \cos^2 au\ du = \frac{u^2}{4} + \frac{u \sin 2au}{4a} + \frac{\cos 2au}{8a^2} + C$

117. $\displaystyle \int \frac{du}{\cos^2 au} = \frac{\tan au}{a} + C$

118. $\displaystyle \int \cos qu \cos pu\ du = \frac{\sin(q-p)u}{2(q-p)} + \frac{\sin(q+p)u}{2(q+p)} + C,\quad q \neq \pm p$

119. $\displaystyle \int \frac{du}{p + q \cos au} = \begin{cases} \dfrac{2}{a\sqrt{p^2-q^2}} \tan^{-1}\sqrt{(p-q)/(p+q)}\ \tan \tfrac{1}{2}au + C, & |p| > |q| \\[3mm] \dfrac{1}{a\sqrt{q^2-p^2}} \ln\left[\dfrac{\tan \tfrac{1}{2}au + \sqrt{(q+p)/(q-p)}}{\tan \tfrac{1}{2}au - \sqrt{(q+p)/(q-p)}}\right] + C, & |p| < |q| \end{cases}$

120. $\displaystyle \int u^m \cos au\ du = \frac{u^m \sin au}{a} + \frac{mu^{m-1}}{a^2} \cos au - \frac{m(m-1)}{a^2} \int u^{m-2} \cos au\ du$

121. $\displaystyle \int \cos^n au\ du = \frac{\sin au \cos^{n-1} au}{an} + \frac{n-1}{n} \int \cos^{n-2} au\ du$

122. $\displaystyle \int \frac{du}{\cos^n au} = \frac{\sin au}{a(n-1)\cos^{n-1} au} + \frac{n-2}{n-1} \int \frac{du}{\cos^{n-2} au}$

123. $\displaystyle \int \sin au \cos au\ du = \frac{\sin^2 au}{2a} + C$

124. $\displaystyle \int \sin pu \cos qu\ du = -\frac{\cos(p-q)u}{2(p-q)} - \frac{\cos(p+q)u}{2(p+q)} + C,\quad p \neq \pm q$

125. $\displaystyle \int \sin^n au \cos au\ du = \frac{\sin^{n+1} au}{(n+1)a} + C,\quad n \neq -1$

126. $\displaystyle \int \cos^n au \sin au\ du = -\frac{\cos^{n+1} au}{(n+1)a} + C,\quad n \neq -1$

127. $\displaystyle \int \sin^2 au \cos^2 au\ du = \frac{u}{8} - \frac{\sin 4au}{32a} + C$ **128.** $\displaystyle \int \frac{du}{\sin au \cos au} = \frac{1}{a} \ln|\tan au| + C$

129. $\displaystyle \int \frac{du}{\cos au(1 \pm \sin au)} = \mp\frac{1}{2a(1 \pm \sin au)} + \frac{1}{2a} \ln\left|\tan\left(\frac{au}{2} + \frac{\pi}{4}\right)\right| + C$

130. $\int \dfrac{du}{\sin au(1 \pm \cos au)} = \pm \dfrac{1}{2a(1 \pm \cos au)} + \dfrac{1}{2a}\ln\left|\tan\dfrac{au}{2}\right| + C$

131. $\int \dfrac{du}{\sin au \pm \cos au} = \dfrac{1}{a\sqrt{2}}\ln\left|\tan\left(\dfrac{au}{2} \pm \dfrac{\pi}{8}\right)\right| + C$

132. $\int \dfrac{\sin au\,du}{\sin au \pm \cos au} = \dfrac{u}{2} \mp \dfrac{1}{2a}\ln|\sin au \pm \cos au| + C$

133. $\int \dfrac{\cos au\,du}{\sin au \pm \cos au} \pm \dfrac{u}{2} + \dfrac{1}{2a}\ln|\sin au \pm \cos au| + C$

134. $\int \dfrac{\sin au\,du}{p + q\cos au} = -\dfrac{1}{aq}\ln|p + q\cos au| + C$ **135.** $\int \dfrac{\cos au\,du}{p + q\sin au} = \dfrac{1}{aq}\ln|p + q\sin au| + C$

136. $\int \sin^m au\cos^n au\,du = \begin{cases} -\dfrac{\sin^{m-1}au\cos^{n+1}au}{a(m+n)} + \dfrac{m-1}{m+n}\displaystyle\int \sin^{m-2}au\cos^n au\,du,\ m \neq -n \\[3mm] \dfrac{\sin^{m+1}au\cos^{n-1}au}{a(m+n)} + \dfrac{n-1}{m+n}\displaystyle\int \sin^m au\cos^{n-2}au\,du,\ m \neq -n \end{cases}$

137. $\int \tan au\,du = -\dfrac{1}{a}\ln|\cos au| = \dfrac{1}{a}\ln|\sec au| + C$ **138.** $\int \tan^2 au\,du = \dfrac{\tan au}{a} - u + C$

139. $\int \tan^n au\sec^2 au\,du = \dfrac{\tan^{n+1}au}{(n+1)a} + C,\ n \neq -1$ **140.** $\int \tan^n au\,du = \dfrac{\tan^{n-1}au}{(n-1)a} - \int \tan^{n-2}au\,du + C,\ n \neq 1$

141. $\int \cot au\,du = \dfrac{1}{a}\ln|\sin au| + C$ **142.** $\int \cot^2 au\,du = -\dfrac{\cot au}{a} - u + C$

143. $\int \cot^n au\csc^2 au\,du = -\dfrac{\cot^{n+1}au}{(n+1)a} + C,\ n \neq -1$ **144.** $\int \cot^n au\,du = -\dfrac{\cot^{n-1}au}{(n-1)a} - \int \cot^{n-2}au\,du,\ n \neq 1$

145. $\int \sec au\,du = \dfrac{1}{a}\ln|\sec au + \tan au| = \dfrac{1}{a}\ln\left|\tan\left(\dfrac{au}{2} + \dfrac{\pi}{4}\right)\right| + C$

146. $\int \sec^2 au\,du = \dfrac{\tan au}{a} + C$

147. $\int \sec^3 au\,du = \dfrac{\sec au\tan au}{2a} + \dfrac{1}{2a}\ln|\sec au + \tan au| + C$

148. $\int \sec^n au\tan au\,du = \dfrac{\sec^n au}{na} + C$

149. $\int \sec^n au\,du = \dfrac{\sec^{n-2}au\tan au}{a(n-1)} + \dfrac{n-2}{n-1}\int \sec^{n-2}au\,du,\ n \neq 1$

150. $\int \csc au\,du = \dfrac{1}{a}\ln|\csc au - \cot au| = \dfrac{1}{a}\ln\left|\tan\dfrac{au}{2}\right| + C$

151. $\int \csc^2 au\,du = -\dfrac{\cot au}{a} + C$ **152.** $\int \csc^n au\cot au\,du = -\dfrac{\csc^n au}{na} + C$

153. $\int \csc^n au\,du = -\dfrac{\csc^{n-2}au\cot au}{a(n-1)} + \dfrac{n-2}{n-1}\int \csc^{n-2}au\,du,\ n \neq 1$

INTEGRALS INVOLVING INVERSE TRIGONOMETRIC FUNCTIONS

154. $\displaystyle\int \sin^{-1}\frac{u}{a}\,du = u\sin^{-1}\frac{u}{a} + \sqrt{a^2 - u^2} + C$

155. $\displaystyle\int u\sin^{-1}\frac{u}{a}\,du = \left(\frac{u^2}{2} - \frac{a^2}{4}\right)\sin^{-1}\frac{u}{a} + \frac{u\sqrt{a^2 - u^2}}{4} + C$

156. $\displaystyle\int \cos^{-1}\frac{u}{a}\,du = u\cos^{-1}\frac{u}{a} - \sqrt{a^2 - u^2} + C$

157. $\displaystyle\int u\cos^{-1}\frac{u}{a}\,du = \left(\frac{u^2}{2} - \frac{a^2}{4}\right)\cos^{-1}\frac{u}{a} - \frac{u\sqrt{a^2 - u^2}}{4} + C$

158. $\displaystyle\int \tan^{-1}\frac{u}{a}\,du = u\tan^{-1}\frac{u}{a} - \frac{a}{2}\ln(u^2 + a^2) + C$ **159.** $\displaystyle\int u\tan^{-1}\frac{u}{a}\,du = \frac{1}{2}(u^2 + a^2)\tan^{-1}\frac{u}{a} - \frac{au}{2} + C$

160. $\displaystyle\int u^m\sin^{-1}\frac{u}{a}\,du = \frac{u^{m+1}}{m+1}\sin^{-1}\frac{u}{a} - \frac{1}{m+1}\int \frac{u^{m+1}}{\sqrt{a^2 - u^2}}\,du$

161. $\displaystyle\int u^m\cos^{-1}\frac{u}{a}\,du = \frac{u^{m+1}}{m+1}\cos^{-1}\frac{u}{a} + \frac{1}{m+1}\int \frac{u^{m+1}}{\sqrt{a^2 - u^2}}\,du$

162. $\displaystyle\int u^m\tan^{-1}\frac{u}{a}\,du = \frac{u^{m+1}}{m+1}\tan^{-1}\frac{u}{a} - \frac{a}{m+1}\int \frac{u^{m+1}}{u^2 + a^2}\,du$

INTEGRALS INVOLVING e^{au}

163. $\displaystyle\int e^{au}\,du = \frac{e^{au}}{a} + C$ **164.** $\displaystyle\int ue^{au}\,du = \frac{e^{au}}{a}\left(u - \frac{1}{a}\right) + C$

165. $\displaystyle\int u^2 e^{au}\,du = \frac{e^{au}}{a}\left(u^2 - \frac{2u}{a} + \frac{2}{a^2}\right) + C$

166. $\displaystyle\int u^n e^{au}\,du = \frac{u^n e^{au}}{a} - \frac{n}{a}\int u^{n-1}e^{au}\,du$

$\displaystyle\qquad\qquad = \frac{e^{au}}{a}\left[u^n - \frac{nu^{n-1}}{a} + \frac{n(n-1)u^{n-2}}{a^2} - \cdots \frac{(-1)^n n!}{a^n}\right]$ if n is a positive integer

167. $\displaystyle\int \frac{du}{p + qe^{au}} = \frac{u}{p} - \frac{1}{ap}\ln|p + qe^{au}| + C$

168. $\displaystyle\int e^{au}\sin bu\,du = \frac{e^{au}(a\sin bu - b\cos bu)}{a^2 + b^2} + C$

169. $\displaystyle\int e^{au}\cos bu\,du = \frac{e^{au}(a\cos bu + b\sin bu)}{a^2 + b^2} + C$

170. $\displaystyle\int ue^{au}\sin bu\,du = \frac{ue^{au}(a\sin bu - b\cos bu)}{a^2 + b^2} - \frac{e^{au}[(a^2 - b^2)\sin bu - 2ab\cos bu]}{(a^2 + b^2)^2} + C$

171. $\displaystyle\int ue^{au}\cos bu\,du = \frac{ue^{au}(a\cos bu + b\sin bu)}{a^2 + b^2} - \frac{e^{au}[(a^2 - b^2)\cos bu + 2ab\sin bu]}{(a^2 + b^2)^2} + C$

172. $\displaystyle\int e^{au}\sin^n bu\,du = \frac{e^{au}\sin^{n-1}bu}{a^2 + n^2b^2}(a\sin bu - nb\cos bu) + \frac{n(n-1)b^2}{a^2 + n^2b^2}\int e^{au}\sin^{n-2}bu\,du$

173. $\int e^{au} \cos^n bu \, du = \dfrac{e^{au} \cos^{n-1} bu}{a^2 + n^2 b^2} \, (a \cos bu + nb \sin bu) + \dfrac{n(n-1)b^2}{a^2 + n^2 b^2} \int e^{au} \cos^{n-2} bu \, du$

INTEGRALS INVOLVING ln u

174. $\int \ln u \, du = u \ln u - u + C$

175. $\int u \ln u \, du = \dfrac{u^2}{2} (\ln u - \tfrac{1}{2}) + C$

176. $\int u^m \ln u \, du = \dfrac{u^{m+1}}{m+1} \left(\ln u - \dfrac{1}{m+1} \right)$ if $m \neq -1$

177. $\int \dfrac{\ln u}{u} \, du = \dfrac{1}{2} \ln^2 u + C$

178. $\int \dfrac{\ln^n u \, du}{u} = \dfrac{\ln^{n+1} u}{n+1} + C$ if $n \neq -1$

179. $\int \dfrac{du}{u \ln u} = \ln |\ln u| + C$

180. $\int \ln^n u \, du = u \ln^n u - n \int \ln^{n-1} u \, du + C$

181. $\int u^m \ln^n u \, du = \dfrac{u^{m+1} \ln^n u}{m+1} - \dfrac{n}{m+1} \int u^m \ln^{n-1} u \, du + C$ if $m \neq -1$

182. $\int \ln(u^2 + a^2) \, du = u \ln(u^2 + a^2) - 2u + 2a \tan^{-1} \dfrac{u}{a} + C$

183. $\int \ln |u^2 - a^2| \, du = u \ln |u^2 - a^2| - 2u + a \ln \left| \dfrac{u+a}{u-a} \right| + C$

INTEGRALS INVOLVING HYPERBOLIC FUNCTIONS

184. $\int \sinh au \, du = \dfrac{\cosh au}{a} + C$

185. $\int u \sinh au \, du = \dfrac{u \cosh au}{a} - \dfrac{\sinh au}{a^2} + C$

186. $\int \cosh au \, du = \dfrac{\sinh au}{a} + C$

187. $\int u \cosh au \, du = \dfrac{u \sinh au}{a} - \dfrac{\cosh au}{a^2} + C$

188. $\int \cosh^2 au \, du = \dfrac{u}{2} + \dfrac{\sinh au \cosh au}{2a} + C$

189. $\int \sinh^2 au \, du = \dfrac{\sinh au \cosh au}{2a} - \dfrac{u}{2} + C$

190. $\int \sinh^n au \, du = \dfrac{\sinh^{n-1} au \cosh au}{an} - \dfrac{n-1}{n} \int \sinh^{n-2} au \, du$

191. $\int \cosh^n au \, du = \dfrac{\cosh^{n-1} au \sinh au}{an} + \dfrac{n-1}{n} \int \cosh^{n-2} au \, du$

192. $\int \sinh au \cosh au \, du = \dfrac{\sinh^2 au}{2a} + C$

193. $\int \sinh pu \cosh qu \, du = \dfrac{\cosh(p+q)u}{2(p+q)} + \dfrac{\cosh(p-q)u}{2(p-q)} + C$

194. $\int \tanh au \, du = \dfrac{1}{a} \ln \cosh au + C$

195. $\int \tanh^2 au \, du = u - \dfrac{\tanh au}{a} + C$

196. $\int \tanh^n au \, du = \dfrac{-\tanh^{n-1} au}{a(n-1)} + \int \tanh^{n-2} au \, du$

197. $\int \coth au \, du = \dfrac{1}{a} \ln |\sinh au| + C$

198. $\displaystyle\int \coth^2 au\,du = u - \frac{\coth au}{a} + C$

199. $\displaystyle\int \text{sech}\,au\,du = \frac{2}{a}\tan^{-1}e^{au} + C$

200. $\displaystyle\int \text{sech}^2 au\,du = \frac{\tanh au}{a} + C$

201. $\displaystyle\int \text{sech}^n au\,du = \frac{\text{sech}^{n-2}au\,\tanh au}{a(n-1)} + \frac{n-2}{n-1}\int \text{sech}^{n-2}au\,du$

202. $\displaystyle\int \text{csch}\,au\,du = \frac{1}{a}\ln\left|\tanh\frac{au}{2}\right| + C$

203. $\displaystyle\int \text{csch}^2 au\,du = -\frac{\coth au}{a} + C$

204. $\displaystyle\int \text{sech}\,u\,\tanh u\,du = -\text{sech}\,u + C$

205. $\displaystyle\int \text{csch}\,u\,\coth u\,du = -\text{csch}\,u + C$

SOME DEFINITE INTEGRALS

Unless otherwise stated, all letters stand for positive numbers.

206. $\displaystyle\int_0^\infty \frac{dx}{x^2 + a^2} = \frac{\pi}{2a}$

207. $\displaystyle\int_0^\infty \frac{x^{p-1}}{1+x}\,dx = \frac{\pi}{\sin p\pi}$

208. $\displaystyle\int_0^a \frac{dx}{\sqrt{a^2 - x^2}} = \frac{\pi}{2}$

209. $\displaystyle\int_0^a \sqrt{a^2 - x^2}\,dx = \frac{\pi a^2}{4}$

210. $\displaystyle\int_0^\pi \sin mx \sin nx\,dx = \begin{cases} 0, & \text{if } m, n \text{ integers and } m \neq n \\ \dfrac{\pi}{2}, & \text{if } m, n \text{ integers and } m = n \end{cases}$

211. $\displaystyle\int_0^\pi \cos mx \cos nx\,dx = \begin{cases} 0, & \text{if } m, n \text{ integers and } m \neq n \\ \dfrac{\pi}{2}, & \text{if } m, n \text{ integers and } m = n \end{cases}$

212. $\displaystyle\int_0^\pi \sin mx \cos nx\,dx = \begin{cases} 0, & \text{if } m, n \text{ integers and } m + n \text{ is odd, or } m = \pm n \\ \dfrac{2m}{(m^2 - n^2)}, & \text{if } m, n \text{ integers and } m + n \text{ is even} \end{cases}$

213. $\displaystyle\int_0^{\pi/2} \sin^2 x\,dx = \int_0^{\pi/2} \cos^2 x\,dx = \frac{\pi}{4}$

214. $\displaystyle\int_0^\infty e^{-ax}\cos bx\,dx = \frac{a}{a^2 + b^2}$

215. $\displaystyle\int_0^\infty e^{-ax}\sin bx\,dx = \frac{b}{a^2 + b^2}$

216. $\displaystyle\int_0^\infty e^{-a^2x^2}\,dx = \frac{\sqrt{\pi}}{2a}$

217. $\displaystyle\int_0^{\pi/2} \sin^{2m} x\,dx = \int_0^{\pi/2} \cos^{2m} x\,dx = \frac{1\cdot 3\cdot 5\cdots(2m-1)}{2\cdot 4\cdot 6\cdots 2m}\frac{\pi}{2}, \quad m = 1, 2, 3, \ldots$

218. $\displaystyle\int_0^{\pi/2} \sin^{2m+1} x\,dx = \int_0^{\pi/2} \cos^{2m+1} x\,dx = \frac{2\cdot 4\cdot 6\cdots 2m}{1\cdot 3\cdot 5\cdots(2m+1)}, \quad m = 1, 2, 3, \ldots$

219. $\displaystyle\int_0^\infty \frac{e^{-x}}{\sqrt{x}}\,dx = \sqrt{\pi}$

220. $\displaystyle\int_0^1 x^m(\ln x)^n\,dx = \frac{(-1)^n n!}{(m+1)^{n+1}}$

Answers to Odd-Numbered Problems and Review Exercises

CHAPTER ONE
PROBLEMS 1.1
1. $-3 < 2$ 3. $\pi/3 > 1$ 5. $\sqrt{8} < 3.1$
9. (a) $\{x : x = 2n \text{ or } x = 5n, \, n \text{ an integer}\}$ (b) $\{x : x = 15n, \, n \text{ an integer}\}$
(c) $\{x : x = 2n \text{ or } x = 3n \text{ or } x = 5n, \, n \text{ an integer}\}$ 13. (a) bounded (b) bounded
(c) bounded below (d) bounded above (e) bounded below (f) bounded
15. (a) A (b) B (c) $\{x : x \text{ a positive integer and } x \neq 3n, \, n \text{ an integer}\}$
(d) $\{x : x = 3n, \, n \text{ a nonnegative integer}\}$ 17. (a) $(-\infty, -1)$ (b) $[5, 10]$
19. $[1, 2] \cup [3, 5]$ 23. (c), (d), (f) are true; (a), (b), (e) are false

PROBLEMS 1.2
1. $(-\infty, 7)$ 3. $[-\frac{1}{2}, 1]$ 5. $(-\frac{11}{5}, -\frac{1}{5})$ 7. $(\frac{3}{2}, \infty)$ 9. $[2, \frac{8}{3})$ 11. $(0, \frac{1}{3})$
13. $(-\infty, 0) \cup (\frac{1}{10}, \infty)$ 15. $(-\infty, -2] \cup [7, \infty)$ 17. (a) 1 (b) 4 (c) -1
25. $(-\infty, -2) \cup (2, \infty)$ 27. $[-7, 1]$ 29. $(-\frac{7}{2}, -\frac{1}{2})$ 31. $[\frac{2}{3}, \frac{14}{3}]$ 33. $(-\infty, \infty)$
35. $(-\infty, -5) \cup (-3, \infty)$ 37. $(-\infty, -3) \cup (2, 7)$ 39. $((-c - b)/a, (c - b)/a)$ if
$c > 0$ and $\emptyset$ if $c \leq 0$ 41. $(-\infty, \infty)$ 43. $(\frac{5}{4}, \infty)$ 45. (a) $|x - 3| < 7$ (b) $|x| \geq 3$
(c) $|x - 8| > 3$ (d) $|x - 5.5| \geq 3.5$ (e) $|x + 3| < 7$ 53. no, $[1.5625, 1.8225] =$
$[(1.25)^2, (1.35)^2]$

PROBLEMS 1.3
1. (a) IV (b) III (c) III (d) I 3. $\sqrt{122}$ 5. $\sqrt{2}/6$ 7. $|b - a|\sqrt{2}$

17. $(x - 1)^2 + y^2 = 1$ **19.** $(x - 1)^2 + (y - 1)^2 = 2$ **21.** $(x + 1)^2 + (y - 4)^2 = 25$
23. $(x - \pi)^2 + (y - 2\pi)^2 = \pi$ **25.** $(x - 3)^2 + (y + 2)^2 = 16$

PROBLEMS 1.4

1. -2 **3.** 1 **5.** $-7/13$ **7.** 0 **9.** -1 **11.** $(d - b)/(c - a)$
13. collinear **15.** collinear **17.** collinear **19.** $1/\sqrt{3}$ **21.** -1 **23.** $-1/\sqrt{3}$
25. 0.0087268677 **27.** 114.5886501 **29.** 0.7535540501 **31.** -2.747477419
33. $45°$ **35.** $150°$ **37.** $87.70938996°$ **39.** $73.61045967°$ **41.** $56.30993247°$
43. $113.49856568°$ **45.** parallel **47.** perpendicular
49. neither parallel nor perpendicular **51.** neither **53.** perpendicular

PROBLEMS 1.5

In Problems 1–11 one point-slope equation is given first (using the first point given). Then the slope-intercept form is given.
1. $y - 2 = 2(x - 1); y = 2x$ **3.** $y - 7 = \frac{1}{2}(x - 3); y = \frac{1}{2}x + \frac{11}{2}$
5. $x = -3$ (line is vertical) **7.** $y + 4 = \frac{11}{5}(x + 2); y = \frac{11}{5}x + \frac{2}{5}$
9. $y + 3 = -\frac{4}{3}(x - 7); y = -\frac{4}{3}x + \frac{19}{3}$ **11.** $y - b = [(d - b)/(c - a)](x - a);$
$y = [(d - b)/(c - a)]x + [(bc - ad)/(c - a)]$ **13.** $y - 1 = -\frac{2}{5}(x + 1); y = -\frac{2}{5}x + \frac{3}{5}$
15. $-3x + y = 1$ **17.** $y - (b/a)x = \beta - (\alpha b/a)$ **19.** none (lines are parallel)
21. all points on the line $2x - 3y = 5$ (lines are the same) **23.** $(\frac{67}{45}, \frac{2}{15})$
25. $y = 1$ **27.** $x + 2y = 4$
31. $2\sqrt{R^2 - (a - b)^2/2}$ provided $|a - b| \leq \sqrt{2}|R|$; otherwise the circles do not intersect
33. $x^2 - 3x + y^2 - 22y = 0$ **35.** $\sqrt{13}/13$ **37.** $\sqrt{61}/5$ **39.** $\sqrt{5}$
43. for example, the "straight line" determined by $(-2, 3)$ and $(5, 4)$ is
$\{(x, y): x \leq -2 \text{ and } y \leq 3\} \cup \{(x, y): -2 \leq x \leq 5 \text{ and } 3 \leq y \leq 4\} \cup \{(x, y): 5 \leq x \text{ and } 4 \leq y\}$

PROBLEMS 1.6

1. $f(0) = 5, f(1) = 5, f(3) = 29, f(-2) = -1$ **3.** $f(0) = 1, f(1) = 2, f(16) = 5, f(25) = 6$
5. yes **7.** yes **9.** no **11.** yes **13.** yes **15.** yes
19. domain $g = \mathbb{R}$, range $g = \mathbb{R}$ **21.** domain $h = \mathbb{R} - \{0\}$, range $h = (0, \infty)$
23. $\mathbb{R} - \{-1\}, \mathbb{R} - \{0\}$ **25.** $[1, \infty), [0, \infty)$ **27.** $\mathbb{R} - \{0\}, (0, \infty)$ **29.** $\mathbb{R}, [0, \infty)$
31. $\mathbb{R}, [0, \infty)$ **33.** $\mathbb{R} - \{0\}, \mathbb{R} - \{5\}$ **35.** $\mathbb{R}, (3, 4]$ **37.** $\mathbb{R}, (0, 1]$ **39.** $\mathbb{R}, [0, 1)$
41. $\mathbb{R}, [0, 1)$ if n even; $\mathbb{R} - \{-1\}, \mathbb{R} - \{1\}$ if n odd **43.** $\mathbb{R}, [-\sqrt[3]{\frac{1}{4}}, \infty)$
45. $(-\infty, 3) - \{0\}, (-\infty, 0) \cup (\frac{1}{3}, \infty)$
47. $f(x + \Delta x) = (x + \Delta x)^2, \dfrac{f(x + \Delta x) - f(x)}{\Delta x} = 2x + \Delta x, \Delta x \neq 0$
49. $f(-5x) = \dfrac{-5x}{(-5x)^5 + 3}, f\left(\dfrac{1}{x}\right) = \dfrac{x^4}{1 + 3x^5}, f(x^{10} - 25) = \dfrac{x^{10} - 25}{(x^{10} - 25)^5 + 3},$
$f(g(t)) = \dfrac{g(t)}{[g(t)]^5 + 3}, f(x + \Delta x) = \dfrac{x + \Delta x}{(x + \Delta x)^5 + 3}$
51. domain $f = \mathbb{R} - \{0\}$, range $f = \{-1, 1\}$
53. $A = A(W) = W(25 - W)$, domain $A = (0, 25)$, range $A = (0, 156.25]$.
55. (a) $T(x) = 6692 + 0.4(x - 28{,}800)$ dollars (c) \$8272
57. $d(t) = \begin{cases} \sqrt{90^2 + (90 - 30t)^2} \text{ ft}, & 0 \leq t < 3 \\ |180 - 30t| \text{ ft}, & 3 \leq t \leq 9 \\ \sqrt{90^2 + (270 - 30t)^2} \text{ ft}, & 9 < t \leq 12 \end{cases}$

59. the conversion of u U.S. dollars into c Canadian dollars is $c = V(u) = (1 + 0.12)u$ and the conversion of d Canadian dollars into t U.S. dollars is $t = H(d) = (1 - 0.12)d$; therefore $H(V(u)) = (1 - 0.12)(1 + 0.12)u = 0.9856u$

PROBLEMS 1.7

For 1–8, answers are given in the order $f + g$, $f - g$, $f \cdot g$, f/g.

1. $-2x - 5$, $\mathbb{R}$; $6x - 5$, $\mathbb{R}$; $-8x^2 + 20x$, $\mathbb{R}$; $(2x - 5)/(-4x)$, $\mathbb{R} - \{0\}$

3. $\sqrt{x + 2} + \sqrt{2 - x}$, $[-2, 2]$; $\sqrt{x + 2} - \sqrt{2 - x}$, $[-2, 2]$; $\sqrt{4 - x^2}$, $[-2, 2]$; $\sqrt{(x + 2)/(2 - x)}$, $[-2, 2)$ **5.** $2 + x^5 - |x|$, $\mathbb{R}$; $x^5 + |x|$, $\mathbb{R}$; $1 + x^5 - |x| - |x|x^5$, $\mathbb{R}$; $(1 + x^5)/(1 - |x|)$, $\mathbb{R} - \{-1, 1\}$ **7.** $\sqrt[5]{x + 2} + \sqrt[4]{x - 3}$, $[3, \infty)$; $\sqrt[5]{x + 2} - \sqrt[4]{x - 3}$, $[3, \infty)$; $\sqrt[5]{x + 2} \sqrt[4]{x - 3}$, $[3, \infty)$; $\sqrt[5]{x + 2}/\sqrt[4]{x - 3}$, $(3, \infty)$

For 9–20, answers are given in the order $f \circ g$ and $g \circ f$.

9. $2x + 1$, $\mathbb{R}$; $2x + 2$, $\mathbb{R}$ **11.** $15x + 11$, $\mathbb{R}$; $15x + 27$, $\mathbb{R}$

13. $(x - 1)/(3x - 1)$, $\mathbb{R} - \{0, \frac{1}{3}\}$; $-2/x$, $\mathbb{R} - \{-2, 0\}$

15. $\sqrt{1 - \sqrt{x - 1}}$, $[1, 2]$; $\sqrt{\sqrt{1 - x} - 1}$, $(-\infty, 0]$

17. $1/\sqrt{x^5 + 1}$, $(-1, \infty)$; $(x + 1)^{-5/2}$, $(-1, \infty)$

19. $\begin{cases} -6x, & x \geq 0 \\ 10x, & x < 0 \end{cases}$, $\mathbb{R}$; $\begin{cases} -3x, & x \geq 0 \\ 10x, & x < 0 \end{cases}$, $\mathbb{R}$

23. four different functions are $x - 5$, $|x - 5|$, $5 - x$, $-|x - 5|$ **25.** $x^3 + 1$

27. $f(x) = \sqrt[3]{x}$, $g(x) = x^2 - 5$

29. domain $k = [0, \infty)$; two solutions are $h(x) = \sqrt{x}$, $g(y) = 1 + y$, $f(z) = z^{5/7}$; $H(x) = 1 + \sqrt{x}$, $G(y) = y^{1/14}$, $F(z) = z^{10}$

PROBLEMS 1.8

1.

(a)   (b)   (c)

3.

(a)   (b)   (c)

(d)

(e)

5.

(a)

(b)

(c)

(d)

(e)

7.

(a)

(b)

(c)

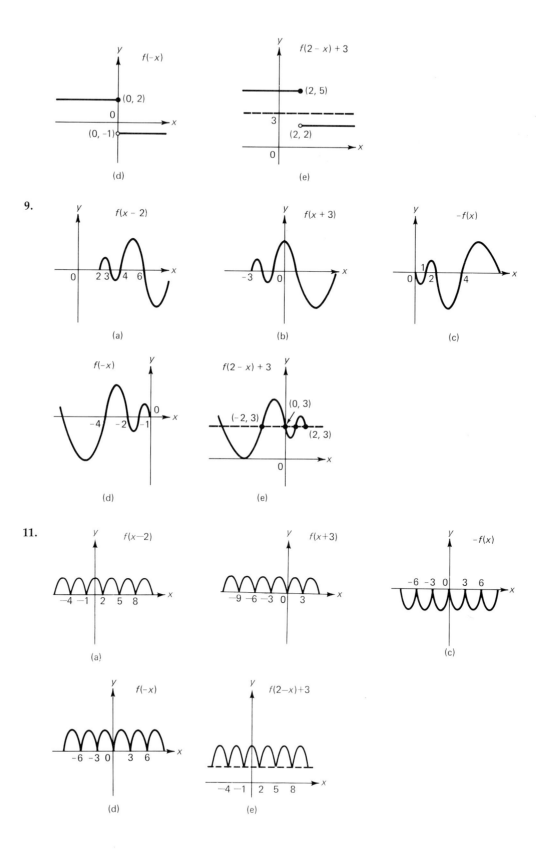

$f(-x)$

(0, 2)

0

(0, –1)

(d)

$f(2 - x) + 3$

(2, 5)

3

(2, 2)

0

(e)

9.

$f(x - 2)$

0   2 3 4 6

(a)

$f(x + 3)$

–3   0

(b)

$-f(x)$

1

0 2   4

(c)

$f(-x)$

–4   –2   –1

0

(d)

$f(2 - x) + 3$

(–2, 3)   (0, 3)

(2, 3)

0

(e)

11.

$f(x-2)$

–4 –1   2   5   8

(a)

$f(x+3)$

–9 –6 –3 0   3

(b)

$-f(x)$

–6 –3 0   3   6

(c)

$f(-x)$

–6 –3 0   3   6

(d)

$f(2-x)+3$

–4 –1   2   5   8

(e)

ΧERCISES FOR CHAPTER ONE

1. $A \cup B = \mathbb{R}; A \cup C = A; B \cup C = B; A \cap B = (-4, 2]; A \cap C = C; B \cap C = C;$
$\overline{A} = (2, \infty); \overline{B} = (-\infty, -4]; \overline{C} = (-\infty, -1) \cup (1, \infty); A - B = (-\infty, -4];$
$B - A = (2, \infty); A - C = (-\infty, -1) \cup (1, 2]; C - A = \emptyset;$
$B - C = (-4, -1) \cup (1, \infty); C - B = \emptyset$ **3.** $(-\infty, -4] \cup [4, \infty)$
5. $(-\infty, -4) \cup (-2, \infty)$ **7.** $\mathbb{R}$ **9.** $(-3, 2) \cup (\frac{11}{3}, \infty)$ **11.** $(1, -3)$
13. $y = \frac{2}{3}x + \frac{11}{3}, 33.7°$ **15.** $y = x - 4, 45°$ **17.** $x = 1, 90°$ **19.** $2x - 5y = 18$
21. $3/\sqrt{10}$ **23.** $y = -\frac{1}{3}x + \frac{11}{3}$ **25.** yes; $\mathbb{R}; [-8, \infty)$ **27.** no **29.** yes; $\mathbb{R}; [\frac{1}{6}, \infty)$
31. yes; $\mathbb{R} - \{-1, 1\}, \mathbb{R}$
33. yes; $f(2) = 0, f(-\sqrt{5}) = 1, f(x + 4) = \sqrt{x^2 + 8x + 12};$
$f(x^3 - 2) = \sqrt{x^6 - 4x^3} = |x|\sqrt{x^4 - 4x}, f(-1/x) = \sqrt{(1/x^2) - 4} = \sqrt{1 - 4x^2}/|x|$
35. $\sqrt{x + 1} + x^3, [-1, \infty); \sqrt{x + 1} - x^3, [-1, \infty); x^3\sqrt{x + 1}, [-1, \infty);$
$x^3/\sqrt{x + 1}, (-1, \infty); \sqrt{x^3 + 1}, [-1, \infty); (x + 1)^{3/2}, [-1, \infty)$ **37.** $\frac{1}{4}x + \frac{3}{2}$

39.

(a)   (b)   (c)

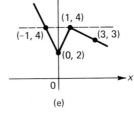

(d)   (e)

CHAPTER TWO
PROBLEMS 2.2

1. (a) (b) $f(3) = 10, f(1) = 8, f(2.5) = 9.5, f(1.5) = 8.5, f(2.1) = 9.1, f(1.9)$
$= 8.9, f(2.01) = 9.01, f(1.99) = 8.99$ (c) 9

3. (a) 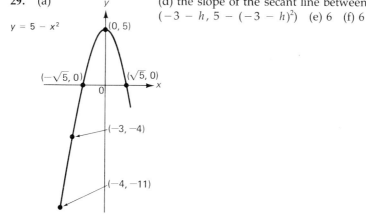 (b) $f(-0.5) = 5.75$, $f(-1.5) = 10.75$, $f(-0.9) = 7.51$, $f(-1.1) = 8.51$, $f(-0.99) = 7.9501$, $f(-1.01) = 8.0501$ (c) 8

5. (a) can't divide by zero (b) $g(x) = x^2 + 2x + 4$ (c) 12 **7.** yes; $\sqrt{2}$ **9.** 45
11. $\frac{1}{2}$ **13.** 0 **15.** 0 **17.** 3 **19.** 2 **21.** doesn't exist **23.** 1 **25.** $\frac{1}{2}$
27. (a) $f(-1) \approx 0.49487$, $f(-3)$ is undefined, $f(-1.5) \approx 0.47730$, $f(-2.5)$ is undefined, $f(-1.9) \approx 0.40625$, $f(-2.1) \approx 0.32774$, $f(-1.99) \approx 0.37647$, $f(-2.01) \approx 0.36877$, $f(-1.999) \approx 0.37306$, $f(-2.001) \approx 0.37229$ (b) ≈ 0.3725 (c) $f(-2) = \sqrt{5}/6 \approx 0.37268$
29. (a) (d) the slope of the secant line between $(-3, -4)$ and $(-3 - h, 5 - (-3 - h)^2)$ (e) 6 (f) 6

31. (a) (b) 0

$y = |3x - 4|$

33. (a) (b) get different answers as 2 is approached from the left and right (see Example 10) (c) 1, -1

$y = \dfrac{|2x - 4|}{2x - 4}$

35. (a) $A(x) = \begin{cases} 7 + |x - 2|, & \text{if } x \neq 2 \\ 8, & \text{if } x = 1 \end{cases}$ (b) $b(z) = 3 - z$, $B(z) = 6 - 2z$

(c) $c(x) = 22/7\pi$ for all x, $C(x) = \begin{cases} 22/7\pi, & \text{if } x \neq -5 \\ 1, & \text{if } x = 1 \end{cases}$ **37.** 0 **39.** 1

41. 0 **43.** does not exist **45.** 0 **47.** -1 **49.** -6 **51.** 0
53. 1 **55.** -7 **57.** 0, 0 **59.** both one-sided limits $= 2$

PROBLEMS 2.3
1. 2 **3.** 3 **5.** 6 **7.** -4 **9.** 6 **11.** -7 **13.** 2 **15.** -34 **17.** 1
19. $\frac{1}{2}$ **21.** $\frac{1}{2}$ **23.** 1 **25.** $\frac{5}{4}$ **27.** $\frac{1}{2}$ **29.** 5 **31.** $\lim_{x \to -3} f(x) = 41$

PROBLEMS 2.4
1. ∞ **3.** ∞ **5.** no limit **7.** ∞ **9.** no limit **11.** 1 **13.** 0 **15.** -1
17. $\frac{2}{3}$ **19.** 0 **21.** 0 **23.** 0 **25.** $\frac{3}{7}$ **27.** ∞ **29.** 1 **31.** $-\infty$
33. $-\infty$ **35.** ∞ **37.** 10,000; 1,000,000; 100,000,000 **41.** (a) -1 (b) 0

PROBLEMS 2.5
1. (a) positive (b) zero (c) negative (d) positive (e) negative
3. negative at x_1; 0 at x_2; positive at x_3; positive at x_4; 0 at x_5; negative at x_6
5. positive at x_1, x_5, x_6, x_{10}, x_{14}; 0 at x_2, x_4, x_7, x_9, x_{11}, x_{13}; negative at x_3, x_8, x_{12}
7. undefined at x_2; positive at x_3
9. positive at x_3; negative at x_1, x_5; undefined at x_2, x_4
11. positive at x_3; negative at x_1, x_5; undefined at x_2, x_4
13. (a) $f(2.5) = 18.75$, $f(2.1) = 13.23$, $f(2.01) = 12.1203$, $f(2.001) = 12.012003$, $f(1.99) = 11.8803$,
$f(1.999) = 11.988003$ (b) 13.5, 12.3, 12.03, 12.003, 11.97, 11.997 (c) $f'(x) = 6x$, $f'(2) = 12$
(d) $y = 12x - 12$ (e)

$y = 3x^2$
(parabola)

$\dfrac{dy}{dx} < 0$ $\dfrac{dy}{dx} > 0$

15. (a) $f(1.5) \approx 6.1237$, $f(1.1) \approx 5.2440$, $f(1.01) \approx 5.0249$, $f(1.001) \approx 5.0025$, $f(0.99) \approx 4.9749$,
$f(0.999) \approx 4.9975$ (b) 2.2474, 2.4404, 2.4938, 2.4994, 2.5100, 2.500
(c) $f'(x) = 5/(2\sqrt{x})$, $f'(1) = \frac{5}{2}$ (d) $y = \frac{5}{2}x + \frac{5}{2}$ (e)

$y = 5\sqrt{x}$

17. $f'(x) = -4$; $y = -4x + 6$ **19.** $f'(x) = 4x - 12$; $y = 4x - 14$
21. $f'(x) = -2x + 3$; $y = 3x + 5$ **23.** $f'(x) = 3x^2$; $y = 3x + 8$
25. $f'(x) = -1/(2\sqrt{x + 3})$; $y = -\frac{1}{6}x - 2$ **27.** $f'(x) = 1/\sqrt{2x}$; $y = \frac{1}{4}x + 2$
29. $f'(x) = 1/(2\sqrt{x}) + \frac{3}{2}\sqrt{x}$; $y = 2x$ **31.** $f'(x) = 4x^3$; $y = 32x - 48$ **37.** $x = \frac{1}{2}$
39. $a = \frac{3}{2}$ **43.** $f'(0) = -1$

PROBLEMS 2.6

1. 9   **3.** $\frac{1}{4}$ **5.** 40   **7.** $v(3) = 420$ ft/sec, $v(10) = 1400$ ft/sec
9. (a) -35 individuals/day   (b) $\sqrt{4000} \approx 63.2$ days   **11.** (a) $80\pi \;\mu\mathrm{m}^2/\mu\mathrm{m}$
(b) $160\pi \;\mu\mathrm{m}^2/\mu\mathrm{m}$ **13.** marginal cost is constant, \$100 per part

PROBLEMS 2.7

1. $(-\infty, \infty)$ **3.** $(0, \infty)$ **5.** $(-\infty, -1), (-1, 1), (1, \infty)$ **7.** $(-\infty, 2), (2, \infty)$
9. $2, -2$ **11.** $f(x) = x[x]$ is continuous on each open interval $(n, n + 1)$ where n is an integer; a bit more can be said; f is continuous on $(-1, 1)$ **13.** 4   **15.**

17. $\frac{1}{3}$ **23.** no on $[-3, 3]$; yes on $[-1, 1]$
25. not continuous   **27.** not continuous   **29.** not continuous   **31.** not continuous
33. (a) $f(x)g(x) = \begin{cases} (x + 3)(x^2 + 6), & x \le 1 \\ (2x + 5)(5x^3 - 1), & x > 1 \end{cases}$ (b) $\lim_{x\to 1-} f(x) = 4$ and $\lim_{x\to 1+} f(x) = 7$
(c) $\lim_{x\to 1-} g(x) = 7$ and $\lim_{x\to 1+} g(x) = 4$ (d) $\lim_{x\to 1-} f(x)g(x) = 28 = \lim_{x\to 1+} f(x)g(x)$
35. if $p(x_0) = q(x_0)$ **9.** $M = \frac{7}{2}, m = 1$ **45.** $F, G,$ and H are continuous everywhere.
47. (a) $[x] + [-x] + 1 = \begin{cases} 1, & \text{if } x \text{ is an integer} \\ 0, & \text{if } x \text{ is not an integer} \end{cases}$ $y = [x] + [-x] + 1$
(b) continuous at each nonintegral real number

PROBLEMS 2.8

In Problems 1–9 any smaller positive values for δ will also work.

1. $\frac{1}{70}, \frac{1}{700}, \delta = \epsilon/7$ **3.** $\frac{1}{50}, \frac{1}{500}, \delta = \epsilon/5$ **5.** $\frac{1}{70}, \frac{1}{700}$ (for $\epsilon = 0.1$, the largest value of δ is 0.016620626; to see this, calculate $\sqrt{9.1}$.), $\delta = \sqrt{9 + \epsilon} - 3$
7. $\frac{1}{11}, \frac{1}{110}$ (If $|x + 1| < \frac{1}{11}$, then $|x| < 1.1$ and $|x(x + 1)| = |x||x + 1| < 1.1/11 = \frac{1}{10}$)
for $\epsilon = 0.1$, the largest value of δ is $\frac{1}{2}(\sqrt{1.4} - 1) \approx 0.0916079783)$, $\delta = (\sqrt{1 + 4\epsilon} - 1)/2$
9. $0.008, 0.0008$ (If $|x - 2| < 0.008$, then
$|x^3 - 8| = |(x^2 + 2x + 4)(x - 2)| < (0.008)[(2.008)^2 + 2(2.008) + 4] \approx 0.096)$ (the largest value of δ is 0.00829885 since $\sqrt[3]{8.1} \approx 2.00829885)$, $\delta = \sqrt[3]{8 + \epsilon} - 2$ **11.** (b) $0 < \delta \le 0.01$
13. (b) $N \ge \sqrt[3]{4000} \approx 15.874$ **27.** let $\delta = 0.01/3$; any smaller (positive) choice is also OK
29. let $\delta = \sqrt{9 + 0.1} - 3$; any smaller positive choice is also OK   **31.** let $\delta = \frac{1}{11}$; any
smaller positive choice is also OK   **33.** let $\delta = \dfrac{1}{10^9 + 10^4}$; any smaller positive choice
is also OK

REVIEW EXERCISES FOR CHAPTER TWO

1. $f(3) = 6, f(1) = 4, f(2.5) = 4.75, f(1.5) = 3.75, f(2.1) = 4.11, f(1.9) = 3.91, f(2.01) = 4.0101,$
$f(1.99) = 3.9901; \lim_{x\to 2}(x^2 - 3x + 6) = 4$ **3.** (a) 0   (b) -108 (c) $\frac{76}{31}$ (d) 0
5. (a) -1 (b) 4   **7.** (a) not defined   (b) 5   (c) $\frac{25}{7}$ (d) 5   (e) -6 (f) $4^5 = 1024$
(g) $\frac{1}{3}$ (h) c/f **9.** $\lim_{x\to 0+} 1/x^3 = \infty$ but $\lim_{x\to 0-} 1/x^3 = -\infty;$
$\lim_{x\to 0+} 1/x^4 = \infty = \lim_{x\to 0-} 1/x^4$ **11.** (a) 0   (b) 0   (c) 0   (d) $\frac{1}{3}$ (e) 0   (f) ∞

13. (a) positive at x_1, x_2, x_6; 0 at x_3, x_5, x_7; negative at x_4, x_8 (b) positive at x_1, x_2, x_{10};
0 at x_3, x_4, x_5, x_7, x_8, x_9; negative at x_6 **15.** $\lim_{x \to -1+} dy/dx = 1$; $\lim_{x \to -1-} dy/dx = -1$
17. $\mu'(3) = \frac{1}{4}$ kg/m   **19.** $C'(x) = 8 - 0.04x$; yes; C' is a decreasing function
21. $\lim_{x \to -2} f(x) = -1$ **23.** $(0, \infty)$ **25.** $(-\infty, 6)$, (6∞)
27. $(-\infty, -2)$, $(-2, 2)$, $(2, \infty)$ **29.** $(-\infty, -3)$, $(-3, \infty)$ **31.** $(-\infty, -3)$, $(-3, 3)$, $(3, \infty)$;
removable discontinuity at -3

CHAPTER THREE
PROBLEMS 3.1
1. $5x^4$ **3.** $6x + 19$ **5.** $5t^4 + 1/(2\sqrt{t})$ **7.** $100z^{99} + 1000z^9$
9. $24r^7 - 48r^5 - 28r^3 + 4r$ **11.** $y = 4x - 3$ **13.** $y = 5x - 5$ **15.** $y = x + 1$
17. $y = 3x - 8$ **21.** $y = \frac{1}{23}x - \frac{116}{23}$ **23.** $x = 0$ **25.** x values are $(-1 \pm \sqrt{5})/2$
27. $y = 6x - 6$; $y = -2x + 2$ **29.** (a) $3000 - 20\sqrt{t} - 30t + 30t^2$
(b) $-(10/\sqrt{t}) - 30 + 60t$ organisms/hr   (c) 205 organisms/hr; 927.5 organisms/hr
33. $4\pi[1 - (11/12)^3]3^2 \approx 25.98$ in.^{3}/in. of radius   **35.** (a) $x = 0$, $x = 2$ (b) $y = 0$, $y = \frac{64}{81}$
37. $y = x$, $y = -6x$ **39.** $y = 8x - 16$, $y = 2x - 1$

PROBLEMS 3.2
1. $6x^2 + 2$ **3.** $\dfrac{(1 + \sqrt{t})3t^2 - t^3/(2\sqrt{t})}{(1 + \sqrt{t})^2} = \dfrac{3t^2 + \frac{5}{2}t^{5/2}}{(1 + \sqrt{t})^2}$
5. $(1 + x + x^5)(-1 + 6x^5) + (2 - x + x^6)(1 + 5x^4)$
7. $\dfrac{(1 + x + x^5)(-1 + 6x^5) - (2 - x + x^6)(1 + 5x^4)}{(1 + x + x^5)^2}$
9. $1/[\sqrt{t}(1 - \sqrt{t})^2]$ **11.** $5v^4 + 7v^{5/2} - \frac{5}{2}v^{3/2} - 2$
13. $\dfrac{(v^3 - \sqrt{v})[2v + 1/\sqrt{v}] - (v^2 + 2\sqrt{v})[3v^2 - 1/(2\sqrt{v})]}{(v^3 - \sqrt{v})^2}$ (not simplified)
15. $\frac{5}{2}v^{3/2}$ **17.** $-\frac{1}{2}r^{-3/2}$ **19.** $-(9t^4 + 2)/[2t^{3/2}(t^4 + 2)^2]$ **21.** $-6/x^7$
23. $-15/(7x^4)$ **25.** $y = 28x - 20$ **27.** $y = -\frac{7}{2}u + \frac{11}{2}$ **29.** $y = 2x - 1$
31. $y = 4x - 4$
35. $(x^2 + 1)(x^3 + 2)(4x^3) + (x^2 + 1)(x^4 + 3)(3x^2) + (x^3 + 2)(x^4 + 3)(2x)$
37. $(f^2g' + f'g^2)/(f + g)^2$ **39.** $dR/dr = -4\alpha l/(0.2)^5 = -12{,}500\alpha l$
41. $(x^7)' = 7x^6$, $(x \cdot x^6)' = 1 \cdot x^6 + x \cdot 6x^5$, $(x^2 \cdot x^5)' = 2x \cdot x^5 + x^2 \cdot 5x^4$,
$(x^3 \cdot x^4)' = 3x^2 \cdot x^4 + x^3 \cdot 4x^3$

PROBLEMS 3.3
1. $3(x + 1)^2$ **3.** $(2/\sqrt{x})(\sqrt{x} + 2)^3$ **5.** $36x^5(1 + x^6)^5$
7. $5(x^2 - 4x + 1)^4(2x - 4)$ **9.** $-6(t + 1)^2/(t - 1)^4$
11. $2(u^5 + u^4 + u^3 + u^2 + u + 1)(5u^4 + 4u^3 + 3u^2 + 2u + 1)$ **13.** $-8y(y^2 + 3)^{-5}$
15. $10x(x^2 + 2)^4(x^4 + 3)^3 + 12x^3(x^4 + 3)^2(x^2 + 2)^5$
17. $(-3t^2 + 2t - 4)/[\sqrt{t^2 + 1}(t + 2)^5]$
19. $\dfrac{\sqrt{u - 2}\{6u(u^2 + 1)^2(u^2 - 1)^2 + 4u(u^2 - 1)(u^2 + 1)^3\} - \frac{1}{2}(u^2 + 1)^3(u^2 - 1)^2(u - 2)^{-1/2}}{u - 2}$
21. $\left[\dfrac{1}{2\sqrt{x + \sqrt{1 + \sqrt{x}}}}\right]\left[1 + \dfrac{1}{2\sqrt{1 + \sqrt{x}}}\left(\dfrac{1}{2\sqrt{x}}\right)\right]$
23. $-5(y^{-2} + y^{-3} + y^{-7})^{-6}(-2y^{-3} - 3y^{-4} - 7y^{-8})$
25. $-\frac{1}{2}(q^{-7} - q^{-3})^{-3/2}(-7q^{-8} + 3q^{-4})$ **27.** 24 kg/m
31. $p'(x) = (x - 17)[2q(x) + (x - 17)q'(x)]$

PROBLEMS 3.4
1. $\frac{2}{5}x^{-3/5} + \frac{2}{3}x^{-2/3}$ **3.** $\frac{10}{3}x(x^2 + 1)^{2/3}$ **5.** $\frac{3}{10}x^2(x^3 + 3)^{-9/10}$ **7.** $\frac{40}{7}t^9(t^{10} - 2)^{-3/7}$
9. $(u^3 + 3u + 1)^{-2/3}(u^2 + 1)$ **11.** $-\frac{28}{17}[(z^2 - 1)/(z^2 + 1)]^{-24/17}[z/(z^2 + 1)^2]$
13. $\frac{5}{3}r(r^2 + 1)^{-1/6}(r - 1)^{1/2} + \frac{1}{2}(r - 1)^{-1/2}(r^2 + 1)^{5/6}$ **15.** $3\sqrt{2}x^{\sqrt{2}-1}$

17. $2\sqrt{3}t(t^2 + 1)^{\sqrt{3}-1}$ **19.** $4(u^{\sqrt{2}} - u^{\sqrt{3}})^3(\sqrt{2}u^{\sqrt{2}-1} - \sqrt{3}u^{\sqrt{3}-1})$

21. $y = 2x; y = -\frac{1}{2}x + \frac{5}{2}$ **23.** $y = 0; x = 1$

27. (a) $V_{rms} \approx 5.7767$ (b) $dV_{rms}/d\rho = -\frac{1}{2}\sqrt{3P/\rho^3}$ (c) $dV_{rms}/d\rho \approx -32.1285$

PROBLEMS 3.5

1. $3\cos 3x$ **3.** $\cos x - x \sin x$,   **5.** $(3x^2 - 2)\cos(x^3 - 2x + 6)$ **7.** $\cos x/(2\sqrt{\sin x})$

9. $2 \sin x \cos x$ **11.** $6x(\sin^2 x^2)\cos x^2$ **13.** 0 **15.** $(3\sin^2 x + \sin^4 x)/\cos^5 x$

17. $x \sec^2 x + \tan x$ **19.** $0 \ (\tan x \cot x = 1)$ **21.** $2 \tan x \sec^2 x$

23. $-\sqrt{x} \csc^2 x + \cot x/(2\sqrt{x})$ **25.** $\frac{3}{2} \tan^{1/2} x \sec^2 x$ **27.** $\sec^3 x + \sec x \tan^2 x$

29. $(\cos x + \sec^2 x)/2\sqrt{\sin x + \tan x}$ **31.** $\frac{1}{2}$ **33.** 16 **35.** 32 **37.** 0 **39.** 0

41. $\frac{3}{4}$ **43.** a/b **45.** $-\frac{1}{2}$ **51.** (a) 1

PROBLEMS 3.6

1. $\dfrac{dy}{dx} = -\dfrac{x^2}{y^2}; \dfrac{dx}{dy} = -\dfrac{y^2}{x^2}$ **3.** $\dfrac{dy}{dx} = -\sqrt{\dfrac{y}{x}}; \dfrac{dx}{dy} = -\sqrt{\dfrac{x}{y}}$

5. $\dfrac{dy}{dx} = -\left(\dfrac{y}{x}\right)^2; \dfrac{dx}{dy} = -\left(\dfrac{x}{y}\right)^2$

7. $\dfrac{dy}{dx} = \dfrac{\frac{1}{2}(x + y)^{-1/2} - \frac{2}{3}x(x^2 + y)^{-2/3}}{\frac{1}{3}(x^2 + y)^{-2/3} - \frac{1}{2}(x + y)^{-1/2}}; \dfrac{dx}{dy} = \dfrac{1}{dy/dx}$ **9.** $\dfrac{dy}{dx} = -\dfrac{x}{y}; \dfrac{dx}{dy} = -\dfrac{y}{x}$

11. $\dfrac{dy}{dx} = \dfrac{2}{15(3xy + 1)^4} - \dfrac{y}{x}; \dfrac{dx}{dy} = \dfrac{1}{dy/dx}$ **13.** $\dfrac{dy}{dx} = \dfrac{x}{y}; \dfrac{dx}{dy} = \dfrac{y}{x}$ **15.** $\dfrac{dy}{dx} = \dfrac{y}{x}; \dfrac{dx}{dy} = \dfrac{x}{y}$

17. $\dfrac{dy}{dx} = -\dfrac{y}{x}; \dfrac{dx}{dy} = -\dfrac{x}{y}$ **19.** $\dfrac{dy}{dx} = \dfrac{-y}{x}; \dfrac{dx}{dy} = \dfrac{-x}{y}$

21. $\dfrac{dy}{dx} = -\left(\dfrac{y}{x}\right)^{15/8}; \dfrac{dx}{dy} = -\left(\dfrac{x}{y}\right)^{15/8}$ **23.** $\dfrac{dy}{dx} = -\dfrac{y}{x}; \dfrac{dx}{dy} = -\dfrac{x}{y}$

25. $\dfrac{dy}{dx} = \dfrac{y - 2xy^3 - 3x^2y^2}{3x^2y^2 + 2x^3y - x}; \dfrac{dx}{dy} = \dfrac{3x^2y^2 + 2x^3y - x}{y - 2xy^3 - 3x^2y^2}$

27. $\dfrac{dy}{dx} = -\left(\dfrac{1}{y}\right)\left(x + \dfrac{1}{\sin(x^2 + y^2)}\right); \dfrac{dx}{dy} = \dfrac{1}{dy/dx}$

29. $\dfrac{dy}{dx} = \dfrac{\cos^2 y \cos(x - y) - \cos x \cos y}{\cos^2 y \cos(x - y) + \sin x \sin y}; \dfrac{dx}{dy} = \dfrac{1}{dy/dx}$

31. vertical tangent at $(0, 1)$; horizontal tangent at $(1, 0)$

33. no vertical or horizontal tangents

35. vertical tangents at $(-a, 0)$, $(a, 0)$; no horizontal tangents

37. no vertical tangents; horizontal tangents at $(\pi/2 + k\pi, (-1)^k)$ where k is an integer

PROBLEMS 3.7

1. $0; 0$ **3.** $8; 0$ **5.** $-\frac{1}{4}x^{-3/2}; \frac{3}{8}x^{-5/2}$ **7.** $-\frac{2}{9}(x + 1)^{-4/3}; \frac{8}{27}(x + 1)^{-7/3}$

9. $-(1 - x^2)^{-3/2}; -3x(1 - x^2)^{-5/2}$ **11.** $r(r - 1)x^{r-2}; r(r - 1)(r - 2)x^{r-3}$

13. $2a; 0$ **15.** $30(x + 1)^{-7}; -210(x + 1)^{-8}$

17. $\dfrac{\sin \sqrt{x}}{4x^{3/2}} - \dfrac{\cos \sqrt{x}}{4x}; (\frac{1}{8}x^{-3/2} - \frac{3}{8}x^{-5/2})\sin \sqrt{x} + \frac{3}{8}x^{-2} \cos \sqrt{x}$

19. $2 \csc^2 x \cot x; -2 \csc^4 x - 4 \csc^2 x \cot^2 x$

21. $\sec^3 x + \sec x \tan^2 x; 5 \sec^3 x \tan x + \sec x \tan^3 x$

25. 50 **27.** $56{,}000$ N

31. $(uv)'' = u''v + 2u'v' + uv''; (uv)''' = u'''v + 3u''v' + 3u'v'' + uv'''$

33. $y'' = -r^2/y^3; y''' = -3r^2x/y^5$

35. $\dfrac{d^n \cos x}{dx^n} = \begin{cases} \cos x, & \text{if } n = 4k \\ -\sin x, & \text{if } n = 4k + 1 \\ -\cos x, & \text{if } n = 4k + 2 \\ \sin x, & \text{if } n = 4k + 3 \end{cases}$

PROBLEMS 3.8

1.

Δx	Δy	dy	Error = $\Delta y - dy$
1	37	27	10
0.5	15.875	13.5	2.375
0.1	2.791	2.7	0.091
0.05	1.372625	1.35	0.022625
0.01	0.270901	0.27	0.000901
0.001	0.027009	0.027	0.000009

3.

Δx	Δy	dy	Error = $\Delta y - dy$
1.0	3.180	3.0	0.180
-1.0	-2.804	-3.0	0.196
0.5	1.5459	1.5	0.0459
-0.5	-1.4521	-1.5	0.0479
0.1	0.30187	0.3	0.00187
-0.1	-0.29812	-0.3	0.00188
0.01	0.030019	0.03	0.000019
-0.01	-0.029981	-0.03	0.000019

5. 0.96 **7.** 0.196 **9.** $\frac{157}{320} = 0.490625$ **11.** 1.6 **13.** 7.98125 **15.** -0.08

17. 339.325 **19.** $\dfrac{1}{\sqrt{2}} + \dfrac{\sqrt{\pi}}{10\sqrt{2}} \approx 0.83244$ **21.** $\tan(11\pi/4) + 0.1 = -0.9$

23. $\csc(5\pi/4) - \pi\sqrt{2}/180 = -\sqrt{2} - \pi\sqrt{2}/180 \approx -1.439$ **25.** 0 **27.** $\frac{1}{3}x^{-2/3}\,dx$

29. $\frac{3}{4}x^2(1 + x^3)^{-3/4}\,dx$ **31.** $-(1/x^2)\,dx$ **33.** $-x^{-1/2}(1 + \sqrt{x})^{-2}\,dx$

35. $(1 + x)^{-1/2}(1 - x)^{-3/2}\,dx$ **37.** $-[2x/(x^2 + 2)^2]\,dx$ **39.** $[-\sin \sqrt{x}/(2\sqrt{x})]\,dx$

41. $(-\cos x/\sin^2 x)\,dx$ **43.** (a) 4000π kg (b) $\pm 40\pi$ kg (c) ± 0.01 ($\pm 1\%$)

45. -0.0053547 m($\approx -0.54\%$) **47.** $\Delta R/R < 0.08$

REVIEW EXERCISES FOR CHAPTER THREE

1. 3 **3.** $-(y + 1)/(x + 1)$ **5.** $6x\sqrt{x + 1} + [(3x^2 + 2)/2\sqrt{x + 1}]$ **7.** $-(y/x)^{1/4}$

9. $\dfrac{1}{3}\left(\dfrac{x + 2}{x - 3}\right)^{-2/3}\left[\dfrac{-5}{(x - 3)^2}\right]$ **11.** $\dfrac{-x^4 + 4x^3 - 4x^2 - 12x - 3}{(x^3 + 5x - 6)^2}$

13. $\dfrac{18x^2(x^3 - 3)^{5/6}}{7(x^3 + 4)^{1/7}} + \dfrac{5x^2(x^3 + 4)^{6/7}}{2(x^3 - 3)^{1/6}}$ **15.** $\dfrac{1 + \frac{1}{2}(x + y)^{-1/2}}{\frac{1}{3}y^{-2/3} - \frac{1}{2}(x + y)^{-1/2}}$

17. $3\sqrt{2}x^2(1 + x^3)^{\sqrt{2}-1}$ **19.** $\dfrac{2x^2(x^5 + 3)^{3/4}(x^3 - 1)^{-1/3} - \frac{15}{4}x^4(x^3 - 1)^{2/3}(x^5 + 3)^{-1/4}}{(x^5 + 3)^{3/2}}$

21. $3x^2 \cos x^3$ **23.** $(1 + \sin^2 x)/\cos^3 x$ **25.** $y = 2x + 2;\ y = -\frac{1}{2}x + \frac{9}{2}$

27. $y = -\frac{183}{8}x + \frac{327}{8};\ y = \frac{8}{183}x + \frac{3286}{183}$ **29.** $y = -(3\sqrt{3}/4\pi)x - (\pi/4\sqrt{3}) + \frac{1}{2};$
$y = (4\pi/3\sqrt{3})x - (4\pi^3/27\sqrt{3}) + \frac{1}{2}$ **31.** $42x^5 - 210x^4 + 6x;\ 210x^4 - 840x^3 + 6$

33. $2(x + 1)^{-3};\ -6(x + 1)^{-4}$ **35.** $-16(3x^2 - 5)/(x^2 + 5)^3;\ 192x(x^2 - 5)/(x^2 + 5)^4$

37. $(2x^{-3} - x^{-1})\cos x + 2x^{-2} \sin x;\ (-6x^{-4} + 3x^{-2})\cos x + (-6x^{-3} + x^{-1})\sin x$

39. $51x^2\,dx$ **41.** $[-11/(x - 7)^2]\,dx$

43. $[(4x^3/3)(x^2 - 3)^{1/2}(x^4 + 5)^{-2/3} + x(x^2 - 3)^{-1/2}(x^4 + 5)^{1/3}]\,dx$

45. $(1/x^2)\sin(1/x)\,dx$ **47.** (a) 31.2 (b) $1199/2400 = 0.49958333 \cdots$

CHAPTER FOUR
PROBLEMS 4.1

1. $-\frac{10}{3}$ **3.** $\frac{8}{3}$ ft/sec **5.** $500\sqrt{1.5^2 - 1}/1.5 = 500\sqrt{5}/3 \approx 372.7$ km/hr

7. $-18^{1.67}(0.01)/[(1.67)(0.3)(18^{0.67})] \approx -0.36$ m^3/sec; decreasing
9. $-11550/\sqrt{9000} \approx -121.75$ ft/sec   **11.** If triangle is not inverted, $dh/dt = 0.08$ ft/min.
if it is inverted, $dh/dt = 0.4$ ft/min.   **13.** 50π cm/sec   **15.** $-\frac{8}{3}$ ft^2/hr   **17.** 5 ft/sec
19. $\frac{11}{2}\pi r^2$ **21.** 3; $75/\sqrt{17}$
23. $dx/dt = (dx/d\theta)(d\theta/dt) = -13 \sin \theta \, d\theta/dt = -100\pi$ cm/min; $dy/dt = 240\pi$ cm/min
25. $12,000\pi$ m/sec (since $l/300 = \tan \theta$ and $d\theta/dt = 10\pi$)   **27.** ≈ 64.78 km/hr
29. $dH/dt = -1500/99^{3/2} \approx -1.52$ m/sec
31. $12,000\pi$ m/sec (since $l/300 = \tan \theta$ and $d\theta/dt = 10\pi$)
33. $dQ/dt = (2.505)(\sec^2(\pi/9))(2^{2.47})\frac{1}{2}(d\theta/dt) \approx 7.8587(d\theta/dt)$ m^3/sec^2

PROBLEMS 4.2

1. $\sqrt{7/3}$ **3.** $c \approx 0.69$ radians ($\sin c = 2/\pi$)   **5.** $c \approx 0.482$ radians ($\cos c = \sqrt{\pi/4}$)
7. 0   **9.** $\pm 1/\sqrt{3}$ **11.** because f is constant, any choice of c from $(-\pi/2, \pi/4)$ works
13. no c satisfies the equation $\frac{2}{3}c^{-1/3} = 0$ **17.** (b) 2   **23.** $0 < b < \frac{1}{2}$
27. no; $\alpha = 1$ produces continuity but not differentiability at $x = 1$,
$\alpha = \frac{2}{3}$ produces $\lim_{x \to 1^-} f'(x) = \lim_{x \to 1^+} f'(x)$ but then f is not continuous at $x = 1$
35. $\epsilon_n = (1/n)^{1/(n-1)}$ **43.** $P(x) = Cx$ for all x

PROBLEMS 4.3

1. (a) increasing on $(-\frac{1}{2}, \infty)$; decreasing on $(-\infty, -\frac{1}{2})$ (b) $(-\frac{1}{2}, -\frac{121}{4})$
(c) y-intercept -30; x-intercepts -6, 5   (d) global minimum $(-\frac{1}{2}, -\frac{121}{4})$; no local maximum
(e)

3. (a) decreasing on $(-\infty, 0)$; increasing on $(0, \infty)$ (b) $(0, 0)$ (c) x-intercepts $\pm\sqrt{2}$;
y-intercept -8 (d) global minimum at $(0, 8)$ (e)

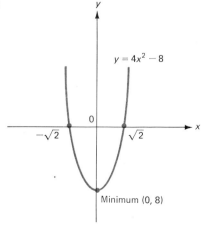

5. (a) increasing on $(-\infty, \infty)$ (b) (0, 0)   (c) x-intercept $\sqrt[3]{3/2}$; y-intercept -3 (d) none
(e)

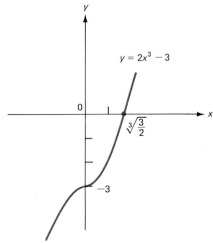

$y = 2x^3 - 3$

7. (a) increasing on $(-\infty, -\frac{2}{3})$ and $(0, \infty)$; decreasing on $(-\frac{2}{3}, 0)$ (b) (0, 0) and $(-\frac{2}{3}, \frac{4}{27})$
(c) x-intercepts 0 and -1; y-intercept 0   (d) local maximum $(-\frac{2}{3}, \frac{4}{27})$; local minimum (0, 0)
(e)

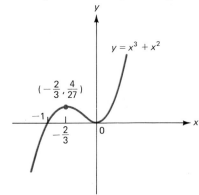

$y = x^3 + x^2$

$\left(-\dfrac{2}{3}, \dfrac{4}{27}\right)$

9. (a) increasing on $(-\infty, 0)$ and $(\frac{2}{3}, \infty)$; decreasing on $(0, \frac{2}{3})$ (b) (0, 0) and $(\frac{2}{3}, -\frac{4}{27})$
(c) x-intercepts 0 and -1; y-intercept 0   (d) local maximum (0, 0); local minimum $(\frac{2}{3}, -\frac{4}{27})$
(e)

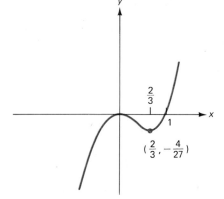

$\left(\dfrac{2}{3}, -\dfrac{4}{27}\right)$

11.   (a) increasing on $(-\infty, -3) \cup (5, \infty)$; decreasing on $(-3, 5)$ (b) $(-3, 106)$, $(5, -150)$
(c) y-intercept 25   (d) local minimum $(5, -150)$; local maximum $(-3, 106)$
(e)

13.   (a) increasing on $(0, \infty)$; decreasing on $(-\infty, 0)$ (b) $(0, 0)$
(c) y-intercept $= x$-intercept $= 0$ (d) global minimum $(0, 0)$; no local maximum
(e)

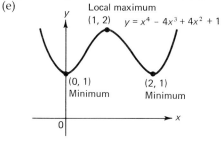

15.   (a) increasing on $(0, 1) \cup (2, \infty)$; decreasing on $(-\infty, 0) \cup (1, 2)$ (b) $(0, 1)$, $(1, 2)$, $(2, 1)$
(c) y-intercept 1; no x-intercept   (d) global minima $(0, 1)$, $(2, 1)$; local maximum $(1, 2)$
(e)

17.   (a) increasing on $(-\infty, \infty)$ (b) $(0, 3)$ (c) y-intercept 3; x-intercept $-\sqrt[3]{3}$
(d) no local maximum or minimum   (e)

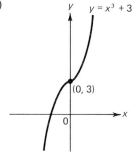

19.   (a) increasing on $(-\infty, \infty)$ (b) $(-1, 0)$ (c) y-intercept 1; x-intercept -1
(d) no local maximum or minimum (e)

21.   (a) increasing on $((2k - 1)\pi, 2k\pi)$, k an integer; decreasing on $(2k\pi, (2k + 1)\pi)$
(b) $(2k\pi, 1)$, $((2k + 1)\pi, -1)$ (c) y-intercept 1; x-intercepts $\pi/2 + k\pi$
(d) global maximum $(2k\pi, 1)$; global minimum $((2k + 1)\pi, -1)$ (e)

23.   (a) increasing on $((4k - 2)\pi, 4k\pi)$; decreasing on $(4k\pi, (4k + 2)\pi)$
(b) $(4k\pi, 1)$, $((4k + 2)\pi, -1)$ (c) y-intercept 1; x-intercepts $(2k + 1)\pi$
(d) global maximum $(4k\pi, 1)$; global minimum $((4k + 2)\pi, -1)$ (e)

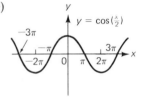

25.   (c) y-intercept 1; x-intercept -1
(d) as $x \rightarrow -1$, $f(x) \rightarrow 0$ and $f'(x) \rightarrow \infty$ so that $x = -1$ is a vertical tangent

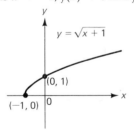

27.   (c) y-intercept 1; no x-intercept
(d) as $x \rightarrow -1^+$, $f(x) \rightarrow \infty$ so that $x = -1$ is a vertical asymptote

29.

31. (a) yes (b) not necessarily (c) yes

35.

p	q	$x = -1$	$x = \dfrac{p - q}{p + q}$	$x = 1$
odd	odd		global min	
even	odd	local max	local min	
odd	even		local max	local min
even	even	global min	local max	global min

PROBLEMS 4.4

1. same graph as for Problem 4.3.1(e) **3.** same graph as for Problem 4.3.11(e) **5.** same graph as for Problem 4.3.17(e)

7.

9.

11.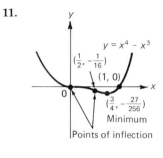

13. same graph as for Problem 4.3.25(d),

15.

17.

19.

21.

23.

25.

27.

29.

31.

33.

35.

37.

39.

41.

45. positive

53.

55.

57.

59.

63.

65. $y'' = -2n(ny - x)(1 - y^2)/(1 - x^2)^2$

PROBLEMS 4.5

In the following answers the maximum value (if any) is given first.

1. -24 at $x = 2$, -30 at $x = 0$ **3.** 890 at $x = 10$, -870 at $x = -10$
5. 1 at $x = 1$, no minimum **7.** 106 at $x = -3$, -150 at $x = 5$
9. 1 at $x = 1$, 0 at $x = 0$ **11.** 2 at $x = 7$, -1 at $x = -2$
13. 250 at $x = 1$, -54 at $x = -1$ **15.** $\frac{1}{2}$ at $x = 3$, $\frac{1}{12}$ at $x = 5$
17. $\frac{4}{5}$ at $x = -3$, 0 at $x = 1$ **19.** 0 at $x = 0$, -4 at $x = -1$
21. $3\frac{1}{2}$ at $x = \frac{1}{2}$, $2\frac{1}{2}$ at $x = 1$ **23.** no maximum, 1 at $x = 0$
25. 20 at $x = 3$, 0 at $x = 0$ **27.** 10 at $x = 3$, -8 at $x = -5$
29. 16 at $x = \pm 2$, 0 at $x = 0$ **31.** 1 at $x = 0$, -1 at $x = \pi/3$
33. 1 at $x = \sqrt{\pi/2}$ and $x = \sqrt{5\pi/2}$, -1 at $x = \sqrt{3\pi/2}$
35. 1 at $x = -3\pi/(2 + 3\pi) \approx -0.82494$ and many others,
-1 at $x = -\pi/(2 + \pi) \approx -0.61102$ and many others
37. no maximum; 1.8 at $x = -\frac{1}{5}$ **39.** 1 at $x = 0$; no minimum
41. *Hint:* If $p(x) \geq 0$ for all x, then n must be even.

PROBLEMS 4.6

1. $L = W = 75$ m **3.** Equilateral triangle $8 \times 8 \times 8$ cm **5.** $\left(\frac{12}{13}, \frac{18}{13}\right)$
7. $(2/\sqrt{29}, 5/\sqrt{29})$ **9.** Maximum $= \pi(35/2\pi)^2 \approx 97.5$ cm^2 when all wire is used for
circle; minimum ≈ 42.88 cm^2 when radius of circle is $35/[2(\pi + 4)] \approx 2.45$ cm **11.** 200
13. choose both to be 10 **15.** closest distance is $30\sqrt{2}$ km after 6 hr; no
17. semicircle is $60/(4 + \pi)$ ft in diameter, rectangle is $30/(4 + \pi)$ ft high **19.** $\frac{40}{7}$ ft
21. breadth $= \frac{1}{2}$ m, depth $= \sqrt{3}/2$ m
23. $\frac{4000}{3}$ pigeons; the integer giving the largest value is 1333
25. $a = -12$; max occurs at each t
27. $\cos \theta = \frac{1}{16}$ and $\theta = 86.4°$ **29.** one-to-one **33.** $\sqrt{8} \approx 2.83$ m
35. use 100 m for N–S and 100 m for E–W fences

PROBLEMS 4.7

1. $7,500$ sets **3.** $\eta(10,000) = 1$ so neither an increase nor a decrease **5.** $4,750$ bottles
7. $\eta(10,000) = 0.2 < 1$, an increase in revenue **9.** 0 sinks **11.** $\eta(2,000) = 0.25 < 1$,
an increase in revenue
13. about $\sqrt{4 \cdot 10^7} \approx 6,325$ trucks per batch and $50,000/\sqrt{4 \cdot 10^7} \approx 7.9 \approx 8$ batches per year
15. $\eta(q) = (q - 5,000)/2q$ **17.** $\$8,000$ in company 1, $\$2,000$ in company 2
23. (a) $q = (\alpha - b)/(2a + 2\beta)$ if $\alpha \geq b$, otherwise $q = 0$ (b) $\eta(q) = -1 + \alpha/(\beta q)$
(c) $q = \alpha/(2\beta)$

PROBLEMS 4.8

1. $[9, 10]$; 9.486832981 **3.** 1.584893192 **5.** 6.192582404; 0.8074175964
7. -2.090520511; 0.2438484828; 7.846672028 **9.** 7.984792473 (only real root)
11. $x_{n+1} = 2x_n - ax_n^2$ converges to $1/a$ provided that $x_0 \in (0, 2/a)$

13. (a) 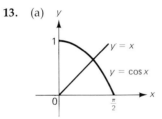   (b) 0.7390851332

15.   hypothesis (i) is violated; $\sin x < x$ for all $x > 0$

REVIEW EXERCISES FOR CHAPTER FOUR

1. $\sqrt[4]{\frac{1}{5}}$ **3.** $-\left(\frac{31}{5}\right)^{5/4}$

5.   **7.**

9.   **11.**

13. same graph as for Problem 4.4.23   **15.**

17.

19.

$y = \cot x$

21. $\sqrt{34}/2$ m/sec   **23.** (a) maximum is 358 at $x = 5$; minimum is -10 at $x = 1$
(b) maximum is $2^{1/3}$ at $x = 4$; minimum is $-2^{1/3}$ at $x = 0$ (c) maximum is 0 at $x = -1$; minimum
is $-\frac{2}{3}$ at $x = 1$ **25.** $(\frac{12}{13}, -\frac{18}{13})$ **27.** 8 weeks (giving 2592 individuals/week)
29. ≈ 1250 units (positive root of $-0.04q^3 + 50q^2 + 1{,}000 = 0$)   **31.** 2.264934401

CHAPTER FIVE
PROBLEMS 5.2

1. $x + C$ **3.** $ax + C$ **5.** $x + x^2/2 + x^3/3 + x^4/4 + x^5/5 + C$
7. $-(2/x^2) + C$ **9.** $\frac{2}{3}x^{3/2} + 2\sqrt{x} + C$ **11.** $\frac{2}{3}x^{3/2} + \frac{2}{5}x^{5/2} + \frac{4}{7}x^{7/4} + C$
13. $\frac{2}{5}x^{5/2} + \frac{2}{3}x^{3/2} - 2x^{1/2} + C$ **15.** $-\frac{289}{4}x^{4/17} + \frac{27}{5}x^{5/9} + C$
17. $\frac{1}{4}\sin x + \frac{5}{4}\cos x + C$ **19.** $y = \frac{2}{3}x^3 + x^2 - \frac{28}{3}$ **21.** $y = \frac{78}{11}x^{11/6} - 3x + \frac{109}{11}$
23. $y = \sin x + 3.5$ **25.** $x = (2.9t^2 + 0.2t + 25)$ m   **27.** $x = (\frac{52}{15}t^{5/2} + 25t + 100)$ m
29. $\frac{3}{2}(100) + \frac{2}{3}(1000) - \frac{2}{5}(10)^{5/2} \approx 690.176$ kg   **31.** \$120,890; \$2,302,223.33
33. unprofitable when $q > 3{,}333$ **35.** $\frac{1}{2}x|x| + C$

PROBLEMS 5.3

1. 20   **3.** 7   **5.** $\frac{61}{20}$ **7.** $\sum_{k=0}^{4} 2^k$ **9.** $\sum_{k=2}^{n} k/(k+1)$ **11.** $\sum_{k=1}^{n} k^{1/k}$
13. $\sum_{k=1}^{8}(-1)^{k+1}x^{5k}$ **15.** $\sum_{k=1}^{8}(2k-1)(2k+1) = \sum_{k=0}^{7}(2k+1)(2k+3)$
17. $32^{-3}\sum_{k=1}^{32} k^2$ **19.** $64^{-4}\sum_{j=1}^{64} j^3$ **21.** $\frac{1}{50}\sum_{k=1}^{50}\sqrt{k/50} = 50^{-3/2}\sum_{k=1}^{50}\sqrt{k}$
23. $\sum_{i=1}^{7}(7-i)^2$ is the odd one   **27.** $\sum_{k=1}^{n} k^3 = \left(\dfrac{n(n+1)}{2}\right)^2$

29. $\displaystyle\sum_{k=2}^{n} \dfrac{1}{k^2 - 1} = \dfrac{3}{4} - \dfrac{n + \frac{1}{2}}{n(n+1)}$

PROBLEMS 5.4

1. $s = \frac{155}{512} \approx 0.3$; $S = \frac{187}{512} \approx 0.37$
5. $28 < A < 84$; $42 < A < 70$; $49 < A < 63$; $A = 56$
7. $12\frac{3}{4} < A < 26\frac{1}{4}$; $16\frac{1}{8} < A < 22\frac{7}{8}$; $17\frac{13}{16} < A < 21\frac{3}{16}$; $A = 19\frac{1}{2}$

In the answers to Problems 9–21, the left-hand endpoint of each subinterval has been chosen.

9. (a) $s_4 = \frac{7}{8}$ (b) $s_8 = \frac{35}{32}$ (c) $s_n = \dfrac{2}{n}\sum_{k=0}^{n-1}\dfrac{1}{2}\left(\dfrac{2k}{n}\right)^2 = \dfrac{4}{n^3}\sum_{k=0}^{n-1}k^2 = \dfrac{2(n-1)(2n-1)}{3n^2}$ (d) $A = \frac{4}{3}$

11. (a) $s_4 = 5^4 \cdot 3^2/4^3 \approx 87.89$ (b) $s_8 = 5^4 \cdot 7^2/(4 \cdot 8^2) \approx 119.63$
(c) $s_n = \dfrac{5}{n}\sum_{k=0}^{n-1}\left(\dfrac{5k}{n}\right)^3 = \left(\dfrac{5}{n}\right)^4\sum_{k=0}^{n-1}k^3 = \dfrac{625}{4}\left(\dfrac{n-1}{n}\right)^2$ (d) $A = 5^4/4 = 156.25$
13. (a) $s_8 = \frac{243}{256} \approx 0.9472$ (b) $s_{16} = \frac{867}{1024} \approx 0.8467$

(c) $s_n = \frac{1}{n}\sum\limits_{k=0}^{n-1}\left[0 - 3\left(-1 + \frac{k}{n}\right)^3\right] = 3\frac{1}{n^4}\left[\frac{n(n+1)}{2}\right]^2 = \frac{3}{4}\left(\frac{n+1}{n}\right)^2$ (d) $A = \frac{3}{4}$

15. (a) $s_4 = 1 - (3\cdot 7)/(6\cdot 4^2) = 0.78125$ (b) $s_8 = 1 - (7\cdot 15)/(6\cdot 8^2) \approx 0.7266$

(c) $s_n = \frac{1}{n}\sum\limits_{k=0}^{n-1}\left[1 - \left(\frac{k}{n}\right)^2\right] = \left(\frac{1}{n}\sum\limits_{k=0}^{n-1}1\right) - \left(\frac{1}{n^3}\sum\limits_{k=0}^{n-1}k^2\right) = 1 - \frac{(n-1)(2n-1)}{6n^2}$ (d) $A = \frac{2}{3}$

17. (a) $s_8 = (15\cdot 55)/(6\cdot 8^2) \approx 2.148$ (b) $s_{16} = (31\cdot 111)/(6\cdot 16^2) \approx 2.240$

(c) $s_n = \frac{1}{n}\sum\limits_{k=0}^{n-1}\left(1 + \frac{k}{n}\right)^2 = 1 + \left(\frac{n-1}{n}\right) + \frac{(n-1)(2n-1)}{6n^2}$ (d) $A = \frac{7}{3}$

19. (a) $s_6 = (5\cdot 29)/(6\cdot 36) \approx 0.6713$ (b) $s_{12} = (11\cdot 59)/(6\cdot 144) \approx 0.7512$

(c) $s_n = \frac{1}{n}\sum\limits_{k=0}^{n-1}\left[\left(\frac{k}{n}\right) + \left(\frac{k}{n}\right)^2\right] = \frac{n-1}{2n} + \frac{(n-1)(2n-1)}{6n^2}$ (d) $A = \frac{5}{6}$

21. (a) $s_4 = (7\cdot 11)/(2\cdot 4^2) = 2.40625$ (b) $s_8 = (15\cdot 23)/(2\cdot 8^2) \approx 2.695$

(c) $s_n = \frac{1}{n}\sum\limits_{k=0}^{n-1}\left[1 + 2\left(\frac{k}{n}\right) + 3\left(\frac{k}{n}\right)^2\right] = 1 + \frac{n-1}{n} + \frac{(n-1)(2n-1)}{2n^2}$ (d) $A = 3$

23. (a) $a, a + \dfrac{b-a}{n}, a + 2\left(\dfrac{b-a}{n}\right), \ldots, a + (n-1)\left(\dfrac{b-a}{n}\right), b$

(b) $x_i^* = a + i\left(\dfrac{b-a}{n}\right) = \dfrac{1}{n}\{(n-i)a + ib\};$

$(x_i^*)^2 = 1/n^2\{(n-i)a + ib\}^2 + 1/n^2\{(n-i)^2a^2 + 2(n-i)iab + i^2b^2\}$

(c) (using right-hand endpoint of each subinterval)

$s_n = \left(\dfrac{b-a}{n}\right)\sum\limits_{k=1}^{n}\left[a + k\left(\dfrac{b-a}{n}\right)\right]^2 = \left(\dfrac{b-a}{n^3}\right)\sum\limits_{k=1}^{n}[(n-k)a + kb]^2$

$\quad = \dfrac{b-a}{6n^2}[(n-1)(2n-1)a^2 + (n^2-1)2ab + (n+1)(2n+1)b^2]$

$\quad = \dfrac{b^3-a^3}{3} + \dfrac{(b-a)^2(b+a)}{2n} + \dfrac{(b-a)^3}{6n^2}$ (d) $A = \dfrac{b^3-a^3}{3}$

25. (a) s_n estimates the area of one-quarter of a circle with radius 1 and centered at the origin

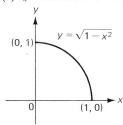

(b) assuming it is already known that the area of a circle with radius r is πr^2,
$\lim_{n\to\infty}(1/n)\sum\limits_{k=1}^{n}\sqrt{1 - (k/n)^2}$ must be $\pi 1^2/4 = \pi/4 \approx 0.785398$

PROBLEMS 5.5

1. 56 **3.** $37\frac{1}{2}$ **5.** $\frac{4}{3}$ **7.** $\frac{4}{3}$ **9.** $(17\cdot 8^3)/3 = 2901\frac{1}{3}$ **11.** $-\frac{1}{4}$ **13.** $\frac{2}{3}$

15. $-152\frac{1}{4}$ **17.** $-6\frac{3}{4}$ **19.** 1 **21.** 10 **23.** $(b^3 - a^3)/3$

29. use fact that $s^{55/3} > s^{17}$ for $23 \le s \le 47$ and Theorem 7 **31.** $\int_1^2 x\,dx$

33. Further hint: $\sqrt{1 + x^3}$ is an increasing function; thus $\sqrt{1 + x^3} \le 3/2\sqrt{2}$ on $[0, \frac{1}{2}]$ and $\sqrt{1 + x^3} \le \sqrt{2}$ on $[\frac{1}{2}, 1]$. Now get overestimates for $\int_0^{1/2}\sqrt{1 + x^3}\,dx$ and $\int_{1/2}^1\sqrt{1 + x^3}\,dx$.

35. $12 < \int < 24$ **37.** $\frac{8}{3} < \int < 8$ **39.** $\frac{99}{100} < \int < 99$ **41.** $\frac{1}{2} < \int < 1$

43. $\pi/6 < \int < \pi/3$ **45.** (a) 64 ft/sec (b) 128 ft (c) $\sqrt{50} \approx 7.07$ sec

PROBLEMS 5.6

1. $\frac{33}{5}$ **3.** $\frac{26}{3}$ **5.** $(c_1/3)(b^3 - a^3) + (c_2/2)(b^2 - a^2) + c_3(b - a)$ **7.** $\frac{333}{4}$ **9.** 0

11. $\frac{80}{3}$ **13.** $\frac{41}{3}$ **15.** $\frac{6}{23} - \frac{1}{3} - \frac{4}{13} + \frac{12}{29} \approx 0.0336$ **17.** $-\frac{1}{101}$ **23.** $2 - \sqrt{2}$

25. 94　**27.** $\frac{32}{3}$　**29.** $\frac{32}{3}$　**31.** $\frac{1056}{5}$　**33.** 8
35. $(b^3 - 3ab^2 + 3a^2b - a^3)/6 = (b - a)^3/6$　**37.** $2\sqrt{2}$
41. $F'(x) = 1/(1 + x^3); F'(2) = \frac{1}{9}$　**43.** $F'(x) = \sqrt{(x - 1)/(x + 1)}; F'(1) = 0$
45. 609.4 m　**47.** (a) 2 kg　(b) $\frac{5}{4}$ m　**49.** \$271.67　**51.** (a) 10^6 m/sec^2　(b) 55 m
53. C　**55.** $\frac{1}{4}$　**57.** $\frac{19}{3}$　**59.** $1 + \frac{1}{2} + \frac{1}{3} + \frac{1}{4} + \frac{1}{5} = \frac{137}{60}$　**61.** $\frac{231}{70} = 3.3$
63. $\frac{2047}{11} + \frac{1}{4608} - \frac{1}{9} \approx 185.98$　**65.** \$45　**67.** -80 ft/sec　**69.** each integral equals 1
71. (a), (c), (e) are true; (b) and (d) are false.
73. one pair of functions which works is $f(x) = x - \frac{1}{2}$ and $g(x) = |x - \frac{1}{2}|$
75. the integrand is not defined at $x = 0$ (the area under the curve between 0 and 1 is infinite)

PROBLEMS 5.7

1. $\frac{1}{6}(1 + x^2)^6 + C$　**3.** 20　**5.** $\frac{3}{2}$　**7.** $\frac{74}{3}$　**9.** $2(10^{3/2} - 1)/27$
11. $\frac{5}{72}(1 + 3x^4)^{6/5} + C$　**13.** $\frac{2}{15}(s^5 + 5s)^{3/2} + C$　**15.** $\sqrt{20} - \sqrt{3}$
17. $-\frac{1}{2}(1 + \sqrt{x})^{-4} + C$　**19.** $-\frac{3}{16}[1 + (1/v^2)]^{8/3} + C$　**21.** $\frac{1}{3}(ax^2 + 2bx + c)^{3/2} + C$
23. $\frac{7}{8}(ax^2 + 2bx + c)^{4/7} + C$　**25.** $[2a^3/3(n + 1)][(1 + \alpha^{n-1})^{3/2} - 1]$　**27.** $(4\sqrt{2}/9)\alpha^{9/2}$
29. 0 (odd function)　**31.** $\frac{1}{3}\sin x^3 + C$　**33.** 0　**35.** $\frac{1}{3}(64 - 7^{3/2})$　**37.** $\frac{7}{2}$
39. $1/2\sqrt{2}$　**41.** (a) $\frac{5}{2}$ kg　(b) $(\frac{13}{2})^{3/2} - 8 \approx 8.57$ m　**43.** $\frac{2}{15}$ m/min　**45.** $\frac{2}{3}$

PROBLEMS 5.8

1. $\frac{1}{6}$　**3.** $\frac{137}{12}$　**5.** $\frac{343}{6}$　**7.** $\frac{355}{3}$　**9.** $\frac{1944}{5}$　**11.** $\frac{128}{3}$　**13.** $\frac{27}{5}$　**15.** $\frac{1}{2}$
17. 108　**19.** 49　**21.** 1　**23.** $\frac{7}{2}$　**25.** $\frac{169}{24}$

PROBLEMS 5.9

1. 98.1 J　**3.** 20 ft·lb　**5.** 45 J　**7.** 2.5 ft
9. $\int_0^{10} g\,\delta\pi r^2 x\,dx \approx 3.85 \times 10^7$ J　**11.** $\int_0^5 g\,\delta\pi r^2 x\,dx \approx 9.63 \times 10^6$ J
13. $(9.81)(1000)(\pi \cdot 5^2 \cdot 10)20 \approx 1.54 \times 10^8$ J (since all the water has to be raised 20 m)
15. 3.54×10^7 J (multiply result of Problem 9 by 0.92)
17. $\int_2^7 \pi g\delta(3 - \frac{3}{7}x)^2 x\,dx \approx 6.209 \times 10^5$ J　**19.** 525 ft·lb　**21.** 449,280 ft·lb
23. 45.38 sec　**25.** Evaluate $|\int_{3.8 \times 10^8 m}^{R} Gm_1m_2/r^2\,dr|$ and see what happens as $R \to \infty$.
The answer is $GM^2 \times 3.23158 \times 10^{-11}$ J.　**27.** 337.5 J
29. $\sqrt{140,000/3} \approx 216$ ft/sec ≈ 147 mi/hr
31. $v_0 = \sqrt{2gd} = \sqrt{2(3.92)(3,430,000)} \approx 5.186$ km/sec ≈ 3.222 mi/sec
33. $x = \frac{1}{2}v^2/g \approx 83.5$ ft ($v = 50$ mi/hr $= \frac{220}{3}$ ft/sec)

PROBLEMS 5.10

9. $G'(x) = (\cos x + \sin x - 10)\cos x \sin x$　**11.** $G'(x) = 5\cos(5x)$
13. $G'(x) = 3x^2\cos(x^3) - 2\cos(2x)$　**15.** $G'(x) = 2\sin(2x)/[(2x)^2 - 1]$

REVIEW EXERCISES FOR CHAPTER FIVE

1. $x^6/6 + C$　**3.** $\frac{9}{2}$　**5.** $(2^{3/2} - 1)/3$　**7.** $-\frac{32}{3}$　**9.** $\frac{1}{320}$　**11.** $\frac{3}{4}$
13. $\frac{233}{6}$　**15.** $\frac{9}{2}$　**17.** $\frac{3}{10}$　**19.** $\frac{1}{6}$　**21.** $\frac{9}{16} = 0.5625$　**23.** $2\sqrt{2}$
25. $F'(x) = x^3/\sqrt{x^2 + 17}, F'(8) = \frac{512}{9}$　**27.** $F'(s) = (1 + s^2)^{2001}, F'(0) = 1$
29. (a) 400 kg　(b) 8 m　**31.** $62.4\pi 6^2 \cdot 12^2/12 \approx 84,687$ ft·lb
33. $v_{final} = g\sqrt{500/(g/2)} \approx 99.045$ m/sec, $k = \frac{1}{2}mv_{final}^2 = m\,4905$ J

CHAPTER SIX
PROBLEMS 6.1

1. $\mathbb{R}; x = \frac{1}{3}(y - 5); dx/dy = \frac{1}{3}$　**3.** $\mathbb{R} - \{0\}; x = 1/y; dx/dy = -1/y^2$
5. $[-\frac{3}{4}, \infty); x = \frac{1}{4}(y^2 - 3); dx/dy = y/2$　**7.** $\mathbb{R}; x = \sqrt[3]{1 - y}; dx/dy = -\frac{1}{3}(1 - y)^{-2/3}$
9. $y = \sin x$ is 1–1 on each interval of the form $[-\pi/2 + k\pi, \pi/2 + k\pi]$ where k is an integer;
$x = \begin{cases} k\pi + \sin^{-1} y, \dagger & \text{if } k \text{ is even} \\ k\pi - \sin^{-1} y, & \text{if } k \text{ is odd} \end{cases}, \frac{dx}{dy} = \frac{(-1)^k}{\sqrt{1 - y^2}}$

$\dagger \sin^{-1} y \in [-\pi/2, \pi/2]$; see Example 9.

11. $y = \cos(x/4)$ is 1–1 on each interval of the form $(4k\pi, 4(k + 1)\pi)$ where k is an integer;
$x = \begin{cases} 4(k\pi + \cos^{-1} y), \dagger & \text{if } k \text{ is even} \\ 4([k + 1]\pi - \cos^{-1} y), & \text{if } k \text{ is odd} \end{cases}$; $\dfrac{dx}{dy} = \dfrac{4(-1)^{k+1}}{\sqrt{1 - y^2}}$

13. $y = \tan x$ is 1–1 on each interval of the form $(-\pi/2 + k\pi, \pi/2 + k\pi)$, where k is an integer. The inverse is denoted $\tan^{-1} x$. That is, $x = \tan^{-1} y$ if $y = \tan x$; $dx/dy = 1/(1 + y^2)$ (if $-\pi/2 < x < \pi/2$)

15. $(-\infty, \frac{5}{2}]$ and $[\frac{5}{2}, \infty)$; $x_1 = (5 - \sqrt{1 + 4y})/2$, $x_2 = (5 + \sqrt{1 + 4y})/2$; $dx_1/dy = -1/\sqrt{1 + 4y}$, $dx_2/dy = 1/\sqrt{1 + 4y}$

17. $\mathbb{R}$; from the cubic formula (which no one really knows but which you can find in the *Handbook of Tables for Mathematics* (Chemical Rubber Co.)),
$x = \sqrt[3]{(y/2) + \sqrt{(y^2/4) + \frac{1}{27}}} + \sqrt[3]{(y/2) - \sqrt{(y^2/4) + \frac{1}{27}}}$; $dx/dy = 1/(1 + 3x^2)$.
Note that y is increasing since $y' = 3x^2 + 1 > 0$

19. 1–1 on $(-\infty, \infty)$; $x = y^{1/101}$; $\dfrac{dx}{dy} = \dfrac{1}{101x^{100}} = \dfrac{1}{101y^{100/101}} = \dfrac{1}{101}y^{-100/101}$

PROBLEMS 6.2

1. 2　**3.** -3　**5.** -4π　**7.** 5　**9.** $-\frac{1}{2}$　**11.** -1　**13.** $\pm\sqrt{6}$　**15.** $\sqrt{2}$
17. e^π　**19.** 4　**21.** $\frac{1}{2}$　**23.** 10^{10-23}　**25.** $\frac{27}{16}$　**27.** 24　**29.** $\sqrt[3]{25}/2$
31. 3　**33.** $\frac{7}{2}\log x$　**35.** $5 \ln x - 18 \ln(1 + x)$

37.

n	$\dfrac{1}{n}$	$1 - \dfrac{1}{n}$	$\left(1 - \dfrac{1}{n}\right)^{-n}$
2	0.5	0.5	4.0
5	0.2	0.8	3.051757813
10	0.1	0.9	2.867971991
1,000	0.001	0.999	2.719642216
10,000	0.0001	0.9999	2.718417754
1,000,000	0.000001	0.999999	2.718283188

41. $999^{1000} > 1000^{999}$ (since $\log 999^{1000} \approx 2999.57 > \log 1000^{999} = 2997$)
43. (a) $L_2/L_1 = 1 + (10 \log 2)/L_1$ if $I_2 = 2I_1$　(b) $I_2/I_1 = I_1/I_0 = I_1 \times 10^{12}$ if $L_2 = 2L_1$
(c) $I_2/I_1 = (I_1/I_0)^5 = I_1 \times 10^{60}$ if $L_2 = 6L_1$　**45.** (a) 1.138828　(b) 0.690731
(c) $(e^4)(e^{0.13}) \approx 62.1779$　(d) $e^{-3}/e^{-0.37} \approx 0.072079$　(e) 123.97, 0.23693
47. (a) $100! \approx 9.324847626 \times 10^{157}$; $200! \approx 7.883293288 \times 10^{374}$
55. if $\log B$ is negative, then step (b) does not necessarily follow from step (a)

PROBLEMS 6.3

1. $1/(x \ln 5)$　**3.** $1/(x \ln 10)$　**5.** $-1/(x \ln 3)$　**7.** $1/(x \ln \pi)$　**9.** $14/(\ln 2)$
11. 3　**13.** 2　**15.** 15

17.　　　**19.**　　　**21.**

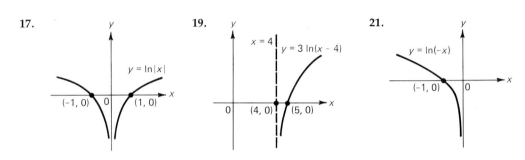

$\dagger \cos^{-1} y \in [0, \pi]$—by analogy with Example 9; also see Section 7.2.

23.

25.

27.

29.

31.

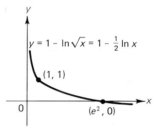

33. $\ln x$ **35.** $[(1/x_1) + (1/x_2) + \cdots + (1/x_n)] \Delta x \approx \int_1^2 (1/t)\, dt = \ln 2$, where $x_i = 1 + i/n$ and $\Delta x = 1/n$ **37.** $W = -500 - 1 \cdot R \cdot 300 \cdot \ln 2 \approx -913.185$ cal
47. $A = e^{1/e} \approx 1.4446679$ (graphs are tangent at $x = e$)

PROBLEMS 6.4

For Problems 1–47, see answer for problem with same number in Section 6.2.
49. $1/(x \ln 5)$ **51.** $1/(x \ln 10)$ **53.** $-1/(x \ln 3)$ **55.** $1/(x \ln \pi)$ **57.** $14/(\ln 2)$
59. 3   **61.** 2   **63.** 15

For Problems 65–79, see answers for Section 6.3, Problems 17–31.

PROBLEMS 6.5

1. $2/(1 + 2x)$ **3.** $5x^4/(1 + x^5)$ **5.** $20x^3 e^{1+5x^4}$ **7.** $\frac{1}{2}\ln|x| + C$ **9.** $\frac{1}{3}e^{x^3} + C$
11. $\frac{1}{3}\ln|1 + x^3| + C$ **13.** $-1/x$ **15.** $\frac{1}{4}\ln(x^4 + 25) + C$ **17.** $3e^{\sqrt[3]{x}} + C$
19. $5x^4$ **21.** $x^{\sqrt{x}-(1/2)}(\frac{1}{2}\ln x + 1)$ **23.** $(x/3)(2\ln x - 1)/(\ln x)^2$
25. $[(2 - x)/(2x^2\sqrt{x} - 1)]e^{\sqrt{x}-1/x}$ **27.** $-\frac{1}{2}\ln^2(1/x) + C = -\frac{1}{2}\ln^2 x + C$ **29.** $\frac{7}{3}$
31. $\ln\frac{9}{7}$ **33.** $3x^2/(x^3 + 3)$ **35.** $1/[x(\ln x - 1)]$ **37.** $1/(x \ln x)$ **39.** $\ln 3$
41. $-1/(4\ln^4 x) + C$ **43.** $\ln|\ln|\ln|x||| + C$ **45.** $(1 + e^x)e^{x+e^x}$

47.

49.

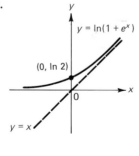

51. $y = e^{-x}\cos x$

53.

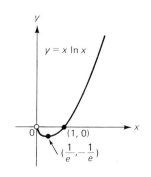

$y = x \ln x$

$(1, 0)$

$\left(\dfrac{1}{e}, -\dfrac{1}{e}\right)$

55.

$y = e^x \ln x$

$(1, 0)$

$\approx (0.591, -0.95)$
Point of inflection

57. $1 + e^{-1}$ **59.** $1 + e^3 - 3 \ln 3$ **61.** $(e - 1)/2$

63. $\sqrt[5]{\dfrac{x^3 - 3}{x^2 + 1}}\left(\dfrac{1}{5}\right)\left(\dfrac{3x^2}{x^3 - 3} - \dfrac{2x}{1 + x^2}\right)$ **65.** $\sqrt{x}\sqrt[3]{x + 2}\sqrt[5]{x - 1}\left[\dfrac{1}{2x} + \dfrac{1}{3(x + 2)} + \dfrac{1}{5(x - 1)}\right]$

69. $2x(\ln 3)(3^{x^2 + 5})$ **71.** $\dfrac{(\log_5 2x)^2}{2 \log_5 e} + C$ **73.** $3^{\sqrt{x}}\left(\dfrac{2}{\ln 3}\right) + C$ **75.** $\dfrac{3(\log_{1/2} x)^2}{2 \log_{1/2} e} + C$

77. $\dfrac{3^{\ln x}}{\ln 3} + C$ **79.** $-\dfrac{1}{2 \log_{14} e}(\log_{14} x)^{-2} + C$ **81.** $(\log_8 e)\left[\dfrac{9 - x^4}{x(x^4 + 3)}\right]$

83. (a) $0.45e^{-0.1} \approx 0.4072$ (b) 1000 units

85. (a) at $x = \sqrt{1.5} \approx 1.225$ m (b) $\frac{3}{2}(1 - e^{-25/3}) \approx 1.49964$ kg

89. if $E(x)$ is not identically equal to 0, then $E(x) = e^{kx}$ where $k = E'(0)$

91. $(-\infty, -1]$ and $[-1, \infty)$; inverse function not obtainable in closed form;
$dx/dy = 1/[(x + 1)e^x] = 1/(y + e^x)$

PROBLEMS 6.6

1. 62,500 after 20 days; 156,250 after 30 days **3.** $250,000e^{0.6} \approx 455,530$ in 1980;
$250,000e^{1.8} \approx 1,512,412$ in 2000 **5.** (a) $-10 + 160e^{-2\alpha} \approx -4.14\,°C$, where $\alpha = 2 \ln \frac{16}{7}$
(b) $(\ln 16)/\alpha \approx 1.6769$ min ≈ 100.62 sec **7.** $[(\ln 0.7)/(\ln 0.5)]5580 \approx 2871$ yr
9. (a) $25(0.6)^{24/10} \approx 7.3367$ kg (b) $[(\ln 0.02)/(\ln 0.6)]10 \approx 76.58$ hr
11. $(\ln 0.5)/(-\alpha) \approx 4.62 \times 10^6$ yr **13.** $\beta = \frac{1}{1500} \ln(845.6/1013.25) \approx -1.205812 \times 10^{-4}$
(a) 625.526 mbar (b) 303.416 mbar (c) 429.033 mbar (d) 348.632 mbar
(e) $-(1/\beta)\ln 1013.25 \approx 57.4$ km **15.** (a) 10,000 (b) 7 (c) 214.15 hr ≈ 8.9 days

PROBLEMS 6.7

1. $2800 **3.** $5000(1 + 0.07/12)^{(12)(8)} \approx \$8739.13$ **5.** 79.1% annually; 81.4% quarterly
7. Choose $5\frac{1}{8}\%$ compounded semiannually. **9.** $i = -\frac{1}{8}\ln \frac{750}{1000} \approx 0.03596 = 3.596\%$
11. $(\ln 3)/15 \approx 0.07324 = 7.324\%$ **13.** $10,000(1.09)^{-5} \approx \$6499.31$
15. $8000e^{-0.4} \approx \$5362.56$ **17.** $t = 1.4$ yr (must solve $[(0.1/\sqrt{0.2t}) - 0.12]e^{\sqrt{0.2t}} = 0.12$;
solution obtained by trial and error)
19. $25,000/[(1.0125^{60} - 1)/0.0125] \approx \$282.25$ quarterly $\approx \$94.08$ monthly
21. $250,000e^{-(0.11)(3)} \approx \$179,730.93$ **23.** (a) $(6000/0.005)[1 - (1/1.005)^{(12)(20)}] \approx \$837,484.63$
(b) yes; buy because present value of investment exceeds purchase price
25. Solve $750,000 = (6000/i)\{1 - [1/(1 + i)]^{240}\}$ to obtain $i \approx 0.00617365 = 0.6174\%$
monthly $= 7.408\%$ annually.
27. $400\left[\dfrac{0.005}{1 - (1/1.005)^{24}}\right] \approx \$17.73$ **29.** $25,000\left[\dfrac{0.00625}{1 - (1/1.00625)^{(12)(25)}}\right] \approx \$184.75$

PROBLEMS 6.8

1. yes; $W = 907$ **3.** yes; $W = 2498$ **5.** yes; $W = 22,994$

REVIEW EXERCISES FOR CHAPTER SIX

1. 2 **3.** $\frac{1}{2}$ **5.** $10^{10^{-9}}$

7. $-\frac{5}{6}[\frac{1}{2}\ln(x^3 - 1) + \frac{4}{3}\ln(x^5 + 1) - \frac{1}{2}\ln(x - 3) - \frac{1}{2}\ln(x - 2)]$

9.

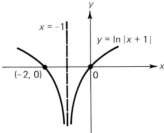

11. $2x/(1 + x^2)$ **13.** $2(x + 1)e^{(x+1)^2}$

15. $\frac{2}{5}e^{\sqrt{x}} + C$ **17.** $e^x/(e^x - 5)$ **19.** $\frac{1}{2}e^{-1/x^2} + C$

21. $\dfrac{\sqrt[3]{x + 3}(x - 5)}{\sqrt{(x + 1)(x + 2)}}\left[\dfrac{1}{3(x + 3)} + \dfrac{1}{x - 5} - \dfrac{1}{2(x + 1)} - \dfrac{1}{(x + 2)}\right]$

23. $[2/(3\log_5 e)](\log_5 x)^{3/2} + C$ **25.** $3x^2$

27.

29.

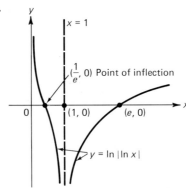

31. $(\ln 2)/0.15 \approx 4.62$ years **33.** 8451.3 years

35. (a) $(\ln 0.25)/(\ln 0.8) \approx 6.212$ weeks (b) $(\ln 0.05)/(\ln 0.8) \approx 13.425$ weeks

37. $10,000e^{(0.06)8} \approx \$16,160.74$ **39.** \$3,778.92 **41.** $[\frac{5}{2}, \infty)$; $x = \frac{1}{2}(5 + y^2)$; $dx/dy = y$

43. $\mathbb{R}$; $x = \ln y - 4$; $dx/dy = 1/y$

45. $y = \sin(3x)$ is 1–1 on each interval of the form $[-(\pi/6) + k(\pi/3), (\pi/6) + k(\pi/3)]$ where k is an integer;

$x = \begin{cases} \frac{1}{3}(k\pi + \sin^{-1} y), & \text{if } k \text{ is even} \\ \frac{1}{3}(k\pi - \sin^{-1} y), & \text{if } k \text{ is odd}; \end{cases}$

$dx/dy = (\frac{1}{3})(-1)^k/\sqrt{1 - y^2}$

47. $(-\infty, -\frac{5}{2}]$ and $[-\frac{5}{2}, \infty)$; $x_1 = (-5 - \sqrt{1 + 4y})/2$, $x_2 = (-5 + \sqrt{1 + 4y})/2$, $dx_1/dy = -1/\sqrt{1 + 4y}$, $dx_2/dy = 1/\sqrt{1 + 4y}$

CHAPTER SEVEN
PROBLEMS 7.1

1. $-\frac{1}{3}\cos 3x + C$ **3.** $\frac{3}{4}$ **5.** $\frac{1}{4}\sin^2 2x + C = -\frac{1}{4}\cos^2 2x + C_1$ **7.** $1/3\sqrt{2}$

9. $\frac{1}{4}\sec 4x + C$ **11.** $\pi/9$ **13.** $-\frac{2}{3}\cos^{3/2} x + C$ **15.** $-\frac{1}{2}\cot x^2 + C$

17. $(-\frac{3}{8})(\cos 2x)^{4/3} + C$ **19.** $\frac{5}{6}(\sec x)^{6/5} + C$ **21.** $-\frac{1}{2}\csc x^2 + C$ **23.** $(2^{5/2} - 1)/5$

25. $\ln |\tan x| + C$ **27.** $2/\sqrt{\csc x} + C = 2\sqrt{\sin x} + C$ **29.** $2\sin\sqrt{x} + C$

31. $\ln(1 + \sin x) + C$ **33.** $\ln |\sin(2 + x)| + C$ **35.** $-\frac{1}{2}e^{-2\tan x} + C$

37. $\ln |\sec x + \tan x| + C$

39. Both answers are essentially the same since $\sin^2 x$ and $-\cos^2 x$ differ only by a constant.

41. 2 **43.** $\pi/2$ **45.** $3\pi/2$ **47.** $\int_0^{\pi/4}(1 - \tan x)\,dx = (\pi/4) - \ln\sqrt{2}$ **49.** $\frac{1}{2}$

51. $1925 + (1000/\pi) \approx \$2243.31$ **53.** (a) $\ln(1 + \sin \theta) + C$ (b) $-\ln |\cos \theta| + C$
(c) $\ln(1 + \sin \theta) - \ln |\cos \theta| + C = \ln |\sec \theta + \tan \theta| + C$
55. (a) $e^x(\sin x + \cos x)$ (b) $e^x(\cos x - \sin x)$ (c) $2e^x \cos x$ (d) $2e^x \sin x$
(e) $\frac{1}{2}e^x(\sin x - \cos x) + C$ (f) $\frac{1}{2}e^x(\sin x + \cos x) + C$

PROBLEMS 7.2

1. $\pi/3$ **3.** $-\pi/6$ **5.** $\pi/6$ **7.** $-\pi/4$ **9.** 0 **11.** $\frac{4}{5}$ **13.** $3/\sqrt{34}$
15. $\sin^{-1} 5$ is undefined. **17.** $\sqrt{1 - x^2}$ **21.** $3/\sqrt{1 - 9x^2}$ **23.** $1/(2\sqrt{x - x^2})$
25. $-(3x^2 + 1)/\sqrt{1 - (x^3 + x)^2}$ **27.** $2/(4 + x^2)$ **29.** $2(x - 4)/(1 + (x - 4)^4)$
31. $\frac{1}{5}\sin^{-1}(5x/3) + C$ **33.** $\frac{1}{2}\tan^{-1}(x^2) + C$ **35.** $\frac{2}{3}\tan^{-1}(x^{3/2}) + C$
37. $1/[x(1 + \ln^2 x)]$ **39.** 0 **41.** $-\dfrac{1}{2}\tan^{-1}\left(\dfrac{e^{-x}}{2}\right) + C$ **43.** $\dfrac{-e^{-x}}{\sin^{-1}(e^{-x})\sqrt{1 - e^{-2x}}}$
45. $2x \cos^{-1}(1 - x) + x^2/\sqrt{2x - x^2}$ **47.** $\tan^{-1}(x - 1) + C$ **49.** $\pi/2$
51. $-\frac{1}{2}[\cos^{-1} x]^2 + C$ **53.** $\tan^{-1}(\sec x) + C$ **55.** $\pi/4$ **57.** $2\sqrt{3} - 2\pi/3$
69. (a) $\pi/3$ (b) $2\pi/3$ (c) $\pi/4$ (d) $\pi/6$ (e) 0 (f) π **71.**

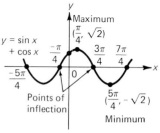

77. To verify an integral formula, it suffices to check that the derivative of the would-be antiderivative actually equals the function being integrated; the derivative results you need are given in Theorems 1, 2, 3 and Problems 65, 70(c), 75(c), plus $(d/dx)|x| = |x|/x$ if $x \neq 0$.
79. $\dfrac{4}{|4x + 2| \sqrt{(4x + 2)^2 - 1}}$ **81.** $-1/(2x\sqrt{x - 1})$, $x > 1$
83. $\frac{1}{3}(\sec^{-1}\sqrt{2} - \sec^{-1} 2) = -\pi/36$ **85.** $-e^x/(1 + e^{2x})$ **87.** $|1/a| \sec^{-1}|x/a| + C$
89. $\sec^{-1} 2 - \sec^{-1} 1 = \pi/3$ **91.** $-\dfrac{\csc^2 x}{|\cot x|\sqrt{\cot^2 x - 1}}$
93. $\dfrac{-1}{x(1 + \ln^2 x)\sqrt{1 - (\cot^{-1}\ln x)^2}}$ **97.** $\pi/4$ for $x > 1$ and $5\pi/4$ for $x < 1$
99. $0.3/1.36 \approx 0.22$ radians per minute, decreasing
101. $3 \cdot 2 \cdot (\pi/9) = 2\pi/3 \approx 2.09$ km/sec **103.** (a) $A(1) = \pi/4 = B(1)$
(b) $A'(x) = -1/(1 + x^2)$; $B'(0)$ is not defined but otherwise $B'(x) = A'(x)$ if $x \neq 0$
(c)

$3\pi/2$

$y = A(x) - B(x)$

105. (a) $f(0) = -\pi/2 = g(0)$ (b) $[-4, 4]$ is maximal domain for f and for g
(c) $f'(x) = (x/|x|)(2/\sqrt{16 - x^2})$, $g'(x) = 2/\sqrt{16 - x^2}$; $f' = g'$ on $(0, 4)$, $f' = -g'$ on $(-4, 0)$, $f'(0)$ does not exist, $g'(0) = \frac{1}{2}$
(d) $f = g$ on $[0, 4]$; note that f is an even function while g is a vertical translation of an odd function; therefore $f \neq g$ throughout $[-4, 0)$

PROBLEMS 7.3

1. $x = \sin t$
3. period $= 16\pi/(\pi - 1)$, frequency $= (\pi - 1)/16\pi$, spring constant $= 5[(\pi - 1)/8]^2$
5. $x(t) = A \cos(\omega t + \delta)$, where $\omega = 2\pi/T$, $A = \sqrt{x_0^2 + (v_0 T/2\pi)^2}$, and $\delta = \tan^{-1}(-v_0 T/2\pi x_0)$

9. $x = A \cos(4t + \delta)$, $y = -2A \sin(4t + \delta)$, where $\delta = \tan^{-1}(-5) \approx -1.3734$ and $A = 1000\sqrt{26} \approx 5099$. The prey will be extinct within $-\frac{1}{4} \tan^{-1}(-5) \approx 0.3434$ weeks.

PROBLEMS 7.4

7. $4 \cosh(4x + 2)$ 9. $-\dfrac{[\operatorname{csch}(\ln x)][\coth(\ln x)]}{x} = -\dfrac{2(1 + x^2)}{(x^2 - 1)^2}$ 11. $2 \tanh \sqrt{x} + C$

13. $\cosh x \cos(\sinh x)$ 15. $2 \tanh x \operatorname{sech}^2 x \, e^{\tanh^2 x}$ 17. $\sinh x + [(\sinh^3 x)/3] + C$

19. $\ln |\sinh x| + C$ 21. $-\cosh(1/x) + C$

23. undefined because $\sin^{-1}(\cosh x)$ is only defined when $x = 0$ (otherwise $\cosh x > 1$)

25. $\frac{1}{2} \ln |1 + \sinh 2x| + C$ 27. $-\frac{3}{11} \operatorname{sech}^{11/3} x + C$

29. $[(\cosh x)/x + (\sinh x)(\ln x)] x^{\cosh x}$ 31. $2a^2 \sinh 1 \approx 2.35a^2$ 45. $\sqrt{29}/5$

47. $\pm\sqrt{55}/8$ 49. $\sqrt{5}/2$

51. $\sinh x = \frac{1}{5}$, $\cosh x = \sqrt{26}/5$, $\tanh x = 1/\sqrt{26}$, $\coth x = \sqrt{26}$, $\operatorname{sech} x = 5/\sqrt{26}$

PROBLEMS 7.5

1. $3/\sqrt{(3x + 2)^2 + 1}$ 3. $1/[x(1 - \ln^2 x)]$, $1/e < x < e$

5. $(\sec \sinh^{-1} x)(\tan \sinh^{-1} x)/\sqrt{x^2 + 1}$ 7. $x/\sqrt{x^2 + 1}$ 9. $1/2\sqrt{(x^2 + 1)} \sinh^{-1} x$

11. $\dfrac{[(\cosh^{-1} x)/\sqrt{x^2 + 1}] - [(\sinh^{-1} x)/\sqrt{x^2 - 1}]}{(\cosh^{-1} x)^2}$, $x > 1$

19. (a) $\ln(4 + \sqrt{17})$;   (b) $\ln(3 + \sqrt{8})$;   (c) $\frac{1}{2} \ln 3$;   (d) $\ln(\sqrt{2} - 1)$

23. $\frac{1}{2} \cosh^{-1}(x^2/2) + C = \frac{1}{2} \ln(x^2 + \sqrt{x^4 - 4}) + C$

25. $\frac{1}{4} \tanh^{-1}(x^2/2) + C = \frac{1}{8} \ln[(2 + x^2)/(2 - x^2)] + C$

27. $\cosh^{-1}(e^x/3) + C = \ln(e^x + \sqrt{e^{2x} - 9}) + C$

29. $\sinh^{-1}(\sin x) + C = \ln(\sin x + \sqrt{1 + \sin^2 x}) + C$

31. $\sinh^{-1}(\frac{1}{2} \sec x) + C = \ln(\sec x + \sqrt{4 + \sec^2 x}) + C$

33. $\cosh^{-1}(2x + 1) + C = \ln(2x + 1 + \sqrt{4x^2 + 4x}) + C$

REVIEW EXERCISES FOR CHAPTER SEVEN

1. period $= \pi/5$

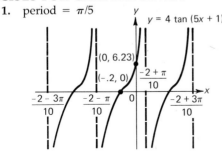

$y = 4 \tan (5x + 1)$

$(0, 6.23)$

$(-.2, 0)$ $\dfrac{-2 + \pi}{10}$

$\dfrac{-2 - 3\pi}{10}$ $\dfrac{-2 - \pi}{10}$ 0 $\dfrac{-2 + 3\pi}{10}$

3. $16x \cos(8x^2 + 2)$ 5. undefined since $e^x + 1 > 1$

7. $\frac{1}{3}[\cos x - (1/\sqrt{1 - x^2})](\sin x + \cos^{-1} x)^{-2/3}$ 9. $-5 \ln (\sqrt{3}/2)$ 11. $2 \sec \sqrt{x} + C$

13. $\frac{3}{2} \tan^{-1} x^2 + C$ 15. $-2x/(x^4 + 1)$ 17. $e^x[(x/\sqrt{1 - x^2}) + \sin^{-1} x + x \sin^{-1} x]$

19. $-\cot e^x + C$ 21. $-\sin x \sec^2(\cos x)$

23. $[\sin^{-1} x \cos x - (\sin x/\sqrt{1 - x^2})]/(\sin^{-1} x)^2$ 25. $(x/2) - [\sin(6x + 4)]/12 + C$

27. (a) $\sqrt{51}/10$ (b) $-\sqrt{15}$ (c) $\sqrt{13}/3$ 29. $(2x + 3) \cosh(x^2 + 3x)$

31. $-3 \operatorname{sech}^3 x \tanh x \, e^{\operatorname{sech}^3 x}$ 33. $\frac{1}{4} \cosh x^4 + C$ 35. $(\sinh x) \cos(\cosh x)$

37. $(\csc^2 x) \operatorname{csch}^2(\cot x)$ 39. $-\frac{3}{4} \operatorname{csch}^{4/3} x + C$ 41. $-(\sin x)/\sqrt{1 + \cos^2 x}$

43. $-1/(2x\sqrt{1 + x})$ 45. $\cosh^{-1}[(\tan x)/2] + C = \ln |\tan x + \sqrt{\tan^2 x - 4}| + C$

47. $\pm\sqrt{3}$

CHAPTER EIGHT
PROBLEMS 8.2

1. $\frac{1}{3} x e^{3x} - \frac{1}{9} e^{3x} + C$ 3. $64e - 128$ 5. $(x^8/8) \ln x - (x^8/64) + C$

7. $(x^2/2) \sin 2x + (x/2) \cos 2x - \frac{1}{4} \sin 2x + C$ 9. $-\frac{4}{3} x(1 - x/2)^{3/2} - \frac{16}{15}(1 - x/2)^{5/2} + C$

11. $(\pi/8) - \frac{1}{4}\ln 2$ **13.** $\frac{3}{4}\sinh 2 - \frac{1}{2}\cosh 2$ **15.** $(\pi/2) - 1$

17. $(x/2)[\sin(\ln x) + \cos(\ln x)] + C$ **19.** $\frac{1}{13}e^{3x}(3\sin 2x - 2\cos 2x) + C$ **21.** $\frac{1}{3}$

23. $(-\sin^3 x \cos x)/4 + (3x/8) - (3\sin x \cos x)/8 + C$

25. $\frac{1}{3}x^3\tan^{-1} x - \frac{1}{6}x^2 + \frac{1}{6}\ln(1 + x^2) + C$ (note that $x^3/(1 + x^2) = x - x/(1 + x^2)$)

27. $\frac{5}{21}\sinh 2x \cosh 5x - \frac{2}{21}\sinh 5x \cosh 2x + C$ **31.** $1 - 6e^{-5}$ kg

33. (a) y (b) $2 - 2/e$

$y = xe^{-x}$

$(1, \frac{1}{e})$ $(2, \frac{2}{e^2})$

0 x

Point of inflection

Maximum

35. $\frac{1}{3}\cos^2 x \sin x + \frac{2}{3}\sin x + C$

37. $\frac{1}{4}\tan x \sec^3 x + \frac{3}{8}\tan x \sec x + \frac{3}{8}\ln|\sec x + \tan x| + C$

39. $(x^3 - 6x)\sin x + (3x^2 - 6)\cos x + C$

41. (a) $T_0(x) = x$, $T_1(x) = -\ln|\cos x|$ (b) $T_n(x) = [1/(n - 1)]\tan^{n-1} x - T_{n-2}(x)$

(c) $T_2(x) = \tan x - x$, $T_3(x) = \frac{1}{2}\tan^2 x + \ln|\cos x|$ **43.** (b) $I_1(x) = \sqrt{x^2 + 1} - 1$

(c) $I_n(x) = (1/n)x^{n-1}\sqrt{x^2 + 1} - [(n - 1)/n]I_{n-2}(x)$

45. $J_n(x) = [2/(2n - k + 2)]x^{n-k+1}\sqrt{x^k + 1} - [2(n - k + 1)/(2n - k + 2)]J_{n-k}(x)$

PROBLEMS 8.3

1. $\frac{4}{15}$ **3.** $\pi/8 + \frac{1}{4}$ **5.** $\frac{3}{8}x + \dfrac{\sin 4x}{8} + \dfrac{\sin 8x}{64} + C$ **7.** $\frac{1}{2}\cos x - \frac{1}{10}\cos 5x + C$

9. $(-1/\sqrt{2})\cos(x/\sqrt{2}) - (1/3\sqrt{2})\cos(3x/\sqrt{2}) + C$ **11.** $\frac{1}{180}\sin 90 x + \frac{1}{220}\sin 110x + C$

13. $\frac{1}{10}(\sec 5x \tan 5x + \ln|\sec 5x + \tan 5x|) + C$ **15.** $3\pi/256$ **17.** $(2\sqrt{2} + 2)/15$

19. $-\frac{1}{3}\cos^3 x + \frac{2}{5}\cos^5 x - \frac{1}{7}\cos^7 x + C$ **21.** $-\frac{1}{2} - 2\ln(1/\sqrt{2}) = -\frac{1}{2} + \ln 2$

23. $\frac{1}{6}\tan^6 x + \frac{1}{4}\tan^4 x + C$ **25.** $\frac{1}{2}\ln 2 - \frac{1}{4}$ **27.** $2\sqrt{\sin x} - \frac{2}{5}\sin^{5/2} x + C$

37. $-\frac{1}{4}\cot^4 x + \frac{1}{2}\cot^2 x + \ln|\sin x| + C$ **39.** $-\frac{1}{2}\csc x \cot x - \frac{1}{2}\ln|\csc x + \cot x| + C$

41. $-\frac{1}{7}\csc^7 x + \frac{1}{5}\csc^5 x + C$ **43.** $-\frac{1}{2}\cot 2x - \frac{1}{6}\cot^3 2x + C$

55. (a) $\int(\cos^2 x - \sin^2 x)\,dx = \frac{1}{2}(\sin x + \cos x)^2 + C$

(b) $\int\cos^2 x\,dx = \frac{1}{2}(x + \sin x \cos x) + C$, $\int\sin^2 x\,dx = \frac{1}{2}(x - \sin x \cos x) + C$

PROBLEMS 8.4

1. $\frac{1}{4}\int_0^{81} e^u u^{-3/4}\,du$ **3.** $\frac{1}{5}\int_1^2 z^{1/2}(z - 1)^{-4/5}\,dz$ **5.** $2\int_0^1 [v/(3 + v)]\,dv$ **7.** $\int_0^{\pi/2}\cos^4\theta\,d\theta$

9. $\int_0^4 (16 - y^2)y^2\,dy$ **11.** $243\int_0^{\pi/2}\cos^3\theta\sin^2\theta\,d\theta$ **13.** $\int_0^{\sinh^{-1}1}\cosh^2 t\,dt$

15. $u = 1/x$ is not differentiable on $(-1, 1)$ (trouble at 0); furthermore this change of variables maps $[-1, 0) \cup (0, 1]$ onto $(-\infty, -1] \cup [1, \infty)$, not onto $[-1, 1]$

17. The correct numerical value is 17,152. The given procedure goes astray because $t = x^2$ does not specify t in terms of a 1-1 function of x. [*Note:* This substitution can be made to work by splitting the interval of integration: $[-1, 7] = [-1, 0] \cup [0, 7]$ and $\int_{-1}^7 f(x)\,dx = \int_{-1}^0 f(x)\,dx + \int_0^7 f(x)\,dx$.]

PROBLEMS 8.5

1. $2 - \sqrt{3}$ **3.** $-(x/2)\sqrt{4 - x^2} + 2\sin^{-1}(x/2) + C$ **5.** $\frac{64}{15} - \frac{11}{5}\sqrt{3}$

7. $\ln|x + \sqrt{x^2 - 4}| + C$ **9.** $4\sqrt{3} - \frac{4}{3} + 2\ln[\sqrt{3}/(2 + \sqrt{3})]$ **11.** $2(\sqrt{2} - 1)$

13. $\frac{1}{2}\tan^{-1}(x/2) + C$ **15.** $\frac{1}{4}[24\ln|x + \sqrt{x^2 + 4}| + 10x\sqrt{x^2 + 4} + x^3\sqrt{x^2 + 4}] + C$

17. $\frac{3}{4} + \ln 2$ **19.** $\pi/(4\sqrt{2})$ **21.** $2\sqrt{55} - \sqrt{7} + \frac{9}{4}\ln[(4 + \sqrt{7})/(8 + \sqrt{55})]$

23. $(2/\sqrt{3}) - (4/\sqrt{15}) + \ln[(4 + \sqrt{15})/(2 + \sqrt{3})]$ **25.** $\frac{1}{9}[(4/\sqrt{7}) - \frac{5}{4}]$

27. $-(x/4)(a^2 - x^2)^{3/2} + (a^2/8)[x\sqrt{a^2 - x^2} + a^2\sin^{-1}(x/|a|)] + C$

29. $\frac{1}{2}[x\sqrt{1 + a^2x^2} + (1/a)\ln|ax + \sqrt{a^2x^2 + 1}|] + C$

31. $\sqrt{5x^2 - 9} - 3\cos^{-1}(3|x|/\sqrt{5}) + C$

33. $\sqrt{9 - 5x^2} - 3 \ln |(3 + \sqrt{9 - 5x^2})/(\sqrt{5}x)| + C$ **35.** $(1/|a|) \sec^{-1}|x/a| + C$

37. $\frac{1}{7}(x^2 - a^2)^{7/2} + (a^2/5)(x^2 - a^2)^{5/2} + C$ **39.** $\frac{1}{3}(a^2 - x^2)^{3/2} - a^2\sqrt{a^2 - x^2} + C$

41. $-\dfrac{\sqrt{a^2 - x^2}}{2a^2x^2} - \dfrac{1}{2a^3} \ln \left|\dfrac{a + \sqrt{a^2 - x^2}}{x}\right| + C$ **43.** $\dfrac{-x}{a^2\sqrt{x^2 - a^2}} + C$

45. $\dfrac{-1}{a} \ln \left|\dfrac{a + \sqrt{x^2 + a^2}}{x}\right| + C$ **47.** $\dfrac{-x}{\sqrt{x^2 + a^2}} + \ln |x + \sqrt{x^2 + a^2}| + C$

49. $x - a \tan^{-1}(x/a) + C$ **51.** $\dfrac{-1}{a^4x} - \dfrac{x}{2a^4(x^2 + a^2)} - \dfrac{3}{2a^5} \tan^{-1}\left(\dfrac{x}{a}\right) + C$

53. $(1/b) \ln |bx + \sqrt{b^2x^2 + a^2}| + C$

55. Note that they have the same derivative and the same value when $x = 1$

57. $25\pi/2\sqrt{2}$ **59.** (a) the part of the unit circle lying in the first quadrant   (b) $\pi/4$

63.

(a) has area $\pi/4 = \frac{1}{4}\pi(1^2) = \int_0^1 \sqrt{2x - x^2}\, dx$

(b) has area $\frac{1}{8}\pi(\sqrt{2})^2 + \frac{1}{2}\cdot 1 \cdot 1 = \pi/4 + \frac{1}{2} = \int_0^1 \sqrt{2 - x^2}\, dx$

(c) has area $\frac{1}{2} = \int_0^1 \sqrt{2 - x^2}\, dx - \int_0^1 \sqrt{2x - x^2}\, dx$

PROBLEMS 8.6

1. $\frac{1}{2} \ln |2x - 5| + C$ **3.** $\frac{1}{3} \ln |3x + 11| + C$ **5.** $\frac{1}{3} \tan^{-1}(x/3) + C$

7. $\frac{1}{8} \ln |(x - 4)/(x + 4)| + C$ **9.** $(1/2\sqrt{3}) \tan^{-1}(x/2\sqrt{3}) + C$ **11.** $x - 2 \ln |x + 1| + C$

13. $\frac{1}{2}x^2 + 6x + 17 \ln |x - 2| + C$ **15.** $\frac{1}{4}x^4 - x^3 + \frac{9}{2}x^2 - 20x + 47 \ln |x + 2| + C$

17. $\frac{1}{4} \ln |x - 2| + \frac{7}{4} \ln |x + 2| + C$ **19.** $\frac{1}{3} \ln |(x - 4)/(x - 1)| + C$

21. $[1/(b - a)] \ln |(x - b)/(x - a)| + C$ **23.** $\frac{7}{5} \ln |x + 3| + \frac{8}{5} \ln |x - 7| + C$

25. $x - 2 \tan^{-1}x + C$ **27.** $x + 3 \ln |x^2 - 6x + 25| - 2 \tan^{-1}[(x - 3)/4] + C$

29. $\frac{1}{2}x^2 + 6x + \frac{1}{2} \ln |x + 1| + \frac{85}{2} \ln |x - 7| + C$ **31.** $\frac{1}{4}x^4 + \frac{1}{2}x^2 + \ln |x + 1| + C$

33. $\frac{1}{2}x^2 - 2x + \frac{3}{2} \ln |x^2 + x + 1| + (11/\sqrt{3}) \tan^{-1}[(2x + 1)/\sqrt{3}] + C$

35. $\frac{1}{3}x - \frac{1}{2} \ln |x^2 + 2x - 5| + [5/(3\sqrt{6})] \ln |(x + 1 - \sqrt{6})/(x + 1 + \sqrt{6})| + C$

37. $1/(\cos x + 2) + C$ **39.** $2 - 2 \ln 2$

PROBLEMS 8.7

1. $\frac{3}{2} \ln 3 - \frac{5}{2} \ln 2$ **3.** $4 \ln |x - 1| - 14 \ln |x - 2| + 11 \ln |x - 3| + C$

5. $\dfrac{x^3}{3} + x + \dfrac{1}{2} \ln \left|\dfrac{x - 1}{x + 1}\right| + C$ **7.** $\dfrac{1}{2}x^2 + 12x + 96 \ln |x - 4| - \dfrac{256}{x - 4} - \dfrac{128}{(x - 4)^2} + C$

9. $\dfrac{1}{4} \ln \left|\dfrac{x + 2}{x}\right| - \dfrac{1}{4x} - \dfrac{1}{4(x + 2)} + C$ **11.** $\ln |x| - \dfrac{1}{x} + \dfrac{2}{3x^3} + C$

13. $-\dfrac{1}{6} \ln \left|\dfrac{x - 1}{x + 1}\right| + \dfrac{1}{12} \ln \left|\dfrac{x - 2}{x + 2}\right| + C$

15. $\frac{1}{2}x^2 - 2x + 20 \ln |x + 3| - \frac{22}{3} \ln |x + 2| + \frac{13}{3} \ln |x - 4| + C$ **17.** $8 \ln \frac{5}{4} - \frac{3}{2}$

19. $2 \ln |x| - \dfrac{3}{4} \ln |x + 1| - \dfrac{3}{2(x - 1)} - \dfrac{5}{4} \ln |x - 1| + C$

21. $\frac{1}{5} \ln |x + 2| - \frac{1}{10} \ln (x^2 + 1) + \frac{2}{5} \tan^{-1}x + C$

23. $\frac{1}{2} \ln(x^2 + 2) + (5/4\sqrt{2}) \tan^{-1}(x/\sqrt{2}) + [1/(x^2 + 2)] + [x/4(x^2 + 2)] + C$

25. $\frac{1}{2} \ln |x^2 - 4| - [2/(x^2 - 4)] + C$ **27.** $\frac{1}{4} \ln |(x^2 - 1)/(x^2 + 1)| + C$

29. $\frac{109}{442} \ln |x^2 - x + \frac{1}{2}| + \frac{261}{221} \tan^{-1}(2x - 1) + \frac{56}{221} \ln |x^2 - 6x + 10| + \frac{1026}{221} \tan^{-1}(x - 3) + C$

31. $\ln |x + (1/x)| - (1/2x^2)$

33. $\frac{1}{2}\ln|x^2 + 4x + 13| + [(47x + 67)/18(x^2 + 4x + 13)] - \frac{61}{54}\tan^{-1}[(x + 2)/3] + C$

35. $\int_0^{\pi/2} \frac{\cos x}{(\sin x + 1)^2 + 1}\,dx = \tan^{-1}(\sin x + 1)\Big|_0^{\pi/2} = \tan^{-1} 2 - \frac{\pi}{4} \approx 0.32175$

37. $\int_0^1 \frac{x^3}{(x^2 + 1)^3}\,dx = \frac{1}{16}$ **39.** $\frac{1}{4}\ln|x^4 - 2x^2| + C$ **41.** $\frac{7}{40}\ln(2x^{60/7} + 5x^{32/7}) + C$

PROBLEMS 8.8

1. $x - 4\sqrt{x + 2} + 4\ln|1 + \sqrt{x + 2}| + C$ (let $u = \sqrt{x + 2}$)

3. $\frac{2}{5}\sqrt{x} - (\frac{8}{25})\ln|4 + 5\sqrt{x}| + C$ (let $u = 5\sqrt{x}$)

5. $\frac{2}{5}(1 + x)^{5/2} - \frac{2}{3}(1 + x)^{3/2} + C$ (let $u = 1 + x$)

7. $\frac{1}{4}[\frac{3}{8}(1 + 2x)^{8/3} - \frac{3}{5}(1 + 2x)^{5/3}]\Big|_0^{13} = 578.7$ $(u = 1 + 2x)$

9. $-\frac{1}{32}[(16/x^3) - 1]^{2/3} + C$ (if $u = (16/x^3) - 1$, then
$$\frac{dx}{x^3(16 - x^3)^{1/3}} = \frac{dx}{x^4[(16 - x^3)/x^3]^{1/3}} = \frac{3x^2\,dx}{3x^6[(16 - x^3)/x^3]^{1/3}} = \frac{-16(1 + u)^{-2}\,du}{3[16/(u + 1)]^2\,u^{1/3}}$$
$= -(1/48)\,u^{-1/3}\,du)$

11. $\frac{8}{3}$ (if $u = 1 + x^3$, $[3x^5/(1 + x^3)^{3/2}]\,dx = (u^{-1/2} - u^{-3/2})\,du)$

13. $\frac{6}{5}(1 + x)^{5/2} - \frac{14}{3}(1 + x)^{3/2} + 8(1 + x)^{1/2} + C$ $(u = 1 + x)$

15. $12(\frac{1}{8} - \frac{1}{7} + \frac{1}{6} - \frac{1}{5} + \frac{1}{4} - \frac{1}{3} + \frac{1}{2} - 1 + \ln 2) \approx 0.7$ $(u = x^{1/12})$

17. $\frac{6}{5}x^{5/6} - 6x^{1/6} + (3/2\sqrt{2})\ln|(x^{1/3} + \sqrt{2}x^{1/6} + 1)/(x^{1/3} - \sqrt{2}x^{1/6} + 1)| +$
$(3/\sqrt{2})[\tan^{-1}(\sqrt{2}x^{1/6} + 1) + \tan^{-1}(\sqrt{2}x^{1/6} - 1)]$ (let $u = x^{1/6}$, use the factorization
$(u^2 + au + 1)(u^2 - au + 1) = u^4 + (2 - a^2)u^2 + 1)$

19. $3(1 - x)^{1/3} + \frac{1}{2^{2/3}}\ln\left|\frac{[2^{1/3} - (1 - x)^{1/3}]^2}{2^{2/3} + 2^{1/3}(1 - x)^{1/3} + (1 - x)^{2/3}}\right| -$
$2^{1/3}\sqrt{3}\tan^{-1}\left[\frac{2(1 - x)^{1/3} + 2^{1/3}}{2^{1/3}\sqrt{3}}\right] + C$ $\left(\text{if } u = (1 - x)^{1/3}, \text{ then } \frac{(1 - x)^{1/3}}{1 + x}\,dx = \frac{-3u^3}{2 - u^3}\,du\right)$

21. $x + 6x^{1/3} + 3\ln|(x^{1/3} - 1)/(x^{1/3} + 1)| + C$ $(u = x^{1/3})$

23. $\frac{1}{27}[\frac{2}{7}u^{7/2} - \frac{8}{5}u^{5/2} + \frac{8}{3}u^{3/2}]\Big|_2^5 = \frac{1}{27}(\frac{190}{21}\sqrt{5} - \frac{128}{105}\sqrt{2}) \approx 0.6854$ $(u = 2 + 3x)$

25. $6[(x^{1/3}/2) + \tan^{-1}x^{1/6} - \frac{1}{2}\ln|x^{1/3} + 1|] + C$ $(u = x^{1/6})$

27. $\frac{-2(1 + \cos x)}{1 + \cos x + \sin x} + C = \frac{-2\sin x}{1 - \cos x + \sin x} + C$ (let $u = \tan(x/2) = \sin x/(1 + \cos x)$)

29. $\int_0^{1/\sqrt{3}}[2/(5 + u^2)]\,du = (2/\sqrt{5})\tan^{-1}(u/\sqrt{5})\Big|_0^{1/\sqrt{3}} = (2/\sqrt{5})\tan^{-1}(1/\sqrt{15})$ (let $u = \tan(x/2)$)

31. $(1/2\sqrt{2})\ln|[\sqrt{2} + (\tan x - 1)]/[\sqrt{2} - (\tan x - 1)]| + C$ (find $\int[1/(\sin u + \cos u)]\,du$ and
then show, by substituting $u = 2x$, that $\int[1/(\sin 2x + \cos 2x)]\,dx = \frac{1}{2}\int[1/(\sin u + \cos u)]\,du$)

33. $2\tan^{-1}[2 + \tan(x/2)] + C$ **35.** $(1/\sqrt{5})\cos^{-1}\{[(5/x) - 2]/3\} + C$

37. $2 - \sqrt{3}$ (let $u = 4 - x^2$)   **39.** $\frac{64}{15} - \frac{11}{5}\sqrt{3}$ (let $u = 4 - x^2$)

41. $\frac{4}{3}(x^2 - 4)^{3/2} + \frac{1}{5}(x^2 - 4)^{5/2} + C$ (let $u = x^2 - 4$)

43. $\sqrt{5x^2 - 9} - 3\tan^{-1}(\sqrt{5x^2 - 9}/3) + C$ (let $u = \sqrt{5x^2 - 9}$, then
$(1/x)\,dx = [(u\,du)/(u^2 + 9)])$

45. $\sqrt{9 - 5x^2} + \frac{3}{2}\ln\left|\frac{\sqrt{9 - 5x^2} - 3}{\sqrt{9 - 5x^2} + 3}\right| + C$ (let $u = \sqrt{9 - 5x^2}$, then $(1/x)\,dx = [u/(u^2 - 9)]\,du$)

47. $\frac{1}{7}(x^2 - a^2)^{7/2} + (a^2/5)(x^2 - a^2)^{5/2} + C$ (let $u = x^2 - a^2$)

49. $\frac{1}{3}(a^2 - x^2)^{3/2} - a^2(a^2 - x^2)^{1/2} + C$ (let $u = a^2 - x^2$)

51. $x - \ln(e^x + 1) + C$ (let $v = e^x$)   **53.** $\ln|x + \sqrt{x^2 + 5}| + C$

55. $\ln|\sqrt{\sec^2 x + A} + \tan x| + (\sqrt{A})\sin^{-1}(\sqrt{A/(A + 1)}\sin x) + C$ (Substitutions:
(1) $u = \sec x$; (2) $v = u^2$; (3) $w^2 = [(v + A)/(v - 1)])$ **57.** $\ln|\sec x + \tan x| + C$

59. $\frac{1}{2}(\sec x \tan x + \ln|\sec x + \tan x|) + C$

PROBLEMS 8.9

1. $\frac{1}{12}\ln|(2x + 3)/(2x - 3)| + C$ **3.** $\frac{1}{6}\tan^{-1}(2x/3) + C$

5. $\frac{1}{2}x\sqrt{16 - 3x^2} + (8/\sqrt{3})\sin^{-1}(\sqrt{3}x/4) + C$ **7.** $\frac{1}{3}\sqrt{16 + 3x^2} + C$

9. $\sqrt{10 + 2x^2} - \sqrt{10}\ln\left|\frac{(\sqrt{10} + \sqrt{10 + 2x^2})}{x}\right| + C$

11. $\frac{1}{3}\cos^{-1}(3/\sqrt{2}x) + C = \frac{1}{3}\sec^{-1}(\sqrt{2}x/3) + C$

13. $\dfrac{-x}{16\sqrt{4x^2 - 16}} + C = \dfrac{-x}{32\sqrt{x^2 - 4}} + C$ **15.** $\dfrac{\sqrt{9x^2 - 1}}{x} + C$

17. $\frac{1}{4}x^2 - \frac{3}{4}x + \frac{9}{8}\ln|2x + 3| + C$ **19.** $-\frac{1}{3}x + \frac{2}{9}\ln|2 + (3/x)| + C$

21. $\tan^3(x/3) - 3\tan(x/3) + x + C$ **23.** $(-\frac{1}{7})(\sin^2 x + \frac{2}{5})\cos^5 x + C$

25. $\frac{1}{6}[\sin(x^3) + \frac{1}{9}\sin(9x^3)] + C$ **27.** $\frac{1}{16}[(8x^2 - 1)\cos^{-1}(2x) - 2x\sqrt{1 - 4x^2}] + C$

29. $[(x^4/4) - 24]\sin^{-1}(x/4) + [(x^3/4) + 6x]\sqrt{1 - (x^2/16)} + C$ **31.** $-(x + 1)e^{-x} + C$

33. $\frac{1}{12}e^{4x^3} + C$

35. $\frac{-1}{12}(3x - 4x^2)^{3/2} - \frac{3}{128}(3 - 8x)\sqrt{3x - 4x^2} + (\frac{27}{512})\sin^{-1}[(8x - 3)/3] + C$

37. $\frac{1}{195}e^{4x^3}(4\cos 7x^3 + 7\sin 7x^3) + C$

PROBLEMS 8.10

In Problems 1–15 the answers are given in the order requested in the text. The last two numbers give the actual error in the trapezoidal and Simpson's estimates, respectively. The calculations were made with a hand calculator with ten decimal place precision.

1. $0.5; 0.5; 0; 0; 0.5; 0; 0$ **3.** $\frac{11}{32} = 0.34375; \frac{1}{3}; \frac{1}{96} = 0.0104167; 0, \frac{1}{3}; -\frac{1}{96}; 0$

5. $6.448104763; 6.389488576; 0.1368343722; 0.0010135879; 6.389056099; 0.0590486641;$
0.0004324773 **7.** $1.218760835; 1.218951005; 0.0003255208; 0.0000012716; 1.218951416;$
$-0.0001905814; -0.0000004111$ **9.** $0.9871158010; 1.000134585; 0.0201863780; 0.0002075329;$
$1.0; -0.012884199; 0.000134585$ **11.** $0.6973999653; 0.6932588905; 0.0207193081;$
$0.0010413912; \ln 2 = 0.6931471806; 0.0042527848; 0.0001117099$

13. $0.9956971321; 0.9999360657; 0.0117435512; 0.0003852527; 1.0; -0.0043028736;$
-0.000063343 **15.** $0.9129205114; 0.9093257681; 0.0040387583; 0.0000124097; 0.9093306736;$
$0.0035898377; -0.0000049056$

In Problems 17–29 the trapezoidal approximation is given first.

17. $0.6590191268; 0.6593301635$ **19.** $1.987795499; 1.994503740$

21. $0.6903257237; 0.8367702680$ **23.** $1.488736680; 1.493674110$

25. $0.9091616587; 0.9096068101$ **27.** $0.9841199229; 0.9838189106$

29. $0.3604819483; 0.3579282259$

31. $|y''| = |3(2x + 3x^4)e^{x^3}| \le 15e$ on $[0, 1]$ and $|y^{(4)}| = |9(9x^8 + 36x^5 + 20x^2)e^{x^3}| \le 585e$. Thus $|\epsilon_8^T| \le 15e/(12 \cdot 8^2) \approx 0.05309$ and $|\epsilon_8^S| \le 585e/(180 \cdot 8^4) \approx 0.00216$

33. $|y''| = |-2 + 4x^2|e^{-x^2} \le 2$ on $[-1, 1]$ and $|y^{(4)}| = |12 - 48x^2 + 16x^4|e^{-x^2} \le 12$. Thus $|\epsilon_{10}^T| \le 2^4[/(12)10^2] \approx 0.01333$ and $|\epsilon_{10}^S| \le 12 \cdot 2^5/(180 \cdot 10^4) \approx 0.00021$

35. $|y''| = |3(4x^2 + 3x^5)e^{x^3}| \le 21e$ on $[0, 1]; |y^{(4)}| = |3(8 + 132x^3 + 144x^6 + 27x^9)e^{x^3}| \le 933e$. Thus $|\epsilon_{10}^T| \le 21e/(12 \cdot 10^2) \approx 0.04757$ and $|\epsilon_{10}^S| \le 933e/(180 \cdot 10^4) \approx 0.00141$

37. $|y^{(4)}| \le 3$ so we need $(3/\sqrt{2\pi})1^5/(180 \cdot n^4) \le 0.005$ (for half the integral—using the hint); $n = 2$ will do; we obtain 0.6830581043 (the "true" value is 0.6826894921 giving an error of 0.0003686122)

39. (a) On $[0, 50]$ need $(3/\sqrt{2\pi})50^5/(180 \cdot n^4) \le 0.05$ or $n \ge 82$; this leads to the estimate $(1/\sqrt{2\pi})\int_{-50}^{50} = (2/\sqrt{2\pi})\int_0^{50} \approx 2(0.4999994266) = 0.9999988532$
(b) $\lim_{N \to \infty}(1/\sqrt{2\pi})\int_{-N}^{N} e^{-x^2/2}\, dx = 1$

41. $|y''| = 2/x^3 \le 2$ on $[1, 2]$; need $2/12n^2 \le 10^{-10}$ or $12n^2 > 2 \cdot 10^{10}$ which implies that $n > 40,824.8$

43. It is not difficult to verify that on $[\frac{1}{2}, 1]$, $|J_{1/2}''(x)| \le 8$. Then, with the trapezoidal rule, we need $8(\frac{1}{2})^3/12n^2 \le 0.01$ or $n > 2.88$. Using $n = 3$ gives $\int_{1/2}^{1} J_{1/2}(x)\, dx \approx 0.3095670957$.

REVIEW EXERCISES FOR CHAPTER EIGHT

1. $\frac{1}{2}\{\sqrt{5} + 4\ln[2/(1 + \sqrt{5})]\}$ **3.** $2\tan^{-1} x + x/(x^2 + 1) + C$ **5.** $\frac{2}{15}$

7. $-e^{-2x}[(x/2) + \frac{1}{4}] + C$ **9.** $\frac{1}{4}\sec^4 x - \sec^2 x - \ln|\cos x| + C$ **11.** $3\ln 2 - \frac{3}{2}$

13. $(-2/\sqrt{3}) + \sqrt{2} + \ln[(2 + \sqrt{3})/(\sqrt{2} + 1)]$ **15.** $\frac{1}{4}e^{2x}(\sin 2x - \cos 2x) + C$

17. $-\frac{1}{4}\cos^4 x + \frac{1}{6}\cos^6 x + C$ **19.** $-\frac{1}{5}\csc^5 x + \frac{1}{3}\csc^3 x + C$

21. $\frac{1}{3}x^3 + x + \ln|x/(x + 1)| + C$ **23.** $x\cosh x - \sinh x + C$ **25.** $-1/(x - 2) + C$

27. $\frac{1}{2}(\sin x + \frac{1}{13}\sin 13x) + C$ **29.** $-\frac{11}{14}\ln|x - 3| + \frac{2}{7}\ln|x + 4| + \frac{3}{2}\ln|x - 5| + C$

31. $\frac{1}{5}(x^2 - 4)^{5/2} + \frac{4}{3}(x^2 - 4)^{3/2} + C$ $(u = x^2 - 4)$ **33.** $\frac{1}{3}\tan^3 x + \tan x + C$

35. $x + 3\ln|x| - (3/x) - (1/2x^2) + C$ **37.** $5\pi/32$ **39.** $\ln(\sqrt{2} + 1)$

41. $\dfrac{1}{2\sqrt{6}}\ln\left|\dfrac{\tan(x/2) + 5 - \sqrt{24}}{\tan(x/2) + 5 + \sqrt{24}}\right| + C = \dfrac{1}{2\sqrt{6}}\ln\left|\dfrac{1 - \cos x + (5 - 2\sqrt{6})\sin x}{1 - \cos x + (5 + 2\sqrt{6})\sin x}\right| + C$

43. $-\frac{3}{2}(1 + x^4)^{-1/3} + C$

45. $\frac{1}{81}\left[-\frac{4}{7}(2 + 3x)^{7/2} + \frac{24}{5}(2 + 3x)^{5/2} - 10(2 + 3x)^{3/2} - 4(2 + 3x)^{1/2}\right] + C$ $(u = 2 + 3x)$

47. $\frac{9}{10}e^{-x}\left[-\cos(x/3) + \frac{1}{3}\sin(x/3)\right] + C$ **49.** $\frac{3}{8}\cosh x \sinh 3x - \frac{1}{8}\sinh x \cosh 3x + C$

51. $-(x^2 + 2x + 2)e^{-x} + C$ **53.** $\sin^2(\sqrt{x}) + C$

55. $\frac{1}{3}\ln|x/(3 + \sqrt{9 - 4x^2})| + C$ (let $u = 3/2x$)

57. $-(\sqrt{x^2 - 16}/x) + \ln|x + \sqrt{x^2 - 16}| + C$ **59.** $\frac{1}{7}\sin^{7/4}(2x^2) + C$ **61.** $\frac{16}{35}$

63. $\frac{1}{12}\sinh(6x) - (x/2) + C$ **65.** $-\frac{1}{39}e^{-3x^3}[3\sin(2x^3) + 2\cos(2x^3)] + C$

67. $\frac{1}{9}[\sec^2(3x) + 2]\tan(3x) + C$

69. $e^{2x}[\frac{3}{5}\cosh 3x - \frac{2}{5}\sinh 3x] + C = \frac{1}{10}e^{5x} + \frac{1}{2}e^{-x} + C$ **71.** $x - \frac{1}{3}\tanh(3x) + C$

73. $(1/\sqrt{7})\tan^{-1}(\sqrt{7}\tan x) + C$

75. $[-1/(1 + \tan x)] + C = [(\sin 2x)/(\cos 2x - \sin 2x - 1)] + C$

77. $(\frac{3}{16}x^2 - \frac{3}{128})\cos 4x + (\frac{1}{4}x^3 - \frac{3}{32}x)\sin 4x + C$ **79.** $\frac{1}{18}\tan^{-1}(x^6/3) + C$

81. $\frac{1}{12}[x^6\sqrt{9 + x^{12}} + 9\ln|x^6 + \sqrt{9 + x^{12}}|] + C$

83. $-\frac{2}{3}\ln|x - 1| + \frac{5}{6}\ln|x^2 + x + 1| + \sqrt{3}\tan^{-1}[(2x + 1)/\sqrt{3}] + C$

85. $[1/(x - 1)] + 2\ln|x - 1| + C$ **87.** $\frac{1}{8}\ln|(x + 4)/(x - 4)| + C$

89. $-\frac{1}{9}\sqrt{25 - 9x^2} + C$ **91.** $(x/2)\sqrt{16 + 9x^2} - \frac{8}{3}\ln|3x + \sqrt{16 + 9x^2}| + C$

93. $(\frac{1}{4}x^4 - \frac{1}{864})\cos^{-1}(3x) - (\frac{1}{48}x^3 + \frac{1}{288}x)\sqrt{1 - 9x^2} + C$

95. $\frac{1}{3}[x^3\tan^{-1}(x^3) - \frac{1}{2}\ln(1 + x^6)] + C$ **97.** 1.383213747 **99.** 0.9253959267

101. 1.005709257

103. $|y''| = |3(2x + 3x^4)e^{x^3}| \leq 15e$ on $[0, 1]$; need $15e/12n^2 \leq 0.01$ or $n \geq 18.4$

105. $|y^{(4)}| = 120/x^6 \leq 120$ on $[1, 2]$; we need $120/2880n^4 \leq 0.0001$ or $n \geq 4.52$.

Using Simpson's rule with $2n = 10$ yields the value 0.5000124699. The actual value is 0.5.

CHAPTER NINE
PROBLEMS 9.1

1. $8\pi/3$ **3.** $4\pi/5$ **5.** $64\pi/3$ **7.** $\pi^2/4$ **9.** $(\pi/2)(e^2 - e^{-2}) = \pi \sinh 2$

11. $\pi^2/4$ **13.** $\pi(e - 2)$ **15.** $\pi/2$ **17.** $(\pi/4)(e^2 - 1)$ **19.** $\pi/2$ **21.** 9π

23. $\pi/12$ **25.** $\pi(9\ln 3 - 4)$ **27.** $3\pi/10$ **29.** $\pi[(\pi/\sqrt{2}) - 2]$ **31.** $\pi(e - 1)$

33. $\frac{4}{3}\pi r^3$ **35.** $\frac{4}{3}\pi ab^2$ **37.** 6π

39. $(\pi/3)(b_1 + b_2 + b_3)|(a_2 - a_1)(b_3 - b_1) - (b_2 - b_1)(a_3 - a_1)|$ **41.** $\pi[4 - (\pi/4)]$

43. $\frac{128}{3}$ **45.** $16\pi/3$ **47.** $16\sqrt{3}/3$ m³   **49.** (a) $\frac{1}{30}$, (b) $\frac{1}{120}$, (c) $\pi/240$

51. $920g\pi \int_0^3 (3 - y)e^{2y}\,dy \approx 2.81 \times 10^6$ joules (assuming that distances are measured in meters)

PROBLEMS 9.2

1. $4\sqrt{10}$ **3.** $\frac{1}{27}(13^{3/2} - 8)$ **5.** $\frac{14}{3}$ **7.** $\frac{123}{32}$

9. $\sqrt{1 + e^4} - \sqrt{1 + e^2} + \frac{1}{2}\ln\left[\dfrac{(\sqrt{1 + e^4} - 1)(\sqrt{1 + e^2} + 1)}{(\sqrt{1 + e^4} + 1)(\sqrt{1 + e^2} - 1)}\right]$

11. (b) look at the definition of the radian   **13.** 6 **15.** $\frac{4}{27}(10^{3/2} - 1)$

17. $\int_0^1 \sqrt{1 + 9x^2 + 9x^4}\,dx$ **19.** $\int_0^1 \sqrt{1 + x^2\cosh^2 x + 2x\cosh x \sinh x + \sinh^2 x}\,dx$

21. $\int_0^4 \sqrt{1 + 4x^2 e^{2x^2}}\,dx$ **23.** $\int_{-1}^1 \sqrt{1 + 4x^2\sinh^2(1 + x^2)}\,dx$

PROBLEMS 9.3

1. $(\pi/6)(17^{3/2} - 5^{3/2})$ **3.** $2\pi\sqrt{5}$ **5.** $\pi\sqrt{1 + \alpha^2}[\alpha(b^2 - a^2) + 2\beta(b - a)]$

7. $(\pi/6)(27 - 5^{3/2})$ **9.** $47\pi/16$

11. $2\pi\displaystyle\int_1^2 \frac{1}{x}\sqrt{1 + \frac{1}{x^4}}\,dx = \pi\int_{1/4}^1 \frac{\sqrt{1 + v^2}}{v}\,dv\left(v = \frac{1}{x^2}\right) = \pi\int_{\tan^{-1}\frac{1}{4}}^{\pi/4} \frac{\sec^3\theta}{\tan\theta}\,d\theta$ $(v = \tan\theta)$

$= \pi\left[\ln\left(\dfrac{\sqrt{17} + 4}{\sqrt{2} + 1}\right) + \sqrt{2} - \dfrac{\sqrt{17}}{4}\right]$

15. (i) if $b > a$, $S = 2\pi b^2 + \dfrac{2\pi b a^2}{\sqrt{b^2 - a^2}} \ln\left|\dfrac{b + \sqrt{b^2 - a^2}}{a}\right|$

(ii) if $a > b$, $S = 2\pi b^2 + \dfrac{2\pi b a^2}{\sqrt{a^2 - b^2}} \sin^{-1}\left(\dfrac{\sqrt{a^2 - b^2}}{a}\right)$ **17.** 24π

PROBLEMS 9.4

1. -18 kg-m; -1.8 m   **3.** 101 kg-m; $\frac{101}{18}$ m   **5.** $\frac{1}{3}; \frac{1}{2}; \frac{2}{3}$ **7.** $\frac{31}{5}; \frac{15}{4}; \frac{124}{75}$
9. $2\pi(2\pi - 1); 4\pi; \pi - \frac{1}{2}$ **11.** $8 \ln 4 - \frac{15}{4}; 4 \ln 4 - 3; (8 \ln 4 - \frac{15}{4})/(4 \ln 4 - 3) \approx 2.88$
13. $(2e^3 + 1)/9; (e^2 + 1)/4; 4(2e^3 + 1)/9(e^2 + 1)$ **15.** $\ln \frac{25}{16}; \ln \frac{8}{5}; \ln \frac{25}{16}/\ln \frac{8}{5} \approx 0.95$

PROBLEMS 9.5

1. $(-1.8, 3.4)$ **3.** $(\frac{101}{18}, \frac{3}{2})$ **5.** $(\frac{13}{24}, \frac{49}{24})$ **7.** $(\frac{4}{7}, \frac{2}{5})$ **9.** $(\frac{493}{185}, \frac{525}{296})$

11. $\left(\dfrac{1 - \ln 2}{\ln 2}, \dfrac{1}{4 \ln 2}\right) \approx (0.44, 0.36)$ **13.** $\left(\dfrac{\ln (9/8)}{\ln (4/3)}, \dfrac{\ln (3/4) + 1/3}{\ln (4/3)}\right) \approx (0.409, 0.159)$

15. $\left(\dfrac{\pi - \sqrt{3}}{9}, \dfrac{\pi}{6\sqrt{3}} + \dfrac{1}{8}\right) \approx (0.157, 0.427)$

17. $\left(\dfrac{2 \ln 2 - \frac{3}{4}}{2 \ln 2 - 1}, \dfrac{(\ln 2)^2 - 2 \ln 2 + 1}{2 \ln 2 - 1}\right) \approx (1.65, 0.24)$

19. $\left(\dfrac{\frac{1}{2} - \ln 2}{1 - \ln 2}, \dfrac{\frac{3}{4} - \ln 2}{1 - \ln 2}\right) \approx (-0.63, 0.19)$ **21.** $(\frac{7}{10}, \frac{9}{16})$ **23.** $(\frac{13}{20}, \frac{77}{50})$ **25.** $(\frac{8}{7}, \frac{289}{11})$

27. $(\frac{213}{85}, \frac{6927}{595})$ **29.** $(4, \frac{1}{2})$ **31.** $\left(\dfrac{\pi\sqrt{2} - 4}{4(\sqrt{2} - 1)}, \dfrac{1}{4(\sqrt{2} - 1)}\right) \approx (0.27, 0.6)$

33. $\left(\dfrac{e^2 + \frac{3}{4} - 2 \ln 2}{e^2 + 1 - 2 \ln 2 - e}, \dfrac{\frac{1}{4}e^4 - (\ln 2)^2 + 2 \ln 2 - 1 - \frac{1}{4}e^2}{e^2 + 1 - 2 \ln 2 - e}\right) \approx (1.58, 2.73)$

37. $(\pi, -\pi/8)$ **43.** $2\pi^2 r^2 a$ **45.** $4\sqrt{2}\pi^2$

PROBLEMS 9.6

1. 120 kg $\cdot$ m^2 **3.** 758 kg $\cdot$ m^2 **5.** (a) $\frac{25}{18}\pi^2$ J (b) $\frac{125}{81}\pi^2$ J   **7.** (a) 3 kg $\cdot$ m^2
(b) 7 kg $\cdot$ m^2 (c) 5 kg $\cdot$ m^2 (d) $\frac{17}{4}$ kg $\cdot$ m^2 **9.** (a) $\frac{1}{2}$ m (b) $\sqrt{\frac{7}{12}}$ m (c) $\sqrt{\frac{5}{12}}$ m (d) $\sqrt{\frac{17}{48}}$ m
11. $800\pi^2(\frac{3}{4}\sqrt{7} + 32 \sin^{-1}\frac{3}{4}) \approx 229{,}940$ J (note that $\sin 4\theta = -3\sqrt{7}/32$ if $\theta = \sin^{-1}\frac{3}{4}$)
13. $I = 16, r = 2/\sqrt{3}$ **15.** $I = \frac{3}{5}, r = \sqrt{\frac{3}{5}}$ **17.** $I = \frac{383}{35}, r = \sqrt{\frac{383}{35}}$
19. $I = (\pi^2/2) - 4, r = \frac{1}{2}\sqrt{\pi^2 - 8}$ **21.** $I = 6, r = \sqrt{3}/(2 \ln 2)$
23. $I = 5(e - 2), r = \sqrt{(e - 2)/(e - 1)}$ **25.** $I = 17\pi/2, r = \sqrt{17}/2$
27. $I = 145\pi/2, r = \sqrt{145}/2$ **29.** $I = \frac{224}{5}, r = 2\sqrt{\frac{3}{5}}$
31. $I = 3[\int_0^{1/\sqrt{2}} 2x^3 \, dx + \int_{1/\sqrt{2}}^1 2x^2\sqrt{1 - x^2} \, dx] = \frac{3}{8} + 3\pi/16$ (we first rotated everything $45°$
clockwise); $r = (\frac{1}{2})\sqrt{1 + 2/\pi}$ **33.** $k = \frac{9}{2}\rho\omega^2$ J

PROBLEMS 9.7

1. $72{,}000(9.81)\pi \approx 2{,}218{,}970$ N   **3.** (a) $P_A = 1000(13.6)(9.81)h = 133{,}416h$ **5.** 975 lb
7. 5297.4 N   **9.** $(1000)(9.81) \int_0^{25/4}(\frac{25}{4} - y)2\sqrt{25y} \, dy = 2{,}554{,}687.4$ N
11. $(1000)(9.81) \int_0^{25/8}(\frac{25}{8} - y)10\sqrt{2y} \, 2dy = 1{,}277{,}343.8$ N   **13.** $3(1030)(9.81) = 30{,}312.9$ N
15. $(1000)(9.81)(3\sqrt{\frac{5}{2}}) \int_0^1(\frac{3}{2} - 3x/2) \, dx = 34{,}900$ N
17. $F = (62.4)(10) \int_0^{5/2}\sqrt{3}(\frac{5}{2} - x) \, 2dx = 3900\sqrt{3}$ lb
19. $(1030)(9.81)(10 - 10e^{-5} - \frac{1}{4} + \frac{1}{4}e^{-10}) \approx 97{,}836$ N
23. $\frac{45}{4}(1000)(9.81) = 110{,}362.5$ N   **25.** $(62.4)(30)(\frac{32}{5}) = 11{,}980.8$ lb
27. $(62.4)(20)(2 \ln 2 - 1) \approx 482$ lb

REVIEW EXERCISES FOR CHAPTER NINE

1. 56π **3.** $\pi/11$ **5.** $\pi^2/8$ **7.** $4\pi^2$ **9.** 144 **11.** $5\sqrt{5}$ **13.** $\frac{2}{3}[2^{3/2} - 1]$
15. $\int_0^{\pi/2}\sqrt{1 + (x \cos x + \sin x)^2} \, dx$
17. $\pi[\ln(\sqrt{1 + e^2} + e) + e\sqrt{1 + e^2} - \ln(\sqrt{2} + 1) - \sqrt{2}]$
19. $\pi\left\{\ln[(\sqrt{17} + 4)/(\sqrt{2} + 1)] + \sqrt{2} - \dfrac{\sqrt{17}}{4}\right\}$ (see answer to Problem 9.3.11)

21. $4\pi[2 + (1/\sqrt{3})\ln(2 + \sqrt{3})]$ **23.** $(\frac{9}{4}, 0)$ **25.** $(\frac{17}{11}, \frac{15}{22})$ **27.** $(\frac{58}{35}, \frac{45}{56})$
29. $(-2, \frac{3}{5})$ **31.** $356 \text{ kg} \cdot \text{m}^2$ **33.** (a) $\frac{1}{2}$ m (b) $\sqrt{\frac{3}{8}}$ m (c) $\sqrt{46}/12$ m
35. $I = \frac{1}{15} \text{ kg} \cdot \text{m}^2$; $r = \sqrt{\frac{2}{15}}$ m   **37.** $250\pi^2/3$ J
39. $\frac{1}{2} \cdot \frac{5}{12} \cdot [\frac{1}{4}(\text{rev/sec})2\pi]^2 = (5\pi^2/96) \text{ g} \cdot \text{cm}^2/\text{sec}^2 \approx 5.14 \times 10^{-8}$ J
41. $I = 101\pi/4$; $r = \sqrt{101}/2$ **43.** 392,400 N
45. $(1000)(9.81) \int_0^2 [(3x^2 + 6x)/4] \, dx = 49,050$ N   **47.** (a) 3525.47 N (b) 13,488.75 N
(c) 28,970.2 N

CHAPTER TEN
PROBLEMS 10.1

1. center : $(0, 0)$
foci : $(0, \pm 3)$
vertices : $(0, \pm 5)$
major axis : line segment between
$(0, -5)$ and $(0, 5)$
minor axis : line segment between
$(-4, 0)$ and $(4, 0)$
eccentricity : $\frac{3}{5}$

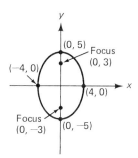

$$1 = \frac{x^2}{16} + \frac{y^2}{25} = \left(\frac{x}{4}\right)^2 + \left(\frac{y}{5}\right)^2$$

Note: In the following answers $\overline{(a, b), (c, d)}$ denotes the line segment between (a, b) and (c, d).

3. center : $(0, 0)$
foci : $(0, \pm 2\sqrt{2})$
vertices : $(0, \pm 3)$
major axis : $\overline{(0, -3)(0, 3)}$
minor axis : $(-1, 0)(1, 0)$
eccentricity : $2\sqrt{2}/3$

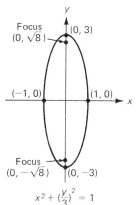

$$x^2 + \left(\frac{y}{3}\right)^2 = 1$$

5. center : $(0, 0)$
foci : $(\pm 2\sqrt{3}, 0)$
vertices : $(\pm 4, 0)$
major axis : $\overline{(-4, 0)(4, 0)}$
minor axis : $(0, -2)(0, 2)$
eccentricity : $\sqrt{3}/2$

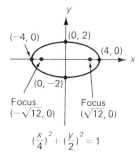

$$\left(\frac{x}{4}\right)^2 + \left(\frac{y}{2}\right)^2 = 1$$

7. center : $(1, -3)$
foci : $(1, -6), (1, 0)$
vertices : $(1, -8), (1, 2)$
major axis : $\overline{(1, -8)(1, 2)}$
minor axis : $(-3, -3)(5, -3)$
eccentricity : $\frac{3}{5}$

9. The graph is a circle, centered
at $(0, 0)$ with radius 1 (the unit circle)

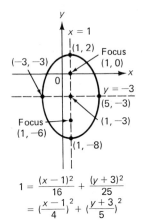

$$1 = \frac{(x-1)^2}{16} + \frac{(y+3)^2}{25}$$
$$= \left(\frac{x-1}{4}\right)^2 + \left(\frac{y+3}{5}\right)^2$$

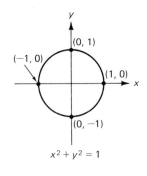

$$x^2 + y^2 = 1$$

11.   center : $(0, 0)$
foci : $(\pm 3\sqrt{3}/2, 0)$
vertices : $(\pm 3, 0)$
major axis : $\overline{(-3, 0)(3, 0)}$
minor axis : $(0, -3/2)(0, 3/2)$
eccentricity : $\sqrt{3}/2$

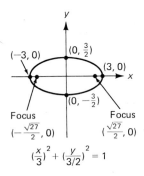

$$\left(\frac{x}{3}\right)^2 + \left(\frac{y}{3/2}\right)^2 = 1$$

13.   center : $(-1, -3)$
foci : $(-1, -3 \pm 2\sqrt{3})$
vertices : $(-1, -7), (-1, 1)$
major axis : $\overline{(-1, -7)(-1, 1)}$
minor axis: $(-3, -3)(1, -3)$
eccentricity : $\sqrt{3}/2$

15.   center : $(-1, 3)$
foci : $(-1, 3 \pm 2\sqrt{3})$
vertices : $(-1, -1), (-1, 7)$
major axis : $\overline{(-1, -1)(-1, 7)}$
minor axis : $(-3, 3)(1, 3)$
eccentricity : $\sqrt{3}/2$

$$4(x + 1)^2 + (y + 3)^2 = 16$$
$$\left(\frac{x+1}{2}\right)^2 + \left(\frac{y+3}{4}\right)^2 = 1$$

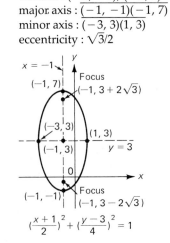

$$\left(\frac{x+1}{2}\right)^2 + \left(\frac{y-3}{4}\right)^2 = 1$$

17. center : $(-2, \frac{1}{4})$
foci : $(-2 \pm \sqrt{\frac{325}{48}}, \frac{1}{4})$
vertices : $(-2 \pm. \sqrt{\frac{65}{6}}, \frac{1}{4})$
major axis : $(-2 - \sqrt{\frac{65}{6}}, \frac{1}{4})(-2 + \sqrt{\frac{65}{6}}, \frac{1}{4})$
minor axis : $(-2, (1 - \sqrt{65})/4)(-2, (1 + \sqrt{65})/4)$
eccentricity : $\sqrt{\frac{5}{8}}$

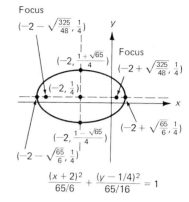

$$\frac{(x+2)^2}{65/6} + \frac{(y-1/4)^2}{65/16} = 1$$

19. $\left(\dfrac{x}{3}\right)^2 + \left(\dfrac{y}{5}\right)^2 = 1$ **21.** $\left(\dfrac{x+1}{3}\right)^2 + \left(\dfrac{y-4}{5}\right)^2 = 1, \left(\dfrac{x+1}{5}\right)^2 + \left(\dfrac{y-4}{3}\right)^2 = 1$

23. $y = (7 - x)/3$ **27.** πab

PROBLEMS 10.2

1. focus : $(0, 4)$
directrix : $y = -4$
axis : $x = 0$
vertex : $(0, 0)$

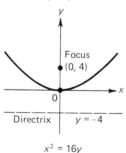

$x^2 = 16y$

3. focus : $(0, -4)$
directrix : $y = 4$
axis : $x = 0$
vertex : $(0, 0)$

$x^2 = -16y$

5. focus : $(0, \frac{3}{8})$
directrix : $y = -\frac{3}{8}$
axis : $x = 0$
vertex : $(0, 0)$

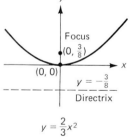

$y = \dfrac{2}{3}x^2$

7. focus : $(0, -\frac{9}{16})$
directrix : $y = \frac{9}{16}$
axis : $x = 0$
vertex : $(0, 0)$

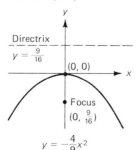

$y = -\dfrac{4}{9}x^2$

9. vertex : $(1, -3)$
focus : $(1, -7)$
directrix : $y = 1$
axis : $x = 1$

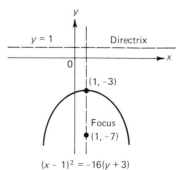

$(x - 1)^2 = -16(y + 3)$

11. vertex : $(0, \frac{9}{4})$
focus : $(0, \frac{5}{4})$
directrix : $y = \frac{13}{4}$
axis : $x = 0$

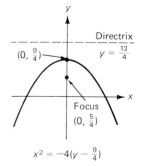

$x^2 = -4(y - \frac{9}{4})$

13.   vertex : $(-1, 0)$
focus : $(-1, -\frac{1}{4})$
directrix : $y = \frac{1}{4}$
axis : $x = -1$

15.   vertex : $(-2, 4)$
focus : $(-2, \frac{15}{4})$
directrix : $y = \frac{17}{4}$
axis : $x = -2$

17.   vertex : $(-2, -4)$
focus : $(-2, -\frac{15}{4})$
directrix : $y = -\frac{17}{4}$
axis : $x = -2$

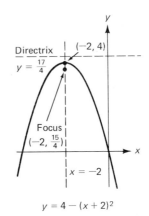

$$y = 4 - (x + 2)^2$$

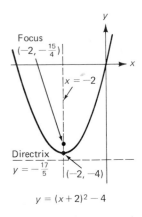

$$y = (x + 2)^2 - 4$$

19. $16y = x^2$ **21.** $-12(x + 2) = (y - 5)^2$

23.   new focus : $(0, -1)$; new directrix : $x = 6$

27.   If the asteroid's position on its orbit looks like

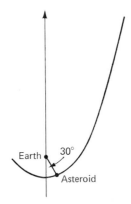

then its closest point to the earth is
$\frac{150{,}000}{2}(1 + \sqrt{3}/2) \approx 139{,}952$ km.

On the other hand, if the problem statement refers to a position like

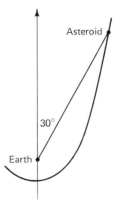

then the closest point is
$\frac{150{,}000}{2}(1 - \sqrt{3}/2) \approx 10{,}048$ km.

29.   6

PROBLEMS 10.3

1.   center : $(0, 0)$
foci : $(\pm\sqrt{41}, 0)$
vertices : $(\pm 4, 0)$
transverse axis : $\overline{(-4, 0)(4, 0)}$
asymptotes : $y = \pm\frac{5}{4}x$

3.   center : $(0, 0)$
foci : $(0, \pm\sqrt{41})$
vertices : $(0, \pm 5)$
transverse axis : $\overline{(0, -5)(0, 5)}$
asymptotes : $y = \pm\frac{5}{4}x$

5. center : $(0, 0)$
foci : $(0, \pm\sqrt{2})$
vertices : $(0, \pm 1)$
transverse axis : $\overline{(0, -1)(0, 1)}$
asymptotes : $y = \pm x$

7. center : $(0, 0)$
foci : $(\pm 3\sqrt{5}/2, 0)$
vertices : $(\pm 3, 0)$
transverse axis : $\overline{(-3, 0)(3, 0)}$
asymptotes : $y = \pm x/2$

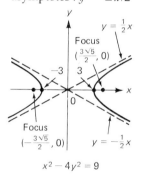

9. center : $(0, 0)$
foci : $(0, \pm 3\sqrt{5}/2)$
vertices : $(0, \pm 3)$
transverse axis : $\overline{(0, -3)(0, 3)}$
asymptotes : $y = \pm 2x$

11. center : $(0, 0)$
foci : $(\pm\sqrt{\frac{10}{3}}, 0)$
vertices : $(\pm\sqrt{2}, 0)$
transverse axis : $\overline{(-\sqrt{2}, 0)(\sqrt{2}, 0)}$
asymptotes : $y = \pm\sqrt{\frac{2}{3}}x$

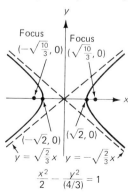

13. center : $(0, 0)$
foci : $(0, \pm\sqrt{\frac{10}{3}})$
vertices : $(0, \pm\sqrt{2})$
transverse axis : $\overline{(0, -\sqrt{2})(0, \sqrt{2})}$
asymptotes : $y = \pm\sqrt{\frac{3}{2}}x$

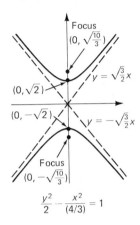

$$\frac{y^2}{2} - \frac{x^2}{(4/3)} = 1$$

15. center : $(1, -2)$
foci : $(1 - \sqrt{5}, -2), (1 + \sqrt{5}, -2)$
vertices : $(-1, -2), (3, -2)$
transverse axis : $\overline{(-1, -2)(3, -2)}$
asymptotes : $y = \pm(\frac{1}{2})(x - 1) - 2$

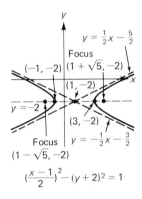

$$\left(\frac{x-1}{2}\right)^2 - (y + 2)^2 = 1$$

17. center : $(-1, -3)$
foci : $(-1 \pm 2\sqrt{5}, -3)$
vertices : $(-3, -3), (1, -3)$
transverse axis : $\overline{(-3, -3)(1, -3)}$
asymptotes : $y = \pm 2(x + 1) - 3$

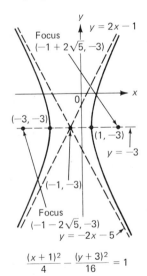

$$\frac{(x + 1)^2}{4} - \frac{(y + 3)^2}{16} = 1$$

19. center : $(4, 2)$
foci : $(4 - \sqrt{\frac{325}{6}}, 2) \ (4 + \sqrt{\frac{325}{6}}, 2)$
vertices : $(4 - \sqrt{\frac{65}{2}}, 2), (4 + \sqrt{\frac{65}{2}}, 2)$
transverse axis : $\overline{(4 - \sqrt{\frac{65}{2}}, 2)(4 + \sqrt{\frac{65}{2}}, 2)}$
asymptotes : $y = \pm\sqrt{\frac{2}{3}}(x - 4) + 2$

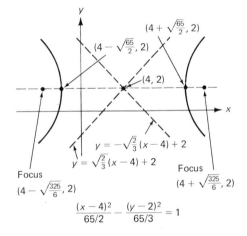

$$\frac{(x - 4)^2}{65/2} - \frac{(y - 2)^2}{65/3} = 1$$

21. $(x/4)^2 - (y/3)^2 = 1$ **23.** $(x/2)^2 - (y/6)^2 = 1$ **27.** $\sqrt{41}/4$ **29.** $\sqrt{2}$
31. $\sqrt{5}/2$ **33.** $\sqrt{\frac{5}{3}}$ **35.** $\sqrt{\frac{5}{3}}$
37. $((x - 4)/2)^2 - ((y + 5)/4)^2 = 1$ or $4x^2 - 32x - y^2 - 10y = -23$
39. $64x^2 - 260x + 300y - 120xy + 475 = 0$ or $120(y - \frac{1}{2}) = 225/(x - \frac{5}{2}) + 64(x - \frac{5}{2})$
41. (c) $y = (\pm 1/\sqrt{3})(x + 2) + 1$

PROBLEMS 10.4

1. ellipse; $3(x')^2 + 5(y')^2 = 45$ **3.** $(\sqrt{3} - \tfrac{3}{2})x' - (1 + 3\sqrt{3}/2)y' = 6$

5. There are two distinct rotations which will align the axis of the parabola $y^2 = -12x$ with the line $y = \sqrt{3}x$: Case (a) $\theta = \pi/3$; the resulting equation is $3(x')^2 + 2\sqrt{3}x'y' + (y')^2 + 24x' - 24\sqrt{3}y' = 0$; Case (b) $\theta = 4\pi/3$; for this rotation of axes we get $3(x')^2 + 2\sqrt{3}x'y' + (y')^2 - 24x' + 24\sqrt{3}y' = 0$

7.   Rotate axes through $\theta = \tfrac{1}{2}\tan^{-1}\tfrac{4}{3} \approx 26.6^0$ to obtain two parallel lines given by $5(x')^2 = 9$.

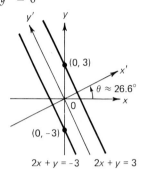

$2x + y = -3$ $2x + y = 3$

9.   Rotate axes through $\theta = \tfrac{1}{2}\tan^{-1}(-\tfrac{2}{3}) \approx -16.8^0$ to obtain the hyperbola $[(\sqrt{13} + 3)/10](x')^2 - [(\sqrt{13} - 3)/10](y')^2 = 1$.

$3x^2 - 2xy - 5 = 0$

11.   Rotate axes through $\theta = \pi/4$ to obtain the hyperbola $[(x')^2/2a] - [(y')^2/2a] = 1$.

$xy = a, \ a > 0$

13. Rotate axes through $\theta = \frac{1}{2} \tan^{-1} \frac{4}{3} \approx 26.6^0$
to obtain the parabola
$[x' + (3/\sqrt{5})]^2 = (8/\sqrt{5})[y' + (\frac{9}{8}\sqrt{5})]$.

$4x^2 + 4xy + y^2 + 20x - 10y = 0$

15. Rotate axes through $\theta = \pi/8$ to obtain the ellipse
$[(3 + \sqrt{2})/8](x')^2 + [(3 - \sqrt{2})/8](y')^2 = 1$.

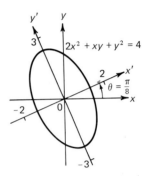

$2x^2 + xy + y^2 = 4$

$\theta = \frac{\pi}{8}$

17. Rotate axes through $\theta = \frac{1}{2} \tan^{-1} 3 \approx 35.8^0$
to obtain the ellipse
$[(4 - \sqrt{10})/36](x')^2 + [(4 + \sqrt{10})/36](y')^2 = 1$.

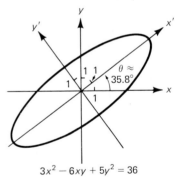

$\theta \approx 35.8°$

$3x^2 - 6xy + 5y^2 = 36$

19. Rotate axes through $\theta = \frac{1}{2} \tan^{-1} \frac{4}{3} \approx 26.6^0$
to obtain the hyperbola
$[4(y' + 2\sqrt{5})^2/35] - [(x' + \sqrt{5})^2/35] = 1$.

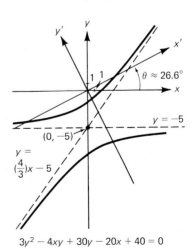

$\theta \approx 26.6°$

$y = -5$

$(0, -5)$

$y = (\frac{4}{3})x - 5$

$3y^2 - 4xy + 30y - 20x + 40 = 0$

REVIEW EXERCISES FOR CHAPTER TEN

1. Ellipse
   center : $(0, 0)$
   foci : $(0, \pm\sqrt{7})$
   vertices : $(0, \underline{\pm 4})$
   major axis : $\overline{(0, -4)(0, 4)}$
   minor axis : $(-3, 0)(3, 0)$
   eccentricity : $\sqrt{7}/4$

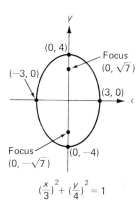

$$\left(\frac{x}{3}\right)^2 + \left(\frac{y}{4}\right)^2 = 1$$

3. Parabola
   focus : $(0, \frac{9}{64})$
   directrix : $y = -\frac{9}{64}$
   axis : $x = 0$
   vertex : $(0, 0)$

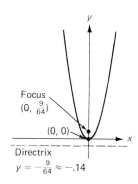

5. Hyperbola
   center : $(0, 0)$
   foci : $(0, \pm 5)$
   vertices : $(0, \pm 3)$
   eccentricity : $\frac{5}{3}$
   transverse axis : $\overline{(0, -3)(0, 3)}$
   asymptotes : $y = \pm\frac{3}{4}x$

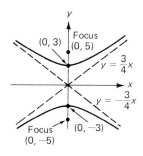

7. Ellipse
   center : $(1, -1)$
   foci : $(1, -1 - \sqrt{5})$, $(1, -1 + \sqrt{5})$
   vertices : $(1, -4)$, $(1, 2)$
   major axis : $\overline{(1, -4)(1, 2)}$
   minor axis : $(-1, -1)(3, -1)$
   eccentricity : $\sqrt{5}/3$

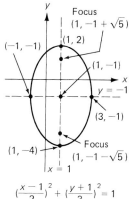

$$\left(\frac{x-1}{2}\right)^2 + \left(\frac{y+1}{3}\right)^2 = 1$$

9. single point $(-2, 5)$

11. two straight lines; i.e., a degenerate hyperbola

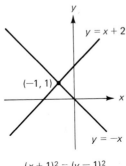

$(x + 1)^2 = (y - 1)^2$

13. Ellipse

center : $(-\frac{1}{2}, -4)$

foci : $(-\frac{1}{2}, -4 - 3/\sqrt{2})$, $(-\frac{1}{2}, -4 + 3/\sqrt{2})$

vertices : $(-\frac{1}{2}, -4 - 3\sqrt{2})$, $(-\frac{1}{2}, -4 + 3\sqrt{2})$

major axis : $\overline{(-\frac{1}{2}, -4 - 3\sqrt{2})(-\frac{1}{2}, -4 + 3\sqrt{2})}$

minor axis : $(-\frac{1}{2} - 3\sqrt{\frac{3}{2}}, -4)(-\frac{1}{2} + 3\sqrt{\frac{3}{2}}, -4)$

eccentricity : $\frac{1}{2}$

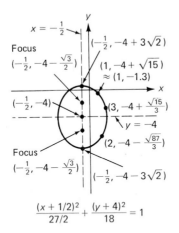

$$\frac{(x + 1/2)^2}{27/2} + \frac{(y + 4)^2}{18} = 1$$

15. unique ellipse: $(x/5)^2 + (y/4)^2 = 1$ **17.** $y^2/4 - x^2/5 = 1$

19. After the rotation, the curve satisfies the equation $5(x')^2 + 3(y')^2 = 20$; the curve is an ellipse. Referring to the (x', y') coordinates, we find

center : $(0, 0)$

foci : $(0, \pm 2\sqrt{\frac{2}{3}})$

vertices : $(0, \pm 2\sqrt{\frac{5}{3}})$

major axis : $\overline{(0, -2\sqrt{\frac{5}{3}})(0, 2\sqrt{\frac{5}{3}})}$

minor axis : $\overline{(-2, 0)(2, 0)}$

eccentricity : $\sqrt{\frac{2}{5}}$

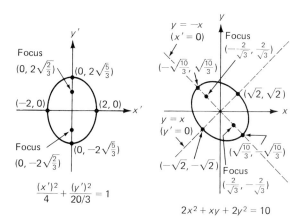

$$\frac{(x')^2}{4} + \frac{(y')^2}{20/3} = 1$$

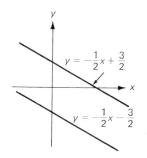

$$2x^2 + xy + 2y^2 = 10$$

21. $9 = x^2 + 4xy + 4y^2 = (x + 2y)^2$;
the graph consists of the two parallel lines with
equations $3 = x + 2y$ and $-3 = x + 2y$. If the
coordinate axes are rotated through the angle
$\frac{1}{2}\cot^{-1}((1 - 4)/4) \approx 63.43°$, the curve satisfies
the equation $9 = 5(x')^2$ or $x' = \pm 3/\sqrt{5}$.

23. Rotate the axes through $\pi/4$ radians; the transformed equation is
$-3 = xy = [(x')^2 - (y')^2]/2$ or $(y')^2/6 - (x')^2/6 = 1$.

	(x', y') coordinates	(x, y) coordinates
center :	$(0, 0)$	$(0, 0)$
foci :	$(0, \pm 2\sqrt{3})$	$(-\sqrt{6}, \sqrt{6}), (\sqrt{6}, -\sqrt{6})$
vertices :	$(0, \pm\sqrt{6})$	$(-\sqrt{3}, \sqrt{3}), (\sqrt{3}, -\sqrt{3})$
transverse axis :	$(0, -\sqrt{6}), (0, \sqrt{6})$	$(-\sqrt{3}, \sqrt{3}), (\sqrt{3}, -\sqrt{3})$
asymptotes :	$y' = \pm x'$	$x = 0, y = 0$
eccentricity :		$\sqrt{2}$

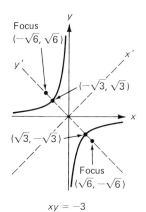

$$xy = -3$$

25. Rotate axes through $\theta = \frac{1}{2}\cot^{-1}((-2-4)/3 \approx 76.717°$; the curve satisfies the transformed equation $(3\sqrt{5}/2 + 1)(x')^2 - (3\sqrt{5}/2 - 1)(y')^2 = 5$; or, using numerical approximations, $(x')^2/1.148 - (y')^2/2.124 = 1$. At this stage, we can identify the curve as a hyperbola opening along the x'-axis.

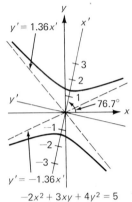

$$-2x^2 + 3xy + 4y^2 = 5$$

$$\left(\frac{3\sqrt{5}}{2} + 1\right)x'^2 - \left(\frac{3\sqrt{5}}{2} - 1\right)y'^2 = 5$$

CHAPTER ELEVEN
PROBLEMS 11.1

1. $(3, 0)$ **3.** $(-5, 0)$ **5.** $(-3\sqrt{3}, -3)$ **7.** $(-5, 0)$ **9.** $(5, 0)$
11. $(\sqrt{2}, -\sqrt{2})$ **13.** $(\sqrt{2}, -\sqrt{2})$ **15.** $(0, -1)$ **17.** $(0, -1)$ **19.** $(3, 0)$
21. $(3, \pi)$ **23.** $(\sqrt{2}, 5\pi/4)$ **25.** $(\sqrt{2}, 3\pi/4)$ **27.** $(1, 3\pi/2)$ **29.** $(4, 2\pi/3)$
31. $(4, 4\pi/3)$ **33.** $(4, 11\pi/6)$ **35.** $(4, 7\pi/6)$

PROBLEMS 11.2

1. circle centered at origin with radius 5   **3.** circle centered at origin with radius 4
5. straight line through origin with slope $-1/\sqrt{3}$

7. circle centered at $(0, \frac{5}{2})$ with radius $\frac{5}{2}$
9. circle centered at $(0, -\frac{5}{2})$ with radius $\frac{5}{2}$ **11.** circle centered at $(\frac{5}{2}, \frac{5}{2})$ with radius $5/\sqrt{2}$

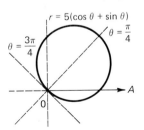

13. circle centered at $(\frac{5}{2}, -\frac{5}{2})$ with radius $5/\sqrt{2}$
15. cardioid, symmetric about $\theta = \pi/2$

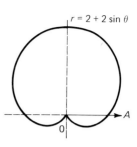

17. cardioid obtained by reflecting curve in Problem 15 about polar axis
19. same graph as in Problem 15   **21.**   same graph as in Problem 17
23. limaçon with loop, symmetric about $\theta = \pi/2$

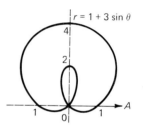

25. limaçon with loop, symmetric about polar axis

27. limaçon with loop, symmetric about   polar axis

29. limaçon with loop, symmetric about $\theta = \pi/2$

31. limaçon without loop, symmetric about polar axis

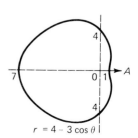

33. limaçon without loop, symmetric about $\theta = \pi/2$

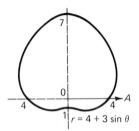

$r = 4 + 3 \sin \theta$

35. four-leafed rose, symmetric about polar axis, pole, and $\theta = \pi/2$

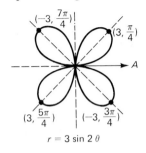

$r = 3 \sin 2\theta$

37. same graph as in Problem 35
39. symmetric about $\theta = \pi/2$

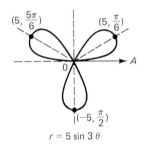

$r = 5 \sin 3\theta$

41. this is the graph of Problem 39 reflected about polar axis
43. eight-leafed rose, symmetric about polar axis, pole, and $\theta = \pi/2$

45. same graph as in Problem 43

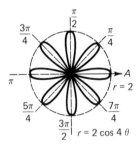

$r = 2 \cos 4\theta$

47.

49.

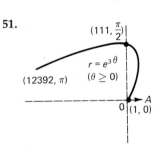

51.

53. lemniscate, symmetric about pole   **55.** lemniscate, symmetric about pole

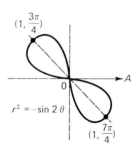

57. lemniscate, symmetric about polar axis, pole, and $\theta = \pi/2$

59. cissoid, symmetric about polar axis

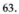

61. conchoid, symmetric about $\theta = \pi/2$

63.

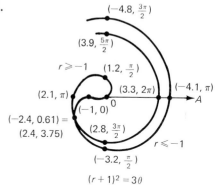

65. two circles (called *osculating circles*) with centers at $(0, \pm\frac{1}{2})$, radii $= \frac{1}{2}$, symmetric about polar axis, pole, and $\theta = \pi/2$

69. $r = -4\cos\theta + 3\sin\theta$

73.   (a) the line $y = 3$ (b) the line $x = -2$ (c) the line $x = \frac{8}{3}$ (d) the line $2y - 3x = 4$
(e) the line $-y + 2x = 4$

PROBLEMS 11.3

1. $(1, \pi/2)$ **3.** $(2 - \sqrt{2}, \pi/4); (2 + \sqrt{2}, 5\pi/4)$; pole   **5.** $(3, \pi/4); (3, 5\pi/4)$
7. $(-2, \pi/2); (-4, \pi)$ **9.** $(\sqrt{3}/2, \pi/6)$; pole   **11.** $(\frac{3}{2}, k\pi/6)$ for $k = 1, 2, 4, 5, 7, 8, 10, 11$
13. $(1, 0)$; pole   **15.** $(1, 1)$; pole   **17.**   pole; $(1/\sqrt{2}, k\pi/8)$ for $k = 1, 3, 5, 7, 9, 11, 13, 15$
19.   none   **25.** $(1, k\pi/8)$ for $k = 1, 3, 5, 7, 9, 11, 13, 15$
27. $(12, \pi/6); (12, 5\pi/6); (4, 7\pi/6); (4, 11\pi/6)$ **29.**   pole; $(1/(2\sqrt{3}), \pi/6), (1/(2\sqrt{3}), 5\pi/6)$
31. $(-1, \pi) = (1, 0)$ **33.** $(4k\pi, 0)$ for $k = 0, 1, 2, 3, \ldots$

PROBLEMS 11.4

1.   0   **3.**   undefined   **5.** $\sqrt{3}/5$ **7.** $-2/\pi$ **9.** $-\sqrt{3}/5$
11. $\tan \beta$ is undefined; $\tan \alpha = 0$ **13.** $\tan \beta = 5/(3\sqrt{3})$; $\tan \alpha = -7\sqrt{3}/3$
15. $\tan \beta = 2$; $\tan \alpha = -\frac{1}{2}$ **17.** $\tan \beta = \frac{1}{3}$; $\tan \alpha = 2$ **19.** $\tan \beta = -1/\sqrt{3}$; $\tan \alpha = 0$
21. $\tan \beta = 1/(2\sqrt{3})$; $\tan \alpha = 3\sqrt{3}/5$
23. $\tan \beta = -\frac{1}{7}$; $\tan \alpha = [\tan(\pi/28) - \frac{1}{7}]/[1 + \frac{1}{7}\tan(\pi/28)] \approx -0.03$
27.   relative maxima of y at $\left(\dfrac{3 + \sqrt{33}}{4}, \cos^{-1}\left(\dfrac{-1 + \sqrt{33}}{8}\right)\right)$,
$\left(\dfrac{3 - \sqrt{33}}{4}, -\cos^{-1}\left(\dfrac{-1 - \sqrt{33}}{8}\right)\right)$; relative minima of y at
$\left(\dfrac{3 - \sqrt{33}}{4}, \cos^{-1}\left(\dfrac{-1 - \sqrt{33}}{8}\right)\right)$, $\left(\dfrac{3 + \sqrt{33}}{4}, -\cos^{-1}\left(\dfrac{-1 + \sqrt{33}}{8}\right)\right)$
29.   relative maxima of y at $(\frac{1}{2}, \pi/6), (\frac{1}{2}, 5\pi/6)$; relative minima of y at $(2, 3\pi/2)$ and pole
33. $\tan^{-1}(\frac{4}{3}) \approx 53°$ **35.** $\tan^{-1}(\sqrt{3}) = 60°$ **37.**   no points of intersection   **39.**   0°
43. $0, \pi$; circle, centered at $(\frac{1}{2}, \pi/2)$
with radius $\frac{1}{2}$
45. $0, \pi/2, \pi, 3\pi/2$; four-leafed rose

$r = \sin \theta$

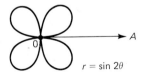

$r = \sin 2\theta$

47. $0, \pi/2, \pi, \pi/2$; four-leafed rose;
the graph of Problem 45 magnified by the factor a

49. $0, \pi/2, \pi, 3\pi/2$; lemniscate

$r^2 = \sin 2\theta$

51. $0, \pi/2, \pi, 3\pi/2$; lemniscate

$r^2 = a \sin 2\theta$

53. $0, \pi/3, 2\pi/3, \pi, 4\pi/3, 5\pi/3$;
three-leafed rose

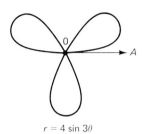

$r = 4 \sin 3\theta$

55. $7\pi/6, 11\pi/6$; cardioid

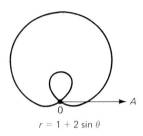

$r = 1 + 2 \sin \theta$

PROBLEMS 11.5

1. $\pi^3/48$ **3.** π **5.** $\pi a^2/20$ **7.** $\pi^5/80$ **9.** $25\pi^9/18$

11. circle; center $(a/2, \pi/2)$,
radius $a/2$, $A = \pi a^2/4$

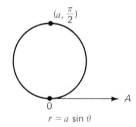

$r = a \sin \theta$

13. circle: center $(a/\sqrt{2}, \pi/4)$,
radius $a/\sqrt{2}$, $A = \pi a^2/2$

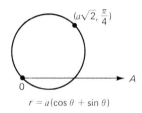

$r = a(\cos \theta + \sin \theta)$

15. circle: center $(a/\sqrt{2}, 3\pi/4)$,
radius $a/\sqrt{2}$, $A = \pi a^2/2$

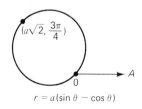

$r = a(\sin \theta - \cos \theta)$

17. limaçon without loop,
symmetric about polar axis, $A = 11\pi$

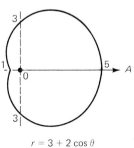

$r = 3 + 2 \cos \theta$

19. limaçon without loop,
symmetric about polar axis, $A = 11\pi$

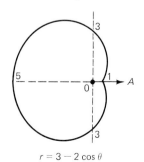

$r = 3 - 2 \cos \theta$

21.   four-leafed rose; $A = \pi a^2/2$

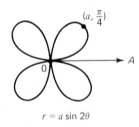

$(a, \frac{\pi}{4})$

0

$r = a \sin 2\theta$

23.   five-leafed rose; $A = \pi a^2/4$

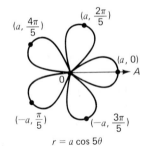

$(a, \frac{2\pi}{5})$

$(a, \frac{4\pi}{5})$

$(a, 0)$

A

0

$(-a, \frac{\pi}{5})$

$(-a, \frac{3\pi}{5})$

$r = a \cos 5\theta$

25.   lemniscate; symmetric about pole, polar axis, and $\theta = \pi/2$; $A = a$

$(-\sqrt{a}, 0)$

$(\sqrt{a}, 0)$

A

0

$r^2 = a \cos 2\theta$

27. $A = 2 \int_{\sin^{-1}(2/3)}^{\pi/2} \frac{1}{2}(2 - 3 \sin \theta)^2 \, d\theta = \frac{17}{4}\pi - \frac{17}{2} \sin^{-1}\left(\frac{2}{3}\right) - 3\sqrt{5} \approx 0.44$

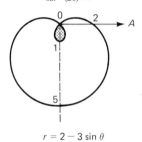

0   2 A

1

5

$r = 2 - 3 \sin \theta$

29. $A = 2 \int_{\pi/2}^{\sin^{-1}(-a/b)} \frac{1}{2}(a + b \sin \theta)^2 \, d\theta = (a^2 + \frac{1}{2}b^2)[\sin^{-1}(-a/b) - \pi/2] - \frac{3}{2}a\sqrt{b^2 - a^2}$

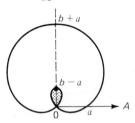

$b + a$

$b - a$

0   a A

$r = a + b \sin \theta$

31. $A = \int_0^{\cos^{-1}(-2/3)}(2 + 3 \cos \theta)^2 \, d\theta - \int_{\cos^{-1}(-2/3)}^{\pi}(2 + 3 \cos \theta)^2 \, d\theta = \frac{17}{2}[2 \cos^{-1}(-\frac{2}{3}) - \pi] + 6\sqrt{5} \approx 25.8$

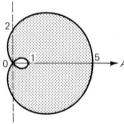

2

0   1   5 A

$r = 2 + 3 \cos \theta$

33. $A = \int_0^\pi \frac{1}{2}(2 + 2\sin\theta)^2\, d\theta - \int_0^\pi \frac{1}{2}(2\sin\theta)^2\, d\theta = 2\pi + 8$

$r = 2 + 2\sin\theta$ (cardioid)
$r = 2\sin\theta$ (circle)

35. $A = \frac{1}{2}\int_\alpha^{\pi/2} a^2 \sin 2\theta\, d\theta - \frac{1}{2}\int_\alpha^{\pi/2}(a\cos\theta)^2\, d\theta + \frac{1}{2}\int_\pi^{3\pi/2} a^2 \sin 2\theta\, d\theta =$
$(a^2/2)[2 - (\pi/4) + \frac{1}{2}\tan^{-1}(\frac{1}{2})]$ [here $\alpha = \tan^{-1}(\frac{1}{2})$]

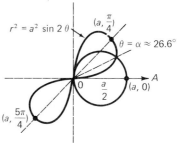

37. $A = 8[\int_0^{\pi/8} \frac{1}{2}(4\sin 2\theta)^2\, d\theta + \int_{\pi/8}^{\pi/4} \frac{1}{2}(4\cos 2\theta)^2\, d\theta] = 8\pi - 16$

39. $A = \pi a^2/4n$. Note that the rose has n petals if n is odd and $2n$ petals if n is even.

41. $9m^2$

REVIEW EXERCISES FOR CHAPTER ELEVEN

1. $(2, 0)$ **3.** $(0, -7)$ **5.** $(0, 1)$ **7.** $(2, 0)$ **9.** $(2, 11\pi/6)$ **11.** $(6\sqrt{2}, 5\pi/4)$

13. circle: radius 8, center $(0, 0)$ **15.** circle: radius 1, center $(1, 0)$

17. limaçon, symmetric about line $\theta = \pi/2$

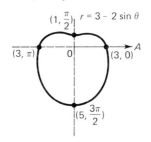

19.   four-leafed rose, symmetric about polar axis, pole, $\theta = \pi/2$

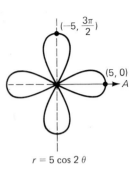

21.   eight-leafed rose, symmetric about pole, polar axis, $\theta = \pi/2$.

23.   logarithmic spiral

25.   straight line $y = 4$ **27.** $(x - 2)^2 + (y - \frac{3}{2})^2 = (\frac{5}{2})^2$
29. $(1, 0)$ **31.**   pole, $(1, \pi/2)$
33.   pole, $(4, 4)$ and, in general, (θ, θ), where $\theta = 2 + 2\sqrt{1 + 2k\pi}$, k a positive integer
35. $(1, k\pi/12)$ for $k = 1, 5, 7, 11, 13, 17, 19, 23$ **37.**   slope $= 0$ (tan β is undefined)
39.   slope $= (4 + \pi)/(4 - \pi)$ (tan $\beta = \pi/4$)
41.   slope is undefined since $\alpha = \pi/2$ (tan $\beta = \sqrt{3}$)
43.   slope $= -1$ (tan β is undefined)
45.   slope $= -\cot(\pi/20) \approx -6.314$ (tan β is undefined)
47. $90°$
49. $\theta = 0, \pi/2, \pi, 3\pi/2$; four-leafed rose

51. $\frac{1}{6}[(\pi/2)^3 - (\pi/6)^3] = 13\pi^3/648$ **53.** $\pi a^2/2$ **55.** 6π
57. $34[\pi/2 - \cos^{-1}(\frac{3}{4})] + 9\sqrt{7}$ **59.** $(\pi/2) - 1$ **61.** $2\pi/5$

CHAPTER TWELVE
PROBLEMS 12.1

1.   0   **3.**   0   **5.** $\frac{3}{2}$ **7.** $\frac{1}{3}$ **9.** $(1 + a)/(1 - a)$ **11.**   0
13. $\sqrt{3}/2$ (first calculate $\lim_{x \to 1^-}[(1 - x^3)/(1 - x^4)] = \frac{3}{4}$. Then
$\lim_{x \to 1^-}\sqrt{(1 - x^3)/(1 - x^4)} = \sqrt{\lim_{x \to 1^-}[(1 - x^3)/(1 - x^4)]} = \sqrt{3}/2$ by Theorem 2.7.2)
15.   0   **17.** e^2 **19.**   0   **21.** $\frac{1}{3}$ **23.** -2 **25.** $1/(2\sqrt{x_0})$ **29.** $\frac{1}{4}$ **35.** Et/L
37. $\int (1/(x - a)^2)\, dx = (-1/(x - a)) + C$

PROBLEMS 12.3

1. $\frac{1}{2}$ **3.**   0   **5.**   0   **7.**   0   **9.**   1   **11.**   0   **13.**   1   **15.**   0   **17.** $+\infty$

19. e **21.** $\frac{3}{2}$

23. 1 (since $\lim_{x\to0^+}(\sin x)^{(\sin x)-x} = \lim_{x\to0^+}[(\sin x)^{\sin x}/(\sin x)^x] =$ $\lim_{x\to0^+}(\sin x)^{\sin x}/\lim_{x\to0^+}(\sin x)^x = 1/1 = 1)$

25. limit doesn't exist since $\lim_{x\to2^+} x^{1/(2-x)} = 0$ and $\lim_{x\to2^-} x^{1/(2-x)} = \infty$ **27.** 1

29. $+\infty$ **31.** $+\infty$ **37.** (a) e^{-k} (b) $+\infty$

PROBLEMS 12.4

1. $\frac{1}{2}$ **3.** diverges   **5.** 0   **7.** diverges   **9.** $\frac{1}{2}\ln\frac{11}{9}$

11. $(1/b)[(\pi/2) - \tan^{-1}(a/b)]$ **13.** diverges   **15.** diverges   **17.** 2   **19.** diverges

21. diverges   **23.** $\frac{3}{4}(3^{2/3} - 1)$ **25.** diverges   **27.** $\frac{1}{2}$ **29.** $\pi/2$ **31.** diverges

33. $\frac{5}{4}\cdot 2^{4/5}$ **35.** diverges   **37.** diverges (note that $1/(1 - \sin x) = (1 + \sin x)/\cos^2 x$

39. diverges   **41.** diverges (since $\int_1^2 \{1/[(x - 2)(x - 4)]\}\,dx$ diverges)

43. $1/\ln 5$ **45.** 1   **47.** $\pi/2$

49. $\pi[\sqrt{2} + \ln(\sqrt{2} + 1)]$ (since $2\pi\int_0^\infty e^{-x}\sqrt{1 + e^{-2x}}\,dx = 2\pi\int_0^1\sqrt{1 + u^2}\,du\ (u = e^{-x}) =$ $2\pi\int_0^{\pi/4}\sec^3\theta\,d\theta\ (u = \tan\theta))$ **55.** use $\sqrt{1 + x^{1/8}}/x^{3/4} > 1/x^{3/4}$

57. for $|x| \ge 1$, $|xe^{-x^2}| > e^{-x^2}$ **63.** $(a + b)/2$ **65.** 0   **67.** $a^2; a$

69. $\sum_{j=1}^N (1/j)$ **73.** $[1/(r - 1)]\,\Gamma\,(2 - r)$

REVIEW EXERCISES FOR CHAPTER TWELVE

1. $\frac{3}{2}$ **3.** no limit   **5.** 0   **7.** $+\infty$ **9.** 1   **11.** $+\infty$ **13.** 1   **15.** $-\frac{1}{2}$

17. $+\infty$ **19.** 0   **21.** $\frac{1}{3}$ **23.** $6/7^4 = \frac{6}{2401}$ **25.** $\frac{1}{4}\ln\frac{13}{12}$ **27.** diverges   **29.** $\frac{2}{3}$

31. diverges   **33.** diverges ($\sqrt{x - 2}$ not defined for $x < 2$)   **35.** diverges

37. diverges   **39.** diverges (since $\int_0^1 \{[1/(x - 2)] - [1/(x - 1)]\}\,dx$ diverges)   **41.** $\pi/3$

43. $(\bar{x}, \bar{y}) = (1, \frac{1}{5})$ **45.** use $1/\ln\sqrt{2 + x} > 1/\sqrt{2 + x}$ **47.** 5

CHAPTER THIRTEEN

PROBLEMS 13.1

1. $(1/\sqrt{2})[1 - (x - \pi/4) - (x - \pi/4)^2/2! + (x - \pi/4)^3/3! + (x - \pi/4)^4/4! - (x - \pi/4)^5/5!$ $- (x - \pi/4)^6/6!]$

3. $1 + (1/e)(x - e) - (1/2e^2)(x - e)^2 + (1/3e^3)(x - e)^3 - (1/4e^4)(x - e)^4 + (1/5e^5)(x - e)^5$

5. $1 - (x - 1) + (x - 1)^2 - (x - 1)^3 + (x - 1)^4$ **7.** $x + \frac{1}{3}x^3$ **9.** $(x - \pi) + \frac{1}{3}(x - \pi)^3$

11. $1 - x^2 + x^4$ **13.** $x + (x^3/3!)$ **15.** $-\frac{1}{2}[x - (\pi/2)]^2$

17. $\dfrac{1}{2} + \dfrac{1}{2^4}x + \dfrac{3}{2^7\cdot 2!}x^2 + \dfrac{3\cdot5}{2^{10}\cdot 3!}x^3 + \dfrac{3\cdot5\cdot7}{2^{13}\cdot 4!}x^4$

19. $1 + \alpha x + (\alpha^2/2!)x^2 + (\alpha^3/3!)x^3 + (\alpha^4/4!)x^4 + (\alpha^5/5!)x^5 + (\alpha^6/6!)x^6$ **21.** $x + x^3/3!$

23. $(a_0 + a_1 + a_2 + a_3) + (a_1 + 2a_2 + 3a_3)(x - 1) + (a_2 + 3a_3)(x - 1)^2 + a_3(x - 1)^3$

25. x^2

PROBLEMS 13.3

1. $(1/7!)(\pi/4)^7 \approx 0.0000366$ **3.** $(120/5!)(\frac{1}{2})^{-6}(\frac{1}{2})^5 = 2$

5. $(1\cdot3\cdot5\cdot7\cdot9\cdot11/64)(\frac{19}{4})^{-13/2}(\frac{1}{4})^6(1/6!) \approx 2.2 \times 10^{-9}$

7. $[2\sec^4(\pi/6) + 4\sec^2(\pi/6)\tan^2(\pi/6)](1/4!)(\pi/6)^4 \approx 0.0167$

9. $(1/5!)(\frac{1}{3})^5(e^{1/9})(\frac{120}{3} + \frac{160}{27} + \frac{32}{243}) \approx 0.0018$

11. $\frac{1}{2} + (\sqrt{3}/2)(0.2) - (\frac{1}{2})(\frac{1}{2})(0.2)^2 - (\sqrt{3}/2)(\frac{1}{6})(0.2)^3 \approx 0.66205$

13. $\tan(\pi/4) + \sec^2(\pi/4)(0.1) + [2\sec^2(\pi/4)\tan(\pi/4)/2](0.1)^2$ $+ [4\sec^2(\pi/4)\tan^2(\pi/4) + 2\sec^4(\pi/4)](0.1)^3/3! \approx 1.2227$

15. $1 + (-1) + \dfrac{1}{2!}(-1)^2 + \dfrac{1}{3!}(-1)^3 + \dfrac{1}{4!}(-1)^4 + \dfrac{1}{5!}(-1)^5 + \dfrac{1}{6!}(-1)^6 \approx 0.36806$

17. $0.5 - \dfrac{(0.5)^2}{2} + \dfrac{(0.5)^3}{3} - \dfrac{(0.5)^4}{4} + \dfrac{(0.5)^5}{5} - \dfrac{(0.5)^6}{6} + \dfrac{(0.5)^7}{7} \approx 0.4058$

19. $1 + 3 + 3^2/2! + \cdots + 3^{12}/12! \approx 20.0852$ **21.** $\frac{1}{2} + (\frac{1}{2})^3/6 \approx 0.5208$

23. $\sin[(\pi/2) + (\pi/18)] \approx 1 - \frac{1}{2}(\pi/18)^2 \approx 0.98477$ **25.** $1 - \frac{1}{2}(0.1) + \frac{3}{8}(0.1)^2 = 0.95375$

27. $\int_0^{1/3} e^{x^2}\, dx \approx \int_0^{1/3} [1 + x^2 + (x^4/2)]\, dx = \frac{1}{3} + \frac{1}{3}(\frac{1}{3})^3 + \frac{1}{10}(\frac{1}{3})^5 \approx 0.3461$; error $< \frac{1}{3}(0.0018) = 0.0006$

REVIEW EXERCISES FOR CHAPTER THIRTEEN

1. $1 + x + (x^2/2!) + (x^3/3!)$
3. $\frac{1}{2} + (\sqrt{3}/2)[x - (\pi/6)] - \frac{1}{4}[x - (\pi/6)]^2 - (\sqrt{3}/12)[x - (\pi/6)]^3$
5. $-[x - (\pi/2)] - \frac{1}{3}[x - (\pi/2)]^3$ **7.** $3 + 2x - x^2 + x^3$ **9.** $(\pi/3)^6/6! \approx 0.00183$
11. $e/7! < 3/7! \approx 0.000595$
13. $f^{(5)}(x) = -8x(15 - 20x^2 + 4x^4)e^{-x^2}$ and crude bound is $(8)(15)/5! = 1$
15. $(1/\sqrt{2}) - (1/\sqrt{2})(-\pi/90) \approx 0.73179$
17. $\ln[(1 + x)/(1 - x)] = 2(x + \frac{1}{3}x^3 + \frac{1}{5}x^5 + \cdots)$ and, with $x = \frac{1}{3}$,
$\ln 2 \approx 2(\frac{1}{3} + \frac{1}{81} + \frac{1}{1215} + \frac{1}{15309}) \approx 0.69313$
19. $0.3 - \frac{1}{3}(0.3)^3 + \frac{1}{5}(0.3)^5 = 0.291486$ (actual error ≈ 0.000029)
21. $\int_0^{1/2} [1 - (x^4/2)]\, dx = \frac{159}{320} = 0.496875$; error $\leq \frac{1}{2}[(1/2^8)/4!] \approx 0.00008$

CHAPTER FOURTEEN
PROBLEMS 14.1

1. $\frac{1}{3}, \frac{1}{9}, \frac{1}{27}, \frac{1}{81}, \frac{1}{243}$ **3.** $\frac{3}{4}, \frac{15}{16}, \frac{63}{64}, \frac{255}{256}, \frac{1023}{1024}$ **5.** $e, e^{1/2}, e^{1/3}, e^{1/4}, e^{1/5}$ **7.** $0, 0, 0, 0, 0$
9. $1, 0, -1, 0, 1$ **11.** 0 **13.** divergent **15.** divergent **17.** $\frac{4}{7}$ **19.** $e^{1/4}$
21. 0 **23.** 0 **25.** divergent **27.** 0 **29.** $a_n = n \cdot 5^{n-1}$ **31.** $a_n = 1 - 1/2^n$
33. $a_n = (-\frac{1}{3})^{n-1}$ **39.** choose a_0 from $[0, 1]$
41. disprove; $p = 1$ and $a_n = 1/(n \ln n)$ is a counterexample

PROBLEMS 14.2

1. $\frac{1}{2}$ **3.** 1 **5.** 1 **7.** 1 **9.** 2 **11.** $(\ln 3)/3 \approx 0.366$ **15.** strictly
increasing **17.** not monotonic **19.** strictly decreasing **21.** strictly decreasing
23. not monotonic **25.** strictly increasing **27.** strictly increasing

PROBLEMS 14.3

1. 364 **3.** $[1 - (-5)^6]/[1 - (-5)] = -2604$ **5.** $(0.3)^2[1 - (-0.3)^7]/1.3 \approx 0.07$
7. $(b^{16} - 1)/(b^{16} + b^{14})$ **9.** $(16\sqrt{2} - 1)/(\sqrt{2} - 1)$ **11.** -3280 **13.** $\frac{4}{3}$ **15.** $\frac{10}{9}$
17. $\pi/(\pi - 1)$ **19.** $1/1.62 \approx 0.61728$ **21.** $\frac{1}{2}$ **23.** $n = 7$ **25.** $n = 459$
27. (b) 12 min

PROBLEMS 14.4

1. $\frac{4}{3}$ **3.** $2 - 1 - \frac{1}{2} = \frac{1}{2}$ **5.** 2 **7.** 125 **9.** $\frac{1}{2}$ **11.** 1
13. $8(3 - 1 - \frac{2}{3}) = \frac{32}{3}$ **15.** $25/6^4 = \frac{25}{1296}$ **17.** $\frac{35}{99}$ **19.** $\frac{71}{99}$
21. $\frac{501}{999} = \frac{167}{333}$ **23.** $\frac{11351}{99900}$ **27.** $\frac{3}{2}$ **29.** $\frac{1}{4}$ **39.** $1\frac{1}{11}$ hr $= 1:05\frac{5}{11}$ P.M.
41. $8 + 8 \cdot 2 \cdot \frac{2}{3} + 8 \cdot 2 \cdot (\frac{2}{3})^2 + 8 \cdot 2 \cdot (\frac{2}{3})^3 + \cdots = 40$ m

PROBLEMS 14.5

Note: C = series converges, D = series diverges
1. C **3.** D **5.** C **7.** C **9.** C **11.** C **13.** C **15.** D **17.** D
19. D **21.** D **23.** C **25.** D **27.** C **29.** D **31.** C **33.** D

PROBLEMS 14.6

Note: C = series converges, D = series diverges
1. D **3.** C **5.** C **7.** D **9.** C **11.** C **13.** D **15.** C **17.** D
19. C **21.** C **23.** C **25.** C

PROBLEMS 14.7

Note: AC = absolutely convergent, CC = conditionally convergent, D = divergent
1. D **3.** CC **5.** D **7.** CC **9.** D **11.** D **13.** AC **15.** CC
17. AC **19.** CC **21.** D **23.** D **25.** CC **27.** D **29.** CC
33. $S_{10} \approx 0.818$ with error ≤ 0.0083, $T_5 \approx 0.819$ with error ≤ 0.0061
35. $S_9 \approx -0.503$ with error ≤ 0.043, $T_3 \approx -0.508$ with error ≤ 0.028

37. $S_{100} \approx 0.555$ with error ≤ 0.1, $T_2 \approx 0.646$ with error ≤ 0.065 (note that $T_9 \approx 0.6003$ with error ≤ 0.0086)   **39.** $1 - \frac{1}{2} - \frac{1}{4} + \frac{1}{3} - \frac{1}{6} - \frac{1}{8} + \frac{1}{5} - \frac{1}{10} - \frac{1}{12} - \frac{1}{14}$
43. $\Sigma a_k = \Sigma 1/k$ is the most obvious example.

PROBLEMS 14.8

1. $6; (-6, 6)$ **3.** $3; (-4, 2)$ **5.** $\frac{1}{3}; (-\frac{1}{3}, \frac{1}{3})$ **7.** $1; [0, 2]$ **9.** $\infty; (-\infty, \infty)$
11. $1; (-1, 1)$ **13.** $\infty; (-\infty, \infty)$ **15.** $1; (-1, 1)$ **17.** $\infty; (-\infty, \infty)$
19. $\frac{1}{2}; [-\frac{1}{2}, \frac{1}{2}]$ **21.** $\frac{5}{2}; (-4, 1)$ **23.** $0; x = 0$ **25.** $1; (-1, 1)$ **27.** $0; x = -1$
29. $3; [-13, -7)$ **31.** $1; (-1, 1)$ **33.** $1; (-1, 1)$

PROBLEMS 14.9

1. $\Sigma_{k=0}^{\infty} (-1)^k x^{2k}$
3. $\pi = 4 \tan^{-1} 1 = 4 \Sigma_{k=0}^{\infty} [(-1)^k/(2k + 1)]$; $T_{10} \approx 3.1371$ with error ≤ 0.00828, $S_{199} \approx 3.1366$ with error ≤ 0.00998
5. $S_3 \approx 0.743$ with error ≤ 0.0046, $T_3 \approx 0.755$ with error ≤ 0.0096
7. $S_2 \approx 0.496875$ with error ≤ 0.000009, $T_3 \approx 0.496795$ with error ≤ 0.000005
9. $S_3 \approx 0.1862$ with error ≤ 0.0038, $T_3 \approx 0.1955$ with error ≤ 0.0074
11. $S_2 \approx 0.7639$ with error ≤ 0.0003, $T_2 \approx 0.7569$ with error ≤ 0.0068
13. $S_1 \approx 0.499783$ with error ≤ 0.0000004, $T_2 \approx 0.4997832$ with error ≤ 0.0000002
15. $\Sigma_{k=0}^{\infty} x^{k+2}/[(k + 2)k!]$ **17.** $\Sigma_{k=1}^{\infty} (-1)^{k+1} x^k/k^2$ **19.** (a) $(-\infty, \infty)$

PROBLEMS 14.10

3. $\Sigma_{k=0}^{\infty} e(x - 1)^k/k!$
5. $\frac{1}{\sqrt{2}} \left[1 - \left(x - \frac{\pi}{4} \right) - \frac{[x - (\pi/4)]^2}{2} + \frac{[x - (\pi/4)]^3}{3!} + \frac{[x - (\pi/4)]^4}{4!} - \cdots \right]$ **7.** $\Sigma_{k=0}^{\infty} (\alpha x)^k/k!$
9. $\Sigma_{k=0}^{\infty} (-1)^k x^{2k+2}/k!$ **11.** $(1/e) \Sigma_{k=0}^{\infty} (x + 1)^k/k!$
13. $(x - 1) \ln x = (x - 1) \sum_{k=1}^{\infty} \frac{(-1)^{k+1}(x - 1)^k}{k} = \sum_{k=1}^{\infty} \frac{(-1)^{k+1}(x - 1)^{k+1}}{k}$; $(0, 2]$
15. $1 + 0 + \frac{[x - (\pi/2)]^2}{2!} + 0$; $(0, \pi)$
17. $2 + \sum_{k=1}^{\infty} \frac{(-1)^{k+1} 1 \cdot 3 \cdot \cdots \cdot (2k - 3)}{(2^{3k-1})k!} (x - 4)^k$; 4 **19.** $\sum_{k=1}^{\infty} \frac{(-1)^{k+1} x^{4k-2}}{(2k - 1)!}$
27. $1 + \frac{1}{4}x^3 - \frac{3}{32}x^6 + \frac{7}{128}x^9 - \frac{77}{2048}x^{12} + \frac{231}{8192}x^{15} - \cdots$
29. $\int_0^{1/4} (1 + \sqrt{x})^{3/5}\, dx = x + \frac{2}{5}x^{3/2} - \frac{3}{50}x^2 + \frac{14}{625}x^{5/2} - \frac{7}{625}x^3 + \frac{102}{15625}x^{7/2} - \cdots \Big|_0^{1/4}$. Using the first five terms (through x^3), we obtain the estimate 0.296775 with error bounded by 0.000051
31. $\mathrm{erfc}(x) = 1 - \frac{2}{\sqrt{\pi}} \sum_{k=0}^{\infty} \frac{(-1)^k x^{2k+1}}{(2k + 1)k!}$;
$\mathrm{erfc}(1) \approx 1 - \frac{2}{\sqrt{\pi}} \left(1 - \frac{1}{3} + \frac{1}{5 \cdot 2} - \frac{1}{7 \cdot 3!} + \frac{1}{9 \cdot 4!} - \frac{1}{11 \cdot 5!} + \frac{1}{13 \cdot 6!} \right) \approx 0.1572858$ with error ≤ 0.0000149,
$\mathrm{erfc}\left(\frac{1}{2}\right) \approx 1 - \frac{2}{\sqrt{\pi}} \left(\frac{1}{2} - \frac{(\frac{1}{2})^3}{3} + \frac{(\frac{1}{2})^5}{5 \cdot 2} - \frac{(\frac{1}{2})^7}{7 \cdot 3!} \right) \approx 0.4795099$ with error ≤ 0.00001

REVIEW EXERCISES FOR CHAPTER FOURTEEN

1. $-1, 0, \frac{1}{3}, \frac{2}{4}, \frac{3}{5}$ **3.** 0 **5.** 0 **7.** e^{-2} **9.** $a_n = (2n - 1)/2^{n+2}$
11. unbounded; not monotonic   **13.** bounded below by $\frac{2}{3}$ and above by 1; strictly increasing
15. bounded above by 1 and below by 0; strictly decreasing
17. bounded below by 0 and above by 1; strictly increasing
19. bounded above by 8 and below by 5; strictly decreasing   **21.** $1{,}398{,}096$
23. $\left(\frac{27}{16}\right) - \left(\frac{8}{75}\right) = \frac{1897}{1200} \approx 1.58$ **25.** $\frac{79}{99}$ **27.** $9\frac{9}{11}$ hr $\approx$ 9:49:05 P.M.
29. convergent   **31.** divergent   **33.** convergent   **35.** divergent
37. convergent   **39.** convergent   **41.** conditionally convergent

43. conditionally convergent   **45.** absolutely convergent

47. conditionally convergent   **49.** divergent   **51.** divergent

53. $T_6 = 1 - (1/2^3) + (1/3^3) - (1/4^3) + (1/5^3) + (1/2)(-1/6^3) \approx 0.9021$ with error ≤ 0.00086; alternatively compute

$S_{10} = 1 - (1/2^3) + (1/3^3) - (1/4^3) + (1/5^3) - (1/6^3) + (1/7^3) - (1/8^3) + (1/9^3) - (1/10^3) \approx$ 0.9011 with error ≤ 0.00075

55. $1 - \frac{1}{2} - \frac{1}{4} + \frac{1}{3} - \frac{1}{6} + \frac{1}{5} - \frac{1}{8} + \frac{1}{7} - \frac{1}{10} - \frac{1}{12}$ **57.** $3; (-3, 3)$

59. $\infty, (-\infty, \infty)$ **61.** $\infty; (-\infty, \infty)$ **63.** $0; x = 0$ **65.** $1; [0, 2)$

67. $3; [-11, -5)$ **69.** $\int_0^{1/2} t^2\, dt = \frac{1}{24}$ with error $\leq \int_0^{1/2} (t^6/3!)\, dt \approx 0.00019$

71. $\int_0^{1/2} (1 - t^4 + t^8)\, dt = \frac{11381}{23040} \approx 0.493967$ with error $\leq \int_0^{1/2} t^{12}\, dt \approx 0.000009$

73. $2\Sigma_{k=0}^{\infty} (x - \ln 2)^k/k!$ **75.** $\Sigma_{k=0}^{\infty} (-1)^k(\alpha x)^{2k+1}/(2k + 1)!$

CHAPTER FIFTEEN
PROBLEMS 15.1

1.

3.

5.
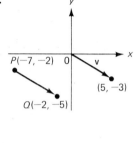

7. $|\mathbf{v}| = 4\sqrt{2}, \theta = \pi/4$

9. $|\mathbf{v}| = 4\sqrt{2}, \theta = 7\pi/4$ **11.** $|\mathbf{v}| = 2, \theta = \pi/6$ **13.** $|\mathbf{v}| = 2, \theta = 2\pi/3$

15. $|\mathbf{v}| = 2, \theta = 4\pi/3$ **17.** $|\mathbf{v}| = \sqrt{89}, \theta = \pi + \tan^{-1}(-\frac{8}{5}) \approx 2.13$ (second quadrant)

19.

21.

23.

25.
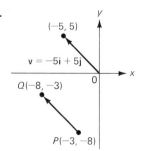

27. (a) (6, 9) (b) (−3, 7) (c) (−7, 1) (d) (39, −22) **33.** $(1/\sqrt{2})\mathbf{i} - (1/\sqrt{2})\mathbf{j}$

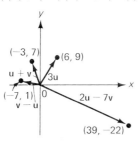

35. $(\frac{3}{5}, -\frac{4}{5})$ **37.** $(1/\sqrt{2}, 1/\sqrt{2})$ if $a > 0$, $(-1/\sqrt{2}, -1/\sqrt{2})$ if $a < 0$
39. $\sin\theta = -3/\sqrt{13}$, $\cos\theta = 2/\sqrt{13}$ **41.** $-(1/\sqrt{2})\mathbf{i} - (1/\sqrt{2})\mathbf{j}$ **43.** $(\frac{3}{5}, -\frac{4}{5})$
45. $\frac{3}{5}\mathbf{i} + \frac{4}{5}\mathbf{j}$
47. (a) $(1/\sqrt{2})\mathbf{i} - (1/\sqrt{2})\mathbf{j}$, (b) $(7/\sqrt{193})\mathbf{i} - (12/\sqrt{193})\mathbf{j}$, (c) $-(2/\sqrt{53})\mathbf{i} + (7/\sqrt{53})\mathbf{j}$
51. $4\mathbf{i} + 4\sqrt{3}\mathbf{j}$ **53.** $4\mathbf{j}$ **55.** $-3\mathbf{i} + 3\sqrt{3}\mathbf{j}$ **57.** $3\sqrt{3}\mathbf{i} - 3\mathbf{j}$

PROBLEMS 15.2

1. 0; 0 **3.** 0; 0 **5.** 20; $\frac{20}{29}$ **7.** -22; $-22/5\sqrt{53}$ **9.** 100; $20/\sqrt{481}$
11. $\mathbf{u} \cdot \mathbf{v} = \alpha\beta - \beta\alpha = 0$
13. parallel **15.** neither **17.** orthogonal **19.** parallel

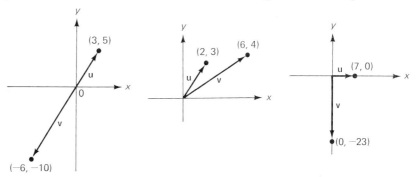

21. (a) $-\frac{3}{4}$ (b) $\frac{4}{3}$ (c) $\frac{1}{7}$ (d) $(-96 + \sqrt{7500})/78 \approx -0.12$ **25.** $\frac{3}{2}\mathbf{i} + \frac{3}{2}\mathbf{j}$ **27.** 0
29. $-\frac{2}{13}\mathbf{i} + \frac{3}{13}\mathbf{j}$ **31.** $\frac{14}{5}\mathbf{i} + \frac{28}{5}\mathbf{j}$ **33.** $-\frac{14}{5}\mathbf{i} + \frac{28}{5}\mathbf{j}$ **35.** $[(\alpha + \beta)/2]\mathbf{i} + [(\alpha + \beta)/2]\mathbf{j}$
37. $[(\alpha - \beta)/2]\mathbf{i} + [(\alpha - \beta)/2]\mathbf{j}$ **39.** $a_1 a_2 + b_1 b_2 > 0$
41. $\text{Proj}_{\overrightarrow{PQ}}\ \overrightarrow{RS} = \frac{51}{25}\mathbf{i} + \frac{68}{25}\mathbf{j}$; $\text{Proj}_{\overrightarrow{RS}}\ \overrightarrow{PQ} = -\frac{17}{26}\mathbf{i} + \frac{85}{26}\mathbf{j}$
47. $52/(5\sqrt{113}) \approx 0.9783$; $61/(\sqrt{34}\sqrt{113}) \approx 0.9841$; $-27/(5\sqrt{34}) \approx -0.9261$
49. when the vectors (a_1, a_2) and (b_1, b_2) have the same or opposite direction **51.** $\sqrt{5}$

PROBLEMS 15.3

7. $-2\mathbf{i} - 5\mathbf{j}$ N; $2\mathbf{i} + 5\mathbf{j}$ N **9.** $-2\mathbf{i} - 3\mathbf{j}$ N; $2\mathbf{i} + 3\mathbf{j}$ N
11. $-2\sqrt{3}\mathbf{i} - 7\mathbf{j}$ lb; $2\sqrt{3}\mathbf{i} + 7\mathbf{j}$ lb
13. $(3/\sqrt{2})\mathbf{i} - [(3/\sqrt{2}) + 2]\mathbf{j}$ N; $-(3/\sqrt{2})\mathbf{i} + [(3/\sqrt{2}) + 2]\mathbf{j}$ N
15. $(7\sqrt{2} - (7\sqrt{3}/2) - \frac{7}{2})(\mathbf{i} + \mathbf{j})$ N; $(-7\sqrt{2} + (7\sqrt{3}/2) + \frac{7}{2})(\mathbf{i} + \mathbf{j})$ N **17.** -12 J
19. $(8\sqrt{3} + 4)$ J **21.** $3\sqrt{2}$ J **23.** $12/\sqrt{13}$ J **25.** $500(\sin 20°/\sin 30°) \approx 342$ N
27. tugboat 1: $(500)(\cos 20°)(750) \approx 352{,}385$ J; tugboat 2: $(342)(\cos 30°)(750) \approx 221{,}149$ J

REVIEW EXERCISES FOR CHAPTER FIFTEEN

1. $|\mathbf{v}| = 3\sqrt{2}$, $\theta = \pi/4$ **3.** $|\mathbf{v}| = 4$, $\theta = 5\pi/3$ **5.** $|\mathbf{v}| = 12\sqrt{2}$, $\theta = 5\pi/4$
7. $2\mathbf{i} + 2\mathbf{j}$ **9.** $4\mathbf{i} + 2\mathbf{j}$ **11.** (a) (10, 5), (b) (5, −3), (c) (−31, 12)
13. $(1/\sqrt{2})\mathbf{i} + (1/\sqrt{2})\mathbf{j}$ **15.** $(2/\sqrt{29})\mathbf{i} + (5/\sqrt{29})\mathbf{j}$ **17.** $\frac{3}{5}\mathbf{i} + \frac{4}{5}\mathbf{j}$

19. $(1/\sqrt{2})\mathbf{i} - (1/\sqrt{2})\mathbf{j}$ if $a > 0$ and $-(1/\sqrt{2})\mathbf{i} + (1/\sqrt{2})\mathbf{j}$ if $a < 0$
21. $-(5/\sqrt{29})\mathbf{i} - (2/\sqrt{29})\mathbf{j}$ **23.** $-(10/\sqrt{149})\mathbf{i} + (7/\sqrt{149})\mathbf{j}$ **25.** $\mathbf{j}$
27. $-\frac{7}{2}\sqrt{3}\mathbf{i} + \frac{7}{2}\mathbf{j}$ **29.** $0; 0$ **31.** $-14, -14/(\sqrt{5}\sqrt{41})$ **33.** neither   **35.** parallel
37. parallel   **39.** $7\mathbf{i} + 7\mathbf{j}$ **41.** $\frac{15}{13}\mathbf{i} + \frac{10}{13}\mathbf{j}$ **43.** $-\frac{3}{2}\mathbf{i} - \frac{7}{2}\mathbf{j}$
45. $\text{Proj}_{\overrightarrow{RS}}\ \overrightarrow{PQ} = -\frac{99}{25}\mathbf{i} + \frac{132}{25}\mathbf{j}$; $\text{Proj}_{\overrightarrow{PQ}}\ \overrightarrow{RS} = -\frac{33}{82}\mathbf{i} - \frac{297}{82}\mathbf{j}$ **47.** $-3\mathbf{i} - 3\mathbf{j}$
49. $-\sqrt{2}\ \mathrm{J}$ **51.** $-11(1 + 3\sqrt{3})\ \mathrm{J}$

CHAPTER SIXTEEN
PROBLEMS 16.1

1. $\mathbb{R} - \{0, 1\}$ **3.** $\mathbb{R} - \{-1, 1\}$ **5.** $(0, 1)$ **7.** $\mathbb{R} - \{n\pi/2: n = 0, \pm 1, \pm 2, \ldots\}$
9. $y^2 = 4x$ **11.** $y = \pm x^{3/2}$ **13.** $y = 2x + 5$ **15.** $x = y^2 + y + 1, y \geq 0$
17. $y = x^3 - 1$ **19.** $y = (\ln x)^2, x > 0$
21. $x^2 + y^2 = e^{2\tan^{-1}(x/y)}$ valid for $-\pi/2 < t < \pi/2$. The graph of this equation is a spiral.
23. $y = x^2, x > 0$ **25.** $(2 - t)\mathbf{i} + (4 + 2t)\mathbf{j}$ **27.** $(3 - 4t)\mathbf{i} + (5 - 12t)\mathbf{j}$
$(1 + t)\mathbf{i} + (6 - 2t)\mathbf{j}$ $(-1 + 4t)\mathbf{i} + (-7 + 12t)\mathbf{j}$
29. $(-2 + 6t)\mathbf{i} + (3 + 4t)\mathbf{j}$ **33.** $(650\sqrt{2}/4)^2 = 52{,}812.5$ ft (using the value $g = 32$ ft/sec^2)
$(4 - 6t)\mathbf{i} + (7 - 4t)\mathbf{j}$
35. $x(t) = 50\sqrt{3}t;\ y(t) = -(9.81t^2/2) - 50t + 150$
37. $x = \frac{1}{2}\alpha - \frac{1}{4}\sin \alpha$ m, $y = \frac{1}{2}\alpha - \frac{1}{4}\cos \alpha$ m
47.

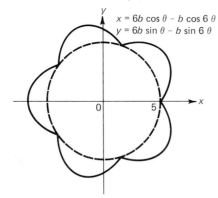

$x = 6b \cos \theta - b \cos 6\theta$
$y = 6b \sin \theta - b \sin 6\theta$

PROBLEMS 16.2

1. $\frac{2}{3}$ **3.** $-1\sqrt{2}$ **5.** $-e^{-4}$ **7.** 0 **9.** $\sqrt{2}$ **11.** 0 **13.** $-1/\sqrt{3}$
15. $-16/\pi^2$ **17.** $y = \frac{2}{3}x + \frac{22}{3}$ **19.** $y = -e^{-4}x + 2e^{-2}$ **21.** $y = \sqrt{2}$
23. none (note that condition (1) does not hold at $t = 0$ and that the Cartesian equation of the curve is the ray $y = x - 3, x \geq -1$, which has neither horizontal nor vertical tangents)
25. V: $((-1)^k, \cos \frac{3}{5}((\pi/2) + k\pi))$ for $k = 0, 1, 2, 3, 4, 5$;
H: $(\sin \frac{3}{5}k\pi, (-1)^k)$ for $k = 0, 1, 2, 3, 4, 5, 6, 7, 8, 9$
27. none   **29.** V: $(1, 0)$; H: none   **31.** none (condition (1) does not hold at $t = 0$)
35. -1 **37.** $(4 + 3\sqrt{3})/(4\sqrt{3} + 3)$ **39.** $\frac{1}{2}$ **41.** 0
43. $y = -x; y = x - 2; y = -(x/2) - \frac{1}{2}$ **45.** when $\cos \alpha = r/s$

PROBLEMS 16.3

1. $\mathbf{f}' = \mathbf{i} - 5t^4\mathbf{j}; \mathbf{f}'' = -20t^3\mathbf{j}$
3. $\mathbf{f}' = (2 \cos 2t)\mathbf{i} - 3(\sin 3t)\mathbf{j}; \mathbf{f}'' = (-4 \sin 2t)\mathbf{i} - (9 \cos 3t)\mathbf{j}$
5. $\mathbf{f}' = (1/t)\mathbf{i} + 3e^{3t}\mathbf{j}; \mathbf{f}'' = -(1/t^2)\mathbf{i} + 9e^{3t}\mathbf{j}$
7. $\mathbf{f}' = (\sec^2 t)\mathbf{i} + (\sec t)(\tan t)\mathbf{j}; \mathbf{f}'' = 2(\sec^2 t)(\tan t)\mathbf{i} + [\sec^3 t + (\sec t)(\tan^2 t)]\mathbf{j}$
9. $\mathbf{f}' = -(\tan t)\mathbf{i} + (\cot t)\mathbf{j}; \mathbf{f}'' = -(\sec^2 t)\mathbf{i} - (\csc^2 t)\mathbf{j}$ **11.** $(2/\sqrt{13})\mathbf{i} + (3/\sqrt{13})\mathbf{j}$
13. $\mathbf{j}$ **15.** $-(1/\sqrt{2})\mathbf{i} + (1/\sqrt{2})\mathbf{j}$ **17.** $\mathbf{i}$ **19.** $(4/\sqrt{97})\mathbf{i} - (9/\sqrt{97})\mathbf{j}$
21. $\frac{8}{3}\mathbf{i} + \frac{32}{5}\mathbf{j}$ **23.** $(2\sqrt{t} + C_1)\mathbf{i} + (\frac{2}{3}t^{3/2} + C_2)\mathbf{j}$ **25.** $(\cosh 1 - 1)\mathbf{i} - (\sinh 1)\mathbf{j}$

27. $(t \ln t - t + C_1)\mathbf{i} + (te^t - e^t + C_2)\mathbf{j}$ **29.** $[(t^4/4) + 2]\mathbf{i} + [5 - (t^6/6)\mathbf{j}$
31. $(\sin t)\mathbf{i} - (\cos t)\mathbf{j}$ **33.** $-(a/\sqrt{a^2 + b^2})\mathbf{i} + (b/\sqrt{a^2 + b^2})\mathbf{j}$ **35.** $(1/\sqrt{2})\mathbf{i} + (1/\sqrt{2})\mathbf{j}$
37. $\dfrac{1}{\sqrt{18 + (9/\sqrt{2})(1 - \sqrt{3})}}\left[\left(-\dfrac{3}{2} - \dfrac{3}{\sqrt{2}}\right)\mathbf{i} + \left(\dfrac{3\sqrt{3}}{2} - \dfrac{3}{\sqrt{2}}\right)\mathbf{j}\right]$

PROBLEMS 16.4

1. $(2 + \sec^2 t)\mathbf{i} - [\sin t + (\sec t)(\tan t)]\mathbf{j}$ **3.** $(3/2\sqrt{t}) - 2t^{-3/2}$ **5.** 0
7. $(3 \sin t \cosh t + 3 \cos t \sinh t)\mathbf{i} + (3 \sin t \sinh t + 3 \cos t \cosh t)\mathbf{j}$
9. $10t^9 + 9t^8 + \{(10t + 9)/[(t + 1)(t^{11} + t^{10})]\}$
11. $\ln t \operatorname{sech}^2 t + [(\tanh t)/t] - 3 \ln t \operatorname{sech} t \tanh t + [(3 \operatorname{sech} t)/t]$
13. $[1/(\sqrt{1 - t^2}) - \sin t]\mathbf{i} + [\cos t - (1/\sqrt{1 - t^2}]\mathbf{j}$
15. $[(\sin^{-1} t)/(1 + t^2)] + [(\tan^{-1} t)/\sqrt{1 - t^2}] - (\cos^{-1} t)\sin t - [(\cos t)/\sqrt{1 - t^2}]$
17. $\mathbf{T}(t) = -(\sin 5t)\mathbf{i} + (\cos 5t)\mathbf{j}$; $\mathbf{T}(\pi/2) = -\mathbf{i}$; $\mathbf{n}(t) = -(\cos 5t) - (\sin 5t)\mathbf{j}$; $\mathbf{n}(\pi/2) = -\mathbf{j}$
19. $\mathbf{T}(t) = (\sin 10t)\mathbf{i} - (\cos 10t)\mathbf{j}$; $\mathbf{T}(\pi) = -\mathbf{j}$; $\mathbf{n}(t) = (\cos 10t)\mathbf{i} + (\sin 10t)\mathbf{j}$; $\mathbf{n}(\pi) = \mathbf{i}$
21. $\mathbf{T}(t) = (1/\sqrt{1 + t^2})\mathbf{i} + (t/\sqrt{1 + t^2})\mathbf{j}$; $\mathbf{T}(1) = (1/\sqrt{2})\mathbf{i} + (1/\sqrt{2})\mathbf{j}$;
$\mathbf{n}(t) = -(t/\sqrt{1 + t^2})\mathbf{i} + (1/\sqrt{1 + t^2})\mathbf{j}$; $\mathbf{n}(1) = -(1/\sqrt{2})\mathbf{i} + (1/\sqrt{2})\mathbf{j}$
23. $\mathbf{T}(t) = -(7/\sqrt{74})\mathbf{i} + (5/\sqrt{74})\mathbf{j}$ (constant); $\mathbf{n}(t) = (5/\sqrt{74})\mathbf{i} + (7/\sqrt{74})\mathbf{j}$ or
$-(5/\sqrt{74})\mathbf{i} - (7/\sqrt{74})\mathbf{j}$
25. $\mathbf{T}(t) = (\operatorname{sech} t)\mathbf{i} + (\tanh t)\mathbf{j}$; $\mathbf{T}(0) = \mathbf{i}$; $\mathbf{n}(t) = -(\tanh t)\mathbf{i} + (\operatorname{sech} t)\mathbf{j}$; $\mathbf{n}(0) = \mathbf{j}$

27. $\mathbf{T}(t) = \sqrt{\dfrac{1 + \sin t}{2}}\,\mathbf{i} - \dfrac{\cos t}{\sqrt{2 + 2\sin t}}\,\mathbf{j}$; $\mathbf{T}(\pi) = \dfrac{1}{\sqrt{2}}\mathbf{i} + \dfrac{1}{\sqrt{2}}\mathbf{j}$;

$\mathbf{n}(t) = \dfrac{\cos t}{\sqrt{2 + 2\sin t}}\,\mathbf{i} + \sqrt{\dfrac{1 + \sin t}{2}}\,\mathbf{j}$; $\mathbf{n}(\pi) = -(1/\sqrt{2})\mathbf{i} + (1/\sqrt{2})\mathbf{j}$
29. $\mathbf{T}(\pi/6) = (3/\sqrt{10})\mathbf{i} - (1/\sqrt{10})\mathbf{j}$; $\mathbf{n}(\pi/6) = -(1/\sqrt{10})\mathbf{i} - (3/\sqrt{10})\mathbf{j}$

PROBLEMS 16.5

1. $\frac{1}{27}(148^{3/2} - 13^{3/2})$ **3.** $8(2^{3/2} - 1)$
5. $\displaystyle\int_{1/\sqrt{5}}^{1} \sqrt{1 + x^2}\, dx = \int_{\tan^{-1}(1/\sqrt{5})}^{\pi/4} \sec^3 \theta\, d\theta = \dfrac{1}{2}\left[\sqrt{2} - \left(\dfrac{\sqrt{6}}{5}\right) + \ln\left(\dfrac{\sqrt{5}(1 + \sqrt{2})}{1 + \sqrt{6}}\right)\right] \approx 0.6861$
7. $\sqrt{2}$ **9.** $8a$ (see Example 3)   **11.** $\frac{2}{27}[13^{3/2} - 8]$ **13.** $|a|\pi$ **15.** $\sqrt{2}(e^3 - 1)$
17. $6\sqrt{2}$ **19.** $\sqrt{2}(e^{\pi/2} - 1)$ **21.** $4\int_0^{\pi/2} \sqrt{a^2 \sin^2 \theta + b^2 \cos^2 \theta}\, d\theta$

PROBLEMS 16.6

1. $\mathbf{f} = 3\{[(s + 2)/2]^{3/2} - 1\}\mathbf{i} + 2\{[(s + 2)/2]^{2/3} - 1\}^{3/2}\mathbf{j}$
3. $\mathbf{f} = \frac{1}{27}\{[(27s + 8)^{2/3} - 4]^{3/2} + 27\}\mathbf{i} + \frac{1}{9}\{(27s + 8)^{2/3} - 13\}\mathbf{j}$
5. $\mathbf{f} = 3 \cos(s/3)\mathbf{i} + 3 \sin(s/3)\mathbf{j}$ **7.** $\mathbf{f} = a \cos(s/a)\mathbf{i} + a \sin(s/a)\mathbf{j}$
9. $\mathbf{f} = (a + b \cos(s/|b|))\mathbf{i} + (c + b \sin (s/|b|))\mathbf{j}$
11. $\mathbf{f} = a[1 - (2s/3a)]^{3/2}\mathbf{i} + b(2s/3a)^{3/2}\mathbf{j}$; here $\theta = \sin^{-1}\sqrt{2s/3a}$

PROBLEMS 16.7

1. $\mathbf{v} = 3\mathbf{j}$, $|\mathbf{v}| = 3$, $\mathbf{a} = -9\mathbf{i}$, $|\mathbf{a}| = 9$ **3.** $\mathbf{v} = -4\sqrt{3}\mathbf{i} - 4\mathbf{j}$; $|\mathbf{v}| = 8$;
$\mathbf{a} = 16\mathbf{i} - 16\sqrt{3}\mathbf{j}$; $|\mathbf{a}| = 32$ **5.** $\mathbf{v} = 4\mathbf{i} + 4\mathbf{j}$; $|\mathbf{v}| = 4\sqrt{2}$; $\mathbf{a} = 4\mathbf{j}$; $|\mathbf{a}| = 4$
7. $\mathbf{v} = -7\mathbf{i} + 5\mathbf{j}$; $|\mathbf{v}| = \sqrt{74}$; $\mathbf{a} = 0$; $|\mathbf{a}| = 0$
9. $\mathbf{v} = 2\mathbf{i}$; $|\mathbf{v}| = 2$; $\mathbf{a} = 2\mathbf{j}$; $|\mathbf{a}| = 2$ **11.** $\mathbf{v} = 2\mathbf{i}$; $|\mathbf{v}| = 2$; $\mathbf{a} = \mathbf{j}$; $|\mathbf{a}| = 1$
13. $\mathbf{v} = [1 + (1/\sqrt{2})]\mathbf{i} - (1/\sqrt{2})\mathbf{j}$; $|\mathbf{v}| = \sqrt{2 + \sqrt{2}}$; $\mathbf{a} = (1/\sqrt{2})\mathbf{i} + (1/\sqrt{2})\mathbf{j}$; $|\mathbf{a}| = 1$
15. (a) $s = \displaystyle\int_0^{t_f} \sqrt{(600\sqrt{2})^2 + (600\sqrt{2} - gt)^2}\, dt = \dfrac{1200^2}{9.81}\left[\dfrac{1}{\sqrt{2}} + \dfrac{1}{4}\ln\left(\dfrac{\sqrt{2} + 1}{\sqrt{2} - 1}\right)\right]$
$\approx 168,483$ m ≈ 168.5 km where $t_f = 1200\sqrt{2}/g$ sec. This can be integrated via the
substitution $600\sqrt{2} - gt = 600\sqrt{2} \tan \theta$.   (b) $(1200)^2/g$ m   (c) $(600)^2/g$ m   (d) 1200 m/sec
17. (a) $s = \int_0^{20/\sqrt{g}} \sqrt{400 + g^2 t^2}\, dt = (200/g)[\sqrt{g}\sqrt{1 + g} + \ln(\sqrt{g} + \sqrt{1 + g})] \approx 247.9$ m
(b) $400/\sqrt{g} \approx 127.7$ m   (c) $\tan^{-1}(-gt_f/20) \approx 107.7°$ (d) $20\sqrt{1 + g}$
19. $20\sqrt{3} - (g/2)(\frac{2}{3})^2 + 4 = 20\sqrt{3} + \frac{4}{9} \approx 32$ ft   **21.** $\ln 2.5 \approx 0.9163$ km/sec
23. $5000g \approx 49,050$ N

PROBLEMS 16.8

1. $\kappa = \frac{1}{2}, \rho = 2$ **3.** $\kappa = 2/5^{3/2}, \rho = 5^{3/2}/2$ **5.** $\kappa = \frac{3}{16}, \rho = \frac{16}{3}$ **7.** $\kappa = 6/\pi, \rho = \pi/6$
9. $\kappa = 2/5^{3/2}, \rho = 5^{3/2}/2$ **11.** $\kappa = 1/2\sqrt{2}, \rho = 2\sqrt{2}$ **13.** $\kappa = 1/2\sqrt{2}, \rho = 2\sqrt{2}$
15. $\kappa = 2a/(1 + b^2)^{3/2}, \rho = (1 + b^2)^{3/2}/2a$ **17.** $\kappa = 1, \rho = 1$
19. $\kappa = 0, \rho$ is undefined   **21.** at the origin   **23.** minimum for $t = \pi/2$, maximum
for $t = \pi/4$ **25.** $13\sqrt{2}/(5\sqrt{5}a)$ **27.** $1/a$ **29.** $3/4a$ **31.** $a_T = 0, a_n = 4$
33. $a_T = (4t/\sqrt{1 + 4t^2}), a_n = (2/\sqrt{1 + 4t^2})$
35. $a_T = (18t^3 - 14t)/\sqrt{9t^4 - 14t^2 + 9}, a_n = 6(1 + t^2)/\sqrt{9t^4 - 14t^2 + 9}$
37. $a_T = (4 + 18t^2)/\sqrt{4 + 9t^2}, a_n = 6t/\sqrt{4 + 9t^2}$
41. $[(10{,}000)(80{,}000)^2/(3600)^2] \cdot (1/\sqrt{2}) \approx 3{,}491{,}885.3$ N
43. $v = 150/\sqrt{2} \approx 106$ km/hr (reduce speed by a factor of $\sqrt{2}$)

REVIEW EXERCISES FOR CHAPTER SIXTEEN

1. $y = 2x$ **3.** $x = [(y/2) + 3]^2$ **5.** $x^2 + y^2 = 1$ **7.** $x = y^3, x \geq 0$
9. 2; V: $(0, 0)$; no H   **11.** $\sqrt{3}$; V: $(1, 0), (-1, 0)$; H: $(0, 1), (0, -1)$
13. undefined; V: $(1, 0)$; no H   **15.** $-4/3\sqrt{3}$; V: $(3, 0), (-3, 0)$; H: $(0, 4), (0, -4)$
17. $1/\sqrt{3}$ **19.** π **21.** $\mathbf{f}' = 2\mathbf{i} - 3t^2\mathbf{j}; \mathbf{f}'' = -6t\mathbf{j}$
23. $\mathbf{f}' = (-5 \sin 5t)\mathbf{i} + (8 \cos 4t)\mathbf{j}; \mathbf{f}'' = (-25 \cos 5t)\mathbf{i} - (32 \sin 4t)\mathbf{j}$
25. $\mathbf{T} = (4/\sqrt{41})\mathbf{i} + (5/\sqrt{41})\mathbf{j}; \mathbf{n} = (-5/\sqrt{41})\mathbf{i} + (4/\sqrt{41})\mathbf{j}$
27. $\mathbf{T} = (\sqrt{3}/2)\mathbf{i} - \frac{1}{2}\mathbf{j}; \mathbf{n} = -\frac{1}{2}\mathbf{i} - (\sqrt{3}/2)\mathbf{j}$
29. $\mathbf{T} = (2/\sqrt{5})\mathbf{i} + (1/\sqrt{5})\mathbf{j}; \mathbf{n} = -(1/\sqrt{5})\mathbf{i} + (2/\sqrt{5})\mathbf{j}$ **31.** $\frac{81}{4}\mathbf{i} + \frac{243}{2}\mathbf{j}$
33. $(\frac{2}{3}t^{3/2} + C_1)\mathbf{i} + (\frac{3}{4}t^{4/3} + C_2)\mathbf{j}$ **35.** $[(t^8/8) - 1]\mathbf{i} + [3 - (t^7/7)]\mathbf{j}$ **37.** $-2 \sin t \cosh t$
39. $\{e^t[(t + 1)^2 - t]/(t + 1)^2\} + [e^{-t}(t^2 + t + 1)/t^2]$ **41.** $\pi/3$ **43.** 16
45. $\frac{1}{3}(29^{3/2} - 5^{3/2})$ **47.** $\mathbf{f} = \frac{3}{4}[(2s + 1)^{2/3} - 1]\mathbf{i} + [(2s + 1)^{2/3} - 1]^{3/2}\mathbf{j}$
49. $\mathbf{f} = 2 \cos(s/2)\mathbf{i} + 2 \sin(s/2)\mathbf{j}$
51. $\mathbf{v} = -\sqrt{3}\mathbf{i} + \mathbf{j}; |\mathbf{v}| = 2; \mathbf{a} = -2\mathbf{i} - 2\sqrt{3}\mathbf{j}; |\mathbf{a}| = 4$
53. $\mathbf{v} = (\ln 2 - 1)\mathbf{i} + 2\mathbf{j}; |\mathbf{v}| = \sqrt{(\ln 2 - 1)^2 + 4}; \mathbf{a} = (\ln^2 2 + 1)\mathbf{i}; |\mathbf{a}| = \ln^2 2 + 1$
55. $\mathbf{v} = 2(\cosh 1)\mathbf{i} + 4\mathbf{j}; |\mathbf{v}| = \sqrt{4 \cosh^2 1 + 16}; \mathbf{a} = 2(\sinh 1)\mathbf{i}; |\mathbf{a}| = 2 \sinh 1$

57. (a) $s = \displaystyle\int_0^{t_f} \sqrt{675 + (15 - gt)^2}\, dt$ m ≈ 87 m, where $t_f = (30 + \sqrt{900 + 12g})/2g \approx 3.155$ sec
(b) $15\sqrt{3}t_f \approx 81.97$ m   (c) $1.5 + 15(15/g) - (g/2)(15/g)^2 \approx 12.97$ m
(d) $\sqrt{675 + (15 - gt_f)^2} \approx 30.49$ m/sec   (e) $\tan^{-1}[(15 - gt_f)/15\sqrt{3}] \approx 148.45°$
59. (a) $(1500)(40) - (9.81)(2500) = 35{,}475$ N   (b) $(1500)(40) - (9.81)(1500) = 45{,}285$ N
61. $\kappa = \rho = 1$ **63.** $\kappa = 36/(97/2)^{3/2}, \rho = (97/2)^{3/2}/36$
65. $\kappa = 16/17^{3/2}, \rho = 17^{3/2}/16$ **67.** $\kappa = 2/17^{3/2}, \rho = 17^{3/2}/2$ **69.** $\kappa = \frac{3}{4}, \rho = \frac{4}{3}$
71. $a_T = 0, a_n = 2$ **73.** $a_T = (6t + 12t^3)/\sqrt{t^2 + t^4}; a_n = 6t^2/\sqrt{t^2 + t^4}$
75. $[1300(175{,}000)^2/(3600)^2] \cdot (1/65) \approx 47{,}261$ N

CHAPTER SEVENTEEN
PROBLEMS 17.1

1.

3.

5.

7.

9.

11.

13.

15.

17. $\sqrt{40}$ **19.** 6   **21.** 5   **23.** $\sqrt{329}$ **25.** $\sqrt{250}$

27. $(x + 1)^2 + (y - 8)^2 + (z + 3)^2 = 5$ **29.** center $(-\frac{3}{2}, \frac{1}{2}, -1)$, $r = 3/\sqrt{2}$

31. $\alpha > 23\frac{1}{4}$ **37.** $(x - \frac{7}{2})^2 + (y - 1)^2 + (z - 2)^2 = \frac{65}{4}$

33. $\overrightarrow{RP} + \overrightarrow{PQ} = \sqrt{14} + \sqrt{126} = \sqrt{224} = \overrightarrow{RQ}$ **39.** $\frac{1}{6}\pi(65)^{3/2}; \frac{4}{3}\pi$

PROBLEMS 17.2

1. $|\mathbf{v}| = 3; 0, 1, 0$ **3.** $|\mathbf{v}| = 14; 0, 0, 1$ **5.** $|\mathbf{v}| = \sqrt{17}; 4/\sqrt{17}, -1/\sqrt{17}, 0$

7. $|\mathbf{v}| = \sqrt{13}; -2/\sqrt{13}, 3/\sqrt{13}, 0$ **9.** $|\mathbf{v}| = \sqrt{3}; 1/\sqrt{3}, -1/\sqrt{3}, 1/\sqrt{3}$

11. $|\mathbf{v}| = \sqrt{3}; -1/\sqrt{3}, 1/\sqrt{3}, 1/\sqrt{3}$ **13.** $|\mathbf{v}| = \sqrt{3}; -1/\sqrt{3}, 1/\sqrt{3}, -1/\sqrt{3}$

15. $|\mathbf{v}| = \sqrt{3}; -1/\sqrt{3}, -1/\sqrt{3}, -1/\sqrt{3}$

17. $|\mathbf{v}| = \sqrt{222}; -7/\sqrt{222}, 2/\sqrt{222}, -13/\sqrt{222}$

19. $|\mathbf{v}| = \sqrt{82}; -3/\sqrt{82}, -3/\sqrt{82}, 8/\sqrt{82}$ **21.** $(1/\sqrt{3})\mathbf{i} + (1/\sqrt{3})\mathbf{j} + (1/\sqrt{3})\mathbf{k}$

23. $\cos^2(\pi/6) + \cos^2(\pi/3) + \cos^2(\pi/4) = \frac{3}{4} + \frac{1}{4} + \frac{1}{2} = 1.5 > 1$ **25.** $-\mathbf{i}$

29. $10\mathbf{i} + 3\mathbf{j} - 7\mathbf{k}$ **31.** $\mathbf{i} - \mathbf{j} - 6\mathbf{k}$ **33.** $-13\mathbf{i} + 28\mathbf{j} + 12\mathbf{k}$ **35.** 25   **37.** 45

39. $\cos^{-1}(-10/\sqrt{59}\sqrt{50}) \approx 1.76$ rad $\approx 100.6°$ **41.** $\frac{25}{38}\mathbf{v} = -\frac{25}{19}\mathbf{i} - \frac{75}{38}\mathbf{j} + \frac{125}{38}\mathbf{k}$

43. $-\frac{1}{5}\mathbf{t} = -\frac{3}{5}\mathbf{i} - \frac{4}{5}\mathbf{j} - \mathbf{k}$ **45.** $\frac{35}{59}\mathbf{w} = \frac{35}{59}\mathbf{i} - \frac{245}{59}\mathbf{j} + \frac{105}{59}\mathbf{k}$

47. $\sqrt{5610}/51 = \sqrt{110/51}$ (Note that the distance is given by $|\overrightarrow{QP} - \text{Proj}_{\overrightarrow{QR}}\overrightarrow{QP}|$. Draw a picture.)

49. $\overrightarrow{PQ} \cdot \overrightarrow{PR} = -6 + 12 - 6 = 0$ **53.** $14\sqrt{3}$ J

61. (a) $\mathbf{u} + \mathbf{v}$ (b) $\mathbf{u} + 2\mathbf{v}$ (c) $\mathbf{u} + \mathbf{v} + \mathbf{w}$ **63.** $(\pm 1/\sqrt{180})(8, 10, 4)$

PROBLEMS 17.3

1. $x\mathbf{i} + y\mathbf{j} + z\mathbf{k} = (2\mathbf{i} + \mathbf{j} + 3\mathbf{k}) + t(-\mathbf{i} + \mathbf{j} - 4\mathbf{k}); x = 2 - t, y = 1 + t, z = 3 - 4t;$ $(x - 2)/(-1) = y - 1 = (z - 3)/(-4)$

3. $x\mathbf{i} + y\mathbf{j} + z\mathbf{k} = (\mathbf{i} + 3\mathbf{j} + 2\mathbf{k}) + t(\mathbf{i} + \mathbf{j} - 4\mathbf{k}); x = 1 + t, y = 3 + t, z = 2 - 4t;$ $x - 1 = y - 3 = (z - 2)/(-4)$

5. $x\mathbf{i} + y\mathbf{j} + z\mathbf{k} = (-4\mathbf{i} + \mathbf{j} + 3\mathbf{k}) + t(-\mathbf{j} - 2\mathbf{k}); x = -4, y = 1 - t, z = 3 - 2t;$ $x = -4$ and $(y - 1)/(-1) = (z - 3)/(-2)$

7. $x\mathbf{i} + y\mathbf{j} + z\mathbf{k} = (\mathbf{i} + 2\mathbf{j} + 3\mathbf{k}) + t(2\mathbf{i} - 2\mathbf{k});\ x = 1 + 2t,\ y = 2,\ z = 3 - 2t;\ y = 2$ and $(x - 1)/2 = (z - 3)/(-2)$

9. $x\mathbf{i} + y\mathbf{j} + z\mathbf{k} = (\mathbf{i} + 2\mathbf{j} + 4\mathbf{k}) + t(3\mathbf{k});\ x = 1,\ y = 2,\ z = 4 + 3t;\ x = 1$ and $y = 2$

11. $x\mathbf{i} + y\mathbf{j} + z\mathbf{k} = (2\mathbf{i} + 2\mathbf{j} + \mathbf{k}) + t(2\mathbf{i} - \mathbf{j} - \mathbf{k});\ x = 2 + 2t,\ y = 2 - t,\ z = 1 - t;$ $(x - 2)/2 = (y - 2)/(-1) = (z - 1)/(-1)$

13. $x\mathbf{i} + y\mathbf{j} + z\mathbf{k} = (\mathbf{i} + 3\mathbf{k}) + t(\mathbf{i} - \mathbf{j});\ x = 1 + t,\ y = -t,\ z = 3;\ x + y = 1$ and $z = 3$

15. $x\mathbf{i} + y\mathbf{j} + z\mathbf{k} = (-\mathbf{i} - 2\mathbf{j} + 5\mathbf{k}) + t(-3\mathbf{j} + 7\mathbf{k});\ x = -1,\ y = -2 - 3t,\ z = 5 + 7t;$ $x = -1$ and $7y + 3z = 1$

17. $x\mathbf{i} + y\mathbf{j} + z\mathbf{k} = (-\mathbf{i} - 3\mathbf{j} + \mathbf{k}) + t(-7\mathbf{j});\ x = -1,\ y = -3 - 7t,\ z = 1;\ x = -1$ and $z = 1$

19. $x\mathbf{i} + y\mathbf{j} + z\mathbf{k} = (a\mathbf{i} + b\mathbf{j} + c\mathbf{k}) + t(d\mathbf{i} + e\mathbf{j});\ x = a + dt,\ y = b + et,\ z = c;$ $ex - dy = ea - db$ and $z = c$

21. $x\mathbf{i} + y\mathbf{j} + z\mathbf{k} = (4\mathbf{i} + \mathbf{j} - 6\mathbf{k}) + t(3\mathbf{i} + 6\mathbf{j} + 2\mathbf{k});\ x = 4 + 3t,\ y = 1 + 6t,$ $z = -6 + 2t;\ (x - 4)/3 = (y - 1)/6 = (z + 6)/2$

27. no point of intersection   29. none   31. $(-3, 2, 7)$ 33. none

35. (a) $\sqrt{186}/3\ (t = \frac{1}{3})$ (b) $\sqrt{1518}/11 = \sqrt{138/11}\ (t = -\frac{4}{11})$; (c) $\sqrt{750}/6 = 5\sqrt{30}/6\ (t = -\frac{1}{6})$

37. the lines do intersect because
$$\begin{vmatrix} 1 & 1 & 2 - 9 \\ 2 & -1 & -1 - (-2) \\ 4 & -2 & 3 - 1 \end{vmatrix} = 0$$

PROBLEMS 17.4

1. $-6\mathbf{i} - 3\mathbf{j}$ 3. $-\mathbf{i} - \mathbf{j} + \mathbf{k}$ 5. $12\mathbf{i} + 8\mathbf{j} - 21\mathbf{k}$ 7. $(bc - ad)\mathbf{j}$

9. $-5\mathbf{i} - \mathbf{j} + 7\mathbf{k}$ 11. $\mathbf{0}$ 13. $42\mathbf{i} + 6\mathbf{j}$ 15. $-9\mathbf{i} + 39\mathbf{j} + 61\mathbf{k}$

17. $-4\mathbf{i} + 8\mathbf{k}$ 19. $\mathbf{0}$

21. $-(9/\sqrt{181})\mathbf{i} - (6/\sqrt{181})\mathbf{j} + (8/\sqrt{181})\mathbf{k};\ (9/\sqrt{181})\mathbf{i} + (6/\sqrt{181})\mathbf{j} - (8/\sqrt{181})\mathbf{k}$

23. $\sqrt{30}/(\sqrt{6}\sqrt{29}) \approx 0.415$ 25. $(x - 1)/2 = (y + 3)/(-26) = (z - 2)/(-22)$

27. $x = -2 + 13t,\ y = 3 + 22t,\ z = 4 - 8t$ 29. $5\sqrt{5}$ 31. $\sqrt{523}$

33. $\sqrt{a^2b^2 + a^2c^2 + b^2c^2}$ 39. $\frac{1}{2}\sqrt{595}$ 43. 23   45. $48/\sqrt{437}$ 47. $23/\sqrt{27}$

PROBLEMS 17.5

1. $x = 0$ (yz-plane))   3. $z = 0$ (xy-plane)   5. $x + z = 4$ 7. $3x - y + 2z = 19$

9. $4x + y - 7z = -15$ 11. $x + y + z = -6$ 13. $20x + 13y - 3z = 58$

15. $x + y + z = 1$ 17. coincident   19. orthogonal   21. orthogonal

23. coincident   25. $x = t,\ y = -11 - 37t,\ z = -2 - 10t$ 27. $33/\sqrt{59}$

29. $11/\sqrt{68}$ 33. $\cos^{-1}(18/\sqrt{26}\sqrt{69}) \approx 1.132 \approx 64.9°$ 37. $x - 22y - 17z = 0$

39. not coplanar   41. $x - 8y + 10z = 0$

PROBLEMS 17.6

1.

$y = \sin x$

3.

$y = \cos z$

5.

$z = x^3$

7. right circular cylinder, radius 1, centered on line $x = 0$, $y = -1$
9. right circular cylinder, radius 2, centered on x-axis **11.** hyperbolic cylinder
13. elliptic cylinder. **15.** hyperbolic cylinder **17.** hyperboloid of one sheet
19. ellipsoid **21.** hyperbolic paraboloid (cross sections parallel to the xz-plane are hyperbolas)
23. hyperboloid of two sheets centered at (2, 0, 0) ($[(x - 2)^2/4] - (y^2/4) - (z^2/4) = 1$); cross sections parallel to the yz-plane are ellipses; surface only defined for $|x - 2| \geq 2$
25. hyperboloid of two sheets (like the surface in Problem 23)
27. hyperboloid of two sheets; cross sections parallel to the xy-plane are ellipses; surface defined for $|z| \geq \sqrt{2}$ **29.** sphere, radius $\sqrt{3}$, center (1, 1, 1)
31. hyperboloid of one sheet, cross sections parallel to the yz-plane are ellipses
33. hyperbolic paraboloid; like the surface in Figure 9 except that (i) the center is $(1, 3, -8)$ instead of the origin and (ii) the line $x = 1$, $z = -8$ instead of the x-axis passes through the center of the surface
35. ellipsoid with equation $(x^2 + z^2)/(a^2) + (y^2/b^2) = 1$

PROBLEMS 17.7

1. $(1/\sqrt{14})\mathbf{i} + (2/\sqrt{14})\mathbf{j} + (3/\sqrt{14})\mathbf{k}$ **3.** $(1/\sqrt{3})\mathbf{i} + (1/\sqrt{3})\mathbf{j} - (1/\sqrt{3})\mathbf{k}$
5. $(-8/\sqrt{65})\mathbf{i} + (1/\sqrt{65})\mathbf{k}$
7. $\int_0^{10} \sqrt{36 + 4t^2}\, dt = 18 \int_0^{\tan^{-1}(10/3)} \sec^3 \theta\, d\theta = 10\sqrt{109} + 9 \ln[(\sqrt{109} + 10)/3]$
9. $\frac{4}{729}(328^{3/2} - 8) \approx 32.55$
11. $\mathbf{v} = -(\sin 1)\mathbf{i} + (\cos 1)\mathbf{j} + 4\mathbf{k}$; $|\mathbf{v}| = \sqrt{17}$; $\mathbf{a} = -(\cos 1)\mathbf{i} - (\sin 1)\mathbf{j} + 12\mathbf{k}$; $|\mathbf{a}| = \sqrt{145}$
13. $\mathbf{v} = \mathbf{j} + \mathbf{k}$; $|\mathbf{v}| = \sqrt{2}$, $\mathbf{a} = \mathbf{i}$; $|\mathbf{a}| = 1$
15. $\mathbf{v} = \frac{1}{2}\mathbf{i} + \frac{1}{3}\mathbf{j} + \frac{1}{4}\mathbf{k}$; $|\mathbf{v}| = \sqrt{61}/12$; $\mathbf{a} = -\frac{1}{4}\mathbf{i} - \frac{2}{9}\mathbf{j} - \frac{3}{16}\mathbf{k}$; $|\mathbf{a}| = \sqrt{3049}/144$
17. $\mathbf{T} = \dfrac{\sqrt{3}a}{2\sqrt{a^2 + 1}}\mathbf{i} - \dfrac{a}{2\sqrt{a^2 + 1}}\mathbf{j} + \dfrac{1}{\sqrt{a^2 + 1}}\mathbf{k}$; $\kappa = \dfrac{a}{a^2 + 1}$; $\mathbf{n} = -\dfrac{1}{2}\mathbf{i} - \dfrac{\sqrt{3}}{2}\mathbf{j}$;

$\mathbf{B} = \dfrac{\sqrt{3}}{2\sqrt{a^2 + 1}}\mathbf{i} - \dfrac{1}{2\sqrt{a^2 + 1}}\mathbf{j} - \dfrac{a}{\sqrt{a^2 + 1}}\mathbf{k}$

19. $\mathbf{T} = \dfrac{-a}{\sqrt{1 + a^2}}\mathbf{i} + \dfrac{1}{\sqrt{1 + a^2}}\mathbf{k}$; $\kappa = \dfrac{b}{1 + a^2}$ $\left(\text{since } \dfrac{d\mathbf{T}}{dt} = -\dfrac{b}{\sqrt{1 + a^2}}\mathbf{j}, \dfrac{ds}{dt} = \sqrt{1 + a^2} \text{ and}\right.$

$\dfrac{d\mathbf{T}}{ds} = \dfrac{d\mathbf{T}/dt}{ds/dt} = \dfrac{-b}{1 + a^2}\mathbf{j}\Big)$; $\mathbf{n} = -\mathbf{j}$; $\mathbf{B} = \dfrac{1}{\sqrt{1 + a^2}}\mathbf{i} + \dfrac{a}{\sqrt{1 + a^2}}\mathbf{k}$
21. $\mathbf{T} = (1/\sqrt{6})\mathbf{i} + (2/\sqrt{6})\mathbf{j} + (1/\sqrt{6})\mathbf{k}$; $\kappa = \sqrt{5}/3$, $\mathbf{n} = -(2/\sqrt{5})\mathbf{i} + (1/\sqrt{5})\mathbf{j}$;
$\mathbf{B} = -1/(\sqrt{30})\mathbf{i} - (2/\sqrt{30})\mathbf{j} + (5/\sqrt{30})\mathbf{k}$
23. $\mathbf{T} = (1/\sqrt{5})\mathbf{j} + (2/\sqrt{5})\mathbf{k}$; $\kappa = \sqrt{20}/25 = 2/5^{3/2}$; $\mathbf{n} = -(2/\sqrt{5})\mathbf{j} + (1/\sqrt{5})\mathbf{k}$; $\mathbf{B} = \mathbf{i}$
31. $\frac{1}{2}$

PROBLEMS 17.8

1. $(1, \sqrt{3}, 5)$ **3.** $(-4, 4\sqrt{3}, 1)$ **5.** $(-3/\sqrt{2}, 3/\sqrt{2}, 2)$ **7.** $(-10, 0, -3)$
9. $(-7/\sqrt{2}, -7/\sqrt{2}, 2)$ **11.** $(1, \pi/2, 0)$ **13.** $(\sqrt{2}, \pi/4, 2)$ **15.** $(4, \pi/3, -5)$
17. $(4, 5\pi/3, 4)$ **19.** $(\sqrt{3}, 0, 1)$ **21.** $(0, 3\sqrt{3}, 3)$ **23.** $(\frac{7}{2}, -\frac{7}{2}, -7/\sqrt{2})$
25. $(\sqrt{3}, 3, -2)$ **27.** $(5\sqrt{3}/4, -\frac{5}{4}, -5\sqrt{3}/2)$ **29.** $(2, \pi/4, \pi/4)$ **31.** $(2, 5\pi/4, \pi/4)$
33. $(2\sqrt{2}, 7\pi/6, \pi/4)$ **35.** $(\sqrt{23}, \cos^{-1}(-2/\sqrt{7}), \cos^{-1}(-4/\sqrt{23}))$
37. cylindrical: $r^2 + z^2 = 25$; spherical: $\rho = 5$ **39.** $x^2 + y^2 - 9y = 0$
41. cylindrical: $r^2 - z^2 = 1$; spherical: $\rho^2(\sin^2 \varphi - \cos^2 \varphi) = 1$ **43.** $z = x^2 + y^2$
45. $z\sqrt{x^2 + y^2} = 1$ **47.** $z = 2xy$ **49.** $\rho = 6 \sin \varphi \cos \theta$

REVIEW EXERCISES FOR CHAPTER SEVENTEEN

1. $\sqrt{68}$ **3.** $\sqrt{216}$ **5.** $(x + 1)^2 + (y - 4)^2 + (z - 2)^2 = 9$
9. $\sqrt{130}$; 0, $3/\sqrt{130}$, $11/\sqrt{130}$ **11.** $\sqrt{53}$; $-4/\sqrt{53}$, $1/\sqrt{53}$, $6/\sqrt{53}$
13. $(2/\sqrt{6})\mathbf{i} - (1/\sqrt{6})\mathbf{j} + (1/\sqrt{6})\mathbf{k}$ **15.** $\mathbf{i} - 14\mathbf{j} + 20\mathbf{k}$ **17.** $\frac{26}{21}\mathbf{i} - \frac{52}{21}\mathbf{j} + \frac{13}{21}\mathbf{k}$
19. 22 **21.** $\cos^{-1}(-9/\sqrt{798}) \approx 108.6°$ **23.** $68/\sqrt{3}$ J
25. $x\mathbf{i} + y\mathbf{j} + z\mathbf{k} = (-4\mathbf{i} + \mathbf{j}) + t(7\mathbf{i} - \mathbf{j} + 7\mathbf{k})$; $x = -4 + 7t$, $y = 1 - t$, $z = 7t$;
$(x + 4)/7 = (y - 1)/(-1) = z/7$

27. $x\mathbf{i} + y\mathbf{j} + z\mathbf{k} = (\mathbf{i} - 2\mathbf{j} - 3\mathbf{k}) + t(5\mathbf{i} - 3\mathbf{j} + 2\mathbf{k}); x = 1 + 5t, y = -2 - 3t,$
$z = -3 + 2t; (x - 1)/5 = (y + 2)/(-3) = (z + 3)/2$ **29.** $\sqrt{165}/3$ **31.** $-7\mathbf{i} - 7\mathbf{k}$
33. $-26\mathbf{i} - 8\mathbf{j} + 7\mathbf{k}$ **35.** $(x + 1)/14 = (y - 2)/(-26) = (z - 4)/(-11)$ **37.** $2\sqrt{22}$
39. $x + z = -1$ **41.** $2x - 3y + 5z = 19$ **43.** $x = \frac{1}{2} - \frac{9}{2}t, y = \frac{7}{2} - \frac{11}{2}t, z = t$
45. $x = \frac{4}{3} - \frac{5}{2}t, y = -4 - \frac{7}{2}t, z = t$ **47.** $\cos^{-1}|-1/\sqrt{207}| \approx 1.501 \approx 86.01°$
49.

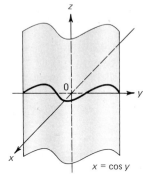

$x = \cos y$

51.

$z = \sqrt[3]{x}$

53. hyperboloid of one sheet; cross sections parallel to xz-plane are ellipses
55. hyperboloid of two sheets; cross sections parallel to xz-plane are ellipses; only defined for $|y| \ge \frac{5}{4}$
57. hyperbolic paraboloid; cross sections parallel to yz-plane are parabolas.
59. $(6/\sqrt{53})\mathbf{i} - (1/\sqrt{53})\mathbf{j} + (4/\sqrt{53})\mathbf{k}$ **61.** $-(3\sqrt{3}/2\sqrt{10})\mathbf{i} - (3/2\sqrt{10})\mathbf{j} + (1/\sqrt{10})\mathbf{k}$
63. $\mathbf{v} = \frac{3}{16}\pi^2\mathbf{i} - (1/\sqrt{2})\mathbf{j} - (1/\sqrt{2})\mathbf{k}; |\mathbf{v}| = \sqrt{1 + (9\pi^4/256)};$
$\mathbf{a} = (3\pi/2)\mathbf{i} - (1/\sqrt{2})\mathbf{j} + (1/\sqrt{2})\mathbf{k}; |\mathbf{a}| = \sqrt{1 + (9\pi^2/4)}$
65. $\mathbf{T} = -(1/\sqrt{5})\mathbf{i} + (\sqrt{3/5})\mathbf{j} + (1/\sqrt{5})\mathbf{k}; \kappa = \frac{2}{5}; \mathbf{n} = -(\sqrt{3}/2)\mathbf{i} - \frac{1}{2}\mathbf{j};$
$\mathbf{B} = (1/2\sqrt{5})\mathbf{i} - (\sqrt{3}/2\sqrt{5})\mathbf{j} + (2/\sqrt{5})\mathbf{k}$
67. $\mathbf{T} = (1/\sqrt{3})\mathbf{i} + (1/\sqrt{3})\mathbf{j} + (1/\sqrt{3})\mathbf{k}; \kappa = \sqrt{2}/3; \mathbf{n} = -(1/\sqrt{2})\mathbf{j} + (1/\sqrt{2})\mathbf{k};$
$\mathbf{B} = (2/\sqrt{6})\mathbf{i} - (1/\sqrt{6})\mathbf{j} - (1/\sqrt{6})\mathbf{k}$ **69.** $(3\sqrt{3}/2, \frac{3}{2}, -1)$ **71.** $(-1, \sqrt{3}, 4)$
73. $(3\sqrt{2}/4, 3\sqrt{6}/4, 3\sqrt{2}/2)$ **75.** $(2, 3\pi/4, 3\pi/4)$
77. cylindrical: $r^2 + z^2 = 25$; spherical: $\rho = 5$
79. cylindrical: $r^2(\cos^2\theta - \sin^2\theta) + z^2 = 1$;
spherical: $\rho^2(\sin^2\varphi\cos^2\theta - \sin^2\varphi\sin^2\theta + \cos^2\varphi) = 1$ **81.** $x^2 + y^2 = 1$

CHAPTER EIGHTEEN
PROBLEMS 18.1

1. $\mathbb{R}^2; [0, \infty)$ **3.** $\{(x, y): y \ne 0\}; \mathbb{R}$ **5.** $\{(x, y): x^2 - 4y^2 \le 1\}; [0, \infty)$ **7.** $\mathbb{R}^2; (0, \infty)$
9. $\{(x, y): x - y \ne (n + \frac{1}{2})\pi, n = 0, \pm1, \pm2, \ldots\}; \mathbb{R}$ **11.** $\{(x, y): |x| \ge |y|$ and
$x \ne -y\}; [0, \infty)$ **13.** $\{(x, y): |x - y| \le 1\}: [0, \pi]$ **15.** $\{(x, y): x \ne -y\}; \mathbb{R}$
17. $\{(x, y): x \ne 0$ and $y \ne 0\}; (-\infty, -2] \cup [2, \infty)$ (The range is obtained by using the fact that if $u = x/(2y)$ and $u > 0$, then $f(x, y) = u + (1/u)$ which has minimum value 2.)
19. $\{(x, y, z): x + y + z \ge 0\}$–this is the half-space "in front of" the plane $x + y + z = 0$; $[0, \infty)$
21. $\{(x, y, z): y^2 < x^2 + z^2\}$–this is the region "outside" the cone $y^2 = x^2 + z^2$; $(0, \infty)$
23. $\{(0, 0, 0)\}; \{0\}$ **25.** $\{(x, y, z): z \ne 0\}; \mathbb{R}$ **27.** $\{(x, y, z): |x + y - z| \le 1\}$–this is the part of $\mathbb{R}^3$ between the planes $x + y - z = -1$ and $x + y - z = 1$; $[-\pi/2, \pi/2]$
29. $\{(x, y, z): y \ne 0$ and $(x + z)/y \ne (n + \frac{1}{2})\pi, n = 0, \pm1, \pm2, \ldots\}; \mathbb{R}$ **31.** $\mathbb{R}^3; (0, \infty)$
33. $\{(x, y, z): x \ne 0, y \ne 0$ and $z \ne 0\}; \mathbb{R} - \{0\}$ **35.** $\mathbb{R}^3; [-3, 3]$ **37.** elliptic paraboloid opening around the y-axis; cross sections parallel to the xz-plane are ellipses
39. hyperbolic paraboloid; cross sections parallel to the xy-plane are hyperbolas
41. two half planes parallel to the xy-plane; $\{(x, y, z): z = 1$ and $x \ge 0\}$ and $\{(x, y, z); z = -1$ and $x \ge 0\}$

43. these are the parallel straight lines (with slopes of -1) $y = -x + (z^2 - 1)$
45. these are concentric ellipses (centered at the origin) with equations $x^2 + 4y^2 = 1 - z^2$, $|z| \le 1$; for $z = 1$ we obtain the single point $(0, 0, 0)$
47. parallel straight lines (with slopes of 1) $y = x - \cos z$
49. for each value of z we get a family of straight lines all of which have a slope of -1: for $z = 0$ we obtain $x + y = n\pi$, n an integer; for $z = 1$ we obtain $x + y = (\pi/4) + n\pi$; for $z = -1$ we obtain $x + y = -(\pi/4) + n\pi$; for $z = \sqrt{3}$ we obtain $x + y = (\pi/3) + n\pi$
51. concentric ellipses (centered at the origin); $a = \sqrt{T - 20}$, $b = \frac{1}{2}\sqrt{T - 20}$, for $T > 20$
53. concentric ellipses (centered at the origin); $a = \sqrt{(P - 100)/2}$, $b = \sqrt{(P - 100)/3}$, for $P > 100$; for $P = 100$ we obtain the single point $(0, 0)$
55. (h)   **57.** (a)   **59.** (e)   **61.** (g)   **63.** (d)

PROBLEMS 18.2

13. along the line $ax + by = 0$, $xy/(x^2 - y^2) = ab/(a^2 - b^2)$; the limit depends on the choice of a and b†   **15.** along the line $ax + by = 0$, $xy^3/(x^4 + y^4) = -a^3b/(a^4 + b^4)$
17. along the line $ax + by = 0$, $(x^2 + y^2)^2/(x^4 + y^4) = (a^2 + b^2)^2/(a^4 + b^4)$
19. along the line $y = kx$, $(ax^2 + by)/(cy^2 + dx) = (ax + bk)/(ck^2x + d) \to bk/d$ as $(x, y) \to (0, 0)$ **21.** along the line $y = kx = z$, $xyz/(x^3 + y^3 + z^3) = k^2/(1 + 2k^3)$
23. 0; since $0 \le (x^2 - y)^2 = x^4 - 2x^2y + y^2$, $2x^2|y| \le x^4 + y^2$ so that $5x^2y^2/(x^4 + y^2) = (\frac{5}{2})|y|2x^2|y|/(x^4 + y^2) \le (\frac{5}{2})|y|$
25. 0; since $0 < |(yx^2 + z^3)/(x^2 + y^2 + z^2)| \le (|y|x^2 + |z|z^2)/(x^2 + y^2 + z^2) \le |y| + |z| \to 0$.   **27.** $-\frac{61}{25}$ **29.** $\ln(4\pi/3)$ **31.** $\sinh 1$ **33.** $\frac{1}{6}$ **35.** $\ln 769$
37. $\mathbb{R}^2 - \{(0, 0)\}$ **39.** $\{(x, y): y \ne 4$ and $x \ne -3\}$ **41.** $\mathbb{R}^2$ **43.** $\{(x, y): xy \ne 1\}$
45. $\{(x, y, z): x > 0)\}$ **47.** $\{(x, y, z): xz > 0\}$
49. $\{(x, y, z): x^2 + y^2 + z^2 < 1\}$ = the "interior" of the unit sphere   **51.** $c = 0$
53. Along the line $y = kx = z$, $(yz - x^2)/(x^2 + y^2 + z^2) = (k^2 - 1)/(1 + 2k^2)$ which shows that $\lim_{(x, y, z) \to (0, 0, 0)} (yz - x^2)/(x^2 + y^2 + z^2)$ does not exist. The function is, therefore, not continuous at the origin.

PROBLEMS 18.3

1. $2xy$; x^2 **3.** $3y^3e^{xy^3}$; $9xy^2e^{xy^3}$ **5.** $4/y^5$; $-20x/y^6$ **7.** $3x^2y^5/(x^3y^5 - 2)$; $5x^3y^4/(x^3y^5 - 2)$ **9.** $\frac{4}{3}(1 + 5y \cos x)(x + 5y \sin x)^{1/3}$; $\frac{20}{3} \sin x(x + 5y \sin x)^{1/3}$
11. $[2y/(x^2 + y^2)](xy/|xy|)$; $[-2x/(x^2 + y^2)](xy/|xy|)$ **13.** 3   **15.** 0   **17.** 1
19. $\frac{48}{169}$ **21.** yz; xz; xy **23.** $1/z$; $1/z$; $-(x + y)/z^2$
25. $3x^2/(x^3 + y^2 + z)$; $2y/(x^3 + y^2 + z)$; $1/(x^3 + y^2 + z)$ **27.** $yz \cos xyz$; $xz \cos xyz$; $xy \cos xyz$ **29.** $\dfrac{\sinh \sqrt{x + 2y + 5z}}{2\sqrt{x + 2y + 5z}}$; $\dfrac{\sinh \sqrt{x + 2y + 5z}}{\sqrt{x + 2y + 5z}}$; $\dfrac{5 \sinh \sqrt{x + 2y + 5z}}{2\sqrt{x + 2y + 5z}}$ **31.** $\frac{1}{3}$
33. $-2 \cos 1$ **35.** $-\frac{1}{2}$ **37.** $-1/(3\sqrt{3})$ **39.** $a/\sqrt{a^2 + b^2 + c^2}$
43. (a) $x = 1$, $(z - 5)/(-12) = (y + 1)/1$; (b) $y = -1$, $(z - 5)/3 = (x - 1)/1$ (The first line is obtained as in $\mathbb{R}^2$ by setting $(z - 5)/[y - (-1)] = -12 = $ the slope.)
45. $y = 1$, $(x - 1)/(-4) = z - 1$
49. $\partial R/\partial x = [50/(1 + 50x + 75y)] + (20/\sqrt{1 + 40x + 125y})$, $\partial R/\partial y = [75/(1 + 50x + 75y)] + (125/2\sqrt{1 + 40x + 125y})$

PROBLEMS 18.4

In Problems 1–11 the answers are given in the order f_{xx}, $f_{xy}(= f_{yx})$, f_{yy}.
1. $2y$; $2x$; 0   **3.** $3y^6e^{xy^3}$; $9y^2(1 + xy^3)e^{xy^3}$; $9xy(2 + 3xy^3)e^{xy^3}$ **5.** 0; $-20/y^6$; $120x/y^7$
7. $\dfrac{-3xy^5(4 + x^3y^5)}{(x^3y^5 - 2)^2}$; $\dfrac{-30x^2y^4}{(x^3y^5 - 2)^2}$; $\dfrac{-5x^3y^3(8 + x^3y^5)}{(x^3y^5 - 2)^2}$
9. $\frac{4}{9}(1 + 5y \cos x)^2(x + 5y \sin x)^{-2/3} - \frac{20}{3} y \sin x(x + 5y \sin x)^{1/3}$; $\frac{20}{9} \sin x(1 + 5y \cos x)(x + 5y \sin x)^{-2/3} + \frac{20}{3} \cos x(x + 5y \sin x)^{1/3}$; $\frac{100}{9} \sin^2 x(x + 5y \sin x)^{-2/3}$

---

†In Problems 18.2.13, 19, 21, it is immediately evident that the limit does not exist because the function is not defined in any neighborhood of the origin.

11. $[-4xy/(x^2 + y^2)^2](xy/|xy|); 2[(x^2 - y^2)/(x^2 + y^2)^2][(xy/|xy|)]; [4xy/(x^2 + y^2)^2][(xy/|xy|)]$

13. $f_{xx} = 0; f_{xy} = z = f_{yx}; f_{xz} = y = f_{zx}; f_{yy} = 0; f_{yz} = x = f_{zy}; f_{zz} = 0$

15. $f_{xx} = 0; f_{xy} = 0 = f_{yx}; f_{xz} = -1/z^2 = f_{zx}; f_{yy} = 0; f_{yz} = -1/z^2 = f_{zy}; f_{zz} = 2(x + y)/z^3$

17. $f_{xx} = \dfrac{-2xyz^3}{(y^2 + x^2z^2)^2}; f_{xy} = \dfrac{x^2z^3 - y^2z}{(y^2 + x^2z^2)^2} = f_{yx}; f_{xz} = \dfrac{y^3 - x^2yz^2}{(y^2 + x^2z^2)^2} = f_{zx}; f_{yy} = \dfrac{2xyz}{(y^2 + x^2z^2)^2};$

$f_{yz} = \dfrac{x^3z^2 - xy^2}{(y^2 + x^2z^2)^2} = f_{zy}; f_{zz} = \dfrac{-2x^3yz}{(y^2 + x^2z^2)^2}$

19. $f_{xx} = 9y^2e^{3xy}\cos z; f_{xy} = 3(1 + 3xy)e^{3xy}\cos z = f_{yx}; f_{xz} = -3ye^{3xy}\sin z = f_{zx};$
$f_{yy} = 9x^2e^{3xy}\cos z; f_{yz} = -3xe^{3xy}\sin z = f_{zy}; f_{zz} = -e^{3xy}\cos z$

21. $f_{xx} = \dfrac{\cosh\sqrt{x + yz}}{4(x + yz)} - \dfrac{\sinh\sqrt{x + yz}}{4(x + yz)^{3/2}}; f_{xy} = zf_{xx} = f_{yx}; f_{xz} = yf_{xx} = f_{zx}; f_{yy} = z^2f_{xx};$

$f_{yz} = \dfrac{\sinh\sqrt{x + yz}}{2\sqrt{x + yz}} + yzf_{xx} = f_{zy}; f_{zz} = y^2f_{xx}$ **23.** (a) 16 (b) 81   **25.** $6y^2$

27. $24/(3x - 2y)^3$ **29.** $9\sin(x + 2y + 3z)$

PROBLEMS 18.5

5. $(e/2)\mathbf{i} + (e/2)\mathbf{j}$ **7.** $[2/(2x - y + 1)]\mathbf{i} + [-1/(2x - y + 1)]\mathbf{j}$ **9.** $-\frac{1}{6}\mathbf{i} + \frac{1}{6}\mathbf{j}$

11. $2x(\sinh y)\mathbf{i} + x^2(\cosh y)\mathbf{j}$ **13.** $\frac{1}{8}\mathbf{i} - \frac{3}{8}\mathbf{j}$

15. $(2xe^{x^2}/3y)\mathbf{i} + [-e^{x^2} + (1 + 2y^2)e^{-y^2}]/3y^2\mathbf{j}$ **17.** $(3\sqrt{2}/4)\mathbf{i} - (\sqrt{6}/4)\mathbf{j} + \sqrt{2}\mathbf{k}$

19. $[\ln y - (z/x)]\mathbf{i} + (x/y)\mathbf{j} - (\ln x)\mathbf{k}$ **21.** $-4e^3\mathbf{i} - 7e^3\mathbf{j} - 13e^3\mathbf{k}$ **23.** $\mathbf{i} - \mathbf{k}$

29. $f(x, y) = (x^2/2) + (y^2/2) + C$

PROBLEMS 18.6

1. $3e^{3t}$ **3.** 1   **5.** $[5(\cos 5t)\cos 3t + 3(\sin 5t)\sin 3t]/(\sin^2 5t + \cos^2 3t)$ **7.** $2t$

9. $(-t - 2)/t^3$ **11.** $[2e^t - (3/t) + 4\sinh t]/(2e^t - 3\ln t + 4\cosh t)$

13. $\partial z/\partial r = -2\sin(r + s)\cos(r + s) + 2\sin(r - s)\cos(r - s);$
$\partial z/\partial s = -2\sin(r + s)\cos(r + s) - 2\sin(r - s)\cos(r - s)$

15. $\partial z/\partial r = -(2s^2/r^3)\cos(s^2/r^2); \partial z/\partial s = (2s/r^2)\cos(s^2/r^2)$

17. $\partial z/\partial r = 2(r - s^2)(2s + r)^3 + 3(r - s^2)^2(2s + r)^2;$
$\partial z/\partial s = -4s(r - s^2)(2s + r)^3 + 6(r - s^2)^2(2s + r)^2$

19. $\partial w/\partial r = s/t; \partial w/\partial s = r/t; \partial w/\partial t = -rs/t^2$

21. $\partial w/\partial r = (4s^3r^3 - 3s^4r^2)\cos[s^3r^3(r - s)]; \partial w/\partial s = (3s^2r^4 - 4s^3r^3)\cos[s^3r^3(r - s)]$

23. $\partial w/\partial r = 3r^2s^3t - 1/(r^2st^3); \partial w/\partial s = 3r^3s^2t - 1/(rs^2t^3); \partial w/\partial t = r^3s^3 - 3/(rst^4)$

25. $\partial w/\partial r = 2r[(s^2 - t^2)/(r^2 + s^2)]^2w; \partial w/\partial s = 2s[(r^2 + t^2)/(r^2 + s^2)]^2w;$
$\partial w/\partial t = 2t[(s^2 - r^2 - 2t^2)/(r^2 + s^2)]w$ **27.** $V_t = 7500\pi$ in^3/min; increasing

29. $P_t = (10R + 75)/1000$ N/cm^2/min; increasing

31. $(2\pi - 38 - 2\sqrt{3})/\sqrt{676 - 240\sqrt{3}} \approx -2.18$ cm/sec; decreasing

33. $(\partial z/\partial x)^2 + (\partial z/\partial y)^2 = [(1/r)(\partial z/\partial \theta)]^2 + (\partial z/\partial r)^2$

35. $\partial w/\partial \rho = (\partial w/\partial x)\sin\varphi\cos\theta + (\partial w/\partial y)\sin\varphi\sin\theta + (\partial w/\partial z)\cos\varphi$
$\partial w/\partial \varphi = (\partial w/\partial x)\rho\cos\varphi\cos\theta + (\partial w/\partial y)\rho\cos\varphi\sin\theta - (\partial w/\partial z)\rho\sin\varphi$
$\partial w/\partial \theta = -(\partial w/\partial x)\rho\sin\varphi\sin\theta + (\partial w/\partial y)\rho\sin\varphi\cos\theta$

PROBLEMS 18.7

1. $x = 1; y = 0, z = 0$ **3.** $z = 1; x = 0, y = 0$

5. $(1/a)(x - a) + (1/b)(y - b) + (1/c)(z - c) = 0; a(x - a) = b(y - b) = c(z - c)$

7. $\frac{1}{4}(x - 4) + \frac{1}{2}(y - 1) + \frac{1}{6}(z - 9) = 0; 4(x - 4) = 2(y - 1) = 6(z - 9)$

9. $2(x - 1) + (y - 2) + (z - 2) = 0; (x - 1)/2 = y - 2 = z - 2$

11. $24(x - 3) - 2(y - 1) + 20(z + 2) = 0; (x - 3)/24 = (y - 1)/(-2) = (z + 2)/20$

13. $(\pi/\sqrt{3})(y - 1) + \sqrt{3}[z - (\pi/3)] = 0; x = \pi/2, (\sqrt{3}/\pi)(y - 1) = (1/\sqrt{3})[z - (\pi/3)]$

15. $(\ln 5)(x - 1) + (\ln 5)(y - 1) + (z - \ln 5) = 0; (x - 1)/(\ln 5) = (y - 1)/(\ln 5) = z - \ln 5$

17. $z = 1 + (x - 1) + 2(y - 1); x - 1 = (y - 1)/2 = (z - 1)/(-1)$

19. $z = 1; x = \pi/8, y = \pi/20$ **21.** $z = -(\pi/4) - \frac{1}{4}(x + 2) - \frac{1}{4}(y - 2);$
$4(x + 2) = 4(y - 2) = z + \pi/4$ **23.** $z = 2 + 2\sqrt{3}(x - \pi/2) - 2\sqrt{3}(y - \pi/6);$
$(x - \pi/2)/2\sqrt{3} = (y - \pi/6)/(-2\sqrt{3}) = (z - 2)/(-1)$ **25.** $(0, 2, 4), (1, 1, 2)$

29. All tangent planes pass through the origin. (Note that the surface cannot pass through the origin since $f(y/x)$ is not defined when $x = 0$.)
33. $[(x - x_0)\mathbf{i} + (y - y_0)\mathbf{j}] \cdot \nabla F(x_0, y_0) = 0$ **35.** $3\mathbf{i} - 8\mathbf{j}; 3(x - 1) - 8(y + 2) = 0$
37. $\mathbf{i} + \mathbf{j}; (x - 1) + y = 0$ **39.** $2\mathbf{i} + 2\mathbf{j}; (x - \pi/4) + y = 0$

PROBLEMS 18.8

1. $9/\sqrt{10}$ **3.** $\sqrt{2}/7$ **5.** $-5/(4\sqrt{13})$ **7.** $(2e^2 + 3e)/\sqrt{2}$ **9.** $2\sqrt{3}$
11. $1/(5\sqrt{6})$ **13.** $61/\sqrt{26}$ **15.** 0   **17.** (a) at the origin
(b) $\nabla T = 200e^{-(x^2 + y^2 + z^2)}(-x\mathbf{i} - y\mathbf{j} - z\mathbf{k})$
19. $(x(t), y(t)) = (ae^{-2t/a^2}, be^{-2t/b^2}); (x/a)^{a^2} = (y/b)^{b^2}$
21. (a) ascend, because $\nabla h \cdot \mathbf{u} > 0$ when $\mathbf{u} = (-1/\sqrt{2})\mathbf{i} - (1/\sqrt{2})\mathbf{j}$
(b) $(-120\mathbf{i} + 40\mathbf{j})/\sqrt{120^2 + 40^2} = (-3\mathbf{i} + \mathbf{j})/\sqrt{10}$; i.e., $\approx 161.6°$ (toward the southeast)
23. $\sqrt{5}$ **25.** $1/\sqrt{2}$

PROBLEMS 18.9

1. $F(x, y, z) = (\alpha/2) \ln(x^2 + y^2 + z^2) = \alpha \ln|\mathbf{x}|$
3. If $k = 1$, let $F(x, y, z) = \alpha|\mathbf{x}|$. If $k = 2$, look at Problem 1. If $k > 2$, a potential function is given by $F(x, y, z) = [-\alpha/(k - 2)](1/|\mathbf{x}|^{k-2})$.

PROBLEMS 18.10

1. $y^3 \Delta x + 3xy^2 \Delta y$ **3.** $(x - y)^{-1/2}(x + y)^{-3/2}(y \Delta x - x \Delta y)$
5. $[1/(2x + 3y)](2 \Delta x + 3 \Delta y)$ **7.** $y^2 z^5 \Delta x + 2xyz^5 \Delta y + 5xy^2 z^4 \Delta z$
9. $[1/(x + 2y + 3z)](\Delta x + 2 \Delta y + 3 \Delta z)$ **11.** $\sinh(xy - z)(y \Delta x + x \Delta y - \Delta z)$
13. (a) $x(\Delta y)^2 + 2y(\Delta x)(\Delta y) + (\Delta x)(\Delta y)^2$
(b) $\Delta f - df = [(0.99)(2.03)^2 - (1)(2)^2] - [2^2(-0.01) + (2)(1)(2)(0.03)] = [0.079691] - [0.08] = -0.000309$ **15.** $\frac{3}{6} + \frac{1}{6}(0.01) - [3(-0.01)/6^2] = 0.5025$ (actual value ≈ 0.50250417)
17. $\sqrt{36}\sqrt[3]{64} + (\frac{1}{2})(\frac{1}{6})(4)(-0.4) + (\frac{1}{3})(6)(\frac{1}{16})(0.08) = 23.87666\ldots$ (actual value ≈ 23.87623437)
19. $\sqrt{\frac{1}{9}} + \frac{1}{27}[4(0.02) - 5(-0.04)] = 0.3437037037\ldots$ (actual value ≈ 0.3435696277)
21. $(3 \cdot 2)/\sqrt{9} + (2/\sqrt{9})(0.02) + (3/\sqrt{9})(-0.03) - (3 \cdot 2/2(9)^{3/2})(-0.05) = 1.9888\ldots$
(actual value ≈ 1.988665097)
23. $8\sqrt{5 - 1^2} + (\sqrt{5 - 1^2})(-0.08) + (8/2\sqrt{5 - 1^2})(0.01) - [(8)(1)/\sqrt{5 - 1^2}](-0.02) = 15.94$
(actual value ≈ 15.93790543)   **25.** (a) $\sqrt{9100} \approx 95.3939$ m
(b) $(1/\sqrt{9100})\{|50 - 110(\frac{1}{2})|[0.3] + [110 - 50(\frac{1}{2})][0.3] + [(50)(110)(\sqrt{3}/2)](\pi/180)\} \approx 0.899768$ m
27. (a) $\frac{8}{3} = 2.6666\ldots$ ohms   (b) $(\frac{8}{3})^2[(0.1/6^2) + (0.03/8^2) + (0.15/12^2)] \approx 0.03049$

PROBLEMS 18.11

1. $x^2 y + y + C$ **3.** not exact   **5.** not exact   **7.** $(x^2/2) - y \sin x + C$
9. $(3x^2/2) \ln x - (3x^2/4) + (x^6/6) - xy + C$ **11.** $\tan x - x + \tan^{-1}(y/x) - e^y + C$
13. not exact   **15.** $x + y + z + C$ **17.** $(xy/z) + (x^3/3) + \cos y + \sin z + C$
19. not exact   **21.** $\mathbf{f}(t) = -\alpha \mathbf{x}/|\mathbf{x}|^4$ and a potential function for $\mathbf{F}$ is $f(\mathbf{x}) = -\alpha/(2|\mathbf{x}|^2)$ $(-\nabla f = \mathbf{F})$ **23.** $\mathbf{F} = -\alpha \mathbf{x}/|\mathbf{x}|$ and $f(\mathbf{x}) = \alpha|\mathbf{x}|$ $(-\nabla f = \mathbf{F})$

PROBLEMS 18.12

1. $(0, 0)$ is a local minimum   **3.** $(-2, 1)$ is a local minimum   **5.** $(-2, 1)$ is a local minimum   **7.** $(\sqrt{5}, 0)$ is a local minimum; $(-\sqrt{5}, 0)$ is a local maximum; $(-1, 2)$ and $(-1, -2)$ are saddle points   **9.** $(\frac{1}{2}, 1)$ and $(-\frac{1}{2}, -1)$ are saddle points; $(-\frac{1}{2}, 1)$ is a local minimum; $(\frac{1}{2}, -1)$ is a local maximum   **11.** $(-2, -2)$ is a local minimum
13. $(0, 0), (0, 4)$ and $(4, 0)$ are saddle points; $(\frac{4}{3}, \frac{4}{3})$ is a local maximum
15. $(1, 1)$ and $(-1, 1)$ are local minima   **17.** $(0, 0)$ is a saddle point. (Note that $D = 0$, but it is evident that f can take positive and negative values near the origin.)
19. $x = y = z = \frac{50}{3}$ **21.** $x = y = z = \frac{50}{3}$ **23.** $9\sqrt{3}$
25. Corners are at $(\pm a/\sqrt{3}, \pm b/\sqrt{3}, \pm c/\sqrt{3})$. (The problem is to maximize $(2x)(2y)(2z)$ with $(x^2/a^2) + (y^2/b^2) + (z^2/c^2) = 1$.)
27. The profit is given by $P(x, y) = 40(8xy + 32x + 40y - 4x^2 - 6y^2) - 10x - 4y$, which is maximized when $x \approx 22$ and $y \approx 18$ ($x = 21.88125, y = 17.9125$).

PROBLEMS 18.13

1. $3/\sqrt{13}$ at the point $(\frac{7}{13}, \frac{17}{13})$ **3.** $5/\sqrt{3}$ at the point $(\frac{8}{3}, \frac{2}{3}, \frac{1}{3})$

5. $|d|/\sqrt{a^2 + b^2 + c^2}$ at the point $(ad/(a^2 + b^2 + c^2), bd/(a^2 + b^2 + c^2), cd/(a^2 + b^2 + c^2))$

Alts—kvs-G—12/12/83

7. minimum is $\frac{15}{32}$ at $(\frac{9}{16}, \frac{1}{8})$; there is no maximum

9. maximum is 8 at (2, 0, 0), (0, 2, 0) and (0, 0, 2);
minimum is -8 at $(-2, 0, 0)$, $(0, -2, 0)$ and $(0, 0, -2)$

11. maximum is $2/\sqrt{3}$ at $(1/\sqrt{3}, 2/\sqrt{3}, 3/\sqrt{3})$ and at 3 other points;
minimum is $-2/\sqrt{3}$ at $(-1/\sqrt{3}, 2/\sqrt{3}, 3/\sqrt{3})$ and at 3 other points. **13.** $\frac{148}{45}$ at $(\frac{74}{45}, \frac{-20}{45}, \frac{28}{45})$

15. maximum is the larger of $|a|$ and $|b|$; minimum is the smaller of $|a|$ and $|b|$

17. 2 at $(\frac{1}{2}, \frac{1}{2}, \frac{1}{2}, \frac{1}{2})$ **21.** maximum is $2/(3\sqrt{3})$ at $(\sqrt{\frac{2}{3}}, \sqrt{\frac{2}{3}}, \sqrt{\frac{1}{3}})$ and at 3 other points;
minimum is $-2/(3\sqrt{3})$ at $(-\sqrt{\frac{2}{3}}, -\sqrt{\frac{2}{3}}, -\sqrt{\frac{1}{3}})$ and at 3 other points

23. height of cone is $12/\sqrt{5}$ m; height of cylinder is $(50/3\pi) - (9/\sqrt{5})$ m

25. base is a square: 2×2 m; height is 3 m **27.** $x/y = \frac{1}{2}$

REVIEW EXERCISES FOR CHAPTER EIGHTEEN

1. $\{(x, y): |x| \geq |y|\}; [0, \infty)$ **3.** $\mathbb{R}^2; [-1, 1]$

5. $\{(x, y, z): x^2 + y^2 + z^2 > 1\}$ = the "exterior" of the unit sphere; $(0, \infty)$

7. parallel straight lines with slopes of -1; $x + y = 1 - z^2$

9. parallel straight lines with slopes of $\frac{1}{3}$; $x - 3y = e^z$ **13.** along $y = kx$
limit is $k/(k^2 - 1)$ **15.** since $0 \leq (x - y^2)^2 = x^2 - 2xy^2 + y^4$, $|2xy^2| \leq x^2 + y^4$ so that
$|4xy^3/(x^2 + y^4)| \leq |4xy^3/2xy^2| = 2|y|$ **17.** $-\frac{1}{4}$ **19.** 5

21. $\{(x, y, z): z < x - y + 4\}$, i.e., the half-space "below" but not including the plane
$x - y - z = -4$ **23.** 0 **25.** $f_x = -\sin(x - 3y)$; $f_y = 3\sin(x - 3y)$

27. $f_x = -y/[(1 + x)^2 + y^2]$; $f_y = (1 + x)/[(1 + x)^2 + y^2]$

29. $f_x = -x(x^2 + y^2 + z^2)^{-3/2}$; $f_y = -y(x^2 + y^2 + z^2)^{-3/2}$; $f_z = -z(x^2 + y^2 + z^2)^{-3/2}$

31. $f_x = \dfrac{1}{z}\sec\left(\dfrac{x - y}{z}\right)\tan\left(\dfrac{x - y}{z}\right)$; $f_y = \dfrac{-1}{z}\sec\left(\dfrac{x - y}{z}\right)\tan\left(\dfrac{x - y}{z}\right)$;

$f_z = \dfrac{y - x}{z^2}\sec\left(\dfrac{x - y}{z}\right)\tan\left(\dfrac{x - y}{z}\right)$

33. $f_x = \dfrac{2x}{y^3 + z^4}$; $f_y = \dfrac{-3y^2(x^2 + z^4)}{(y^3 + z^4)^2}$; $f_z = \dfrac{4z^3(y^3 - x^2)}{(y^3 + z^4)^2}$

35. $f_x = \dfrac{1}{y + z}f$; $f_y = \dfrac{w - x}{(y + z)^2}f$; $f_z = \dfrac{w - x}{(y + z)^2}f$; $f_w = \dfrac{-1}{y + z}f$

37. $f_{xx} = \dfrac{2xy}{(x^2 + y^2)^2}$; $f_{xy} = \dfrac{y^2 - x^2}{(x^2 + y^2)^2} = f_{yx}$; $f_{yy} = \dfrac{-2xy}{(x^2 + y^2)^2}$

39. $f_{xx} = \dfrac{4y}{(x - y)^3}$; $f_{xy} = \dfrac{-2(x + y)}{(x - y)^3} = f_{yx}$; $f_{yy} = \dfrac{4x}{(x - y)^3}$

41. Let $R = \sqrt{1 - x^2 - y^2 - z^2}$, then $f_{xx} = \dfrac{1 + 2x^2 - y^2 - z^2}{R^5}$; $f_{xy} = \dfrac{3xy}{R^5} = f_{yx}$;

$f_{xz} = \dfrac{3xz}{R^5} = f_{zx}$; $f_{yy} = \dfrac{1 - x^2 + 2y^2 - z^2}{R^5}$; $f_{yz} = \dfrac{3yz}{R^5} = f_{zy}$; $f_{zz} = \dfrac{1 - x^2 - y^2 + 2z^2}{R^5}$

43. $f_{zxx} = 0 = f_{zxxy} = f_{zxxyz}$ **45.** $\frac{1}{2}i - \frac{1}{2}j$ **47.** $-\frac{1}{2}i + j$ **49.** $\frac{1}{12}i - \frac{1}{12}j - \frac{1}{48}k$

51. $-3j - 4k$ **53.** $(t^2 + 2t)/(1 + t)^2\sqrt{1 - t^4/(1 + t)^2}$ **55.** $2r/(r - s)^2$

57. $[\cos(e^{r+s} - e^{r-s})][e^{r+s} + e^{r-s}]$ **59.** $\partial w/\partial s = 4r^4s^3t^2 + 2r^4st^{-2}$;
$\partial w/\partial t = 2r^4s^4t - 2r^4s^2t^{-3}$ **61.** $x + y + z = 3$; $x = y = z$ **63.** $3x - y + 5z = 15$;
$(x + 1)/3 = (y - 2)/(-1) = (z - 4)/5$ **65.** $-3x + 6y - 2z = 18$;
$(x + 2)/(-3) = (y - 1)/6 = (z + 3)/(-2)$ **67.** $-3/\sqrt{2}$ **69.** $-1/(2\sqrt{13})$

71. $9/(\sqrt{14})(6^{3/2})$ **73.** $-1/|x|^3$ is a potential function **75.** $3x^2y^2\,\Delta x + 2x^3y\,\Delta y$

77. $\Delta x/2\sqrt{(x + 1)(y - 1)} - (\Delta y)\sqrt{x + 1}/2(y - 1)^{3/2}$

79. $[1/(x - y + 4z)](\Delta x - \Delta y + 4\,\Delta z)$

81. $\frac{4}{7} + (0.03/7) - \frac{4}{49}(-0.03) \approx 0.578163$ (actual value ≈ 0.578192)

83. $54 + \frac{27}{4}(-0.03) + 9(0.05 - 0.03) = 53.9775$ (actual value ≈ 53.97654)
85. $xy^2 + x + C$ **87.** not exact   **89.** $xy^3z^5 + \frac{3}{2}x^2 - \frac{8}{3}y^{3/2} - \cos z + C$
91. $(0, 0)$ is a local minimum
93. $(\sqrt{2}, 1/\sqrt{2})$ is a local minimum; $(\sqrt{2}, -1/\sqrt{2})$ and $(-\sqrt{2}, 1/\sqrt{2})$ are saddle points;
$(-\sqrt{2}, -1/\sqrt{2})$ is a local maximum.   **95.** $(2^{-1/5}, 2^{3/10})$ and $(2^{-1/5}, -2^{3/10})$ are local minima.
97. $8/\sqrt{11}$ at $(\frac{14}{11}, \frac{-3}{11}, \frac{20}{11})$ **99.** max $V = \frac{1}{2}(\frac{10}{3})^{3/2}$ if base is a square $\sqrt{\frac{10}{3}} \times \sqrt{\frac{10}{3}}$
and height $= \frac{1}{2}\sqrt{\frac{10}{3}}$ **101.** vertices are $(\pm 6/\sqrt{3}, \pm 2/\sqrt{3}, \pm 3/\sqrt{3})$;
dimensions are $12/\sqrt{3} \times 4/\sqrt{3} \times 6/\sqrt{3}$

CHAPTER NINETEEN
PROBLEMS 19.1
1. $\frac{45}{2}$ **3.** 0   **5.** 16   **7.** 39   **9.** 0   **11.** $\frac{15}{2}$ **13.** 10   **15.** $0 \le I \le 6$
17. $-\sqrt{2}\pi/3 \le I \le \sqrt{2}\pi/3$ (These are crude bounds obtained from $|x - y| \le \sqrt{2}$ and
$1/(4 - x^2 - y^2) \le \frac{1}{3}$ on the unit disk.); by symmetry, $I = 0$.   **19.** $0 \le I \le \frac{1}{4} \ln 2$

PROBLEMS 19.2
1. $\frac{2}{3}$ **3.** $e^{-5} - e^{-1} - e^{-2} + e^2$ **5.** -31 **7.** $\frac{162}{5}$ **9.** $\frac{16}{3}$ **11.** $\frac{20}{3}$
13. $\frac{1}{2}(e^{19} - e^{17} - e^3 + e)$ **15.** $\frac{1}{3} + \pi/16$ **17.** $\int_0^{1/2} \int_x^{1-x} (x + 2y)\, dy\, dx = \frac{7}{24}$
19. $\int_0^{1/\sqrt{2}} \int_{x^2}^{1-x^2} (x^2 + y)\, dy\, dx = \sqrt{2}/5$
21. $\int_1^2 \int_1^y (y/\sqrt{x^2 + y^2})\, dx\, dy = \int_1^2 \int_x^2 (y/\sqrt{x^2 + y^2})\, dy\, dx = \int_1^2 (\sqrt{x^2 + 4} - x\sqrt{2})\, dx$
$= 1/\sqrt{2} - \frac{1}{2}\sqrt{5} + 2 \ln[(2 + 2\sqrt{2})/(1 + \sqrt{5})]$
23. $\int_0^\infty \int_0^\infty (x + y)e^{-(x+y)}\, dy\, dx = \int_0^\infty (1 + y)e^{-y}\, dy = 2$ **25.** (a) $\int_{-5}^8 \int_0^4 (x + y)\, dx\, dy$
(b) 182   (c) region is a rectangle   **27.** $\int_0^1 \int_y^1 dx\, dy$ (b) $\frac{1}{2}$
(c) triangle with vertices at $(0, 0)$, $(1, 0)$ and $(1, 1)$ **29.** $\int_0^1 \int_0^{\cos^{-1}x} y\, dy\, dx$ (b) $(\pi/2) - 1$
(c) region in first quadrant bounded by x-axis, y-axis, and curve $y = \cos^{-1} x$
(vertices are $(0, 0)$, $(\pi/2, 0)$, $(1, 0)$)   **31.** (a) $\int_0^1 \int_{y^3}^{y^2} (1 + y^6)\, dx\, dy$ (b) $\frac{17}{180}$
(c) region bounded by $y = x^{1/2}$ and $y = x^{1/3}$ (the curves meet at $(0, 0)$ and $(1, 1)$)
33. (a) $\int_0^\infty \int_0^y (1 + y^2)^{-7/5}\, dx\, dy$ (b) $\frac{5}{4}$
(c) region is the "triangular" part of the first quadrant above the line $y = x$
35. $\frac{9}{2}$ **37.** $\frac{128}{3}$ **39.** 48π **41.** $2\int_0^2 \int_0^x \sqrt{2 - x}\, dy\, dx = 32\sqrt{2}/15$
43. $\int_{(-1-\sqrt{13})/2}^{(-1+\sqrt{13})/2} \int_{x^2}^{3-x} 2\sqrt{y - x^2}\, dy\, dx = \frac{4}{3} \int_{(-1-\sqrt{13})/2}^{(-1+\sqrt{13})/2} (3 - x - x^2)^{3/2}\, dx = 169\pi/32$
(Note: Limits on x were obtained by first setting $z = 0$ to obtain $y = x^2$ and then equating
this with the line $y = -x + 3$.)
45. $\int_{-\sqrt{3}}^{-1} \int_0^{3-x^2} dy\, dx + \int_{-1}^1 \int_{x^3+1}^{3-x} dy\, dx = 2\sqrt{3} + \frac{2}{3}$;
also $\int_{-1}^1 \int_0^{x^3+1} dy\, dx + \int_1^{\sqrt{3}} \int_0^{3-x^2} dy\, dx = 2\sqrt{3} - \frac{2}{3}$
49. The surface $z = \sqrt{x^2 + y^2}$ is a cone; the region in the first quadrant bounded by the
curves $y = \sqrt{x}$ and $y = x^2$ is the directrix of a cylinder whose generatrix is the z-axis; the
solid in question is bounded below by the xy-plane, above by the cone, and whose "walls"
are the sides of the cylinder.

PROBLEMS 19.3
1. $\frac{16}{3}$; $(\frac{51}{32}, 0)$ (See Problem 19.2.9)
3. $(\sqrt{3} - 1)/12$; $(\frac{1}{2} + (\pi/6) - [\pi/6(\sqrt{3} - 1)], \frac{1}{3} - [\pi/18(\sqrt{3} - 1)])$
5. $3e^3 - 3e + e^{-1} - e^{-3}$; $\left(\dfrac{10e^3 - 10e - 2e^{-1} + 2e^{-3}}{3e^3 - 3e + e^{-1} - e^{-3}}, \dfrac{6e^3 - 12e + 2e^{-1} - 4e^{-3}}{3e^3 - 3e + e^{-1} - e^{-3}}\right)$
7. $(\pi/16) + \frac{1}{3}$; $(62/5(16 + 3\pi), (16 + 15\pi)/5(16 + 3\pi))$ **9.** $\frac{5}{24}$; $(\frac{11}{20}, \frac{1}{5})$ **11.** $\frac{1}{2}$; $(1, 1)$
13. $4\sqrt{2}\pi^2$ **15.** $6\pi^2 a^2 b$

PROBLEMS 19.4
1. $\frac{2}{3}\pi a^3$ **3.** $\int_0^\pi \int_0^{2\cos\theta} (3 - r)r\, dr\, d\theta = 3\pi$ **5.** $\frac{32}{15}\pi^6 a^5$ **7.** $(4^2)(3\pi/2) = 24\pi$
9. 11π **11.** $2\int_0^{\pi/2} \int_0^{2\sqrt{\sin 2\theta}} r\, dr\, d\theta = 6$ **17.** 6
19. $4\int_0^{\pi/2} \int_0^2 [r - (r^2/2)]r\, dr\, d\theta = 4\pi/3$ **21.** $\pi(1 - e^{-1})$
23. area $= 19\pi/2$; centroid $= (0, \frac{37}{38})_{\text{rectangular}} = (\frac{37}{38}, \pi/2)_{\text{polar}}$

PROBLEMS 19.5

1. $2\sqrt{6}$ **3.** $(\pi/2)\sqrt{1 + a^2 + b^2}$ **5.** $2\int_0^1 \int_1^2 \sqrt{1 + \frac{4}{9}x^{-2/3}}\, dy\, dx = \frac{2}{27}(13^{3/2} - 8)$
(Note that $\int_{-1}^1 \int_1^2 \sqrt{1 + \frac{4}{9}x^{-2/3}}\, dy\, dx$ is *wrong* since the integrand is not defined at $x = 0$; as in Chapter 12, this improper integral is treated by breaking it into two parts.)
7.   132   **9.** $\int_0^1 \int_0^2 \sqrt{1 + 9x}\, dy\, dx = \frac{4}{27}(10^{3/2} - 1)$ (This is the surface area for $z \geq -1$.)
11. $2\pi a^2$ **13.**   16
15. $\int_{-4}^4 \int_{-\sqrt{16-y^2}}^{\sqrt{16-y^2}} \sqrt{1 + (x^2/4) + (y^2/4)}\, dx\, dy = \frac{1}{2}\int_0^{2\pi} \int_0^4 r\sqrt{4 + r^2}\, dr\, d\theta = (\pi/3)(20^{3/2} - 8)$
17. $\int_{-1}^1 \int_{-\sqrt{1-y^2}}^{\sqrt{1-y^2}} \sqrt{1 + 9x^4 + 9y^4}\, dx\, dy$
19. $\int_0^2 \int_x^{4-x} \sqrt{1 + 1/2(1 + x + y)}\, dy\, dx = (\int_0^2 \int_0^y + \int_2^4 \int_0^{4-y}) \sqrt{1 + 1/2(1 + x + y)}\, dx\, dy$
21. $2\int_{-a}^a \int_{-b\sqrt{1-x^2/a^2}}^{b\sqrt{1-x^2/a^2}} \sqrt{\dfrac{1 + (c^2/a^2 - 1)(x^2/a^2) + (c^2/b^2 - 1)(y^2/b^2)}{1 - (x^2/a^2) - (y^2/b^2)}}\, dy\, dx$; see Problem 9.3.15

PROBLEMS 19.6

1. $\frac{1}{8}$ **3.** $(a_2 - a_1)(b_2 - b_1)(c_2 - c_1)$ **5.** $6(\frac{31}{5} + \frac{127}{7})$
7. $\int_{-\pi/2}^{\pi/2} \int_0^1 \int_{-\sqrt{1-r^2}}^{\sqrt{1-r^2}} z^2\, r\, dz\, dr\, d\theta = 2\pi/15$ **9.** (a) $\int_0^1 \int_0^{\sqrt{1-z^2}} \int_z^{\sqrt{1-y^2}} yz\, dx\, dy\, dz$
(b) $\int_0^1 \int_0^x \int_0^{\sqrt{1-x^2}} yz\, dy\, dz\, dx$ **11.** $\frac{1}{6}$ **13.** $\int_0^3 \int_0^{\sqrt{9-x^2}} \int_0^{4-x} dy\, dz\, dx = 9(\pi - 1)$
15. $\int_{-\sqrt{12}}^{\sqrt{12}} \int_{-\sqrt{12-x^2}}^{\sqrt{12-x^2}} \int_2^{\sqrt{16-x^2-y^2}} dz\, dy\, dx = 40\pi/3$ **17.** $\frac{4}{3}\pi abc$ **19.** $\mu = \frac{1}{24};\ (\frac{2}{5}, \frac{1}{5}, \frac{1}{5})$
21. $\mu = \frac{207}{8};\ (\frac{108}{115}, \frac{176}{115}, 18\pi/23 - \frac{72}{115})$
23. $\mu = (4\pi/15)abc(\alpha a^2 + \beta b^2 + \gamma c^2);\ (0, 0, 0)$ **25.** $(a/4, b/4, c/4)$ **27.** $(0, \frac{15}{7}, \frac{20}{7})$
29. $\mu = \int_0^1 \int_0^y \int_0^{xy} (1 + 2z)\, dz\, dx\, dy = \frac{13}{72}$ (a) $(\bar{x}, \bar{y}, \bar{z}) = (\frac{258}{455}, \frac{372}{455}, \frac{7}{26})$
(b) $\bar{x} \approx 0.57,\ \bar{y} \approx 0.82,\ \bar{z} \approx 0.27$, and $(\bar{x})(\bar{y}) \approx 0.46$ so that $0 < \bar{z} < \bar{x}\bar{y},\ 0 < \bar{x} < \bar{y}$, and $0 < \bar{y} < 1$

PROBLEMS 19.7

1. $\int_0^\pi \int_0^{2\cos\theta} \int_{-\sqrt{4-r^2}}^{\sqrt{4-r^2}} r\, dz\, dr\, d\theta = \frac{16}{3}\left[\pi - \frac{4}{3}\right]$
3. $\mu = \int_0^\pi \int_0^{2\cos\theta} (2\alpha/3)r(4 - r^2)^{3/2}\, dr\, d\theta = \frac{64}{15}\alpha(\pi - \frac{16}{15});\ (32/[21(\pi - \frac{16}{15})], 0, 0)$
5. $\mu = \int_0^{2\pi} \int_0^2 \int_0^{4-r^2} \alpha rz\, dz\, dr\, d\theta = \frac{32}{3}\alpha\pi;\ (0, 0, 2)$
7. $V = \int_0^\pi \int_0^{\sin\theta} \int_{r^2}^{r\sin\theta} r\, dz\, dr\, d\theta = \pi/32;\ (0, \frac{1}{2}, \frac{5}{12})$
9. $\mu = \int_0^{2\pi} \int_0^2 \int_{z/4}^z \alpha r^2\, dr\, dz\, d\theta = \frac{21}{8}\alpha\pi;\ (0, 0, \frac{8}{5})$ **11.** $\int_0^{\pi/2} \int_0^1 \int_r^{\sqrt{1-r^2}} rz^3\, dz\, dr\, d\theta = 0$
13. $V = \int_0^2 \int_0^{2\pi} \int_{\pi/4}^{3\pi/4} \rho^2 \sin\varphi\, d\varphi\, d\theta\, d\rho = (16\sqrt{2}/3)\pi$
15. $\int_0^{2\pi} \int_0^\pi \int_a^b (\alpha/\rho)\, \rho^2 \sin\varphi\, d\rho\, d\varphi\, d\theta = 2\pi\alpha(b^2 - a^2)$. Here the density is given by α/ρ, where α is a constant of proportionality.
17. $V = \int_0^1 \int_0^\pi \int_0^{\pi/6} \rho^2 \sin\varphi\, d\theta\, d\varphi\, d\rho = \pi/9$
19. $I = \int_0^3 \int_0^{2\pi} \int_0^\pi \rho^3 \cdot \rho^2 \sin\varphi\, d\varphi\, d\theta\, d\rho = 486\pi$
21. $I = \int_0^a \int_0^{\pi/2} \int_0^{\pi/2} \dfrac{(\rho \cos\varphi)^3}{\rho \sin\varphi} \cdot \rho^2 \sin\varphi\, d\varphi\, d\theta\, d\rho = \pi a^5/15$

REVIEW EXERCISES FOR CHAPTER NINETEEN

1. $\frac{4}{3}$ **3.** $\frac{67}{3}$ **5.** -24 **7.** $\frac{1}{3} + \pi/16$ **9.** $-\pi \leq I \leq \pi$

11. $\int_0^3 \int_0^{\sqrt{9-y^2}} (9 - y^2)^{3/2}\, dx\, dy = \frac{648}{5}$ **13.** 6 (See Problem 19.6.12)

15. $400\pi/3$ (See Problem 19.6.17)   **17.** $(0, 0)$ **19.** $\mu = \frac{5}{2}$; $(\frac{28}{75}, \frac{19}{15})$

21. 6π **23.** $\int_0^{2\pi} \int_0^4 \int_{r^2/4}^r r\, dz\, dr\, d\theta = 32\pi/3$ **25.** $\sqrt{35}/12$ **27.** 32π

29. $[(96)(31)/5] + [(127)(24)/7] = 36072/35$ **31.** 1 (See Problem 19.6.12)

33. $\frac{11}{10}$ **35.** $V = \frac{128}{15}$; $(\frac{12}{7}, 0, \frac{16}{7})$

37. $V = \frac{2}{3}a^3(\pi - \frac{4}{3})$; $(0, 4a/5(\pi - 4/3), 0)$

(Note that the equation of the cylinder is $r = a \sin \theta$.)

39. $\int_0^1 \int_{x^2}^x \int_{-\sqrt{z-x^2}}^{\sqrt{z-x^2}} dy\, dz\, dx = \int_0^1 \frac{4}{3}(x - x^2)^{3/2}\, dx = \pi/32$

41. $\int_{\pi/4}^{3\pi/4} \int_0^{2\pi} \int_0^3 \rho^2 \sin \varphi\, d\rho\, d\theta\, d\varphi = 18\sqrt{2}\pi$ **43.** $\int_0^\pi \int_0^{\pi/3} \int_0^2 \rho^2 \sin \varphi\, d\rho\, d\theta\, d\varphi = 16\pi/9$

CHAPTER TWENTY

PROBLEMS 20.1

1. $2(x + y)(\mathbf{i} + \mathbf{j})$ **3.** $\sin(x - y)(-\mathbf{i} + \mathbf{j})$ **5.** $(x/\sqrt{x^2 + y^3})\mathbf{i} + (3y^2/2\sqrt{x^2 + y^3})\mathbf{j}$

7. $-y \sec^2(y - x)\mathbf{i} + [\tan(y - x) + y \sec^2(y - x)]\mathbf{j}$

9. $\sec(x + 3y) \tan(x + 3y)(\mathbf{i} + 3\mathbf{j})$ **11.** $[4xy/(x^2 + y^2)^2](y\mathbf{i} - x\mathbf{j})$

13. $yz\mathbf{i} + xz\mathbf{j} + xy\mathbf{k}$ **15.** $\left(\frac{1}{3xy}\right)\left[\left(\frac{x^2 + y^2 - z^2}{x}\right)\mathbf{i} - \left(\frac{x^2 + y^2 + z^2}{y}\right)\mathbf{j} + 2z\mathbf{k}\right]$

17. $y^2\mathbf{i} + 2y(x + z^3)\mathbf{j} + 3y^2z^2\mathbf{k}$ **19.** $\sin y \ln z\, \mathbf{i} + x \cos y \ln z\, \mathbf{j} + (x/z) \sin y\, \mathbf{k}$

21. $(\cosh z - y \cos x)\mathbf{i} - \sin x\mathbf{j} + x \sinh z\mathbf{k}$ **23.** $-\omega y\mathbf{i} + \omega x\mathbf{j}$

PROBLEMS 20.2

1. $\mathbf{x} = (t, 2t)$; $I = \int_0^2 9t^2\, dt = 24$ **3.** $\mathbf{x} = (t, 2t - 4)$; $I = \int_1^2 (2t^2 - 2t - 8)\, dt = -\frac{19}{3}$

5. $\mathbf{x} = (\cos \theta, \sin \theta)$; $I = \int_0^{2\pi} [(\cos \theta)(\sin \theta)(-\sin \theta) + (\sin \theta - \cos \theta)\cos \theta]\, d\theta = -\pi$

7. $\int_0^1 0\, dt + \int_0^1 (t - 1)\, dt + \int_0^1 t^2\, dt = -\frac{5}{6}$ **9.** $\mathbf{x} = (3 \cos \theta, -\sin \theta)$;

$I = \int_{\pi/2}^{3\pi/2} [(9 \cos^2 \theta - 2 \sin \theta)(-3 \sin \theta) + (-\sin^2 \theta)(-\cos \theta)]\, d\theta = 3\pi - \frac{2}{3}$

11. $\int_0^1 e^t\, dt + \int_0^1 (-e + e^{1-2t})\, dt + \int_1^0 e^{-t}\, dt = \frac{1}{2}(e + e^{-1}) - 2 = \cosh 1 - 2$

13. $\mathbf{x} = (2 + 2t, 1 + 5t)$; $I = \int_0^1 \left\{ \frac{1 + 5t}{[2(1 + t)]^2} \cdot 2 + \frac{2(1 + t)}{[1 + 5t]^2} \cdot 5 \right\} dt = \frac{5}{2} \ln 2 + \frac{2}{5} \ln 6 + \frac{1}{3}$

15. $W = \int_0^{\pi/2} [\sin^3 t \cos t + e^{2t} \sin t]\, dt = \frac{9}{20} + (2e^\pi/5) \approx 9.71$ J

17. $W = -\int_0^{2\pi} [(\cos \theta \sin \theta)(-\sin \theta) + (2 \cos^3 \theta - \sin \theta)(\cos \theta)]\, d\theta = -3\pi/2$ J

19. $W = \int_0^1 0\, dt + \int_0^1 t^2\, dt + \int_1^0 (2t^2 + t^2)\, dt = -\frac{2}{3}$ J

21. $\mathbf{F} = \nabla(x^2y + y)$; $I = 14$ **23.** $\mathbf{F} = \nabla x \sin(x + y)$; $I = \pi/6$

25. $\mathbf{F} = \nabla x^2 \cos y$; $I = \pi^2/4$ **27.** $\mathbf{F} = \nabla xe^y$; $I = 5e^7$

PROBLEMS 20.3

1. 2   **3.** $\int_0^1 \int_0^{1-y} 2e^x \sin y\, dx\, dy = \cos 1 - \sin 1 + e - 2$ **5.** $2(e^2 - 1)(1 - \cos 1)$

7. $\iint_{\text{disk}} (-4y) = 0$ **9.** $(\pi/8) - (\pi/3) + (\pi/3\sqrt{2}) = \pi[(4\sqrt{2} - 5)/24]$

11. $\frac{242}{5} - \frac{26}{3} = \frac{596}{15}$ **13.** $\iint_\Omega (b - a)\, dA = (b - a)(\text{area of } \Omega)$

15. 0 (exact differential with $f = -2x^2/\sqrt{1 + y^2}$)   **17.** Use (14). Note that the line from (a_1, b_1) to (a_2, b_2) is parametrized as $\mathbf{x} = (a_1 + t(a_2 - a_1), b_1 + t(b_2 - b_1))$ for $0 \leq t \leq 1$. This yields area as $\frac{1}{2} |(a_1 + a_2)(b_2 - b_1) + (a_2 + a_3)(b_3 - b_2) + (a_3 + a_1)(b_1 - b_3)|$; (also see Problem 1.3.11)

19. $\frac{1}{2} |(a_1 + a_2)(b_2 - b_1) + (a_2 + a_3)(b_3 - b_2) + (a_3 + a_4)(b_4 - b_3) + (a_4 + a_1)(b_1 - b_4)|$ See answer to Problem 17.)   **21.** (a) $2x - 2y$ (b) 0 (c) 0 (d) 0

23. (a) $3(x^2 - y^2)$ (b) 0 (c) 0 (d) 0   **25.** (a) -2 (b) -2π (c) 0 (d) 0

27. curl $\mathbf{F} = 0$ **29.** curl $\mathbf{F}$ is an odd function   **31.** curl $\mathbf{F} = 0$ **33.** div $\mathbf{F} = 0$

PROBLEMS 20.4

1. $\int_0^1 [(t)(1) + (t^2)(2t) + (t^3)(3t^2)]\, dt = \frac{3}{2}$

3. $\int_0^{\pi/2}[(\cos^2 t)(-\sin t) + (\sin^2 t)(\cos t) + (t^2)(1)]\, dt = \pi^3/24$ **5.** $\mathbf{F} = \nabla xy^2z^4; I = 12$

7. $\mathbf{F} = \nabla(\ln|x + 2y + 3z| - \frac{3}{2}x^2 + \frac{1}{3}y^3); I = \ln 20 + \frac{9}{2}$

PROBLEMS 20.5

1. $(5^{3/2} - 1)/6$ **3.** $(9\sqrt{5}/16) - (\frac{1}{32})\ln(2 + \sqrt{5})$ **5.** $\frac{14}{3}$ **7.** 0   **9.** $-7\sqrt{14}/108$

11. $76\sqrt{2}\pi$ **13.** $64 + 32\sqrt{21}/3$; note that a tetrahedron has four faces. The integral over the sloping face is $32\sqrt{21}/3$. The other three integrals are 0, $\frac{64}{3}$; and $\frac{128}{3}$.

15. 0   **17.** 54π **19.** $\sqrt{3}/12$ **23.** $-\frac{32}{3}$

25. 2π **27.** 0   **29.** $-\pi/3$ **33.** $\frac{419}{3360}$

PROBLEMS 20.6

1. div $\mathbf{F} = 2(x + y + z)$; curl $\mathbf{F} = \mathbf{0}$ **3.** div $\mathbf{F} = 0$; curl $\mathbf{F} = \mathbf{0}$

5. div $\mathbf{F} = x + y + z$; curl $\mathbf{F} = -y\mathbf{i} - z\mathbf{j} - x\mathbf{k}$

7. div $\mathbf{F} = 0$; curl $\mathbf{F} = x(e^{xy} - e^{xz})\mathbf{i} + y(e^{yz} - e^{xy})\mathbf{j} + z(e^{xz} - e^{yz})\mathbf{k}$

9. div $\mathbf{F} = (1/y) + (1/z) + (1/x)$; curl $\mathbf{F} = (y/z^2)\mathbf{i} + (z/x^2)\mathbf{j} + (x/y^2)\mathbf{k}$ **19.** 0   **21.** 0

29. (b) $\mathbf{F} = \nabla(2x^2yz)$

PROBLEMS 20.7

1. -3π **3.** -6 **5.** 108π **7.** $21 - \sin 3$

11. the integral of $\mathbf{F}$ along the straight line segment from $(6, 0, 0)$ to $(0, 3, 0)$ is -18, from $(0, 3, 0)$ to $(0, 0, 2)$ is -4, from $(0, 0, 2)$ to $(6, 0, 0)$ is -24; the integral of $\mathbf{F}$ along the triangle is -46 **13.** the integral equals -18π

PROBLEMS 20.8

1. 4π **3.** 36π **5.** 0   **7.** 2   **9.** 108π **11.** $\frac{1}{6}$ **13.** $\frac{1}{2}$ **15.** $\frac{184}{35}$

PROBLEMS 20.9

1. $[\partial(x, y)/\partial(u, v)] = -2$ **3.** $[\partial(x, y)/\partial(u, v)] = 4(u^2 + v^2)$

5. $[\partial(x, y)/\partial(u, v)] = -2$ **7.** $[\partial(x, y)/\partial(u, v)] = -(a^2 + b^2)$

9. $[\partial(x, y)/\partial(u, v)] = (1 - uv)e^{u+v}$ **11.** $[\partial(x, y)/\partial(u, v)] = [1/(u + v)][(1/v) - (1/u)]$

13. $[\partial(x, y)/\partial(u, v)] = \sec u \csc v(1 + uv \cot u \tan v)$ **15.** $[\partial(x, y, z)/\partial(u, v, w)] = 0$

17. $[\partial(x, y, z)/\partial(u, v, w)] = 2(u^2v - uv^2 + v^2w - vw^2 + w^2u - wu^2)$

19. $[\partial(x, y, z)/\partial(u, v, w)] = e^{u+v+w}$

21. $\int_0^1 \int_y^1 xy\, dx\, dy = \int_{-1/2}^0 \int_{-v}^{1+v}(u^2 - v^2)(2)\, du\, dv = \frac{1}{8}$ **23.** 2π **25.** $\frac{128}{15}$ **27.** $4\pi abc/3$

29. $\frac{594}{5}$

REVIEW EXERCISES FOR CHAPTER TWENTY

1. $3(x + y)^2(\mathbf{i} + \mathbf{j})$ **3.** $\frac{1}{2}\sqrt{y/x}\,\mathbf{i} + \frac{1}{2}\sqrt{x/y}\,\mathbf{j}$ **5.** $2x\mathbf{i} + 2y\mathbf{j} + 2z\mathbf{k}$

9. $\mathbf{F} = \nabla((x^3/3) + (y^3/3)); I = \frac{2}{3}$ **11.** $-\frac{1}{2}$ **13.** $\frac{5}{3}$ **15.** (a) $\mathbf{F} = \nabla(xe^{xy})$ (b) 1

17. $\int_0^1 \int_x^1 (y^2 - x^2)\, dy\, dx = \frac{1}{6}$ (using Green's theorem)

19. $\int_{-1}^1 \int_{x^2}^1 (x/\sqrt{1 + x^2})\, dy\, dx = 0$ **21.** (a) curl $\mathbf{F} = \mathbf{0}$ (b) 0 (c) div $\mathbf{F} = y^2 + x^2$ (d) 8π

23. curl$[(\cos x^2)\mathbf{i} + e^y\mathbf{j}] = \mathbf{0}$ **25.** $\frac{3}{2}$ **27.** $-\frac{3}{5}$ **29.** $(\frac{91}{48})\sqrt{5} + (\frac{61}{96})\ln(2 + \sqrt{5})$

31. 0   **33.** $567\sqrt{2}\pi/4$ **35.** $\alpha\pi^2/2$ (α is proportionality constant)

37. 4π **39.** π **41.** div $\mathbf{F} = 3$; curl $\mathbf{F} = \mathbf{i} + \mathbf{j} + \mathbf{k}$

43. div $\mathbf{F} = (1/x) + (1/y) + (1/z)$; curl $\mathbf{F} = \mathbf{0}$

45. div $\mathbf{F} = -\sin z$; curl $\mathbf{F} = (-\sin x + \sin y)\mathbf{k}$ **47.** 0   **49.** 0

51. $4(a + b + c)\pi/3$ **53.** 0   **55.** $\frac{5}{4}$ if the unit cube is $(0, 1)^3$; 0 if it is $(-\frac{1}{2}, \frac{1}{2})^3$

57. $[\partial(x, y)/\partial(u, v)] = 18u^2v^2$ **59.** $[\partial(x, y)/\partial(u, v)] = (uv - 1)e^{u+v}$

61. $[\partial(x, y)/\partial(u, v)] = v[\sec^2(uv)][u \sec^2 u - \tan u]$ **63.** $[\partial(x, y, z)/\partial(u, v, w)] = 2uvw$

65. $\int_2^3 \int_{-v}^v e^{(u/v)}\frac{1}{2}\, du\, dv = \frac{5}{4}(e - e^{-1}) = \frac{5}{2}\sinh 1$

CHAPTER TWENTY-ONE

PROBLEMS 21.1

1. first   **3.** third   **5.** fourth

PROBLEMS 21.2

1. $y = -(7x^2/2) + C$ **3.** $\sin t + \ln|\csc x + \cot x| = 1 + \ln(\csc 3 + \cot 3)$
5. $y^{-4} - 2y^{-2} = -4x^{-1} - 6x^{-2} + C$ **7.** $x = -\ln(\cos t + e^{-1} - 1)$
9. $x = e^{t-(\sin 2t)/2}$ **11.** $y = \tan(\frac{1}{3}x^3 + C)$ **13.** ≈ 4486
15. 14,549,000 in 1978; 15,953,000 in 1983; 17,493,000 in 1988 **17.** $y = Cx^{-k}$
19. $(2(25^2) - (\frac{16}{15})a^2)\sqrt{a}/\sqrt{2g} \approx 365.1$ sec where $a = \sqrt{25^2 - 1^2}$ is the initial height of the water level above the hole

PROBLEMS 21.3

1. $y = 2x^2 + C$ **3.** $x = \frac{1}{2}(\sin t - \cos t) + \frac{3}{2}e^{-t}$ **5.** $y = x^x(1 + Ce^{-x})$
7. $x = (bt + C)e^{at}$ **9.** $y = \frac{1}{4} + \frac{1}{2}x \tan 2x + C \sec 2x$ **11.** $y = \frac{1}{2} + [1/2(x^2 + 1)]$
13. $x = (t^2 + 1)/2 + [7/2(t^2 + 1)^2]$ **15.** $x = ye^y + Cy$
17. (a) $A'(t) + \alpha A = c$, where α is the constant of proportionality
(b) $A(t) = (c/\alpha)(1 - e^{-\alpha t})$; (c) c/α **19.** $I = 2(1 - e^{-25t})$
21. $Q = \frac{1}{680}(3 \sin 60t + 5 \cos 60t - 5e^{-100t})$
23. $I(0.1) = (e^{-10}/100) + 600\pi(1 - e^{-10})/[(100)^2 + (120\pi)^2]$
25. $Q(0.000001) = \frac{1}{2000}(1 - e^{-1})$
27. $I(t) = \{E_0C/[1 + (\omega RC)^2]\}[RC\omega^2 \cos \omega t - \omega \sin \omega t + (1/RC) e^{-t/RC}]$;
$Q(t) = \{E_0C/[1 + (\omega RC)^2]\}[\cos \omega t - e^{-t/RC} + \omega RC \sin \omega t]$

PROBLEMS 21.4

1. $y = c_1e^{2x} + c_2e^{-2x}$ **3.** $y = c_1e^{-x} \sin x + c_2e^{-x} \cos x$ **5.** $y = 4e^{-x/4} \sin(x/4)$
7. $y = 5e^{-2x} - 4e^{-3x}$ **9.** $y = (1 + \frac{9}{2}x)e^{-5x/2}$ **11.** $y = -\frac{1}{5}e^{3x} - \frac{4}{5}e^{-2x}$
13. $y = \sin 2x - \frac{3}{2} \cos 2x$ **15.** $y \equiv 1$ **17.** $y = c_1e^{6x} + c_2e^{7x}$ **19.** $y = -\frac{1}{2}e^x + \frac{5}{2}e^{-x}$

PROBLEMS 21.5

1. $y = \frac{4}{5} \cos x + c_1 \sin\sqrt{6}x + c_2 \cos\sqrt{6}x$ **3.** $y = \frac{1}{10}e^{4x} + c_1e^{-x} + c_2e^{-2x}$
5. $y = \frac{4}{3}x^3 - 4x^2 + 8x - 6 + 9e^{-x}$ **7.** $y = (\frac{2}{3}x^3 + 1 - 3x)e^{3x}$
9. $y = (x^2 + c_1 + c_2x)e^x$ **11.** $y = [(4x + 1 - 2\pi)/4] \sin 4x$
13. $y = e^{2x}((-\frac{1}{80}) \cos x + (\frac{1}{40}) \sin x) + (x/4)e^{-2x} \cos x + e^{-2x}(c_1 \cos x + c_2 \sin x)$ (rewrite $\sinh 2x$ as $(e^{2x} - e^{-2x})/2$)
15. $y = -(1/2\sqrt{2}) x(\cos \sqrt{2}x)e^x + c_1e^x \sin \sqrt{2}x + c_2e^x \cos \sqrt{2}x$ **17.** $y = 3 - 2x \sin x$
19. $y = -\frac{1}{3}x \cos 3x + \frac{1}{9}x^2 - \frac{2}{81}$ **21.** 5
23. $Q(t) = \frac{1}{14,430}(-48 \cos 30t + 45 \sin 30t + 74e^{-20t} + 26e^{-5t})C$
25. $I_{\text{steady state}} = \cos t + 2 \sin t$ **27.** $I_{\text{transient}} = e^{-t}(-\cos t - 3 \sin t)$

PROBLEMS 21.6

3. $V = 4\sqrt{g/2}$ cm/sec (since $x = c_1 \cos\sqrt{(g/32)}t + c_2 \sin\sqrt{(g/32)}t$ with $x(0) = 16$ and $x'(0) = 0$) **5.** period $2\pi/\sqrt{5\pi g}$ sec; $x(t) = [(1/5\pi) - 1] \cos\sqrt{5\pi g}t$
7. $k = (100\pi^2 + 1)/160$
9. $x(t) = e^{-\sqrt{17}t}(c_1 = \cos 9t + c_2 = \sin 9t) + [(47 \sin 2t - 2\sqrt{17} \cos 2t)/22770]$

REVIEW EXERCISES FOR CHAPTER TWENTY-ONE

1. $y = \frac{3}{2}x^2 + C$ **3.** $x = -\ln(e^{-3} - \sin t)$ **5.** $y = 0.7e^{-3x} + 0.3 \cos x + 0.1 \sin x$
7. $x = (2e^{-3} - \frac{1}{4} + \frac{1}{4}t^4)e^{3t}$ **9.** $y = c_1e^x + c_2e^{4x}$ **11.** $y = c_1e^{3x} + c_2e^{-3x}$
13. $y = (c_1 + c_2x)e^{-3x}$ **15.** $y = e^x \sin x$ **17.** $y = \frac{2}{3} \sin x + c_1 \cos 2x + c_2 \sin 2x$
19. $y = c_1e^{-x/2} \sin(\sqrt{3}/2)x + c_2e^{-x/2} \cos(\sqrt{3}/2)x - (x/\sqrt{3})[\cos(\sqrt{3}/2)x]e^{-x/2}$
21. $y = x^3 - 7x + c_1 \cos x + c_2 \sin x$ **23.** $y = (3 - 11x + x^2/2)e^{4x}$
25. $y = x + (\frac{1}{2})e^x - (x/2)\cos x + c_1 \cos x + c_2 \sin x$
27. $y \equiv 0$ (this is a harmonic oscillator which never gets going)
29. period $= 2\pi\sqrt{60/g}$; frequency $= (\sqrt{g/60}/2\pi)$;
amplitude $= \sqrt{10^2 + (-5)^2} \cdot 10/(g/6) \approx 10.076$ cm **31.** $\mu = \sqrt{4(10)(g/6)} \approx 80.87$

APPENDIX 1

PROBLEMS A1.1

1. $5\pi/6$ **3.** $5\pi/3$ **5.** $4\pi/5$ **7.** $15°$ **9.** $22.5°$ **11.** $-60°$ **13.** $90°$

PROBLEMS A1.2

In Problems 1–15 the value of $\sin\theta$ is given first.

1. $0, 1$ **3.** $-\frac{1}{2}, -\sqrt{3}/2$ **5.** $(\sqrt{2}/4)(1 + \sqrt{3}), (\sqrt{2}/4)(\sqrt{3} - 1)$

7. $-\sqrt{2 - \sqrt{3}}/2, -\sqrt{2 + \sqrt{3}}/2$ **9.** $-\sqrt{2 - \sqrt{3}}/2, \sqrt{2 + \sqrt{3}}/2$

11. $\sqrt{2 - \sqrt{2 + \sqrt{2}}}/2, \sqrt{2 + \sqrt{2 + \sqrt{2}}}/2$ **13.** $\sqrt{2 + \sqrt{2}}/2, \sqrt{2 - \sqrt{2}}/2$

15. $-\sqrt{2 - \sqrt{2 + \sqrt{3}}}/2, \sqrt{2 + \sqrt{2 + \sqrt{3}}}/2$

29. amplitude = 2

31. amplitude = 4

33.

35. period = $2\pi/3$, amplitude = 3

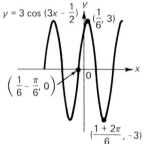

PROBLEMS A1.3

In Problem 1–15 the answers are given in the order $\tan x$, $\cot x$, $\sec x$, $\csc x$.

1. 0, undefined, 1, undefined **3.** $1/\sqrt{3}, \sqrt{3}, -2/\sqrt{3}, -2$

5. $2 + \sqrt{3} = \sqrt{(2 + \sqrt{3})/(2 - \sqrt{3})}, 2 - \sqrt{3}, 2/\sqrt{2 - \sqrt{3}}, 2/\sqrt{2 + \sqrt{3}}$

7. $2 - \sqrt{3}, 2 + \sqrt{3}, -2/\sqrt{2 + \sqrt{3}}, -2/\sqrt{2 - \sqrt{3}}$

9. $-2 + \sqrt{3}, -2 - \sqrt{3}, 2/\sqrt{2 + \sqrt{3}}, -2/\sqrt{2 - \sqrt{3}}$

11. $\sqrt{2 - \sqrt{2 + \sqrt{2}}}/\sqrt{2 + \sqrt{2 + \sqrt{2}}}, \sqrt{2 + \sqrt{2 + \sqrt{2}}}/\sqrt{2 - \sqrt{2 + \sqrt{2}}},$
$2/\sqrt{2 + \sqrt{2 + \sqrt{2}}}, 2/\sqrt{2 - \sqrt{2 + \sqrt{2}}}$

13. $\sqrt{2} + 1 = \sqrt{2 + \sqrt{2}}/\sqrt{2 - \sqrt{2}}, \sqrt{2} - 1, 2/\sqrt{2 - \sqrt{2}}, 2/\sqrt{2 + \sqrt{2}}$

15. $-\sqrt{2 - \sqrt{2 + \sqrt{3}}}/\sqrt{2 + \sqrt{2 + \sqrt{3}}}, -\sqrt{2 + \sqrt{2 + \sqrt{3}}}/\sqrt{2 - \sqrt{2 + \sqrt{3}}},$
$2/\sqrt{2 + \sqrt{2 + \sqrt{3}}}, -2/\sqrt{2 - \sqrt{2 + \sqrt{3}}}$

17. period = 6π

19.   period $= \pi/3$

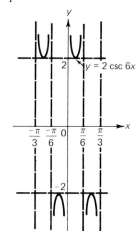

$y = 2 \csc 6x$

21.   period $= 2\pi/5$

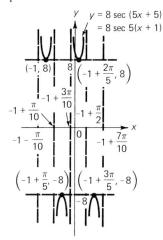

$y = 8 \sec (5x + 5)$
$= 8 \sec 5(x + 1)$

PROBLEMS A1.4

1. $\sin \theta = \frac{4}{11}\sqrt{6}$, $\cos \theta = \frac{5}{11}$, $\tan \theta = 4\sqrt{6}/5$, $\csc \theta = 11/4\sqrt{6}$, $\sec \theta = \frac{11}{5}$, $\cot \theta = 5/4\sqrt{6}$

3. $\sin \theta = -\sqrt{3}/2$, $\cos \theta = \frac{1}{2}$, $\tan \theta = -\sqrt{3}$, $\csc \theta = -2/\sqrt{3}$, $\sec \theta = 2$, $\cot \theta = -1/\sqrt{3}$

5. $\sin \theta = \frac{1}{5}$, $\cos \theta = -2\sqrt{6}/5$, $\tan \theta = -1/2\sqrt{6}$, $\csc \theta = 5$, $\sec \theta = -5/2\sqrt{6}$, $\cot \theta = -2\sqrt{6}$

7. $\sin \theta = -\frac{2}{3}$, $\cos \theta = \sqrt{5}/3$, $\tan \theta = -2/\sqrt{5}$, $\csc \theta = -\frac{3}{2}$, $\sec \theta = 3/\sqrt{5}$, $\cot \theta = -\sqrt{5}/2$

9. $\sin \theta = -10/\sqrt{101}$, $\cos \theta = -1/\sqrt{101}$, $\tan \theta = 10$, $\csc \theta = -\sqrt{101}/10$, $\sec \theta = -\sqrt{101}$, $\cot \theta = \frac{1}{10}$

APPENDIX 2

1.   *Hint:* Verify the identity $[n^2(n + 1)^2/4] + (n + 1)^3 = (n + 1)^2(n + 2)^2/4$.

3.   *Hint:* Apply the product rule to $xP_n(x) + b$.

5.   *Hint:* Pick one element from the set and color it orange. Now count how many subsets contain this special orange element; count how many subsets do not contain it. Can there be any other subsets?

APPENDIX 3

1. $D = -2$; $x_1 = 3$, $x_2 = 0$ **3.** $D = 0$; infinite number—the entire line $x_1 + 2x_2 = 3$

5. $D = 0$; no solutions   **7.** $D = -29$; $x_1 = x_2 = 0$ **9.** -56 **11.**   140   **13.**   96

15. -120 **17.** -132 **19.** -398 **23.** $x_1 = -3$, $x_2 = 0$, $x_3 = 5$

25. $x_1 = -\frac{26}{9}$, $x_2 = -\frac{1}{2}$, $x_3 = \frac{19}{9}$ **27.** $x_1 = -2$, $x_2 = 0$, $x_3 = 1$, $x_4 = 5$

APPENDIX 4

1.   10   **3.**   36   **5.**   165   **7.**   1

11. $a^7 + 7a^6b + 21a^5b^2 + 35a^4b^3 + 35a^3b^4 + 21a^2b^5 + 7ab^6 + b^7$

13. $x^5 - 10x^4y + 40x^3y^2 - 80x^2y^3 + 80xy^4 - 32y^5$

15. $a^5x^5 - 5a^4bx^4y + 10a^3b^2x^3y^2 - 10a^2b^3x^2y^3 + 5ab^4xy^4 - b^5y^5$

17. $\frac{1}{32}x^5 + \frac{5}{48}x^4y + \frac{5}{36}x^3y^2 + \frac{5}{54}x^2y^3 + \frac{5}{162}xy^4 + \frac{1}{243}y^5$ **19.**   165

APPENDIX 5

7.   Let $f(x) = x + 1$ and $g(x) = x$ (more complicated examples can be found).

9.   Let $f(x) = x$ and $g(x) = 1/x$.   **11.**   Let $f(x) = g(x) = x$. Note that ∞/∞ is not defined.

APPENDIX 6

1. $9 - 7i$ **3.**   2   **5.** $-27 + 5i$ **7.** $5\sqrt{2}e^{\pi i/4}$ **9.** $3\sqrt{2}e^{-\pi i/4}$ **11.** $6e^{\pi i/6}$

13. $8e^{-\pi i/6}$ **15.** $2e^{-2\pi i/3}$ **17.** -2 **19.** $-[1/(2\sqrt{2})] - (1/2\sqrt{2})i$

21. $-2\sqrt{3} + 2i$ **23.** $-\frac{3}{2} - (3\sqrt{3}/2)i$ **25.** $\cos 1 + i \sin 1 \approx 0.54 + 0.84i$

27. $4 - 6i$ **29.** $7i$ **31.** $2e^{-\pi i/7}$ **33.** $3e^{4\pi i/11}$

39.   circle of radius a centered at z_0 if $a > 0$, single point z_0 if $a = 0$, empty set if $a < 0$

Index

A4
B5
C6
D7
E8